Developmental Mathematics

eText Reference

Kirk Trigsted
University of Idaho

Kevin Bodden
Lewis & Clark Community College

Randy Gallaher
Lewis & Clark Community College

PEARSON

Boston Columbus Indianapolis New York San Francisco Upper Saddle River
Amsterdam Cape Town Dubai London Madrid Milan Munich Paris Montreal Toronto
Delhi Mexico São Paulo Sydney Hong Kong Seoul Singapore Taipei Tokyo

Editorial Director, Mathematics: Christine Hoag
Editor in Chief: Michael Hirsch
Senior Acquisitions Editor: Dawn Giovanniello
Editorial Assistant: Ashley Yee
Executive Content Editor: Kari Heen
Content Editor: Katherine Minton
Design Manager: Andrea Nix
Art Director: Heather Scott
Senior Production Supervisor: Ron Hampton
Math Media Producer: Audra Walsh
Content Development Manager, MathXL: Rebecca Williams
Marketing Manager: Rachel Ross
Associate Marketing Manager: Alica Frankel
Market Development Manager: Dona Kenly
Procurement Specialist: Carol Melville

1 2 3 4 5 6—V003—16 15 14 13

PEARSON

ISBN-13: 978-0-321-88023-9
ISBN-10: 0-321-88023-4

Contents

Preface

Introduction

This *Developmental Mathematics eText Reference* contains the pages of Kirk Trigsted, Kevin Bodden, and Randy Gallaher's *Developmental Mathematics eText* in a portable, spiral-bound format. The structure of the eText Reference helps students organize their notes by providing them with space to summarize the videos and animations. Students can also use it to review the eText's material anytime, anywhere.

A Note to Students

This textbook was created for you! Unlike a traditional text, we wanted to create content that gives you, the reader, the ability to be an active participant in your learning. The eText was specifically designed to be read online. The eText pages have large, readable fonts and were designed without the need to scroll. Throughout the material, we have carefully placed thousands of hyperlinks to definitions, previous modules, interactive videos, animations, and other important content. Many of the videos and animations allow you to actively participate and interact. Take some time to "click around" and get comfortable with the navigation of the eText and explore its many features.

Before you attempt each homework assignment, read the appropriate topic(s) of the eText. At the beginning of each topic (starting in Module 1), you will encounter a feature called Things to Know. This feature includes all the prerequisite objectives from previous topics that you will need to successfully learn the material presented in the new topic. If you do not have a basic understanding of these objectives, click on the desired hyperlinks and rework through the objectives, taking advantage of any videos or animations.

An additional feature of the eText is the inclusion of an audio icon. By clicking on this icon, you can listen to the text as you read. While you read through the pages of the eText, use the margins of this *Developmental Mathematics eText Reference* to take notes, summarize key points, and list helpful tips and reminders. An additional option, if made available by your instructor, is to use the *Guided Notebook* to guide your note taking as you work through the eText. The *Guided Notebook* provides more specific direction on how to proceed through the material while providing more space for note taking.

Try testing yourself by working through the corresponding You Try It exercises. Remember, you learn math by *doing* math! The more time you spend working through the videos, animations, and exercises, the more you will understand. If your instructor assigns homework in MyMathLab or MathXL, rework the exercises until you get them right. Be sure to go back and read the eText at anytime while you are working on your homework. This text caters to your educational needs. We hope you enjoy the experience!

A Note to Instructors

Today's students have grown up in a technological world where everything is "clickable." We have taught with MyMathLab for many years and have experienced firsthand how fewer and fewer students have been referring to their traditional textbooks, opting instead to use electronic resources. As the use of technology plays an ever increasing role in how we are teaching our students, it seems only natural to have a textbook that mirrors the way our students are learning. We are excited to have written a text that was conceived from the ground up to be used as an online interactive tool. Unlike traditional printed textbooks, this eText was specifically designed for students to read online while working in MyMathLab. Therefore, we wrote this text entirely from an online perspective, with MyMathLab and its existing functionality specifically in mind. Every hyperlink, video, and animation has been strategically integrated within the context of each page to engage the student and maximize his or her learning experience. All of the interactive media was designed so students can actively participate while they learn math. Many of the interactive videos and animations require student interaction, giving specific feedback for incorrect responses.

We are proponents of students learning terms and definitions. Therefore, we have created hyperlinks throughout the text to the definitions of important mathematical terms. We have also inserted a tremendous amount of "just-in-time review" throughout the text by creating links to prerequisite topics. Students have the ability to reference these review materials with just a click of the mouse. You will see that the exercise sets are concise and nonrepetitive. Since MyMathLab will be used as the main assessment engine, there is no need for a repetitive exercise set in the hardcopy version of the text. Every exercise is available for you to assign within MyMathLab or MathXL. For the first time, instructors can assess reading assignments. We have created five conceptual reading assessment questions for every section of the eText, giving the student specific feedback for both correct and incorrect responses. Each feedback message directs them back to the appropriate location for review. Our hope is that every student who uses the eText will have a positive learning experience.

Acknowledgments

First of all, we want to thank our wives for their loving support, encouragement, and sacrifices. Thanks to our children for being patient when we needed to work and for reminding us when we needed to take breaks. We could not have completed this project without generous understanding and forgiveness from each of you.

Writing this textbook has been one of the most difficult and rewarding experiences of our lives. We truly could not have accomplished our goal without the support and contributions of so many wonderful, talented people. These extraordinary talents deserve every accolade we can give them. From Pearson, we would like to thank Dawn Giovanniello, our editor and friend, for believing in us; Greg Tobin, Chris Hoag, and Michael Hirsch for their continued support; Heather Scott and Andrea Nix for their design brilliance; Dona Kenly, Rachel Ross, and Alicia Frankel for their marketing expertise; Tracy Menoza and Ruth Berry for taking care of all the details; Eileen Moore, Rebecca Williams, and the rest of the MathXL development team for an amazing job; Brian Morris and the art team for their great eye for detail; Revathi Viswanathan of PreMedia Global; and Ron Hampton, our production manager, for all his support.

Along the way, we had the help of so many people from around the world and from many different walks of life. Alice Champlin and Anthony T. J. Kilian at Magnitude Entertainment for creating quality videos and animations; Helen Medley for creating the answers; and Kim Johnson for her work on the *Guided Notebook* and Sue Glascoe for creating the PowerPoints.

The following list attempts to recognize all of the reviewers. Please accept our deepest apologies if we have inadvertently omitted anyone. We have benefited greatly from your honest feedback, constructive criticism, and thoughtful suggestions. Each of you has helped us create a resource we truly believe will be effective in helping students learn mathematics. We are deeply grateful and we genuinely thank you all from the bottom of our hearts.

Kevin Bodden
Randy Gallaher

Ali M. Ahmad, Dona Ana Community College
Khadija Ahmen, Monroe County Community College
Mary Jo Anhalt, Bakersfield College, Delano Center
Hamid Attarzadeh, Jefferson Community and Technical College
Tony Ayers, Collin College

Sam Bazzi, Henry Ford Community College
Disa Beaty, Rose State College
David M. Behrman, Somerset Community College
Charlotte Ellen Bell, Richland College
David Bell, Florida Community College at Jacksonville
Marion G. Ben-Jacob, Ph.D., Mercy College
Annette Benbow, Tarrant County College
Rosanne B. Benn, Prince George's Community College
Tom Blackburn, Northeastern Illinois University
Gregory Bloxom, Pensacola Junior College
Amberlee Bosse, University of Phoenix
Joe Brenkert, Front Range Community College
Gail Brooks, McLennan Community College
Shirley Brown, Weatherford College
Hien Bui, Hillsborough Community College
Debby Casson, University of New Mexico
Mark Chapman, Baker College
Shawn Clift, Eastern Kentucky University
Jeremy Coffelt, Blinn College
Arunas Dagys, Saint Xavier University
Marlene Dean, Oxnard College
Donna Densmore, Bossier Parish Community College
Jill M. DeWitt, Baker College of Muskegon
Karen Egedy, Baton Rouge Community College
Nancy Fardelius Fees, Northwest College
Matthew Flacche, Camden County College
Irina Golub, Saint Paul College
Edna Greenwood, Tarrant County College
Susan Grody, Broward College
Ryan E. Grossman, Ivy Tech Community College
Shawna Haider, Salt Lake Community College
Steven Hair, Pennsylvania State University
Professor Richard Halkyard, Gateway Community College
Amadou Hama, Kennedy-King College, Chicago
Amanda Hanley, Cuyahoga Community College
Thomas Hartfield, Gainesville State College
Al Hemenway, Los Angeles Mission College
Gloria P. Hernandez, Louisiana State University at Eunice
Stephanie Houdek, St. Cloud State University
Teresa Houston, East Mississippi Community College
Elizabeth Howell, North Central Texas College
Mark Hunter, Cosumnes River College
Sally Jackman, Hopkinsville Community College
Sharon Jackson, Brookhaven College
Marilyn Jacobi, Gateway Community College
Yvette Janecek, Blinn College
Brian Karasek, Glendale Community College
Mike Kirby, Tidewater Community College
Cameron S. Kishel, Columbus State Community College
Daniela Kojouharov, Tarrant County College, SE
Carla K. Kulinsky, Salt Lake Community College
Rita LaChance, University of Maine–Augusta
Dr. Marsha Lake, Brevard Community College (FL)–Titusville Campus
Xiaodan Leng, Pasadena City College
Joyce Lindstrom, St. Charles Community College
Charyl A. Link, Kansas City Kansas Community College
Annette Magyar, Southwestern Michigan College
Frank Marfai
Ilva Mariani, Cerritos College
Stacy Martig, St. Cloud State University
Mike May, SJ, Saint Louis University
Jeremy McClure, Heartland Community College

Dr. Kimberley McHale, Heartland Community College
Stacey Moore, Wallace State Community College–Hanceville, AL
Kathrine Murray, Tyler Junior College
Dr. Said Ngobi, Victor Valley College
Denise Nunley, Glendale Community College
Kathleen Offenholley, Borough of Manhattan Community College
Staci Osborn, Cuyahoga Community College
Louise W. Pack, Rowan-Cabarrus Community College
Karen D. Pain, Palm Beach State College
Linda Parrish, Brevard Community College
Margaret Patin, Vernon College
Kathleen Faulds Peters, Manchester Community College
Allison Pfaff, Harrisburg Area Community College
Evelyn Porter, Utah Valley University
Tom Pulver, Waubonsee Community College
Carolynn Reed, Austin Community College
George Reed, Angelina College
Pat Rhodes, Treasure Valley Community College
Natalie Rivera, Estrella Mountain Community College
Lisa Rombes, Washtenaw Community College
Andrew S. H. Russell, Queensborough Community College, CUNY
Jeffrey Saikali, San Diego Miramar College
Togba Sapolucia, Houston Community College
Nicole Saporito, Luzerne County Community College
Mario Scribner, Tidewater Community College
Karra Seggerman, Heartland Community College
Lisa Sheppard, Lorain County Community College
Jeanette Shotwell, Central Texas College
Joseph A. Spadaro, Gateway Community College
Marie E. St. James, St. Clair County Community College
Brad Stetson, Schoolcraft College
Sean K. Stewart, Owens Community College
Susan Tummers Stocum, El Camino College
Kenneth Takvorian, Mount Wachusett Community College
Janet E. Teeguarden, Ivy Tech Community College
Mary Ann Teel, University of North Texas
Diane Valade, Piedmont Virginia Community College
Phil Veer, Johnson County Community College
Dr. Mary Wagner-Krankel, St. Mary's University
Tyler Wallace, Big Bend Community College
Denise Walsh, Quinebaug Valley Community College
Christina Weston, Nashville State Community College
Yong Wilson, Cleveland State Community College
Claire Winkel, Pima Community College

MODULE ONE

Whole Numbers

MODULE ONE CONTENTS

1.1 Study Tips for This Course

OBJECTIVES

1 Prepare for This Course

2 Study for Optimal Success

3 Use the eText Effectively

4 Get Ready for an Exam

OBJECTIVE 1 PREPARE FOR THIS COURSE

The path to success in a math course starts even before the first day of class. Careful preparation can help you get off on the right foot. The following tips can help you have the best start to your course.

1. **Select a section that is right for you.** Most colleges offer courses at different times and in various formats. When selecting a course, don't always choose a section that seems most convenient. A longer class that meets once a week may sound good, but if you need more sessions to process information, it might be better to choose a section that meets several times each week for shorter periods of time. Also, online or web-enhanced courses tend to require more self-discipline and motivation. Be sure to talk with your advisor to discuss which class types best fit your learning style.

2. **Have a positive attitude.** One of the most important keys to success in a course is a positive attitude. You may or may not enjoy math. Either way, you have signed up for the course and want to complete it to the best of your ability. Thinking positively can help you get past stumbling blocks and keep you on track to complete your assignments.
3. **Know your course details ahead of time.** Before the first day of class, make sure you know where your class is located and what time your class begins. Plan to arrive early on the first day in case you underestimate the time it will take to get to your class. If your college posts syllabi online, download the syllabus and review the course requirements and policies.
4. **Develop a weekly schedule.** Create a weekly schedule that includes your work hours and class times. Add study time for each of your classes and stick to your schedule. Having set times to study will help you avoid cramming for exams or waiting until the last minute to work on homework or quizzes. Be sure to schedule some down time with family and friends to avoid burning out. Use this **template** to build your weekly schedule.
5. **Don't overload yourself.** Many instructors expect students to work at least 2 hours outside of class for every 1 hour in class. For a 15-hour course load, this means you will spend at least 45 hours per week on school. Doctors also recommend 8 hours of sleep each night. This takes up another 56 hours per week. This leaves you 67 hours per week for work, family, friends, household duties, etc. While this may seem like a lot of time, it goes by quickly. You will need to balance the demands for your time without overloading yourself. Taking on too much can lead to problems in all areas of your life.
6. **Get to know your classmates.** Introduce yourself to other students in your class and try to form a study group. Working together on homework can be an excellent learning aid. Explaining a concept to another student solidifies your understanding and may also help the other student understand more by getting a different perspective.

OBJECTIVE 2 STUDY FOR OPTIMAL SUCCESS

Doing well in your course will require a significant amount of studying outside class. In fact, a large portion of your learning will take place after you leave the classroom. Managing your time wisely and using all available resources will help you achieve success in your course. The following tips can help you reach this goal.

1. **Select an appropriate study area.** Having a designated quiet area for studying can help you stay organized and focused. Choose a study location that will allow you to concentrate and avoid frequent distractions.
2. **Study daily (or close to it).** Letting homework pile up can be a huge problem. It is easy to put off doing your assignments and think that you can catch up on the weekend. However, in most cases this doesn't happen; you just get further behind and more confused. Plan to work on math daily, or as close to it as possible. Even if you can spend only 15 or 20 minutes on math, doing so will help keep the material from piling up. Daily work also helps you learn the material in smaller doses, keeping new concepts fresh in your mind. Waiting until the last minute to cram for an exam or rushing through a homework assignment to get it turned in on time will not help you learn the material well enough to retain what you need.
3. **Read the eText.** Just as a novel builds the story from one chapter to the next, the material in each module of your eText builds on previous skills and concepts covered in earlier modules. Use the Things to Know feature at the start of each topic to go back and review important concepts needed for the topic. When reading the eText, do not just focus on the examples. Also read the text *between* the examples. Often this is where concepts are explained in simpler terms, which helps with your overall understanding of

the material. This includes popup boxes that illustrate or extend concepts further. Do not ignore the videos and animations. These support the eText by providing additional detailed examples. As you read the eText and work the exercises, write down any questions that arise so you can ask your instructor. If you rely only on your memory for this, you may forget some of the questions you wanted to ask once you get to class.

4. **Take notes even outside of class.** You should take notes on the material as you read your eText. Read each topic prior to class and outline the topic in your notebook. If taking notes is difficult for you, use the Guided Notebook that accompanies the eText to help structure your notes. That way you can use class time for filling in the gaps in your notes instead of rushing to write down everything that is written on the board. Take notes even when doing homework. If you use the available help resources, take notes on specific tips or techniques that helped you, and write down those exercises that were particularly difficult to complete. Keep all your notes and work together in a three-ring binder so they can be easily reviewed when needed.

5. **Create a glossary of important terms.** Understanding the mathematical terminology is an important part of any math course. Also this makes it easier to read the eText. As you come across a new term or phrase, write it on a note card along with its definition. You can then review the terms and phrases periodically to make sure that you still understand each one. Throughout the eText, key terms are clickable so you can quickly find the definition of a term you may have forgotten.

6. **Do your homework.** This should go without saying, but it is often overlooked. Homework is not meant to be busy work. Rather, it is an important practice designed to help you understand the concepts covered. Practice will improve your skills, make you feel more comfortable with the material, and build your confidence. If you are having difficulty with a certain type of exercise, do more of that type until you can do them easily. Utilize the available help features within each homework assignment. However, do not just mimic a procedure. Take time to understand why each step in a solution was carried out.

   Complete your homework immediately after reading the eText. This way, the information will still be fresh in your mind. Try each homework problem. As you read through a problem, it may appear to be simple. However, when working on your own, you may find that this problem is suddenly more difficult. It is better to find this out on homework than on an exam. Do not lose points needlessly. Have all assignments completed and turned in on time so you don't lose valuable points. In fact, try to complete assignments early in case you run into technical issues or other difficulties.

7. **Check your work.** When working homework problems, allow yourself plenty of paper to show every step of a problem in an organized manner. This will reduce the number of careless mistakes. In addition, when you look back at the problem later in the course, you will be able to follow your work more easily. If you work a problem incorrectly, do not erase your work. Just put an X through the work so you can still review it later. Your instructor or anyone helping you can review your work to see where your mistakes are being made. This is very helpful for getting you past difficult spots. When two different approaches are available for solving a problem, use one approach to solve the problem and the second approach to check your answer. If your two answers do not agree, you know that you have made an error. When your instructor returns an assignment or quiz, be sure to review it and understand your mistakes. Do not simply discard old papers without first making an effort to learn from your mistakes.

8. **Aim for perfection.** Keep in mind that practice does not make perfect. Only *perfect practice* makes perfect. Make sure to always check your answers to see if they are correct. If you have to continually rework exercises to get the right answer, seek help from your instructor or tutoring center. If you practice something the wrong way, you will only strengthen your bad habits. Continually working a problem incorrectly increases the chance that you will make the same mistake on an exam. Find out what you are

doing wrong so that you can complete the exercise correctly every time. This is what is meant by 'perfect practice.' If you have attempted a problem several times and carefully checked your work, but still can't get the correct answer, you should seek help immediately to find your mistake so your practice can be perfect.

9. **Avoid absences.** Although situations may arise that cause you to miss a class, plan to attend every class. If you miss a class, it is your responsibility to find out what was covered. As soon as possible, contact your instructor or another student in the class to find out what was covered in your absence. Find someone in the class who takes good notes and ask if you can copy them to review the covered material.
10. **Get help right away when you need it.** Although this is the last tip, it may be the best advice you can receive. Often students are afraid to ask for help, which only leads to greater frustration and more mistakes. Without getting proper help at the right time, you might continue to miss questions because you do not understand the concepts and are unable to identify your mistakes. Follow these suggestions throughout your course:
    - Use the resources provided with the eText, including tutorials and examples.
    - Visit your college's math resource center or tutoring service.
    - Form study groups with other students in your class.
    - Contact your instructor. Ask questions in class, utilize available office hours, or even send an email. This includes questions on technology used in the course such as graphing calculators or online homework.

    Help is never far away, but you need to seek it. The sooner you reach out for help, the sooner help can be given.

OBJECTIVE 3 USE THE ETEXT EFFECTIVELY

The eText was written to be fully integrated with MyMathLab so that you can benefit from a wide variety of resources. Using the eText effectively is the first step in the following four-step process:

(eText Screens 1.1-1–1.1-16)

Your instructor may or may not use all four steps. However, the general approach remains the same. Read and interact first, then practice and get help as needed.

My animation summary

The eText has a number of features in it to help you in your course. These include

- Things to Know links
- You Try It links
- Animations
- Videos
- Interactive Videos
- Popup definitions
- Audio links
- Sticky notes
- Highlighter

Watch this **video animation** to see a tour of the various components of the eText and how you can use them to learn the course content most effectively. Then watch this **video** that shows you how to efficiently navigate through the eText.

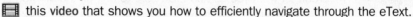

My video summary

OBJECTIVE 4 GET READY FOR AN EXAM

For many students, taking a math exam creates an enormous amount of anxiety and stress. It doesn't need to be that way. If you have studied properly and practiced prior to the exam date, you should feel confident that you are well prepared for the exam.

The following tips can help you as you get ready to take an exam.

1. **Relax.** Some students create anxiety and stress for themselves simply because they believe that they are not good at math and never will be good at math. These students are capable of doing the math, but the anxiety and stress interfere with their concentration. As a result, they often do poorly. If you have properly prepared for the exam, you have a good chance of doing well. Take some deep breaths prior to the start of the exam to help calm your nerves. Then trust the skills you have developed through studying the material.
2. **Prepare well in advance.** Do not try to cram for an exam. This just causes you to confuse concepts and miss questions that you might have answered correctly with proper studying. Each time that you study math, you are preparing for the exam. Remember this as you work through assignments. Review topics immediately if you struggle with them. At least several days before the exam, you should increase the amount of review and study to help prepare you for exam day.
3. **Review the main objectives.** Take a few minutes to review the objectives in the eText for each topic covered by the exam. This can help you identify important concepts that are likely to appear on the exam.
4. **Take a practice test.** Students often do well on homework and quizzes but "freeze up" on exams. To prevent this, be sure to take a practice exam on the material. Take the practice exam in a similar setting as you will have for the real exam. This will help make you more comfortable on test day. Be sure to review the practice test to identify weak areas that need more review prior to the graded exam.
5. **Be comfortable and alert.** On the night before the exam, be sure you get a good night's sleep. Being well rested will help you work more effectively. On the day of the exam, be sure to eat well-balanced meals so that you have the energy necessary to focus. Avoid heavy foods that can make you sluggish or sleepy. Dress comfortably, but bring a light

jacket in case the testing room is cold. Use the bathroom before taking the exam so you don't waste exam time.

6. **Read directions carefully.** Make sure to read the exam directions completely before starting the exam. Otherwise, you might miss important information. Pay attention to the point value on test questions. The instructor might assign more points for more difficult problems, or all problems may be worth the same number of points. Knowing the point distribution will affect your test-taking strategy. Read each question carefully to be sure that you understand what is being asked. Check to see that your answer does indeed answer the question asked.
7. **Manage time wisely.** While taking the test, work all the problems you are confident about first. Do not get bogged down on one question. If there is a problem that you do not know how to work, skip it and move on to problems you can work. Then, return to that problem later on if time permits. This is particularly helpful if all the questions are worth the same number of points. However, be sure to answer every question on the test. Leaving a question blank automatically means zero points for that question. Do as much work for each question as possible, even if it is just a step or two. Your instructor may give partial credit for what you do know. Be sure to use all the allowed time for the exam. If you have answered all the questions, go back and check your work. Leaving early without checking over all your answers can be costly. The exam is not a race to see who finishes first. Don't worry if others turn in their test early. This doesn't necessarily mean they did well on the exam. Half their test may be blank.
8. **Know your technology.** If you are allowed to use a calculator or other technology during the exam, be sure you are familiar with the technology and how to use its features. Because button sequences may vary from calculator to calculator, using a different calculator for the first time on an exam can be problematic. The exam is not the time to learn how to use it. Practice using the technology as you study for the exam. Ask your instructor for help if you don't know how to use certain features.
9. **Don't toss old exams.** When an exam is returned, make sure you review it to see what mistakes you made, if any, and learn from them. If you have a final exam, old tests are often helpful to study from because they indicate what the instructor feels is important for you to know.

1.1 Exercises

1. What is your instructor's name?
2. What are your instructor's office hours?
3. Where is your instructor's office? Do you know how to get there?
4. What is your instructor's email address? How else can you reach him or her?
5. Does your college have a tutoring center? If so, what hours is it open?
6. Have you exchanged contact information with at least one other student in your class?
7. Where will you access the eText and online homework?
8. How will your grade in the course be determined?
9. Can you use a graphing calculator or other technology in your course?
10. How many credit hours are you taking? How many hours per week do you plan to work?

11. Have you filled out a weekly schedule? Did you allow yourself study time and free time?
12. What online help tools are available for your course in MyMathLab?
13. What are some suggested study tips?
14. What are some suggested test-taking tips?
15. Using the eText effectively is the first step in what four-step process?
16. What does the  icon in the eText mean?
17. What does the  icon in the eText mean?
18. What does the  icon in the eText mean?
19. What does the  icon in the eText mean?
20. What does the  icon in the eText mean?
21. What does the  icon in the eText mean?
22. Why is it important to have a positive attitude when taking your math course?
23. Explain the statement "Perfect practice makes perfect."

Use the video in Objective 3 to help you answer these remaining questions.

24. Explain the purpose of "Things to Know" at the beginning of each topic.
25. Explain why some terms in the eText are highlighted links.
26. Explain why three types of examples are used in the eText.

(eText Screens 1.2-1–1.2-29)

1.2 Introduction to Whole Numbers

OBJECTIVES

1 Identify the Place Value of a Digit in a Whole Number

2 Write Whole Numbers in Standard Form and Word Form

3 Change Whole Numbers from Standard Form to Expanded Form

4 Use Inequality Symbols to Compare Whole Numbers

5 Round Whole Numbers

6 Read Tables and Bar Graphs Involving Whole Numbers

OBJECTIVE 1 IDENTIFY THE PLACE VALUE OF A DIGIT IN A WHOLE NUMBER

The numbers $0, 1, 2, 3, 4, 5, 6, 7, 8, 9, 10, 11, \ldots$ form a group called the **whole numbers**. This list of numbers continues without end. While zero (0) is the smallest whole number, there is no largest whole number.

The symbols $0, 1, 2, 3, 4, 5, 6, 7, 8$, and 9 are called **digits**. When a whole number is written using digits, it is written in **standard form**. Whole numbers with many digits are written using commas to separate the digits into groups of three, called **periods**. The names of the first five periods from right to left are *ones*, *thousands*, *millions*, *billions*, and *trillions*.

The **place value** of a digit is determined by its position in the number. For example, the U.S. Census Bureau projects that the world population in 2020 will be 7,584,821,144. The digit 8 appears twice in this number, but each 8 represents a different amount. Figure 1 shows the projected 2020 world population on a **place-value chart**. Looking from left to right, we see that the place value of the first 8 is ten-millions, and the place value of the second 8 is hundred-thousands.

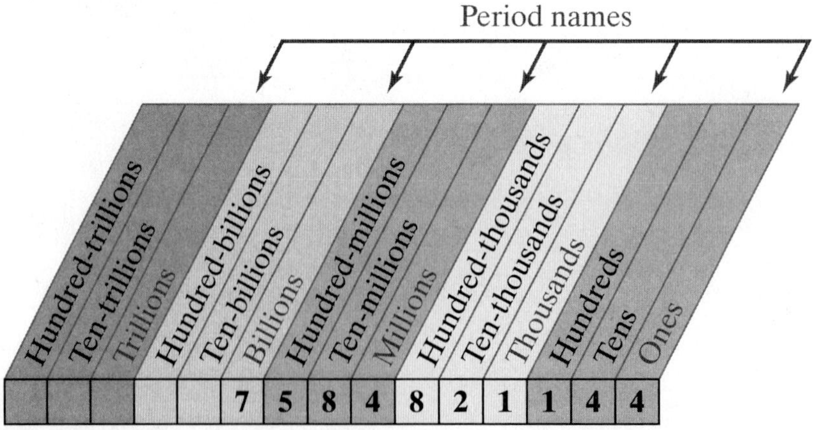

Figure 1 Place-Value Chart

Example 1 Identifying Place Value

Identify the **place value** of the **digit** 6 in each **whole number**.

a. 87,962 **b.** 506,721 **c.** 160,942,328

(eText Screens 1.2-1–1.2-29)

Solutions

a. View this **popup box** to see 87,962 on a place-value chart. The digit 6 is in the tens place.

b. View this **popup box** to see 506,721 on a place-value chart. The digit 6 is in the thousands place.

c. View this **popup box** to see 160,942,328 on a place-value chart. The digit 6 is in the ten-millions place.

You Try It **Work through this You Try It problem.**

Work Exercises 1–4 in this textbook or in the MyMathLab Study Plan.

OBJECTIVE 2 WRITE WHOLE NUMBERS IN STANDARD FORM AND WORD FORM

Whole numbers can be written in **standard form** or **word form**. For example, the word form for 57 is *fifty-seven*, and the standard form for *two hundred eighteen* is 218.

> **Writing Whole Numbers in Word Form**
>
> To write a whole number in word form, work from left to right. Write the word form of the number in each **period**, followed by the name of the period. (Normally, the ones period name is not written.) Skip periods that contain all zeros (000).

Let's look at the word form of 7,584,821,144, which is *seven billion, five hundred eighty-four million, eight hundred twenty-one thousand, one hundred forty-four*.

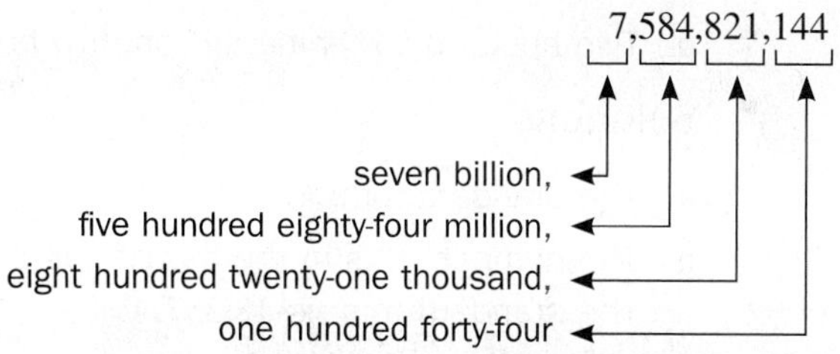

Example 2 Writing Whole Numbers in Word Form

My video summary

Write each whole number in word form.

a. 894 **b.** 23,715 **c.** 6,047,829 **d.** 164,302,000

Solutions

a. The three **digits** are in the ones period. The word form is *eight hundred ninety-four*.

b. From left to right, the first two digits are in the *thousands* period, and the last three digits are in the *ones* period. The word form is *twenty-three thousand, seven hundred fifteen*.

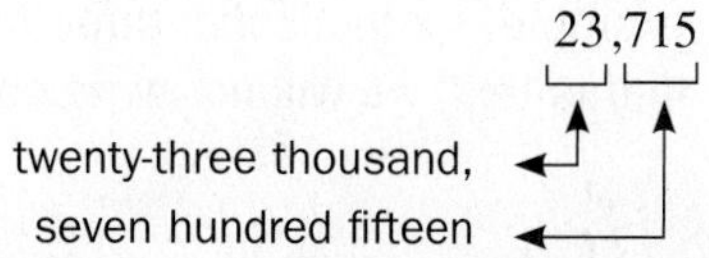

c. From left to right, the first digit is in the *millions* period, the next three digits are in the *thousands* period, and the last three digits are in the *ones* period. Try to finish writing the word form on your own. View the **answer**, or watch this **video** for detailed solutions to parts a–d.

d. Try to write this word form on your own. View the **answer**, or watch this **video** for detailed solutions to parts a–d.

You Try It **Work through this You Try It problem.**

Work Exercises 5–10 in this textbook or in the MyMathLab Study Plan.

CAUTION

The word "and" should *not* be used when speaking or writing **whole numbers**. For example, 125 is "one hundred twenty-five," *not* "one hundred and twenty-five."

Writing Whole Numbers in Standard Form

To write a **whole number** in standard form, write the number in each **period** using **digits** and separate the periods by commas. Fill in any skipped periods with zeros (000).

Example 3 Writing Whole Numbers in Standard Form

Write each whole number in standard form.

a. Six hundred twelve

b. Eighteen thousand, two hundred fifty-seven

c. One billion, three hundred seventy-two million, five hundred thirteen thousand, six hundred ninety-eight

d. Two hundred sixty-nine million, five hundred thousand

Solutions

a. The standard form is 612.

b. The number 18 is in the thousands period, and the number 257 is in the ones period. The standard form is 18,257.

c. The number 1 is in the billions period, 372 is in the millions period, 513 is in the thousands period, and 698 is in the ones period. The standard form is 1,372,513,698.

d. The number 269 is in the millions period, and the number 500 is in the thousands period. Because no number is mentioned for the ones period, we fill it with all zeros (000). The standard form is 269,500,000.

You Try It **Work through this You Try It problem.**

Work Exercises 11–16 in this textbook or in the MyMathLab Study Plan.

CAUTION

A four-digit number may be written in standard form either with or without a comma. For example, *six thousand, three hundred twenty-five* can be written as 6,325 or 6325. Typically in this text, we will not write commas in four-digit numbers.

OBJECTIVE 3 CHANGE WHOLE NUMBERS FROM STANDARD FORM TO EXPANDED FORM

The **expanded form** of a whole number is written as the addition of each **digit's** place value. For example, the expanded form of 2743 is

$$\boxed{2 \text{ thousands}} + \boxed{7 \text{ hundreds}} + \boxed{4 \text{ tens}} + \boxed{3 \text{ ones}}$$
$$\downarrow \quad \downarrow \quad \downarrow \quad \downarrow$$
$$2000 + 700 + 40 + 3$$

Example 4 Writing Whole Numbers in Expanded Form

Write each whole number in expanded form.

a. 32,589 **b.** 4,520,708

Solutions

a. $32{,}589 = 30{,}000 + 2{,}000 + 500 + 80 + 9$

b. Try to write this expanded form on your own, then check your **answer**.

You Try It **Work through this You Try It problem.**

Work Exercises 17–22 in this textbook or in the MyMathLab Study Plan.

OBJECTIVE 4 USE INEQUALITY SYMBOLS TO COMPARE WHOLE NUMBERS

Every **whole number** can be represented visually with a point on a **number line.** Figure 2 shows a number line with marks equally spaced to represent whole numbers. The point for the smallest whole number 0 is called the **origin**. The arrowhead on the right indicates that the whole numbers continue forever and that there is no largest whole number. The lack of an arrowhead on the left indicates that 0 is the smallest whole number.

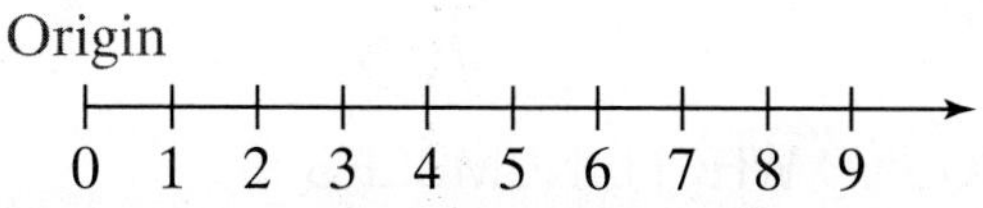

Figure 2
Number Line

We **graph**, or **plot**, a whole number by placing a solid circle (•) at its location on the number line. For example, the whole numbers 4 and 7 are graphed in Figure 3.

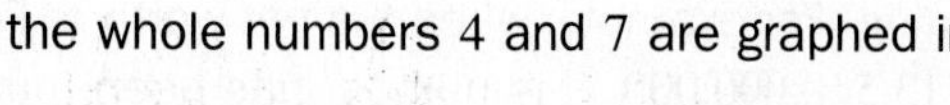

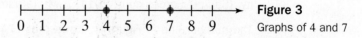

Figure 3
Graphs of 4 and 7

The whole numbers increase in value as we move from left to right on the number line. Because 7 is to the right of 4, we say that 7 *is greater than* 4. We can also say that 4 *is less than* 7. We can write such comparison statements using **inequality symbols**.

Inequality Symbols

The symbol $>$ means *is greater than*.
The symbol $<$ means *is less than*.

$7 > 4$ is read as "seven is greater than four."

$4 < 7$ is read as "four is less than seven."

Example 5 Using Inequality Symbols to Compare Whole Numbers

Fill in the blank with the appropriate inequality symbol, $<$ or $>$, to make a true comparison statement.

a. 6____9 **b.** 15____12

Solutions

a. Because 6 is to the left of 9 on the number line, we have $6 < 9$.

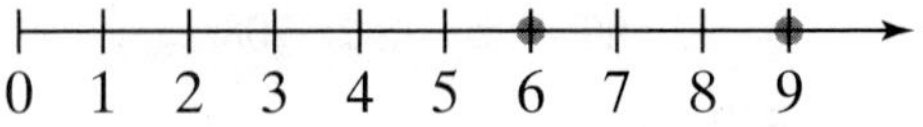

b. Because 15 is to the right of 12 on the number line, we have $15 > 12$.

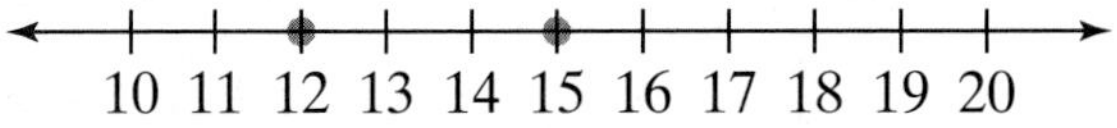

You Try It **Work through this You Try It problem.**

Work Exercises 23–26 in this textbook or in the MyMathLab Study Plan.

TIP When writing comparison statements, remember that the inequality symbol always points to the smaller number.

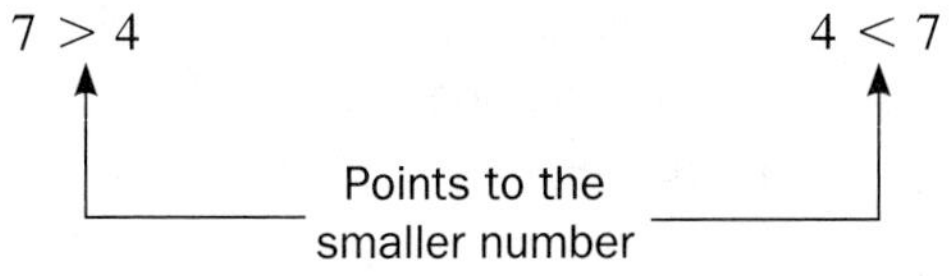

OBJECTIVE 5 ROUND WHOLE NUMBERS

When we approximate a **whole number** we **round** the whole number to a desired **place value.** For example, when the net worth of Facebook creator Mark Zuckerberg is reported as \$13,500,000,000, this number has been rounded to the nearest hundred million for simplicity. (*Source:* Forbes.com, 2011) A whole number can be rounded to any given place value.

The number line in Figure 4 shows that 27 is closer to 30 than 20, so 27 rounded to the nearest ten is 30. We say that 27 **rounds up** to 30.

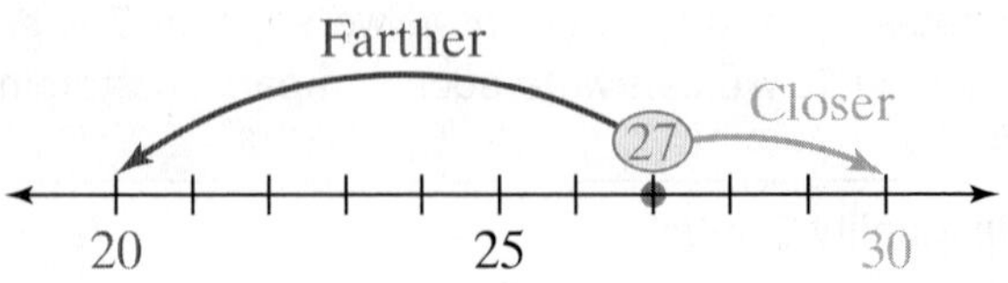

Figure 4
27 rounds up to 30.

The number line in Figure 5 shows that 438 is closer to 400 than 500, so 438 rounded to the nearest hundred is 400. We say that 438 **rounds down** to 400.

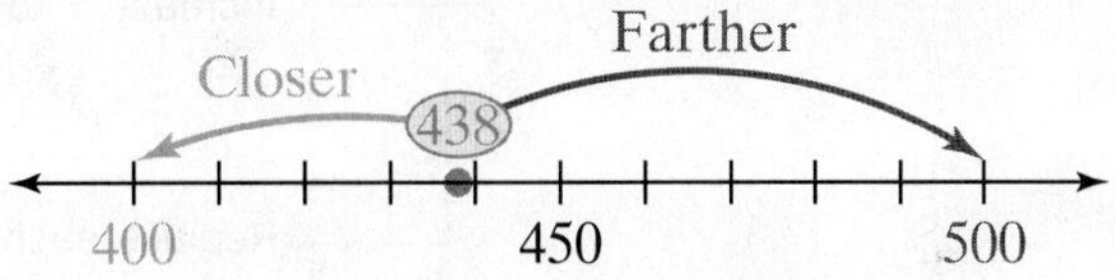

Figure 5
438 rounds down to 400.

We can round a whole number to a given **place value** using the following steps:

> **Rounding a Whole Number to a Given Place Value**
>
> **Step 1.** Find the **digit** to the right of the given place value (the place value that you are rounding to).
>
> **Step 2.** If the digit from Step 1 is 5 or greater, then increase the digit in the given place value by 1 and replace each digit to its right with a 0. This is called **rounding up**.
>
> **Step 3.** If the digit from Step 1 is less than 5, then keep the digit in the given place value and replace each digit to its right with a 0. This is called **rounding down**.

Example 6 Rounding Whole Numbers to a Given Place Value

My video summary

Round each whole number to the given place value.

a. 643 to the nearest ten

b. 3765 to the nearest hundred

c. 314,861 to the nearest ten-thousand

d. 29,534 to the nearest thousand

Solutions We follow the **steps for rounding whole numbers**.

a. The **digit** 4 is in the tens **place**, where we want to round. The digit 3 is in the ones place, which is to the immediate right of the tens place.

643
↗ ↖
Tens Ones

Because 3 is less than 5, we "round down" by keeping the 4 in the tens place and replacing the 3 in the ones place with a 0.

Keep the 4
640
Replace with 0

So to the nearest ten, 643 rounds to 640.

b. The **digit** 7 is in the hundreds **place**, where we want to round. The digit 6 is in the tens place, which is to the immediate right of the hundreds place.

3765
↗ ↖
Hundreds Tens

Because 6 is greater than 5, we "round up" by increasing the hundreds digit from 7 to 8 and replacing the tens and ones digits each with 0.

Increase 7 to 8

3800

Replace each digit with 0

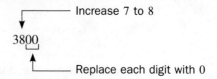

So to the nearest hundred, 3765 rounds to 3800.

c.–d. Try rounding these two numbers on your own. Check the **answers**, or watch this **video** for complete solutions to all four parts.

You Try It **Work through this You Try It problem.**

Work Exercises 27–34 in this textbook or in the MyMathLab Study Plan.

OBJECTIVE 6 READ TABLES AND BAR GRAPHS INVOLVING WHOLE NUMBERS

Tables and **graphs** such as **bar graphs** are often used to organize and display information. For example, Table 1 shows the number of video game consoles sold in 2010 by model and world region.

Table 1

2010 Video Game Consoles Sold				
Model	**Americas**	**Japan**	**EMEAA**	**World Total**
DS	9,758,315	2,887,216	8,404,433	21,049,964
PSP	2,266,282	2,883,770	4,215,118	9,365,170
PS2	1,867,225	89,949	2,634,606	4,591,760
PS3	5,542,514	1,553,638	7,354,707	14,450,859
Wii	9,064,682	1,661,630	7,342,873	18,069,185
X360	8,308,470	211,585	5,086,583	13,606,638
Total	36,807,488	9,287,788	35,038,320	81,133,596

(*Source:* vgchartz.com)

Note: *The acronym EMEAA stands for Europe, Middle East, Africa, and Asia. In this table, it represents all regions of the world other than the Americas and Japan.*

Table 1 displays a lot of information. For example, reading across the row marked "Wii," we see that 9,064,682 Wii consoles were sold in the Americas, 1,661,630 Wii consoles were sold in Japan, and 7,342,873 Wii consoles were sold in EMEAA for a total of 18,069,185 Wii consoles sold worldwide during 2010.

Example 7 Video Games

Use **Table 1** to answer each question.

a. How many X360 consoles were sold in Japan in 2010?

b. Which model had the least number of consoles sold in the Americas?

c. In which region were the most PS3 consoles sold?

Solutions

a. View the **table**. Look across the row marked "X360" until the "Japan" column is reached. The number of X360 consoles sold in Japan in 2010 was 211,585.

b. View the **table**. Looking down the "Americas" column, we see the smallest number is 1,867,225, which is the number of PS2 consoles sold. The PS2 model had the least number of consoles sold in the Americas.

c. View the **table**. Looking across the "PS3" row, we see the largest number (other than the total) is 7,354,707, which is the number of PS3 consoles sold in the EMEAA column. The most PS3 consoles were sold in the EMEAA region.

You Try It **Work through this You Try It problem.**

Work Exercises 35 and 36 in this textbook or in the MyMathLab Study Plan.

A **bar graph** uses **vertical** or **horizontal** bars to display information visually. Each bar represents a different category for data. Longer bars indicate larger values and shorter bars indicate smaller values, so bar graphs are useful for making comparisons among the categories.

Example 8 Hybrid Car Gas Mileage

The bar graph in Figure 6 shows the gas mileage (combined city and highway) for selected 2012 hybrid cars. Use the graph to answer the following questions.

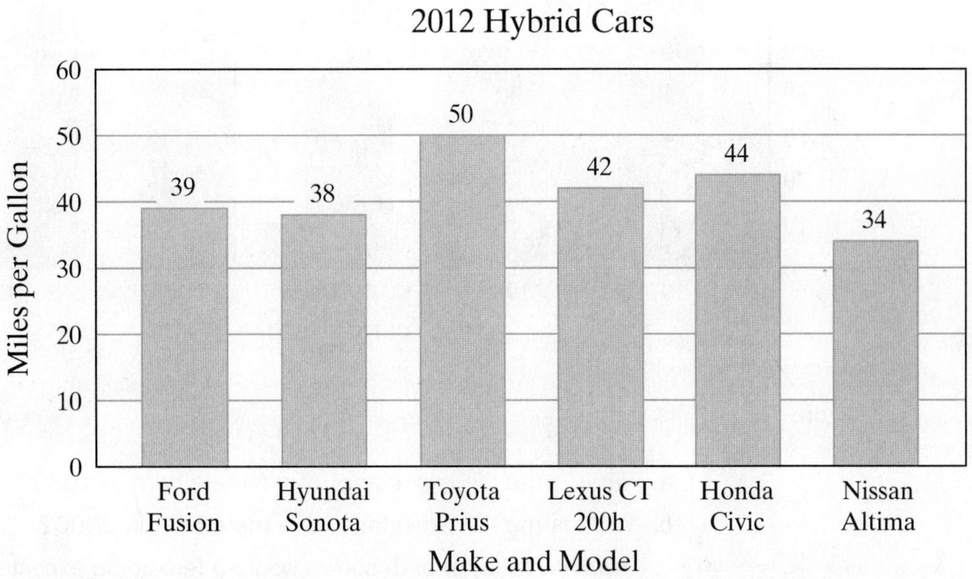

Source: HybridCars.com

Figure 6

(*eText Screens 1.2-1–1.2-29*)

a. What is the gas mileage of the Hyundai Sonata?

b. Does a Ford Fusion or Honda Civic get more miles per gallon?

c. Which car gets the highest gas mileage? What is its gas mileage?

Solutions

a. View the **graph**. Look at the bar labeled "Hyundai Sonata." It shows that gas mileage for a Hyundai Sonata is 38 miles per gallon.

b. View the **graph**. Look at the bars labeled "Ford Fusion" and "Honda Civic." The bar for Honda Civic is taller than the bar for Ford Fusion (and $44 > 39$), so a Honda Civic gets more miles per gallon.

c. View the **graph**. The tallest bar indicates the highest gas mileage, so the Toyota Prius gets the highest gas mileage at 50 miles per gallon.

You Try It **Work through this You Try It problem.**

Work Exercises 37 and 38 in this textbook or in the MyMathLab Study Plan.

Sometimes a bar graph will contain more than one bar for each category. In Example 9, we see a **double bar graph**.

Example 9 Life Expectancy

The double bar graph in Figure 7 shows the life expectancy of females and males born in the given years. Use the graph to answer the following questions.

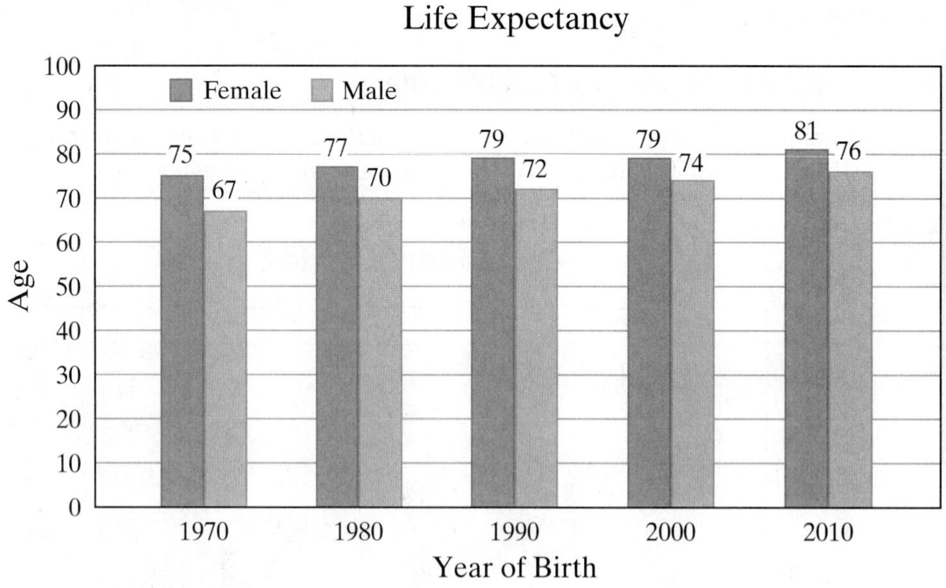

Source: U.S. Census Bureau

Figure 7

a. What is the life expectancy of a female born in 1980?

b. What is the life expectancy of a male born in 2000?

c. In what year(s) of birth shown would a female be expected to live to the age of 79?

(eText Screens 1.2-1–1.2-29)

My video summary

Solutions

a. View the **graph**. Look at the bars labeled with the year "1980." The bar to the left is for females. It shows that a female born in 1980 is expected to live to the age of 77.

b.–c. Try answering these questions on your own. View the **answers**, or watch this **video** for complete solutions to all three parts.

You Try It **Work through this You Try It problem.**

Work Exercises 39 and 40 in this textbook or in the MyMathLab Study Plan.

1.2 Exercises

In Exercises 1–4, identify the place value of the digit 3 in each whole number.

You Try It 1. 539 2. 2,308 3. 356,980 4. 13,275,000

In Exercises 5–10, write each whole number in word form.

You Try It 5. 903 6. 8,245 7. 12,934

8. 654,038 9. 7,320,695 10. 249,581,000,000

In Exercises 11–16, write each whole number in standard form.

You Try It 11. Four hundred sixty-three

12. Thirty-six thousand, eight hundred twenty-seven

13. Eleven billion, six hundred fifty-five million, seven hundred twelve thousand, nine hundred thirty-one

14. Five hundred four million, seven hundred sixty-one thousand, eight hundred ninety-two

15. Nine hundred twenty-eight million, three hundred fifty thousand

16. Two trillion, six hundred seventy-five billion

In Exercises 17–22, write each whole number in expanded form.

You Try It 17. 748 18. 6524 19. 52,806

20. 959,874 21. 2,706,540 22. 160,879,326

In Exercises 23–26, fill in the blank with the appropriate inequality symbol, $<$ or $>$, to make a true comparison statement.

You Try It 23. 12____8 24. 36____41

25. 715____751 26. 1123____967

In Exercises 27–34, round each whole number to the given place value.

You Try It 27. 564 to the nearest ten

28. 2862 to the nearest hundred

29. 6809 to the nearest thousand

30. 38,497 to the nearest thousand

31. 23,899 to the nearest ten-thousand

32. 4,953,120 to the nearest hundred-thousand

33. 56,140,875 to the nearest million

34. 99,950 to the nearest hundred

In Exercises 35–40, use the given table or bar graph to answer each question.

You Try It 35. **Olympic Medal Count** The following table shows the nations whose athletes won the most medals during the 2010 Winter Olympics Games in Vancouver, British Columbia, Canada.

2010 Winter Olympics Medal Count				
Nation	**Gold**	**Silver**	**Bronze**	**Total**
United States	9	15	13	37
Germany	10	13	7	30
Canada	14	7	5	26
Norway	9	8	6	23
Austria	4	6	6	16
Russia	3	5	7	15
South Korea	6	6	2	14
China	5	2	4	11
Sweden	5	2	4	11
France	2	3	6	11
Switzerland	6	0	3	9
Netherlands	4	1	3	8

Source: International Olympic Committee

a. How many gold medals were won by athletes from China?
b. How many bronze medals were won by athletes from Austria?
c. Which nation(s) took home 6 silver medals?
d. Which nation(s) took home more gold medals than the United States?

(*eText Screens 1.2-1–1.2-29*)

36. **Planets** The following table provides information about the eight planets of our solar system.

The Eight Planets of Our Solar System			
Planet	**Diameter (in Miles)**	**Average Distance from Sun (in Miles)**	**Number of Moons**
Mercury	3031	36,000,000	0
Venus	7521	67,200,000	0
Earth	7926	93,000,000	1
Mars	4222	141,600,000	2
Jupiter	88,729	483,600,000	18 named (and many smaller ones)
Saturn	74,600	886,700,000	18+
Uranus	32,600	1,784,000,000	15
Neptune	30,200	2,794,400,000	2

Source: EnchantedLearning.com

a. What is Jupiter's average distance from the Sun?
b. Which planet(s) have 2 moons?
c. Which planet has the largest diameter?
d. Which planet(s) have a smaller diameter than Earth?

You Try It 37. **Male Heights** The following bar graph gives the average height (in centimeters) for males by age.

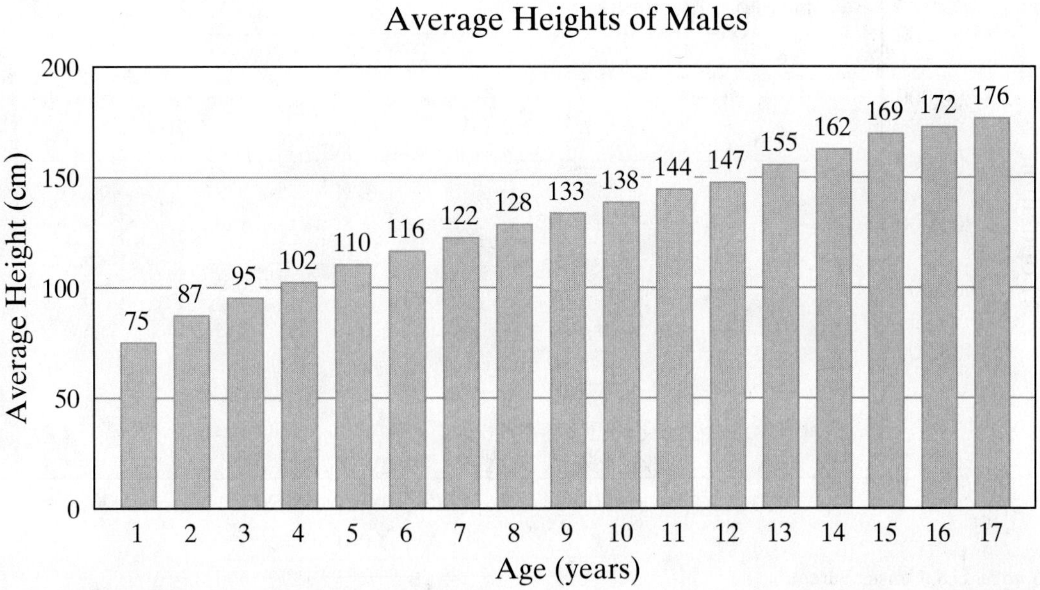

Source: Centers for Disease Control and Prevention

(eText Screens 1.2-1–1.2-29)

a. What is the average height of a 3-year-old boy?
b. What is the average height of an 8-year-old male?
c. At what age is the average boy 110 cm tall?
d. At what age is the average man 162 cm tall?

38. **Space Shuttle Program** A June 2011 CBS News survey asked 1045 Americans: "How do you feel about the U.S. space shuttle program coming to an end?" The results are shown in the following bar graph.

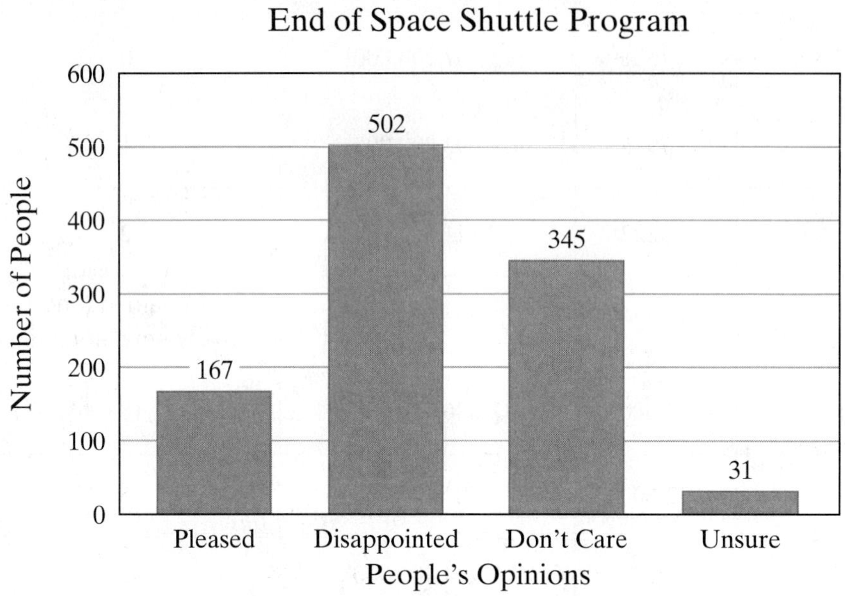

Source: PollingReport.com

a. How many people responded that they did not care?
b. Were more people pleased or disappointed that the space shuttle program was ending?
c. What opinion was given by 31 Americans?

You Try It 39. **Population** The following double bar graph gives the populations of Hawaii and Alaska in the given years.

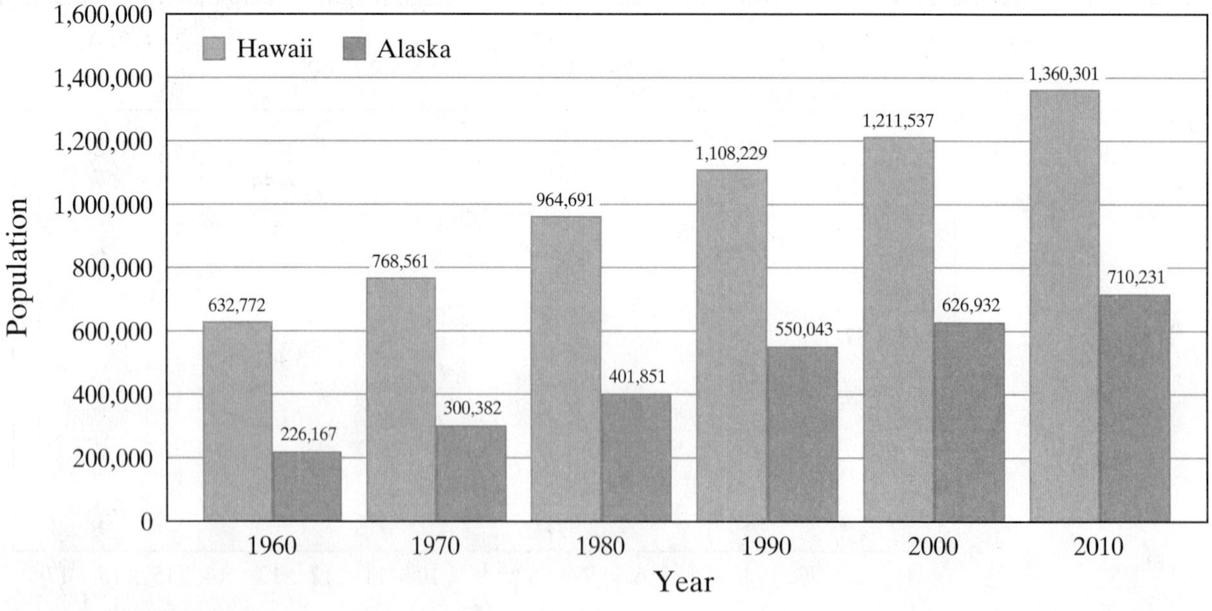

Source: U.S. Census Bureau

a. What was the population of Alaska in 1970?
b. What was the population of Hawaii in 2010?
c. In what year was the population of Alaska 300,382?
d. In what year was the population of Hawaii 964,691?

40. **Super Bowl Champions** The following double bar graph gives information about the NFL teams with the most Super Bowl appearances and wins through 2011.

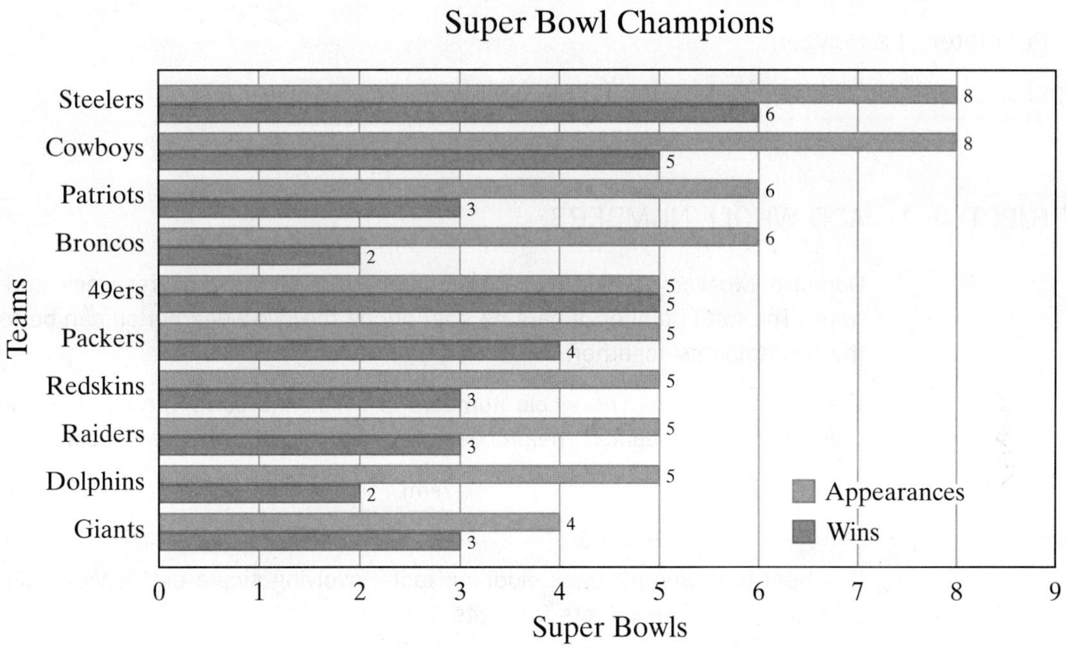

a. How many Super Bowl appearances have the Dolphins had? How many Super Bowl wins?
b. How many Super Bowl appearances have the Broncos had? How many Super Bowl wins?
c. Which teams have won the Super Bowl 5 times?
d. What teams have had more Super Bowl appearances than the 49ers?

1.3 Adding and Subtracting Whole Numbers; Perimeter

THINGS TO KNOW

Before working through this topic, be sure you are familiar with the following concepts:

		VIDEO	ANIMATION	INTERACTIVE
You Try It	**1.** Identify the Place Value of a Digit in a Whole Number (Topic 1.2, **Objective 1**)			
You Try It	**2.** Use Inequality Symbols to Compare Whole Numbers (Topic 1.2, **Objective 4**)			
You Try It	**3.** Round Whole Numbers (Topic 1.2, **Objective 5**)	■		
You Try It	**4.** Read Tables and Bar Graphs Involving Whole Numbers (Topic 1.2, **Objective 6**)	■		

OBJECTIVES

1 Add Whole Numbers

2 Use Special Properties of Addition

3 Add Several Whole Numbers

4 Subtract Whole Numbers

5 Estimate Sums and Differences of Whole Numbers

6 Find the Perimeter of a Polygon

7 Solve Applications by Adding or Subtracting Whole Numbers

OBJECTIVE 1 ADD WHOLE NUMBERS

During a two-week period, a car salesman sold 5 cars the first week and 3 the second week. The total number of cars he sold during the two-week period can be found by adding the two amounts together. $5 + 3 = 8$ cars sold.

The result of adding two **whole numbers** is called the **sum**. The numbers being added are called **terms**, or **addends**. The process of finding the sum is called **addition**.

$$\overbrace{5}^{\text{Term}} + \overbrace{3}^{\text{Term}} = \overbrace{8}^{\text{Sum}}$$

It is best to memorize basic addition facts involving single digits. View this **addition table** to review these basic facts for digits $0–9$.

To add whole numbers with more than one digit, we add **digits** with corresponding **place values**, working from right to left. That is, we add the digits in the ones place, then add the digits in the tens place, then the digits in the hundreds place, and so on. When adding, it is often helpful to write the numbers vertically and line up the place values.

Example 1 Adding Two Whole Numbers

Add: $217 + 62$

Solution Begin by writing the numbers vertically and lining up the **place values**.

$$\begin{array}{r} 217 \\ +\ \ 62 \\ \hline \end{array}$$

Ones (right digit), Tens (middle digit), Hundreds (left digit)

Next add the **digits** in the ones place, then add the digits in the tens place and add the digits in the hundreds place. Note that there is no hundreds place in 62, so we use a zero (0) as a placeholder and add.

$$\begin{array}{r} 217 \\ +\ \ 62 \\ \hline 279 \end{array}$$

- 7 ones $+$ 2 ones $=$ 9 ones
- 1 ten $+$ 6 tens $=$ 7 tens
- 2 hundreds $+$ 0 hundreds $=$ 2 hundreds

So, $217 + 62 = 279$.

You Try It **Work through this You Try It problem.**

Work Exercises 1–6 in this textbook or in the MyMathLab Study Plan.

Example 2 Adding Two Whole Numbers

My video summary

Add: $5203 + 1781$

Solution Begin by writing the numbers vertically and lining up the **place values**.

$$\begin{array}{r} 5203 \\ +1781 \\ \hline \end{array}$$

Finish this problem on your own. Check your **answer**, or watch this **video** for the complete solution.

You Try It **Work through this You Try It problem.**

Work Exercises 7–11 in this textbook or in the MyMathLab Study Plan.

Recall that each place value can contain only one **digit**, 0–9. If we add digits for a given place value and the result is larger than 9, we must *carry*. **Carrying** means to move a digit from one column of numbers to a column with a higher place value. Consider the following example.

$$\begin{array}{r} 38 \\ +14 \\ \hline \end{array}$$

To **add** these numbers, we start by adding the digits in the ones place. However, $8 + 4 = 12$, which is more than 9. Because $12 = 12$ ones $= 1$ ten $+ 2$ ones, we put a 2 in the ones place of our answer and **carry** the 1 ten to the tens place.

$$\begin{array}{r} {\scriptstyle 1} \;\; \\ 3\,8 \\ +1\,4 \\ \hline 2 \end{array} \quad \begin{array}{l} \leftarrow \text{Carry the 1 ten} \\ \\ \\ \leftarrow \text{2 ones} \end{array}$$

We now add the digits in the tens place. $1 + 3 + 1 = 5$ tens, so we place a 5 in the tens place of our answer.

$$\begin{array}{r} {\scriptstyle 1} \;\; \\ 38 \\ +14 \\ \hline 52 \end{array}$$

Example 3 Adding Two Whole Numbers with Carrying

Add: $256 + 78$

Solution Begin by writing the numbers vertically and lining up the **place values**.

$$\begin{array}{r} 256 \\ +\ 78 \\ \hline \end{array}$$

(eText Screens 1.3-1–1.3-42)

Next add the **digits** in the ones place, then add the digits in the tens place, etc.

$$\begin{array}{r} {}^{1} \\ 256 \\ +\ 78 \\ \hline 4 \end{array}$$

← Carry 1 ten

6 ones + 8 ones = 14 ones
= 1 ten + 4 ones

← 4 ones

Carry 1 hundred →

$$\begin{array}{r} {}^{1\,1} \\ 256 \\ +\ 78 \\ \hline 34 \end{array}$$

1 ten + 5 tens + 7 tens = 13 tens
= 1 hundred + 3 tens

↑ 3 tens

$$\begin{array}{r} {}^{1\,1} \\ 256 \\ +\ 78 \\ \hline 334 \end{array}$$

1 hundred + 2 hundreds = 3 hundreds

↑ 3 hundreds

So, 256 + 78 = 334.

You Try It **Work through this You Try It problem.**

Work Exercises 12–17 in this textbook or in the MyMathLab Study Plan.

Example 4 Adding Two Whole Numbers with Carrying

My video summary

Add: 237,603 + 91,829

Solution Begin by writing the numbers vertically and lining up the **place values**.

$$\begin{array}{r} 237{,}603 \\ +\ 91{,}829 \\ \hline \end{array}$$

Finish this problem on your own. Check your **answer**, or watch this **video** for the complete solution.

You Try It **Work through this You Try It problem.**

Work Exercises 18–22 in this textbook or in the MyMathLab Study Plan.

OBJECTIVE 2 USE SPECIAL PROPERTIES OF ADDITION

To help with adding, we review and name three special properties.

If 0 is added to any given number, then the **sum** is the given number. For example, $5 + 0 = 5$ and $0 + 9 = 9$. This is called the **identity property of addition.**

Identity Property of Addition

For any number, a,

$$a + 0 = a \quad \text{and} \quad 0 + a = a.$$

The **commutative property of addition** states that the *order* of the **terms** in **addition** does not affect the sum. For example, $6 + 9 = 15$ and $9 + 6 = 15$, so $6 + 9 = 9 + 6$.

> **Commutative Property of Addition**
> For any numbers a and b,
> $$a + b = b + a.$$

The **associative property of addition** states that changing the grouping of terms does not affect the sum. For example, $(3 + 5) + 2 = 8 + 2 = 10$ and $3 + (5 + 2) = 3 + 7 = 10$, so $(3 + 5) + 2 = 3 + (5 + 2)$.

> **Associative Property of Addition**
> For any numbers, a, b, and c,
> $$(a + b) + c = a + (b + c).$$

Example 5 Using the Commutative and Associative Properties of Addition

a. Use the **commutative property of addition** to complete the statement.

$4 + 9 =$ ________

b. Use the **associative property of addition** to complete the statement.

$5 + (3 + 6) =$ ________

Solutions

a. The commutative property of addition says that the order of the **terms** does not matter. Therefore, we can write

$$4 + 9 = 9 + 4.$$

b. The associative property of addition says that the grouping of the terms does not matter. Therefore, we can write

$$5 + (3 + 6) = (5 + 3) + 6.$$

CAUTION

Notice that for part b the order of the terms stayed the same because we were only applying the associative property. The associative property involves the *grouping* of terms, not the order.

You Try It **Work through this You Try It problem.**

Work Exercises 23–26 in this textbook or in the MyMathLab Study Plan.

OBJECTIVE 3 ADD SEVERAL WHOLE NUMBERS

We can use the **commutative** and **associative** properties to simplify the addition of several numbers. Together they tell us that when adding several numbers, we can add the numbers in any order or grouping that we prefer. We can follow our usual approach of writing the numbers vertically, lining up **place values**, and adding the digits in each place value.

Sometimes the addition of several numbers can be simplified by looking for **sums** that are multiples of 10 (e.g. 10, 20, 30, etc.). View this **popup** for an illustration.

Example 6 Adding Several Whole Numbers

Add: $167 + 26 + 43$

Solution Begin by writing the numbers vertically and lining up the place values.

$$\begin{array}{r} 167 \\ 26 \\ +\ 43 \\ \hline \end{array}$$

Add the **digits** in the ones place.

$$\begin{array}{r} {\scriptstyle 1} \\ 167 \\ 26 \\ +\ 43 \\ \hline 6 \end{array}$$

1 ← Carry 1 ten

$\underbrace{7 + 6 + 3}_{10} = 10 + 6 = 16$ ones $= 1$ ten $+ 6$ ones

6 ← 6 ones

Add the **digits** in the tens place.

$$\begin{array}{r} {\scriptstyle 11} \\ 167 \\ 26 \\ +\ 43 \\ \hline 36 \end{array}$$

Carry 1 hundred → 11

$1 + \overbrace{6 + 2 + 4}^{10}$ with $\underbrace{1 + 6 + 2}_{}$… : $1 + 6 + 2 + 4 = 10 + 3 = 13$ tens $= 1$ hundred $+ 3$ tens

3 ← 3 tens

Add the digits in the hundreds place.

$$\begin{array}{r} {\scriptstyle 11} \\ 167 \\ 26 \\ +\ 43 \\ \hline 236 \end{array}$$

$1 + 1 = 2$ hundreds

2 ← 2 hundreds

So, $167 + 26 + 43 = 236$.

Example 7 Adding Several Whole Numbers

My video summary

Add: $142 + 83 + 2061 + 17$

Solution Try adding these numbers on your own. Check your **answer**, or watch this **video** for a fully worked solution.

You Try It **Work through this You Try It problem.**

Work Exercises 27–36 in this textbook or in the MyMathLab Study Plan.

(*eText Screens 1.3-1–1.3-42*)

OBJECTIVE 4 SUBTRACT WHOLE NUMBERS

The opposite of **addition** is **subtraction**. When subtracting, we take one number away from another to find the **difference** of the two numbers. The **subtrahend** is the number being subtracted (think *subtra*hend → *subtra*cted) and the **minuend** is the number being subtracted *from*.

If a jogger has 2 energy packs and buys 4 more, she will have a total of $2 + 4 = 6$ energy packs.

If she then runs a marathon and uses 4 energy packs, she will have 2 energy packs left at the end of the race.

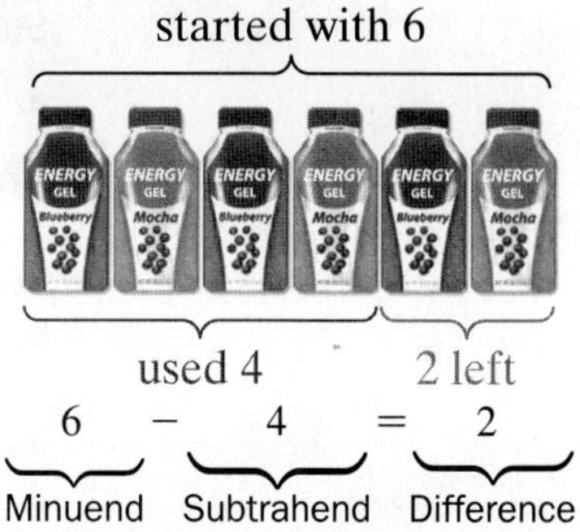

We define **subtraction** in terms of **addition**. That is, we would say $6 - 4 = 2$ because $2 + 4 = 6$. This means we can always check the result of subtraction by adding, as shown below.

Subtraction	Check
Minuend	Difference
− Subtrahend	+ Subtrahend
Difference	Minuend

If adding the **difference** to the **subtrahend** results in the original **minuend**, the result checks.

Example 8 Subtracting Whole Numbers

Subtract. Check by adding.

a. $12 - 8$ **b.** $9 - 6$

Solutions

a. $\underbrace{12}_{\text{Minuend}} - \underbrace{8}_{\text{Subtrahend}} = \underbrace{4}_{\text{Difference}}$ Check: $\underbrace{4}_{\text{Difference}} + \underbrace{8}_{\text{Subtrahend}} = \underbrace{12}_{\text{Minuend}}$

b. $9 - 6 = 3$ Check: $3 + 6 = 9$

You Try It **Work through this You Try It problem.**

Work Exercises 37 and 38 in this textbook or in the MyMathLab Study Plan.

(eText Screens 1.3-1–1.3-42)

Because subtraction can be defined in terms of **addition**, we can use the **identity property of addition** to note two special properties of subtraction.

> **Identity Properties of Subtraction**
> For any number a,
>
> $$a - 0 = a \quad \text{because} \quad a + 0 = a$$
>
> and
>
> $$a - a = 0 \quad \text{because} \quad 0 + a = a.$$

CAUTION Unlike addition, there is no commutative or associative property for subtraction.

Example 9 Using the Identity Properties of Subtraction

Subtract. Check by adding.

a. $10 - 0$ **b.** $14 - 14$

Solutions

a. $10 - 0 = 10$ because $10 + 0 = 10.$

b. $14 - 14 = 0$ because $0 + 14 = 14.$

You Try It **Work through this You Try It problem.**

Work Exercises 39 and 40 in this textbook or in the MyMathLab Study Plan.

As with **addition**, our general approach to **subtraction** is to line up the numbers vertically then begin by subtracting the ones, then subtracting the tens, and so on.

Example 10 Subtracting Whole Numbers

Subtract. Check by adding.

a. $67 - 35$ **b.** $428 - 205$

Solutions

a. Begin by writing the numbers vertically and lining up the **place values**.

$$\begin{array}{r} 67 \\ \underline{-35} \end{array} \begin{array}{l} \leftarrow \text{Minuend} \\ \leftarrow \text{Subtrahend} \end{array}$$

Next subtract the **digits** in the ones place, then subtract the digits in the tens place. We can check by adding.

$$\begin{array}{r} 67 \\ \underline{-35} \\ 32 \end{array} \begin{array}{l} \leftarrow \text{Minuend} \\ \leftarrow \text{Subtrahend} \\ \leftarrow \text{Difference} \end{array}$$

7 ones − 5 ones = 2 ones
6 tens − 3 tens = 3 tens

Check:

$$\begin{array}{r} 32 \\ \underline{+35} \\ 67 \end{array} \begin{array}{l} \leftarrow \text{Difference} \\ \leftarrow \text{Subtrahend} \\ \leftarrow \text{Minuend} \end{array}$$

The result checks, so $67 - 35 = 32$.

My video summary

b. Try working this problem on your own. Check your **answer**, or watch this **video** for a fully worked solution to part b.

You Try It **Work through this You Try It problem.**

Work Exercises 41–45 in this textbook or in the MyMathLab Study Plan.

Sometimes **subtraction** in a **place value** is not possible because the number being subtracted (**subtrahend**) is larger than the number it is being subtracted from (**minuend**). Consider the following.

$$\begin{array}{r} 52 \\ -38 \\ \hline \end{array}$$

We cannot subtract in the ones column, $2 - 8$, because 8 is more than 2. However, we can **borrow** from the tens column in order to do the subtraction. Because 1 ten $= 10$ ones, we can subtract 1 (or borrow) from the tens and increase the ones by 10.

1 ten = 10 ones

Borrow 1 from the tens. (5 tens − 1 ten = 4 tens) → $\begin{array}{r} \overset{4}{\cancel{5}}\,\overset{12}{\cancel{2}} \\ -3\,8 \\ \hline \end{array}$ ← Increase the ones by 10. (2 ones + 10 ones = 12 ones)

We can now subtract $12 - 8 = 4$ in the ones column and subtract $4 - 3 = 1$ in the tens column.

$$\begin{array}{r} \overset{4}{\cancel{5}}\,\overset{12}{\cancel{2}} \\ -3\,8 \\ \hline 1\,4 \end{array} \quad \text{Original minuend} \qquad \text{Check} \quad \begin{array}{r} \overset{1}{1}\,4 \\ +3\,8 \\ \hline 5\,2 \end{array}$$

The result of the check is the original **minuend**, so $52 - 38 = 14$.

Example 11 Subtracting Whole Numbers with Borrowing

My interactive video summary

Subtract. Check by adding.

a. $234 - 68$ **b.** $500 - 269$ **c.** $3270 - 1433$

Solutions

a. Begin by subtracting the ones.

$$\begin{array}{r} 234 \\ -\ 68 \\ \hline \end{array} \xrightarrow{\text{Borrow from the tens}} \begin{array}{r} 2\,\overset{2}{\cancel{3}}\,\overset{14}{\cancel{4}} \\ -\ \ 6\,8 \\ \hline 6 \end{array}$$

Next subtract the tens.

$$\begin{array}{r} 2\,\overset{2}{\cancel{3}}\,\overset{14}{\cancel{4}} \\ -\ \ 6\,8 \\ \hline 6 \end{array} \xrightarrow{\text{Borrow from the hundreds}} \begin{array}{r} \overset{1}{\cancel{2}}\,\overset{\overset{12}{\cancel{2}}}{\cancel{3}}\,\overset{14}{\cancel{4}} \\ -\ \ 6\,8 \\ \hline 6\,6 \end{array}$$

Lastly, subtract the hundreds.

$$\begin{array}{r} \overset{1}{\cancel{2}}\,\overset{\overset{12}{\cancel{2}}}{\cancel{3}}\,\overset{14}{\cancel{4}} \\ -\;\;6\,8 \\ \hline 1\,6\,6 \end{array}$$

So, $234 - 68 = 166$. View the **check**.

b. Begin by subtracting the ones. Notice that we will need to **borrow** in order to do this **subtraction**. However, there are no tens in the **minuend** to borrow from. Therefore, we first borrow 1 hundred and increase the tens by 10.

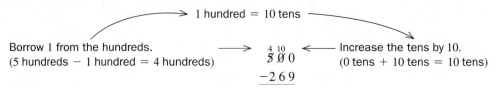

1 hundred = 10 tens

Borrow 1 from the hundreds. (5 hundreds − 1 hundred = 4 hundreds) → $\begin{array}{r} \overset{4}{\cancel{5}}\,\overset{10}{\cancel{0}}\,0 \\ -2\,6\,9 \\ \hline \end{array}$ ← Increase the tens by 10. (0 tens + 10 tens = 10 tens)

Then we can **borrow** 1 ten and increase the ones by 10.

1 ten = 10 ones

Borrow 1 from the tens. (10 tens − 1 ten = 9 tens) → $\begin{array}{r} \overset{4}{\cancel{5}}\,\overset{\overset{9}{\cancel{10}}}{\cancel{0}}\,\overset{10}{\cancel{0}} \\ -2\,6\,9 \\ \hline \end{array}$ ← Increase the ones by 10. (0 ones + 10 ones = 10 ones)

Finish working this problem on your own. View this **popup** to compare your answer, or see the complete solution in this **interactive video**.

c. Try to work this problem on your own. Check your **answer**, or watch this **interactive video** for fully worked solutions to all three parts.

You Try It **Work through this You Try It problem.**

Work Exercises 46–60 in this textbook or in the MyMathLab Study Plan.

OBJECTIVE 5 ESTIMATE SUMS AND DIFFERENCES OF WHOLE NUMBERS

Sometimes we want to estimate a **sum** or **difference** so that we can check if an answer is reasonable or because we need only an approximate answer. Before continuing with this objective, you may want to review rounding whole numbers in **Topic 1.2**.

To estimate a sum or difference, first **round** each number to the appropriate **place value** and then carry out the desired operation.

Example 12 Estimating Sums of Whole Numbers

Round each value to the nearest ten. Then estimate the sum.

$42 + 379 + 7 + 51$

Solution

$$\begin{array}{r} 42 \\ 379 \\ 7 \\ +\;51 \\ \hline \end{array} \xrightarrow{\text{Round to nearest 10}} \begin{array}{r} 40 \\ 380 \\ 10 \\ +\;50 \\ \hline 480 \end{array} \leftarrow \text{Estimated sum}$$

The estimated sum is 480. This is close to the actual sum, which is 479.

(eText Screens 1.3-1–1.3-42)

You Try It **Work through this You Try It problem.**

Work Exercises 61, 63, and 65 in this textbook or in the MyMathLab Study Plan.

Example 13 Estimating Differences of Whole Numbers

My video summary

Round each value to the nearest thousand, then estimate the **difference**.

$$\begin{array}{r} 4207 \\ -\ 2869 \\ \hline \end{array}$$

Solution Work through the following, or watch this **video** for the solution.

$$\begin{array}{r} 4207 \\ -\ 2869 \\ \hline \end{array} \xrightarrow{\text{Round to nearest 1000}} \begin{array}{r} 4000 \\ -\ 3000 \\ \hline 1000 \end{array} \leftarrow \text{Estimated difference}$$

The estimated difference is 1000. The actual difference is 1338.

You Try It **Work through this You Try It problem.**

Work Exercises 62, 64, and 66 in this textbook or in the MyMathLab Study Plan.

OBJECTIVE 6 FIND THE PERIMETER OF A POLYGON

A **polygon** is a closed figure made up of connected line segments. Each line segment is called a **side** of the polygon. The following are examples of polygons.

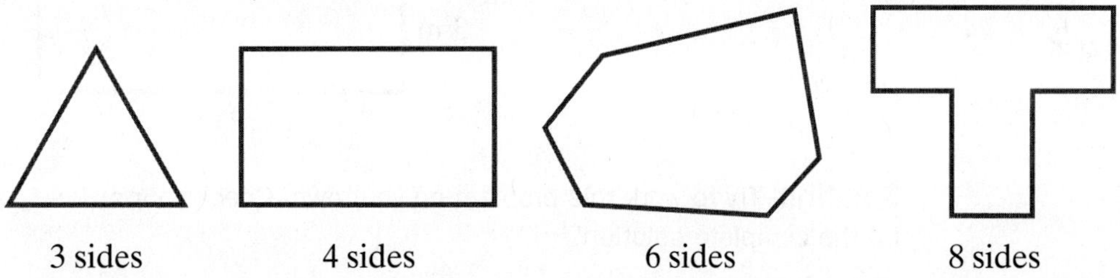

Figure 8 Polygons

The **perimeter** of a polygon is the distance around that polygon. Perimeter is measured in units of length or distance such as inches (in.), feet (ft), yards (yd), miles (mi), millimeters (mm), centimeters (cm), and kilometers (km). We find the perimeter by **adding** the lengths of the sides that make up the polygon.

Example 14 Finding the Perimeter of a Polygon

Find the **perimeter** of each **polygon**.

a.

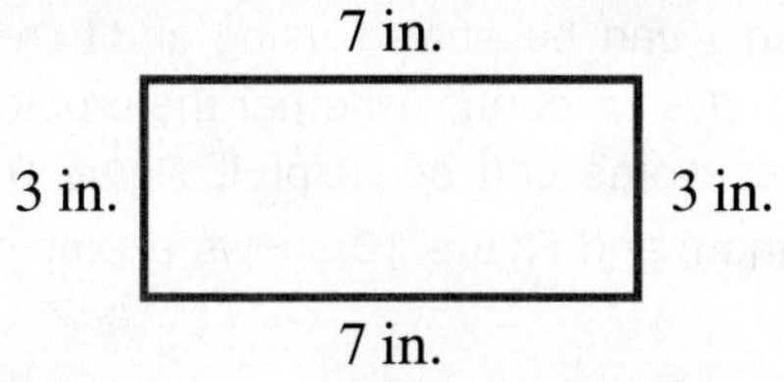

b.

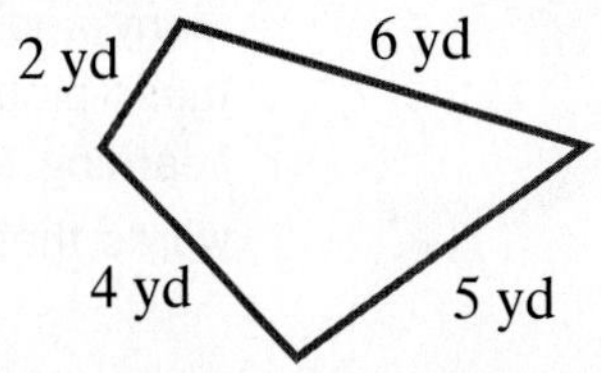

(eText Screens 1.3-1–1.3-42)

Solutions

a. We find the **perimeter** by adding up the lengths of the **sides** that make up the **polygon**.

$$\underbrace{3\text{ in.} + 7\text{ in.}}_{10\text{ in.}} + \underbrace{3\text{ in.} + 7\text{ in.}}_{10\text{ in.}} = 20\text{ in.}$$

The perimeter is 20 inches.

b. Add the lengths of the sides of the polygon. We can leave off the units when performing calculations, but then we must remember to write the units with the final answer.

$$2 + 6 + 4 + 5 = 17$$

The lengths of the sides are in yards, so the perimeter is 17 yards.

You Try It **Work through this You Try It problem.**

Work Exercises 67–70 in this textbook or in the MyMathLab Study Plan.

CAUTION Because perimeter is a length (or distance), be sure to include units when giving the perimeter.

Example 15 Finding the Perimeter of a Polygon

My video summary

Find the **perimeter** of the **polygon**.

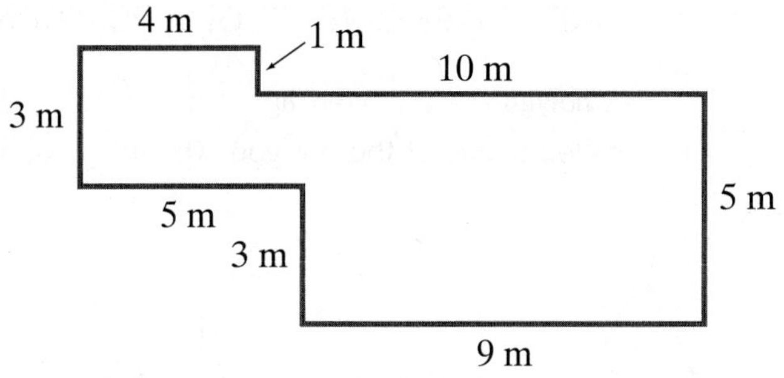

Solution Try to work this problem on your own. Check your **answer**, or watch this **video** for the complete solution.

You Try It **Work through this You Try It problem.**

Work Exercises 71 and 72 in this textbook or in the MyMathLab Study Plan.

OBJECTIVE 7 SOLVE APPLICATIONS BY ADDING OR SUBTRACTING WHOLE NUMBERS

Many application problems can be solved using **addition** and **subtraction**. When solving these problems, we first have to identify whether the problem involves a **sum** or **difference**. Learning to recognize key words can be helpful. **Figure 9** shows some examples of key words that indicate addition, and **Figure 10** shows examples for subtraction.

Key Word	Word Statement	Mathematical Expression
Sum	The *sum* of 2 and 6	$2 + 6$
Increased by	3 *increased by* 5	$3 + 5$
Added to	11 *added to* 8	$8 + 11$
More than	7 *more than* 15	$15 + 7$

Figure 9
Key words Meaning Addition

Key Word	Word Statement	Mathematical Expression
Difference	The *difference* of 12 and 3	$12 - 3$
Decreased by	15 *decreased by* 5	$15 - 5$
Subtracted from	7 *subtracted from* 20	$20 - 7$
Less than	9 *less than* 34	$34 - 9$

Figure 10
Key words Meaning Subtraction

Example 16 Translating Phrases Involving Addition or Subtraction

Translate to an addition or subtraction problem, then simplify.

a. Find 27 subtracted from 42.

b. Find 35 more than 19.

Solutions

a. The key words "subtracted from" mean **subtraction.** Remember that the order of the numbers matters with subtraction. What is being subtracted? 27 is being subtracted *from* 42. This means we can write

$$\underbrace{42}_{\text{Subtract } \textit{from} \text{ this value.}} - \underbrace{27}_{\text{This is what is subtracted.}} = 15.$$

View this **popup** for the subtraction details.

b. The key words "more than" mean addition. What is being added? 35 and 19.

$$35 + 19 = 54$$

View this **popup** for the addition details.

You Try It **Work through this You Try It problem.**

Work Exercises 73–76 in this textbook or in the MyMathLab Study Plan.

Once we identify the type of problem, we translate the word statement into a mathematical statement and carry out the **addition** or **subtraction**. Remember that the order of the numbers does not matter with addition but does matter with subtraction.

Example 17 Stock Market

On July 6, 2010, the Dow Jones Industrial Average closed at 9744 points. One year later, on July 6, 2011, it closed at 12,626 points. (*Source:* money.cnn.com)

a. **Round** each closing amount to the nearest thousand and estimate the difference between these amounts.

b. Find the exact **difference** between these closing amounts.

Solutions The key word ***difference*** means we want to subtract.

a.
$$\begin{array}{r} 12{,}626 \\ -\ \ 9\,744 \\ \hline \end{array} \xrightarrow{\text{Round to nearest thousand}} \begin{array}{r} 13{,}000 \\ -\ 10{,}000 \\ \hline 3\,000 \end{array}$$

The estimated difference in closing amounts is 3000 points.

b.
$$\begin{array}{r} 12{,}626 \\ -\ \ 9\,744 \\ \hline \end{array} \qquad \begin{array}{r} \scriptstyle 11\ 15 \\ \scriptstyle 0\ \not{1}\ \not{5}\ 12 \\ \not{1}\,\not{2},\not{6}\,\not{2}\,6 \\ -\ \ \ 9\,7\,4\,4 \\ \hline 2\,8\,8\,2 \end{array}$$

The exact difference between the amounts is 2882 points.

You Try It **Work through this You Try It problem.**

Work Exercises 77–80 in this textbook or in the MyMathLab Study Plan.

Example 18 Fencing in Horses

My video summary

John wants to fence off part of his land so his horses can graze. How much fencing will he need?

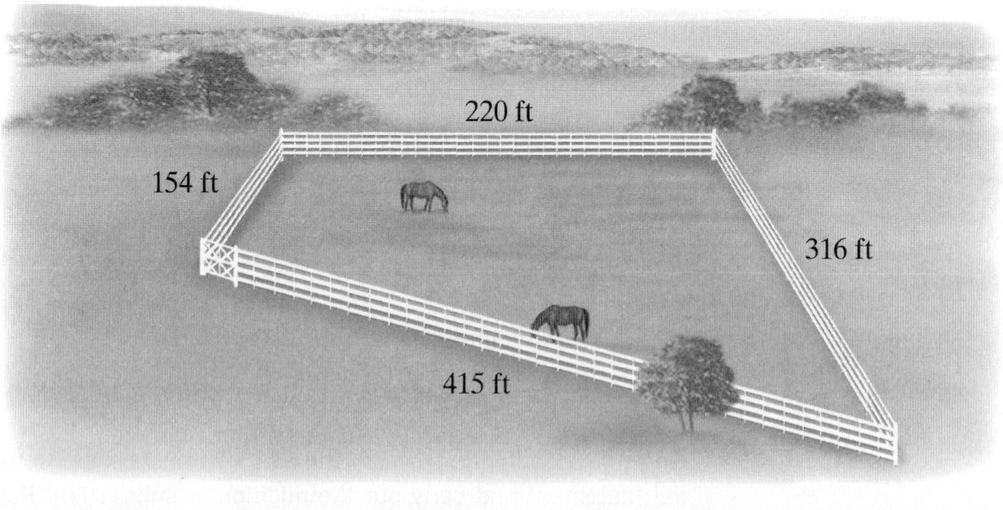

a. **Round** each value to the nearest hundred to estimate the total fencing needed.

b. Find the exact amount of fencing needed.

Solutions Because the fencing goes around the land being fenced in, we are finding the **perimeter**. Add up the lengths of all the sides to find the perimeter.

Try to solve this problem on your own. Check your **answer**, or watch this **video** for a complete solution to both parts.

You Try It **Work through this You Try It problem.**

Work Exercises 81 and 82 in this textbook or in the MyMathLab Study Plan.

1.3 Exercises

In Exercises 1–24, add the numbers.

You Try It

1. $9 + 3$
2. $14 + 5$
3. $22 + 37$
4. $\begin{array}{r} 406 \\ +\ 81 \\ \hline \end{array}$
5. $\begin{array}{r} 567 \\ +\ 200 \\ \hline \end{array}$
6. $\begin{array}{r} 532 \\ +\ 114 \\ \hline \end{array}$

You Try It

7. $\begin{array}{r} 6713 \\ +\ 225 \\ \hline \end{array}$
8. $\begin{array}{r} 1261 \\ +\ 3024 \\ \hline \end{array}$
9. $\begin{array}{r} 10{,}305 \\ +\ 2\,004 \\ \hline \end{array}$
10. $82{,}166 + 3701$
11. $14{,}296 + 43{,}502$
12. $8 + 9$
13. $24 + 7$
14. $68 + 25$
15. $\begin{array}{r} 575 \\ +\ 77 \\ \hline \end{array}$
16. $\begin{array}{r} 388 \\ +\ 422 \\ \hline \end{array}$
17. $\begin{array}{r} 893 \\ +\ 236 \\ \hline \end{array}$

You Try It

18. $\begin{array}{r} 4806 \\ +\ 837 \\ \hline \end{array}$
19. $\begin{array}{r} 2993 \\ +\ 4532 \\ \hline \end{array}$
20. $\begin{array}{r} 18{,}729 \\ +\ 4\,265 \\ \hline \end{array}$
21. $94{,}218 + 5307$
22. $80{,}504 + 17{,}199$
23. Use the *commutative property of addition* to rewrite the statement: $14 + 10 =$ __________
24. Use the *associative property of addition* to rewrite the statement: $7 + (3 + 15) =$ ______________
25. Use the *commutative property of addition* to rewrite the statement: $(2 + 16) + 8 =$ __________
26. Use the *associative property of addition* to rewrite the statement by inserting grouping symbols: $22 + 7 + 15 =$ ____________

In Exercises 27–36, add the numbers.

You Try It

27. $4 + 12 + 6$
28. $18 + 7 + 2 + 13$
29. $5 + 4 + 3 + 9 + 6 + 5$
30. $26 + 12 + 15 + 37$
31. $6 + 47 + 32 + 14 + 13$
32. $151 + 25 + 19$

33. 247 + 158 + 313   34. 4219 + 53 + 321 + 207

35. 6231 + 5922 + 316 + 587   36. 8644 + 7135 + 2015 + 4261

In Exercises 37–60, subtract. Check by adding.

You Try It 37. 8 − 5   38. 19 − 7   39. 19 − 0

You Try It 40. 32 − 32   41. 65 − 32   42. 59 − 20

43. 246 − 25   44. 784 − 530   45. 492 − 131

46. 34 − 19   47. 46 − 38   48. 708 − 25

You Try It 49. 352 − 27   50. 537 − 59   51. 134 − 36

52. 448 − 162   53. 600 − 143   54. 974 − 254

55. 831 − 127   56. 4179 − 343   57. 7258 − 1306

58. 24,719 − 12,455   59. 56,422 − 38,106   60. 40,000 − 6727

In Exercises 61 and 62, estimate the sum or difference by rounding each number to the nearest ten.

You Try It 61. 43 + 38 + 22 + 59   62. 98 − 53

In Exercises 63 and 64, estimate the sum or difference by rounding each number to the nearest hundred.

You Try It 63. 2637 + 452 + 3520   64. 5719 − 481

In Exercises 65 and 66, estimate the sum or difference by rounding each number to the nearest thousand.

65. 43,290 + 250,637 + 95,445   66. 734,012 − 238,929

In Exercises 67–72, find the perimeter of each polygon.

You Try It 67.

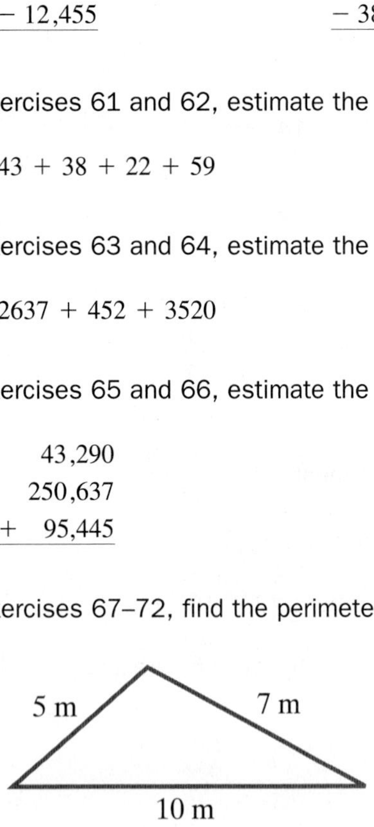

68.

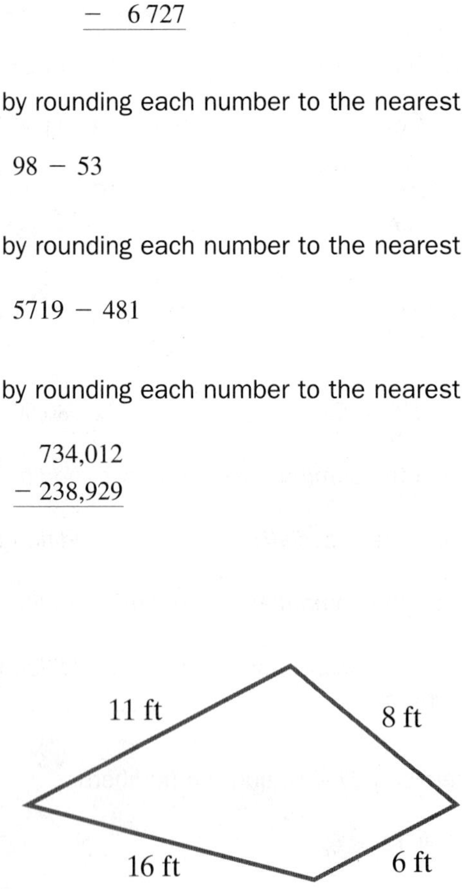

69.
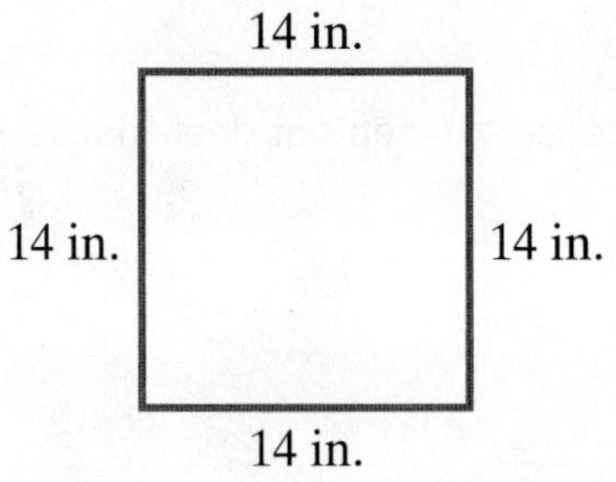

70.
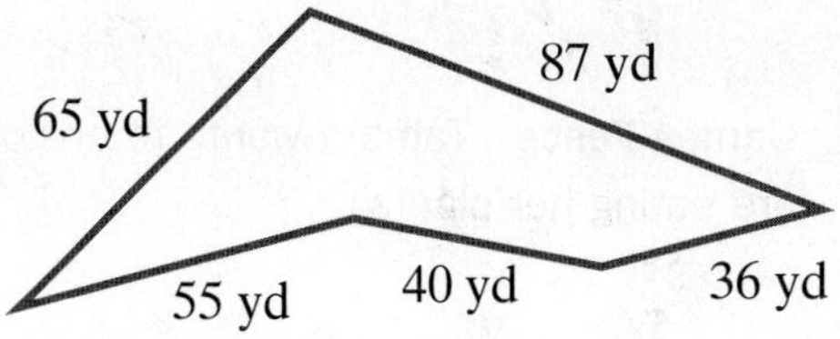

You Try It 71.
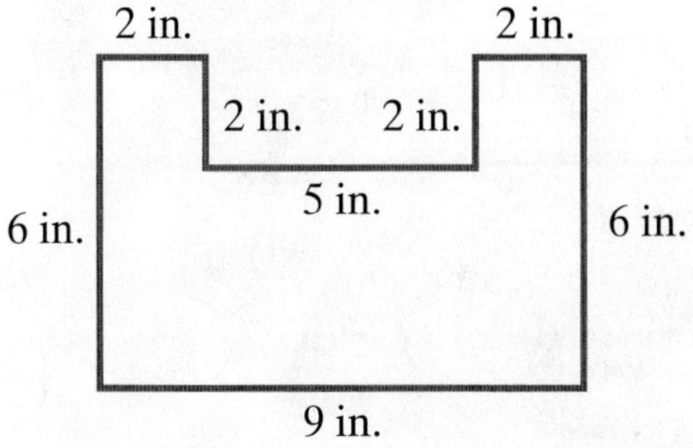

72.
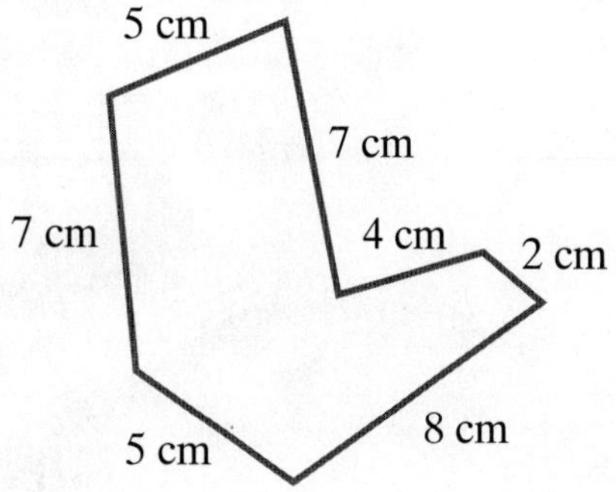

You Try It 73. Find the sum of 126 and 77.

74. Find 53 subtracted from 286.

75. Find the difference of 372 and 181.

76. Find the total of 784 and 605.

You Try It 77. **Game Show** A family competing on the game show *Family Game Night* collects cash cards worth \$575, \$1120, \$380, and \$765. Find the total value of their cash cards.

a. Estimate the total value by rounding each number to the nearest hundred and then finding the sum.
b. Find the exact total value of the cash cards.

78. **Population** The population of a city was 293,953 in the 2000 census and rose to 567,582 in the 2010 census. Find the difference between the two counts to determine the population increase.

a. Estimate the population increase by rounding the counts to the nearest thousand and then finding the difference.
b. Find the exact population increase.

79. **Video Stores** On January 1, 2010, a video rental company operated 3,525 stores. The following year on January 1, 2011, the company operated 1,716 stores. Find the decrease in the number of stores operated by the rental company.

a. Estimate the decrease by rounding each number to the nearest thousand and then finding the difference.
b. Find the exact decrease in the number of stores operated by the company.

80. **Seating Room** A pizzeria can seat 165 customers in its main dining room. There are also two banquet rooms than can hold 85 and 110 customers. What is the total capacity of the pizzeria?

a. Estimate the total capacity by rounding each number to the nearest ten and then finding the sum.
b. Find the exact total capacity of the pizzeria.

You Try It 81. **Garden Fence** Tamara wants to enclose her garden with a fence to keep out deer that are eating her plants.

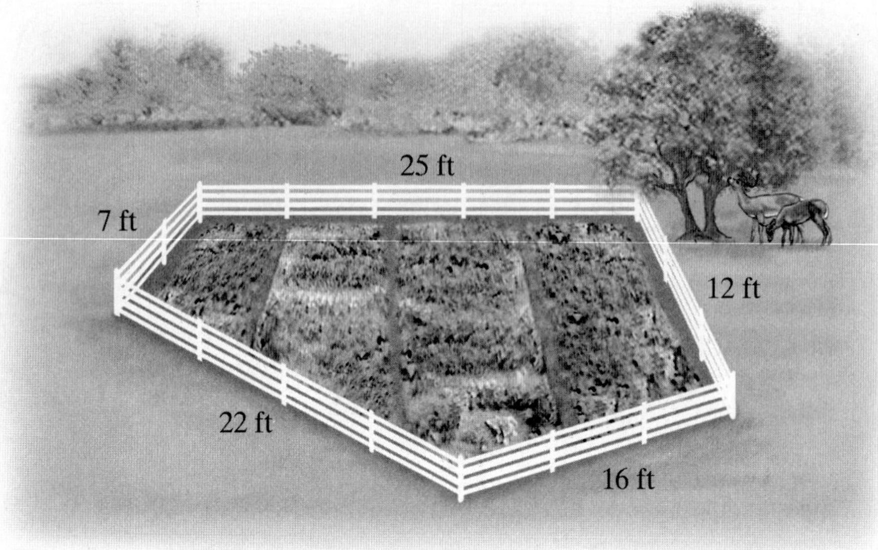

a. Round each value to the nearest ten to estimate the total fencing needed.
b. Find the exact amount of fencing needed.

82. **Crowd Control** Security personnel at a music venue need to put up a barricade around the stage to keep back the crowd. Find the total length of the barricade.

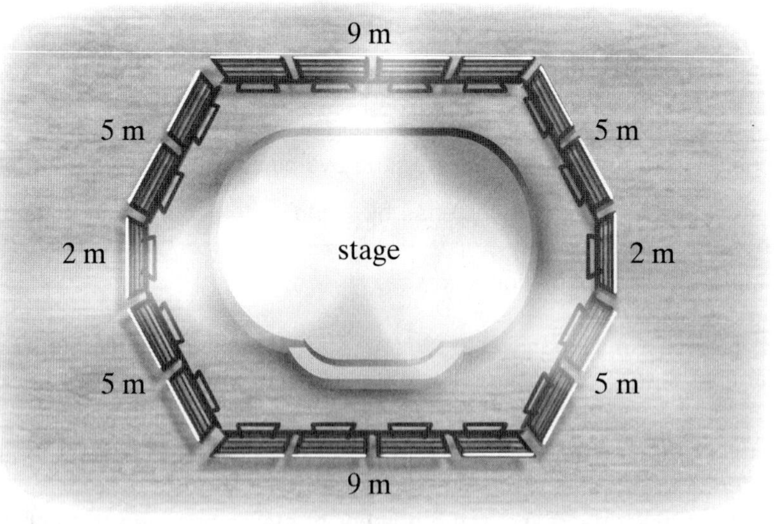

1.4 Multiplying Whole Numbers; Area

THINGS TO KNOW

Before working through this topic, be sure you are familiar with the following concepts:

VIDEO ANIMATION INTERACTIVE

You Try It
1. Round Whole Numbers (Topic 1.2, **Objective 5**)

You Try It
2. Add Whole Numbers (Topic 1.3, **Objective 1**)

You Try It
3. Estimate Sums and Differences of Whole Numbers (Topic 1.3, **Objective 5**)

You Try It
4. Find the Perimeter of a Polygon (Topic 1.3, **Objective 6**)

You Try It
5. Solve Applications by Adding or Subtracting Whole Numbers (Topic 1.3, **Objective 7**)

OBJECTIVES

1 Use Special Properties of Multiplication

2 Use the Distributive Property

3 Multiply Whole Numbers

4 Estimate Products of Whole Numbers

5 Find the Area of a Rectangle

6 Solve Applications by Multiplying Whole Numbers

OBJECTIVE 1 USE SPECIAL PROPERTIES OF MULTIPLICATION

Hannah purchases 4 six-packs of cola. We can find the total number of colas she purchased by adding $6 + 6 + 6 + 6 = 24$ colas. Notice that the **terms** in the **addition** are all the same. We can simplify this **repeated addition** by introducing **multiplication**, which is another way of writing repeated addition.

(*eText Screens 1.4-1–1.4-32*)

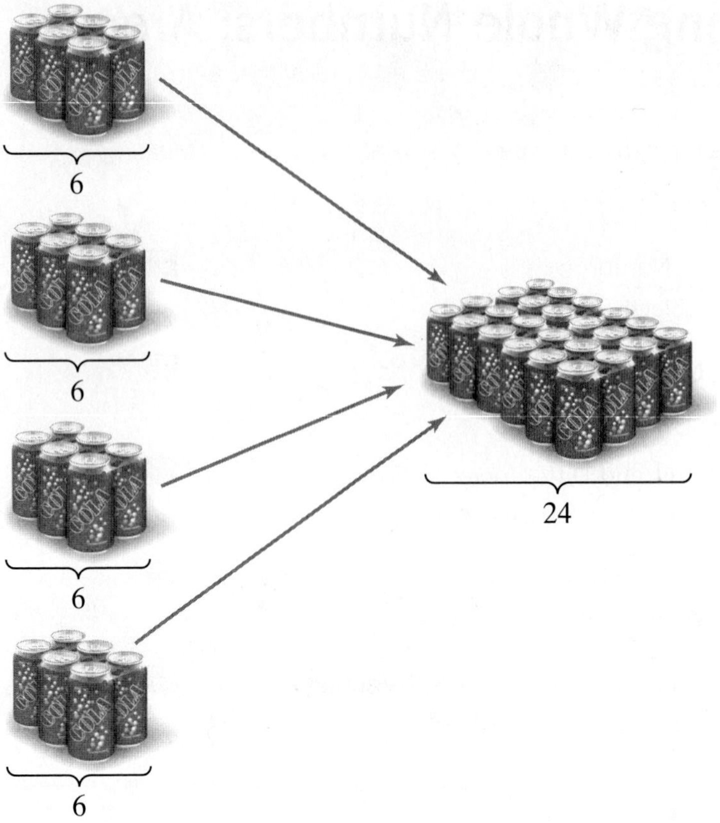

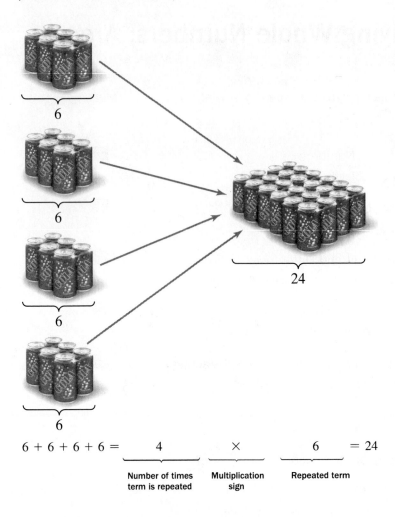

$$6 + 6 + 6 + 6 = \underbrace{4}_{\text{Number of times term is repeated}} \underbrace{\times}_{\text{Multiplication sign}} \underbrace{6}_{\text{Repeated term}} = 24$$

The numbers 4 and 6 are called **factors**. The result of the multiplication, 24, is called the **product**. We read 4×6 as "four times six," which means we are adding 6 four times. The product of two **whole number** factors is called a **multiple** of each factor. For example, 24 is a multiple of 4 and 6 because $4 \times 6 = 24$.

In our example, we used a cross, $\times$, to indicate **multiplication**. We could also have used a raised dot, $\cdot$, or parentheses, ( ). Although less common, an asterisk,*, is sometimes used for multiplication (e.g., in some calculator displays).

$$4 \times 6 = 24 \text{ or } 4 \cdot 6 = 24 \text{ or } 4(6) = 24 \text{ or } (4)(6) = 24 \text{ or } 4 * 6 = 24$$

Just like with **addition**, it is important to memorize basic multiplication facts involving single **digits**. View this **multiplication table** to review these basic facts.

Before we move on to multiplying **whole numbers** with more than one digit, we should review and name several properties of multiplication.

The **commutative property of multiplication** states that the *order* of the **factors** in multiplication does not affect the **product**. For example, $4 \cdot 6 = 24$ and $6 \cdot 4 = 24$, so $4 \cdot 6 = 6 \cdot 4$.

Commutative Property of Multiplication

For any numbers a and b,

$$a \cdot b = b \cdot a.$$

(eText Screens 1.4-1–1.4-32)

The **associative property of multiplication** states that changing the grouping of factors does not affect the product. For example, $(3 \cdot 2) \cdot 4 = 6 \cdot 4 = 24$ and $3 \cdot (2 \cdot 4) = 3 \cdot 8 = 24$, so $(3 \cdot 2) \cdot 4 = 3 \cdot (2 \cdot 4)$.

Associative Property of Multiplication

For any numbers, a, b, and c,

$$(a \cdot b) \cdot c = a \cdot (b \cdot c).$$

Example 1 Using the Commutative and Associative Properties of Multiplication

Use the given property to rewrite each statement.

a. Associative property of multiplication: $(4 \cdot 7) \cdot 9 =$ ______

b. **Commutative property of multiplication:** $7 \cdot 3 =$ ______

Solutions

a. Using the associative property of multiplication, we can change the grouping of the **factors**: $(4 \cdot 7) \cdot 9 = 4 \cdot (7 \cdot 9)$. Notice that we only changed the grouping, not the order.

b. The commutative property of multiplication allows us to change the order of the factors: $7 \cdot 3 = 3 \cdot 7$.

You Try It **Work through this You Try It problem.**

Work Exercises 1 and 2 in this textbook or in the MyMathLab Study Plan.

The **product** of any number and 0 is equal to 0. For example, $0 \cdot 3 = 0$ and $7 \cdot 0 = 0$. This is called the **multiplication property of 0**.

Multiplication Property of Zero

For any number, a,

$$0 \cdot a = 0 \quad \text{and} \quad a \cdot 0 = 0.$$

If 1 is multiplied by any number, then the product is the original number. For example, $7 \cdot 1 = 7$ and $1 \cdot 4 = 4$. This is called the **identity property of multiplication** or **multiplicative identity property**.

Identity Property of Multiplication

For any number, a,

$$a \cdot 1 = a \quad \text{and} \quad 1 \cdot a = a.$$

Example 2 Using Multiplication Properties

Multiply.

a. $46 \cdot 0 =$ ______

b. $1 \cdot 18 =$ ______

(eText Screens 1.4-1–1.4-32)

Solutions

a. Using the multiplication property of zero, the product of any number and 0 is equal to 0. So, $46 \cdot 0 = 0$.

b. Using the **identity property of multiplication**, the product of any number and 1 is equal to that number. So, $1 \cdot 18 = 18$.

You Try It **Work through this You Try It problem.**

Work Exercises 3 and 4 in this textbook or in the MyMathLab Study Plan.

OBJECTIVE 2 USE THE DISTRIBUTIVE PROPERTY

Another important property is the **distributive property**, which states that **multiplication** *distributes* over **addition** (or **subtraction**). For example, $4(2 + 5) = 4 \cdot 2 + 4 \cdot 5 = 8 + 20 = 28$.

> **Distributive Property**
>
> For any numbers a, b, and c,
>
> $$a \cdot (b + c) = a \cdot b + a \cdot c \quad \text{or} \quad a \cdot (b - c) = a \cdot b - a \cdot c.$$

Because multiplication is **commutative**, we can also write the distributive property as

$$(b + c) \cdot a = b \cdot a + c \cdot a.$$

A similar result is also true for subtraction. This means that multiplying a number by a sum or difference is the same as first multiplying each **term** by the number and then adding or subtracting the results.

CAUTION

Be careful not to confuse the distributive property with the **associative property of multiplication**. We can distribute over a sum but not a product. That is,

$$a \cdot (b \cdot c) \neq (a \cdot b) \cdot (a \cdot c).$$

Example 3 Using the Distributive Property

My video summary

Rewrite each statement using the **distributive property**.

a. $4(6 + 12)$ **b.** $(8 - 3)5$

Solutions Work through the following, or watch this **video** for detailed solutions to both parts.

a. The distributive property tells us that multiplying a number by a **sum** is the same as first **multiplying** each term by the number and then summing the results. So, we can rewrite the statement $4(6 + 12)$ by multiplying each term in the parentheses by 4.

$$4(6 + 12) = 4 \cdot 6 + 4 \cdot 12$$

b. Rewrite the statement $(8 - 3)5$ by multiplying each term in the parentheses by 5 and then subtracting the results.

$$(8 - 3)5 = 8 \cdot 5 - 3 \cdot 5$$

You Try It **Work through this You Try It problem.**

Work Exercises 5–8 in this textbook or in the MyMathLab Study Plan.

OBJECTIVE 3 MULTIPLY WHOLE NUMBERS

We can use the **distributive property** to multiply numbers with more than one **digit**. For example,

$$4(17) = \underbrace{4(10 + 7)}_{\text{Expand}} = \underbrace{4 \cdot 10 + 4 \cdot 7}_{\text{Distributive property}} = 40 + 28 = 68$$

Here we applied the distributive property after **expanding** 17. We can use a similar approach when multiplying vertically.

We can start by multiplying the 7 ones by 4.

$$\begin{array}{r} 17 \\ \times\ 4 \\ \hline 28 \end{array} \qquad \begin{aligned} 4(7 \text{ ones}) &= 28 \text{ ones} \\ &= 2 \text{ tens} + 8 \text{ ones} \end{aligned}$$

Notice that we put a 2 in the tens place and an 8 in the ones place because the product, 28, contains 2 tens and 8 ones. Next we multiply the 1 ten by 4.

$$\begin{array}{r} 17 \\ \times\ 4 \\ \hline 28 \\ 4 \\ \hline \end{array} \qquad 4(1 \text{ ten}) = 4 \text{ tens}$$

4 ← Put a 4 in the tens place.

$$\begin{array}{r} 17 \\ \times\ 4 \\ \hline 28 \\ 40 \\ \hline \end{array}$$

40 ← We could also have included a 0 in the ones place.

Adding the two **products** gives us the same result we obtained earlier.

$$\begin{array}{r} 17 \\ \times\ 4 \\ \hline 28 \\ 40 \\ \hline 68 \end{array}$$

We can make this process more compact, as shown below, so that only one line is needed.

$$\begin{array}{r} {\scriptstyle 2} \\ 17 \\ \times\ 4 \\ \hline 8 \end{array} \qquad \begin{aligned} 4(7 \text{ ones}) &= 28 \text{ ones} \\ &= 2 \text{ tens} + 8 \text{ ones} \end{aligned}$$

Carry 2 tens

8 ← 8 ones

$$\begin{array}{r} 17 \\ \times\ 4 \\ \hline 68 \end{array} \qquad 4(1 \text{ ten}) = \text{④ tens}$$

② +

2 tens + 4 tens = 6 tens

My video summary Watch this **video** for a detailed explanation.

(eText Screens 1.4-1–1.4-32)

Example 4 Multiplying Whole Numbers

Multiply:

a. 243×8 **b.** 1472×6

Solutions

a.

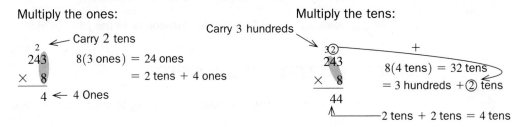

Multiply the ones:

$$\begin{array}{r} {}^{2} \\ 243 \\ \times 8 \\ \hline 4 \end{array}$$

Carry 2 tens

8(3 ones) = 24 ones
= 2 tens + 4 ones

4 Ones

Multiply the tens:

Carry 3 hundreds

$$\begin{array}{r} {}^{3\,2} \\ 243 \\ \times 8 \\ \hline 44 \end{array}$$

8(4 tens) = 32 tens
= 3 hundreds + ② tens

2 tens + 2 tens = 4 tens

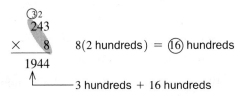

Multiply the hundreds:

$$\begin{array}{r} {}^{3\,2} \\ 243 \\ \times 8 \\ \hline 1944 \end{array}$$

8(2 hundreds) = ⑯ hundreds

3 hundreds + 16 hundreds

So, $243 \times 8 = 1944$.

My video summary

b. Try working this problem on your own. Check your **answer**, or watch this **video** for a fully worked solution to part b.

You Try It **Work through this You Try It problem.**

Work Exercises 9–12 in this textbook or in the MyMathLab Study Plan.

In the previous example, we multiplied by a single digit and were able to write the answer in a single line. If we multiply by a number with more than one digit, we will need a line for each digit along with a line to **sum** the results for each digit. Example 5 illustrates this idea.

Example 5 Multiplying Whole Numbers

Multiply:

a. $\begin{array}{r} 78 \\ \times 34 \\ \hline \end{array}$ **b.** $\begin{array}{r} 254 \\ \times 429 \\ \hline \end{array}$ **c.** $\begin{array}{r} 3402 \\ \times 326 \\ \hline \end{array}$

Solutions

a. Begin by multiplying 78×4. Because 4 is in the ones **place**, we write the result starting in the ones place. 8 times 4 is 32. We write the 2 in the ones place and carry the 3. 7 times 4 is 28 plus the 3 carried equals 31. So our first line in the solution should be 312.

$$\begin{array}{r} {}^{3} \\ 78 \\ \times\ 34 \\ \hline 312 \end{array} \qquad 78 \times 4 = 312$$

Next we multiply 78×3. Because 3 is in the tens place, we put a 0 in the ones place and write the result starting in the tens place. 8 times 3 is 24. We write the 4 in the tens place and carry the 2. 7 times 3 is 21 plus the carried 2 is 23. The second line of the solution should be 2340.

$$\begin{array}{r} {}^{2} \\ 78 \\ \times\ 34 \\ \hline 312 \\ 2340 \end{array}$$

$78 \times 3 = 234$ — Put a 0 in the ones place.

Add the results by adding the two solution lines.

$$\begin{array}{r} 78 \\ \times\ 34 \\ \hline 312 \\ 2340 \\ \hline 2652 \end{array} \quad 2340 + 312 = 2652$$

So, $78 \times 34 = 2652$.

b. Begin by **multiplying** 254×9. Because 9 is in the ones **place**, we write the result starting in the ones place. 9×4 is 36. Write the 6 in the ones place and carry the 3. Next, 5×9 is 45 plus the carried 3 is 48. We write the 8 in the tens place and carry the 4. 2×9 is 18 plus the carried 4 is 22. We write 2286 in the first solution line.

$$\begin{array}{r} {}^{4\,3} \\ 254 \\ \times\ 429 \\ \hline 2286 \end{array} \quad 254 \times 9 = 2286$$

Next we multiply 254×2. Because 2 is in the tens **place**, we put a 0 in the ones place and write the result starting in the tens place. 4×2 is 8. Write the 8 in the tens place. 5×2 is 10, so we write a 0 in the hundreds place and carry the 1. 2×2 is 4 plus the carried 1 equals 5. We have 5080 in the second solution line.

$$\begin{array}{r} {}^{1} \\ 254 \\ \times\ 429 \\ \hline 2286 \\ 5080 \end{array}$$

$254 \times 2 = 508$ — Put a 0 in the ones place.

Next we multiply 254×4. Because 4 is in the hundreds place, we put 0's in the ones place and tens place, and write the result starting in the hundreds place. 4×4 is 16. Write the 6 in the hundreds place and carry the 1. 5×4 is 20 plus the carried 1 is 21. Write the 1 in the thousands place and carry the 2. 2×4 is 8 plus the carried 2 is 10. So the third solution line is 101600.

$$\begin{array}{r} {}^{2\,1} \\ 254 \\ \times\ 429 \\ \hline 2286 \\ 5080 \\ 101600 \end{array}$$

$254 \times 4 = 1016$ — Put a 0 in the ones place and the tens place.

Add the three results (add the three solution lines).

$$\begin{array}{r} 254 \\ \times \quad 429 \\ \hline 2\,286 \\ 5\,080 \\ 101{,}600 \\ \hline 108{,}966 \end{array} \qquad 2286 + 5080 + 101{,}600 = 108{,}966$$

So, $254 \times 429 = 108{,}966$.

My video summary

c. Try working this problem on your own. Check your **answer**, or watch this **video** for a fully worked solution to part c.

You Try It **Work through this You Try It problem.**

Work Exercises 13–23 in this textbook or in the MyMathLab Study Plan.

OBJECTIVE 4 ESTIMATE PRODUCTS OF WHOLE NUMBERS

As with **addition** and **subtraction**, we may only need an estimate of the **product** when we **multiply**. If so, we can **round** each **factor** to a given **place value** and then carry out the multiplication.

Numbers that end in 0 are **multiples** of 10. Multiplying by multiples of 10, such as 10, 100, or 1000, is relatively easy. Notice the pattern in the following multiplications.

$$\begin{aligned} 12 \times 10 &= 120 \\ 12 \times 100 &= 1200 \\ 12 \times 1000 &= 12{,}000 \\ &\vdots \end{aligned}$$

To multiply a number by 10, just add 1 zero to the end of the number. To multiply a number by 100, add 2 zeros to the end of the number. To multiply a number by 1000, add 3 zeros to the end of the number. This pattern continues without end.

Multiplying by other multiples of 10, such as 30, 300, or 3000, follows a similar approach.

$$\begin{aligned} 12 \times 3 &= 36 \\ 12 \times 30 &= 360 \\ 12 \times 300 &= 3600 \\ 12 \times 3000 &= 36{,}000 \\ &\vdots \end{aligned}$$

In each case above, the first two **digits** of the **product** are the same and the number of **trailing zeros** (zeros at the end of the number) in the product is the same as the number of trailing zeros in the second **factor**. So, we could have simply **multiplied** $12 \times 3 = 36$ and added the number of trailing zeros to the result.

This idea extends to cases where both factors have trailing zeros. For example, to multiply 1200×3000, we can multiply $12 \times 3 = 36$ and then add 5 zeros to the product. We add 5 zeros because there are 2 trailing zeros in the first factor and 3 in the second. So there are $2 + 3 = 5$ total trailing zeros, which means that $1200 \times 3000 = 3{,}600{,}000$.

This technique will be useful when estimating products of **whole numbers**.

Example 6 Estimating Products of Whole Numbers

Estimate the product 3782×739 by first **rounding** the factors as indicated.

a. Nearest ten **b.** Nearest hundred

Solutions

a. $\begin{array}{r} 3782 \\ \times\ 739 \\ \hline \end{array} \xrightarrow{\text{Round to the nearest ten}} \begin{array}{r} 3780 \\ \times\ 740 \\ \hline \end{array}$

There are a total of two **trailing zeros**. Multiply 378×74 and add two zeros to the product.

$$\begin{array}{r} {\scriptstyle 5\,5} \\ {\scriptstyle \not 3\,\not 3} \\ 378 \\ \times\quad 74 \\ \hline 1512 \\ 26460 \\ \hline 27972 \end{array} \quad \begin{array}{l} \leftarrow \text{carries from } 378 \times 70 \\ \leftarrow \text{carries from } 378 \times 4 \\ \\ \\ \\ \\ \\ \\ \end{array}$$

So, $3780 \times 740 = 2{,}797{,}200$.

b. $\begin{array}{r} 3782 \\ \times\ 739 \\ \hline \end{array} \xrightarrow{\text{Round to the nearest hundred}} \begin{array}{r} 3800 \\ \times\ 700 \\ \hline \end{array}$

There are a total of 4 trailing zeros. Multiply 38×7 and add 4 zeros to the **product**.

$$\begin{array}{r} {\scriptstyle 5} \\ 38 \\ \times\quad 7 \\ \hline 266 \end{array}$$

So, $3800 \times 700 = 2{,}660{,}000$.

You Try It **Work through this You Try It problem.**

Work Exercises 24–27 in this textbook or in the MyMathLab Study Plan.

Example 7 Estimating Products of Whole Numbers

My video summary

Estimate the **product** $26{,}159 \times 8733$ by first **rounding** the **factors** to the nearest thousand.

Solution Try working this problem on your own. Check your **answer**, or watch this **video** for a fully worked solution.

You Try It **Work through this You Try It problem.**

Work Exercises 28 and 29 in this textbook or in the MyMathLab Study Plan.

(*eText Screens 1.4-1–1.4-32*)

OBJECTIVE 5 FIND THE AREA OF A RECTANGLE

The **area** of a figure is the amount of surface contained within the **sides** of the figure. Surface is two-dimensional (involving two dimensions such as length and width), so area is measured in units that are also two-dimensional, **square units**. For example, a **square inch** (sq in.) is a square with 1-inch sides. A **square centimeter** (sq cm) is a square with 1-centimeter sides.

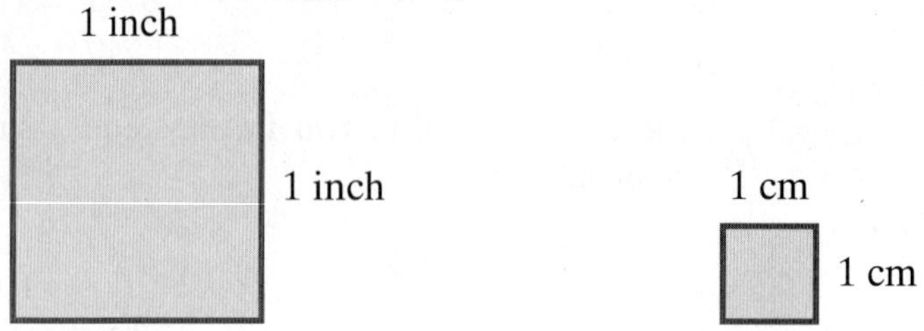

1 square inch (sq in.)   1 square centimeter (sq cm)

To find the area of a figure, we find the number of square units required to cover the figure. For now we will look only at the area of a rectangular region. However, we will always measure area in *square units* regardless of the shape of the figure.

The following rectangular region contains 4 rows, each of which contains 5 squares. Counting the number of squares, there are 20 square units in the rectangle. Therefore, we say the **area** of the rectangle is 20 square units.

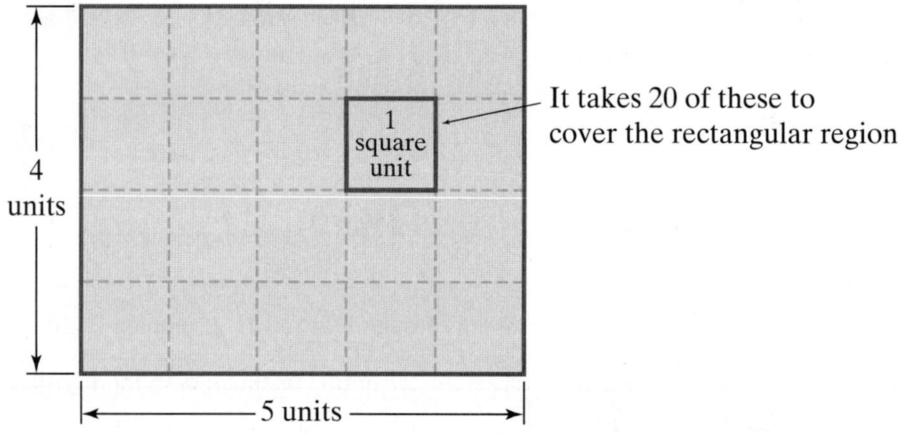

We could also find the area of the rectangle by multiplying the length and width. That is, $A = l \cdot w$.

$$5 \text{ units} \times 4 \text{ units} = 20 \text{ square units}$$

Remember that **perimeter** is one-dimensional (length), so it has single units such as feet, while area is two-dimensional (length and width) so it has square units such as sq ft. Watch this **animation** for an illustration of the difference between perimeter and area.

My animation summary

(eText Screens 1.4-1–1.4-32)

Example 8 Finding the Area of a Rectangle

Find the area of each rectangle.

a.

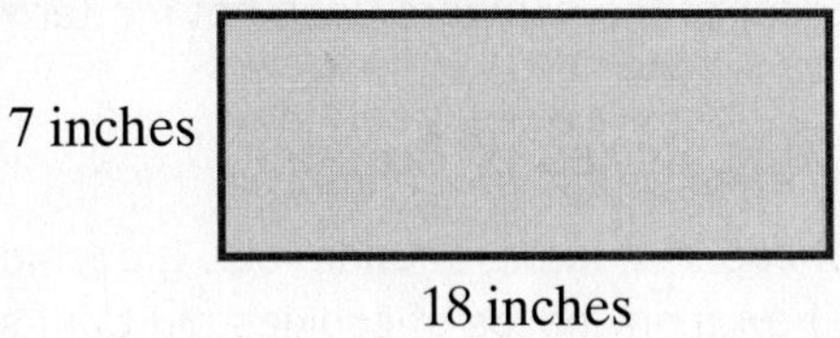

b.

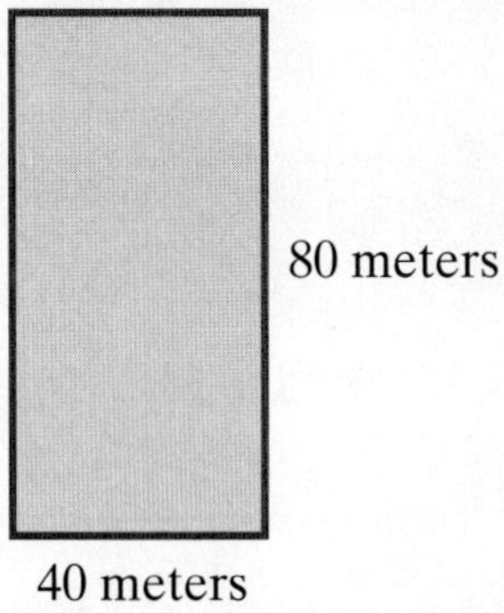

Solutions

a. We find the area by multiplying the length and the width.

$$\begin{array}{r} \scriptstyle 5 \\ 18 \\ \times\ \ 7 \\ \hline 126 \end{array}$$

The area of the rectangle is 126 square inches.

My video summary

b. Watch this **video** to confirm that the area is 3200 square meters.

You Try It **Work through this You Try It problem.**

Work Exercises 30–32 in this textbook or in the MyMathLab Study Plan.

OBJECTIVE 6 SOLVE APPLICATIONS BY MULTIPLYING WHOLE NUMBERS

As with **addition** and **subtraction**, many application problems can be solved by multiplying **whole numbers**. When solving these problems, we want to learn to recognize key words that indicate **multiplication**. Figure 11 shows some examples of key words that mean multiplication.

Key Word	Word Phrase	Mathematical Expression
Product	The *product* of 5 and 12	5(12) [or 5×12, $5 \cdot 12$, $5 * 12$]
Times	9 *times* 6	9(6)
Twice	*Twice* 18	2(18)

Figure 11 Key Words Meaning Multiplication

(eText Screens 1.4-1–1.4-32)

To solve application problems using multiplication, we translate the word statement into a mathematical statement and carry out the multiplication.

CAUTION

Remember that the order of the numbers does not matter with multiplication.

Example 9 Counting Boxes of Cookies

Carli sold 47 cases of cookies during a fundraiser. If each case contains 12 boxes of cookies, how many boxes of cookies did Carli sell?

Solution We **multiply** the number of cases by the number of boxes in each case. Carli sold $47 \times 12 = 564$ boxes of cookies.

Example 10 Counting Calories

My video summary

A 12-ounce can of Pepsi® contains 150 calories. How many total calories are there in a 24-pack of Pepsi? (*Source:* www.pepsicobeveragefacts.com)

Solution Multiply the calories in one can by 24 to find the total calories on your own. Check your **answer**, or watch this **video** for the solution.

You Try It **Work through this You Try It problem.**

Work Exercises 33–36 in this textbook or in the MyMathLab Study Plan.

Example 11 Estimating the Value of Gold

On July 14, 2011, the price for 1 troy ounce of gold at the New York Mercantile was \$1595. At this price, estimate the value of 32 troy ounces, **rounding** each value to the nearest ten. (*Source:* money.cnn.com)

Solution To estimate the total value, we multiply the price of 1 troy ounce by the number of ounces after rounding each value to the nearest ten.

$$\begin{array}{r} \$1595 \\ \times \quad 32 \\ \hline \end{array} \xrightarrow{\text{Round to nearest ten}} \begin{array}{r} \$1600 \leftarrow \text{Price of 1 ounce} \\ \times \quad 30 \leftarrow \text{Ounces} \\ \hline \end{array}$$

There are three **trailing zeros** so we can multiply 16×3 and add three zeros to the **product**.

$16 \times 3 = 48$, so the estimated value is $1600 \times 30 = \$48{,}000$.

Example 12 Amount of Carpet Needed

My video summary

Tobin wants to put new carpet in her family room. The room is rectangular in shape with a length of 26 ft and a width of 16 ft.

a. Find the **area** of the room to determine how much carpet Tobin needs to buy.

b. How much will she spend on the carpet if each square foot costs \$4?

(eText Screens 1.4-1–1.4-32)

Solutions

a. To find the area of the room, we multiply the length and the width.

$$\underbrace{(26\text{ feet})}_{\text{Length}}\underbrace{(16\text{ feet})}_{\text{Width}} = 416\text{ square feet}$$

The area of the room is 416 square feet.

b. Try to find the amount she spends by multiplying the price of 1 square foot by the total number of square feet. Check your **answer**, or watch this **video** for a complete solution to both parts.

You Try It **Work through this You Try It problem.**

Work Exercises 37–40 in this textbook or in the MyMathLab Study Plan.

Example 13 Travel Cost

My video summary

A youth group is planning to send 18 members to a rally in Washington, D.C. If each member attending the rally must pay $532 in airfare and $220 for food and lodging, what is the total cost of the trip for the entire group?

Solution To find the total cost of the trip for the entire youth group, we need to add the total cost of airfare and the total cost of food and lodging. Try to work this problem on your own. Check your **answer**, or watch this **video** for a fully worked solution.

You Try It **Work through this You Try It problem.**

Work Exercises 41 and 42 in this textbook or in the MyMathLab Study Plan.

1.4 Exercises

You Try It

1. Use the given property to rewrite each statement.

   a. Commutative property of multiplication: $9 \cdot 5 =$ ______

   b. Associative property of multiplication: $5 \cdot (6 \cdot 3) =$ ______

2. Use the properties of multiplication to show that $(3 \cdot 5) \cdot 7 = 5 \cdot (7 \cdot 3)$.

In Exercises 3 and 4, use properties of multiplication to multiply.

You Try It 3. **a.** $19 \cdot 1$ **b.** $0 \cdot 15$

4. **a.** $1 \cdot 0$ **b.** $34 \cdot 0$

In Exercises 5–8, rewrite each statement using the distributive property.

You Try It 5. $7(8 + 1) =$ ______ 6. $(12 - 4)5 =$ ______

7. $7(20 - 9) =$ ______ 8. $(2 + 5)8 =$ ______

In Exercises 9–23, multiply.

You Try It

9. 25×6

10. 327×6

11. 2835×9

12. $14{,}205 \times 7$

13. $\begin{array}{r} 62 \\ \times\ 29 \\ \hline \end{array}$

14. $\begin{array}{r} 47 \\ \times\ 96 \\ \hline \end{array}$

You Try It

15. $\begin{array}{r} 381 \\ \times\ 24 \\ \hline \end{array}$

16. $\begin{array}{r} 600 \\ \times\ 73 \\ \hline \end{array}$

17. $\begin{array}{r} 1285 \\ \times\ 58 \\ \hline \end{array}$

18. $\begin{array}{r} 328 \\ \times\ 129 \\ \hline \end{array}$

19. $\begin{array}{r} 652 \\ \times\ 111 \\ \hline \end{array}$

20. $\begin{array}{r} 283 \\ \times\ 367 \\ \hline \end{array}$

21. $\begin{array}{r} 2747 \\ \times\ 274 \\ \hline \end{array}$

22. $\begin{array}{r} 3218 \\ \times\ 2651 \\ \hline \end{array}$

23. $\begin{array}{r} 20{,}015 \\ \times\ 42 \\ \hline \end{array}$

In Exercises 24–29, estimate each product by first rounding the factors as indicated.

You Try It

24. 126×57 (nearest ten)

25. 1184×241 (nearest ten)

26. 6265×3718 (nearest hundred)

27. $15{,}419 \times 2874$ (nearest hundred)

You Try It

28. $52{,}114 \times 4739$ (nearest thousand)

29. $322{,}871 \times 42{,}851$ (nearest thousand)

In Exercises 30–32, find the area of the figure.

You Try It

30.

2 inches

5 inches

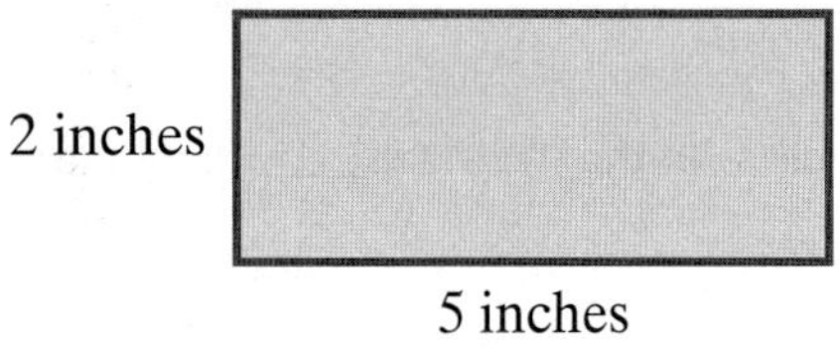

31.

4 ft

4 ft

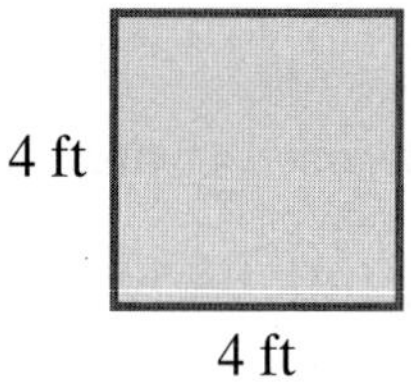

32.

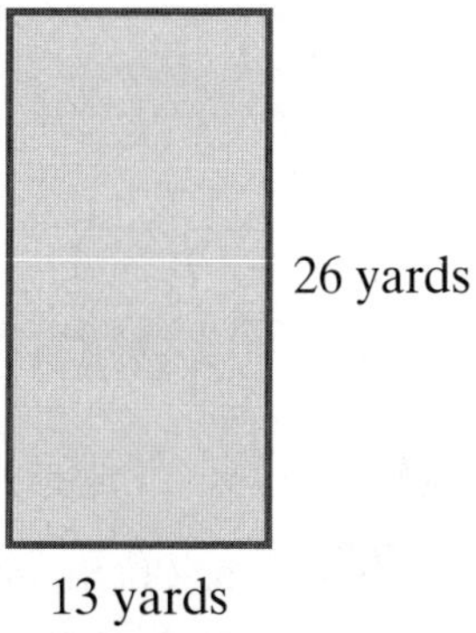

26 yards

13 yards

You Try It 33. **Calorie Intake**   Joey Chestnut won the 2011 Nathan's Famous International Hot Dog Eating Contest consuming 62 HDBs (hot dog and bun) in 10 minutes. If each HDB has 270 calories, how many total calories did Joey consume?

34. **Pain Reliever**   A case of pain reliever contains 24 bottles. If each bottle contains 100 tablets, how many tablets are there in a case?

35. **Data Storage**   A single layer Blu-ray disc can hold 27 GB (gigabytes) of data. How much data will fit on 25 such discs?

36. **Hamburger Calories**   A White Castle cheeseburger contains 170 calories. How many calories are in a sack of 20 White Castle cheeseburgers?

You Try It 37. **Membership Fee**   A fitness club has a monthly fee of $56. What is its annual (twelve months) fee?

38. **Mexico Cruise**   A family of twelve goes on a 5-day cruise to Mexico. If the cost for the cruise (including insurance) is $576 per person, what is the total cost for the family?

39. **Tuition Cost** For the 2011–2012 academic year, undergraduate evening students enrolling in University College in Arts & Sciences at Washington University paid tuition of $585 per credit hour.

    a. Estimate the tuition cost of 32 credit hours by first rounding each value to the nearest ten.

    b. Find the exact tuition cost of 32 credit hours during the 2011–2012 academic year.

40. **Cost of Platinum** On August 12, 2011, the price for 1 troy ounce of platinum at the New York Mercantile was $1797. (*Source:* money.cnn.com)

    a. At this price, estimate the value of 45 troy ounces of platinum by first rounding each value to the nearest ten.

    b. Find the exact value of 45 troy ounces of platinum.

You Try It

41. **Concert Tickets** Tasha and her five friends bought tickets for a Lil' Wayne concert at the AT&T Center in San Antonio, TX. Tickets were $80, but there was also a service fee of $14 per ticket. How much did they spend for the six tickets?

42. **Copy Paper** A case of copier paper costs $47 plus $4 tax. What is the total cost for 27 cases?

1.5 Dividing Whole Numbers

THINGS TO KNOW

Before working through this topic, be sure you are familiar with the following concepts:

VIDEO ANIMATION INTERACTIVE

You Try It 1. Round Whole Numbers (Topic 1.2, **Objective 5**)

You Try It 2. Read Tables and Bar Graphs Involving Whole Numbers (Topic 1.2, **Objective 6**)

You Try It 3. Add Whole Numbers (Topic 1.3, **Objective 1**)

You Try It 4. Subtract Whole Numbers (Topic 1.3, **Objective 4**)

You Try It 5. Multiply Whole Numbers (Topic 1.4, **Objective 3**)

OBJECTIVES

1 Divide Whole Numbers

2 Estimate Quotients of Whole Numbers

3 Solve Applications by Dividing Whole Numbers

(eText Screens 1.5-1–1.5-31)

OBJECTIVE 1 DIVIDE WHOLE NUMBERS

To **divide** a quantity means to separate it into groups of equal size. For example, consider a soft drink producer with 20 bottles of root beer to be packaged into cartons that hold 4 bottles each. To find the resulting number of cartons, we divide the 20 bottles into groups of 4. Figure 12 shows that the result is 5 groups (or cartons) of 4 bottles. We say that *20 divided by 4 equals 5.*

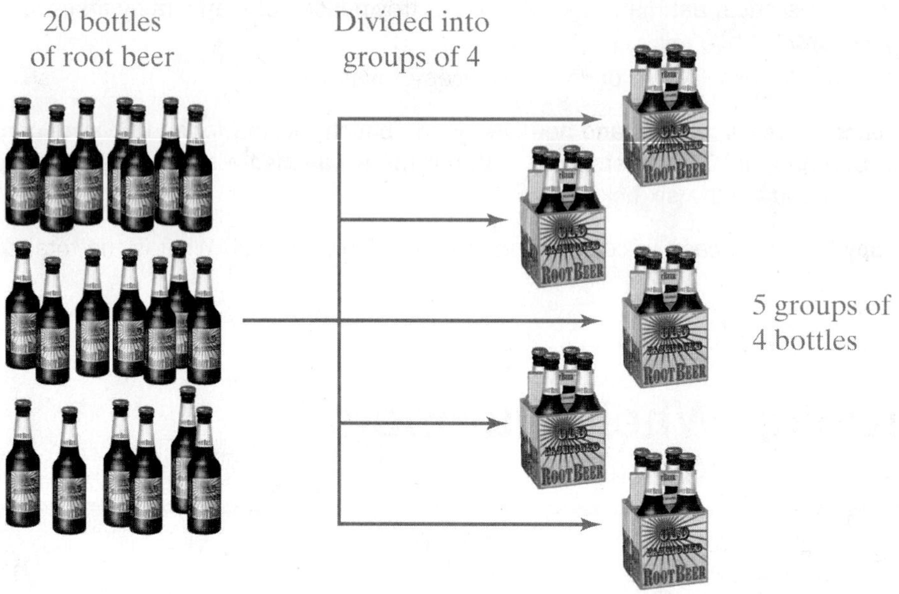

Figure 12

This division problem can be written using a **division symbol** $\div$, a **forward slash** $/$, a **long division symbol** $\overline{)\ \ }$, or a **fraction bar** $-$. The quantity being divided (20) is called the **dividend**, whereas the quantity we divide by (4) is the **divisor**. The result of division (5) is called the **quotient**.

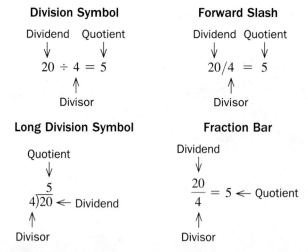

Division Symbol

Dividend ↓ Quotient ↓

$$20 \div 4 = 5$$

↑ Divisor

Forward Slash

Dividend ↓ Quotient ↓

$$20/4 = 5$$

↑ Divisor

Long Division Symbol

Quotient ↓

$$4\overline{)20}^{\,5} \leftarrow \text{Dividend}$$

↑ Divisor

Fraction Bar

Dividend ↓

$$\frac{20}{4} = 5 \leftarrow \text{Quotient}$$

↑ Divisor

Just as we defined **subtraction** in terms of **addition** in **Topic 1.3**, we can define **division** in terms of **multiplication**. We can say $20 \div 4 = 5$ because $5 \cdot 4 = 20$. This means we can check the result of division by multiplying:

$$\text{Division: } \quad \text{Dividend} \div \text{Divisor} = \text{Quotient}$$
$$\text{Check: } \quad \text{Quotient} \cdot \text{Divisor} = \text{Dividend}$$

If multiplying the **quotient** by the **divisor** results in the **dividend**, the result checks.

(eText Screens 1.5-1–1.5-31)

Example 1 Dividing Whole Numbers

Find each quotient. Check by multiplying.

a. $45 \div 5$ **b.** $21/7$ **c.** $\dfrac{32}{8}$ **d.** $9\overline{)72}$

Solutions

a. $45 \div 5 = 9$ because $9 \cdot 5 = 45$.

b. $21/7 = 3$ because $3 \cdot 7 = 21$.

c.–d. Try to find these quotients on your own. See the **answers** to parts c and d.

You Try It **Work through this You Try It problem.**

Work Exercises 1–4 in this textbook or in the MyMathLab Study Plan.

Example 2 Quotients Involving 1

Find each **quotient**. Check by multiplying.

a. $5 \div 1$ **b.** $\dfrac{13}{1}$ **c.** $8/8$ **d.** $10\overline{)10}$

Solutions

a. $5 \div 1 = 5$ because $5 \cdot 1 = 5$.

b. $\dfrac{13}{1} = 13$ because $13 \cdot 1 = 13$.

c. $8/8 = 1$ because $1 \cdot 8 = 8$.

d. $\begin{array}{r}1\\10\overline{)10}\end{array}$ because $1 \cdot 10 = 10$.

Parts a and b of Example 2 illustrate that any number divided by 1 results in that same number. Parts c and d illustrate that any number, except 0, divided by itself is 1. These are the **division properties of 1.**

> **Division Properties of 1**
>
> **1.** For any number, a,
>
> $$a \div 1 = a.$$
>
> **2.** For any number, a except 0,
>
> $$a \div a = 1.$$

Example 3 Quotients Involving 0

My video summary

Find each **quotient**. Check by multiplying.

a. $0 \div 16$ **b.** $\dfrac{7}{0}$

(eText Screens 1.5-1–1.5-31)

Solutions Work through the following, or watch this **video** for detailed solutions.

a. $0 \div 16 = 0$ because $0 \cdot 16 = 0$.

b. To determine the quotient of $\frac{7}{0}$, we must find a number that we can multiply by 0 to result in 7. That is, if $\frac{7}{0} = \square$, then $0 \cdot \square = 7$.

But recall the **multiplication property of zero** from **Topic 1.4.** Any number multiplied by 0 results in 0, which means we can never get the desired result of 7. We say that $\frac{7}{0}$ is **undefined.**

Part a of Example 3 illustrates that 0 divided any other number results in 0. Part b illustrates that division by 0 is not possible. These are the **division properties of 0.**

> **Division Properties of 0**
> For any number, a except 0,
>
> $0 \div a = 0$ and $a \div 0$ is undefined.

My video summary

CAUTION $0 \div 0$ does not equal 1. We say that $0 \div 0$ is **indeterminate.** Watch this **video** for the details.

You Try It **Work through this You Try It problem.**

Work Exercises 5–8 in this textbook or in the MyMathLab Study Plan.

When a **dividend** is large, we can divide using **long division**, which involves a repeated four-step process: **divide**, **multiply**, **subtract**, and **drop down** a **digit**. We review this process in the next example.

Example 4 Using Long Division

Divide $956 \div 4$. Check the result.

Solution We write the problem using long division notation: $4\overline{)956}$.

Divide 9 by 4 to find the digit in the hundreds **place** of the quotient.

$$\begin{array}{r} 2 \\ 4\overline{)956} \end{array}$$

Ask: "How many times will 4 go into 9?" Answer: "$9 \div 4$ is about 2." Place this 2 above the 9 (in the hundreds place).

Multiply the 2 by 4 to get 8. Write this result beneath the 9 in the hundreds **place**, then **subtract.**

$$\begin{array}{rl} 2 & \\ 4\overline{)956} & \\ \underline{-8} & \quad 2 \cdot 4 = 8 \\ 1 & \quad 9 - 8 = 1 \end{array}$$

Note that 1 is less than the divisor, 4.

Drop down the next **digit**, 5, from the tens place in the **dividend** and repeat the process.

$$\begin{array}{r} 2 \\ 4\overline{)956} \\ \underline{-8\downarrow} \\ 15 \end{array}$$

Divide 15 by 4 to find the digit in the tens place of the **quotient** (3). **Multiply** the result by 4 (aligning this **product** with the 15), and **subtract**.

$$\begin{array}{rl} 23 & \\ 4\overline{)956} & \text{Ask: "How many times will 4 go into 15?" Answer: "}15 \div 4\text{ is about }3.\text{"} \\ \underline{-8} & \text{Place this 3 above the 5 (in the tens place).} \\ 15 & \\ \underline{-12} & \text{Multiply: } 3 \cdot 4 = 12 \\ 3 & \text{Subtract: } 15 - 12 = 3 \end{array}$$

Drop down the last **digit**, 6, from the ones **place** and repeat the process one more time. **Divide** 36 by 4 to find the digit in the ones place of the **quotient** (9). **Multiply** the result by 4 (aligning this product with the 36), then **subtract**.

$$\begin{array}{rl} 239 & \text{Ask: "How many times will 4 go into 36?" Answer: "}36 \div 4\text{ is }9.\text{"} \\ 4\overline{)956} & \text{Place this 9 above the 6 (in the ones place).} \\ \underline{-8} & \\ 15 & \\ \underline{-12\downarrow} & \\ 36 & \\ \underline{-36} & \text{Multiply: } \quad 9 \cdot 4 = 36 \\ 0 & \text{Subtract: } 36 - 36 = 0 \end{array}$$

To check, we see if the quotient multiplied by the **divisor** equals the **dividend**.

$$\begin{array}{rl} {}^{1\,3} & \\ 239 & \leftarrow \text{Quotient} \\ \underline{\times\ \ 4} & \leftarrow \text{Divisor} \\ 956 & \leftarrow \text{Dividend} \end{array}$$

The result checks, so $956 \div 4 = 239$.

Note: Because the final subtraction in Example 4 resulted in 0, we say that the divisor, 4, **divides exactly** into the dividend, 956.

You Try It **Work through this You Try It problem.**

Work Exercises 9–11 in this textbook or in the MyMathLab Study Plan.

We can see why the **long division** process works by thinking about the **place value** of each **digit**. Let's look at Example 4 in more detail.

(eText Screens 1.5-1–1.5-31)

Hundreds Tens Ones

$$\begin{array}{r} 239 \\ 4\overline{)956} \\ -800 \\ \hline 156 \\ -120 \\ \hline 36 \\ -36 \\ \hline 0 \end{array}$$

Here we are really subtracting $200 \cdot 4 = 800$ from 956. This is why it is important to write the 2 in the hundreds place.

Here we are really subtracting $30 \cdot 4 = 120$ from 156. This is why it is important to write the 3 in the tens place.

Here we are subtracting $9 \cdot 4 = 36$ from 36. This is why the 9 goes in the ones place.

Typically, we leave off the **trailing zeros** to simplify the process.

Sometimes a quantity cannot be divided into a whole number of equal-sized groups. For example, suppose the soft drink producer in the opening example decided to package the 20 bottles of root beer into cartons that hold 6 bottles each, instead of 4. Figure 13 shows that the result is 3 cartons of 6 bottles, with 2 bottles left over.

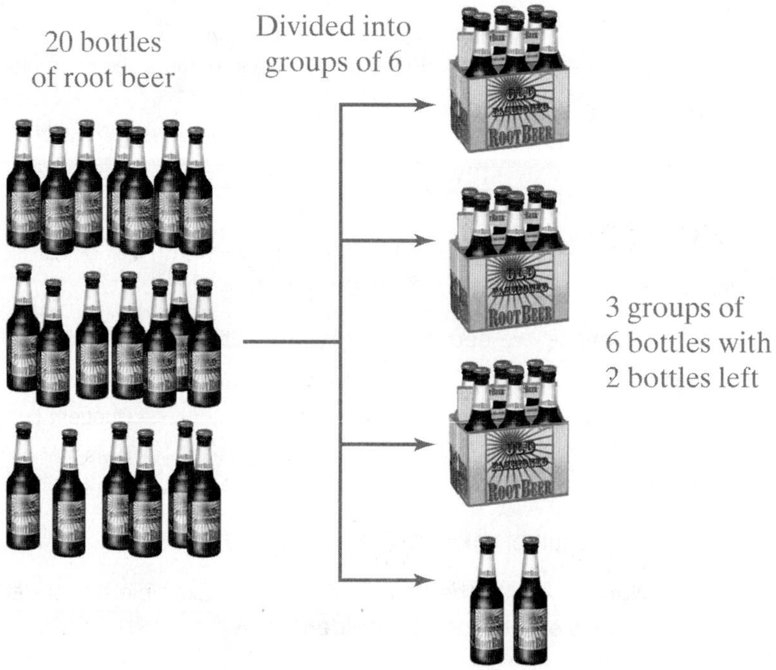

Figure 13

The amount left over after division is called the **remainder.** We say that *20 divided by 6 equals 3 with remainder 2*, or $20 \div 6 = 3 \text{ R } 2$. We can check such division problems using multiplication and addition in the following form:

$$\text{Quotient} \cdot \text{Divisor} + \text{Remainder} = \text{Dividend.}$$

Example 5 Division with a Remainder

Divide. Check the result.

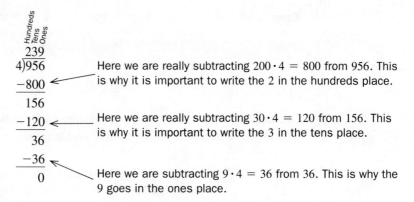

a. $6401 \div 7.$ **b.** $16{,}542 \div 8$

Solutions

a. Because the **digit** 6 (in the thousands **place**) cannot be divided by 7, we begin by dividing the first two digits 64 by 7.

$7\overline{)6401}$ with 9 above	Divide: $64 \div 7$ is about 9. Place this 9 above the 4.
-63	Multiply: $9 \cdot 7 = 63$. Write this 63 beneath the 64.
1	Subtract: $64 - 63 = 1$
$7\overline{)6401}$ with 91 above; $-63\downarrow$	Drop down the next digit, 0.
10	Divide: $10 \div 7$ is about 1. Place this 1 above the 0.
-7	Multiply: $1 \cdot 7 = 7$. Write this 7 beneath the 10.
3	Subtract: $10 - 7 = 3$
$7\overline{)6401}$ with 914 above; -63, 10, $-7\downarrow$	Drop down the last digit, 1, from the ones place.
31	Divide: $31 \div 7$ is about 4. Place this 4 above the 1.
-28	Multiply: $4 \cdot 7 = 28$. Write this 28 beneath the 31.
3	Subtract: $31 - 28 = 3$

At this point, there are no more **digits** to drop down and 7 will not divide into 3, so 3 is the **remainder**.

To check, we see if the quotient multiplied by the **divisor**, plus the remainder, equals the **dividend**.

$$\begin{array}{rl} \overset{2}{914} & \leftarrow \text{Quotient} \\ \times\ \ 7 & \leftarrow \text{Divisor} \\ \hline 6398 & \\ +3 & \leftarrow \text{Remainder} \\ \hline 6401 & \leftarrow \text{Dividend} \end{array}$$

The result checks, so $6401 \div 7 = 914\text{ R }3$.

My video summary

b. Try to work this problem on your own. View the **answer**, or watch this **video** for a complete solution to part b.

You Try It **Work through this You Try It problem.**

Work Exercises 12–14 in this textbook or in the MyMathLab Study Plan.

When the **divisor** has more than one **digit**, we can follow the same process of divide, multiply, subtract, and drop down a digit. To make the division step easier, we often estimate by **rounding**. Sometimes our estimates will be wrong. When that happens, we need to make an adjustment. We demonstrate this in the next example.

Example 6 Dividing by a Two-Digit Number

Divide $18\overline{)1325}$. Check the result.

Solution The divisor 18 cannot divide into the first digit 1 nor the first two digits 13, so we begin by dividing the first three digits 132 by 18.

Divide 132 by 18 to find the digit in the tens **place** of the quotient.

$$\begin{array}{r} 6 \\ 18\overline{)1325} \end{array}$$

Ask: "How many times will 18 go into 132?" To help answer this question, round 18 to 20. Estimate $132 \div 20$ by thinking about $13 \div 2$, which is about 6. Place this 6 in the tens place above the 2.

Multiply the 6 by 18 to get 108. Write this **product** beneath the 132, then **subtract**.

$$\begin{array}{r} 6 \\ 18\overline{)1325} \\ -108 \\ \hline 24 \end{array}$$

$6 \cdot 18 = 108$

$132 - 108 = 24$ Notice that 24 is greater than the divisor, 18. This means our estimate for the quotient, 6, is too small. We must increase it by 1 to 7 and try again.

$$\begin{array}{r} 7 \\ 18\overline{)1325} \\ -126 \\ \hline 6 \end{array}$$

$7 \cdot 18 = 126$

$132 - 126 = 6$ Notice $6 < 18$, so our new estimate works!

Drop down the last **digit**, 5, from the ones place and repeat the process. **Divide** 65 by 18 to find the digit in the ones place of the **quotient**.

$$\begin{array}{r} 7 \\ 18\overline{)1325} \\ -126\downarrow \\ \hline 65 \end{array}$$

Ask: "How many times will 18 go into 65?" Again, rounding 18 to 20, we estimate $65 \div 20$ by thinking about $6 \div 2$, which is 3. Place this 3 in the ones place above the 5.

Multiply the 3 by 18 to get 54. **Subtract**.

$$\begin{array}{r} 73 \\ 18\overline{)1325} \\ -126 \\ \hline 65 \\ -54 \\ \hline 11 \end{array}$$

$3 \cdot 18 = 54$

$65 - 54 = 11$ Notice $11 < 18$.

There are no more **digits** to drop down and 18 will not divide into 11, so 11 is the **remainder**.

Thus, $1325 \div 18 = 73 \text{ R } 11$. View the **check**.

CAUTION

When using **long division**, if a multiplication step results in a **product** larger than the dividend for that step, then your estimate for the division step is too big. Likewise, if a subtraction step results in a **difference** larger than the **divisor**, then your estimate for the division step is too small. In either case, you must adjust the estimate and try again.

You Try It **Work through this You Try It problem.**

Work Exercises 15–18 in this textbook or in the MyMathLab Study Plan.

(eText Screens 1.5-1–1.5-31)

Example 7 Dividing Whole Numbers

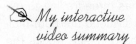

My interactive video summary

Divide. Check the result.

a. $896 \div 43$ **b.** $18{,}719/508$ **c.** $\dfrac{49{,}910}{62}$

Solutions Work through the following, or watch this **interactive video** for detailed solutions.

a.

$$\begin{array}{r@{\quad}l}
20 & \\
43\overline{)896} & \\
-86\downarrow & 2\cdot 43 = 86 \\
\hline
36 & 89 - 86 = 3.\ \text{Drop down the 6.} \\
-0 & 0\cdot 43 = 0.\ \text{(43 cannot divide into 36.)} \\
\hline
36 & 36 - 0 = 36.\ \text{This is the remainder.}
\end{array}$$

So, $896 \div 43 = 20\text{ R }36$. View the **check**.

b.

$$\begin{array}{r@{\quad}l}
36 & \\
508\overline{)18719} & \\
-1524\downarrow & 3\cdot 508 = 1524 \\
\hline
3479 & 1871 - 1524 = 347.\ \text{Drop down the 9.} \\
-3048 & 6\cdot 508 = 3048 \\
\hline
431 & 3479 - 3048 = 431.\ \text{This is the remainder.}
\end{array}$$

So, $18{,}719/508 = 36\text{ R }431$. View the **check**.

c.

$$\begin{array}{r@{\quad}l}
805 & \\
62\overline{)49910} & \\
-496\downarrow & 8\cdot 62 = 496 \\
\hline
31 & 499 - 496 = 3.\ \text{Drop down the 1.} \\
-0\downarrow & 0\cdot 62 = 0.\ \text{(62 cannot divide into 31.)} \\
\hline
310 & 31 - 0 = 31.\ \text{Drop down the 0.} \\
-310 & 5\cdot 62 = 310 \\
\hline
0 & 310 - 310 = 0.\ \text{There is no remainder.}
\end{array}$$

So, $\dfrac{49{,}910}{62} = 805$. View the **check**.

You Try It **Work through this You Try It problem.**

Work Exercises 19–24 in this textbook or in the MyMathLab Study Plan.

OBJECTIVE 2 ESTIMATE QUOTIENTS OF WHOLE NUMBERS

As with the other **operations**, we sometimes need only an estimate of the **quotient** when dividing rather than an exact answer. To find such an estimate, we can **round** the **divisor** and/or **dividend** to desired **place values** and then divide using the rounded values.

(eText Screens 1.5-1–1.5-31)

In **Topic 1.4**, we learned that numbers with a **trailing zero** are **multiples** of 10. If a number has two trailing zeros, then it is also a multiple of 100. If a number has three trailing zeros, then it is also a multiple of 1000, and so on. When the divisor and dividend are both multiples of 10, 100, 1000, and so on, we can use a shortcut to find the quotient. Notice the pattern in the following divisions:

$$70 \div 10 = 7$$
$$700 \div 100 = 7$$
$$7000 \div 1000 = 7$$
$$\vdots$$

To divide a multiple of 10 by 10, remove one trailing zero. To divide a multiple of 100 by 100, remove two trailing zeros. To divide a multiple of 1000 by 1000, remove three trailing zeros. And so on.

Division involving other multiples with **trailing zeros** is similar. Consider the following:

$$28 \div 2 = 14$$
$$280 \div 20 = 14$$
$$2800 \div 200 = 14$$
$$\vdots$$

Notice that the **quotient** is the same for each of these division problems. Notice also that except for the equal number of trailing zeros in the **dividend** and **divisor**, the nonzero numbers in both the dividend and the divisor are the same in each case. We could have removed the equal number of trailing zeros and found the quotient simply by using the remaining **digits**.

A Division Shortcut

In a division problem, an equal number of trailing zeros can be removed from the dividend and divisor without changing the quotient.

Example 8 Using a Division Shortcut

Use the division shortcut to find each quotient.

a. $120 \div 30$ **b.** $\dfrac{43{,}000}{100}$ **c.** $50\overline{)2100}$

Solutions

a. The **dividend**, 120, and divisor, 30, each have one **trailing zero**. We can remove the trailing zero from each number and divide:

$$120 \div 30 = 12 \div 3 = 4$$

Remove one trailing zero from the dividend and one from the divisor.

b. The dividend, 43,000, has three trailing zeros, but the divisor, 100, has only two. So, we can remove two trailing zeros from each and divide.

$$\frac{43{,}000}{100} = \frac{430}{1} = 430$$

c. The dividend, 2100, has two trailing zeros, while the divisor, 50, has one. We remove one trailing zero from each and divide.

$$50\overline{)2100} \Leftrightarrow \begin{array}{r} 42 \\ 5\overline{)210} \\ -20\downarrow \\ \hline 10 \\ -10 \\ \hline 0 \end{array} \quad \begin{array}{l} \\ \\ 4 \cdot 5 = 20 \\ 21 - 20 = 1. \text{ Bring down the } 0. \\ 2 \cdot 5 = 10 \\ 10 - 10 = 0. \end{array}$$

So, $2100 \div 50 = 42$.

You Try It **Work through this You Try It problem.**

Work Exercises 25–28 in this textbook or in the MyMathLab Study Plan.

CAUTION While a **quotient** will be the same when using the **division shortcut**, a **remainder** will not. For example, $270 \div 40 = 6\text{ R }30$, but $27 \div 4 = 6\text{ R }3$.

Note that a remainder will always contain the same number of trailing zeros that the dividend and **divisor** have in common.

The division shortcut is helpful when estimating quotients. We can approximate both the **dividend** and the divisor so that they divide more easily. For example, to approximate $389 \div 48$, we notice that 389 is close to 400 and 48 is close to 50. We choose 400 and 50 because they divide more easily:

$$400 \div 50 = 40 \div 5 = 8.$$

So, our estimate for the quotient $389 \div 48$ is 8.

If we calculate $389 \div 48$ exactly, the result is $8\text{ R }5$, which is close to our estimate.

CAUTION As a general rule, when using this estimation method for division, either increase both the dividend and divisor to convenient numbers that can divide easily, or decrease them both. This practice will usually give better estimates. Increasing one and decreasing the other will lead to poor estimates.

Example 9 Estimating Quotients

Estimate each **quotient**.

a. $56{,}478 \div 82$ **b.** $715{,}809 \div 887$

Solutions

a. Notice that the first two **digits** in the **dividend** are 56 and the first digit in the **divisor** is 8. Because 56 is divisible by 8, we round 56,478 down to 56,000, and we round 82 down to 80.

Approximate the dividend — Drop one trailing zero

$$56{,}478 \div 82 \qquad 56{,}000 \div 80 = 5600 \div 8 = 700$$

Approximate the divisor

The estimated quotient is 700.

(eText Screens 1.5-1–1.5-31)

My video summary

b. Try to estimate this quotient on your own. View the **answer**, or watch this **video** for a complete solution to part b.

You Try It **Work through this You Try It problem.**

Work Exercises 29–32 in this textbook or in the MyMathLab Study Plan.

OBJECTIVE 3 SOLVE APPLICATIONS BY DIVIDING WHOLE NUMBERS

Many application problems can be solved using **division**. These problems typically involve separating a quantity into equal groups. Recognizing key words that indicate a **quotient** can be essential to solving such problems. Figure 14 provides some examples of key words or phrases that mean division.

Key Word or Phrase	Word Phrase	Mathematical Expression
Quotient	The *quotient* of 56 and 4	$56 \div 4 \left(\text{or } \frac{56}{4}, 56/4, 4\overline{)56}\right)$
Divide	*Divide* 100 by 25	$100 \div 25$
Divided by	21 *divided by* 7	$21 \div 7$
Goes into	5 *goes into* 30	$30 \div 5$
Shared evenly among	$100 *shared evenly among* 4 people	$100 \div 4$
Split equally between	12 pounds *split equally between* 2 brothers	$12 \div 2$
Per	320 miles *per* 16 gallons	$320 \div 16$

Figure 14

To solve applications involving division, translate the words into a mathematical statement and carry out the division.

Example 10 Paying for a Party

To throw a party, six friends will rent a hall for $450. If this rent is shared evenly among the friends, what is each friend's share of the rent?

Solution The key phrase *shared evenly among* tells us to **divide** the cost by the number of friends.

Total rent ÷ Number of friends = Each friend's share

$$450 \div 6 = 75$$

View this **popup box** to see the work. Each friend's share of the rent is $75.

You Try It **Work through this You Try It problem.**

Work Exercises 33–36 in this textbook or in the MyMathLab Study Plan.

Example 11 Chartering Busses

My video summary

An executive secretary must charter busses to transport 965 workers from a conference hotel to a restaurant across town. If each bus carries 56 people, how many busses are needed? How many seats will be empty? (*Source:* newcharterbus.com)

Solution Work through the following, or watch this **video** for a detailed solution.

To find the number of busses needed, we **divide** the number of workers by the number of people who can fit on each bus.

Number of workers	÷	People per bus	=	Number of busses
↓		↓		↓
965	÷	56	=	17 R 13

There are enough workers to fill 17 busses completely, with 13 workers left over. If the executive secretary hires only 17 busses, then the 13 remaining workers will have no transportation across town. Therefore, 18 busses will be needed.

Hiring the 18th bus means that there are now empty seats. To find how many, subtract: $56 - 13 = 43$. There will be 43 empty seats with 18 busses.

You Try It **Work through this You Try It problem.**

Work Exercises 37–40 in this textbook or in the MyMathLab Study Plan.

Example 12 Estimating Debt per Citizen

During July 2011, the New York state debt was \$283,477,740,095 when the population of New York was 19,466,875. Estimate the debt per citizen of New York at that time. (*Source:* U.S. Debt Clock.org)

Solution The key phrase *debt per citizen* tells us to **divide** the amount of debt by the number of citizens: $283{,}477{,}740{,}095 \div 19{,}466{,}875$. To estimate, we notice that debt of 283,477,740,095 is close to 300,000,000,000 and the population of 19,466,875 is close to 20,000,000. (Observe that we obtained both of these estimates by rounding up.) Using the **division shortcut** and then dividing, we get

Drop seven trailing zeros

$$300{,}000{,}000{,}000 \div 20{,}000{,}000 = 30{,}000 \div 2 = 15{,}000.$$

The New York state debt was about \$15,000 per citizen in July 2011.

You Try It **Work through this You Try It problem.**

Work Exercises 41 and 42 in this textbook or in the MyMathLab Study Plan.

1.5 Exercises

In Exercises 1–8, find each quotient. Check by multiplying.

You Try It 1. $54 \div 9$ 2. $40/10$ 3. $\frac{56}{8}$ 4. $2\overline{)16}$

You Try It 5. $12/12$ 6. $15 \div 1$ 7. $\frac{0}{6}$ 8. $45 \div 0$

In Exercises 9–24, divide. Check the result.

You Try It 9. $72 \div 3$ 10. $\frac{534}{6}$ 11. $8\overline{)1976}$

12. $7\overline{)257}$ 13. $4306/9$ 14. $56{,}321 \div 4$

You Try It 15. $16\overline{)1184}$ 16. $948 \div 21$ 17. $2275/52$

18. $\frac{9076}{39}$ 19. $17\overline{)4095}$ 20. $16{,}632 \div 54$

You Try It 21. $84\overline{)20{,}486}$ 22. $13{,}608 \div 324$ 23. $127\overline{)89{,}670}$

24. $\frac{89{,}670}{287}$

You Try It

In Exercises 25–28, use the division shortcut to find each quotient.

25. $3700 \div 100$ 26. $\frac{24{,}000}{8000}$

27. $6300/70$ 28. $40\overline{)2600}$

In Exercises 29–32, estimate each quotient.

29. $79{,}512 \div 789$ 30. $45{,}235 \div 513$

You Try It 31. $324{,}677 \div 82$ 32. $1{,}794{,}658 \div 1983$

In Exercises 33–42, solve each application.

33. **Inheritance** In her will, a grandmother leaves \$325,000 to be shared equally among her 5 grandchildren. How much inheritance will each grandchild receive?

You Try It 34. **Water Cooler** An office water cooler holds 960 ounces of water. How many 8-ounce cups can be filled from the cooler?

35. **Car Payments** To purchase a car, Reagan borrowed \$17,616 that will be paid back in 48 equal monthly payments. What is her monthly payment?

36. **Gas Mileage** Karen's hybrid SUV gets 32 miles per gallon. How many gallons of gas will Karen need for an 800-mile trip?

37. **Ordering Pizzas** Beverly needs to order pizzas to provide lunch for a group of 435 students. If one pizza will feed 3 people, how many pizzas should Beverly order?

You Try It

38. **Delivering Concrete** If a concrete truck can hold 9 cubic yards of concrete, how many trips must it make to deliver 70 cubic yards of concrete to a construction site?

39. **Doggie Treats** A kennel owner has 75 treats to distribute evenly among 14 dogs being boarded at the kennel. How many treats will each dog get? How many treats will the owner have left?

40. **Cake Reception** Rena expects 240 guests to attend a cake reception. If a sheet cake will serve 36 people, how many sheet cakes should Rena order for the reception? If each guest has one serving of cake, how many servings will be left over?

You Try It

41. **Earnings per Game** During the 2010–11 NBA season, Kobe Bryant's salary from the LA Lakers was $24,806,250. If the team played in 82 games that season, estimate Bryant's earnings per game. **Hint:** 24 is divisible by 8. (*Source:* InsideHoops.com)

42. **Google** In August 2004, Google raised approximately $1,666,000,000 in capital with its initial public offering of stock for $85 per share. Estimate the number of shares sold. **Hint:** 16 is divisible by 8. (*Source:* Google, Inc.)

1.6 Exponents and Order of Operations

THINGS TO KNOW

Before working through this topic, be sure you are familiar with the following concepts:

	VIDEO	ANIMATION	INTERACTIVE
You Try It 1. Add Whole Numbers (Topic 1.3, **Objective 1**)	▤		
You Try It 2. Subtract Whole Numbers (Topic 1.3, **Objective 4**)	▤		
You Try It 3. Multiply Whole Numbers (Topic 1.4, **Objective 3**)	▤		
You Try It 4. Divide Whole Numbers (Topic 1.5, **Objective 1**)	▤		▶

OBJECTIVES

1 Use Exponential Notation

2 Evaluate Exponential Expressions

3 Simplify Expressions Using the Order of Operations

4 Find the Average of a List of Numbers

OBJECTIVE 1 USE EXPONENTIAL NOTATION

In **Topic 1.4**, we learned that **multiplication** is repeated addition. Similarly, exponent notation is used to represent repeated multiplication.

Consider the **product** $2 \cdot 2 \cdot 2 \cdot 2$. The **factor** 2 is repeated 4 times. We can write this product as the **exponential expression** 2^4. The number 2 is called the **base** and indicates the factor that is being repeated. The number 4 is called the **exponent**, or **power**, and tells us how many times the factor is repeated. In this case, the exponent is 4 because the factor 2 is repeated 4 times.

$$\underbrace{2 \cdot 2 \cdot 2 \cdot 2}_{\text{4 factors of 2}} = \underset{\uparrow \atop \text{base}}{2}^{4} \leftarrow \text{exponent}$$

We read 2^4 as "2 raised to the fourth power." The following gives some examples of how we can read exponential expressions.

The exponential expression . . .	can be read as . . .
4^2	"four to the second power," or "four squared"
8^3	"eight to the third power," or "eight cubed"
7^5	"seven to the fifth power"

Example 1 Using Exponential Notation

My video summary

Write each of the following **products** using exponential notation.

a. $5 \cdot 5 \cdot 5$ **b.** $3 \cdot 3 \cdot 3 \cdot 3 \cdot 3 \cdot 3 \cdot 3$

Solutions

a. The **factor** 5 is being repeated, so the **base** is 5. The factor is being repeated three times, so the exponent is 3.

$$5 \cdot 5 \cdot 5 = 5^3$$

b. The factor 3 is being repeated, so the base is 3. The factor is being repeated seven times, so the exponent is 7. Write the exponential expression, then check your **answer**. Or watch this **video** for the solutions to both parts.

You Try It **Work through this You Try It problem.**

Work Exercises 1 and 2 in this textbook or in the MyMathLab Study Plan.

We can still use **exponent notation** if we have different **factors** that are repeated. For example, suppose we have the product

$$2 \cdot 2 \cdot 2 \cdot 7 \cdot 7 \cdot 7 \cdot 7.$$

Because there are three factors of 2 and four factors of 7, we can write this **product** as

$$2^3 \cdot 7^4.$$

Notice that we have two different **bases** and each has its own **exponent**.

Example 2 Using Exponential Notation

Write the following product using exponential notation.

$$4 \cdot 4 \cdot 4 \cdot 8 \cdot 8 \cdot 15 \cdot 15 \cdot 15 \cdot 15$$

Solution There are three factors of 4, two factors of 8, and four factors of 15. So, we have a base 4 with an exponent of 3, a base 8 with an exponent of 2, and a base 15 with an exponent of 4.

$$4 \cdot 4 \cdot 4 \cdot 8 \cdot 8 \cdot 15 \cdot 15 \cdot 15 \cdot 15 = 4^3 \cdot 8^2 \cdot 15^4$$

You Try It **Work through this You Try It problem.**

Work Exercises 3 and 4 in this textbook or in the MyMathLab Study Plan.

Recall that when we found **area** in **Topic 1.4**, we used *square units*. We can abbreviate these units by using an **exponent**. For example, "square feet" can be abbreviated as "ft^2" and "square centimeters" can be abbreviated as "cm^2". We can see this by multiplying the units as we multiply numbers. View this **popup** for an example.

OBJECTIVE 2 EVALUATE EXPONENTIAL EXPRESSIONS

A **numeric expression** is a combination of numbers and arithmetic operations such as $4 + 8$ or 3^2. We **evaluate** a numeric expression by simplifying the expression to a single numeric value. To evaluate **exponential expressions** we write the expression as a **product** and then carry out the multiplication. For example, $3^4 = 3 \cdot 3 \cdot 3 \cdot 3 = 81$.

When we raise a number to the first **power**, the exponent is 1. Any number raised to the first power is equal to itself. For example, $2^1 = 2$, $3^1 = 3$, and so on. If no **exponent** is written, it is assumed to be an exponent of 1.

Example 3 Evaluate Exponential Expressions

Evaluate each exponential expression.

a. 2^6 **b.** 10^4 **c.** 7^1

Solutions

a. $2^6 = 2 \cdot 2 \cdot 2 \cdot 2 \cdot 2 \cdot 2 = 64$

b. $10^4 = 10 \cdot 10 \cdot 10 \cdot 10 = 10{,}000$

c. $7^1 = 7$

You Try It **Work through this You Try It problem.**

Work Exercises 5–10 in this textbook or in the MyMathLab Study Plan.

CAUTION Although **exponent notation** represents repeated multiplication, we do not multiply the base and the **exponent**. That is, 2^6 does not equal $2 \cdot 6$ ($2^6 = 64$ while $2 \cdot 6 = 12$).

Example 4 Evaluate Exponential Expressions

Evaluate each exponential expression.

a. $4 \cdot 5^3$ **b.** $3^2 \cdot 10^3$

Solutions

a. $4 \cdot 5^3 = 4 \cdot 5 \cdot 5 \cdot 5 = 500$

My video summary

b. Try working this problem on your own. View the **answer**, or watch this **video** for the solutions to both parts.

You Try It **Work through this You Try It problem.**

Work Exercises 11–14 in this textbook or in the MyMathLab Study Plan.

OBJECTIVE 3 SIMPLIFY EXPRESSIONS USING THE ORDER OF OPERATIONS

The numeric expression $4 + 6 \cdot 8$ contains two **arithmetic operations: addition** and **multiplication**. Depending on which operation is performed first, a different answer will result when we simplify.

Add first, then multiply: $4 + 6 \cdot 8 = 10 \cdot 8 = 80$

Multiply first, then add: $4 + 6 \cdot 8 = 4 + 48 = 52$

Different Results

For this reason, mathematicians have agreed on a specific **order of operations**.

Order of Operations

1. **Parentheses (or other grouping symbols)** Evaluate operations within parentheses (or other **grouping symbols**) first, starting with the innermost set and working out.
2. **Exponents** Work from left to right and evaluate any **exponential expressions** as they occur.
3. **Multiplication and Division** Work from left to right and perform any multiplication or **division** operations as they occur.
4. **Addition and Subtraction** Work from left to right and perform any addition or **subtraction** operations as they occur.

Because we perform **multiplication** before addition, the correct **simplification** for our numeric expression above is $4 + 6 \cdot 8 = 52$, not 80.

View this **popup** for a tip on remembering the **order of operations**.

Example 5 Simplifying Expressions Using the Order of Operations

Simplify each expression.

a. $7 + 5^2$ **b.** $70 \div 5 - 3 \cdot 2$ **c.** $10^2 \div 5 - 4 \cdot 2$

Solutions For each expression, we follow the order of operations.

a. The two **operations** are **addition** and **exponents**. Exponents have priority over addition, so we evaluate the exponent first.

Begin with the original expression: $7 + 5^2$

Evaluate 5^2: $= 7 + 25$

Add: $= 32$

b. We have three operations: **division, subtraction**, and **multiplication**. Remember that multiplication and division have equal priority and both have priority over subtraction. Working from left to right, we would do the division first followed by the multiplication. The subtraction would be done last.

Begin with the original expression: $70 \div 5 - 3 \cdot 2$

Divide $70 \div 5$: $= 14 - 3 \cdot 2$

Multiply $3 \cdot 2$: $= 14 - 6$

Subtract: $= 8$

My video summary

c. There are four **operations** in this problem. Can you identify each one? Try working this problem on your own. Check your **answer**, or watch this **video** for the complete solution to part c.

You Try It **Work through this You Try It problem.**

Work Exercises 15–24 in this textbook or in the MyMathLab Study Plan.

TIP In the **order of operations**, multiplication and division are given the same priority and are performed from left to right, in order, as *either* occurs. It is incorrect to do all the multiplication first and then all the **division**. View this **popup** for an illustration.

Similarly, **addition** and subtraction have equal priority and are performed from left to right, in order, as *either* occurs. View this **popup** for an illustration.

Example 6 Using Order of Operations to Simplify Numeric Expressions

My video summary

Simplify each expression.

a. $22 + 5 - 3 + 6 - 4$ **b.** $48 \div 3 \cdot 2 - 5$

Solutions For each expression, follow the **order of operations**. Try to do these on your own. Check your **answers**, or watch the **video** for complete solutions to both parts.

You Try It **Work through this You Try It problem.**

Work Exercises 25–28 in this textbook or in the MyMathLab Study Plan.

Example 7 Using Order of Operations to Simplify Numeric Expressions

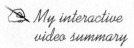

My interactive video summary

Simplify each expression.

a. $(4^2 - 14) \cdot 3^2$ **b.** $3 \cdot (7 - 5)^2 + 4$ **c.** $28 \div 7 + [6 - 2^2]^3$

Solutions For each expression, we follow the order of operations.

a. Begin with the original expression: $(4^2 - 14) \cdot 3^2$

Evaluate 4^2 inside parentheses: $= (16 - 14) \cdot 3^2$

Evaluate $16 - 14$ inside parentheses: $= 2 \cdot 3^2$

Evaluate 3^2: $= 2 \cdot 9$

Multiply $2 \cdot 9$: $= 18$

(eText Screens 1.6-1–1.6-21)

b.–c. Try working these problems on your own. Check your **answers** to parts b and c or watch this **interactive video** for complete solutions to all three parts.

You Try It **Work through this You Try It problem.**

Work Exercises 29–36 in this textbook or in the MyMathLab Study Plan.

Parentheses, ( ), brackets, [ ], and braces, { }, are the most common **grouping symbols.** However, other grouping symbols do exist, such as the fraction bar, ———. A fraction bar separates the expression into two parts: a top expression and a bottom expression. The top expression is grouped together and the bottom expression is grouped together. View an **illustration.**

We must **simplify** expressions separately in the top and bottom before dividing.

Example 8 Using Order of Operations with Special Grouping Symbols

Simplify: $\dfrac{5 \cdot 7 - 2^3}{4(7 - 5) + 1}$

Solution The **fraction bar** is a **grouping symbol.** Simplify the top and bottom separately before dividing. Let's begin by simplifying the top expression.

Evaluate the exponent: $= \dfrac{5 \cdot 7 - 8}{4(7 - 5) + 1} \quad \leftarrow 2^3 = 8$

Multiply: $= \dfrac{35 - 8}{4(7 - 5) + 1} \quad \leftarrow 5 \cdot 7 = 35$

Subtract: $= \dfrac{27}{4(7 - 5) + 1} \quad \leftarrow 35 - 8 = 27$

Now we simplify the bottom expression and then divide.

Subtract inside parentheses: $= \dfrac{27}{4(2) + 1} \quad \leftarrow 7 - 5 = 2$

Multiply: $= \dfrac{27}{8 + 1} \quad \leftarrow 4(2) = 8$

Add: $= \dfrac{27}{9} \quad \leftarrow 8 + 1 = 9$

Divide: $= 3$

You Try It **Work through this You Try It problem.**

Work Exercises 37–40 in this textbook or in the MyMathLab Study Plan.

CAUTION

Grouping the top together and grouping the bottom together is particularly important when using a calculator to **evaluate** an expression. A calculator will strictly follow **order of operations** and cannot guess what you meant to enter. View this **example.**

(eText Screens 1.6-1–1.6-21)

If an expression contains one set of **grouping symbols** inside another, we begin from the inside with the innermost grouping symbols and work our way outward. If writing an expression with grouping symbols, it is helpful to use different types of grouping symbols if more than one set is required.

Example 9 Using Order of Operations with Multiple Grouping Symbols

Simplify the expression.

$$25 - [4^2 - (10 - 3)] + 7 \cdot 6$$

Solution We have more than one pair of **grouping symbols**. So, we start simplifying by working on the innermost grouping symbols.

Begin with the original expression: $25 - \overbrace{[4^2 - \underbrace{(10 - 3)}_{\text{Innermost grouping}}]}^{\text{Outermost grouping}} + 7 \cdot 6$

Evaluate $(10 - 3)$: $= 25 - [4^2 - 7] + 7 \cdot 6$

Evaluate 4^2: $= 25 - [16 - 7] + 7 \cdot 6$

Subtract $16 - 7$: $= 25 - 9 + 7 \cdot 6$

Multiply $7 \cdot 6$: $= 25 - 9 + 42$

Subtract $25 - 9$: $= 16 + 42$

Add $16 + 42$: $= 58$

You Try It **Work through this You Try It problem.**

Work Exercises 41–44 in this textbook or in the MyMathLab Study Plan.

OBJECTIVE 4 FIND THE AVERAGE OF A LIST OF NUMBERS

To find the **average** of a list of numbers, we **add** up the numbers and **divide** the **sum** by the number of **addends**.

> **Average**
>
> The **average** of a list of numbers is the sum of the numbers divided by the number of addends.

When calculating an average, we must add first and then divide, which seems to contradict the **order of operations**. However, if we put the sum inside **grouping symbols**, such as parentheses, this order makes sense because we always work with grouping symbols first.

To find the average of a list of numbers such as 4, 7, 12, 15, and 17, we write

$$(4 + 7 + 12 + 15 + 17) \div 5.$$

We can write this expression using a **fraction bar** as well because it is also a grouping symbol.

$$\frac{4 + 7 + 12 + 15 + 17}{5} \quad \begin{array}{l}\leftarrow \text{Sum of the addends} \\ \leftarrow \text{Number of addends}\end{array}$$

Example 10 Finding the Average of a List of Numbers

Find the average of 7, 15, 8, 23, 47, and 14.

Solution There are 6 numbers, so to find the average, we add up the numbers and divide by 6.

$$\frac{7 + 15 + 8 + 23 + 47 + 14}{6} = \frac{114}{6} \quad \begin{array}{l}\leftarrow \text{Sum of the addends} \\ \leftarrow \text{Number of addends}\end{array}$$
$$= 19$$

The average of the numbers is 19.

You Try It **Work through this You Try It problem.**

Work Exercises 45 and 46 in this textbook or in the MyMathLab Study Plan.

Example 11 Finding the Average of a List of Numbers

My video summary

Based on data from hotels.com, the four most expensive U.S. cities for hotel rooms during 2010 are given in the following table.

City	Daily Rate
New York	\$194
Honolulu	\$159
Boston	\$155
Santa Barbara	\$144

Find the average of these four values. (*Source:* hotels.com)

Solution Find the average of these values on your own. Check your answer, or watch this video for the complete solution.

You Try It **Work through this You Try It problem.**

Work Exercises 47–49 in this textbook or in the MyMathLab Study Plan.

1.6 Exercises

In Exercises 1–4, write the product using exponential notation.

You Try It 1. $8 \cdot 8 \cdot 8 \cdot 8 \cdot 8$

2. $15 \cdot 15 \cdot 15 \cdot 15 \cdot 15 \cdot 15 \cdot 15 \cdot 15$

You Try It 3. $7 \cdot 7 \cdot 9 \cdot 9 \cdot 9 \cdot 9$

4. $4 \cdot 3 \cdot 3 \cdot 3 \cdot 10 \cdot 10 \cdot 10 \cdot 10$

In Exercises 5–14, evaluate the exponential expression.

You Try It 5. 4^2

6. 3^5

7. 1^8

8. 10^5

9. 0^7

10. 17^1

You Try It 11. $2 \cdot 6^2$

12. $3^3 \cdot 5^2$

13. $2^3 \cdot 10^4$

14. $0 \cdot 7^3$

In Exercises 15–38, simplify each expression using the order of operations.

You Try It 15. $20 - 4 \cdot 3$

16. $20 - 3^2$

17. $38 + \frac{12}{2}$

18. $7^2 + 24 \div 6$

19. $7 \cdot 3 - 1^5 \cdot 4$

20. $12 \cdot 10^3 + 36 \div 4$

21. $40 - 5 \times 2^3$

22. $15 \cdot 4 - 0 \div 3$

23. $8^2 \div 4 + 16 \cdot 3$

24. $8 \times 2 + 4^3 \cdot 1$

25. $30 - 7 + 3 - 4$

26. $6 \cdot 12 \div 2 \cdot 4$

You Try It 27. $25 - 3 - 8 + 6$

28. $24 \div 6 \cdot 2 + 7$

29. $6 + (5 - 3)^2$

30. $2^3 \cdot (7 + 3) - 2$

31. $(8 + 4^2) + \frac{28}{4}$

32. $31 - [3^2 + 2]$

You Try It 33. $24 \div 3 + [7 - 2^2]^3$

34. $(13 - 9)^2 \times 5 + 2^5$

35. $(18 + 9) \times (6 \div 2)$

36. $5^2 \cdot 3 - 6 \cdot (8 + 4)$

37. $\frac{12 + 8}{7 - 3}$

38. $\frac{(4 + 3)(6 - 2)}{2^2 + 5 \cdot 0}$

You Try It 39. $\frac{10^2 - 5 \cdot 8}{3(7 - 5)}$

40. $\frac{4^3 + 1^5}{7 \times 4 \div 2 - 1}$

41. $5 - [10 - (5 - 2)^2]$

42. $(12 \div 4) + [(7 + 5) \cdot 4]$

You Try It 43. $28 \div (4^2 + [2 + 3]^2 - 3^3)$

44. $\{4 + 3[18 - 2(4 + 1)]^2\} + 34$

In Exercises 45 and 46, find the average of the set of numbers.

You Try It 45. 7, 10, 19, 23, and 16

46. 28, 43, 55, 79, 80, 93, and 112

You Try It 47. **Bid Analyst** Roisin is a price analyst for a food distributor. One week she put together bids for 693 cases, 504 cases, 845 cases, 377 cases, and 591 cases. What was the average number of cases per bid?

(eText Screens 1.7-1–1.7-26)

48. **Harry Potter** The following table shows the ticket revenue (in $ millions) for the opening weekend of each movie in the Harry Potter series. What is the average opening weekend revenue for the series? (*Source:* http://www.the-numbers.com)

Movie	Opening Weekend Revenue ($ millions)	Movie	Opening Weekend Revenue ($ millions)
Sorcerer's Stone	90	*Order of the Phoenix*	77
Chamber of Secrets	88	*Half-Blood Prince*	78
Prisoner of Azkaban	94	*Deathly Hallows: Part I*	125
Goblet of Fire	103	*Deathly Hallows: Part II*	169

49. **Shuttle Missions** The following list shows the number of missions for each of the five shuttles used by NASA between 1981 and 2011. (*Source:* www.nasa.gov)

Shuttle	Missions
Columbia	28
Challenger	10
Endeavor	25
Atlantis	33
Discovery	39

What is the average number of missions for the five shuttles?

1.7 Introduction to Variables, Algebraic Expressions, and Equations

THINGS TO KNOW

Before working through this topic, be sure you are familiar with the following concepts:	VIDEO	ANIMATION	INTERACTIVE
You Try It — 1. Solve Applications by Adding or Subtracting Whole Numbers (Topic 1.3, **Objective 7**)	✓		
You Try It — 2. Solve Applications by Multiplying Whole Numbers (Topic 1.4, **Objective 6**)	✓		
You Try It — 3. Solve Applications by Dividing Whole Numbers (Topic 1.5, **Objective 3**)	✓		
You Try It — 4. Simplify Expressions Using the Order of Operations (Topic 1.6, **Objective 3**)	✓		✓

(eText Screens 1.7-1–1.7-26)

OBJECTIVES

1 Evaluate Algebraic Expressions

2 Distinguish Between Expressions and Equations

3 Determine If a Value Is a Solution to an Equation

4 Translate Word Phrases into Algebraic Expressions

5 Translate Sentences into Equations

OBJECTIVE 1 EVALUATE ALGEBRAIC EXPRESSIONS

A community college charges $150 per credit hour for courses. To find the cost of a course, we multiply 150 by the number of credit hours.

Rule for finding the cost of a course: $150 \cdot$ (number of credit hours in the course)

How much would a 4-credit-hour prealgebra course cost?

Cost per credit hour → $150 \cdot 4$ ← Number of credit hours

$150 \cdot 4$ The course would cost $600.

How much would a 2-credit-hour weight training course cost?

Cost per credit hour → $150 \cdot 2$ ← Number of credit hours

$150 \cdot 2$ The course would cost $300.

We can use a table to track the cost of various courses, depending on their credit hours.

Course	Rule: $150 \cdot$ (number of credit hours)	Cost of Course
Prealgebra	$150 \cdot 4$	$600
Weight Training	$150 \cdot 2$	$300
Computer Aided Drafting	$150 \cdot 5$	$750
Non-Western Cultures	$150 \cdot 3$	$450
Study Skills	$150 \cdot 1$	$150

Our rule for finding the cost of a course can be written more simply as follows:

$150 \cdot x$

↑ The x represents "the number of credit hours."

Because the number of credit hours per course may vary, we use a letter, x, to represent a general number of credit hours. The x is called a *variable*.

A **variable** is a symbol (usually a letter) that is used in place of a numeric value that can change, or *vary*, depending on the situation. A **constant** has a value that never changes. For example, the 150 in our rule for finding the cost of a course is a constant.

(eText Screens 1.7-1–1.7-26)

An **algebraic expression** is a variable or combination of variables, constants, **operations**, and **grouping symbols**. So $150 \cdot x$ is an algebraic expression. Other examples of algebraic expressions include

$$x,\quad 3z,\quad xy,\quad y^2,\quad a+4,\quad x+y,\quad 4x+7,\quad 5(a-b),\quad \text{and}\quad \frac{w-8}{2}.$$

Note: If two variables or a number and a variable are written directly next to each other with no operation written between them, then the operation is understood to be **multiplication**. So, $3z$ means $3 \cdot z$, xy means $x \cdot y$, and $4x + 7$ means $4 \cdot x + 7$. Our rule for finding the cost of a course can be written as $150x$.

An **algebraic expression** will have a different value depending on the specific values of the **variables**. For example, $150x$ has a value of 450 when $x = 3$, but it has a value of 750 when $x = 5$. When we find these specific values of an expression, we are *evaluating the algebraic expression*.

Evaluate Algebraic Expressions

To **evaluate an algebraic expression**, **substitute** the given values for the variables and **simplify** the resulting **numeric expression** using the **order of operations**.

Example 1 Evaluating Algebraic Expressions

My video summary

Evaluate each algebraic expression for the given value of the variable.

a. $n + 4$ for $n = 3$ **b.** $7a$ for $a = 9$

c. $x - 12$ for $x = 15$ **d.** $\frac{y}{4}$ for $y = 20$

Solutions

a. Begin with the original algebraic expression: $n + 4$

Substitute 3 for n: $= 3 + 4$

Add: $= 7$

The value of the expression $n + 4$ is 7 when n is 3.

b. Begin with the original algebraic expression: $7a$

Substitute 9 for a: $= 7(9)$

Multiply: $= 63$

The value of $7a$ is 63 when a is 9.

c.–d. Try to **evaluate** these **expressions** on your own, then check your **answers**. Or, watch this **video** for complete solutions to all four parts.

You Try It **Work through this You Try It problem.**

Work Exercises 1–4 in this textbook or in the MyMathLab Study Plan.

Example 2 Evaluating Algebraic Expressions

Evaluate each expression for the given values of the **variables**.

a. $25 - 3m$ for $m = 6$ **b.** $4p^2 + 10$ for $p = 3$.

Solutions

a. Begin with the original expression: $25 - 3m$

Substitute 6 for m: $= 25 - 3(6)$

Multiply: $= 25 - 18$ ← $3 \cdot 6 = 18$

Subtract: $= 7$

The value of the expression $25 - 3m$ is 7 when $m = 6$.

b. Begin with the original expression: $4p^2 + 10$

Substitute 3 for p: $= 4(3)^2 + 10$

Evaluate the exponent: $= 4(9) + 10$ ← $3 \cdot 3 = 9$

Multiply: $= 36 + 10$ ← $4 \cdot 9 = 36$

Add: $= 46$

The value of the expression $4p^2 + 10$ is 46 when $p = 3$.

CAUTION

When **substituting** a given value for a **variable**, it is often helpful to write the value within parentheses to avoid error, particularly when **multiplication** or **exponents** are involved.

You Try It **Work through this You Try It problem.**

Work Exercises 5–10 in this textbook or in the MyMathLab Study Plan.

Given a value for each variable, we can **evaluate** an **algebraic expression** containing more than one variable.

Example 3 Evaluating Algebraic Expressions

My interactive video summary

Evaluate each algebraic expression for the given value of the variable.

a. $x^2 + y^3$ for $x = 5$ and $y = 2$

b. $5(m - n) - 14$ for $m = 21$ and $n = 13$

c. $\dfrac{w^2 - 5z}{z}$ for $w = 8$ and $z = 4$

d. $8x - 2xy$ for $x = 11$ and $y = 3$

Solutions

a. Begin with the original expression: $x^2 + y^3$

Substitute 5 for x and 2 for y: $= (5)^2 + (2)^3$

Evaluate the exponents: $= 25 + 8$ ← $5 \cdot 5 = 25; 2 \cdot 2 \cdot 2 = 8$

Add: $= 33$

The value of the **expression** $x^2 + y^3$ is 33 when $x = 5$ and $y = 2$.

b. Begin with the original expression: $5(m - n) - 14$

Substitute 21 for m and 13 for n: $= 5(21 - 13) - 14$

Subtract within the parentheses: $= 5(8) - 14$ ← $21 - 13 = 8$

Multiply: $= 40 - 14$ ← $5(8) = 40$

Subtract: $= 26$

The value of the expression $5(m - n) - 14$ is 26 when $m = 21$ and $n = 13$.

(eText Screens 1.7-1–1.7-26)

c.–d. Try to **evaluate** these expressions on your own. Check the **answers**, or watch this **interactive video** for complete solutions to all four parts.

You Try It **Work through this You Try It problem.**

Work Exercises 11–20 in this textbook or in the MyMathLab Study Plan.

OBJECTIVE 2 DISTINGUISH BETWEEN EXPRESSIONS AND EQUATIONS

An **equation** is a statement that two quantities are equal. For example, the equation

$$\underbrace{18 - 13}_{\text{Quantity}} \overset{\text{Equal sign}}{=} \underbrace{5}_{\text{Quantity}}$$

means that the quantities $18-13$ and 5 are equal. This equation is a **numeric equation** because it contains only **numeric expressions** and no **variables**. When an equation contains one or more variables, it is called an **algebraic equation**. An algebraic equation indicates that two **algebraic expressions** are equal. Following is an example of an algebraic equation.

$$\underbrace{2x + 6}_{\text{Algebraic Expression}} \overset{\text{Equal sign}}{=} \underbrace{x + 18}_{\text{Algebraic Expression}}$$

CAUTION

An *equation* contains an equal sign $(=)$, while an *expression* does not. Typically, we will use the word "equation" to mean "algebraic equation."

Example 4 Distinguishing Between Expressions and Equations

Determine whether each of the following is an **expression** or an **equation**.

a. $5x - 8 = 2x + 10$ **b.** $3x^2 - 7x + 12$

c. $6x + 8y - 3z$ **d.** $4(10) - 3(5) = 3^2 + 4^2$

Solutions

a. $5x - 8 = 2x + 10$ has an equal sign, so it is an equation.

b. $3x^2 - 7x + 12$ has no equal sign, so this is an expression.

c. $6x + 8y - 3z$ has no equal sign, so this is an expression.

d. $4(10) - 3(5) = 3^2 + 4^2$ has an equal sign, so it is an equation.

You Try It **Work through this You Try It problem.**

Work Exercises 21–26 in this textbook or in the MyMathLab Study Plan.

OBJECTIVE 3 DETERMINE IF A VALUE IS A SOLUTION TO AN EQUATION

Consider the following two **numeric equations:**

$$2(3) = 6 \qquad 8 - 8 = 16$$

True False

The first equation is a **true equation** because its statement is true: $2(3)$ does equal 6. However, the second equation is a **false equation** because its statement is false: $8 - 8$ does *not* equal 16 (because $8 - 8$ results in 0, not 16).

When working with an **algebraic equation**, typically our goal is to **solve** it, or find all values for the **variable** that make a true equation. Any value that makes the equation true is called a **solution** of the equation.

> **Definition** Solve an Equation
>
> To **solve an equation** means to find all values for the variable that make the equation true.

> **Definition** Solution
>
> A **solution** of an equation is a value that, when **substituted** for the variable, makes the equation true.

For example, 3 is a **solution** of the **equation** $x + 6 = 9$ because substituting 3 for x results in the true statement $3 + 6 = 9$. On the other hand, 2 is *not* a solution of $x + 6 = 9$ because substituting 2 for x results in the false statement $2 + 6 = 9$.

To determine if a given value is a solution to an equation, **substitute** the given value for the **variable**, and **simplify** both sides of the equation. If the resulting statement is true, then the value is a solution. If the resulting statement is false, then the value is not a solution.

Example 5 Determining If a Given Value Is a Solution to an Equation

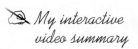

My interactive video summary

Determine if the given value of the variable is a solution to the equation.

a. $6x - 5 = 43;\ 8$ **b.** $4a = a + 8;\ 3$

c. $3n - 7 = 2n + 5;\ 10$ **d.** $\dfrac{y + 11}{2} = 2(y - 1);\ 5$

Solutions

a. Begin with the original equation: $6x - 5 = 43$

Substitute 8 for x: $6(8) - 5 \stackrel{?}{=} 43$

Multiply: $48 - 5 \stackrel{?}{=} 43$

Subtract: $43 = 43$ True

The final statement is true, so 8 is a solution to the equation.

(eText Screens 1.7-1–1.7-26)

b. Begin with the original equation: $4a = a + 8$

Substitute 3 for a: $4(3) \stackrel{?}{=} 3 + 8$

Multiply or add: $12 = 11$ False

The final statement is false, so 3 is not a solution to the equation.

c.–d. Try to do parts c and d on your own, then view the **answers**. Watch this **interactive video** to see complete solutions to all four parts.

You Try It **Work through this You Try It problem.**

Work Exercises 27–38 in this textbook or in the MyMathLab Study Plan.

OBJECTIVE 4 TRANSLATE WORD PHRASES INTO ALGEBRAIC EXPRESSIONS

Problem solving often involves translating word phrases into **algebraic expressions.** We have seen before that many key words or phrases translate into the **arithmetic operations.** Some of these key words or phrases, along with their respective operations, are listed in Table 2 for your review.

Addition	Subtraction	Multiplication	Division
Add	Subtract	Multiply	Divide
Added to	Subtracted from	Times	Divided by
Plus	Minus	Product	Quotient
Sum	Difference	Of	Per
Increased by	Decreased by	Double	Ratio
More than	Less	Triple	Goes into
Total	Less than	Twice	Shared evenly

Table 2

In earlier topics, we translated word statements into **numeric expressions.** We now use **variables** to represent unknown values and translate word statements into algebraic expressions.

Consider the following word phrase:

The sum of a number and 15

(a number: Unknown value; 15: Known value)

To translate this phrase into an **algebraic expression**, we use the variable x to represent the unknown value.

The sum of a number and 15

$x + 15$

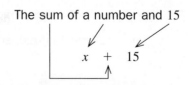

So, the phrase *"the sum of a number and 15"* translates into the algebraic expression $x + 15$.

Example 6 Writing Word Phrases as Algebraic Expressions

My video summary

Write each word phrase as an algebraic expression. Use x to represent the unknown number.

a. 12 increased by a number

b. The product of 6 and a number

c. 25 less than a number

d. The quotient of 30 and a number

Solutions Work through the following, or watch this **video**.

a. Looking at **Table 2**, the phrase "increased by" means **addition.** What is being added? 12 is increased by an unknown number x, so x is being added to 12.

12 increased by a number

$12 + x$

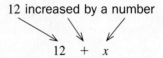

So, "12 increased by a number" is written as $12 + x$.

b. "Product" means **multiplication.** What is being multiplied? 6 and an unknown number x are being multiplied.

The product of 6 and a number

$6 \cdot x$

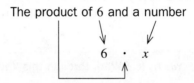

"The product of 6 and a number" translates into $6 \cdot x$ or $6x$.

c. "Less than" indicates **subtraction.** What is being subtracted? 25 is being subtracted from an unknown number x.

25 less than a number

$x - 25$

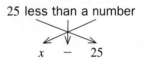

"25 less than a number" is the expression $x - 25$.

d. "Quotient" implies **division.** What is being divided? 30 is being divided by the unknown number x.

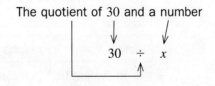

The quotient of 30 and a number

$30 \div x$

"The quotient of 30 and a number" translates into $30 \div x$, or $\frac{30}{x}$.

CAUTION

When a word phrase indicates subtraction, pay particular attention to which quantity is the **minuend** and which is the **subtrahend.** Likewise, when a word phrase indicates division, pay special attention to which quantity is the **dividend** and which is the **divisor.**

(eText Screens 1.7-1–1.7-26)

Sometimes a word phrase indicates more than one **operation**.

Example 7 Writing Word Phrases as Algebraic Expressions

Write each word phrase as an algebraic expression. Use x to represent the unknown number.

a. The **sum** of 4 and triple a number

b. 8 times the **difference** of a number and 9

Solutions

a. The word "sum" means **addition**. What is being added? 4 is being added with "triple a number." The word "triple" means to multiply by 3. What is being multiplied by 3? An unknown number x.

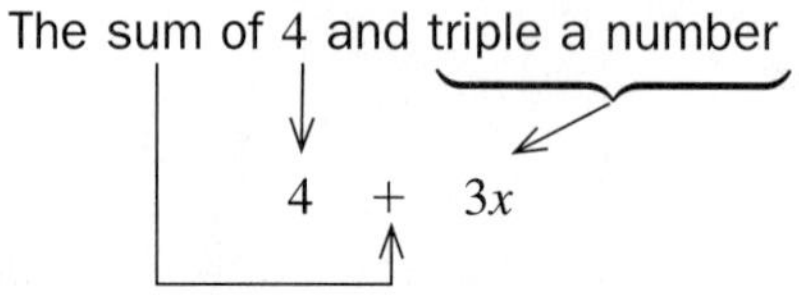

So, "the sum of 4 and triple a number" is written as $4 + 3x$.

My video summary

b. Try to translate this word phrase on your own. View the **answer**, or watch this **video** for the complete solution to part b.

You Try It **Work through this You Try It problem.**

Work Exercises 39–48 in this textbook or in the MyMathLab Study Plan.

OBJECTIVE 5 TRANSLATE SENTENCES INTO EQUATIONS

Just as we can translate word phrases to **algebraic expressions**, we can also translate sentences into **equations**.

Like the **key words** that translate into **arithmetic operations**, there are key words that translate into an equal sign $(=)$. Some of these words are shown in Table 3.

Key Words That Translate into an Equal Sign			
is	was	will be	gives
yields	results in	equals	is equal to
is equivalent to	is the same as	amounts to	makes

Table 3

Consider the following sentence:

The **product** of 3 and a number is 21.

The key word "product" implies multiplication, and the key word "is" indicates an equal sign. Letting x represent the unknown number, we translate:

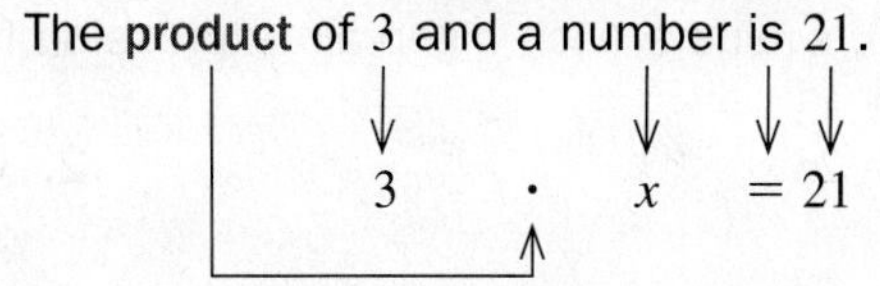

The equation is written as $3x = 21$.

Example 8 Translating Sentences into Equations

Translate each sentence into an **equation**. Let x represent the unknown number.

a. Ten more than a number is 42.

b. A number divided by 5 will be 18.

c. Four times a number, decreased by 27, gives the number.

d. The **sum** of a number and 6 is the same as the **difference** of 12 and the number.

Solutions

a. The phrase "more than" means **addition**. "Ten more than a number" translates to the **algebraic expression** $x + 10$. "Is" indicates an equal sign.

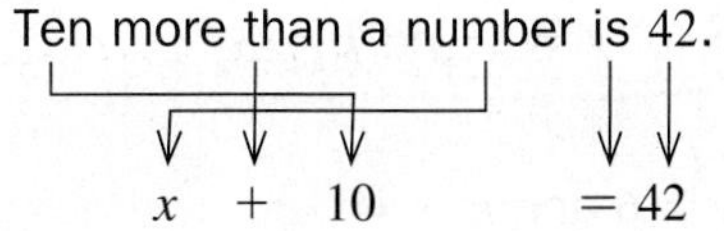

The equation is written as $x + 10 = 42$.

b. "Divided by" indicates **division**. "A number divided by 5" translates to the expression $\frac{x}{5}$. "Will be" indicates an equal sign.

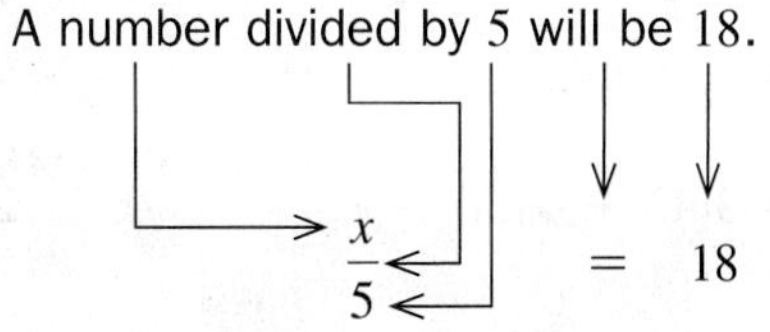

The equation is $\frac{x}{5} = 18$.

My video summary

c.–d. Try to translate these sentences on your own, then check your **answers**. Watch this **video** for the complete solutions to parts c and d.

You Try It **Work through this You Try It problem.**

Work Exercises 49–58 in this textbook or in the MyMathLab Study Plan.

1.7 Exercises

In Exercises 1–20, evaluate each algebraic expression for the given values of the variables.

You Try It

1. $x + 9$ for $x = 25$
2. $y - 8$ for $y = 14$
3. $13m$ for $m = 4$
4. $\frac{n}{3}$ for $n = 72$

You Try It

5. $7x + 1$ for $x = 8$
6. $64 - w \div 2$ for $w = 42$
7. $6z^3$ for $z = 2$
8. $5a^2 - 14$ for $a = 6$
9. $\frac{2p + 6}{3}$ for $p = 9$
10. $4(t - 7) + 3t$ for $t = 9$

You Try It

11. $7r - 3s$ for $r = 11$ and $s = 10$
12. $9(x - y) + 5$ for $x = 13$ and $y = 8$
13. $a^3 + b^2$ for $a = 4$ and $b = 3$
14. $7mn - 3n$ for $m = 2$ and $n = 5$
15. $(b - 2c)^3$ for $b = 9$ and $c = 4$
16. $\frac{7w - z^2}{2z}$ for $w = 9$ and $z = 3$
17. $2x^2 + 3xy - y^2$ for $x = 1$ and $y = 3$
18. $4w(5x - y)$ for $w = 3$, $x = 2$, and $y = 6$
19. $\frac{2m - 3n}{2n + m}$ for $m = 6$ and $n = 4$
20. $\frac{5xy}{z}$ for $x = 4$, $y = 3$, and $z = 6$

In Exercises 21–26, determine whether the statement is an expression or an equation.

21. $4x^2 + x - 9$
22. $2(x + 4) = 3x - 5$

You Try It

23. $2x - 3y = 18$
24. $17w - 12x + 5y$
25. $8(2)^3 - 13 = 3(13 + 4)$
26. $x(x - 4) = 5$

In Exercises 27–38, determine if the given value of the variable is a solution to the equation.

You Try It

27. $x + 4 = 15$; 9
28. $3y - 4 = 20$; 8
29. $7a = 3a + 20$; 4
30. $4n - 3 = 15 - 2n$; 3
31. $10(z + 3) = 50$; 2
32. $3t + 8 = 2(t + 4)$; 0
33. $20 = 80a$; 4
34. $\frac{m + 3}{4} = 5$; 9
35. $\frac{k + 7}{3} = 3(k - 1)$; 2
36. $3(2t + 5) - 13 = 6t + 2$; 10
37. $2n^2 = 16$; 4
38. $x(x - 7) = 18$; 9

(eText Screens 1.7-1–1.7-26)

In Exercises 39–48, write each word phrase as an algebraic expression. Use x to represent the unknown number.

39. The difference of a number and 12

40. The sum of a number and 57

41. 80 decreased by a number

42. The product of 12 and a number

43. The quotient of a number and 32

44. 7 more than the product of 6 and a number

45. 25 less than a number

46. The quotient of a number and 5, minus 16

47. 4 times a number, increased by 10

48. Double a number, minus 45

In Exercises 49–58, translate each sentence into an equation. Use x to represent the unknown number.

You Try It 49. Twelve less than a number is 65.

50. Twenty-one more than a number is 84.

51. The quotient of a number and 4 results in 25.

52. The product of 8 and a number yields 496.

53. Adding 2 to the product of 6 and a number gives 50.

54. A number divided by 7, minus 12, results in 11.

55. A number subtracted from 63 makes 41.

56. A number increased by 49 equals 83.

57. Twice the sum of a number and 1 is the same as the number plus 8.

58. Six times a number, increased by 21, amounts to 9 times the number, decreased by 3.

MODULE TWO

Integers and Introduction to Solving Equations

MODULE TWO CONTENTS

2.1 Introduction to Integers

THINGS TO KNOW

Before working through this topic, be sure you are familiar with the following concepts:

VIDEO ANIMATION INTERACTIVE

You Try It 1. Use Inequality Symbols to Compare Whole Numbers (Topic 1.2, **Objective 4**)

You Try It 2. Read Tables and Bar Graphs Involving Whole Numbers (Topic 1.2, **Objective 6**)

OBJECTIVES

1 Graph Integers on a Number Line

2 Use Inequality Symbols to Compare Integers

3 Find the Absolute Value of a Number

4 Find the Opposite of a Number

5 Solve Applications Involving Integers

OBJECTIVE 1 GRAPH INTEGERS ON A NUMBER LINE

A **whole number** can be either 0 or greater than 0. When a number is greater than 0, it is called a **positive number**. We can use positive numbers to describe real-world situations such as a temperature of 5° *above zero*, a $100 *deposit* into a checking account, or a 10-yard *gain* on a football field. We can write positive numbers with or without a **positive sign**, +. Table 1 shows each situation with its representative positive number.

Situation	Number Form	Word Form
5° above zero	+5 or 5	positive five
$100 deposit	+100 or 100	positive one hundred
10-yard gain	+10 or 10	positive ten

Table 1

Many situations cannot be described by numbers that are 0 or positive. For example, consider a temperature of 5° *below zero*, a $100 *withdrawal* from a checking account, or a 10-yard *loss* on a football field. When a number is less than 0, it is called a **negative number.** We write negative numbers with a **negative sign**, −. Table 2 shows each situation with its representative negative number.

Situation	Number Form	Word Form
5° below zero	−5	negative five
$100 withdrawal	−100	negative one hundred
10-yard loss	−10	negative ten

Table 2

Every **positive whole number** has a related **negative number.** For example, 3 is related to −3, and 50 is related to −50. If the positive whole numbers, their related negative numbers, and zero are all grouped together, they form the *integers*.

Definition Integers

The **integers** are

$$\ldots, -5, -4, -3, -2, -1, 0, 1, 2, 3, 4, 5, \ldots$$

The ellipsis (. . .) at each end means that the list of integers continues without end in both the negative and positive directions.

CAUTION Zero is neither positive nor negative.

In **Topic 1.2**, we visualized **whole numbers** using a number line. We can extend the number line to the left of 0, the **origin**, to represent the negative numbers. **Figure 1** shows the number line with equally spaced marks that represent the integers.

(eText Screens 2.1-1–2.1-20)

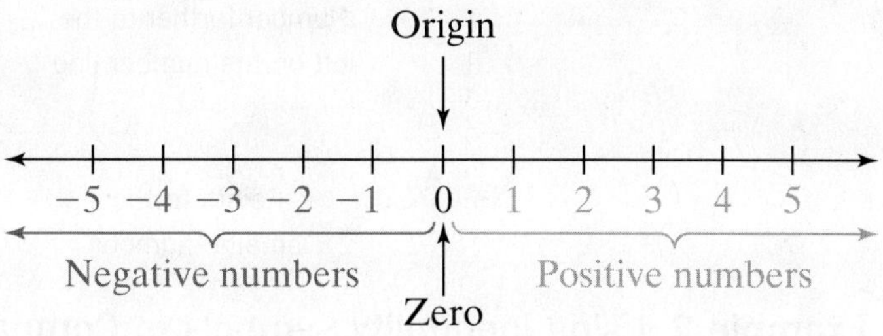

Figure 1 The Number Line

We can **graph**, or **plot**, an **integer** by placing a solid circle (•) at its location on the **number line**.

Example 1 Graphing Integers on the Number Line

My video summary

Graph 0, -4, 3, and -2 on the number line.

Solution To graph each integer, we place a solid circle at its location on the number line. See Figure 2. Watch this **video** for a detailed solution.

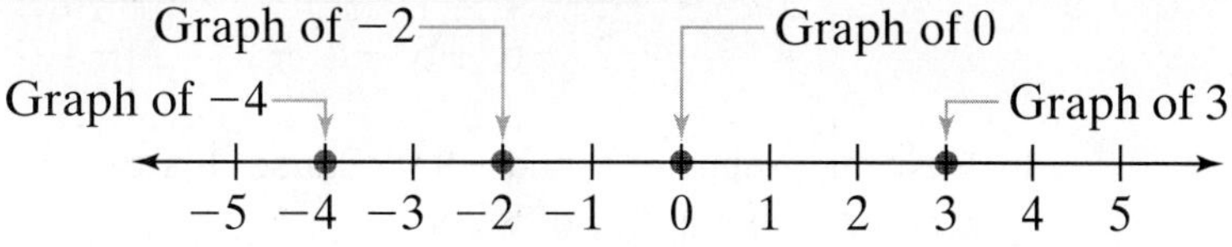

Figure 2

You Try It **Work through this You Try It problem.**

Work Exercises 1–4 in this textbook or in the MyMathLab Study Plan.

OBJECTIVE 2 USE INEQUALITY SYMBOLS TO COMPARE INTEGERS

We can use the **number line** to compare **integers.** As we move from left to right on the number line, the numbers increase in value, or get larger. Likewise, as we move from right to left, the numbers decrease in value, or get smaller.

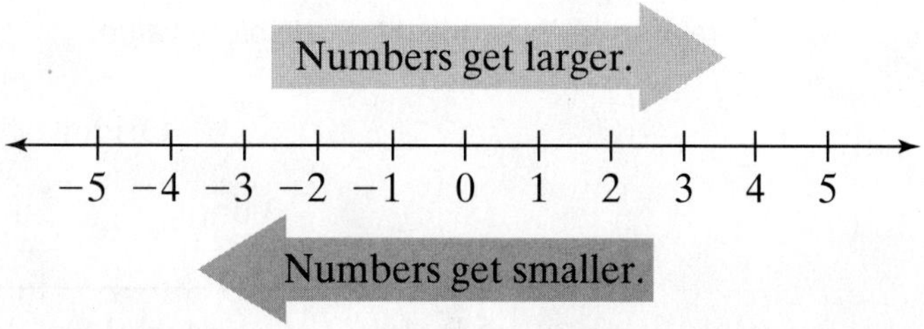

We can use **inequality symbols** to write comparison statements for integers. For example, because -4 is to the left of -2 on the number line, we say that -4 *is less than* -2, or $-4 < -2$. We can also say that -2 *is greater than* -4, or $-2 > -4$.

CAUTION Remember that the inequality symbol will always point to the smaller number, which is the number farther to the left on the number line.

(eText Screens 2.1-1–2.1-20)

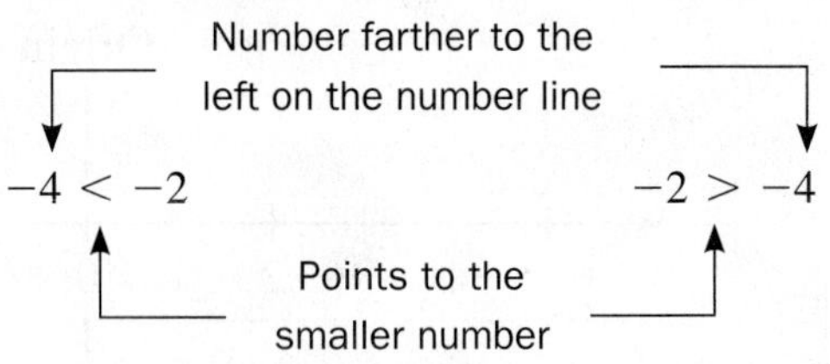

Example 2 Using Inequality Symbols to Compare Integers

Fill in the blank with an **inequality symbol**, $<$ or $>$, to make a true comparison statement.

a. -6____1 **b.** 0____-8 **c.** -5____-7

Solutions

a. The graphs of -6 and 1 are shown on the **number line** below. Because -6 is to the left of 1, -6 *is less than* 1. So, $-6 < 1$.

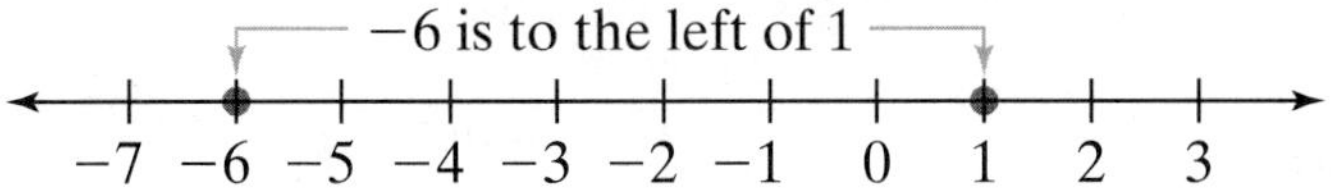

b. View the **graphs of** 0 **and** -8. Because 0 is to the right of -8 on the number line, 0 *is greater than* -8. So, $0 > -8$.

c. View the **graphs of** -5 **and** -7. Because -5 is to the right of -7 on the number line, 5 *is greater than* -7. So, $-5 > -7$.

You Try It **Work through this You Try It problem.**

Work Exercises 5–10 in this textbook or in the MyMathLab Study Plan.

OBJECTIVE 3 FIND THE ABSOLUTE VALUE OF A NUMBER

Figure 3 shows the graphs of -3 and 3 on the **number line**. Notice that the distance from 0 to 3 is 3 units and that the distance from 0 to -3 is also 3 units. These distances are related to the concept of *absolute value*.

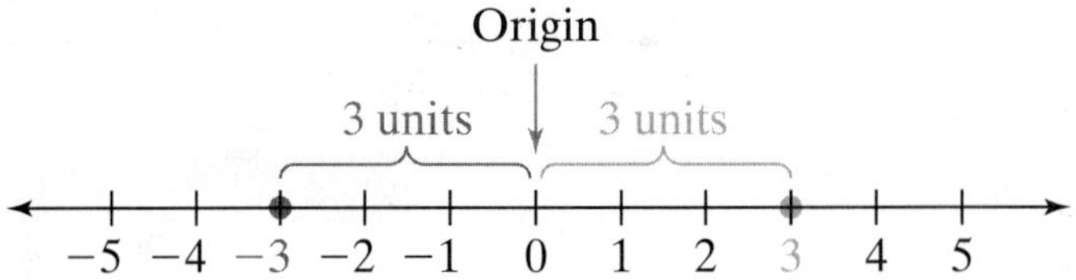

Figure 3

> **Definition Absolute Value**
>
> The **absolute value** of a number a, written as $|a|$, is the distance from 0 to a on the number line.

We read the notation $|3|$ as "the absolute value of three," and we read $|-3|$ as "the absolute value of negative three." Because both 3 and -3 are 3 units from 0 on the number line, both absolute values equal 3. So, $|3| = 3$ and $|-3| = 3$.

CAUTION

Absolute value is an expression of distance, so it cannot be **negative**. The absolute value of a number will always be **positive** or 0.

Example 3 Finding Absolute Values of Numbers

My video summary

Find each **absolute value**.

a. $|5|$ **b.** $|-4|$ **c.** $|0|$

Solutions Work through the following, or watch this **video** for detailed solutions to all three parts.

a. The distance from 0 to 5 on the **number line** is 5 units. See Figure 4. So, $|5| = 5$.

b. The distance from 0 to -4 on the number line is 4 units. See Figure 4. So, $|-4| = 4$.

c. The distance from 0 to 0 on the number line is 0 units. So, $|0| = 0$.

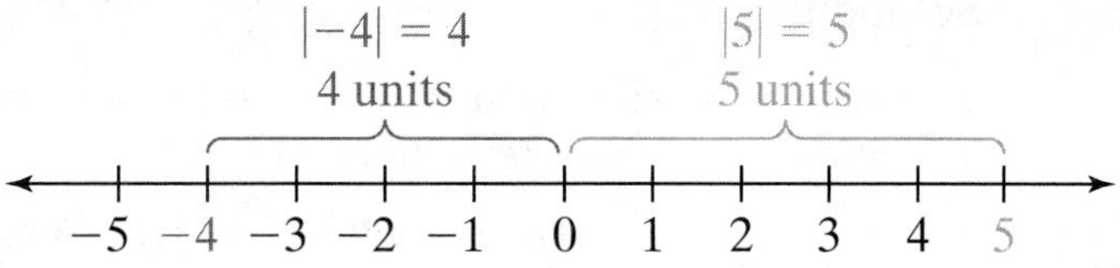

Figure 4

You Try It **Work through this You Try It problem.**

Work Exercises 11–15 in this textbook or in the MyMathLab Study Plan.

OBJECTIVE 4 FIND THE OPPOSITE OF A NUMBER

Look again at **Figure 3**. Notice that 3 and -3 are the same distance from 0 on the **number line** (or have the same **absolute value**), but they appear in opposite directions. Because of this, we say that 3 and -3 are *opposites*.

> **Definition Opposites**
>
> Two numbers are **opposites** if their graphs are located the same distance from 0 on the number line but lie on opposite sides of 0.

So, the opposite of a **positive number** is the **negative number** with the same absolute value, and the opposite of a negative number is the positive number with the same absolute value. Zero is its own opposite. See **Figure 5**.

(eText Screens 2.1-1–2.1-20)

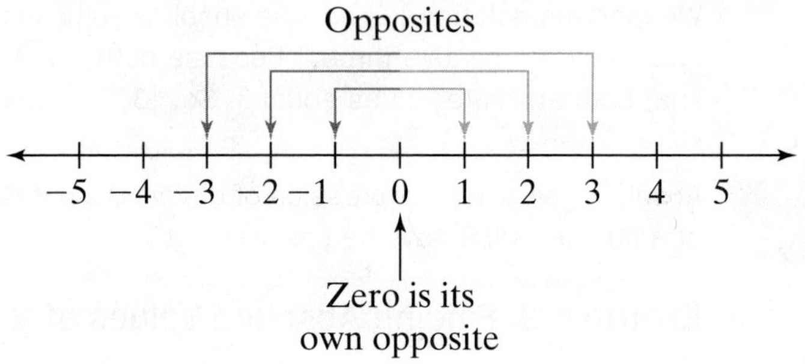

Figure 5

> **Finding the Opposite of a Number**
> To find the **opposite** of a number, change its **sign**.

Example 4 Finding Opposites of Numbers

Find the opposite of each number.

a. 16 **b.** -9 **c.** 0

Solutions

a. Because 16 is a **positive number**, we find its opposite by changing the sign to negative. The opposite of 16 is -16.

b. Because -9 is a **negative number**, we find its opposite by changing the sign to positive. The opposite of -9 is 9.

c. Because 0 is neither positive nor negative, it is its own opposite. The opposite of 0 is 0.

You Try It **Work through this You Try It problem.**

Work Exercises 16–20 in this textbook or in the MyMathLab Study Plan.

When two numbers are **opposites**, each one is the opposite of the other. For example, the opposite of 3 is -3, and the opposite of -3 is 3. We often read a negative sign $(-)$ as "the opposite of." For example, we can read -3 as "the opposite of 3," and we can read $-(-3)$ as "the opposite of negative three." The opposite of -3 is 3, so $-(-3) = 3$. This provides a rule that is true for all numbers.

> **Double-Negative Rule**
>
> If a is a number, then $-(-a) = a$.

A number containing more than one **sign** can be simplified to an equivalent number with no more than one sign.

Example 5 Simplifying Expressions with Multiple Signs

Simplify each expression.

a. $-(-20)$ **b.** $-|-8|$ **c.** $-|13|$ **d.** $|-(-6)|$

Solutions

My video summary

a. We can read $-(-20)$ as "the opposite of -20," which is 20. So, $-(-20) = 20$.

$$\underbrace{\text{The opposite of}}\ \underbrace{-20} \;\rightarrow\; -(-20) = 20$$

b. We can read $-|-8|$ as "the **opposite** of the **absolute value** of -8." Because the absolute value of -8 is 8, we have the opposite of 8, or -8. So, $-|-8| = -8$.

$$\underbrace{\text{The opposite of}}\ \underbrace{\text{the absolute value of } -8} \;\rightarrow\; -|-8| = -8$$

c.–d. Try **simplifying** these **expressions** on your own. View the **answers**, or watch this **video** for complete solutions to all four parts.

You Try It **Work through this You Try It problem.**

Work Exercises 21–25 in this textbook or in the MyMathLab Study Plan.

OBJECTIVE 5 SOLVE APPLICATIONS INVOLVING INTEGERS

We now look at some applications involving **integers**.

Example 6 Using Integers to Represent Real-World Situations

In each situation, write the given number as an integer.

a. The TauTona gold mine in South Africa is currently the world's deepest mine at nearly 4 kilometers below the surface. (*Source: National Geographic*)

b. On September 13, 1922, the highest temperature ever recorded on Earth occurred in El Azizia, Libya, at 136° above 0° Fahrenheit. (*Source:* National Climatic Data Center)

Solutions

a. If 0 represents ground level, then a depth of 4 kilometers below the surface is represented by -4.

b. If 0 represents $0°$ Fahrenheit, then a temperature of $136°$ Fahrenheit is represented by $+136$ or 136.

You Try It **Work through this You Try It problem.**

Work Exercises 26–31 in this textbook or in the MyMathLab Study Plan.

(eText Screens 2.1-1–2.1-20)

Example 7 Comparing Integers in an Application

In Major League Baseball, **integers** are used to find a team's rank in its division by representing the number of games that the team is behind the front-runner. At the end of the 2010 season, the Cincinnati Reds led the National League Central Division (NLC). The other teams in the NLC (and games behind) are the Chicago Cubs (-16), Houston Astros (-15), Milwaukee Brewers (-14), Pittsburgh Pirates (-34), and St. Louis Cardinals (-5). (*Source:* baseball-reference.com)

Use the information to rank the teams from best to worst in the NLC that season.

Rank	NLC Team	Games Behind the Front-runner
1	Cincinnati Reds	0
2		
3		
4		
5		
6		

Solution Writing the integers in order from largest to smallest, we get -5, -14, -15, -16, and -34. This represents the ranking of fewer games behind (best) to more games behind (worst). We write these numbers in the table along with the corresponding team.

Rank	NLC Team	Games Behind
1	Cincinnati Reds	0
2	St. Louis Cardinals	-5
3	Milwaukee Brewers	-14
4	Houston Astros	-15
5	Chicago Cubs	-16
6	Pittsburgh Pirates	-34

You Try It **Work through this You Try It problem.**

Work Exercises 32 and 33 in this textbook or in the MyMathLab Study Plan.

2.1 Exercises

In Exercises 1–4, graph each list of integers on the number line.

You Try It 1. $0, 2, 5, 8$ 2. $-1, -3, -5, -7$ 3. $2, -5, -8, 9$ 4. $-10, 15, -16, 20$

In Exercises 5–10, fill in the blank with the correct inequality symbol, $<$ or $>$, to make a true comparison statement.

You Try It 5. -8____2 6. -3____-6 7. 0____-6

8. 10____-32 9. -18____18 10. -21____-16

In Exercises 11–15, find each absolute value.

You Try It 11. $|6|$ 12. $|-2|$ 13. $|0|$ 14. $|-34|$ 15. $|78|$

In Exercises 16–20, find the opposite of each number.

You Try It 16. 11 17. -4 18. -34 19. 85 20. 0

In Exercises 21–25, simplify each expression.

You Try It 21. $-(-13)$ 22. $-(19)$ 23. $-|-10|$ 24. $-|3|$ 25. $|-(-12)|$

In Exercises 26–31, write the given number in each situation as an integer.

You Try It 26. **NASDAQ** On September 1, 2011, the NASDAQ Composite Index closed with a loss for the day of about $33. (*Source:* TD Ameritrade)

27. **Elevation** The highest point in the U.S. is Mount McKinley, Alaska, with an elevation of 20,320 feet above sea level. (*Source:* U.S. Geological Survey)

28. **Record Cold** On July 21, 1983, the lowest temperature ever recorded on Earth occurred at Vostok, Antarctica, at $129°$ below $0°$ Fahrenheit. (*Source:* National Climatic Data Center)

29. **Profits** During the first quarter of 2011, Ford Motor Company earned an average profit of $2806 for each car it produced in North America. (*Source: The Wall Street Journal*)

30. **Shipwreck Discovery** On September 1, 1985, the wreck of the RMS *Titanic* was discovered by Robert Ballard at a depth of about 12,500 feet below sea level. (*Source: National Geographic*)

31. **Deepest Lake** The world's deepest lake is Lake Baikal in southern Russia. Its floor is 5314 feet below the water's surface. (*Source:* geology.com)

You Try It 32. **Golf Match** In golf, *par* is the expected number of strokes needed to complete a hole or a round of holes. *Below par* means that a player took fewer strokes than expected, and *above par* means that a player took more strokes than expected. The player finishing with the fewest number of strokes wins. At the end of a match, the results for six players are as follows: Murphy, 1 above par; Johnson, 2 below par; Kinley, 3 below par; Renda, par; Forbes, 5 below par; and Finley, 3 above par.

(eText Screens 2.2-1–2.2-20)

Write each player's score as an integer. Then rank the players from best to worst for this match, and complete the table.

Rank	Player	Score (in integer form)
1		
2		
3		
4		
5		
6		

For Exercise 33, you may want to review how to read bar graphs in **Topic 1.2**.

33. **Water Level** For a given week, the following bar graph shows the water level of a river as compared to flood stage (0 feet). Use integers to fill in the table first and then answer the questions.

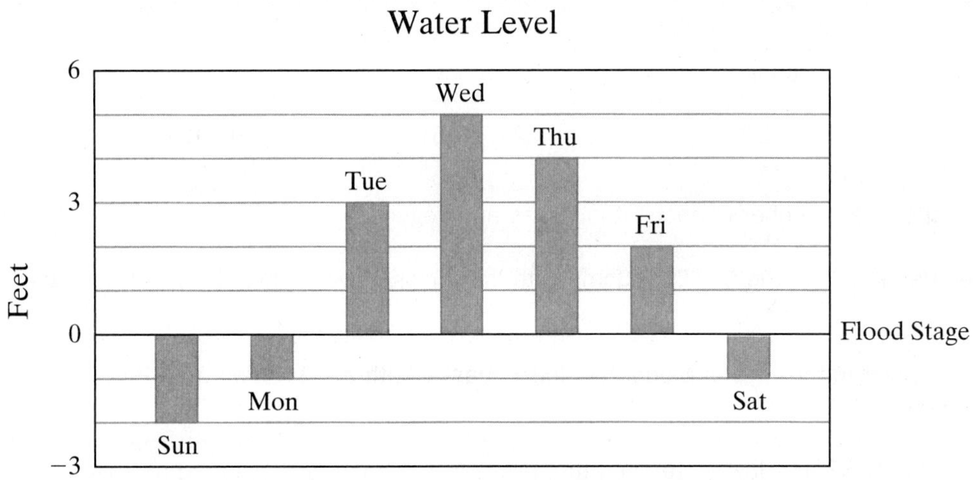

Day	Water Level (feet)
Sunday	
Monday	
Tuesday	
Wednesday	
Thursday	
Friday	
Saturday	

a. On what day(s) was the river above flood stage?

b. On what day was the river at its highest level? What was the highest water level?

c. On what day was the river at its lowest level? What was the lowest water level?

d. List the two days on which the water levels were opposite integers. What are the opposite integers?

2.2 Adding Integers

THINGS TO KNOW

Before working through this topic, be sure you are familiar with the following concepts: VIDEO ANIMATION INTERACTIVE

You Try It **1.** Add Whole Numbers (Topic 1.3, **Objective 1**)

You Try It **2.** Graph Integers on a Number Line (Topic 2.1, **Objective 1**)

(eText Screens 2.2-1–2.2-20)

You Try It — **3.** Find the Absolute Value of a Number (Topic 2.1, **Objective 3**)

You Try It — **4.** Find the Opposite of a Number (Topic 2.1, **Objective 4**)

OBJECTIVES

1 Add Two Integers with the Same Sign

2 Add Two Integers with Different Signs

3 Add More Than Two Integers

4 Solve Applications Involving Addition of Integers

OBJECTIVE 1 ADD TWO INTEGERS WITH THE SAME SIGN

When adding **integers**, it can be helpful to visualize the **addition** by using a **number line.** To begin to add 2 and 3, we locate the first number 2 on the number line. Recall that the **sign** of the number tells us on which side of zero the number lies: left of zero (**negative**) or right of zero (**positive**).

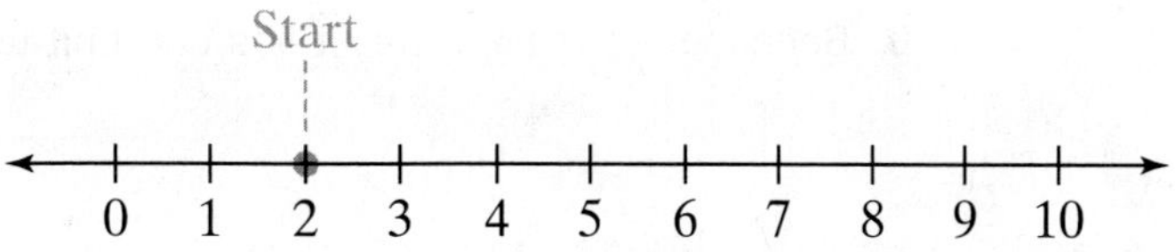

When adding, the sign of the second number tells us which direction to move: left (negative) or right (positive). Our second number, 3, is positive so we will move to the right. We then move along the number line the same number of units as the **absolute value** of the second number. Because the absolute value of 3 is 3, we move 3 units to the right and land on 5. Therefore,

$$2 + 3 = 5.$$

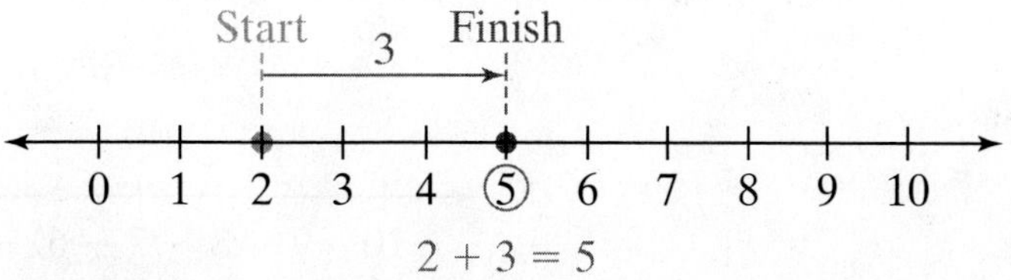

Example 1 Adding Two Integers with the Same Sign

Add.

a. $3 + 4$ **b.** $-1 + (-5)$

Solutions

a. Because 3 is positive, it lies three units to the right of zero on the number line.

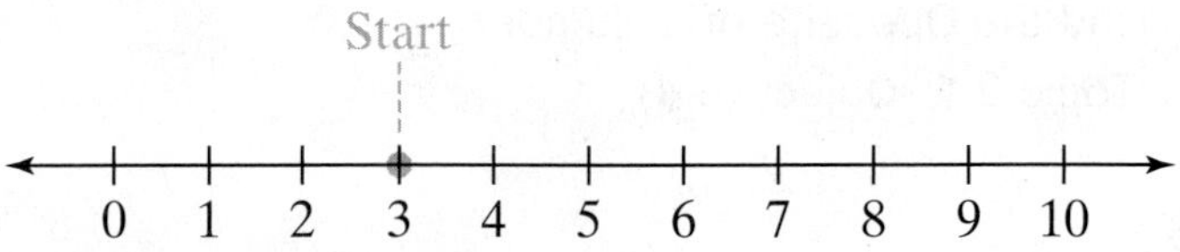

Next we find the absolute value of the second number.

$$|4| = 4$$

Starting at the first number, we move the number of units indicated by the absolute value of the second number, 4 units. Because the second number is positive, we move 4 units to the *right*.

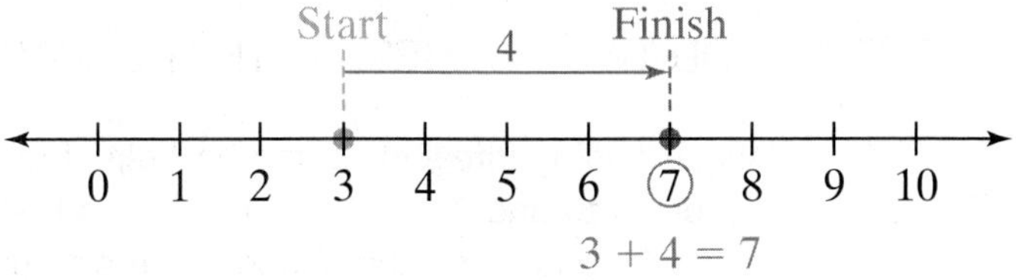

So, $3 + 4 = 7$.

b. Because -1 is **negative**, it lies one unit to the left of zero on the **number line**.

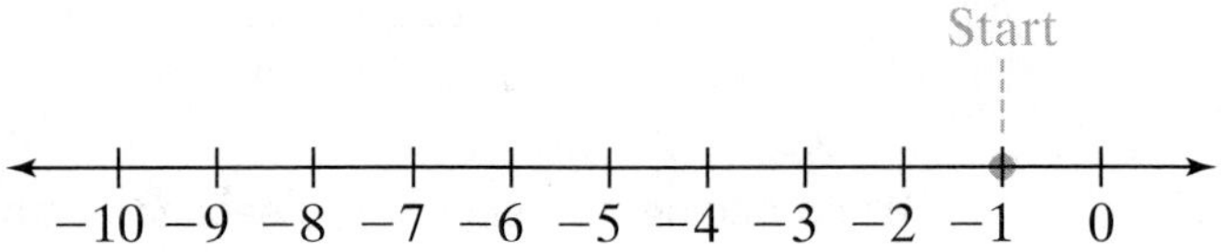

Next we find the **absolute value** of the second number.

$$|-5| = 5$$

Starting at the first number, we move the number of units indicated by the absolute value of the second number, 5 units. Because the second number is negative, we move 5 units to the *left*.

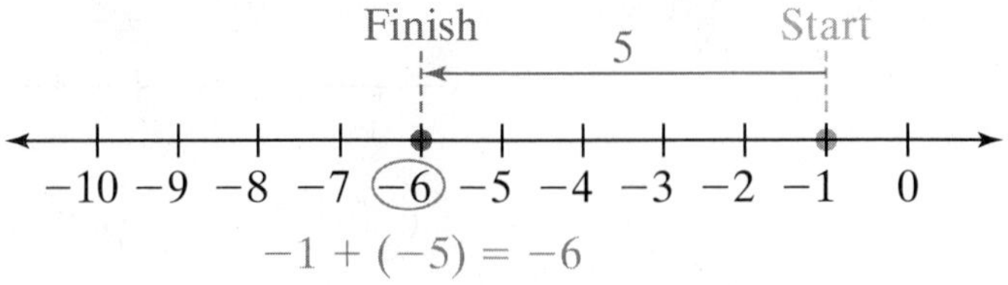

So, $-1 + (-5) = -6$.

You Try It **Work through this You Try It problem.**

Work Exercises 1–4 in this textbook or in the MyMathLab Study Plan.

Drawing **number lines** each time we want to add two numbers can be time consuming. Instead, we can replace the number line approach with the following procedure.

> **Adding Two Numbers with the Same Sign**
>
> If two numbers being added have the same **sign** (both **positive** or both **negative**), then we add as follows:
>
> **Step 1.** Find the **absolute value** of each term.
>
> **Step 2.** Add the two results from Step 1.
>
> **Step 3.** Use the common sign as the sign of the **sum**.

Example 2 Adding Two Integers with the Same Sign

Add: $-2 + (-8)$

Solution First, we find the absolute values, $|-2| = 2$ and $|-8| = 8$. Then we add the results, $2 + 8 = 10$. Because the common sign is negative, the sum will be negative. Thus, $-2 + (-8) = -10$.

You Try It **Work through this You Try It problem.**

Work Exercise 5 in this textbook or in the MyMathLab Study Plan.

Example 3 Adding Two Integers with the Same Sign

Add: $5 + 3$

Solution First, we find the absolute values, $|5| = 5$ and $|3| = 3$. Then we add the results, $5 + 3 = 8$. Because the common sign is positive, the sum will be positive. Thus, $5 + 3 = 8$.

You Try It **Work through this You Try It problem.**

Work Exercise 6 in this textbook or in the MyMathLab Study Plan.

Example 4 Adding Two Integers with the Same Sign

My interactive video summary

Add.

a. $16 + 8$ **b.** $-17 + (-39)$

Solutions Try working these problems on your own. Check your **answers**, or watch this **interactive video** for complete solutions to both parts.

You Try It **Work through this You Try It problem.**

Work Exercises 7–14 in this textbook or in the MyMathLab Study Plan.

TIP Another way to think about adding **signed** numbers is to consider money gained or lost. View this **popup** to see an example.

(eText Screens 2.2-1–2.2-20)

OBJECTIVE 2 ADD TWO INTEGERS WITH DIFFERENT SIGNS

Sometimes we must add two numbers with different **signs**. For example, if the value of one share of Apple Inc. stock increased $\$7$ one day and decreased $\$4$ the next, then the share increased in value by $\$3$ over the two days, or $7 + (-4) = 3$.

When **adding** two numbers with the same sign, we saw that the sign of the **sum** was the same as the common sign. However, when adding two numbers with different signs, the sum may be **positive**, **negative**, or zero. We can visualize such addition by using the **number line** approach.

Example 5 Adding Two Integers with Different Signs

Add.

a. $2 + (-5)$ **b.** $-3 + (7)$

Solutions

a. Because 2 is positive, it lies two units to the right of zero on the number line.

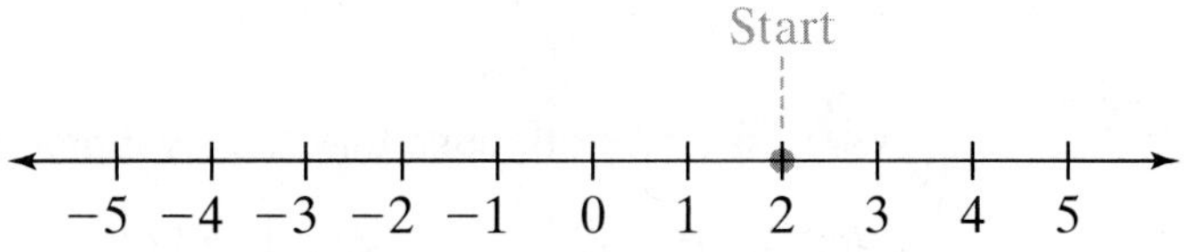

Next we find the **absolute value** of the second number.

$$|-5| = 5$$

Starting at the first number, we move the number of units indicated by the absolute value of the second number, 5 units. Because the second number is negative, we move 5 units to the *left*.

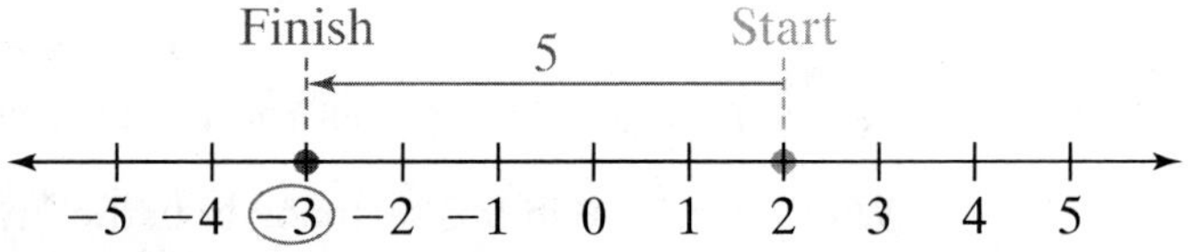

So, $2 + (-5) = -3$.

My video summary

b. Try adding the numbers on your own. Check your **answer**, or watch this **video** for a complete solution to part b.

You Try It **Work through this You Try It problem.**

Work Exercises 15 and 16 in this textbook or in the MyMathLab Study Plan.

Again, using **number lines** can be cumbersome, so we replace this number line approach with the following procedure.

Adding Two Numbers with Different Signs

If two numbers being added have different **signs** (one **positive** and one **negative**), then we add as follows:

Step 1. Find the **absolute value** of each **term**.

Step 2. Subtract the smaller result from the larger result from Step 1.

Step 3. Use the sign of the term with the larger absolute value as the sign of the **sum**.

Example 6 Adding Two Integers with Different Signs

Add.

a. $12 + (-7)$ **b.** $8 + (-19)$

Solutions

a. First, we find the absolute value of each term.

$$|12| = 12 \quad \text{and} \quad |-7| = 7$$

The terms have different signs, so we subtract the smaller result from the larger result.

$$12 - 7 = 5$$

The larger absolute value result belongs to the positive number, so the sum will be positive.

$$12 + (-7) = 5$$

b. First, we find the **absolute value** of each term.

$$|8| = 8 \quad \text{and} \quad |-19| = 19$$

The terms have different **signs**, so we **subtract** the smaller result from the larger result.

$$19 - 8 = 11$$

The larger absolute value result belongs to the **negative** number, so the **sum** will be negative.

$$8 + (-19) = -11$$

You Try It **Work through this You Try It problem.**

Work Exercises 17–20 in this textbook or in the MyMathLab Study Plan.

Example 7 Adding Two Integers with Different Signs

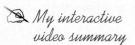

My interactive video summary

Add.

a. $-48 + 73$ **b.** $-35 + 35$ **c.** $28 + (-94)$

(eText Screens 2.2-1–2.2-20)

Solutions Try working these problems on your own. Check your **answers**, or watch this **interactive video** for complete solutions to all three parts.

You Try It **Work through this You Try It problem.**

Work Exercises 21–28 in this textbook or in the MyMathLab Study Plan.

In Example 7b, the two numbers are **opposites** with a sum of 0. This result shows why we refer to opposites as **additive inverses.**

> **Adding a Number and Its Opposite**
>
> The sum of a number a and its opposite, or additive inverse, $-a$, is 0.
>
> $$a + (-a) = 0$$

We now combine our procedures for adding two numbers.

> **Adding Two Numbers**
>
> 1. If the **signs** of the two **terms** are the same (both **positive** or both **negative**), add their **absolute values** and use the common sign as the sign of the **sum.**
> 2. If the signs of the two terms are different (one positive and one negative), subtract the smaller absolute value from the larger absolute value and use the sign of the term with the larger absolute value as the sign of the sum.

Example 8 Adding Two Integers

My interactive video summary

Add.

a. $24 + (-37)$ **b.** $44 + 79$ **c.** $-27 + (-115)$ **d.** $-274 + (513)$

Solutions

a. Find the absolute value of each number.

$$|24| = 24 \quad \text{and} \quad |-37| = 37$$

The terms have different signs, so **subtract** the smaller result from the larger result.

$$37 - 24 = 13$$

Because the larger absolute value result belongs to the negative number, the sum will be negative.

$$24 + (-37) = -13$$

b. Because both numbers are positive, we **add** the absolute values and keep the sum positive.

$$|44| = 44 \quad \text{and} \quad |79| = 79$$

So,

$$44 + 79 = 123.$$

c.–d. Try to work these problems on your own. Start by finding the **absolute value** of each number. Add or subtract the absolute values as required, then determine the **sign** of the result. Check your **answers**, or watch this **interactive video** for the complete solutions to all four parts.

You Try It **Work through this You Try It problem.**

Work Exercises 29–38 in this textbook or in the MyMathLab Study Plan.

OBJECTIVE 3 ADD MORE THAN TWO INTEGERS

If we need to add more than two **integers**, we can follow the **order of operations** and perform the **additions** in order from left to right.

Example 9 Adding More Than Two Integers

Add: $15 + (-23) + 19$

Solution

Begin with the original expression:	$15 + (-23) + 19$
Add $15 + (-23)$:	$= -8 + 19$
Add $-8 + 19$:	$= 11$

Recall that the **commutative** and **associative** properties of addition allow us to add numbers in any order. This means we could add all the **positive** integers and add all the **negative** integers, then add the two **sums**. We would get the same result as we would if adding in order from left to right.

Example 10 Adding More Than Two Integers

My video summary

Add: $-16 + 49 + 26 + (-35)$

Solution Work through the following, or watch this **video** for the complete solution.

Add the positive integers:	$49 + 26 = 75$
Add the negative integers:	$-16 + (-35) = -51$
Add the two sums:	$75 + (-51) = 24$

View this **popup** to see a solution working the additions from left to right.

You Try It **Work through this You Try It problem.**

Work Exercises 39–46 in this textbook or in the MyMathLab Study Plan.

OBJECTIVE 4 SOLVE APPLICATIONS INVOLVING ADDITION OF INTEGERS

Addition of **integers** can be used to solve a variety of application problems. Remember to get in the habit of looking for key words that indicate different **mathematical operations**. Review some terms that mean addition in this **popup**.

(eText Screens 2.2-1–2.2-20)

Example 11 Translating Phrases Involving Addition

My video summary

Translate each sentence into an addition statement and **simplify**.

a. The sum of -49 and -16 **b.** 12 more than -27

Solutions

a. The key word *sum* indicates addition. Because the two numbers have opposite signs, we add the absolute values and keep the common sign.

$$|-49| = 49 \quad \text{and} \quad |-16| = 16$$
$$49 + 16 = 65$$

The common sign is negative, so our result must be negative. The sum of -49 and -16 is -65.

b. The key words *more than* tell us we are adding 12 to -27. Work this problem on your own. Check your **answer**, or watch this **video** for solutions to both parts.

You Try It **Work through this You Try It problem.**

Work Exercises 47–49 in this textbook or in the MyMathLab Study Plan.

Example 12 Change in Temperature

At 6 a.m. on February 1, 2011, the temperature in Cheyenne, WY, was $-18°F$. By noon, the temperature had increased by $23°F$. What was the temperature at noon? (*Source:* www.weatherunderground.com)

Solution The key words *increased by* indicate addition. To find the temperature at noon, we add -18 and 23.

$$-18 + 23 = 5$$

The temperature at noon was $5°F$.

Example 13 Change in Stock Market

My video summary

The following shows the point change in the Dow Jones Industrial Average (DJIA) for 5 days in 2011. What was the total point change for the 5 days? (*Source:* finance.yahoo.com)

July 19	July 20	July 21	July 22	July 25
201	−12	157	−44	−87

Solution Try to solve this problem on your own. Check your **answer**, or watch this **video** for a complete solution.

You Try It **Work through this You Try It problem.**

Work Exercises 50–55 in this textbook or in the MyMathLab Study Plan.

(eText Screens 2.2-1–2.2-20)

2.2 Exercises

In Exercises 1–4, add using a number line.

You Try It 1. $7 + 4$ 2. $8 + (-5)$ 3. $-7 + 3$ 4. $-4 + (-7)$

In Exercises 5–46, add the integers.

You Try It 5. $-7 + (-3)$ 6. $5 + 9$ 7. $7 + 15$ 8. $-3 + (-8)$

9. $8 + 10$ 10. $-11 + (-6)$ 11. $15 + 123$ 12. $-102 + (-17)$

You Try It

13. $422 + 253$ 14. $-121 + (-219)$ 15. $-7 + 5$ 16. $8 + (-14)$

You Try It

17. $8 + (-5)$ 18. $-7 + (12)$ 19. $-9 + (-14)$ 20. $-21 + 7$

You Try It

21. $14 + (-22)$ 22. $31 + (-21)$ 23. $-52 + 52$ 24. $-19 + 16$

You Try It

25. $-62 + 63$ 26. $26 + (-27)$ 27. $37 + (-37)$ 28. $-17 + 26$

29. $32 + 85$ 30. $-27 + 76$ 31. $24 + (-27)$ 32. $74 + (-114)$

You Try It

33. $(-33) + (-26)$ 34. $-19 + 0$ 35. $-17 + 17$ 36. $352 + (-228)$

37. $(-218) + (-202)$ 38. $0 + 184$ 39. $(-18) + 12 + 6$ 40. $17 + 18 + (-39)$

You Try It 41. $(-31) + (-19) + 28$ 42. $35 + (-12) + (-17)$ 43. $(-22) + 18 + (-17) + 11$

44. $25 + (-12) + (-23) + 14$ 45. $(-13) + (-15) + (-18) + 52$ 46. $18 + 45 + 30 + (-100)$

In Exercises 47–49, translate each phrase into an expression involving addition, then simplify.

You Try It 47. -53 added to -26

48. The sum of 43 and 28

49. -31 increased by 19

In Exercises 50–55, solve each application problem by adding integers.

You Try It 50. **Checking Account** Ferdinand has \$78 in his checking account. If he deposits a check for \$109, how much will be in his account?

51. **Temperature** The temperature at 5 a.m. in Chicago was -6°F. By 3 p.m. the temperature had risen 24 degrees. What was the temperature at 3 p.m.?

52. **Grocery Bill** At the grocery store, Hollynn spent \$10 on vegetables, \$8 on fruit, \$22 on meat, and \$18 on other items. What was her total bill?

53. **Credit Card Balance** The following shows a summary of Payton's credit card activity for the past week. Add the values in the *Amount* column to find her current balance.

Activity	Amount
Previous balance	$\$514$
Payment	$-\$140$
Tuition installment	$\$360$
Returned books	$-\$95$

54. **Card Flipping** Sam is playing a card flipping game with his roommate. Each player starts with 30 cards and take turns flipping until a match occurs at which point he either wins cards or loses cards. The following table shows Sam's result for three turns. How many cards does Sam have after the third turn?

Turn 1	Turn 2	Turn 3
-8	$+12$	-9

55. **Stock Value** A certain stock index was at 680 points on Monday morning. The following table shows the point change for the week. What was the value of the stock index at the end of the week?

Monday	Tuesday	Wednesday	Thursday	Friday
-70	-50	130	80	-110

2.3 Subtracting Integers

THINGS TO KNOW

Before working through this topic, be sure you are familiar with the following concepts:

		VIDEO	ANIMATION	INTERACTIVE
You Try It	1. Subtract Whole Numbers (Topic 1.3, **Objective 4**)	▤		▤
You Try It	2. Find the Opposite of a Number (Topic 2.1, **Objective 4**)	▤		
You Try It	3. Add Two Integers with the Same Sign (Topic 2.2, **Objective 1**)			▤
You Try It	4. Add Two Integers with Different Signs (Topic 2.2, **Objective 2**)	▤		▤
You Try It	5. Solve Applications Involving Addition of Integers (Topic 2.2, **Objective 4**)	▤		

(eText Screens 2.3-1–2.3-13)

OBJECTIVES

1 Subtract Integers

2 Add and Subtract Integers

3 Solve Applications Involving Subtraction of Integers

OBJECTIVE 1 SUBTRACT INTEGERS

In **Topic 1.3**, we subtracted whole numbers such as $12 - 7 = 5$. In **Topic 2.2**, we added integers such as $12 + (-7) = 5$. In **Topic 2.1**, we learned about **opposites**. Using all three concepts together, we can define **subtraction** in terms of **addition**. Consider the following.

$12 - 7 = 5$	$12 + (-7) = 5$
Subtracting Whole Numbers	Adding Integers

Notice that subtracting 7 from 12 is equivalent to adding -7 to 12, and -7 is the opposite of 7.

Subtraction				Addition			
12	$-$	7	$=$	12	$+$	(-7)	$= 5$
1st number		2nd number		1st number		Opposite of 2nd number	

This illustrates a result that is true in general. We can subtract a number by adding its opposite.

> **Subtracting Two Numbers**
>
> If a and b represent two numbers, then
>
> $$a - b = a + (-b).$$

Example 1 Subtracting Integers by Adding the Opposite

My video summary

Subtract.

a. $19 - 7$ **b.** $-6 - 9$

Solutions For each subtraction, we add the opposite.

a. Change to add the opposite of 7: $19 - 7 = 19 + (-7)$

We now choose the appropriate **addition rule**. The **signs** are different. $|19| = 19$ and $|-7| = 7$. Because 19 has the larger **absolute value**, this sum is **positive**. We subtract 7 from 19 to find the result, $19 - 7 = 12$. The **difference** is 12.

b. Change to add the **opposite** of 9: $-6 - 9 = -6 + (-9)$

The signs are the same, so add the absolute values of the numbers and keep the common sign. Try finishing this problem on your own. Check your **answer**, or watch this **video** for the complete solutions to both parts.

You Try It **Work through this You Try It problem.**

Work Exercises 1–6 in this textbook or in the MyMathLab Study Plan.

(eText Screens 2.3-1–2.3-13)

Before working through Example 2, you might wish to review the **Double-Negative Rule**. Recall that a number containing more than one sign can be simplified to a number containing at most one sign.

Example 2 Subtracting Integers Using the Double-Negative Rule

My video summary

Subtract.

a. $-7 - (-6)$ **b.** $10 - (-16)$ **c.** $25 - (-25)$

Solutions

a. Change to add the opposite of -6: $-7 - (-6) = -7 + (-(-6))$ ← $-(-6) = 6$

Double-negative rule: $= -7 + 6$

We now choose the appropriate **addition rule**. The signs are different. $|-7| = 7$ and $|6| = 6$. We subtract 6 from 7 to get $7 - 6 = 1$. Because -7 has the larger **absolute value**, our result will be **negative**. The difference is $-7 - (-6) = -1$.

b.–c. Try working these problems on your own. Check your **answers**, or watch this **video** for complete solutions to all three parts.

You Try It **Work through this You Try It problem.**

Work Exercises 7–16 in this textbook or in the MyMathLab Study Plan.

OBJECTIVE 2 ADD AND SUBTRACT INTEGERS

If an **expression** contains both addition and subtraction, we change each subtraction to addition and find the sum.

$$7 + 23 - 18 + 4 - (-9) = 7 + 23 + (-18) + 4 + 9$$

Opposite of 18 → (-18); Opposite of -9 → 9; Change to addition

TIP Remember that the **commutative** and **associative** properties of addition allow us to add numbers in any order.

Example 3 Adding and Subtracting Integers

Simplify: $5 - 8 + (-3) - (-6)$

Solution

Change subtractions to adding the opposite: $= 5 + (-8) + (-3) + 6$ (Opposite of 8: -8; Opposite of -6: 6)

Add $5 + (-8)$: $= -3 + (-3) + 6$

Add $-3 + (-3)$: $= -6 + (6)$

Add $-6 + 6$: $= 0$

View this **popup** to see an alternate solution.

Example 4 Adding and Subtracting Integers

My video summary

Simplify: $-12 + 24 - (-15) - 8 + 10$

Solution

Change to adding the opposite:

$$-12 + 24 - (-15) - 8 + 10 = -12 + 24 + 15 + (-8) + 10$$

Finish **simplifying** this expression on your own. Check your **answer**, or watch this **video** for a complete solution.

You Try It **Work through this You Try It problem.**

Work Exercises 17–28 in this textbook or in the MyMathLab Study Plan.

OBJECTIVE 3 SOLVE APPLICATIONS INVOLVING SUBTRACTION OF INTEGERS

Subtraction of integers can also be used to solve application problems. Review some key words that mean subtraction in this **popup**.

CAUTION Remember that *order matters* with subtraction.
For example, $7 - 5 = 2$ and $5 - 7 = -2$, but $2 \neq -2$.

Example 5 Translating Phrases Involving Subtraction

My video summary

For each phrase, write an expression involving subtraction and **simplify**.

a. The **difference** of -4 and 20

b. -18 subtracted from 35

Solutions

a. To find the difference of two numbers, we subtract the second number from the first.

Write a subtraction problem:	$-4 - 20$
Change to adding the opposite:	$= -4 + (-20)$ ← -20 is the opposite of 20
Add:	$= -24$

The difference of -4 and 20 is -24.

b. The key words *subtracted from* tell us which number is first in the subtraction. To subtract -18 *from* 35, we write $35 - (-18)$. Simplify this on your own. Check your **answer**, or watch this **video** for solutions to both parts.

You Try It **Work through this You Try It problem.**

Work Exercises 29–32 in this textbook or in the MyMathLab Study Plan.

(eText Screens 2.3-1–2.3-13)

Example 6 Solving Applications Involving Subtraction of Integers

Brandi overdrew her checking account so her balance was $-\$24$. The bank charged her a \$45 overdraft fee and deducted that amount from her account. After the fee, what was her balance?

Solution To find Brandi's balance after the fee, we subtract the \$45 fee *from* her balance before the fee.

Write a subtraction problem: $-24 - 45$

Change to adding the opposite: $= -24 + (-45) \leftarrow -45$ is the opposite of 45

Add: $= -69$

After the fee, Brandi's account balance is $-\$69$.

You Try It Work through this You Try It problem.

Work Exercises 33–36 in this textbook or in the MyMathLab Study Plan.

Example 7 Checking Account Balance

My video summary

Isabella has \$425 in her checking account. She withdraws \$275 for her share of rent and withdraws \$78 to pay her cell phone bill. She then deposits a payroll check in the amount of \$317. How much money is now in her checking account?

Solution Each withdrawal amount must be subtracted from her balance, while the deposit must be added. To find the amount in her checking account, we simplify

$$425 - 275 - 78 + 317.$$

Try to finish this problem on your own. Check your **answer**, or watch this **video** for a complete solution.

You Try It Work through this You Try It problem.

Work Exercises 37–39 in this textbook or in the MyMathLab Study Plan.

2.3 Exercises

In Exercises 1–16, subtract.

You Try It

1. $8 - 3$ 2. $-4 - 5$ 3. $26 - 15$ 4. $34 - 62$

5. $-18 - 12$ 6. $-30 - 143$ 7. $4 - (-6)$ 8. $-3 - (-9)$

You Try It

9. $-5 - (-14)$ 10. $-17 - (-11)$ 11. $34 - (-16)$ 12. $22 - (-31)$

13. $127 - (-46)$ 14. $-201 - (-70)$ 15. $-38 - (-38)$ 16. $54 - (-54)$

(eText Screens 2.3-1–2.3-13)

In Exercises 17–28, add and subtract as indicated.

You Try It 17. $7 - 4 + 8$ 18. $-8 + 25 - 17$ 19. $15 - 6 - 9$

20. $38 - 12 - 10$ 21. $12 - (-7) - 3 + 11$ 22. $5 - 11 + 4 - 6$

23. $18 - 4 - (-8) + 14$ 24. $-22 + 19 - (-16) - 13$ 25. $48 - (-27) - 18 + 4$

26. $31 + 7 - 20 - (-6) + 18$ 27. $10 - 46 + 18 - 36 - (-18)$ 28. $-7 - 12 + 34 - (-18) + 29$

In Exercises 29–32, write an expression involving subtraction and simplify.

You Try It 29. 12 subtracted from -38

30. -73 decreased by 35

31. The difference of -16 and -37

32. 26 less than 91.

In Exercises 33–39, solve the given application problem.

You Try It 33. **Temperature** The record high temperature in Punxsutawney, PA (home to the famous groundhog Punxsutawney Phil) on February 2nd is 54°F, and the record low temperature is -17°F. What is the difference between these high and low temperatures? (*Source:* www.weather.com)

34. **Elevation** The lowest point in Australia is Lake Eyre, at an elevation of -15 m. The highest point is Mt. Kosciusko, at an elevation of 2230 m. What is the difference in elevation between Mt. Kosciusko and Lake Eyre? (*Source:* www.australianexplorer.com)

35. **Golf Scores** During one round of golf, Chris scored a 3, or 3 strokes over par, and Shane scored a -5, or 5 strokes under par. What was the difference between Chris's score and Shane's score?

36. **Goal Difference** In soccer, the goal difference (GD) for a team is found by subtracting *goals against* (GA) from *goals for* (GF). The following table shows goals for and goals against for three MLS teams in the 2010 season. Find the goal difference for each team.

Team	GF	GA
L.A. Galaxy	44	26
Kansas City Wizards	36	35
D.C. United	21	47

You Try It 37. **Savings Account** Russell opens a savings account with \$500. He later deposits a check for \$237 but has to withdraw \$186 for car repairs the next day. He then withdraws \$200 to put into his checking account. At the end of the year, he is credited \$2 in interest. How much money is in Russell's savings account at the end of the year?

38. **Credit Downgrade** On August 5, 2011, the U.S. credit rating was downgraded from AAA to AA+. On that day, the Dow Jones Industrial Average (DJIA) closed at 11,445 points. In the four trading days following the downgrade, the DJIA lost 635 points, gained 430 points, fell 520 points, and increased 423 points. What was the closing value on the fourth day? (*Source:* finance.yahoo.com)

39. **Cell Phone Minutes** Andre has a pay-as-you-go cell phone with a current balance of 74 minutes. He makes two calls, one for 27 minutes and one for 36 minutes, then purchases 100 additional minutes. If he talks to his mother for 52 minutes, how many minutes does he have left?

2.4 Multiplying and Dividing Integers

THINGS TO KNOW

Before working through this topic, be sure you are familiar with the following concepts:

		VIDEO	ANIMATION	INTERACTIVE
You Try It	1. Multiply Whole Numbers (Topic 1.4, **Objective 3**)			
You Try It	2. Solve Applications by Multiplying Whole Numbers (Topic 1.4, **Objective 6**)			
You Try It	3. Divide Whole Numbers (Topic 1.5, **Objective 1**)			
You Try It	4. Solve Applications by Dividing Whole Numbers (Topic 1.5, **Objective 3**)			
You Try It	5. Evaluate Exponential Expressions (Topic 1.6, **Objective 2**)			
You Try It	6. Find the Absolute Value of a Number (Topic 2.1, **Objective 3**)			
You Try It	7. Add and Subtract Integers (Topic 2.3, **Objective 2**)			

OBJECTIVES

1 Multiply Two Integers

2 Multiply More than Two Integers

3 Evaluate Exponential Expressions Involving an Integer Base

4 Divide Integers

5 Solve Applications by Multiplying or Dividing Integers

(eText Screens 2.4-1–2.4-23)

OBJECTIVE 1 MULTIPLY TWO INTEGERS

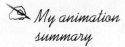

My animation summary

Multiplying **integers** is similar to multiplying **whole numbers** except the **product** will sometimes be **negative**. The **sign** of the product depends on the signs of the **factors** being multiplied. Work through this **animation** (or continue reading below) to develop *rules for multiplying two numbers*.

Look at the pattern that follows when we multiply 5 by different factors as they decrease:

The first factor decreases by 1 each time.

$$\left.\begin{array}{l} 4\cdot 5 = 20 \\ 3\cdot 5 = 15 \\ 2\cdot 5 = 10 \\ 1\cdot 5 = 5 \\ 0\cdot 5 = 0 \end{array}\right\}$$

The product decreases by 5 each time.

We can continue decreasing the first factor by 1 and product by 5 to continue the pattern:

Decrease the first factor by 1 each time.

$$\left.\begin{array}{l} -1\cdot 5 = -5 \\ -2\cdot 5 = -10 \\ -3\cdot 5 = -15 \\ -4\cdot 5 = -20 \end{array}\right\}$$

Decrease the product by 5 each time.

View this **popup** to see each product above confirmed by repeated **addition**. This pattern shows that *multiplying a negative number and a positive number results in a negative product*.

Multiplying a Positive Number and a Negative Number

To multiply a **positive number** and a **negative number**, multiply their **absolute values** and make the **product** negative.

Example 1 Multiplying a Positive Integer and a Negative Integer

Multiply.

a. $8(-9)$ **b.** $(-6)(2)$

Solutions

a. $|8| = 8$ and $|-9| = 9$. We are multiplying a positive **integer** and a negative integer, so we find $8\cdot 9 = 72$ and make this product negative. So, $8(-9) = -72$.

b. $|-6| = 6$ and $|2| = 2$. Because $6\cdot 2 = 12$, we get $(-6)(2) = -12$.

You Try It **Work through this You Try It problem.**

Work Exercises 1 and 2 in this textbook or in the MyMathLab Study Plan.

(eText Screens 2.4-1–2.4-23)

What if both **factors** being multiplied are **negative**? Look at the following pattern when we multiply -5 by different factors as they decrease:

$$\begin{matrix}\text{The first factor}\\\text{decreases by}\\\text{1 each time.}\end{matrix}\left\{\begin{aligned} 4\cdot(-5) &= -20\\ 3\cdot(-5) &= -15\\ 2\cdot(-5) &= -10\\ 1\cdot(-5) &= -5\\ 0\cdot(-5) &= 0\end{aligned}\right\}\begin{matrix}\text{The product}\\\text{increases by}\\\text{5 each time.}\end{matrix}$$

View this **popup** to see each product above confirmed by repeated **addition.** We can continue decreasing the first factor by 1 and increasing the **product** by 5 to continue the pattern:

$$\begin{matrix}\text{Decrease the}\\\text{first factor by}\\\text{1 each time.}\end{matrix}\left\{\begin{aligned} -1\cdot(-5) &= 5\\ -2\cdot(-5) &= 10\\ -3\cdot(-5) &= 15\\ -4\cdot(-5) &= 20\end{aligned}\right\}\begin{matrix}\text{Increase the}\\\text{product by}\\\text{5 each time.}\end{matrix}$$

This pattern shows that *multiplying two negative numbers results in a positive product.*

Multiplying Two Negative Numbers

To multiply two negative numbers, multiply their **absolute values** and make the product positive.

Example 2 Multiplying Two Negative Integers

Multiply.

a. $(-6)(-8)$ **b.** $(-4)(-7)$

Solutions

a. $|-6| = 6$ and $|-8| = 8$. We are multiplying two negative **integers**, so we find $6\cdot 8 = 48$ and make this **product** positive. So, $(-6)(-8) = 48$.

b. $|-4| = 4$ and $|-7| = 7$. Because $4\cdot 7 = 28$, we get $(-4)(-7) = 28$.

You Try It **Work through this You Try It problem.**

Work Exercises 3 and 4 in this textbook or in the MyMathLab Study Plan.

We now summarize the rules for all cases of multiplying two numbers.

Multiplying Two Numbers

Multiply the **absolute values** of the two **factors** to get the absolute value of the product. Determine the **sign** of the product using the following rules:

1. If the signs of the two factors are the same (both **positive** or both **negative**), then the product is positive.
2. If the signs of the two factors are different (one positive and one negative), then the product is negative.
3. If either factor is 0, then the product is 0 (neither positive nor negative).

(eText Screens 2.4-1–2.4-23)

Example 3 Multiplying Two Integers

My video summary

Multiply.

a. $(-8)(11)$ **b.** $(-6)(0)$ **c.** $15 \cdot 7$ **d.** $-13 \cdot (-12)$

Solutions Try finding these products on your own. View the **answers**, or work through this **video** for complete solutions.

You Try It **Work through this You Try It problem.**

Work Exercises 5–10 in this textbook or in the MyMathLab Study Plan.

OBJECTIVE 2 MULTIPLY MORE THAN TWO INTEGERS

If multiplying more than two **integers**, we can multiply in order from left to right.

Example 4 Multiplying More Than Two Integers

Multiply $-2(4)(-3)$.

Solution

Begin with the original expression: $\underbrace{-2(4)}(-3)$

Multiply the first two factors, $-2(4)$: $= \underbrace{-8(-3)}$

Multiply $-8(-3)$: $= 24$

In **Topic 1.4**, we named several **properties of multiplication** that are true for all numbers including integers. If desired, because of the **commutative** and **associative** properties of multiplication, we can change the order and grouping of the **factors** being multiplied. For example, in Example 4, we could have multiplied the two **negative** factors first and then multiplied the result by the **positive** factor as follows:

Begin with the original expression: $-2(4)(-3)$

Multiply the two negative factors, $-2(-3)$: $= (6)(4)$

Multiply $(6)(4)$: $= 24$

Often changing the order or grouping is helpful when multiplying more than two integers.

Example 5 Multiplying More Than Two Negative Integers

Multiply.

a. $-5(-2)(-7)$ **b.** $-3(-2)(-5)(-4)$ **c.** $-4(-3)(-2)(-6)(-1)$

Solutions

a. Begin with the original expression: $\underbrace{-5(-2)}(-7)$

Multiply the first two factors, $-5(-2)$: $= \underbrace{10(-7)}$

Multiply $10(-7)$: $= -70$

b. Begin with the original expression: $-3(-2)(-5)(-4)$

Multiply the first two factors, $-3(-2)$; multiply the last two factors, $(-5)(-4)$: $= 6(20)$

Multiply $6(20)$: $= 120$

My video summary

c. Try finding this **product** on your own. View the **answer**, or watch this **video** for a complete solution to part c.

You Try It **Work through this You Try It problem.**

Work Exercises 11–14 in this textbook or in the MyMathLab Study Plan.

Part a of Example 5 shows that multiplying three **negative** integers results in a negative **product**. Part b shows that multiplying four negative **integers** results in a **positive** product. Part c shows that multiplying five negative integers results in a negative product. This leads us to the following rule.

Multiplying an Even or Odd Number of Negative Factors

The product of an **even number** of negative numbers is positive.

The product of an **odd number** of negative numbers is negative.

When multiplying more than two **nonzero** factors, we can determine the **sign** of the product by counting the number of **negative signs.** An odd number of negative signs leads to a negative product. An even number of negative signs leads to a positive product.

CAUTION

If any **factor** being multiplied is zero (0), then the product is zero (neither positive nor negative).

Example 6 Multiplying More Than Two Integers

My interactive video summary

Multiply.

a. $4(9)(-2)$ **b.** $3(-8)(2)(-7)$ **c.** $4(-8)(-7)(-3)$

Solutions

a. None of the three **factors** being multiplied is zero, and only one of the factors is **negative** (an odd number of **negative signs**). So, the **product** is negative.

$$4(9)(-2) = -72$$

b. The four factors being multiplied are all **nonzero**. Two of the factors are negative (an even number of negative signs), so the product is **positive**. Try to finish this multiplication on your own. View the **answer**, or watch this **interactive video** for complete solutions to all three parts.

c. Try finding this product on your own. View the **answer**, or watch this **interactive video** for complete solutions to all three parts.

You Try It **Work through this You Try It problem.**

Work Exercises 15–18 in this textbook or in the MyMathLab Study Plan.

(eText Screens 2.4-1–2.4-23)

OBJECTIVE 3 EVALUATE EXPONENTIAL EXPRESSIONS INVOLVING AN INTEGER BASE

Recall from **Topic 1.6** that an **exponent** or **power** tells how many times a **base** is multiplied by itself.

$$\overset{\text{Exponent}}{\downarrow}$$
$$\underset{\substack{\uparrow \\ \text{Base}}}{2}^4 = \underbrace{2 \cdot 2 \cdot 2 \cdot 2}_{\text{4 factors of 2}} = 16$$

Sometimes the base of an **exponential expression** will be **negative**.

$$\overset{\text{Exponent}}{\downarrow}$$
$$\underset{\substack{\uparrow \\ \text{Base}}}{(-3)}^2 = \underbrace{(-3)(-3)}_{\text{2 factors of } -3} = 9$$

Example 7 Evaluating Exponential Expressions with a Negative Base

Evaluate each exponential expression.

a. $(-5)^2$ **b.** $(-5)^3$ **c.** $(-5)^4$

Solutions

a. $(-5)^2 = (-5)(-5) = 25$

b. $(-5)^3 = (-5)(-5)(-5) = -125$

c. $(-5)^4 = (-5)(-5)(-5)(-5) = 625$

In **Example 7** parts a and c, notice that the negative **base** is raised to an **even power** and each result is **positive**. In part b, however, the negative base is raised to an **odd power** and the result is **negative**. This leads to the following rule.

> **Even and Odd Powers of a Negative Base**
>
> If a negative base is raised to an even power, then the result is positive.
>
> If a negative base is raised to an odd power, then the result is negative.

Consider the following two **exponential expressions:**

$$(-4)^2 \quad \text{and} \quad -4^2$$

Although these two expressions look similar, they are quite different. The first expression has parentheses around the -4, but the second expression does not. If **grouping symbols** enclose the negative with the number, then the negative sign is a part of the base. We read $(-4)^2$ as "the square of negative four" or "negative four squared."

$$\overset{\text{Exponent}}{\downarrow}$$
$$\underbrace{(-4)}_{\substack{\text{Base} \\ \text{is } -4}}{}^2 = \underbrace{(-4)(-4)}_{\text{2 factors of } -4} = 16$$

(eText Screens 2.4-1–2.4-23)

If there are no **grouping symbols**, then the **negative sign** is not part of the **base**. We read -4^2 as "the opposite of the square of four" or "the opposite of four squared."

Exponent ↓ Take the opposite ↓

$$\underbrace{-4^2}_{\text{Base is 4}} = -\underbrace{(4 \cdot 4)}_{\text{2 factors of 4}} = -16$$

So, $(-4)^2 = 16$ and $-4^2 = -16$.

CAUTION: The base of an exponential expression only includes the negative sign when the negative sign is enclosed within grouping symbols.

Example 8 Evaluating Exponential Expressions Involving Integer Bases

My video summary

Evaluate each exponential expression.

a. $(-8)^2$ **b.** -2^4 **c.** -10^2 **d.** $(-4)^3$ **e.** -6^3

Solutions

a. $(-8)^2 = (-8)(-8) = 64$

b. $-2^4 = -(2 \cdot 2 \cdot 2 \cdot 2) = -16$

c.–e. Try to evaluate these exponential expressions on your own. View the **answers**, or work through this **video** for complete solutions to all five parts.

You Try It **Work through this You Try It problem.**

Work Exercises 19–26 in this textbook or in the MyMathLab Study Plan.

OBJECTIVE 4 DIVIDE INTEGERS

In **Topic 1.5**, we saw that **division** can be defined in terms of **multiplication**. For example,

$$\underset{\text{Dividend}}{35} \div \underset{\text{Divisor}}{5} = \underset{\text{Quotient}}{7} \quad \text{because} \quad \underset{\text{Quotient}}{7} \cdot \underset{\text{Divisor}}{5} = \underset{\text{Dividend}}{35}.$$

We can use this same idea when **negative numbers** are involved in division.

Consider $(-35) \div 5$. The **quotient** is the number that when multiplied by 5 results in -35.

Think: $(?) \cdot 5 = -35$ Answer: -7

So, $(-35) \div 5 = -7$ because $(-7) \cdot 5 = -35$. This example demonstrates that *dividing a negative number by a positive number results in a negative quotient.*

Next, consider $24 \div (-6)$.

Think: $(?)(-6) = 24$ Answer: -4

So, $24 \div (-6) = -4$ because $(-4)(-6) = 24$. This example demonstrates that *dividing a positive number by a negative number results in a negative quotient.*

Finally, consider $(-18) \div (-9)$.

Think: $(?)(-9) = -18$ Answer: 2

So, $(-18) \div (-9) = 2$ because $(2)(-9) = -18$. This example demonstrates that *dividing a negative number by a negative number results in a positive quotient*.

The rules for dividing two numbers are similar to the rules for multiplying two numbers.

Dividing Two Numbers

Divide the **absolute values** of the numbers to get the absolute value of the **quotient**. Determine the **sign** of the quotient using the following rules:

1. If the signs of the two numbers are the same (both **positive** or both **negative**), then the quotient is positive.
2. If the signs of the two numbers are different (one positive and one negative), then the quotient is negative.
3. If the **dividend** is 0 and the **divisor** is **nonzero**, then the quotient is 0.
4. If the dividend is nonzero and the divisor is 0, the quotient is **undefined**.

CAUTION Recall from **Topic 1.5** that $0 \div 0$ is **indeterminate**. Watch this **video** to review why.

Example 9 Dividing Integers

My video summary

Divide.

a. $(-15) \div (-3)$ b. $\dfrac{-56}{7}$ c. $72/(-8)$ d. $\dfrac{0}{-12}$

Solutions

a. $|-15| = 15$ and $|-3| = 3$. We are dividing two negative **integers**, so we find $15 \div 3 = 5$ and make this **quotient** positive. So, $(-15) \div (-3) = 5$.

b. $|-56| = 56$ and $|7| = 7$. We are dividing a **negative** integer by a **positive** integer, so we find $\dfrac{56}{7} = 8$ and make this quotient negative. So, $\dfrac{-56}{7} = -8$.

c.–d. Try finding these quotients on your own. View the **answers**, or work through this **video** for complete solutions to all four parts.

You Try It **Work through this You Try It problem.**

Work Exercises 27–36 in this textbook or in the MyMathLab Study Plan.

OBJECTIVE 5 SOLVE APPLICATIONS BY MULTIPLYING OR DIVIDING INTEGERS

We now look at some applications involving **multiplication** and **division** of **integers**. Review some **key words that mean multiplication** and **key words that mean division**.

Example 10 Translating Sentences Involving Multiplication or Division

Translate each sentence into a multiplication or division statement and solve.

a. Find the **product** of -9 and -6. b. Divide -112 by 4.

(eText Screens 2.4-1–2.4-23)

Solutions

My video summary

a. To find the product of two numbers, we multiply them.

Write a multiplication problem: $(-9)(-6)$

We are multiplying two negative integers, so we find $9 \cdot 6 = 54$ and make this product positive. So, $(-9)(-6) = 54$. Thus, the product of -9 and -6 is 54.

b. We want to divide -112 *by* 4, so we must simplify $-112 \div 4$. Simplify this on your own. Check your **answer**, or watch this **video** for solutions to both parts.

You Try It **Work through this You Try It problem.**

Work Exercises 37–44 in this textbook or in the MyMathLab Study Plan.

Example 11 Credit Card Debt

Each month, Taima spends $75 more than she earns by making charges on her credit card. Use an integer to describe Taima's financial condition after 12 months.

Solution Taima is going in debt at a rate of $75 per month, represented by the integer −$75 per month. To find her financial condition after 12 months, we multiply:

$$(-75)(12).$$

We are multiplying a negative integer by a positive integer, so we find $75 \cdot 12 = 900$ and make this product negative. So, $(-75)(12) - -900$.

In 12 months, Taima will be in debt by $900, which is represented by the integer −$900.

You Try It **Work through this You Try It problem.**

Work Exercises 45 and 46 in this textbook or in the MyMathLab Study Plan.

Example 12 Digging a Well

My video summary

To dig a well for water, a pile driver is used to hammer a pipe into the ground. Each strike of the pile driver sinks the pipe 2 inches deeper (-2 inches). How many strikes of the pile driver are needed for the pipe to reach water at a depth of 4800 inches below the surface (-4800 inches)?

Solution We know the water is at -4800 inches, and we know that the pipe moves -2 inches per strike of the pile driver. To determine the number of strikes needed to reach the water, we divide:

$$(-4800) \div (-2).$$

Try to perform this division on your own to solve the problem. Check your **answer**, or watch this **video** for a complete solution.

You Try It **Work through this You Try It problem.**

Work Exercises 47 and 48 in this textbook or in the MyMathLab Study Plan.

2.4 Exercises

In Exercises 1–18, multiply.

You Try It

1. $4(-7)$
2. $(-3)(8)$
3. $(-5)(-9)$
4. $(-2)(-6)$

You Try It

5. $-10(4)$
6. $0(-7)$

You Try It

7. $-6(-12)$
8. $11(9)$
9. $(-14)(-15)$
10. $(13)(-8)$

You Try It

11. $-3(-4)(-5)$
12. $(-2)(-5)(-3)(-6)$
13. $-1(-2)(-4)(-7)(-3)$
14. $(-3)(-2)(-1)(-8)$

You Try It

15. $(-5)(-3)(4)$
16. $3(7)(-6)$
17. $(-2)(3)(-9)(-1)$
18. $6(0)(-2)(5)$

In Exercises 19–26, evaluate the exponential expression.

You Try It

19. $(-6)^2$
20. -9^2
21. -3^4
22. $(-2)^4$
23. $(-10)^3$
24. -5^3
25. $(-1)^6$
26. -1^{10}

In Exercises 27–36, divide.

You Try It

27. $\dfrac{45}{-5}$
28. $\dfrac{-63}{-9}$
29. $(-18) \div 6$
30. $0 \div (-10)$
31. $55 \div 5$
32. $(-48) \div (-12)$
33. $\dfrac{-4}{0}$
34. $\dfrac{-52}{4}$
35. $(-49)/7$
36. $54/(-9)$

In Exercises 37–44, translate each sentence into a multiplication or division statement and solve.

You Try It

37. Double -16.
38. What is the quotient of -36 and -3?
39. Divide -234 by 9.
40. Find the product of -17 and -8.
41. What is 32 times -6?
42. How many times does -12 go into -72?
43. Multiply -25 by -6.
44. Find 252 divided by -14.

In Exercises 45–48, solve each application.

You Try It

45. **Falling Temperature** The temperature in a town drops 4 degrees per hour for 9 consecutive hours. Use an integer to describe the overall change in temperature for the town.

(eText Screens 2.5-1–2.5-19)

46. **Lost Yardage** During one possession, a football team lost 10 yards (-10) on each of three consecutive plays. What was the team's total change in yardage on these three plays?

You Try It 47. **Scuba Diving** A scuba diver descends slowly beneath the water of a lake at a rate of 15 feet per minute (-15 feet per minute). How long will it take her to reach the floor of the lake if it is 120 feet deep (-120 feet)?

48. **Business Losses** A company is equally owned by four partners. If the company loses \$200,000 in one year, use an integer to describe the loss for each individual partner.

2.5 Order of Operations

THINGS TO KNOW

Before working through this topic, be sure you are familiar with the following concepts:

		VIDEO	ANIMATION	INTERACTIVE
You Try It	**1.** Simplify Expressions Using the Order of Operations (Topic 1.6, **Objective 3**)			
You Try It	**2.** Add and Subtract Integers (Topic 2.3, **Objective 2**)			
You Try It	**3.** Multiply Integers (Topic 2.4, **Objective 1**)			
You Try It	**4.** Divide Integers (Topic 2.4, **Objective 4**)			

OBJECTIVES

1 Use the Order of Operations with Integers

2 Evaluate Algebraic Expressions Using Integers

3 Translate Word Phrases Using Integers

OBJECTIVE 1 USE THE ORDER OF OPERATIONS WITH INTEGERS

In **Topic 1.6**, we simplified **numeric expressions** involving **whole numbers.** Here we review the order of operations and simplify numeric expressions involving **integers.**

Order of Operations

1. **Parentheses (or other grouping symbols)** Evaluate operations within parentheses (or other **grouping symbols**) first, starting with the innermost set and working out.
2. **Exponents** Work from left to right and evaluate any **exponential expressions** as they occur.
3. **Multiplication and Division** Work from left to right and perform any multiplication or **division** operations as they occur.
4. **Addition and Subtraction** Work from left to right and perform any addition or **subtraction** operations as they occur.

CAUTION

Remember that multiplication and division have the same priority. Likewise, addition and subtraction have the same priority.

Example 1 Using the Order of Operations with Integers (Two Operations)

Simplify.

a. $8 - 4 + 3$ **b.** $-2 \cdot 4^2$ **c.** $6 + (-15) \div 3$

Solutions For each expression, follow the **order of operations**.

a. The expression has a subtraction and an addition. We perform these in order from left to right.

Begin with the original expression: $8 - 4 + 3$

Subtract $8 - 4$: $= 4 + 3$

Add $4 + 3$: $= 7$

b. The expression has an **exponent** and a multiplication. We do the exponent first.

Begin with the original expression: $-2 \cdot 4^2$

Evaluate 4^2: $= -2 \cdot 16 \leftarrow 4^2 = 4 \cdot 4 = 16$

Multiply $-2 \cdot 16$: $= -32$

c. The expression has a division and an addition. We do the division first.

Begin with the original expression: $6 + (-15) \div 3$

Divide $-15 \div 3$: $= 6 + (-5)$

Add $6 + (-5)$: $= 1$

You Try It **Work through this You Try It problem.**

Work Exercises 1–8 in this textbook or in the MyMathLab Study Plan.

Example 2 Using the Order of Operations with Integers (Two Operations)

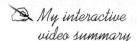

My interactive video summary

Simplify.

a. $\dfrac{4-10}{-2}$ b. $13 - 2^2$ c. $-80 \div (-10)\cdot(-4)$

Solutions Try working these problems on your own. Check your **answers**, or watch this **interactive video** for complete solutions to all three parts.

You Try It Work through this You Try It problem.

Work Exercises 9–16 in this textbook or in the MyMathLab Study Plan.

Example 3 Using the Order of Operations with Integers (Three Operations)

Simplify.

a. $\dfrac{-2-7}{7-4}$ b. $(-3)^2 - 4^3$ c. $|2-5| + 3$

Solutions For each expression, follow the **order of operations.**

a. Begin with the original expression: $\dfrac{-2-7}{7-4}$

Subtract $-2-7$: $= \dfrac{-9}{7-4}$

Subtract $7-4$: $= \dfrac{-9}{3}$

Divide $\dfrac{-9}{3}$: $= -3$

b. Begin with the original expression: $(-3)^2 - 4^3$

Evaluate $(-3)^2$: $= 9 - 4^3 \leftarrow (-3)^2 = (-3)(-3) = 9$

Evaluate 4^3: $= 9 - 64 \leftarrow 4^3 = 4\cdot4\cdot4 = 64$

Subtract $9-64$: $= -55$

c. Begin with the original expression: $|2-5| + 3$

Subtract $2-5$: $= |-3| + 3 \leftarrow$ Work inside **grouping symbols** first

Evaluate $|-3|$: $= 3 + 3 \quad \leftarrow |-3| = 3$

Add $3+3$: $= 6$

You Try It Work through this You Try It problem.

Work Exercises 17–24 in this textbook or in the MyMathLab Study Plan.

(eText Screens 2.5-1–2.5-19)

Example 4 Using the Order of Operations with Integers (Three Operations)

My interactive video summary

Simplify.

a. $12 - (-2) \cdot |-4|$ b. $(7 - 4)(3 - 8)$

c. $20 - (6 \cdot 4 - 1)$ d. $-4 - 12 \div 2 \times (-3)$

Solutions Try working these problems on your own. Check your **answers**, or watch this **interactive video** for complete solutions to all four parts.

You Try It **Work through this You Try It problem.**

Work Exercises 25–32 in this textbook or in the MyMathLab Study Plan.

Example 5 Using the Order of Operations with Integers (Grouping Symbols)

Simplify.

a. $-3[7 - 2(4 - 9)^2]$ b. $8 - |6 - 9|^3 - 7$

Solutions For each expression, follow the **order of operations**.

a. Original expression: $-3[7 - 2(4 - 9)^2]$ ← Work innermost grouping symbols first

Subtract $4 - 9$: $= -3[7 - 2(-5)^2]$

Evaluate $(-5)^2$: $= -3[7 - 2(25)]$ $\leftarrow (-5)^2 = (-5)(-5) = 25$

Multiply $2(25)$: $= -3[7 - 50]$

Subtract $7 - 50$: $= -3[-43]$

Multiply $-3(-43)$: $= 129$

b. Original expression: $8 - |6 - 9|^3 - 7$

Subtract $6 - 9$: $= 8 - |-3|^3 - 7$

Absolute value $|-3|$: $= 8 - 3^3 - 7$ $\leftarrow |-3| = 3$

Evaluate 3^3: $= 8 - 27 - 7$ $\leftarrow 3^3 = 27$

Subtract $8 - 27$: $= -19 - 7$

Subtract $-19 - 7$: $= -26$

You Try It **Work through this You Try It problem.**

Work Exercises 33–35 in this textbook or in the MyMathLab Study Plan.

Example 6 Using Order of Operations with Integers (Grouping Symbols)

My interactive video summary

Simplify.

a. $|(-3)^2 - 4| - 2^3 + 6$ b. $-4 \cdot 5^3 - (4 - 2 \cdot 3)$

Solutions Try working these problems on your own. Check your **answers**, or watch this **interactive video** for complete solutions to both parts.

You Try It **Work through this You Try It problem.**

Work Exercises 36–38 in this textbook or in the MyMathLab Study Plan.

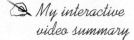

(eText Screens 2.5-1–2.5-19)

OBJECTIVE 2 EVALUATE ALGEBRAIC EXPRESSIONS USING INTEGERS

In **Topic 1.7**, we evaluated **algebraic expressions** using **whole numbers**. We now review **evaluating algebraic expressions** using **integers**.

Example 7 Evaluating Algebraic Expressions Using Integers

My video summary

Evaluate each algebraic expression for the given value of the **variable**.

a. $x + 6$ for $x = -2$ **b.** $9 - m$ for $m = 17$

c. $-5z$ for $z = -4$ **d.** $\frac{p}{-2}$ for $p = 18$

Solutions

a. Begin with the original algebraic expression: $x + 6$

Substitute -2 for x: $= -2 + 6$

Add: $= 4$

So, the value of the expression $x + 6$ is 4 when $x = -2$.

b. Begin with the original algebraic expression: $9 - m$

Substitute 17 for m: $= 9 - 17$

Subtract: $= -8$

So, the value of the expression $9 - m$ is -8 when $m = 17$.

c.–d. Try to **evaluate** these **expressions** on your own, then check your **answers** to parts c and d. Or, watch this **video** for complete solutions to all four parts.

You Try It **Work through this You Try It problem.**

Work Exercises 39–46 in this textbook or in the MyMathLab Study Plan.

Example 8 Evaluating Algebraic Expressions

My video summary

Evaluate $3 - 2x^2$ for the given values of the **variable**.

a. $x = -1$ **b.** $x = 0$ **c.** $x = 4$

Solutions

a. Begin with the original expression: $3 - 2x^2$

Substitute -1 for x: $= 3 - 2(-1)^2$

Evaluate $(-1)^2$: $= 3 - 2(1)$ ← $(-1)^2 = (-1)(-1) = 1$

Multiply $2 \cdot 1$: $= 3 - 2$

Subtract: $= 1$

The value of the expression $3 - 2x^2$ is 1 when $x = -1$.

b. Begin with the original expression: $3 - 2x^2$

Substitute 0 for x: $= 3 - 2(0)^2$

Finish **evaluating** this **expression** on your own, then check your **answer** to part b. Or, watch this **video** for complete solutions to all three parts.

c. Try to evaluate this expression on your own. Check your **answer** to part c, or watch this **video** for complete solutions to all three parts.

You Try It **Work through this You Try It problem.**

Work Exercises 47–50 in this textbook or in the MyMathLab Study Plan.

Example 9 Evaluating Algebraic Expressions

My interactive video summary

Evaluate each algebraic expression for the given values of the variables.

a. $3x - y$ for $x = 2$ and $y = -8$

b. $5(x + y) - z$ for $x = -5$, $y = 7$, and $z = 10$

c. $-2m^2 + n^3$ for $m = -3$ and $n = 4$

d. $b^2 - 4ac$ for $a = 7$, $b = 2$, and $c = -1$

Solutions

a. Begin with the original expression: $3x - y$

Substitute 2 for x and -8 for y: $= 3(2) - (-8)$

Multiply $3(2)$: $= 6 - (-8)$

Subtract: $= 14$ ← $6 - (-8) = 6 + 8 = 14$

The value of the **expression** $3x - y$ is 14 when $x = 2$ and $y = -8$.

b. Begin with the original expression: $5(x + y) - z$

Substitute -5 for x, 7 for y, and 10 for z: $= 5(-5 + 7) - 10$

Add $-5 + 7$: $= 5(2) - 10$ ← Work inside parentheses first

Multiply $5(2)$: $= 10 - 10$

Subtract: $= 0$

The value of the expression $5(x + y) - z$ is 0 when $x = -5$, $y = 7$, and $z = 10$.

c.–d. Try to **evaluate** these expressions on your own. Check your **answers** to parts c and d, or watch this **interactive video** for complete solutions to all four parts.

You Try It **Work through this You Try It problem.**

Work Exercises 51–58 in this textbook or in the MyMathLab Study Plan.

OBJECTIVE 3 TRANSLATE WORD PHRASES USING INTEGERS

In **Topic 1.7**, we also used whole numbers to translate word phrases into **algebraic expressions** involving several operations. We revisit this topic again to include integers. You may wish to review the key words and phrases from **Table 2** in Topic 1.7.

(eText Screens 2.5-1–2.5-19)

Example 10 Writing Word Phrases as Algebraic Expressions

My video summary

Write each word phrase as an algebraic expression. Use x to represent the unknown number.

a. 9 more than the product of -4 and a number

b. The sum of -8 and a number, divided by the difference of the number and 11

c. 6 less than twice the sum of a number and -15

d. A number divided by -3, increased by twice the number

Solutions

a. "More than" indicates **addition**. What is being added? 9 is being added to the product of -4 and a number. "Product" means **multiplication**. What is being multiplied? -4 is being multiplied by the unknown number x. This translates as $-4x$.

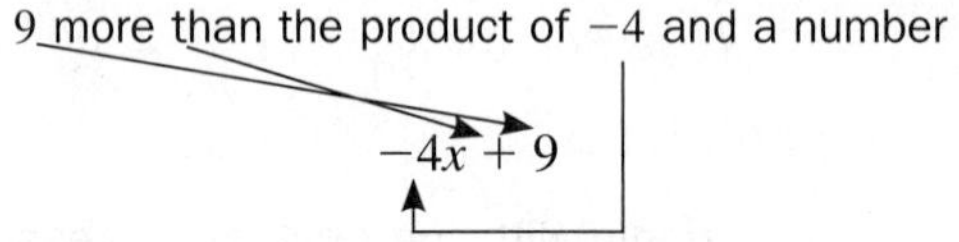

So, we translate "9 more than the product of -4 and a number" as the expression $-4x + 9$.

b. The key words *divided by* mean **division**. What is being divided? The comma indicates a separation of quantities so "the sum of -8 and a number" is being divided by "the difference of the number and 11." The key word "sum" means addition, so we translate "the sum of -8 and a number" as $-8 + x$. The key word "difference" means **subtraction**, so we translate "the difference of the number and 11" as $x - 11$.

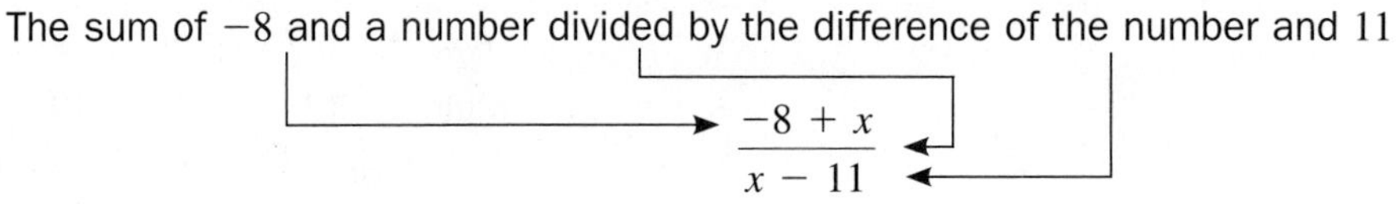

So, we can translate "The sum of -8 and a number, divided by the difference of the number and 11" as

$$\frac{-8 + x}{x - 11}.$$

Using parentheses, we could also write $(-8 + x)/(x - 11)$.

c.–d. Try translating these two expressions on your own. Check your **answers**, or watch this video for complete solutions to all four parts.

You Try It **Work through this You Try It problem.**

Work Exercises 59–68 in this textbook or in the MyMathLab Study Plan.

CAUTION Watch the wording carefully when translating. Sometimes phrases that sound similar will actually have entirely different translations. For example, see this **popup** for an explanation of the difference between "less than a number" and "less a number."

2.5 Exercises

In Exercises 1–38, simplify the expression using the order of operations.

You Try It 1. $5 \cdot 3^2$ 2. $-4 + 3(-2)$ 3. $14 - 8 + 7$ 4. $(-2)^4 - 15$

5. $|-7| + 8$ 6. $(-6)(-3) - 7$ 7. $-2(5 - 9)$ 8. $-(-2)^3$

You Try It 9. $-6 \cdot 8 \div 4$ 10. $1^5 - 4$ 11. $\dfrac{-14 - 6}{-2}$ 12. $\dfrac{36}{8 - 12}$

13. $-3^2 - 3^2$ 14. $100 \div (-20) \cdot 5$ 15. $10 - (-5)^3$ 16. $(7 + (-5))^3$

You Try It 17. $-1(4 - 11) + 6$ 18. $5(-6) - (-2)$ 19. $-12 + 3 \cdot 4^3$

20. $|-5 - 9| - 12$ 21. $8 \cdot (-2)^3 + 13$ 22. $\dfrac{17 - 5}{6 - 3}$

23. $(14 - 20)(15 - 8)$ 24. $(3 \cdot 15 - 10) - 7$ 25. $5^2 - 0 \cdot 3$

You Try It 26. $(10 - 4)^3 \div (-3 + 1)$ 27. $(-7 + 17)^2 \div (6 - 5)^3$ 28. $-4 \cdot |-5| - 2(-8)$

29. $-40 \div (-4) \div (-2) \cdot 6$ 30. $(12 - 3^2)^3$ 31. $(18 \div 3) - (35 \div 7)$

You Try It 32. $4(7 + (-5)) - (-8 + 3)$ 33. $-2[8 - 3(9 - 14)^2]$ 34. $16 - |4 - 16|^2 + 3$

You Try It 35. $\dfrac{-2^3 - 7}{|7 - 10|}$ 36. $|2^3 - 10| - 4^2 + 11$ 37. $-2^4 \cdot 6 - (4 \cdot 8 - 5^2)$

38. $\dfrac{(-5)(-4) + (-2)(6)}{|3^2 - 7|}$

In Exercises 39–58, evaluate each expression for the given values of the variables.

You Try It 39. $y - 3$ for $y = -4$ 40. $b + 10$ for $b = -7$

41. $\dfrac{x}{-3}$ for $x = 27$ 42. $-2z$ for $z = 6$

43. $7 - x$ for $x = 20$ 44. $-4 + m$ for $m = 11$

45. $\dfrac{24}{p}$ for $p = -2$ 46. $3s$ for $s = -8$

You Try It 47. $7 - 3x^2$ **a.** for $x = -4$ **b.** for $x = 0$ **c.** for $x = 6$

48. $|x + 4| - 6$ **a.** for $x = 2$ **b.** for $x = 0$ **c.** for $x = -12$

49. $2x^3 + 7$ **a.** for $x = -3$ **b.** for $x = 0$ **c.** for $x = 5$

50. $4x^2 - 5x$ **a.** for $x = 4$ **b.** for $x = 0$ **c.** for $x = -2$

You Try It 51. $x - 4y$ for $x = 2$ and $y = -4$ 52. $-2m^2 - m + n^2$ for $m = -4$ and $n = 2$

53. $(x - h)^2 + k$ for $x = 10$, $h = -1$, and $k = 3$ 54. $5x - 2y + z$ for $x = 8$, $y = -2$, and $z = -5$

(eText Screens 2.6-1–2.6-20)

55. $|x^2 - 3z|$ for $x = -7$ and $z = -12$

56. $4z - 3(x - y)$ for $x = 0$, $y = -7$, and $z = 23$

57. $\dfrac{a^2}{2b - c}$ for $a = 8$, $b = -5$, and $c = 6$

58. $mx + b$ for $m = 2$, $b = -3$, and $x = -6$

In Exercises 59–68, write each word phrase as an algebraic expression. Use x to represent the unknown number in each phrase.

You Try It

59. -9 times the difference of -4 and a number

60. 10 less than the product of -3 and a number

61. The difference of -15 and 20 times a number

62. 8 more than the quotient of -7 and a number

63. The sum of a number and 7, divided by -3

64. The sum of -4 and triple a number, divided by the sum of the number and 5

65. The product of a number and 9 more than the number

66. 3 more than the product of -5 and the difference between a number and 6

67. Twice a number decreased by the number divided by -5

68. 10 minus a number, increased by the square of the number

2.6 Solving Equations: The Addition and Multiplication Properties

THINGS TO KNOW

Before working through this topic, be sure you are familiar with the following concepts:

		VIDEO	ANIMATION	INTERACTIVE
You Try It	1. Determine If a Value Is a Solution to an Equation (Topic 1.7, **Objective 3**)			
You Try It	2. Add and Subtract Integers (Topic 2.3, **Objective 2**)			
You Try It	3. Multiply Two Integers (Topic 2.4, **Objective 1**)			
You Try It	4. Divide Integers (Topic 2.4, **Objective 4**)			
You Try It	5. Evaluate Algebraic Expressions Using Integers (Topic 2.5, **Objective 2**)			

(eText Screens 2.6-1–2.6-20)

OBJECTIVES

1 Determine If an Integer Is a Solution to an Equation

2 Use the Addition Property of Equality to Solve Equations

3 Use the Multiplication Property of Equality to Solve Equations

OBJECTIVE 1 DETERMINE IF AN INTEGER IS A SOLUTION TO AN EQUATION

Recall from **Topic 1.7** that a **solution** of an **algebraic equation** is a value that, when **substituted** for the variable, makes the equation true. In **Topic 1.7**, we determined if **whole-number** values were **solutions** to equations. However, some equations have solutions that are **negative**. For example, -7 is a solution to the equation $-2x = 14$ because substituting -7 for x results in the true statement $-2(-7) = 14$.

We can determine if a given **integer** is a solution to an equation by substituting the integer for the **variable** and **simplifying** both sides of the equation. If the resulting statement is true, then the value is a solution. If the resulting statement is false, then the value is not a solution.

Example 1 Determining If an Integer Is a Solution to an Equation

My video summary

Determine if the given value of the variable is a solution to the equation.

a. $3x + 5 = -16;\ -7$ **b.** $-3n = 20 - n;\ -5$

Solutions We substitute the given integer value for the variable and simplify. If the resulting statement is true, then the value is a solution to the equation.

a. Begin with the original equation: $3x + 5 = -16$

Substitute -7 for x: $3(-7) + 5 \stackrel{?}{=} -16$

Multiply: $-21 + 5 \stackrel{?}{=} -16$

Add: $-16 = -16$ True

The final statement is true, so -7 *is* a solution to the equation.

b. Try to work this problem on your own. View the **answer**, or watch this **video** to see complete solutions to both parts.

You Try It **Work through this You Try It problem.**

Work Exercises 1–6 in this textbook or in the MyMathLab Study Plan.

OBJECTIVE 2 USE THE ADDITION PROPERTY OF EQUALITY TO SOLVE EQUATIONS

When two or more **equations** have the exact same **solutions**, they are called **equivalent equations.** For example, the three equations $3x + 2 = 17$, $3x = 15$, and $x = 5$ are all equivalent equations because all three equations have the same solution, 5. We can tell that 5 is the solution of these three equations because 5 makes all three equations true:

$3x + 2 = 17$	$3x = 15$	$x = 5$
$3(5) + 2 \stackrel{?}{=} 17$	$3(5) \stackrel{?}{=} 15$	$5 \stackrel{?}{=} 5$
$15 + 2 \stackrel{?}{=} 17$	$15 = 15$ True	$5 = 5$ True
$17 = 17$ True		

(eText Screens 2.6-1–2.6-20)

We know how to determine if a given value is a solution to an equation, but we need an efficient way to find the right value to check. The key is to look for simpler equivalent equations until we find one that ends with an **isolated variable** of the form

$$variable = value \quad \text{or} \quad value = variable.$$

Some *properties of equality* will help. We start with the **addition property of equality.**

> **Addition Property of Equality**
>
> Let a, b, and c be numbers or **algebraic expressions.** Then,
>
> $$a = b \quad \text{and} \quad a + c = b + c$$
>
> are equivalent equations.

The **addition property of equality** means that we can add the same value to both sides of an **equation** without changing its **solution.**

To better understand the addition property of equality, think of an equation as a balanced scale. Figure 6a illustrates the **numeric equation** $20 = 20$. If we add 4 to the left side of this equation, then the two sides are no longer equal and the scale will not balance $(20 + 4 \neq 20)$. See Figure 6b. To maintain equality and balance, we must also add 4 to the right side of the equation $(20 + 4 = 20 + 4)$. See Figure 6c.

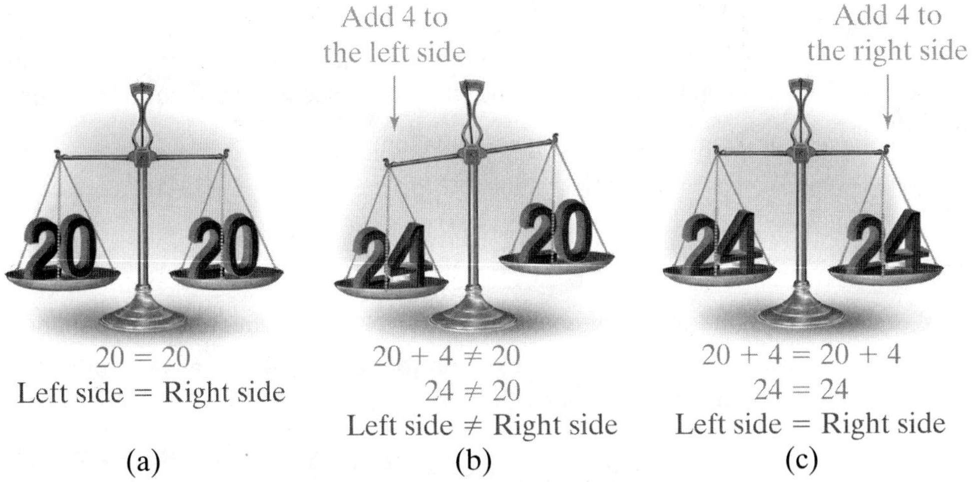

Figure 6

Example 2 Using the Addition Property to Solve an Equation

Solve the equation $x - 5 = 7$.

Solution To solve, we **isolate the variable** x on one side of the equal sign. To isolate x, we must remove the "-5" from the left side of the equation. We can do this by **adding** 5 to both sides.

Begin with the original equation:	$x - 5 = 7$
Add 5 to both sides:	$x - 5 + 5 = 7 + 5$
Simplify $-5 + 5$; simplify $7 + 5$:	$x + 0 = 12$ ← $\boxed{x - 5 + 5 = x + [(-5) + 5] = x + 0}$
Simplify $x + 0$:	$x = 12$

↑ $x + 0 = x$ **(Identity Property of Addition)**

Check Begin with the original equation: $x - 5 = 7$

Substitute 12 for x: $12 - 5 \stackrel{?}{=} 7$

Simplify: $7 = 7$ True

The final statement is true, so 12 checks. Thus, 12 is the **solution** of the equation.

You Try It **Work through this You Try It problem.**

Work Exercise 7 in this textbook or in the MyMathLab Study Plan.

Recall that **subtraction** is defined in terms of **addition** $[a - c = a + (-c)]$. Therefore, the **addition property of equality** will allow us to also subtract the same value from both sides of an **equation** without changing its **solution**. So,

$a = b$ and $a - c = b - c$ are **equivalent equations.**

Some textbooks refer to this as the ***subtraction property of equality***, but we include it as part of the addition property of equality.

My video animation summary

Watch the first part of this **video animation** to further visualize the addition property of equality.

Example 3 Using the Addition Property to Solve an Equation

Solve $24 = x + 10$.

Solution To isolate x, we remove the "$+10$" from the right side of the equation by subtracting 10 from both sides.

Begin with the original equation: $24 = x + 10$

Subtract 10 from both sides: $24 - 10 = x + 10 - 10$

Simplify $24 - 10$; simplify $10 - 10$: $14 = x + 0$

Simplify $x + 0$: $14 = x$

$x + 0 = x$ **(Identity Property of Addition)**

View the **check**. The solution is 14.

You Try It **Work through this You Try It problem.**

Work Exercise 8 in this textbook or in the MyMathLab Study Plan.

TIP When **solving an equation**, you may **isolate the variable** on either side of the equal sign. Choose the more convenient side.

Recall that sometimes a **solution** of an equation will be a **negative number** (as we saw in **Example 1a**).

Example 4 Using the Addition Property to Solve an Equation

Solve each equation.

a. $n - 3 = -7$

b. $2 = y + 9$

c. $-4 + m = -13$

My video summary

Solutions

a. To isolate n, we remove the "-3" from the left side of the equation by **adding** 3 to both sides.

Begin with the original equation:	$n - 3 = -7$
Add 3 to both sides:	$n - 3 + 3 = -7 + 3$
Simplify $-3 + 3$; simplify $-7 + 3$:	$n + 0 = -4$
Simplify $n + 0$:	$n = -4$

$n - 3 + 3$
$= n + [(-3) + 3]$
$= n + 0$

$n + 0 = n$ **(Identity Property of Addition)**

View the **check**. The solution is -4.

b. To isolate y, we remove the "$+9$" from the right side of the equation by **subtracting** 9 from both sides. Try doing this on your own. View the **answer**, or watch this **video** for complete solutions to all three parts.

c. Try solving this equation on your own. View the **answer**, or work through this **video** for complete solutions to all three parts.

You Try It **Work through this You Try It problem.**

Work Exercises 9–21 in this textbook or in the MyMathLab Study Plan.

OBJECTIVE 3 USE THE MULTIPLICATION PROPERTY OF EQUALITY TO SOLVE EQUATIONS

Not all **equations** can be **solved** using the **addition property of equality**. For example, the equation $\frac{x}{2} = 3$ cannot be solved by **adding** or **subtracting** a value from both sides. Instead, we need the **multiplication property of equality**.

> **Multiplication Property of Equality**
>
> Let a, b, and c be numbers or **algebraic expressions** with $c \neq 0$. Then,
>
> $$a = b \quad \text{and} \quad a \cdot c = b \cdot c$$
>
> are **equivalent equations**.

Figure 7a shows the **numeric equation** $5 = 5$ as a balanced scale. **Figure 7b** shows that, if we multiply the left side of the equation by 2, then the two sides are no longer equal and the scale will not balance $(2 \cdot 5 \neq 5)$. **Figure 7c** shows that we must also multiply the right side of the equation by 2 to maintain equality and balance $(2 \cdot 5 = 2 \cdot 5)$.

(eText Screens 2.6-1–2.6-20)

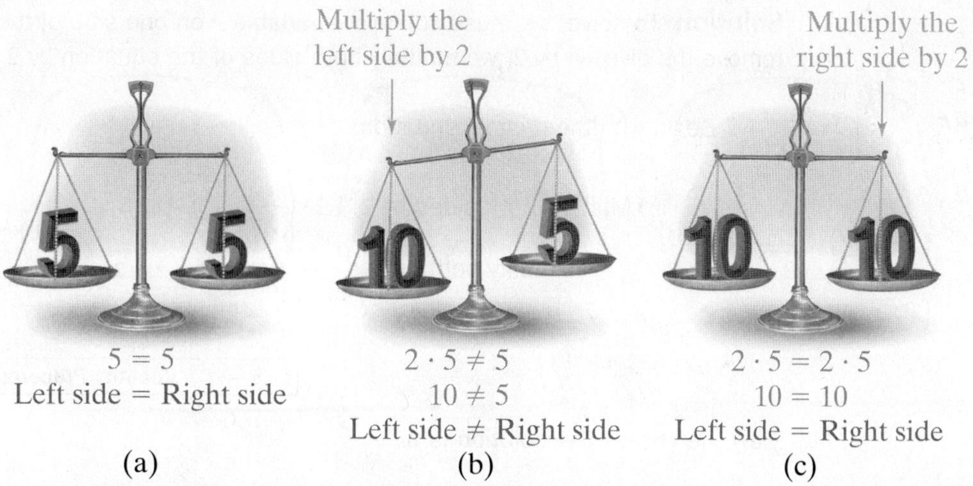

Figure 7

The **multiplication property of equality** means that we can multiply both sides of an **equation** by the same value without changing its **solution.**

Before using this property to solve an equation, consider the following problem involving **multiplication** and **division:** $3 \cdot \left(\frac{6}{3}\right)$.

Following the correct **order of operations**, we first divide within the **grouping symbols**. Then, we multiply the result by the remaining factor, 3.

$$3 \cdot \left(\frac{6}{3}\right) = 3(2) = 6$$

Now let's look at a similar problem: $\left(\frac{3}{3}\right) \cdot 6$.

Again, we follow the correct order of operations, dividing first within the grouping symbols and then multiplying the result by the remaining factor, 6.

$$\left(\frac{3}{3}\right) \cdot 6 = (1) \cdot 6 = 6$$

For both problems we obtained a result of 6, which means $3 \cdot \left(\frac{6}{3}\right) = \left(\frac{3}{3}\right) \cdot 6 = 6$. This is always true.

> **A Helpful Fact When Multiplying and Dividing**
>
> If a and b are numbers or **algebraic expressions**, then, $a \cdot \left(\frac{b}{a}\right) = \left(\frac{a}{a}\right) \cdot b = b$.

This fact is helpful when using the **multiplication property of equality** to solve equations.

Example 5 Using the Multiplication Property to Solve an Equation

Solve the equation $\frac{x}{2} = 3$.

Solution To solve, we must **isolate the variable** x on one side of the equal sign. To remove the **division** by 2, we **multiply** both sides of the equation by 2.

Begin with the original equation: $\frac{x}{2} = 3$

Multiply both sides by 2: $2 \cdot \left(\frac{x}{2}\right) = 2 \cdot (3)$

Simplify both sides: $1 \cdot x = 6$ ← $2 \cdot \left(\frac{x}{2}\right) = \left(\frac{2}{2}\right) \cdot x = 1 \cdot x$

Simplify $1 \cdot x$: $x = 6$

↑ $1 \cdot x = x$ (Identity Property of Multiplication)

View the **check**. The solution is 6.

You Try It **Work through this You Try It problem.**

Work Exercise 22 in this textbook or in the MyMathLab Study Plan.

Recall that **division** can be defined in terms of **multiplication**. For example, $\frac{20}{4} = 5$ because $4 \cdot 5 = 20$. Therefore, the **multiplication property of equality** will allow us to also divide both sides of an **equation** by the same **nonzero** value without changing its **solution**. So,

$a = b$ and $\frac{a}{c} = \frac{b}{c}, c \neq 0$, are **equivalent equations.**

Some textbooks refer to this as the ***division property of equality***, but we include it as part of the multiplication property of equality.

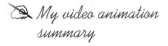

My video animation summary

Watch the second part of this **video animation** to further visualize the multiplication property of equality.

Example 6 Using the Multiplication Property to Solve an Equation

Solve $4x = 36$.

Solution To remove the **multiplication** by 4 to **isolate** x, we **divide** both sides of the **equation** by 4.

Begin with the original equation: $4x = 36$

Divide both sides by 4: $\frac{4x}{4} = \frac{36}{4}$

Simplify both sides: $1 \cdot x = 9$ ← $\frac{4x}{4} = \left(\frac{4}{4}\right)x = 1 \cdot x$

Simplify $1 \cdot x$: $x = 9$

↑ $1 \cdot x = x$ (Identity Property of Multiplication)

View the **check**. The **solution** is 9.

You Try It **Work through this You Try It problem.**

Work Exercise 23 in this textbook or in the MyMathLab Study Plan.

Sometimes we may need to multiply or divide both sides of an equation by a **negative number** in order to **isolate the variable.**

Example 7 Using the Multiplication Property to Solve an Equation

My video summary

Solve each equation.

a. $-3z = 21$ **b.** $-8 = \frac{w}{-5}$ **c.** $-96 = 16m$ **d.** $\frac{t}{-7} = 0$

Solutions

a. To isolate z, we **divide** both sides of the equation by -3.

Begin with the original equation: $-3z = 21$

Divide both sides by -3: $\frac{-3z}{-3} = \frac{21}{-3}$

Simplify both sides: $1 \cdot z = -7$ ← $\frac{-3z}{-3} = \left(\frac{-3}{-3}\right)z = 1 \cdot z$

Simplify $1 \cdot z$: $z = -7$

↑ $1 \cdot z = z$ (**Identity Property of Multiplication**)

View the **check**. The solution is -7.

b. To isolate w, we **multiply** both sides of the equation by -5.

Begin with the original equation: $-8 = \frac{w}{-5}$

Multiply both sides by -5: $-5(-8) = -5\left(\frac{w}{-5}\right)$

Simplify both sides: $40 = 1 \cdot w$ ← $-5\left(\frac{w}{-5}\right) = \left(\frac{-5}{-5}\right)w = 1 \cdot w$

Simplify $1 \cdot w$: $40 = w$

↑ $1 \cdot w = w$ (**Identity Property of Multiplication**)

View the **check**. The solution is 40.

c.–d. Try solving these **equations** on your own. View the **answers**, or work through this **video** for complete solutions to all four parts.

You Try It **Work through this You Try It problem.**

Work Exercises 24–36 in this textbook or in the MyMathLab Study Plan.

2.6 Exercises

In Exercises 1–6, determine if the given integer value for the variable is a solution to the equation.

You Try It

1. $5x + 12 = 32; 4$
2. $40 = 6n - 7; 8$
3. $7 - a = 2a + 11; -4$
4. $-8y = 35 - y; -5$
5. $3(t + 8) = -2t + 19; -1$
6. $-9k + 16 = -7k - 4; -11$

(eText Screens 2.6-1–2.6-20)

In Exercises 7–36, solve each equation. Remember to check your answers.

You Try It 7. $x - 3 = 4$

8. $18 = y + 6$

9. $m - 3 = -9$

10. $8 = w + 10$

You Try It 11. $-12 + n = -20$

12. $x - 10 = 0$

13. $-5 = -1 + x$

14. $k + 12 = 12$

15. $-11 = t - 16$

16. $m + 6 = 0$

17. $3 + y = 12$

18. $n - 4 = -4$

19. $x + 124 = 76$

20. $t - 742 = -587$

21. $-73 + p = -128$

You Try It 22. $\frac{w}{8} = 4$

23. $5x = 45$

24. $-9n = -63$

You Try It 25. $-3 = \frac{m}{6}$

26. $\frac{a}{-2} = -20$

27. $8t = -24$

28. $\frac{k}{-3} = 12$

29. $-5h = 0$

30. $\frac{x}{13} = -13$

31. $-7y = 105$

32. $-18m = -18$

33. $\frac{n}{9} = 0$

34. $\frac{x}{-125} = -375$

35. $112k = 1568$

36. $\frac{x}{200} = -325$

MODULE THREE

Solving Equations and Problem Solving

MODULE THREE CONTENTS

3.1 Simplifying Algebraic Expressions

THINGS TO KNOW

Before working through this topic, be sure you are familiar with the following concepts:

		VIDEO	ANIMATION	INTERACTIVE
You Try It	1. Add Two Integers with the Same Sign (Topic 2.2, **Objective 1**)			
You Try It	2. Add Two Integers with Different Signs (Topic 2.2, **Objective 2**)			
You Try It	3. Subtract Integers (Topic 2.3, **Objective 1**)			
You Try It	4. Multiply Two Integers (Topic 2.4, **Objective 1**)			
You Try It	5. Divide Integers (Topic 2.4, **Objective 4**)			

OBJECTIVES

1 Identify Terms, Coefficients, and Like Terms

2 Combine Like Terms

3 Multiply Algebraic Expressions

4 Multiply Expressions Using the Distributive Property

5 Simplify Algebraic Expressions

OBJECTIVE 1 IDENTIFY TERMS, COEFFICIENTS, AND LIKE TERMS

In **Topic 1.7**, we learned that an **algebraic expression** is a **variable** or combination of variables, **constants**, **operations**, and **grouping symbols**. Some examples of algebraic expressions are

$$x, \quad 2y, \quad n + 8, \quad 6ab - 5c, \quad 3x^2 + 7x + 12, \quad \text{and} \quad 8w^2 - 4wz + 9z^2.$$

In **Topic 1.3**, we learned that numbers being added in a **numeric expression** are called *terms*. The quantities that are added in an algebraic expression are also called **terms**.

> **Definition Term**
>
> The **terms** of an algebraic expression are the quantities being added. A term that contains one or more variables is a **variable term**. A term without any variables is a **constant term**.

For example, the expression $3x^2 + 7x + 12$ has three terms: $3x^2$, $7x$, and 12. The $3x^2$ and $7x$ are *variable terms*, while the 12 is a *constant term*. The plus signs $(+)$ separate the terms.

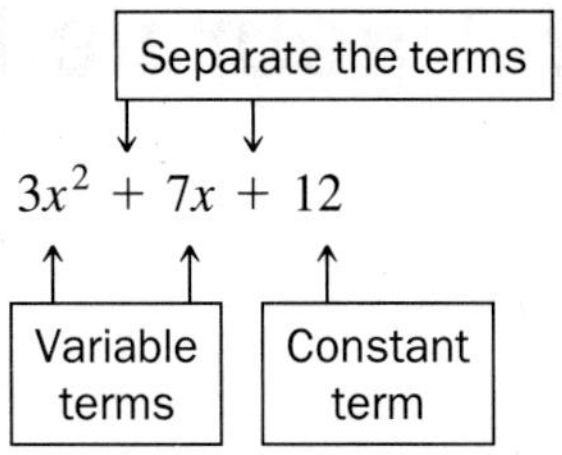

Because **subtraction** is defined by adding the **opposite**, a subtracted quantity is a "negative" **term**. For example, in the **expression** $6ab - 5c$, the terms are $6ab$ and $-5c$ because $6ab - 5c = 6ab + (-5c)$.

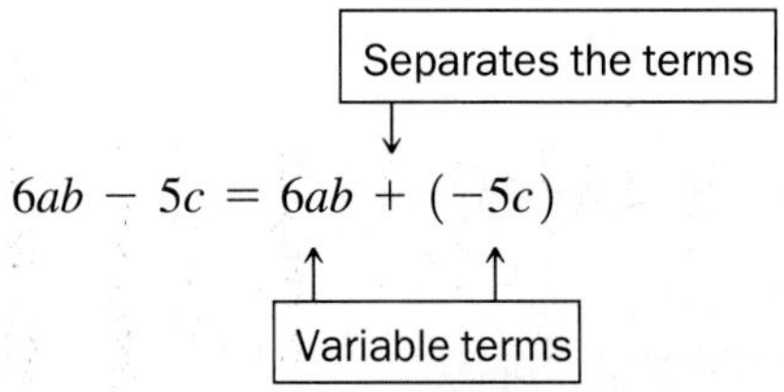

Notice that the **variable term** $6ab$ has a numeric **factor**, 6, and **variable factors**, a and b. The numeric factor of a term is called the *coefficient* or *numerical coefficient*.

> **Definition Coefficient (or Numerical Coefficient)**
>
> A **coefficient**, or **numerical coefficient**, is the numeric factor of a term.

Every term in an algebraic expression has a coefficient. For example, the expression $8w^2 - 4wz + 9z^2$ has three terms, so it has three coefficients. Similarly, $6ab - 5c$ has two terms, so it has two coefficients.

Expression	Terms	Coefficients
$8w^2 - 4wz + 9z^2$ or $8w^2 + (-4wz) + 9z^2$	$8w^2, -4wz, 9z^2$	$8, -4, 9$
$6ab - 5c$ or $6ab + (-5c)$	$6ab, -5c$	$6, -5$

When a **variable term** has no written **coefficient**, a 1 is understood to be the numerical coefficient. For example, in the expression $n + 8$, the coefficient of the variable term n is 1 because $n = 1n$.

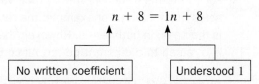

$$n + 8 = 1n + 8$$

No written coefficient (points to n) — Understood 1 (points to $1n$)

The coefficient of a **constant term** is the **constant** itself.

Example 1 Identifying Terms and Coefficients

My video summary

List the **terms** of each **expression**. Then identify the coefficient of each term.

a. $12m - 7n$ **b.** $x^2 - 13x + 42$ **c.** $7x^3 - x^2 + x - 16$

Solutions

a. $12m - 7n = \underbrace{12m}_{\text{Term 1}} + \underbrace{(-7n)}_{\text{Term 2}}$

Term	Coefficient
$12m$	12
$-7n$	-7

b. $x^2 - 13x + 42 = \underbrace{1x^2}_{\text{Term 1}} + \underbrace{(-13x)}_{\text{Term 2}} + \underbrace{42}_{\text{Term 3}}$ (Understood 1 points to the 1 in $1x^2$)

Term	Coefficient
x^2 or $1x^2$	1
$-13x$	-13
42	42

c. Try to list the **terms** and **coefficients** on your own. View this **popup** for a complete solution to part c, or watch this **video** for complete solutions to all three parts.

CAUTION

Pay attention to the **signs** when listing terms and coefficients.

You Try It **Work through this You Try It problem.**

Work Exercises 1–6 in this textbook or in the MyMathLab Study Plan.

(eText Screens 3.1-1–3.1-19)

Consider the **expression** $4x + 7x$. Notice that the two terms $4x$ and $7x$ have the same **variable factor**, x. Terms with the exact same variable **factors** are called *like terms*.

> **Definition Like Terms**
>
> Two terms are **like terms** if they contain the exact same variable factors. The coefficients may differ.

For two terms to be *like* terms, their **variables** must match, and any **exponents** on the variables must match. For example, the terms $2x^2$ and $4x^2$ are **like terms** because the x^2 factor is the same in both terms. However, the terms $5n^3$ and $7n^2$ are not like terms because each n is raised to a different power. All **constant terms** are like terms.

Example 2 Identifying Like Terms

My video summary

Identify the like terms in each algebraic expression.

a. $3x^2 + 7x - 2 + 5x^2 - x + 4$ **b.** $m^2 - 2mn + 6n^2 - mn + 9m^2$

Solutions

a. $3x^2$ and $5x^2$ are like terms because they have the same variable factor, x^2. Similarly, $7x$ and $-x$ are like terms with the same variable factor of x. Lastly, -2 and 4 are like terms because they are both **constants**.

b. Try to identify these like terms on your own. View this **popup** for a complete solution to part b, or watch this **video** for complete solutions to both parts.

You Try It **Work through this You Try It problem.**

Work Exercises 7–10 in this textbook or in the MyMathLab Study Plan.

OBJECTIVE 2 COMBINE LIKE TERMS

In **Topic 1.4**, we used the **distributive property** to distribute multiplication over addition or subtraction. When an **expression** contains **like terms**, we can use the distributive property in reverse to **combine like terms**. For example, in the expression $4x + 7x$, we can add the terms $4x$ and $7x$ as follows:

Add coefficients

$$4x + 7x = \overbrace{(4 + 7)}x = 11x$$

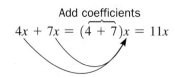

When we combine the like terms of an expression, we **simplify the expression** by writing it with as few **terms** and **operations** as possible.

Example 3 Combining Like Terms

Simplify each expression by combining like terms.

a. $12y - 10y$ **b.** $4xy + xy + 19xy$ **c.** $x^2 - 17x^2 + 6y^2 - 5y^2$

Solutions

a. $12y$ and $-10y$ are like terms because they have the same **variable factor**, y.

Begin with the original expression: $12y - 10y$

Reverse the distributive property: $= (12 - 10)y$

Subtract $12 - 10$: $= 2y$

b. $4xy$, xy, and $19xy$ are **like terms**. The **coefficient** of the term xy is understood to be 1.

Begin with the original expression: $4xy + xy + 19xy$

Write in the understood 1: $= 4xy + 1xy + 19xy$

Reverse the distributive property: $= (4 + 1 + 19)xy$

Add $4 + 1 + 19$: $= 24xy$

My video summary

c. x^2 and $-17x^2$ are like terms, and $6y^2$ and $-5y^2$ are like terms. Try to simplify this **expression** on your own by combining each pair of like terms. View the **answer**, or watch this **video** for a complete solution to part c.

You Try It **Work through this You Try It problem.**

Work Exercises 11–14 in this textbook or in the MyMathLab Study Plan.

Sometimes, it is necessary to rearrange the **terms** in an expression so that the like terms are grouped together. Doing this is called **collecting like terms**. Collecting like terms is possible because of the **commutative** and **associative properties of addition**.

Example 4 Combining Like Terms

Simplify each **expression** by **combining like terms**.

a. $8x - 6 - 13x + 14$ **b.** $2t^2 + 14 - 5t - 3t^2 + 18t$

Solutions

a. $8x$ and $-13x$ are like terms, as are -6 and 14.

Begin with the original expression: $8x - 6 - 13x + 14$

Change subtractions to add the opposite: $= 8x + (-6) + (-13x) + 14$

Rearrange to **collect like terms**: $= 8x + (-13x) + (-6) + 14$

Combine like terms: $= -5x + 8$

My video summary

b. $2t^2$ and $-3t^2$ are **like terms**, as are $-5t$ and $18t$. The **constant term** 14 has no like terms in this expression.

Begin with the original expression: $2t^2 + 14 - 5t - 3t^2 + 18t$

Change subtractions to add the opposite: $= 2t^2 + 14 + (-5t) + (-3t^2) + 18t$

Rearrange to collect like terms: $= 2t^2 + (-3t^2) + (-5t) + 18t + 14$

Try to finish this problem on your own. View the **answer**, or watch this **video** for a complete solution to part b.

(eText Screens 3.1-1–3.1-19)

Example 5 Combining Like Terms

My interactive video summary

Simplify each algebraic expression by combining like terms.

a. $5x - 2x$ b. $6x^2 - 12x - 3x^2 + 4x$

c. $3z - 2z^2 + 7z^2$ d. $6x^2 + 2x + 4x + 3$

e. $-3x + 5 - y + x - 8$

Solutions Try to simplify these expressions on your own. Check your **answers**, or watch this **interactive video** for complete solutions to all five parts.

You Try It **Work through this You Try It problem.**

Work Exercises 15–26 in this textbook or in the MyMathLab Study Plan.

OBJECTIVE 3 MULTIPLY ALGEBRAIC EXPRESSIONS

In **Topic 1.4**, we used the **associative property of multiplication** to change the grouping of **factors** during multiplication. For example, $4 \cdot (5 \cdot 6) = (4 \cdot 5) \cdot 6$. We can also use the associative property to multiply **algebraic expressions** such as $8(2x)$:

$$8(2x) = 8 \cdot (2 \cdot x) = \underbrace{(8 \cdot 2)}_{\text{Regroup}} \cdot x = \underbrace{(16)}_{\text{Multiply}} \cdot x = 16x$$

Understood multiplication

So, $8(2x)$ *simplifies* to $16x$.

Example 6 Multiplying Algebraic Expressions

Simplify each expression by multiplying.

a. $-9(5m)$ b. $4(-6y^2)$ c. $-13(-7a)$

Solutions

a. Begin with the original expression: $-9(5m)$

Use the associative property to regroup factors: $= (-9 \cdot 5)m$

Multiply $-9 \cdot 5$: $= -45m$

b. Begin with the original expression: $4(-6y^2)$

Use the associative property to regroup factors: $= [4(-6)]y^2$

Multiply $4(-6)$: $= -24y^2$

c. Begin with the original expression: $-13(-7a)$

Use the associative property to regroup factors: $= [-13(-7)]a$

Multiply $-13(-7)$: $= 91a$

You Try It **Work through this You Try It problem.**

Work Exercises 27–34 in this textbook or in the MyMathLab Study Plan.

OBJECTIVE 4 MULTIPLY EXPRESSIONS USING THE DISTRIBUTIVE PROPERTY

Earlier in this topic, we used the **distributive property** to **combine like terms.** We can also use the distributive property to multiply out **expressions** such as $4(x + 7)$. In this expression, we are unable to combine the terms within the parentheses because x and 7 are not **like terms**. However, we can still multiply by distributing multiplication over addition as follows:

Distributive Property

$$4(x + 7) = 4 \cdot (x + 7) = 4 \cdot x + 4 \cdot 7$$
$$= 4x + 28$$

Understood multiplication

So, $4(x + 7)$ multiplies to $4x + 28$.

Recall that multiplication also distributes over subtraction, so we can multiply out expressions such as $3(y - 8)$:

Distributive Property

$$3(y - 8) = 3 \cdot y - 3 \cdot 8 = 3y - 24$$

So, $3(y - 8)$ multiplies to $3y - 24$.

Example 7 Using the Distributive Property

My video summary

Multiply using the **distributive property**.

a. $9(2x + 5)$ **b.** $-8(10x + 2y)$ **c.** $-10(3m - 4)$

Solutions

a. Begin with the original expression: $9(2x + 5)$

Distribute the 9: $= 9 \cdot (2x) + 9 \cdot 5$

Simplify the multiplications: $= 18x + 45$

$9 \cdot (2x) = (9 \cdot 2)x = 18x$

b. Begin with the original expression: $-8(10x + 2y)$

Distribute the -8: $= -8 \cdot (10x) + (-8) \cdot (2y)$

Simplify the multiplications: $= -80x + (-16y)$

Change "$+(-16y)$" to "$-16y$": $= -80x - 16y$

$-8 \cdot (10x) = (-8 \cdot 10)x = -80x$
$-8 \cdot (2y) = (-8 \cdot 2)y = -16y$

c. Try to work this problem on your own. View this **popup** for a complete solution to part c, or watch this **video** for complete solutions to all three parts.

You Try It **Work through this You Try It problem.**

Work Exercises 35–42 in this textbook or in the MyMathLab Study Plan.

(eText Screens 3.1-1–3.1-19)

OBJECTIVE 5 SIMPLIFY ALGEBRAIC EXPRESSIONS

Sometimes, **simplifying** an **algebraic expression** requires several steps. For example, we may need to multiply using the **distributive property** first and then **combine like terms.**

Example 8 Simplifying Algebraic Expressions

My interactive video summary

Simplify each algebraic expression.

a. $7(x - 2) + 9$

b. $7(n + 13) - (5n + 21)$

c. $-2(2y - 11) + 9(y - 1)$

d. $35 - 10(3t + 8)$

Solutions

a. Begin with the original expression: $7(x - 2) + 9$

Distribute the 7: $= 7x - 7 \cdot 2 + 9$

Multiply $7 \cdot 2$: $= 7x - 14 + 9$

Combine like terms: $= 7x - 5$

b. We can think of $-(5n + 21)$ as $-1(5n + 21)$.

Begin with the original expression: $7(n + 13) - (5n + 21)$

Rewrite $-(5n + 21)$ as $-1(5n + 21)$: $= 7(n + 13) - 1(5n + 21)$

Distribute the 7; distribute the -1: $= 7n + 7 \cdot 13 + (-1)(5n) + (-1)(21)$

Perform the multiplications: $= 7n + 91 + (-5n) + (-21)$

Rearrange terms to collect like terms: $= 7n + (-5n) + 91 + (-21)$

Combine like terms: $= 2n + 70$

c.–d. Try to simplify these expressions on your own. View the **answers** to parts c and d, or watch this **interactive video** for complete solutions to all four parts.

You Try It **Work through this You Try It problem.**

Work Exercises 43–50 in this textbook or in the MyMathLab Study Plan.

3.1 Exercises

In Exercises 1–6, list the terms of each expression. Then identify the coefficient of each term.

You Try It

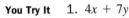

1. $4x + 7y$
2. $18m - 5$
3. $9z^2 - 8z + 6$
4. $t^2 - t - 12$
5. $w^3 - 4w^2 - 6w + 5$
6. $2a^2 + 3ab$

In Exercises 7–10, identify the like terms in each algebraic expression.

You Try It

7. $3x + 9y - x + 8y$
8. $x^2 + 5x + 2 - 8x + 6 + 3x^2$
9. $7m - 12 + 4n + 8 - m$
10. $2p^2 - 3pq + 8q^2 + p^2 - 6pq$

(eText Screens 3.1-1–3.1-19)

In Exercises 11–26, simplify each expression by combining like terms.

You Try It 11. $8z + 15z$

12. $-6x^2 + 11x^2$

13. $ab - 5ab + 13ab$

14. $5m^2 + 4m^2 + 7n^2 - 10n^2$

You Try It 15. $7p^2 - 2p - 9p + 20$

16. $6y - 25 - 8y + 38$

17. $8m + 3n + 7m$

18. $a + 5b + a + 4c$

19. $t^2 + 19t - 8t^2 + t$

20. $-3a^2 + a^2 - 8a^2$

21. $4w - 3w^2 + 10w^2$

22. $-4x + 13 - 3y + 7x - 18$

23. $7k^2 - 12 + 6k - 4k^2 - 19k$

24. $6w - 28 + 32 - 6w$

25. $-ab + 23ab - 15ab + 17$

26. $x - 5x + x^2 - 6x^2$

In Exercises 27–34, simplify each expression by multiplying.

You Try It 27. $7(8x)$

28. $-3(6y)$

29. $12(-5n)$

30. $-2(-11t)$

31. $13(7z^2)$

32. $4(-9x^3)$

33. $-14(8ab)$

34. $-10(-m)$

In Exercises 35–42, multiply using the distributive property.

You Try It 35. $7(x + 3)$

36. $2(y - 6)$

37. $8(10w + 1)$

38. $5(7t - 9)$

39. $-6(2x + 7)$

40. $-9(-3m + 5n)$

41. $-4(5y - 8)$

42. $3(2x + 8y - 5z)$

In Exercises 43–50, simplify each algebraic expression.

You Try It 43. $4(x + 8) - 9$

44. $34 + 8(y - 7)$

45. $6(2y - 1) + 3y$

46. $y + 5(y - 8)$

47. $-2(w - 6) + 7w + 1$

48. $-3(2x + 5) + 6x$

49. $7(x + 2) - (3n + 8)$

50. $-4(y - 5) + 2(3y - 1)$

(eText Screens 3.2-1–3.2-15)

3.2 Revisiting the Properties of Equality

THINGS TO KNOW

Before working through this topic, be sure you are familiar with the following concepts:	VIDEO	ANIMATION	INTERACTIVE
You Try It — **1.** Determine If an Integer Is a Solution to an Equation (Topic 2.6, **Objective 1**)	✓		
You Try It — **2.** Use the Addition Property of Equality to Solve Equations (Topic 2.6, **Objective 2**)	✓	✓	
You Try It — **3.** Use the Multiplication Property of Equality to Solve Equations (Topic 2.6, **Objective 3**)	✓	✓	
You Try It — **4.** Identify Terms, Coefficients, and Like Terms (Topic 3.1, **Objective 1**)	✓		
You Try It — **5.** Combine Like Terms (Topic 3.1, **Objective 2**)	✓		

OBJECTIVES

1 Solve Equations Using the Addition Property of Equality

2 Solve Equations Using the Multiplication Property of Equality

3 Solve Equations Using Both Properties of Equality

OBJECTIVE 1 SOLVE EQUATIONS USING THE ADDITION PROPERTY OF EQUALITY

In **Topic 2.6**, we learned to **solve** some **equations.** Now that we know how to **combine like terms**, we can solve even more equations.

Recall that the **addition property of equality** allows the same quantity to be added to or subtracted from both sides of an equation without changing the **solution.** Often we use this property to **isolate the variable** in the form

$$\textit{variable} = \textit{value} \quad \text{or} \quad \textit{value} = \textit{variable}.$$

We review this process in Example 1.

Example 1 Solving an Equation Using the Addition Property of Equality

Solve the equation $x - 6 = -15$.

Solution

Begin with the original equation:	$x - 6 = -15$	
Add 6 to both sides:	$x - 6 + 6 = -15 + 6$	← Addition property of equality
Simplify both sides:	$x = -9$	
	↑ Isolated variable	

View the **check**. The solution is -9.

You Try It **Work through this You Try It problem.**

Work Exercises 1–4 in this textbook or in the MyMathLab Study Plan.

Sometimes, the quantity to be added to or subtracted from both sides of an equation is a **variable term**. For example, if an **equation** has variable terms on both sides of the equal sign $(=)$, then we can use the **addition property of equality** to remove one of the variable terms from one side of the equation so that the resulting **equivalent equation** contains only one variable term. We demonstrate this process in Example 2.

Example 2 Solving Equations Using the Addition Property of Equality

My video summary

Solve each equation.

a. $3x = 7 + 2x$ **b.** $-4n + 5 = -3n$

Solutions

a. To solve, we **isolate the variable** x on one side of the equal sign. The equation has variable terms on both sides, $3x$ and $2x$, so we subtract $2x$ from both sides to remove it from the right side.

Begin with the original equation: $3x = 7 + 2x$

Subtract $2x$ from both sides: $3x - 2x = 7 + 2x - 2x$ ← Addition property of equality

Combine like terms: $1x = 7 + 0x$

Simplify: $x = 7$ ← $1x = x; 7 + 0x = 7 + 0 = 7$

↑ Isolated variable

View the **check**. The solution is 7.

b. Notice that adding the **variable term** $4n$ to both sides of the **equation** will remove the $4n$ from the left side of the equation, leaving the **constant term** 5 on the left side and **isolating the variable** n on the right side. Try doing this to solve the equation on your own. View the **answer**, or watch this **video** for complete solutions to both parts.

You Try It **Work through this You Try It problem.**

Work Exercises 5–8 in this textbook or in the MyMathLab Study Plan.

To solve some equations, we must apply the **addition property of equality** twice in order to get the **variable terms** and **constant terms** on opposite sides of the equal sign.

Example 3 Solving Equations Using the Addition Property of Equality Twice

My video summary

Solve $6y - 7 = 5y + 2$.

Solution To solve, we **isolate the variable** y on one side of the equal sign. To do this, we begin by subtracting $5y$ from both sides of the equation.

(eText Screens 3.2-1–3.2-15)

Begin with the original equation: $6y - 7 = 5y + 2$

Subtract $5y$ from both sides: $6y - 7 - 5y = 5y + 2 - 5y$ ← Addition property of equality

Collect like terms on both sides: $\underbrace{6y - 5y} - 7 = \underbrace{5y - 5y} + 2$

Combine like terms: $1y \quad - 7 = \quad 0y \quad + 2$

Simplify: $y - 7 = 2$ ← $1y = y; 0y + 2 = 0 + 2 = 2$

This last equation now has a form like the one we solved in **Example 1**. Try to finish solving this equation on your own. View the **answer**, or watch this **video** for a complete solution.

You Try It **Work through this You Try It problem.**

Work Exercises 9–12 in this textbook or in the MyMathLab Study Plan.

OBJECTIVE 2 SOLVE EQUATIONS USING THE MULTIPLICATION PROPERTY OF EQUALITY

Recall from **Topic 2.6** that the **multiplication property of equality** allows us to multiply or divide both sides of an equation by the same **nonzero** quantity.

Example 4 Solving Equations Using the Multiplication Property of Equality

Solve each equation.

a. $\dfrac{x}{-3} = 12$ **b.** $4y = -32$

Solutions

a. To **isolate the variable** x, **multiply** both sides of the equation by -3.

Begin with the original equation: $\dfrac{x}{-3} = 12$

Multiply both sides by -3: $-3\left(\dfrac{x}{-3}\right) = -3(12)$ ← Multiplication property of equality

Simplify both sides: $x = -36$

$-3\left(\dfrac{x}{-3}\right) = \left(\dfrac{-3}{-3}\right)x = 1x = x$

View the **check**. The **solution** is -36.

b. To **isolate the variable** y, **divide** both sides of the equation by 4.

Begin with the original equation: $4y = -32$

Divide both sides by 4: $\dfrac{4y}{4} = \dfrac{-32}{4}$ ← Multiplication property of equality

Simplify both sides: $y = -8$

$\dfrac{4y}{4} = \left(\dfrac{4}{4}\right)y = 1y = y$

View the **check**. The **solution** is -8.

You Try It **Work through this You Try It problem.**

Work Exercises 13–20 in this textbook or in the MyMathLab Study Plan.

OBJECTIVE 3 SOLVE EQUATIONS USING BOTH PROPERTIES OF EQUALITY

To **solve** some **equations**, we must apply both **properties of equality**. For these equations, we use the **addition property** to write the **variable terms** on one side of the equal sign and the **constant terms** on the other side. Then, we use the **multiplication property** to **isolate the variable.**

Example 5 Solving Equations Using Both Properties of Equality

My interactive video summary

Solve each equation.

a. $9x - 4 = 68$ **b.** $\dfrac{t}{3} + 5 = -1$

c. $2 - 13n = -63$ **d.** $10 = -3x - 17$

Solutions

a. The only variable term in the equation, $9x$, is on the left side of the equal sign. To get the $9x$ by itself, add 4 to both sides.

Begin with the original equation: $9x - 4 = 68$

Add 4 to both sides: $9x - 4 + 4 = 68 + 4$ ← Addition property of equality

Simplify both sides: $9x = 72$

Divide both sides by 9: $\dfrac{9x}{9} = \dfrac{72}{9}$ ← Multiplication property of equality

Simplify: $x = 8$

$\dfrac{9x}{9} = \left(\dfrac{9}{9}\right)x = 1x = x$

View the **check**. The **solution** is 8.

b. The only **variable term** in the equation, $\dfrac{t}{3}$, is on the left side of the equal sign. To get the $\dfrac{t}{3}$ by itself, subtract 5 from both sides.

Begin with the original equation: $\dfrac{t}{3} + 5 = -1$

Subtract 5 from both sides: $\dfrac{t}{3} + 5 - 5 = -1 - 5$ ← Addition property of equality

Simplify both sides: $\dfrac{t}{3} = -6$

Multiply both sides by 3: $3\left(\dfrac{t}{3}\right) = 3(-6)$ ← Multiplication property of equality

Simplify: $t = -18$ ← $3\left(\dfrac{t}{3}\right) = \left(\dfrac{3}{3}\right)t = 1t = t$

View the **check**. The **solution** is -18.

c.–d. Try solving these two equations on your own. View the **answers**, or watch this **interactive video** for complete solutions to all four parts.

You Try It **Work through this You Try It problem.**

Work Exercises 21–36 in this textbook or in the MyMathLab Study Plan.

Example 6 Solving Equations Using Both Properties of Equality

Solve each equation.

a. $6x = -2x + 40$ **b.** $15n + 42 = 8n$

Solutions

a. There are **variable terms** on both sides of the equation, but the only **constant term** is on the right side. Let's move all the variables to the left side by adding $2x$ to both sides.

Begin with the original equation: $6x = -2x + 40$

Add $2x$ to both sides: $6x + 2x = -2x + 40 + 2x$ ← Addition property of equality

Collect like terms on the right side: $\underbrace{6x + 2x} = \underbrace{-2x + 2x} + 40$

Combine like terms: $8x = 0x + 40$

Simplify: $8x = 40$ ← $0x + 40 = 0 + 40 = 40$

Divide both sides by 8: $\dfrac{8x}{8} = \dfrac{40}{8}$ ← Multiplication property of equality

Simplify: $x = 5$ ← $\dfrac{8x}{8} = \left(\dfrac{8}{8}\right)x = 1x = x$

View the **check**. The **solution** is 5.

My video summary

b. There are **variable terms** on both sides of the equation, but the only **constant term** is on the left side. So, we can move all the variables to the right side by subtracting $15n$ from both sides.

Try **solving** this equation on your own. View the **answer**, or watch this **video** for a complete solution to part b.

You Try It **Work through this You Try It problem.**

Work Exercises 37–42 in this textbook or in the MyMathLab Study Plan.

Example 7 Solving Equations Using Both Properties of Equality

My interactive video summary

Solve each equation.

a. $5x - 8 = 3x + 12$ **b.** $16 - 5t = 19 - 2t$

Solutions

a. This equation has both variable terms and constant terms on both sides of the equal sign. We must move all variable terms to one side and the constant terms to the other side. Which sides we choose does not matter. Let's move the variable terms to the left side and the constant terms to the right side.

Begin with the original equation: $5x - 8 = 3x + 12$

Subtract $3x$ from both sides: $5x - 8 - 3x = 3x + 12 - 3x$ ← Addition property of equality

Collect like terms on both side: $\underbrace{5x - 3x} - 8 = \underbrace{3x - 3x} + 12$

Combine like terms: $2x - 8 = 0x + 12$ ← $0x + 12 = 0 + 12 = 12$

Simplify: $2x - 8 = 12$

This last **equation** now has a form like the ones we solved in **Example 5**. Try to finish **solving** it on your own. View the **answer**, or watch this **interactive video** for complete solutions to both parts.

b. Try solving this equation on your own. View the **answer**, or watch this **interactive video** for complete solutions to both parts.

You Try It **Work through this You Try It problem.**

Work Exercises 43–50 in this textbook or in the MyMathLab Study Plan.

3.2 Exercises

In Exercises 1–12, use the addition property of equality to solve each equation.

You Try It

1. $x + 21 = 16$
2. $y - 12 = -3$
3. $-10 = t - 4$
4. $35 = w + 7$

You Try It

5. $10n = 3 + 9n$
6. $-4z = 10 - 5z$
7. $m + 18 = 2m$
8. $25 - 8k = -7k$

You Try It

9. $2x + 5 = x + 9$
10. $6h - 15 = 5h + 2$
11. $24 - 3p = 19 - 2p$
12. $11a - 13 = 12a + 7$

In Exercises 13–20, use the multiplication property of equality to solve each equation.

You Try It

13. $\dfrac{x}{4} = 20$
14. $\dfrac{w}{-1} = -2$
15. $\dfrac{k}{-6} = 3$
16. $\dfrac{n}{2} = -8$

17. $-3y = -12$

18. $4p = 16$

19. $-x = 9$

20. $15a = -45$

In Exercises 21–50, use both properties of equality to solve each equation.

You Try It

21. $4x - 3 = 25$

22. $3x + 1 = 19$

23. $7y + 8 = -48$

24. $2n - 7 = -21$

25. $\frac{x}{2} - 3 = 5$

26. $\frac{t}{4} + 1 = -1$

27. $\frac{k}{5} - 6 = -9$

28. $\frac{m}{9} + 4 = 1$

29. $-2y + 1 = 13$

30. $-6x - 17 = -71$

31. $-n - 3 = -9$

32. $4 - x = 15$

33. $16 = 3k - 8$

34. $-15 = 2t + 13$

35. $27 = 3 - 4m$

36. $-2 = 8 - h$

You Try It

37. $9w = 7w + 16$

38. $2p = 42 - p$

39. $10u - 91 = 3u$

40. $21y + 112 = 17y$

41. $136 - 5x = -13x$

42. $-9a = -12a - 48$

You Try It

43. $3x + 1 = x + 9$

44. $5n - 4 = -3n + 20$

45. $8m + 5 = 5m - 1$

46. $3t + 7 = 17 - 2t$

47. $-2p + 2 = p + 5$

48. $10x - 9 = 4x - 9$

49. $11 - 5u = -9 - u$

50. $-9y - 17 = -13y - 13$

3.3 Solving Linear Equations in One Variable

THINGS TO KNOW

Before working through this topic, be sure you are familiar with the following concepts:

	VIDEO	ANIMATION	INTERACTIVE
You Try It 1. Multiply Expressions Using the Distributive Property (Topic 3.1, **Objective 4**)	✓		
You Try It 2. Simplify Algebraic Expressions (Topic 3.1, **Objective 5**)			✓
You Try It 3. Solve Equations Using the Addition Property of Equality (Topic 3.2, **Objective 1**)	✓	✓	

You Try It

4. Solve Equations Using the Multiplication Property of Equality (Topic 3.2, **Objective 2**)

You Try It

5. Solve Equations Using Both Properties of Equality (Topic 3.2, **Objective 3**)

OBJECTIVES

1 Identify Linear Equations in One Variable

2 Solve Linear Equations Involving Non-Simplified Expressions

3 Solve Linear Equations Involving Grouping Symbols

OBJECTIVE 1 IDENTIFY LINEAR EQUATIONS IN ONE VARIABLE

The equations $4x - 2 = 6$ and $-z + 5 = 8$ are both examples of *linear equations in one variable*. Recall that an *equation* contains an equal sign $(=)$, while an *expression* does not.

Linear equations are also called **first-degree equations** because the **exponent** on the variable is **understood to be 1**. If an **equation** contains a variable raised to an exponent other than 1, or if a variable appears beneath a **fraction bar**, then the equation is **nonlinear.**

Definition Linear Equation in One Variable

A **linear equation in one variable** is an equation that can be written in the form $ax + b = c$, where x is a variable and a, b, and c are any numbers such that $a \neq 0$.

TIP

It is important to understand that an equation does not have to be written in the form $ax + b = c$ to be linear. Rather, the equation must *be able to be written* in the form $ax + b = c$. For example, the equation $x + x + 3 = 7$ is a linear equation in one variable because it can be written as $2x + 3 = 7$, which has the form $ax + b = c$. Also keep in mind that the **variable** in the equation does not have to be x. Any variable can be used, but it must be the *only* variable used.

Example 1 Identifying Linear Equations in One Variable

My video summary

Determine if each is a **linear equation in one variable.**

If not, state why.

a. $-2x + 5 - 7$

b. $3y + 1 = 7$

c. $x^3 - 5x = 14$

d. $7a - 3b = 27$

e. $\dfrac{3}{x + 2} = 7$

f. $9x + 20 = 7 - 4x$

Solutions

a. $-2x + 5 - 7$ is not a linear equation in one variable because it is an algebraic expression, not an **equation**.

b. $3y + 1 = 7$ is a linear equation in one **variable.** Only one variable occurs, and the **exponent** on the variable is 1.

(eText Screens 3.3-1–3.3-12)

c. $x^3 - 5x = 14$ is an equation in one variable, but it is nonlinear because x^3 has an exponent other than 1.

d.–f. Try working the last three parts on your own. Check your **answers**, or watch this **video** for the complete solutions to all six parts.

You Try It **Work through this You Try It problem.**

Work Exercises 1–8 in this textbook or in the MyMathLab Study Plan.

OBJECTIVE 2 SOLVE LINEAR EQUATIONS INVOLVING NON-SIMPLIFIED EXPRESSIONS

When **solving** a **linear equation** in one variable with **non-simplified expressions** on one or both sides, we simplify each side first by **combining like terms**. Then, we continue solving the equation by using the **properties of equality** as in **Topic 3.2**.

Example 2 Solving Linear Equations Containing Non-Simplified Expressions

Solve: $7x - 8 + 3x = 5x - 13$

Solution Start by simplifying the expression on the left side of the equation. The right side of the equation is already simplified.

Begin with the original equation: $7x - 8 + 3x = 5x - 13$

Combine like terms on the left: $10x - 8 = 5x - 13 \leftarrow 7x + 3x = 10x$

Now apply the addition and multiplication properties of equality to finish solving the equation.

Subtract $5x$ from both sides: $10x - 8 - 5x = 5x - 13 - 5x$ ← Addition property of equality

Simplify: $5x - 8 = 0x - 13$ ← $10x - 5x = 5x$; $5x - 5x = 0x$

$5x - 8 = -13$

Add 8 to both sides: $5x - 8 + 8 = -13 + 8$ ← Addition property of equality

Simplify: $5x = -5$

Divide both sides by 5: $\dfrac{5x}{5} = \dfrac{-5}{5}$ ← Multiplication property of equality

Simplify: $x = -1$ ← $\dfrac{5x}{5} = \left(\dfrac{5}{5}\right)x = 1x = x$

Check Begin with the original equation: $7x - 8 + 3x = 5x - 13$

Substitute -1 for x: $7(-1) - 8 + 3(-1) \stackrel{?}{=} 5(-1) - 13$

Multiply: $-7 - 8 + (-3) \stackrel{?}{=} -5 - 13$

Subtract $-7 - 8$ on the left; subtract $-5 - 13$ on the right: $-15 + (-3) \stackrel{?}{=} -18$

Add $-15 + (-3)$: $-18 = -18$ True

The resulting statement is true, so -1 checks. The **solution** to the equation is -1.

(eText Screens 3.3-1–3.3-12)

You Try It Work through this You Try It problem.

Work Exercises 9–12 in this textbook or in the MyMathLab Study Plan.

CAUTION

Remember to always check your potential solution in the *original* equation, not the simplified version.

Example 3 Solving Linear Equations Containing Non-Simplified Expressions

My interactive video summary

Solve:

a. $6x + 5 - 4x = 8 - x + 9$ **b.** $3 + 6x + 15 = -7x - 2 + 3x$

Solutions

a. Start by **simplifying** the expressions on each side of the equation.

Begin with the original equation: $6x + 5 - 4x = 8 - x + 9$

Combine like terms on the left: $2x + 5 = 8 - x + 9 \leftarrow 6x - 4x = 2x$

Combine like terms on the right: $2x + 5 = 17 - x \leftarrow 8 + 9 = 17$

Now apply the addition and multiplication **properties of equality** to finish solving the equation. Check your **answer**, or watch this **interactive video** for the complete solutions to both parts.

b. Try solving this equation on your own. Check your **answer**, or watch this **interactive video** for the fully worked solutions to both parts.

You Try It Work through this You Try It problem.

Work Exercises 13–22 in this textbook or in the MyMathLab Study Plan.

OBJECTIVE 3 SOLVE LINEAR EQUATIONS INVOLVING GROUPING SYMBOLS

If a linear equation contains **grouping symbols**, we must first remove the grouping symbols by using the **distributive property**. We then **combine like terms** and apply the addition and multiplication **properties of equality** as appropriate to **isolate the variable.**

Example 4 Solving Linear Equations Involving Grouping Symbols

Solve:

a. $3(x + 4) = 6$ **b.** $5(3 - 2x) + 4 = 2x - 5$

Solutions

a. Start by using the distributive property to remove the grouping symbols.

Begin with the original equation: $3(x + 4) = 6$

Apply the distributive property: $3x + 12 = 6 \leftarrow$ $\boxed{3(x + 4) = 3 \cdot x + 3 \cdot 4 = 3x + 12}$

Subtract 12 from both sides: $3x + 12 - 12 = 6 - 12 \leftarrow$ Addition property of equality

Simplify: $3x = -6$

Divide both sides by 3: $\frac{3x}{3} = \frac{-6}{3}$ ← Multiplication property of equality

Simplify: $x = -2$

View the **check**. The solution is -2.

b. Start by using the **distributive property** to remove the **grouping symbols**.

My video summary

Begin with the original equation: $5(3 - 2x) + 4 = 2x - 5$

Apply the distributive property: $15 - 10x + 4 = 2x - 5$ ← $5(3 - 2x) = 5 \cdot 3 - 5 \cdot 2x = 15 - 10x$

Finish solving this equation on your own. Check your **answer**, or watch this **video** for the fully worked solution.

You Try It **Work through this You Try It problem.**

Work Exercises 23–32 in this textbook or in the MyMathLab Study Plan.

The following general strategy may be used for solving linear equations in one variable.

A General Strategy for Solving Linear Equations in One Variable

Step 1. Remove grouping symbols using the distributive property.

Step 2. Simplify each side of the equation by **combining like terms.**

Step 3. Use the **addition property of equality** to collect all **variable terms** on one side of the equation and all constant terms on the other side.

Step 4. Use the **multiplication property of equality** to **isolate the variable.**

Step 5. Check the result in the original equation.

Example 5 Solving Linear Equations Involving Grouping Symbols

Solve:

My interactive video summary

a. $5(x + 7) = 3(x - 5)$ **b.** $3(2x - 1) - 7 = 10 - 2(x - 2)$

Solutions Try solving these equations on your own using the **general strategy for solving linear equations in one variable.** Start by using the distributive property to remove grouping symbols, then simplify each side by **combining like terms.** Check your **answers,** or watch this **interactive video** for the fully worked solutions to both parts.

You Try It **Work through this You Try It problem.**

Work Exercises 33–42 in this textbook or in the MyMathLab Study Plan.

(eText Screens 3.3-1–3.3-12)

3.3 Exercises

In Exercises 1–8, determine if each is a linear equation in one variable. If not, state why not.

You Try It

1. $3 + 2x - 1$
2. $5y + 1 = 2y$
3. $7 = 3 - 2x$
4. $5x + 3 = 0$
5. $x^2 + x + 1 = 7$
6. $3a - 2 = 5 - 2a$
7. $5u + 7v + 1 = 3 - 2u$
8. $\dfrac{2}{x-1} + 3 = 5 - x$

In Exercises 9–38, solve each linear equation.

You Try It

9. $8 - 3x + 3 = 2x + 1$
10. $5 - 2x = 2x - 3 - 6x$
11. $5 - 3x + 1 = 5x - 2$
12. $2y + 4 = 3y + 25 - 4y$

You Try It

13. $2x + 35 - 3x = 2 - 6x + 3$
14. $3y - 14 - 5y = 4y - 2 - 12y$
15. $2 - 5y + 5 = 3y + 10 - 5y$
16. $5a - 2 - 3a = 2a + 4 + 3a$
17. $3x + 18 + 2x - 1 = 59 - 4x + 39$
18. $2 - 5x + 3 = 5 - 2x + 7 - 10x$
19. $5n - 3 + 2n + 1 = 8 + 3n + 2$
20. $3x - 14 - 5x + 2 = 27 - 2x - 3 - 3x$
21. $5 + 3x + 1 - x = -6 + 3x + 7 - 2x$
22. $8 + 5x - 1 - 3x = 2x + 3 + 3x + 1 - 4x$

You Try It

23. $5(2x - 3) = 15$
24. $5x = 7(2 + x)$
25. $2(x + 1) - 3 = 3x - 19$
26. $3u - 5 = 2(3 - u) - 1$
27. $-3(m - 9) = 2m - 8$
28. $7 - 3x = 3(x + 1) - 2$
29. $2(w - 2) + 7 = 3 - 2w$
30. $14 - (2m + 7) = 4 + 2m - 13$
31. $2(3x + 8) + 9 = 7 + 4x$
32. $6(x + 1) + x + 2 = 2 + 5x$

You Try It

33. $3(2x + 1) = 5(x + 1)$
34. $2(b + 1) + b = 2(b - 1) - 1$
35. $4(x - 9) - 3x = 3(x - 2) - 4$
36. $4(x + 4) + x = 3(x + 3) + 1$
37. $-2n + 4(3 - n) = 2(3n + 42)$
38. $15(x - 2) = 10(x + 1) - 15$
39. $3(4x + 1) - 4 = 3(5x + 1) - 4x$
40. $4(5 - 2x) - 1 = 6(4 - x) - 3x - 5$
41. $3(x + 11) - 2(x + 2) = 2(x - 5) + 3(x - 7)$
42. $3(2h + 1) - 2(h + 3) = 2(h - 3) - 1$

(eText Screens 3.4-1–3.4-30)

3.4 Using Linear Equations to Solve Problems

THINGS TO KNOW

Before working through this topic, be sure you are familiar with the following concepts: VIDEO ANIMATION INTERACTIVE

You Try It 1. Translate Sentences into Equations (Topic 1.7, **Objective 5**)

You Try It 2. Translate Word Phrases Using Integers (Topic 2.5, **Objective 3**)

You Try It 3. Solve Equations Using Both Properties of Equality (Topic 3.2, **Objective 3**)

You Try It 4. Solve Linear Equations Involving Non-Simplified Expressions (Topic 3.3, **Objective 2**)

You Try It 5. Solve Linear Equations Involving Grouping Symbols (Topic 3.3, **Objective 3**)

OBJECTIVES

1 Translate Word Statements into Equations

2 Solve Problems Involving Numbers

3 Solve Problems Involving Perimeter and Area

4 Solve Applications Using Linear Equations

OBJECTIVE 1 TRANSLATE WORD STATEMENTS INTO EQUATIONS

In **Topic 1.7**, we translated sentences into **equations** using whole numbers. We revisit this topic again to include **integers** and develop a problem-solving strategy for applications. You may wish to review some **key words for operations** and some **key words that translate to an equal sign** from Topic 1.7.

CAUTION

Remember to watch the wording! The same words in a different order can have different translations. For example, view this **popup** to see the difference between "twice the sum of a number and 9" and "the sum of twice a number and 9."

Example 1 Translating Word Statements into Equations

My video summary

Write each word statement as an equation. Use x to represent the unknown number. Do not solve.

a. 14 more than a number is -20.

b. The difference of a number and 7 equals twice the number increased by 5.

c. The product of -4 and the sum of a number and 7 is the same as the number decreased by 10.

d. The quotient of a number and 5 results in -3 plus eight times the number.

Solutions

a. "More than" indicates **addition**. What is being added? 14 is being added to the unknown number x. The key word *is* translates to an equal sign $(=)$.

14 more than a number is -20

$$x + 14 = -20$$

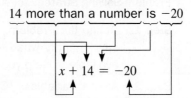

We can translate "14 more than a number is -20" as the equation $x + 14 = -20$.

b. *Difference* means **subtraction**. What is being subtracted? 7 is being subtracted *from* an unknown number x. We can translate this as $x - 7$. The key word *equals* translates to an equal sign. "Increased by" means addition. What is being added? Twice the unknown number is being added to 5. We can translate this as $2x + 5$.

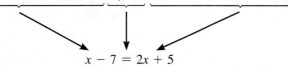

The difference of a number and 7 equals twice the number increased by 5.

$$x - 7 = 2x + 5$$

So, we can translate "The difference of a number and 7 equals twice the number increased by 5" as the equation $x - 7 = 2x + 5$.

c.–d. Try translating these two word statements on your own. Check your **answers** to parts c and d, or watch this **video** for the complete solutions to all four parts.

You Try It **Work through this You Try It problem.**

Work Exercises 1–12 in this textbook or in the MyMathLab Study Plan.

OBJECTIVE 2 SOLVE PROBLEMS INVOLVING NUMBERS

In real-world problem solving, we are generally given a description of the problem and asked to find a solution. Just as a **model** can be used to illustrate something that cannot be seen easily (for example, a DNA model), a **mathematical model** uses the language of mathematics to describe a problem. Typically, a model is an **equation** that describes a relationship within an application.

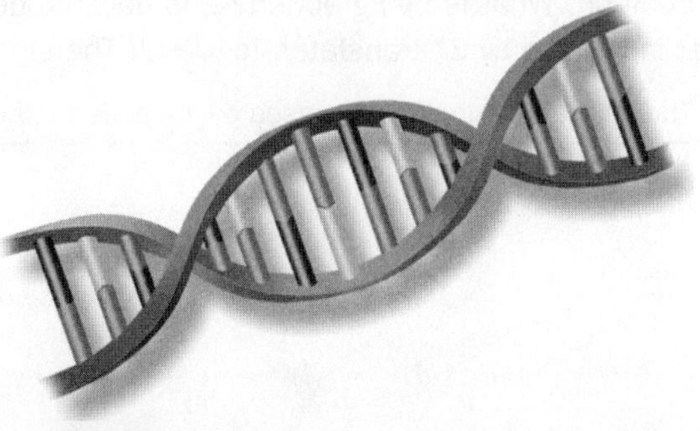

Figure 1 Model of DNA

The area of a rectangle is given by the product of its length and width.

Area = Length × Width

Figure 2 Mathematical model for area

(eText Screens 3.4-1–3.4-30)

We have practiced translating word phrases and statements into **algebraic expressions** and equations. This will prepare us for applications in which we must first translate a problem description into an equation and then use the equation to solve the problem.

The following problem-solving strategy will be used throughout this course to translate, model, and solve applications involving **linear equations.**

Problem-Solving Strategy for Applications of Linear Equations

Step 1. **Define the Problem.** Read the problem carefully, or multiple times if necessary. Identify what you are trying to find and determine what information is available to help you find it.

Step 2. **Assign Variables.** Choose a **variable** to assign to an unknown quantity in the problem. For example, use p for price. If other unknown quantities exist, express them in terms of the selected variable.

Step 3. **Translate into an Equation.** Use the relationships among the known and unknown quantities to form an **equation**.

Step 4. **Solve the Equation.** Determine the value of the variable and use the result to find any other unknown quantities in the problem.

Step 5. **Check the Reasonableness of Your Answer.** Check to see if your answer makes sense within the context of the problem. If not, check your work for errors and try again.

Step 6. **Answer the Question.** Write a clear statement that answers the question(s) posed.

We begin by using **linear equations** to solve problems involving numbers. If necessary, review the **Strategy for Solving Linear Equations in One Variable**.

Example 2 Solving Problems Involving Numbers

Three times a number, decreased by 4, equals the number increased by 2. Find the number.

Solution Follow the **problem-solving strategy**.

Step 1. We want to find "the number."

Step 2. Let x be the unknown number.

Step 3. "Times" means multiplication. What is being multiplied? Three is multiplied by the number, so "Three times a number" translates to $3x$. The key words *decreased by* mean subtraction. What is being subtracted? 4 is being subtracted from $3x$, so "Three times a number, decreased by 4" translates to $3x - 4$. The key word *equals* translates to an equal sign ($=$). The key words *increased by* mean addition. What is being added? 2 is being added to the number, so "the number increased by 2" translates to $x + 2$. The equation is $3x - 4 = x + 2$.

$$\underbrace{\text{Three times a number decreased by 4}}_{3x - 4}\ \underbrace{\text{equals}}_{=}\ \underbrace{\text{the number increased by 2}}_{x + 2}$$

Step 4. Write the equation: $3x - 4 = x + 2$

Add 4 to both sides: $3x - 4 + 4 = x + 2 + 4$ ← Addition property of equality

Simplify: $3x = x + 6$

Subtract x from both sides: $3x - x = x + 6 - x$ ← Addition property of equality

Simplify: $2x = 6$

Divide both sides by 2: $\frac{2x}{2} = \frac{6}{2}$ ← Multiplication property of equality

Simplify: $x = 3$

Step 5. Check the number 3 in the original sentence to see if it is a reasonable solution.
"3 times 3, decreased by 4" translates to $3(3) - 4$.

Simplify: $3(3) - 4 = 9 - 4 = 5$.

"3 increased by 2" translates to $3 + 2$.

Simplify: $3 + 2 = 5$.

Both phrases simplify to the same result, 5, so 3 checks.

Step 6. The number is 3.

You Try It **Work through this You Try It problem.**

Work Exercises 13–16 in this textbook or in the MyMathLab Study Plan.

Example 3 Solving Problems Involving Numbers

My video summary

The sum of -6 and twice a number is equal to 5 times the sum of the number and 3. Find the number.

Solution

Step 1. We want to find "the number."

Step 2. Let x be the unknown number.

Step 3. The key word *sum* means addition. What is being added? -6 is being added with twice a number, x. Notice the word "and" separates the quantities being added. "Twice" means to multiply by 2, so we can translate "twice a number" as $2x$. "The sum of -6 and twice a number" translates to $-6 + 2x$.

The key words *is equal to* translate to an equal sign $(=)$.

"5 times" means to multiply by 5. What gets multiplied? 5 is multiplied by the *sum* of the number and 3. Because the entire sum is multiplied by 5, we must put the sum in parentheses. The phrase "5 times the sum of the number and 3" translates to $5(x + 3)$. The equation is $-6 + 2x = 5(x + 3)$.

The sum of −6 and twice a number	is equal to	5 times the sum of the number and 3.
$-6 + 2x$	$=$	$5(x + 3)$

Continue following the **problem-solving strategy** and try to finish this problem on your own. Check your **answer**, or watch this **video** for the fully worked solution.

You Try It **Work through this You Try It problem.**

Work Exercises 17–20 in this textbook or in the MyMathLab Study Plan.

(eText Screens 3.4-1–3.4-30)

Example 4 Solving Problems Involving Numbers

My video summary

Four times the difference of three times a number and 2, increased by 5 is the same as seven times the difference of the number and 4. Find the number.

Solution Try solving this problem on your own using the **problem-solving strategy**. Check your **answer**, or watch this **video** for a detailed solution.

You Try It **Work through this You Try It problem.**

Work Exercises 21–24 in this textbook or in the MyMathLab Study Plan.

OBJECTIVE 3 SOLVE PROBLEMS INVOLVING PERIMETER AND AREA

My animation summary

We introduced **perimeter** in **Topic 1.3** and **area** in **Topic 1.4**. We now look at applications of linear equations involving perimeter and area. Review this **animation** on perimeter and area before proceeding to Example 5.

Example 5 Perimeter of a Rectangle

The length of a rectangle is 3 feet shorter than twice its width. Find the dimensions of the rectangle if the perimeter is 36 feet.

Solution Follow the **problem-solving strategy**.

Step 1. We know the perimeter is 36 feet and the length is 3 feet shorter than twice the width. We want to find the length and width (the dimensions) of the rectangle.

Step 2. Let w represent the width of the rectangle. Then the length can be expressed as $2w - 3$ (3 feet shorter than twice the width).

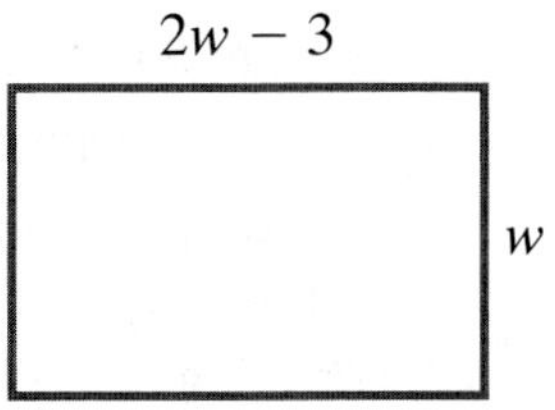

Step 3. The perimeter is 36 feet. We add the lengths of the sides to get the perimeter, 36.

$$w + (2w - 3) + w + (2w - 3) = 36$$

Step 4.

Write the equation:	$w + (2w - 3) + w + (2w - 3) = 36$
Remove grouping symbols:	$w + 2w - 3 + w + 2w - 3 = 36$
Combine like terms:	$6w - 6 = 36$
Add 6 to both sides:	$6w - 6 + 6 = 36 + 6$ ← Addition property of equality
Simplify:	$6w = 42$
Divide both sides by 6:	$\frac{6w}{6} = \frac{42}{6}$ ← Multiplication property of equality
Simplify:	$w = 7$

If $w = 7$, then $2(7) - 3 = 14 - 3 = 11$.

Step 5. We have found that the length is 11 feet and the width is 7 feet. This would give a perimeter of $7 + 11 + 7 + 11 = 36$ feet, so our result seems reasonable.

Step 6. The width of the rectangle is 7 feet and the length is 11 feet.

Example 6 Poster Border

My video summary

The length of a rectangular poster is 17 inches longer than its width. If it takes 134 inches of trim to put a border around the poster, what are the poster's dimensions (length and width)?

Solution

Step 1. We have a rectangular poster with a length that is 17 inches longer than its width. We know the perimeter is 134 inches because it takes that much trim to put a border around the poster. We want to find the length and width of the poster.

Step 2. Let $w =$ the width of the poster. Then the length is given by $w + 17$ (17 inches longer than the width).

Step 3. The perimeter is 134 inches. We find the perimeter by adding the lengths of the sides of the rectangle.

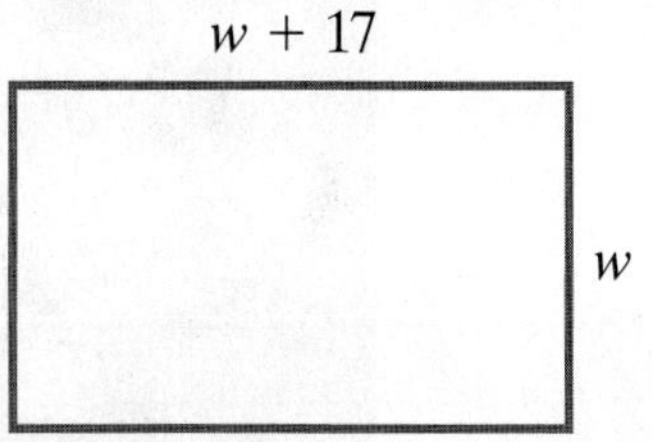

$$(w + 17) + w + (w + 17) + w = 134$$

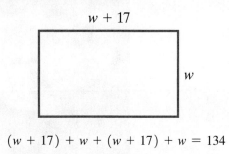

Continue following the **problem-solving strategy** and finish the problem on your own. Check your **answer**, or watch this **video** for the fully worked solution.

You Try It **Work through this You Try It problem.**

Work Exercises 25–31 in this textbook or in the MyMathLab Study Plan.

Example 7 Area of a Rectangle

My video summary

If the area of the rectangle below is 92 square feet, find the value of x.

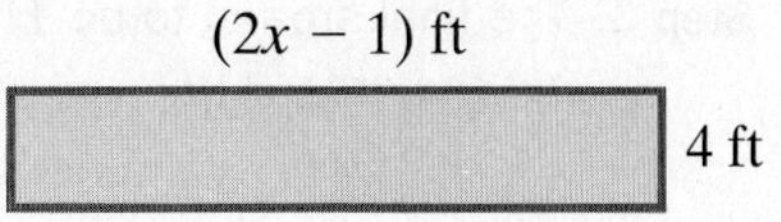

Solution

Step 1. We know the area of the rectangle is 92 square feet, the width is 4 feet, and the length is $2x - 1$ feet. We want to find the value of x.

Step 2. The variable x is given in the problem statement.

Step 3. The area is 92 square feet. Because area equals length times width, we multiply the length and width to get the area, 92.

(eText Screens 3.4-1–3.4-30)

$$4(2x - 1) = 92$$

Width ↗ Length ↖ Area ↖

Continue following the **problem-solving strategy** as you solve this equation and try to finish the problem on your own. Check your **answer**, or watch this **video** for the fully worked solution.

Example 8 Trimming a Picture

My video summary

A photograph is 10 cm wide and 15 cm long. To fit in a picture frame, the picture length must be trimmed the same amount on both sides. If the final area is 110 square centimeters, how much should be trimmed from each side?

Solution

Step 1. We know the original length is 15 cm, the original width is 10 cm, and the final area must be 110 sq cm. We want to find the value of x, the amount from each side of the length that must be trimmed.

Step 2. The variable x is given in the diagram as the amount that needs to be trimmed from each side of the length. This means the new length will be $15 - 2x$ (subtract an x from each side).

Step 3. The final area is to be 110 square feet. We multiply the new length and width to get the area, 110.

$$10(15 - 2x) = 110$$

Finish solving this problem on your own using the **problem-solving strategy**. Check your **answer**, or watch this **video** for a detailed solution.

You Try It **Work through this You Try It problem.**

Work Exercises 32–35 in this textbook or in the MyMathLab Study Plan.

(eText Screens 3.4-1–3.4-30)

OBJECTIVE 4 SOLVE APPLICATIONS USING LINEAR EQUATIONS

We now look at applications that can be solved by translating a situation into a **linear equation in one variable**.

Example 9 Blood Drive

During a blood drive, 12 more pints were collected on the first day than on the second day. If the total number of pints collected over the two days was 58, how many pints were collected each day?

Solution Follow the **problem-solving strategy**.

Step 1. We know that blood was collected over two days. The total number of pints for the two days was 58 pints, and there were 12 more pints collected on the first day than the second. We want to find the number of pints collected each day.

Step 2. Let p represent the number of pints collected on the second day. Then the number of pints collected on the first day can be expressed as $p + 12$ (12 more than the second day).

Step 3. The two-day total is 58 pints. We add the number of pints from each day to get the total, 58.

$$\underbrace{(p + 12)}_{\text{Pints on 1st day}} + \underbrace{p}_{\text{Pints on 2nd day}} = \underbrace{58}_{\text{Total pints}}$$

Step 4.

Write the equation: $(p + 12) + p = 58$

Remove grouping symbols: $p + 12 + p = 58$

Combine like terms: $2p + 12 = 58$

Subtract 12 from both sides: $2p + 12 - 12 = 58 - 12$ ← Addition property of equality

Simplify: $2p = 46$

Divide both sides by 2: $\frac{2p}{2} = \frac{46}{2}$ ← Multiplication property of equality

Simplify: $p = 23$

If $p = 23$, then $23 + 12 = 35$.

35 pints were collected on the first day and 23 pints were collected on the second day.

Step 5. We have found that 35 pints were collected on the first day and 23 pints on the second. Because 35 is 12 more than 23 and $35 + 23 = 58$, our result checks.

Step 6. There were 35 pints collected on the first day of the blood drive and 23 pints collected on the second day.

You Try It **Work through this You Try It problem.**

Work Exercises 36–39 in this textbook or in the MyMathLab Study Plan.

Example 10 Football Yards

My video summary

During Week 2 of the 2011 NFL season, Cam Newton had a total of 485 yards passing and running. If he passed (threw) for 8 yards more than 8 times his number of running yards, for how many yards did he pass? For how many yards did he run? (*Source:* espn.com)

(eText Screens 3.4-1–3.4-30)

Solution

Step 1. We know the total number of yards was 485 and that Cam passed for 8 more yards than 8 times his number running. We want to find the number of passing yards and the number of running yards.

Step 2. Let $y =$ the number of running yards. Then we can express the number of passing yards as $8y + 8$ (8 more than 8 times the running yards).

Step 3. The total number of yards is 485. We can add the running yards and the passing yards to get the total yards, 485.

$$\underbrace{y}_{\text{Running yards}} + \underbrace{(8y + 8)}_{\text{Passing yards}} = \underbrace{485}_{\text{Total yards}}$$

Continue following the **problem-solving strategy** as you solve this equation and try to finish the problem on your own. Check your **answer**, or watch this **video** for the fully worked solution.

You Try It **Work through this You Try It problem.**

Work Exercises 40–45 in this textbook or in the MyMathLab Study Plan.

3.4 Exercises

In Exercises 1–12, write each word statement as an equation. Use x to represent the unknown number. Do not solve.

You Try It

1. 12 less than twice a number equals the number.
2. The difference between 17 and a number is 4 more than twice the number.
3. Three times the sum of a number and 5 is the same as four times the number divided by 7.
4. 5 more than a number divided by 3 results in twice the number, decreased by 2.
5. The sum of 7 and the difference of 8 minus a number is 15.
6. -12 divided by the sum of twice a number and 3 is the same as the number.
7. The sum of a number and 3 multiplied by the sum of the number and 7 is equivalent to 17.
8. The quotient of the difference of 6 and a number and the sum of twice the number and 5, yields eleven times the number added to 3.
9. Half the sum of a number plus 7 results in 4 less than twice the number.
10. The difference between the sum of three times a number and 7 and twice the number results in the number increased by 11.
11. The product of 17 and the difference between a number and 4 yields 3 times the number plus 7.
12. A number increased by four times the number and decreased by twice the number equals three times the number, decreased by 9.

(eText Screens 3.4-1–3.4-30)

In Exercises 13–24, use the problem-solving strategy to solve each problem and find the unknown number.

You Try It 13. Five times a number increased by two yields seven times the number decreased by four.

14. Three added to the difference of a number and 1 equals 11.

15. Two times a number added to 3 times a number is the same as four times the number minus 5.

16. Seven times a number decreased by 4 yields four times the number increased by 11.

You Try It 17. Three times the sum of twice a number and 3 is the same as 5 times the number and 1.

18. 5 less than six times a number equals the sum of four times the number and 3.

19. The difference of five times a number and 4 results in three times the difference of twice the number and 1.

20. Four times the sum of -2 and a number yields the difference between 6 and three times the number.

You Try It 21. Twice the difference of three times a number and 1 equals 5 times the sum of the number and 1.

22. Five times the difference of twice a number and 2, increased by 8, equals seven times the sum of the number and 3, decreased by 11.

23. The difference of twice a number and 3, increased by 7, is the same as seven times the difference of the number and 2, decreased by twice the number.

24. The quotient of the difference of twice a number and 3, divided by 5, is the same as three times the difference of the number and 1, decreased by 8.

In Exercises 25–45, use the problem-solving strategy to solve each problem.

You Try It 25. The length of a rectangle is one more than twice its width. Its perimeter is 26 inches. Find its dimensions (length and width).

26. The lengths of the sides of a triangle are $x + 2$, $2x + 1$, and $3x$. If its perimeter is 21, find the lengths of the sides.

27. **Garden Fence** The length of a rectangular garden is 8 ft longer than its width. The garden has a wall along one length. If 26 ft of fencing would be required to finish enclosing the garden, what are its dimensions (length and width)?

(eText Screens 3.4-1–3.4-30)

28. A triangle has two sides that are the same length, the length of the third side is 17 in., and its perimeter is 43 in. How long are the sides with the same length?

29. **Pool Fence** Valerie needs to put a fence around her pool and deck. The area to be fenced is rectangular in shape such that the length is 8 feet longer than the width. If she needs a total of 120 feet of fencing, what are the dimensions (length and width) of the area to be fenced?

30. **Christmas Lights** If it takes Joe 48 ft of Christmas lights to outline his garage (except along the ground), how tall is Joe's garage door, x?

31. The rectangle and the triangle have the same perimeters. Find the dimensions of each figure.

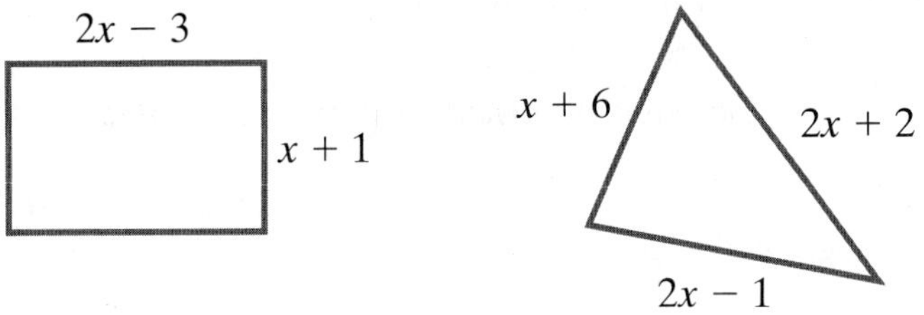

You Try It 32. The rectangle has an area of 105 sq ft. Find the value of x.

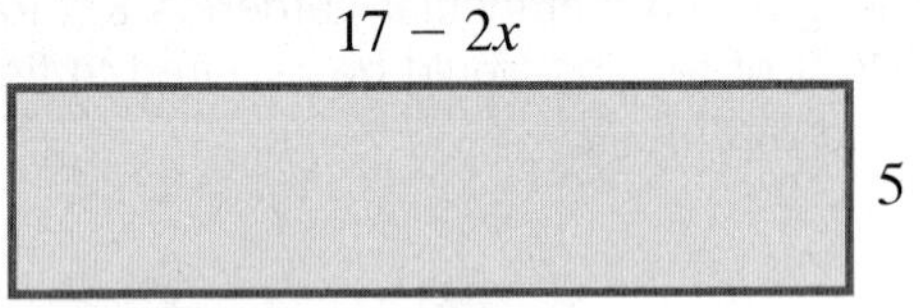

33. The rectangle has an area of 153 sq in. Find the value of x.

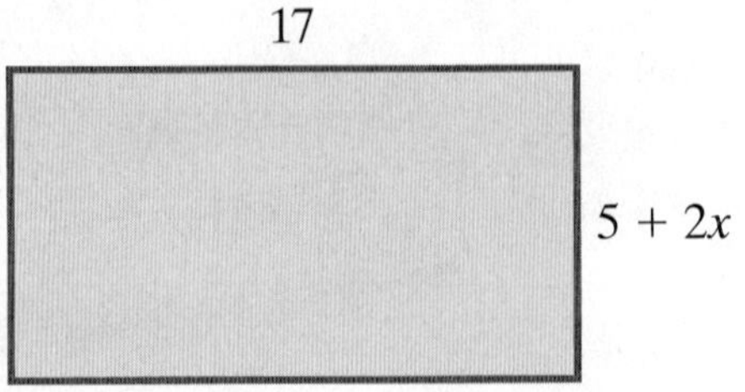

34. **Air Filter** An air filter for a furnace has a width of 20 inches and a length of $(4x + 37)$ inches. Find the value of x such that the filter has an area of 500 square inches.

35. The outer rectangle is 12 units by 10 units and the inner rectangle has an area of 60 sq units. Find x.

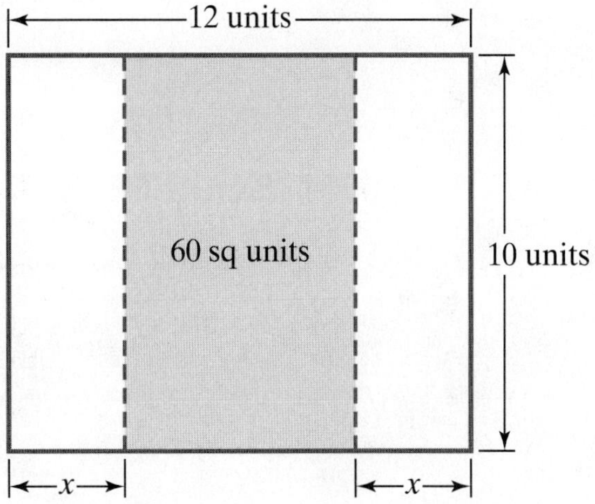

You Try It 36. **Tropical Storms and Hurricanes** Between 1851 and 2010, there have been 192 more tropical storms and hurricanes in the U.S. during September than during August. If the total number for the two months is 904, how many tropical storms and hurricanes occurred in each month?

37. **Cutting a Board** A 95-cm board is cut into two pieces. If one piece is 37 cm shorter than the other, how long are the two pieces?

38. **The Lion King** In its opening weekend in 2011, *The Lion King 3D* made $10 million less than the original when it opened in 1994. If the total amount for the two opening weekends was $70 million, how much did *The Lion King* make during each opening weekend? (*Source:* movies.yahoo.com)

39. **Deep Space Rocket** In 2011, NASA unveiled plans for a deep space rocket that could carry 5 tons less cargo than 3 times the cargo carried by the now-retired space shuttles. If the deep space rocket and space shuttle could carry a combined cargo of 95 tons, how much cargo could each carry? (*Source:* www.nasa.gov)

You Try It 40. **Fantasy Football** During one week on Jesica's Fantasy Football team, Miles Austin earned 8 more points than twice the number of points earned by Jake Cutler. If they combined for 44 points, how many points did each player earn for Jesica?

41. **Golf Score** After a round of golf, Keegan's score was 9 strokes more than three times Brennan's score. If their combined score was 1, what was each player's score?

42. **Video Game System** An electronics store sold a total of 20 video games that cost either $40 or $50. If the number of $50 games sold was one less than twice the number of $40 games sold, how many of each were sold?

43. **Red Bull® Calories** A small can of Red Bull contains 2 more calories than 4 times the number of calories in a Red Bull shot. If a small can of Red Bull and a Red Bull shot contain a total of 137 calories, how many calories are in each drink? (*Source:* www.redbull.com)

44. **Chocolate Spending** Per capita spending on chocolate in Norway during 2011 was $23 less than four times the per capita spending in the U.S. If the combined per capita spending was $267, what was the per capita amount for each country? (*Source:* usatoday.com)

45. **Cutting a Wire** A wire that is 65 feet long is cut into three pieces. One piece is 2 feet longer than the shortest piece, and another piece is 5 feet shorter than twice the shortest piece. How long is each piece?

MODULE FOUR

Fractions and Mixed Numbers

MODULE FOUR CONTENTS

4.1 Introduction to Fractions and Mixed Numbers

THINGS TO KNOW

Before working through this topic, be sure you are familiar with the following concepts:

		VIDEO	ANIMATION	INTERACTIVE
You Try It	1. Graph Integers on a Number Line (Topic 2.1, **Objective 1**)	▣		

OBJECTIVES

1 Identify the Numerator and Denominator of a Fraction

2 Represent Information with Fractions and Mixed Numbers

3 Graph Fractions and Mixed Numbers on a Number Line

4 Write Improper Fractions as Mixed Numbers or Whole Numbers

5 Write Mixed Numbers as Improper Fractions

OBJECTIVE 1 IDENTIFY THE NUMERATOR AND DENOMINATOR OF A FRACTION

To count whole items such as students, calculators, cell phones, and so forth, we use **whole numbers.** To describe part of a whole item, we can use *fractions*. For example, consider a chocolate bar that is divided into 12 pieces of equal size. If 5 of these pieces are eaten for a snack, then five-twelfths, or $\frac{5}{12}$, of the chocolate bar has been eaten. See **Figure 1.**

(*eText Screens 4.1-1–4.1-31)*

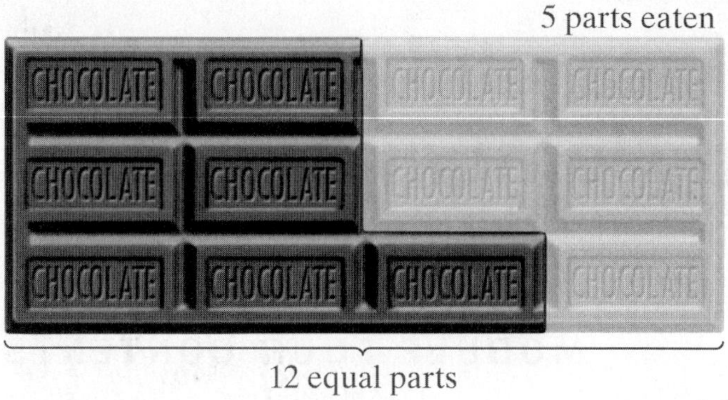

Figure 1 $\frac{5}{12}$ of the chocolate bar has been eaten.

The parts of fractions have names. For the fraction $\frac{5}{12}$, the 12 is called the *denominator* and refers to the number of equal parts within the whole item. The 5 is called the *numerator* and refers to the number of equal parts of interest, in this case the 5 parts eaten. The two parts of a fraction are separated by a *fraction bar.*

$$\frac{5}{12} \begin{array}{l} \leftarrow \text{Numerator} \\ \leftarrow \text{Fraction bar} \\ \leftarrow \text{Denominator} \end{array}$$

Definition Fraction

A **fraction** is a number of the form $\frac{a}{b}$, where a and b are numbers and $b \neq 0$. a is called the **numerator**, and b is called the **denominator**. The division line that separates the numerator and denominator is called the **fraction bar**.

TIP Fractions can be **positive** or **negative**. Some fractions have numerators and/or denominators that are **algebraic expressions.**

Example 1 Identifying Numerators and Denominators

Identify the **numerator** and **denominator** of each fraction.

a. $\frac{13}{15}$ **b.** $\frac{-5}{8}$ **c.** $\frac{2x}{3y}$

Solutions

a. $\frac{13}{15}$ ← Numerator / ← Denominator — The numerator is 13 and the denominator is 15.

b. $\frac{-5}{8}$ ← Numerator / ← Denominator — The numerator is -5 and the denominator is 8.

c. $\frac{2x}{3y}$ ← Numerator / ← Denominator — The numerator is $2x$ and the denominator is $3y$.

You Try It **Work through this You Try It problem.**

Work Exercises 1–4 in this textbook or in the MyMathLab Study Plan.

(eText Screens 4.1-1–4.1-31)

Note: For the remainder of this topic, we will focus only on positive fractions and mixed numbers.

OBJECTIVE 2 REPRESENT INFORMATION WITH FRACTIONS AND MIXED NUMBERS

Figure 2 shows a circle divided into 8 equal parts with 7 of the parts shaded. Therefore, the **fraction** of the circle shaded is seven-eighths, or $\frac{7}{8}$. (7 ← Shaded parts; 8 ← Equal parts)

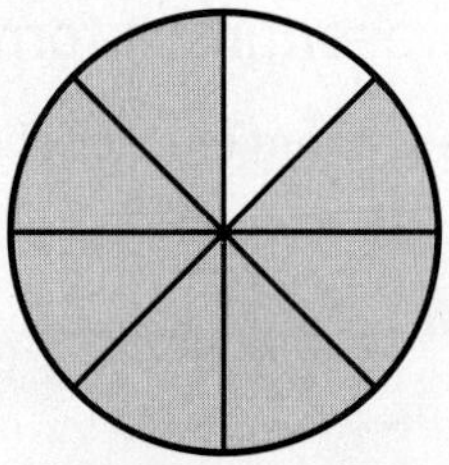

Figure 2

$\frac{7}{8}$ of the circle is shaded.

Example 2 Using Fractions to Represent the Shaded Portion of a Figure

Write a fraction to represent the shaded portion of each figure.

a.

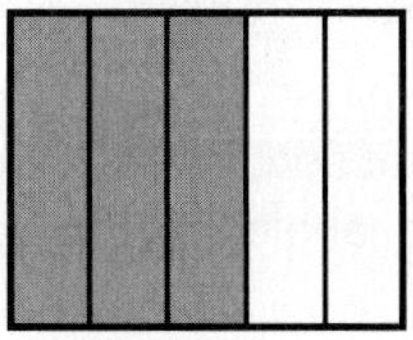

b.

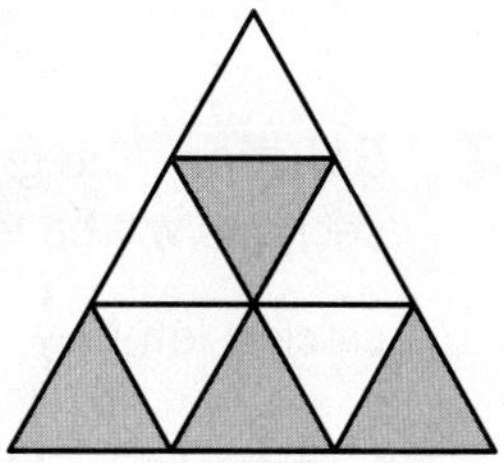

Solutions

a. 3 of the 5 equal parts are shaded, so the fraction is $\frac{3}{5}$. (3 ← Shaded parts; 5 ← Equal parts)

b. 4 of the 9 equal parts are shaded, so the fraction is $\frac{4}{9}$. (4 ← Shaded parts; 9 ← Equal parts)

You Try It **Work through this You Try It problem.**

Work Exercises 5–8 in this textbook or in the MyMathLab Study Plan.

Example 3 Representing Information with Fractions

Graduated cylinders like the one shown below are used in science to measure liquids. What **fraction** of the graduated cylinder shown is filled?

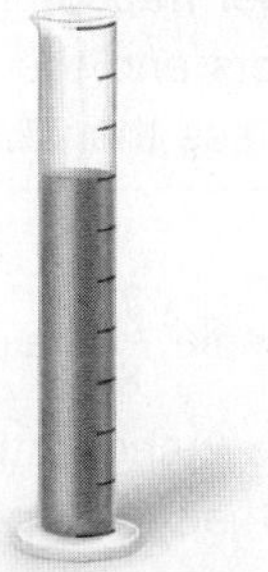

(eText Screens 4.1-1–4.1-31)

Solution The markings divide the graduated cylinder into 10 equal parts. The cylinder has 7 parts filled. View this **popup box** for a labeled figure. The fraction of the cylinder that is filled is $\frac{7}{10}$. ($7 \leftarrow$ Parts filled; $10 \leftarrow$ Equal parts)

You Try It **Work through this You Try It problem.**

Work Exercises 9–12 in this textbook or in the MyMathLab Study Plan.

Example 4 Representing Information with Fractions

If one section of the graham cracker shown is eaten, what **fraction** of the graham cracker will be left?

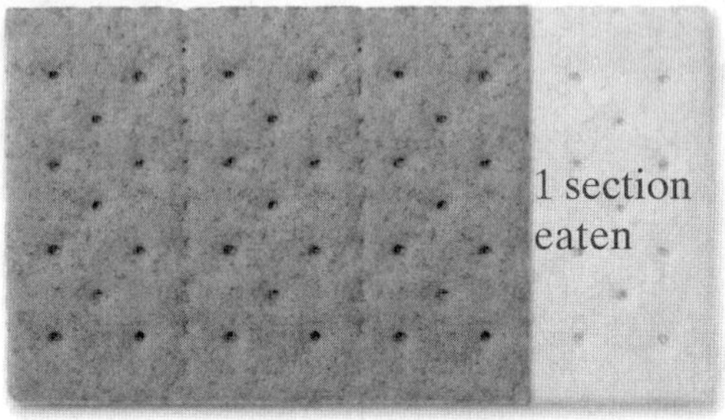

Solution The graham cracker has 4 equal sections. If 1 section is eaten, then $4 - 1 = 3$ sections will be left. View this **popup** for a labeled figure. So, the fraction of the graham cracker left is $\frac{3}{4}$. ($3 \leftarrow$ Sections left; $4 \leftarrow$ Equal sections)

Example 5 Representing Information with Fractions

My video summary

A prealgebra class has 15 male students and 13 female students. What **fraction** of the class is male?

Solution Try to answer this question on your own. View the **answer**, or watch this **video** for a complete solution.

You Try It **Work through this You Try It problem.**

Work Exercises 13–20 in this textbook or in the MyMathLab Study Plan.

Each fraction we have used so far to represent information has been a *proper fraction*. Remember, for the following definitions we are considering only *positive* fractions.

> **Definition Proper Fraction**
>
> A **proper fraction** is a fraction in which the **numerator** and **denominator** are both **whole numbers** and the numerator is smaller than the denominator. A proper fraction has a value less than 1.

For example, $\frac{7}{8}$ is a proper fraction because the numerator 7 is smaller than the denominator 8. **Figure 2** shows that the value of the fraction is less than 1 because the shaded portion is less than the whole circle.

What if the numerator is greater than or equal to the denominator, such as $\frac{8}{8}$ or $\frac{11}{8}$? These types of fractions are called *improper fractions*.

Definition Improper Fraction

An **improper fraction** is a fraction in which the **numerator** and **denominator** are both whole numbers and the numerator is greater than or equal to the denominator. An improper fraction has a value greater than or equal to 1.

Figure 3 shows a circle divided into 8 equal parts with all 8 parts shaded. Therefore, we can say that the fraction of the circle shaded is eight-eighths, or $\frac{8}{8}$. ← Shaded parts / ← Equal parts

Also, we can say that the whole circle, or 1 circle, is shaded. So, $\frac{8}{8} = 1$.

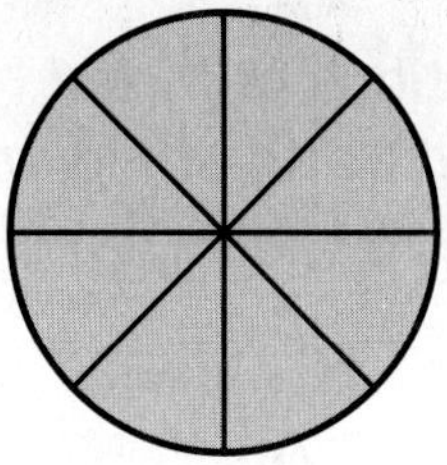

Figure 3

$\frac{8}{8}$ of the circle, or 1 whole circle, is shaded.

Figure 4 shows two circles. Each circle is divided into 8 equal parts, and a total of 11 parts are shaded. The shaded portion represents the **improper fraction** $\frac{11}{8}$. ← Shaded parts / ← Equal parts per circle

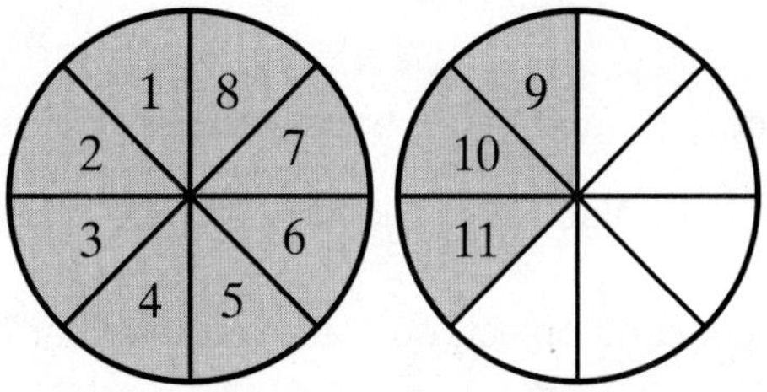

Figure 4

The shaded portion represents the fraction $\frac{11}{8}$.

Example 6 Identifying Proper and Improper Fractions

Identify each fraction as a **proper fraction** or an improper fraction.

a. $\frac{14}{9}$ **b.** $\frac{3}{5}$ **c.** $\frac{10}{10}$

Solutions

a. The **numerator** 14 is greater than the **denominator** 9, so $\frac{14}{9}$ is an improper fraction.

b. The numerator 3 is less than the denominator 5, so $\frac{3}{5}$ is a **proper fraction.**

c. The numerator and denominator are equal, so $\frac{10}{10}$ is an **improper fraction.**

You Try It **Work through this You Try It problem.**

Work Exercises 21–24 in this textbook or in the MyMathLab Study Plan.

(eText Screens 4.1-1–4.1-31)

Look again at the shaded portion of **Figure 4**, which represents $\frac{11}{8}$. Another way to describe **Figure 4** is to say that the shaded portion consists of one whole circle and three-eighths of the second circle, or *one and three-eighths* circles, which is a *mixed number*. We write "one and three-eighths" as $1\frac{3}{8}$, which means $1 + \frac{3}{8}$, but the + symbol is not written.

> **Definition Mixed Number**
>
> A **mixed number** is a number of the form $a\frac{b}{c}$, where a is a nonzero **whole number** and $\frac{b}{c}$ is a proper fraction. $a\frac{b}{c}$ means $a + \frac{b}{c}$, but the + symbol is not written. a is the **whole-number part** and $\frac{b}{c}$ is the **fraction part** of the mixed number.

Example 7 Using Fractions or Mixed Numbers to Represent the Shaded Parts of Figures

My video summary

For parts a and b, write both an **improper fraction** and a **mixed number** to represent the shaded portion.

a.

b.

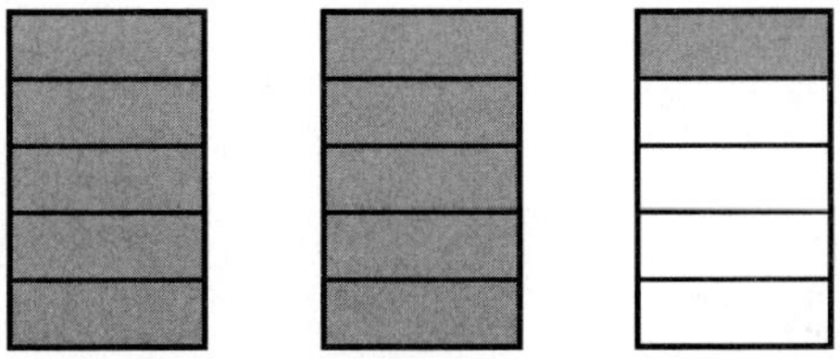

Solutions

a. Each whole object is divided into 4 equal parts. A total of 7 parts are shaded, so the improper fraction is $\frac{7}{4}$. Also, the shaded portions represent one whole object and three-fourths of a second object, so the mixed number is $1\frac{3}{4}$.

b. Try to work this problem on your own. View the **answer**, or watch this **video** for complete solutions to both parts.

You Try It **Work through this You Try It problem.**

Work Exercises 25–28 in this textbook or in the MyMathLab Study Plan.

Example 8 Representing Information with Mixed Numbers

Write a **mixed number** to represent the length in inches of the red ribbon shown in the figure.

My video summary

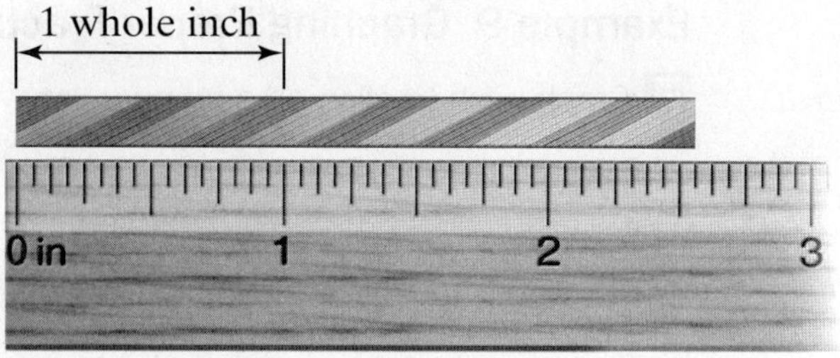

Solution In the figure, the ruler and ribbon are placed next to each other so that the ribbon starts at the mark for 0 inches on the ruler. We can determine the length of the ribbon by looking at the mark on the ruler where the ribbon ends. Because the ribbon ends between the marks for 2 and 3 whole inches, the length is a mixed number with a whole part of 2. That is, the length is 2 and a fraction inches.

Try to determine the fraction part of the mixed number and answer the question on your own. View the **answer**, or watch this **video** for a complete solution.

You Try It **Work through this You Try It problem.**

Work Exercises 29 and 30 in this textbook or in the MyMathLab Study Plan.

OBJECTIVE 3 GRAPH FRACTIONS AND MIXED NUMBERS ON A NUMBER LINE

In **Topic 2.1**, we learned to graph **integers** on a **number line**. As with integers, we can graph, or plot, a **fraction** or **mixed number** by plotting a solid circle (¥) at its location on the number line. To find the location of a fraction, think of the distance between two **consecutive integers**, called the **unit distance**, as a "whole object." For example, to graph $\frac{5}{8}$, divide the distance between 0 and 1 into 8 equal parts. Each part represents one-eighth, or $\frac{1}{8}$, of the unit distance.

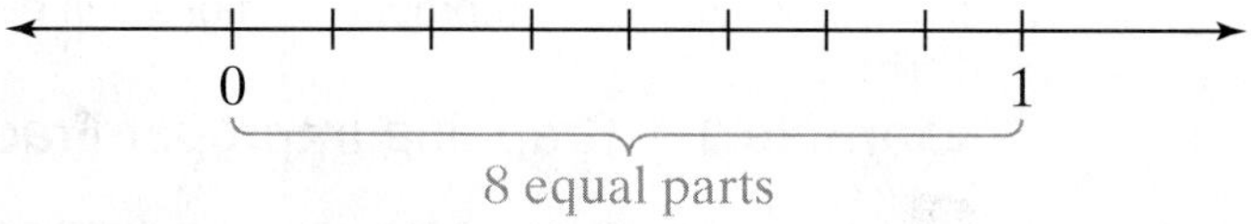

Starting at 0, move 5 places to the right. This is the location of $\frac{5}{8}$. Graph it by placing a solid circle at this location.

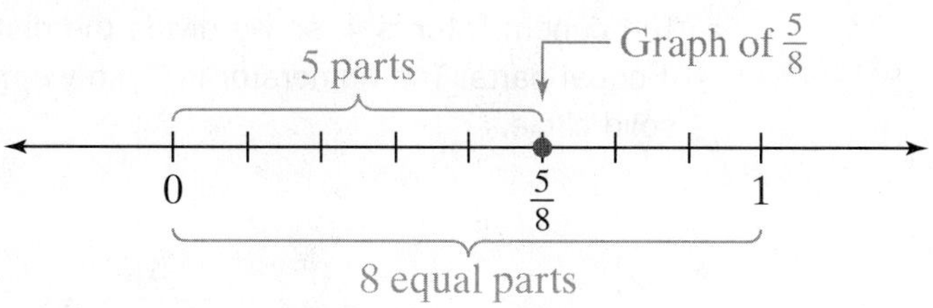

When graphing fractions on a number line, the **denominator** tells us how many equal parts are within the **unit distance**. The **numerator** tells us the position of the fraction within these parts.

Example 9 Graphing Proper Fractions on a Number Line

My video summary

Graph each **fraction** on a **number line**.

a. $\frac{2}{3}$ **b.** $\frac{1}{5}$ **c.** $\frac{5}{6}$

Solutions

a. The **denominator** is 3, so we divide the distance between 0 and 1 into 3 equal parts. The **numerator** is 2, so we move 2 places to the right of 0 and plot a solid circle.

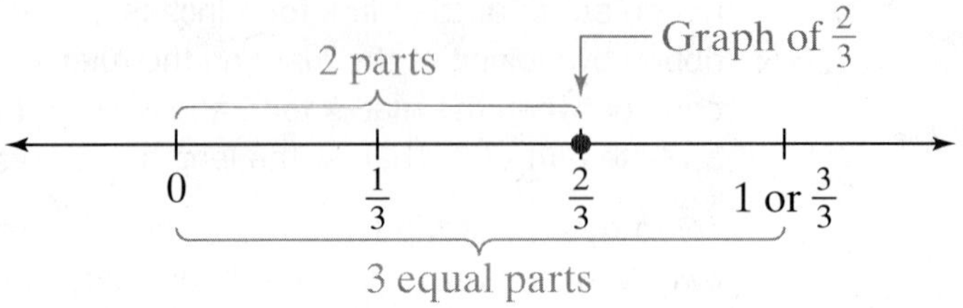

b. The denominator is 5, so divide the distance between 0 and 1 into 5 equal parts. Because the numerator is 1, move 1 place to the right of 0 to plot the solid circle. Try to finish graphing this fraction on your own. View the **answer** to part b, or watch this **video** for complete solutions to all three parts.

c. Try to graph this **fraction** on your own. View the **answer** to part c, or watch this **video** for complete solutions to all three parts.

You Try It **Work through this You Try It problem.**

Work Exercises 31–33 in this textbook or in the MyMathLab Study Plan.

The fractions in Example 9 are all **proper fractions**, so their graphs are between 0 and 1 on the **number line**. Recall that **improper fractions** have values greater than or equal to 1. We graph improper fractions as we graph proper fractions: the denominator tells us how many equal parts are within the **unit distance**, and the numerator tells us the position within the parts. The graphs of improper fractions will be at 1 or to its right.

Example 10 Graphing Improper Fractions on a Number Line

My video summary

Graph each improper fraction on a number line.

a. $\frac{7}{4}$ **b.** $\frac{5}{2}$ **c.** $\frac{3}{3}$

Solutions

a. The denominator is 4, so we divide the distance between **consecutive integers** into 4 equal parts. The numerator is 7, so we move 7 places to the right of 0 and plot a solid circle.

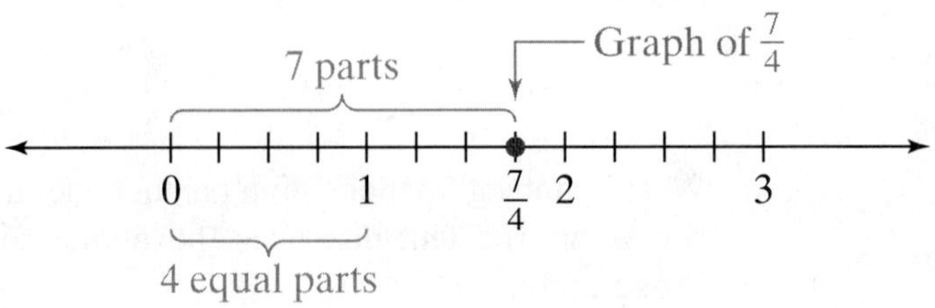

b. The **denominator** is 2, so we divide the distance between **consecutive integers** into 2 equal parts. Because the **numerator** is 5, move 5 places to the right of 0 to plot the solid circle. Try to finish graphing this fraction on your own. View the **answer** to part b, or watch this **video** for complete solutions to all three parts.

c. Try to graph this **fraction** on your own. View the **answer** to part c, or watch this **video** for complete solutions to all three parts.

You Try It **Work through this You Try It problem.**

Work Exercises 34–36 in this textbook or in the MyMathLab Study Plan.

To graph a **mixed number**, first locate the consecutive integers between which the mixed number lies. Then divide the distance between the integers into equal parts according to the denominator, and locate the position within these parts using the numerator.

Example 11 Graphing Mixed Numbers on a Number Line

My video summary

Graph each **mixed number** on a **number line**.

a. $1\frac{3}{5}$ **b.** $2\frac{1}{4}$

Solutions

a. $1\frac{3}{5}$ is between the **consecutive integers** 1 and 2. The **denominator** is 5, so we divide the distance between 1 and 2 into 5 equal parts. The **numerator** is 3, so we move 3 places to the right of 1 and plot a solid circle.

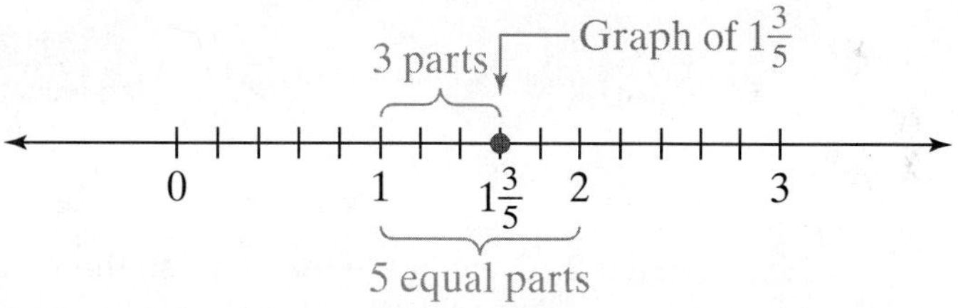

b. Try to graph this mixed number on your own. View the **answer**, or watch this **video** for complete solutions to both parts.

You Try It **Work through this You Try It problem.**

Work Exercises 37 and 38 in this textbook or in the MyMathLab Study Plan.

OBJECTIVE 4 WRITE IMPROPER FRACTIONS AS MIXED NUMBERS OR WHOLE NUMBERS

In **Figure 4**, we see that the **improper fraction** $\frac{11}{8}$ can be written as the **mixed number** $1\frac{3}{8}$.

This means that $\frac{11}{8} = 1\frac{3}{8}$. An improper fraction can *always* be changed into either a mixed number or a **whole number**.

(*eText Screens 4.1-1–4.1-31*)

To write an improper fraction as a mixed number or whole number, divide.

Writing an Improper Fraction as a Mixed Number or Whole Number

Step 1. Divide the **numerator** by the **denominator**.

Step 2. If a **remainder** occurs in Step 1, use the **quotient** as the **whole-number part** of the mixed number. Place the remainder over the original denominator to form the **fraction part** of the mixed number. The mixed number will take the form

$$\text{Quotient}\,\frac{\text{Remainder}}{\text{Original Denominator}}.$$

If no remainder occurs in Step 1, then the result is a whole number, not a mixed number.

Example 12 Writing Improper Fractions as Mixed Numbers or Whole Numbers

Write each **improper fraction** as a **mixed number** or **whole number**.

a. $\frac{74}{9}$ **b.** $\frac{126}{7}$ **c.** $\frac{185}{12}$

Solutions We follow the **steps** for writing improper fractions as mixed numbers or whole numbers.

a. The **numerator** of the improper fraction is 74 and the **denominator** is 9.

$$\frac{74 \leftarrow \text{Numerator}}{9 \leftarrow \text{Denominator}}$$

Step 1. Divide.

$$\begin{array}{r} 8 \leftarrow \text{Quotient} \\ \text{Denominator} \rightarrow 9\overline{)74} \leftarrow \text{Numerator} \\ \underline{72} \\ 2 \leftarrow \text{Remainder} \end{array}$$

Step 2. There is a **remainder**, so the result is a mixed number.

$$\underset{\uparrow\ \text{Quotient}}{8}\frac{2 \leftarrow \text{Remainder}}{9 \leftarrow \text{Original denominator}}$$

The improper fraction $\frac{74}{9}$ equals the mixed number $8\frac{2}{9}$.

b. The **numerator** of the **improper fraction** is 126 and the **denominator** is 7.

$$\frac{126 \leftarrow \text{Numerator}}{7 \leftarrow \text{Denominator}}$$

Step 1. Divide.

$$\begin{array}{r} 18 \leftarrow \text{Quotient} \\ \text{Denominator} \rightarrow 7\overline{)126} \leftarrow \text{Numerator} \\ \underline{7} \\ 56 \\ \underline{56} \\ 0 \leftarrow \text{No remainder} \end{array}$$

Step 2. There is no **remainder**, so the result is a **whole number**.

The improper fraction $\frac{126}{7}$ equals the whole number 18.

My video summary

c. Try working this problem on your own. View the **answer**, or watch this **video** for a complete solution to part c.

You Try It **Work through this You Try It problem.**

Work Exercises 39–46 in this textbook or in the MyMathLab Study Plan.

OBJECTIVE 5 WRITE MIXED NUMBERS AS IMPROPER FRACTIONS

Just as an **improper fraction** can be written as either a **mixed number** or a **whole number**, a mixed number can be written as an improper fraction.

Writing a Mixed Number as an Improper Fraction

Step 1. Multiply the **whole-number part** by the **denominator** of the **fraction part**.

Step 2. Add the **product** from Step 1 to the **numerator** of the fraction part.

Step 3. Write the **sum** from Step 2 over the denominator of the fraction part.

Example 13 Writing Mixed Numbers as Improper Fractions

Write each mixed number as an improper fraction.

a. $3\frac{2}{5}$ **b.** $12\frac{9}{16}$

Solutions We follow the **steps** for writing **mixed numbers** as **improper fractions**.

a. The **whole-number part** is 3, the **numerator** of the **fraction part** is 2, and the **denominator** of the fraction part is 5.

Whole-number part ↓ $3\frac{2}{5}$ ← Numerator, ← Denominator

Step 1. Multiply: $3 \cdot 5 = 15$

Step 2. Add: $15 + 2 = 17$

Step 3. Write the 17 over the 5: $\frac{17}{5}$

So, the mixed number $3\frac{2}{5}$ equals the improper fraction $\frac{17}{5}$.

My video summary

b. Try working this problem on your own. View the **answer**, or watch this **video** for a complete solution to part b.

You Try It **Work through this You Try It problem.**

Work Exercises 47–52 in this textbook or in the MyMathLab Study Plan.

(eText Screens 4.1-1–4.1-31)

4.1 Exercises

In Exercises 1–4, identify the numerator and denominator of each fraction.

You Try It

1. $\frac{7}{12}$
2. $\frac{25}{6}$
3. $\frac{-19}{32}$
4. $\frac{4x}{5y}$

In Exercises 5–8, write a fraction to represent the shaded portion of each figure.

You Try It

5. 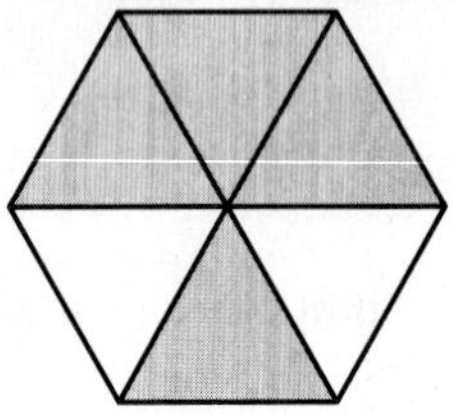

6. 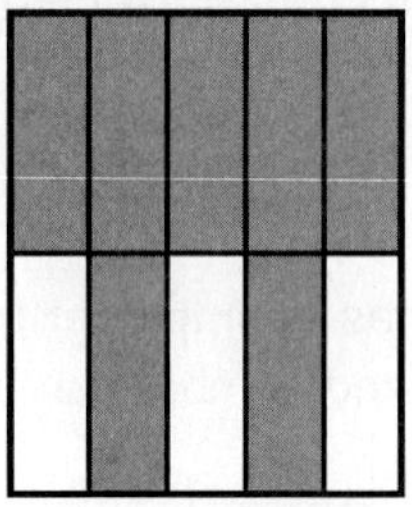

7. 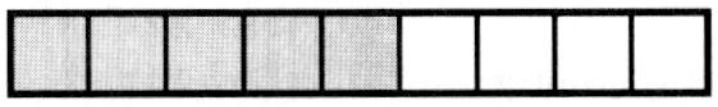

8. 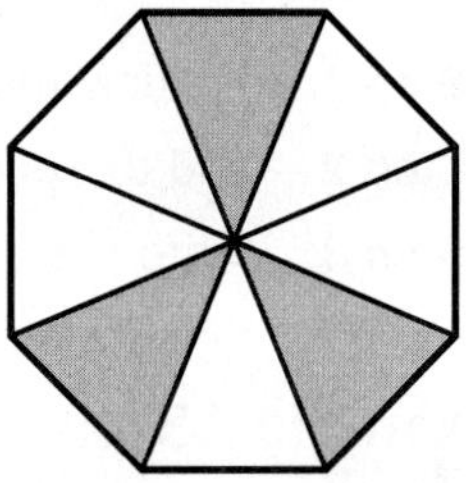

In Exercises 9–20, write a fraction to represent the given information.

You Try It

9. What fraction of the graduated cylinder is filled with water?

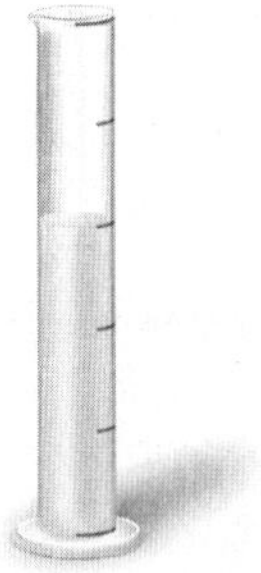

10. What fraction of the fuel tank holds fuel?

11. What fraction of an inch is the length of ribbon?

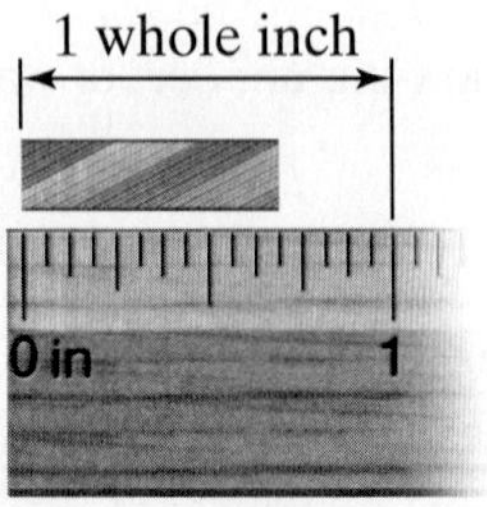

12. What fraction of the syringe contains fluid?

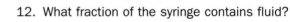

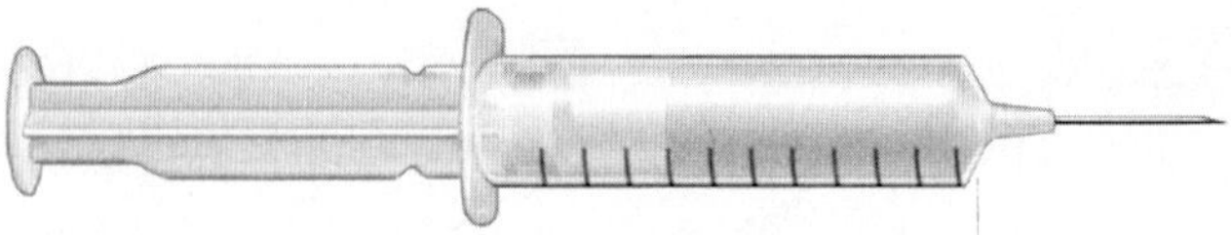

You Try It 13. **Eating Cheesecake** If you eat one slice of the cheesecake shown below, what fraction of the cheesecake will be left?

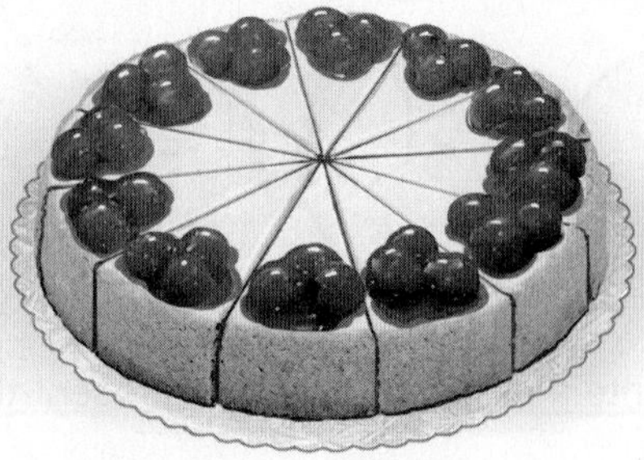

14. **Animal Shelter** One week, an animal shelter rescued 5 cats and 16 dogs. What fraction of rescued animals were cats?

15. **Music Selection** The songs on Steve's iPod are as follows: 25 rock, 14 rap, 34 pop, and 16 country. What fraction of Steve's iPod songs is rock?

16. **Left-handed Presidents** As of 2011, forty-three different individuals have served as president of the United States. Eight of the presidents have been left handed. What fraction of presidents has been left handed? What fraction of presidents has been right handed? (*Source:* Infoplease.com)

    Note: Typically, Grover Cleveland is counted twice as president because he served two nonconsecutive terms. We have counted him only once for this exercise.

17. **Broken Eggs** Upon inspecting a carton of 12 eggs, a shopper discovered that 3 of the eggs were cracked. What fraction of eggs in the carton was cracked? What fraction of eggs in the carton was not cracked?

18. **House of Representatives** At the first session of the 112th congress, the U.S. House of Representatives consisted of 242 Republicans, 192 Democrats, and 1 vacant seat. What fraction of the U.S. House of Representatives was Republican? What fraction was vacant? (*Source:* clerk.house.gov)

19. **Free Throws** During the 2010–11 NBA season, Dirk Nowitzki of the Dallas Mavericks attempted 443 free-throw shots and made 395 of them. What fraction of free-throw shots did Dirk make? What fraction did he miss? (*Source:* basketball-reference.com)

20. **Votes for Test Day** To choose a test date, a prealgebra instructor let her students vote for either Wednesday or Friday. If 13 students voted for Wednesday and 27 voted for Friday, what fraction of students voted for Wednesday? What fraction voted for Friday?

In Exercises 21–24, identify each fraction as a proper fraction or an improper fraction.

You Try It 21. $\frac{9}{10}$ 22. $\frac{23}{8}$ 23. $\frac{7}{7}$ 24. $\frac{12}{4}$

In Exercises 25–28, write both an improper fraction and a mixed number or whole number to represent the shaded portion in each exercise.

You Try It 25.

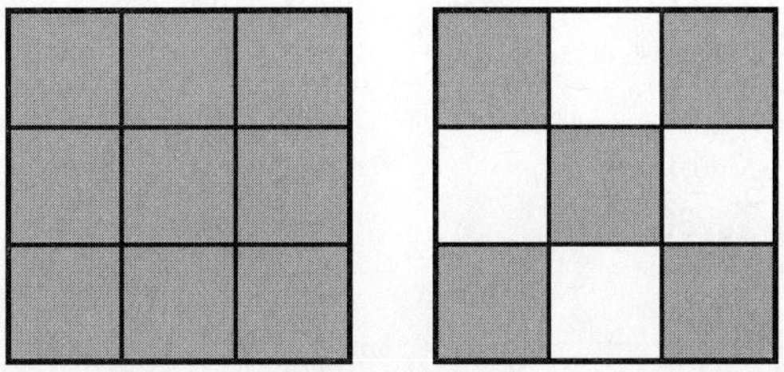

26.

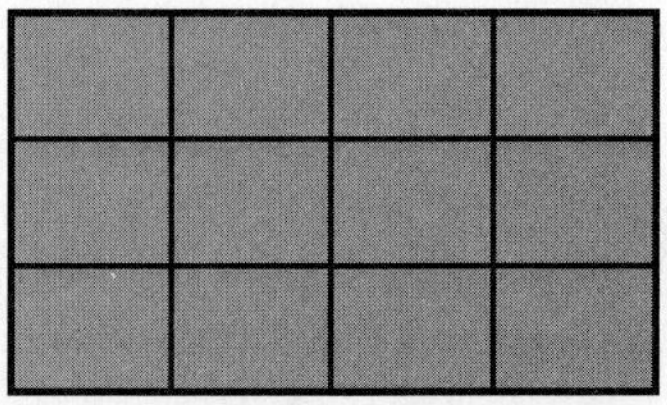

27. 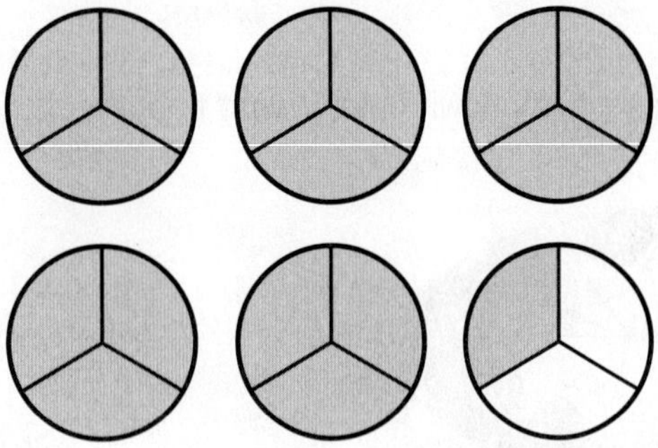

28. 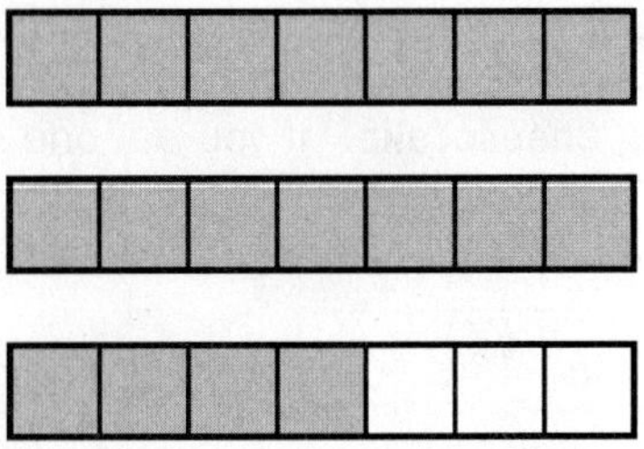

In Exercises 29 and 30, write a mixed number to represent the measurement shown.

You Try It 29. The length of red ribbon shown

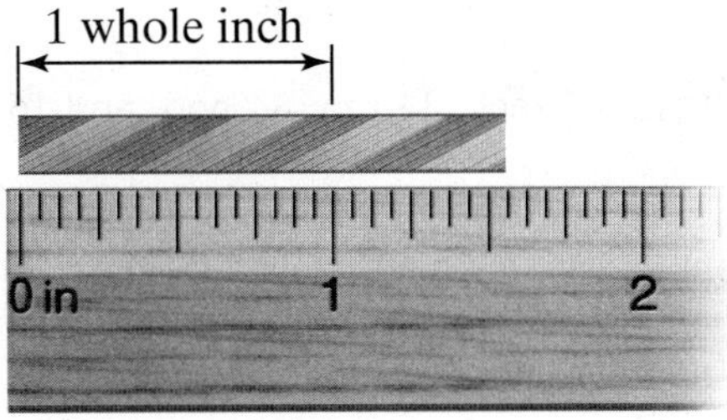

30. The amount of vegetable oil shown .

In Exercises 31–38, graph each fraction or mixed number on a number line.

You Try It 31. $\frac{3}{4}$ 32. $\frac{1}{2}$ 33. $\frac{5}{7}$

You Try It 34. $\frac{8}{5}$ 35. $\frac{7}{3}$ 36. $\frac{6}{6}$

You Try It 37. $1\frac{4}{5}$ 38. $2\frac{3}{8}$

In Exercises 39–46, write each improper fraction as a mixed number or whole number.

You Try It 39. $\frac{8}{3}$ 40. $\frac{32}{5}$ 41. $\frac{117}{9}$ 42. $\frac{55}{12}$

43. $\frac{522}{18}$ 44. $\frac{191}{25}$ 45. $\frac{149}{32}$ 46. $\frac{237}{116}$

In Exercises 47–52, write each mixed number as an improper fraction.

You Try It 47. $5\frac{1}{4}$ 48. $6\frac{3}{8}$ 49. $9\frac{7}{10}$

50. $7\frac{9}{16}$ 51. $14\frac{21}{25}$ 52. $3\frac{121}{150}$

4.2 Factors and Simplest Form

THINGS TO KNOW

Before working through this topic, be sure you are familiar with the following concepts:

		VIDEO	ANIMATION	INTERACTIVE
You Try It	1. Multiply Whole Numbers (Topic 1.4, **Objective 3**)	■		
You Try It	2. Divide Whole Numbers (Topic 1.5, **Objective 1**)	■		
You Try It	3. Use Exponential Notation (Topic 1.6, **Objective 1**)	■		
You Try It	4. Multiply Two Integers (Topic 2.4, **Objective 1**)	■	■	
You Try It	5. Identify the Numerator and Denominator of a Fraction (Topic 4.1, **Objective 1**)			

OBJECTIVES

1 List the Factors of a Number

2 Find the Prime Factorization of a Number

3 Find the Greatest Common Factor

4 Determine If Two Fractions are Equivalent

5 Write Equivalent Fractions

6 Write Fractions in Simplest Form

OBJECTIVE 1 LIST THE FACTORS OF A NUMBER

In **Topic 1.4**, we saw that 4 and 6 were called **factors** of 24 because $4 \cdot 6 = 24$. A factor of a number **divides the number evenly** (that is, there is no remainder), and we say the number is **divisible** by the factor. For example, 24 is divisible by 6 because $24 \div 6 = 4$ has no remainder. Similarly, 24 is divisible by 4 because $24 \div 4 = 6$ has no remainder. We would say 24 is *not* divisible by 5 because $24 \div 5 = 4$ R4 has a remainder. Because 24 is not divisible by 5, we know that 5 is not a factor of 24.

(eText Screens 4.2-1–4.2-34)

To **factor** a nonzero **whole number** means to write it as the **product** of two or more whole numbers. For example, we can factor 24 by writing it as $4 \cdot 6$.

$$\underset{\text{Factor}}{4} \cdot \underset{\text{Factor}}{6} = \underset{\text{Product}}{24}$$

Because 24 can also be written as $1 \cdot 24$, $2 \cdot 12$, or $3 \cdot 8$, the numbers 1, 2, 3, 8, 12, and 24 are also factors of 24. There are no other factors of 24, so we can say that the factors of 24 are 1, 2, 3, 4, 6, 8, 12, and 24.

Some special rules can be used to determine if certain numbers are **factors**. We call these **divisibility tests**. The following are divisibility tests for 2, 3, 5, and 10.

Divisibility Tests for 2, 3, 5, and 10

- Every even number is **divisible** by 2. A number is even if it ends with 2, 4, 6, 8, or 0.
- A number is divisible by 3 if the sum of its **digits** is divisible by 3.
- A number is divisible by 5 if its last digit is 0 or 5.
- A number is divisible by 10 if its last digit is 0.

View this **popup** to see divisibility tests for 4, 6, 8, and 9.

We find the list of factors for a number by first finding pairs of factors, then writing the factors in increasing order.

Example 1 Listing the Factors of a Number

My interactive video summary

List the factors for each number.

a. 36 **b.** 64 **c.** 120

Solutions

a. We look for pairs of **factors** that multiply to 36. Starting with 1, we check each **positive** whole number to see if it **divides evenly** into 36, using the **divisibility tests** if appropriate. Every number is divisible by 1, and a number divided by 1 is just that number. So, 1 and the number form the first pair of factors.

Factor	Factor
1	36

Because 36 is even, we know it is **divisible** by 2. $36 \div 2 = 18$, so 2 and 18 form another pair of factors.

Factor	Factor
1	36
2	18

The sum of the **digits** in 36 is $3 + 6 = 9$, which is divisible by 3. Therefore, 36 is divisible by 3. $36 \div 3 = 12$, so 3 and 12 form another pair of factors.

Factor	Factor
1	36
2	18
3	12

Continuing on, $36 \div 4 = 9$, so 4 and 9 form another pair of factors.

Factor	Factor
1	36
2	18
3	12
4	9

5 is not a **factor** of 36 because the last **digit** in 36 is 6, not 0 or 5. 6 is a factor of 36 because both 2 and 3 are factors. $36 \div 6 = 6$, so 6 and 6 form another pair of factors.

Factor	Factor
1	36
2	18
3	12
4	9
6	6

Notice that as we increase the factor on the left, the factor on the right decreases. Using this approach, we continue listing pairs of factors until the two factors are the same (as in this case), or there are no factors possible between the last pair in the table.

The factors of 36, in increasing order, are 1, 2, 3, 4, 6, 9, 12, 18, and 36.

b. $64 \div 1 = 64$, so 1 and 64 are a pair of **factors.**

64 is even, so it is **divisible** by 2. $64 \div 2 = 32$, so 2 and 32 are a pair of factors.

Adding the digits, we get $6 + 4 = 10$, which is not divisible by 3, so 3 is not a factor.

$64 \div 4 = 16$. So 4 and 16 are a pair of factors.

Continue finding pairs of factors on your own, then list all the factors of 64. Check your **answer**, or watch this **interactive video** for complete solutions to all three parts.

c. Try to list the factors for 120 on your own. Check your **answer**, or watch this **interactive video** for complete solutions to all three parts.

You Try It **Work through this You Try It problem.**

Work Exercises 1–10 in this textbook or in the MyMathLab Study Plan.

OBJECTIVE 2 FIND THE PRIME FACTORIZATION OF A NUMBER

Some numbers, such as 7, have only two **factors**, 1 and itself. Such numbers are called *prime numbers*.

> **Definition Prime Number**
>
> A **prime number** is a **whole number** greater than 1 whose only whole number factors are 1 and itself.

The first ten prime numbers are 2, 3, 5, 7, 11, 13, 17, 19, 23, and 29.

The number 24 is not a prime number because it has factors other than 1 and 24. Instead, it is a *composite number*.

(eText Screens 4.2-1–4.2-34)

> **Definition Composite Number**
>
> A **composite number** is a whole number greater than 1 that is not prime.

CAUTION The numbers 0 and 1 are restricted from the definitions of both prime and composite numbers. This means 0 and 1 are neither prime nor composite.

Every composite number can be **factored** into the **product** of prime numbers. For example, the composite number 30 can be written as

$$\underbrace{30}_{\text{Composite number}} = \underbrace{2 \cdot 3 \cdot 5.}_{\text{Product of prime factors}}$$

The product of prime factors is called the **prime factorization** of the **composite number**. Every composite number has a unique prime factorization.

> **Finding the Prime Factorization of a Composite Number**
>
> **Step 1.** Excluding 1 and the number itself, find any two **factors** whose product is the given composite number.
>
> **Step 2.** If either factor from Step 1 is not a **prime** number, continue by factoring each composite factor. Continue this process until every factor is a prime number.

TIP If there are repeated factors in the final prime factorization, write the factorization using **exponential notation.**

CAUTION Finding the **prime factorization** is not the same as finding the factors of a number. The prime factorization is a product, such as $2 \cdot 3 \cdot 5$, whereas the factors of a number are given as a list, such as 1, 2, 3, 5, 6, 10, 15, and 30.

The ability to factor composite numbers is necessary to simplify **fractions**, as we will see later in this topic, and when performing certain operations on fractions.

One way to find the **prime factorization** of a **composite number** is by using a **factor tree.** This method is shown in the following example.

Example 2 Finding the Prime Factorization of a Composite Number

Find the prime factorization of each composite number.

a. 42 **b.** 60 **c.** 252

Solutions

We follow the **steps** for finding the prime factorization of a composite number.

a. Step 1. We know from our multiplication facts that $6 \cdot 7 = 42$, so let's use those two **factors** first and organize the factorization with a factor tree.

42
/ \
Composite → 6 · 7 ← Prime

Step 2. The factor 7 is prime, but the factor 6 is composite, so we continue to factor: $6 = 2 \cdot 3$.

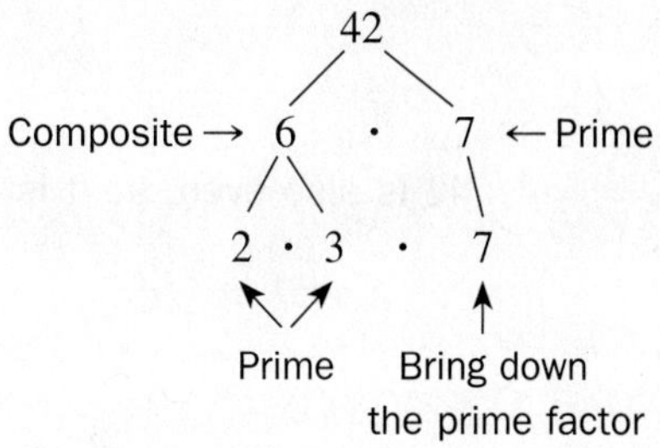

Now all of the factors are **prime numbers.** Remember that the **commutative property of multiplication** tells us that the order of the **factors** is not important when multiplying. However, it is common practice when performing a prime factorization to place the factors in increasing order from smallest to largest, so we will do the same in this text. The prime factorization of 42 is

$$42 = 2 \cdot 3 \cdot 7.$$

My video summary

b. Step 1. We know from our multiplication facts that 6×10 is 60, so we begin with the factors 6 and 10 and organize the factorization with a **factor tree.**

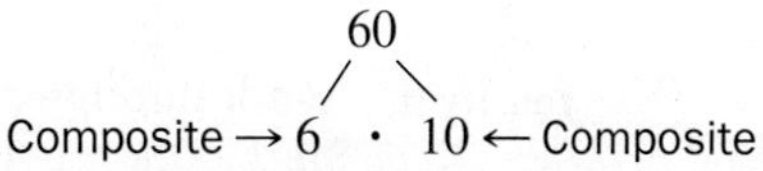

Step 2. The factors 6 and 10 are both composite numbers, so we continue by factoring 6 and 10 separately. Finish the prime factorization on your own. View this **popup** to check your answer, or watch this **video** for a complete solution to part b.

My video summary

c. Try to do this **prime factorization** on your own. Check your **answer**, or watch this **video** for a complete solution to part c.

You Try It **Work through this You Try It problem.**

Work Exercises 11–16 in this textbook or in the MyMathLab Study Plan.

Note: The pair of factors we begin with in our **prime factorization** does not matter. The end result will be the same. View this **popup** to see the factor tree for Example 2b starting with the factors 4 and 15.

Another approach to finding the prime factorization of a number is to divide by the smallest **prime** factor that evenly divides the number. This approach is illustrated in the following example.

Example 3 Finding the Prime Factorization of a Composite Number

Find the prime factorization of each **composite number.**

a. 84 **b.** 300

(eText Screens 4.2-1–4.2-34)

Solutions

a. Because 84 is an even number, it is **divisible** by 2.

$$2\overline{)84}^{\,42}$$

42 is also even, so it is also divisible by 2.

$$\begin{array}{r} 21 \\ 2\overline{)42} \\ \uparrow 2\overline{)84} \end{array}$$

The **sum** of the digits of 21 is $2 + 1 = 3$, which is divisible by 3. So, 21 is divisible by 3.

$$\begin{array}{r} 7 \\ 3\overline{)21} \\ \uparrow 2\overline{)42} \\ \uparrow 2\overline{)84} \end{array}$$

The **quotient** 7 is prime, so we finish our factorization. We have the prime factors of 2, 2, 3, and 7, so the prime factorization of 84 is $2 \cdot 2 \cdot 3 \cdot 7$ or $2^2 \cdot 3 \cdot 7$.

My video summary

b. Try to do this **prime factorization** on your own. Check your **answer**, or watch this **video** for a complete solution to part b.

You Try It **Work through this You Try It problem.**

Work Exercises 17–24 in this textbook or in the MyMathLab Study Plan.

OBJECTIVE 3 FIND THE GREATEST COMMON FACTOR

Consider the following **prime factorizations** for 30 and 42.

$$30 = 2 \cdot 3 \cdot 5 \qquad 42 = 2 \cdot 3 \cdot 7$$

Notice that 2 and 3 are **factors** of both 30 and 42. We call these **common factors** because they are common to both numbers.

Now consider the list of factors for 30 and 42.

Factors of 30	Factors of 42
1, 2, 3, 5, 6, 10, 15, 30	1, 2, 3, 6, 7, 14, 21, 42

In addition to the common **prime** factors 2 and 3, we see that 6 is also a common factor of 30 and 42. In fact, 6 is the largest common factor, so we call this the **greatest common factor (GCF).** Multiplying the two common prime factors gives us this greatest common factor, $2 \cdot 3 = 6$.

TIP For large numbers, prime factorizations can help us determine the GCF.

(eText Screens 4.2-1–4.2-34)

Example 4 Finding the Greatest Common Factor of a List of Numbers

Find the **GCF** for each list of numbers.

a. 40, 60 **b.** 108, 180 **c.** 252, 420, 980

Solutions

a. Start by writing the **prime factorization** of each number.

$$40 = 2 \cdot 2 \cdot 2 \cdot 5 = 2^3 \cdot 5$$
$$60 = 2 \cdot 2 \cdot 3 \cdot 5 = 2^2 \cdot 3 \cdot 5$$

We see that there are **common factors** of 2, 2, and 5. Multiplying, we find that $2 \cdot 2 \cdot 5 = 2^2 \cdot 5 = 20$ is the greatest common factor of 40 and 60.

b. Start by writing the prime factorization of each number.

$$108 = 2 \cdot 2 \cdot 3 \cdot 3 \cdot 3 = 2^2 \cdot 3^3$$
$$180 = 2 \cdot 2 \cdot 3 \cdot 3 \cdot 5 = 2^2 \cdot 3^2 \cdot 5$$

We see that there are common factors of 2, 2, 3, and 3. Multiplying, we find that $2 \cdot 2 \cdot 3 \cdot 3 = 2^2 \cdot 3^2 = 36$ is the greatest common factor of 108 and 180.

My video summary

c. Try to find the GCF for this list of numbers on your own. Check your **answer**, or watch this **video** for a complete solution to part c.

You Try It **Work through this You Try It problem.**

Work Exercises 25–28 in this textbook or in the MyMathLab Study Plan.

Notice in Example 4 that, in exponent form, the **GCF** contained each *different* **factor** using the *smallest* **exponent** that occurred on the factor in any of the prime factorizations. This idea extends to cases where terms involve **variable factors** as well. For example, the terms x^5yz^2 and x^4y^2 have common variable factors of x and y. The smallest exponent on x is 4 and the smallest exponent on y is 1. So the GCF of x^5yz^2 and x^4y^2 is x^4y.

Greatest Common Factor

The **greatest common factor (GCF)** of a list of terms is the largest factor common to all terms in the list. The GCF can include both numeric factors such as 2 and variable factors such as x.

We summarize the procedure for finding the GCF below.

Finding the Greatest Common Factor

Step 1. Write each term as the **product** of prime factors and variable factors in exponent form.

Step 2. Identify each different **common prime factor** or variable factor.

Step 3. Multiply the common factors using the *smallest* exponent that occurs on the factor in any term.

(eText Screens 4.2-1–4.2-34)

Example 5 Finding the Greatest Common Factor

Find the GCF.

a. 225, 945 **b.** $4a^2b^4c$, $6a^7b^3$ **c.** $12m^2n^5$, $8m^3n^3$, $20m^4n^3$

Solutions

For each part, we follow the **steps** for finding the **GCF**.

a. Step 1.

Smallest exponent

$$225 = 3\cdot3\cdot\quad 5\cdot5 \quad = 3^{2}\cdot5^{2}$$
$$945 = 3\cdot3\cdot3\cdot5\cdot\quad 7 = 3^{3}\cdot5^{1}\cdot7^{1}$$

Step 2. We have **common factors** of 3 and 5. The prime factor 7 is not common to both terms, so it is not included in the GCF.

Step 3. The smallest **exponent** on 3 is 2 and the smallest exponent on 5 is 1. Therefore, we have

$$\text{GCF} = 3^2\cdot5^1 = 9\cdot5 = 45.$$

b. Step 1.

$$4a^2b^4c = 2^2\cdot \qquad a^2\cdot b^4\cdot c^1$$
$$6a^7b^3 = 2^1\cdot3^1\cdot a^7\cdot b^3$$

Step 2. We have common factors of 2, a, and b. The prime factor 3 and the variable factor c are not common to both terms, so they are not included in the GCF.

Using the smallest exponent on the common factors, finish writing the GCF on your own. View this **popup** to check your answer.

My video summary

c. Try to find the **GCF** for this part on your own. Check your **answer**, or watch this **video** for a complete solution to part c.

You Try It **Work through this You Try It problem.**

Work Exercises 29–37 in this textbook or in the MyMathLab Study Plan.

OBJECTIVE 4 DETERMINE IF TWO FRACTIONS ARE EQUIVALENT

Look at the **fractions** illustrated in **Figure 5**. The rectangle in Figure 5(a) is divided into three equal parts with two parts shaded, so $\frac{2}{3}$ of the rectangle is shaded. The identical rectangle in Figure 5(b) is divided into six equal parts with four parts shaded, so $\frac{4}{6}$ of the rectangle is shaded. Likewise, in Figure 5(c), $\frac{6}{9}$ of the rectangle is shaded. Notice that the amount of shaded area is the same in each of the three rectangles.

This means $\frac{2}{3} = \frac{4}{6} = \frac{6}{9}$.

(eText Screens 4.2-1–4.2-34)

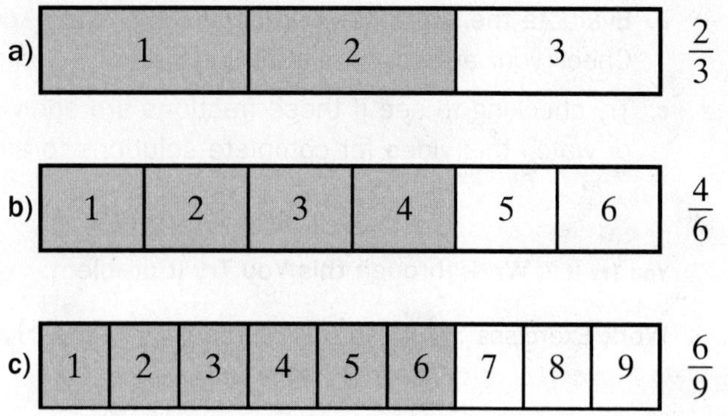

Figure 5

The three **fractions** illustrated in Figure 5 are *equivalent fractions*.

> **Definition Equivalent Fractions**
>
> Two or more fractions are **equivalent fractions** if they represent the same value.

So, from Figure 5, we can say that $\frac{2}{3} = \frac{4}{6}$. If we multiply the **numerator** of one fraction by the **denominator** of the other, we get equal **products**.

$$6 \cdot 2 = 12 \qquad 3 \cdot 4 = 12$$

We call $6 \cdot 2$ and $3 \cdot 4$ **cross products** because we obtained the products by multiplying *across* the equal sign.

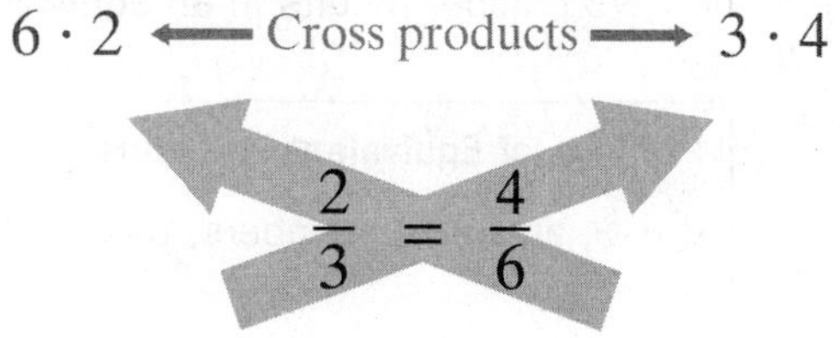

TIP For equivalent fractions, the cross products will always be equal.

Example 6 Determining If Two Fractions are Equivalent

My video summary

Determine if each pair of fractions is **equivalent**.

a. $\frac{3}{7}$ and $\frac{27}{63}$ **b.** $\frac{5}{9}$ and $\frac{60}{72}$ **c.** $\frac{-3}{5}$ and $\frac{27}{-45}$

Solutions

a. Check the cross products.

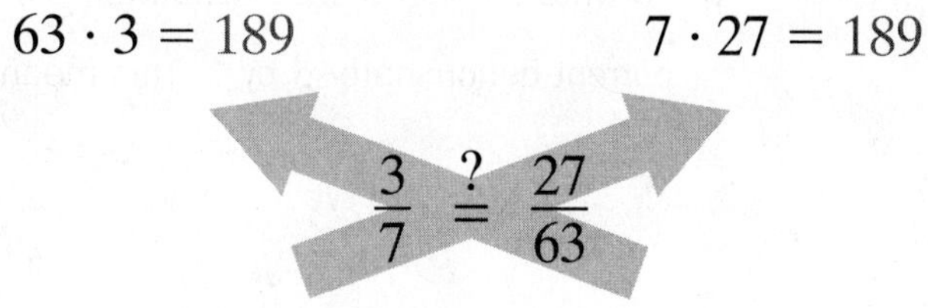

63 · 3 = 189
7 · 27 = 189
$\frac{3}{7} \stackrel{?}{=} \frac{27}{63}$

Because the two **cross products** are equal (see the **details**), the fractions are equivalent.

b. Evaluate the two cross products, $72 \cdot 5$ and $9 \cdot 60$, to see if the results are equal. Check your **answer**, or watch this **video** for complete solutions to all three parts.

c. Try checking to see if these fractions are equivalent on your own. Check your **answer**, or watch this **video** for complete solutions to all three parts.

You Try It Work through this You Try It problem.

Work Exercises 38–46 in this textbook or in the MyMathLab Study Plan.

In the fraction $\frac{-3}{5}$, the **numerator** is negative and the **denominator** is positive, while in $\frac{27}{-45}$, the numerator is positive and the denominator is negative. However, in part c of Example 6, we saw that $\frac{-3}{5}$ and $\frac{27}{-45}$ were **equivalent fractions.** This illustrates the general result that

$$-\frac{a}{b} = \frac{-a}{b} = \frac{a}{-b}$$

provided $b \neq 0$. In other words, if a fraction is negative, we can put the negative sign in either the numerator or denominator (but not both) and have an equivalent fraction.

OBJECTIVE 5 WRITE EQUIVALENT FRACTIONS

Multiplying or dividing both the **numerator** and **denominator** of a fraction by the same nonzero number results in an **equivalent fraction.**

Property of Equivalent Fractions

If a, b, and c are numbers, then

$$\frac{a}{b} = \frac{a \cdot c}{b \cdot c} \quad \text{and} \quad \frac{a}{b} = \frac{a \div c}{b \div c}$$

as long as b and c are not equal to 0.

Example 7 Finding Equivalent Fractions

My video summary

Write the equivalent fraction with the given **numerator** or **denominator**.

a. $\frac{4}{5}$; denominator 35 **b.** $\frac{30}{54}$; numerator 5

Solutions

a. To write an **equivalent fraction** for $\frac{4}{5}$ with a denominator of 35, we must multiply the current denominator 5 by 7. This means we must also multiply the current numerator 4 by 7:

$$\frac{4}{5} = \frac{4 \cdot 7}{5 \cdot 7} = \frac{28}{35}$$

We can check that $\frac{4}{5}$ and $\frac{28}{35}$ are equivalent by using **cross products.**

$$35 \cdot 4 = 140 \qquad\qquad 5 \cdot 28 = 140$$

$$\frac{4}{5} \stackrel{?}{=} \frac{28}{35}$$

Because the cross products are equal, the two fractions are equivalent.

b. To write an **equivalent fraction** for $\frac{30}{54}$ with a numerator of 5, we must divide the current numerator 30 by 6. This means we must also divide the current denominator 54 by 6. Try to find this equivalent fraction on your own. Check your **answer**, or watch this **video** for fully worked solutions to both parts.

You Try It **Work through this You Try It problem.**

Work Exercises 47–58 in this textbook or in the MyMathLab Study Plan.

OBJECTIVE 6 WRITE FRACTIONS IN SIMPLEST FORM

A **fraction** can be written in infinitely many equivalent forms. Only one of these is the *simplest form*. For example, of all the **equivalent fractions** $\frac{2}{3} = \frac{4}{6} = \frac{6}{9}$ in **Figure 5**, the fraction $\frac{2}{3}$ is in simplest form.

> **Definition Simplest Form for Fractions**
>
> A fraction is in **simplest form**, or **lowest terms**, when the **numerator** and **denominator** have no **common factors** other than 1. A fraction in simplest form is called a **simplified fraction.**

The fraction $\frac{8}{12}$ is not in simplest form because $8 = 2 \times 4$ and $12 = 3 \times 4$ share the common factor 4, which is their **greatest common factor**. Using the **property of equivalent fractions,** we can divide both the numerator and denominator by 4:

$$\frac{8}{12} = \frac{8 \div 4}{12 \div 4} = \frac{2}{3}$$

Because 1 is the only common factor of 2 and 3, the fraction $\frac{2}{3}$ is in simplest form. In general, we can simplify a fraction by dividing both the numerator and denominator by their greatest common factor.

(eText Screens 4.2-1–4.2-34)

Example 8 Simplifying Fractions Using the GCF

Write each fraction in **simplest form** by dividing the numerator and denominator by the **GCF**.

a. $\frac{30}{42}$ **b.** $\frac{18}{35}$ **c.** $\frac{16}{80}$ **d.** $\frac{210}{63}$

Solutions

a. List the **factors** of 30 and 42 and determine the GCF.

Factors of 30: 1, 2, 3, 5, 6, 10, 15, 30
Factors of 42: 1, 2, 3, 6, 7, 14, 21, 42

The greatest common factor of 30 and 42 is 6. Divide both the numerator and denominator by 6.

$$\frac{30 \div 6}{42 \div 6} = \frac{5}{7}$$

The fraction $\frac{30}{42}$ simplifies to $\frac{5}{7}$.

b. List the factors of 18 and 35 and determine the GCF.

Factors of 18: 1, 2, 3, 6, 9, 18
Factors of 35: 1, 5, 7, 35

Because the GCF is 1, the fraction $\frac{18}{35}$ is already written in simplest form.

My video summary

c.–d. Try to simplify these fractions on your own. Check your **answers**, or watch this **video** for fully worked solutions to parts c and d.

You Try It **Work through this You Try It problem.**

Work Exercises 59–66 in this textbook or in the MyMathLab Study Plan.

We can also use **prime factorization** to simplify fractions by first writing both the **numerator** and **denominator** in terms of their prime factorizations. We then divide both the numerator and denominator by each **common factor**. We show this process by placing a slash mark through each common factor and replacing it with a 1, the result of dividing the common factor by itself (review the **division properties of** 1). For example,

Divide $2/2 = 1$ in the numerator

$$\frac{4}{6} = \frac{2 \cdot 2}{2 \cdot 3} = \frac{\overset{1}{\cancel{2}} \cdot 2}{\underset{1}{\cancel{2}} \cdot 3} = \frac{1 \cdot 2}{1 \cdot 3} = \frac{2}{3}$$

Divide $2/2 = 1$ in the denominator

Simplifying Fractions

If a, b, and c are numbers, then $\frac{a \cdot c}{b \cdot c} = \frac{a \cdot \overset{1}{\cancel{c}}}{b \cdot \underset{1}{\cancel{c}}} = \frac{a \cdot 1}{b \cdot 1} = \frac{a}{b}$ for $b \neq 0$ and $c \neq 0$.

(eText Screens 4.2-1–4.2-34)

TIP Replacing each divided-out factor with a 1 can lead to unnecessary clutter. It is, therefore, okay to leave the 1's unwritten. If every factor in the numerator divides out, however, be sure to include a 1 in the numerator of the **simplified fraction**.

In general, we can use the following steps to simplify fractions.

Writing a Fraction in Simplest Form

Step 1. Write the **prime factorizations** of the numerator and denominator.

Step 2. Divide both the numerator and denominator by each **common factor**.

Step 3. Separately in the numerator and denominator, multiply any factors that are not divided out to obtain the simplified fraction. If all factors in the numerator divide out, the numerator will be 1.

My video summary

Before proceeding, you may want to watch this **video** for a review of prime factorization.

Example 9 Simplifying Fractions

Write each fraction in simplest form.

a. $\frac{12}{45}$ b. $-\frac{44}{110}$ c. $-\frac{18}{108}$ d. $\frac{55}{63}$

Solutions

a. **Factor** the numerator and denominator: $\frac{12}{45} = \frac{2\cdot 2\cdot 3}{3\cdot 3\cdot 5}$

Divide out the common factor 3: $= \frac{2\cdot 2\cdot \cancel{3}^1}{_1\cancel{3}\cdot 3\cdot 5}$

Multiply the remaining factors: $= \frac{2\cdot 2\cdot 1}{1\cdot 3\cdot 5} = \frac{4}{15}$

The fraction $\frac{12}{45}$ simplifies to $\frac{4}{15}$.

b. We can **simplify** negative fractions in the same manner that we simplify positive fractions.

Factor the numerator and denominator: $-\frac{44}{110} = -\frac{2\cdot 2\cdot 11}{2\cdot 5\cdot 11}$

Divide out the common factor 2: $= -\frac{\cancel{2}\cdot 2\cdot 11}{\cancel{2}\cdot 5\cdot 11}$

Divide out the common factor 11: $= -\frac{\cancel{2}\cdot 2\cdot \cancel{11}}{\cancel{2}\cdot 5\cdot \cancel{11}}$

Multiply the remaining factors: $= -\frac{2}{5}$

The fraction $-\frac{44}{110}$ simplifies to $-\frac{2}{5}$.

(eText Screens 4.2-1–4.2-34)

My video summary **c.–d.** Try to simplify these fractions on your own. Check your **answers**, or watch this **video** for fully worked solutions to parts c and d.

You Try It **Work through this You Try It problem.**

Work Exercises 67–76 in this textbook or in the MyMathLab Study Plan.

We can use this same approach to simplify fractions involving **variable factors** by remembering that **exponent notation** represents repeated multiplication. For example, $x^2 = x \cdot x$ or $y^4 = y \cdot y \cdot y \cdot y$. If a variable appears in the denominator of a fraction, we will assume the variable is never 0 because the denominator of a fraction cannot be 0.

Example 10 Simplifying Fractions

My video summary Write each fraction in simplest form.

a. $\dfrac{30x^4}{9x^2}$ **b.** $\dfrac{42x^2y^3}{14x^3y}$

Solutions

a. Factor the numerator and denominator: $\dfrac{30x^4}{9x^2} = \dfrac{2 \cdot 3 \cdot 5 \cdot x \cdot x \cdot x \cdot x}{3 \cdot 3 \cdot x \cdot x}$

Divide out the common factors: $= \dfrac{2 \cdot \cancel{3} \cdot 5 \cdot \cancel{x} \cdot \cancel{x} \cdot x \cdot x}{\cancel{3} \cdot 3 \cdot \cancel{x} \cdot \cancel{x}}$

Multiply the remaining factors: $= \dfrac{2 \cdot 5 \cdot x \cdot x}{3}$

$= \dfrac{10x^2}{3}$

The fraction $\dfrac{30x^4}{9x^2}$ simplifies to $\dfrac{10x^2}{3}$.

b. Try to simplify this fraction on your own. Check your **answer**, or watch this **video** for fully worked solutions to both parts.

You Try It **Work through this You Try It problem.**

Work Exercises 77–86 in this textbook or in the MyMathLab Study Plan.

CAUTION When simplifying fractions, we can divide out **common factors** of the **numerator** and **denominator**. However, it is incorrect to divide out **terms**.

Correct Simplification	**Incorrect Simplification**
$\dfrac{2 \cdot \cancel{7}}{3 \cdot \cancel{7}} = \dfrac{2}{3}$	$\dfrac{2 + \cancel{7}}{3 + \cancel{7}} = \dfrac{2}{3}$ Note: $\dfrac{2+7}{3+7} = \dfrac{9}{10}$, not $\dfrac{2}{3}$

(eText Screens 4.2-1–4.2-34)

4.2 Exercises

In Exercises 1–10, list the factors for each number.

You Try It

1. 12
2. 30
3. 19
4. 16
5. 72
6. 90
7. 165
8. 66
9. 24
10. 144

In Exercises 11–24, find the prime factorization of each composite number. If the number is a prime number, identify it as such.

You Try It

11. 6
12. 36
13. 41
14. 350
15. 104
16. 819
17. 75
18. 128
19. 144
20. 216

You Try It

21. 315
22. 620
23. 396
24. 1287

In Exercises 25–37, find the GCF for each group of terms.

You Try It

25. 6 and 15
26. 18 and 78
27. 15 and 90
28. 27, 90, and 378
29. 700 and 280
30. $22x^2$ and $143x$

You Try It

31. $6a^2$, $9a^3$, and $12a^2$
32. $6xz$, $15x^2y$, and $40xyz$
33. 42, 189, and 252
34. a^6b^3 and a^3b^5
35. m^3n, $4m^5n^3$, $18m^2n^2$
36. $36a^3b^3c$, $70a^2bc^4$, and $94abc^3$
37. $114x^2yz$, $174xz$, and $198xyz^2$

In Exercises 38–46, determine if the two fractions are equivalent.

You Try It

38. $\frac{10}{15}$ and $\frac{2}{3}$
39. $\frac{12}{15}$ and $\frac{4}{5}$
40. $\frac{4}{3}$ and $\frac{9}{12}$
41. $\frac{15}{4}$ and $\frac{30}{8}$
42. $\frac{-12}{18}$ and $\frac{2}{-3}$
43. $\frac{6}{9}$ and $\frac{11}{14}$
44. $\frac{9}{19}$ and $\frac{6}{16}$
45. $\frac{12}{18}$ and $\frac{18}{27}$
46. $\frac{2}{-7}$ and $\frac{-3}{8}$

In Exercises 47–58, write an equivalent fraction satisfying the stated condition.

You Try It

47. $\frac{5}{7}$; numerator $= 40$
48. $\frac{16}{24}$; denominator $= 3$
49. $\frac{17}{51}$; numerator $= 1$
50. $\frac{9}{6}$; denominator $= 24$
51. $\frac{16}{24}$; denominator $= 12$
52. $\frac{22}{11}$; numerator $= 66$

53. $\frac{1}{7}$; numerator $= 4$ 54. $\frac{24}{18}$; denominator $= 3$ 55. 3; numerator $= 6$

56. $\frac{72}{144}$; denominator $= 2$ 57. $\frac{3x}{10}$; denominator $= 40$ 58. $\frac{36}{80x}$; numerator $= 9$

In Exercises 59–86, write each fraction in simplest form using the method of your choice.

You Try It 59. $\frac{3}{6}$ 60. $\frac{15}{12}$ 61. $\frac{49}{42}$ 62. $\frac{52}{64}$

63. $\frac{38}{81}$ 64. $\frac{27}{72}$ 65. $\frac{14}{32}$ 66. $\frac{80}{400}$

You Try It 67. $-\frac{27}{81}$ 68. $-\frac{24}{36}$ 69. $\frac{51}{17}$ 70. $-\frac{180}{123}$

71. $\frac{105}{405}$ 72. $-\frac{114}{180}$ 73. $-\frac{210}{315}$ 74. $\frac{45}{56}$

You Try It 75. $\frac{132}{187}$ 76. $\frac{90}{540}$ 77. $-\frac{30x^2}{16x}$ 78. $\frac{98y}{21xy}$

79. $\frac{55x^2y}{121xy^2}$ 80. $\frac{205x}{215y}$ 81. $\frac{51x^2y^2z}{34xy^2z^2}$ 82. $\frac{58x^2}{87x^3}$

83. $\frac{385xy}{210z}$ 84. $\frac{20x^3y}{4xy}$ 85. $-\frac{9z}{90x^2z}$ 86. $-\frac{42x^2y}{7x^2y}$

4.3 Multiplying and Dividing Fractions

THINGS TO KNOW

Before working through this topic, be sure you are familiar with the following concepts:

	VIDEO	ANIMATION	INTERACTIVE
You Try It **1.** Find the Area of a Rectangle (Topic 1.4, **Objective 5**)			
You Try It **2.** Multiply Two Integers (Topic 2.4, **Objective 1**)			
You Try It **3.** Evaluate Exponential Expressions Involving an Integer Base (Topic 2.4, **Objective 3**)			
You Try It **4.** Divide Integers (Topic 2.4, **Objective 4**)			

You Try It

5. Identify the Numerator and Denominator of a Fraction (Topic 4.1, **Objective 1**)

You Try It

6. Write Fractions in Simplest Form (Topic 4.2, **Objective 6**)

OBJECTIVES

1 Multiply Fractions

2 Evaluate Exponential Expressions Involving Fraction Bases

3 Find Reciprocals

4 Divide Fractions

5 Simplify Expressions by Multiplying and Dividing Fractions

6 Solve Applications by Multiplying or Dividing Fractions

7 Find the Area of a Triangle

OBJECTIVE 1 MULTIPLY FRACTIONS

Following an office party, $\frac{1}{2}$ of a sheet cake remained as leftovers. If three coworkers each take $\frac{1}{3}$ of the leftover cake, what **fraction** of a whole sheet cake does each coworker get? Each coworker gets $\frac{1}{3}$ of $\frac{1}{2}$ sheet cake. Figure 6 shows that each coworker's share amounts to $\frac{1}{6}$ of a whole sheet cake.

Figure 6

The word "of" indicates multiplication, so

$$\frac{1}{3} \text{ of } \frac{1}{2} \quad \text{means} \quad \frac{1}{3} \cdot \frac{1}{2}.$$

The result is $\frac{1}{6}$ of a whole sheet cake, so $\frac{1}{3} \cdot \frac{1}{2} = \frac{1}{6}$.

(eText Screens 4.3-1–4.3-39)

The sheet cake example shows that to multiply **fractions**, we multiply the **numerators** and multiply the **denominators**. If necessary, we then **simplify** the result.

Multiplying Fractions

If a, b, c, and d are numbers and b and d are not equal to 0, then $\frac{a}{b} \cdot \frac{c}{d} = \frac{a \cdot c}{b \cdot d}$.

Example 1 Multiplying Fractions

Multiply.

a. $\frac{2}{3} \cdot \frac{4}{5}$ **b.** $\frac{5}{8} \cdot \frac{7}{9}$

Solutions

a. $\frac{2}{3} \cdot \frac{4}{5} = \frac{\overbrace{2 \cdot 4}^{\text{Multiply numerators}}}{\underbrace{3 \cdot 5}_{\text{Multiply denominators}}} = \frac{8}{15}$

Because 8 and 15 have no **common factors**, our answer is written in **simplest form**.

My video summary

b. Try working this problem on your own. View the **answer**, or watch this **video** for a detailed solution to part b.

You Try It **Work through this You Try It problem.**

Work Exercises 1 and 2 in this textbook or in the MyMathLab Study Plan.

Now let's multiply fractions when **simplifying** is necessary. Consider $\frac{9}{10} \cdot \frac{5}{6}$.

$$\frac{9}{10} \cdot \frac{5}{6} = \frac{\overbrace{9 \cdot 5}^{\text{Multiply numerators}}}{\underbrace{10 \cdot 6}_{\text{Multiply denominators}}} = \frac{45}{60}$$

Because $\frac{45}{60}$ is not in **simplest form**, we need to simplify it. As we did in **Topic 4.2**, we can write the **prime factorizations** of the numerator and denominator and divide out the **common factors**:

Write the prime factorizations: $\frac{45}{60} = \frac{3 \cdot 3 \cdot 5}{2 \cdot 2 \cdot 3 \cdot 5}$

Divide out common factors: $= \frac{\overset{1}{\cancel{3}} \cdot 3 \cdot \overset{1}{\cancel{5}}}{2 \cdot 2 \cdot \underset{1}{\cancel{3}} \cdot \underset{1}{\cancel{5}}}$

Multiply the remaining factors: $= \frac{1 \cdot 3 \cdot 1}{2 \cdot 2 \cdot 1 \cdot 1} = \frac{3}{4}$ ← Simplest form

So, $\frac{9}{10} \cdot \frac{5}{6} = \frac{3}{4}$.

(eText Screens 4.3-1–4.3-39)

Typically, it is quicker to find the **prime factorizations** of the original numerators and denominators instead of multiplying them out. If all of the **common factors** are divided out, then the resulting fraction will be in simplest form. To see this process, let's again consider $\frac{9}{10}\cdot\frac{5}{6}$:

Multiply numerators; multiply denominators: $\frac{9}{10}\cdot\frac{5}{6}=\frac{9\cdot 5}{10\cdot 6}$

Write the prime factorizations of 9, 5, 10, and 6: $=\frac{3\cdot 3\cdot 5}{2\cdot 5\cdot 2\cdot 3}$ Note that 5 is prime.

Divide out common factors: $=\frac{\overset{1}{\cancel{3}}\cdot 3\cdot \overset{1}{\cancel{5}}}{2\cdot \underset{1}{\cancel{5}}\cdot 2\cdot \underset{1}{\cancel{3}}}$

Multiply the remaining factors: $=\frac{1\cdot 3\cdot 1}{2\cdot 1\cdot 2\cdot 1}=\frac{3}{4}$ ← Simplest form

This process will be our usual approach when multiplying fractions.

Example 2 Multiplying Fractions

My interactive video summary

Multiply and **simplify**.

a. $\frac{2}{7}\cdot\frac{3}{4}$ b. $\frac{3}{50}\cdot\frac{10}{21}$ c. $24\cdot\frac{5}{8}$ d. $\frac{7}{15}\cdot\frac{55}{14}$

Solutions

a. Write down the original problem: $\frac{2}{7}\cdot\frac{3}{4}$

Multiply numerators; multiply denominators: $=\frac{2\cdot 3}{7\cdot 4}$

Find the **prime factorization** of each factor: $=\frac{2\cdot 3}{7\cdot 2\cdot 2}$

Divide out the **common factor** 2: $=\frac{\cancel{2}\cdot 3}{7\cdot \cancel{2}\cdot 2}$

Multiply the remaining factors: $=\frac{3}{7\cdot 2}=\frac{3}{14}$

So, $\frac{2}{7}\cdot\frac{3}{4}=\frac{3}{14}$.

b. Write the original problem: $\frac{3}{50}\cdot\frac{10}{21}$

Multiply numerators; multiply denominators: $=\frac{3\cdot 10}{50\cdot 21}$

Find the **prime factorization** of each factor: $=\frac{3\cdot 2\cdot 5}{2\cdot 5\cdot 5\cdot 3\cdot 7}$

Divide out the **common factors**: $=\frac{\cancel{3}\cdot \cancel{2}\cdot \cancel{5}}{\cancel{2}\cdot \cancel{5}\cdot 5\cdot \cancel{3}\cdot 7}$

Multiply the remaining factors: $=\frac{1}{5\cdot 7}=\frac{1}{35}$

Every factor divides out, so we have 1 in the numerator.

So, $\frac{3}{50}\cdot\frac{10}{21}=\frac{1}{35}$.

(eText Screens 4.3-1–4.3-39)

c. To multiply $24 \cdot \frac{5}{8}$, we think of 24 as a fraction with a denominator of 1.

$$24 \cdot \frac{5}{8} = \frac{24}{1} \cdot \frac{5}{8}$$

Try to finish this problem on your own. View the **answer**, or watch this **interactive video** for complete solutions to all four parts.

d. Try working this problem on your own. View the **answer**, or watch this **interactive video** for complete solutions to all four parts.

TIP When multiplying fractions, if a number in a numerator has a factor in common with a number in a denominator, then this **common factor** can be divided out of both numbers first before multiplying the fractions. This generally results in fewer shown steps. For example, see these **alternative solutions** to Examples 2a and 2b.

You Try It **Work through this You Try It problem.**

Work Exercises 3–10 in this textbook or in the MyMathLab Study Plan.

In the next example, we multiply fractions involving **negative signs.** Before proceeding, you may want to review the rules for **multiplying any two numbers.**

Example 3 Multiplying Fractions Involving Negative Numbers

Multiply and **simplify.**

a. $-\frac{3}{7} \cdot \frac{2}{5}$ b. $\left(-\frac{14}{15}\right)\left(-\frac{12}{35}\right)$

Solutions

a. Recall that the **product** of a **negative number** and a **positive number** is negative, so our product will be negative.

$$-\frac{3}{7} \cdot \frac{2}{5} = -\frac{\overbrace{3 \cdot 2}^{\text{Multiply numerators}}}{\underbrace{7 \cdot 5}_{\text{Multiply denominators}}} = -\frac{6}{35}$$

My video summary

b. Recall that the product of two negative numbers is positive. Try working this problem on your own. View the **answer**, or watch this **video** for a detailed solution to part b.

You Try It **Work through this You Try It problem.**

Work Exercises 11–16 in this textbook or in the MyMathLab Study Plan.

As before, if a **variable** appears in the **denominator**, we will assume that it does not cause any denominators to equal 0.

Example 4 Multiplying Fractions Involving Variables

Multiply and **simplify**. Assume variables do not cause any denominators to equal 0.

a. $\frac{x}{y} \cdot \frac{y^2}{x^3}$ **b.** $\frac{6m}{4n} \cdot \frac{5}{2m}$

Solutions

a. Write down the original problem: $\frac{x}{y} \cdot \frac{y^2}{x^3}$

Multiply numerators; multiply denominators: $= \frac{x \cdot y^2}{y \cdot x^3}$

Write y^2 as $y \cdot y$ and x^3 as $x \cdot x \cdot x$: $= \frac{x \cdot y \cdot y}{y \cdot x \cdot x \cdot x}$

Divide out the **common factors**: $= \frac{\cancel{x} \cdot \cancel{y} \cdot y}{\cancel{y} \cdot \cancel{x} \cdot x \cdot x}$

Multiply the remaining factors: $= \frac{y}{x \cdot x} = \frac{y}{x^2}$

So, $\frac{x}{y} \cdot \frac{y^2}{x^3} = \frac{y}{x^2}$.

My video summary

b. Try working this problem on your own. View the **answer**, or watch this **video** for a complete solution to part b.

You Try It **Work through this You Try It problem.**

Work Exercises 17–20 in this textbook or in the MyMathLab Study Plan.

OBJECTIVE 2 EVALUATE EXPONENTIAL EXPRESSIONS INVOLVING FRACTION BASES

In **Topic 2.4,** we evaluated **exponential expressions** with **integer** bases. The **base** of an exponential expression can also be a fraction.

Exponent

$$\left(\frac{1}{2}\right)^3 = \underbrace{\frac{1}{2} \cdot \frac{1}{2} \cdot \frac{1}{2}}_{3 \text{ factors of } \frac{1}{2}} = \frac{1 \cdot 1 \cdot 1}{2 \cdot 2 \cdot 2} = \frac{1}{8}$$

Base

Example 5 Evaluating Exponential Expressions Involving Fraction Bases

My video summary

Evaluate each exponential expression.

a. $\left(\frac{2}{3}\right)^4$ **b.** $\left(-\frac{5}{8}\right)^2$

Solutions

a. $\left(\frac{2}{3}\right)^4 = \frac{2}{3} \cdot \frac{2}{3} \cdot \frac{2}{3} \cdot \frac{2}{3} = \frac{2 \cdot 2 \cdot 2 \cdot 2}{3 \cdot 3 \cdot 3 \cdot 3} = \frac{16}{81}$

b. Try to evaluate this expression on your own. View the **answer**, or watch this **video** for complete solutions to both parts.

You Try It **Work through this You Try It problem.**

Work Exercises 21–26 in this textbook or in the MyMathLab Study Plan.

OBJECTIVE 3 FIND RECIPROCALS

The number 0 is called the **additive identity** because if 0 is added to any number, the ***sum*** is the same identical number. For example, $0 + 6 = 6$. Similarly, the number 1 is called the **multiplicative identity** because if 1 is multiplied by any number, the ***product*** is the same identical number. For example, $1 \cdot 8 = 8$.

In **Topic 2.2**, we learned that the **sum** of two **opposites** is 0, and we referred to the opposites as *additive inverses* because their sum is the additive identity. For example, the additive inverse of -3 is 3 because $-3 + 3 = 0$, the additive identity.

In the same way, two numbers are *multiplicative inverses* if their product is the multiplicative identity, 1. Another name for multiplicative inverse is *reciprocal*.

Definition **Reciprocal or Multiplicative Inverse**

Two **nonzero** numbers are **reciprocals**, or **multiplicative inverses**, if their product is 1.

For example, the reciprocal of $\frac{2}{3}$ is $\frac{3}{2}$ because $\frac{2}{3} \cdot \frac{3}{2} = \frac{2 \cdot 3}{3 \cdot 2} = \frac{\overset{1}{\cancel{2}} \cdot \overset{1}{\cancel{3}}}{\underset{1}{\cancel{3}} \cdot \underset{1}{\cancel{2}}} = \frac{1 \cdot 1}{1 \cdot 1} = \frac{1}{1} = 1$.

Similarly, the reciprocal of 5 is $\frac{1}{5}$ because $5 \cdot \frac{1}{5} = 1$, and the reciprocal of $-\frac{7}{6}$ is $-\frac{6}{7}$ because $\left(-\frac{7}{6}\right)\left(-\frac{6}{7}\right) = 1$.

We can find the **reciprocal** of a **nonzero** fraction by switching its **numerator** and **denominator**.

Finding Reciprocals

If a and b are numbers not equal to 0, then the reciprocal of the fraction $\frac{a}{b}$ is $\frac{b}{a}$.

Once again, we will assume any **variable** appearing in a denominator does not cause the denominator to equal 0.

Example 6 Finding Reciprocals

Find the reciprocal of each number or expression.

a. $\frac{5}{8}$ **b.** $-\frac{11}{6}$ **c.** 12 **d.** $\frac{3x}{y^2}$

(eText Screens 4.3-1–4.3-39)

Solutions

My video summary

a. To find the reciprocal, switch the numerator 5 and denominator 8. The reciprocal of $\frac{5}{8}$ is $\frac{8}{5}$.

b. The reciprocal of a negative number is also negative. Switch the numerator 11 with the denominator 6. The reciprocal of $-\frac{11}{6}$ is $-\frac{6}{11}$.

c. Think of 12 as a fraction with a **denominator** of 1:

$$12 = \frac{12}{1}.$$

Try to finish finding the **reciprocal** on your own. View the **answer**, or watch this **video** for detailed solutions to all four parts.

d. Try finding this reciprocal on your own. View the **answer**, or watch this **video** for detailed solutions to all four parts.

You Try It **Work through this You Try It problem.**

Work Exercises 27–30 in this textbook or in the MyMathLab Study Plan.

CAUTION The number 0 does not have a reciprocal. Do you see **why**?

OBJECTIVE 4 DIVIDE FRACTIONS

In **Topic 1.5**, we saw that a division problem such as $12 \div 4$ means, "How many groups of size 4 are in 12?" Because there are 3 groups of 4 in 12, we know that $12 \div 4 = 3$. See Figure 7.

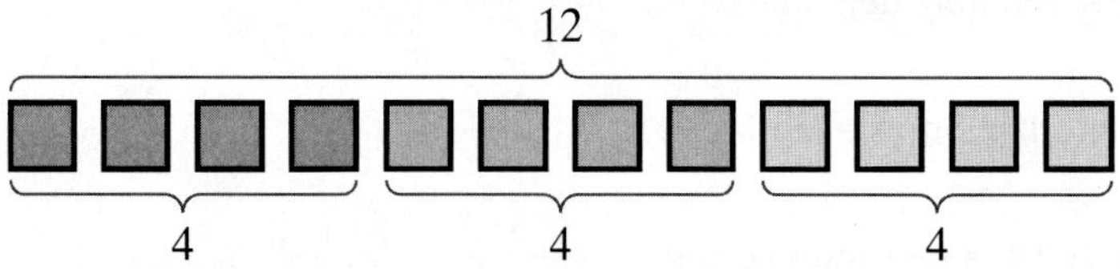

There are 3 groups of 4 in 12, so $12 \div 4 = 3$.

Figure 7

When we divide fractions, we can think in the same way. For example, $\frac{2}{3} \div \frac{1}{6}$ means, "How many groups of size $\frac{1}{6}$ are in $\frac{2}{3}$?" Figure 8 illustrates the answer.

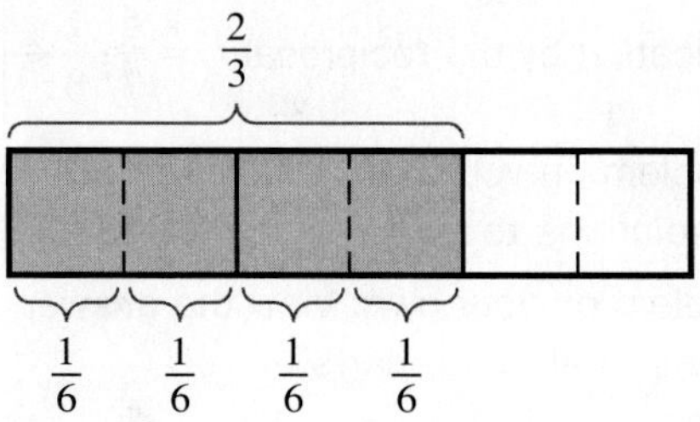

There are 4 groups of $\frac{1}{6}$ in $\frac{2}{3}$, so $\frac{2}{3} \div \frac{1}{6} = 4$.

Figure 8

(eText Screens 4.3-1–4.3-39)

To divide two fractions, we can **multiply** the first fraction by the **reciprocal** of the second fraction.

Dividing Fractions

If $a, b, c,$ and d are numbers such that b, c, and d are not equal to 0, then

$$\frac{a}{b} \div \frac{c}{d} = \frac{a}{b} \cdot \frac{d}{c} = \frac{a \cdot d}{b \cdot c}.$$

Multiply by the reciprocal

Multiply by reciprocal

For example, $\frac{2}{3} \div \frac{1}{6} = \frac{2}{3} \cdot \frac{6}{1} = \frac{2 \cdot 6}{3 \cdot 1} = \frac{2 \cdot \overset{2}{\cancel{6}}}{\underset{1}{\cancel{3}} \cdot 1} = \frac{2 \cdot 2}{1 \cdot 1} = \frac{4}{1} = 4.$

Example 7 Dividing Fractions

My interactive video summary

Divide and **simplify**.

a. $\frac{5}{6} \div \frac{3}{2}$ b. $\frac{6}{7} \div 8$ c. $\frac{9}{4} \div \frac{20}{12}$

Solutions

a. Write the original problem: $\frac{5}{6} \div \frac{3}{2}$

Change to multiplication by the **reciprocal**: $= \frac{5}{6} \cdot \frac{2}{3}$ ← The reciprocal of $\frac{3}{2}$ is $\frac{2}{3}$.

Multiply numerators; multiply denominators: $= \frac{5 \cdot 2}{6 \cdot 3}$

Find the **prime factorization** of each factor: $= \frac{5 \cdot 2}{2 \cdot 3 \cdot 3}$

Divide out the **common factors**: $= \frac{5 \cdot \cancel{2}}{\cancel{2} \cdot 3 \cdot 3}$

Multiply the remaining factors: $= \frac{5}{3 \cdot 3} = \frac{5}{9}$

b. Write the original problem: $\frac{6}{7} \div 8$

Change to multiplication by the reciprocal: $= \frac{6}{7} \cdot \frac{1}{8}$ ← The reciprocal of 8 is $\frac{1}{8}$.

Try to finish this problem on your own. Check the **answer**, or watch this **interactive video** for complete solutions to all three parts.

c. Try working this problem on your own. View the **answer**, or watch this **interactive video** for complete solutions to all three parts.

You Try It **Work through this You Try It problem.**

Work Exercises 31–34 in this textbook or in the MyMathLab Study Plan.

CAUTION When **dividing fractions**, it is incorrect to divide out **common factors** before the problem has been changed to multiplication.

Incorrect	Correct
~~$\frac{5}{6} \div \frac{3}{2} = \frac{5}{\cancel{6}_{2}} \div \frac{\cancel{3}^{1}}{2}$~~	$\frac{5}{6} \div \frac{3}{2} = \frac{5}{6} \cdot \frac{2}{3} = \frac{5}{\cancel{6}_{3}} \cdot \frac{\cancel{2}^{1}}{3}$

Example 8 Dividing Fractions Involving Negatives

Divide and **simplify**.

a. $9 \div \left(-\frac{3}{8}\right)$ **b.** $\left(-\frac{8}{21}\right) \div \left(-\frac{6}{14}\right)$

Solutions

a. Recall that, when dividing two numbers with different **signs**, the **quotient** is **negative**. So, our quotient will be negative.

Write the original problem: $9 \div \left(-\frac{3}{8}\right)$

Change to multiplication by the **reciprocal**: $= \frac{9}{1} \cdot \left(-\frac{8}{3}\right)$ ← The reciprocal of $-\frac{3}{8}$ is $-\frac{8}{3}$.

Multiply numerators; multiply denominators: $= -\frac{9 \cdot 8}{1 \cdot 3}$

Divide out **common factors**: $= -\frac{\cancel{9}^{3} \cdot 8}{1 \cdot \cancel{3}_{1}}$

Multiply the remaining factors: $= -\frac{3 \cdot 8}{1 \cdot 1} = -\frac{24}{1} = -24$

My video summary

b. Recall that, when dividing a negative number by a negative number, the quotient will be positive. Try dividing these fractions on your own. Check the **answer**, or watch this **video** for a detailed solution to part b.

You Try It **Work through this You Try It problem.**

Work Exercises 35–38 in this textbook or in the MyMathLab Study Plan.

Example 9 Dividing Fractions Involving Variables

Divide and **simplify**. Assume **variables** do not cause any denominators to equal 0.

a. $\frac{4x}{y^2} \div \frac{18x^3}{y^3}$ **b.** $\frac{25t}{6r} \div (15rt)$

Solutions

a. Write down the original problem: $\frac{4x}{y^2} \div \frac{18x^3}{y^3}$

Change to multiplication by the **reciprocal**: $= \frac{4x}{y^2} \cdot \frac{y^3}{18x^3}$ ← The reciprocal of $\frac{18x^3}{y^3}$ is $\frac{y^3}{18x^3}$.

Multiply numerators; multiply denominators: $= \dfrac{4x \cdot y^3}{y^2 \cdot 18x^3}$

Write the **prime factorizations:** $= \dfrac{2 \cdot 2 \cdot x \cdot y \cdot y \cdot y}{y \cdot y \cdot 2 \cdot 3 \cdot 3 \cdot x \cdot x \cdot x}$

Divide out the **common factors:** $= \dfrac{\cancel{2} \cdot 2 \cdot \cancel{x} \cdot \cancel{y} \cdot \cancel{y} \cdot y}{\cancel{y} \cdot \cancel{y} \cdot \cancel{2} \cdot 3 \cdot 3 \cdot \cancel{x} \cdot x \cdot x}$

Multiply the remaining factors: $= \dfrac{2 \cdot y}{3 \cdot 3 \cdot x \cdot x} = \dfrac{2y}{9x^2}$

So, $\dfrac{4x}{y^2} \div \dfrac{18x^3}{y^3} = \dfrac{2y}{9x^2}$.

My video summary

b. Think of $15rt$ with a **denominator** of 1: $15rt = \dfrac{15rt}{1}$.

Change the division to multiplication by the reciprocal: $\dfrac{25t}{6r} \div (15rt) = \dfrac{25t}{6r} \div \dfrac{15rt}{1} = \dfrac{25t}{6r} \cdot \dfrac{1}{15rt}$ ← The reciprocal of $\dfrac{15rt}{1}$ is $\dfrac{1}{15rt}$.

Try to finish this problem on your own. View the **answer**, or watch this **video** for a complete solution to part b.

You Try It **Work through this You Try It problem.**

Work Exercises 39–44 in this textbook or in the MyMathLab Study Plan.

OBJECTIVE 5 SIMPLIFY EXPRESSIONS BY MULTIPLYING AND DIVIDING FRACTIONS

If an **expression** contains both multiplication and division of fractions, we change each division to multiplication by the **reciprocal** and find the **product**.

Example 10 Multiplying and Dividing Fractions

Multiply and **divide** to **simplify** each expression. Assume **variables** do not cause any denominators to equal 0.

a. $\dfrac{5}{9} \div \dfrac{7}{12} \cdot \dfrac{3}{10}$ **b.** $\left(-\dfrac{35p}{6q}\right)\left(-\dfrac{q^2}{55p}\right) \div \left(-\dfrac{14pq}{11}\right)$

Solutions

a. In this problem, we have both division and multiplication. We change the division to multiplication by the reciprocal and then simplify.

Write down the original problem: $\dfrac{5}{9} \div \dfrac{7}{12} \cdot \dfrac{3}{10}$

Change to multiplication by the reciprocal: $= \dfrac{5}{9} \cdot \dfrac{12}{7} \cdot \dfrac{3}{10}$ ← The reciprocal of $\dfrac{7}{12}$ is $\dfrac{12}{7}$.

Multiply numerators; multiply denominators: $= \dfrac{5 \cdot 12 \cdot 3}{9 \cdot 7 \cdot 10}$

Write the **prime factorizations:** $= \dfrac{5 \cdot 2 \cdot 2 \cdot 3 \cdot 3}{3 \cdot 3 \cdot 7 \cdot 2 \cdot 5}$

Divide out the **common factors:** $= \dfrac{\cancel{5}\cdot\cancel{2}\cdot 2\cdot\cancel{3}\cdot\cancel{3}}{\cancel{3}\cdot\cancel{3}\cdot 7\cdot\cancel{2}\cdot\cancel{5}}$

Multiply the remaining factors: $= \dfrac{2}{7}$

So, $\dfrac{5}{9} \div \dfrac{7}{12}\cdot\dfrac{3}{10} = \dfrac{2}{7}$.

My video summary

b. The reciprocal of $-\dfrac{14pq}{11}$ is $-\dfrac{11}{14pq}$. Try working this problem on your own. View the **answer**, or watch this **video** for a complete solution to part b.

You Try It **Work through this You Try It problem.**

Work Exercises 45–50 in this textbook or in the MyMathLab Study Plan.

OBJECTIVE 6 SOLVE APPLICATIONS BY MULTIPLYING OR DIVIDING FRACTIONS

When working with fractions, we have seen that the keyword *of* means to multiply. Typically, when an application requires us to find a part of something, we will multiply by a fraction.

Example 11 A Scholarship

Haley received a "five-eighths scholarship," which pays $\dfrac{5}{8}$ of the tuition and fees to her local community college. If tuition and fees total $2632, what dollar amount will Haley's scholarship cover? How much additional money will she need in order to pay the rest of the tuition and fees?

Solution Haley's scholarship covers $\dfrac{5}{8}$ of the \$2632 in tuition and fees. To determine the dollar amount covered by the scholarship, we multiply:

$$\frac{5}{8}\text{ of } \$2632 \quad\text{means}\quad \frac{5}{8}\cdot 2632.$$

$$\frac{5}{8}\cdot 2632 = \frac{5\cdot 2632}{8} = \frac{5\cdot \overset{329}{\cancel{2632}}}{\underset{1}{\cancel{8}}} = 1645$$

So Haley's scholarship covers \$1645 of the tuition.

Haley will need \$2632 − \$1645 = \$987 of additional money to pay the remaining tuition and fees.

Example 12 Breaking a Filibuster

My video summary

A *filibuster* is an action, such as long speechmaking, that is sometimes used in the U.S. Senate to prevent votes. At least $\dfrac{3}{5}$ of the votes in the Senate are needed to break (or stop) a filibuster. If all 100 senators vote, how many votes are needed to stop a filibuster?

Solution We need to find how many votes are needed to break a filibuster, which is $\dfrac{3}{5}$ of 100 votes. Try working this problem on your own. View the **answer**, or watch this **video** for a complete solution.

(eText Screens 4.3-1–4.3-39)

You Try It **Work through this You Try It problem.**

Work Exercises 51 and 52 in this textbook or in the MyMathLab Study Plan.

When an application requires us to make groups of equal size, we divide.

Example 13 Grilling Burgers

My video summary

To grill burgers for a holiday gathering, Otis purchased 12 pounds of ground beef to make into patties. If each burger patty is to weigh $\frac{3}{8}$ pound, how many burgers can Otis make?

Solution We need to find how many burgers Otis can make. We divide the total amount of meat he has by the amount of meat needed for each burger.

$$\begin{array}{ccc} \text{Total amount of meat} & \div & \text{Meat needed for each burger} \\ \downarrow & & \downarrow \\ 12 \text{ pounds} & \div & \frac{3}{8} \text{ pound per burger} \end{array}$$

Try to finish this problem on your own. View the **answer**, or watch this **video** for a complete solution.

You Try It **Work through this You Try It problem.**

Work Exercises 53–58 in this textbook or in the MyMathLab Study Plan.

OBJECTIVE 7 FIND THE AREA OF A TRIANGLE

In **Topic 1.4**, we learned how to find the **area of a rectangle**. Now we learn to find the area of a *triangle*. Recall that **area** is the amount of surface contained within the **sides** of a figure with two dimensions, so area is measured in **square units**. To find area, we find the number of square units needed to cover the surface.

Each corner point of a triangle is called a **vertex**. The distance from a vertex to the opposite side, called the **base**, gives the **height** of the triangle. The height must be **perpendicular** to the base, which means it forms a 90-degree angle with the base. Figure 9 shows some examples of triangles with a vertex, base, and height labeled.

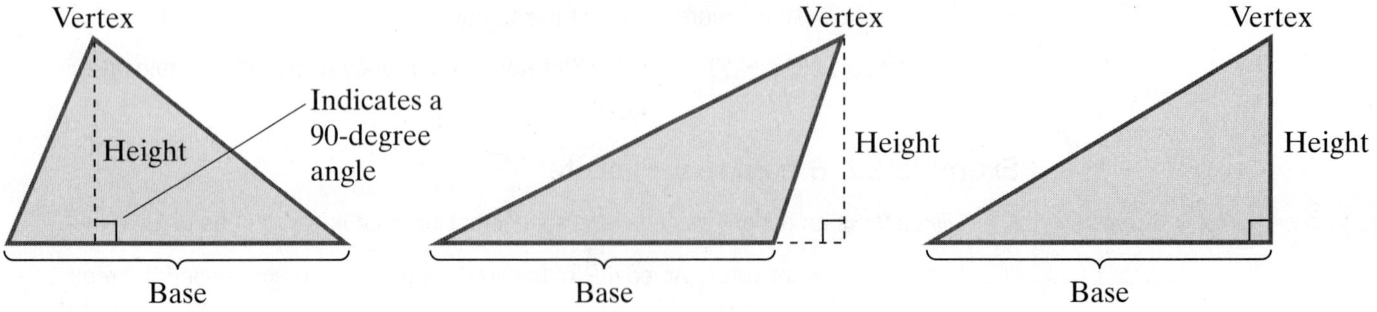

Figure 9

(eText Screens 4.3-1–4.3-39)

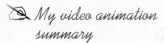

My video animation summary

Watch this **video animation** or read the following discussion to develop a formula for finding the **area** of a triangle.

Figure 10a shows a triangle with base b and height h. If a copy of the triangle is split (Figure 10b) and the parts rotated, a rectangle can be formed (Figure 10c) that has twice the area of the original triangle.

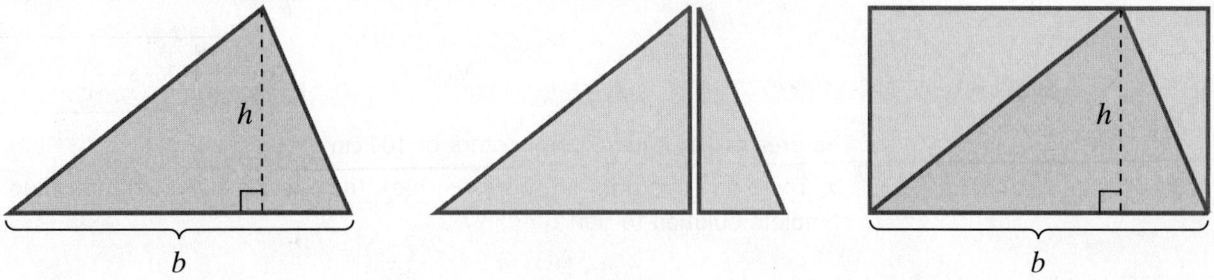

(a) (b) (c)

Figure 10

Because the **area of the rectangle** in Figure 10c is $b \cdot h$, the triangle in Figure 10a has half that area, or $\frac{1}{2} \cdot b \cdot h$.

Area of a Triangle

If a triangle has base b and height h, then its area A is given by the formula

$$A = \frac{1}{2}(\text{base})(\text{height}) \quad \text{or} \quad A = \frac{1}{2}bh.$$

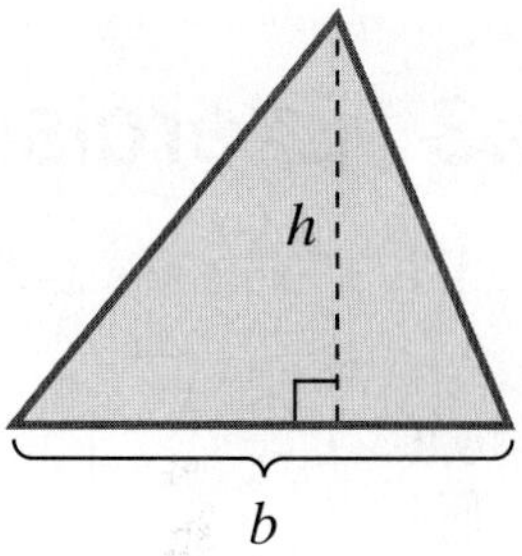

Example 14 Finding the Area of a Triangle

Find the **area of each triangle**.

a.

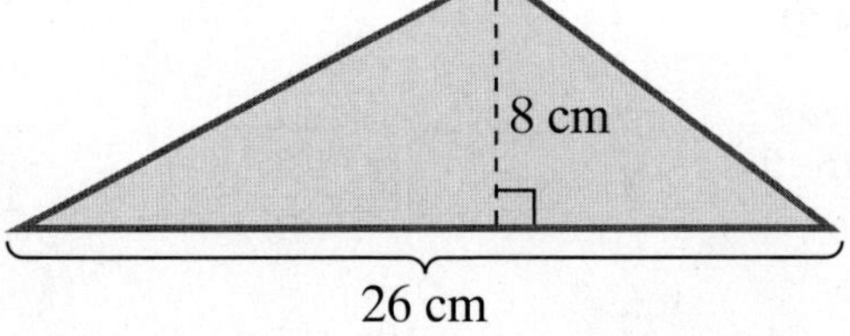

b.

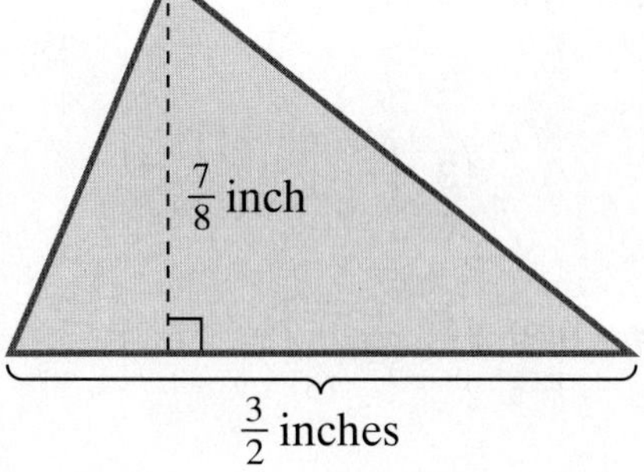

(eText Screens 4.3-1–4.3-39)

Solutions

a. The base of this triangle is 26 centimeters, and the height is 8 centimeters.

Write the area formula for triangles: $A = \frac{1}{2}bh$

Substitute 26 for b and 8 for h: $= \frac{1}{2}\cdot 26\cdot 8$

Multiply: $= 104 \leftarrow$ $\boxed{\frac{1}{\cancel{2}_1}\cdot \overset{13}{\cancel{26}}\cdot 8 = 1\cdot 13\cdot 8 = 104}$

The area is 104 square centimeters or 104 cm^2.

My video summary

b. Try to find this area on your own. View the **answer**, or watch this **video** for a complete solution to part b.

CAUTION Remember that **area** is measured in *square units* regardless of the shape of the figure.

You Try It **Work through this You Try It problem.**

Work Exercises 59–62 in this textbook or in the MyMathLab Study Plan.

4.3 Exercises

In Exercises 1–20, multiply and simplify. Assume variables do not cause any denominators to equal 0.

You Try It

1. $\frac{5}{9}\cdot\frac{4}{7}$
2. $\frac{1}{10}\cdot\frac{11}{15}$
3. $\frac{2}{13}\cdot\frac{1}{8}$
4. $\frac{8}{9}\cdot\frac{6}{11}$

You Try It

5. $\frac{15}{16}\cdot\frac{4}{5}$
6. $\frac{6}{15}\cdot\frac{5}{16}$
7. $\frac{42}{9}\cdot\frac{3}{7}$
8. $\frac{1}{6}\cdot 78$

You Try It

9. $21\cdot\frac{2}{3}$
10. $\frac{70}{45}\cdot\frac{15}{21}$
11. $-\frac{5}{6}\cdot\frac{1}{2}$
12. $\frac{3}{12}\cdot\left(-\frac{8}{9}\right)$
13. $\left(-\frac{7}{10}\right)\left(-\frac{15}{14}\right)$
14. $-\frac{16}{6}\cdot\frac{9}{8}$
15. $-48\left(-\frac{5}{16}\right)$
16. $\frac{10}{39}\left(-\frac{13}{40}\right)$

You Try It

17. $\frac{x}{y^2}\cdot\frac{y^3}{x^3}$
18. $\frac{mn}{32}\cdot\frac{24m}{n}$
19. $-\frac{10p}{21q}\cdot\frac{14q^2}{15p}$
20. $\left(-\frac{20xy}{26y^2}\right)\left(-\frac{13xy}{55x^2y}\right)$

In Exercises 21–26, evaluate each exponential expression.

You Try It

21. $\left(\frac{4}{7}\right)^2$
22. $\left(\frac{2}{5}\right)^3$
23. $\left(-\frac{8}{9}\right)^2$
24. $\left(-\frac{1}{4}\right)^3$
25. $\left(\frac{5}{3}\right)^2$
26. $\left(\frac{3}{10}\right)^4$

(eText Screens 4.3-1–4.3-39)

In Exercises 27–30, find the reciprocal of each number or expression. Assume variables do not cause any denominators to equal 0.

You Try It 27. $\frac{7}{4}$ 28. $-\frac{3}{16}$ 29. 10 30. $\frac{4y}{5x^2}$

In Exercises 31–44, divide and simplify. Assume variables do not cause any denominators to equal 0.

You Try It 31. $\frac{5}{12} \div \frac{4}{3}$ 32. $\frac{3}{8} \div 18$ 33. $\frac{15}{6} \div \frac{35}{4}$ 34. $40 \div \frac{5}{4}$

You Try It 35. $28 \div \left(-\frac{7}{3}\right)$ 36. $\left(-\frac{48}{81}\right) \div \left(-\frac{32}{72}\right)$ 37. $-\frac{13}{3} \div \frac{39}{6}$ 38. $\frac{12}{20} \div \frac{9}{15}$

You Try It 39. $\frac{5}{7} \div \frac{x}{3}$ 40. $\left(-\frac{5}{x}\right) \div \left(-\frac{35}{x}\right)$ 41. $\frac{x^2y}{w} \div \frac{xy^2}{w^3}$ 42. $-\frac{9}{10} \div \frac{25m}{9}$

43. $\frac{16p}{15q} \div (-20pq)$ 44. $\frac{3x}{70xy^2} \div \frac{1}{63y}$

In Exercises 45–50, multiply and divide to simplify each expression. Assume variables do not cause any denominators to equal 0.

45. $\frac{4}{7} \div \frac{6}{5} \cdot \frac{14}{15}$ 46. $\frac{3}{8} \cdot \frac{12}{3} \div \frac{27}{2}$ 47. $\frac{18}{55} \div \left(-\frac{1}{5}\right) \div \left(-\frac{12}{11}\right)$

You Try It 48. $\frac{5x}{3y^3} \div \frac{105xy}{6y^2} \cdot \frac{4y^2}{7}$ 49. $-10p \div \frac{64pq}{15q} \div \left(-\frac{25q}{8p}\right)$ 50. $\left(-\frac{25}{18m}\right)\left(-\frac{27n}{15m}\right) \div \left(-\frac{n}{10mn}\right)$

In Exercises 51–58, solve each application.

You Try It 51. **Crop Farming** Leo farms 576 acres of land. One year, he planted soybeans on $\frac{5}{16}$ of his land. How many acres did he plant with soybeans?

52. **Overriding a Veto** If the U.S. president vetoes a bill, at least $\frac{2}{3}$ of members voting in the U.S. House of Representatives must vote to override the veto for it to become law anyway. If all 435 members of the House vote, how many votes are required to override a veto? (Note: A two-thirds vote is also required by the U.S. Senate.)

You Try It 53. **Cereal Servings** According to the nutrition facts on a box of Cinnamon Toast Crunch®, the serving size for the cereal is $\frac{3}{4}$ cup. How many servings of the cereal can be placed in a container that holds 24 cups?

54. **Machine Shop** With each turn, a nut moves $\frac{3}{32}$ inch on a bolt. How many turns will it take for the nut to move $\frac{3}{4}$ inch?

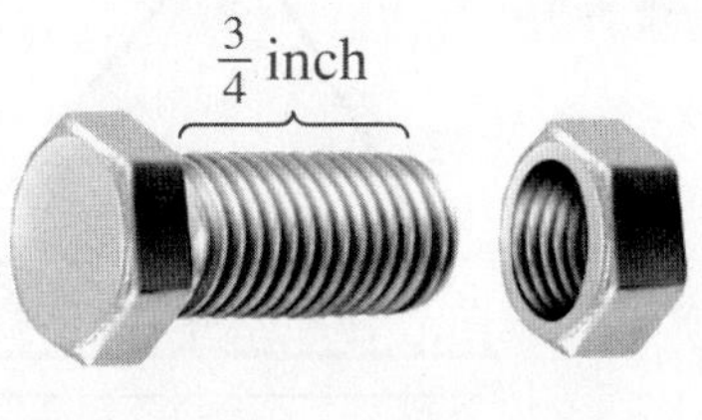

55. **Filling a Prescription** A small child has been prescribed an antibiotic as follows: $\frac{1}{2}$ teaspoon to be given 4 times a day for 10 days. How many teaspoons of the antibiotic are needed to fill the prescription?

56. **Measuring Thickness** If a stack of 75 business cards is $\frac{5}{8}$ inch thick, how thick is one business card?

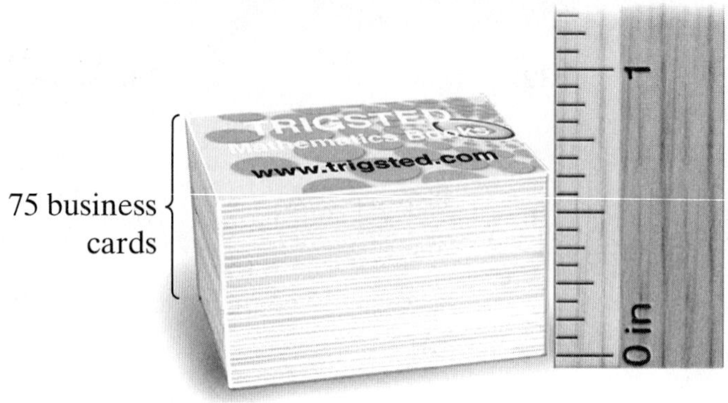

57. **Area** Find the area of the rectangle shown.

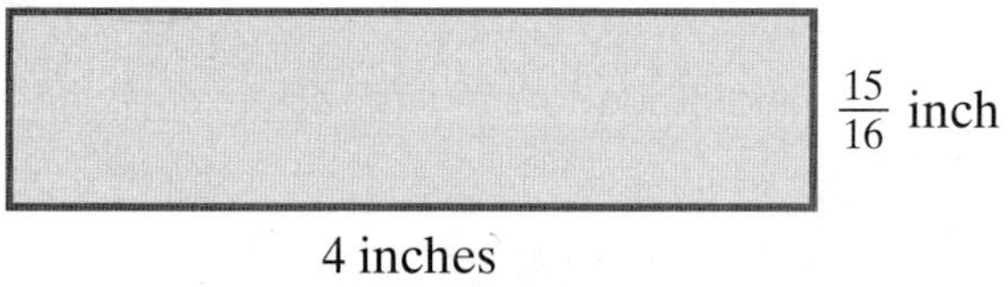

58. **Solar Panel** The solar panel on the top of a garden lamp is the shape of a square measuring $\frac{11}{16}$ inch on each side. What is the area of the solar panel?

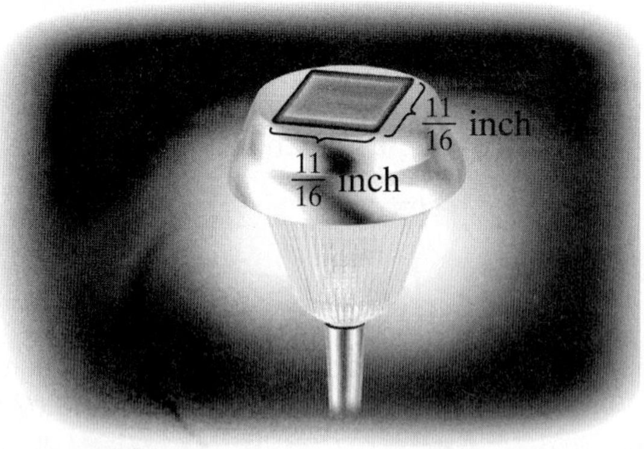

In Exercises 59–62, find the area of each triangle.

You Try It

59.

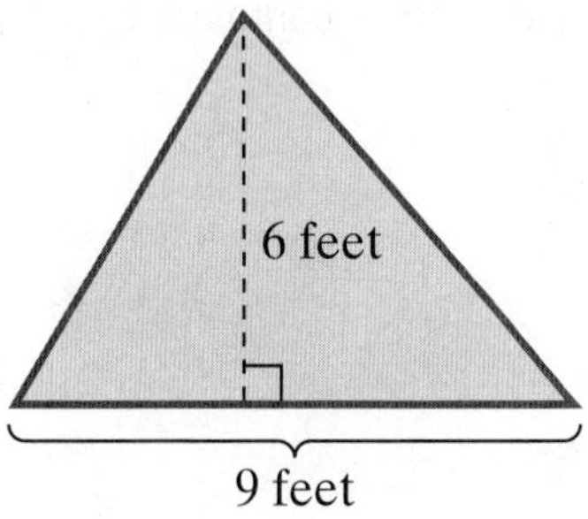

60.

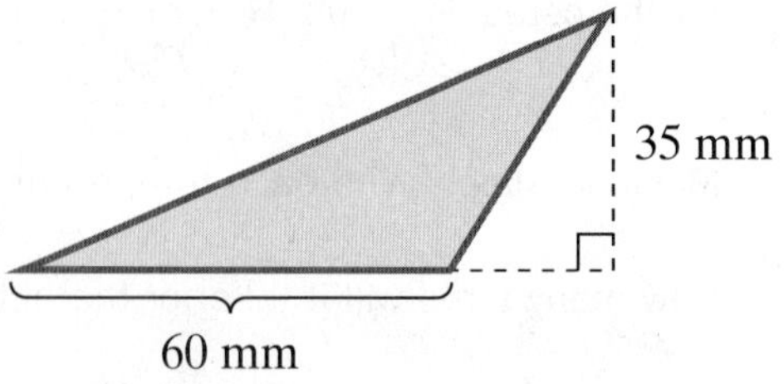

(eText Screens 4.4-1–4.4-33)

61.

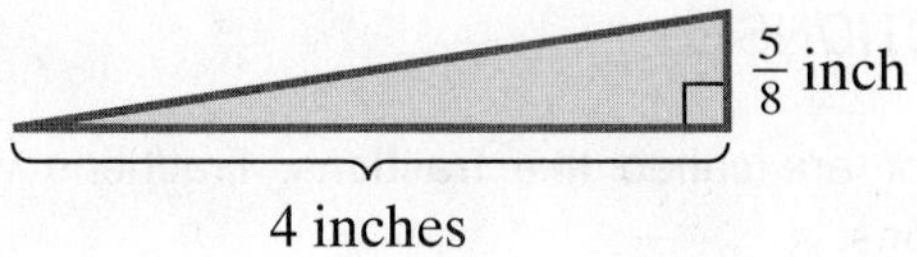

62.

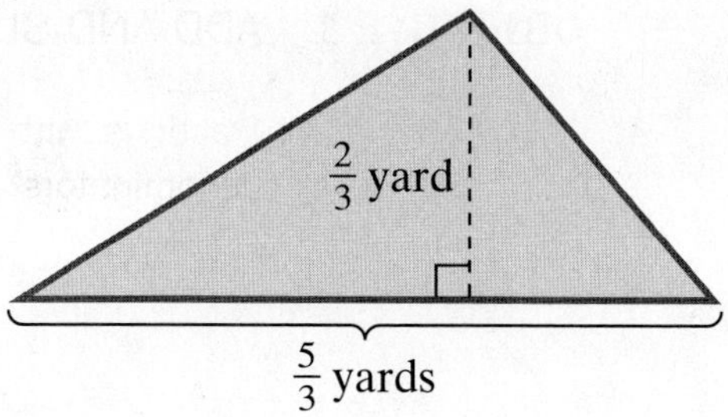

4.4 Adding and Subtracting Fractions

THINGS TO KNOW

Before working through this topic, be sure you are familiar with the following concepts:

		VIDEO	ANIMATION	INTERACTIVE
You Try It	1. Multiply Two Integers (Topic 2.4, **Objective 1**)	✓	✓	
You Try It	2. Divide Integers (Topic 2.4, **Objective 4**)	✓		
You Try It	3. Combine Like Terms (Topic 3.1, **Objective 2**)	✓		✓
You Try It	4. Write Improper Fractions as Mixed Numbers or Whole Numbers (Topic 4.1, **Objective 4**)	✓		
You Try It	5. Find the Prime Factorization of a Number (Topic 4.2, **Objective 2**)	✓		
You Try It	6. Write Equivalent Fractions (Topic 4.2, **Objective 5**)	✓		
You Try It	7. Write Fractions in Simplest Form (Topic 4.2, **Objective 6**)	✓		

OBJECTIVES

1 Add and Subtract Like Fractions

2 Find the Least Common Denominator for Unlike Fractions

3 Add and Subtract Unlike Fractions

4 Add and Subtract Fractions with Variables

5 Solve Applications by Adding or Subtracting Fractions

(eText Screens 4.4-1–4.4-33)

OBJECTIVE 1 ADD AND SUBTRACT LIKE FRACTIONS

Fractions with the same denominator are called **like fractions**. Fractions with different denominators are called **unlike fractions**.

Like Fractions	Unlike Fractions
$\frac{2}{3}, \frac{7}{3}$	$\frac{4}{5}, \frac{8}{9}$
Same denominator	Different denominators

Just as we could combine only **like terms** in **Topic 3.1**, we can add only like fractions. View this **popup** to help see why this is the case.

The following illustration shows how we can add two like fractions.

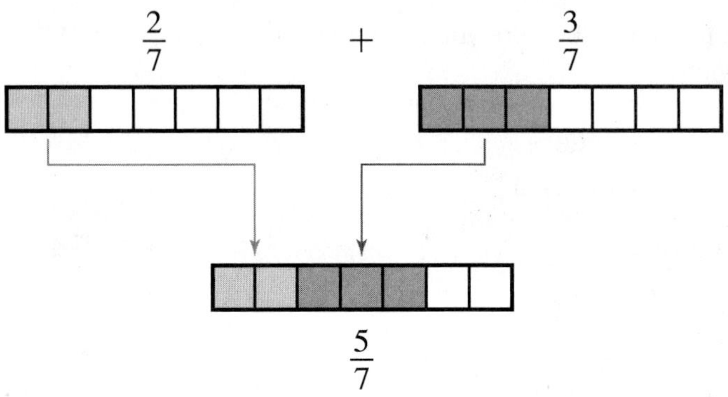

Figure 11

From Figure 11 we see that $\frac{2}{7} + \frac{3}{7} = \frac{5}{7}$. The **numerator** of the **sum** is the sum of the two numerators, $2 + 3 = 5$. The **denominator** of the sum is the common denominator, 7. This result is true in general.

> **Adding Like Fractions**
>
> If a, b, and c are numbers, and $c \neq 0$, then
>
> $$\frac{a}{c} + \frac{b}{c} = \frac{a+b}{c}.$$

So, to add fractions with the same, or common, denominator, we add the numerators and keep the common denominator.

Example 1 Adding Like Fractions

Add.

a. $\frac{3}{11} + \frac{5}{11}$ **b.** $\frac{4}{5} + \frac{3}{5}$

Solutions

a. The fractions have the same **denominator**, 11. Add the **numerators** and keep the common denominator.

$$\frac{3}{11} + \frac{5}{11} = \frac{3+5}{11} = \frac{8}{11} \begin{array}{l} \leftarrow \text{Add the numerators} \\ \leftarrow \text{Keep the common denominator} \end{array}$$

b. The fractions have the same denominator, 5. Add the numerators and keep the common denominator.

$$\frac{4}{5} + \frac{3}{5} = \frac{4+3}{5} = \frac{7}{5} \begin{array}{l}\leftarrow \text{Add the numerators}\\ \leftarrow \text{Keep the common denominator}\end{array}$$

You Try It **Work through this You Try It problem.**

Work Exercises 1–4 in this textbook or in the MyMathLab Study Plan.

Recall from **Topic 4.2** that when a fraction is negative, the **negative sign** can go in either the numerator or denominator. That is,

$$-\frac{a}{b} = \frac{-a}{b} = \frac{a}{-b}$$

provided $b \neq 0$. We can use this result to add signed fractions.

Example 2 Adding Like Fractions Involving Negative Signs

My video summary

Add.

a. $-\frac{3}{7} + \frac{4}{7}$ **b.** $-\frac{4}{9} + \left(-\frac{13}{9}\right)$ **c.** $-\frac{10}{3} + \frac{8}{3}$

Solutions

a. $-\frac{3}{7} + \frac{4}{7} = \frac{-3}{7} + \frac{4}{7}$ ← Write $-\frac{3}{7}$ as $\frac{-3}{7}$

$= \frac{-3+4}{7}$ ← Add the numerators; ← Keep the common denominator

$= \frac{1}{7}$

b. $-\frac{4}{9} + \left(-\frac{13}{9}\right) = \frac{-4}{9} + \frac{(-13)}{9}$ ← Write $-\frac{4}{9}$ as $\frac{-4}{9}$ and $-\frac{13}{9}$ as $\frac{-13}{9}$

Finish adding these fractions. Check your **answer**, or watch this **video** for complete solutions to all three parts.

c. Try adding these fractions on your own. Check your **answer**, or watch this **video** for complete solutions to all three parts.

You Try It **Work through this You Try It problem.**

Work Exercises 5–8 in this textbook or in the MyMathLab Study Plan.

CAUTION Remember to write your results in **simplest form**.

Example 3 Adding Like Fractions

Add. Simplify if necessary.

a. $-\frac{5}{6} + \frac{13}{6}$ **b.** $-\frac{5}{8} + \left(-\frac{11}{8}\right)$ **c.** $\frac{3}{20} + \frac{9}{20}$

(eText Screens 4.4-1–4.4-33)

My video summary

Solutions Try working these problems on your own. Remember to write your answers in **lowest terms**. Check your **answers**, or watch this **video** for complete solutions to all three parts.

You Try It **Work through this You Try It problem.**

Work Exercises 9–12 in this textbook or in the MyMathLab Study Plan.

To subtract **like fractions**, we follow an approach much like that for addition. The **numerator** of the **difference** is the difference of the two numerators and the **denominator** of the difference is the common denominator. For example,

$$\frac{14}{5} - \frac{8}{5} = \frac{14 - 8}{5} \begin{array}{l}\leftarrow \text{Subtract the numerators}\\ \leftarrow \text{Keep the common denominator}\end{array}$$

$$= \frac{6}{5}$$

This result is true in general.

> **Subtracting Like Fractions**
>
> If a, b, and c are numbers, and $c \neq 0$, then
>
> $$\frac{a}{c} - \frac{b}{c} = \frac{a - b}{c}.$$

So, to subtract fractions with the same, or common, denominator, we subtract the numerators and keep the common denominator.

Example 4 Subtracting Like Fractions

My interactive video summary

Subtract. Simplify if necessary.

a. $\frac{7}{3} - \frac{5}{3}$ **b.** $-\frac{17}{6} - \frac{4}{6}$ **c.** $\frac{6}{7} - \frac{20}{7}$ **d.** $-\frac{3}{14} - \left(-\frac{5}{14}\right)$

Solutions

a. The fractions have the same **denominator**, 3. Subtract the **numerators** and keep the common denominator.

$$\frac{7}{3} - \frac{5}{3} = \frac{7 - 5}{3} = \frac{2}{3} \begin{array}{l}\leftarrow \text{Subtract the numerators}\\ \leftarrow \text{Keep the common denominator}\end{array}$$

b. $-\frac{17}{6} - \frac{4}{6} = \frac{-17}{6} - \frac{4}{6}$ ← Change $-\frac{17}{6}$ to $\frac{-17}{6}$

Subtract the numerators and keep the common denominator. Remember to **simplify** your result. Check your **answer**, or watch this **interactive video** for complete solutions to all four parts.

c.–d. Try subtracting these fractions on your own. Check your **answers**, or watch this **interactive video** for complete solutions to all four parts.

You Try It **Work through this You Try It problem.**

Work Exercises 13–20 in this textbook or in the MyMathLab Study Plan.

OBJECTIVE 2 FIND THE LEAST COMMON DENOMINATOR FOR UNLIKE FRACTIONS

Recall that fractions can be added or subtracted only if they have the same **denominator**. Therefore, when we wish to add or subtract fractions with different denominators, we must first write each one as an **equivalent fraction** with a **common denominator**. Though any common denominator can be used for this purpose (such as multiplying all the denominators), we typically use the *least common denominator*. We do this to keep the numbers as small as possible and to avoid more difficult computations.

> **Definition Least Common Denominator (LCD)**
>
> The **least common denominator (LCD)** of a group of fractions is the smallest number that is **divisible** by all the denominators in the group.

For example, the LCD of $\frac{1}{3}$ and $\frac{1}{4}$ is 12 because 12 is the smallest number that is divisible by both 3 and 4. See the **details**.

The least common denominator is also called the **least common multiple (LCM)** of the denominators. We can find the LCD by listing multiples of each denominator until we find the smallest number that is a **multiple** of all the denominators.

TIP Always check first to see if the largest denominator among the unlike fractions is a multiple of all the other denominators. For example, the LCD of $\frac{1}{3}$ and $\frac{1}{6}$ is 6 because 6 is a multiple of 3.

Example 5 Finding a Least Common Denominator by Inspection

Find the **LCD** of the given fractions by inspection.

a. $\frac{5}{12}$ and $\frac{17}{30}$ **b.** $\frac{3}{5}$ and $\frac{49}{40}$

Solutions

a. We start by checking if 30 is a **multiple** of 12. Because 12 does not **divide evenly** into 30, 30 is not a multiple of 12.

Next, we list multiples of 12 and 30. For multiples of 12, we get $12 \cdot 1 = 12$, $12 \cdot 2 = 24$, $12 \cdot 3 = 36$, and so on. For multiples of 30, we get $30 \cdot 1 = 30$, $30 \cdot 2 = 60$, $30 \cdot 3 = 90$, and so on.

$$\text{Multiples of 12: } 12, 24, 36, 48, 60, 72, 84, \ldots$$
$$\text{Multiples of 30: } 30, 60, 90, 120, \ldots$$

We see that 60 is the **least common multiple** of 12 and 30. Therefore, 60 is the LCD of $\frac{5}{12}$ and $\frac{17}{30}$.

b. Because 40 is a multiple of 5 $(40 = 5 \cdot 8)$, the LCD of $\frac{3}{5}$ and $\frac{49}{40}$ is 40.

You Try It **Work through this You Try It problem.**

Work Exercises 21–26 in this textbook or in the MyMathLab Study Plan.

(eText Screens 4.4-1–4.4-33)

Sometimes finding the LCD by inspection is difficult, especially if the denominators are large. The following steps can help us deal with those situations.

Steps for Finding the LCD of a Group of Fractions by Prime Factorization

Step 1. Write out the **prime factorization** of each denominator in the group.

Step 2. Write down each distinct **prime** factor the greatest number of times it appears in any one factorization.

Step 3. Multiply the **factors** listed in Step 2 to find the LCD.

Example 6 Finding a Least Common Denominator by Prime Factorization

My interactive video summary

Find the **LCD** of the given fractions by prime factorization.

a. $\frac{5}{18}$ and $\frac{7}{24}$ **b.** $\frac{11}{75}$ and $\frac{13}{120}$ **c.** $\frac{5}{6}, \frac{1}{21},$ and $\frac{23}{98}$

Solutions We use the **steps** for finding the LCD.

a. Step 1. Write out the prime factorizations of 18 and 24:

$$18 = 2 \cdot 3 \cdot 3$$
$$24 = 2 \cdot 2 \cdot 2 \cdot 3$$

Step 2. There are two distinct prime factors: 2 and 3. The greatest number of 2's in one factorization is three, so we include three 2's for the **LCD**. The greatest number of 3's in one factorization is two, so we include two 3's in the list for the LCD.

$$\text{LCD} = 2 \cdot 2 \cdot 2 \cdot 3 \cdot 3$$

Step 3. Multiply. $\text{LCD} = 2 \cdot 2 \cdot 2 \cdot 3 \cdot 3 = 72$

So, 72 is the LCD of $\frac{5}{18}$ and $\frac{7}{24}$.

b. Step 1. The **prime factorizations** of 75 and 120 are

$$75 = 3 \cdot 5 \cdot 5$$
$$120 = 2 \cdot 2 \cdot 2 \cdot 3 \cdot 5$$

Step 2. There are three distinct **prime** factors: 2, 3, and 5. The greatest number of 2's in one factorization is three; the greatest number of 3's in one factorization is one; the greatest number of 5's in one factorization is two. Try to finish finding the LCD on your own. View the **answer**, or watch this **interactive video** to view complete solutions for all three parts.

c. Try finding this **LCD** on your own. View the **answer**, or watch this **interactive video** for complete solutions to all three parts.

You Try It **Work through this You Try It problem.**

Work Exercises 27–32 in this textbook or in the MyMathLab Study Plan.

(eText Screens 4.4-1–4.4-33)

OBJECTIVE 3 ADD AND SUBTRACT UNLIKE FRACTIONS

Earlier in this topic we saw that to add or subtract fractions with a **common denominator**, we add or subtract the **numerators** and write the result over the common denominator.

Adding and Subtracting Fractions

If $a, b,$ and c are numbers and $c \neq 0$, then

$$\frac{a}{c} + \frac{b}{c} = \frac{a+b}{c} \quad \text{and} \quad \frac{a}{c} - \frac{b}{c} = \frac{a-b}{c}.$$

My animation summary

To add fractions with unlike denominators, we first write **equivalent fractions** using the least common denominator. Then we add the **like fractions** as before. Watch this **animation** to see an overview of adding and subtracting fractions with unlike denominators, then continue reading.

Suppose we want to add

$$\frac{1}{4} + \frac{1}{3}.$$

The fractions $\frac{1}{4}$ and $\frac{1}{3}$ have an **LCD** of 12. So we start by writing equivalent fractions with a denominator of 12 using the **property of equivalent fractions.**

$$\frac{1}{4} = \frac{1 \cdot 3}{4 \cdot 3} = \frac{3}{12} \quad \text{and} \quad \frac{1}{3} = \frac{1 \cdot 4}{3 \cdot 4} = \frac{4}{12}$$

So, $\frac{1}{4} + \frac{1}{3} = \frac{3}{12} + \frac{4}{12} = \frac{3+4}{12} = \frac{7}{12}$. This is illustrated in the following figure.

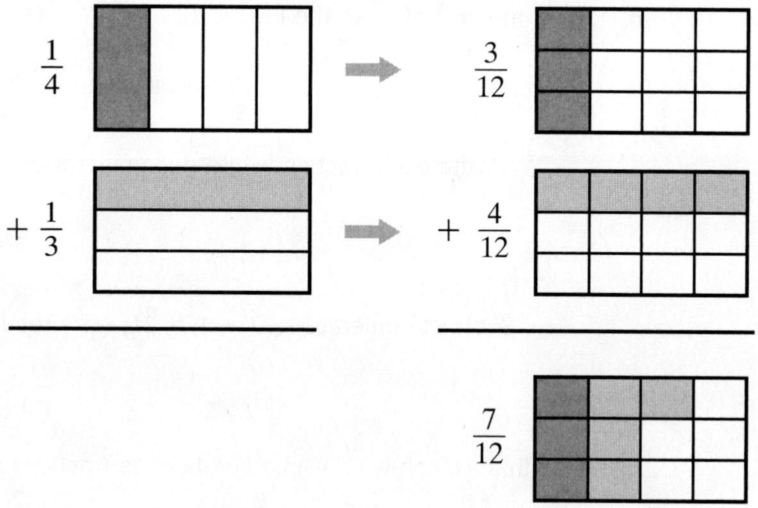

Figure 12

The following steps will help us add and subtract any two fractions.

(eText Screens 4.4-1–4.4-33)

Steps for Adding or Subtracting Fractions

Step 1. If necessary, find the **LCD** of the fractions.

Step 2. Write each fraction as an **equivalent fraction** with the LCD as its denominator.

Step 3. Add or subtract the **numerators** of the **like fractions** and write the result over the LCD.

Step 4. **Simplify** if necessary.

Example 7 Adding and Subtracting Fractions

My interactive video summary

Add or subtract. Simplify if necessary.

a. $\frac{1}{9}+\frac{5}{12}$ **b.** $\frac{11}{6}-\frac{1}{12}$ **c.** $\frac{3}{7}-2$

d. $-\frac{11}{14}+\frac{5}{24}$ **e.** $\frac{7}{48}-\frac{8}{15}$ **f.** $\frac{10}{3}-\frac{7}{4}+\frac{1}{2}$

Solutions

a. The **LCD** is 36 (see the **details**).

Begin with the original problem: $\frac{1}{9}+\frac{5}{12}$

Write each fraction with a denominator of LCD $=36$: $=\frac{1\cdot 4}{9\cdot 4}+\frac{5\cdot 3}{12\cdot 3}$

$=\frac{4}{36}+\frac{15}{36}$

Add numerators $4+15=19$; write result over 36: $=\frac{19}{36}$

b. 12 is a **multiple** of 6, so the LCD is 12.

Begin with the original problem: $\frac{11}{6}-\frac{1}{12}$

Write each fraction with a denominator of LCD $=12$: $=\frac{11\cdot 2}{6\cdot 2}-\frac{1}{12}$

$=\frac{22}{12}-\frac{1}{12}$

Subtract numerators $22-1=21$; write result over 12: $=\frac{21}{12}$

Simplify: $=\frac{7\cdot \cancel{3}}{4\cdot \cancel{3}}=\frac{7}{4}$

c. Recall that we can write **whole numbers** as **fractions** by writing the whole number over 1. So, we can write 2 as $\frac{2}{1}$. The **LCD** of $\frac{3}{7}$ and $\frac{2}{1}$ is 7 because 7 is a multiple of 1 $(1\cdot 7=7)$.

Begin with the original problem: $\frac{3}{7}-2$

Rewrite 2 as $\frac{2}{1}$: $=\frac{3}{7}-\frac{2}{1}$

Write each fraction with a denominator of LCD $= 7$: $= \frac{3}{7} - \frac{2 \cdot 7}{1 \cdot 7}$

$= \frac{3}{7} - \frac{14}{7}$

Subtract numerators $3 - 14 = -11$; write result over 7: $= \frac{-11}{7}$ or $-\frac{11}{7}$

d.–f. Try working these problems on your own. View the **answers**, or watch this **interactive video** for complete solutions to all six parts.

You Try It **Work through this You Try It problem.**

Work Exercises 33–44 in this textbook or in the MyMathLab Study Plan.

OBJECTIVE 4 ADD AND SUBTRACT FRACTIONS WITH VARIABLES

We can also add and subtract fractions involving **variables.** When there are variables in the numerators, remember that we can add and subtract only **like terms.**

Example 8 Adding and Subtracting Fractions with Variables in the Numerator

Add or subtract as indicated. **Simplify** if necessary.

a. $\frac{3x}{2} + \frac{5}{2}$ **b.** $\frac{2m}{15} - \frac{3m}{10}$

Solutions

a. The fractions have the same denominator 2, so we can add the numerators and keep the **common denominator.**

Begin with the original problem: $\frac{3x}{2} + \frac{5}{2}$

Add numerators; keep common denominator: $= \frac{3x + 5}{2}$

Because $3x$ and 5 are not like terms, we cannot combine them. There are no **common factors** to divide out, so our result is simplified.

My video summary

b. The denominators are not the same, so we need to find the **LCD.**

$$15 = 3 \cdot 5$$
$$10 = 2 \cdot 5$$

The LCD is $2 \cdot 3 \cdot 5 = 30$.

Begin with the original problem: $\frac{2m}{15} - \frac{3m}{10}$

Write each fraction with a denominator of LCD $= 30$: $= \frac{2m \cdot 2}{15 \cdot 2} - \frac{3m \cdot 3}{10 \cdot 3}$

$= \frac{4m}{30} - \frac{9m}{30}$

(eText Screens 4.4-1–4.4-33)

Subtract the numerators and keep the common denominator. Simplify your result. Check your **answer**, or watch this **video** for a complete solution to part b.

You Try It **Work through this You Try It problem.**

Work Exercises 45–48 in this textbook or in the MyMathLab Study Plan.

Remember that when a variable occurs in a **denominator**, we will assume the denominator will not equal 0. So, when finding an LCD involving variables, we follow an approach similar to that for numbers. We include each unique **variable** factor the largest number of times it occurs, as illustrated in part b of the next example and in Example 10.

Example 9 Adding and Subtracting Fractions with Variables in the Denominator

Add or subtract as indicated. **Simplify** if necessary.

a. $\frac{10}{3x} + \frac{2}{3x}$ **b.** $\frac{7}{6y^2} - \frac{2}{3y^2}$

Solutions

a. The fractions have the same denominator $3x$, so we can add the numerators and keep the **common denominator.**

Begin with the original problem: $\frac{10}{3x} + \frac{2}{3x}$

Add numerators; keep common denominator: $= \frac{10 + 2}{3x}$

Simplify: $= \frac{12}{3x}$

Divide out common factors: $= \frac{2 \cdot 2 \cdot \cancel{3}}{\cancel{3} \cdot x}$

Multiply remaining factors: $= \frac{4}{x}$

My video summary

b. The denominators are not the same, so we need to find a **common denominator,** or the **LCD**.

$$6y^2 = 2 \cdot 3 \cdot y \cdot y$$
$$3y^2 = 3 \cdot y \cdot y$$

There are unique factors of 2, 3, and y. The factors 2 and 3 occur at most once in any factorization, while the factor y occurs at most twice. So, we include one factor of 2, one factor of 3, and two factors of y in the LCD. The LCD is $2 \cdot 3 \cdot y \cdot y = 6y^2$.

Begin with the original problem: $\frac{7}{6y^2} - \frac{2}{3y^2}$

Write each fraction with a denominator of LCD $= 6y^2$: $= \frac{7}{6y^2} - \frac{2 \cdot 2}{3y^2 \cdot 2}$

$= \frac{7}{6y^2} - \frac{4}{6y^2}$

Subtract the **numerators** and keep the common denominator. Simplify your result. Check your **answer**, or watch this **video** for a complete solution to part b.

(eText Screens 4.4-1–4.4-33)

You Try It **Work through this You Try It problem.**

Work Exercises 49–52 in this textbook or in the MyMathLab Study Plan.

Example 10 Adding and Subtracting Fractions with Variables

Add or subtract as indicated. **Simplify** if necessary.

a. $\dfrac{7}{10x^2} + \dfrac{5}{2x}$ **b.** $\dfrac{1}{x} - \dfrac{2}{5}$

Solutions

a. The denominators are not the same, so we need to find the **LCD**.

$$10x^2 = 2 \cdot 5 \cdot x \cdot x$$
$$2x = 2 \cdot x$$

There are unique **factors** of 2, 5, and x. The factors 2 and 5 occur at most once in any factorization, while the factor x occurs at most twice. So we include one factor of 2, one factor of 5, and two factors of x in the LCD. The LCD is $2 \cdot 5 \cdot x \cdot x = 10x^2$.

Begin with the original problem: $\dfrac{7}{10x^2} + \dfrac{5}{2x}$

Write each fraction with a denominator of LCD $= 10x^2$: $= \dfrac{7}{10x^2} + \dfrac{5 \cdot 5x}{2x \cdot 5x}$

$= \dfrac{7}{10x^2} + \dfrac{25x}{10x^2} \leftarrow 2x \cdot 5x = 2 \cdot 5 \cdot x \cdot x = 10x^2$

Add numerators; keep common denominator: $= \dfrac{7 + 25x}{10x^2}$ or $\dfrac{25x + 7}{10x^2}$

b. The denominators are not the same so we need to find the **LCD**. In this case, the LCD is $5 \cdot x$ or $5x$, the product of the two denominators.

Begin with the original problem: $\dfrac{1}{x} - \dfrac{2}{5}$

Write each fraction with a denominator of LCD $= 5x$: $= \dfrac{1 \cdot 5}{x \cdot 5} - \dfrac{2 \cdot x}{5 \cdot x}$

$= \dfrac{5}{5x} - \dfrac{2x}{5x}$

Subtract numerators; keep common denominator: $= \dfrac{5 - 2x}{5x}$ or $\dfrac{-2x + 5}{5x}$

You Try It **Work through this You Try It problem.**

Work Exercises 53–56 in this textbook or in the MyMathLab Study Plan.

Example 11 Adding and Subtracting Fractions with Variables

Add or subtract as indicated. Simplify if necessary.

a. $\dfrac{5y}{3} - \dfrac{4}{3} + \dfrac{2y}{3}$ **b.** $\dfrac{a}{2} - \dfrac{6a}{7} + \dfrac{5}{4}$ **c.** $\dfrac{2}{3} - \dfrac{1}{6y} + \dfrac{3}{5}$

(eText Screens 4.4-1–4.4-33)

My interactive video summary

Solutions Try working these problems on your own. Remember that we need a **common denominator** to add or subtract fractions, and that we can add or subtract only **like terms.** Check your **answers**, or watch this **interactive video** for complete solutions to all three parts.

You Try It **Work through this You Try It problem.**

Work Exercises 57–60 in this textbook or in the MyMathLab Study Plan.

OBJECTIVE 5 SOLVE APPLICATIONS BY ADDING OR SUBTRACTING FRACTIONS

We can often solve applied problems by adding or subtracting fractions.

Example 12 Engagement Month

A survey asked couples the month in which they became engaged. Of those surveyed, $\frac{8}{25}$ became engaged in either July or August. If $\frac{7}{50}$ became engaged in August, what fraction became engaged in July? (*Source:* bridalguide.com)

Solution We can find the fraction of couples that became engaged in July by subtracting the fraction engaged in August from the two-month total.

Because the denominators are different, we need a **common denominator.** 50 is a multiple of 25, so the **LCD** is 50.

$$\frac{8}{25} - \frac{7}{50} = \frac{8 \cdot 2}{25 \cdot 2} - \frac{7}{50}$$
$$= \frac{16}{50} - \frac{7}{50} = \frac{9}{50}$$

$\frac{9}{50}$ of the couples in the survey became engaged in July.

You Try It **Work through this You Try It problem.**

Work Exercises 61–64 in this textbook or in the MyMathLab Study Plan.

Example 13 Cheesy Goodness

My video summary

A recipe calls for $\frac{4}{3}$ cup of cheddar cheese, $\frac{3}{5}$ cup of provolone cheese, and $\frac{1}{2}$ cup of pepper jack cheese. In total, how many cups of cheese are needed for the recipe?

Solution Add the three fractions together to find the total number of cups of cheese needed. Check your **answer**, or watch this **video** for a complete solution.

You Try It **Work through this You Try It problem.**

Work Exercises 65–70 in this textbook or in the MyMathLab Study Plan.

4.4 Exercises

In Exercises 1–20, add or subtract as indicated. Simplify if necessary.

You Try It 1. $\frac{2}{9}+\frac{5}{9}$ 2. $\frac{5}{7}+\frac{4}{7}$ 3. $\frac{2}{15}+\frac{14}{15}$ 4. $\frac{4}{5}+\frac{17}{5}$

You Try It 5. $-\frac{2}{11}+\frac{5}{11}$ 6. $-\frac{7}{5}+\frac{1}{5}$ 7. $-\frac{3}{7}+\frac{4}{7}$ 8. $-\frac{5}{3}+\left(-\frac{8}{3}\right)$

You Try It 9. $-\frac{3}{13}+\frac{16}{13}$ 10. $-\frac{7}{8}+\frac{11}{8}$ 11. $-\frac{2}{9}+\left(-\frac{16}{9}\right)$ 12. $\frac{7}{15}+\frac{13}{15}$

You Try It 13. $\frac{6}{7}-\frac{1}{7}$ 14. $\frac{4}{15}-\frac{11}{15}$ 15. $\frac{8}{15}-\frac{14}{15}$ 16. $-\frac{11}{5}-\frac{4}{5}$

17. $\frac{22}{13}-\frac{9}{13}$ 18. $\frac{15}{14}-\frac{23}{14}$ 19. $\frac{21}{10}-\frac{19}{10}$ 20. $-\frac{3}{10}-\left(-\frac{7}{10}\right)$

In Exercises 21–26, find the LCD of the given fractions by inspection.

You Try It 21. $\frac{9}{14}$ and $\frac{5}{6}$ 22. $\frac{4}{7}$ and $\frac{15}{28}$ 23. $\frac{3}{10}$ and $\frac{7}{8}$

24. $\frac{5}{7}$ and $\frac{7}{5}$ 25. $\frac{5}{17}$ and 3 26. $\frac{4}{45}$ and $\frac{35}{12}$

In Exercises 27–32, find the LCD of the given fractions by prime factorization.

You Try It 27. $\frac{5}{18}$ and $\frac{11}{12}$ 28. $\frac{11}{6}$ and $\frac{11}{35}$ 29. $\frac{14}{45}$ and $\frac{175}{54}$

30. $\frac{59}{140}$ and $\frac{117}{350}$ 31. $\frac{1}{10}, \frac{3}{20}$ and $\frac{7}{30}$ 32. $\frac{5}{6}, \frac{22}{15}$ and $\frac{11}{35}$

In Exercises 33–59, add or subtract as indicated. Simplify if necessary.

You Try It 33. $\frac{5}{8}+\frac{7}{12}$ 34. $\frac{2}{9}-\frac{5}{12}$ 35. $\frac{5}{8}-1$ 36. $\frac{7}{6}-\frac{7}{15}$

37. $\frac{7}{12}+\frac{11}{21}$ 38. $4+\frac{3}{10}$ 39. $\frac{13}{5}-\frac{4}{15}$ 40. $-\frac{5}{6}-\frac{5}{4}$

41. $-\frac{4}{5}+\frac{7}{10}$ 42. $\frac{4}{3}-\frac{6}{5}+\frac{2}{7}$ 43. $\frac{1}{2}+\frac{1}{3}+\frac{1}{5}$ 44. $\frac{1}{3}+\frac{1}{2}-\frac{1}{4}$

You Try It 45. $\frac{8x}{25}+\frac{3}{25}$ 46. $\frac{2y}{5}-\frac{11y}{6}$ 47. $\frac{7z}{15}+\frac{5z}{18}$ 48. $\frac{5}{21}-\frac{3z}{14}$

You Try It 49. $\frac{5}{2a}-\frac{6}{7a}$ 50. $\frac{5}{8x^2}-\frac{3}{20x^2}$ 51. $\frac{11}{7z}+\frac{3}{7z}$ 52. $\frac{2}{9b^3}+\frac{4}{15b^3}$

You Try It 53. $\frac{3}{x} - \frac{5}{12}$

54. $\frac{1}{2x} + \frac{1}{3x}$

55. $\frac{2x}{15} + \frac{5}{21x}$

56. $\frac{5}{18x} - \frac{2}{15x^2}$

You Try It 57. $\frac{a}{6} + \frac{4}{3} + \frac{a}{2}$

58. $\frac{5}{6y} + \frac{7}{3} - \frac{5}{2}$

59. $\frac{3x}{4} + \frac{2}{3} + \frac{x}{12}$

60. $\frac{7x}{5} - \frac{3x}{4} - \frac{9}{10}$

You Try It 61. **Sharing Candy** Logan gave $\frac{1}{4}$ of her candy to Sophie and $\frac{1}{3}$ to Caty. How much of her candy did she give away?

62. **Teenager Time** On average, $\frac{3}{4}$ of a teenager's day is spent at school or sleeping. If $\frac{21}{50}$ is spent sleeping, what fraction is spent in school?

63. **Cosmetic Surgery** When asked which cosmetic surgery would be their first choice, $\frac{4}{25}$ of those surveyed would opt for Lasik surgery, while $\frac{7}{50}$ would opt for liposuction. What total fraction would opt for either Lasik surgery or liposuction? (*Source:* usatoday.com)

64. **Grade Weight** In a statistics course, $\frac{4}{5}$ of a student's grade is from exams and quizzes. If $\frac{1}{6}$ of the grade is from quizzes, what fraction is from exams?

You Try It 65. **Nut Mixture** A snack mix consists of $\frac{3}{4}$ cup peanuts, $\frac{2}{3}$ cup almonds, and $\frac{1}{2}$ cup cashews. How many cups of nuts are in the mix?

66. **Jogging Distance** Angie jogged $8\frac{1}{2}$ miles on Monday, $13\frac{2}{5}$ miles on Wednesday, $10\frac{1}{3}$ miles on Friday, and $18\frac{3}{4}$ miles on Sunday. How many miles did Angie jog in total?
Hint: Change each mixed number to an improper fraction, then add.

67. **Scariest Movie** In a recent survey, $\frac{3}{25}$ of those surveyed said *A Nightmare on Elm Street* is the scariest movie of all time, while $\frac{1}{10}$ said *Halloween* was the scariest. What fraction selected a movie other than these two as the scariest movie of all time? (*Source:* redbox.com)

In Exercises 68–70, use the following chart to answer the questions. (*Source:* mashable.com)

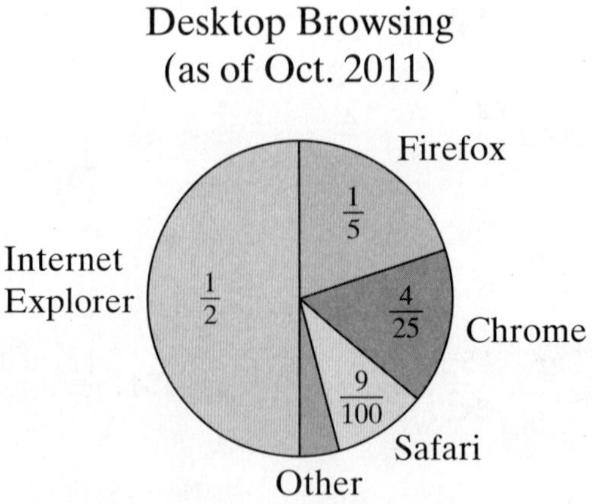

(eText Screens 4.5-1–4.5-21)

68. What fraction of desktop browsing is done using Firefox or Safari?

69. What is the difference in the fraction of desktop browsing done using Firefox rather than Chrome?

70. What fraction of desktop browsing is done using a browser other than Internet Explorer, Firefox, Chrome, or Safari?

4.5 Complex Fractions and Review of Order of Operations

THINGS TO KNOW

Before working through this topic, be sure you are familiar with the following concepts:

		VIDEO	ANIMATION	INTERACTIVE
You Try It	1. Use the Order of Operations with Integers (Topic 2.5, **Objective 1**)			
You Try It	2. Evaluate Algebraic Expressions Using Integers (Topic 2.5, **Objective 2**)			
You Try It	3. Identify the Numerator and Denominator of a Fraction (Topic 4.1, **Objective 1**)			
You Try It	4. Multiply Fractions (Topic 4.3, **Objective 1**)			
You Try It	5. Evaluate Exponential Expressions Involving Fraction Bases (Topic 4.3, **Objective 2**)			
You Try It	6. Divide Fractions (Topic 4.3, **Objective 4**)			
You Try It	7. Find the Least Common Denominator for Unlike Fractions (Topic 4.4, **Objective 2**)			
You Try It	8. Add and Subtract Unlike Fractions (Topic 4.4, **Objective 3**)			
You Try It	9. Add and Subtract Fractions with Variables (Topic 4.4, **Objective 4**)			

OBJECTIVES

1 Simplify Complex Fractions

2 Use the Order of Operations with Fractions

3 Evaluate Algebraic Expressions Using Fractions

(eText Screens 4.5-1–4.5-21)

OBJECTIVE 1 SIMPLIFY COMPLEX FRACTIONS

Sometimes, the **numerator** or **denominator** of a **fraction** will contain one or more fractions. We begin this topic by learning to **simplify** such *complex fractions*.

> **Definition Complex Fraction**
>
> A **complex fraction** is a fraction in which the numerator and/or denominator contain(s) fractions.

Some examples of complex fractions are

$$\frac{\frac{3}{8}}{\frac{5}{4}},\quad \frac{\frac{5}{6}-\frac{3}{4}}{\frac{7}{9}+\frac{1}{6}},\quad \frac{\frac{7}{10}}{\frac{11}{15}-2},\quad \frac{\frac{5}{8}+\frac{17}{12}}{7},\quad \text{and}\quad \frac{\frac{x}{y}}{\frac{wx^2}{y^3}}.$$

The fractions within the numerator and denominator are called **minor fractions.** The numerator and denominator of the complex fraction are separated by the **main fraction bar.**

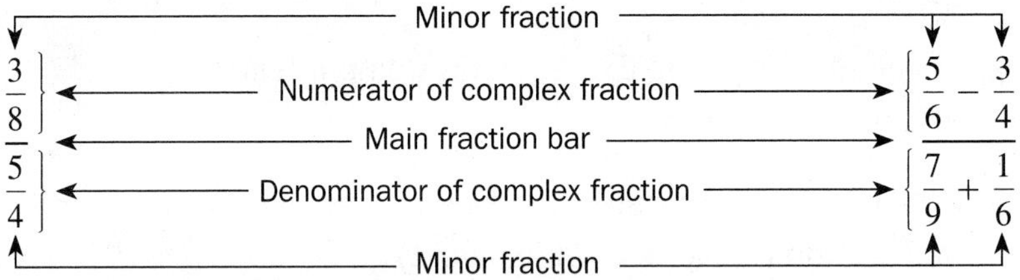

To **simplify a complex fraction**, we rewrite it as an **equivalent fraction** in which the **numerator** and **denominator** contain no **minor fractions** and have no **common factors** other than 1. We show two methods for simplifying complex fractions, called *Method 1* and *Method 2*. With Method 1, we recognize that the **main fraction bar** is a **division** symbol. So, we divide the numerator by the denominator.

Example 1 Simplifying a Complex Fraction

Simplify the complex fraction.

$$\frac{\frac{3}{8}}{\frac{5}{4}}$$

Solution We divide the minor fraction from the numerator by the minor fraction from the denominator. To do this, we multiply the numerator by the reciprocal of the denominator.

Division

Write the **complex fraction** as division: $\frac{\frac{3}{8}}{\frac{5}{4}} = \frac{3}{8} \div \frac{5}{4}$

Change to multiplication by the **reciprocal**: $= \frac{3}{8} \cdot \frac{4}{5}$ ← The reciprocal of $\frac{5}{4}$ is $\frac{4}{5}$.

Multiply **numerators**; multiply **denominators**: $= \frac{3 \cdot 4}{8 \cdot 5}$

Divide out the **common factor** 4: $= \frac{3 \cdot \cancel{4}}{{}_2\cancel{8} \cdot 5}$

Multiply the remaining factors: $= \frac{3}{2 \cdot 5} = \frac{3}{10}$

You Try It **Work through this You Try It problem.**

Work Exercise 1 in this textbook or in the MyMathLab Study Plan.

In addition to being a division symbol, the **main fraction bar** of a **complex fraction** is also a **grouping symbol.** If the **numerator** and **denominator** are not each written as a single **fraction**, then they must be simplified before dividing.

> **Method 1 for Simplifying Complex Fractions**
>
> **Step 1.** Simplify the expression in the numerator into a single **minor fraction.**
>
> **Step 2.** Simplify the expression in the denominator into a single minor fraction.
>
> **Step 3.** Divide the minor fraction in the numerator by the minor fraction in the denominator. To do this, multiply the minor fraction in the numerator by the reciprocal of the minor fraction in the denominator. Simplify if possible.

Example 2 Simplifying a Complex Fraction Using Method 1

My video summary

Use Method I to simplify the complex fraction.

$$\frac{\frac{5}{6} - \frac{3}{4}}{\frac{7}{9} + \frac{1}{4}}$$

Solution

Step 1. In the **numerator**, the **LCD** for the **minor fractions** $\frac{5}{6}$ and $\frac{3}{4}$ is 12.

$$\frac{5}{6} - \frac{3}{4} = \frac{5}{6} \cdot \frac{2}{2} - \frac{3}{4} \cdot \frac{3}{3} = \frac{10}{12} - \frac{9}{12} = \frac{10 - 9}{12} = \frac{1}{12}$$

Step 2. In the **denominator**, the LCD for $\frac{7}{9}$ and $\frac{1}{4}$ is 36.

$$\frac{7}{9} + \frac{1}{4} = \frac{7}{9} \cdot \frac{4}{4} + \frac{1}{4} \cdot \frac{9}{9} = \frac{28}{36} + \frac{9}{36} = \frac{28 + 9}{36} = \frac{37}{36}$$

Step 3. Substitute the fractions from Steps 1 and 2, for the numerator and denominator of the **complex fraction**, and then divide.

$$\frac{\frac{5}{6} - \frac{3}{4}}{\frac{7}{9} + \frac{1}{4}} = \frac{\frac{1}{12}}{\frac{37}{36}} = \frac{1}{12} \div \frac{37}{36}$$

Try to complete the division to finish this problem on your own. Check the **answer**, or watch this **video** for a complete solution.

(eText Screens 4.5-1–4.5-21)

You Try It **Work through this You Try It problem.**

Work Exercises 2 and 3 in this textbook or in the MyMathLab Study Plan.

In **Topic 4.2**, we saw that multiplying both the **numerator** and the **denominator** of a fraction by the same nonzero number results in an **equivalent fraction**. For example,

$$\frac{2}{3} = \frac{5 \cdot 2}{5 \cdot 3} = \frac{10}{15} \text{ because } \frac{5}{5} = 1.$$

With Method 2 for simplifying **complex fractions**, we use this same concept. We multiply the numerator and denominator of the complex fraction by the **LCD** of its **minor fractions**.

> **Method 2 for Simplifying Complex Fractions**
>
> **Step 1.** Determine the LCD of all the minor fractions within the complex fraction.
>
> **Step 2.** Multiply the numerator and denominator of the complex fraction by the LCD from Step 1.
>
> **Step 3.** Simplify.

Let's take another look at the complex fraction from **Example 1**.

Example 3 Simplifying a Complex Fraction Using Method 2

Use Method 2 to simplify the complex fraction.

$$\frac{\frac{3}{8}}{\frac{5}{4}}$$

Solution

Step 1. The **LCD** for the two **minor fractions** $\frac{3}{8}$ and $\frac{5}{4}$ is 8.

Step 2. Multiply the **numerator** and **denominator** by 8: $\frac{\frac{3}{8}}{\frac{5}{4}} = \frac{8 \cdot \frac{3}{8}}{8 \cdot \frac{5}{4}}$

Step 3. Divide out **common factors**: $= \frac{\cancel{8} \cdot \frac{3}{\cancel{8}}}{{}_{2}\cancel{8} \cdot \frac{5}{\cancel{4}}}$

Multiply the remaining factors: $= \frac{3}{2 \cdot 5}$

$$= \frac{3}{10}$$

Notice that this is the same result we obtained in Example 1.

You Try It **Work through this You Try It problem.**

Work Exercise 4 in this textbook or in the MyMathLab Study Plan.

For comparison, let's revisit the **complex fraction** from **Example 2.**

Example 4 Simplifying a Complex Fraction Using Method 2

My video summary

Use **Method 2** to simplify the complex fraction.

$$\frac{\frac{5}{6}-\frac{3}{4}}{\frac{7}{9}+\frac{1}{4}}$$

Solution

Step 1. The **denominators** of all the **minor fractions** are 6, 4, 9, and 4, so the **LCD** is 36.

Step 2. Multiply both the **numerator** and the denominator by 36:

$$\frac{\frac{5}{6}-\frac{3}{4}}{\frac{7}{9}+\frac{1}{4}}=\frac{36\left(\frac{5}{6}-\frac{3}{4}\right)}{36\left(\frac{7}{9}+\frac{1}{4}\right)}$$

Step 3. Use the **distributive property:** $=\dfrac{36\cdot\frac{5}{6}-36\cdot\frac{3}{4}}{36\cdot\frac{7}{9}+36\cdot\frac{1}{4}}$

Simplify the expressions in the numerator and denominator to finish simplifying the complex fraction. Check the **answer**, or watch this **video** for a complete solution.

You Try It **Work through this You Try It problem.**

Work Exercises 5 and 6 in this textbook or in the MyMathLab Study Plan.

As you have seen in Examples 1–4, **Method 1** and **Method 2** give the same simplification. You may prefer one method over the other, but we suggest that you practice using both methods for a while to discover when one method might be better suited than the other.

Example 5 Simplifying Complex Fractions

My interactive video summary

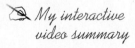

Use **Method 1** or **Method 2** to simplify each complex fraction. Assume **variables** do not cause any **denominator** to equal 0.

a. $\dfrac{\frac{7}{10}}{\frac{11}{15}-2}$ **b.** $\dfrac{\frac{5}{8}+\frac{17}{12}}{7}$ **c.** $\dfrac{\frac{x}{y}}{\frac{wx^2}{y^3}}$

Solutions Try simplifying these complex fractions on your own. Check the **answers**, or watch this **interactive video** for complete solutions.

(eText Screens 4.5-1–4.5-21)

You Try It **Work through this You Try It problem.**

Work Exercises 7–15 in this textbook or in the MyMathLab Study Plan.

OBJECTIVE 2 USE THE ORDER OF OPERATIONS WITH FRACTIONS

In **Topic 2.5**, we simplified **numeric expressions** involving **integers**. Next, we **simplify** numeric expressions involving **fractions**. Let's begin by reviewing the order of operations.

> **Order of Operations**
>
> 1. **Parentheses (or other grouping symbols)** Evaluate operations within parentheses (or other **grouping symbols**) first, starting with the innermost set and working out.
> 2. **Exponents** Work from left to right and evaluate any **exponential expressions** as they occur.
> 3. **Multiplication and Division** Work from left to right and perform any **multiplication** or **division** operations as they occur.
> 4. **Addition and Subtraction** Work from left to right and perform any **addition** or **subtraction** operations as they occur.

CAUTION When simplifying expressions involving fractions, remember to always write the final result in **lowest terms**.

Example 6 Using the Order of Operations with Fractions

Simplify each expression.

a. $\dfrac{1}{2}+\dfrac{3}{4}\cdot\dfrac{1}{6}$ **b.** $\left(-\dfrac{5}{4}\right)^2-\dfrac{7}{6}$

Solutions

a. Begin with the original expression: $\dfrac{1}{2}+\dfrac{3}{4}\cdot\dfrac{1}{6}$

Multiply $\dfrac{3}{4}\cdot\dfrac{1}{6}$: $=\dfrac{1}{2}+\dfrac{1}{8} \leftarrow \boxed{\dfrac{3}{4}\cdot\dfrac{1}{6}=\dfrac{3\cdot1}{4\cdot6}=\dfrac{\overset{1}{\cancel{3}}\cdot1}{4\cdot\underset{2}{\cancel{6}}}=\dfrac{1\cdot1}{4\cdot2}=\dfrac{1}{8}}$

8 is a **multiple** of 2, so the **LCD** is 8.

Write $\dfrac{1}{2}$ with a denominator of 8: $=\dfrac{1\cdot4}{2\cdot4}+\dfrac{1}{8}$

$=\dfrac{4}{8}+\dfrac{1}{8}$

Add: $=\dfrac{5}{8} \leftarrow \boxed{\dfrac{4}{8}+\dfrac{1}{8}=\dfrac{4+1}{8}=\dfrac{5}{8}}$

So, $\dfrac{1}{2}+\dfrac{3}{4}\cdot\dfrac{1}{6}=\dfrac{5}{8}$.

My video summary **b.** Write the original problem: $\left(-\frac{5}{4}\right)^2 - \frac{7}{6}$

Evaluate the exponent, $\left(-\frac{5}{4}\right)^2$: $= \frac{25}{16} - \frac{7}{6} \leftarrow \boxed{\left(-\frac{5}{4}\right)^2 = \left(-\frac{5}{4}\right)\left(-\frac{5}{4}\right) = \frac{5\cdot 5}{4\cdot 4} = \frac{25}{16}}$

Try to subtract these fractions on your own to finish this problem. Check the **answer**, or watch this **video** for a complete solution to part b.

You Try It **Work through this You Try It problem.**

Work Exercises 16–21 in this textbook or in the MyMathLab Study Plan.

Example 7 Using the Order of Operations with Fractions

Simplify each expression.

a. $\frac{11}{10} - \frac{3}{10}\left(\frac{2}{3} - \frac{1}{6}\right)$ **b.** $\left(\frac{5}{6} - \frac{4}{9}\right)\left(\frac{1}{3} + \frac{7}{6}\right)$

Solutions

a. Begin within the parentheses and subtract $\frac{2}{3} - \frac{1}{6}$. Because 6 is a **multiple** of 3, the **LCD** is 6.

Subtract: $\frac{2}{3} - \frac{1}{6} = \frac{2\cdot 2}{3\cdot 2} - \frac{1}{6} = \frac{4}{6} - \frac{1}{6} = \frac{4-1}{6} = \frac{3}{6} = \frac{1}{2}$

So, we have $\frac{11}{10} - \frac{3}{10}\left(\frac{2}{3} - \frac{1}{6}\right) = \frac{11}{10} - \frac{3}{10}\left(\frac{1}{2}\right)$

Multiply $\frac{3}{10}\cdot\frac{1}{2}$: $= \frac{11}{10} - \frac{3}{20} \leftarrow \boxed{\frac{3}{10}\cdot\frac{1}{2} = \frac{3\cdot 1}{10\cdot 2} = \frac{3}{20}}$

Because 20 is a multiple of 10, the LCD is 20.

Subtract: $= \frac{11\cdot 2}{10\cdot 2} - \frac{3}{20} = \frac{22}{20} - \frac{3}{20} = \frac{22-3}{20} = \frac{19}{20}$

So, $\frac{11}{10} - \frac{3}{10}\left(\frac{2}{3} - \frac{1}{6}\right) = \frac{19}{20}$.

My video summary **b.** Try to **simplify** this expression on your own by first performing the **operations** within the parentheses, and then multiplying. Check the **answer**, or watch this **video** for a detailed solution.

You Try It **Work through this You Try It problem.**

Work Exercises 22–27 in this textbook or in the MyMathLab Study Plan.

OBJECTIVE 3 EVALUATE ALGEBRAIC EXPRESSIONS USING FRACTIONS

We **evaluated algebraic expressions** using **whole numbers** in **Topic 1.7** and **integers** in **Topic 2.5**. Next, we evaluate algebraic expressions using **fractions.** As before, we **substitute** the given values for the **variables** and **simplify** using the **order of operations.**

(eText Screens 4.5-1–4.5-21)

Example 8 Evaluating Algebraic Expressions Using Fractions

My video summary

Evaluate $4p - q^2$ for $p = \frac{7}{12}$ and $q = -\frac{2}{3}$.

Solution Work through the following, or watch this **video** for a detailed solution.

Substitute $\frac{7}{12}$ for p and $-\frac{2}{3}$ for q: $4p - q^2 = 4\left(\frac{7}{12}\right) - \left(-\frac{2}{3}\right)^2$

Evaluate the exponent: $= 4\left(\frac{7}{12}\right) - \frac{4}{9}$ ← $\left(-\frac{2}{3}\right)^2 = \left(-\frac{2}{3}\right)\left(-\frac{2}{3}\right) = \frac{4}{9}$

Multiply: $= \frac{7}{3} - \frac{4}{9}$ ← $4\left(\frac{7}{12}\right) = \frac{4 \cdot 7}{12} = \frac{{}^{1}\cancel{4} \cdot 7}{\cancel{12}_{3}} = \frac{7}{3}$

Write $\frac{7}{3}$ with a denominator of 9: $= \frac{21}{9} - \frac{4}{9}$ ← LCD = 9

Subtract: $= \frac{17}{9}$ or $1\frac{8}{9}$

The value of $4p - q^2$ is $\frac{17}{9}$ or $1\frac{8}{9}$ when $p = \frac{7}{12}$ and $q = -\frac{2}{3}$.

CAUTION When **substituting** a fraction as the **base** of an **exponential expression**, be sure to place it within parentheses or other **grouping symbols** to avoid sign errors.

You Try It **Work through this You Try It problem.**

Work Exercises 28–33 in this textbook or in the MyMathLab Study Plan.

Example 9 Evaluating Algebraic Expressions Using Fractions

My interactive video summary

Evaluate $\frac{x + y}{z}$ for $x = -\frac{3}{5}$, $y = \frac{5}{6}$ and $z = \frac{7}{9}$.

Solution

Begin with the original expression: $\frac{x + y}{z}$

Substitute $-\frac{3}{5}$ for x, $\frac{5}{6}$ for y, and $\frac{7}{9}$ for z: $= \frac{-\frac{3}{5} + \frac{5}{6}}{\frac{7}{9}}$

Try to finish simplifying this expression on your own, using either **Method 1** or **Method 2** for simplifying complex fractions. View the **answer**, or watch this **interactive video** for a complete solution.

You Try It **Work through this You Try It problem.**

Work Exercises 34 and 35 in this textbook or in the MyMathLab Study Plan.

4.5 Exercises

In Exercises 1–3, use Method 1 to simplify each complex fraction.

You Try It

1. $\dfrac{\frac{5}{6}}{\frac{15}{16}}$

2. $\dfrac{\frac{1}{6}+\frac{1}{9}}{-\frac{2}{3}}$

3. $\dfrac{\frac{9}{16}-\frac{1}{8}}{\frac{9}{16}+\frac{3}{4}}$

In Exercises 4–6, use Method 2 to simplify each complex fraction.

You Try It

4. $\dfrac{\frac{7}{12}}{\frac{11}{18}}$

5. $\dfrac{\frac{5}{6}}{\frac{9}{10}+\frac{3}{5}}$

6. $\dfrac{\frac{3}{4}-\frac{1}{3}}{\frac{2}{3}+\frac{1}{6}}$

In Exercises 7–15, use Method 1 or Method 2 to simplify each complex fraction. Assume variables do not cause any denominators to equal 0.

You Try It

7. $\dfrac{-\frac{10}{21}}{-\frac{5}{14}}$

8. $\dfrac{-\frac{15}{16}}{24}$

9. $\dfrac{\frac{3}{5}-\frac{9}{10}}{\frac{5}{4}}$

10. $\dfrac{\frac{9}{20}}{\frac{7}{25}-4}$

11. $\dfrac{\frac{5}{6}-\frac{2}{3}}{\frac{7}{9}+\frac{1}{4}}$

12. $\dfrac{\frac{1}{5}+\frac{1}{2}}{\frac{4}{5}-\frac{1}{10}}$

13. $\dfrac{\frac{8x}{yz}}{\frac{4x^2}{3z}}$

14. $\dfrac{\frac{11}{5m}}{\frac{1}{3}-\frac{16}{15}}$

15. $\dfrac{\frac{3}{x}}{5+\frac{1}{x}}$

In Exercises 16–27, simplify each expression.

You Try It

16. $\frac{3}{8}+\frac{1}{4}\cdot\frac{6}{5}$

17. $\frac{7}{9}-6\div\frac{9}{2}$

18. $\left(\frac{3}{7}\right)^2-\frac{9}{14}$

19. $\frac{2}{9}\cdot\frac{3}{8}+\frac{5}{6}\cdot\frac{3}{10}$

20. $\left(-\frac{3}{2}\right)^2-\left(\frac{7}{4}\right)^2$

21. $\left(1-\frac{5}{8}\right)^2$

You Try It

22. $\frac{15}{16}\left(\frac{2}{3}-\frac{4}{15}\right)$

23. $\frac{9}{8}\div\left(\frac{1}{2}-\frac{5}{4}\right)$

24. $\frac{8}{9}-\frac{1}{3}\left(\frac{5}{9}-\frac{1}{3}\right)$

25. $\left(\frac{20}{13}-\frac{15}{13}\right)\left(\frac{7}{10}+\frac{3}{5}\right)$

26. $28\left(\frac{1}{4}\right)^2-\frac{11}{12}$

27. $2\left(\frac{1}{3}\right)^3-\frac{2}{9}\cdot\frac{3}{4}$

In Exercises 28–35, evaluate each expression for the given values of the variables.

You Try It 28. $\frac{8}{15}t$ for $t = \frac{5}{2}$

29. $m + n$ for $m = \frac{7}{6}$ and $n = -\frac{1}{4}$

30. $a^2 - b$ for $a = -\frac{3}{5}$ and $b = -\frac{2}{5}$

31. $2x + y^2$ for $x = -\frac{7}{16}$ and $y = \frac{1}{8}$

32. $p(1 - p)$ for $p = \frac{4}{5}$

33. $\frac{1 - k}{1 + k}$ for $k = \frac{7}{12}$

You Try It 34. $\frac{x - y}{z}$ for $x = \frac{5}{6}, y = \frac{1}{4}$, and $z = -\frac{1}{2}$

35. $\frac{p \cdot q}{n}$ for $p = \frac{21}{25}, q = \frac{4}{25}$, and $n = 14$

4.6 Operations on Mixed Numbers

THINGS TO KNOW

Before working through this topic, be sure you are familiar with the following concepts:

		VIDEO	ANIMATION	INTERACTIVE
You Try It	1. Write Improper Fractions as Mixed Numbers or Whole Numbers (Topic 4.1, **Objective 4**)	✓		
You Try It	2. Multiply Fractions (Topic 4.3, **Objective 1**)	✓		✓
You Try It	3. Divide Fractions (Topic 4.3, **Objective 4**)	✓		✓
You Try It	4. Find the Least Common Denominator for Unlike Fractions (Topic 4.4, **Objective 2**)			✓
You Try It	5. Add and Subtract Unlike Fractions (Topic 4.4, **Objective 3**)			✓
You Try It	6. Solve Applications by Adding or Subtracting Fractions (Topic 4.4, **Objective 5**)	✓		

OBJECTIVES

1 Multiply Mixed Numbers

2 Divide Mixed Numbers

3 Add Mixed Numbers

4 Subtract Mixed Numbers

5 Solve Applications Involving Mixed Numbers

OBJECTIVE 1 MULTIPLY MIXED NUMBERS

To multiply **mixed numbers**, we first convert each mixed number into an **improper fraction** and then follow the procedures for **multiplying** fractions.

Example 1 Multiplying a Mixed Number and a Proper Fraction

Multiply. Write the answer as a mixed number if possible.

a. $\frac{4}{7} \cdot 3\frac{2}{5}$ **b.** $\frac{2}{7} \cdot 4\frac{3}{8}$

Solutions

a. Begin with the original problem: $\frac{4}{7} \cdot 3\frac{2}{5}$

Convert mixed numbers to improper fractions: $= \frac{4}{7} \cdot \frac{17}{5} \leftarrow 3\frac{2}{5} = \frac{5 \cdot 3 + 2}{5} = \frac{17}{5}$

Multiply numerators; multiply denominators: $= \frac{4 \cdot 17}{7 \cdot 5}$

There are no **common factors** to **divide out**, so we multiply the factors and write the result as a mixed number.

Multiply remaining factors: $= \frac{68}{35}$ $\quad 35\overline{)68}$, $1 \leftarrow$ Quotient, -35, $33 \leftarrow$ Remainder

Convert back to a mixed number: $= 1\frac{33}{35}$ ($33 \leftarrow$ Remainder, $35 \leftarrow$ Denominator, $1 \uparrow$ Quotient)

b. Begin with the original problem: $\frac{2}{7} \cdot 4\frac{3}{8}$

Convert mixed numbers to **improper fractions**: $= \frac{2}{7} \cdot \frac{35}{8} \leftarrow 4\frac{3}{8} = \frac{8 \cdot 4 + 3}{8} = \frac{35}{8}$

Multiply numerators; multiply denominators: $= \frac{2 \cdot 35}{7 \cdot 8}$

Divide out common factors: $= \frac{\overset{1}{\cancel{2}} \cdot \overset{5}{\cancel{35}}}{\underset{1}{\cancel{7}} \cdot \underset{4}{\cancel{8}}}$

Multiply remaining factors: $= \frac{1 \cdot 5}{1 \cdot 4} = \frac{5}{4}$

Convert back to a **mixed number**: $= 1\frac{1}{4}$

You Try It **Work through this You Try It problem.**

Work Exercises 1–4 in this textbook or in the MyMathLab Study Plan.

(eText Screens 4.6-1–4.6-29)

Example 2 Multiplying Mixed Numbers

My video summary

Multiply. Write the answer as a **mixed number** if possible.

a. $5 \cdot 3\frac{1}{4}$ **b.** $6\frac{2}{5} \cdot 4\frac{4}{9}$

Solutions

a. Recall that we can write the **whole number** 5 as $\frac{5}{1}$.

Begin with the original problem: $5 \cdot 3\frac{1}{4}$

Convert mixed numbers to **improper fractions:** $= \frac{5}{1} \cdot \frac{13}{4} \leftarrow 3\frac{1}{4} = \frac{4 \cdot 3 + 1}{4} = \frac{13}{4}$

Finish multiplying this problem on your own, then check your **answer**. Watch this **video** to see complete solutions to both parts.

b. Write each **mixed number** as an **improper fraction**, then multiply. Remember to **simplify** and write your answer as a mixed number if possible. Check your **answer**, or watch this **video** for complete solutions to both parts.

You Try It **Work through this You Try It problem.**

Work Exercises 5–10 in this textbook or in the MyMathLab Study Plan.

TIP When all the numbers being multiplied are **whole numbers** or mixed numbers, it is a good idea to check your answers by estimating the product. View this **popup** to see the estimates for Example 2.

OBJECTIVE 2 DIVIDE MIXED NUMBERS

As with multiplication, to divide **mixed numbers** we first convert each mixed number into an **improper fraction**. We then follow the usual procedure for **dividing** fractions.

Example 3 Dividing a Mixed Number and a Proper Fraction

Divide. Write the answer as a mixed number if possible.

a. $2\frac{3}{4} \div \frac{8}{9}$ **b.** $\frac{3}{5} \div 5\frac{1}{7}$

Solutions

a. Begin with the original problem: $2\frac{3}{4} \div \frac{8}{9}$

Convert mixed numbers to improper fractions: $= \frac{11}{4} \div \frac{8}{9} \leftarrow 2\frac{3}{4} = \frac{4 \cdot 2 + 3}{4} = \frac{11}{4}$

Change to multiplication by the **reciprocal:** $= \frac{11}{4} \cdot \frac{9}{8}$

Multiply numerators; multiply denominators: $= \frac{11 \cdot 9}{4 \cdot 8}$

There are no **common factors** to **divide out**, so we multiply the numerators and multiply the denominators, then write the result as a mixed number.

Multiply remaining factors: $= \frac{99}{32}$

Convert back to a mixed number: $= 3\frac{3}{32}$

b. Begin with the original problem: $\frac{3}{5} \div 5\frac{1}{7}$

Convert mixed numbers to **improper fractions:** $= \frac{3}{5} \div \frac{36}{7} \leftarrow 5\frac{1}{7} = \frac{7 \cdot 5 + 1}{7} = \frac{36}{7}$

Change to multiplication by the **reciprocal:** $= \frac{3}{5} \cdot \frac{7}{36}$

Divide out common factors: $= \frac{\overset{1}{\cancel{3}}}{5} \cdot \frac{7}{\underset{12}{\cancel{36}}}$

Multiply remaining factors: $= \frac{1 \cdot 7}{5 \cdot 12} = \frac{7}{60}$

You Try It **Work through this You Try It problem.**

Work Exercises 11–14 in this textbook or in the MyMathLab Study Plan.

Example 4 Dividing Mixed Numbers

My video summary

Divide. Write the answer as a **mixed number** if possible.

a. $4\frac{5}{9} \div 2\frac{1}{3}$ b. $7\frac{3}{5} \div 3$

Solutions

a. Begin with the original problem: $4\frac{5}{9} \div 2\frac{1}{3}$

Convert mixed numbers to **improper fractions:** $= \frac{41}{9} \div \frac{7}{3} \leftarrow$ $4\frac{5}{9} = \frac{9 \cdot 4 + 5}{9} = \frac{41}{9}$; $2\frac{1}{3} = \frac{3 \cdot 2 + 1}{3} = \frac{7}{3}$

Change to multiplication by the **reciprocal:** $= \frac{41}{9} \cdot \frac{3}{7}$

Finish multiplying this problem on your own, then check your **answer.** Watch this **video** to see complete solutions to both parts.

b. Write each **mixed number** as an improper fraction, then multiply. Remember that we can write whole numbers as fractions with a denominator of 1. Check your **answer**, or watch this **video** for complete solutions to both parts.

You Try It **Work through this You Try It problem.**

Work Exercises 15–20 in this textbook or in the MyMathLab Study Plan.

OBJECTIVE 3 ADD MIXED NUMBERS

To add two **mixed numbers**, first we could convert each mixed number into an improper fraction, just as we did to multiply and divide. However, because the equivalent improper fractions can involve very large **numerators**, it is usually easier to add the similar parts. This means that we can add the **fraction parts** and then add the **whole-number parts**.

Example 5 Adding Mixed Numbers

My video summary

Add. Write the answer as a mixed number if possible.

a. $6\frac{1}{3} + 2\frac{5}{12}$ **b.** $4\frac{1}{6} + 5\frac{3}{8}$

Solutions Remember that to add the fraction parts, we need a common denominator. So, we first write equivalent fraction parts for each mixed number using the **LCD** as the **common denominators**. Then we add the fraction parts, followed by the whole-number parts. For simplicity, we will add vertically so that we can line up the fraction parts and the whole-number parts.

a. We have denominators of 3 and 12. Because 12 is a **multiple** of 3, the **LCD** of the fraction parts is 12.

Whole-number parts

$$\begin{array}{rcl} 6\frac{1}{3} & = & 6\frac{4}{12} \leftarrow \boxed{\frac{1}{3} = \frac{1\cdot 4}{3\cdot 4} = \frac{4}{12}} \\ +\,2\frac{5}{12} & = & +\,2\frac{5}{12} \\ \hline & & 8\frac{9}{12} \leftarrow \frac{4+5}{12} \end{array}$$

Proper-fraction parts

$8 \uparrow$ $6 + 2$

As always, we want to make sure the **fraction part** of our answer is in **simplest form**. Notice that $\frac{9}{12}$ can be simplified by dividing out a **common factor** of 3.

$$\frac{9}{12} = \frac{\cancel{3}\cdot 3}{\cancel{3}\cdot 4} = \frac{3}{4}$$

Therefore, $6\frac{1}{3} + 2\frac{5}{12} = 8\frac{3}{4}$.

b. We have denominators of 6 and 8. The **LCD** of the fraction parts is 24.

Whole-number parts

$$\begin{array}{rcl} 4\frac{1}{6} & = & 4\frac{4}{24} \leftarrow \boxed{\frac{1}{6} = \frac{1\cdot 4}{6\cdot 4} = \frac{4}{24}} \\ +\,5\frac{3}{8} & = & +\,5\frac{9}{24} \leftarrow \boxed{\frac{3}{8} = \frac{3\cdot 3}{8\cdot 3} = \frac{9}{24}} \\ \hline \end{array}$$

Proper-fraction parts

Finish adding on your own. Check your **answer**, or watch this **video** to see the solutions to both parts worked out in detail.

(eText Screens 4.6-1–4.6-29)

You Try It **Work through this You Try It problem.**

Work Exercises 21–26 in this textbook or in the MyMathLab Study Plan.

Sometimes when we add the fraction parts of two **mixed numbers**, the result is an **improper fraction**. If this occurs, we must convert the improper-fraction part into a mixed number and "carry" the whole-number part of the result to the whole-number parts of the original mixed numbers before finding the final sum.

For example, consider the number $4\frac{5}{3}$. This number is technically *not* a mixed number because the fraction part is not a **proper fraction**. We can convert this number into a mixed number as follows:

$$4\frac{5}{3} = 4 + \frac{5}{3} = 4 + 1\frac{2}{3} = 5\frac{2}{3}$$

Example 6 Adding Mixed Numbers with Carrying

Add. Write your answer as a mixed number if possible.

$$8\frac{7}{10} + 3\frac{14}{15}$$

Solution First, we write the problems vertically and convert the fraction parts to have **common denominators**. Next, we add the fraction parts, followed by the whole-number parts.

The **LCD** of the fraction parts is 30.

$$\begin{array}{rcrl} 8\frac{7}{10} & = & 8\frac{21}{30} & \leftarrow \boxed{\frac{7}{10} = \frac{7\cdot 3}{10\cdot 3} = \frac{21}{30}} \\ +3\frac{14}{15} & = & +\,3\frac{28}{30} & \leftarrow \boxed{\frac{14}{15} = \frac{14\cdot 2}{15\cdot 2} = \frac{28}{30}} \\ \hline & & 11\frac{49}{30} & \end{array}$$

Because $\frac{49}{30}$ is an **improper fraction**, we must convert $11\frac{49}{30}$ into a **mixed number:**

$$11\frac{49}{30} = 11 + \frac{49}{30} = 11 + 1\frac{19}{30} = 12\frac{19}{30}$$

You Try It **Work through this You Try It problem.**

Work Exercises 27–31 in this textbook or in the MyMathLab Study Plan.

Example 7 Adding Several Mixed Numbers

Add. Write your answer as a mixed number if possible.

$$2\frac{3}{4} + 7 + 1\frac{4}{5}$$

(eText Screens 4.6-1–4.6-29)

✍ *My video summary*

Solution The **LCD** of the fraction parts is 20. Convert the **fraction parts** to get a common denominator, then add the fraction parts and the whole-number parts.

$$2\frac{3}{4} = 2\frac{15}{20} \leftarrow \boxed{\frac{3}{4} = \frac{3\cdot 5}{4\cdot 5} = \frac{15}{20}}$$

$$7 = 7$$

$$+\,1\frac{4}{5} = +\,1\frac{16}{20} \leftarrow \boxed{\frac{4}{5} = \frac{4\cdot 4}{5\cdot 4} = \frac{16}{20}}$$

Finish solving this problem on your own. Check your **answer**, or watch this **video** for a fully worked solution.

You Try It **Work through this You Try It problem.**

Work Exercises 32–34 in this textbook or in the MyMathLab Study Plan.

TIP Remember, when all the numbers are whole numbers or mixed numbers, it is a good idea to check your answers by estimating. View this **popup** to see the estimate for the sum in Example 7.

OBJECTIVE 4 SUBTRACT MIXED NUMBERS

To subtract two **mixed numbers**, we follow the same approach as for adding mixed numbers. We obtain a **common denominator** for the fraction parts, then subtract the **fraction parts** and subtract the **whole-number parts**.

Example 8 Subtracting Mixed Numbers

Subtract. Write answers as a mixed number if possible.

a. $5\frac{1}{3} - 2\frac{2}{9}$ **b.** $10\frac{5}{6} - 3\frac{8}{15}$

Solutions Just as with addition, we will need to have a common denominator to subtract the fraction parts. So, we first write **equivalent fraction** parts for each mixed number using the **LCD** as the common denominator. Then we subtract the fraction parts, followed by the whole-number parts. For simplicity, we will again subtract vertically so that we can line up the fraction parts and the whole-number parts.

a. We have denominators of 3 and 9. Because 9 is a **multiple** of 3, the **LCD** of the fraction parts is 9.

Whole-number parts

$$5\frac{1}{3} = 5\frac{3}{9}$$

Proper-fraction parts

$$-2\frac{2}{9} = -2\frac{2}{9}$$

$$3\frac{1}{9} \leftarrow \frac{3-2}{9}$$

$$5 - 2$$

Therefore, $5\frac{1}{3} - 2\frac{2}{9} = 3\frac{1}{9}$.

My video summary

b. We have denominators of 6 and 15. The LCD of the **fraction parts** is 30.

Whole-number parts

$$10\frac{5}{6} = 10\frac{25}{30}$$

Proper-fraction parts

$$-3\frac{8}{15} = -3\frac{16}{30}$$

Finish subtracting on your own. As always, we want to make sure that the **fraction part** of our answer is in **simplest form**. Check your **answer**, or watch this **video** to see a complete solution to part b.

You Try It **Work through this You Try It problem.**

Work Exercises 35–41 in this textbook or in the MyMathLab Study Plan.

When subtracting mixed numbers, sometimes the fraction part being subtracted is larger than the fraction part it is being subtracted from. When this occurs, we must "borrow" from the **whole-number part**.

Example 9 Subtracting Mixed Numbers with Borrowing

My video summary

Subtract. Write answers as a mixed number if possible.

a. $7\frac{3}{8} - \frac{5}{8}$ **b.** $4\frac{2}{5} - 1\frac{3}{4}$

Solutions

a. Writing the subtraction vertically, we get

$$\begin{array}{r} 7\frac{3}{8} \\ -\ \frac{5}{8} \\ \hline \end{array}$$

To subtract the fraction parts, we need to compute $\frac{3}{8} - \frac{5}{8}$.

However, because 5 is bigger than 3, we must borrow:

$$7\frac{3}{8} = 7 + \frac{3}{8} = \underbrace{6 + 1 + \frac{3}{8}}_{\text{Borrow 1 from the whole number}} = 6 + 1\frac{3}{8} = 6 + \underbrace{\frac{11}{8}}_{\frac{8\cdot 1+3}{8}} = 6\frac{11}{8}$$

The problem becomes

$$\begin{array}{r} 7\frac{3}{8} \\ -\ \frac{5}{8} \\ \hline \end{array} \longrightarrow \begin{array}{r} 6\frac{11}{8} \\ -\ \frac{5}{8} \\ \hline 6\frac{6}{8} \end{array}$$

$\boxed{6 - 0 = 6}$

Notice that $\frac{6}{8}$ can be simplified by **dividing out** a **common factor** of 2.

$$\frac{6}{8} = \frac{\not{2}\cdot 3}{\not{2}\cdot 4} = \frac{3}{4}$$

Therefore, $7\frac{3}{8} - \frac{5}{8} = 6\frac{3}{4}$.

My video summary **b.** We have denominators of 5 and 4. The **LCD** of the **fraction parts** is 20.

$$\begin{array}{rcr} 4\frac{2}{5} & = & 4\frac{8}{20} \\ -\ 1\frac{3}{4} & = & -\ 1\frac{15}{20} \\ \hline \end{array}$$

Notice that to subtract $\frac{8}{20} - \frac{15}{20}$ we must borrow because 15 is larger than 8. To do so, we can rewrite the first mixed number as follows:

$$4\frac{8}{20} = 4 + \frac{8}{20} = 3 + 1\frac{8}{20} = 3 + \frac{28}{20} = 3\frac{28}{20}$$

Finish subtracting on your own. Check your **answer**, or watch this **video** to see a complete solution to part b.

You Try It **Work through this You Try It problem.**

Work Exercises 42–50 in this textbook or in the MyMathLab Study Plan.

Example 10 Adding and Subtracting Mixed Numbers

My video summary Simplify: $8\frac{1}{3} + 1\frac{4}{9} - 3\frac{5}{18}$

Solution From the **order of operations**, we perform addition and subtraction from left to right, so our first operation is to add $8\frac{1}{3} + 1\frac{4}{9}$. The **LCD** of the fraction parts is 9.

$$\begin{array}{rcr} 8\frac{1}{3} & = & 8\frac{3}{9} \\ +\ 1\frac{4}{9} & = & +\ 1\frac{4}{9} \\ \hline & & 9\frac{7}{9} \end{array}$$

Finish solving this problem on your own by subtracting $9\frac{7}{9} - 3\frac{5}{18}$. Check your **answer**, or watch this **video** for a fully worked solution.

You Try It **Work through this You Try It problem.**

Work Exercises 51 and 52 in this textbook or in the MyMathLab Study Plan.

OBJECTIVE 5 SOLVE APPLICATIONS INVOLVING MIXED NUMBERS

We can use operations on mixed numbers to solve real-life application problems.

Example 11 iPhone Dimensions

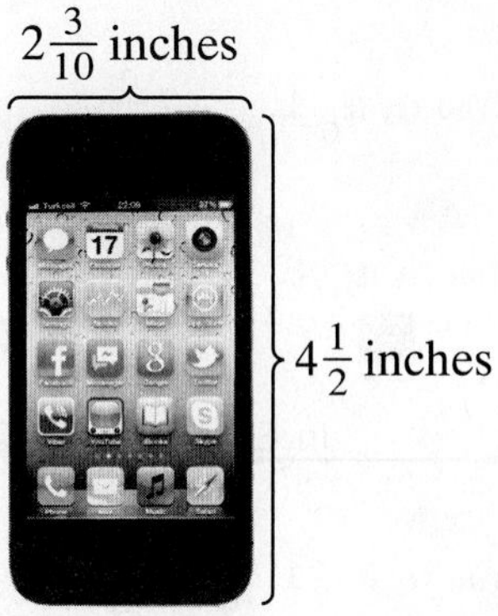

The iPhone 4 is rectangular in shape and measures $4\frac{1}{2}$ inches long by $2\frac{3}{10}$ inches wide. Find its **perimeter**.
(*Source:* www.apple.com)

Solution Recall from **Topic 1.3** that the perimeter of a **polygon** is found by adding up the lengths of the sides of the polygon. So, the perimeter of the iPhone 4 (in inches) is given by

$$4\frac{1}{2} + 2\frac{3}{10} + 4\frac{1}{2} + 2\frac{3}{10}.$$

Begin by adding $4\frac{1}{2} + 2\frac{3}{10}$.

$$\begin{array}{rcr} 4\frac{1}{2} & = & 4\frac{5}{10} \\ +\,2\frac{3}{10} & = & +\,2\frac{3}{10} \\ \hline & & 6\frac{8}{10} \text{ or } 6\frac{4}{5} \end{array}$$

Notice that the first two numbers and the last two numbers are the same.

$$\underbrace{4\frac{1}{2} + 2\frac{3}{10}}_{6\frac{4}{5}} + \underbrace{4\frac{1}{2} + 2\frac{3}{10}}_{6\frac{4}{5}}$$

So, we can find the **perimeter** by adding $6\frac{4}{5} + 6\frac{4}{5}$, or by multiplying $6\frac{4}{5}$ by 2.

Multiply $6\frac{4}{5}$ by 2: $2 \cdot 6\frac{4}{5}$

Convert to **improper fractions**: $= \frac{2}{1} \cdot \frac{34}{5} \leftarrow 6\frac{4}{5} = \frac{5 \cdot 6 + 4}{5} = \frac{34}{5}$

Multiply numerators; multiply denominators: $= \frac{68}{5}$

Convert back to a **mixed number**: $= 13\frac{3}{5}$

The perimeter of the iPhone 4 is $13\frac{3}{5}$ inches.

You Try It **Work through this You Try It problem.**

Work Exercises 53–60 in this textbook or in the MyMathLab Study Plan.

CAUTION Remember to include units in your answer when appropriate.

4.6 Exercises

In Exercises 1–10, multiply and simplify. Write your answer as a mixed number if possible.

You Try It 1. $\frac{1}{2}\cdot 7\frac{3}{10}$ 2. $8\frac{2}{7}\cdot\frac{3}{4}$ 3. $\frac{4}{9}\cdot 6\frac{3}{5}$ 4. $\frac{3}{5}\cdot 2\frac{7}{12}$ 5. $3\frac{1}{2}\cdot 2\frac{1}{3}$

You Try It 6. $4\frac{3}{5}\cdot 6\frac{2}{3}$ 7. $3\frac{2}{7}\cdot 1\frac{5}{6}$ 8. $1\frac{1}{4}\cdot 2\frac{2}{11}$ 9. $4\cdot 5\frac{3}{7}$ 10. $10\cdot 4\frac{2}{5}$

In Exercises 11–20, divide and simplify. Write your answer as a mixed number if possible.

You Try It 11. $\frac{8}{9}\div 2\frac{1}{4}$ 12. $4\frac{7}{8}\div\frac{2}{3}$ 13. $\frac{5}{6}\div 1\frac{1}{2}$ 14. $3\frac{13}{20}\div\frac{3}{10}$

You Try It 15. $6\frac{1}{3}\div 3\frac{1}{5}$ 16. $4\frac{3}{8}\div 7\frac{5}{9}$ 17. $2\frac{1}{6}\div 5\frac{7}{12}$ 18. $8\frac{5}{12}\div 4\frac{2}{3}$

19. $6\frac{3}{4}\div 8\frac{1}{3}$ 20. $3\frac{4}{5}\div 2\frac{1}{10}$

In Exercises 21–34, add and simplify. Write your answer as a mixed number if possible.

You Try It 21. $\frac{3}{7}+3\frac{1}{7}$ 22. $1\frac{1}{10}+\frac{3}{10}$ 23. $4\frac{1}{12}+3\frac{1}{4}$ 24. $4+3\frac{1}{6}$

25. $2\frac{3}{5}+7\frac{1}{10}$ 26. $3\frac{2}{9}+4\frac{5}{12}$ 27. $\frac{3}{8}+5\frac{7}{8}$ 28. $\frac{4}{5}+2\frac{3}{10}$

You Try It 29. $3\frac{8}{15}+2\frac{4}{5}$ 30. $8\frac{3}{16}+3\frac{7}{8}$ 31. $4\frac{3}{5}+2\frac{7}{8}$ 32. $4\frac{1}{2}+6+1\frac{1}{3}$

You Try It 33. $2\frac{4}{5}+3\frac{9}{10}+7$ 34. $5\frac{2}{7}+4\frac{1}{4}+9\frac{3}{5}$

In Exercises 35–50, subtract and simplify. Write your answer as a mixed number if possible.

You Try It 35. $5\frac{7}{9}-\frac{5}{9}$ 36. $8\frac{7}{10}-\frac{3}{10}$ 37. $7\frac{11}{12}-2\frac{2}{3}$ 38. $16\frac{4}{5}-9\frac{7}{10}$

39. $10\frac{7}{8}-4\frac{1}{6}$ 40. $7\frac{11}{12}-1\frac{5}{12}$ 41. $7\frac{9}{16}-3\frac{7}{20}$ 42. $16\frac{3}{10}-3\frac{7}{10}$

You Try It 43. $5\frac{1}{3}-\frac{7}{9}$ 44. $4\frac{3}{8}-2\frac{1}{2}$ 45. $7-3\frac{1}{5}$ 46. $10\frac{1}{20}-6\frac{1}{4}$

47. $3\frac{2}{9}-2\frac{5}{6}$ 48. $11\frac{1}{3}-7\frac{5}{6}$ 49. $8\frac{1}{6}-5\frac{7}{10}$ 50. $9\frac{5}{6}-3\frac{14}{15}$

You Try It 51. $3\frac{2}{5}+2\frac{1}{4}-4\frac{1}{10}$ 52. $7\frac{1}{6}-3\frac{1}{2}+5\frac{2}{9}$

You Try It 53. **Wood Dimensions** A standard 2×8 board actually has a rectangular cross section that measures $1\frac{1}{2}$ inches by $7\frac{1}{4}$ inches. Find the perimeter of the cross section.

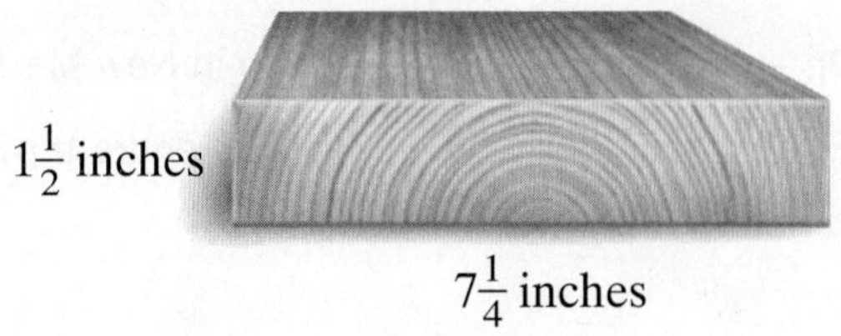

54. **Snickerdoodles** One batch of snickerdoodle cookies calls for $2\frac{3}{4}$ cups of flour. How many cups of flour would be needed to make 5 batches of cookies?

55. **Inflatable Movie Screen** A rectangular inflatable movie screen has a length of $10\frac{5}{12}$ feet and a width of $6\frac{1}{2}$ feet. What is the area of the screen?

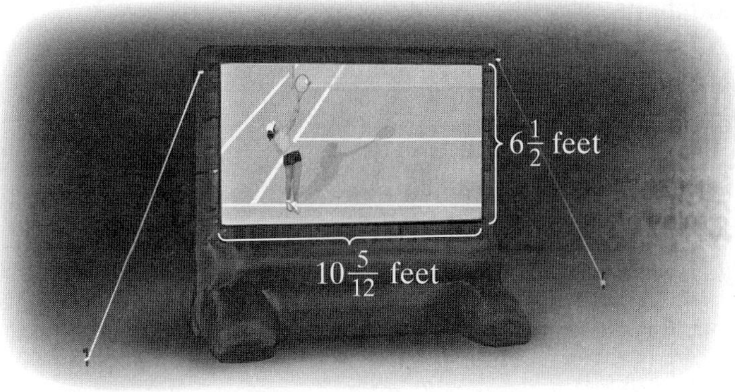

56. **Sports Drink** A sports drink bottle contains $2\frac{1}{2}$ cups of liquid. If a basketball player drinks $1\frac{1}{3}$ cups before a game, how much liquid is left in the bottle?

57. **Board Length** A board has a length of $21\frac{7}{8}$ feet and must be divided into 5 pieces of equal length. How long will each piece be?

58. **Fish Pond** A triangular fish pond has a base of $3\frac{1}{6}$ meters and a height of $1\frac{1}{4}$ meters. What is the area of the surface of the pond?

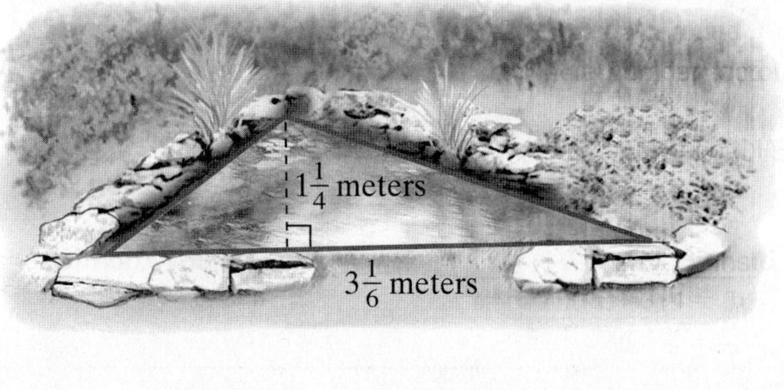

59. **Carpeting Rooms** Alisha is carpeting three rooms in her condo. She needs $18\frac{1}{4}$ square yards of carpet for a bedroom, $31\frac{5}{8}$ square yards for a family room, and $4\frac{1}{9}$ square yards for a bathroom. How many total square yards of carpet will she need?

60. **Backpack Trek** On an eight-day backpacking trip in New Mexico, Nicholas walked an average of $10\frac{2}{3}$ miles per day. How many miles did he walk for the entire trip?

4.7 **Solving Equations Containing Fractions**

THINGS TO KNOW

Before working through this topic, be sure you are familiar with the following concepts:

		VIDEO	ANIMATION	INTERACTIVE
You Try It	1. Determine If an Integer Is a Solution to an Equation (Topic 2.6, **Objective 1**)	▪		
You Try It	2. Solve Equations Using the Addition Property of Equality (Topic 3.2, **Objective 1**)	▪		
You Try It	3. Solve Equations Using the Multiplication Property of Equality (Topic 3.2, **Objective 2**)	▪		
You Try It	4. Solve Equations Using Both Properties of Equality (Topic 3.2, **Objective 3**)	▪		▪
You Try It	5. Multiply Fractions (Topic 4.3, **Objective 1**)	▪		▪
You Try It	6. Divide Fractions (Topic 4.3, **Objective 4**)	▪		
You Try It	7. Find the Least Common Denominator for Unlike Fractions (Topic 4.4, **Objective 2**)			▪
You Try It	8. Add and Subtract Unlike Fractions (Topic 4.4, **Objective 3**)			▪
You Try It	9. Add and Subtract Fractions with Variables (Topic 4.4, **Objective 4**)	▪		

OBJECTIVES

1 Determine If a Fraction Is a Solution to an Equation

2 Use the Properties of Equality to Solve Linear Equations Involving Fractions

3 Solve Linear Equations by Clearing Fractions

(eText Screens 4.7-1–4.7-23)

OBJECTIVE 1 DETERMINE IF A FRACTION IS A SOLUTION TO AN EQUATION

Recall that a **solution** of an **algebraic equation** is a value that, when **substituted** for the **variable**, makes the equation true. In **Topic 1.7**, we determined when **whole numbers** were **solutions** to equations, and in **Topic 2.6**, we determined when **integers** were solutions to equations. A **fraction** can also be a solution to an equation. For example, $\frac{1}{2}$ is a solution to $10x = 5$ because substituting $\frac{1}{2}$ for x results in the true statement

$$10\left(\frac{1}{2}\right) = 5.$$

As with whole numbers and integers, to determine if a given fraction is a solution to an equation, substitute the fraction for the variable and **simplify** both sides of the equation. If the resulting statement is true, then the fraction is a solution. If the resulting statement is false, then the fraction is not a solution.

Example 1 Determining If a Fraction Is a Solution to an Equation

My video summary

Determine if the given fraction is a **solution** to the equation.

a. $3x - 1 = 9x - 5;\ \frac{2}{3}$ **b.** $-8y = 9 - 4y;\ -\frac{3}{4}$

Solutions We **substitute** the given fraction for the variable and **simplify**. If the resulting statement is true, then the value is a solution to the equation.

a. Begin with the original equation: $3x - 1 = 9x - 5$

Substitute $\frac{2}{3}$ for x: $3\left(\frac{2}{3}\right) - 1 \stackrel{?}{=} 9\left(\frac{2}{3}\right) - 5$

Multiply: $2 - 1 \stackrel{?}{=} 6 - 5$ ←

$$3 \cdot \frac{2}{3} = \frac{\cancel{3}}{1} \cdot \frac{2}{\cancel{3}} = \frac{2}{1} = 2;$$
$$9 \cdot \frac{2}{3} = \frac{{}^{3}\cancel{9}}{1} \cdot \frac{2}{\cancel{3}} = \frac{3 \cdot 2}{1} = 6$$

Subtract: $1 = 1$ True

The final statement is true, so $\frac{2}{3}$ is a solution to the equation.

b. Try to work this problem on your own. View the **answer**, or watch this **video** to see a complete solution to both parts.

You Try It **Work through this You Try It problem.**

Work Exercises 1–4 in this textbook or in the MyMathLab Study Plan.

Example 2 Determining If a Value Is a Solution to an Equation Involving Fractions

My video summary

Determine if the given value is a **solution** to the equation.

a. $\frac{2}{5}x - \frac{3}{10} = \frac{1}{2};\ 2$

b. $\frac{5}{8}n + \frac{3}{4} = -\frac{5}{16}n;\ -\frac{4}{5}$

(eText Screens 4.7-1–4.7-23)

Solutions Work through the following, or watch this **video** for detailed solutions to both parts.

a. Begin with the original equation: $\frac{2}{5}x - \frac{3}{10} = \frac{1}{2}$

Substitute 2 for x: $\frac{2}{5}(2) - \frac{3}{10} \stackrel{?}{=} \frac{1}{2}$

Multiply: $\frac{4}{5} - \frac{3}{10} \stackrel{?}{=} \frac{1}{2} \leftarrow$ $\boxed{\frac{2}{5}\cdot 2 = \frac{2}{5}\cdot\frac{2}{1} = \frac{2\cdot 2}{5\cdot 1} = \frac{4}{5}}$

Write $\frac{4}{5}$ with a denominator of 10: $\frac{8}{10} - \frac{3}{10} \stackrel{?}{=} \frac{1}{2} \leftarrow$ $\boxed{\text{LCD} = 10; \frac{4}{5} = \frac{4\cdot 2}{5\cdot 2} = \frac{8}{10}}$

Subtract: $\frac{5}{10} \stackrel{?}{=} \frac{1}{2}$

Simplify: $\frac{1}{2} = \frac{1}{2}$ True

The final statement is true, so 2 is a solution to the equation.

b. Begin with the original equation: $\frac{5}{8}n + \frac{3}{4} = -\frac{5}{16}n$

Substitute $-\frac{4}{5}$ for n: $\frac{5}{8}\left(-\frac{4}{5}\right) + \frac{3}{4} \stackrel{?}{=} -\frac{5}{16}\left(-\frac{4}{5}\right)$

Multiply: $-\frac{1}{2} + \frac{3}{4} \stackrel{?}{=} \frac{1}{4} \leftarrow$ $\boxed{\begin{array}{l}\frac{5}{8}\left(-\frac{4}{5}\right) = -\frac{{}^{1}\cancel{5}\cdot\cancel{4}^{1}}{{}_{2}\cancel{8}\cdot\cancel{5}_{1}} = -\frac{1}{2}; \\ -\frac{5}{16}\left(-\frac{4}{5}\right) = \frac{{}^{1}\cancel{5}\cdot\cancel{4}^{1}}{{}_{4}\cancel{16}\cdot\cancel{5}_{1}} = \frac{1}{4}\end{array}}$

Write $-\frac{1}{2}$ with a denominator of 4: $-\frac{2}{4} + \frac{3}{4} \stackrel{?}{=} \frac{1}{4} \leftarrow$ $\boxed{\text{LCD} = 4; -\frac{1}{2} = -\frac{1\cdot 2}{2\cdot 2} = -\frac{2}{4}}$

Add: $\frac{1}{4} = \frac{1}{4}$ True

The final statement is true, so $-\frac{4}{5}$ is a solution to the equation.

You Try It **Work through this You Try It problem.**

Work Exercises 5–8 in this textbook or in the MyMathLab Study Plan.

OBJECTIVE 2 USE THE PROPERTIES OF EQUALITY TO SOLVE LINEAR EQUATIONS INVOLVING FRACTIONS

In **Topic 2.6** and again in **Module 3**, we used the **addition property of equality** and the **multiplication property of equality** to solve **linear equations** involving **integers**. Next, we use these properties to **solve** linear equations involving **fractions**.

Recall that the addition property allows the same quantity to be added to or subtracted from both sides of an equation, while the multiplication property allows both sides of the equation to be multiplied or divided by the same **nonzero** quantity. We find the **solution** to the equation by **isolating the variable** in the form

$$\textit{variable} = \textit{value} \quad \text{or} \quad \textit{value} = \textit{variable}.$$

Example 3 Using the Addition Property to Solve Equations Involving Fractions

Solve each equation.

a. $x - \frac{3}{5} = \frac{1}{5}$ **b.** $\frac{9}{14} = m + \frac{3}{14}$

Solutions

a. To **isolate** x on the left side of the equal sign, we can add $\frac{3}{5}$ to both sides.

Begin with the original equation: $x - \frac{3}{5} = \frac{1}{5}$

Add $\frac{3}{5}$ to both sides: $x - \frac{3}{5} + \frac{3}{5} = \frac{1}{5} + \frac{3}{5}$ ← **Addition property of equality**

Simplify both sides: $x = \frac{4}{5}$ ← **Isolated variable**

View the **check**. The solution is $\frac{4}{5}$.

b. To **isolate** m on the right side of the equal sign, we can subtract $\frac{3}{14}$ from both sides of the equation.

Begin with the original equation: $\frac{9}{14} = m + \frac{3}{14}$

Subtract $\frac{3}{14}$ from both sides: $\frac{9}{14} - \frac{3}{14} = m + \frac{3}{14} - \frac{3}{14}$ ← **Addition property of equality**

Simplify both sides: $\frac{6}{14} = m$

Isolated variable

Write $\frac{6}{14}$ in **lowest terms**: $\frac{3}{7} = m$

$\frac{6}{14} = \frac{2 \cdot 3}{2 \cdot 7} = \frac{\cancel{2} \cdot 3}{\cancel{2} \cdot 7} = \frac{3}{7}$

View the **check**. The **solution** is $\frac{3}{7}$.

You Try It **Work through this You Try It problem.**

Work Exercises 9 and 10 in this textbook or in the MyMathLab Study Plan.

Example 4 Using the Addition Property to Solve Equations Involving Fractions

Solve each equation.

a. $y + \frac{5}{4} = \frac{11}{12}$ **b.** $\frac{2}{15} = x - \frac{7}{10}$

(eText Screens 4.7-1–4.7-23)

Solutions

a. To **isolate** y on the left side of the equal sign, we can subtract $\frac{5}{4}$ from both sides of the equation.

Begin with the original equation: $y + \frac{5}{4} = \frac{11}{12}$

Subtract $\frac{5}{4}$ from both sides: $y + \frac{5}{4} - \frac{5}{4} = \frac{11}{12} - \frac{5}{4}$ ← **Addition property of equality**

The **LCD** of the fractions on the right side is 12.

Write $\frac{5}{4}$ as $\frac{15}{12}$ on the right: $y + \frac{5}{4} - \frac{5}{4} = \frac{11}{12} - \frac{15}{12}$ ← $\frac{5}{4} = \frac{5 \cdot 3}{4 \cdot 3} = \frac{15}{12}$

Simplify both sides: $y = -\frac{4}{12}$

Write fraction in **lowest terms**: $y = -\frac{1}{3}$ ← $-\frac{4}{12} = -\frac{4}{4 \cdot 3} = -\frac{\cancel{4}}{\cancel{4} \cdot 3} = -\frac{1}{3}$

Isolated variable (→ y)

View the **check**. The **solution** is $-\frac{1}{3}$.

My video summary

b. Try solving this equation on your own. View the **answer**, or watch this **video** for a complete solution to part b.

You Try It **Work through this You Try It problem.**

Work Exercises 11–16 in this textbook or in the MyMathLab Study Plan.

Recall that the **product** of a **nonzero** number and its **reciprocal** is 1. For example,

$$\frac{5}{8} \cdot \frac{8}{5} = 1, \quad \left(-\frac{2}{3}\right)\left(-\frac{3}{2}\right) = 1, \quad \text{and} \quad \frac{1}{7} \cdot 7 = 1.$$

We can use reciprocals with the **multiplication property of equality** to solve equations.

Example 5 Using the Multiplication Property to Solve Equations Involving Fractions

Solve. $\frac{3}{4}x = 15$

Solution To **isolate** x on the left side of the equation, we can multiply both sides by the reciprocal of $\frac{3}{4}$, which is $\frac{4}{3}$.

Begin with the original equation: $\frac{3}{4}x = 15$

Multiply both sides by $\frac{4}{3}$: $\frac{4}{3} \cdot \frac{3}{4}x = \frac{4}{3} \cdot 15$ ← **Multiplication property of equality**

On each side, divide out **common factors**: $\frac{\cancel{4} \cdot \cancel{3}}{\cancel{3} \cdot \cancel{4}}x = \frac{4 \cdot \cancel{15}^{5}}{\cancel{3}}$

Simplify each side: $\quad 1\cdot x = 4\cdot 5$

$$x = 20$$

View the **check**. The **solution** is 20.

Example 6 Using the Multiplication Property to Solve Equations Involving Fractions

My interactive video summary

Solve each equation.

a. $35 = -\frac{5}{7}k$ **b.** $\frac{w}{6} = \frac{5}{9}$ **c.** $20n = -\frac{4}{5}$ **d.** $60 = 8p$

Solutions

a. To **isolate** k on the right side of the equation, we can multiply both sides by the **reciprocal** of $-\frac{5}{7}$, which is $-\frac{7}{5}$.

Begin with the original equation: $\quad 35 = -\frac{5}{7}k$

Multiply both sides by $-\frac{7}{5}$: $\quad -\frac{7}{5}\cdot 35 = -\frac{7}{5}\cdot\left(-\frac{5}{7}k\right)$ ← Multiplication property of equality

On the left side, the multiplication involves a negative number and a positive number, so the product will be negative. On the right side, the multiplication involves two negatives, so the result will be positive.

On each side, divide out **common factors**: $\quad -\frac{7\cdot \cancel{35}^{7}}{\cancel{5}} = \left(-\frac{\cancel{7}}{\cancel{5}}\cdot -\frac{\cancel{5}}{\cancel{7}}\right)k$

Simplify each side: $\quad -7\cdot 7 = 1\cdot k$

$$-49 = k$$

View the **check**. The solution is -49.

b. Note that $\frac{w}{6}$ is equivalent to $\frac{1}{6}\cdot w$. To **isolate** w on the left side of the equation, we can multiply both sides by the **reciprocal** of $\frac{1}{6}$, which is 6.

Begin with the original equation: $\quad \frac{w}{6} = \frac{5}{9}$

Multiply both sides by 6: $\quad 6\cdot\frac{w}{6} = 6\cdot\frac{5}{9}$ ← Multiplication property of equality

Try to finish solving this equation on your own. View the **answer** to part b, or watch this **interactive video** for complete solutions to all four parts.

c. To **isolate** n on the left side of the equation, we can multiply both sides by the **reciprocal** of 20, which is $\frac{1}{20}$. Try to solve this equation on your own. View the **answer** to part c, or watch this **interactive video** for complete solutions to all four parts.

d. Try to solve this equation on your own. View the **answer** to part d, or watch this **interactive video** for complete solutions to all four parts.

(eText Screens 4.7-1–4.7-23)

You Try It **Work through this You Try It problem.**

Work Exercises 17–24 in this textbook or in the MyMathLab Study Plan.

To **solve** some **equations**, we need to use both **properties of equality.**

Example 7 Using Both Properties of Equality to Solve Equations Involving Fractions

My video summary

Solve. $\frac{3}{10}x + \frac{7}{5} = \frac{1}{5}$

Solution The only **variable term,** $\frac{3}{10}x$, is on the left side of the equation. To get the $\frac{3}{10}x$ by itself, we can subtract $\frac{7}{5}$ from both sides.

Begin with the original equation: $\frac{3}{10}x + \frac{7}{5} = \frac{1}{5}$

Subtract $\frac{7}{5}$ from both sides: $\frac{3}{10}x + \frac{7}{5} - \frac{7}{5} = \frac{1}{5} - \frac{7}{5}$ ← Addition property of equality

Simplify both sides: $\frac{3}{10}x = -\frac{6}{5}$

Next, to **isolate** x, we can multiply both sides by $\frac{10}{3}$, the **reciprocal** of $\frac{3}{10}$. Try to finish solving this equation on your own. View the **answer**, or watch this **video** for a complete solution.

You Try It **Work through this You Try It problem.**

Work Exercises 25–30 in this textbook or in the MyMathLab Study Plan.

OBJECTIVE 3 SOLVE LINEAR EQUATIONS BY CLEARING FRACTIONS

If we multiply both sides of an **equation** by the **LCD** of all the **fractions** in the equation, we can **clear the fractions** from the equation. Typically, this action results in an **equivalent equation** that is less tedious to **solve**. Let's revisit the equations from **Example 4** and solve them by first clearing the fractions.

Example 8 Solving Equations by Clearing Fractions

Solve each equation by first clearing the fractions.

a. $y + \frac{5}{4} = \frac{11}{12}$

b. $\frac{2}{15} = x - \frac{7}{10}$

(eText Screens 4.7-1–4.7-23)

Solutions

a. The fractions in this equation are $\frac{5}{4}$ and $\frac{11}{12}$. Because 12 is a multiple of 4, the LCD is 12.

Begin with the original equation: $y + \frac{5}{4} = \frac{11}{12}$

Multiply both sides by the LCD, 12: $12\left(y + \frac{5}{4}\right) = 12\left(\frac{11}{12}\right)$ ← **Multiplication property of equality**

Distribute on the left: $12y + 12 \cdot \frac{5}{4} = 12 \cdot \frac{11}{12}$

Simplify on both sides: $12y + 15 = 11$ ← $\overset{3}{\cancel{12}} \cdot \frac{5}{\cancel{4}} = 15;\ \cancel{12} \cdot \frac{11}{\cancel{12}} = 11$

We now have an equation to solve that has no fractions.

Subtract 15 from both sides: $12y + 15 - 15 = 11 - 15$ ← **Addition property of equality**

Simplify: $12y = -4$

Divide both sides by 12: $\frac{12y}{12} = \frac{-4}{12}$ ← **Multiplication property of equality**

Simplify on both sides: $y = -\frac{1}{3}$ ← $\frac{\cancel{12}y}{\cancel{12}} = y;\ -\frac{\cancel{4}}{\cancel{12}_3} = -\frac{1}{3}$

Note that this is the same result from **Example 4a**.

The **solution** is $-\frac{1}{3}$. View the **check**.

My video summary

b. The fractions in this equation are $\frac{2}{15}$ and $-\frac{7}{10}$. The **LCD** is 30. Try to clear the fractions in this equation on your own. Remember to multiply both sides by the LCD, 30. Then finish solving the equation. Compare the **answer** to the one found in Example 4b. Watch this **video** for a complete solution to part b.

You Try It **Work through this You Try It problem.**

Work Exercises 31 and 32 in this textbook or in the MyMathLab Study Plan.

In **Topic 3.3**, we developed a **general strategy for solving linear equations in one variable.** Let's now add another step for clearing fractions.

A General Strategy for Solving Linear Equations in One Variable

Step 1. If the equation contains **fractions**, multiply both sides by the **LCD** of all the fractions.

Step 2. Remove grouping symbols using the **distributive property**.

Step 3. Simplify each side of the equation by **combining like terms.**

Step 4. Use the **addition property of equality** to collect all **variable terms** on one side of the equation and all **constant terms** on the other side.

Step 5. Use the **multiplication property of equality** to **isolate the variable.**

Step 6. Check the result in the original equation.

(eText Screens 4.7-1–4.7-23)

Example 9 Solving Equations by Clearing Fractions

Solve each equation by first clearing the fractions.

a. $\frac{x}{6} - \frac{x}{4} = -3$ **b.** $\frac{n}{5} = \frac{n}{15} + \frac{2}{3}$

Solutions

a. We follow the **general strategy for solving linear equations in one variable.**

Begin with the original equation: $\frac{x}{6} - \frac{x}{4} = -3$

Step 1. The **LCD** of the fractions is 12.

Multiply both sides by the LCD, 12: $12\left(\frac{x}{6} - \frac{x}{4}\right) = 12(-3)$

Step 2. Use the **distributive property:** $12 \cdot \frac{x}{6} - 12 \cdot \frac{x}{4} = 12(-3)$

Step 3. Simplify on both sides: ${}^{2}\cancel{12} \cdot \frac{x}{\cancel{6}} - {}^{3}\cancel{12} \cdot \frac{x}{\cancel{4}} = 12(-3)$

$2x - 3x = -36$

Step 4. All of the **variable terms** are already on the left side, and the only **constant term** is already on the right side.

Combine like terms: $-x = -36$

Step 5. Divide both sides by -1: $\frac{-x}{-1} = \frac{-36}{-1}$

Simplify: $x = 36$

Step 6. View the **check**. The **solution** is 36.

My video summary

b. Try to solve this equation on your own by clearing the fractions. View the **answer**, or watch this **video** for a complete solution to part b.

You Try It **Work through this You Try It problem.**

Work Exercises 33–40 in this textbook or in the MyMathLab Study Plan.

4.7 Exercises

In Exercises 1–8, determine if the given value is a solution to the equation.

You Try It

1. $8x + 7 = 12x + 4$; $\frac{3}{4}$
2. $15n + 2 = 5n - 2$; $-\frac{1}{5}$
3. $-4y = -6y - 7$; $-\frac{7}{2}$
4. $-2(k - 1) = 4k - 3$; $\frac{5}{6}$

You Try It

5. $\frac{7}{16}x - \frac{1}{8} = \frac{1}{4}x$; 2
6. $\frac{t}{3} + 4 = \frac{t}{2} + 10$; -36
7. $\frac{1}{2}x - \frac{1}{3} = \frac{1}{9}$; $\frac{8}{9}$
8. $\frac{7}{6}m - \frac{1}{4} = -\frac{2}{3}$; $-\frac{5}{14}$

(eText Screens 4.7-1–4.7-23)

In Exercises 9–30, use the properties of equality to solve each equation.

You Try It 9. $x - \frac{8}{17} = -\frac{2}{17}$

10. $y + \frac{11}{4} = \frac{9}{4}$

11. $w + \frac{1}{3} = \frac{7}{9}$

12. $\frac{7}{3} + a = \frac{14}{15}$

You Try It 13. $\frac{2}{3} = t + \frac{4}{5}$

14. $m - \frac{3}{10} = -\frac{5}{8}$

15. $\frac{5}{6} + n = \frac{5}{9} - \frac{1}{3}$

16. $\frac{5}{8} + \frac{7}{12} = 2 + y$

You Try It 17. $\frac{2}{3}x = 12$

18. $\frac{7}{4}p = -56$

19. $-\frac{1}{5}y = -\frac{3}{20}$

20. $\frac{n}{12} = \frac{3}{20}$

21. $-16 = 56t$

22. $-\frac{13}{40}x = \frac{39}{20}$

23. $8n = \frac{10}{21}$

24. $\frac{5}{16} = -\frac{7}{8}w$

You Try It 25. $\frac{5}{6}x - \frac{4}{3} = \frac{1}{3}$

26. $5 = 9 - \frac{6}{5}x$

27. $15x - 7 = 2$

28. $\frac{1}{4}y + \frac{1}{2} = -\frac{1}{8}$

29. $-\frac{2}{7}x + \frac{1}{10} = \frac{3}{5}$

30. $\frac{x}{10} + \frac{4}{5} = \frac{17}{20}$

In Exercises 31–40, solve each equation by first clearing the fractions.

You Try It 31. $x - \frac{7}{16} = -\frac{5}{8}$

32. $\frac{1}{4} = n - \frac{7}{10}$

33. $\frac{2}{3}p + 1 = \frac{1}{2}$

34. $\frac{9}{16} - \frac{5}{8}t = \frac{3}{16}$

You Try It 35. $\frac{x}{5} - \frac{x}{3} = 6$

36. $\frac{n}{12} = \frac{n}{4} - \frac{4}{3}$

37. $\frac{2}{9}w = 1 + \frac{1}{6}w$

38. $\frac{13}{10}k + \frac{9}{5} = \frac{1}{15}$

39. $\frac{3}{7}y + \frac{5}{14} = \frac{1}{2}y$

40. $\frac{1}{4} + \frac{1}{3}x = \frac{1}{2}x + \frac{5}{6}$

MODULE FIVE

Decimals

MODULE FIVE CONTENTS

5.1 Introduction to Decimals

THINGS TO KNOW

Before working through this topic, be sure you are familiar with the following concepts:

		VIDEO	ANIMATION	INTERACTIVE
You Try It	1. Identify Place Value of a Digit in a Whole Number (Topic 1.2, **Objective 1**)			
You Try It	2. Write Whole Numbers in Standard Form and Word Form (Topic 1.2, **Objective 2**)	▣		
You Try It	3. Round Whole Numbers (Topic 1.2, **Objective 5**)	▣		
You Try It	4. Use Inequality Symbols to Compare Integers (Topic 2.1, **Objective 2**)			
You Try It	5. Graph Fractions and Mixed Numbers on a Number Line (Topic 4.1, **Objective 3**)	▣		
You Try It	6. Write Fractions in Simplest Form (Topic 4.2, **Objective 6**)	▣		

OBJECTIVES

1 Identify the Place Value of a Digit in a Decimal Number

2 Write Decimals in Word Form

3 Change Decimals from Words to Standard Form

(*eText Screens 5.1-1–5.1-27*)

4 Change Decimals into Fractions or Mixed Numbers

5 Graph Decimals on a Number Line

6 Compare Decimals

7 Round Decimals to a Given Place Value

OBJECTIVE 1 IDENTIFY THE PLACE VALUE OF A DIGIT IN A DECIMAL NUMBER

As with **fractions**, a **decimal number**, or simply a **decimal**, can be used to describe a part of a whole. As we saw with whole numbers in **Topic 1.2**, the position of each **digit** in a decimal number determines its **place value**. For example, the per capita consumption of energy drinks in the United States in 2010 was 12.891596 cups. (*Source:* Mintel) The figure below shows this number on a **place-value chart**. In the chart, the digit 8 represents a value of eight-tenths, or $\frac{8}{10}$. The digit 6 represents a value of six-millionths, or $\frac{6}{1{,}000{,}000}$.

Digits appearing in front of the **decimal point** form the **whole-number part** of the decimal number. Digits appearing behind the decimal point form the **decimal part** (or **fractional part**) of the number. In the number 12.891596, for example, the "12" is the whole-number part and the "891596" is the decimal part.

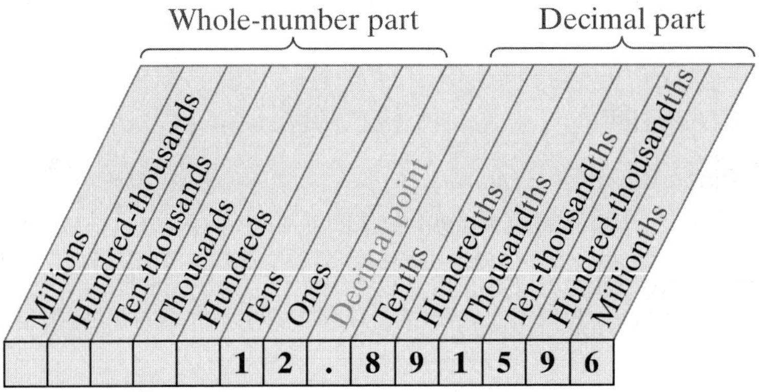

Place-Value Chart

CAUTION **Place values** that occur in front of the **decimal point** end with "s" and represent whole-number values, but place values that occur behind the decimal point end with "ths" and represent fractional values.

Example 1 Identifying Place Value in Decimals

Identify the place value of the **digit** 4 in each **decimal number**.

a. 32.10483 **b.** 29.415 **c.** 6.8421 **d.** 42.063

Solutions

a. View this **popup** to see 32.10483 on a place-value chart. The digit 4 is in the thousandths place.

b. View this **popup** to see 29.415 on a place-value chart. The digit 4 is in the tenths place.

(eText Screens 5.1-1–5.1-27)

c. View this **popup** to see 6.8421 on a place-value chart. The digit 4 is in the hundredths place.

d. View this **popup** to see 42.063 on a place-value chart. The digit 4 is in the tens place.

You Try It **Work through this You Try It problem.**

Work Exercises 1–8 in this textbook or in the MyMathLab Study Plan.

OBJECTIVE 2 WRITE DECIMALS IN WORD FORM

We read and write **decimal numbers** in a manner similar to the way we read and write **whole numbers**. For example, the whole number 187 is written in words as *one hundred eighty-seven*. If these same three **digits** fall directly behind a **decimal point**, then we write the decimal number in the same way, but we follow it with the **place value** name of the digit in the last position. For the decimal number 0.187, the last digit 7 is in the thousandths place. So, 0.187 is written as *one hundred eighty-seven thousandths*.

TIP Typically, when a decimal has no **whole-number part**, the digit 0 is written in front of the decimal point as the whole-number part. This 0 is not absolutely required, but it makes reading the decimal easier.

> **Writing a Decimal Number in Words**
>
> **Step 1.** Write the whole-number part in words.
>
> **Step 2.** Write the word "and" for the decimal point.
>
> **Step 3.** Write the **decimal part** as if it were a whole number and follow it with the place value name of the last digit.

The **decimal number** 349.561 is written as *three hundred forty-nine and five hundred sixty-one thousandths*.

349 · 561

↑ ↑ ↑

Three hundred forty-nine and five hundred sixty-one thousandths

Example 2 Writing Decimal Numbers in Word Form

My video summary

Write each decimal number in words.

a. 63.52 **b.** 259.3184 **c.** 0.436 **d.** 400.036

Solutions

a. The **whole-number part** is 63, written in words as *sixty-three*. The **decimal part** is .52. Without the **decimal point**, the two **digits** would form the number 52, written in words as *fifty-two*. The last digit, 2, is in the hundredths **place**, so .52 is *fifty-two hundredths*. Including the word "and" for the decimal point, we write 63.52 in words as *sixty-three and fifty-two hundredths*.

(*eText Screens 5.1-1–5.1-27*)

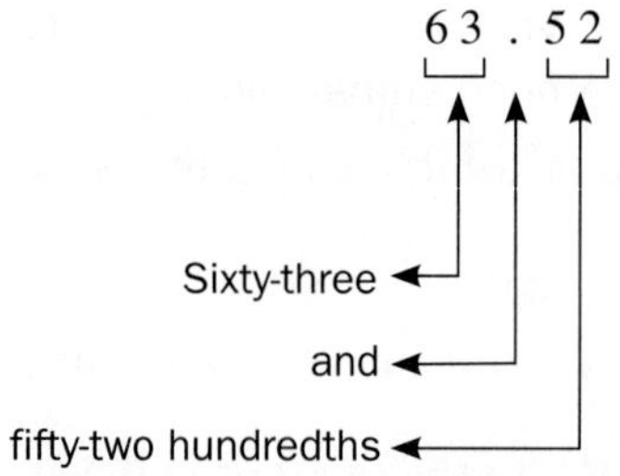

b. The **whole-number part** is 259. The last **digit** of the **decimal part** .3184 is in the ten-thousandths place. The number is *two hundred fifty-nine and three thousand, one hundred eighty-four ten-thousandths*.

c. This number has no whole-number part. The last digit of the decimal part is in the *thousandths* place. Try to finish writing the word form on your own. View the **answer**, or watch this **video** for detailed solutions to all four parts.

d. Try to write this word form on your own. View the **answer**, or watch this **video** for detailed solutions to all four parts.

You Try It **Work through this You Try It problem.**

Work Exercises 9–16 in this textbook or in the MyMathLab Study Plan.

CAUTION

There is a big difference between the values of 0.436 and 400.036 from Example 2, but when reading or writing these numbers, the only difference is the word "and." Proper use of the word "and" is important!

OBJECTIVE 3 CHANGE DECIMALS FROM WORDS TO STANDARD FORM

Decimals expressed using **digits**, such as 1.79, are written in **standard form**. To change a decimal from word form to standard form, recognize that the **whole-number part** comes *before* the word "and," and the **decimal part** comes *afterward*. So, we write the whole-number part, then a decimal point, and finally the decimal part. To be sure the last digit of the decimal part falls in the correct **place**, we may need to include 0's as the beginning digits of the decimal part.

Changing a Decimal from Words to Standard Form

Step 1. Form the whole-number part, which comes before the word "and."

Step 2. Write a decimal point in place of the word "and."

Step 3. Form the decimal part as if it were a whole number, making sure the last digit falls in the correct place. Insert 0's, if necessary, as the beginning digits.

Example 3 Changing Decimals from Words to Standard Form

My video summary

Change each decimal to standard form.

a. Eight and fifteen hundredths

b. Four hundred sixty-three and five hundred seventy-one ten-thousandths

c. Two hundred eight thousandths

d. Two hundred and eight thousandths

(eText Screens 5.1-1–5.1-27)

Solutions

a. The word "and" separates the **whole-number part** from the **decimal part.** The word "hundredths" tells us that the last **digit** of the decimal part is in the hundredths **place,** which is the second place behind the **decimal point.** So, the whole-number part is 8 and the decimal part is .15. The standard form is 8.15.

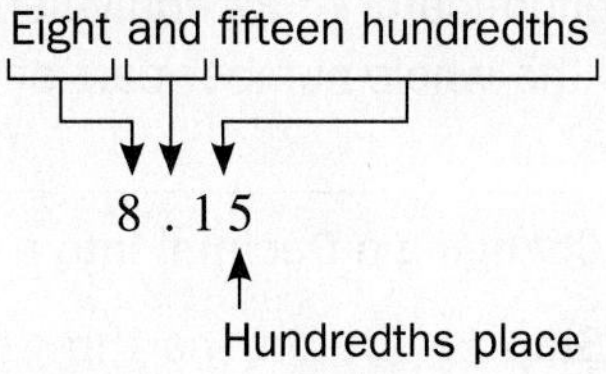

b. The whole-number part is 463. The word "ten-thousandths" tells us that the last digit of the decimal part is in the ten-thousandths place, which is the fourth place behind the decimal point. Because 571 has only three digits, we must place a 0 between it and the decimal point to form the decimal part .0571. The standard form is 463.0571.

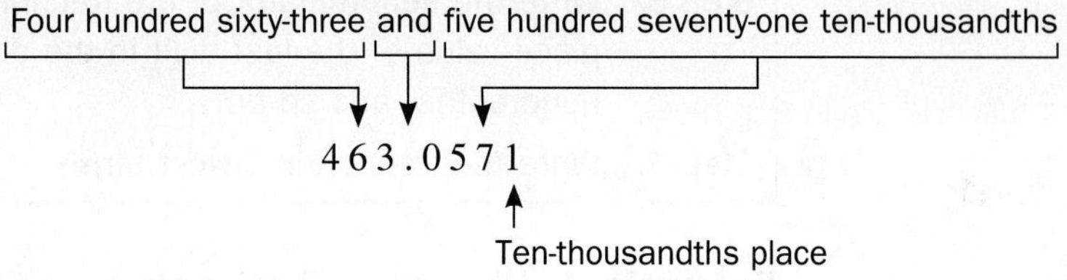

c. The word "thousandths" tells us that the last digit of the decimal part is in the thousandths place, which is the third place behind the decimal point. The word "and" does not appear in the written phrase, so there is no whole-number part (or the whole-number part is zero). Try to finish writing the standard form on your own. View the **answer**, or watch this **video** for detailed solutions to all four parts.

d. Try to write this standard form on your own. View the **answer**, or watch this **video** for detailed solutions to all four parts.

You Try It **Work through this You Try It problem.**

Work Exercises 17–24 in this textbook or in the MyMathLab Study Plan.

OBJECTIVE 4 CHANGE DECIMALS INTO FRACTIONS OR MIXED NUMBERS

Being familiar with the **place-value system** is key to changing a **decimal** into a **fraction** or **mixed number.** Consider the decimal 0.3 and the fraction $\frac{3}{10}$. Both of these numbers are read as *three tenths* because they are equal in value.

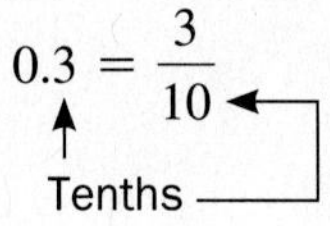

Similarly, the decimal 2.47 and the mixed number $2\frac{47}{100}$ are both read as *two and forty-seven hundredths* because they are equal in value.

(eText Screens 5.1-1–5.1-27)

$$2.47 = 2\frac{47}{100}$$

Whole-number part

Hundredths

We can see that the **place value** of the last **digit** in the **decimal number** determines the denominator of its equivalent fraction. Also, any **whole-number part** of the decimal number is the whole-number part of its equivalent **mixed number**.

> **Changing a Decimal into a Fraction or Mixed Number**
>
> **Step 1.** Determine the whole-number part of the decimal number, and write it as the whole-number part of a mixed number. (Note: If there is no whole-number part, then the decimal number will become a **proper fraction**.)
>
> **Step 2.** Write the digits from the **decimal part** as the **numerator** of the fraction. Exclude any leading zeros.
>
> **Step 3.** Write the **denominator** of the fraction as 10, 100, 1000, etc., depending on the place value of the last digit in the decimal number. Use 10 for tenths, 100 for hundredths, and so on.
>
> **Step 4.** Write the fraction in **lowest terms**.

Example 4 Changing Decimals into Fractions or Mixed Numbers

My video summary

Change each decimal into a fraction or mixed number. **Simplify** if necessary.

a. 0.69 **b.** 3.8 **c.** 0.375 **d.** 9.005

Solutions We follow the **steps** for changing a decimal into a fraction or mixed number.

a. There is no **whole-number part**, so this decimal will change to a **proper fraction**. Write the 69 from the **decimal part** in the **numerator** of the fraction. The last **digit** is in the hundredths place, so the **denominator** of the fraction is 100.

$$0.69 = \frac{69}{100}$$

Hundredths

Because 69 and 100 have no **common factors** other than 1, the fraction $\frac{69}{100}$ is in **lowest terms**.

b. The whole-number part is 3. Write the 8 from the decimal part in the numerator of the **fraction**. The last digit is in the tenths place, so the denominator of the fraction is 10.

$$3.8 = 3\frac{8}{10}$$

Tenths

Whole-number part

The fraction part simplifies: $\frac{8 \div 2}{10 \div 2} = \frac{4}{5}$. So, $3.8 = 3\frac{8}{10} = 3\frac{4}{5}$.

c. There is no whole-number part, so this decimal will change to a **proper fraction**. Write the 375 from the decimal part in the numerator of the fraction. The last digit is in the thousandths place, so the denominator of the fraction is 1000.

$$0.375 = \frac{375}{1000}$$

Thousandths

Try to finish this problem on your own by writing this fraction in **lowest terms**. Check the **answer**, or watch this **video** for complete solutions to all four parts.

d. Try working this problem on your own. View the **answer**, or watch this **video** for complete solutions to all four parts.

You Try It **Work through this You Try It problem.**

Work Exercises 25–32 in this textbook or in the MyMathLab Study Plan.

OBJECTIVE 5 GRAPH DECIMALS ON A NUMBER LINE

In **Topic 4.1**, we graphed **fractions** and **mixed numbers** on a **number line**. We can graph **decimals** in a similar way. For example, consider 0.3. The last digit has a tenths **place value**, so we divide the distance between 0 and 1 into 10 equal parts. Each part represents one-tenth, or 0.1, of the **unit distance**. Starting at 0 and moving 3 places to the right gives the location of 0.3. We graph it by placing a solid circle at this location as shown in Figure 1.

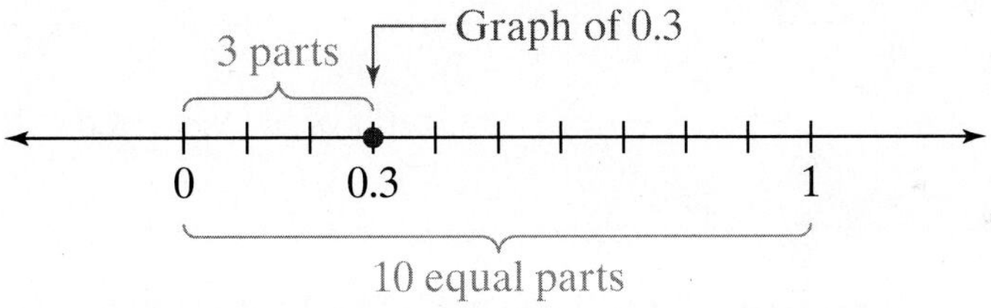

Figure 1

We can graph -0.3 in a similar way. Divide the distance between -1 and 0 into 10 equal parts. Starting at 0 and moving 3 places to the *left* gives the location of -0.3. We graph it by placing a solid circle at its location as shown in Figure 2.

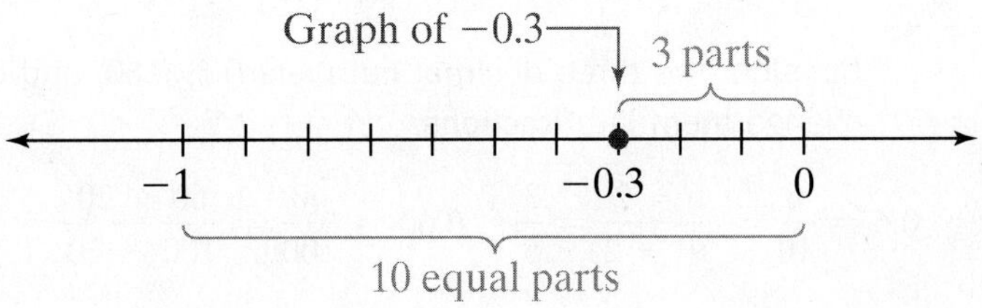

Figure 2

Next, consider 2.47. The last digit has a hundredths **place value,** so we divide the distance between 2 and 3 into 100 equal parts. Each part represents one-hundredth, or 0.01, of the **unit distance.** Starting at 2 and moving 47 places to the right gives the location of 2.47. We place a solid circle at this location to graph 2.47 as shown in **Figure 3**.

(eText Screens 5.1-1–5.1-27)

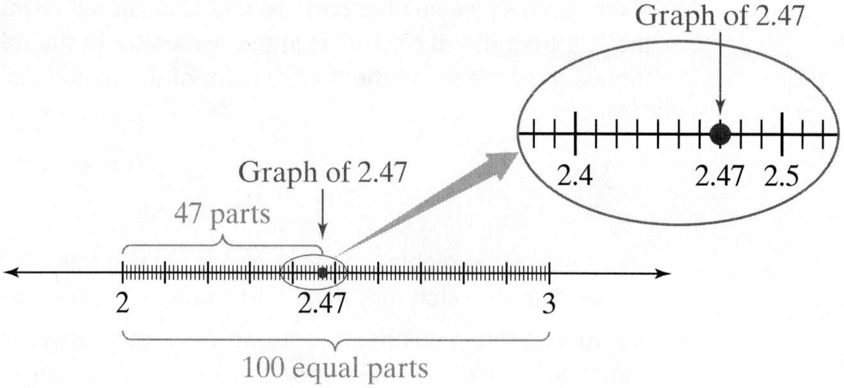

Figure 3

TIP Typically, when graphing a number with two or more decimal places, rather than trying to accurately divide the unit distance into 100 or more equal parts, we estimate the number's location on the number line. We demonstrate this in Example 5c and d.

Example 5 Graphing Decimals on a Number Line

My video summary

Graph each **decimal** on a **number line**.

a. 0.6 **b.** −0.7 **c.** 1.76 **d.** −1.33

Solutions The decimals are plotted on the number line in Figure 4. Watch this **video** for complete solutions.

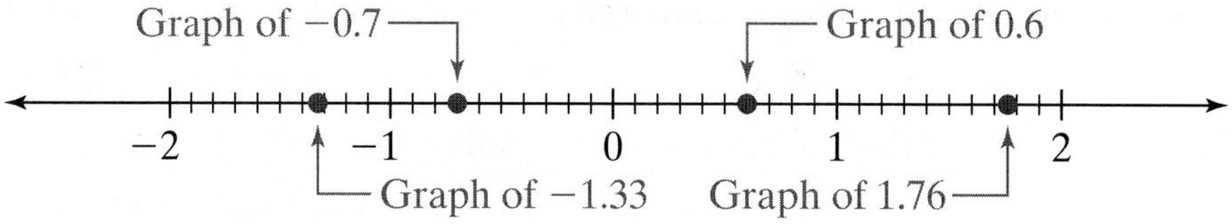

Figure 4

You Try It **Work through this You Try It problem.**

Work Exercises 33–40 in this textbook or in the MyMathLab Study Plan.

OBJECTIVE 6 COMPARE DECIMALS

Consider the three **decimal numbers** 0.6, 0.60, and 0.600. Look at what happens when we change them into **fractions**.

$$0.6 = \frac{6}{10} = \frac{6 \div 2}{10 \div 2} = \frac{3}{5}, \quad 0.60 = \frac{60}{100} = \frac{60 \div 20}{100 \div 20} = \frac{3}{5}, \quad 0.600 = \frac{600}{1000} = \frac{600 \div 200}{1000 \div 200} = \frac{3}{5}$$

All three decimals result in the same simplified fraction, so

$$0.6 = 0.60 = 0.600.$$

This illustrates that a 0 **digit** can be written behind the last digit to the right of a **decimal point** without changing the value of the number.

CAUTION A 0 digit that is written in front of the last digit to the right of a decimal point changes the value of the number. For example, 0.6 and 0.06 have different values. That is, $0.6 \neq 0.06$.

(eText Screens 5.1-1–5.1-27)

When we compared **integers** in **Topic 2.1**, we learned that numbers increase in value as we move from left to right on a **number line.** Similarly, numbers decrease in value as we move from right to left on a number line. Looking back at the decimals we graphed on the number line in **Figure 4**, we can see that 0.6 *is less than* 1.76, or $0.6 < 1.76$, and -1.33 *is less than* -0.7, or $-1.33 < -0.7$. Equivalently, we can say that 1.76 *is greater than* 0.6, or $1.76 > 0.6$, and -0.7 *is greater than* -1.33, or $-0.7 > -1.33$.

The following rules will help us compare two **decimals.**

Comparing Two Decimals

1. If both decimals are **positive**, work from left to right comparing the **digits** in each **place value**. When two different digits are encountered, the decimal having the larger digit is the larger positive number.
2. If both decimals are **negative**, work from left to right comparing the digits in each place value. When two different digits are encountered, the decimal having the larger digit is the smaller negative number.
3. If one decimal is positive and the other is negative, then the positive decimal is larger.

TIP If one number has more decimal places than the other, fill in each "missing" place value with a 0 digit to make comparisons easier.

Example 6 Comparing Decimals

My video summary

Fill in the blank with $<$, $>$, or $=$ to make a true comparison statement.

a. 0.531 ____ 0.529 **b.** 9.42 ____ 9.426 **c.** -0.65 ____ -0.69

d. -5.7 ____ -5.07 **e.** -0.7 ____ 0.3 **f.** 2.49 ____ 2.4900

Solutions We follow the **rules for comparing two decimals.**

a. We compare **digits** from left to right until we find two that are different.

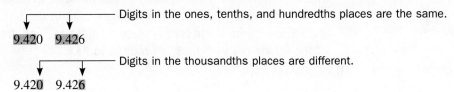

Digits in the tenths places are the same.

$0.531 \quad 0.529$

Digits in the hundredths places are different.

$0.531 \quad 0.529$

Comparing the digits in the hundredths **places**, we have $3 > 2$. Because both decimals are **positive**, this means $0.531 > 0.529$.

b. Notice that 9.426 has three decimal places while 9.42 has only two. Therefore, we write in a 0 digit at the end of 9.42 to get 9.420.

Digits in the ones, tenths, and hundredths places are the same.

$9.420 \quad 9.426$

Digits in the thousandths places are different.

$9.420 \quad 9.426$

Both decimals are positive, and $0 < 6$, so $9.42 < 9.426$.

c. We are comparing two **negative** decimals.

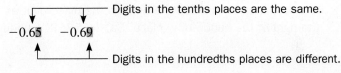

Digits in the tenths places are the same.

$-0.65 \quad -0.69$

Digits in the hundredths places are different.

Comparing the digits in the hundredths **places**, we have $5 < 9$. Because both decimals are negative, the larger number actually becomes the smaller value. This means $-0.65 > -0.69$.

d.–f. Try working these problems on your own. View the **answers**, or watch this **video** for complete solutions to all six parts.

You Try It **Work through this You Try It problem.**

Work Exercises 41–50 in this textbook or in the MyMathLab Study Plan.

OBJECTIVE 7 ROUND DECIMALS TO A GIVEN PLACE VALUE

In **Topic 1.2**, we rounded whole numbers. We round **decimals** to the right of the decimal point in a similar way. The only difference is that we drop all of the **digits** to the right of the **place** being rounded to, instead of replacing them with 0 digits.

Rounding a Decimal to the Right of the Decimal Point

Step 1. Find the digit to the right of the given place value (the place value that you are rounding to).

Step 2. If the digit from Step 1 is 5 or greater, then increase the digit in the given place value by 1 and drop all digits to its right.

If the digit from Step 1 is less than 5, then drop all digits to the right of the given place value.

Example 7 Rounding Decimals

My video summary

Round each decimal to the given place value.

a. 23.769 to the nearest tenth

b. -435.8648 to the nearest hundredth

c. 18.369532 to the nearest thousandth

d. 361.208 to the nearest whole number

Solutions We follow the **steps for rounding a decimal.**

a. 2 3 . 7 6 9
↗ ↖
Tenths Hundredths

The digit 7 is in the tenths place, where we want to round. The digit 6 is in the hundredths place, which is to the immediate right of the tenths place. Because 6 is larger than 5, we increase the tenths place from 7 to 8, and drop all of the digits to the right of the tenths place. So, 23.769 rounds to 23.8.

b. -435.8648
↗ ↖
Hundredths Thousandths

The digit 6 is in the hundredths place, where we want to round. The digit 4 is in the thousandths place, which is to the immediate right of the hundredths place. Because 4 is less than 5, we keep the hundredths place digit as 6 and drop all of the digits to the right of the hundredths place. So, -435.8648 rounds to -435.86.

(eText Screens 5.1-1–5.1-27)

c.–d. Try rounding these decimals on your own. View the **answers**, or watch this **video** for complete solutions to all four parts.

You Try It **Work through this You Try It problem.**

Work Exercises 51–60 in this textbook or in the MyMathLab Study Plan.

CAUTION

When rounding, pay close attention to the given place value (the place value that you are rounding to). Do not confuse tens with tenths, hundreds with hundredths, and so forth. For example, if 759.248 is rounded to the nearest *hundredth*, the result is 759.25. However, if 759.248 is rounded to the nearest *hundred*, the result is 800.

5.1 Exercises

In Exercises 1–8, identify the place value of the digit 3 in each decimal number.

You Try It

1. 257.361
2. 65.013476
3. 134.761
4. 8.045939
5. 53,942.810
6. 456.8372
7. 17,653.029
8. 0.00132

In Exercises 9–16, write each decimal number in words.

You Try It

9. 0.63
10. 8.6
11. 43.59
12. 42.079
13. 7516.534
14. 865.10227
15. 0.0513
16. 500.0013

In Exercises 17–24, change each decimal to standard form.

You Try It

17. twelve and eight tenths
18. seventeen and thirty-two hundredths
19. three hundred fifteen and thirteen thousandths
20. five thousand eight hundred twelve ten-thousandths
21. twenty-one and five hundredths
22. one hundred eighty-two and six thousand five hundred forty-eight ten-thousandths
23. eight hundred and seventy-five thousandths
24. eight hundred seventy-five thousandths

In Exercises 25–32, change each decimal into a proper fraction or mixed number. Simplify if necessary.

You Try It

25. 0.57
26. 0.4
27. 19.7
28. 9.6
29. 0.625
30. 14.56
31. 2.004
32. 0.072

(eText Screens 5.2-1–5.2-25)

In Exercises 33–40, graph each decimal on a number line.

You Try It

33. 0.1
34. −0.7
35. −2.3
36. 1.8
37. 0.83
38. −0.27
39. 3.26
40. −1.18

In Exercises 41–50, fill in the blank with <, >, or = to make a true comparison statement.

You Try It

41. 0.5 _____ 0.6
42. −0.7 _____ −0.9
43. 0.642 _____ 0.638
44. −0.26 _____ −0.23
45. 7.834 _____ 7.85
46. −1.4 _____ −1.37
47. 0.395 _____ 0.39461
48. 0.024 _____ −0.919
49. 5.4 _____ 5.400
50. 12.003 _____ 12.3

In Exercises 51–60, round each decimal to the given place value.

You Try It

51. 0.6714, nearest tenth
52. −359.6847, nearest hundredth
53. 24.90568, nearest thousandth
54. −64.257, nearest whole number
55. −0.09624, nearest thousandth
56. 23.96801, nearest tenth
57. 132.598, nearest hundredth
58. 0.051172, nearest ten-thousandth
59. −237.1842, nearest ten
60. 514.0263, nearest hundred

5.2 Adding and Subtracting Decimals

THINGS TO KNOW

Before working through this topic, be sure you are familiar with the following concepts:

	VIDEO	ANIMATION	INTERACTIVE
You Try It — **1.** Add Whole Numbers (Topic 1.3, **Objective 1**)	✓		
You Try It — **2.** Subtract Whole Numbers (Topic 1.3, **Objective 4**)	✓		✓
You Try It — **3.** Estimate Sums and Differences of Whole Numbers (Topic 1.3, **Objective 5**)	✓		
You Try It — **4.** Solve Applications by Adding or Subtracting Whole Numbers (Topic 1.3, **Objective 7**)	✓		

(eText Screens 5.2-1–5.2-25)

You Try It 5. Add and Subtract Integers (Topic 2.3, **Objective 2**)

You Try It 6. Combine Like Terms (Topic 3.1, **Objective 2**)

OBJECTIVES

1 Add Positive Decimals

2 Subtract Positive Decimals

3 Estimate Sums and Differences of Decimals

4 Add and Subtract Negative Decimals

5 Add and Subtract Variable Expressions Involving Decimals

6 Solve Applications by Adding or Subtracting Decimals

OBJECTIVE 1 ADD POSITIVE DECIMALS

Addition of **decimals** is similar to addition of **whole numbers.** We line up the digits with corresponding **place values** by writing the numbers vertically and lining up the **decimal points.** We then add **digits** with corresponding place values, working from right to left.

Adding Decimals

Step 1. Write the numbers vertically, lining up the decimal points.

Step 2. Add digits in the corresponding place values, working from right to left.

Step 3. Write the decimal point in the answer so that it lines up with the decimal points in the problem.

TIP We can add 0's after the last digit to the right of the decimal point to help line up place values.

Example 1 Adding Decimals

Add.

a. $42.37 + 23.41$ **b.** $817.0316 + 41.95$

Solutions We follow the **steps for adding decimals.**

a. Write the numbers vertically, lining up the **decimal points.**

$$\begin{array}{r} 42.37 \\ +23.41 \\ \hline \end{array}$$

↑ Align decimal points

Add **digits** with corresponding **place values**, working from right to left. Place a decimal point in the sum so it lines up with the decimal points in the problem.

(eText Screens 5.2-1–5.2-25)

$$\begin{array}{r} 42.37 \\ +23.41 \\ \hline 65.78 \end{array}$$

↑
Line up decimal point in the sum with the other decimal points

b. Write the numbers vertically lining up the **decimal points.**

$$\begin{array}{r} 817.0316 \\ +\ 41.9500 \\ \hline \end{array} \leftarrow \text{Add 0's as placeholders}$$

↑
Align decimal points

Add **digits** with corresponding **place values**, working from right to left. Place a decimal point in the sum so it lines up with the decimal points in the problem.

$$\begin{array}{r} 817.0316 \\ +41.9500 \\ \hline 858.9816 \end{array}$$

↑
Line up decimal point in the sum with the other decimal points

You Try It **Work through this You Try It problem.**

Work Exercises 1–6 in this textbook or in the MyMathLab Study Plan.

CAUTION Why is it important to line up decimal points? View this **popup** for an illustration.

Just as when adding whole numbers, sometimes it is necessary to **carry** when adding **decimals.**

Example 2 Adding Decimals with Carrying

My video summary

Add.

a. $57.62 + 28.741$ **b.** $3.019 + 0.085$

Solutions We follow the **steps for adding decimals.**

a.
$$\begin{array}{r} {\scriptstyle 1\,1} \\ 57.620 \\ +28.741 \\ \hline 86.361 \end{array} \leftarrow \text{Add 0 as a placeholder}$$

↑
Line up decimal points

b. Try **adding** these decimals on your own. Check your **answer**, or watch this **video** for the complete solutions to both parts.

You Try It **Work through this You Try It problem.**

Work Exercises 7–12 in this textbook or in the MyMathLab Study Plan.

(eText Screens 5.2-1–5.2-25)

The **decimal point** in a **whole number** is positioned after the last **digit**. For example, we could write 35 as 35. by showing the decimal point after the last digit. We could also add 0's after the decimal point as placeholders, such as 35.0 or 35.000, if necessary. Example 3 illustrates this concept further.

Example 3 Adding Whole Numbers and Decimals

Add. $26 + 17.38$

Solution Add decimal point after last digit

$$\begin{array}{r} \overset{1}{2}6.00 \\ +17.38 \\ \hline 43.38 \end{array} \quad \leftarrow \text{Add 0's as placeholders}$$

Line up decimal points

You Try It **Work through this You Try It problem.**

Work Exercises 13 and 14 in this textbook or in the MyMathLab Study Plan.

We can add several **decimals** using the same approach.

Example 4 Adding Several Decimals

My video summary

Add. $14.07 + 263.4 + 8.125$

Solution We follow the **steps for adding decimals.**

$$\begin{array}{r} 14.070 \\ 263.400 \\ +8.125 \\ \hline \end{array} \quad \begin{array}{l} \leftarrow \text{Add 0's as placeholders} \\ \leftarrow \\ \\ \end{array}$$

Line up decimal points

Try finishing this problem on your own. Check your **answer**, or watch this **video** for the complete solution.

You Try It **Work through this You Try It problem.**

Work Exercises 15–20 in this textbook or in the MyMathLab Study Plan.

OBJECTIVE 2 SUBTRACT POSITIVE DECIMALS

Subtraction of **decimals** follows the same approach as addition. We line up the **place values** by writing the numbers vertically and lining up the **decimal points**. Then we subtract **digits** with corresponding place values, working from right to left.

Subtracting Decimals

Step 1. Write the numbers vertically, lining up the decimal points.

Step 2. Subtract digits in the corresponding place values, working from right to left.

Step 3. Write the decimal point in the answer so that it lines up with the decimal points in the problem.

(eText Screens 5.2-1–5.2-25)

Example 5 Subtracting Decimals

Subtract. 128.73 − 15.21

Solution We follow the steps for subtracting decimals.

$$\begin{array}{r} 128.73 \\ -\ 15.21 \\ \hline 113.52 \end{array}$$

↑ Line up decimal points

You Try It **Work through this You Try It problem.**

Work Exercises 21–25 in this textbook or in the MyMathLab Study Plan.

TIP Remember that you can check your answer to subtraction by adding the **difference** to the **subtrahend** and see if the **sum** is the original **minuend**. View this **popup** to see the check for Example 5.

As with **whole numbers**, sometimes we need to **borrow** while subtracting **decimals**. We also continue to use 0's as placeholders after the last **digit** to the right of the **decimal point** to help line up **place values**.

Example 6 Subtracting Decimals with Borrowing

My interactive video summary

Subtract.

a. 94.02 − 35.64 **b.** 7.1371 − 0.95 **c.** 139 − 47.205

Solutions We follow the **steps for subtracting decimals**.

a.

$$\begin{array}{r} \scriptstyle 8\ \overset{13}{\cancel{3}}\ \ \overset{9}{\cancel{10}}\ 12 \\ \cancel{9}\cancel{4}.\cancel{0}\ \cancel{2} \\ -35.6\ 4 \\ \hline 58.3\ 8 \end{array}$$

↑ Line up decimal points

View the **check**.

b. Write the problem vertically, lining up the **decimal points** and inserting 0's as placeholders where needed.

$$\begin{array}{r} 7.1371 \\ -0.9500 \\ \hline \end{array}$$

↑ Line up decimal points

Try finishing this problem on your own. Remember to **borrow**, if necessary. Check your **answer**, or watch this **interactive video** for the complete solutions to all three parts.

c. Try **subtracting** these decimals on your own. Check your **answer**, or watch this **interactive video** for the complete solutions to all three parts.

You Try It **Work through this You Try It problem.**

Work Exercises 26–32 in the MyMathLab Study Plan.

(eText Screens 5.2-1–5.2-25)

OBJECTIVE 3 ESTIMATE SUMS AND DIFFERENCES OF DECIMALS

When working with **mixed numbers**, we saw that **estimates** could be used to help us check if our answer was reasonable. We can also use estimates to check the reasonableness of our answers when adding and subtracting **decimals**.

Example 7 Estimating Sums and Differences of Decimals

Estimate the **sum** or **difference** by first rounding each number to the nearest ten.

a. $739.85 - 264.361$ **b.** $28.6 + 154.608 + 77.39$

Solutions We follow the **steps for adding decimals**.

a. Start by writing the problem vertically, lining up the **decimal points**.

$$\begin{array}{r} 739.850 \\ -264.361 \\ \hline \end{array} \quad \begin{array}{l} \leftarrow \text{Add 0 as a placeholder} \\ \\ \end{array}$$

↑ Line up decimal points

Round each value to the nearest ten.

$$\begin{array}{r} 739.850 \\ -264.361 \\ \hline \end{array} \xrightarrow{\text{Round to the nearest ten}} \begin{array}{r} 740 \\ -260 \\ \hline \end{array}$$

Carry out the subtraction.

$$\begin{array}{r} \scriptstyle 6\ 14 \\ \cancel{7}\cancel{4}0 \\ -260 \\ \hline 480 \end{array}$$

So, $739.85 - 264.361$ is about 480. This is close to the exact value of 475.489.

$$\begin{array}{r} \scriptstyle 6\ 13\ \ 7\ \overset{14}{\cancel{4}}\,10 \\ \cancel{7}\cancel{3}9.\cancel{8}\cancel{5}\cancel{0} \\ -264.361 \\ \hline 475.489 \end{array}$$

My video summary

b. Try estimating this sum on your own. Check your **answer**, or watch this **video** for the complete solution.

You Try It **Work through this You Try It problem.**

Work Exercises 33–38 in this textbook or in the MyMathLab Study Plan.

OBJECTIVE 4 ADD AND SUBTRACT NEGATIVE DECIMALS

We can add and subtract negative **decimals** in the same way that we added and subtracted **integers**. We can change each subtraction to addition of the **opposite** and follow the appropriate addition procedure. Review the steps for **adding two numbers with the same sign** and **adding two numbers with different signs** from Topic 2.2. Remember to line up **decimal points** when adding or subtracting decimals.

(eText Screens 5.2-1–5.2-25)

Example 8 Adding and Subtracting Negative Decimals

My interactive video summary

Add or subtract as indicated.

a. $-37.48 + 74.63$ **b.** $-14.92 - 4.037$

c. $21 - 36.84$ **d.** $-46.72 - (-67.306)$

Solutions Change each subtraction to addition of the opposite, then follow the appropriate addition procedure.

a. We are adding two numbers with different signs. First, we find the **absolute value** of each term.

$$|-37.48| = 37.48 \text{ and } |74.63| = 74.63$$

Then subtract the smaller result from the larger result.

$$\begin{array}{r} 74.63 \\ -37.48 \end{array} \qquad \begin{array}{r} {\scriptsize 6\ 14\ \ 5\ 13} \\ \cancel{7}\cancel{4}.\cancel{6}\cancel{3} \\ -37.48 \\ \hline 37.15 \end{array}$$

↑
Line up decimal points

The larger **absolute value** result belongs to the positive number, so the sum will be positive.

$$-37.48 + 74.63 = 37.15$$

b. Change the subtraction to addition of the **opposite**.

$$-14.92 - 4.037 = -14.92 + (-4.037)$$

We are adding two numbers with the same **sign**. Add the absolute values and use the common sign as the sign of the sum.

$$|-14.92| = 14.92 \text{ and } |-4.037| = 4.037$$

$$\begin{array}{r} 14.920 \\ +\ \ 4.037 \\ \hline \end{array} \leftarrow \text{Add 0 as a placeholder}$$

↑
Line up decimal points

Try finishing this problem on your own. Check your **answer**, or watch this **interactive video** for the complete solutions to all four parts.

c.–d. Try working these problems on your own. Check your **answers**, or watch this **interactive video** for the complete solutions to all four parts.

You Try It **Work through this You Try It problem.**

Work Exercises 39–50 in this textbook or in the MyMathLab Study Plan.

OBJECTIVE 5 ADD AND SUBTRACT VARIABLE EXPRESSIONS INVOLVING DECIMALS

We can also simplify **algebraic expressions** involving **decimals** by **combining like terms**.

Example 9 Combining Like Terms Involving Decimals

My video summary

Simplify each expression by combining like terms.

a. $5.8 - 3.2x + 4.7x - 9.5$ **b.** $6.5y + 18.72 - 7.05y + 4.02y - 11$

Solutions

a. 5.8 and -9.5 are like terms, as are $-3.2x$ and $4.7x$.

Begin with the original expression: $5.8 - 3.2x + 4.7x - 9.5$

Change subtractions to add the opposite: $= 5.8 + (-3.2x) + 4.7x + (-9.5)$

Rearrange to **collect like terms**: $= 5.8 + (-9.5) + (-3.2x) + 4.7x$

Combine like terms: $= -3.7 + 1.5x$ or $1.5x - 3.7$

$5.8 + (-9.5) = -3.7$

$-3.2x + 4.7x$
$= (-3.2 + 4.7)x$
$= 1.5x$

b. $6.5y$, $-7.05y$, and $4.02y$ are like terms, and 18.72 and -11 are **like terms**. Try to simplify this **expression** on your own by combining like terms. View the **answer**, or watch this **video** for the complete solutions to both parts.

You Try It **Work through this You Try It problem.**

Work Exercises 51–55 in this textbook or in the MyMathLab Study Plan.

OBJECTIVE 6 SOLVE APPLICATIONS BY ADDING OR SUBTRACTING DECIMALS

Many real-world applications can be solved by adding or subtracting **decimals**.

Example 10 Data Download Speed

A speed test conducted by Engadget in February 2011 compared 3G download speeds of the Verizon iPhone 4 and the AT&T iPhone 4. The average download speed for the AT&T iPhone 4 was 2.702 Mbps (megabytes per second), while the average download speed for the Verizon iPhone 4 was 1.01 Mbps. What is the difference in download speed between the AT&T iPhone 4 and the Verizon iPhone 4? (*Source:* osxdaily.com)

Solution To find the difference in download speeds, we subtract $2.702 - 1.01$.

$$\begin{array}{r} 2.\overset{6}{\not{7}}\overset{10}{\not{0}}2 \\ -1.010 \\ \hline 1.692 \end{array}$$

← Add 0 as a placeholder

↑ Line up decimal points

The difference in download speeds is 1.692 Mbps. The AT&T iPhone 4 can download data at a speed 1.692 Mbps faster than the Verizon iPhone 4.

Example 11 Restaurant Check Total

My video summary

TJ and some friends went out to dinner at a popular barbecue restaurant. The bill subtotal was \$179.28 including tax. Because there were more than 6 people in their party, a tip was automatically added. If the added tip was \$32.27, what was the final amount of the bill?

(eText Screens 5.2-1–5.2-25)

Solution To find the final amount of the bill, we add 179.28 + 32.27. Add these values on your own and check your **answer**. Watch this **video** for the complete solution.

You Try It **Work through this You Try It problem.**

Work Exercises 56–62 in this textbook or in the MyMathLab Study Plan.

5.2 Exercises

In Exercises 1–20, add as indicated.

You Try It

1. 4.82 + 3.15
2. 32.21 + 17.36
3. 25.63 + 114.05
4. 82.1632 + 13.73
5. 321.67 + 142.315
6. 121.007 + 217.9923
7. 9.73 + 8.09
8. 59.26 + 20.793
9. 122.6397 + 217.223
10. 222.8888 + 28.091
11. 157.69 + 382.2727
12. 387.463 + 259.537
13. 22.748 + 35
14. 19 + 42.86
15. 12.241 + 18 + 31.7683
16. 52.13 + 23.005 + 17.0199
17. 22.1 + 17.3421 + 14.387
18. 11.343 + 16.27 + 42.3127
19. 15.09 + 11.903 + 22.0089
20. 15.17 + 12.249 + 220.4371 + 112.3

In Exercises 21–32, subtract as indicated.

21. 7.69 − 2.04
22. 12.748 − 9.13
23. 147.393 − 25.172
24. 258.486 − 124.27
25. 146.2743 − 23.252
26. 8.04 − 2.95
27. 18.93 − 10.98
28. 54.26 − 21.368
29. 6 − 0.87
30. 212.756 − 109.583
31. 58 − 13.72
32. 43.107 − 17.25

In Exercises 33–35, estimate the sum or difference by first rounding each number to the nearest hundred.

33. 122.365 + 178.29
34. 184.75 − 121.986
35. 615.217 − 428.4091

In Exercises 36–38, estimate the sum or difference by first rounding each number to the nearest ten.

36. 232.7852 − 19.52
37. 18.17 + 123.484 + 25.3
38. 27.625 + 187.13 + 53

In Exercises 39–50, add or subtract as indicated.

39. −17.48 + 42.3
40. −18.75 − 31.24
41. 105 − 217.63
42. −48.264 + 22.75
43. −48.235 − (−631.177)
44. 15.862 − 22.68
45. −19 − (14.22)
46. −122.72 − 215.364
47. 25.8132 − (−16.297)
48. −192.47 + 225.261
49. 19 − 32.72
50. −25 − 17.618

In Exercises 51–55, simplify the expression by combining like terms.

You Try It 51. $3.27 - 16.531x - 1.95$

52. $5.713x + 2.852 - 3.84x$

53. $6.72 + 18.214y - 12.715y - 8.818$

54. $14.6218 + 17.32 - 2.459z - 41.6 - 5.82z$

55. $61.83b - 17.549 - 32.1b + 41.623 + 14.291b$

You Try It 56. **File Size** Daniel downloads three video files to his computer. The file sizes are 25.131 Mb, 17.64 Mb, and 28.035 Mb. What is the total size of the three files?

57. **Data Download Speed** A speed test conducted by Metrico Wireless in October 2011 compared download speeds of the Sprint iPhone 4S and the AT&T iPhone 4S. The maximum download speed for the AT&T iPhone 4S was 6.407 Mbps (megabytes per second), while the maximum download speed for the Sprint iPhone 4S was 1.767 Mbps. What is the difference in download speed between the AT&T iPhone 4S and the Sprint iPhone 4S? (*Source:* www.cnn.com)

58. **Garden Size** Find the perimeter of a rectangular garden whose length is 12.257 ft and whose width is 8.59 ft.

59. **Online Order Total** Justin placed an order at an online retailer. The following amounts were shown in his emailed receipt:

Subtotal:	\$94.97
Percent Off (15.0%):	\$14.25
Shipping (Free, excluding surcharge):	\$0.00
Tax:	\$3.41

Determine the total amount for Justin's order.

60. **Change Due** Jeff purchases three items at a retail store. The total cost (including tax) was \$17.38. If Jeff pays with a \$20 bill, how much change should he receive back?

61. **Sale Price** A backpacking tent originally cost \$229.99. It is placed on clearance for \$106.78. How much lower is the clearance price?

62. **Stock Fluctuation** On a certain Monday, the price for one share of Google stock increased \$3.56. On Tuesday the price decreased \$5.09, and on Wednesday the price decreased \$10.60. What was the total change in price for the three days?

5.3 Multiplying Decimals; Circumference

THINGS TO KNOW

Before working through this topic, be sure you are familiar with the following concepts:

VIDEO ANIMATION INTERACTIVE

You Try It 1. Identify the Place Value of a Digit in a Whole Number (Topic 1.2, **Objective 1**)

You Try It 2. Multiply Two Integers (Topic 2.4, **Objective 1**)

(eText Screens 5.3-1–5.3-23)

You Try It 3. Solve Applications by Multiplying or Dividing Integers (Topic 2.4, **Objective 5**)

You Try It 4. Multiply Fractions (Topic 4.3, **Objective 1**)

You Try It 5. Change Decimals into Fractions or Mixed Numbers (Topic 5.1, **Objective 4**)

You Try It 6. Round Decimals to a Given Place Value (Topic 5.1, **Objective 7**)

OBJECTIVES

1 Multiply Decimals

2 Multiply Decimals by Powers of 10

3 Estimate Products of Decimals

4 Evaluate Exponential Expressions Involving Decimal Bases

5 Find the Circumference of a Circle

6 Solve Applications by Multiplying Decimals

OBJECTIVE 1 MULTIPLY DECIMALS

We multiply **decimals** in the same way that we multiply **whole numbers**. The only difference is that the product will contain a **decimal point**. We multiply the numbers as if there were no decimal points and then determine the proper location of the decimal point in the product. Consider the following:

$$\underbrace{0.7}_{\substack{\uparrow\\ \text{1 decimal place}}} \times \underbrace{0.12}_{\substack{\uparrow\\ \text{2 decimal places}}} = \frac{7}{10} \times \frac{12}{100} = \frac{7 \cdot 12}{10 \cdot 100} = \frac{84}{1000} = \underbrace{0.084}_{\substack{\uparrow\\ \text{3 decimal places}}}$$

In the **factors** of the original problem, notice that there are a total of three **digits** to the right of decimal points, $1 + 2 = 3$. The final **product** also has three digits to the right of the decimal point.

$$\underbrace{1.24}_{\substack{\uparrow\\ \text{2 decimal places}}} \times \underbrace{0.003}_{\substack{\uparrow\\ \text{3 decimal places}}} = \frac{124}{100} \times \frac{3}{1000} = \frac{124 \cdot 3}{100 \cdot 1000} = \frac{372}{100{,}000} = \underbrace{0.00372}_{\substack{\uparrow\\ \text{5 decimal places}}}$$

In this case, there are a total of five digits to the right of decimal points, $2 + 3 = 5$. The final product also has five digits to the right of the decimal point. This illustrates a general result. The total number of digits to the right of the decimal points in the factors will equal the number of digits to the right of the decimal point in the product. Instead of writing the **decimals as fractions** and multiplying, we can use this idea to simplify the multiplication process.

Steps for Multiplying Decimals

Step 1. Multiply the numbers as though no **decimal points** were present.

Step 2. Place the decimal point in the product so that the number of **digits** to its right is the same as the total number of digits to the right of the decimal points in the **factors**.

Example 1 Multiplying Decimals

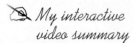

My interactive video summary

Multiply.

a. 4.2×7.6 **b.** 9.53×0.64 **c.** 12×1.006

Solutions We follow the steps for multiplying decimals.

a. Multiply the numbers as though no decimal points were present.

$$\begin{array}{r} 42 \\ \times\ 76 \\ \hline 252 \\ 2940 \\ \hline 3192 \end{array} \qquad \begin{array}{rl} 4.2 & \leftarrow 1 \text{ decimal place} \\ \times\ 7.6 & \leftarrow 1 \text{ decimal place} \\ \hline 252 & \\ 2940 & \\ \hline 31.92 & \leftarrow 2 \text{ decimal places} \end{array}$$

The factors have a total of $1 + 1 = 2$ digits behind their decimal points, so the product will have 2 digits behind its decimal point.

b. Start by multiplying the numbers as though no **decimal points** were present. The **factors** have a total of $2 + 2 = 4$ digits behind the decimal points, so the product will have 4 digits behind its decimal point.

$$\begin{array}{rl} 9.53 & \leftarrow 2 \text{ decimal places} \\ \times\ 0.64 & \leftarrow 2 \text{ decimal places} \\ \hline & \leftarrow 4 \text{ decimal places} \end{array}$$

Finish multiplying on your own and check your **answer**. Watch this **interactive video** to see complete solutions to all three parts.

c. Try working this problem on your own. Check your **answer**, or watch this **interactive video** for detailed solutions to all three parts.

You Try It **Work through this You Try It problem.**

Work Exercises 1–12 in this textbook or in the MyMathLab Study Plan.

When multiplying **signed** decimals, remember that the **product of two numbers** with the same sign is positive and the product of two numbers with different signs is negative.

Example 2 Multiplying Signed Decimals

Multiply.

a. $(5.78)(-3.2)$ **b.** $(-4.75)(-2.013)$

(eText Screens 5.3-1–5.3-23)

Solutions We follow the **steps for multiplying decimals.**

a. The two numbers have different **signs**. Multiply the **absolute values** of the numbers and make the product negative.

$$|5.78| = 5.78 \qquad |-3.2| = 3.2$$

The **factors** have a total of $1 + 2 = 3$ digits behind the decimal points, so the product will have 3 digits behind its decimal point.

$$\begin{array}{rl} 5.78 & \leftarrow 2 \text{ decimal places} \\ \times \quad 3.2 & \leftarrow 1 \text{ decimal place} \\ \hline 1156 & \\ 17340 & \\ \hline 18.496 & \leftarrow 3 \text{ decimal places} \end{array}$$

Therefore, $(5.78)(-3.2) = -18.496$.

My video summary

b. The two numbers have the same sign so the product will be positive. Multiply these numbers on your own and check your **answer**. Or watch this **video** to see the complete solution to part b.

You Try It **Work through this You Try It problem.**

Work Exercises 13–20 in this textbook or in the MyMathLab Study Plan.

OBJECTIVE 2 MULTIPLY DECIMALS BY POWERS OF 10

In **Topic 1.2**, we saw that the position of a **digit** in a number determines its **place value**. These place values were **powers of 10** such as 10, 100, 1000, and so on. Because of this, some patterns arise when multiplying a decimal by a power of 10.

Consider the following pattern for powers of 10 that are greater than or equal to 1:

$$64.2713 \times 10 = 642.713$$

1 zero — Decimal point moves 1 place to the *right*.

$$64.2713 \times 100 = 6427.13$$

2 zeros — Decimal point moves 2 places to the *right*.

$$64.2713 \times 1000 = 64{,}271.3$$

3 zeros — Decimal point moves 3 places to the *right*.

Notice that when multiplying by a power of 10 greater than or equal to 1, we simply move the **decimal point** to the *right* the same number of places as there are zeros in the power of 10.

Multiplying Decimals by a Power of 10 Greater Than or Equal to 1

To multiply a number by a **power of 10** greater than or equal to 1, such as 10, 100, 1000, and so on, move the **decimal point** in the number to the *right* the same number of places as there are zeros in the power of 10.

A similar pattern exists for powers of 10 that are less than 1, such as 0.1, 0.01, 0.001, and so on. Consider the following pattern for powers of 10 that are less than 1:

$$64.2713 \times 0.1 = 6.42713$$

↑ 1 decimal place — Decimal point moves 1 place to the *left*.

$$64.2713 \times 0.01 = 0.642713$$

↑ 2 decimal places — Decimal point moves 2 places to the *left*.

$$64.2713 \times 0.001 = 0.0642713$$

↑ 3 decimal places — Decimal point moves 3 places to the *left*.

Notice that when multiplying by a power of 10 less than 1, we simply move the decimal point to the *left* the same number of places as there are **decimal places** in the power of 10.

Multiplying Decimals by a Power of 10 less Than 1

To multiply a number by a **power of** 10 less than 1, such as 0.1, 0.01, 0.001, and so on, move the **decimal point** in the number to the *left* the same number of places as there are **decimal places** in the power of 10.

Example 3 Multiplying Decimals by Powers of 10

Multiply.

a. 5.378×100 **b.** -18.735×0.1

c. 58.49×0.0001 **d.** $(-98.17)(-1000)$

Solutions

a. We are multiplying by a power of 10 that is greater than 1. There are 2 zeros in the power of 10, so we move the decimal point to the *right* 2 places.

$$5.378 \times 100 = 537.8$$

↑ 2 zeros — Decimal point moves 2 places to the *right*.

b. We are multiplying by a **power of 10** that is less than 1. There is 1 decimal place in the power of 10 so we move the decimal point to the *left* 1 place. We are multiplying two numbers with different **signs** so the result will be negative.

$$-18.735 \times 0.1 = -1.8735$$

↑ 1 decimal place — Decimal point moves 1 place to the *left*.

My video summary

c.–d. Try working these problems on your own. Check your **answers**, or watch this **video** for detailed solutions to both parts.

You Try It **Work through this You Try It problem.**

Work Exercises 21–30 in this textbook or in the MyMathLab Study Plan.

(eText Screens 5.3-1–5.3-23)

OBJECTIVE 3 ESTIMATE PRODUCTS OF DECIMALS

In **Topic 5.2**, we rounded decimals to estimate **sums** and **differences**. We can do the same when multiplying decimals to check the reasonableness of our **products**. In particular, the estimate can help us see if the **decimal point** has been placed properly in the product.

Example 4 Estimating Products of Decimals

Estimate the product by first rounding each number to the nearest ten.

a. 12.78×45.96 **b.** 276.82×93.51

Solutions

a. Round each number to the nearest ten.

$$\begin{array}{r} 12.78 \\ \times 45.96 \\ \hline \end{array} \xrightarrow{\text{Round to the nearest ten}} \begin{array}{r} 10 \\ \times 50 \\ \hline \end{array}$$

Now carry out the multiplication.

$$\begin{array}{r} 10 \\ \times 50 \\ \hline 00 \\ 500 \\ \hline 500 \end{array}$$

As an alternative, we could have noted that there are two trailing zeros, computed $1 \times 5 = 5$, and then put the two trailing zeros back in the result.

$$10 \times 50 = 500$$

The estimated product is 500. This is reasonably close to the actual value of 587.3688 (see the **details**).

My video summary

b. Try estimating this product on your own. Check your **answer**, or watch this **video** for the complete solution to part b.

You Try It **Work through this You Try It problem.**

Work Exercises 31–35 in this textbook or in the MyMathLab Study Plan.

OBJECTIVE 4 EVALUATE EXPONENTIAL EXPRESSIONS INVOLVING DECIMAL BASES

As with **integers** and **fractions**, the **base** of an **exponential expression** can also be a **decimal**.

$$\overset{\text{Exponent}}{(3.5)^3} = \underbrace{(3.5)(3.5)(3.5)}_{3 \text{ Factors of } 3.5}$$

(Base: 3.5)

To evaluate an exponential expression with a decimal base, write the expression in **expanded form** and perform the multiplications.

Example 5 Evaluating Exponential Expressions Involving Decimal Bases

Evaluate each exponential expression.

a. $(4.3)^2$ **b.** $(-2.7)^3$

(eText Screens 5.3-1–5.3-23)

Solutions

a. $(4.3)^2 = 4.3 \times 4.3$

$$\begin{array}{rl} 4.3 & \leftarrow 1 \text{ decimal place} \\ \times \quad 4.3 & \leftarrow 1 \text{ decimal place} \\ \hline 1\,2\,9 & \\ 17\,2\,0 & \leftarrow \text{Insert 0 as a placeholder} \\ \hline 18.4\,9 & \leftarrow 1 + 1 = 2 \text{ decimal places} \end{array}$$

So, $(4.3)^2 = 18.49$.

b. $(-2.7)^3 = -2.7 \times -2.7 \times -2.7$

My video summary

Start by multiplying the **absolute values** of the first two factors. Because the two factors have the same sign, the **product** will be positive.

$$|-2.7| = 2.7 \rightarrow \quad \begin{array}{rl} 2.7 & \leftarrow 1 \text{ decimal place} \\ \times \;\; 2.7 & \leftarrow 1 \text{ decimal place} \\ \hline 1\,8\,9 & \\ 5\,4\,0 & \leftarrow \text{Insert 0 as a placeholder} \\ \hline 7.2\,9 & \leftarrow 1 + 1 = 2 \text{ decimal places} \end{array}$$

$$(-2.7)^3 = \underbrace{-2.7 \times -2.7}_{7.29} \times -2.7$$

$$= 7.29 \times -2.7$$

Finish this problem by multiplying 7.29×-2.7. View the **answer**, or watch this **video** for the complete solution to part b.

You Try It **Work through this You Try It problem.**

Work Exercises 36–41 in this textbook or in the MyMathLab Study Plan.

OBJECTIVE 5 FIND THE CIRCUMFERENCE OF A CIRCLE

In **Topic 1.3**, we learned that the **perimeter** of a **polygon** was the distance around the polygon. We found the perimeter by adding the lengths of the sides that made up the polygon.

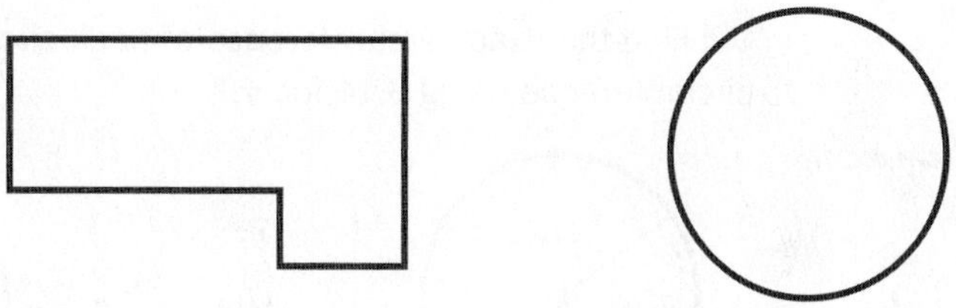

Polygon Circle

The distance around a circle is called the **circumference** of the circle. A circle has no sides, so we need to find this distance in another way. The **radius** of a circle is the distance from the center of the circle to the edge of the circle. The circumference of a circle is related to the radius through a constant denoted by the Greek letter π (spelled "pi" and pronounced "pie").

(eText Screens 5.3-1–5.3-23)

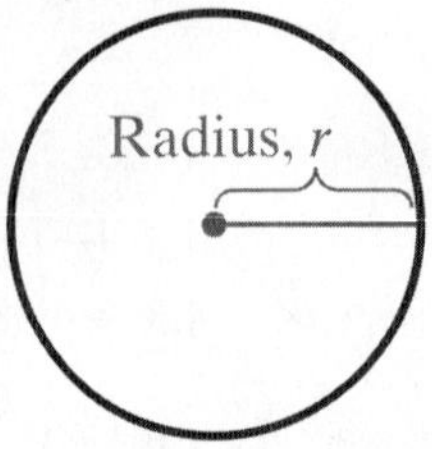

$$\text{Circumference} = 2 \cdot \pi \cdot \text{radius}$$
$$C = 2\pi r$$

The **diameter** of a circle (distance across the circle through its center) is equal to twice the **radius** of the circle. That is, $\text{diameter} = 2 \cdot \text{radius}$ or $d = 2r$. Therefore, we can also relate the circumference of a circle to its diameter.

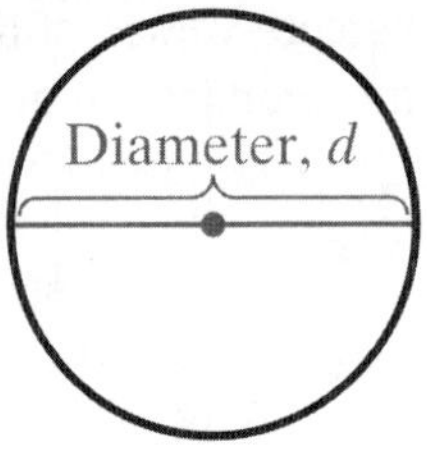

$$\text{Circumference} = \pi \cdot \text{diameter}$$
$$C = \pi d$$

$$\begin{aligned} C &= 2\pi r \\ &= \pi \cdot 2r \\ &= \pi \cdot d \\ &= \pi d \end{aligned}$$

TIP We often approximate the value of π with the decimal 3.14. In some cases, the fraction $\frac{22}{7}$ is also used to approximate the value of π.

Circumference of a Circle

The **circumference** of a circle is the distance around the circle. Its value is found by using either

$$C = 2\pi r \quad \text{or} \quad C = \pi d,$$

where r is the radius of the circle and d is the diameter. The constant π is often approximated by the decimal 3.14 $\left(\text{or in some cases, the fraction } \frac{22}{7}\right)$.

My video animation summary

Watch this **animation** to see an illustration on the computation of π.

Example 6 Finding the Circumference of a Circle

Find the exact **circumference** of each circle in terms of π. Then approximate the circumference using 3.14 for π.

My video summary

a.

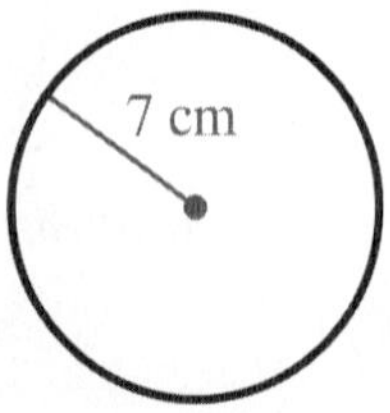

b.

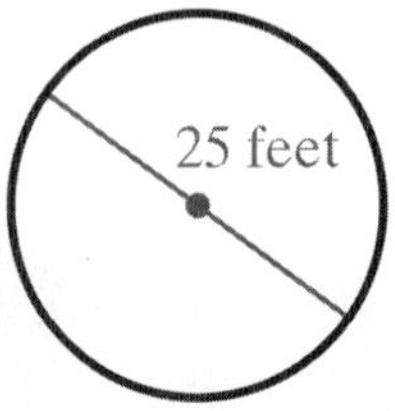

Solutions

a. From the figure, we see that the **radius** is 7 cm. Using $C = 2\pi r$, we substitute 7 for r.

$$C = 2\pi r = 2\pi(7) = 14\pi$$

The exact circumference is 14π cm. (Remember to include units.) Using 3.14 for π, we get

≈ means "is approximately equal to" →

$$\begin{aligned} C &= 14\pi \\ &\approx 14(3.14) \\ &= 43.96 \text{ cm} \end{aligned}$$

$$\begin{array}{r} 3.14 \\ \times \quad 14 \\ \hline 12\,56 \\ 31\,40 \\ \hline 43.96 \end{array}$$

The circumference is approximately 43.96 cm.

b. Here, we are given the **diameter**, so we can find the **circumference** by using $C = \pi d$. Try doing this on your own, along with the estimation. Check your **answer**, or watch this **video** for the complete solutions to both parts.

You Try It **Work through this You Try It problem.**

Work Exercises 42–45 in this textbook or in the MyMathLab Study Plan.

OBJECTIVE 6 SOLVE APPLICATIONS BY MULTIPLYING DECIMALS

A number of real-life application problems can be solved by **multiplying decimals**.

Example 7 Nip and Tuck

According to the American Society of Plastic Surgeons, 2.4 million cosmetic procedures were performed on 30–39-year-olds in 2010. Write this number in **expanded form**. (*Source:* www.plasticsurgery.org)

Solution To write 2.4 million in expanded form, we multiply 2.4 by 1,000,000. Notice that 1,000,000 is a power of 10 that is greater than 1. There are 6 zeros, so we move the **decimal point** in 2.4 six places to the *right*.

$$2.4 \times 1{,}000{,}000 = 2{,}400{,}000.$$

6 zeros ↑ — Decimal point moves 6 places to the *right*.

So, 2.4 million in expanded form is 2,400,000.

You Try It **Work through this You Try It problem.**

Work Exercises 46 and 47 in this textbook or in the MyMathLab Study Plan.

Example 8 Filling a Gas Tank

My video summary

Before returning a rental car, Leighton fills the gas tank. If the gasoline costs \$3.179 per gallon and it takes 8.5 gallons to fill the tank, how much did it cost Leighton to fill the tank? Round the answer to two **decimal places**.

Solution The cost to fill the tank can be found by multiplying the price per gallon by the number of gallons. That is, we need to compute 3.179×8.5. There are a total of four decimal places $(3 + 1 = 4)$, so the product will have four decimal places. Because money is usually rounded to the nearest cent, we round the product to two decimal places. Do the multiplication and rounding on your own. Check your **answer**, or watch this **video** for the complete solution.

(eText Screens 5.3-1–5.3-23)

You Try It **Work through this You Try It problem.**

Work Exercises 48–53 in this textbook or in the MyMathLab Study Plan.

5.3 Exercises

In Exercises 1–30, multiply as indicated.

You Try It

1. 5.1×6.8
2. 3.72×12.51
3. 12.79×8.3
4. 9.31×15.621
5. 7.29×42
6. 1.03×21.971
7. 6.22×5.002
8. 24.99×1.085
9. 258.2×14.013

You Try It

10. 0.367×18
11. 7.003×8.004
12. 2.0901×15.2
13. $(-27.9)(3.85)$
14. $(7.862)(-2.74)$
15. $(-48)(2.83)$
16. -1.318×-73.9
17. -3.71×-5.01
18. $(-13.71)(51)$

You Try It

19. 0.53×0.27
20. -14.2×0.045
21. 51.17×10
22. 171.8×0.01
23. $(27.284)(1000)$
24. $(15.32)(-0.1)$
25. 15.2×100
26. $(-28.113)(100)$
27. 154.72×0.001
28. 0.573×0.01
29. -0.35×1000
30. $(132)(0.01)$

In Exercises 31–33, estimate the product by first rounding each value to the nearest ten.

You Try It

31. 17.6×122.981
32. 28×14.99
33. 153.71×19.402

In Exercises 34 and 35, estimate the product by first rounding each value to the nearest hundred.

34. 312.71×559.218
35. 89.51×317.621

In Exercises 36–41, evaluate each exponential expression.

You Try It

36. 4.6^2
37. $(-3.47)^2$
38. -2.9^2
39. 4.7^3
40. -5.1^3
41. $(3.2)^4$

In Exercises 42–45, find the exact circumference of each circle in terms of π. Then approximate the circumference using 3.14 for π.

You Try It

42.

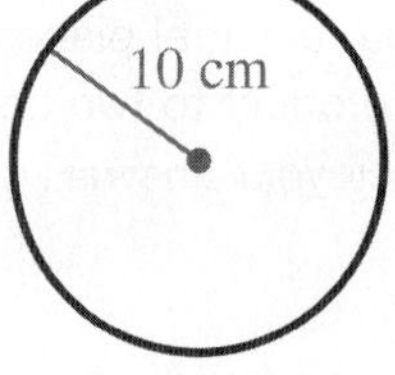

43.

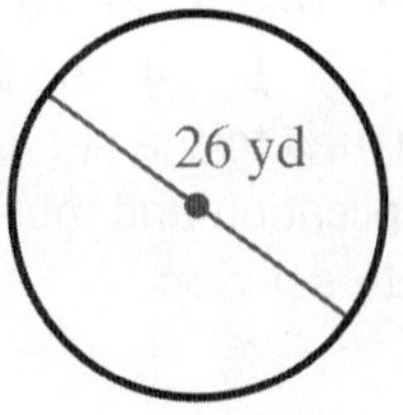

44.

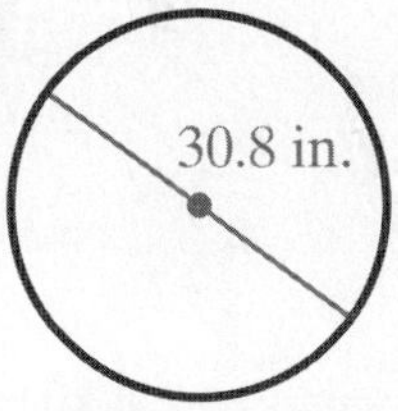

45.

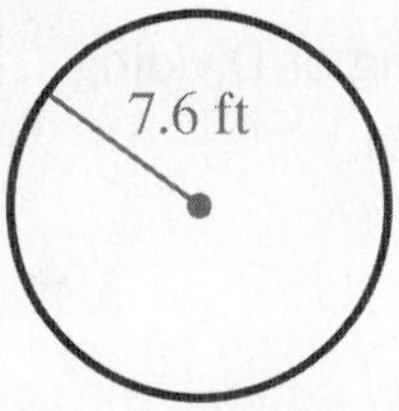

You Try It 46. **Football Attendance** Attendance at college football games in 2010 totaled 49.671 million. Write this number in expanded form. (*Source:* www.ncaa.org)

47. **Coffee Consumption** In 2009, the average amount of coffee consumed per person in the U.S. was 23.3 gallons. How many total gallons were consumed by 100 people? (*Source:* U.S. Statistical Abstract 2012)

You Try It 48. **Road Trip** Colton is planning a road trip during spring break. The gas tank in his car holds 12.4 gallons of gas, and his car averages 31.5 miles per gallon. How many miles can Colton travel on one tank of gas?

49. **Exchange Rate** On November 22, 2011, the exchange rate between euros (EUR) and U.S. dollars (USD) was 1 EUR = 1.3525 USD. What is the equivalent of 25.8 euros in U.S. dollars? Do not round. (*Source:* x-rates.com)

50. **Tablet Screen Area** A wireless tablet screen has a length of 8.75 inches and a width of 7.02 inches. What is the area of the tablet screen?

51. **Sales Tax** Cadence buys a new tent for an upcoming camping trip. To find the sales tax, she needs to multiply by 0.085. What must she pay in sales tax for the tent if the selling price is $269? Round the answer to two decimal places.

52. **Taco Cost** Taco Bell offers a 12-pack of hard-shell tacos for $10. How much would it cost to buy the tacos separately if each taco costs $0.89?

53. **Unit Conversion** An inch is equivalent to 2.54 centimeters. How many centimeters are in 12 inches (1 foot)?

5.4 Dividing Decimals

THINGS TO KNOW

Before working through this topic, be sure you are familiar with the following concepts:

	VIDEO	ANIMATION	INTERACTIVE
1. Divide Whole Numbers (Topic 1.5, **Objective 1**) — You Try It	▤		▤
2. Estimate Quotients of Whole Numbers (Topic 1.5, **Objective 2**) — You Try It	▤		
3. Solve Applications by Dividing Whole Numbers (Topic 1.5, **Objective 3**) — You Try It	▤		
4. Divide Integers (Topic 2.4, **Objective 4**) — You Try It	▤		

(eText Screens 5.4-1–5.4-29)

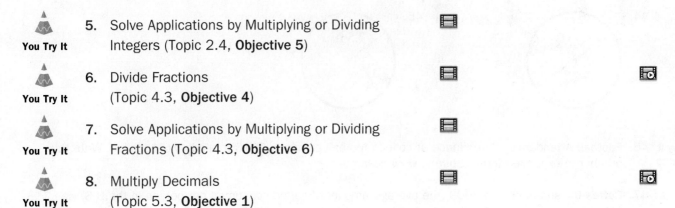

You Try It 5. Solve Applications by Multiplying or Dividing Integers (Topic 2.4, **Objective 5**)

You Try It 6. Divide Fractions (Topic 4.3, **Objective 4**)

You Try It 7. Solve Applications by Multiplying or Dividing Fractions (Topic 4.3, **Objective 6**)

You Try It 8. Multiply Decimals (Topic 5.3, **Objective 1**)

OBJECTIVES

1 Divide Decimals

2 Divide Decimals by Powers of 10

3 Estimate Quotients of Decimals

4 Solve Applications by Dividing Decimals

OBJECTIVE 1 DIVIDE DECIMALS

Division involving **decimal numbers** is similar to division involving **whole numbers** except that we must place a **decimal point** in the **quotient**. Consider the following in which a decimal is divided by a whole number:

$$0.34 \div 2 = \frac{34}{100} \div 2 = \frac{34}{100} \cdot \frac{1}{2} = \frac{34 \cdot 1}{100 \cdot 2} = \frac{\overset{17}{\cancel{34}} \cdot 1}{100 \cdot \underset{1}{\cancel{2}}} = \frac{17}{100} = 0.17$$

So, $0.34 \div 2 = 0.17$. Compare this to $34 \div 2 = 17$. Do you see the similarity? The only difference is the location of the decimal point (and the 0 in front of the decimal point that makes the decimal easier to read).

Writing our decimal division problem in **long-division** form, the decimal points in the **dividend** and quotient align. This always happens when a decimal is divided by a whole number. We use this idea to develop a long-division process for dividing a decimal by a whole number.

```
            ┌──── Decimal points align
            ▼
           0.17 ← Quotient
Divisor → 2)0.34 ←Dividend
```

> **Dividing a Decimal by a Whole Number**
>
> **Step 1.** Write the problem in **long-division** form and place a **decimal point** in the **quotient** directly above the decimal point in the **dividend**.
>
> **Step 2.** Divide as if dividing whole numbers. If necessary, write 0 **digits** behind the last digit of the dividend to continue the division process.

TIP Remember to check your answer. Multiplying the quotient and **divisor** should result in the dividend.

(eText Screens 5.4-1–5.4-29)

Example 1 Dividing Decimals by Whole Numbers

Divide. Check the result.

a. $29.4 \div 6$ **b.** $-73.36 \div 14$

Solutions

a. We write the problem in long-division form and place the decimal point in the quotient directly above the decimal point in the dividend.

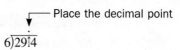

Place the decimal point

$$6\overline{)29.4}$$

Because the **digit** 2 (in the tens **place**) cannot be divided by 6, we begin by dividing the first two digits 29 by 6.

$$\begin{array}{r} 4. \\ 6\overline{)29.4} \\ -24 \\ \hline 5 \end{array}$$

Divide: $29 \div 6$ is about 4. Place this 4 above the 9.

Multiply: $4 \cdot 6 = 24$

Subtract: $29 - 24 = 5$

$$\begin{array}{r} 4.9 \\ 6\overline{)29.4} \\ -24\downarrow \\ \hline 5\,4 \\ -5\,4 \\ \hline 0 \end{array}$$

Drop down the next digit, 4.

Divide: $54 \div 6$ is 9. Place this 9 above the 4.

Multiply: $9 \cdot 6 = 54$

Subtract: $54 - 54 = 0$

We can check this result by seeing if the **quotient** multiplied by the **divisor** equals the **dividend**. The result checks (view the **details**), so $29.4 \div 6 = 4.9$.

My video summary

b. Recall that dividing a **negative** number by a **positive** number gives a negative **quotient**. We divide the **absolute values**, and make the quotient negative. (Review the rules for **dividing two numbers.**)

$|-73.36| = 73.36$ and $|14| = 14$, so we divide $73.36 \div 14$ and make the quotient negative.

Place the decimal point

$$14\overline{)73.36}$$

Try to finish this problem on your own. View the **answer**, or watch this **video** for a complete solution to part b.

You Try It **Work through this You Try It problem.**

Work Exercises 1–4 in this textbook or in the MyMathLab Study Plan.

Sometimes, we must insert one or more 0's behind the last **digit** of the dividend to continue the division process.

Example 2 Dividing Decimals Involving Insertion of 0's in the Dividend

Divide. Check the result.

a. $22.8 \div 5$ **b.** $8.7 \div 12$

(eText Screens 5.4-1–5.4-29)

Solutions

a. Write the problem in **long-division** form and place the **decimal point** in the quotient directly above the decimal point in the **dividend**.

$5\overline{)22.8}$ (decimal point placed above)	Place the decimal point
$\begin{array}{r} 4. \\ 5\overline{)22.8} \end{array}$	Divide: $22 \div 5$ is about 4. Place this 4 above the 2 in the ones place.
-20	Multiply: $4 \cdot 5 = 20$
2	Subtract: $22 - 20 = 2$
$\begin{array}{r} 4.5 \\ 5\overline{)22.8} \\ -20\downarrow \end{array}$	Drop down the next digit, 8.
28	Divide: $28 \div 5$ is about 5. Place this 5 above the 8 in the tenths place.
-25	Multiply: $5 \cdot 5 = 25$
3	Subtract: $28 - 25 = 3$
$\begin{array}{r} 4.56 \\ 5\overline{)22.80} \\ -20 \\ 28 \\ -25\downarrow \end{array}$	Insert a 0 digit in the hundredths place behind the 8, then drop it down.
30	Divide: $30 \div 5$ is 6. Place this 6 above the inserted 0.
-30	Multiply: $6 \cdot 5 = 30$
0	Subtract: $30 - 30 = 0$

So, $22.8 \div 5 = 4.56$. View the **check**.

My video summary

b. Try to work this problem on your own. Hint: You will need to insert two 0 **digits** behind the 7 in the dividend. View the **answer**, or watch this **video** for a complete solution to part b.

You Try It **Work through this You Try It problem.**

Work Exercises 5–8 in this textbook or in the MyMathLab Study Plan.

In the first two examples, each quotient is a *terminating decimal*. A **terminating decimal** is a **decimal number** that ends or has a fixed number of **digits** behind the **decimal point**. For example, the number 4.56 is a terminating decimal because it has exactly two digits behind the decimal point.

Sometimes the result of dividing two decimal numbers is a *repeating decimal*. A **repeating decimal** is a decimal in which a fixed number of digits form a pattern that repeats without end. We indicate a repeating decimal by placing a bar over the pattern of digits that repeat or by writing three dots (. . .), called an *ellipsis*, behind the established pattern. For example, both expressions below indicate the same repeating decimal in which the digit 7 repeats forever.

$4.5\overline{7}$ ← Bar over the 7 indicates the 7 repeats forever   or $4.5777\ldots$ (Ellipsis indicates the 7 repeats forever)

(eText Screens 5.4-1–5.4-29)

When a repeating pattern is established during the division process, we can write the **quotient** using one of these forms. Typically, we will use the bar notation rather than the ellipsis notation.

CAUTION When writing a repeating decimal, do not use both a bar and an ellipsis at the same time. Use one or the other, but not both, to indicate the repeating pattern of digits.

Example 3 Dividing Decimals Involving Repeating Decimals

Divide.

a. $7.4 \div 3$ **b.** $58.6 \div (-11)$

Solutions

a. We write the problem in **long-division** form and place the **decimal point** in the **quotient** directly above the decimal point in the **dividend**. During the division process, we add as many 0 digits as necessary to result in a **terminating** or **repeating decimal.**

Repeating pattern

$2.4666...$	
$3\overline{)7.4000}$	
-6	$2 \cdot 3 = 6$
14	$7 - 6 = 1$. Drop down the 4.
-12	$4 \cdot 3 = 12$
20	$14 - 12 = 2$. Insert a 0 digit and drop it down.
-18	$6 \cdot 3 = 18$
20	$20 - 18 = 2$. Insert a 0 digit and drop it down.
-18	$6 \cdot 3 = 18$
20	$20 - 18 = 2$. Insert a 0 digit and drop it down.
-18	$6 \cdot 3 = 18$
2	$20 - 18 = 2$

The 6 pattern keeps repeating, so $7.4 \div 3 = 2.4\overline{6}$ (or $2.4666\ldots$).

My video summary

b. In this problem, we are dividing a **positive** number by a **negative** number, which gives a negative **quotient**. Try working this problem on your own. View the **answer**, or watch this **video** for a complete solution to part b.

You Try It **Work through this You Try It problem.**

Work Exercises 9–12 in this textbook or in the MyMathLab Study Plan.

CAUTION When using the bar notation to write a **repeating decimal**, use as few digits as possible to show the repeating pattern.

Correct	Incorrect	Incorrect
$0.\overline{24}$	~~$0.24\overline{24}$~~	~~$0.\overline{2424}$~~

In **Topic 2.4**, we divided **integers** that always resulted in integer quotients. However, the quotient of two integers will not always be an integer. In those cases, we can write the quotient as either a **terminating** or repeating decimal. To divide, we recognize that an unwritten **decimal point** is understood to be located behind the last **digit** of an integer (or behind the ones **place**).

(eText Screens 5.4-1–5.4-29)

Example 4 Dividing Integers with Decimal Quotients

Divide.

a. $-13 \div 8$

b. $-82 \div (-33)$

Solutions

a. We are dividing a **negative** integer by a **positive** integer, so we divide the **absolute values** and make the **quotient** negative.

$$|-13| = 13 \quad \text{and} \quad |8| = 8$$

Putting the problem in **long-division** form, we write in the understood **decimal point** both behind the last digit of the **dividend** 13 and again directly above it in the quotient.

$$8\overline{)13.}$$

Write in the understood decimal point and then place it in the quotient.

During the division process, we include as many 0 digits behind the decimal point as necessary to result in a **terminating** or **repeating decimal.**

$$\begin{array}{r} 1.625 \\ 8\overline{)13.000} \end{array}$$

-8	$1 \cdot 8 = 8$
50	$13 - 8 = 5$. Insert a 0 digit and drop it down.
-48	$6 \cdot 8 = 48$
20	$50 - 48 = 2$. Insert a 0 digit and drop it down.
-16	$2 \cdot 8 = 16$
40	$20 - 16 = 4$. Insert a 0 digit and drop it down.
-40	$5 \cdot 8 = 40$
0	$40 - 40 = 0$. The decimal terminates.

The decimal **terminates,** so $-13 \div 8 = -1.625$.

My video summary

b. In this problem, we are dividing a **negative** integer by a negative integer so the **quotient** will be **positive.** Try working this problem on your own. View the **answer,** or watch this **video** for a complete solution to part b.

You Try It **Work through this You Try It problem.**

Work Exercises 13–16 in this textbook or in the MyMathLab Study Plan.

In Examples 1–4, the **divisors** are **whole numbers** or **integers.** Next, we look at dividing when the divisor is a **decimal.** Consider the problem $2.34 \div 0.6$. Let's begin by writing the problem in fraction form:

$$2.34 \div 0.6 = \frac{2.34}{0.6}.$$

Using the **equivalent property of fractions** from **Topic 4.2,** we can multiply both the numerator and denominator by the same nonzero number without changing the value of the fraction. Let's multiply both the **numerator** and **denominator** by 10. Remember that multiplying a decimal number by 10 simply moves the **decimal point** one place to the right. So,

$$2.34 \div 0.6 = \frac{2.34}{0.6} = \frac{2.34 \cdot 10}{0.6 \cdot 10} = \frac{23.4}{6} = 23.4 \div 6.$$

This illustrates that the division problems $2.34 \div 0.6$ and $23.4 \div 6$ are equivalent. With the divisor being a whole number, we could use the process for **dividing a decimal by a whole number** to finish this problem.

In our example above, the decimal points in the **divisor** and **dividend** are both moved one place to the right, which results in the divisor becoming a **whole number**. This leads to the following process for dividing by decimals.

> **Dividing Decimals**
>
> **Step 1.** Move the decimal point in the divisor to the right until it is behind the last **digit**.
>
> **Step 2.** Move the decimal point in the dividend to the right the same number of places as the decimal point was moved in Step 1. Add zeros as placeholders if necessary.
>
> **Step 3.** Divide. Place the decimal point in the **quotient** directly above the location of the moved decimal point in the dividend.

Example 5 Dividing by Decimals

My interactive video summary

Divide.

a. $7.836 \div 0.8$

b. $-5.24/0.9$

c. $\dfrac{-0.105}{-0.875}$

d. $15.9 \div 0.33$

Solutions

a. We follow the **steps for dividing decimals.** The **divisor** has one **digit** behind the decimal point. So, we move the decimal points in both the divisor and **dividend** one place to the right, making the divisor a whole number.

```
                        9.7 9 5
0.8)7.836   ⇒   8.)7 8.3 6 0  ← Insert a 0 digit
                  −7 2↓
                     6 3
                   −5 6↓
                       7 6
                     −7 2↓
                         4 0
                       −4 0
                           0  ← Terminates
```

The quotient is 9.795.

b. Recall that dividing a **negative** number by a **positive** number gives a negative **quotient**. We divide the **absolute values** and make the quotient negative. Because $|-5.24| = 5.24$ and $|0.9| = 0.9$, we divide $5.24/0.9$ and make the quotient negative.

$$0.9\overline{)5.24} \Rightarrow \begin{array}{r} 5.8222\ldots \\ 9.\overline{)52.4000} \\ \underline{-45} \\ 74 \\ \underline{-72} \\ 20 \\ \underline{-18} \\ 20 \\ \underline{-18} \\ 20 \\ \underline{-18} \\ 2 \end{array}$$

← Insert three 0 digits

Repeating pattern (20, −18, 20, −18, 20, −18, 2)

$5.24/0.9 = 5.8\overline{2}$, so $-5.24/0.9 = -5.8\overline{2}$ (or $-5.8222\ldots$).

c.–d. Try working these problems on your own. View the **answers**, or watch this **interactive video** for detailed solutions to all four parts.

You Try It **Work through this You Try It problem.**

Work Exercises 17–24 in this textbook or in the MyMathLab Study Plan.

Suppose we wanted to **round** the quotient 9.795 from **Example 5a** to the nearest tenth. Because the digit 9 in the hundredths place is larger than 5, we would increase the tenths place digit from 7 to 8, and drop all of the digits to the right of the tenths place.

$$9.795 \approx 9.8$$

"is approximately equal to" (points to $\approx$); Tenths (points to 7); Hundredths (points to 9)

Notice that we did not need to know the digit in the thousandths place in order to round the quotient to the nearest tenth. We needed to know only the digit in the hundredths place, which is to the immediate right of the tenths place.

TIP If we know that we will be rounding a quotient to a given **place value**, then we only need to carry out the division process until we have the digit to the immediate right of that place value. We can then stop the division process and round.

Example 6 Dividing Decimals with Rounding

Divide $53.6 \div 0.83$. **Round** the **quotient** to the nearest hundredth.

(eText Screens 5.4-1–5.4-29)

My video summary **Solution** Continue working below, or watch this **video** for a complete solution.

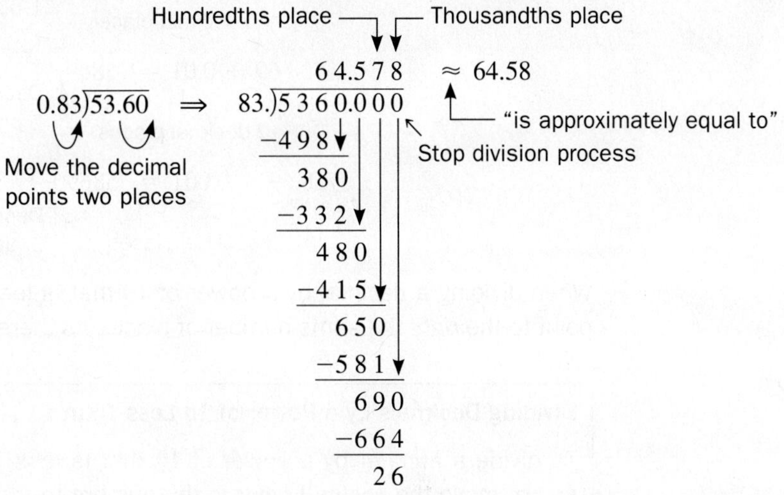

You Try It Work through this You Try It problem.

Work Exercises 25–32 in this textbook or in the MyMathLab Study Plan.

OBJECTIVE 2 DIVIDE DECIMALS BY POWERS OF 10

In **Topic 5.3**, we learned that multiplying a decimal by a **power of 10** simply moves the **decimal point** in the decimal number. Similar results occur when dividing decimals by a power of 10. Consider the following pattern when dividing decimals by powers of 10 that are greater than or equal to 1:

$758.69 \div 10 = 75.869$ — 1 zero; Decimal point moves 1 place to the *left*.

$758.69 \div 100 = 7.5869$ — 2 zeros; Decimal point moves 2 places to the *left*.

$758.69 \div 1000 = 0.75869$ — 3 zeros; Decimal point moves 3 places to the *left*.

Notice that when dividing a decimal by a power of 10 that is greater than or equal to 1, we simply move the decimal point to the *left* the same number of places as there are zeros in the power of 10.

> **Dividing Decimals by a Power of 10 Greater Than or Equal to 1**
>
> To divide a number by a **power of 10** greater than or equal to 1, such as 10, 100, 1000, and so on, move the **decimal point** in the number to the *left* the same number of places as there are zeros in the power of 10.

A similar pattern exists when dividing by powers of 10 that are smaller than 1, such as 0.1, 0.01, 0.001, and so on. Consider the following pattern.

(eText Screens 5.4-1–5.4-29)

$758.69 \div 0.1 = 7586.9$ — 1 decimal place; Decimal point moves 1 place to the *right*.

$758.69 \div 0.01 = 75869.$ — 2 decimal places; Decimal point moves 2 places to the *right*.

$758.69 \div 0.001 = 758690.$ — 3 decimal places; Decimal point moves 3 places to the *right*.

When dividing a decimal by a power of 10 that is less than 1, we simply move the decimal point to the *right* the same number of places as there are **decimal places** in the power of 10.

Dividing Decimals by a Power of 10 Less Than 1

To divide a number by a **power of 10** that is less than 1, such as 0.1, 0.01, 0.001, and so on, move the **decimal point** in the number to the *right* the same number of places as there are **decimal places** in the power of 10.

Example 7 Dividing Decimals by Powers of 10

My video summary

Divide.

a. $287.349 \div 100$

b. $92.594/(-0.001)$

c. $\dfrac{-781.93}{-10{,}000}$

d. $\dfrac{3.91}{-0.00001}$

Solutions

a. The **divisor** is a power of 10 that is greater than 1. Because there are 2 zeros in the power of 10, we move the decimal point 2 places to the *left*.

$287.349 \div 100 = 2.87349$ — 2 zeros; Decimal point moves 2 places to the *left*.

b. Note that we are dividing a **positive** number by a **negative** number, so the **quotient** will be negative. Because $|92.594| = 92.594$ and $|-0.001| = 0.001$, we divide $92.594/0.001$ and make the quotient negative.

The **divisor** is a **power of** 10 that is less than 1. Because there are 3 **decimal places** in the power of 10, we move the decimal point to the *right* 3 places.

$92.594/0.001 = 92594.$ — 3 decimal places; Decimal point moves 3 places to the *right*.

So, $92.594/(-0.001) = -92{,}594$.

c.–d. Try working these problems on your own. Check your **answers,** or watch this **video** for detailed solutions to all four parts.

You Try It **Work through this You Try It problem.**

Work Exercises 33–40 in this textbook or in the MyMathLab Study Plan.

OBJECTIVE 3 ESTIMATE QUOTIENTS OF DECIMALS

As with the other **arithmetic operations**, we can estimate when dividing **decimals** to check the reasonableness of our **quotient**. For division, we approximate both the **dividend** and the **divisor** with numbers that can be divided more easily. Recall that this is the same approach we used to estimate quotients of **whole numbers** back in **Topic 1.5.** To obtain better estimates, remember to either increase both the dividend and divisor, or decrease them both, if possible.

Example 8 Estimating Quotients of Decimals

Estimate each quotient by approximating the dividend and divisor with numbers that can be divided more easily.

a. $26.74 \div 2.8$ **b.** $394.632 \div 48.6$

Solutions

a. We want to approximate both the dividend and divisor with numbers that will divide easily. Notice that the dividend is close to 27 and the divisor is close to 3. Also, 3 divides evenly into 27.

Approximate the dividend

$$26.74 \div 2.8 \qquad 27 \div 3 = 9$$

Approximate the divisor

The estimated quotient is 9. This is reasonably close to the actual quotient of 9.55. View the **details.**

Note: A second reasonable estimate for $26.74 \div 2.8$ can be found by rounding 26.74 to 30 and by rounding 2.8 to 3. This would give the reasonable estimate $30 \div 3 = 10$.

My video summary

b. Try to estimate this quotient on your own. View the **answer**, or watch this **video** for a complete solution to part b.

You Try It **Work through this You Try It problem.**

Work Exercises 41–44 in this textbook or in the MyMathLab Study Plan.

OBJECTIVE 4 SOLVE APPLICATIONS BY DIVIDING DECIMALS

Many real-life application problems can be solved by **dividing decimals.**

Example 9 Monthly Car Payment

My video summary

To purchase a used car, Reagan borrowed \$12,856.58. If the loan must be paid back in 48 equal monthly payments, what will be the monthly payment? Round to the nearest cent (hundredths place).

Solution To find the monthly payment, we divide the loan amount by the number of monthly payments.

$$\$12{,}856.58 \div 48$$

(eText Screens 5.4-1–5.4-29)

Try to perform this division and rounding on your own to answer the question. View the **answer**, or watch this **video** for the complete solution.

You Try It **Work through this You Try It problem.**

Work Exercises 45–52 in this textbook or in the MyMathLab Study Plan.

5.4 Exercises

In Exercises 1–8, divide.

You Try It

1. $4.77 \div 9$
2. $89.6 \div 7$
3. $-37.6 \div 8$
4. $-4.65 \div (-15)$
5. $8.25/6$
6. $\dfrac{17.4}{-24}$
7. $-39.7 \div 5$
8. $18.3 \div 16$

In Exercises 9–24, divide. Use bar notation to write repeating decimals.

You Try It

9. $6.4 \div 9$
10. $1.84/6$
11. $\dfrac{-3.5}{33}$
12. $19.6 \div 11$
13. $25 \div 9$
14. $-85/33$
15. $124 \div (-16)$
16. $-500 \div (-111)$
17. $1.7 \div 0.9$
18. $9.512 \div 0.4$
19. $0.259 \div 0.14$
20. $42.56 \div 0.88$
21. $-2.68/1.2$
22. $\dfrac{1.566}{-0.027}$
23. $1.525/(-0.6)$
24. $\dfrac{-2.07}{-0.375}$

In Exercises 25–32, divide. Round the quotients as indicated.

You Try It

25. $0.873 \div 0.51$; hundredths place
26. $9.452/(-7.2)$; thousandths place
27. $63.5 \div 0.49$; whole number
28. $254.9/0.6$; tenths place
29. $\dfrac{-1.293}{17}$; thousandths place
30. $\dfrac{-87.34}{-9}$; hundredths place
31. $56 \div 3.25$; tenths place
32. $0.0259/0.012$; ten-thousandths place

In Exercises 33–40, divide.

You Try It

33. $78.913 \div 100$
34. $\dfrac{812.439}{0.01}$
35. $-9.41 \div 0.1$
36. $\dfrac{-32.062}{-1000}$
37. $8.64/(-0.001)$
38. $5.978/(-10)$
39. $\dfrac{31.94}{10{,}000}$
40. $\dfrac{578.65}{0.0001}$

In Exercises 41–44, estimate each quotient by approximating the dividend and divisor with numbers that can be divided more easily.

You Try It 41. $\frac{7892.22}{38.7}$

42. $\frac{422.809}{71.3}$

43. $\frac{21.63}{1.89}$

44. $\frac{3871.83}{47.51}$

In Exercises 45–52, solve each application problem.

You Try It 45. **Selling Gold** In December 2011, the price of gold was $1742.50 per ounce. If Sarah sold some gold for $4774.45, how many ounces of gold did she sell? (*Source:* monex.com)

46. **Wall Street Wins the Lottery** In November 2011, three colleagues from a wealth-management firm in Connecticut, who jointly purchased a Powerball lottery ticket, won a $254.2 million jackpot. If the money was split equally, how much did each person receive? Round to the nearest whole million. (*Source: Wall Street Journal*)

47. **Unit Conversions** A kilogram is equivalent to 2.2 pounds. If Gerard weighs 165 pounds, what is his weight in kilograms?

48. **Earnings per Hour** If Brendan earned $339.08 for working 24.5 hours one week, find his earnings per hour.

49. **Gas Mileage** To measure her car's gas mileage, Bianca keeps track of her mileage and gasoline consumption. If she has traveled 433.1 miles since her last refill and it takes 15.25 gallons of gasoline to fill up her tank, determine the miles per gallon for Bianca's car.

50. **Allergy Shots** If an allergy patient is to receive 0.25 milliliter of serum per injection, how many injections can be administered from a 15-milliliter vial of the serum?

51. **Storing Digital Photos** Paige has a 16-gigabyte memory stick for storing digital photos. If each photo uses 0.005 gigabyte of memory, how many photos can Paige store on the memory stick?

52. **Sliced Bread** A bread slicing machine cuts loaves of bread into pieces that are 0.625 inch thick. How many slices will be cut from an 10-inch loaf of bread?

5.5 Fractions, Decimals, and Order of Operations

THINGS TO KNOW

Before working through this topic, be sure you are familiar with the following concepts:

	VIDEO	ANIMATION	INTERACTIVE
You Try It 1. Use the Order of Operations with Fractions (Topic 4.5, **Objective 2**)	✓		
You Try It 2. Evaluate Algebraic Expressions Using Fractions (Topic 4.5, **Objective 3**)	✓		✓
You Try It 3. Compare Decimals (Topic 5.1, **Objective 6**)	✓		

(eText Screens 5.5-1–5.5-26)

You Try It **4.** Add Positive Decimals (Topic 5.2, **Objective 1**)

You Try It **5.** Subtract Positive Decimals (Topic 5.2, **Objective 2**)

You Try It **6.** Add and Subtract Negative Decimals (Topic 5.2, **Objective 4**)

You Try It **7.** Multiply Decimals (Topic 5.3, **Objective 1**)

You Try It **8.** Find the Circumference of a Circle (Topic 5.3, **Objective 5**)

You Try It **9.** Divide Decimals (Topic 5.4, **Objective 1**)

OBJECTIVES

1 Change Fractions or Mixed Numbers into Decimals

2 Compare Fractions and Decimals

3 Use the Order of Operations with Decimals

4 Evaluate Algebraic Expressions Using Decimals

5 Find the Area of a Circle

OBJECTIVE 1 CHANGE FRACTIONS OR MIXED NUMBERS INTO DECIMALS

When a **fraction** and a **decimal** represent the same value, such as $\frac{1}{2}$ and 0.5, they are called **equivalent**. Recall that a **fraction bar** is a division symbol.

$$\text{Fraction bar} \rightarrow \frac{1}{2} \begin{array}{l} \leftarrow \text{Numerator} \\ \leftarrow \text{“divided by”} \\ \leftarrow \text{Denominator} \end{array}$$

Therefore, to change a fraction into an equivalent decimal, we divide the **numerator** by the **denominator**. To do this, we use the method from **Topic 5.4** for **dividing a decimal by a whole number.**

> **Changing a Fraction into a Decimal**
>
> To change a fraction into a decimal, divide the numerator by the denominator.

TIP The result will be either a **terminating** or **repeating decimal.**

Example 1 Changing a Fraction into a Decimal

Change $\frac{3}{4}$ into an equivalent decimal.

(eText Screens 5.5-1–5.5-26)

My video summary

Solution We divide the **numerator** by the **denominator.** Using **long division,** we write in the understood **decimal points.**

$$4.\overline{)3.}$$

↑ Write in the understood decimal points.

During the division process, we include as many 0 digits behind the decimal point as necessary to result in a **terminating** or **repeating decimal.**

$$\begin{array}{r} 0.75 \\ 4.\overline{)3.00} \\ -28\downarrow \\ \hline 20 \\ -20 \\ \hline 0 \end{array}$$

Insert 0 digits as needed to continue dividing.
$7 \cdot 4 = 28$
$30 - 28 = 2$. Insert a 0 digit and drop it down.
$5 \cdot 4 = 20$
$20 - 20 = 0$. The decimal terminates.

So, the fraction $\frac{3}{4}$ is equivalent to the decimal 0.75, or $\frac{3}{4} = 0.75$.

You Try It **Work through this You Try It problem.**

Work Exercises 1 and 2 in this textbook or in the MyMathLab Study Plan.

If a **fraction** is **negative**, then its **decimal** equivalent is negative also. We can change a negative fraction into a decimal as if it were positive and then add on the **negative sign** to the decimal form.

Example 2 Changing a Fraction into a Decimal

My video summary

Change $-\frac{11}{16}$ into an equivalent decimal.

Solution Dividing 11 by 16 results in the decimal 0.6875. So, $-\frac{11}{16} = -0.6875$. See the **details**, or watch this **video** for a complete solution.

You Try It **Work through this You Try It problem.**

Work Exercises 3 and 4 in this textbook or in the MyMathLab Study Plan.

As we have stated before, some fractions change into **repeating decimals.** As we did when dividing decimals, we will use the **bar notation** rather than the **ellipsis notation** to write repeating decimals.

Example 3 Changing Fractions into Repeating Decimals

Change each fraction into an equivalent decimal.

a. $\frac{5}{6}$ **b.** $\frac{83}{99}$

(eText Screens 5.5-1–5.5-26)

My video summary

Solutions Work through the following, or watch this **video** for solutions to both parts. For each fraction, pay close attention to the pattern of **digits** that repeat to form the **repeating decimal**.

a. Repeating pattern

$$6\overline{)5.0000} = 0.8333\ldots$$

$6\overline{)5.0000}$	Insert 0 digits as needed to continue dividing.
-48	$8 \cdot 6 = 48$
20	$50 - 48 = 2$. Insert a 0 digit and drop it down.
-18	$3 \cdot 6 = 18$
20	$20 - 18 = 2$. Insert a 0 digit and drop it down.
-18	$3 \cdot 6 = 18$
20	$20 - 18 = 2$. Insert a 0 digit and drop it down.
-18	$3 \cdot 6 = 18$
2	$20 - 18 = 2$

The digit 3 keeps repeating, so $\frac{5}{6} = 0.8\overline{3}$ (or $0.8333\ldots$).

b. Repeating pattern

$$99\overline{)83.0000} = 0.8383\ldots$$

$99\overline{)83.0000}$	Insert 0 digits as needed to continue dividing.
-792	$8 \cdot 99 = 792$
380	$830 - 792 = 38$. Insert a 0 digit and drop it down.
-297	$3 \cdot 99 = 297$
830	$380 - 297 = 83$. Insert a 0 digit and drop it down.
-792	$8 \cdot 99 = 792$
380	$830 - 792 = 38$. Insert a 0 digit and drop it down.
-297	$3 \cdot 99 = 297$
83	$380 - 297 = 83$.

The two-**digit** pattern "83" keeps **repeating**, so $\frac{83}{99} = 0.\overline{83}$ (or $0.838383\ldots$).

You Try It **Work through this You Try It problem.**

Work Exercises 5–8 in this textbook or in the MyMathLab Study Plan.

CAUTION

Notice the similarity between the two different repeating decimals $0.8\overline{3}$ and $0.\overline{83}$ that we found in Example 3. Clear placement of the bar above the repeating pattern is essential when using **bar notation**.

Example 4 Changing Fractions into Decimals

My video summary

Write each **fraction** as a **decimal**.

a. $\frac{9}{11}$ **b.** $\frac{7}{20}$ **c.** $\frac{19}{8}$

Solutions Try changing these fractions into decimals on your own. View the **answers**, or watch this **video** for complete solutions.

(eText Screens 5.5-1–5.5-26)

You Try It **Work through this You Try It problem.**

Work Exercises 9–16 in this textbook or in the MyMathLab Study Plan.

TIP The decimal equivalent of a **proper fraction** will have 0 as the only **digit** in front of the **decimal point** because its **absolute value** is less than 1. The decimal equivalent of an **improper fraction** will have a nonzero digit in front of the decimal point because its absolute value is greater than or equal to 1.

To write a **mixed number** as a **decimal number**, we recognize that the **whole-number part** of the mixed number will also be the whole-number part of the decimal number (in front of the decimal point). Changing the **fraction part** of the mixed number into a decimal gives the **decimal part** of the number (behind the decimal point).

Example 5 Changing Mixed Numbers into Decimals

My video summary

Change each **mixed number** into an equivalent **decimal**.

a. $3\frac{18}{25}$ b. $-5\frac{2}{3}$

Solutions Work through the following, or watch this **video** for solutions to both parts.

a. Change $\frac{18}{25}$ into a decimal, then include the whole-number part, 3.

$$\frac{18}{25} \Rightarrow \begin{array}{r} 0.72 \\ 25\overline{)18.00} \\ -17\,5 \\ \hline 50 \\ -50 \\ \hline 0 \end{array} \qquad \text{So, } 3\frac{18}{25} = 3.72.$$

b. Changing $\frac{2}{3}$ into a decimal gives $0.\overline{6}$. So, $-5\frac{2}{3} = -5.\overline{6}$.

You Try It **Work through this You Try It problem.**

Work Exercises 17–20 in this textbook or in the MyMathLab Study Plan.

Recall that the **decimal** equivalent of a **fraction** will be either a **terminating** or **repeating decimal**. Sometimes, the pattern of **digits** that repeat can be long.

$$\frac{12}{37} = 0.\overline{324} \quad \leftarrow \text{Repeating pattern of 3 digits}$$

$$\frac{51}{101} = 0.\overline{5148} \quad \leftarrow \text{Repeating pattern of 4 digits}$$

$$\frac{13}{21} = 0.\overline{619047} \quad \leftarrow \text{Repeating pattern of 6 digits}$$

For this reason, among others, we may want to **round the decimal** to a given **place value**. This rounded decimal will give a **decimal approximation** for a fraction.

(eText Screens 5.5-1–5.5-26)

TIP To approximate a fraction to a given decimal place value, carry out the division process until the digit to the immediate right of that place value is known, then round.

Example 6 Finding Decimal Approximations

Find a decimal approximation for each fraction, rounded to the given place value.

a. $\frac{3}{7}$; hundredths **b.** $\frac{24}{19}$; thousandths

Solutions

a. Because we are **rounding** to the hundredths **place**, we carry out the division process until we know the **digit** in the thousandths place, which is the place value to the immediate right of the hundredths place. At that point, we stop the division process and round.

Hundredths place → 2; Thousandths place → 8

$$\begin{array}{r} 0.428 \\ 7\overline{)3.000} \\ -28 \\ \hline 20 \\ -14 \\ \hline 60 \\ -56 \\ \hline 4 \end{array} \leftarrow \text{Stop division process}$$

The digit 8 in the thousandths place is larger than 5, so we increase the digit in the hundredths place to 3, and drop all digits to its right. Thus, $\frac{3}{7} \approx 0.43$.

My video summary

b. $\frac{24}{19}$ is an improper fraction, so its decimal approximation will have a nonzero digit in front of the decimal point. Try to divide 24 by 19 on your own, carrying out the division process until a digit is found in the ten-thousandths place. Then stop the division process and round.

View the **answer**, or watch this **video** for a complete solution to part b.

You Try It **Work through this You Try It problem.**

Work Exercises 21–28 in this textbook or in the MyMathLab Study Plan.

OBJECTIVE 2 COMPARE FRACTIONS AND DECIMALS

By changing **fractions** into **decimals**, we can then compare fractions with decimals (and other fractions) using the **rules for comparing two decimals** from **Topic 5.1**.

Example 7 Comparing Fractions and Decimals

Fill in the blank with $<$, $>$, or $=$ to make a true comparison statement.

a. $\frac{3}{8}$ ____ 0.38 **b.** $\frac{5}{9}$ ____ 0.5 **c.** $\frac{11}{5}$ ____ 2.2

d. $\frac{7}{12}$ ____ $\frac{4}{7}$ **e.** $6\frac{7}{40}$ ____ 6.175 **f.** $0.\overline{56}$ ____ $\frac{17}{30}$

(eText Screens 5.5-1–5.5-26)

Solutions

My interactive video summary

a. First, we change $\frac{3}{8}$ into a decimal. Then, we compare the decimals.

$$\frac{3}{8} \Rightarrow \begin{array}{r} 0.375 \\ 8\overline{)3.000} \\ \underline{-24} \\ 60 \\ \underline{-56} \\ 40 \\ \underline{-40} \\ 0 \end{array} \qquad \text{So, } \frac{3}{8} = 0.375.$$

We compare **digits** of the **decimals** from left to right until we find two that are different.

Digits in the tenths places are the same.

$0.375 \quad 0.38$

Digits in the hundredths places are different.

Comparing the digits in the hundredths **places**, we have $7 < 8$.

This means $0.375 < 0.38$, which then means $\frac{3}{8} < 0.38$.

b. The decimal equivalent of $\frac{5}{9}$ is $0.\overline{5}$ (or $0.555\ldots$). View the **details.**

To make the comparison easier, we write the **repeating decimal** using **ellipsis notation,** and we fill in zeros behind the **terminating decimal.**

Digits in the tenths places are the same.

$0.555\ldots \quad 0.500$

Digits in the hundredths places are different.

Comparing the digits in the hundredths places, we have $5 > 0$.

This means $0.555\ldots > 0.500$, which means $\frac{5}{9} > 0.5$.

c. First, we change $\frac{11}{5}$ into a **decimal**. Then we compare the decimals.

$$\frac{11}{5} \Rightarrow \begin{array}{r} 2.2 \\ 5\overline{)11.0} \\ \underline{-10} \\ 10 \\ \underline{-10} \\ 0 \end{array} \qquad \text{Therefore, we see that } \frac{11}{5} = 2.2.$$

d.–f. Try working these problems on your own. View the **answers,** or work through this **interactive video** for complete solutions to all six parts.

You Try It **Work through this You Try It problem.**

Work Exercises 29–36 in this textbook or in the MyMathLab Study Plan.

(eText Screens 5.5-1–5.5-26)

Example 8 Arranging Numbers in Order

Write the numbers in order from smallest to largest.

$\frac{18}{25}, 0.685, \frac{5}{8}$

Solution First, change the two **fractions** into **decimal** form: $\frac{18}{25} = 0.72$ and $\frac{5}{8} = 0.625$.

View the **details.** The decimals in order are 0.625, 0.685, 0.72, so the original numbers in order are $\frac{5}{8}, 0.685, \frac{18}{25}$.

You Try It **Work through this You Try It problem.**

Work Exercises 37–40 in this textbook or in the MyMathLab Study Plan.

OBJECTIVE 3 USE THE ORDER OF OPERATIONS WITH DECIMALS

In **Topic 4.5**, we simplified **numeric expressions** involving **fractions.** Next, we **simplify** numeric expressions involving **decimals.** Let's begin by reviewing the order of operations.

Order of Operations

1. **Parentheses (or other grouping symbols)** Evaluate operations within parentheses (or other **grouping symbols**) first, starting with the innermost set and working out.
2. **Exponents** Work from left to right and evaluate any **exponential expressions** as they occur.
3. **Multiplication and Division** Work from left to right and perform any **multiplication** or **division** operations as they occur.
4. **Addition and Subtraction** Work from left to right and perform any **addition** or **subtraction** operations as they occur.

Example 9 Using the Order of Operations with Decimals

Simplify each expression using the **order of operations.**

a. $0.37(9.4 - 2.6)$ **b.** $(2.5)^2 + (1.4)^2$

Solutions

a. Begin with the original expression: $0.37(9.4 - 2.6)$

Subtract within the parentheses: $= 0.37(6.8)$

Multiply: $= 2.516$

$$\begin{array}{r} \overset{8}{\not 9}.\overset{14}{\not 4} \\ -2.6 \\ \hline 6.8 \end{array}$$

$$\begin{array}{rl} 0.37 & \leftarrow \text{2 decimal places} \\ \times\ 6.8 & \leftarrow \text{1 decimal place} \\ \hline 296 & \\ 2220 & \\ \hline 2.516 & \leftarrow \text{3 decimal places} \end{array}$$

My video summary

b. In this expression, there are two exponents and addition. Remember, we must evaluate both exponents before we add. Try to simplify this expression on your own. Check the **answer**, or watch this **video** for a complete solution to part b.

You Try It **Work through this You Try It problem.**

Work Exercises 41–44 in this textbook or in the MyMathLab Study Plan.

(eText Screens 5.5-1–5.5-26)

Example 10 Using the Order of Operations with Decimals

Simplify each expression using the **order of operations.**

a. $\frac{3}{5}(8.75) - 3.98$ **b.** $\frac{5 - (1.7)^2}{6(0.7) + 5.8}$

Solutions

a. Begin with the original expression: $\frac{3}{5}(8.75) - 3.98$

Change $\frac{3}{5}$ to a decimal: $= (0.6)(8.75) - 3.98$

Multiply: $= 5.25 - 3.98$

Subtract: $= 1.27$

$\frac{3}{5} \Rightarrow 5\overline{)3.0}$ quotient 0.6; $-3\,0$; 0

$8.75 \leftarrow$ 2 decimal places
$\times\ 0.6 \leftarrow$ 1 decimal place
$5.250 \leftarrow$ 3 decimal places

$\overset{4\ 11\ 15}{5.2\,5}$
$-3.9\,8$
$1.2\,7$

My video summary

b. Try to simplify this expression on your own by first simplifying the numerator and denominator, and then dividing. Check the **answer**, or watch this **video** for a detailed solution.

You Try It **Work through this You Try It problem.**

Work Exercises 45–48 in this textbook or in the MyMathLab Study Plan.

OBJECTIVE 4 EVALUATE ALGEBRAIC EXPRESSIONS USING DECIMALS

We have **evaluated algebraic expressions** using **whole numbers** in **Topic 1.7**, **integers** in **Topic 2.5**, and **fractions** in **Topic 4.5**. Next, we evaluate using **decimals**. We **substitute** the given values for the **variables** and **simplify** using the **order of operations**.

Example 11 Evaluating Algebraic Expressions Using Decimals

My video summary

Evaluate $\frac{6x - 2y}{5}$ for $x = 1.7$ and $y = 1.28$.

Solution Work through the following, or watch this **video** for a detailed solution.

Substitute 1.7 for x and 1.28 for y: $\frac{6x - 2y}{5} = \frac{6(1.7) - 2(1.28)}{5}$

Perform the two multiplications: $= \frac{10.2 - 2.56}{5}$

Subtract: $= \frac{7.64}{5}$

Divide: $= 1.528$

The value of $\frac{6x - 2y}{5}$ is 1.528 when $x = 1.7$ and $y = 1.28$.

You Try It **Work through this You Try It problem.**

Work Exercises 49–56 in this textbook or in the MyMathLab Study Plan.

(eText Screens 5.5-1–5.5-26)

OBJECTIVE 5 FIND THE AREA OF A CIRCLE

Recall that the **area** of a two-dimensional figure is the amount of surface contained within the **sides** of the figure. In **Topic 1.4**, we learned how to find the **area of a rectangle**, and in **Topic 4.3**, we learned how to find the **area of a triangle**. Now we learn to find the area of a *circle*.

As with **circumference**, the area of a circle is related to the **radius** of the circle and the constant π.

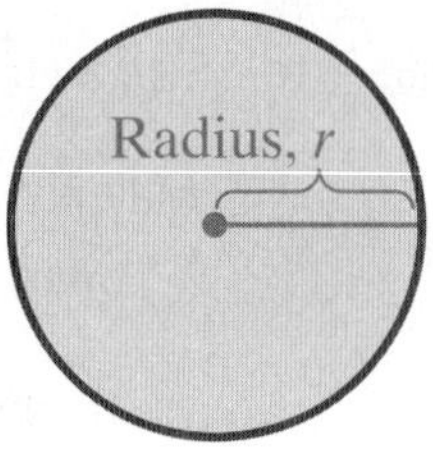

Area $= \pi \cdot (\text{radius})^2$

$A = \pi r^2$

Area of a Circle

The **area** of a circle is the amount of surface contained within the circle. Its value is found by

$$A = \pi r^2,$$

where r is the radius of the circle. The constant π is often approximated by the decimal 3.14 (or in some cases, the fraction $\frac{22}{7}$).

TIP Recall that area is measured in **square units** regardless of the shape of the figure.

Example 12 Finding the Area of a Circle

Find the exact area of each circle in terms of π. Then approximate the area using 3.14 for π.

a.

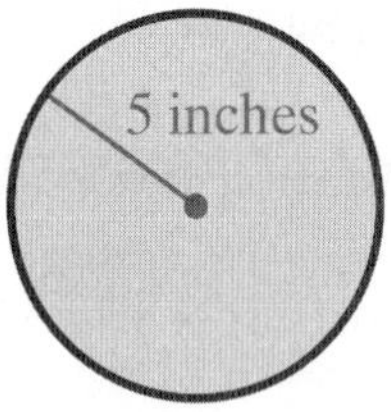

b.

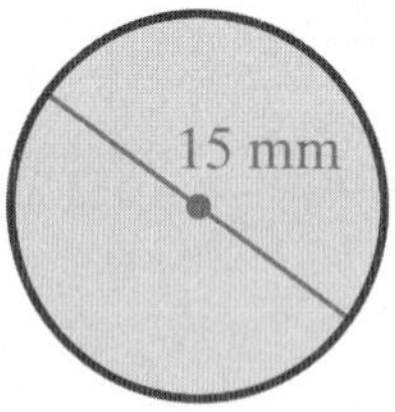

Solutions

a. From the figure, we see that the **radius** is 5 inches.

Exact Area: Using $A = \pi r^2$, we substitute 5 for r.

$$A = \pi r^2 = \pi(5)^2 = 25\pi$$

The exact area is 25π square inches, or $25\pi \text{ in}^2$.

Approximate Area: Using 3.14 for π, we get

"is approximately equal to"

$$A = 25\pi \approx 25(3.14) = 78.50$$

$$\begin{array}{r} 3.14 \\ \times\ 25 \\ \hline 1570 \\ 6280 \\ \hline 78.50 \end{array}$$

The area is approximately 78.50 square inches, or 78.50 in^2.

My video summary

b. For this circle, we have the **diameter**. To find the **area** by using $A = \pi r^2$, we need to know the **radius**. We can find the radius by dividing the diameter by 2, so

$$r = \frac{15}{2} = 7.5 \text{ mm}.$$

Try to find the area on your own. Remember to find both the exact area and the approximation using 3.14 for π. View the **answer**, or watch this **video** for the complete solution to part b.

You Try It **Work through this You Try It problem.**

Work Exercises 57–60 in this textbook or in the MyMathLab Study Plan.

CAUTION

The area of a circle is found using the *radius* of the circle, not the *diameter*. If the diameter is provided, we must first determine the radius from it and then use the radius to find the area.

5.5 Exercises

In Exercises 1–20, change each fraction or mixed number into an equivalent decimal. To write repeating decimals, use bar notation.

You Try It
1. $\frac{4}{5}$
2. $\frac{21}{25}$
3. $-\frac{7}{8}$
4. $-\frac{13}{20}$

You Try It
5. $\frac{5}{9}$
6. $-\frac{1}{12}$
7. $\frac{8}{11}$
8. $-\frac{14}{33}$

You Try It
9. $\frac{9}{16}$
10. $\frac{16}{99}$
11. $-\frac{15}{4}$
12. $\frac{9}{24}$

13. $-\frac{25}{18}$
14. $-\frac{17}{36}$
15. $\frac{31}{45}$
16. $\frac{47}{32}$

You Try It
17. $6\frac{1}{4}$
18. $2\frac{5}{12}$
19. $-4\frac{39}{50}$
20. $-1\frac{4}{15}$

In Exercises 21–28, find a decimal approximation for each fraction or mixed number, rounded to the given place value.

You Try It
21. $\frac{9}{13}$; tenths
22. $\frac{22}{7}$; hundredths
23. $-\frac{8}{17}$; thousandths

24. $\frac{20}{21}$; hundredths
25. $4\frac{2}{15}$; hundredths
26. $\frac{73}{60}$; tenths

27. $\frac{14}{3}$; thousandths
28. $\frac{16}{27}$; ten-thousandths

(eText Screens 5.5-1–5.5-26)

In Exercises 29–36, fill in the blank with $<$, $>$, or $=$ to make a true comparison statement.

You Try It 29. $\frac{1}{4}$ ____ 0.27

30. $\frac{2}{3}$ ____ 0.6

31. $\frac{11}{20}$ ____ 0.55

32. $\frac{8}{15}$ ____ $\frac{13}{25}$

33. $0.\overline{24}$ ____ $\frac{8}{33}$

34. $\frac{4}{9}$ ____ $\frac{5}{11}$

35. $5\frac{7}{8}$ ____ 5.9

36. $\frac{7}{9}$ ____ $\frac{17}{22}$

In Exercises 37–40, write the given list of numbers in order from smallest to largest.

You Try It 37. $\frac{1}{3}, 0.325, \frac{5}{16}$

38. $1\frac{4}{5}, 1.88, 1.\overline{7}$

39. $8.56, \frac{17}{2}, 8.\overline{5}$

40. $\frac{4}{3}, \frac{11}{8}, \frac{33}{25}$

In Exercises 41–48, simplify each expression using the order of operations.

You Try It 41. $0.54(8.2 - 7.3)$

42. $-6(0.35) + 8.9$

43. $(1.3)^2 - (1.2)^2$

44. $(5 - 1.8)^2$

You Try It 45. $\frac{3}{4}(5.42) + 8.63$

46. $-\frac{5}{8}(5.3 - 3.9)$

47. $1.125 + \frac{7}{2}(8.25 - 4.61)$

48. $\dfrac{8.65 - 1.2(3.5)}{7.44 + (1.6)^2}$

In Exercises 49–56, evaluate each expression for the given value(s) of the variable(s).

You Try It 49. $3.45x$ for $x = 16.2$

50. $\dfrac{m + n}{2}$ for $m = 2.57$ and $n = -0.96$

51. $a^2 - b$ for $a = -0.6$ and $b = -0.4$

52. $2.5x + 6y$ for $x = 8.4$ and $y = 3.9$

53. $np(1 - p)$ for $n = 30$ and $p = 0.55$

54. $\frac{1}{2}bh$ for $b = 10.6$ and $h = 8.4$

55. $\dfrac{t}{8}$ for $t = 14.9$

56. $\dfrac{x - m}{s}$ for $x = 18.9$, $m = 17.4$, and $s = 2.5$

In Exercises 57–60, find the exact area of each circle in terms of π. Then approximate the area using 3.14 for π.

You Try It 57.

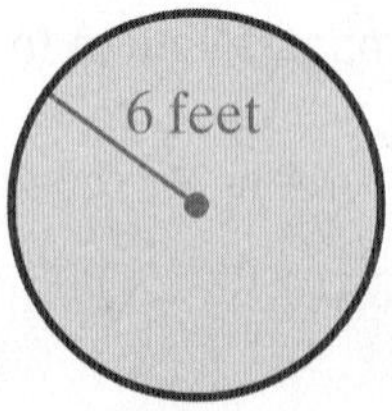

58.

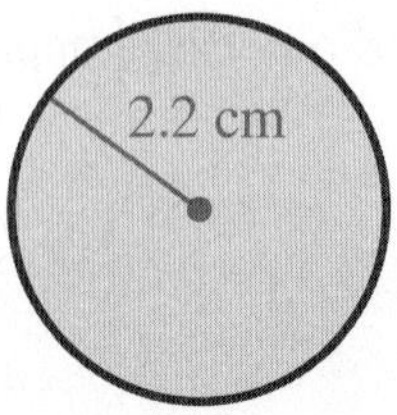

59.

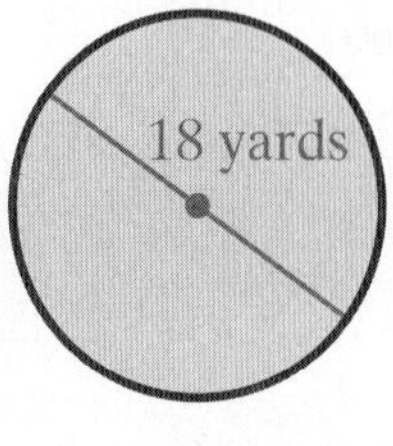

60.

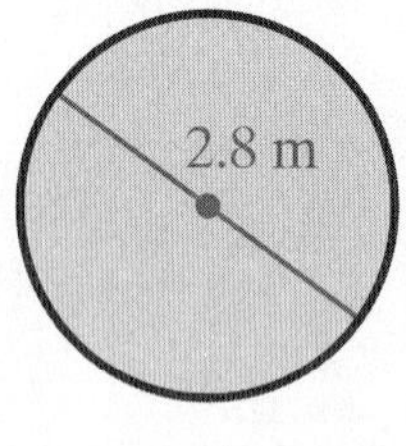

5.6 Solving Equations Containing Decimals

THINGS TO KNOW

Before working through this topic, be sure you are familiar with the following concepts:

VIDEO ANIMATION INTERACTIVE

You Try It 1. Determine If a Fraction Is a Solution to an Equation (Topic 4.7, **Objective 1**)

You Try It 2. Use the Properties of Equality to Solve Linear Equations Involving Fractions (Topic 4.7, **Objective 2**)

You Try It 3. Solve Linear Equations by Clearing Fractions (Topic 4.7, **Objective 3**)

You Try It 4. Add Positive Decimals (Topic 5.2, **Objective 1**)

You Try It 5. Subtract Positive Decimals (Topic 5.2, **Objective 2**)

You Try It 6. Add and Subtract Negative Decimals (Topic 5.2, **Objective 4**)

You Try It 7. Multiply Decimals (Topic 5.3, **Objective 1**)

You Try It 8. Divide Decimals (Topic 5.4, **Objective 1**)

OBJECTIVES

1 Determine If a Decimal Is a Solution to an Equation

2 Use the Properties of Equality to Solve Linear Equations Involving Decimals

3 Solve Linear Equations by Clearing Decimals

4 Use Linear Equations to Solve Applications Involving Decimals

OBJECTIVE 1 DETERMINE IF A DECIMAL IS A SOLUTION TO AN EQUATION

We have seen previously that a **Solution** of an **algebraic equation** is a value that, when **substituted** for the variable, makes the equation true. We checked if whole numbers were solutions in **Topic 1.7**, integers in **Topic 2.6**, and **fractions** in **Topic 4.7**. A **decimal number** can also be a solution to an equation. For example, 9.3 is a solution to $x - 1.5 = 7.8$ because substituting 9.3 for x results in the true statement, $9.3 - 1.5 = 7.8$.

To determine if a given decimal is a solution to an equation, substitute the decimal for the variable and **simplify** both sides of the equation. If the resulting statement is true, then the decimal is a solution. If the resulting statement is false, then the decimal is not a solution.

(eText Screens 5.6-1–5.6-18)

Example 1 Determining If a Decimal Is a Solution to an Equation

My video summary

Determine if the given decimal is a **solution** to the equation.

a. $8x - 7 = 35.4;\ 5.3$ **b.** $1.5y + 4.8 = -3 + 2y;\ -2.6$

Solutions We **substitute** the given decimal for the variable and **simplify**. If the resulting statement is true, then the value is a solution to the equation. If the resulting statement is false, then the value is not a solution to the equation.

a. Begin with the original equation: $8x - 7 = 35.4$

Substitute 5.3 for x: $8(5.3) - 7 \stackrel{?}{=} 35.4$

Multiply: $42.4 - 7 \stackrel{?}{=} 35.4$ ←

Subtract: $35.4 = 35.4$ True

$$\begin{array}{r} {}^{2} \\ 5.3 \\ \times\ \ 8 \\ \hline 42.4 \end{array}$$

The final statement is true, so 5.3 *is* a solution to the equation.

b. Try to work this problem on your own. View the **answer**, or watch this **video** to see a complete solution to both parts.

You Try It **Work through this You Try It problem.**

Work Exercises 1–6 in this textbook or in the MyMathLab Study Plan.

OBJECTIVE 2 USE THE PROPERTIES OF EQUALITY TO SOLVE LINEAR EQUATIONS INVOLVING DECIMALS

We developed a **strategy for solving linear equations** in one variable in **Topic 3.3**. We then applied this same strategy in **Topic 4.7** to equations involving **fractions**. In this topic, we now use the **addition** and **multiplication** properties of equality to solve equations involving **decimals**.

Example 2 Using the Addition Property to Solve Equations Involving Decimals

Solve each equation.

a. $x - 3.8 = 7$ **b.** $y + 4.3 = -2.3$

Solutions

a. To **isolate** the variable x, we can add 3.8 to both sides of the equation.

Begin with the original equation: $x - 3.8 = 7$

Add 3.8 to both sides: $x - 3.8 + 3.8 = 7 + 3.8$ ← Addition property of equality

Simplify both sides: $x = 10.8$

View the **check**. The solution is 10.8.

My video summary

b. Try solving this equation on your own. View the **answer**, or watch this **video** for the complete solution.

You Try It **Work through this You Try It problem.**

Work Exercises 7–12 in this textbook or in the MyMathLab Study Plan.

(eText Screens 5.6-1–5.6-18)

Example 3 Using the Multiplication Property to Solve Equations Involving Decimals

Solve each equation.

a. $\frac{x}{3} = -4.7$ **b.** $-2.4y = -16.8$

Solutions

a. To **isolate** the variable x, we multiply both sides of the equation by 3.

Begin with the original equation: $\frac{x}{3} = -4.7$

Multiply both sides by 3: $3 \cdot \frac{x}{3} = 3 \cdot (-4.7)$ ← Multiplication property of equality

Simplify both sides: $x = -14.1$ ← $3 \cdot \frac{x}{3} = \frac{3x}{3} = \frac{\cancel{3}x}{\cancel{3}} = x$

View the **check**. The solution is -14.1.

My video summary

b. Try solving this equation on your own. View the **answer**, or watch this **video** for the complete solution.

You Try It **Work through this You Try It problem.**

Work Exercises 13–18 in this textbook or in the MyMathLab Study Plan.

As we have seen previously, we sometimes need both **properties of equality** to **solve** an **equation**,

Example 4 Using Both Properties of Equality to Solve Equations Involving Decimals

Solve each equation.

a. $3x + 1.9 = 15.4$ **b.** $8 = 5.8 - 1.1y$

Solutions

a. The only **variable term**, $3x$, appears on the left. We can **isolate** this term by subtracting 1.9 from both sides of the equation.

Begin with the original equation: $3x + 1.9 = 15.4$

Subtract 1.9 from both sides: $3x + 1.9 - 1.9 = 15.4 - 1.9$ ← Addition property of equality

Simplify both sides: $3x = 13.5$

Next, we isolate the variable by dividing both sides of the equation by 3.

Divide both sides by 3: $\frac{3x}{3} = \frac{13.5}{3}$ ← Multiplication property of equality

Simplify both sides: $x = 4.5$ ←

$$\begin{array}{r} 4.5 \\ 3\overline{)13.5} \\ \underline{12} \\ 1\,5 \\ \underline{1\,5} \\ 0 \end{array}$$

View the **check**. The solution is 4.5.

(eText Screens 5.6-1–5.6-18)

My video summary

b. The only **variable term**, $-1.1y$, appears on the right. We can **isolate** this term by subtracting 5.8 from both sides of the equation. Try solving this equation on your own. View the **answer**, or watch this **video** for the complete solution.

You Try It **Work through this You Try It problem.**

Work Exercises 19–24 in this textbook or in the MyMathLab Study Plan.

Example 5 Using Both Properties of Equality to Solve Equations Involving Decimals

My interactive video summary

Solve each equation.

a. $14.5x + 3 = 5x - 12.2$ b. $7z - 4.04 = 1.3z + 14.2$

Solutions

a. There are variable terms and **constant terms** on both sides of the equation. We start by moving the variable terms to one side of the equation and the constants to the other.

Begin with the original equation:	$14.5x + 3 = 5x - 12.2$	
Subtract $5x$ from both sides:	$14.5x + 3 - 5x = 5x - 12.2 - 5x$	← Addition property of equality
Simplify:	$9.5x + 3 = -12.2$	
Subtract 3 from both sides:	$9.5x + 3 - 3 = -12.2 - 3$	← Addition property of equality
Simplify:	$9.5x = -15.2$	

Next we isolate the variable by dividing both sides of the equation by 9.5. Finish working this problem on your own. Check your **answer**, or watch this **interactive video** for detailed solutions to both parts.

b. Try solving this equation on your own. View the **answer**, or watch this **interactive video** for the complete solution to both parts.

You Try It **Work through this You Try It problem.**

Work Exercises 25–30 in this textbook or in the MyMathLab Study Plan.

If an equation contains **grouping symbols**, remember to remove the grouping symbols first by using the **distributive property**.

Example 6 Using the Distributive Property to Solve Equations Involving Decimals

My interactive video summary

Solve each equation.

a. $3(x - 2.7) = 17.4$

b. $1.8(b + 10) = 0.92 - b$

c. $2(m - 3.7) = 4(m + 5.2) - 44$

Solutions In each problem, use the **distributive property** first to remove **grouping symbols.**

a. Begin with the original equation: $3(x - 2.7) = 17.4$

Apply the distributive property: $3x - 8.1 = 17.4 \leftarrow 3(x - 2.7) = 3 \cdot x - 3 \cdot 2.7 = 3x - 8.1$

Add 8.1 to both sides: $3x - 8.1 + 8.1 = 17.4 + 8.1$

Simplify: $3x = 25.5$

Divide both sides by 3: $\dfrac{3x}{3} = \dfrac{25.5}{3}$

Simplify: $x = 8.5$

$$\begin{array}{r} 8.5 \\ 3\overline{)25.5} \\ \underline{24} \\ 1\,5 \\ \underline{1\,5} \\ 0 \end{array} \qquad \begin{array}{r} \overset{2}{2}.7 \\ \underline{\times\ \ 3} \\ 8.1 \end{array}$$

View the **check**. The solution is 8.5.

b. Begin with the original equation: $1.8(b + 10) = 0.92 - b$

Apply the distributive property: $1.8b + 18 = 0.92 - b$

Finish solving this equation on your own using the **properties of equality**. View the **answer**, or watch this **interactive video** for the complete solution to all three parts.

c. Try solving this equation on your own. View the **answer**, or watch this **interactive video** for the complete solution to all three parts.

You Try It **Work through this You Try It problem.**

Work Exercises 31–36 in this textbook or in the MyMathLab Study Plan.

OBJECTIVE 3 SOLVE LINEAR EQUATIONS BY CLEARING DECIMALS

In **Topic 4.7**, we learned how to solve a linear equation containing fractions by first **clearing the fractions.** When an equation contains **decimals**, it is often helpful to clear them as we do with fractions. To **clear the decimals**, multiply both sides of the equation by an appropriate **power of 10**, such as 10, 100, or 1000. Determine which power of 10 to use by looking at the **constants** in the equation and choosing the constant with the greatest number of **decimal places.** We count those decimal places and then raise 10 to that power.

Example 7 Solving Equations by Clearing Decimals

Solve each equation by first clearing the decimals.

a. $4x - 3.9 = -14.3$

b. $3.7y + 4.25 = 9.8$

Solutions

a. The constant 4 has 0 decimal places, while the constants -3.9 and -14.3 have one decimal place. So the greatest number of decimal places is one. To clear the decimals, we multiply both sides of the equation by $10^1 = 10$.

$$4x - \underbrace{3.9}_{\text{one decimal place}} = \underbrace{-14.3}_{\text{one decimal place}}$$

(eText Screens 5.6-1–5.6-18)

Begin with the original equation: $4x - 3.9 = -14.3$

Multiply both sides by 10: $10(4x - 3.9) = 10(-14.3)$

Apply the **distributive property**: $40x - 39 = -143$

Add 39 to both sides: $40x - 39 + 39 = -143 + 39$

Simplify: $40x = -104$

Divide both sides by 40: $\frac{40x}{40} = \frac{-104}{40}$

Simplify: $x = -2.6$

$$\begin{array}{r} 2.6 \\ 40\overline{)104.0} \\ \underline{80} \\ 240 \\ \underline{240} \\ 0 \end{array}$$

View the **check**. The solution is -2.6.

My video summary

b. The **constants** 3.7 and 9.8 have one decimal place and the constant 4.25 has two decimal places. So the greatest number of decimal places is two. To clear the decimals, multiply both sides of the equation by $10^2 = 100$.

$$\underbrace{3.7}_{\text{one decimal place}}y + \underbrace{4.25}_{\text{two decimal places}} = \underbrace{9.8}_{\text{one decimal place}}$$

Finish solving this equation on your own. View the **answer**, or watch this **video** for the complete solution to part b.

You Try It **Work through this You Try It problem.**

Work Exercises 37–42 in this textbook or in the MyMathLab Study Plan.

We make one more adjustment to our **strategy for solving linear equations in one variable** by clearing **decimals** along with **fractions** in Step 1. Remember that clearing fractions or decimals is optional but usually makes equations easier to solve. If a particular step does not apply to a given equation, then skip it.

A General Strategy for Solving Linear Equations in One Variable

Step 1. If the equation contains fractions, multiply both sides by the **LCD** of all the fractions.
If the equation contains decimals, multiply both sides by an appropriate **power of 10**.

Step 2. Remove grouping symbols using the **distributive property**.

Step 3. Simplify each side of the equation by **combining like terms**.

Step 4. Use the **addition property of equality** to collect all **variable terms** on one side of the equation and all **constant terms** on the other side.

Step 5. Use the **multiplication property of equality** to **isolate the variable**.

Step 6. Check the result in the original equation.

OBJECTIVE 4 USE LINEAR EQUATIONS TO SOLVE APPLICATIONS INVOLVING DECIMALS

We can use equations containing decimals to solve application problems. We can use our **problem-solving strategy** for applications of linear equations presented in **Topic 3.4**.

Example 8 Moving Costs

Kennedy needs to rent a truck to move from her parent's house to her apartment at college. The cost to rent a 14-ft truck for one day is $29.95 plus $0.79 per mile. If x represents the number of miles driven, solve the following equation to determine how many miles Kennedy drove with the truck. (*Source:* www.uhaul.com)

$$29.95 + 0.79x = 216.39$$

Solution We can apply our **strategy for solving linear equations in one variable.**

Begin with the original equation:	$29.95 + 0.79x = 216.39$
Subtract 29.95 from both sides:	$29.95 + 0.79x - 29.95 = 216.39 - 29.95$
Simplify:	$0.79x = 186.44$
Divide both sides by 0.79:	$\dfrac{0.79x}{0.79} = \dfrac{186.44}{0.79}$
Simplify:	$x = 236$

$$\begin{array}{r} 236. \\ 0.79\overline{)186.44} \\ \underline{158} \\ 28\,4 \\ \underline{23\,7} \\ 474 \\ \underline{474} \\ 0 \end{array}$$

Kennedy drove the truck 236 miles.

You Try It **Work through this You Try It problem.**

Work Exercises 43 and 44 in this textbook or in the MyMathLab Study Plan.

Example 9 iTunes® Purchases

My video summary

Quan uses iTunes® to purchase some apps for $0.99 each and some songs for $1.29 each. If he purchased 15 more songs than apps and spent a total of $37.59, how many of each did he buy?

Solution

Step 1. We know the total amount spent and that Quan purchased 15 more songs than apps. We also know that each app cost $0.99 and each song cost $1.29.

Step 2. Let a = the number of apps purchased. Then we can express the number of songs purchased as $a + 15$ (15 more than the number of apps).

Step 3. The amount spent on apps is found by multiplying the price per app by the number of apps, $0.99a$. Likewise, the amount spent on songs is found by multiplying the price per song by the number of songs, $1.29(a + 15)$. We can get the total spent by adding the amount spent on apps to the amount spent on songs.

$$\underbrace{0.99a}_{\text{Amount spent on apps}} + \underbrace{1.29(a + 15)}_{\text{Amount spent on songs}} = \underbrace{37.59}_{\text{Total spent}}$$

Finish solving this problem on your own. Check your **answer**, or watch this **video** for the fully worked solution.

You Try It **Work through this You Try It problem.**

Work Exercises 45–48 in this textbook or in the MyMathLab Study Plan.

5.6 Exercises

In Exercises 1–6, determine if the given value is a solution to the given equation.

You Try It

1. $2x - 5 = 21.4$; 13.2
2. $3.2y + 2.1 = 11$; 2.5
3. $5.1x + 3 = 26.1 - 2.9x$; 4
4. $2.8y + 3.7 = -6.1$; -3.5
5. $-5.2x - 5.04 = 14.2$; -3.7
6. $5x - 3.7 = 7.62 - 3.4x$; 4.6

In Exercises 7–36, solve each equation.

You Try It

7. $x - 2.5 = 6.3$
8. $y + 2.9 = 4.8$
9. $z - 3.7 = 5.3$
10. $w + 11.3 = 8.3$
11. $5.2 - m = 4.8$
12. $b + 13.7 = 4.9$

You Try It

13. $\frac{x}{5} = 2.6$
14. $2.4x = 3.6$
15. $5.2y = 26$
16. $-3.5y = 9.1$
17. $-1.7x = -2.38$
18. $\frac{b}{2.6} = -5.8$

You Try It

19. $5.4x + 2.3 = 10.4$
20. $6y - 14.5 = 5.3$
21. $2.5x + 3.5 = -3$
22. $7.24 + 5.4y = -12.2$
23. $-2.2x - 10 = 2.1$
24. $4.2 - 3.5x = -9.8$

You Try It

25. $5.6x + 2.7 = 2.4x + 10.7$
26. $-1.8x - 5.6 = -8x + 13$
27. $16.3 + 2.5x = 1.6 - 4.5x$
28. $8.8y + 3.6 = 3.3y - 3$
29. $3.8y + 5.07 = 6.5y - 3.3$
30. $-4 - 7.1m = 34.88 - 12.5m$

You Try It

31. $3(x - 2.4) = 8.1$
32. $2.2(b + 3) = 6.82 - 2b$
33. $1.5(m + 2.2) = 2m + 1.3$
34. $3(x + 1.7) = 2(x - 1.2) + 4.2$
35. $-1.2(x + 3) = 2.2(x - 4.5) + 1.2$
36. $2.1(b - 3.4) = -4.3(b + 3.75) - 2.44$

In Exercises 37–42, solve each equation by first clearing decimals.

You Try It

37. $2.2x + 1.7 = 13.8$
38. $3.7y + 10 = 6.2y$
39. $5.7 - 1.6b = 1.7$
40. $1.4m - 2.24 = 2.1$
41. $1.22y + 5.15 = -4$
42. $4.3x - 2.65 = 6.8x - 21.15$

You Try It

43. **Cell Phone Plan** Raevyn signs up for a voice plan with her new smartphone. The plan costs \$59.99 per month and provides her with 900 anytime minutes. Additional minutes cost \$0.40 each. One month her bill was for \$158.79. If x represents the number of minutes used, solve the following equation to find the number of minutes Raevyn talked on her smartphone.

$$0.4(x - 900) + 59.99 = 158.79$$

(eText Screens 5.6-1–5.6-18)

44. **Weekly Wages** Bradi earns \$9.25 per hour for the first 40 hours that she works per week. For overtime hours (hours over 40 hours), she earns \$13.88 per hour. One week, she earned \$474.10. If x represents the number of hours she worked, solve the following equation to determine the number of hours Bradi worked that week.

$$9.25(40) + 13.88(x - 40) = 474.10$$

You Try It 45. **Burning Calories** Marcus can burn 632 calories per hour swimming and 905 calories per hour jogging. If he burned 4568 calories one week and ran 2.5 hours more than he swam, how long did he spend on each type of exercise?

46. **Taping a Window** To paint a wall in his apartment, Naheem needs to put tape around a rectangular window. If the length is 1.25 feet longer than the width and the perimeter is 7.5 feet, what are the dimensions of the window?

47. **Rental Charge** For a 1-day rental, E-Z Rental charges \$40 plus \$0.35 per mile to rent a truck. To rent the same truck at Regal Rentals would cost \$20.50 plus \$0.65 per mile. Determine the number of miles when the two charges will be the same for a 1-day rental.

48. **Photo Book** Mindy created a photo book at www.snapfish.com. The price of the book was \$27.99 for the first 20 pages plus \$1.99 for each additional 2 pages. If she spent \$39.93 on the book, how many pages did it contain? (*Source:* www.snapfish.com)

MODULE SIX

Ratios and Proportions

MODULE SIX CONTENTS

6.1 Ratios, Rates, and Unit Prices

THINGS TO KNOW

Before working through this topic, be sure you are familiar with the following concepts:

		VIDEO	ANIMATION	INTERACTIVE
You Try It	1. Divide Whole Numbers (Topic 1.5, **Objective 1**)	✓		✓
You Try It	2. Identify the Numerator and Denominator of a Fraction (Topic 4.1, **Objective 1**)			
You Try It	3. Write Fractions in Simplest Form (Topic 4.2, **Objective 6**)	✓		
You Try It	4. Divide Fractions (Topic 4.3, **Objective 4**)	✓		✓
You Try It	5. Divide Mixed Numbers (Topic 4.6, **Objective 2**)	✓		
You Try It	6. Multiply Decimals by Powers of 10 (Topic 5.3, **Objective 2**)	✓		

OBJECTIVES

1 Write Two Quantities as a Ratio

2 Write Two Quantities as a Rate

3 Find a Unit Rate

4 Compare Unit Prices

(eText Screens 6.1-1–6.1-25)

OBJECTIVE 1 WRITE TWO QUANTITIES AS A RATIO

If there are 7 males and 15 females in an algebra class, then we can say that the *ratio* of males to females is 7 to 15. A **ratio** is a comparison of two quantities, usually in the form of a **quotient**. There are three common notations for expressing ratios. One is to use the word "to," as we just saw, but ratios may also be expressed with *fraction notation* or *colon notation*. So, we might write the ratio of 7 to 15 as

$$7 \text{ to } 15 \qquad \text{or} \qquad \underset{\substack{\uparrow \\ \text{Fraction notation}}}{\frac{7}{15}} \qquad \text{or} \qquad \underset{\substack{\uparrow \\ \text{Colon notation}}}{7{:}15.}$$

In each case, we read the ratio as "7 to 15." In this text, we will typically use fraction notation when writing a ratio, and we will write the **fraction** in **simplified form**. If the ratio is an **improper fraction**, we will *not* write it as a **mixed number**.

CAUTION Pay attention to the wording because the order of the quantities in a ratio is important. For example, the ratio of 7 to 15 is $\frac{7}{15}$ and *not* $\cancel{\frac{15}{7}}$. To write a ratio as a fraction, remember that the first number of the ratio is the **numerator** and the second number is the **denominator**.

Example 1 Writing Ratios

Write each **ratio** in fraction notation.

a. 5 to 24 **b.** 14:9

Solutions

a. To write a ratio as a fraction, the first number of the ratio is the **numerator** and the second number is the **denominator**.

The ratio is $\frac{5}{24}$.

b. Here, the ratio is given in colon notation. The first number is the numerator and the second number is the denominator.

The ratio is $\frac{14}{9}$.

You Try It **Work through this You Try It problem.**

Work Exercises 1–4 in this textbook or in the MyMathLab Study Plan.

TIP Remember that if a ratio is an **improper fraction**, we leave it that way. We do not change it to a **mixed number** because a ratio is a comparison of two quantities and a single mixed number does not show the relationship between the two quantities.

Ratios behave like **fractions,** which is why fraction notation is often used. We simplify ratios the same way that we simplify fractions, by dividing out **common factors**. For example, the ratio of 12 to 28 is equivalent to a ratio of 3 to 7. We can see this by using fraction notation and writing the fraction in **simplest form**:

$$\frac{12}{28} = \frac{3 \cdot 4}{7 \cdot 4} = \frac{3 \cdot \cancel{4}}{7 \cdot \cancel{4}} = \frac{3}{7}.$$

(eText Screens 6.1-1–6.1-25)

In addition to dividing out common factors, we can divide out **common units**, if any. For example, in the following expression, the common units 'feet' divide out.

$$\frac{7 \text{ feet}}{5 \text{ feet}} = \frac{7\ \cancel{\text{feet}}}{5\ \cancel{\text{feet}}} = \frac{7}{5}.$$

> **Simplifying a Ratio**
>
> To **simplify a ratio**, write the fraction notation in simplest form by dividing out all common factors and common units.

Example 2 Writing and Simplifying Ratios

My video summary

Write each ratio in simplest form.

a. 24 to 10 **b.** 30:63

Solutions

a. The ratio in **fraction notation** is $\frac{24}{10}$. Next, we simplify the ratio by **dividing out** common factors.

$$\frac{24}{10} = \frac{2 \cdot 12}{2 \cdot 5} = \frac{\cancel{2} \cdot 12}{\cancel{2} \cdot 5} = \frac{12}{5}$$

b. The **ratio** in fraction notation is $\frac{30}{63}$. Simplify this ratio by dividing out common factors.

View the **answer**, or watch this **video** for complete solutions to both parts.

You Try It **Work through this You Try It problem.**

Work Exercises 5–10 in this textbook or in the MyMathLab Study Plan.

Example 3 Writing and Simplifying Ratios with Common Units

My video summary

a. Amazon's Kindle Fire tablet has a length of 190 mm and a width of 120 mm. Write the ratio of length to width as a fraction in simplest form. (*Source:* amazon.com)

b. A dietician recommends to a patient that he consume 125 grams of protein, 140 grams of fat, and 50 grams of carbohydrates each day. Write the ratio of carbohydrates to protein in simplest form.

Solutions

a. We want the ratio of length to width. Because length is given first, it goes in the **numerator** of our ratio and width goes in the **denominator**.

$$\frac{\text{length}}{\text{width}} = \frac{190 \text{ mm}}{120 \text{ mm}} = \frac{19 \cdot 10 \text{ mm}}{12 \cdot 10 \text{ mm}} = \frac{19 \cdot \cancel{10}\ \cancel{\text{mm}}}{12 \cdot \cancel{10}\ \cancel{\text{mm}}} = \frac{19}{12}$$

Divide out the common factors

Divide out the common units

b. We want the **ratio** of carbohydrates to protein. Because carbohydrates is given first, it goes in the numerator of our ratio and protein goes in the denominator. The ratio is $\frac{50\text{ g}}{125\text{ g}}$. Simplify this ratio on your own. View the **answer**, or watch this **video** for complete solutions to both parts.

You Try It **Work through this You Try It problem.**

Work Exercises 11–16 in this textbook or in the MyMathLab Study Plan.

We want to express ratios using two **whole numbers**. If a ratio involves **decimals**, we use the **property of equivalent fractions** to first clear the decimals by multiplying the **numerator** and **denominator** by an appropriate **power of 10**. We determine the power of 10 by determining the largest number of **decimal places** in either the numerator or denominator. Consider the following:

$$\frac{2.4}{8.35}\begin{array}{l}\leftarrow \text{One decimal place}\\ \leftarrow \text{Two decimal places}\end{array}$$

To clear the decimals, we would multiply both the numerator and denominator by 100 because the largest number of decimal places is 2 and $10^2 = 100$.

$$\frac{2.4}{8.35} = \frac{2.4 \cdot 100}{8.35 \cdot 100} = \frac{240}{835}$$

$2.4 \cdot 100 = 240$

$8.35 \cdot 100 = 835$

Moves the decimal point to the right two decimal places

We then simplify the **ratio** as in previous examples.

If a ratio involves **fractions** or **mixed numbers**, we divide as we did in **Topic 4.3** and **Topic 4.6**, respectively, writing the final ratio as a ratio of two **whole numbers**.

Example 4 Writing and Simplifying Ratios Involving Decimals, Fractions, and Mixed Numbers

My interactive video summary

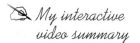

Write each ratio in simplest form.

a. 4.2 to 2.8 **b.** $\frac{3}{4}$ to $\frac{1}{6}$ **c.** $3\frac{1}{4}$ to $2\frac{1}{2}$

Solutions

a. The ratio is $\frac{4.2}{2.8}$. The largest number of **decimal places** in either the numerator or denominator is one. We clear the decimals by multiplying both the numerator and denominator by $10^1 = 10$, then **simplify**.

$$\frac{4.2}{2.8} = \underbrace{\frac{4.2 \cdot 10}{2.8 \cdot 10}}_{\text{Clear decimals}} = \frac{42}{28} = \frac{2 \cdot 3 \cdot 7}{2 \cdot 2 \cdot 7} = \underbrace{\frac{\cancel{2} \cdot 3 \cdot \cancel{7}}{\cancel{2} \cdot 2 \cdot \cancel{7}}}_{\text{Divide out the common factors}} = \frac{3}{2} \leftarrow \text{Ratio of two whole numbers}$$

b. The **ratio** is $\frac{\frac{3}{4}}{\frac{1}{6}}$. Recall that $\frac{\frac{3}{4}}{\frac{1}{6}} = \frac{3}{4} \div \frac{1}{6}$, so we simplify by dividing the fractions.

Change to multiplication by the **reciprocal**: $= \frac{3}{4} \cdot \frac{6}{1}$ ← The reciprocal of $\frac{1}{6}$ is $\frac{6}{1}$

Multiply numerators; multiply denominators: $= \frac{3 \cdot 6}{4 \cdot 1}$

Finish simplifying this ratio on your own. View the **answer**, or watch this **interactive video** for complete solutions to all three parts.

c. The ratio is $\frac{3\frac{1}{4}}{2\frac{1}{2}}$.

$$3\frac{1}{4} = \frac{4 \cdot 3 + 1}{4} = \frac{13}{4} \qquad 2\frac{1}{2} = \frac{2 \cdot 2 + 1}{2} = \frac{5}{2}$$

$$\frac{3\frac{1}{4}}{2\frac{1}{2}} = 3\frac{1}{4} \div 2\frac{1}{2} = \frac{13}{4} \div \frac{5}{2}$$

Finish simplifying this ratio on your own. View the **answer**, or watch this **interactive video** for complete solutions to all three parts.

You Try It **Work through this You Try It problem.**

Work Exercises 17–24 in this textbook or in the MyMathLab Study Plan.

Ratios involving **decimals** or **fractions** also appear in applications.

Example 5 Golden Ratio

My video summary

According to ancient Greeks, the **golden ratio** is supposed to be the most pleasing to the eye. It can be roughly approximated by the fraction $\frac{81}{50}$ and appears frequently in nature, art, and architecture. For the same reasons, the ratio 2:1 is considered to be the least pleasing to the eye.

a. A dollar bill is approximately 6.2 inches long and 2.6 inches wide. Find the ratio of length to width for a dollar bill.

b. A credit card is approximately 3.3 inches long and 2.1 inches wide. Find the ratio of length to width for a credit card.

c. Which of the two ratios is closer to the golden ratio?

Solutions Try working these problems on your own. View the **answers**, or watch this **video** for complete solutions.

You Try It **Work through this You Try It problem.**

Work Exercises 25 and 26 in this textbook or in the MyMathLab Study Plan.

(eText Screens 6.1-1–6.1-25)

OBJECTIVE 2 WRITE TWO QUANTITIES AS A RATE

Rates are special types of **ratios** in which quantities of different types are compared. For example, consider a cyclist who can bike 9 miles in 30 minutes. Writing this rate as a **fraction**, we have

$$\frac{9 \text{ miles}}{30 \text{ minutes}}, \text{ which simplifies to } \frac{3 \text{ miles}}{10 \text{ minutes}}. \quad \leftarrow \text{Different units}$$

Notice that because the two quantities have different units, the units *do not* **divide out.**

Recall that the word *to* can be used in a ratio to separate the quantities being compared. When working with rates, we often see other words used to separate the two quantities. Some examples are given below.

per	20 miles *per* gallon
for	10 cookies *for* \$3.00
in	26.2 miles in 5 hours

Example 6 Writing and Simplifying Rates

a. Applying a 30-4-4 fertilizer grade requires 9 lb of fertilizer for every 3000 square feet of lawn. Write this rate as a fraction in **simplest form.**

b. Dionna earned \$315 for working 36 hours at her summer job. Write this rate as a fraction in simplest form.

Solutions

a. We want the **ratio** of fertilizer amount to lawn size. There are different units, so the ratio is also a rate. Because the amount of fertilizer is given first, it goes in the **numerator** of our **rate** and lawn size goes in the **denominator.**

Divide out the common factors

$$\frac{\text{fertilizer}}{\text{lawn size}} = \frac{9 \text{ lb}}{3000 \text{ ft}^2} = \frac{3 \cdot 3 \text{ lb}}{1000 \cdot 3 \text{ ft}^2} = \frac{3 \cdot \cancel{3} \text{ lb}}{1000 \cdot \cancel{3} \text{ ft}^2} = \frac{3 \text{ lb}}{1000 \text{ ft}^2}$$

Keep the different units

The rate in simplest form is $\frac{3 \text{ lb}}{1000 \text{ ft}^2}$ or, 3 lb per 1000 ft^2.

b. We want the **ratio** of money earned to hours worked. Notice the different units, which means that we have a **rate**. The amount of money earned is given first, so it goes in the numerator of our rate. Hours worked goes in the denominator.

Divide out the common factors

$$\frac{\text{money earned}}{\text{hours worked}} = \frac{315 \text{ dollars}}{36 \text{ hours}} = \frac{35 \cdot 9 \text{ dollars}}{4 \cdot 9 \text{ hours}} = \frac{35 \cdot \cancel{9} \text{ dollars}}{4 \cdot \cancel{9} \text{ hours}} = \frac{35 \text{ dollars}}{4 \text{ hours}}$$

Keep the different units

The rate in simplest form is $\frac{35 \text{ dollars}}{4 \text{ hours}}$.

(eText Screens 6.1-1–6.1-25)

Example 7 Writing and Simplifying Rates

My video summary

a. Tanner typed a 2000-word paper in 45 minutes. Write his typing **rate** as a **simplified fraction.**

b. During the 2011 season, Adrian Gonzalez of the Boston Red Sox hit 27 home runs in 630 at-bats. Write his home run to at-bat rate as a simplified fraction. (*Source:* espn.com)

Solutions Try writing these simplified rates on your own. View the **answers**, or watch this **video** for complete solutions to both parts.

You Try It **Work through this You Try It problem.**

Work Exercises 27–34 in this textbook or in the MyMathLab Study Plan.

OBJECTIVE 3 FIND A UNIT RATE

A **unit rate** is a **rate** that has a denominator of 1 unit. A familiar unit rate is "miles per hour." For example, the speed 65 miles per hour can be written as $\frac{65 \text{ miles}}{1 \text{ hour}}$.

> **Converting Rates to Unit Rates**
>
> To convert a rate to a unit rate, divide the quantity in the **numerator** by the quantity in the **denominator**.

Example 8 Converting Rates to Unit Rates

a. A jet travels 960 miles in 2 hours. What is the jet's unit rate in miles per hour?

b. A movie rental chain hires a store manager at a salary of $32,500 per year. Given that there are 52 weeks in a year, find the manager's unit rate in dollars per week.

Solutions

a. $\frac{960 \text{ miles}}{2 \text{ hours}} = \frac{960 \div 2 \text{ miles}}{2 \div 2 \text{ hours}} = \frac{480 \text{ miles}}{1 \text{ hour}} = 480 \text{ miles per hour}$

b. $\frac{\$32{,}500}{52 \text{ weeks}} = \frac{\$32{,}500 \div 52}{52 \div 52 \text{ weeks}} = \frac{\$625}{1 \text{ week}} = \$625 \text{ per week}$

You Try It **Work through this You Try It problem.**

Work Exercises 35–38 in this textbook or in the MyMathLab Study Plan.

TIP Because the key word *per* indicates **division**, we can replace it with a forward slash to show the division. For example, we can write "480 miles per hour" as "480 miles/hour," or "$625 per week" as "$625/week."

Although we typically write **ratios** and **rates** using **whole numbers**, it is possible to have a **unit rate** with a **numerator** that is a **fraction** or **decimal**.

Example 9 Converting Rates to Unit Rates Involving Fractions or Decimals

My video summary

a. Using her DSL modem, Tameka can download a 360 MB file in 16 seconds. Find her modem's unit rate in MB per second.

b. The U.S. government often reports statistics as "per capita" (per person). If the population of the U.S. was 305,000,000 in 2009 and 1,982,500,000 pounds of peanuts were consumed during that year, find the unit rate of peanut consumption per person for 2009. In other words, find the amount of peanuts eaten per capita in the U.S. for 2009. (*Source:* U.S. Department of Agriculture)

Solutions Try writing these unit rates on your own. View the **answers**, or watch this **video** for complete solutions to both parts.

You Try It **Work through this You Try It problem.**

Work Exercises 39–42 in this textbook or in the MyMathLab Study Plan.

OBJECTIVE 4 COMPARE UNIT PRICES

Unit rates are used in many fields such as sports and business in order to perform decision-making comparisons. When the unit rate involves a "cost per unit," we call this a **unit price**.

> **Unit Price**
>
> A **unit price** is found by dividing the total cost by the number of units.
>
> $$\text{Unit price} = \frac{\text{Total cost}}{\text{Number of units}}$$

Example 10 Finding a Unit Price

a. Piper used 60 MB of data on her smartphone beyond what is covered in her monthly plan. The cost for this additional data was \$15. Find the unit price in dollars per MB.

b. If a 12-ounce can of soda costs 75¢, what is the unit price in ¢ per ounce?

Solutions

a. $\frac{\$15}{60\,\text{MB}} = \frac{15 \div 60 \text{ dollars}}{60 \div 60\,\text{MB}} = \frac{0.25 \text{ dollars}}{1\,\text{MB}} = \$0.25 \text{ per MB}$

b. $\frac{75¢}{12 \text{ ounces}} = \frac{75¢ \div 12}{12 \div 12 \text{ ounces}} = \frac{6.25¢}{1 \text{ ounce}} = 6.25¢ \text{ per ounce}$

You Try It **Work through this You Try It problem.**

Work Exercises 43–46 in this textbook or in the MyMathLab Study Plan.

Unit prices can be used to compare options so that you can determine which is the better deal. In doing so, we assume there is no waste. That is, the entire product will be consumed.

(eText Screens 6.1-1–6.1-25)

Example 11 Comparing Unit Prices

A box of 3 microwave popcorn bags costs \$3.18, while a box of 5 bags costs \$5.19. Determine each unit price and state which is the better deal. Round unit prices to two **decimal places** if necessary.

Solution

$$\text{3-pack: unit price} = \frac{\text{total cost}}{\text{number of units}} = \frac{\$3.18}{\text{3 bags}} = \$1.06 \text{ per bag} \qquad 3\overline{)3.18}\ \ 1.06$$

$$\text{5-pack: unit price} = \frac{\text{total cost}}{\text{number of units}} = \frac{\$5.19}{\text{5 bags}} \approx \$1.04 \text{ per bag} \qquad 5\overline{)5.190}\ \ 1.038 \approx 1.04$$

"is approximately equal to"

The box of 5 bags has a lower unit price, so it is the better deal.

Example 12 Comparing Grocery Store Prices

My video summary

Grocery stores will often display **unit prices** on price labels so that shoppers can make comparisons quickly. The following price labels show that a 40-oz jar of a certain brand of creamy peanut butter sells for \$6.38 and a 56-oz jar of the same type of peanut butter sells for \$9.98 (the unit prices have been hidden). Find the unit prices and compare them to determine which is the better deal. Round to two **decimal places** if necessary.

Solution Find the unit prices and make the comparison on your own to determine which is the better deal. View the **answer**, or watch this **video** for a complete solution.

You Try It **Work through this You Try It problem.**

Work Exercises 47–50 in this textbook or in the MyMathLab Study Plan.

6.1 Exercises

In Exercises 1–4, write each ratio in fraction notation.

You Try It 1. 6 to 17   2. 9:14   3. 15 to 11   4. 37:36

In Exercises 5–10, write each ratio in simplest form.

You Try It 5. 6 to 20   6. 14:8   7. 22 to 12

8. 13 to 39   9. 36:18   10. 14:52

You Try It 11. A garden is 8 ft long and 5 ft wide. Find the ratio of width to length.

(eText Screens 6.1-1–6.1-25)

12. A doctor recommends that a patient eat 125 g of protein for each 30 g of carbohydrates. What is her recommended ratio of protein to carbohydrates?

13. In the 2011 UEFA Champions League, Barcelona had 4 goals scored against them and scored 20 goals. What was the club's ratio of "goals against" to "goals for"? (*Source:* www.uefa.com)

14. Anna earns $2500 per month and spends $775 on debt payments such as her credit card and school loan. What is her ratio of debt to income?

15. A small college has 49 math majors, 91 engineering majors, and 168 business majors in its graduate programs. What is the ratio of engineering majors to math majors?

16. A bag of mixed nuts contains 18 oz of peanuts, 12 oz of walnuts, and 8 oz of cashews. What is the ratio of walnuts to cashews?

In Exercises 17–34, write each ratio or rate in simplest form.

You Try It 17. 2.6 to 1.4 18. $\frac{3}{8}:\frac{5}{12}$ 19. $2\frac{1}{4}$ to $1\frac{3}{8}$ 20. 5.7:1.26

21. 5.85 to 8 22. $\frac{5}{3}$ to $\frac{7}{9}$ 23. $3\frac{2}{3}:5\frac{1}{2}$ 24. 2.25 to $1\frac{1}{4}$

You Try It 25. The Great Pyramid of Giza has sides of length 756 feet and a height of approximately 480 feet. Find the simplified ratio of length to height. How close is this ratio to the golden ratio, which is approximated by the fraction $\frac{81}{50}$? Round your answer to two decimal places. (*Source:* pbs.org)

26. The CN Tower in Toronto is approximately 554 meters tall and has an observation deck at an approximate height of 342 meters. Find the simplified ratio of tower height to deck height. How close is this ratio to the golden ratio, which is approximated by the fraction $\frac{81}{50}$? Round your answer to two decimal places. (*Source:* goldennumber.com)

You Try It 27. Willie earned $816 for 40 hours of work.

28. A trip of 290 miles uses 19 gallons of gasoline.

29. A weight of 771 lb rests on a 12 square foot platform.

30. 80 ounces of split peas cost 50 cents.

31. A brand of cheese crackers contains 150 calories in 28 crackers.

32. A typist can type 1610 words in 25 minutes.

33. A basketball team scores 27 points in 15 minutes.

34. Charney's utility bill shows a usage of 845 kilowatt-hours for 26 days.

In Exercises 35–42, find the indicated unit rate.

You Try It 35. A 2012 Chevy Volt can travel 360 miles on 9 gallons of gasoline. What is the Volt's unit rate in miles per gallon? (*Source:* motortrend.com)

36. Khiry earned $4420 for 13 weeks of summer work. What was his unit rate in dollars per week?

37. Amanda types a 1575-word paper in 35 minutes. What is her unit rate in words per minute?

38. Boyd runs 12 miles in 60 minutes. What is his unit rate in minutes per mile?

You Try It 39. Using his cable modem, Patrick can download a 486-MB file in 20 seconds. Find his modem's unit rate in MB per second.

40. Find the per capita consumption of whole milk in the U.S. in 2011 if 1,767,000,000 gallons were consumed and the population was 310,000,000. (*Source:* U.S. Department of Agriculture)

41. During the first 14 games of the 2011 NFL season, the New Orleans Saints had a total of 6494 yards on offense. Determine the number of yards per game rounded to one decimal place. (*Source:* neworleanssaints.com)

42. On Black Friday 2011, the average Apple Store sold 178 iPads® during a 12-hour period. Find the unit rate in iPads per hour rounded to one decimal place.

In Exercises 43–46, find the indicated unit price.

You Try It 43. A store charges $2.36 for 4 cucumbers. What is the unit price per cucumber?

44. Shirlene pays $4.48 for a 32-ounce bag of potato chips. What is the unit price per ounce?

45. Bianca pays $8.50 for a box of 50 envelopes. What is the unit price per envelope?

46. A box of six microwave popcorn bags costs $4.88. What is the unit price per bag?

You Try It 47. **Bulk Batteries** A pack of 16 Energizer® batteries costs $9.97, while a pack of 20 costs $12.38. Find the unit prices and compare them to determine which is the better deal. Round to two decimal places if necessary.

48. **Grocery Shopping** A grocery store sells a 20-oz bottle of ketchup for $1.73 and a 32-oz bottle for $2.60. Find the unit prices and compare them to determine which is the better deal. Round to two decimal places if necessary.

49. **Grocery Shopping** A grocery store sells a 5-oz can of solid white albacore tuna for $1.48 and a 20-oz 4-pack for $7.63. Find the unit prices and compare them to determine which is the better deal. Round to two decimal places if necessary.

50. **Tanning Packages** A tanning salon offers 12 tans for $50 or 17 tans for $68. Find the unit prices and compare them to determine which is the better deal. Round to two decimal places if necessary.

6.2 Proportions

THINGS TO KNOW

Before working through this topic, be sure you are familiar with the following concepts:

VIDEO | ANIMATION | INTERACTIVE

You Try It 1. Use the Multiplication Property of Equality to Solve Equations (Topic 2.6, **Objective 3**)

You Try It 2. Identify the Numerator and Denominator of a Fraction (Topic 4.1, **Objective 1**)

(eText Screens 6.2-1–6.2-20)

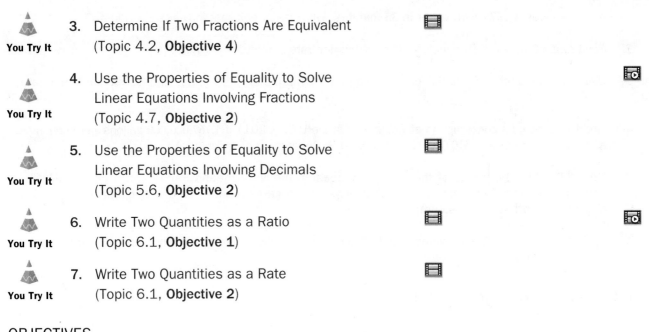

You Try It 3. Determine If Two Fractions Are Equivalent (Topic 4.2, **Objective 4**)

You Try It 4. Use the Properties of Equality to Solve Linear Equations Involving Fractions (Topic 4.7, **Objective 2**)

You Try It 5. Use the Properties of Equality to Solve Linear Equations Involving Decimals (Topic 5.6, **Objective 2**)

You Try It 6. Write Two Quantities as a Ratio (Topic 6.1, **Objective 1**)

You Try It 7. Write Two Quantities as a Rate (Topic 6.1, **Objective 2**)

OBJECTIVES

1 Write Proportions

2 Determine Whether Proportions Are True or False

3 Solve Proportions

OBJECTIVE 1 WRITE PROPORTIONS

An **equation** stating that two **ratios** (or **rates**) are equal is called a *proportion*. For example,

$$\frac{\$18}{2\text{ hours}} = \frac{\$27}{3\text{ hours}}$$

is a proportion stating that the pay rate of $18 for 2 hours of work is equivalent to the pay rate of $27 for 3 hours of work. We read this proportion as

$18 *is to* 2 hours *as* $27 *is to* 3 hours.

> **Definition Proportion**
> A **proportion** is an equation stating that two ratios (or rates) are equal, such as $\frac{a}{b} = \frac{c}{d}$, where $b \neq 0$ and $d \neq 0$.

$$\frac{a}{b} = \frac{c}{d}$$

a is to b as c is to d.

Example 1 Writing Proportions

Write each sentence as a **proportion**.

a. 2 is to 12 as 8 is to 48.

b. 424 miles is to 16 gallons as 318 miles is to 12 gallons.

c. 6 cups is to 4 cups as 1.5 cups is to 1 cup.

My video summary

Solutions Work through the following, or watch this **video** for complete solutions to all three parts.

a. 2 is to 12 as 8 is to 48.

$$\frac{2}{12} = \frac{8}{48}$$

b. 424 miles is to 16 gallons as 318 miles is to 12 gallons.

$$\frac{424 \text{ miles}}{16 \text{ gallons}} = \frac{318 \text{ miles}}{12 \text{ gallons}}$$

c. Recall that, if comparing quantities with the same units, then the units can be divided out of the **ratio**. So, the **proportion** is

$$\frac{6 \cancel{\text{cups}}}{4 \cancel{\text{cups}}} = \frac{1.5 \cancel{\text{cups}}}{1 \cancel{\text{cup}}} \text{ or simply } \frac{6}{4} = \frac{1.5}{1}.$$

You Try It **Work through this You Try It problem.**

Work Exercises 1–6 in this textbook or in the MyMathLab Study Plan.

TIP When writing proportions, divide out common units within each ratio if possible. If the units within the **numerator** and **denominator** of each ratio are not the same, then they should be included in your proportion.

OBJECTIVE 2 DETERMINE WHETHER PROPORTIONS ARE TRUE OR FALSE

In **Topic 1.7**, we saw that **equations** may be true or false. In the same way, **proportions** may be true or false. One way to determine whether a proportion is true or false is to write both **ratios** in **simplest form** and compare the results.

Example 2 Using Simplest Form to Determine If Proportions Are True

Write each ratio in simplest form to determine whether each proportion is true or false.

a. $\frac{15}{20} = \frac{18}{24}$ **b.** $\frac{45}{18} = \frac{24}{9}$

Solutions

a. Write each ratio in simplest form.

$$\frac{15}{20} = \frac{15 \div 5}{20 \div 5} = \frac{3}{4} \quad \text{and} \quad \frac{18}{24} = \frac{18 \div 6}{24 \div 6} = \frac{3}{4}$$

Both ratios simplify to $\frac{3}{4}$, so the proportion is *true*.

b. Write each **ratio** in **simplest form**.

$$\frac{45}{18} = \frac{45 \div 9}{18 \div 9} = \frac{5}{2} \quad \text{and} \quad \frac{24}{9} = \frac{24 \div 3}{9 \div 3} = \frac{8}{3}$$

The two ratios simplify to different results $\left(\frac{5}{2} \neq \frac{8}{3}\right)$, so the **proportion** is *false*.

(eText Screens 6.2-1–6.2-20)

You Try It **Work through this You Try It problem.**

Work Exercises 7–10 in this textbook or in the MyMathLab Study Plan.

In **Topic 4.2**, we used **cross products** to determine if two **fractions** were **equivalent.** We also can use cross products to determine whether a proportion is true or false. Let's find the cross products for the proportions from Example 2.

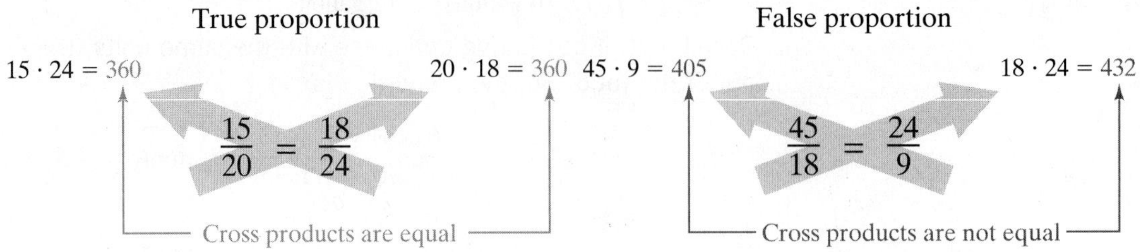

Notice that the **cross products** of the true **proportion** from Example 2a are equal $(15 \cdot 24 = 20 \cdot 18)$, while the cross products of the false proportion from Example 2b are not equal $(45 \cdot 9 \neq 18 \cdot 24)$. This leads to the following rule:

Using Cross Products to Determine Whether Proportions Are True or False

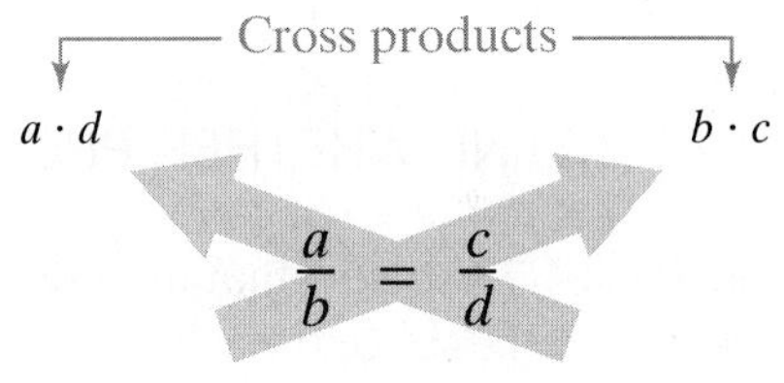

If the cross products of a proportion are *equal*, then the proportion is *true*.

If the cross products are *not equal*, the proportion is *false*.

Example 3 Using Cross Products to Determine If Proportions Are True

My video summary

Use cross products to determine whether each proportion is true or false.

a. $\frac{8}{50} = \frac{12}{75}$ **b.** $\frac{4}{18} = \frac{9}{27}$

Solutions

a. Check the **cross products.**

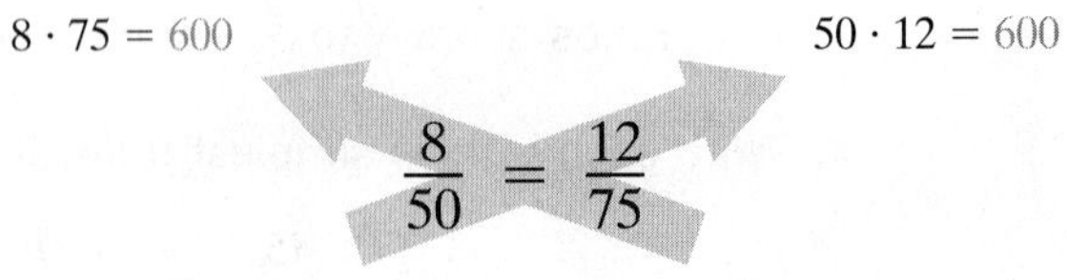

The cross products are *equal* $(600 = 600)$, so the **proportion** is *true*.

b. Try working this problem on your own. View the **solution** to this part, or watch this **video** for complete solutions to both parts.

You Try It **Work through this You Try It problem.**

Work Exercises 11–14 in this textbook or in the MyMathLab Study Plan.

The quantities within a proportion are not always **whole numbers.** Sometimes they are **decimals, fractions, mixed numbers,** and so forth.

Example 4 Determining If Proportions Involving Decimals and Fractions Are True

Use cross products to determine whether each proportion is true or false.

a. $\frac{3.5}{8} = \frac{5.4}{12}$ **b.** $\frac{\frac{3}{4}}{2\frac{1}{2}} = \frac{\frac{1}{2}}{1\frac{2}{3}}$

Solutions

a. Check the **cross products.**

$$3.5 \cdot 12 = 42 \qquad \frac{3.5}{8} = \frac{5.4}{12} \qquad 8 \cdot 5.4 = 43.2$$

The cross products are *not equal* $(42 \neq 43.2)$, so the **proportion** is *false*.

My video summary

b. Check the cross products.

$$\frac{3}{4} \cdot 1\frac{2}{3} \qquad \frac{\frac{3}{4}}{2\frac{1}{2}} = \frac{\frac{1}{2}}{1\frac{2}{3}} \qquad 2\frac{1}{2} \cdot \frac{1}{2}$$

Try to perform the multiplication for each cross product on your own, then compare the results to determine whether the proportion is true. View the **answer**, or watch this **video** for a complete solution to part b.

You Try It **Work through this You Try It problem.**

Work Exercises 15–18 in this textbook or in the MyMathLab Study Plan.

(eText Screens 6.2-1–6.2-20)

OBJECTIVE 3 SOLVE PROPORTIONS

If one of the four quantities within a **proportion** is unknown, we can find it by using **cross products.** When we find an unknown quantity that makes a proportion true, we **solve the proportion**.

Consider the following:

$$\frac{x}{4} = \frac{3}{2}$$

For this proportion to be true, the cross products must be equal. By forming an equation from the cross products and solving, we can find the unknown quantity, x.

Cross products

$2 \cdot x$ $\qquad$ $4 \cdot 3$

$$\frac{x}{4} = \frac{3}{2}$$

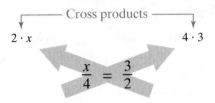

Set cross products equal: $2 \cdot x = 4 \cdot 3$

Simplify both sides: $2x = 12$

Divide both sides by 2: $\dfrac{2x}{2} = \dfrac{12}{2}$

Simplify: $x = 6$

$\dfrac{2x}{2} = x$

We can check this result by substituting 6 for x in the original **proportion** to see if a true proportion results.

$6 \cdot 2 = 12$ $\qquad$ $4 \cdot 3 = 12$

$$\frac{6}{4} \stackrel{?}{=} \frac{3}{2}$$

The cross products are equal $(12 = 12)$, so 6 makes the proportion true. Therefore, 6 is the **solution** to the proportion.

The preceding example leads us to the following process for **solving proportions.**

> **Solving a Proportion**
>
> **Step 1.** Form an equation by setting the **cross products** equal to each other.
>
> **Step 2.** Use the **multiplication property of equality** to **solve** the equation from Step 1.
>
> **Step 3.** Check the result from Step 2 by substituting it into the original proportion and confirming that the cross products are equal.

Example 5 Solving Proportions

Solve each **proportion**.

a. $\dfrac{75}{x} = \dfrac{3}{8}$ $\qquad$ **b.** $\dfrac{21}{14} = \dfrac{18}{y}$

Solutions

a. We begin by finding the **cross products** and setting them equal to each other.

$75 \cdot 8$ $\quad$ $3 \cdot x$

$$\frac{75}{x} = \frac{3}{8}$$

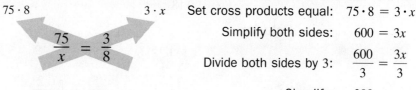

Set cross products equal:	$75 \cdot 8 = 3 \cdot x$
Simplify both sides:	$600 = 3x$
Divide both sides by 3:	$\frac{600}{3} = \frac{3x}{3}$
Simplify:	$200 = x$

View the **check**. The solution is 200.

My video summary

b. Try solving this proportion on your own. View the **answer**, or watch this **video** for a complete solution to part b.

You Try It **Work through this You Try It problem.**

Work Exercises 19–24 in this textbook or in the MyMathLab Study Plan.

TIP When solving proportions, **ratios** may be simplified before the cross products are found. For example, in Example 5b, $\frac{21}{14}$ can be simplified to $\frac{3}{2}$ before finding the cross products. The **solution** of the proportion will be the same either way.

Example 6 Solving Proportions Involving Negatives

Solve each **proportion**.

a. $\frac{t}{-15} = \frac{8}{-12}$ $\qquad$ b. $\frac{-10}{7} = \frac{n}{28}$

Solutions

a. $-12 \cdot t$ $\quad$ $-15 \cdot 8$

$$\frac{t}{-15} = \frac{8}{-12}$$

Set cross products equal:	$-12 \cdot t = -15 \cdot 8$
Simplify both sides:	$-12t = -120$
Divide both sides by -12:	$\frac{-12t}{-12} = \frac{-120}{-12}$
Simplify:	$t = 10$

View the **check**. The solution is 10.

Note: To solve this proportion, as described in the previous tip, we could have simplified $\frac{8}{-12}$ before finding the cross product. View this **alternate solution** to Example 6a.

My video summary

b. Try solving this proportion on your own. View the **answer**, or watch this **video** for a complete solution to part b.

You Try It **Work through this You Try It problem.**

Work Exercises 25–28 in this textbook or in the MyMathLab Study Plan.

Now let's practice solving proportions that include **decimals, fractions,** or **mixed numbers.**

Example 7 Solving a Proportion Involving Decimals

My video summary

Solve $\frac{0.5}{k} = \frac{0.2}{0.18}$.

Solution

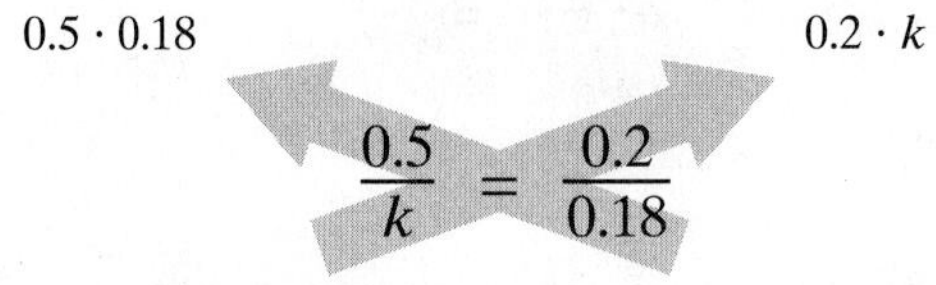

Set cross products equal: $0.5 \cdot 0.18 = 0.2 \cdot k$

Simplify both sides: $0.09 = 0.2k$

Divide both sides by 0.2: $\frac{0.09}{0.2} = \frac{0.2k}{0.2}$

Simplify: $0.45 = k$

$0.2\overline{)0.09} \Rightarrow 02.\overline{)00.90}$

$$\begin{array}{r} 0.45 \\ 02.\overline{)00.90} \\ -8 \\ \hline 10 \\ -10 \\ \hline 0 \end{array}$$

View the **check**. The solution is 0.45.

You Try It **Work through this You Try It problem.**

Work Exercises 29–32 in this textbook or in the MyMathLab Study Plan.

Example 8 Solving Proportions Involving Fractions and Mixed Numbers

Solve each **proportion**. Write the **solution** as a simplified **proper fraction** or **mixed number** when applicable.

a. $\dfrac{\frac{1}{2}}{\frac{3}{4}} = \dfrac{x}{\frac{1}{6}}$ **b.** $\dfrac{m}{4\frac{1}{6}} = \dfrac{1\frac{4}{5}}{2\frac{3}{16}}$

Solutions

a.

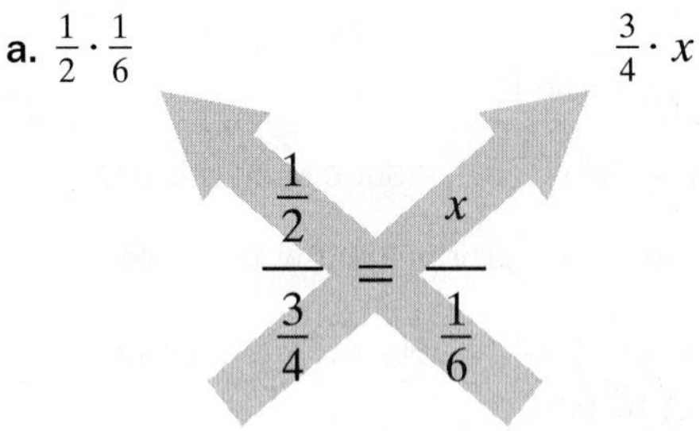

Set cross products equal: $\frac{1}{2} \cdot \frac{1}{6} = \frac{3}{4} \cdot x$

Simplify both sides: $\frac{1}{12} = \frac{3}{4}x$

Multiply both sides by $\frac{4}{3}$: $\frac{4}{3} \cdot \frac{1}{12} = \frac{4}{3} \cdot \frac{3}{4}x$

Simplify: $\frac{1}{9} = x$

$\frac{4}{3} \cdot \frac{1}{12} = \frac{\cancel{4} \cdot 1}{3 \cdot \cancel{12}_3} = \frac{1}{9}$

View the **check**. The solution is $\frac{1}{9}$.

(eText Screens 6.2-1–6.2-20)

My video summary

b. Try solving this **proportion** on your own. To multiply **mixed numbers**, remember to first write each one as an **improper fraction**. View the **answer**, or watch this **video** for a complete solution to part b.

You Try It **Work through this You Try It problem.**

Work Exercises 33–36 in this textbook or in the MyMathLab Study Plan.

Sometimes, we may want to **round** the **solution** of a proportion to a given **place value**.

Example 9 Solving Proportions with Rounding

My video summary

Solve $\frac{6}{7} = \frac{y}{4.5}$. Round the solution to the hundredths place.

Solution Watch this **video** or work through the following for a detailed solution.

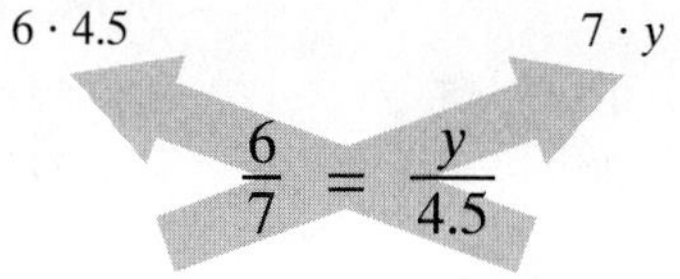

Set cross products equal: $6 \cdot 4.5 = 7 \cdot y$

Simplify both sides: $27 = 7y$

Divide both sides by 7: $\frac{27}{7} = \frac{7y}{7}$

Simplify: $3.86 \approx y$

$$\begin{array}{r} 3.857 \\ 7\overline{)27.000} \\ -21 \\ \hline 6\,0 \\ -5\,6 \\ \hline 40 \\ -35 \\ \hline 50 \\ -49 \\ \hline 1 \end{array}$$

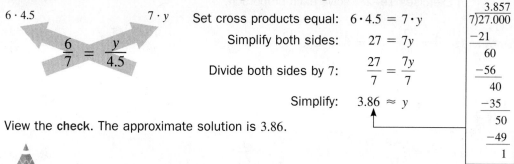

View the **check**. The approximate solution is 3.86.

You Try It **Work through this You Try It problem.**

Work Exercises 37–40 in this textbook or in the MyMathLab Study Plan.

6.2 Exercises

In Exercises 1–6, write each sentence as a proportion.

You Try It

1. 10 is to 15 as 30 is to 45.
2. 3.6 is to 10 as 4.5 is to 12.5.
3. 480 miles is to 15 gallons as 256 miles is to 8 gallons.
4. 9 field goals is to 12 attempts as 75 field goals is to 100 attempts.
5. 60 meters is to 75 meters as 4 meters is to 5 meters.
6. $\frac{3}{4}$ cup is to $1\frac{1}{2}$ cups as $2\frac{1}{4}$ cups is to $4\frac{1}{2}$ cups.

(eText Screens 6.2-1–6.2-20)

In Exercises 7–10, write each ratio in simplest form to determine whether each proportion is true or false.

You Try It 7. $\frac{5}{8} = \frac{65}{104}$ 8. $\frac{10}{15} = \frac{35}{45}$ 9. $\frac{12}{20} = \frac{180}{300}$ 10. $\frac{-32}{12} = \frac{56}{-21}$

In Exercises 11–18, use cross products to determine whether each proportion is true or false.

You Try It 11. $\frac{7}{8} = \frac{20}{24}$ 12. $\frac{7}{9} = \frac{21}{27}$ 13. $\frac{-25}{200} = \frac{-7}{56}$ 14. $\frac{13}{162} = \frac{1}{14}$

You Try It 15. $\frac{0.1}{-0.4} = \frac{-0.3}{0.7}$ 16. $\frac{0.75}{0.5} = \frac{2.7}{1.8}$ 17. $\frac{\frac{2}{7}}{\frac{4}{3}} = \frac{\frac{1}{8}}{\frac{7}{6}}$ 18. $\frac{3\frac{1}{4}}{1\frac{3}{8}} = \frac{2\frac{3}{5}}{1\frac{1}{10}}$

In Exercises 19–28, solve each proportion.

You Try It 19. $\frac{x}{8} = \frac{7}{32}$ 20. $\frac{15}{y} = \frac{1}{6}$ 21. $\frac{10}{8} = \frac{45}{n}$ 22. $\frac{52}{72} = \frac{k}{54}$

You Try It 23. $\frac{m}{9} = \frac{56}{168}$ 24. $\frac{68}{80} = \frac{p}{100}$ 25. $\frac{-7}{a} = \frac{49}{-14}$ 26. $\frac{8}{6} = \frac{-48}{h}$

27. $\frac{x}{18} = \frac{-6}{-27}$ 28. $\frac{120}{-48} = \frac{f}{64}$

In Exercises 29–32, solve each proportion. Write the solution in decimal form when applicable.

You Try It 29. $\frac{0.8}{0.5} = \frac{z}{4.5}$ 30. $\frac{2}{0.6} = \frac{0.3}{w}$ 31. $\frac{y}{0.625} = \frac{1.5}{1.25}$ 32. $\frac{10.8}{-2.4} = \frac{t}{4.5}$

In Exercises 33–36, solve each proportion. Write the solution as a simplified proper fraction or mixed number when applicable.

You Try It 33. $\frac{\frac{3}{10}}{12} = \frac{n}{15}$ 34. $\frac{\frac{3}{4}}{\frac{1}{2}} = \frac{\frac{5}{6}}{x}$ 35. $\frac{3\frac{3}{4}}{m} = \frac{1\frac{7}{8}}{4\frac{1}{10}}$ 36. $\frac{a}{2\frac{1}{2}} = \frac{4\frac{1}{4}}{3\frac{3}{16}}$

In Exercises 37–40, solve each proportion. Round the solution to the given place value.

You Try It 37. $\frac{29}{y} = \frac{7}{5}$, tenths place 38. $\frac{k}{4.6} = \frac{3.3}{5.2}$, hundredths place

39. $\frac{11}{5.4} = \frac{9.1}{n}$, thousandths place 40. $\frac{15}{-13} = \frac{p}{23}$, hundredths place

6.3 Proportions and Problem Solving

THINGS TO KNOW

Before working through this topic, be sure you are familiar with the following concepts:

VIDEO ANIMATION INTERACTIVE

You Try It 1. Write Two Quantities as a Ratio (Topic 6.1, **Objective 1**)

You Try It 2. Write Two Quantities as a Rate (Topic 6.1, **Objective 2**)

You Try It 3. Write Proportions (Topic 6.2, **Objective 1**)

You Try It 4. Solve Proportions (Topic 6.2, **Objective 3**)

OBJECTIVE

1 Use Proportions to Solve Applications

OBJECTIVE 1 USE PROPORTIONS TO SOLVE APPLICATIONS

Proportions can be used to solve a wide variety of application problems from daily life and from fields of study such as architecture, business, science, medicine, home economics, and so forth. If a **ratio** (or **rate**) between two quantities is known, then we can use a proportion to find an unknown quantity.

Below, we refine our **problem-solving strategy** from **Topic 3.4** for use with proportions.

Problem-Solving Strategy for Applications of Proportions

Step 1. **Define the Problem.** Read the problem carefully, or multiple times if necessary. Identify what you are trying to find and determine what information is available to help you find it.

Step 2. **Assign Variables.** Choose a **variable** for the unknown quantity in the problem.

Step 3. **Translate into a Proportion.** Use the relationships among the known and unknown quantities to write a proportion.

Step 4. **Solve the Proportion.** Determine the value of the variable.

Step 5. **Check the Reasonableness of Your Answer.** Check to see if your answer makes sense within the context of the problem. If not, check your work for errors and try again.

Step 6. **Answer the Question.** Write a clear statement that answers the question(s) posed.

Example 1 Text Messaging

In 2011, researchers reported that the average 18–24-year-old sends and receives 110 text messages per day. At this rate, how many texts does the average 18–24-year-old send and receive in a 30-day month? (*Source:* Pew Research Center, April–May 2011)

(eText Screens 6.3-1–6.3-16)

Solution We follow the **problem-solving strategy for applications of proportions.**

Step 1. We know that the average 18–24-year-old sends and receives 110 texts per day. We want to find the number of texts sent and received in a 30-day month.

Step 2. Let t = the number of texts sent/received in a 30-day month.

Step 3. We write a **proportion** by setting the **rates** for 1 day and 1 month equal. We must be sure to keep like quantities in similar positions in the proportion. Writing in the units can help with this.

Daily rate — Monthly Rate

$$\text{Number of texts} \rightarrow \frac{110 \text{ texts}}{1 \text{ day}} = \frac{t \text{ texts}}{30 \text{ days}} \leftarrow \text{Number of texts}$$

Number of days → (denominators) ← Number of days

Step 4. We find the cross products and set them equal to each other. For this process, we can work without the units.

$$110 \cdot 30 \qquad \frac{110}{1} = \frac{t}{30} \qquad 1 \cdot t$$

Set cross products equal: $110 \cdot 30 = 1 \cdot t$

Simplify both sides: $3300 = t$

Step 5. We can use 3300 texts in the monthly rate of our proportion and cross multiply to check.

$$110 \cdot 30 = 3300 \qquad \frac{110}{1} \stackrel{?}{=} \frac{3300}{30} \qquad 1 \cdot 3300 = 3300$$

The cross products are equal, so 3300 is a reasonable solution and checks.

Step 6. The average 18–24-year-old sends and receives 3300 texts in a 30-day month.

You Try It **Work through this You Try It problem.**

Work Exercises 1 and 2 in this textbook or in the MyMathLab Study Plan.

Example 2 Estimating the Bass Population in a Lake

My video summary

To estimate the population of largemouth bass in a small lake, a conservation agent catches 120 bass, tags them, and releases them back into the lake. Some time later, the agent returns, catches a sample of 90 bass, and finds that 8 of them are tagged. Estimate the number of largemouth bass in the entire population.

Solution We follow the **problem-solving strategy**.

Step 1. Out of the total number of largemouth bass in the lake, we know that 120 bass were tagged by the agent. In the sample of 90 bass, the agent found that 8 of them were tagged. We want to find an estimate of the total number of largemouth bass in the lake.

Step 2. Let n = the number of largemouth bass in the entire population.

Step 3. For our estimate, we assume that the **ratio** of tagged bass to total bass is the same for the entire population of bass in the lake as it is in the sample of 90 bass caught by the agent. This leads to the following **proportion**:

$$\begin{array}{rcccl} & \text{Population} & & \text{Sample} & \\ \text{Tagged bass} \longrightarrow & 120 & = & 8 & \longleftarrow \text{Tagged bass} \\ \text{Total number of bass} \longrightarrow & n & & 90 & \longleftarrow \text{Total number of bass} \end{array}$$

Try to solve this proportion on your own. View the **answer**, or watch this **video** for a complete solution.

You Try It **Work through this You Try It problem.**

Work Exercises 3–7 in this textbook or in the MyMathLab Study Plan.

Examples 1 and 2 worked with whole numbers only. However, as we learned in Topic 6.2, sometimes we must solve proportions that include decimals, which we will see in the next example.

Example 3 Finding a Distance from a Map

My video summary

On the given map showing part of the state of Illinois, an actual distance of 50 miles is represented by 2.5 centimeters. If the distance between Springfield and Chicago is 10.6 centimeters on the map, find the actual distance between the two cities.

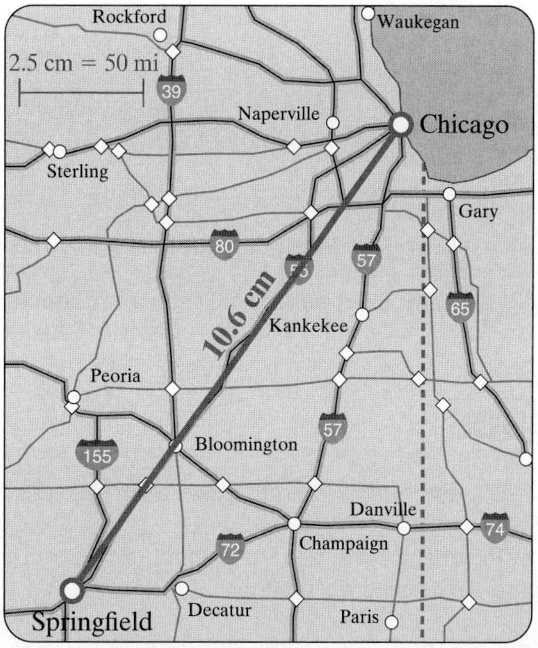

Solution We follow the **problem-solving strategy**.

Step 1. We know that an actual distance of 50 miles is represented by 2.5 cm on the map. We want to find the actual distance between Springfield and Chicago, which is represented by 10.6 cm on the map.

Step 2. Let x = the actual distance between Springfield and Chicago.

Step 3. Write a **proportion** that compares map distances to the actual distances.

$$\begin{array}{rcccl} & \text{Given scale for map} & & \text{Springfield to Chicago} & \\ \text{Map distance} \longrightarrow & \dfrac{2.5 \text{ cm}}{50 \text{ miles}} & = & \dfrac{10.6 \text{ cm}}{x \text{ miles}} & \longleftarrow \text{Map distance} \\ \text{Actual distance} \longrightarrow & & & & \longleftarrow \text{Actual distance} \end{array}$$

Try to solve this proportion on your own. View the **answer**, or watch this **video** for a complete solution.

(eText Screens 6.3-1–6.3-16)

You Try It **Work through this You Try It problem.**

Work Exercises 8–10 in this textbook or in the MyMathLab Study Plan.

The next example also involves solving proportions when decimals are included in the ratios. However, in this case, we will be rounding our answer and therefore want to find an approximate, not exact, solution.

Example 4 Driving a Hybrid Car

My video summary

In his new hybrid car, Malik can travel 194 miles on 4.5 gallons of gas. How far can Malik travel on 12 gallons of gas? Round to the nearest mile.

Solution We follow the **problem-solving strategy**.

Step 1. We know that Malik travels 194 miles on 4.5 gallons of gas. We want to find the number of miles he can travel on 12 gallons of gas.

Step 2. Let m = the number of miles Malik can travel on 12 gallons of gas.

Step 3. Write a **proportion** that compares the miles per gallon of each trip.

$$\begin{array}{rccl} & \text{1st trip} & & \text{2nd trip} \\ \text{Number of miles} \rightarrow & \dfrac{194 \text{ miles}}{4.5 \text{ gallons}} & = & \dfrac{m \text{ miles}}{12 \text{ gallons}} \leftarrow \text{Number of miles} \\ \text{Number of gallons} \rightarrow & & & \leftarrow \text{Number of gallons} \end{array}$$

Try to solve this proportion on your own. View the **answer**, or watch this **video** for a complete solution.

You Try It **Work through this You Try It problem.**

Work Exercises 11–20 in this textbook or in the MyMathLab Study Plan.

6.3 Exercises

In Exercises 1–20, use a proportion to solve each application.

You Try It

1. **Printing a Document** If a color laser printer can print 25 pages per minute, how many minutes will it take to print a 375-page document?

2. **Student–Teacher Ratio** An elementary school maintains a student-to-teacher ratio of 21 to 1. If the school enrolls 840 students, how many teachers does the school employ? Give the answer to the nearest whole number.

You Try It

3. **Estimating Fish Population** To determine the number of fish in a lake, researchers caught and tagged 200 of them. In a later sample of 250 fish, the researchers found that 16 were tagged. Estimate the total number of fish in the lake.

4. **Too Much Email?** In a 2011 survey of U.S. consumers, 3 out of 4 reported that they receive more email than they can read. If 2224 U.S. consumers were included in the survey, how many consumers reported that they receive more email than they can read? (*Source:* Epsilon Targeting, 12/01/2011)

5. **Burning Calories** The readout on Jamarie's elliptical machine shows that he burned 375 calories during a 25-minute workout. How many calories will he burn during a 40-minute workout? (*Source:* livestrong.com)

(eText Screens 6.3-1–6.3-16)

6. **Piano Lessons** If Parker pays $140 for 8 piano lessons, what must she pay for 52 lessons?

7. **Painting a Wall** Kamara used 2 gallons of paint to cover 750 square feet of wall space. How many gallons of paint will she need to cover another 3000 square feet?

You Try It

8. **Finding Distance from a Map** The given map shows part of the state of Tennessee. On the map, an actual distance of 30 miles is represented by 1 inch. If the distance between Memphis and Nashville is 7 inches on the map, find the actual distance between the two cities.

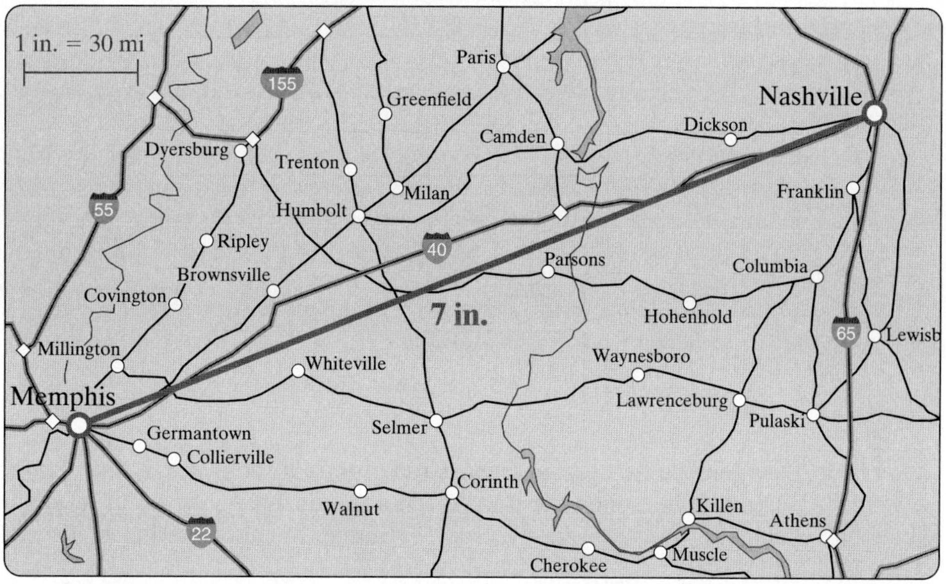

9. **Finding Distance from a Map** The given map shows part of the National Mall in Washington, DC. A distance of 2 centimeters on the map represents an actual distance of 500 feet. If the actual distance from the Lincoln Memorial to the Washington Monument is 4240 feet, what is the distance on this map? Round to the nearest tenth of a centimeter.

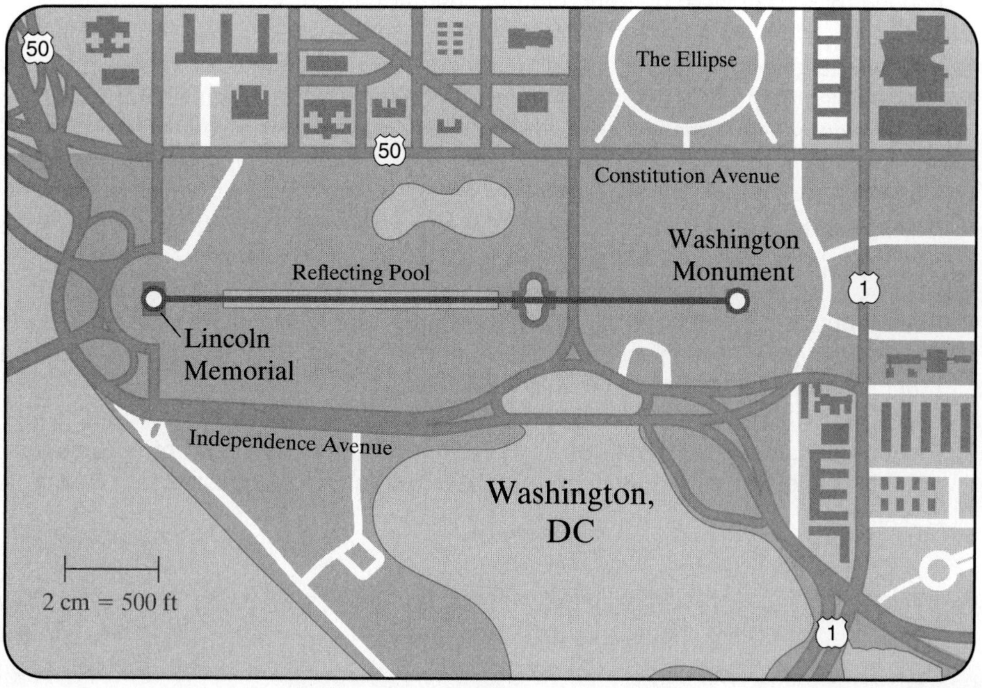

10. **Scale Drawing** For a scale drawing of a bathroom, $\frac{1}{4}$ inch represents an actual length of 1 foot. If the bathroom measures $1\frac{1}{2}$ inches by 2 inches in the drawing, what are its actual dimensions (length and width)?

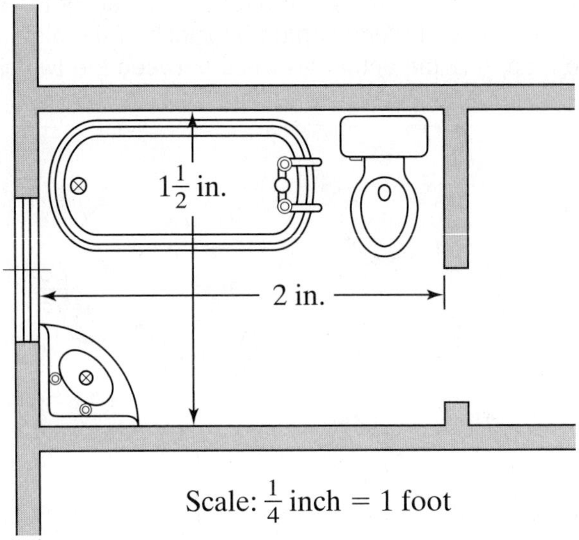

You Try It 11. **Watching Television** In 2011, researchers reported that the average American (age 15 years or older) watches 2.7 hours of television per day. At this rate, how many hours of television does the average American watch in a year (365 days)? (*Source:* Bureau of Labor Statistics, Time Use Survey, June 22, 2011)

12. **Pet Ownership** According to the American Pet Products Association, 72.9 million U.S. households have pets. Out of every 10 pet-owning households, 4 own more than one pet. How many U.S. households own more than one pet? (*Source:* APPA, 2011–2012 National Pet Owners Survey)

13. **Gas Mileage** Carson can drive 329 miles on 12 gallons of gas. How many gallons of gas would be needed for a 228-mile trip? Round to the nearest tenth of a gallon.

14. **Waist-to-Height Ratio** Medical Research has shown that the ratio of a person's waist size to height can be an indicator of heart disease. A healthy man should have a maximum waist-to-height ratio of 0.55 to 1. To be healthy, what should a man's maximum waist size be if his height is 72 inches? (*Source:* WebMD.com)

15. **Hours Worked** If Kelsey earned $79.56 by working 6 hours one day, how much would she earn by working 8 hours in one day?

16. **Peanut Butter** If a 16.3-ounce jar of Peter Pan® creamy peanut butter contains enough peanut butter for 14 servings, how many servings are in a 28-ounce jar? Round to the nearest whole number. (*Source:* peterpanpb.com)

17. **Fertilizing a Lawn** If it takes 20 pounds of fertilizer to treat a lawn of 3000 square feet, how many pounds of the fertilizer are needed to treat a lawn of 11,000 square feet? Round to the nearest tenth of a pound.

18. **Baking Cookies** A cookie recipe calls for $2\frac{1}{4}$ cups of flour and $1\frac{1}{2}$ cups of sugar. Paula has plenty of flour but only $1\frac{1}{4}$ cups of sugar. How much flour should she use with the $1\frac{1}{4}$ cups of sugar to keep the ingredients in proportion?

19. **Download Speed** If it takes Caleb 34 seconds to download a 10-megabyte file from the Internet to his computer, how long will it take him to download an 18-megabyte file? Round to the nearest second.

20. **Cancer Treatment** Xeloda® is an oral chemotherapy medication sometimes used in combination with other drugs to treat breast cancer or colorectal cancer. The recommended dosage is 2500 milligrams per square meter of body surface area. Calculate the recommended dosage for a breast cancer patient who has a body surface area of 1.72 square meters. (*Source:* drugs.com)

6.4 Congruent and Similar Triangles

THINGS TO KNOW

Before working through this topic, be sure you are familiar with the following concepts:

		VIDEO	ANIMATION	INTERACTIVE
You Try It	1. Write Two Quantities as a Ratio (Topic 6.1, **Objective 1**)	✓		✓
You Try It	2. Write Proportions (Topic 6.2, **Objective 1**)	✓		
You Try It	3. Solve Proportions (Topic 6.2, **Objective 3**)	✓		

OBJECTIVES

1 Identify the Corresponding Parts of Congruent Triangles

2 Determine Whether Two Triangles Are Congruent

3 Identify the Corresponding Parts of Similar Triangles

4 Find Unknown Lengths of Sides in Similar Triangles

5 Solve Applications Involving Similar Triangles

OBJECTIVE 1 IDENTIFY THE CORRESPONDING PARTS OF CONGRUENT TRIANGLES

Recall that each corner point of a triangle is called a **vertex.** A triangle has three **vertices** (plural of vertex). Typically, uppercase letters are used to label the vertices so that we can easily tell them apart. The triangle in Figure 1 has vertices P, Q, and R. Often, we name a triangle by using these vertex labels. For example, we might name Figure 1 as ΔPQR, which reads as "triangle PQR." (Note: We could also name Figure 1 as ΔQRP, ΔRPQ, ΔPRQ, ΔRQP, and ΔQPR.)

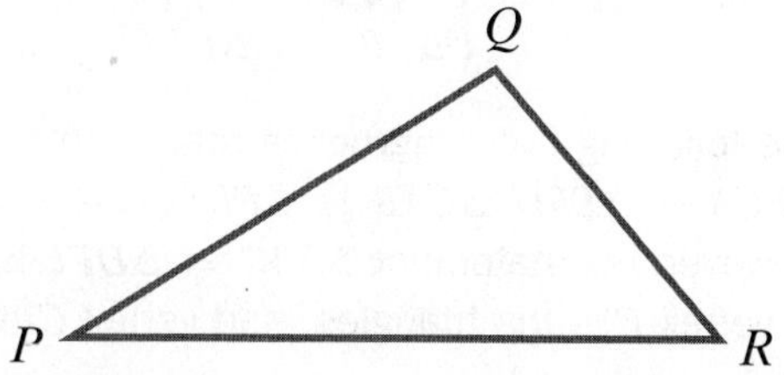

Figure 1

(eText Screens 6.4-1–6.4-37)

The vertex labels also can be used to name the **angles** and **sides** of a triangle. For example, $\angle P$ reads as "angle P," which is the angle with vertex P, and $\overline{PQ}$ reads as "side PQ," which is the side extending between P and Q.

My video animation summary

Two triangles are **congruent** if they have the exact same shape and size. Watch the first part of this **video animation**, or continue reading, to learn more about congruent triangles. Consider the two triangles in Figure 2.

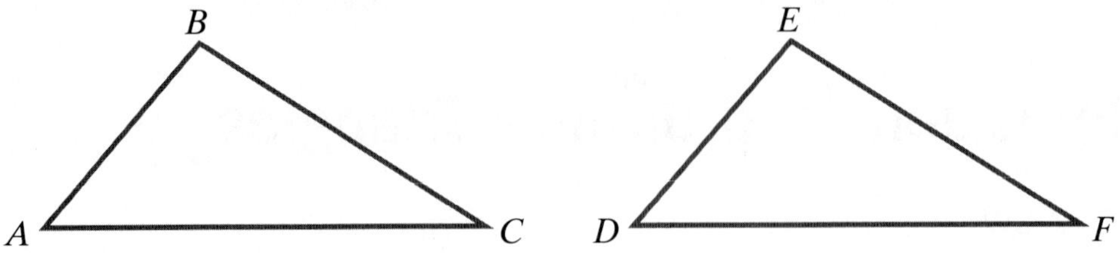

Figure 2

If we were to slide ΔABC on top of ΔDEF, they would match exactly. The two triangles have the exact same shape and size, so ΔABC *is congruent to* ΔDEF. Using the **congruence symbol** ($\cong$), we write this as

$$\Delta ABC \cong \Delta DEF.$$

"is congruent to"

Because $\angle A$ matches up with $\angle D$, these angles are called **corresponding angles.** Similarly, $\angle B$ and $\angle E$ are corresponding angles, and $\angle C$ and $\angle F$ are corresponding angles. Because $\overline{AB}$ matches up with $\overline{DE}$, they are called **corresponding sides**. Similarly, $\overline{BC}$ and $\overline{EF}$ are corresponding sides, and $\overline{AC}$ and $\overline{DF}$ are corresponding sides.

TIP Each pair of **corresponding sides** is located opposite a pair of **corresponding angles.** For example in Figure 2, $\angle A$ corresponds to $\angle D$, and the sides opposite these angles, $\overline{BC}$ and $\overline{EF}$, also correspond.

Each pair of corresponding angles has the same measure. To state this, we use the letter m as notation to mean "the measure of":

$$m\angle A = m\angle D \qquad m\angle B = m\angle E \qquad m\angle C = m\angle F$$

"The measure of $\angle A$ equals the measure of $\angle D$."

Likewise, each pair of corresponding sides has the same length. We indicate "the length of" a side by writing it without the bar on top:

$$AB = DE \qquad BC = EF \qquad AC = DF$$

"The length of $\overline{AB}$ equals the length of $\overline{DE}$."

When stating that two triangles are **congruent**, we must take care to write corresponding **vertices** in the same positions.

$$\Delta ABC \cong \Delta DEF$$

Vertex A corresponds to vertex D.
Vertex B corresponds to vertex E.
Vertex C corresponds to vertex F.

The following five **congruence** statements could also be used for the triangles in Figure 2: $\Delta BCA \cong \Delta EFD$, $\Delta CAB \cong \Delta FDE$, $\Delta ACB \cong \Delta DFE$, $\Delta CBA \cong \Delta FED$, and $\Delta BAC \cong \Delta EDF$. However, the statement $\Delta ABC \cong \Delta DFE$ is incorrect because vertex B does not correspond to vertex F in the triangles, and vertex C does not correspond to vertex E.

(eText Screens 6.4-1–6.4-37)

Example 1 Identifying Corresponding Parts of Congruent Triangles

My video summary

The two triangles shown below are **congruent**.

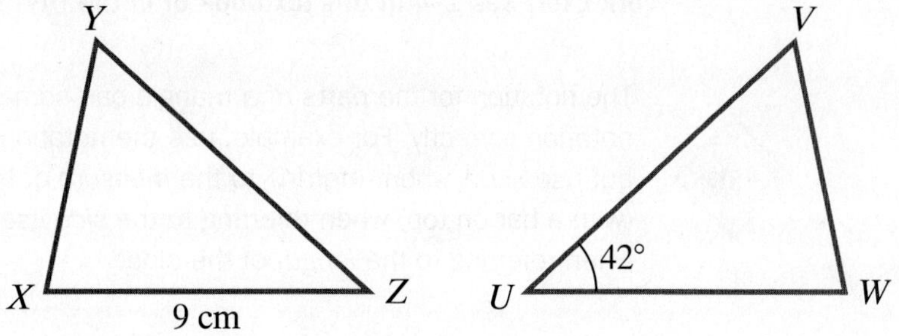

a. List the **corresponding angles** and **corresponding sides**.

b. Complete the sentence $\Delta XYZ \cong \Delta$____.

c. Find the measure of angle Z.

d. Find the length of $\overline{UW}$.

Solutions Work through the following, or watch this **video** for complete solutions to all four parts.

a. If we were to slide ΔXYZ on top of ΔUVW, they would *not* match up. However, if we were to first reflect ΔXYZ (or flip it over) before sliding it on top of ΔUVW, then they *would* match up.

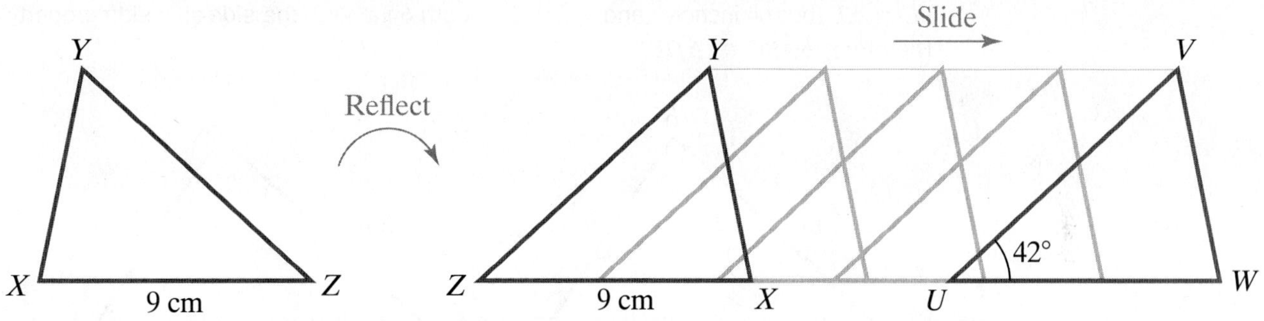

The **corresponding angles** and **corresponding sides** are

$\angle Z$ corresponds to $\angle U$	$\overline{YX}$ corresponds to $\overline{VW}$
$\angle Y$ corresponds to $\angle V$	$\overline{ZX}$ corresponds to $\overline{UW}$
$\angle X$ corresponds to $\angle W$	$\overline{ZY}$ corresponds to $\overline{UV}$

b. To complete the sentence $\Delta XYZ \cong \Delta$____, we must write the corresponding **vertices** in the same order as XYZ.

Vertex X corresponds to vertex W.
Vertex Y corresponds to vertex V.
Vertex Z corresponds to vertex U.

$\Delta XYZ \cong \Delta WVU$

c. We know $\angle Z$ corresponds to $\angle U$, so they have the same measure. Looking at ΔUVW, we see $m\angle U = 42°$, so $m\angle Z = 42°$.

d. We know $\overline{ZX}$ corresponds to $\overline{UW}$, so they are the same length. Looking at ΔZYX, we see $ZX = 9$ cm, so $UW = 9$ cm.

(eText Screens 6.4-1–6.4-37)

You Try It **Work through this You Try It problem.**

Work Exercises 1–4 in this textbook or in the MyMathLab Study Plan.

CAUTION The notation for the parts of a triangle can sometimes be confusing. Take care to use the notation correctly. For example, use the notation $\angle A$ when referring to the *angle itself*, but use $m\angle A$ when referring to the *measure* of the angle. Similarly, use the notation $\overline{AB}$ (with a bar on top) when referring to the *side itself*, but use AB (without a bar on top) when referring to the *length* of the side.

OBJECTIVE 2 DETERMINE WHETHER TWO TRIANGLES ARE CONGRUENT

Sometimes, we may need to determine whether two triangles are **congruent**. We will use three different properties of congruence for this: (1) the *side-side-side (SSS) property*, (2) the *side-angle-side (SAS) property*, and (3) the *angle-side-angle (ASA) property*.

> **Side-Side-Side (SSS) Property of Congruence**
>
> If the lengths of the three sides of one triangle equal the lengths of the **corresponding sides** of another triangle, then the two triangles are congruent.

For example, look at the two triangles in Figure 3. Because $AB = DE$ (both 5 inches), $BC = EF$ (both 4 inches), and $AC = DF$ (both 6 inches), the side-side-side property applies. Therefore, $\Delta ABC \cong \Delta DEF$.

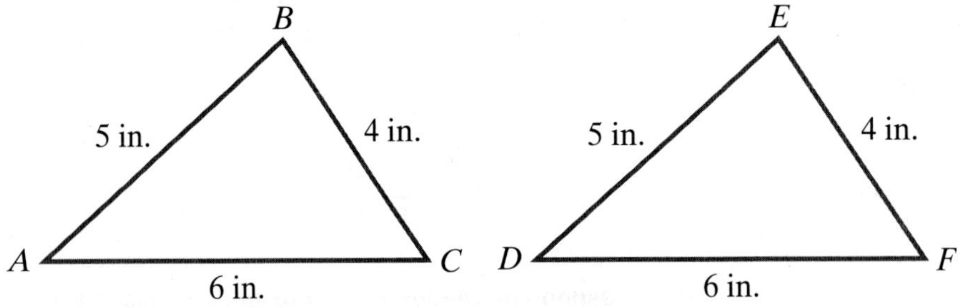

Figure 3

> **Side-Angle-Side (SAS) Property of Congruence**
>
> If the lengths of two sides of one triangle equal the lengths of the **corresponding sides** of another triangle and if the **corresponding angles** between each pair of sides have the same measure, then the two triangles are **congruent**.

For example, look at the two triangles in **Figure 4**. Because $AB = DE$ (both 7.3 m), $m\angle A = m\angle D$ (both 30°), and $AC = DF$ (both 5.2 m), the side-angle-side property applies. Therefore, $\Delta ABC \cong \Delta DEF$.

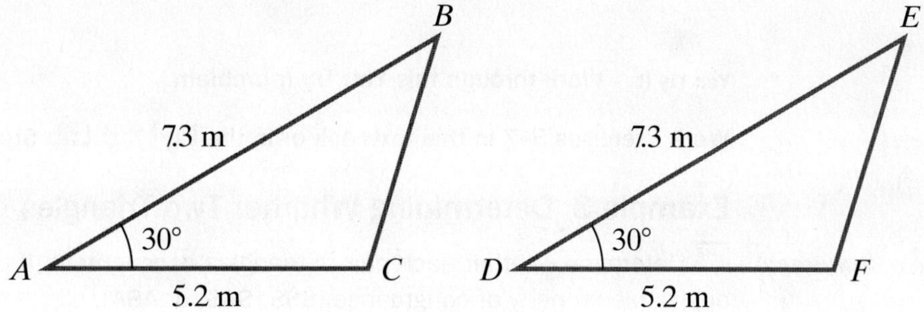

Figure 4

> **Angle-Side-Angle (ASA) Property of Congruence**
>
> If two angles of one triangle have the same measures as the **corresponding angles** of another triangle and if the **corresponding sides** between each pair of angles have equal lengths, then the two triangles are **congruent**.

For example, look at the two triangles in Figure 5. Because $m\angle A = m\angle D$ (both 46°), $AC = DF$ (both 10 cm), and $m\angle C = m\angle F$ (both 34°), the angle-side-angle property applies. So, $\Delta ABC \cong \Delta DEF$.

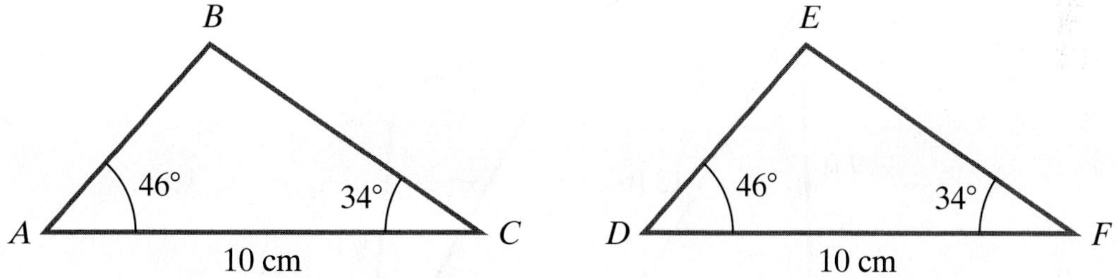

Figure 5

Example 2 Explaining Why Triangles Are Congruent

For each pair of triangles, state the property (**SSS**, **SAS**, or **ASA**) that explains why the triangles are **congruent**.

a.

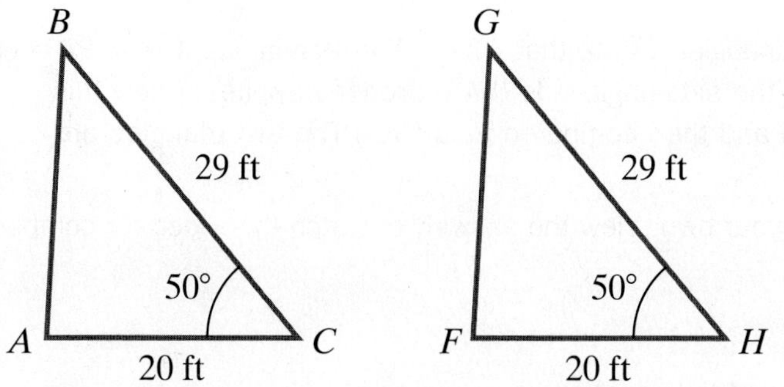

b.

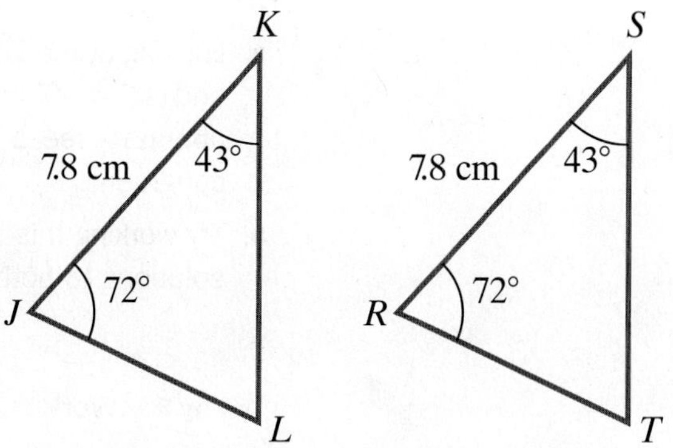

Solutions

a. Looking at the labels on the triangles, we can see that $AC = FH = 20$ ft, $m\angle C = m\angle H = 50°$, and $BC = GH = 29$ ft. So by the side-angle-side (SAS) property, the triangles must be congruent.

b. Because $m\angle J = m\angle R = 72°$, $JK = RS = 7.8$ cm, and $m\angle K = m\angle S = 43°$, the angle-side-angle (ASA) property applies. Thus, the triangles are congruent.

(eText Screens 6.4-1–6.4-37)

You Try It **Work through this You Try It problem.**

Work Exercises 5–7 in this textbook or in the MyMathLab Study Plan.

Example 3 Determining Whether Two Triangles Are Congruent

My video summary

Determine whether each pair of triangles is **congruent**. If congruent, state the applicable property of congruence (**SSS**, **SAS**, or **ASA**).

a.

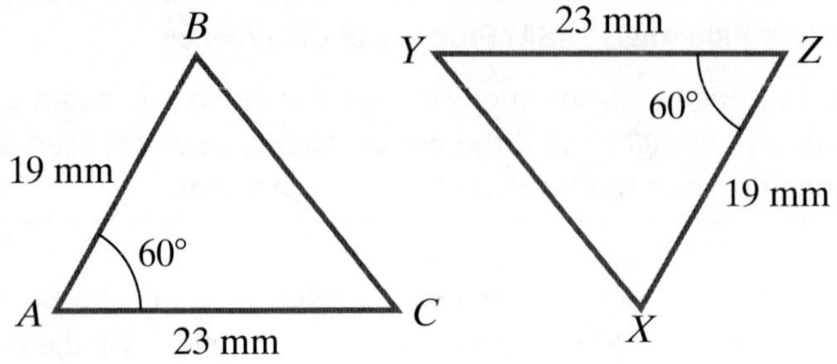

b.

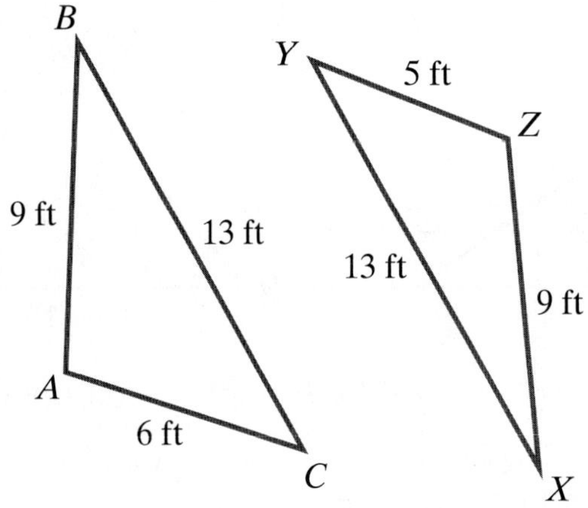

Solutions

a. Looking at the labels on the triangles, notice that $AB = ZX = 19 \text{ mm}$, $m\angle A = m\angle Z = 60°$, and $AC = ZY = 23 \text{ mm}$, so the side-angle-side (SAS) property applies. (View this **popup** to see ΔXYZ rotated and then compared to ΔABC.) The two triangles are congruent.

b. Try working this problem on your own. View the **answer**, or watch this **video** for complete solutions to both parts.

You Try It **Work through this You Try It problem.**

Work Exercises 8–13 in this textbook or in the MyMathLab Study Plan.

OBJECTIVE 3 IDENTIFY THE CORRESPONDING PARTS OF SIMILAR TRIANGLES

We know that two triangles are **congruent** if they have the same shape *and* size. Two triangles are **similar** if they have the same shape but not necessarily the same size. Watch the second part of this **video animation**, or continue reading, to learn more about similar triangles. The two triangles shown in **Figure 6** are similar. They have the same shape, but they are different in size.

(eText Screens 6.4-1–6.4-37)

My video animation summary

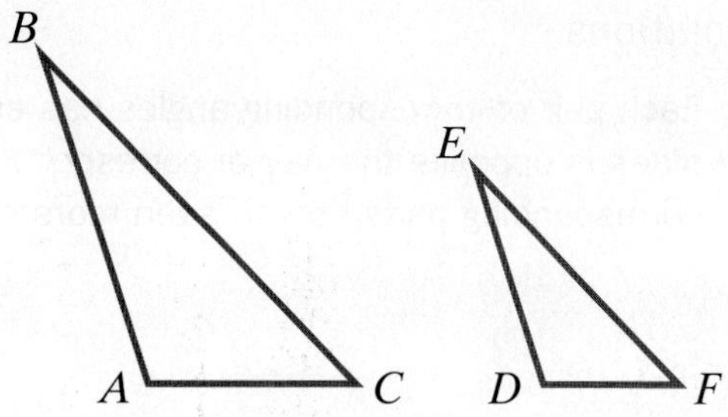

Figure 6

In similar triangles, **corresponding angles** are located in common positions relative to the shape of the triangles. In Figure 6, $\angle A$ corresponds to $\angle D$, $\angle B$ corresponds to $\angle E$, and $\angle C$ corresponds to $\angle F$. As with congruent triangles, each pair of corresponding angles in similar triangles has equal measure.

Corresponding sides are located opposite to the corresponding angles. In Figure 6, $\overline{AB}$ corresponds to $\overline{DE}$, $\overline{BC}$ corresponds to $\overline{EF}$, and $\overline{AC}$ corresponds to $\overline{DF}$. The lengths of the corresponding sides are not always equal, but they are **proportional**:

$$\frac{AB}{DE} = \frac{BC}{EF} = \frac{AC}{DF}$$

We can use the **similarity symbol** $(\sim)$ to state that two triangles are **similar**. For example,

$$\Delta ABC \sim \Delta DEF.$$

"is similar to"

As with **congruence**, we must take care to write corresponding **vertices** in the same positions.

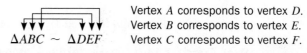

$\Delta ABC \sim \Delta DEF$

Vertex A corresponds to vertex D.
Vertex B corresponds to vertex E.
Vertex C corresponds to vertex F.

The following five similarity statements could also be used for the triangles in Figure 6: $\Delta BCA \sim \Delta EFD$, $\Delta CAB \sim \Delta FDE$, $\Delta ACB \sim \Delta DFE$, $\Delta CBA \sim \Delta FED$, and $\Delta BAC \sim \Delta EDF$. However, the statement $\Delta ABC \sim \Delta DFE$ is incorrect because vertex B does not correspond to vertex F in the triangles, and vertex C does not correspond to vertex E.

Example 4 Identifying Corresponding Parts of Similar Triangles

My video summary

The two triangles shown below are similar.

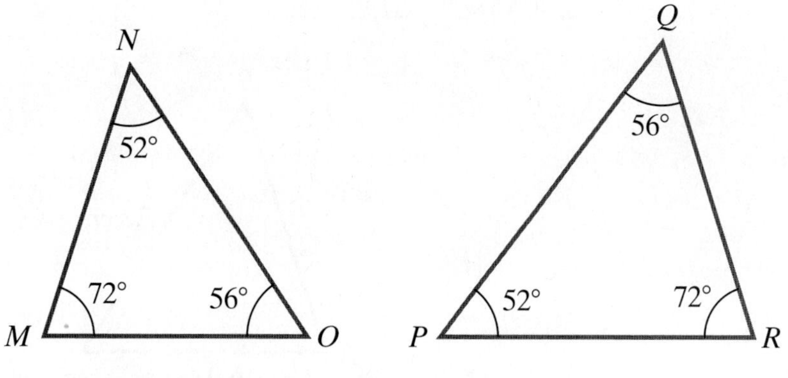

a. List the **corresponding angles** and **corresponding sides**.

b. Complete the sentence $\Delta MNO \sim \Delta$____.

(eText Screens 6.4-1–6.4-37)

Solutions

a. Each pair of corresponding angles has equal measure, and each pair of corresponding sides is opposite the pair of corresponding angles. If we rotate ΔMNO, then the corresponding parts can be seen more easily.

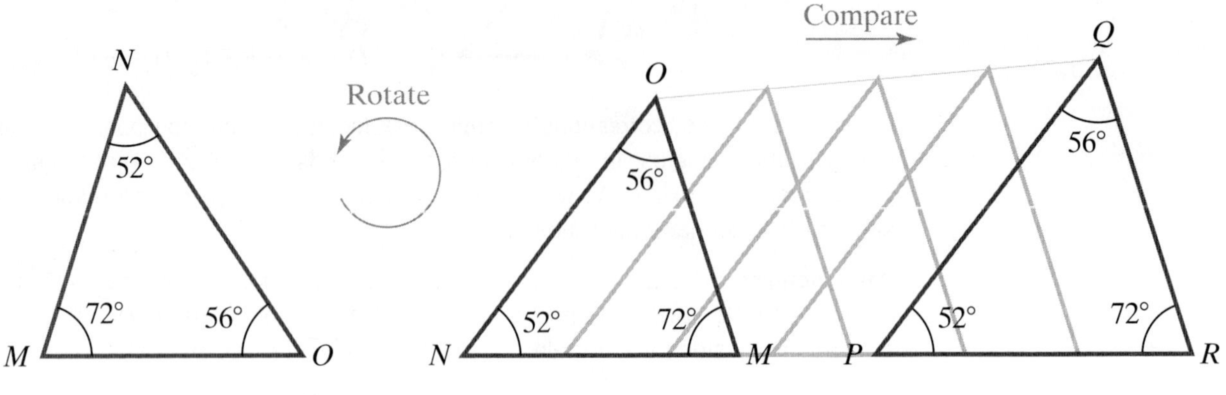

The corresponding angles and corresponding sides are

$\angle N$ corresponds to $\angle P$	$\overline{OM}$ corresponds to $\overline{QR}$
$\angle O$ corresponds to $\angle Q$	$\overline{NM}$ corresponds to $\overline{PR}$
$\angle M$ corresponds to $\angle R$	$\overline{NO}$ corresponds to $\overline{PQ}$

b. Try to finish this part on your own. View the **answer**, or watch this **video** for complete solutions to both parts.

You Try It **Work through this You Try It problem.**

Work Exercises 14–17 in this textbook or in the MyMathLab Study Plan.

OBJECTIVE 4 FIND UNKNOWN LENGTHS OF SIDES IN SIMILAR TRIANGLES

Because the lengths of **corresponding sides** of **similar triangles** are **proportional**, we can use **proportions** to find unknown lengths by setting **ratios** of corresponding sides equal.

Example 5 Finding Lengths of Unknown Sides in Similar Triangles

a. If $\Delta ABC \sim \Delta DEF$, find x.

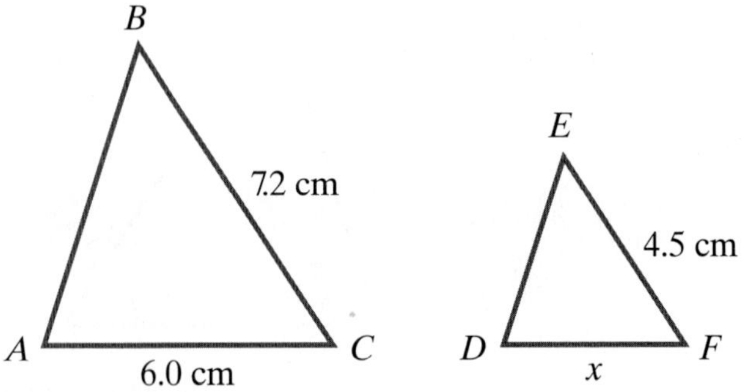

b. If $\Delta XYZ \sim \Delta PQR$, find n.

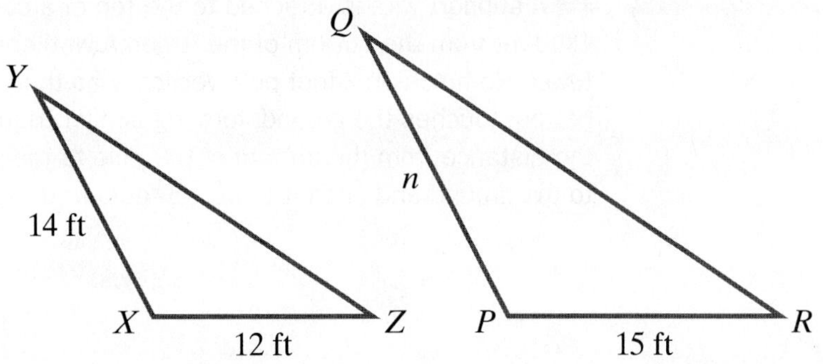

Solutions

a. Because the triangles are similar, the lengths of their corresponding sides are proportional. So, $\frac{BC}{EF} = \frac{AC}{DF}$. In the figure, we are given that $BC = 7.2\text{ cm}$, $EF = 4.5\text{ cm}$, $AC = 6.0\text{ cm}$, and $DF = x$. This gives the following proportion:

$$\begin{matrix} BC \rightarrow \\ EF \rightarrow \end{matrix} \frac{7.2\text{ cm}}{4.5\text{ cm}} = \frac{6.0\text{ cm}}{x} \begin{matrix} \leftarrow AC \\ \leftarrow DF \end{matrix}$$

$7.2 \cdot x$ $\qquad$ $4.5 \cdot 6.0$

$$\frac{7.2}{4.5} = \frac{6.0}{x}$$

$$\begin{array}{r} 4.5 \leftarrow 1 \text{ decimal place} \\ \times\ 6.0 \leftarrow 1 \text{ decimal place} \\ \hline 00 \\ 2700 \\ \hline 27.00 \leftarrow 2 \text{ decimal places} \end{array}$$

Set cross products equal: $7.2 \cdot x = 4.5 \cdot 6.0$

Simplify both sides: $7.2x = 27$

Divide both sides by 7.2: $\frac{7.2x}{7.2} = \frac{27}{7.2}$

Simplify: $x = 3.75$

$$7.2\overline{)27.0} \Rightarrow 72.\overline{)270.00} \quad \begin{array}{r} 3.75 \\ -216 \\ \hline 54\,0 \\ -50\,4 \\ \hline 3\,60 \\ -3\,60 \\ \hline 0 \end{array}$$

The unknown length x is 3.75 cm.

My video summary

b. Try to work this problem on your own. View the **answer**, or watch this **video** for a complete solution to part b.

You Try It **Work through this You Try It problem.**

Work Exercises 18–23 in this textbook or in the MyMathLab Study Plan.

CAUTION Be sure to include units in final answers whenever units are included in problems.

OBJECTIVE 5 SOLVE APPLICATIONS INVOLVING SIMILAR TRIANGLES

Similar triangles can be used in applications to find lengths that are difficult to measure physically. In Example 6, we use similar triangles to determine the height of a communication tower.

(eText Screens 6.4-1–6.4-37)

Example 6 Finding the Height of a Communication Tower

My video summary

A support wire is attached to the top of a communication tower and to the ground 180 feet from the bottom of the tower. A worker wants to determine the height of the tower. He holds an 8-foot pole vertically so that its top touches the support wire and its bottom touches the ground, forming similar triangles (see the diagram). He then measures the distance from the bottom of the pole to the point where the support wire is attached to the ground and finds it to be 4.5 feet. Find the height of the tower.

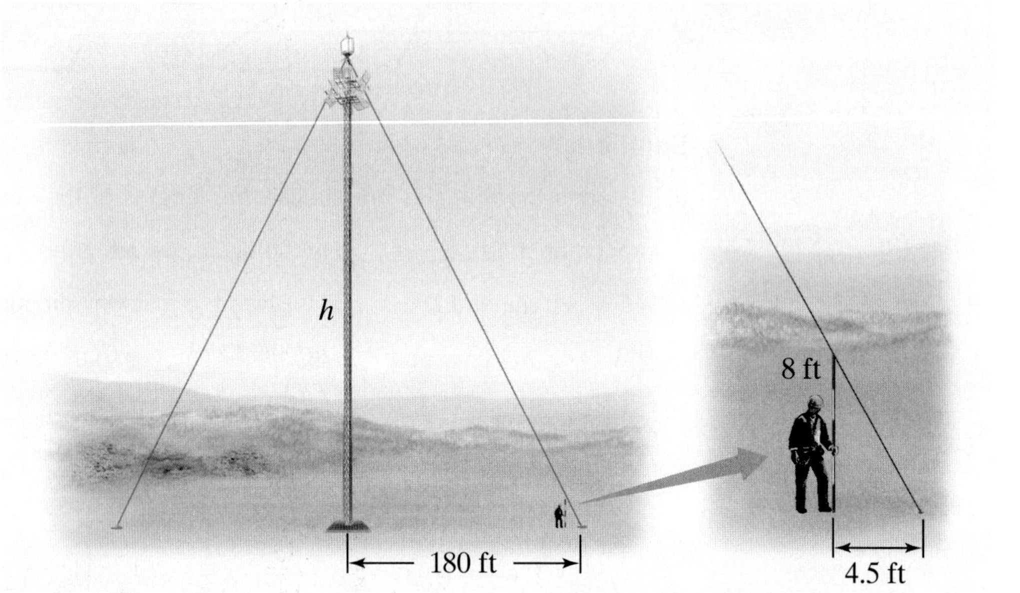

Solution The triangle formed by the pole, ground, and support wire is similar to the triangle formed by the tower, ground, and support wire. So, the lengths of the corresponding sides are proportional. The height of the tower, h, corresponds to the height of the pole, 8 feet, and the 180-foot distance corresponds to the 4.5-foot distance. This gives the proportion

$$\frac{h}{8\text{ ft}} = \frac{180\text{ ft}}{4.5\text{ ft}}.$$

Try to solve this proportion on your own. View the **answer**, or watch this **video** for a complete solution.

You Try It **Work through this You Try It problem.**

Work Exercises 24–29 in this textbook or in the MyMathLab Study Plan.

6.4 Exercises

In Exercises 1–4, each pair of triangles is congruent.

You Try It 1.

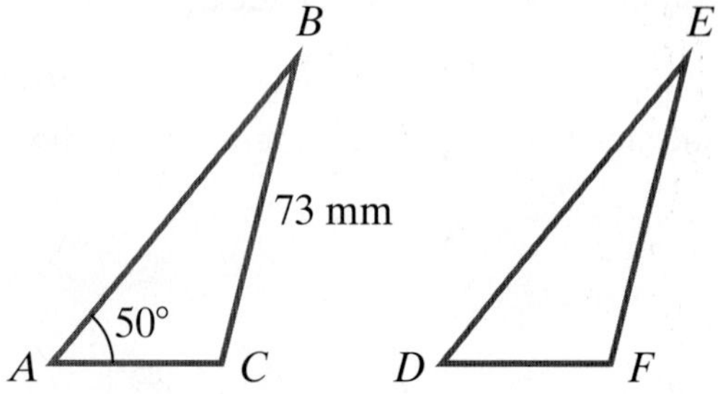

a. List the corresponding angles and sides.

b. Complete the sentence: $\Delta ABC \cong \Delta$____.

c. Find $m\angle D$.

d. Find EF.

(eText Screens 6.4-1–6.4-37)

2.

a. List the corresponding angles and sides.

b. Complete the sentence: $\Delta JKL \cong \Delta$____.

c. Find $m\angle F$.

d. Find KL.

3.

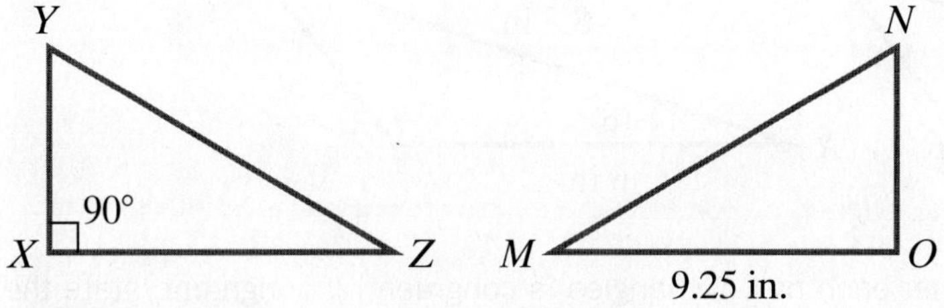

a. List the corresponding angles and sides.

b. Complete the sentence: $\Delta XYZ \cong \Delta$____.

c. Which angle of ΔMNO measures $90°$?

d. Which side of ΔXYZ is 9.25 inches long?

4.

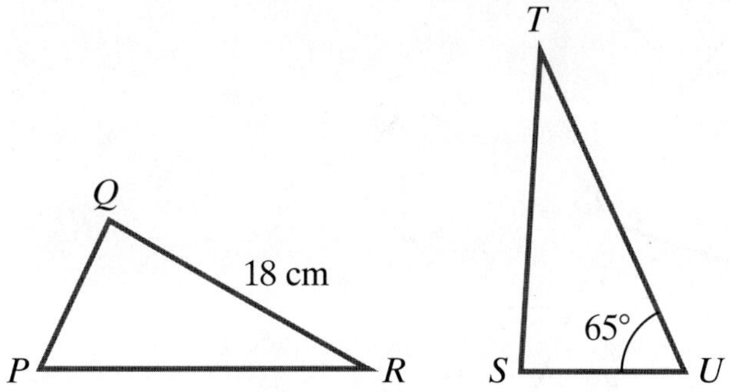

a. List the corresponding angles and sides.

b. Complete the sentence: $\Delta PQR \cong \Delta$____.

c. Which angle of ΔPQR measures $65°$?

d. Which side of ΔSTU is 18 cm long?

In Exercises 5–7, for each pair of triangles, state the property (SSS, SAS, or ASA) that explains why the triangles are congruent.

You Try It 5.

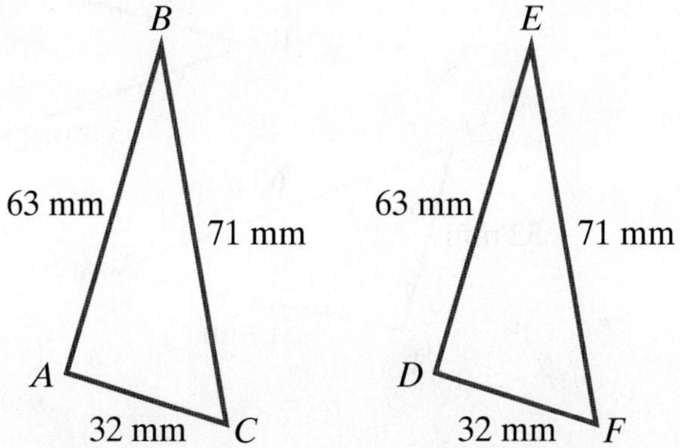

6.

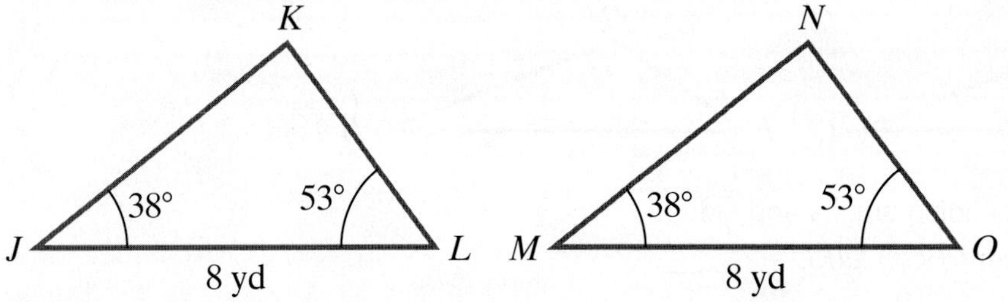

7.

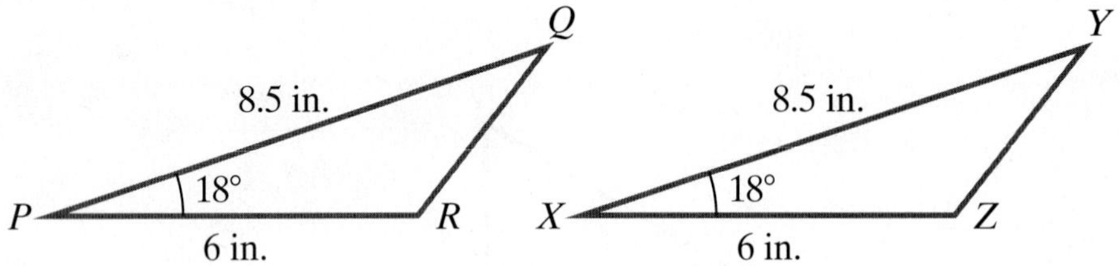

In Exercises 8–13, determine whether each pair of triangles is congruent. If congruent, state the property of congruence that applies (SSS, SAS, or ASA).

You Try It

8.

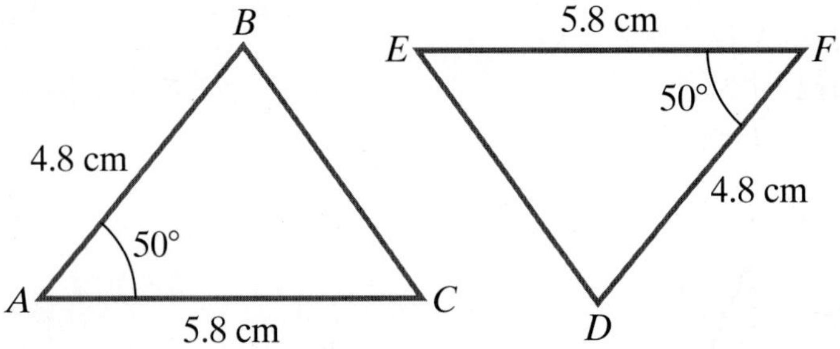

9.

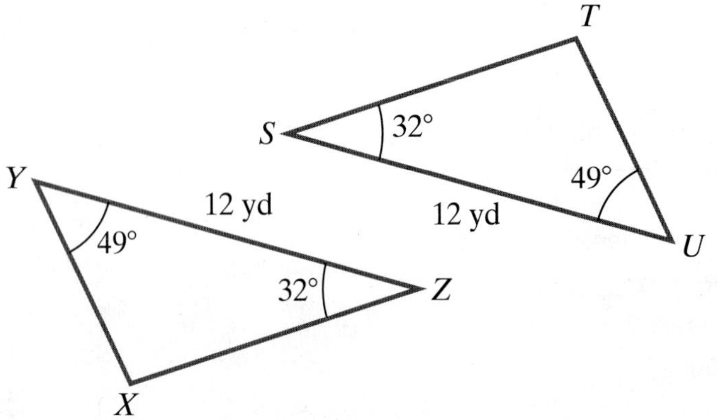

10.

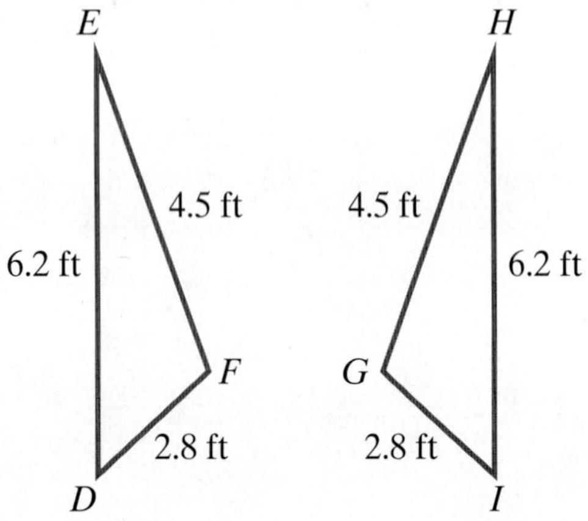

11.

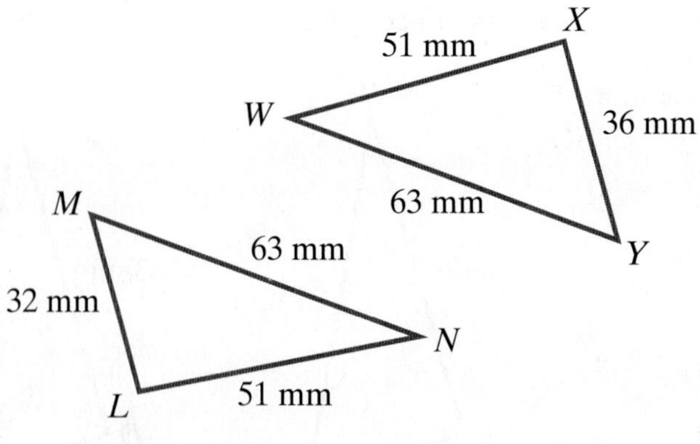

12.

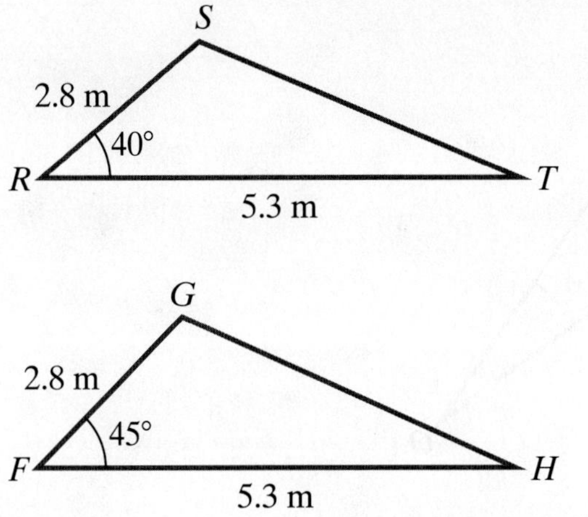

13.

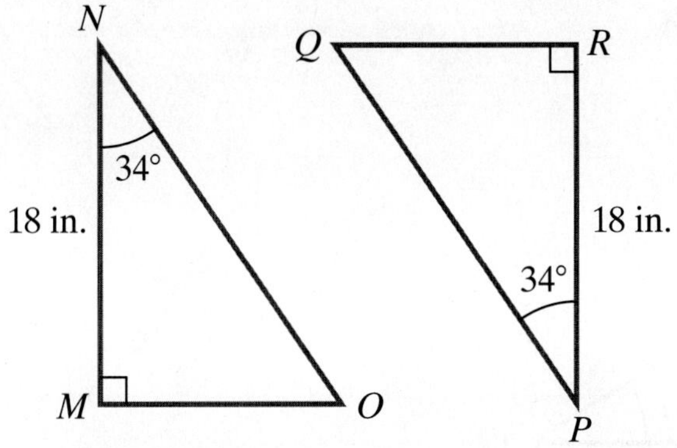

In Exercises 14–17, the given triangles are similar.

You Try It 14.

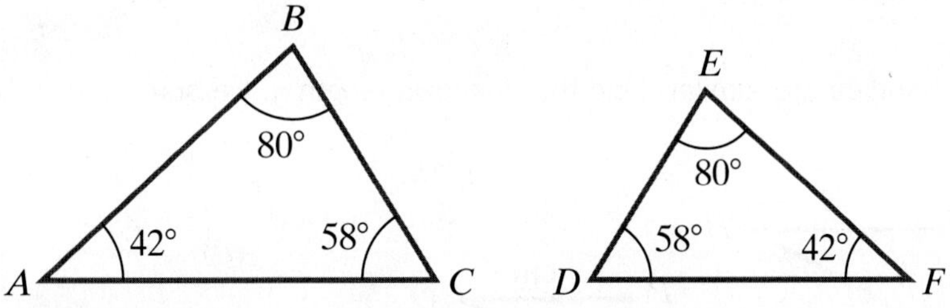

a. List the corresponding angles and sides.

b. Complete the sentence: $\Delta ABC \sim \Delta$____.

15.

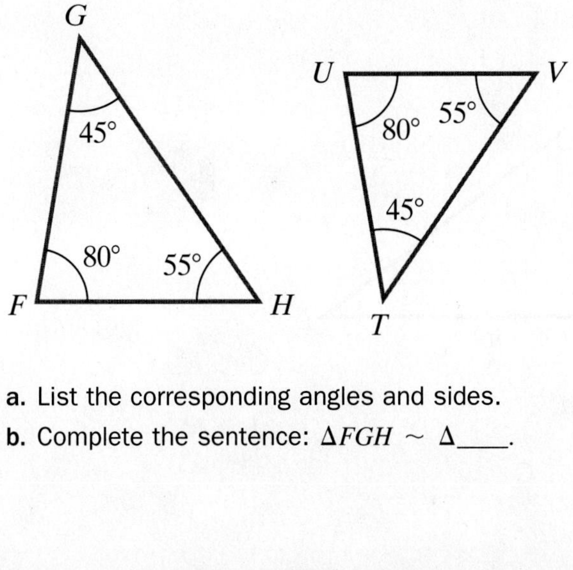

a. List the corresponding angles and sides.

b. Complete the sentence: $\Delta FGH \sim \Delta$____.

(eText Screens 6.4-1–6.4-37)

16.

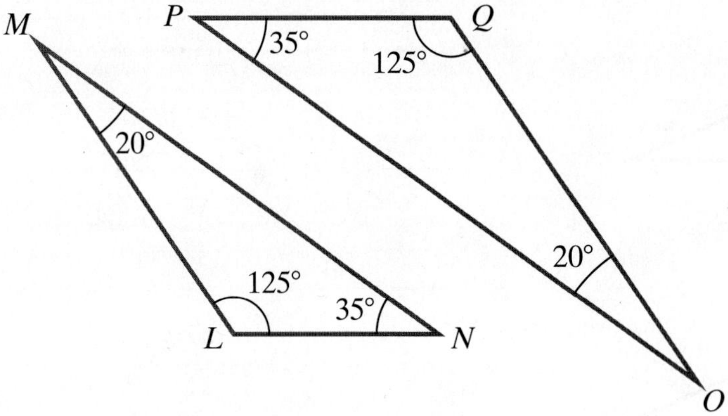

a. List the corresponding angles and sides.

b. Complete the sentence: $\Delta LMN \sim \Delta$____.

17.

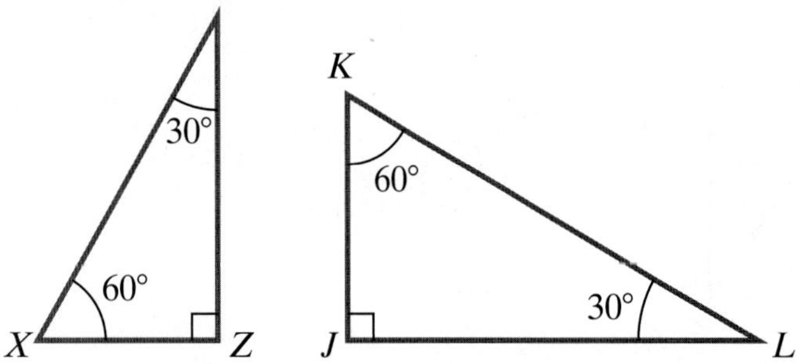

a. List the corresponding angles and sides.

b. Complete the sentence: $\Delta XYZ \sim \Delta$____.

In Exercises 18–23, the given triangles are similar. Find the unknown length n. Be sure to include units in your answers.

You Try It 18.

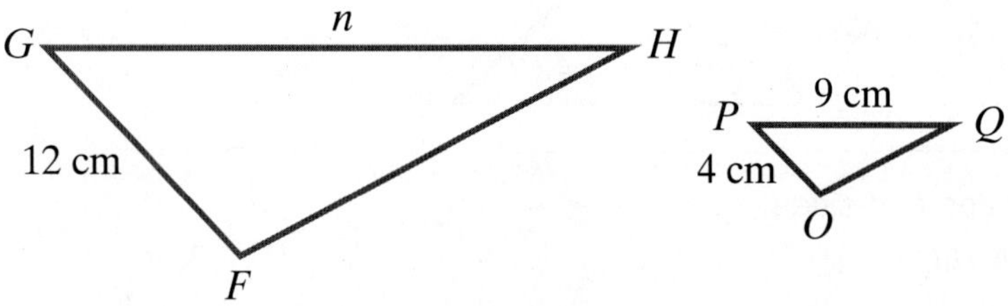

19.

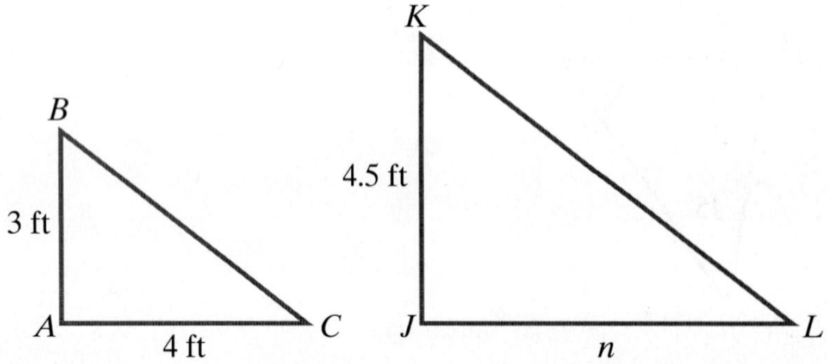

20.

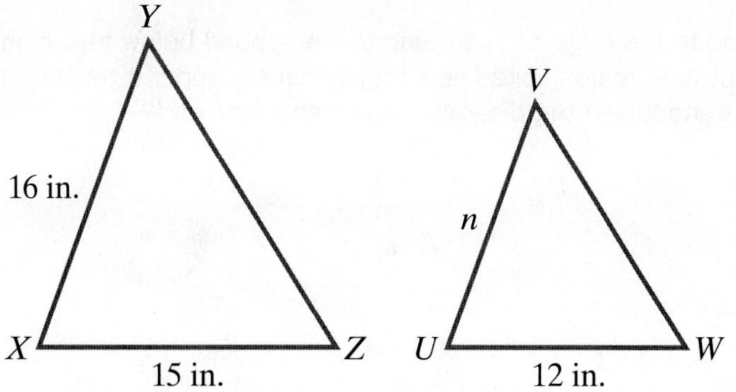

21.

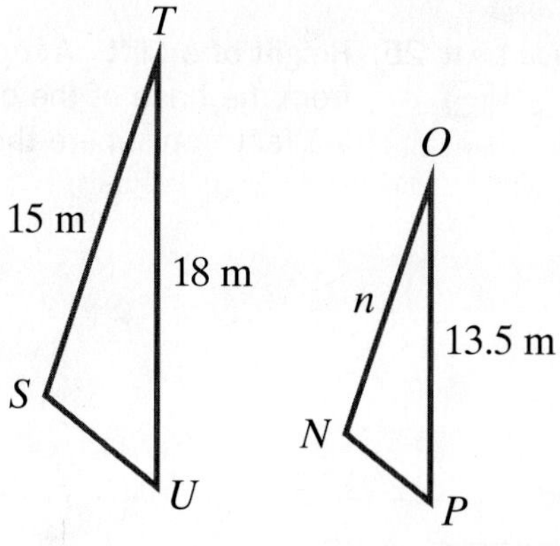

22. Hint: $\Delta ABC \sim \Delta FED$

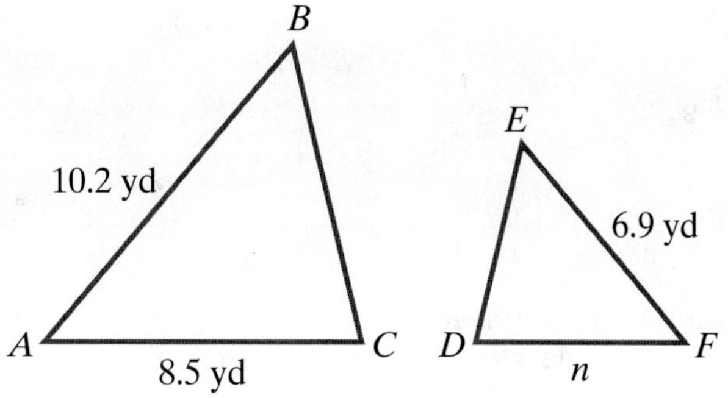

23. Hint: $\Delta JKL \sim \Delta NML$

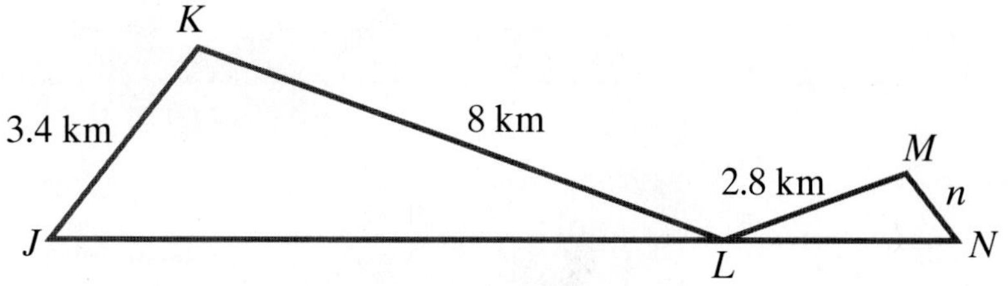

In Exercises 24–29, use similar triangles to solve each application.

24. **Height of a Tree** A hiker wants to determine the height of a tree. She measures the tree's shadow to be 25.3 meters long. Her own shadow at the same time is 2.2 meters long. If the hiker is 1.7 meters tall, how tall is the tree? Round to the nearest tenth of a meter.

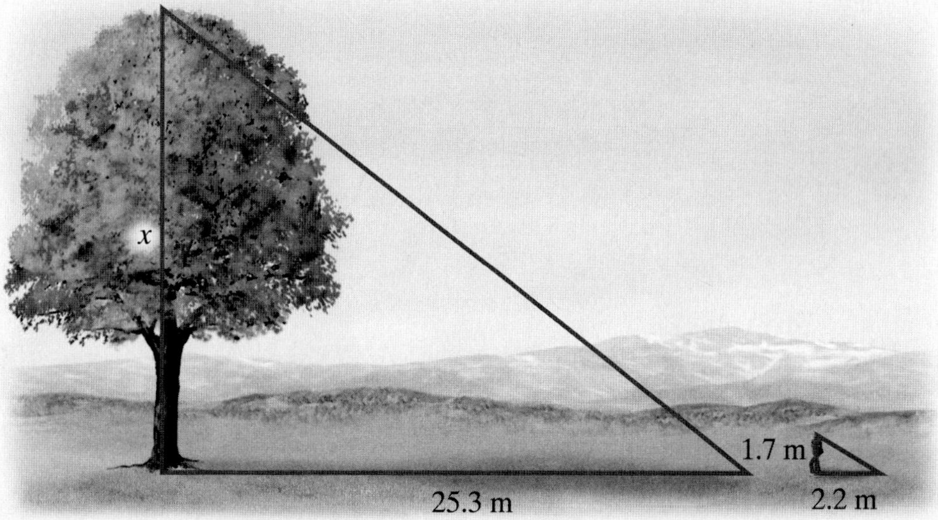

You Try It 25. **Height of a Cliff** A rope is fastened to the edge of a cliff and to the ground below at a point 120 feet from the base of the cliff. If a 5-foot pole is positioned vertically under the rope, it touches the ground 7.5 feet from where the rope is fastened. See the diagram. How high is the cliff?

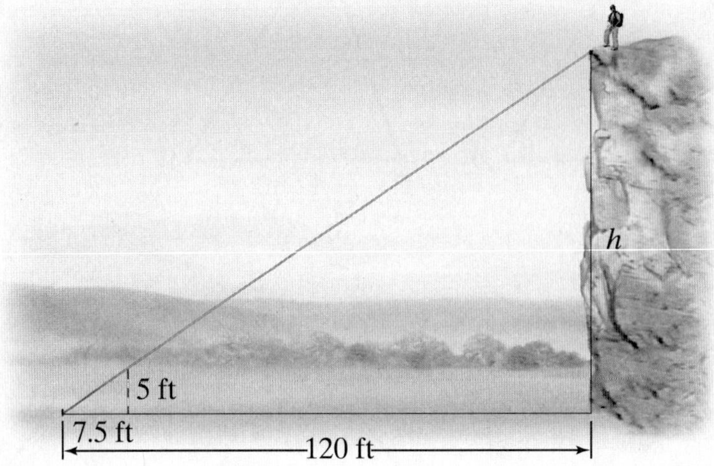

26. **Width of a River** A surveyor needs to determine the width of a river, but he does not want to get wet. He lays out the similar triangles shown in the diagram. What is the width of the river?

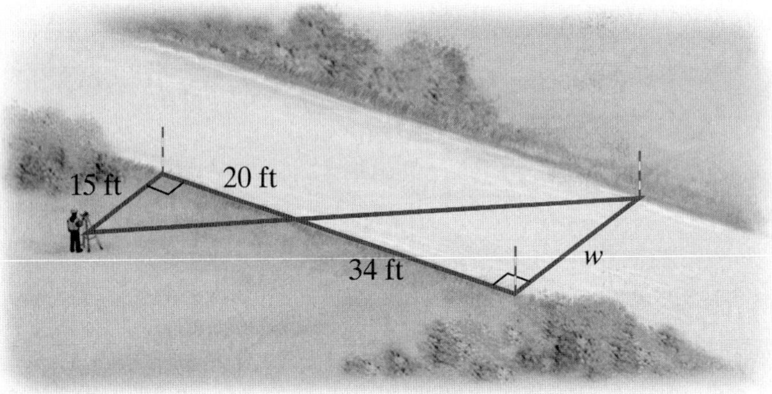

27. **Distance Across a Lake** To find the distance across a lake, a surveyor lays out the similar triangles shown in the diagram. What is the distance d across the lake?

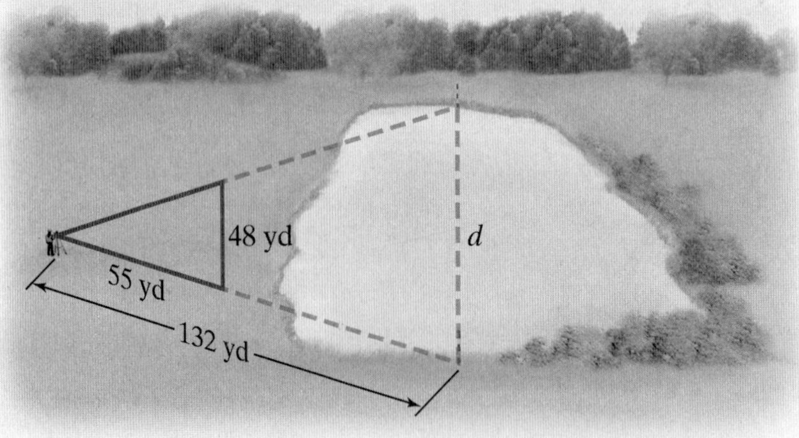

28. **Height of a Building** To find the height of a building, a 6-foot-tall man places a mirror on the ground 80 feet from the building. He then positions himself 10 feet farther where he can see the top of the building reflected in the mirror. See the diagram. Because of reflection properties of mirrors, $\Delta ABC \sim \Delta DEC$. Find the height of the building.

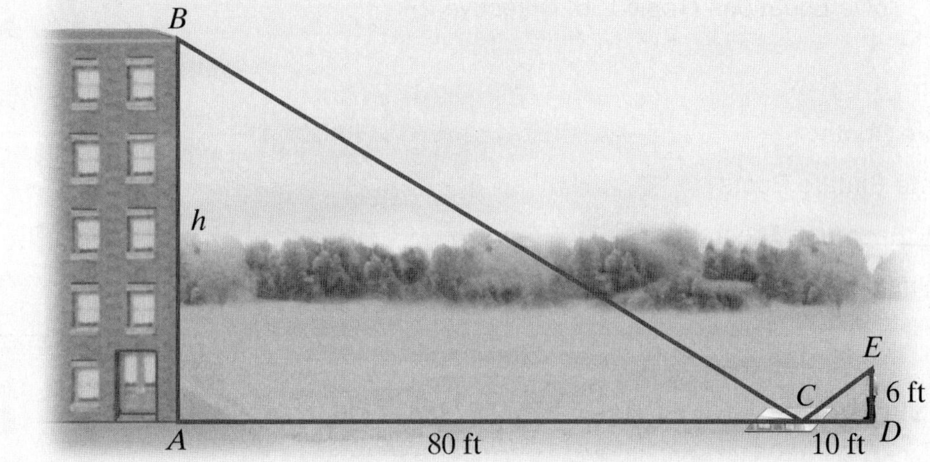

29. **Tennis Anyone?** A tennis player serves a ball from behind the base line 39 feet from the net. At what height must he hit the ball for it to just clear the 3-foot net and land 21 feet past the net (on the service line)? Assume the ball travels in a straight line as shown in the figure. Round the height to the nearest tenth of a foot.

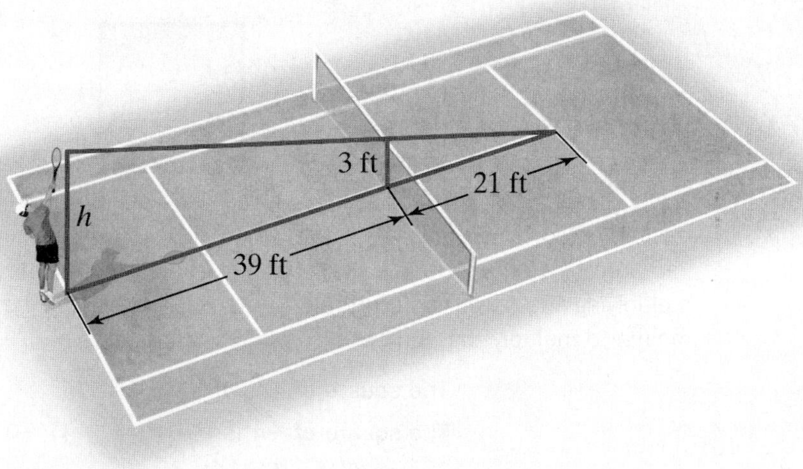

6.5 Square Roots and the Pythagorean Theorem

THINGS TO KNOW

Before working through this topic, be sure you are familiar with the following concepts:

		VIDEO	ANIMATION	INTERACTIVE
You Try It	1. Use Inequality Symbols to Compare Whole Numbers (Topic 1.2, **Objective 4**)			
You Try It	2. Find the Area of a Rectangle (Topic 1.4, **Objective 5**)	✓	✓	

You Try It **3.** Evaluate Exponential Expressions (Topic 1.6, **Objective 2**)

You Try It **4.** Use the Addition Property of Equality to Solve Equations (Topic 2.6, **Objective 2**)

OBJECTIVES

1 Find Square Roots

2 Approximate Square Roots

3 Use the Pythagorean Theorem

4 Solve Applications Using the Pythagorean Theorem

OBJECTIVE 1 FIND SQUARE ROOTS

In this topic, we continue our work with triangles by exploring right triangles and the Pythagorean Theorem. First, we need to study square roots.

We saw in **Topic 1.4** that the **area** A of a rectangle can be found by **multiplying** the rectangle's length l and width w. That is, $A = l \cdot w$. A square is a special case of a rectangle in which the length and width are the same.

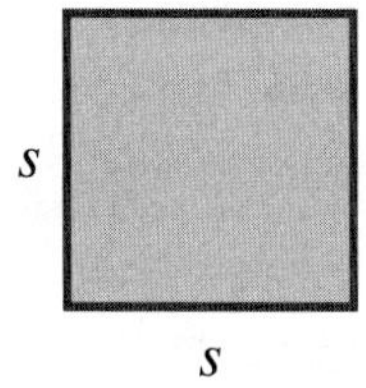

The area of a square is given by $A = s \cdot s = s^2$. That is, the area of a square is found by multiplying the length of one side by itself. This is why we say that *squaring* a number means to multiply the number by itself. For example,

The square of 4 is $4^2 = 4 \cdot 4 = 16.$

The square of -4 is $(-4)^2 = (-4)(-4) = 16.$

The square of $\frac{1}{3}$ is $\left(\frac{1}{3}\right)^2 = \frac{1}{3} \cdot \frac{1}{3} = \frac{1}{9}.$

Given the length of the sides of a square, we find the **area** by **squaring** the length (**Figure 7a**). But suppose we have a square with a given area and want to determine the length of each side (**Figure 7b**)? We need to find a number that we can multiply by itself to get the given area. This number is called the **square root** of the area. Because $6 \cdot 6 = 36$, the length of each side in Figure 7b would be 6 ft and we would say that 6 is the square root of 36.

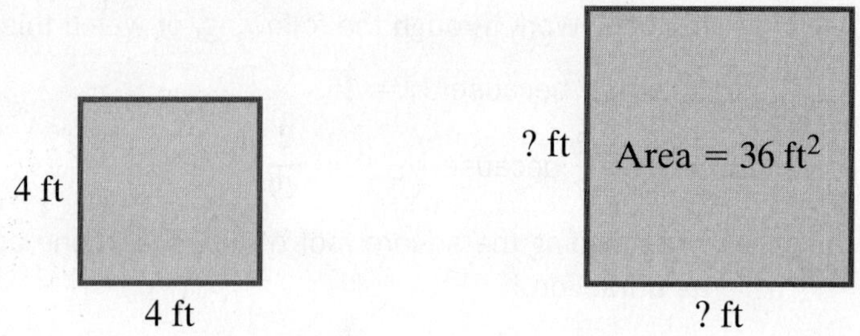

$$\text{Area} = (4 \text{ ft})(4 \text{ ft})$$
$$= 16 \text{ ft}^2$$

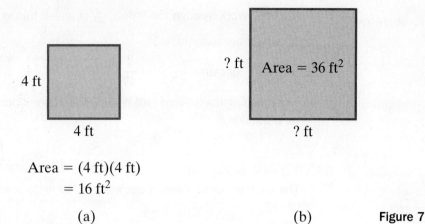

(a) (b) **Figure 7**

CAUTION Remember that the length of a side is given in single units, like feet, while area is given in square units, like square feet (ft^2).

A square root of a **positive number** a is a number b that, when squared, results in a. So, b is a square root of a if $b^2 = a$.

Every positive number has two **square roots**: one positive and one negative. For example, 4 and -4 are both square roots of 16 because

$$4^2 = 16 \quad \text{and} \quad (-4)^2 = 16.$$

4 is the **positive** (or **principal**) **square root** of 16, and -4 is the **negative square root** of 16. For the remainder of this text, we will only work with positive (or principal) square roots.

CAUTION Notice that we gave the positive square root when finding the length of the side of a square. This is because length cannot be negative.

We use the **radical sign** $\sqrt{\ }$ to denote positive square roots. For example, $\sqrt{16}$ represents the positive square root of 16, and $\sqrt{36}$ represents the positive square root of 36. We can **evaluate** these square roots as follows:

$$\sqrt{16} = 4 \quad \text{because} \quad 4^2 = 16.$$
$$\sqrt{36} = 6 \quad \text{because} \quad 6^2 = 36.$$

Definition Square Root

For $a > 0$, $\sqrt{a}$ is the **positive** or **principal square root** of a.

$$\sqrt{a} = b \text{ only if } b^2 = a \text{ and } b > 0$$

Also, $\sqrt{0} = 0$.

Example 1 Finding Square Roots

Evaluate each **square root**.

a. $\sqrt{25}$ **b.** $\sqrt{\dfrac{4}{49}}$ **c.** $\sqrt{0.04}$

(eText Screens 6.5-1–6.5-23)

My video summary

Solutions Work through the following, or watch this **video** for the complete solutions.

a. $\sqrt{25} = 5$ because $5^2 = 25$.

b. $\sqrt{\dfrac{4}{49}} = \dfrac{2}{7}$ because $\left(\dfrac{2}{7}\right)^2 = \dfrac{4}{49}$

c. When finding the square root of a **decimal**, one approach is first to write the decimal as a fraction.

$$\sqrt{0.04} = \sqrt{\frac{4}{100}} = \frac{2}{10} \text{ because } \left(\frac{2}{10}\right)^2 = \frac{4}{100}$$

The original number was a decimal, so we write our final answer as a decimal: $\dfrac{2}{10} = 0.2$, so $\sqrt{0.04} = 0.2$.

You Try It **Work through this You Try It problem.**

Work Exercises 1–6 in this textbook or in the MyMathLab Study Plan.

OBJECTIVE 2 APPROXIMATE SQUARE ROOTS

So far, we have only taken **square roots** of *perfect squares*. A number whose square root is a **whole number** or a **fraction** is called a **perfect square**. View this **popup** for a list of the first ten whole numbers that are perfect squares.

We now look at square roots of numbers that are not perfect squares.

Consider $\sqrt{13}$. Because 13 is not a perfect square, $\sqrt{13}$ will not be a whole number or a fraction. However, we can **estimate** its value by using a calculator or a table (see **Appendix A**). Before doing this, think about what might be a reasonable result. Notice that 13 is between the perfect squares 9 and 16. That is,

$$9 < 13 < 16.$$

So, the square root of 13 will be between the principal square roots of 9 and 16.

$$\sqrt{9} < \sqrt{13} < \sqrt{16} \quad \text{or} \quad 3 < \sqrt{13} < 4$$

We can see this visually on the following **number line**.

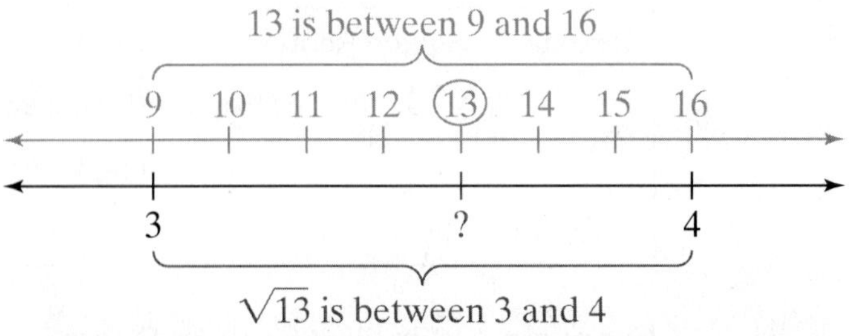

Because 13 is closer to 16 than to 9, we might expect that $\sqrt{13}$ will be closer to 4 than 3.

Figure 8 shows the TI-84 Plus calculator display for $\sqrt{13}$. Rounding to the nearest thousandth, we have $\sqrt{13} \approx 3.606$. This approximation is between 3 and 4 as expected (and closer to 4 than to 3).

(eText Screens 6.5-1–6.5-23)

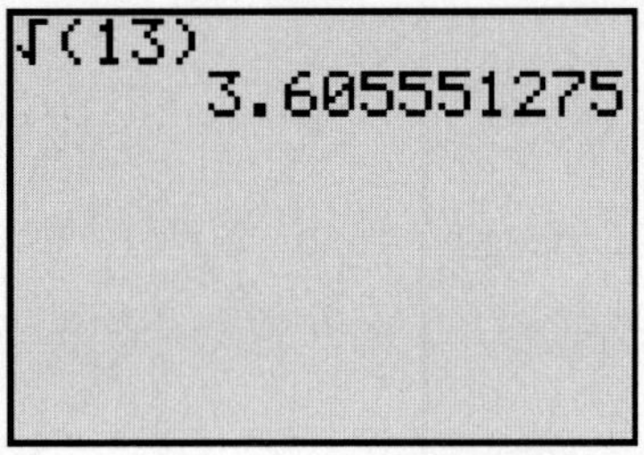

Figure 8
TI-84 Plus Display Approximating $\sqrt{13}$.

Example 2 Approximating Square Roots

My video summary

For each part, determine the two **whole numbers** that the **square root** lies between. Then use a calculator to approximate each square root. Round answers to the nearest thousandth.

a. $\sqrt{57}$ **b.** $\sqrt{\dfrac{39}{7}}$ **c.** $\sqrt{19.5}$

Solutions

a. 57 is between the **perfect squares** 49 and 64. Because $49 = 7^2$ and $64 = 8^2$, $\sqrt{57}$ is between $\sqrt{49} = 7$ and $\sqrt{64} = 8$. Using a calculator, we find

$$\sqrt{57} \approx 7.550.$$

↑ is approximately equal to

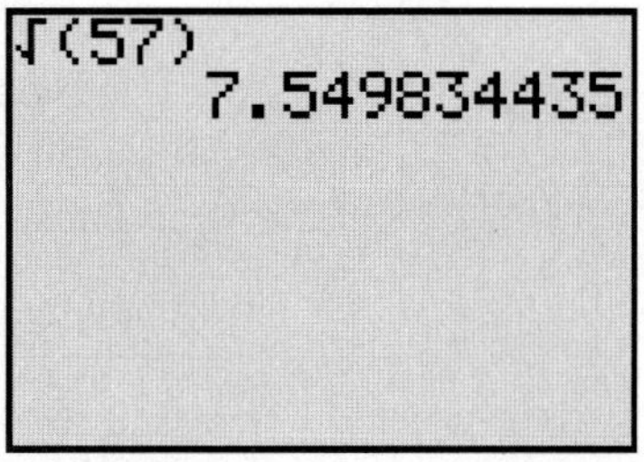

b. Writing $\frac{39}{7}$ as a **mixed number**, we get $\frac{39}{7} = 5\frac{4}{7}$, which is between the perfect squares 4 and 9. Finish working this problem on your own and check your **answer**. Or watch this **video** for complete solutions to all three parts.

c. Try working this problem on your own. Check your **answer**, or watch this **video** for complete solutions to all three parts.

You Try It **Work through this You Try It problem.**

Work Exercises 7–12 in this textbook or in the MyMathLab Study Plan.

TIP To compute a **square root** on a calculator, look for the key with the **radical sign**, $\sqrt{\ }$ .On some models, the square root feature may be a 2nd function on the square key. In addition, some calculators require you to press the square root key before the number, while others require you to press it after the number. Check your instruction manual to be sure.

OBJECTIVE 3 USE THE PYTHAGOREAN THEOREM

Square roots are frequently used when working with *right triangles.* **Right triangles** are triangles with a 90° angle, or **right angle**. The **hypotenuse** of a right triangle is the side opposite the right angle and is the longest of the three sides. The other two sides are called the **legs** of the triangle.

(eText Screens 6.5-1–6.5-23)

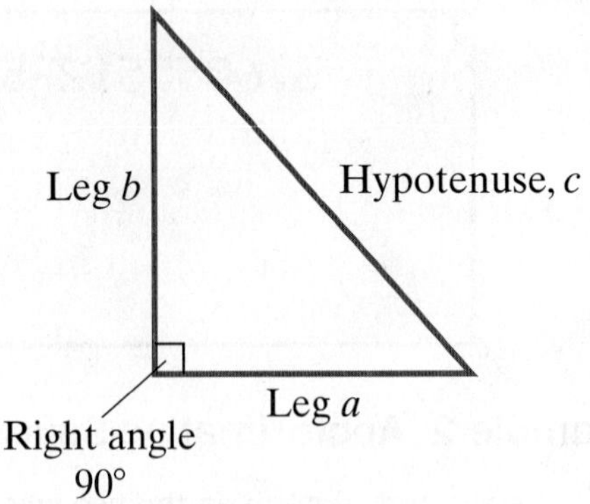

Figure 9

TIP Recall from **Topic 4.3** that we use a small box to indicate a 90° angle.

For any right triangle, the following is true.

Pythagorean Theorem

If a and b are the lengths of the **legs** of a **right triangle**, and c is the length of the **hypotenuse**, then

$$a^2 + b^2 = c^2.$$

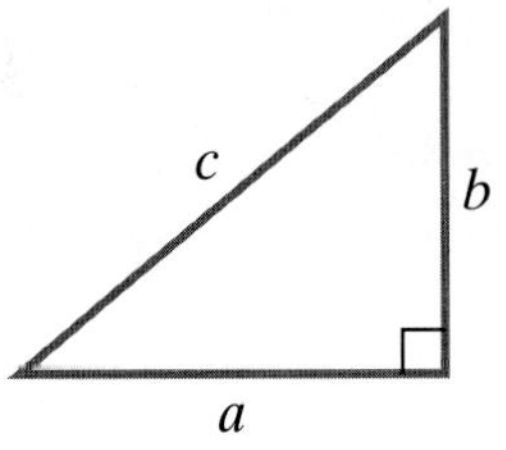

In words, this means that the sum of the squares of the leg lengths in a right triangle is equal to the square of the length of the hypotenuse.

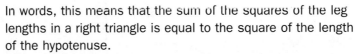

My video animation summary

View this **video animation** for an illustration of the Pythagorean Theorem.

Given the lengths of two sides of a right triangle, we can use the Pythagorean Theorem to find the length of the missing side.

Example 3 Using the Pythagorean Theorem to Find the Length of a Side

My video summary

Find the length of the missing side for each right triangle.

a.

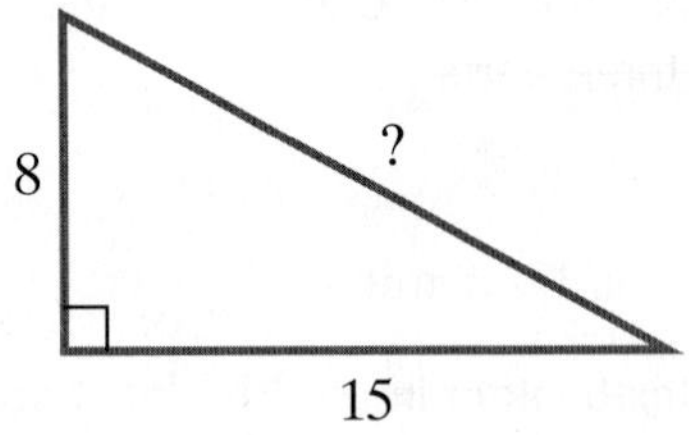

b.

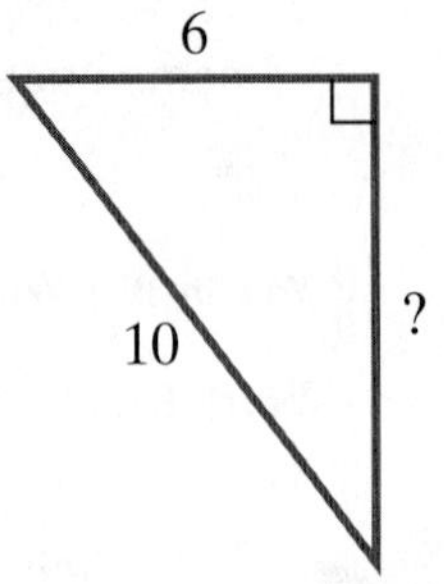

Solutions

a. In this triangle, the missing side is the **hypotenuse**. We are given that the legs have lengths of 8 units and 15 units. Using the **Pythagorean Theorem**, we let $a = 8$, $b = 15$, and we want to find c.

Pythagorean Theorem:	$a^2 + b^2 = c^2$
Substitute 8 for a and 15 for b:	$(8)^2 + (15)^2 = c^2$
Evaluate 8^2 and 15^2:	$64 + 225 = c^2$
Add:	$289 = c^2$

To find the **solutions** to this equation, we want to find values that when **squared** will equal 289. Note that $(17)(17) = 289$ and $(-17)(-17) = 289$, so both 17 and -17 are solutions to the equation $c^2 = 289$. These are the positive and negative **square roots** of 289. However, c is a *length*, so its value must be **non-negative**. We discard the negative solution.

$$\begin{aligned} c &= \sqrt{289} \leftarrow (17)(17) = 289 \\ &= 17 \end{aligned}$$

The length of the missing side is 17 units.

b. In this triangle, the missing side is one of the **legs**. We are given that the other leg has length 6 units and the **hypotenuse** has length 10 units. Using the **Pythagorean Theorem**, we let $a = 6$, $c = 10$, and we want to find b.

Pythagorean Theorem: $a^2 + b^2 = c^2$

Substitute 6 for a and 10 for c: $(6)^2 + (b)^2 = (10)^2$

Try finishing this problem on your own. Check your **answer**, or watch this **video** for complete solutions to both parts.

You Try It **Work through this You Try It problem.**

Work Exercises 13–15 in this textbook or in the MyMathLab Study Plan.

Example 4 Using the Pythagorean Theorem to Approximate the Length of a Side

My video summary

Find the exact length of the missing side of each right triangle. Then approximate the length to the nearest tenth.

a.

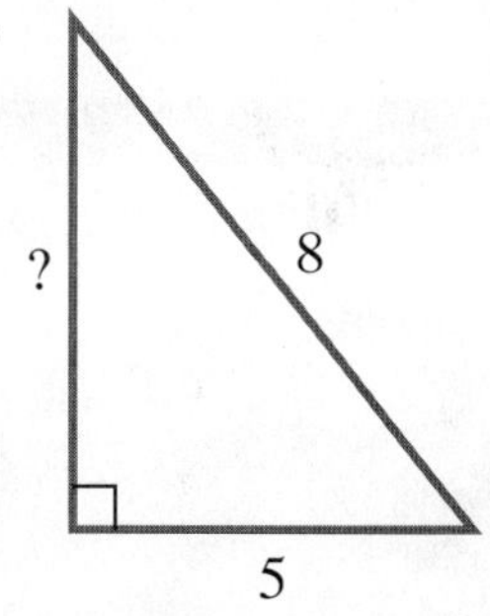

b.

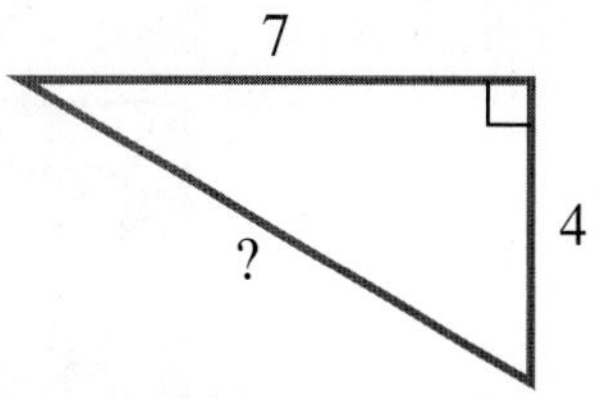

Solutions

a. In this triangle, the missing side is one of the **legs**. We are given that the other leg has length 5 units and the **hypotenuse** has length 8 units. Using the **Pythagorean Theorem**, we let $a = 5$, $c = 8$, and we want to find b.

Pythagorean Theorem: $a^2 + b^2 = c^2$

Substitute 5 for a and 8 for c: $(5)^2 + b^2 = (8)^2$

Evaluate 5^2 and 8^2: $25 + b^2 = 64$

Subtract 25 from both sides: $b^2 = 39$

Use the positive square root of 39: $b = \sqrt{39}$

39 is not a **perfect square**, so the exact length is given using the **radical sign**. The exact length of the missing side is $\sqrt{39}$ units.

(eText Screens 6.5-1–6.5-23)

We can use the square root feature of a calculator, $\sqrt{\ }$, to approximate the length. The length of the missing side is approximately 6.2 units.

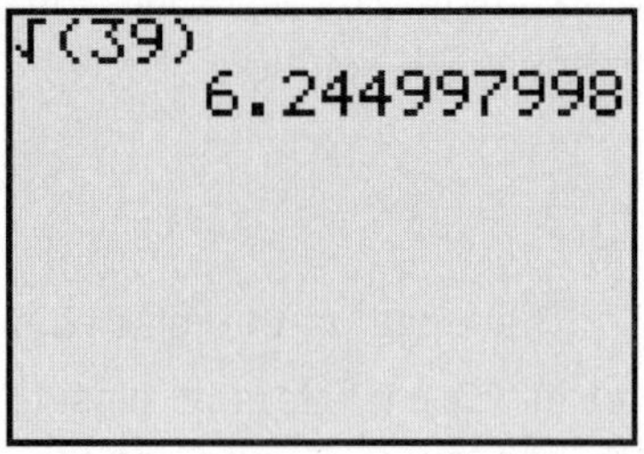

b. Try working this problem on your own. Check your **answer**, or watch this **video** for complete solutions to both parts.

You Try It **Work through this You Try It problem.**

Work Exercises 16–18 in this textbook or in the MyMathLab Study Plan.

OBJECTIVE 4 SOLVE APPLICATIONS USING THE PYTHAGOREAN THEOREM

Some applications can be solved by using a **right triangle** to represent a situation.

Example 5 TV Dimensions

My video summary

Television manufacturers give the screen size of a television by reporting the length of its diagonal. If the length of the screen is 40 inches and its width is 25 inches, approximate the screen size to the nearest tenth of an inch.

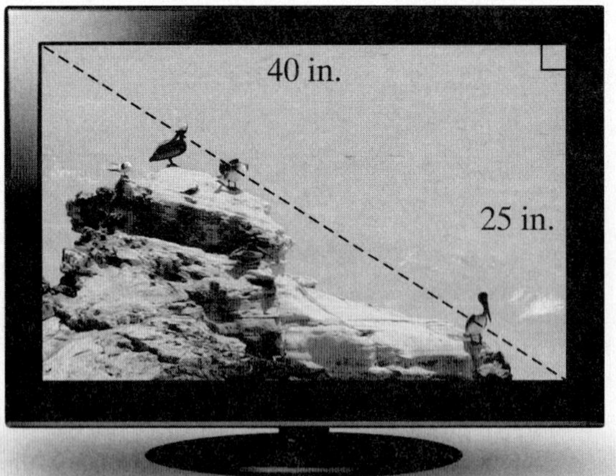

Solution Work through the following, or watch this **video** for the complete solution.

From the diagram, we can see that the length, width, and diagonal form a right triangle with the diagonal being the **hypotenuse** of the triangle. We can use the **Pythagorean Theorem** with $a = 40$, $b = 25$, and solve for the screen size, c.

Pythagorean Theorem: $a^2 + b^2 = c^2$

Substitute 40 for a and 25 for b: $(40)^2 + (25)^2 = c^2$

Evaluate 40^2 and 25^2: $1600 + 625 = c^2$

(*eText Screens 6.5-1–6.5-23*)

Add: $2225 = c^2$

Use the positive square root of 2225: $c = \sqrt{2225} \approx 47.2$

The screen size is approximately 47.2 inches.

You Try It **Work through this You Try It problem.**

Work Exercises 19–22 in this textbook or in the MyMathLab Study Plan.

6.5 Exercises

In Exercises 1–6, evaluate each square root.

You Try It

1. $\sqrt{49}$
2. $\sqrt{144}$
3. $\sqrt{0.25}$
4. $\sqrt{1.96}$
5. $\sqrt{\dfrac{9}{25}}$
6. $\sqrt{\dfrac{81}{16}}$

In Exercises 7–12, use a calculator to approximate each square root. Round to three decimal places. Check that the answer is reasonable.

You Try It

7. $\sqrt{78}$
8. $\sqrt{214}$
9. $\sqrt{\dfrac{75}{8}}$
10. $\sqrt{12\dfrac{3}{7}}$
11. $\sqrt{23.7}$
12. $\sqrt{62.9}$

In Exercises 13–15, use the Pythagorean Theorem to find the length of the missing side.

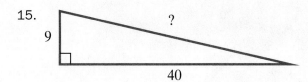

You Try It

13.

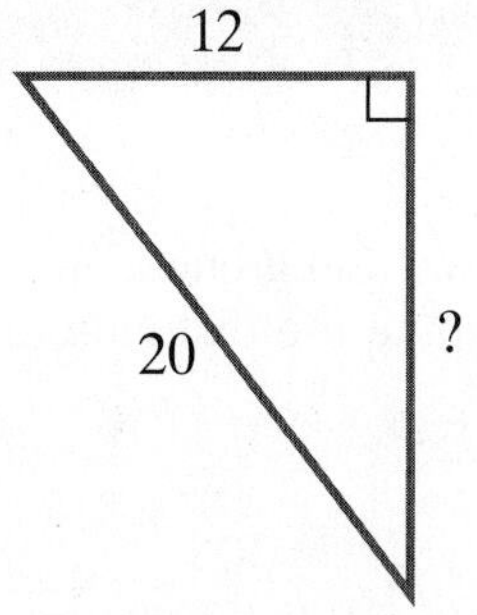

14.

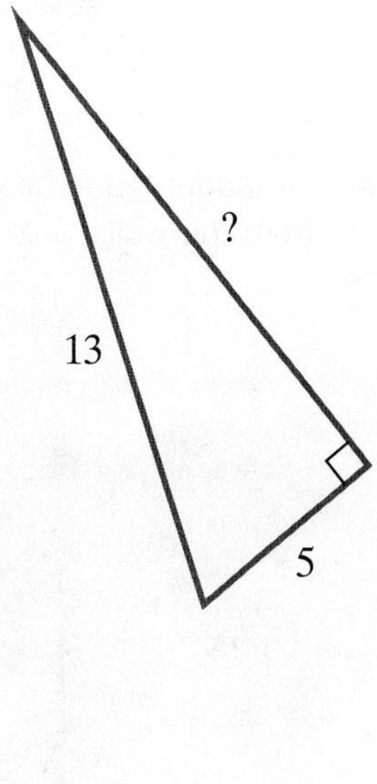

15.

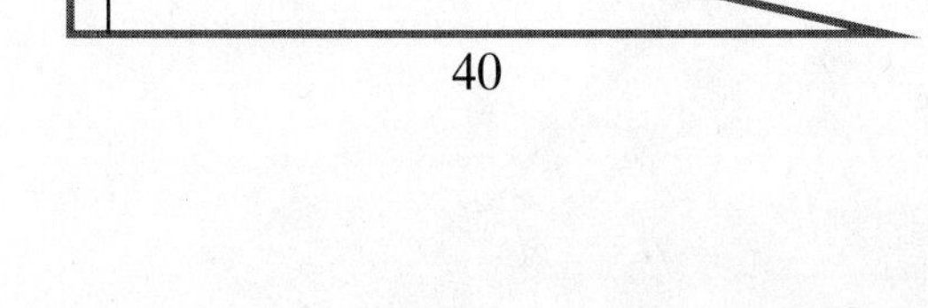

(eText Screens 6.5-1–6.5-23)

In Exercises 16–18, use the Pythagorean Theorem to find the exact length of the missing side. Then approximate the length to the nearest hundredth.

You Try It 16.

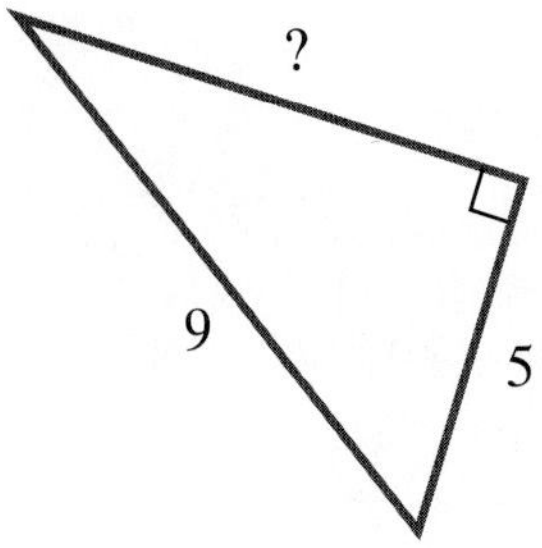

17.

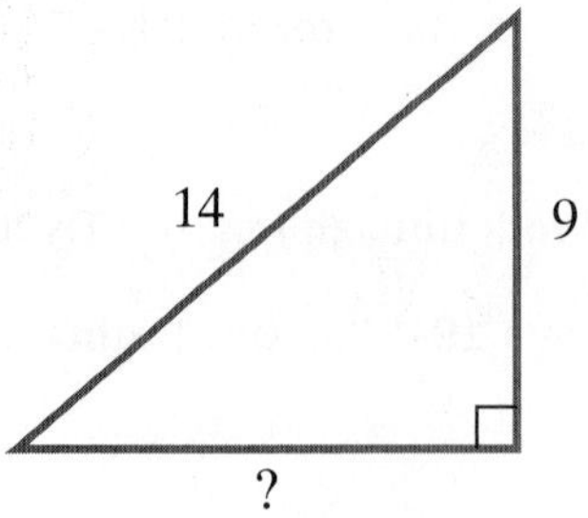

18.

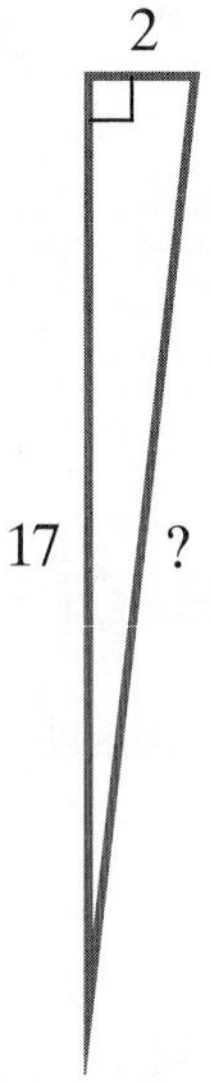

You Try It 19. **HD Television** A 54.6-inch television has a screen width of 28.1 inches. Approximate the screen length to the nearest tenth of an inch.

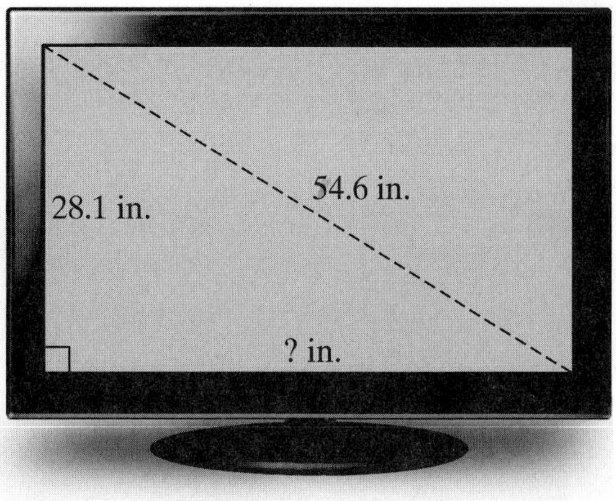

20. **Leaning Ladder** A 10-ft ladder leans against a wall at a point 7 feet above the ground. How far is the bottom of the ladder from the wall? Give an exact answer, then approximate the distance to the nearest hundredth of a foot.

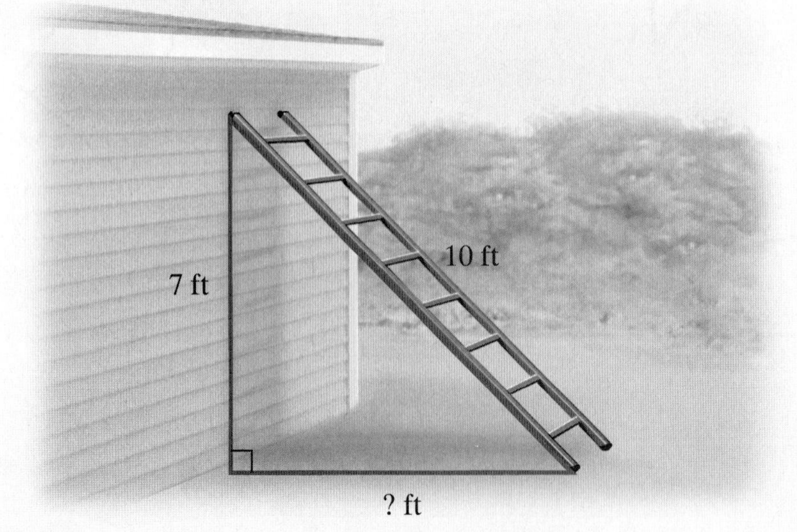

21. **Parasailing** At the beach, Zoey decides to go parasailing. If she is at a height of 150 feet and 530 feet behind the boat, how long is the anchor rope? Give an exact answer, then approximate the distance to the nearest foot.

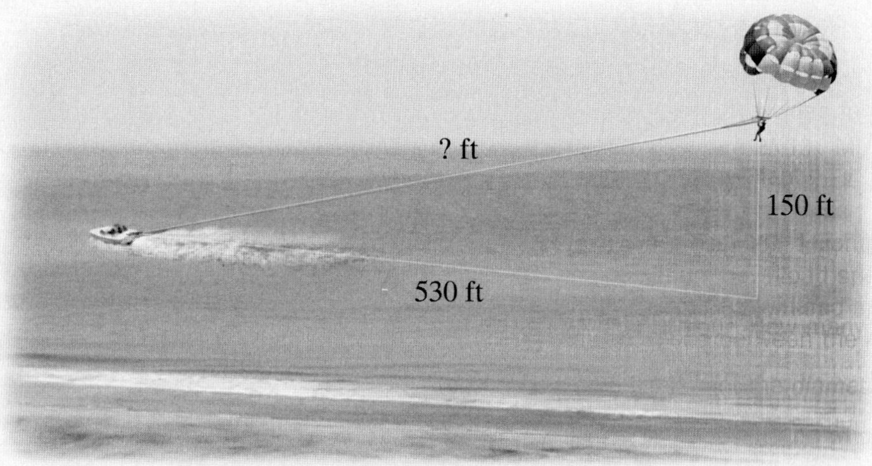

22. **Ramp Design** The Americans with Disabilities Act guidelines state that a wheelchair ramp cannot be more than 30 inches high. If a ramp has a horizontal projection of 750 inches and a height of 30 inches, how long is the surface of the ramp? Give the exact answer, then approximate the length to the nearest inch. (*Source:* www.ada.gov)

MODULE SEVEN

Percent

MODULE SEVEN CONTENTS

7.1 Percents, Decimals, and Fractions

THINGS TO KNOW

Before working through this topic, be sure you are familiar with the following concepts:

		VIDEO	ANIMATION	INTERACTIVE
You Try It	1. Write Fractions in Simplest Form (Topic 4.2, **Objective 6**)	▤		
You Try It	2. Change Decimals into Fractions or Mixed Numbers (Topic 5.1, **Objective 4**)	▤		
You Try It	3. Round Decimals to a Given Place Value (Topic 5.1, **Objective 7**)	▤		
You Try It	4. Multiply Decimals by Powers of 10 (Topic 5.3, **Objective 2**)	▤		
You Try It	5. Divide Decimals by Powers of 10 (Topic 5.4, **Objective 2**)	▤		
You Try It	6. Change Fractions or Mixed Numbers into Decimals (Topic 5.5, **Objective 1**)	▤		

OBJECTIVES

1 Write Percents as Decimals

2 Write Decimals as Percents

(eText Screens 7.1-1–7.1-25)

3 Write Percents as Fractions

4 Write Fractions as Percents

5 Perform Conversions Among Percents, Decimals, and Fractions

OBJECTIVE 1 WRITE PERCENTS AS DECIMALS

The word **percent** means "per hundred" or "out of 100." A percent is denoted by the **percent symbol** %. For example, if you received a score of 87% on a test, then you earned 87 points out of every 100 points possible on the test.

In the figure below, 87 out of 100 squares are shaded. Therefore, 87% of the squares are shaded.

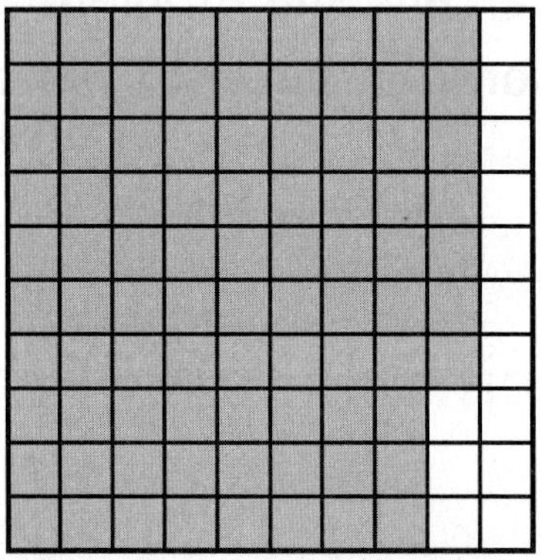

87 out of 100 squares are shaded, or 87% of the squares are shaded.

Because 87% and $\frac{87}{100}$ both mean "87 out of 100," we can say that $87\% = \frac{87}{100}$. Because $\frac{87}{100} = 0.87$ as well, we can say that $87\% = 0.87$. This illustrates that a percent can be written as an equivalent **fraction** or **decimal**. Changing among equivalent **percents**, fractions, and decimals is an important step in solving percent problems.

Because $100\% = 1$, multiplying or dividing by 100% is equivalent to multiplying or dividing by 1. This means that a value does not change if it is multiplied or divided by 100%. Let's look at how to write percents as decimals.

Writing a Percent as a Decimal

To write a percent as a decimal, divide the percent by 100% and write the answer in decimal form.

TIP Recall from **Topic 5.4** that when a number is divided by a **power of** 10 greater than 1, the result is that the **decimal point** in the number moves to the left the same number of places as there are zeros in the power of 10. So when a number is divided by 100, its decimal point moves 2 places to the left.

Example 1 Writing a Percent as a Decimal

Write 38.5% as a **decimal**.

Solution

Write the original percent: 38.5%

Divide by 100%: $= \frac{38.5\%}{100\%}$

Divide out the % symbol: $= \frac{38.5\%}{100\%} = \frac{38.5}{100}$

Divide 38.5 by 100: $= 0.385$

$\frac{38.5}{100} = 0.385$ — 2 zeros; Decimal point moves 2 places to the *left*

So, 0.385 is the equivalent decimal of 38.5%.

You Try It **Work through this You Try It problem.**

Work Exercise 1 in this textbook or in the MyMathLab Study Plan.

Dividing by 100% has two results:

1. The percent symbol (%) is removed.
2. The **decimal point** is moved two places to the left.

Knowing this can be helpful when changing from a **percent** to a decimal.

TIP If a **percent** is less than 100%, then its equivalent **decimal** will be less than 1. Similarly, if a percent is larger than 100%, then its equivalent decimal form will be larger than 1.

Example 2 Writing Percents as Decimals

My video summary

Write each percent as a decimal.

a. 63% **b.** 7.25% **c.** 248% **d.** 0.5%

Solutions To **write each percent as a decimal**, divide by 100%. Doing this removes the percent symbol and moves the decimal point two places to the left.

a. $63\% = \underbrace{\frac{63\%}{100\%}}_{\text{Divide by }100\%} = \frac{63}{100} = 0.63$

Percent symbol is removed; Decimal point moves 2 places to the *left*

So, 0.63 is the equivalent decimal of 63%.

b. $7.25\% = \underbrace{\frac{7.25\%}{100\%}}_{\text{Divide by }100\%} = \frac{7.25}{100} = 0.0725$

Percent symbol is removed; Decimal point moves 2 places to the *left*

So, 0.0725 is the equivalent decimal of 7.25%.

c.–d. Try finding these equivalent **decimals** on your own. View the **answers** to parts c and d, or watch this **video** for complete solutions to all four parts.

You Try It **Work through this You Try It problem.**

Work Exercises 2–8 in this textbook or in the MyMathLab Study Plan.

(eText Screens 7.1-1–7.1-25)

OBJECTIVE 2 WRITE DECIMALS AS PERCENTS

Sometimes, we need to write a **decimal** as a **percent**.

> **Writing a Decimal as a Percent**
>
> To write a decimal as a percent, multiply the decimal by 100%.

TIP When a number is multiplied by 100, the **decimal point** in the number moves 2 places to the right. See **Topic 5.3** for a review.

Example 3 Writing a Decimal as a Percent

Write 0.542 as a percent.

Solution

Write the original decimal:	0.542
Multiply by 100%:	$= 0.542(100\%)$
	$= (0.542 \cdot 100)\%$
Multiply 0.542 by 100:	$= 54.2\%$

$0.542 \cdot 100 = 54.2$

2 zeros — Decimal point moves 2 places to the *right*

So, 54.2% is the equivalent percent for 0.542.

You Try It **Work through this You Try It problem.**

Work Exercise 9 in this textbook or in the MyMathLab Study Plan.

Multiplying by 100% has two results:

1. The **decimal point** is moved two places to the right.
2. The percent symbol $(\%)$ is attached.

TIP If a **decimal** is less than 1, then its equivalent **percent** will be less than 100%. If a decimal is larger than 1, then its equivalent percent will be larger than 100%.

Example 4 Writing Decimals as Percents

My video summary

Write each decimal as a percent.

a. 0.48 **b.** 2.3 **c.** 0.0675 **d.** 0.009

Solutions To **write each decimal as a percent**, we multiply by 100%. This moves the **decimal point** two places to the right and attaches the percent symbol.

a. $0.48 = \underbrace{0.48(100\%)}_{\text{Multiply by }100\%} = 48.\%$

Percent symbol is attached

Decimal point moves 2 places to the *right*

So, 48% is the equivalent percent of 0.48.

Multiply by 100% — Percent symbol is attached

b. $2.3 = 2.3(100\%) = 230.\%$

Decimal point moves 2 places to the *right*

So, 230% is the equivalent percent of 2.3.

c.–d. Try finding these equivalent **percents** on your own. View the **answers** to parts c and d, or watch this **video** for complete solutions to all four parts.

You Try It **Work through this You Try It problem.**

Work Exercises 10–16 in this textbook or in the MyMathLab Study Plan.

CAUTION Be careful to move the **decimal point** in the correct direction when changing between decimals and percents. If changing from a percent to a decimal, then the decimal point moves to the *left*. If changing from a decimal to a percent, then the decimal point moves to the *right*.

OBJECTIVE 3 WRITE PERCENTS AS FRACTIONS

To write a **percent** as a **fraction** or **mixed number**, we use a process similar to that for **writing a percent as a decimal.** We divide by 100%, but we **simplify** the result as a fraction or mixed number instead of a decimal.

> **Writing a Percent as a Fraction or Mixed Number**
>
> To write a percent as a fraction or mixed number, divide the percent by 100% and write the answer as a simplified fraction or mixed number.

Example 5 Writing a Percent as a Fraction

Write 12% as a fraction.

Solution

Write the original percent: 12%

Divide by 100%: $= \dfrac{12\%}{100\%}$

Divide out the % symbol: $= \dfrac{12\not\%}{100\not\%} = \dfrac{12}{100}$

Simplify the fraction: $= \dfrac{12 \div 4}{100 \div 4} = \dfrac{3}{25}$

So, $\dfrac{3}{25}$ is the equivalent fraction of 12%.

You Try It **Work through this You Try It problem.**

Work Exercises 17–19 in this textbook or in the MyMathLab Study Plan.

(eText Screens 7.1-1–7.1-25)

Dividing by 100% leads to the following:

1. The percent symbol (%) is removed.
2. The number in the percent remains as the **numerator** of a **fraction** with 100 as the **denominator**.
3. The fraction is **simplified**.

TIP If a **percent** is less than 100%, then its equivalent fraction will be a **proper fraction**. If a percent is larger than 100%, then its equivalent fraction will be an **improper fraction** or **mixed number**.

Example 6 Writing Percents as Fractions or Mixed Numbers

My video summary

Write each percent as a simplified fraction or mixed number.

a. 8.6% **b.** 0.25% **c.** 320%

Solutions We write each percent over 100% and simplify.

a. Write the original percent: 8.6%

Divide by 100%: $= \dfrac{8.6\%}{100\%}$

Divide out the % symbol: $= \dfrac{8.6\%}{100\%} = \dfrac{8.6}{100}$

Recall that we can clear **decimals** from a **fraction** by multiplying the **numerator** and **denominator** by an appropriate **power of** 10. Here, we multiply the numerator and denominator by 10 because the largest number of decimal places in either the numerator or denominator is 1 and $10^1 = 10$.

Clear the decimals: $= \dfrac{8.6 \cdot 10}{100 \cdot 10} = \dfrac{86}{1000}$

Simplify the fraction: $= \dfrac{86 \div 2}{1000 \div 2} = \dfrac{43}{500}$

So, $\dfrac{43}{500}$ is the equivalent fraction of 8.6%.

b. $0.25\% = \overbrace{\dfrac{0.25\%}{100\%}}^{\text{Divide by } 100\%} = \dfrac{0.25}{100}$ Try to finish this problem on your own.

Remember to multiply the **numerator** and **denominator** by the appropriate **power of** 10 to remove the **decimal point** from the numerator. Then **simplify**. View the **answer** to part b, or watch this **video** for complete solutions to all three parts.

c. Try to work this problem on your own. View the **answer** to part c, or watch this **video** for complete solutions to all three parts.

You Try It **Work through this You Try It problem.**

Work Exercises 20–24 in this textbook or in the MyMathLab Study Plan.

(eText Screens 7.1-1–7.1-25)

Sometimes, **fractions** are contained within a **percent**. For example, consider 8.6% from **Example 6a**. Because $0.6 = \frac{6}{10} = \frac{3}{5}$, we can use the fraction in place of the decimal and write this percent as $8\frac{3}{5}\%$. So, $8.6\% = 8\frac{3}{5}\%$. Note that $8\frac{3}{5}\%$ is in percent form, not fraction form. In the solution to Example 6a, we saw $8.6\% = \frac{43}{500}$. This also means $8\frac{3}{5}\% = \frac{43}{500}$. In the next example, we change such percents to fractions.

Example 7 Writing Percents as Fractions or Mixed Numbers

My interactive video summary

Write each **percent** as a **simplified fraction** or **mixed number**.

a. $\frac{3}{8}\%$ **b.** $4\frac{1}{6}\%$ **c.** $\frac{20}{3}\%$

Solutions To find each fraction, we divide the percent by 100% and **simplify**.

Divide by 100% — Multiply by the reciprocal of 100

$$\text{a. } \frac{3}{8}\% = \frac{\frac{3}{8}\%}{100\%} = \frac{\frac{3}{8}}{100} = \frac{3}{8} \div 100 = \frac{3}{8}\cdot\frac{1}{100} = \frac{3\cdot 1}{8\cdot 100} = \frac{3}{800}$$

So, the equivalent fraction of $\frac{3}{8}\%$ is $\frac{3}{800}$.

Divide by 100%

$$\text{b. } 4\frac{1}{6}\% = \frac{4\frac{1}{6}\%}{100\%} = \frac{4\frac{1}{6}}{100} = 4\frac{1}{6} \div 100$$

Try to finish this problem on your own by performing the division. View the **answer** to part b, or watch this **interactive video** for complete solutions to all three parts.

c. Try working this problem on your own. View the **answer** to part c, or watch this **interactive video** for complete solutions to all three parts.

You Try It **Work through this You Try It problem.**

Work Exercises 25–30 in this textbook or in the MyMathLab Study Plan.

CAUTION

Do not confuse a percent, such as $6\frac{1}{4}\%$, as being a fraction or a mixed number. It is not a fraction, just as 6.25% is not a decimal. The percent symbol is very important and must not be overlooked.

OBJECTIVE 4 WRITE FRACTIONS AS PERCENTS

Now let's write **fractions** or **mixed numbers** as **percents**.

Writing a Fraction as a Percent

To write a fraction or mixed number as a percent, multiply it by 100% and **simplify**.

(eText Screens 7.1-1–7.1-25)

TIP The equivalent percent of a **proper fraction** will be less than 100%. The equivalent percent of an **improper fraction** or mixed number will be larger than 100%.

Example 8 Writing Fractions or Mixed Numbers as Percents

My interactive video summary

Write each fraction or mixed number as a percent.

a. $\frac{3}{4}$ **b.** $\frac{5}{9}$ **c.** $\frac{19}{16}$ **d.** $5\frac{3}{8}$

Solutions To write each fraction as a percent, we multiply by 100%.

a. $\frac{3}{4} = \overbrace{\frac{3}{4} \cdot 100\%}^{\text{Multiply by }100\%} = \left(\frac{3}{4} \cdot \frac{100}{1}\right)\% = \left(\frac{3 \cdot \cancel{100}^{25}}{{}_{1}\cancel{4} \cdot 1}\right)\% = 75\%$ ← Percent symbol is attached

So, the equivalent percent of $\frac{3}{4}$ is 75%.

b. $\frac{5}{9} = \overbrace{\frac{5}{9} \cdot 100\%}^{\text{Multiply by }100\%} = \left(\frac{5}{9} \cdot \frac{100}{1}\right)\% = \frac{500}{9}\% = 55\frac{5}{9}\% \text{ or } 55.\overline{5}\%$

So, the equivalent **percent** of $\frac{5}{9}$ is $55\frac{5}{9}\%$ or $55.\overline{5}\%$.

c.–d. Try finding these percents on your own. View the **answers** to parts c and d, or watch this **interactive video** for detailed solutions to all four parts.

Note: We can write the final percent by using either a fraction or a decimal.

You Try It **Work through this You Try It problem.**

Work Exercises 31–38 in this textbook or in the MyMathLab Study Plan.

Sometimes, we might want to approximate **percents** by **rounding** to a given **place value**.

Example 9 Writing Fractions as Approximate Percents

Write $\frac{7}{12}$ as a percent rounded to the nearest hundredth of a percent.

Solution

$$\frac{7}{12} = \overbrace{\frac{7}{12} \cdot 100\%}^{\text{Multiply by }100\%} = 58.\overline{3}\% \approx 58.33\%$$

↑ "is approximately equal to"

You Try It **Work through this You Try It problem.**

Work Exercises 39–42 in this textbook or in the MyMathLab Study Plan.

OBJECTIVE 5 PERFORM CONVERSIONS AMONG PERCENTS, DECIMALS, AND FRACTIONS

In **Module 5**, we learned how to **change decimals into fractions** and to **change fractions into decimals.** Adding these skills to those we have learned in this topic, we can perform conversions among **percents, decimals,** and **fractions.**

Example 10 Converting Among Percents, Decimals, and Fractions

My interactive video summary

Complete the chart with appropriate percents, decimals, and fractions.

	Percent	Decimal	Fraction
a.	28%		
b.			$\frac{17}{40}$
c.		0.875	
d.			$\frac{13}{12}$

Solutions

a. To change 28% to a decimal, divide it by 100% and write the answer in decimal form.

$$28\% = \overbrace{\frac{28\%}{100\%}}^{\text{Divide by }100\%} = \frac{28}{100} = 0.28$$

Percent symbol is removed; Decimal point moves 2 places to the *left*

To change 28% to a **fraction**, divide it by 100% and write the answer as a **simplified fraction.**

$$28\% = \overbrace{\frac{28\%}{100\%}}^{\text{Divide by }100\%} = \frac{28}{100} = \frac{28 \div 4}{100 \div 4} = \frac{7}{25}$$

(Simplify)

b. To write $\frac{17}{40}$ as a **percent**, multiply by 100%.

$$\frac{17}{40} = \overbrace{\frac{17}{40} \cdot 100\%}^{\text{Multiply by }100\%} = \left(\frac{17}{\cancel{40}_{2}} \cdot \frac{\cancel{100}^{5}}{1}\right)\% = \frac{85}{2}\% = 42\frac{1}{2}\% \text{ or } 42.5\%$$

To write $\frac{17}{40}$ as a **decimal**, divide 17 by 40.

$$\frac{17}{40} \Rightarrow 40\overline{)17.000}$$

```
     0.425
40)17.000
   16 0
    1 00
      80
     200
     200
       0 ← Terminates
```

(eText Screens 7.1-1–7.1-25)

c.–d. Try to complete the chart on your own. View the **answers**, or watch this **interactive video** for fully worked solutions to all four parts.

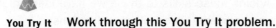

You Try It **Work through this You Try It problem.**

Work Exercises 43–48 in this textbook or in the MyMathLab Study Plan.

7.1 Exercises

In Exercises 1–8, write each percent as a decimal.

You Try It 1. 94.2% 2. 52% 3. 2% 4. 8.3%

You Try It 5. 235% 6. 400% 7. 0.4% 8. 19.25%

In Exercises 9–16, write each decimal as a percent.

You Try It 9. 0.735 10. 0.09 11. 0.064 12. 3

You Try It 13. 0.91 14. 0.0058 15. 1.24 16. 0.2

In Exercises 17–30, write the percent as a simplified fraction or mixed number.

You Try It 17. 13% 18. 52% 19. 65% 20. 0.6%

You Try It 21. 7.2% 22. 13.36% 23. 240% 24. 325%

You Try It 25. $\frac{3}{4}\%$ 26. $1\frac{3}{8}\%$ 27. $5\frac{2}{9}\%$ 28. $96\frac{4}{5}\%$

29. $\frac{16}{3}\%$ 30. $\frac{20}{17}\%$

In Exercises 31–38, write each fraction or mixed number as a percent. Write each percent using both fractions and decimals.

You Try It 31. $\frac{3}{5}$ 32. $\frac{1}{8}$ 33. $\frac{1}{15}$ 34. $\frac{13}{125}$

35. $\frac{53}{40}$ 36. $\frac{29}{24}$ 37. $7\frac{13}{16}$ 38. $8\frac{2}{3}$

In Exercises 39–42, write each fraction as a percent rounded to the nearest hundredth of a percent.

You Try It 39. $\frac{9}{11}$ 40. $\frac{7}{18}$ 41. $\frac{19}{7}$ 42. $\frac{1}{130}$

(eText Screens 7.2-1–7.2-13)

In Exercises 43–48, perform conversions between percents, decimals, and fractions to complete the chart. Write any repeating decimals with as few digits as possible.

	Percent	Decimal	Fraction
You Try It 43.	2.5%		
44.			$\frac{17}{125}$
45.		0.48	
46.	180%		
47.			$\frac{22}{15}$
48.		0.054	

7.2 Solving Percent Problems with Equations

THINGS TO KNOW

Before working through this topic, be sure you are familiar with the following concepts:

		VIDEO	ANIMATION	INTERACTIVE
You Try It	1. Solve Equations Using the Multiplication Property of Equality (Topic 3.2, **Objective 2**)	✓	✓	
You Try It	2. Translate Word Statements into Equations (Topic 3.4, **Objective 1**)	✓		
You Try It	3. Use the Properties of Equality to Solve Linear Equations Involving Decimals (Topic 5.6, **Objective 2**)	✓		✓
You Try It	4. Write Percents as Decimals (Topic 7.1, **Objective 1**)	✓		

OBJECTIVES

1 Translate Word Statements into Percent Equations

2 Solve Percent Equations

OBJECTIVE 1 TRANSLATE WORD STATEMENTS INTO PERCENT EQUATIONS

We have already seen how recognizing key words in a statement is helpful when translating statements to equations. In this topic, we focus primarily on the key word *of*, which translates to **multiplication** (·), and the key word *is*, which translates to an equal sign (=). In addition, recall that we use **variables**, such as x, to represent unknown quantities. Sometimes, the key word *what* will be used to identify the unknown quantity, such as *what number* or *what percent*.

We can use this information to solve percent problems by first translating the problem into a **percent equation**, which appears in its general form below.

Percent Equation

$$\text{Percent} \cdot \text{Base} = \text{Amount}$$

When we translate **percent** problems into **equations**, we can use the following clues to help identify the different quantities in the **percent equation**.

Quantity	Clue
Percent	Followed by the word *percent* or by a % symbol
Base	Represents one whole and typically follows the key word *of*
Amount	The part compared to the whole, normally isolated from the **percent** and the base on the other side of the word *is* (or other key word meaning ***equals***)

For example, the sentence "seventy percent of fifty is thirty-five" translates into $70\% \cdot 50 = 35$.

70% of 50 is 35

$$70\% \cdot 50 = 35$$

Percent (70%), Base (50), Amount (35)

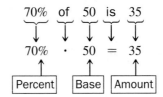

In the sentence, 70% is the *percent*, 50 is the *base*, and 35 is the *amount*.

Example 1 Writing Percent Problems as Equations

Translate each word statement into a **percent equation**. Do not solve.

a. 20% of 190 is what number?

b. What percent of 16 is 40?

c. 65% of what number is 79.3?

Solutions

a. Translate into an equation: 20% of 190 is what number ← The amount is unknown

$$20\% \cdot 190 = x$$

Percent (20%), Base (190), Amount (x)

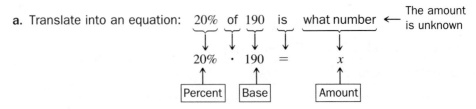

b. Translate into an equation: What percent of 16 is 40 ← The percent is unknown

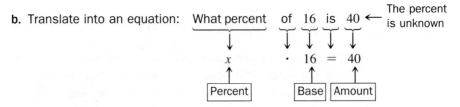

$$x \cdot 16 = 40$$

Percent (x), Base (16), Amount (40)

c. Translate into an equation: 65% of what number is 79.3 ← The base is unknown

$$\underbrace{65\%}_{\text{Percent}} \cdot \underbrace{x}_{\text{Base}} = \underbrace{79.3}_{\text{Amount}}$$

You Try It **Work through this You Try It problem.**

Work Exercises 1–6 in this textbook or in the MyMathLab Study Plan.

Keep in mind that the side of the **percent equation** where the amount is located does not matter. That is,

$$\text{Percent} \cdot \text{Base} = \text{Amount} \quad \text{and} \quad \text{Amount} = \text{Percent} \cdot \text{Base}$$

are equivalent forms of the percent equation. In either case, we use the key words in the problem statement to identify the different parts of the percent equation.

Example 2 Writing Percent Problems as Equations

My interactive video summary

Translate each word statement into a **percent equation**. Do not solve.

a. 105 is 140% of what number?

b. 18 is what percent of 34?

c. What number is $7\frac{1}{2}\%$ of 46?

Solutions

a. Translate into an equation: 105 is 140% of what number ← The base is unknown

$$\underbrace{105}_{\text{Amount}} = \underbrace{140\%}_{\text{Percent}} \cdot \underbrace{x}_{\text{Base}}$$

b.–c. Try translating these statements on your own. Check your **answers**, or watch this **interactive video** for the complete solutions to all three parts.

You Try It **Work through this You Try It problem.**

Work Exercises 7–12 in this textbook or in the MyMathLab Study Plan.

OBJECTIVE 2 SOLVE PERCENT EQUATIONS

Once a percent problem has been translated into a **percent equation**, we can solve the equation for an unknown quantity by using the **multiplication property of equality**. The percent equation contains three quantities (**percent**, **base**, and **amount**), so given any two of the quantities, we can solve the equation for the third.

CAUTION When using the **percent equation**, the percent must be **written as a decimal** or **fraction** in order to solve the problem. Typically, we will use decimal form.

Example 3 Solving a Percent Equation for the Amount

a. What number is 60% of 55?

b. $3\frac{1}{2}\%$ of 90 is what number?

Solutions Using the percent equation, we translate each sentence into an equation and solve.

a. Translate into an equation: What number is 60% of 55

$$\underbrace{\text{What number}}_{x}\ \underbrace{\text{is}}_{=}\ \underbrace{60\%}_{60\%}\ \underbrace{\text{of}}_{\cdot}\ \underbrace{55}_{55}$$

Convert 60% to a decimal: $x = 0.60(55)$

Multiply 0.60×55: $x = 33$

$$\begin{array}{r} 55 \\ \times\ 0.6 \\ \hline 33.0 \end{array}$$

So, 33 is 60% of 55.

b. Translate into an equation: $3\frac{1}{2}\%$ of 90 is what number

$$\underbrace{3\frac{1}{2}\%}_{3\frac{1}{2}\%}\ \underbrace{\text{of}}_{\cdot}\ \underbrace{90}_{90}\ \underbrace{\text{is}}_{=}\ \underbrace{\text{what number}}_{x}$$

Convert $3\frac{1}{2}\%$ to a decimal: $0.035(90) = x$ ← $3\frac{1}{2}\% = 3.5\% = 0.035$

Multiply 0.035×90: $3.15 = x$

$$\begin{array}{r} \overset{4}{0.035} \\ \times\ \ 90 \\ \hline 3.150 \end{array}$$

So, 3.15 is $3\frac{1}{2}\%$ of 90.

You Try It **Work through this You Try It problem.**

Work Exercises 13–16 in this textbook or in the MyMathLab Study Plan.

CAUTION When solving the **percent equation** for the percent, the solution will be in **decimal** form. Remember to convert the decimal to a percent in your final answer.

Example 4 Solving a Percent Equation for the Percent

a. 85 is what percent of 50?

b. What percent of 140 is 49?

Solutions Using the percent equation, we translate each sentence into an equation and solve.

a. Translate into an equation: 85 is what percent of 50

$$\underbrace{85}_{85}\ \underbrace{\text{is}}_{=}\ \underbrace{\text{what percent}}_{x}\ \underbrace{\text{of}}_{\cdot}\ \underbrace{50}_{50}$$

Divide both sides by 50: $\frac{85}{50} = \frac{x \cdot 50}{50}$

Simplify: $1.7 = x$

$$\begin{array}{r} 1.7 \\ 50\overline{)85.0} \\ -50 \\ \hline 35\ 0 \\ -35\ 0 \\ \hline 0 \end{array}$$

Change to % form: $170\% = x$ ← $1.7 \times 100\% = 170\%$

So, 85 is 170% of 50.

(eText Screens 7.2-1–7.2-13)

My video summary

b. Translate into an equation: What percent of 140 is 49

$$x \cdot 140 = 49$$

Try to finish the problem on your own. View the **answer**, or watch this **video** for a fully worked solution to part b.

You Try It **Work through this You Try It problem.**

Work Exercises 17–20 in this textbook or in the MyMathLab Study Plan.

In our next example, we solve a percent equation for the base.

Example 5 Solving a Percent Equation for the Base

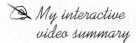

My interactive video summary

a. 42% of what number is 63?

b. 14.5 is 20% of what number?

Solutions

a. Translate into an equation: 42% of what number is 63

$$42\% \cdot x = 63$$

Try finishing this problem on your own. View the **answer**, or watch this **interactive video** for fully worked solutions to both parts.

b. Try to solve this problem on your own. View the **answer**, or watch this **interactive video** for fully worked solutions to both parts.

You Try It **Work through this You Try It problem.**

Work Exercises 21–24 in this textbook or in the MyMathLab Study Plan.

7.2 Exercises

In Exercises 1–12, translate each word statement into a percent equation. Do not solve.

You Try It

1. 40% of what number is 25?
2. What percent of 26 is 52?
3. 22.5% of 160 is what number?
4. What percent of 65 is 32?
5. 12% of what number is 17?
6. 18% of 18 is what number?

You Try It

7. What number is 15% of 36?
8. 17 is what percent of 68?
9. 22 is 130% of what number?
10. 112 is what percent of 72?
11. 11% of 94 is what number?
12. 40 is $10\frac{1}{2}\%$ of what number?

(eText Screens 7.3-1–7.3-15)

In Exercises 13–24, use an equation to solve each percent problem.

You Try It 13. 60% of 35 is what number?

14. What number is 17% of 200?

15. 115% of 12 is what number?

16. 14.5% of 40 is what number?

You Try It 17. What percent of 24 is 36?

18. 27 is what percent of 150?

19. What percent of 32 is 6.4?

20. 35 is what percent of 75?

You Try It 21. 22 is 40% of what number?

22. 121% of what number is 121?

23. 17% of what number is 5.1?

24. 23.1 is 42% of what number?

7.3 Solving Percent Problems with Proportions

THINGS TO KNOW

Before working through this topic, be sure you are familiar with the following concepts:

		VIDEO	ANIMATION	INTERACTIVE
You Try It	1. Solve Equations Using the Multiplication Property of Equality (Topic 3.2, **Objective 2**)	✓	✓	
You Try It	2. Translate Word Statements into Equations (Topic 3.4, **Objective 1**)	✓		
You Try It	3. Use the Properties of Equality to Solve Linear Equations Involving Decimals (Topic 5.6, **Objective 2**)	✓		✓
You Try It	4. Solve Proportions (Topic 6.2, **Objective 3**)	✓		
You Try It	5. Write Percents as Fractions (Topic 7.1, **Objective 3**)	✓		
You Try It	6. Translate Word Statements into Percent Equations (Topic 7.2, **Objective 1**)	✓		✓
You Try It	7. Solve Percent Equations (Topic 7.2, **Objective 2**)	✓		✓

OBJECTIVES

1 Write Percent Problems as Proportions

2 Solve Percent Problems Using Proportions

(eText Screens 7.3-1–7.3-15)

OBJECTIVE 1 WRITE PERCENT PROBLEMS AS PROPORTIONS

In Topic 7.2, we solved percent problems by using the **percent equation.** In this topic, we solve percent problems using **proportions.** Recall that a proportion is an equation stating that two ratios are equal to each other. Here, our proportion will involve the ratio of an **amount** to a **base** and an equivalent number of parts out of 100.

Consider the following percent equation:

$$\underbrace{9}_{\text{Amount}} = \underbrace{25\%}_{\text{Percent}} \cdot \underbrace{36}_{\text{Base}}$$

Dividing both sides of the equation by the **base** gives

$$\frac{9}{36} = 25\%,$$

which can be rewritten as

$$\frac{9}{36} = \frac{25}{100}. \leftarrow \boxed{\text{Change percent to fraction form}}$$

This illustrates the general form of our **percent proportion**.

Percent Proportion

$$\begin{array}{r} \text{The part compared to the whole} \rightarrow \\ \text{The number after } \textit{of} \rightarrow \end{array} \frac{\text{Amount}}{\text{Base}} = \frac{\text{Percent Number}}{100} \begin{array}{l} \leftarrow \text{The number before \%} \\ \leftarrow \text{Always } 100 \end{array}$$

To translate a percent problem into a **percent proportion**, we first need to identify the **amount**, the **base**, and the **percent number**. In a percent problem, we will be given two of these quantities and must use a variable to represent the unknown quantity. Typically, we will use the letter a to represent the *amount,* the letter b to represent the *base,* and the letter p to represent the *percent number*. This gives us the following alternate form of the percent proportion:

$$\begin{array}{r} \text{Amount} \rightarrow \\ \text{Base} \rightarrow \end{array} \frac{a}{b} = \frac{p}{100} \begin{array}{l} \leftarrow \text{Percent number} \\ \leftarrow \text{Always } 100 \end{array}$$

Example 1 Writing Percent Problems as Proportions

Translate each word statement into a percent proportion. Do not solve.

a. 12% of 75 is what number?

b. What percent of 50 is 18?

c. 30% of what number is 62.5?

Solutions

a. We start by identifying the **amount**, the **base**, and the **percent number**.

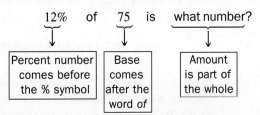

12% of 75 is what number?

- 12% → Percent number comes before the % symbol
- 75 → Base comes after the word *of*
- what number? → Amount is part of the whole

(eText Screens 7.3-1–7.3-15)

Next, we put in the known quantities in our **percent proportion** and a **variable** for the unknown.

$$\begin{matrix}\text{Amount} \rightarrow \\ \text{Base} \rightarrow\end{matrix} \frac{a}{75} = \frac{12}{100} \begin{matrix}\leftarrow \text{Percent number} \\ \leftarrow \text{Always } 100\end{matrix}$$

b. Identify the different quantities in the **percent** problem.

What percent of 50 is 18?

- What percent → Percent number
- 50 → Base
- 18 → Amount

Translate into a **percent proportion.**

$$\begin{matrix}\text{Amount} \rightarrow \\ \text{Base} \rightarrow\end{matrix} \frac{18}{50} = \frac{p}{100} \begin{matrix}\leftarrow \text{Percent number} \\ \leftarrow \text{Always } 100\end{matrix}$$

My video summary

c. Try writing this percent proportion on your own. Check your **answer**, or watch this **video** for the complete solution.

You Try It **Work through this You Try It problem.**

Work Exercises 1–6 in this textbook or in the MyMathLab Study Plan.

Example 2 Writing Percent Problems as Proportions

My interactive video summary

Translate each word statement into a **percent proportion.** Do not solve.

a. What number is 9.5% of 77?

b. 230 is 175% of what number?

c. 5 is what percent of 14?

Solutions

a. Identify the different quantities in the percent problem.

What number is 9.5% of 77?

- What number → Amount
- 9.5% → Percent Number
- 77 → Base

Translate into a percent proportion.

$$\begin{matrix}\text{Amount} \rightarrow \\ \text{Base} \rightarrow\end{matrix} \frac{a}{77} = \frac{9.5}{100} \begin{matrix}\leftarrow \text{Percent number} \\ \leftarrow \text{Always } 100\end{matrix}$$

b.–c. Try translating these statements on your own. Check your **answers,** or watch this **interactive video** for the complete solutions to all three parts.

You Try It **Work through this You Try It problem.**

Work Exercises 7–12 in this textbook or in the MyMathLab Study Plan.

OBJECTIVE 2 SOLVE PERCENT PROBLEMS USING PROPORTIONS

As with the **percent equation**, the **percent proportion** contains three quantities (**percent number, base, and amount**), so given any two of the quantities, we can solve the equation for the third. We solve these proportions the same way we did in **Topic 6.2**—by finding the

cross products and then solving for the variable. Let's start with solving a percent proportion for the amount.

Example 3 Solving a Percent Proportion for the Amount

a. What number is 45% of 80?

b. 7.5% of 140 is what number?

Solutions We translate each sentence into a percent proportion and solve.

a. Identify the different quantities in the **percent** problem.

What number is 45% of 80?

What number → Amount; 45% → Percent Number; 80 → Base

Translate into a percent proportion.

$$\text{Amount} \rightarrow \frac{a}{80} = \frac{45}{100} \begin{array}{l}\leftarrow \text{Percent number} \\ \leftarrow \text{Always } 100\end{array} \quad (\text{Base} \rightarrow 80)$$

We next find the **cross products** and set them equal to each other.

Set cross products equal: $100 \cdot a = 80 \cdot 45$

Simplify both sides: $100a = 3600$

Divide both sides by 100: $\dfrac{100a}{100} = \dfrac{3600}{100}$

Simplify: $a = 36$

So, 45% of 80 is 36.

TIP Recall that when solving **proportions**, we can simplify the **ratios** before finding the cross products. View this **alternate solution** to Example 3a.

b. Identify the different quantities in the percent problem.

7.5% of 140 is what number?

7.5% → Percent Number; 140 → Base; what number → Amount

Translate into a **percent proportion**.

$$\text{Amount} \rightarrow \frac{a}{140} = \frac{7.5}{100} \begin{array}{l}\leftarrow \text{Percent number} \\ \leftarrow \text{Always } 100\end{array} \quad (\text{Base} \rightarrow 140)$$

Solve the **proportion**.

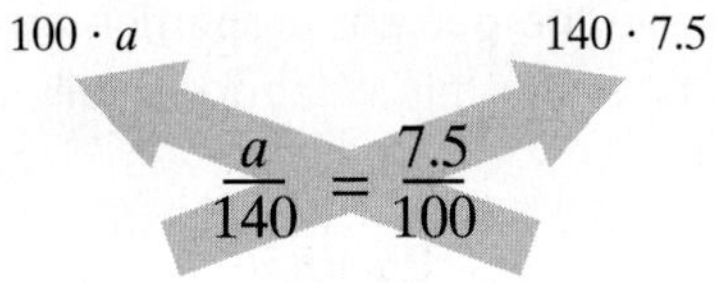

Set cross products equal: $100 \cdot a = 140 \cdot 7.5$

Simplify both sides: $100a = 1050$

Divide both sides by 100: $\dfrac{100a}{100} = \dfrac{1050}{100}$

Simplify: $a = 10.5$

So, 7.5% of 140 is 10.5.

(eText Screens 7.3-1–7.3-15)

You Try It **Work through this You Try It problem.**

Work Exercises 13–16 in this textbook or in the MyMathLab Study Plan.

In the next example, we solve percent proportions for the percent.

Example 4 Solving a Percent Proportion for the Percent

a. 90 is what percent of 200?

b. What percent of 72 is 93.6?

Solutions

a. Identify the different quantities in the percent problem.

90 is what percent of 200?

90 → Amount; what percent → Percent Number; 200 → Base

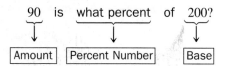

Translate into a percent proportion.

$$\frac{90}{200} = \frac{p}{100}$$

Amount → 90; Base → 200; p ← Percent number; 100 ← Always 100

Solve the **proportion**.

$100 \cdot 90$ $\quad$ $200 \cdot p$

$$\frac{90}{200} = \frac{p}{100}$$

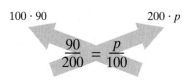

Set cross products equal: $100 \cdot 90 = 200 \cdot p$

Simplify both sides: $9000 = 200p$

Divide both sides by 200: $\frac{9000}{200} = \frac{200p}{200}$

Simplify: $45 = p$

So, 90 is 45% of 200.

CAUTION Remember that the **percent number** is already in percent form except for the % symbol. When solving a **percent proportion** for the **percent**, keep the percent number as is and add the % symbol.

My video summary

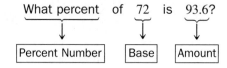

b. Identify the different quantities in the **percent** problem.

What percent of 72 is 93.6?

What percent → Percent Number; 72 → Base; 93.6 → Amount

Try to write and solve the **percent proportion** on your own to finish the problem. View the **answer**, or watch this **video** for a fully worked solution to part b.

You Try It **Work through this You Try It problem.**

Work Exercises 17–20 in this textbook or in the MyMathLab Study Plan.

In Example 5, we solve percent proportions for the base.

Example 5 Solving a Percent Proportion for the Base

My interactive video summary

a. 52% of what number is 117?

b. 57.5 is 125% of what number?

Solutions

a. Identify the different quantities in the **percent** problem.

52% (Percent Number) of what number (Base) is 117? (Amount)

Translate into a **percent proportion.**

$$\begin{array}{r} \text{Amount} \rightarrow \\ \text{Base} \rightarrow \end{array} \frac{117}{b} = \frac{52}{100} \begin{array}{l} \leftarrow \text{Percent number} \\ \leftarrow \text{Always } 100 \end{array}$$

Try to finish working this problem on your own. View the **answer**, or watch this **interactive video** for fully worked solutions to both parts.

b. Try to write and solve a **percent proportion** for this problem on your own. View the **answer**, or watch this **interactive video** for fully worked solutions to both parts.

You Try It **Work through this You Try It problem.**

Work Exercises 21–24 in this textbook or in the MyMathLab Study Plan.

7.3 Exercises

In Exercises 1–12, translate each word statement into a percent proportion. Do not solve.

You Try It

1. 30 is 22% of what number?
2. 27% of 18 is what number?
3. What number is 19% of 130?
4. 59% of 42 is what number?
5. 17.5 is what percent of 80?
6. 14 is 120% of what number?

You Try It

7. 16.5% of 110 is what number?
8. What percent of 34 is 78?
9. 42% of what number is 18?
10. What percent of 72 is 48?
11. 18.2 is what percent of 72.6?
12. 152% of what number is 16?

In Exercises 13–24, use a percent proportion to solve the percent problem.

You Try It

13. What number is 25% of 42?
14. 12% of 75 is what number?
15. What number is $22\frac{1}{2}\%$ of 80?
16. 16% of 45 is what number?

You Try It

17. 36 is what percent of 60?
18. 6.9 is what percent of 46?

19. What percent of 50 is 80?

20. What percent of 64 is 24?

You Try It 21. 16 is 20% of what number?

22. 12% of what number is 5.4?

23. 110% of what number is 44?

24. $8\frac{1}{4}\%$ of what number is 3.63?

7.4 Applications of Percent

THINGS TO KNOW

Before working through this topic, be sure you are familiar with the following concepts:

VIDEO ANIMATION INTERACTIVE

You Try It 1. Solve Applications Using Linear Equations (Topic 3.4, **Objective 4**)

You Try It 2. Write Percents as Decimals (Topic 7.1, **Objective 1**)

You Try It 3. Write Percents as Fractions (Topic 7.1, **Objective 3**)

You Try It 4. Translate Word Statements into Percent Equations (Topic 7.2, **Objective 1**)

You Try It 5. Solve Percent Equations (Topic 7.2, **Objective 2**)

You Try It 6. Write Percent Problems as Proportions (Topic 7.3, **Objective 1**)

You Try It 7. Solve Percent Problems Using Proportions (Topic 7.3, **Objective 2**)

OBJECTIVES

1 Solve Applications Involving Percents

2 Solve Applications Involving a Percent Increase

3 Solve Applications Involving a Percent Decrease

OBJECTIVE 1 SOLVE APPLICATIONS INVOLVING PERCENTS

Percents appear quite often in real-world situations. Using the problem-solving strategies for applications of linear equations and proportions, we can apply these steps when solving applications involving **percent equations** or **percent proportions**.

Example 1 Healthy Diet

As part of a healthy diet, Amanda keeps her fat consumption to 20% of her daily calories. If she consumes 1875 calories in a day, how many of her calories are from fat?

(eText Screens 7.4-1–7.4-21)

Solution We will use a percent equation to solve this problem. Follow the **problem-solving strategy** for applications of linear equations.

Step 1. We know Amanda consumes 1875 calories in a day and that 20% of her calories are from fat. We want to find the daily amount of calories from fat. So, in words, we have the following percent problem:

20% of 1875 calories is the amount of calories from fat.

Step 2. Let $a =$ the amount of calories from fat.

Step 3. Translate the percent problem using the **percent equation**.

20%	of	1875	is	the amount of calories from fat ← The amount is unknown
↓	↓	↓	↓	↓
20%	·	1875	=	a
↑		↑		↑
Percent		Base		Amount

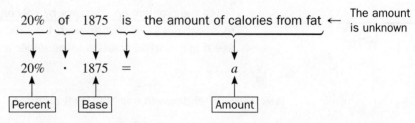

Step 4. Write the equation: $0.20(1875) = a \leftarrow 20\% = 0.20$

Multiply: $375 = a$

Step 5. $20\% = \frac{1}{5}$ and $1875 \approx 2000$. **Estimating**, we get $\frac{1}{5}(2000) = 400$, which is close to 375. This means that our result is reasonable.

Step 6. 375 of Amanda's calories are from fat.

You Try It **Work through this You Try It problem.**

Work Exercises 1–5 in this textbook or in the MyMathLab Study Plan.

My video summary

TIP We solved Example 1 by using a **percent equation.** However, we could also have used a **percent proportion.** Watch this **video** to see this alternate solution.

Example 2 Households with Televisions

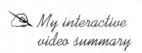

My interactive video summary

Approximately 54 million U.S. households with a television tuned in to the 2012 Super Bowl between the New York Giants and the New England Patriots. If this was 47% of the total number of U.S. households with televisions, how many U.S. households have a television? Round to the nearest million. (*Source:* The Nielsen Company)

Solution Follow the **problem-solving strategy** for applications of linear equations.

Step 1. We know that 54 million U.S. households watched the 2012 Super Bowl and that this was 47% of the total number of U.S. households with televisions. We want to find the total number of U.S. households with a television. So, in words, we have the following percent problem:

47% of the total number of U.S. households with televisions is 54 million households.

Step 2. Let $h =$ the total number of U.S. households with televisions.

Step 3. Translate the percent problem using the **percent equation**.

47%	of	the total number of households with televisions	is	54 million households	← The base is unknown
↓	↓	↓	↓	↓	
47%	$\cdot$	h	$=$	54	
↑		↑		↑	
Percent		Base		Amount	

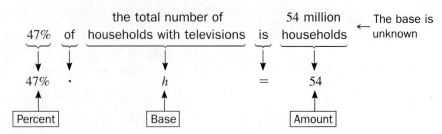

Step 4. Write the equation: $0.47h = 54 \leftarrow 47\% = 0.47$

Finish the problem on your own by solving the equation for h. Be sure to check the reasonableness of your answer. View the **answer**, or watch this **interactive video** to see a fully worked solution using either a percent equation or **percent proportion**.

You Try It **Work through this You Try It problem.**

Work Exercises 6–10 in this textbook or in the MyMathLab Study Plan.

Example 3 College Expenses

My interactive video summary

The estimated total cost of attendance for an in-state freshman at Arizona State University was $25,600 for the 2011–2012 academic year. Of this, $9728 was for tuition and fees. What percent of the total cost of attendance was for tuition and fees? (*Source:* https://students.asu.edu)

Solution

Step 1. We know that the total cost of attendance was $25,600 and of that, $9728 was for tuition and fees. We want to determine what **percent** of the total cost is tuition and fees. So, in words, we have the following percent problem:

$9728 *is what percent of* $25,600.

Step 2. Let $p =$ the percent of total cost that is tuition and fees.

Step 3. Translate the percent problem using the **percent equation**.

9728	is	what percent	of	25,600	← The percent is unknown
↓	↓	↓	↓	↓	
9728	$=$	p	$\cdot$	$25{,}600$	
↑		↑		↑	
Amount		Percent		Base	

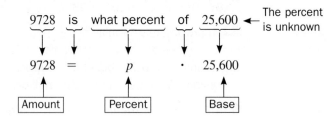

Step 4. Write the equation: $9728 = 25{,}600p$

Finish the problem on your own by solving the equation for p. Be sure to check the reasonableness of your answer. The answer will be in decimal form, so remember to **change the decimal to a percent** when writing your result. View the **answer**, or watch this **interactive video** to see a fully worked solution using either a percent equation or **percent proportion**.

You Try It **Work through this You Try It problem.**

Work Exercises 11–15 in this textbook or in the MyMathLab Study Plan.

(eText Screens 7.4-1–7.4-21)

OBJECTIVE 2 SOLVE APPLICATIONS INVOLVING A PERCENT INCREASE

In many applications, we are interested in showing how much an amount has increased or decreased. We can use **percents** to give a relative size to the change by considering the *percent increase* or *percent decrease* as compared to some original amount. We look at these two types of applications next.

Finding a percent increase is really just an application of our general **percent equation.** Consider the following:

$$\boxed{\text{Percent}} \cdot \boxed{\text{Base}} = \boxed{\text{Amount}}$$

$$\downarrow \qquad \downarrow \qquad \downarrow$$

$$\text{Percent increase} \cdot \text{Original value} = \text{Amount of increase}$$

We can solve this equation for the percent increase by dividing both sides of the equation by the original value. This gives

$$\text{Percent increase} = \frac{\text{Amount of increase}}{\text{Original value}}.$$

So, a **percent increase** is the amount of increase expressed as a percentage of the original value. The **amount of increase** can be found by subtracting the original value from the new value.

$$\text{Amount of increase} = \text{New value} - \text{Original value}$$

TIP One way to remember this is to consider that the new value will always be larger than the original value in a percent increase problem. New value $>$ Original value.

For example, in the fall of 2011, Netflix announced that it would raise the price of its bundle plan (streaming Internet and DVDs) from \$10 per month to \$16 per month. Notice that the new value is greater than the original value. The **amount of increase** is found by subtracting the original price, \$10, from the new price, \$16. (*Source:* nytimes.com)

$$\text{Amount of increase} = \$16 - \$10 = \$6$$

The **percent increase** is then found by dividing the amount of increase by the original price.

$$\text{Percent increase} = \frac{\text{Amount of increase}}{\text{Original price}} = \frac{\$6}{\$10} = \frac{\cancel{\$}6}{\cancel{\$}10} = 0.6$$

↑ Divide out the common units

Because the percent increase is written in percent form, we need to change this answer from a **decimal to a percent.**

$$0.6 = 0.6 \cdot 100\% = 60\%$$

↑ Multiply by 100%

Therefore, the bundle price change by Netflix is a 60% increase.

Example 4 Cost of Commercials

My video summary

The cost of a 30-second commercial during the 2012 Super Bowl was \$3.5 million, up from \$3.0 million in 2011. Find the percent increase. Round to the nearest tenth of a percent. (*Source:* forbes.com)

(eText Screens 7.4-1–7.4-21)

Solution Work through the following, or watch this **video** for the complete solution. We start by finding the **amount of increase**. Subtract the original price (the price in 2011) from the new price (the price in 2012).

$$\text{Amount of increase} = \$3.5 \text{ million} - \$3.0 \text{ million} = \$0.5 \text{ million}$$

We then find the **percent increase** by dividing the amount of increase by the original price (the price in 2011).

$$\text{Percent increase} = \frac{\text{Amount of increase}}{\text{Original price}} = \frac{\$0.5 \text{ million}}{\$3.0 \text{ million}} = \underbrace{\frac{\$0.5 \cancel{\text{ million}}}{\$3.0 \cancel{\text{ million}}}}_{\text{Divide out the common units}} = \frac{0.5}{3.0} = 0.16666\ldots$$

CAUTION Make sure to use the *original* value when finding the percent increase, not the new value.

Multiply the result by 100% and round to one **decimal place**.

$$0.16666\ldots = 16.666\ldots\% \approx 16.7\%$$

Move the decimal point 2 places to the right; Add the % symbol; Round to the nearest tenth of a percent

So, the price of a 30-second commercial during the Super Bowl increased by approximately 16.7% from 2011 to 2012.

You Try It **Work through this You Try It problem.**

Work Exercises 16–21 in this textbook or in the MyMathLab Study Plan.

CAUTION When rounding **percents**, make sure you round *after* you change the **decimal to a percent.** Notice that if we had rounded first in Example 4 and then changed to a percent, our result would have been

$$\frac{0.5}{3.0} = 0.16666\ldots \approx 0.2 = 20\%.$$

OBJECTIVE 3 SOLVE APPLICATIONS INVOLVING A PERCENT DECREASE

As with percent increase, finding a percent decrease is just an application of our general **percent equation**. Consider the following:

$$\boxed{\text{Percent}} \cdot \boxed{\text{Base}} = \boxed{\text{Amount}}$$

$$\downarrow \qquad \downarrow \qquad \downarrow$$

$$\text{Percent decrease} \cdot \text{Original value} = \text{Amount of decrease}$$

We can solve this equation for the percent decrease by dividing both sides of the equation by the original value. This gives

$$\text{Percent decrease} = \frac{\text{Amount of decrease}}{\text{Original value}}.$$

So, a **percent decrease** is the amount of decrease expressed as a percentage of the original value. The **amount of decrease** can be found by subtracting the new value from the original value.

$$\text{Amount of decrease} = \text{Original value} - \text{New value}$$

CAUTION Notice that to find the **amount of increase**, we subtract the original value from the new value, but for the amount of decrease, we subtract the new value from the original value. Also, in a **percent increase**, the new value is greater than the original value, but in a percent decrease, the new value is less than the original value.

For example, in the fall of 2011, Netflix announced that it would lower the price of its DVD-only plan from $10 per month to $8 per month. (*Source:* nytimes.com)

The **amount of decrease** is found by subtracting the new price, $8, from the original price, $10.

$$\text{Amount of decrease} = \$10 - \$8 = \$2$$

The **percent decrease** is then found by dividing the amount of decrease by the original price.

$$\text{Percent decrease} = \frac{\text{Amount of decrease}}{\text{Original price}} = \frac{\$2}{\$10} = \frac{\cancel{\$}2}{\cancel{\$}10} = 0.2$$

↑ Divide out the common units

Change from a **decimal to a percent.**

$$0.2 = 0.2 \cdot 100\% = 20\%$$

↑ Multiply by 100%

So, the DVD-only price change by Netflix was a 20% decrease.

Example 5 Hours Worked in a Week

My interactive video summary

The average number of hours worked per week in the U.S. was 40.2 in 2009 but was only 34.6 in 2011. Find the percent decrease. Round to the nearest percent. (*Source:* usatoday.com)

Solution Work through the following, or watch this **video** for the complete solution.

We start by finding the **amount of decrease**. Subtract the new value (the average number of hours worked per week in 2011) from the original value (the average number of hours worked per week in 2009).

$$\text{Amount of decrease} = 40.2 \text{ hours} - 34.6 \text{ hours} = 5.6 \text{ hours}$$

We then find the **percent decrease** by dividing the amount of decrease by the original value (average hours worked in 2009).

$$\text{Percent decrease} = \frac{\text{Amount of decrease}}{\text{Original value}} = \frac{5.6 \text{ hours}}{40.2 \text{ hours}} = \frac{5.6 \cancel{\text{hours}}}{40.2 \cancel{\text{hours}}} = \frac{5.6}{40.2} = 0.139303\ldots$$

Divide out the common units

CAUTION Make sure to use the *original* value when finding the percent decrease, not the new value.

Multiply the result by 100% and **round** to the nearest whole number.

$$0.139303\ldots = 13.9303\ldots\% \approx 14\%$$

Move the decimal point 2 places to the right — Add the % symbol — Round to the nearest whole percent

So, the average number of hours worked each week decreased by approximately 14% from 2009 to 2011.

You Try It **Work through this You Try It problem.**

Work Exercises 22–27 in this textbook or in the MyMathLab Study Plan.

7.4 Exercises

In Exercises 1–27, use percents, percent equations, and/or percent proportions to solve the following applications.

You Try It

1. **Tablet Computers** In January 2012, 19% of adults in the U.S. (aged 18 years or older) owned a tablet computer. If there were 236 million adults in the U.S. at the time, how many owned a tablet computer? Round to the nearest million. (*Source:* msnbc.com)

2. **Juice Content** A bottle of a cranberry–apple juice cocktail is 15% juice. How many ounces of juice are in a 64 oz bottle of the juice cocktail?

3. **Fund-raising** As part of a fund-raising program, Chelsea can earn a 4% rebate for purchasing gift cards to a local restaurant. If she purchases $160 in gift cards, how much is her rebate?

4. **Response Rate** A researcher sends out a survey to 640 subjects by mail and receives completed surveys from 15% of the subjects. How many surveys did the researcher receive?

5. **Video Game Ratings** In 2010, 55% of the Entertainment Software Rating Board's 1628 ratings for video games were E (for everyone). How many video games received the E rating? (*Source:* gamespot.com)

You Try It

6. **Student Population** 47% of undergraduate students at a university are women. If there are 3807 female undergraduate students, how many total undergraduate students does the university have?

7. **Recycled Paper** A paper manufacturer uses recycled material for 30% of the total weight of a box of new paper. If a box of new paper contains 10.8 pounds of recycled material, how much does the box of new paper weigh?

8. **Mobile Phone Profits** In the final quarter of 2011, Samsung received 16% of the total profit for all mobile phone manufacturers. If the total profit for Samsung was $2.4 billion in that quarter, what was the total profit received for all mobile phone manufacturers for all carriers? (*Source:* news.cnet.com)

9. **Facebook Revenue** In 2011, the application publisher Zynga accounted for 12% of Facebook's revenue, or $445 million. What was Facebook's total revenue in 2011? Round to the nearest million.

10. **Completion Rate** A quarterback made 375 pass completions during one football season. If this was 60% of his attempts, how many passes did he attempt?

You Try It 11. **New Year's Resolution** In a survey, 1200 people were asked to list their New Year's resolutions. Of those surveyed, 516 said that they planned to lose weight. What percent of those surveyed planned to lose weight? (*Source:* usatoday.com)

12. **Course Grade** A course syllabus states that the course will have a total of 650 points, which will be used to determine a student's grade. If a student earns 546 points, what is the student's course grade as a percent?

13. **iPhone Sales** For the final quarter of 2011, it was anticipated that Apple would sell 30 million iPhones. If Apple actually sold 37 million iPhones that quarter, what percent of the anticipated sales were sold? Round to the nearest whole percent. (*Source:* news.cnet.com)

14. **Quality Control** In a sample of 120 electronic circuits, a quality-control inspector finds 3 defective circuits. What percent of the sample is defective?

15. **Nutrition Data** The following nutrition label gives the total calories for a serving of chicken noodle soup, and the calories from fat. Use the label to determine the percent of calories from fat.

Nutrition Facts

Serving Size 1 cup (8 fl oz)(241.0 g)

Amount Per Serving	
Calories 60	**Calories from Fat** 21
	% Daily Value
Total Fat 2.3g	**4%**
Saturated Fat 0.6g	**3%**
Trans Fat 0.6g	
Polyunsaturated Fat 0.6g	
Monounsaturated Fat 1.0g	

16. **Deductible Increase** To lower his insurance rates, Damon increases his deductible from \$250 to \$350. What is the percent increase in his deductible?

17. **Tuition Increase** A community college raises its tuition from \$80 per credit hour to \$92 per credit hour. What is the percent increase?

You Try It 18. **Franchise Increase** In 2011, Subway had 199 franchises in China. This number is expected to increase to 500 by 2015. What is this percent increase? Round to the nearest whole percent. (*Source: The Wall Street Journal*)

19. **Job Growth** In 2011, there were approximately 2,740,000 registered nurses in the U.S. The Bureau of Labor Statistics expects this number to grow to 3,452,000 by 2020. What is this percent increase? Round to the nearest whole percent. (*Source:* bls.gov)

20. **Stamp Prices** On January 22, 2012, the U.S. Postal Service implemented a postage increase for all mail classes. The cost to mail a postcard increased from 29¢ to 32¢. What is this percent increase? Round to the nearest tenth of a percent. (*Source:* stamps.com)

21. **Gasoline Prices** The average price of regular unleaded gasoline in the U.S. on January 3, 2011 was \$3.33 per gallon. This rose to \$3.78 per gallon on January 2, 2012. What was the percent increase in average gasoline price during this period of time? Round to the nearest tenth of a percent. (*Source:* eia.gov)

(*eText Screens 7.5-1–7.5-17*)

You Try It 22. **Diet Soft Drink** Jayla switches from drinking regular soda to drinking diet soda. If her regular soda contains 125 mg of sodium per bottle and the diet soda contains 85 mg of sodium, find her percent decrease in sodium by switching from regular to diet soda.

23. **Class Withdrawals** Students enrolled in a prealgebra class can withdraw within the first two weeks without receiving a grade. If there were 28 students registered for the class on the first day and 21 remaining at the end of the second week, what was the percent decrease in students for the first two weeks?

24. **Housing Crash** A house was bought in 2005 for \$155,000. In 2012, it was valued at \$138,000. Find the percent decrease in value of the house. Round to the nearest whole percent.

25. **Kids Meals** In 2012, McDonald's required all of its restaurants to sell healthier versions of kid's meals. The meals featured smaller portions of french fries, added apple slices, and offered low-fat milk as a drink alternative. If the number of calories of one kid's meal was reduced from 520 to 410, what was the percent decrease? Round to the nearest whole percent. (*Source:* nytimes.com)

26. **Falling Stocks** During a bad economy, a stock index decreased from 11,896 points to 11,384 points in a single day. What was the percent decrease for that day? Round to the nearest tenth of a percent.

27. **Pizza Size** Holly orders a large 16-inch pizza, but decides that is too much food and changes her order to a medium 14-inch pizza. If the area of the large pizza is 64π square inches and the area of the medium pizza is 49π square inches, what is the percent decrease in area? Round to the nearest tenth of a percent.

7.5 Percent and Problem Solving: Sales Tax, Commission, and Discount

THINGS TO KNOW

Before working through this topic, be sure you are familiar with the following concepts:

		VIDEO	ANIMATION	INTERACTIVE
You Try It	1. Write Percents as Decimals (Topic 7.1, **Objective 1**)	✓		
You Try It	2. Write Decimals as Percents (Topic 7.1, **Objective 2**)	✓		
You Try It	3. Translate Word Statements into Percent Equations (Topic 7.2, **Objective 1**)			✓
You Try It	4. Solve Percent Equations (Topic 7.2, **Objective 2**)	✓		✓
You Try It	5. Solve Percent Problems Using Proportions (Topic 7.3 **Objective 2**)	✓		✓
You Try It	6. Solve Applications Involving Percents (Topic 7.4, **Objective 1**)	✓		

(eText Screens 7.5-1–7.5-17)

OBJECTIVES

1 Compute Sales Tax, Overall Price, and Tax Rate

2 Compute Commission and Commission Rate

3 Compute Discount, Sale Price, and Discount Rate

OBJECTIVE 1 COMPUTE SALES TAX, OVERALL PRICE, AND TAX RATE

Many applications of **percent** relate to buying and selling merchandise. In this topic, we will look at three such applications: *sales tax*, *commission*, and *discount*.

Most state and local governments charge a **sales tax** when merchandise is purchased. Typically, the sales tax is a percent of the **purchase price**. The percent is called the **tax rate**. We can use the **percent equation** to create a formula for computing sales tax where the sales tax is the ***amount***, the tax rate is the *percent*, and the purchase price is the ***base***.

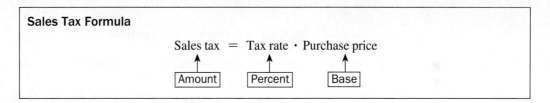

Sales Tax Formula

$$\underset{\text{Amount}}{\text{Sales tax}} = \underset{\text{Percent}}{\text{Tax rate}} \cdot \underset{\text{Base}}{\text{Purchase price}}$$

Usually the merchant collects the sales tax at the time of purchase, so the **overall price** paid to the merchant is the purchase price plus the sales tax.

Computing the Overall Price

$$\text{Overall price} = \text{Purchase price} + \text{Sales tax}$$

Example 1 Computing Sales Tax and Overall Price

Peg plans to buy a 3-D television priced at $\$1299$. The **tax rate** is 7%.

a. How much will Peg have to pay in **sales tax**?

b. What will be Peg's **overall price** for the television?

Solutions

a. Recognize that the **purchase price** is $\$1299$ and the tax rate is 7%. Substitute these into the **sales tax formula**.

Sales tax = Tax rate · Purchase price

Substitute: $= 7\% \cdot \$1299$

Write 7% as a decimal: $= 0.07 \cdot \$1299$

Multiply: $= \$90.93$

$$7\% = \frac{7\%}{100\%} = 0.07$$

Peg will have to pay $\$90.93$ in sales tax.

b. Overall price = Purchase price + Sales tax

$= \$1299 + \$90.93$

$= \$1389.93$

Peg's overall price for the television will be $\$1389.93$.

(eText Screens 7.5-1–7.5-17)

You Try It **Work through this You Try It problem.**

Work Exercise 1 in this textbook or in the MyMathLab Study Plan.

In **Example 1**, the **sales tax** worked out evenly to the nearest cent (hundredths **place**). Sometimes is necessary to **round**.

Example 2 Computing Sales Tax and Overall Price

My video summary

A pair of basketball shoes is priced at $189.79. If the **tax rate** is 6.5%, find the sales tax and the **overall price**. Round to the nearest cent.

Solution The **purchase price** is $189.79 and the tax rate is 6.5%. First, we find the sales tax by using the **sales tax formula** and then rounding to the nearest cent. Then, we **compute the overall price** by adding the sales tax to $189.79. Try to work this problem on your own. Then check the **answer**, or watch this **video** for a complete solution.

You Try It **Work through this You Try It problem.**

Work Exercises 2 and 3 in this textbook or in the MyMathLab Study Plan.

Sometimes, we want to find the **tax rate**. The next two examples address this type of situation.

Example 3 Finding the Tax Rate

The **sales tax** on a case of energy drinks is $1.95. If the **purchase price** is $32.50, find the tax rate.

Solution Let r represent the tax rate. The purchase price is $32.50 and the sales tax is $1.95. We substitute these values into the **sales tax formula** and **solve** for r:

$$\text{Sales tax} = \text{Tax rate} \cdot \text{Purchase price}$$

Substitute: $$1.95 = r \cdot 32.50$$

Divide both sides by 32.50: $$\frac{1.95}{32.50} = \frac{r \cdot 32.50}{32.50}$$

Simplify: $$0.06 = r$$

Write 0.06 as a percent: $$6\% = r$$

$0.06 = 0.06 \cdot 100\% = 006.\% = 6\%$

The tax rate is 6%.

You Try It **Work through this You Try It problem.**

My video summary

Work Exercises 4 and 5 in this textbook or in the MyMathLab Study Plan.

Example 4 Finding the Tax Rate

Sam bought a new coat priced at $159.99. When he was paying for the coat, the cashier asked Sam for $170.39. What was the **tax rate**? **Round** to the nearest tenth of a **percent**.

Solution To find the tax rate, we first must find the **sales tax paid**. The **purchase price** of the coat is $159.99 and the overall price is $170.39, so

$$\text{Sales tax} = \$170.39 - \$159.99 = \$10.40.$$

Let r represent the tax rate. Use the **sales tax formula** to finish finding the tax rate on your own. View the **answer**, or watch this **video** for a complete solution.

You Try It **Work through this You Try It problem.**

Work Exercises 6–8 in this textbook or in the MyMathLab Study Plan.

OBJECTIVE 2 COMPUTE COMMISSION AND COMMISSION RATE

People who work in sales often are paid on **commission**, which means that they earn a **percent** of their **sales**. The percent is called the **commission rate**. We can use the **percent equation** to create a formula for computing commission.

Commission Formula

$$\underbrace{\text{Commission}}_{\text{Amount}} = \underbrace{\text{Commission rate}}_{\text{Percent}} \cdot \underbrace{\text{Sales}}_{\text{Base}}$$

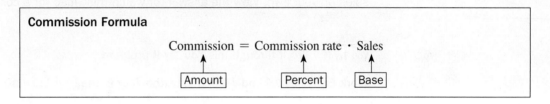

We can use the commission formula to solve for any unknown quantity as long as the other two values are given. The next three examples illustrate this point.

Example 5 Finding the Commission

Sherry is a realtor who just sold a house for $235,000. If her commission rate is 1.5% of the selling price, compute Sherry's **commission**.

Solution We substitute the **commission rate** of 1.5% and sales of $235,000 into the **commission formula**.

	Commission	= Commission rate	· Sales
Substitute:		= 1.5%	· \$235,000
Write 1.5% as a decimal:		= 0.015	· \$235,000 ← $1.5\% = 0.015$
Multiply:		= \$3525	

Sherry's commission is $3525.

You Try It **Work through this You Try It problem.**

Work Exercises 9–11 in this textbook or in the MyMathLab Study Plan.

Example 6 Finding the Commission Rate

My video summary

A jewelry salesperson earned $300 in **commission** by selling a wedding ring for $7500. What was the **commission rate**?

Solution Let r represent the commission rate. The commission is $300 and sales are $7500. Substitute these values into the **commission formula**:

	Commission	= Commission rate	· Sales
Substitute:	300	= r	· 7500

(eText Screens 7.5-1–7.5-17)

Try to **solve** for r on your own. Remember to write the commission rate as a **percent**. View the **answer**, or watch this **video** for a complete solution.

You Try It **Work through this You Try It problem.**

Work Exercises 12–14 in this textbook or in the MyMathLab Study Plan.

Example 7 Finding the Sales

My video summary

Lindsay, a furniture saleswoman, earns a 2.5% commission rate on her total sales each week. If she earned \$338.10 in commission one week, what were Lindsay's sales that week?

Solution The **commission rate** is 2.5%, and the **commission** is \$338.10. Try to find the sales on your own. View the **answer**, or watch this **video** for a complete solution.

You Try It **Work through this You Try It problem.**

Work Exercises 15 and 16 in this textbook or in the MyMathLab Study Plan.

OBJECTIVE 3 COMPUTE DISCOUNT, SALE PRICE, AND DISCOUNT RATE

During a sale, a store decreases the price of merchandise. The **amount** by which the price is decreased is called a **discount**. Typically, the discount is a **percent** of the **original price**. This percent is called the **discount rate**. We can use the **percent equation** to create a formula for computing discounts.

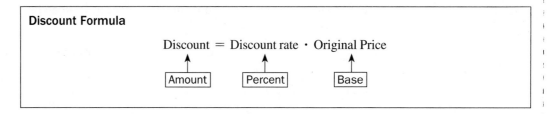

Discount Formula

$$\text{Discount} = \text{Discount rate} \cdot \text{Original Price}$$

Discount → Amount; Discount rate → Percent; Original Price → Base

To find the **sale price**, we subtract the discount from the original price.

Sale Price Formula

$$\text{Sale price} = \text{Original price} - \text{Discount}$$

In the following examples, we use these two formulas to find unknown quantities.

Example 8 Computing the Discount and Sale Price

My video summary

A department store is having a sale on men's clothing. Karl wants to buy a new suit originally priced at \$595.89. A sign reads, "Take 35% off the price marked on any suit." Find the **discount** and the **sale price** for Karl's suit. **Round** to the nearest cent.

(eText Screens 7.5-1–7.5-17)

Solution First, we find the discount and then subtract it from the **original price** to find the sale price. To compute the discount, substitute the original price of $595.89 and the **discount rate** of 35% into the **discount formula**:

Discount	=	Discount rate	·	Original price	
Substitute:	=	35%	·	$595.89	
Write 35% as a decimal:	=	0.35	·	$595.89	35% = 0.35
Multiply:	=	$208.5615			
Round to the cent (hundredth):	≈	$208.56			

≈ "is approximately equal to"

Finish finding the sale price on your own. View the **answer**, or watch this **video** for a complete solution.

You Try It **Work through this You Try It problem.**

Work Exercises 17–20 in this textbook or in the MyMathLab Study Plan.

Example 9 Computing the Discount Rate

My video summary

A cordless power saw that normally sells for $125 is on sale for $87.50. Find the **discount rate**.

Solution To find the discount rate, we first must find the **discount**. The **original price** of the saw is $125 and the **sale price** is $87.50, so

$$\text{Discount} = \$125 - \$87.50 = \$37.50.$$

Let r represent the discount rate. Use the **discount formula** to finish finding the discount rate on your own. View the **answer**, or watch this video for a complete solution.

You Try It **Work through this You Try It problem.**

Work Exercises 21–24 in this textbook or in the MyMathLab Study Plan.

7.5 Exercises

In Exercises 1–8, solve each sales tax problem.

You Try It 1. Kirsten is buying a tablet computer priced at $549. The tax rate is 8%.

   a. How much will Kirsten pay in sales tax?

   b. What is the overall price for Kirsten's tablet?

2. The purchase price of a designer dress is $859. If the tax rate is 6.5%, find the sales tax and overall price. Round to the nearest cent.

You Try It 3. Carmen bought a pair of jeans for $49.95. If the tax rate is 8.25%, find the sales tax and overall price. Round to the nearest cent.

You Try It 4. Regina purchased $150 in groceries. If the sales tax was $3.75, what was the tax rate?

5. A one-day ticket to an amusement park costs \$40, plus \$2.94 in sales tax. What is the tax rate?

You Try It 6. Ebony bought a leather sofa priced at \$1250. Including sales tax, she paid \$1325 for the sofa. What was the tax rate?

7. The purchase price for an air hockey table is \$640. Including sales tax, the overall price is \$683.20. What is the tax rate?

8. The sales tax on a game system is \$18.45. If the tax rate is 7.5%, find the purchase price.

In Exercises 9–16, solve each commission problem.

You Try It 9. A real estate company earns a commission rate of 6% on properties that it lists and sells. If the company lists and sells a house for \$150,000, what is the commission?

10. A used car salesman earns an 8.5% commission rate on all sales. Find his commission if he sells a car for \$12,800.

11. For conducting an estate auction, an auctioneer earns a 15% commission rate on gross sales. If the gross sales are \$29,876, find the auctioneer's commission.

You Try It 12. If an appliance salesperson earns a \$180 commission by selling a front-load washer/dryer for \$1500, what is her commission rate?

13. If a stock broker earns a \$1500 commission by handling \$300,000 worth of stock transactions, what is his commission rate?

14. A travel agent sold a cruise vacation to a couple for \$1460 and earned a commission of \$163.52. What was the agent's commission rate?

You Try It 15. A beauty consultant is paid a 12.5% commission rate for selling cosmetics. What would her total sales need to be in order to earn a commission of \$500?

16. A salesperson who is paid a 4.2% commission rate earned \$612.15 in commission. What were his total sales?

In Exercises 17–24, solve each discount problem.

You Try It 17. A shoe store is having a sale for 25% off original prices. A pair of running shoes has an original price of \$86.

a. What is the discount?

b. What is the sale price?

18. A smartphone originally priced at \$259.99 is on sale for 30% off. Find the discount and the sale price. Round to the nearest cent.

19. Following a holiday season, holiday decorations are put on sale for 75% off. Find the discount and the sale price for a decoration with an original price of \$35.99. Round to the nearest cent.

20. Missy used a coupon for 20% off the price of her meal at a buffet restaurant and received a \$3 discount. What was the original price of Missy's meal before the discount?

21. If Dayton receives a \$13 discount on a college sweatshirt originally priced at \$40, what is the discount rate?

22. At Quick Change Auto Service, an oil change normally costs \$35. To grow business, Quick Change offers a \$10 discount to new customers. What is the discount rate?

You Try It 23. A television originally priced at \$650 is on sale for \$546. Find the discount rate.

24. An office chair that normally sells for \$159 is on sale for \$79. Find the discount rate. Round to the nearest tenth of a percent.

7.6 Percent and Problem Solving: Interest

THINGS TO KNOW

Before working through this topic, be sure you are familiar with the following concepts:

		VIDEO	ANIMATION	INTERACTIVE
You Try It	1. Use the Order of Operations with Decimals (Topic 5.5, **Objective 3**)	✓		
You Try It	2. Evaluate Algebraic Expressions Using Decimals (Topic 5.5, **Objective 4**)	✓		
You Try It	3. Write Percents as Decimals (Topic 7.1, **Objective 1**)	✓		
You Try It	4. Write Decimals as Percents (Topic 7.1, **Objective 2**)	✓		
You Try It	5. Solve Percent Equations (Topic 7.2, **Objective 2**)	✓		✓

OBJECTIVES

1 Compute Simple Interest

2 Compute Compound Interest

OBJECTIVE 1 COMPUTE SIMPLE INTEREST

When we borrow money, we pay an extra amount of money called **interest** for this privilege. Similarly, if we lend money to someone else, the payment we receive for this generosity is also called **interest**. So, when we borrow money, we must *pay* interest; when we lend or invest money, we *earn* interest. The amount of money that is borrowed or invested is called the **principal**. Typically, the amount of interest paid or earned is a **percent** of the principal over a given time interval. The percent is called the **interest rate**. There are two primary types of interest: *simple interest* and *compound interest*.

When interest is paid or earned on the original principal only, it is called **simple interest**. We compute simple interest using the **simple interest formula**:

Simple Interest Formula

$$\text{Simple interest} = \text{Principal} \cdot \text{Rate} \cdot \text{Time}$$

$$I = P \cdot r \cdot t$$

TIP When entering the interest rate into the simple interest formula, remember to **write the percent as a decimal.** For example, we would enter 0.08 for an interest rate of 8%.

Example 1 Earning Simple Interest

If $2500 is invested for 3 years in a **certificate of deposit**, or CD, that pays **simple interest** at an **interest rate** of 5% per year, how much **interest** will be earned?

Solution The **principal** is $P = \$2500$. The interest rate is $r = 5\%$ per year, or 0.05. The time is $t = 3$ years. To find the simple interest I, we substitute these values into the **simple interest formula.**

Write the formula: $I = P \cdot r \cdot t$

Substitute: $= \$2500 \cdot 0.05 \cdot 3$

Multiply: $= \$375$

The simple interest is $375.

You Try It **Work through this You Try It problem.**

Work Exercises 1 and 2 in this textbook or in the MyMathLab Study Plan.

CAUTION The interest rate r is given as a percent per unit of time. The time t must have matching units. For example, if the interest rate is 7% per year, then time must be in years. If the interest rate is 2% per month, then time must be in months. Standard practice is to use annual (yearly) interest rates.

TIP If no unit of time is given with the interest rate, it is understood to be an annual rate.

Example 2 Paying Simple Interest

My video summary

Kendall borrowed $1200 for 9 months at a **simple interest** rate of 8% per year. How much **interest** will Kendall have to pay?

Solution The **interest rate** is per *year*, but the time is in *months*. So, we must change one or the other units so that they match. We choose to change 9 months to years:

$$t = 9 \text{ months} = \frac{9}{12} \text{ year} = \frac{3}{4} \text{ year} \qquad \boxed{\frac{9}{12} = \frac{3 \cdot \cancel{3}}{4 \cdot \cancel{3}} = \frac{3}{4}}$$

Substitute this time along with the **principal** $P = \$1200$ and the rate $r = 8\%$ per year, or 0.08, into the **simple interest formula** to find the interest on your own. View the **answer**, or watch this **video** for a complete solution.

You Try It **Work through this You Try It problem.**

Work Exercises 3 and 4 in this textbook or in the MyMathLab Study Plan.

When paying back a loan, a borrower must pay back the original **principal** plus the **interest.** In **Example 2**, Kendall will have to pay back the original $1200 that he borrowed plus the $72 in interest. So, Kendall will pay back an *overall amount* of $\$1200 + \$72 = \$1272$.

Similarly, when an investment ends, the investor gets back the original principal *plus* the interest. In **Example 1**, the investor will get back the original $2500 invested plus the $375 in interest. So, the investor will get back an *overall amount* of $\$2500 + \$375 = \$2875$.

Notice that whether borrowing or investing, the **overall amount** of money involved results from adding the principal and the interest.

Finding the Overall Amount for a Loan or Investment

$$\text{Overall amount} = \text{Principal} + \text{Interest}$$
$$A = P + I$$

Example 3 Finding the Overall Amount for a Loan

To buy textbooks, a college student borrows $450 at 12% **simple interest** for 4 months. Find the **overall amount** that the student must repay at the end of the 4-month period.

Solution No unit of time is stated for the interest rate, so it is understood to be an annual rate. To match this, we change 4 months to years.

$$t = 4 \text{ months} = \frac{4}{12} \text{ year} = \frac{1}{3} \text{ year} \qquad \boxed{\frac{4}{12} = \frac{1 \cdot \cancel{4}}{3 \cdot \cancel{4}} = \frac{1}{3}}$$

The **principal** is $P = \$450$ and the interest rate $r = 12\%$ per year, or 0.12. We can find the simple interest using the **simple interest formula**:

$$I = Prt = \$450 \cdot 0.12 \cdot \frac{1}{3} = \$18$$

We know the $P = \$450$ and $I = \$18$, so we can find the overall amount:

$$A = P + I = \$450 + \$18 = \$468$$

The student must repay an overall amount of $468.

Example 4 Finding the Overall Amount for an Investment

My video summary

If Martha invests $15,000 at a **simple interest** rate of 6.2% for 1.5 years, what is the **overall amount** of money that she will have at the end of the 1.5-year period?

Solution Try working this problem on your own. View the **answer**, or watch this **video** for a complete solution.

You Try It **Work through this You Try It problem.**

Work Exercises 5–8 in this textbook or in the MyMathLab Study Plan.

OBJECTIVE 2 COMPUTE COMPOUND INTEREST

A second, more commonly used type of **interest** is **compound interest**. The difference between **simple interest** and compound interest is that simple interest is computed on just the **principal**, while compound interest is computed on both the principal and the previously paid interest.

My animation summary

To see how compound interest works, watch this **animation**. Then skip forward to Example 5 or continue reading.

Suppose that $4000 is invested in a savings account that pays 5% interest for one year. The interest earned at the end of that year is

$$I = Prt = \$4000 \cdot 0.05 \cdot 1 = \$200.$$

Adding this interest to the original principal, we now have $\$4000 + \$200 = \$4200$ in the savings account. If we keep it all in the savings account for another year at the same interest rate, the principal for the second year will be \$4200. So, for the second year, we will receive interest on both the interest from the first year and the original principal. The interest earned at the end of the second year is

$$I = Prt = \$4200 \cdot 0.05 \cdot 1 = \$210.$$

Adding in this interest, we now have $\$4200 + \$210 = \$4410$ in the savings account.

Comparing this \$4410 to the original **principal** \$4000, a total of \$410 has been earned in **compound interest.**

If we compute **simple interest** on \$4000 at 5% for 2 years, the **interest** earned would be $I = Prt = \$4000 \cdot 0.05 \cdot 2 = \$400$. So, by using compound interest, we earned \$10 more than we would have earned if we had used simple interest.

When investing, compound interest earns more than simple interest because compound interest earns interest on interest. Similarly, when borrowing, compound interest costs more than simple interest.

In the example just described, the compound interest was computed at the end of each year, so the interest was compounded **annually.** We can also compute the interest over other time intervals, such as **semiannually** (twice per year), **quarterly** (4 times per year), **monthly** (12 times per year), or **daily** (365 times per year).

Computing compound interest in this way can take a long time. The **compound interest formula** can be used to quickly find the overall amount of money in an account.

Compound Interest Formula

The **overall amount** A in an account is given by the formula

$$A = P\left(1 + \frac{r}{n}\right)^{n \cdot t},$$

where P is the **principal,** r is the annual **interest rate** written in decimal form, t is the length of time in years, and n is the number times compounded per year.

CAUTION Recognize that this formula gives the *overall amount*, not the *interest*.

To find out how much **interest** has been earned, we can subtract the principal from the overall amount.

Finding Interest

$$\text{Interest} = \text{Overall amount} - \text{Principal}$$
$$I = A - P$$

Example 5 Computing Compound Interest

\$3250 is invested at 4% **interest** compounded semiannually for 3 years.

a. Find the **overall amount.** Round to the nearest cent.

b. Find the **compound interest.** Round to the nearest cent.

(eText Screens 7.6-1–7.6-15)

Solutions

a. The **principal** is $P = \$3250$, the rate is $r = 4\%$ or 0.04, and the time is $t = 3$ years. "Compounded semiannually" means twice per year, so $n = 2$. To find the overall amount A, we substitute these values into the **compound interest formula.**

$$\text{Write the formula:} \quad A = P\left(1 + \frac{r}{n}\right)^{n \cdot t}$$

$$\text{Substitute:} \quad = \$3250\left(1 + \frac{0.04}{2}\right)^{2 \cdot 3}$$

$$\text{Simplify within the parentheses:} \quad = \$3250(1.02)^6 \leftarrow \boxed{1 + \frac{0.04}{2} = 1 + 0.02 = 1.02}$$

$$\text{Evaluate the exponent, then multiply:} \quad = \$3660.027862608$$

$$\text{Round to the cent (hundredth):} \quad \approx \$3660.03$$

The overall amount is $3660.03.

b. From part a, we know that $A = \$3660.03$ and $P = \$3250$, so we can find the **compound interest** earned:

$$I = A - P = \$3660.03 - \$3200 = \$460.03.$$

The compound interest earned was $460.03.

Example 6 Computing Compound Interest

My video summary

Kejwan put $8500 into a savings account that pays 3.5% interest compounded quarterly. He plans to grow the money in the account for 5 years.

a. How much money will Kejwan have in the account after 5 years? Round to the nearest cent.

b. How much compound interest will Kejwan have earned over the 5 years? Round to the nearest cent.

Solution Try working this problem on your own. View the **answer**, or watch this **video** for a complete solution.

You Try It **Work through this You Try It problem.**

Work Exercises 9–14 in this textbook or in the MyMathLab Study Plan.

7.6 Exercises

In Exercises 1–8, solve each simple interest problem. Round to the nearest cent as necessary.

You Try It 1. An accountant invests $7500 for 2 years in a certificate of deposit (CD) that pays a simple interest rate of 4% per year. How much interest will be earned?

2. A retiree invests $1250 for 1.5 years in a savings plan that pays simple interest of 2.6% per year. How much interest will be earned?

You Try It 3. If Lexi borrows $950 for 10 months at a simple interest rate of 6% per year, how much interest will she have to pay for this loan?

4. If a company borrows $175,000 for 4 years at a simple interest rate of 8.5% per year, find the interest paid on the loan.

(eText Screens 7.6-1–7.6-15)

You Try It 5. A student takes out a short-term loan to pay for tuition, books, and supplies. If the simple interest loan is \$3500 at a rate of 5% per year for 6 months, find the overall amount that the student must pay back at the end of the 6-month period.

6. To build a house, a construction company borrows \$95,000 for 9 months at a simple interest rate of 6.5% per year. Find the interest paid on the loan and the overall amount paid back.

7. If Damon invests \$800 for 1 year in a savings account that pays a simple interest rate of 1.75% annually, then what is the overall amount of money that he will have in the account?

8. If Pearl makes an investment of \$50,000 at a simple interest rate of 5.8% for 8 years, then what is the overall amount of money that she will have at the end of the 8-year period?

In Exercises 9–14, solve each compound interest problem. Round to the nearest cent as necessary.

You Try It 9. \$9000 is invested at 6% compounded annually for 3 years.

a. Find the overall amount.

b. Find the compound interest.

10. \$3600 is invested at 9% compounded semiannually for 5 years.

a. Find the overall amount.

b. Find the compound interest.

11. \$4825 is invested at 4.5% compounded monthly for 6 years.

a. Find the overall amount.

b. Find the compound interest.

12. \$12,000 is invested at 7.3% compounded daily for 4 years.

a. Find the overall amount.

b. Find the compound interest.

13. After receiving an inheritance of \$18,000, a woman deposits the money in an account paying 6.3% compounded monthly. How much money will be in the account after 3 years?

14. On the day a child was born, his grandparents started a college fund for him by investing \$5000 in an account paying 7% interest compounded quarterly. How much money will be in the college fund when the child turns 18 years old? How much interest will have been earned?

MODULE EIGHT

Geometry and Measurement

MODULE EIGHT CONTENTS

8.1 Lines and Angles

THINGS TO KNOW

Before working through this topic, be sure you are familiar with the following concepts:	VIDEO	ANIMATION	INTERACTIVE
You Try It 1. Identify the Corresponding Parts of Congruent Triangles (Topic 6.4, **Objective 1**)	▣		
You Try It 2. Identify the Corresponding Parts of Similar Triangles (Topic 6.4, **Objective 3**)	▣		

OBJECTIVES

1 Identify and Name Lines, Segments, Rays, and Angles

2 Classify Angles as Acute, Obtuse, Right, or Straight

3 Identify Complementary and Supplementary Angles and Find Their Measures

4 Find an Unknown Measure of an Angle Using Given Information

---

OBJECTIVE 1 IDENTIFY AND NAME LINES, SEGMENTS, RAYS, AND ANGLES

A **line** is a straight row of points that extend forever in two directions. When we draw a line, we place arrowheads at each end to indicate that the line continues without end in both directions. We can name a line using any two points on the line. For example, we might name the line in Figure 1a as $\overleftrightarrow{AB}$, which reads as "line AB." We can also name the line as $\overleftrightarrow{BA}$. A line also may be named with a single variable, usually a lowercase letter. For example, the line in Figure 1b is line l.

(eText Screens 8.1-1–8.1-31)

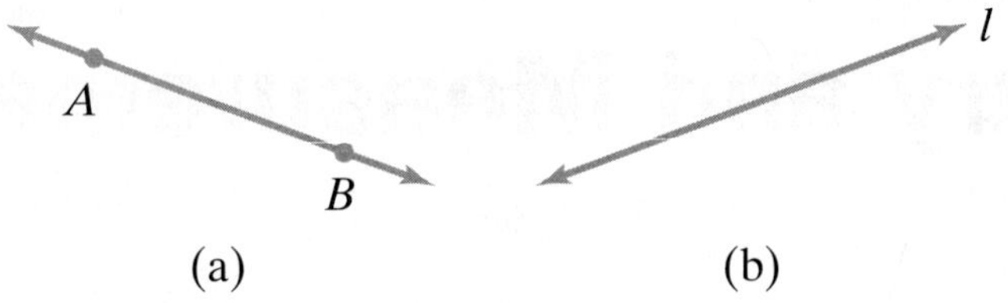

Figure 1

A **line segment**, or simply a **segment**, is a section of a line that has two endpoints. We use the endpoints to name the segment. For example, we can name the segment in Figure 2 as $\overline{CD}$, which reads "segment CD." We can also name the segment as $\overline{DC}$.

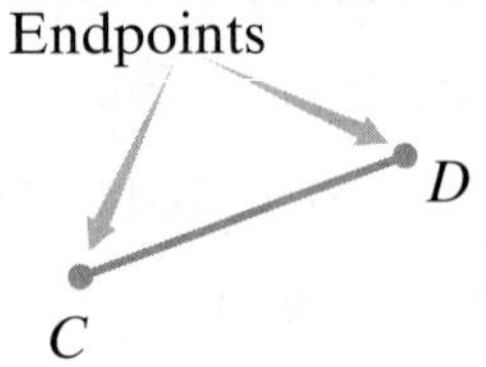

Figure 2

CAUTION Pay attention to the difference in the symbols used for naming **lines** and **segments.** The symbol above the points naming a line has arrowheads on each side ($\overleftrightarrow{AB}$). The symbol above the endpoints naming a segment has no arrowheads ($\overline{CD}$),

A **ray** is a section of a line that has one endpoint and continues without end in one direction. We name a ray using its endpoint and one other point on the ray. For example, we name the ray in Figure 3 as $\overrightarrow{EF}$, which reads "ray EF."

Figure 3

CAUTION The symbol above the points naming a ray has only one arrowhead. The side of the symbol *without* the arrowhead goes above the endpoint of the ray, while the side of the symbol *with* the arrowhead goes above the other point naming the ray. The ray in Figure 3 is correctly named $\overrightarrow{EF}$ because E is the endpoint of the ray and F is another point on the ray. $\overrightarrow{FE}$ does not correctly name Figure 3 because $\overrightarrow{FE}$ names a ray with endpoint F and another point E. The ray in Figure 4 is named $\overrightarrow{FE}$.

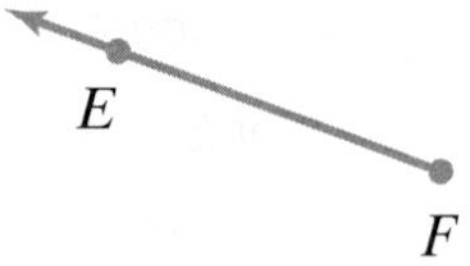

Figure 4

Example 1 Identifying Lines, Segments, and Rays

Identify each figure as a **line**, **segment**, or **ray**. Then use the correct symbol to name the figure.

a.

b.

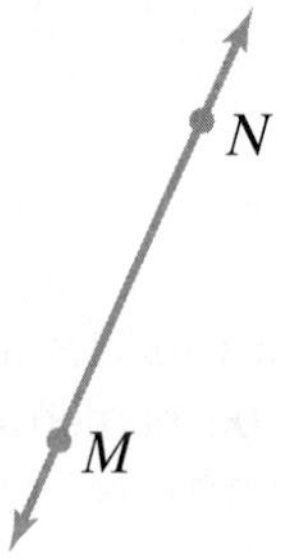

c.

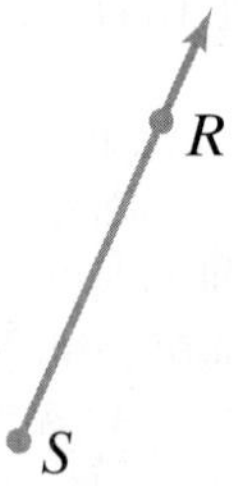

(eText Screens 8.1-1–8.1-31)

Solutions

a. This figure is a section of a line with two endpoints, so it a line segment. It can be named as either $\overline{JK}$ or $\overline{KJ}$.

b. This figure is a straight row of points that continues without end in both directions, so it is a line. It can be named as either $\overleftrightarrow{MN}$ or $\overleftrightarrow{NM}$.

c. This figure is a section of a line with one endpoint and continues without end in one direction, so it is a ray. It can be named as $\overrightarrow{SR}$.

You Try It **Work through this You Try It problem.**

Work Exercises 1–6 in this textbook or in the MyMathLab Study Plan.

Two **rays** that have the same endpoint form an **angle**. The common endpoint is called the **vertex**, and the two rays are called the **sides** of the angle. For example, the vertex of the angle in Figure 5 is B, and the sides are $\overrightarrow{BA}$ and $\overrightarrow{BC}$.

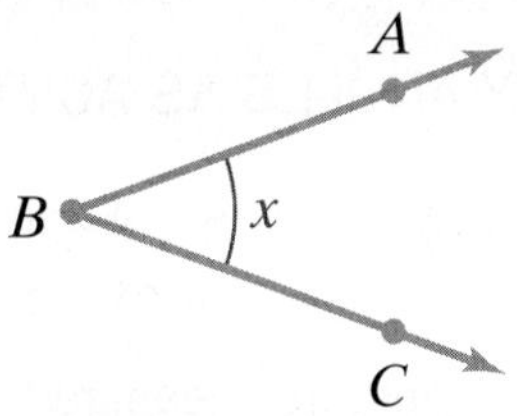

Figure 5

We can name an angle in several ways:

1. Write the angle symbol $\angle$ followed by a point from one side, the vertex, and a point from the other side. The vertex must be in the middle. For example, we can name the angle in Figure 5 as $\angle ABC$ or $\angle CBA$.
2. Write the angle symbol $\angle$ followed by the vertex. For example, the angle in Figure 5 can be named $\angle B$.
3. Write the angle symbol $\angle$ and a variable or number to label the angle. For example, the angle in Figure 5 can be named as $\angle x$.

CAUTION If two or more **angles** share the same **vertex**, then the vertex alone cannot be used to name an angle. For example, we cannot use $\angle B$ to name an angle in Figure 6 because it would be unclear if $\angle B$ means $\angle ABD$, $\angle ABC$, or $\angle CBD$.

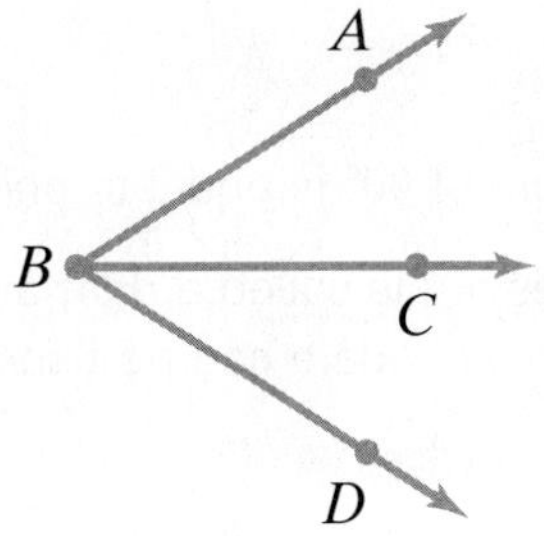

Figure 6

Example 2 Naming Angles

My video summary

Using the figure below, name each angle in two other ways.

a. $\angle 1$ **b.** $\angle TQR$ **c.** $\angle 2$

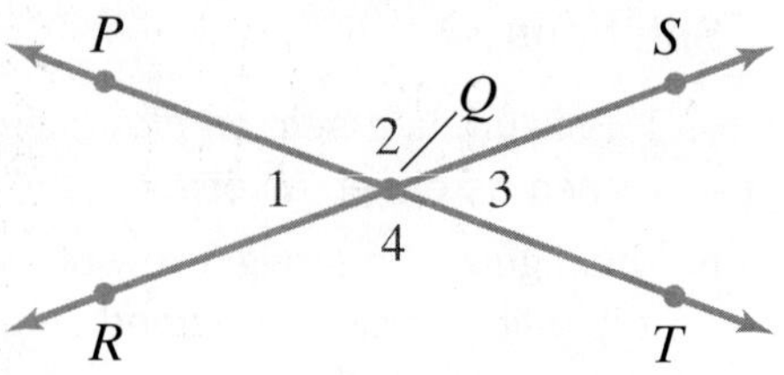

Solutions View the answers below, or watch this **video** for complete solutions.

a. Two other names for $\angle 1$ are $\angle PQR$ and $\angle RQP$.

b. $\angle TQR$ can also be named as $\angle RQT$ and $\angle 4$.

c. $\angle PQS$ and $\angle SQP$ are other names for $\angle 2$.

You Try It **Work through this You Try It problem.**

Work Exercises 7–10 in this textbook or in the MyMathLab Study Plan.

OBJECTIVE 2 CLASSIFY ANGLES AS ACUTE, OBTUSE, RIGHT, OR STRAIGHT

Recall from **Topic 6.4** that the letter m is used with an angle as a symbol representing *the measure of an angle*. For example, $m\angle ABC$ reads as "the measure of $\angle ABC$."

Typically, **angles** are measured in **degrees**. The symbol used to represent degrees is a small raised circle, °. A measure of 360 degrees, or 360°, is equal to one full revolution as shown in Figure 7.

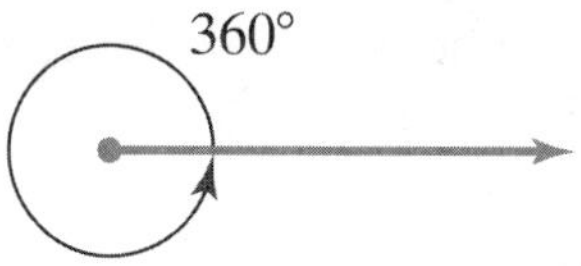

Figure 7

A measure of 180° is equal to one-half of a full revolution $\left(\frac{1}{2}\cdot 360° = 180°\right)$. An angle that measures 180° is called a **straight angle**. For example, $\angle EFG$ in Figure 8 is a straight angle.

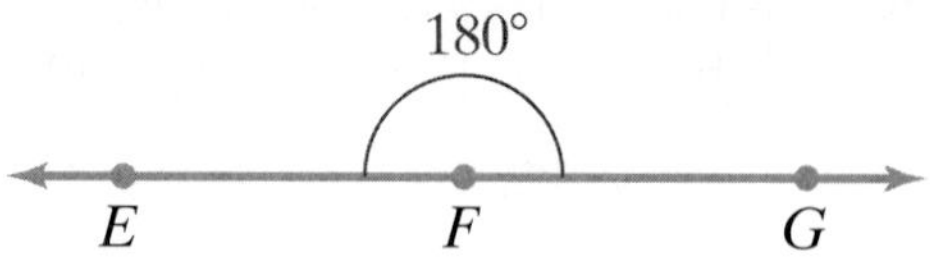

Figure 8

A measure of 90° is equal to one-fourth of a full revolution $\left(\frac{1}{4}\cdot 360° = 90°\right)$. An **angle** that measures 90° is called a **right angle**. For example, $\angle LMN$ in Figure 9 is a right angle. Note that a small square can be used as a symbol to label a right angle.

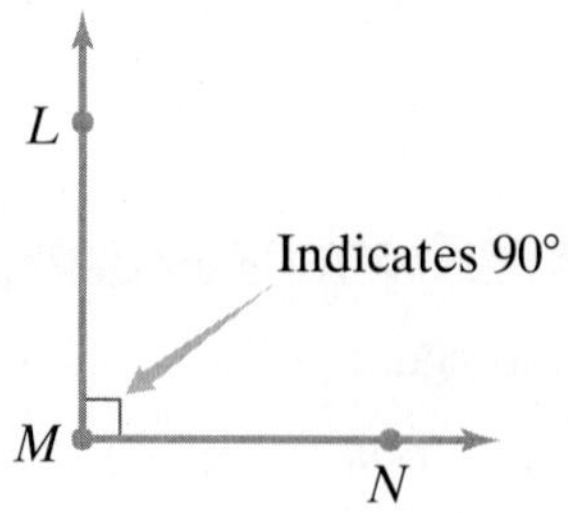

Figure 9

An **acute angle** is an angle that measures between $0°$ and $90°$. An **obtuse angle** measures between $90°$ and $180°$. In Figure 10, $\angle QRS$ is an acute angle and $\angle TUV$ is an obtuse angle.

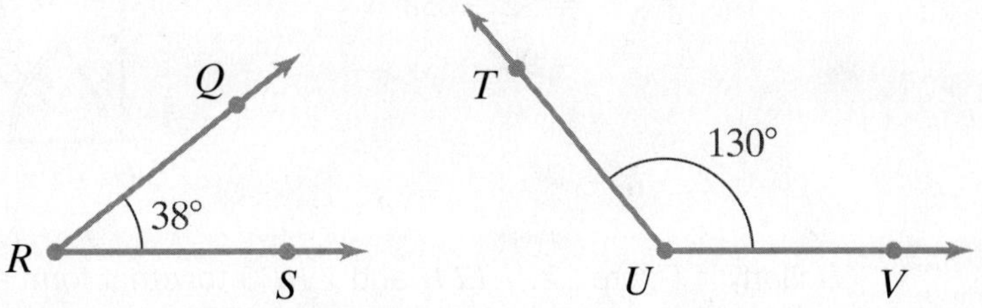

Figure 10

Example 3 Classifying Angles

My video summary

Classify each angle as **acute**, **obtuse**, **right**, or **straight**.

a.

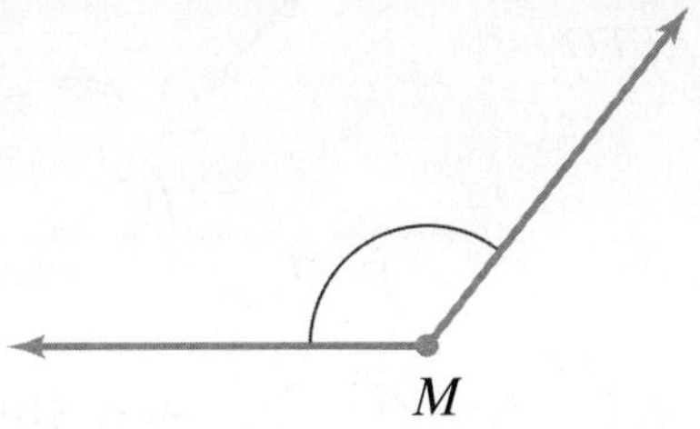

b.

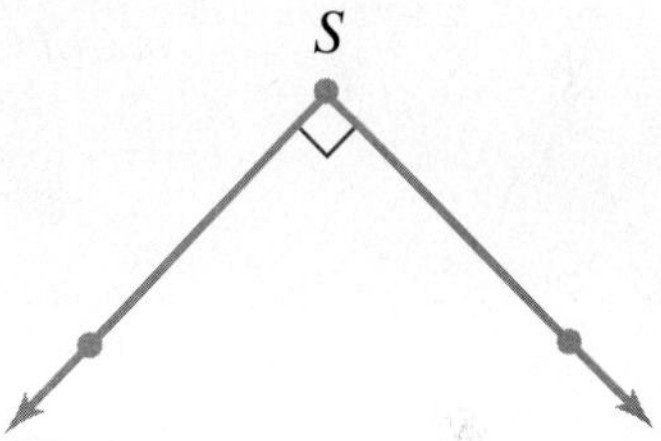

c.

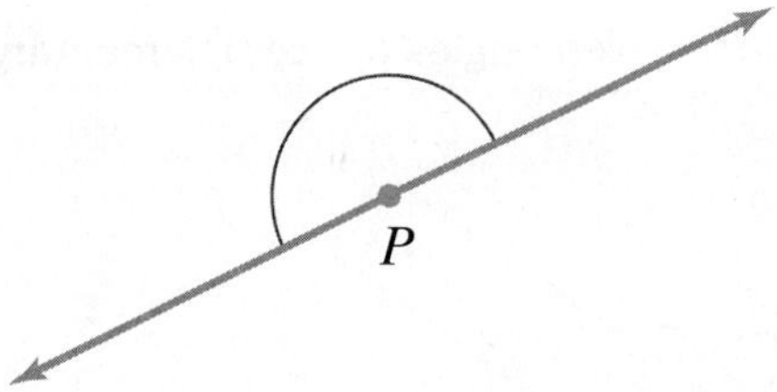

d.

Solutions Continue reading for the answers, or watch this **video** for complete solutions.

a. $\angle M$ is an obtuse angle. Its measure is between $90°$ and $180°$.

b. $\angle S$ is a right angle. It measures $90°$ as indicated by the square symbol.

c. $\angle P$ is a straight angle. Its measure is $180°$.

d. $\angle D$ is an acute angle. Its measure is between $0°$ and $90°$.

You Try It **Work through this You Try It problem.**

Work Exercises 11–18 in this textbook or in the MyMathLab Study Plan.

OBJECTIVE 3 IDENTIFY COMPLEMENTARY AND SUPPLEMENTARY ANGLES AND FIND THEIR MEASURES

In **Figure 11**, notice that $\angle ABD$ and $\angle DBC$ together form a **right angle**, $\angle ABC$. Two angles are **complementary angles** if the **sum** of their measures is $90°$. We say that $\angle ABD$ and $\angle DBC$ are **complements** of each other.

$$m\angle ABC = m\angle ABD + m\angle DBC$$
$$= 32^\circ + 58^\circ$$
$$= 90^\circ$$

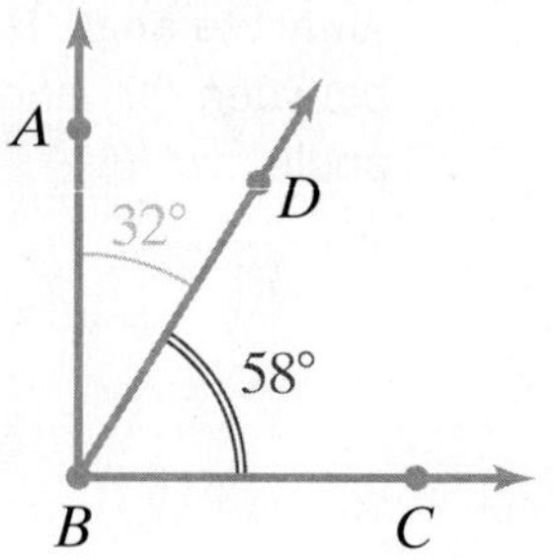

Figure 11

Similarly in Figure 12, $\angle EFH$ and $\angle HFG$ together form a **straight angle**, $\angle EFG$. Two **angles** are **supplementary angles** if the sum of their measures is 180°. We say that $\angle EFH$ and $\angle HFG$ are **supplements** of each other.

$$m\angle EFG = m\angle EFH + m\angle HFG$$
$$= 45^\circ + 135^\circ$$
$$= 180^\circ$$

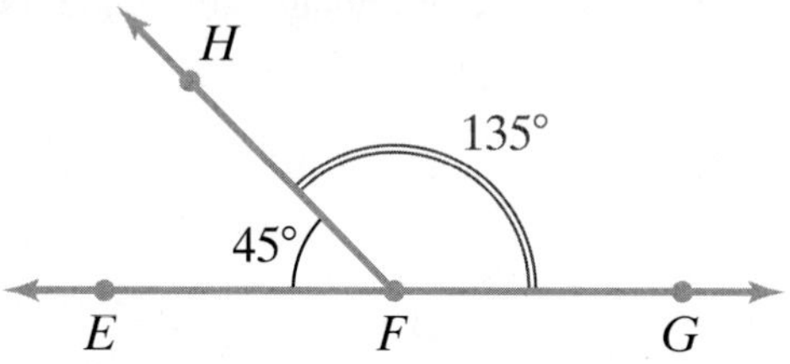

Figure 12

Example 4 Identifying Complementary and Supplementary Angles

My video summary

Determine whether the given **angles** are **complementary angles, supplementary angles**, or neither.

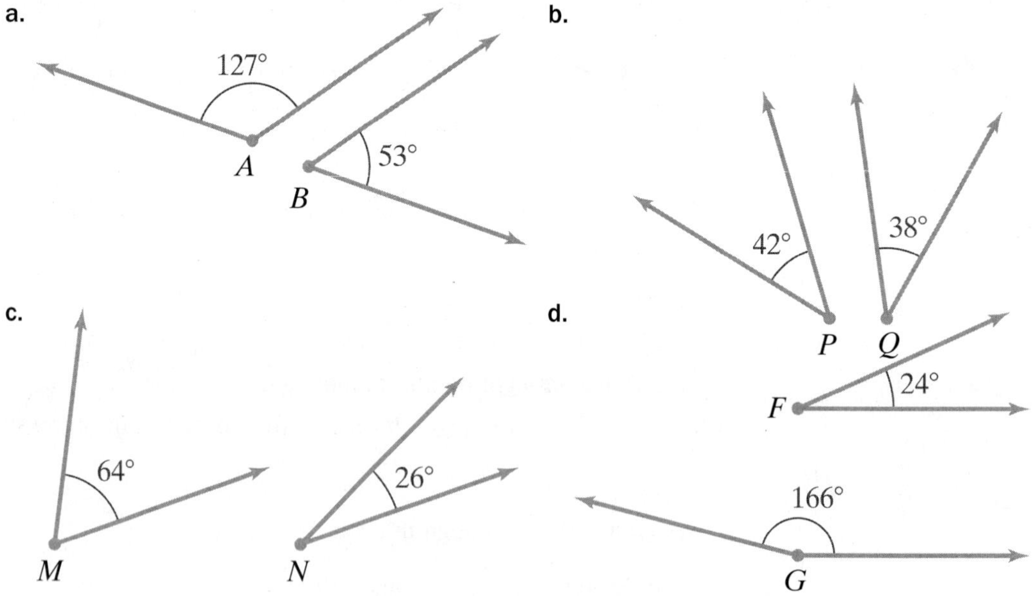

Solutions We check to see if the **sum** of the angle measures is 90°, 180°, or neither.

a. $127^\circ + 53^\circ = 180^\circ$, so $\angle A$ and $\angle B$ are supplementary angles.

b. $42^\circ + 38^\circ = 80^\circ$. This sum is not 90° or 180°, so $\angle P$ and $\angle Q$ are neither complementary nor supplementary angles.

c.–d. Try working these problems on your own. View the **answers**, or watch this **video** for complete solutions to all four parts.

You Try It **Work through this You Try It problem.**

Work Exercises 19–24 in this textbook or in the MyMathLab Study Plan.

Example 5 Finding the Measures of Complementary and Supplementary Angles

Find the measure of each angle described.

a. The **complement** of an $18°$ angle.

b. The **supplement** of a $23°$ angle.

c. The complement of a $51°$ angle.

d. The supplement of a $104°$ angle.

Solutions

a. The measures of complementary angles add to $90°$, so the complement of an angle measuring $18°$ has a measure of $90° - 18° = 72°$.

b. Supplementary angles have measures that add to $180°$. Thus, the supplement of a $23°$ angle has a measure of $180° - 23° = 157°$.

My video summary

c.–d. Try working these problems on your own. View the **answers**, or watch this **video** for complete solutions to parts c and d.

You Try It **Work through this You Try It problem.**

Work Exercises 25–28 in this textbook or in the MyMathLab Study Plan.

OBJECTIVE 4 FIND AN UNKNOWN MEASURE OF AN ANGLE USING GIVEN INFORMATION

Two **angles** that have the same **vertex** and a common **side** are **adjacent angles** provided that they do not overlap. For example, $\angle ABC$ and $\angle CBD$ in Figure 13 are adjacent angles because they have the same vertex, B, and a common side, $\overrightarrow{BC}$. The two adjacent angles together form a third angle, $\angle ABD$. If we know the measures of two of these angles, we can add or subtract to find the measure of the third angle.

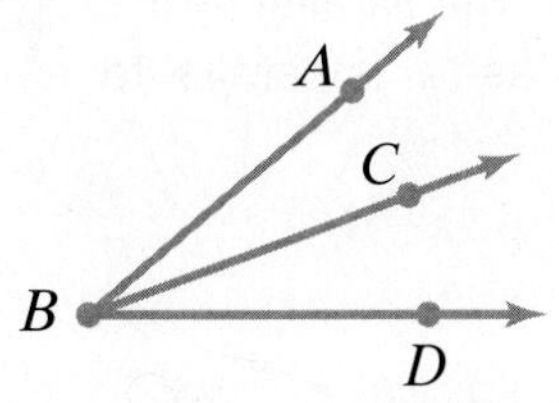

Figure 13

Example 6 Finding Unknown Measures of Angles

Find the measure of each indicated angle.

a. $\angle LMO$ **b.** $\angle x$

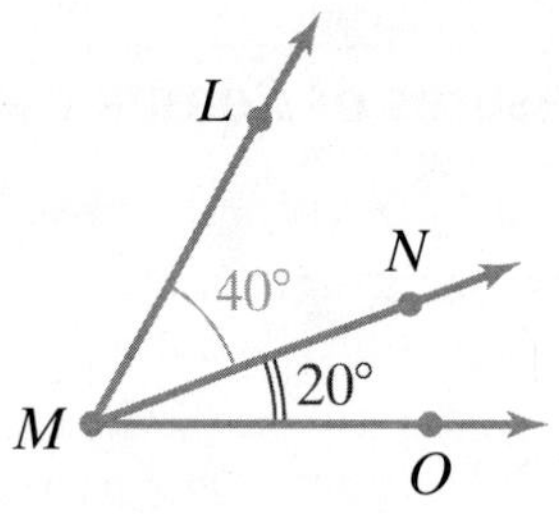

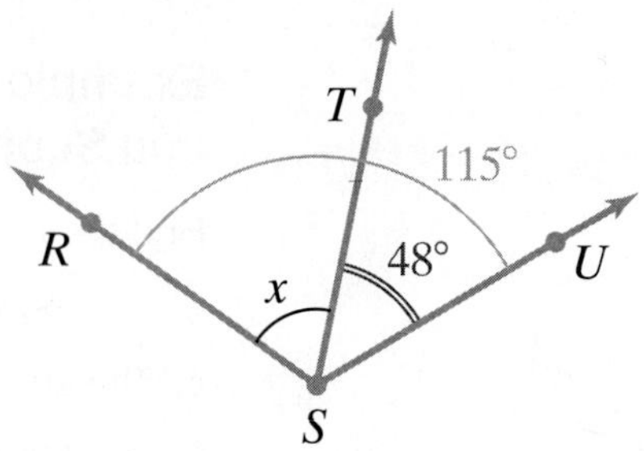

Solutions

a. $\angle LMN$ and $\angle NMO$ are **adjacent angles** that together form $\angle LMO$. From the figure, we know $m\angle LMN = 40°$ and $m\angle NMO = 20°$. We can add to find the measure of $\angle LMO$:

$$m\angle LMO = m\angle LMN + m\angle NMO = 40° + 20° = 60°$$

My video summary

b. $\angle RST$ and $\angle TSU$ are adjacent angles that together form $\angle RSU$. Try to use the given measures to find the measure of $\angle x$ on your own. View the **answer**, or watch this **video** for a complete solution to part b.

You Try It **Work through this You Try It problem.**

Work Exercises 29–32 in this textbook or in the MyMathLab Study Plan.

A **plane** is a two-dimensional flat surface that extends forever in all four directions.

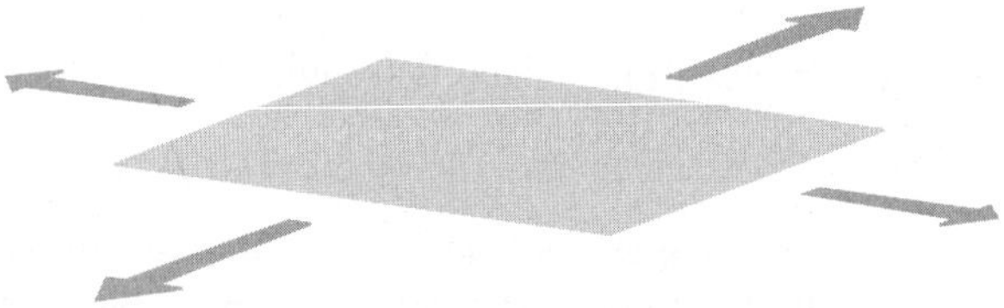

When two different **lines** lie on the same **plane**, they must be either *intersecting lines* or *parallel lines*. **Intersecting lines** cross (or touch) at a single point, whereas **parallel lines** never cross (or touch). In Figure 14, lines k and l are intersecting lines, while lines m and n are parallel. The parallel symbol, $\|$, can be used to indicate two parallel lines. For example, $m \| n$ reads as "m is parallel to n."

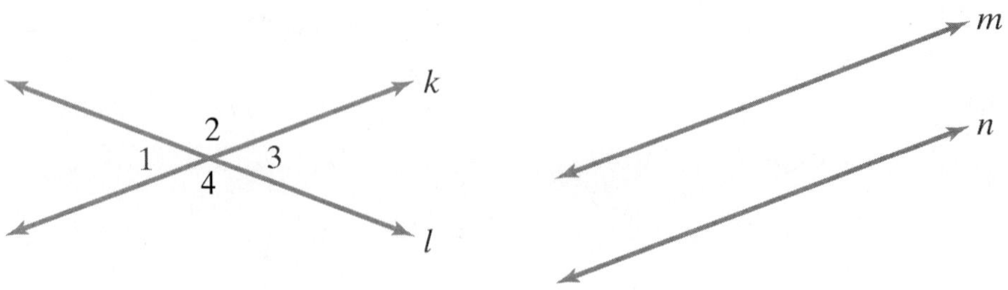

(a) Intersecting lines (b) Parallel lines **Figure 14**

When two lines intersect, they form four **angles**. Two angles that are opposite each other (not adjacent) are called **vertical angles**. In Figure 14a, $\angle 1$ and $\angle 3$ are vertical angles. Similarly, $\angle 2$ and $\angle 4$ are vertical angles. The measures of two vertical angles are equal. So, $m\angle 1 = m\angle 3$ and $m\angle 2 = m\angle 4$.

Any two **adjacent angles** formed by two intersecting lines are **supplementary angles**. Each of the following pairs of angles from Figure 14a are supplementary: $\angle 1$ and $\angle 2$, $\angle 2$ and $\angle 3$, $\angle 3$ and $\angle 4$, and $\angle 4$ and $\angle 1$.

Example 7 Finding Measures of Angles Involving Intersecting Lines

In the figure, the measure of $\angle a$ is $70°$. Find the measures of $\angle b$, $\angle c$, and $\angle d$.

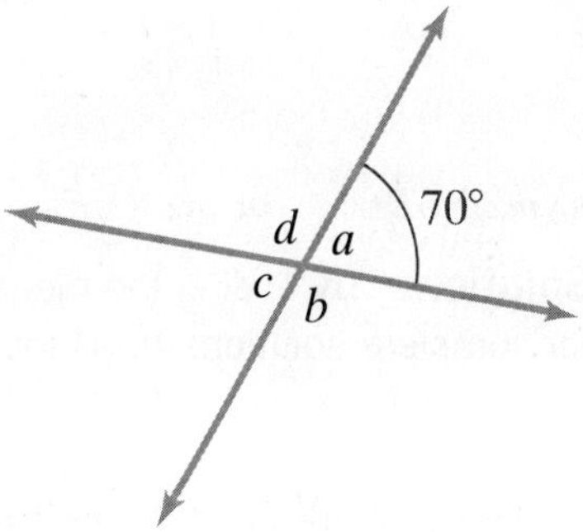

Solution Because $\angle a$ and $\angle c$ are **vertical angles**, their measures are equal. Therefore, $m\angle c = 70°$.

Because $\angle a$ and $\angle b$ are **adjacent angles** that are **supplementary**, their measures add to $180°$. Thus, $m\angle b = 180° - 70° = 110°$.

Finally, because $\angle b$ and $\angle d$ are vertical angles, their measures are equal. So, $m\angle d = 110°$.

You Try It **Work through this You Try It problem.**

Work Exercises 33 and 34 in this textbook or in the MyMathLab Study Plan.

CAUTION

Identifying **vertical angles** can be confusing when **angles** are formed by three or more lines **intersecting** at the same point. To help with this, remember that a pair of vertical angles must be formed by the same two lines. For example, in Figure 15, $\angle 1$ and $\angle 4$ are vertical angles. Both are formed by the same lines, m and n. However, $\angle 1$ and $\angle 3$ are *not* vertical angles because they are formed by different pairs of lines. $\angle 1$ is formed by lines m and n, while $\angle 3$ is formed by lines l and m.

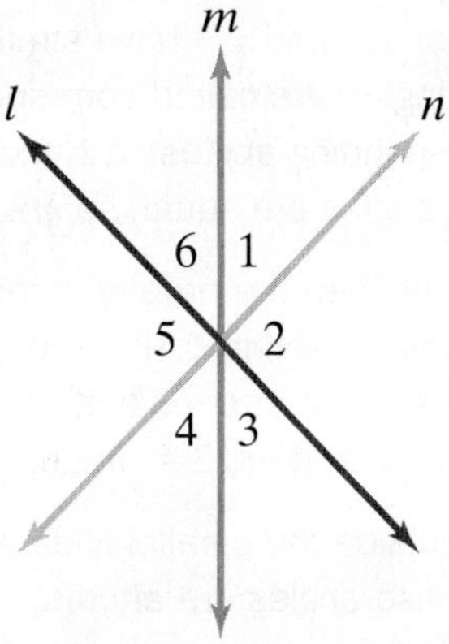

Figure 15

Example 8 Finding Measures of Angles Involving Intersecting Lines

My video summary

In the figure, $m\angle BGC = 65°$ and $m\angle CGD = 77°$. Find the following measures.

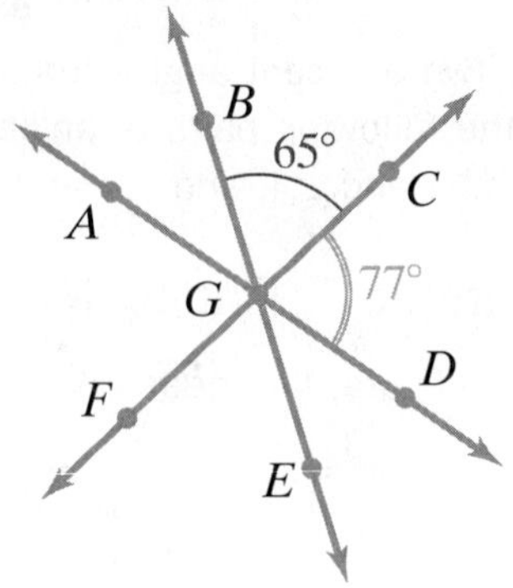

a. $m\angle FGE$ **b.** $m\angle AGF$ **c.** $m\angle DGE$ **d.** $m\angle AGB$

Solutions Try finding the measures on your own. View the **answers**, or watch this **video** for complete solutions to all four parts.

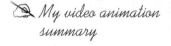

You Try It **Work through this You Try It problem.**

Work Exercises 35 and 36 in this textbook or in the MyMathLab Study Plan.

A **transversal line** is a line that intersects two or more **lines** at different points. In Figure 16, line t is a transversal intersecting two **parallel lines,** l and m. When a transversal intersects with two parallel lines, eight **angles** are formed that have special relationships.

My video animation summary

Continue reading to learn more about these relationships, or watch this **video animation** and then skip forward to **Example 9**.

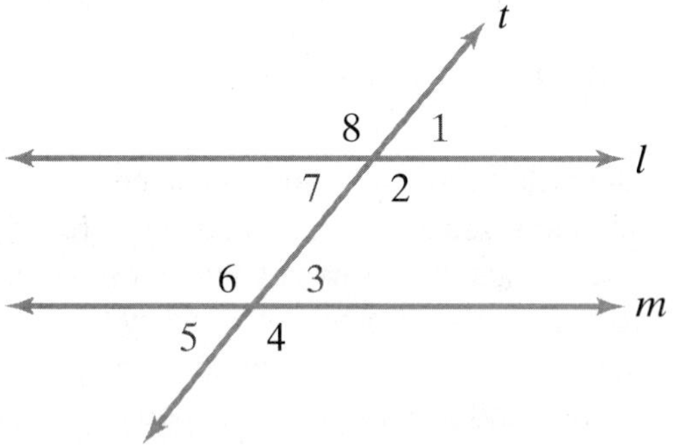

Figure 16

Notice that $\angle 1$ and $\angle 3$ have similar positions along the transversal and parallel lines. Such pairs of angles are called **corresponding angles.** Similarly, the following pairs of angles also are corresponding angles: $\angle 2$ and $\angle 4$, $\angle 5$ and $\angle 7$, and $\angle 6$ and $\angle 8$. The measures of corresponding angles are equal, so $m\angle 1 = m\angle 3$, $m\angle 2 = m\angle 4$, $m\angle 5 = m\angle 7$, and $m\angle 6 = m\angle 8$.

Looking between the **parallel lines** l and m in Figure 16, $\angle 7$ and $\angle 3$ are on opposite sides of the **transversal line** t. These angles are called **alternate interior angles.** Similarly, $\angle 6$ and $\angle 2$ are alternate interior angles. The measures of alternate interior angles are equal, so $m\angle 3 = m\angle 7$ and $m\angle 2 = m\angle 6$.

Looking outside the parallel lines in Figure 16, $\angle 1$ and $\angle 5$ are on opposite sides of the transversal. These angles are **alternate exterior angles.** Similarly, $\angle 8$ and $\angle 4$ are alternate exterior angles. The measures of alternate exterior angles are equal, so $m\angle 1 = m\angle 5$ and $m\angle 8 = m\angle 4$.

Measures of Angles Formed by a Transversal and Parallel Lines

When a transversal line intersects two parallel lines, then

1. The measures of **corresponding angles** are equal.
2. The measures of alternate interior angles are equal.
3. The measures of alternate exterior angles are equal.

If the measure of any one of the angles formed by a transversal intersecting two parallel lines is known, then the measures of the other seven angles can be found.

Example 9 Finding Measures of Angles Involving a Transversal and Parallel Lines

My video summary

In the figure, the measure of $\angle 3$ is $42°$. Given $k \| l$, find the measures of the other seven labeled **angles**.

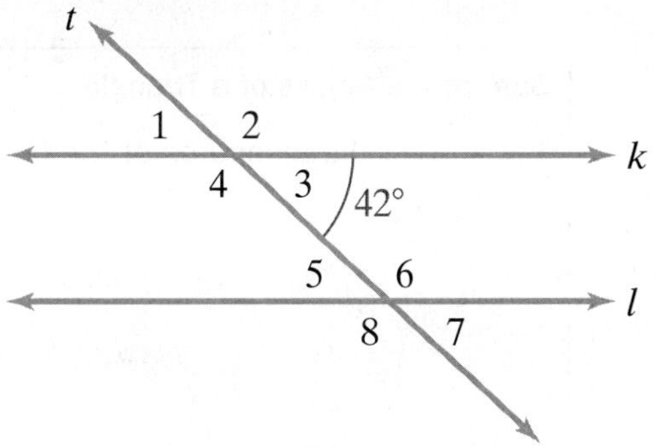

Solution Work through the following, or watch this **video** for a complete solution.

$\angle 3$ and $\angle 5$ are **alternate interior angles**, so $m\angle 5 = 42°$.

$\angle 3$ and $\angle 7$ are **corresponding angles**, so $m\angle 7 = 42°$.

$\angle 3$ and $\angle 1$ are **vertical angles**, so $m\angle 1 = 42°$.

$\angle 1$ and $\angle 2$ together form a **straight angle**, which means they are **supplementary angles.** Thus, $m\angle 2 = 180° - 42° = 138°$.

$\angle 2$ and $\angle 4$ are vertical angles, so $m\angle 4 = 138°$.

$\angle 2$ and $\angle 6$ are corresponding angles, so $m\angle 6 = 138°$.

$\angle 2$ and $\angle 8$ are **alternate exterior angles**, so $m\angle 8 = 138°$.

You Try It **Work through this You Try It problem.**

Work Exercises 37 and 38 in this textbook or in the MyMathLab Study Plan.

TIP Often when working with **parallel lines** that are intersected by a **transversal**, we can find the measures of **angles** in more than one way. For instance, in Example 9, we found the measure of $\angle 7$ by recognizing that $\angle 7$ and $\angle 3$ were **corresponding angles.** We could also have found the measure of $\angle 7$ by recognizing that $\angle 7$ and $\angle 5$ were **vertical angles.**

(eText Screens 8.1-1–8.1-31)

To conclude this topic, we examine the angles of a triangle. Figure 17a shows a triangle with angles x, y, and z. If the triangle is split apart (Figure 17b) and the three angles are arranged to fit together (Figure 17c), a **straight angle** is formed. This shows that the measures of the three angles of a triangle add to $180°$.

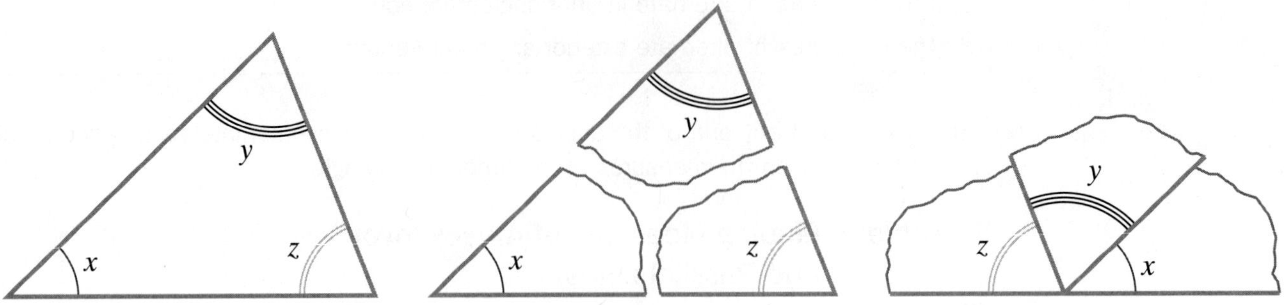

Figure 17

Sum of the Angles of a Triangle

The **sum** of the measures of the three **angles** in a triangle is $180°$.

For ΔABC,

$$m\angle A + m\angle B + m\angle C = 180°.$$

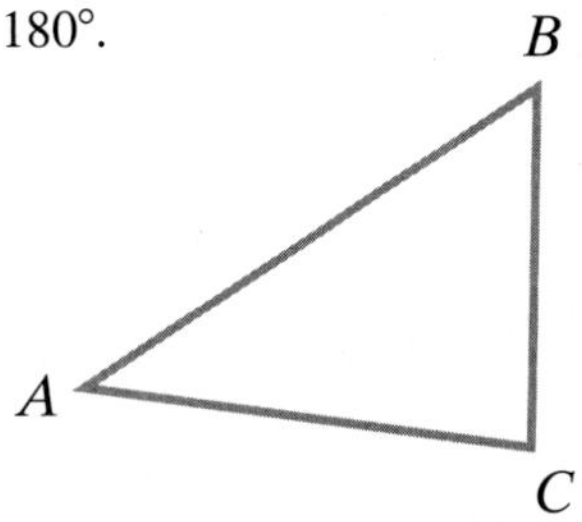

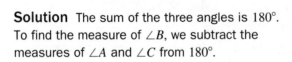

Example 10 Finding the Measure of an Angle in a Triangle

In the triangle shown, the measure of $\angle A$ is $72°$, and the measure of $\angle C$ is $34°$. Find the measure of $\angle B$.

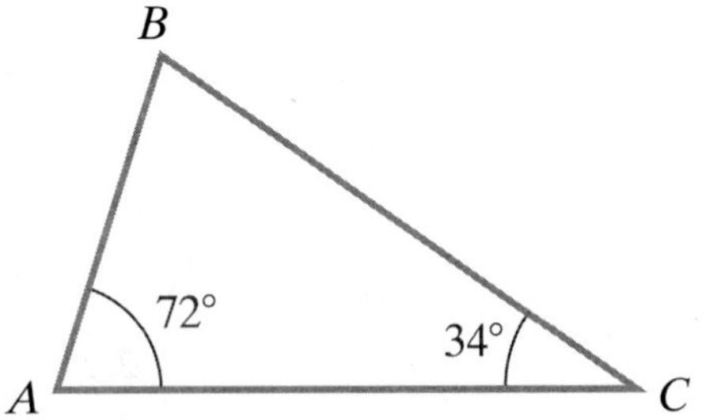

Solution The sum of the three angles is $180°$. To find the measure of $\angle B$, we subtract the measures of $\angle A$ and $\angle C$ from $180°$.

$$\begin{aligned} m\angle B &= 180° - m\angle A - m\angle C \\ &= 180° - 72° - 34° \\ &= 74° \end{aligned}$$

You Try It **Work through this You Try It problem.**

Work Exercises 39 and 40 in this textbook or in the MyMathLab Study Plan.

(eText Screens 8.1-1–8.1-31)

8.1 Exercises

In Exercises 1–6, identify the figure as a line, segment, or ray. Then use the correct symbol to name the figure.

You Try It

1. 

2. 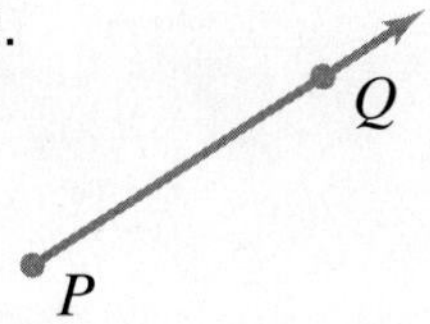

3. 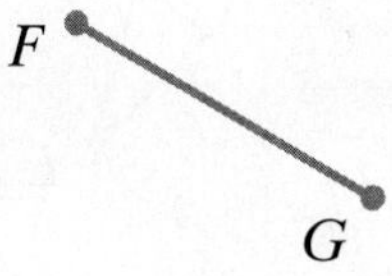

4. 

5. 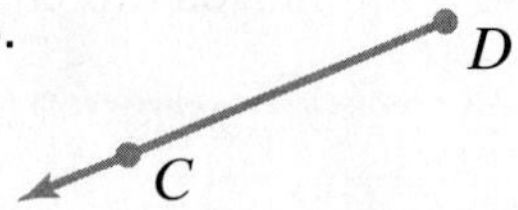

6. 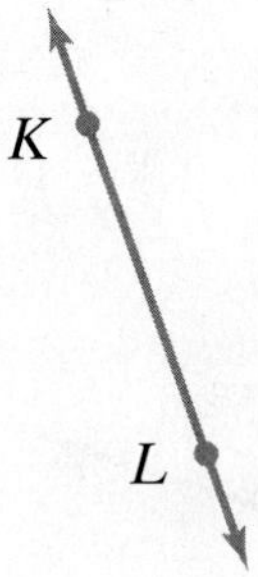

In Exercises 7–10, using the appropriate figure, name the given angle in two other ways.

You Try It

7. $\angle y$ 8. $\angle MON$ 9. $\angle 5$ 10. $\angle AGF$

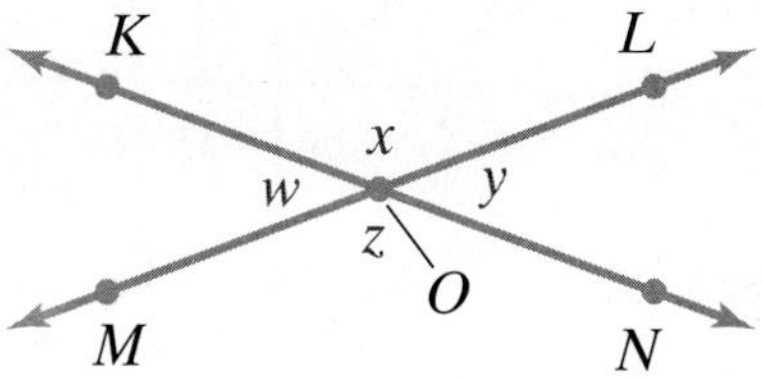

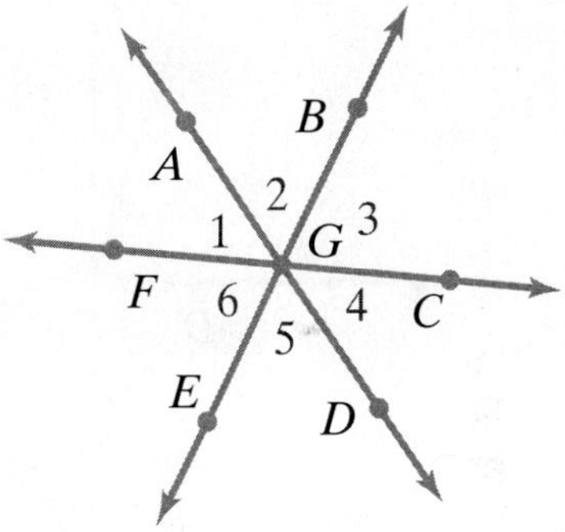

In Exercises 11–18, classify the angle as acute, obtuse, right, or straight.

You Try It

11. 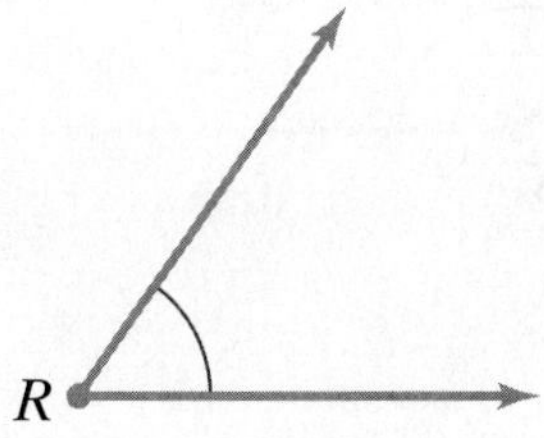

12. 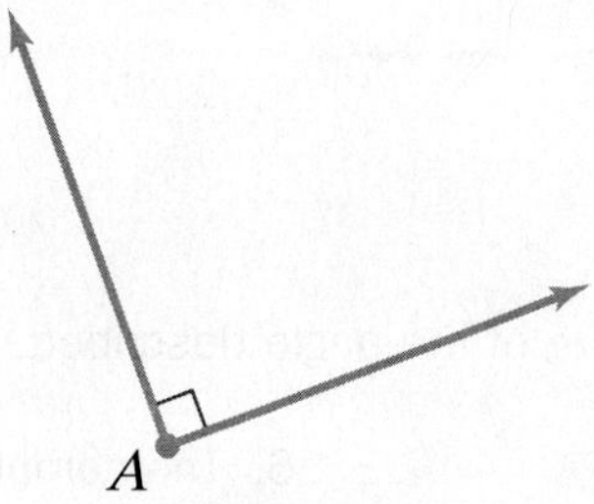

13. 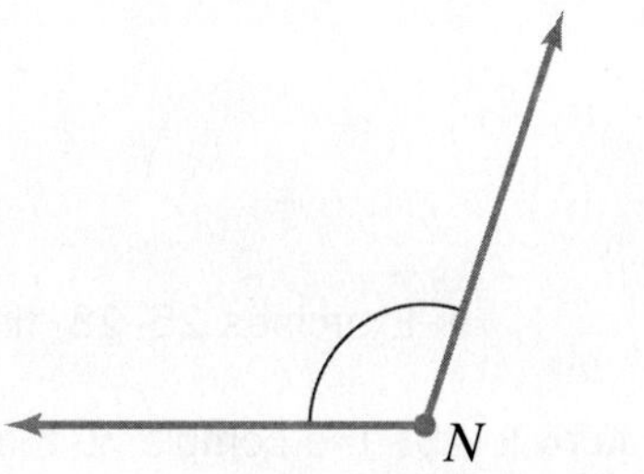

14. 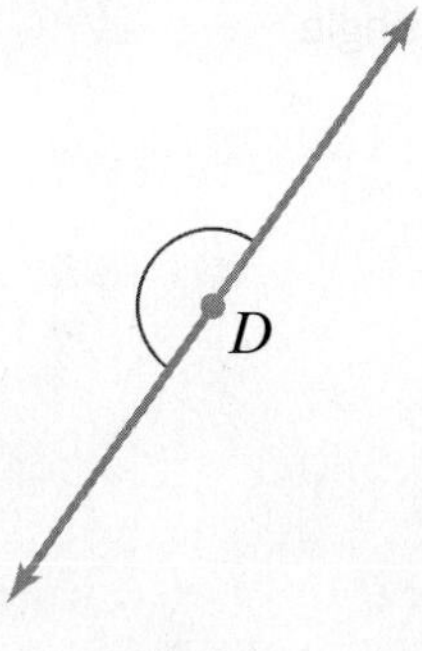

15. 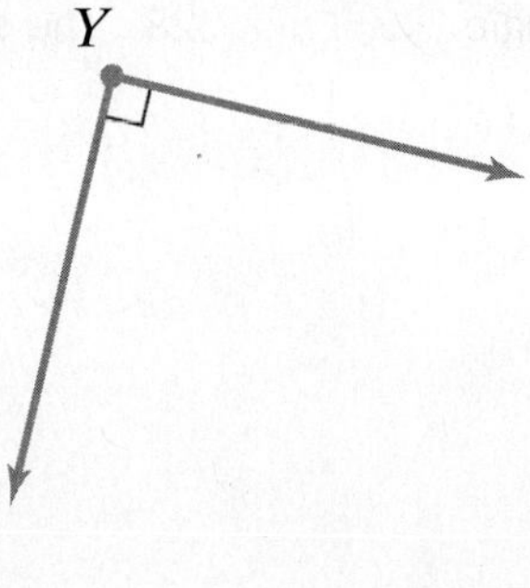

16. 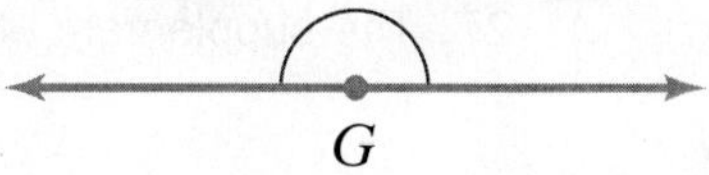

17.

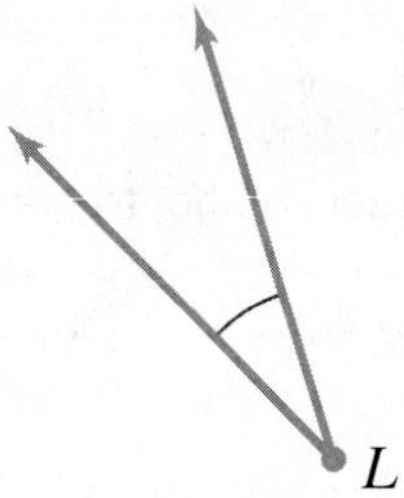

18.

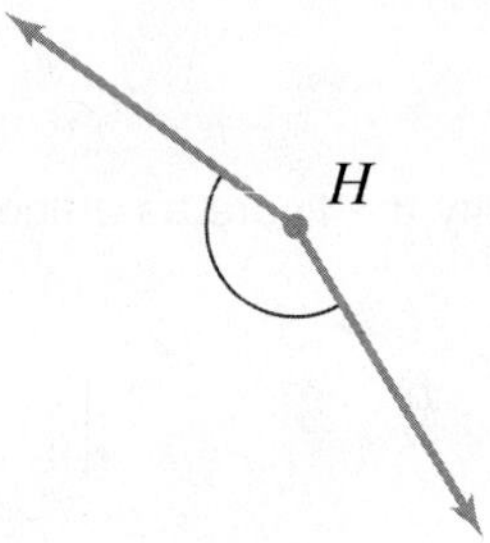

In Exercises 19–24, determine whether the given angles are complementary, supplementary, or neither.

You Try It 19.

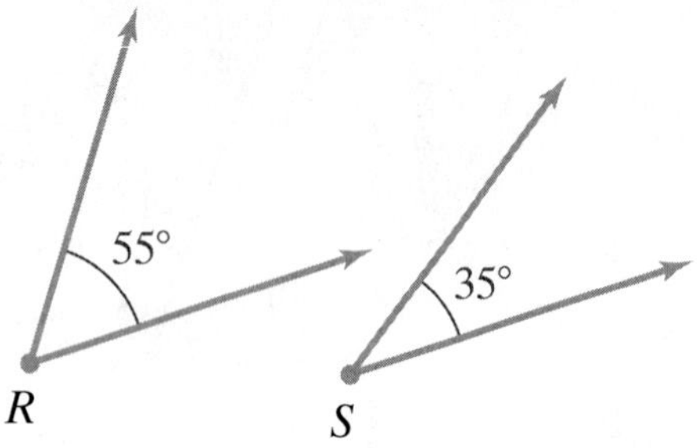

20.

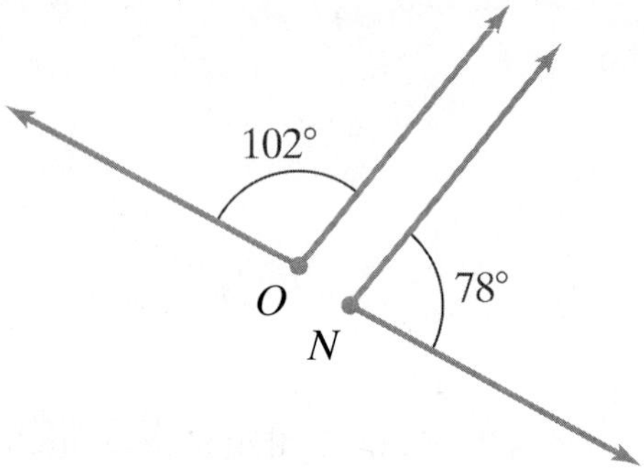

21.

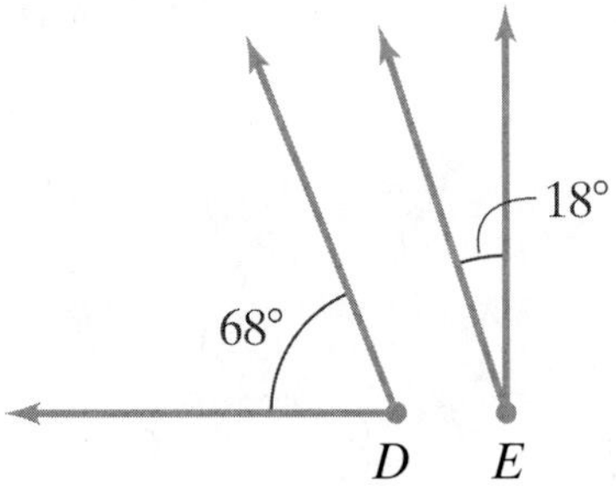

22.

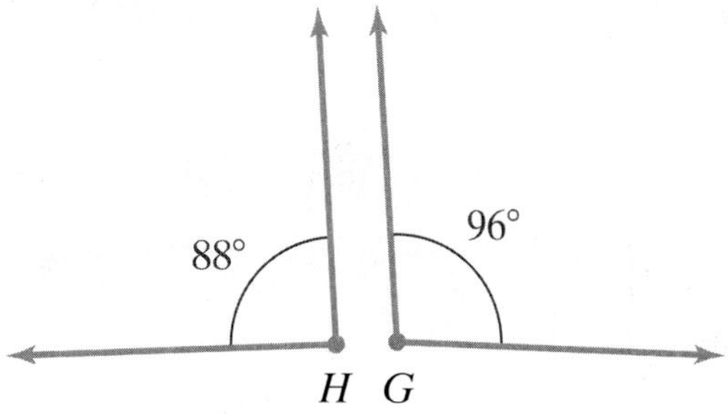

23.

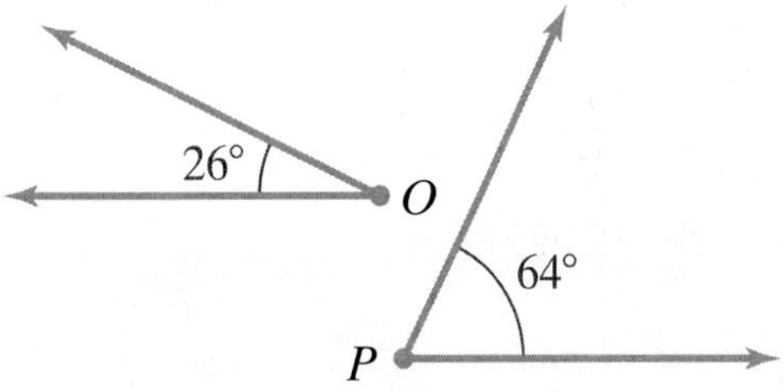

24.

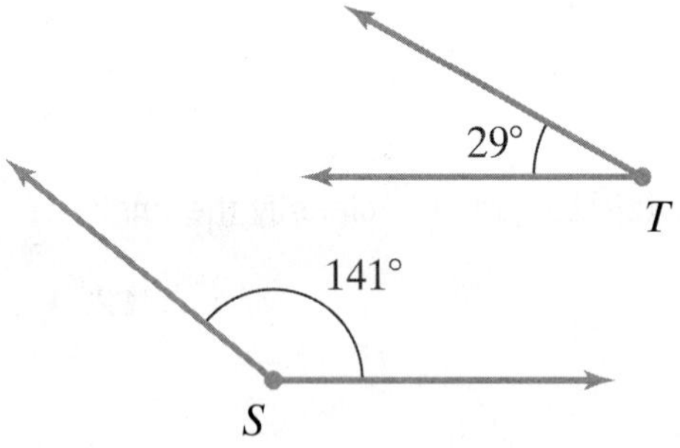

In Exercises 25–28, find the measure of the angle described.

You Try It 25. The complement of a 23° angle

26. The complement of a 56° angle

27. The supplement of a 41° angle

28. The supplement of a 116° angle

In Exercises 29–32, find the measure of the indicated angle.

You Try It 29. $\angle ABD$

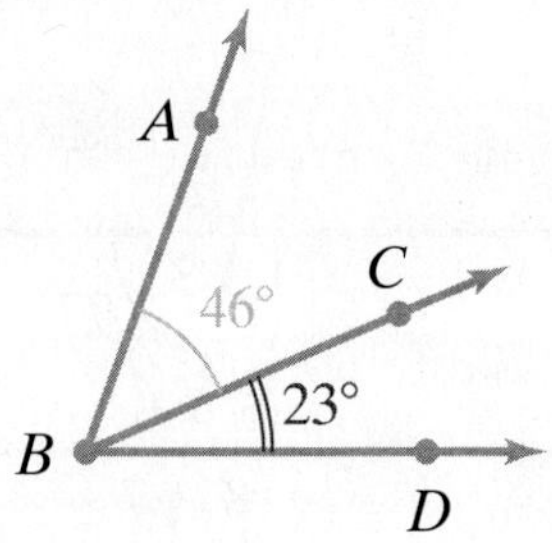

30. $\angle RSU$

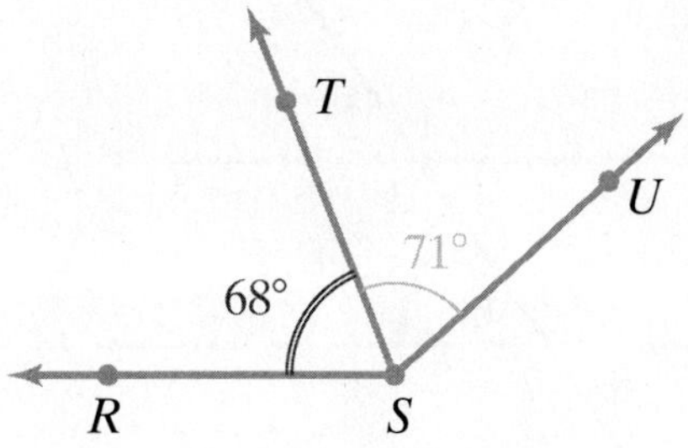

31. $\angle PNO$

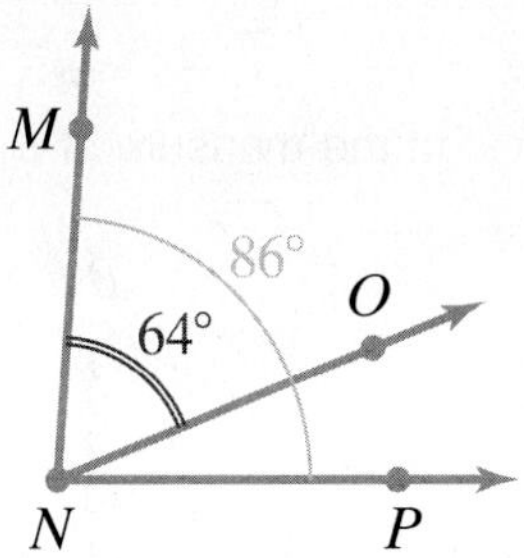

32. $\angle EFG$

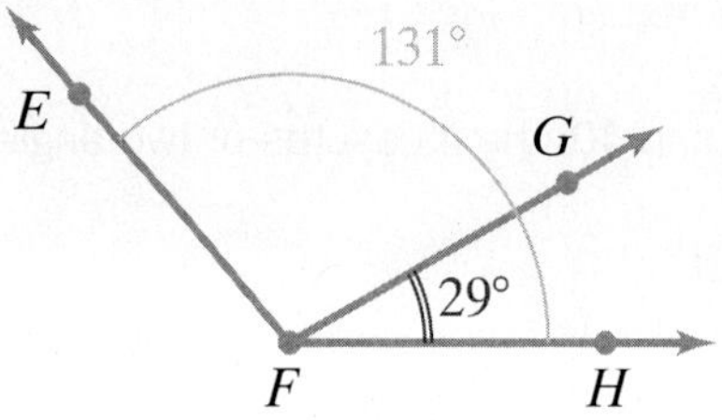

In Exercises 33 and 34, the measure of $\angle a$ is given. Find the measures of $\angle b$, $\angle c$, and $\angle d$.

You Try It 33.

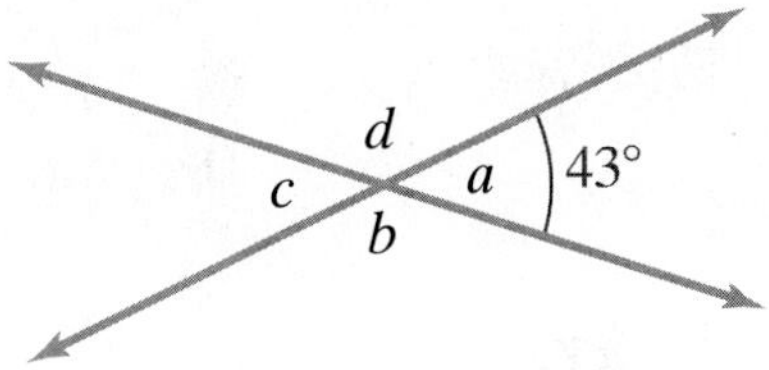

34.

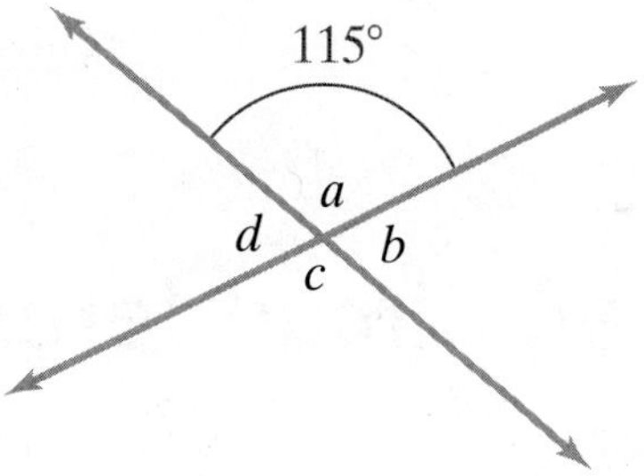

In Exercises 35 and 36, find the measure of the indicated angle.

You Try It 35. **a.** $\angle FGE$

b. $\angle AGF$

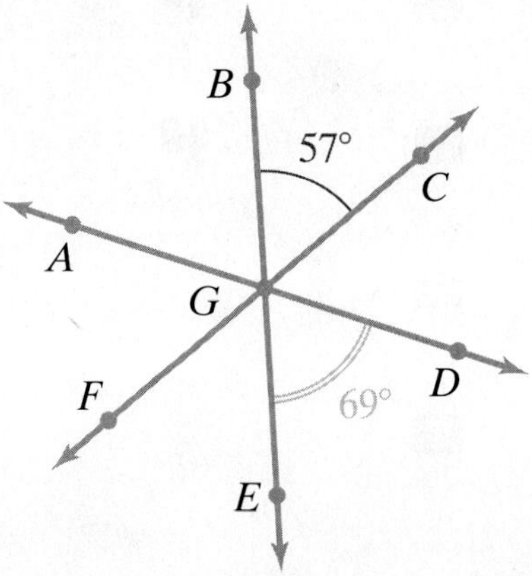

36. **a.** $\angle LOM$

b. $\angle KOL$

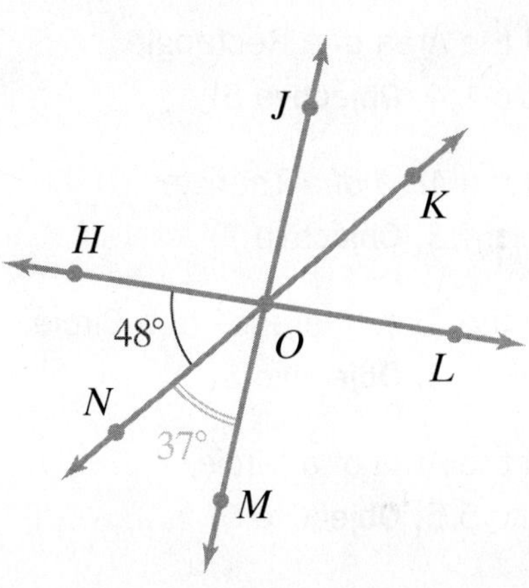

In Exercises 37 and 38, the measure of one angle is shown in the figure. Given $k \parallel l$, find the measures of the other seven labeled angles.

You Try It 37.

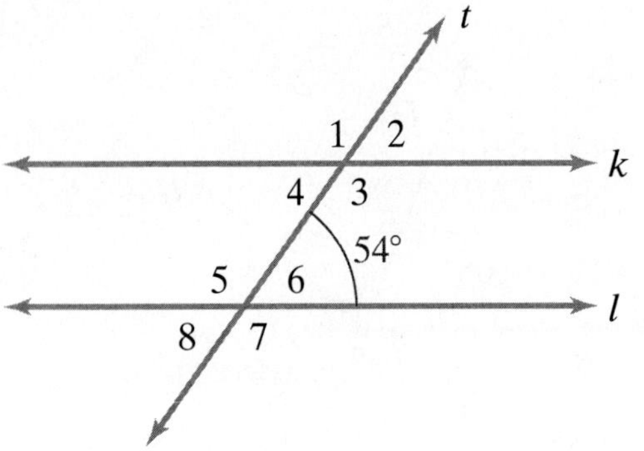

38.

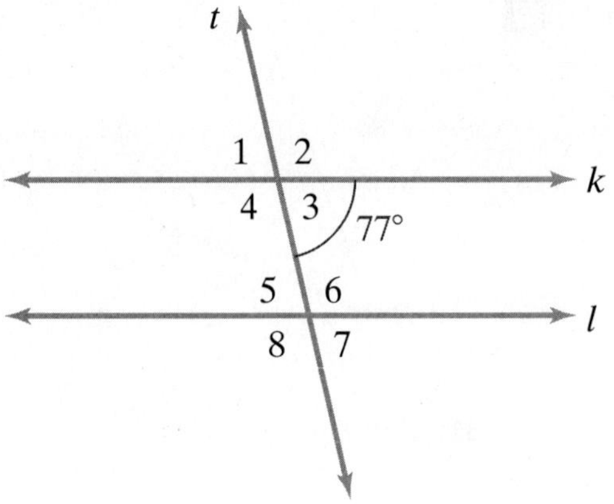

In Exercises 39 and 40, the measures of two angles of a triangle are shown. Find the measure of the third angle.

You Try It 39.

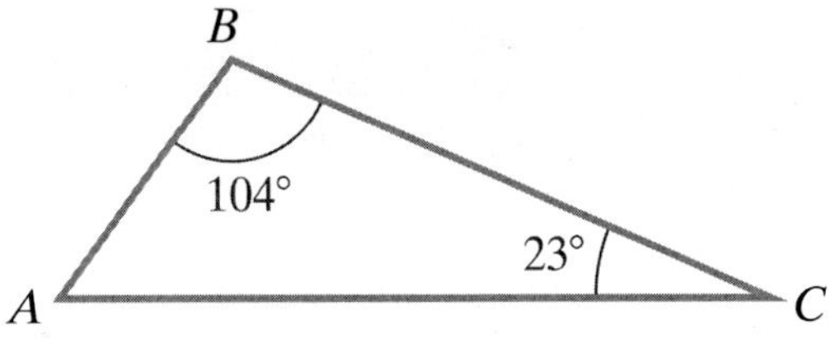

40.

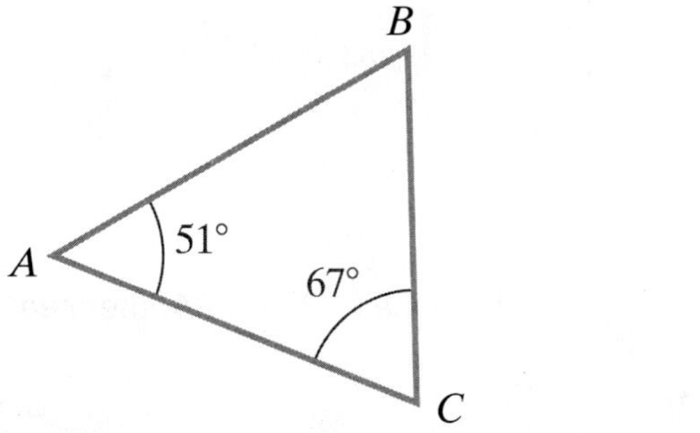

8.2 Perimeter, Circumference, and Area

THINGS TO KNOW

Before working through this topic, be sure you are familiar with the following concepts:

VIDEO ANIMATION INTERACTIVE

You Try It **1.** Find the Perimeter of a Polygon (Topic 1.3, **Objective 6**)

You Try It **2.** Find the Area of a Rectangle (Topic 1.4, **Objective 5**)

You Try It **3.** Find the Area of a Triangle (Topic 4.3, **Objective 7**)

You Try It **4.** Find the Circumference of a Circle (Topic 5.3, **Objective 5**)

You Try It **5.** Find the Area of a Circle (Topic 5.5, **Objective 5**)

(eText Screens 8.2-1–8.2-47)

OBJECTIVES

1 Find the Perimeter of Common Polygons

2 Find the Area of Common Polygons

3 Find the Circumference and Area of Circles

4 Find the Perimeter and Area of Figures Formed from Two or More Common Polygons

5 Solve Applications Involving Perimeter, Circumference, or Area

OBJECTIVE 1 FIND THE PERIMETER OF COMMON POLYGONS

In **Topic 1.3**, we found the **perimeter** of a **polygon** by adding up the lengths of the sides of the polygon. For common polygons, special **formulas** can be used to find the perimeter.

A **rectangle** is a four-sided polygon in which the sides meet at **right angles**. Consider the following general rectangle.

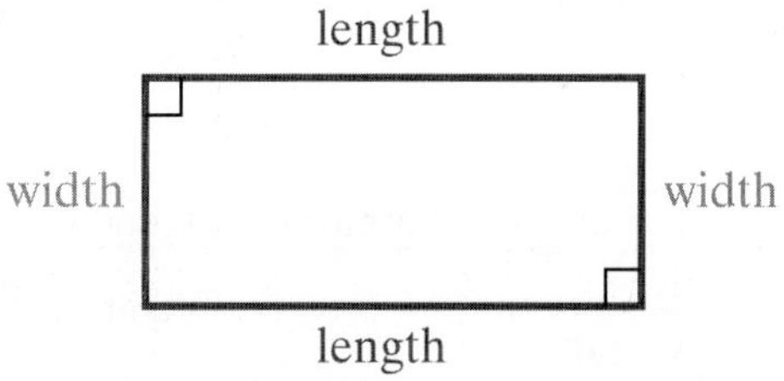

To find the perimeter of the rectangle, we add up the lengths of the sides.

$$\text{Perimeter} = \text{length} + \text{width} + \text{length} + \text{width}$$

In general, we find the perimeter of a rectangle by adding two lengths and two widths. That is,

$$\text{Perimeter} = 2 \cdot \text{length} + 2 \cdot \text{width} \quad \text{or} \quad P = 2l + 2w.$$

CAUTION Remember to include units when finding perimeters. The units for perimeter are the same as the units for the lengths of the sides of the polygon.

Perimeter of a Rectangle

If a rectangle has length l and width w, then its perimeter P is given by the formula

$$\textbf{Perimeter} = 2 \cdot \textbf{length} + 2 \cdot \textbf{width} \quad \text{or} \quad P = 2l + 2w.$$

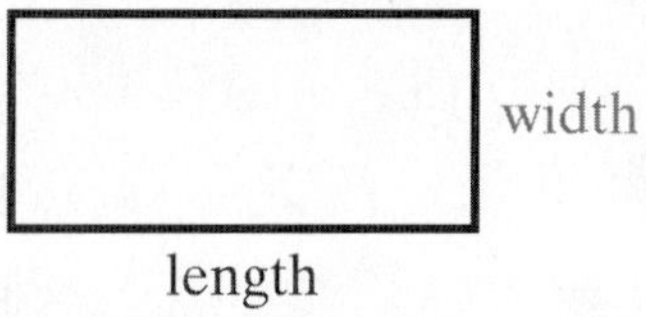

Example 1 Finding the Perimeter of a Rectangle

Find the perimeter of a **rectangle** with length 12 feet and width 7 feet.

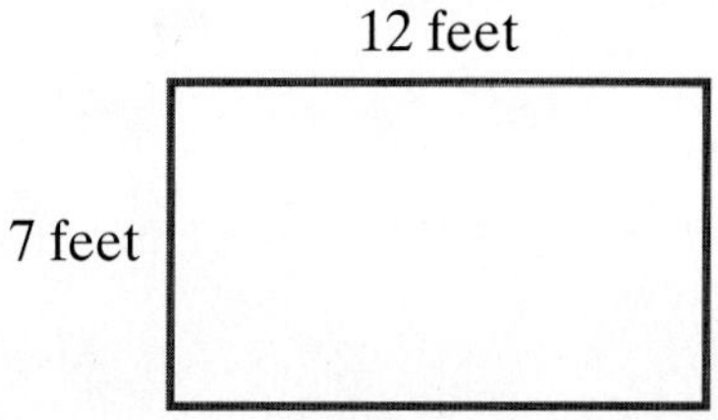

Solution Start with the perimeter formula: $P = 2l + 2w$

Substitute 12 ft for l and 7 ft for w: $= 2(12\text{ ft}) + 2(7\text{ ft})$

Multiply: $= 24\text{ ft} + 14\text{ ft}$

Add: $= 38\text{ ft}$

The perimeter of the rectangle is 38 feet.

You Try It **Work through this You Try It problem.**

Work Exercises 1 and 2 in this textbook or in the MyMathLab Study Plan.

A **square** is a special **rectangle** in which the lengths of all the sides are the same.

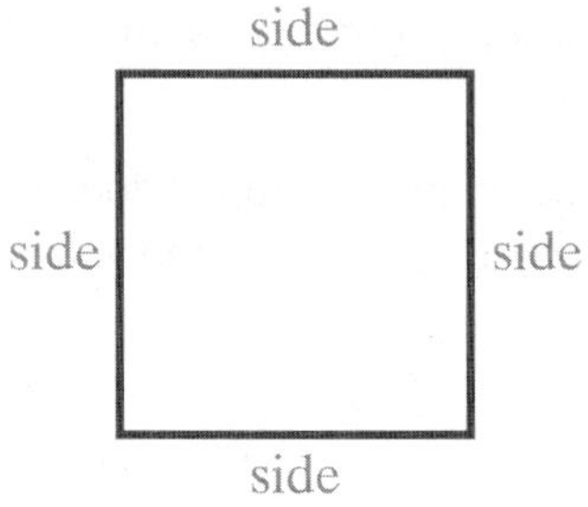

To find the **perimeter** of the square, we add up the lengths of the sides.

$$\text{Perimeter} = \text{side} + \text{side} + \text{side} + \text{side}$$

Generally, we find the perimeter of a square by adding the length of a side of the square four times. That is,

$$P = 4 \cdot \text{side} \quad \text{or} \quad P = 4s.$$

A square is an example of a **regular polygon**. In a regular polygon, the lengths of all the sides are the same. We can find the perimeter of a regular polygon by multiplying the length of one side of the **polygon** by the number of sides.

Perimeter of a Square

If a square has sides of length s, then its perimeter P is given by the formula

$$\text{Perimeter} = 4 \cdot \text{side} \quad \text{or} \quad P = 4s.$$

Example 2 Finding the Perimeter of a Square

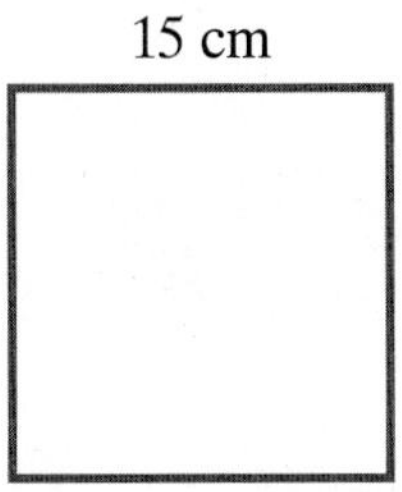

Find the perimeter of a **square** whose sides are length 15 cm.

Solution Start with the perimeter formula: $P = 4s$

Substitute 15 cm for s: $= 4(15\text{ cm})$

Multiply: $= 60\text{ cm}$

The perimeter of the square is 60 centimeters.

You Try It **Work through this You Try It problem.**

Work Exercises 3 and 4 in this textbook or in the MyMathLab Study Plan.

A **parallelogram** is a four-sided **polygon** with two pairs of parallel sides. Opposite sides in a parallelogram have the same length.

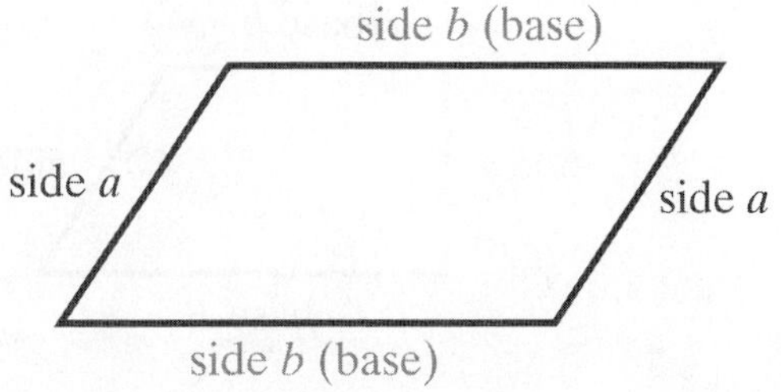

As with other polygons, we find the **perimeter** by adding up the lengths of the sides.

$$\text{Perimeter} = \text{side } a + \text{side } b + \text{side } a + \text{side } b$$

In general, we find the perimeter of a parallelogram by adding the length of side a twice and the length of side b twice. That is,

$$\text{Perimeter} = 2 \cdot (\text{side } a) + 2 \cdot (\text{side } b) \quad \text{or} \quad P = 2a + 2b.$$

TIP The formula for the perimeter of a parallelogram looks much like the formula for the **perimeter of a rectangle**. This is because a rectangle is a special parallelogram in which the sides meet at **right angles**. What does this tell you about **squares**?

Perimeter of a Parallelogram

If a parallelogram has parallel sides of length a and parallel base sides of length b, then its perimeter P is given by the formula

$$\text{Perimeter} = 2 \cdot (\text{side } a) + 2 \cdot (\text{side } b)$$

or

$$P = 2a + 2b.$$

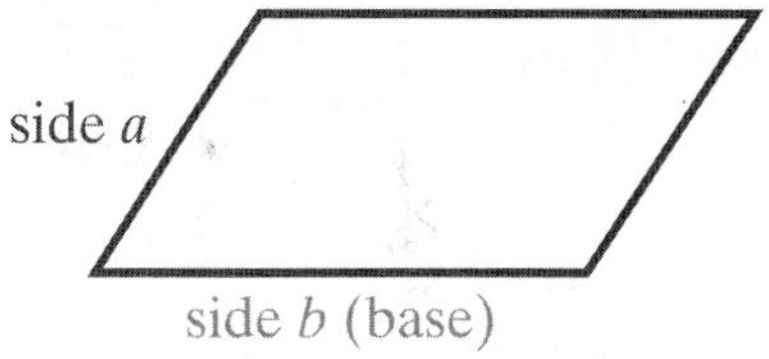

Example 3 Finding the Perimeter of a Parallelogram

Find the **perimeter** of a **parallelogram** with one side of length 23 yd and an adjacent side of length 14 yd.

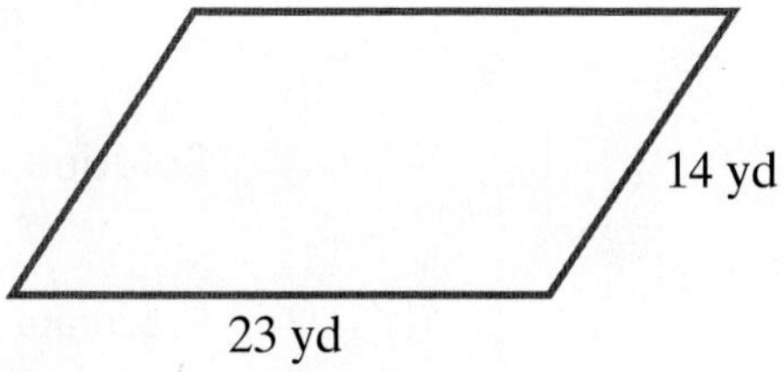

Solution

Start with the perimeter formula:	$P = 2a + 2b$
Substitute 14 yd for a and 23 yd for b:	$= 2(14\text{ yd}) + 2(23\text{ yd})$
Multiply:	$= 28\text{ yd} + 46\text{ yd}$
Add:	$= 74\text{ yd}$

The perimeter of the parallelogram is 74 yards.

(eText Screens 8.2-1–8.2-47)

You Try It **Work through this You Try It problem.**

Work Exercises 5 and 6 in this textbook or in the MyMathLab Study Plan.

A **trapezoid** is a four-sided **polygon** with one pair of parallel sides called *bases*. Some examples are shown below.

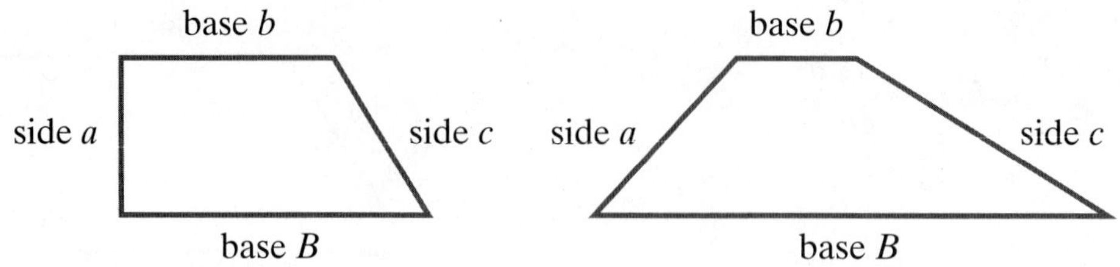

Again, we find the **perimeter** by adding up the lengths of the sides.

$$\text{Perimeter} = \text{side } a + \text{base } b + \text{side } c + \text{Base } B$$

Perimeter of a Trapezoid

If a trapezoid has sides of length a and c and bases b (small base) and B (large base), then its perimeter P is given by the formula

$$\textbf{Perimeter} = \text{side } \boldsymbol{a} + \text{base } \boldsymbol{b} + \text{side } \boldsymbol{c} + \text{Base } \boldsymbol{B}$$

or

$$P = a + b + c + B.$$

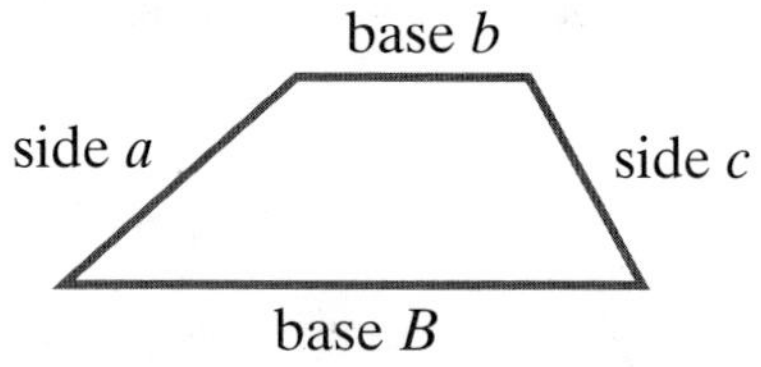

Example 4 Finding the Perimeter of a Trapezoid

Find the perimeter of the **trapezoid** shown below.

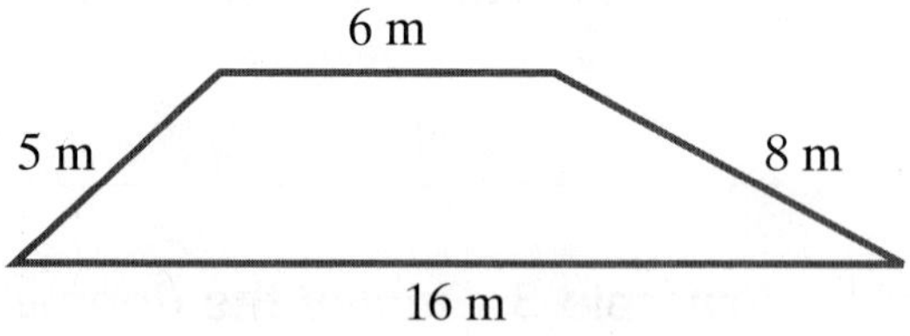

Solution Add up the lengths of the four sides.

Start with the perimeter formula:	$P = a + b + c + B$
Substitute 5 m for a, 6 m for b, 8 m for c, and 16 m for B:	$= 5\text{ m} + 6\text{ m} + 8\text{ m} + 16\text{ m}$
Add:	$= 35\text{ m}$

The perimeter of the trapezoid is 35 meters.

You Try It **Work through this You Try It problem.**

Work Exercises 7 and 8 in this textbook or in the MyMathLab Study Plan.

(*eText Screens 8.2-1–8.2-47*)

A **triangle** is a three-sided **polygon**. Some examples are shown below.

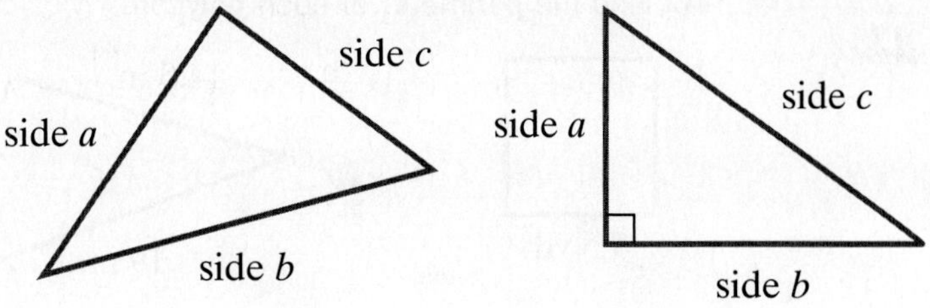

We find the **perimeter** by adding up the lengths of the sides.

$$\text{Perimeter} = \text{side } a + \text{side } b + \text{side } c$$

> **Perimeter of a Triangle**
>
> If a triangle has sides of length a, b, and c, then its perimeter P is given by the formula
>
> 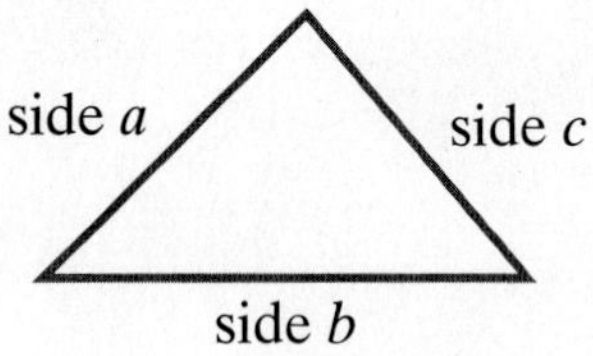
>
>
> $$\text{Perimeter} = \text{side } \boldsymbol{a} + \text{side } \boldsymbol{b} + \text{side } \boldsymbol{c}$$
>
> or
>
> $$P = a + b + c.$$

Example 5 Finding the Perimeter of a Triangle

Find the perimeter of the **triangle** shown below.

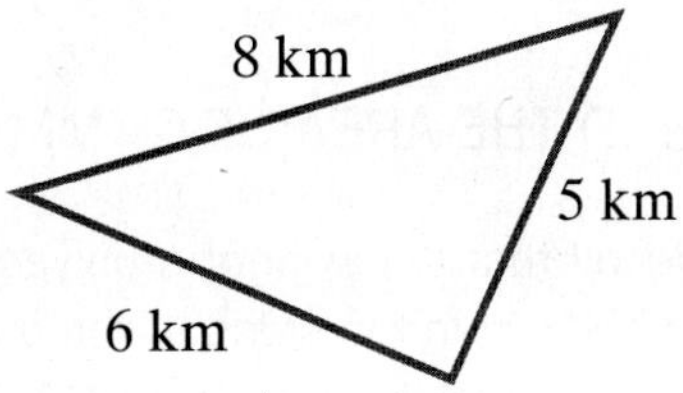

Solution Add up the lengths of the three sides.

Start with the perimeter formula:	$P = a + b + c$
Substitute 6 km for a, 5 km for b, and 8 km for c:	$= 6\text{ km} + 5\text{ km} + 8\text{ km}$
Add:	$= 19\text{ km}$

The perimeter of the triangle is 19 kilometers.

You Try It **Work through this You Try It problem.**

Work Exercises 9 and 10 in this textbook or in the MyMathLab Study Plan.

TIP If all the units are the same, then the units may be left off when doing computations. However, be sure to include units in the final answer.

(eText Screens 8.2-1–8.2-47)

Example 6 Finding the Perimeter of Common Polygons

My interactive video summary

Find the **perimeter** of each **polygon**.

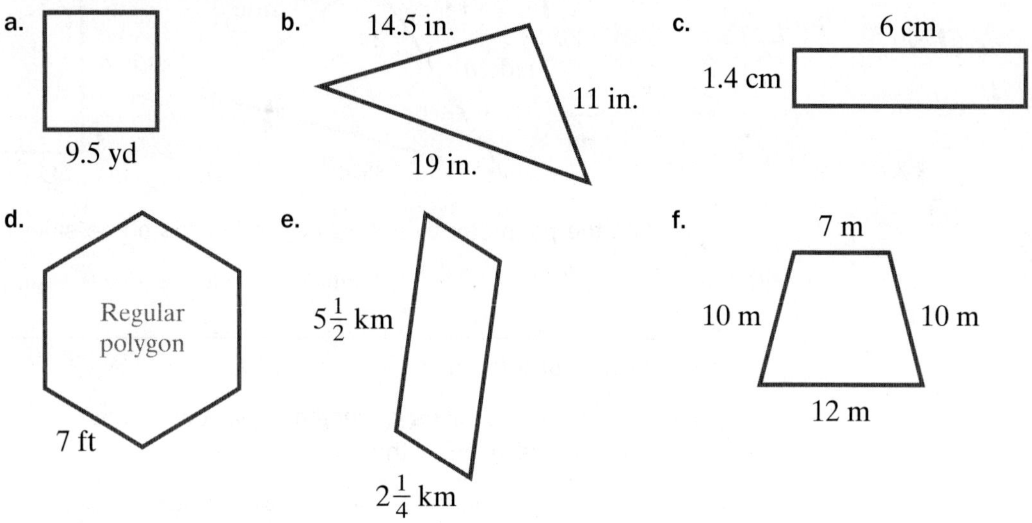

Solutions Use the appropriate **perimeter formula** for each polygon to find the perimeter. Check your **answers**, or watch this **interactive video** to see complete solutions for each part.

You Try It **Work through this You Try It problem.**

Work Exercises 11–16 in this textbook or in the MyMathLab Study Plan.

OBJECTIVE 2 FIND THE AREA OF COMMON POLYGONS

Recall that the **area** of a **polygon** is the amount of surface, measured in *square units*, contained within the sides of the polygon. We found the **area of a rectangle** in **Topic 1.4** and the **area of a triangle** in **Topic 4.3**. We review these formulas below, along with area formulas for other common polygons.

Area of a Rectangle

If a rectangle has length l and width w, then its area A is given by the formula

$$\textbf{Area} = \textbf{length} \cdot \textbf{width}$$

or

$$A = l \cdot w.$$

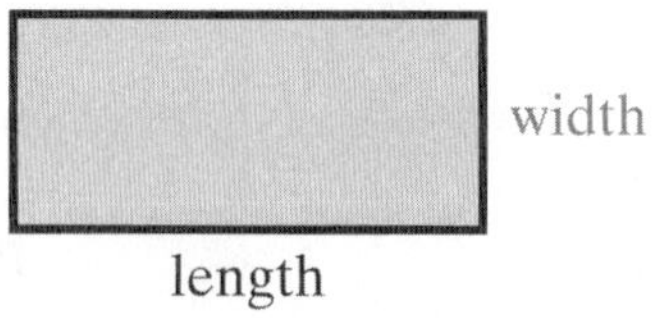

Area of a Triangle

If a triangle has base b and corresponding height h, then its area A is given by the formula

$$\textbf{Area} = \frac{1}{2}(\textbf{base})(\textbf{height}) \quad \text{or} \quad A = \frac{1}{2}bh.$$

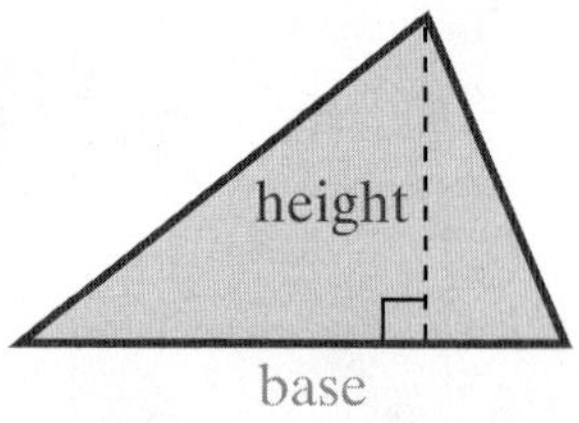

My video animation summary

Watch this **video animation** to see the development of the formula for the area of a triangle.

CAUTION

Remember to include units when finding **area**. The units for area are *square units* such as square feet (ft^2) or square meters (m^2).

Example 7 Finding the Area of Rectangles and Triangles

Find the area of each **polygon**.

a.

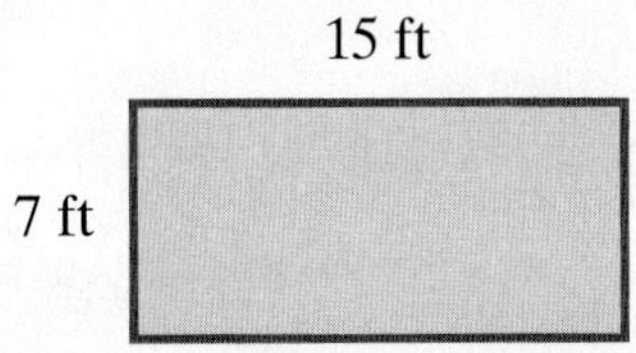

b.

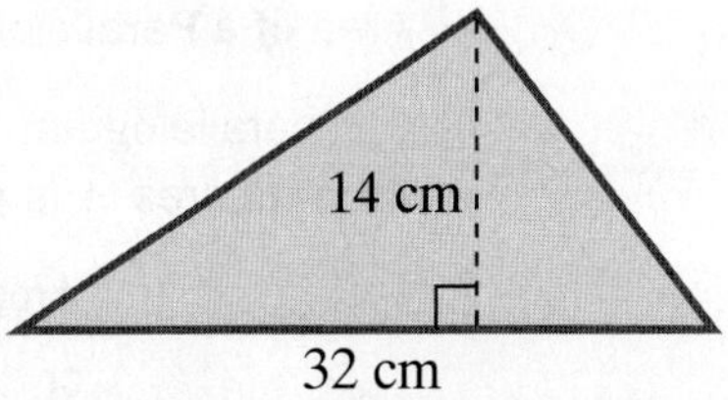

My video summary

Solutions Try to find the area of these figures on your own. Check your **answers**, or watch this **video** to see complete solutions for both parts.

You Try It Work through this You Try It problem.

Work Exercises 17–20 in this textbook or in the MyMathLab Study Plan.

Recall that a square is a special **rectangle** in which the lengths of all the sides are the same.

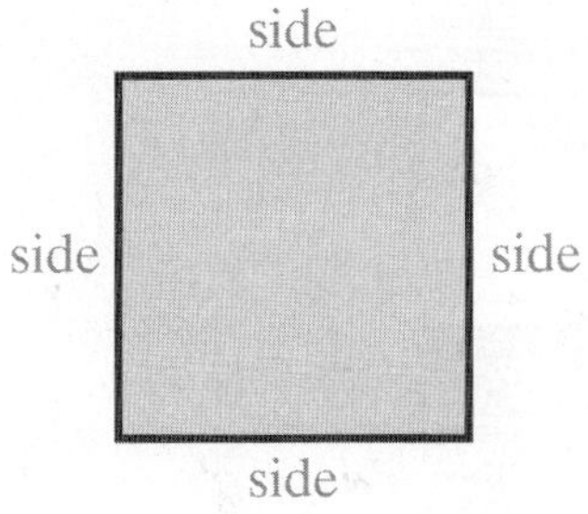

To find the **area** of the square, we multiply the length and width. Because these are the same, we get

$$\text{Area} = \text{side} \cdot \text{side} \quad \text{or} \quad A = s \cdot s = s^2.$$

Area of a Square

If a square has sides of length s, then its area A is given by the formula

$$\text{Area} = \text{side} \cdot \text{side} \quad \text{or} \quad A = s \cdot s = s^2.$$

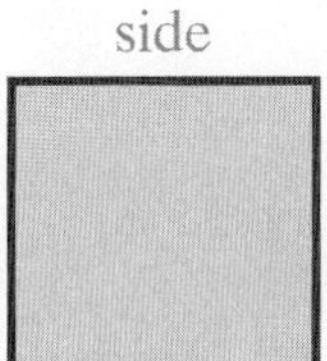

Example 8 Finding the Area of a Square

Find the **area of a square** whose sides are length 8 in.

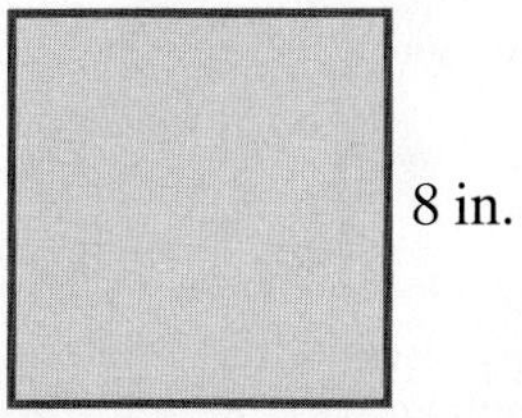

Solution Start with the area formula: $A = s^2$

Substitute 8 in. for s: $= (8 \text{ in.})^2$

Simplify: $= 64$ square inches

The area of the square is 64 square inches.

(eText Screens 8.2-1–8.2-47)

You Try It **Work through this You Try It problem.**

Work Exercises 21 and 22 in this textbook or in the MyMathLab Study Plan.

My video animation summary

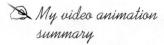

Two additional common **polygons** are **parallelograms** and **trapezoids**. Watch this **video animation** to see how area formulas are developed for these polygons.

Area of a Parallelogram

If a parallelogram has base of length b and height h, then its area A is given by the formula

$$\textbf{Area} = (\textbf{base}) \cdot (\textbf{height})$$

or

$$A = bh.$$

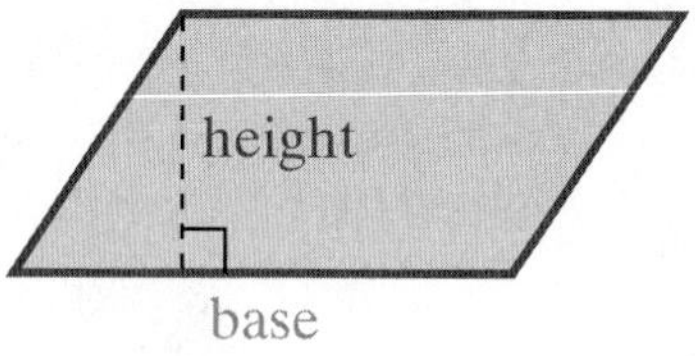

Area of a Trapezoid

If a trapezoid has height h and bases b and B, then its area A is given by the formula

$$\textbf{Area} = \frac{1}{2} \cdot \textbf{height} \cdot (\textbf{base} + \textbf{Base})$$

or

$$A = \frac{1}{2}h(b + B).$$

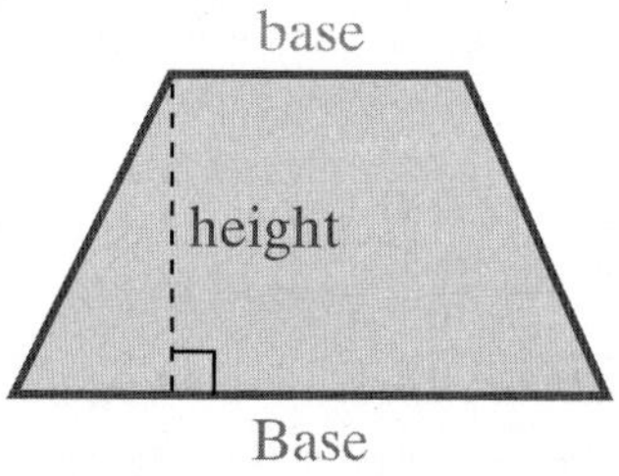

Example 9 Finding the Area of Parallelograms and Trapezoids

Find the **area** of each **polygon**.

a.

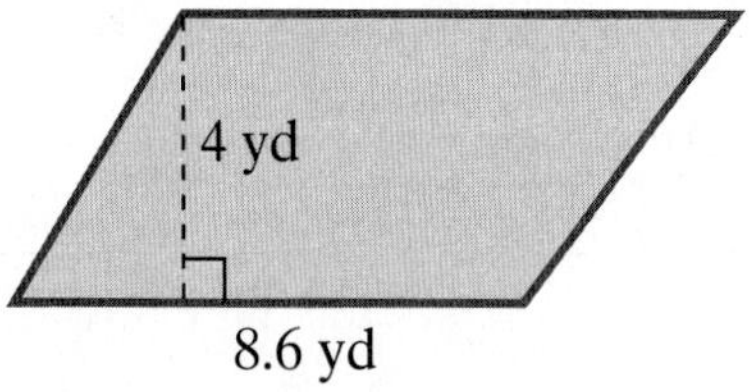

b.

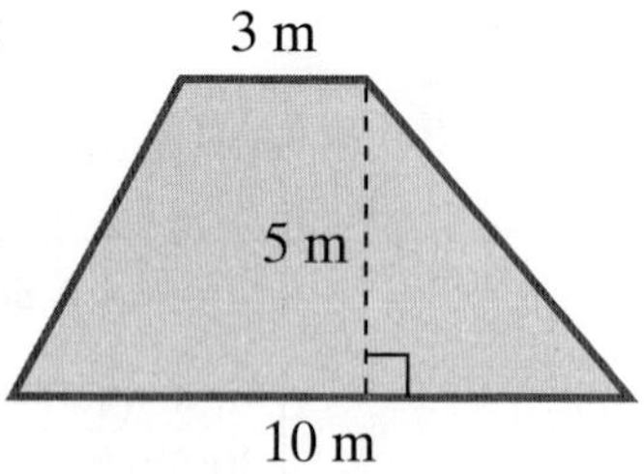

Solutions

a. This polygon is a **parallelogram** with a base of 8.6 yards and height of 4 yards.

Start with the area formula:	$A = bh$	
Substitute 8.6 yd for b and 4 yd for h:	$= (8.6\text{ yd})(4\text{ yd})$	← $\begin{array}{r} {}^{2} \\ 8.6 \\ \times\ 4 \\ \hline 34.4 \end{array}$
Multiply:	$= 34.4$ square yards	

The area of the parallelogram is 34.4 square yards.

b. This polygon is a **trapezoid** with a height of 5 meters, and bases of 3 meters and 10 meters.

Start with the area formula: $A = \frac{1}{2}h(b + B)$

Substitute 5 m for h, 3 m for b, and 10 m for B: $= \frac{1}{2}(5\text{ m})(3\text{ m} + 10\text{ m})$

Add: $= \frac{1}{2}(5\text{ m})(13\text{ m})$

Multiply: $= \frac{65}{2}$ or 32.5 m^2

The area of the trapezoid is 32.5 square meters.

You Try It **Work through this You Try It problem.**

Work Exercises 23–26 in this textbook or in the MyMathLab Study Plan.

TIP As with **perimeter**, if all the units are the same, then the units may be left off when doing computations. However, be sure to include units in the final answer. Remember that area has *square* units.

Example 10 Finding the Area of Common Polygons

My interactive video summary

Find the area of each **polygon**.

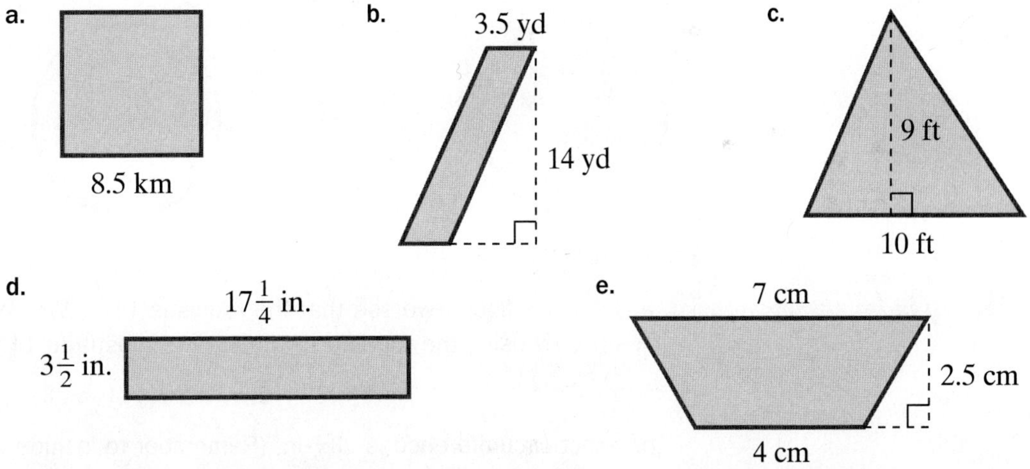

Solutions Use the appropriate **area formula** for each polygon to find the area. Check your **answers**, or watch this **interactive video** to see complete solutions for each part.

You Try It **Work through this You Try It problem.**

Work Exercises 27–31 in this textbook or in the MyMathLab Study Plan.

OBJECTIVE 3 FIND THE CIRCUMFERENCE AND AREA OF CIRCLES

We introduced a formula for the **circumference** of a circle in **Topic 5.3**, and a formula for the area of a circle in **Topic 5.5**. Let's review those formulas now.

(eText Screens 8.2-1–8.2-47)

Circumference of a Circle

The **circumference** of a circle is the distance around the circle. Its value is found by using either

$$C = 2\pi r \quad \text{or} \quad C = \pi d$$

where r is the **radius** of the circle and d is the **diameter**.

Area of a Circle

The **area** of a circle is the amount of surface contained within the circle. Its value is found by

$$A = \pi r^2$$

where r is the radius of the circle.

TIP

Recall the following:

1. we often approximate the value of π with the decimal 3.14,
2. the radius of a circle is equal to half the diameter of the circle $\left(r = \frac{d}{2}\right)$.

Example 11 Finding the Circumference and Area of a Circle

Find the exact **circumference** and **area** of each circle in terms of π. Then approximate the circumference and area using 3.14 for π.

a.

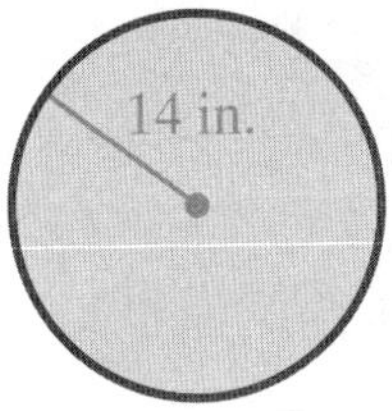

b.

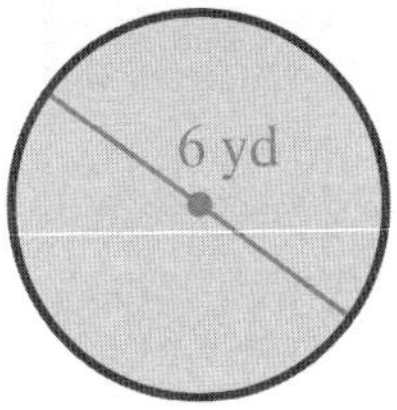

Solutions

My video summary

a. From the figure, we see that the **radius** is 14 in. We can get the exact circumference by using the formula $C = 2\pi r$. We substitute 14 for r to get

$$C = 2\pi r = 2\pi(14) = 28\pi.$$

The exact circumference is 28π in. (Remember to include units.)

We can find the exact area by using the formula $A = \pi r^2$. **Substitute** 14 for r to get

$$A = \pi r^2 = \pi(14)^2 = 196\pi.$$

The exact area is 196π square inches. (Remember that area is given in square units.)

Using 3.14 for π, we can get approximate values for both the **circumference** and area by computing the following:

$$\begin{array}{r} 3.14 \\ \times\ 28 \\ \hline \end{array} \qquad \begin{array}{r} 3.14 \\ \times\ 196 \\ \hline \end{array}$$

Approximate circumference → ← Approximate area

Find the approximate values on your own. Check your **answers**, or watch this **video** for the complete solution to part a.

My video summary **b.** Try working this problem on your own. Check your **answers**, or watch this **video** for the complete solution to part b.

You Try It **Work through this You Try It problem.**

Work Exercises 32–35 in this textbook or in the MyMathLab Study Plan.

CAUTION Remember that π has no units so it does not affect the units of our results.

OBJECTIVE 4 FIND THE PERIMETER AND AREA OF FIGURES FORMED FROM TWO OR MORE COMMON POLYGONS

In **Topic 1.3**, we found the **perimeter** of **polygons** when the lengths of all the sides were given. Here we look at situations where not all the lengths are directly given. In such cases, we look for smaller common polygons that are used to form the figure and use them to find any missing lengths.

Example 12 Finding the Perimeter of a Figure Formed from Two or More Common Polygons

My video summary Find the perimeter of the polygon shown below.

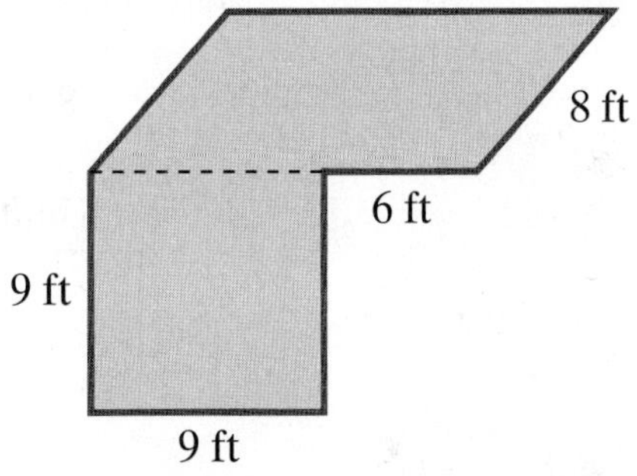

Solution Looking at the figure, we see that it is made up of a **parallelogram** and a **square**.

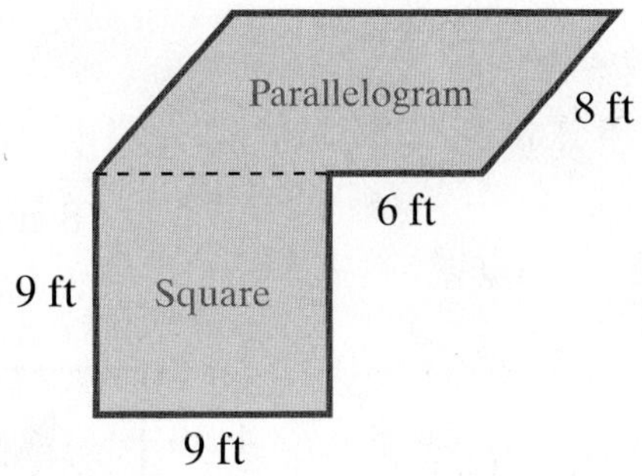

Recall that all four sides of a square have the same length and that opposite sides in a parallelogram have the same length. We can use this information to help us determine the missing lengths.

(eText Screens 8.2-1–8.2-47)

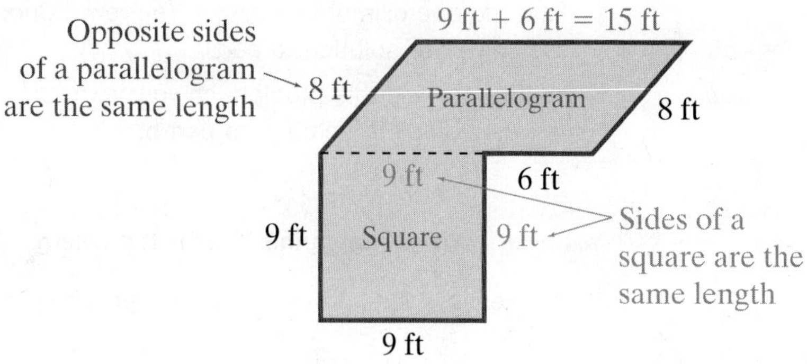

Now that we know the lengths of all the sides, we can find the **perimeter** of the figure. Try to finish this on your own. Check your **answer**, or watch this **video** for the complete solution.

As with perimeter, sometimes we can find the **area** of more complicated figures by breaking the figure down into smaller pieces. We find the area of the figure by finding the area of each piece that makes up the figure and then adding these individual areas together.

Example 13 Finding the Area of a Figure Formed from Two or More Common Polygons

My video summary

Find the area of the **polygon** shown below.

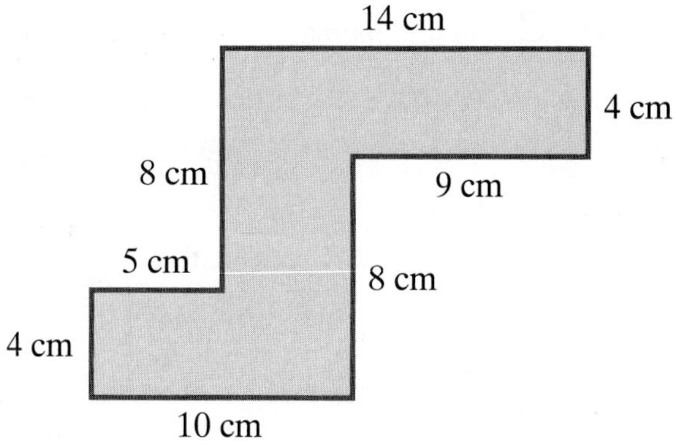

Solution Start by dividing up the polygon into three rectangular regions. One possibility is next.

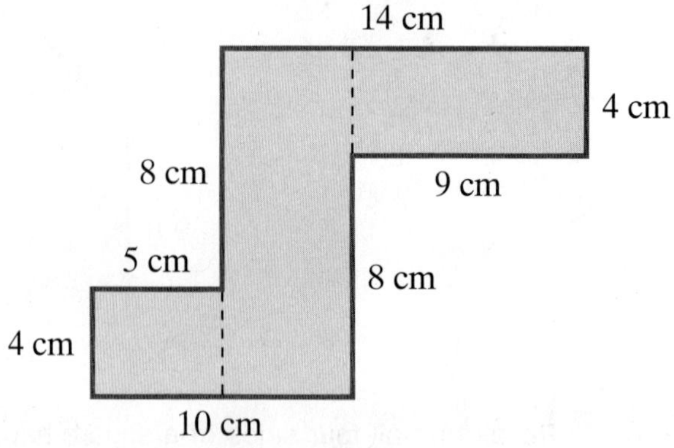

Next we need to determine the length and width of each **rectangle**.

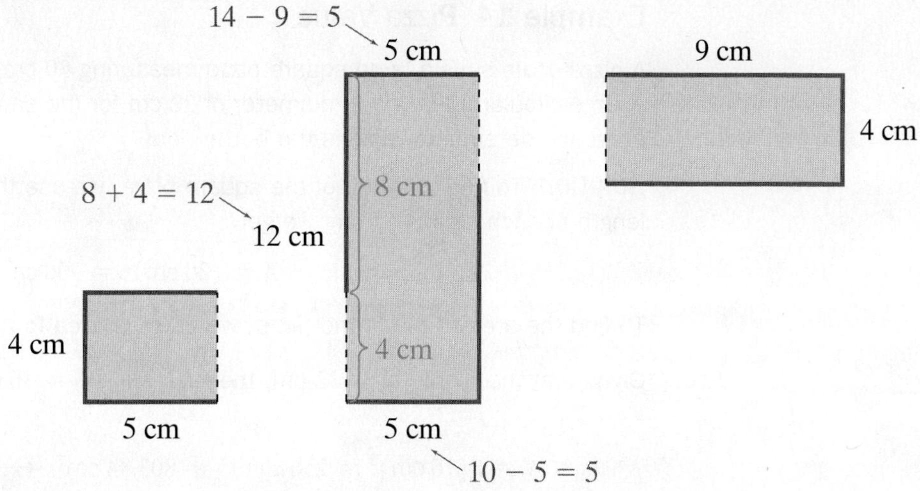

Notice that we had to add two sides of the original **polygon** to get the length of the middle rectangle and subtract to get its width.

Find the **area** of each rectangle, then add the individual areas to find the area of the polygon. Check your **answer**, or watch this **video** for the complete solution.

You Try It **Work through this You Try It problem.**

Work Exercises 36–43 in this textbook or in the MyMathLab Study Plan.

OBJECTIVE 5 SOLVE APPLICATIONS INVOLVING PERIMETER, CIRCUMFERENCE, OR AREA

We can use formulas for **perimeter, circumference**, and **area** to solve a variety of real-world applications. The following table summarizes the formulas we have seen so far.

Figure	Formulas	Figure	Formulas
Square 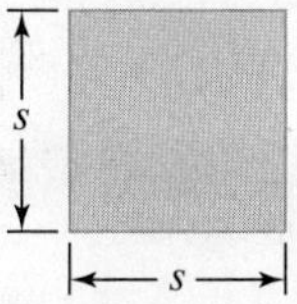	Perimeter: $P = 4s$ Area: $A = s^2$	Triangle 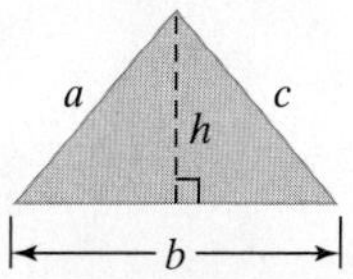	Perimeter: $P = a + b + c$ Area: $A = \frac{1}{2}bh$
Rectangle 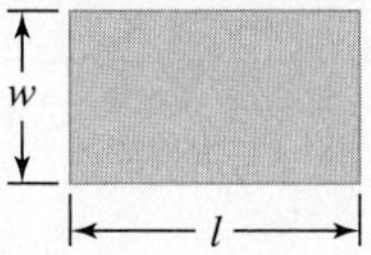	Perimeter: $P = 2l + 2w$ Area: $A = lw$	Trapezoid 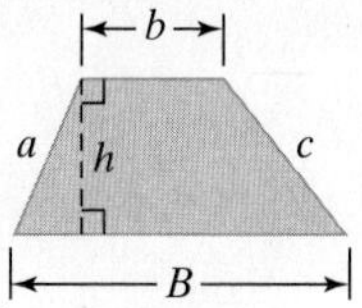	Perimeter: $P = a + b + c + B$ Area: $A = \frac{1}{2}h(b + B)$
Parallelogram 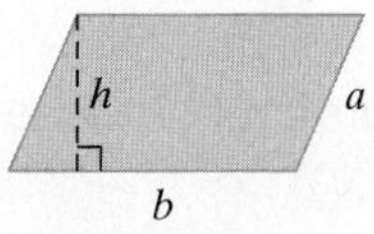	Perimeter: $P = 2a + 2b$ Area: $A = bh$	Circle 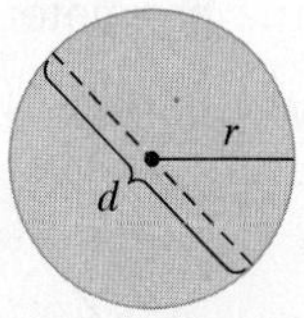	Circumference: $C = 2\pi r = \pi d$ Area: $A = \pi r^2$

(eText Screens 8.2-1–8.2-47)

Example 14 Pizza Value

A pizza chain sells a large square pizza measuring 30 cm on each side. A competitor sells a large circular pizza with a diameter of 32 cm for the same price. Find the area of each pizza and determine which is the better deal.

Solution To find the area of the square pizza, we use the formula $A = s^2$. Given that the length of each side is 30 cm, we get

$$A = (30\text{ cm})^2 = 900\text{ cm}^2.$$

To find the area of the round pizza, we use the **area formula** $A = \pi r^2$.

Given that the diameter is 32 cm, the radius is $\frac{32}{2} = 16$ cm. Thus, we have

$$A = \pi(16\text{ cm})^2 = 256\pi\text{ cm}^2 \approx 803.84\text{ cm}^2.$$

Use 3.14 for π to get an approximate value

The square pizza is the better value because it has more pizza for the same price.

You Try It **Work through this You Try It problem.**

Work Exercises 44–48 in this textbook or in the MyMathLab Study Plan.

Example 15 Replacing a Countertop

My video summary

Sandra wants to put a new granite countertop on her kitchen island. Use the diagram below to find the **perimeter** and **area** of her countertop. If the countertop costs \$30 per square foot, how much will it cost?

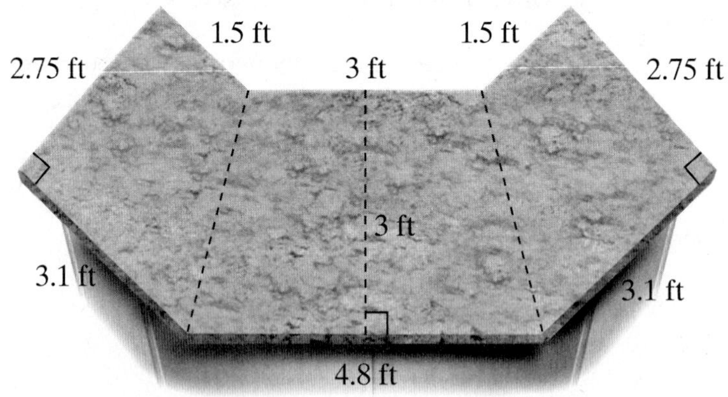

Solution The perimeter can be found by adding up the lengths of the sides.

$$\begin{aligned} P &= \overbrace{2.75\text{ ft} + 1.5\text{ ft}} + \overbrace{3\text{ ft} + 1.5\text{ ft}} + \overbrace{2.75\text{ ft} + 3.1\text{ ft}} + \overbrace{4.8\text{ ft} + 3.1\text{ ft}} \\ &= 4.25\text{ ft} + 4.5\text{ ft} + 5.85\text{ ft} + 7.9\text{ ft} \\ &= 8.75\text{ ft} + 13.75\text{ ft} \\ &= 22.5\text{ ft} \end{aligned}$$

The perimeter of the countertop is 22.5 feet.

We start finding the area by dividing the figure up into three **trapezoids**.

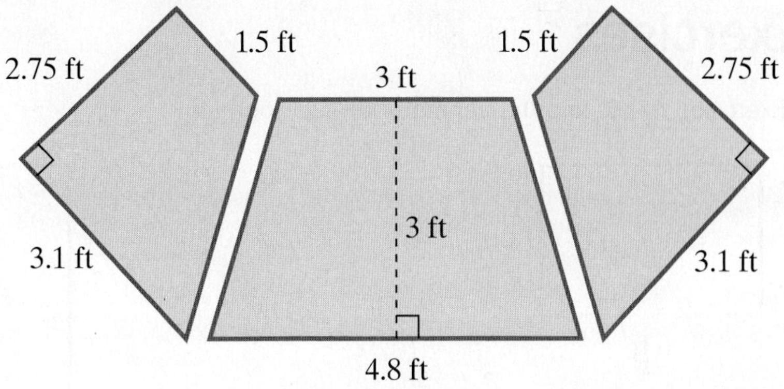

We can find the area of the countertop by adding up the individual areas. Then multiply the total area by the price of the granite to determine Sandra's cost. Find the total area and cost of the countertop on your own. Check your **answers**, or watch this **video** for the complete solution.

You Try It **Work through this You Try It problem.**

Work Exercises 49–53 in this textbook or in the MyMathLab Study Plan.

Example 16 Painting a Door

My video summary

Jerianne wants to paint the inside of a door. The door is rectangular in shape and contains a small rectangular window (which will not be painted). If Jerianne wants to use painter's tape around the window, how much tape will she need? What is the area of the door that will be painted?

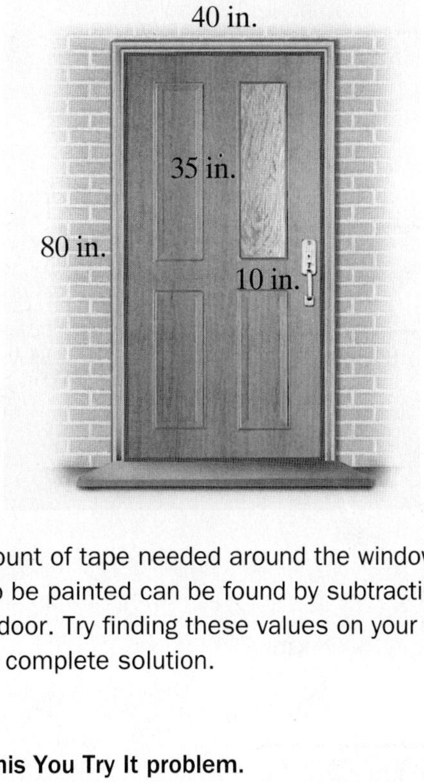

Solution To find the amount of tape needed around the window, we find the **perimeter** of the window. The **area** to be painted can be found by subtracting the area of the window from the total area of the door. Try finding these values on your own. Check your **answers**, or watch this **video** for the complete solution.

You Try It **Work through this You Try It problem.**

Work Exercises 54–56 in this textbook or in the MyMathLab Study Plan.

8.2 Exercises

In Exercises 1–16, find the perimeter of each polygon.

You Try It 1.

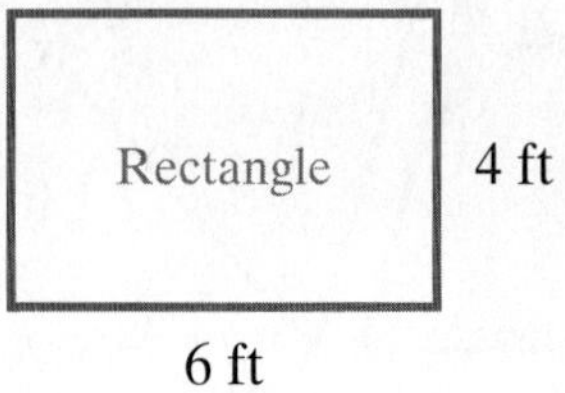

2.

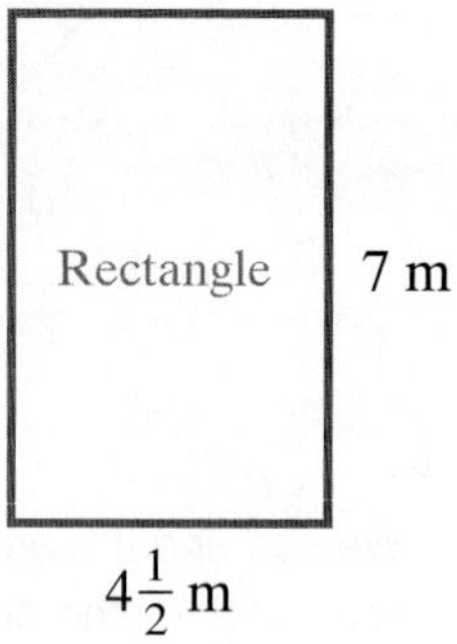

You Try It 3.

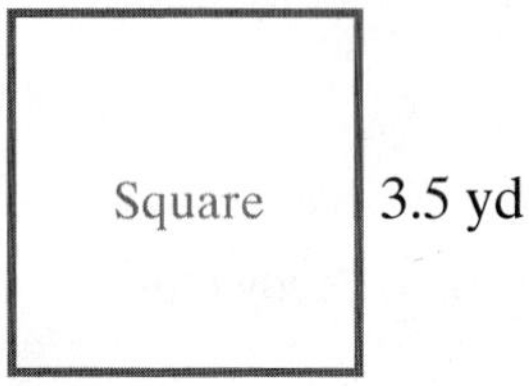

4.

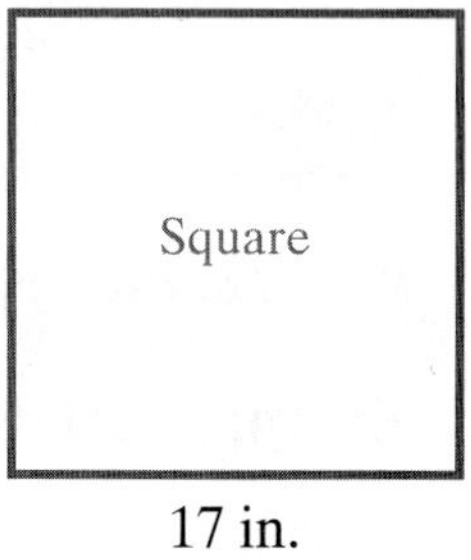

You Try It 5.

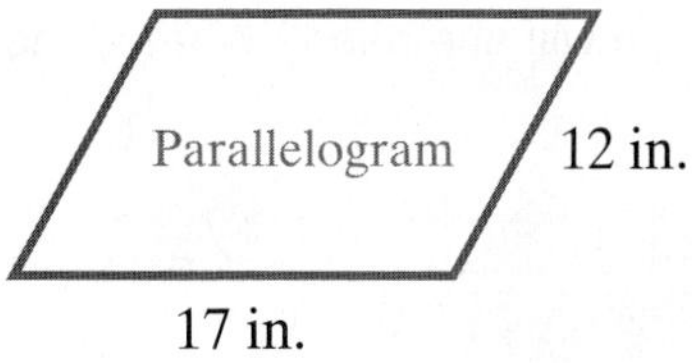

6.

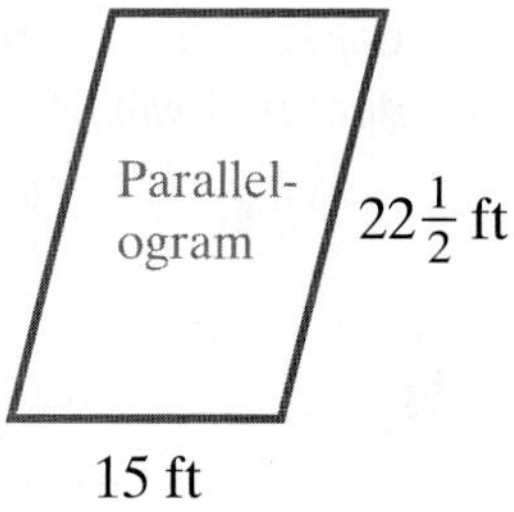

You Try It 7.

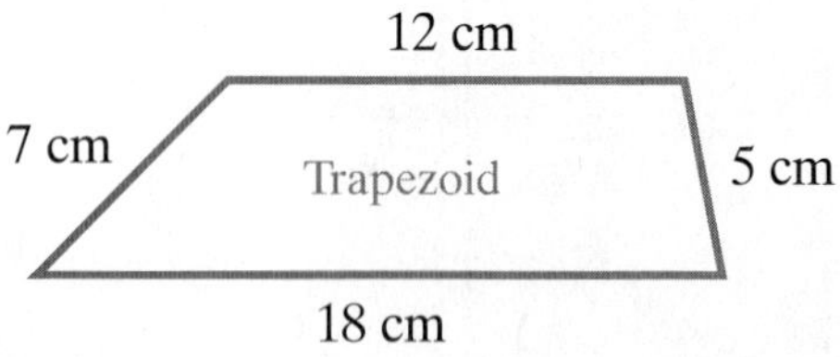

8.

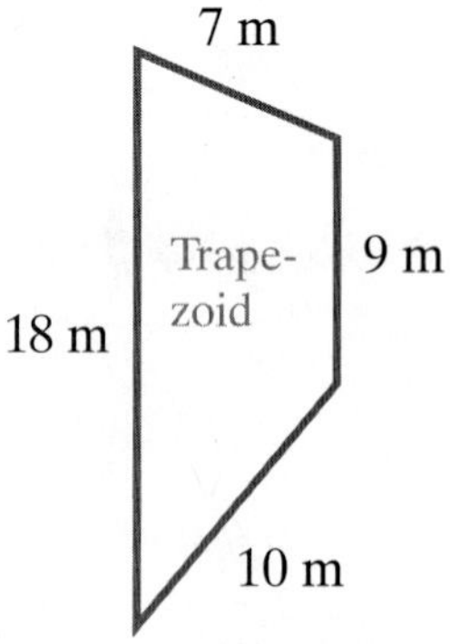

You Try It 9.

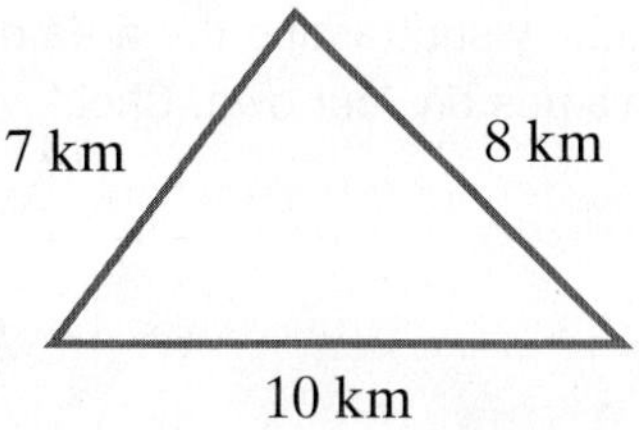

10.

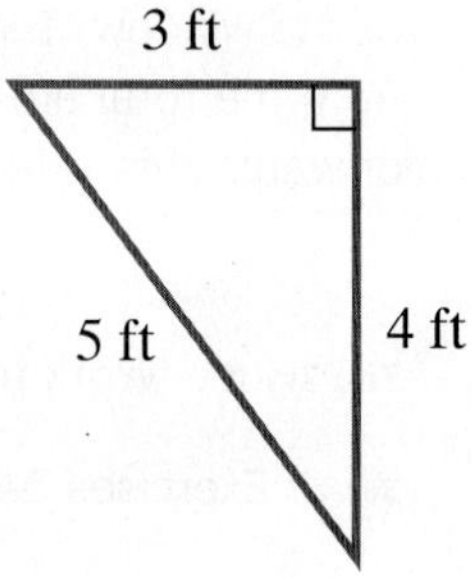

(eText Screens 8.2-1–8.2-47)

You Try It 11.

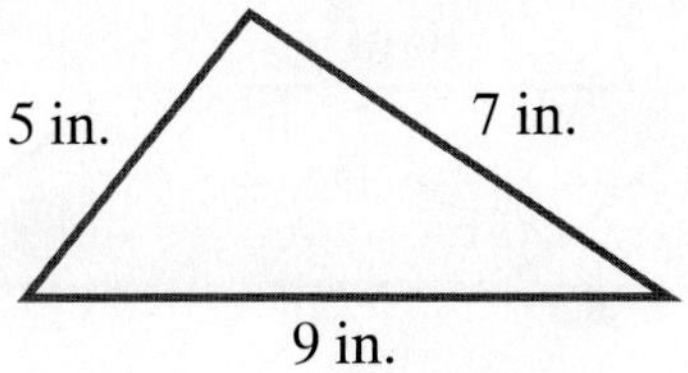

12.

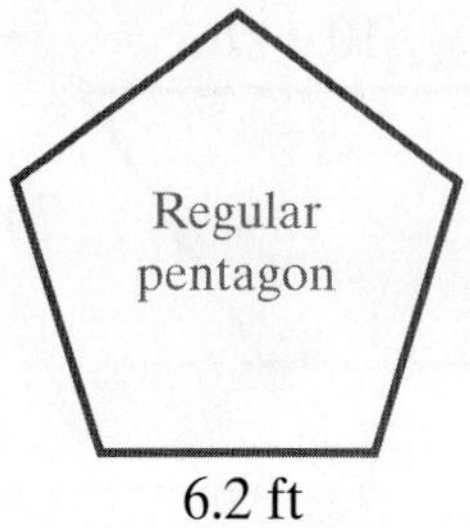

13.

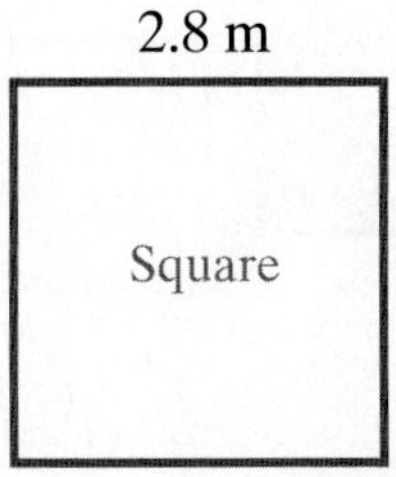

14.

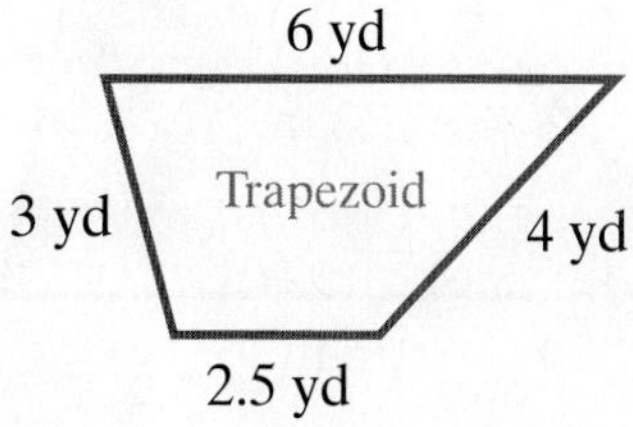

15.

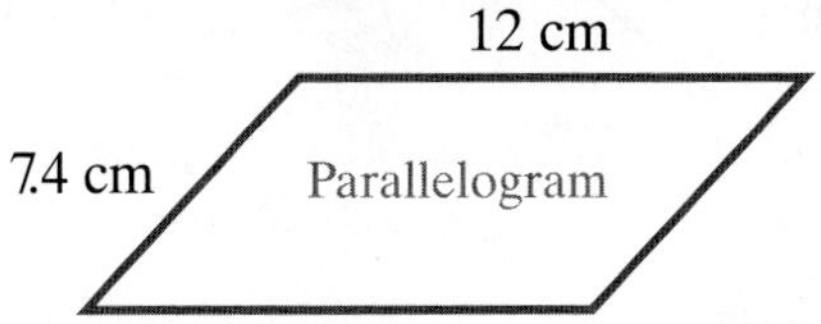

16.

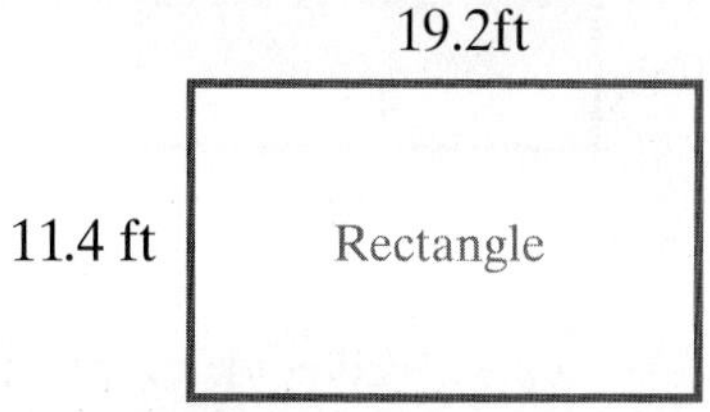

In Exercises 17–31, find the area of each polygon.

You Try It 17.

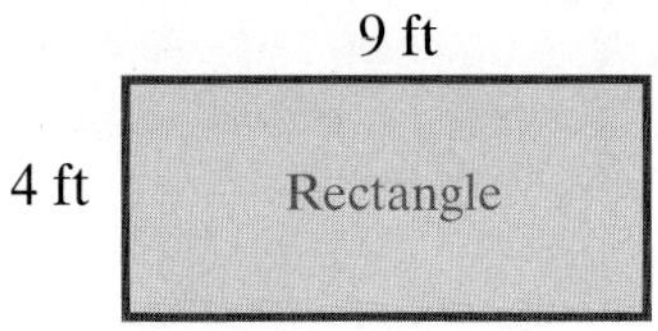

18.

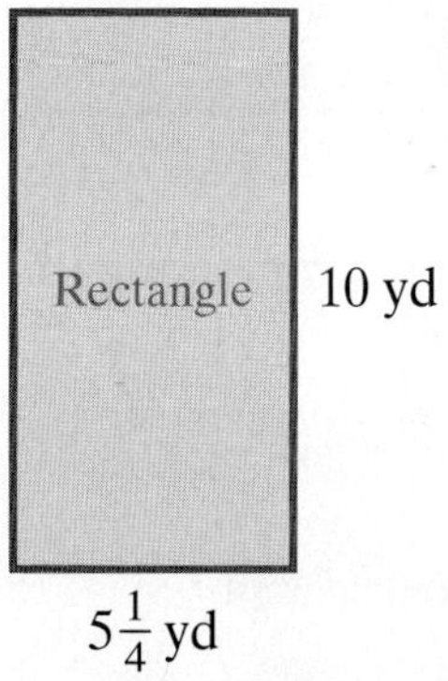

19.

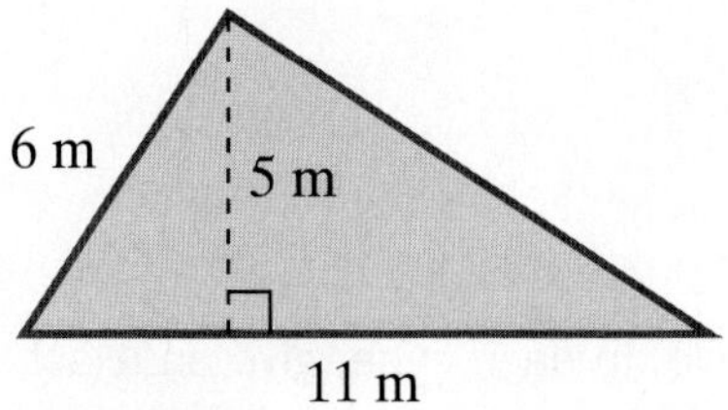

20.

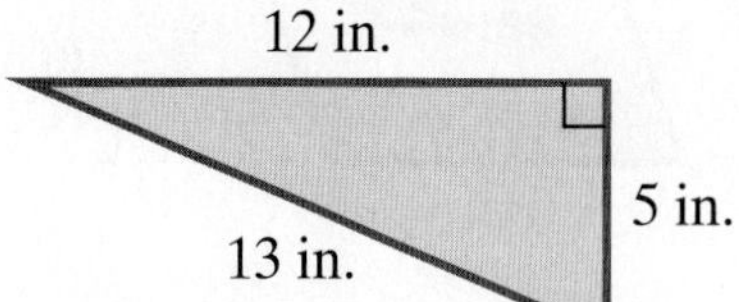

You Try It 21.

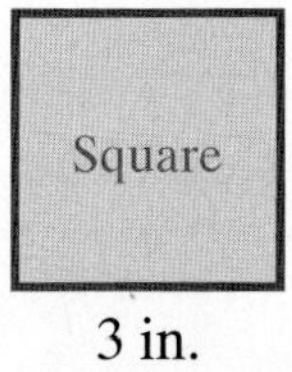

22.

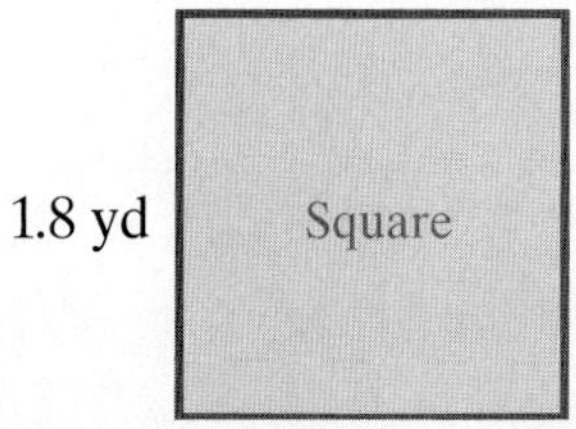

You Try It 23.

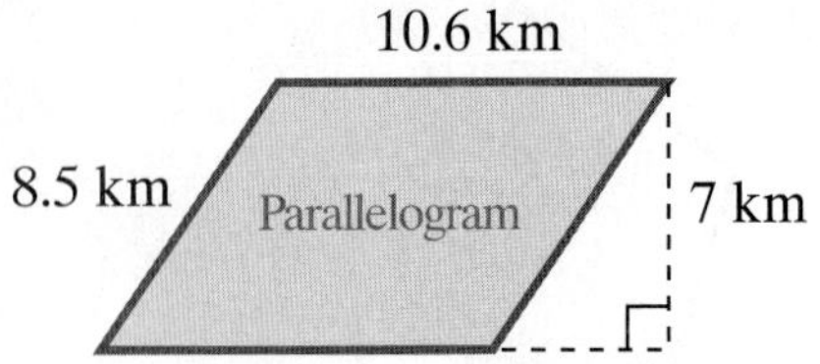

24.

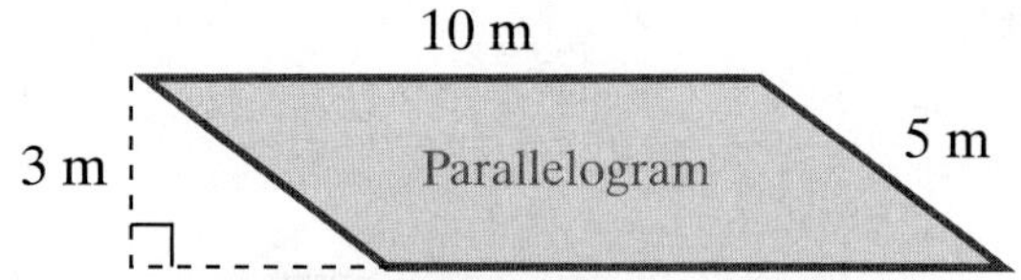

25.

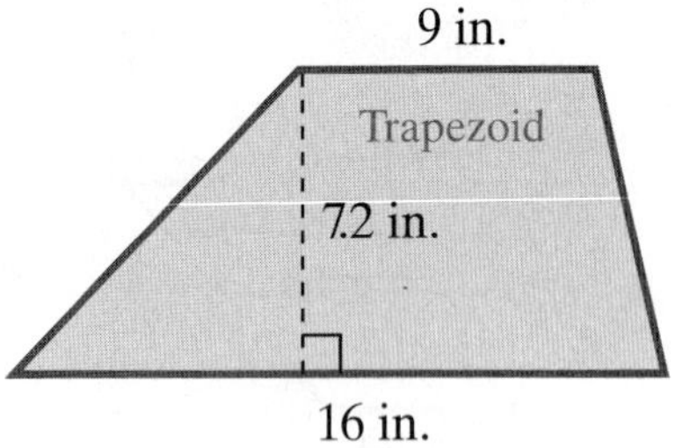

26.

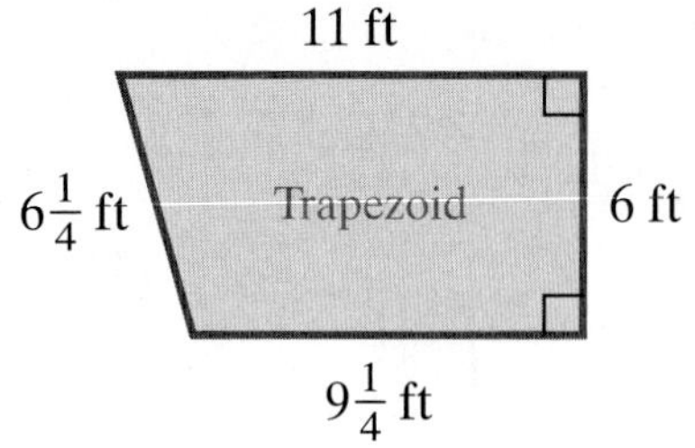

You Try It 27.

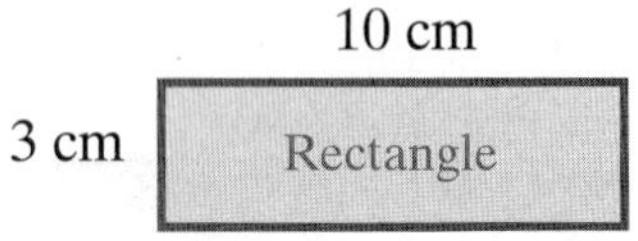

28.

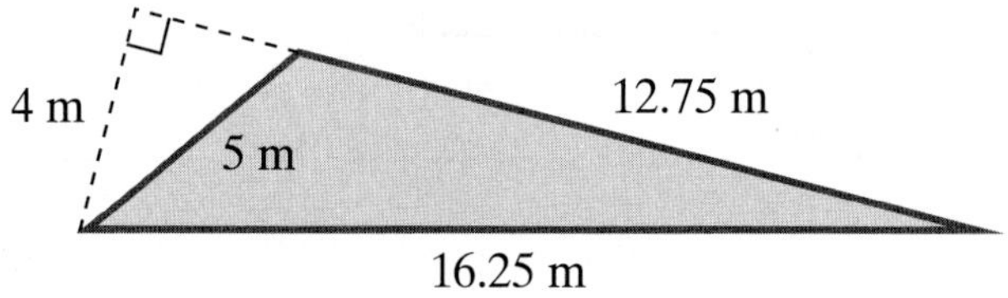

29.

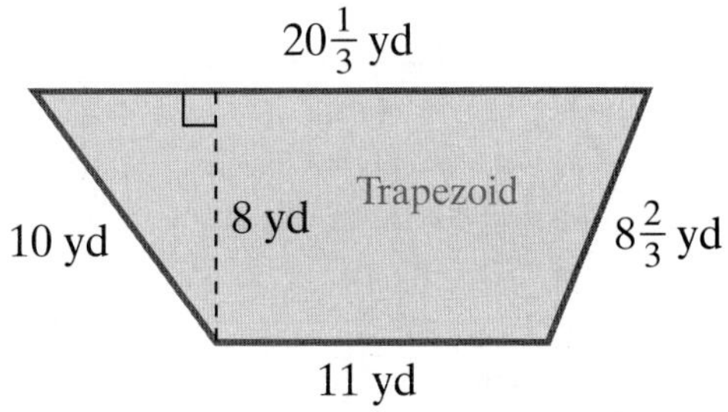

30.

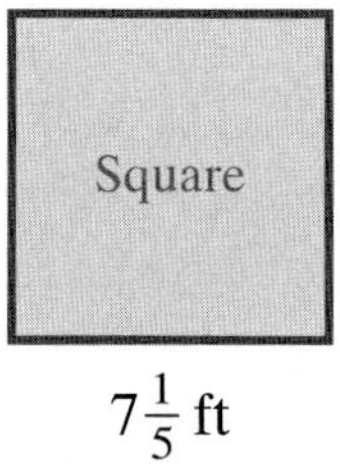

31.

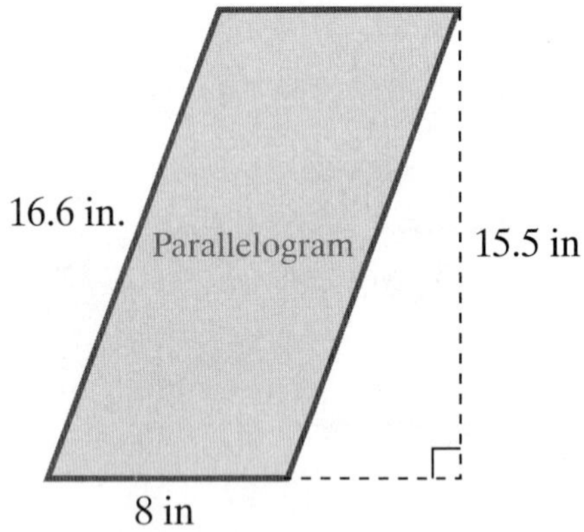

In Exercises 32–35, find the circumference and area of each circle. In each case, give an exact answer (in terms of π) and an approximation to two decimal places using 3.14 for π.

You Try It 32.

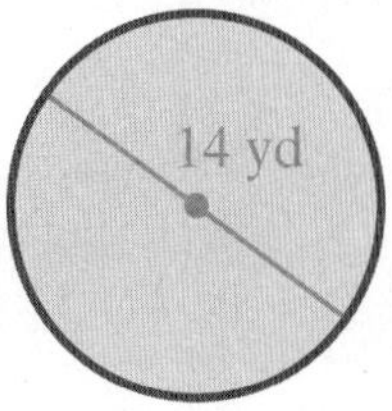

33.

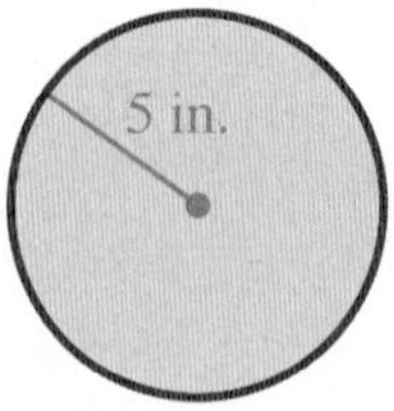

34.

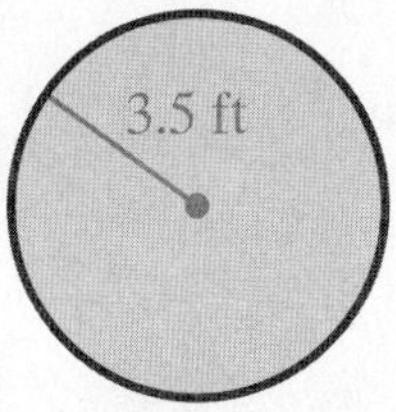

35.

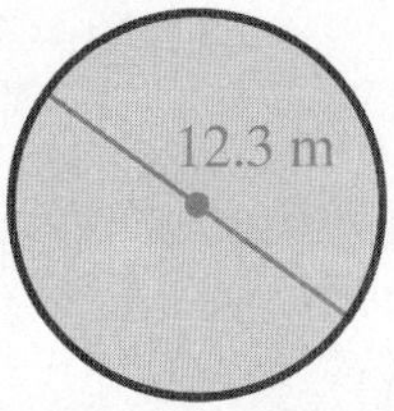

In Exercises 36–43, find the perimeter and area of each figure. If necessary, round answers to two decimal places.

You Try It 36.

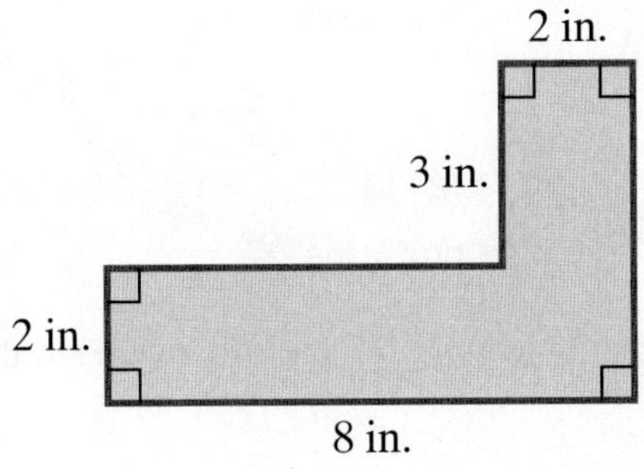

37.

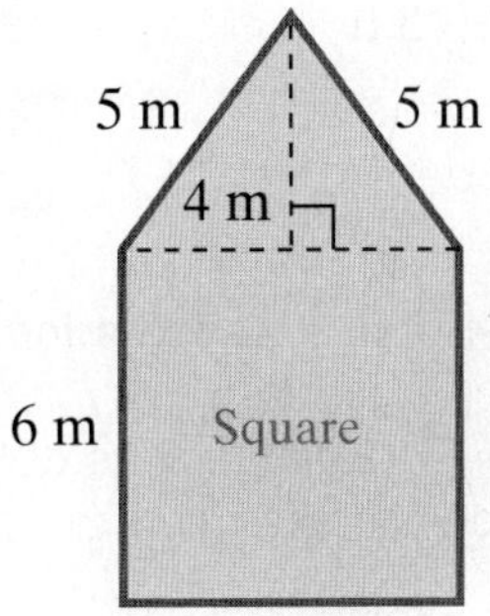

38.

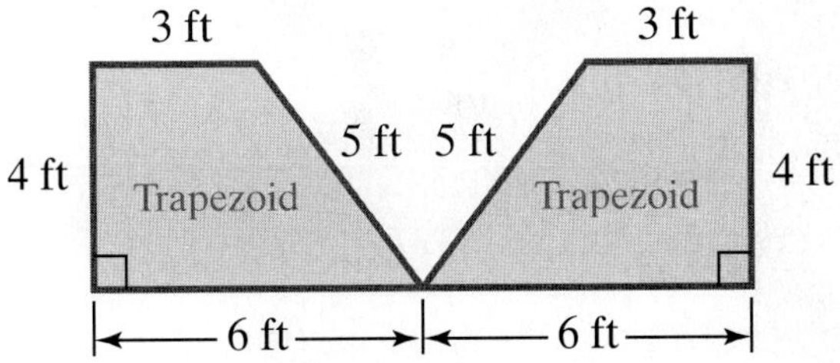

39.

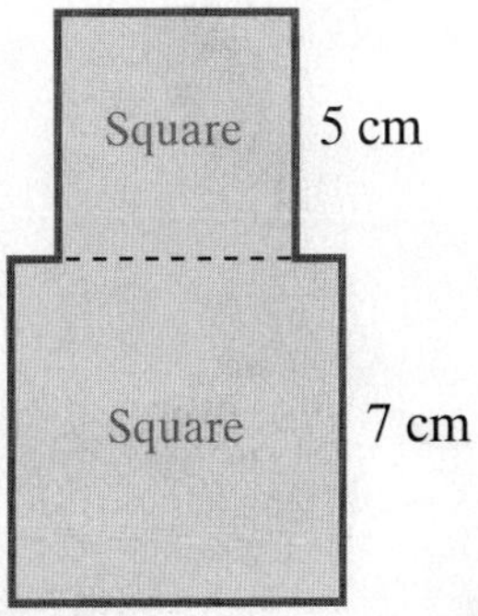

40.

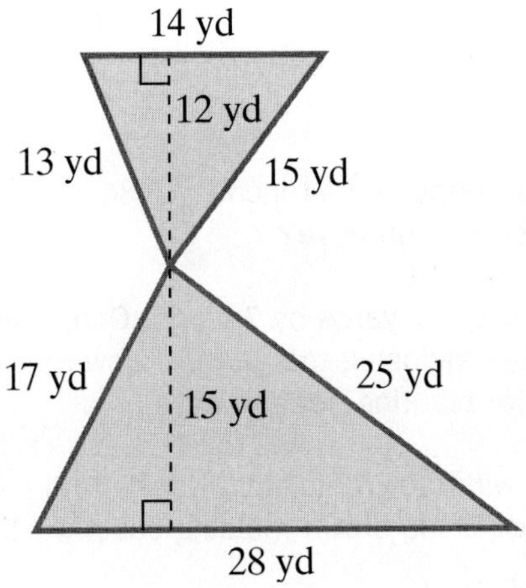

41.

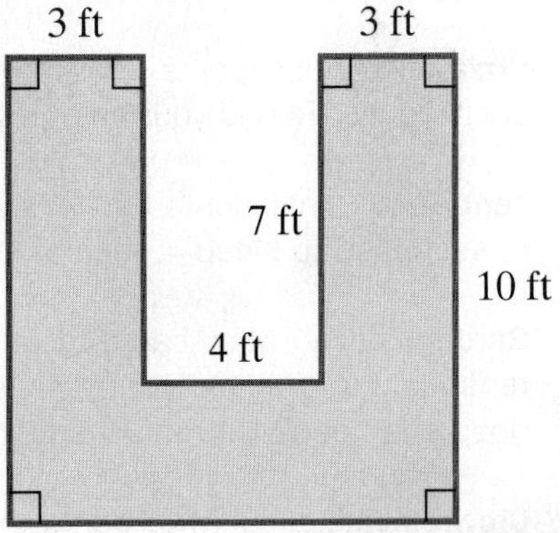

42.

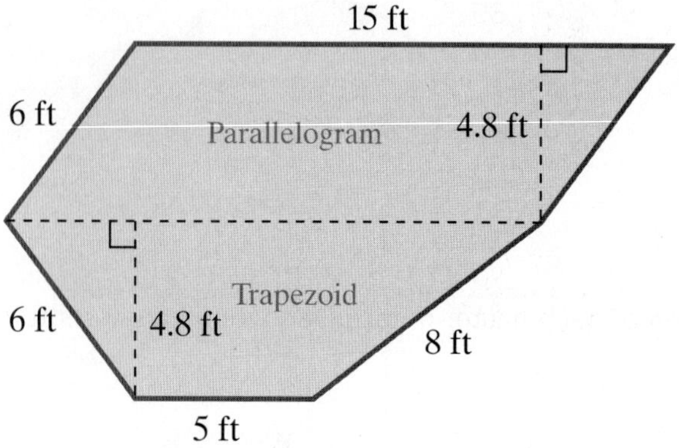

43.

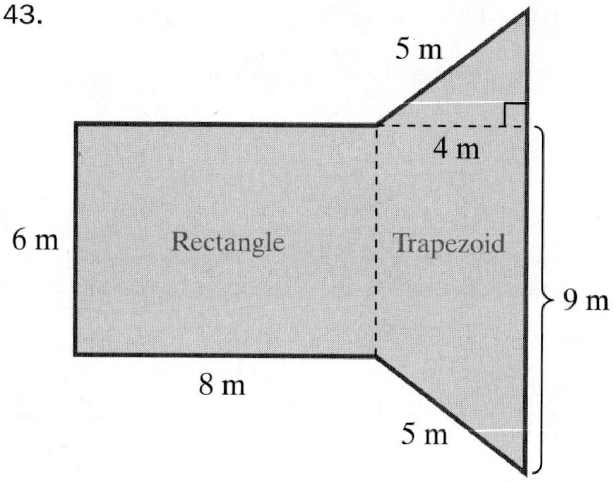

In Exercises 44–56, use formulas for perimeter and area to solve the problems.

You Try It 44. **Home Plate** The dimensions of home plate in baseball are given in the figure below. Find the area. (*Source:* infoplease.com)

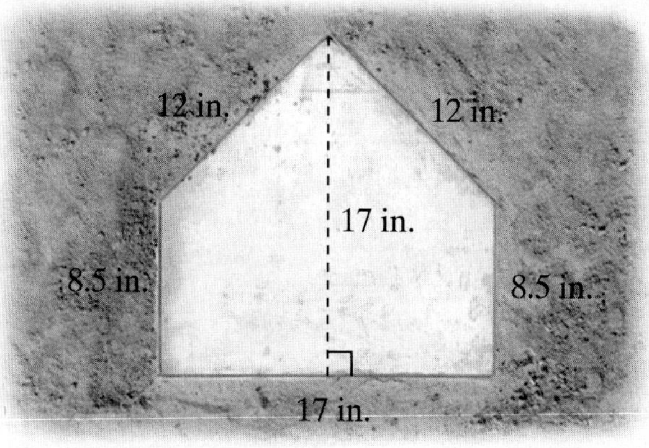

45. **Pizza Slice** Susie cuts a 10-inch diameter pizza into 6 equal wedges. Find the perimeter and the area of each wedge. Round your answers to 2 decimal places.

46. **Tent Size** A half-dome tent has a rectangular bottom that measures 94 inches by 86 inches. If the tent is supposed to sleep 4 people, how much area would each person have?

47. **Garden Fence** Mary has a rectangular garden that measures 12 yards by 7 yards. One of the long sides runs along a wall, and Mary wishes to use fencing to finish enclosing the garden. How much fencing does she need? How many square yards does she have for planting her garden?

48. **Crown Molding** Shayna wants to border her dining room with crown molding. If the room measures 12 feet by 15 feet, how much crown molding will she need? If the crown molding costs $3.20 per foot, what will be her total cost for the job?

You Try It 49. **Stop Sign** A standard U.S. Stop sign is a regular octagon (8-sides) that measures 31 cm on a side and 75 cm across opposite flats. Find the area of the sign. How much reflective tape would it take to add a reflective border along the edges of the sign? (*Source:* dimensionsguide.com)

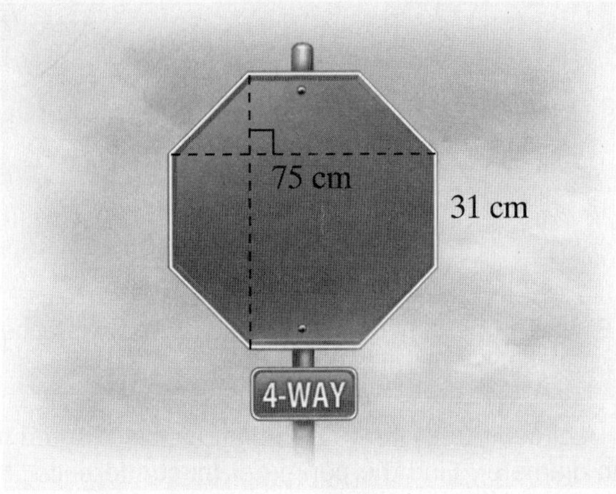

50. **Running Track** A running track consists of a rectangle capped by two semicircles (half a circle). If the rectangle measures 120 yards by 50 yards and Pete runs one lap around the track, how far did he run? Round your answer to the nearest tenth.

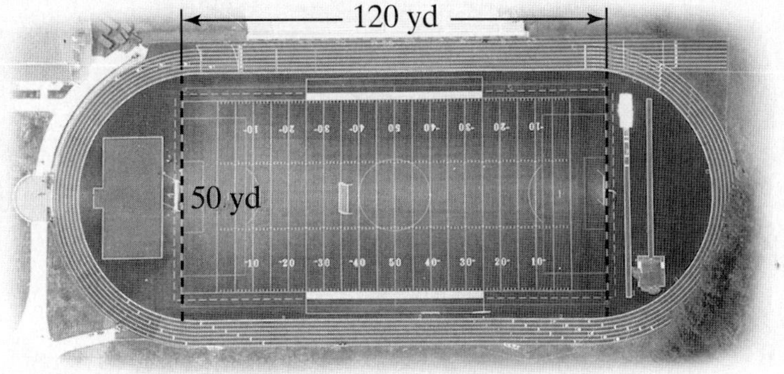

51. **Window Dimensions** A window consists of a rectangle and a semicircle (half a circle). The dimensions of the rectangle are $4\frac{1}{2}$ ft by 3 ft. Find the area of the window rounded to two decimal places.

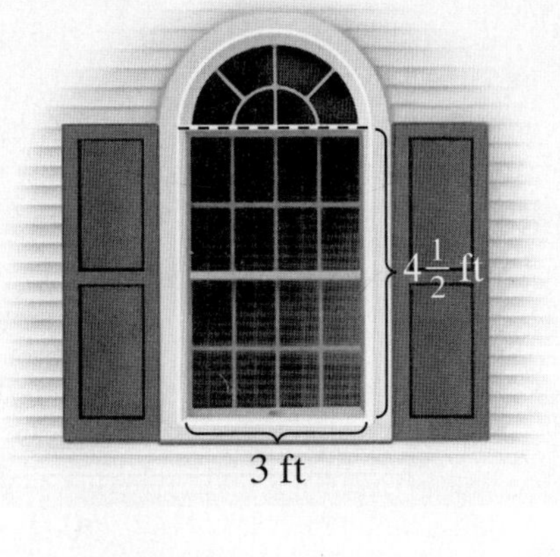

52. **Painting a Barn** John has to paint the back wall of his family's barn (see figure). If a gallon of paint covers 400 sq ft, how many gallons of paint will he need?

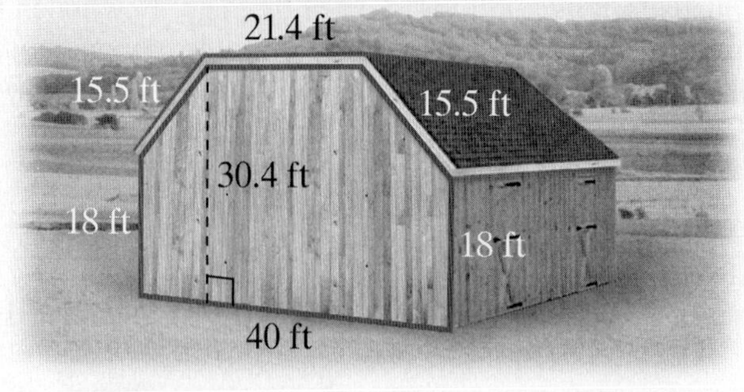

53. **Soccer Penalty Box** The penalty box on a soccer field has a perimeter of 124 yards. If the box is 18 yards wide, what is the area of the box?

You Try It 54. **Flower Garden** A circular tree of 6 feet in diameter is planted in the middle of a circular flower garden that is 15 feet in diameter. Find the perimeter (circumference) of the flower garden and find the area of the garden that contains flowers.

55. **Paving a Parking Lot** A restaurant owner wants to pave his parking lot. Use the diagram below to determine the area to be paved.

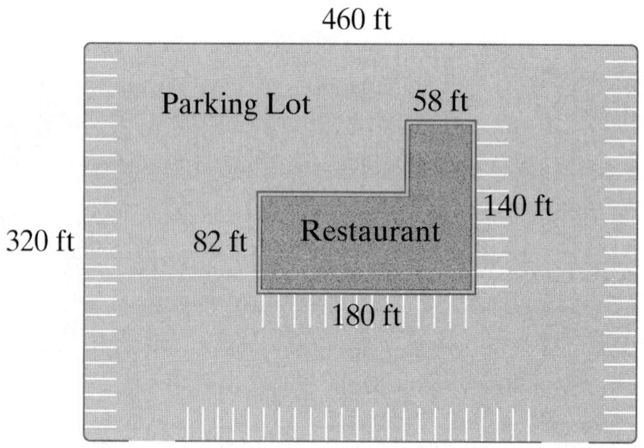

56. **Grass Coverage** Garrett has a rectangular backyard measuring 40 feet by 50 feet. Part of the yard is covered by a circular pool 20 feet in diameter, and the rest is covered with grass. What percent of the yard is covered with grass? Round your answer to the nearest whole percent.

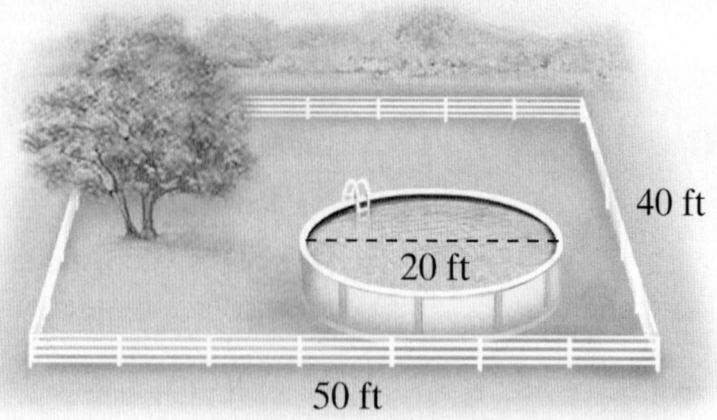

8.3 Volume and Surface Area

THINGS TO KNOW

Before working through this topic, be sure you are familiar with the following concepts:

VIDEO ANIMATION INTERACTIVE

You Try It 1. Find the Area of a Rectangle (Topic 1.4, **Objective 5**)

You Try It 2. Find the Area of a Triangle (Topic 4.3, **Objective 7**)

You Try It 3. Find the Circumference of a Circle (Topic 5.3, **Objective 5**)

You Try It 4. Find the Area of a Circle (Topic 5.5, **Objective 5**)

You Try It 5. Find the Area of Common Polygons (Topic 8.2, **Objective 2**)

OBJECTIVES

1 Find the Volume of Common Solids

2 Find the Surface Area of Common Solids

3 Solve Applications Involving Volume or Surface Area

OBJECTIVE 1 FIND THE VOLUME OF COMMON SOLIDS

In **Topic 8.2**, we found the **perimeter** and **area** of *plane figures* such as rectangles, triangles, and circles. **Plane figures** are two-dimensional (flat) figures that have length and width but zero thickness. In this topic, we turn our attention to *solids*. **Solids** are three-dimensional figures that have length, width, and thickness (or height). Some household examples of solids are boxes, cans, and balls.

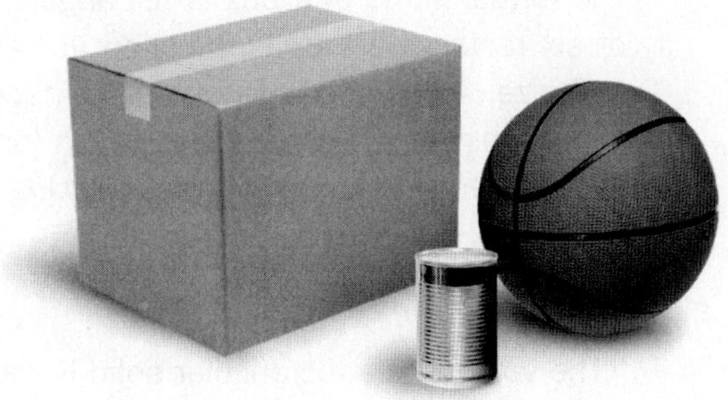

The **volume** of a **solid** is the amount of the space within the solid. Space is three dimensional, so volume is measured in units that are also three dimensional, **cubic units**. For example, a **cubic inch** (cu in., or in^3) is a cube with 1-inch edges. A **cubic centimeter** (cu cm, or cm^3) is a cube with 1-centimeter edges.

(eText Screens 8.3-1–8.3-49)

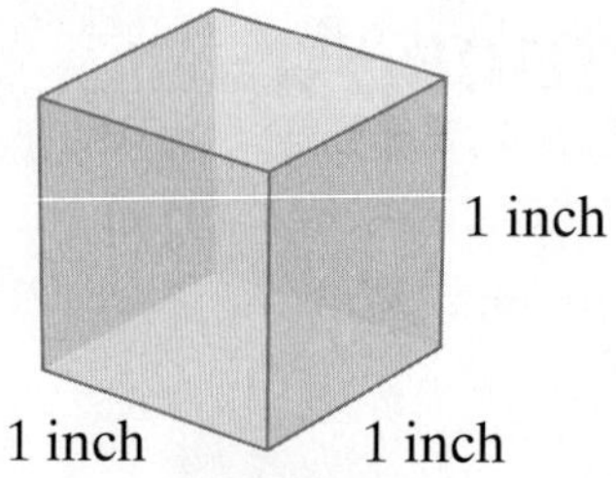

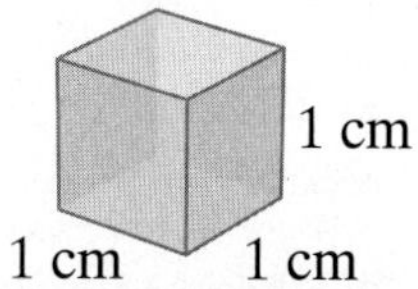

To find the volume of a solid, we find the number of cubic units required to fill up the solid. Continue reading to learn more about the volume of solids, or watch the first part of this **video animation** and then skip forward to **Example 1**.

My video animation summary

To fill the box in Figure 18 with cubes, consider one level at a time. There are 3 rows, with 4 cubes in each row, totaling 12 cubes per level. Because there are 2 levels, we need a total of 24 cubes to fill the box. Therefore, we say the **volume** of the box is 24 cubic units.

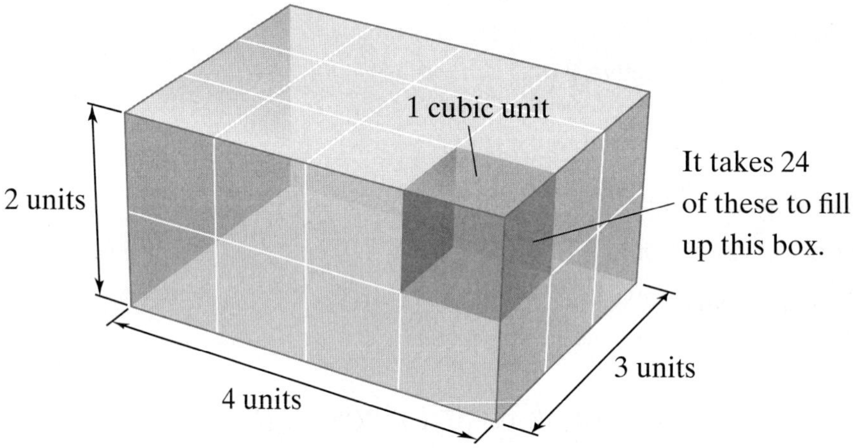

Figure 18

TIP The volume of this box is the result of multiplying the **area** of the bottom rectangle, called the **base**, by the height:

$$\underbrace{4 \text{ units} \cdot 3 \text{ units}}_{\text{Area of base}} \cdot 2 \text{ units} = 24 \text{ cubic units}$$

The formal name of a box is *rectangular solid*. A **rectangular solid** is a **solid** that consists of six rectangular surfaces called its **faces**. Opposite faces have the exact same shape and size. A **line segment** where two faces touch (or intersect) is an **edge** of the solid. A corner point where three faces touch (or intersect) is a **vertex**. We can find the **volume** of rectangular solids by using the following **formula**:

Volume of a Rectangular Solid

The volume of a rectangular solid is found by multiplying the length times the width times the height.

$$V = \underbrace{l \cdot w}_{\text{Area of base}} \cdot h$$

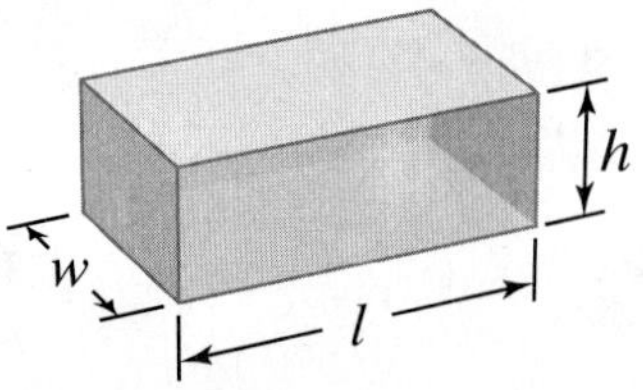

Example 1 Finding the Volume of Rectangular Solids

Find the volume of each rectangular solid.

a.

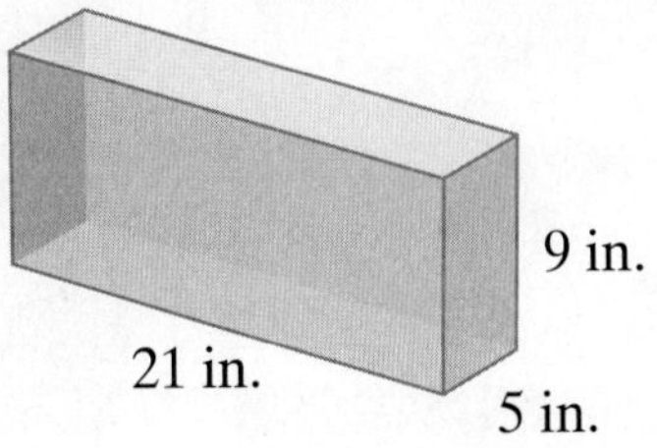

b.

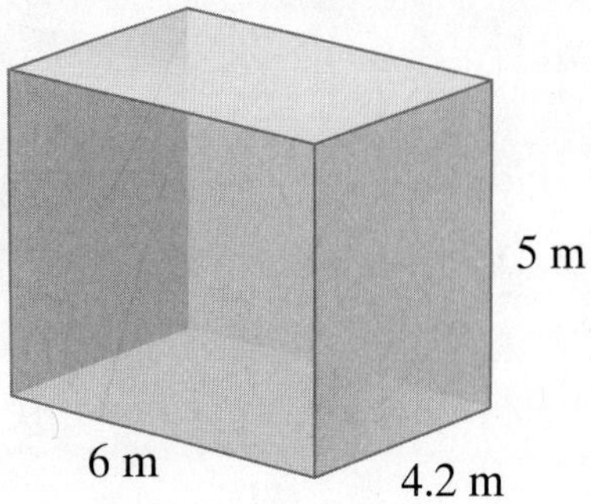

Solutions

a. We use the **formula** $V = lwh$ for the **volume** of a **rectangular solid** with $l = 21$ in., $w = 5$ in., and $h = 9$ in.

Write the formula: $V = lwh$

Substitute for l, w, and h: $= (21\text{ in.})(5\text{ in.})(9\text{ in.})$

Multiply: $= 945\text{ in}^3$

The volume is 945 cubic inches.

My video summary

b. Try to find this volume on your own by substituting $l = 6$ m, $w = 4.2$ m, and $h = 5$ m into the formula. View the **answer**, or watch this **video** for a complete solution to part b.

You Try It **Work through this You Try It problem.**

Work Exercises 1–4 in this textbook or in the MyMathLab Study Plan.

A **cube** is a **rectangular solid** for which all six **faces** are squares of equal size. Because the length, width, and height are all equal, the **volume** of a cube can be found using the following formula.

Volume of a Cube

The volume of a cube is found by **cubing** its side length, s.

$$V = s^3$$

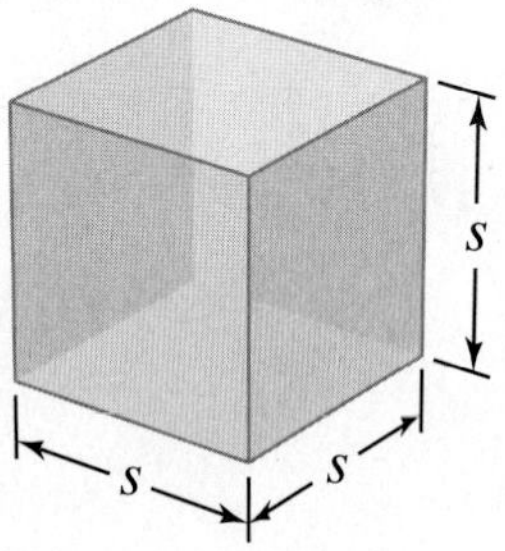

(eText Screens 8.3-1–8.3-49)

Example 2 Finding the Volume of Cubes

My video summary

Find the volume of each cube.

a.

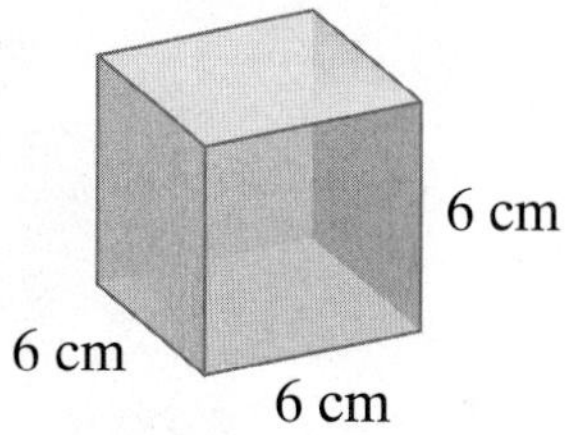

b.

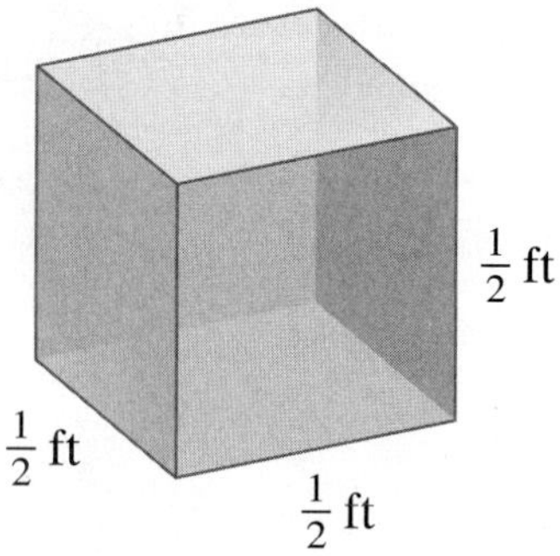

Solutions

a. We use the **formula** $V = s^3$ for the **volume** of a **cube** with $s = 6$ cm.

Write the formula: $V = s^3$

Substitute $s = 6$ cm: $= (6\text{ cm})^3$

Evaluate: $= 216\text{ cm}^3$

The volume is 216 cubic centimeters.

b. Substitute $s = \frac{1}{2}$ ft into the formula, and try to find the volume on your own. Check your **answer**, or watch this **video** for complete solutions to both parts.

You Try It **Work through this You Try It problem.**

Work Exercises 5 and 6 in this textbook or in the MyMathLab Study Plan.

The formal name of a can is a *right circular cylinder*. A **right circular cylinder** is a **solid** that consists of two circles of equal size and a curved rectangular side. When positioned upright, the circles align one directly above the other so that each is **perpendicular** to the height.

As with a **rectangular solid**, we can find the **volume** of a right circular cylinder by multiplying the **area** of the bottom circle, called the **base**, by the height.

Volume of a Right Circular Cylinder

The volume of a right circular cylinder is found by multiplying the area of the circular base, πr^2, by the height, h.

$$V = \underbrace{\pi \cdot r^2}_{\text{Area of base}} \cdot h,$$

where r is the **radius**, and h is the height.

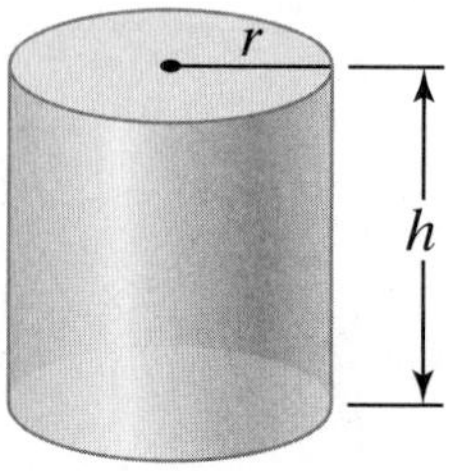

As with **circumference** and area of circles, we can either find the exact volume of a cylinder in terms of π, or we can approximate the volume by using 3.14 for π.

CAUTION The volume of a **solid** is measured in *cubic units* regardless of shape of the solid.

Example 3 Finding the Volume of Right Circular Cylinders

Find the exact **volume** of each **right circular cylinder** in terms of π. Then approximate the volume using 3.14 for π.

a.

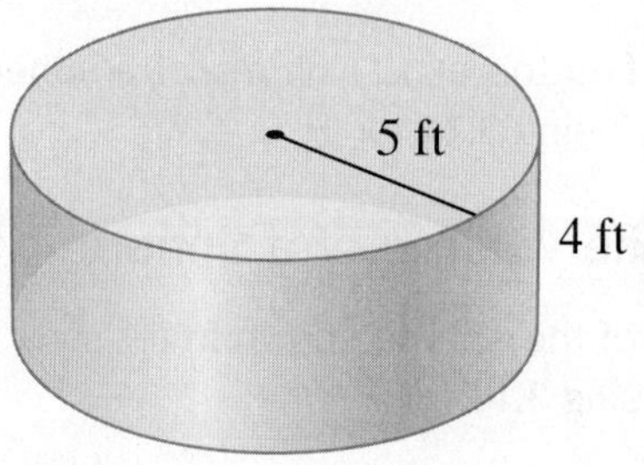

b.

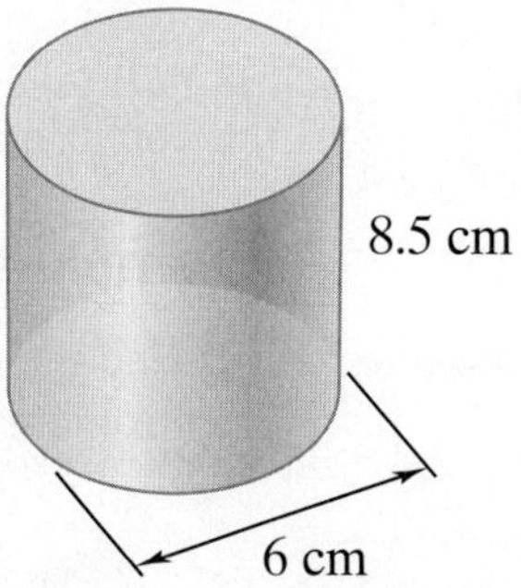

Solutions

a. From the figure, we see the **radius** is 5 feet and the height is 4 feet.
Exact Volume: Using the **formula** $V = \pi r^2 h$, we substitute in $r = 5$ ft and $h = 4$ ft.

$$V = \pi r^2 h = \pi(5\text{ ft})^2(4\text{ ft}) = 100\pi\text{ ft}^3$$

The exact volume is 100π cubic feet.

Approximate Volume: Using 3.14 for π, we get

"is approximately equal to"

$$V = 100\pi \approx 100(3.14) = 314\text{ ft}^3$$

The approximate volume is 314 cubic feet.

My video summary

b. From the figure, we see that the **diameter** is 6 cm and the height is 8.5 cm. To use the **formula** $V = \pi r^2 h$, we need the **radius**. Because the radius is half the diameter, the radius is $r = d \div 2 = 6 \div 2 = 3$ cm.

Now that we have the radius, try to finish finding the exact **volume** and approximate volume on your own. Check your **answer**, or watch this **video** for a complete solution to part b.

You Try It **Work through this You Try It problem.**

Work Exercises 7–10 in this textbook or in the MyMathLab Study Plan.

CAUTION The formula for finding the volume of a cylinder requires the *radius* of the cylinder. Do not substitute in the *diameter* by mistake. If the diameter is given, begin by finding the radius, which is half of the diameter.

The technical name for a ball is *sphere*. A **sphere** is made up of all points in three dimensions that are located at an equal distance from a fixed point called the **center.** A **radius** of a sphere extends from the center to the surface. A **diameter** extends from one side of the sphere to the other, through the center.

Volume of a Sphere

The volume of a sphere is found by the formula

$$V = \frac{4}{3}\pi r^3,$$

where r is the radius.

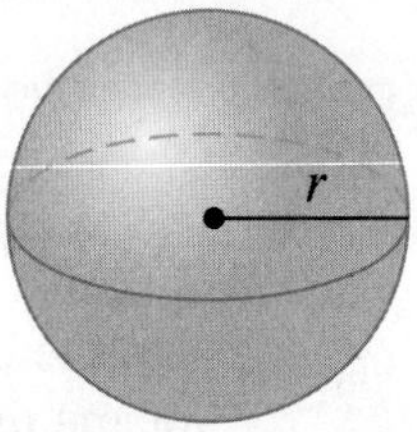

As with **cylinders**, we can find the exact **volume** of a **sphere** in terms of π, or we can find an approximate volume by using 3.14 for π.

Example 4 Finding the Volume of a Sphere

My video summary

Find the exact volume of the sphere shown in terms of π. Then approximate the volume using 3.14 for π.

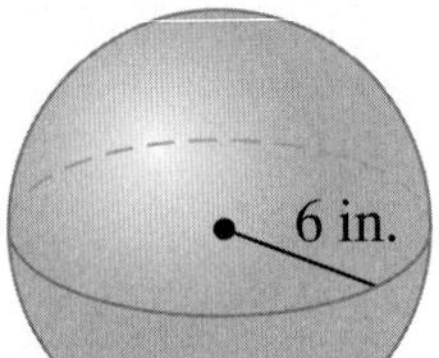

Solution From the figure, we see the **radius** is 6 inches.

Exact Volume: Write the formula: $V = \frac{4}{3}\pi r^3$

Substitute 6 in. for r: $= \frac{4}{3}\pi(6\text{ in.})^3$

Evaluate the exponent: $= \frac{4}{3}\pi \cdot 216\text{ in}^3$

Multiply: $= 288\pi\text{ in}^3 \leftarrow$ $\boxed{\frac{4}{3}\pi \cdot 216 = \frac{4\cdot \cancel{216}^{72}}{{}_1\cancel{3}}\pi = 288\pi}$

The exact volume is 288π cubic inches.

Approximate Volume: Using 3.14 for π, try to find the approximate volume on your own. Check your **answer**, or watch this **video** for a complete solution.

You Try It **Work through this You Try It problem.**

Work Exercises 11 and 12 in this textbook or in the MyMathLab Study Plan.

A **right cone**, or cone, contains a circular base and a curved surface with one **vertex** called the **apex**. If positioned upward, the apex (or tip) aligns directly above the center of the circle so that the circle is **perpendicular** to the height.

The **volume** of a cone is one-third of the volume of a **cylinder** with the same base and height. This fact leads to the following formula for the volume of a cone.

Volume of a Cone

The volume of a cone is found by the formula

$$V = \frac{1}{3}\cdot \underbrace{\pi \cdot r^2}_{\text{Area of base}} \cdot h,$$

where r is the **radius**, and h is the height.

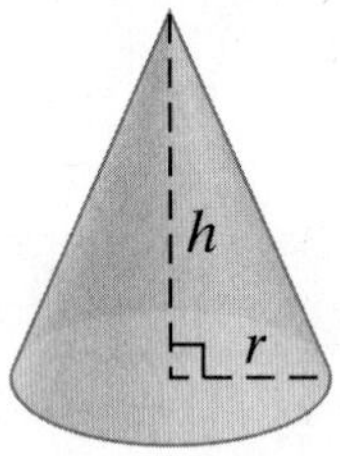

(eText Screens 8.3-1–8.3-49)

As with cylinders and spheres, we can find the exact volume of cones in terms of π, or we can approximate the volume by using 3.14 for π.

Example 5 Finding the Volume of a Cone

My video summary

Find the exact **volume** of the **cone** shown in terms of π. Then approximate the volume using 3.14 for π.

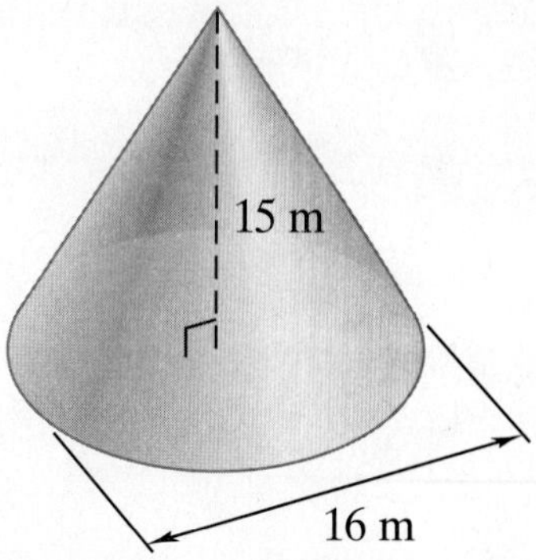

Solution The **diameter** is 16 m, so the **radius** is $16 \div 2 = 8$ m. The height is 15 m.

Exact Volume: Using $V = \frac{1}{3}\pi r^2 h$, substitute $r = 8$ m and $h = 15$ m.

$$V = \frac{1}{3}\pi r^2 h = \frac{1}{3}\pi(8\text{ m})^2(15\text{ m}) = \frac{1}{\cancel{3}_1}\pi(64\text{ m}^2)(\cancel{15}^{5}\text{ m}) = 320\pi\text{ m}^3$$

The exact volume is 320π cubic meters.

Approximate Volume: Use 3.14 for π to approximate the volume on your own. Check your **answer**, or watch this **video** for a complete solution.

You Try It **Work through this You Try It problem.**

Work Exercises 13 and 14 in this textbook or in the MyMathLab Study Plan.

A **pyramid** is like a **cone** except its base is a **polygon** instead of a circle. The sides of a pyramid are triangles. A **regular pyramid** has a base that is a **regular polygon** and is **perpendicular** to the height. Its triangle sides are **congruent**. In this text, we consider only regular pyramids with square bases, **square pyramids**.

The **volume** of a pyramid is one-third of the **area** of the base, times the height.

Volume of a Pyramid

The volume of a pyramid is found by the formula

$$V = \frac{1}{3}\cdot B \cdot h,$$

where B is the area of the base, and h is the height.

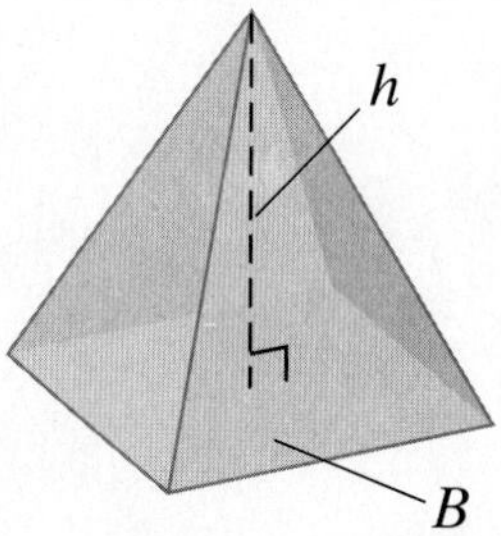

Example 6 Finding the Volume of a Pyramid

My video summary

Find the volume of the pyramid.

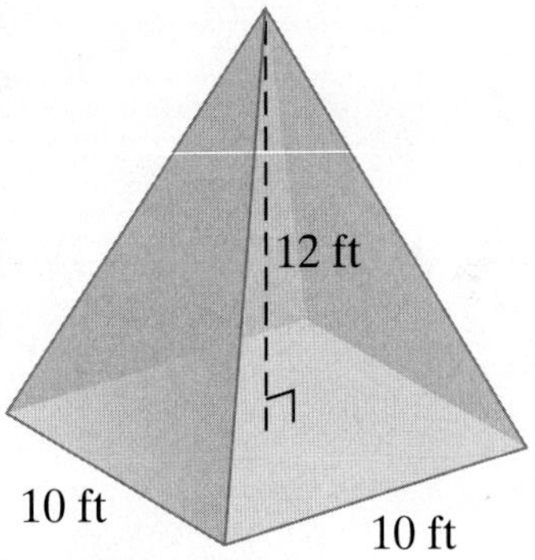

Solution Read below, or watch this **video** for a complete solution.

The base of the **pyramid** is a square 10 ft by 10 ft. So, the area of the base is

$$B = (10\text{ ft})^2 = 100\text{ ft}^2.$$

Using $V = \frac{1}{3}Bh$, we substitute $B = 100\text{ ft}^2$ and $h = 12\text{ ft}$.

$$V = \frac{1}{3}Bh = \frac{1}{3}(100\text{ ft}^2)(12\text{ ft}) = \frac{1}{\cancel{3}_1}(100\text{ ft}^2)(\cancel{12}^{4}\text{ ft}) = 400\text{ ft}^3$$

The **volume** of the pyramid is 400 cubic feet.

You Try It **Work through this You Try It problem.**

Work Exercises 15 and 16 in this textbook or in the MyMathLab Study Plan.

OBJECTIVE 2 FIND THE SURFACE AREA OF COMMON SOLIDS

The **surface area** of a **solid** is the total **area** of the surface on all sides of the solid. Like the area of **plane figures**, surface area is measured in **square units**. To find the surface area of a solid, we find the number of square units required to cover the surface of the solid.

My video animation summary

Continue reading to learn more about the surface area of solids, or watch the second part of this **video animation** and then skip forward to **Example 7**.

To cover the box in Figure 19a with squares, consider each **face** separately.

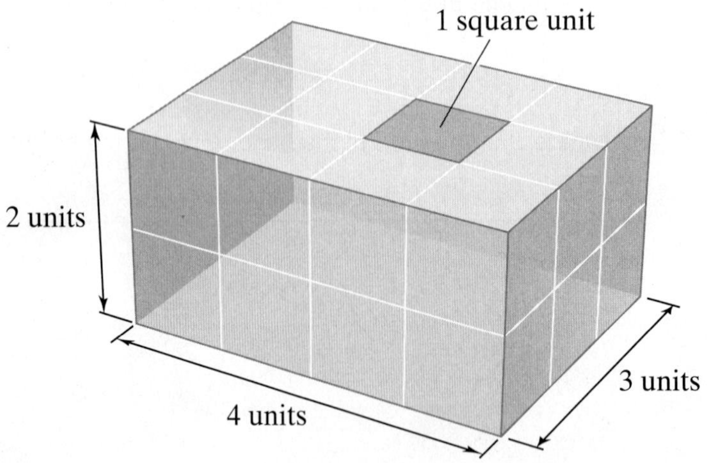

Figure 19a

Figure 19b shows the six faces of the box separated. We can find the number of square units needed to cover each face and then add them together to find the total surface area.

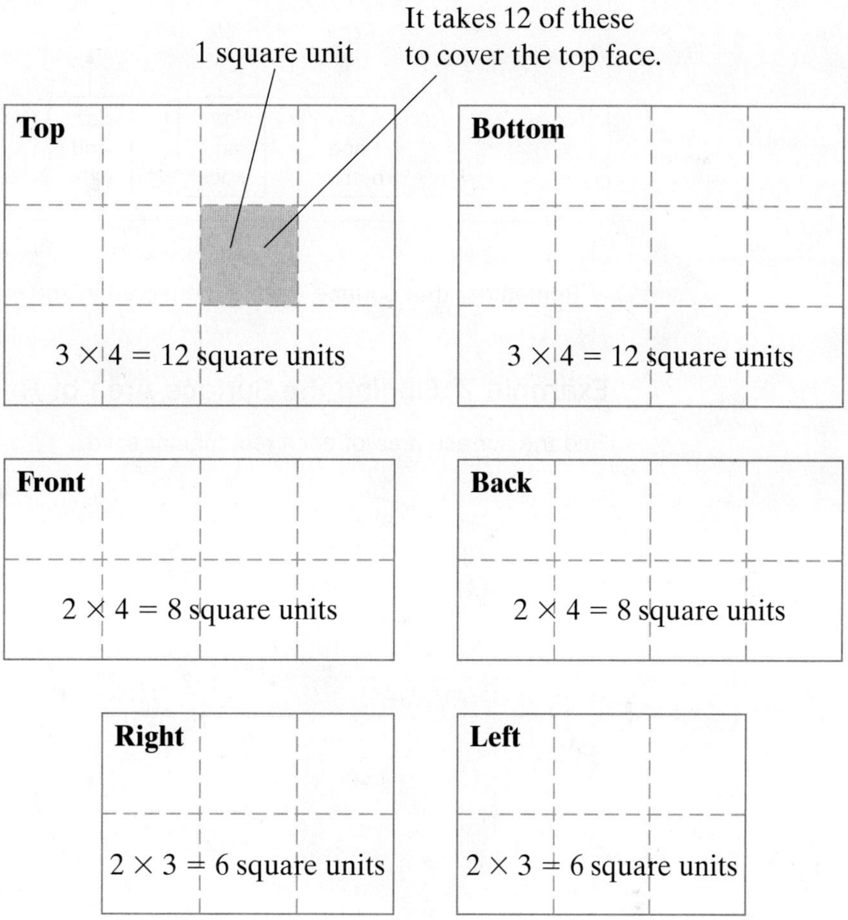

Figure 19b

Top and Bottom To cover the top **face**, we need 3 rows of 4 squares, or 12 square units. The bottom face is the same size as the top, so we also need 12 square units to cover it.

Front and Back To cover the front face, we need 2 rows of 4 squares, or 8 square units. The back face is the same size as the front, so we need another 8 square units for it.

Right and Left To cover the right face, we need 2 rows of 3 squares, or 6 square units. The left face is the same size as the right, so we also need 6 square units for the left.

All together, we need $8 + 8 + 12 + 12 + 6 + 6 = 52$ square units to cover the surface of all six faces. So, the **surface area** of the box is 52 square units.

Notice that once we have the **area** of the top **face**, we can double it to obtain the combined area of the top and bottom faces. Similarly, we can double the area of the front face to find the combined area of the front and back faces, and we can double the area of the left face to find the combined area of the left and right faces. This leads to the following **formula** for finding the **surface area** of **rectangular solids**.

(eText Screens 8.3-1–8.3-49)

Surface Area of a Rectangular Solid

If a rectangular solid has length l, width w, and height h, then its surface area SA is given by the formula

$$SA = \underbrace{2lw}_{\text{Top and bottom}} + \underbrace{2lh}_{\text{Front and back}} + \underbrace{2wh}_{\text{Left and right}}.$$

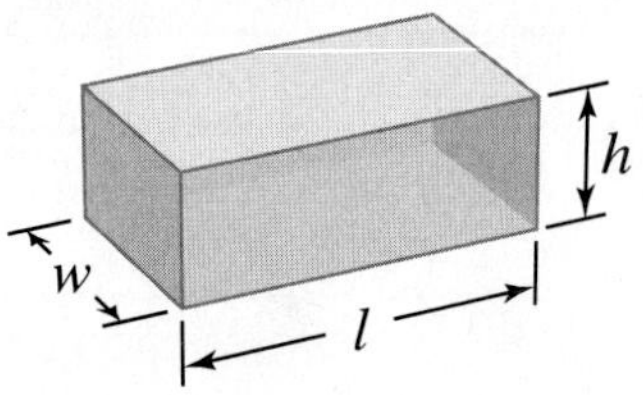

CAUTION Remember that surface area is measured in *square units*.

Example 7 Finding the Surface Area of Rectangular Solids

Find the **surface area** of each **rectangular solid**.

a.

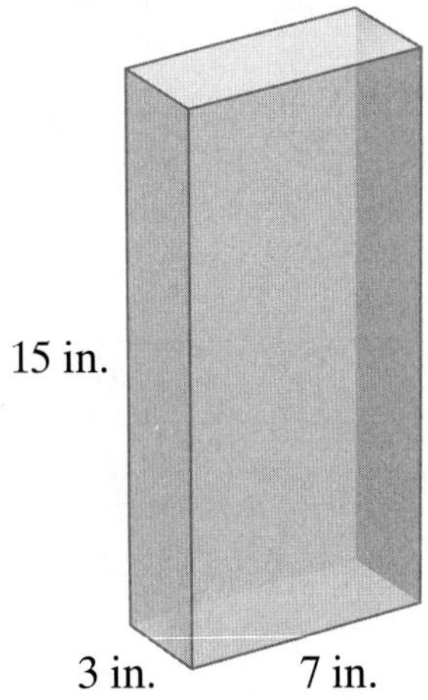

b.

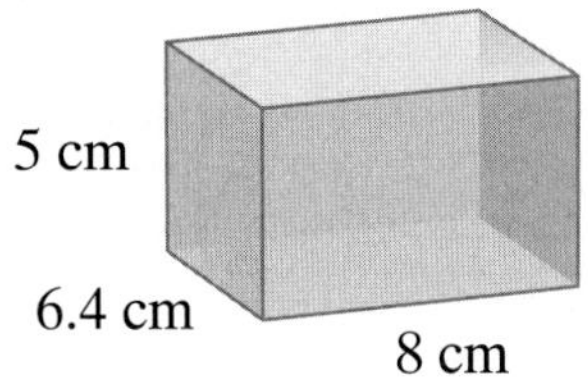

Solutions

a. From the figure, we see the length is 7 inches, the width is 3 inches, and the height is 15 inches. We use the **formula** $SA = 2lw + 2lh + 2wh$ with $l = 7$ in., $w = 3$ in., and $h = 15$ in.

Write the formula:	$SA = 2lw + 2lh + 2wh$
Substitute:	$= 2(7\text{ in.})(3\text{ in.}) + 2(7\text{ in.})(15\text{ in.}) + 2(3\text{ in.})(15\text{ in.})$
Multiply:	$= 42\text{ in}^2 + 210\text{ in}^2 + 90\text{ in}^2$
Add:	$= 342\text{ in}^2$

The surface area is 342 square inches.

My video summary

b. Try to find this **surface area** on your own by substituting $l = 8$ cm, $w = 6.4$ cm, and $h = 5$ cm into the **formula** $SA = 2lw + 2lh + 2wh$. View the **answer**, or watch this **video** for a complete solution to part b.

You Try It **Work through this You Try It problem.**

Work Exercises 17–20 in this textbook or in the MyMathLab Study Plan.

Because the six **faces** of a **cube** are all squares of the same size, we can multiply the **area** of one square by 6 to obtain the surface area.

Surface Area of a Cube

If a cube has side length s, then its surface area SA is given by the formula

$$SA = 6 \cdot \underbrace{s^2}_{\text{Area of one face}}.$$

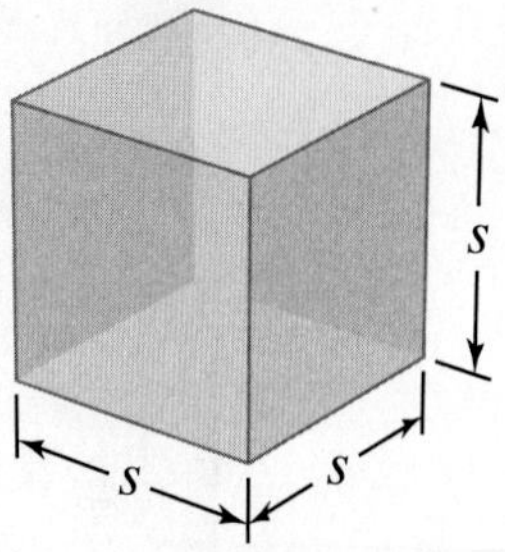

Example 8 Finding the Surface Area of a Cube

Find the **surface area** of the cube.

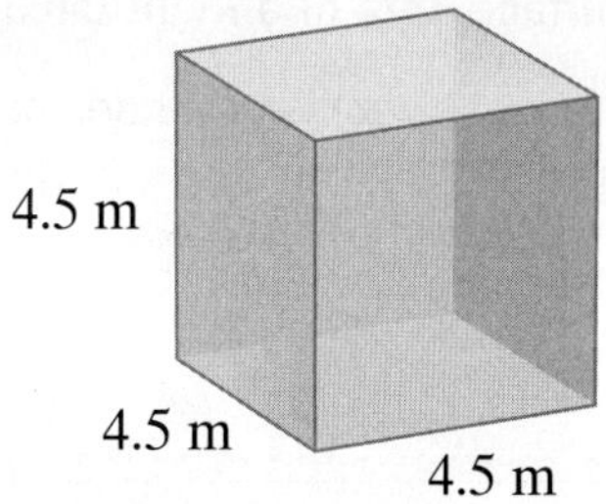

Solution We use the **formula** $SA = 6s^2$ with $s = 4.5$ m.

$$SA = 6s^2 = 6(4.5\text{ m})^2 = 6(20.25\text{ m}^2) = 121.5\text{ m}^2$$

The surface area of the cube is 121.5 square meters.

You Try It **Work through this You Try It problem.**

Work Exercises 21 and 22 in this textbook or in the MyMathLab Study Plan.

Figure 20 shows that the surface of a **right circular cylinder** consists of two circles and a rectangle. We can find the **surface area** of the cylinder by adding the **area** of the two circles and the area of the rectangle.

Circles The area of the top and bottom circles are the same, πr^2. So, the area of the two circles combined is $2\pi r^2$.

Rectangle The length of the rectangle is equal to the **circumference** of the cylinder, which is $2\pi r$. The width of the rectangle is the height of the cylinder, h. So, the area of the rectangle is $2\pi r \cdot h$.

Adding the area of the circles to the area of the rectangle gives the **formula** for the surface area of right circular cylinders, $SA = 2\pi r^2 + 2\pi rh$.

(eText Screens 8.3-1–8.3-49)

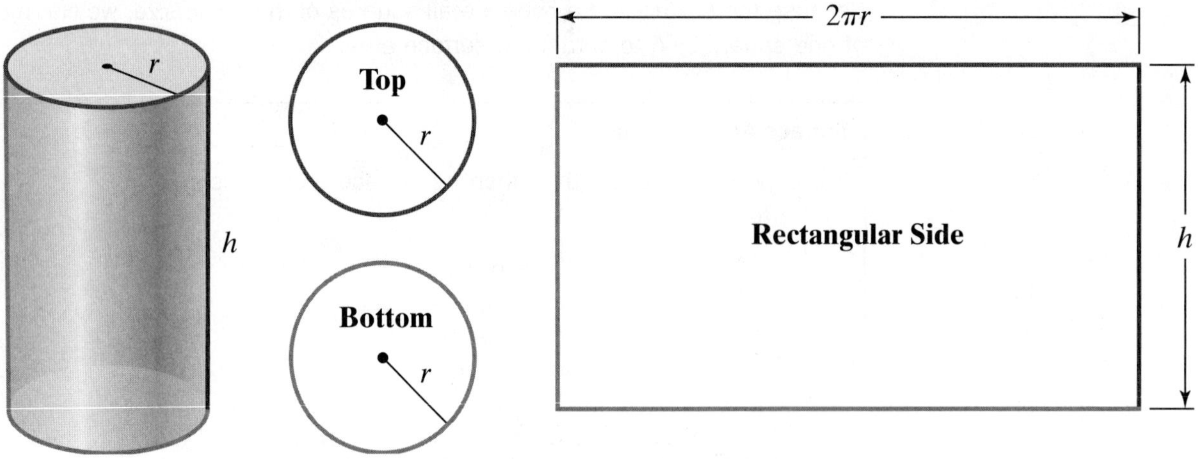

Figure 20

> **Surface Area of a Right Circular Cylinder**
>
> If a right circular cylinder has radius r and height h, its surface area SA is given by the formula
>
> $$SA = \underbrace{2\pi r^2}_{\text{Circles}} + \underbrace{2\pi rh}_{\text{Rectangle}}.$$
>
> 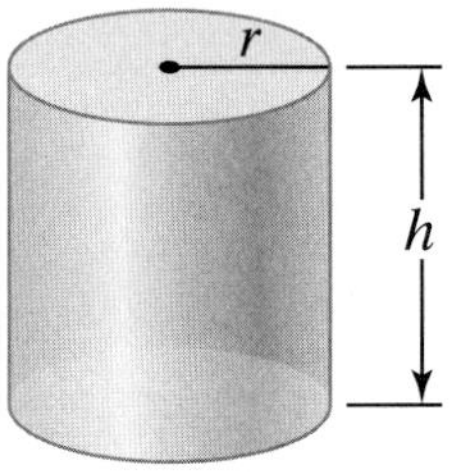
>

As with volume, we can find the **surface area** of a cylinder exactly in terms of π, or we can find it approximately by using 3.14 for π.

Example 9 Finding the Surface Area of a Right Circular Cylinder

My video summary

For the **right circular cylinder** shown, find the exact surface area in terms of π. Then approximate the surface area using 3.14 for π.

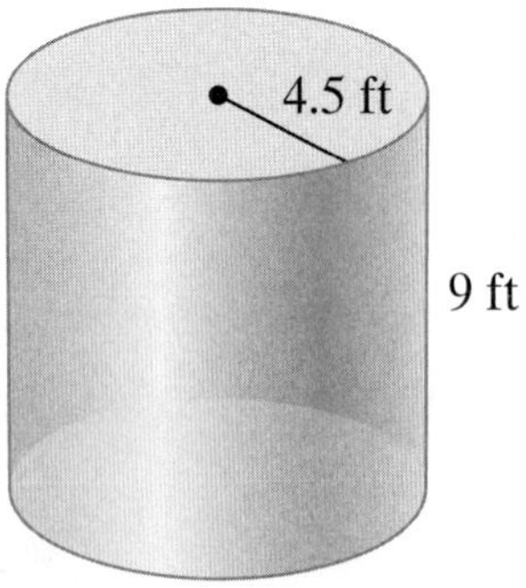

Solution We see the **radius** is 4.5 feet and the height is 9 feet.

Exact Surface Area: Using $SA = 2\pi r^2 + 2\pi rh$, substitute $r = 4.5$ ft and $h = 9$ ft.

$$\begin{aligned} SA = 2\pi r^2 + 2\pi rh &= 2\pi(4.5\text{ ft})^2 + 2\pi(4.5\text{ ft})(9\text{ ft}) \\ &= 40.5\pi\text{ ft}^2 + 81\pi\text{ ft}^2 \\ &= 121.5\pi\text{ ft}^2 \end{aligned}$$

The exact surface area is 121.5π square feet.

Approximate Surface Area: Using 3.14 for π, find the approximate surface area on your own. Check your **answer**, or watch this **video** for a complete solution.

You Try It **Work through this You Try It problem.**

Work Exercises 23–26 in this textbook or in the MyMathLab Study Plan.

We can use the following **formulas** to find the **surface area of spheres, cones,** and **pyramids.**

Surface Area of a Sphere

The surface area of a sphere is found by the formula

$$SA = 4\pi r^2,$$

where r is the radius.

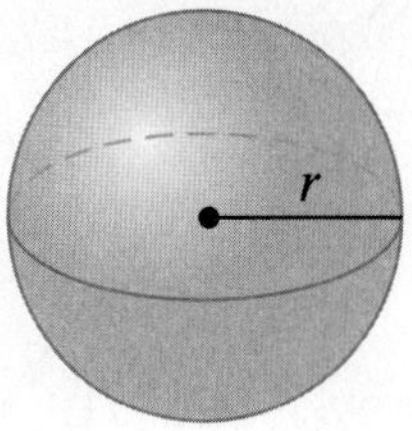

Surface Area of a Right Cone

The surface area of a cone is given by the formula

$$SA = \pi rs + \pi r^2,$$

where r is the **radius,** and $s = \sqrt{r^2 + h^2}$ is the **slant height.**

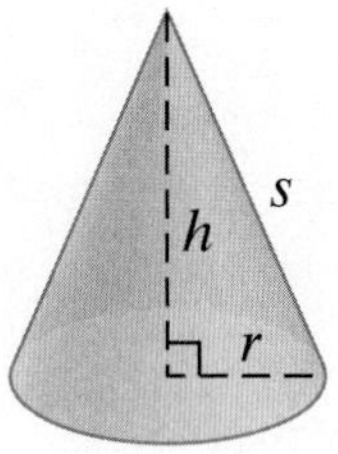

CAUTION If the slant height is not given, then we must find it by using the radius and height: $s = \sqrt{r^2 + h^2}$.

Surface Area of a Regular Pyramid

The surface area of a **regular pyramid** is found by the formula

$$SA = B + \frac{1}{2}Pl,$$

where B is the area of the base, P is the **perimeter** of the base, and l is the **slant height.**

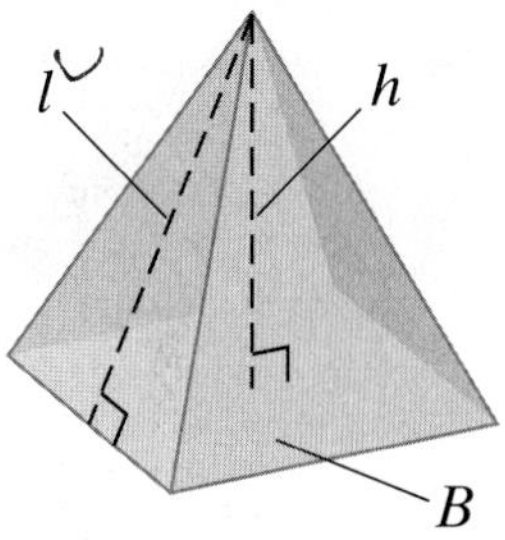

Remember, we will be considering **square pyramids** only in this text.

Example 10 Finding the Surface Area of Common Solids

My interactive video summary

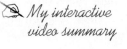

Find the **surface area** of each **solid.** When appropriate, give both an exact answer in terms of π and an approximation using 3.14 for π.

a.

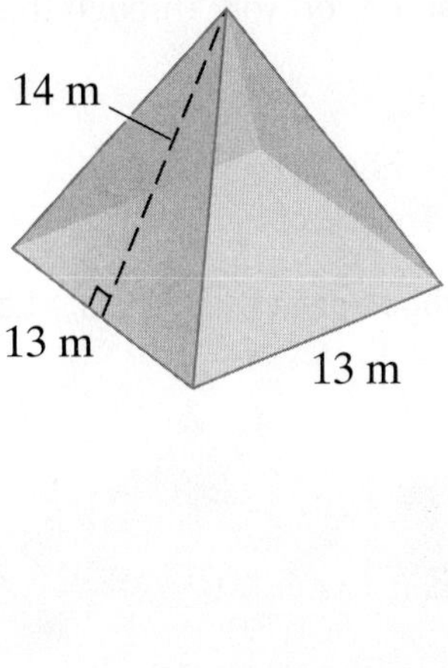

b.

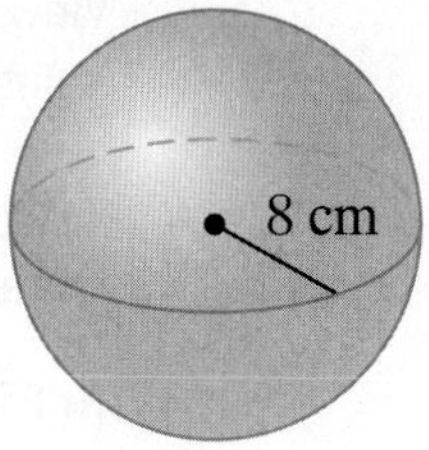

c.

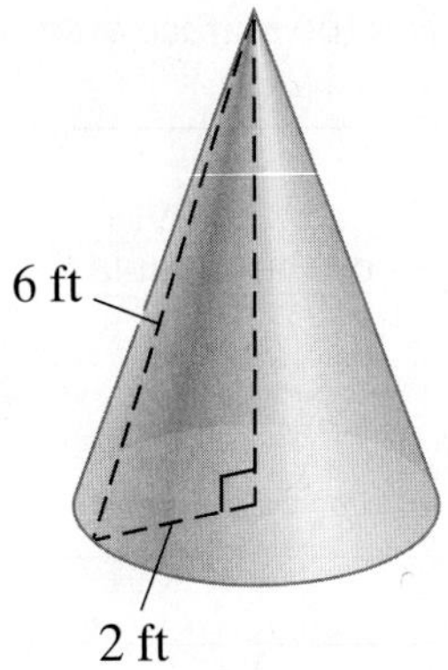

d.

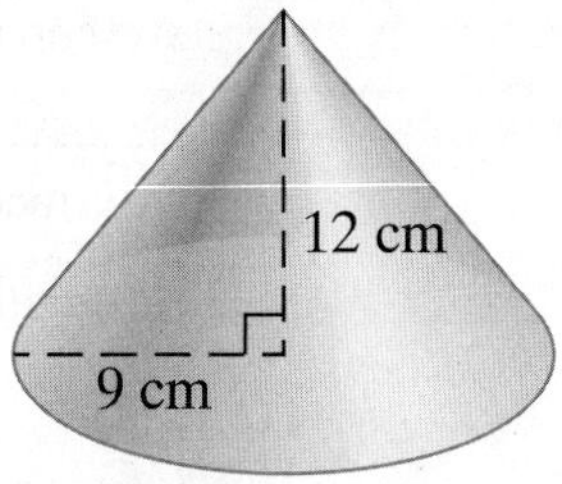

Solutions

a. We see this **square pyramid** has a base measuring 13 m by 13 m and a slant height of 14 m. To find the **surface area**, we begin by finding the **area** of the base, B, and the **perimeter** of the base, P. Each side of the square base is 13 m, so

$$B = (13\text{ m})^2 = 169\text{ m}^2 \quad \text{and} \quad P = 4(13\text{ m}) = 52\text{ m}.$$

Next, using $SA = B + \frac{1}{2}Pl$, we substitute $B = 169\text{ m}^2$, $P = 52\text{ m}$, and $l = 14\text{ m}$:

$$\begin{aligned} SA = B + \frac{1}{2}Pl &= 169\text{ m}^2 + \frac{1}{2}(52\text{ m})(14\text{ m}) \\ &= 169\text{ m}^2 + 364\text{ m}^2 \\ &= 533\text{ m}^2 \end{aligned}$$

The surface area of the pyramid is 533 square meters.

b. From the figure, we see a **sphere** with a radius of 8 cm. We need to find both the exact **surface area** and an approximation.

Exact Surface Area: Using $SA = 4\pi r^2$, we substitute in $r = 8$ cm.

$$SA = 4\pi r^2 = 4\pi(8\text{ cm})^2 = 4\pi(64\text{ cm}^2) = 256\pi\text{ cm}^2$$

The exact surface area is 256π square centimeters.

Approximate Surface Area: Using 3.14 for π, we get

$$SA = 256\pi \approx 256(3.14) = 803.84\text{ cm}^2$$

The surface area is approximately 803.84 square centimeters.

c.–d. Try finding the exact and approximate surface areas of these two **right cones** on your own by using the **formula** $SA = \pi rs + \pi r^2$. Notice that the **slant height**, s, is given in part c, but not in part d. In part d, you first will need to find s by using $s = \sqrt{r^2 + h^2}$. View the **answers** for parts c and d, or work through this **interactive video** for complete solution to all four parts.

You Try It **Work through this You Try It problem.**

Work Exercises 27–32 in this textbook or in the MyMathLab Study Plan.

(eText Screens 8.3-1–8.3-49)

OBJECTIVE 3 SOLVE APPLICATIONS INVOLVING VOLUME OR SURFACE AREA

The following table gives a summary of the **formulas** we have seen for finding the **volume** and **surface area** of common **solids**. We can use these formulas to solve applications.

Figure	Formulas	Figure	Formulas
Cube 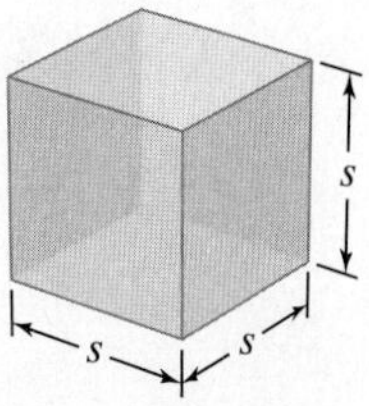	Volume: $V = s^3$ Surface Area: $SA = 6s^2$	Sphere 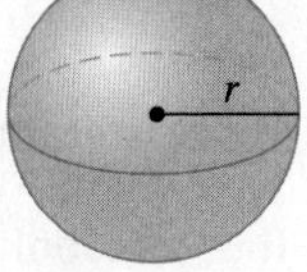	Volume: $V = \frac{4}{3}\pi r^3$ Surface Area: $SA = 4\pi r^2$
Rectangular Solid 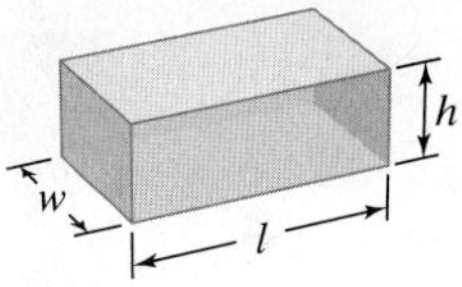	Volume: $V = lwh$ Surface Area: $A = 2lw + 2lh + 2wh$	Right Cone 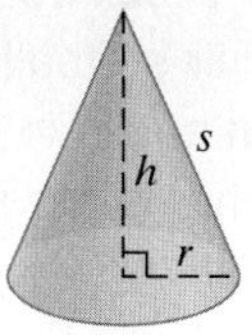	Volume: $V = \frac{1}{3}\pi r^2 h$ Surface Area: $SA = \pi rs + \pi r^2 = \pi r\sqrt{r^2 + h^2} + \pi r^2$
Right Circular Cylinder 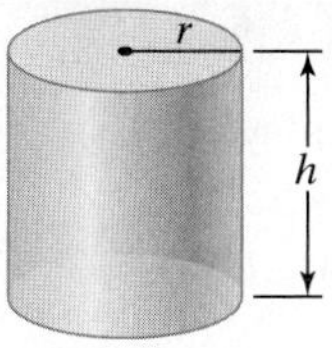	Volume: $V = \pi r^2 h$ Surface Area: $SA = 2\pi r^2 + 2\pi rh$	Pyramid 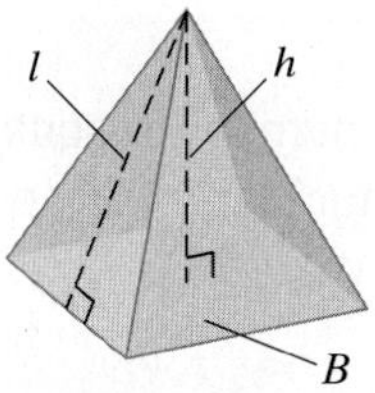	Volume: $V = \frac{1}{3}Bh$, where B is the area of the base. Surface Area (of a Regular Pyramid): $SA = B + \frac{1}{2}Pl$, where P is the perimeter of the base.

Example 11 Volume of a Fish Tank

The fish tank shown is 50 cm long, 30 cm wide, and 40 cm deep. How many cubic centimeters of water can the tank hold?

(eText Screens 8.3-1–8.3-49)

Solution To find the number of cubic centimeters the tank can hold means to find the tank's **volume.** The tank is a **rectangular solid** with $l = 50$ cm, $w = 30$ cm, and $h = 40$ cm. We substitute these values into the **formula** $V = lwh$

$$\text{Write the volume formula: } V = lwh$$
$$\text{Substitute: } = (50\text{ cm})(30\text{ cm})(40\text{ cm})$$
$$\text{Multiply: } = 60{,}000\text{ cm}^3$$

The tank can hold 60,000 cubic centimeters of water.

You Try It **Work through this You Try It problem.**

Work Exercises 33–36 in this textbook or in the MyMathLab Study Plan.

Example 12 Souvenir Display Case

My video summary

A company makes clear **cubes** from plastic sheeting as a display case for souvenir softballs. If the cubes measure 7.6 centimeters on each side, how many square centimeters of plastic sheeting are needed to make one display case?

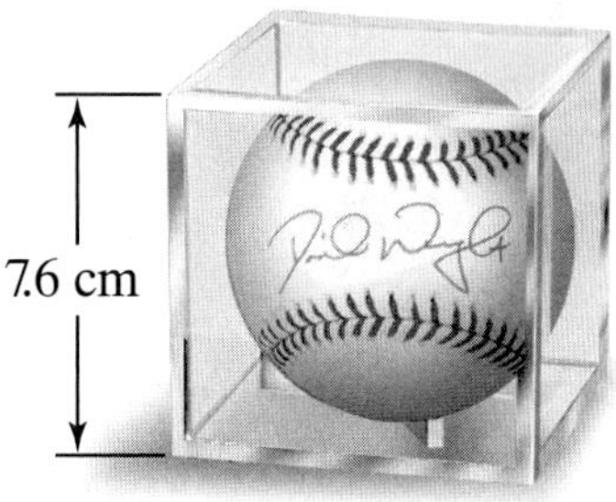

Solution To find the number of square centimeters needed to make a display case means to find the surface area of the cube. Try to find the surface area on your own to answer the question. View the **answer**, or watch this **video** for a complete solution.

You Try It **Work through this You Try It problem.**

Work Exercises 37 and 38 in this textbook or in the MyMathLab Study Plan.

Example 13 Installing a Pool Filter

My video summary

Kenny is installing the filter shown on his new pool. The filter is a sphere with a diameter of 22 inches. The instructions state to fill the filter three-fourths full with sand for proper filtration. How many cubic inches of sand should Kenny put into the filter? Give an approximate answer using 3.14 for π.

(eText Screens 8.3-1–8.3-49)

Solution Read below, or watch this **video** for a complete solution.

To fill the filter three-fourths full with sand means that three-fourths of the filter's volume should contain sand. We need to find three-fourths of the filter's volume.

We begin by finding the volume of the filter, which is a sphere with a 22-inch diameter. To use the **formula** $V = \frac{4}{3}\pi r^3$, we first find the radius of the sphere: $r = \frac{d}{2} = \frac{22}{2} = 11$ inches. Therefore, the volume is

$$V = \frac{4}{3}\pi(11)^3 = \frac{4}{3}\pi(1331) = \frac{4 \cdot 1331}{3}\pi = \frac{5324}{3}\pi \text{ in}^3.$$

To find three-fourths of this volume, multiply by $\frac{3}{4}$:

$$\frac{3}{4}\cdot\left(\frac{5324}{3}\pi \text{ in}^3\right) = \frac{\cancel{3}}{\underset{1}{\cancel{4}}}\cdot\frac{\overset{1331}{\cancel{5324}}}{\cancel{3}}\pi \text{ in}^3 = 1331\pi \text{ in}^3.$$

Using 3.14 for π, we get

$$1331\pi \text{ in}^3 \approx 1331(3.14) \text{ in}^3 = 4179.34 \text{ in}^3.$$

Kenny should put about 4179.34 cubic inches of sand in the filter.

You Try It **Work through this You Try It problem.**

Work Exercises 39–50 in this textbook or in the MyMathLab Study Plan.

8.3 Exercises

In Exercises 1–6, find the volume of each rectangular solid or cube.

You Try It

1.

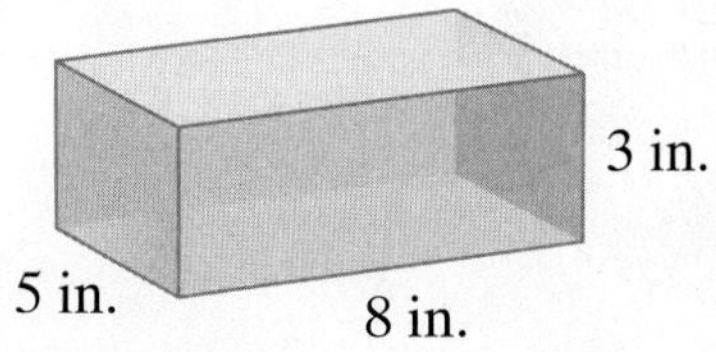

2.

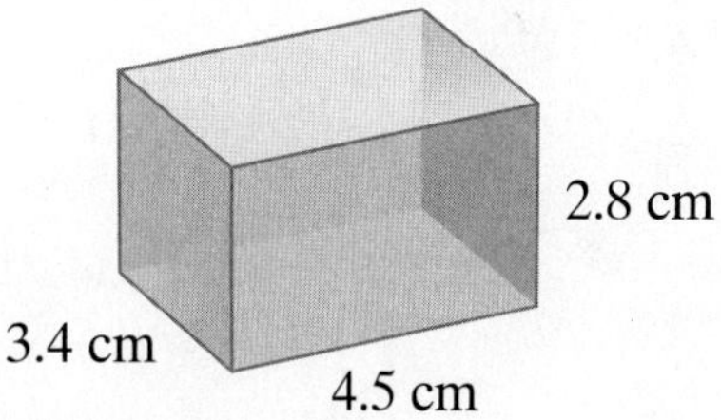

3.

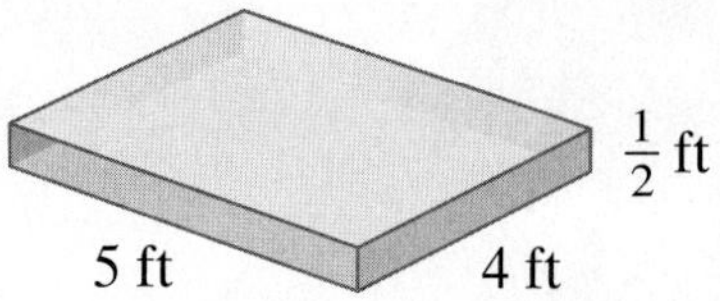

4.

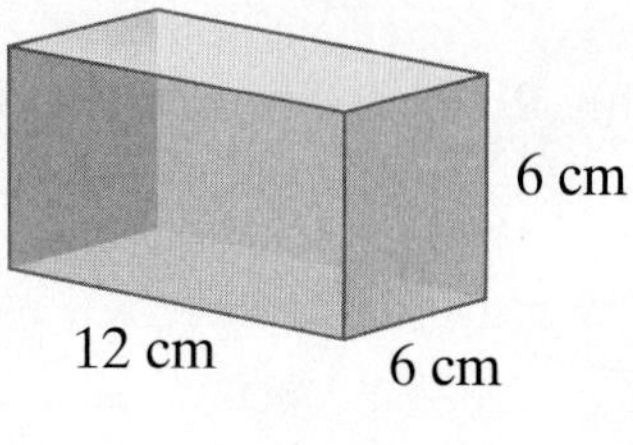

(eText Screens 8.3-1–8.3-49)

You Try It 5.

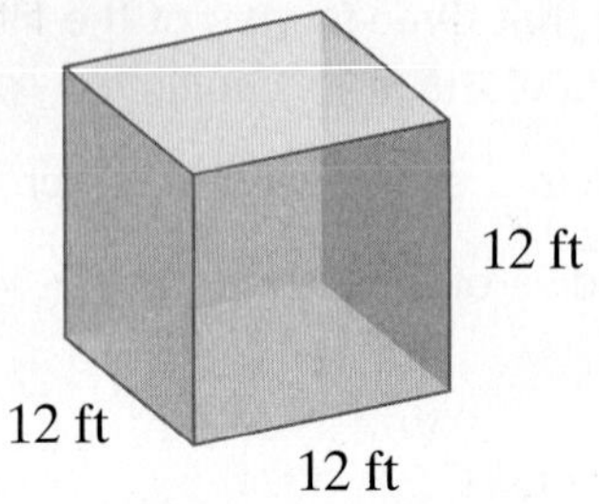

6.

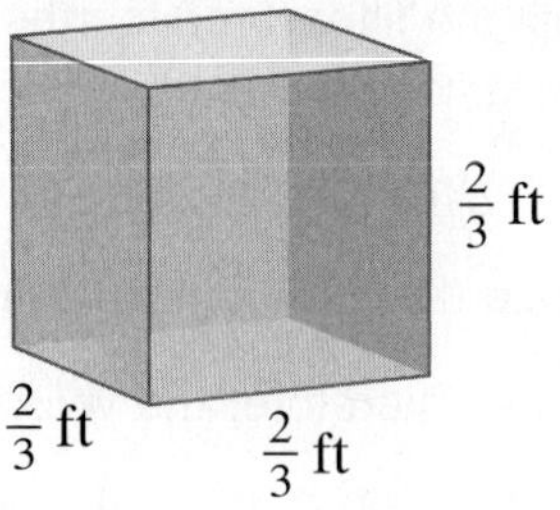

You Try It In Exercises 7–14, find the exact volume of each solid in terms of π. Then approximate the volume using 3.14 for π.

7.

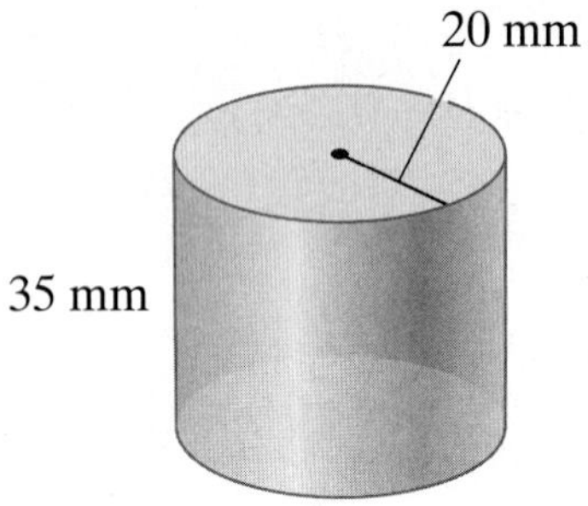

8.

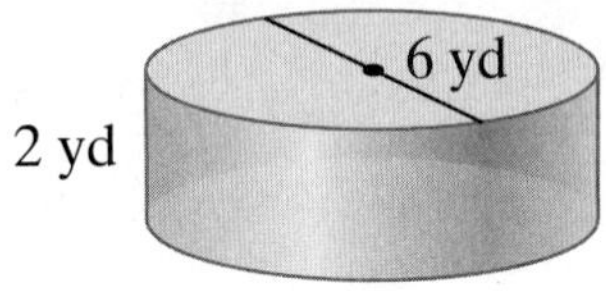

9.

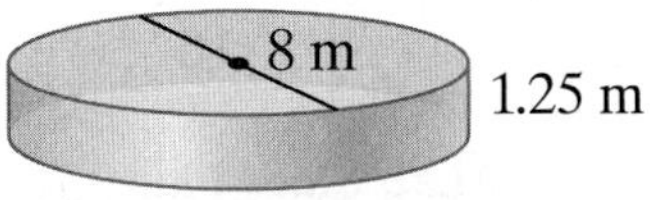

10.

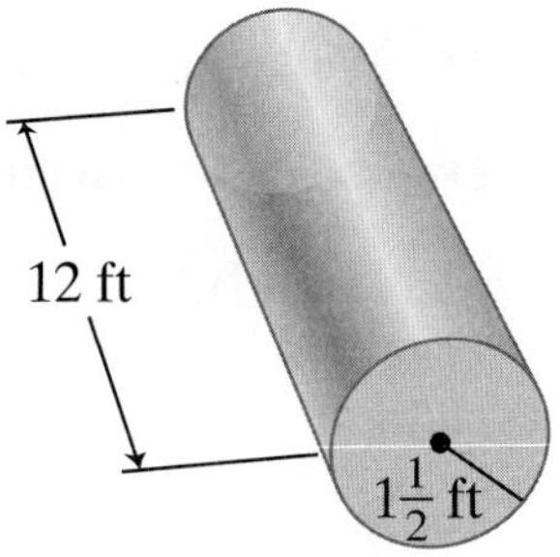

You Try It 11.

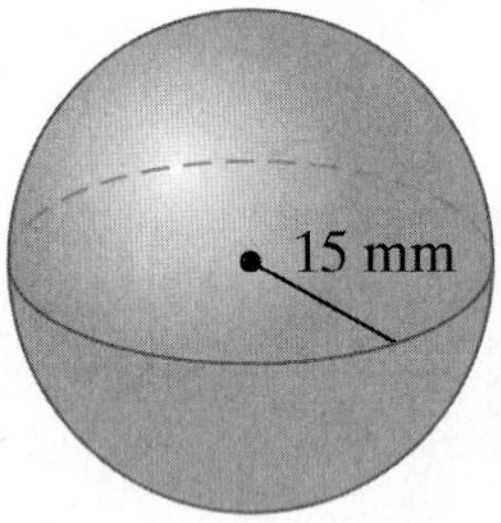

12.

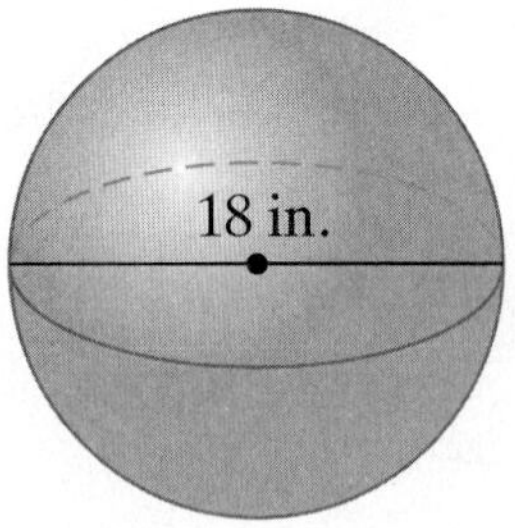

You Try It 13.

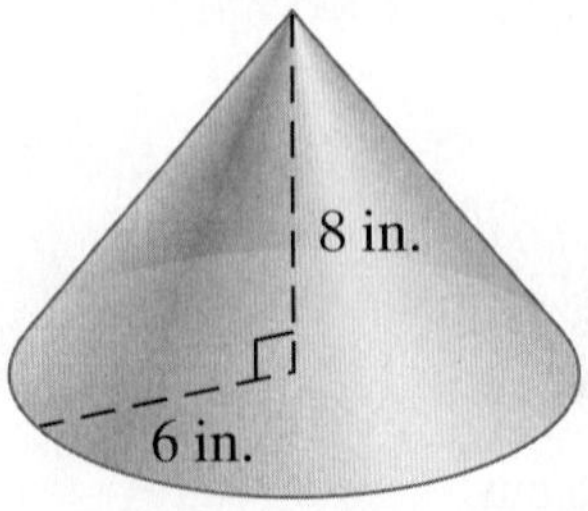

14.

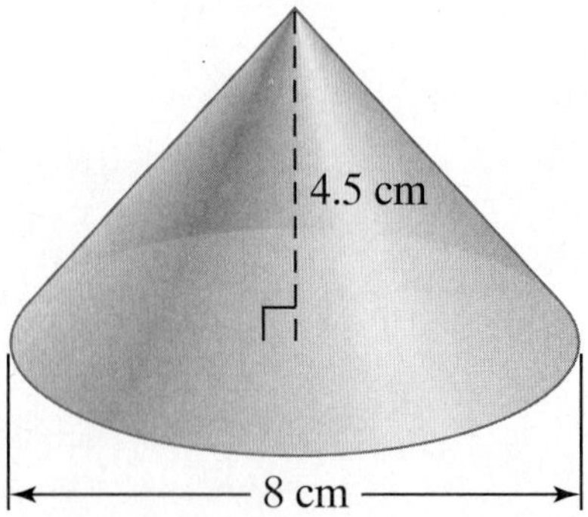

In Exercises 15 and 16, find the volume of each pyramid.

You Try It 15.

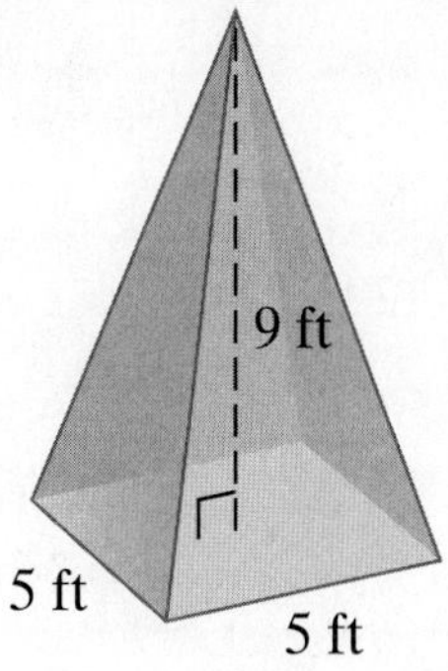

16.

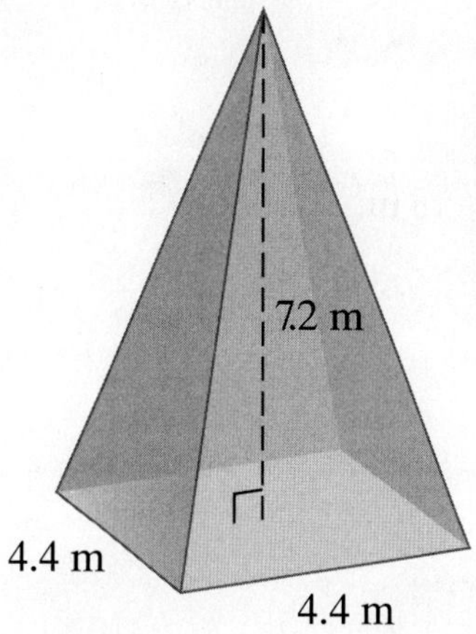

In Exercises 17–22, find the surface area of each rectangular solid or cube.

You Try It 17.

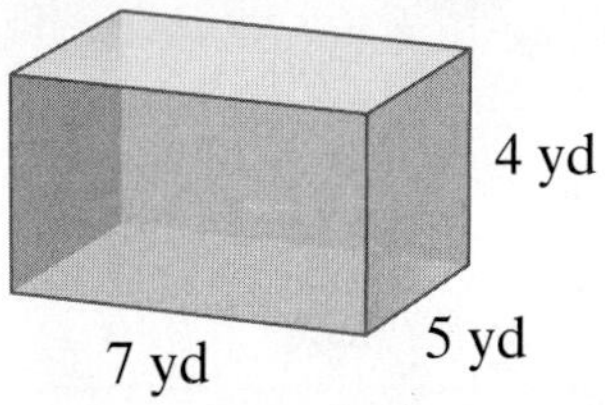

18.

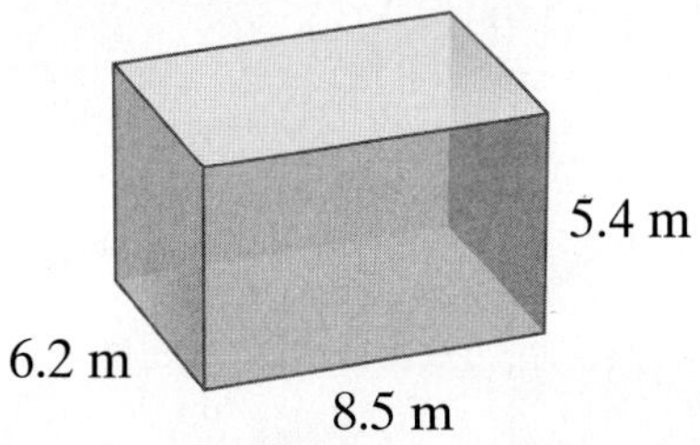

19.

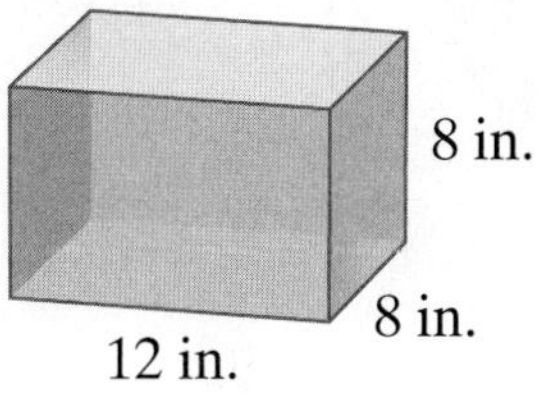

20.

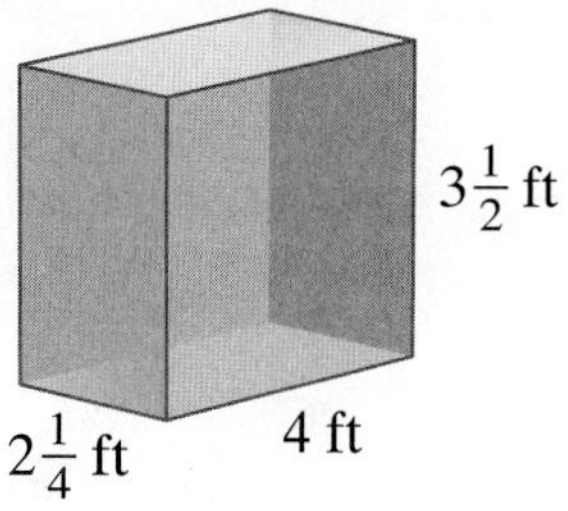

You Try It 21.

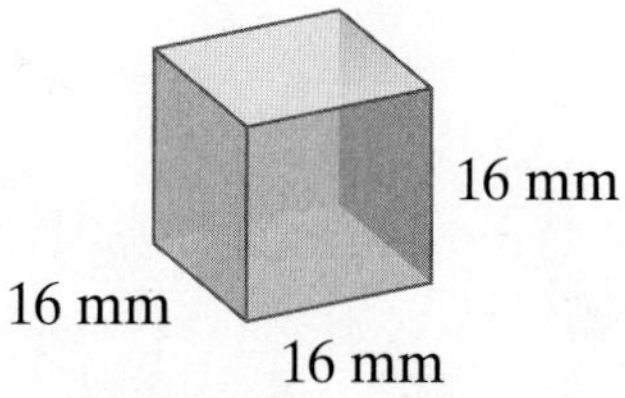

22.

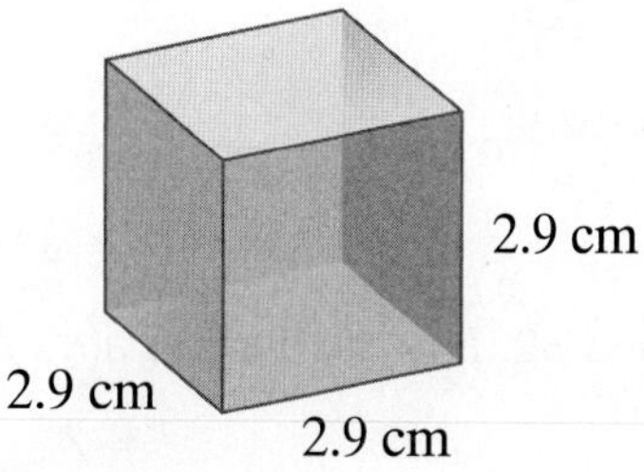

(eText Screens 8.3-1–8.3-49)

In Exercises 23–30, find the exact surface area of each solid in terms of π. Then approximate the surface area using 3.14 for π.

You Try It 23.

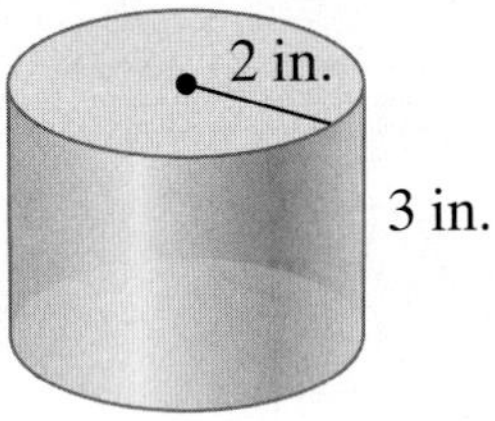

24.

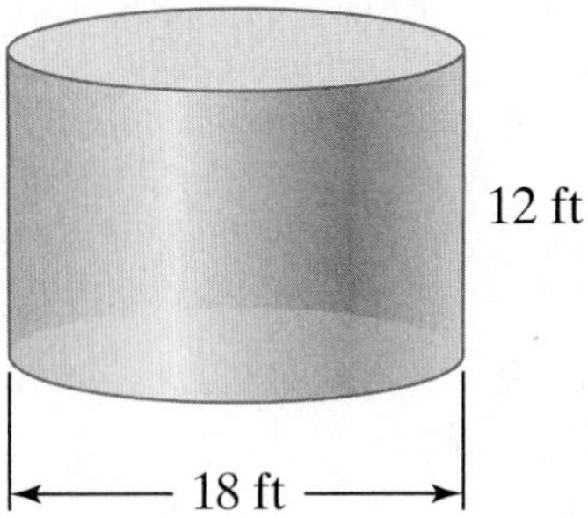

25.

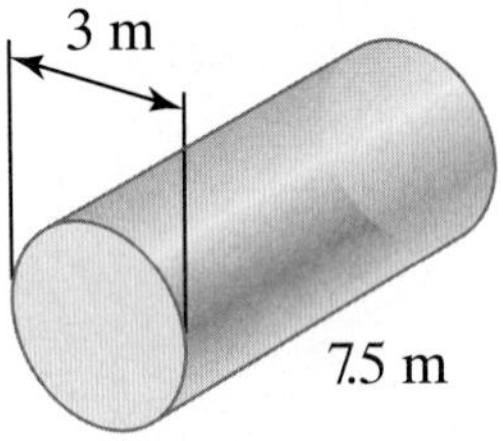

26.

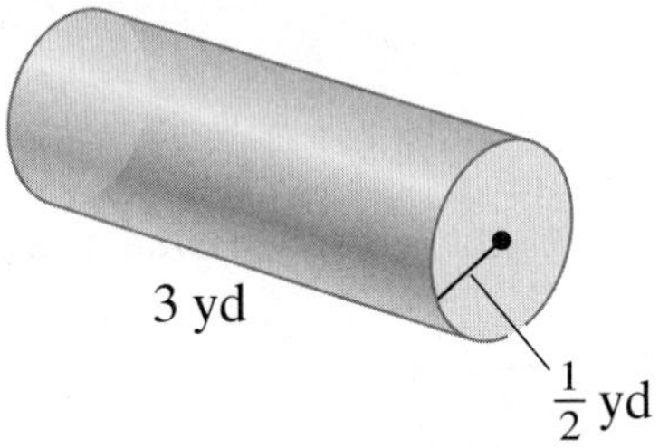

You Try It 27.

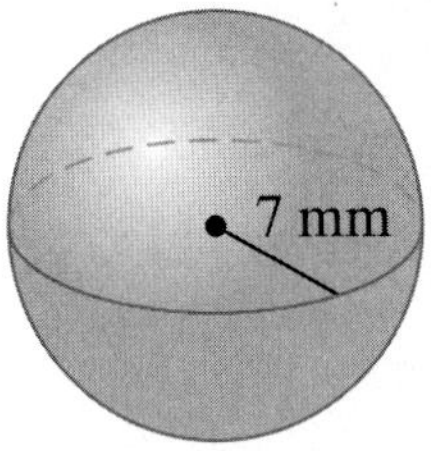

28.

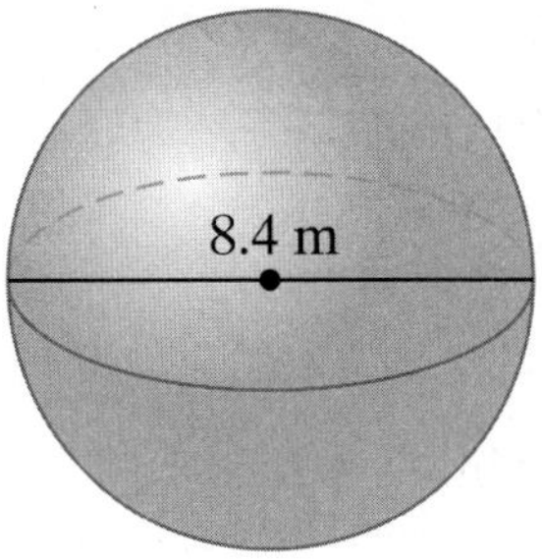

29.

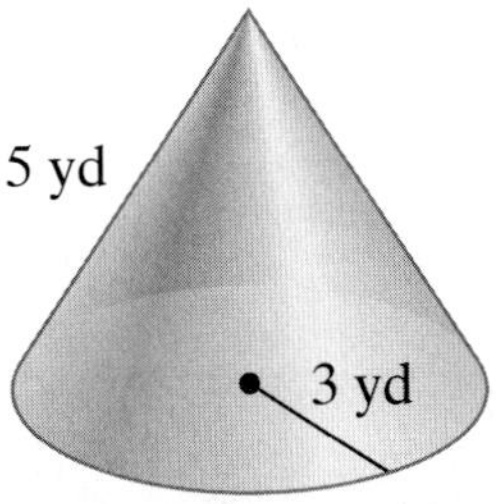

30.

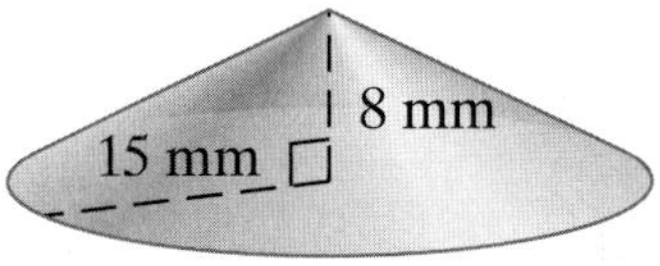

In Exercises 31 and 32, find the surface area of each pyramid.

31.

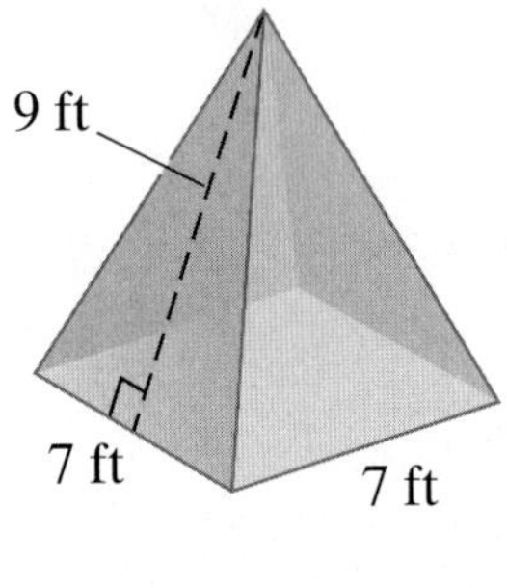

32.

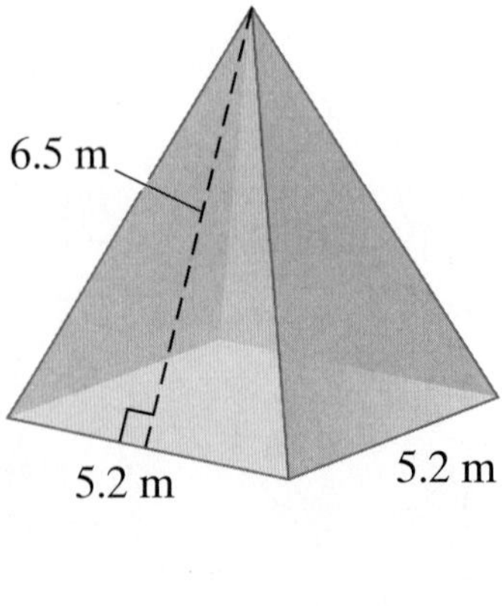

In Exercises 33–40, solve each application.

You Try It 33. **Carry-on Luggage** The largest piece of carry-on luggage allowed on several major airlines measures 23 in. by 10 in. by 16 in., as shown. Find the volume of a piece of luggage with these dimensions. (*Source*: luggagesource.com)

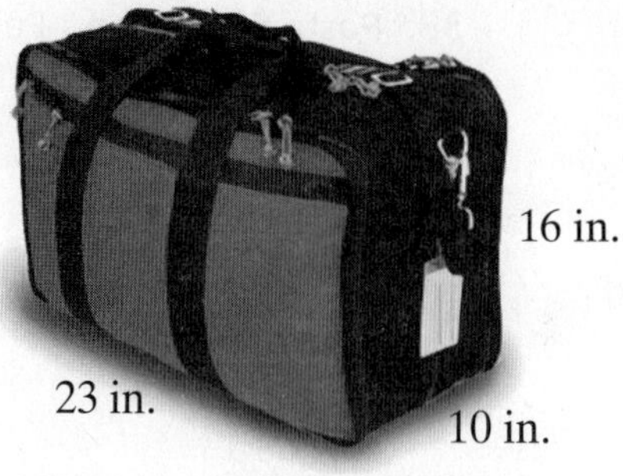

34. **Volume of a Tablet Computer** The tablet computer shown is approximately a rectangular solid. What is the volume of the tablet computer?

35. **Cord of Firewood** A cord of firewood is a stack 8 feet long, 4 feet wide, and 4 feet tall, as shown. Find the volume of a cord of wood.

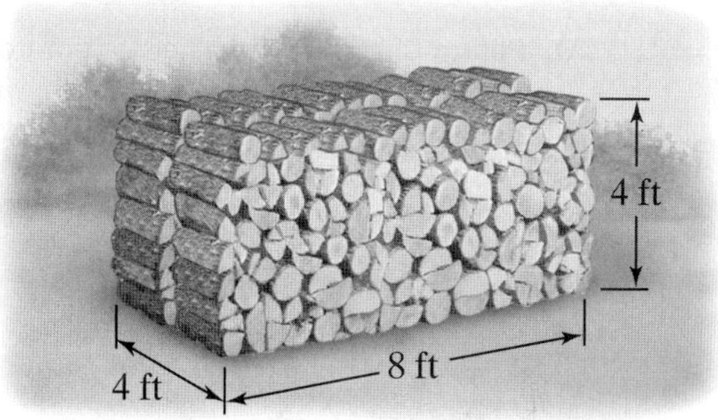

36. **Rain Barrel** A rain barrel has the shape of a right circular cylinder with a diameter of 2 feet and a height of 3 feet, as shown. How many cubic feet of rainwater will the barrel hold? Use 3.14 for π to approximate the answer to two decimal places.

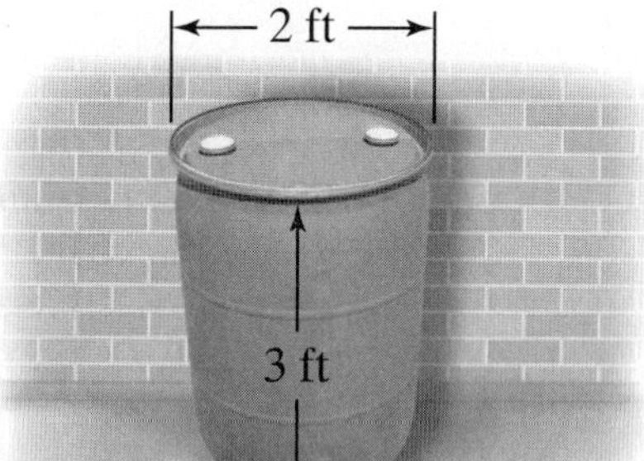

You Try It 37. **Furniture Upholstery** A furniture company makes footstools in the shape of a cube, as shown. If leather upholstery covers all six sides, how many square centimeters of leather are needed to cover one stool?

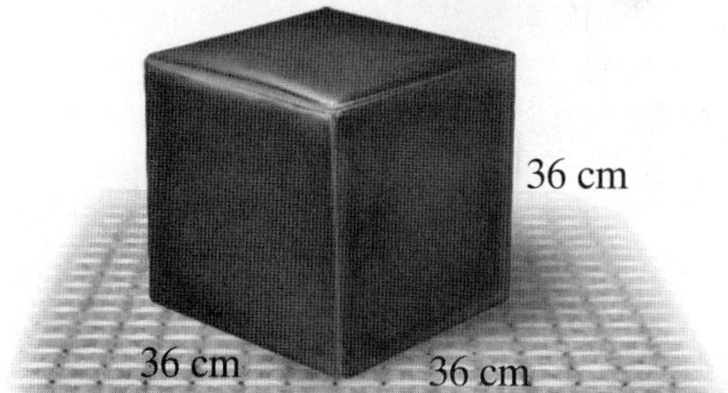

38. **Pasta Sauce Can** Pasta sauce is packaged in a metal can with a diameter of 8.5 cm and a height of 14 cm. How many square centimeters of metal are needed to make this can? Use 3.14 for π to approximate the answer to two decimal places.

You Try It 39. **Water Tower** The reservoir at the top of a water tower has the shape of a sphere with a radius of 15 meters, as shown. If the reservoir is filled to two-thirds capacity, how many cubic meters of water are in the tower? Use 3.14 for π to approximate the answer to two decimal places.

40. **Potato Chips** A container for potato chips is a circular cylinder with diameter 7.4 cm and height 26.5 cm, as shown. If the container is 90% full when opened, what is the volume of chips in the container? Use 3.14 for π to approximate the answer to two decimal places.

41. **Basketball** The diameter of an NBA basketball is 9.4 inches. What is the volume of the basketball? What is the surface area? Use 3.14 for π to approximate each answer to two decimal places.

42. **Popcorn Box** A popcorn box measures 6 inches long, 2.5 inches wide, and 8 inches tall, as shown. Find the volume and surface area of the box.

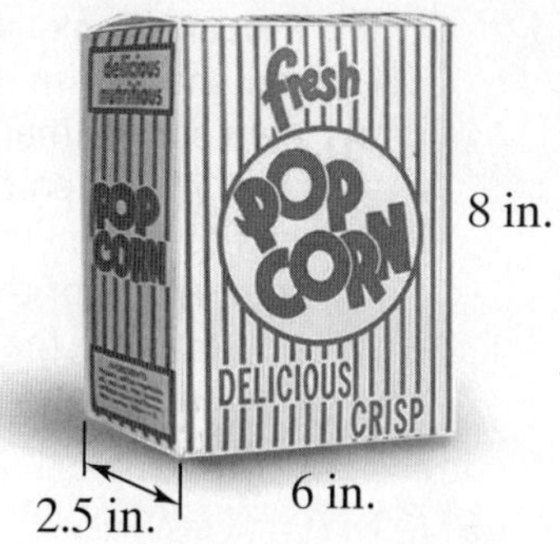

43. **Swimming Pool** An above-ground pool is shaped like a right circular cylinder with a diameter of 24 ft and a depth of 4 ft. How many cubic feet of water can the pool hold? Use 3.14 for π to approximate the answer to two decimal places.

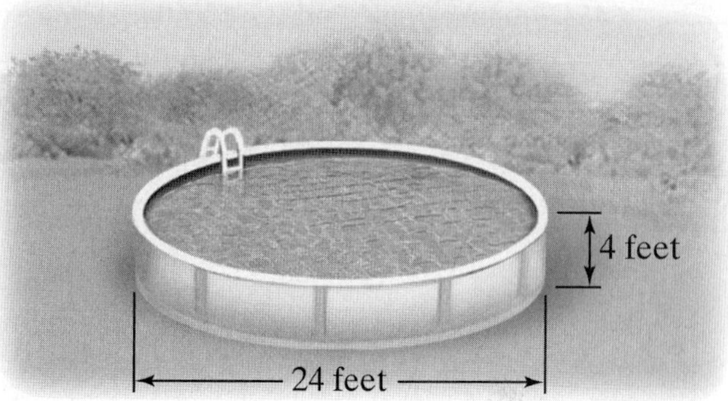

44. **Ancient Pyramid** The Great Pyramid of Khufu has a square base measuring 230 meters on each side. Though shorter now, the original height was 146 meters. Originally, what was the volume of the ancient pyramid? (*Source: International Dictionary of Historic Places*)

45. **Hanging Basket** A hanging basket has the shape of a hemisphere (half a sphere) with a diameter of 14 inches, as shown. How many cubic inches of potting soil can the basket hold? Use 3.14 for π to approximate the answer to two decimal places.

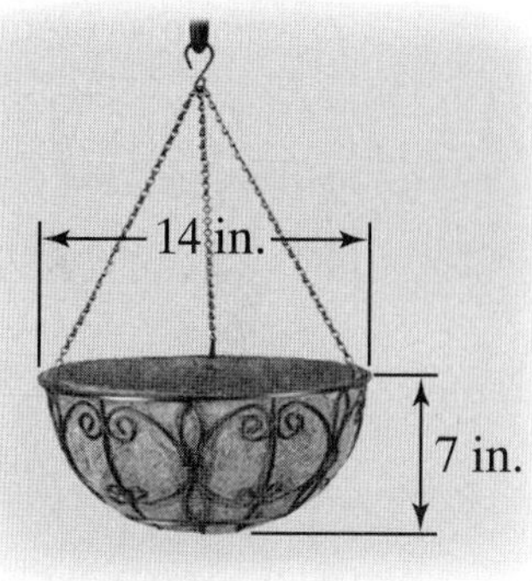

46. **Dessert Dish** A restaurant serves dessert in a cone-shaped dish with a diameter of 8 cm. If the dish is 6.5 cm deep, as shown, how many cubic centimeters of dessert can the dish hold? Use 3.14 for π to approximate the answer to two decimal places.

47. **Propane Tank** A propane gas tank has the shape of a right circular cylinder with a hemisphere (half a sphere) on each end. If the cylinder part of the tank is 4 feet long and the diameter is 3 feet, as shown, find the volume of the tank. Use 3.14 for π to approximate the answer to two decimal places.

48. **Propane Tank, Part 2** Find the surface area of the tank shown in Exercise 47.

49. **Grain Bin** The grain bin shown has the shape of a cylinder with a cone-shaped roof. How many cubic feet of grain will fit in the bin if it is filled to the peak of the cone? Use 3.14 for π to approximate the answer to two decimal places.

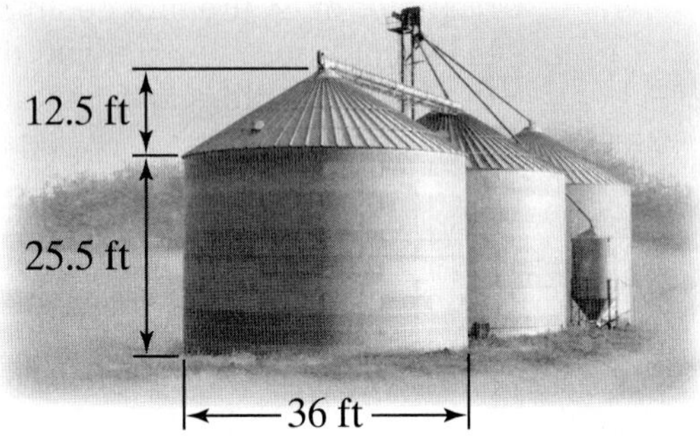

50. **Ice Cream** A spherical scoop of ice cream with a diameter of 6 cm is placed on top of an empty sugar cone with a diameter of 4 cm and a depth of 12 cm, as shown. Can the volume of ice cream fit completely inside the cone? If so, what volume will remain empty inside the cone? If not, what volume of ice cream will be left over? Use 3.14 for π to approximate the answer to two decimal places.

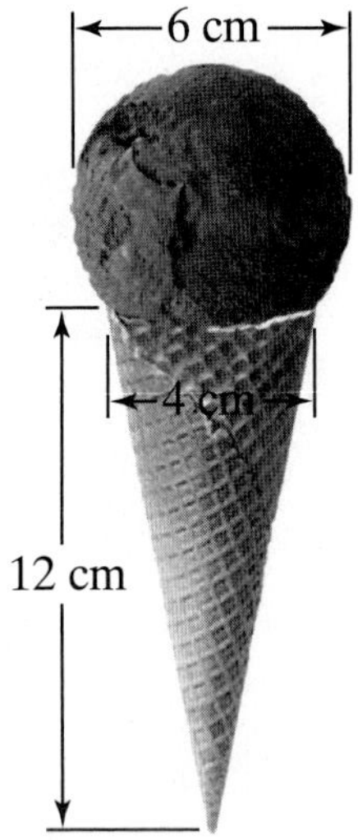

8.4 Linear Measurement

THINGS TO KNOW

Before working through this topic, be sure you are familiar with the following concepts:

	VIDEO	ANIMATION	INTERACTIVE
You Try It 1. Write Fractions in Simplest Form (Topic 4.2, **Objective 6**)	▤		
You Try It 2. Multiply Fractions (Topic 4.3, **Objective 1**)	▤		▤

You Try It 3. Multiply Decimals by Powers of 10 (Topic 5.3, **Objective 2**)

You Try It 4. Divide Decimals by Powers of 10 (Topic 5.4, **Objective 2**)

You Try It 5. Write Two Quantities as a Rate (Topic 6.1, **Objective 2**)

OBJECTIVES

1 Make Conversions Within American Units of Length

2 Make Conversions Involving Mixed Units of Length

3 Make Conversions Within Metric Units of Length

4 Make Conversions Between American and Metric Units of Length

OBJECTIVE 1 MAKE CONVERSIONS WITHIN AMERICAN UNITS OF LENGTH

In the United States, there are two systems of measurement used. These are the **U.S. Customary Units system** (also known as **American Units**) and the **metric system** (also known as the **International System of Units**). In science, medicine, and most manufacturing settings, the metric system is used. Some examples of these units are meters, kilograms, and liters. In the U.S., we still use American Units for most everyday measurements. Some examples of these units are inches, ounces, and gallons.

In this topic we discuss units of *length*. The **length** of an object is the distance from one end of the object to another along its largest dimension. The U.S. system of measurement uses **inch**, **foot**, **yard**, and **mile** to measure length.

The following box shows how these U.S. units of length are related.

U.S. Customary (American) Units of Length

$$12 \text{ inches (in.)} = 1 \text{ foot (ft)}$$
$$3 \text{ feet} = 1 \text{ yard (yd)}$$
$$5280 \text{ feet} = 1 \text{ mile (mi)}$$

We can convert between the different units by multiplying a given value by an appropriate *unit fraction* (this process is referred to as *unit analysis* or *dimensional analysis*). A **unit fraction** is a fraction that is equivalent to 1. For example, because 12 in. = 1 ft, we have $\frac{12 \text{ in.}}{1 \text{ ft}} = 1$. Likewise, $\frac{1 \text{ ft}}{12 \text{ in.}} = 1$. The **fractions** $\frac{12 \text{ in.}}{1 \text{ ft}}$ and $\frac{1 \text{ ft}}{12 \text{ in.}}$ are unit fractions and can be used to convert between inches and feet.

(eText Screens 8.4-1–8.4-28)

One inch is about the length of the head of a toothbrush.

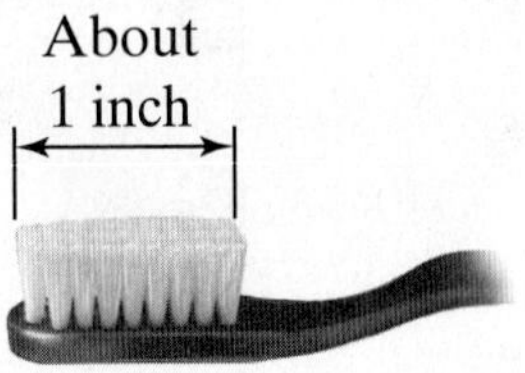

One yard is about the length of a guitar.

TIP Inches, in., are the only units to use a period in its abbreviation. This is to avoid confusion with 'in' the preposition.

The following summary gives the unit fractions that can be used to convert between various units of length in the American system of measurement.

Unit Fractions for American Units of Length

$$\frac{12 \text{ in.}}{1 \text{ ft}} = 1 \text{ (feet to inches)} \qquad \frac{1 \text{ ft}}{12 \text{ in.}} = 1 \text{ (inches to feet)}$$

$$\frac{3 \text{ ft}}{1 \text{ yd}} = 1 \text{ (yards to feet)} \qquad \frac{1 \text{ yd}}{3 \text{ ft}} = 1 \text{ (feet to yards)}$$

$$\frac{5280 \text{ ft}}{1 \text{ mi}} = 1 \text{ (miles to feet)} \qquad \frac{1 \text{ mi}}{5280 \text{ ft}} = 1 \text{ (feet to miles)}$$

Because a unit fraction is equal to 1, and multiplying by 1 does not change the value of a number, multiplying by a unit fraction gives us an equivalent value but in different units.

TIP To determine which **unit fraction** to use in a conversion, use the fraction that has the given units in the **denominator** and the desired units in the **numerator**.

Example 1 Converting Between American Units of Length

a. Convert 72 inches to feet.

b. Convert 4 yards to feet.

Solutions

a. Because we are given inches (in.) and want to convert to feet (ft), we use the unit fraction

$$\frac{1 \text{ ft}}{12 \text{ in.}}. \begin{matrix} \leftarrow \text{Desired units} \\ \leftarrow \text{Given units} \end{matrix}$$

Multiply the given value by the unit fraction and **simplify** by **dividing out** common **factors** and units.

Multiply by 1

$$72 \text{ in.} = \frac{72 \text{ in.}}{1} \cdot \boxed{\frac{1 \text{ ft}}{12 \text{ in.}}}$$

$$= \frac{\overset{6}{\cancel{72}} \cancel{\text{in.}}}{1} \cdot \frac{1 \text{ ft}}{\underset{1}{\cancel{12}} \cancel{\text{in.}}}$$

$$= 6 \cdot 1 \text{ ft}$$

$$= 6 \text{ ft}$$

Notice that the given units divide out, leaving the desired units

So, 72 in. = 6 ft.

b. Because we are given yards (yd) and want to convert to feet (ft), we use the **unit fraction**

$$\frac{3 \text{ ft}}{1 \text{ yd}} \begin{array}{l} \leftarrow \text{Desired units} \\ \leftarrow \text{Given units} \end{array}.$$

Multiply the given value by the unit fraction and **simplify** by **dividing out** common **factors** and units.

Multiply by 1

$$4 \text{ yd} = \frac{4 \text{ yd}}{1} \cdot \boxed{\frac{3 \text{ ft}}{1 \text{ yd}}}$$

$$= \frac{4 \cancel{\text{yd}}}{1} \cdot \frac{3 \text{ ft}}{1 \cancel{\text{yd}}}$$

$$= 4 \cdot 3 \text{ ft}$$

$$= 12 \text{ ft}$$

Notice that the given units divide out, leaving the desired units

So, 4 yd = 12 ft.

You Try It **Work through this You Try It problem.**

Work Exercises 1–6 in this textbook or in the MyMathLab Study Plan.

Example 2 Converting Between American Units of Length

My video summary

a. Convert 22 feet to yards. **b.** Convert 2.4 miles to feet.

Solutions

a. Because we are given feet (ft) and want to convert to yards (yd), we use the **unit fraction**

$$\frac{1 \text{ yd}}{3 \text{ ft}} \begin{array}{l} \leftarrow \text{Desired units} \\ \leftarrow \text{Given units} \end{array}.$$

Multiply the given value by the unit fraction and **simplify** by **dividing out** common **factors** and units.

$$22 \text{ ft} = \frac{22 \text{ ft}}{1} \cdot \frac{1 \text{ yd}}{3 \text{ ft}}$$

$$= \frac{22 \cancel{\text{ft}}}{1} \cdot \frac{1 \text{ yd}}{3 \cancel{\text{ft}}}$$

$$= \frac{22}{3} \text{yd}$$

$$= 7\frac{1}{3} \text{yd}$$

So, $22\text{ ft} = 7\frac{1}{3}\text{ yd}$.

b. Because we are given miles (mi) and want to convert to feet (ft), we use the **unit fraction**

$$\frac{5280\text{ ft}}{1\text{ mi}}. \quad \begin{matrix}\leftarrow \text{Desired units} \\ \leftarrow \text{Given units}\end{matrix}$$

Finish doing this conversion on your own. Check your **answer**, or watch this **video** for complete solutions to both parts.

You Try It **Work through this You Try It problem.**

Work Exercises 7–13 in this textbook or in the MyMathLab Study Plan.

In some cases, it may be necessary to use more than one unit fraction when doing a conversion.

Example 3 Converting Between American Units Using More Than One Unit Fraction

My video summary

Convert 78 inches to yards.

Solutions Work through the following, or watch this **video** for the complete solution. We are given inches (in.) and want to convert to yards (yd). We do this by first converting from inches to feet (ft), then from feet to yards using two unit fractions:

$$\text{Given units} \rightarrow \frac{1\text{ ft}}{12\text{ in.}} \cdot \frac{1\text{ yd}}{3\text{ ft}} \leftarrow \text{Desired units}$$

Multiply the given value by the two **unit fractions** and simplify.

$$78\text{ in.} = \frac{78\text{ in.}}{1} \cdot \boxed{\frac{1\text{ ft}}{12\text{ in.}} \cdot \frac{1\text{ yd}}{3\text{ ft}}} \quad \leftarrow \text{Multiply by 1}$$

$$= \frac{\overset{13}{\cancel{\overset{26}{\cancel{78}}}}\ \cancel{\text{in.}}}{1} \cdot \frac{1\ \cancel{\text{ft}}}{\underset{6}{\cancel{12}}\ \cancel{\text{in.}}} \cdot \frac{1\text{ yd}}{\underset{1}{\cancel{3}}\ \cancel{\text{ft}}}$$

$$= \frac{13}{6}\text{ yd}$$

$$= 2\frac{1}{6}\text{ yd}$$

So, $78\text{ in.} = 2\frac{1}{6}\text{ yd}$.

You Try It **Work through this You Try It problem.**

Work Exercises 14–17 in this textbook or in the MyMathLab Study Plan.

OBJECTIVE 2 MAKE CONVERSIONS INVOLVING MIXED UNITS OF LENGTH

In **Example 2** and **Example 3**, the converted value included a fraction or decimal. In some cases, it is more helpful to use mixed units to express a measurement.

For example, it might seem awkward to say that a person is $5\frac{5}{6}$ feet tall. Instead of using a **mixed number** with single units, we can convert the **fractional part** to a **whole number** using smaller units. In this case, we choose the unit of length that is smaller than a foot, which is an inch.

$$\frac{5}{6}\text{ ft} = \frac{5}{6}\text{ ft}\cdot\underbrace{\frac{12\text{ in.}}{1\text{ ft}}}_{\substack{\text{Unit}\\ \text{fraction} = 1}} = \frac{5}{\cancel{6}_{1}}\cancel{\text{ft}}\cdot\frac{\cancel{12}^{2}\text{ in.}}{1\cancel{\text{ft}}} = 10\text{ in.}$$

So, rather than say a person is $5\frac{5}{6}$ feet tall, we would say that the person is 5 feet and 10 inches tall, or 5 ft 10 in. tall.

Example 4 Converting Between American Units of Length Using Mixed Units

My video summary

a. Convert: 44 feet = ______ yd ______ ft

b. Convert: 3 ft 9 in. = ______ inches

Solutions

a. Work through the following, or watch this **video** for complete solutions to both parts.

We are given feet (ft) and want to convert to yards (yd).

Because 1 yard = 3 feet, we divide the number of feet by 3 to determine the number of whole yards. The remainder is the remaining number of feet.

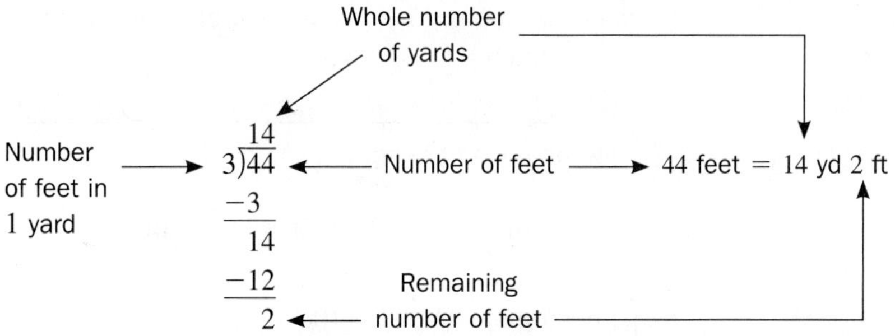

So, 44 feet = 14 yd 2 ft.

b. Try making this conversion on your own. Check your **answer**, or watch this **video** for a complete solution to part b.

You Try It **Work through this You Try It problem.**

Work Exercises 18–26 in this textbook or in the MyMathLab Study Plan.

OBJECTIVE 3 MAKE CONVERSIONS WITHIN METRIC UNITS OF LENGTH

The metric system of measurement was developed in the late 1700s by French scientists who desired to have a standard set of measurements. The system established *base units* and formed larger and smaller units by using **powers of** 10.

Base Unit

A **base unit** is a unit of measurement that has other units named in terms of it.

In the metric system, the base unit of length is the **meter**. The word 'meter' comes from the Greek word 'metron' which means 'a measure.' Its actual value has changed over the years, but the current standard is that 1 meter is the length travelled by light in a vacuum during 1/299792458 of a second. A meter is slightly larger than a yard and is roughly the height of a doorknob or countertop. We use the symbol **m** as an abbreviation for meters.

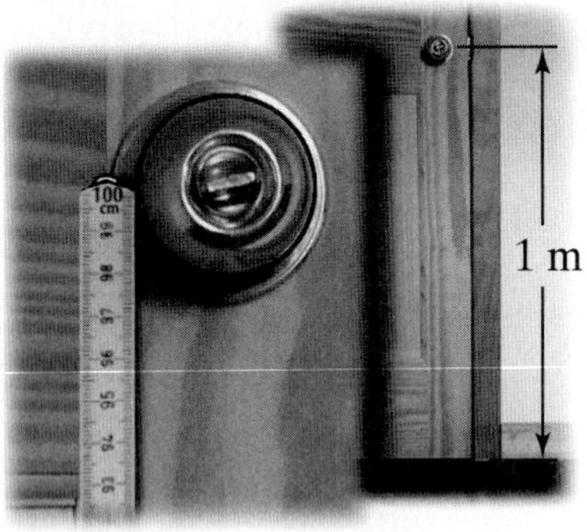

The metric system uses prefixes to create larger and smaller units. The following table summarizes the most common metric prefixes and their meaning.

Prefix	kilo	hecto	deka	base	deci	centi	milli
Meaning	1000	100	10	1	$\frac{1}{10}$	$\frac{1}{100}$	$\frac{1}{1000}$

← Larger   Smaller →

Putting these prefixes in front of the word *meter* allows us to create longer or shorter measurements. The following table shows the common metric units of **length** and their abbreviations.

Metric Units of Length

1 kilometer (km) $= 1000$ meters (m)	$1\text{ m} = \frac{1}{1000}\text{ km}$ or 0.001 km
1 hectometer (hm) $= 100\text{ m}$	$1\text{ m} = \frac{1}{100}\text{ hm}$ or 0.01 hm
1 dekameter (dam) $= 10\text{ m}$	$1\text{ m} = \frac{1}{10}\text{ dam}$ or 0.1 dam
1 meter (m) $= 1\text{ m}$	$1\text{ m} = 1\text{ m}$
1 decimeter (dm) $= \frac{1}{10}\text{ m}$ or 0.1 m	$1\text{ m} = 10\text{ dm}$
1 centimeter (cm) $= \frac{1}{100}\text{ m}$ or 0.01 m	$1\text{ m} = 100\text{ cm}$
1 millimeter (mm) $= \frac{1}{1000}\text{ m}$ or 0.001 m	$1\text{ m} = 1000\text{ mm}$

TIP These same prefixes will be used for metric units of weight and capacity, which we cover in Topics 8.5 and 8.6.

(eText Screens 8.4-1–8.4-28)

To convert units of **length** within the metric system, we use **unit fractions** just as we did with American units. For example, to convert 3408 meters to kilometers, we multiply by the unit fraction $\frac{1\text{ km}}{1000\text{ m}}$ ← Desired units / ← Given units because 1 km = 1000 m.

$$3408\text{ m} = \frac{3408\text{ m}}{1} \cdot \frac{1\text{ km}}{1000\text{ m}} = \frac{3408}{1000}\text{ km} = 3.408\text{ km}$$

Dividing by 1000 moves the decimal point 3 places to the left.

So, 3408 m = 3.408 km.

The fact that the metric system is based on **powers of** 10 gives it a distinct advantage over American units when it comes to converting units. Consider the following metric chart for metric units of length. Here we list the common metric units of length from largest to smallest.

Metric Chart for Units of Length

km hm dam m dm cm mm

In the example just shown, we divided by 1000, which moved the **decimal point** three places to the left. Notice in the metric chart for length that kilometers (km) is three positions (or units) to the left of meters (m).

TIP When converting between metric units, the **decimal point** moves the same number of places, and in the same direction, that it takes go from the given units to the desired units in the **metric chart for units of length**.

To convert from 3408 meters to kilometers, we start at the given units, m, and move to the desired units, km. This requires a move of 3 units to the *left*.

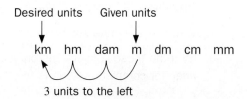

Desired units (km) Given units (m)

km hm dam m dm cm mm

3 units to the left

So, we move the decimal point in our original number 3 places to the left.

3408 m = 3.408 km

3 places to the left

From this point forward, we will use a metric chart and move the decimal point when doing conversions in metric units.

Example 5 Converting Between Metric Units of Length

My interactive video summary

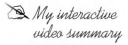

Perform each of the following conversions.

a. 5 kilometers to meters

b. 142.8 centimeters to meters

c. 650 millimeters to meters

d. 7.3 centimeters to millimeters

(eText Screens 8.4-1–8.4-28)

Solutions

a. We are given kilometers (km) and want to convert to meters (m). Looking at the **metric chart for units of length**, we move from km to m, which is 3 units to the *right*.

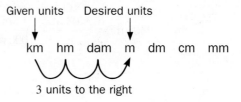

Given units Desired units

km hm dam m dm cm mm

3 units to the right

Therefore, we move the **decimal point** in our original number 3 places to the right.

$$5.000 \text{ km} = 5000 \text{ m}$$

3 places to the right

So, $5 \text{ km} = 5000 \text{ m}$.

View this **alternate solution** to see the conversion worked with a **unit fraction**.

b. Here, we are given centimeters (cm) and want to convert to meters (m). Moving from cm to m on the **metric chart for length** requires a move of 2 units to the *left*.

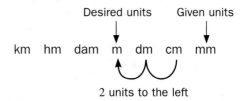

Desired units Given units

km hm dam m dm cm mm

2 units to the left

Move the **decimal point** 2 places to the left in the original number.

$$142.8 \text{ cm} = 1.428 \text{ m}$$

2 places to the left

Therefore, $142.8 \text{ cm} = 1.428 \text{ m}$.

View this **alternate solution** to see the conversion worked with a **unit fraction**.

c.–d. Try to do these conversions on your own. View the **answers**, or watch this **interactive video** for complete solutions to all four parts.

You Try It **Work through this You Try It problem.**

Work Exercises 27–46 in this textbook or in the MyMathLab Study Plan.

OBJECTIVE 4 MAKE CONVERSIONS BETWEEN AMERICAN AND METRIC UNITS OF LENGTH

Sometimes, for consistency, it is necessary to convert between American units and metric units of length. To do so, we need a key conversion that relates the two systems. It turns out that there is an exact conversion between the two systems by means of inches and centimeters.

$$1 \text{ in.} = 2.54 \text{ cm} \leftarrow \text{1 inch is exactly 2.54 centimeters}$$

However, converting through inches and centimeters can be tedious, so it is better to have more direct conversions between units. The following summary shows how we can convert between the two systems. Note that the conversions are all approximations except for the exact relationship between inches and centimeters.

American and Metric Conversion Facts for Length

$$1 \text{ in.} = 2.54 \text{ cm}$$

$1 \text{ ft} \approx 0.305 \text{ m}$	$1 \text{ m} \approx 3.281 \text{ ft}$
$1 \text{ yd} \approx 0.914 \text{ m}$	$1 \text{ m} \approx 1.094 \text{ yd}$
$1 \text{ mi} \approx 1.609 \text{ km}$	$1 \text{ km} \approx 0.621 \text{ mi}$

These conversion facts lead to the following **unit fractions**. Again note that all but one of these (inches to centimeters) are approximations.

(Approximate) Unit Fractions for American and Metric Units of Length

$\frac{2.54 \text{ cm}}{1 \text{ in.}} = 1$ (inches to centimeters)	$\frac{0.394 \text{ in.}}{1 \text{ cm}} \approx 1$ (centimeters to inches)
$\frac{0.305 \text{ m}}{1 \text{ ft}} \approx 1$ (feet to meters)	$\frac{3.281 \text{ ft}}{1 \text{ m}} \approx 1$ (meters to feet)
$\frac{0.914 \text{ m}}{1 \text{ yd}} \approx 1$ (yards to meters)	$\frac{1.094 \text{ yd}}{1 \text{ m}} \approx 1$ (meters to yards)
$\frac{1.609 \text{ km}}{1 \text{ mi}} \approx 1$ (miles to kilometers)	$\frac{0.621 \text{ mi}}{1 \text{ km}} \approx 1$ (kilometers to miles)

We have written the unit fractions with a **denominator** of 1. For example, we convert from centimeters to inches using the approximate unit fraction $\frac{0.394 \text{ in.}}{1 \text{ cm}}$ rather than the exact unit fraction $\frac{1 \text{ in.}}{2.54 \text{ cm}}$. This is to simplify computations because it is usually easier to multiply by a decimal than divide. Remember to select the unit fraction that has the given units in the denominator and the desired units in the **numerator**.

Example 6 Converting Between American and Metric Units of Length

a. Convert 13 inches to centimeters.

b. Convert 15 kilometers to miles. Round to the nearest tenth.

Solutions

a. Because we are given inches (in.) and want to convert to centimeters (cm), we use the **unit fraction**

$$\frac{2.54 \text{ cm}}{1 \text{ in.}}. \quad \begin{matrix} \leftarrow \text{Desired units} \\ \leftarrow \text{Given units} \end{matrix}$$

Multiply the given value by the unit fraction and simplify.

$$\begin{aligned} 13 \text{ in.} &= \frac{13 \text{ in.}}{1} \cdot \frac{2.54 \text{ cm}}{1 \text{ in.}} \\ &= \frac{13 \cancel{\text{in.}}}{1} \cdot \frac{2.54 \text{ cm}}{1 \cancel{\text{in.}}} \\ &= 33.02 \text{ cm} \end{aligned}$$

$$\begin{array}{r} 2.54 \\ \times \quad 13 \\ \hline 7\,62 \\ 25\,40 \\ \hline 33.02 \end{array}$$

So, 13 in. $= 33.02$ cm. Notice that this is an exact result because we used an exact unit fraction.

b. Because we are given kilometers (km) and want to convert to miles (mi), we use the **unit fraction**

$$\frac{0.621 \text{ mi}}{1 \text{ km}}. \quad \begin{matrix} \leftarrow \text{Desired units} \\ \leftarrow \text{Given units} \end{matrix}$$

Unlike part a, this will be an approximate unit fraction, so the result will not be exact. Try making this conversion on your own. Check your **answer**, or watch this **video** for complete solutions to both parts.

You Try It **Work through this You Try It problem.**

Work Exercises 47–54 in this textbook or in the MyMathLab Study Plan.

As we have seen previously, we sometimes need to use more than one conversion to change units.

Example 7 Converting Between American and Metric Units of Length

My video summary

a. Convert 7 kilometers to feet. Round to the nearest **whole number.**

b. Convert 22 meters to inches. Round to the nearest tenth.

Solutions

a. We are given kilometers (km) and want to convert to feet (ft). We do this by first converting from kilometers to miles (mi), then from miles to feet using two unit fractions:

$$\text{Given units} \rightarrow \frac{0.621 \text{ mi}}{1 \text{ km}} \cdot \frac{5280 \text{ ft}}{1 \text{ mi}} \leftarrow \text{Desired units}$$

Multiply the given value by the two **unit fractions** and simplify.

$$\begin{aligned} 7 \text{ km} &\approx \frac{7 \text{ km}}{1} \cdot \boxed{\frac{0.621 \text{ mi}}{1 \text{ km}} \cdot \frac{5280 \text{ ft}}{1 \text{ mi}}} \quad \text{(Approximately equal to; Multiply by 1)} \\ &= \frac{7 \cancel{\text{km}}}{1} \cdot \frac{0.621 \cancel{\text{mi}}}{1 \cancel{\text{km}}} \cdot \frac{5280 \text{ ft}}{1 \cancel{\text{mi}}} \\ &= 4.347 \cdot 5280 \text{ ft} \\ &\approx 22{,}952 \text{ ft} \quad \text{(Rounded to the nearest whole number)} \end{aligned}$$

$$\begin{array}{r} 5\,280 \\ \times\ 4.347 \\ \hline 36\,960 \\ 211\,200 \\ 1584\,000 \\ 21120\,000 \\ \hline 22952.160 \end{array}$$

So, 7 km $\approx$ 22,952 ft.

View this popup to see an **alternate approach.**

b. We are given meters (m) and want to convert to inches (in.). We do this by first converting from meters to feet (ft), then from feet to inches using two unit fractions:

$$\text{Given units} \rightarrow \frac{3.281 \text{ ft}}{1 \text{ m}} \cdot \frac{12 \text{ in.}}{1 \text{ ft}} \leftarrow \text{Desired units}$$

Try finishing this conversion on your own. Check your **answer**, or watch this **video** for complete solutions to both parts.

(eText Screens 8.4-1–8.4-28)

You Try It **Work through this You Try It problem.**

Work Exercises 55–62 in this textbook or in the MyMathLab Study Plan.

8.4 Exercises

In Exercises 1–26, perform the indicated conversion within American units of length.

You Try It 1. 5 yards to feet

2. 42 feet to yards

3. 60 inches to feet

4. 12 feet to inches

5. 15,840 feet to miles

6. 4 miles to feet

You Try It 7. 8.5 yards to feet

8. $3\frac{1}{4}$ feet to inches

9. 28 feet to yards

10. 34 inches to feet

11. 3.5 miles to feet

12. 6000 feet to miles

13. A lacrosse field is 110 yards long. How long is it in feet?

You Try It 14. 42 inches to yards

15. 5000 yards to miles

16. 2.4 yards to inches

17. $8\frac{1}{4}$ miles to yards

18. 76 in. = ____ ft ____ in.

You Try It 19. 38 ft = ____ yd ____ ft

20. 5 yd 1 ft = _____ ft

21. 3 ft 5 in. = _____ in.

22. 142 in. = ____ yd _____ ft _____ in.

23. 6 yd 2 ft 8 in. = _______ in.

24. A refrigerator is 5 ft 10 in. tall. How many inches is this?

25. At its widest point, the Lincoln Memorial is 201 ft 10 in. wide. What is this measurement in meters? Round to the nearest tenth. (*Source:* nps.gov)

26. A trash dumpster has a length of 68 in. Write this length in terms of feet and inches.

In Exercises 27–46, perform the indicated conversion within metric units of length.

You Try It 27. 12 meters to centimeters

28. 4.3 kilometers to meters

29. 654 millimeters to meters

30. 22.7 centimeters to millimeters

31. 842 centimeters to meters

32. 16.2 millimeters to centimeters

33. 0.41 meters to centimeters

34. 0.036 meters to millimeters

35. 5400 meters to kilometers

36. 1.9 centimeters to millimeters

37. 3025 millimeters to meters

38. 0.012 kilometers to centimeters

39. 14 kilometers to hectometers

40. 8 centimeters to decimeters

41. 13.9 meters to dekameters

42. 174 millimeters to decimeters

43. 830 dekameters to kilometers

44. 516 decimeters to dekameters

45. Janet ran a 12-kilometer race. How long was the race in meters?

46. A cell phone is 125.8 millimeters long. How long is it in centimeters?

In Exercises 47–62, perform the indicated conversion between American and metric units of length. Round answers to the nearest tenth.

You Try It 47. Convert 55 inches to centimeters.

48. Convert 83 centimeters to inches.

49. Convert 14.5 feet to meters.

50. Convert 30 kilometers to miles.

51. Convert 16 yards to meters.

52. Convert 6.4 meters to feet.

53. Convert 26.2 miles to kilometers.

54. Convert 20 meters to yards.

You Try It 55. Convert 8 feet to centimeters.

56. Convert 9.5 meters to inches.

57. Convert 140 centimeters to yards.

58. Convert 1700 yards to kilometers.

59. A laptop computer is 38 centimeters wide. How wide is it in inches? Round to the nearest tenth.

60. A signal tower is 85 meters tall. How tall is it in yards? Round to the nearest tenth.

61. A telephone handset measures 6.5 inches by 1.5 inches. What are the dimensions in centimeters? Round to the nearest tenth.

62. A passport book is 125 millimeters long. How long is it in inches? Round to the nearest tenth.

8.5 Weight and Mass

THINGS TO KNOW

Before working through this topic, be sure you are familiar with the following concepts:

VIDEO ANIMATION INTERACTIVE

You Try It 1. Multiply Fractions (Topic 4.3, **Objective 1**)

You Try It 2. Multiply Decimals by Powers of 10 (Topic 5.3, **Objective 2**)

You Try It 3. Divide Decimals by Powers of 10 (Topic 5.4, **Objective 2**)

You Try It 4. Write Two Quantities as a Rate (Topic 6.1, **Objective 2**)

You Try It 5. Make Conversions Within American Units of Length (Topic 8.4, **Objective 1**)

You Try It 6. Make Conversions Involving Mixed Units of Length (Topic 8.4, **Objective 2**)

You Try It 7. Make Conversions Within Metric Units of Length (Topic 8.4, **Objective 3**)

You Try It 8. Make Conversions Between American and Metric Units of Length (Topic 8.4, **Objective 4**)

OBJECTIVES

1 Make Conversions Within American Units of Weight

2 Make Conversions Involving Mixed Units of Weight

3 Make Conversions Within Metric Units of Mass

4 Make Conversions Between American and Metric Units of Weight and Mass

OBJECTIVE 1 MAKE CONVERSIONS WITHIN AMERICAN UNITS OF WEIGHT

In this topic, we look at measurements of **weight** and **mass**. The terms *weight* and *mass* are often used interchangeably, but there is a difference. **Weight** is a measure related to the pull of Earth's gravity on an object, while **mass** is a measure related to the amount of matter within an object. The farther an object moves from the center of Earth, the less *weight* it will have, but the *mass* of the object will not change. For example, if you were to travel to the International Space Station, your *mass* would not change, but you would weigh less than you do here on Earth.

TIP Though in science it is important to distinguish an object's weight from its mass, for most practical purposes here on earth, we can use the terms to mean the same. In this text, we will associate the term *weight* with American units and the term *mass* with metric units.

(eText Screens 8.5-1–8.5-21)

The most commonly used American units of weight are ***ounces***, ***pounds***, and ***tons***. An average adult mouse weighs about 1 ounce. Flour is often purchased in 5-pound bags. An adult male African elephant will weigh between 6 and 7 tons.

1 ounce

5 pounds

6 to 7 tons

These units are related as follows:

American Units of Weight

$$16 \text{ ounces (oz)} = 1 \text{ pound (lb)} \qquad 2000 \text{ pounds} = 1 \text{ ton}$$

To make conversions among units of weight and mass, we can use the same process we used in **Topic 8.4** for units of length. We multiply by appropriate **unit fractions**.

Unit Fractions for American Units of Weight

$$\frac{16 \text{ oz}}{1 \text{ lb}} = 1 \quad \frac{1 \text{ lb}}{16 \text{ oz}} = 1 \quad \frac{2000 \text{ lb}}{1 \text{ ton}} = 1 \quad \frac{1 \text{ ton}}{2000 \text{ lb}} = 1$$

Recall that when we multiply a quantity by a **unit fraction**, we are really multiplying by 1. Therefore, the result is the same quantity, just expressed in different units.

TIP Remember to choose the unit fraction with the desired units (units being converted to) in the **numerator** and the original units in the **denominator**. This way, the original units divide out, leaving the desired units.

Example 1 Converting Tons to Pounds

Convert 9 tons to pounds.

Solution To convert from tons to pounds, multiply by $\frac{2000 \text{ lb}}{1 \text{ ton}}$. ← Desired units / ← Original units

Multiply by 1

$$9 \text{ tons} = \frac{9 \text{ tons}}{1} \cdot \frac{2000 \text{ lb}}{1 \text{ ton}} = \frac{9 \cancel{\text{tons}}}{1} \cdot \frac{2000 \text{ lb}}{1 \cancel{\text{ton}}} = 9(2000 \text{ lb}) = 18{,}000 \text{ lb}$$

So, 9 tons is the same weight as 18,000 pounds.

You Try It **Work through this You Try It problem.**

Work Exercises 1 and 2 in this textbook or in the MyMathLab Study Plan.

(eText Screens 8.5-1–8.5-21)

Example 2 Converting Ounces to Pounds

My video summary

Convert 224 ounces to pounds.

Solution To convert from ounces to pounds, multiply by $\frac{1 \text{ lb}}{16 \text{ oz}}$. ← Desired units / ← Original units

Multiply by 1

$$224 \text{ oz} = \frac{224 \text{ oz}}{1} \cdot \frac{1 \text{ lb}}{16 \text{ oz}}$$

Try to finish this conversion on your own. View the **answer**, or watch this **video** for a complete solution.

You Try It **Work through this You Try It problem.**

Work Exercises 3 and 4 in this textbook or in the MyMathLab Study Plan.

Sometimes weight conversions involve **fractions**, **mixed numbers**, or **decimals**.

Example 3 Converting Between American Units of Weight

My interactive video summary

Make each of the following conversions.

a. 16,500 pounds to tons **b.** 4.2 pounds to ounces

c. 54 ounces to pounds **d.** $3\frac{1}{8}$ tons to pounds

Solutions In each case, we multiply by the appropriate **unit fraction**.

a. To convert from pounds to tons, multiply by $\frac{1 \text{ ton}}{2000 \text{ lb}}$. ← Desired units / ← Original units

$$16{,}500 \text{ lb} = \frac{16{,}500 \text{ lb}}{1} \cdot \frac{1 \text{ ton}}{2000 \text{ lb}} = \frac{16{,}500 \text{ ton}}{2000} = 8\frac{1}{4} \text{ tons, or } 8.25 \text{ tons}$$

Thus, 16,500 pounds is equivalent to $8\frac{1}{4}$ tons, or 8.25 tons.

b. To convert from pounds to ounces, multiply by $\frac{16 \text{ oz}}{1 \text{ lb}}$. ← Desired units / ← Original units

$$4.2 \text{ lb} = \frac{4.2 \text{ lb}}{1} \cdot \frac{16 \text{ oz}}{1 \text{ lb}} = 4.2(16 \text{ oz}) = 67.2 \text{ oz}$$

Thus, 4.2 pounds is equivalent to 67.2 oz.

c.–d. Try to make these conversions on your own. View the **answers**, or watch this **interactive video** for complete solutions to all four parts.

You Try It **Work through this You Try It problem.**

Work Exercises 5–14 in this textbook or in the MyMathLab Study Plan.

(eText Screens 8.5-1–8.5-21)

OBJECTIVE 2 MAKE CONVERSIONS INVOLVING MIXED UNITS OF WEIGHT

Sometimes, weights are expressed using more than one unit. For example, we might say that a newborn baby weighs 8 pounds 10 ounces.

Example 4 Converting Mixed Units

Express 8 lb 10 oz in ounces only.

Solution We convert the 8 pounds to ounces and add the result to the 10 ounces.

To convert from pounds to ounces, multiply by $\frac{16 \text{ oz}}{1 \text{ lb}}$. ← Desired units / ← Original units

$$8 \text{ lb} = \frac{8 \cancel{\text{lb}}}{1} \cdot \frac{16 \text{ oz}}{1 \cancel{\text{lb}}} = 8(16 \text{ oz}) = 128 \text{ oz}$$

Therefore, $8 \text{ lb } 10 \text{ oz} = 128 \text{ oz} + 10 \text{ oz} = 138 \text{ oz}$.

8 lb 10 oz is equivalent to 138 oz.

You Try It **Work through this You Try It problem.**

Work Exercises 15 and 16 in this textbook or in the MyMathLab Study Plan.

Example 5 Converting Mixed Units

My video summary

Express 195 ounces in mixed units of pounds and ounces.

Solution Read below, or watch this **video** for a complete solution.

To convert from ounces to mixed units (pounds and ounces), we divide to find the whole number of pounds in 195 ounces. The remainder is the number of ounces. We divide because $195 \text{ oz} = \frac{195 \cancel{\text{oz}}}{1} \cdot \frac{1 \text{ lb}}{16 \cancel{\text{oz}}} = \frac{195 \text{ lb}}{16}$.

$$\begin{array}{r} 12 \\ 16\overline{)195} \\ \underline{-16} \\ 35 \\ \underline{-32} \\ 3 \end{array}$$

12 → 12 lb 3 oz ← Remainder 3

Therefore, 195 ounces is equivalent to 12 pounds 3 ounces.

You Try It **Work through this You Try It problem.**

Work Exercises 17–20 in this textbook or in the MyMathLab Study Plan.

OBJECTIVE 3 MAKE CONVERSIONS WITHIN METRIC UNITS OF MASS

In the metric system, the basic unit of mass is the **gram**. When metric units were originally developed in the late 1700s, a gram was defined as the mass of 1 cubic centimeter (cm^3) of water. In the late 1800s, the definition was refined to base the **gram** on the mass of a specific object kept in Sèvres, France, called the international prototype.

(eText Screens 8.5-1–8.5-21)

For practical purposes, you can think of a mass of 1 gram as roughly equivalent to the weight of a large paperclip.

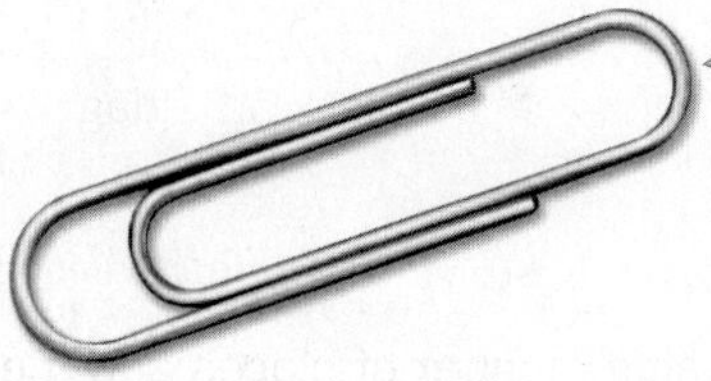

The table below shows the metric units of **mass**. Notice that the prefixes for these units are the same as those we saw in **Topic 8.4** for metric units of length.

Metric Units of Mass

1 kilogram (kg) $= 1000$ grams (g)	$1\text{ g} = \frac{1}{1000}\text{ kg}$ or 0.001 kg
1 hectogram (hg) $= 100\text{ g}$	$1\text{ g} = \frac{1}{100}\text{ hg}$ or 0.01 hg
1 dekagram (dag) $= 10\text{ g}$	$1\text{ g} = \frac{1}{10}\text{ dag}$ or 0.1 dag
1 gram (g) $= 1\text{ g}$	$1\text{ g} = 1\text{ g}$
1 decigram (dg) $= \frac{1}{10}\text{ g}$ or 0.1 g	$1\text{ g} = 10\text{ dg}$
1 centigram (cg) $= \frac{1}{100}\text{ g}$ or 0.01 g	$1\text{ g} = 100\text{ cg}$
1 milligram (mg) $= \frac{1}{1000}\text{ g}$ or 0.001 g	$1\text{ g} = 1000\text{ mg}$

To convert units of **mass** within the metric system, we can use our familiar procedure of multiplying by an appropriate unit fraction. For example, to convert 2681 milligrams to grams, we can multiply by the **unit fraction** $\frac{1\text{ g}}{1000\text{ mg}}$ (because $1\text{ g} = 1000\text{ mg}$):

$$2681\text{ mg} = \frac{2681\text{ mg}}{1}\cdot\frac{1\text{ g}}{1000\text{ mg}} = \frac{2681\text{ g}}{1000} = 2.681\text{ g}.$$

Dividing by 1000 moves the decimal point 3 places to the left.

Therefore, $2681\text{ mg} = 2.681\text{ g}$.

TIP The metric units of mass are all **powers of** 10 of the gram. Therefore, conversions within these units result in a movement of the **decimal point**.

As we did for metric units of length, we can create a metric chart for the units of mass. Moving left to right in this chart, we list the metric units of mass from largest to smallest.

Metric Chart for Units of Mass

kg hg dag g dg cg mg

(eText Screens 8.5-1–8.5-21)

To convert from 2681 milligrams to grams, start at mg on the chart (the original units) and move to g (desired units), which requires a move of 3 units to the *left*.

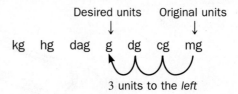

Desired units ↓ (g) Original units ↓ (mg)

kg hg dag g dg cg mg

3 units to the *left*

This is the same number of places and the direction that the **decimal point** will move.

$$2681.\text{ mg} = 2.681\text{ g}$$

3 places to the *left*

The most commonly used metric units of **mass** are *kilogram*, *gram*, and *milligram*, so we focus our examples and exercises more heavily on these units.

Example 6 Converting Between Metric Units of Mass

My interactive video summary

Make each of the following conversions.

a. 8.5 kilograms to grams

b. 12,300 grams to kilograms

c. 0.2 gram to milligrams

d. 79.25 milligrams to centigrams

Solutions

a. We are converting from kilograms (kg) to grams (g). On the metric chart, we move from kg to g, which is 3 units to the *right*.

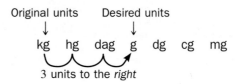

Original units ↓ (kg) Desired units ↓ (g)

kg hg dag g dg cg mg

3 units to the *right*

So, we move the **decimal point** 3 places to the right.

$$8.500\text{ kg} = 8500\text{ g}$$

3 places to the *right*

Therefore, 8.5 kg = 8500 g.

View this **alternate solution** to see this conversion worked with a **unit fraction**.

b. Here, we are converting grams (g) to kilograms (kg). Moving from g to kg on the metric chart requires a move of 3 units to the *left*.

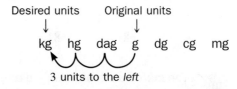

Desired units ↓ (kg) Original units ↓ (g)

kg hg dag g dg cg mg

3 units to the *left*

Moving the **decimal point** 3 places to the left gives

$$12{,}300\text{ g} = 12.3\text{ kg}.$$

3 places to the *left*

Thus, 12,300 g = 12.3 kg. View this **alternate solution**.

c.–d. Try to make these conversions on your own. View the **answers**, or watch this **interactive video** for complete solutions to all four parts.

(eText Screens 8.5-1–8.5-21)

You Try It **Work through this You Try It problem.**

Work Exercises 21–36 in this textbook or in the MyMathLab Study Plan.

OBJECTIVE 4 MAKE CONVERSIONS BETWEEN AMERICAN AND METRIC UNITS OF WEIGHT AND MASS

As with units of length, it is sometimes necessary to convert between American units of weight and metric units of mass. The table below gives approximate equivalences that can be used to make conversions.

American and Metric Conversion Facts

$1 \text{ kg} \approx 2.20 \text{ lb} \qquad 1 \text{ g} \approx 0.035 \text{ oz} \qquad 1 \text{ lb} \approx 0.45 \text{ kg} \qquad 1 \text{ oz} \approx 28.35 \text{ g}$

The approximate equivalences lead to the following *approximate* **unit fractions**, among others.

Approximate Unit Fractions for American and Metric Units of Weight and Mass

$$\frac{2.20 \text{ lb}}{1 \text{ kg}} \approx 1 \qquad \frac{0.035 \text{ oz}}{1 \text{ g}} \approx 1 \qquad \frac{0.45 \text{ kg}}{1 \text{ lb}} \approx 1 \qquad \frac{28.35 \text{ g}}{1 \text{ oz}} \approx 1$$

TIP To help simplify computations when making conversions between American and metric units, multiply by the appropriate unit fraction that has 1 unit in the denominator. For example, to convert from pounds to kilograms, multiply by $\frac{0.45 \text{ kg}}{1 \text{ lb}}$ rather than $\frac{1 \text{ kg}}{2.20 \text{ lb}}$.

Example 7 Converting Pounds to Kilograms

Convert 5.7 pounds to kilograms. Round to the nearest tenth.

Solution To convert from pounds to kilograms, multiply by $\frac{0.45 \text{ kg}}{1 \text{ lb}}$. ← Desired units (numerator) ← Original units (denominator)

$$5.7 \text{ lb} \approx \frac{5.7 \cancel{\text{lb}}}{1} \cdot \frac{0.45 \text{ kg}}{1 \cancel{\text{lb}}} = 5.7(0.45 \text{ kg}) \approx 2.6 \text{ kg}$$

Round to the nearest tenth.

$$\begin{array}{r} 5.7 \\ \times\ 0.45 \\ \hline 285 \\ 2280 \\ \hline 2.565 \end{array}$$

5.7 ← 1 decimal place
× 0.45 ← 2 decimal places
2.565 ← 3 decimal places

So, 5.7 pounds is approximately equivalent to 2.6 kilograms.

You Try It **Work through this You Try It problem.**

Work Exercises 37 and 38 in this textbook or in the MyMathLab Study Plan.

CAUTION Because these conversion facts are approximations, computations can sometimes vary depending on which conversion fact is used. For this reason, we will **round** our results.

(eText Screens 8.5-1–8.5-21)

Example 8 Converting Between American and Metric Units of Weight and Mass

My video summary

Convert 12.5 ounces to grams. Round to the nearest tenth.

Solution Try to make this conversion on your own. View the **answer**, or watch this **video** for a complete solution.

You Try It **Work through this You Try It problem.**

Work Exercises 39–50 in this textbook or in the MyMathLab Study Plan.

8.5 Exercises

In Exercises 1–20, make the indicated conversion within American units of weight.

You Try It 1. 7 pounds to ounces

2. 8 tons to pounds

You Try It 3. 224 ounces to pounds

4. 34,000 pounds to tons

You Try It 5. $5\frac{1}{4}$ pounds to ounces

6. $7\frac{3}{4}$ tons to pounds

7. 3600 pounds to tons

8. 52 pounds to ounces

9. 9.4 pounds to ounces

10. 200 ounces to pounds

11. 8.6 tons to pounds

12. 900 pounds to tons

13. A child's bowling ball weighs 128 ounces. What is this weight in pounds?

14. A truck has a hauling capacity of 4.5 tons. What is the hauling capacity in pounds?

15. Express 12 lb 3 oz in ounces only.

You Try It 16. Express 4 tons 1325 lb in pounds only.

You Try It 17. Express 236 ounces in mixed units of pounds and ounces.

18. Express 13,250 pounds in mixed units of tons and pounds.

19. A cat weighs 13 pounds 6 ounces. Express this weight in ounces only.

20. Darrell purchased a 35-ounce bag of grass seed. Express this weight in mixed units of pounds and ounces.

In Exercises 21–36, make the indicated conversion within metric units of mass.

You Try It 21. 16,000 milligrams to grams

22. 17 grams to milligrams

23. 12,000 grams to kilograms

24. 7.25 kilograms to grams

25. 6520 milligrams to grams

26. 2300 grams to kilograms

27. 84 grams to kilograms

28. 0.0258 kilograms to grams

29. 250 milligrams to grams

30. 0.56 grams to milligrams

31. 4.9 hectograms to grams

32. 7600 grams to decigrams

33. 0.5 centigrams to milligrams

34. 540 dekagrams to kilograms

35. A capsule contains 500 milligrams of a steroid. Convert this amount to grams.

36. A bag contains 1.81 kilograms of sugar. Convert this amount to grams.

In Exercises 37–50, make the indicated conversion between American and metric units of weight and mass. Round answers to the nearest tenth.

You Try It 37. 8.2 pounds to kilograms

38. 250 grams to ounces

You Try It 39. 125 kilograms to pounds

40. 14 ounces to grams

41. 187 pounds to kilograms

42. 17.25 grams to ounces

43. 2.5 ounces to grams

44. 0.65 kilograms to pounds

45. 8 pounds to grams

46. 12 kilograms to ounces

47. A smart phone weighs 4.9 ounces. What is the mass of the smart phone in grams? Round to the nearest tenth.

48. A sign in an elevator indicates a weight limit of 3500 pounds. Convert this limit to kilograms.

49. A cereal box contains 510 grams of puffed rice. Express this amount of puffed rice in ounces.

50. A bag contains 32 kilograms of horse feed. How many pounds of horse feed are in the bag?

8.6 Capacity

THINGS TO KNOW

Before working through this topic, be sure you are familiar with the following concepts:

	VIDEO	ANIMATION	INTERACTIVE
You Try It 1. Multiply Fractions (Topic 4.3, **Objective 1**)	✓		✓
You Try It 2. Multiply Decimals by Powers of 10 (Topic 5.3, **Objective 2**)	✓		
You Try It 3. Divide Decimals by Powers of 10 (Topic 5.4, **Objective 2**)	✓		

(eText Screens 8.6-1–8.6-19)

You Try It 4. Write Two Quantities as a Rate (Topic 6.1, **Objective 2**)

You Try It 5. Make Conversions Within American Units of Weight (Topic 8.5, **Objective 1**)

You Try It 6. Make Conversions Involving Mixed Units of Weight (Topic 8.5, **Objective 2**)

You Try It 7. Make Conversions Within Metric Units of Mass (Topic 8.5, **Objective 3**)

You Try It 8. Make Conversions Between American and Metric Units of Weight and Mass (Topic 8.5, **Objective 4**)

OBJECTIVES

1 Make Conversions Within American Units of Capacity

2 Make Conversions Involving Mixed Units of Capacity

3 Make Conversions Within Metric Units of Capacity

4 Make Conversions Between American and Metric Units of Capacity

OBJECTIVE 1 MAKE CONVERSIONS WITHIN AMERICAN UNITS OF CAPACITY

In this topic, we look at measurements of *capacity*. **Capacity** is a measure of **volume** that is often used when measuring liquids. For example, the amount of fuel in a car's tank, the amount of milk in a pitcher, and the amount of vaccine in a syringe are all measured by units of capacity.

The most commonly used American units of capacity are ***fluid ounces***, ***cups***, ***pints***, ***quarts***, and ***gallons***. A can of soft drink may contain 12 fluid ounces. Drinks are often served in 8-ounce glasses, which is equivalent to 1 cup. Milk can be purchased in pints, quarts, or gallons.

12 fluid ounces 1 cup 1 pint 1 quart 1 gallon

The American units of **capacity** are related as follows:

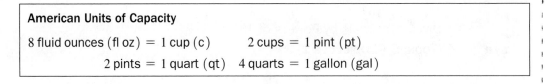

American Units of Capacity

8 fluid ounces (fl oz) = 1 cup (c) 2 cups = 1 pint (pt)

2 pints = 1 quart (qt) 4 quarts = 1 gallon (gal)

(eText Screens 8.6-1–8.6-19)

As in **Topic 8.4** and **Topic 8.5**, to convert among units of capacity, we can multiply by appropriate **unit fractions.**

Unit Fractions for American Units of Capacity

$\frac{8\text{ fl oz}}{1\text{ c}} = 1$ or $\frac{1\text{ c}}{8\text{ fl oz}} = 1$ $\qquad$ $\frac{2\text{ c}}{1\text{ pt}} = 1$ or $\frac{1\text{ pt}}{2\text{ c}} = 1$

$\frac{2\text{ pt}}{1\text{ qt}} = 1$ or $\frac{1\text{ qt}}{2\text{ pt}} = 1$ $\qquad$ $\frac{4\text{ qt}}{1\text{ gal}} = 1$ or $\frac{1\text{ gal}}{4\text{ qt}} = 1$

TIP Remember, when multiplying by a unit fraction, the desired units (units being converted to) should be in the **numerator** of the unit fraction, and the original units should be in the **denominator**. This way, the original units divide out, leaving the desired units.

Example 1 Converting Between American Units of Capacity

My video summary

Make each indicated conversion.

a. 5 gallons to quarts $\qquad$ **b.** 96 fluid ounces to cups

Solutions

a. To convert from gallons to quarts, multiply by $\frac{4\text{ qt}}{1\text{ gal}}$. ← Desired units / ← Original units

Multiply by 1

$$5\text{ gal} = \frac{5\text{ gal}}{1}\cdot\frac{4\text{ qt}}{1\text{ gal}} = \frac{5\cancel{\text{ gal}}}{1}\cdot\frac{4\text{ qt}}{1\cancel{\text{ gal}}} = 5(4\text{ qt}) = 20\text{ qt}$$

So, 20 quarts and 5 gallons are equal in capacity.

b. To convert from fluid ounces to cups, multiply by $\frac{1\text{ c}}{8\text{ fl oz}}$. ← Desired units / ← Original units

Try to finish this problem on your own. View the **answer**, or watch this **video** for complete solutions to both parts.

You Try It **Work through this You Try It problem.**

Work Exercises 1–8 in this textbook or in the MyMathLab Study Plan.

Sometimes measures of **capacity** involve **fractions, mixed numbers,** or **decimals.**

Example 2 Converting Between American Units of Capacity

My video summary

Make each indicated conversion.

a. $3\frac{1}{4}$ cups to pints $\qquad$ **b.** 8.25 quarts to pints

Solutions In each case, we multiply by the appropriate **unit fraction.**

a. To convert from cups to pints, multiply by $\frac{1\text{ pt}}{2\text{ c}}$. ← Desired units / ← Original units

(eText Screens 8.6-1–8.6-19)

$$3\frac{1}{4}\text{ c} = \frac{13}{4}\text{ c} = \frac{13\ \cancel{\text{c}}}{4}\cdot\frac{1\text{ pt}}{2\ \cancel{\text{c}}} = \frac{13\text{ pt}}{8} = 1\frac{5}{8}\text{ pt, or } 1.625\text{ pt}$$

Thus, a measure of $3\frac{1}{4}$ cups is equivalent to $1\frac{5}{8}$ pints, or 1.625 pints.

b. Try to complete this conversion on your own. View the **answer**, or watch this **video** for complete solutions to both parts.

You Try It **Work through this You Try It problem.**

Work Exercises 9–14 in this textbook or in the MyMathLab Study Plan.

Sometimes, we need to use more than one **unit fraction** when making conversions.

Example 3 Using More Than One Unit Fraction

My video summary

Convert 3 gallons to cups.

Solution Work through the following, or watch this **video** for a complete solution. Our **table of American units of capacity** does not provide a direct conversion fraction for gallons to cups. However, we can multiply using a sequence of unit fractions that will move us from the original units to the desired units.

Change gal to qt	Change qt to pt	Change pt to c
↓	↓	↓

$$\text{Original units} \rightarrow \frac{4\text{ qt}}{1\text{ gal}}\cdot\frac{2\text{ pt}}{1\text{ qt}}\cdot\frac{2\text{ c}}{1\text{ pt}} \leftarrow \text{Desired units}$$

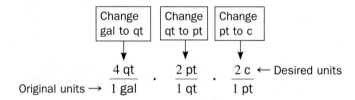

Multiplying the original value by the sequence gives

$$3\text{ gal} = \frac{3\ \cancel{\text{gal}}}{1}\cdot\frac{4\ \cancel{\text{qt}}}{1\ \cancel{\text{gal}}}\cdot\frac{2\ \cancel{\text{pt}}}{1\ \cancel{\text{qt}}}\cdot\frac{2\text{ c}}{1\ \cancel{\text{pt}}} = 3(4)(2)(2\text{ c}) = 48\text{ c}.$$

So, 3 gallons is equivalent to 48 cups.

You Try It **Work through this You Try It problem.**

Work Exercises 15–24 in this textbook or in the MyMathLab Study Plan.

OBJECTIVE 2 MAKE CONVERSIONS INVOLVING MIXED UNITS OF CAPACITY

As with units of length and weight, **capacity** can be expressed using more than one unit.

Example 4 Converting Mixed Units

Convert as indicated.

a. 6 gal 3 qt = _____ qt **b.** 30 fl oz = _____ c _____ fl oz

Solutions

a. We convert 6 gallons to quarts and then add the result to the 3 quarts.

To convert from gallons to quarts, multiply by $\frac{4 \text{ qt}}{1 \text{ gal}}$. ← Desired units / ← Original units

$$6 \text{ gal} = \frac{6 \cancel{\text{gal}}}{1} \cdot \frac{4 \text{ qt}}{1 \cancel{\text{gal}}} = 6(4 \text{ qt}) = 24 \text{ qt}$$

Therefore, 6 gal 3 qt = 24 qt + 3 qt = 27 qt.

6 gal 3 qt is equivalent to 27 qt.

My video summary

b. Try to complete this conversion on your own. View the **answer**, or watch this **video** for a complete solution to part b.

You Try It **Work through this You Try It problem.**

Work Exercises 25–30 in this textbook or in the MyMathLab Study Plan.

OBJECTIVE 3 MAKE CONVERSIONS WITHIN METRIC UNITS OF CAPACITY

In the metric system, the basic unit of **capacity** is the *liter*. A **liter** is the capacity of a cube that measures 10 centimeters on each side.

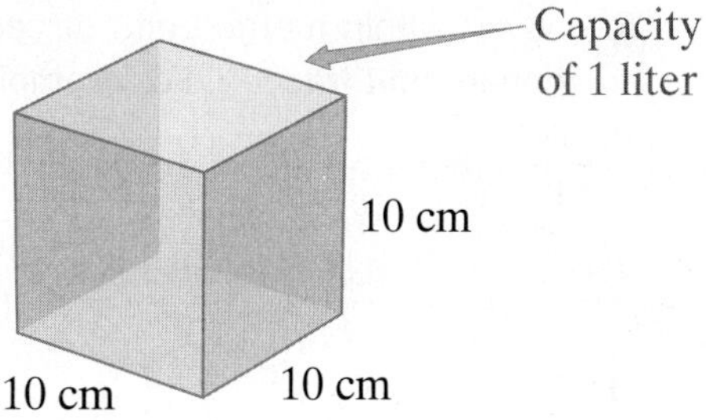

A liter is slightly larger than a quart. Soft drinks are commonly packaged in 1- or 2-liter bottles. A teaspoon holds 5 milliliters.

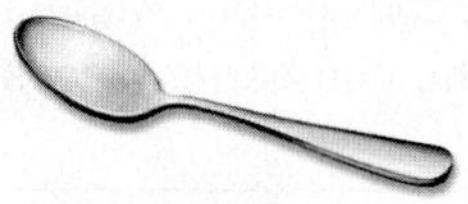

Holds 5 milliliters

Holds 2 liters

(*eText Screens 8.6-1–8.6-19*)

The table below shows the metric units of **capacity**.

Metric Units of Capacity

1 kiloliter (kL) $= 1000$ liters (L)	$1\text{ L} = \frac{1}{1000}\text{ kL}$ or 0.001 kL
1 hectoliter (hL) $= 100\text{ L}$	$1\text{ L} = \frac{1}{100}\text{ hL}$ or 0.01 hL
1 dekaliter (daL) $= 10\text{ L}$	$1\text{ L} = \frac{1}{10}\text{ daL}$ or 0.1 daL
1 liter (L) $= 1\text{ L}$	$1\text{ L} = 1\text{ L}$
1 deciliter (dL) $= \frac{1}{10}\text{ L}$ or 0.1 L	$1\text{ L} = 10\text{ dL}$
1 centiliter (cL) $= \frac{1}{100}\text{ L}$ or 0.01 L	$1\text{ L} = 100\text{ cL}$
1 milliliter (mL) $= \frac{1}{1000}\text{ L}$ or 0.001 L	$1\text{ L} = 1000\text{ mL}$

Notice once again that the prefixes for these units are the same as those we saw for metric units of **length** and **mass**. The most commonly used metric units of **capacity** are the *liter* and *milliliter*.

To convert within metric units of capacity, we can use the procedure of multiplying by an appropriate **unit fraction**. For example, to convert 6.4 liters to milliliters, we can multiply by $\frac{1000\text{ mL}}{1\text{ L}}$ (because $1\text{ L} = 1000\text{ mL}$):

$$6.4\text{ L} = \frac{6.4\not{\text{L}}}{1} \cdot \frac{1000\text{ mL}}{1\not{\text{L}}} = 6.4(1000\text{ mL}) = 6400\text{ mL}$$

Multiplying by 1000 moves the decimal point 3 places to the right.

Therefore, $6.4\text{ L} = 6400\text{ mL}$.

TIP As with metric units of length and mass, the metric units of capacity are all **powers of** 10 of the liter. So, conversions within these units result in a movement of the **decimal point**.

Below is a metric chart with the units of **capacity** listed in order from largest to smallest. As before, when making conversions within these units, we can use the chart to determine both the direction and the number of places to move the **decimal point**.

Metric Chart for Units of Capacity

kL hL daL L dL cL mL

To convert from 6.4 liters to milliliters, start at L on the chart (the original units) and move to mL (desired units), which requires a move of 3 units to the *right*.

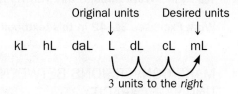

Original units ↓ (L) Desired units ↓ (mL)

kL hL daL L dL cL mL

3 units to the *right*

This is the same number of places and the direction that the decimal point will move.

$$6.400 \text{ L} = 6400 \text{ mL}$$

3 places to the *right*

Example 5 Converting Metric Units of Capacity

My interactive video summary

Make each of the following conversions.

a. 7250 milliliters to liters

b. 24.3 deciliters to milliliters

c. 0.15 liters to milliliters

d. 525 centiliters to liters

Solutions

a. We are converting from milliliters (mL) to liters (L). On the **metric chart**, we move from mL to L, which is 3 units to the *left*.

Desired units ↓ (L) Original units ↓ (mL)

kL hL daL L dL cL mL

3 units to the *left*

So, we move the **decimal point** 3 places to the left.

$$7250. \text{ mL} = 7.250 \text{ L}$$

3 places to the *left*

Therefore, 7250 mL = 7.25 L.

View this **alternate solution** to see this conversion worked with a **unit fraction**.

b. Here we are converting deciliters (dL) to milliliters (mL). Moving from dL to mL on the **metric chart** requires a move of 2 units to the *right*.

Original units ↓ (dL) Desired units ↓ (mL)

kL hL daL L dL cL mL

2 units to the *right*

Moving the **decimal point** 2 places to the right gives

$$24.30 \text{ dL} = 2430 \text{ mL}.$$

2 places to the *right*

Thus, 24.3 dL = 2430 mL.

View this **alternate solution** to see this conversion worked with **unit fractions.**

c.–d. Try to make these conversions on your own. View the **answers**, or work through this **interactive video** for complete solutions to all four parts.

(eText Screens 8.6-1–8.6-19)

You Try It **Work through this You Try It problem.**

Work Exercises 31–42 in this textbook or in the MyMathLab Study Plan.

OBJECTIVE 4 MAKE CONVERSIONS BETWEEN AMERICAN AND METRIC UNITS OF CAPACITY

Sometimes it is necessary to convert between American and metric units of **capacity**. The table below gives approximate equivalences that can be used to make conversions.

American and Metric Conversion Facts for Capacity

1 fl oz ≈ 29.57 mL	1 L ≈ 33.81 fl oz
1 c ≈ 237 mL	1 L ≈ 4.23 c
1 pt ≈ 0.47 L	1 L ≈ 2.11 pt
1 qt ≈ 0.95 L	1 L ≈ 1.06 qt
1 gal ≈ 3.79 L	1 L ≈ 0.26 gal

The approximate conversion facts lead to the following *approximate* **unit fractions.**

Approximate Unit Fractions for American and Metric Units of Capacity

$$\frac{29.57\text{ mL}}{1\text{ fl oz}} \approx 1 \qquad \frac{33.81\text{ fl oz}}{1\text{ L}} \approx 1 \qquad \frac{237\text{ mL}}{1\text{ c}} \approx 1 \qquad \frac{4.23\text{ c}}{1\text{ L}} \approx 1$$

$$\frac{0.47\text{ L}}{1\text{ pt}} \approx 1 \qquad \frac{2.11\text{ pt}}{1\text{ L}} \approx 1 \qquad \frac{0.95\text{ L}}{1\text{ qt}} \approx 1 \qquad \frac{1.06\text{ qt}}{1\text{ L}} \approx 1$$

$$\frac{3.79\text{ L}}{1\text{ gal}} \approx 1 \qquad \frac{0.26\text{ gal}}{1\text{ L}} \approx 1$$

Example 6 Converting Between American and Metric Units of Capacity

Make each indicated conversion. Round to the nearest tenth.

a. 3.5 quarts to liters **b.** 16 fluid ounces to milliliters

Solutions

a. To convert from quarts to liters, multiply by $\frac{0.95\text{ L}}{1\text{ qt}}$. ← Desired units / ← Original units

$$3.5\text{ qt} \approx \frac{3.5\cancel{\text{qt}}}{1} \cdot \frac{0.95\text{ L}}{1\cancel{\text{qt}}} = 3.5(0.95\text{ L}) \approx 3.3\text{ L}$$

Round

```
   3.5   ← 1 decimal place
 × 0.95  ← 2 decimal places
   175
  3150
 3.325   ← 3 decimal places
```

So, 3.5 quarts is approximately equivalent to 3.3 liters.

(eText Screens 8.6-1–8.6-19)

b. Try to make this conversion on your own. View the **answer**, or watch this **video** for a complete solution to part b.

You Try It **Work through this You Try It problem.**

Work Exercises 43–58 in this textbook or in the MyMathLab Study Plan.

8.6 Exercises

In Exercises 1–30, make the indicated conversion within American units of capacity.

You Try It

1. 32 fluid ounces to cups
2. 18 pints to cups
3. 25 gallons to quarts
4. 64 pints to quarts
5. 6 cups to fluid ounces
6. 23 quarts to pints
7. 124 quarts to gallons
8. 14 cups to pints

You Try It

9. $\frac{3}{4}$ cup to fluid ounces
10. 17 fluid ounces to cups
11. 3 pints to quarts
12. 6.5 cups to pints
13. 19 quarts to gallons
14. $5\frac{3}{8}$ gallons to quarts

You Try It

15. 15 gallons to pints
16. 96 fluid ounces to pints
17. 8.5 gallons to cups
18. 132 cups to quarts
19. 288 fluid ounces to quarts
20. 4 gallons to fluid ounces
21. A container holds 64 fluid ounces of orange juice. How many pints of orange juice is this?
22. A recipe calls for 3 cups of milk. Convert this to pints.
23. One day, John drank $\frac{1}{2}$ gallon of water. How many quarts of water did John drink?
24. A can holds 2.75 gallons of gasoline. Convert this to fluid ounces.

You Try It

25. 12 gal 2 qt = ______ qt
26. 10 qt 1 pt = ______ pt
27. 86 qt = ____ gal ____ qt
28. 140 fl oz = ____ c____ fl oz
29. Karen used 4 gallons 3 quarts of paint to paint a large room. Express this amount in quarts only.
30. A bottle contains 28 fluid ounces of sports drink. Convert this amount to mixed units of cups and fluid ounces.

(*eText Screens 8.7-1–8.7-15*)

In Exercises 31–42, make the indicated conversion within metric units of capacity.

You Try It 31. 21 liters to milliliters

32. 3200 milliliters to liters

33. 52 milliliters to liters

34. 8.39 liters to milliliters

35. 67.4 centiliters to liters

36. 0.5 centiliter to milliliter

37. 875 milliliters to deciliters

38. 2.65 kiloliters to liters

39. 9.1 dekaliters to liters

40. 19,340 liters to milliliters

41. An IV bag contains 750 milliliters of sodium chloride. How many liters is this?

42. A container holds 3.25 liters of sweet tea. Express this amount in milliliters.

In Exercises 43–58, make the indicated conversion between American and metric units of capacity. Round answers to the nearest tenth.

You Try It 43. 45 fluid ounces to liters

44. 2 liters to fluid ounces

45. 6.5 cups to milliliters

46. 750 liters to gallons

47. 25 pints to liters

48. 3.5 liters to cups

49. 15 quarts to liters

50. 13.5 liters to pints

51. 16.25 gallons to liters

52. 60 liters to quarts

53. 250 fluid ounces to liters

54. 3 cups to liters

55. A bottle contains 2 liters of soft drink. Convert this amount to quarts.

56. Ethan purchased 18.5 gallons of gasoline. How many liters of gasoline did he buy? Round to the nearest tenth.

57. A tank has a capacity of 190,000 liters. What is the capacity of the tank in gallons?

58. A syringe contains 1.75 fluid ounces of fluid. Convert this to milliliters. Round to the nearest tenth.

8.7 Time and Temperature

THINGS TO KNOW

Before working through this topic, be sure you are familiar with the following concepts:

	VIDEO	ANIMATION	INTERACTIVE
You Try It 1. Write Fractions in Simplest Form (Topic 4.2, **Objective 6**)	✓		
You Try It 2. Multiply Fractions (Topic 4.3, **Objective 1**)	✓		✓

(eText Screens 8.7-1–8.7-15)

You Try It 3. Write Two Quantities as a Rate (Topic 6.1, **Objective 2**)

You Try It 4. Make Conversions Within American Units of Length (Topic 8.4, **Objective 1**)

You Try It 5. Make Conversions Within American Units of Weight (Topic 8.5, **Objective 1**)

You Try It 6. Make Conversions Within American Units of Capacity (Topic 8.6, **Objective 1**)

OBJECTIVES

1 Make Conversions Within Units of Time

2 Make Conversions Between Fahrenheit and Celsius Temperatures

OBJECTIVE 1 MAKE CONVERSIONS WITHIN UNITS OF TIME

We finish this module by looking at two more sets of conversions: time and temperature. Unlike the other measurements we have seen, *time* is measured in the same units in both the American system and the metric system. Some common measures of time are **second**, **minute**, **hour**, **day**, and **week**. The following shows how these units are related.

Units of Time

$$60 \text{ seconds (s or sec)} = 1 \text{ minute (min)}$$
$$60 \text{ minutes} = 1 \text{ hour (hr)}$$
$$24 \text{ hours} = 1 \text{ day (d)}$$
$$7 \text{ days} = 1 \text{ week (wk)}$$

The following summary gives the unit fractions that can be used to convert between various units of time.

Unit Fractions for American Units of Time

$\frac{1 \text{ min}}{60 \text{ sec}} = 1$ (seconds to minutes) $\qquad \frac{60 \text{ sec}}{1 \text{ min}} = 1$ (minutes to seconds)

$\frac{1 \text{ hr}}{60 \text{ min}} = 1$ (minutes to hours) $\qquad \frac{60 \text{ min}}{1 \text{ hr}} = 1$ (hours to minutes)

$\frac{1 \text{ day}}{24 \text{ hr}} = 1$ (hours to days) $\qquad \frac{24 \text{ hr}}{1 \text{ d}} = 1$ (days to hours)

$\frac{1 \text{ wk}}{7 \text{ d}} = 1$ (days to weeks) $\qquad \frac{7 \text{ d}}{1 \text{ wk}} = 1$ (weeks to days)

(eText Screens 8.7-1–8.7-15)

Example 1 Converting Between Units of Time

a. Convert 3 days to hours.

b. Convert 240 seconds to minutes.

Solutions

a. Because we are given days (d) and want to convert to hours (h), we use the **unit fraction**

$$\frac{24\text{ h}}{1\text{ d}}. \quad \begin{matrix}\leftarrow \text{Desired units}\\ \leftarrow \text{Given units}\end{matrix}$$

Multiply the given value by the unit fraction and **simplify** by **dividing out** common **factors** or units.

$$\begin{aligned} 3\text{ d} &= \frac{3\text{ d}}{1}\cdot\boxed{\frac{24\text{ h}}{1\text{ d}}} \quad \leftarrow \text{Multiply by } 1\\ &= \frac{3\,\cancel{\text{d}}}{1}\cdot\frac{24\text{ h}}{1\,\cancel{\text{d}}}\\ &= 3\cdot 24\text{ h}\\ &= 72\text{ h}\end{aligned}$$

So, 3 days = 72 hours.

b. Because we are given seconds (s) and want to convert to minutes (min), we use the **unit fraction**

$$\frac{1\text{ min}}{60\text{ s}}. \quad \begin{matrix}\leftarrow \text{Desired units}\\ \leftarrow \text{Given units}\end{matrix}$$

Multiply the given value by the unit fraction and **simplify** by **dividing out** common **factors** or units.

$$\begin{aligned} 240\text{ s} &= \frac{240\text{ s}}{1}\cdot\boxed{\frac{1\text{ min}}{60\text{ s}}} \quad \leftarrow \text{Multiply by } 1\\ &= \frac{\overset{4}{\cancel{240}}\,\cancel{\text{s}}}{1}\cdot\frac{1\text{ min}}{\underset{1}{\cancel{60}}\,\cancel{\text{s}}}\\ &= 4\text{ min}\end{aligned}$$

So, 240 seconds = 4 minutes.

You Try It **Work through this You Try It problem.**

Work Exercises 1–8 in this textbook or in the MyMathLab Study Plan.

Example 2 Converting Between Units of Time

a. Convert 240 days to weeks.

b. Convert 3.5 hours to minutes.

My video summary

Solutions

a. Because we are given days (d) and want to convert to weeks (wk), we use the **unit fraction**

$$\frac{1\ \text{wk}}{7\ \text{d}}. \quad \begin{array}{l}\leftarrow \text{Desired units}\\ \leftarrow \text{Given units}\end{array}$$

Multiply the given value by the unit fraction and **simplify** by **dividing out** common **factors** or units.

$$\begin{aligned} 240\ \text{d} &= \frac{240\ \text{d}}{1} \cdot \frac{1\ \text{wk}}{7\ \text{d}} \\ &= \frac{240\ \not{\text{d}}}{1} \cdot \frac{1\ \text{wk}}{7\ \not{\text{d}}} \\ &= \frac{240}{7}\ \text{wk} \\ &= 34\frac{2}{7}\ \text{wk} \end{aligned}$$

$$\begin{array}{r} 34 \\ 7\overline{)240} \\ -\ 210 \\ \hline 30 \\ -\ 28 \\ \hline 2 \end{array}$$

So, 240 days $= 34\frac{2}{7}$ weeks. We could also say 240 days = 34 weeks 2 days.

b. Because we are given hours (h) and want to convert to minutes (min), we use the **unit fraction**

$$\frac{60\ \text{min}}{1\ \text{h}}. \quad \begin{array}{l}\leftarrow \text{Desired units}\\ \leftarrow \text{Given units}\end{array}$$

Finish doing this on your own. Check your **answer**, or watch this **video** for complete solutions to both parts.

You Try It **Work through this You Try It problem.**

Work Exercises 9–16 in this textbook or in the MyMathLab Study Plan.

As with other units, it may be necessary to use more than one unit fraction when doing a conversion.

Example 3 Converting Time Units Using More Than One Unit Fraction

My video summary

Convert 3600 minutes to days.

Solutions Work through the following, or watch this **video** for the complete solution. We are given minutes (min) and want to convert to days (d). We do this by first converting from minutes to hours (h), then from hours to days using two unit fractions:

$$\text{Given units} \rightarrow \frac{1\ \text{h}}{60\ \text{min}} \cdot \frac{1\ \text{d}}{24\ \text{h}} \leftarrow \text{Desired units}$$

(eText Screens 8.7-1–8.7-15)

Multiply the given value by the two **unit fractions** and simplify.

$$3600\text{ min} = \frac{3600\text{ min}}{1} \cdot \boxed{\frac{1\text{ h}}{60\text{ min}} \cdot \frac{1\text{ d}}{24\text{ h}}} \quad \leftarrow \text{Multiply by 1}$$

$$= \frac{\overset{5}{\cancel{\overset{60}{\cancel{3600}}}}\,\cancel{\text{min}}}{1} \cdot \frac{\cancel{1}\,\cancel{\text{h}}}{\underset{1}{\cancel{60}}\,\cancel{\text{min}}} \cdot \frac{1\text{ d}}{\underset{2}{\cancel{24}}\,\cancel{\text{h}}}$$

$$= \frac{5}{2}\text{ d}$$

$$= 2\frac{1}{2}\text{ days}$$

So, 3600 minutes $= 2\frac{1}{2}$ days.

You Try It **Work through this You Try It problem.**

Work Exercises 17–24 in this textbook or in the MyMathLab Study Plan.

OBJECTIVE 2 MAKE CONVERSIONS BETWEEN FAHRENHEIT AND CELSIUS TEMPERATURES

In American units, we measure temperature on the Fahrenheit scale. For example, the temperature of a room might be 77°F (read "seventy-seven degrees Fahrenheit"). The raised circle indicates "degrees" and the "F" indicates we are using the **Fahrenheit scale.** Using this scale, we would say that water freezes at 32°F (thirty-two degrees Fahrenheit) and boils at 212°F (two hundred, twelve degrees Fahrenheit).

In metric units, we measure temperature on the **Celsius scale.** For example, the temperature of the room mentioned above might be 25°C (twenty-five degrees Celsius). The following illustrates a few temperatures using both scales.

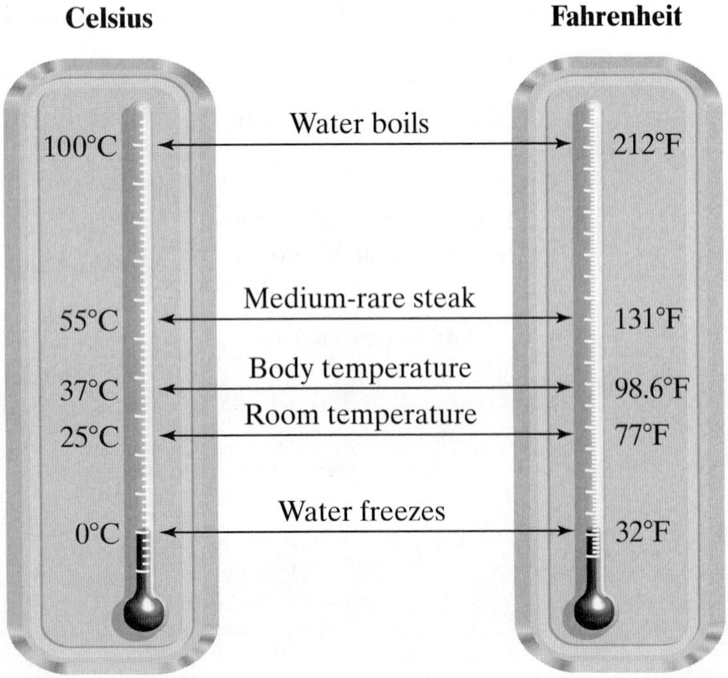

Note: The temperatures for which water freezes or boils are exact.

(eText Screens 8.7-1–8.7-15)

The Celsius scale is similar to the Centigrade scale. View this **popup** to see the difference.

We can convert from one temperature scale to the other by using the following **formulas**.

Temperature Conversion Formulas

Celsius to Fahrenheit: $F = \frac{9}{5}C + 32$

Fahrenheit to Celsius: $C = \frac{5}{9}(F - 32)$

Example 4 Converting Between Fahrenheit and Celsius Temperatures

My video summary

a. Convert $18°\text{C}$ to Fahrenheit.

b. Convert $15°\text{F}$ to Celsius. Round to the nearest tenth of a degree if necessary.

Solutions

a. Because we are given degrees Celsius (°C) and want to convert to degrees Fahrenheit, we use the formula

$$F = \frac{9}{5}C + 32.$$

Substitute 18 for C and simplify to find the value for F.

$$\begin{aligned} F &= \frac{9}{5}(18) + 32 \\ &= \frac{162}{5} + 32 \\ &= 32.4 + 32 \\ &= 64.4 \end{aligned}$$

So, $18°\text{C} = 64.4°\text{F}$.

b. **Substitute** 15 for F in the correct **temperature conversion formula** and solve for C. Try making this conversion on your own. Check your **answer**, or watch this **video** for complete solutions to both parts.

You Try It **Work through this You Try It problem.**

Work Exercises 25–36 in this textbook or in the MyMathLab Study Plan.

8.7 Exercises

In Exercises 1–20, make the indicated conversion within units of time.

You Try It

1. Convert 6 weeks to days.
2. Convert 63 days to weeks.
3. Convert 12 days to hours.
4. Convert 120 hours to days.
5. Convert 28 minutes to seconds.
6. Convert 315 seconds to minutes.

(*eText Screens 8.7-1–8.7-15*)

7. Convert 30 hours to minutes.

8. Convert 472 minutes to hours.

You Try It 9. Convert 12.5 weeks to days.

10. Convert 275 seconds to minutes.

11. Convert 4.5 hours to minutes.

12. Convert 34 days to weeks.

13. Convert $7\frac{1}{3}$ days to hours.

14. Convert $14\frac{1}{2}$ minutes to seconds.

15. Convert 320 minutes to hours.

16. Convert 430 hours to days.

You Try It 17. Convert 3.2 hours to seconds.

18. Convert 2.5 weeks to hours.

19. Convert 5000 minutes to days.

20. Convert 420 hours to weeks.

21. A day on a planet is the time it takes the planet to rotate on its axis. If a day on Jupiter is 9.8 hours, how long is this in minutes?

22. In 2011, Geoffrey Mutai ran the Boston Marathon in the fastest time ever for a marathon. His winning time was 2 h 3 min 2 s. How many seconds did it take him to run the marathon?

23. The 2012 Iditarod red lantern, given to the last musher to finish, was awarded to Jan Steves. Her final time for the race was 14 days, 11 hours, 57 minutes and 11 seconds. What was her time in hours? Round to the nearest tenth. (*Source:* Iditarod.com)

24. If a posted speed limit is $\frac{80\text{ km}}{\text{h}}$, what is the speed limit in $\frac{\text{m}}{\text{s}}$ (meters per second)? Round to the nearest whole number.

In Exercises 25–36, make the indicated temperature conversion. Round answers to one decimal place if necessary.

You Try It 25. Convert 45°C to Fahrenheit.

26. Convert 59°F to Celsius.

27. Convert −25°C to Fahrenheit.

28. Convert −20°F to Celsius.

29. Convert 98°F to Celsius.

30. Convert 87°C to Fahrenheit.

31. Convert 54.5°F to Celsius.

32. Convert 42.6°C to Fahrenheit.

33. Magic Ice Istanbul is the world's first ice museum in a warm climate. The temperature inside is kept at −5°C. What is this temperature on the Fahrenheit scale? (*Source:* istanbulview.com)

34. The melting point of aluminum is 660°C. What is the melting point in Fahrenheit? (*Source:* engineeringtoolbox.com)

35. An oral body temperature above 100°F is generally considered to be a fever. What is body temperature of 102°F on the Celsius scale?

36. The average low temperature in August in San Diego, CA, is 67°F and the average high temperature is 77°F. Convert these temperatures to Celsius. (*Source:* weather.com)

MODULE NINE

STATISTICS

MODULE NINE CONTENTS

9.1 Mean, Median, and Mode

THINGS TO KNOW

Before working through this topic, be sure you are familiar with the following concepts:

VIDEO ANIMATION INTERACTIVE

You Try It 1. Find the Average of a List of Numbers (Topic 1.6, **Objective 4**)

You Try It 2. Add Positive Decimals (Topic 5.2, **Objective 1**)

You Try It 3. Add and Subtract Negative Decimals (Topic 5.2, **Objective 4**)

You Try It 4. Divide Decimals (Topic 5.4, **Objective 1**)

OBJECTIVES

1 Find the Mean

2 Find the Median

3 Find the Mode

OBJECTIVE 1 FIND THE MEAN

In this topic, we look at some additional applications of **decimals** from statistics.

When given a list of data values, it is convenient to summarize the data with a single "middle" number called a **measure of central tendency.** For example, your GPA (grade point average) is a single number that summarizes all your grades at a given school.

(eText Screens 9.1-1–9.1-16)

The three most common measures of central tendency are the *mean*, the *median*, and the *mode*.

The **mean**, denoted by $\bar{x}$ (read "x bar"), is often referred to as the **average** and is computed by adding up all the data values and then dividing by the number of values. Recall that we learned how to find the average of a list of numbers in **Topic 1.6**.

Example 1 Finding the Mean

Valerie keeps track of her gasoline purchases to help her budget monthly expenses. The following data shows the price per gallon (in dollars) for regular unleaded gasoline on her last 10 fill-ups.

2.97	3.00	3.08	3.10	3.04
2.96	3.06	3.07	2.95	3.07

Find the mean (average) price per gallon for her last 10 fill-ups.

Solutions To find the **mean**, $\bar{x}$, we first add the data values.

$$\begin{aligned}&\underbrace{2.97 + 3.00} + \underbrace{3.08 + 3.10} + \underbrace{3.04 + 2.96} + \underbrace{3.06 + 3.07} + \underbrace{2.95 + 3.07}\\ &= 5.97 + 6.18 + 6 + 6.13 + 6.02\\ &= 12.15 + 12.13 + 6.02\\ &= 24.28 + 6.02\\ &= 30.3\end{aligned}$$

Add pairs of values working left to right

Next, we divide the **sum** by the number of values, 10.

$$\bar{x} = \frac{\overbrace{30.3}^{\text{Sum}}}{\underbrace{10}_{\text{Number of values}}} = 3.03$$

The mean (**average**) price per gallon is $3.03.

Example 2 Blowing the Curve

My video summary

If the **average** of an exam is below 70, a physics teacher will curve the grades by adding enough points to each exam to bring the average up to 70. The following are the exam scores for 7 students in one of her physics classes:

76 64 57 66 65 77 55

a. Find the average (**mean**) exam score and round to 1 decimal place. Will the teacher curve this exam?

b. Suppose that the class actually has 8 students and the last student scored a 100 on the exam. Find the average of the eight exam scores. Will the teacher curve this exam?

(eText Screens 9.1-1–9.1-16)

Solutions Work through the following, or watch this **video** for the complete solution.

a. To find the average (mean), $\bar{x}$, add the data values and divide by 7, the number of values.

$$\bar{x} = \frac{76 + 64 + 57 + 66 + 65 + 77 + 55}{7}$$

$$= \frac{460}{7} \approx 65.7$$

↑ approximately equal to

$$\begin{array}{r} 65.7... \\ 7\overline{)460.0} \\ -42 \\ \hline 40 \\ 35 \\ \hline 50 \\ 49 \\ \hline 1 \end{array}$$

The average grade was approximately 65.7. Since this is less than 70, the teacher will curve the exam.

b. We add the score of 100 and calculate the **mean** of the 8 values.

$$\bar{x} = \frac{\overbrace{460}^{\text{Sum of first seven values}} + 100}{\underbrace{8}_{7 + 1 = 8 \text{ values}}} = \frac{560}{8} = 70$$

The average of the 8 grades is 70. Since this is not below 70, the teacher will not curve the exam. The grade of 100 pulled the mean up enough to "blow the curve."

You Try It **Work through this You Try It problem.**

Work Exercises 1–4 in this textbook or in the MyMathLab Study Plan.

OBJECTIVE 2 FIND THE MEDIAN

Because the mean uses all the data values and gives each one equal weight, the **mean** can be affected greatly by **extreme values** (values far away from the rest of the data). We saw the effect of this in **Example 2b**. Next, we look at a **measure of central tendency** that is not greatly affected by extreme values, called the **median**.

Much like a highway median is in the middle of the road, the median of an **ordered list of data** is located in the middle of the data. By "ordered," we mean that the list of data is written from smallest to largest in value. If the number of data values in the ordered list is odd, the median is the middle value. If the number of values is even, the median is the **average** of the two middle values.

Odd number of values	Even number of values
4 7 9 10 14	3 6 11 15 20 23
↑ (at 9) Median is the middle value	↑ (at 11 15) Median is the average of the two middle values

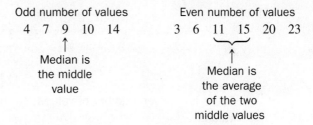

The median is not as affected by extreme values because it is based more on the position of data in the ordered list rather than on actual values. The median is a value such that half of the values in the ordered list are smaller than the median and half are larger.

(eText Screens 9.1-1–9.1-16)

Consider the **median** of the physics exam scores from **Example 2** with and without the score of 100. The **ordered** lists and corresponding medians are given below.

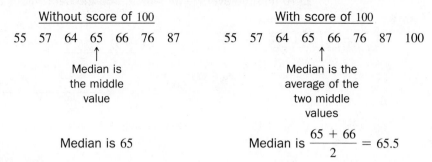

Without score of 100	With score of 100
55 57 64 65 66 76 87	55 57 64 65 66 76 87 100
↑ Median is the middle value (65)	↑ Median is the average of the two middle values (65, 66)
Median is 65	Median is $\frac{65+66}{2} = 65.5$

Notice how including the score of 100 did not greatly affect the median.

Example 3 Finding the Median

My video summary

The following data show the number of apps downloaded on 8 randomly selected iPads®.

48 38 43 34 27 22 37 46

a. Find the median.

b. How many values are below the median? How many are above the median?

c. Suppose the value 48 was mistakenly recorded as 84. Find the median using this value.

d. How did changing the value in part c affect the value of the median?

Solutions Work through the following, or watch this **video** for the complete solution.

a. To find the **median**, begin by **ordering** the data.

22 27 34 $\underbrace{37\quad 38}_{\text{Two middle values}}$ 43 46 48

There is an even number of values, so the median is the **average** of the two middle values.

$$\text{median} = \frac{37+38}{2} = \frac{75}{2} = 37.5$$

The median number of apps downloaded is 37.5.

b. There are four values below the median (22 27 34 37).
There are four values above the median (38 43 46 48).

c. **Order** the data, replacing 48 with 84.

22 27 34 $\underbrace{37\quad 38}_{\text{Two middle values}}$ 43 46 $\overbrace{84}^{\text{Incorrect value}}$

There is an even number of values, so the **median** is the **average** of the two middle values.

$$\text{median} = \frac{37+38}{2} = \frac{75}{2} = 37.5$$

Using the incorrect value, the median number of apps downloaded is 37.5.

d. Using the incorrect value, the median remained the same.

(eText Screens 9.1-1–9.1-16)

You Try It **Work through this You Try It problem.**

Work Exercises 5–10 in this textbook or in the MyMathLab Study Plan.

CAUTION

Before finding the median, the data values must be ordered, which means to write the values from smallest to largest.

OBJECTIVE 3 FIND THE MODE

A third **measure of central tendency** is the **mode**. The mode is the value that appears in the data set most often. It is possible for a data set to have no mode, one mode, or more than one mode. Like the **median**, the mode is not greatly affected by a few **extreme values**.

Example 4 Finding the Mode

My video summary

The following data are the snowfall readings (in inches) at Boston's Logan International Airport for Boston's top twelve winter storms. (*Source:* cbsboston.com, February 1, 2011)

27.5	27.1	26.3	25.4	22.5	21.4
19.8	19.4	18.7	18.2	18.2	18.2

a. Find the mode.

b. Suppose the value 27.5 was mistakenly recorded as 72.5. Find the mode using this value.

c. How did changing the value in part b affect the value of the mode?

Solutions Work through the following, or watch this **video** for the complete solution.

a. To find the mode, locate the value that occurs most often.

27.5	27.1	26.3	25.4	22.5	21.4
19.8	19.4	18.7	18.2	18.2	18.2

The mode is 18.2 because it occurs most often (3 times).

b. Replace 27.5 with 72.5 and locate the value that occurs most often.

Incorrect value					
72.5	27.1	26.3	25.4	22.5	21.4
19.8	19.4	18.7	18.2	18.2	18.2

Using the incorrect value, the **mode** is 18.2.

c. Using the incorrect value, the mode remained the same.

You Try It **Work through this You Try It problem.**

Work Exercises 11–15 in this textbook or in the MyMathLab Study Plan.

TIP

Of the three **measures of central tendency**, the mode is the only one that must always be one of the data values.

View this **popup** to learn more about the mode as a measure of central tendency.

9.1 Exercises

In Exercises 1–4, find the mean for each list of data.

1. 21, 24, 32, 19, 25, 23

2. 4.4 cm, 6.5 cm, 2.6 cm, 8.1 cm, 4.1 cm

3. 534, 413, 201, 397, 369, 294

4. $1.52, $0.85, $3.28, $5.11, $3.08, $2.78

In Exercises 5–10, find the median for each list of data.

5. 17, 40, 16, 47, 83

6. 28, 14, 40, 40, 35, 23

7. 10.4, 1.6, 3.5, 7.4, 9.1, 3.8

8. 6.5 h, 4.3 h, 8.2 h, 17.4 h, 10.3 h, 12.7 h, 8.2 h

9. 120 mi, 214 mi, 186 mi, 312 mi, 249 mi

10. 350, 198, 412, 376, 343, 581

In Exercises 11–15, find the mode for each list of data.

11. 12, 6, 8, 4, 4, 6, 4

12. 0.09, 0.12, 0.03, 0.07, 0.09, 0.04, 0.1

13. 2.5, 2.3, 2.3, 3.0, 2.2, 2.3, 2.4

14. 10, 13, 4, 8, 12, 10, 4, 6

15. 58 m, 60 m, 29 m, 75 m, 80 m, 63 m, 20 m

16. **Texting Habits** The following data represent the average number of texts per day for 12 random 18–24-year-olds.

57	56	75	82	64	46
82	44	111	49	75	112

Find the mean, median, and mode for the data. Round answers to one decimal place if necessary.

17. **Chicago Snowfall** The following data are the snowfall readings (in inches) for Chicago's top ten winter storms as of February 2, 2011. (*Source:* National Weather Service)

23	21.6	20.2	19.2	18.8
16.2	15	14.9	14.9	14.3

a. Find the mean, median, and mode for the data.

b. Suppose the value 14.3 was actually recorded as 41.3. Compute the mean, median, and mode again using this value. Comment on any differences from the results in part a.

(eText Screens 9.2-1–9.2-25)

18. **Basketball Players** The following data are the heights (in inches) for a sample of 12 basketball players.

72	78	78	76	82	75
78	79	75	80	80	83

a. Find the mean, median, and mode for the data. Round answers to one decimal place if necessary.

b. Suppose the value 83 was actually recorded as 38. Compute the mean, median, and mode again using this value. Comment on any differences from the results in part a.

19. **Exam Scores** The following data are the final exam scores for an online algebra course.

88	80	73	95	98	81	86	93
95	90	90	96	90	92	88	87

a. Find the mean, median, and mode for the data. Round answers to one decimal place if necessary.

b. A student taking the exam late scored a 97. Add this value to the data set and find the new mean, median, and mode.

20. **Online Friends** The following data are the number of Facebook friends for 9 random Facebook users.

75 192 135 240 196 60 24 271 100

a. Find the mean, median, and mode for the data. Round answers to one decimal place if necessary.

b. A new user opens a Facebook account starting with 10 friends. Add this data value to the data set and find the new mean, median, and mode for the data.

9.2 Histograms

THINGS TO KNOW

Before working through this topic, be sure you are familiar with the following concepts:

		VIDEO	ANIMATION	INTERACTIVE
You Try It	1. Read Tables and Bar Graphs Involving Whole Numbers (Topic 1.2, **Objective 6**)			
You Try It	2. Add Positive Decimals (Topic 5.2, **Objective 1**)			
You Try It	3. Change Fractions or Mixed Numbers into Decimals (Topic 5.5, **Objective 1**)			
You Try It	4. Write Decimals as Percents (Topic 7.1, **Objective 2**)			
You Try It	5. Write Fractions as Percents (Topic 7.1, **Objective 4**)			

OBJECTIVES

1 Read a Histogram

2 Construct a Frequency Table

3 Construct a Histogram

(eText Screens 9.2-1–9.2-25)

OBJECTIVE 1 READ A HISTOGRAM

A **histogram** is a special type of **bar graph** in which the bars touch, and the width of the bars has meaning. As with a bar graph, each bar in a histogram represents a **class**, or category. However, in a histogram, each class is really a range of values called a **class interval.** The height of each bar is determined by the number of data values that fall within the corresponding class interval. This is called the **class frequency.** There are no gaps between the bars of a histogram, unless a class has a frequency of 0, and there is no overlap in the class intervals. Consider the histogram in Figure 1 for the final exam scores of 25 prealgebra students.

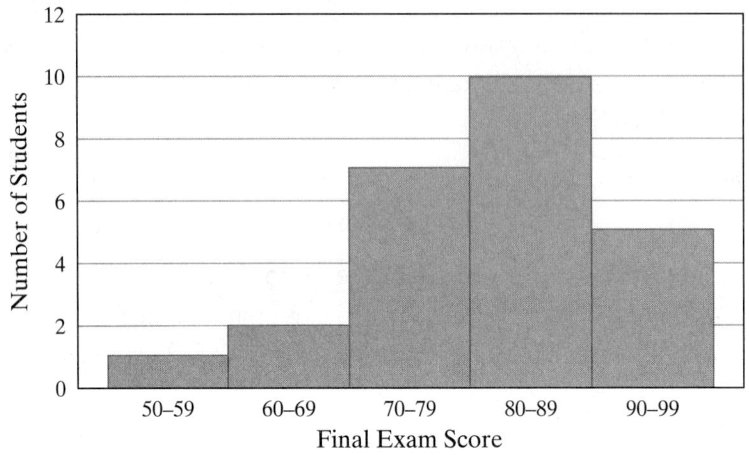

Figure 1 Frequency Histogram

The histogram in Figure 1 has five classes. The class intervals are 50–59, 60–69, 70–79, 80–89, and 90–99. The height of the bar for the class interval 70–79 is 7, so the class interval has a frequency of 7.

For large sets of data, it is common to use *relative* frequencies rather than frequencies when constructing a **histogram**. The **relative frequency** of a class is the ***percent*** of values (written as a decimal) falling within the corresponding class interval. The relative frequency is found by dividing the class frequency by the total number of data values. For example, there are 25 data values in Figure 1, so the relative frequency for the interval 70–79 is $\frac{7}{25} = 0.28$. Figure 2 shows the relative frequency histogram for the data in Figure 1. Notice that the shape of the histogram is the same in both cases.

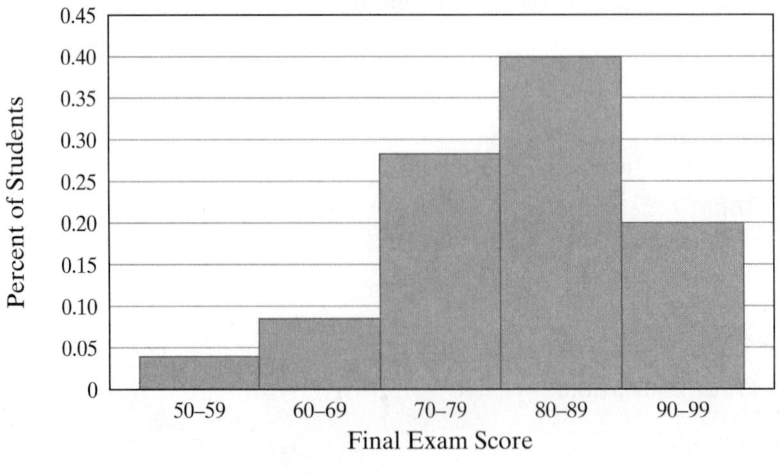

Figure 2 Relative Frequency Histogram

(eText Screens 9.2-1–9.2-25)

Example 1 Reading a Frequency Histogram

The following **histogram** shows the heights, in inches, for the members of a men's college basketball team. Use the histogram to answer the following questions.

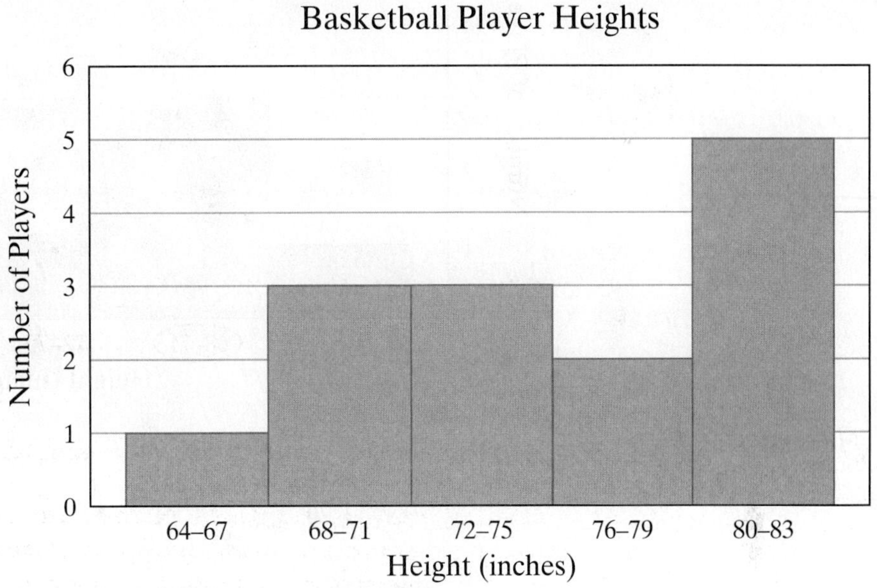

a. Which height range has the most number of players?

b. How many players are between 68 and 71 inches tall?

c. How many players are at least 76 inches tall?

Solutions

a. The height range with the largest number of players is the **class** with the tallest bar. The tallest bar corresponds to the class 80–83, so the height range 80 inches to 83 inches has the largest number of players.

b. We look for the bar that corresponds to the class 68–71. The height of this bar is 3, so there are three players on the team with heights between 68 inches and 71 inches.

c. Notice that there are actually two bars which meet the condition "at least 76 inches tall". Players falling in the interval 76–79 or the interval 80–84 would have heights of at least 76 inches. We add the two frequencies together to get the total number of players who are at least 76 inches tall.

(eText Screens 9.2-1–9.2-25)

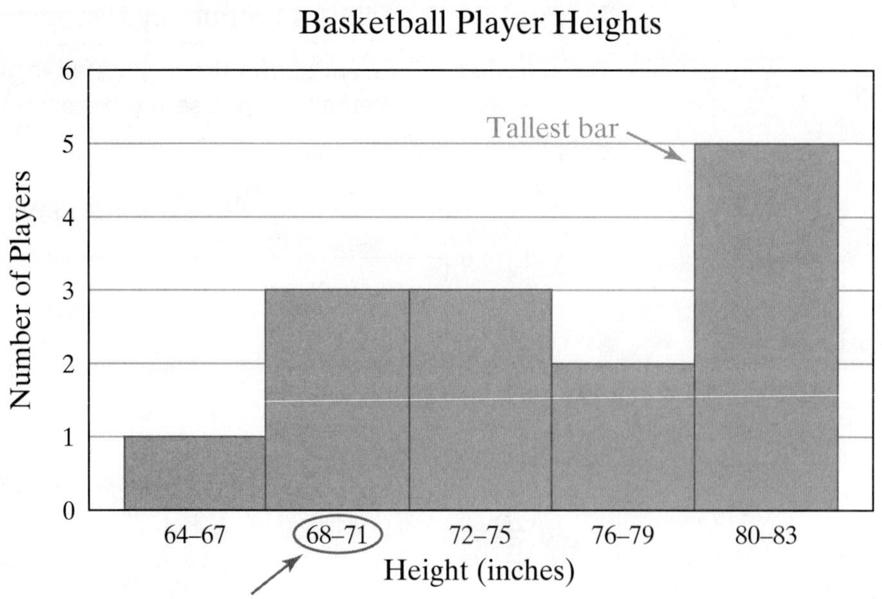

80–83 inches

$$2 + 5 = 7$$

76–79 inches

There are 7 members of the team with a height of at least 76 inches.

You Try It **Work through this You Try It problem.**

Work Exercises 1 and 2 in this textbook or in the MyMathLab Study Plan.

Example 2 Reading a Relative Frequency Histogram

My video summary

The following **histogram** shows the speeds, in miles per hour, for a sample of cars along a stretch of highway. Use the histogram to answer the following questions.

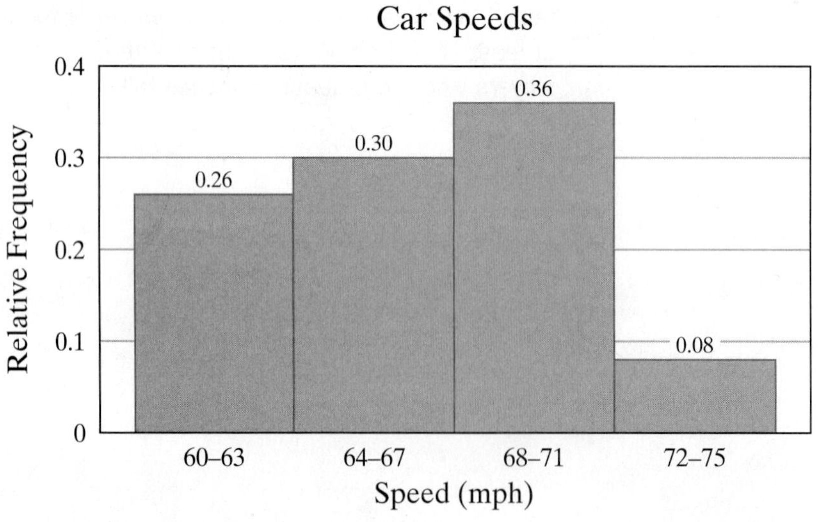

(eText Screens 9.2-1–9.2-25)

a. What percent of cars were driving between 68 and 71 miles per hour?

b. What percent of cars were driving less than 68 miles per hour?

c. Which speed interval had the least number of drivers?

Solutions Try answering these questions on your own using the given histogram. Check your **answers**, or watch this **video** for the complete solutions.

You Try It **Work through this You Try It problem.**

Work Exercises 3 and 4 in this textbook or in the MyMathLab Study Plan.

OBJECTIVE 2 CONSTRUCT A FREQUENCY TABLE

To create a **histogram**, we first create a **frequency table** in which we organize the data by tallying the number of values that fall within given **class intervals.**

Example 3 Constructing a Frequency Table

My video summary

The following data are the number of minutes of television watched on a weekday morning by a sample of 25 college students.

15	30	40	10	65	90	25	40	50
60	45	30	30	75	15	80	30	55
20	20	45	70	60	5	35		

Construct a frequency table for the data using the given class intervals.

Television Time (minutes)	Tally	Frequency
0–19		
20–39		
40–59		
60–79		
80–99		

Solution Starting with the first row, go through the list of data and tally in which class interval each data value lies. Count up the tally marks for each class interval to get the class frequency, also known simply as the **frequency**. Watch this **video** for the complete solution.

Television Time (minutes)	Tally	Frequency
0–19	||||	4
20–39	~~||||~~ |||	8
40–59	~~||||~~ |	6
60–79	~~||||~~	5
80–99	||	2

(eText Screens 9.2-1–9.2-25)

Alternately, we could have constructed a **relative frequency** table as shown in this **popup**.

You Try It **Work through this You Try It problem.**

Work Exercises 5 and 6 in this textbook or in the MyMathLab Study Plan.

OBJECTIVE 3 CONSTRUCT A HISTOGRAM

Before data can be organized in a **frequency table** and a **histogram** constructed, it is necessary to determine the number of **classes** and the **class intervals.** This process is beyond the scope of this text, so for now the class intervals will be provided. To construct a histogram, we first complete the frequency table or **relative frequency table**, then we use the table to construct the individual bars of the histogram.

Example 4 Constructing a Frequency Histogram

The following data are the commute distances, in miles, to work for a sample of 30 adults.

5	10	20	35	5	23	14	45	4	8
47	13	11	2	30	12	40	10	16	13
7	4	17	10	20	24	25	30	15	21

Construct a frequency histogram using the given class intervals.

Commute Distance (miles)	Tally	Frequency
1–8		
9–16		
17–24		
25–32		
33–40		
41–48		

Solution Starting with the first row, go through the list of data and tally in which **class interval** each data value lies. Count up the tally marks for each interval to get the class frequency.

Commute Distance (miles)	Tally	Frequency
1–8	𝍸 ||	7
9–16	𝍸 𝍸	10
17–24	𝍸 |	6
25–32	|||	3
33–40	||	2
41–48	||	2

(eText Screens 9.2-1–9.2-25)

For each class interval, construct a bar using the class interval for the width and the **frequency** for the height. Remember to make the bars of the **histogram** touch, unless a class has a frequency of 0.

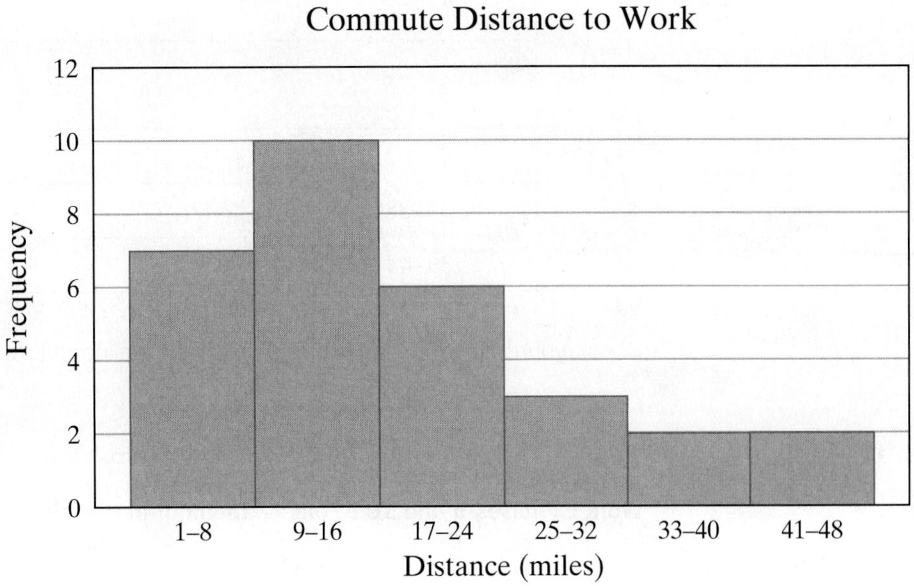

You Try It **Work through this You Try It problem.**

Work Exercises 7 and 8 in this textbook or in the MyMathLab Study Plan.

Example 5 Constructing a Relative Frequency Histogram

My video summary

The following data are the highway gas mileages, in miles per gallon, for a sample of 2012 model year compact cars. (*Source:* fueleconomy.gov)

31	31	34	22	29	35	21	36	30	18
28	44	32	27	42	32	29	33	36	40

Construct a relative frequency **histogram** for the data using the given **class intervals**.

Gas Mileage (mpg)	Tally	Frequency	Relative Frequency
15–20			
21–26			
27–32			
33–38			
39–44			

Solution Begin by completing the table below, then use the relative frequencies and class intervals to create a histogram (see **Figure 2** for an illustration.)

(eText Screens 9.2-1–9.2-25)

Gas Mileage (mpg)	Tally	Frequency	Relative Frequency
15–20	|	1	$\frac{1}{20} = 0.05$
21–26	||	2	$\frac{2}{20} = 0.10$
27–32	𝍸 ||||		
33–38	𝍸		
39–44	|||		

Check your **answer**, or watch this **video** for the complete solution.

You Try It **Work through this You Try It problem.**

Work Exercises 9 and 10 in this textbook or in the MyMathLab Study Plan.

9.2 Exercises

In Exercises 1–4, use the given histogram to answer the questions.

You Try It 1. **Football Player Weight** The following histogram shows the weights, in pounds, for the members of a college football team. Use the histogram to answer the questions.

College Football Player Weights

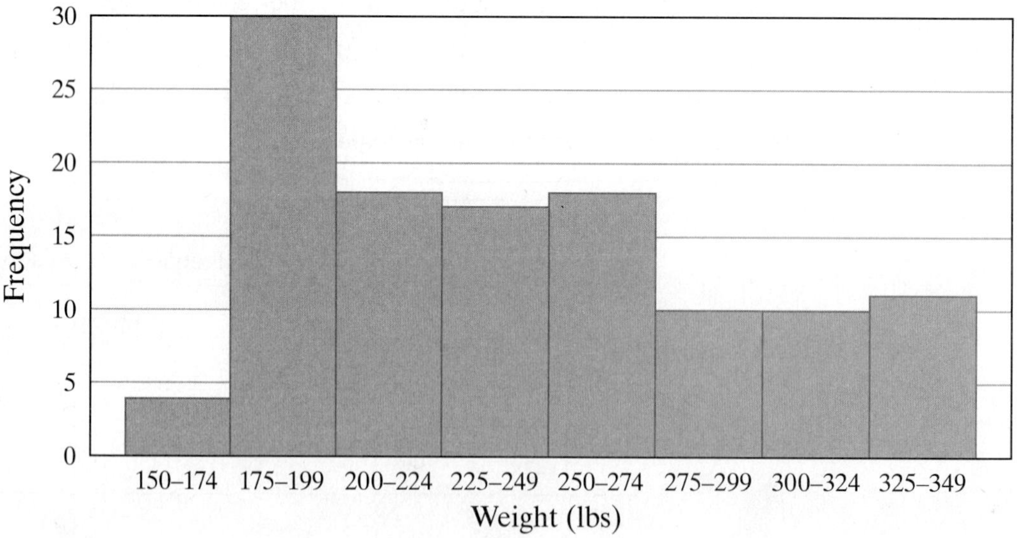

a. Which weight range has the greatest number of players?

b. Which weight range has the least number of players?

c. How many players weighed between 250 and 274 pounds?

d. How many players weighed less than 200 pounds?

e. How many players weighed at least 275 pounds?

2. **High Temperature** The following histogram shows the high temperatures, in degrees Fahrenheit, for March 2012 in Phoenix, AZ. Use the histogram to answer the questions.

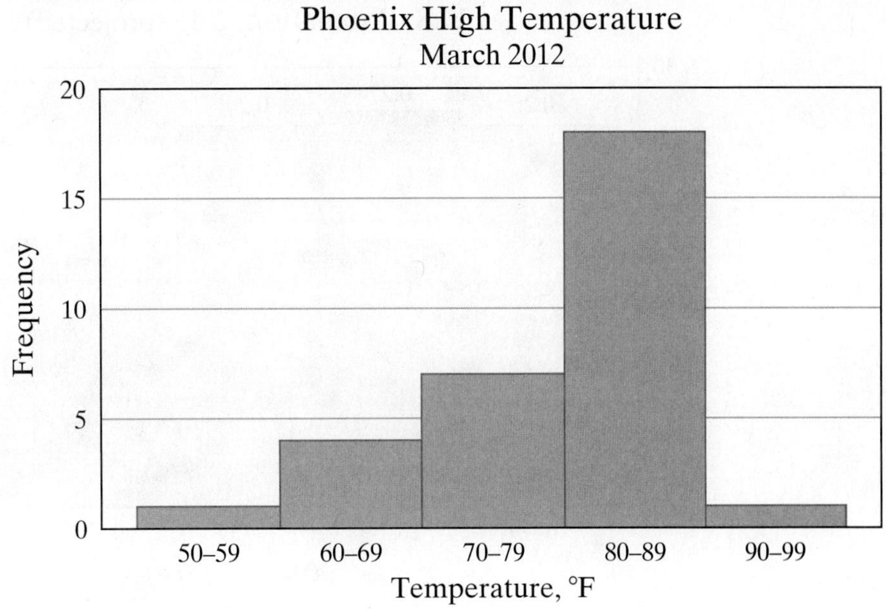

a. How many days had a high temperature between 70°F and 79°F?
b. How many days had a high temperature less than 90°F?
c. How many days had a high temperature of at least 70°F?
d. Which temperature range was most frequent?

You Try It

3. **ACT Scores** The following histogram shows the distribution of ACT Math scores in 2011 for the state of Illinois. Use the histogram to answer the questions. (*Source:* act.org)

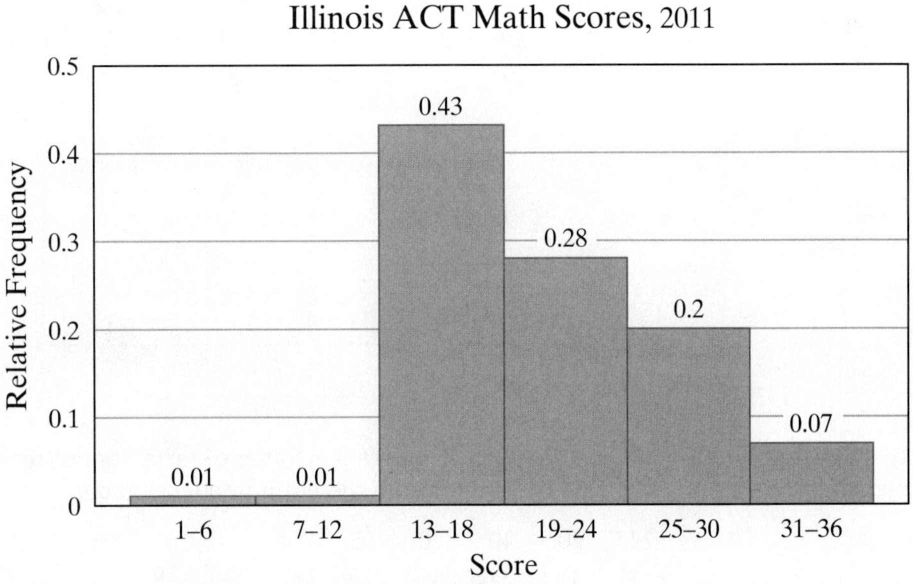

a. What percent of students scored between 19 and 24?
b. Which interval had the largest number of scores?
c. What percent of students scored a 25 or higher?

(*eText Screens 9.2-1–9.2-25*)

4. **U.S. Population** The following histogram shows the projected age distribution in the U.S. for the year 2015. Use the histogram to answer the questions. (*Source:* U.S. Census Bureau)

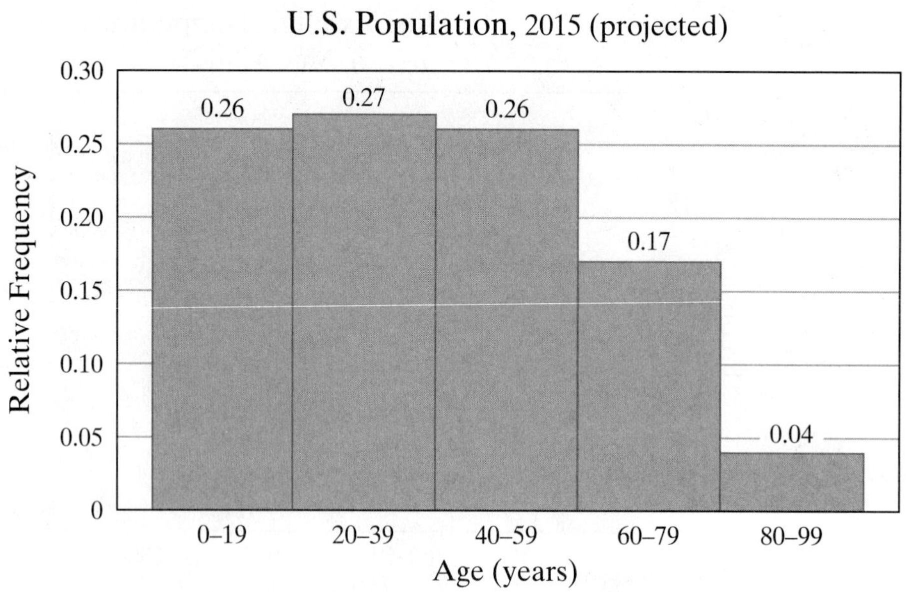

a. What percent of the population is projected to be between 60 and 79 years of age in 2015?

b. Which age interval is projected to be the largest?

c. What percent of the population is projected to be less than 40 years of age in 2015?

In Exercises 5 and 6, use the given data to complete the frequency table.

You Try It 5. **Unemployment Length** The following data are the number of weeks a sample of unemployed persons has been unemployed. Use the data to complete the given frequency table. (*Source:* Monthly Labor Review, March 2012)

3	10	12	20	23	28	1	12	7	35
29	17	33	4	9	27	30	2	28	36

Weeks unemployed	Tally	Frequency
Less than 5		
5–14		
15–26		
27 or more		

6. **Text Messages** The following data are the average number of texts sent or received per day by a sample of teenagers. Use the data to complete the given frequency table.

15	20	40	70	105	174	0	62	206	12
80	31	234	255	310	181	230	8	300	415
16	46	93	25	120					

(eText Screens 9.2-1–9.2-25)

Daily Texts	Tally	Frequency
0–49		
50–149		
150–249		
250–349		
350 or more		

In Exercises 7 and 8, construct a frequency histogram for the given data.

You Try It 7. **Commute Times** The following data are the commute times, in minutes, to work for a sample of 28 adults.

34	28	18	24	21	19	30	19	24	20
31	28	29	22	28	36	22	29	24	28
16	17	21	28	26	21	23	27		

Construct a frequency histogram using the given class intervals.

Commute Time (minutes)	Tally	Frequency
15–18		
19–22		
23–26		
27–30		
31–34		
35–38		

8. **Hours Worked** The following data are the number of hours worked per week by a sample of employed full-time community college students.

11	15	32	27	12	22	13	13	31	33
25	18	19	18	27	29	19	17	14	8
26	25	34	28	16	10	24	27	12	10

Construct a frequency histogram for the data using the given class intervals.

Weekly hours worked	Tally	Frequency
6–10		
11–15		
16–20		
21–25		
26–30		
31–35		

In Exercises 9 and 10, construct a relative frequency histogram for the given data.

You Try It

9. **Gas Mileage** The following data are the highway gas mileages, in miles per gallon, for a sample of 2012 model year mid-size cars. (*Source:* fueleconomy.gov)

31	31	34	22	29	35	21	36	30	18
28	44	32	27	42	32	29	33	36	40

Construct a relative frequency histogram for the data using the given class intervals.

Gas Mileage (mpg)	Tally	Frequency	Relative Frequency
15–20			
21–26			
27–32			
33–38			
39–44			

10. **Luggage Weight** The following data are the weights, in pounds, for a sample of 25 suitcases on an airline flight.

28	35	37	41	32	36	34	37	39	41
36	33	42	38	38	40	34	29	38	38
40	36	36	31	45					

Use the given class intervals to construct a relative frequency histogram for the data.

Weight (lbs)	Tally	Frequency	Relative Frequency
25–28			
29–32			
33–36			
37–40			
41–44			
45–48			

9.3 Counting

THINGS TO KNOW

Before working through this topic, be sure you are familiar with the following concepts:

	VIDEO	ANIMATION	INTERACTIVE
You Try It 1. Multiply Whole Numbers (Topic 1.4, **Objective 1**)	▣		
You Try It 2. Evaluate Exponential Expressions (Topic 1.6, **Objective 2**)	▣		

(eText Screens 9.3-1–9.3-13)

OBJECTIVES

1 Use a Tree Diagram to Count Outcomes

2 Use the Fundamental Counting Principle to Count Outcomes

OBJECTIVE 1 USE A TREE DIAGRAM TO COUNT OUTCOMES

Figure 3

Consider the advertisement shown in Figure 3. How many possibilities are there for a sub? Each different sub is considered an **outcome** and is the result of selecting a meat, a cheese, a topping, and a condiment. In this topic, we consider two ways to determine the number of possible outcomes when performing a series of tasks. First we look at **tree diagrams.**

To create a tree diagram, you create a **branch** for each possible choice of the first task. From each of these values, create a branch for each possible choice of the second task. And so on for each additional task. The different outcomes are found by following each unique path in the diagram. For example, the following illustrates how to construct a tree diagram for the possible outcomes when tossing two coins.

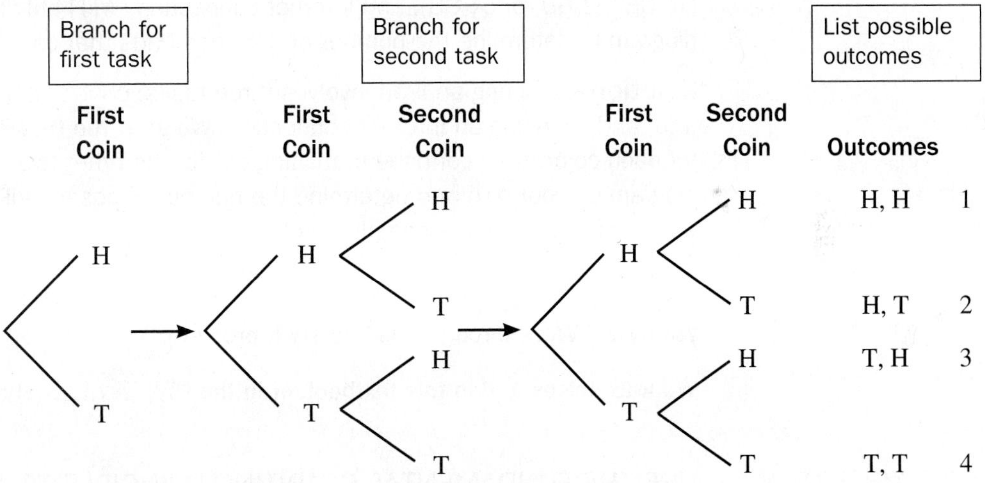

There are four possible outcomes when tossing two coins.

Example 1 Using a Tree Diagram to Count the Number of Lunches

For lunch, Lindsay needs to select a soup and sandwich from the following menu.

Soup	Sandwich
Chicken Noodle	Turkey
Broccoli Cheese	Ham
Vegetable Beef	Roast Beef
	Veggie

(eText Screens 9.3-1–9.3-13)

Draw a **tree diagram** to determine the number of possible soup and sandwich combinations.

Solution Selecting a lunch involves two tasks: choosing a soup and choosing a sandwich. We start the tree diagram by creating a branch for each soup type. We then continue branching out for the choice of sandwich.

From the diagram, we count 12 different soup and sandwich combinations that are possible.

We could also have started by choosing the sandwich. The 12 outcomes would be the same.

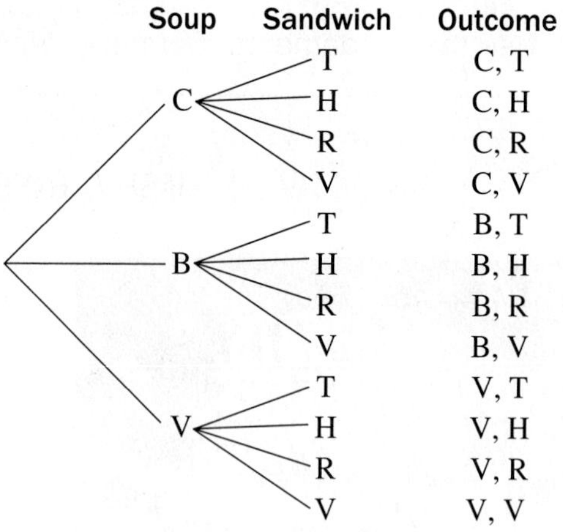

You Try It **Work through this You Try It problem.**

Work Exercises 1–3 in this textbook or in the MyMathLab Study Plan.

Example 2 Using a Tree Diagram to Count the Number of iPads®

My video summary

A customer purchasing an iPad has a choice of color (black or white), storage size (16 GB, 32 GB, or 64 GB), and Internet connection (Wi-Fi only or Wi-Fi + 4G). Draw a tree diagram to determine the number of different iPads that can be purchased.

Solution Selecting an iPad involves three tasks: choosing a color, choosing a storage size, and choosing an Internet connection. We start the **tree diagram** by creating a branch for each color, then continue branching out for the other two choices. Try to draw the diagram on your own and determine the number of possible iPads. View the **answer**, or watch this **video** for the complete solution.

You Try It **Work through this You Try It problem.**

Work Exercises 4–6 in this textbook or in the MyMathLab Study Plan.

OBJECTIVE 2 USE THE FUNDAMENTAL COUNTING PRINCIPLE TO COUNT OUTCOMES

Tree diagrams provide a visual display of possible **outcomes**. However, they can be large and tedious depending on the number of tasks and the number of choices for each task.

In Example 2 we had three tasks. There were two choices for the first task (choosing a color), 3 choices for the second (choosing a storage size), and 2 for the third (choosing a Internet connection). Multiplying the number of possible choices for each task together gives

$$2 \cdot 3 \cdot 2 = 12.$$

Notice that this is equal to the total number of different iPads that could be ordered. This illustrates a general result known as the **fundamental counting principle**.

Fundamental Counting Principle

The number of outcomes for a series of tasks can be found by multiplying the number of ways each task can be completed.

For example, if there are n_1 ways to do task 1 and n_2 ways to do task 2, then there are $n_1 \cdot n_2$ ways to complete the two tasks. This result extends to any number of tasks.

TIP Note that the fundamental counting principle only gives the number of outcomes, not the actual outcomes. One advantage of a tree diagram is that the actual outcomes can all be determined from the diagram.

Example 3 Using the Fundamental Counting Principle to Count the Number of Sandwiches

Using the information from **Figure 3**, determine the number of possible sandwiches if you select one meat, one cheese, one topping, and one condiment.

Solution Creating a sandwich involves four tasks: choosing a meat, choosing a cheese, choosing a topping, and choosing a condiment. There are 11 types of meat, 4 types of cheese, 10 types of toppings, and 11 types of condiments. Using the **fundamental counting principle**, we multiply the number of choices for each task together.

$$\begin{aligned} 11 \cdot 4 \cdot 10 \cdot 11 &= 44 \cdot 10 \cdot 11 \\ &= 440 \cdot 11 \\ &= 4840 \end{aligned}$$

There are 4840 possible sandwiches.

You Try It **Work through this You Try It problem.**

Work Exercises 7–10 in this textbook or in the MyMathLab Study Plan.

Example 4 Using the Fundamental Counting Principle to Count the Number of Computers

A customer ordering a computer online can select between 3 processors (3.60 GHz, 3.50 GHz, 3.10 GHz), 3 hard drive sizes (250 GB, 750 GB, 1 TB), 2 video cards (1 GB, 2 GB), 4 optical drives (DVD player, DVD writer, Blu-ray player, Blu-ray writer), and 2 batteries (6 cell, 9 cell). How many different ways could the customer order a computer?

Solution There are five tasks to complete to order a computer. Using the **fundamental counting principle**, multiply the number of ways to perform each task to determine the total number of ways a computer could be ordered.

$$\underbrace{\quad\;\;}_{\text{processors}} \cdot \underbrace{\quad\;\;}_{\text{hard drives}} \cdot \underbrace{\quad\;\;}_{\text{video cards}} \cdot \underbrace{\quad\;\;}_{\text{optical drives}} \cdot \underbrace{\quad\;\;}_{\text{batteries}}$$

Check your **answer**, or view this **popup** for the solution.

You Try It **Work through this You Try It problem.**

Work Exercises 11–14 in this textbook or in the MyMathLab Study Plan.

(eText Screens 9.3-1–9.3-13)

Example 5 Using the Fundamental Counting Principle to Count the Number of Passwords

My video summary

Use the **fundamental counting principle** to determine the number of possible five-character passwords with the following conditions:

- the first three characters must be letters
- the letters O and L cannot be used
- the last two characters must be digits
- uppercase and lowercase letters are treated as the same
- no letter may be repeated

Solution There are five tasks to complete to choose a password (three letters and two digits). Try to determine the number of passwords on your own. Check your **answer**, or watch this **video** for the complete solution.

You Try It **Work through this You Try It problem.**

Work Exercises 15–18 in this textbook or in the MyMathLab Study Plan.

9.3 Exercises

In Exercises 1–6, use a tree diagram to count the number of outcomes.

You Try It

1. **Chicken Nugget Orders** Chicken Palace offers three sizes of nuggets—4 piece, 6 piece, and 10 piece, and four types of sauces—BBQ, Salsa, Honey Mustard, and Ranch. Assuming a customer always orders a sauce, draw a tree diagram and use it to determine the number of possible nugget orders.

2. **Twister®** On a single turn in the game of Twister, players must move one of four body parts—Right foot (RF), Left foot (LF), Right hand (RH), or Left hand (LH), to one of four colors—Red (R), Yellow (Y), Blue (B), Green (G). Use a tree diagram to list all the possible moves on a single turn.

3. **Painting Rooms** Sally wants to paint her bedroom and her living room and has four colors to choose from for each room (pink, yellow, light blue, and light green). Draw a tree diagram to determine the different ways she can paint the two rooms. How many ways can she paint the rooms so that they are different colors?

You Try It

4. **Buying a Car** Geoff is buying a new car and needs to decide on a color (silver, red, or blue), whether or not to have a sunroof, and the type of interior (cloth or leather). Draw a tree diagram to determine the number of possible different cars from which Geoff may choose.

5. **Lunch Combos** Bob's Burger Bungalow offers the following on its lunch menu:

   Entrée: hamburger (H), fish sandwich (F), chicken sandwich (C)
   Drink: soda (S), water (W), tea (T)
   Side: french fries (FF), onion rings (O)

   A combo meal is an entrée, drink, and a side. Use a tree diagram to list all the possible combo meals.

6. **Tossing Coins** Phil tosses a coin three times. Draw a tree diagram to determine all the possible outcomes. How many ways can he toss three coins and get exactly 2 heads?

(eText Screens 9.4-1–9.4-13)

In Exercises 7–18, use the fundamental counting principle to count the number of outcomes.

You Try It 7. **Team Captain** A baseball team has 17 members and must select a captain and an assistant. How many ways can this be done?

8. **Dressing Up** Tom has 3 shirts, 4 pants, and 7 ties that he can wear on a date. How many outfits can he choose from?

9. **Pizza Parlor** Pepi's Pizza Palace offers a $4 medium pizza. If you can choose from one of seven toppings, one of three cheeses, one of two kinds of sauce, and one of four types of crust, how many different pizzas are possible?

10. **Radio Station Call Sign** A radio station is to be assigned a four-letter call sign. If the first letter must be K or W, how many four-letter call signs are possible?

You Try It 11. **Car Purchase** Darius wants to purchase a new car. He must choose one of 7 colors, one of 3 transmission types, one of 6 trim packages, one of 2 interiors, and one of 3 wheel packages. How many different cars are possible?

12. **Waiting in Line** How many different ways can 8 people stand in line at a grocery store?

13. **Quiz Guessing** A six-question multiple choice quiz has 4 answers per question. Carolyn decides to randomly select which choice will be correct for each question. How many different answer keys are possible?

14. **Making Music** How many ten-note 'songs' can be made from the notes A, B, C, D, E, F, G?

You Try It 15. **License Plates** Keegan is getting a vanity license plate and must select five characters. The first three must be a letter, and the last two must be numbers. If the letters O, I, and Z, and the number 0 cannot be used, how many vanity plates are possible?

16. **Word Game** Jesse wants to make a four-letter word in a Hangman game. If he has 8 distinct letters to choose from, how many four-letter words can he make (assuming all letter combinations are words)?

17. **Password Selection** Logan needs a 4-letter password for a computer program. The letters I, O, and Z cannot be used, and two consecutive letters cannot be the same. How many passwords are possible if there is no difference between lowercase and uppercase letters?

18. **Sorority Officers** Payton's sorority has 7 seniors, 5 juniors, 6 sophomores, and 4 freshmen. They want to choose a president, a vice-president, and social secretary. The president must be a senior and the vice-president cannot be a freshman. How many different ways can they pick the officers?

9.4 Probability

THINGS TO KNOW

Before working through this topic, be sure you are familiar with the following concepts:

	VIDEO	ANIMATION	INTERACTIVE
You Try It 1. Add Several Whole Numbers (Topic 1.3, **Objective 3**)	▤		
You Try It 2. Change Fractions or Mixed Numbers into Decimals (Topic 5.5, **Objective 1**)	▤		

(eText Screens 9.4-1–9.4-13)

OBJECTIVES

1 Estimate the Probability of an Event

2 Compute the Probability of an Event for Equally Likely Outcomes

OBJECTIVE 1 ESTIMATE THE PROBABILITY OF AN EVENT

When events in life are uncertain, we often assign a chance or likelihood to the event. We call this the **probability** of the event. For example, we might say there is an 80% chance of rain, or the probability of rain is 0.80. In this topic, we will look at two ways of assigning probabilities: *empirically* (using observed data) and *theoretically* (using equally likely outcomes). A third approach, not covered in this text, subjectively assigns probabilities based on an educated guess from personal experience.

A **probability experiment**, or simply an **experiment**, is any task or process in which the result is not known in advance. Some examples are flipping a coin, drawing a name from a hat, or rolling a pair of dice. Each instance of the experiment is called a **trial**. As we saw in the counting problems of the previous topic, each possible result of an experiment is called an **outcome**. An **event** is a combination of possible outcomes. For example, if we toss two coins the four possible outcomes are

HH, HT, TH, and TT, with H for heads and T for tails.

One event might be *two heads*, an outcome of HH. Another event might be *one head*, with two possible outcomes HT and TH since both outcomes contain one head.

Probability can be expressed as a fraction, a decimal, or a percent. In fraction or decimal form, all probabilities must be between 0 and 1 (inclusive). A probability of 0 means an event will not happen and a probability of 1 means that the event will definitely happen. The closer a probability is to 0, the less likely the event. Similarly, the closer the probability is to 1, the more likely the event.

When we estimate the probability of an event, we are looking for the **empirical probability.** This is found by dividing the number of times the event occurs (or is observed) by the number of trials of the experiment.

Estimated Probability of an Event (Empirical Probability)

$$\text{estimated probability of an event} = \frac{\text{number of times the event is observed}}{\text{number of trials of the experiment}}$$

Example 1 Estimating the Probability of Making a Free Throw

A basketball player has made 28 of his last 32 free throws. Estimate the **probability** that he will make his next free throw.

Solution To estimate the probability that the basketball player will make his next free throw, we divide the number of free throws made by the number attempted.

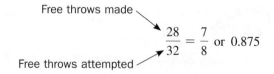

Free throws made → 28; Free throws attempted → 32

$$\frac{28}{32} = \frac{7}{8} \text{ or } 0.875$$

(eText Screens 9.4-1–9.4-13)

The estimated probability that the basketball player will make his next free throw is 0.875 or 87.5%.

You Try It **Work through this You Try It problem.**

Work Exercises 1–4 in this textbook or in the MyMathLab Study Plan.

Example 2 Estimating the Probability of a Sum

My video summary

Mia rolls two six-sided dice 20 times. For each roll, she sums the two values.

6	9	3	4	9	8	4	7	7	4
7	7	10	10	7	6	5	7	7	5

Estimate the **probability** that rolling two six-sided dice results in

a. a sum of 2.

b. a sum of 9 or more.

Solutions

a. We **estimate** the probability of obtaining a **sum** of 2 by counting the number of observations with a sum of 2 and dividing by the number of observations. Since none of the observed sums equals 2, our estimated probability is

Number of sums equal to 2 → $\frac{0}{20} = 0$ ← Number of observations

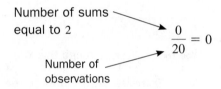

b. A sum of 9 or more means the sum could be 9, 10, 11, or 12. Note that the highest possible sum would be $6 + 6 = 12$. Estimate the probability of getting a sum of 9 or more by counting the number of **outcomes** with a sum of 9 or more and divide by the number of **trials**. Check your **answer**, or watch this **video** for the complete solution to both parts.

You Try It **Work through this You Try It problem.**

Work Exercises 5–8 in this textbook or in the MyMathLab Study Plan.

OBJECTIVE 2 COMPUTE THE PROBABILITY OF AN EVENT FOR EQUALLY LIKELY OUTCOMES

Empirical probability relies on observed **outcomes** to estimate the **probability** of an **event**. On the other hand, **theoretical probability** takes *all* outcomes into account when finding the probability of an event.

The simplest case of theoretical probability involves **equally likely outcomes**, which means that each outcome has the exact same chance of occurring. This is what is referred to when we say a coin or die is *fair*. A *fair* coin is one in which both sides (heads and tails) have the same chance of occurring. Likewise, a *fair* six-sided die (singular for dice) is one in which all six sides have the same chance of occurring.

For equally likely outcomes, we find the probability of an event by counting the number of outcomes that satisfy the event and dividing by the total number of possible outcomes.

(eText Screens 9.4-1–9.4-13)

Probability of an Event for Equally Likely Outcomes (Theoretical Probability)

If all outcomes are equally likely, then the probability of an event is given by

$$\text{probability of an event} = \frac{\text{number of outcomes satisfying the event}}{\text{number of possible outcomes}}.$$

TIP To compute theoretical probabilities, we must know all the outcomes. Sometimes a **tree diagram** is helpful in determining all the possible outcomes of an **experiment**.

Example 3 Computing the Probability of a Yellow Marble

A bag of marbles contains 7 yellow marbles, 10 blue marbles, 6 red marbles, and 2 green marbles. What is the probability that a marble selected at random is yellow?

Solution There are 7 marbles in the bag that correspond to the event "yellow marble," and there are $7 + 10 + 6 + 2 = 25$ marbles in the bag. Only one marble is drawn, so each marble is a possible **outcome**. The **probability** that a randomly selected marble is yellow is

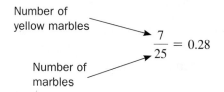

Number of yellow marbles →

$$\frac{7}{25} = 0.28$$

Number of marbles →

You Try It Work through this You Try It problem.

Work Exercises 9–12 in this textbook or in the MyMathLab Study Plan.

Example 4 Computing the Probability of a Sum

My video summary

Find the **probability** of each of the following **events** when rolling a pair of six-sided dice.

a. The sum is 2.

b. The sum is 9 or more.

Solutions

a. We start by determining all the possible **outcomes** for rolling two six-sided dice. From the **fundamental counting principle**, we know that there are $6 \cdot 6 = 36$ possible outcomes. The following diagram shows all 36 possibilities.

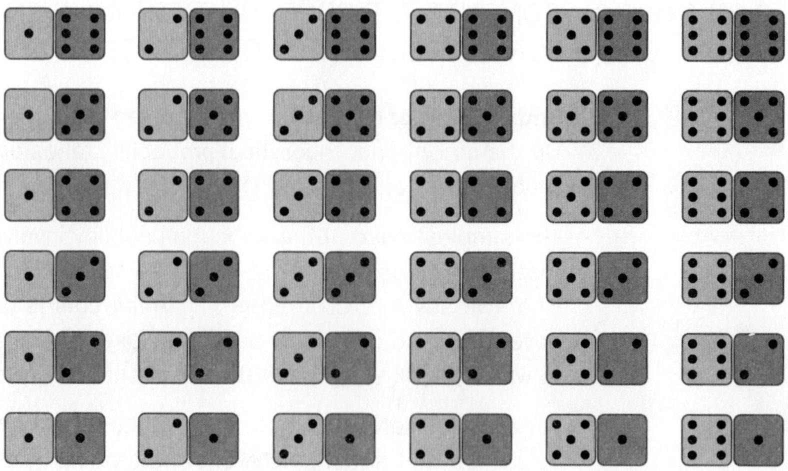

(eText Screens 9.4-1–9.4-13)

There is only one possible outcome with a sum of 2. That is when a 1 appears on both dice. Therefore,

$$\text{probability of a sum of } 2 = \frac{1}{36} \approx 0.028 \text{ or } 2.8\%.$$

b. Use the **diagram** from part a to find the number of **outcomes** with a sum of 9 or more, then divide by the number of outcomes. Check your **answer**, or watch this video for the complete solution to both parts.

You Try It **Work through this You Try It problem.**

Work Exercises 13–16 in this textbook or in the MyMathLab Study Plan.

9.4 Exercises

In Exercises 1–8, estimate the probability of an event (empirical probability).

You Try It 1. **Batting Average** If a baseball player made a hit 4 times in his last 14 at-bats, estimate the probability that the player will get a hit on his next at-bat.

2. **Quality Control** A quality control technician collects a sample of 50 electrical components produced from a certain machine and finds 3 defective components. Use this information to estimate the probability that a randomly selected electrical component produced from this machine is defective.

3. **Gender** A prealgebra class at a college contains 12 male students and 16 female students. Estimate the probability that a randomly selected student at the college is male.

4. **Side Effect** In a clinical trial for a new diabetes medication, 14 of 394 patients receiving the medication reported having headaches as a side effect. Estimate the probability that someone using the medication will experience headaches as a side effect. Round your answer to three decimal places.

You Try It 5. **Smartphone Demographics** A survey of 688 adult smartphone owners found the following:

Education Level	n
Did not finish high school	40
High school graduate (or GED)	169
Some college	171
College graduate	308

Estimate the probability that a smartphone owner has an education level of at least some college.

(*eText Screens 9.4-1–9.4-13*)

6. **e-Reading** A survey of those who own an e-book reader found the following regarding how often they read an e-book on their reader.

How frequently the e-reader is used to read e-books	n
Daily, or almost daily	302
A few times each week	93
A few times each month	61
A few times each year	42
Not at all	24

Estimate the probability that someone who owns an e-book reader uses it to read e-books less than a few times each month. Round your answer to three decimal places if necessary.

7. **Flipping Coins** You toss three fair coins 500 times and obtain the following results for the number of heads.

Number of heads	n
0 heads	115
1 head	143
2 heads	162
3 heads	80

Estimate the probability that when the three coins are tossed, they will all be the same.

8. **Age of Best Actor Winner** The following are the ages of the best actor winner from 2003 to 2012.

29	43	37	38	45
50	48	60	50	39

Estimate the probability that the best actor in a given year will be 40 years or older.

In Exercises 9–16, compute the probability of an event for equally likely outcomes (theoretical probability).

You Try It 9. **M&M Color** Suppose a bag of M&M candies contains 8 brown, 7 red, 8 yellow, 9 green, 11 orange, and 12 blue. If one candy is selected from the bag at random, what is the probability that the candy will be green?

10. **Soft Drinks** A cooler contains 4 cans of root beer, 8 cans of regular cola, 3 cans of diet cola, 4 cans of grape, and 6 cans of orange. If a can is selected at random from the cooler, what is the probability of selecting a grape soda?

11. **Drawing a Card** In a standard deck of 52 cards, there are four suits (diamonds, hearts, clubs, spades) of 13 cards each (2, 3, 4, 5, 6, 7, 8, 9, 10, jack, queen, king, ace). If a single card is drawn from a standard deck, what is the probability of drawing a diamond?

12. **Selecting a Book** A math teacher has a bookshelf containing 10 algebra books, 4 trigonometry books, 3 calculus books, and 5 statistics books. If she selects a book at random, what is the probability that the she selects a trigonometry book?

(eText Screens 9.4-1–9.4-13)

You Try It 13. **Number of Tails** If five fair coins are tossed, what is the probability that there are at least 3 tails?

14. **Even Roll** If a fair twenty-sided die is rolled, what is the probability of obtaining an even number greater than 10?

15. **Marble Color** A bag of marbles contains 5 yellow marbles, 12 blue marbles, 8 red marbles, and 6 green marbles. What is the probability that a marble selected at random is yellow or red?

16. **Roulette Table** An American roulette wheel has slots numbered 1–36 (half red, half black) along with slots for 0 and 00 (in green). Use the following illustration of a roulette betting table to determine the probability of obtaining a red number that is less than 20 on a single spin of the wheel.

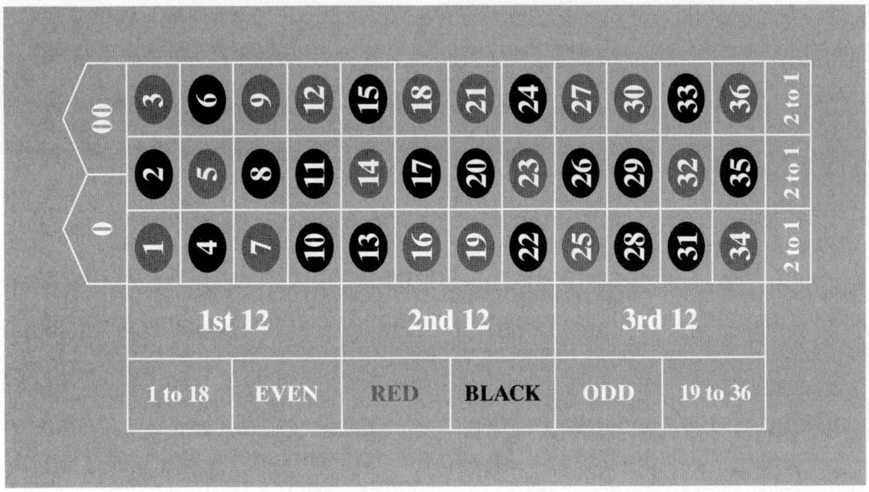

MODULE TEN

Real Numbers and Algebraic Expressions

MODULE TEN CONTENTS

10.1 The Real Number System

THINGS TO KNOW

Before working through this topic, be sure you are familiar with the following concepts:

		VIDEO	ANIMATION	INTERACTIVE
You Try It	1. Write Fractions in Simplest Form (Topic 4.2, **Objective 6**)	▣		
You Try It	2. Perform Conversions Among Percents, Decimals and Fractions (Topic 7.1, **Objective 5**)			▣

OBJECTIVES

1 Classify Real Numbers

2 Plot Real Numbers on a Number Line

3 Find the Opposite of a Real Number

4 Find the Absolute Value of a Real Number

5 Use Inequality Symbols to Order Real Numbers

6 Translate Word Statements Involving Inequalities

OBJECTIVE 1 CLASSIFY REAL NUMBERS

A **set** is a collection of objects. Each object in a set is called an **element** or a **member** of the set. Typically, capital letters name sets, and braces { } group the list of elements in a set. For example, the days of the week form a set. If we name this set D, we write:

$$D = \{\text{Sunday, Monday, Tuesday, Wednesday, Thursday, Friday, Saturday}\}.$$

If S represents the set of single-digit numbers, then

$$S = \{0, 1, 2, 3, 4, 5, 6, 7, 8, 9\}.$$

A set that contains no elements is called an **empty set** or **null set**. We represent this set with a pair of **empty braces** $\{\,\}$ or the **null symbol** $\varnothing$. For example, let A represent the set of all days of the week that begin with the letter B. Because there are no such days, we write $A = \{\,\}$ or $A = \varnothing$.

The symbol $\{\varnothing\}$ is not the empty set. It is the set containing the element $\varnothing$. Be sure to use the symbols $\{\;\}$ or $\varnothing$ to represent the empty set, not $\{\varnothing\}$.

Both D and S are **finite sets**, meaning that each set has a fixed number of elements (7 days, 10 single-digit numbers). When sets are **infinite**, they have an unlimited number of elements. For example, the set of positive **even numbers** is written as

$$E = \{2, 4, 6, 8, \ldots\}.$$

The three dots, . . . , are called an *ellipsis* and indicate the given pattern continues. We encounter a variety of numbers every day. A plane might carry 120 passengers,and a netbook may cost \$349.99. Your friend may pay $\frac{1}{2}$ the rent for your apartment.

The numbers 120, 349.99, and $\frac{1}{2}$ are examples of *real numbers*. Let's look at the smaller sets of numbers that make up the set of real numbers.

When counting books on a shelf, we count them as 1, 2, 3, 4, and so on. These values form the set of *natural numbers*, or *counting numbers*, represented by the symbol $\mathbb{N}$.

Definition   Natural Numbers

A **natural number**, or **counting number**, is an element of the set $\mathbb{N} = \{1, 2, 3, 4, 5, \ldots\}$.

Adding the element 0 to the set of natural numbers gives us the set of *whole numbers*, represented by the symbol $\mathbb{W}$.

Definition   Whole Numbers

A **whole number** is an element of the set $\mathbb{W} = \{0, 1, 2, 3, 4, 5, \ldots\}$.

Notice that every natural number is also a whole number because all natural numbers are included within the set of whole numbers.

Adding the negative numbers, $-1, -2, -3, -4, \ldots$ to the set of whole numbers forms the set of *integers*, represented by $\mathbb{Z}$.

Definition   Integers

An **integer** is an element of the set $\mathbb{Z} = \{\ldots, -4, -3, -2, -1, 0, 1, 2, 3, 4, \ldots\}$.

Notice that every whole number is also an integer. What does this tell us about **natural numbers**? View the **answer**.

Integers are still "whole" in the sense that they are not **fractions.** Adding fractions to the set of integers gives us the set of *rational numbers*, represented by $\mathbb{Q}$.

A **rational number** can be expressed as the **quotient** of two **integers** p and q, written as $\frac{p}{q}$, when $q \neq 0$. $\frac{p}{q}$ is called the **fraction form** of the rational number such that p is the **numerator** and q is the **denominator**. Examples of rational numbers include $\frac{3}{4}$, $-\frac{7}{8}$, and $\frac{13}{11}$.

> **Definition Rational Numbers**
>
> A **rational number** is a number that can be written as a fraction, $\frac{p}{q}$, where p and q are integers and $q \neq 0$.

Rational numbers can be written in **decimal form**. When a **fraction** is written as a **decimal**, the result is either a **terminating decimal**, such as $\frac{3}{4} = 0.75$ and $-\frac{7}{8} = -0.875$, or a **repeating decimal**, such as $\frac{5}{6} = 0.8333\ldots = 0.8\overline{3}$ and $\frac{13}{11} = 1.181818\ldots = 1.\overline{18}$.

Every **integer** is also a **rational number** because we can write the integer with a denominator of 1. For example, $5 = \frac{5}{1}$, $0 = \frac{0}{1}$, and $-8 = \frac{-8}{1}$. What does this tell us about **whole numbers** and **natural numbers**? View the **answer**.

Not all numbers can be written as the **quotient** of two **integers**. Such numbers belong to the set of *irrational numbers*.

> **Definition Irrational Numbers**
>
> An **irrational number** is a number that cannot be written as a fraction, $\frac{p}{q}$, where p and q are integers. In its decimal form, an irrational number does not repeat or terminate.

Square roots and other numbers expressed with **radicals** are often **irrational numbers**. For example, $\sqrt{3} = 1.7320508\ldots$ and $\sqrt{2} = 1.41421356\ldots$ have decimal forms that continue indefinitely without a repeating pattern, so each is an irrational number. The number $\pi = 3.14159265\ldots$ is another example of an irrational number.

CAUTION: Not all numbers expressed with radicals are irrational numbers. For example, $\sqrt{9} = 3$ and $\sqrt{225} = 15$ are rational numbers. In fact, $\sqrt{9} = 3$ and $\sqrt{225} = 15$ are also natural numbers, whole numbers, and integers.

Combining the set of **rational numbers** with the set of irrational numbers forms the set of *real numbers*, represented by $\mathbb{R}$.

> **Definition Real Numbers**
>
> A **real number** is any number that is either rational or irrational.

(eText Screens 10.1-1–10.1-29)

Figure 1 shows the following relationships involving the set of real numbers:

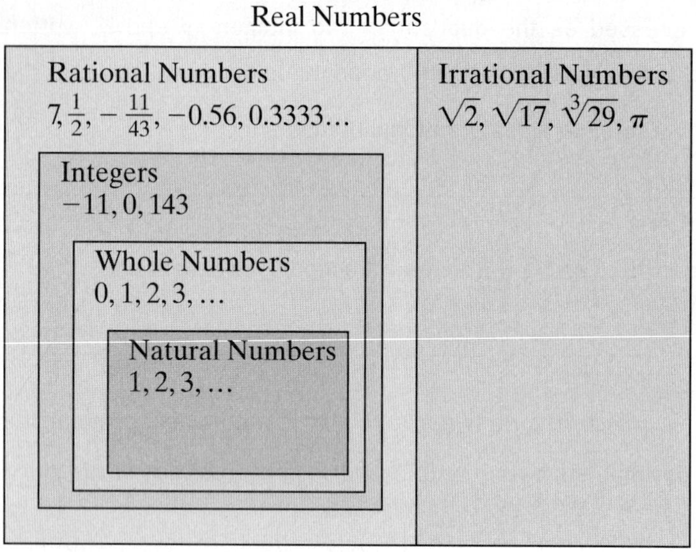

- Every real number is either rational or irrational.
- All natural numbers are whole numbers.
- All whole numbers are integers.
- All integers are rational numbers.

Figure 1 The set of real numbers

Example 1 Classifying Real Numbers

Classify each **real number** as a **natural number**, **whole number**, **integer**, **rational number**, and/or **irrational number**. Each number may belong to more than one **set**.

a. 8 **b.** -4.8 **c.** $\sqrt{10}$ **d.** -7

e. $-\frac{4}{7}$ **f.** $\sqrt{25}$ **g.** 0 **h.** $3.\overline{45}$

Solutions

a. Because 8 is one of the natural (or counting) numbers, we can use **Figure 1** to classify it as a natural number, a whole number, an integer, and a rational number.

b. Because -4.8 is **negative**, it cannot be a natural number or a whole number. Also, because it contains a fractional portion $(.8)$, it cannot be an integer. Notice that -4.8 is a **terminating decimal**. Therefore, we can classify -4.8 as a rational number.

c. $\sqrt{10} = 3.1622776\ldots$ is not a terminating or **repeating decimal**, so it cannot be a rational number. Therefore, $\sqrt{10}$ is an irrational number.

My video summary

d–h. Try classifying these numbers on your own. Check your **answers**, or watch this **video** for the complete solutions.

Before classifying a **real number**, check to see if it can be **simplified**. For example, the **rational number** $\frac{6}{2}$ can be simplified as $\frac{6}{2} = 3$, so it is also an **integer**, a **whole number**, and a **natural number**.

You Try It **Work through this You Try It problem.**

Work Exercises 1–8 in this textbook or in the MyMathLab Study Plan.

OBJECTIVE 2 PLOT REAL NUMBERS ON A NUMBER LINE

The **real number line**, also known simply as the number line, is a **graph** that represents the **set** of all **real numbers.** Every point on the number line corresponds to exactly one real number, and every real number corresponds to exactly one point on the number line. The point is called the **graph** of that real number, and the real number corresponding to a point is called the **coordinate** of that point.

The point 0 is called the **origin** of the number line. Numbers located to the left of the origin are **negative numbers**, and numbers located to the right of the origin are **positive numbers.** The number 0 is neither negative nor positive. Figure 2 shows us that any two **consecutive integers** are located one unit apart. The word *consecutive* means that the integers follow in order, one after another, without interruption.

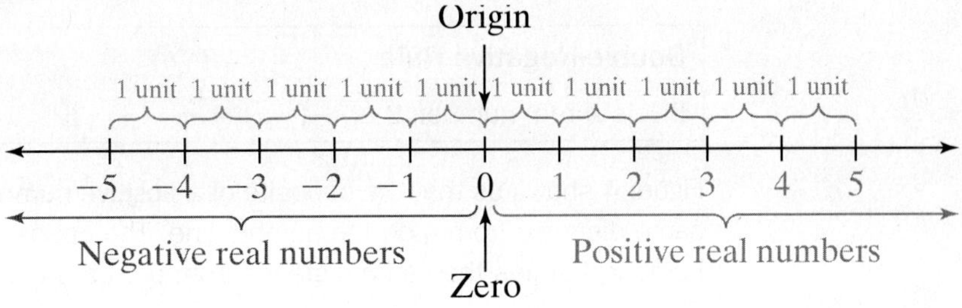

Figure 2 The real number line

We **plot**, or **graph**, a real number by placing a solid circle (●) at its location on the **number line.**

Example 2 Plotting Real Numbers on the Number Line

Plot the following **set** of numbers on the **number line**.

$$\left\{-3, -\frac{3}{2}, 1, 2.25\right\}$$

Solution We plot each number by placing a solid circle at its location on the number line. See Figure 3.

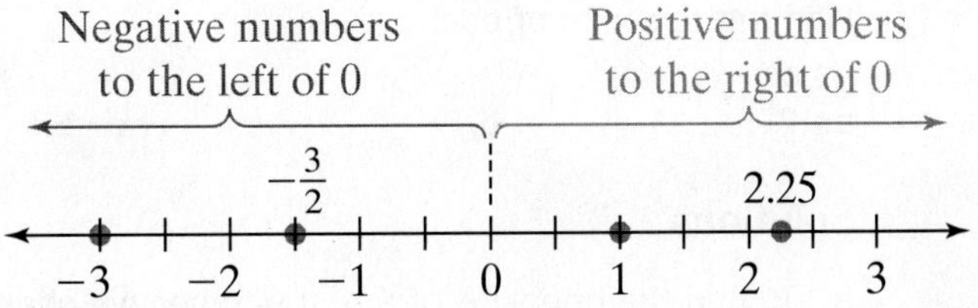

Figure 3

Notice that the sign of each number indicates whether the number lies on the left or right side of 0. Because $-\frac{3}{2} = -1\frac{1}{2}$, its point lies midway between -2 and -1. However, because $2.25 = 2\frac{1}{4}$, its point lies one-fourth of the way between 2 and 3.

You Try It **Work through this You Try It problem.**

Work Exercises 9 and 10 in this textbook or in the MyMathLab Study Plan.

OBJECTIVE 3 FIND THE OPPOSITE OF A REAL NUMBER

For every **real number** other than 0 on the **number line**, there is a corresponding number located the same distance away from 0 on the number line but in the opposite direction. These pairs of numbers are called *opposites*.

> **Definition Opposites**
>
> Two numbers are **opposites**, if they are located the same distance away from 0 on the number line but lie on opposite sides of 0.

We use a negative sign $(-)$ to represent "the opposite of." So, -8 reads as "the opposite of eight," and $-(-8)$ reads as "the opposite of negative eight." The opposite of negative eight equals eight, so, $-(-8) = 8$. This rule is true for all real numbers.

> **Double-Negative Rule**
>
> If a is a real number, then $-(-a) = a$.

Figure 4 shows us that the opposite of a **positive number** is the **negative number** that lies the same distance from 0 on the number line. The opposite of a negative number is the positive number that lies the same distance from 0 on the number line. Zero is its own opposite.

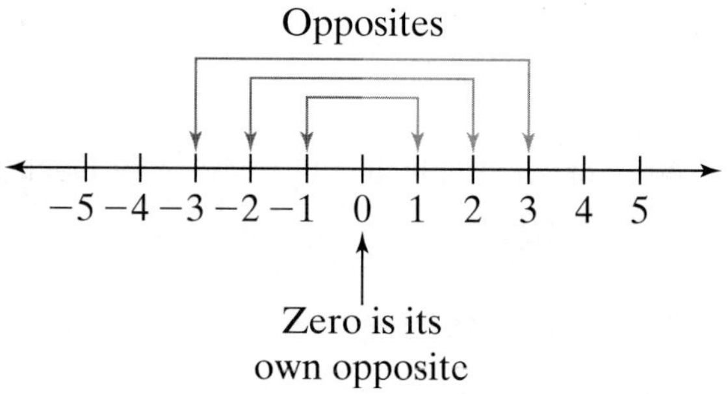

Figure 4

> **Finding the Opposite of a Real Number**
>
> To find the opposite of a **real number**, change its **sign**.

Example 3 Finding Opposites of Real Numbers

Find the **opposite** of each real number.

a. 22 **b.** $-\frac{4}{5}$ **c.** 6.4 **d.** 0

Solutions

a. To find the opposite of a real number, we change its sign. Because 22 is **positive**, the opposite of 22 is -22.

b. Because $-\frac{4}{5}$ is **negative**, the opposite of $-\frac{4}{5}$ is $\frac{4}{5}$.

c. The opposite of 6.4 is -6.4.

d. Because 0 is its own opposite, the opposite of 0 is 0.

You Try It **Work through this You Try It problem.**

Work Exercises 11–15 in this textbook or in the MyMathLab Study Plan.

OBJECTIVE 4 FIND THE ABSOLUTE VALUE OF A REAL NUMBER

As defined earlier in this topic and illustrated in **Figure 2**, the distance from 0 to 1 on a **real number line** is the **unit distance** for the number line. The concept of unit distance is related to the concept of *absolute value*.

> **Definition Absolute Value**
>
> The **absolute value** of a **real number** a, written as $|a|$, is the distance from 0 to a on the number line.

The symbol $|4|$ reads as "the absolute value of four," but $|-4|$ reads as "the absolute value of negative four." Because both 4 and -4 are 4 units from 0 on the number line, both absolute values equal 4. So, $|4| = 4$ and $|-4| = 4$. See Figure 5.

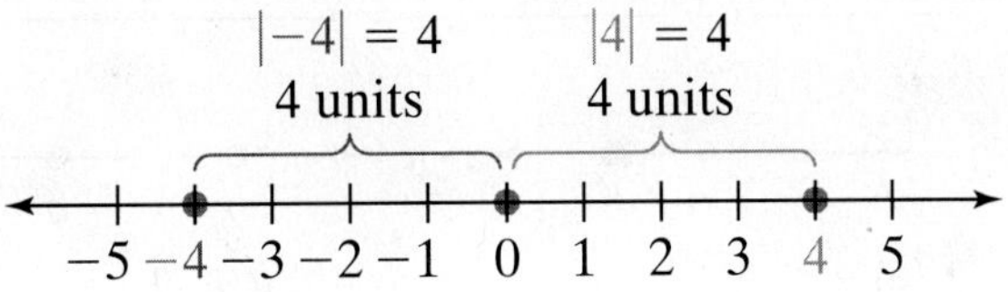

Figure 5

CAUTION: Because distance cannot be negative, the absolute value of a real number will always be **non-negative.**

What does Figure 5 tell us about the absolute value of opposites? See the **answer**.

Example 4 Finding Absolute Values

Find each **absolute value**.

a. $|3|$ **b.** $|-5|$ **c.** $|-1.5|$ **d.** $\left|-\frac{7}{2}\right|$ **e.** $|0|$

Solutions See Figure 6 for an illustration of parts a and b.

a. $|3| = 3$
The absolute value of 3 is 3 because 3 lies 3 units away from 0 on the **number line.**

b. $|-5| = 5$
The absolute value of -5 is 5 because -5 lies 5 units away from 0 on the number line.

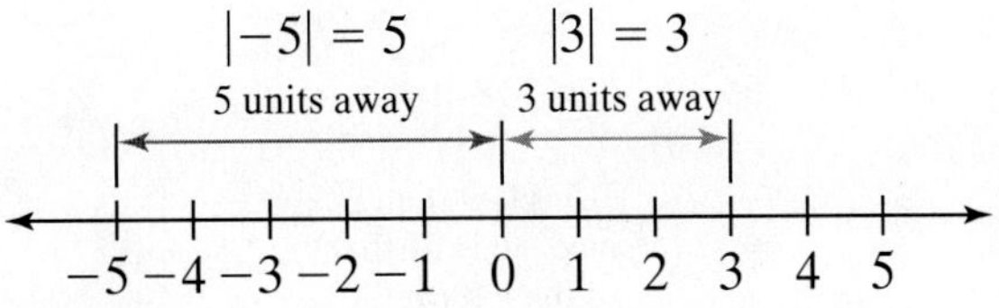

Figure 6

My video summary

c–e. Try to find each absolute value on your own. Check your **answers**, or watch this **video** for the complete solutions.

You Try It **Work through this You Try It problem.**

Work Exercises 16–19 in this textbook or in the MyMathLab Study Plan.

OBJECTIVE 5 USE INEQUALITY SYMBOLS TO ORDER REAL NUMBERS

To find the **order** of two **real numbers** a and b means to determine if the first number a is smaller than, larger than, or equal to the second number b. We can find the *order* of any two real numbers by comparing their locations on the **number line**. Because numbers on the number line increase as we read from left to right, as we move further to the right, the larger the number.

Order of Real Numbers

1. If a is located to the left of b on the real number line, as shown in Figure 7(a), then a ***is less than*** b, written as $a < b$.
2. If a is located to the right of b, as shown in Figure 7(b), then a ***is greater than*** b, written as $a > b$.
3. If a and b are located at the same position on the number line, as shown in Figure 7(c), then a ***is equal to*** b, written as $a = b$.

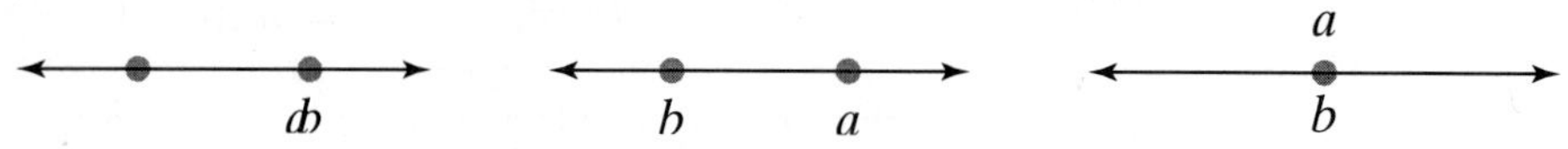
b
a
a
b

(a) $a < b$ (b) $a > b$ (c) $a = b$ **Figure 7**

The symbol for equality is the **equal sign**, =. Symbols for less than, <, and greater than, >, are called **inequality symbols.** We use these symbols to order real numbers. For example, because -5 is located to the left of -2 on the **number line**, as shown in Figure 8, we say that -5 *is less than* -2 and write $-5 < -2$. We say that -2 *is greater than* -5 and write $-2 > -5$.

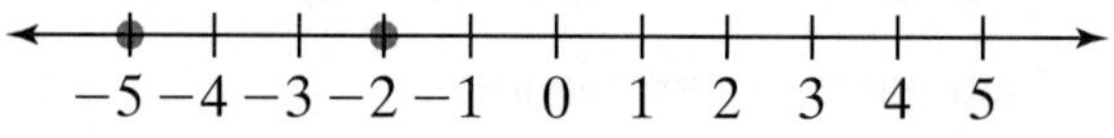
−5 −4 −3 −2 −1 0 1 2 3 4 5

Figure 8

When using the symbols < or > to **compare** or order numbers, the symbol should point to the smaller of the two numbers, as shown in Figure 9.

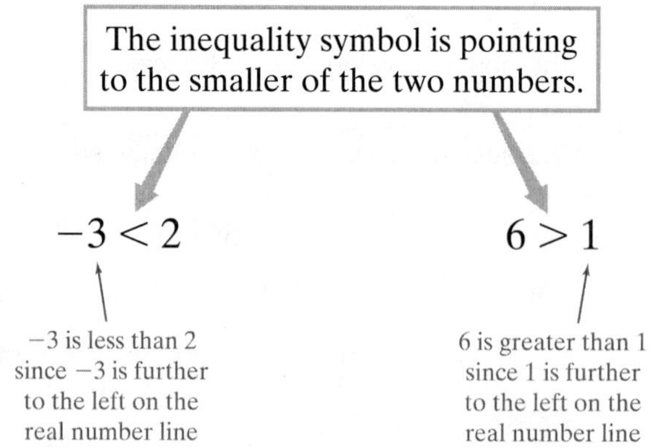
The inequality symbol is pointing to the smaller of the two numbers.
−3 < 2
6 > 1
−3 is less than 2 since −3 is further to the left on the real number line
6 is greater than 1 since 1 is further to the left on the real number line

Figure 9

Figure 10 shows that 0.5 and $\frac{1}{2}$ share the same position on the number line. So, we say that 0.5 *equals* $\frac{1}{2}$ and write $0.5 = \frac{1}{2}$.

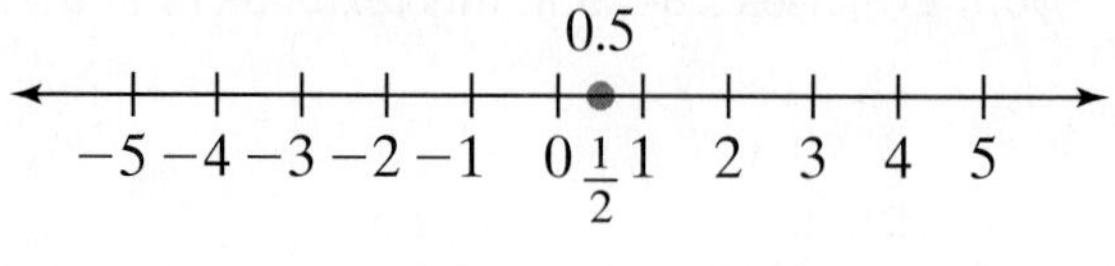
0.5
−5 −4 −3 −2 −1 0 1/2 1 2 3 4 5

Figure 10

Example 5 Ordering Real Numbers Using Equality and Inequality Symbols

Fill in the blank with the correct symbol, $<$, $>$, or $=$, to make a true statement.

a. 0 ____ 3 **b.** -3.7 ____ -1.5 **c.** $-\frac{5}{4}$ ____ -1.25 **d.** $\frac{4}{5}$ ____ $\frac{5}{9}$

Solutions

a. View 0 **and** 3 **plotted** on the **number line.** Because 0 is to the left of 3 on the number line, 0 *is less than* 3. So, we write $0 < 3$.

b. View -3.7 **and** -1.5 **plotted** on the number line. Because -3.7 is to the left of -1.5 on the number line, -3.7 *is less than* -1.5. So, we write $-3.7 < -1.5$.

c. View $-\frac{5}{4}$ **and** -1.25 **plotted** on the number line. The two numbers share the same position, so they are equal, or $-\frac{5}{4} = -1.25$.

d. To compare $\frac{4}{5}$ and $\frac{5}{9}$ more easily, first we convert both **fractions to decimals:** $\frac{4}{5} = 0.8$ and $\frac{5}{9} = 0.\overline{5}$. On the number line, 0.8 is to the right of $0.\overline{5}$, so $\frac{4}{5}$ *is greater than* $\frac{5}{9}$, or $\frac{4}{5} > \frac{5}{9}$. View the **numbers plotted** on the number line.

You Try It **Work through this You Try It problem.**

Work Exercises 20–28 in this textbook or in the MyMathLab Study Plan.

In addition to $<$ and $>$, three other **inequality symbols** can be used to **compare** real numbers: *less than or equal to* ($\leq$), *greater than or equal to* ($\geq$), and *not equal to* ($\neq$). Notice that $\leq$ and $\geq$ allow for equality as part of the comparison. For this reason, inequalities of the forms $a \leq b$ and $a \geq b$ are **non-strict inequalities.** Similarly, inequalities of the forms $a < b, a > b$, and $a \neq b$ are **strict inequalities** because they do not allow for equality.

For non-strict inequalities, we need to remember that the inequality contains two parts ($<$ or $>$, and equality). In order for the inequality to be true, only one of the two parts must be a true statement.

$$a \leq b \begin{cases} \text{True if } a < b & (4 \leq 8 \text{ is true because } 4 < 8 \text{ is true}) \\ \text{True if } a = b & \left(3.5 \leq \frac{7}{2} \text{ is true because } 3.5 = \frac{7}{2} \text{ is true}\right) \end{cases}$$

$$a \geq b \begin{cases} \text{True if } a > b & (9 \geq 6 \text{ is true because } 9 > 6 \text{ is true}) \\ \text{True if } a = b & \left(2.5 \geq \frac{5}{2} \text{ is true because } 2.5 = \frac{5}{2} \text{ is true}\right) \end{cases}$$

Example 6 Using Inequality Symbols

Determine if each statement is true or false.

a. $\frac{7}{10} \leq 0.7$ **b.** $-8 \geq 4$ **c.** $-2 \geq -4$

d. $\frac{7}{3} \leq 1.\overline{3}$ **e.** $-\frac{9}{4} \neq -2.75$

(eText Screens 10.1-1–10.1-29)

Solutions

a. In order for the statement to be true, we need either $\frac{7}{10} < 0.7$ or $\frac{7}{10} = 0.7$ to be true. The comparison may be easier if we first convert the **fraction to a decimal.** In decimal form, $\frac{7}{10} = 0.7$ So, the statement $\frac{7}{10} \leq 0.7$ is true because $\frac{7}{10} = 0.7$.

b. In order for the statement to be true, we need either $-8 > 4$ or $-8 = 4$ to be true. Because -8 lies to the left of 4 on the **real number line**, we have $-8 < 4$. So, the statement $-8 \geq 4$ is false.

My video summary

c–e. Try to answer these problems on your own. Check your **answers** or watch this **video** for the complete solutions.

You Try It **Work through this You Try It problem.**

Work Exercises 29–33 in this textbook or in the MyMathLab Study Plan.

OBJECTIVE 6 TRANSLATE WORD STATEMENTS INVOLVING INEQUALITIES

When solving application problems, understanding the language of mathematics is important in order to translate word statements into **mathematical expressions.** When working with inequalities, it will be important to use key words to distinguish between the various **inequality symbols.** There are many key words or phrases that describe inequality symbols. Figure 11 lists some examples.

Inequality	Key Words	Word Statement	Mathematical Expression
$<$	is less than, is fewer than	Six is fewer than eight	$6 < 8$
$>$	is more than, is greater than	10 is more than two	$10 > 2$
$\leq$	is less than or equal to, at most, no more than	Seven is less than or equal to nine; Four is no more than five	$7 \leq 9$; $4 \leq 5$
$\geq$	is greater than or equal to, at least, no less than	15 is greater than or equal to 13; Eight is no less than three	$15 \geq 13$; $8 \geq 3$
$\neq$	not equal to, different than	12 is not equal to 15	$12 \neq 15$

Figure 11 Key words for inequalities

Example 7 Translating Word Statements Involving Inequalities

Write a mathematical expression for each word phrase.

a. Fourteen is greater than ten

b. Twenty-four is no more than thirty

c. Nine is not equal to eighteen

Solutions Try translating the statements on your own, then check your **answers.**

You Try It **Work through this You Try It problem.**

Work Exercises 34–38 in this textbook or in the MyMathLab Study Plan.

Example 8 Translating Word Statements Involving Inequalities

Use **real numbers** and write an inequality that represents the given comparison.

a. Orchid Island Gourmet Orange Juice sells for $\$6$, which is more than Florida's Natural Premium Orange Juice that sells for $\$3$.

b. In 2008 there were 4983 identity thefts reported in Colorado, which is different than the 4433 reported in Missouri during the same year. (*Source: census.gov/compendia/*)

c. On May 20, 2010, the Dow Jones Industrial Average closed at 10,068.01 which was less than on May 19, 2010, when it closed at 10,444.37. (*Source: dowjonesclose.com/*)

Solutions

a. The key words "is more than" indicate we should use the **inequality symbol** '>'. We represent the cost of the Orchid Island juice with the **integer** 6 and the Florida's Natural juice with the integer 3.

$$\underbrace{\text{cost of Orchid Island juice}}_{6} \quad \underbrace{\text{is more than}}_{>} \quad \underbrace{\text{cost of Florida Natural juice}}_{3}$$

My video summary

b–c. Try translating these statements on your own. Check your **answers**, or watch this **video** for the solutions.

You Try It **Work through this You Try It problem.**

Work Exercises 39–42 in this textbook or in the MyMathLab Study Plan.

10.1 Exercises

In Exercises 1–8, classify each real number as a natural number, whole number, integer, rational number, and/or irrational number. Each number may belong to more than one set.

You Try It

1. -15 **2.** 9.5 **3.** $-\sqrt{11}$ **4.** $\frac{2}{5}$

5. $-\frac{24}{3}$ **6.** 18 **7.** $4.\overline{35}$ **8.** $\sqrt{36}$

In Exercises 9 and 10, plot each set of numbers on a number line.

You Try It

9. $\left\{-3, 2.5, -\frac{7}{4}, 4\right\}$ **10.** $\left\{\frac{3}{4}, -4.\overline{2}, 0, 5, -3\right\}$

In Exercises 11–15, find the opposite of each real number.

You Try It

11. -17 **12.** 29 **13.** 12.45 **14.** $-\frac{7}{12}$ **15.** $4.\overline{27}$

In Exercises 16–19, find each absolute value.

You Try It

16. $|32|$ **17.** $|-12.5|$ **18.** $|-9|$ **19.** $\left|\frac{15}{28}\right|$

In Exercises 20–28, fill in the blank with the correct symbol, $<$, $>$, or $=$, to make a true statement.

You Try It

20. -6.9____-3.2

21. 5____0

22. $-\frac{1}{2}$____$\frac{1}{3}$

23. $\frac{9}{4}$____2.25

24. 15____30

25. $-3\frac{3}{4}$____$-\frac{15}{4}$

26. 1.4____$-\frac{1}{5}$

27. $\frac{7}{3}$____$\frac{9}{4}$

28. -4.35____$1\frac{1}{3}$

In Exercises 29–33, determine if each statement is true or false.

You Try It

29. $-3 \leq -7$

30. $\frac{22}{5} \leq 4.6$

31. $-2 \neq 2$

32. $8.4 \geq 8.40$

33. $6.40 \geq 6\frac{4}{5}$

In Exercises 34–38, write a mathematical expression for each word phrase.

You Try It

34. Seventeen is greater than twelve

35. Forty-five is less than or equal to fifty

36. Eight is at least three

37. Four is different than five

38. Thirty is greater than or equal to fifteen

In Exercises 39–42, use real numbers and write an inequality that compares the given numbers.

You Try It

39. In September 2009, 54% of teens reported texting friends on a daily basis. This is greater than the 38% who reported daily texting in February 2008. (*Source: pewinternet.org/*)

40. The average price per gallon for unleaded gas in the U.S. on May 17, 2010 was \$2.864, which was different than the average price of \$2.309 on May 17, 2009. (*Source: eia.doe.gov*)

41. The average high temperature for Phoenix, AZ in January is 66°F, which is less than the average high temperature of 70°F for February. (*Source: phoenix.about.com/*)

42. Between 1990 and 2000, the population of Philadelphia, PA decreased by 68,000 while the population of Chicago, IL increased by 112,000. (*Source: infoplease.com/*)

10.2 Adding and Subtracting Real Numbers

THINGS TO KNOW

Before working through this topic, be sure you are familiar with the following concepts:

VIDEO ANIMATION INTERACTIVE

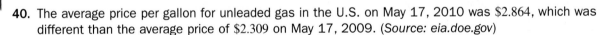

You Try It

1. Write Fractions in Simplest Form (Topic 4.2, **Objective 6**)

You Try It

2. Add and Subtract Unlike Fractions (Topic 4.4, **Objective 3**)

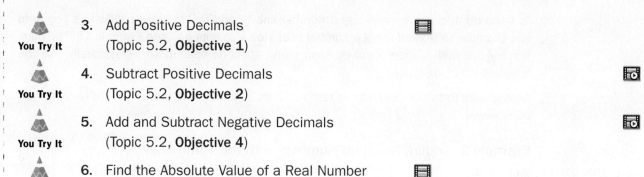

You Try It 3. Add Positive Decimals (Topic 5.2, **Objective 1**)

You Try It 4. Subtract Positive Decimals (Topic 5.2, **Objective 2**)

You Try It 5. Add and Subtract Negative Decimals (Topic 5.2, **Objective 4**)

You Try It 6. Find the Absolute Value of a Real Number (Topic 10.1, **Objective 4**)

OBJECTIVES

1 Add Two Real Numbers with the Same Sign
2 Add Two Real Numbers with Different Signs
3 Subtract Real Numbers
4 Translate Word Statements Involving Addition or Subtraction
5 Solve Applications Involving Addition or Subtraction of Real Numbers

OBJECTIVE 1 ADD TWO REAL NUMBERS WITH THE SAME SIGN

The result of adding two **real numbers** is called the **sum** of the numbers. The numbers being added are called **terms**, or **addends**. How we add terms depends on their **signs** (**positive** or **negative**). First, let's add two real numbers with the same sign.

Adding Two Real Numbers with the Same Sign

If two real numbers being added have the same sign (both positive or both negative), then we add as follows:

Step 1. Find the **absolute value** of each term.

Step 2. Add the two results from Step 1.

Step 3. Use the common sign as the sign of the **sum**.

Example 1 Adding Two Real Numbers with the Same Sign

Add.

a. $2 + 5$ **b.** $-4 + (-3)$

Solutions

a. First, we find the absolute values, $|2| = 2$ and $|5| = 5$. Then we add the results, $2 + 5 = 7$. Because the common sign is positive, the sum will be positive. Thus, $2 + 5 = 7$.

b. First, we find the absolute values, $|-4| = 4$ and $|-3| = 3$. Then, we add the results, $4 + 3 = 7$. Because the common sign is negative, the sum will be negative. So, $-4 + (-3) = -7$.

You Try It **Work through this You Try It problem.**

Work Exercises 1–4 in this textbook or in the MyMathLab Study Plan.

(eText Screens 10.2-1–10.2-26)

To visualize addition, we can use a **number line** to add **real numbers**. View this **popup** to see Example 1a worked using a number line. View this **popup** to see Example 1b. The number line method is more involved than using absolute values, so we generally use the absolute value approach.

Another way to think about adding real numbers is to consider money gained or lost. View an **example**.

Example 2 Adding Two Real Numbers with the Same Sign

Add.

a. $-3.65 + (-7.45)$ **b.** $\frac{4}{5} + \frac{13}{5}$ **c.** $-\frac{3}{5} + \left(-\frac{7}{2}\right)$ **d.** $-3\frac{1}{3} + \left(-5\frac{1}{4}\right)$

Solutions

a. We find the absolute values, $|-3.65| = 3.65$ and $|-7.45| = 7.45$. Both numbers are negative, so the sum will be negative. Because $3.65 + 7.45 = 11.10$, we get $-3.65 + (-7.45) = -11.10$.

b. We find the **absolute values,** $\left|\frac{4}{5}\right| = \frac{4}{5}$ and $\left|\frac{13}{5}\right| = \frac{13}{5}$. Both numbers are **positive,** so the **sum** will be positive.

$$\underbrace{\frac{4}{5} + \frac{13}{5}}_{\text{Common denominator}} = \frac{\overbrace{4 + 13}^{\text{Add the numerators}}}{\underbrace{5}_{\text{Keep the common denominator}}} = \frac{17}{5}$$

The sum is $\frac{17}{5}$.

My video summary

c. Because both numbers are **negative,** we add the absolute values and make the sum negative.

$$\left|-\frac{3}{5}\right| = \frac{3}{5} \quad \text{and} \quad \left|-\frac{7}{2}\right| = \frac{7}{2}$$

To add $\frac{3}{5}$ and $\frac{7}{2}$, we need a **common denominator.**

$$\underbrace{\frac{3}{5} + \frac{7}{2}}_{\text{LCD} = 10} = \frac{3}{5}\cdot\frac{2}{2} + \frac{7}{2}\cdot\frac{5}{5} = \frac{6}{10} + \frac{35}{10}$$

Finish adding the numbers on your own. Check your **answer,** or watch this **video** for the complete solution.

My video summary

d. Try adding the numbers on your own. Check your **answer,** or watch this **video** for the complete solution.

You Try It **Work through this You Try It problem.**

Work Exercises 5–10 in this textbook or in the MyMathLab Study Plan.

(eText Screens 10.2-1–10.2-26)

OBJECTIVE 2 ADD TWO REAL NUMBERS WITH DIFFERENT SIGNS

Sometimes we must add two numbers with different **signs**. For example, if the value of one share of Apple Inc. stock increased $\$8$ one day and decreased $\$3$ the next, then the share increased in value by $\$5$ over the two days, or $8 + (-3) = 5$.

When adding two **real numbers** with different signs, we can use **absolute value**.

> **Adding Two Real Numbers with Different Signs**
>
> If two real numbers being added have different signs (one **positive** and one **negative**), then we add as follows:
>
> **Step 1.** Find the absolute value of each **term**.
>
> **Step 2.** Subtract the smaller result from the larger result from Step 1.
>
> **Step 3.** Use the sign of the term with the larger absolute value as the sign of the **sum**.

Example 3 Adding Two Real Numbers with Different Signs

Add.

a. $7 + (-4)$ **b.** $-6 + 4$ **c.** $6 + (-6)$

Solutions

a. First, we find the absolute value of each **term**.

$$|7| = 7 \quad \text{and} \quad |-4| = 4$$

The terms have different signs so we subtract the smaller result from the larger result.

$$7 - 4 = 3$$

The larger absolute value result belongs to the positive number, so the sum will be positive.

$$7 + (-4) = 3$$

b. Find the **absolute value** of each number.

$$|-6| = 6 \quad \text{and} \quad |4| = 4$$

Subtract the smaller result from the larger result.

$$6 - 4 = 2$$

Because the larger absolute value result belongs to the **negative** number, the **sum** will be negative.

$$-6 + 4 = -2$$

c. Find the absolute values.

$$|6| = 6 \quad \text{and} \quad |-6| = 6$$

The numbers have the same absolute value, so the **difference** is 0.

$$6 - 6 = 0$$

Therefore,

$$6 + (-6) = 0.$$

You Try It **Work through this You Try It problem.**

Work Exercises 11–16 in this textbook or in the MyMathLab Study Plan.

We can use a **number line** to visualize the addition of real numbers with different signs. View this **popup** to see Example 3a worked using a number line. View this **popup** to see Example 3b.

In Example 3c, the two numbers are **opposites** with a sum of 0. This result shows why we refer to opposites as **additive inverses**.

Adding a Real Number and Its Opposite

The sum of a real number a and its opposite, or **additive inverse**, $-a$, is 0.

$$a + (-a) = 0$$

We now summarize our procedures for adding two **real numbers**.

Adding Two Real Numbers

Step 1. If the **signs** of the two **terms** are the same (both **positive** or both **negative**), add their **absolute values** and use the common sign as the sign of the **sum**.

Step 2. If the signs of the two terms are different (one positive and one negative), subtract the smaller absolute value from the larger absolute value and use the sign of the term with the larger absolute value as the sign of the sum.

Example 4 Adding Two Real Numbers

Add.

a. $-12 + (-9)$ **b.** $7 + (-18)$ **c.** $\frac{4}{3} + \frac{5}{6}$ **d.** $-5.7 + 12.3$

Solutions

a. Because both numbers are negative, we add the absolute values and make the sum negative.

$$|-12| = 12 \quad \text{and} \quad |-9| = 9$$

$12 + 9 = 21$, so $-12 + (-9) = -21$.

b. Find the absolute value of each number.

$$|7| = 7 \quad \text{and} \quad |-18| = 18$$

Subtract the smaller result from the larger result.

$$18 - 7 = 11$$

Because the larger absolute value result belongs to the negative number, the sum will be negative.

$$7 + (-18) = -11$$

My video summary

c–d. Try to work these problems on your own. Start by finding the absolute value of each number. Add or subtract the absolute values as required, then determine the sign of the result. Check your **answers**, or watch this **video** for the complete solutions.

You Try It **Work through this You Try It problem.**

Work Exercises 17–20 in this textbook or in the MyMathLab Study Plan.

OBJECTIVE 3 SUBTRACT REAL NUMBERS

The next **arithmetic operation** is **subtraction**. The result of subtracting two **real numbers** is called the **difference** of the numbers. The number being subtracted is called the **subtrahend**, and the number being subtracted from is called the **minuend**.

$$\underset{\text{Minuend}}{23} - \underset{\text{Subtrahend}}{7} = \underset{\text{Difference}}{16}$$

Using the concept of **opposite**, we can define subtraction in terms of **addition**. We subtract a number by adding its opposite.

Subtracting Two Real Numbers

If a and b represent two real numbers, then

$$a - b = a + (-b).$$

Example 5 Subtracting Two Real Numbers

My interactive video summary

Subtract.

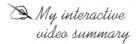

a. $15 - 6$ b. $9 - 17$ c. $-4.5 - 3.2$ d. $\frac{7}{3} - \frac{2}{5}$

e. $-4.9 - (-2.5)$ f. $7\frac{3}{4} - \left(-2\frac{1}{5}\right)$ g. $4 - (-4)$

Solutions For each **subtraction**, we add the opposite.

a. Add the opposite of 6: $15 - 6 = 15 + (-6)$
We now choose the appropriate **addition rule**. The **signs** are different. $|15| = 15$ and $|-6| = 6$. Because 15 has the larger **absolute value**, this **sum** is **positive**. We subtract 6 from 15 to find the result, $15 - 6 = 9$. The **difference** is 9.

b. Add the **opposite** of 17: $9 - 17 = 9 + (-17)$
The **signs** are different. $|9| = 9$ and $|-17| = 17$. Because -17 has the larger **absolute value**, this **sum** is **negative**. We subtract 9 from 17 and make the result negative. $17 - 9 = 8$, so $9 + (-17) = -8$. The **difference** is -8.

c. Add the opposite of 3.2: $-4.5 - 3.2 = -4.5 + (-3.2)$
Both -4.5 and -3.2 are negative, so the sum is negative. $|-4.5| = 4.5$ and $|-3.2| = 3.2$. Adding gives $4.5 + 3.2 = 7.7$, so $-4.5 + (-3.2) = -7.7$. The difference is -7.7.

d. Add the opposite of $\frac{2}{5}$: $\frac{7}{3} - \frac{2}{5} = \frac{7}{3} + \left(-\frac{2}{5}\right)$

The signs are different. $\left|\frac{7}{3}\right| = \frac{7}{3}$ and $\left|-\frac{2}{5}\right| = \frac{2}{5}$. The sum is positive because $\frac{7}{3}$ has the larger absolute value. We subtract $\frac{2}{5}$ from $\frac{7}{3}$ to find the result.

$$\underbrace{\frac{7}{3} - \frac{2}{5}}_{\text{LCD} = 15} = \frac{7}{3}\cdot\frac{5}{5} - \frac{2}{5}\cdot\frac{3}{3} = \frac{35}{15} - \frac{6}{15} = \underbrace{\frac{35-6}{15}}_{\text{Combine the numerators}} = \frac{29}{15}$$

The difference is $\frac{29}{15}$.

e–g. Try to work these problems on your own. Check your **answers**, or watch this **interactive video** for the complete solutions.

You Try It **Work through this You Try It problem.**

Work Exercises 21–30 in this textbook or in the MyMathLab Study Plan.

OBJECTIVE 4 TRANSLATE WORD STATEMENTS INVOLVING ADDITION OR SUBTRACTION

When solving application problems, we use the language of mathematics to translate word statements into **mathematical expressions**. There are many key words and phrases that indicate **addition**. Figure 12 shows some examples.

Key Words	Word Phrase	Mathematical Expression
Sum	The *sum* of -2 and 5	$-2 + 5$
Increased by	4 *increased by* 7	$4 + 7$
Added to	3 *added to* 8	$8 + 3$
More than	6 *more than* -4	$-4 + 6$

Figure 12 Key words meaning addition

Example 6 Translating Word Statements Involving Addition

Write a mathematical expression for each word phrase.

a. Five more than -8

b. 8.4 increased by 0.17

c. The sum of -4 and -10

d. 15 added to -30

Solutions

a. The key words "more than" mean addition. What is being added? 5 and -8.

Addition

Five more than -8

The two quantities being added

So, "five more than -8" is written as $-8 + 5$.

b. The key words "increased by" mean **addition**. What is being added? 0.17 is being added to 8.4.

Addition

8.4 increased by 0.17

The two quantities being added

So, "8.4 increased by 0.17" translates to $8.4 + 0.17$.

c. The keyword "**sum**" indicates addition. -4 and -10 are being added. The key word "and" separates the two quantities and can be translated to a $+$ sign.

Addition

$$\text{The } \overbrace{\text{sum of}} \underbrace{-4} \text{ and } \underbrace{-10}$$

The two quantities being added

So, "the sum of -4 and -10" is written as $-4 + (-10)$.

d. The key words are "added to." The numbers -30 and 15 are being added.

Addition

$$\underbrace{15} \overbrace{\text{added to}} \underbrace{-30}$$

The two quantities being added

So, "15 added to -30" translates to $-30 + 15$.

You Try It **Work through this You Try It problem.**

Work Exercises 31–36 in this textbook or in the MyMathLab Study Plan.

As with **addition**, many keywords or phrases indicate **subtraction**. Figure 13 shows some examples.

Key Word	Word Phrase	Mathematical Expression
Difference	The ***difference*** of 9 and 7	$9 - 7$
Decreased by	18 ***decreased by*** 4	$18 - 4$
Subtracted from	6 ***subtracted from*** 3	$3 - 6$
Less than	10 ***less than*** 25	$25 - 10$

Figure 13 Key words for subtraction

Unlike addition, the order of numbers in subtraction matters, so we have to pay close attention to wording. For example, $10 - 2$ (the difference of 10 and 2) is not the same as $2 - 10$. (the difference of 2 and 10)

Example 7 Translating Word Statements Involving Subtraction

Write a **mathematical expression** for each word phrase.

a. Fifteen subtracted from 22

b. The **difference** of 7 and 12

c. Eight decreased by 11

d. 20 less than the difference of 4 and 9

e. The **sum** of 8 and 13, decreased by 5

(eText Screens 10.2-1–10.2-26)

Solutions

a. The key words "subtracted from" mean **subtraction**. What is being subtracted? 15 is being subtracted *from* 22.

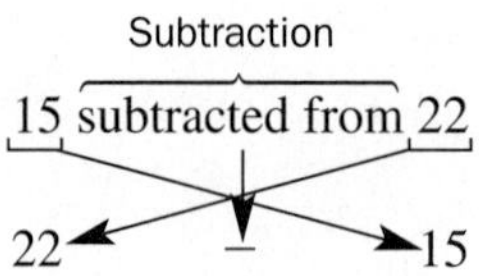

So, "fifteen subtracted from 22" is written as $22 - 15$.

b. "**Difference**" means subtraction. 12 is being subtracted from 7. Note in **Figure 13** that with a difference, the second number is subtracted from the first.

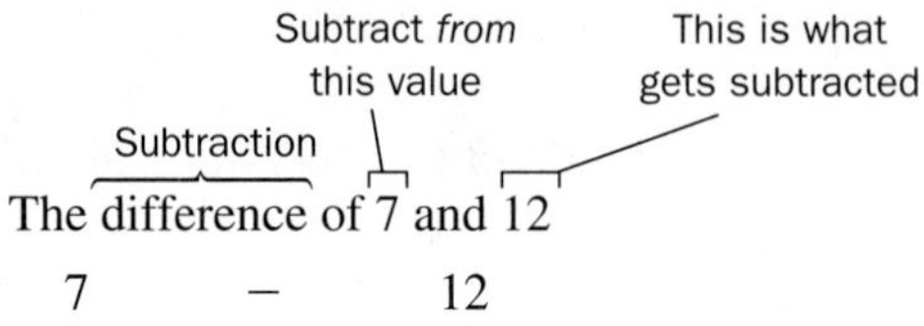

"The difference of 7 and 12" translates to $7 - 12$.

c. "Decreased by" indicates **subtraction**, and the numbers 8 and 11 are being subtracted. Because 8 is being decreased, we write this value first.

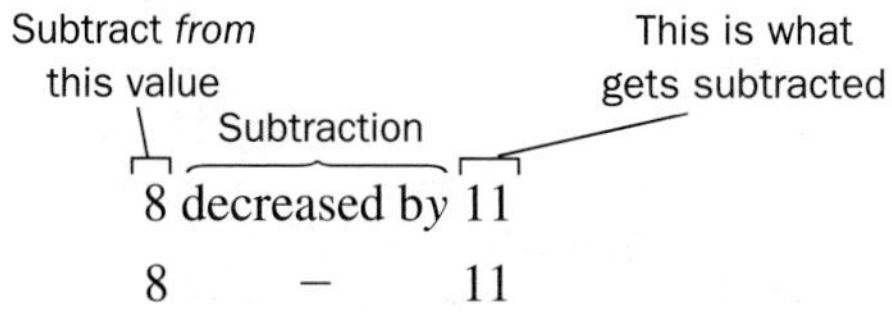

"Eight decreased by 11" translates to $8 - 11$.

My video summary

d. The key words "less than" and "difference" suggest two subtractions. Try to complete this translation on your own. Check your **answer**, or watch this **video** for a complete solution.

e. This problem involves both **addition** and subtraction. Try to do this translation on your own. Check your **answer**, or watch this **video** for a complete solution.

You Try It **Work through this You Try It problem.**

Work Exercises 37–42 in this textbook or in the MyMathLab Study Plan.

OBJECTIVE 5 SOLVE APPLICATIONS INVOLVING ADDITION OR SUBTRACTION OF REAL NUMBERS

Addition and subtraction can be used to solve a variety of application problems that involve finding a **sum** or **difference**. Looking for key words is important because they will tell us which **operation** to perform and on which quantities.

Example 8 Solving Applications Involving Addition or Subtraction

On May 20, 2010 the Dow Jones Industrial Average lost 376.36 points. The next day it gained 125.38 points. What was the total result for the two days? (*Source: Yahoo! Finance*)

Solution The loss of 376.36 points can be expressed as -376.36 while the gain of 125.38 points can be expressed as $+125.38$. To find the total result, we **add**.

$$\underbrace{-376.36}_{\text{May 20}} + \underbrace{125.38}_{\text{May 21}}$$

The numbers have different signs so we find the **absolute value** of each number and subtract the smaller result from the larger.

$$|-376.36| = 376.36 \quad \text{and} \quad |125.38| = 125.38$$

$$376.36 - 125.38 = 250.98$$

Because the larger absolute value result belongs to the **negative** number, the sum will be negative.

$$-376.36 + 125.38 = -250.98$$

The total result for the two days was a loss of 250.98 points.

Example 9 Solving Applications Involving Addition or Subtraction

My video summary

The record high temperature in Alaska was 100°F recorded in 1915 at Fort Yukon. The record low was -80°F recorded in 1971 at Prospect Creek Camp. What is the difference between these record high and low temperatures? (*Source: National Climatic Data Center*)

Solution Try to solve this problem on your own. Check your **answer**, or watch this **video** for a complete solution.

You Try It **Work through this You Try It problem.**

Work Exercises 43–46 in this textbook or in the MyMathLab Study Plan.

10.2 Exercises

In Exercises 1–20, add the real numbers.

You Try It

1. $7 + 8$ **2.** $-2 + (-3)$ **3.** $-23 + (-48)$ **4.** $29 + 85$

You Try It

5. $-2.78 + (-1.25)$ **6.** $15.06 + 0.78$ **7.** $-\frac{4}{3} + \left(-\frac{7}{3}\right)$ **8.** $\frac{3}{4} + \frac{7}{8}$

9. $-\frac{2}{3} + \left(-\frac{12}{5}\right)$ **10.** $\left(-2\frac{1}{3}\right) + \left(-3\frac{1}{2}\right)$ **11.** $6 + (-2)$ **12.** $(-5) + 8$

You Try It

13. $-3 + 1$ **14.** $7 + (-9)$ **15.** $-12 + 12$ **16.** $(-2.79) + 4.21$

You Try It

17. $\left(-\frac{13}{4}\right) + \left(\frac{13}{4}\right)$ **18.** $\left(-\frac{2}{7}\right) + \left(-\frac{7}{2}\right)$ **19.** $\left(-\frac{7}{3}\right) + \left(2\frac{1}{3}\right)$ **20.** $(3.87) + (-10.4)$

In Exercises 21–30, subtract the real numbers.

You Try It

21. $15 - 9$ **22.** $8 - 11$ **23.** $\frac{7}{3} - \frac{3}{7}$ **24.** $-3.1 - 5.9$

25. $-7-(-3)$ **26.** $-1\frac{2}{3}-3\frac{1}{4}$ **27.** $-\frac{2}{9}-\frac{5}{18}$ **28.** $2.1-1.2$

29. $\frac{5}{8}-\left(-\frac{3}{4}\right)$ **30.** $-2.1-(-3.2)$

In Exercises 31–36, write a mathematical expression for each word phrase.

You Try It

31. 5 added to -7

32. -3 increased by 12

33. 8.1, added to 2.1 more than 3.7

34. The sum of $-\frac{3}{5}$ and $\frac{5}{3}$

35. Twelve more than -15

36. $2\frac{3}{4}$ increased by $\frac{1}{2}$

In Exercises 37–42, write a mathematical expression for each word phrase.

You Try It

37. The difference of 12 and -13

38. -18 decreased by 15

39. 5 more than the difference of 2 and 7

40. 8 subtracted from 5, decreased by 3

41. Two-thirds subtracted from $\frac{5}{9}$

42. 34 less than the sum of -3 and 19

In Exercises 43–46, solve the given application problem.

You Try It

43. On May 18, 2010 the Dow Jones Industrial Average lost 114.88 points. The next day it lost 66.58 points. What was the total result for the two days? Express this as a real number. (*Source: Yahoo! Finance*)

44. For the 2008 tax year, the IRS received 131,543,000 individual tax returns. Of those, 90,639,000 were filed electronically. How many individual tax returns were sent by mail for 2008? (*Source: irs.gov*)

45. The lowest point in the Atlantic Ocean is the Milwaukee Deep in the Puerto Rico Trench, at an elevation of -8605 m. The lowest point in the Pacific Ocean is the Challenger Deep in the Mariana Trench, at an elevation of $-10{,}924$ m. What is the difference in elevation between the Milwaukee Deep and the Challenger Deep? (*Source: www.worldatlas.com*)

46. The record high temperature in Montana was 47.2°C recorded in 1937 at Medicine Lake. The record low was -56.7°C recorded in 1954 at Rogers Pass. What is the difference between these record high and low temperatures? (*Source: National Climatic Data Center*)

10.3 Multiplying and Dividing Real Numbers

THINGS TO KNOW

Before working through this topic, be sure you are familiar with the following concepts:	VIDEO	ANIMATION	INTERACTIVE
You Try It **1.** Write Fractions in Simplest Form (Topic 4.2, **Objective 6**)	✓		
You Try It **2.** Multiply Fractions (Topic 4.3, **Objective 1**)	✓		✓
You Try It **3.** Divide Fractions (Topic 4.3, **Objective 4**)	✓		✓

(eText Screens 10.3-1–10.3-25)

You Try It — 4. Multiply Decimals (Topic 5.3, **Objective 1**)

You Try It — 5. Divide Decimals (Topic 5.4, **Objective 1**)

You Try It — 6. Find the Absolute Value of a Real Number (Topic 10.1, **Objective 4**)

OBJECTIVES

1 Multiply Real Numbers

2 Divide Real Numbers

3 Translate Word Statements Involving Multiplication or Division

4 Solve Applications Involving Multiplication or Division

OBJECTIVE 1 MULTIPLY REAL NUMBERS

Multiplication is simply repeated **addition**. For example, $4(6)$ means that 6 is added four times, and $5(-6)$ means that -6 is added five times. Figure 14 displays this process visually.

$$4(6) = 6 + 6 + 6 + 6 = 24$$
$$5(-6) = (-6) + (-6) + (-6) + (-6) + (-6) = -30$$

Figure 14

The result of multiplying two **real numbers** is called the **product** of the numbers. The numbers being multiplied are called **factors**. The **sign** of a product depends on the signs of the factors.

The following gives a summary for multiplying two real numbers.

> **Multiplying Two Real Numbers**
>
> Multiply the **absolute values** of the two **factors** to get the absolute value of the **product**. Determine the sign of the product using the following rules:
>
> **1.** If the signs of the two factors are the same (both **positive** or both **negative**), then the product is positive.
>
> **2.** If the signs of the two factors are different (one positive and one negative), then the product is negative.
>
> **3.** The product of a real number and 0 is equal to 0 (**Multiplication Property of Zero**).
>
> View this **illustration** of how the factor signs affect the sign of the product.

Example 1 Multiplying Two Real Numbers

Multiply.

a. $5 \cdot 13$ **b.** $6(-7)$ **c.** 0×15 **d.** $(-4)(-12)$

Solutions

a. $|5| = 5$ and $|13| = 13$. The **absolute value** of the **product** is $5 \cdot 13 = 65$. Since 5 and 13 have the same **sign**, the product is positive. So, $5 \cdot 13 = 65$.

b. $|6| = 6$ and $|-7| = 7$. The absolute value of the product is $6 \cdot 7 = 42$. One **factor** is positive and the other is negative, so the product is negative. So, $6(-7) = -42$.

c. One of the factors is 0, so the product is 0.

d. $|-4| = 4$ and $|-12| = 12$. The absolute value of the product is $4 \cdot 12 = 48$. Since -4 and -12 have the same sign, the product is positive. So, $(-4)(-12) = 48$.

You Try It **Work through this You Try It problem.**

Work Exercises 1–8 in this textbook or in the MyMathLab Study Plan.

You may wish to review multiplying **fractions** in **Topic 4.3** before working through Example 2.

Example 2 Multiplying Two Real Numbers

Multiply.

a. $\left(-\frac{3}{4}\right)\left(-\frac{7}{9}\right)$ **b.** $5 \cdot \frac{3}{10}$ **c.** $\frac{3}{8} \times 0$ **d.** $\left(-\frac{2}{3}\right)\left(\frac{6}{14}\right)$

Solutions

a. $\left|-\frac{3}{4}\right| = \frac{3}{4}$ and $\left|-\frac{7}{9}\right| = \frac{7}{9}$. The absolute value of the product is

$$\frac{3}{4} \cdot \frac{7}{9} = \underbrace{\frac{\cancel{3}^1}{4} \cdot \frac{7}{\cancel{9}^3}}_{\text{Simplify}} = \frac{1 \cdot 7}{4 \cdot 3} = \frac{7}{12}.$$

Because $-\frac{3}{4}$ and $-\frac{7}{9}$ have the same **sign**, the **product** is **positive**. So,

$$\left(-\frac{3}{4}\right)\left(-\frac{7}{9}\right) = \frac{7}{12}.$$

My video summary

b–d. Try to finish these problems on your own. Remember to write **fractions** in **lowest terms**. Check your **answers**, or watch this **video** for the complete solutions.

You Try It **Work through this You Try It problem.**

Work Exercises 9–14 in this textbook or in the MyMathLab Study Plan.

Example 3 Multiplying Two Real Numbers

Multiply.

a. $(1.4)(-3.5)$ **b.** $(8.32)(0)$ **c.** $-\frac{3}{5} \times 6\frac{1}{3}$ **d.** $(4)(5.8)$

Solutions

a. The **factors** have different signs, so the product is **negative**.
Since $(1.4)(3.5) = 4.9$ (View the **details**), the product is $(1.4)(-3.5) = -4.9$.

b. Since one of the factors is 0, the product is 0. So, $(8.32)(0) = 0$.

My video summary

c–d. Try to work these problems on your own. Remember to write **mixed numbers** as **improper fractions** before multiplying, and write fractions in lowest terms. Check your **answers**, or watch this **video** for the complete solutions.

You Try It **Work through this You Try It problem.**

Work Exercises 15–20 in this textbook or in the MyMathLab Study Plan.

(eText Screens 10.3-1–10.3-25)

OBJECTIVE 2 DIVIDE REAL NUMBERS

Our next **arithmetic operation** is **division**. The result of dividing two **real numbers** is called the **quotient** of the numbers. The number being divided is called the **dividend**, and we divide by a number called the **divisor**.

Dividend		Divisor		Quotient
↓		↓		↓
20	÷	4	=	5

$$\begin{matrix}\text{Dividend} \rightarrow \\ \text{Divisor} \rightarrow\end{matrix} \frac{20}{4} = 5 \leftarrow \text{Quotient}$$

In Topic 10.2, we saw that **subtraction** is defined in terms of **addition**. We subtract a number by adding its **opposite**, or **additive inverse**.

$$a - b = a + (-b)$$

Similarly, we can define division in terms of **multiplication**. Consider the following:

$$10 \div 5 = \frac{10}{5} = \frac{10 \cdot 1}{1 \cdot 5} = \frac{10}{1} \cdot \frac{1}{5} = 10 \cdot \frac{1}{5}$$

The number $\frac{1}{5}$ is called the **reciprocal**, or **multiplicative inverse**, of 5. Two numbers are reciprocals if their **product** is 1. So, 5 and $\frac{1}{5}$ are reciprocals of each other because $5 \cdot \frac{1}{5} = 1$.

> **Definition Reciprocals (or Multiplicative Inverses)**
>
> Two numbers are **reciprocals**, or **multiplicative inverses**, if their product is 1.

The number 0 does not have a reciprocal. Find out **why**.

Every nonzero **real number** b has a **reciprocal**, given by $\frac{1}{b}$. We can find the reciprocal of a **rational number** by inverting or "flipping" the **fraction**. View some **examples**.

The reciprocal of a number will have the same **sign** as the number. For example, the reciprocal of $\frac{3}{4}$ is $\frac{4}{3}$, and the reciprocal of $-\frac{2}{3}$ is $-\frac{3}{2}$.

To divide by a number, we multiply by its reciprocal.

> **Definition Division of Two Real Numbers**
>
> If a and b represent two **real numbers** and $b \neq 0$, then
>
> $$a \div b = \frac{a}{b} = a \cdot \frac{1}{b}.$$

Example 4 Dividing Two Real Numbers

Divide.

a. $\frac{-60}{4}$ **b.** $\frac{3}{25} \div \left(-\frac{9}{20}\right)$

Solutions

a. $\frac{-60}{4} = \underbrace{-60 \cdot \frac{1}{4}}_{\text{Change to multiplication by the reciprocal}} = -15$ ← the reciprocal of 4 is $\frac{1}{4}$

(*eText Screens 10.3-1–10.3-25*)

b. $\frac{3}{25} \div \left(-\frac{9}{20}\right) = \underbrace{\frac{3}{25} \cdot \left(-\frac{20}{9}\right)}_{\substack{\text{Change to}\\ \text{multiplication}\\ \text{by the reciprocal}}}$

$\left|\frac{3}{25}\right| = \frac{3}{25}$ and $\left|-\frac{20}{9}\right| = \frac{20}{9}$, so the **absolute value** of the **product** is

$$\frac{3}{25} \cdot \frac{20}{9} = \underbrace{\frac{\cancel{3}^1}{\cancel{25}^5} \cdot \frac{\cancel{20}^4}{\cancel{9}^3}}_{\text{Simplify}} = \frac{1 \cdot 4}{5 \cdot 3} = \frac{4}{15}.$$

Since $\frac{3}{25}$ and $-\frac{20}{9}$ have different **signs**, the product is **negative**.

$$\frac{3}{25} \div \left(-\frac{9}{20}\right) = -\frac{4}{15}$$

We can use other techniques to **divide** real numbers in special cases. For example, dividing two **integers** is the same process as simplifying a **fraction** to **lowest terms**. Also, when dividing **decimals**, we can use **long division** and then determine the appropriate sign for the **quotient**.

CAUTION

Because division is defined in terms of **multiplication**, the rules for finding the sign of a quotient are the same as the rules for finding the sign of a product.

Dividing Two Real Numbers Using Absolute Value

Divide the **absolute values** of the two **real numbers** to find the absolute value of the **quotient**. Determine the **sign** of the quotient by the following rules:

1. If the signs of the two numbers are the same (both **positive** or both **negative**), then their quotient is positive.
2. If the signs of the two numbers are different (one positive and one negative), then their quotient is negative.
3. The quotient of 0 and any nonzero real number is 0. So, $\frac{0}{b} = 0$ for $b \neq 0$. (**See why.**)
4. Division by zero, or $\frac{a}{0}$, is undefined.

Example 5 Dividing Two Real Numbers

Divide.

a. $(-8) \div (-36)$ **b.** $\frac{0}{5}$ **c.** $\frac{12}{35} \div \left(-\frac{27}{14}\right)$ **d.** $15 \div \left(\frac{3}{4}\right)$

Solutions

a. The two numbers are the same sign so the quotient is positive. Dividing the **absolute values**, we get

$$\frac{8}{36} = \frac{2 \cdot 4}{9 \cdot 4} = \frac{2 \cdot \cancel{4}}{9 \cdot \cancel{4}} = \frac{2}{9}$$

The quotient is $\frac{2}{9}$.

(eText Screens 10.3-1–10.3-25)

b. Since the quotient of zero and a nonzero real number is $0, \frac{0}{5} = 0$.

My video summary

c. The two numbers have different **signs** so we divide the **absolute values** of the numbers and make the **quotient** negative. To divide, we multiply the first **fraction** by the **reciprocal** of the second fraction.

$$\frac{12}{35} \div \frac{27}{14} = \frac{12}{35} \cdot \frac{14}{27}$$

Try to finish this problem on your own. Check your **answer**, or watch this **video** for the complete solution.

My video summary

d. Try to work this problem on your own. Check your **answer**, or watch this **video** for the complete solution.

You Try It **Work through this You Try It problem.**

Work Exercises 21–29 in this textbook or in the MyMathLab Study Plan.

Example 6 Dividing Two Real Numbers

Divide.

a. $\frac{48.6}{-3}$ **b.** $-7\frac{2}{5} \div (-3)$ **c.** $\frac{-59.4}{4.5}$ **d.** $6\frac{5}{8} \div 2\frac{1}{4}$

Solutions

a. The two numbers have different **signs** so we divide the **absolute values** of the numbers and make the quotient **negative**. (View the **details**.)

$$3\overline{)48.6} = 16.2$$

The quotient is -16.2.

b. The two numbers have the same **sign** (both negative) so the **quotient** will be positive. We change the **mixed number** to an **improper fraction** and then divide the **absolute values** of the two numbers.

$$7\frac{2}{5} \div 3 = \frac{37}{5} \div 3 = \underbrace{\frac{37}{5} \cdot \frac{1}{3}}_{\text{Change to multiplication by the reciprocal}} = \frac{37}{15} = 2\frac{7}{15}$$

So, $-7\frac{2}{5} \div (-3) = 2\frac{7}{15}$.

My video summary

c–d. Try to complete these problems on your own. Check your **answers**, or watch this **video** for the complete solutions.

You Try It **Work through this You Try It problem.**

Work Exercises 30–34 in this textbook or in the MyMathLab Study Plan.

CAUTION If a quotient is **negative**, it can be written in one of three ways: $-\frac{a}{b} = \frac{-a}{b} = \frac{a}{-b}$.

(eText Screens 10.3-1–10.3-25)

OBJECTIVE 3 TRANSLATE WORD STATEMENTS INVOLVING MULTIPLICATION OR DIVISION

We have already seen many key words and phrases that indicate **addition** or **subtraction.** Figure 15 shows some examples of key words and phrases for **multiplication.**

Key Word	Word Phrase	Mathematical Expression
Product	The *product* of 7 and 8	$7(8)$ $[\text{or } 7 \times 8, 7 \cdot 8, 7*8]$
Times	-7 *times* -9	$(-7)(-9)$
Of (with fractions)	One-third *of* 27	$\frac{1}{3}(27)$
Of (with percents)	15% *of* 200	$0.15(200)$
Twice	*Twice* 7	$2(7)$

Figure 15 Key words meaning multiplication

Just as there are many keywords to indicate multiplication, Figure 15 illustrates different **mathematical symbols** that can be used to represent multiplication. The most common symbols are the cross, $\times$, and the dot, $\cdot$. **Grouping symbols** such as parentheses can also be used to indicate multiplication. (Note: The asterisk, *, is also used to display multiplication, primarily on calculator view screens. View an **example**.)

Example 7 Translating Word Statements Involving Multiplication

Write a **mathematical expression** for each word phrase.

a. The product of 3 and -6

b. 30% of 50

c. Three times the sum of 10 and 4

d. Three-fourths of 20, increased by 7

e. The difference of 2 and the product of 8 and 15

f. 3 increased by 15, times 4

Solutions

a. The key word "**product**" indicates **multiplication.** What is being multiplied? 3 and -6. Notice that the key word "and" separates the quantities that are being multiplied.

Multiplication

The product of 3 and -6

The two quantities being multiplied

"The product of 3 and -6" is written as $3(-6)$.

b. The key word "of" indicates **multiplication** because it follows a percent.

30% of 50

$\downarrow$ $\downarrow$ $\downarrow$

Write the percent as a decimal → 0.30×50

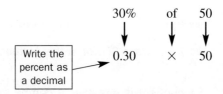

So, "30% of 50" is written as 0.30×50 or $0.30(50)$.

c. "Times" means multiplication. Three and the **sum** are being multiplied. "Sum" indicates **addition**, and the numbers 10 and 4 are being added.

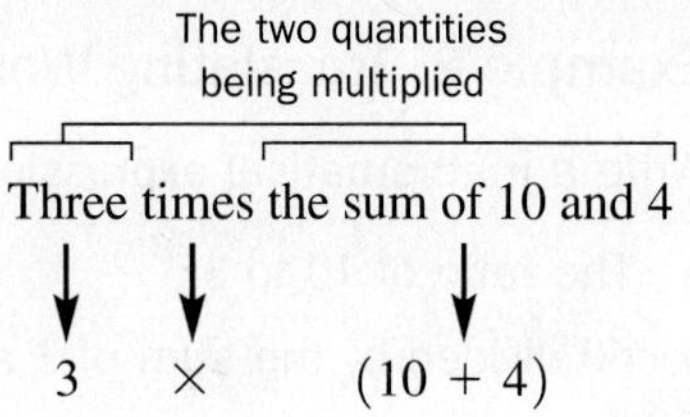

The sum is treated as a single quantity, so we put it within **grouping symbols**. "Three times the sum of 10 and 4" translates to $3 \times (10 + 4)$ or $3(10 + 4)$.

d. The comma separates the phrase into two parts: "Three-fourths of 20" and "increased by 7." The key word "of" indicates **multiplication**, and the numbers $\frac{3}{4}$ and 20 are being multiplied. We stop at the comma when determining the second **factor**. The phrase "increased by" indicates **addition**. We add 7 to everything before the comma: "three-fourths of 20."

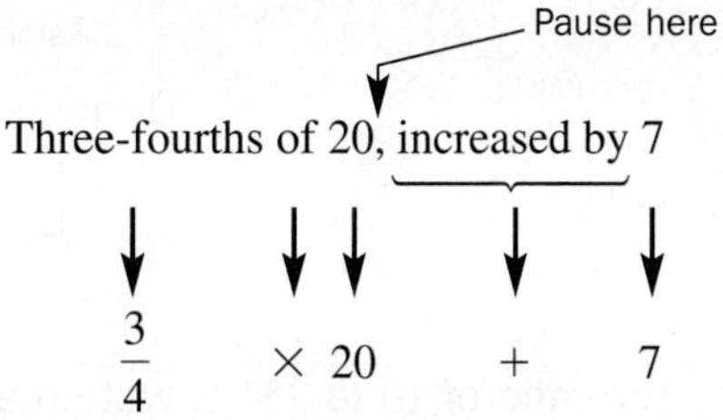

"Three-fourths of 20, increased by 7" translates to $\left(\frac{3}{4} \times 20\right) + 7$ or $\frac{3}{4}(20) + 7$.

My video summary

e. Try to translate this phrase on your own. Check your **answer**, or watch this **video** for the complete solution.

f. Try to translate this phrase on your own. Remember that the comma is used to separate phrases into parts. Check your **answer**, or watch this **video** for the complete solution.

You Try It **Work through this You Try It problem.**

Work Exercises 35–41 in this textbook or in the MyMathLab Study Plan.

How would the result in Example 7d change if the phrase did not contain a comma? **Find out.**

As with **multiplication**, there are many key words and phrases that indicate **division**. Figure 16 shows some examples.

Key Word	Word Phrase	Mathematical Expression
Quotient	The *quotient* of 12 and 6	$\frac{12}{6}$ [or $12 \div 6$, $12/6$]
Divided by	24 *divided by* −3	$24 \div (-3)$
Per	3 tutors *per* 50 students	$\frac{3 \text{ tutors}}{50 \text{ students}}$
Ratio	The *ratio* of 4 to 9	$\frac{4}{9}$

Figure 16 Key words meaning division

(eText Screens 10.3-1–10.3-25)

When translating word statements involving division, typically the **numerator** is the first number or expression given and the **denominator** is the second number or expression.

Example 8 Translating Word Statements Involving Division

Write a **mathematical expression** for each word phrase.

a. The ratio of 10 to 35

b. 60 divided by the sum of 3 and 7

c. The quotient of 20 and 4

d. The difference of 12 and 7, divided by the difference of 8 and -3

Solutions

a. The key word "ratio" indicates **division.** What is being divided? The first number, 10, is being divided by the second number, 35. Remember that the order of the given numbers is important.

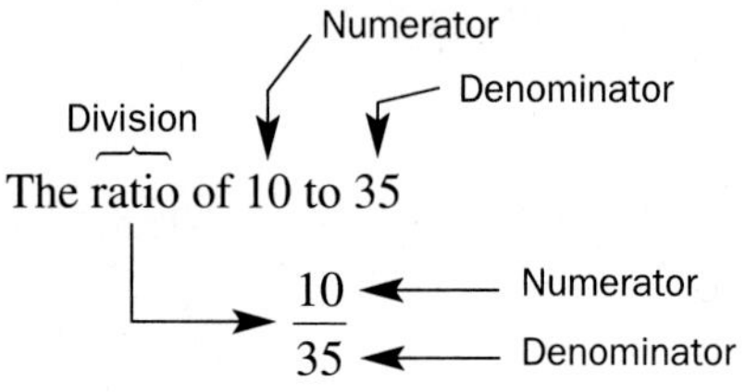

"The ratio of 10 to 35" is written as $\frac{10}{35}$.

b. "Divided by" means **division.** 60 is being divided by the **sum,** and "sum" means addition. What is being added? 3 and 7 are being added.

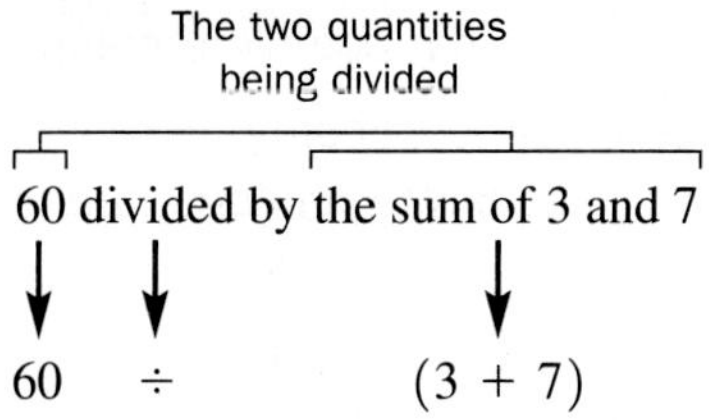

"60 divided by the sum of 3 and 7" translates to $60 \div (3 + 7)$ or $\frac{60}{3 + 7}$.

Note that the sum is grouped as a single quantity either by using a fraction bar or **grouping symbols.**

My video summary

c–d. Try to translate these phrases on your own. Check your **answers,** or watch this **video** for the complete solutions.

You Try It **Work through this You Try It problem.**

Work Exercises 42–48 in this textbook or in the MyMathLab Study Plan.

How would the result in Example 8d have changed if the phrase did not contain a comma? **Find out.**

(eText Screens 10.3-1–10.3-25)

OBJECTIVE 4 SOLVE APPLICATIONS INVOLVING MULTIPLICATION OR DIVISION

As with **addition** and **subtraction**, **multiplication** and **division** occur frequently in application problems.

Example 9 Solving Applications Involving Multiplication or Division

For their 8 home games in the 2009 regular season, the Denver Broncos had a total attendance of 600,928. Assuming equal attendance at all home games, how many people attended each home game? (*Source: espn.go.com*)

Solution The number attending each home game can be found by dividing the total attendance number by the number of home games.

$$\frac{600{,}928}{8} = 75{,}116$$

There were 75,116 people in attendance at each home game (See the **work**).

Example 10 Solving Applications Involving Multiplication or Division

The amount of acid in a solution can be found by multiplying the volume of solution by the percent of acid in the solution (written in decimal form). How much acid is in 20 liters of a 3% solution?

Solution We find the amount of acid by multiplying the solution volume by the percent of the solution. In decimal form, 3% is written as 0.03.

$$20(0.03) = 0.6$$

The solution contains 0.6 liter of acid.

You Try It **Work through this You Try It problem.**

Work Exercises 49–54 in this textbook or in the MyMathLab Study Plan.

10.3 Exercises

In Exercises 1–20, multiply the real numbers.

You Try It

1. $(3)(-2)$ **2.** 4×5 **3.** $(-6)(-1)$ **4.** $9 \cdot 17$

5. $14(-5)$ **6.** $27 \cdot 0$ **7.** $(-18)(-52)$ **8.** $(-7)(-7)$

You Try It

9. $\left(-\frac{3}{2}\right)\left(-\frac{8}{9}\right)$ **10.** $\frac{3}{5}\left(-\frac{2}{7}\right)$ **11.** $0 \cdot \frac{4}{25}$ **12.** $(-7)\left(\frac{5}{14}\right)$

13. $\frac{4}{3} \times 8$ **14.** $\frac{7}{12} \cdot \frac{12}{5}$ **15.** $(5)(-2.6)$ **16.** $(-3.5)(4.2)$

You Try It

17. $(-4.59)(0)$ **18.** $(-2.3)(-10.5)$ **19.** $8 \times 2\frac{1}{4}$ **20.** $\left(-3\frac{2}{9}\right)(6)$

(eText Screens 10.3-1–10.3-25)

In Exercises 21–34, divide the real numbers.

21. $\frac{42}{-6}$ **22.** $160 \div (-20)$ **23.** $\frac{-64}{-4}$ **24.** $\frac{0}{-8}$

You Try It

25. $-\frac{5}{8} \div \left(-\frac{3}{2}\right)$ **26.** $-15 \div \frac{5}{2}$ **27.** $\frac{1}{3} \div \frac{1}{4}$ **28.** $0 \div \left(-\frac{2}{5}\right)$

29. $\frac{\frac{5}{12}}{\frac{3}{10}}$ **30.** $\frac{-5}{0.4}$ **31.** $\frac{-0.6}{-4}$

You Try It

32. $\frac{47.25}{-6.3}$ **33.** $-13 \div 4\frac{1}{3}$ **34.** $\left(3\frac{1}{3}\right) \div \left(-1\frac{1}{2}\right)$

In Exercises 35–48, write a mathematical expression for each word phrase.

You Try It

35. 22% of 18 **36.** The product of 5 and −20

37. Two-thirds of −20 **38.** 3 more than twice 0.37

39. 19 decreased by 12, times $\frac{7}{8}$ **40.** 19, decreased by 12 times $\frac{7}{8}$

41. Eighteen more than the product of 0.7 and the difference of 20 and 4

You Try It

42. The quotient of 24 and −7 **43.** 5 chaperones per 60 students

44. The ratio of 7 to 12 **45.** 3.76 divided by 5

46. 26% of 20, divided by 8 **47.** 30 increased by 12, divided by $\frac{7}{8}$

48. The difference of 5 and 9, divided by the difference of 10 and 4

In Exercises 49–54, solve each application by using multiplication or division.

You Try It

49. In 2010, first-class postage for a standard postcard was \$0.28. At this price, how much would it cost to mail 9 postcards? (*Source: usps.com*)

50. A serving of chicken rice soup contains 1.5 grams of fat. If a can of soup contains 2.5 servings, how many total grams of fat are in the can?

51. For 80 home games in the 2009 season, the Chicago Cubs had a total attendance of about 3,168,800. Assuming equal attendance at all home games, how many people attended each home game? (*Source: espn.go.com*)

52. A farmer decides to retire and distribute his 357 acres of land equally among his 5 children. How many acres will each child receive?

53. How much acid is in 40 liters of a 30% solution?

54. How much acid is in 7.5 gallons of a 7% solution?

(eText Screens 10.4-1–10.4-17)

10.4 Exponents and Order of Operations

THINGS TO KNOW

Before working through this topic, be sure you are familiar with the following concepts:

		VIDEO	ANIMATION	INTERACTIVE
You Try It	1. Add Two Real Numbers with Different Signs (Topic 10.2, **Objective 2**)	▣		
You Try It	2. Subtract Real Numbers (Topic 10.2, **Objective 3**)			▣
You Try It	3. Multiply Real Numbers (Topic 10.3, **Objective 1**)	▣		
You Try It	4. Divide Real Numbers (Topic 10.3, **Objective 2**)	▣		

OBJECTIVES

1 Evaluate Exponential Expressions

2 Use the Order of Operations to Evaluate Numeric Expressions

---

OBJECTIVE 1 EVALUATE EXPONENTIAL EXPRESSIONS

In **Topic 10.3**, we learned that multiplication is repeated addition. Similarly, we can represent repeated multiplication by using *exponents*.

Consider the **product** $2 \cdot 2 \cdot 2 \cdot 2 \cdot 2$. The **factor** 2 is repeated 5 times. We can write this product as the *exponential expression* 2^5. The number 5 is called the **exponent**, or **power**, and indicates that the factor 2 is repeated 5 times. The number 2 is called the **base** and indicates the factor that is being repeated.

$$\underbrace{2 \cdot 2 \cdot 2 \cdot 2 \cdot 2}_{5 \text{ factors of } 2} = 2^5 \quad \leftarrow \text{Exponent; } 2 \text{ is the Base}$$

We read 2^5 as "2 raised to the fifth power."

> **Definition Exponential Expression**
>
> If a is a **real number** and n is a **natural number**, then the **exponential expression** a^n represents the product of n factors of a.
>
> $$a^n = \underbrace{a \cdot a \cdot a \cdot \ldots \cdot a}_{n \text{ factors of } a}$$
>
> a is the **base**, and n is the **exponent** or **power**. We read a^n as "a raised to the nth power" or "a to the nth."

When we raise a number to the first **power**, the exponent is 1. Any real number raised to the first power is equal to itself. For example, $2^1 = 2$, $3^1 = 3$, and so on. If no exponent is written, it is assumed to be an **exponent of** 1.

(eText Screens 10.4-1–10.4-17)

We **evaluate** numeric **exponential expressions** by multiplying the **factors**. For example, $2^5 = 2 \cdot 2 \cdot 2 \cdot 2 \cdot 2 = 32$.

Example 1 Evaluate Exponential Expressions

Evaluate each exponential expression.

a. 4^3 **b.** $\left(\frac{2}{3}\right)^4$ **c.** $(0.3)^2$

Solutions

a. 4^3 means to multiply 3 factors of 4, so $4^3 = 4 \cdot 4 \cdot 4 = 64$.

b. $\left(\frac{2}{3}\right)^4$ means to multiply 4 factors of $\frac{2}{3}$, so $\left(\frac{2}{3}\right)^4 = \frac{2}{3} \cdot \frac{2}{3} \cdot \frac{2}{3} \cdot \frac{2}{3} = \frac{16}{81}$.

c. $(0.3)^2 = (0.3)(0.3) = 0.09$

You Try It **Work through this You Try It problem.**

Work Exercises 1–5 in this textbook or in the MyMathLab Study Plan.

When a negative **sign** is involved, we must see if the negative sign is part of the **base**. For example, consider $(-3)^2$. The parentheses around -3 indicate that the negative sign is part of the base.

Exponent

$$\underbrace{(-3)}_{\text{Base is } -3}{}^{2} = \overbrace{(-3)(-3)}^{\text{Two factors of } -3} = 9.$$

In the expression -3^2, the negative sign is not part of the base. Instead, we must find the **opposite** of 3^2.

Exponent

$$-\underbrace{3}_{\text{Base is } 3}{}^{2} = -\overbrace{(3 \cdot 3)}^{\text{Two factors of } 3} = -9.$$

Take the opposite

Example 2 Evaluate Exponential Expressions

Evaluate each exponential expression.

My video summary

a. $(-5)^3$ **b.** -4^2 **c.** $(-2)^4$

Solutions

a. The parentheses around the -5 indicate that the base is -5, and the negative sign is part of the base. So, $(-5)^3 = (-5)(-5)(-5) = -125$.

b. Because there are no parentheses around the -4, the negative **sign** is not part of the **base**. Instead, we find the **opposite** of 4^2. So, $-4^2 = -(4 \cdot 4) = -16$.

c. Try to evaluate this expression on your own. Check your **answer**, or watch this **video** for the complete solutions to all three parts.

You Try It **Work through this You Try It problem.**

Work Exercises 6–12 in this textbook or in the MyMathLab Study Plan.

(eText Screens 10.4-1–10.4-17)

When the base of an exponential expression is negative, how do we use the **exponent** to determine the sign of the result? **Find out.**

OBJECTIVE 2 USE THE ORDER OF OPERATIONS TO EVALUATE NUMERIC EXPRESSIONS

The **numeric expression** $2 + 3 \cdot 4$ contains two **operations**: addition and multiplication. Depending on which operation is performed first, a different answer will result.

Add first, then multiply: $2 + 3 \cdot 4 = \quad 5 \cdot 4 = 20$

Different Results

Multiply first, then add: $2 + 3 \cdot 4 = 2 + 12 = 14$

For this reason, mathematicians have agreed on a specific **order of operations**.

Order of Operations

1. **Parentheses (or other grouping symbols)** Evaluate operations within parentheses (or other **grouping symbols**) first, starting with the innermost set and working out.
2. **Exponents** Work from left to right and evaluate any **exponential expressions** as they occur.
3. **Multiplication and Division** Work from left to right and perform any **multiplication** or **division** operations as they occur.
4. **Addition and Subtraction** Work from left to right and perform any **addition** or **subtraction** operations as they occur.

Because we perform multiplication before addition, the correct **evaluation** for our numeric expression above is $2 + 3 \cdot 4 = 14$, not 20.

View this **tip** on remembering the order of operations.

Example 3 Using Order of Operations to Evaluate Numeric Expressions

Simplify each expression.

a. $10 - 4^2$ **b.** $12 \div 4 + 8$

Solutions For each expression, we follow the **order of operations**.

a. The two **operations** are subtraction and exponents. Exponents have priority over subtraction, so we evaluate the exponent first.

Begin with the original expression: $10 - 4^2$

Evaluate 4^2: $= 10 - 16$

Subtract: $= -6 \leftarrow 10 - 16 = 10 + (-16)$

b. The two operations are division and addition. Because division has priority over addition, we divide first.

Begin with the original expression: $12 \div 4 + 8$

Divide $12 \div 4$: $= 3 + 8$

Add: $= 11$

You Try It **Work through this You Try It problem.**

Work Exercises 13–20 in this textbook or in the MyMathLab Study Plan.

(eText Screens 10.4-1–10.4-17)

In the **order of operations,** multiplication and division are given the same priority and are performed from left to right as *either* occurs. It is incorrect to do all the multiplication first and then all the division. View this **illustration**.

Similarly, addition and subtraction are given equal priority and are performed from left to right as *either* occurs. It is incorrect to do all the addition first and then all the subtraction. View this **illustration.**

Example 4 Using Order of Operations to Evaluate Numeric Expressions

Simplify each expression.

My video summary

a. $15 - 3 + 6 - 8 + 7$ **b.** $3 \cdot 15 \div 5 \cdot 6 \div 2$

Solutions For each expression, follow the order of operations. Try to do these on your own. Check your **answers**, or watch the **video** for complete solutions.

Example 5 Using Order of Operations to Evaluate Numeric Expressions

Simplify each expression.

a. $5 + (4 - 2)^2 - 3^2$ **b.** $[5 - 9]^2 + 12 \div 4$

Solutions For each expression, we follow the order of operations.

a. Begin with the original expression: $5 + (4 - 2)^2 - 3^2$

Evaluate $4 - 2$ inside parentheses: $= 5 + (2)^2 - 3^2$

Evaluate 2^2 and 3^2: $= 5 + 4 - 9$

Add $5 + 4$: $= 9 - 9$

Subtract $9 - 9$: $= 0$

b. Begin with the original expression: $[5 - 9]^2 + 12 \div 4$

Evaluate $5 - 9$ inside brackets: $= [-4]^2 + 12 \div 4$

Evaluate $[-4]^2$: $= 16 + 12 \div 4$

Divide $12 \div 4$: $= 16 + 3$

Add $16 + 3$: $= 19$

You Try It **Work through this You Try It problem.**

Work Exercises 21–27 in this textbook or in the MyMathLab Study Plan.

Because **grouping symbols** have the highest priority in **order of operations**, we can use them to control the order in which computations are completed. This can be especially helpful if using a calculator to do computations.

Example 6 Using Order of Operations to Evaluate Numeric Expressions

My video summary

Simplify each expression.

a. $(-5 + 8) \cdot 3$ **b.** $(10 - 4)^2$ **c.** $12 \div (4 + 8)$

Solutions Try to do these problems on your own. Check your **answers**, or watch the **video** for complete solutions. Compare your results in parts b and c to the results from **Example 3**. Notice how the parentheses affect which operation is performed first and the final result.

You Try It **Work through this You Try It problem.**

Work Exercises 28–30 in this textbook or in the MyMathLab Study Plan.

(eText Screens 10.4-1–10.4-17)

Parentheses, ( ), brackets, [ ], and braces, { }, are the most common **grouping symbols.** Fraction bars, ——, however, and **absolute value** symbols, | |, are also treated as grouping symbols. A fraction bar separates the expression into two parts: the **numerator** and **denominator.** The numerator is grouped together and the denominator is grouped together. View this **illustration.**

CAUTION Grouping the numerator together and grouping the denominator together is particularly important when using a calculator to **evaluate** an expression. A calculator will strictly follow order of operations and cannot guess what you meant to enter. View this **calculator example.**

We must **simplify expressions** separately in the numerator and denominator of a **fraction** before dividing. Similarly, we simplify expressions within absolute value symbols before evaluating the absolute value.

$$3^2 - 5\cdot 2 + |3^2 - 8 \div 4|$$

Absolute value is treated as a grouping symbol so we simplify inside the absolute value first.

CAUTION Once the expression within the absolute value is simplified, the absolute value can be evaluated with the same priority as **exponents.** For example, in the expression $-5\cdot|-6|$, we find the absolute value first and then perform the multiplication.

Original expression: $-5\cdot|-6|$

Evaluate $|-6|$: $= -5\cdot 6$

Multiply $-5\cdot 6$: $= -30$

Example 7 Using Order of Operations with Special Grouping Symbols

Simplify each expression.

a. $\dfrac{-2(3) + 6^2}{(-4)^2 - 1}$ b. $|7^2 - 5(3)| \div 2 + 8$

Solutions

a. The **fraction bar** is a **grouping symbol**. We begin by simplifying the numerator and denominator separately before dividing.

Begin with the original expression: $\dfrac{-2(3) + 6^2}{(-4)^2 - 1}$

Evaluate the exponents: $= \dfrac{-2(3) + 36}{16 - 1}$ $\leftarrow 6^2 = 36$, $\leftarrow (-4)^2 = 16$

Multiply in the numerator: $= \dfrac{-6 + 36}{16 - 1}$

Add and subtract: $= \dfrac{30}{15}$ $\leftarrow -6 + 36 = 30$, $\leftarrow 16 - 1 = 15$

Divide: $= 2$

My video summary

b. The **absolute value** symbol is a grouping symbol. First, we simplify within the absolute value symbols. Try to do this on your own. Check your **answer**, or watch the **video** for a complete solution.

You Try It **Work through this You Try It problem.**

Work Exercises 31–36 in this textbook or in the MyMathLab Study Plan.

(eText Screens 10.4-1–10.4-17)

Some **numeric expressions** contain **nested grouping symbols** (grouping symbols within grouping symbols). In such situations, we begin with the innermost set of grouping symbols and work our way outward. To avoid confusion, we can use different types of grouping symbols if more than one set is required.

Example 8 Using Order of Operations with Nested Grouping Symbols

Simplify each expression.

a. $[2^3 - 3(5 - 7)^2] \div 6 - 9$ **b.** $\dfrac{|-5^2 + 2^3| - 10}{4^2 - 6 \cdot 5}$

Solutions

a. Begin with the original expression: $[2^3 - 3(\underbrace{5 - 7}_{\text{Innermost grouping 1st}})^2] \div 6 - 9$

Evaluate $(5 - 7)$: $= \underbrace{[2^3 - 3(-2)^2]}_{\text{Outermost grouping 2nd}} \div 6 - 9$

Evaluate exponents 2^3 and $(-2)^2$: $= [8 - 3(4)] \div 6 - 9$

Multiply $3(4)$: $= [8 - 12] \div 6 - 9$

Subtract $8 - 12$: $= (-4) \div 6 - 9$

Divide $-4 \div 6$: $= -\dfrac{2}{3} - 9 \leftarrow$ $\boxed{\dfrac{-4}{6} = \dfrac{-2 \cdot \cancel{2}}{3 \cdot \cancel{2}} = \dfrac{-2}{3} = -\dfrac{2}{3}}$

Subtract $-\dfrac{2}{3} - 9$: $= -\dfrac{29}{3} \leftarrow$ $\boxed{-\frac{2}{3} - 9 = -\frac{2}{3} - \frac{27}{3} = -\frac{2}{3} + \left(-\frac{27}{3}\right) = -\frac{29}{3}}$

My video summary

b. The **absolute value** bars act as grouping symbols. First, we simplify within the absolute value symbols. Try to do this problem on your own, then check your **answer**. Watch this **video** for a complete solution.

You Try It **Work through this You Try It problem.**

Work Exercises 37–40 in this textbook or in the MyMathLab Study Plan.

Example 9 Using Order of Operations to Evaluate Numeric Expressions

My interactive video summary

Simplify each expression.

a. $\dfrac{3}{10} \cdot \dfrac{5}{2} - \dfrac{1}{2}$ **b.** $36 \div \dfrac{8}{3^2 - 5} + (-2)^3$ **c.** $\dfrac{\left|\dfrac{1}{3} - \dfrac{3}{5}\right|}{4} \div \dfrac{1}{2} - 1$

Solutions For each expression, follow the **order of operations**. Check your **answers**, or watch the **interactive video** for the complete solutions.

You Try It **Work through this You Try It problem.**

Work Exercises 41–48 in this textbook or in the MyMathLab Study Plan.

10.4 Exercises

In Exercises 1–12, evaluate each exponential expression.

You Try It

1. 2^8 **2.** $(0.2)^5$ **3.** $\left(\frac{3}{4}\right)^4$ **4.** $\left(\frac{3}{2}\right)^4$ **5.** 1^9 **6.** $(-3)^4$

You Try It

7. -3^4 **8.** $-\left(\frac{3}{5}\right)^3$ **9.** $-\left(-\frac{3}{5}\right)^3$ **10.** $-(0.1)^4$ **11.** $-(-0.1)^4$ **12.** 0^3

In Exercises 13–48, simplify each expression using the order of operations.

You Try It

13. $3\cdot 2 + 4$ **14.** $3 + 2\cdot 4$ **15.** $3 - 2^2$ **16.** $6 \div 3 - 1$

17. $\frac{1}{2} \div \frac{1}{3} + \frac{1}{4}$ **18.** $-6 + 3 - 1$ **19.** $6 \div 2\cdot\frac{1}{3}$ **20.** $(3 + 2)\cdot 4$

You Try It

21. $20 - 4\cdot 9 + 12$ **22.** $70 - 42 \div 7 + 9$ **23.** $8 \div 4 + 3^2$ **24.** $2(-3 + 5)^2 - (8 - 5)^2$

25. $(2\cdot 3)^2 + 2\cdot 3^2$ **26.** $2(9 - 5)^3$ **27.** $7\cdot 2 + 3^2$ **28.** $2^2 - 5\cdot 2 + (2^2 - 5)\cdot 2$

You Try It

29. $6 \div (3 - 1) + 6 \div 3 - 1$ **30.** $(40 \div 2)^2 - 40 \div 2^2$

31. $\dfrac{2 + 5\cdot 3}{2\cdot 5 + 3}$ **32.** $|-2|^3 - |3|^2$

You Try It

33. $\dfrac{|6 - 11| + 4}{4^2 - 2(5)}$ **34.** $\dfrac{5^2 - 3^2}{5 - 3}$ **35.** $\dfrac{3(15)-5}{(-2)^2 + 3}$ **36.** $\dfrac{2 - 7\cdot 1}{3\cdot 2 - 1} \div \dfrac{2\cdot 7 + 1}{3 + 2\cdot 1}$

You Try It

37. $[5 + 3(10 - 7)^2]$ **38.** $\dfrac{(3 + 2^3)^2 - 11}{7(4 - 5)-3}$

39. $|1 - 3|^3\cdot\left[6\cdot\frac{2}{3} + 4\left(1 + \frac{3}{4}\right)\right]$ **40.** $\left\{\left[(-2)^3 + \frac{2}{3}\cdot 6\right] \div 12\right\}^5$

You Try It

41. $12 \div 4 \div 3$ **42.** $24 \div (6\cdot 2)$ **43.** $12 \div 4\cdot 3$ **44.** $24 \div (6 \div 2)$

45. $1 + 2\cdot 3 - 4 \div 5$ **46.** $7 - 2(3 - 1)$ **47.** $3^2 + 2(3 - 5)$ **48.** $\dfrac{\left(\frac{3}{2} + 1\right) \div \frac{1}{2}}{2^6 - 8\cdot 4}$

10.5 Variables and Properties of Real Numbers

THINGS TO KNOW

Before working through this topic, be sure you are familiar with the following concepts:

		VIDEO	ANIMATION	INTERACTIVE
You Try It	1. Find the Opposite of a Real Number (Topic 10.1, **Objective 3**)			
You Try It	2. Add Two Real Numbers with Different Signs (Topic 10.2, **Objective 2**)	✓		
You Try It	3. Subtract Real Numbers (Topic 10.2, **Objective 3**)			✓
You Try It	4. Multiply Real Numbers (Topic 10.3, **Objective 1**)	✓		
You Try It	5. Use the Order of Operations to Evaluate Numeric Expressions (Topic 10.4, **Objective 2**)	✓		✓

OBJECTIVES

1 Evaluate Algebraic Expressions

2 Use the Commutative and Associative Properties

3 Use the Distributive Property

4 Use the Identity and Inverse Properties

OBJECTIVE 1 EVALUATE ALGEBRAIC EXPRESSIONS

So far, we have focused only on **numeric expressions.** Now we expand our discussion to include *algebraic expressions*.

A **variable** is a symbol (usually a letter) that represents a changing value. For example, suppose the letter x represents the number of iTunes downloads purchased by a user. This might be 20 for one user and 2000 for another. A **constant** has a value that never changes. For example, 3 is a constant because the value of 3 is always the same.

An **algebraic expression** is a **variable** or a combination of variables, **constants, operations,** and **grouping symbols.** Examples of algebraic expressions include

$$\underbrace{x}_{\text{A number}},\quad \underbrace{5z}_{\text{Five times a number}},\quad \underbrace{x^2}_{\text{The square of a number}},\quad \underbrace{w+7}_{\text{A number increased by 7}},\quad \underbrace{5a-3}_{\text{Three less than five times a number}},\quad \text{and}\quad \underbrace{2(x+y)}_{\text{Twice the sum of two numbers}}.$$

If a constant appears next to a variable, then we multiply the number and the variable. For example, $5z = 5 \cdot z$ and $5a - 3 = 5 \cdot a - 3$.

Algebraic expressions can describe quantities in a general way. For example, if the common price for an iPhone app is $\$0.99$ (*Source:* pinchmedia.com), then the algebraic expression

$0.99x$ gives us the cost for purchasing x apps at this price. To determine the cost for a specific number of apps, we replace the variable with a number and then **simplify** the resulting **numeric expression**. This process is called **evaluating** the expression for a given value of the variable.

For example, to find the cost for purchasing 6 apps, we substitute 6 for x and simplify the resulting numeric expression, or **evaluate** the algebraic expression $0.99x$ for $x = 6$:

$$0.99x = 0.99(6) = 5.94$$

The value of the algebraic expression $0.99x$ is 5.94 when $x = 6$, so 6 apps cost \$5.94. When the value of the **variable** changes, the value of the algebraic expression will change as well.

Evaluate Algebraic Expressions

To **evaluate an algebraic expression**, substitute the given values for the variables and simplify the resulting numeric expression using the **order of operations**.

Example 1 Evaluating Algebraic Expressions

Evaluate each **algebraic expression** for the given values of the variables. Remember to follow **order of operations** when simplifying.

a. $3x + 7$ for $x = 5$

b. $a^2 - 3$ for $a = -4$

Solutions

a. Begin with the original algebraic expression: $3x + 7$

Substitute 5 for x: $= 3(5) + 7$

Multiply: $= 15 + 7$

Add: $= 22$

b. Begin with the original algebraic expression: $a^2 - 3$

Substitute -4 for a: $= (-4)^2 - 3$

Evaluate the exponent: $= 16 - 3$

Subtract: $= 13$

You Try It **Work through this You Try It problem.**

Work Exercises 1–5 in this textbook or in the MyMathLab Study Plan.

When substituting a given value for a variable, write the value within parentheses to help avoid mathematical mistakes.

In Example 1, notice that we put parentheses around given values for a variable during the **substitution** process. View this **popup** to see how the problem would be different if we had not used parentheses in part (b).

Example 2 Evaluating Algebraic Expressions

Evaluate each **algebraic expression** for the given values of the **variables**. Remember to follow **order of operations** when simplifying.

a. $5x^2 + 9$ for $x = 7$

b. $-2(m + 3) - 5$ for $m = 8$

Solutions

a. Begin with the original algebraic expression: $5x^2 + 9$

Substitute 7 for x: $= 5(7)^2 + 9$

Evaluate the exponent: $= 5(49) + 9$

Multiply: $= 245 + 9$

Add: $= 254$

My video summary

b. Begin with the original algebraic expression: $-2(m + 3) - 5$

Substitute 8 for m: $= -2(8 + 3) - 5$

Try to finish **simplifying** on your own. Check your **answer**, or watch this **video** for the complete solution.

You Try It **Work through this You Try It problem.**

Work Exercises 6–10 in this textbook or in the MyMathLab Study Plan.

Example 3 Evaluating Algebraic Expressions

Evaluate each **algebraic expression** for the given values of the **variables**.

a. $\dfrac{x^2 + 6}{5x - 2}$ for $x = 2$

b. $|3y - 4| + 7y - 1$ for $y = -3$

Solutions

a. Begin with the original algebraic expression: $\dfrac{x^2 + 6}{5x - 2}$

Substitute 2 for x: $= \dfrac{(2)^2 + 6}{5(2) - 2}$

Evaluate the exponent: $= \dfrac{4 + 6}{5(2) - 2}$

Multiply: $= \dfrac{4 + 6}{10 - 2}$

Add and subtract: $= \dfrac{10}{8}$

Simplify: $= \dfrac{5}{4}$

My video summary

b. Try evaluating this expression on your own using the **order of operations**. Check your **answer**, or watch this **video** for the complete solution.

You Try It **Work through this You Try It problem.**

Work Exercises 11–16 in this textbook or in the MyMathLab Study Plan.

Given values for the **variables**, we can evaluate **algebraic expressions** involving more than one variable.

Example 4 Evaluating Algebraic Expressions

Evaluate each algebraic expression for the given values of the variables.

a. $12a + 7b$ for $a = -4$ and $b = 12$

b. $x^2 - 2xy + 3y^2$ for $x = 3$ and $y = -1$

(eText Screens 10.5-1–10.5-27)

Solutions

a. Begin with the original expression: $12a + 7b$

Substitute -4 for a and 12 for b: $= 12(-4) + 7(12)$

Multiply: $= -48 + 84$

Add: $= 36$

My video summary

b. Try to **simplify** on your own by following the **order of operations**. Check your **answer**, or watch this **video** for the complete solutions.

You Try It **Work through this You Try It problem.**

Work Exercises 17–24 in this textbook or in the MyMathLab Study Plan.

OBJECTIVE 2 USE THE COMMUTATIVE AND ASSOCIATIVE PROPERTIES

In this module, we have reviewed how to simplify **numeric expressions** involving **addition**, **subtraction**, **multiplication**, and **division**. To help with this process, let's review and name several properties of **real numbers**.

The **commutative property of addition** states that the *order* of the **terms** in addition does not affect the **sum**. For example, $4 + 7 = 11$ and $7 + 4 = 11$, so $4 + 7 = 7 + 4$.

Commutative Property of Addition

If a and b are real numbers, then $a + b = b + a$.

The **commutative property of multiplication** states that the order of the **factors** does not affect the **product**. For example, $8(-4) = -32$ and $(-4)(8) = -32$, so $8(-4) = (-4)(8)$.

Commutative Property of Multiplication

If a and b are real numbers, then $a \cdot b = b \cdot a$.

CAUTION

Subtraction and division **do not** have **commutative properties**. For example, $9 - 2 = 7$ but $2 - 9 = -7$. Similarly, $6 \div 3 = 2$ but $3 \div 6 = 0.5$.

Example 5 Using the Commutative Properties

Use the given property to rewrite each statement. Do not simplify.

a. **Commutative property of multiplication:** $-2(6) =$ ______

b. **Commutative property of addition:** $5.03 + 9.2 =$ ______

Solutions

a. Using the commutative property of multiplication, we can change the order of the **factors**: $-2(6) = 6(-2)$.

b. The commutative property of addition allows us to change the order of the **terms**. $5.03 + 9.2 = 9.2 + 5.03$.

You Try It **Work through this You Try It problem.**

Work Exercises 25–30 in this textbook or in the MyMathLab Study Plan.

(eText Screens 10.5-1–10.5-27)

The **associative property of addition** states that regrouping terms does not affect the **sum**. For example, $(7 + 2) + 5 = 9 + 5 = 14$ and $7 + (2 + 5) = 7 + 7 = 14$, so $(7 + 2) + 5 = 7 + (2 + 5)$.

Associative Property of Addition

If a, b, and c are real numbers, then $(a + b) + c = a + (b + c)$.

The **associative property of multiplication** states that regrouping factors does not affect the product. For example, $(-10 \cdot 4) \cdot 6 = (-40) \cdot 6 = -240$ and $-10 \cdot (4 \cdot 6) = -10 \cdot (24) = -240$, so $(-10 \cdot 4) \cdot 6 = -10 \cdot (4 \cdot 6)$.

Associative Property of Multiplication

If a, b, and c are real numbers, then $(a \cdot b) \cdot c = a \cdot (b \cdot c)$.

CAUTION

Subtraction and division **do not** have associative properties. For example, $(7 - 4) - 2 = 3 - 2 = 1$ but $7 - (4 - 2) = 7 - 2 = 5$. Similarly, $(80 \div 8) \div 2 = 10 \div 2 = 5$ but $80 \div (8 \div 2) = 80 \div 4 = 20$.

Example 6 Using the Associative Properties

Use the given property to rewrite each statement. Do not simplify.

a. **Associative property of addition:** $\left(\frac{2}{3} + \frac{1}{6}\right) + \frac{5}{6} =$ ______

b. **Associative property of multiplication:** $5 \cdot (2 \cdot 13) =$ ______

Solutions

a. Using the associative property of addition, we can change the grouping of the terms: $\left(\frac{2}{3} + \frac{1}{6}\right) + \frac{5}{6} = \frac{2}{3} + \left(\frac{1}{6} + \frac{5}{6}\right)$.

b. The associative property of multiplication allows us to change the grouping of the factors: $5 \cdot (2 \cdot 13) = (5 \cdot 2) \cdot 13$.

You Try It **Work through this You Try It problem.**

Work Exercises 31–38 in this textbook or in the MyMathLab Study Plan.

The commutative and associative properties will be helpful when we later simplify expressions and solve equations.

Example 7 Using the Commutative and Associative Properties

Use the commutative and associative properties to simplify each expression.

a. $(3 + x) + 7$

b. $(8y)\left(\frac{1}{2}\right)$

(eText Screens 10.5-1–10.5-27)

Solutions

a. Commutative property of addition: $(3+x)+7 = \overbrace{(x+3)}^{\text{Changed order of the terms}} + 7$

Associative property of addition: $= \overbrace{x+(3+7)}^{\text{Changed grouping}}$

Add 3 and 7: $= x + \overbrace{10}^{3+7=10}$

b. Commutative property of multiplication: $(8y)\left(\frac{1}{2}\right) = \overbrace{\frac{1}{2}(8y)}^{\text{Changed order of the terms}}$

Associative property of multiplication: $= \overbrace{\left(\frac{1}{2}\cdot 8\right)y}^{\text{Changed grouping}}$

Multiply $\frac{1}{2}$ and 8: $= \overbrace{4y}^{\frac{1}{2}\cdot 8 = 4}$

You Try It **Work through this You Try It problem.**

Work Exercises 39–42 in this textbook or in the MyMathLab Study Plan.

OBJECTIVE 3 USE THE DISTRIBUTIVE PROPERTY

Another important property is the **distributive property**, which states that **multiplication** *distributes* over **addition** (or **subtraction**). This means that multiplying a number by a sum is **equivalent** to first multiplying each term by the number and then summing the results. For example, compare $5(3+4) = 5(7) = 35$ with $5\cdot 3 + 5\cdot 4 = 15 + 20 = 35$. Because both expressions simplify to 35, they are equivalent: $5(3+4) = 5\cdot 3 + 5\cdot 4$.

Distributive Property

If a, b, and c are **real numbers**, then $a(b+c) = ab + ac$.

Because multiplication is **commutative,** we can also write the distributive property as

$$(b+c)a = ba + ca.$$

In **Topic 10.2**, we defined subtraction by using addition. So the distributive property also applies to subtraction.

$$a(b-c) = ab - ac$$

The **distributive property** extends to sums (or differences) involving more than two **terms.**

$$a(b+c+d) = ab + ac + ad$$

See **why this works.**

The distributive property will be helpful because it allows us to remove **grouping symbols** so we can combine terms, then regroup in a simpler form.

Example 8 Using the Distributive Property

Use the **distributive property** to remove parentheses and write the **product** as a **sum**. Simplify if possible.

a. $9(x + 2)$ **b.** $(7x - 5) \cdot 3$

Solutions

a. Apply the distributive property: $9(x + 2) = 9 \cdot x + 9 \cdot 2$

Find each product: $= 9x + 18$

b. Apply the **distributive property**: $(7x - 5) \cdot 3 = 7x \cdot 3 - (5) \cdot 3$

Commutative property of multiplication: $= 3 \cdot 7x - (5) \cdot 3$

Find each product: $= 21x - 15$

You Try It **Work through this You Try It problem.**

Work Exercises 43–50 in this textbook or in the MyMathLab Study Plan.

In **Topic 10.1**, we used a negative **sign** to indicate the **opposite** of a real number. We can also find a number's opposite if we multiply the number by -1. For example, the opposite of 8 is $-1 \cdot 8 = -8$, and the opposite of -3 is $-1(-3) = -(-3) = 3$.

The **expression** $-1 \cdot (3z - 4)$ could also be written as $-(3z - 4)$. This simply means we want to find the opposite of the expression inside the parentheses. The result $-3z + 4$ is found by writing the opposite of each **term** within the parentheses. This leads us to the following rule.

> **Opposite of an Expression**
>
> If a negative sign appears in front of an expression within parentheses, remove the parentheses and then change the sign of each term in the expression. The result is the **opposite of the expression**.

Example 9 Using the Distributive Property

Use the **distributive property** to remove parentheses and write the **product** as a **sum**. Simplify if possible.

a. $2(4y + 3z - 5)$ **b.** $-6(3y - 8)$ **c.** $-(2a - 7b + 8)$

Solutions

a. Begin with the original expression: $2(4y + 3z - 5)$

Apply the distributive property: $= 2 \cdot 4y + 2 \cdot 3z - 2 \cdot 5$

Associative property of multiplication: $= (2 \cdot 4)y + (2 \cdot 3)z - (2 \cdot 5)$

Find each product: $= 8y + 6z - 10$

My video summary

b–c. Try to work these problems on your own. Check your **answers**, or watch this **video** for the complete solutions.

You Try It **Work through this You Try It problem.**

Work Exercises 51–60 in this textbook or in the MyMathLab Study Plan.

Example 10 Using the Distributive Property

Use the **distributive property** to write each sum as a product.

My video summary

a. $9 \cdot x + 9 \cdot 4$ b. $4x + 4y$

Solutions Remove the common **factor** from each **term**, then write the remaining expression in parentheses. Work through the following, or watch the **video** for the solutions.

a. Begin with the original expression: $9 \cdot x + 9 \cdot 4$

Remove the common factor: $= (9 \cdot x + 9 \cdot 4)$

Remaining expression in parentheses: $= 9(x + 4)$

b. Begin with the original expression: $4x + 4y$

Remove the common factor: $= (4x + 4y)$

Remaining expression in parentheses: $= 4(x + y)$

You Try It **Work through this You Try It problem.**

Work Exercises 61–64 in this textbook or in the MyMathLab Study Plan.

OBJECTIVE 4 USE THE IDENTITY AND INVERSE PROPERTIES

We complete our discussion of the properties of **real numbers** with the *identity* and *inverse properties*.

If 0 is added to any real number, then the **sum** is the original number. For example, $4 + 0 = 4$ and $0 + 7 = 7$. This is called the **additive identity**.

Identity Property of Addition

If a is a real number, then $a + 0 = a$ and $0 + a = a$.

Similarly, if 1 is multiplied by a real number, then the **product** is the original number. For example, $5 \cdot 1 = 5$ and $1 \cdot 9 = 9$. This is called the **multiplicative identity**.

Identity Property of Multiplication

If a is a real number, then $a \cdot 1 = a$ and $1 \cdot a = a$.

The **opposite** of a real number is also called the **additive inverse** of the number because the sum of a number and its inverse is zero. For example, -6 is the opposite of 6, and -6 is the additive inverse of 6 because $6 + (-6) = 0$.

Inverse Property of Addition

If a is a real number, then there is a **unique** real number $-a$ such that

$$a + (-a) = -a + a = 0.$$

(eText Screens 10.5-1–10.5-27)

Similarly, the **reciprocal** of a **real number** is called the **multiplicative inverse** of the number because the product of a number and its reciprocal is 1. For example, the reciprocal, or multiplicative inverse, of $\frac{5}{3}$ is $\frac{3}{5}$ because $\frac{5}{3}\cdot\frac{3}{5} = 1$.

Inverse Property of Multiplication

If a is a nonzero real number, then there is a unique real number $\frac{1}{a}$ such that

$$a\cdot\frac{1}{a} = \frac{1}{a}\cdot a = 1.$$

Example 11 Using the Identity and Inverse Properties

Identify the property of real numbers illustrated in each statement.

a. $-4\cdot 1 = -4$

b. $(-5 + 5) + x = 0 + x$

c. $0 + y = y$

d. $\frac{1}{2}\cdot 2x = x$

Solutions Try to identify each property on your own. Check your **answers**.

You Try It **Work through this You Try It problem.**

Work Exercises 65–68 in this textbook or in the MyMathLab Study Plan.

10.5 Exercises

In Exercises 1–16, evaluate each algebraic expression for the given value of the variable.

You Try It

1. $2x + 6$ for $x = 7$

2. $8 - 5x$ for $x = -3$

3. $4w^2$ for $w = -6$

4. $\frac{3}{5}x - 7$ for $x = 10$

5. $\frac{3}{4}x - 9$ for $x = -\frac{5}{3}$

6. $\frac{2.5y + 3.8}{4}$ for $y = 1.4$

You Try It

7. $3a^2 - 10$ for $a = 4$

8. $-2(n + 4) + 9$ for $n = \frac{1}{2}$

9. $\frac{4b + 3}{5}$ for $b = -2$

10. $4|5 - x| + 1$ for $x = -8$

11. $-z^2 + 3z$ for $z = -1$

12. $5 - 4y + 2y^2$ for $y = 3$

You Try It

13. $\frac{2x + 6}{x - 5}$ for $x = 10$

14. $\frac{7 - 3x}{x^2 + 1}$ for $x = -2$

15. $|2a - 3| + a - 8$ for $a = -5$

16. $\frac{2(c + 3)}{c - 3}$ for $c = \frac{2}{3}$

In Exercises 17–24, evaluate each algebraic expression for the given values of the variables.

You Try It

17. $2x - 3y$ for $x = -4$ and $y = 5$

18. $20a + 9b$ for $a = -\frac{1}{4}$ and $b = \frac{2}{3}$

19. $m^2 - 4mn - 12n^2$ for $m = 1$ and $n = -3$

20. $a^2 - b^2$ for $a = -3$ and $b = -5$

21. $|3x^2 - y|$ for $x = 3$ and $y = -2$

22. $mx + b$ for $m = \frac{1}{5}, x = 15$, and $b = -4$

23. $b^2 - 4ac$ for $a = 1, b = -3$, and $c = -6$

24. $\frac{y_2 - y_1}{x_2 - x_1}$ for $x_1 = 9, x_2 = -3, y_1 = -4$, and $y_2 = 2$

In Exercises 25–30, use a commutative property to complete each statement. Do not simplify.

You Try It

25. $5 + 9 =$ ______

26. $-6 \cdot y =$ ______

27. $11 + x =$ ______

28. $\frac{1}{5}x =$ ______

29. $8.3(-3) =$ ______

30. $m + (-8) =$ ______

In Exercises 31–38, use an associative property to complete each statement. Do not simplify.

You Try It

31. $(3 + 6) + 15 =$ ______

32. $-3(4 \cdot 17) =$ ______

33. $(7y) \cdot 4 =$ ______

34. $\frac{3}{4} + \left(x + \frac{5}{9}\right) =$ ______

35. $(1.45 + a) + b =$ ______

36. $0.05(20x) =$ ______

37. $-6 + (6 + c) =$ ______

38. $\frac{3}{5}\left(\frac{5}{3}w\right) =$ ______

In Exercises 39–42, use the commutative and associative properties to simplify each expression.

You Try It

39. $-2 + (7 + x)$

40. $-4(2m)$

41. $(3z) \cdot 8$

42. $\left(\frac{1}{3} + y\right) + \frac{2}{5}$

In Exercises 43–60, use the distributive property to remove parentheses and write the product as a sum. Simplify if possible.

You Try It

43. $3(a + b)$

44. $10(x - 4)$

45. $4(2y - 5)$

46. $3.2(5x + 2)$

47. $(8x + 3) \cdot 4$

48. $-1 \cdot (2z + 7)$

49. $(3x + 2y)(-4)$

50. $\frac{2}{3}(3x + 9)$

You Try It

51. $-(7 - n)$

52. $0.5(4x - 1)$

53. $(3x - 16) \cdot \frac{1}{4}$

54. $-4\left(\frac{3}{2}m - \frac{1}{10}\right)$

55. $0.2(1.5 + 3.4x)$

56. $-(3a - 2b - 4)$

57. $12(x + 4 - 5y)$

58. $-2(4w - 9z)$

59. $3(5x + y - 7)$

60. $(4r + s + 5)(-0.4)$

In Exercises 61–64, use the distributive property to write each sum as a product.

You Try It

61. $4 \cdot x + 4 \cdot 1$

62. $10a + 10b$

63. $\frac{1}{2} \cdot y - \frac{1}{2} \cdot 6$

64. $(-5)m + (-5)(2)$

In Exercises 65–68, identify the property of real numbers illustrated in each statement.

You Try It

65. $20 \cdot 1 = 20$

66. $x + 14 + (-14) = x$

67. $\left(\frac{1}{3} \cdot 3\right)x = 1x$

68. $0 + 9 = 9$

10.6 Simplifying Algebraic Expressions

THINGS TO KNOW

Before working through this topic, be sure you are familiar with the following concepts:

		VIDEO	ANIMATION	INTERACTIVE
You Try It	1. Translate Word Statements Involving Addition or Subtraction (Topic 10.2, **Objective 4**)	✓		
You Try It	2. Translate Word Statements Involving Multiplication or Division (Topic 10.3, **Objective 3**)	✓		
You Try It	3. Use the Order of Operations to Evaluate Numeric Expressions (Topic 10.4, **Objective 2**)	✓		✓

OBJECTIVES

1 Identify Terms, Coefficients, and Like Terms of an Algebraic Expression

2 Simplify Algebraic Expressions

3 Write Word Statements as Algebraic Expressions

4 Solve Applied Problems Involving Algebraic Expressions

OBJECTIVE 1 IDENTIFY TERMS, COEFFICIENTS, AND LIKE TERMS OF AN ALGEBRAIC EXPRESSION

In **Topic 10.2**, we learned that numbers being added are called *terms*. Similarly, the **terms** of an **algebraic expression** are the quantities being added. For example, in the expression $9x^2 + 6x + 4$, the terms are $9x^2, 6x$, and 4. Because **subtraction** is defined by adding the opposite, subtracted quantities are "negative" terms. In the expression $7x - 3y$, the terms are $7x$ and $-3y$ because $7x - 3y = \underbrace{7x}_{\text{Term}} + \underbrace{(-3y)}_{\text{Term}}$.

> **Definition Term**
>
> The **terms** of an algebraic expression are the quantities being added. Terms containing variable factors are called **variable terms**, while terms without variables are called **constant terms**.

The numeric factor of a term is called the **coefficient** of the term. Every term in an algebraic expression has a coefficient. The expression $9x^2 + 6x + 4$ has three terms, so it has 3 coefficients. Similarly, the expression $7x - 3y$ has two terms, so it has 2 coefficients.

Expression	Coefficients
$9x^2 + 6x + 4$	9, 6, 4
$7x - 3y$	7, −3

Definition Coefficient (or Numerical Coefficient)

A **coefficient**, or **numerical coefficient**, is the numeric factor of a term.

CAUTION Remember the coefficient of a term includes the sign. For example, $4x - 8y = 4x + (-8y)$, so the coefficients are 4 and -8.

Example 1 Identifying Terms and Coefficients

Determine the number of **terms** in each expression and list the **coefficients** for each term.

a. $3x^2 + 7x - 3$ **b.** $4x^3 - \frac{3}{2}x^2 + x - 1$ **c.** $3x^2 - 2.3x + x - \frac{3}{4}$

Solutions

a. $3x^2 + 7x - 3 = \underbrace{3x^2}_{\text{Term 1}} + \underbrace{7x}_{\text{Term 2}} + \underbrace{(-3)}_{\text{Term 3}}$

The expression has three terms, so there are three coefficients.

Term	Coefficient
$3x^2$	3
$7x$	7
-3	-3

b. $4x^3 - \frac{3}{2}x^2 + x - 1 = \underbrace{4x^3}_{\text{Term 1}} + \underbrace{\left(-\frac{3}{2}x^2\right)}_{\text{Term 2}} + \underbrace{(1x)}_{\text{Term 3}} + \underbrace{(-1)}_{\text{Term 4}}$

The expression has four **terms**, so there are four **coefficients**.

Term	Coefficient
$4x^3$	4
$-\frac{3}{2}x^2$	$-\frac{3}{2}$
x	1
-1	-1

My video summary

c. Try this problem on your own. Check your **answer**, or watch this **video** for the complete solution.

You Try It **Work through this You Try It problem.**

Work Exercises 1–6 in this textbook or in the MyMathLab Study Plan.

Two terms are **like terms** if their **variable factors** are exactly the same. For example, the terms $5x^2y$ and $4x^2y$ are **like terms** because they have the same variable factor, x^2y. The terms $6a^2b$ and $9ab^2$ are not like terms because the variable factors are not exactly the same. All **constants** are like terms.

(eText Screens 10.6-1–10.6-21)

Example 2 Identifying Like Terms

Identify the like terms in each **algebraic expression**.

a. $5x^2 + 3x - 6 + 4x^2 - 7x + 10$

b. $3.5a^2 + 2.1ab + 6.9b^2 - ab + 8a^2$

Solutions

a. $5x^2$ and $4x^2$ are like terms because they have the same variable factor, x^2. Similarly, $3x$ and $-7x$ are like terms with the same variable factor of x. Lastly, -6 and 10 are like terms because they are both constants.

My video summary

b. Compare the variable factors to identify like terms. Check your **answer**, or watch this **video** for the complete solution.

You Try It **Work through this You Try It problem.**

Work Exercises 7–10 in this textbook or in the MyMathLab Study Plan.

OBJECTIVE 2 SIMPLIFY ALGEBRAIC EXPRESSIONS

When an algebraic expression contains **like terms**, we **simplify the expression** by combining the like terms. To **combine like terms**, we use the **distributive property** in reverse.

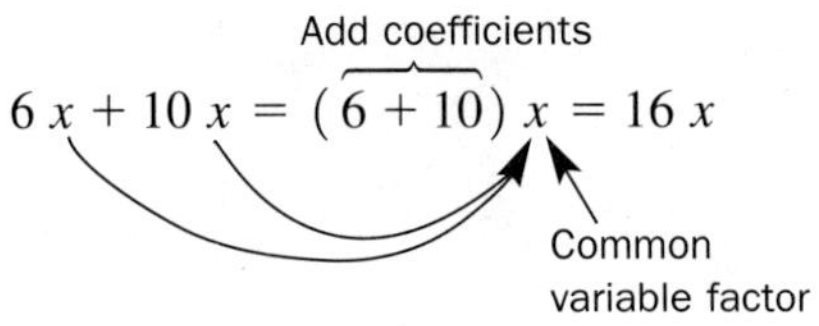

We can also use the **commutative** and **associative** properties of addition to reorder terms in an **algebraic expression** so like terms are grouped, or collected, together.

Example 3 Combining Like Terms

My interactive video summary

Simplify each algebraic expression by combining like terms.

a. $5x - 2x$ **b.** $6x^2 - 12x - 3x^2 + 4x$ **c.** $3z - 2z^2 + 7z^2$

d. $6x^2 + 2x + 4x + 3$ **e.** $-3x + 5 - y + x - 8$

Solutions

a. $5x$ and $-2x$ are like terms because they have the same **variable factor** x. To combine like terms, we reverse the distributive property.

Begin with the original expression:	$5x - 2x$
Reverse the distributive property:	$= (5 - 2)x$
Subtract 5 and 2:	$= 3x$

b. $6x^2$ and $-3x^2$ are **like terms**, as are $-12x$ and $4x$.

Begin with the original expression:	$6x^2 - 12x - 3x^2 + 4x$
Rearrange to **collect like terms**:	$= 6x^2 - 3x^2 - 12x + 4x$
Reverse the **distributive property**:	$= (6 - 3)x^2 + (-12 + 4)x$

Add and subtract: $= 3x^2 + (-8)x$

$= 3x^2 - 8x$

c.–e. Try to **simplify** the remaining expressions on your own. View the **answers**, or watch this **interactive video** for the complete solutions to all five parts.

You Try It **Work through this You Try It problem.**

Work Exercises 11–23 in this textbook or in the MyMathLab Study Plan.

Before we **combine like terms**, often we must first remove **grouping symbols** using the distributive property. In general, an algebraic expression is **simplified** if all grouping symbols are removed and like terms have been combined.

Simplifying an Algebraic Expression

1. Remove grouping symbols using the distributive property.
2. Combine like terms by using the distributive property in reverse.

Example 4 Simplifying Algebraic Expressions

My interactive video summary

Simplify each algebraic expression.

a. $3(x - 4) + 2$ **b.** $8(x + 6) + 7x$

c. $5(x - 6) - 3(x - 7)$ **d.** $2(5z + 1) - (3z - 2)$

Solutions

a. Use the **distributive property** to remove the parentheses, then **combine like terms**.

Begin with the original expression: $3(x - 4) + 2$

Distribute the 3: $= 3x - 3(4) + 2$

Multiply $3 \cdot 4$: $= 3x - 12 + 2$

Combine like terms: $= 3x - 10 \leftarrow -12 + 2 = -10$

b. Begin with the original expression: $8(x + 6) + 7x$

Distribute the 8: $= 8x + 8(6) + 7x$

Multiply $8 \cdot 6$: $= 8x + 48 + 7x$

Combine like terms: $= 15x + 48 \leftarrow 8x + 7x = 15x$

c.–d. Try to simplify these expressions on your own. Each contains two sets of grouping symbols. Use the distributive property to remove both sets, then collect and combine like terms. Check your **answers**, or watch this **interactive video** for the complete solutions to all four parts.

You Try It **Work through this You Try It problem.**

Work Exercises 24–32 in this textbook or in the MyMathLab Study Plan.

OBJECTIVE 3 WRITE WORD STATEMENTS AS ALGEBRAIC EXPRESSIONS

In algebra, problem solving often involves writing word statements as **algebraic expressions.** When translating word statements, look for key words or phrases that translate into **arithmetic operations.** Table 2 provides a review of some phrases and their corresponding operations.

Addition	Subtraction	Multiplication	Division
add	subtract	multiply	divide
plus	minus	times	divided by
sum	difference	product	quotient
increased by	decreased by	of	per
more than	less	double	ratio
total	less than	triple	
		twice	

Table 2

In Topics 10.2 and 10.3, we translated word statements into **numeric expressions.** Now we can use **variables** to represent any unknown values within verbal descriptions. Consider the following:

$$\text{The sum of } \underbrace{\text{a number}}_{\text{Unknown value}} \text{ and } \underbrace{10}_{\text{Known value}}$$

When translating this word statement, we use the variable x to represent the unknown value.

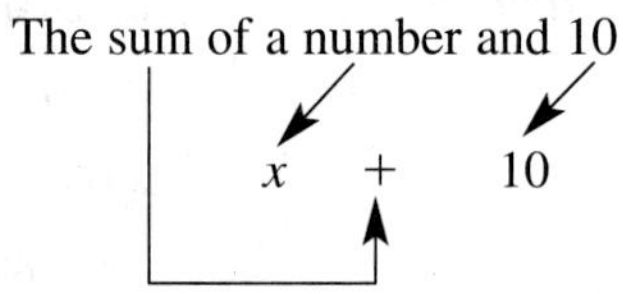

Example 5 Writing Word Statements as Algebraic Expressions

My video summary

Write each word statement as an **algebraic expression.** Use x to represent the unknown number.

a. Twenty decreased by a number

b. The product of sixteen and a number

c. Five more than twice a number

d. Three-fourths of the square of a number

e. The quotient of 12 and a number, increased by the number

f. The sum of a number and 4, divided by the difference of the number and 9.

Solutions

a. Looking at **Table 2**, the phrase "decreased by" indicates **subtraction.** What is being subtracted? 20 is decreased by an unknown number x, so x is being subtracted from 20. "Twenty decreased by a number" is written as $20 - x$.

b. The word "product" means **multiplication.** 16 and an unknown number x are being multiplied. "The product of sixteen and a number" translates to $16x$.

c. "More than" indicates **addition.** 5 and twice x are being added, which means 5 and $2x$. "Five more than twice a number x" is written as $5 + 2x$.

d.–f. Try to finish these problems on your own. View the **answers**, or watch this **video** for the complete solutions.

You Try It **Work through this You Try It problem.**

Work Exercises 33–38 in this textbook or in the MyMathLab Study Plan.

Example 6 Writing Word Statements as Algebraic Expressions

a. The longest side of a triangle is four units longer than five times the length of the shortest side. Express the length of the longest side in terms of the shortest side, a.

b. Michelle invests d dollars in one account and \$6750 less than this amount in the second account. Express the amount she invests in the second account in terms of d.

c. The state of Texas has 10 fewer institutes of higher education than twice the number in Virginia. If we let $n =$ the number of institutes in Virginia, express the number in Texas, in terms of n. (*Source: Statistical Abstract, 2010*)

Solutions

a. The key word '*times*' means **multiplication**, while the key words 'longer than' imply **addition**. Letting $a =$ the length of the shortest side, we get

$$\underbrace{\text{four units}}_{4}\ \underbrace{\text{longer than}}_{+}\ \underbrace{\text{five}}_{5}\ \underbrace{\text{times}}_{\cdot}\ \underbrace{\text{the length of the shortest side}}_{a}$$

The length of the longest side is $4 + 5a$, or $5a + 4$, units.

b. The key words 'less than' mean **subtraction**. What is being subtracted? 6750 is being subtracted *from* the amount in the first account.

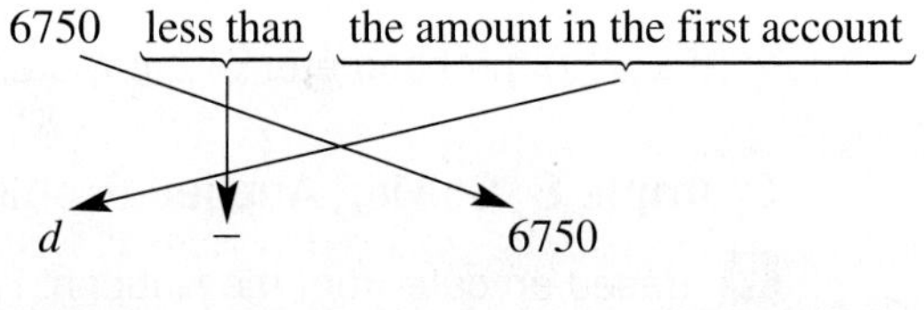

The amount invested in the second account is $d - 6750$ dollars.

My video summary

c. Try to work this problem on your own. View the **answer**, or watch this **video** for the complete solution.

You Try It **Work through this You Try It problem.**

Work Exercises 39–42 in this textbook or in the MyMathLab Study Plan.

OBJECTIVE 4 SOLVE APPLIED PROBLEMS INVOLVING ALGEBRAIC EXPRESSIONS

Simplifying algebraic expressions can be helpful when solving applied problems.

(eText Screens 10.6-1–10.6-21)

Example 7 Solving Applied Problems Involving Algebraic Expressions

The perimeter of a rectangle is the sum of the lengths of the sides of the rectangle. Use the following rectangle to answer the questions.

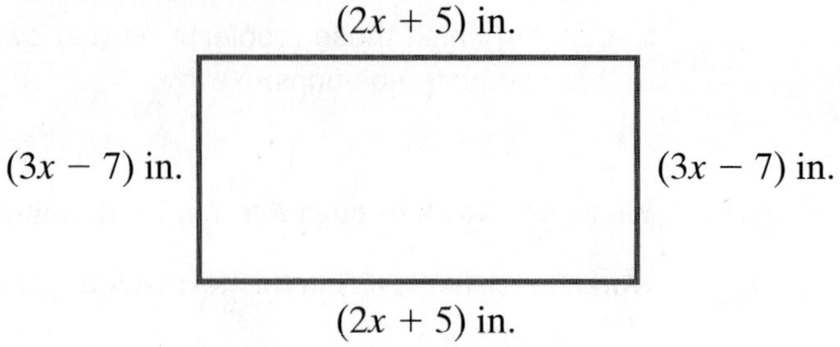

a. Write a **simplified algebraic expression** that represents the perimeter of the rectangle.

b. Use your result from part (a) to find the perimeter if $x = 7$.

Solutions

a. To find the perimeter, we add the lengths of the four sides.

Add the lengths of the four sides: $(2x + 5) + (3x - 7) + (2x + 5) + (3x - 7)$

Remove grouping symbols: $= 2x + 5 + 3x - 7 + 2x + 5 + 3x - 7$

Rearrange to collect like terms: $= \underbrace{2x + 3x + 2x + 3x} \underbrace{+ 5 - 7 + 5 - 7}$

Combine like terms: $= 10x + (-4)$

Simplify: $= 10x - 4$

The perimeter of the rectangle is $(10x - 4)$ in.

b. **Evaluate** the expression $10x - 4$ for $x = 7$.

Begin with the original algebraic expression: $10x - 4$

Substitute 7 for x: $= 10(7) - 4$

Multiply: $= 70 - 4$

Subtract: $= 66$

If $x = 7$, the perimeter of the rectangle would be 66 inches.

Example 8 Solving Applied Problems Involving Algebraic Expressions

My video summary

Based on data from the National Fire Protection Association, the number of residential property fires, in thousands, is given by $111x^2 - 1366x + 6959$, where $x =$ the number of years after 2000. The number of vehicle fires, in thousands, is given by $-52.5x + 775$. (*Source: nfpa.org*)

a. Write a **simplified algebraic expression** for the difference between the number of residential property fires and the number of vehicle fires.

b. Use your result from part (a) to estimate the difference in 2010.

Solutions Try to work this problem on your own. Check your **answers**, or watch this **video** for the complete solutions.

You Try It **Work through this You Try It problem.**

Work Exercises 43–46 in this textbook or in the MyMathLab Study Plan.

10.6 Exercises

In Exercises 1–6, determine the number of terms in the expression and then list the coefficients for each term.

You Try It

1. $3x - 5$

2. $5x^2 - 6x + 4$

3. $-\frac{3}{4}x^2 - 9$

4. $w^3 - 5w^2 + w - 10$

5. $a^4 - \frac{1}{2}a^2 + 8$

6. $3x^2 - x - \frac{4}{3}$

In Exercises 7–10, identify the like terms in each algebraic expression.

You Try It

7. $2a - b + 3 + 2b - 4a$

8. $x^2 - 4x + 2y + 3x^2 - y$

9. $6.3m - 4.5n^2 + 3.1mn + 2.2n^2 - 8m$

10. $5m - 3 + 6n + 4 - m$

In Exercises 11–32, simplify each algebraic expression by collecting and combining like terms.

You Try It

11. $-6z + 13z$

12. $8y + \frac{5}{3}y$

13. $5x^2 - 3x + 7x - 8$

14. $ab - 4.3ab + 7.8ab - 4.1$

15. $7x + 4x^2 + 6x^2$

16. $14 - \frac{1}{2}y + \frac{2}{3}y^2 - \frac{1}{4}y^2$

17. $8a - 2b + 14a$

18. $6m - 20 - 12m + 50$

19. $4.7m + 6.9 - 4.7m + 5.4$

20. $4x + y - 3 + x^2 - 6y + 7$

21. $-2c^2 + 0.7c^2 - 0.05c^2$

22. $\frac{x}{3} + \frac{3}{4} + \frac{x}{6} - \frac{5}{12}$

23. $3x^2 - 5x + 6 - 7x^2 + 4x - 8$

24. $6(x + 3) - 4$

You Try It

25. $-(4x - 6) + 9$

26. $5(2t - 11) - 4t$

27. $-3(1 - 5x) + 4x - 5$

28. $\frac{4}{3}\left(9x + \frac{15}{28}\right) - x + \frac{3}{7}$

29. $-5(w + 3) + 2(6w - 1)$

30. $3.5(y^2 + 4y - 9) + 1.8y^2 + 6.3$

31. $4(x^2 - 3x + 5) + 6(3x^2 + 5x - 2)$

32. $3(y^2 - 5) - 7(1 + 4y - 3y^2)$

In Exercises 33–38, write each word phrase as an algebraic expression. Use x to represent the unknown value.

You Try It

33. The sum of a number and eight

34. Twice the difference of a number and seventeen

35. The quotient of ten and a number, increased by two

36. A number decreased by twenty-five

37. One more than three-fifths of a number

38. The product of a number and three, minus the product of nine and the number

You Try It

39. A board is cut into two pieces. The longer piece is 18 cm longer than three times the shorter piece. If $s =$ the length of the shorter piece, express the length of the longer piece in terms of s.

40. Stephanie has d dollars to invest and splits the money between two accounts. If she invests $3000 in one account, write an algebraic expression in terms of d for the amount she invests in the second account.

41. China is building a network of global navigation satellites to be completed in 2020. The total number of satellites in the network will be one less than three times the number expected to be operational in 2012. Let $n =$ the number of expected satellites in the network in 2012. Express the number of satellites in the completed network in terms of n. (*Source:* www.gpsdaily.com)

42. In 2009, the percentage of 12 year-olds who owned a cell phone was 5 points higher than three times the percentage in 2004. If p is the percentage in 2004, write an algebraic expression in terms of p for the percentage in 2009.

You Try It

43. The perimeter of a triangle is the sum of the lengths of the sides of the triangle. Use the following triangle to answer the questions.

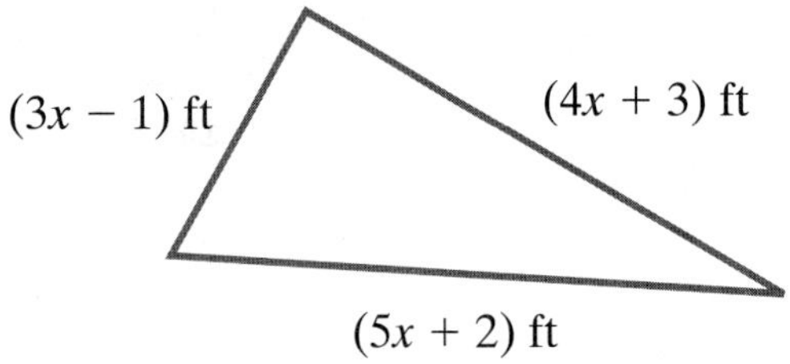

a. Write a simplified algebraic expression that represents the perimeter of the rectangle.

b. Use your result from part (a) to find the perimeter if $x = 5$.

44. A wire with total length $(5x + 4.6)$ meters is cut into two pieces such that one piece measures $(3x - 2.5)$ meters.

a. Write a simplified algebraic expression that represents the length of the second piece.

b. Use your result from part (a) to find the length of the second piece if $x = 4.5$.

45. Based on data from the U.S. Department of Agriculture, the average annual per capita consumption of non-alcoholic beverages other than tap water (in gallons) is given by $0.44x + 150$, where $x =$ the number of years after 2000. The average annual per capita consumption of alcoholic beverages (in gallons) is given by $0.05x + 25$. (*Source: census.gov/compendia/statab/2010*)

a. Write a simplified algebraic expression that represents the total average per capita beverage consumption.

b. Use your result from part (a) to estimate the total average per capita beverage consumption in 2012.

46. To build a bookcase, Brennan needs two side boards and four shelf boards. Side boards have length $(7x - 2)$ inches and shelf boards have length $(3x + 6)$ inches.

a. Write a simplified algebraic expression that represents the total length of wood that Brennan would need to build a bookcase.

b. Use your result from part (a) to find the total length if $x = 6$.

MODULE ELEVEN

Linear Equations and Inequalities in One Variable

MODULE ELEVEN CONTENTS

11.1 The Addition and Multiplication Properties of Equality

THINGS TO KNOW

Before working through this topic, be sure you are familiar with the following concepts:

	VIDEO	ANIMATION	INTERACTIVE
You Try It 1. Use the Order of Operations to Evaluate Numeric Expressions (Topic 10.4, **Objective 2**)	✓		✓
You Try It 2. Evaluate Algebraic Expressions (Topic 10.5, **Objective 1**)	✓		
You Try It 3. Simplify Algebraic Expressions (Topic 10.6, **Objective 2**)			✓

OBJECTIVES

1 Identify Linear Equations in One Variable

2 Determine If a Given Value Is a Solution to an Equation

3 Solve Linear Equations Using the Addition Property of Equality

4 Solve Linear Equations Using the Multiplication Property of Equality

5 Solve Linear Equations Using Both Properties of Equality

(eText Screens 11.1-1–11.1-24)

OBJECTIVE 1 IDENTIFY LINEAR EQUATIONS IN ONE VARIABLE

An **equation** is a statement that two quantities are equal. For example, the numeric equation

$$\underbrace{12+4}_{\text{Quantity}} = \underbrace{16}_{\text{Quantity}}$$

means that the quantities $12 + 4$ and 16 are equal. An equation that contains one or more **variables** is called an **algebraic equation**. An algebraic equation indicates that two **algebraic expressions** are equal. Algebraic equations contain an equality symbol ($=$), whereas algebraic expressions do not.

For example, $5x + 3$ and $4x + 6$ are both algebraic *expressions*, but

$$\underbrace{5x+3}_{\text{Algebraic expression}} \quad \underbrace{=}_{\text{Equality symbol}} \quad \underbrace{4x+6}_{\text{Algebraic expression}}$$

My interactive video summary

is an algebraic *equation*. Work through this **interactive video** to practice distinguishing between expressions and equations.

An **equation in one variable** contains a single variable. Each of the following is an example of an equation in one variable:

$$\underbrace{2x+3=-7,}_{\text{One variable} \to x} \quad \underbrace{a^2-3a+8=0,}_{\text{One variable} \to a} \quad \text{and} \quad \underbrace{z-\frac{2}{5}=\frac{3}{4}z}_{\text{One variable} \to z}$$

Note: Even if the same variable appears multiple times, the equation is still considered to be "in one variable."

The equations $2x + 3 = -7$ and $z - \frac{2}{5} = \frac{3}{4}z$ are both examples of *linear equations in one variable*.

> **Definition Linear Equation in One Variable**
>
> A **linear equation in one variable** is an equation that can be written in the form $ax + b = c$, where a, b, and c are real numbers and $a \neq 0$.

Linear equations are also **first-degree equations** because the **exponent** on the variable is **understood to be** 1. Until we learn how to simplify equations, this is how we will identify linear equations. If an equation contains a variable raised to an exponent other than 1, then the equation is *nonlinear*. View this **popup** to see other characteristics of **nonlinear equations**.

My interactive video summary

Watch this **interactive video** to practice distinguishing between linear and nonlinear equations. It is important to understand that an equation does not have to be in the form $ax + b = c$ to be considered linear. Rather, the equation must be able *to be written in the form* $ax + b = c$.

Example 1 Identifying Linear Equations in One Variable

Determine if each is a linear equation in one variable. If not, state why.

a. $4x + 3 - 2x$ **b.** $4x + 2 = 3x - 1$

c. $x^2 + 3x = 5$ **d.** $2x + 3y = 6$

Solutions

a. $4x + 3 - 2x$ is not a **linear equation in one variable** because it is not an **equation.**

b. $4x + 2 = 3x - 1$ is a linear equation in one variable.

c. $x^2 + 3x = 5$ is an equation in one variable, but it is nonlinear because x^2 has an **exponent** other than 1.

d. $2x + 3y = 6$ is not a linear equation in one variable because it has more than one **variable.**

You Try It **Work through this You Try It problem.**

Work Exercises 1–6 in this textbook or in the MyMathLab Study Plan.

OBJECTIVE 2 DETERMINE IF A GIVEN VALUE IS A SOLUTION TO AN EQUATION

We can check the truth of an **equation** by simplifying both sides of the equation and comparing the results. Consider the following two equations:

$$-3 + (-2)^2 = -7 + 1 \qquad |10 - 26| + 4 = 3^2 + 11$$

Simplify both sides of each equation and compare the results to determine if the statement is true. View this **popup** to see the steps. Simplifying the first equation results in a false statement, so the equation is not true. Simplifying the second equation results in a true statement, so the equation is true.

When we *solve* an **algebraic equation**, we look for all values for the **variable(s)** that make the equation true. Values that result in a true statement are called *solutions* of the equation.

Definition Solve

To **solve** an equation means to find its **solution set**, or the set of all values that make the equation true.

Definition Solution

A **solution** is a value that, when substituted for a variable, makes the equation true.

CAUTION Remember that we *simplify* expressions (as in **Topic 10.6**) but we *solve* equations.

To determine if a given value is a **solution** to an equation, we **substitute** the value for the variable, **simplify** both sides of the equation, and compare the results to see if the statement is true. For example, consider the equation $5x + 2 = x - 6$. You can **view the checks** to see if either 3 or -2 is a solution.

CAUTION When substituting values for a variable, it is best to use parentheses around the substituted value to avoid mistakes in performing operations.

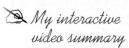

My interactive video summary

Example 2 Determining If a Given Value Is a Solution to an Equation

Determine if the given value is a solution to the equation.

a. $-3x + 5 = 8 - 4x;\ x = 3$ **b.** $2(y + 1) + 8 = 7 - y;\ y = -3$

c. $|a - 6| - 1 = 9 + a^2;\ a = -2$ **d.** $\frac{3}{5}w - \frac{1}{2} = -\frac{3}{10}w;\ w = \frac{5}{9}$

(eText Screens 11.1-1–11.1-24)

Solutions First, we substitute the given value for the **variable** and simplify. If the resulting statement is true, then the value is a solution to the equation.

a. Begin with the original equation: $-3x + 5 = 8 - 4x$

Substitute 3 for x: $-3(3) + 5 \stackrel{?}{=} 8 - 4(3)$

Simplify: $-9 + 5 \stackrel{?}{=} \; = 8 - 12$

Compare: $-4 = -4$ True

The final statement is true, so $x = 3$ is a solution to the equation.

b. Begin with the original equation: $2(y + 1) + 8 = 7 - y$

Substitute -3 for y: $2((-3) + 1) + 8 \stackrel{?}{=} 7 - (-3)$

Simplify: $2(-2) + 8 \stackrel{?}{=} 7 + 3$

$-4 + 8 \stackrel{?}{=} 7 + 3$

Compare: $4 = 10$ False

The final statement is false, so $y = -3$ is not a solution to the equation.

c.–d. Try completing parts c and d on your own, then view the **answers**. Watch this **interactive video** to see the complete solutions to all four parts.

You Try It **Work through this You Try It problem.**

Work Exercises 7–12 in this textbook or in the MyMathLab Study Plan.

OBJECTIVE 3 SOLVE LINEAR EQUATIONS USING THE ADDITION PROPERTY OF EQUALITY

All the solutions to an equation form the **solution set** of the equation. Two or more equations with the same solution set are called **equivalent equations.** When solving a **linear equation in one variable,** we look for simpler equivalent equations until we find one that ends with an **isolated variable** of the form:

$$\textit{variable} = \textit{value} \quad \text{or} \quad \textit{value} = \textit{variable}.$$

To find simpler equivalent equations, we use the *properties of equality* to add, subtract, multiply, and/or divide both sides of an equation by the same quantity without changing its solution set.

Consider the true statement, $15 = 15$. Figure 1a shows that we can think of an equation as a balance scale in which both sides are equal. Notice that if we add 5 to the left side of the equation, the two sides are no longer equal (Figure 1b), and the equality statement is no longer true. To maintain equality (balance), we also need to add 5 to the right side of the equation (Figure 1c).

Figure 1

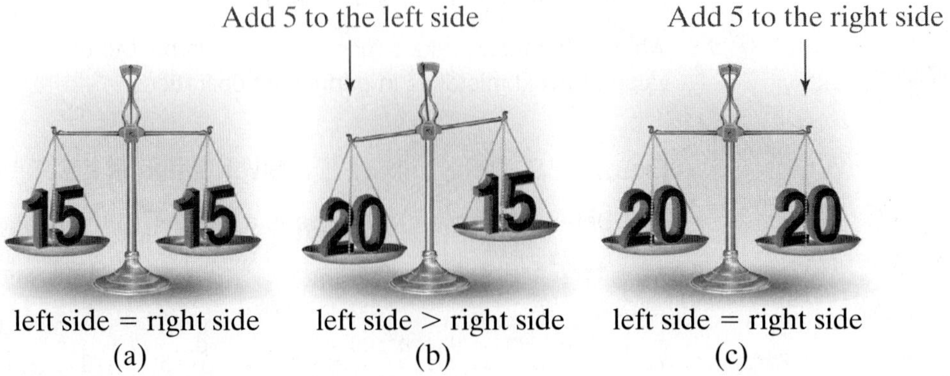

(eText Screens 11.1-1–11.1-24)

Figure 1 illustrates the addition property of equality.

> **Addition Property of Equality**
>
> Let a, b, and c be real numbers or **algebraic expressions**. Then,
>
> $$a = b \quad \text{and} \quad a + c = b + c$$
>
> are equivalent equations.

Using the addition property of equality, we can add the same value to both sides of an **equation** without changing its **solution set**. Because **subtraction** is defined in terms of addition, this property also holds true for subtraction:

$$a = b \quad \text{and} \quad a - c = b - c \quad \text{are equivalent equations.}$$

Example 3 Solving Linear Equations Using the Addition Property of Equality

Solve.

a. $x - 5 = 3$ **b.** $y + \frac{2}{3} = \frac{1}{5}$

Solutions

a. To solve the equation, we want to get the variable x on one side of the equation.

Begin with the original equation: $x - 5 = 3$

Add 5 to both sides: $x \underbrace{- 5 + 5}_{0} = 3 + 5$

Simplify: $x = 8$

To check, we **substitute** 8 for x in the original equation to see if a true statement results. View the **check**.

b. To **solve** the equation, we want to get the variable y by itself on one side of the equation.

Begin with the original equation: $y + \frac{2}{3} = \frac{1}{5}$

Subtract $\frac{2}{3}$ from both sides: $y + \underbrace{\frac{2}{3} - \frac{2}{3}}_{0} = \frac{1}{5} - \frac{2}{3}$

Simplify: $y = \frac{1}{5} - \frac{2}{3}$

Write each fraction with a denominator of LCD = 15: $y = \frac{1 \cdot 3}{5 \cdot 3} - \frac{2 \cdot 5}{3 \cdot 5}$

$y = \frac{3}{15} - \frac{10}{15}$

Subtract: $y = -\frac{7}{15}$

Check this result by **substituting** $-\frac{7}{15}$ for y in the original equation. View the **check**.

You Try It **Work through this You Try It problem.**

Work Exercises 13–16 in this textbook or in the MyMathLab Study Plan.

(*eText Screens 11.1-1–11.1-24*)

The **addition property of equality** can be applied to **variable terms** as well as **constants.** Example 4 illustrates this idea.

Example 4 Solving Linear Equations Using the Addition Property

Solve.

a. $3z - 4.5 = 4z$ **b.** $12 - 7x = -6x$

Solutions

a. To solve the equation we want to get the variable z by itself on one side of the equation.

Begin with the original equation: $3z - 4.5 = 4z$

Subtract $3z$ from both sides: $\underbrace{3z - 3z}_{0} - 4.5 = 4z - 3z$

Simplify: $-4.5 = z$

The solution set is $\{-4.5\}$. View the **check**.

b. To solve the equation we want to get the variable x by itself on one side of the equation. Add $7x$ to both sides and **simplify**. View the **solution and check**.

You Try It **Work through this You Try It problem.**

Work Exercises 17–20 in this textbook or in the MyMathLab Study Plan.

We can also apply the **addition property of equality** to both **constants** and **variable terms** within the same equation.

Example 5 Solving Linear Equations Using the Addition Property

My video summary

Solve.

a. $5x - 3 = 6x + 2$ **b.** $-5w + 27 = 13 - 4w$

Solutions Try solving these two equations on your own. Apply the addition property of equality twice within each equation: once for the constant and once for the variable term. View the **answers**, or watch this **video** for complete solutions.

You Try It **Work through this You Try It problem.**

Work Exercises 21–25 in this textbook or in the MyMathLab Study Plan.

OBJECTIVE 4 SOLVE LINEAR EQUATIONS USING THE MULTIPLICATION PROPERTY OF EQUALITY

Not all equations can be solved using the **addition property of equality**. For example, the equation $\frac{x}{5} = -3$ cannot be solved by adding or subtracting the same value on both sides. Instead, we apply the **multiplication property of equality.**

Multiplication Property of Equality

Let a, b, and c be real numbers or **algebraic expressions with** $c \neq 0$. Then,

$$a = b \quad \text{and} \quad a \cdot c = b \cdot c$$

are equivalent equations.

(eText Screens 11.1-1–11.1-24)

Using the multiplication property of equality, we can multiply both sides of an **equation** by the same **nonzero** value without changing the **solution set** of the equation. Because **division** is defined in terms of multiplication, this property also holds true for division:

$$a = b \quad \text{and} \quad \frac{a}{c} = \frac{b}{c} \quad \text{are equivalent equations.}$$

Multiplying or dividing both sides of an equation by zero is not allowed. Do you see **why**?

Example 6 Solving Linear Equations Using the Multiplication Property of Equality

Solve.

a. $\frac{x}{5} = -3$ **b.** $-4x = -32$

Solutions

a. We want to **isolate** the variable x on one side of the equation. While we have the variable *term* by itself, we do not have the *variable* by itself. To do this, we need to multiply both sides of the equation by 5.

Begin with the original equation: $\frac{x}{5} = -3$

Multipy both sides by 5: $\underbrace{5 \cdot \frac{x}{5}}_{5 \cdot \frac{1}{5} = 1} = 5 \cdot (-3)$

Simplify: $x = -15$

The solution set is $\{-15\}$. View the **check**.

b. Again we have the **variable term** by itself but not the variable. To isolate the variable, we divide both sides of the equation by -4.

Begin with the original equation: $-4x = -32$

Divide both sides by -4: $\frac{-4x}{-4} = \frac{-32}{-4}$ ← $\left(\frac{-4}{-4} = 1\right)$

Simplify: $x = 8$

The solution set is $\{8\}$. View the **check**.

You Try It **Work through this You Try It problem.**

Work Exercises 26–29 in this textbook or in the MyMathLab Study Plan.

Example 7 Solving Linear Equations Using the Multiplication Property of Equality

My video summary

Solve.

a. $\frac{4}{3}x = 52$ **b.** $2.2x = 6.93$

Solutions Try solving these equations on your own. Apply the **multiplication property of equality** to isolate the variable. View the **answers**, or watch this **video** for complete solutions.

You Try It **Work through this You Try It problem.**

Work Exercises 30–36 in this textbook or in the MyMathLab Study Plan.

OBJECTIVE 5 SOLVE LINEAR EQUATIONS USING BOTH PROPERTIES OF EQUALITY

To solve some equations, we must use both the addition and multiplication **properties of equality**.

First, we use the addition property to get the **variable terms** on one side of the equation and the **constants** on the other. Then we use the multiplication property to isolate the variable.

Example 8 Solving Linear Equations in One Variable

My video summary

Use the properties of equality to solve each equation.

a. $\frac{2}{5}x - 4 = 4$ **b.** $7n = 3n - 8$

Solutions Work through each problem, or watch this **video** to see the complete solutions to both parts.

a. Begin with the original equation: $\frac{2}{5}x - 4 = 4$

Add 4 to both sides: $\frac{2}{5}x - 4 + 4 = 4 + 4$ ← Addition property of equality

Simplify: $\frac{2}{5}x = 8$

Multiply both sides by $\frac{5}{2}$: $\frac{5}{2}\cdot\frac{2}{5}x = \frac{5}{2}\cdot 8$ ← Multiplication property of equality

Simplify: $x = 20$

The solution set is $\{20\}$. **View the check** of this **solution** in the original equation.

b. Begin with the original equation: $7n = 3n - 8$

Subtract $3n$ from both sides: $7n - 3n = 3n - 8 - 3n$

Simplify: $4n = -8$

Divide both sides by 4: $\frac{4n}{4} = \frac{-8}{4}$

Simplify: $n = -2$

The solution set is $\{-2\}$. View the **check**.

You Try It **Work through this You Try It problem.**

Work Exercises 37–45 in this textbook or in the MyMathLab Study Plan.

Some equations have **algebraic expressions** on both sides of the equality symbol. We again use the **properties of equality** to move all variables to one side and constants to the other. The only difference here is that we need to apply the **addition property of equality** twice: once for the **constant** and once for the **variable terms**.

Example 9 Solving Linear Equations with Variables on Both Sides

Use the properties of equality to solve each equation.

a. $7y + 4 = 2y - 6$ **b.** $2x - 14.5 = 0.5x + 50$

(eText Screens 11.1-1–11.1-24)

Solutions

a. Begin with the original equation: $7y + 4 = 2y - 6$

Subtract 4 from both sides: $7y + 4 - 4 = 2y - 6 - 4$ ← Addition property of equality

Simplify: $7y = 2y - 10$

Subtract $2y$ from both sides: $7y - 2y = 2y - 10 - 2y$ ← Addition property of equality

Simplify: $5y = -10$

Divide both sides by 5: $\frac{5y}{5} = \frac{-10}{5}$ ← Multiplication property of equality

Simplify: $y = -2$

The solution set is $\{-2\}$. Check this **solution** in the original equation.

My video summary

b. Try solving this equation on your own. Be sure to pay attention to decimal places when combining **coefficients**. View the **answer**, or watch this **video** for the complete solutions to both parts.

You Try It **Work through this You Try It problem.**

Work Exercises 46–52 in this textbook or in the MyMathLab Study Plan.

11.1 Exercises

In Exercises 1–6, determine if each is a linear equation in one variable. If not, state why.

You Try It

1. $3x - 2(x + 1)$ **2.** $3x = 2(x + 1)$ **3.** $3x^{-1} = 2(x + 1)$

4. $3x^2 = 2(x^2 + 1)$ **5.** $3x = 2(y + 1)$ **6.** $1.3x - 2 + 5x = x - 8.6$

In Exercises 7–12, determine if the given value is a solution to the equation.

You Try It

7. $2x - 5 = 13 - 4x;\ x = 3$ **8.** $5 - 2(y - 1) = 3y + (y - 5);\ y = 5$

9. $z^2 + 2z + 1 = (-1 - z)^2;\ z = -4$ **10.** $\frac{1}{2}w - \frac{1}{3} = \frac{1}{6}w;\ w = 6$

11. $\frac{2}{3}x - \frac{3}{4}(x + 1) = \frac{1}{6}x - \frac{21}{16};\ x = \frac{9}{4}$ **12.** $4.5m - 5 = 3.1(m + 2.6);\ m = 2.5$

In Exercises 13–25, solve each linear equation using the addition property of equality.

You Try It

13. $a - 3 = 2$ **14.** $c + 4 = 7$ **15.** $y - 2 = -5$

You Try It

16. $z + \frac{2}{3} = \frac{5}{4}$ **17.** $10x = 9x + 7$ **18.** $4.8 + 1.5x = 2.5x$

19. $7x + \frac{3}{4} = 6x$ **20.** $6 - 5x = -4x$ **21.** $4x + 2 = 3x + 5$

(eText Screens 11.2-1–11.2-21)

22. $4 + 3c = 4c - 3$ **23.** $-1 - 6m = 4 - 7m$ **24.** $3z + 2.7 = 8.1 + 2z$

You Try It

25. $\frac{1}{8} - \frac{5}{6}w = \frac{1}{6}w + \frac{9}{8}$

In Exercises 26–36, solve each linear equation using the multiplication property of equality.

26. $5x = 10$ **27.** $\frac{x}{3} = 9$ **28.** $21 = -3x$

You Try It

29. $\frac{m}{-2} = -16$ **30.** $\frac{3}{4}x = 27$ **31.** $-\frac{1}{4}z = -3.5$

32. $3.2y = 2.4$ **33.** $\frac{2}{5}x = \frac{1}{3}$ **34.** $\frac{3}{2}z = -\frac{9}{20}$

You Try It

35. $17 = 5w$ **36.** $\frac{n}{-2} = -\frac{3}{8}$

In Exercises 37–52, solve using both the addition and multiplication properties of equality.

37. $5x + 2 = 27$ **38.** $2 + 4x = x$ **39.** $7 - 3x = -8$

You Try It

40. $-12 = 3y + 10$ **41.** $5z = 16 - 3z$ **42.** $-4n + 7 = -6n$

43. $3.7x - 4.1 = 3.2x$ **44.** $\frac{2}{3}w - 1 = 5$ **45.** $\frac{2}{5}y + \frac{1}{2} = \frac{5}{2}$

46. $2x + 1 = 5x - 8$ **47.** $8 + b = 2 - 2b$ **48.** $5 - 4n = 6n + 1$

You Try It

49. $2z + 6 = 5z + 6$ **50.** $3x - 2.9 = 5.1 - 5x$ **51.** $\frac{2}{3}y + 11 = -\frac{1}{2}y + 8$

52. $0.1x - 14.5 = 0.04x + 10.7$

11.2 Solving Linear Equations in One Variable

THINGS TO KNOW

Before working through this topic, be sure you are familiar with the following concepts:

VIDEO ANIMATION INTERACTIVE

You Try It

1. Use the Order of Operations to Evaluate Numeric Expressions (Topic 10.4, **Objective 2**)

You Try It

2. Evaluate Algebraic Expressions (Topic 10.5, **Objective 1**)

You Try It

3. Simplify Algebraic Expressions (Topic 10.6, **Objective 2**)

You Try It

4. Determine If a Given Value Is a Solution to an Equation (Topic 11.1, **Objective 2**)

You Try It

5. Solve Linear Equations Using Both Properties of Equality (Topic 11.1, **Objective 5**)

(eText Screens 11.2-1–11.2-21)

OBJECTIVES

1 Solve Linear Equations Containing Non-Simplified Expressions

2 Solve Linear Equations Containing Fractions

3 Solve Linear Equations Containing Decimals; Apply a General Strategy

4 Identify Contradictions and Identities

5 Use Linear Equations to Solve Application Problems

OBJECTIVE 1 SOLVE LINEAR EQUATIONS CONTAINING NON-SIMPLIFIED EXPRESSIONS

When **solving** a **linear equation** with **non-simplified expressions** on one or both sides, we simplify each side first before using the **properties of equality.**

Example 1 Solving Linear Equations Containing Non-Simplified Expressions

Solve: $4x + 7 - 2x = 5 - 3x - 3$

Solution Begin with the original equation: $4x + 7 - 2x = 5 - 3x - 3$

Combine like terms: $2x + 7 = 2 - 3x$

Add $3x$ to both sides: $2x + 7 + 3x = 2 - 3x + 3x$ ← Addition property of equality

Simplify: $5x + 7 = 2$

Subtract 7 from both sides: $5x + 7 - 7 = 2 - 7$ ← Addition property of equality

Simplify: $5x = -5$

Divide both sides by 5: $\dfrac{5x}{5} = \dfrac{-5}{5}$ ← Multiplication property of equality

Simplify: $x = -1$

The **solution set** is $\{-1\}$. View the **check.**

You Try It **Work through this You Try It problem.**

Work Exercises 1–5 in this textbook or in the MyMathLab Study Plan.

If an equation contains **grouping symbols**, we remove the grouping symbols by applying the **distributive property**, **simplify** both sides, and then use the **properties of equality** to **solve** the equation.

Example 2 Solving Linear Equations Containing Non-Simplified Expressions and Grouping Symbols

My video summary

Solve: $7 - 2(4z - 3) = 3z + 1$

Solution

Begin with the original equation: $7 - 2(4z - 3) = 3z + 1$

Use the **distributive property:** $7 - 8z + 6 = 3z + 1$ ← $-2(4z - 3) = -2 \cdot 4z - 2 \cdot (-3)$

Combine like terms: $13 - 8z = 3z + 1$

Both sides of the equation are now **simplified expressions.** Finish solving the equation on your own using the **properties of equality**. Check your **answer**, or watch this **video** for the complete solution.

You Try It **Work through this You Try It problem.**

Work Exercises 6–10 in this textbook or in the MyMathLab Study Plan.

Example 3 Solving Equations Containing Non-Simplified Expressions

My video summary

Solve: $2(3x - 1) - 5x = 3 - (3x + 1)$

Solution Begin with the original equation: $2(3x - 1) - 5x = 3 - (3x + 1)$

Use the **distributive property**: $6x - 2 - 5x = 3 - 3x - 1$

The **grouping symbols** have been removed from both sides of the equation. Simplify each side by **combining like terms**, then finish solving by using the properties of equality. Check your **answer**, or watch this **video** for a fully worked solution.

You Try It **Work through this You Try It problem.**

Work Exercises 11–18 in this textbook or in the MyMathLab Study Plan.

OBJECTIVE 2 SOLVE LINEAR EQUATIONS CONTAINING FRACTIONS

When an equation contains **fractions**, we can make the calculations more manageable if we remove the fractions before **combining like terms** or applying the **properties of equality.** To do this, we multiply both sides of the equation by an appropriate **common multiple** of all the **denominators,** usually the **least common denominator (LCD)** of all the fractions.

Example 4 Solving Equations Containing Fractions

My video summary

Solve: $y + \frac{2}{3} = \frac{1}{5}$

Solution The equation contains fractions with the denominators 3 and 5, so the LCD is 15. We can **clear the fractions** by multiplying both sides of the equation by 15.

Begin with the original equation: $y + \frac{2}{3} = \frac{1}{5}$

Multiply both sides by 15: $15\left(y + \frac{2}{3}\right) = 15\left(\frac{1}{5}\right)$

Use the **distributive property**: $15(y) + 15\left(\frac{2}{3}\right) = 15\left(\frac{1}{5}\right)$

Clear the fractions: $15y + 10 = 3$

Subtract 10 from both sides: $15y + 10 - 10 = 3 - 10$

Simplify: $15y = -7$

Divide both sides by 15: $\frac{15y}{15} = \frac{-7}{15}$

Simplify: $y = -\frac{7}{15}$

The **solution set** is $\left\{-\frac{7}{15}\right\}$. This is the same result that we obtained in Example 3 of **Topic 11.1** where we solved the same equation without clearing fractions first.

Example 5 Solving Equations Containing Fractions

Solve: $\frac{w+3}{2} - 4 = w + \frac{1}{3}$

Solution The denominators in this equation are 2 and 3, so the **LCD** is 6. We can **clear the fractions** by multiplying both sides of the equation by 6.

Begin with the original equation: $\frac{w+3}{2} - 4 = w + \frac{1}{3}$

Multiply both sides by 6: $6\left(\frac{w+3}{2} - 4\right) = 6\left(w + \frac{1}{3}\right)$

Use the **distributive property**: $6\left(\frac{w+3}{2}\right) - 6(4) = 6(w) + 6\left(\frac{1}{3}\right)$

Clear the fractions: $3(w+3) - 24 = 6w + 2$ ← $6\left(\frac{w+3}{2}\right) = \overset{3}{\cancel{6}}\left(\frac{w+3}{\underset{1}{\cancel{2}}}\right) = 3(w+3)$

Use the distributive property: $3w + 9 - 24 = 6w + 2$ ← $3(w) + 3(3) = 3w + 9$

Simplify: $3w - 15 = 6w + 2$

My video summary

We now have **simplified expressions** on each side of the equation. Finish solving the equation on your own. View the **answer**, or watch this **video** for a complete solution.

You Try It **Work through this You Try It problem.**

Work Exercises 19–26 in this textbook or in the MyMathLab Study Plan.

Note: When solving an equation that contains fractions, it is usually helpful to clear the fractions first. However, clearing fractions first is not always necessary, as shown in this **popup**.

Example 6 Solving Linear Equations Containing Fractions

My video summary

Solve: $\frac{5x}{2} - \frac{7}{8} = \frac{3}{4}x - \frac{11}{8}$

Solution This equation contains fractions with the denominators 2, 4, and 8, so the **LCD** is 8. Begin by multiplying both sides of the equation by 8 to **clear the fractions.** Try solving this equation on your own. Check your **answer** or watch this **video** for a fully worked solution.

You Try It **Work through this You Try It problem.**

Work Exercises 27–32 in this textbook or in the MyMathLab Study Plan.

The advantage of clearing fractions is that we can work with integers which are easier to use when **performing operations.** However, equations containing fractions can still be solved without clearing the fractions. View this **popup** to see Example 6 solved without clearing fractions.

(eText Screens 11.2-1–11.2-21)

OBJECTIVE 3 SOLVE LINEAR EQUATIONS CONTAINING DECIMALS; APPLY A GENERAL STRATEGY

When an equation contains **decimals**, it is often helpful to clear them as we do with fractions. To **clear decimals**, multiply both sides of the equation by an appropriate **power of** 10. Determine which power of 10 to use by looking at the **constants** in the equation and choosing the constant with the greatest number of decimal places. We count those decimal places and then raise 10 to that power. Multiplying both sides of the equation by that power of 10 will usually clear all the decimals.

Example 7 Solving Linear Equations Containing Decimals

Solve: $1.4x - 3.8 = 6$

Solution The **coefficients** 1.4 and -3.8 have one decimal place, so the greatest number of decimal places is one. To clear the decimals, we multiply both sides of the equation by $10^1 = 10$.

$$\underbrace{1.4}_{\text{One decimal place}}x - \underbrace{3.8}_{\text{One decimal place}} = 6$$

Begin with the original equation:	$1.4x - 3.8 = 6$
Multiply both sides by 10:	$10(1.4x - 3.8) = 10(6)$
Apply the **distributive property:**	$14x - 38 = 60$
Add 38 to both sides:	$14x - 38 + 38 = 60 + 38$
Simplify:	$14x = 98$
Divide both sides by 14:	$\dfrac{14x}{14} = \dfrac{98}{14}$
Simplify:	$x = 7$

The **solution set** is $\{7\}$. View the **check**.

Example 8 Solving Linear Equations Containing Decimals

My video summary

Solve: $0.1x + 0.03(7 - x) = 0.05(7)$

Solution The **coefficient** 0.1 has one decimal place. The **factors** 0.03 and 0.05 have two decimal places, so the greatest number of decimal places is two.

$$\underbrace{0.1}_{\text{One decimal place}}x + \underbrace{0.03}_{\text{Two decimal places}}(7 - x) = \underbrace{0.05}_{\text{Two decimal places}}(7)$$

To **clear the decimals**, we multiply both sides of the equation by $10^2 = 100$.

Begin with the original equation:	$0.1x + 0.03(7 - x) = 0.05(7)$
Multiply both sides by 100:	$100[0.1x + 0.03(7 - x)] = 100[0.05(7)]$
Distribute:	$100[0.1x] + 100[0.03(7 - x)] = 100[0.05(7)]$
Multiply:	$10x + 3(7 - x) = 5(7)$

CAUTION Notice that we only multiplied 100 by the first factor in each term.

(eText Screens 11.2-1–11.2-21)

Finish solving the equation, then check your **answer**. Watch this **video** for a complete solution.

You Try It **Work through this You Try It problem.**

Work Exercises 33–40 in this textbook or in the MyMathLab Study Plan.

Our work so far leads us to the following general strategy for solving linear equations in one variable. If a particular step does not apply to a given equation, then skip it. Remember that clearing fractions or decimals is optional but usually makes equations easier to solve.

A General Strategy for Solving Linear Equations in One Variable

Step 1. **Clear all fractions** by multiplying both sides of the equation by the **LCD**. **Clear all decimals** by multiplying both sides of the equation by the appropriate **power of** 10.

Step 2. Remove grouping symbols using the **distributive property**.

Step 3. Simplify each side of the equation by **combining like terms**.

Step 4. Use the **addition property of equality** to collect all **variable terms** on one side of the equation and all **constant terms** on the other side.

Step 5. Use the **multiplication property of equality** to **isolate the variable**.

Step 6. Check that the result **satisfies** the original equation.

OBJECTIVE 4 IDENTIFY CONTRADICTIONS AND IDENTITIES

So far, we have solved equations with exactly one **solution**. However, not every **linear equation in one variable** has a single solution. There are two other cases—no solution and a solution set of all real numbers.

Consider the equation $x = x + 1$. No matter what value is **substituted** for x, the resulting value on the right side will always be one greater than the value on the left side. Therefore, the equation can never be true. Trying to solve the equation, we get

$$\begin{aligned} x &= x + 1 \\ x - x &= x + 1 - x \\ 0 &= 1 \end{aligned}$$

← Variables are gone and statement is false

No variable terms remain, and a false statement results. We call such an equation a **contradiction**, and it has *no solution*. Its **solution set** is the empty or **null set**, denoted by $\{\ \}$ or $\varnothing$, respectively.

Now consider the equation $(x + 3) + (x - 8) = 2x - 5$. The expression on the left side of the equation **simplifies** to the expression on the right side. No matter what value we **substitute** for x, the resulting values on both the left and right sides will always be the same. Therefore, the equation is always true regardless of the value for the variable. Trying to solve this equation, we get

$$\begin{aligned} (x + 3) + (x - 8) &= 2x - 5 \\ x + 3 + x - 8 &= 2x - 5 \\ 2x - 5 &= 2x - 5 \\ 2x - 5 - 2x &= 2x - 5 - 2x \\ -5 &= -5 \end{aligned}$$

← Variables are gone and statement is true

No **variable terms** remain, and a true statement results. We call such an equation an **identity**, and its **solution set** is the set of all real numbers, denoted by $\mathbb{R}$ or $\{x|x \text{ is a real number}\}$.

(eText Screens 11.2-1–11.2-21)

Example 9 shows how to identify **contradictions** and **identities** when solving equations.

Example 9 Identifying Contradictions and Identities

Determine if the equation is a **contradiction** or an **identity**. State the solution set.

a. $3x + 2(x - 4) = 5x + 7$ **b.** $3(x - 4) = x + 2(x - 6)$

Solutions

a. Begin with the original equation: $3x + 2(x - 4) = 5x + 7$

Use the **distributive property**: $3x + 2x - 8 = 5x + 7$

Combine like terms: $5x - 8 = 5x + 7$

Subtract $5x$ from both sides: $5x - 8 - 5x = 5x + 7 - 5x$

Simplify: $-8 = 7$ False

All the **variable terms** drop out, leaving a false statement. Therefore, the equation is a contradiction and has no solution. Its solution set is $\{\ \}$ or $\emptyset$.

My video summary

b. Try to complete this problem on your own to verify that it is an identity. View this **popup** for the steps, or watch this **video** for the complete solution.

You Try It **Work through this You Try It problem.**

Work Exercises 41–46 in this textbook or in the MyMathLab Study Plan.

OBJECTIVE 5 USE LINEAR EQUATIONS TO SOLVE APPLICATION PROBLEMS

We now look at how we can use linear equations to solve application problems. At this point you may wish to review the **Strategy for Solving Linear Equations in One Variable.**

Example 10 Body Surface Area of Infants and Children

The body surface area and weight of well proportioned infants and children are related by the equation

$$30S = W + 4$$

where S = the body surface area in square meters and W = weight in kilograms. Find the body surface area of a well-proportioned child that weighs 18 kg. *(Source: The Internet Journal of Anesthesiology, Vol. 2, No. 2)*

Solution We are given the child's weight, so we begin by **substituting** 18 for W.

$$30S = (18) + 4$$

Simplifying on the right-hand side gives

$$30S = 22.$$

To solve for S, we apply the **multiplication property of equality** and divide both sides of the equation by 30.

$$\frac{30S}{30} = \frac{22}{30}$$

$$S = \frac{22}{30} = \frac{11}{15} \approx 0.733$$

The child's body surface area is approximately 0.733 square meters.

(eText Screens 11.2-1–11.2-21)

Example 11 Red Meat vs Poultry

My video summary

In the U.S., the average pounds of red meat eaten, M, is related to the average pounds of poultry eaten, P, by the equation

$$100M = 14{,}000 - 42P.$$

Determine the average amount of poultry eaten if the average amount of red meat eaten is 100.1 pounds. *(Source: U.S. Department of Agriculture)*

Solution Substitute 100.1 for M in the equation, then solve the equation for P. Check your answer, or watch this video for the complete solution.

You Try It Work through this You Try It problem.

Work Exercises 47–49 in this textbook or in the MyMathLab Study Plan.

11.2 Exercises

In Exercises 1–46, solve the linear equations.

You Try It

1. $2x + 3 - 5x = 7 + x$

2. $3 + 5x - 2 = 2x + 3 - x$

3. $3y + 7 + 2y - 3 = 4y + 1$

4. $y + 5 + 2y + 4 = 3 + 3y + 2 - 4y$

You Try It

5. $3z - 2 - 6z + 1 = 4 - 3z - 7 + z$

6. $2(x + 1) + 3x = x - 8$

7. $3x + 7(1 - x) = 15$

8. $-2(x - 1) + 5x = -2 + x + 7$

9. $2x + 6 = 3(x + 2)$

10. $2x - 6 + 3x = 2x + 5(x - 2)$

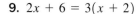

You Try It

11. $7 + 2(3 - 4z) = 7z + 2(3z - 4)$

12. $2(3x - 5) + 1 = (2x - 1) - 3(x - 2)$

13. $(4x - 1) + 3(2x + 8) = 5x - 2$

14. $(x + 3) - 2(2 - 3x) = 4(x + 4) - 2$

15. $3(2x - 1) - 6(x + 1) = 2x + 3$

16. $4(5 - 2x) + 1 = 2(x + 3) - 5x$

17. $2[3(x + 2) - 2] = x - 2$

18. $2x - (x + 1) = -3 + 5(x - 6)$

You Try It

19. $2y + \frac{1}{5} = \frac{4}{3}$

20. $\frac{y}{3} + \frac{1}{2} = 3y$

21. $\frac{1}{3}(y + 1) = \frac{2}{3}y$

22. $3 - \frac{2z}{5} = 5 - \frac{2z}{3}$

23. $2x + \frac{1}{2} = \frac{x}{3} + 3$

24. $\frac{x + 1}{2} + 3 = \frac{x + 4}{5}$

You Try It

25. $2w + \frac{7}{3} = \frac{w + 2}{3} - 2$

26. $2 + \frac{3y}{5} = \frac{2}{5} + 3y$

27. $\frac{a}{2} + \frac{2}{3} = \frac{3a}{4} + \frac{4}{5}$

28. $\frac{2x}{3} + \frac{1}{3} + 2 = \frac{3x}{4} - \frac{3}{2}$

29. $\frac{2x + 2}{5} + 1 = \frac{5x + 1}{3} - 4$

30. $\frac{5x - 2}{3} + \frac{x + 1}{5} = 7$

31. $\left(\frac{x + 1}{2} + 1\right) + 1 = \frac{3x + 3}{4}$

32. $3\left[2 - \frac{z + 2}{5}\right] + 1 = 5\left(\frac{z - 1}{3} + 1\right)$

33. $1.5x + 2.4 = 3.2x - 1$

34. $2.5x + 1.2 = 6.8$

You Try It **35.** $2.57y + 3.21 = 1.5y$

36. $5.2z - 7.85 = 9.726$

37. $2.26 - 1.12x = 5x - 3.554$

38. $2.3(5 + 2x) - 2.5 = 9$

39. $0.24x - 0.66 = 1.35x - 0.65x - 2.806$

40. $1.5x - 3.5 + 2x = 3.2 + 1.9x - 11.7$

41. $3(x + 2) - 2x + 4 = x + 5$

42. $5(x + 2) - 2x = 3(x + 1) + 7$

You Try It

43. $-3(x + 1) + x = 2(x + 2) - 7 - 4x$

44. $(2x + 5) + (x + 1) = 3(x - 1) + 7$

45. $2(x + 3) - 2 = 5(x + 1) - 3x$

46. $2\left(\frac{x + 1}{3} - 1\right) - x = -\frac{x + 1}{3}$

You Try It

47. Sliders The cost, C (in dollars), and the number of mini-burgers (or *sliders*), N, are related by the following equation:

$$40C = 27N + 2.$$

Find the number of sliders that can be purchased for \$14.90.

48. Chirp Rate Temperature, T (in °F) and the *chirp rate of crickets*, C, (in chirps per minute) have an actual mathematical relationship as shown in the following equation:

$$T = 50 + \frac{C - 40}{4}$$

While sitting outside last night, I heard crickets chirping. If the temperature was about 52 degrees, how many chirps did I hear per minute?

49. Child Height The following equation shows the approximate linear relationship between the age (in years), a, of a child and the child's height, h (in inches):

$$h = 2.5433a + 28.465$$

If Payton, is 11.5 years old, approximately how tall might she be?
(Source: The Merck Manual of Diagnosis and Therapy, 15th ed)

11.3 Introduction to Problem Solving

THINGS TO KNOW

Before working through this topic, be sure you are familiar with the following concepts:

	VIDEO	ANIMATION	INTERACTIVE
You Try It — 1. Write Word Statements as Algebraic Expressions (Topic 10.6, **Objective 3**)	▤		
You Try It — 2. Solve Linear Equations Using Both Properties of Equality (Topic 11.1, **Objective 5**)	▤		
You Try It — 3. Solve Linear Equations Containing Non-Simplified Expressions (Topic 11.2, **Objective 1**)	▤		
You Try It — 4. Solve Linear Equations Containing Decimals; Apply a General Strategy (Topic 11.2, **Objective 3**)	▤		

(eText Screens 11.3-1–11.3-21)

OBJECTIVES

1 Translate Sentences into Equations
2 Use the Problem-Solving Strategy to Solve Direct Translation Problems
3 Solve Problems Involving Related Quantities
4 Solve Problems Involving Consecutive Integers
5 Solve Problems Involving Value

OBJECTIVE 1 TRANSLATE SENTENCES INTO EQUATIONS

In **Topic 10.6**, we translated *word phrases* into **algebraic expressions.** We can also translate *sentences* into **equations.** Like the **key words** that translate into **arithmetic operations,** there are key words that translate into an equal sign ($=$). See Table 1.

Table 1

Key Words That Translate to an Equal Sign			
is	was	will be	gives
yields	results in	equals	is equal to
is equivalent to	is the same as		

Consider the following sentence:

The **product** of 5 and a number is 45.

The key word "product" indicates multiplication, and the key word "is" indicates an equal sign. Letting x represent the unknown number, we can translate:

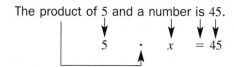

The product of 5 and a number is 45.

$$5 \cdot x = 45$$

The equation is $5x = 45$.

Example 1 Translating Sentences into Equations

Translate each sentence into an **equation**. Use x to represent each unknown number.

a. Fifty-two less than a number results in -21.

b. Three-fourths of a number, increased by 8, gives the number.

c. The **difference** of 15 and a number is the same as the **sum** of the number and 1.

d. If the sum of a number and 4 is multiplied by 2, the result will be 2 less than the **product** of 4 and the number.

Solutions

a. The phrase "less than" indicates subtraction. "Fifty-two less than a number" translates into the **algebraic expression** $x - 52$. "Results in" indicates an equal sign.

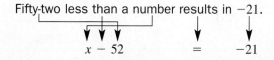

Fifty-two less than a number results in -21.

$$x - 52 = -21$$

The equation is $x - 52 = -21$.

b. "Increased by" indicates addition. Three-fourths of x and 8 are the **terms** being added, which means $\frac{3}{4}x + 8$. The word "gives" means an equal sign, and "the number" means x.

Three-fourths of a number, increased by 8, gives the number.

$$\frac{3}{4} \quad \cdot \quad x \quad + \quad 8 \quad = \quad x$$

The equation is $\frac{3}{4}x + 8 = x$.

My video summary

c.–d. Try to translate these sentences on your own, then check your **answers**. Watch this **video** for the complete solutions.

You Try It **Work through this You Try It problem.**

Work Exercises 1–8 in this textbook or in the MyMathLab Study Plan.

OBJECTIVE 2 USE THE PROBLEM-SOLVING STRATEGY TO SOLVE DIRECT TRANSLATION PROBLEMS

A **mathematical model** uses the language of mathematics to describe a problem. Typically the model is an equation that describes a relationship within an application.

As you proceed through this course, you will learn how to model applied situations by using equations. Let's explore the following problem-solving strategy when translating, modeling, and solving applied problems involving linear equations.

Problem-Solving Strategy for Applications of Linear Equations

Step 1. **Define the Problem.** Read the problem carefully, or multiple times if necessary. Identify what you are trying to find and determine what information is available to help you find it.

Step 2. **Assign Variables.** Choose a variable to assign to an unknown quantity in the problem. For example, use p for price. If other unknown quantities exist, express them in terms of the selected variable.

Step 3. **Translate into an Equation.** Use the relationships among the known and unknown quantities to form an equation.

Step 4. **Solve the Equation.** Determine the value of the variable and use the result to find any other unknown quantities in the problem.

Step 5. **Check the Reasonableness of Your Answer.** Check to see if your answer makes sense within the context of the problem. If not, check your work for errors and try again.

Step 6. **Answer the Question.** Write a clear statement that answers the question(s) posed.

We can use this **problem-solving strategy** to solve problems involving **direct translation**.

Example 2 Solving Direct-Translation Problems

Five times a number, increased by 17, is the same as 11 subtracted from the number. Find the number.

Solution Follow the problem-solving strategy.

Step 1. We must find "the number."

Step 2. Let x be the unknown number.

Step 3. "Five times a number, increased by 17" translates to $5x + 17$. The word phrase "is the same as" translates to an equal sign ($=$). The phrase "11 subtracted from the number" translates to $x - 11$.

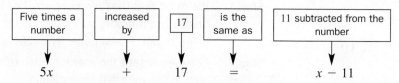

Five times a number	increased by	17	is the same as	11 subtracted from the number
↓	↓	↓	↓	↓
$5x$	$+$	17	$=$	$x - 11$

The equation is $5x + 17 = x - 11$.

Step 4.

Write the equation:	$5x + 17 = x - 11$
Subtract x from both sides:	$5x - x + 17 = x - x - 11$
Simplify:	$4x + 17 = -11$
Subtract 17 from both sides:	$4x + 17 - 17 = -11 - 17$
Simplify:	$4x = -28$
Divide both sides by 4:	$\dfrac{4x}{4} = \dfrac{-28}{4}$
Simplify:	$x = -7$

Step 5. Check the number -7 in the original sentence to see if it is a reasonable solution.

"5 times -7, increased by 17" translates to $5(-7) + 17$.

Simplify: $5(-7) + 17 = -35 + 17 = -18$.

"11 subtracted from -7" translates to $-7 - 11$.

Simplify: $-7 - 11 = -18$.

Both phrases simplify to the same result, -18, so -7 checks.

Step 6. The number is -7.

You Try It **Work through this You Try It problem.**

Work Exercises 9–12 in this textbook or in the MyMathLab Study Plan.

Example 3 Solving Direct-Translation Problems

My video summary

Four times the difference of twice a number and 5 results in the number increased by 50. Find the number.

Solution Using the problem-solving strategy, try to solve this problem on your own. View the answer, or watch this video for a detailed solution.

You Try It **Work through this You Try It problem.**

Work Exercises 13–16 in this textbook or in the MyMathLab Study Plan.

(eText Screens 11.3-1–11.3-21)

OBJECTIVE 3 SOLVE PROBLEMS INVOLVING RELATED QUANTITIES

For some problems, we need to find two or more quantities that are related in some way.

Example 4 Storage Capacity

The storage capacity of Deon's external hard drive is 32 times that of his jump drive, a small portable memory device. Together, his two devices have 264 gigabytes of memory. What is the memory size of each device?

Solution Follow the **problem-solving strategy**.

Step 1. We must find the "memory size" or number of gigabytes that each device holds. The memory size of the external hard drive is 32 times that of the jump drive, and the total capacity is 264 gigabytes.

Step 2. Let g represent the memory size of the jump drive. Then $32g$ is the memory size of the external hard drive.

Step 3. The total size is 264 gigabytes. We add the sizes of the jump drive and external hard drive to get 264.

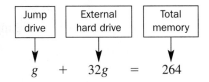

Jump drive		External hard drive		Total memory
↓		↓		↓
g	$+$	$32g$	$=$	264

Step 4.

Write the equation: $g + 32g = 264$

Combine like terms: $33g = 264$

Divide both sides by 33: $\dfrac{33g}{33} = \dfrac{264}{33}$

Simplify: $g = 8$

If $g = 8$, then $32g = 32(8) = 256$.

Step 5. The jump drive holds 8 gigabytes of memory, and the external hard drive holds 256 gigabytes. Because 256 is 32 times 8 and the sum of 8 and 256 is 264, these results make sense.

Step 6. Deon's jump drive has 8 gigabytes of memory, and his external hard drive has 256 gigabytes of memory.

You Try It **Work through this You Try It problem.**

Work Exercises 17–20 in this textbook or in the MyMathLab Study Plan.

Example 5 Movie Running Times

My video summary

Disney's *Toy Story* is 11 minutes shorter than its sequel *Toy Story 2*. *Toy Story 3* is 17 minutes longer than *Toy Story 2*. If the total running time for the three movies is 282 minutes, find the running time of each movie. (*Source*: Disney)

Solution

Step 1. We need to find the running time of each movie. The relationship between the quantities is that *Toy Story* is 11 minutes shorter than *Toy Story 2*, and *Toy Story 3* is 17 minutes longer than *Toy Story 2*. Also, the total running time of the three movies is 282 minutes.

Step 2. Let t represent the running time of *Toy Story 2*. Then $t - 11$ is the running time for *Toy Story*, and $t + 17$ is the running time for *Toy Story 3*.

Step 3. The sum of the three individual running times equals the total running time of 282 minutes.

Toy Story		*Toy Story 2*		*Toy Story 3*		Total Running Time
↓		↓		↓		↓
$(t - 11)$	$+$	t	$+$	$(t + 17)$	$=$	282

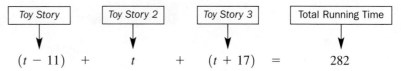

Complete the remaining steps of the **problem-solving strategy** to finish this problem on your own. View the **answer**, or watch this **video** for a complete solution.

You Try It **Work through this You Try It problem.**

Work Exercises 21 and 22 in this textbook or in the MyMathLab Study Plan.

OBJECTIVE 4 SOLVE PROBLEMS INVOLVING CONSECUTIVE INTEGERS

Integers that appear next to each other in an ordered list of all integers are called **consecutive integers**. For example, 7, 8, and 9 are three consecutive integers. Similarly, 10, 12, and 14 are **consecutive even integers**, and 9, 11, and 13 are **consecutive odd integers.**

The **difference** of two consecutive integers is 1 (for example, $9 - 8 = 1$), and the difference of two consecutive even integers or two consecutive odd integers is 2 (for example, $10 - 8 = 2$ and $9 - 7 = 2$). Using these facts, we can establish the general relationships shown in Table 2.

	Example	**General Relationship**
Consecutive integers	7, 8, 9 (+1, +1)	$x, x + 1, x + 2$, where x is an integer (+1, +1)
Consecutive even integers	10, 12, 14 (+2, +2)	$x, x + 2, x + 4$, where x is an even integer (+2, +2)
Consecutive odd integers	9, 11, 13 (+2, +2)	$x, x + 2, x + 4$, where x is an odd integer (+2, +2)

Table 2

CAUTION: A common error is to use x, $x + 1$, and $x + 3$ to represent consecutive odd integers (because 1 and 3 are odd). This is wrong! Remember consecutive odd integers are 2 units apart.

Example 6 Solving a Consecutive Integer Problem

The sum of two **consecutive integers** is 79. Find the two integers.

Solution

Step 1. We are looking for two consecutive integers with a sum of 79.

Step 2. Let x be the first integer. Then $x + 1$ is the next consecutive integer.

Step 3. The sum is 79, so we add the two integers to equal 79.

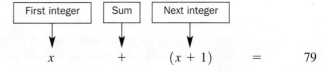

First integer	Sum	Next integer		
↓	↓	↓		
x	$+$	$(x + 1)$	$=$	79

Step 4.

Write the equation:	$x + (x + 1) = 79$
Combine like terms:	$2x + 1 = 79$
Subtract 1 from both sides:	$2x + 1 - 1 = 79 - 1$
Simplify:	$2x = 78$
Divide both sides by 2:	$\frac{2x}{2} = \frac{78}{2}$
Simplify:	$x = 39$

If $x = 39$, then $x + 1 = 39 + 1 = 40$.

Step 5. Because 39 and 40 are consecutive integers and their sum is 79, the answers make sense.

Step 6. The two consecutive integers are 39 and 40.

You Try It **Work through this You Try It problem.**

Work Exercises 23 and 24 in this textbook or in the MyMathLab Study Plan.

Example 7 Solving a Consecutive Integer Problem

My video summary

Three **consecutive even integers** add to 432. Find the three integers.

Solution Try to solve this problem on your own. View the **answer**, or watch this **video** for a detailed solution.

You Try It **Work through this You Try It problem.**

Work Exercises 25–28 in this textbook or in the MyMathLab Study Plan.

OBJECTIVE 5 SOLVE PROBLEMS INVOLVING VALUE

Many applications involve finding the cost or value of a set of items. For example, suppose you purchase 3 pastries that cost 69¢ each. Your total cost for the three pastries will be $(\$0.69)(3) = \$2.07$. The following equation helps us to solve this problem:

$$\text{Total value} = (\text{Value per item})(\text{Number of items})$$

CAUTION

The "value" in the above equation can represent cost, profit, revenue, earnings, and so forth.

Example 8 Cell Phone Plan

My video summary

Ethan's cell phone plan costs $34.99 per month for the first 700 minutes, plus $0.35 for each additional minute. If Ethan's bill is $57.39, how many minutes did he use?

Solution

Step 1. We need to find the number of minutes Ethan used on his cell phone. We know that the first 700 minutes are included in the monthly cost of $34.99 and that additional minutes cost $0.35 each. We also know the total cost was $57.39.

Step 2. Let m be the total number of minutes Ethan used. Then $m - 700$ is the number of "additional minutes" above 700.

(eText Screens 11.3-1–11.3-21)

Step 3. The cost of the first 700 minutes is $34.99. The cost of the additional minutes is $0.35(m - 700)$. Adding these two amounts results in the overall bill amount of $57.39. So, we have the equation:

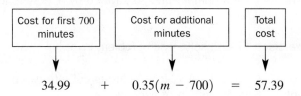

Cost for first 700 minutes | Cost for additional minutes | Total cost

$$34.99 \quad + \quad 0.35(m - 700) \quad = \quad 57.39$$

Finish solving the equation to find the number of minutes Ethan used. Compare your **answer**, or watch this **video** for a fully worked solution.

You Try It **Work through this You Try It problem.**

Work Exercises 29–32 in this textbook or in the MyMathLab Study Plan.

11.3 Exercises

In Exercises 1–8, translate each sentence into an equation. Use x to represent each unknown number. Do not solve.

1. Thirteen more than a number is 38.

2. The sum of twice a number and 16 is equal to three times the number.

3. Subtracting 5 from the quotient of a number and 4 gives 31.

You Try It

4. The product of 7 and the sum of a number and 4 results in the difference of 4 and the number.

5. The sum of a number and 2.7, divided by 6, yields the quotient of the number and 9.

6. If -7 is decreased by a number, the result will be three-fourths of the number.

7. The ratio of a number and 2 is equivalent to the difference of the number and 19.

8. Three times the sum of a number and 8, decreased by 10, is the same as five times the difference of the number and 1, increased by 14.

In Exercises 9–16, solve each problem involving direct translation.

9. The quotient of a number and 3, decreased by 1, is equal to 37. Find the number.

10. Nineteen less than triple a number is -53. Find the number.

You Try It

11. The sum of a number and 11 is the same as the difference of 17 and the number. Find the number.

12. The product 8 and a number, increased by 84, is equivalent to the product of 12 and the sum of the number increased by 4. Find the number.

13. If -11 is added to a number, the sum will be 5 times the number. Find the number.

14. Five-eighths more than three-fourths of a number results in two times the number.

15. Eight times the difference of twice a number and 9 yields the number increased by 93. Find the number.

You Try It

16. The sum of a number and 80, divided by 6, results in one-half of the number. Find the number.

In Exercises 17–22, solve each problem involving related quantities.

17. Cutting a Board A 72-inch board is cut into two pieces so that one piece is four times as long as the other. Find the length of each piece.

You Try It

18. Inheritance An inheritance of $35,000 will be divided between sisters Hannah and Taylor. If Hannah is to receive $3000 less than Taylor, how much does each sister receive?

19. Skyscrapers The Burj Khalifa (in Dubai, United Arab Emirates) is 320 meters taller than Taipei 101 (in Taipei, Taiwan). If the height of the Burj Khalifa is subtracted from twice the height of Taipei 101, the result is 188 meters. How tall is each skyscraper? (*Source*: Infoplease.com)

20. Movie Classics The running time for *Gone with the Wind* is 36 minutes more than twice the running time for *The Wizard of Oz*. If the running time for *Gone with the Wind* is subtracted from triple the running time for *The Wizard of Oz*, the result is 65 minutes. Find the running times for each movie. (*Source*: The Internet Movie Database, IMDb.com)

21. Most Valuable Player During his career, Michael Jordan was named the NBA's MVP one time less than Kareem Abdul-Jabbar. Bill Russell was MVP two more than half the times of Abdul-Jabbar. Together, the three men were named MVP 16 times. How many times were each of the men named MVP? (*Source*: NBA.com)

You Try It

22. Grand Slam Tennis During his tennis career in singles play, Pete Sampras won 3 fewer Australian Open titles than U.S. Open titles and 2 more Wimbledon titles than U.S. Open titles. If he won 14 of these titles total, how many times did he win each one? (*Source:* petesampras.com)

In Exercises 23–28, solve each problem involving consecutive integers.

23. The sum of two consecutive integers is 691. Find the integers.

You Try It

24. The sum of three consecutive integers is 258. Find the integers.

25. The sum of two consecutive odd integers is 116. Find the integers.

26. The sum of three consecutive even integers is 762. Find the integers.

You Try It

27. An Open Book The page numbers showing on an open book are consecutive integers with a sum of 317. Find the two page numbers.

28. Office Numbers The numbers on the doors of four adjacent offices are consecutive odd integers. If their sum is 880, find the four office numbers.

In Exercises 29–32, solve each problem involving value.

29. Selling Aluminum Doug collects and recycles aluminum cans by selling them to a local salvage yard. If he earns $136.30 by selling aluminum at a price of $0.58 per pound, how many pounds of aluminum did Doug sell?

You Try It

30. Car Rental A car rental cost $80 per day, for the first 250 miles, plus $0.20 for each additional mile. If Sonya's rental cost was $116 for a one-day rental, how many miles did she drive?

(eText Screens 11.4-1–11.4-25)

31. Working Two Jobs John works two jobs. As a security guard he earns $9.50 per hour. As a landscaper he earns $13.00 per hour. One week John worked a total of 40 hours and earned $432.50. How many hours did he work at each job?

32. Concession Stand A basketball concession stand sells sodas for $1.50 each and bags of popcorn for $0.75 each. One night, 36 more sodas were sold than bags of popcorn. If the concession stand made $153 that night, how many sodas and how many bags of popcorn were sold?

11.4 Formulas

THINGS TO KNOW

Before working through this topic, be sure you are familiar with the following concepts:

		VIDEO	ANIMATION	INTERACTIVE
You Try It	1. Evaluate Algebraic Expressions (Topic 10.5, **Objective 1**)	✓		
You Try It	2. Solve Linear Equations Using Both Properties of Equality (Topic 11.1, **Objective 5**)	✓		
You Try It	3. Solve Linear Equations Containing Non-Simplified Expressions (Topic 11.2, **Objective 1**)	✓		
You Try It	4. Solve Linear Equations Containing Fractions (Topic 11.2, **Objective 2**)	✓		
You Try It	5. Use Linear Equations to Solve Application Problems (Topic 11.2, **Objective 5**)	✓		

OBJECTIVES

1 Evaluate a Formula

2 Find the Value of a Non-isolated Variable in a Formula

3 Solve a Formula for a Given Variable

4 Use Geometric Formulas to Solve Applications

OBJECTIVE 1 EVALUATE A FORMULA

Definition Formula

A **formula** is an **equation** that describes the relationship between two or more **variables**.

Typically formulas apply to physical or financial situations that relate quantities such as length, area, volume, time, speed, money, interest rates, and so on. For example, $d = rt$ is the **distance formula.** It relates three variables: distance d, rate (or speed) r, and time t. If we know values for r and t, we can **evaluate the formula** to find the value of d.

(eText Screens 11.4-1–11.4-25)

Example 1 Evaluating the Distance Formula

A car travels at an average speed (rate) of 55 miles per hour for 3 hours. How far does the car travel?

Solution Using the distance formula, we have $r = \dfrac{55 \text{ mi}}{\text{h}}$ and $t = 3 \text{ h}$ and need to find d.

Write the distance formula: $d = rt$

Substitute $r = \dfrac{55 \text{ mi}}{\text{h}}$ and $t = 4\text{h}$: $d = \dfrac{55 \text{ mi}}{\text{h}} \cdot 4\text{h}$

Simplify: $d = \dfrac{55 \text{ mi}}{\cancel{\text{h}}} \cdot \dfrac{4\cancel{\text{h}}}{1} = 220 \text{ mi}$

The car traveled 220 miles.

CAUTION When evaluating formulas, the units must be consistent. For example, when evaluating the distance formula, if the unit for rate is meters per second, then the unit for time must be seconds, and the resulting unit for distance will be meters.

You Try It **Work through this You Try It problem.**

Work Exercises 1 and 2 in this textbook or in the MyMathLab Study Plan.

The **Devine Formula** is a popular **formula** in the healthcare profession that uses a person's height to compute his or her ideal body weight. The formula is $w = 110 + 5.06(h - 60)$ for men and $w = 100.1 + 5.06(h - 60)$ for women, where w is ideal body weight, in pounds, and h is height, in inches.

Note: This formula is only used for people over 60 inches, or 5 feet, tall.

Example 2 Computing Ideal Body Weight

My video summary

Evaluate the Devine Formula to find the ideal body weight of each person described.

a. A man 72 inches tall

b. A woman 66 inches tall

Solutions

a. Write the Devine Formula (for men): $w = 110 + 5.06(h - 60)$

Substitute $h = 72$: $w = 110 + 5.06(72 - 60)$

Subtract within parentheses: $w = 110 + 5.06(12)$

Multiply: $w = 110 + 60.72$

Add: $w = 170.72$

The ideal body weight of a man 72 inches tall is 170.72 pounds.

b. Try to evaluate the Devine Formula (for women) on your own, then check your **answer**. Watch this **video** for complete solutions to both parts.

You Try It **Work through this You Try It problem.**

Work Exercises 3–8 in this textbook or in the MyMathLab Study Plan.

Many **formulas** come from geometry. To review the formulas for **area** and **perimeter**, launch this **popup box**. To review formulas for volume, launch this **popup box**.

Example 3 Evaluating Geometry Formulas

a. The top of a stainless steel sink is shaped like a square with each side measuring $15\frac{3}{4}$ inches long. How many inches of aluminum molding will be required to surround the outside of the sink? Hint: Use the **formula** for the **perimeter of a square**.

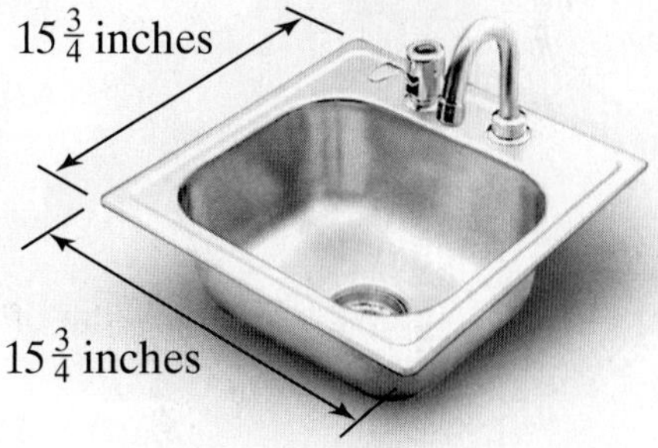

b. A yield sign has the shape of a triangle with a base of 3 feet and a height of 2.6 feet. Find the area of the sign. Hint: Use the formula for the **area of a triangle**.

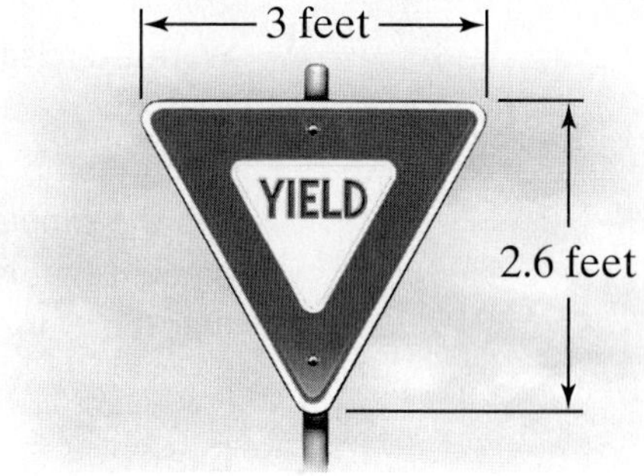

Solutions

a. We must find the **perimeter** of the sink. The **formula** for the perimeter of a square is $P = 4s$. We are given $s = 15\frac{3}{4}$ inches.

Write the formula: $P = 4s$

Substitute $s = 15\frac{3}{4}$ in: $P = 4\left(15\frac{3}{4}\text{ in}\right)$

Simplify: $P = 4\left(\frac{63}{4}\text{ in}\right) = 63\text{ in}$

63 inches of aluminum molding are needed.

My video summary

b. Try working this problem on your own by using the formula $A = \frac{1}{2}bh$. View the **answer**, or watch this **video** for a detailed solution.

You Try It **Work through this You Try It problem.**

Work Exercises 9–12 in this textbook or in the MyMathLab Study Plan.

OBJECTIVE 2 FIND THE VALUE OF A NON-ISOLATED VARIABLE IN A FORMULA

Most **formulas** are solved for one of their variables. This means that one variable is **isolated** on one side of the equation. For example, the formula $d = rt$ is solved for d. We have **evaluated formulas** to find values of isolated variables. Now let's find values of non-isolated variables.

Example 4 Finding the Value of a Non-isolated Variable

The **perimeter** of a rectangle is given by the formula $P = 2l + 2w$. If $P = 84$ cm and $l = 26$ cm, find w.

Solution We want to find the width w of a rectangle with perimeter $P = 84$ cm and length $l = 26$ cm.

Write the original formula: $P = 2l + 2w$

Substitute $P = 84$ cm and $l = 26$ cm: $84\text{ cm} = 2(26\text{ cm}) + 2w$

(eText Screens 11.4-1–11.4-25)

Simplify:	$84 \text{ cm} = 52 \text{ cm} + 2w$
Subtract 52 cm from both sides:	$84 \text{ cm} - 52 \text{ cm} = 52 \text{ cm} + 2w - 52 \text{ cm}$
Simplify:	$32 \text{ cm} = 2w$
Divide both sides by 2:	$\dfrac{32 \text{ cm}}{2} = \dfrac{2w}{2}$
Simplify:	$16 \text{ cm} = w$

The width of the rectangle is 16 cm.

You Try It **Work through this You Try It problem.**

Work Exercises 13–16 in this textbook or in the MyMathLab Study Plan.

A common banking formula, $I = Prt$, is used for computing **simple interest**. In this formula, I is the *simple interest* earned on an investment or paid for a loan; P is the *principal*, or the amount that is invested or borrowed; r is the *interest rate* in decimal form; and t is the *time* that the money is invested or borrowed.

Example 5 Using the Simple Interest Formula

My video summary

Paige has invested \$15,000 in a **certificate of deposit** (CD) that pays 4% simple interest annually. If she earns \$750 in interest when the CD matures, how long has Paige invested the money?

Solution We know $P = \$15{,}000$, $r = 4\% = 0.04$, and $I = \$750$. To find t, we substitute these values into the formula $I = Prt$ and solve for t. Try to finish this problem on your own. View the **answer**, or watch this **video** for the complete solution.

You Try It **Work through this You Try It problem.**

Work Exercise 17 in this textbook or in the MyMathLab Study Plan.

A **formula** that relates the two most common measures of temperature is $F = \frac{9}{5}C + 32$, where F represents degrees Fahrenheit and C represents degrees Celsius.

Example 6 Finding Equivalent Temperatures

My video summary

During the month of February, the average high temperature in Montreal, QC is $-4.6°C$ while the average high temperature in Phoenix, AZ is $70.7°F$. (*Source:* World Weather Information Service)

a. What is the equivalent Fahrenheit temperature in Montreal?

b. What is the equivalent Celsius temperature in Phoenix?

Solutions Try working these problems on your own, then check your **answers**. Watch this **video** for complete solutions.

You Try It **Work through this You Try It problem.**

Work Exercises 18–22 in this textbook or in the MyMathLab Study Plan.

(eText Screens 11.4-1–11.4-25)

OBJECTIVE 3 SOLVE A FORMULA FOR A GIVEN VARIABLE

Sometimes it is helpful to **solve a formula for a given variable** in terms of the other **variables.** This means that we write the **formula** with the desired variable **isolated** on one side of the equal sign, resulting in all other variables and **constants** on the other side of the equal sign. For example, the distance formula $d = rt$ can be solved for t:

Write the distance formula: $d = rt$

Divide both sides by r: $\frac{d}{r} = \frac{rt}{r}$

Simplify: $\frac{d}{r} = t$

To solve a formula for a given variable, we follow the **general strategy for solving linear equations** from **Topic 11.2**.

CAUTION: When we solve a formula for a given variable, the appearance of the formula changes, but the relationships among the variables do not.

Example 7 Solving a Formula for a Given Variable

Solve each formula for the given variable.

a. Selling price: $S = C + M$ for M

b. Area of a triangle: $A = \frac{1}{2}bh$ for b

c. **Perimeter** of a rectangle: $P = 2l + 2w$ for l

Solutions

a. Write the original formula: $S = C + M$

Subtract C from both sides: $S - C = C + M - C$

Simplify: $S - C = M$

The formula $S - C = M$, or $M = S - C$, is solved for M.

b. Write the original formula: $A = \frac{1}{2}bh$

Multiply both sides by 2 to clear the fraction: $2(A) = 2\left(\frac{1}{2}bh\right)$

Simplify: $2A = bh$

Divide both sides by h: $\frac{2A}{h} = \frac{bh}{h}$

Simplify: $\frac{2A}{h} = b$

The formula $b = \frac{2A}{h}$ is solved for the variable b.

My video summary

c. Try to solve this formula for l on your own. View the **answer**, or watch this **video** to see a fully worked solution.

CAUTION: For a formula to be solved for a given variable, the given variable must be isolated on one side of the equation and must be the only variable of its type in the equation.

You Try It **Work through this You Try It problem.**

Work exercises 23–30 in this textbook or in the MyMathLab Study Plan.

(eText Screens 11.4-1–11.4-25)

OBJECTIVE 4 USE GEOMETRIC FORMULAS TO SOLVE APPLICATIONS

We have seen that **formulas** can be very useful when solving applications. Some applications require work beyond a single formula. Let's look at some applied problems that involve geometric formulas.

Example 8 Filling a Swimming Pool

An above-ground pool is shaped like a **circular cylinder** with a **diameter** of 28 ft and a depth of 4.5 ft. If $1 \text{ ft}^3 \approx 7.5$ gal, how many gallons of water will the pool hold? Use $\pi \approx 3.14$ and round to the nearest thousand gallons.

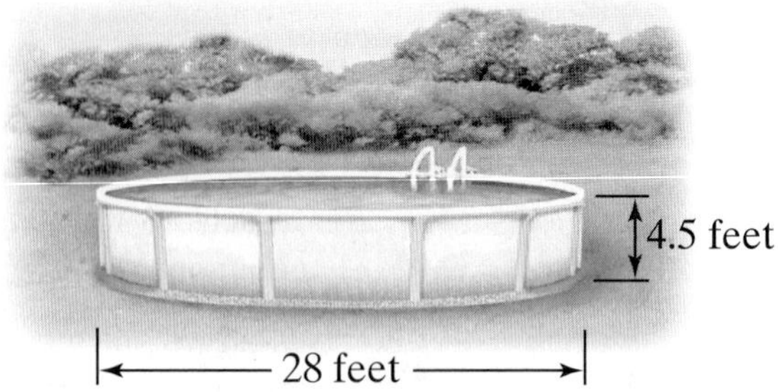

Solution We must find the **volume** of the pool in cubic feet and convert to gallons. The formula for the volume of a cylinder is $V = \pi r^2 h$. The diameter is 28 ft, so the **radius** is half of the diameter, or 14 ft.

Write the volume formula: $V = \pi r^2 h$

Substitute $r = 14$ ft, $h = 4.5$ ft, and $\pi \approx 3.14$: $V \approx (3.14)(14 \text{ ft})^2(4.5 \text{ ft})$

Evaluate the exponent: $V \approx (3.14)(196 \text{ ft}^2)(4.5 \text{ ft})$

Multiply: $V \approx 2769.48 \text{ ft}^3$

To find the number of gallons, we multiply the volume by $\frac{7.5 \text{ gal}}{1 \text{ ft}^3}$:

$$2769.48 \text{ ft}^3 \cdot \frac{7.5 \text{ gal}}{1 \text{ ft}^3} \approx 20{,}771.1 \text{ gal}$$

Rounding to the nearest thousand gallons, the pool will hold approximately 21,000 gallons.

You Try It **Work through this You Try It problem.**

Work Exercises 31 and 32 in this textbook or in the MyMathLab Study Plan.

Example 9 Finding the Cost of a New Floor

My video summary

Terrence wants to have a new floor installed in his living room, which measures 20 ft by 15 ft. Extending out 3 ft from one wall is a fireplace in the shape of a **trapezoid** with base lengths of 4 ft and 8 ft.

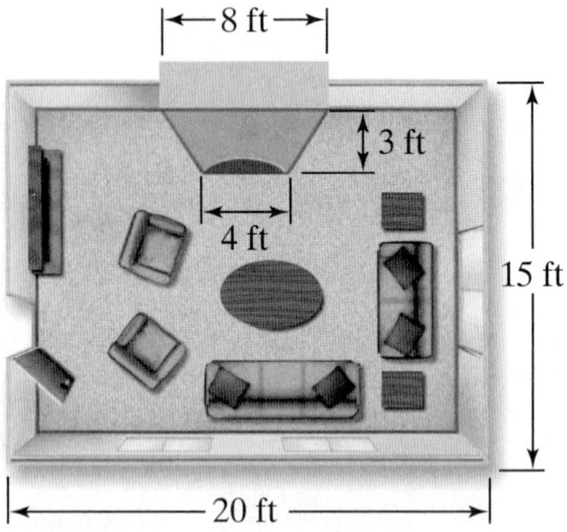

a. Find the **area** that needs flooring.

b. If the flooring costs \$5.29 per square foot, how much will Terrence pay for the new floor? (Assume there is no wasted flooring.)

(*eText Screens 11.4-1–11.4-25*)

Solutions

a. The **area** that needs new flooring is the area of the room (rectangle) minus the area of the fireplace (trapezoid). The area formula for a rectangle is $A = lw$, and the area formula for a trapezoid is $A = \frac{1}{2}h(a + b)$, so

$$\text{Area needing flooring} = \underbrace{lw}_{\text{Area of Rectangle}} - \underbrace{\frac{1}{2}h(a + b)}_{\text{Area of Trapezoid}}.$$

Substitute $l = 20$ ft, $w = 15$ ft, $h = 3$ ft, $a = 4$ ft, and $b = 8$ ft to finish part a, then complete all of part b on your own. View the **answers**, or watch this **video** for detailed solutions.

You Try It **Work through this You Try It problem.**

Work Exercises 33 and 34 in this textbook or in the MyMathLab Study Plan.

Example 10 Constructing a Roundabout Intersection

My video summary

The roundabout intersection with the top view shown in the figure will be constructed using concrete pavement 9 inches thick. How many cubic yards of concrete will be needed for the roundabout? Use $\pi \approx 3.14$.

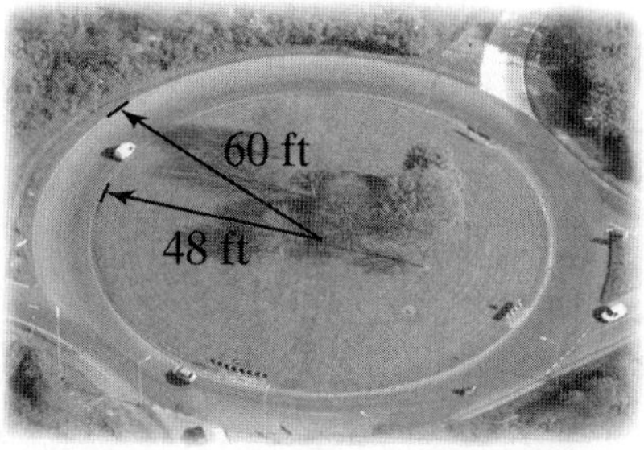

Solution We must find the **volume** of concrete needed, in cubic yards. This problem involves finding the volume of two **cylinders**: an outer cylinder with a **radius** of 60 feet and an inner cylinder with a radius of 48 feet. Both cylinders have heights of 9 inches. The volume of concrete needed is the difference in the volumes of these two cylinders:

$$\text{Volume of concrete} = \text{Volume of outer cylinder} - \text{Volume of inner cylinder}$$

Try to finish solving this problem on your own. Remember to use consistent units when computing the volumes. (Hint: 1 yd $= 3$ ft $= 36$ in.) View the **answer**, or watch this **video** for a detailed solution.

You Try It **Work through this You Try It problem.**

Work Exercises 35 and 36 in this textbook or in the MyMathLab Study Plan.

11.4 Exercises

In Exercises 1 and 2, use the distance formula $d = rt$.

You Try It

1. An airplane travels for 2.5 hours at an average rate of 140 miles per hour. How far does the plane travel?

2. If a cheetah runs at a rate of 16 meters per second for 7 seconds, how far will it travel?

In Exercises 3 and 4, use the appropriate **Devine Formula** to compute the ideal body weight of each person described.

You Try It

3. A man 70 inches tall

4. A woman 62 inches tall

In Exercises 5–8, use the formula provided to answer each question.

5. Good Tip $T = 0.2C$ is a formula for computing a 20% tip T in a restaurant for a meal that costs C dollars. Find the tip for a meal costing \$75.

6. Retail Price The formula $R = W + pW$ gives the retail price R of an item with a wholesale price W that is marked up p percent (in decimal form). Determine the retail price of a pair of jeans with a wholesale price of \$35.40 that is marked up 40%.

7. High Blood Pressure The formula $p = 1.068a - 18.39$ can be used to find the percent p of men at age a, in years, with high blood pressure. What percent of 35 year-old men have high blood pressure? Remember to convert from decimals to percent for your answer. Round to the nearest tenth of a percent. (*Source:* Based on data from the British Heart Foundation)

8. Horse Power The formula $H = \dfrac{VIE}{746}$ gives the horsepower H produced by an electrical motor with voltage V in volts, current I in amps, and percent efficiency E in decimal form. What is the horsepower of a 240-volt motor pulling 40 amps and having 86% efficiency? Remember to convert percents to decimals in your calculations. Round H to the nearest tenth.

In Exercises 9–12, evaluate the appropriate geometry formula to answer each question.

You Try It

9. Road Work Ahead The roadwork ahead sign shown has the shape of a square (rotated to look like a diamond). A construction worker wants to outline the edge of the sign with reflective tape. What length of reflective tape will be needed?

10. NBA Basketball Court The dimensions of an official NBA basketball court are shown in the figure. Find both the perimeter and area of the court. (*Source:* NBA.com)

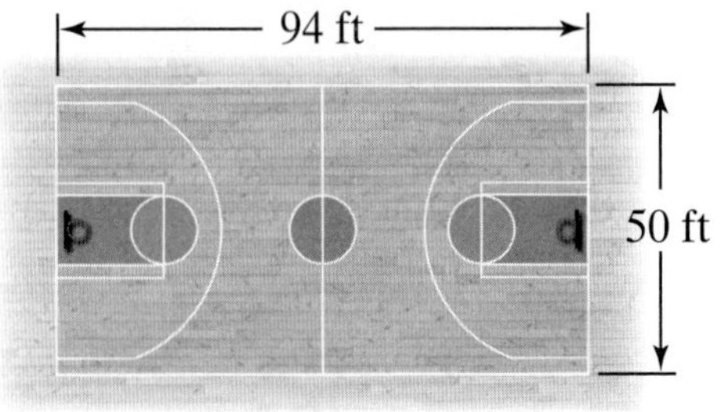

11. Table Top The top of a multi-purpose table is the shape of the trapezoid shown. Find both the perimeter and area of the table top.

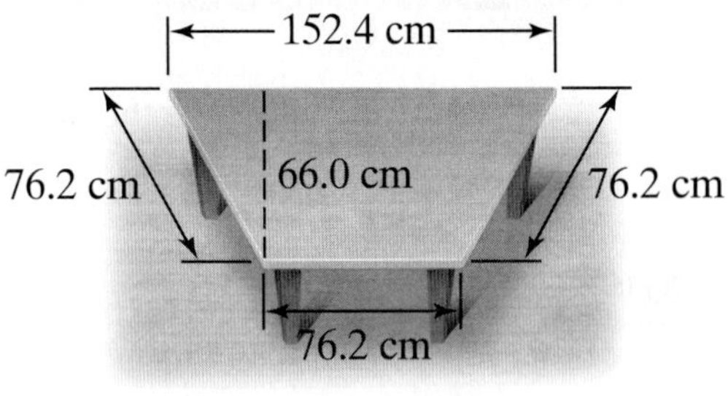

(eText Screens 11.4-1–11.4-25)

12. Volume of a Printer The laser printer shown is approximately a rectangular solid. What is the volume of the printer?

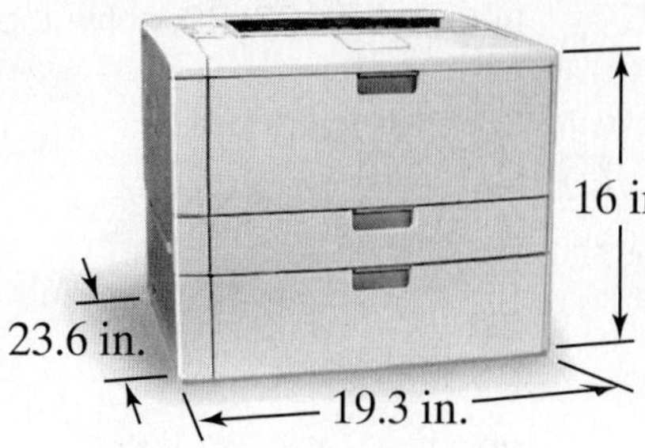

In Exercises 13–16, find the value of the unknown variable.

13. Profit Find R if $P = R - C$, $P = \$2150$, and $C = \$1525$.

You Try It

14. Area of a Rectangle Find the value of w if $A = lw$, $A = 48\text{ m}^2$, and $l = 4\text{ m}$.

15. Retail Price Find the value of W if $R = W + pW$, $R = \$31.85$, and $p = 30\% = 0.3$.

16. Area of a Trapezoid Find the value of a if $A = \frac{1}{2}h(a + b)$, $A = 210\text{ mm}^2$, $h = 14\text{ mm}$, and $b = 18\text{ mm}$.

In Exercises 17–22, use the appropriate formula to answer each question.

17. Simple Interest Principal How much principal should be invested in a savings account paying 6% simple interest if you want to earn \$450 in interest in 2 years.

You Try It

18. Record Heat The warmest temperature officially recorded on Earth is 136°F, which occurred on September 13, 1922 in Al 'Aziziyah, Libya. (*Source:* World Meteorological Organization) What is the equivalent temperature in Celsius? Round to the nearest tenth.

19. Time for a Trip How long will it take Cory to drive 210 miles if he averages 60 miles per hour on the trip?

20. NBA Ball Size The circumference of an official NBA basketball is 29.5 inches. Find the diameter of the ball. Use $\pi \approx 3.14$. Round to the nearest tenth. (*Source:* NBA.com)

21. Sail Height A triangular sail for a boat has an area of 120 ft^2. If the base of the sail is 16 ft, find the height.

22. Popcorn Box The popcorn box shown has a volume of 123 in^3. What is the height of the box?

(eText Screens 11.4-1–11.4-25)

In Exercises 23–30, solve the formula for the given variable.

You Try It

23. $C = 2\pi r$ for r

24. $P = a + b + c$ for a

25. $Ax + By = C$ for y

26. $A = P + Prt$ for t

27. $E = I(r + R)$ for R

28. $A = \frac{1}{2}h(a + b)$ for a

29. $F = \frac{9}{5}C + 32$ for C

30. $E = \frac{I - P}{I}$ for P

In Exercises 31–32, solve each application. When necessary, use $\pi \approx 3.14$.

You Try It

31. **Rain Barrel** The rain barrel shown is a right circular cylinder. If $1 \text{ ft}^3 \approx 7.5$ gallons, how many gallons of rainwater can the barrel hold? Round to the nearest tenth.

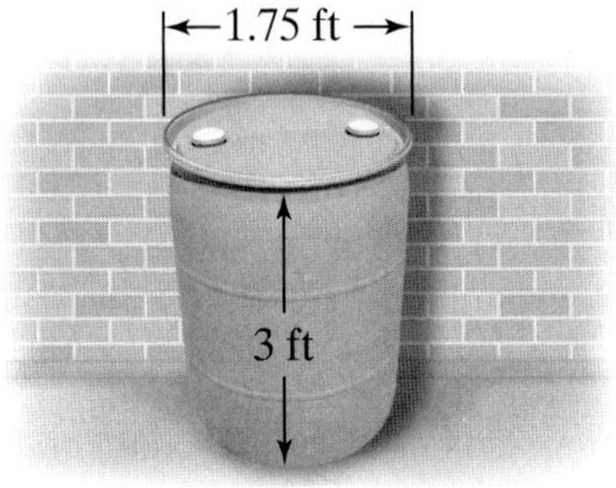

32. **Martini Glass** The martini glass shown has the shape of cone (on a stem). If $1 \text{ cm}^3 = 1 \text{ mL}$, how many milliliters of liquid can the glass hold? Round to the nearest tenth.

You Try It

33. **Carpeting a Bedroom** Lauren's bedroom measures 12 feet by 15 feet, with a closet that measures 6 feet by 3 feet. If carpet costs \$4.75 per square foot, how much will it cost to carpet Lauren's room and closet?

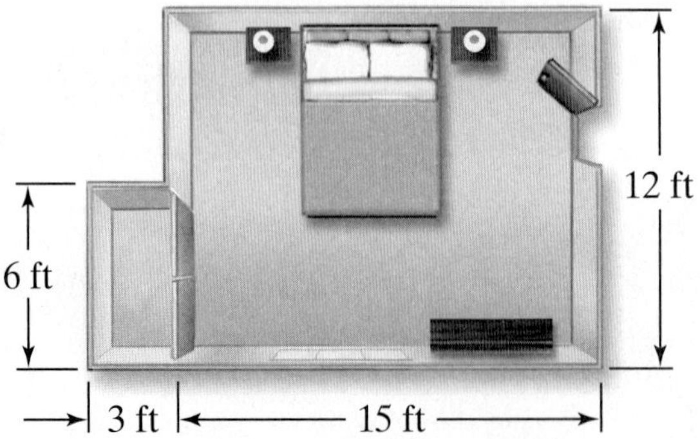

34. Sodding a Lawn If sod costs $0.45 per square foot, how much will it cost to sod the lawn shown in the figure? Hint: Sod is not needed where the house and driveway are located.

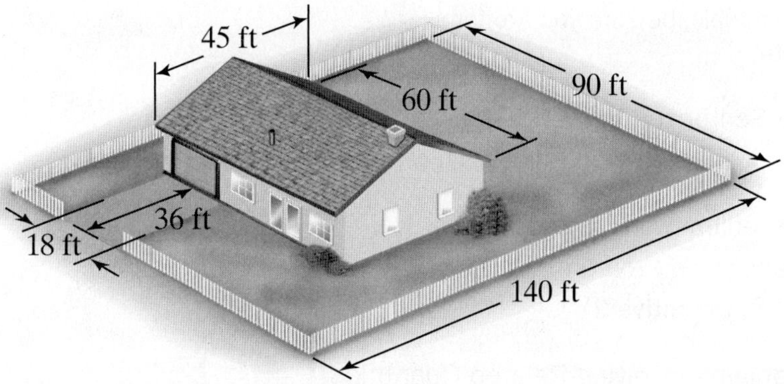

You Try It

35. Building a Sidewalk A concrete sidewalk 5 feet wide and 6 inches thick will be built around the square playground shown. How many cubic yards of concrete will be needed for the sidewalk?

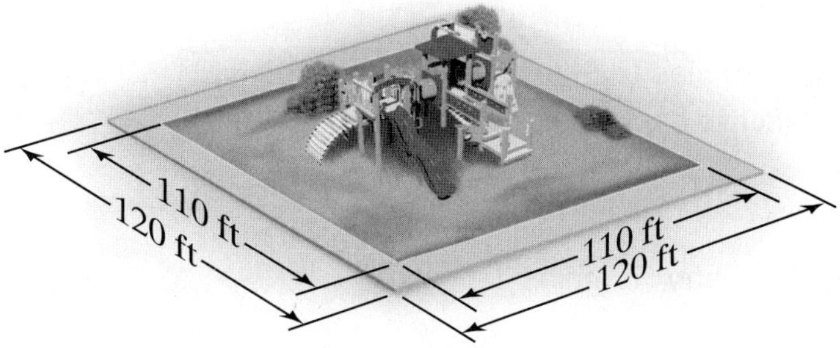

36. Resurfacing a Driveway A semicircular driveway 10 feet wide will be resurfaced with asphalt. See the figure. If the 2-inch thick asphalt costs $1.95 per square foot, find the cost of the driveway.

(eText Screens 11.5-1–11.5-22)

11.5 Geometry and Uniform Motion Problem Solving

THINGS TO KNOW

Before working through this topic, be sure you are familiar with the following concepts:

		VIDEO	ANIMATION	INTERACTIVE
You Try It	1. Translate Sentences into Equations (Topic 11.3, **Objective 1**)	▤		
You Try It	2. Use the Problem-Solving Strategy to Solve Direct Translation Problems (Topic 11.3, **Objective 2**)	▤		
You Try It	3. Solve Problems Involving Related Quantities (Topic 11.3, **Objective 3**)	▤		
You Try It	4. Evaluate a Formula (Topic 11.4, **Objective 1**)	▤		
You Try It	5. Solve a Formula for a Given Variable (Topic 11.4, **Objective 3**)	▤		
You Try It	6. Use Geometric Formulas to Solve Applications (Topic 11.4, **Objective 4**)	▤		

OBJECTIVES

1 Solve Problems Involving Geometry Formulas

2 Solve Problems Involving Angles

3 Solve Problems Involving Uniform Motion

OBJECTIVE 1 SOLVE PROBLEMS INVOLVING GEOMETRY FORMULAS

We begin by looking at applications involving basic geometry formulas. **Review** some of the common formulas. In the following examples, the dimensions are given in terms of a common variable, which may or may not be one of the dimensions. Once a value for the variable is known, the individual dimensions can be found.

Example 1 Miniature Golf

A green on a miniature golf course has a rectangular boundary. The length of the boundary is six feet longer than twelve times its width. If the perimeter is 103 feet, what are the dimensions of the green?

Solution We use the **problem-solving steps.**

Step 1. We must find both the length and the width of the miniature golf course green. The perimeter is 103 feet, and the length is 6 feet longer than 12 times the width. We will use the formula for the **perimeter of a rectangle** to help us solve this problem.

Step 2. If we let w represent the width of the green, then the length is $12w + 6$.

Step 3. Use the formula for the perimeter of a rectangle.

$$2l + 2w = P$$

Substituting 103 for P (the perimeter) and $12w + 6$ for l we have

$$2(12w + 6) + 2w = 103$$

Step 4. Solve the equation for w and then use the result to find the length.

Begin with the original equation:	$2(12w + 6) + 2w = 103$
Distribute:	$24w + 12 + 2w = 103$
Combine like terms:	$26w + 12 = 103$
Subtract 12 from both sides:	$26w + 12 - 12 = 103 - 12$
Simplify:	$26w = 91$
Divide both sides by 26:	$\frac{26w}{26} = \frac{91}{26}$
Simplify:	$w = \frac{7}{2}$ or 3.5

Step 5. The width of the green is 3.5 feet. Substituting 3.5 for w, we get $l = 12(3.5) + 6 = 48$ feet. This gives a perimeter of $P = 2(48) + 2(3.5) = 103$ feet, as given in the original problem.

Step 6. The width of the green is 3.5 feet, and the length is 48 feet.

Example 2 Canoe Paddle

My video summary

The blade of a canoe paddle is in the shape of an **isosceles triangle** so that two sides have the same length. The two common sides are each 4 inches longer than twice the length of the third side. If the **perimeter** is 48 inches, find the lengths of the sides of the blade.

Solution

Step 1. We need to find the lengths of the sides of a canoe paddle blade. We know that the blade is an isosceles triangle, the common side is 4 inches longer than twice the length of the third side, and the perimeter of the blade is 48 inches. We will use the perimeter formula for a triangle to solve this problem.

Step 2. If we let a represent the length of the third side, then the length of each common side is $2a + 4$.

Step 3. Start with the formula for the perimeter of a triangle.

$$a + b + c = P$$

Substituting 48 for P (the perimeter) and $2a + 4$ for b and c (the two common sides), we have the equation:

$$a + (2a + 4) + (2a + 4) = 48$$

Step 4. Finish solving the equation to find the length of the third side, then use that result to find the length of the common side. When finished, check your **answer**, or watch this **video** for a fully worked solution.

Example 3 Fresno Favorite

My video summary

The Triangle Drive In has been a local favorite for hamburgers in Fresno, CA since 1963. It is located on a triangular-shaped lot. One side of the lot is 4 meters longer than the shortest side, and the third side is 32 meters less than twice the length of the shortest side. If the perimeter of the lot is 180 meters, find the length of each side.

Solution Follow the **problem-solving steps** to solve this problem on your own. View the **answer**, or watch this **video** for a fully worked solution.

You Try It **Work through this You Try It problem.**

Work Exercises 1–5 in this textbook or in the MyMathLab Study Plan.

Example 4 Bathtub Surround

My video summary

A bathtub is surrounded on three sides by a vinyl wall enclosure. The height of the enclosure is 20 inches less than three times the width, and the length is 5 inches less than twice the width. If the sum of the length, width, and height is 185 inches, what is the **volume** of the enclosed space?

Solution

Step 1. We need to find the volume of the enclosed space. We know that the height is 20 inches less than three times the width, the length is 5 inches less than twice the width, and the sum of the length, width, and height is 185 inches. We will need to use the **formula** for the volume of a **rectangular solid** to compute the volume.

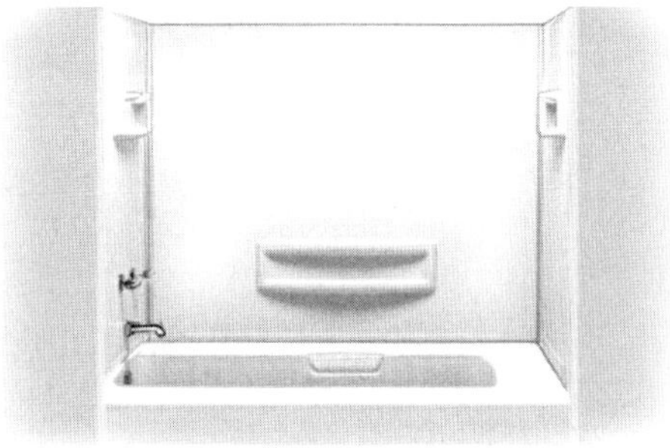

Step 2. If we let w represent the width of the enclosure, then the height is $h = 3w - 20$ and the length is $l = 2w - 5$.

Step 3. Start with the **sum** of the three dimensions:

$$l + w + h = 185 \quad \leftarrow \text{Given sum}$$

Substituting $3w - 20$ for h and $2w - 5$ for l, we have the equation:

$$(2w - 5) + w + (3w - 20) = 185$$

Step 4. Finish solving the equation to find the width of the enclosure. Use the width to find the height and length and then the volume of the enclosed space. When finished, view the **answer**, or watch this **video** for a fully worked solution.

You Try It **Work through this You Try It problem.**

Work Exercises 6 and 7 in this textbook or in the MyMathLab Study Plan.

OBJECTIVE 2 SOLVE PROBLEMS INVOLVING ANGLES

Geometry problems sometimes involve *complementary* and *supplementary* **angles**. The **measures** of two **complementary angles** add to 90°. The measures of two **supplementary angles** add to 180°. Thus, if an angle measures x degrees, its **complement** is given by $90 - x$ and its **supplement** is given by $180 - x$. Complementary angles form **right angles**, while supplementary angles form a straight line.

Example 5 Solve Problems Involving Complementary Angles

Find the measure of each **complementary angle** in the following figure.

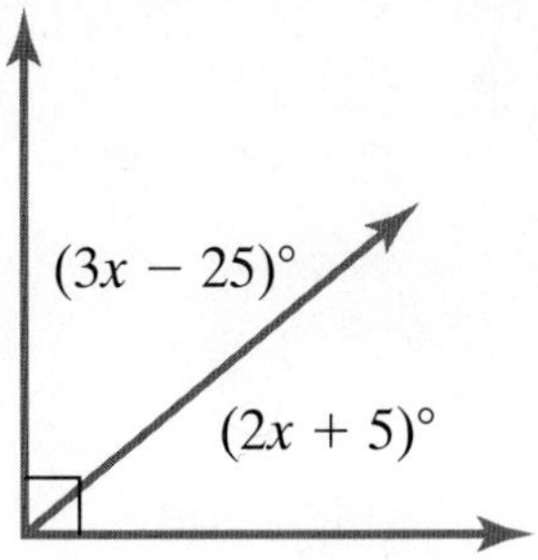

Solution We use the **problem-solving steps.**

Step 1. We must find the measure of each angle. We know that one angle measure is $3x - 25$ degrees, the other is $2x + 5$ degrees, and the sum of the measures of complementary angles is $90°$.

Step 2. The variable x is the unknown quantity. Each angle measure is given in terms of x.

Step 3. Because the sum of the measures of complementary angles is $90°$, we write

$$\overbrace{(3x - 25)}^{\text{1st angle}} + \overbrace{(2x + 5)}^{\text{2nd angle}} = 90.$$

Step 4. Solve the equation for x and then use the result to find the **measure** of each angle.

Begin with the original equation:	$(3x - 25) + (2x + 5) = 90$
Remove grouping symbols:	$3x - 25 + 2x + 5 = 90$
Combine like terms:	$5x - 20 = 90$
Add 20 to both sides:	$5x - 20 + 20 = 90 + 20$
Simplify:	$5x = 110$
Divide both sides by 5:	$\frac{5x}{5} = \frac{110}{5}$
Simplify:	$x = 22$

Step 5. Substituting 22 for x, we get $3(22) - 25 = 41°$ for the measure of one angle and $2(22) + 5 = 49°$ for the measure of the second. The sum of the two angle measures is $90°$, as required. $(41° + 49° = 90°)$

Step 6. The angle measures are $41°$ and $49°$.

You Try It **Work through this You Try It problem.**

Work Exercises 8 and 9 in this textbook or in the MyMathLab Study Plan.

Example 6 Solve Problems Involving Supplementary Angles

My video summary

Find the measure of each **supplementary angle** in the following figure.

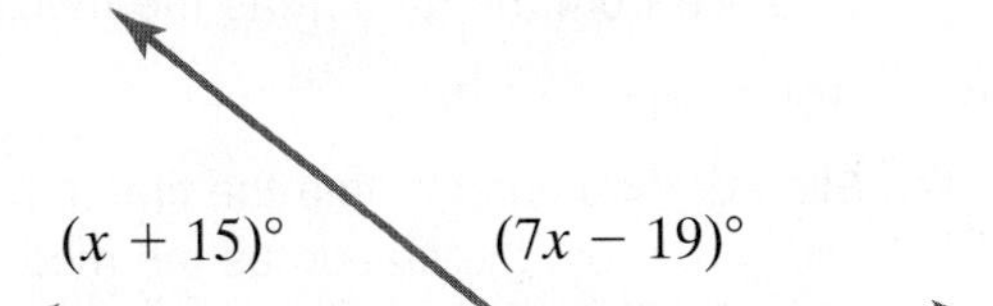

Solution

Step 1. We want to find the measure of each angle. We know that one angle measure is $x + 15$ degrees, the other is $7x - 19$ degrees, and the sum of the measures of supplementary angles is $180°$.

Step 2. In the diagram, the variable x is the unknown quantity. Each angle measure is given in terms of x.

Step 3. Because the sum of the measures of supplementary angles is $180°$, we can write

$$\overbrace{(x + 15)}^{\text{1st angle}} + \overbrace{(7x - 19)}^{\text{2nd angle}} = 180.$$

Step 4. Finish solving the equation to find the angle measures. When finished, view the **answer**, or watch this **video** for a fully worked solution.

You Try It **Work through this You Try It problem.**

Work Exercises 10 and 11 in this textbook or in the MyMathLab Study Plan.

A well-known property of triangles is that the **sum** of the **measures** of the three **angles** equals $180°$. We can use this fact as a **formula** to solve problems involving triangles.

Example 7 Triangle Park

My video summary

Triangle Park in Lexington, KY, has a roughly triangular shape such that the smallest angle measures 10 degrees less than the middle-sized angle. The largest angle measures 30 degrees less than twice the middle-sized angle. Find the measures of all three angles.

Solution Work through the **problem-solving steps** to solve this problem on your own. View the **answer**, or watch this **video** for a fully worked solution.

You Try It **Work through this You Try It problem.**

Work Exercises 12–16 in this textbook or in the MyMathLab Study Plan.

OBJECTIVE 3 SOLVE PROBLEMS INVOLVING UNIFORM MOTION

Some problems involve two **rates**. This means that there will also be two distances and two times. These are called **uniform motion** problems. In these types of problems, the **equations** formed will typically set two quantities equal to each other (e.g. distance traveled) or add two quantities together (e.g. time of travel).

Example 8 High Speed Trains

In January 2010, the U.S. government announced plans for the development of high-speed rail projects. A medium-fast passenger train leaves a station traveling 100 mph. Two hours later, a high-speed passenger train leaves the same station traveling 180 mph on a different track. How long will it take the high-speed train to be the same distance from the station as the medium-fast passenger train?

Solution

Step 1. We need to find the time it takes the high-speed train to be the same distance from the station as the medium-fast train. The medium-fast train travels at 100 mph, and the high-speed train travels at 180 mph. The medium-fast train travels 2 hours more than the high-speed train. We also know that the distance traveled by each train will be the same.

Step 2. If we let $t =$ the travel time for the high-speed train, then the travel time for the medium-fast train will be $t + 2$ (two hours more).

Step 3. Because the distance traveled is the same, we can write

$$\text{distance}_{\text{med-fast}} = \text{distance}_{\text{high-speed}}$$

Using the **distance formula**, $d = rt$, we write

$$(\text{rate})(\text{time})_{\text{med-fast}} = (\text{rate})(\text{time})_{\text{high-speed}}$$

Substituting the given rates and our **expressions** for the times from Step 2, we get

$$(100)(t + 2) = (180)(t)$$

Sometimes it is helpful to first summarize the given information in a table. View a **table** for this example.

Step 4. Solve the equation for t, the travel time for the high-speed train.

Begin with the original equation:	$(100)(t + 2) = (180)(t)$
Distribute:	$100t + 200 = 180t$
Subtract $100t$ from both sides:	$100t + 200 - 100t = 180t - 100t$
Simplify:	$200 = 80t$
Divide both sides by 80:	$\frac{200}{80} = \frac{80t}{80}$
Simplify:	$2.5 = t$

Step 5. Substituting 2.5 for t, we get $(2.5) + 2 = 4.5$ hours for the travel time of the medium-fast train.

$$\left.\begin{aligned} 100(4.5) &= 450 \text{ miles} \\ 180(2.5) &= 450 \text{ miles} \end{aligned}\right\} \text{Same distance}$$

My animation summary

The distance traveled by each train is the same, as described in the original problem.

Step 6. The high-speed train will be the same distance away from the station after 2.5 hours.

Watch this **animation** for an example where we add distances.

Example 9 Travel Mileage

Brennan provides in-home healthcare in a rural county and gets reimbursed for mileage. On one particular day he spent 4 hours driving to visit patients. His average speed is 50 mph on the highway but then slows to 30 mph when driving through towns. If he traveled five times as far on the highway as through towns, how far did he travel that day?

My video summary

Solution

Step 1. We must find the total distance Brennan traveled on the day in question. We know that the total time is four hours, his rate on the highway is 50 mph, and his rate through towns is 30 mph. We also know that the distance he travels on the highway is five times the distance traveled through towns.

Step 2. If we let d = distance traveled in towns, then the distance traveled on the highway is $5d$ (five times the distance in towns).

Step 3. The total travel time is the sum of the time spent driving on the highway and the time spent driving through towns.

$$\text{time}_{\text{highway}} + \text{time}_{\text{town}} = \text{time}_{\text{total}}$$

Solving the **distance formula** for t gives us $t = \frac{d}{r}$, or $\text{time} = \frac{\text{distance}}{\text{rate}}$. So we rewrite our equation as

$$\left(\frac{\text{distance}}{\text{rate}}\right)_{\text{highway}} + \left(\frac{\text{distance}}{\text{rate}}\right)_{\text{town}} = \text{time}_{\text{total}}$$

Substituting the given rates, total time, and our expressions for the distances, we get

$$\frac{5d}{50} + \frac{d}{30} = 4.$$

View a **table summary**.

Step 4. Solve the equation for d (the distance traveled in towns) on your own and then answer the question. Remember that his total distance is the sum of his distance on the highway and his distance in towns. When finished, view the **answer**, or watch this **video** for a fully worked solution.

You Try It **Work through this You Try It problem.**

Work Exercises 17–20 in this textbook or in the MyMathLab Study Plan.

CAUTION When working with the distance formula, the units for distance, rate, and time must be consistent. For example, if the unit for rate is miles per hour, then the unit for time must be hours and the unit for distance must be miles.

11.5 Exercises

You Try It

1. **Banquet Table** The length of a banquet table is 2 feet longer than 5 times its width. If the perimeter is 52 feet, what are the dimensions of the table?

2. **Solar Panel** A rectangular solar panel has a length that is 12 inches shorter than 3 times its width. If the perimeter of the panel is 136 inches, what are the dimensions of the panel?

3. **Lacrosse Field** The length of a lacrosse field is 10 yards less than twice its width, and the perimeter is 340 yards. The defensive area of the field is $\frac{7}{22}$ of the total field area. Find the defensive area of the lacrosse field.

4. **Patio Garden** A man has a rectangular garden. One length of the garden lies along a patio wall. However, the rest of the garden is enclosed by 36 feet of fencing. If the length of the garden is twice its width, what is the area of the garden?

5. **Triangular Deck** A triangular-shaped deck has one side that is 5 feet longer than the shortest side and a third side that is 5 feet shorter than twice the length of the shorter side. If the perimeter of the deck is 60 feet, what are the lengths of the three sides?

You Try It

6. **Brick Size** The width of a brick is half the length, which is 1 inch less than four times the height. If the sum of the three dimensions is 14.25 inches, find the volume of the brick.

7. **Mailing a Package** The sum of the length, width, and height of a rectangular package is 60 inches. If the length is 2 inches longer than three times the height and the width is 2 inches shorter than twice the height, what is the volume of the package?

You Try It

8. **Complementary Angles** Find the measures of the angles in the following diagram.

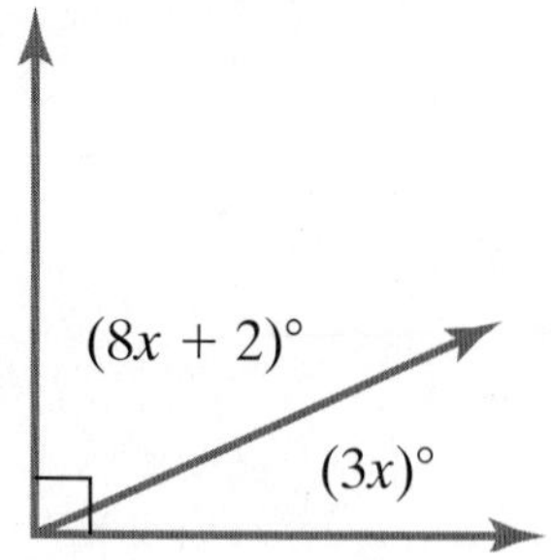

9. **Complementary Angles** Two complementary angles are $(2x - 2)^\circ$ and $(6x + 4)^\circ$. Find their measures.

You Try It

10. **Supplementary Angles** Find the measures of the angles in the following diagram.

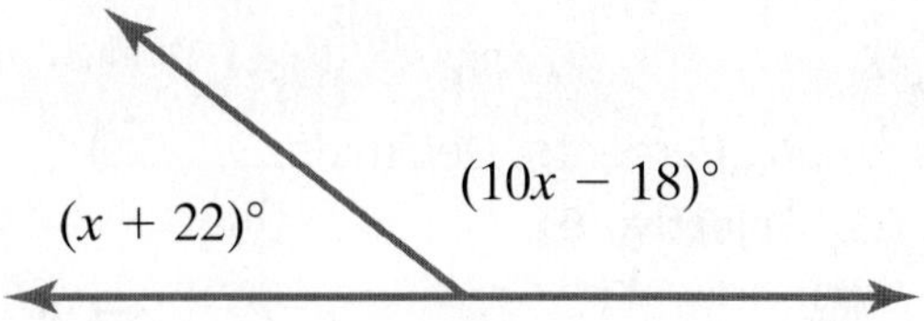

11. **Supplementary Angles** Angle A has the property that five times its complement is equal to twice its supplement. Find the measure of angle A.

You Try It

12. **Triangle Angles** The largest angle of a triangle is 5 times the size of the smallest, and the middle angle is 3 times the size of the smallest. What are the three angle measures?

13. **Isosceles Triangle** An isosceles triangle contains two equal angles. Each of these angles is five degrees larger than twice the smallest angle. What are the measures of the three angles?

14. **Triangular Sunshade** The smallest angle of a triangular sunshade measures 10 degrees less than the middle-sized angle. The largest angle measures 10 degrees more than twice the middle-sized angle. What are the angle measures?

15. **Support Cable** A platform on the back of an RV is anchored by two support cables. Each cable is attached to the front of the platform and the back wall of the RV. See the diagram below. If angle B measures 30° less than twice the measure of angle A, what is the measure of angle B (the angle between the cable and the wall)?

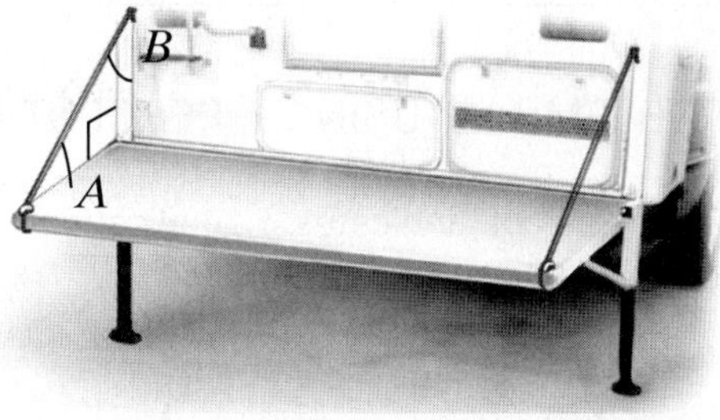

16. **Triangular Flag** The three angles of a triangular flag are x, twice the complement of x, and one-fourth the supplement of x. Find the angle measures.

You Try It

17. **Jogging Time** David leaves his house, jogging down the street at 6 mph. Once David has gone $\frac{1}{4}$ mile, Jacob comes out and follows him at a rate of 6.5 mph. How long will it take Jacob to catch David?

18. **Hiking Speed** Landon can climb a certain hill at a rate that is 2.5 mph slower than his rate coming down the hill. If it takes him 2 hours to climb the hill and 45 minutes to come down the hill, what is his rate coming down?

19. **Road Trip** Perry goes for a 3-hour trip through towns and on country roads. If he averages 55 mph on country roads and 35 mph through towns, and if he travels four times as far on country roads as he does through towns, what is the total length of his trip?

20. **Biathlon** Francesca plans to compete in a biathlon that involves running and cycling. During one training session, she covered a total distance of 70 miles in three hours. If she ran at a rate of 6 mph and cycled at a rate of 30 mph, how long did she spend cycling?

11.6 Percent and Mixture Problem Solving

THINGS TO KNOW

Before working through this topic, be sure you are familiar with the following concepts:

		VIDEO	ANIMATION	INTERACTIVE
You Try It	1. Perform Conversions Among Percents, Decimals, and Fractions (Topic 7.1, **Objective 5**)	▤		
You Try It	2. Solve Linear Equations Containing Decimals; Apply a General Strategy (Topic 11.2, **Objective 3**)	▤		
You Try It	3. Use the Problem-Solving Strategy to Solve Direct Translation Problems (Topic 11.3, **Objective 2**)	▤		
You Try It	4. Solve Problems Involving Related Quantities (Topic 11.3, **Objective 3**)	▤		

OBJECTIVES

1 Solve Problems by Using a Percent Equation

2 Solve Percent Problems Involving Discounts, Markups, and Sales Tax

3 Solve Percent of Change Problems

4 Solve Mixture Problems

OBJECTIVE 1 SOLVE PROBLEMS BY USING A PERCENT EQUATION

Recall that the word "of" indicates multiplication. For example, the sentence "thirty percent of sixty is eighteen" translates into $30\% \cdot 60 = 18$ or $0.3 \cdot 60 = 18$.

30%	of	60	is	18
↓	↓	↓	↓	↓
30%	·	60	=	18
↓	↓	↓	↓	↓
Change 30% to decimal form. → 0.3	·	60	=	18
↑ Percent		↑ Base		↑ Amount

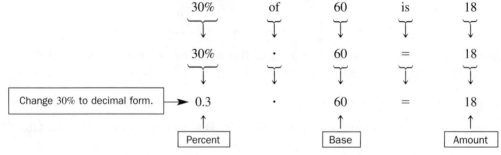

In the sentence, 30% is the *percent*, 60 is the *base*, and 18 is the *amount*. Using these labels, we can write the **general equation for percents.**

General Equation for Percents

$$\text{Percent} \cdot \text{Base} = \text{Amount}$$

When we translate **percent** problems into **equations**, the word *percent* or the % symbol identifies the percent. The **base** represents one whole and typically follows the word *of*. The

amount is the part compared to the whole and is normally isolated from the percent and the base on one side of the word *is* (or a different **key word meaning equals**).

CAUTION When using the **general equation for percents**, the percent must be **written as a decimal** or **fraction** in order to solve the problem. Typically, we will use decimal form.

Example 1 Using the Percent Equation

My video summary

Use equations to solve each percent problem.

a. 32 is 40% of what number?
b. 145% of 78 is what number?
c. 8.2 is what percent of 12.5?

Solutions Using the general equation for percents, we translate each sentence into an equation and solve.

a. Translate into an equation: 32 is 40% of what number

$$\underbrace{32}_{32} \; \underbrace{\text{is}}_{=} \; \underbrace{40\%}_{40\%} \; \underbrace{\text{of}}_{\cdot} \; \underbrace{\text{what number}}_{x}$$

$$32 = 40\% \cdot x$$

Convert 40% to a decimal: $32 = 0.4 \cdot x$

Divide both sides by 0.4: $\frac{32}{0.4} = \frac{0.4 \cdot x}{0.4}$

Simplify: $80 = x$

So, 32 is 40% of 80.

b. Translate into an equation: 145% of 78 is what number

$$\underbrace{145\%}_{145\%} \; \underbrace{\text{of}}_{\cdot} \; \underbrace{78}_{78} \; \underbrace{\text{is}}_{=} \; \underbrace{\text{what number}}_{x}$$

$$145\% \cdot 78 = x$$

Try to solve this equation on your own to finish the problem. View the **answer**, or watch this **video** for fully-worked solutions to all three parts.

c. Try solving this problem on your own, then check your **answer**. Watch this **video** for fully worked solutions to all three parts.

You Try It **Work through this You Try It problem.**

Work Exercises 1–6 in this textbook or in the MyMathLab Study Plan.

Example 2 Bleach Concentration

6% of a 128 fluid-ounce bottle of bleach is sodium hypochlorite. How many fluid ounces of sodium hypochlorite are in the bottle?

Solution We follow the **problem-solving steps.**

Step 1. We need to find the *amount* of sodium hypochlorite in the bottle of bleach. We know the *percent*, 6%, and the *base*, 128 fluid ounces.

Step 2. Let x be the number of fluid ounces of sodium hypochlorite.

Step 3. Translate into an equation: 6% of 128 fl oz is sodium hypochlorite

$$\underbrace{6\%}_{6\%} \; \underbrace{\text{of}}_{\cdot} \; \underbrace{128\text{ fl oz}}_{128} \; \underbrace{\text{is}}_{=} \; \underbrace{\text{sodium hypochlorite}}_{x}$$

$$\underset{\text{Percent}}{6\%} \cdot \underset{\text{Base}}{128} = \underset{\text{Amount}}{x}$$

Step 4. Change 6% to decimal form: $0.06 \cdot 128 = x$

Multiply: $7.68 = x$

Step 5. To check, we divide the amount, 7.68 by the base, 128, to see if we get the correct percent, 6%: $7.68 \div 128 = 0.06 = 6\%$. The result checks.

Step 6. The bottle contains 7.68 fluid ounces of sodium hypochlorite.

You Try It **Work through this You Try It problem.**

Work Exercises 7–10 in this textbook or in the MyMathLab Study Plan.

OBJECTIVE 2 SOLVE PERCENT PROBLEMS INVOLVING DISCOUNTS, MARKUPS, AND SALES TAX

During a sale, a store may take a certain percent off the price of merchandise. We can calculate the **amount** of this **discount** by using a **percent equation**. We then *subtract* the discount from the original price to find the new price.

Computing Discounts

$$\text{Discount} = \text{Percent} \cdot \text{Original price}$$
$$\text{New price} = \text{Original price} - \text{Discount}$$

Example 3 Going Out of Business Sale

A furniture store is going out of business and cuts all prices by 55%. What is the sale price of a sofa with an original price of $1199?

Solution Compute the discount and subtract it from the original price.

$$\text{Discount} = \text{Percent} \cdot \text{Original price} = 55\% \cdot \$1199 = 0.55 \cdot \$1199 = \$659.45$$

$$\text{Sale price} = \text{Original price} - \text{Discount} = \$1199 - \$659.45 = \$539.55$$

The sale price of the sofa is $539.55.

You Try It **Work through this You Try It problem.**

Work Exercises 11–13 in this textbook or in the MyMathLab Study Plan.

Markups work like discounts, except that they are *added* to the original price.

Computing Markups

$$\text{Markup} = \text{Percent} \cdot \text{Original price}$$
$$\text{New price} = \text{Original price} + \text{Markup}$$

Example 4 College Book Store

My video summary

A college book store sells all textbooks at a 30% **markup** over its cost. If the price marked on a biology textbook is $124.28, what was the cost of the book to the store? Round to the nearest cent.

Solution Follow the **problem-solving strategy**.

Step 1. We need to find the original price of the book to the bookstore. We know the markup is 30% and the selling price is $124.28.

Step 2. Let x be the original price of the book to the bookstore.

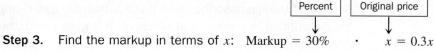

Step 3. Find the markup in terms of x:

	Percent		Original price
Markup =	30%	$\cdot$	$x = 0.3x$

	New price		Original price		Markup
Form an equation:	$\$124.28$	$=$	x	$+$	$0.3x$

Try to finish this problem by completing the **problem-solving steps**. View the **answer**, or watch this **video** for a complete solution.

You Try It **Work through this You Try It problem.**

Work Exercises 14 and 15 in this textbook or in the MyMathLab Study Plan.

Sales tax is handled like markups. Multiply the tax rate (a percent) by the purchase price to determine the sales tax amount. Then add the sales tax to the purchase price to get the overall price.

> **Computing Sales Tax**
>
> $$\text{Sales tax} = \text{Tax rate} \cdot \text{Purchase price}$$
> $$\text{Overall price} = \text{Purchase price} + \text{Sales tax}$$

Example 5 Buying Jeans

Charlotte bought a pair of jeans priced at $51.99. When sales tax was added, she paid an overall price of $55.37. What was the tax rate? Round to the nearest tenth of a percent.

Solution Follow the **problem-solving steps**.

Step 1. We need to find the tax rate, a percent. We know that the purchase price is $51.99 and the overall (total) price is $55.37.

Step 2. Let x be the tax rate.

Step 3. Find the sales tax in terms of x:

	Tax rate		Purchase price
Sales tax =	x	$\cdot$	$51.99 = 51.99x$

	Overall price	Purchase price		Sales tax
Form an equation:	$55.37 =$	51.99	$+$	$51.99x$

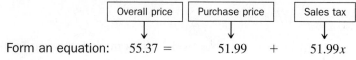

Step 4. Subtract 51.99 from both sides: $3.38 = 51.99x$

Divide both sides by 51.99: $0.065 \approx x$

Change to a percent: $6.5\% \approx x$

Step 5. View the check.

Step 6. The tax rate is 6.5%.

You Try It **Work through this You Try It problem.**

Work Exercises 16–18 in this textbook or in the MyMathLab Study Plan.

OBJECTIVE 3 SOLVE PERCENT OF CHANGE PROBLEMS

Percent of change describes how much a quantity has changed. If the quantity goes up, the description is a **percent of increase.** If it goes down, the description is a **percent of decrease.**

When computing the **amount** of change from the **percent** of change, the original amount is the **base.** We compute the new amount by adding or subtracting the amount of change to the original amount, depending if the quantity is increasing or decreasing.

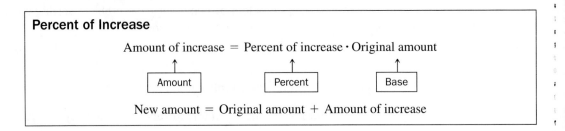

Percent of Increase

$$\text{Amount of increase} = \text{Percent of increase} \cdot \text{Original amount}$$

Amount ↑ — Percent ↑ — Base ↑

$$\text{New amount} = \text{Original amount} + \text{Amount of increase}$$

Example 6 Enrollment Increase

Last year, $16{,}528$ students attended City Community College. This year, enrollment increased by 3.2%. How many students attend City Community College this year? Round to the nearest whole student.

Solution Follow the **problem-solving steps.**

Step 1. We must find the current enrollment of City Community College. Last year's enrollment was $16{,}528$, and the percent of increase is 3.2%.

Step 2. Let x be the **amount of increase** in enrollment. The current enrollment is $16{,}528 + x$.

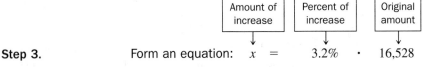

Amount of increase ↓ — Percent of increase ↓ — Original amount ↓

Step 3. Form an equation: $x = 3.2\% \cdot 16{,}528$

Step 4. Change 3.2% to decimal form: $x = 0.032 \cdot 16{,}528$

Multiply: $x \approx 529$

So, $16{,}528 + x = 16{,}528 + 529 = 17{,}057$ students.

Step 5. View the check.

Step 6. This year's enrollment at City Community College is $17{,}057$ students.

You Try It **Work through this You Try It problem.**

Work Exercises 19–21 in this textbook or in the MyMathLab Study Plan.

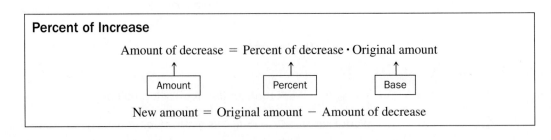

Percent of Increase

$$\text{Amount of decrease} = \text{Percent of decrease} \cdot \text{Original amount}$$

Amount ↑ — Percent ↑ — Base ↑

$$\text{New amount} = \text{Original amount} - \text{Amount of decrease}$$

(*eText Screens 11.6-1–11.6-23*)

Example 7 General Motors Reorganization

My video summary

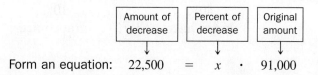

Prior to reorganization in 2010, General Motors (GM) had 91,000 U.S. employees. After the reorganization, GM had 68,500 U.S. employees. By what percent did the number of U.S. employees decrease? (*Source*: General Motors)

Solution

Step 1. We must find the **percent of decrease**. The original number of U.S. employees was 91,000 (before reorganization), and the new number was 68,500 (afterwards).

Step 2. Let x be the percent of decrease in the number of U.S. employees.

Step 3. The **amount of decrease** is $91{,}000 - 68{,}500 = 22{,}500$ employees.

Amount of decrease	Percent of decrease	Original amount
↓	↓	↓

Form an equation: $22{,}500 = x \cdot 91{,}000$

Try to finish this problem by completing the **problem-solving steps**. View the **answer**, or watch this **video** for a complete solution.

You Try It **Work through this You Try It problem.**

Work Exercises 22–24 in this textbook or in the MyMathLab Study Plan.

CAUTION When calculating the **percent of change**, be sure to use the original amount as the **base**, not the new amount.

OBJECTIVE 4 SOLVE MIXTURE PROBLEMS

Suppose that a 50-pound bag of lawn fertilizer has a 12% nitrogen **concentration**. This means that out of all the components in the fertilizer, 12% is pure nitrogen. The 50-pound bag contains a total of $0.12(50) = 6$ pounds of pure nitrogen.

We can find the amount of a particular component in a **mixture** by multiplying its concentration (in decimal form) by the amount of mixture. The *amount of component* is the **amount** in the **general equation for percents**, the *concentration* is the **percent**, and the *amount of mixture* is the **base**.

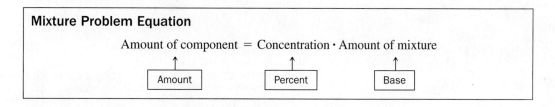

Mixture Problem Equation

$$\text{Amount of component} = \text{Concentration} \cdot \text{Amount of mixture}$$

Amount	Percent	Base

My animation summary

We can also write this in the following equivalent form:

$$\text{Concentration} = \frac{\text{Amount of component}}{\text{Amount of mixture}}$$

With mixture problems, remember that the total *amount* of a component does not change when two or more substances are mixed together, but the *concentration* of that component might change. Watch this **animation** for an illustration.

(eText Screens 11.6-1–11.6-23)

Example 8 Organic Juice Mix

My video summary

An organic cranberry-grape juice is 40% grape juice, while an organic fruit cocktail juice is 10% grape juice. If 8 ounces of the cranberry-grape juice are mixed with 22 ounces of the fruit cocktail juice, what is the mixed juice's **concentration** of grape juice?

Solution Follow the **problem-solving steps.**

Step 1. 8 ounces of cranberry-grape juice (40% grape) are mixed with 22 ounces of fruit cocktail juice (10% grape) to form $8 + 22 = 30$ ounces of a third juice. We must find the concentration of grape juice in the mix. See Figure 2.

Figure 2

Step 2. Let x be the concentration (percent) of grape juice in the mix.

Step 3. The **amount** of grape juice in the cranberry-grape juice is

Concentration		Base		Amount of grape juice
↓		↓		↓
40%	·	8	=	$0.4 \cdot 8 = 3.2$ ounces

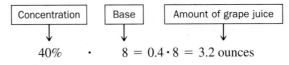

The amount of grape juice in the fruit cocktail juice is

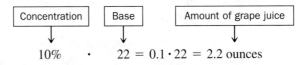

Concentration		Base		Amount of grape juice
↓		↓		↓
10%	·	22	=	$0.1 \cdot 22 = 2.2$ ounces

The amount of grape juice in the mixture is

Concentration		Base		Amount of grape juice
↓		↓		↓
x	·	30	=	$30x$

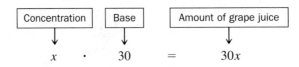

Now we can write the equation:

Amount of grape juice from cranberry-grape		Amount of grape juice from fruit cocktail		Amount of grape juice from mixture
↓		↓		↓
3.2 ounces	+	2.2 ounces	=	$30x$

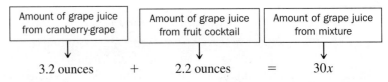

Try to finish this problem on your own. View the **answer**, or watch this **video** to see a detailed solution.

You Try It **Work through this You Try It problem.**

Work Exercises 25 and 26 in this textbook or in the MyMathLab Study Plan.

(eText Screens 11.6-1–11.6-23)

Example 9 Alcohol Concentration

My video summary

How many milliliters of a 25% alcohol solution must be mixed with 10 mL of a 60% alcohol solution to result in a mixture that is 30% alcohol?

Solution

Step 1. We mix the two solutions together to result in a third solution. We know that an unknown amount of a 25% alcohol solution will be mixed with 10 mL of a 60% alcohol solution to result in a 30% alcohol solution.

Step 2. Let x be the unknown amount of the 25% alcohol solution (in mL). When we mix this amount with 10 mL of a 60% alcohol solution, the resulting amount of a 30% alcohol solution is $x + 10$ (in mL). See Figure 3.

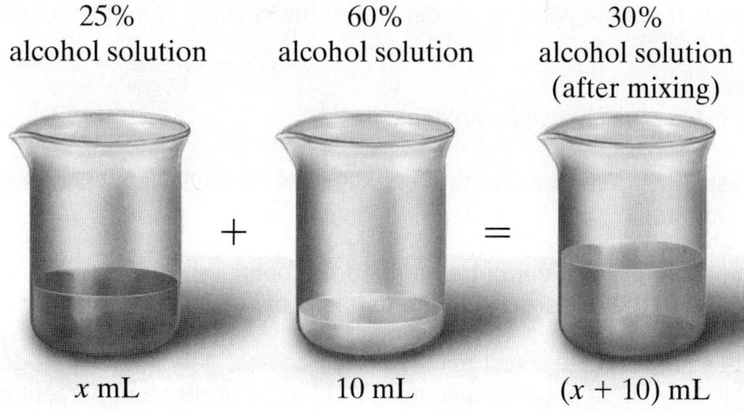

Figure 3

Step 3.

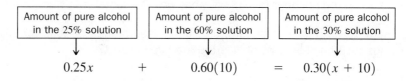

Amount of pure alcohol in the 25% solution		Amount of pure alcohol in the 60% solution		Amount of pure alcohol in the 30% solution
$0.25x$	$+$	$0.60(10)$	$=$	$0.30(x + 10)$

Try to finish this problem on your own. View the **answer**, or watch this **video** to see a detailed solution.

You Try It **Work through this You Try It problem.**

Work Exercises 27–30 in this textbook or in the MyMathLab Study Plan.

CAUTION The **concentration** of a mixture must always be between the concentrations of the two mixed solutions. Do you see **why**?

11.6 Exercises

In Exercises 1–6, use an equation to solve each percent problem.

You Try It

1. 21 is 75% of what number?
2. What number is 38% of 155?
3. 52.7 is what percent of 62?
4. What number is 214% of 86.5?
5. 2.5% of what number is 91?
6. What percent of 55 is 88?

In Exercises 7–10, use a percent equation to solve each percent problem.

You Try It

7. **Sculpture** A 240-pound sculpture is made by using a cast that is 90% copper and 10% tin. How many pounds of copper are in the sculpture?

8. **Ethanol Production** Twenty-six percent of all U.S. grain crops grown in 2008 were used to produce ethanol, which amounts to 104 million tons of grain. (*Source*: Earth Policy Institute) How much grain was grown in the U.S. in 2008?

9. **Mobile Devices** In March 2010, a total of 234 million people age 13 or older in the U.S. used mobile devices. Of these, 51.2 million people used devices manufactured by Motorola. (*Source*: comScore.com) What percent of the population used Motorola devices? Round to the nearest tenth of a percent.

10. **Bartending** A Harvey Wallbanger cocktail is made by mixing 1 ounce of vodka, $\frac{1}{2}$ ounce of Galliano, and 4 ounces of orange juice. What percent of the cocktail is orange juice? Round to the nearest tenth of a percent.

In Exercises 11–18, solve each problem.

You Try It

11. **Mattress Sale** A mattress that normally sells for $499.90 is on sale for 30% off. What is the sale price of the mattress?

12. **Clearance Sale** To clear out a brand of discontinued power tools, a home supply store marks down the tools by 25%. If the sale price for a cordless power drill is $82.35, what was the original price?

13. **Caribbean Cruise** As a last minute deal, Don and Mary booked a 7-day Caribbean cruise for a total of $737. If the normal price for a couple is $1340, what discount percent did Don and Mary receive?

You Try It

14. **Seafood Restaurant** An oceanside restaurant prices its fresh fish using a 90% markup over its cost. If it sells fresh red snapper for $19.95, what was the restaurant's cost for this fish?

15. **Fundraiser** An athletic booster club purchased bottles of soda for $0.80 each and sold them during games for $2.00 each. What was the markup percent?

16. **Buying a Computer** The purchase price of a laptop computer is $529.50. If the sales tax rate is 7.5%, find the overall price of the laptop. Round to the nearest cent.

You Try It

17. **Buying Jewelry** Trevor bought a diamond pendant with a purchase price of $324. When sales tax was added, he paid an overall price of $349.92. What was the tax rate?

18. **Buying a Game System** Destiny bought a game system for an overall price of $239.03. If the tax rate is 6%, what was the purchase price of the game system?

In Exercises 19–24, solve each percent of change problem.

You Try It

19. **Tuition Increase** A state university's board of directors approved a 7.5% tuition increase for next year. If the current annual tuition is $9485, what will the tuition be next year? Round to the nearest whole dollar.

20. **Population Increase** According to the U.S. Census Bureau, the population of the Palm Coast metro area in Florida increased by 84% to 92,000 people between 2000 and 2010. What was the population of the Palm Coast metro area in 2000?

21. **Life Expectancy** The life expectancy for a newborn child in the U.S. increased from 73.7 years in 1980 to 78.3 years in 2010. What is the percent of increase for this time period? Round to the nearest tenth of a percent. (*Source*: U.S. Census Bureau)

22. **Daycare Enrollment** Following an investigation into finance fraud, a daycare center's enrollment decreased by 30% to 56 students. What was the daycare center's enrollment before the investigation?

You Try It

23. GM Plant Closings Before the 2010 reorganization, GM had 47 U.S. plants. After the reorganization, GM had 34 U.S. plants. By what percent did the number of U.S. plants decrease? Round to the nearest tenth of a percent. (*Source*: General Motors)

24. The Biggest Loser During Season 8 of NBC's reality show *The Biggest Loser*, the winner Danny Cahill lost 55.6% of his body weight. If he weighed 430 pounds at the start of the show, what was his weight at the end of the show?

In Exercises 25–30, solve each mixture problem.

You Try It

25. Mixing Alcohol Suppose 8 pints of a 12% alcohol solution is mixed with 2 pints of a 60% alcohol solution. What is the concentration of alcohol in the new 10-pint mixture?

26. Mixing Fertilizer The granular fertilizer 12-12-12 is composed of 12% nitrogen, 12% phosphate, and 12% potassium. Similarly, the fertilizer 16-20-0 is composed of 16% nitrogen, 20% phosphate, and 0% potassium. If a gardener mixes a 10 pound bag of 12-12-12 with a 15 pound bag of 16-20-0, what are the concentrations of nitrogen, phosphate, and potassium in the mixture? Express the answers as percents.

You Try It

27. Mixing Cranberry Juice How much of a 75% cranberry juice drink must be mixed with 3 liters of a 20% cranberry juice drink to result in a mixture that is 50% cranberry juice?

28. Salad Dressing A cook needs to mix a regular ranch dressing containing 46% fat with a light ranch dressing containing 26% fat to result in 5 cups of a dressing containing 32% fat. How many cups of each kind of dressing should be mixed?

29. Diluting Acid How many cups of pure water should be mixed with 1.5 cups of a 30% acid solution to dilute it into a 2% acid solution?

30. Fuel Mixture A two-cycle engine mechanic has 2 gallons of a fuel mixture that is 96% gasoline and 4% oil. How many gallons of gasoline should be added to form a fuel mixture that is 99% gasoline and 1% oil?

11.7 Linear Inequalities in One Variable

THINGS TO KNOW

Before working through this topic, be sure you are familiar with the following concepts:

		VIDEO	ANIMATION	INTERACTIVE
You Try It	1. Use the Order of Operations to Evaluate Numeric Expressions (Topic 10.4, **Objective 2**)	✓		✓
You Try It	2. Evaluate Algebraic Expressions (Topic 10.5, **Objective 1**)	✓		
You Try It	3. Simplify Algebraic Expressions (Topic 10.6, **Objective 2**)			✓
You Try It	4. Solve Linear Equations Containing Non-simplified Expressions (Topic 11.2, **Objective 1**)	✓		
You Try It	5. Use the Problem-Solving Strategy to Solve Direct Translation Problems (Topic 11.3, **Objective 2**)	✓		
You Try It	6. Solve Problems Involving Geometry Formulas (Topic 11.5, **Objective 1**)	✓		

OBJECTIVES

1 Write the Solution Set of an Inequality in Set-Builder Notation

2 Graph the Solution Set of an Inequality on a Number Line

3 Use Interval Notation to Express the Solution Set of an Inequality

4 Solve Linear Inequalities in One Variable

5 Solve Three-Part Inequalities

6 Use Linear Inequalities to Solve Application Problems

OBJECTIVE 1 WRITE THE SOLUTION SET OF AN INEQUALITY IN SET-BUILDER NOTATION

An equal sign is used in **equations** to show that two quantities are equal. We use **inequality symbols** in **inequalities** to show that two quantities are unequal.

My interactive video summary

Watch this **interactive video** to practice distinguishing between equations and inequalities.

While most equations have a *finite* number of solutions, most **inequalities** have an *infinite* number of solutions. For example, the inequality $x > -2$ has infinitely many solutions because there are infinitely many values of x that are greater than -2. View this **popup** for an explanation of the difference between "finite" and "infinite."

The collection of all solutions to an inequality forms the **solution set** of the inequality. When a solution set contains an infinite number of values, we cannot list each solution, so we use **set-builder notation** to express the solution set. The set of all numbers greater than -2 is written in set-builder notation as follows.

$$\{x \mid x > -2\}$$

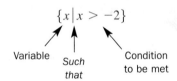

Variable — Such that — Condition to be met

The variable appears to the left of the vertical bar. The vertical bar means "such that," and the expression to the right of the vertical bar is the condition that must be met for a number to be part of the solution set.

Example 1 Writing the Solution Set of an Inequality in Set-Builder Notation

Write the **solution set** of each inequality in **set-builder notation**.

a. $x < 6$ **b.** $y \geq -3$ **c.** $2 < m \leq 9$

Solutions

a. To be part of the solution set, the value for x must be less than 6. We write the solution set as $\{x|x < 6\}$ and read this as "the set of all values for x such that x is less than 6."

b. We write the solution set as $\{y|y \geq -3\}$ and read this as "the set of all values for y such that y is greater than or equal to -3."

c. To be part of the solution set, the value for m must be greater than 2 but less than or equal to 9. We write the solution set as $\{m|2 < m \leq 9\}$ and read this as "the set of all values for m such that 2 is less than m and m is less than or equal to 9."

You Try It **Work through this You Try It problem.**

Work Exercises 1–4 in this textbook or in the MyMathLab Study Plan.

OBJECTIVE 2 GRAPH THE SOLUTION SET OF AN INEQUALITY ON A NUMBER LINE

To graph the **solution set** on a number line, we use an open circle (○) to indicate that a value *is not* included in the solution set and a closed circle (●) to indicate that a value *is* included in the solution set. For example, the graph of $\{x|x > -2\}$ is shown below.

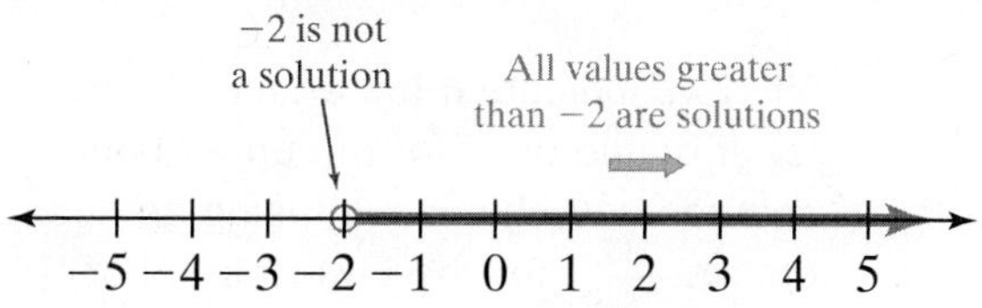

Example 2 Graphing the Solution Set of an Inequality on a Number Line

Graph each solution set on a number line.

a. $\{x|x \le 0\}$ b. $\{x|-2 \le x < 4\}$ c. $\{x|x > -1\}$

d. $\{x|3 < x < 7\}$ e. $\{x|-1 \le x \le 5\}$ f. $\{x|x \text{ is any real number}\}$

Solutions

a. We translate this solution set as "the set of all values for x such that x is less than or equal to 0." Because the inequality is **non-strict**, we place a closed circle at 0 to show that 0 is a solution. Then we shade the number line to the left to show that all values less than 0 are also solutions.

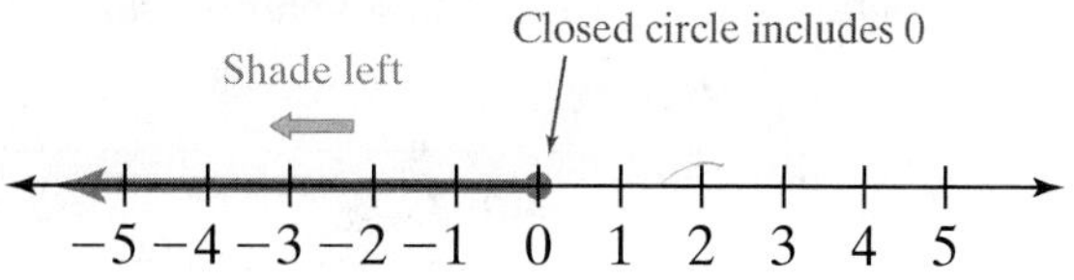

b. The solution set reads as "the set of all values for x such that -2 is less than or equal to x and x is less than 4." The inequality on the left is non-strict, so we place a closed circle at -2 to show that -2 is a solution. The inequality on the right is **strict**, so we place an open circle at 4 to show that 4 is not a solution and then shade the number line between the two circles to indicate that all values between -2 and 4 are also solutions.

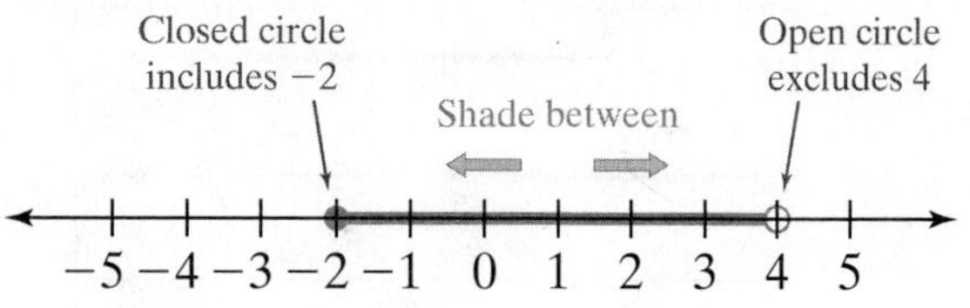

c.–f. View this **popup** for the remaining solution sets and their graphs.

You Try It **Work through this You Try It problem.**

Work Exercises 5–9 in this textbook or in the MyMathLab Study Plan.

(eText Screens 11.7-1–11.7-29)

OBJECTIVE 3 USE INTERVAL NOTATION TO EXPRESS THE SOLUTION SET OF AN INEQUALITY

We can also use **interval notation** to express the **solution set** of an inequality. First, each solution set has a **lower bound** and an **upper bound**, separated by a comma, which make up the **endpoints** of the interval. An interval has the form:

lower bound, upper bound

Then we indicate if the endpoints are included in the interval. A parenthesis— '(' for the lower bound or ')' for the upper bound — shows that the endpoint is not included in the solution set. This is like using an open circle when graphing on a number line. A square bracket— '[' for the lower bound or ']' for the upper bound — shows that the endpoint is included in the solution set. This is like using a closed circle when graphing on a number line.

In **Example 2b**, the solution set $\{x|-2 \leq x < 4\}$ is written as $[-2, 4)$ in interval notation. We use a square bracket on -2 because it *is* included in the solution set, and we use a parenthesis on 4 because it *is not* included in the solution set.

CAUTION A parenthesis is always used for $-\infty$ and ∞ because these are infinity symbols, not numbers. They indicate that the interval is **unbounded** in a particular direction.

My interactive video summary

Watch this **interactive video** to help you determine the lower and upper bounds of an interval or solution set.

Table 1 summarizes the three ways of expressing intervals used in this text: **number line graph**, **interval notation**, and **set-builder notation**. Typically we use interval notation and graphs when expressing **solution sets** for inequalities.

Graph	Interval Notation	Set-Builder Notation
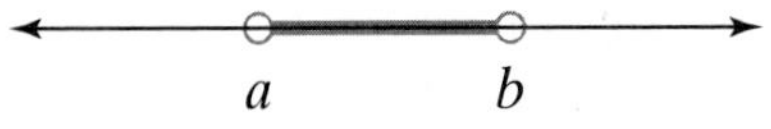	(a, b)	$\{x \mid a < x < b\}$
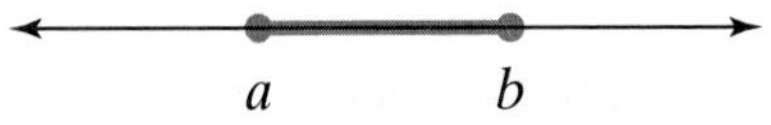	$[a, b,]$	$\{x \mid a \leq x \leq b\}$
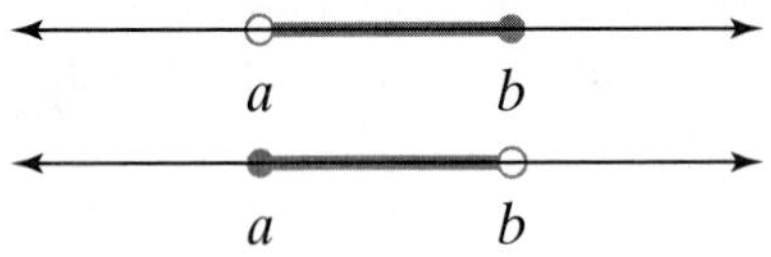	$(a, b]$ $[a, b)$	$\{x \mid a < x \leq b\}$ $\{x \mid a \leq x < b\}$
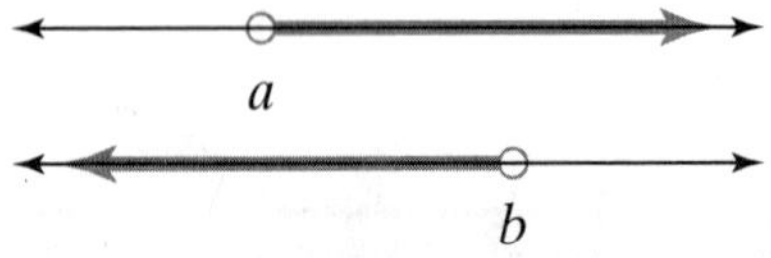	(a, ∞) $(-\infty, b)$	$\{x \mid x > a\}$ $\{x \mid x < b\}$
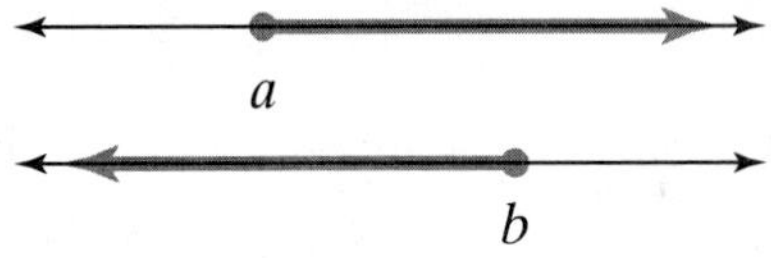	$[a, \infty)$ $(-\infty, b]$	$\{x \mid x \geq a\}$ $\{x \mid x \leq b\}$

Table 1

(eText Screens 11.7-1–11.7-29)

Example 3 Using Interval Notation to Express the Solution Set of an Inequality

Write each **solution set** using **interval notation**.

a. $\{x \mid x \geq -2\}$ **b.** $\{x \mid 0 < x \leq 6\}$ **c.** x is less than 4

d. x is between -1 and 5, **inclusive**

e. $\{x \mid x \text{ is any real number}\}$ **f.** $\{x \mid 8 > x \geq -3\}$

Solutions

a. This solution set has a **lower bound** of -2 and no **upper bound**. The interval notation is $[-2, \infty)$. We use a square bracket on -2 to show that -2 is included in the solution set. We use a parenthesis on ∞ to show that there is no upper bound.

b. There is a lower bound of 0 and an upper bound of 6. The interval notation is $(0, 6]$. We use a parenthesis on 0, the lower bound, to show that 0 is not included in the solution set, and a square bracket on 6, the upper bound, to show that 6 is included in the solution set.

My video summary

c.–f. Try working these parts on your own. View the **answers**, or watch this **video** for complete solutions.

You Try It **Work through this You Try It problem.**

Work Exercises 10–14 in this textbook or in the MyMathLab Study Plan.

CAUTION When writing a solution set in **interval notation**, the values should increase from left to right.

OBJECTIVE 4 SOLVE LINEAR INEQUALITIES IN ONE VARIABLE

The inequality $3x + 2 \leq -6$ is a **linear inequality in one variable**.

> **Definition Linear Inequality in One Variable**
>
> A **linear inequality in one variable** is an inequality that can be written in the form $ax + b < c$, where a, b, and c are real numbers and $a \neq 0$.
>
> **Note:** The inequality symbol $<$ can be replaced with $>, \leq, \geq,$ or $\neq$.

Why do we require $a \neq 0$ in our definition? View this **popup** for an explanation.

Solving linear inequalities is similar to solving linear equations. Our goal is to **isolate** the variable on either side of the inequality symbol, then form the **solution set** based on the resulting inequality.

Similar to solving equations, we have an **addition property of inequality**.

> **Addition Property of Inequality**
>
> Let a, b, and c be real numbers.
>
> If $a < b$, then $a + c < b + c$ and $a - c < b - c$.
>
> Adding or subtracting the same quantity from both sides of an inequality results in an **equivalent inequality**.
>
> **Note:** The inequality symbol $<$ can be replaced with $>, \leq, \geq,$ or $\neq$.

Example 4 Solving a Linear Inequality in One Variable Using the Addition Property of Inequality

Solve each inequality using the **addition property of inequality**. Write the **solution set** in **interval notation** and graph it on a number line.

a. $x + 5 > 4$ **b.** $y - 3 \leq 1$

Solutions

a. Our approach is similar to the one used when solving linear equations. First, we **isolate** the variable on one side of the inequality symbol.

Begin with the original inequality: $x + 5 > 4$

Subtract 5 from both sides: $x + 5 - 5 > 4 - 5$

Simplify: $x > -1$

The graph of the solution set appears on the following number line:

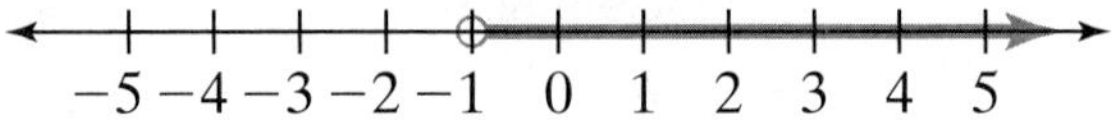

In interval notation, the solution set is $(-1, \infty)$.

Check With an infinite number of values in the solution set, we cannot check every solution. So, we pick one **test value** from the solution set as a "check." Let's choose $x = 0$ as our test value and **substitute** it for x in the original inequality to see if a true statement results.

Begin with the original inequality: $x + 5 > 4$

Substitute 0 for x: $0 + 5 \overset{?}{>} 4$

Simplify: $5 > 4$ True

Because the resulting statement, $5 > 4$, is true, $x = 0$ is a solution to the inequality.

CAUTION Checking one solution, or even several solutions, does not guarantee that the solution set of the inequality is correct. It only guarantees that the tested values are solutions. However, it is still good practice to check one or two test values.

My video summary

b. Apply the **addition property of inequality**. View the **answer**, or watch this **video** for a complete solution.

You Try It **Work through this You Try It problem.**

Work Exercises 15–18 in this textbook or in the MyMathLab Study Plan.

Equations and **inequalities** have similar addition properties. However, there is an important difference between the two when we multiply or divide. Consider the following examples.

True statement:	$5 > 3$		True statement:	$5 > 3$	
Multiply both sides by 4:	$4(5) > 4(3)$		Multiply both sides by −4:	$-4(5) > -4(3)$	
Simplify:	$20 > 12$	True	Simplify:	$-20 > -12$	False

When we multiply (or divide) both sides of a true inequality statement by a positive number, the resulting inequality is also a true statement. However, when we multiply (or divide) both sides of the same true inequality statement by a **negative** number, the resulting inequality is a false statement.

(eText Screens 11.7-1–11.7-29)

In order to make the second inequality true, we must switch the direction of the inequality. View this **illustration**.

$-20 > -12$ False $\qquad$ $-20 < -12$ True

Switch the inequality directions

This leads us to the **multiplication property of inequality**.

> **Multiplication Property of Inequality**
>
> Let a, b, and c be real numbers.
>
> If $a < b$ and $c > 0$, then $ac < bc$ and $\frac{a}{c} < \frac{b}{c}$.
>
> Multiplying or dividing both sides of an inequality by a positive number c results in an equivalent inequality.
>
> If $a < b$ and $c < 0$, then $ac > bc$ and $\frac{a}{c} > \frac{b}{c}$.
>
> Multiplying or dividing both sides of an inequality by a negative number c, and switching the direction of the inequality, results in an equivalent inequality.
>
> **Note:** The inequality symbol $<$ can be replaced with $>, \leq, \geq,$ or $\neq$.

Example 5 Solving a Linear Inequality in One Variable Using the Multiplication Property of Inequality

Solve each inequality. Graph the solution set on a number line and write the solution set in **interval notation**.

a. $-5x < 15$ $\qquad$ **b.** $\frac{m}{3} \geq 2$

Solutions

a. Apply the **multiplication property of inequality** to isolate the variable.

Begin with the original inequality: $-5x < 15$

Divide both sides by -5: $\frac{-5x}{-5} > \frac{15}{-5}$ ← Reverse the direction of the inequality

Simplify: $x > -3$

The graph of the solution set appears on the following number line:

My video summary

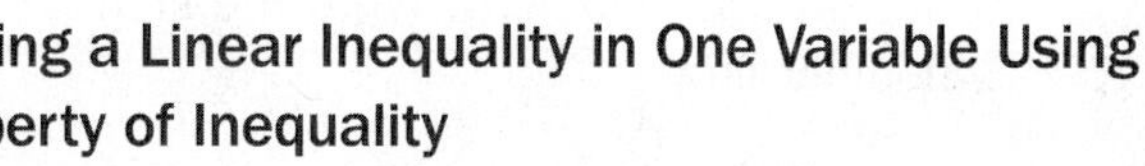

In interval notation, the solution set is $(-3, \infty)$.

b. Apply the **multiplication property of inequality**. View the **answer**, or watch this **video** for a complete solution.

You Try It **Work through this You Try It problem.**

Work Exercises 19–22 in this textbook or in the MyMathLab Study Plan.

(eText Screens 11.7-1–11.7-29)

Our approach to solving **linear inequalities in one variable** is similar to our approach to **solving linear equations** in one variable. Let's apply the following guidelines.

Guidelines for Solving Linear Inequalities in One Variable

Step 1. Clear all fractions from the inequality by multiplying both sides by the **LCD**. (If LCD < 0, reverse the direction of the inequality)
Clear all decimals by multiplying both sides by the appropriate power of 10.

Step 2. Remove grouping symbols using the **distributive property**.

Step 3. Simplify each side of the inequality by **combining like terms.**

Step 4. Use the **addition property of inequality** to move all variable terms to one side of the inequality and all constant terms to the other side.

Step 5. Use the **multiplication property of inequality** to isolate the variable.

Step 6. Write the solution set in **interval notation** (or **set-builder notation**) and graph it on a number line.

View this summary of the **inequality properties**.

Example 6 Solving a Linear Inequality in One Variable

Solve each inequality. Write the **solution set** in **interval notation** and graph it on a number line.

a. $-3x + 2 \le 8$ **b.** $6x - 3 > 4x + 9$

Solutions

a. Follow the **guidelines** for solving linear inequalities in one variable.

Begin with the original inequality: $-3x + 2 \le 8$

Subtract 2 from both sides: $-3x + 2 - 2 \le 8 - 2$

Simplify: $-3x \le 6$

Divide both sides by -3: $\dfrac{-3x}{-3} \ge \dfrac{6}{-3}$ ← Remember to switch the direction

Simplify: $x \ge -2$

The graph of the solution set appears on the following number line:

My video summary

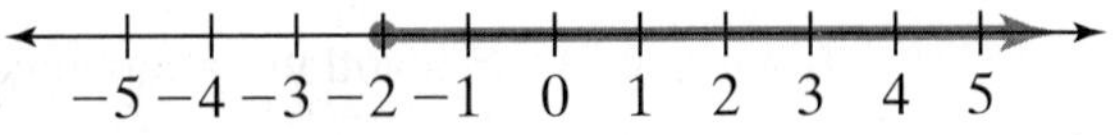

In interval notation, the solution set is $[-2, \infty)$.

b. Try solving this inequality on your own. View the **answer**, or watch this **video** for a complete solution.

You Try It **Work through this You Try It problem.**

Work Exercises 23–26 in this textbook or in the MyMathLab Study Plan.

(eText Screens 11.7-1–11.7-29)

Example 7 Solving a Linear Inequality in One Variable

Solve the inequality $5x - 4 < 2(x + 3) + x$. Write the solution set in **interval notation**, and graph it on a number line.

Solution Follow the **guidelines** for solving linear inequalities in one variable.

Begin with the original inequality:	$5x - 4 < 2(x + 3) + x$
Distribute:	$5x - 4 < 2x + 6 + x$
Combine like terms:	$5x - 4 < 3x + 6$
Subtract $3x$ from both sides:	$5x - 4 - 3x < 3x + 6 - 3x$
Simplify:	$2x - 4 < 6$
Add 4 to both sides:	$2x - 4 + 4 < 6 + 4$
Simplify:	$2x < 10$
Divide both sides by 2:	$\dfrac{2x}{2} < \dfrac{10}{2}$
Simplify:	$x < 5$

The graph of the solution set:

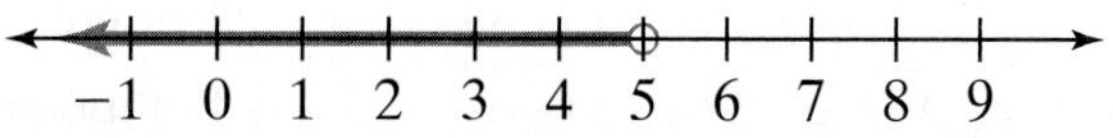

In interval notation, the solution set is $(-\infty, 5)$.

Example 8 Solving a Linear Inequality in One Variable

My video summary

Solve the inequality $4 + 2(3 - x) > 3(2x + 7) + 5$. Write the solution set in **set-builder notation** and graph it on a number line.

Solution Try to solve this inequality on your own. View the **answer**, or watch this **video** to see the complete solution.

You Try It **Work through this You Try It problem.**

Work Exercises 27–30 in this textbook or in the MyMathLab Study Plan.

Example 9 Solving a Linear Inequality Containing Fractions

Solve the inequality $\frac{n}{3} - 4 > -\frac{n}{6} + 1$ and write the solution set in **interval notation.**

Solution View the **answer**, or watch this **video** to see the detailed solution.

You Try It **Work through this You Try It problem.**

Work Exercises 31–34 in this textbook or in the MyMathLab Study Plan.

CAUTION When solving inequalities involving fractions, some students like to clear fractions right away. This is helpful, but not required. The same solution set will result whether or not fractions are cleared immediately.

As with equations, a linear inequality can have no solution or all real numbers as its solution set. Remember that a **contradiction** has no solution, and an **identity** has the set of all real numbers as its solution set.

Example 10 Solving Special Cases of Linear Inequalities

Solve the following inequalities. Write each solution set in **interval notation**.

a. $10 - 2(x + 1) > -5x + 3(x + 8)$ **b.** $2(5 - x) - 2 < 3(x + 3) - 5x$

Solutions

a. We use the **properties of inequalities** to isolate the variable on one side of the inequality.

Begin with the original ineqality:	$10 - 2(x + 1) > -5x + 3(x + 8)$
Use the distributive property:	$10 - 2x - 2 > -5x + 3x + 24$
Simplify:	$-2x + 8 > -2x + 24$
Add $2x$ to both sides:	$-2x + 8 + 2x > -2x + 24 + 2x$
Simplify:	$8 > 24$° ° False

Because the final statement is a contradiction, the inequality has no solution. The solution set is the empty { } or null set, ∅.

My video summary

b. Watch the **video** to see the full solution process and confirm that the result is an identity. The solution set is all real numbers.

You Try It **Work through this You Try It problem.**

Work Exercises 35 and 36 in this textbook or in the MyMathLab Study Plan.

OBJECTIVE 5 SOLVE THREE-PART INEQUALITIES

Consider the following application: A Bing internet search in June 2010 showed the range in price of iPad apps from $0.99 to $14.99. We can use the **three-part inequality** $0.99 \leq P \leq 14.99$ to show this relationship. Let P be the price of an iPad app.

To solve a three-part inequality, we use the **properties of inequalities** to isolate the variable between the two inequality symbols. Remember that *what we do to one part, we must do to all three parts* in order to write an equivalent inequality.

Example 11 Solving a Three-Part Inequality

Solve the inequality $4 \leq 3x - 2 < 7$. Write this solution set in **interval notation** and then graph it on a number line.

Solution

Begin with the original inequality:	$4 \leq 3x - 2 < 7$
Add 2 to all three parts:	$4 + 2 \leq 3x - 2 + 2 < 7 + 2$
Simplify:	$6 \leq 3x < 9$
Divide all three parts by 3:	$\frac{6}{3} \leq \frac{3x}{3} < \frac{9}{3}$
Simplify:	$2 \leq x < 3$ ← Variable isolated between inequality symbols

(eText Screens 11.7-1–11.7-29)

The graph of this solution set is shown below:

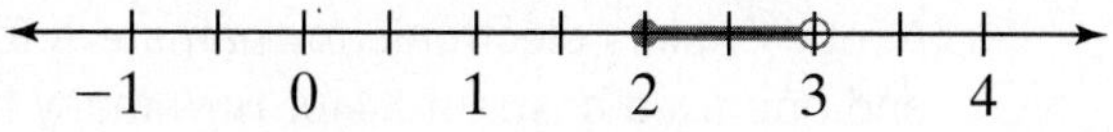

The solution set in interval notation is $[2, 3)$.

CAUTION When solving **three-part inequalities**, remember that what you do to one part of the inequality must be done to **all three parts** in order to obtain an equivalent inequality.

You Try It **Work through this You Try It problem.**

Work Exercises 37–40 in this textbook or in the MyMathLab Study Plan.

Example 12 Solving a Three-Part Inequality

My video summary

Solve the inequality $-1 < \frac{2x+1}{3} < 1$ and write its solution set in **interval notation**.

Solution Begin with the original inequality: $-1 < \frac{2x+1}{3} < 1$

Multiply all parts by 3 to clear the fraction: $3(-1) < \cancel{3} \cdot \frac{2x+1}{\cancel{3}} < 3(1)$

Simplify: $-3 < 2x + 1 < 3$

Finish solving the inequality on your own. View the **answer**, or watch this **video** for the complete solution.

You Try It **Work through this You Try It problem.**

Work Exercises 41–44 in this textbook or in the MyMathLab Study Plan.

OBJECTIVE 6 USE LINEAR INEQUALITIES TO SOLVE APPLICATION PROBLEMS

At this point, you may wish to review the **strategy** for solving application problems with linear equations. We present a variation of this strategy here using **mathematical models** for linear inequalities.

Strategy for Solving Application Problems Involving Linear Inequalities

Step 1. **Define the Problem.** Read the problem carefully, or multiple times if necessary. Identify what you are trying to find and determine what information is available to help you find it.

Step 2. **Assign Variables.** Choose a **variable** to assign to an unknown quantity in the problem. For example, use p for price. If other unknown quantities exist, express them in terms of the selected variable.

Step 3. **Translate into an Inequality.** Use the relationships among the known and unknown quantities to form an **inequality**.

Step 4. **Solve the Inequality.** Determine the **solution set** of the inequality.

Step 5. **Check the Reasonableness of Your Answer.** Check to see if your results make sense within the context of the problem. If not, check your work for errors and try again.

Step 6. **Answer the Question.** Write a clear statement that answers the question(s) posed.

(eText Screens 11.7-1–11.7-29)

Example 13 Making a Profit

A retailer sells electronic toy hamsters for $10. He pays $2 for each hamster wholesale and has fixed costs of $440. How many hamsters must he sell in order to make a **profit**? Solve the inequality $R > C$ with R as his **revenue** and C as his **cost**.

Solution

Step 1. We want to find the number of electronic hamsters that must be sold in order to make a profit. The retail price is $10 each, and the wholesale cost to the retailer is $2 each. The fixed costs are $440.

Step 2. Let the variable x represent the number of hamsters that are sold. The retailer's revenue, R, is $10x$ because the number sold times the price equals the revenue. The retailer's cost, C, is the sum of the total wholesale cost and fixed costs. Therefore, C is $2x + 440$.

Step 3. We can write the following inequality:

$$\overbrace{10x}^{\text{Revenue, } R} > \overbrace{2x + 440}^{\text{Cost, } C}$$

Step 4. Begin with the original inequality: $10x > 2x + 440$

Subtract $2x$ from both sides: $10x - 2x > 2x + 440 - 2x$

Simplify: $8x > 440$

Divide both sides by 8: $\frac{8x}{8} > \frac{440}{8}$

Simplify: $x > 55$

Step 5. Because $55(10) = 550$ and $2(55) + 440 = 550$, the retailer's cost and revenue are equal if 55 hamsters are sold. To make a profit, the retailer needs to sell more than 55 electronic hamsters, so this answer is reasonable.

Step 6. The retailer needs to sell more than 55 (at least 56) electronic hamsters to make a profit.

You Try It **Work through this You Try It problem.**

Work Exercises 45 and 46 in this textbook or in the MyMathLab Study Plan.

Example 14 Incredible Birthday Party

My video summary

Lesley is planning a birthday party for her son at Incredible Pizza. The party will cost $43 plus $15 for each guest. If she does not want to spend more than $300, what is the largest number of guests that can attend the party?

Solution

Step 1. We want to find the number of guests that can attend the birthday party. We know that the cost is $43 plus $15 per guest and Lesley does not want to spend more than $300.

Step 2. Let the variable x represent the number of guests. The cost of the party is $43 + 15x$ because there is a fixed cost of $43 and a per guest cost of $15.

Step 3. Because Lesley has a maximum amount she wants to spend, the inequality will involve a "less than or equal to" symbol.

$$\overbrace{43 + 15x}^{\text{Party cost}} \leq \overbrace{300}^{\text{Maximum amount}}$$

(*eText Screens 11.7-1–11.7-29*)

Watch this **video** to see the rest of the solution completed in detail. There can be at most 17 guests at the party.

CAUTION Because it is not feasible for a fractional guest to attend, we must round down the number of guests.

You Try It **Work through this You Try It problem.**

Work Exercises 47 and 48 in this textbook or in the MyMathLab Study Plan.

11.7 Exercises

In Exercises 1–4, write the solution set in set-builder notation.

You Try It

1. $x < -3$

2. $-1 < y < 5$

3. z is any real non-negative number

4. x is smaller than 8 and greater than 5

In Exercises 5–9, graph the solution set on a number line.

You Try It

5. $\{x|x > 0\}$

6. $\{x|-3 \le x < 2\}$

7. $\{x|x \le 5\}$

8. $\{x|2 \le x \le 5\}$

9. $\{x|-3 < x < 1\}$

In Exercises 10–14, write the solution set in interval notation.

You Try It

10. $\{x|x \ge -4\}$

11. $\{y|-5 < y < -2\}$

12. $\{y|0 < y\}$

13. t is larger than 3 but not larger than 10

14. w is nonnegative

In Exercises 15–30, solve the inequality. Graph the solution set and write it in interval notation.

You Try It

15. $x + 3 < 5$

16. $x - 1 \ge 4$

17. $x + 3 \le 2x + 1$

18. $3x + 7 > 2x - 5$

You Try It

19. $5x > 10$

20. $\frac{m}{-4} \ge 1$

21. $\frac{y}{2} \le -3$

22. $-6x < 24$

You Try It

23. $2x + 1 > 5x + 7$

24. $15 - x < 3x - 1$

25. $3x - 2 < 8 + x$

26. $5x - 2 < 3x + 8$

27. $4x - 5 > 3(x + 2) + 1$

28. $2x - 1 \ge 5(x + 1) + 3$

You Try It

29. $5 + (x - 3) > 3 - (x - 5)$

30. $2 + 3(4 + x) < 2(7x + 3) - 3$

(eText Screens 11.7-1–11.7-29)

In Exercises 31–36, solve the inequality. Write the solution set in interval notation.

You Try It

31. $\frac{m}{5} - 2 + m > -\frac{m}{2} + \frac{7}{5}$

32. $\frac{3}{5}x + 1 - 2x \leq \frac{x}{2} - \frac{9}{10}$

33. $2x + \frac{1}{2} > \frac{2}{3}x - 1$

34. $2.5x - 4 \geq 3.1 + 2.25x$

You Try It

35. $5x - (2x - 5) \geq 3x + 7$

36. $3(x - 1) + 2x < 2 + 5(x + 1)$

In Exercises 37–44, solve the inequality. Graph the solution set and write it in interval notation.

37. $-2 < 2x + 1 < 7$

38. $-4 < 1 - 3x < -1$

You Try It

39. $1 < 5x + 6 \leq 16$

40. $6.4 \leq 3.5x - 2 < 14.1$

41. $1 \leq \frac{2x + 1}{5} < 3$

42. $-1 \leq \frac{3 - 3x}{-3} \leq 3$

You Try It

43. $-4 \leq \frac{3(x - 1) + 4}{2} < 2$

44. $2 < \frac{8 - 2x}{5} < 4$

In Exercises 45–48, solve each application problem.

You Try It

45. Flea Market Nancy sells handcrafted bracelets at a flea market for \$7. If her monthly fixed costs are \$675 and each bracelet costs her \$2.75 to make, how many bracelets must she sell in a month to make a profit?

46. Car Rental A rental car company offers a rental plan of \$75 per day, plus \$0.12 per mile. An alternate rental plan is for \$60 per day, plus \$0.20 per mile. For what mileage is the first plan cheaper than the second?

You Try It

47. Wedding Reception Beverly is planning her wedding reception at the Missouri Botanical Gardens. The rental cost is \$2000 plus \$76 per person for catering. How many guests can she invite if her reception budget is \$10,000? (*Source:* www.mobot.org)

48. Wireless Plan Suppose Verizon Wireless offers a Nationwide 450 Talk & Test monthly plan that includes 450 anytime minutes and unlimited nights and weekends for \$60. Each additional anytime minute (or fraction of a minute) costs the user \$0.45. If Tori subscribes to this plan, how many anytime minutes can she use each month while keeping her total cost to no more than \$100 (before taxes)? (*Source:* www.verizonwireless.com)

11.8 Compound Inequalities; Absolute Value Equations and Inequalities

THINGS TO KNOW

Before working through this topic, be sure you are familiar with the following concepts:

		VIDEO	ANIMATION	INTERACTIVE
You Try It	1. Find the Absolute Value of a Real Number (Topic 10.1, **Objective 4**)	▤		
You Try It	2. Solve Linear Equations Using Both Properties of Equality (Topic 11.1, **Objective 5**)	▤		
You Try It	3. Solve Linear Inequalities in One Variable (Topic 11.7, **Objective 4**)	▤		

OBJECTIVES

1 Find the Union and Intersection of Two Sets

2 Solve Compound Linear Inequalities in One Variable

3 Solve Absolute Value Equations

4 Solve Absolute Value Inequalities

OBJECTIVE 1 FIND THE UNION AND INTERSECTION OF TWO SETS

The words *and* and *or* are sometimes used when working with **sets** of numbers.

Intersection

For any two sets A and B, the **intersection** of A and B is given by $A \cap B$ and represents the **elements** that are in set A **and** in set B.

$$A \cap B = \{x \mid x \text{ is an element of } A \text{ and an element of } B\}$$

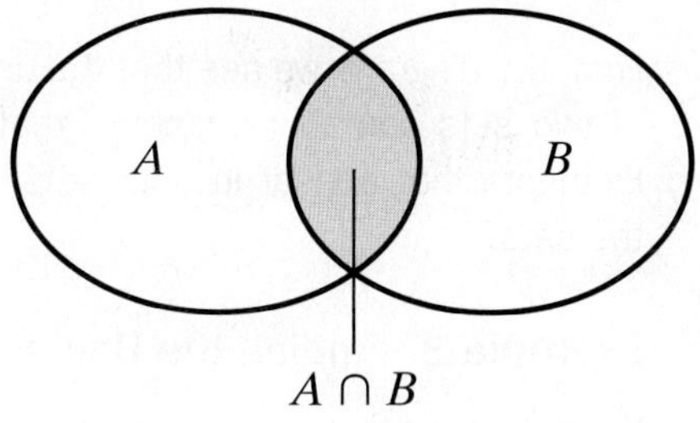

From the diagram, we see that the intersection of two sets is the overlap of the sets. Much like the intersection of two roads is the region common to both roads, the intersection of two sets is the set of elements that are common to both sets. If there is no overlap, the intersection is the **empty set**, and we write $A \cap B = \{\ \}$ or $A \cap B = \varnothing$.

(eText Screens 11.8-1–11.8-31)

Example 1 Finding the Intersection of Two Sets

Let $A = \{1, 3, 4, 5, 7, 10, 12\}$ and $B = \{2, 4, 6, 8, 10, 12\}$. Find $A \cap B$, the **intersection** of the two **sets**.

Solution The set $A \cap B$ is the set of elements that are in both A and B. Both sets contain the numbers 4, 10, and 12. So,

$$A \cap B = \{1, 3, 4, 5, 7, 10, 12\} \cap \{2, 4, 6, 8, 10, 12\} = \{4, 10, 12\}$$

You Try It **Work through this You Try It problem.**

Work Exercises 1 and 2 in this textbook or in the MyMathLab Study Plan.

Example 2 Finding the Intersection of Two Sets

My video summary

Let $A = \{x|x > -2\}$ and $B = \{x|x \leq 5\}$. Find $A \cap B$, the intersection of the two sets.

Solution View the **answer**, or watch the **video** for the solution.

You Try It **Work through this You Try It problem.**

Work Exercise 5 in this textbook or in the MyMathLab Study Plan.

Union

For any two **sets** A and B, the **union** of A and B is given by $A \cup B$ and represents the **elements** that are in set A **or** in set B.

$$A \cup B = \{x|x \text{ is an element of } A \text{ or an element of } B\}$$

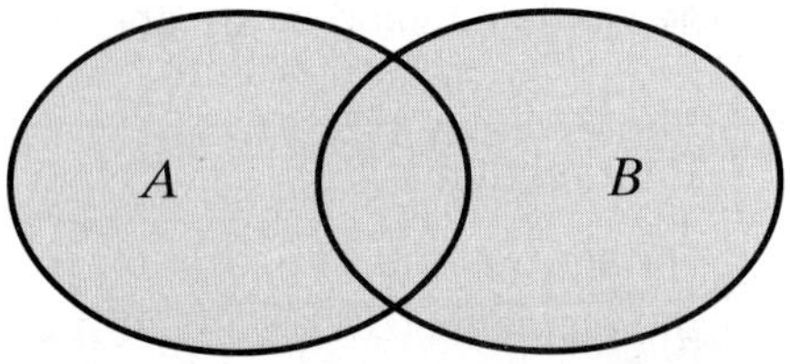

$A \leqslant B$

From the diagram, we see that the union of two sets is the combination of the sets. The union of two sets contains elements that are just in set A, just in set B, or in both A and B. Elements that appear in both sets A and B are only listed once when writing the union of the sets.

Example 3 Finding the Union of Two Sets

Let $A = \{1, 3, 4, 5, 7, 10, 12\}$ and $B = \{2, 4, 6, 8, 10, 12\}$. Find $A \cup B$, the **union** of the two sets.

Solution The union is the set of all unique values that are in either set A or in set B.

$$A \cup B = \{1, 2, 3, 4, 5, 6, 7, 8, 10, 12\}$$

Notice that 4, 10, and 12 occurred in both sets, but were only listed once in the union.

(eText Screens 11.8-1–11.8-31)

You Try It **Work through this You Try It problem.**

Work Exercises 3 and 4 in this textbook or in the MyMathLab Study Plan.

Example 4 Finding the Union of Two Sets

My video summary

Let $A = \{x | x < -2\}$ and $B = \{x | x \geq 5\}$. Find $A \cup B$, the union of the two sets.

Solution View the **answer**, or watch the **video** for the solution.

You Try It **Work through this You Try It problem.**

Work Exercise 6 in this textbook or in the MyMathLab Study Plan.

Example 5 Finding the Intersection of Intervals

My video summary

Find the **intersection** of the following intervals and graph the set on a number line.

a. $[0, \infty) \cap (-\infty, 5]$ **b.** $\left((-\infty, -2) \cup (-2, \infty)\right) \cap [-4, \infty)$

Solutions View the **answers**, or watch the **video** for the solutions.

You Try It **Work through this You Try It problem.**

Work Exercises 7 and 8 in this textbook or in the MyMathLab Study Plan.

OBJECTIVE 2 SOLVE COMPOUND LINEAR INEQUALITIES IN ONE VARIABLE

A **compound inequality** consists of two **inequalities** that are joined together using the words *and* or *or*. Figure 4 shows two examples of compound inequalities.

(a) $x + 2 < 5$ and $3x \geq -6$

(b) $x + 3 \leq 1$ or $2x - 5 > 7$

Figure 4
Examples of compound inequalities

The word *and* indicates **intersection**. A number is a **solution** to a compound inequality using *and* if it is a solution to *both* inequalities. For example, $x = 2$ is a solution to the compound inequality in Figure 4(a) because it is a solution to both $x + 2 < 5$ *and* $3x \geq -6$. View the **solution process**.

The word *or* indicates **union**. A number is a solution to a compound inequality using *or* if it is a solution to *either* inequality. For example, $x = 8$ is a solution to the compound inequality in Figure 4(b) because it is a solution to at least one of the inequalities. View the **solution process**.

My interactive video summary

For more practice on checking solutions to compound inequalities, work through this **interactive video**.

We now present a general strategy for solving compound inequalities.

(eText Screens 11.8-1–11.8-31)

> **Guidelines for Solving Compound Linear Inequalities**
>
> **Step 1.** Solve each **inequality** separately.
>
> **Step 2.** Graph each **solution set** on a number line.
>
> **Step 3.** For **compound inequalities** using *and*, the solution set is the **intersection** of the individual solution sets.
>
> For compound inequalities using *or*, the solution set is the **union** of the individual solution sets.

Example 6 Solving a Compound Linear Inequality Using *and*

My video summary

Solve $3x - 5 < -2$ and $4x + 11 \geq 3$. Graph the solution set and then write it in **interval notation.**

Solution Let's follow the guidelines for solving **compound linear inequalities.**

Step 1. Solve each **inequality** separately.

Original inequality:	$3x - 5 < -2$	and	Original inequality:	$4x + 11 \geq 3$
Add 5 to both sides:	$3x - 5 + 5 < -2 + 5$		Subtract 11 from both sides:	$4x + 11 - 11 \geq 3 - 11$
Simplify:	$3x < 3$		Simplify:	$4x \geq -8$
Divide both sides by 3:	$\frac{3x}{3} < \frac{3}{3}$		Divide both sides by 4:	$\frac{4x}{4} \geq \frac{-8}{4}$
Simplify:	$x < 1$		Simplify:	$x \geq -2$

Step 2. Graph each solution set on a number line.

$\{x \mid x < 1\}$

$\{x \mid x \geq -2\}$

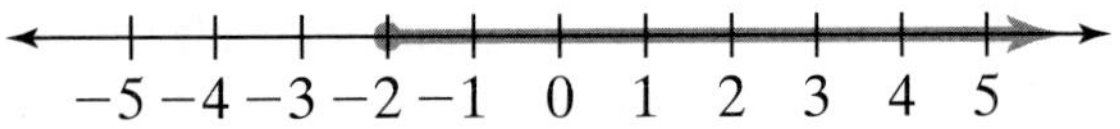

Continue to follow the solution process as shown here and/or watch this **video** to see the solution worked out in detail.

Step 3. Since the **compound inequality** uses *and*, the **solution set** is the **intersection** of the two graphs. We look for all x-values that are common to both solution sets. The first solution set includes all values of x that are less than 1. The second solution set includes all values of x that are greater than or equal to -2. The x-values common to both solution sets include all x-values that are greater than or equal to -2 and less than 1, or $\{x \mid -2 \leq x < 1\}$. The following graph shows this solution set:

$\{x \mid -2 \leq x < 1\}$

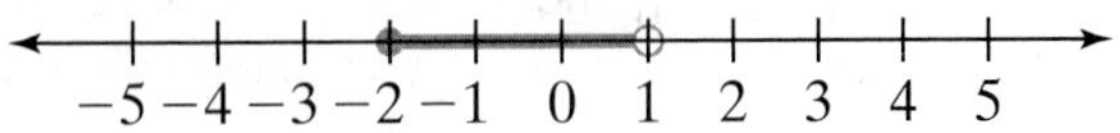

The solution set of the compound inequality, written in **interval notation**, is $[-2, 1)$.

You Try It **Work through this You Try It problem.**

Work Exercises 9 and 10 in this textbook or in the MyMathLab Study Plan.

Example 7 Solving a Compound Linear Inequality Using *or*

My video summary

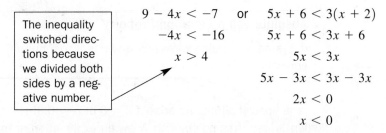

Solve $9 - 4x < -7$ or $5x + 6 < 3(x + 2)$. Graph the solution set and then write it in interval notation.

Solution

Step 1. Solve each **inequality** separately.

$$\begin{aligned} 9 - 4x &< -7 \\ -4x &< -16 \\ x &> 4 \end{aligned} \qquad \text{or} \qquad \begin{aligned} 5x + 6 &< 3(x + 2) \\ 5x + 6 &< 3x + 6 \\ 5x &< 3x \\ 5x - 3x &< 3x - 3x \\ 2x &< 0 \\ x &< 0 \end{aligned}$$

The inequality switched directions because we divided both sides by a negative number.

Step 2–3. Graph each **solution set** on a number line. Since the **compound inequality** uses *or*, the solution set is the **union** of the two graphs. We look for all x values that appear in either of the two solution sets. Complete the solution; then check your **answer**. Or, watch the **video** for a detailed solution.

You Try It **Work through this You Try It problem.**

Work Exercises 11 and 12 in this textbook or in the MyMathLab Study Plan.

Like other inequalities, a **compound inequality** can have no solution or the set of all **real numbers** as its solution set. For example, the solution set of the compound inequality

$$2x - 3 \le -1 \quad \text{and} \quad x - 7 \ge -3$$

is the **null set**, $\varnothing$ (see **why**), whereas the solution set of the compound inequality

$$10x + 7 > 2 \quad \text{or} \quad 3x - 6 \le 9$$

is the set of all real numbers, $\mathbb{R}$ or $(-\infty, \infty)$ in **interval notation** (see **why**).

You Try It **Work through this You Try It problem.**

Work Exercises 13–16 in this textbook or in the MyMathLab Study Plan.

OBJECTIVE 3 SOLVE ABSOLUTE VALUE EQUATIONS

My video summary

Before reading further, you may want to review **absolute value** in **Topic 10.1**. Consider how to solve the equation $|x| = 5$. First, we look for values of x that are 5 units away from zero on a number line. The two numbers that are 5 units away from zero on a number line are $x = -5$ and $x = 5$, as shown in Figure 5.

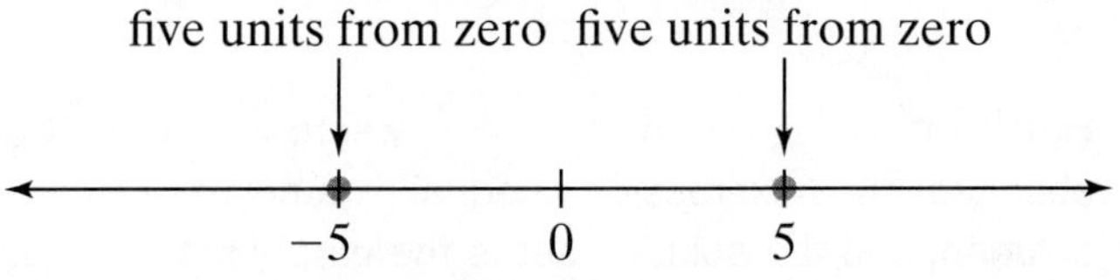

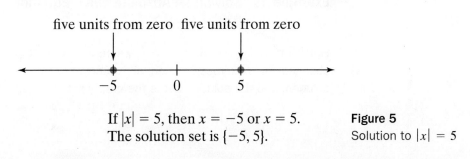

If $|x| = 5$, then $x = -5$ or $x = 5$.
The solution set is $\{-5, 5\}$.

Figure 5
Solution to $|x| = 5$

(eText Screens 11.8-1–11.8-31)

How do we solve the equation $|x + 2| = 5$? First, we consider the expression inside the absolute value bars as a single quantity. To solve the equation, we look for values of x that make that quantity equal to -5 or 5. So, we write the following:

$$x + 2 = 5 \quad \text{or} \quad x + 2 = -5$$

This idea can be applied to all **algebraic expressions**, as shown in the following box.

Absolute Value Equation Property

If u is an algebraic expression and c is a **real number** such that $c > 0$ then $|u| = c$ is equivalent to $u = -c$ or $u = c$.

Some special situations arise if $c = 0$ or $c < 0$. If $c = 0$, then $|u| = 0$ and $u = 0$. If $c < 0$, then $|u| = c$ has no solution. View an **explanation** of these results.

Example 8 Solving an Absolute Value Equation

Solve: $|m + 4| = 8$

Solution Using the form $|u| = c$, where $u = m + 4$ and $c = 8$, we look for values of m such that $m + 4$ is 8 units away from zero on a number line. The absolute value equation property shows that $|m + 4| = 8$ is equivalent to

$$m + 4 = -8 \quad \text{or} \quad m + 4 = 8$$

Now we can solve the two **equations** separately and combine the two **solution sets.**

$$\begin{aligned} m + 4 &= -8 \quad \text{or} \quad & m + 4 &= 8 \\ m &= -12 & m &= 4 \end{aligned}$$

The solution set is $\{-12, 4\}$. View the check of this **answer**.

Example 9 Solving an Absolute Value Equation

My video summary

Solve: $|1 - 3x| = 4$

Solution View the **answer**, or watch this **video** for a detailed solution.

You Try It **Work through this You Try It problem.**

Work Exercises 17–20 in this textbook or in the MyMathLab Study Plan.

Example 10 Solving an Absolute Value Equation

My video summary

Solve: $|2x - 5| = 0$

Solution View the **answer**, or watch this **video** for a detailed solution.

Example 11 Solving an Absolute Value Equation

Solve: $|3x + 7| = -4$

Solution Using the form $|u| = c$, where $u = 3x + 7$ and $c = -4$, we see that the absolute value expression is equal to a negative number. This means that there is no solution, and the solution set is the **empty set**, $\{\ \}$.

(eText Screens 11.8-1–11.8-31)

You Try It **Work through this You Try It problem.**

Work Exercises 21 and 22 in this textbook or in the MyMathLab Study Plan.

CAUTION Care must be taken to not just blindly apply the absolute value equation property. View this **popup** to see what would happen in Example 11 if we mistakenly applied the absolute value equation property when $c < 0$.

When applying the absolute value equation property, remember to write the absolute value equation in the form $|u| = c$. If the equation is not in this form, then we must isolate the absolute value expression first before applying the property.

Now we present a general strategy for solving **absolute value** equations.

Strategy for Solving Absolute Value Equations

Step 1. Isolate the absolute value expression on one side of the equation to obtain the form $|u| = c$.

Step 2. Apply the **absolute value equation property**.

If $c > 0$, write $|u| = c$ as $u = -c$ or $u = c$.

If $c = 0$, write $|u| = c$ as $u = c$.

If $c < 0$, the equation has no solution.

Step 3. Solve any equations from Step 2 to find the solution set.

Step 4. Check your answer within the original absolute value equation to confirm the solution set.

Example 12 Solving an Absolute Value Equation

My video summary

Solve: $2|w - 1| + 3 = 11$

Solution Following the **strategy for solving absolute value equations**, first we isolate the **absolute value** expression on one side of the equation.

Start with the original equation: $2|w - 1| + 3 = 11$

Subtract 3 from both sides: $2|w - 1| + 3 - 3 = 11 - 3$

Simplify: $2|w - 1| = 8$

Divide both sides by 2: $\dfrac{2|w - 1|}{2} = \dfrac{8}{2}$

Simplify: $|w - 1| = 4$

Finish solving the equation. View the **answer**, or watch the **video** to see the full solution.

Example 13 Solving an Absolute Value Equation

My video summary

Solve: $-3|2 - m| + 8 = 2$

Solution View the **answer**, or watch the **video** to see the full solution.

You Try It **Work through this You Try It problem.**

Work Exercises 23 and 24 in this textbook or in the MyMathLab Study Plan.

(eText Screens 11.8-1–11.8-31)

OBJECTIVE 4 SOLVE ABSOLUTE VALUE INEQUALITIES

If $|u|$ and c are real numbers, then the **trichotomy property** tells us that exactly one of the following is true:

$$|u| = c, |u| < c, \quad \text{or} \quad |u| > c$$

Now that we know how to solve absolute value equations of the form $|u| = c$, let's turn our attention to solving absolute value inequalities of the forms $|u| < c$ or $|u| > c$.

When solving $|x| = 5$, we looked for values of x that were 5 units away from zero on a number line. To solve the inequality $|x| < 5$, we look for values of x that are *less than* 5 units away from zero on a number line. Similarly, to solve the inequality $|x| > 5$, we look for values of x that are *greater than* 5 units away from zero on a number line. The solutions to these two inequalities are illustrated in Figures 6 and 7, respectively.

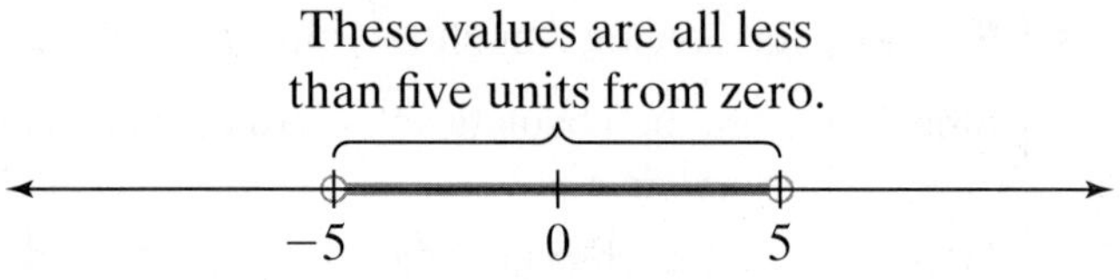
These values are all less
than five units from zero.
−5
0
5

If $|x| < 5$, then $-5 < x < 5$.
The solution set is $\{x|-5 < x < 5\}$ in set-builder notation or $(-5, 5)$ in interval notation.

Figure 6
Solution to $|x| < 5$

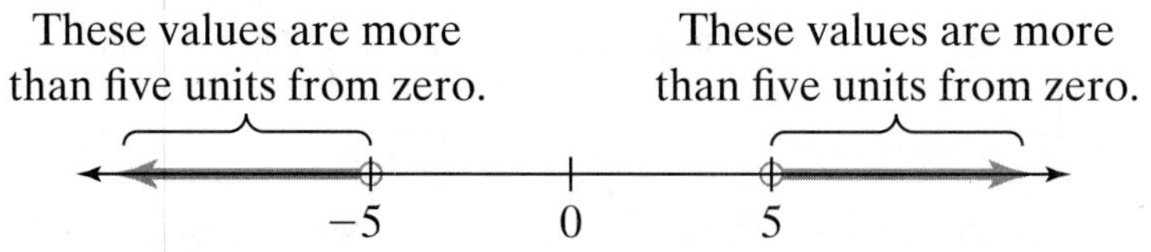
These values are more
than five units from zero.
These values are more
than five units from zero.
−5
0
5

If $|x| > 5$, then $x < -5$ or $x > 5$.
The solution set is $\{x|x < -5 \text{ or } x > 5\}$ in set-builder notation or $(-\infty, -5) \cup (5, \infty)$ in interval notation.

Figure 7
Solution to $|x| > 5$

Figures 6 and 7 help us see that absolute value inequalities can be converted into alternate forms that do not include absolute value expressions. We present these alternate forms in the following summary.

Absolute Value Inequality Property

If u is an algebraic expression and c is a real number such that $c > 0$, then

1. $|u| < c$ is equivalent to $-c < u < c$

and

2. $|u| > c$ is equivalent to $u < -c \quad \text{or} \quad u > c$

Similar forms exist for the non-strict cases $|u| \leq c$ and $|u| \geq c$.

As with absolute value equations, special cases arise when working with absolute value inequalities. Two special cases are when $|u| < 0$ and $|u| \geq 0$. View this **popup** to see the **solution sets** of these two cases.

(eText Screens 11.8-1–11.8-31)

Example 14 Solving an Absolute Value Inequality

My video summary

Solve: $|2m - 1| \le 5$

Solution Using the form $|u| \le c$, where $u = 2m - 1$ and $c = 5$, we look for values of m such that $2m - 1$ is less than or equal to 5 units away from zero on a number line. Next we use the **absolute value inequality property** to write $|2m - 1| \le 5$ is equivalent to $-5 \le 2m - 1 \le 5$.

Now solve the resulting **three-part inequality**. View this **popup** to compare your answer and see a check, or watch this **video** to confirm your results.

You Try It **Work through this You Try It problem.**

Work Exercises 25–30 in this textbook or in the MyMathLab Study Plan.

Example 15 Solving an Absolute Value Inequality

My video summary

Solve: $|5x + 1| > 3$

Solution View this **popup** to compare your answer and see a check, or watch this **video** to confirm your results.

CAUTION

In Example 15, $|5x + 1| > 3$ is *not* equivalent to $-3 > 5x + 1 > 3$. A **three-part inequality** must be true from far left to far right: $-3 > 3$ is a false statement. Another common error in this type of problem is to write $5x + 1 > -3$ for the first inequality, instead of $5x + 1 < -3$. Think carefully about the meaning of the inequality before writing it.

You Try It **Work through this You Try It problem.**

Work Exercises 31–36 in this textbook or in the MyMathLab Study Plan.

As with absolute value equations, not every absolute value inequality will be written in a form that will allow us to use one of the **absolute value inequality properties** immediately. The following general strategy for solving absolute value inequalities can be used as a guide.

Strategy for Solving Absolute Value Inequalities

Step 1. Isolate the absolute value expression to obtain one of the following forms: $|u| < c$, $|u| > c$, $|u| \le c$, or $|u| \ge c$.

Step 2. Apply the absolute value inequality property.

Step 3. Solve the resulting compound inequality.

Example 16 Solving an Absolute Value Inequality

My video summary

Solve: $|4x - 3| + 2 \le 7$

Solution Following the **strategy for solving absolute value inequalities**, first we isolate the absolute value expression.

Write the original inequality: $|4x - 3| + 2 \le 7$

Subtract 2 from both sides: $|4x - 3| \le 5$

(eText Screens 11.8-1–11.8-31)

Now we have the form $|u| \leq c$, where $u = 4x - 3$ and $c = 5$. We look for values of x such that $4x - 3$ is less than or equal to 5 units away from zero on a number line. Using the **absolute value inequality property**, $|4x - 3| \leq 5$ is equivalent to $-5 \leq 4x - 3 \leq 5$.

We find the solution set by solving the **three-part inequality**:

Write the three-part inequality: $-5 \leq 4x - 3 \leq 5$

Add 3 to all three parts: $-2 \leq 4x \leq 8$

Divide all three parts by 4 and simplify: $-\dfrac{1}{2} \leq x \leq 2$

The solution set is $\left\{x \middle| -\dfrac{1}{2} \leq x \leq 2\right\}$ in **set-builder notation** or $\left[-\dfrac{1}{2}, 2\right]$ in **interval notation**. View the **check** for this solution set. The following shows the graph of the solution set on a number line:

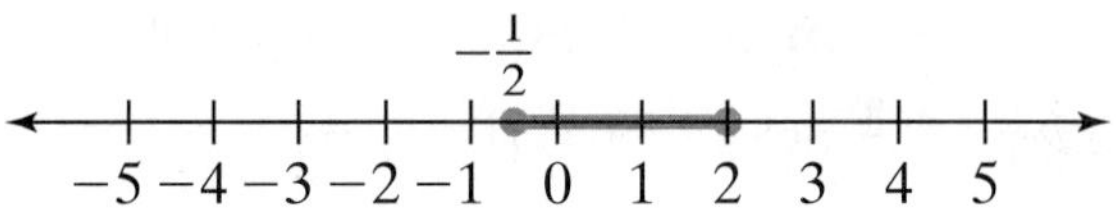

Example 17 Solving an Absolute Value Inequality

My interactive video summary

Solve: $5|1 - 2x| - 3 > 12$

Solution Follow the **strategy for solving absolute value inequalities**; then view this **popup** to check your solution set, or work through this **interactive video**.

You Try It **Work through this You Try It problem.**

Work Exercises 37–42 in this textbook or in the MyMathLab Study Plan.

11.8 Exercises

In Exercises 1–4, find the indicated set given $A = \{-2, 0, 1, 5, 6, 9, 15\}$, $B = \{-6, -4, -2, 0, 2, 4, 6\}$, and $C = \{-1, 3, 7, 11, 20\}$.

You Try It **1.** $A \cap B$

2. $A \cap C$

You Try It **3.** $A \cup B$

4. $A \cup B \cup C$

You Try It **5.** Find $A \cap B$ if $A = \{x|x < 9\}$ and $B = \{x|x \geq 2\}$.

You Try It **6.** Find $A \cup B$ if $A = \{x|x \leq -3\}$ and $B = \{x|x \geq 5\}$.

You Try It **7.** Find the intersection: $(-8, 5] \cap (-12, 3)$

8. Find the intersection: $((-\infty, 0) \cup (0, \infty)) \cap [-3, \infty)$

(eText Screens 11.8-1–11.8-31)

In Exercises 9–16, solve the compound inequality. Write your answer in interval notation or state that there is no solution.

You Try It

9. $2x - 3 \le 5$ and $5x + 1 > 6$

10. $\dfrac{n-1}{3} \ge 1$ and $\dfrac{4n-2}{2} \le 9$

You Try It

11. $4k + 2 \le -10$ or $3k - 4 > 8$

12. $2x + 1 > 5$ or $5 - 3x < 11$

You Try It

13. $2y - 1 \le 7$ or $3y - 5 > 1$

14. $3b - 5 \le 1$ and $1 - b < -3$

15. $x + 1 < -5$ and $2x - 3 < -11$

16. $0.3x + 3 \le x + 0.2$ or $4(2x - 5) > 3x + 5$

In Exercises 17–24, solve each absolute value equation.

You Try It

17. $|x| = 3$

18. $|x - 1| = 5$

19. $|6x + 7| = 3$

20. $\left|\dfrac{3x+1}{5}\right| = \dfrac{2}{3}$

You Try It

21. $|7x + 3| = 0$

22. $|x - 2| = -4$

You Try It

23. $3|4x - 3| + 1 = 10$

24. $-2|3x + 2| + 7 = 5$

In Exercises 25–30, solve each absolute value inequality. Write your answer in interval notation.

You Try It

25. $|9x| \le 3$

26. $|x - 2| < 3$

27. $|3 - 4x| < 11$

28. $\left|\dfrac{2x-5}{3}\right| \le \dfrac{1}{5}$

29. $|3x - 7| \le 0$

30. $|x - 1| < -2$

In Exercises 31–36, solve each absolute value inequality. Write your answer in interval notation.

You Try It

31. $|2x| > 6$

32. $|x + 4| > 5$

33. $\left|\dfrac{2x-6}{7}\right| \ge \dfrac{3}{14}$

34. $|2 - 5x| \ge 3$

35. $|9x - 4| > 0$

36. $|3x + 1| \ge -5$

In Exercises 37–42, solve each absolute value inequality. Write your answer in interval notation.

37. $|4x - 3| + 2 \le 9$

38. $|1 - 2x| - 4 > 3$

39. $3|x + 5| - 1 \ge 8$

You Try It

40. $-|3x + 2| + 5 > -6$

41. $-4|5x + 2| - 3 < 5$

42. $2|3 - 4x| + 1 < 4$

MODULE TWELVE

Graphs of Linear Equations and Inequalities in Two Variables

MODULE TWELVE CONTENTS

12.1 The Rectangular Coordinate System

THINGS TO KNOW

Before working through this topic, be sure you are familiar with the following concepts:

VIDEO ANIMATION INTERACTIVE

You Try It 1. Plot Real Numbers on a Number Line (Topic 10.1, **Objective 2**)

OBJECTIVES

1 Read Line Graphs

2 Identify Points in the Rectangular Coordinate System

3 Plot Ordered Pairs in the Rectangular Coordinate System

4 Create Scatter Plots

OBJECTIVE 1 READ LINE GRAPHS

Graphs are often used to show relationships between two **variables**. Figure 1 shows the selling price of MasterCard, Inc. stock at the start of each month from May 2006 (when it began trading publicly) until June 2010. The graph shows us how the value of the stock has changed over time.

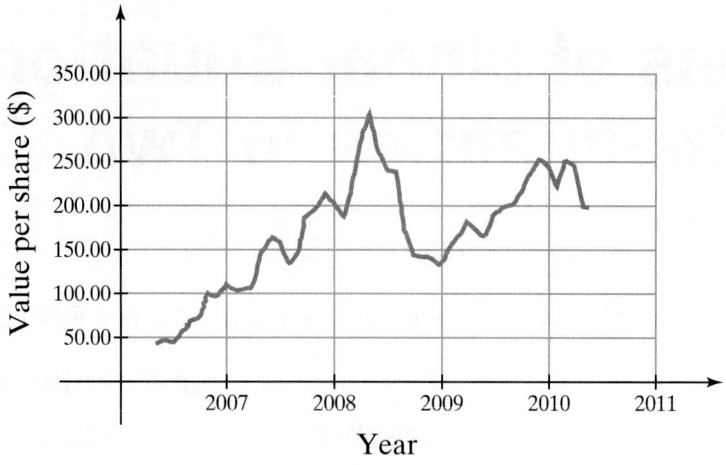

Figure 1 MasterCard, Inc. stock value
Source: TD Ameritrade

Figure 1 is an example of a *line graph*. **Line graphs** consist of a series of points (•) that are connected by **line segments.**

Example 1 Finding Temperatures on a Line Graph

My video summary

The following **line graph** shows the average daily temperature in St. Louis, MO, for each month.

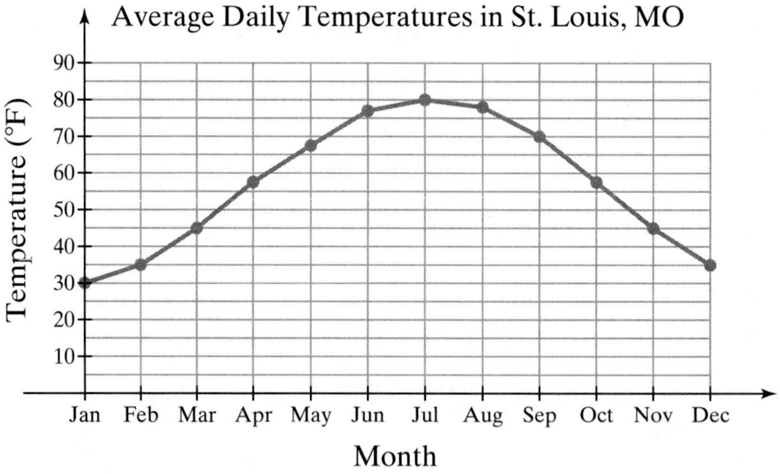

Source: National Climatic Data Center

a. What is the average daily temperature in February?

b. What is the average daily temperature in November?

c. In what month is the average daily temperature $70°F$?

d. Which month has the highest average daily temperature? What is the average daily temperature for that month?

e. In what months are the average daily temperatures above $65°F$?

(eText Screens 12.1-1–12.1-17)

Solutions

a. The red point on the graph below corresponds to February. Note that the point for February is about halfway between the tick marks for 30°F and 40°F, which means that the average daily temperature for February is about 35°F.

b. The green point on the graph below corresponds to November. Because that point lies about halfway between the tick marks for 40°F and 50°F, the average daily temperature is about 45°F.

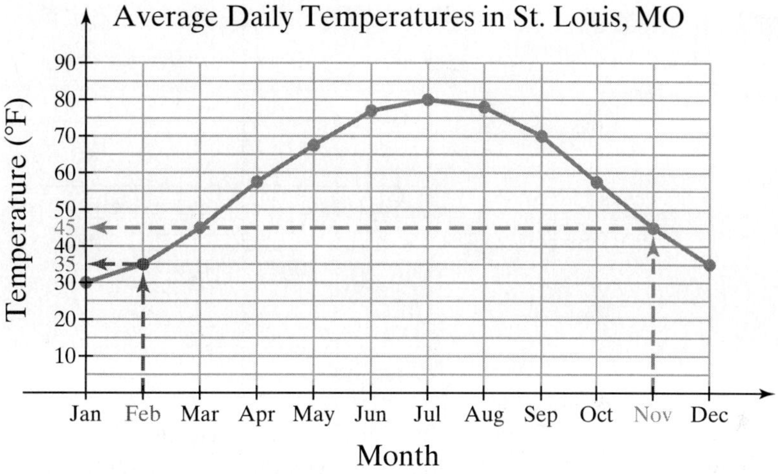

Source: National Climatic Data Center

c.–e. Try answering these questions on your own. View the **answers**, or watch this **video** for detailed solutions to all five parts.

You Try It **Work through this You Try It problem.**

Work Exercises 1 and 2 in this textbook or in the MyMathLab Study Plan.

OBJECTIVE 2 IDENTIFY POINTS IN THE RECTANGULAR COORDINATE SYSTEM

My animation summary

Line graphs like **Figure 1** use points in two dimensions to show how two **variables** are related. We now extend this method of identifying points by defining the *rectangular coordinate system*. Watch this **animation** for an overview.

The **rectangular coordinate system**, also known as the **Cartesian coordinate system** in honor of its inventor **René Descartes**, consists of two **perpendicular** real **number lines** called the **coordinate axes.** The horizontal axis is called the ***x*-axis,** and the vertical axis is called the ***y*-axis.** The two axes intersect at a point called the **origin.**

The **plane** represented by this system is called the **coordinate plane,** also known as the **Cartesian plane** or ***xy*-plane.** The axes divide the plane into four regions called **quadrants.** The upper-right region is *Quadrant I,* the upper-left region is *Quadrant II,* the lower-left region is *Quadrant III,* and the lower-right region is *Quadrant IV.* See Figure 2.

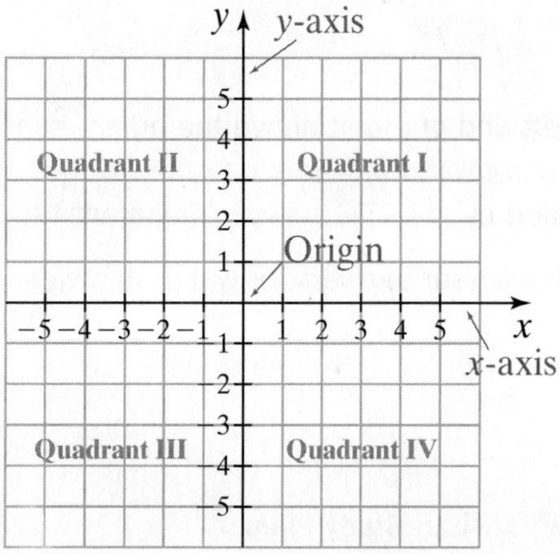

Figure 2
The rectangular coordinate system

(eText Screens 12.1-1–12.1-17)

Each position, or **point**, on the **coordinate plane** can be identified using an **ordered pair** of numbers in the form (x, y). For example, the point shown in Figure 3 is identified by the ordered pair $(3, 2)$. The first number 3 is called the **x-coordinate** or **abscissa** and indicates the point is located 3 units to the right of the origin. The second number 2 is called the **y-coordinate** or **ordinate** and indicates the point is located 2 units above the **origin.**

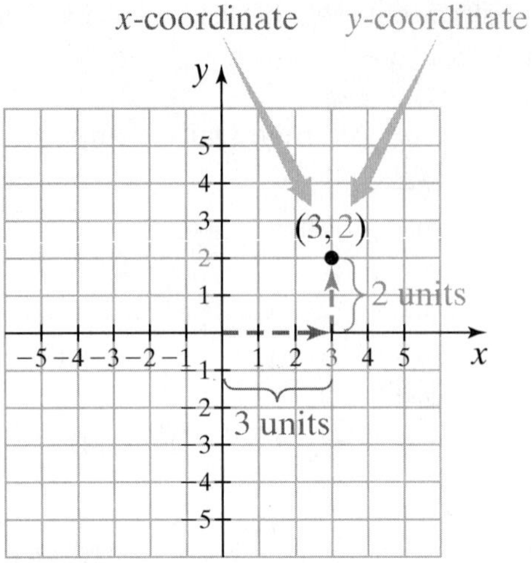

Figure 3
The ordered pair $(3, 2)$

In Figure 3, both the x-coordinate and y-coordinate are positive $(x > 0, y > 0)$ because the point is located in Quadrant I. Launch this **popup box** to determine the signs for x- and y-coordinates of points that lie in the other three **quadrants** or on the **axes.**

Example 2 Identifying Points

My video summary

Use an **ordered pair** to identify each point on the **coordinate plane** shown. State the **quadrant** or **axis** where each point lies.

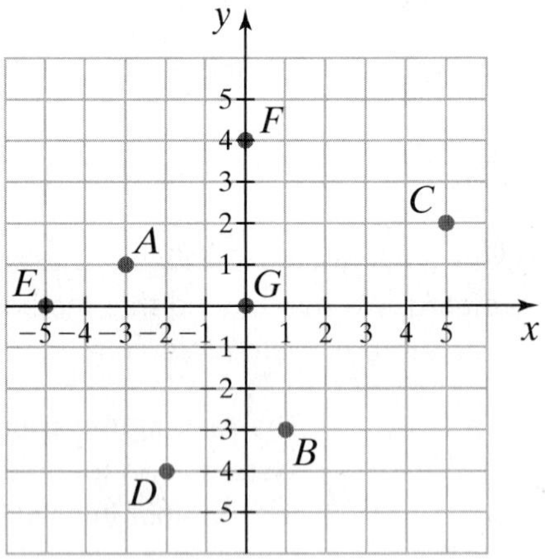

Solution Point A is positioned three units to the left and one unit above the **origin**, so it is identified by the ordered pair $(-3, 1)$. It lies in Quadrant II. Point B is one unit to the right and three units below the origin, so it is identified by $(1, -3)$. It lies in Quadrant IV.

Try to finish identifying these points on your own. Check your **answers**, or watch this **video** for a detailed solution.

You Try It **Work through this You Try It problem.**

Work Exercises 3 and 4 in this textbook or in the MyMathLab Study Plan.

(*eText Screens 12.1-1–12.1-17*)

The order of the pair of **coordinates** in an **ordered pair** is just as important as the coordinates themselves. For example, compare the ordered pairs and the positions of points A and B in **Example 2**. Both ordered pairs include the numbers 1 and -3, but they are in reverse order: $A(-3, 1)$ versus $B(1, -3)$. Point A lies in Quadrant II, whereas point B lies in Quadrant IV.

OBJECTIVE 3 PLOT ORDERED PAIRS IN THE RECTANGULAR COORDINATE SYSTEM

We **plot**, or **graph**, an **ordered pair** by placing a point (•) at its location on the **coordinate plane**.

Example 3 Plotting Ordered Pairs

My video summary

Plot each ordered pair on the coordinate plane. State the **quadrant** or **axis** where each point lies.

a. $(2, 4)$ **b.** $(4, -5)$ **c.** $(0, -2)$

d. $(-3, -4)$ **e.** $\left(-\frac{7}{2}, \frac{5}{2}\right)$ **f.** $(1.5, 0)$

Solutions

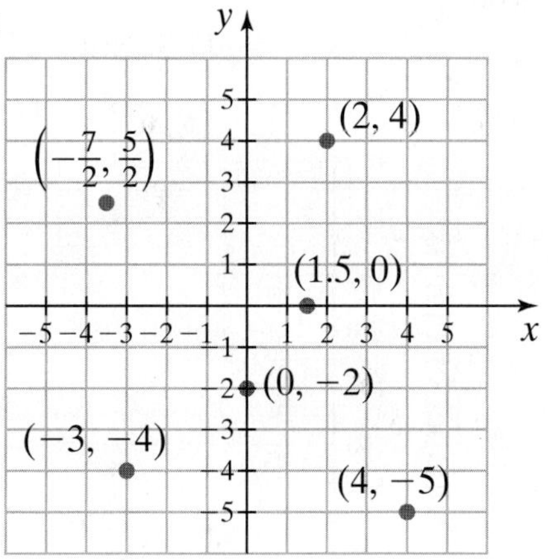

Point $(2, 4)$ lies in Quadrant I.

Point $(4, -5)$ lies in Quadrant IV.

Point $(0, -2)$ lies on the y-axis.

Point $(-3, -4)$ lies in Quadrant III.

Point $\left(-\frac{7}{2}, \frac{5}{2}\right)$ lies in Quadrant II.

Point $(1.5, 0)$ lies on the x-axis.

Watch this **video** for a detailed solution.

You Try It **Work through this You Try It problem.**

Work Exercises 5–10 in this textbook or in the MyMathLab Study Plan.

OBJECTIVE 4 CREATE SCATTER PLOTS

Ordered pairs can be used to study relationships between variables. For example, a doctor studying height and weight can use ordered pairs of the form (*height*, *weight*) to record data for each patient. Data represented as ordered pairs are called **paired data**. A **scatter plot** is created by graphing paired data as points on the **coordinate plane**. Scatter plots can reveal patterns in the paired data.

Example 4 Ethanol Industry

The table to the right shows the number of U.S. ethanol plants operating in the month of January for the years 2000–2010. List ordered pairs in the form (*year, number of plants*). Create a scatter plot of the paired data. Do the paired data show a trend? If so, what is the trend?

Year	Number of Plants
2000	54
2001	56
2002	61
2003	68
2004	72
2005	81
2006	95
2007	110
2008	139
2009	170
2010	189

Source: Renewable Fuels Association

Solution The ordered pairs are $(2000, 54)$, $(2001, 56)$, $(2002, 61)$, $(2003, 68)$, $(2004, 72)$, $(2005, 81)$, $(2006, 95)$, $(2007, 110)$, $(2008, 139)$, $(2009, 170)$, and $(2010, 189)$.

Because the paired data consist of all positive numbers, we need only **Quadrant I** of the **coordinate plane**. We mark the given years along the x-axis. For the y-axis, we let the grid marks represent multiples of 10 ethanol plants. (**Note:** We labeled only the multiples of 20 to avoid clutter.) Plotting the ordered pairs gives us the graph shown.

The scatter plot shows that the number of U.S. ethanol plants has grown steadily from 2000 to 2010, with faster growth after 2005.

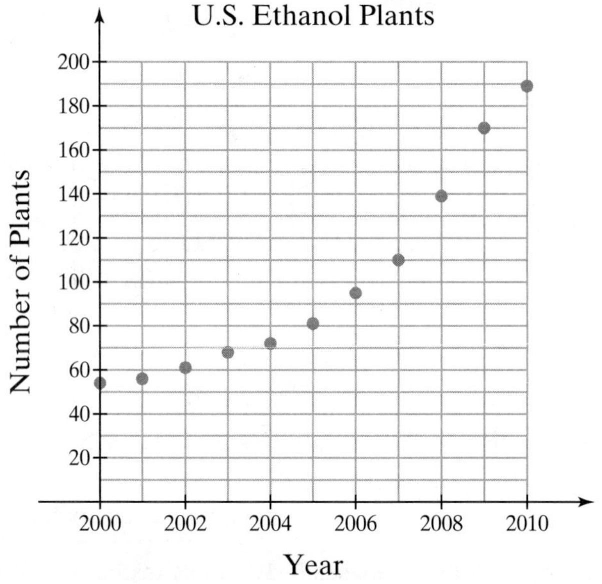

You Try It **Work through this You Try It problem.**

Work Exercises 11–14 in this textbook or in the MyMathLab Study Plan.

CAUTION **Scatter plots** and **line graphs** typically involve only Quadrant I because real data primarily involve non-negative numbers.

12.1 Exercises

In Exercises 1 and 2, use the line graph provided to answer the questions.

You Try It **1. Twisters** The following line graph shows the average number of tornadoes in Oklahoma each month.

a. To the nearest whole number, what is the average number of tornadoes in March?

b. What month has an average of about 8 tornadoes?

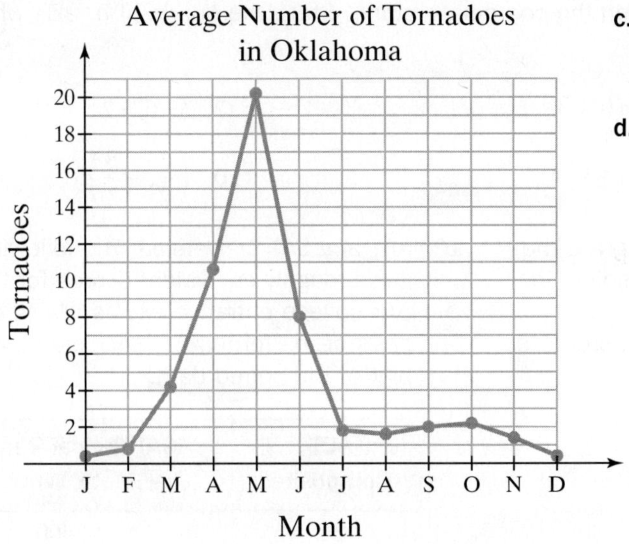

Source: National Weather Service, NOAA

c. Which month has the highest average number of tornadoes? To the nearest whole number, what is the average number of tornadoes this month?

d. In what months are the average numbers of tornadoes below 1?

2. Florida The following line graph shows the population of Florida since 1900.

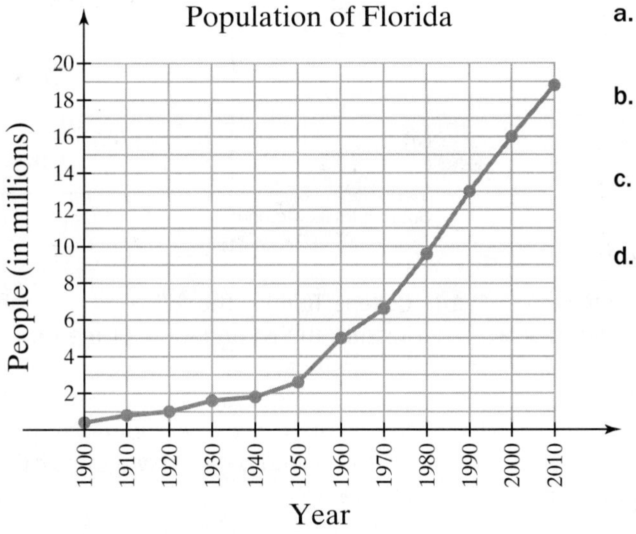

Source: U.S. Census Bureau

a. To the nearest million, what was the population of Florida in 1960?

b. To the nearest million, what was the population in 2000?

c. In what year was the population about 13 million people?

d. Estimate the increase in population from 1900 to 2010.

In Exercises 3 and 4, use an ordered pair to identify each point on the given coordinate plane. State the quadrant or axis where each point lies.

You Try It

3.

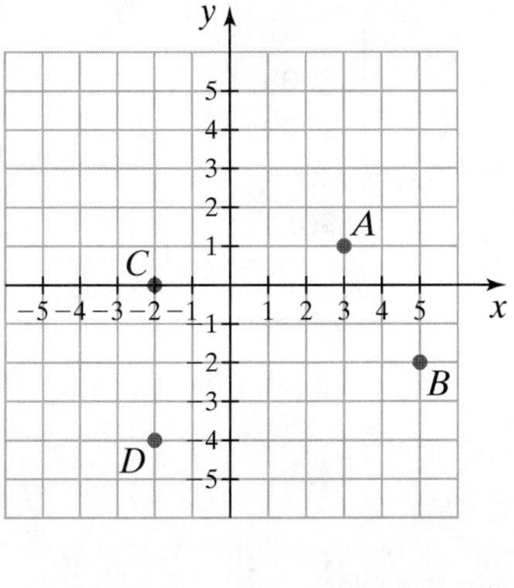

4.

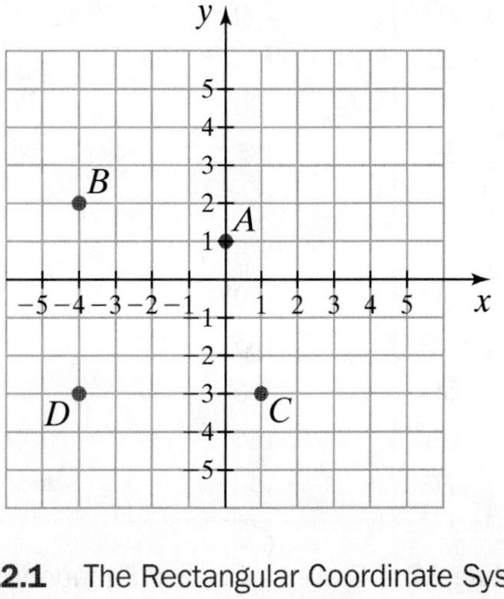

In Exercises 5–10, plot each ordered pair on the coordinate plane. State the quadrant or axis where each point lies.

You Try It

5. $(-4, 1)$ **6.** $(0, -3)$ **7.** $(-6, -2)$

8. $(2, 5)$ **9.** $(3.5, 0)$ **10.** $\left(\frac{1}{2}, -\frac{5}{2}\right)$

11. Price of Gasoline The following table gives the average price per gallon of regular gasoline in the U.S. for the years 2000–2009. Using ordered pairs of the form (*year*, *price*), create a scatter plot of the paired data.

Year	Price per gallon (dollars)
2000	1.48
2001	1.42
2002	1.35
2003	1.56
2004	1.85
2005	2.27
2006	2.57
2007	2.80
2008	3.25
2009	2.35

Source: Energy Information Administration

12. ACT and SAT Conversion The following table gives roughly equivalent scores for the ACT and SAT college entrance exams. Using ordered pairs of the form (*ACT*, *SAT*), create a scatter plot of the paired data.

ACT Composite	SAT Critical Reading, Math, Writing
36	2400
33	2190
30	1980
27	1820
24	1650
21	1500
18	1290
15	1060
12	780

Source: InLikeMe.com

You Try It

13. Maximum Heart Rate *Maximum heart rate* is the highest number of times a person's heart can beat in one minute of exercise. The following table gives the average maximum heart rate for ages 20–70. Using ordered pairs of the form (*age*, *maximum heart rate*), create a scatter plot of the paired data. Do the paired data show a trend? If so, what is the trend?

Age	Average Maximum Heart Rate (beats per minute)
20	200
25	195
30	190
35	185
40	180
45	175
50	170
55	165
60	160
65	155
70	150

Source: American Heart Association

14. College Tuition The following table shows annual in-state tuition and fees for the Connecticut State University system for the years 2005–2011. Using ordered pairs of the form (*year*, *tuition & fees*), create a scatter plot of the paired data. Do the paired data show a trend? If so, what is the trend?

Year	Tuition & Fees (dollars)
2005	5,611
2006	5,936
2007	6,284
2008	6,736
2009	7,180
2010	7,566
2011	8,043

Source: State of Connecticut Department of Higher Education

(eText Screens 12.2-1–12.2-30)

12.2 Graphing Linear Equations in Two Variables

THINGS TO KNOW

Before working through this topic, be sure you are familiar with the following concepts:

		VIDEO	ANIMATION	INTERACTIVE
You Try It	1. Determine If a Given Value Is a Solution to an Equation (Topic 11.1, **Objective 2**)			
You Try It	2. Solve Linear Equations Using Both Properties of Equality (Topic 11.1, **Objective 5**)			
You Try It	3. Plot Ordered Pairs in the Rectangular Coordinate System (Topic 12.1, **Objective 3**)			

OBJECTIVES

1 Determine If an Ordered Pair Is a Solution to an Equation

2 Determine the Unknown Coordinate of an Ordered Pair Solution

3 Graph Linear Equations by Plotting Points

4 Find x- and y-Intercepts

5 Graph Linear Equations Using Intercepts

6 Use Linear Equations to Model Data

7 Graph Horizontal and Vertical Lines

OBJECTIVE 1 DETERMINE IF AN ORDERED PAIR IS A SOLUTION TO AN EQUATION

In **Topic 11.1** we learned that a **solution** to an **equation in one variable** is a value that, when substituted for the **variable**, makes the equation true. In this topic, we expand on this idea and consider equations that contain two variables. For example, we may wish to find solutions to the equation $x + y = 4$.

Solutions to equations in two variables require *two* values—one value for each variable. For the equation $x + y = 4$, $x = 1$ and $y = 3$ form a solution because a true statement results if we **substitute** the values for the variables.

Begin with the original equation: $x + y = 4$

Substitute 1 for x and 3 for y: $1 + 3 \stackrel{?}{=} 4$

Simplify: $4 = 4$ True

We can write this solution as the **ordered pair** $(1, 3)$, where the first number is the x-value and the second number is the y-value.

Solution to an Equation in Two Variables

A **solution to an equation in two variables** is an ordered pair of values that, when substituted for the variables, makes the equation true.

(eText Screens 12.2-1–12.2-30)

Example 1 Determining If an Ordered Pair Is a Solution to an Equation

Determine if each **ordered pair** is a **solution** to the equation $x + 2y = 8$.

a. $(-2, 5)$ **b.** $(2, 6)$ **c.** $\left(-11, \frac{3}{2}\right)$ **d.** $(0, 4)$

Solutions

a. Substitute -2 for x and 5 for y. Work through the simplification to determine if the ordered pair is a solution.

Begin with the original equation: $x + 2y = 8$

Substitute -2 for x and 5 for y: $(-2) + 2(5) \stackrel{?}{=} 8$

Simplify: $-2 + 10 \stackrel{?}{=} 8$

$8 = 8$ True

The final statement is true, so $(-2, 5)$ is a solution to the equation.

b. Begin with the original equation: $x + 2y = 8$

Substitute 2 for x and 6 for y: $(2) + 2(6) \stackrel{?}{=} 8$

Simplify: $2 + 12 \stackrel{?}{=} 8$

$14 = 8$ False

The final statement is not true, so $(2, 6)$ is not a solution to the equation.

My video summary

c–d. Substitute the values for the variables and work through the simplification to determine if each ordered pair is a solution. View the **answers**, or watch this **video** to see the details.

You Try It **Work through this You Try It problem.**

Work Exercises 1–3 in this textbook or in the MyMathLab Study Plan.

OBJECTIVE 2 DETERMINE THE UNKNOWN COORDINATE OF AN ORDERED PAIR SOLUTION

In Example 1, we determined if a given ordered pair was a **solution** to an equation in two variables. But how do we find such ordered pair solutions if we know only one **coordinate**? Substituting the given coordinate for the corresponding variable will result in an equation in one variable that we can solve as we did in **Topic 11.1** or **Topic 11.2**. The resulting values for both variables together form an **ordered pair solution** to the equation.

Example 2 Determining Unknown Coordinates

My interactive video summary

Find the unknown coordinate so that each ordered pair **satisfies** $2x - 3y = 15$.

a. $(6, ?)$ **b.** $(?, 7)$ **c.** $\left(-\frac{5}{2}, ?\right)$

Solutions

a. To find the y-coordinate when the x-coordinate is 6, we substitute 6 for x and solve for y.

Begin with the original equation: $2x - 3y = 15$

Substitute 6 for x: $2(6) - 3y = 15$

Simplify: $12 - 3y = 15$ ← Linear equation in one variable

Subtract 12 from both sides: $-3y = 3$

Divide both sides by -3: $y = -1$ ← Missing coordinate

We combine the two coordinates to form the solution. The ordered pair $(6, -1)$ is a solution to the equation $2x - 3y = 15$.

b. Begin with the original equation: $2x - 3y = 15$

Substitute 7 for y: $2x - 3(7) = 15$

Simplify: $2x - 21 = 15$ ← linear equation in one variable

Finish solving for x to find the x-coordinate when the y-coordinate is 7. Once completed, view the **answer** or watch this **interactive video** to check your work.

c. Substitute the x-coordinate into the equation and solve for y to determine the y-coordinate. Once finished, view the **answer** or watch this **interactive video** to check your work for all parts.

You Try It **Work through this You Try It problem.**

Work Exercises 4–6 in this textbook or in the MyMathLab Study Plan.

OBJECTIVE 3 GRAPH LINEAR EQUATIONS BY PLOTTING POINTS

From **Example 1**, we see that an equation in two variables can have more than one solution. The ordered pairs $(-2, 5)$ and $(0, 4)$ are two of the solutions to the equation $x + 2y = 8$. Building on **Example 2**, we can find additional solutions to this equation by selecting different values for x and determining the corresponding values for y. Or, we can select values for y and determine the corresponding values for x.

x	y
2	?
?	2
8	?
?	−1

Substitute given values →

$(2) + 2y = 8$
$x + 2(2) = 8$
$(8) + 2y = 8$
$x + 2(-1) = 8$

Solve for the remaining variable →

x	y
2	3
4	2
8	0
10	−1

Solutions →

$(2, 3)$
$(4, 2)$
$(8, 0)$
$(10, -1)$

The **graph of an equation in two variables** includes all points whose **coordinates** are solutions to the equation. So a graph is a visual display of the solution set for an equation. To make such a graph, we can **plot** several points that **satisfy** the equation. Then we connect the points with a line or smooth curve.

In Figure 4a, we plot each of the **ordered pair solutions** we have found for $x + 2y = 8$.

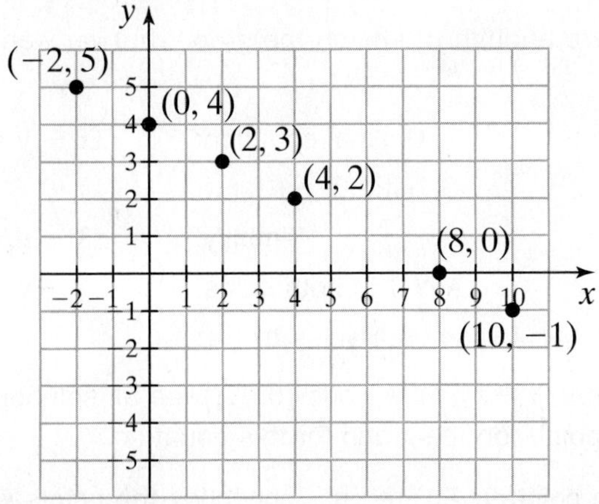

Figure 4a

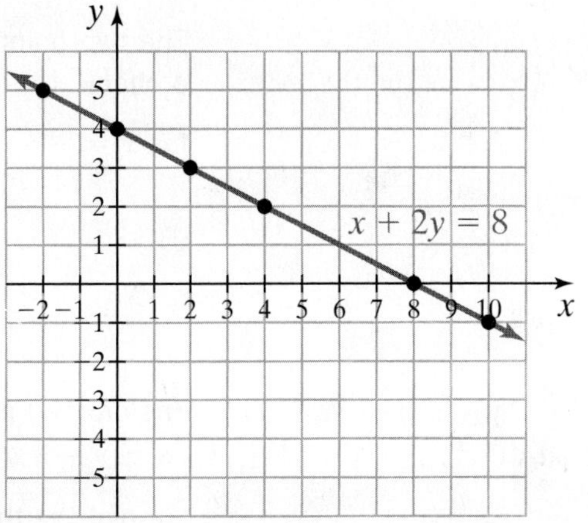

Figure 4b

There is a pattern to the plotted points. They all appear to lie on the same line. Connecting the points with a line as in Figure 4b completes the **graph** of the equation $x + 2y = 8$. For every point on this line, the coordinates of the point give an ordered pair solution to the equation.

The equations from **Example 1**, $x + 2y = 8$, and from **Example 2**, $2x - 3y = 15$, are both examples of **linear equations in two variables**. The graphs of these equations are always straight lines.

Definition Linear Equation in Two Variables (Standard Form)

A **linear equation in two variables** is an equation that can be written in the standard form $Ax + By = C$, where A, B, and C are real numbers, and A and B are not both equal to 0.

A line can be completely determined by two points. So, to graph a linear equation in two variables, we find two **ordered pair solutions**, plot the corresponding points, and connect the points with a line. In practice, we will generally find a third solution to serve as a **check**. If the corresponding third point also lies on the **graph**, we can feel comfortable with our solution.

Example 3 Graphing Linear Equations by Plotting Points

Graph $3x - y = 2$ by plotting points.

Solution First, we determine some **ordered pair solutions** to the equation. Select a value for one of the variables and then solve for the remaining variable.

Let $x = 1$:	Original equation:	$3x - y = 2$
	Substitute 1 for x:	$3(1) - y = 2$
	Simplify:	$3 - y = 2$
	Subtract 3 from both sides:	$-y = -1$
	Divide both sides by -1:	$y = 1$
Let $y = -2$:	Original equation:	$3x - y = 2$
	Substitute -2 for y:	$3x - (-2) = 2$
	Simplify:	$3x + 2 = 2$
	Subtract 2 from both sides:	$3x = 0$
	Divide both sides by 3:	$x = 0$

When $x = 1$, we get $y = 1$ and when $y = -2$ we get $x = 0$. So, the ordered pairs $(1, 1)$ and $(0, -2)$ are solutions to the equation and are points on the graph.

The two points are enough to sketch the graph, but we want to find a third point as a **check**.

Let $x = -1$:	Original equation:	$3x - y = 2$
	Substitute -1 for x:	$3(-1) - y = 2$
	Simplify:	$-3 - y = 2$
	Add 3 to both sides:	$-y = 5$
	Divide both sides by -1:	$y = -5$

The ordered pairs $(-1, -5)$, $(1, 1)$, and $(0, -2)$ are all solutions to the equation $3x - y = 2$ and points on the graph for this equation.

We plot the three points in Figure 5a. Connecting the points with a line, the complete graph of $3x - y = 2$ is given in Figure 5b.

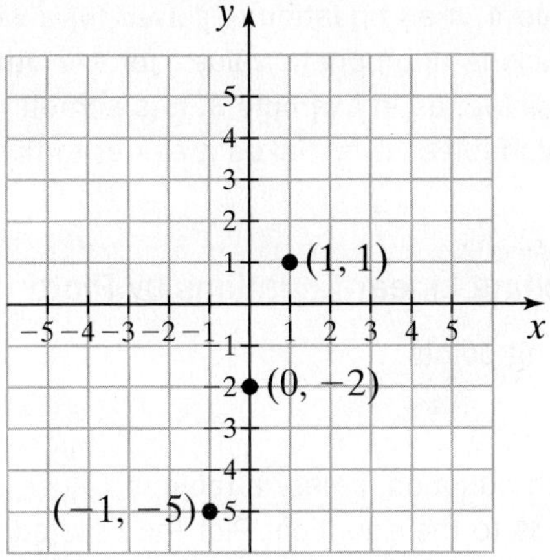

Figure 5a

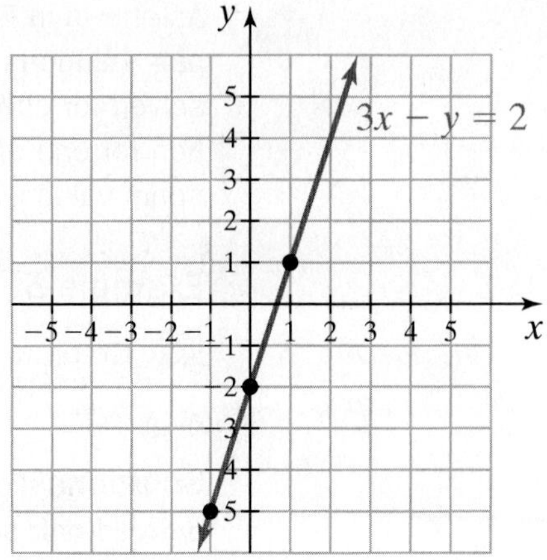

Figure 5b

You Try It Work through this You Try It problem.

Work Exercises 7 and 8 in this textbook or in the MyMathLab Study Plan.

Example 4 Graphing Linear Equations by Plotting Points

Graph $y = -\frac{1}{3}x + 2$ by plotting points.

Solution The equation is solved for y. We can create a table of values by selecting appropriate values for x and determining the corresponding values for y. Because the **coefficient** on x has a **denominator** of 3, we will select three distinct values for x that are **multiples** of 3. This will allow us to **clear the fraction** in the equation and avoid fractions as coordinates in our solutions.

Using $x = -3$, $x = 0$, and $x = 3$, the resulting **ordered pair solutions** are $(-3, 3)$, $(0, 2)$, and $(3, 1)$, respectively. View the **work**.

The points are plotted in **Figure 6a** and the complete graph is shown in **Figure 6b**.

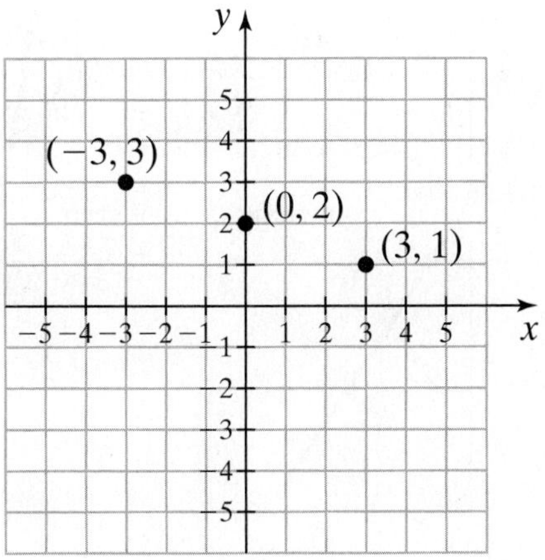

Figure 6a

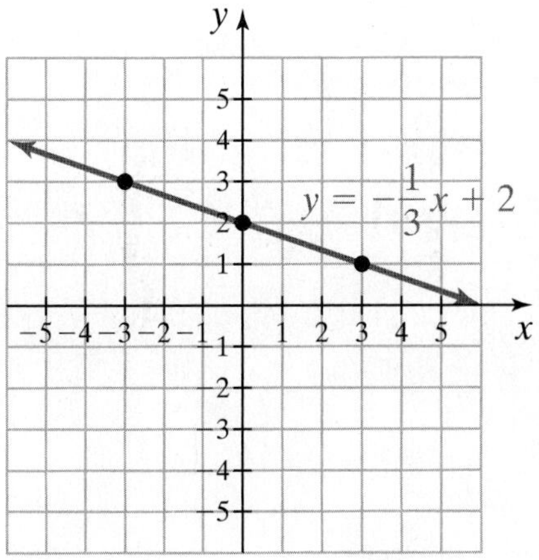

Figure 6b

You Try It Work through this You Try It problem.

Work Exercises 9 and 10 in this textbook or in the MyMathLab Study Plan.

(eText Screens 12.2-1–12.2-30)

As shown in Example 4, if an equation is **solved for a variable**, we can easily find ordered pair solutions by picking appropriate values for the other variable. If an equation is not solved for either variable, as in **Example 3**, it is sometimes helpful to first solve the equation for one of the variables. This makes it easier to find **integer** solutions and avoid fractional values.

Example 5 Graphing Linear Equations by Plotting Points

My interactive video summary

Graph by plotting points.

a. $y = 2x$ b. $3x + 2y = 5$

Solutions For each equation, create a **table of values** so you have at least three distinct **ordered pair solutions** to the equation. Plot the ordered pairs and connect the points with a straight line. Check your **answers**, or watch this **interactive video** for complete solutions.

You Try It **Work through this You Try It problem.**

Work Exercises 11–16 in this textbook or in the MyMathLab Study Plan.

OBJECTIVE 4 FIND x- AND y-INTERCEPTS

Look at **Figure 4b**. Notice that the graph of $x + 2y = 8$ crosses the y-axis at the point $(0, 4)$ and also crosses the x-axis at the point $(8, 0)$. The points where a graph crosses or touches the **axes** are called its **intercepts**. The graph of $x + 2y = 8$ has two intercepts: $(0, 4)$ and $(8, 0)$. If a graph never crosses an axis, then it will have no intercepts.

A **y-intercept** is the **y-coordinate** of a point where a graph crosses or touches the **y-axis**. Its corresponding ordered pair would be $(0, y)$. An **x-intercept** is the **x-coordinate** of a point where a graph crosses or touches the **x-axis**. Its corresponding ordered pair would be $(x, 0)$. So for the graph of $x + 2y = 8$, the y-intercept is 4 and the x-intercept is 8.

Figure 7 illustrates examples of x- and y-intercepts.

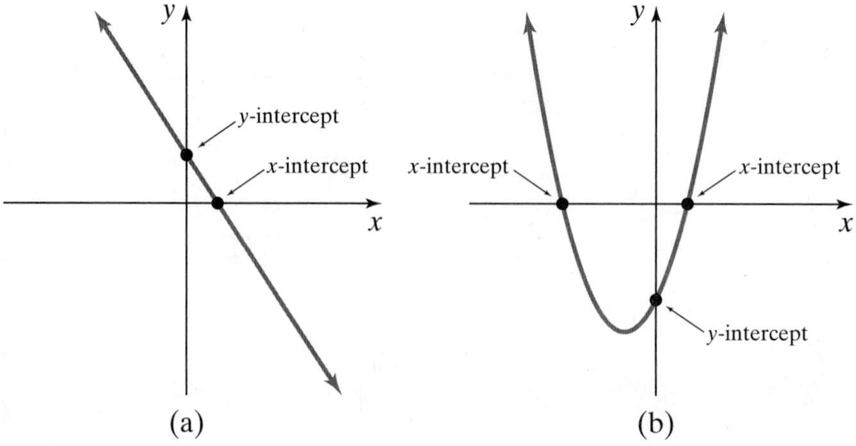

Figure 7 x- and y-Intercepts

Note: When the type of intercept is not specified, we must list the intercept as an ordered pair. If the type of intercept is specified, then we can list only the coordinate of the intercept.

(eText Screens 12.2-1–12.2-30)

Example 6 Finding x- and y-Intercepts

Find the **intercepts** of the graph shown in Figure 8. What are the **x-intercepts**? What are the **y-intercepts**?

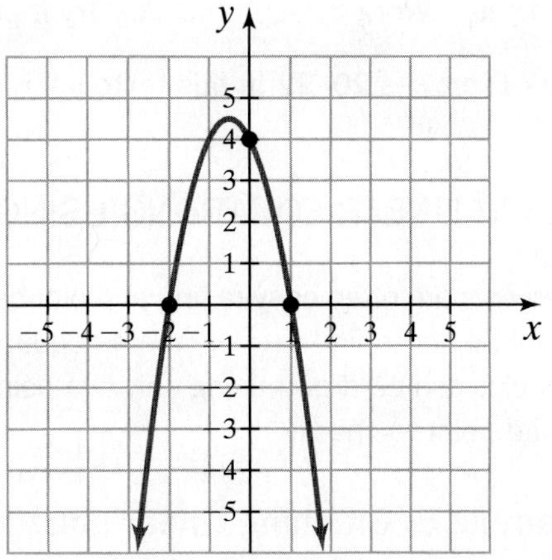

Figure 8

Solution The graph crosses the y-axis at $(0, 4)$ and crosses the x-axis at $(-2, 0)$ and $(1, 0)$. The x-intercepts are -2 and 1. The y-intercept is 4. The intercepts are $(0, 4)$, $(-2, 0)$, and $(1, 0)$.

You Try It **Work through this You Try It problem.**

Work Exercises 17–19 in this textbook or in the MyMathLab Study Plan.

To find **x-intercepts** of the graph of an equation, we let $y = 0$ and solve for x because every point on the x-axis has a y-coordinate of 0. To find **y-intercepts**, we let $x = 0$ and solve for y because every point on the y-axis has an x-coordinate of 0.

Finding x- and y-Intercepts of a Graph Given an Equation

- To find an x-intercept, let $y = 0$ and solve for x.
- To find a y-intercept, let $x = 0$ and solve for y.

Example 7 Finding x- and y-Intercepts

Find the x- and y-intercepts for the graph of each equation.

a. $2x + y = 4$ **b.** $4x = 3y + 8$

Solutions

a. To find the **x-intercept**, we let $y = 0$ and solve for x. To find the **y-intercept**, we let $x = 0$ and solve for y.

x-intercept:

Let $y = 0$: $2x + (0) = 4$

Simplify: $2x = 4$

Divide both sides by 2: $x = 2$

The x-intercept is 2.

y-intercept:

Let $x = 0$: $2(0) + y = 4$

Simplify: $y = 4$

The y-intercept is 4.

My video summary

b. Try finding the intercepts for this equation on your own. Check your **answers**, or watch this **video** for the complete solution.

You Try It **Work through this You Try It problem.**

Work Exercises 20–22 in this textbook or in the MyMathLab Study Plan.

OBJECTIVE 5 GRAPH LINEAR EQUATIONS USING INTERCEPTS

Intercepts are often easy to find and plot because one of the **coordinates** is 0. This makes them useful points to find when graphing equations. Because the graph of a **linear equation in two variables** can be drawn using only two points, we can use the x- and y-intercepts together with a third point to check.

Example 8 Graphing Linear Equations Using Intercepts

Graph $3x - 2y = 6$ using intercepts.

Solution To find the ***x*-intercept**, we let $y = 0$ and solve for x. To find the ***y*-intercept**, we let $x = 0$ and solve for y. As a check, we randomly let $x = 4$ and find the corresponding y-value.

***x*-intercept:** Let $y = 0$: $3x - 2(0) = 6$

Simplify: $3x = 6$

Divide both sides by 3: $x = 2$

The x-intercept is 2, so the corresponding point is $(2, 0)$.

***y*-intercept:** Let $x = 0$: $3(0) - 2y = 6$

Simplify: $-2y = 6$

Divide both sides by -2: $y = -3$

The ***y*-intercept** is -3, so the corresponding point is $(0, -3)$.

Check point: Let $x = 4$: $3(4) - 2y = 6$

Simplify: $12 - 2y = 6$

Subtract 12 from both sides: $-2y = -6$

Divide both sides by -2: $y = 3$

The corresponding check point is $(4, 3)$.

We plot the three points and connect the points with a straight line. The resulting graph is shown in Figure 9.

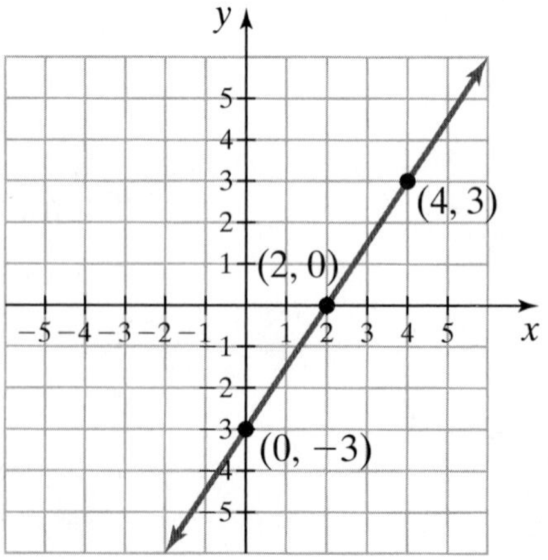

Figure 9

You Try It **Work through this You Try It problem.**

Work Exercises 23–30 in this textbook or in the MyMathLab Study Plan.

(*eText Screens 12.2-1–12.2-30*)

The graph of a **linear equation** always has at least one intercept, but it may have two. When the graph has only one **intercept**, we need to find at least one more point in order to graph the equation.

Example 9 Graphing Linear Equations Using Intercepts

My video summary

Graph $2x = 5y$ using intercepts.

Solution Find the **x-intercept** by letting $y = 0$ and solving for x. Find the **y-intercept** by letting $x = 0$ and solving for y.

Let $y = 0$:	$2x = 5(0)$	Let $x = 0$:	$2(0) = 5y$
Simplify:	$2x = 0$	Simplify:	$0 = 5y$
Divide by 2:	$x = 0$	Divide by 5:	$0 = y$
The x-intercept is 0.		The y-intercept is 0.	

The graph has only one intercept: $(0, 0)$. Select two more values for x and find the corresponding y-values to plot two additional points on the graph. Use the points to graph the equation. Check your **answer**, or watch this **video** for a detailed solution.

You Try It **Work through this You Try It problem.**

Work Exercises 31 and 32 in this textbook or in the MyMathLab Study Plan.

Compare the equations in **Examples 8** and 9. How can we tell when the graph of a linear equation in two variables will go through the origin? **Find out.**

OBJECTIVE 6 USE LINEAR EQUATIONS TO MODEL DATA

Linear equations can be used to **model** real-world applications. In these situations, the **coordinates** of the points on the graph have meaning in the context of the situation.

Example 10 Disappearing Drive-Ins

My video summary

The number of U.S. drive-in theaters can be modeled by the linear equation $y = -7.5x + 435$, where x is the number of years after 2000.
(*Source:* United Drive-In Theater Owners Association, 2009)

a. Sketch the graph of the equation for the year 2000 and beyond.

b. Find the missing coordinate for the ordered pair solution $(?, 390)$.

c. Interpret the point from part (b).

d. Find and interpret the y-intercept.

e. What does the x-intercept represent in this problem?

Solutions Try to work this problem on your own using earlier examples to guide you. View the **answers**, or watch this **video** for the complete solution.

You Try It **Work through this You Try It problem.**

Work Exercises 37 and 38 in this textbook or in the MyMathLab Study Plan.

(eText Screens 12.2-1–12.2-30)

OBJECTIVE 7 GRAPH HORIZONTAL AND VERTICAL LINES

The equations $x = a$ and $y = b$ are considered **linear equations in two variables** because they can be written in standard form. The equation $x = a$ can be written as $x + 0y = a$ and its graph is a **vertical line**. The equation $y = b$ can be written as $0x + y = b$ and its graph is a **horizontal line**.

Example 11 Graphing Horizontal Lines

Graph $y = 4$.

Solution The equation $y = 4$ can be written as $0x + y = 4$ to show that it is a linear equation in two variables. For any value **substituted** for x in the equation, y will always equal 4. Therefore, every point on the graph of the equation will have a y-coordinate of 4. If we choose x-values of $-2, 0$, and 3, the corresponding ordered pair solutions are $(-2, 4)$, $(0, 4)$, and $(3, 4)$. The resulting graph, shown in Figure 10, is a horizontal line with **y-intercept** 4 and no **x-intercept** because the graph does not cross the x-axis.

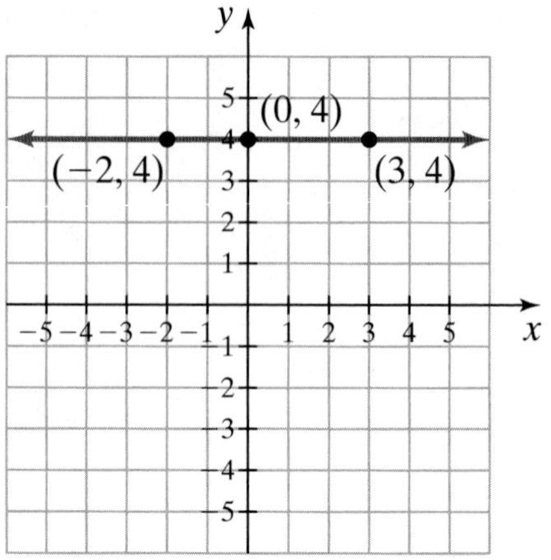

Figure 10

Example 12 Graphing Vertical Lines

My video summary

Graph $x = -3$.

Solution The equation $x = -3$ can be written as $x + 0y = -3$ to show that it is a **linear equation in two variables**. For any value **substituted** for y in the equation, x will always equal -3. Therefore, every point on the graph of the equation will have an **x-coordinate** of -3. Choose three values for y and find the corresponding points. Plot the points and connect them with a line. View the **answer**, or watch this **video** to see a detailed solution.

You Try It **Work through this You Try It problem.**

Work Exercises 33–36 in this textbook or in the MyMathLab Study Plan.

12.2 Exercises

In Exercises 1–3, determine if the given ordered pairs are solutions to the equations.

You Try It

1. $4x + 3y = 17$

a. $(3, 0)$

b. $(2, 3)$

c. $\left(-\frac{1}{4}, 6\right)$

2. $y = \frac{3}{4}x - 2$

a. $(-8, -8)$

b. $(4, -1)$

c. $\left(2, -\frac{1}{2}\right)$

3. $2x = 3y + 2$

a. $\left(\frac{1}{2}, 1\right)$

b. $(-2, -2)$

c. $\left(-\frac{2}{3}, 0\right)$

In Exercises 4–6, find the unknown coordinate so that each ordered pair satisfies the given equation.

4. $5x + 4y = 10$

You Try It

a. $(2, ?)$

b. $\left(?, -\frac{5}{4}\right)$

5. $-3x = 4y + 3$

a. $\left(?, -\frac{3}{4}\right)$

b. $(-2, ?)$

6. $y = -\frac{1}{3}x + 4$

a. $(6, ?)$

b. $(?, -5)$

In Exercises 7–16, graph each equation by plotting points.

You Try It

7. $2x + y = 6$

8. $4x - y = 2$

9. $y = \frac{3}{4}x - 4$

10. $y = -\frac{2}{3}x + 3$

You Try It

11. $y = 2x$

12. $y = -\frac{3}{2}x$

You Try It

13. $3x - 2y = 5$

14. $2x + 5y = 3$

15. $1.2x - 0.5y = 2.4$

16. $1.6x + 1.2y = 2.4$

In Exercises 17–19, find the x- and y-intercepts of each graph.

You Try It

17.

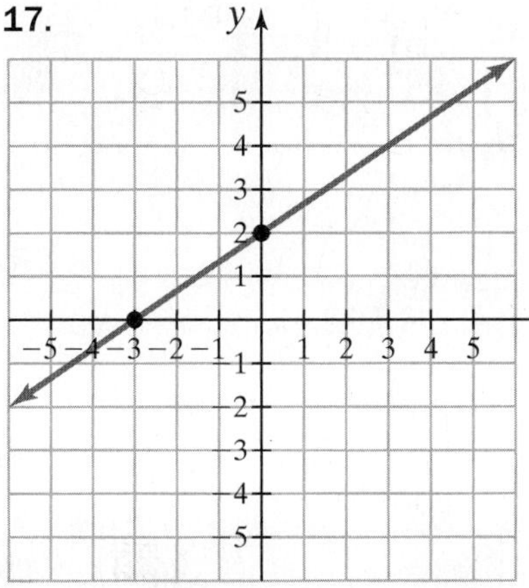

18.

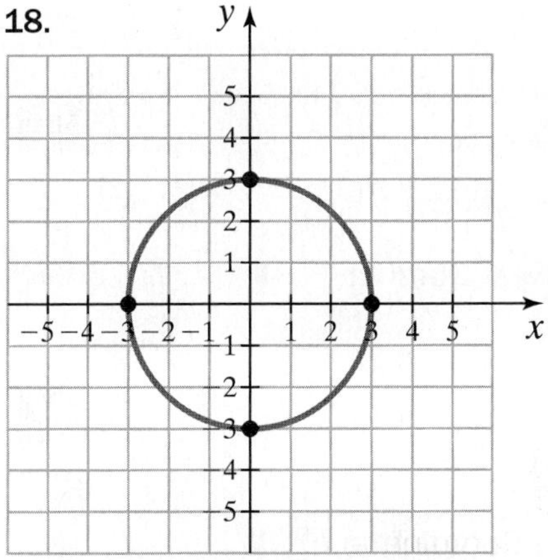

19.

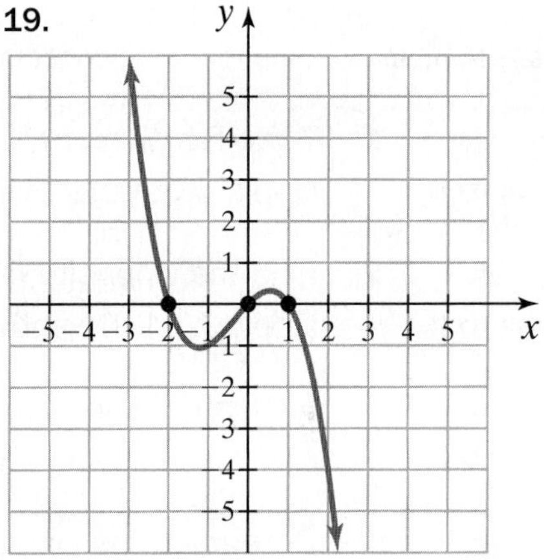

In Exercises 20–22, find the x- and y-intercepts of the graph of the given equation.

20. $3x + y = -2$

21. $y = 4x - 2$

22. $y = 0.4x + 1.4$

You Try It

In Exercises 23–32, graph the linear equation using intercepts.

23. $4x + y = 8$

24. $x - 2y = -6$

25. $-9x + 3y = 9$

You Try It

26. $y = x - 3$

27. $y = -x - \frac{5}{2}$

28. $y = 3.5x + 1.4$

29. $x + 3y = \frac{3}{4}$

30. $y = -\frac{3}{5}x - 2$

31. $y = -2x$

32. $3y = 4x$

You Try It

In Exercises 33–36, graph the given line.

33. $y = 3$

34. $x = -2$

35. $x - 2.5 = 0$

36. $y + 2 = 0$

You Try It

(eText Screens 12.3-1–12.3-28)

You Try It

37. **Car Loan** Natalie took out a loan for a 2011 Ford Fusion with 0% financing and a monthly payment of \$335. The balance remaining on the loan, y, is given by the linear equation $y = -335x + 24{,}120$, where x is the number of months that Natalie has made payments.
   a. Sketch the graph of the equation.
   b. Find the missing coordinate for the ordered pair solution $(25, ?)$.
   c. Interpret the point from part (b).
   d. Find and interpret the y-intercept.
   e. What does the x-intercept represent in this problem?

38. **Video Stores** The number of Blockbuster video stores in the U.S. can be approximated by the linear equation $y = -318x + 4935$, where x is the number of years after 2005. (*Source:* www.blockbuster.com)
   a. Sketch the graph of the equation.
   b. Find the missing coordinate for the ordered pair solution $(8, ?)$.
   c. Interpret the point from part (b).
   d. Find and interpret the y-intercept.
   e. What does the x-intercept represent in this problem?

12.3 Slope

THINGS TO KNOW

Before working through this topic, be sure you are familiar with the following concepts:

	VIDEO	ANIMATION	INTERACTIVE
You Try It 1. Write Fractions in Simplest Form (Topic 4.2, **Objective 6**)	▤		
You Try It 2. Find the Opposite of a Real Number (Topic 10.1, **Objective 3**)			
You Try It 3. Evaluate a Formula (Topic 11.4, **Objective 1**)	▤		
You Try It 4. Determine the Unknown Coordinate of an Ordered Pair Solution (Topic 12.2, **Objective 2**)			▤
You Try It 5. Graph Linear Equations by Plotting Points (Topic 12.2, **Objective 3**)			▤
You Try It 6. Graph Horizontal and Vertical Lines (Topic 12.2, **Objective 7**)	▤		

OBJECTIVES

1 Find the Slope of a Line Given Two Points
2 Find the Slopes of Horizontal and Vertical Lines
3 Graph a Line Using the Slope and a Point
4 Find and Use the Slopes of Parallel and Perpendicular Lines
5 Use Slope in Applications

OBJECTIVE 1 FIND THE SLOPE OF A LINE GIVEN TWO POINTS

In **Topic 12.2**, we graphed **linear equations in two variables** by **plotting points**. Although the graph of every linear equation is a straight line, there can be many differences between the graphs. A key feature of a line is its **slant** or **steepness**. For example, looking from left to right at the graphs in Figure 11, the lines in (a) and (b) slant upward (or rise), whereas the lines in (c) and (d) slant downward (or fall). Also, the line in (a) has a steeper upward slant than the line in (b), and the line in (c) has a steeper downward slant than the line in (d).

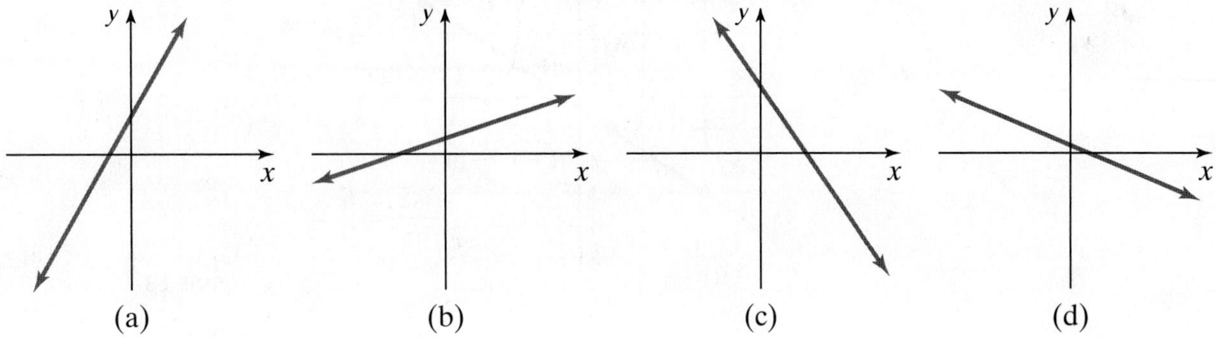

Figure 11

We measure the slant or steepness of a line using *slope*.

> **Definition Slope**
>
> The **slope** of a line is the **ratio** of the vertical change in y, or **rise**, to the horizontal change in x, or **run**.
>
> $$\text{slope} = \frac{\text{vertical change}}{\text{horizontal change}} = \frac{\text{change in } y}{\text{change in } x} = \frac{\text{rise}}{\text{run}}$$

The line graphed in Figure 12 goes through the points $(2, 1)$ and $(5, 3)$. The **rise** of the line is the **difference** between the two y-coordinates: $3 - 1 = 2$, and the **run** of the line is the difference between the two corresponding x-coordinates: $5 - 2 = 3$. The **slope** of the line is the ratio of rise to run:

$$\text{slope} = \frac{\text{rise}}{\text{run}} = \frac{2}{3}.$$

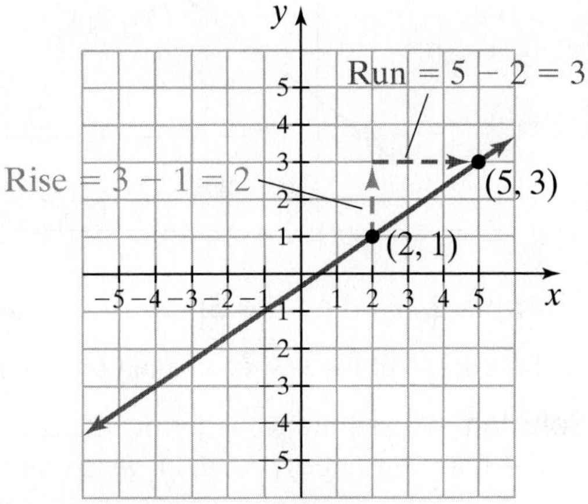

Figure 12

A slope of $\frac{2}{3}$ means that for every horizontal change of 3 units, there is a corresponding vertical change of 2 units. Because the slope is the same everywhere on a line, it does not matter which two points on the line are used to find its slope, as shown in Figure 13.

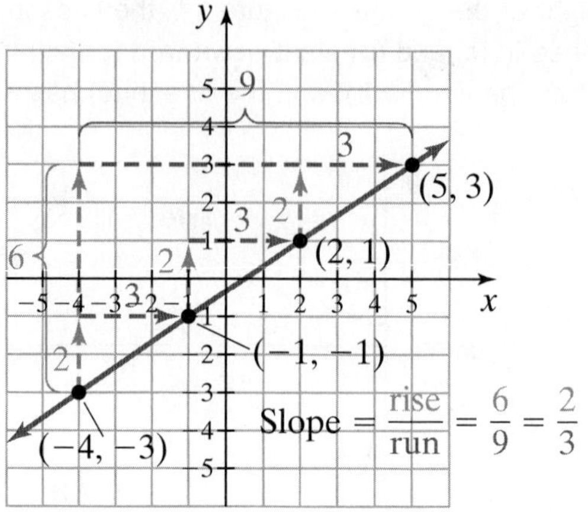

Figure 13

Slope is traditionally identified with the letter m. (Read **why**.) We can write a **formula** for finding slope using **subscript notation** to identify the two general points on the line, (x_1, y_1) and (x_2, y_2). We read x_1 as "x sub one" and y_2 as "y sub two."

Slope Formula

Given two points, (x_1, y_1) and (x_2, y_2), on the graph of a line, the **slope** m of the line containing the two points is given by the formula

$$m = \frac{\text{change in } y}{\text{change in } x} = \frac{\text{rise}}{\text{run}} = \frac{y_2 - y_1}{x_2 - x_1},$$

where $x_1 \neq x_2$.

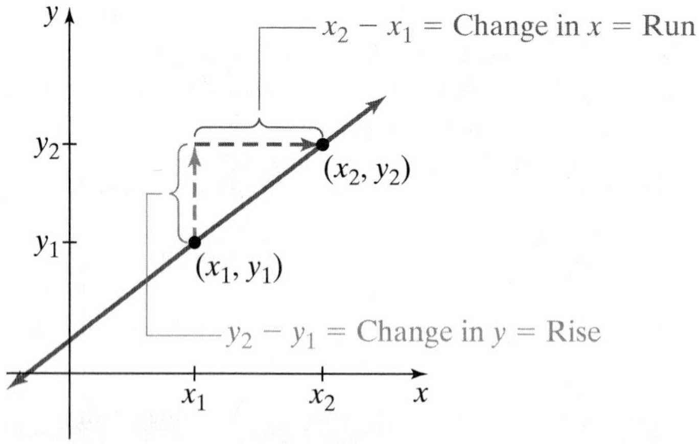

Example 1 Finding the Slope of a Line Given Two Points

Find the **slope** of the line containing the points $(-2, -4)$ and $(0, 1)$.

Solution We use the **slope formula**. Let $(x_1, y_1) = (-2, -4)$, which means $x_1 = -2$ and $y_1 = -4$, and let $(x_2, y_2) = (0, 1)$, which means $x_2 = 0$ and $y_2 = 1$.

$$\text{Write the slope formula:} \quad m = \frac{y_2 - y_1}{x_2 - x_1}$$

$$\text{Substitute } x_1 = -2, y_1 = -4, x_2 = 0, \text{ and } y_2 = 1: \quad = \frac{1 - (-4)}{0 - (-2)}$$

$$\text{Simplify:} \quad = \frac{1 + 4}{0 + 2}$$

$$= \frac{5}{2}$$

The slope of the line is $\frac{5}{2}$. This means that for every horizontal change, or **run**, of 2 units, there is a corresponding vertical change, or **rise**, of 5 units. The slope of this line is illustrated in Figure 14.

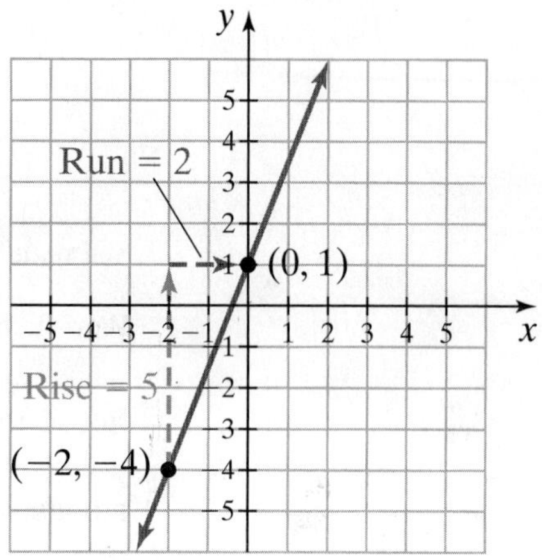

Figure 14

You Try It **Work through this You Try It problem.**

Work Exercises 1 and 2 in this textbook or in the MyMathLab Study Plan.

CAUTION

It does not matter which point is called (x_1, y_1) and which point is called (x_2, y_2). Just be consistent. Whichever y-coordinate goes first in the **numerator** of the **slope formula**, the corresponding x-coordinate must go first in the **denominator**. This **alternate solution** shows that letting $(x_1, y_1) = (0, 1)$ and $(x_2, y_2) = (-2, -4)$ in **Example 1** gives the same result $m = \frac{5}{2}$.

Example 2 Finding the Slope of a Line Given Two Points

Find the **slope** of the line containing the points $(-2, 4)$ and $(1, -3)$.

Solution We use the slope formula. Let $(x_1, y_1) = (-2, 4)$ and $(x_2, y_2) = (1, -3)$.

$$\text{Write the slope formula:} \quad m = \frac{y_2 - y_1}{x_2 - x_1}$$

$$\text{Substitute } x_1 = -2, y_1 = 4, x_2 = 1, \text{ and } y_2 = -3: \quad = \frac{-3 - 4}{1 - (-2)}$$

$$\text{Simplify:} \quad = \frac{-3 + (-4)}{1 + 2}$$

$$= \frac{-7}{3} = -\frac{7}{3}$$

The slope of the line is $-\frac{7}{3}$. For every horizontal change of 3 units, there is a corresponding vertical change of -7 units. This slope is illustrated in Figure 15.

(eText Screens 12.3-1–12.3-28)

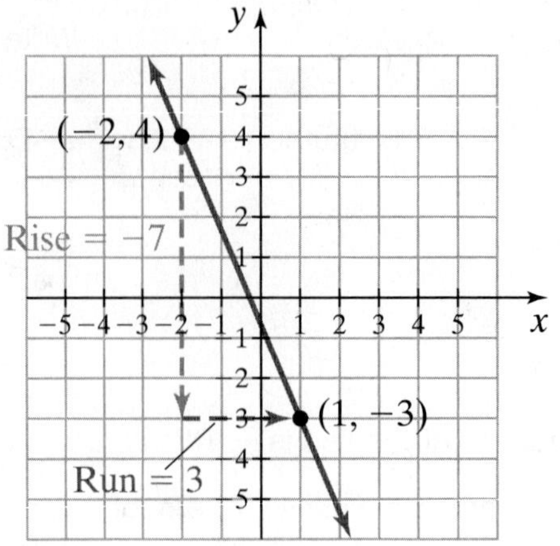

Figure 15

You Try It **Work through this You Try It problem.**

Work Exercises 3 and 4 in this textbook or in the MyMathLab Study Plan.

Notice that the line in **Figure 14** slants upward, or rises, from left to right and has a positive **slope**, whereas the line in **Figure 15** slants downward, or falls, from left to right and has a negative slope.

> **Positive versus Negative Slope**
>
> A line that slants upward, or rises, from left to right has a positive slope.
> A line that slants downward, or falls, from left to right has a negative slope.

Example 3 Finding the Slope of a Line Given Two Points

My video summary

Find the slope of the line containing the given points. Simplify if possible.

a. $(-6, -1)$ and $(4, 5)$ **b.** $(1, 5)$ and $(3, -1)$

Solutions Use the **slope formula**.

a. Let $(x_1, y_1) = (-6, -1)$ and $(x_2, y_2) = (4, 5)$.

Write the slope formula: $m = \dfrac{y_2 - y_1}{x_2 - x_1}$

Substitute $x_1 = -6, y_1 = -1, x_2 = 4$, and $y_2 = 5$: $= \dfrac{5 - (-1)}{4 - (-6)}$

Simplify: $= \dfrac{5 + 1}{4 + 6} = \dfrac{6}{10} = \dfrac{3}{5}$

See this **visual representation** of this slope.

b. Try to find this slope on your own, then check your **answer**.
Watch this **video** for detailed solutions to both parts of this example.

You Try It **Work through this You Try It problem.**

Work Exercises 5–8 in this textbook or in the MyMathLab Study Plan.

(eText Screens 12.3-1–12.3-28)

OBJECTIVE 2 FIND THE SLOPES OF HORIZONTAL AND VERTICAL LINES

Let's look at the **slopes** of lines that do not slant upward or downward from left to right.

Example 4 Finding the Slope of a Line Given Two Points

My video summary

Find the slope of the line containing the given points. Simplify if possible.

a. $(-3, 2)$ and $(1, 2)$ **b.** $(4, 2)$ and $(4, -5)$

Solutions Use the **slope formula.**

a. Let $(x_1, y_1) = (-3, 2)$ and $(x_2, y_2) = (1, 2)$.

Write the slope formula: $m = \dfrac{y_2 - y_1}{x_2 - x_1}$

Substitute $x_1 = -3, y_1 = 2, x_2 = 1$, and $y_2 = 2$: $= \dfrac{2 - 2}{1 - (-3)}$

Simplify: $= \dfrac{0}{4}$

$= 0$

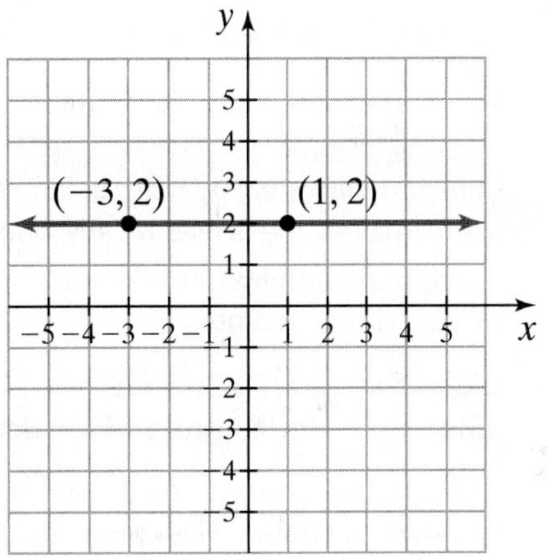

Figure 16 Slope 0

The slope of the line is 0. Figure 16 shows that the line through these points is horizontal. It does not slant upward or downward.

b. Let $(x_1, y_1) = (4, 2)$ and $(x_2, y_2) = (4, -5)$.

Write the slope formula: $m = \dfrac{y_2 - y_1}{x_2 - x_1}$

Substitute $x_1 = 4, y_1 = 2, x_2 = 4$, and $y_2 = -5$: $= \dfrac{-5 - 2}{4 - 4}$

Simplify: $= \dfrac{-7}{0}$

Division by 0 is undefined, so this line has an *undefined slope*. Figure 17 shows that the graph of the line through these points is vertical. It does not slant upward nor downward.

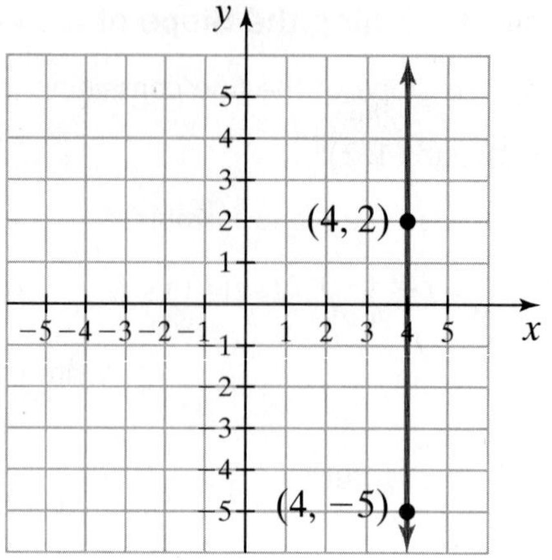

Figure 17
Undefined slope

Watch this **video** for detailed solutions to both parts of this example.

You Try It **Work through this You Try It problem.**

Work Exercises 9 and 10 in this textbook or in the MyMathLab Study Plan.

Horizontal lines have no vertical change. This makes the **numerator** in the slope formula 0, which makes the slope of any horizontal line 0. Similarly, **vertical lines** have no horizontal change. This makes the **denominator** of the slope formula 0, which makes the slope of any vertical line undefined.

Slopes of Horizontal and Vertical Lines

All horizontal lines (which have equations of the form $y = b$) have **slope 0**.
All vertical lines (which have equations of the form $x = a$) have **undefined slope**.

CAUTION Avoid using the term "no slope." Technically, "no slope" means undefined slope (vertical line), but it can easily be confused with zero slope (horizontal line). Therefore, it is better to clearly state "zero slope" or "undefined slope" and avoid "no slope."

Figure 18 summarizes the relationship between the slope and the graph of a linear equation.

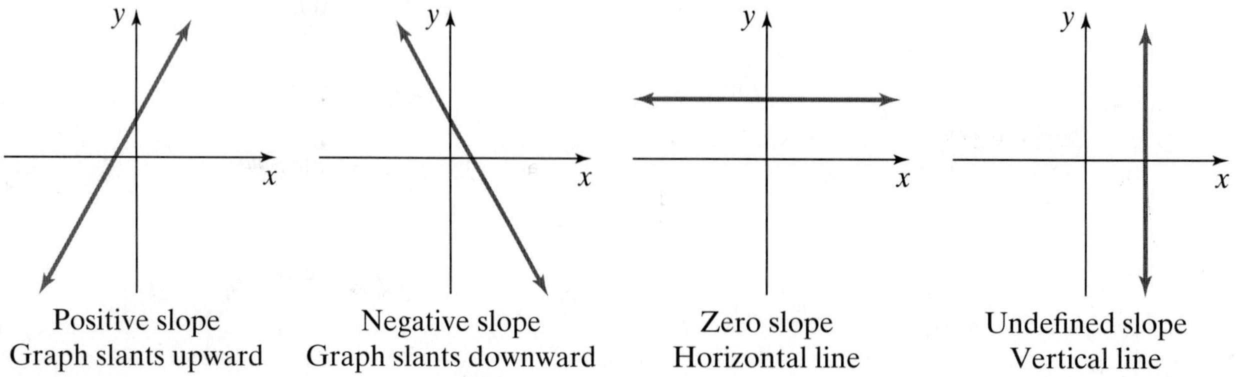

Figure 18 Types of lines

OBJECTIVE 3 GRAPH A LINE USING THE SLOPE AND A POINT

In **Topic 12.2**, we found that only two points are needed to sketch the graph of a line. Given one point on a line and the **slope** of the line, we can also sketch the graph. We **plot** the given point and then use the **rise** and the **run** to find a second point.

Example 5 Graphing a Line Using the Slope and a Point

Graph the line that has slope $m = \frac{3}{2}$ and passes through the point $(1, -2)$.

Solution Note that the slope is positive, so the line will slant upward from left to right. The slope is $m = \frac{3}{2}$, so rise $= 3$ and run $= 2$. Plot the given point $(1, -2)$. From this point, move up (rise) 3 units and move right (run) 2 units. This brings us to a second point on the line, $(3, 1)$. Draw the line between $(1, -2)$ and $(3, 1)$, as shown in Figure 19.

Note: If desired, additional points on the line can be plotted using the slope. Launch this **popup** box to see.

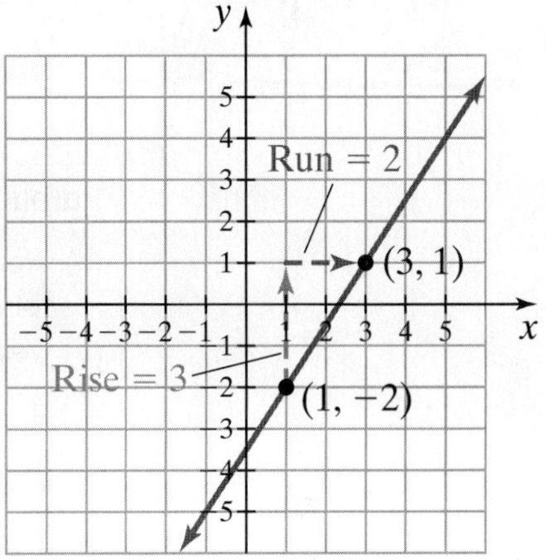

Figure 19

You Try It **Work through this You Try It problem.**

Work Exercises 11 and 12 in this textbook or in the MyMathLab Study Plan.

Example 6 Graphing a Line Using the Slope and a Point

My video summary

Graph the line that has slope $m = -3$ and passes through the point $(2, -1)$.

Solution The slope is negative, so the line will slant downward from left to right. The slope is $m = -3 = \frac{-3}{1} = \frac{3}{-1}$. We can use either rise $= -3$ and run $= 1$, or rise $= 3$ and run $= -1$. Try to graph this line on your own. View the **answer**, or watch this **video** for a detailed solution.

You Try It **Work through this You Try It problem.**

Work Exercises 13–16 in this textbook or in the MyMathLab Study Plan.

OBJECTIVE 4 FIND AND USE THE SLOPES OF PARALLEL AND PERPENDICULAR LINES

Parallel lines are two lines in the **coordinate plane** that never touch (have no points in common). The distance between two parallel lines remains the same across the plane. In Figure 20, lines l_1 and l_2 are parallel. Notice that both lines have the same **slope** $m_1 = m_2 = \frac{2}{3}$. If the slope of either line changed, then the two lines would no longer be parallel.

Notice also that l_1 and l_2 have different y-intercepts. If two lines have equal slopes and different y-intercepts, then the lines are parallel.

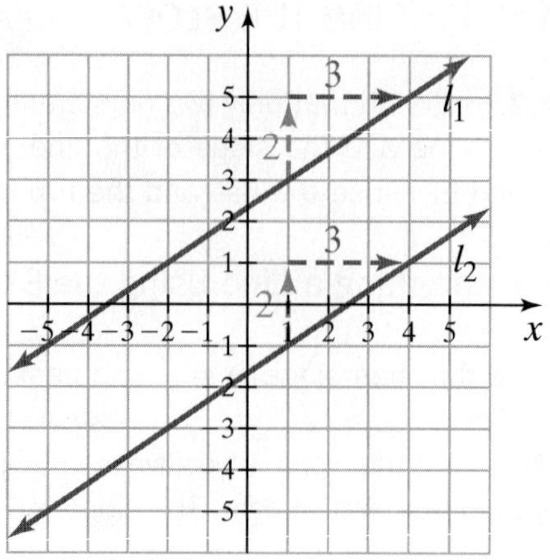

Figure 20 Parallel lines

Parallel Lines

(a) Non-vertical lines are parallel **if and only if** they have equal slopes and different y-intercepts.

(b) Vertical lines are parallel if they have different **x-intercepts.**

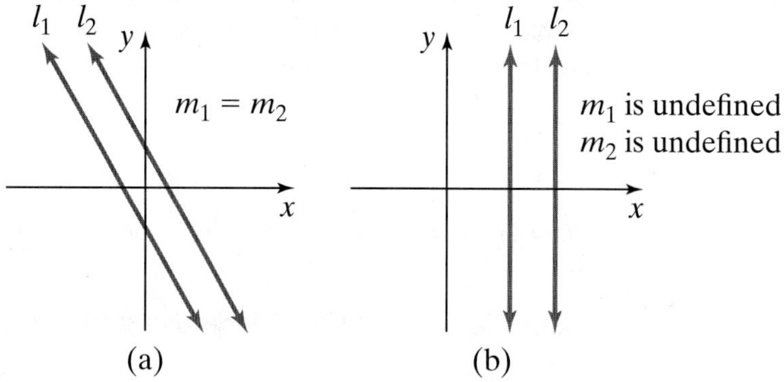

Two lines that **intersect** at a 90° (or right) angle are called **perpendicular lines.** In Figure 21, lines l_1 and l_2 are perpendicular. Notice that l_1 has **slope** $m_1 = 2$, whereas l_2 has slope $m_2 = -\frac{1}{2}$. If the slope of either line were changed, then the two lines would no longer be perpendicular. Notice also that the **product** of the slopes is $m_1 \cdot m_2 = 2\left(-\frac{1}{2}\right) = -1$. This is true for the slopes of perpendicular lines. We say that the slopes of perpendicular lines are **opposite reciprocals**, which means that the slopes have opposite signs and their **absolute values** are **reciprocals.**

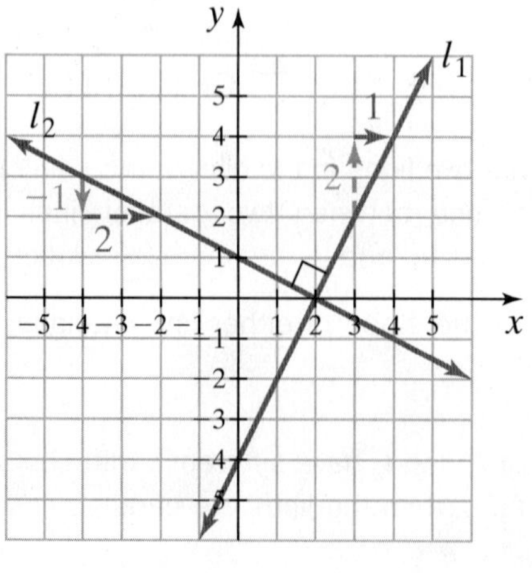

Figure 21 Perpendicular lines

Perpendicular Lines

(a) Two non-vertical lines are perpendicular if and only if the product of their slopes is -1. Their slopes are opposite reciprocals.

(b) A vertical line and a horizontal line are perpendicular to each other.

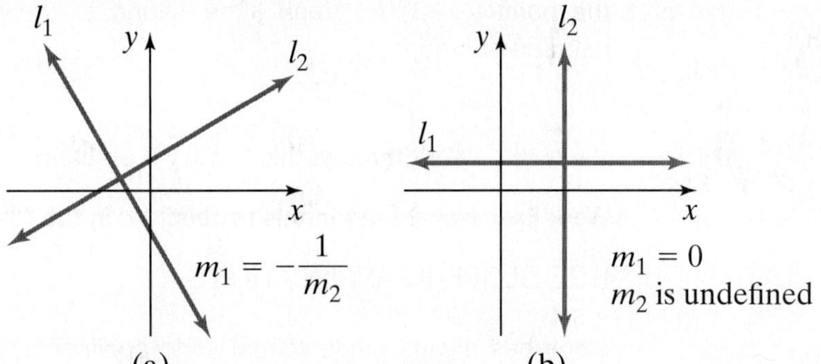

(a) (b)

Example 7 Finding Slopes of Parallel and Perpendicular Lines

Line l_1 has a slope of $m_1 = -\frac{4}{5}$.

a. If line l_2 is **parallel** to line l_1, what is its slope?

b. If line l_3 is **perpendicular** to line l_1, what is its slope?

Solutions

a. Parallel lines have equal slopes, so the slope of line l_2 is equal to the slope of line l_1. Line l_2 has a slope of $m_2 = -\frac{4}{5}$.

b. Perpendicular lines have slopes that are **opposite reciprocals**, so the slope of line l_3 is the opposite reciprocal of $m_1 = -\frac{4}{5}$. Changing the sign and finding the reciprocal, we get $\frac{5}{4}$. Line l_3 has a slope of $m_3 = \frac{5}{4}$. Note that $m_1 \cdot m_3 = \left(-\frac{4}{5}\right)\left(\frac{5}{4}\right) = -1$.

You Try It **Work through this You Try It problem.**

Work Exercises 17–20 in this textbook or in the MyMathLab Study Plan.

Example 8 Graphing Parallel and Perpendicular Lines

My video summary

Figure 22 shows the graph of a line l_1.

a. Graph a line l_2 that is **parallel** to l_1 and passes through the point $(3, -2)$.

b. Graph a line l_3 that is **perpendicular** to l_1 and passes through the point $(3, -2)$.

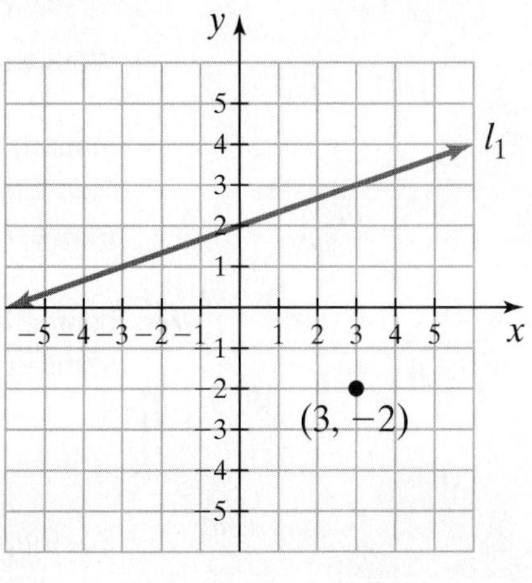

Figure 22

(eText Screens 12.3-1–12.3-28)

Solutions From looking at the graph, we can see that the slope of l_1 is $m_1 = \frac{1}{3}$. Because l_2 is parallel to l_1, its slope will be the same, $m_2 = \frac{1}{3}$. Because l_3 is perpendicular to l_1, its slope will be $m_3 = -3$, the **opposite reciprocal** of $m_1 = \frac{1}{3}$. Use these slopes with the point $(3,-2)$ to graph lines l_2 and l_3. View the **answers**, or watch this **video** for detailed solutions.

You Try It **Work through this You Try It problem.**

Work Exercises 21–24 in this textbook or in the MyMathLab Study Plan.

OBJECTIVE 5 USE SLOPE IN APPLICATIONS

Slope has a wide range of real-world applications. For example, in road construction, the slope of a road in percent form is called its **grade**. A road with a 6% grade means that the slope is $\frac{6}{100}$. For every 100 feet of horizontal change, the road rises 6 feet vertically. The slope of a roof is called its **pitch**. The pitch of the following roof is $\frac{4}{9}$.

$\frac{6}{100} = 6\%$ grade

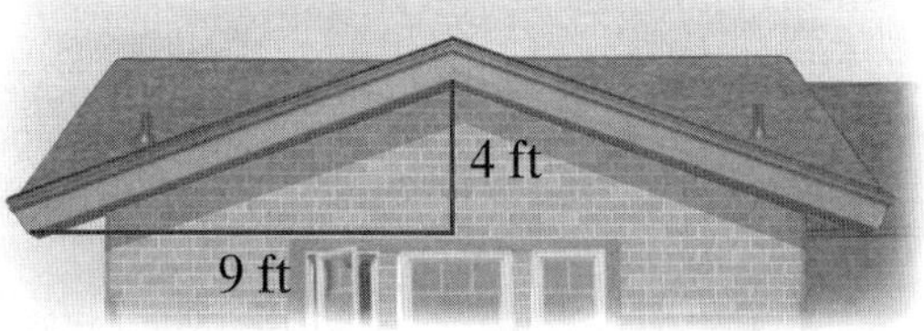

Example 9 Finding the Grade of a Wheelchair Ramp

A standard wheelchair ramp should rise no more than 1 foot vertically for every 12 feet horizontally. Find the grade of this ramp. Round to the nearest tenth of a percent. (*Source:* Americans with Disabilities Act Accessibility Guidelines (ADAAG))

Solution The grade of the ramp is its slope expressed as a percent.

$$\text{grade} = m = \frac{\text{rise}}{\text{run}} = \frac{1}{12} \approx 0.083 \text{ or } 8.3\%$$

The grade of a wheelchair ramp should be about 8.3%.

You Try It **Work through this You Try It problem.**

Work Exercises 25 and 26 in this textbook or in the MyMathLab Study Plan.

In many applications, the **slope** of a line is called the **average rate of change** because we can measure the change in one variable with respect to another variable. For example, suppose that a car's gas tank is full with $x_1 = 16$ gallons of gasoline when the odometer reads $y_1 = 35{,}652$ miles. Later, when the tank is one-fourth full with $x_2 = 4$ gallons of gasoline, the odometer reads $y_2 = 35{,}976$ miles. We can use slope to measure the average rate of change in miles driven with respect to gallons of gasoline used:

$$m = \frac{y_2 - y_1}{x_2 - x_1} = \frac{\text{change in miles}}{\text{change in gasoline}} = \frac{35{,}976 - 35{,}652}{16 - 4} = \frac{324}{12} = 27$$

So, the car traveled 27 miles per 1 gallon of gasoline used, on average.

(eText Screens 12.3-1–12.3-28)

Example 10 Using Slope as the Average Rate of Change

My video summary

The average tuition and fees for U.S. public two-year colleges were $\$2130$ in 1999. The average tuition and fees were $\$2540$ in 2009. Find and interpret the slope of the line connecting the points $(1999, 2130)$ and $(2009, 2540)$. (*Source:* College Board, *Trends in College Pricing 2009*)

Solution Let $(x_1, y_1) = (1999, 2130)$ and $(x_2, y_2) = (2009, 2540)$. Use the **slope formula** to find $m = 41$. Between 1999 and 2009, the average tuition and fees for two-year colleges increased by $\$41$ per year. Watch this **video** for a detailed solution.

You Try It **Work through this You Try It problem.**

Work Exercises 27–30 in this textbook or in the MyMathLab Study Plan.

12.3 Exercises

In Exercises 1–10, find the slope of the line containing the given points. Simplify if possible.

You Try It

1. $(-1, 4)$ and $(3, 7)$

2. $(1, -2)$ and $(3, 2)$

3. $(-3, 5)$ and $(2, -1)$

4. $(-1, 7)$ and $(5, 3)$

You Try It

5. $(5, 0)$ and $(-1, -3)$

6. $(2, 13)$ and $(8, 5)$

You Try It

7.

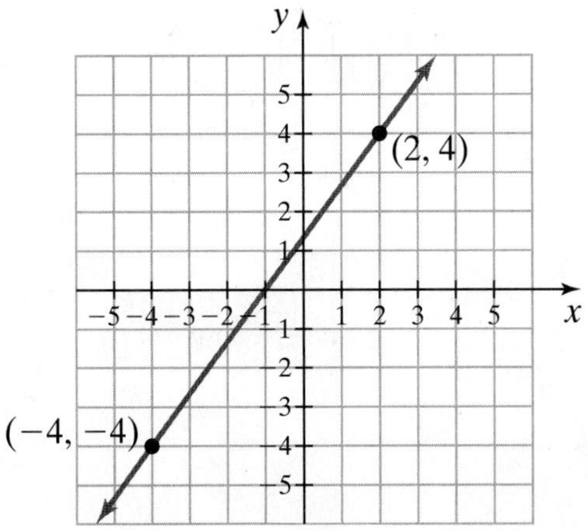

8.

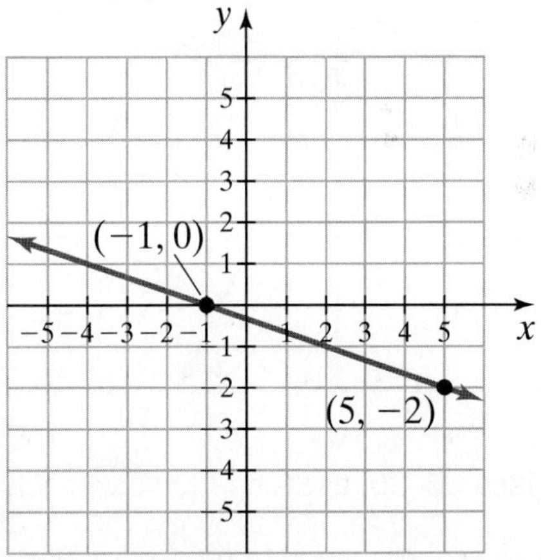

You Try It

9. $(4, -2)$ and $(-1, -2)$

10. $(5, -3)$ and $(5, 4)$

In Exercises 11–16, graph the line given the slope and a point on the line.

You Try It

11. $m = \frac{1}{3}; (-2, -2)$

12. $m = 2; (0, 4)$

13. $m = -\frac{5}{2}; (3, 1)$

You Try It

14. $m = -\frac{3}{4}; (0, 0)$

15. $m = 0; (4, -3)$

16. Undefined slope; $(-2, 5)$

In Exercises 17–20, for the given slope of line l_1, **(a)** find the slope of a line l_2 parallel to l_1, and **(b)** find the slope of a line l_3 perpendicular to l_1.

You Try It

17. $m = \frac{2}{3}$ **18.** $m = -\frac{7}{4}$ **19.** $m = 6$ **20.** $m = 0$

In Exercises 21 and 22, graph a line l_2 that is parallel to the given line l_1 and passes through the given point.

You Try It

21.

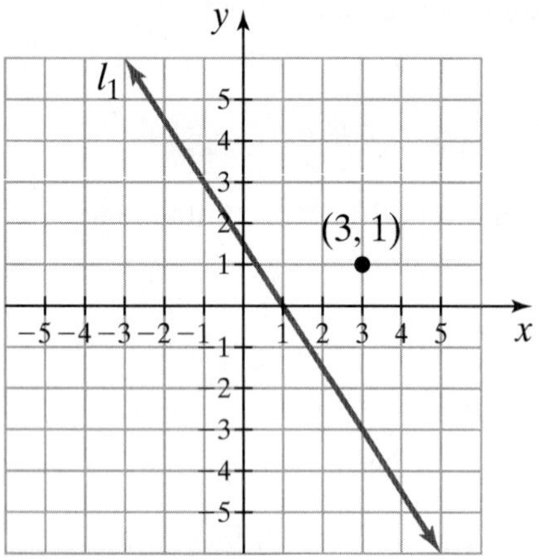

22.

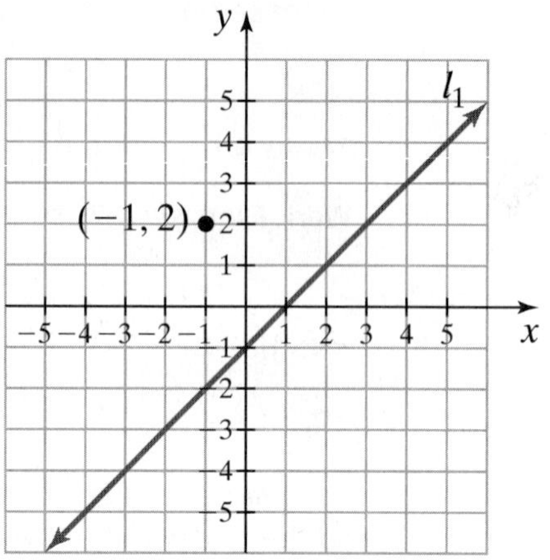

In Exercises 23 and 24, graph a line l_3 that is perpendicular to the given line l_1 and passes through the given point.

23.

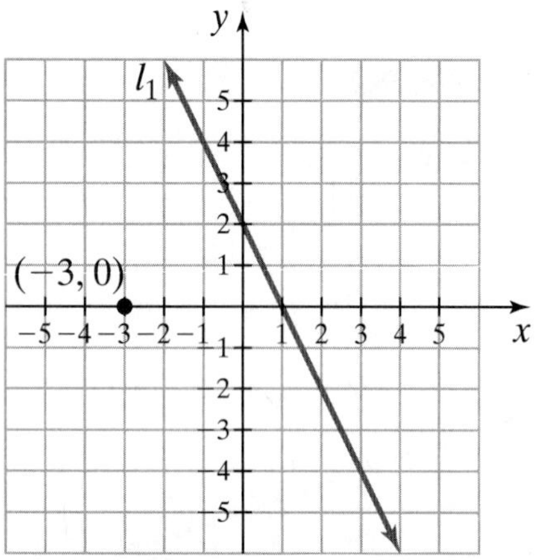

24.

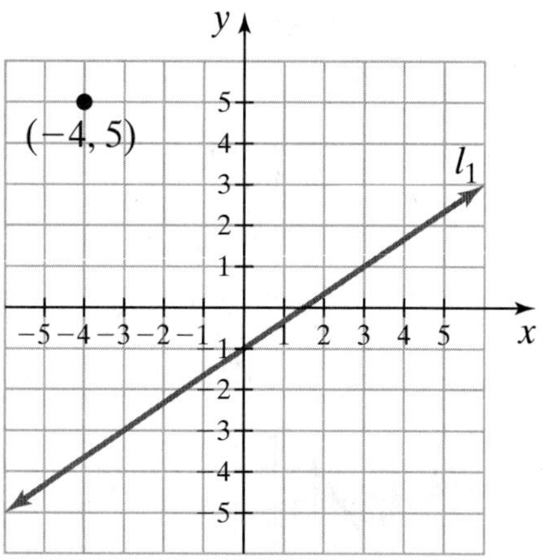

In Exercises 25–30, use slope to solve each application.

You Try It

25. Grade of a Road Find the grade of the road shown, written as a percent.

(eText Screens 12.4-1–12.4-29)

26. Pitch of a Roof Find the pitch of the roof shown, written as a fraction in simplest form.

You Try It

27. Tuition and Fees The average tuition and fees for U.S. public four-year colleges and universities were $3930 in 1994. The average tuition and fees were $7020 in 2009. Find and interpret the slope of the line connecting the points $(1994, 3930)$ and $(2009, 7020)$. (*Source:* College Board, *Trends in College Pricing 2009*)

28. Car Value In 2005, Julie purchased a new SUV for $32,000. In 2010, she sold the SUV for $18,000. Find and interpret the slope of the line connecting the points $(2005, 32000)$ and $(2010, 18000)$.

29. Autism Prevalence In 2000, there were 93,650 cases of autism in the U.S. among people ages 3 to 22. In 2008, there were 337,795 cases of autism among this same U.S. age group. Find and interpret the slope of the line connecting the points $(2000, 93650)$ and $(2008, 337795)$. (*Source:* ThoughtfulHouse.org)

30. Video MP4 Players An 8-gigabyte video MP4 player costs $79.99 in a store. A 16-gigabyte video MP4 player of the same brand cost $99.99 in the same store. Find and interpret the slope of the line connecting the points $(8, 79.99)$ and $(16, 99.99)$.

12.4 Equations of Lines

THINGS TO KNOW

Before working through this topic, be sure you are familiar with the following concepts:

		VIDEO	ANIMATION	INTERACTIVE
You Try It	1. Solve Linear Equations Using Both Properties of Equality (Topic 11.1, **Objective 5**)			
You Try It	2. Solve a Formula for a Given Variable (Topic 11.4, **Objective 3**)			
You Try It	3. Find the Slope of a Line Given Two Points (Topic 12.3, **Objective 1**)			
You Try It	4. Graph a Line Using the Slope and a Point (Topic 12.3, **Objective 3**)			

OBJECTIVES

1 Determine the Slope and y-Intercept from a Linear Equation

2 Use the Slope-Intercept Form to Graph a Linear Equation

3 Write the Equation of a Line Given Its Slope and y-Intercept

4 Write the Equation of a Line Given Its Slope and a Point on the Line

5 Write the Equation of a Line Given Two Points

(eText Screens 12.4-1–12.4-29)

6 Determine the Relationship Between Two Lines

7 Write the Equation of a Line Parallel or Perpendicular to a Given Line

8 Use Linear Equations to Solve Applications

OBJECTIVE 1 DETERMINE THE SLOPE AND y-INTERCEPT FROM A LINEAR EQUATION

In **Topic 12.2**, we graphed the **linear equation** $x + 2y = 8$ by **plotting points** as shown in Figure 23a. In Topic 12.3, we learned how to use the graph to determine the **slope** of the line $m = \frac{\text{rise}}{\text{run}} = -\frac{1}{2}$, as shown in Figure 23b. Notice that the graph crosses the y-axis at the point $(0, 4)$, so the **y-intercept** is 4.

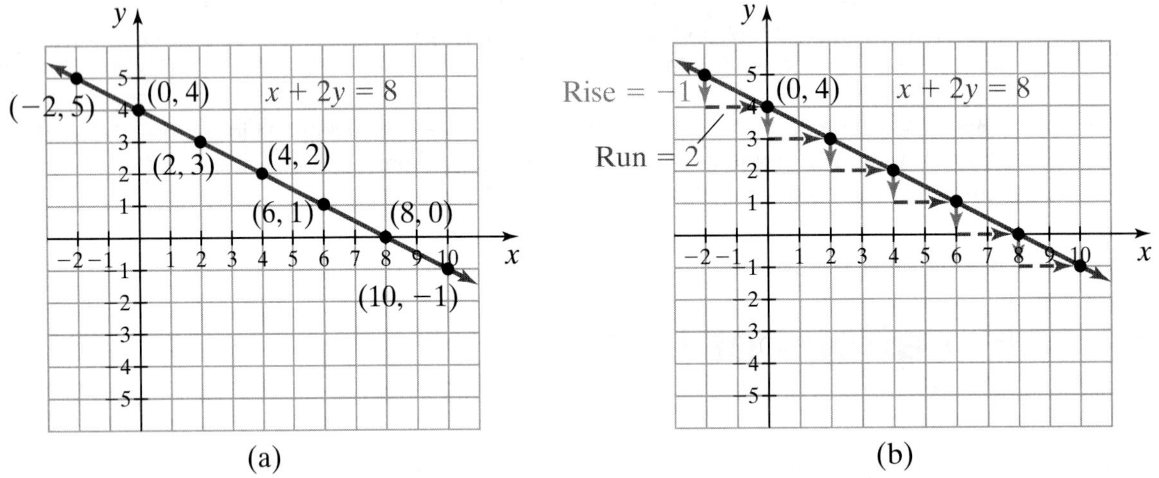

Figure 23 Graph of $x + 2y = 8$

We can determine the **slope** and **y-intercept** directly from the equation by first solving the equation for y.

Original equation:	$x + 2y = 8$
Subtract x from both sides:	$2y = -x + 8$
Divide both sides by 2:	$\frac{2y}{2} = \frac{-x}{2} + \frac{8}{2}$
Simplify:	$y = -\frac{1}{2}x + 4$

Notice that, in this form, the slope, $m = -\frac{1}{2}$, is the **coefficient** of x and the y-intercept, $b = 4$, is the **constant term**. This result is true in general.

Slope-Intercept Form

A **linear equation in two variables** of the form

$$y = mx + b$$

is written in **slope-intercept form**, where m is the slope of the line and b is the y-intercept.

Be careful not to assume that the coefficient of x is always the slope. This is true when the equation is in slope-intercept form but is not true in general.

Example 1 Determining the Slope and y-Intercept from a Linear Equation

Find the **slope** and **y-intercept** of the given line.

a. $2x + y = 3$ **b.** $4x - 3y = 6$

Solutions

a. Put the equation in slope-intercept form by solving for y.

Begin with the original equation: $2x + y = 3$

Subtract $2x$ from both sides: $y = -2x + 3 \leftarrow y = mx + b$

Comparing this result to the slope-intercept form, we see that the coefficient of x is -2 and the constant term is 3. Therefore, the slope is $m = -2$, and the y-intercept is $b = 3$.

b. Begin with the original equation: $4x - 3y = 6$

Subtract $4x$ from both sides: $-3y = -4x + 6$

Divide both sides by -3: $\dfrac{-3y}{-3} = \dfrac{-4x}{-3} + \dfrac{6}{-3}$

Simplify: $y = \dfrac{4}{3}x - 2 \leftarrow y = mx + b$

Now determine the slope and y-intercept on your own, then view the **answers**.

You Try It **Work through this You Try It problem.**

Work Exercises 1–4 in this textbook or in the MyMathLab Study Plan.

It is possible that either m or b (or both) are equal to 0, as shown in Example 2.

Example 2 Determining the Slope and y-Intercept from a Linear Equation

Find the **slope** and **y-intercept** of the given line.

My video summary

a. $4x - 10y = 0$ **b.** $y = 4$

Solutions Write each equation in the form $y = mx + b$ to determine the slope and y-intercept, then view the **answers**. For a detailed solution, watch this **video**.

You Try It **Work through this You Try It problem.**

Work Exercises 5–8 in this textbook or in the MyMathLab Study Plan.

Not every linear equation in two variables can be written in **slope-intercept form**. Think about what type of line cannot have its equation written in this form. View this **example** of such a line.

OBJECTIVE 2 USE THE SLOPE-INTERCEPT FORM TO GRAPH A LINEAR EQUATION

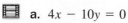

Recall that the **y-intercept**, b, corresponds to the point $(0, b)$ on the graph. Given the equation of a nonvertical line, we can use the **slope-intercept form** to determine the slope

(eText Screens 12.4-1–12.4-29)

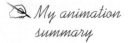

My animation summary

of the line and the point on the line corresponding to the y-intercept. Watch this **animation**, which illustrates the concept, and then work through Example 3.

Example 3 Graphing a Linear Equation Using Slope-Intercept Form

Graph the equation $y = \frac{3}{5}x - 2$ using the slope and y-intercept.

Solution The equation is written in slope-intercept form. The slope is $m = \frac{3}{5}$, and the y-intercept is $b = -2$. We plot the point $(0, -2)$ and use the slope to obtain a second point on the graph. Then we can complete the graph by connecting the points with a straight line.

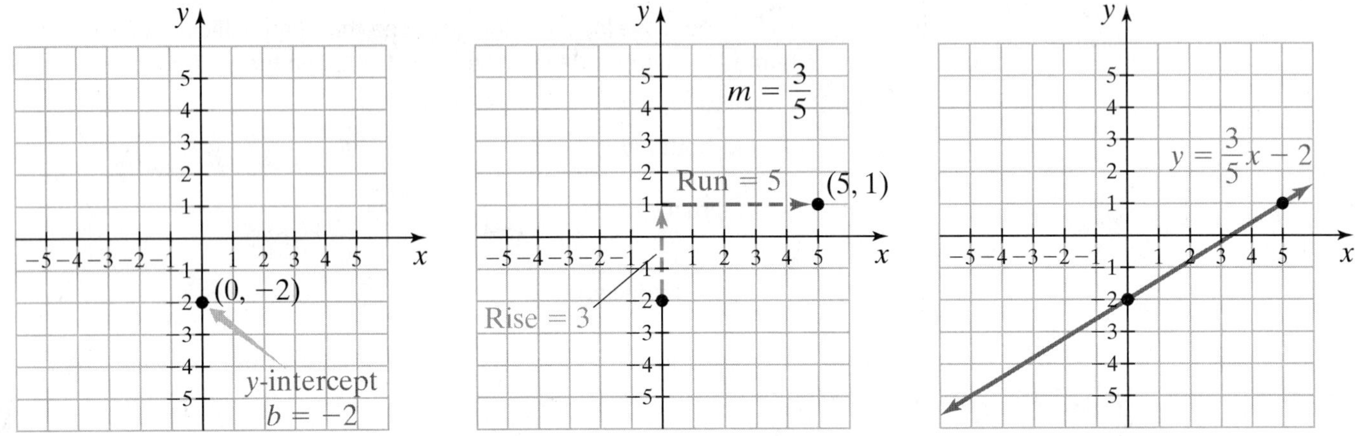

(a) (b) (c)

You Try It Work through this You Try It problem.

Work Exercises 9–11 in this textbook or in the MyMathLab Study Plan.

Example 4 Graphing a Linear Equation Using Slope-Intercept Form

My video summary

Graph the equation $2x + 3y = 9$ using the **slope** and **y-intercept**.

Solution We start by writing the equation in **slope-intercept form**, $y = -\frac{2}{3}x + 3$. The slope is $m = -\frac{2}{3}$ and the y-intercept is $b = 3$, so the point $(0, 3)$ is on the graph of the equation. Use this point and the slope to sketch the graph of the equation. Compare your graph to the **answer**, or watch this **video** for a detailed solution.

You Try It Work through this You Try It problem.

Work Exercises 12–14 in this textbook or in the MyMathLab Study Plan.

OBJECTIVE 3 WRITE THE EQUATION OF A LINE GIVEN ITS SLOPE AND y-INTERCEPT

Given the **slope** and **y-intercept** of a line, we can write the equation of the line by **substituting** the corresponding values into the **slope-intercept form**.

Example 5 Writing an Equation of a Line Given Its Slope and y-Intercept

Write an equation of the line with the given slope and y-intercept.

a. slope -4; y-intercept 3 **b.** slope $\frac{2}{5}$; y-intercept -7

Solutions

a. We have $m = -4$ and $b = 3$. Substitute these values into the slope-intercept form.

$$y = mx + b \qquad y = -4x + 3$$

The equation of the line is $y = -4x + 3$.

b. We have $m = \frac{2}{5}$ and $b = -7$. Substitute these values into the slope-intercept form.

$$y = mx + b \qquad y = \frac{2}{5}x + (-7)$$

The equation of the line is $y = \frac{2}{5}x - 7$.

You Try It **Work through this You Try It problem.**

Work Exercises 15–20 in this textbook or in the MyMathLab Study Plan.

OBJECTIVE 4 WRITE THE EQUATION OF A LINE GIVEN ITS SLOPE AND A POINT ON THE LINE

Given the **slope** of a nonvertical line and a point on the line (other than the **y-intercept**), we can still use the **slope-intercept form** to write the equation of the line. The given point (x, y) is an **ordered pair solution** to the equation, so must satisfy the equation. We use these **coordinates** and the given slope to solve for the y-intercept, b.

Example 6 Writing the Equation of a Line Given Its Slope and a Point on the Line

Write the equation of the line that has slope 2 and passes through the point $(4,7)$.

Solution We are given $m = 2$ and $(x, y) = (4, 7)$. Substitute these values into the slope-intercept form and solve for b.

Slope-intercept form:	$y = mx + b$
Substitute 2 for m, 4 for x, and 7 for y:	$7 = (2)(4) + b$
Simplify:	$7 = 8 + b$
Subtract 8 from both sides:	$-1 = b$

Now use the slope and y-intercept to write the equation $y = 2x - 1$.

We can use an alternate form to write the equation of a line given the **slope** and a point on the line. This form is called the **point-slope form**.

(eText Screens 12.4-1–12.4-29)

Point-Slope Form

Given the slope m of a line and a point (x_1, y_1) on the line, the **point-slope form** of the equation of the line is given by

$$y - y_1 = m(x - x_1).$$

Example 7 Writing the Equation of a Line Given Its Slope and a Point on the Line

My interactive video summary

Use the point-slope form to determine the equation of the line that has slope $-\frac{3}{4}$ and passes through the point $(2, -5)$. Write the equation in **slope-intercept form.**

Solution We are given $m = -\frac{3}{4}$ and $(x_1, y_1) = (2, -5)$. **Substitute** these values into the **point-slope form.**

Point-slope form:	$y - y_1 = m(x - x_1)$
Substitute $-\frac{3}{4}$ for m, 2 for x_1, and -5 for y_1:	$y - (-5) = \left(-\frac{3}{4}\right)(x - (2))$
Simplify:	$y + 5 = -\frac{3}{4}(x - 2)$

Finish solving the equation for y and write it in **slope-intercept form.** Check your **answer**, or watch this **interactive video** to see a detailed solution using either point-slope form or slope-intercept form.

You Try It **Work through this You Try It problem.**

Work Exercises 21–26 in this textbook or in the MyMathLab Study Plan.

Whether we use the slope-intercept form or the point-slope form, the work is basically the same to determine the equation of the line. To see **Example 6** worked out using the **point-slope form**, view this popup.

Example 8 Writing the Equation of a Line Given Its Slope and a Point on the Line

Write the equation of a line that passes through the point $(-3, 2)$ and has the given **slope**.

a. $m = 0$

b. undefined slope

Solutions

a. **Horizontal lines** have a slope of 0, and the equation of a horizontal line is given by $y = b$ where b is the **y-intercept.** Recall that every point on the graph of a horizontal line has the same **y-coordinate**, b. The line passes through the point $(-3, 2)$, which has a y-coordinate of 2. Therefore, the equation of the line passing through $(-3, 2)$ with slope 0 is $y = 2$.

b. Similarly, **vertical lines** have undefined slope, and the equation of a vertical line is given by $x = a$ where a is the **x-intercept.** Recall that every point on the graph of a vertical line has the same **x-coordinate**, a. The line passes through the point $(-3, 2)$, which has an x-coordinate of -3. Therefore, the equation of the line passing through $(-3, 2)$ with undefined slope is $x = -3$.

You Try It **Work through this You Try It problem.**

Work Exercises 27 and 28 in this textbook or in the MyMathLab Study Plan.

OBJECTIVE 5 WRITE THE EQUATION OF A LINE GIVEN TWO POINTS

Recall that two points determine a line. Given two points on a line, we can find the equation of the line by first finding the slope of the line and then using the slope and either given point to determine the equation.

Example 9 Writing the Equation of a Line Given Two Points

My video summary

Write the equation of the line passing through the points $(-4, 1)$ and $(2, 4)$. Write your answer in **slope-intercept form**.

Solution We are given two points, but we don't know the slope. We can compute the slope by using the **slope formula**:

$$m = \frac{4 - 1}{2 - (-4)} = \frac{4 - 1}{2 + 4} = \frac{3}{6} = \frac{1}{2}.$$

Now that we know the slope, we can use it together with *either* of the two given points to write the equation of the line. Remember that you can use either the slope-intercept form or the **point-slope form** to determine the equation. View the **answer** or watch this **video** to see a detailed solution.

You Try It **Work through this You Try It problem.**

Work Exercises 29–32 in this textbook or in the MyMathLab Study Plan.

We summarize the slope formula and the different forms for equations of lines in Table 1.

$m = \dfrac{y_2 - y_1}{x_2 - x_1}$	**Slope** Average rate of change
$y - y_1 = m(x - x_1)$	**Point-Slope Form** Slope is m, and (x_1, y_1) is a point on the line.
$y = mx + b$	**Slope-Intercept Form** Slope is m, and y-intercept is b.
$Ax + By = C$	**Standard Form** A, B, and C are real numbers, with A and B not both zero.
$y = b$	**Horizontal Line** Slope is zero, and y-intercept is b.
$x = a$	**Vertical Line** Slope is undefined, and x-intercept is a.

Table 1 Slope and equations of lines

OBJECTIVE 6 DETERMINE THE RELATIONSHIP BETWEEN TWO LINES

Slopes and **y-intercepts** can be used to determine the relationship between two lines. This will be helpful when we solve systems of linear equations in Module 16.

Recall that **parallel lines** have the same slope but different y-intercepts, while perpendicular lines have **opposite-reciprocal** slopes (product of the slopes is -1). **Coinciding lines** have the same slope *and* the same y-intercept. Coinciding lines appear to be the same line because their graphs lie on top of each other. Two lines with different slopes will intersect. Two intersecting lines that are not perpendicular are called **only intersecting** lines.

In Table 2 we summarize how the slope and y-intercept allow us to determine the relationship between two lines.

	Parallel Lines	Coinciding Lines	Intersecting and Perpendicular Lines	Only Intersecting Lines (not Perpendicular)
Slopes are	Same	Same	opposite reciprocals $\left(m_1 \cdot m_2 = -1 \text{ or } m_1 = -\dfrac{1}{m_2}\right)$	Different (but $m_1 \cdot m_2 \neq -1$)
y-Intercepts are	Different	Same	Same or Different	Same or Different

Table 2 Relationship between two lines

Example 10 Determining the Relationship Between Two Lines

My interactive video summary

For each pair of lines, determine if the lines are **parallel, perpendicular, coinciding,** or **only intersecting.**

a. $3y = -2x + 7$
$3x - 2y = 8$

b. $y = -3x + 1$
$6x + 2y = 2$

c. $4x - 5y = 15$
$y = \dfrac{4}{5}x + 1$

d. $3x - 4y = 2$
$x + 2y = -12$

Solutions

a. Begin by writing each equation in **slope-intercept** form.

Original equation: $3y = -2x + 7$

Divide both sides by 3: $y = -\dfrac{2}{3}x + \dfrac{7}{3} \leftarrow y = mx + b$

Original equation: $3x - 2y = 8$

Subtract $3x$ from both sides: $-2y = -3x + 8$

Divide both sides by -2: $y = \dfrac{3}{2}x - 4 \leftarrow y = mx + b$

In slope-intercept form, we see that the two lines have **opposite-reciprocal** slopes, $\left(-\dfrac{2}{3}\right)\left(\dfrac{3}{2}\right) = -1$, so the lines are perpendicular.

b–d. View the **answers** or watch this **interactive video** for a more detailed solution to all parts.

You Try It **Work through this You Try It problem.**

Work Exercises 33–38 in this textbook or in the MyMathLab Study Plan.

OBJECTIVE 7 WRITE THE EQUATION OF A LINE PARALLEL OR PERPENDICULAR TO A GIVEN LINE

When writing the equation of a **parallel line** or **perpendicular line**, we must remember that parallel lines have the same slope and perpendicular lines have **opposite-reciprocal** slopes.

Example 11 Writing the Equation of a Perpendicular Line or a Parallel Line

Write the equation of the line that passes through the point $(6, -5)$ and is

a. perpendicular to $6x - 2y = -1$. **b.** parallel to $y = -2x + 4$.

Solutions

a. We are given a point on the graph, but we do not know the slope. However, we know that the line is perpendicular to $6x - 2y = -1$, which gives us information about the slope. Perpendicular lines have opposite-reciprocal slopes, so we first need to determine the slope of the given line.

Original equation: $6x - 2y = -1$

Subtract $6x$ from each side: $-2y = -6x - 1$

Divide both sides by -2: $y = 3x + \frac{1}{2} \leftarrow \frac{-6}{-2} = 3; \frac{-1}{-2} = \frac{1}{2}$

The slope of the given line is 3, or $\frac{3}{1}$. The opposite reciprocal is $-\frac{1}{3}$, so the slope of our line is $-\frac{1}{3}$. Now that we know the slope and a point on the graph, we can use the **slope-intercept form** or **point-slope form** to determine the equation.

Slope-intercept form: $y = mx + b$

Substitute $-\frac{1}{3}$ for m, 6 for x, -5 for y: $-5 = \left(-\frac{1}{3}\right)(6) + b$

Simplify: $-5 = -2 + b$

Add 2 to both sides: $-3 = b$

The equation of the line is $y = -\frac{1}{3}x - 3$.

(View the **alternate steps** using point-slope form.)

My video summary

b. The equation is already written in slope-intercept form. Determine the slope of the given line. Then use this slope and the given point to determine the equation of the parallel line. See the answer in this **popup**, or see the complete solution in this **video**.

You Try It **Work through this You Try It problem.**

Work Exercises 39–42 in this textbook or in the MyMathLab Study Plan.

Table 3 summarizes how to write the equation of a line from given information.

(*eText Screens 12.4-1–12.4-29*)

If you are given ...	Then do this ...
Slope and **y-intercept**	Plug values for m and b directly into the **slope-intercept form** to write the equation.
Slope and a point	Use the slope-intercept form to find b and then write the equation in slope-intercept form.
Two points	Find the slope. Then use this slope and either point in the slope-intercept form to find b.
Point and slope $= 0$ (horizontal line)	Use the y-coordinate of the point to write $y = b$.
Point and undefined slope (vertical line)	Use the x-coordinate of the point to write $x = a$.
Point and **parallel** line in equation form	Find the slope of the given line. Use this slope and the point in the slope-intercept form to find b.
Point and **perpendicular** line in equation form	Find the slope of the given line. Find its **opposite reciprocal** to find the slope of the perpendicular line. Using this as the slope and the given point, find b by using the slope-intercept form.

Table 3 Writing equations of lines from given information

Table 3 emphasizes the use of the slope-intercept form because it is generally more useful (view this **popup** for an explanation). Remember that given the slope and a point, we could also use the **point-slope form** if desired.

OBJECTIVE 8 USE LINEAR EQUATIONS TO SOLVE APPLICATIONS

Many real-life applications can be **modeled** using **linear equations**. Work through Example 12, or watch the **video** solution.

Example 12 Football Attendance

My video summary

If attendance at professional football games is 17 million in a given year, then the corresponding attendance at college football games is 31 million. Increasing attendance at professional football games to 25 million increases attendance at college football games to 55 million. (*Source: Statistical Abstract, 2010*)

a. Assume that the relationship between professional football attendance (in millions) and college football attendance (in millions) is linear. Find the equation of the line that describes this relationship. Write your answer in **slope-intercept form**.

b. Use your equation from part (a) to estimate the attendance at college football games if the attendance at professional football games is 21 million.

Solutions

a. We start by writing two **ordered pairs** of the form $(x, y) =$ (**pro atd, college atd**), where each coordinate represents millions of people. From the problem statement, the first ordered pair is $(17, 31)$ and the second is $(25, 55)$.

Next we use the two ordered pairs to determine the slope.

$$m = \frac{55 - 31}{25 - 17} = \frac{24}{8} = 3$$

We then use the slope-intercept form to determine the y-intercept, b. Either of the two ordered pairs can be used. We will use $(17, 31)$.

Slope-intercept form:	$y = mx + b$
Substitute 3 for m, 17 for x, and 31 for y:	$31 = (3)(17) + b$
Simplify:	$31 = 51 + b$
Subtract 51 from both sides:	$-20 = b$

Using the slope $m = 3$ and y-intercept $b = -20$, the equation is $y = 3x - 20$.

b. To estimate the attendance at college football games if the attendance at professional games is 21 million, we substitute 21 for x and solve for y.

Slope-intercept form:	$y = 3x - 20$
Substitute 21 for x:	$y = 3(21) - 20$
Simplify:	$y = 63 - 20$
	$y = 43$

The estimated attendance at college football games is 43 million.

You Try It **Work through this You Try It problem.**

Work Exercises 43–45 in this textbook or in the MyMathLab Study Plan.

12.4 Exercises

You Try It

In Exercises 1–8, determine the slope and y-intercept for each equation.

1. $3x + y = 9$ **2.** $-4x + y = 7$ **3.** $6x + 5y = -4$ **4.** $2x - 5y = 3$

5. $5x + 4y = 0$ **6.** $3x - 8y = 0$ **7.** $y = 10$ **8.** $x = -2$

You Try It

In Exercises 9–14, graph the equation using the slope and y-intercept.

You Try It

9. $y = 3x + 2$ **10.** $y = \frac{4}{3}x - 1$ **11.** $y = -\frac{1}{2}x$

12. $2x + 3y = 4$ **13.** $5x - 3y = -12$ **14.** $1.2x + 1.5y = 4.5$

You Try It

In Exercises 15–20, write the equation of the line with the given slope and y-intercept.

You Try It

15. $m = 4, b = 8$ **16.** $m = -2, b = -7.2$ **17.** slope $= \frac{1}{3}$, y-intercept $= 4$

18. slope $= \frac{3}{4}$, y-intercept $= \frac{1}{2}$ **19.** slope $= 0$, y-intercept $= \frac{5}{4}$ **20.** slope $= -3.5$, y-intercept $= 0$

In Exercises 21–26, write the equation of the line that has the given slope and passes through the given point. Write the equation in slope-intercept form.

You Try It

21. $m = -3; (2, 5)$ **22.** $m = \frac{3}{5}; (10, -3)$ **23.** slope $= -\frac{2}{3}; (-3, 11)$

24. slope $= 3.5; (-5, 2.4)$

25. slope $= \frac{2}{3}; (6, 4)$

26. $m = -\frac{1}{4}; \left(\frac{8}{3}, -\frac{5}{2}\right)$

27. $m = 0; (4, 1)$

28. undefined slope; $(-5, 9)$

You Try It

In Exercises 29–32, write the equation of the line passing through the given points. Write your answer in slope-intercept form.

You Try It

29. Passing through $(2, 6)$ and $(4, 9)$

30. Passing through $(-1, 2)$ and $(2, -7)$

31. Passing through $(0, 0)$ and $\left(-3, \frac{2}{3}\right)$

32. Passing through $(3, 5)$ and $(-6, 5)$

In Exercises 33–38, determine if the two lines are parallel, perpendicular, coinciding, or only intersecting.

You Try It

33. $y = -2x + 9$
$8x + 4y = 1$

34. $5x - 3y = 4$
$3x - 5y = -7$

35. $-8x + 2y = 13$
$x + 4y = -3$

36. $3x - 4y = 28$
$y = \frac{3}{4}x - 7$

37. $y = -6$
$-2y = -6$

38. $x = 4$
$y = 4$

You Try It

39. Write the equation of the line perpendicular to $y = \frac{1}{6}x + 4$ that passes through the point $(-3, 7)$.

40. Write the equation of the line perpendicular to $7x - 5y = -2$ that passes through the point $(-7, 4)$.

41. Write the equation of the line parallel to $y = \frac{1}{6}x + 4$ that passes through the point $(-8, 1)$.

42. Write the equation of the line parallel to $7x - 5y = -2$ that passes through the point $(10, -2)$.

You Try It

43. Mortgage Rates If the average interest rate for a 15-year fixed rate mortgage is 5.00%, then the corresponding 30-year fixed rate is 5.85%. If the 15-year rate decreases to 4.00%, then the 30-year rate decreases to 4.55%. (*Source: mortgagenewsdaily.com*).

a. Assume that the relationship between the 15-year fixed rate (%) and the 30-year fixed rate (%) is linear. Find the equation of the line that describes this relationship. Write your answer in slope-intercept form.

b. Use your equation from part (a) to estimate the 30-year fixed rate if the 15-year fixed rate is 4.8%.

44. Hard Drive Size Jasmine is building a new computer. She notices that for \$55, she can get 250 GB of hard drive space, but for \$75, she can get 500 GB of space.

a. Assume that the relationship between price (\$) and hard drive space (GB) is linear. Find the equation of the line that describes this relationship. Write your answer in slope-intercept form.

b. Use your equation from part (a) to estimate the hard drive space that Jasmine can get for \$85.

45. e-Reader Sales An e-reader manufacturer expects to sell 50,000 units each year if the selling price is \$250. If the selling price is reduced to \$180, then the manufacturer expects to sell 140,000 units each year.

a. Assume that the relationship between selling price (\$) and units sold (1000s) is linear. Find the equation of the line that describes this relationship. Write your answer in slope-intercept form.

12.5 Linear Inequalities in Two Variables

THINGS TO KNOW

Before working through this topic, be sure you are familiar with the following concepts:

VIDEO ANIMATION INTERACTIVE

1. Graph the Solution Set of an Inequality on a Number Line (Topic 11.7, **Objective 2**) — You Try It
2. Solve Linear Inequalities in One Variable (Topic 11.7, **Objective 4**) — You Try It
3. Use Linear Inequalities to Solve Application Problems (Topic 11.7, **Objective 6**) — You Try It
4. Identify Points in the Rectangular Coordinate System (Topic 12.1, **Objective 2**) — You Try It
5. Graph Horizontal and Vertical Lines (Topic 12.2, **Objective 7**) — You Try It
6. Use the Slope-Intercept Form to Graph a Linear Equation (Topic 12.4, **Objective 2**) — You Try It
7. Use Linear Equations to Solve Applications (Topic 12.4, **Objective 8**) — You Try It

OBJECTIVES

1 Determine If an Ordered Pair Is a Solution to a Linear Inequality in Two Variables

2 Graph a Linear Inequality in Two Variables

3 Solve Applications Involving Linear Inequalities in Two Variables

OBJECTIVE 1 DETERMINE IF AN ORDERED PAIR IS A SOLUTION TO A LINEAR INEQUALITY IN TWO VARIABLES

In **Topic 11.7**, we solved **linear inequalities in one variable**. The **solution set** for such an inequality is the set of all values that make the inequality true. Typically, we graph the solution set of a linear inequality in one variable on a **number line**.

In this topic, we solve **linear inequalities in two variables.** A linear inequality in two variables looks like a **linear equation in two variables** except that an **inequality symbol** replaces the equal sign.

Definition Linear Inequality in Two Variables

A **linear inequality in two variables** is an inequality that can be written in the form $Ax + By < C$, where A, B, and C are real numbers, and A and B are not both equal to zero.

Note: The inequality symbol "$<$" can be replaced with $>$, $\leq$, or $\geq$.

(eText Screens 12.5-1–12.5-16)

An **ordered pair** is a **solution to a linear inequality in two variables** if, when substituted for the variables, it makes the inequality true.

Example 1 Determining If an Ordered Pair Is a Solution to a Linear Inequality in Two Variables

My video summary

Determine if the given **ordered pair** is a **solution to the inequality** $2x - 3y < 6$.

a. $(-1, -2)$ **b.** $(4, -1)$ **c.** $(6, 2)$

Solutions We substitute the x- and y-coordinates for the **variables** and simplify. If the resulting statement is true, then the ordered pair is a solution to the inequality.

a. Begin with the original inequality: $2x - 3y < 6$

Substitute -1 for x and -2 for y: $2(-1) - 3(-2) \overset{?}{<} 6$

Simplify: $-2 + 6 \overset{?}{<} 6$

$4 < 6$ True

The final statement is true, so $(-1, -2)$ is a solution to the inequality.

b–c. Try to determine if these ordered pairs are solutions on your own. View the **answers**, or watch this **video** for detailed solutions to all three parts.

You Try It **Work through this You Try It problem.**

Work Exercises 1–4 in this textbook or in the MyMathLab Study Plan.

OBJECTIVE 2 GRAPH A LINEAR INEQUALITY IN TWO VARIABLES

Let's look for all solutions to a **linear inequality in two variables**. First, we focus on the **linear equation** related to the inequality. To find solutions to $x + y > 2$ or $x + y < 2$, we would look at the equation $x + y = 2$. For an **ordered pair** to **satisfy** this equation, the sum of its **coordinates** must be 2. So, $(-2, 4)$, $(1, 1)$, $(2, 0)$, and $(4, -2)$ are all examples of solutions to the equation. (View the **details**.) Plotting and connecting these points gives the line shown in Figure 24.

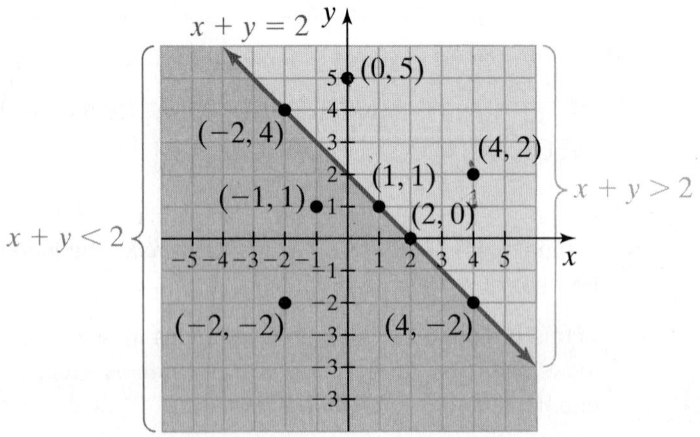

Figure 24

Now let's look to see which **ordered pairs** satisfy the inequalities $x + y > 2$ or $x + y < 2$. Notice that the line in **Figure 24** divides the **coordinate plane** into two **half-planes**, an *upper half-plane* and a *lower half-plane*.

Choose any point that lies in the upper half-plane, such as $(4, 2)$. The sum of its **coordinates** is $4 + 2 = 6$, which is larger than 2. Choose any other point in this region, such as $(0, 5)$. The sum of its coordinates is $0 + 5 = 5$, which is also larger than 2.

(eText Screens 12.5-1–12.5-16)

If we choose any point in the upper half-plane, the sum of its coordinates will be larger than 2. This means that any ordered pair from the upper half-plane is a **solution to the inequality** $x + y > 2$. The area shaded blue in **Figure 24** represents the set of all **ordered pair solutions** to the inequality $x + y > 2$.

Repeat this process for the lower half-plane to see that the area shaded pink in **Figure 24** represents the set of all ordered pair solutions to the inequality $x + y < 2$. (Launch this **popup box** to see the process.)

The linear equation $x + y = 2$ acts as a **boundary line** that separates the solutions of the two linear inequalities $x + y > 2$ and $x + y < 2$.

Based on this information, we can define **steps for graphing linear inequalities in two variables.**

Steps for Graphing Linear Inequalities in Two Variables

Step 1. Find the **boundary line** for the inequality by replacing the **inequality symbol** with an equal sign and graphing the resulting equation. If the inequality is **strict**, graph the boundary using a dashed line. If the inequality is **non-strict**, graph the boundary using a solid line.

Step 2. Choose a **test point** that does not belong to the boundary line and determine if it is a solution to the inequality.

Step 3. If the test point is a solution to the inequality, then shade the **half-plane** that contains the test point. If the test point is not a solution to the inequality, then shade the half-plane that does not contain the test point. The shaded area represents the set of all **ordered pair solutions** to the inequality.

Note: When graphing a **linear inequality in two variables** on the **coordinate plane**, using a dashed line is similar to using an open circle when graphing a **linear inequality in one variable** on a **number line**. Using a solid line is similar to using a solid circle. Do you see **why**?

Example 2 Graphing a Linear Inequality in Two Variables

My interactive video summary

Graph each inequality.

a. $3x - 4y \leq 8$ **b.** $y > 3x$ **c.** $y < -2$

Solutions

a. We follow the **three-step process.**

Step 1. The **boundary line** is $3x - 4y = 8$. We use a solid line as the graph because the inequality is **non-strict**. See Figure 25(a).

Step 2. We choose the **test point** $(0, 0)$ and check to see if it **satisfies** the inequality.

Begin with the original inequality: $3x - 4y \leq 8$

Substitute $x = 0$ and $y = 0$: $3(0) - 4(0) \overset{?}{\leq} 8$

Simplify: $0 \leq 8$ True

The point $(0, 0)$ is a solution to the inequality.

Step 3. Because the test point is a solution to the inequality, we shade the **half-plane** that contains $(0, 0)$. See Figure 25(b).
The shaded region, including the boundary line, represents all ordered pair solutions to the inequality $3x - 4y \leq 8$.

(eText Screens 12.5-1–12.5-16)

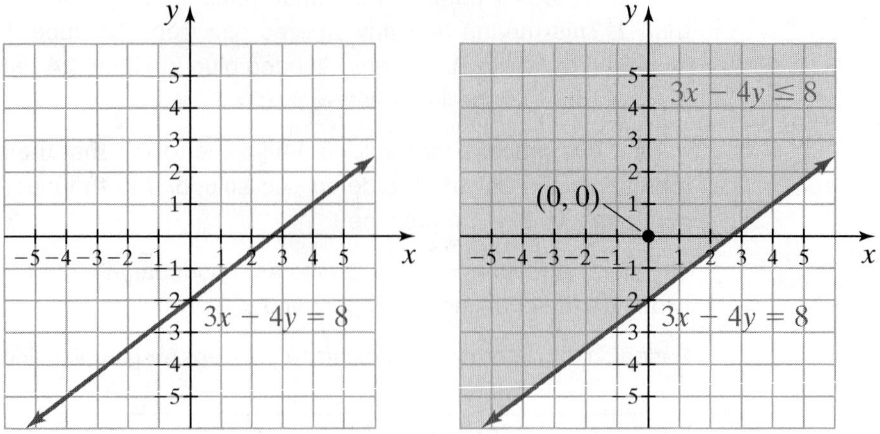

(a) (b) **Figure 25**

b. For $y > 3x$, the **boundary line** is $y = 3x$. Because the inequality is **strict**, we graph the boundary line using a dashed line. Choose a **test point** and complete the graph. See the **answer**, or watch this **interactive video** for a detailed solution.

c. Try to graph this inequality $y < -2$ on your own. Note that boundary line is the **horizontal line** $y = -2$. See the **answer**, or watch this **interactive video** for a detailed solution.

CAUTION A test point cannot belong to the boundary line. Do you see **why**?

You Try It **Work through this You Try It problem.**

Work Exercises 5–12 in this textbook or in the MyMathLab Study Plan.

OBJECTIVE 3 SOLVE APPLICATIONS INVOLVING LINEAR INEQUALITIES IN TWO VARIABLES

Sometimes real-life applications can be **modeled** by **linear inequalities in two variables.**

Example 3 A Piggy Bank

My video summary

A piggy bank contains only nickels and dimes with a total value of less than $9. Let n = the number of nickels and d = the number of dimes.

a. Write an inequality describing the possible numbers of coins in the bank.

b. Graph the inequality. Because n and d must be **whole numbers**, restrict the graph to **Quadrant I.**

c. Could the piggy bank contain 90 nickels and 60 dimes?

Solutions

a. The value of n nickels, in cents, is $5n$, and the value of d dimes is $10d$, so the total value of the coins in the bank is $5n + 10d$. The total value is less than $9, or 900¢, so the inequality is

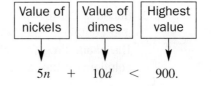

Value of nickels	Value of dimes	Highest value
↓	↓	↓

$$5n \quad + \quad 10d \quad < \quad 900.$$

b. Our two variables are n and d, so label the horizontal axis n and the vertical axis d. Follow the **steps for graphing linear inequalities in two variables** to obtain the graph in **Figure 26**. Watch this **video** for a detailed solution.

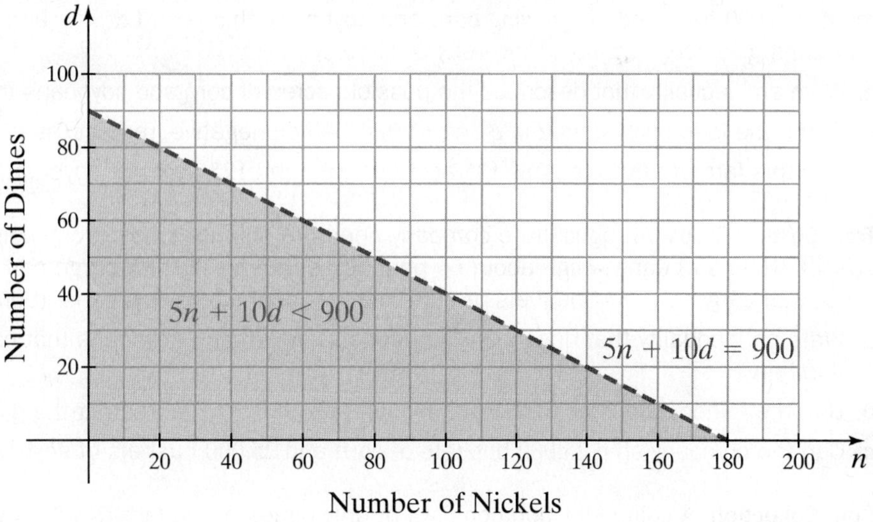

Figure 26

CAUTION The number of nickels and dimes must be **whole numbers**, so only those points in the shaded region that have whole number coordinates are acceptable solutions to this application.

c. Substitute $(n, d) = (90, 60)$ into the inequality to see if a true statement results.

Write the inequality:	$5n + 10d < 900$
Substitute $n = 90$ and $d = 60$:	$5(90) + 10(60) \overset{?}{<} 900$
Simplify:	$1050 < 900$ False

The piggy bank cannot contain 90 nickels and 60 dimes.

You Try It **Work through this You Try It problem.**

Work Exercises 13–16 in this textbook or in the MyMathLab Study Plan.

12.5 Exercises

In Exercises 1–4, determine if each ordered pair is a solution to the given inequality.

You Try It

1. $2x - y > 5$
a. $(6, 8)$
b. $(3, -4)$
c. $(3, 4)$

2. $4x + 3y \le 24$
a. $(8, -2)$
b. $(-5, 10)$
c. $(3, 4)$

3. $y < \frac{4}{3}x - 2$
a. $(-6, -5)$
b. $(3, 0)$
c. $\left(\frac{3}{2}, \frac{1}{2}\right)$

4. $3x - 5y \ge 45$
a. $(7, -5)$
b. $(-2, -10)$
c. $(7.5, -4.5)$

In Exercises 5–12, graph each inequality.

You Try It

5. $3x + y \le 6$ **6.** $5x - 3y > 15$ **7.** $-x + y \ge 0$ **8.** $x + 2y < 0$

9. $y < -\frac{3}{2}x + 4$ **10.** $y \ge -2x$ **11.** $y > -1$ **12.** $x \le 3$

(*eText Screens 12.5-1–12.5-16*)

In Exercises 13–16, solve each application by using an inequality.

You Try It

13. Growing Crops In Illinois the annual cost to grow corn is about $450 per acre, and the annual cost to grow soybeans is about $300 per acre (not counting the cost of land). Suppose a small farmer has at most $90,000 to spend on growing corn and soybeans this year. Let c = acres of corn and s = acres of soybeans. (*Source: Farmdoc*, University of Illinois)

a. Write an inequality that describes the possible acres of corn and soybeans this farmer can afford to grow.

b. Graph the inequality. Because c and s must be non-negative, restrict the graph to Quadrant I.

c. Can this farmer afford to grow 125 acres of corn and 125 acres of soybeans?

14. Transporting Crops An agriculture company needs to ship its crops. Corn weighs about 50 pounds per bushel, while soybeans weigh about 60 pounds per bushel. A river barge can safely carry up to 3,000,000 pounds of cargo. Let c = bushels of corn and s = bushels of soybeans. (*Source:* powderandbulk.com)

a. Write an inequality that describes the bushels of corn and soybeans that can be shipped together on this barge.

b. Graph the inequality. Because c and s must be non-negative, restrict the graph to Quadrant I.

c. Can the company ship 25,000 bushels of corn and 25,000 bushels of soybeans on this barge?

15. Coin Collection A coin collection consists of only dimes and quarters with a value of more than $12. Let d = the number of dimes and q = the number of quarters.

a. Write an inequality that describes the possible numbers of dimes and quarters in the collection.

b. Graph the inequality. Because d and q must be whole numbers, restrict the graph to Quadrant I.

c. Could this collection consist of 35 dimes and 45 quarters?

16. Car Rental While on a business trip, Bruce pays $50 per day to rent a car, plus $0.20 per mile driven. Bruce's company will reimburse him at most $200 for the entire rental. Let d = number of days and m = number of miles.

a. Write an inequality that describes Bruce's car rental limitations.

b. Graph the inequality. Because d and m must be non-negative, restrict the graph to Quadrant I.

c. If Bruce keeps the car for 3 days and drives 400 miles, will he be within his budget?

MODULE THIRTEEN

Systems of Linear Equations and Inequalities

MODULE THIRTEEN CONTENTS

13.1 Solving Systems of Linear Equations by Graphing

THINGS TO KNOW

Before working through this topic, be sure you are familiar with the following concepts:

		VIDEO	ANIMATION	INTERACTIVE
You Try It	1. Determine If an Ordered Pair Is a Solution to an Equation (Topic 12.2, **Objective 1**)	✓		
You Try It	2. Graph a Line Using the Slope and a Point (Topic 12.3, **Objective 3**)	✓		
You Try It	3. Use the Slope-Intercept Form to Graph a Linear Equation (Topic 12.4, **Objective 2**)	✓	✓	
You Try It	4. Determine the Relationship between Two Lines (Topic 12.4, **Objective 6**)			✓

OBJECTIVES

1 Determine If an Ordered Pair Is a Solution to a System of Linear Equations in Two Variables

2 Determine the Number of Solutions to a System Without Graphing

3 Solve Systems of Linear Equations by Graphing

OBJECTIVE 1 DETERMINE IF AN ORDERED PAIR IS A SOLUTION TO A SYSTEM OF LINEAR EQUATIONS IN TWO VARIABLES

Figure 1 shows that the approximate revenue for digital music is projected to increase steadily through 2014, while revenue for physical music (CDs and vinyl) is projected to decrease. The **x-coordinate** of the point where the two lines **intersect** gives the year when revenues are expected to be equal for both music types, and the **y-coordinate** of the intersection point gives the expected revenue when the values are equal for both music types.

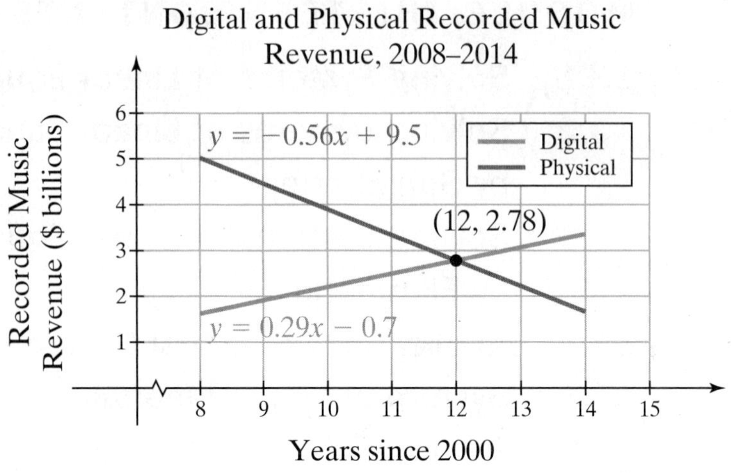

Figure 1

Source: Forrester Research Internet Music Forecast, 12/09

Figure 1 illustrates the graph of a **system of linear equations in two variables**

Definition System of Linear Equations in Two Variables

A system of linear equations in two variables is a collection of two or more **linear equations in two variables** considered together.

Note: Although a system of linear equations in two variables can contain more than two equations, we will only consider cases involving two equations in this text.

The system graphed in Figure 1 is given by the equations $\begin{cases} y = 0.29x - 0.7 \\ y = -0.56x + 9.5 \end{cases}$.

The following examples also represent systems of linear equations in two variables.

$$\begin{cases} 5x - y = -9 \\ 3x + 4y = 15 \end{cases} \qquad \begin{cases} \frac{1}{3}a - \frac{2}{5}b = 6 \\ -\frac{1}{6}a + \frac{4}{3}b = \frac{3}{7} \end{cases} \qquad \begin{cases} 10p + 3q = 28 \\ q = -5 \end{cases}$$

Within a given system of linear equations in two variables, the same two **variables** will be used. For example, the variables x and y are used in both equations of the first system, while the variables a and b are used in the second system.

From **Topic 12.2** we know that an **ordered pair** is a **solution** to an equation in two variables if a true statement results when **substituting** the values for the variables. Each ordered pair solution is a point on the graph of the equation. The **intersection point** in Figure 1, $(12, 2.78)$, lies on the graph of both equations, so it is a solution to both equations. Substituting 12 for x and 2.78 for y in either equation results in a true statement (view the **checks**). The ordered pair $(12, 2.78)$ is an example of a **solution to a system of linear equations.**

(eText Screens 13.1-1–13.1-17)

> **Definition Solution to a System of Linear Equations in Two Variables**
>
> A **solution to a system of linear equations** in two variables is an ordered pair that, when substituted for the variables, makes all equations in the system true.

CAUTION

To determine if an ordered pair is a solution to a system, we check to see if the ordered pair makes *both* equations true when substituted for the variables. It is not enough to check only one equation in the system.

Example 1 Determining If an Ordered Pair Is a Solution to a System of Linear Equations in Two Variables

My interactive video summary

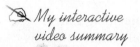

Determine if each ordered pair is a solution to the following system:

$$\begin{cases} 2x + 3y = 12 \\ x + 2y = 7 \end{cases}$$

a. $(-3, 6)$ **b.** $(3, 2)$

Solutions Check each ordered pair to see if it makes both equations true. If both equations are true, then the ordered pair is a **solution** to the system. If either equation is false, then the ordered pair is not a solution to the system.

a.		First Equation	Second Equation
	Begin with the original equations:	$2x + 3y = 12$	$x + 2y = 7$
	Substitute -3 for x and 6 for y in each equation:	$2(-3) + 3(6) \stackrel{?}{=} 12$	$(-3) + 2(6) \stackrel{?}{=} 7$

Finish simplifying both equations to see if $(-3, 6)$ is a solution to the system. View the **answer**, or watch this **interactive video** for the full solution.

b. Substitute 3 for x and 2 for y in each equation. Simplify to see if the resulting equations are true. View the **answer**, or watch this **interactive video** for the full solution.

You Try It **Work through this You Try It problem.**

Work Exercises 1–5 in this textbook or in the MyMathLab Study Plan.

OBJECTIVE 2 DETERMINE THE NUMBER OF SOLUTIONS TO A SYSTEM WITHOUT GRAPHING

Solutions to **systems of linear equations** are the **intersection points** of the graphs of the equations. When two linear equations are graphed, one of three possible outcomes will occur.

1. The two lines intersect at one point. See Figure 2(a). The system has one solution.
2. The two lines are **parallel** and do not intersect at all. See Figure 2(b). The system has no solution.
3. The two lines **coincide** and have an **infinite** number of intersection points. See Figure 2(c). The system has an infinite number of solutions.

(eText Screens 13.1-1–13.1-17)

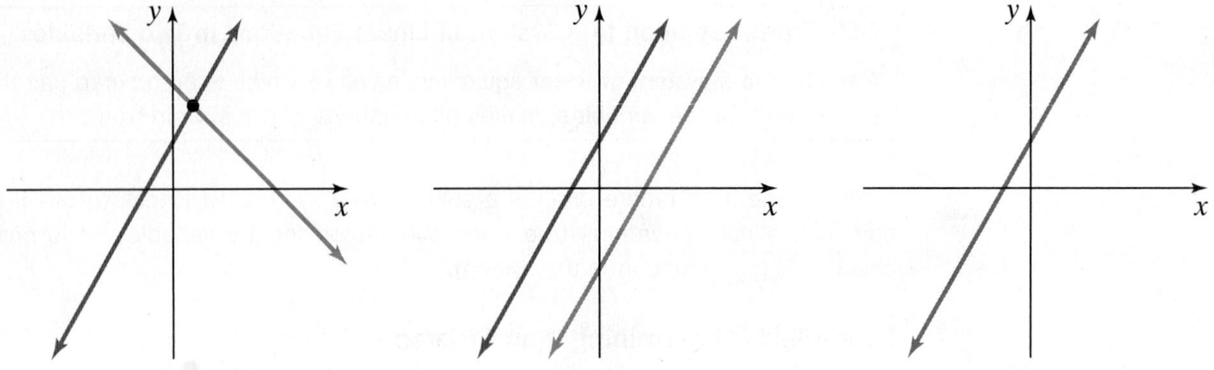

(a)	(b)	(c)
Intersecting Lines	**Parallel Lines**	**Coinciding Lines**
One solution	No solution	Infinitely many solutions
Two lines are different, having one common point.	Two lines are different, having no common points.	Two lines are the same, having infinitely many common points.

Figure 2

A system with at least one solution is **consistent**. A system without a solution is **inconsistent**. When the equations in a system are coinciding lines, the system is **dependent**. When the equations in a system are different lines, the system is **independent**.

In **Topic 12.4** we saw how to use the **slope** and ***y*-intercept** to determine the relationship between two lines. We can use this same information to find the number of **solutions** to a system of linear equations without actually graphing the system.

Intersecting lines (different slopes)	→	One solution
Parallel lines (same slope, different y-intercepts)	→	No solution
Coinciding lines (same slope, same y-intercept)	→	Infinite number of solutions

Example 2 Determining the Number of Solutions to a System Without Graphing

Determine the number of solutions to each system without graphing.

a. $\begin{cases} y = 3x - 4 \\ 6x + 3y = 8 \end{cases}$ **b.** $\begin{cases} 2x - 4y = \dfrac{8}{3} \\ 3x - 6y = 4 \end{cases}$ **c.** $\begin{cases} 5x - 2y = 3 \\ -\dfrac{5}{2}x + y = 7 \end{cases}$

Solutions

a. Write each equation in **slope-intercept form** so that the slope is easy to find.

$$\begin{cases} y = 3x - 4 \\ 6x + 3y = 8 \end{cases}$$

First equation: $y = 3x - 4$ ← Slope-intercept form

Second Equation: $6x + 3y = 8$

Subtract $6x$ from both sides: $3y = -6x + 8$

Divide both sides by 3: $y = \dfrac{-6x + 8}{3} \leftarrow \boxed{\dfrac{-6x + 8}{3} = \dfrac{-6x}{3} + \dfrac{8}{3}}$

Simplify: $y = -2x + \dfrac{8}{3} \leftarrow$ Slope-intercept form

The two **slopes**, 3 and -2, are different. Therefore, the lines will **intersect**, and the system will have one solution.

b. The equations in slope-intercept form are both $y = \dfrac{1}{2}x - \dfrac{2}{3}$ (view the **details**). This means that the graphs of the equations are coinciding lines and there are an infinite number of solutions. All **ordered pairs** that satisfy the equation $y = \dfrac{1}{2}x - \dfrac{2}{3}$ are solutions to the system.

My video summary

c. Watch this **video** to confirm that the system will have no solution because the two lines are **parallel**.

You Try It **Work through this You Try It problem.**

Work Exercises 6–11 in this textbook or in the MyMathLab Study Plan.

OBJECTIVE 3 SOLVE SYSTEMS OF LINEAR EQUATIONS BY GRAPHING

In this module, we look at three methods for solving **systems of linear equations** in two variables: *graphing, substitution,* and *elimination*.

If the system has *no solution* or an *infinite number of solutions*, then there is no need to graph the system since we already know the solution set. We will look at these special cases more in the next two topics. For now, we will focus on **consistent systems** with one solution (intersecting lines).

To solve systems of linear equations by graphing, we use a three-step process.

Solving Systems of Linear Equations in Two Variables by Graphing

Step 1. Graph the two equations on the same set of **axes**.

Step 2. If the lines intersect, find the **coordinates** of the **intersection point**. The ordered pair is the **solution to the system**.

Step 3. Check the ordered-pair solution in <u>both</u> of the original equations.

Note: If the lines are **parallel**, then the system has no solution. If the lines **coincide**, then the system has infinitely many solutions.

CAUTION: It may not be possible to identify the exact intersection point by graphing, so it is essential to check the ordered-pair solution in both equations.

Example 3 Solving Systems of Linear Equations by Graphing

Solve the following system by graphing:

$$\begin{cases} y = 2x + 1 \\ y = -x + 4 \end{cases}$$

(eText Screens 13.1-1–13.1-17)

Solution Note that the **slopes,** 2 and -1, are different, so there will be one solution to the system. We follow the **three-step process.**

Step 1. Graph each line.

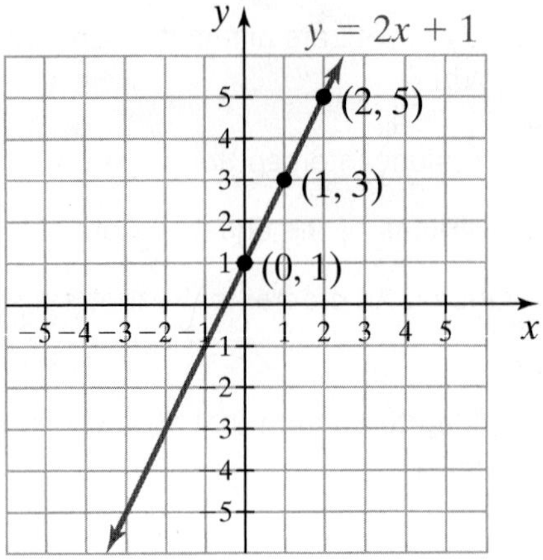

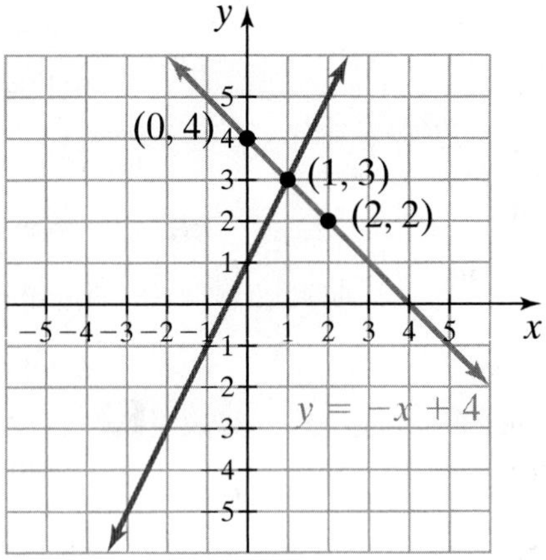

Step 2. Find the **coordinates** of the intersection point.

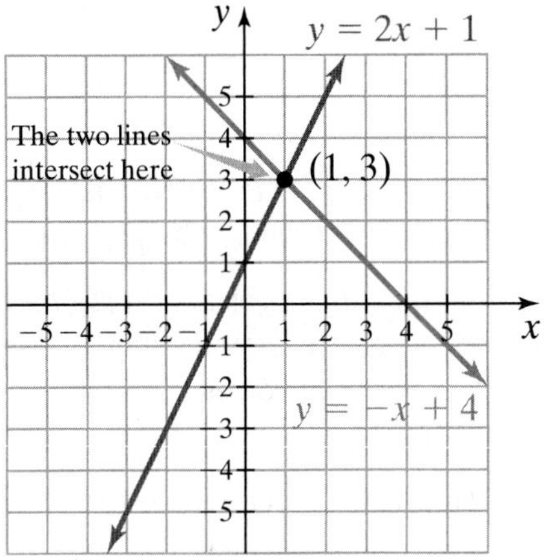

Figure 3

Graph of system $\begin{cases} y = 2x + 1 \\ y = -x + 4 \end{cases}$

The **intersection point** is $(1, 3)$.

Step 3. Check $(1, 3)$ in both equations to see if it is the **solution** to the system.

	First Equation	Second Equation
Original equation:	$y = 2x + 1$	$y = -x + 4$
Substitute 1 for x and 3 for y:	$(3) \stackrel{?}{=} 2(1) + 1$	$(3) \stackrel{?}{=} -(1) + 4$
Simplify:	$3 = 3$ True	$3 = 3$ True

The **ordered pair** checks in both equations, so $(1, 3)$ is the solution to the system.

Example 4 Solving Systems of Linear Equations by Graphing

My video summary

Solve the following system by graphing:

$$\begin{cases} 3x + y = -2 \\ x + y = 2 \end{cases}$$

(eText Screens 13.1-1–13.1-17)

Solution We follow the **three-step process.**

Step 1. Graph each line, as shown in Figure 4.

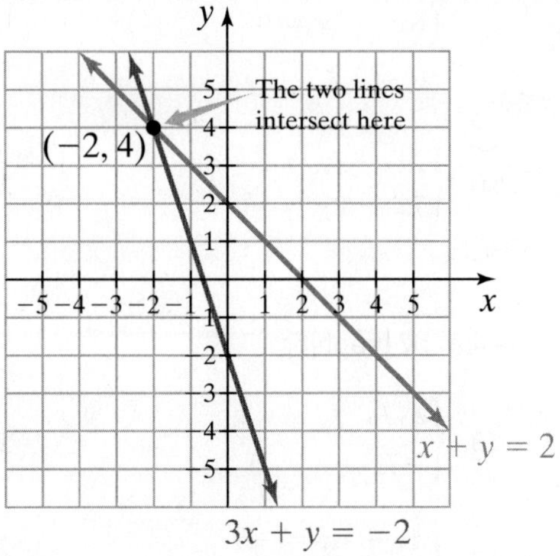

Figure 4

Graph of system $\begin{cases} 3x + y = -2 \\ x + y = 2 \end{cases}$

Step 2. The **intersection point** is $(-2, 4)$.

Step 3. Check $(-2, 4)$ in both equations to see if it is the **solution** to the system.

Watch this **video** to see the fully worked solution.

You Try It **Work through this You Try It problem.**

Work Exercises 12–21 in this textbook or in the MyMathLab Study Plan.

13.1 Exercises

In Exercises 1–5, determine if each ordered pair is a solution to the given system.

You Try It

1. $\begin{cases} 3x + 2y = 1 \\ 5x + 3y = 1 \end{cases}$ **a.** $(-1, 2)$ **b.** $(5, -7)$

2. $\begin{cases} 6x - y = 1 \\ -x + 4y = 19 \end{cases}$ **a.** $(1, 5)$ **b.** $(3, 17)$

3. $\begin{cases} 0.7x - 0.2y = 0.8 \\ 0.4x + 0.8y = 3.2 \end{cases}$ **a.** $(6, 14)$ **b.** $(2, 3)$

4. $\begin{cases} x + 2y = 6.3 \\ 5x - 7y = -4 \end{cases}$ **a.** $\left(\frac{2}{5}, -\frac{1}{3}\right)$ **b.** $(1.3, 2.5)$

5. $\begin{cases} \frac{2}{3}x - y = -2 \\ \frac{5}{4}x + \frac{1}{2}y = \frac{23}{4} \end{cases}$ **a.** $(3, 4)$ **b.** $(2, -3)$

In Exercises 6–11, determine the number of solutions to the given system without graphing.

You Try It

6. $\begin{cases} y = 7x + 2 \\ y = 2x + 7 \end{cases}$

7. $\begin{cases} y = 3x + 7 \\ 6x - 2y = 9 \end{cases}$

8. $\begin{cases} 2x - 5y = 15 \\ y = \frac{2}{5}x - 3 \end{cases}$

9. $\begin{cases} 5x - 8y = 4 \\ 2y - 1.25x = 1 \end{cases}$

10. $\begin{cases} 2x + 7y = 1 \\ 7x + 2y = 1 \end{cases}$

11. $\begin{cases} y = \frac{3}{4}x - 1 \\ x = \frac{4}{3}y + \frac{4}{3} \end{cases}$

In Exercises 12–21, solve each system by graphing.

You Try It

12. $\begin{cases} y = 3x - 1 \\ y = 2x + 1 \end{cases}$

13. $\begin{cases} 4x + 2y = -8 \\ -7x + y = 5 \end{cases}$

14. $\begin{cases} 3x - 4y = 0 \\ 9x - 4y = 24 \end{cases}$

15. $\begin{cases} 2x - 6y = -6 \\ 2x + 3y = 12 \end{cases}$

16. $\begin{cases} 2x - 3y = -9 \\ x + 0.5y = -2.5 \end{cases}$

17. $\begin{cases} x + 0.5y = 6 \\ 1.8x + 1.5y = 12 \end{cases}$

18. $\begin{cases} 9x - 2y - 8 = 0 \\ y = 2x + 1 \end{cases}$

19. $\begin{cases} 4x - \frac{5}{3}y = -15 \\ -5y = x + 20 \end{cases}$

20. $\begin{cases} 7x + \frac{7}{2}y = 28 \\ \frac{3}{2}x + \frac{5}{4}y = 5 \end{cases}$

21. $\begin{cases} 3x - \frac{7}{3}y = \frac{14}{3} \\ 6x - \frac{7}{2}y = \frac{35}{2} \end{cases}$

13.2 Solving Systems of Linear Equations by Substitution

THINGS TO KNOW

Before working through this topic, be sure you are familiar with the following concepts:

VIDEO | ANIMATION | INTERACTIVE

You Try It

1. Determine If an Ordered Pair Is a Solution to a System of Linear Equations in Two Variables (Topic 13.1, **Objective 1**)

You Try It

2. Determine the Number of Solutions to a System Without Graphing (Topic 13.1, **Objective 2**)

You Try It

3. Solve Systems of Linear Equations by Graphing (Topic 13.1, **Objective 3**)

(eText Screens 13.2-1–13.2-17)

OBJECTIVES

1 Solve Systems of Linear Equations by Substitution

2 Solve Special Systems by Substitution

OBJECTIVE 1 SOLVE SYSTEMS OF LINEAR EQUATIONS BY SUBSTITUTION

In this topic, we present the first of two algebraic methods for solving systems of linear equations in two variables. Consider the following system:

$$\begin{cases} 2x + 3y = 14 \\ y = x - 2 \end{cases}$$

Recall that when solving a system, we are looking for the **intersection point(s)** of the two lines in the system. At these points, both equations have the same x-values and the same y-values. If the second equation says that y equals $x - 2$, this means that y must equal $x - 2$ in the first equation as well.

$$2x + 3y = 14 \qquad\qquad y = x - 2$$

y must equal $x - 2$ in both equations

Replacing y in the first equation by $x - 2$ gives us an equation we can solve for x.

First equation:	$2x + 3y = 14$
Replace y with $x - 2$:	$2x + 3(x - 2) = 14$
Distribute:	$2x + 3x - 6 = 14$
Simplify:	$5x - 6 = 14$
Add 6 to both sides:	$5x = 20$
Divide both sides by 5:	$x = 4$

This value for x can then be used to determine the corresponding value for y. We can choose either original equation to find y. Here we substitute 4 for x in the second equation.

Second equation:	$y = x - 2$
Replace x with 4:	$y = (4) - 2$
Simplify:	$y = 2$

The **solution** to this system is the ordered pair $(4, 2)$.

The **substitution method** involves solving one of the equations for one **variable**, substituting the resulting expression into the other equation, and then solving for the remaining variable. The substitution method can be summarized in four steps.

Solving Systems of Linear Equations in Two Variables by Substitution

Step 1. Choose an equation and solve for one variable in terms of the other variable.

Step 2. Substitute the expression from Step 1 into the other equation.

Step 3. Solve the equation in one variable from Step 2.

Step 4. Substitute the solution from Step 3 into one of the original equations to find the value of the other variable.

CAUTION It is always a good idea to check the ordered-pair solution by **substituting** the x- and y-values into *both* original equations to see if both equations result in true statements.

(eText Screens 13.2-1–13.2-17)

Example 1 Solving Systems of Linear Equations by Substitution

Use the **substitution method** to solve the following system:

$$\begin{cases} 4x + 2y = 10 \\ y = 3x - 10 \end{cases}$$

Solution

Step 1. The second equation, $y = 3x - 10$, is already solved for y.

Step 2. Substitute $3x - 10$ for y in the first equation.

First equation: $4x + 2y = 10$

Substitute $3x - 10$ for y: $4x + 2\overbrace{(3x - 10)}^{y} = 10$

Step 3. Solve for x.

Equation from Step 2: $4x + 2(3x - 10) = 10$

Distribute: $4x + 6x - 20 = 10$

Simplify: $10x - 20 = 10$

Add 20 to both sides: $10x = 30$

Divide both sides by 10: $x = \frac{30}{10} = 3$

Step 4. Find y by substituting $x = 3$ into one of the original equations.

Original second equation: $y = 3x - 10$

Substitute 3 for x: $y = 3(3) - 10$

Multiply: $y = 9 - 10$

Simplify: $y = -1$

The solution to this system is the **ordered pair** $(3, -1)$. Check this solution in both original equations. View this **popup** to see the check.

Figure 5 shows the graph of the system in Example 1. We see that the solution $(3, -1)$ is the **intersection point** of the two lines in the system.

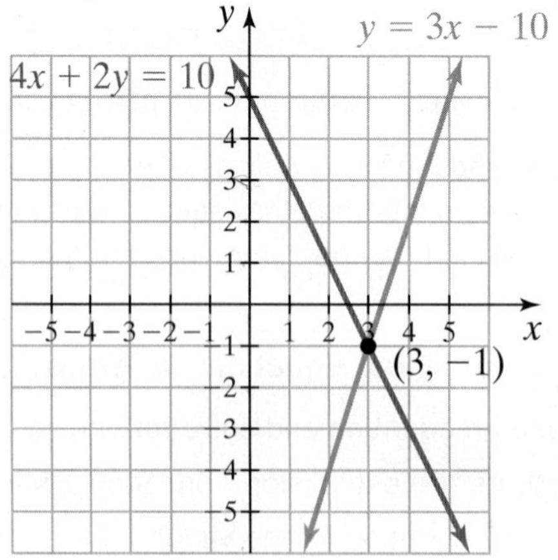

Figure 5

Graph of system $\begin{cases} 4x + 2y = 10 \\ y = 3x - 10 \end{cases}$

You Try It **Work through this You Try It problem.**

Work Exercises 1–8 in this textbook or in the MyMathLab Study Plan.

(eText Screens 13.2-1–13.2-17)

Example 2 Solving Systems of Linear Equations by Substitution

My video summary

Use the **substitution method** to solve the following system:

$$\begin{cases} 3x + y = -6 \\ 2x - \dfrac{1}{3}y = 2 \end{cases}$$

Solution

Step 1. Choose an equation and solve for one of the variables. If possible, choose a variable that has a **coefficient** of 1. In this example, it is easiest to solve the first equation for y.

Begin with the first equation: $3x + y = -6$

Subtract $3x$ from both sides: $y = -3x - 6$

Step 2. Substitute $-3x - 6$ for y in the second equation.

Second equation: $2x - \frac{1}{3}y = 2$

Substitute $-3x - 6$ for y: $2x - \frac{1}{3}\overbrace{(-3x - 6)}^{y} = 2$

Step 3. Solve for x.

Equation from Step 2: $2x - \frac{1}{3}\overbrace{(-3x - 6)}^{y} = 2$

Distribute: $2x + x + 2 = 2$ ← $-\frac{1}{3}(-3x) - \frac{1}{3}(-6) = x + 2$

Simplify: $3x + 2 = 2$

Subtract 2 from both sides: $3x = 0$

Divide both sides by 3: $x = \frac{0}{3} = 0$

Use this value for x to find the value for y and complete the solution to the system. View the **answer**, or watch this **video** for a complete solution.

Example 3 Solving Systems of Linear Equations by Substitution

My video summary

Solve the following system:

$$\begin{cases} 4x + 3y = 7 \\ x + 9y = -1 \end{cases}$$

Solution Try solving this system on your own. View the **answer**, or watch this **video** for the fully worked solution.

You Try It **Work through this You Try It problem.**

Work Exercises 9–16 in this textbook or in the MyMathLab Study Plan.

(eText Screens 13.2-1–13.2-17)

Example 4 Solving Systems of Linear Equations by Substitution

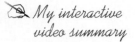

My interactive video summary

Use the **substitution method** to solve the following system:

$$\begin{cases} 6x - 3y = -33 \\ 2x + 4y = 4 \end{cases}$$

Solution In this system, we do not see any variables with a **coefficient** of 1 or −1. We can select either equation and solve for either variable. Notice that if we divide both sides of the first equation by 3, we get

$$\frac{6x}{3} - \frac{3y}{3} = \frac{-33}{3} \rightarrow 2x - y = -11.$$

And if we divide both sides of the second equation by 2, we get

$$\frac{2x}{2} + \frac{4y}{2} = \frac{4}{2} \rightarrow x + 2y = 2.$$

Try solving this system on your own. View the **answer**, or watch this **interactive video** for a fully worked solution.

You Try It **Work through this You Try It problem.**

Work Exercises 17–20 in this textbook or in the MyMathLab Study Plan.

OBJECTIVE 2 SOLVE SPECIAL SYSTEMS BY SUBSTITUTION

Recall that a **system of linear equations in two variables** will always result in one of three possible situations. We see these situations graphically in Figure 6.

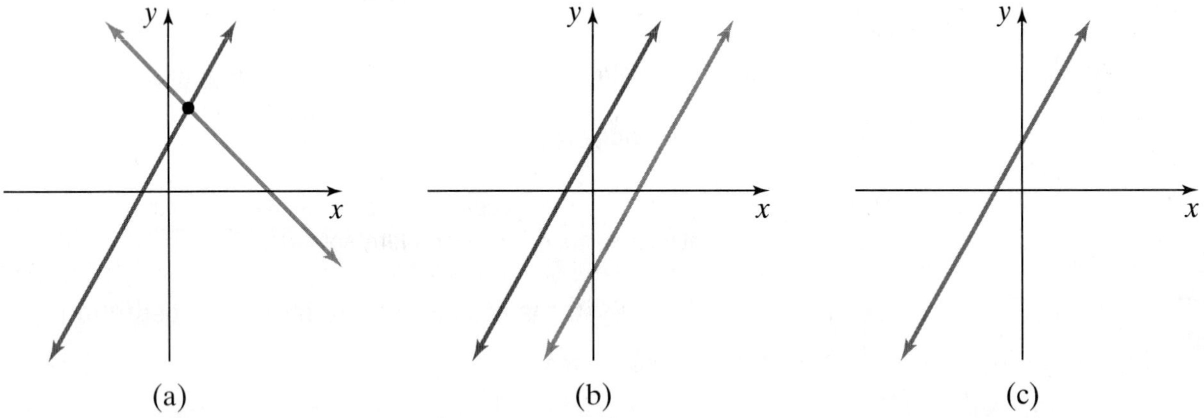

(a)
Intersecting Lines
One solution
Consistent, Independent
Two lines are different, having one common point. The lines have different slopes.

(b)
Parallel Lines
No solution
Inconsistent, Independent
Two lines are different, having no common points. The lines have the same slope but different y-intercepts.

(c)
Coinciding Lines
Infinitely many solutions
Consistent, Dependent
Two lines are the same, having infinitely many common points. The lines have the same slope and same y-intercept.

Figure 6

When solving a system algebraically, we recognize these three situations as follows:

1. **A value for both variables exists.** This is the "one solution" found for **intersecting lines.** The system is **independent** and **consistent.**
2. **A contradiction occurs.** This is the "no solution" found for **parallel lines.** The system is independent and **inconsistent.**
3. **An identity results.** This is the "infinite solutions" found for **coinciding lines.** The system is **dependent** and consistent.

So far we have focused on solving systems with one solution. We now consider systems with no solution or an infinite number of solutions.

Example 5 Solving a Dependent System by Substitution

Use the **substitution method** to solve the following system:

$$\begin{cases} 2x + 10y = 8 \\ x + 5y = 4 \end{cases}$$

Solution

Step 1. Solve the second equation for x.

Begin with the second equation: $x + 5y = 4$

Subtract $5y$ from both sides: $x = -5y + 4$

Step 2. Substitute $-5y + 4$ for x in the first equation.

First equation: $2x + 10y = 8$

Substitute $-5y + 4$ for x: $2(-5y + 4) + 10y = 8$

Step 3. Solve for y.

Distribute: $-10y + 8 + 10y = 8$

Combine like terms: $8 = 8$ True

Our equation in Step 3 simplifies to $8 = 8$, which is an **identity**, so the equation is true for any value of y. For each y, there is a corresponding value for x such that $x = -5y + 4$. So, the system has an infinite number of solutions and is a **dependent system**. Any ordered pair that is a solution to one of the equations is a solution to both equations.

Figure 7 shows that the graphs of the two equations in the system are **coinciding lines.** The **solution to the system** is the set of all ordered pairs that lie on the graph of either equation.

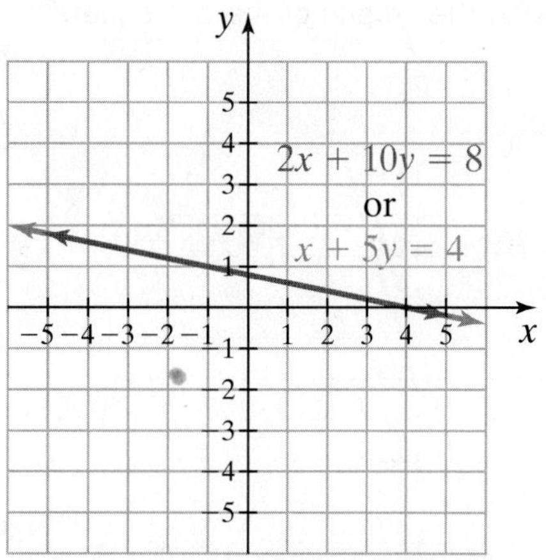

Figure 7

Graph of system $\begin{cases} 2x + 10y = 8 \\ x + 5y = 4 \end{cases}$

(eText Screens 13.2-1–13.2-17)

Because there are an **infinite** number of solutions to **dependent systems**, it is not possible to write all the solutions. Instead, we can use **set-builder notation**, including one of the equations from the system (or any **equivalent equation**). The solution set for the system in **Example 5** can be written as $\{(x, y) \mid x + 5y = 4\}$. We read this as "the set of all ordered pairs (x, y), such that $x + 5y = 4$." We can also write the solution set equivalently as $\{(x, y) \mid 2x + 10y = 8\}$ by using the first equation, or as $\left\{(x, y) \,\middle|\, y = -\frac{1}{5}x + \frac{4}{5}\right\}$ by writing either equation in **slope-intercept form**. For convenience, we will agree to use slope-intercept form when writing the equation of the common line in a dependent system.

CAUTION

Although dependent systems have an infinite number of ordered-pair solutions, not all ordered pairs are solutions. The solution set consists only of those ordered pairs whose points lie on the coinciding lines.

Example 6 Solving an Inconsistent System by Substitution

Use the **substitution method** to solve the following system:

$$\begin{cases} 3x - y = -1 \\ -12x + 4y = 8 \end{cases}$$

Solution

Step 1. Solve the first equation for y.

First equation: $3x - y = -1$

Subtract $3x$ from both sides: $-y = -3x - 1$

Multiply both sides by -1: $y = 3x + 1$

Step 2. Substitute $3x + 1$ for y in the second equation.

Second equation: $-12x + 4y = 8$

Substitute $3x + 1$ for y: $-12x + 4(3x + 1) = 8$

Step 3. Solve for x.

Distribute: $-12x + 12x + 4 = 8$

Combine like terms: $4 = 8$ False

The equation in Step 3 simplifies to $4 = 8$, which is a **contradiction**. This means that the equation is never true. Therefore, the system has no solution and is **inconsistent**. The solution set is $\varnothing$ or $\{\ \}$.

Figure 8 shows that the graphs of the two equations in the system are **parallel lines**.

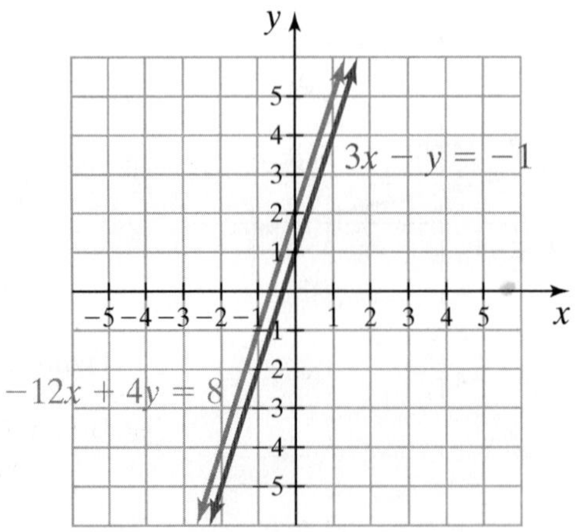

Figure 8

Graph of system $\begin{cases} 3x - y = -1 \\ -12x + 4y = 8 \end{cases}$

(eText Screens 13.2-1–13.2-17)

You Try It Work through this You Try It problem.

Work Exercises 21–24 in this textbook or in the MyMathLab Study Plan.

Example 7 Solving Special Systems by Substitution

My interactive video summary

Use the **substitution method** to solve the following systems:

a. $\begin{cases} \frac{1}{4}x + y = 5 \\ x + 4y = 8 \end{cases}$ b. $\begin{cases} -2.4x + 1.5y = -3 \\ 0.8x - 0.5y = 1 \end{cases}$

Solutions Try solving these systems on your own. View the **answers**, or watch this **interactive video** for fully worked solutions.

You Try It Work through this You Try It problem.

Work Exercises 25–28 in this textbook or in the MyMathLab Study Plan.

13.2 Exercises

In Exercises 1–20, solve each system by substitution.

You Try It

1. $\begin{cases} y = 3x - 2 \\ 3x + 5y = 44 \end{cases}$

2. $\begin{cases} 2x - 3y = 17 \\ x = 2y + 10 \end{cases}$

3. $\begin{cases} x = 4y - 3 \\ 5x - 3y = 2 \end{cases}$

4. $\begin{cases} 5x - 2y = -1 \\ y = 3x \end{cases}$

5. $\begin{cases} 6x + y = 9 \\ y = 3 - 3x \end{cases}$

6. $\begin{cases} y = 2x - 5 \\ y = 5x - 2 \end{cases}$

7. $\begin{cases} \frac{2}{3}x + \frac{5}{2}y = -3 \\ x = \frac{2}{3}y + \frac{13}{3} \end{cases}$

8. $\begin{cases} y = \frac{3}{7}x + \frac{1}{7} \\ \frac{1}{2}x + \frac{3}{8}y = \frac{11}{8} \end{cases}$

You Try It

9. $\begin{cases} 2x + y = 7 \\ 4x - 7y = 5 \end{cases}$

10. $\begin{cases} x + y = 5 \\ 3x - y = -13 \end{cases}$

11. $\begin{cases} x - 4y = 6 \\ 7x - 5y = -4 \end{cases}$

12. $\begin{cases} 5x - 3y = -18 \\ -2x + y = 7 \end{cases}$

13. $\begin{cases} 5x - 2y = 0 \\ 3x + y = 0 \end{cases}$

14. $\begin{cases} x - 2y = 7 \\ 3x + 2y = -3 \end{cases}$

15. $\begin{cases} 2x + \frac{3}{2}y = 1 \\ x + 3y = 7 \end{cases}$

16. $\begin{cases} 1.5x + y = 1.1 \\ 2.4x - 3y = 8.4 \end{cases}$

You Try It

17. $\begin{cases} 2x + 4y = 6 \\ 5x + 2y = -1 \end{cases}$

18. $\begin{cases} 3x + 6y = \frac{19}{2} \\ \frac{8}{5}x + 3y = 5 \end{cases}$

19. $\begin{cases} 1.7x + 3.2y = 0.22 \\ 2.1x - 2.1y = 6.93 \end{cases}$

20. $\begin{cases} \frac{1}{2}x + \frac{5}{3}y = \frac{2}{3} \\ 3x - 2y = \frac{8}{5} \end{cases}$

(*eText Screens 13.3-1–13.3-17*)

In Exercises 21–28, solve each special system by substitution. If the system is dependent, write the solution using set-builder notation with the common line expressed in slope-intercept form.

You Try It **21.** $\begin{cases} 6x + 2y = 18 \\ y = 8 - 3x \end{cases}$

22. $\begin{cases} 2x + 6y = 18 \\ x + 3y = 9 \end{cases}$

23. $\begin{cases} 5x - 15y = -10 \\ -x + 3y = 2 \end{cases}$

24. $\begin{cases} -2x + y = 11 \\ 6x - 3y = 5 \end{cases}$

You Try It **25.** $\begin{cases} x - 3y = -1 \\ -2x + 6y = 2 \end{cases}$

26. $\begin{cases} y = 4x + 3 \\ y = 4x - 3 \end{cases}$

27. $\begin{cases} 3.4x - 5.1y = -1.7 \\ -2.4x + 3.6y = 1.2 \end{cases}$

28. $\begin{cases} \frac{2}{3}x + \frac{7}{3}y = \frac{13}{3} \\ \frac{4}{5}x + \frac{14}{5}y = \frac{18}{5} \end{cases}$

13.3 Solving Systems of Linear Equations by Elimination

THINGS TO KNOW

Before working through this topic, be sure you are familiar with the following concepts:

	VIDEO	ANIMATION	INTERACTIVE
You Try It 1. Determine If an Ordered Pair Is a Solution to a System of Linear Equations in Two Variables (Topic 13.1, **Objective 1**)			✓
You Try It 2. Determine the Number of Solutions to a System Without Graphing (Topic 13.1, **Objective 2**)	✓		
You Try It 3. Solve Systems of Linear Equations by Graphing (Topic 13.1, **Objective 3**)	✓		
You Try It 4. Solve Systems of Linear Equations by Substitution (Topic 13.2, **Objective 1**)	✓		
You Try It 5. Solve Special Systems by Substitution (Topic 13.2, **Objective 2**)			✓

OBJECTIVES

1 Solve Systems of Linear Equations by Elimination

2 Solve Special Systems by Elimination

OBJECTIVE 1 SOLVE SYSTEMS OF LINEAR EQUATIONS BY ELIMINATION

The **elimination method** for solving a **system of linear equations** in two variables involves adding the two equations together in a way that will *eliminate* one of the **variables.** This algebraic method is based on the following logic.

> **Logic for the Elimination Method**
>
> If $A = B$ and $C = D$, then $A + C = B + D$.

This means that if two true equations are added, then the result will be a third true equation. Because equations are added, the elimination method is also known as the **addition method**.

Example 1 Solving Systems of Linear Equations by Elimination

Solve the following system.

$$\begin{cases} x + y = 8 \\ x - y = -2 \end{cases}$$

Solution Notice that the first equation includes "$+y$" while the second equation includes "$-y$." Let's see what happens when we add these two equations together.

$$\begin{array}{r} x + y = 8 \\ x - y = -2 \\ \hline 2x + 0 = 6 \end{array}$$

The y-variable is eliminated. By solving the resulting equation for x, we can find the **x-coordinate** of the **solution to the system.**

Rewrite the equation:	$2x + 0 = 6$
Simplify:	$2x = 6$
Divide both sides by 2:	$x = 3$

The value of x is 3. We can now substitute $x = 3$ into one of the original equations to find the value of y.

Begin with the first original equation:	$x + y = 8$
Substitute 3 for x:	$3 + y = 8$
Subtract 3 from both sides:	$3 + y - 3 = 8 - 3$
Simplify:	$y = 5$

The solution is $(3, 5)$. Now we can check $(3, 5)$ in both of the original equations.

	First Equation	Second Equation
Begin with each original equation:	$x + y = 8$	$x - y = -2$
Substitute $x = 3$ and $y = 5$:	$3 + 5 \stackrel{?}{=} 8$	$3 - 5 \stackrel{?}{=} -2$
Simplify:	$8 = 8$ True	$-2 = -2$ True

The ordered pair checks, so $(3, 5)$ is the solution to the system.

Figure 9 shows the graph of the system. We see that the solution $(3, 5)$ is the **intersection point** of the two lines in the system.

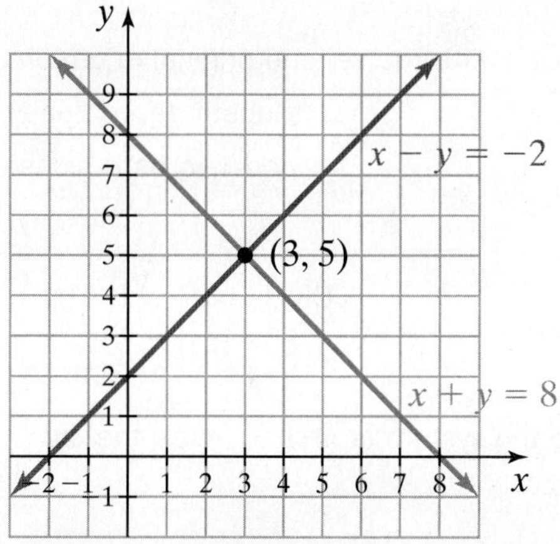

Figure 9

Graph of system $\begin{cases} x + y = 8 \\ x - y = -2 \end{cases}$

You Try It **Work through this You Try It problem.**

Work Exercises 1–6 in this textbook or in the MyMathLab Study Plan.

In **Example 1**, the y-variable was eliminated simply by adding the two equations together. Unfortunately, a variable will not always be eliminated so easily. Consider the following system:

$$\begin{cases} x - y = -4 \\ x + 2y = 5 \end{cases}$$

Adding these two equations as a first step does not eliminate a variable:

$$\begin{array}{r} x - y = -4 \\ \underline{x + 2y = 5} \\ 2x + y = 1 \end{array}$$

In cases like this, we must write **equivalent equations** in the system before adding so that a variable will be eliminated.

To eliminate a variable, the **coefficients** of the variable in the two equations must be **opposites**. We can make this happen by multiplying one or both of the equations by an appropriate nonzero **constant**.

Example 2 Solving Systems of Linear Equations by Elimination

Solve the following system.

$$\begin{cases} x - y = -4 \\ x + 2y = 5 \end{cases}$$

Solution Multiplying both sides of the first equation by 2 will make the coefficient of its y-variable -2, which is the **opposite** of the coefficient of the y-variable in the second equation. Add the result to the second equation in order to eliminate the y-variable.

$$x - y = -4 \xrightarrow{\text{Multiply by 2}} 2(x - y) = 2(-4) \xrightarrow{\text{Simplify}} 2x - 2y = -8$$

$$x + 2y = 5 \xrightarrow{\text{No change}} x + 2y = 5 \xrightarrow{\text{No change}} \underline{x + 2y = 5}$$

Add the new equations: $3x = -3$

Solve for x by dividing both sides by 3: $x = -1$

The value of x is -1. Substitute $x = -1$ into one of the original equations to find the value of y.

Begin with the second original equation: $x + 2y = 5$

Substitute -1 for x: $-1 + 2y = 5$

Add 1 to both sides: $-1 + 2y + 1 = 5 + 1$

Simplify: $2y = 6$

Divide both sides by 2: $\frac{2y}{2} = \frac{6}{2}$

Simplify: $y = 3$

The **solution to the system** is $(-1, 3)$. View the **check**.

(eText Screens 13.3-1–13.3-17)

You Try It **Work through this You Try It problem.**

Work Exercises 7–9 in this textbook or in the MyMathLab Study Plan.

The elimination method can be summarized in five steps.

Solving Systems of Linear Equations in Two Variables by Elimination

Step 1. Choose a **variable** to eliminate.

Step 2. If necessary, multiply one or both equations by an appropriate nonzero **constant** so that the sum of the **coefficients** of one of the variables is zero.

Step 3. Add the two equations from Step 2 together to obtain an equation in one variable.

Step 4. Solve the equation in one variable from Step 3.

Step 5. Substitute the value found in Step 4 into one of the original equations to solve for the other variable.

CAUTION It is good practice to check the solution. Do this by substituting the ordered pair into the original equation that was not used in Step 5.

Example 3 Solving Systems of Linear Equations by Elimination

My video summary

Use the **elimination method** to solve the following system.

$$\begin{cases} x - 3y = -9 \\ 5x + 4y = -7 \end{cases}$$

Solution

Step 1. We choose to eliminate the x-variable because the **coefficients** of the x-variables will be **opposites** if we multiply the first equation by -5.

Step 2. Multiply both sides of the first equation by -5.

$$x - 3y = -9 \xrightarrow{\text{Multiply by } -5} -5(x - 3y) = -5(-9) \xrightarrow{\text{Simplify}} -5x + 15y = 45$$

$$5x + 4y = -7 \xrightarrow{\text{No change}} 5x + 4y = -7 \xrightarrow{\text{No change}} 5x + 4y = -7$$

Step 3. Add the two new equations together.

$$\begin{array}{r} -5x + 15y = 45 \\ 5x + 4y = -7 \\ \hline 19y = 38 \end{array}$$

Step 4. Solve the resulting equation for y.

Rewrite the resulting equation from Step 3: $19y = 38$

Divide both sides by 19: $y = 2$

Step 5. Find x by substituting $y = 2$ into one of the original equations.

Begin with the original first equation: $x - 3y = -9$

Substitute 2 for y: $x - 3(2) = -9$

Simplify: $x - 6 = -9$

Add 6 to both sides: $x - 6 + 6 = -9 + 6$

Simplify: $x = -3$

The **solution to the system** is $(-3, 2)$

(eText Screens 13.3-1–13.3-17)

Since we used the original first equation in Step 5, we use the original second equation to check the final answer. View the **check**, or watch this **video** for a fully worked solution.

You Try It **Work through this You Try It problem.**

Work Exercises 10–12 in this textbook or in the MyMathLab Study Plan.

In **Example 3**, we had to multiply one equation by a nonzero constant to eliminate a variable. To solve some systems, we must multiply both equations by nonzero constants. Consider the next example.

Example 4 Solving Systems of Linear Equations by Elimination

My video summary

Use the **elimination method** to solve the following system.

$$\begin{cases} -3x + 4y = 7 \\ 5x + 6y = 1 \end{cases}$$

Solution

Step 1. For this problem, we choose to eliminate the x-variable. The **coefficients** of the x-variables will be **opposites** if we multiply the first equation by 5 and the second equation by 3.

Step 2. Multiply both sides of the first equation by 5, and multiply both sides of the second equation by 3.

$$-3x + 4y = 7 \xrightarrow{\text{Multiply by 5}} 5(-3x + 4y) = 5(7) \xrightarrow{\text{Simplify}} -15x + 20y = 35$$

$$5x + 6y = 1 \xrightarrow{\text{Multiply by 3}} 3(5x + 6y) = 3(1) \xrightarrow{\text{Simplify}} 15x + 18y = 3$$

Step 3. Add the two new equations together.

$$\begin{aligned} -15x + 20y &= 35 \\ 15x + 18y &= 3 \\ \hline 38y &= 38 \end{aligned}$$

Step 4. Solve the resulting equation for y.

Rewrite the resulting equation from Step 3: $38y = 38$

Divide both sides by 38: $y = 1$

Step 5. Try to finish solving this problem on your own. See the **answer**, or watch this **video** for the complete solution.

Example 5 Solving Systems of Linear Equations by Elimination

My video summary

Use the **elimination method** to solve the following system.

$$\begin{cases} 5x - 6y = 20 \\ 4x + 9y = 16 \end{cases}$$

Solution Try solving this system on your own. View the **answer**, or watch this **video** for a fully worked solution.

You Try It **Work through this You Try It problem.**

Work Exercises 13–18 in this textbook or in the MyMathLab Study Plan.

If the equations in a system include fractions, then it is usually helpful to **clear the fractions** before proceeding. It may also be helpful to **clear any decimals** from a system.

Example 6 Solving Systems of Linear Equations Involving Fractions

My video summary

Use the **elimination method** to solve the following system.

$$\begin{cases} x - \frac{3}{5}y = \frac{4}{5} \\ \frac{1}{2}x + 3y = -\frac{9}{5} \end{cases}$$

Solution We begin by clearing the fractions from the equations. Multiply both sides of the first equation by the **LCD** 5, and multiply both sides of the second equation by the LCD 10.

$$x - \frac{3}{5}y = \frac{4}{5} \xrightarrow{\text{Multiply by 5}} 5\left(x - \frac{3}{5}y\right) = 5\left(\frac{4}{5}\right) \xrightarrow{\text{Simplify}} 5x - 3y = 4$$

$$\frac{1}{2}x + 3y = -\frac{9}{5} \xrightarrow{\text{Multiply by 10}} 10\left(\frac{1}{2}x + 3y\right) = 10\left(-\frac{9}{5}\right) \xrightarrow{\text{Simplify}} 5x + 30y = -18$$

Try to finish this problem on your own by solving the resulting system.

$$\begin{cases} 5x - 3y = 4 \\ 5x + 30y = -18 \end{cases}$$

See the **answer**, or watch this **video** for a fully worked solution.

You Try It **Work through this You Try It problem.**

Work Exercises 19–22 in this textbook or in the MyMathLab Study Plan.

OBJECTIVE 2 SOLVE SPECIAL SYSTEMS BY ELIMINATION

Now we look at **inconsistent** and **dependent systems**. Like the **substitution method**, when solving by **elimination**, an inconsistent system will lead to a **contradiction** and a dependent system will lead to an **identity**.

Example 7 Solving Inconsistent and Dependent Systems by Elimination

Use the elimination method to solve each system.

a. $\begin{cases} 3x + y = 6 \\ 6x + 2y = 4 \end{cases}$ **b.** $\begin{cases} 2x - 8y = 6 \\ 3x - 12y = 9 \end{cases}$

Solutions

a. Step 1. For this problem, we choose to eliminate the x-variable. The **coefficients** of the x-variables will be **opposites** if we multiply the first equation by -2.

Step 2. Multiply both sides of the first equation by -2.

$$3x + y = 6 \xrightarrow{\text{Multiply by } -2} -2(3x + y) = -2(6) \xrightarrow{\text{Simplify}} -6x - 2y = -12$$

$$6x + 2y = 4 \xrightarrow{\text{No change}} 6x + 2y = 4 \xrightarrow{\text{No change}} 6x + 2y = 4$$

Step 3. Add the two new equations together.

$$\begin{array}{r} -6x - 2y = -12 \\ 6x + 2y = 4 \\ \hline 0 = -8 \end{array} \quad \text{False}$$

Both variables are eliminated, and we are left with the **contradiction** $0 = -8$. The system has no solution and is **inconsistent**. The solution set is $\varnothing$ or $\{\}$. Figure 10 shows that the graphs of the two linear equations in this system are **parallel**.

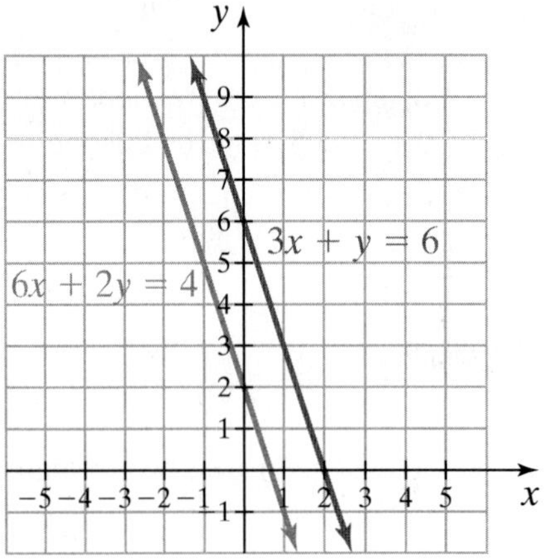

Figure 10

Graph of system $\begin{cases} 3x + y = 6 \\ 6x + 2y = 4 \end{cases}$

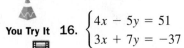

My video summary

b. Try to solve this system on your own. View the **answer**, or watch this **video** for a complete solution.

You Try It **Work through this You Try It problem.**

Work Exercises 23–26 in this textbook or in the MyMathLab Study Plan.

13.3 Exercises

In Exercises 1–22, solve each system by elimination.

You Try It **1.** $\begin{cases} x + y = 8 \\ x - y = 4 \end{cases}$ **2.** $\begin{cases} x + 4y = 1 \\ -x + 2y = 5 \end{cases}$ **3.** $\begin{cases} x - 5y = -3 \\ x + 5y = 17 \end{cases}$

4. $\begin{cases} -4x + 3y = -4 \\ 4x + 5y = -28 \end{cases}$ **5.** $\begin{cases} -3x + 4y = -15 \\ 2x - 4y = 10 \end{cases}$ **6.** $\begin{cases} 10x - 9y = 11 \\ 2x + 9y = -5 \end{cases}$

You Try It **7.** $\begin{cases} x + 3y = 4 \\ 2x - y = -6 \end{cases}$ **8.** $\begin{cases} 2x - 5y = 13 \\ 6x - 5y = 29 \end{cases}$ **9.** $\begin{cases} 8x - 3y = -17 \\ 2x + y = 1 \end{cases}$

You Try It **10.** $\begin{cases} 4x - y = 10 \\ x - 3y = -25 \end{cases}$ **11.** $\begin{cases} 10x - 3y = 45 \\ 2x - 5y = -13 \end{cases}$ **12.** $\begin{cases} 7x - 2y = -22 \\ 5x + 6y = 14 \end{cases}$

13. $\begin{cases} 2x + 5y = 3 \\ 3x + 2y = -12 \end{cases}$ **14.** $\begin{cases} 3x + 5y = -20 \\ -5x + 4y = -16 \end{cases}$ **15.** $\begin{cases} 2x + 3y = 13 \\ 5x + 2y = -6 \end{cases}$

You Try It **16.** $\begin{cases} 4x - 5y = 51 \\ 3x + 7y = -37 \end{cases}$ **17.** $\begin{cases} 9x - 2y = 0 \\ 5x + 3y = 0 \end{cases}$ **18.** $\begin{cases} 4x + 9y = 1 \\ 6x + 6y = -1 \end{cases}$

(eText Screens 13.4-1–13.4-24)

You Try It

19. $\begin{cases} \frac{1}{4}x - y = \frac{9}{8} \\ x + \frac{3}{5}y = -\frac{1}{10} \end{cases}$

20. $\begin{cases} \frac{1}{3}x + \frac{1}{2}y = -\frac{1}{3} \\ \frac{1}{2}x + \frac{1}{3}y = -\frac{4}{3} \end{cases}$

21. $\begin{cases} 0.3x + 0.2y = 1 \\ 0.1x - 0.3y = 1.8 \end{cases}$

22. $\begin{cases} 0.5x - 0.4y = 1.5 \\ 0.1x - 0.2y = 0.3 \end{cases}$

In Exercises 23–26, solve each special system by elimination. If the system is dependent, write the solution using set-builder notation with the common line expressed in slope-intercept form.

You Try It

23. $\begin{cases} 2x - 6y = 12 \\ -3x + 9y = -18 \end{cases}$

24. $\begin{cases} -x - 5y = 7 \\ 8x + 40y = -25 \end{cases}$

25. $\begin{cases} 6x - 10y = 40 \\ -9x + 15y = 45 \end{cases}$

26. $\begin{cases} 6x - 12y = 9 \\ -4x + 8y = -6 \end{cases}$

13.4 Applications of Linear Systems

THINGS TO KNOW

Before working through this topic, be sure you are familiar with the following concepts:

VIDEO ANIMATION INTERACTIVE

You Try It 1. Solve Problems Involving Related Quantities (Topic 11.3, **Objective 3**)

You Try It 2. Solve Problems Using Geometry Formulas (Topic 11.5, **Objective 1**)

You Try It 3. Solve Problems Involving Angles (Topic 11.5, **Objective 2**)

You Try It 4. Solve Problems Involving Uniform Motion (Topic 11.5, **Objective 3**)

You Try It 5. Solve Mixture Problems (Topic 11.6, **Objective 4**)

You Try It 6. Solve Systems of Linear Equations by Substitution (Topic 13.2, **Objective 1**)

You Try It 7. Solve Systems of Linear Equations by Elimination (Topic 13.3, **Objective 1**)

OBJECTIVES

1 Solve Related Quantity Applications Using Systems

2 Solve Geometry Applications Using Systems

3 Solve Uniform Motion Applications Using Systems

4 Solve Mixture Applications Using Systems

(*eText Screens 13.4-1–13.4-24*)

OBJECTIVE 1 SOLVE RELATED QUANTITY APPLICATIONS USING SYSTEMS

Sometimes we can use an **equation in one variable** to solve application problems such as those in **Topic 11.3**. However, it is often easier to use two variables and create a **system of linear equations**. The following steps are based on the **problem-solving strategy** for applications of linear equations from Topic 11.3.

Problem-Solving Strategy for Applications Using Systems of Linear Equations

Step 1. **Define the Problem.** Read the problem carefully; multiple times if necessary. Identify what you need to find and determine what information is available to help you find it.

Step 2. **Assign Variables.** Choose **variables** that describe each unknown quantity.

Step 3. **Translate into a System of Equations.** Use the relationships among the known and unknown quantities to form a system of equations.

Step 4. **Solve the System.** Use **graphing**, **substitution**, or **elimination** to solve the system.

Step 5. **Check the Reasonableness of Your Answers.** Check to see if your answers make sense within the context of the problem. If not, check your work for errors and try again.

Step 6. **Answer the Question.** Write a clear statement that answers the question(s).

Example 1 Storage Capacity

The storage capacity of Deon's external hard drive is 32 times that of his jump drive, a small portable memory device. Together, his two devices have 264 gigabytes of memory. What is the memory size of each device?

Solution Follow the **problem-solving strategy** for applications using systems.

Step 1. We must find the "memory size" or number of gigabytes that each device holds. The memory size of the external hard drive is 32 times that of the jump drive, and the total capacity is 264 gigabytes.

Step 2. Let e = the memory size of the external hard drive and j = the memory size of the jump drive. Note that we use the descriptive variables e for external and j for jump because they are easier to remember than other letters such as x or y.

Step 3. Because the memory of the external hard drive is 32 times that of the jump drive, we have the equation $e = 32j$.

Also, the two devices have a total of 264 gigabytes of memory, so we have the equation $e + j = 264$.

The two equations together give the following **system**:

$$\begin{cases} e = 32j \\ e + j = 264 \end{cases}$$

Step 4. Because the first equation is solved for e, we will solve the system using **substitution**.

Rewrite the second equation: $e + j = 264$

Substitute $32j$ for e: $(32j) + j = 264$

Simplify: $33j = 264$

Divide both sides by 33: $j = 8$

To find e, substitute 8 for j in the first equation.

Rewrite the first equation: $e = 32j$

Substitute 8 for j: $e = 32(8)$

Simplify: $e = 256$

Step 5. The jump drive holds 8 gigabytes of memory, and the external hard drive holds 256 gigabytes. Because 256 is 32 times 8 and the sum of 8 and 256 is 264, these results make sense.

Step 6. Deon's jump drive holds 8 gigabytes of memory, and his external hard drive holds 256 gigabytes of memory.

Did the problem in Example 1 look familiar? Look back to the solution shown in **Example 4 in Topic 11.3**. Now compare that solution process with the one just shown in Example 1. Which approach do you prefer? Algebra problems often can be solved in more than one way!

Example 2 Ages of a Brother and Sister

My interactive video summary

The **sum** of the ages of Ben and his younger sister Annie is 18 years. The **difference** of their ages is 4 years. What is the age of each child?

Solution We follow the **problem-solving strategy** for applications using systems.

Step 1. We need to find the ages of Ben and Annie. We know that the sum of their ages is 18 years and the difference of their ages is 4 years.

Step 2. Let B = Ben's age and A = Annie's age.

Step 3. The sum of their ages is 18 years, so we have the equation $B + A = 18$.

Also, the difference of their ages is 4 years. Because Ben is the older child, we have the equation $B - A = 4$.

The two equations together give the following **system**:

$$\begin{cases} B + A = 18 \\ B - A = 4 \end{cases}$$

Step 4. Finish solving this system of equations on your own. View the **answer**, or watch this **interactive video** for a fully worked solution.

You Try It **Work through this You Try It problem.**

Work Exercises 1–6 in this textbook or in the MyMathLab Study Plan.

OBJECTIVE 2 SOLVE GEOMETRY APPLICATIONS USING SYSTEMS

In **Topic 11.5**, we solved geometry problems using **formulas**. We can also use **systems of equations** to solve such applications.

Example 3 A Calculator Display Panel

My interactive video summary

The display panel of a graphing calculator has the shape of a rectangle with a **perimeter** of 264 millimeters. If the length of the display panel is 18 millimeters longer than the width, find its dimensions.

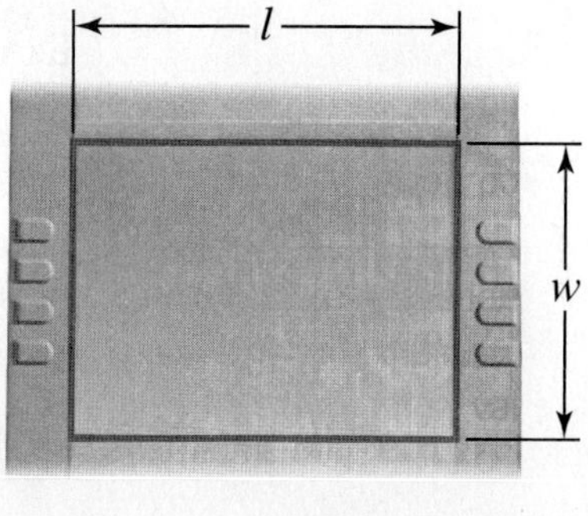

(eText Screens 13.4-1–13.4-24)

Solution

Step 1. We need to find the length and width of the display panel. We know that the perimeter is 264 mm and the length is 18 mm longer than the width.

Step 2. Let l = length and w = width.

Step 3. The perimeter of a rectangle is given by the formula $P = 2l + 2w$, so we have the equation $2l + 2w = 264$.

Also, the length is 18 mm longer than the width, so we have the equation $l = w + 18$.

The two equations together give the following system:

$$\begin{cases} 2l + 2w = 264 \\ l = w + 18 \end{cases}$$

Step 4. Finish solving this system of equations on your own. See the **answer**, or watch this **interactive video** for a fully worked solution.

You Try It **Work through this You Try It problem.**

Work Exercises 7–9 in this textbook or in the MyMathLab Study Plan.

Geometry problems sometimes involve *complementary* and *supplementary* angles. The measures of two **complementary angles** add to 90°. The measures of two **supplementary angles** add to 180°. Complementary angles form a **right angle**, and supplementary angles form a **straight angle**.

Example 4 Supplementary Angles

My interactive video summary

Find the measures of two **supplementary angles** if the measure of the larger angle is 20 degrees less than three times the measure of the smaller angle.

Solution Follow the **problem-solving strategy** for applications using systems and try to solve this problem on your own. View the **answer**, or watch this **interactive video** for a detailed solution.

You Try It **Work through this You Try It problem.**

Work Exercises 10–12 in this textbook or in the MyMathLab Study Plan.

OBJECTIVE 3 SOLVE UNIFORM MOTION APPLICATIONS USING SYSTEMS

My interactive video summary

In **Topic 11.5** we solved **uniform motion** applications involving only one variable. We can use a **system of equations** to solve similar types of problems when there are two variables.

Example 5 Finding Different Times

Shawn is training for the Dirty Duo running-and-bicycling race. During a three-hour training session, his total distance cycling and running was 33 miles. If he cycled at a rate of 18 miles per hour and ran at a rate of 6 miles per hour, how much time did he spend doing each activity?

(eText Screens 13.4-1–13.4-24)

Solution We apply our **problem-solving strategy** for applications using systems.

Step 1. We want to find the time spent cycling and the time spent running. We know the total distance is 33 miles and the total time is 3 hours. We also know that Shawn can cycle 18 miles in one hour and run 6 miles in one hour.

Step 2. We know the time traveled and the distance traveled but not the individual times. We can define two variables to represent the times.

x = time spent cycling

y = time spent running

Step 3. This is a **uniform motion** problem, so we will need the **distance formula,** $d = r \cdot t$. The distance traveled for each activity is given by the product of the person's speed and time traveled. Because the total distance is 33 miles, we can write the following equation:

$$\underbrace{18x}_{\text{Distance cycling}} + \underbrace{6y}_{\text{Distance running}} = \underbrace{33}_{\text{Total distance}}$$

And, because the total time was three hours, we can write the following second equation:

$$x + y = 3$$

Writing the two equations together gives the system:

$$\begin{cases} 18x + 6y = 33 \\ x + y = 3 \end{cases}$$

Step 4. Solve this system by **substitution** or **elimination**. Complete Steps 5 and 6 on your own. View the **answer**, or watch this **interactive video** for a detailed solution.

You Try It **Work through this You Try It problem.**

Work Exercises 13 and 14 in this textbook or in the MyMathLab Study Plan.

Some **uniform motion** problems involve one motion that works with or against another. Walking up the down escalator or paddling upstream are examples of two motions working against each other. Kicking a football with the wind or walking in the direction of motion on a moving sidewalk are examples of two motions working together. For these problems, it is important to remember that when motions work together the **rates** are added, but when they work against each other, the rates are subtracted.

Example 6 Finding Different Rates

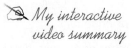

My interactive video summary

A jet plane travels 1950 miles in 3.9 hours going with the wind. On the return trip, the plane must fly into the wind and the travel time increases to 5 hours. Find the speed of the jet plane in still air and the speed of the wind. Assume the wind speed is the same for both trips.

Solution We apply our **problem-solving strategy** for applications using systems.

Step 1. We want to find the speed of the plane in still air and the speed of the wind. We know that the time of travel is 3.9 hours with the wind and 5 hours against (or into) the wind. We also know that the plane travels 1950 miles each way.

(eText Screens 13.4-1–13.4-24)

Step 2. We know the time traveled and the distance traveled but not the speeds. We can define two variables to represent the speeds.

$$r = \text{speed of the plane in still air}$$
$$w = \text{wind speed}$$

Step 3. This is a **uniform motion** problem, so we will need the **distance formula**, $d = r \cdot t$. The distance traveled by the plane is given by the product of the plane's speed and time traveled. When flying with the wind, the plane's speed is found by adding the speed of the plane in still air to the wind speed. This allows us to write the following equation:

$$\underbrace{1950}_{\substack{\text{Distance}\\\text{plane}\\\text{traveled}}} = \underbrace{(r + w)}_{\substack{\text{Plane}\\\text{speed with}\\\text{the wind}}} \cdot \underbrace{(3.9)}_{\substack{\text{Time of}\\\text{travel with}\\\text{the wind}}}$$

We also know that when flying into (or against) the wind, we find the plane's speed by subtracting the speed of the wind from the speed of the plane in still air. The distance traveled is the same, so we get the following second equation:

$$\underbrace{1950}_{\substack{\text{Distance}\\\text{plane}\\\text{traveled on}\\\text{return trip}}} = \underbrace{(r - w)}_{\substack{\text{Plane}\\\text{speed}\\\text{against}\\\text{the wind}}} \cdot \underbrace{(5)}_{\substack{\text{Time of}\\\text{travel}\\\text{against}\\\text{the wind}}}$$

Writing the two equations together gives the system:

$$\begin{cases} 1950 = (r + w)(3.9) \\ 1950 = (r - w)(5) \end{cases}$$

Step 4. Solve the system by **substitution** or **elimination**. Complete Steps 5 and 6 on your own. View the **answer**, or watch this **interactive video** for a detailed solution.

You Try It **Work through this You Try It problem.**

Work Exercises 15 and 16 in this textbook or in the MyMathLab Study Plan.

OBJECTIVE 4 SOLVE MIXTURE APPLICATIONS USING SYSTEMS

Example 7 Mailing Packages

My interactive video summary

A shipping company delivered 160 packages one day. The cost of regular delivery is \$6.50, and the cost for express delivery is \$17.50. Total shipping **revenue** for the day was \$1513. How many of each kind of delivery were made?

Solution We apply our **problem-solving strategy** for systems of linear equations.

Step 1. We want to find the number of regular deliveries and the number of express deliveries. We know that the total number of deliveries was 160 with a total revenue of \$1513. We also know that regular deliveries cost \$6.50 and express deliveries cost \$17.50.

Step 2. We know the total number of deliveries and total revenue, but not the number of each delivery type. We define two variables to represent the number of each type of delivery.

$$R = \text{number of regular deliveries in one day}$$
$$E = \text{number of express deliveries in one day}$$

(eText Screens 13.4-1–13.4-24)

Step 3. For mixture problems, *totals* are good places to start when trying to write equations in a system. The first sentence tells us about the *total* number of deliveries.

A shipping company delivered 160 packages one day.

The total number of deliveries is the sum of the numbers for each delivery type. Therefore, our first equation is

$$\underbrace{R}_{\text{Regular deliveries}} + \underbrace{E}_{\text{Express deliveries}} = \underbrace{160}_{\text{Total deliveries}}$$

The second and third sentences tell us about the *total* **revenue.**

The cost of regular delivery is \$6.50, and the cost for express delivery is \$17.50. Total shipping revenue for the day was \$1513.

Revenue is found by multiplying the number of packages delivered by the charge per package. The total revenue is the sum of the revenue from each delivery type. This will be our second equation.

$$\underbrace{6.50R}_{\text{Regular delivery revenue}} + \underbrace{17.50E}_{\text{Express delivery revenue}} = \underbrace{1513}_{\text{Total revenue}}$$

Writing the two equations together gives the system:

$$\begin{cases} R + E = 160 \\ 6.50R + 17.50E = 1513 \end{cases}$$

Step 4. Solve this system by **substitution** or **elimination**. Complete Steps 5 and 6 on your own. View the **answer**, or watch this **interactive video** for a detailed solution.

You Try It **Work through this You Try It problem.**

Work Exercises 17–19 in this textbook or in the MyMathLab Study Plan.

Example 8 Mixing Solutions

My interactive video summary

A chemist needs eight liters of a 50% alcohol solution but only has a 30% solution and an 80% solution available. How many liters of each solution should be mixed to form the needed solution?

Solution We apply our **problem-solving strategy** for applications using systems.

Step 1. We want to find the number of liters of a 30% alcohol solution and the number of liters of an 80% alcohol solution that must be mixed to form 8 liters of a 50% alcohol solution.

Step 2. We know the total number of liters needed for the mixture, but not the number of liters for each type of solution. We define two variables to represent the number of liters of each type.

x = liters of 30% alcohol solution

y = liters of 80% alcohol solution

Step 3. Since we know the total number of liters, our first equation deals with the number of liters.

$$x + y = 8$$

The second equation deals with the percents, but we will consider the amount of alcohol contained in each solution. The total amount of alcohol in the mixture will equal the sum of the alcohol from each solution. If we multiply each percent by the number of liters of that solution type, we can find the number of liters of alcohol in each solution. In words we have

$$\underbrace{(\text{percent})(\text{liters of }30\%)}_{\text{Alcohol in }30\%\text{ solution}} + \underbrace{(\text{percent})(\text{liters of }80\%)}_{\text{Alcohol in }80\%\text{ solution}} = \underbrace{(\text{percent})(\text{total liters of }50\%)}_{\text{Alcohol in }50\%\text{ mixture}}$$

Using the given percents and the defined variables, we get our second equation:

$$0.3(x) + 0.8(y) = 0.5(8)$$

or

$$0.3x + 0.8y = 4$$

So the two equations in our system are

$$\begin{cases} x + \quad y = 8 \\ 0.3x + 0.8y = 4 \end{cases}.$$

Step 4. Solve this system by **substitution** or **elimination**. Complete Steps 5 and 6 on your own. View the **answer**, or watch this **interactive video** for a detailed solution.

You Try It **Work through this You Try It problem.**

Work Exercises 20–22 in this textbook or in the MyMathLab Study Plan.

Example 9 Counting Calories

My interactive video summary

Logan and Payton went to Culver's for lunch. Logan ate two Butterburgers with cheese and a small order of fries for a total of 1801 calories. Payton ate one Butterburger with cheese and two small orders of fries for a total of 1313 calories. How many calories are in a Culver's Butterburger with cheese? How many calories are in a small order of fries? (*Source*: Culver's Nutritional Guide, September 2010)

Solution Apply the **problem-solving strategy** for applications using systems and try solving this problem on your own. View the **answer**, or watch this **interactive video** for the complete solution using either **substitution** or **elimination**.

You Try It **Work through this You Try It problem.**

Work Exercises 23–25 in this textbook or in the MyMathLab Study Plan.

13.4 Exercises

In Exercises 1–25, use a system of linear equations to solve each application.

You Try It

1. **Two Brothers** Clayton is 2 years younger than his brother, Josh. If the sum of their ages is 26, how old is each boy?

2. **Number Problem** The sum of two numbers is 121, while the difference of the two numbers is 15. What are the two numbers?

3. **Weights of Friends** Isaiah weighs 20 pounds more than his friend, Geoff. If the sum of their weights is 340 pounds, how much does each man weigh?

4. **Coin Collection** Susan has a collection of 50 nickels and dimes. If the number of nickels is four times the number of dimes, how many nickels and how many dimes does she have?

5. **Heights of Landmarks** The height of the St. Louis Arch is 19 feet more than twice the height of the Statue of Liberty (from the base of its pedestal). If the difference in their heights is 324.5 feet, find the heights of the two landmarks. (*Sources*: gatewayarch.com and statueofliberty.org)

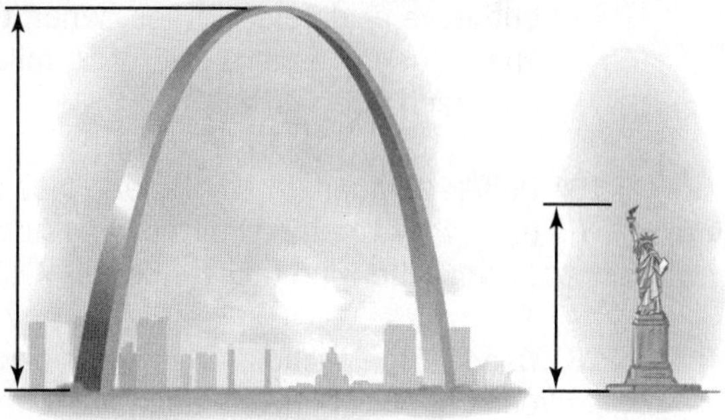

6. **River Lengths** The Missouri River is 200 miles longer than the Mississippi River. If the sum of their lengths is 4880 miles, what is the length of each river? (*Source*: U.S. Geological Survey)

You Try It

7. **TV Dimensions** The flat screen of a 3D television is 49 centimeters longer than it is wide. If the perimeter of the screen is 398 centimeters, find its dimensions.

8. **Tennis Court** A doubles tennis court has a perimeter of 228 feet. If six times the length of the court equals thirteen times the width, what are its dimensions?

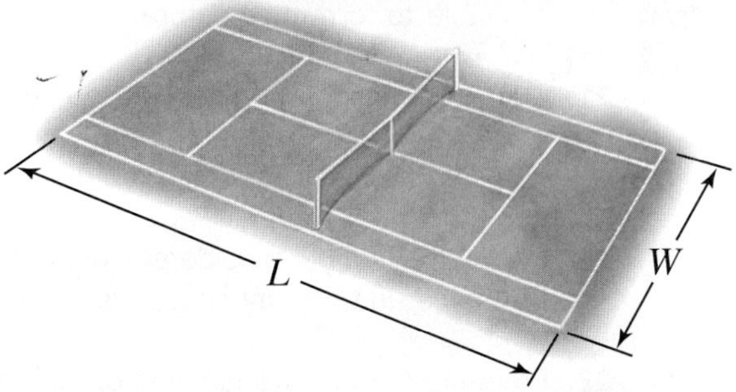

9. **Isosceles Triangle** Recall that an isosceles triangle has two sides with equal lengths. The perimeter of an isosceles triangle is 80 centimeters. If the non-equal side is 10 centimeters shorter than each of the two equal sides, find the lengths of the three sides.

You Try It

10. **Supplementary Angles** Two angles are supplementary. The measure of the larger angle is 8 degrees more than three times the measure of the smaller angle. Find the measure of each angle.

11. **Complementary Angles** Two angles are complementary. The measure of the larger angle is 35 degrees less than four times the measure of smaller angle. Find the measure of each angle.

12. **Right Triangle** Recall that a right triangle contains a 90-degree angle. Also recall that the measures of the three angles in any triangle add to 180 degrees. In the figure shown, the measure of angle y is 6 degrees larger than twice the measure of angle x. Find the measures of angles x and y.

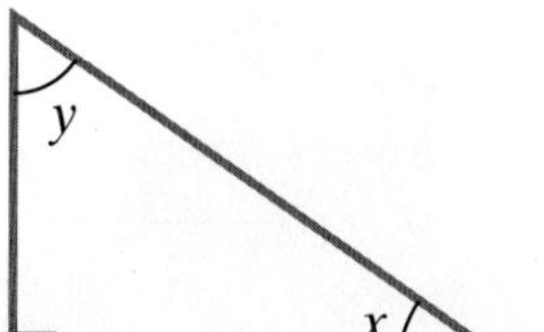

(eText Screens 13.4-1–13.4-24)

You Try It

13. Aquathon Nikola competed in an aquathon (swimming and running) competition. He swam at a rate of 3 km/hr and ran at a rate of 10 km/hr for a total distance traveled of 11.5 km. If he completed the race in 1.5 hours, how long did he take to complete each part of the race?

14. Traveling Apart Two friends, Thaddeus and Ian, start at the same location and drive in opposite directions, but leave at different times. When they are 620 miles apart, their combined travel time is 10 hours. If Thaddeus drives at a rate of 60 miles per hour and Ian drives at a rate of 65 miles per hour, how long had each been driving?

You Try It

15. Puddle Jumper A small plane flies roundtrip between Wichita, KS, and Columbus, OH. The trip is 784 miles each way. The trip with the wind (the wind going in the same direction) takes 3.5 hours, while the trip against the wind takes 4 hours. What is the speed of the plane in still air? What is the speed of the wind?

16. Canoe Trip Nicholas paddles his canoe downstream from the Lodge to Big Bend in 4 hours and then back upstream to the Lodge in 6 hrs. If the distance from the Lodge to Big Bend is 18 miles, find Nicholas' speed in still water and the speed of the current.

You Try It

17. Soda Mix Elodie is buying cans of soda for the school picnic. Her choices are Coke at $0.50 per can and Dr. Pepper at $0.55 per can. She wants to provide a mixture of choices but has a limited budget. If she buys 271 cans and spends $141, how many cans of each flavor did she buy?

18. Pancake Breakfast A youth club held a pancake breakfast to raise money for a trip. Tickets were $3 for children and $5.50 for adults. If 279 tickets were sold and the group took in $1322, find the number of adults and the number of children that attended the breakfast.

19. Trail Mix Chloe purchased some dried mango and dried kiwi to make some trail mix. The dried mango cost $3.86 per pound and the dried kiwi cost $4.60 per pound. If she spent $17.29 for 4 pounds of dried fruit, how many pounds of each fruit did she have?

You Try It

20. Acid Solution A chemist needs 3 liters of a 12% acid solution. He has a 10% solution and a 20% solution available to form the mixture. How much of each should be used to form the 12% solution?

21. Gold Jewelry A jeweler wants to make gold chains that are 60% gold. She has a supply of 18-carat gold (75% gold) and 12-carat gold (50% gold) that can be melted to form the desired alloy. How much 18-carat gold and how much 12-carat gold should be used to form 350 ounces of the 60% alloy?

22. Fruit Juice To be considered "nectar" a drink must contain 40% fruit juice. Find how much pure (100%) fruit juice and how much fruit drink (10% juice) must be mixed to make 9 gallons of "nectar."

You Try It

23. Counting Carbs Two glasses of milk and 3 snack bars have a total of 61 carbs, and 1 glass of milk and 5 snack bars have a total of 76 carbs. How many carbs are in one glass of milk? How many carbs are in one snack bar?

24. Pastry Purchase Bill and Mary Ann went to the Viola bakery. Bill bought 5 danishes and 7 filled donuts for $13.30. Mary Ann bought 6 of each for $13.08. What is the price of each type of pastry?

25. Counting Calories At In-N-Out Burger, 6 cheeseburgers w/onion and 3 orders of french fries contain 4080 calories. 3 cheeseburgers w/onion and 2 orders of french fries contain 2240 calories. How many calories are in a cheeseburger w/onion? How many calories are in one order of french fries? (*Source*: In-N-Out Burger Nutritional Facts, September 2010)

(*eText Screens 13.5-1–13.5-21*)

13.5 Systems of Linear Inequalities

THINGS TO KNOW

Before working through this topic, be sure you are familiar with the following concepts:

		VIDEO	ANIMATION	INTERACTIVE
You Try It	1. Determine If an Ordered Pair Is a Solution to a Linear Inequality in Two Variables (Topic 12.5, **Objective 1**)	✓		
You Try It	2. Graph a Linear Inequality in Two Variables (Topic 12.5, **Objective 2**)			✓
You Try It	3. Solve Applications Involving Linear Inequalities in Two Variables (Topic 12.5, **Objective 3**)	✓		

OBJECTIVES

1 Determine If an Ordered Pair Is a Solution to a System of Linear Inequalities in Two Variables

2 Graph Systems of Linear Inequalities

3 Solve Applications Involving Systems of Linear Inequalities

OBJECTIVE 1 DETERMINE IF AN ORDERED PAIR IS A SOLUTION TO A SYSTEM OF LINEAR INEQUALITIES IN TWO VARIABLES

In **Topic 12.5**, we learned about **linear inequalities in two variables.** In this topic, we explore *systems of linear inequalities in two variables.*

> **Definition System of Linear Inequalities in Two Variables**
>
> A **system of linear inequalities in two variables** is a collection of two or more linear inequalities in two variables considered together.

The following three examples all represent systems of linear inequalities in two variables.

$$\begin{cases} 2x - 3y < 6 \\ 3x + 2y > 12 \end{cases} \qquad \begin{cases} q \geq -\dfrac{5}{2}p + 6 \\ q \leq 3 \end{cases} \qquad \begin{cases} y < -x + 5 \\ x \geq 0 \\ y \geq 0 \end{cases}$$

Notice that a system of linear inequalities looks like a **system of linear equations,** except **inequality symbols** replace the equal signs.

> **Definition Solution to a System of Linear Inequalities in Two Variables**
>
> A **solution to a system of linear inequalities** in two variables is an **ordered pair** that, when substituted for the variables, makes all inequalities in the system true.

(*eText Screens 13.5-1–13.5-21*)

Example 1 Determining If an Ordered Pair Is a Solution to a System of Linear Inequalities in Two Variables

My video summary

Determine if each ordered pair is a **solution** to the following **system of inequalities.**

$$\begin{cases} 2x + y \geq -3 \\ x - 4y \leq 12 \end{cases}$$

a. $(4, 2)$ **b.** $(2, -5)$ **c.** $(0, -3)$

Solutions We substitute the ***x*- and *y*-coordinates** for the variables in each inequality and **simplify**. If the resulting statements are both true, then the ordered pair is a solution to the system.

a.

	First inequality		Second inequality	
Write the original inequalities:	$2x + y \geq -3$		$x - 4y \leq 12$	
Substitute 4 for x and 2 for y:	$2(4) + 2 \overset{?}{\geq} -3$		$4 - 4(2) \overset{?}{\leq} 12$	
Multiply:	$8 + 2 \overset{?}{\geq} -3$		$4 - 8 \overset{?}{\leq} 12$	
Add or subtract:	$10 \geq -3$	True	$-4 \leq 12$	True

The ordered pair makes both inequalities true, so $(4, 2)$ is a solution to the system of inequalities.

b.–c. Try working through these parts on your own, then check your **answers**. Watch this **video** for detailed solutions to all three parts.

You Try It **Work through this You Try It problem.**

Work Exercises 1–3 in this textbook or in the MyMathLab Study Plan.

OBJECTIVE 2 GRAPH SYSTEMS OF LINEAR INEQUALITIES

The **graph of a system of linear inequalities in two variables** is the **intersection** of the graphs of each inequality in the system. This graph represents the set of all **solutions to the system of inequalities.**

My animation summary

To graph a system of linear inequalities, graph each inequality in the system and find the region they all have in common, if any. This common region is called the **solution region**. Watch this **animation** for an overview of graphing systems of linear inequalities, using the system $\begin{cases} 2x - 3y \leq 9 \\ 2x - y > -1 \end{cases}$.

Steps for Graphing Systems of Linear Inequalities

Step 1. Use the **steps for graphing linear inequalities in two variables** from **Topic 12.5** to graph each linear inequality in the system on the same **coordinate plane.**

Step 2. Determine the region where the shaded areas overlap, if any. This region represents the set of all solutions to the system of inequalities.

CAUTION Remember to use a dashed **boundary line** when the inequality is **strict** and a solid boundary line when the inequality is **non-strict.**

(eText Screens 13.5-1–13.5-21)

Note: For the remainder of this topic, we will not find or label the exact **intersection point** of boundary lines. If desired, however, these points could be found by using the **substitution** or **elimination methods** for solving **systems of linear equations**.

Example 2 Graphing Systems of Linear Inequalities

My video summary

Graph the system of linear inequalities from **Example 1**.

$$\begin{cases} 2x + y \geq -3 \\ x - 4y \leq 12 \end{cases}$$

Solution Follow the **steps for graphing systems of linear inequalities**.

Step 1. **Graph $2x + y \geq -3$.** The inequality is **non-strict**, so we use a solid line to graph its **boundary line** $2x + y = -3$. We use the **test point** $(0, 0)$. Do you see **why**? Because $2(0) + 0 \geq -3$ is true, we shade the **half-plane** that contains $(0, 0)$. See **Figure 11(a)**.

Graph $x - 4y \leq 12$. The inequality is non-strict, so we use a solid line to graph its boundary line $x - 4y = 12$. The test point $(0, 0)$ satisfies the inequality because $0 - 4(0) \leq 12$ is true, so we shade the half-plane that contains $(0, 0)$. See **Figure 11(b)**.

Step 2. The graph of the system of inequalities is the area where the two inequalities overlap. This is the darkest shaded region in **Figure 11(c)**. Any point that falls in this darkest region is a **solution to the system**.

Watch this **video** for a detailed solution.

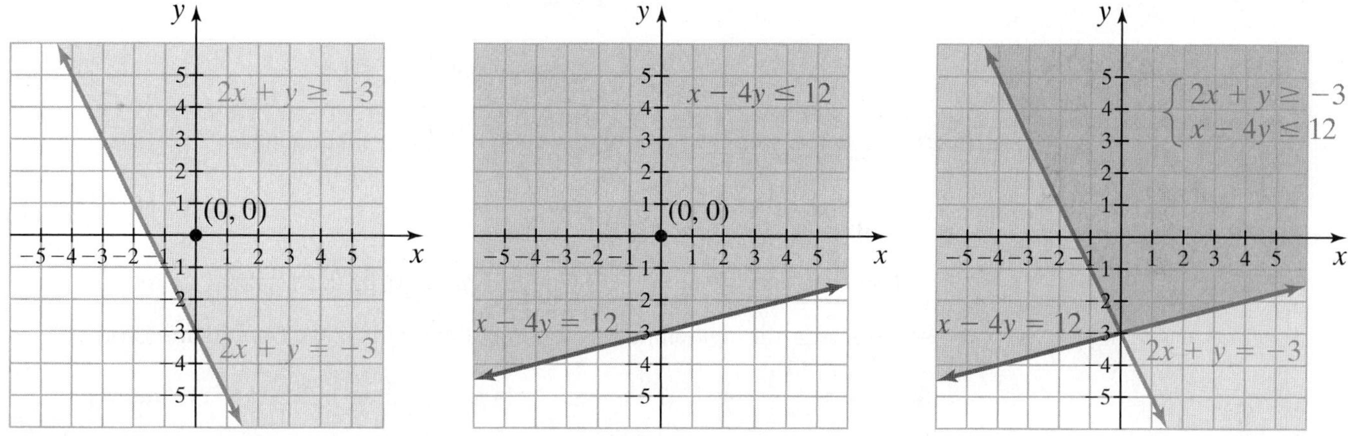

(a) Graph of the linear inequality $2x + y \geq -3$

(b) Graph of the linear inequality $x - 4y \leq 12$

(c) Graph of the system $\begin{cases} 2x + y \geq -3 \\ x - 4y \leq 12 \end{cases}$

Figure 11

CAUTION: Only the points that fall in the darkest-shaded region of Figure 11(c) are **solutions to the system** in **Example 2**. This includes the points on the adjacent sections of the two solid boundary lines $2x + y = -3$ and $x - 4y = 12$, including their intersection point.

You Try It **Work through this You Try It problem.**

Work Exercises 4–6 in this textbook or in the MyMathLab Study Plan.

(eText Screens 13.5-1–13.5-21)

Example 3 Graphing Systems of Linear Inequalities

My video summary

Graph the system of linear inequalities.

$$\begin{cases} x + y < 4 \\ x - 2y < -2 \end{cases}$$

Solution Both inequalities in this system are **strict**, so the **boundary lines** $x + y = 4$ and $x - 2y = -2$ are dashed lines. **Graph the inequalities** on the same **coordinate plane** and find the region they have in common. View the **graph**, or watch this **video** for a detailed solution.

CAUTION The points on the two dashed boundary lines $x + y = 4$ and $x - 2y = -2$, including the intersection point, are not solutions to this system.

You Try It **Work through this You Try It problem.**

Work Exercises 7–9 in this textbook or in the MyMathLab Study Plan.

A system of linear inequalities typically has an infinite number of solutions, but a system of inequalities with no solution, called an **inconsistent system of inequalities**, is also possible.

Example 4 Identifying Inconsistent Systems of Linear Inequalities

My video summary

Graph the system of linear inequalities.

$$\begin{cases} y \le -\dfrac{1}{3}x - 3 \\[2ex] y > -\dfrac{1}{3}x + 2 \end{cases}$$

Solution Follow the **steps for graphing systems of linear inequalities**.

Step 1. **Graph $y \le -\frac{1}{3}x - 3$.** The inequality is **non-strict**, so we use a solid line to graph its **boundary line** $y = -\frac{1}{3}x - 3$. The **test point** $(0, 0)$ does not **satisfy** the inequality, so we shade the **half-plane** that does not contain $(0, 0)$. See **Figure 12**(a).

Graph $y > -\frac{1}{3}x + 2$. The inequality is strict, so we use a dashed line to graph its boundary line $y = -\frac{1}{3}x + 2$. The test point $(0, 0)$ does not satisfy the inequality, so we shade the half-plane that does not contain $(0, 0)$. See **Figure 12**(b).

Step 2. The boundary lines $y = -\frac{1}{3}x - 3$ and $y = -\frac{1}{3}x + 2$ are **parallel**, and the shaded regions are on opposite sides of these parallel lines, so there is no shared region. See **Figure 12**(c). There are no ordered pairs that satisfy both inequalities, so the system has no solution and is **inconsistent**. We can use the null symbol $\varnothing$ or empty set $\{\ \}$ to indicate this.

Watch this **video** for a more detailed solution.

(eText Screens 13.5-1–13.5-21)

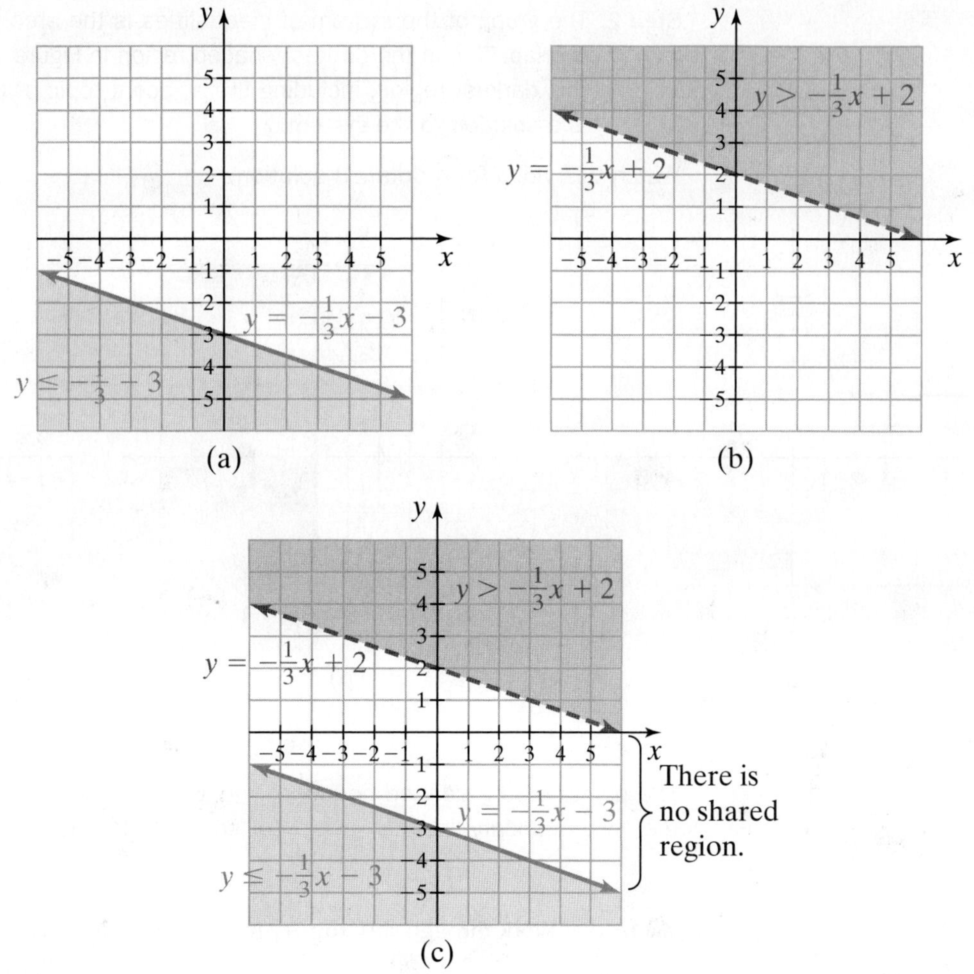

Figure 12

You Try It **Work through this You Try It problem.**

Work Exercises 10–12 in this textbook or in the MyMathLab Study Plan.

Some systems of linear inequalities involve **horizontal** or **vertical** boundary lines.

Example 5 Graphing Systems of Linear Inequalities

My video summary

Graph the system of linear inequalities.

$$\begin{cases} x - 3y > 6 \\ x \geq 1 \end{cases}$$

Solution Follow the **steps for graphing systems of linear inequalities.**

Step 1. **Graph $x - 3y > 6$.** The inequality is **strict**, so we use a dashed line to graph its **boundary line** $x - 3y = 6$. The **test point** $(0, 0)$ does not **satisfy** the inequality because $0 - 3(0) > 6$ is false, so we shade the **half-plane** that does not contain $(0, 0)$. See **Figure 13(a)**.

Graph $x \geq 1$. The inequality is **non-strict**, so we use a solid line to graph its vertical boundary line $x = 1$. The test point $(0, 0)$ does not satisfy the inequality because $0 \geq 1$ is false, so we shade the half-plane that does not contain $(0, 0)$. See **Figure 13(b)**.

(eText Screens 13.5-1–13.5-21)

Step 2. The graph of the system of inequalities is the area where the two inequalities overlap. This is the darkest shaded region in Figure 13(c). Any point that falls in this darkest region, including the adjacent topic of the solid boundary line $x = 1$, is a **solution to the system.**

Watch this **video** for a detailed solution.

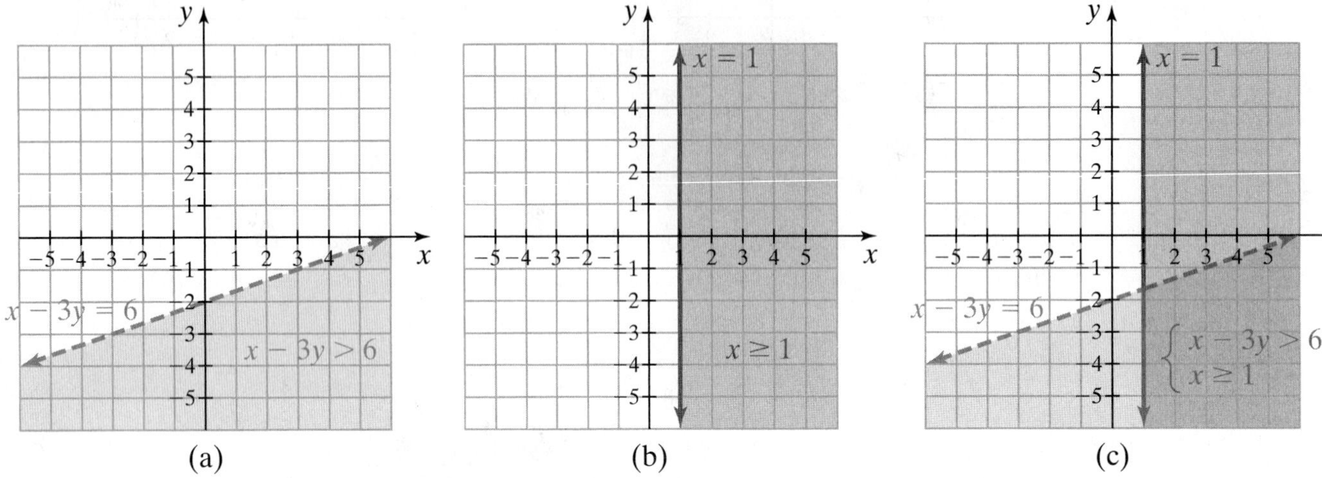

Figure 13

CAUTION The points on the dashed boundary line $x - 3y = 6$, including the intersection point with the solid boundary line $x = 1$, are not solutions to this system.

You Try It **Work through this You Try It problem.**

Work Exercises 13–15 in this textbook or in the MyMathLab Study Plan.

Consider the system $\begin{cases} x \geq 0 \\ y \geq 0 \end{cases}$. What is its **solution region**? See this **explanation**.

Sometimes a **system of linear inequalities** contains more than two inequalities.

Example 6 Graphing Systems with More Than Two Linear Inequalities

My video summary

Graph the system of linear inequalities.

$$\begin{cases} x + y < 4 \\ y > 0 \\ 2x - y > -4 \end{cases}$$

Solution Follow the **steps for graphing systems of linear inequalities.**

Step 1. Graph $x + y < 4$. The inequality is **strict**, so we use a dashed line to graph its **boundary line** $x + y = 4$. Check a convenient **test point** to confirm that we shade the **half-plane** below this boundary line. View this **graph**.

Graph $y > 0$. The inequality is strict, so we use a dashed line to graph its horizontal boundary line $y = 0$. Test a convenient point to confirm that we shade above this boundary line. (Note that the test point in this case cannot be $(0, 0)$.) View this **graph**.

Graph $2x - y > -4$. The inequality is strict, so we use a dashed line to graph its boundary line $2x - y = -4$. Test a convenient point to confirm that the half plane below this boundary line is shaded. View this **graph**.

Step 2. The graph of the system of inequalities is the area where the three inequalities overlap. This is the shaded region in Figure 14. Any point that falls in this shaded region is a **solution to the system.**

Watch this **video** for a detailed solution.

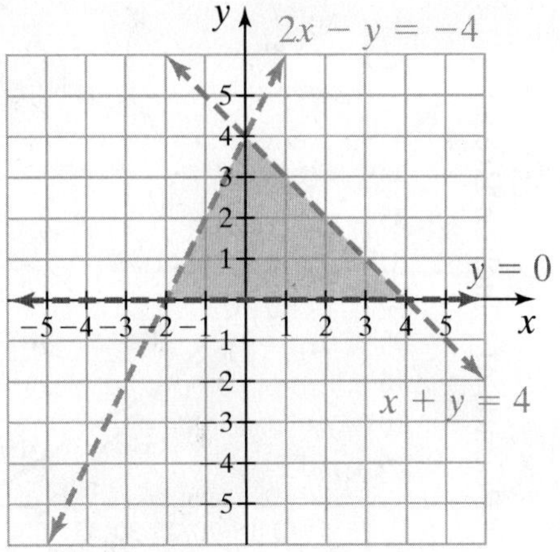

Figure 14

Graph of $\begin{cases} x + y < 4 \\ y > 0 \\ 2x - y > -4 \end{cases}$

CAUTION The points on the dashed boundary lines, including the intersection points, are not solutions to this system.

You Try It **Work through this You Try It problem.**

Work Exercises 16–18 in this textbook or in the MyMathLab Study Plan.

OBJECTIVE 3 SOLVE APPLICATIONS INVOLVING SYSTEMS OF LINEAR INEQUALITIES

Let's look at a real life application that is **modeled** by a **system of linear inequalities.**

Example 7 Planning a Barbeque

My video summary

Savannah is planning a barbeque for her family and friends. She will spend $150 or less to buy hamburger patties that cost $3 per pound and boneless chicken breast that cost $5 per pound. To limit waste, she will purchase at most 40 pounds of meat all together. Also, the amount of hamburger and chicken purchased must be **non-negative.** A system of linear inequalities that models this situation is

$$\begin{cases} 3h + 5c \le 150 \\ h + c \le 40 \\ h \ge 0 \\ c \ge 0 \end{cases}$$

where h = pounds of hamburger patties and c = pounds of chicken breast.

a. Graph the system of linear inequalities.

b. Can Savannah purchase 20 pounds of hamburger patties and 15 pounds of chicken breast for the barbeque?

c. Can Savannah purchase 10 pounds of hamburger patties and 30 pounds of chicken breast for the barbeque?

Solutions

a. Label the horizontal axis h and the vertical axis c. The inequalities $h \geq 0$ and $c \geq 0$ restrict the graph to **Quadrant I**. Follow the **steps for graphing systems of linear inequalities** to result in the graph in Figure 15. Watch this **video** for a detailed solution.

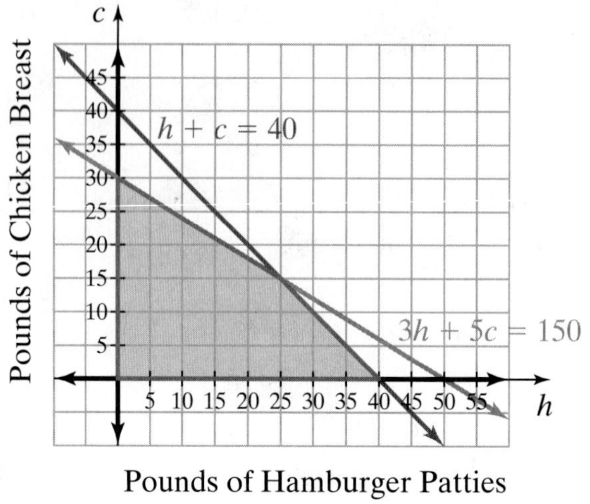

Figure 15

b. Yes. The ordered pair $(h, c) = (20, 15)$ lies within the shaded region in Figure 15, so Savannah can purchase 20 pounds of hamburger patties and 15 pounds of chicken.

c. No. The ordered pair $(h, c) = (10, 30)$ lies outside of the shaded region in Figure 15, so Savannah cannot purchase 10 pounds of hamburger patties and 30 pounds of chicken.

You Try It **Work through this You Try It problem.**

Work Exercises 19–22 in this textbook or in the MyMathLab Study Plan.

13.5 Exercises

In Exercises 1–3, determine if each ordered pair is a solution to the given system of linear inequalities in two variables.

You Try It

1. $\begin{cases} x + y \leq 10 \\ x - y \geq -2 \end{cases}$

a. $(1, -5)$
b. $(4, 6)$
c. $(-3, 7)$

2. $\begin{cases} 2x - 3y > 6 \\ 5x + 2y < -1 \end{cases}$

a. $(2, -2)$
b. $(1, -3)$
c. $(0, -4)$

3. $\begin{cases} 2x - 5y \leq 10 \\ x - y > -6 \\ 3x + 4y < 12 \end{cases}$

a. $(-1, -2)$
b. $(3, 2)$
c. $(-5, -4)$

In Exercises 4–18, graph each system of linear inequalities in two variables.

You Try It

4. $\begin{cases} y > 2x - 1 \\ y \leq x + 2 \end{cases}$

5. $\begin{cases} x + y \geq 2 \\ x - y \leq 4 \end{cases}$

6. $\begin{cases} 3x - 2y \leq -2 \\ x + 2y > 6 \end{cases}$

You Try It

7. $\begin{cases} 3x + y < 2 \\ x + 2y < 8 \end{cases}$

8. $\begin{cases} y < 2x \\ y > -3x \end{cases}$

9. $\begin{cases} 3x - 2y \leq -2 \\ x + 2y \geq 6 \end{cases}$

You Try It

10. $\begin{cases} y < x - 2 \\ y \geq x + 4 \end{cases}$

11. $\begin{cases} 2x + y \leq 4 \\ y \geq -2x - 1 \end{cases}$

12. $\begin{cases} 3x - 4y > 12 \\ 6x - 8y < -32 \end{cases}$

You Try It

13. $\begin{cases} 4x - y > 2 \\ y \geq -1 \end{cases}$

14. $\begin{cases} x \geq -1 \\ y \leq 3 \end{cases}$

15. $\begin{cases} \frac{1}{2}x - \frac{1}{3}y \leq 1 \\ x \geq -1 \end{cases}$

You Try It

16. $\begin{cases} x + y \leq 5 \\ x \geq 0 \\ y \geq 0 \end{cases}$

17. $\begin{cases} x + y > -2 \\ x - y < 2 \\ y < 2 \end{cases}$

18. $\begin{cases} x + y < 6 \\ 3x - y > -2 \\ x \geq 0 \\ y \geq 0 \end{cases}$

In Exercises 19–22, solve each application by using a system of linear inequalities in two variables.

You Try It

19. Investing Judy has $50,000 to invest. Her financial advisor recommends that she place at least $30,000 in mutual funds and no more than $20,000 in bonds. A system of linear inequalities that models this situation is

$$\begin{cases} m + b \leq 50{,}000 \\ m \geq 30{,}000 \\ b \leq 20{,}000 \\ b \geq 0 \end{cases}$$

where m = the dollar amount invested in mutual funds and b = the dollar amount invested in bonds.

a. Graph the system of linear inequalities.
b. Can Judy invest $20,000 in mutual funds and $20,000 in bonds?
c. Can Judy invest $38,000 in mutual funds and $12,000 in bonds?

20. Getting Cash For an upcoming shopping trip, Elaine will withdraw cash from her checking account which contains $100. She will want the cash in only five-dollar and ten-dollar bills, and she will carry no more than 12 bills all together. A system of linear inequalities that models this situation is

$$\begin{cases} 5f + 10t \leq 100 \\ f + t \leq 12 \\ f \geq 0 \\ t \geq 0 \end{cases}$$

where f = the number $5 bills and t = the number of $10 bills Elaine receives.

a. Graph the system of linear inequalities.
b. Can Elaine receive 4 five-dollar bills and 8 ten-dollar bills?
c. Can Elaine receive 7 five-dollar bills and 3 ten-dollar bills?

21. Growing Crops In Illinois, the annual cost to grow corn is about $450 per acre, and the annual cost to grow soybeans is about $300 per acre (not counting the cost of land). A small farmer has at most 150 acres for growing corn and soybeans. His budget for growing these crops is at most $60,000. A system of linear inequalities that models this situation is

$$\begin{cases} c + s \leq 150 \\ 450c + 300s \leq 60{,}000 \\ c \geq 0 \\ s \geq 0 \end{cases}$$

where c = acres of corn and s = acres of soybeans. (*Source: Farmdoc*, University of Illinois)

a. Graph the system of linear inequalities.
b. Can this farmer grow 100 acres of corn and 50 acres of soybeans?
c. Can this farmer grow 120 acres of corn and 30 acres of soybeans?

22. **Dimensions of a Garden** Austin is planning to build a rectangular garden. Its perimeter will be at most 36 feet, and its length will be no more than twice its width. A system of linear inequalities that models this situation is

$$\begin{cases} 2l + 2w \leq 36 \\ l \leq 2w \\ l \geq w \end{cases}$$

where l = the length and w = the width of the garden.

a. Graph the system of linear inequalities.

b. Can Austin's garden be 15 feet long and 10 feet wide?

c. Can Austin's garden be 10 feet long and 8 feet wide?

13.6 Systems of Linear Equations in Three Variables

THINGS TO KNOW

Before working through this topic, be sure you are familiar with the following concepts:

		VIDEO	ANIMATION	INTERACTIVE
You Try It	1. Determine If an Ordered Pair Is a Solution To a System of Linear Equations in Two Variables (Topic 13.1, **Objective 1**)			✓
You Try It	2. Solve Systems of Linear Equations by Elimination (Topic 13.3, **Objective 1**)	✓		
You Try It	3. Solve Special Systems by Elimination (Topic 13.3, **Objective 2**)	✓		
You Try It	4. Solve Related Quantity Applications Using Systems (Topic 13.4, **Objective 1**)			✓
You Try It	5. Solve Mixture Applications Using Systems (Topic 13.4, **Objective 4**)			✓

OBJECTIVES

1 Determine If an Ordered Triple Is a Solution to a System of Linear Equations in Three Variables

2 Solve Systems of Linear Equations in Three Variables

3 Use Systems of Linear Equations in Three Variables to Solve Application Problems

OBJECTIVE 1 DETERMINE IF AN ORDERED TRIPLE IS A SOLUTION TO A SYSTEM OF LINEAR EQUATIONS IN THREE VARIABLES

An equation such as $4x + 2y - 7z = 12$ is called a **linear equation in three variables** because there are three **variables** and each variable term is **linear**. Comparing this equation to our definitions for **linear equations in one variable** and **linear equations in two variables** leads us to the following definition.

(eText Screens 13.6-1–13.6-33)

> **Definition Linear Equation in Three Variables**
>
> A **linear equation in three variables** is an equation that can be written in the form $Ax + By + Cz = D$, where A, B, C, and D are **real numbers**, and A, B, and C are not all equal to 0.

A solution to a linear equation in three variables is an **ordered triple** (x, y, z) that makes the equation true. As with the two-variable case, there are infinitely many solutions to these equations. The graph of the **solution set** to a linear equation in two variables is a **line** in two-dimensional space. With three variables, the graph of the solution set becomes a **plane** in three-dimensional space.

In **Topic 13.1**, we saw that an ordered pair was an **ordered-pair solution** to a **system of linear equations in two variables** if it was a solution to both equations in the system. Similarly, an **ordered triple** is an **ordered-triple solution** to a **system of linear equations in three variables** if it is a solution to all three equations in the system.

> **Definition System of Linear Equations in Three Variables**
>
> A **system of linear equations in three variables** is a collection of **linear equations in three variables** considered together. A **solution to a system** of linear equations in three variables is an **ordered triple** that satisfies all equations in the system.

Example 1 Determining If an Ordered Triple Is a Solution to a System

My interactive video summary

Determine if each ordered triple is a **solution** to the given system:

$$\begin{cases} 3x + y - 2z = 4 \\ 2x - 2y + 3z = 9 \\ x + y - z = 5 \end{cases}$$

a. $(3, 9, 7)$ **b.** $(2, -4, -1)$

Solutions

a. To determine if the **ordered triple** $(3, 9, 7)$ is a **solution** to the system, substitute the values for the variables into each equation to see if a true statement results. To be a solution to the system, the ordered triple must be a solution to all of the equations in the system.

	First Equation	Second Equation	Third Equation
Original equation:	$3x + y - 2z = 4$	$2x - 2y + 3z = 9$	$x + y - z = 5$
Substitute:	$3(3) + (9) - 2(7) \stackrel{?}{=} 4$	$2(3) - 2(9) + 3(7) \stackrel{?}{=} 9$	$(3) + (9) - (7) \stackrel{?}{=} 5$
Simplify:	$9 + 9 - 14 \stackrel{?}{=} 4$	$6 - 18 + 21 \stackrel{?}{=} 9$	$3 + 9 - 7 \stackrel{?}{=} 5$
	$4 = 4$ True	$9 = 9$ True	$5 = 5$ True

Since the ordered triple makes all three equations true, $(3, 9, 7)$ is a solution to the system.

b. Check the ordered triple $(2, -4, -1)$ on your own. View the **answer**, or watch this **interactive video** for a detailed solution to both parts.

You Try It **Work through this You Try It problem.**

Work Exercises 1 and 2 in this textbook or in the MyMathLab Study Plan.

(eText Screens 13.6-1–13.6-33)

OBJECTIVE 2 SOLVE SYSTEMS OF LINEAR EQUATIONS IN THREE VARIABLES

Graphically, we can visualize **solutions to systems of linear equations in two variables** as **intersection points** of **lines**. These systems could have one solution (**intersecting lines**), no solution (**parallel lines**), or infinitely many solutions (**coinciding lines**). These same situations apply when solving **systems of linear equations in three variables**, but graphically, we view the solutions as intersection points of **planes**. Figure 16 illustrates the possibilities for systems of linear equations in three variables.

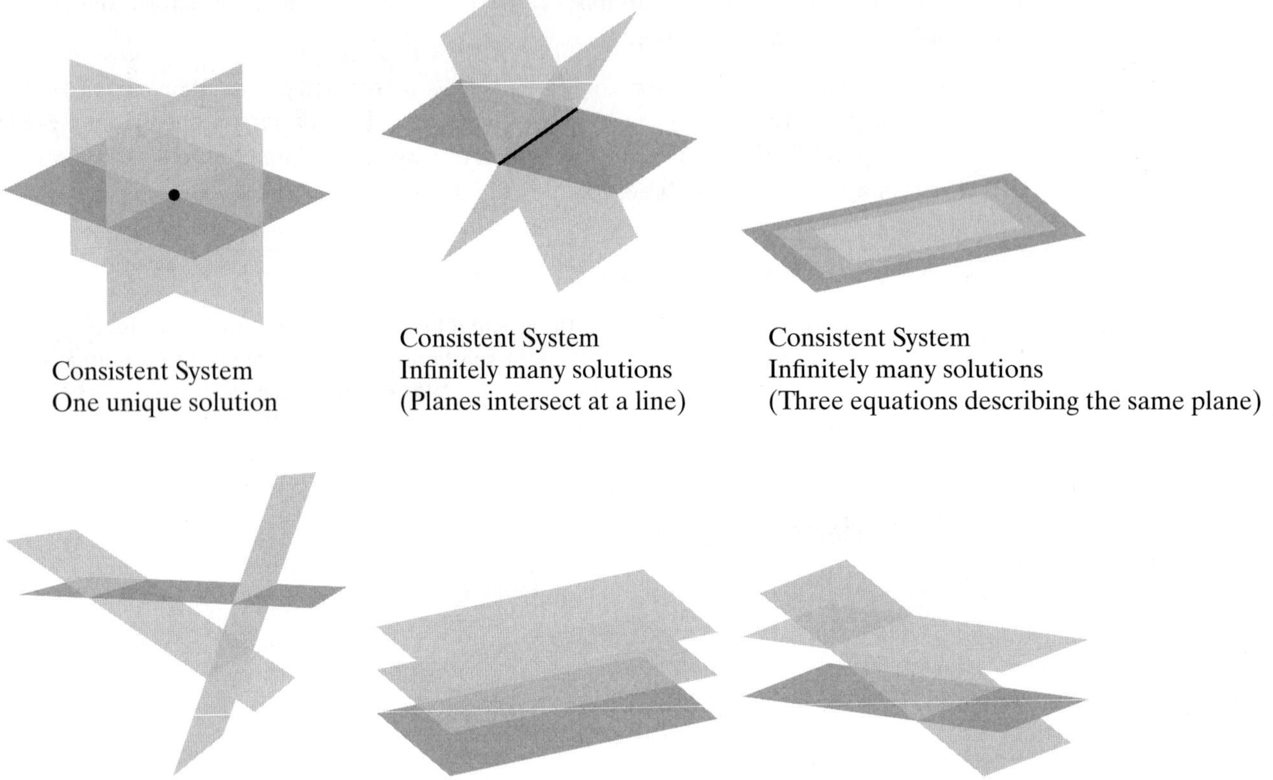

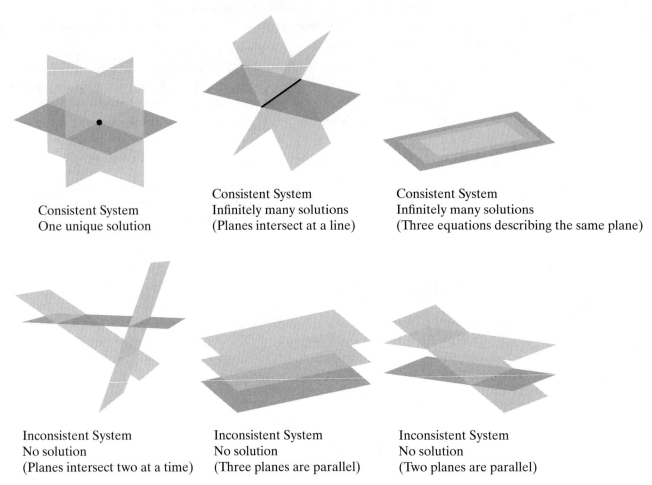

Consistent System
One unique solution

Consistent System
Infinitely many solutions
(Planes intersect at a line)

Consistent System
Infinitely many solutions
(Three equations describing the same plane)

Inconsistent System
No solution
(Planes intersect two at a time)

Inconsistent System
No solution
(Three planes are parallel)

Inconsistent System
No solution
(Two planes are parallel)

Figure 16

To solve **systems of linear equations in three variables** by graphing would require us to graph **planes** in three dimensions, which is not a practical task. Instead, we will focus on solving systems of linear equations in three variables by applying the **elimination method** used for solving systems of linear equations in two variables.

When solving systems of **linear equations in two variables** by elimination, the goal is to reduce the system of two equations in two variables to a single **linear equation in one variable**. We then solve this equation for the remaining variable, substitute this value into either of the original equations, and solve for the variable that had been eliminated. See **Figure 17**. This process of substituting back into an original equation to find values of remaining variables is called **back substitution**.

(eText Screens 13.6-1–13.6-33)

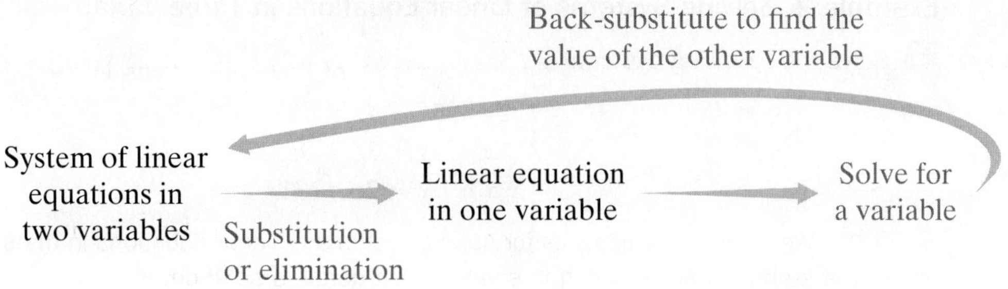

Figure 17 Solving a system of linear equations in two variables

A similar process is used to solve a **system of linear equations in three variables**. The goal is to reduce the system of three equations in three variables down to a system of two equations in two variables. At that point, we can reduce the two-equation system to a single equation in one variable and easily solve that equation. Using **back substitution**, we can find the values of the other two variables. See Figure 18.

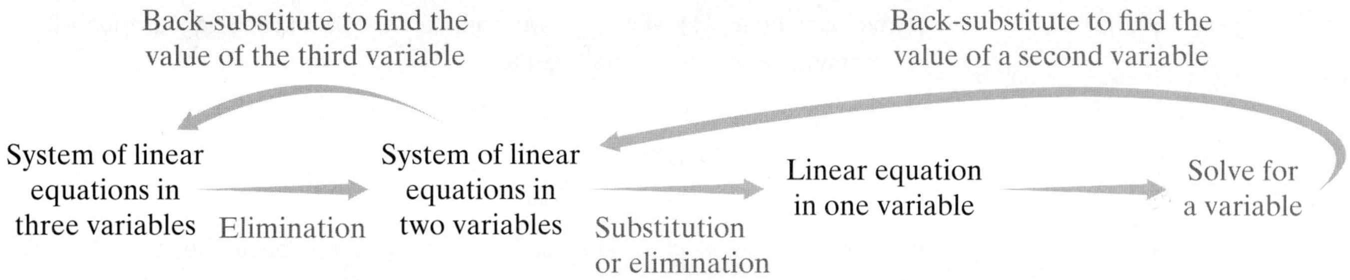

Figure 18 Solving a system of linear equations in three variables

Before looking at an example, we present some **guidelines** for solving systems of linear equations in three variables by elimination.

Guidelines for Solving a System of Linear Equations in Three Variables by Elimination

Step 1. **Write each equation in standard form.** Write each equation in the form $Ax + By + Cz = D$, lining up the variable terms. Number the equations to keep track of them.

Step 2. **Eliminate a variable from one pair of equations.** Use the **elimination method** to eliminate a variable from any two of the original three equations, leaving one equation in two variables.

Step 3. **Eliminate the same variable again.** Use a different pair of the original equations and eliminate the same variable again, leaving one equation in two variables.

Step 4. **Solve the system of linear equations in two variables.** Use the resulting equations from Steps 2 and 3 to create and solve the corresponding system of linear equations in two variables by **substitution** or **elimination**.

Step 5. **Use back substitution to find the value of the third variable.** Substitute the results from Step 4 into any of the original equations to find the value of the remaining variable.

Step 6. **Check the solution.** Check the proposed solution in each equation of the system and write the solution set.

(eText Screens 13.6-1–13.6-33)

Example 2 Solving Systems of Linear Equations in Three Variables

My video summary

Solve the following system:

$$\begin{cases} 2x + 3y + 4z = 12 \\ x - 2y + 3z = 0 \\ -x + y - 2z = -1 \end{cases}$$

Solution We follow our **guidelines** for solving systems of linear equations in three variables. Follow the steps below, or watch this **video** for a detailed solution.

Step 1. The equations are already in **standard form** and all the variables are lined up. We rewrite the system and number each equation.

$$\begin{cases} 2x + 3y + 4z = 12 & (1) \\ x - 2y + 3z = 0 & (2) \\ -x + y - 2z = -1 & (3) \end{cases}$$

Step 2. We can eliminate any of the variables. For convenience, we will eliminate the variable x from equations (1) and (2). We can do this by multiplying equation (2) by -2 and adding the equations together.

$$\begin{aligned} (1)&: & 2x + 3y + 4z &= 12 \\ \text{Multiply } (2) \text{ by } -2&: & -2x + 4y - 6z &= 0 \\ \text{Add}&: & 7y - 2z &= 12 \quad (4) \end{aligned}$$

How do you know which variable to eliminate? View this **popup box** to find out.

Step 3. We need to eliminate the same variable, x, from a different pair of equations. We will use equations (2) and (3) for the second pair. Since the **coefficients** of x in these equations have the same **absolute value**, but opposite signs, we can eliminate the variable by simply adding the equations.

$$\begin{aligned} (2)&: & x - 2y + 3z &= 0 \\ (3)&: & -x + y - 2z &= -1 \\ \text{Add}&: & -y + z &= -1 \quad (5) \end{aligned}$$

Step 4. Combining equations (4) and (5), we form a system of linear equations in two variables.

$$\begin{cases} 7y - 2z = 12 & (4) \\ -y + z = -1 & (5) \end{cases}$$

To solve this system, we use the **elimination method** by multiplying equation (5) by 7 and adding the result to equation (4) to eliminate the variable y.

$$\begin{aligned} (4)&: & 7y - 2z &= 12 \\ \text{Multiply } (5) \text{ by } 7&: & -7y + 7z &= -7 \\ \text{Add}&: & 5z &= 5 \\ \text{Divide by } 5&: & z &= 1 \end{aligned}$$

Since $z = 1$, we **back-substitute** this into (5) to solve for y.

$$\begin{aligned} (5)&: & -y + z &= -1 \\ \text{Substitute } 1 \text{ for } z&: & -y + (1) &= -1 \\ \text{Simplify}&: & -y + 1 &= -1 \\ \text{Subtract } 1&: & -y &= -2 \\ \text{Divide by } -1&: & y &= 2 \end{aligned}$$

(eText Screens 13.6-1–13.6-33)

When **back-substituting**, we can use either of the equations, but often one equation is preferred over another. Can you see why we chose to use equation (5) instead of equation (4)? See an **explanation.**

Step 5. Substitute 2 for y and 1 for z in any of the original equations, (1), (2), or (3), and solve for x. We will back-substitute using equation (2).

(2):	$x - 2y + 3z = 0$
Substitute 2 for y and 1 for z:	$x - 2(2) + 3(1) = 0$
Multiply:	$x - 4 + 3 = 0$
Simplify:	$x - 1 = 0$
Add 1:	$x = 1$

The solution to the system is the **ordered triple** $(1, 2, 1)$.

Step 6. View this **popup box** for the check.

You Try It **Work through this You Try It problem.**

Work Exercises 3–5 in this textbook or in the MyMathLab Study Plan.

Example 3 Solving Systems of Linear Equations in Three Variables with Missing Terms

My video summary

Solve the following system:

$$\begin{cases} 2x + y = 13 \\ 3x - 2y + z = 8 \\ x + 2y - 3z = 5 \end{cases}$$

Solution We follow our **guidelines** for solving a system of linear equations in three variables.

Step 1. First, we rewrite and number each equation, lining up the variables. Notice that the first equation has a variable term missing, so we put a gap in its place.

$$\begin{cases} 2x + y \qquad = 13 & (1) \\ 3x - 2y + z = 8 & (2) \\ x + 2y - 3z = 5 & (3) \end{cases}$$

Step 2. We can eliminate any of the variables, but we notice that one equation already has z eliminated. By selecting z as the variable to eliminate, we can move directly to Step 3.

Step 3. Looking at equations (2) and (3), we might be tempted to add these equations to eliminate y. However, in Step 2, we selected z as the variable to eliminate, so we need to eliminate z again in this step.

Try to finish solving this system on your own. View the **answer**, or watch this **video** for a detailed solution.

You Try It **Work through this You Try It problem**

Work Exercises 9–11 in this textbook or in the MyMathLab Study Plan.

(eText Screens 13.6-1–13.6-33)

Example 4 Solving Systems of Linear Equations in Three Variables Involving Fractions

My video summary

Solve the following system:

$$\begin{cases} \frac{1}{2}x + \quad y + \frac{2}{3}z = 2 \\ \frac{3}{4}x + \frac{5}{2}y - \ 2z = -7 \\ \quad x + 4y + 2z = 4 \end{cases}$$

Solution If an equation in the system contains fractions, then it is often helpful to **clear the fractions** first. After doing this, follow the **guidelines** for solving a system of linear equations in three variables to solve this system on your own. View the **answer**, or watch this **video** for a detailed solution.

You Try It **Work through this You Try It problem.**

Work Exercises 6–8 in this textbook or in the MyMathLab Study Plan.

Systems of linear equations in three variables may be **inconsistent** or may include **dependent equations**. If, in our solution process, we find a **contradiction**, then the system is inconsistent and has no solution. In the two-variable case, this occurred if we had **parallel lines** because the lines had no points in common.

For the three-variable case, the system will be inconsistent if all three **planes** have no points in common. This would happen if all the planes were parallel, but would also happen if two or none of the planes are parallel, as shown in Figure 19.

Inconsistent System
No solution
(Planes intersect two at a time)

Inconsistent System
No solution
(Three planes are parallel)

Inconsistent System
No solution
(Two planes are parallel)

Figure 19

When solving **systems of linear equations in two variables**, encountering an **identity** meant the system was **dependent** and had an infinite number of **ordered-pair solutions**. This happened when the lines were **coinciding**.

However, unlike the two-variable case, identifying dependent systems in three variables takes a bit more work because we must consider all possible pairings of the equations in the system. Obtaining an identity with one pairing is not sufficient to say the system is dependent. This idea is illustrated in the following example.

Example 5 Solving Systems of Linear Equations in Three Variables with No Solution

Solve the following system:

$$\begin{cases} x - y + 2z = 5 \\ 3x - 3y + 6z = 15 \\ -2x + 2y - 4z = 7 \end{cases}$$

Solution We follow our **guidelines** for solving a system of linear equations in three variables.

Step 1. The equations are already in standard form with all the variables lined up. We rewrite the system and number each equation.

$$\begin{cases} x - y + 2z = 5 & (1) \\ 3x - 3y + 6z = 15 & (2) \\ -2x + 2y - 4z = 7 & (3) \end{cases}$$

Step 2. For convenience, we will eliminate the variable x from equations (1) and (2). To do this, we multiply equation (1) by -3 and add the equations together.

$$\begin{array}{rr} \text{Multiply } (1) \text{ by } -3: & -3x + 3y - 6z = -15 \\ (2): & \underline{3x - 3y + 6z = 15} \\ \text{Add:} & 0 = 0 \quad \text{True} \end{array}$$

The last line is an **identity**. In the two-variable case, we would stop and say that the system had an infinite number of solutions. However, in systems of three variables, this is not necessarily the case.

Step 3. Let's continue our process and eliminate the variable x from the pairing of equations (1) and (3). To do this, we multiply equation (1) by 2 and add the equations.

$$\begin{array}{rr} \text{Multiply } (1) \text{ by } 2: & 2x - 2y + 4z = 10 \\ (3): & \underline{-2x + 2y - 4z = 7} \\ \text{Add:} & 0 = 17 \quad \text{False} \end{array}$$

The last line is a **contradiction,** so the system is **inconsistent** and has no solution. View this **popup box** to find out why this occurs.

You Try It **Work through this You Try It problem.**

Work Exercises 14 and 15 in this textbook or in the MyMathLab Study Plan.

My video summary

As we saw in Example 5, finding an **identity** in the three-variable case does not necessarily mean the system is **dependent** and has an infinite number of solutions. Watch this **video** for a more detailed explanation.

A dependent system of linear equations in three variables can occur in two ways: (1) all three equations describe the same plane, or (2) the three planes intersect in a line. These are shown again in **Figure 20.**

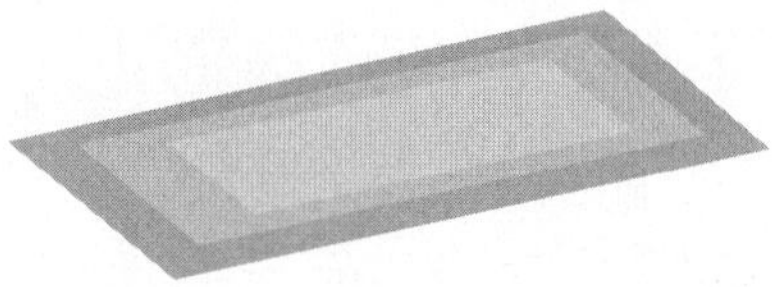

Consistent System
Infinitely many solutions
(Three equations describing the same plane)

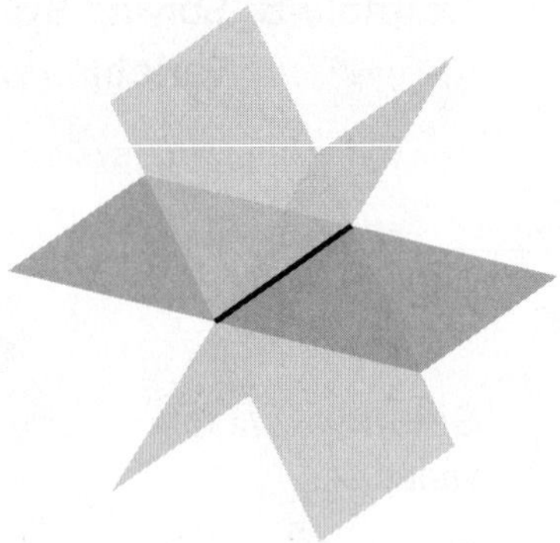

Consistent System
Infinitely many solutions
(Planes intersect at a line)

Figure 20 Dependent Systems

Example 6 Solving Systems of Dependent Linear Equations in Three Variables

Solve the following system.

$$\begin{cases} x - y + 2z = 5 \\ 3x - 3y + 6z = 15 \\ -2x + 2y - 4z = -10 \end{cases}$$

Solution

$$\begin{cases} x - y + 2z = 5 & (1) \\ 3x - 3y + 6z = 15 & (2) \\ -2x + 2y - 4z = -10 & (3) \end{cases}$$

For this system, notice that multiplying equation (1) by 3 gives us equation (2) and multiplying equation (1) by -2 gives us equation (3). Therefore, all three equations describe the same **plane** and the system is dependent.

Using **set-builder notation**, the solution set for this system can be written as $\{(x, y, z) | x - y + 2z = 5\}$. View this **popup box** for other variations.

You Try It **Work through this You Try It problem.**

Work Exercises 12 and 13 in this textbook or in the MyMathLab Study Plan.

When a dependent system of equations in three variable results from the three planes intersecting in a line, we write the solution set using an **ordered-triple notation.** Using this notation, we write two of the coordinates in the ordered triple in terms of the third. This is shown in Example 7.

Example 7 Solving Systems of Dependent Linear Equations in Three Variables

Solve the following system.

$$\begin{cases} 4x + y - 5z = -1 & (1) \\ -2x - y + z = 3 & (2) \\ x + y + z = -4 & (3) \end{cases}$$

My video summary

Solution Eliminating y using equations (1) and (2) results in the equation $2x - 4z = 2$. Eliminating y again using equations (1) and (3) results in the equation $3x - 6z = 3$. Combining these two equations gives a two-variable system:

$$\begin{cases} 2x - 4z = 2 & (4) \\ 3x - 6z = 3 & (5) \end{cases}$$

If we divide equation (4) by 2 and divide equation (5) by 3, then we see that the two equations in the system are equivalent. Therefore, this two-variable system has an infinite number of solutions. This indicates that the three planes described by the equations in the original system intersect in a line. Solving either equation (4) or (5) for x gives the relationship $x = 2z + 1$. When we substitute this result into any of the three original equations and solve for y, we have $y = -3z - 5$. Since we have written two of the variables in terms of the third, the **ordered-triple solution** is $(2z + 1, -3z - 5, z)$. Watch this **video** to see the complete solution.

You Try It **Work through this You Try It problem.**

Work Exercises 16 and 17 in this textbook or in the MyMathLab Study Plan.

Which notation should be used to express the solution of a dependent system? View this **suggestion**.

CAUTION Having an infinite number of ordered-triple solutions does not mean that all ordered triples are solutions to a system of linear equations in three variables. View this **explanation**.

OBJECTIVE 3 USE SYSTEMS OF LINEAR EQUATIONS IN THREE VARIABLES TO SOLVE APPLICATION PROBLEMS

Building on the **problem-solving strategy** using systems of equations from Topic 13.4, we now look at applications involving systems of linear equations in three variables.

Example 8 Real-Time Strategy Game

My video summary

While playing a real-time strategy game, Joel created military units to defend his town: warriors, skirmishers, and archers. Warriors require 20 units of food and 50 units of gold. Skirmishers require 25 units of food and 35 units of wood. Archers require 32 units of wood and 32 units of gold. If Joel used 506 units of gold, 606 units of wood, and 350 units of food to create the units, how many of each type of military unit did he create?

Solution

Step 1. Define the Problem. We want to find the number of each type of military unit created. There are three types of units: warriors, skirmishers, and archers. Each unit requires a certain amount of gold, wood, and food. We know how much of each resource is needed for each unit, and we know the total amount of each resource that is used. We summarize this information in the following table.

	Each Warrior	Each Skirmisher	Each Archer	Total
Units of Gold	50	0	32	506
Units of Wood	0	35	32	606
Units of Food	20	25	0	350

Step 2. Assign Variables. Let W, S, and A represent the number of warrior, skirmisher, and archer units, respectively.

Step 3. Translate into a System of Equations. We need to translate the given information into three equations to form a system. When writing a system of equations, totals are often a good place to start. We know the total units of gold, wood, and food used, and we know how much of each resource is required by the individual units. Therefore, we can write one equation based on total gold, a second equation based on total wood, and a third equation based on total food. A total of 506 units of gold were used. Since warriors require 50 units of gold and archers require 32 units of gold, we write the following equation.

$$\overbrace{50W}^{\text{Gold for warriors}} + \overbrace{32A}^{\text{Gold for archers}} = \overbrace{506}^{\text{Total gold}}$$

Note that there is no variable term involving S because skirmishers do not require gold.

A total of 606 units of wood were used. Since skirmishers require 35 units of wood and archers require 32 units of wood, we write a second equation:

$$\overbrace{35S}^{\text{Wood for skirmishers}} + \overbrace{32A}^{\text{Wood for archers}} = \overbrace{606}^{\text{Total wood}}$$

Again notice that there is a variable term missing because warriors do not require wood.

A total of 350 units of food were used. Since warriors require 20 units of food and skirmishers require 25 units of food, we write a third equation:

$$\overbrace{20W}^{\text{Food for warriors}} + \overbrace{25S}^{\text{Food for skirmishers}} = \overbrace{350}^{\text{Total food}}$$

Step 4. Solve the System. Using the three equations, we form the system

$$\begin{cases} 50W \phantom{{}+25S} + 32A = 506 \\ \phantom{50W+{}} 35S + 32A = 606 \\ 20W + 25S \phantom{{}+32A} = 350 \end{cases}$$

Now solve this system using the **elimination method**. Remember to check the reasonableness of your answer. View the **answer**, or watch this **video** for a detailed solution.

You Try It **Work through this You Try It problem.**

Work Exercises 18 and 19 in this textbook or in the MyMathLab Study Plan.

Example 9 Buy Clothes Online

My video summary

Wendy ordered 30 T-shirts online for her three children. Small T-shirts cost \$4 each, medium T-shirts cost \$5 each, and large T-shirts are \$6 each. She spent \$40 more for the large T-shirts than for the small T-shirts. Wendy's total bill was \$154. How many T-shirts of each size did she buy?

Solution Following the **problem-solving strategy**, we begin by defining some variables. Let S, M, and L represent the number of small, medium, and large T-shirts, respectively. With the given information, we can write and solve a system of three linear equations. One equation uses the total number of shirts purchased; a second equation uses the total amount spent, and a third equation uses the fact that Wendy spent \$40 more on large T-shirts than on small T-shirts. Try to finish this problem on your own. View the **answer**, or watch this **video** for a detailed solution.

You Try It **Work through this You Try It problem.**

Work Exercises 20–23 in this textbook or in the MyMathLab Study Plan.

13.6 Exercises

In Exercises 1 and 2, two ordered triples are given. Determine if each ordered triple is a solution to the given system.

You Try It

1. $(-1, 1, -2), (1, -1, 2)$

$$\begin{cases} x + y + z = -2 \\ -x - 3y - 2z = 2 \\ 2x - 2y + 5z = -14 \end{cases}$$

2. $(2, -1, 4), (-2, 1, -4)$

$$\begin{cases} \frac{1}{2}x + 3y - z = 6 \\ -x + y + \frac{1}{4}z = 2 \\ x - 4y + z = -10 \end{cases}$$

In Exercises 3–11, solve each system of linear equations.

You Try It

3. $$\begin{cases} x + y + z = 4 \\ 2x - y - 2z = -10 \\ -x - y + 3z = 8 \end{cases}$$

4. $$\begin{cases} x - 2y + z = 6 \\ 2x + y - 3z = -3 \\ x - 3y + 3z = 10 \end{cases}$$

5. $$\begin{cases} x - 2y + 2z = 2 \\ 3x + 2y - 2z = -1 \\ x - y - 2z = 0 \end{cases}$$

You Try It

6. $$\begin{cases} x - \frac{1}{2}y + \frac{1}{2}z = -3 \\ x + y - z = 0 \\ -3x - 3y + 4z = 1 \end{cases}$$

7. $$\begin{cases} \frac{1}{3}x - \frac{2}{3}y + z = 0 \\ \frac{1}{2}x - \frac{3}{4}y + z = -\frac{1}{2} \\ -2x - y + z = 7 \end{cases}$$

8. $$\begin{cases} x + y + 10z = 3 \\ \frac{1}{2}x - y + z = -\frac{5}{6} \\ -2x + 3y - 5z = \frac{7}{3} \end{cases}$$

You Try It

9. $$\begin{cases} -4x + 5y + 9z = -9 \\ x - 2y + z = 0 \\ 2y - 8z = 8 \end{cases}$$

10. $$\begin{cases} 2x + 2y + z = 9 \\ x + z = 4 \\ 4y - 3z = 17 \end{cases}$$

11. $$\begin{cases} x - y = 7 \\ y - z = 2 \\ x + z = 1 \end{cases}$$

In Exercises 12–17, determine if the system has no solution or infinitely many solutions. If the system has infinitely many solutions, describe the solution with the equation of a plane or an ordered triple in terms of one variable.

You Try It

12. $$\begin{cases} 2x + 6y - 4z = 8 \\ -x - 3y + 2z = -4 \\ x + 3y - 2z = 4 \end{cases}$$

13. $$\begin{cases} 2x - y + z = -6 \\ x - \frac{1}{2}y + \frac{1}{2}z = -3 \\ 4x - 2y + 2z = -12 \end{cases}$$

You Try It

14. $$\begin{cases} x - 4y + 2z = 7 \\ \frac{1}{2}x - 2y + z = 1 \\ -3x + y - 4z = 9 \end{cases}$$

15. $$\begin{cases} 4x - y + z = 8 \\ x + y + 3z = 2 \\ 3x - 2y - 2z = 5 \end{cases}$$

You Try It

16. $\begin{cases} x + 2y - z = 11 \\ x + 3y - 2z = 14 \\ 3x + 7y - 4z = 36 \end{cases}$

17. $\begin{cases} x - y + z = 5 \\ 2x + 3y - 3z = -5 \\ 3x + 2y - 2z = 0 \end{cases}$

In Exercises 18–23, solve using a system of linear equations in three variables.

You Try It

18. **Real-Time Strategy Game** While playing a real-time strategy game, Arvin created military units for a battle: long swordsmen, spearmen, and crossbowmen. Long swordsmen require 60 units of food and 20 units of gold. Spearmen require 35 units of food and 25 units of wood. Crossbowmen require 25 units of wood and 45 units of gold. If Arvin used 1975 units of gold, 1375 units of wood, and 1900 units of food to create the units, how many of each type of military unit did he create?

You Try It

19. **Concession Stand** The concession stand at a school basketball tournament sells hot dogs, hamburgers, and chicken sandwiches. During one game, the stand sold 16 hot dogs, 14 hamburgers, and 8 chicken sandwiches for a total of \$89.00. During a second game, the stand sold 10 hot dogs, 13 hamburgers, and 5 chicken sandwiches for a total of \$66.25. During a third game, the stand sold 4 hot dogs, 7 hamburgers, and 7 chicken sandwiches for a total of \$49.75. Determine the price of each product.

20. **Ordering Pizza** Ben ordered 35 pizzas for an office party. He ordered three types: cheese, supreme, and pepperoni. Cheese pizza costs \$9 each, pepperoni pizza costs \$12 each, and supreme pizza costs \$15 each. He spent exactly twice as much on the pepperoni pizzas as he did on the cheese pizzas. If Ben spent \$420, how many pizzas of each type did he buy?

21. **Theater Tickets** On opening night of the play *The Music Man*, 1010 tickets were sold for a total of \$10,300. Adult tickets cost \$12 each, children's tickets cost \$10 each, and senior citizen tickets cost \$7 each. If the total number of adult and children tickets sold exceeded twice the number of senior citizen tickets sold by 170 tickets, then how many tickets of each type were sold?

22. **NCAA Basketball** Tyler Hansbrough was the leading scorer of the 2009 NCAA Basketball champions, the North Carolina Tar Heels. Hansbrough scored a total of 722 points during the 2009 season. He made 26 more one-point free throws than two-point field goals, and his number of two-point field goals was two less than 25 times his number of three-point field goals. How many free throws, two-point field goals, and three-point field goals did Tyler Hansbrough make during the 2009 season? (*Source:* espn.com)

23. **Facebook Users** The number of new Facebook users, y (in millions), between September 2008 and March 2009 can be modeled by the equation $y = ax^2 + bx + c$, where x represents the age of the user. Using the ordered-pair solutions $(15, 1)$, $(35, 7)$, and $(55, 3)$, create a system of linear equations in three variables for a, b, and c. Do this by substituting each ordered-pair solution into the model, creating an equation in three variables. Solve the resulting system to find the coefficients of the model. Then use the model to predict the number of new Facebook users who were 25 years old. (*Source:* www.facebook.com)

MODULE FOURTEEN

Exponents and Polynomials

MODULE FOURTEEN CONTENTS

14.1 Exponents

THINGS TO KNOW

Before working through this topic, be sure you are familiar with the following concepts:

	VIDEO	ANIMATION	INTERACTIVE
You Try It 1. Evaluate Exponential Expressions (Topic 10.4, **Objective 1**)	🎞		

OBJECTIVES

1 Simplify Exponential Expressions Using the Product Rule

2 Simplify Exponential Expressions Using the Quotient Rule

3 Use the Zero-Power Rule

4 Use the Power-to-Power Rule

5 Use the Product-to-Power Rule

6 Use the Quotient-to-Power Rule

7 Simplify Exponential Expressions Using a Combination of Rules

OBJECTIVE 1 SIMPLIFY EXPONENTIAL EXPRESSIONS USING THE PRODUCT RULE

An **exponential expression** is a **constant** or **algebraic expression** that is raised to a **power**. The constant or algebraic expression makes up the **base**, and the power is the **exponent** on the base. In **Topic 10.4**, we learned that an **exponent** can be used to show repeated multiplication. For example, $3 \cdot 3 \cdot 3 \cdot 3$ can be written as 3^4.

(eText Screens 14.1-1–14.1-23)

$$\underbrace{3\cdot 3\cdot 3\cdot 3}_{\text{4 factors of 3}} = \underset{\substack{\uparrow\\ \text{Base}}}{3}^{4} \leftarrow \text{Exponent}$$

The exponent 4 indicates that there are four **factors** of the base 3.

The same is true when the base is a **variable** or an algebraic expression.

$$\underbrace{x\cdot x\cdot x\cdot x\cdot x}_{\text{5 factors of } x} = \underset{\substack{\uparrow\\ \text{Base}}}{x}^{5} \leftarrow \text{Exponent} \qquad \underbrace{(2y)\cdot(2y)\cdot(2y)}_{\text{3 factors of } 2y} = \underset{\substack{\uparrow\\ \text{Base}}}{(2y)}^{3} \leftarrow \text{Exponent}$$

The exponent 5 means there are five factors of the base x, and the exponent 3 means there are three factors of the base $2y$.

When an algebraic expression such as $2y$ is the base of an exponential expression, we must use parentheses (or other **grouping symbols**) to show that the entire expression $2y$ is raised to the third power, not just the variable y.

$$(2y)\cdot(2y)\cdot(2y) = (2y)^3 \neq 2y^3 = 2\cdot y\cdot y\cdot y$$

If we multiply **exponential expressions** with the same **base**, such as $x^3\cdot x^4$, we write

$$\begin{aligned} x^3\cdot x^4 &= \underbrace{(x\cdot x\cdot x)}_{\text{3 factors of } x} \cdot \underbrace{(x\cdot x\cdot x\cdot x)}_{\text{4 factors of } x} \\ &= \underbrace{x\cdot x\cdot x\cdot x\cdot x\cdot x\cdot x}_{\text{7 factors of } x} \\ &= x^7. \end{aligned}$$

Multiplying three **factors** of x by four factors of x means we are really multiplying seven factors of x, which is the sum of the two initial exponents 3 and 4:

$$x^3\cdot x^4 = x^{3+4} = x^7$$

This result suggests the following rule.

> **The Product Rule for Exponents**
>
> When multiplying exponential expressions with the same base, add the exponents and keep the common base.
>
> $$a^m\cdot a^n = a^{m+n}$$

CAUTION: When a factor has no written **exponent**, it is understood to be 1. For example, $x = x^1$.

Example 1 Using the Product Rule for Exponents

My video summary

Use the **product rule** to simplify each expression.

a. $5^4\cdot 5^6$ **b.** $x^5\cdot x^7$ **c.** $y^3\cdot y$ **d.** $b^3\cdot b^5\cdot b^4$

Solutions

a. The two exponential expressions 5^4 and 5^6 have the same base 5, so we add the exponents:

$$5^4\cdot 5^6 = 5^{4+6} = 5^{10}$$

Add the exponents.

Keep the common base.

b. $x^5 \cdot x^7 = x^{5+7} = x^{12}$

c.–d. Try to simplify these expressions on your own. View the **answers**, or watch this **video** for detailed solutions to all four parts of this example.

You Try It **Work through this You Try It problem.**

Work Exercises 1–4 in this textbook or in the MyMathLab Study Plan.

CAUTION

When using the product rule for exponents, do not multiply the bases. Instead, keep the common base and add the exponents. For example,

Incorrect	**Correct**
~~$2^8 \cdot 2^6 = 4^{14}$~~	$2^8 \cdot 2^6 = 2^{14}$

When the **exponential expressions** being multiplied involve more than one base, we first use the **commutative** and **associative** properties of multiplication to group **like bases** together and then apply the **product rule.**

Example 2 Using the Product Rule for Exponents

My video summary

Simplify using the product rule.

a. $(4x^2)(7x^3)$ **b.** $(m^4n^2)(m^3n^6)$ **c.** $(-3a^5b^3)(-8a^2b)$

Solutions

a. Begin with the original expression: $(4x^2)(7x^3)$

Rearrange factors to group like bases: $= 4 \cdot 7 \cdot x^2 \cdot x^3$

Multiply constants; apply product rule: $= 28 \cdot x^{2+3}$

Simplify: $= 28x^5$

b. Begin with the original expression: $(m^4n^2)(m^3n^6)$

Rearrange factors to group like bases: $= m^4 \cdot m^3 \cdot n^2 \cdot n^6$

Apply the product rule: $= m^{4+3} \cdot n^{2+6}$

Simplify: $= m^7n^8$

c. Try to simplify this expression on your own. View the **answer**, or watch this **video** for detailed solutions to all three parts of this example.

You Try It **Work through this You Try It problem.**

Work Exercises 5–8 in this textbook or in the MyMathLab Study Plan.

OBJECTIVE 2 SIMPLIFY EXPONENTIAL EXPRESSIONS USING THE QUOTIENT RULE

When dividing **exponential expressions** with the same nonzero **base**, such as $\frac{x^5}{x^2}$ with $x \neq 0$, we can expand each expression and divide out common factors:

Expand: $\frac{x^5}{x^2} = \frac{x \cdot x \cdot x \cdot x \cdot x}{x \cdot x}$ $\begin{array}{l} \leftarrow \text{5 factors of } x \\ \leftarrow \text{2 factors of } x \end{array}$

Divide out common factor: $= \frac{\cancel{x} \cdot \cancel{x} \cdot x \cdot x \cdot x}{\cancel{x} \cdot \cancel{x}}$

Simplify: $= x \cdot x \cdot x$

$= x^3$

Notice that both factors of x from the **denominator** divide out, leaving three remaining factors of x in the **numerator**. This result is the same if we subtract the initial exponents 5 and 2:

$$\frac{x^5}{x^2} = x^{5-2} = x^3$$

As long as the base is not 0, this result is true in general.

The Quotient Rule for Exponents

When dividing **exponential expressions** with the same nonzero **base**, subtract the **denominator** exponent from the **numerator** exponent and keep the common base.

$$\frac{a^m}{a^n} = a^{m-n} \quad (a \neq 0)$$

Example 3 Using the Quotient Rule for Exponents

My video summary

Use the quotient rule to simplify each expression.

a. $\frac{t^9}{t^5}$ b. $\frac{7^5}{7^3}$ c. $\frac{y^{24}}{y^{15}}$ d. $\frac{(-4)^{14}}{(-4)^{11}}$

Solutions

a. The two exponential expressions t^9 and t^5 have the same base t, so we subtract the exponents:

Subtract the exponents.

$$\frac{t^9}{t^5} = t^{9-5} = t^4$$

Keep the common base.

b. $\frac{7^5}{7^3} = 7^{5-3} = 7^2 = 49$

c.–d. Try to simplify these expressions on your own. View the **answers**, or watch this **video** for detailed solutions to all four parts of this example.

You Try It **Work through this You Try It problem.**

Work Exercises 9–12 in this textbook or in the MyMathLab Study Plan.

When the **exponential expressions** being divided involve more than one **base**, we group **like bases** into individual **quotients**. Then we apply the **quotient rule** to each quotient and simplify.

Example 4 Using the Quotient Rule for Exponents

Simplify using the quotient rule.

a. $\frac{15x^6}{3x^2}$ b. $\frac{a^4b^9c^5}{a^2b^3c}$ c. $\frac{4m^6n^7}{12m^5n^2}$

My video summary

Solutions

a. Begin with the original expression: $\dfrac{15x^6}{3x^2}$

Group like bases into individual quotients: $= \dfrac{15}{3} \cdot \dfrac{x^6}{x^2}$

Divide the constants; apply the quotient rule: $= 5 \cdot x^{6-2}$

Simplify: $= 5x^4$

b. Begin with the original expression: $\dfrac{a^4b^9c^5}{a^2b^3c}$

Group like bases into individual quotients: $= \dfrac{a^4}{a^2} \cdot \dfrac{b^9}{b^3} \cdot \dfrac{c^5}{c^1}$ ← Understood exponent of 1

Apply the quotient rule to each quotient: $= a^{4-2} \cdot b^{9-3} \cdot c^{5-1}$

Simplify: $= a^2b^6c^4$

c. Try to simplify this expression on your own. View the **answer**, or watch this **video** for detailed solutions to all three parts of this example.

You Try It **Work through this You Try It problem.**

Work Exercises 13–16 in this textbook or in the MyMathLab Study Plan.

OBJECTIVE 3 USE THE ZERO-POWER RULE

If we multiply 2^3 by 2^0, we can use the **product rule** and write

$$2^3 \cdot 2^0 = 2^{3+0} = 2^3.$$

From the **multiplicative identity property** we know that $2^3 \cdot 1 = 2^3$. Therefore, it makes sense that $2^0 = 1$ because $2^3 \cdot 2^0 = 2^3 = 2^3 \cdot 1$. As long as the **base** is not 0, this result is true in general.

The Zero-Power Rule

A nonzero base raised to the 0 power equals 1.

$$a^0 = 1 \quad (a \neq 0)$$

The zero-power rule can also be derived from the **quotient rule**. View this **popup box** to see how.

Example 5 Using the Zero-Power Rule

Simplify using the zero-power rule.

a. 6^0 **b.** $(-3)^0$ **c.** -3^0 **d.** $(2x)^0$ **e.** $2x^0$

Solutions

a. The base 6 is nonzero, so $6^0 = 1$.

b. The parentheses indicate that the base is -3, which is nonzero, so $(-3)^0 = 1$.

c. There are no parentheses, so the base is 3, not -3. We need to find the "opposite of 3^0." So, $-3^0 = -(3^0) = -1$.

d. The parentheses indicate that the base is $2x$, which is nonzero provided that $x \neq 0$, so $(2x)^0 = 1$.

e. There are no parentheses, so the base is x, not $2x$. We have "2 times x^0." So, assuming $x \neq 0$, $2x^0 = 2 \cdot 1 = 2$.

You Try It **Work through this You Try It problem.**

Work Exercises 17–22 in this textbook or in the MyMathLab Study Plan.

OBJECTIVE 4 USE THE POWER-TO-POWER RULE

An **exponential expression** itself can be raised to a power, such as $(x^2)^3$. We can simplify this expression by expanding and then using the **product rule**.

$$(x^2)^3 = \underbrace{x^2 \cdot x^2 \cdot x^2}_{\text{3 factors of } x^2} = \underbrace{x^{2+2+2}}_{\text{Product rule}} = x^6$$

Notice the final exponent 6 is the result of multiplying the two original exponents 2 and 3, so $(x^2)^3 = x^{2 \cdot 3} = x^6$. This result is true in general.

> **The Power-to-Power Rule**
>
> When an exponential expression is raised to a power, multiply the exponents.
>
> $$(a^m)^n = a^{m \cdot n}$$

Example 6 Using the Power-to-Power Rule

My video summary

Simplify using the **power-to-power rule**.

a. $(y^5)^6$ **b.** $[(-2)^3]^5$

Solutions

a. Because we are raising the exponential expression, y^5, to the sixth power, we multiply the exponents:

Multiply the exponents.

$$(y^5)^6 = y^{5 \cdot 6} = y^{30}$$

Keep the original base.

b. Try to simplify this expression on your own. View the **answer**, or watch this **video** for detailed solutions to both parts of this example.

You Try It **Work through this You Try It problem.**

Work Exercises 23–26 in this textbook or in the MyMathLab Study Plan.

OBJECTIVE 5 USE THE PRODUCT-TO-POWER RULE

When the **base** of an **exponential expression** is a **product**, such as $(xy)^4$, we can expand the expression and use the **commutative** and **associative** properties of multiplication to regroup like factors.

$$\text{Expand: } (xy)^4 = (xy)\cdot(xy)\cdot(xy)\cdot(xy) \leftarrow \boxed{\text{4 factors of } xy}$$

$$\text{Group common factors: } = \underbrace{x\cdot x\cdot x\cdot x}_{\text{4 factors of } x} \cdot \underbrace{y\cdot y\cdot y\cdot y}_{\text{4 factors of } y}$$

$$\text{Rewrite with exponents: } = x^4y^4$$

The final result is that each factor of the original base is raised to the exponent: $(xy)^4 = x^4y^4$. This result is true in general.

The Product-to-Power Rule

When raising a product to a power, raise each factor of the base to the common exponent.

$$(ab)^n = a^nb^n$$

Example 7 Using the Product-to-Power Rule

Simplify using the **product-to-power rule**.

a. $(mn)^8$ **b.** $(x^2y)^5$ **c.** $(3y)^4$ **d.** $(-4p^5q^3)^2$

Solutions

a. The **base** is the **product** mn.

Apply the product-to-power rule: $(mn)^8 = m^8n^8$

b. The base is the product x^2y.

Apply the product-to-power rule: $(x^2y)^5 = (x^2)^5\cdot y^5$

Apply the power-to-power rule: $= x^{2\cdot 5}\cdot y^5$

Simplify: $= x^{10}y^5$

My video summary

c.–d. Try to simplify these expressions on your own. View the **answers**, or watch this **video** for detailed solutions to parts c and d.

You Try It **Work through this You Try It problem.**

Work Exercises 27–32 in this textbook or in the MyMathLab Study Plan.

OBJECTIVE 6 USE THE QUOTIENT-TO-POWER RULE

When the **base** of an **exponential expression** is a **quotient**, such as $\left(\frac{x}{y}\right)^4$ with $y \neq 0$, we can expand the expression and multiply the resulting fractions.

$$\text{Expand: } \left(\frac{x}{y}\right)^4 = \frac{x}{y}\cdot\frac{x}{y}\cdot\frac{x}{y}\cdot\frac{x}{y} \leftarrow \boxed{\text{4 factors of } \frac{x}{y}}$$

$$\text{Multiply the fractions: } = \frac{x\cdot x\cdot x\cdot x}{y\cdot y\cdot y\cdot y} \begin{array}{l}\leftarrow \boxed{\text{4 factors of } x \text{ in numerator}}\\ \leftarrow \boxed{\text{4 factors of } y \text{ in denominator}}\end{array}$$

$$\text{Rewrite with exponents: } = \frac{x^4}{y^4}$$

The final result is that the **numerator** and **denominator** are each raised to the **exponent**: $\left(\frac{x}{y}\right)^4 = \frac{x^4}{y^4}$. This result is true in general.

The Quotient-to-Power Rule

When raising a quotient to a power, raise both the numerator and denominator to the common exponent.

$$\left(\frac{a}{b}\right)^n = \frac{a^n}{b^n} \quad (b \neq 0)$$

Example 8 Using the Quotient-to-Power Rule

Simplify using the **quotient-to-power rule**.

a. $\left(\frac{m}{n}\right)^9$ b. $\left(\frac{x^2}{y^5}\right)^4$ c. $\left(\frac{x}{2}\right)^5$ d. $\left(\frac{3x^2}{5y^4}\right)^3$

Solutions

a. The **base** is the **quotient** $\frac{m}{n}$.

Apply the quotient-to-power rule: $\left(\frac{m}{n}\right)^9 = \frac{m^9}{n^9}$

b. The base is the quotient $\frac{x^2}{y^5}$.

Apply the quotient-to-power rule: $\left(\frac{x^2}{y^5}\right)^4 = \frac{(x^2)^4}{(y^5)^4}$

Apply the power-to-power rule: $= \frac{x^{2\cdot 4}}{y^{5\cdot 4}}$

Simplify: $= \frac{x^8}{y^{20}}$

My video summary

c.–d. Try to simplify these expressions on your own. View the **answers**, or watch this **video** for detailed solutions to parts c and d.

You Try It **Work through this You Try It problem.**

Work Exercises 33–38 in this textbook or in the MyMathLab Study Plan.

OBJECTIVE 7 SIMPLIFY EXPONENTIAL EXPRESSIONS USING A COMBINATION OF RULES

In some of the earlier examples, more than one rule for exponents was used to simplify the **exponential expressions**. To be considered **simplified**, an exponential expression must meet the following conditions:

- No parentheses or other **grouping symbols** are present.
- No zero **exponents** are present.
- No **powers** are raised to powers.
- Each **base** occurs only once.

Following is a summary of the exponent rules we have seen so far. We can combine these rules to simplify exponential expressions.

Rules for Exponents	
Product Rule	$a^m \cdot a^n = a^{m+n}$
Quotient Rule	$\dfrac{a^m}{a^n} = a^{m-n} \quad (a \neq 0)$
Zero-Power Rule	$a^0 = 1 \quad (a \neq 0)$
Power-to-Power Rule	$(a^m)^n = a^{m \cdot n}$
Product-to-Power Rule	$(ab)^n = a^n b^n$
Quotient-to-Power Rule	$\left(\dfrac{a}{b}\right)^n = \dfrac{a^n}{b^n} \quad (b \neq 0)$

Example 9 Using a Combination of Exponent Rules

My interactive video summary

Simplify using the **rules for exponents**.

a. $(c^3)^5(c^2)^6$

b. $\left(\dfrac{15x^8y^5}{3x^6y}\right)^2$

c. $(-2w^3z^2)(-2wz^2)^4$

d. $\dfrac{(4m^2n^0)(2n^3)^2}{8mn^5}$

Solutions

a. Apply the power-to-power rule: $(c^3)^5(c^2)^6 = c^{3\cdot 5} \cdot c^{2\cdot 6}$

Multiply: $= c^{15} \cdot c^{12}$

Apply the product rule: $= c^{15+12}$

Add: $= c^{27}$

b. Within the parentheses, group like bases: $\left(\dfrac{15x^8y^5}{3x^6y}\right)^2 = \left(\dfrac{15}{3} \cdot \dfrac{x^8}{x^6} \cdot \dfrac{y^5}{y^1}\right)^2$

Divide constants; apply the quotient rule: $= (5 \cdot x^{8-6} \cdot y^{5-1})^2$

Simplify: $= (5x^2y^4)^2$

Apply the product-to-power rule: $= 5^2(x^2)^2(y^4)^2$

Apply the power-to-power rule: $= 5^2 \cdot x^{2\cdot 2} \cdot y^{4\cdot 2}$

Simplify: $= 25x^4y^8$

c.–d. Try to simplify these expressions on your own. View the **answers**, or work through this **interactive video** for complete solutions to all four parts.

You Try It **Work through this You Try It problem.**

Work Exercises 39–44 in this textbook or in the MyMathLab Study Plan.

Note: There is often more than one way to simplify an exponential expression. Do not feel that you need to always work problems using the exact same steps as shown in the examples. Just make sure that your solutions meet the **conditions** for simplified exponential expressions.

14.1 Exercises

In Exercises 1–8, simplify using the product rule.

You Try It

1. $7^4 \cdot 7^2$
2. $x^8 \cdot x^5$
3. $w^5 \cdot w$
4. $t^6 \cdot t^2 \cdot t^5$

You Try It

5. $(6a^3)(5a^7)$
6. $(x^2y^7)(x^4y^3)$
7. $(-2m^9n)(-9m^4n^5)$
8. $(-6p^2)(5p^4)(2p^3)$

In Exercises 9–16, simplify using the quotient rule.

You Try It

9. $\dfrac{b^{12}}{b^8}$
10. $\dfrac{5^9}{5^7}$
11. $\dfrac{(-10)^7}{(-10)^4}$
12. $\dfrac{y^{17}}{y^{16}}$

You Try It

13. $\dfrac{24q^{15}}{6q^3}$
14. $\dfrac{60a^4b^{13}}{12ab^9}$
15. $\dfrac{x^7y^9z^8}{x^5y^3z^7}$
16. $\dfrac{10m^{21}n^9}{15m^{13}n^2}$

In Exercises 17–22, simplify using the zero-power rule.

You Try It

17. 4^0
18. $(-8)^0$
19. -12^0
20. $6m^0$
21. $(3x)^0$
22. $3p^0 + 2q^0$

In Exercises 23–26, simplify using the power-to-power rule.

You Try It

23. $(x^3)^8$
24. $(y^9)^4$
25. $[(-2)^4]^3$
26. $[(m^4)^5]^3$

In Exercises 27–32, simplify using the product-to-power rule.

You Try It

27. $(pq)^6$
28. $(2m)^5$
29. $(a^3b)^7$
30. $(-5t^6)^4$
31. $(-7r^3s^9)^2$
32. $(2x^3y^2z)^4$

In Exercises 33–38, simplify using the quotient-to-power rule.

You Try It

33. $\left(\dfrac{a}{b}\right)^8$
34. $\left(\dfrac{3}{5}\right)^4$
35. $\left(\dfrac{t}{4}\right)^3$
36. $\left(\dfrac{m^4}{n^3}\right)^5$
37. $\left(\dfrac{u^5}{2}\right)^7$
38. $\left(\dfrac{3p^4}{4q^5}\right)^3$

In Exercises 39–44, simplify using the rules for exponents.

You Try It

39. $(x^2)^4(x^3)^6$
40. $(m^3n^2)^3(m^2n^5)^2$
41. $\left(\dfrac{16p^9q^5}{8p^3q^5}\right)^2$
42. $\dfrac{(18c^0d^{10})(4c^5d^2)}{12cd^5}$
43. $\left(\dfrac{x^2}{y}\right)^3\left(\dfrac{x^3}{y^2}\right)^2$
44. $\left(\dfrac{-2ac^2}{b^3}\right)^2\left(\dfrac{9a^5b^8}{c^3}\right)$

14.2 Introduction to Polynomials

THINGS TO KNOW

Before working through this topic, be sure you are familiar with the following concepts:

		VIDEO	ANIMATION	INTERACTIVE
You Try It	1. Evaluate Algebraic Expressions (Topic 10.5, **Objective 1**)	✓		
You Try It	2. Identify Terms, Coefficients, and Like Terms of an Algebraic Expression (Topic 10.6, **Objective 1**)	✓		
You Try It	3. Simplify Algebraic Expressions (Topic 10.6, **Objective 2**)			✓
You Try It	4. Simplify Exponential Expressions Using a Combination of Rules (Topic 14.1, **Objective 7**)			✓

OBJECTIVES

1 Classify Polynomials

2 Determine the Degree and Coefficient of a Monomial

3 Determine the Degree and Leading Coefficient of a Polynomial

4 Evaluate a Polynomial for a Given Value

5 Simplify Polynomials by Combining Like Terms

OBJECTIVE 1 CLASSIFY POLYNOMIALS

Recall that the **terms** of an algebraic expression are the quantities being added. A term can be a **constant**, a **variable**, or the product of a constant and one or more variables raised to **powers**. For example, the terms of the expression $3x^5 - 2x^2 + 7$ are $3x^5$, $-2x^2$, and 7.

If a term contains a single numeric **factor** and if none of the **variable factors** can be combined using the **rules for exponents**, then it is called a **simplified term**. Examples of simplified terms include

$$5x^3, -7y^2z, \frac{2}{3}b^2, \text{ and } 8.$$

The terms $-4m^2m^5$ and $\frac{12y^3}{20}$ are not simplified. Do you see **why**?

Some **simplified terms** are also *monomials*.

> **Definition Monomial**
>
> A **monomial** is a simplified term in which all variables are raised to non-negative **integer** powers and no variables appear in any **denominator**.

My interactive video summary

Note that a monomial can be a constant, such as 7. Some examples of monomials are $7x^4$, $-\frac{3}{4}x^3y^2$, and 9. The terms $2y^{-3}$, $6x^{1/4}$, and $\frac{2}{x}$ are not monomials. Do you see **why**? Watch this **interactive video** to determine if a given term is a monomial.

We studied **algebraic expressions** in **Topic 10.5** and **Topic 10.6**. We now look at special kinds of algebraic expressions called *polynomials*.

> **Definition Polynomial**
>
> A **polynomial** in x is a **monomial** or a **finite** sum of monomials of the form ax^n, where a is any real number and n is any whole number.

For example,

$$10x^7 + 6x^5 - 3x^2 - 5x + 4$$

is a polynomial in x. However, the expressions

$$5x^2 + 3x - 7x^{-4} \quad \text{and} \quad -3x^5 + \frac{2}{x^2} - 6x$$

are not polynomials. Do you see **why**?

The monomials that make up a polynomial are called the **terms of the polynomial**. A polynomial is a **simplified polynomial** if all of its **terms** are simplified and none of its terms are **like terms**. Polynomials can be defined using variables other than x and may have terms with more than one variable. For now, we will consider **polynomials in one variable**. In Topic 14.8, we visit polynomials in several variables.

We can classify simplified polynomials by the number of terms in the polynomial.

There are special names for **polynomials** with 1, 2, or 3 **terms**. If there are more than three terms, we use the general name, *polynomial*.

Polynomial	Number of terms	Name	Hint
$4x$	1	Monomial	← "Mono-" means "one"
$3x^4 + 2$	2	Binomial	← "Bi-" means "two"
$-2x^4 - 3x + 1$	3	Trinomial	← "Tri-" means "three"
$2x^3 - 8x^2 + 5x - 9$	4 (or more)	Polynomial	← "Poly-" means "many"

Table 1

Example 1 Classifying Polynomials

Classify each polynomial as a monomial, binomial, trinomial, or none of these.

a. $5x - 7$ **b.** $\frac{1}{3}x^2$ **c.** $5x^3 - 7x^2 + 4x + 1$ **d.** $-2x^3 - 5x^2 + 8x$

Solutions

a.–d. Count the number of terms, then use Table 1 to help you classify the expressions on your own. View the **answers**.

You Try It **Work through this You Try It problem.**

Work Exercises 1–6 in this textbook or in the MyMathLab Study Plan.

OBJECTIVE 2 DETERMINE THE DEGREE AND COEFFICIENT OF A MONOMIAL

Every **monomial** has both a *degree* and a *coefficient*.

> **Definition Degree of a Monomial**
>
> The **degree of a monomial** is the *sum* of the **exponents** on the **variables**.

For example, the degree of $-5x^2$ is 2, and the degree of $\frac{5}{8}x^3y^4$ is $3 + 4 = 7$.

CAUTION The degree of any constant term is 0 because there is no **variable factor**.

> **Definition Coefficient of a Monomial**
>
> The **coefficient of a monomial** is the **constant** factor.

For example, the coefficient of $-5x^2$ is -5, and the coefficient of $\frac{5}{8}x^3y^4$ is $\frac{5}{8}$.

Example 2 Determining the Coefficient and Degree of a Monomial

Determine the coefficient and degree of each monomial.

a. $4.6x^3$ **b.** $7x$ **c.** x^2y^4 **d.** 12 **e.** $\frac{3}{4}x^2yz^3$ **f.** $-2xyz^7$

Solutions

a. The **coefficient** of $4.6x^3$ is the **constant** factor 4.6. Because the **exponent** is 3, the **degree** is 3.

b. The coefficient of $7x$ is 7. The exponent of x is **understood to be** 1, so the degree is 1.

c. Because $x^2y^4 = 1x^2y^4$, the coefficient is 1. The degree is the sum of the exponents on the variables: $2 + 4 = 6$.

My video summary

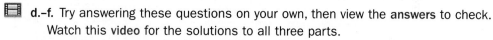

d.–f. Try answering these questions on your own, then view the **answers** to check. Watch this **video** for the solutions to all three parts.

You Try It **Work through this You Try It problem.**

Work Exercises 7–12 in this textbook or in the MyMathLab Study Plan.

OBJECTIVE 3 DETERMINE THE DEGREE AND LEADING COEFFICIENT OF A POLYNOMIAL

Because a **polynomial** is made up of **monomials**, every polynomial also has a degree.

> **Definition Degree of a Polynomial**
>
> The **degree of a polynomial** is the largest degree of its terms.

For example, $4x + 5x^3 - x^2 + 6$ has a degree of 3, $-3x^5 - 2x^3 + 5x^2 - 7$ has a degree of 5, and 15 has a degree of 0 (think $15 = 15(1) = 15x^0$).

(eText Screens 14.2-1–14.2-15)

The polynomial $-3x^5 - 2x^3 + 5x^2 - 7$ is an example of a polynomial written in **descending order.** This means that the terms are listed so that the first term has the largest degree and the **exponents** on the variable decrease from left to right. This is called the **standard form** for polynomials. When a polynomial is in standard form, we can find its *leading coefficient*.

> **Definition Leading Coefficient of a Polynomial in One Variable**
>
> When a polynomial in one variable is written in **standard form**, the **coefficient** of the first term (the term with the highest degree) is called the **leading coefficient.**

Example 3 Determining the Degree and Leading Coefficient of a Polynomial

Write each polynomial in **standard form**. Then find its **degree** and **leading coefficient.**

a. $4.2m - 3m^2 + 1.8 - 7m^3$ **b.** $\frac{2}{3}x^3 - 3x^2 + 5 - x^4 + \frac{1}{4}x$

Solutions

a. Start by determining the degree of each term.

Term	Degree	
$4.2m$	1	← $m = m^1$
$-3m^2$	2	
1.8	0	← $1.8 = 1.8m^0$
$-7m^3$	3	

Now write the polynomial so that the terms descend in order from the largest degree, 3, to the smallest degree, 0 (the constant term): $-7m^3 - 3m^2 + 4.2m + 1.8$. The degree of the polynomial is 3, the largest degree. The leading coefficient is -7, the coefficient of the first term when the polynomial is written in standard form.

My video summary

b. Try working this problem on your own. View the **answer**, or watch this **video** for a detailed solution.

You Try It **Work through this You Try It problem.**

Work Exercises 13–18 in this textbook or in the MyMathLab Study Plan.

OBJECTIVE 4 EVALUATE A POLYNOMIAL FOR A GIVEN VALUE

We **evaluate** polynomials exactly the same way we evaluated **algebraic expressions** in **Topic 10.5.** We substitute the given value for the **variable** and simplify.

Example 4 Evaluating Polynomials

Evaluate the **polynomial** $x^3 + 3x^2 + 4x - 5$ for the given values of x.

a. $x = -2$ **b.** $x = 0$ **c.** $x = 2$ **d.** $x = \frac{5}{2}$

(eText Screens 14.2-1–14.2-15)

Solutions

a. Begin with the polynomial: $x^3 + 3x^2 + 4x - 5$

Substitute -2 for x: $(-2)^3 + 3(-2)^2 + 4(-2) - 5$ ← Put values in parentheses when substituting

Simplify the exponents: $= -8 + 3(4) + 4(-2) - 5$

Simplify the multiplication: $= -8 + 12 - 8 - 5$

Add and subtract: $= -9$

The value of the polynomial is -9 when x is -2.

b. Begin with the polynomial: $x^3 + 3x^2 + 4x - 5$

Substitute 0 for x: $(0)^3 + 3(0)^2 + 4(0) - 5$ ← Put values in parentheses when substituting

Simplify the exponents: $= 0 + 3(0) + 4(0) - 5$

Simplify the multiplication: $= 0 + 0 + 0 - 5$

Add and subtract: $= -5$

The value of the polynomial is -5 when x is 0.

My video summary

c.–d. Try to evaluate the **polynomial** for the remaining values on your own. Then view the **answers**, or watch this **video** to see the complete solutions.

You Try It **Work through this You Try It problem.**

Work Exercises 19–25 in this textbook or in the MyMathLab Study Plan.

OBJECTIVE 5 SIMPLIFY POLYNOMIALS BY COMBINING LIKE TERMS

We learned how to simplify **algebraic expressions** in **Topic 10.6**. This often involved **combining like terms**. Similarly, we can simplify a **polynomial** with **like terms** by combining the like terms. This will be a necessary step as we perform operations on polynomials.

Example 5 Simplifying Polynomials

My interactive video summary

Simplify each polynomial by combining like terms.

a. $3x^2 + 8x - 4x + 2$

b. $2.3x - 3 - 5x + 8.4$

c. $2x + 3x^2 - 6 + x^2 - 2x + 9$

d. $\frac{2}{3}x^2 + \frac{1}{5}x - \frac{1}{10}x - \frac{1}{6}x^2 + \frac{1}{4}$

e. $6x^3 + x^2 - 7$

Solutions

a. Begin with the original expression: $3x^2 + 8x - 4x + 2$

Reverse the **distributive property**: $= 3x^2 + (8 - 4)x + 2$

Subtract: $= 3x^2 + 4x + 2$

b. Begin with the original expression: $2.3x - 3 - 5x + 8.4$

Rearrange to **collect like terms**: $= 2.3x - 5x - 3 + 8.4$

Reverse the distributive property: $= (2.3 - 5)x + (-3 + 8.4)$

Add or subtract: $= -2.7x + 5.4$

(eText Screens 14.2-1–14.2-15)

c.–e. Try to simplify these polynomials on your own. Then view the **answers**, or watch this **interactive video** to see the complete solution for any of the five parts.

You Try It **Work through this You Try It problem.**

Work Exercises 26–34 in this textbook or in the MyMathLab Study Plan.

14.2 Exercises

In Exercises 1–6, classify each polynomial as a monomial, binomial, trinomial, or none of these.

You Try It

1. $3x^2 - 5x + 1$

2. $-\frac{3}{5}x^2$

3. $x^5 - 7x^3 + x^2 - 4$

4. $5x^3 + 2x$

5. $2.4x^3 - 0.5x^2 + 0.25x + 1.35$

6. $\frac{1}{2}x^4 - \frac{4}{5}x + 1$

In Exercises 7–12, determine the coefficient and degree of each monomial.

You Try It

7. $4x^3y$

8. 8.1

9. x^4y^3

10. $-7xy^2z^2$

11. $-9x$

12. $\frac{2}{3}x^2y^4z$

In Exercises 13–18, write each polynomial in standard form. Then find its degree and leading coefficient.

You Try It

13. $4 + 5x$

14. $\frac{2}{3}y - 12y^2 - 5$

15. $9 - 3x$

16. $m - 8 - m^7$

17. $4.1x^3 - 6.7x + 3.8 + x^2$

18. $7a - a^5 + 4a^2 + 3 - 8a^2$

In Exercises 19–25, evaluate each polynomial for the given value.

You Try It

19. $y + 4$ when $y = 3$.

20. $2x - 5$ when $x = -6$.

21. $5x^2 - 9x - 7$ when $x = 0$.

22. $x^2 - 7x - 12$ when $x = -1$

23. $3.5x^2 + 2.4$ when $x = 1.2$.

24. $x^2 + \frac{2}{5}x - 2$ when $x = \frac{3}{2}$

25. $-2x^3 + 6x^2 - 5x - 11$ for $x = -2$.

In Exercises 26–34, simplify each polynomial by combining like terms.

You Try It

26. $6x - 3x$

27. $2x + 5 - 7x$

28. $5x^2 - 9x + 3$

29. $14x + 3x - x + 1$

30. $-4x^2 + 9x + 7x^2 - 7$

31. $0.2x^2 - 4.1 + 3.2x + 1.1x^2 - 5$

32. $\frac{1}{3}x - \frac{2}{5} + 2x^2 - \frac{3}{2}x - \frac{4}{5}$

33. $x^3 - 2x + 3 + 6x^3 + 2x - 11$

34. $3x^4 + x^2 - 8 - 7x^4 + 3x - 2x^2$

14.3 Adding and Subtracting Polynomials

THINGS TO KNOW

Before working through this topic, be sure you are familiar with the following concepts:

		VIDEO	ANIMATION	INTERACTIVE
You Try It	**1.** Find the Opposite of a Real Number (Topic 10.1, **Objective 3**)			
You Try It	**2.** Identify Terms, Coefficients, and Like Terms of an Algebraic Expression (Topic 10.6, **Objective 1**)	✓		
You Try It	**3.** Simplify Algebraic Expressions (Topic 10.6, **Objective 2**)			✓

OBJECTIVES

1 Add Polynomials

2 Find the Opposite of a Polynomial

3 Subtract Polynomials

OBJECTIVE 1 ADD POLYNOMIALS

In **Topic 10.6**, we learned how to simplify **algebraic expressions.** This process often involves combining like terms.

To add **polynomials**, we remove all **grouping symbols**, use the **commutative** and **associative properties** of addition to rearrange the terms so that **like terms** are grouped, and combine all like terms.

Adding Polynomials

To add polynomials, remove all grouping symbols and combine like terms.

Example 1 Adding Polynomials

Add $(2x + 8) + (7x - 3)$.

Solution

Begin with the original expression: $(2x + 8) + (7x - 3)$

Remove the grouping symbols: $= 2x + 8 + 7x - 3$

Rearrange to group like terms: $= 2x + 7x + 8 - 3$

Combine like terms: $= 9x + 5$

So, $(2x + 8) + (7x - 3) = 9x + 5$.

We can also add **polynomials** vertically. To do this, line up **like terms** in columns and then combine like terms. We repeat **Example 1** to show this method:

$$\begin{array}{r} 2x + 8 \\ +\ \underline{7x - 3} \\ 9x + 5 \end{array}$$

You Try It **Work through this You Try It problem.**

Work Exercises 1 and 2 in this textbook or in the MyMathLab Study Plan.

CAUTION Remember, when a **term** has no **coefficient** shown, it is understood to be 1.

Example 2 Adding Polynomials

My interactive video summary

Add.

a. $(y^2 + 3y + 7) + (y^2 - 3y - 2)$
b. $(10p^3 + 7p - 13) + (5p^2 - 4p)$
c. $(3m^3 + m^2 - 8) + (2m^3 - 4m^2 + 3m) + (5m^2 + 4)$

Solutions We follow the procedure for **adding polynomials**.

a. Begin with the original expression: $(y^2 + 3y + 7) + (y^2 - 3y - 2)$
Remove the grouping symbols: $= y^2 + 3y + 7 + y^2 - 3y - 2$
Rearrange to group like terms: $= y^2 + y^2 + 3y - 3y + 7 - 2$
Combine like terms: $= 2y^2 + 5$

Therefore, $(y^2 + 3y + 7) + (y^2 - 3y - 2) = 2y^2 + 5$.

b. Begin with the original expression: $(10p^3 + 7p - 13) + (5p^2 - 4p)$
Remove the grouping symbols: $= 10p^3 + 7p - 13 + 5p^2 - 4p$
Rearrange to group like terms: $= 10p^3 + 5p^2 + 7p - 4p - 13$

Finish this problem on your own. Check your **answer**, or watch this **interactive video** for the complete solutions to all three parts.

c. Try adding these polynomials on your own. Check your **answer**, or watch this **interactive video** to see the complete solutions to all three parts.

Now try reworking **Example 2** by adding vertically. See this **popup box** for the solutions. Which method do you prefer?

You Try It **Work through this You Try It problem.**

Work Exercises 3–8 in this textbook or in the MyMathLab Study Plan.

OBJECTIVE 2 FIND THE OPPOSITE OF A POLYNOMIAL

In **Topic 10.1** we learned that a negative sign can be used to represent the "opposite" of a **real number**. For example, -5 is "the opposite of 5," and $-(-7)$ is "the opposite of negative 7." We find the opposite of a real number by changing its sign.

In the same way, $-(x^2 - 5x + 7)$ is "the opposite of the **polynomial** $x^2 - 5x + 7$," and we can find the *opposite of a polynomial* by changing the sign of each of its **terms**. So, the opposite of $x^2 - 5x + 7$, or $-(x^2 - 5x + 7)$, is $-x^2 + 5x - 7$.

Opposite Polynomials

To find the **opposite of a polynomial**, change the sign of each term.

(eText Screens 14.3-1–14.3-11)

TIP Finding the opposite of a polynomial, such as $-(x^2 - 5x + 7)$, can be thought of as **distributing** the negative sign through the polynomial.

$$-(x^2 - 5x + 7) = -x^2 - (-5x) + (-7) = -x^2 + 5x - 7$$

Example 3 Finding Opposite Polynomials

Find the opposite of each polynomial.

a. $x^2 + 6x + 8$ **b.** $8y - 27$ **c.** $-m^3 - 5m^2 + m + 7$

Solutions To find the opposite polynomial, change the sign of each **term**.

a. $-(x^2 + 6x + 8) = -x^2 - 6x - 8$

b. $-(8y - 27) = -8y + 27$

c. $-(-m^3 - 5m^2 + m + 7) = m^3 + 5m^2 - m - 7$

You Try It **Work through this You Try It problem.**

Work Exercises 9–12 in this textbook or in the MyMathLab Study Plan.

OBJECTIVE 3 SUBTRACT POLYNOMIALS

In **Topic 10.2**, we learned to subtract a real number by adding its opposite. In the same way, we **subtract** a polynomial by adding its **opposite polynomial**.

> **Subtracting Polynomials**
>
> To subtract a polynomial, add its opposite polynomial.

Example 4 Subtracting Polynomials

My video summary

Subtract.

a. $(9x + 13) - (6x - 4)$ **b.** $(3a^2 + 5a - 8) - (-2a^2 + a - 7)$

Solutions We subtract by adding the opposite polynomial.

a. Begin with the original expression: $(9x + 13) - (6x - 4)$

Add the opposite polynomial: $= (9x + 13) + (-6x + 4)$

Remove the grouping symbols: $= 9x + 13 - 6x + 4$

Rearrange to group like terms: $= 9x - 6x + 13 + 4$

Combine like terms: $= 3x + 17$

So, $(9x + 13) - (6x - 4) = 3x + 17$.

b. Begin with the original expression: $(3a^2 + 5a - 8) - (-2a^2 + a - 7)$

Add the opposite polynomial: $= (3a^2 + 5a - 8) + (+2a^2 - a + 7)$

Try finishing this problem on your own. Check your **answer**, or watch this **video** for detailed solutions to both parts.

(eText Screens 14.3-1–14.3-11)

As with addition, polynomials can be subtracted vertically. We repeat **Example 4a** to show this method:

$$\begin{array}{r} 9x + 13 \\ -\ \underline{(6x - 4)} \end{array} \xrightarrow{\text{Add the opposite polynomial}} \begin{array}{r} 9x + 13 \\ +\ \underline{-6x + \ \ 4} \\ 3x + 17 \end{array}$$

Now try reworking **Example 4b** by subtracting vertically. See this **popup box** for the solution.

You Try It **Work through this You Try It problem.**

Work Exercises 13–20 in this textbook or in the MyMathLab Study Plan.

14.3 Exercises

In Exercises 1–8, add.

You Try It

1. $(2x + 5) + (4x + 9)$

2. $(5y^2 - 3y) + (2y^2 - y)$

3. $(2m^2 - 5m - 13) + (m^2 + 5m + 23)$

4. $(7n^2 + 3n - 6) + (-2n^2 + 5)$

You Try It

5. $(12p^3 - 3p^2 - 7p - 5) + (-4p^3 + p^2 + 2p - 6)$

6. $(2a^2 - 11a + 9) + (8a^2 - 13) + (2a - 3)$

7. $(0.4w^2 - 5.6w + 3.1) + (1.3w^2 + 2.1w - 4.0)$

8. $\left(\frac{3}{5}x - \frac{1}{3}\right) + \left(\frac{1}{10}x + \frac{5}{6}\right)$

In Exercises 9–12, find the opposite of each polynomial.

You Try It

9. $7x^2 - 24$

10. $y^3 - 8y^2 - 17$

11. $-8z^3 + 3z^2 + 4z - 13$

12. $-\frac{7}{9}t^2 - \frac{4}{3}t + \frac{5}{6}$

In Exercises 13–20, subtract.

You Try It

13. $(5y + 8) - (2y - 14)$

14. $(7p^2 + 3p + 17) - (5p + 12)$

15. $(2t^2 + 13t - 17) - (-3t^2 + 11t - 13)$

16. $(-6z^2 + 3z - 14) - (8z^2 + z)$

17. $(-7x^3 + 6x^2 + x - 15) - (-2x^3 - 6x^2 + 3x - 15)$

18. $(m^2 - 9m + 1) - (3m^2 - 7) - (-2m + 3)$

19. $(5.1n^2 - 3.6n + 7.8) - (2.6n^2 + 1.9)$

20. $\left(\frac{7}{8}a - \frac{1}{3}\right) - \left(\frac{3}{8}a - \frac{1}{2}\right)$

14.4 Multiplying Polynomials

THINGS TO KNOW

Before working through this topic, be sure you are familiar with the following concepts:

		VIDEO	ANIMATION	INTERACTIVE
You Try It	1. Simplify Algebraic Expressions (Topic 10.6, **Objective 2**)	▤		
You Try It	2. Simplify Exponential Expressions Using a Combination of Rules (Topic 14.1, **Objective 7**)			▣

OBJECTIVES

1 Multiply Monomials

2 Multiply a Polynomial by a Monomial

3 Multiply Two Binomials

4 Multiply Two or More Polynomials

OBJECTIVE 1 MULTIPLY MONOMIALS

To multiply **monomials,** we use the **commutative** and **associative** properties to group **factors** with **like bases** and then apply the **product rule for exponents**. This is similar to the way we simplified some exponential expressions in **Topic 14.1**.

Multiplying Monomials

Rearrange the factors to group the **coefficients** and to group like bases. Then, multiply the coefficients and apply the product rule for exponents.

Example 1 Multiplying Monomials

Multiply.

a. $(6x^5)(7x^2)$ **b.** $\left(-\frac{3}{4}x^2\right)\left(-\frac{2}{9}x^8\right)$ **c.** $(3x^2)(-0.2x^3)$

Solutions

a.

Begin with the original expression:	$(6x^5)(7x^2)$
Rearrange factors to group coefficients and like bases:	$= (6 \cdot 7)(x^5x^2)$
Multiply coefficients; apply product rule for exponents:	$= 42x^{5+2}$
Simplify:	$= 42x^7$

b.–c. Try to work these problems on your own. View this **popup** for the complete solutions to both parts.

You Try It **Work through this You Try It problem.**

Work Exercises 1–6 in this textbook or in the MyMathLab Study Plan.

OBJECTIVE 2 MULTIPLY A POLYNOMIAL BY A MONOMIAL

To multiply a **monomial** and a **polynomial** with more than one **term**, we use the **distributive property**.

> **Multiplying Polynomials by Monomials**
>
> To multiply a polynomial by a monomial, use the distributive property to multiply each term of the polynomial by the monomial. Then simplify using the method for **multiplying monomials**.

If m is a monomial and $p_1, p_2, p_3, \ldots, p_n$ are the terms of a polynomial, then

$$m \cdot (p_1 + p_2 + p_3 + \ldots + p_n) = m \cdot p_1 + m \cdot p_2 + m \cdot p_3 + \ldots + m \cdot p_n$$

Example 2 Multiplying a Polynomial by a Monomial

Multiply.

a. $3x(4x - 5)$ **b.** $-4x^2(3x^2 + x - 7)$

Solutions

a. Begin with the original expression: $3x(4x - 5)$

Distribute the monomial $3x$: $= 3x \cdot 4x - 3x \cdot 5$

Rearrange the factors: $= 3 \cdot 4 \cdot x \cdot x - 3 \cdot 5 \cdot x$

Multiply coefficients; apply product rule for exponents: $= 12x^{1+1} - 15x$

Simplify: $= 12x^2 - 15x$

b. Begin with the original expression: $-4x^2(3x^2 + x - 7)$

Distribute the monomial $-4x^2$: $= -4x^2 \cdot 3x^2 + (-4x^2) \cdot x - (-4x^2) \cdot 7$

Try to finish this problem on your own. View this **popup** for the complete solution.

Example 3 Multiplying a Polynomial by a Monomial

My video summary

Multiply.

a. $\frac{1}{2}x^2(4x^2 - 6x + 2)$ **b.** $0.25x^3(6x^3 - 10x^2 + 4x - 7)$

Solutions Try to work these problems on your own. View the **answers**, or watch this **video** for the complete solutions to both parts.

You Try It **Work through this You Try It problem.**

Work Exercises 7–12 in this textbook or in the MyMathLab Study Plan.

(eText Screens 14.4-1–14.4-15)

OBJECTIVE 3 MULTIPLY TWO BINOMIALS

To multiply two binomials, we can use the **distributive property** twice. We can distribute the first binomial to each **term** in the second binomial. We can then distribute each term from the second binomial through the first binomial (from the back).

$$(a + b)(c + d) = (a + b)c + (a + b)d = ac + bc + ad + bd$$

The end result is that each term in the first binomial gets multiplied by each term in the second binomial.

$$(a + b)(c + d) = ac + ad + bc + bd$$

> **Multiplying Two Binomials**
>
> To multiply two binomials, multiply each term of the first binomial by each term of the second binomial. To simplify, **combine like terms**, if any.

Example 4 Multiplying Two Binomials

Multiply using the distributive property twice.

a. $(x + 3)(x + 2)$ **b.** $(x + 6)(x - 2)$ **c.** $(x - 4)(x - 5)$

Solutions

a. Multiply each **term** in the first **binomial** by each term in the second binomial.

Begin with the original expression: $(x + 3)(x + 2)$

Multiply x by x and 2;
multiply 3 by x and 2: $= x \cdot x + x \cdot 2 + 3 \cdot x + 3 \cdot 2$

Simplify: $= x^2 + 2x + 3x + 6$ ← $\boxed{x \cdot x = x^2}$

Combine like terms: $= x^2 + 5x + 6$ ← $\boxed{2x + 3x = 5x}$

b. Multiply each term in the first binomial by each term in the second.

Begin with the original expression: $(x + 6)(x - 2)$

Multiply x by x and -2;
multiply 6 by x and -2: $= x \cdot x + x \cdot (-2) + 6 \cdot x + 6 \cdot (-2)$

Simplify: $= x^2 - 2x + 6x - 12$ ← $\boxed{x \cdot (-2) = -2x, 6 \cdot (-2) = -12}$

Combine like terms: $= x^2 + 4x - 12$ ← $\boxed{-2x + 6x = 4x}$

My video summary

c. Try to work this problem on your own. View the **answer**, or watch this **video** for the complete solution.

CAUTION: Remember that a term contains the sign of its **coefficient**. For example, in the expression $3x^2 - 7x$ the terms are $3x^2$ and $-7x$.

The process used in Example 4 follows a specific order for multiplying **binomials**. The **acronym** FOIL summarizes the process and reminds us to multiply the two First terms, the

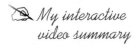

My animation summary

two Outside terms, the two Inside terms, and the two Last terms. Watch this **animation**, which illustrates how to use the **FOIL method** to multiply two binomials.

> **The FOIL Method**
>
> FOIL reminds us to multiply the two First terms, the two Outside terms, the two Inside terms, and the two Last terms.

CAUTION: After using FOIL, be sure to simplify by combining any like terms.

Example 5 Multiplying Two Binomials

My interactive video summary

Multiply using the FOIL method.

a. $(x-4)(2x+3)$ **b.** $\left(\frac{1}{2}x-6\right)(3x-4)$ **c.** $(5x+7)(4x+3)$

Solutions

a. The two first terms are x and $2x$. The two outside terms are x and 3. The two inside terms are -4 and $2x$. The two last terms are -4 and 3.

Outside, First, Inside, Last

$$\begin{aligned}(x-4)(2x+3) &= \overbrace{x\cdot 2x}^{\text{First}} + \overbrace{x\cdot 3}^{\text{Outside}} + \overbrace{(-4)\cdot 2x}^{\text{Inside}} + \overbrace{(-4)\cdot 3}^{\text{Last}}\\ &= 2x^2 + 3x + (-8x) + (-12)\\ &= 2x^2 + 3x - 8x - 12\\ &= 2x^2 - 5x - 12 \quad \leftarrow \boxed{3x - 8x = -5x}\end{aligned}$$

b. Apply the **FOIL method.**

$$\left(\frac{1}{2}x-6\right)(3x-4) = \overbrace{\frac{1}{2}x\cdot 3x}^{\text{First}} + \overbrace{\frac{1}{2}x\cdot(-4)}^{\text{Outside}} + \overbrace{(-6)\cdot 3x}^{\text{Inside}} + \overbrace{(-6)\cdot(-4)}^{\text{Last}}$$

Try finishing this problem on your own. View the **answer**, or watch this **interactive video** to see the complete solution.

c. Try working this problem on your own. View the **answer**, or watch this **interactive video** to see the complete solution.

You Try It **Work through this You Try It problem.**

Work Exercises 13–20 in this textbook or in the MyMathLab Study Plan.

CAUTION: The FOIL method can be used only when multiplying two binomials.

We can multiply two binomials vertically, just as we were able to add or subtract vertically. View this **popup** to see Example 5c worked in a vertical format.

OBJECTIVE 4 MULTIPLY TWO OR MORE POLYNOMIALS

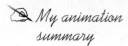

My animation summary

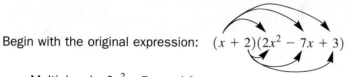

We can expand our work in Objective 3 to multiply any two **polynomials.** Watch this **animation** for an explanation.

> **Multiplying Two or More Polynomials**
>
> To multiply two polynomials, multiply each term of the first polynomial by each term of the second polynomial. To simplify, **combine like terms,** if any. If there are more than two polynomials, then we use this procedure to multiply two polynomials at a time.

Example 6 Multiplying Two Polynomials

Multiply: **a.** $(x + 2)(2x^2 - 7x + 3)$ **b.** $(y^2 + 2y - 9)(2y^2 - 4y + 7)$

Solutions

a. We multiply each term in the **binomial** by each term in the **trinomial.**

Begin with the original expression:	$(x + 2)(2x^2 - 7x + 3)$
Multiply x by $2x^2, -7x$, and 3;	
Multiply 2 by $2x^2, -7x$, and 3:	$= x \cdot 2x^2 + x \cdot (-7x) + x \cdot 3 + 2 \cdot 2x^2 + 2 \cdot (-7x) + 2 \cdot 3$
Simplify:	$= 2x^3 - 7x^2 + 3x + 4x^2 - 14x + 6$
Combine like terms:	$= 2x^3 - 3x^2 - 11x + 6$

View this **popup** to see the problem solved by multiplying vertically.

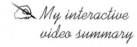

My video summary

b. Try to solve this problem on your own. View the **answer,** or watch this **video** to see the detailed solution using the vertical format. View this **popup** to see the solution using the **distributive property.**

You Try It **Work through this You Try It problem.**

Work Exercises 21–27 in this textbook or in the MyMathLab Study Plan.

To multiply three or more **polynomials,** we multiply two polynomials at a time.

Example 7 Multiplying Three or More Polynomials

My interactive video summary

Multiply: **a.** $-4x(2x - 1)(x + 3)$ **b.** $(x - 1)(x + 3)(3x - 2)$

Solutions

a. Because multiplication is **commutative,** we can perform the multiplications in any order we wish. For this problem, we will first multiply the two **binomials** and then distribute the **monomial** through the resulting **product.** Work through the steps below, or watch this **interactive video** to see the complete solutions to both parts of this example.

Begin with the original expression:	$-4x(2x - 1)(x + 3)$
Multiply $2x - 1$ and $x + 3$:	$= -4x(2x \cdot x + 2x \cdot 3 - 1 \cdot x - 1 \cdot 3)$ ← FOIL
	$= -4x(2x^2 + 6x - x - 3)$
Combine like terms:	$= -4x(2x^2 + 5x - 3)$
Distribute $-4x$:	$= (-4x) \cdot 2x^2 + (-4x) \cdot 5x + (-4x) \cdot (-3)$
Simplify:	$= -8x^3 + (-20x^2) + 12x$
	$= -8x^3 - 20x^2 + 12x$

(eText Screens 14.4-1–14.4-15)

b. Work this problem by first multiplying two of the **binomials** and then **combining like terms.** Then multiply the resulting product by the remaining binomial. View the **answer**, or watch this **interactive video** for the complete solutions to both parts.

You Try It **Work through this You Try It problem.**

Work Exercises 28–30 in this textbook or in the MyMathLab Study Plan.

14.4 Exercises

In Exercises 1–6, multiply each pair of monomials.

You Try It

1. $(3x^5)(2x^3)$
2. $(5x^4)(-x^5)$
3. $(-3y^5)(-2y^9)$
4. $(3.2w^3)(2.5w^2)$
5. $\left(\frac{2}{3}x\right)\left(\frac{6}{5}x^4\right)$
6. $(3x^2y^3)(5x^4y^2)$

In Exercises 7–12, multiply.

You Try It

7. $5x(3x + 2)$
8. $-8x(4x - 9)$
9. $3x(x^2 - x + 1)$
10. $-\frac{1}{2}x^2(2x^2 + 4x + 6)$
11. $-3x^3(x^3 + 2x - 7)$
12. $2x^2(3x^5 - 2x^3 + 7x - 4)$

In Exercises 13–16, multiply using the distributive property twice.

13. $(x + 2)(x + 5)$
14. $(w - 3)(2w - 5)$
15. $\left(y + \frac{3}{5}\right)\left(y - \frac{4}{3}\right)$
16. $(2x^2 - 3)(3x + 2)$

In Exercises 17–20, multiply using the FOIL method.

You Try It

17. $(x - 2)(x + 9)$
18. $(2m + 1)(m + 3)$
19. $(7z + 6)(8z - 3)$
20. $(2x^2 - 3)(3x - 2)$

In Exercises 21–30, multiply the polynomials and simplify by combining like terms.

You Try It

21. $(x - 3)(x^2 + 3x + 9)$
22. $(m - 1)(3m^2 + 2m - 9)$
23. $(2x + 1)(x^2 - x + 1)$
24. $(3x + 2)(4x^3 - 2x + 11)$
25. $(x^2 + 3)(2x^3 + x - 3)$
26. $(x^2 + 3x + 1)(2x^2 + x - 3)$
27. $(x^3 - x + 1)(x^2 + x - 1)$
28. $(2x)(-3x)(x^2 + x + 1)$

You Try It

29. $(5x)(3x - 2)(2x + 3)$
30. $(x + 4)(2x - 1)(3x - 2)$

14.5 Special Products

THINGS TO KNOW

Before working through this topic, be sure you are familiar with the following concepts:

		VIDEO	ANIMATION	INTERACTIVE
You Try It	1. Simplify Algebraic Expressions (Topic 10.6, **Objective 2**)			✓
You Try It	2. Simplify Exponential Expressions Using a Combination of Rules (Topic 14.1, **Objective 7**)			✓
You Try It	3. Multiply a Polynomial by a Monomial (Topic 14.4, **Objective 2**)	✓		
You Try It	4. Multiply Two Binomials (Topic 14.4, **Objective 3**)	✓		✓

OBJECTIVES

1 Square a Binomial Sum

2 Square a Binomial Difference

3 Multiply the Sum and Difference of Two Terms

OBJECTIVE 1 SQUARE A BINOMIAL SUM

In this topic, we look at three special **products** involving **binomials**.

A **binomial sum** is a binomial in which the two **terms** are added such as $A + B$. We can use the **FOIL method** to find a general result for the square of a binomial sum, $(A + B)^2$.

$$(A + B)^2 = (A + B)(A + B) = \overbrace{A \cdot A}^{\text{F}} + \overbrace{A \cdot B}^{\text{O}} + \overbrace{B \cdot A}^{\text{I}} + \overbrace{B \cdot B}^{\text{L}}$$
$$= A^2 + AB + AB + B^2$$
$$= A^2 + 2AB + B^2$$

This result is our first special product rule.

The Square of a Binomial Sum Rule

$$\underbrace{(A + B)^2}_{\text{The square of a binomial sum}} \underbrace{=}_{\text{equals}} \underbrace{A^2}_{\text{the square of the first term}} \underbrace{+}_{\text{plus}} \underbrace{2AB}_{\text{2 times the product of the two terms}} \underbrace{+}_{\text{plus}} \underbrace{B^2}_{\text{the square of the second term.}}$$

CAUTION Note that the square of a **binomial sum** is not the sum of the squares: $(A + B)^2 \neq A^2 + B^2$. See this **popup** for a geometric explanation of the square of a binomial sum.

Example 1 Squaring a Binomial Sum

Multiply.

a. $(x+7)^2$ **b.** $(0.2m+1)^2$ **c.** $\left(z^2+\frac{1}{4}\right)^2$ **d.** $\left(10y+\frac{2}{5}\right)^2$

Solutions In each case, use the **square of a binomial sum rule.**

a. Square of a binomial sum rule: $(A+B)^2 = A^2 + 2AB + B^2$

Substitute x for A and 7 for B: $(x+7)^2 = (x)^2 + 2(x)(7) + (7)^2$

Simplify: $= x^2 + 14x + 49$

b. Square of a binomial sum rule: $(A+B)^2 = A^2 + 2AB + B^2$

Substitute $0.2m$ for A and 1 for B: $(0.2m+1)^2 = (0.2m)^2 + 2(0.2m)(1) + (1)^2$

Simplify: $= 0.04m^2 + 0.4m + 1$

My video summary

c.–d. Try working these parts on your own. View the **answers**, or watch this **video** for the complete solutions.

You Try It **Work through this You Try It problem.**

Work Exercises 1–8 in this textbook or in the MyMathLab Study Plan.

OBJECTIVE 2 SQUARE A BINOMIAL DIFFERENCE

A **binomial difference** is a binomial in which one **term** is subtracted from the other such as $A - B$. We can again use the **FOIL method** to find a general result for the square of a binomial difference, $(A-B)^2$.

$$(A-B)^2 = (A-B)(A-B) = \overbrace{A\cdot A}^{\text{F}} + \overbrace{A\cdot(-B)}^{\text{O}} + \overbrace{(-B)\cdot A}^{\text{I}} + \overbrace{(-B)\cdot(-B)}^{\text{L}}$$
$$= A^2 + (-AB) + (-AB) + B^2$$
$$= A^2 - AB - AB + B^2$$
$$= A^2 - 2AB + B^2$$

This result is our second special product rule.

The Square of a Binomial Difference Rule

$$\underbrace{(A-B)^2}_{\text{The square of a binomial difference}} \underbrace{=}_{\text{equals}} \underbrace{A^2}_{\text{the square of the first term}} \underbrace{-}_{\text{minus}} \underbrace{2AB}_{\text{2 times the product of the two terms}} \underbrace{+}_{\text{plus}} \underbrace{B^2}_{\text{the square of the second term.}}$$

CAUTION Note that the square of a **binomial difference** is not the difference of the squares: $(A-B)^2 \neq A^2 - B^2$.

Example 2 Squaring a Binomial Difference

Multiply.

a. $(x-3)^2$ **b.** $\left(2z-\dfrac{1}{6}\right)^2$ **c.** $(w^3-0.7)^2$ **d.** $(5p-1.2)^2$

Solutions In each case, use the **square of a binomial difference rule.**

a. Square of a binomial difference rule: $(A-B)^2 = A^2 - 2AB + B^2$

Substitute x for A and 3 for B: $(x-3)^2 = (x)^2 - 2(x)(3) + (3)^2$

Simplify: $= x^2 - 6x + 9$

b. Square of a binomial difference rule: $(A-B)^2 = A^2 - 2AB + B^2$

Substitute $2z$ for A and $\frac{1}{6}$ for B: $\left(2z-\dfrac{1}{6}\right)^2 = (2z)^2 - 2(2z)\left(\dfrac{1}{6}\right) + \left(\dfrac{1}{6}\right)^2$

Simplify: $= 4z^2 - \dfrac{2}{3}z + \dfrac{1}{36}$

My video summary

c.–d. Try working these parts on your own. View the **answers**, or watch this **video** for the complete solutions.

You Try It **Work through this You Try It problem.**

Work Exercises 9–16 in this textbook or in the MyMathLab Study Plan.

Notice that the first two special product rules are very similar:

Middle term has the same sign as in the binomial

$$(A+B)^2 = A^2 + 2AB + B^2 \qquad (A-B)^2 = A^2 - 2AB + B^2$$

Last term is always positive

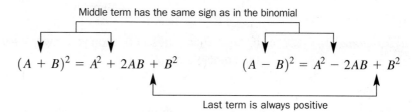

The **trinomials** $A^2 + 2AB + B^2$ and $A^2 - 2AB + B^2$ are called **perfect square trinomials** because they come from squaring a **binomial**. Recognizing perfect square trinomials will be helpful when we get to Module 15.

OBJECTIVE 3 MULTIPLY THE SUM AND DIFFERENCE OF TWO TERMS

When a **binomial sum** and a **binomial difference** are made from the same two terms, they are called **conjugates** of each other. For example, the binomial sum $x+5$ and the binomial difference $x-5$ are **conjugates** of each other.

We can use the **FOIL method** to find a general result for the **product** of conjugates, $(A+B)(A-B)$.

$$\begin{aligned}(A+B)(A-B) &= \overbrace{A\cdot A}^{\text{F}} + \overbrace{A\cdot(-B)}^{\text{O}} + \overbrace{B\cdot A}^{\text{I}} + \overbrace{B\cdot(-B)}^{\text{L}}\\ &= A^2 - AB + AB - B^2\\ &= A^2 - B^2\end{aligned}$$

So, the product of the sum and difference of two terms (product of conjugates) equals the difference of the squares of the two terms.

(eText Screens 14.5-1–14.5-12)

This result is our third special product rule.

The Sum and Difference of Two Terms Rule (Product of Conjugates Rule)

$$\underbrace{(A+B)(A-B)}_{\text{The product of the sum and difference of two terms}} \underbrace{=}_{\text{equals}} \underbrace{A^2}_{\text{the square of the first term}} \underbrace{-}_{\text{minus}} \underbrace{B^2}_{\text{the square of the second term.}}$$

Example 3 Multiplying the Sum and Difference of Two Terms

Multiply.

a. $(x+4)(x-4)$ b. $\left(5y+\frac{1}{2}\right)\left(5y-\frac{1}{2}\right)$

c. $(8-x)(8+x)$ d. $(3z^2+0.5)(3z^2-0.5)$

Solution In each case, use the **sum and difference of two terms rule** (product of **conjugates**).

a. Sum and difference of two terms rule: $(A+B)(A-B) = A^2 - B^2$

Substitute x for A and 4 for B: $(x+4)(x-4) = (x)^2 - (4)^2$

Simplify: $= x^2 - 16$

b. Sum and difference of two terms rule: $(A+B)(A-B) = A^2 - B^2$

Substitute $5y$ for A and $\frac{1}{2}$ for B: $\left(5y+\frac{1}{2}\right)\left(5y-\frac{1}{2}\right) = (5y)^2 - \left(\frac{1}{2}\right)^2$

Simplify: $= 25y^2 - \frac{1}{4}$

My video summary

c.–d. Try working these parts on your own. View the **answers**, or watch this **video** for the complete solutions.

You Try It **Work through this You Try It problem.**

Work Exercises 17–24 in this textbook or in the MyMathLab Study Plan.

In summary, the results from Examples 1–3 give us the following **special product rules for binomials.**

Special Product Rules for Binomials

The square of a binomial sum	$(A+B)^2 = A^2 + 2AB + B^2$	Perfect square trinomial
The square of a binomial difference	$(A-B)^2 = A^2 - 2AB + B^2$	Perfect square trinomial
The product of the sum and difference of two terms (product of conjugates)	$(A+B)(A-B) = A^2 - B^2$	Difference of two squares

Recognizing these special binomial sums and differences will allow us to find their **products** more quickly by using the special product rules.

14.5 Exercises

Multiply.

You Try It

1. $(y + 2)^2$
2. $\left(n + \frac{4}{3}\right)^2$
3. $(x + 3.5)^2$
4. $(2x + 5)^2$
5. $\left(\frac{3}{4}z + 1\right)^2$
6. $(3.2m + 7)^2$
7. $\left(6x + \frac{3}{5}\right)^2$
8. $(y^3 + 0.8)^2$

You Try It

9. $(x - 6)^2$
10. $\left(y - \frac{3}{7}\right)^2$
11. $(z^2 - 4.2)^2$
12. $(4x - 1)^2$
13. $(8 - x)^2$
14. $(1.5 - 3m)^2$
15. $\left(\frac{5}{2}z - 4\right)^2$
16. $(1.2x - 2.5)^2$

You Try It

17. $(x + 9)(x - 9)$
18. $(y - 2)(y + 2)$
19. $(3z + 4)(3z - 4)$
20. $\left(w + \frac{1}{5}\right)\left(w - \frac{1}{5}\right)$
21. $(2.1 + 3y)(2.1 - 3y)$
22. $\left(\frac{5}{3}v + 2\right)\left(\frac{5}{3}v - 2\right)$
23. $(4x^2 + 7)(4x^2 - 7)$
24. $(3.5 - x^2)(3.5 + x^2)$

14.6 Negative Exponents and Scientific Notation

THINGS TO KNOW

Before working through this topic, be sure you are familiar with the following concepts:

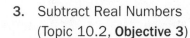

	VIDEO	ANIMATION	INTERACTIVE
You Try It 1. Add Two Real Numbers with the Same Sign (Topic 10.2, **Objective 1**)	✓		
You Try It 2. Add Two Real Numbers with Different Signs (Topic 10.2, **Objective 2**)	✓		
You Try It 3. Subtract Real Numbers (Topic 10.2, **Objective 3**)			✓

You Try It

4. Multiply Real Numbers
(Topic 10.3, **Objective 1**)

You Try It

5. Divide Real Numbers
(Topic 10.3, **Objective 2**)

You Try It

6. Simplify Exponential Expressions Using a Combination of Rules (Topic 14.1, **Objective 7**)

OBJECTIVES

1 Use the Negative-Power Rule

2 Simplify Expressions Containing Negative Exponents Using a Combination of Rules

3 Convert a Number from Standard Form to Scientific Notation

4 Convert a Number from Scientific Notation to Standard Form

5 Multiply and Divide with Scientific Notation

OBJECTIVE 1 USE THE NEGATIVE-POWER RULE

So far, we have encountered only **exponential expressions** with non-negative exponents. But what if an exponential expression contains a negative exponent, such as x^{-2}? We can use the **quotient rule** to understand the meaning of such an expression. Consider $\frac{x^2}{x^5}$, where $x \neq 0$. We can find this quotient using two different methods.

Method 1: Apply the **quotient rule for exponents:** $\frac{x^2}{x^5} = x^{2-5} = x^{-3}$

Subtract the denominator exponent from the numerator exponent.

Keep the common base.

Method 2: Expand the numerator and denominator: $\frac{x^2}{x^5} = \frac{x \cdot x}{x \cdot x \cdot x \cdot x \cdot x}$ ← 2 factors of x / ← 5 factors of x

Divide out common factors: $= \frac{\cancel{x} \cdot \cancel{x}}{\cancel{x} \cdot \cancel{x} \cdot x \cdot x \cdot x}$

Simplify: $= \frac{1}{x \cdot x \cdot x} = \frac{1}{x^3}$

Because $\frac{x^2}{x^5}$ equals both x^{-3} and $\frac{1}{x^3}$, this means $x^{-3} = \frac{1}{x^3}$. So, x^{-3} is the **reciprocal** of x^3.

Definition Negative Exponent

A negative exponent can be expressed with a positive exponent as follows:

$$a^{-n} = \frac{1}{a^n} \quad (a \neq 0)$$

(eText Screens 14.6-1–14.6-21)

Example 1 Using the Definition of a Negative Exponent

Write each expression with positive exponents. Then simplify if possible.

a. x^{-4} b. 2^{-3} c. $7x^{-3}$ d. $(-2)^{-4}$ e. -3^{-2} f. $2^{-1} + 3^{-1}$

Solutions In each case, we use the definition of a negative exponent to rewrite the expression with all positive exponents. Then we simplify if necessary.

a. $x^{-4} = \dfrac{1}{x^4}$

b. $2^{-3} = \dfrac{1}{2^3} = \dfrac{1}{2 \cdot 2 \cdot 2} = \dfrac{1}{8}$

c. There are no parentheses, so only x is the **base** of the exponential expression.

$$7x^{-3} = 7 \cdot \frac{1}{x^3} = \frac{7}{x^3}$$

My video summary

d.–f. Try to simplify these expressions on your own. View the **answers**, or watch this **video** for detailed solutions to parts d through f.

You Try It **Work through this You Try It problem.**

Work Exercises 1–8 in this textbook or in the MyMathLab Study Plan.

CAUTION

A **negative exponent** does not mean the value of the expression is negative. For example, $2^{-1} = \frac{1}{2}$, which is positive.

Sometimes, a negative exponent appears in a **denominator**. Consider the following:

$$\frac{1}{x^{-3}} = 1 \div x^{-3} = 1 \div \frac{1}{x^3} = 1 \cdot \frac{x^3}{1} = x^3$$

As long as the **base** is not 0, this result is true in general.

The Negative-Power Rule

To remove a negative exponent, switch the location of the base (**numerator** or denominator) and change the exponent to be positive.

$$a^{-n} = \frac{1}{a^n} \quad \text{and} \quad \frac{1}{a^{-n}} = a^n \quad (a \neq 0)$$

Example 2 Using the Negative-Power Rule

Write each expression with positive exponents. Then simplify if possible.

a. $\dfrac{1}{y^{-5}}$ b. $\dfrac{1}{6^{-2}}$ c. $\dfrac{3}{4t^{-7}}$ d. $\dfrac{-8}{q^{-11}}$ e. $\dfrac{m^{-9}}{n^{-4}}$ f. $\dfrac{5^{-3}}{2^{-4}}$

Solutions In each case, we use the **negative-power rule** to rewrite the expression with all positive exponents. Then we simplify if necessary.

a. The **base** y is raised to the -5 power. Switch the location of y from the **denominator** to the **numerator** and change its exponent from -5 to 5.

$$\frac{1}{y^{-5}} = y^5$$

Exponent is now positive

Base is now in the numerator

b. The base 6 is raised to the -2 power. Switch the location of 6 from the denominator to the numerator and change the exponent from -2 to 2. Then simplify.

$$\frac{1}{6^{-2}} = 6^2 = 36$$

c. There are no parentheses, so the **base** is only t, which is raised to the -7 power. Switch the location of t from the **denominator** to the **numerator**, and change the exponent from -7 to 7.

$$\frac{3}{4t^{-7}} = \frac{3t^7}{4}$$

Note that the 3 and 4 do not switch locations because they are not affected by the negative exponent.

My video summary

d.–f. Try to simplify these expressions on your own. View the **answers**, or watch this **video** for detailed solutions to parts d through f.

You Try It **Work through this You Try It problem.**

Work Exercises 9–16 in this textbook or in the MyMathLab Study Plan.

OBJECTIVE 2 SIMPLIFY EXPRESSIONS CONTAINING NEGATIVE EXPONENTS USING A COMBINATION OF RULES

An **exponential expression** is not simplified if it contains negative exponents. Adding this to the conditions from **Topic 14.1** gives the following complete list of requirements for **simplified exponential expressions:**

- No parentheses or other **grouping symbols** are present.
- No zero or **negative exponents** are present.
- No **powers** are raised to powers.
- Each **base** occurs only once.

The following table summarizes all of our rules for exponents.

Rules for Exponents

Product Rule	$a^m \cdot a^n = a^{m+n}$
Quotient Rule	$\frac{a^m}{a^n} = a^{m-n} \quad (a \neq 0)$
Zero-Power Rule	$a^0 = 1 \quad (a \neq 0)$
Power-to-Power Rule	$(a^m)^n = a^{m \cdot n}$
Product-to-Power Rule	$(ab)^n = a^n b^n$
Quotient-to-Power Rule	$\left(\frac{a}{b}\right)^n = \frac{a^n}{b^n} \quad (b \neq 0)$
Negative-Power Rule	$a^{-n} = \frac{1}{a^n}$ and $\frac{1}{a^{-n}} = a^n \quad (a \neq 0)$

We can combine the **rules for exponents** to simplify exponential expressions.

(eText Screens 14.6-1–14.6-21)

CAUTION: Remember, there is typically more than one path that leads to a correct, simplified exponential expression. Just make sure your path correctly applies the rules for exponents.

Example 3 Using a Combination of Exponent Rules

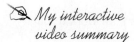

My interactive video summary

Simplify.

a. $(9x^{-5})(7x^2)$ b. $(p^{-4})^2$ c. $\frac{52m^{-4}}{13m^{-10}}$ d. $(w^{-1}z^3)^{-4}$

Solutions

a. Begin with the original expression: $(9x^{-5})(7x^2)$

Rearrange factors to group like bases: $= 9 \cdot 7 \cdot x^{-5} \cdot x^2$

Multiply constants; apply **product rule:** $= 63 \cdot x^{-5+2}$

Simplify: $= 63x^{-3}$

Apply the **negative-power rule:** $= \frac{63}{x^3}$ ← Switch location of base x from numerator to denominator, and change exponent from -3 to 3.

b. Begin with the original expression: $(p^{-4})^2$

Apply the **power-to-power rule:** $= p^{(-4)(2)}$

Simplify: $= p^{-8}$

Apply the **negative-power rule:** $= \frac{1}{p^8}$

c.–d. Try to simplify these expressions on your own. View the **answers**, or work through this **interactive video** for complete solutions to all four parts.

You Try It **Work through this You Try It problem.**

Work Exercises 17–22 in this textbook or in the MyMathLab Study Plan.

Example 4 Using a Combination of Exponent Rules

My interactive video summary

Simplify.

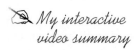

a. $\frac{(3xz)^{-2}}{(2yz)^{-3}}$ b. $\left(\frac{10}{x}\right)^{-3}$ c. $\frac{(2a^5b^{-6})^3}{4a^{-1}b^5}$ d. $\left(\frac{-5xy^{-3}}{x^{-2}y^5}\right)^4$

Solutions

a. Begin with the original expression: $\frac{(3xz)^{-2}}{(2yz)^{-3}}$

Apply the **negative-power rule:** $= \frac{(2yz)^3}{(3xz)^2}$

Apply the **product-to-power rule:** $= \frac{8y^3z^3}{9x^2z^2}$ ← $(2yz)^3 = 2^3y^3z^3 = 8y^3z^3$; ← $(3xz)^2 = 3^2x^2z^2 = 9x^2z^2$

Apply the **quotient rule:** $= \frac{8y^3z}{9x^2}$ ← $\frac{z^3}{z^2} = z^{3-2} = z^1 = z$

b. Begin with the original expression: $\left(\frac{10}{x}\right)^{-3}$

Apply the **quotient-to-power rule:** $= \frac{10^{-3}}{x^{-3}}$

(eText Screens 14.6-1–14.6-21)

Apply the **negative-power rule:** $= \dfrac{x^3}{10^3}$

Simplify: $= \dfrac{x^3}{1000}$

c.–d. Try to simplify these expressions on your own. View the **answers**, or work through this **interactive video** for complete solutions to all four parts.

You Try It **Work through this You Try It problem.**

Work Exercises 23–28 in this textbook or in the MyMathLab Study Plan.

OBJECTIVE 3 CONVERT A NUMBER FROM STANDARD FORM TO SCIENTIFIC NOTATION

Scientists frequently work with very large or very small numbers. For example, the distance from the Sun to the center of the Milky Way is about 19,200,000,000,000 miles, and the mass of a neutron is about 0.000000000000000000000001675 grams.

Calculating with such numbers in **standard form** can be difficult. To make the numbers more manageable, we can use **scientific notation.**

> **Scientific Notation**
>
> A number is written in **scientific notation** if it has the form
>
> $$a \times 10^n,$$
>
> where a is a **real number**, such that $1 \leq |a| < 10$, and n is an **integer**.

Note: When writing scientific notation, typically we use a times sign ($\times$) instead of a dot ($\cdot$) to indicate multiplication.

In scientific notation, we write the distance from the Sun to the center of the Milky Way as 1.92×10^{13} miles and the mass of a neutron as 1.675×10^{-24} grams. Notice that for the very large number, the **exponent** on 10 is positive, while for the very small number, the exponent on 10 is negative.

The following procedure can be used to convert a **real number** from **standard form** to **scientific notation.**

> **Converting from Standard Form to Scientific Notation**
>
> **Step 1.** Write the real number factor a by moving the **decimal point** so that $|a|$ is greater than or equal to 1 but less than 10. To do this, place the decimal point to the right of the first nonzero **digit**.
>
> **Step 2.** Multiply the number by 10^n, where $|n|$ is the number of places that the decimal point moves. If the decimal point moves to the left, then $n > 0$ (n is positive). If the decimal point moves to the right, then $n < 0$ (n is negative). Remove any zeros lying to the right of the last nonzero digit or to the left of the first nonzero digit.

Example 5 Converting from Standard Form to Scientific Notation

My video summary Write each number in scientific notation.

a. 56,800,000,000,000,000 **b.** 0.0000000467 **c.** 0.00009012 **d.** 200,000,000

Solutions

a. Move the **decimal point** to the right of the 5, the first nonzero **digit**.

56,800,000,000,000,000.

16 places

The decimal point has moved sixteen places to the left, so the exponent on the 10 will be 16. Remove all of the zeros behind the last nonzero digit 8.

$$56{,}800{,}000{,}000{,}000{,}000 = 5.68 \times 10^{16}$$

b. Move the **decimal point** to the right of the 4, the first nonzero **digit**.

0.0000000467

8 places

The decimal point has moved eight places to the right, so the exponent on the 10 will be -8. Remove all of the zeros in front of the 4.

$$0.0000000467 = 4.67 \times 10^{-8}$$

c.–d. Try to convert these numbers to scientific notation on your own. View the **answers**, or watch this **video** for detailed solutions to all four parts.

You Try It **Work through this You Try It problem.**

Work Exercises 29–36 in this textbook or in the MyMathLab Study Plan.

OBJECTIVE 4 CONVERT A NUMBER FROM SCIENTIFIC NOTATION TO STANDARD FORM

Converting from Scientific Notation to Standard Form

1. Remove the exponential factor, 10^n.
2. Move the **decimal point** $|n|$ places, inserting zero placeholders as needed. If $n > 0$ (n is positive), move the decimal point to the right. If $n < 0$ (n is negative), move the decimal point to the left.

CAUTION: Remember, if the **exponent** on 10 is positive, then the absolute value of the number is larger than 1. If the exponent on 10 is negative, then the absolute value of the number is smaller than 1.

Example 6 Converting from Scientific Notation to Standard Form

My video summary Write each number in **standard form**.

a. 4.98×10^{-5} **b.** 9.4×10^{7} **c.** -3.015×10^{9} **d.** 1.203×10^{-4}

Solutions

a. The exponent is -5, so we move the decimal point five places to the left, filling in zero placeholders as needed.

00004.98

5 places

So, $4.98 \times 10^{-5} = 0.0000498$.

b. The exponent is 7, so we move the decimal point seven places to the right, filling in zero placeholders as needed.

$$9.4000000 \quad \text{(7 places)}$$

So, $9.4 \times 10^7 = 94{,}000{,}000$.

c.–d. Try to convert these numbers to standard form on your own. View the **answers**, or watch this **video** for detailed solutions to all four parts.

You Try It **Work through this You Try It problem.**

Work Exercises 37–44 in this textbook or in the MyMathLab Study Plan.

OBJECTIVE 5 MULTIPLY AND DIVIDE WITH SCIENTIFIC NOTATION

We use **rules for exponents** to multiply or divide numbers written in **scientific notation**.

Example 7 Multiplying and Dividing with Scientific Notation

Perform the indicated operations. Write your results in scientific notation.

a. $(1.8 \times 10^5)(3 \times 10^8)$ **b.** $\dfrac{2.16 \times 10^{12}}{4.5 \times 10^3}$

My video summary

c. $(-7.4 \times 10^9)(6.5 \times 10^{-4})$ **d.** $\dfrac{5.7 \times 10^{-3}}{7.5 \times 10^{-7}}$

Solutions

a.

Begin with original expression: $(1.8 \times 10^5)(3 \times 10^8)$

Regroup factors: $= (1.8 \times 3) \times (10^5 \times 10^8)$

Multiply numeric factors; apply **product rule:** $= 5.4 \times 10^{5+8}$

Simplify: $= 5.4 \times 10^{13}$ ← This answer is in scientific notation.

So, $(1.8 \times 10^5)(3 \times 10^8) = 5.4 \times 10^{13}$.

b.

Begin with original expression: $\dfrac{2.16 \times 10^{12}}{4.5 \times 10^3}$

Regroup factors: $= \left(\dfrac{2.16}{4.5}\right) \times \left(\dfrac{10^{12}}{10^3}\right)$

Divide numeric factors; apply **quotient rule:** $= 0.48 \times 10^9$ ← Not scientific notation because 0.48 is not between 1 and 10.

Write 0.48 in scientific notation: $= (4.8 \times 10^{-1}) \times 10^9$

Regroup factors: $= 4.8 \times (10^{-1} \times 10^9)$

Apply the **product rule:** $= 4.8 \times 10^8$

So, $\dfrac{2.16 \times 10^{12}}{4.5 \times 10^3} = 4.8 \times 10^8$.

c.–d. Try to work these problems on your own. View the **answers**, or watch this **video** for detailed solutions to parts c and d.

You Try It **Work through this You Try It problem.**

Work Exercises 45–48 in this textbook or in the MyMathLab Study Plan.

(eText Screens 14.6-1–14.6-21)

14.6 Exercises

In Exercises 1–16, write each expression with positive exponents. Then simplify if possible.

You Try It

1. y^{-3} 2. 5^{-3} 3. $6w^{-5}$ 4. $-3t^{-8}$

5. $(-8)^2$ 6. -10^{-4} 7. $2^{-3} + 4^{-1}$ 8. $3^{-1} \cdot 5^{-2}$

You Try It

9. $\dfrac{1}{z^{-5}}$ 10. $\dfrac{1}{4^{-3}}$ 11. $\dfrac{1}{(-3)^{-4}}$ 12. $\dfrac{5}{7x^{-4}}$

13. $\dfrac{-12}{t^{-3}}$ 14. $\dfrac{a^{-8}}{b^{-15}}$ 15. $\dfrac{9^{-2}}{2^{-5}}$ 16. $\dfrac{x^{-2}y^7}{z^{-5}}$

In Exercises 17–28, simplify each exponential expression using the rules for exponents.

You Try It

17. $(8x^{-7})(4x^3)$ 18. $(q^{-6})^4$ 19. $\dfrac{45t^{-2}}{9t^{-8}}$ 20. $\dfrac{-24x^{-7}y^5}{4x^{-3}y^2}$

21. $(6x^{-2}y^5)^{-3}$ 22. $\dfrac{5r^5s^0t^4}{-3r^3s^{-4}t^9}$ 23. $\dfrac{(5ac)^{-4}}{(3bc)^{-2}}$ 24. $\left(\dfrac{x}{2}\right)^{-5}$

You Try It

25. $\left(\dfrac{p^2}{q^7}\right)^{-8}$ 26. $(2x^4)^{-3}(-3y^{-7})$ 27. $\dfrac{(3a^5b^{-6})^4}{9a^{-7}b^{-1}}$ 28. $\left(\dfrac{8m^{-4}n^5}{m^3n^{-1}}\right)^3$

In Exercises 29–36, write each number in scientific notation.

You Try It

29. 89,000,000,000 30. 1,025,000,000,000,000

31. 0.000000406 32. −0.000005

33. **Microns** A micron is a metric unit for measuring very small lengths or distances. There are 25,400 microns in 1 inch. Write 25,400 in scientific notation.

34. **X-rays** An X-ray has a wavelength of 0.000000724 millimeters. Write 0.000000724 in scientific notation.

35. **Response Time** A computer's RAM response time is 0.000000005 second. Write 0.000000005 in scientific notation.

36. **National Debt** On November 21, 2010, the U.S. national debt was approximately $13,800,000,000,000. Write 13,800,000,000,000 in scientific notation. (*Source*: usdebtclock.org)

In Exercises 37–44, write each number in standard form.

You Try It

37. 2.97×10^{-6} 38. 9.63×10^{13}

39. 8×10^{-10} 40. -6.045×10^8

41. **Light-Year** Astronomers measure large distances in light-years. One light-year is the distance light can travel in one year, which is approximately 5.878×10^{12} miles. Write 5.878×10^{12} in standard form.

42. **Chickenpox** A chickenpox virus measures 1.6×10^{-4} millimeters in diameter. Write 1.6×10^{-4} in standard form.

43. **Dust Mites** A dust mite weighs 5.3×10^{-5} grams. Write 5.3×10^{-5} in standard form.

44. **Google Search** In February 2010, an average of 1.21×10^{8} searches per hour were conducted using the Google search engine. Write 1.21×10^{8} in standard form. (*Source*: searchengineland.com)

You Try It

In Exercises 45–48, perform the indicated operations. Write your results in scientific notation.

45. $(4.6 \times 10^{6})(1.5 \times 10^{9})$

46. $(9.2 \times 10^{-5})(6.8 \times 10^{7})$

47. $\dfrac{7.2 \times 10^{9}}{2.4 \times 10^{4}}$

48. $\dfrac{5.2 \times 10^{5}}{6.5 \times 10^{-3}}$

14.7 Dividing Polynomials

THINGS TO KNOW

Before working through this topic, be sure you are familiar with the following concepts:

VIDEO ANIMATION INTERACTIVE

You Try It 1. Simplify Exponential Expressions Using the Quotient Rule (Topic 14.1, **Objective 2**)

You Try It 2. Add Polynomials (Topic 14.3, **Objective 1**)

You Try It 3. Subtract Polynomials (Topic 14.3, **Objective 3**)

You Try It 4. Multiply Monomials (Topic 14.4, **Objective 1**)

You Try It 5. Multiply a Polynomial by a Monomial (Topic 14.4, **Objective 2**)

You Try It 6. Multiply Two Binomials (Topic 14.4, **Objective 3**)

You Try It 7. Multiply Two or More Polynomials (Topic 14.4, **Objective 4**)

OBJECTIVES

1 Divide Monomials

2 Divide a Polynomial by a Monomial

3 Divide Polynomials Using Long Division

(eText Screens 14.7-1–14.7-16)

OBJECTIVE 1 DIVIDE MONOMIALS

In **Topic 14.1**, we used the **quotient rule for exponents** to simplify **exponential expressions.** We grouped **like bases** into individual **quotients** and then applied the quotient rule to simplify. Dividing monomials is a similar process.

> **Dividing Monomials**
>
> Group the **coefficients** and group like bases into individual quotients. Then divide the coefficients and apply the quotient rule for exponents.

Example 1 Dividing Monomials

Divide.

a. $\dfrac{32x^7}{4x^3}$ **b.** $\dfrac{9y^4}{45y^4}$ **c.** $\dfrac{60y}{5y^4}$

Solutions

a. Begin with the original expression: $\dfrac{32x^7}{4x^3}$

Group like bases into individual quotients: $= \dfrac{32}{4} \cdot \dfrac{x^7}{x^3}$

Divide the coefficients; apply the quotient rule: $= 8 \cdot x^{7-3}$

Simplify: $= 8x^4$

b. Begin with the original expression: $\dfrac{9y^4}{45y^4}$

Group **like bases** into individual **quotients**: $= \dfrac{9}{45} \cdot \dfrac{y^4}{y^4}$

Divide the **coefficients**; apply the **quotient rule**: $= \dfrac{1}{5} \cdot y^0 \leftarrow \boxed{4 - 4}$

Apply the **zero-power rule**: $= \dfrac{1}{5} \leftarrow \boxed{\tfrac{1}{5} \cdot y^0 = \tfrac{1}{5} \cdot 1 = \tfrac{1}{5}}$

c. Begin with the original expression: $\dfrac{60y}{5y^4}$

Group like bases into individual quotients: $= \dfrac{60}{5} \cdot \dfrac{y^1}{y^4} \leftarrow \boxed{y = y^1}$

Divide the coefficients; apply the quotient rule: $= 12 \cdot y^{-3} \leftarrow \boxed{1 - 4}$

Apply the **negative-power rule**: $= \dfrac{12}{y^3}$

You Try It **Work through this You Try It problem.**

Work Exercises 1–4 in this textbook or in the MyMathLab Study Plan.

(eText Screens 14.7-1–14.7-16)

OBJECTIVE 2 DIVIDE A POLYNOMIAL BY A MONOMIAL

From **Topic 14.4**, we learned how to multiply a **polynomial** by a **monomial**. Using the **distributive property**, we multiplied each **term** of the polynomial by the monomial and then simplified the expression. Dividing a polynomial by a monomial is a similar process. We divide each term of the polynomial by the monomial.

Dividing Polynomials by Monomials

To divide a polynomial by a monomial, divide each term of the polynomial by the monomial and simplify using the method for **dividing monomials.**

We can check our division answers by multiplying. For example, $\frac{10}{2} = 5$ checks because $2 \cdot 5 = 10$. In general, if $\frac{\text{dividend}}{\text{divisor}} = \text{quotient}$, then $(\text{divisor})(\text{quotient}) = \text{dividend}$.

Example 2 Dividing a Polynomial by a Monomial

Divide.

a. $\dfrac{12x^3 - 28x^2}{4x^2}$ b. $(9m^5 - 15m^4 + 18m^3) \div 3m^3$

Solutions

a. Begin with the original expression: $\dfrac{12x^3 - 28x^2}{4x^2}$

Divide each term of the **polynomial** by the **monomial**: $= \dfrac{12x^3}{4x^2} - \dfrac{28x^2}{4x^2}$

Use the method for **dividing monomials**: $= \dfrac{12}{4}x^{3-2} - \dfrac{28}{4}x^{2-2}$

Simplify: $= 3x^1 - 7x^0$

$= 3x - 7$

Check: If $\dfrac{12x^3 - 28x^2}{4x^2} = 3x - 7$, then $4x^2(3x - 7)$ must equal $12x^3 - 28x^2$. We multiply to check.

$$4x^2(3x - 7) = 4x^2 \cdot 3x - 4x^2 \cdot 7 = 12x^3 - 28x^2$$

So, our **quotient** $3x - 7$ checks.

My video summary

b. Try to work this problem on your own. View the **answer**, or watch this **video** for a complete solution to part b.

You Try It **Work through this You Try It problem.**

Work Exercises 5–8 in this textbook or in the MyMathLab Study Plan.

Example 3 Dividing a Polynomial by a Monomial

Divide $\dfrac{54t^3 - 12t^2 - 24t}{6t^2}$.

(eText Screens 14.7-1–14.7-16)

Solution

Begin with the original expression: $\dfrac{54t^3 - 12t^2 - 24t}{6t^2}$

Divide each term of the **polynomial** by the **monomial**: $= \dfrac{54t^3}{6t^2} - \dfrac{12t^2}{6t^2} - \dfrac{24t}{6t^2}$

Use the method for **dividing monomials**: $= \dfrac{54}{6}t^{3-2} - \dfrac{12}{6}t^{2-2} - \dfrac{24}{6}t^{1-2}$

Simplify: $= 9t^1 - 2t^0 - 4t^{-1}$

Apply the **negative-power rule**: $= 9t - 2 - \dfrac{4}{t}$

Check: $6t^2\left(9t - 2 - \dfrac{4}{t}\right) = 6t^2 \cdot 9t - 6t^2 \cdot 2 - 6t^2 \cdot \dfrac{4}{t}$

$= 54t^3 - 12t^2 - 24t$

Take note that $9t - 2 - \dfrac{4}{t}$ is not a polynomial because of the term $-\dfrac{4}{t}$. This illustrates that the **quotient** of two polynomials is not always a polynomial.

You Try It **Work through this You Try It problem.**

Work Exercises 9 and 10 in this textbook or in the MyMathLab Study Plan.

OBJECTIVE 3 DIVIDE POLYNOMIALS USING LONG DIVISION

When dividing a **polynomial** by a polynomial, we use **long division**. **Polynomial long division** follows the same approach as long division for **real numbers**. View this **popup box** to review an example using long division to divide real numbers.

We can think of long division as a four-step process: **divide, multiply, subtract**, and **drop**.

CAUTION Before performing polynomial long division, write the **dividend** and **divisor** in **descending order**. If any **powers** are missing, then insert them with a **coefficient** of 0 as a placeholder.

Process for Polynomial Long Division

1. **Divide** the first term of the dividend by the first term of the divisor.
2. **Multiply** the result of the division by the divisor. Write this result under the dividend, making sure to line up like terms.
3. **Subtract** the result of the multiplication from the dividend.
4. **Drop** down the next term from the original dividend to form a reduced polynomial.

Repeat these four steps using the reduced polynomial. Continue repeating until the **remainder** can no longer be divided, which occurs when the **degree** of the remainder is less than the degree of the divisor.

The final result can be written in the form

$$\frac{\text{dividend}}{\text{divisor}} = \text{quotient} + \frac{\text{remainder}}{\text{divisor}}.$$

(eText Screens 14.7-1–14.7-16)

We can check this result by noting that

$$\text{dividend} = (\text{divisor})(\text{quotient}) + \text{remainder}.$$

We demonstrate **polynomial long division** in the following examples.

Example 4 Dividing Polynomials Using Long Division

Divide $(2x^2 + x - 15) \div (x + 3)$.

Solution The **dividend** is $2x^2 + x - 15$ and the **divisor** is $x + 3$. Both are in **descending order**. We apply the long division process of divide, multiply, subtract, and drop until the **remainder** can no longer be divided.

Divide the first **term** of the dividend, $2x^2$, by the first term of the divisor, x.

Because $\dfrac{2x^2}{x} = 2x$, $2x$ is the first term in the **quotient**.

$$\begin{array}{r} 2x \phantom{{}+x-15} \\ x+3\overline{)2x^2 + x - 15} \end{array}$$

Multiply the result of the division, $2x$, by the divisor, $x + 3$, to get $2x(x + 3) = 2x^2 + 6x$. Write this result under the dividend, making sure to line up **like terms**.

$$\begin{array}{r} 2x \phantom{{}+x-15} \\ x+3\overline{)2x^2 + x - 15} \\ 2x^2 + 6x \phantom{{}-15} \end{array}$$

Subtract the result of the multiplication, $2x^2 + 6x$, from the first two terms of the dividend. To do the subtraction, we change the sign of each term in $(2x^2 + 6x)$ and add.

$$\begin{array}{r} 2x \phantom{{}+x-15} \\ x+3\overline{)2x^2 + x - 15} \\ \underline{-(2x^2 + 6x)} \phantom{{}-15} \end{array} \quad \rightarrow \quad \begin{array}{r} 2x \phantom{{}+x-15} \\ x+3\overline{)2x^2 + x - 15} \\ \text{Add} \rightarrow \underline{-2x^2 - 6x} \phantom{{}-15} \\ -5x \phantom{{}-15} \end{array}$$

Drop down the next term, -15, and repeat the process.

$$\begin{array}{r} 2x \phantom{{}+x-15} \\ x+3\overline{)2x^2 + x - 15} \\ \underline{-2x^2 - 6x} \phantom{{}-15} \\ -5x - 15 \end{array}$$

Divide $-5x$ by x. Because $\dfrac{-5x}{x} = -5$, the next term in the **quotient** is -5.

$$\begin{array}{r} 2x - 5 \phantom{{}-15} \\ x+3\overline{)2x^2 + x - 15} \\ \underline{-2x^2 - 6x} \phantom{{}-15} \\ -5x - 15 \end{array}$$

Multiply -5 by $x + 3$ to get $-5(x + 3) = -5x - 15$.

$$\begin{array}{r} 2x - 5 \phantom{{}-15} \\ x+3\overline{)2x^2 + x - 15} \\ \underline{-2x^2 - 6x} \phantom{{}-15} \\ -5x - 15 \\ -5x - 15 \end{array}$$

Subtract $-5x - 15$ from $-5x - 15$. We do this by changing the sign of each term in $(-5x - 15)$ and adding.

$$
\begin{array}{r}
2x - 5 \\
x + 3\overline{)2x^2 + x - 15} \\
\underline{-2x^2 - 6x} \\
-5x - 15 \\
\underline{-(-5x - 15)}
\end{array}
\quad \rightarrow \quad
\begin{array}{r}
2x - 5 \\
x + 3\overline{)2x^2 + x - 15} \\
\underline{-2x^2 - 6x} \\
-5x - 15 \\
\text{Add} \rightarrow \underline{5x + 15} \\
0 \leftarrow \text{Remainder}
\end{array}
$$

The remainder is 0. So, $(2x^2 + x - 15) \div (x + 3) = 2x - 5$.

Check We must show that dividend $=$ (divisor)(quotient) $+$ remainder.

$$\underbrace{(x+3)}_{\text{Divisor}}\underbrace{(2x-5)}_{\text{Quotient}} + \underbrace{(0)}_{\text{Remainder}} = 2x^2 - 5x + 6x - 15 + 0 = \underbrace{2x^2 + x - 15}_{\text{Dividend}}$$

Our result checks, so the division is correct.

You Try It **Work through this You Try It problem.**

Work Exercises 11–14 in this textbook or in the MyMathLab Study Plan.

Example 5 Dividing Polynomials Using Long Division

My video summary

Divide $\dfrac{x^2 + 26x - 6x^3 - 12}{2x - 3}$.

Solution First, we rewrite the **dividend** in **descending order:** $-6x^3 + x^2 + 26x - 12$. The **divisor** is $2x - 3$. We apply the **long division process** of divide, multiply, subtract, and drop until the **remainder** can no longer be divided.

Divide the first **term** of the dividend, $-6x^3$, by the first term of the divisor, $2x$. Because $\dfrac{-6x^3}{2x} = -3x^2$, $-3x^2$ is the first term in the **quotient**.

$$
\begin{array}{r}
-3x^2 \qquad\qquad\quad \\
2x - 3\overline{)-6x^3 + x^2 + 26x - 12}
\end{array}
$$

Multiply the result of the division, $-3x^2$, by the divisor, $2x - 3$, to get $-3x^2(2x - 3) = -6x^3 + 9x^2$. Write this result under the dividend, making sure to line up **like terms.**

$$
\begin{array}{r}
-3x^2 \qquad\qquad\quad \\
2x - 3\overline{)-6x^3 + x^2 + 26x - 12} \\
-6x^3 + 9x^2 \qquad\qquad\quad
\end{array}
$$

Subtract the result of the multiplication, $-6x^3 + 9x^2$, from the first two terms of the dividend. To do the subtraction, we change the sign of each term in $(-6x^3 + 9x^2)$ and add.

$$
\begin{array}{r}
-3x^2 \qquad\qquad\quad \\
2x - 3\overline{)-6x^3 + x^2 + 26x - 12} \\
\underline{-(-6x^3 + 9x^2)} \qquad\qquad\quad
\end{array}
\quad \rightarrow \quad
\begin{array}{r}
-3x^2 \qquad\qquad\quad \\
2x - 3\overline{)-6x^3 + x^2 + 26x - 12} \\
\text{Add} \rightarrow \underline{+6x^3 - 9x^2} \qquad\qquad\quad \\
-8x^2 \qquad\qquad\quad
\end{array}
$$

(eText Screens 14.7-1–14.7-16)

Drop down the next term, $26x$, and repeat the process.

$$\begin{array}{r} -3x^2 \\ 2x - 3\overline{)-6x^3 + x^2 + 26x - 12} \\ \underline{+6x^3 - 9x^2} \\ -8x^2 + 26x \end{array}$$

Try finishing the problem on your own. See the **answer**, or watch this **video** for the complete solution.

Check Show that dividend = (divisor)(quotient) + remainder. View the **check**.

You Try It **Work through this You Try It problem.**

Work Exercises 15–18 in this textbook or in the MyMathLab Study Plan.

Example 6 Dividing Polynomials Using Long Division

My video summary

Divide $\dfrac{3t^3 - 11t - 12}{t + 4}$.

Solution The **dividend** is $3t^3 - 11t - 12$, and the **divisor** is $t + 4$. There is no t^2-term in the dividend, so we insert one with a **coefficient** of 0 as a placeholder.

$$t + 4\overline{)3t^3 + 0t^2 - 11t - 12}$$

Try to complete this problem on your own. See the **answer**, or watch this **video** for a detailed solution.

You Try It **Work through this You Try It problem.**

Work Exercises 19 and 20 in this textbook or in the MyMathLab Study Plan.

14.7 Exercises

In Exercises 1–4, divide the monomials.

You Try It

1. $\dfrac{54x^{11}}{9x^3}$
2. $\dfrac{-108z^{36}}{27z^{12}}$
3. $\dfrac{5m^3}{30m^3}$
4. $\dfrac{72t^4}{8t^9}$

In Exercises 5–10, divide the polynomial by the monomial.

You Try It

5. $\dfrac{14x^4 - 35x^3}{7x^3}$
6. $\dfrac{12m^4 - 16m^3 + 32m^2}{4m}$
7. $(48y^7 - 24y^5 + 6y^2) \div 6y^2$
8. $\dfrac{-10p^5 + 75p^4}{-5p^4}$

You Try It

9. $\dfrac{18a^3 - 63a^2 - 27a}{9a^2}$
10. $\dfrac{8w^3 - 14w + 6}{-2w^2}$

(eText Screens 14.8-1–14.8-13)

In Exercises 11–20, divide the polynomials using polynomial long division.

You Try It 11. $\dfrac{x^2 + 18x + 72}{x + 6}$

12. $\dfrac{6t^2 - t - 43}{2t + 5}$

13. $\dfrac{x^3 - 2x^2 - 8x + 21}{x + 3}$

14. $\dfrac{6a^3 - 5a^2 - 11a + 13}{3a - 1}$

You Try It 15. $\dfrac{2x^2 + x^3 - 9x - 4}{x + 5}$

16. $\dfrac{20 - 11x - 3x^2}{4 - 3x}$

17. $\dfrac{30 + 5x - 6x^2}{5 - 2x}$

18. $\dfrac{x^4 - 2x^3 - 13x^2 + 24x - 36}{x - 6}$

You Try It 19. $\dfrac{4m^3 + 5m - 7}{m + 2}$

20. $\dfrac{9x^4 - 2x^2 + 2x + 9}{x^2 + 5}$

14.8 Polynomials in Several Variables

THINGS TO KNOW

Before working through this topic, be sure you are familiar with the following concepts:

		VIDEO	ANIMATION	INTERACTIVE
You Try It	1. Evaluate Algebraic Expressions (Topic 10.5, **Objective 1**)	✓		
You Try It	2. Determine the Degree and Leading Coefficient of a Polynomial (Topic 14.2, **Objective 3**)	✓		
You Try It	3. Add Polynomials (Topic 14.3, **Objective 1**)			✓
You Try It	4. Subtract Polynomials (Topic 14.3, **Objective 3**)	✓		
You Try It	5. Multiply a Polynomial by a Monomial (Topic 14.4, **Objective 2**)	✓		
You Try It	6. Multiply Two Binomials (Topic 14.4, **Objective 3**)	✓	✓	✓
You Try It	7. Multiply Two or More Polynomials (Topic 14.4, **Objective 4**)	✓	✓	

OBJECTIVES

1 Determine the Degree of a Polynomial in Several Variables

2 Evaluate Polynomials in Several Variables

(eText Screens 14.8-1–14.8-13)

3 Add or Subtract Polynomials in Several Variables

4 Multiply Polynomials in Several Variables

OBJECTIVE 1 DETERMINE THE DEGREE OF A POLYNOMIAL IN SEVERAL VARIABLES

So far in this module, we have focused on **polynomials in one variable.** A polynomial containing two or more variables, such as the following example, is called a **polynomial in several variables.**

$$-2x^2y^2z^4 + 7xyz^3 + 5xz^2$$

> **Definition** **Polynomial in Several Variables**
>
> A **polynomial in several variables** is a polynomial containing two or more variables.

Our discussion of polynomials in one variable extends to polynomials in several variables as well.

Recall that the degree of a monomial is the sum of the exponents on the variables, and that the degree of a polynomial is the largest degree of any of its terms.

Example 1 Determining the Degree of a Polynomial in Several Variables

Determine the **coefficient** and **degree** of each **term**, then find the **degree of the polynomial.**

a. $2x^3y - 7x^2y^3 + xy^2$ **b.** $3x^2yz^3 - 4xy^3z + xy^2z^4$

Solution

a.

Term	Coefficient	Degree	
$2x^3y$	2	$3 + 1 = 4$	← $x^3y = x^3y^1$
$-7x^2y^3$	-7	$2 + 3 = 5$	
xy^2	1	$1 + 2 = 3$	← $xy^2 = x^1y^2$

The degree of the polynomial is 5.

b. Try working this problem on your own. View this **popup** to check your answer.

You Try It **Work through this You Try It problem.**

Work Exercises 1–3 in this textbook or in the MyMathLab Study Plan.

OBJECTIVE 2 EVALUATE POLYNOMIALS IN SEVERAL VARIABLES

We **evaluate** polynomials in several variables the same way we evaluated **algebraic expressions** in **Topic 10.5**. We substitute the given values for the **variables** and simplify.

Example 2 Evaluating Polynomials in Several Variables

a. Evaluate $3x^2y - 2xy^3 + 5$ for $x = -2$ and $y = 3$.

b. Evaluate $-a^3bc^2 + 5a^2b^2c - 2ab$ for $a = 2$, $b = -1$ and $c = 4$.

(eText Screens 14.8-1–14.8-13)

Solution

a. Begin with the polynomial:	$3x^2y - 2xy^3 + 5$	
Substitute -2 for x and 3 for y:	$3(-2)^2(3) - 2(-2)(3)^3 + 5$	← Put values in parentheses when substituting
Simplify the exponents:	$= 3(4)(3) - 2(-2)(27) + 5$	
Simplify the multiplication:	$= 36 - (-108) + 5$	
	$= 36 + 108 + 5$	← $-(-108) = +108$
Add:	$= 149$	

The value of the polynomial is 149 when x is -2 and $y = 3$.

My video summary

b. Try to evaluate this **polynomial** for the given values on your own. Then view the **answer**, or watch this **video** to see the complete solution.

You Try It **Work through this You Try It problem.**

Work Exercises 4–6 in this textbook or in the MyMathLab Study Plan.

OBJECTIVE 3 ADD OR SUBTRACT POLYNOMIALS IN SEVERAL VARIABLES

We add **polynomials in several variables** by removing **grouping symbols** and **combining like terms**. This is the same way we added polynomials in one variable in **Topic 14.3**.

We subtract polynomials in several variables just like we subtract polynomials in one variable—by adding its **opposite**.

Example 3 Adding or Subtracting Polynomials in Several Variables

Add or Subtract as indicated.

a. $(2x^2 + 3xy - 7y^2) + (4x^2 - xy + 11y^2)$

b. $(4a^2 - 3ab + 2b^2) - (6a^2 - 5ab + 7b^2)$

c. $(7x^4 + 3x^3y^3 - 2xy^3 + 5) + (2x^4 - x^3y^3 + 8xy^3 - 10)$

d. $(10x^3y + 2x^2y^2 - 5xy^3 - 8) - (6x^3y + x^2y^2 - 3xy^3)$

Solution

a. Write the original expression:	$(2x^2 + 3xy - 7y^2) + (4x^2 - xy + 11y^2)$
Remove grouping symbols:	$= 2x^2 + 3xy - 7y^2 + 4x^2 - xy + 11y^2$
Rearrange terms:	$= 2x^2 + 4x^2 + 3xy - xy - 7y^2 + 11y^2$
Combine like terms:	$= 6x^2 + 2xy + 4y^2$
b. Write the original expression:	$(4a^2 - 3ab + 2b^2) - (6a^2 - 5ab + 7b^2)$
Change to add the opposite:	$= (4a^2 - 3ab + 2b^2) + (-6a^2 + 5ab - 7b^2)$
Remove grouping symbols:	$= 4a^2 - 3ab + 2b^2 - 6a^2 + 5ab - 7b^2$
Rearrange terms:	$= 4a^2 - 6a^2 - 3ab + 5ab + 2b^2 - 7b^2$
Combine like terms:	$= -2a^2 + 2ab - 5b^2$

My video summary

c.–d. Try working these parts on your own. View the **answers**, or watch this **video** for the complete solutions to both parts.

You Try It **Work through this You Try It problem.**

Work Exercises 7–16 in this textbook or in the MyMathLab Study Plan.

(eText Screens 14.8-1–14.8-13)

OBJECTIVE 4 MULTIPLY POLYNOMIALS IN SEVERAL VARIABLES

As with addition and subtraction, we multiply polynomials in several variables just like we multiply polynomials in one variable. The **FOIL method** and the **special product rules** from **Topic 14.5** can also be used to simplify the multiplication of these polynomials.

Example 4 Multiplying a Polynomial by a Monomial

Multiply $5xy^2(4x^2 - 3xy + 2y^2)$.

Solution

Begin with the original expression: $5xy^2(4x^2 - 3xy + 2y^2)$

Distribute the monomial $5xy^2$: $= 5xy^2 \cdot 4x^2 - 5xy^2 \cdot 3xy + 5xy^2 \cdot 2y^2$

Rearrange the factors: $= 5 \cdot 4 \cdot x \cdot x^2 \cdot y^2 - 5 \cdot 3 \cdot x \cdot x \cdot y^2 \cdot y + 5 \cdot 2 \cdot x \cdot y^2 \cdot y^2$

Multiply the coefficients and apply the product rule for exponents: $= 20x^{1+2}y^2 - 15x^{1+1}y^{2+1} + 10xy^{2+2}$

Simiplify: $= 20x^3y^2 - 15x^2y^3 + 10xy^4$

You Try It **Work through this You Try It problem.**

Work Exercises 17–20 in this textbook or in the MyMathLab Study Plan.

Example 5 Multiplying Two Binomials in Several Variables

Multiply $(3x - 2y)(4x + 3y)$.

Solution Apply the **FOIL method.**

$$\begin{aligned}(3x - 2y)(4x + 3y) &= \overbrace{3x \cdot 4x}^{\text{First}} + \overbrace{3x \cdot 3y}^{\text{Outside}} + \overbrace{(-2y) \cdot 4x}^{\text{Inside}} + \overbrace{(-2y) \cdot 3y}^{\text{Last}} \\ &= 3 \cdot 4 \cdot x \cdot x + 3 \cdot 3 \cdot x \cdot y - 2 \cdot 4 \cdot x \cdot y - 2 \cdot 3 \cdot y \cdot y \\ &= 12x^{1+1} + 9xy - 8xy - 6y^{1+1} \\ &= 12x^2 + 9xy - 8xy - 6y^2 \\ &= 12x^2 + xy - 6y^2 \quad \leftarrow \boxed{\text{Combine like terms.}}\end{aligned}$$

You Try It **Work through this You Try It problem.**

Work Exercises 21–24 in this textbook or in the MyMathLab Study Plan.

Example 6 Using Special Product Rules with Polynomials in Several Variables

My interactive video summary

Multiply.

a. $(6x^2 + 5y)^2$ **b.** $(4x^3 - 9y^2)^2$ **c.** $(2x^2y - 7)(2x^2y + 7)$

Solutions

a. Square of a binomial sum rule: $(A + B)^2 = A^2 + 2AB + B^2$

Substitute $6x^2$ for A and $5y$ for B: $(6x^2 + 5y)^2 = (6x^2)^2 + 2(6x^2)(5y) + (5y)^2$

Simplify: $= 6^2(x^2)^2 + 2(6 \cdot 5)(x^2y) + 5^2y^2$

$= 36x^4 + 60x^2y + 25y^2$

(eText Screens 14.8-1–14.8-13)

b.–c. Try working these problems on your own by applying a **special product rule for binomials.** View the **answers,** or watch this **interactive video** to see solutions for all three parts.

You Try It **Work through this You Try It problem.**

Work Exercises 25–30 in this textbook or in the MyMathLab Study Plan.

Example 7 Multiplying Two Polynomials in Several Variables

Multiply $(x + 2y)(x^2 - 4xy + y^2)$.

Solution We multiply each term in the **binomial** by each term in the **trinomial.**

Begin with the original expression: $(x + 2y)(x^2 - 4xy + y^2)$

Multiply x by x^2, $-4xy$, and y^2;
Multiply $2y$ by x^2, $-4xy$, and y^2: $= x \cdot x^2 + x \cdot (-4xy) + x \cdot y^2 + 2y \cdot x^2 + 2y \cdot (-4xy) + 2y \cdot y^2$

Simplify: $= x^3 - 4x^2y + xy^2 + 2x^2y - 8xy^2 + 2y^3$

Combine like terms: $= x^3 - 2x^2y - 7xy^2 + 2y^3$

You Try It **Work through this You Try It problem.**

Work Exercises 31–34 in this textbook or in the MyMathLab Study Plan.

CAUTION Remember that multiplication is **commutative** when identifying like terms. For example, $-4x^2y$ and $2yx^2$ are like terms because $2yx^2 = 2x^2y$ so the terms have the same variable factor, x^2y.

14.8 Exercises

In Exercises 1–3, determine the coefficient and degree of each term, then find the degree of the polynomial.

You Try It **1.** $3x^2y^5 - 4xy^3 + 12y^2 - 5$

2. $-\frac{1}{2}xy^3 + xy + 3x^4y$

3. $6x^2yz^2 - 2.5xy^2z^2 + 7y^3z^5 - 4.3$

In Exercises 4–6, evaluate the polynomial for the given values of the variables.

You Try It **4.** Evaluate $x^2 - 3xy + 2y^2$ for $x = 5$ and $y = 4$.

5. Evaluate $xy^3 + y^3 - 2xy + 1$ for $x = -4$ and $y = 2$.

6. Evaluate $2x^2y + 3xy - y^2$ for $x = 3$ and $y = -2$.

In Exercises 7–16, add or subtract as indicated.

You Try It **7.** $(-12a + 11b) - (6a - 7b)$

8. $(3x + 2y) + (8x - 7y)$

9. $(2m^2 + n) - (4m^2 - 16mn - 5n)$

10. $(a^3 + b^3) + (2a^3 - 3a^2b + 9b^3)$

11. $(x - y) - (4x + 2xy - 3y)$

12. $(x^2 + 2xy - 5y^2) + (3x^2 - 4xy + 6y^2)$

13. $(5x^3 + xy - 7) + (-2x^3 - 4xy + 3)$

14. $(3a^4b^2 + 7a^2b^2 - 5ab) - (2a^4b^2 - 8a^2b^2 + ab)$

15. $(mn^2 + n) - (m + 3n) + (2m - 7mn^2)$

16. $(5a + 7b - c) + (2b - 13c) - (a - b + 9c)$

In Exercises 17–34, multiply. Simplify by combining like terms.

You Try It **17.** $3x(x^2y - 3xy)$

18. $\frac{1}{2}y^2(3x^2 - y^2)$

19. $3x^2y(x + y)$

20. $-4y^3(4xy + 15y^2)$

You Try It **21.** $(5x - 3y)(3x + 5y)$

22. $(x + 7y)(3x + 8y)$

23. $(0.2a^2 + 3b)(3a - 0.4b^2)$

24. $\left(\frac{1}{3}x - 3y\right)\left(\frac{2}{3}x + 2y\right)$

You Try It **25.** $(a + 3b)^2$

26. $(3a^2 - b^2)(3a^2 + b^2)$

27. $(4m - n)^2$

28. $\left(\frac{2}{5}xy + 3\right)^2$

29. $(5x - 2y)^2$

30. $(x^2y^3 - 4xy^2)(x^2y^3 + 4xy^2)$

31. $(2x - y)(x^2 + 3xy - y^2)$

32. $(x + 5y)(x^2 - 5xy + 25y^2)$

You Try It **33.** $-2x(x - y)(3x^2 + y)$

34. $(3a - b)(9a^2 - 6ab + b^2)$

MODULE FIFTEEN

Factoring Polynomials

MODULE FIFTEEN CONTENTS

15.1 Greatest Common Factor and Factoring by Grouping

THINGS TO KNOW

Before working through this topic, be sure you are familiar with the following concepts:

		VIDEO	ANIMATION	INTERACTIVE
You Try It	1. Use the Distributive Property (Topic 10.5, **Objective 3**)	▣		
You Try It	2. Identify Terms, Coefficients, and Like Terms of an Algebraic Expression (Topic 10.6, **Objective 1**)	▣		
You Try It	3. Multiply a Polynomial by a Monomial (Topic 14.4, **Objective 2**)	▣		
You Try It	4. Multiply Two Binomials (Topic 14.4, **Objective 3**)	▣	▣	▣
You Try It	5. Divide a Polynomial by a Monomial (Topic 14.7, **Objective 2**)	▣		

(eText Screens 15.1-1–15.1-23)

OBJECTIVES

1 Find the Greatest Common Factor of a Group of Integers

2 Find the Greatest Common Factor of a Group of Monomials

3 Factor Out the Greatest Common Factor from a Polynomial

4 Factor by Grouping

OBJECTIVE 1 FIND THE GREATEST COMMON FACTOR OF A GROUP OF INTEGERS

Factoring a **polynomial** is the reverse process of **multiplying polynomials.** When a polynomial is written as an **equivalent expression** that is a **product** of polynomials, we say that the polynomial has been **factored** or written in **factored form.** In **Example 5a of Topic 14.4**, we multiplied the **binomials** $x - 4$ and $2x + 3$ to result in the **trinomial** $2x^2 - 5x - 12$. Reversing the process, we factor $2x^2 - 5x - 12$ into the product $(x - 4)(2x + 3)$.

Multiply

$$(x - 4)(2x + 3) = 2x^2 - 5x - 12$$

Factoring

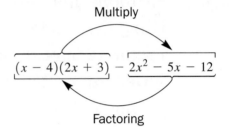

Therefore, $x - 4$ and $2x + 3$ are **factors** of $2x^2 - 5x - 12$.

CAUTION The word *factor* can be used as a noun or verb. As a noun, a *factor* is one of the numbers or expressions being multiplied to form a product. As a verb, *factor* means to rewrite the polynomial as a product.

When factoring a **polynomial**, first we find the *greatest common factor* or GCF of its **terms.** In this topic, we always **factor over the integers**, which means that all **coefficients** in both the original polynomial and its factors will be **integers.**

The **greatest common factor (GCF) of a group of integers** is the largest integer that divides evenly into each integer in the group. For example, the GCF of 12 and 18 is 6 because 6 is the largest integer that divides both 12 and 18 evenly ($12 \div 6 = 2$ and $18 \div 6 = 3$). When the GCF is not easy to find, we use the following process:

Finding the GCF of a Group of Integers

Step 1. Write out the **prime factorization** for each integer in the group.

Step 2. Write down the prime factors common to all the integers in the group.

Step 3. Multiply the **common prime factors** listed in Step 2 to find the GCF. If there are no common prime factors in Step 2, then the GCF is 1.

Example 1 Finding the GCF of a Group of Integers

Find the GCF of each group of integers.

a. 36 and 60 **b.** 28 and 45 **c.** 75, 90, and 105

Solutions We follow the steps for **finding the GCF of a group of integers.**

a. Step 1. $36 = \underbrace{2 \cdot 2 \cdot 3} \cdot 3$ and $60 = \underbrace{2 \cdot 2 \cdot 3} \cdot 5$

Common Prime Factors

Step 2. There are three **common prime factors:** $2, 2,$ and 3.

Step 3. The **GCF** is $2 \cdot 2 \cdot 3 = 12$.

b. Step 1. $28 = 2 \cdot 2 \cdot 7$ and $45 = 3 \cdot 3 \cdot 5$

Step 2. There are no common prime factors.

Step 3. The GCF is 1.

My video summary

c. Try to find this GCF on your own. View the **answer**, or watch this **video** for a detailed solution to part c.

You Try It **Work through this You Try It problem.**

Work Exercises 1–4 in this textbook or in the MyMathLab Study Plan.

OBJECTIVE 2 FIND THE GREATEST COMMON FACTOR OF A GROUP OF MONOMIALS

The **greatest common factor (GCF) of a group of monomials** is the **monomial** with the largest **coefficient** and highest **degree** that divides each monomial evenly.

Consider the **exponential expressions** $x^3, x^4,$ and x^5, each of which is a monomial with a coefficient of 1.

$$x^3 = \underbrace{x \cdot x \cdot x}, \quad x^4 = \underbrace{x \cdot x \cdot x} \cdot x, \quad \text{and} \quad x^5 = \underbrace{x \cdot x \cdot x} \cdot x \cdot x$$

Common Factors Common Factors

The expressions have three factors of the **base** x in common, so the greatest common *factor* of the expressions is $x \cdot x \cdot x = x^3$. Notice that this GCF is the expression with the lowest **power**. This is true in general.

> **Common Variable Factors for a GCF**
>
> If a variable is a **common factor** of a group of monomials, then the lowest power on that variable in the group will be a factor of the GCF.

Example 2 Finding the GCF of a Group of Exponential Expressions

Find the GCF of each group of exponential expressions.

a. x^4 and x^7 **b.** $y^3, y^6,$ and y^9 **c.** $w^6z^2, w^3z^5,$ and w^5z^4

Solutions

a. The variable x is a common factor of both expressions. The lowest power is 4, so the GCF is x^4.

b. The variable y is a **common factor** of all three expressions. The smallest **power** is 3, so the GCF is y^3.

c. The variables w and z are common factors of all three expressions. The lowest power of w is 3 and the lowest power of z is 2, so the GCF is w^3z^2.

You Try It **Work through this You Try It problem.**

Work Exercises 5–8 in this textbook or in the MyMathLab Study Plan.

We can use the following process to find the **GCF of a group of monomials.**

Finding the GCF of a Group of Monomials

Step 1. Find the GCF of the coefficients.

Step 2. Find the lowest power for each **common variable factor.**

Step 3. The GCF is the **monomial** with the coefficient from Step 1 and the variable factor(s) from Step 2.

Example 3 Finding the GCF of a Group of Monomials

Find the GCF of each group of monomials.

a. $14x^6$ and $21x^8$

b. $6a^2$, $10ab$, and $14b^2$

c. $40x^5y^6$, $-48x^9y$, and $24x^2y^4$

d. $14m^3n^2$, $6m^5n$, and $9m^4$

Solutions

a. We follow the steps for **finding the GCF of a group of monomials.**

Step 1. The **prime factorizations** of the **coefficients** 14 and 21 are

$$14 = 2 \cdot 7 \text{ and } 21 = 3 \cdot 7.$$

So, the **GCF** for the coefficients is 7.

Step 2. The variable x is a **common factor** of both expressions, and the lowest power is 6. So, x^6 is a factor of the GCF.

Step 3. Combining Steps 1 and 2, the GCF of the group of monomials is $7x^6$.

b. Step 1. The prime factorizations of the coefficients 6, 10, and 14 are

$$6 = 2 \cdot 3, 10 = 2 \cdot 5, \text{ and } 14 = 2 \cdot 7.$$

The GCF for the coefficients is 2.

My video summary

Step 2. There is no common variable factor for all three monomials. (Note: $6a^2$ does not have a factor of b, and $14b^2$ does not have a factor of a.) So, no variables are included in the GCF.

Step 3. The GCF for the group of monomials is 2.

c.–d. Try finding these GCFs on your own, then check your **answers.** Watch this **video** for complete solutions to parts c and d.

You Try It **Work through this You Try It problem.**

Work Exercises 9–14 in this textbook or in the MyMathLab Study Plan.

OBJECTIVE 3 FACTOR OUT THE GREATEST COMMON FACTOR FROM A POLYNOMIAL

The **greatest common factor (GCF) of a polynomial** is the **expression** with the largest **coefficient** and highest **degree** that divides each **term** of the **polynomial** evenly.

(eText Screens 15.1-1–15.1-23)

Consider the **binomial** $6x^2 + 10x$. The **GCF** of the two **monomials** $6x^2$ and $10x$ is $2x$, so $2x$ is the GCF of $6x^2 + 10x$. Once we know the GCF, we can **factor** it out of the binomial. We write each term of the binomial as a **product** that includes the **factor** $2x$.

$$6x^2 + 10x = 2x \cdot 3x + 2x \cdot 5$$

Reversing the **distributive property**, we factor $2x$ out of each term.

$$\underbrace{2x}\cdot 3x + \underbrace{2x}\cdot 5 = 2x(3x + 5)$$

So, $6x^2 + 10x = 2x(3x + 5)$.

Notice that the final result is the GCF times a **binomial factor**. This is expected because the original expression is a binomial. When the GCF is factored from a polynomial, the **polynomial factor** of the resulting product will have the same number of terms as the original polynomial.

Generalizing our process for this example gives the following steps for factoring out the GCF from a polynomial.

Factoring Out the GCF from a Polynomial

Step 1. Find the **GCF** of all **terms** in the **polynomial**.

Step 2. Write each term as a **product** that includes the GCF.

Step 3. Use the **distributive property** in reverse to factor out the GCF.

Step 4. Check the answer. The terms of the **polynomial factor** should have no more **common factors**. Multiplying the answer back out should give the original polynomial.

Example 4 Factoring Out the GCF from a Binomial

Factor out the GCF from each binomial.

a. $6x + 12$ **b.** $w^5 + w^4$ **c.** $8y^3 - 12y^2$

Solutions We follow our four-step process.

a. Step 1. The two terms are $6x$ and 12. The largest **integer** that divides evenly into the two **coefficients** 6 and 12 is 6. There is no common variable factor, so the GCF is 6.

Step 2. Write each term as the product of 6 and another factor.

$$6x + 12 = 6 \cdot x + 6 \cdot 2$$

Step 3. Factor out 6 using the distributive property in reverse.

$$6 \cdot x + 6 \cdot 2 = 6(x + 2)$$

Step 4. The **terms** of the **binomial factor** $x + 2$ have no **common factors**, so our **GCF** should be correct. Multiplying the final answer,

$$6(x + 2) = 6 \cdot x + 6 \cdot 2 = 6x + 12.$$

This is the original binomial, so our answer checks.

So, $6x + 12 = 6(x + 2)$.

My video summary

b. **Step 1.** The two terms are w^5 and w^4. Both **coefficients** are understood to be 1, so the coefficient of the GCF is 1. The variable w is a common factor of both terms, and the lowest **power** is 4. So the GCF is w^4.

Step 2. Write each term as the product of w^4 and another factor.

$$w^5 + \underbrace{w^4} = w^4 \cdot w + \underbrace{w^4 \cdot 1}$$

Because this term is the GCF, w^4, write it as $w^4 \cdot 1$.

Try to complete Steps 3 and 4 on your own to finish the problem. View the **answer**, or watch this **video** for fully worked solutions to parts b and c.

c. Try factoring out the GCF on your own. View the **answer**, or watch this **video** for complete solutions to parts b and c.

CAUTION

When checking answers after factoring out the **GCF**, multiplying is not enough. We must also make sure that the **terms** of the **polynomial factor** have no more **common factors**. For example, consider $6x + 12 = 2(3x + 6)$. The GCF has not been factored out of the binomial correctly because the terms $3x$ and 6 still have a common factor of 3. However, multiplying the right side results in the left side, so the problem appears to check. From Example 4a, we know the correct answer is $6x + 12 = 6(x + 2)$.

You Try It **Work through this You Try It problem.**

Work Exercises 15–18 in this textbook or in the MyMathLab Study Plan.

Example 5 Factoring Out the GCF from a Polynomial

Factor out the GCF from each **polynomial**.

a. $9p^5 + 18p^4 + 54p^3$ **b.** $10a^4b^6 - 15a^3b^7 + 35a^2b^8$

Solutions We follow our **four-step process.**

a. **Step 1.** The three terms are $9p^5, 18p^4$, and $54p^3$. The largest **integer** that divides evenly into the three **coefficients** 9, 18, and 54 is 9. The variable p is in all three terms and p^3 has the lowest **power**. So, the GCF is $9p^3$.

Step 2. We write each term as the product of $9p^3$ and another factor.

$$9p^5 + 18p^4 + 54p^3 = 9p^3 \cdot p^2 + 9p^3 \cdot 2p + 9p^3 \cdot 6$$

Step 3. Factor out $9p^3$ using the **distributive property** in reverse.

$$9p^3 \cdot p^2 + 9p^3 \cdot 2p + 9p^3 \cdot 6 = 9p^3(p^2 + 2p + 6)$$

Step 4. View the **check**.

The answer checks, so $9p^5 + 18p^4 + 54p^3 = 9p^3(p^2 + 2p + 6)$.

My video summary

b. Try factoring out the GCF on your own. View the **answer**, or watch this **video** for a complete solution to part b.

You Try It **Work through this You Try It problem.**

Work Exercises 19–22 in this textbook or in the MyMathLab Study Plan.

When the **leading coefficient** of a polynomial is negative, we may wish to factor out the negative sign with the GCF.

Example 6 Factoring Out a Negative Sign with the GCF

Factor out the negative sign with the GCF. $-8x^3 + 28x^2 - 20x$

Solution We again follow our **four-step process.**

Step 1. The three terms are $-8x^3, 28x^2$, and $-20x$. The largest **integer** that divides evenly into $-8, 28$, and -20 is 4. The variable x is in all three terms, and the lowest **power** is 1. So, x is a factor of the GCF. Because the leading coefficient is negative, we include a negative sign as part of the GCF. Therefore, we use $-4x$ as the GCF.

Step 2. We write each term as the product of $-4x$ and another factor. Because the second term $28x^2$ has a positive **coefficient**, both factors must be negative.

$$-8x^3 + 28x^2 - 20x = (-4x) \cdot 2x^2 + (-4x) \cdot (-7x) + (-4x) \cdot 5$$

Step 3. Factor out $-4x$ using the **distributive property** in reverse.

$$(-4x) \cdot 2x^2 + (-4x) \cdot (-7x) + (-4x) \cdot 5 = -4x(2x^2 - 7x + 5)$$

Step 4. View the **check**.

The answer checks, so $-8x^3 + 28x^2 - 20x = -4x(2x^2 - 7x + 5)$.

You Try It **Work through this You Try It problem.**

Work Exercises 23–26 in this textbook or in the MyMathLab Study Plan.

In each example so far, the **GCF** has been a **monomial**. In the next example, we factor out a **binomial** as the GCF.

Example 7 Factoring Out a Binomial as the GCF

Factor out the common **binomial factor** as the GCF.

a. $4x(y + 5) + 11(y + 5)$ b. $7x(x + y) - (x + y)$

Solutions

a. Treating $4x(y + 5)$ and $11(y + 5)$ as **terms**, the only **common factor** is the binomial factor $y + 5$. So, the GCF is $y + 5$. Each term is already written as a product of $y + 5$, so we factor it out:

$$4x\underline{(y + 5)} + 11\underline{(y + 5)} = (y + 5)(4x + 11)$$

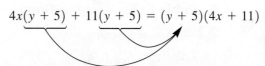

So, $4x(y + 5) + 11(y + 5) = (y + 5)(4x + 11)$.

My video summary

b. Try factoring out the common binomial factor as the GCF on your own. View the **answer**, or watch this **video** for a complete solution.

You Try It **Work through this You Try It problem.**

Work Exercises 27–30 in this textbook or in the MyMathLab Study Plan.

OBJECTIVE 4 FACTOR BY GROUPING

Suppose the **polynomial** $4x(y + 5) + 11(y + 5)$ from **Example 7a** had been simplified so that its **terms** did not contain the common **binomial factor** $y + 5$. Can we **factor** the simplified polynomial $4xy + 20x + 11y + 55$? The answer is yes. To do this, we use a method called **factoring by grouping.** Let's examine this method using the polynomial

$$4xy + 20x + 11y + 55.$$

Looking at all four terms in the polynomial, there are no **common factors** other than 1. However, if we group the first two terms together and consider them by themselves, they have a common factor of $4x$. **Factoring out** $4x$ from these two terms results in

$$4xy + 20x = 4x(y + 5).$$

Similarly, if we group the last two terms, they have a common factor of 11. Factoring out the 11 gives us

$$11y + 55 = 11(y + 5).$$

Using the grouped pairs, we arrive at the original polynomial in **Example 7a.** We can then factor out the common binomial factor to complete the process.

$$\begin{aligned} 4xy + 20x + 11y + 55 &= (4xy + 20x) + (11y + 55) \\ &= 4x(y + 5) + 11(y + 5) \\ &= (y + 5)(4x + 11) \end{aligned}$$

We summarize how to *factor by grouping* with the following steps:

Factoring a Polynomial by Grouping

Step 1. Group **terms** with a **common factor.** It may be necessary to rearrange the terms.

Step 2. For each group, **factor** out the **greatest common factor.**

Step 3. Factor out the common **polynomial factor,** if there is one.

Step 4. Check your answer by multiplying out the factors.

CAUTION If no arrangement of terms leads to a common polynomial factor for Step 3, then the polynomial cannot be factored by grouping.

Example 8 Factoring by Grouping

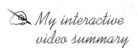

My interactive video summary

Factor by grouping.

a. $2x^2 - 6x + xy - 3y$

b. $5xy + 6 + 5x + 6y$

c. $3m^2 + 3m - 2mn - 2n$

d. $4w^3 - 14w^2 - 10w + 35$

Solutions

a. The first two terms have a common factor of $2x$, and the last two terms have a common factor of y. We follow our process for factoring by grouping.

Begin with the original polynomial expression:	$2x^2 - 6x + xy - 3y$
Group the first two terms and last two terms:	$= (2x^2 - 6x) + (xy - 3y)$
Factor out the GCF from each group:	$= 2x(x - 3) + y(x - 3)$
Factor out the common binomial factor:	$= (x - 3)(2x + y)$

This answer **checks,** so $2x^2 - 6x + xy - 3y = (x - 3)(2x + y)$.

b. For $5xy + 6 + 5x + 6y$, the first two terms and the last two terms have no common factors other than 1, so we must rearrange the terms to group terms with common factors. Notice that $5xy$ and $5x$ have a common factor of $5x$ and that $6y$ and 6 have a common factor of 6.

Begin with the original polynomial expression:	$5xy + 6 + 5x + 6y$
Rearrange to group terms with common factors:	$= (5xy + 5x) + (6y + 6)$
Factor out the GCF from each group:	$= 5x(y + 1) + 6(y + 1)$
Factor out the common binomial factor:	$= (y + 1)(5x + 6)$

This answer **checks**, so $5xy + 6 + 5x + 6y = (y + 1)(5x + 6)$.

c.–d. Try factoring these polynomials on your own. View the **answers**, or watch this **interactive video** for complete solutions to all four parts.

CAUTION If minus signs are involved when factoring by grouping, pay close attention to the placement of grouping symbols. For example, consider $wx - wy - xz + yz$. It is incorrect to write the grouping as $(wx - wy) - (xz + yz)$ because $-(xz + yz) = -xz - yz$, not $-xz + yz$. It is correct to write the grouping as $(wx - wy) + (-xz + yz)$ or $(wx - wy) - (xz - yz)$.

You Try It **Work through this You Try It problem.**

Work Exercises 31–40 in this textbook or in the MyMathLab Study Plan.

15.1 Exercises

In Exercises 1–4, find the GCF of each group of integers.

You Try It **1.** 21 and 35

2. 48 and 80

3. 45, 63, and 90

4. 70, −105, and 175

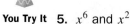

In Exercises 5–14, find the GCF of each group of monomials.

You Try It **5.** x^6 and x^2

6. w^4, w^7, and w^{10}

7. a^3b^9 and a^5b

8. x^4y^7, x^8y^6, and $x^{12}y^5$

You Try It **9.** $12x^8$ and $30x^5$

10. $32y^2$ and $27y^5$

11. $10p^2, 25pq$, and $15q^2$

12. $8x^3, 16x^5$, and $4x^7$

13. $54m^3n^4, -36mn^5$, and $72m^8n^6$

14. $24a^2b^4, 15ab^4$, and $25b^5$

In Exercises 15–22, factor out the GCF from each polynomial.

You Try It **15.** $30x + 12$

16. $y^7 + y^5$

17. $20w^4 - 28w^3$

18. $27x^2y + 63xy$

You Try It **19.** $4y^5 - 16y^4 + 28y^3$

20. $x^2y^6 - 5x^3y^5 + 9x^4y^4$

21. $25ab^5 - 10a^2b^4 + 5a^3b^3$

22. $6m^4 - 10m^3 + 12m^2 - 14m$

In Exercises 23–26, factor out the negative sign with the GCF.

You Try It **23.** $-8x^2 + 24x$

24. $-10m^2n - 35mn^2$

25. $-9x^2 - 27x + 12$

26. $-3x^3 + 15x^2 - 21x$

In Exercises 27–30, factor out the common binomial factor as the GCF.

27. $5x(y - 4) + 8(y - 4)$

28. $2x(x + 8) - (x + 8)$

You Try It **29.** $m(m - 2n) - 3n(m - 2n)$

30. $z^2(w^2 + 4) + 7(w^2 + 4)$

In Exercises 31–40, factor by grouping.

31. $5xy + 15x + 6y + 18$

32. $wz + 10w - 2z - 20$

You Try It **33.** $9ab - 12a - 6b + 8$

34. $4x^2 + 5xy + 8x + 10y$

35. $6x^2 - 11xy - 6x + 11y$

36. $5x^2 - 3y + 5xy - 3x$

37. $x^3 + 6x^2 + 7x + 42$

38. $x^3 - x^2 + 13x - 13$

39. $x^3 - 5x^2 + x - 5$

40. $m^2 + 7n + mn + 7m$

15.2 Factoring Trinomials of the Form $x^2 + bx + c$

THINGS TO KNOW

Before working through this topic, be sure you are familiar with the following concepts:

	VIDEO	ANIMATION	INTERACTIVE
You Try It **1.** Identify Terms, Coefficients, and Like Terms of an Algebraic Expression (Topic 10.6, **Objective 1**)	✓		
You Try It **2.** Determine the Degree and Leading Coefficient of a Polynomial (Topic 14.2, **Objective 3**)	✓		
You Try It **3.** Multiply Two Binomials (Topic 14.4, **Objective 3**)	✓	✓	✓
You Try It **4.** Factor Out the Greatest Common Factor from a Polynomial (Topic 15.1, **Objective 3**)	✓		

OBJECTIVES

1 Factor Trinomials of the Form $x^2 + bx + c$

2 Factor Trinomials of the Form $x^2 + bxy + cy^2$

3 Factor Trinomials of the Form $x^2 + bx + c$ after Factoring Out the GCF

(eText Screens 15.2-1–15.2-15)

OBJECTIVE 1 FACTOR TRINOMIALS OF THE FORM $x^2 + bx + c$

In this topic, we continue learning how to **factor polynomials** by focusing on **trinomials.** From **Topic 14.4**, we know the **product** of two **binomials** is often a trinomial. Since factoring is the reverse of multiplication, trinomials typically factor into the product of two binomials.

My animation summary

Let's look at trinomials with a **leading coefficient** of 1 and a **degree** of 2. These trinomials have the form, $x^2 + bx + c$. Begin by working through this **animation** about factoring trinomials.

Using **FOIL**, pay close attention to the multiplication process of two binomials. For example, look at $(x + 3)(x + 4)$.

$$(x + 3)(x + 4) = \underbrace{x^2}_{\text{F}} + \underbrace{4x}_{\text{O}} + \underbrace{3x}_{\text{I}} + \underbrace{12}_{\text{L}} = x^2 + 7x + 12$$

The product is a trinomial of the form $x^2 + bx + c$. Study the **terms** of the trinomial and the terms of each **binomial factor**, and then answer the following questions:

- Where does the term x^2 in the trinomial come from? View the **answer.**
- Where does the term 12 in the trinomial come from? View the **answer.**
- Where does the term $7x$ in the trinomial come from? View the **answer.**

Reversing our multiplication, the **trinomial** $x^2 + 7x + 12$ factors into the **product** $(x + 3)(x + 4)$. The numbers 3 and 4 are the "Last" **terms** in the **binomial factors** because 3 and 4 multiply to 12 and add to 7.

$$\underset{\substack{\uparrow \\ x \cdot x}}{x^2} + \underset{\substack{\uparrow \\ 3x + 4x}}{7x} + \underset{\substack{\uparrow \\ 3 \cdot 4}}{12} = (x + 3)(x + 4)$$

The relationship between the terms of the trinomial and the terms of the binomial factors leads us to the following steps for *factoring trinomials of the form* $x^2 + bx + c$. Remember that we will be **factoring over integers.**

> **Factoring Trinomials of the Form $x^2 + bx + c$**
>
> **Step 1.** Find two **integers,** n_1 and n_2, whose product is the **constant** term c and whose **sum** is the **coefficient** b. So, $n_1 \cdot n_2 = c$ and $n_1 + n_2 = b$.
>
> **Step 2.** Write the trinomial in the **factored form** $(x + n_1)(x + n_2)$.
>
> **Step 3.** Check the answer by multiplying out the factored form.

Example 1 Factoring Trinomials of the Form $x^2 + bx + c$

Factor each trinomial.

a. $x^2 + 11x + 18$ **b.** $x^2 + 13x + 30$

Solutions Follow the **three-step** process.

a. Step 1. We need to find two **integers** whose **product** is 18 and whose **sum** is 11. Because 18 and 11 are both positive, we begin by listing the pairs of **positive factors** for 18:

Positive Factors of 18	Sum of Factors	
1, 18	$1 + 18 = 19$	
2, 9	$2 + 9 = 11$	← This is the pair.
3, 6	$3 + 6 = 9$	

Step 2. Because $2 \cdot 9 = 18$ and $2 + 9 = 11$, we write

$$x^2 + 11x + 18 = (x + 2)(x + 9).$$

Step 3. Using **FOIL**, we check our answer by multiplying:

$$(x + 2)(x + 9) = \underbrace{x^2}_{F} + \underbrace{9x}_{O} + \underbrace{2x}_{I} + \underbrace{18}_{L}$$
$$= x^2 + 11x + 18 \leftarrow \text{Original trinomial}$$

Our result checks, so $x^2 + 11x + 18 = (x + 2)(x + 9)$.

Note: Because multiplication is **commutative**, the order in which we list the factors in the product does not matter. For example, we could also write this answer as $x^2 + 11x + 18 = (x + 9)(x + 2)$.

My video summary

b. We need to find two **integers** whose **product** is 30 and whose **sum** is 13. Try to find this pair of integers and finish factoring the trinomial on your own. Remember to check your answer by multiplying out the **factors**. View the **answer**, or watch this **video** for a fully worked solution to part b.

You Try It **Work through this You Try It problem.**

Work Exercises 1–3 in this textbook or in the MyMathLab Study Plan.

Recall that a **prime number** is a **whole number** greater than 1 whose only whole number factors are 1 and itself. For example, the first ten prime numbers are 2, 3, 5, 7, 11, 13, 17, 19, 23, and 29. As with numbers, there are also *prime polynomials*.

Definition Prime Polynomial

A polynomial is a **prime polynomial** if its only **factors over the integers** are 1 and itself.

For example, the **binomial** $4x + 3$ is a prime polynomial because it cannot be written as the product of any two factors other than 1 and itself. The binomial $4x + 8$ is not prime because it can be factored into $4(x + 2)$.

CAUTION

When deciding if a polynomial is prime, consider only factors over the integers. For example, even though $4x + 3 = 4(x + 0.75)$, the binomial $4x + 3$ is prime. Because 0.75 is not an integer, $4(x + 0.75)$ is not factored over the integers.

Example 2 Recognizing a Prime Trinomial of the Form $x^2 + bx + c$

Factor $x^2 + 14x + 20$.

Solution

Step 1. We need to find two **integers** whose **product** is 20 and whose **sum** is 14. Because 20 and 14 are both positive, we list the pairs of **positive factors** for 20:

Positive Factors of 20	Sum of Factors
1, 20	$1 + 20 = 21$
2, 10	$2 + 10 = 12$
4, 5	$4 + 5 = 9$

Step 2. Because none of the three possible pairs of factors have a sum of 14, we cannot factor this **trinomial** into the form $(x + n_1)(x + n_2)$. Therefore, $x^2 + 14x + 20$ is a **prime polynomial**.

(eText Screens 15.2-1–15.2-15)

You Try It **Work through this You Try It problem.**

Work Exercise 4 in this textbook or in the MyMathLab Study Plan.

In all of the examples so far involving $x^2 + bx + c$, the **coefficients** b and c have both been positive integers, but these coefficients can also be negative. When this happens, we must consider the sign of the **factors** of c as we look for the pair of factors whose sum is b.

Example 3 Factoring Trinomials of the Form $x^2 + bx + c$

My interactive video summary

Factor.

a. $x^2 - 13x + 40$ **b.** $m^2 - 5m - 36$ **c.** $w^2 + 7w - 60$

Solutions Follow the **three-step process.**

a. Step 1. We need to find two **integers** whose **product** is 40 and whose **sum** is -13. The only way that the product can be positive and the sum can be negative is if both integers are negative. We begin by listing the pairs of **negative factors** for 40:

Negative Factors of 40	Sum of Factors	
$-1, -40$	$(-1) + (-40) = -41$	
$-2, -20$	$(-2) + (-20) = -22$	
$-4, -10$	$(-4) + (-10) = -14$	
$-5, -8$	$(-5) + (-8) = -13$	← This is the pair.

Step 2. Because $(-5)(-8) = 40$ and $(-5) + (-8) = -13$, we write

$$x^2 - 13x + 40 = (x - 5)(x - 8).$$

Step 3. Multiply to check the answer. View the **check.**

Our result checks, so $x^2 - 13x + 40 = (x - 5)(x - 8)$.

b. We need to find two **integers** whose **product** is -36 and whose **sum** is -5. For the product to be negative, one of the integers must be positive and the other must be negative. For the sum to be negative, the integer with the larger **absolute value** must be negative. We list such pairs of **factors** for -36:

Factors of -36	Sum of Factors
$1, -36$	$(1) + (-36) = -35$
$2, -18$	$(2) + (-18) = -16$
$3, -12$	$(3) + (-12) = -9$
$4, -9$	$(4) + (-9) = -5$
$6, -6$	$(6) + (-6) = 0$

Use this information to finish factoring the **trinomial** on your own. View the **answer,** or work through this **interactive video** for the complete solution.

c. Try factoring this trinomial on your own. View the **answer,** or work through this **interactive video** for the complete solution.

You Try It **Work through this You Try It problem.**

Work Exercises 5–16 in this textbook or in the MyMathLab Study Plan.

OBJECTIVE 2 FACTOR TRINOMIALS OF THE FORM $x^2 + bxy + cy^2$

We can factor a **trinomial** in two variables using the same approach as that for factoring a trinomial in one variable. For example, we factor a trinomial of the form $x^2 + bxy + cy^2$ just as we would factor $x^2 + bx + c$. The difference is that the second **term** of each **binomial factor** must contain the y variable. So, we find two **integers**, n_1 and n_2, whose **product** is the **coefficient** c and whose **sum** is the coefficient b. Then we write

$$x^2 + bxy + cy^2 = (x + n_1y)(x + n_2y).$$

Example 4 Factoring a Trinomial of the Form $x^2 + bxy + cy^2$

Factor.

a. $x^2 + 10xy + 24y^2$ **b.** $m^2 + 22mn - 48n^2$

Solutions

a. We need to find two integers whose product is 24 and whose sum is 10. Both 24 and 10 are positive, so we consider the pairs of **positive factors** for 24.

Positive Factors of 24	Sum of Factors	
$1, 24$	$1 + 24 = 25$	
$2, 12$	$2 + 12 = 14$	
$3, 8$	$3 + 8 = 11$	
$4, 6$	$4 + 6 = 10$	← This is the pair.

Because $4 \cdot 6 = 24$ and $4 + 6 = 10$, we write

$$x^2 + 10xy + 24y^2 = (x + 4y)(x + 6y).$$

Check the answer by multiplying. View the **check**.

My video summary

b. Try factoring this **trinomial** on your own. View the **answer**, or watch this **video** for a detailed solution.

You Try It **Work through this You Try It problem.**

Work Exercises 17–20 in this textbook or in the MyMathLab Study Plan.

OBJECTIVE 3 FACTOR TRINOMIALS OF THE FORM $x^2 + bx + c$ AFTER FACTORING OUT THE GCF

A **polynomial** is **factored completely** if it is written as the **product** of all **prime polynomials.** For example, we factored each **trinomial** in **Example 3** as a product of prime **binomials**, so we factored each trinomial completely.

In **Topic 15.1**, we learned how to factor out the **greatest common factor** from a polynomial. If a trinomial has a **common factor** other than 1, we factor out the GCF first.

Example 5 Factoring Trinomials with a Common Factor

Factor completely.

a. $4x^2 - 28x - 32$ **b.** $2y^3 - 36y^2 + 64y$

Solutions

a. The three **terms** have a GCF of 4. First, we factor out the 4, then we factor the remaining trinomial.

Begin with the original trinomial:	$4x^2 - 28x - 32$
Write each term as a product of the GCF 4:	$= 4 \cdot x^2 - 4 \cdot 7x - 4 \cdot 8$
Factor out the 4:	$= 4(x^2 - 7x - 8)$

The trinomial $x^2 - 7x - 8$ is of the form $x^2 + bx + c$, so we use the **three-step process.**

We need to find two **integers** whose **product** is -8 and whose **sum** is -7. For the product to be negative, one of the integers must be positive and the other must be negative. For the sum to be negative, the integer with the larger **absolute value** must be negative. We list such pairs of **factors** for -8:

Factors of -8	Sum of Factors	
$1, -8$	$1 + (-8) = -7$	← This is the pair.
$2, -4$	$2 + (-4) = -2$	

Because $1(-8) = -8$ and $1 + (-8) = -7$, we write

$$x^2 - 7x - 8 = (x + 1)(x - 8).$$

Therefore, $4x^2 - 28x - 32 = 4(x + 1)(x - 8)$.
Check the answer by multiplying. View the **check**.

My video summary

b. Try factoring this **trinomial** on your own. Remember to factor out the **GCF** first. View the **answer**, or watch this **video** for a detailed solution.

You Try It **Work through this You Try It problem.**

Work Exercise 21–26 in this textbook or in the MyMathLab Study Plan.

A trinomial will sometimes have a **leading coefficient** of -1, such as $-x^2 + 7x + 60$. Because it is much easier to factor a trinomial when the leading coefficient is 1, we often begin by factoring out -1.

Example 6 Factoring a Trinomial with a Leading Coefficient of −1

My video summary

Factor $-x^2 + 3x + 10$.

Solution The leading coefficient is -1, so let's begin by factoring it out.

Begin with the original trinomial:	$-x^2 + 3x + 10$
Write each term as a product of -1:	$= (-1)x^2 + (-1)(-3x) + (-1)(-10)$
Factor out the -1:	$= -1(x^2 - 3x - 10)$
	$= -(x^2 - 3x - 10)$

The trinomial $x^2 - 3x - 10$ is of the form $x^2 + bx + c$. Try to finish factoring this trinomial on your own. View the **answer**, or watch this **video** for a detailed solution.

You Try It **Work through this You Try It problem.**

Work Exercises 27–30 in this textbook or in the MyMathLab Study Plan.

15.2 Exercises

In Exercises 1–20, factor each trinomial, or state that the trinomial is prime.

You Try It **1.** $x^2 + 10x + 21$

2. $y^2 + 16y + 28$

3. $w^2 + 17w + 72$

4. $z^2 + 12z + 16$

You Try It **5.** $n^2 - 9n + 14$

6. $k^2 - 4k - 5$

7. $p^2 + 3p - 54$

8. $x^2 - 24x + 44$

9. $y^2 + 10y - 36$

10. $m^2 - 3m - 70$

11. $x^2 - 24x + 108$

12. $n^2 + 18n - 40$

13. $x^2 - 31x - 66$

14. $t^2 - 10t + 25$

15. $w^2 + 4w - 96$

16. $q^2 + 20q + 36$

You Try It **17.** $x^2 + 6xy + 8y^2$

18. $p^2 + 3pq - 54q^2$

19. $a^2 - 11ab + 24b^2$

20. $m^2 - mn - 90n^2$

In Exercises 21–30, factor completely.

You Try It **21.** $5x^2 + 50x + 80$

22. $t^4 - 17t^3 + 60t^2$

23. $2x^3 - 12x^2 - 270x$

24. $8y^2z + 16yz - 280z$

25. $20m^2 - 520m + 500$

26. $2x^3 + 18x^2y + 40xy^2$

You Try It **27.** $-x^2 + 2x + 3$

28. $-y^2 - 10y + 144$

29. $-x^2 + 7xy - 10y^2$

30. $-4t^3 + 32t^2 - 60t$

15.3 Factoring Trinomials of the Form $ax^2 + bx + c$ Using Trial and Error

THINGS TO KNOW

Before working through this topic, be sure you are familiar with the following concepts:

	VIDEO	ANIMATION	INTERACTIVE
You Try It **1.** Identify Terms, Coefficients, and Like Terms of an Algebraic Expression (Topic 10.6, **Objective 1**)	▤		
You Try It **2.** Determine the Degree and Leading Coefficient of a Polynomial (Topic 14.2, **Objective 3**)	▤		

You Try It — **3.** Multiply Two Binomials (Topic 14.4, **Objective 3**)

You Try It — **4.** Factor Out the Greatest Common Factor from a Polynomial (Topic 15.1, **Objective 3**)

OBJECTIVES

1 Factor Trinomials of the Form $ax^2 + bx + c$ Using Trial and Error

2 Factor Trinomials of the Form $ax^2 + bxy + cy^2$ Using Trial and Error

OBJECTIVE 1 FACTOR TRINOMIALS OF THE FORM $ax^2 + bx + c$ USING TRIAL AND ERROR

My animation summary

If the **leading coefficient** of a **trinomial** is not 1, we must consider the coefficient as we look for the **binomial factors**. Work through this **animation**.

Once again, we can use **FOIL** to see how to factor this type of trinomial. For example, look at the **product** $(2x + 3)(2x + 5)$.

$$(2x+3)(2x+5) = \overbrace{2x \cdot 2x}^{\text{F}} + \overbrace{2x \cdot 5}^{\text{O}} + \overbrace{3 \cdot 2x}^{\text{I}} + \overbrace{3 \cdot 5}^{\text{L}} = 4x^2 + 16x + 15$$

The product results in a trinomial of the form $ax^2 + bx + c$ with $a \neq 1$. Study the **terms** of the trinomial and the terms of each binomial factor, and then answer the following questions:

- Where does the first term, $4x^2$, in the trinomial come from? View the **answer**.
- Where does the last term, 15, in the trinomial come from? View the **answer**.
- Where does the middle term, $16x$, in the trinomial come from? View the **answer**.

Reversing the multiplication, the **trinomial** $4x^2 + 16x + 15$ factors into the product $(2x + 3)(2x + 5)$. Using FOIL, the "first" **terms** in our **factors** must multiply to $4x^2$. The "last" terms must multiply to 15. And, the "outside" and "inside" **products** must add to $16x$.

$$\underset{\underset{2x\cdot 2x}{\uparrow}}{4x^2} + \underset{\underset{2x\cdot 5 + 3\cdot 2x}{\uparrow}}{16x} + \underset{\underset{3\cdot 5}{\uparrow}}{15} = (2x+3)(2x+5)$$

If a trinomial of the form $ax^2 + bx + c$ can be factored, it will factor to the form $(m_1x + n_1)(m_2x + n_2)$. We can find the **integers** m_1, m_2, n_1, and n_2 by using the following *trial and error strategy*.

Trial-and-Error Strategy for Factoring Trinomials of the Form $ax^2 + bx + c$

Step 1. Find all pairs of factors for the **leading coefficient** a.

Step 2. Find all pairs of factors for the **constant** term c.

Step 3. By trial and error, check different combinations of factors from Step 1 and factors from Step 2 in the form $(\square x + \square)(\square x + \square)$ until the correct middle term bx is found by adding the "outside" and "inside" products. If no such combination of factors exists, the trinomial is **prime**.

Step 4. Check your answer by multiplying out the **factored form**.

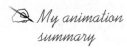

Example 1 Factoring Trinomials of the Form $ax^2 + bx + c$ Using Trial and Error

Factor $3x^2 + 7x + 2$.

Solution We follow the **trial-and-error strategy**. Because all of the **coefficients** are positive, we need to check only combinations of **positive factors**.

Step 1. The **leading coefficient** a is 3. The only pair of positive factors is $3 \cdot 1$.

Step 2. The **constant** term c is 2. The only pair of positive factors is $2 \cdot 1$.

Step 3. Using the form $(\square x + \square)(\square x + \square)$, we try different combinations of the possible factors until we find one that gives the middle term $7x$.

Trial 1. $(3x + 2)(1x + 1)$

$$\begin{array}{r} 2x \\ +3x \\ \hline 5x \end{array} \leftarrow \text{Not } 7x$$

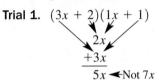

The "outside" **product** is $3x \cdot 1 = 3x$ and the "inside" product is $2 \cdot 1x = 2x$, giving a **sum** of $3x + 2x = 5x$. This is not the correct combination.

Trial 2. $(3x + 1)(1x + 2)$

$$\begin{array}{r} x \\ +6x \\ \hline 7x \end{array} \leftarrow \text{Correct}$$

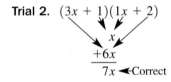

The "outside" product is $3x \cdot 2 = 6x$ and the "inside" product is $1 \cdot 1x = x$, giving a sum of $6x + x = 7x$. This is the correct combination.

Step 4. Check the answer by multiplying out $(3x + 1)(x + 2)$. View the **check**.

The answer checks, so $3x^2 + 7x + 2 = (3x + 1)(x + 2)$.

CAUTION Remember, because multiplication is **commutative**, the order in which the **binomial factors** are listed does not matter. So, we could also write the answer to Example 1 as $3x^2 + 7x + 2 = (x + 2)(3x + 1)$.

In Example 1, there were only two possible combinations of factors to check. Often there are many more combinations to consider.

Example 2 Factoring Trinomials of the Form $ax^2 + bx + c$ Using Trial and Error

My video summary

Factor $5x^2 + 17x + 6$.

Solution We follow the **trial-and-error strategy**. Because all of the **coefficients** are positive, we need to check only combinations of **positive factors**.

Step 1. The **leading coefficient** a is 5. The only pair of positive factors is $5 \cdot 1$.

Step 2. The constant term c is 6. The pairs of positive factors are $1 \cdot 6$ and $2 \cdot 3$.

Step 3. We use the form $(\square x + \square)(\square x + \square)$ with different combinations of the factors from Step 1 and Step 2 until we find one that gives the middle term $17x$.

Trial 1. $(5x + 6)(1x + 1)$

$$\begin{array}{r} 6x \\ +5x \\ \hline 11x \end{array} \leftarrow \text{Not } 17x$$

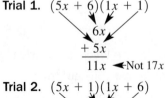

The "outside" **product** is $5x \cdot 1 = 5x$ and the "inside" product is $6 \cdot 1x = 6x$, giving a **sum** of $5x + 6x = 11x$. This is not the correct combination.

Trial 2. $(5x + 1)(1x + 6)$

$$\begin{array}{r} x \\ +30x \\ \hline 31x \end{array} \leftarrow \text{Not } 17x$$

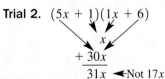

The "outside" product is $5x \cdot 6 = 30x$ and the "inside" product is $1 \cdot 1x = x$, giving a sum of $30x + x = 31x$. This is not the correct combination.

Trial 3. $(5x + 3)(1x + 2)$

$$\begin{array}{r} 3x \\ + 10x \\ \hline 13x \end{array} \leftarrow \text{Not } 17x$$

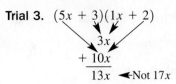

The "outside" product is $5x \cdot 2 = 10x$ and the "inside" product is $3 \cdot 1x = 3x$, giving a sum of $10x + 3x = 13x$. This is not the correct combination.

Trial 4. $(5x + 2)(1x + 3)$

$$\begin{array}{r} 2x \\ + 15x \\ \hline 17x \end{array} \leftarrow \text{Correct}$$

The "outside" product is $5x \cdot 3 = 15x$ and the "inside" product is $2 \cdot 1x = 2x$, giving a sum of $15x + 2x = 17x$. This is the correct combination.

Step 4. Check the answer by multiplying out $(5x + 2)(x + 3)$. View the **check**, or watch this **video** for a complete solution.

The answer checks, so $5x^2 + 17x + 6 = (5x + 2)(x + 3)$.

You Try It **Work through this You Try It problem.**

Work Exercises 1–4 in this textbook or in the MyMathLab Study Plan.

Now let's try factoring **trinomials** with negative **coefficients**.

Example 3 Factoring Trinomials of the Form $ax^2 + bx + c$ Using Trial and Error

My interactive video summary

Factor.

a. $4x^2 - 5x - 6$

b. $12n^2 - 16n + 5$

Solutions

a. We follow the **trial-and-error strategy**.

Step 1. The leading coefficient a is 4. For convenience, we will consider only the **positive factors** of 4, which are $4 \cdot 1$ and $2 \cdot 2$.

Step 2. The constant term c is -6. In this case, we must consider the negative sign. The pairs of factors are $1 \cdot (-6)$, $-1 \cdot 6$, $2 \cdot (-3)$, and $-2 \cdot 3$.

Step 3. We use the form $(\square x + \square)(\square x + \square)$ with different combinations of the factors from Step 1 and Step 2 until we find one that gives the middle term $-5x$. Try to complete this problem on your own. Compare your **answer**, or work through this **interactive video** for complete solutions to parts a and b.

b. Try factoring this trinomial on your own, then compare your **answer**. Work through this **interactive video** for complete solutions to parts a and b.

You Try It **Work through this You Try It problem.**

Work Exercises 5–8 in this textbook or in the MyMathLab Study Plan.

CAUTION

If the **terms** of a **polynomial** have no **common factor** (other than 1), then the terms of its **polynomial factors** will have no common factors either. For example, the terms of the trinomial $4x^2 - 5x - 6$ have no common factors, which means the terms of its binomial factors will have no common factors either. Therefore, some potential **binomial factors**, such as $4x + 6$ and $2x - 2$, can be disregarded because they contain a common factor.

When a trinomial is **prime**, no combination of binomial factors results in the correct middle term.

Example 4 Recognizing a Prime Trinomial of the Form $ax^2 + bx + c$

Factor $2y^2 - 19y + 15$.

Solution We follow the **trial-and-error strategy**.

Step 1. The **leading coefficient** a is 2. The **positive factors** are $2 \cdot 1$.

Step 2. The constant term c is 15. Because the middle **coefficient** b is negative and comes from adding the "outside" and "inside" **products**, we need to consider only the pairs of **negative factors** for 15, which are $-1 \cdot (-15)$ and $-3 \cdot (-5)$.

Step 3. Using the form $(\square x + \square)(\square x + \square)$, we try different combinations of the factors until we find one that gives the middle term $-19y$.

Trial 1. $(2y - 1)(1y - 15)$

$-y + (-30y) = -31y$ ← Not $-19y$

Trial 2. $(2y - 15)(1y - 1)$

$-15y + (-2y) = -17y$ ← Not $-19y$

Trial 3. $(2y - 3)(1y - 5)$

$-3y + (-10y) = -13y$ ← Not $-19y$

Trial 4. $(2y - 5)(1y - 3)$

$-5y + (-6y) = -11y$ ← Not $-19y$

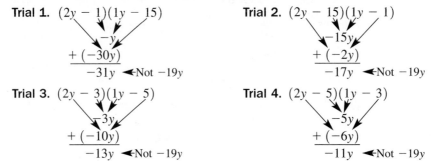

We have checked all the possible combinations, and none result in the middle term $-19y$. This means $2y^2 - 19y + 15$ is **prime**.

You Try It **Work through this You Try It problem.**

Work Exercises 9–26 in this textbook or in the MyMathLab Study Plan.

OBJECTIVE 2 FACTOR TRINOMIALS OF THE FORM $ax^2 + bxy + cy^2$ USING TRIAL AND ERROR

We can factor a **trinomial** of the form $ax^2 + bxy + cy^2$ just as we would factor $ax^2 + bx + c$, except that the second term of each **binomial factor** must contain the variable y. So, we use **trial and error** to check different combinations of factors for a and factors for c in the form $(\square x + \square y)(\square x + \square y)$ until the correct middle term bxy is found.

Example 5 Factoring a Trinomial of the Form $ax^2 + bxy + cy^2$

Factor.

a. $6x^2 + 17xy - 3y^2$ **b.** $2m^2 + 11mn + 12n^2$

Solutions

a. The pairs of **positive factors** of the **leading coefficient** 6 are $6 \cdot 1$ and $3 \cdot 2$.

The **coefficient** of the last **term** is -3, and its **factors** are $1 \cdot (-3)$ and $-1 \cdot 3$.

Using the form $(\square x + \square y)(\square x + \square y)$, we try different combinations of the factors until we find one that gives the middle term $17xy$.

Trial 1. $(6x + 1y)(1x - 3y)$

$xy + (-18xy) = -17xy$ ← Wrong sign

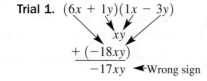

The "outside" **product** is $6x \cdot (-3y) = -18xy$ and the "inside" product is $1y \cdot 1x = xy$, giving a **sum** of $-18xy + xy = -17xy$. This is not the correct combination. Only the sign is wrong, so switch the signs in the binomial factors.

(eText Screens 15.1-1–15.1-23)

Consider the **binomial** $6x^2 + 10x$. The **GCF** of the two **monomials** $6x^2$ and $10x$ is $2x$, so $2x$ is the GCF of $6x^2 + 10x$. Once we know the GCF, we can **factor** it out of the binomial. We write each term of the binomial as a **product** that includes the **factor** $2x$.

$$6x^2 + 10x = 2x \cdot 3x + 2x \cdot 5$$

Reversing the **distributive property**, we factor $2x$ out of each term.

$$\underbrace{2x}\cdot 3x + \underbrace{2x}\cdot 5 = 2x(3x + 5)$$

So, $6x^2 + 10x = 2x(3x + 5)$.

Notice that the final result is the GCF times a **binomial factor**. This is expected because the original expression is a binomial. When the GCF is factored from a polynomial, the **polynomial factor** of the resulting product will have the same number of terms as the original polynomial.

Generalizing our process for this example gives the following steps for factoring out the GCF from a polynomial.

Factoring Out the GCF from a Polynomial

Step 1. Find the **GCF** of all **terms** in the **polynomial**.

Step 2. Write each term as a **product** that includes the GCF.

Step 3. Use the **distributive property** in reverse to factor out the GCF.

Step 4. Check the answer. The terms of the **polynomial factor** should have no more **common factors**. Multiplying the answer back out should give the original polynomial.

Example 4 Factoring Out the GCF from a Binomial

Factor out the GCF from each binomial.

a. $6x + 12$ **b.** $w^5 + w^4$ **c.** $8y^3 - 12y^2$

Solutions We follow our four-step process.

a. Step 1. The two terms are $6x$ and 12. The largest **integer** that divides evenly into the two **coefficients** 6 and 12 is 6. There is no common variable factor, so the GCF is 6.

Step 2. Write each term as the product of 6 and another factor.

$$6x + 12 = 6 \cdot x + 6 \cdot 2$$

Step 3. Factor out 6 using the distributive property in reverse.

$$6 \cdot x + 6 \cdot 2 = 6(x + 2)$$

Step 4. The **terms** of the **binomial factor** $x + 2$ have no **common factors**, so our **GCF** should be correct. Multiplying the final answer,

$$6(x + 2) = 6 \cdot x + 6 \cdot 2 = 6x + 12.$$

This is the original binomial, so our answer checks.

So, $6x + 12 = 6(x + 2)$.

My video summary

b. **Step 1.** The two terms are w^5 and w^4. Both **coefficients** are understood to be 1, so the coefficient of the GCF is 1. The variable w is a common factor of both terms, and the lowest **power** is 4. So the GCF is w^4.

Step 2. Write each term as the product of w^4 and another factor.

$$w^5 + \underbrace{w^4} = w^4 \cdot w + \underbrace{w^4 \cdot 1}$$

Because this term is the GCF, w^4, write it as $w^4 \cdot 1$.

Try to complete Steps 3 and 4 on your own to finish the problem. View the **answer**, or watch this **video** for fully worked solutions to parts b and c.

c. Try factoring out the GCF on your own. View the **answer**, or watch this **video** for complete solutions to parts b and c.

CAUTION

When checking answers after factoring out the **GCF**, multiplying is not enough. We must also make sure that the **terms** of the **polynomial factor** have no more **common factors**. For example, consider $6x + 12 = 2(3x + 6)$. The GCF has not been factored out of the binomial correctly because the terms $3x$ and 6 still have a common factor of 3. However, multiplying the right side results in the left side, so the problem appears to check. From Example 4a, we know the correct answer is $6x + 12 = 6(x + 2)$.

You Try It **Work through this You Try It problem.**

Work Exercises 15–18 in this textbook or in the MyMathLab Study Plan.

Example 5 Factoring Out the GCF from a Polynomial

Factor out the GCF from each **polynomial**.

a. $9p^5 + 18p^4 + 54p^3$ b. $10a^4b^6 - 15a^3b^7 + 35a^2b^8$

Solutions We follow our **four-step process.**

a. **Step 1.** The three terms are $9p^5, 18p^4$, and $54p^3$. The largest **integer** that divides evenly into the three **coefficients** 9, 18, and 54 is 9. The variable p is in all three terms and p^3 has the lowest **power**. So, the GCF is $9p^3$.

Step 2. We write each term as the product of $9p^3$ and another factor.

$$9p^5 + 18p^4 + 54p^3 = 9p^3 \cdot p^2 + 9p^3 \cdot 2p + 9p^3 \cdot 6$$

Step 3. Factor out $9p^3$ using the **distributive property** in reverse.

$$9p^3 \cdot p^2 + 9p^3 \cdot 2p + 9p^3 \cdot 6 = 9p^3(p^2 + 2p + 6)$$

Step 4. View the **check**.

The answer checks, so $9p^5 + 18p^4 + 54p^3 = 9p^3(p^2 + 2p + 6)$.

My video summary

b. Try factoring out the GCF on your own. View the **answer,** or watch this **video** for a complete solution to part b.

You Try It **Work through this You Try It problem.**

Work Exercises 19–22 in this textbook or in the MyMathLab Study Plan.

When the **leading coefficient** of a polynomial is negative, we may wish to factor out the negative sign with the GCF.

(eText Screens 15.3-1–15.3-14)

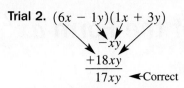

Trial 2. $(6x - 1y)(1x + 3y)$

$$\begin{array}{r} -xy \\ +18xy \\ \hline 17xy \end{array} \leftarrow \text{Correct}$$

The "outside" **product** is $6x \cdot 3y = 18xy$ and the "inside" product is $-1y \cdot 1x = -xy$, giving a **sum** of $18xy - xy = 17xy$. This is the correct combination.

Because we have found the correct factored form, we do not need to test any more combinations. Check the answer by multiplying out $(6x - y)(x + 3y)$. View the **check**.

The answer does check, so $6x^2 + 17xy - 3y^2 = (6x - y)(x + 3y)$.

My video summary

b. Try factoring this trinomial on your own, then compare your **answer**. Work through this **video** for a complete solution.

You Try It **Work through this You Try It problem.**

Work Exercises 27–30 in this textbook or in the MyMathLab Study Plan.

15.3 Exercises

In Exercises 1–30, factor each trinomial, or state that the trinomial is prime.

You Try It

1. $3x^2 + 16x + 5$

2. $2y^2 + 7y + 6$

3. $5x^2 + 22x + 8$

4. $4w^2 + 20w + 21$

You Try It

5. $6n^2 - 17n + 7$

6. $5m^2 + 13m - 6$

7. $12t^2 + 5t - 7$

8. $8x^2 - 6x - 9$

You Try It

9. $3x^2 - 4x + 8$

10. $10m^2 - 23m + 12$

11. $20p^2 - 8p - 1$

12. $9y^2 - 24y + 16$

13. $2w^2 + 23w - 39$

14. $8x^2 + 15x + 9$

15. $15t^2 - 49t + 24$

16. $14p^2 + 28p - 36$

17. $4z^2 - 21z - 18$

18. $5p^2 + 8p - 21$

19. $12x^2 + 16x - 3$

20. $2q^2 + 15q + 7$

21. $7y^2 - y - 8$

22. $6x^2 - 25x + 4$

23. $2w^2 - 15w + 27$

24. $22y^2 + 43y - 30$

25. $24m^2 - 26m - 15$

26. $16x^2 + 32x + 15$

You Try It

27. $9x^2 - 21xy - 8y^2$

28. $10p^2 + 7pq - 3q^2$

29. $2a^2 - 27ab + 70b^2$

30. $3m^2 + 17mn + 20n^2$

15.4 Factoring Trinomials of the Form $ax^2 + bx + c$ Using the ac Method

THINGS TO KNOW

Before working through this topic, be sure you are familiar with the following concepts:

		VIDEO	ANIMATION	INTERACTIVE
You Try It	1. Identify Terms, Coefficients, and Like Terms of an Algebraic Expression (Topic 10.6, **Objective 1**)	✓		
You Try It	2. Determine the Degree and Leading Coefficient of a Polynomial (Topic 14.2, **Objective 3**)	✓		
You Try It	3. Multiply Two Binomials (Topic 14.4, **Objective 3**)	✓	✓	✓
You Try It	4. Factor Out the Greatest Common Factor from a Polynomial (Topic 15.1, **Objective 3**)	✓		
You Try It	5. Factor by Grouping (Topic 15.1, **Objective 4**)			✓
You Try It	6. Factor Trinomials of the Form $ax^2 + bx + c$ Using Trial and Error (Topic 15.3, **Objective 1**)	✓	✓	✓

OBJECTIVES

1 Factor Trinomials of the Form $ax^2 + bx + c$ Using the ac Method

2 Factor Trinomials of the Form $ax^2 + bxy + cy^2$ Using the ac Method

3 Factor Trinomials of the Form $ax^2 + bx + c$ After Factoring Out the GCF

OBJECTIVE 1 FACTOR TRINOMIALS OF THE FORM $ax^2 + bx + c$ USING THE ac METHOD

The **trial-and-error method** for factoring **trinomials** of the form $ax^2 + bx + c$ can be time consuming. In this topic, we learn a second method, known as the ***ac* method**, for factoring these trinomials. We continue to **factor over the integers.**

The *ac* Method for Factoring Trinomials of the Form $ax^2 + bx + c$

Step 1. Multiply $a \cdot c$.

Step 2. Find two **integers**, n_1 and n_2, whose **product** is ac and whose **sum** is b. So, $n_1 \cdot n_2 = ac$ and $n_1 + n_2 = b$. If no such pair of integers exists, the trinomial is **prime**.

Step 3. Rewrite the middle term as the sum of two terms using the integers found in Step 2. So, $ax^2 + bx + c = ax^2 + n_1x + n_2x + c$.

Step 4. **Factor by grouping.**

Step 5. Check your answer by multiplying out the **factored form**.

(eText Screens 15.4-1–15.4-12)

My animation video summary

Watch this **animation** to see how the *ac* method works.

The *ac* method is also known as the **grouping method** or the **expansion method**. It is favored by some over the trial-and-error method because its approach is more systematic, but it can still be time consuming. We recommend that you practice using both methods and then choose the one that you like best.

Example 1 Factoring Trinomials of the Form $ax^2 + bx + c$ Using the *ac* Method

Factor $3x^2 + 14x + 8$ using the ***ac* method**.

My video summary

Solution For $3x^2 + 14x + 8$, we have $a = 3$, $b = 14$, and $c = 8$.

Step 1. $a \cdot c = 3 \cdot 8 = 24$.

Step 2. We must find two **integers** whose **product** is $ac = 24$ and whose **sum** is $b = 14$. Because the product and sum are both positive, we consider only the **positive factors** of 24.

Positive Factors of 24	Sum of Factors
1, 24	$1 + 24 = 25$
2, 12	$2 + 12 = 14$ ← This is the pair.
3, 8	$3 + 8 = 11$
4, 6	$4 + 6 = 10$

From the list above, 2 and 12 are the integers we need.

Step 3. $3x^2 + 14x + 8 = 3x^2 + 2x + 12x + 8$

Step 4. Begin with the new polynomial from Step 3: $3x^2 + 2x + 12x + 8$

Group the first two terms and last two terms: $= (3x^2 + 2x) + (12x + 8)$

Factor out the GCF from each group: $= x(3x + 2) + 4(3x + 2)$

Factor out the common binomial factor: $= (3x + 2)(x + 4)$

Step 5. Check the answer by multiplying out $(3x + 2)(x + 4)$. View the **check**, or watch this **video** for a complete solution.

The answer checks, so $3x^2 + 14x + 8 = (3x + 2)(x + 4)$.

You Try It **Work through this You Try It problem.**

Work Exercises 1 and 2 in this textbook or in the MyMathLab Study Plan.

Example 2 Factoring Trinomials of the Form $ax^2 + bx + c$ Using the *ac* Method

My video summary

Factor $2x^2 - 3x - 20$ using the ***ac* method**.

Solution For $2x^2 - 3x - 20$, we have $a = 2, b = -3$, and $c = -20$.

Step 1. $a \cdot c = 2 \cdot (-20) = -40$.

Step 2. We must find two **integers** whose **product** is $ac = -40$ and whose **sum** is $b = -3$. Because the product is negative, one of the integers must be positive and the other must be negative. For the sum to be negative, the integer with the larger **absolute value** must be negative. We list such pairs of factors for -40.

Factors of -40	Sum of Factors	
$1, -40$	$1 + (-40) = -39$	
$2, -20$	$2 + (-20) = -18$	
$4, -10$	$4 + (-10) = -6$	
$5, -8$	$5 + (-8) = -3$	$\leftarrow$ This is the pair.

From the list above, 5 and -8 are the integers we need.

Step 3. $2x^2 - 3x - 20 = 2x^2 + 5x - 8x - 20$

Step 4. **Factor by grouping** to complete this problem on your own. Remember to check your answer by multiplying out your final result. Compare your **answer**, or watch this **video** for a complete solution.

You Try It **Work through this You Try It problem.**

Work Exercises 3–6 in this textbook or in the MyMathLab Study Plan.

Example 3 Factoring Trinomials of the Form $ax^2 + bx + c$ Using the ac Method

My interactive video summary

Factor each **trinomial** using the ac **method**. If the trinomial is **prime**, state this as your answer.

a. $2x^2 + 9x - 18$ **b.** $6x^2 - 23x + 20$ **c.** $5x^2 + x + 6$

Solutions Try factoring each of these trinomials on your own, and then compare your **answers**. Work through this **interactive video** for detailed solutions to all three parts.

You Try It **Work through this You Try It problem.**

Work Exercises 7–20 in this textbook or in the MyMathLab Study Plan.

OBJECTIVE 2 FACTOR TRINOMIALS OF THE FORM $ax^2 + bxy + cy^2$ USING THE ac METHOD

In **Topic 15.3**, we used **trial and error** to factor **trinomials** of the form $ax^2 + bxy + cy^2$. We can also use the ac **method** to factor such trinomials, but we must be careful to include the second variable throughout the process.

Example 4 Factoring a Trinomial of the Form $ax^2 + bxy + cy^2$

Factor $2p^2 + 7pq - 15q^2$ using the ac method.

Solution For $2p^2 + 7pq - 15q^2$, we have $a = 2, b = 7$, and $c = -15$.

Step 1. $a \cdot c = 2 \cdot (-15) = -30$.

Step 2. We must find two **integers** whose **product** is $ac = -30$ and whose **sum** is $b = 7$. The product is negative, so one integer must be positive and the other must be negative. The sum is positive, so the integer with the larger **absolute value** must be positive. We list such pairs of factors for -30.

(eText Screens 15.4-1–15.4-12)

My video summary

Factors of −30	Sum of Factors
$-1, 30$	$-1 + 30 = 29$
$-2, 15$	$-2 + 15 = 13$
$-3, 10$	$-3 + 10 = 7$ ← This is the pair.
$-5, 6$	$-5 + 6 = 1$

From the list above, -3 and 10 are the integers we need.

Step 3. $2p^2 + 7pq - 15q^2 = 2p^2 - 3pq + 10pq - 15q^2$

Step 4. **Factor by grouping** to complete this problem on your own. Remember to check your answer by multiplying out your final result. Compare your **answer**, or watch this **video** for a complete solution.

You Try It **Work through this You Try It problem.**

Work Exercises 21–24 in this textbook or in the MyMathLab Study Plan.

OBJECTIVE 3 FACTOR TRINOMIALS OF THE FORM $ax^2 + bx + c$ AFTER FACTORING OUT THE GCF

When a **trinomial** has a **common factor** other than 1, we factor out the **GCF** first.

Example 5 Factoring Trinomials with a Common Factor

My interactive video summary

Factor completely: $24t^5 - 52t^4 - 20t^3$

Solution The **terms** in this trinomial have a GCF of $4t^3$. We first factor out $4t^3$ and then factor the remaining trinomial.

Begin with the original trinomial:	$24t^5 - 52t^4 - 20t^3$
Write each term as a product of the GCF $4t^3$:	$= 4t^3 \cdot 6t^2 - 4t^3 \cdot 13t - 4t^3 \cdot 5$
Factor out the $4t^3$:	$= 4t^3(6t^2 - 13t - 5)$

Try to finish the problem on your own by factoring $6t^2 - 13t - 5$. You can use your preferred method, the ***ac* method** or **trial and error**. Compare your **answer**, or watch this **interactive video** for the complete solution.

You Try It **Work through this You Try It problem.**

Work Exercises 25 and 26 in this textbook or in the MyMathLab Study Plan.

Consider the **trinomial** $-2x^2 + 9x + 35$. Notice that the three **terms** do not have a **common factor**. However, when a trinomial has a negative number for a **leading coefficient**, we typically begin by factoring out a -1 as we did in **Example 6 of Topic 15.2**. We do this because it is much easier to factor a trinomial when the leading coefficient is positive.

Example 6 Factoring Trinomials with a Common Factor

My interactive video summary

Factor completely. $-2x^2 + 9x + 35$

Solution Because the leading coefficient is negative, we factor out a -1.

$$-2x^2 + 9x + 35 = -1(2x^2 - 9x - 35) = -(2x^2 - 9x - 35)$$

(eText Screens 15.4-1–15.4-12)

Finish the problem on your own by factoring $2x^2 - 9x - 35$. You can use either the ***ac* method** or **trial-and-error**. View the **answer**, or watch this **interactive video** for the complete solution.

You Try It **Work through this You Try It problem.**

Work Exercises 27–30 in this textbook or in the MyMathLab Study Plan.

15.4 Exercises

In Exercises 1–30, factor each trinomial completely, or state that the trinomial is prime.

You Try It

1. $5x^2 + 17x + 6$

2. $2y^2 + 27y + 13$

You Try It

3. $4t^2 + 7t - 15$

4. $7m^2 - 8m - 12$

5. $3p^2 - 26p + 16$

6. $12n^2 + 28n - 5$

You Try It

7. $2w^2 - 17w + 33$

8. $4x^2 + 3x + 2$

9. $10y^2 + 47y - 15$

10. $16a^2 + 32a - 9$

11. $18q^2 - 13q + 2$

12. $2m^2 - m - 21$

13. $20x^2 + 29x + 5$

14. $15n^2 - 38n + 24$

15. $8t^2 + 6t - 5$

16. $5y^2 + 11y - 12$

17. $6z^2 - 19z - 20$

18. $21t^2 + 8t - 5$

19. $8n^2 - 51n - 56$

20. $22x^2 + 81x + 14$

You Try It

21. $4x^2 + 11xy + 6y^2$

22. $3m^2 - 23mn + 40n^2$

23. $14a^2 + 15ab - 9b^2$

24. $3p^2 - 11pq - 20q^2$

You Try It

25. $10x^4 + 44x^3 - 30x^2$

26. $40y^2 - 285y + 35$

27. $-3m^2 - 10m + 8$

28. $-20n^2 - 26n - 8$

You Try It

29. $-36a^2b + 147ab - 12ab^2$

30. $24x^2 - 4xy - 60y^2$

15.5 Factoring Special Forms

THINGS TO KNOW

Before working through this topic, be sure you are familiar with the following concepts:

		VIDEO	ANIMATION	INTERACTIVE
You Try It	1. Square a Binomial Sum (Topic 14.5, **Objective 1**)	■		
You Try It	2. Square a Binomial Difference (Topic 14.5, **Objective 2**)	■		
You Try It	3. Multiply the Sum and Difference of Two Terms (Topic 14.5, **Objective 3**)	■		
You Try It	4. Factor Out the Greatest Common Factor from a Polynomial (Topic 15.1, **Objective 3**)	■		

OBJECTIVES

1 Factor the Difference of Two Squares

2 Factor Perfect Square Trinomials

3 Factor the Sum or Difference of Two Cubes

OBJECTIVE 1 FACTOR THE DIFFERENCE OF TWO SQUARES

In this topic, we focus on the *form*. For example, the **standard form** for a **linear equation in two variables** is $Ax + By = C$, where A and B are not both zero. The equations $y = -3x$ and $4 + 5y = 8x$ may look different and have different **coefficients**, but they are both linear equations in two variables because they can be written in the same standard form, $Ax + By = C$.

Equation	Standard Form	Coefficients
$y = -3x$	$3x + y = 0$	$A = 3, B = 1, C = 0$
$4 + 5y = 8x$	$8x - 5y = 4$	$A = 8, B = -5, C = 4$

Focusing on *form* is particularly helpful with **factoring**. In **Topic 14.5**, we used **special product rules** to multiply two **binomials.** In that topic, recognizing how two binomial factors fit a special form helped us to quickly find the product. Working in reverse, we can use the results of special binomial products to help us factor certain expressions.

Recall from **Topic 14.5** that the product of a binomial and its **conjugate** (the sum and difference of two terms) is the difference of the **squares** of the **terms** in the binomial. So, $(A + B)(A - B) = A^2 - B^2$. We can use this result to form a rule for factoring the **difference of two squares.**

Factoring the Difference of Two Squares

If A and B are **real numbers, variables,** or **algebraic expressions,** then the difference of their squares can be factored into the product of the sum and difference of the two quantities.

$$A^2 - B^2 = (A + B)(A - B)$$

Note: Because multiplication is **commutative,** we could also write $A^2 - B^2 = (A - B)(A + B)$.

CAUTION Remember to focus on *form*. The quantities A and B could be numbers, variables, or algebraic expressions.

Example 1 Factoring the Difference of Two Squares

Factor each expression completely.

a. $x^2 - 9$ **b.** $16 - y^2$

Solutions

a. Start by rewriting the expression as the difference of two squares.

$$x^2 - 9 = x^2 - (3)^2$$

Here, we have $A = x$ (a **variable**) and $B = 3$ (a **real number**). Applying the **difference of two squares rule,** we get

$$x^2 - 9 = \underbrace{x^2 - 3^2}_{A^2 - B^2} = \underbrace{(x + 3)}_{(A + B)}\underbrace{(x - 3)}_{(A - B)}.$$

b. Start by rewriting the expression as the **difference of two squares.**

$$16 - y^2 = (4)^2 - y^2$$

Here we have $A = 4$ (a real number) and $B = y$ (a variable). Applying the difference of two squares rule, we get

$$16 - y^2 = \underbrace{(4)^2 - y^2}_{A^2 - B^2} = \underbrace{(4 + y)}_{(A + B)}\underbrace{(4 - y)}_{(A - B)}.$$

You Try It **Work through this You Try It problem.**

Work Exercises 1–5 in this textbook or in the MyMathLab Study Plan.

To use the **difference of two squares rule,** both terms must be **perfect squares.** An integer is a perfect square if it is the square of another integer. View this **popup** to review some **perfect square** integers. A rational number is a perfect square if it can be written as the square of another rational number. For example, $\frac{4}{9}$ is a perfect square because we can write $\frac{4}{9} = \left(\frac{2}{3}\right)^2$.

A **monomial** with a perfect square **coefficient** and variables raised to even **powers** is also a perfect square. For example, $9x^2$ and $4x^2y^6$ are perfect squares.

An **algebraic expression** is a perfect square if it is raised to an even power. For example, $(5n)^2$ is a perfect square.

Example 2 Factoring the Difference of Two Squares

Factor each expression completely.

a. $z^2 - \dfrac{25}{16}$ **b.** $36x^2 - 25$ **c.** $4 - 49n^6$ **d.** $81m^2 - n^2$

Solutions

a. Start by rewriting the expression as the **difference of two squares.**

$$z^2 - \frac{25}{16} = z^2 - \left(\frac{5}{4}\right)^2$$

Here, we have $A = z$ (a variable) and $B = \dfrac{5}{4}$ (a real number). Applying the **difference of two squares rule,** we get

$$z^2 - \frac{25}{16} = \underbrace{z^2 - \left(\frac{5}{4}\right)^2}_{A^2 - B^2} = \underbrace{\left(z + \frac{5}{4}\right)}_{(A + B)}\underbrace{\left(z - \frac{5}{4}\right)}_{(A - B)}.$$

b. Start by rewriting the expression as the difference of two squares.

$$36x^2 - 25 = 6^2x^2 - 5^2 = (6x)^2 - 5^2$$

Here we have $A = 6x$ (an **algebraic expression**) and $B = 5$ (a real number). Apply the difference of two squares rule, then view the **answer** to see if you are correct.

My video summary **c.–d.** Try factoring these two expressions on your own using the **difference of two squares rule.** View the **answers**, or watch this **video** for detailed solutions to parts c and d.

You Try It **Work through this You Try It problem.**

Work Exercises 6–10 in this textbook or in the MyMathLab Study Plan.

Example 3 Factoring the Difference of Two Squares with a Greatest Common Factor

Factor each expression completely.

a. $3x^2 - 75$ **b.** $36x^3 - 64x$

Solutions

a. Start by factoring out the **greatest common factor**, 3.

$$\begin{aligned} 3x^2 - 75 &= 3 \cdot x^2 - 3 \cdot 25 \\ &= 3(x^2 - 25) \end{aligned}$$

Next, rewrite the expression in parentheses as the **difference of two squares.**

$$3(x^2 - 25) = 3(x^2 - 5^2)$$

Here, we have $A = x$ and $B = 5$. Applying the **difference of two squares rule,** we get

$$3\overbrace{(x^2 - 5^2)}^{A^2 - B^2} = 3\underbrace{(x + 5)}_{(A + B)}\underbrace{(x - 5)}_{(A - B)}.$$

So, $3x^2 - 75 = 3(x + 5)(x - 5)$.

(eText Screens 15.5-1–15.5-19)

My video summary

b. Try factoring this expression on your own. Begin by looking for a **greatest common factor**, then use the difference of two squares rule. View the **answer**, or watch this **video** for a detailed solution to part b.

You Try It **Work through this You Try It problem.**

Work Exercises 11–16 in this textbook or in the MyMathLab Study Plan.

CAUTION The *sum* of two **perfect squares** cannot be factored using real numbers other than to factor out the greatest common factor.

When factoring, we want to make sure we always factor *completely*. This means we should always check each **factor** to see if it can be factored further.

Example 4 Factoring the Difference of Two Squares More than Once

My video summary

Factor completely.

$16x^4 - 81$

Solution Work through the following, or watch this **video** solution.
Start by rewriting the expression as the **difference of two squares.**

$$16x^4 - 81 = (4x^2)^2 - (9)^2$$

Here, we have $A = 4x^2$ and $B = 9$. Applying the **difference of two squares rule**, we get

$$16x^4 - 81 = \underbrace{(4x^2)^2 - (9)^2}_{A^2 - B^2} = \underbrace{(4x^2 + 9)}_{(A + B)}\underbrace{(4x^2 - 9)}_{(A - B)}.$$

Notice that the factor $(4x^2 - 9)$ can be written as $(2x)^2 - 3^2$, so it too is the **difference of two squares** and can be factored further. The factor $(4x^2 + 9)$ is the sum of two squares, so it cannot be factored further.

$$4x^2 - 9 = \underbrace{(2x)^2 - (3)^2}_{A^2 - B^2} = \underbrace{(2x + 3)}_{(A + B)}\underbrace{(2x - 3)}_{(A - B)}$$

We write

$$16x^4 - 81 = (4x^2 + 9)\overbrace{(2x + 3)(2x - 3)}^{4x^2 - 9}.$$

This polynomial is now factored completely.

You Try It **Work through this You Try It problem.**

Work Exercises 17–22 in this textbook or in the MyMathLab Study Plan.

OBJECTIVE 2 FACTOR PERFECT SQUARE TRINOMIALS

In **Topic 14.5**, we had two **special product rules** for squaring the sum or difference of two terms. The results of these products were called **perfect square trinomials.** We can use these special product rules to factor perfect square trinomials by reversing the process.

Factoring Perfect Square Trinomials

If A and B are **real numbers, variables,** or **algebraic expressions,** then

$$A^2 - 2AB + B^2 = (A - B)(A - B) = (A - B)^2 \quad \text{and}$$
$$A^2 + 2AB + B^2 = (A + B)(A + B) = (A + B)^2.$$

Again, we focus on the form of the **trinomial.** To be a perfect square trinomial, the first and last **terms** must be **perfect squares,** and the middle term must be twice the product of the two quantities being squared or the opposite of the product. To **factor** a perfect square trinomial, we first identify the quantities being squared in the first and last terms and then apply the appropriate rule.

Example 5 Factoring Perfect Square Trinomials

Factor each expression completely.

a. $x^2 + 6x + 9$ **b.** $y^2 - 10y + 25$

Solutions

a. The first **term** is a **perfect square,** $x^2 = (x)^2$, and the last term is also a perfect square, $9 = 3^2$. Because $6x = 2(x)(3)$, we have a **perfect square trinomial** with a positive middle term.

$$x^2 + 6x + 9 = \underbrace{(x)^2 + 2(x)(3) + 3^2}_{A^2 + 2\cdot A\cdot B + B^2} = \underbrace{(x + 3)^2}_{(A + B)^2}$$

b. The first term is a perfect square, $y^2 = (y)^2$, and the last term is also a perfect square, $25 = 5^2$. Because $10y = 2(y)(5)$, we have a perfect square trinomial with a negative middle term.

$$y^2 - 10y + 25 = \underbrace{(y)^2 - 2(y)(5) + 5^2}_{A^2 - 2\cdot A\cdot B + B^2} = \underbrace{(y - 5)^2}_{(A - B)^2}$$

You Try It **Work through this You Try It problem.**

Work Exercises 23 and 24 in this textbook or in the MyMathLab Study Plan.

Example 6 Factoring Perfect Square Trinomials

Factor each expression completely.

a. $4x^2 + 12x + 9$ **b.** $25y^2 - 60y + 36$

Solutions

a. We write $4x^2 = 2^2x^2 = (2x)^2$ and $9 = 3^2$. Because $12x = 2(2x)(3)$, we have a **perfect square trinomial** with a positive middle term.

$$4x^2 + 12x + 9 = \underbrace{(2x)^2 + 2(2x)(3) + 3^2}_{A^2 + 2\cdot A\cdot B + B^2} = \underbrace{(2x + 3)^2}_{(A + B)^2}$$

My video summary

b. Try to do this factoring on your own. When finished, view the **answer,** or watch this **video** for a detailed solution to part b.

(eText Screens 15.5-1–15.5-19)

You Try It **Work through this You Try It problem.**

Work Exercises 25–27 in this textbook or in the MyMathLab Study Plan.

Perfect square trinomials may contain several variables as shown in Example 7a.

Example 7 Factoring Perfect Square Trinomials

My interactive video summary

Factor each expression completely.

a. $16x^2 + 24xy + 9y^2$ **b.** $m^4 - 12m^2 + 36$

Solutions

a. Write the first and last **terms** as **perfect squares** to determine A and B. Now complete the factorization on your own. View the **answer**, or watch this **interactive video** for a detailed solution.

b. Using **rules for exponents**, we can write $m^4 = (m^2)^2$. Now complete the factorization on your own. View the **answer**, or watch this **interactive video** for a detailed solution.

You Try It **Work through this You Try It problem.**

Work Exercises 28–34 in this textbook or in the MyMathLab Study Plan.

OBJECTIVE 3 FACTOR THE SUM OR DIFFERENCE OF TWO CUBES

Like the difference of two **squares**, the difference of two **cubes** is the result of a special product rule. Consider the following example:

$$\underbrace{(A - B)}\underbrace{(A^2 + AB + B^2)}$$

$$(x - 3)(x^2 + 3x + 9)$$

Distribute: $= x \cdot x^2 + x \cdot 3x + x \cdot 9 - 3 \cdot x^2 - 3 \cdot 3x - 3 \cdot 9$

Simplify: $= x^3 + 3x^2 + 9x - 3x^2 - 9x - 27$

Collect like terms: $= x^3 + 3x^2 - 3x^2 + 9x - 9x - 27$

Combine like terms: $= x^3 - 27$

$= \underbrace{x^3}_{A^3} - \underbrace{3^3}_{B^3}$ ← Difference of two cubes

However, unlike the sum of two squares, the sum of two cubes *can* be factored. These two new factor rules are given next.

Factoring the Sum and Difference of Two Cubes

If A and B are **real numbers, variables,** or **algebraic expressions,** then the sum or difference of their cubes can be factored as follows:

$$A^3 + B^3 = (A + B)(A^2 - AB + B^2)$$
$$A^3 - B^3 = (A - B)(A^2 + AB + B^2)$$

We can check these rules using two special products. View this **popup** to see the steps.

In order to use these rules, both terms in the expression must be **perfect cubes.** An integer is a perfect cube if it is the cube of another **integer.** View this **popup** to review some **perfect cube** integers. A **monomial** with a perfect cube coefficient and variables raised to powers that are multiples of 3 is also considered a perfect cube (Example: $27x^3$). An algebraic expression is a perfect cube if it is raised to a power that is a multiple of 3 (Example: $(2m - 1)^3$).

Notice that the form of the results can help us remember how to factor the sum or difference of two cubes. See the **details.**

To factor the sum or difference of two cubes, first we identify the quantities being cubed and then apply the appropriate rule.

Example 8 Factoring the Sum or Difference of Two Cubes

Factor each expression completely.

a. $x^3 + 64$ **b.** $z^3 - 8$

Solutions

a. The first term is a **perfect cube,** $x^3 = (x)^3$, and the second term is also a perfect cube, $64 = 4^3$. We have the **sum of two cubes,** so we apply the rule for $A^3 + B^3$.

$$A^3 + B^3 = (A + B)(A^2 - AB + B^2)$$
$$x^3 + 64 = (x)^3 + (4)^3 = (x + 4)(x^2 - 4x + 4^2)$$
$$= (x + 4)(x^2 - 4x + 16)$$

b. The first term is a perfect cube, $z^3 = (z)^3$, and the second term is also a perfect cube, $8 = 2^3$. We have the **difference of two cubes,** so we apply the rule for $A^3 - B^3$.

$$A^3 - B^3 = (A - B)(A^2 + AB + B^2)$$
$$z^3 - 8 = (z)^3 - (2)^3 = (z - 2)(z^2 + 2z + 2^2)$$
$$= (z - 2)(z^2 + 2z + 4)$$

You Try It **Work through this You Try It problem.**

Work Exercises 35 and 36 in this textbook or in the MyMathLab Study Plan.

Example 9 Factoring the Sum or Difference of Two Cubes

Factor each expression completely.

a. $125y^3 - 1$ **b.** $128z^3 + 54y^3$ **c.** $8x^3y^2 + y^5$

Solutions

a. Because $125y^3 = (5y)^3$ and $1 = 1^3$, we have the **difference of two cubes.** Apply the rule for $A^3 - B^3$.

$$A^3 - B^3 = (A - B)(A^2 + AB + B^2)$$
$$125y^3 - 1 = (5y)^3 - (1)^3 = (5y - 1)((5y)^2 + (5y)(1) + (1)^2)$$
$$= (5y - 1)(25y^2 + 5y + 1)$$

b. First, factor out the **GCF,** 2. Write the remaining two **terms** as **perfect cubes** to determine the quantities $A = 4z$ and $B = 3y$. Try to complete the factorization on your own. When finished, view the rest of the **solution.**

My video summary **c.** First, factor out the GCF. Using **rules for exponents**, we write $y^5 = y^2 \cdot y^3$. Now complete this factorization on your own. Check your **answer**, or watch this **video** for a detailed solution.

You Try It **Work through this You Try It problem.**

Work Exercises 37–43 in this textbook or in the MyMathLab Study Plan.

15.5 Exercises

In Exercises 1–43, factor each expression completely. If a polynomial is prime, then state this as your answer.

You Try It

1. $y^2 - 16$ **2.** $49 - z^2$ **3.** $x^2 - y^2$ **4.** $-p^2 + q^2$

You Try It

5. $x^{10} - 4$ **6.** $x^2 - \frac{49}{9}$ **7.** $9y^2 - 1$ **8.** $x^2y^2 - 25$

9. $16 - 25m^2$ **10.** $4x^2y^4 - \frac{9}{4}z^8$ **11.** $12x^2 - 27$ **12.** $50 - 8y^2$

You Try It

13. $20x^2 + 45$ **14.** $8a^2 - 18b^2$ **15.** $3x^2y - 12y$ **16.** $a^3b - ab^3$

You Try It

17. $x^4 - 1$ **18.** $81x^4 - 16$ **19.** $x^4 - y^4$ **20.** $m^8 - 16n^4$

You Try It

21. $5x^8 - 5$ **22.** $y^{16} - z^{16}$ **23.** $x^2 - 2x + 1$ **24.** $m^2 + 16m + 64$

You Try It

25. $4x^2 + 28x + 49$ **26.** $9h^2 - 30h + 25$ **27.** $\frac{1}{4}x^2 + x + 4$ **28.** $x^2 + 2xy + y^2$

You Try It

29. $m^2 - 14mn + 49n^2$ **30.** $4a^2 - 20ab + 25b^2$ **31.** $25xy^2 + 70xy + 49x$ **32.** $y^4 - 2y^2 + 1$

You Try It

33. $z^6 + 10z^3 + 25$ **34.** $16a^4 - 72a^2b^2 + 81b^4$ **35.** $y^3 - 125$ **36.** $z^3 + 27$

You Try It

37. $8x^3 - 1$ **38.** $5t^3 + 40$ **39.** $64p^3 + 121q^3$ **40.** $3z^5 - 24z^2$

41. $x^3y^3 - 27$ **42.** $a^3b^3 + 8c^3$ **43.** $x^9 - 1$

15.6 A General Factoring Strategy

THINGS TO KNOW

Before working through this topic, be sure you are familiar with the following concepts:

VIDEO ANIMATION INTERACTIVE

You Try It 1. Factor Out the Greatest Common Factor from a Polynomial (Topic 15.1, **Objective 3**)

You Try It 2. Factor by Grouping (Topic 15.1, **Objective 4**)

You Try It 3. Factor Trinomials of the Form $x^2 + bx + c$ (Topic 15.2, **Objective 1**)

You Try It 4. Factor Trinomials of the Form $ax^2 + bx + c$ Using Trial and Error (Topic 15.3, **Objective 1**)

You Try It 5. Factor Trinomials of the Form $ax^2 + bx + c$ Using the ac Method (Topic 15.4, **Objective 1**)

OBJECTIVE

1 Factor Polynomials Completely

OBJECTIVE 1 FACTOR POLYNOMIALS COMPLETELY

Now we know several techniques and rules for factoring **polynomials**. We can use this knowledge to create a general strategy for **factoring polynomials completely**. A polynomial is factored completely if all its polynomial **factors**, other than **monomials**, are **prime**.

General Strategy for Factoring Polynomials Completely

Step 1. If necessary, factor out the **greatest common factor**. If the **leading coefficient** is negative, factor out a **common factor** with a negative coefficient.

Step 2. Select a strategy based on the number of **terms**.

a. If there are two terms, try to use one of the following special factor rules:

Difference of two squares: $A^2 - B^2 = (A + B)(A - B)$

Sum of two cubes: $A^3 + B^3 = (A + B)(A^2 - AB + B^2)$

Difference of two cubes: $A^3 - B^3 = (A - B)(A^2 + AB + B^2)$

b. If there are three terms, see if the **trinomial** is a **perfect square trinomial**. If so, factor using one of these special factor rules:

$$A^2 + 2AB + B^2 = (A + B)^2$$
$$A^2 - 2AB + B^2 = (A - B)^2$$

If the trinomial is not a perfect square, factor using **trial and error** or the ***ac* method**.

c. If there are four or more terms, try **factoring by grouping**.

continued

Step 3.	Check to see if any factors can be factored further. Check each factor, other than **monomial factors**, to make sure they are **prime**.
Step 4.	Check your answer by multiplying. Multiply out your result to see if it equals the original expression.

Example 1 Factoring a Polynomial Completely

My interactive video summary

Factor each expression completely.

a. $w^2 - w - 20$ **b.** $4y^4 - 32y$ **c.** $x^2 - 14x + 49$ **d.** $3z^3 - 15z^2 - 42z$

Solutions

a. Follow the **general factoring strategy**.

Step 1. Looking at each **term**, we see there is no common factor.

Step 2. The expression has three terms. The first and last terms are not perfect squares, so this is not one of our special forms.

Because the **leading coefficient** of the trinomial factor is 1, we need two **factors** whose product is the **constant**, -20, and whose sum is the middle **coefficient** -1. Because the constant is negative, the two factors have opposite signs. And because the middle term is negative, the factor with the larger absolute value must be negative.

Factor 1	Factor 2	Sum	
1	-20	-19	
2	-10	-8	
4	-5	-1	← This is what we need.

The required factors are 4 and -5. We can factor the trinomial as $w^2 - w - 20 = (w + 4)(w - 5)$.

Step 3. All the factors are **prime**, so our factorization is complete.

Step 4. Check by multiplying.

$$\begin{aligned}(w + 4)(w - 5) &= w \cdot w + w(-5) + 4 \cdot w + 4(-5)\\ &= w^2 - 5w + 4w - 20\\ &= w^2 - w - 20 \checkmark\end{aligned}$$

b. Follow the **general factoring strategy**.

Step 1. Looking at each **term**, we can factor out a **common factor** of $4y$.

$$4y^4 - 32y = 4y \cdot y^3 - 4y \cdot 8 = 4y(y^3 - 8)$$

Step 2. The expression in parentheses, $y^3 - 8$, has two terms that are both cubes $(8 = 2^3)$, so this is one of our special forms. We factor this expression using the **difference of two cubes** rule with $A = y$ and $B = 2$.

$$\begin{aligned}y^3 - 8 &= y^3 - 2^3\\ &= (y - 2)(y^2 + 2 \cdot y + 2^2)\\ &= (y - 2)(y^2 + 2y + 4)\end{aligned}$$

Thus, we have $4y^4 - 32y = 4y(y - 2)(y^2 + 2y + 4)$.

Step 3. All the factors, other than the **monomial factor**, are **prime**, so our factorization is complete.

Step 4. Check by multiplying.

$$\begin{aligned} 4y(y-2)(y^2+2y+4) &= (4y\cdot y - 4y\cdot 2)(y^2+2y+4) \\ &= (4y^2-8y)(y^2+2y+4) \\ \text{Distribute}\rightarrow &= 4y^2\cdot y^2 + 4y^2\cdot 2y + 4y^2\cdot 4 - 8y\cdot y^2 - 8y\cdot 2y - 8y\cdot 4 \\ \text{Simplify}\rightarrow &= 4y^4 + 8y^3 + 16y^2 - 8y^3 - 16y^2 - 32y \\ \text{Combine like terms}\rightarrow &= 4y^4 - 32y \ \checkmark \end{aligned}$$

c.–d. Follow the **general factoring strategy** to factor these trinomials on your own. Once finished, view the **answers**. If you need help, watch this **interactive video** for complete solutions to all four parts.

You Try It **Work through this You Try It problem.**

Work Exercises 1–8 in this textbook or in the MyMathLab Study Plan.

Example 2 Factoring a Polynomial Completely

My interactive video summary

Factor each expression completely.

a. $2x^3 - 5x^2 - 8x + 20$ **b.** $3a^2 - 10a - 8$ **c.** $3z^2 + z - 1$

Solutions

a. Use the **general factoring strategy**. Other than 1 or -1, there is no **common factor**. There are four **terms**, so we consider **factoring by grouping**. We try grouping the first two terms and the last two terms.

$$\underbrace{2x^3 - 5x^2}\ \underbrace{-\ 8x + 20}$$

From the first two terms, we can factor out x^2. From the last two terms, we can factor out -4.

$$\begin{aligned} 2x^3 - 5x^2 - 8x + 20 &= (2x^3 - 5x^2) + (-8x + 20) \\ &= x^2(2x-5) - 4(2x-5) \end{aligned}$$

Try to complete this factorization on your own. View the **answer**, or watch this **interactive video** for the full solution.

b.–c. Use the **general factoring strategy**. Other than 1 or -1, there is no **common factor**. There are three **terms** in both cases, but neither is a **perfect square trinomial**. Try factoring on your own using **trial and error** or the ***ac* method**. View the **answers**, or watch this **interactive video** for complete solutions to all three parts.

You Try It **Work through this You Try It problem.**

Work Exercises 9–16 in this textbook or in the MyMathLab Study Plan.

Example 3 Factoring a Polynomial Completely

Factor each expression completely.

a. $10x^2 + 11xy - 6y^2$ **b.** $2p^2 - 32pq + 128q^2$

c. $7x^2z - 14x$ **d.** $-3y^4z - 24yz^4$

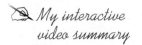

My interactive video summary

Solutions

a. Other than 1 or -1, there is no common factor. We then try to factor the expression using trial and error, or the ac method. We will use the ac method.

$$ac = (10)(-6) = -60$$

We need to find two **integers** whose **product** is -60 and whose **sum** is 11. For the product to be negative, one of the integers must be positive and the other must be negative. For the sum to be positive, the integer with the larger **absolute value** must be positive. We list such pairs of **factors** for -60:

Factor 1	Factor 2	Sum	
-1	60	59	
-2	30	28	
-3	20	17	
-4	15	11	← This is what we need
-5	12	7	
-6	10	4	

Because -4 and 15 multiply to -60 and add to 11, the integers we need are -4 and 15. We use these values to split up the middle term.

$$10x^2 + 11xy - 6y^2 = 10x^2 - 4xy + 15xy - 6y^2$$

We now **factor by grouping.**

Original expression:	$10x^2 + 11xy - 6y^2$
Split up middle term:	$= 10x^2 - 4xy + 15xy - 6y^2$
Group first two terms and last two:	$= (10x^2 - 4xy) + (15xy - 6y^2)$
Factor $2x$ from first two and $3y$ from last two:	$= 2x(5x - 2y) + 3y(5x - 2y)$
Factor out $(5x - 2y)$:	$= (5x - 2y)(2x + 3y)$

Check the answer by multiplying. View the **check.**

b.–d. Try factoring on your own using the **general factoring strategy.** View the **answers,** or watch this **interactive video** for complete solutions to all four parts.

You Try It **Work through this You Try It problem.**

Work Exercises 17–24 in this textbook or in the MyMathLab Study Plan.

15.6 Exercises

In Exercises 1–30, factor completely. If a polynomial is prime, then state this as your answer.

You Try It

1. $3x^2 - 12$

2. $2y^3 - 10y^2 - 28y$

3. $a^3 + 8b^3$

4. $y^2 + 3y + 3$

5. $x^2 + x - 6$

6. $w^2 + 12w + 20$

7. $5z^2 - 30z + 45$

8. $-4m^4 + 4m$

(eText Screens 15.7-1–15.7-17)

You Try It **9.** $4x^2 - 20x + 25$

10. $2x^3 + 5x^2 - 8x - 20$

11. $5x^3 + 14x^2 - 3x$

12. $6x^2 - x - 7$

13. $3w^2 - 5w + 4$

14. $9t^2 - 15t - 36$

15. $9x^2 + 12x + 4$

16. $3x^4 + 3x^3 - x^2 - x$

17. $-6p^2 + 9pq + 6q^2$

18. $5y^2z^3 - 20z$

You Try It **19.** $9y + 18y^2$

20. $6xz^2 - 14xz - 40x$

21. $25m^2 + 40mn + 16n^2$

22. $6x^2 - 5xy - 6y^2$

23. $b^3 + 2ab^2 - a^2b - 2a^3$

24. $-6b^2 + 3ab + a - 2b$

25. $x^2 - x - 20$

26. $1 - a^4$

27. $48y^3 + 6y^2 - 9y$

28. $r^3(t^3 - 8) + 8(t^3 - 8)$

29. $p^2 - 10p + 9$

30. $y^5 - y^3 + y^2 - 1$

15.7 Solving Polynomial Equations by Factoring

THINGS TO KNOW

Before working through this topic, be sure that you are familiar with the following concepts:

	VIDEO	ANIMATION	INTERACTIVE
You Try It 1. Solve Linear Equations Using Both Properties of Equality (Topic 11.1, **Objective 5**)	✓		
You Try It 2. Factor Polynomials Completely (Topic 15.6, **Objective 1**)			✓

OBJECTIVES

1 Solve Quadratic Equations by Factoring

2 Solve Polynomial Equations by Factoring

OBJECTIVE 1 SOLVE QUADRATIC EQUATIONS BY FACTORING

A **polynomial equation** results when we set two **polynomials** equal to each other. Some examples of polynomial equations are

$$2x - 7 = 4, \quad 3x^2 + 5x = x - 2, \quad \text{and} \quad 2x^3 + 7 = 3x^2 - x.$$

A polynomial equation is in **standard form** if one side equals zero and the other side is a simplified polynomial written in **descending order**.

The **standard forms** of the above polynomial equations are

$$2x - 11 = 0, \quad 3x^2 + 4x + 2 = 0, \quad \text{and} \quad 2x^3 - 3x^2 + x + 7 = 0.$$

The **degree of a polynomial equation** in standard form is the same as the highest **degree** of any of its terms. Notice that a polynomial of degree 1, such as $2x - 7 = 4$, is a **linear equation**. We learned how to solve these types of equations in **Topic 11.1** and **Topic 11.2**.

We now look at how to solve polynomial equations of degree 2. These types of equations are called **quadratic equations**. A quadratic equation in one variable is any equation that can be written in the standard form, $ax^2 + bx + c = 0$.

Definition Quadratic Equation

A **quadratic equation** in standard form is written as

$$ax^2 + bx + c = 0,$$

where a, b, and c are real numbers and $a \neq 0$.

Some examples of quadratic equations are

$$5x^2 - 7x + 4 = 0, \quad 3x^2 + 5x = x - 2, \quad 2x^2 + 4 = 9, \quad \text{and} \quad 7x^2 = x.$$

The first equation above is written in standard form. The **standard forms** of the remaining equations are

$$3x^2 + 4x + 2 = 0, \quad 2x^2 - 5 = 0, \quad \text{and} \quad 7x^2 - x = 0.$$

To solve quadratic equations, we can use the **factoring techniques** discussed in this module together with the **zero product property**.

Zero Product Property

If A and B are **real numbers** or **algebraic expressions** and $A \cdot B = 0$, then $A = 0$ or $B = 0$.

Recall that **factors** are quantities that are multiplied together to form a **product**. The **zero product property** tells us that if the product of two factors equals zero, then at least one of the factors must equal zero. This property extends to any number of factors.

CAUTION Note that the zero product property only works when the product equals zero. For instance, if we have $A \cdot B = 18$, it is *not* true that one or both of the factors must be equal to 18. In fact, we can multiply $2 \cdot 9$ to get 18 (or even $3 \cdot 6$), and neither factor is equal to 18. View this **popup** to see another example.

Example 1 Solving a Quadratic Equation by Factoring

Solve each equation.

a. $(x + 10)(x - 3) = 0$ **b.** $x(3x + 5) = 0$

Solutions

a. We have a product of two **factors** that is equal to zero. Using the **zero product property**, we set each factor equal to zero and solve the resulting equations.

$$(x + 10)(x - 3) = 0$$
$$x + 10 = 0 \quad \text{or} \quad x - 3 = 0$$
$$x = -10 \qquad\qquad x = 3$$

To check that these values are **solutions**, substitute them back into the original equation to see if a true statement results.

Check $x = -10$:

$$(x + 10)(x - 3) = 0$$
$$(-10 + 10)(-10 - 3) \stackrel{?}{=} 0$$
$$(0)(-13) \stackrel{?}{=} 0$$
$$0 = 0 \quad \text{True}$$

Check $x = 3$:

$$(x + 10)(x - 3) = 0$$
$$(3 + 10)(3 - 3) \stackrel{?}{=} 0$$
$$(13)(0) \stackrel{?}{=} 0$$
$$0 = 0 \quad \text{True}$$

Both values check, so the **solution set** is $\{-10, 3\}$.

My video summary

b. Set each factor equal to zero and solve the resulting equations. View the **answer**, or watch this **video** for the complete solution.

You Try It **Work through this You Try It problem.**

Work Exercises 1–4 in this textbook or in the MyMathLab Study Plan.

In Example 1, we solved equations in which the quadratic expression was given in factored form. This will not always be the case. In general, we can use the following steps to solve a quadratic equation, or other polynomial equations, by factoring.

Solving Polynomial Equations by Factoring

Step 1. Write the equation in **standard form** so that one side is zero and the other side is a simplified polynomial written in **descending order**.

Step 2. Factor the polynomial **completely**.

Step 3. Set each **distinct factor** with a variable equal to zero, and solve the resulting equations.

Step 4. Check each **solution** in the original equation.

Example 2 Solving Quadratic Equations by Factoring

Solve each equation by factoring.

a. $z^2 + 4z - 12 = 0$ **b.** $-4x^2 + 28x - 40 = 0$

Solutions We follow the **four-step process** for solving polynomial equations by factoring.

a. Step 1. The equation is already in **standard form**, so we move on to Step 2.

Step 2. Using the **general strategy for factoring**, we need two numbers whose product is the constant -12 and whose sum is the middle coefficient 4. Because the product is negative, the two numbers have opposite signs. Because the sum is positive, the number with the larger **absolute value** is positive.

Factor 1	Factor 2	Sum	
-1	12	11	
-2	6	4	← This is the desired sum
-3	4	1	

The required numbers are -2 and 6, which can be used to factor the polynomial.

$$z^2 + 4z - 12 = 0$$
$$(z - 2)(z + 6) = 0$$

Step 3. Set each factor equal to zero and solve the equations.

$$z - 2 = 0 \quad \text{or} \quad z + 6 = 0$$
$$z = 2 \qquad\qquad z = -6$$

Step 4. Check these values on your own to confirm that the **solution set** is $\{-6, 2\}$.

My video summary

b. The equation is in standard form. We factor the expression on the left by first factoring out the **greatest common factor**, -4.

$$-4x^2 + 28x - 40 = 0$$
$$-4(x^2 - 7x + 10) = 0$$

Now **factor** $x^2 - 7x + 10$. Set each variable factor equal to 0 and solve the resulting equations to form the solution set. View the **answer**, or watch this **video** for the complete solution.

You Try It **Work through this You Try It problem.**

Work Exercises 5–10 in this textbook or in the MyMathLab Study Plan.

CAUTION

Note in Example 2b that the factor -4 is a constant and will never equal 0. When solving equations by factoring, we set only factors that contain a **variable** equal to 0.

Example 3 Solving Quadratic Equations by Factoring

Solve each equation by factoring.

a. $9w^2 + 64 = 48w$ **b.** $4m^2 = 49$ **c.** $3x(x - 2) = 2 - x$

Solutions We follow the **four-step process** for solving polynomial equations by factoring.

a. Step 1. Write the equation in **standard form** by subtracting $48w$ from both sides.

Original equation: $9w^2 + 64 = 48w$

Subtract $48w$ from both sides: $9w^2 - 48w + 64 = 0$

Step 2. The **quadratic trinomial** on the left is a **perfect square trinomial** with a negative middle term. Factoring gives us the following:

$$9w^2 - 48w + 64 = 0$$
$$(3w)^2 - 2(3w)(8) + 8^2 = 0 \leftarrow A^2 - 2AB + B^2$$
$$(3w - 8)^2 = 0 \leftarrow (A - B)^2$$
$$\text{or}$$
$$(3w - 8)(3w - 8) = 0$$

Step 3. Set each **distinct factor** equal to zero and solve the equations. The factor $(3w - 8)$ occurs twice, but we only set it equal to 0 once.

$$3w - 8 = 0$$
$$3w = 8$$
$$w = \frac{8}{3}$$

Step 4. Check this value on your own to confirm that the **solution set** is $\left\{\frac{8}{3}\right\}$.

(eText Screens 15.7-1–15.7-17)

CAUTION: When writing a solution set, we include only distinct solutions. This is why in Step 3 we set only each *distinct* variable factor equal to zero.

My video summary

b. Write the equation in **standard form** by subtracting 49 from both sides.

Begin with the original equation: $4m^2 = 49$

Subtract 49 from both sides: $4m^2 - 49 = 0$

The left side is now the **difference of two squares,** $4m^2 - 49 = (2m)^2 - 7^2$. Factor this special form and set each variable factor equal to 0. Solve the resulting equations to form the solution set. View the **answer**, or watch this **video** for the complete solution.

c. To write this equation in **standard form**, we first simplify the left side of the equation, then move all terms to one side so the other side is 0.

Begin with the original equation: $3x(x - 2) = 2 - x$

Distribute $3x$ on the left: $3x^2 - 6x = 2 - x$

Add x to both sides: $3x^2 - 5x = 2$

Subtract 2 from both sides: $3x^2 - 5x - 2 = 0$

Now factor the left side. Set each **variable factor** equal to 0 and solve the resulting equations to form the solution set. View the **answer**, or watch this **video** for the complete solution.

You Try It **Work through this You Try It problem.**

Work Exercises 11–18 in this textbook or in the MyMathLab Study Plan.

Example 4 Solving Quadratic Equations by Factoring

Solve each equation by factoring.

My interactive video summary

a. $(x + 2)(x - 5) = 18$ **b.** $(x + 3)(3x - 5) = 5(x + 1) - 10$

Solutions We follow the **four-step process** for solving polynomial equations by factoring.

a. Begin by simplifying both sides; then write the equation in **standard form.**

Begin with the original equation: $(x + 2)(x - 5) = 18$

Expand the left side: $x^2 - 5x + 2x - 10 = 18$

Simplify: $x^2 - 3x - 10 = 18$

Subtract 18 from both sides: $x^2 - 3x - 28 = 0$

Solve the equation on your own. View the **answer**, or watch this **interactive video** for detailed solutions to both parts.

b. Begin by simplifying both sides; then write the equation in standard form. Solve the equation on your own, then view the **answer**. Or watch this **interactive video** for detailed solutions to both parts.

You Try It **Work through this You Try It problem.**

Work Exercises 19–24 in this textbook or in the MyMathLab Study Plan.

CAUTION: Not every **quadratic equation** can be solved by factoring. We will look at additional methods for solving quadratic equations in Module 19

OBJECTIVE 2 SOLVE POLYNOMIAL EQUATIONS BY FACTORING

The **zero product property** can be extended to any number of factors. This allows us to solve some **polynomial equations** of degree larger than 2 by factoring.

Example 5 Solving Polynomial Equations by Factoring

Solve each equation by factoring.

a. $(x + 7)(2x - 1)(5x + 4) = 0$ **b.** $24x^3 + 8x^2 = 100x^2 - 28x$

c. $z^3 + z^2 = z + 1$

Solutions

a. The equation has 0 on the right and the polynomial on the left is given in factored form. Therefore, we can apply the zero product property directly. We set each **variable factor** equal to 0 and solve the resulting equations.

$$(x + 7)(2x - 1)(5x + 4) = 0$$

$$x + 7 = 0 \quad \text{or} \quad 2x - 1 = 0 \quad \text{or} \quad 5x + 4 = 0$$

$$x = -7 \qquad 2x = 1 \qquad 5x = -4$$

$$x = \frac{1}{2} \qquad x = -\frac{4}{5}$$

Check these values on your own to confirm that the **solution set** is $\left\{-7, -\frac{4}{5}, \frac{1}{2}\right\}$.

b. Write the equation in **standard form.**

My video summary

Begin with the original equation: $24x^3 + 8x^2 = 100x^2 - 28x$

Subtract $100x^2$ from both sides: $24x^3 - 92x^2 = -28x$

Add $28x$ to both sides: $24x^3 - 92x^2 + 28x = 0$

Now factor the left side. Begin by factoring out the GCF, $4x$.

$$24x^3 - 92x^2 + 28x = 4x(6x^2 - 23x + 7)$$

Complete the factoring, then apply the **zero product property** and finish solving the equation. View the **answer**, or watch this **video** for the complete solution to part b.

c. Write the equation in standard form by subtracting $z + 1$ from both sides.

Begin with the original equation: $z^3 + z^2 = z + 1$

Subtract $z + 1$ from both sides: $z^3 + z^2 - (z + 1) = 0$

Take the opposite of $(z + 1)$: $z^3 + z^2 - z - 1 = 0$

To factor the left side, notice that there are four terms so try **factoring by grouping.** Once factored, set each **variable factor** equal to zero, and solve the resulting equations on your own. View the **answer**, or watch this **video** for a detailed solution to part c.

You Try It **Work through this You Try It problem.**

Work Exercises 25–34 in this textbook or in the MyMathLab Study Plan.

Example 6 Solving Polynomial Equations by Factoring

Solve by factoring:

$$(2x - 9)(3x^2 - 16x - 12) = 0$$

(eText Screens 15.7-1–15.7-17)

Solution

Begin with the original equation: $(2x - 9)(3x^2 - 16x - 12) = 0$

Factor the trinomial: $(2x - 9)(x - 6)(3x + 2) = 0 \leftarrow$ See the details

Set each **variable factor** equal to 0 and solve the resulting equations.

$$2x - 9 = 0 \quad \text{or} \quad x - 6 = 0 \quad \text{or} \quad 3x + 2 = 0$$
$$2x = 9 \qquad x = 6 \qquad 3x = -2$$
$$x = \frac{9}{2} \qquad\qquad x = -\frac{2}{3}$$

Check these values on your own to confirm that the **solution set** is $\left\{-\frac{2}{3}, \frac{9}{2}, 6\right\}$.

You Try It **Work through this You Try It problem.**

Work Exercises 35–40 in this textbook or in the MyMathLab Study Plan.

Note that in Example 6, we did not begin by multiplying the factors on the left together as we did in **Example 4a**. This was because the right side of the equation was already 0 as required by the **zero product property**.

15.7 Exercises

In Exercises 1–4, solve each quadratic equation using the zero product property.

You Try It

1. $(x - 1)(x + 2) = 0$

2. $2x(2x - 5) = 0$

3. $(7x + 1)(3x - 1) = 0$

4. $5(7 - x)(x + 9) = 0$

In Exercises 5–24, use factoring techniques to solve each quadratic equation.

You Try It

5. $x^2 - 5x + 6 = 0$

6. $y^2 + 3y - 10 = 0$

7. $2x^2 + 10x + 12 = 0$

8. $w^2 - w - 6 = 0$

9. $z^2 + 9z + 14 = 0$

10. $5x^2 + 10x = 0$

11. $3z^2 + 5z + 1 = 3$

12. $4x^2 + 1 = 4x$

13. $2x^2 = 50$

14. $3q^2 + 5q - 2 = 2q - 2$

You Try It

15. $9x(x + 3) = 3x - 16$

16. $x(x + 3) = 4x + 2$

17. $2x(x + 1) = 3 - 3x$

18. $7x(x + 3) = 2x + 6$

You Try It

19. $(x - 4)(x + 2) = 7$

20. $(x + 3)(x - 2) = x + 3$

(eText Screens 15.8-1–15.8-18)

21. $(x-4)(x+6)=7x$

22. $(3x-2)(x+2)=-4$

23. $(8x+5)(2x-5)=10x-50$

24. $(x+10)(x-1)=2(x-5)-12$

In Exercises 25–28, solve each polynomial equation using the zero product property.

25. $(x-1)(x+12)(x-3)=0$

26. $x(4x+1)(x-2)=0$

27. $(x+2)(x-8)(7-x)=0$

28. $(x-4)(3x+5)(x+6)(x-10)=0$

In Exercises 29–40, use factoring techniques to solve each polynomial equation.

29. $x^3+x^2=2x$

30. $6y^3+7y^2=8y^2+5y$

31. $x^2(7x+3)-4(7x+3)=0$

32. $z^3+4=z^2+4z$

You Try It **33.** $12x^3-75x=25-4x^2$

34. $8h^3+41=14$

35. $(x-2)(x^2-4x+3)=0$

36. $(2x+1)(3x^2+8x-3)=0$

You Try It **37.** $(x^2-3x-10)(3x+5)=0$

38. $(x^2-9)(x^2-25)=0$

39. $(x^2-1)(x^2-4x+4)=0$

40. $x^2+4=5x^2$

15.8 Applications of Quadratic Equations

THINGS TO KNOW

Before working through this topic, be sure that you are familiar with the following concepts:

		VIDEO	ANIMATION	INTERACTIVE
You Try It	1. Solve Linear Equations Using Both Properties of Equality (Topic 11.1, **Objective 5**)	■		
You Try It	2. Use Linear Equations to Solve Application Problems (Topic 11.2, **Objective 5**)	■		
You Try It	3. Factor Polynomials Completely (Topic 15.6, **Objective 1**)			■
You Try It	4. Solve Quadratic Equations by Factoring (Topic 15.7, **Objective 1**)	■		

(eText Screens 15.8-1–15.8-18)

OBJECTIVES

1 Solve Application Problems Involving Consecutive Numbers

2 Solve Application Problems Involving Geometric Figures

3 Solve Application Problems Using the Pythagorean Theorem

4 Solve Application Problems Involving Quadratic Models

OBJECTIVE 1 SOLVE APPLICATION PROBLEMS INVOLVING CONSECUTIVE NUMBERS

Solving some real-world applications will require us to solve a quadratic equation. To solve these types of applications, we follow the same **problem-solving strategy** used for **linear equations.**

CAUTION

Recall that we need to pay attention to **feasible solutions** when working with applications. Based on the problem's context, not every **solution** to an equation will be a solution to the problem. We need to discard any solutions that do not make sense.

Example 1 House Numbers

The house numbers on the west side of a street are consecutive positive odd **integers.** The product of the house numbers for two next-door neighbors on the west side of the street is 575. Find the house numbers.

Solution

Step 1. Define the Problem. We know the house numbers on the west side of a street are consecutive positive odd numbers and that the product of the house numbers for two next-door neighbors on that side of the street is 575. We want to find the house numbers.

Step 2. Assign Variables. Let x = the first odd house number. Then $x + 2$ is the next consecutive odd house number.

Step 3. Translate into an Equation. To write an equation, we write the product of the two house numbers and set it equal to 575.

$$x(x + 2) = 575$$

Step 4. Solve the Equation.

Begin with the original equation: $x(x + 2) = 575$

Distribute on the left: $x^2 + 2x = 575$

Subtract 575 from both sides: $x^2 + 2x - 575 = 0$

Factor the left side: $(x - 23)(x + 25) = 0$ ← See the details

Set each **variable factor** equal to zero and solve the resulting equations.

$$x - 23 = 0 \quad \text{or} \quad x + 25 = 0$$
$$x = 23 \qquad\qquad x = -25$$

Step 5. Check the Reasonableness of Your Answer. Because the house numbers are positive odd integers, we discard the negative solution. The only reasonable solution is $x = 23$.

Step 6. Answer the Question. The house numbers are 23 and $23 + 2 = 25$.

(eText Screens 15.8-1–15.8-18)

You Try It **Work through this You Try It problem.**

Work Exercises 1–5 in this textbook or in the MyMathLab Study Plan.

OBJECTIVE 2 SOLVE APPLICATION PROBLEMS INVOLVING GEOMETRIC FIGURES

Let's look at an application that involves a quadratic equation and a geometric figure.

Example 2 Swimming Pool Border

My video summary

A swimming pool is 20 feet wide and 30 feet long. A sidewalk border around the pool has uniform width and an **area** that is equal to the area of the pool. See Figure 1. Find the width of the border.

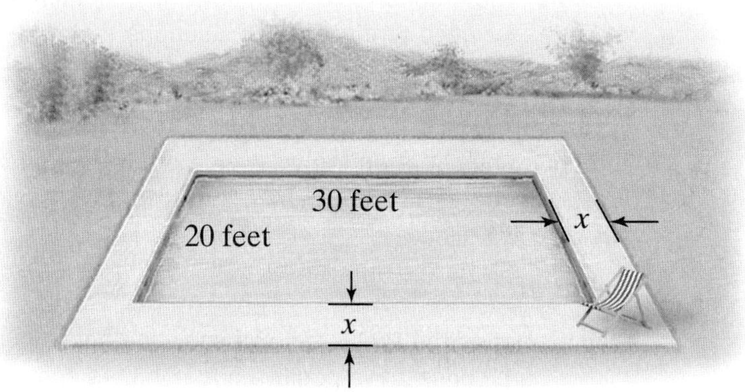

Figure 1

Solution Follow the **problem-solving strategy**.

Step 1. The area of the pool is $20(30) = 600$ sq ft, which is also the area of the sidewalk border. We want to find the width of the border.

Step 2. From the figure, the width of the sidewalk border is x. The combined width of the pool and border is $20 + 2x$ because we are adding one border width to each side of the pool. The combined length is $30 + 2x$ for the same reason. The area of the pool and the area of the border are both 600 ft^2, so the combined area is 1200 ft^2.

Step 3. We can use the formula for the **area of a rectangle** to write an equation for the total combined area.

Begin with the rectangle area formula: $A = lw$

Substitute: $1200 = (20 + 2x)(30 + 2x)$

Finish the problem on your own by solving the equation to find the width of the sidewalk border, x. View the **answer**, or watch this **video** for a detailed solution.

You Try It **Work through this You Try It problem.**

Work Exercises 6–10 in this textbook or in the MyMathLab Study Plan.

OBJECTIVE 3 SOLVE APPLICATION PROBLEMS USING THE PYTHAGOREAN THEOREM

In **Topic 11.4**, we explored **common formulas** for geometric figures such as triangles. **Right triangles** are triangles with a 90° angle, or **right angle**. The **hypotenuse** of a right triangle is the side opposite the right angle and is the longest of the three sides. The other two sides are called the **legs** of the triangle.

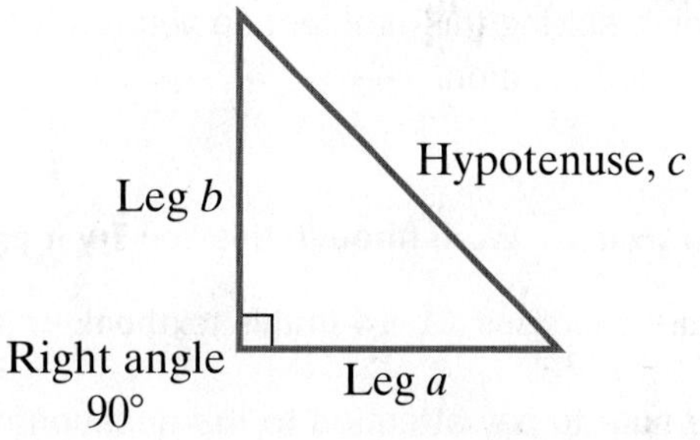

Figure 2

Pythagorean Theorem

For **right triangles**, the sum of the **squares** of the lengths of the **legs** of the triangle equals the square of the length of the **hypotenuse**.

$$a^2 + b^2 = c^2$$

Example 3 Support Wire

My video summary

A wire is attached to a cell phone tower for support. See Figure 3. The length of the wire is 40 meters less than twice the height of the tower. The wire is fixed to the ground at a distance that is 40 meters more than the height of the tower. Find the length of the wire.

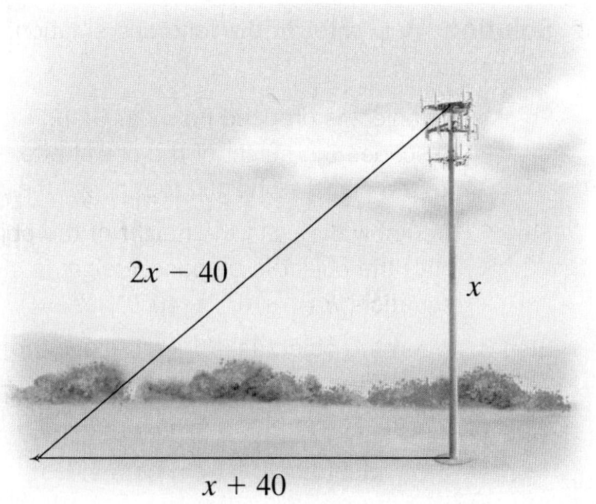

Figure 3

Solution Follow the **problem-solving strategy**.

Step 1. Figure 3 shows us that we can use a **right triangle** to describe this situation. One **leg** is the height of the tower, and the other leg is the distance from the tower where the support wire is fixed to the ground. The length of the wire is the **hypotenuse**.

Step 2. The figure is labeled with the height of the tower as x and the distance from the tower to where the wire is fixed to the ground as $x + 40$. The length of the wire is represented by $2x - 40$.

(eText Screens 15.8-1–15.8-18)

Step 3. We can use the **Pythagorean Theorem** to find the height of the tower.

Pythagorean Theorem: $a^2 + b^2 = c^2$

Substitute expressions for the lengths: $(x + 40)^2 + (x)^2 = (2x - 40)^2$

Step 4. Expand using special products: $x^2 + 80x + 1600 + x^2 = 4x^2 - 160x + 1600$

Simplify the left side: $2x^2 + 80x + 1600 = 4x^2 - 160x + 1600$

Finish solving this problem on your own. View the **answer**, or watch this **video** for a detailed solution.

You Try It **Work through this You Try It problem.**

Work Exercises 11–14 in this textbook or in the MyMathLab Study Plan.

CAUTION Be sure to pay attention to the question being asked. When solving application problems, solving for a variable is not always enough. Sometimes, as in Example 3, we will need to do some computation using the value of the variable in order to answer the question.

OBJECTIVE 4 SOLVE APPLICATION PROBLEMS INVOLVING QUADRATIC MODELS

Now we can solve application problems that are **modeled** by quadratic equations.

Example 4 Falling Object

My video summary

The Grand Canyon Skywalk sits 4000 ft above the Colorado River. If an object is dropped from the observation deck, its height h, in feet after t seconds, is given by

$$h = -16t^2 + 4000.$$

How long will it take for the object to be 400 feet above the Colorado River?

Solution Work through the following solution, or watch this **video** to see the solution worked out.

Step 1. An object is dropped from the Grand Canyon Skywalk. Given the equation that describes the height of the object after some time, we must find the time it takes for the object to be 400 feet above the river.

Step 2. We know that h is the height of the object in feet and t is the time in seconds after the object is dropped. Height is given in terms of time through the quadratic equation, $h = -16t^2 + 4000$.

Step 3. When the object is 400 feet above the Colorado River, $h = 400$. Substituting this value into the equation, we write $400 = -16t^2 + 4000$.

Step 4. Begin with the original equation: $400 = -16t^2 + 4000$

Subtract 400 from both sides: $0 = -16t^2 + 3600$

Factor out the **GCF**: $0 = -16(t^2 - 225)$

Factor the **difference of two squares**: $0 = -16(t + 15)(t - 15)$

Set each **variable factor** equal to zero and solve the resulting equations.

$$t + 15 = 0 \quad \text{or} \quad t - 15 = 0$$
$$t = -15 \qquad\qquad t = 15$$

(eText Screens 15.8-1–15.8-18)

Step 5. Because the time for the object to fall cannot be negative, we discard the negative solution. The only reasonable solution is $t = 15$ seconds.

Step 6. The object will be 400 feet above the Colorado River 15 seconds after it is dropped.

You Try It **Work through this You Try It problem.**

Work Exercises 15–17 in this textbook or in the MyMathLab Study Plan.

Example 5 Broadband Access

My video summary

For household incomes under \$100,000, the relationship between the percentage of households with home broadband access and the annual household income can be approximated by the model,

$$y = -0.01x^2 + 1.7x + 9.5.$$

Here x is the annual household income (in \$1000s) and y is the percentage of households with home broadband access. Use the model to estimate the annual household income if 75.5 percent of such households have home broadband access.

(*Source:* Pew Research Center's Internet & American Life Project, Aug. 9–Sept. 13, 2010)

Solution Follow the **problem-solving strategy**. We are given the relationship between household income and percent with broadband access. We want to determine the household income when 75.5 percent have broadband access. Substitute 75.5 for y and solve the resulting equation for x. Keep in mind that the model is for household incomes *under* \$100,000. View the **answer**, or watch this **video** for the complete solution.

You Try It **Work through this You Try It problem.**

Work Exercises 18–21 in this textbook or in the MyMathLab Study Plan.

15.8 Exercises

In Exercises 1–21, solve each application problem.

You Try It

1. **Room Numbers** The room numbers on one side of a school corridor are consecutive positive even integers. The product of the room numbers for two adjoining rooms on that side is 6560. Find the room numbers.

2. **Flooring Repair** To repair his kitchen subfloor, John cut a rectangular piece of plywood such that the width and length were two consecutive odd integers. If the product of the length and width is 323, what were the dimensions of the piece of plywood?

3. **Consecutive Integers** Find three consecutive integers whose product is 161 larger than the cube of the smallest integer.

4. **Multiples** Find two consecutive positive multiples of 3 such that the sum of their squares is 225.

5. **Mobile Access** According to the 2010 Digital Community College survey, the percent of community colleges that provide access to grades through mobile devices (such as smart phones) and the percent that allow students to register for classes through such devices are consecutive integers. If more allowed students to register than access to grades, and the product of the two percentages is 380, find the percentages.

(eText Screens 15.8-1–15.8-18)

6. **Garden Path** Aliesha has a rectangular plot of land where she can plant a garden. She wants to have a path around the garden with a uniform width of x. If the plot of land measures 30 feet by 45 feet and Aliesha has a usable garden area of 700 square feet, what is the width of the path?

7. **Picture Frame** A portrait has dimensions 20 inches by 30 inches and is surrounded by a frame of uniform width. If the total area (portrait plus frame) is 1064 square inches, find the width of the frame.

8. **Rectangular Table Top** A rectangular table top has a perimeter of 38 inches and an area of 84 square inches. Find its dimensions.

9. **Triangular Sail** The height of a triangular sail is 4 yards less than twice the length of its base. If the area of the sail is 48 square yards, find the dimensions of the sail.

10. **Banner Size** The length of a rectangular banner is 5 feet longer than its width. If the area is 36 square feet, find the dimensions.

11. **Dining Canopy** A dining canopy has corner posts that are anchored by support lines. The length of each support line is 5 feet longer than the height of the post, and the support line is anchored at a distance that is 5 feet less than twice the height of the post. Determine the height of the corner post.

You Try It

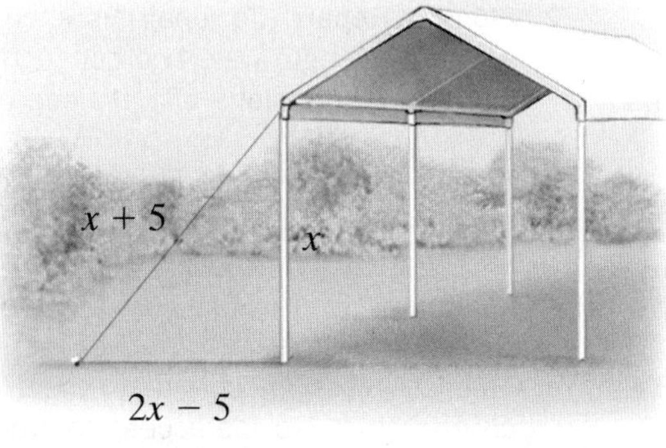

12. **Modern Architecture** An art studio purchased a triangular plot of land in downtown San Diego and constructed a new building whose base is in the shape of a right triangle. The two legs have lengths of $x - 2$ feet and $x + 5$ feet. The hypotenuse is $2x - 27$ feet. Find the lengths of the three sides.

13. **Desktop Organizer** A dorm room desktop organizer is roughly in the shape of a right triangle. The height and base are the solutions to the equation $x^2 + 200 = 28x + 8$. Find the hypotenuse, h.

14. **Bike Ride** At 8 AM, Todd rides his bike due north from campus at 10 mph. One hour later, Chad rides his bike due east from the same point, also at 10 mph. When will they be 50 miles apart?

You Try It

15. **Falling Object** The Infinity Room at Wisconsin's House on the Rock is an observation deck that extends 200 feet out above a scenic valley. The deck sits 150 feet above the valley floor. If an object is dropped from the observation deck, its height h in feet, after t seconds, is given by

$$h = -16t^2 + 150.$$

How long will it take for the object to be 6 feet above the valley floor?

16. **Projectile Motion** A cannonball is fired across a field. Its height above the ground after t seconds is given by $-16t^2 + 40t$. How long does it take for the ball to reach its maximum height of 25 feet?

17. **Projectile Motion** Cierra stands on the edge of her dorm roof and throws a ball into the air. The ball's height above the ground after t seconds is given by

$$h = -16t^2 + 32t + 48.$$

When will the ball hit the ground?

18. **Facebook Growth** The number of Facebook users worldwide can be approximated by the model $N = 24t^2 - 240t + 600$, where N is the number of users in millions and t is the number of years after 2000. Based on this model, predict when the number of Facebook users worldwide will reach 2400 million (i.e. 2.4 billion). (*Source:* internetworldstats.com)

19. **Population** A councilwoman conducts a study and finds the population of her city can be estimated by the equation $p = x^2 - 5x + 1$, where p is thousands of people and x is the number of years after 2000. In what year did the city have a population of 7000?

20. **Wrench Cost** Eduardo notices that the number of oil filter wrenches n that he sells at his hardware store each week is related to the price p (in dollars) by the model,

$$n = -p^2 - 2p + 263.$$

What should Eduardo charge for the oil filter wrench if he wants to sell 200 per week?

21. **Rental Cost** The manager of a 50-unit apartment complex is trying to decide what rent to charge. He knows that all units will be rented if he charges \$600 per month, but for every increase of \$25 over the \$600, there will be a unit vacant. Find the number of occupied units if his total monthly rental income is \$33,600 and the rent per unit is less than \$1000.

MODULE SIXTEEN

Rational Expressions and Equations

MODULE SIXTEEN CONTENTS

16.1 Simplifying Rational Expressions

THINGS TO KNOW

Before working through this topic, be sure you are familiar with the following concepts:

		VIDEO	ANIMATION	INTERACTIVE
You Try It	1. Write Fractions in Simplest Form (Topic 4.2, **Objective 6**)	✓		
You Try It	2. Evaluate Algebraic Expressions (Topic 10.5, **Objective 1**)	✓		
You Try It	3. Factor Polynomials Completely (Topic 15.6, **Objective 1**)			✓

OBJECTIVES

1 Evaluate Rational Expressions

2 Find Restricted Values for Rational Expressions

3 Simplify Rational Expressions

OBJECTIVE 1 EVALUATE RATIONAL EXPRESSIONS

Recall from **Topic 10.1** that a number is a **rational number** if it can be written as a **fraction** $\frac{p}{q}$, where p and q are **integers** and $q \neq 0$. A *rational expression* has a similar definition.

(eText Screens 16.1-1–16.1-20)

> **Definition Rational Expression**
>
> A **rational expression** is an **expression** that can be written as the quotient $\frac{P}{Q}$ of two **polynomials** P and Q as long as $Q \neq 0$.

Some examples of rational expressions are

$$\frac{x-2}{5}, \quad \frac{x+4}{x-7}, \quad \frac{x^2+2x-8}{x^2-x-20}, \quad \text{and} \quad \frac{x+y}{x-y}.$$

The first three expressions are *rational expressions in one variable*, x. The last expression is a *rational expression in two variables*, x and y.

Given values for the variables, we can **evaluate** rational expressions. Using the same process for evaluating expressions in **Topic 10.5**, we substitute the given value for the **variable** and then **simplify**.

Example 1 Evaluating a Rational Expression

Evaluate $\frac{x+8}{x-2}$ for the given value of x.

a. $x = 4$ **b.** $x = -6$

Solutions Substitute the given value for the **variable** and **simplify**.

a. Substitute 4 for x in the expression: $\frac{x+8}{x-2} = \frac{4+8}{4-2}$

Add and subtract: $= \frac{12}{2}$

Simplify the fraction: $= 6$

The value of $\frac{x+8}{x-2}$ is 6 when $x = 4$.

b. Substitute -6 for x in the expression: $\frac{x+8}{x-2} = \frac{-6+8}{-6-2}$

Add and subtract: $= \frac{2}{-8}$

Simplify the fraction: $= -\frac{1}{4}$ ← $\frac{2}{-8} = \frac{1 \cdot 2}{-4 \cdot 2} = \frac{1 \cdot \cancel{2}}{-4 \cdot \cancel{2}} = -\frac{1}{4}$

The value of $\frac{x+8}{x-2}$ is $-\frac{1}{4}$ when $x = -6$.

You Try It **Work through this You Try It problem.**

Work Exercises 1–6 in this textbook or in the MyMathLab Study Plan.

(eText Screens 16.1-1–16.1-20)

Example 2 Evaluating a Rational Expression

My video summary

Evaluate $\frac{x^2 - y}{9x + 5y}$ for $x = 3$ and $y = -1$.

Solution Substitute $x = 3$ and $y = -1$: $\frac{x^2 - y}{9x + 5y} = \frac{(3)^2 - (-1)}{9(3) + 5(-1)}$

Try to finish simplifying this expression on your own by using the correct order of operations. Check your **answer**, or watch this **video** for a complete solution.

You Try It **Work through this You Try It problem.**

Work Exercises 7–10 in this textbook or in the MyMathLab Study Plan.

OBJECTIVE 2 FIND RESTRICTED VALUES FOR RATIONAL EXPRESSIONS

In the definition of a **rational expression**, the statement $Q \neq 0$ means that the **polynomial** Q in the **denominator** cannot equal zero. For example, if we **evaluate** $\frac{x + 8}{x - 2}$ from **Example 1** for $x = 2$, the result is $\frac{2 + 8}{2 - 2} = \frac{10}{0}$, which is **undefined**. So, we must *restrict* $\frac{x + 8}{x - 2}$ to values of x such that $x \neq 2$. We call $x = 2$ a *restricted value* for $\frac{x + 8}{x - 2}$.

Definition Restricted Value

A **restricted value** for an **algebraic expression** is a value that makes the expression undefined.

When working with rational expressions in one variable, values of the variable that cause the denominator to equal zero are restricted values.

Finding Restricted Values for Rational Expressions in One Variable

To find the restricted values for a rational expression in one variable, set the denominator equal to zero. Then, solve the resulting **equation** and restrict the **solutions**. If the equation has no real-number solution, then the expression has no restricted values.

Example 3 Finding Restricted Values for Rational Expressions

Find any **restricted values** for each **rational expression**.

a. $\frac{3x + 5}{3x - 2}$ b. $\frac{x^2 + 2x - 35}{x^2 + x - 30}$

Solutions For each rational expression, we **set** the **denominator** equal to zero and solve the resulting **equation**. The **solutions** are the restricted values.

a. Set the denominator equal to 0: $3x - 2 = 0$

Add 2 to both sides: $3x = 2$

Divide both sides by 3: $x = \frac{2}{3}$

The restricted value for $\frac{3x + 5}{3x - 2}$ is $\frac{2}{3}$. This means that the rational expression is undefined when $x = \frac{2}{3}$.

My video summary

b. Set the denominator equal to 0: $x^2 + x - 30 = 0$

Try to solve this equation on your own by factoring. The solutions will be the restricted values for the rational expression. See the **answer**, or watch this **video** for a complete solution.

You Try It **Work through this You Try It problem.**

Work Exercises 11–18 in this textbook or in the MyMathLab Study Plan.

Example 4 Finding Restricted Values for Rational Expressions

Find any **restricted values** for each **rational expression**.

a. $\frac{2x + 9}{4}$ **b.** $\frac{2x}{x^2 + 1}$

Solutions

a. The **denominator** is a **constant**, 4, which is not equal to 0. This means there are no restricted values for the rational expression $\frac{2x + 9}{4}$.

b. Any restricted value would be a solution to the equation $x^2 + 1 = 0$, but $x^2 + 1$ is a **prime polynomial** and cannot be factored. For any real number that is substituted for x, the expression x^2 will have a non-negative value. Adding 1 will then produce a positive value, which means $x^2 + 1$ cannot equal 0. This means there are no restricted values for $\frac{2x}{x^2 + 1}$.

You Try It **Work through this You Try It problem.**

Work Exercises 19 and 20 in this textbook or in the MyMathLab Study Plan.

OBJECTIVE 3 SIMPLIFY RATIONAL EXPRESSIONS

A fraction is written in **lowest terms**, or **simplest form**, if the **numerator** and **denominator** have no **common factors** other than 1. For example, the fraction $\frac{5}{8}$ is in lowest terms because 5 and 8 have no common factors (except 1). However, the fraction $\frac{6}{10}$ is not in lowest terms because 6 and 10 have a common factor of 2.

Recall that multiplying or dividing both the numerator and denominator of a fraction by the same nonzero value results in an **equivalent fraction**. Therefore, we can **simplify a fraction** by dividing both the numerator and the denominator by their **greatest common factor**. To do this, we can write both the numerator and denominator in terms of their **prime factorizations**. We then **divide out** each common factor by dividing both the numerator and denominator by the common factor. This process is shown by placing slash marks through each common factor and then replacing the factor with 1, the result of dividing the common factor by itself.

(eText Screens 16.1-1–16.1-20)

Separately, in the numerator and denominator, we multiply the 's by any remaining factors to result in the simplified fraction.

$$\frac{6}{10} = \frac{2 \cdot 3}{2 \cdot 5} = \frac{\overset{1}{\cancel{2}} \cdot 3}{\underset{1}{\cancel{2}} \cdot 5} = \frac{1 \cdot 3}{1 \cdot 5} = \frac{3}{5}.$$

We simplify **rational expressions** in the same way that we **simplify fractions**. We **divide out** the **common factors.**

> **Simplification Principle for Rational Expressions**
>
> If P, Q, and R are **polynomials**, then $\dfrac{P \cdot R}{Q \cdot R} = \dfrac{P \cdot \overset{1}{\cancel{R}}}{Q \cdot \underset{1}{\cancel{R}}} = \dfrac{P \cdot 1}{Q \cdot 1} = \dfrac{P}{Q}$ for $Q \neq 0$ and $R \neq 0$.

We can use the following steps to simplify rational expressions.

> **Simplifying Rational Expressions**
>
> **Step 1.** **Factor** the **numerator** and **denominator** completely.
>
> **Step 2.** Divide out each common factor of the numerator and denominator.
>
> **Step 3.** Separately, in the numerator and denominator, multiply any factors that were not divided out to obtain the simplified rational expression. If all **factors** in the numerator divide out, the numerator will be 1.

In most cases, we will not multiply out the polynomial factors in a simplified rational expression. Instead, we will leave them in **factored form.**

Example 5 Simplifying a Rational Expression

Simplify $\dfrac{2x^2 - 6x}{7x - 21}$.

Solution We follow the **three-step process.**

Factor the numerator and denominator: $\dfrac{2x^2 - 6x}{7x - 21} = \dfrac{2x(x - 3)}{7(x - 3)}$

Divide out the common factor $x - 3$: $= \dfrac{2x\overset{1}{\cancel{(x - 3)}}}{7\underset{1}{\cancel{(x - 3)}}}$

Write the simplified rational expression: $= \dfrac{2x}{7}$ for $x \neq 3$

The original expression $\dfrac{2x^2 - 6x}{7x - 21}$ and the simplified expression $\dfrac{2x}{7}$ are equal only when $x \neq 3$. This is why we write "for $x \neq 3$." Do you see **why**? Typically, we will not list such restricted values when simplifying rational expressions.

You Try It **Work through this You Try It problem.**

Work Exercises 21 and 22 in this textbook or in the MyMathLab Study Plan.

(eText Screens 16.1-1–16.1-20)

Example 6 Simplifying a Rational Expression

Simplify $\dfrac{5x}{x^2 + 5x}$.

Solution

Factor the denominator: $\dfrac{5x}{x^2 + 5x} = \dfrac{5x}{x(x + 5)}$

Divide out the common factor x: $= \dfrac{5\overset{1}{\cancel{x}}}{\underset{1}{\cancel{x}}(x + 5)}$

Write the simplified rational expression: $= \dfrac{5}{x + 5}$

You Try It **Work through this You Try It problem.**

Work Exercises 23 and 24 in this textbook or in the MyMathLab Study Plan.

CAUTION Only **common factors** of an expression can be divided out. It is incorrect to divide out **terms**. In Example 6, $5x$ is a term of the original denominator, and 5 is a term of the simplified denominator. Neither can be divided out.

Incorrect

$$\frac{5x}{x^2 + 5x} = \frac{\overset{1}{\cancel{5x}}}{x^2 + \underset{1}{\cancel{5x}}} = \frac{1}{x^2 + 1}$$

Incorrect

$$\frac{5}{x + 5} = \frac{\overset{1}{\cancel{5}}}{x + \underset{1}{\cancel{5}}} = \frac{1}{x + 1}$$

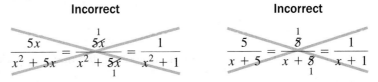

CAUTION When simplifying a **rational expression**, replacing each divided-out **factor** with a 1 can cause unnecessary clutter. Therefore, it is fine to leave out the 1's. However, if every factor in the numerator divides out, be sure to include a 1 in the numerator of the simplified rational expression.

Example 7 Simplifying a Rational Expression

My video summary

Simplify $\dfrac{y^2 + 2y - 24}{y^2 + 4y - 32}$.

Solution Follow the **three-step process**.

Begin with the original expression: $\dfrac{y^2 + 2y - 24}{y^2 + 4y - 32}$

Factor the numerator and denominator: $= \dfrac{(y - 4)(y + 6)}{(y - 4)(y + 8)}$

Try to finish simplifying this rational expression on your own. View the **answer**, or watch this **video** for a complete solution.

You Try It **Work through this You Try It problem.**

Work Exercises 25–28 in this textbook or in the MyMathLab Study Plan.

Example 8 Simplifying a Rational Expression

My video summary

Simplify $\dfrac{2m^2 + m - 15}{2m^3 - 5m^2 - 18m + 45}$.

Solution Follow the **three-step process**, and try to simplify this **rational expression** on your own. View the **answer**, or watch this **video** for a complete solution.

You Try It **Work through this You Try It problem.**

Work Exercises 29 and 30 in this textbook or in the MyMathLab Study Plan.

Example 9 Simplifying a Rational Expression in Two Variables

My video summary

Simplify $\dfrac{x^2 - xy - 12y^2}{2x^2 + 7xy + 3y^2}$.

Solution

Begin with the original expression: $\dfrac{x^2 - xy - 12y^2}{2x^2 + 7xy + 3y^2}$

Factor the numerator and denominator: $= \dfrac{(x - 4y)(x + 3y)}{(2x + y)(x + 3y)}$

Try to finish this problem on your own. View the **answer**, or watch this **video** for a complete solution.

You Try It **Work through this You Try It problem.**

Work Exercises 31 and 32 in this textbook or in the MyMathLab Study Plan.

The **commutative property of addition** states that a sum is not affected by the order of the **terms**. So, $a + b = b + a$. Sometimes, when factoring the numerator and denominator of a **rational expression**, the terms of **common factors** may be arranged in a different order. We can still **divide out** these common factors.

Example 10 Simplifying a Rational Expression

Simplify $\dfrac{w^2 - y^2}{2xy + 2xw}$.

Solution

Begin with the original expression: $\dfrac{w^2 - y^2}{2xy + 2xw}$

Factor the numerator and denominator: $= \dfrac{(w + y)(w - y)}{2x(y + w)}$

Divide out the common factors $w + y = y + w$: $= \dfrac{\cancel{(w + y)}(w - y)}{2x\cancel{(y + w)}}$

Write the simplified rational expression: $= \dfrac{w - y}{2x}$

You Try It **Work through this You Try It problem.**

Work Exercises 33 and 34 in this textbook or in the MyMathLab Study Plan.

(eText Screens 16.1-1–16.1-20)

If the numerator and denominator of a **rational expression** have factors that are **opposite polynomials**, we can simplify by first factoring -1 from one of them.

Example 11 Simplifying a Rational Expression Involving Opposites

Simplify $\dfrac{3x - 10}{10 - 3x}$.

Solution We recognize that $3x - 10$ and $10 - 3x$ are opposite polynomials, so we factor -1 from $10 - 3x$:

$$10 - 3x = -1(-10 + 3x) = -1(3x - 10)$$

This gives

$$\frac{3x - 10}{10 - 3x} = \frac{3x - 10}{-1(3x - 10)} = \frac{\cancel{3x - 10}}{-1\cancel{(3x - 10)}} = \frac{1}{-1} = -1.$$

You Try It **Work through this You Try It problem.**

Work Exercises 35 and 36 in this textbook or in the MyMathLab Study Plan.

Example 11 shows us that **opposite polynomials** in the numerator and denominator of a rational expression will **divide out**, leaving a factor of -1.

Example 12 Simplifying a Rational Expression Involving Opposites

My video summary

Simplify $\dfrac{2x^2 - 27x + 70}{49 - 4x^2}$.

Solution

Begin with the original expression: $\dfrac{2x^2 - 27x + 70}{49 - 4x^2}$

Factor the numerator and denominator: $= \dfrac{(2x - 7)(x - 10)}{(7 - 2x)(7 + 2x)}$

Notice that $2x - 7$ and $7 - 2x$ are opposite polynomials. Finish simplifying the **rational expression** on your own. View the **answer**, or watch this **video** for a complete solution.

You Try It **Work through this You Try It problem.**

Work Exercises 37–40 in this textbook or in the MyMathLab Study Plan.

16.1 Exercises

In Exercises 1–10, evaluate each rational expression for the given value(s) of the variable(s).

You Try It

1. $\dfrac{x + 5}{x - 1}$ for $x = 3$

2. $\dfrac{2p + 1}{3p - 1}$ for $p = 4$

3. $\dfrac{w + 6}{w + 8}$ for $w = -4$

4. $\dfrac{y^2 - 1}{y^3 - 1}$ for $y = 3$

(eText Screens 16.1-1–16.1-20)

5. $\dfrac{x^2 - 5x - 14}{5 - x}$ for $x = 3$

6. $\dfrac{2t^3 - 7}{t^2 + 5}$ for $t = -1$

You Try It

7. $\dfrac{y^2 - x}{5x - 2y}$ for $x = 4$ and $y = -6$

8. $\dfrac{w + 2z}{w - 2z}$ for $w = -10$ and $z = 3$

9. $\dfrac{m^2 - n^2}{m^3 + n^3}$ for $m = 5$ and $n = 3$

10. $\dfrac{x^2 + xy}{5 - y}$ for $x = -3$ and $y = 8$

In Exercises 11–20, find any restricted values for the given rational expression.

You Try It

11. $\dfrac{3}{x - 8}$

12. $\dfrac{y - 4}{y + 7}$

13. $\dfrac{3t + 1}{7t - 4}$

14. $\dfrac{2m - 3}{5m}$

15. $\dfrac{p + 6}{(p - 3)(p + 2)}$

16. $\dfrac{x^2 - 16}{x^2 - x - 20}$

17. $\dfrac{2y^2 - 11y - 6}{3y^2 - 13y - 30}$

18. $\dfrac{n^3 - 3n^2 - 4n}{n^3 + 5n^2 - n - 5}$

You Try It

19. $\dfrac{2t - 7}{3}$

20. $\dfrac{5x - 1}{x^2 + 3}$

In Exercises 21–40, simplify each rational expression. If the expression is already in simplest form, state this in your answer.

You Try It

21. $\dfrac{3m - 9}{5m - 15}$

22. $\dfrac{5x^2 - 10x}{13x - 26}$

You Try It

23. $\dfrac{6y}{y^2 + 6y}$

24. $\dfrac{4n + 24}{9n^3 + 54n^2}$

You Try It

25. $\dfrac{z^2 - 2z - 15}{z^2 + 10z + 21}$

26. $\dfrac{m^2 + 4m - 117}{m^2 + 18m + 80}$

27. $\dfrac{t^2 + 5t - 24}{2t^2 - 11t + 15}$

28. $\dfrac{6w^2 + 5w - 4}{3w^2 + 13w + 12}$

You Try It

29. $\dfrac{2x^2 + 11x + 12}{2x^3 + 3x^2 - 32x - 48}$

30. $\dfrac{12a^3 + 4a^2 + 9a + 3}{3a^2 - 11a - 4}$

You Try It

31. $\dfrac{x^2 - 7xy - 8y^2}{x^2 - 3xy - 4y^2}$

32. $\dfrac{4p^2 + 5pq - 6q^2}{3p^2 + 5pq - 2q^2}$

You Try It **33.** $\dfrac{12 + 5x}{5x^2 + 22x + 24}$

34. $\dfrac{a^2 - 49b^2}{56bc + 8ac}$

You Try It **35.** $\dfrac{2m - 9}{9 - 2m}$

36. $\dfrac{3x + 8}{8 - 3x}$

You Try It **37.** $\dfrac{w^2 - 6w - 16}{64 - w^2}$

38. $\dfrac{2 - z}{z^2 + z - 6}$

39. $\dfrac{n^3 - 5n^2 + 7n - 35}{n^2 - 10n + 25}$

40. $\dfrac{y^2 - 81}{-y - 9}$

16.2 Multiplying and Dividing Rational Expressions

THINGS TO KNOW

Before working through this topic, be sure you are familiar with the following concepts:

	VIDEO	ANIMATION	INTERACTIVE
You Try It **1.** Simplify Exponential Expressions Using the Quotient Rule (Topic 14.1, **Objective 2**)	✓		
You Try It **2.** Factor Polynomials Completely (Topic 15.6, **Objective 1**)			✓
You Try It **3.** Simplify Rational Expressions (Topic 16.1, **Objective 3**)	✓		

OBJECTIVES

1 Multiply Rational Expressions

2 Divide Rational Expressions

OBJECTIVE 1 MULTIPLY RATIONAL EXPRESSIONS

Recall that to multiply **rational numbers** (fractions), we multiply straight across the **numerators**, multiply straight across the **denominators**, and **simplify**. If $\frac{p}{q}$ and $\frac{r}{s}$ are rational numbers, then

$$\frac{p}{q} \cdot \frac{r}{s} = \frac{p \cdot r}{q \cdot s} = \frac{pr}{qs}.$$

For example,

$$\frac{3}{14} \cdot \frac{8}{9} = \frac{3 \cdot 8}{14 \cdot 9} = \frac{24}{126}.$$

(eText Screens 16.2-1–16.2-22)

You may wish to review multiplying fractions in **Topic 4.3**. Recall that we write fractions in simplified form by **dividing out** common factors from the numerator and denominator.

$$\frac{3}{14}\cdot\frac{8}{9}=\frac{24}{126}=\frac{4\cdot 6}{21\cdot 6}=\frac{4\cdot \cancel{6}^{1}}{21\cdot \cancel{6}_{1}}=\frac{4}{21}$$

It is often easier to simplify fractions by first writing the numerator and the denominator as a product of **prime factors**, then divide out any common factors.

$$\frac{3}{14}\cdot\frac{10}{9}=\underbrace{\frac{3}{2\cdot 7}\cdot\frac{2\cdot 5}{3\cdot 3}}_{\text{Factor into prime factors}}=\underbrace{\frac{\cancel{3}^{1}}{\cancel{2}_{1}\cdot 7}\cdot\frac{\cancel{2}^{1}\cdot 5}{3\cdot \cancel{3}_{1}}}_{\text{Divide out common factors}}=\underbrace{\frac{5}{7\cdot 3}}_{\text{Multiply remaining factors}}=\frac{5}{21}$$

We follow the same approach when multiplying **rational expressions**. We factor all numerators and denominators **completely, divide out** any **common factors**, multiply remaining factors in the numerators, and multiply remaining factors in the denominators. So, if $\frac{P}{Q}$ and $\frac{R}{S}$ are rational expressions, then

$$\frac{P}{Q}\cdot\frac{R}{S}=\frac{PR}{QS}.$$

We keep the result in **factored form** just as we did when simplifying rational expressions.

Multiplying Rational Expressions

Step 1. Factor each numerator and denominator completely into **prime factors**.

Step 2. Divide out common factors.

Step 3. Multiply remaining factors in the numerators and multiply remaining factors in the denominators.

CAUTION Factoring is a critical step in the multiplication process. If necessary, review the **factoring techniques** discussed in Module 15.

Example 1 Multiplying Rational Expressions

Multiply $\frac{5x^2}{2y}\cdot\frac{6y^2}{25x^3}$.

Solution We follow the **three-step process** for multiplying **rational expressions**.

Begin with the original expression: $\frac{5x^2}{2y}\cdot\frac{6y^2}{25x^3}$

Factor numerators and denominators: $=\frac{5\cdot x\cdot x}{2\cdot y}\cdot\frac{2\cdot 3\cdot y\cdot y}{5\cdot 5\cdot x\cdot x\cdot x}$

Divide out common factors: $=\frac{\cancel{5}\cdot\cancel{x}\cdot\cancel{x}}{\cancel{2}\cdot\cancel{y}}\cdot\frac{\cancel{2}\cdot 3\cdot\cancel{y}\cdot y}{5\cdot\cancel{5}\cdot\cancel{x}\cdot\cancel{x}\cdot x}$

Multiply remaining factors: $=\frac{3\cdot y}{5\cdot x}=\frac{3y}{5x}$

You Try It **Work through this You Try It problem.**

Work Exercises 1–4 in this textbook or in the MyMathLab Study Plan.

(eText Screens 16.2-1–16.2-22)

Notice that we could also have worked Example 1 by using the **rules for exponents** (view the **details**).

Example 2 Multiplying Rational Expressions

Multiply $\dfrac{3x-6}{2x}\cdot\dfrac{8}{5x-10}$.

Solution We follow the **three-step process** for multiplying **rational expressions.**

Begin with the original expression: $\dfrac{3x-6}{2x}\cdot\dfrac{8}{5x-10}$

Factor numerators and denominators: $=\dfrac{3(x-2)}{2\cdot x}\cdot\dfrac{2\cdot 4}{5(x-2)}$

Divide out common factors: $=\dfrac{3\cancel{(x-2)}}{\cancel{2}\cdot x}\cdot\dfrac{\cancel{2}\cdot 4}{5\cancel{(x-2)}}$

Multiply remaining factors: $=\dfrac{3\cdot 4}{x\cdot 5}=\dfrac{12}{5x}$

You Try It **Work through this You Try It problem.**

Work Exercises 5–8 in this textbook or in the MyMathLab Study Plan.

Example 3 Multiplying Rational Expressions

My video summary

Multiply $\dfrac{x^2-4}{x^2+2x-35}\cdot\dfrac{x^2-25}{x+2}$.

Solution Follow the **three-step process** for multiplying **rational expressions.**

Begin with the original expression: $\dfrac{x^2-4}{x^2+2x-35}\cdot\dfrac{x^2-25}{x+2}$

Factor numerators and denominators: $=\dfrac{(x+2)(x-2)}{(x+7)(x-5)}\cdot\dfrac{(x+5)(x-5)}{(x+2)}$

Continue by dividing out **common factors**, multiplying remaining factors in the numerator, and multiplying remaining factors in the denominator. Remember that even though we are multiplying numerators and denominators, we typically will leave them in **factored form.** View the **answer**, or watch this **video** for a complete solution.

You Try It **Work through this You Try It problem.**

Work Exercises 9–14 in this textbook or in the MyMathLab Study Plan.

Example 4 Multiplying Rational Expressions

My interactive video summary

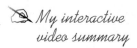

Multiply $\dfrac{2x^2+3x-2}{3x^2-2x-1}\cdot\dfrac{3x^2+4x+1}{2x^2+x-1}$.

Solution Work through this **interactive video** to complete the solution by factoring the expressions and **dividing out** common factors.

(eText Screens 16.2-1–16.2-22)

You Try It **Work through this You Try It problem.**

Work Exercises 15–18 in this textbook or in the MyMathLab Study Plan.

Common factors are not limited to constants or **binomial factors.** It is possible to have other common **polynomial factors,** as shown in the next example.

Example 5 Multiplying Rational Expressions

Multiply $\frac{3x^2 + 9x + 27}{x - 1} \cdot \frac{x + 3}{x^3 - 27}$.

Solution Follow the **three-step process** for multiplying **rational expressions.**

Begin with the original expression: $\frac{3x^2 + 9x + 27}{x - 1} \cdot \frac{x + 3}{x^3 - 27}$

Factor numerators and denominators: $= \frac{3(x^2 + 3x + 9)}{(x - 1)} \cdot \frac{(x + 3)}{(x - 3)(x^2 + 3x + 9)}$

Notice there is a common **trinomial factor** in a numerator and denominator. This trinomial factor happens to be **prime.** So, these factors **divide out** just like any other common factor.

Divide out common factors: $\frac{3\cancel{(x^2 + 3x + 9)}}{(x - 1)} \cdot \frac{(x + 3)}{(x - 3)\cancel{(x^2 + 3x + 9)}}$

Multiply remaining factors: $= \frac{3(x + 3)}{(x - 1)(x - 3)}$

You Try It **Work through this You Try It problem.**

Work Exercises 19–22 in this textbook or in the MyMathLab Study Plan.

Sometimes **factors** in the numerator and denominator are **opposites (or additive inverses).** The **quotient** of opposites equals -1. This can be seen by factoring out -1 from either the numerator or denominator and dividing out the **common factors.** For example,

$$\frac{4x - 7}{7 - 4x} = \frac{4x - 7}{-1(-7 + 4x)} = \frac{4x - 7}{-1(4x - 7)} = \frac{\cancel{4x - 7}}{-1\cancel{(4x - 7)}} = \frac{1}{-1} = -1.$$

Usually, we want the factors to have positive **leading coefficients.** If a leading coefficient is negative, we can factor out a **negative constant.**

Example 6 Multiplying Rational Expressions

My video summary

Multiply $\frac{3x^2 + 10x - 8}{2x - 3x^2} \cdot \frac{4x + 1}{x + 4}$.

Solution Multiply the expressions on your own. Remember to factor out a negative constant if the leading coefficient is negative. View the **answer,** or watch this **video** for a complete solution.

You Try It **Work through this You Try It problem.**

Work Exercises 23–26 in this textbook or in the MyMathLab Study Plan.

(eText Screens 16.2-1–16.2-22)

As we saw when factoring in **Topic 15.2**, rational expressions may contain more than one **variable**.

Example 7 Multiplying Rational Expressions

My video summary

Multiply $\dfrac{x^2 + xy}{3x + y} \cdot \dfrac{3x^2 + 7xy + 2y^2}{x^2 - y^2}$.

Solution

Remember to factor out the **GCF** first. For **polynomials** of the form $ax^2 + bxy + cy^2$, remember that the second **term** in each **binomial factor** must contain the variable y.

Begin with the original expression: $\dfrac{x^2 + xy}{3x + y} \cdot \dfrac{3x^2 + 7xy + 2y^2}{x^2 - y^2}$

Factor the numerators and denominators: $= \dfrac{x(x + y)}{3x + y} \cdot \dfrac{(3x + y)(x + 2y)}{(x + y)(x - y)}$

Finish the multiplication on your own. View the **answer**, or watch this **video** for a complete solution.

You Try It **Work through this You Try It problem.**

Work Exercises 27–30 in this textbook or in the MyMathLab Study Plan.

OBJECTIVE 2 DIVIDE RATIONAL EXPRESSIONS

To divide **rational numbers**, recall that if $\frac{p}{q}$ and $\frac{r}{s}$ are rational numbers such that q, r, and s are not zero, then

$$\frac{p}{q} \div \frac{r}{s} = \frac{p}{q} \cdot \frac{s}{r} = \frac{ps}{qr}$$

The numbers $\frac{r}{s}$ and $\frac{s}{r}$ are **reciprocals** of each other.

For example,

$$\frac{4}{15} \div \frac{6}{25} = \underbrace{\frac{4}{15} \cdot \frac{25}{6}}_{\text{Change to multiplication by the reciprocal}} = \underbrace{\frac{4 \cdot 25}{15 \cdot 6}}_{\text{Multiply numerators and multiply denominators}} = \underbrace{\frac{2 \cdot 2 \cdot 5 \cdot 5}{3 \cdot 5 \cdot 2 \cdot 3}}_{\text{Factor into prime factors}} = \underbrace{\frac{\cancel{2} \cdot 2 \cdot \cancel{5} \cdot 5}{3 \cdot \cancel{5} \cdot \cancel{2} \cdot 3}}_{\text{Divide out common factors}} = \underbrace{\frac{2 \cdot 5}{3 \cdot 3}}_{\text{Multiply remaining factors}} = \frac{10}{9}.$$

Refer to **Topic 4.3** to review dividing rational numbers in fraction form.

We follow the same approach when dividing **rational expressions**. To find the **quotient** of two rational expressions, we find the **product** of the first rational expression and the reciprocal of the second rational expression. So, if $\frac{P}{Q}$ and $\frac{R}{S}$ are rational expressions, then

$$\frac{P}{Q} \div \frac{R}{S} = \frac{P}{Q} \cdot \frac{S}{R} = \frac{PS}{QR}.$$

Dividing Rational Expressions

Step 1. Change the division to multiplication and replace the **divisor** by its **reciprocal.**

Step 2. Multiply the expressions.

Example 8 Dividing Rational Expressions

Divide each rational expression.

a. $\dfrac{6x^5}{9y^3} \div \dfrac{5x^4}{3y^2}$ b. $\dfrac{(x+2)(x-1)}{(3x-5)} \div \dfrac{(x-1)(x+4)}{(2x+3)}$

Solutions

a. We follow the **two-step process** for dividing **rational expressions.**

Begin with the original expression: $\dfrac{6x^5}{9y^3} \div \dfrac{5x^4}{3y^2}$

Change to multiplication by the reciprocal: $= \dfrac{6x^5}{9y^3} \cdot \dfrac{3y^2}{5x^4}$ ← $\dfrac{3y^2}{5x^4}$ is the reciprocal of $\dfrac{5x^4}{3y^2}$

Factor: $= \dfrac{2 \cdot 3 \cdot x \cdot x \cdot x \cdot x \cdot x}{3 \cdot 3 \cdot y \cdot y \cdot y} \cdot \dfrac{3 \cdot y \cdot y}{5 \cdot x \cdot x \cdot x \cdot x}$

Divide out common factors: $= \dfrac{2 \cdot \cancel{3} \cdot \cancel{x \cdot x \cdot x \cdot x} \cdot x}{\cancel{3} \cdot 3 \cdot \cancel{y \cdot y} \cdot y} \cdot \dfrac{\cancel{3 \cdot y \cdot y}}{5 \cdot \cancel{x \cdot x \cdot x \cdot x}}$

Multiply remaining factors: $= \dfrac{2 \cdot x}{y \cdot 5} = \dfrac{2x}{5y}$

b. Begin with the original expression: $\dfrac{(x+2)(x-1)}{(3x-5)} \div \dfrac{(x-1)(x+4)}{(2x+3)}$

Change to multiplication by the reciprocal: $= \dfrac{(x+2)(x-1)}{(3x-5)} \cdot \dfrac{(2x+3)}{(x-1)(x+4)}$

Divide out common factors: $= \dfrac{(x+2)\cancel{(x-1)}}{(3x-5)} \cdot \dfrac{(2x+3)}{\cancel{(x-1)}(x+4)}$

Multiply remaining factors: $= \dfrac{(x+2)(2x+3)}{(3x-5)(x+4)}$

You Try It **Work through this You Try It problem.**

Work Exercises 31–34 in this textbook or in the MyMathLab Study Plan.

Example 9 Dividing Rational Expressions

My video summary

Divide $\dfrac{9y^2-81}{4y^2} \div \dfrac{y+3}{8}$.

Solution

Begin with the original expression: $\dfrac{9y^2-81}{4y^2} \div \dfrac{y+3}{8}$

Change to multiplication by the reciprocal: $= \dfrac{9y^2-81}{4y^2} \cdot \dfrac{8}{y+3}$

Factor out the GCF: $= \dfrac{9(y^2 - 9)}{4y^2} \cdot \dfrac{8}{y + 3}$

Factor difference of two squares: $= \dfrac{9(y - 3)(y + 3)}{4y^2} \cdot \dfrac{8}{y + 3}$

Divide out common factors and finish the problem on your own. View the **answer**, or watch this **video** for the complete solution.

You Try It **Work through this You Try It problem.**

Work Exercises 35–40 in this textbook or in the MyMathLab Study Plan.

Example 10 Dividing Rational Expressions

My video summary

Divide $\dfrac{2x^2 + 21x + 40}{3x^2 + 23x - 8} \div \dfrac{4x^2 + 16x + 15}{x + 2}$.

Solution Begin with the original expression: $\dfrac{2x^2 + 21x + 40}{3x^2 + 23x - 8} \div \dfrac{4x^2 + 16x + 15}{x + 2}$

Change to multiplication by the **reciprocal**: $= \dfrac{2x^2 + 21x + 40}{3x^2 + 23x - 8} \cdot \dfrac{x + 2}{4x^2 + 16x + 15}$

Factor the numerators and denominators and then multiply. View the **answer**, or watch this **video** for a complete solution.

You Try It **Work through this You Try It problem.**

Work Exercises 41–43 in this textbook or in the MyMathLab Study Plan.

Example 11 Dividing Rational Expressions

My interactive video summary

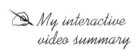

Divide $\dfrac{x^3 - 8}{2x^2 - x - 6} \div \dfrac{x^2 + 2x + 4}{6x^2 + 11x + 3}$.

Solution Work through this **interactive video** to complete the solution, or view the **answer**.

You Try It **Work through this You Try It problem.**

Work Exercises 44–46 in this textbook or in the MyMathLab Study Plan.

As with multiplication, when dividing **rational expressions**, we may have more than one variable as shown in the next example.

Example 12 Dividing Rational Expressions

Divide $\dfrac{x^3 - 8y^3}{3x + y} \div \dfrac{4x - 8y}{6x^2 + 17xy + 5y^2}$.

My video summary

Solution

Begin with the original expression: $\dfrac{x^3 - 8y^3}{3x + y} \div \dfrac{4x - 8y}{6x^2 + 17xy + 5y^2}$

Change to multiplication: $= \dfrac{x^3 - 8y^3}{3x + y} \cdot \dfrac{6x^2 + 17xy + 5y^2}{4x - 8y}$

Factor: $= \dfrac{(x - 2y)(x^2 + 2xy + 4y^2)}{3x + y} \cdot \dfrac{(3x + y)(2x + 5y)}{4(x - 2y)}$

Finish the problem on your own. View the **answer**, or watch the **video** for a complete solution.

You Try It **Work through this You Try It problem.**

Work Exercises 47 and 48 in this textbook or in the MyMathLab Study Plan.

If we need to multiply or divide more than two rational expressions, first we change any divisions to multiplication using the appropriate **reciprocal**. Then we follow the **three-step process** for multiplying rational expressions.

Example 13 Multiplying and Dividing Rational Expressions

My video summary

Perform the indicated operations.

$$\frac{x^2 + 2x - 15}{x^2 + 2x - 8} \cdot \frac{x^2 + 3x + 2}{x^2 + 4x - 21} \div \frac{x + 2}{x^2 + 9x + 14}$$

Solution Try this problem on your own. Remember to first change all divisions to multiplication using the **reciprocal** of the **divisor**. View the **answer**, or watch the **video** to see a complete solution.

You Try It **Work through this You Try It problem.**

Work Exercises 49–54 in this textbook or in the MyMathLab Study Plan.

16.2 Exercises

In Exercises 1–30, multiply the rational expressions.

You Try It

1. $\dfrac{6a^3}{2b^2} \cdot \dfrac{5b^3}{3a^4}$

2. $\dfrac{14x^2y}{3y^2x} \cdot \dfrac{9x^3y^2}{7x^2y^3}$

3. $\dfrac{2xyz}{5y^2z^2} \cdot \dfrac{15xz^2}{6x^2y}$

4. $\dfrac{-3x^2z}{2y^3z^2} \cdot \dfrac{4xy^2}{z^3}$

You Try It

5. $\dfrac{5x}{3x + 6} \cdot \dfrac{x + 2}{7}$

6. $\dfrac{3m + 9}{10m} \cdot \dfrac{5m^2}{2m + 6}$

7. $\dfrac{6x^2}{2x - 2} \cdot \dfrac{x^2 - x}{5x^3}$

8. $\dfrac{2xy + y}{3y^2z} \cdot \dfrac{9yz^2}{2xz + z}$

You Try It

9. $\dfrac{x+1}{x^2-4}\cdot\dfrac{x-2}{x^2-1}$

10. $\dfrac{x-6}{x^2+3x-10}\cdot\dfrac{2x-4}{x^2-36}$

11. $\dfrac{3r^2+12r}{r-1}\cdot\dfrac{r^2-5r+4}{r+2}$

12. $\dfrac{x^2-3x}{x^2+x}\cdot\dfrac{x^2+3x+2}{x^2-2x-3}$

13. $\dfrac{x^2+4x+3}{x^2+5x+6}\cdot\dfrac{x+2}{x-1}$

14. $\dfrac{x^2+12x+36}{x+1}\cdot\dfrac{4x+3}{x^2+2x-24}$

You Try It

15. $\dfrac{x^2-1}{3x^2+11x+6}\cdot\dfrac{3x^2+7x-6}{x^2-2x+1}$

16. $\dfrac{x^2+3x}{x^2-4x-5}\cdot\dfrac{x^2-25}{2x^2+x}$

17. $\dfrac{3b^2+4b-4}{b^2+b-2}\cdot\dfrac{b^2-1}{3b^2+b-2}$

18. $\dfrac{4x^2+20x-24}{6x^2+7x-3}\cdot\dfrac{4x^2+20x+21}{9x^2+51x-18}$

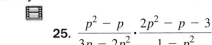

You Try It

19. $\dfrac{x^3-8}{x^2+4x+3}\cdot\dfrac{x^2+3x}{x^3+2x^2+4x}$

20. $\dfrac{x+10}{2x^2-4x+8}\cdot\dfrac{3x^3+24}{x-5}$

21. $\dfrac{5x^2-5x+5}{10x^2+20x}\cdot\dfrac{x^2-2x}{x^3+1}$

22. $\dfrac{2n^3+2n^2+n+1}{n^2+5n+6}\cdot\dfrac{n^2-2n}{n^3+1}$

You Try It

23. $\dfrac{4-x^2}{x^2-x-6}\cdot\dfrac{5x^2-14x-3}{2x^2-3x-2}$

24. $\dfrac{12-2x-2x^2}{2y^2}\cdot\dfrac{x+3y}{-x^2+xy-y^2}$

25. $\dfrac{p^2-p}{3p-2p^2}\cdot\dfrac{2p^2-p-3}{1-p^2}$

26. $\dfrac{3-8x-3x^2}{9-x^2}\cdot\dfrac{x^2-x-6}{9x^2-1}$

You Try It

27. $\dfrac{x^2+2xy-3y^2}{3x^2+2xy-y^2}\cdot\dfrac{x^2+3xy+2y^2}{x^2+xy-2y^2}$

28. $\dfrac{a^3+a^3b+a+b}{2a^3+2a}\cdot\dfrac{18a^2}{6a^2-6b^2}$

29. $\dfrac{x^2-y^2}{2x^2+xy-y^2}\cdot\dfrac{2xy-y^2}{x^3-y^3}$

30. $\dfrac{2x^2+xy-y^2}{2x^2-3xy+y^2}\cdot\dfrac{x^2-y^2}{x^2+2xy+y^2}$

In Exercises 31–48, divide the rational expressions.

You Try It

31. $\dfrac{4y^3}{3x^2}\div\dfrac{8y^2}{15x^3}$

32. $\dfrac{5x}{6}\div\dfrac{10x+15}{6x+18}$

33. $\dfrac{3x+2}{(x-2)(x+1)}\div\dfrac{2x+3}{(x-2)(x+3)}$

34. $\dfrac{x(x+1)}{(x-1)(x+2)}\div\dfrac{2x+2}{(x-1)(x+3)}$

You Try It

35. $\dfrac{2x^2}{3x^2-3}\div\dfrac{14}{x+1}$

36. $\dfrac{3m}{4m-12}\div\dfrac{m^2+m}{6-2m}$

37. $\dfrac{x-2}{3x-9}\div\dfrac{x^2-2x}{x-3}$

38. $\dfrac{6y^2}{3y^2+5y+2}\div\dfrac{10y^2-6y}{2y^2-3y-5}$

39. $\dfrac{x^2-4}{x^3-27}\div\dfrac{x^3-8}{x^2-9}$

40. $\dfrac{5x-x^2}{x^3-125}\div\dfrac{x}{x^2+4x+8}$

You Try It

41. $\frac{x^2 + x - 6}{x + 5} \div \frac{x - 2}{x^2 + 4x - 5}$

42. $\frac{x^2 - 12x + 36}{x^2 - x - 42} \div \frac{x^2 - 36}{4}$

43. $\frac{2x^2 + x - 3}{x^2 + 3x + 2} \div \frac{x^2 + 4x - 5}{3x^2 + 5x + 2}$

44. $\frac{x^2 - 2x - 3}{2x^2 + x - 1} \div \frac{3x^2 - 7x - 6}{6x^2 + x - 2}$

You Try It **45.** $\frac{3h^2 - 3h - 60}{7h^2 - 37h + 10} \div \frac{12h^2 + 20h + 3}{2h^2 - h - 6}$

46. $\frac{10x^2 + 3x - 1}{3x + 2} \div \frac{2x + 1}{9x^2 + 3x - 2}$

You Try It

47. $\frac{x^2 - y^2}{4x + 2y} \div \frac{x^2 - 3xy - 4y^2}{2x^2 - 7xy - 4y^2}$

48. $\frac{x^3 - y^3}{x^2 - xy + y^2} \div \frac{x^2 - y^2}{x^3 + y^3}$

In Exercises 49–54, perform the indicated operations.

You Try It

49. $\frac{5xy}{x^3} \cdot \frac{xy^3}{8y} \div \frac{2y^3}{3x^4y}$

50. $\frac{7}{x^2} \div \frac{9x^2y}{x^3} \cdot \frac{4x}{x^8}$

51. $\frac{x^2 - 2x - 3}{x - 1} \cdot \frac{x + 4}{x^2 + 2x - 15} \cdot \frac{x^2 + 4x - 5}{x^2 - x - 2}$

52. $\frac{x^2 + 6x + 8}{x^2 + 6x + 5} \div \frac{x^2 - 5x - 14}{x^2 + 3x - 10} \cdot \frac{x^2 - 6x - 7}{x^2 + x - 6}$

53. $\frac{x^2 + 5x + 4}{x - 3} \cdot \frac{x^2 - 4x + 3}{x^2 + 6x + 5} \div \frac{x^2 - 3x + 2}{x + 5}$

54. $\frac{x^2 + x - 2}{x^2 - 5x + 6} \div \frac{x^2 + 4x + 3}{x^2 + 2x - 15} \div \frac{(x + 2)^2}{x^2 + 3x - 10}$

16.3 Least Common Denominators

THINGS TO KNOW

Before working through this topic, be sure you are familiar with the following concepts:

	VIDEO	ANIMATION	INTERACTIVE
You Try It **1.** Factor Polynomials Completely (Topic 15.6, **Objective 1**)			
You Try It **2.** Simplify Rational Expressions (Topic 16.1, **Objective 3**)			

OBJECTIVES

1 Find the Least Common Denominator of Rational Expressions

2 Write Equivalent Rational Expressions

(eText Screens 16.3-1–16.3-15)

OBJECTIVE 1 FIND THE LEAST COMMON DENOMINATOR OF RATIONAL EXPRESSIONS

Recall that when adding or subtracting **fractions**, we need a **common denominator**. Typically, we use the least common denominator (**LCD**). The same is true when adding or subtracting rational expressions, which we will visit in the next topic.

To find the LCD of rational expressions, first we **factor** each denominator. Then we include each **unique factor** the largest number of times that it occurs in any denominator. We can illustrate this idea by considering **rational numbers**.

To find the LCD of $\frac{4}{15}$, $\frac{7}{18}$, and $\frac{3}{8}$, first factor each denominator into its **prime factors**.

$$15 = 3 \cdot 5$$
$$18 = 2 \cdot 3 \cdot 3$$
$$8 = 2 \cdot 2 \cdot 2$$

Next, form the LCD by including each unique factor the largest number of times that it occurs in any denominator. The factor 2 occurs at most three times in any denominator, the factor 3 occurs at most two times, and the factor 5 occurs at most once. So, our LCD is

$$\text{LCD} = 2 \cdot 2 \cdot 2 \cdot 3 \cdot 3 \cdot 5 = 2^3 \cdot 3^2 \cdot 5 = 360.$$

View this **popup** for a summary of the steps to find the LCD of rational numbers.

We follow a similar process to find the **LCD** of **rational expressions**.

Finding the Least Common Denominator (LCD) of Rational Expressions

Step 1. **Factor** each **denominator** completely.

Step 2. List each **unique factor** from any denominator.

Step 3. The least common denominator is the product of the unique factors, each raised to a **power** equivalent to the largest number of times that the factor occurs in any denominator.

Don't forget numerical factors when forming the LCD.

Example 1 Finding the Least Common Denominator of Rational Expressions

Find the LCD of the rational expressions.

a. $\frac{7}{10x^3}, \frac{3}{5x^2}$ **b.** $\frac{x+2}{3x}, \frac{x-1}{2x^2+6x}$

Solutions

a. Follow the **three-step process.**

Step 1. $10x^3 = 2 \cdot 5 \cdot x \cdot x \cdot x$

$5x^2 = 5 \cdot x \cdot x$

Step 2. We have **unique factors** of 2, 5, and x. Make sure to include each factor the largest number of times that it occurs in any **denominator.**

Step 3. The factors 2 and 5 occur at most once, and the factor x occurs at most three times.

$\text{LCD} = 2 \cdot 5 \cdot x \cdot x \cdot x = 10x^3$

b. Follow the three-step process.

Step 1. $3x = 3 \cdot x$

$2x^2 + 6x = 2 \cdot x(x + 3)$

Step 2. We have unique factors of 3, x, 2, and $(x + 3)$. Notice that we must list both x and $x + 3$ as factors because x is not a factor of $x + 3$.

Step 3. The factors 3, x, 2, and $(x + 3)$ each occur at most once.

$\text{LCD} = 3 \cdot 2 \cdot x \cdot (x + 3) = 6x(x + 3)$

You Try It **Work through this You Try It problem.**

Work Exercises 1–6 in this textbook or in the MyMathLab Study Plan.

Example 2 Finding the Least Common Denominator of Rational Expressions

Find the LCD of the rational expressions.

a. $\dfrac{z^2}{6 - z}, \dfrac{9}{2z - 12}$ **b.** $\dfrac{y + 2}{y^2 + 2y - 3}, \dfrac{2y}{y^2 + 5y + 6}$

Solutions

a. Follow the **three-step process**.

Step 1. $6 - z = -1 \cdot (z - 6)$

$2z - 12 = 2 \cdot (z - 6)$

Step 2. We have unique factors of -1, $(z - 6)$, and 2. Notice that we factored out a -1 in the first denominator. We will do this when the **leading coefficient** is negative to avoid including **opposite factors** in the **LCD**. In this case, we include -1 and $z - 6$ instead of $z - 6$ and $6 - z$.

Determine the largest number of times that each unique factor occurs and form the LCD. Then view the **answer**.

My video summary

b. Follow the three-step process.

Step 1. $y^2 + 2y - 3 = (y + 3)(y - 1)$

$y^2 + 5y + 6 = (y + 3)(y + 2)$

Determine the unique factors and the largest number of times that each occurs. Then form the LCD. View the **answer**, or watch this **video** for a complete solution.

You Try It **Work through this You Try It problem.**

Work Exercises 7–12 in this textbook or in the MyMathLab Study Plan.

Example 3 Finding the Least Common Denominator of Rational Expressions

My interactive video summary

Find the **LCD** of the **rational expressions**.

a. $\dfrac{4x}{10x^2 - 7x - 12}, \dfrac{2x - 3}{5x^2 - 11x - 12}$

b. $\dfrac{10 - x}{6x^2 + 5x + 1}, \dfrac{-4}{9x^2 + 6x + 1}, \dfrac{x^2 - 7x}{10x^2 - x - 3}$

(eText Screens 16.3-1–16.3-15)

Solutions Try to find the LCDs on your own. View the **answers**, or watch this **interactive video** for complete solutions to both parts.

You Try It **Work through this You Try It problem.**

Work Exercises 13–17 in this textbook or in the MyMathLab Study Plan.

OBJECTIVE 2 WRITE EQUIVALENT RATIONAL EXPRESSIONS

Let's look at how to write an **equivalent rational expression** using the **LCD**. The key to doing this lies with our previous work on dividing out **common factors** in **Topic 16.1.** Recall that

$$\frac{\text{common factor}}{\text{common factor}} = 1.$$

We can therefore **divide out** common factors that occur in the **numerator** and denominator because the quotient of the common factors equals 1.

When writing **equivalent fractions,** we are actually working the other way. We multiply the original expression by 1, but in such a way as to obtain the desired LCD. To do this, we first ask the question,

"What should we multiply the original denominator by to get the LCD?"

Whatever the answer is, we multiply both the numerator and denominator by this answer.

For example, to write $\frac{3}{5}$ as an equivalent fraction with a denominator of 20, we ask, "What do we multiply 5 by to get 20?" The answer is 4, so we multiply both the numerator and denominator by 4 to get $\frac{3}{5} = \frac{3}{5} \cdot 1 = \frac{3}{5} \cdot \frac{4}{4} = \frac{12}{20}$. Notice that $\frac{4}{4} = 1$, so we have not changed the value, just the way it looks.

To work with **rational expressions,** we ask the same question: *"What do we multiply the original denominator by to get the LCD?"* Then we multiply both the **numerator** and **denominator** by the answer.

> **Writing Equivalent Rational Expressions**
>
> To write an equivalent rational expression with a desired denominator:
>
> **Step 1.** Factor the given denominator and the desired denominator.
>
> **Step 2.** Determine the missing factor(s) from the given denominator that must be multiplied to get the desired denominator.
>
> **Step 3.** Multiply the numerator and denominator of the given rational expression by the missing factor(s) from Step 2.

CAUTION From the commutative property of multiplication, the order of the factors in either the numerator or denominator does not matter. For example,

$$\frac{(x+1)(x-4)}{(x+3)(2x-1)} = \frac{(x-4)(x+1)}{(2x-1)(x+3)}.$$

(eText Screens 16.3-1–16.3-15)

Example 4 Writing Equivalent Rational Expressions

Write each rational expression as an equivalent **rational expression** with the desired denominator.

a. $\dfrac{3}{2x} = \dfrac{}{10x^3}$ b. $\dfrac{x+2}{3x+15} = \dfrac{}{3(x-1)(x+5)}$

Solutions We follow the **three-step process**.

a. **Step 1.** $2x = 2 \cdot x$

$10x^3 = 2 \cdot 5 \cdot x \cdot x \cdot x$

Step 2. The given denominator is missing $5 \cdot x \cdot x$. What do we multiply $2x$ by to get $10x^3$? The answer is $5 \cdot x \cdot x = 5x^2$.

Step 3. $\dfrac{3}{2x} = \dfrac{3}{2x} \cdot \underbrace{\dfrac{5x^2}{5x^2}}_{1} = \dfrac{15x^2}{10x^3}$

So, $\dfrac{3}{2x} = \dfrac{15x^2}{10x^3}$.

b. **Step 1.** $3x + 15 = 3 \cdot (x + 5)$

$3(x - 1)(x + 5)$ is already factored.

Step 2. The given denominator is missing $(x - 1)$. What do we multiply $3x + 15$ by to get $3(x - 1)(x + 5)$? The answer is $(x - 1)$.

Step 3. $\dfrac{x+2}{3x+15} = \dfrac{x+2}{3(x+5)} \cdot \underbrace{\dfrac{(x-1)}{(x-1)}}_{1} = \dfrac{(x+2)(x-1)}{3(x+5)(x-1)}$ or $\dfrac{(x+2)(x-1)}{\underbrace{3(x-1)(x+5)}_{\text{Commutative property of multiplication}}}$

So, $\dfrac{x+2}{3x+15} = \dfrac{(x+2)(x-1)}{3(x-1)(x+5)}$.

You Try It **Work through this You Try It problem.**

Work Exercises 18–23 in this textbook or in the MyMathLab Study Plan.

Example 5 Writing Equivalent Rational Expressions

Write each **rational expression** as an equivalent rational expression with the desired **denominator**.

a. $\dfrac{-7}{1-4y} = \dfrac{}{8y^2 - 2y}$ b. $\dfrac{5z}{z^2 + z - 6} = \dfrac{}{(z-4)(z-2)(z+3)}$

Solutions

a. **Step 1.** $1 - 4y = -1 \cdot (4y - 1)$

$8y^2 - 2y = 2 \cdot y \cdot (4y - 1)$

Step 2. What do we multiply $1 - 4y$ by to get $8y^2 - 2y$? The answer is $-2 \cdot y = -2y$. Notice that we need to use -2 so that $-1 \cdot -2 = 2$.

Multiply the numerator and denominator of the given rational expression by $-2y$ on your own. Then view the **answer**.

(eText Screens 16.3-1–16.3-15)

My video summary **b. Step 1.** $z^2 + z - 6 = (z + 3)(z - 2)$

$(z - 4)(z - 2)(z + 3)$ is already factored.

Try finishing this problem on your own. Determine what is missing from the given denominator, then multiply both the numerator and denominator of the given rational expression by this quantity. View the **answer**, or watch this **video** for a complete solution.

You Try It **Work through this You Try It problem.**

Work Exercises 24–28 in this textbook or in the MyMathLab Study Plan.

16.3 Exercises

In Exercises 1–17, find the LCD for the given rational expressions.

You Try It

1. $\frac{5}{6x^5}, \frac{4}{9x^2}$

2. $\frac{6}{5a^3b^4}, \frac{7}{15a^2b^8}$

3. $\frac{3}{7a^3c^3}, \frac{5}{14b^5c^2}$

4. $\frac{x+1}{4x^2}, \frac{x-2}{3x^2+9x}$

5. $\frac{3x-5}{6x^2-3x}, \frac{2x+1}{12x^2+4x}$

6. $\frac{5}{3x^3+9x^2+6x}, \frac{7x}{2x^3-2x}$

You Try It

7. $\frac{x}{3-x}, \frac{2x+1}{x^2-x-6}$

8. $\frac{3x+1}{x^2-4x-5}, \frac{x+7}{2+x-x^2}$

9. $\frac{2x}{3+2x-x^2}, \frac{5x-1}{9-x^2}$

10. $\frac{y-5}{y^2-y-2}, \frac{y+5}{y^2+y-6}$

11. $\frac{2y-1}{y^2-9}, \frac{1-5y}{y^2+6y+9}$

12. $\frac{2x}{4-x^2}, \frac{1-x}{6+x-x^2}$

You Try It

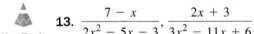

13. $\frac{7-x}{2x^2-5x-3}, \frac{2x+3}{3x^2-11x+6}$

14. $\frac{3x^2+2x+1}{10x^2-15x}, \frac{2x^2-7x}{6x^2-13x+6}$

15. $\frac{5x+1}{6x^3+26x^2-20x}, \frac{5x-1}{6x^4+33x^3+15x^2}$

16. $\frac{7+x}{4-x^2}, \frac{6+x}{x^2+x-2}, \frac{5+x}{x^2-x-2}$

17. $\frac{x+1}{2x^2+x-3}, \frac{x+2}{3x^2+13x-10}, \frac{x+3}{6x^2+5x-6}$

In Exercises 18–28, write each rational expression as an equivalent rational expression with the desired denominator.

You Try It

18. $\frac{2}{3x} = \frac{}{15x^4}$

19. $\frac{3}{5x^2y^3} = \frac{}{15x^4y^5}$

20. $\frac{2b}{3a^2c^3} = \frac{}{21a^5b^3c^3}$

21. $\frac{x+1}{2x+6} = \frac{}{2(x-1)(x+3)}$

22. $\dfrac{7}{x^2 - x - 6} = \dfrac{}{(x+2)(x-2)(x+3)(x-3)}$

23. $\dfrac{3x - 5}{2x + 3} = \dfrac{}{(2x+3)(x+2)}$

You Try It

24. $\dfrac{5}{2 - x}, \dfrac{}{(x^2 - 4)(x^2 + x)}$

25. $\dfrac{x + 1}{9x - 3x^2} = \dfrac{}{3x(x-1)(x-2)(x-3)}$

26. $\dfrac{2x + 1}{x^2 + x - 2} = \dfrac{}{(x-1)(x+1)(x+2)}$

27. $\dfrac{3x - 5}{2x^2 - 4x - 6} = \dfrac{}{10(x-3)(x-2)(x+1)}$

28. $\dfrac{6 - z}{z^2 + 8z + 15} = \dfrac{}{(z^2 + 7z + 12)(z^2 + 7z + 10)}$

16.4 Adding and Subtracting Rational Expressions

THINGS TO KNOW

Before working through this topic, be sure you are familiar with the following concepts:

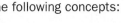

		VIDEO	ANIMATION	INTERACTIVE
You Try It	1. Add Polynomials (Topic 14.3, **Objective 1**)			
You Try It	2. Subtract Polynomials (Topic 14.3, **Objective 3**)			
You Try It	3. Factor Polynomials Completely (Topic 15.6, **Objective 1**)			
You Try It	4. Simplify Rational Expressions (Topic 16.1, **Objective 3**)			

OBJECTIVES

1 Add and Subtract Rational Expressions with Common Denominators

2 Add and Subtract Rational Expressions with Unlike Denominators

OBJECTIVE 1 ADD AND SUBTRACT RATIONAL EXPRESSIONS WITH COMMON DENOMINATORS

Recall that when adding or subtracting **fractions**, we first check for a **common denominator**. Then we add or subtract the **numerators** and keep the common denominator. To add or subtract **rational numbers** with common **denominators**, we follow this general approach:

$$\frac{p}{q} + \frac{r}{q} = \frac{p + r}{q} \quad \text{or} \quad \frac{p}{q} - \frac{r}{q} = \frac{p - r}{q}.$$

(eText Screens 16.4-1–16.4-23)

For example,

$$\frac{4}{5}+\frac{3}{5}=\frac{4+3}{5}=\frac{7}{5} \quad \text{or} \quad \frac{11}{3}-\frac{7}{3}=\frac{11-7}{3}=\frac{4}{3}.$$

The same is true when working with **rational expressions**. Once we have a common denominator, we add or subtract the numerators and keep the common denominator.

Adding and Subtracting Rational Expressions with Common Denominators

If $\frac{P}{Q}$ and $\frac{R}{Q}$ are rational expressions, then

$$\frac{P}{Q}+\frac{R}{Q}=\frac{P+R}{Q} \quad \text{and} \quad \frac{P}{Q}-\frac{R}{Q}=\frac{P-R}{Q}.$$

Example 1 Adding and Subtracting Rational Expressions with Common Denominators

Add or subtract.

a. $\frac{4z}{3}+\frac{5z}{3}$ b. $\frac{3r}{7s^2}-\frac{2r}{7s^2}$

Solutions

a. The **rational expressions** have **common denominators**, so we write the sum of the **numerators** over the common denominator and **simplify**.

Begin with the original expression: $\frac{4z}{3}+\frac{5z}{3}$

Add the numerators: $=\frac{4z+5z}{3}$

Combine like terms: $=\frac{9z}{3}$

Divide out common factor of 3: $=\frac{{}^{3}\cancel{9}z}{\cancel{3}_{1}}$

$=3z$

b. The **rational expressions** have a **common denominator**, so we write the difference of the **numerators** over the common denominator and **simplify**.

Begin with the original expression: $\frac{3r}{7s^2}-\frac{2r}{7s^2}$

Subtract the numerators: $=\frac{3r-2r}{7s^2}$

Simplify: $=\frac{r}{7s^2}$

You Try It **Work through this You Try It problem.**

Work Exercises 1–5 in this textbook or in the MyMathLab Study Plan.

(eText Screens 16.4-1–16.4-23)

When subtracting rational expressions, it is a good idea to use **grouping symbols** around each numerator that involves more than one **term**. This will help remind us to subtract each term in the second numerator. We see an illustration of this point in the next example.

Example 2 Adding and Subtracting Rational Expressions with Common Denominators

Add or subtract.

a. $\dfrac{9x}{x-4}+\dfrac{7x-2}{x-4}$ b. $\dfrac{5y+1}{y-2}-\dfrac{2y+3}{y-2}$

Solutions

a. The **rational expressions** have **common denominators**, so we write the sum of the **numerators** over the common denominator and **simplify**.

Begin with the original expression: $\dfrac{9x}{x-4}+\dfrac{7x-2}{x-4}$

Add the numerators: $=\dfrac{9x+7x-2}{x-4}$

Finish simplifying. Check the resulting rational expression to see if there are any **common factors** that can be **divided out**. Then, view the **answer**.

b. The **rational expressions** have **common denominators**, so we write the difference of the **numerators** over the common denominator and **simplify**.

Begin with the original expression: $\dfrac{5y+1}{y-2}-\dfrac{2y+3}{y-2}$

Subtract the numerators: $=\dfrac{(5y+1)-(2y+3)}{y-2}$

Use the distributive property: $=\dfrac{5y+1-2y-3}{y-2}$

Finish simplifying. Check the resulting rational expression to see if there are any **common factors** that can be **divided out**. Then, view the **answer**.

You Try It **Work through this You Try It problem.**

Work Exercises 6–10 in this textbook or in the MyMathLab Study Plan.

Example 3 Adding and Subtracting Rational Expressions with Common Denominators

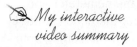

My interactive video summary

Add or subtract.

a. $\dfrac{4}{x^2+2x-8}+\dfrac{x}{x^2+2x-8}$ b. $\dfrac{x}{x+2}-\dfrac{x-3}{x+2}$ c. $\dfrac{x^2-2}{x-5}-\dfrac{4x+3}{x-5}$

Solutions Try to perform the operations on your own. Remember to check your result for any **common factors**. View the **answers**, or watch this **interactive video** for complete solutions to all three parts.

(eText Screens 16.4-1–16.4-23)

You Try It **Work through this You Try It problem.**

Work Exercises 11–15 in this textbook or in the MyMathLab Study Plan.

OBJECTIVE 2 ADD AND SUBTRACT RATIONAL EXPRESSIONS WITH UNLIKE DENOMINATORS

To add or subtract **rational expressions** with unlike **denominators**, first we write each fraction as an **equivalent expression** using the **LCD**. Then we can add or subtract the numerators, keeping the common denominator.

Adding and Subtracting Rational Expressions with Unlike Denominators

Step 1. Find the **LCD** for all expressions being added or subtracted.

Step 2. Write **equivalent expressions** for each term using the LCD as the denominator.

Step 3. Add/subtract the numerators, but keep the denominator the same (the LCD).

Step 4. Simplify if possible.

CAUTION We focus on the LCD because it is the easiest **common denominator** to use. Any common denominator will allow us to add or subtract rational expressions.

Example 4 Adding and Subtracting Rational Expressions

Perform the indicated operations and simplify.

a. $\frac{7}{6x} + \frac{3}{2x^3}$ **b.** $\frac{3x}{x-3} - \frac{x-2}{x+3}$

Solutions We follow the **four-step process.**

a. Step 1.

$$6x = 2 \cdot 3 \cdot x$$
$$2x^3 = 2 \cdot x \cdot x \cdot x$$

We have **unique factors** of 2, 3, and x. Make sure to include each factor the largest number of times that it occurs in any **denominator**.

$$\text{LCD} = 2 \cdot 3 \cdot x \cdot x \cdot x = 6x^3$$

Step 2. In the first expression, we multiply the denominator by x^2 to get the **LCD**, so we obtain an **equivalent expression** by multiplying both the numerator and denominator by x^2. In the second expression, we multiply the denominator by 3 to get the LCD, so we obtain an equivalent expression by multiplying both the numerator and denominator by 3.

The overall expression becomes

$$\frac{7}{6x} + \frac{3}{2x^3} = \frac{7}{6x} \cdot \frac{x^2}{x^2} + \frac{3}{2x^3} \cdot \frac{3}{3}$$
$$= \frac{7x^2}{6x^3} + \frac{9}{6x^3}$$

Step 3. With a **common denominator**, we can now add:

$$\frac{7x^2}{6x^3} + \frac{9}{6x^3} = \frac{7x^2 + 9}{6x^3}$$

Step 4. The expression cannot be simplified further, so $\frac{7}{6x} + \frac{3}{2x^3} = \frac{7x^2 + 9}{6x^3}$.

My video summary

b. Step 1. The **denominators** cannot be factored further.

We have **unique factors** of $x - 3$ and $x + 3$. Make sure to include each factor the largest number of times that it occurs in any denominator.

$$\text{LCD} = (x - 3)(x + 3)$$

Step 2. In the first expression, we multiply the denominator by $x + 3$ to get the **LCD** and obtain an **equivalent expression** by multiplying both the **numerator** and denominator by $x + 3$. In the second expression, we multiply the denominator by $x - 3$ to get the LCD and obtain an equivalent expression by multiplying both the numerator and denominator by $x - 3$.

So, we write

$$\frac{3x}{x - 3} - \frac{x - 2}{x + 3} = \frac{3x}{x - 3} \cdot \frac{x + 3}{x + 3} - \frac{x - 2}{x + 3} \cdot \frac{x - 3}{x - 3}$$

$$= \frac{3x(x + 3)}{(x - 3)(x + 3)} - \frac{(x - 2)(x - 3)}{(x - 3)(x + 3)}.$$

Carry out the subtraction on your own and **simplify** if necessary. View the **answer**, or watch this **video** for a complete solution.

You Try It **Work through this You Try It problem.**

Work Exercises 16–20 in this textbook or in the MyMathLab Study Plan.

Example 5 Adding and Subtracting Rational Expressions

Perform the indicated operations and simplify.

a. $\frac{z + 2}{3z} - \frac{5}{3z + 12}$

b. $\frac{5}{4m - 12} + \frac{3}{2m}$

Solutions

a. Step 1.

$$3z = 3 \cdot z$$

$$3z + 12 = 3(z + 4)$$

We have **unique factors** of $3, z,$ and $(z + 4)$. Make sure to include each factor the largest number of times that it occurs in any **denominator**.

$$\text{LCD} = 3 \cdot z \cdot (z + 4) = 3z(z + 4).$$

Step 2. In the first expression, we multiply the denominator by $z + 4$ to get the LCD and obtain an **equivalent expression** by multiplying both the numerator and denominator by $z + 4$. In the second expression, we multiply the denominator by z to get the LCD, so we obtain an equivalent expression by multiplying both the numerator and denominator by z.

The overall expression becomes

$$\frac{z+2}{3z} - \frac{5}{3z+12} = \frac{z+2}{3z}\cdot\frac{z+4}{z+4} - \frac{5}{3(z+4)}\cdot\frac{z}{z}$$
$$= \frac{z^2+6z+8}{3z(z+4)} - \frac{5z}{3z(z+4)}$$

Step 3. With a **common denominator**, we can now subtract:

Subtract: $= \dfrac{z^2+6z+8-5z}{3z(z+4)}$

Simplify: $= \dfrac{z^2+z+8}{3z(z+4)}$

Step 4. The expression cannot be simplified further, so

$$\frac{z+2}{3z} - \frac{5}{3z+12} = \frac{z^2+z+8}{3z(z+4)}.$$

My video summary

b. Write equivalent expressions using the LCD and carry out the addition on your own. **Simply** if necessary, then view the **answer** or watch this **video** for a complete solution.

You Try It **Work through this You Try It problem.**

Work Exercises 21–25 in this textbook or in the MyMathLab Study Plan.

Example 6 Adding and Subtracting Rational Expressions

Perform the indicated operations and simplify.

a. $2 + \dfrac{4}{x-5}$ **b.** $\dfrac{x^2-2}{x^2+6x+8} - \dfrac{x-3}{x+4}$

Solutions Follow the **four-step process.**

a. Step 1. Note that the first term is a **rational expression** since we can write it as $2 = \frac{2}{1}$. The **LCD** is $x - 5$.

Step 2. In the first expression, we multiply the denominator by $x - 5$ to get the LCD and obtain an **equivalent expression** by multiplying both the numerator and denominator by $x - 5$.

We write

$$2 + \frac{4}{x-5} = \frac{2}{1} + \frac{4}{x-5}$$
$$= \frac{2}{1}\cdot\frac{x-5}{x-5} + \frac{4}{x-5}$$
$$= \frac{2(x-5)}{x-5} + \frac{4}{x-5}.$$

Step 3. With a **common denominator**, we can now add:

Add: $= \dfrac{2(x-5)+4}{x-5}$

Distribute: $= \dfrac{2x-10+4}{x-5}$

$$\text{Simplify:} \quad = \frac{2x - 6}{x - 5}$$

$$= \frac{2(x - 3)}{x - 5}$$

Step 4. The expression cannot be simplified further, so

$$2 + \frac{4}{x - 5} = \frac{2(x - 3)}{x - 5}.$$

My video summary

b. Step 1. Factor the denominators and determine the **LCD**. The second denominator is **prime**, but the first denominator can be factored as $x^2 + 6x + 8 = (x + 4)(x + 2)$

The LCD is $(x + 4)(x + 2)$.

Step 2. In the first expression, we already have the **LCD**. In the second expression, we multiply the denominator by $x + 2$ to get the LCD and obtain an **equivalent expression** by multiplying both the **numerator** and **denominator** by $x + 2$.

Doing this gives

$$\frac{x^2 - 2}{x^2 + 6x + 8} - \frac{x - 3}{x + 4} = \frac{x^2 - 2}{(x + 4)(x + 2)} - \frac{x - 3}{x + 4} \cdot \frac{x + 2}{x + 2}$$

$$= \frac{x^2 - 2}{(x + 4)(x + 2)} - \frac{(x - 3)(x + 2)}{(x + 4)(x + 2)}.$$

Carry out the subtraction on your own and **simplify** if necessary. View the **answer**, or watch this **video** for a complete solution.

You Try It **Work through this You Try It problem.**

Work Exercises 26–30 in this textbook or in the MyMathLab Study Plan.

Example 7 Adding and Subtracting Rational Expressions with Unlike Denominators

My interactive video summary

Perform the indicated operations and simplify.

a. $\dfrac{x + 7}{x^2 - 9} + \dfrac{3}{x + 3}$ b. $\dfrac{x + 1}{2x^2 + 5x - 3} - \dfrac{x}{2x^2 + 3x - 2}$

Solutions Perform the operations on your own and **simplify**. View the **answers**, or watch this **interactive video** to see complete solutions.

You Try It **Work through this You Try It problem.**

Work Exercises 31–38 in this textbook or in the MyMathLab Study Plan.

If we are adding or subtracting **rational expressions** whose denominators are **opposites**, we can find a **common denominator** by multiplying the numerator and denominator of either rational expression (but not both) by -1.

(eText Screens 16.4-1–16.4-23)

Example 8 Adding and Subtracting Rational Expressions with Unlike Denominators

Perform the indicated operation and **simplify**.

$$\frac{2y}{y-5}+\frac{y-1}{5-y}$$

Solution

The denominators are opposites: $\frac{2y}{y-5}+\frac{y-1}{5-y}$

Multiply the numerator and denominator of the second rational expression by -1: $=\frac{2y}{y-5}+\frac{(-1)}{(-1)}\cdot\frac{y-1}{5-y}$

Distribute: $=\frac{2y}{y-5}+\frac{-y+1}{-5+y}$

Rewrite the denominator in second expression: $=\frac{2y}{y-5}+\frac{-y+1}{y-5}$

Add: $=\frac{2y-y+1}{y-5}$

Simplify: $=\frac{y+1}{y-5}$

You Try It **Work through this You Try It problem.**

Work Exercises 39 and 40 in this textbook or in the MyMathLab Study Plan.

Example 9 Adding and Subtracting Rational Expressions with Unlike Denominators

Perform the indicated operations and **simplify**.

$$\frac{x+1}{x^2-6x+9}+\frac{3}{x-3}-\frac{6}{x^2-9}$$

Solution Following the **four-step process**, we start by finding the **LCD** for all the terms.
$x^2-6x+9=(x-3)(x-3)$; $x-3$ is prime; $x^2-9=(x+3)(x-3)$

The LCD is $(x-3)(x-3)(x+3)$.

Next, we rewrite each term as an **equivalent expression** using the LCD.

$$\frac{x+1}{x^2-6x+9}+\frac{3}{x-3}-\frac{6}{x^2-9}$$

$$=\frac{x+1}{(x-3)(x-3)}\cdot\frac{x+3}{x+3}+\frac{3}{x-3}\cdot\frac{(x-3)(x+3)}{(x-3)(x+3)}-\frac{6}{(x+3)(x-3)}\cdot\frac{x-3}{x-3}$$

$$=\frac{(x+1)(x+3)}{(x-3)(x-3)(x+3)}+\frac{3(x-3)(x+3)}{(x-3)(x-3)(x+3)}-\frac{6(x-3)}{(x-3)(x-3)(x+3)}$$

Now we combine the three **numerators** and keep the **common denominator**.

$$=\frac{(x+1)(x+3)+3(x-3)(x+3)-6(x-3)}{(x-3)(x-3)(x+3)}$$

(eText Screens 16.4-1–16.4-23)

Finish **simplifying** the expression on your own. View the **answer**, or watch this **video** for a complete solution.

You Try It **Work through this You Try It problem.**

Work Exercises 41–47 in this textbook or in the MyMathLab Study Plan.

16.4 Exercises

In Exercises 1–47, add or subtract as indicated.

You Try It

1. $\frac{4x}{7} + \frac{2x}{7}$

2. $\frac{8}{7y^3} + \frac{6}{7y^3}$

3. $\frac{a^2 + 1}{5c^2} + \frac{b^2 + 2}{5c^2}$

4. $\frac{2x}{8y^3 + 2} - \frac{x}{8y^3 + 2}$

5. $\frac{5x + 1}{9a^3} - \frac{2x + 1}{9a^3}$

6. $\frac{x}{x + 5} + \frac{6}{x + 5}$

You Try It

7. $\frac{x + 1}{2x + 3} + \frac{2x + 1}{2x + 3}$

8. $\frac{5x + 1}{3x + 2} + \frac{x + 3}{3x + 2}$

9. $\frac{2x - 3}{3x} - \frac{2x + 9}{3x}$

You Try It

10. $\frac{7x + 3}{5x - 1} - \frac{2x + 5}{5x - 1}$

11. $\frac{2}{x^2 + 4x + 3} + \frac{2x}{x^2 + 4x + 3}$

12. $\frac{x^2}{2x + 1} + \frac{x^2 + x}{2x + 1}$

13. $\frac{x^2 + 9}{x^2 - 9} + \frac{6x}{x^2 - 9}$

14. $\frac{x}{x^2 - 4} - \frac{2}{x^2 - 4}$

15. $\frac{x^2 + x + 6}{x^2 - x - 6} - \frac{3x + 9}{x^2 - x - 6}$

You Try It

16. $\frac{x + 5}{5x^2} + \frac{3}{10x}$

17. $\frac{2}{3x} + \frac{3}{2x}$

18. $\frac{5x}{x + 2} + \frac{7}{2x + 1}$

19. $\frac{x + 2}{x - 2} - \frac{x - 2}{x + 2}$

20. $\frac{2x + 1}{x} - \frac{2x}{x + 1}$

21. $\frac{x + 1}{2x} + \frac{3x}{5x - 10}$

You Try It

22. $\frac{2m}{3m + 3} + \frac{m + 1}{2m}$

23. $\frac{x + 2}{2x} - \frac{3}{2x + 10}$

24. $\frac{3}{2x + 2} - \frac{2}{3x + 3}$

You Try It

25. $\frac{x}{x^2 - 1} - \frac{1}{x}$

26. $\frac{3x}{x + 1} + 2$

27. $\frac{2(x + 11)}{2x^2 - x - 10} + \frac{x + 4}{x + 2}$

28. $\frac{3x}{x + 1} - 2$

29. $\frac{x^2 - 3}{x^2 - 5x + 6} - \frac{x - 3}{x - 2}$

30. $\frac{x}{2} - \frac{2}{x}$

You Try It

31. $\frac{x - 1}{x^2 - 4} + \frac{5}{x - 2}$

32. $\frac{4}{x^2 - 4} + \frac{2x}{x + 2}$

33. $\frac{1 - 12x}{2x^2 - 5x - 3} + \frac{x + 2}{x - 3}$

34. $\frac{3x - 4}{x + 3} - \frac{6 - 24x}{x^2 - 9}$

35. $\frac{2x}{x^2 - x - 2} - \frac{x}{x^2 - 1}$

36. $\frac{2x - 1}{x + 1} + \frac{3(3x + 1)}{x^2 - 1}$

37. $\frac{8}{x - 2} - \frac{3x + 26}{x^2 - 4}$

38. $\frac{6x}{2x^2 - 8xy + 4y^2} - \frac{3y}{x^2 - 3xy + 2y^2}$

You Try It

39. $\dfrac{3x}{x-1}+\dfrac{2x+3}{1-x}$

40. $\dfrac{x^2}{x-3}+\dfrac{9}{3-x}$

41. $\dfrac{1}{x}+\dfrac{1}{x+1}-\dfrac{1}{x-1}$

You Try It

42. $\dfrac{1}{x^2+x}+\dfrac{1}{x^2-x}+\dfrac{1}{x^2-1}$

43. $\dfrac{x-3}{x^2+12x+36}+\dfrac{1}{x+6}-\dfrac{2x+3}{2x^2+3x-54}$

44. $\dfrac{1}{x+1}+\dfrac{2}{x+2}-\dfrac{3}{x+3}$

45. $\dfrac{1}{x+2}+\dfrac{1}{x-2}-\dfrac{4}{x^2-4}$

46. $\dfrac{5}{y}-\dfrac{5}{y-3}+\dfrac{16}{(y-3)^2}$

47. $\left(\dfrac{1}{2}+\dfrac{3}{x}\right)-\left(\dfrac{1}{2}-\dfrac{1}{x}\right)$

16.5 Complex Rational Expressions

THINGS TO KNOW

Before working through this topic, be sure you are familiar with the following concepts:

		VIDEO	ANIMATION	INTERACTIVE
You Try It	1. Use the Negative-Power Rule (Topic 14.6, **Objective 1**)	✓		
You Try It	2. Simplify Rational Expressions (Topic 16.1, **Objective 3**)	✓		
You Try It	3. Multiply Rational Expressions (Topic 16.2, **Objective 1**)	✓		✓
You Try It	4. Divide Rational Expressions (Topic 16.2, **Objective 2**)	✓		✓
You Try It	5. Find the Least Common Denominator of Rational Expressions (Topic 16.3, **Objective 1**)	✓		✓
You Try It	6. Add and Subtract Rational Expressions with Unlike Denominators (Topic 16.4, **Objective 2**)	✓		✓

OBJECTIVES

1 Simplify Complex Rational Expressions by First Simplifying the Numerator and Denominator

2 Simplify Complex Rational Expressions by Multiplying by a Common Denominator

OBJECTIVE 1 SIMPLIFY COMPLEX RATIONAL EXPRESSIONS BY FIRST SIMPLIFYING THE NUMERATOR AND DENOMINATOR

Sometimes the **numerator** and/or **denominator** of a **rational expression** will contain one or more rational expressions. In this topic, we learn to simplify such *complex rational expressions*.

(eText Screens 16.5-1–16.5-18)

> **Definition Complex Rational Expression**
>
> A **complex rational expression,** or **complex fraction,** is a rational expression in which the numerator and/or denominator contain(s) rational expressions.

Some examples of **complex rational expressions** are

$$\frac{\frac{2}{9x}}{\frac{5}{6xy}}, \quad \frac{\frac{x+2}{x-4}}{\frac{x^2-4}{x+1}}, \quad \frac{\frac{1}{y}+\frac{y+1}{y-1}}{\frac{1}{y}-\frac{y-1}{y+1}}, \quad \frac{\frac{1}{x}+\frac{1}{y}}{z}, \text{ and } \frac{z^{-2}+z^{-1}}{z^{-1}-z^{-2}}.$$

The rational expressions within the numerator and denominator are called **minor rational expressions** or **minor fractions.** The numerator and denominator of the complex rational expression are separated by the **main fraction bar.**

$$\frac{\frac{x+2}{x-4}}{\frac{x^2-4}{x+1}}$$

Minor rational expressions: $\frac{x+2}{x-4}$ and $\frac{x^2-4}{x+1}$

$\frac{x+2}{x-4}$ ← Numerator of the complex rational expression

← Main fraction bar

$\frac{x^2-4}{x+1}$ ← Denominator of the complex rational expression

To **simplify a complex rational expression**, we rewrite it in the form $\frac{P}{Q}$, $Q \neq 0$, where P and Q are **polynomials** with no **common factors.** In this topic, we show two methods that can be used to **simplify complex rational expressions.** We call these *Method I* and *Method II.* Method I results from recognizing that the **main fraction bar** is a division symbol. So, we divide the **numerator** by the **denominator.**

Example 1 Simplifying a Complex Rational Expression

Simplify $\frac{\frac{2}{9x}}{\frac{5}{6xy}}$.

Solution We **divide** the **minor rational expression** from the **numerator** by the minor rational expression from the **denominator.** To do this, we multiply the numerator by the **reciprocal** of the denominator.

Begin with the original expression: $\frac{\frac{2}{9x}}{\frac{5}{6xy}} = \frac{2}{9x} \div \frac{5}{6xy}$

Change to multiplication by the reciprocal: $= \frac{2}{9x} \cdot \frac{6xy}{5}$

Factor: $= \frac{2}{3 \cdot 3 \cdot x} \cdot \frac{2 \cdot 3 \cdot x \cdot y}{5}$

Divide out common factors: $= \frac{2}{\cancel{3} \cdot 3 \cdot \cancel{x}} \cdot \frac{2 \cdot \cancel{3} \cdot \cancel{x} \cdot y}{5}$

Multiply the remaining factors: $= \frac{4y}{15}$

(eText Screens 16.5-1–16.5-18)

You Try It **Work through this You Try It problem.**

Work Exercises 1–3 in this textbook or in the MyMathLab® Study Plan.

In addition to acting as a division symbol, the **main fraction bar** of a **complex fraction** also serves as a **grouping symbol**. If the **numerator** and **denominator** are not each written as single **rational expressions**, then they must be simplified as such before dividing.

Method I for Simplifying Complex Rational Expressions

Step 1. Simplify the expression in the numerator into a single rational expression.

Step 2. Simplify the expression in the denominator into a single rational expression.

Step 3. Divide the expression in the numerator by the expression in the denominator. To do this, multiply the **minor rational expression** in the numerator by the **reciprocal** of the minor rational expression in the denominator. Simplify if possible.

Example 2 Simplifying a Complex Rational Expression Using Method I

Use **Method I** to simplify each complex rational expression.

a. $\dfrac{\dfrac{1}{3}-\dfrac{1}{x}}{\dfrac{1}{9}-\dfrac{1}{x^2}}$ **b.** $\dfrac{4-\dfrac{5}{x-1}}{\dfrac{6}{x-1}-7}$

Solutions

a. Step 1. The **LCD** for the **minor fractions** $\frac{1}{3}$ and $-\frac{1}{x}$ in the **numerator** is $3x$.

$$\frac{1}{3}-\frac{1}{x}=\frac{1}{3}\cdot\frac{x}{x}-\frac{1}{x}\cdot\frac{3}{3}=\frac{x}{3x}-\frac{3}{3x}=\frac{x-3}{3x}$$

Step 2. The LCD for the minor fractions $\frac{1}{9}$ and $-\frac{1}{x^2}$ in the **denominator** is $9x^2$.

$$\frac{1}{9}-\frac{1}{x^2}=\frac{1}{9}\cdot\frac{x^2}{x^2}-\frac{1}{x^2}\cdot\frac{9}{9}=\frac{x^2}{9x^2}-\frac{9}{9x^2}=\frac{x^2-9}{9x^2}$$

Step 3. Substitute the simplified **rational expressions** from Steps 1 and 2 for the numerator and denominator of the **complex fraction**, and then divide.

$$\frac{\dfrac{1}{3}-\dfrac{1}{x}}{\dfrac{1}{9}-\dfrac{1}{x^2}}=\frac{\dfrac{x-3}{3x}}{\dfrac{x^2-9}{9x^2}}=\frac{x-3}{3x}\div\frac{x^2-9}{9x^2}$$

Change to multiplication by the reciprocal: $=\dfrac{x-3}{3x}\cdot\dfrac{9x^2}{x^2-9}$

Factor: $=\dfrac{x-3}{3x}\cdot\dfrac{3\cdot3\cdot x\cdot x}{(x+3)(x-3)}$

Divide out common factors: $= \dfrac{\cancel{x-3}}{\cancel{3}\cancel{x}} \cdot \dfrac{\cancel{3}\cdot 3\cdot \cancel{x}\cdot x}{(x+3)\cancel{(x-3)}}$

Multiply remaining factors: $= \dfrac{3x}{x+3}$

My video summary

b. Step 1. The **LCD** for the **numerator** is $x - 1$.

$$4 - \frac{5}{x-1} = 4\cdot\frac{x-1}{x-1} - \frac{5}{x-1} = \frac{4(x-1)-5}{x-1} = \frac{4x-4-5}{x-1} = \frac{4x-9}{x-1}$$

Step 2. The LCD for the **denominator** is $x - 1$.

$$\frac{6}{x-1} - 7 = \frac{6}{x-1} - 7\cdot\frac{x-1}{x-1} = \frac{6-7(x-1)}{x-1} = \frac{6-7x+7}{x-1} = \frac{13-7x}{x-1}$$

Step 3. Try to finish simplifying the complex rational expression on your own. View the **answer,** or watch this **video** for the complete solution to part b.

You Try It **Work through this You Try It problem.**

Work Exercises 4–6 in this textbook or in the MyMathLab® Study Plan.

OBJECTIVE 2 SIMPLIFY COMPLEX RATIONAL EXPRESSIONS BY MULTIPLYING BY A COMMON DENOMINATOR

In **Topic 16.3**, we saw that multiplying the **numerator** and **denominator** of a **rational expression** by the same nonzero **expression** resulted in an equivalent expression. For example, $\frac{3}{4} = \frac{3}{4}\cdot\frac{7}{7} = \frac{21}{28}$ because $\frac{7}{7} = 1$.

The second method for simplifying **complex rational expressions** is based on this same concept. Let's take another look at the complex fraction from **Example 1**.

Example 3 Simplifying a Complex Rational Expression

Simplify $\dfrac{\frac{2}{9x}}{\frac{5}{6xy}}$.

Solution Let's multiply the numerator and denominator of the complex rational expression by $18xy$, the **LCD** of $\frac{2}{9x}$ and $\frac{5}{6xy}$.

Begin with the original expression: $\dfrac{\frac{2}{9x}}{\frac{5}{6xy}}$

Multiply the numerator and denominator by $18xy$: $= \dfrac{\frac{2}{9x}}{\frac{5}{6xy}} \cdot \dfrac{18xy}{18xy}$

Divide out common factors: $= \dfrac{\dfrac{2}{\cancel{9x}} \cdot \overset{2}{\cancel{18}}\cancel{x}y}{\dfrac{5}{\cancel{6xy}} \cdot \overset{3}{\cancel{18}}\cancel{xy}}$

Rewrite the remaining factors: $= \dfrac{2 \cdot 2 \cdot y}{5 \cdot 3}$

Multiply: $= \dfrac{4y}{15}$

You Try It **Work through this You Try It problem.**

Work Exercises 7–9 in this textbook or in the MyMathLab® Study Plan.

In **Example 3**, we multiplied the **numerator** and **denominator** of the **complex rational expression** by the **LCD** of its **minor rational expressions.** Multiplying by the LCD is the foundation of Method II for simplifying complex rational expressions.

> **Method II for Simplifying Complex Rational Expressions**
>
> **Step 1.** Determine the LCD of all the minor rational expressions within the complex rational expression.
>
> **Step 2.** Multiply the numerator and denominator of the complex rational expression by the LCD from Step 1.
>
> **Step 3.** Simplify.

For comparison, let's revisit the complex rational expressions from **Example 2.**

Example 4 Simplifying a Complex Rational Expression Using Method II

Use **Method II** to simplify each complex rational expression.

a. $\dfrac{\dfrac{1}{3} - \dfrac{1}{x}}{\dfrac{1}{9} - \dfrac{1}{x^2}}$ **b.** $\dfrac{4 - \dfrac{5}{x-1}}{\dfrac{6}{x-1} - 7}$

Solutions

a. Step 1. The **denominators** of all the **minor fractions** are $3, x, 9$, and x^2, so the **LCD** is $9x^2$.

Step 2. Multiply the **numerator** and denominator by $9x^2$.

$$\frac{\dfrac{1}{3} - \dfrac{1}{x}}{\dfrac{1}{9} - \dfrac{1}{x^2}} \cdot \frac{9x^2}{9x^2}$$

Step 3. Use the distributive property: $= \dfrac{\dfrac{1}{3} \cdot 9x^2 - \dfrac{1}{x} \cdot 9x^2}{\dfrac{1}{9} \cdot 9x^2 - \dfrac{1}{x^2} \cdot 9x^2}$

Divide out common factors: $= \dfrac{\frac{1}{\cancel{3}}\cdot \overset{3}{\cancel{9}}x^2 - \frac{1}{\cancel{x}}\cdot 9\overset{x}{\cancel{x^2}}}{\frac{1}{\cancel{9}}\cdot \cancel{9}x^2 - \frac{1}{\cancel{x^2}}\cdot 9\cancel{x^2}}$

Simplify: $= \dfrac{3x^2 - 9x}{x^2 - 9}$

Factor each polynomial: $= \dfrac{3x(x-3)}{(x+3)(x-3)}$

Divide out common factors: $= \dfrac{3x\cancel{(x-3)}}{(x+3)\cancel{(x-3)}}$

Simplify: $= \dfrac{3x}{x+3}$

My video summary

b. Step 1. The **LCD** for all the **minor fractions** is $x - 1$.

Step 2. Multiply the **numerator** and **denominator** by $x - 1$.

$$\frac{4 - \frac{5}{x-1}}{\frac{6}{x-1} - 7}\cdot\frac{x-1}{x-1}$$

Step 3. Try to finish simplifying the complex rational expression on your own. View the **answer**, or watch this **video** for the complete solution.

You Try It **Work through this You Try It problem.**

Work Exercises 10–12 in this textbook or in the MyMathLab® Study Plan.

As you have seen in Examples 1–4, **Method I** and **Method II** both give the same simplification. Have you developed a preference for one method over the other? Whether you have or not, we suggest that you practice using both methods for a while to discover when one method might be better suited than the other.

Example 5 Simplifying a Complex Rational Expression

My interactive video summary

Simplify the **complex rational expression** using Method I or Method II.

$$\frac{\frac{5}{n-2} - \frac{3}{n}}{\frac{6}{n^2 - 2n} + \frac{2}{n}}$$

Solution Try simplifying on your own. View the **answer**, or watch this **interactive video** to see the complete solution.

You Try It **Work through this You Try It problem.**

Work Exercises 13–24 in this textbook or in the MyMathLab® Study Plan.

(eText Screens 16.5-1–16.5-18)

Complex rational expressions can be written using **negative exponents.** For example, the expression $\frac{3^{-1} - x^{-1}}{3^{-2} - x^{-2}}$ is equivalent to the complex fraction $\frac{\frac{1}{3} - \frac{1}{x}}{\frac{1}{9} - \frac{1}{x^2}}$ from **Example 2.**

When a complex rational expression contains negative exponents, we can rewrite it as an equivalent expression with positive exponents by using the **negative power rule.** Then we simplify it by using Method I or Method II.

Example 6 Simplifying a Complex Rational Expression Containing Negative Exponents

My interactive video summary

Simplify the complex rational expression.

$$\frac{1 - 9y^{-1} + 14y^{-2}}{1 + 3y^{-1} - 10y^{-2}}$$

Solution Use the **negative power rule** to rewrite the expression with positive **exponents.**

$$\frac{1 - 9y^{-1} + 14y^{-2}}{1 + 3y^{-1} - 10y^{-2}} = \frac{1 - \frac{9}{y} + \frac{14}{y^2}}{1 + \frac{3}{y} - \frac{10}{y^2}}$$

Try to finish simplifying the **complex rational expression** on your own using **Method I** or **Method II.** View the **answer,** or watch this **interactive video** for the complete solution.

You Try It **Work through this You Try It problem.**

Work Exercises 25 and 26 in this textbook or in the MyMathLab® Study Plan.

16.5 Exercises

In Exercises 1–6, use Method I to simplify each complex rational expression.

You Try It

1. $\frac{\frac{14}{5x}}{\frac{7}{15x}}$

2. $\frac{\frac{x+7}{2}}{\frac{3x-1}{10}}$

3. $\frac{\frac{5x+5}{x^2-9}}{\frac{x^2-1}{x+3}}$

You Try It

4. $\frac{\frac{5}{6} + \frac{3}{4}}{\frac{8}{3} - \frac{5}{9}}$

5. $\frac{\frac{3}{x} + 2}{\frac{9}{x^2} - 4}$

6. $\frac{\frac{9}{x-4} + \frac{6}{x-5}}{\frac{5x-23}{x^2-9x+20}}$

In Exercises 7–12, use Method II to simplify each complex rational expression.

You Try It

7. $\frac{\frac{x+2}{15x}}{\frac{2x-1}{10x}}$

8. $\frac{\frac{8x}{x+y}}{\frac{2x^3}{y}}$

9. $\frac{\frac{3}{x^2-x-2}}{\frac{x+4}{x-2}}$

(eText Screens 16.6-1–16.6-20)

You Try It

10. $\dfrac{2-\dfrac{a}{b}}{\dfrac{a^2}{b^2}-4}$

11. $\dfrac{3-\dfrac{7}{x-2}}{\dfrac{8}{x-2}-5}$

12. $\dfrac{\dfrac{x+5}{x}-\dfrac{8}{x-1}}{\dfrac{x+1}{x}+\dfrac{x+1}{x-1}}$

In Exercises 13–26, simplify each complex rational expression using Method I or Method II.

13. $\dfrac{7+\dfrac{1}{x}}{7-\dfrac{1}{x}}$

14. $\dfrac{\dfrac{1}{x}+3}{\dfrac{1}{x^2}-9}$

15. $\dfrac{\dfrac{x+2}{x-6}-\dfrac{x+12}{x+5}}{x+82}$

16. $\dfrac{\dfrac{6}{x}+\dfrac{5}{y}}{\dfrac{5}{x}-\dfrac{6}{y}}$

17. $\dfrac{\dfrac{1}{16}-\dfrac{1}{x^2}}{\dfrac{1}{4}+\dfrac{1}{x}}$

18. $\dfrac{\dfrac{8}{x+4}-\dfrac{2}{x+7}}{\dfrac{x+8}{x+4}}$

You Try It

19. $\dfrac{\dfrac{2}{x+13}+\dfrac{1}{x-5}}{2-\dfrac{x+25}{x+13}}$

20. $\dfrac{\dfrac{x-5}{x+5}+\dfrac{x-5}{x-9}}{1+\dfrac{x+5}{x-9}}$

21. $\dfrac{1+\dfrac{4}{x}-\dfrac{5}{x^2}}{1-\dfrac{2}{x}-\dfrac{35}{x^2}}$

22. $\dfrac{\dfrac{x-4}{x^2-25}}{1+\dfrac{1}{x-5}}$

23. $\dfrac{\dfrac{x+3}{x-3}-\dfrac{x-3}{x+3}}{\dfrac{x-3}{x+3}+\dfrac{x+3}{x-3}}$

24. $\dfrac{\dfrac{3}{a^2}-\dfrac{1}{ab}-\dfrac{2}{b^2}}{\dfrac{2}{a^2}-\dfrac{5}{ab}+\dfrac{3}{b^2}}$

You Try It

25. $\dfrac{6x^{-1}+6y^{-1}}{xy^{-1}-x^{-1}y}$

26. $\dfrac{7x^{-1}-3y^{-1}}{49x^{-2}-9y^{-2}}$

16.6 Solving Rational Equations

THINGS TO KNOW

Before working through this topic, be sure you are familiar with the following concepts:

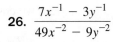

	VIDEO	ANIMATION	INTERACTIVE
You Try It **1.** Solve Linear Equations Containing Fractions (Topic 11.2, **Objective 2**)	▤		
You Try It **2.** Solve a Formula for a Given Variable (Topic 11.4, **Objective 3**)	▤		
You Try It **3.** Factor Polynomials Completely (Topic 15.6, **Objective 1**)			▶
You Try It **4.** Solve Polynomial Equations by Factoring (Topic 15.7, **Objective 2**)	▤		

(eText Screens 16.6-1–16.6-20)

You Try It 5. Evaluate Rational Expressions (Topic 16.1, **Objective 1**)

You Try It 6. Find Restricted Values for Rational Expressions (Topic 16.1, **Objective 2**)

You Try It 7. Find the Least Common Denominator of Rational Expressions (Topic 16.3, **Objective 1**)

OBJECTIVES

1 Identify Rational Equations

2 Solve Rational Equations

3 Identify and Solve Proportions

4 Solve a Formula Containing Rational Expressions for a Given Variable

OBJECTIVE 1 IDENTIFY RATIONAL EQUATIONS

As we have seen before, we make a clear distinction between an **equation** and an **expression**. Recall that we ***simplify*** expressions and ***solve*** equations.

We have previously defined **linear equations** and **polynomial equations**. Now, let's define a *rational equation*.

Definition Rational Equation

A **rational equation** is a statement in which two **rational expressions** are set equal to each other.

Some examples of rational equations are

$$\underbrace{\frac{2x}{x-1}}_{\text{Rational expression}} \underbrace{=}_{\text{Equal sign}} \underbrace{\frac{2x+3}{x}}_{\text{Rational expression}} \quad \text{and} \quad \underbrace{\frac{1}{x-2}+\frac{1}{x+2}}_{\text{Rational expression}} \underbrace{=}_{\text{Equal sign}} \underbrace{\frac{4}{x^2-4}}_{\text{Rational expression}}.$$

Example 1 Identifying Rational Equations

Determine if each statement is a rational equation. If not, state why.

a. $\frac{x-4}{x}+\frac{4}{x+5}=\frac{6}{x}$

b. $\frac{5}{y}+\frac{7}{y+2}$

c. $\frac{\sqrt{k+1}}{k+3}=\frac{k-5}{k+4}$

d. $5n^{-1}=3n^{-2}$

Solutions

a. Both $\frac{x-4}{x}+\frac{4}{x+5}$ and $\frac{6}{x}$ are **rational expressions**. They are set equal to each other, so the statement is a **rational equation**.

b. This statement $\frac{5}{y}+\frac{7}{y+2}$ does not contain an equal sign, so it is not a rational equation. Instead, it is a rational *expression*.

c. This statement is not a rational equation because $\frac{\sqrt{k+1}}{k+3}$ is not a rational expression.

Note that the **numerator** is not a **polynomial** because there is a **variable** under a **radical**.

d. Note that $5n^{-1} = \frac{5}{n}$ and $3n^{-2} = \frac{3}{n^2}$, so they are both rational expressions. They are set equal to each other, so this statement is a rational equation.

You Try It **Work through this You Try It problem.**

Work Exercises 1–4 in this textbook or in the MyMathLab Study Plan.

OBJECTIVE 2 SOLVE RATIONAL EQUATIONS

In **Topic 11.2**, we learned to **clear fractions** from **equations** by multiplying both sides of the equation by the **LCD**. We then solved the resulting equation. This idea is the key to solving **rational equations**.

Example 2 Solving Rational Equations

Solve.

a. $\frac{1}{2}x + \frac{2}{3} = \frac{3}{4}$ **b.** $\frac{1}{x} + \frac{1}{2} = \frac{1}{3}$

Solutions

a. The LCD of the fractions $\frac{1}{2}, \frac{2}{3}$, and $\frac{3}{4}$ is 12.

Write the original equation: $\frac{1}{2}x + \frac{2}{3} = \frac{3}{4}$

Multiply both sides by the LCD: $12\left(\frac{1}{2}x + \frac{2}{3}\right) = 12\left(\frac{3}{4}\right)$

Distribute: $12\left(\frac{1}{2}x\right) + 12\left(\frac{2}{3}\right) = 12\left(\frac{3}{4}\right)$

Simplify: $6x + 8 = 9$

Subtract 8 from both sides: $6x = 1$

Divide both sides by 6: $x = \frac{1}{6}$

View this **popup box** to see that this answer checks. The **solution set** is $\left\{\frac{1}{6}\right\}$.

My video summary

b. The **LCD** of $\frac{1}{x}, \frac{1}{2}$, and $\frac{1}{3}$ is $6x$.

Write the original equation: $\frac{1}{x} + \frac{1}{2} = \frac{1}{3}$

Multiply both sides by the LCD: $6x\left(\frac{1}{x} + \frac{1}{2}\right) = 6x\left(\frac{1}{3}\right)$

Distribute: $6x \cdot \frac{1}{x} + 6x \cdot \frac{1}{2} = 6x \cdot \frac{1}{3}$

Divide out common factors: $6\cancel{x}\cdot\frac{1}{\cancel{x}} + \overset{3}{\cancel{6}}x\cdot\frac{1}{\underset{1}{\cancel{2}}} = \overset{2}{\cancel{6}}x\cdot\frac{1}{\underset{1}{\cancel{3}}}$

Simplify: $6 + 3x = 2x$

Finish solving this equation on your own. Remember to check your result. View the **answer**, or watch this **video** for a complete solution to part b.

You Try It **Work through this You Try It problem.**

Work Exercises 5–8 in this textbook or in the MyMathLab Study Plan.

Recall that values that cause the **denominator** of a **rational expression** to equal zero are **restricted values.** Restricted values cannot be **solutions** of **rational equations.** However, sometimes when we multiply both sides of a rational equation by the **LCD**, a solution of the resulting **linear** or **polynomial equation** will be a restricted value of the original rational equation. Such "solutions" are called **extraneous solutions**, and they must be excluded from the **solution set.**

For example, let's solve the equation $\frac{x+1}{x-2} = \frac{3}{x-2}$. Notice that $x = 2$ is a restricted value for these rational expressions. Also, the LCD of the rational expressions is $x - 2$.

Multiply both sides by the LCD: $(x-2)\left(\frac{x+1}{x-2}\right) = (x-2)\left(\frac{3}{x-2}\right)$

Divide out common factors: $\cancel{(x-2)}\cdot\frac{x+1}{\cancel{(x-2)}} = \cancel{(x-2)}\cdot\frac{3}{\cancel{(x-2)}}$

Simplify: $x + 1 = 3$

Subtract 1 from both sides: $x = 2$

The resulting solution is the restricted value, which means it does not make the original equation true. So, we must discard this result as an extraneous solution. This means that the equation $\frac{x+1}{x-2} = \frac{3}{x-2}$ has no solution.

The following steps can be used to solve rational equations.

Solving Rational Equations

Step 1. List all **restricted values.**

Step 2. Determine the **LCD** of all **denominators** in the equation.

Step 3. Multiply both sides of the equation by the LCD.

Step 4. Solve the resulting **polynomial equation.**

Step 5. Discard any restricted values and check the remaining **solutions** in the original equation.

Example 3 Solving Rational Equations

Solve $\frac{2}{x} - \frac{x-3}{2x} = 3$.

(eText Screens 16.6-1–16.6-20)

Solution We follow the **five-step process.**

Step 1. To find the **restricted values**, we look for values that make any **denominator** equal to zero.

$$\frac{2}{x} - \frac{x-3}{2x} = \frac{3}{1}$$

Examining the variable factors in the denominators, we see that the only restricted value is 0.

Step 2. There are unique **factors** of 2 and x in the denominators, each occurring at most once. The **LCD** is $2x$.

Step 3. Multiply both sides of the equation by the LCD.

Multiply both sides by the LCD: $2x\left(\frac{2}{x} - \frac{x-3}{2x}\right) = 2x(3)$

Distribute: $2x \cdot \frac{2}{x} - 2x \cdot \frac{x-3}{2x} = 2x \cdot 3$

Divide out common factors: $2\cancel{x} \cdot \frac{2}{\cancel{x}} - \cancel{2x} \cdot \frac{x-3}{\cancel{2x}} = 2x \cdot 3$

Multiply: $4 - (x-3) = 6x$

Step 4. Solve the resulting equation.

Distribute: $4 - x + 3 = 6x$

Simplify: $7 - x = 6x$

Add x to both sides: $7 = 7x$

Divide both sides by 7: $1 = x$

Step 5. The potential solution $x = 1$ is not a **restricted value**, so we do not discard it. View this **popup box** to see the check of $x = 1$ in the original equation. The **solution set** is $\{1\}$.

You Try It **Work through this You Try It problem.**

Work Exercises 9 and 10 in this textbook or in the MyMathLab Study Plan.

Example 4 Solving Rational Equations

My video summary

Solve $\frac{4}{5} - \frac{3}{x-3} = \frac{1}{x}$.

Solution We follow the **five-step process.**

Step 1. To find the **restricted values**, we look for values that make any **denominator** equal to zero. Set each denominator with a **variable** equal to zero and solve.

$$x - 3 = 0 \quad \text{or} \quad x = 0$$
$$x = 3$$

The restricted values are 0 and 3.

Step 2. There are unique **factors** of 5, x, and $x - 3$ in the denominators, each occurring at most once. The **LCD** is $5x(x-3)$.

Step 3. Multiply both sides of the equation by the LCD.

Multiply both sides by the LCD: $5x(x-3)\left(\frac{4}{5}-\frac{3}{x-3}\right)=5x(x-3)\left(\frac{1}{x}\right)$

Distribute: $5x(x-3)\cdot\frac{4}{5}-5x(x-3)\cdot\frac{3}{x-3}=5x(x-3)\cdot\frac{1}{x}$

Divide out common factors: $\cancel{5}x(x-3)\cdot\frac{4}{\cancel{5}}-5x\cancel{(x-3)}\cdot\frac{3}{\cancel{x-3}}=5\cancel{x}(x-3)\cdot\frac{1}{\cancel{x}}$

Multiply: $4x(x-3)-15x=5(x-3)$

Try to finish solving the equation on your own. View the **answer**, or watch this **video** for a complete solution.

You Try It **Work through this You Try It problem.**

Work Exercises 11 and 12 in this textbook or in the MyMathLab Study Plan.

Example 5 Solving Rational Equations

My video summary

Solve $\frac{m}{m+2}+\frac{5}{m-2}=\frac{20}{m^2-4}$.

Solution We follow the **five-step process.**

Step 1. To find the **restricted values**, look for values that make any denominator equal to zero. First, factor the denominators.

$$\frac{m}{m+2}+\frac{5}{m-2}=\frac{20}{(m-2)(m+2)}$$

Now set each **variable factor** equal to zero and solve.

$$m+2=0 \quad \text{or} \quad m-2=0$$
$$m=-2 \qquad\qquad m=2$$

The restricted values are -2 and 2.

Step 2. We have unique **factors** of $m-2$ and $m+2$ in the denominators, each occurring at most once. The **LCD** is $(m-2)(m+2)$.

Step 3. Multiply both sides of the equation by the **LCD**, distribute, and **divide out** common factors.

$$(m-2)(m+2)\left(\frac{m}{m+2}+\frac{5}{m-2}\right)=(m-2)(m+2)\cdot\frac{20}{(m-2)(m+2)}$$

$$(m-2)(m+2)\cdot\frac{m}{m+2}+(m-2)(m+2)\cdot\frac{5}{m-2}=(m-2)(m+2)\cdot\frac{20}{(m-2)(m+2)}$$

$$(m-2)\cancel{(m+2)}\cdot\frac{m}{\cancel{m+2}}+\cancel{(m-2)}(m+2)\cdot\frac{5}{\cancel{m-2}}=\cancel{(m-2)(m+2)}\cdot\frac{20}{\cancel{(m-2)(m+2)}}$$

$$m(m-2)+5(m+2)=20$$

Finish solving the equation on your own, making sure to check for **restricted values.** View the **answer**, or watch this **video** for a complete solution.

You Try It **Work through this You Try It problem.**

Work Exercises 13 and 14 in this textbook or in the MyMathLab Study Plan.

(eText Screens 16.6-1–16.6-20)

Example 6 Solving Rational Equations

My video summary

Solve $\dfrac{2}{x-3} - \dfrac{4}{x^2 - 2x - 3} = \dfrac{1}{x+1}$.

Solution Following the **five-step process**, we start by finding the **restricted values.** Factoring the denominators gives

$$\frac{2}{x-3} - \frac{4}{(x-3)(x+1)} = \frac{1}{x+1}.$$

Set each **variable factor** in the denominators equal to zero and solve.

$$x - 3 = 0 \quad \text{or} \quad x + 1 = 0$$
$$x = 3 \qquad\qquad x = -1$$

The restricted values are -1 and 3.

The **LCD** is $(x-3)(x+1)$. Multiplying both sides of the equation by the LCD gives the **polynomial equation**

$$2(x+1) - 4 = x - 3.$$

Solving the polynomial equation leads to $x = -1$. However, $x = -1$ is a restricted value, so it must be discarded. Since no other possible **solution** exists, the equation has no solution. The **solution set** is $\{\ \}$ or $\varnothing$. Watch this **video** for a fully worked solution.

You Try It **Work through this You Try It problem.**

Work Exercises 15 and 16 in this textbook or in the MyMathLab Study Plan.

OBJECTIVE 3 IDENTIFY AND SOLVE PROPORTIONS

A **ratio** is the **quotient** of two numbers or **algebraic expressions**, such as with **slope** $\left(\dfrac{\text{rise}}{\text{run}}\right)$. A **rational number** is the quotient of two **integers**, so rational numbers are ratios. Likewise, a **rational expression** is the quotient of two **polynomials**, so rational expressions are ratios.

A **proportion** is a statement that two ratios are equal, such as $\dfrac{P}{Q} = \dfrac{R}{S}$. When a **rational equation** is a proportion, we can solve it by first **cross multiplying** to get $P \cdot S = Q \cdot R$ and then solving the resulting **polynomial equation**.

Cross Multiplying

$$\frac{P}{Q} \times \frac{R}{S}$$

$$P \cdot S = Q \cdot R$$

Note that any "solution" that is a **restricted value** in the original equation must be discarded. Proportions can be used to solve many types of applications, as we will see later in Topic 16.7.

Example 7 Solving Proportions

Solve.

a. $\dfrac{8}{x+3} = \dfrac{5}{x}$ **b.** $\dfrac{x}{6} = \dfrac{2}{x-1}$

(eText Screens 16.6-1–16.6-20)

Solutions

a. This **rational equation** has the form $\frac{P}{Q} = \frac{R}{S}$, so it is a **proportion.**

First, we identify the **restricted values** by setting each denominator equal to zero and solving.

$$x + 3 = 0 \quad \text{or} \quad x = 0$$
$$x = -3$$

We have two restricted values, -3 and 0.

Next, we **cross multiply** and solve the resulting **polynomial equation.**

Begin with the original equation:	$\frac{8}{x+3} = \frac{5}{x}$
Cross multiply:	$8x = 5(x + 3)$
Distribute:	$8x = 5x + 15$
Subtract $5x$ from both sides:	$3x = 15$
Divide both sides by 3:	$x = 5$

Because 5 is not a restricted value, the solution set is $\{5\}$. View this **popup box** to see that this result checks.

My video summary

b. This **rational equation** has the form $\frac{P}{Q} = \frac{R}{S}$, so it is a **proportion.**

The denominator $x - 1$ is the only one that contains a variable, so the only **restricted value** is 1.

Begin with the original equation:	$\frac{x}{6} = \frac{2}{x-1}$
Cross multiply:	$x(x - 1) = 6(2)$

Try to solve this resulting equation on your own. Be sure to check for **extraneous solutions.** View the **answer,** or watch this **video** for a complete solution.

You Try It **Work through this You Try It problem.**

Work Exercises 17–20 in this textbook or in the MyMathLab Study Plan.

CAUTION

Only proportions can be solved by cross multiplying. If a rational equation is not a proportion, do not try to cross multiply.

OBJECTIVE 4 SOLVE A FORMULA CONTAINING RATIONAL EXPRESSIONS FOR A GIVEN VARIABLE

Recall that a **formula** is an **equation** that describes the relationship between two or more **variables.** In **Topic 11.4,** we learned to **solve a formula** for a given variable. If a formula contains rational expressions, we can solve it for a given variable by using the **steps for solving a rational equation.**

(eText Screens 16.6-1–16.6-20)

Example 8 Solving a Formula for a Given Variable

Solve each formula for the given variable.

a. $I = \dfrac{E}{r + R}$ for R **b.** $\dfrac{1}{f} = \dfrac{1}{c} + \dfrac{1}{d}$ for d

Solutions

a. Write the original formula: $I = \dfrac{E}{r + R}$

Multiply both sides by $r + R$, the LCD: $I(r + R) = \left(\dfrac{E}{r + R}\right)(r + R)$

Distribute; divide out common factor: $Ir + IR = E$ ← $\left(\dfrac{E}{\cancel{r + R}}\right)(\cancel{r + R}) = E$

Subtract Ir from both sides: $IR = E - Ir$

Divide both sides by I: $R = \dfrac{E - Ir}{I}$

Solving the formula $I = \dfrac{E}{r + R}$ for R results in $R = \dfrac{E - Ir}{I}$.

My video summary

b. Multiplying both sides of the **formula** by its **LCD** cdf results in

$$cd = df + cf.$$

To **solve** for d, we must get both **terms** that contain d on the same side of the equal sign. We subtract df from both sides to result in

$$cd - df = cf.$$

To be solved for d, the formula can only contain one d, so we **factor** it out.

$$d(c - f) = cf$$

Dividing both sides by $c - f$ gives

$$d = \frac{cf}{c - f},$$

which is the desired result. Watch this **video** for a fully worked solution.

You Try It **Work through this You Try It problem.**

Work Exercises 21–28 in this textbook or in the MyMathLab Study Plan.

CAUTION When solving a formula for a variable that occurs more than once, you will typically need to move all terms containing the desired variable to the same side of the equal sign and then factor out the variable.

16.6 Exercises

In Exercises 1–4, determine if the statement is a rational equation. If not, state why.

You Try It

1. $\dfrac{x - 3}{x^2 + 1} = \dfrac{\sqrt{x^2 - 4x + 1}}{x - 5}$

2. $4y^{-2} = 15 - 17y^{-1}$

(eText Screens 16.6-1–16.6-20)

3. $\dfrac{x^2 + 3x - 4}{2x + 5} = \dfrac{x - 2}{3x + 1}$

4. $\dfrac{x^3 - 4x^2 + 5x}{2x - 9}$

In Exercises 5–20, solve the rational equation. If there is no solution, state this as your answer.

You Try It

5. $\dfrac{1}{5}x - \dfrac{5}{6} = \dfrac{2}{3}$

6. $\dfrac{n - 7}{8} = \dfrac{5n}{3} - \dfrac{7}{12}$

7. $\dfrac{1}{y} + \dfrac{1}{6} = \dfrac{1}{4}$

8. $\dfrac{5}{m} - \dfrac{3}{10} = \dfrac{3}{2m} - \dfrac{9}{4}$

You Try It

9. $\dfrac{4}{3x} - \dfrac{x + 1}{x} = \dfrac{2}{5}$

10. $\dfrac{x - 1}{x} = \dfrac{3}{4} - \dfrac{3}{2x}$

You Try It

11. $\dfrac{3}{z} - \dfrac{9}{z - 5} = \dfrac{1}{4}$

12. $\dfrac{5}{3p} + \dfrac{2p}{p - 5} = -\dfrac{4}{3}$

You Try It

13. $\dfrac{b}{b - 5} + \dfrac{2}{b + 5} = \dfrac{50}{b^2 - 25}$

14. $\dfrac{3}{3m + 2} - \dfrac{2}{m - 6} = \dfrac{1}{3m^2 - 16m - 12}$

You Try It

15. $\dfrac{1}{n + 4} + \dfrac{3}{n + 5} = \dfrac{1}{n^2 + 9n + 20}$

16. $\dfrac{6}{y^2 - 25} = \dfrac{3}{y^2 - 5y}$

You Try It

17. $\dfrac{3}{t + 2} = \dfrac{6}{t + 16}$

18. $\dfrac{k}{7} = \dfrac{5}{k - 2}$

19. $\dfrac{x + 2}{x - 5} = \dfrac{x - 6}{x + 1}$

20. $\dfrac{x}{3x + 2} = \dfrac{x + 5}{3x - 1}$

In Exercises 21–28, solve the formula for the given variable.

You Try It

21. $E = \dfrac{RC}{T}$ for C

22. $L = \dfrac{h}{mv}$ for v

23. $z = \dfrac{x - m}{s}$ for s

24. $P = \dfrac{A}{1 + rt}$ for t

25. $m = \dfrac{y_2 - y_1}{x_2 - x_1}$ for x_2

26. $\dfrac{F}{G} = \dfrac{m_1 m_2}{r^2}$ for G

27. $R = \dfrac{(E - B) + D}{B}$ for B

28. $\dfrac{1}{A} - \dfrac{1}{B} = \dfrac{1}{C}$ for A

(eText Screens 16.7-1–16.7-26)

16.7 Applications of Rational Equations

THINGS TO KNOW

Before working through this topic, be sure you are familiar with the following concepts: VIDEO ANIMATION INTERACTIVE

You Try It 1. Use the Problem-Solving Strategy to Solve Direct Translation Problems (Topic 11.3, **Objective 2**)

You Try It 2. Find the Value of a Non-isolated Variable in a Formula (Topic 11.4, **Objective 2**)

You Try It 3. Solve Problems Involving Uniform Motion (Topic 11.5, **Objective 3**)

You Try It 4. Solve Rational Equations (Topic 16.6, **Objective 2**)

You Try It 5. Identify and Solve Proportions (Topic 16.6, **Objective 3**)

OBJECTIVES

1 Use Proportions to Solve Problems

2 Use Formulas Containing Rational Expressions to Solve Problems

3 Solve Uniform Motion Problems Involving Rational Equations

4 Solve Problems Involving Rate of Work

OBJECTIVE 1 USE PROPORTIONS TO SOLVE PROBLEMS

Recall from **Topic 16.6** that a **proportion** is a statement that two **ratios** are equal, such as $\frac{P}{Q} = \frac{R}{S}$. Writing proportions is a powerful tool in solving application problems. Given a ratio (or **rate**) of two quantities, a proportion can be used to determine an unknown quantity.

Example 1 Defective Lightbulbs

A quality-control inspector examined a sample of 200 lightbulbs and found 18 of them to be defective. At this ratio, how many defective bulbs can the inspector expect in a shipment of 22,000 light bulbs?

Solution We follow our usual **problem-solving strategy**.

Step 1. We want to find the number of defective lightbulbs expected in a shipment of 22,000 lightbulbs. We know that there were 18 defective bulbs in a sample of 200 bulbs.

Step 2. Let $x =$ the number of defective bulbs expected in the shipment.

Step 3. The ratio of defective bulbs to total bulbs in the shipment is assumed to be the same as the ratio of defective bulbs to total bulbs in the sample. We can set these two ratios equal to each other to form a proportion.

(eText Screens 16.7-1–16.7-26)

$$\begin{array}{cc} \text{Shipment} & \text{Sample} \\ \downarrow & \downarrow \end{array}$$

$$\frac{\text{Defective Bulbs}}{\text{Total Bulbs}} = \frac{\text{Defective Bulbs}}{\text{Total Bulbs}}$$

$$\frac{x}{22{,}000} = \frac{18}{200}$$

Step 4. Because we have a **proportion**, we can **cross multiply** to solve.

$$\text{Cross multiply:} \quad \frac{x}{22{,}000} \times \frac{18}{200}$$

$$200x = 22{,}000(18)$$

$$\text{Simplify:} \quad 200x = 396{,}000$$

$$\text{Divide both sides by 200:} \quad x = 1980$$

Step 5. There are 110 times as many lightbulbs in the shipment as in the sample $(110 \cdot 200 = 22{,}000)$. Therefore, there also should be 110 times as many defects in the shipment as in the sample. Because $110 \cdot 18 = 1980$, our result is reasonable and checks.

Step 6. The inspector can expect 1980 defective bulbs in a shipment of 22,000 lightbulbs.

You Try It **Work through this You Try It problem.**

Work Exercises 1–4 in this textbook or in the MyMathLab Study Plan.

Example 2 Planting Grass Seed

My video summary

A landscaper plants grass seed at a general **rate** of 7 pounds for every 1000 square feet. If the landscaper has 25 pounds of grass seed on hand, how many additional pounds of grass seed will he need to purchase for a job to plant grass on a 45,000 square-foot yard?

Solution We follow our usual **problem-solving strategy**.

Step 1. We need to find how many pounds of grass seed the landscaper must purchase to plant a 45,000 ft^2 yard. We know 7 pounds of seed will plant 1000 ft^2, and we know the landscaper already has 25 pounds of seed on hand.

Step 2. Let $x =$ the pounds of additional seed purchased. Then $x + 25 =$ the total pounds of seed needed for this job.

Step 3. The rate of *pounds of seed* to *square feet of yard* for this job is the same as the general rate, so we can form a proportion.

$$\begin{array}{cc} \text{General rate} & \text{Job rate} \\ \downarrow & \downarrow \end{array}$$

$$\frac{\text{Pounds}}{\text{Square feet}} = \frac{\text{Pounds}}{\text{Square feet}}$$

$$\frac{7}{1000} = \frac{x + 25}{45{,}000}$$

Step 4. **Cross multiply** to solve the **proportion**.

Try to finish solving this problem on your own. View the **answer**, or watch this **video** for the complete solution.

You Try It **Work through this You Try It problem.**

Work Exercises 5 and 6 in this textbook or in the MyMathLab Study Plan.

(eText Screens 16.7-1–16.7-26)

Similar triangles have the same shape but not necessarily the same size. For similar triangles, **corresponding angles** are equal, and **corresponding sides** have lengths that are **proportional.** So, we can use proportions to find unknown lengths in similar triangles.

Example 3 Similar Triangles

Find the unknown length n for the following similar triangles.

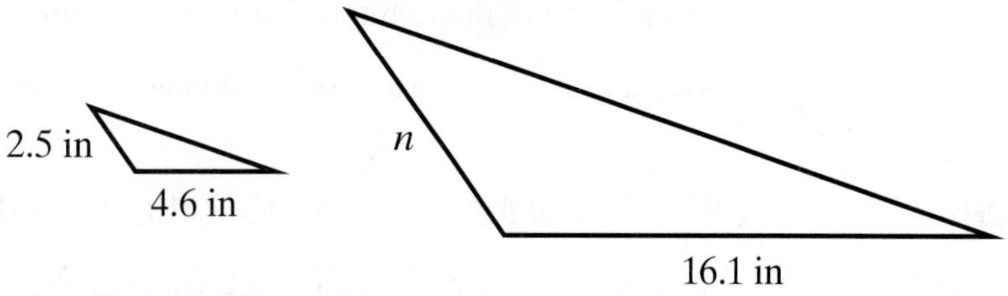

Solution Follow the **problem-solving strategy**. Since the triangles are **similar**, their **corresponding sides** are **proportional**. Side n on the larger triangle corresponds to the 2.5-inch side on the smaller triangle. The 16.1-inch side on the larger triangle corresponds to the 4.6-inch side on the smaller triangle. So, we have the following proportion:

$$\frac{n}{2.5} = \frac{16.1}{4.6}$$

Cross multiply: $4.6n = 2.5(16.1)$

Simplify: $4.6n = 40.25$

Divide both sides by 4.6: $\frac{4.6n}{4.6} = \frac{40.25}{4.6}$

Simplify: $n = 8.75$

The unknown side length n is 8.75 inches long.

You Try It **Work through this You Try It problem.**

Work Exercises 7 and 8 in this textbook or in the MyMathLab Study Plan.

Example 4 Height of a Tree

My video summary

A forest ranger wants to determine the height of a tree. She measures the tree's shadow as 84 feet long. Her own shadow at the same time is 7.5 feet long. If she is 5.5 feet tall, how tall is the tree?

(eText Screens 16.7-1–16.7-26)

Solution Follow the **problem-solving strategy**. Since the shadows were measured at the same time, the **ratio** of the height-to-shadow length of the tree is **proportional** to the ratio of height-to-shadow length of the forest ranger. Write the proportion, and then use it to finish solving this problem. View the **answer**, or watch this **video** for a complete solution.

You Try It **Work through this You Try It problem.**

Work Exercises 9 and 10 in this textbook or in the MyMathLab Study Plan.

OBJECTIVE 2 USE FORMULAS CONTAINING RATIONAL EXPRESSIONS TO SOLVE PROBLEMS

In **Topic 16.6**, we saw that **formulas** often involve **rational expressions**. These formulas can be used to solve application problems.

Example 5 Resistance in Parallel Circuits

In electronics, the total resistance R of a circuit containing two resistors in parallel is given by the formula $\frac{1}{R} = \frac{1}{R_1} + \frac{1}{R_2}$, where R_1 and R_2 are the two individual resistances. If the total resistance is 10 ohms and one resistor has twice the resistance of the other, find the resistance of each circuit.

Solution

Step 1. We want to find the individual resistances for a circuit containing two resistors in parallel. We know that the total resistance is 10 ohms and one resistor has twice the resistance of the other.

Step 2. We are given the resistance of one resistor in terms of the other. If we let $x =$ the resistance of the first resistor, then $2x =$ the resistance of the second.

Step 3. Using the given formula, we substitute in the total resistance and expressions for the individual resistances. This gives the equation

$$\frac{1}{10} = \frac{1}{x} + \frac{1}{2x}.$$

Step 4. The **LCD** is $10x$. Note that the only restricted value is $x = 0$.

Begin with the original equation: $\frac{1}{10} = \frac{1}{x} + \frac{1}{2x}$

Multiply both sides by the LCD: $10x\left(\frac{1}{10}\right) = 10x\left(\frac{1}{x} + \frac{1}{2x}\right)$

Distribute: $10x \cdot \frac{1}{10} = 10x \cdot \frac{1}{x} + 10x \cdot \frac{1}{2x}$

Simplify: $\cancel{10}x \cdot \frac{1}{\cancel{10}} = 10\cancel{x} \cdot \frac{1}{\cancel{x}} + \overset{5}{\cancel{10x}} \cdot \frac{1}{\cancel{2x}}$

$$x \cdot 1 = 10 \cdot 1 + 5 \cdot 1$$
$$x = 10 + 5$$
$$x = 15$$

Step 5. We can check our answer using our equation $\frac{1}{10} = \frac{1}{x} + \frac{1}{2x}$.

$$\frac{1}{15} + \frac{1}{2(15)} = \frac{1}{15} + \frac{1}{30} = \frac{2}{30} + \frac{1}{30} = \frac{3}{30} = \frac{1}{10}$$

The total resistance is 10 ohms, so this result is reasonable and it checks.

Step 6. The resistances are 15 ohms and $2(15) = 30$ ohms.

You Try It **Work through this You Try It problem.**

Work Exercises 11 and 12 in this textbook or in the MyMathLab Study Plan.

OBJECTIVE 3 SOLVE UNIFORM MOTION PROBLEMS INVOLVING RATIONAL EQUATIONS

In **Topic 11.5**, we studied **uniform motion problems** and used the **formula** $d = rt$. Solving this formula for t gives $t = \frac{d}{r}$, or $\text{time} = \frac{\text{distance}}{\text{rate}}$. We now use this **rational equation** to solve problems.

Example 6 Boat Speed

My video summary

Emalie can travel 16 miles upriver in the same amount of time it takes her to travel 24 miles downriver. If the speed of the current is 4 mph, how fast can her boat travel in still water?

Solution Follow the **problem-solving strategy**.

Step 1. We want to find the speed of Emalie's boat in still water. We know she can travel 16 miles upriver in the same amount of time she can travel 24 miles downriver, and we know the speed of the current is 4 mph.

Step 2. Let $r =$ the speed of Emalie's boat in still water. Then her speed upriver is $r - 4$ (because she goes against the current) and her speed downriver is $r + 4$ (because she goes with the current).

Step 3. We know the amount of time traveled is the same in either direction.

$$\text{time}_{\text{upriver}} = \text{time}_{\text{downriver}}$$

Solving the **distance formula** for t gives us $t = \frac{d}{r}$, or $\text{time} = \frac{\text{distance}}{\text{rate}}$. So, we rewrite our equation as follows:

$$\left(\frac{\text{distance}}{\text{rate}}\right)_{\text{upriver}} = \left(\frac{\text{distance}}{\text{rate}}\right)_{\text{downriver}}$$

Substituting the given distances and our expressions for rates from Step 2, we get the equation

$$\frac{16}{r-4} = \frac{24}{r+4}.$$

Try to solve this equation to finish this problem on your own. View the **answer**, or watch this **video** for a complete solution.

You Try It Work through this You Try It problem.

Work Exercises 13 and 14 in this textbook or in the MyMathLab Study Plan.

Example 7 Train Speed

My video summary

Fatima rode an express train 223.6 miles from Boston to New York City and then rode a passenger train 218.4 miles from New York City to Washington, DC. If the express train travels 30 miles per hour faster than the passenger train and her total trip took 6.5 hours, what was the average speed of the express train?

Solution Follow the **problem-solving strategy**.

Step 1. We want to find the average speed of the express train. We know the express train traveled 30 miles per hour faster than the passenger train. We also know that the distances traveled on the express and passenger trains were 223.6 and 218.4 miles, respectively. Finally, we know the time of the trip was 6.5 hours.

Step 2. Let r = the average speed of the express train. Then $r - 30$ = the average speed of the passenger train.

Step 3. The total time of Fatima's trip must be the **sum** of her time on the express train and her time on the passenger train.

$$\text{time}_{\text{express}} + \text{time}_{\text{passenger}} = \text{time}_{\text{total}}$$

Solving the **distance formula** for t gives us $t = \frac{d}{r}$, or $\text{time} = \frac{\text{distance}}{\text{rate}}$.

So, we rewrite our equation as

$$\left(\frac{\text{distance}}{\text{rate}}\right)_{\text{express}} + \left(\frac{\text{distance}}{\text{rate}}\right)_{\text{passenger}} = \text{time}_{\text{total}}.$$

Substituting the known distances, the expressions for train speeds from Step 2, and the total time, we get the equation

$$\frac{223.6}{r} + \frac{218.4}{r-30} = 6.5.$$

Step 4. The **LCD** is $r(r-30)$. Multiply both sides of the equation by the LCD to obtain a **polynomial equation**.

Write the original equation: $$\frac{223.6}{r} + \frac{218.4}{r-30} = 6.5$$

Multiply by the LCD: $$r(r-30)\left(\frac{223.6}{r} + \frac{218.4}{r-30}\right) = r(r-30)(6.5)$$

Distribute: $$r(r-30)\cdot\frac{223.6}{r} + r(r-30)\cdot\frac{218.4}{r-30} = r(r-30)(6.5)$$

Divide out common factors: $$\cancel{r}(r-30)\cdot\frac{223.6}{\cancel{r}} + r\cancel{(r-30)}\cdot\frac{218.4}{\cancel{r-30}} = r(r-30)(6.5)$$

Multiply: $$223.6(r-30) + 218.4r = r(r-30)(6.5)$$

Finish solving the equation on your own to answer the question. View the **answer**, or watch this **video** for a complete solution.

(eText Screens 16.7-1–16.7-26)

You Try It **Work through this You Try It problem.**

Work Exercises 15 and 16 in this textbook or in the MyMathLab Study Plan.

OBJECTIVE 4 SOLVE PROBLEMS INVOLVING RATE OF WORK

For work problems involving multiple workers, such as people, copiers, pumps, etc., it is often helpful to consider **rate of work**.

Rate of Work

The **rate of work** is the number of jobs that can be completed in a given unit of time.

If one job can be completed in t units of time, then the rate of work is given by $\frac{1}{t}$.

We can add **rates** but not times. For example, if it takes Avril 4 hours to paint a room and Anisa 2 hours to paint the same room, it would not take $4 + 2 = 6$ hours for them to paint the room together. It cannot take longer for two people to do a job than it would take either one alone. When dealing with two workers, we use the following formula:

$$\underbrace{\frac{1}{t_1}}_{\text{Work rate of 1st worker}} + \underbrace{\frac{1}{t_2}}_{\text{Work rate of 2nd worker}} = \underbrace{\frac{1}{t}}_{\text{Work rate together}}$$

Here, t_1 and t_2 are the individual times to complete one job, and t is the time to complete the job when working together.

Example 8 Painting a Room

My video summary

Avril can paint a room in 4 hours if she works alone. Anisa can paint the same room in 2 hours if she works alone. How long will it take the two women to paint the room if they work together?

Solution Follow the **problem-solving strategy**.

Step 1. We want to find how long it will take Avril and Anisa to paint the room if they work together. Working alone, Avril can paint the room in 4 hours, while Anisa can paint the room in 2 hours.

Step 2. Let $x =$ the time to paint the room when working together.

Step 3. Avril can paint the room in 4 hours, so her **rate of work** is $\frac{1}{4}$ room per hour. Anisa can paint the room in 2 hours, so her rate of work is $\frac{1}{2}$ room per hour. Together, they can paint the room in x hours, so their combined rate of work is $\frac{1}{x}$ room per hour. The **sum** of the two individual work rates results in the combined work rate:

$$\underbrace{\frac{1}{4}}_{\text{Avril's rate}} + \underbrace{\frac{1}{2}}_{\text{Anisa's rate}} = \underbrace{\frac{1}{x}}_{\text{Combined rate}}.$$

Step 4. The **LCD** is $4x$.

Write the original equation: $\frac{1}{4} + \frac{1}{2} = \frac{1}{x}$

Multiply both sides by the LCD: $4x\left(\frac{1}{4} + \frac{1}{2}\right) = 4x\left(\frac{1}{x}\right)$

Distribute: $4x \cdot \frac{1}{4} + 4x \cdot \frac{1}{2} = 4x \cdot \frac{1}{x}$

Divide out common factors: $\cancel{4x} \cdot \frac{1}{\cancel{4}} + \overset{2}{\cancel{4}}x \cdot \frac{1}{\cancel{2}} = 4\cancel{x} \cdot \frac{1}{\cancel{x}}$

Simplify: $x + 2x = 4$

Combine like terms: $3x = 4$

Divide both sides by 3: $x = \frac{4}{3}$

Step 5. The result $\frac{4}{3}$ hours is less than the 2 hours needed for the faster worker alone, so the result is reasonable.

Step 6. Together, Avril and Anisa can paint the room in $\frac{4}{3}$ hours, or 1 hour and 20 minutes.

You Try It **Work through this You Try It problem.**

Work Exercises 17 and 18 in this textbook or in the MyMathLab Study Plan.

Example 9 Emptying a Pool

My video summary

A small pump takes 8 more hours than a larger pump to empty a pool. Together the pumps can empty the pool in 3 hours. How long will it take the larger pump to empty the pool if it works alone?

Solution Follow the **problem-solving strategy**.

Step 1. We want to find the time it will take the larger pump to empty the pool by itself. We know that the two pumps take 3 hours to empty the pool and the smaller pump takes 8 more hours than the larger pump to empty the pool by itself.

Step 2. Let x = the time for the larger pump to empty the pool. Then $x + 8$ = the time for the smaller pump to empty the pool.

Step 3. Since we are combining **rates of work**, we get

$$\underbrace{\frac{1}{t_1}}_{\text{Larger pump rate}} + \underbrace{\frac{1}{t_2}}_{\text{Smaller pump rate}} = \underbrace{\frac{1}{t}}_{\text{Combined rate}}.$$

Substituting in the given combined time, and expressions for individual times, we get

$$\underbrace{\frac{1}{x}}_{\text{Larger pump rate}} + \underbrace{\frac{1}{x+8}}_{\text{Smaller pump rate}} = \underbrace{\frac{1}{3}}_{\text{Combined rate}}.$$

Step 4. The **LCD** is $3x(x + 8)$. Multiply both sides of the equation by the LCD and solve the resulting **polynomial equation** to finish this problem on your own. View the **answer**, or watch this **video** for a complete solution.

You Try It **Work through this You Try It problem.**

Work Exercises 19 and 20 in this textbook or in the MyMathLab Study Plan.

Example 10 Filling a Pond

My animation summary

A garden hose can fill a pond in 2 hours whereas an outlet pipe can drain the pond in 10 hours. If the outlet pipe is accidentally left open, how long would it take to fill the pond?

Solution Watch this **animation** or continue reading for a complete solution.

We want to find the time required to fill the pond if the hose is running and the outlet pipe is open. Let t = the time required to fill the pond.

We are combining **rates of work**, but the rates are not working together. Rates that work towards completion of a task are positive actions and therefore are positive values. Rates that work against the completion of the task are negative actions and therefore are negative values.

$$\underbrace{\frac{1}{t_1}}_{\text{Garden hose rate}} + \underbrace{\frac{-1}{t_2}}_{\text{Outlet pipe rate}} = \underbrace{\frac{1}{t}}_{\text{Combined rate}}$$

Substituting in the given individual times, we get

$$\underbrace{\frac{1}{2}}_{\text{Garden hose rate}} + \underbrace{\frac{-1}{10}}_{\text{Outlet pipe rate}} = \underbrace{\frac{1}{t}}_{\text{Combined rate}}$$

or

$$\underbrace{\frac{1}{2}}_{\text{Garden hose rate}} - \underbrace{\frac{1}{10}}_{\text{Outlet pipe rate}} = \underbrace{\frac{1}{t}}_{\text{Combined rate}}.$$

Because the outlet pipe is working against the garden hose, we end up subtracting the rates. Note that this is similar to the situation when a boat is traveling upstream or downstream. When going downstream (with the current) we added the rates. However, when going upstream (against the current) we subtracted the rates because the current was working against the boat.

The **LCD** is $10t$. Multiply both sides of the equation by the LCD and solve the resulting equation for t.

Original equation: $\frac{1}{2} - \frac{1}{10} = \frac{1}{t}$

Multiply by the LCD: $10t\left(\frac{1}{2} - \frac{1}{10}\right) = 10t\left(\frac{1}{t}\right)$

Distribute: $5t - t = 10$

Simplify: $4t = 10$

Divide both sides by 4: $t = \frac{10}{4} = 2.5$

It will take 2.5 hours to fill the pond with the outlet pipe open.

(eText Screens 16.7-1–16.7-26)

You Try It **Work through this You Try It problem.**

Work Exercises 21 and 22 in this textbook or in the MyMathLab Study Plan.

16.7 Exercises

In Exercises 1–6, use a proportion to solve the problem.

You Try It

1. **Defective Cell Phones** A quality control inspector finds 8 defective units in a sample of 50 cellular phones. At this ratio, how many defective units can the inspector expect to find in a batch of 1500 such phones?

2. **Students and Teachers** If a high school employs 5 teachers for every 76 students, how many students are in the school if it employs 115 teachers?

3. **Ice Cream** If $\frac{1}{2}$ cup of chocolate chip cookie dough ice cream contains 15 grams of fat, how many grams of fat are in a quart of the ice cream? **Hint:** 1 quart = 4 cups.

4. **Identity Theft** In 2008, there were 75 identity theft complaints per 100,000 residents in the state of Missouri. How many identity theft complaints were made for the entire state if the population of Missouri was 5,900,000?

You Try It

5. **Planting Corn** To plant a 250-acre field, a farmer used 3500 pounds of corn seed. If the farmer still has 500 pounds of corn seed on hand, how many more pounds of corn seed will he need to buy in order to plant an additional 180-acre field?

6. **Time to Complete an Order** Olivia works at a craft store and can make 2 stone crafts in 45 minutes. If she has been working for 30 minutes on an order for 7 stone crafts, how much longer does she need to work to complete the order?

In Exercises 7 and 8, find the unknown length n for the given similar triangles.

You Try It

7.

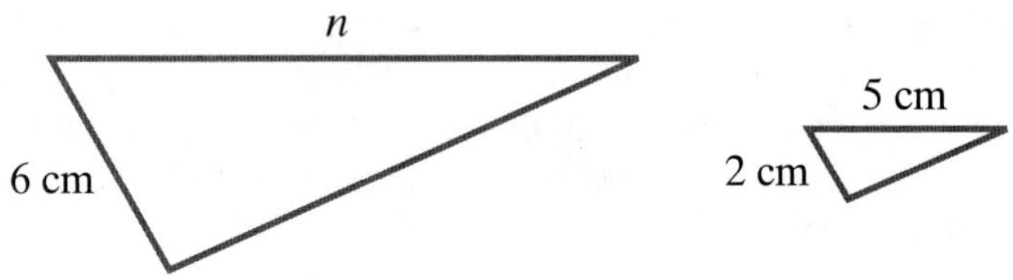

8.

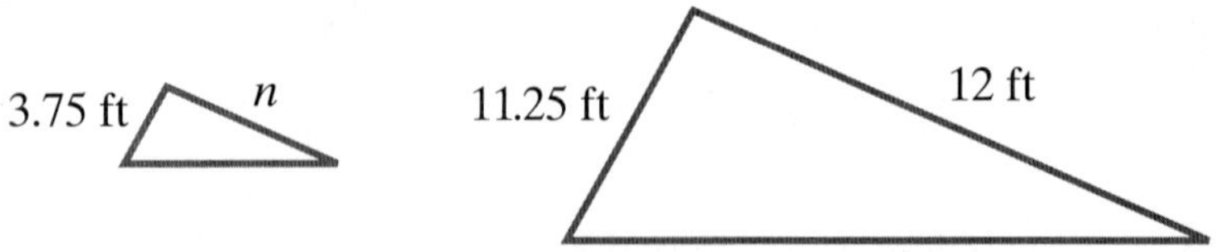

In Exercises 9–22, solve the problem.

You Try It

9. **Height of a Lighthouse** Elena wants to determine the height of a lighthouse. The lighthouse casts an 84 meter shadow. If Elena is 1.8 meters tall and casts a 2.4-meter shadow when standing next to the lighthouse, how tall is the lighthouse?

10. Tennis Anyone? At what height must a tennis ball be served in order to just clear the net and land 20 feet past the net? Assume the ball travels a straight path as shown in the figure.

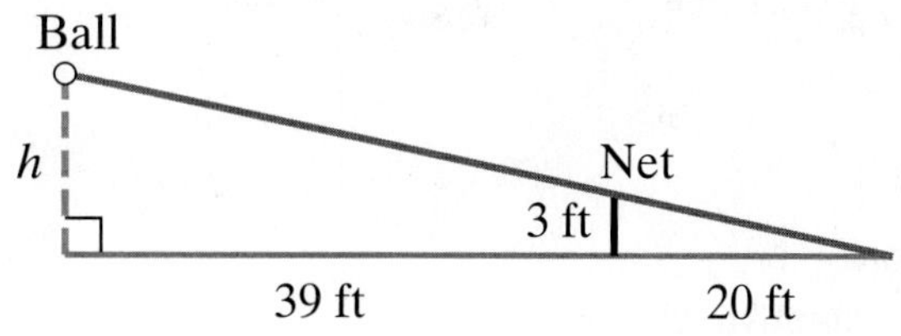

You Try It

11. Resistors in Parallel If the total resistance of two circuits in parallel is 9 ohms and one circuit has three times the resistance of the other, find the resistance of each circuit. Use $\frac{1}{R} = \frac{1}{R_1} + \frac{1}{R_2}$, where R_1 and R_2 are the two individual resistances.

12. Focal Length The focal length f of a lens is given by the equation $\frac{1}{f} = \frac{1}{x_o} + \frac{1}{x_i}$, where x_o is the distance from the object to the lens, and x_i is the distance from the lens to the image of the object. If the focal length of a lens is 12 cm and an object is three times as far from the lens as its image, then how far is the object from the lens?

You Try It

13. Current Speed On a float trip, Kim traveled 30 miles downstream in the same amount of time that it would take her to row 12 miles upstream. If her speed in still water is 7 mph, find the speed of the current.

14. Plane Speed A plane can fly 500 miles against the wind in the same amount of time that it can fly 725 miles with the wind. If the wind speed is 45 mph, find the speed of the plane in still air.

You Try It

15. Walking Speed For exercise, Mae jogs 10 miles then walks an additional 2 miles to cool down. If her jogging speed is 4.5 miles per hour faster than her walking speed and her total exercise time is 3 hours, what is her walking speed?

16. Moving Sidewalk Choi can walk 108 m in 72 seconds. Standing on a moving sidewalk at Thailand's Suvarnabhumi International Airport, she can travel 108 m in 40 seconds. How long will it take her to travel 108 m if she walks on the moving sidewalk? Round your answer to the nearest tenth.

You Try It

17. Making Copies Beverly can copy final exams in 30 minutes using a new copy machine. Using an old copy machine, the same job takes 45 minutes. If both copy machines are used, how long will it take to copy the final exams?

18. Mowing the Lawn By himself, Chris can mow his lawn in 60 minutes. If his daughter Claudia helps, they can mow the lawn together in 40 minutes. How long would it take Claudia to mow the lawn by herself?

You Try It

19. Staining a Deck It takes Shawn 2 more hours to stain a deck than Michelle. Together, it takes them 2.4 hours to complete the work. How long would it take Shawn to stain the deck by himself?

20. Lining Fields Together, Sari and Rilee can mark the lines on the soccer fields at a recreation center 18 minutes faster than if Rilee lines the fields on her own. If it takes Sari 56 minutes to line the fields by herself, how long would it take Rilee to do the job alone?

You Try It

21. Filling a Hot Tub A garden hose can fill a hot tub in 180 minutes, whereas the drain on a hot tub can empty it in 300 minutes. If the drain is accidentally left open, how long will it take to fill the tub?

22. Emptying a Boat's Bilge A leak in the hull will fill a boat's bilge in 8 hours. The boat's bilge pump can empty a full bilge in 3 hours. How long will it take the pump to empty a full bilge if the boat is leaking?

16.8 Variation

THINGS TO KNOW

Before working through this topic, be sure that you are familiar with the following concepts:

		VIDEO	ANIMATION	INTERACTIVE
You Try It	1. Evaluate a Formula (Topic 11.4, **Objective 1**)			
You Try It	2. Find the Value of a Non-isolated Variable in a Formula (Topic 11.4, **Objective 2**)			

OBJECTIVES

1 Solve Problems Involving Direct Variation

2 Solve Problems Involving Inverse Variation

OBJECTIVE 1 SOLVE PROBLEMS INVOLVING DIRECT VARIATION

Often in application problems, we need to know how one quantity varies with respect to other quantities. A **variation equation** allows us to describe how one quantity changes with respect to one or more additional quantities. We will examine two types of variation, *direct* variation and *inverse* variation.

Direct variation means that one **variable** is a constant multiple of another variable.

Definition Direct Variation

For an equation of the form

$$y = kx,$$

we say that y **varies directly** with x, or y is **directly proportional** to x. The nonzero **constant** k is called the **constant of variation** or the **proportionality constant.**

A variation equation such as $y = kx$ is also called a **model** that describes the relationship between the variables x and y.

Example 1 Direct Variation

Suppose y varies directly with x, and $y = 20$ when $x = 8$.

a. Find the equation that relates x and y.

b. Find y when $x = 12$.

Solutions

a. We know that y **varies directly** with x, so the equation has the form $y = kx$. We use the fact that $y = 20$ when $x = 8$ to find k.

Write the direct variation equation: $y = kx$

Substitute $y = 20$ and $x = 8$: $20 = k(8)$

Divide both sides by 8: $\frac{20}{8} = \frac{k(8)}{8}$

Simplify: $2.5 = k$

The **constant of variation** is 2.5, so the equation is $y = 2.5x$.

b. We now know that y and x are related by the equation $y = 2.5x$. To find y when $x = 12$, substitute 12 for x in the equation and simplify:

$$y = 2.5(12) = 30$$

Therefore, $y = 30$ when $x = 12$.

You Try It **Work through this You Try It problem.**

Work Exercises 1 and 2 in this textbook or in the MyMathLab Study Plan.

In Example 1, the **model** $y = kx$ is a **linear equation**. However, direct variation equations will not always be linear. For example, if we say that y varies directly with the **cube** of x, then the equation will be $y = kx^3$, where k is the constant of variation.

Example 2 Direct Variation

My video summary

Suppose y **varies directly** with the cube of x, and $y = 375$ when $x = 5$.

a. Find the equation that relates x and y.

b. Find y when $x = 2$.

Solutions Because y varies directly with the cube of x, or x^3, the variation equation has the form $y = kx^3$. We can use the fact that $y = 375$ when $x = 5$ to find k, and then the equation. Try to complete this problem on your own. View the **answer**, or watch this **video** for a complete solution.

You Try It **Work through this You Try It problem.**

Work Exercises 3–6 in this textbook or in the MyMathLab Study Plan.

For direct variation, the **ratio** of the two quantities is constant (the **constant of variation**). For example, consider $y = kx$ and $y = kx^3$.

$$\underbrace{y = kx}_{\substack{y \text{ varies}\\ \text{directly}\\ \text{with } x}} \rightarrow \underbrace{\frac{y}{x}}_{\substack{\text{Ratio of the}\\ \text{quantities}\\ y \text{ and } x}} = \underbrace{k}_{\substack{\text{Constant of}\\ \text{variation}}} \qquad \underbrace{y = kx^3}_{\substack{y \text{ varies directly}\\ \text{with the cube}\\ \text{of } x}} \rightarrow \underbrace{\frac{y}{x^3}}_{\substack{\text{Ratio of the}\\ \text{quantities}\\ y \text{ and } x^3}} = \underbrace{k}_{\substack{\text{Constant}\\ \text{of variation}}}$$

Problems involving variation can generally be solved using the following guidelines.

Solving Variation Problems

Step 1. Translate the problem into an equation that models the situation.

Step 2. Substitute given values for the variables into the equation and solve for the **constant of variation,** k.

Step 3. Substitute the value for k into the equation to form the general **model**.

Step 4. Use the general model to answer the question posed in the problem.

(eText Screens 16.8-1–16.8-16)

Example 3 Kinetic Energy

The kinetic energy of an object in motion varies directly with the square of its speed. If a van traveling at a speed of 30 meters per second has 945,000 joules of kinetic energy, how much kinetic energy does it have if it is traveling at a speed of 20 meters per second?

Solution Follow the **guidelines for solving variation problems.**

Step 1. We are told that the kinetic energy of an object in motion varies **directly** with the square of its speed. If we let K = kinetic energy and s = speed, we can translate the problem statement into the model

$$K = ks^2,$$

where k is the **constant of variation.**

Step 2. To determine the value of k, we use the fact that the kinetic energy is 945,000 joules when the velocity is 30 meters per second.

Substitute 945,000 for K and 30 for s: $945{,}000 = k(30)^2$

Simplify 30^2: $945{,}000 = 900k$

Divide both sides by 900: $1050 = k$

Step 3. The constant of variation is 1050, so the general model is $K = 1050s^2$.

Step 4. We want to determine the kinetic energy of the van if its speed is 20 meters per second. Substituting 20 for s, we find

$$K = 1050(20)^2 = 1050(400) = 420{,}000.$$

The van will have 420,000 joules of kinetic energy if it is traveling at a speed of 20 meters per second.

You Try It **Work through this You Try It problem.**

Work Exercises 7 and 8 in this textbook or in the MyMathLab Study Plan.

Example 4 Measuring Leanness

My video summary

The Ponderal Index measure of leanness states that weight varies directly with the cube of height. If a "normal" person who is 1.2 m tall weighs 21.6 kg, how much will a "normal" person weigh if they are 1.8 m tall?

Solution Follow the **guidelines for solving variation problems.**

Step 1. We are told that weight varies **directly** with the cube of height. If we let w = weight and h = height, we can translate the problem statement into the **model**

$$w = kh^3,$$

where k is the **constant of variation.**

Step 2. To determine the value of k, we use the fact that a normal person who is 1.2 meters tall has a weight of 21.6 kg.

Substitute 1.2 for h and 21.6 for w: $21.6 = k(1.2)^3$

Simplify $(1.2)^3$: $21.6 = 1.728k$

Divide both sides by 1.728: $12.5 = k$

Step 3. The constant of variation is 12.5, so the general **model** is $w = 12.5h^3$.

(eText Screens 16.8-1–16.8-16)

Use the general model to determine the weight of a normal person who is 1.8 m tall. Check your **answer**, or watch this **video** for a detailed solution.

You Try It **Work through this You Try It problem.**

Work Exercises 9 and 10 in this textbook or in the MyMathLab Study Plan.

OBJECTIVE 2 SOLVE PROBLEMS INVOLVING INVERSE VARIATION

Inverse variation means that one variable is a constant multiple of the **reciprocal** of another variable.

> **Definition Inverse Variation**
>
> For equations of the form
>
> $$y = \frac{k}{x} \quad \text{or} \quad y = k \cdot \frac{1}{x},$$
>
> we say that y **varies inversely** with x, or y is **inversely proportional** to x. The constant k is called the **constant of variation.**

For inverse variation, the **product** of the two quantities is constant (the constant of variation).
For example, consider $y = \frac{k}{x}$ and $y = \frac{k}{x^2}$.

$$\underbrace{y = \frac{k}{x}}_{\substack{y \text{ varies} \\ \text{inversely} \\ \text{with } x}} \rightarrow \underbrace{xy}_{\substack{\text{Product of the} \\ \text{quantities} \\ y \text{ and } x}} = \underbrace{k}_{\substack{\text{Constant} \\ \text{of} \\ \text{variation}}} \qquad \underbrace{y = \frac{k}{x^2}}_{\substack{y \text{ varies} \\ \text{inversely with} \\ \text{the square of } x}} \rightarrow \underbrace{x^2 y}_{\substack{\text{Product of} \\ \text{the quantities} \\ y \text{ and } x^2}} = \underbrace{k}_{\substack{\text{Constant} \\ \text{of} \\ \text{variation}}}$$

Example 5 Inverse Variation

My video summary Suppose y **varies inversely** with x, and $y = 72$ when $x = 50$.

a. Find the equation that relates x and y.

b. Find y when $x = 45$.

Solutions

a. Because y varies inversely with x, the equation has the form $y = \frac{k}{x}$. We use the fact that $y = 72$ when $x = 50$ to find k.

Write the inverse variation equation: $y = \frac{k}{x}$

Substitute $y = 72$ and $x = 50$: $72 = \frac{k}{50}$

Multiply both sides by 50: $3600 = k$

The **constant of variation** is 3600, so the equation is $y = \frac{3600}{x}$.

b. Try to work this part on your own. View the **answer**, or watch this **video** for a complete solution to both parts.

You Try It **Work through this You Try It problem.**

Work Exercises 11–14 in this textbook or in the MyMathLab Study Plan.

Problems involving **inverse variation** can be solved using the same **guidelines for solving variation problems.**

Example 6 Density of an Object

For a given mass, the density of an object is **inversely proportional** to its volume. If 50 cubic centimeters (cm^3) of an object with a density of $28\ \text{g/cm}^3$ is compressed to $40\ \text{cm}^3$, what would be its new density?

Solution

Step 1. We are told that the density of an object with a given mass varies inversely with its volume. If we let $D =$ density and $V =$ volume, we can translate the problem statement into the **model**

$$D = \frac{k}{V},$$

where k is the **constant of variation.**

Step 2. To determine the value of k, we use the fact that the density is $28\ \text{g/cm}^3$ when the volume is $50\ \text{cm}^3$.

Substitute 28 for D and 50 for V: $\quad 28 = \dfrac{k}{50}$

Multiply both sides by 50: $\quad 1400 = k$

Step 3. The constant of variation is 1400, so the general model is $D = \dfrac{1400}{V}$.

Step 4. We want to determine the density of the object if the volume is compressed to $40\ \text{cm}^3$. Substituting 40 for V, we find

$$D = \frac{1400}{40} = 35\ \text{g/cm}^3.$$

The density of the compressed object would be $35\ \text{g/cm}^3$.

You Try It **Work through this You Try It problem.**

Work Exercises 15 and 16 in this textbook or in the MyMathLab Study Plan.

Example 7 Shutter Speed

My video summary

The shutter speed, S, of a camera **varies inversely** as the square of the aperture setting, f. If the shutter speed is 125 for an aperture of 5.6, what is the shutter speed if the aperture is 1.4?

(eText Screens 16.8-1–16.8-16)

Solution Follow the **guidelines for solving variation problems.**

Step 1. We are told that shutter speed varies inversely with the square of the aperture setting. Letting S = shutter speed and f = aperture setting, we can translate the problem statement into the **model**

$$S = \frac{k}{f^2},$$

where k is the **constant of variation.**

Step 2. To determine the value of k, we use the fact that an aperture setting of 5.6 corresponds to a shutter speed of 125.

Substitute 5.6 for f and 125 for S: $125 = \frac{k}{(5.6)^2}$

Simplify $(5.6)^2$: $125 = \frac{k}{31.36}$

Multiply both sides by 31.36: $3920 = k$

Step 3. The constant of variation is 3920, so the general model is $S = \frac{3920}{f^2}$.

Use the general model to determine the shutter speed for an aperture setting of 1.4. Check your **answer**, or watch this **video** for a detailed solution.

You Try It **Work through this You Try It problem.**

Work Exercises 17 and 18 in this textbook or in the MyMathLab Study Plan.

16.8 Exercises

In Exercises 1–18, solve each variation problem.

You Try It

1. Suppose y varies directly with x, and $y = 14$ when $x = 4$.
   a. Find the equation that relates x and y.
   b. Find y when $x = 7$.

2. Suppose S is directly proportional to t, and $S = 900$ when $t = 25$.
   a. Find the equation that relates t and S.
   b. Find S when $t = 4$.

You Try It

3. Suppose y varies directly with the square of x, and $y = 216$ when $x = 3$.
   a. Find the equation that relates x and y.
   b. Find y when $x = 5$.

4. Suppose M is directly proportional to the square of r, and $M = 14$ when $r = \frac{1}{6}$.
   a. Find the equation that relates r and M.
   b. Find M when $r = \frac{2}{3}$.

5. Suppose p is directly proportional to the cube of n, and $p = 135$ when $n = 6$.
   a. Find the equation that relates n and p.
   b. Find p when $n = 10$.

6. Suppose A varies directly with the cube of m, and $A = 48$ when $m = 4$.

a. Find the equation that relates m and A.

b. Find A when $m = 8$.

You Try It

7. Scuba Diving The water pressure on a scuba diver is directly proportional to the depth of the diver. If the pressure on a diver is 13.5 psi when she is 30 feet below the surface, how far below the surface will she be when the pressure is 18 psi?

8. Pendulum Length The length of a simple pendulum varies directly with the square of its period. If a pendulum of length 2.25 meters has a period of 3 seconds, how long is a pendulum with a period of 8 seconds?

You Try It

9. Water Flow Rate For a fixed water flow rate, the amount of water that can be pumped through a pipe varies directly as the square of the diameter of the pipe. In one hour, a pipe with an 8-inch diameter can pump 400 gallons of water. Assuming the same water flow rate, how much water could be pumped through a pipe that is 12 inches in diameter?

10. Falling Distance The distance an object falls varies directly with the square of the time it spends falling. If a ball falls 19.6 meters after falling for 2 seconds, how far will it fall after 9 seconds?

You Try It

11. Suppose y varies inversely with x, and $y = 24$ when $x = 25$.

a. Find the equation that relates x and y.

b. Find y when $x = 30$.

12. Suppose p is inversely proportional to q, and $p = 3.5$ when $q = 4.6$.

a. Find the equation that relates q and p.

b. Find p when $q = 1.4$.

13. Suppose n varies inversely with the square of m, and $n = 6.4$ when $m = 9$.

a. Find the equation that relates m and n.

b. Find n when $m = 4$.

14. Suppose d is inversely proportional to the cube of t, and $d = 76.8$ when $t = 2.5$.

a. Find the equation that relates t and d.

b. Find d when $t = 5$.

You Try It

15. Car Depreciation The value of a car is inversely proportional to its age. If a car is worth \$8100 when it is 4 years old, how old will it be when it is worth \$3600?

16. Electric Resistance For a given voltage, the resistance of a circuit is inversely related to its current. If a circuit has a resistance of 5 ohms and a current of 12 amps, what is the resistance if the current is 20 amps?

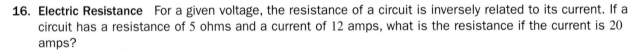

You Try It

17. Weight of an Object The weight of an object within Earth's atmosphere varies inversely with the square of the distance of the object from Earth's center. If a low Earth orbit satellite weighs 100 kg on Earth's surface (6400 km), how much will it weigh in its orbit 800 km above Earth?

18. Light Intensity The intensity of a light varies inversely as the square of the distance from the light source. If the intensity from a light source 3 feet away is 8 lumens, what is the intensity at a distance of 2 feet?

MODULE SEVENTEEN

Introduction to Functions

MODULE SEVENTEEN CONTENTS

17.1 Relations and Functions

THINGS TO KNOW

Before working through this topic, be sure you are familiar with the following concepts:

VIDEO ANIMATION INTERACTIVE

You Try It 1. Solve a Formula for a Given Variable (Topic 11.4, **Objective 3**)

You Try It 2. Write the Solution Set of an Inequality in Set-Builder Notation (Topic 11.7, **Objective 1**)

You Try It 3. Use Interval Notation to Express the Solution Set of an Inequality (Topic 11.7, **Objective 3**)

OBJECTIVES

1 Identify Independent and Dependent Variables

2 Find the Domain and Range of a Relation

3 Determine If Relations Are Functions

4 Determine If Graphs Are Functions

5 Solve Application Problems Involving Relations and Functions

---

OBJECTIVE 1 IDENTIFY INDEPENDENT AND DEPENDENT VARIABLES

We have seen that applying math to everyday life often involves situations in which one quantity is related to another. For example, the **cost** to fill a gas tank is related to the number of gallons purchased. The amount of **simple interest** paid on a loan is related to the amount owed. The total cost to manufacture 3D televisions is related to the number of televisions produced, and so on.

(eText Screens 17.1-1–17.1-22)

In **Topic 11.4** we solved **formulas** for a given **variable**. In doing so, we had to express a relationship between the given variable and any remaining variables. When we solve for a variable, that variable is called the **dependent variable** because its value *depends on* the value(s) of the remaining variable(s). Any remaining variables are called **independent variables** because we are free to select their values.

If the average price per gallon of regular unleaded gas is \$3.24 on a given day, then two gallons would cost $\$3.24(2) = \$6.38$, three gallons would cost $\$3.24(3) = \$9.72$, and so on. We can model the situation with the **equation**

$$y = 3.24x,$$

where y = cost in dollars and x = gallons of gas purchased.

Since the equation is solved for y, we identify y (cost) as the **dependent variable** and x (gallons of gas) as the **independent variable.** When an equation involving x and y is not solved for either variable, like the linear equation $2x + 3y = 12$, we will call x the independent variable and y the dependent variable.

Example 1 Identifying Independent and Dependent Variables

For each of the following equations, identify the **dependent variable** and the **independent variable(s)**.

a. $y = 3x + 5$ **b.** $w = ab + 3c^2$ **c.** $3x^2 + 9y = 12$

Solutions

a. Since the equation is solved for y, we identify y as the dependent variable. The remaining variable, x, is the independent variable.

b. Since the equation is solved for w, we identify w as the dependent variable. The remaining variables, a, b, and c, are independent variables.

c. Since the equation is not solved for either variable, we identify x as the independent variable and y as the dependent variable.

You Try It **Work through this You Try It problem.**

Work Exercises 1–4 in this textbook or in the MyMathLab Study Plan.

In Example 1 we see that it is possible to have more than one independent variable in an equation. However, we will limit our discussion mainly to situations involving one dependent and one independent variable. Our gasoline cost **model,** $y = 3.24x$, is an example of such a situation.

OBJECTIVE 2 FIND THE DOMAIN AND RANGE OF A RELATION

A **relation** is a correspondence between two sets of numbers that can be represented by a set of **ordered pairs.**

Definition

A **relation** is a set of ordered pairs.

(*eText Screens 17.1-1–17.1-22*)

In **Topic 12.2** we learned that **equations in two variables** define a set of **ordered pair solutions**. When writing ordered pair solutions, we write the value of the **independent variable** first, followed by the value of the **dependent variable**. For example, if 1 gallon of regular unleaded gas is purchased, the total cost is $3.24(1) = \$3.24$, which gives the ordered pair $(1, 3.24)$. If 2 gallons are purchased, the total cost is $3.24(2) = \$6.48$, which gives the ordered pair $(2, 6.48)$. We can create more ordered pair solutions by following this process. See a **table of values**.

Since a **graph** is a visual representation of the ordered pair solutions to an equation, we consider equations and graphs to be relations because they define sets of ordered pairs. A relation shows the connection between the set of values for the independent variable, called the **domain** (or *input values*), and the set of values for the dependent variable, called the **range** (or *output values*).

At this point, you may wish to review **set-builder notation** and **interval notation** in **Topic 11.7**.

Definitions

The **domain of a relation** is the set of all values for the **independent variable**. These are the first coordinates in the set of **ordered pairs** and are also known as *input values*.

The **range of a relation** is the set of all values for the **dependent variable**. These are the second coordinates in the set of ordered pairs and are also known as *output values*.

Example 2 Finding the Domain and Range of a Relation

My interactive video summary

Find the domain and range of each **relation**.

a. $\{(-5, 7), (3, 5), (6, 7), (12, -4)\}$

b.

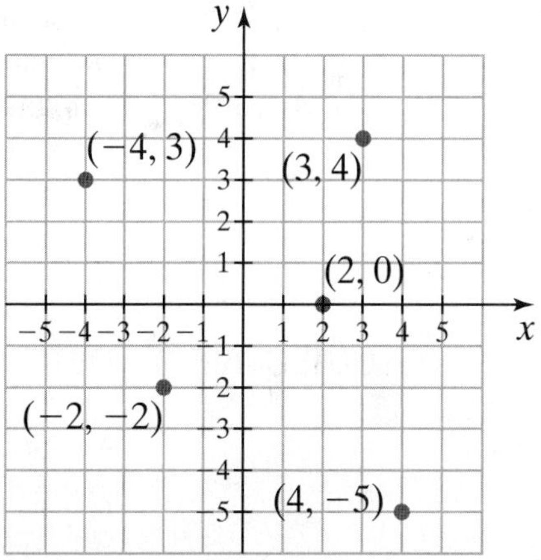

Solutions For parts a and b, identify the first **coordinates** of each **ordered pair** to find the **domain** and the second coordinates of each ordered pair to find the **range**. View the **answers**, or watch this **interactive video** for detailed solutions.

You Try It **Work through this You Try It problem.**

Work Exercises 5 and 6 in this textbook or in the MyMathLab Study Plan.

My interactive video summary

Example 3 Finding the Domain and Range of a Relation

Find the domain and range of each **relation**.

a.

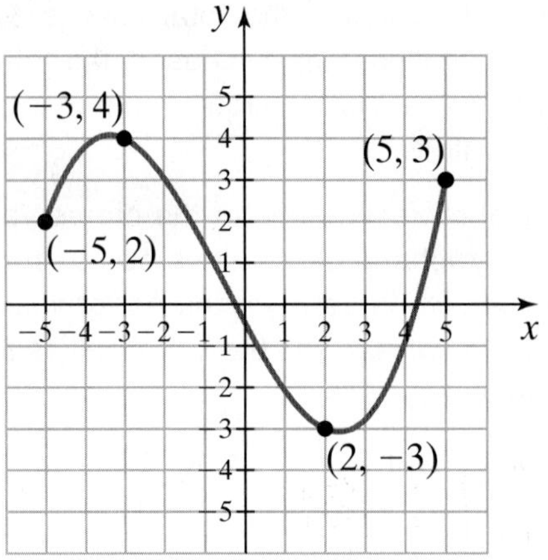

b.

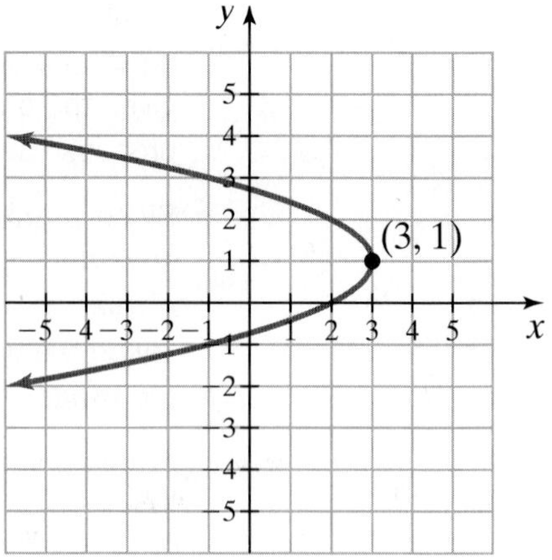

c. $y = |x - 1|$

Solutions Try to find the **domain** and **range** of each relation on your own. View the **answers**, or watch this **interactive video** for detailed solutions.

You Try It **Work through this You Try It problem.**

Work Exercises 7 and 8 in this textbook or in the MyMathLab Study Plan.

When working with application problems, we often need to restrict the domain to use only those values that make sense within the context of the situation. This restricted domain is called the **feasible domain**. The feasible domain is the set of values for the **independent variable** that make sense, or are *feasible*, in the context of the application. For example, in our gasoline cost model, $y = 3.24x$, the domain of the equation is all **real numbers**. However, it doesn't make sense to use negative numbers in the domain since x represents the number of gallons of gas purchased. Therefore, the **feasible domain** would be all real numbers greater than or equal to 0, written as $\{x|x \geq 0\}$ or $[0, \infty)$.

OBJECTIVE 3 DETERMINE IF RELATIONS ARE FUNCTIONS

A **relation** *relates* one set of numbers, the **domain**, to another, the **range**. When each value in the domain corresponds to (is paired with) exactly one value in the range, we have a special type of relation called a **function**.

> **Definition**
>
> A **function** is a special type of relation in which each value in the domain corresponds to exactly one value in the range.

Given a set of ordered pairs, we can determine if the relation is a function by looking at the **x-coordinates**. If no x-coordinate is repeated, then the relation is a function because each **input value** corresponds to exactly one **output value**. If the same x-coordinate corresponds to two or more different **y-coordinates**, then the relation is not a function.

(eText Screens 17.1-1–17.1-22)

Given an equation, we can test input values to see if, when substituted into the equation, there is more than one output value. In order for an equation to be a function, each input value must correspond to one and only one output value.

Example 4 Determining If Relations Are Functions

My interactive video summary

Determine if each of the following relations is a **function**.

a. $\{(-3, 6), (2, 5), (0, 6), (17, -9)\}$

b. $\{(4, 5), (7, -3), (4, 10), (-6, 1)\}$

c. $\{(-2, 3), (0, 3), (4, 3), (6, 3), (8, 3)\}$

d. $|y - 5| = x + 3$

e. $y = x^2 - 3x + 2$

f. $4x - 8y = 24$

Solutions Try to determine which relations are functions on your own. View the **answers**, or watch this **interactive video** for detailed explanations.

You Try It **Work through this You Try It problem.**

Work Exercises 9–16 in this textbook or in the MyMathLab Study Plan.

OBJECTIVE 4 DETERMINE IF GRAPHS ARE FUNCTIONS

If a **relation** appears as a graph, we can determine if the relation is a **function** by using the **vertical line test.**

My animation summary

Vertical Line Test

If a vertical line intersects (crosses or touches) the graph of a relation at more than one point, then the relation is not a function. If every vertical line intersects the graph of a relation at no more than one point, then the relation is a function.

Why does the vertical line test work? Watch this **animation** to find out.

Example 5 Determining If Graphs Are Functions

My animation summary

Use the vertical line test to determine if each graph is a function.

a.

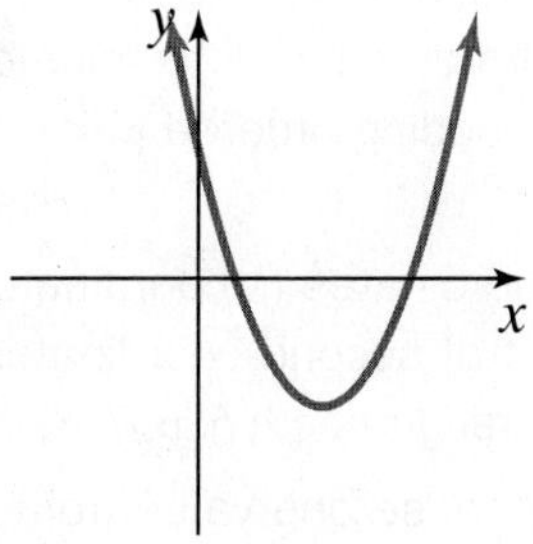

b.

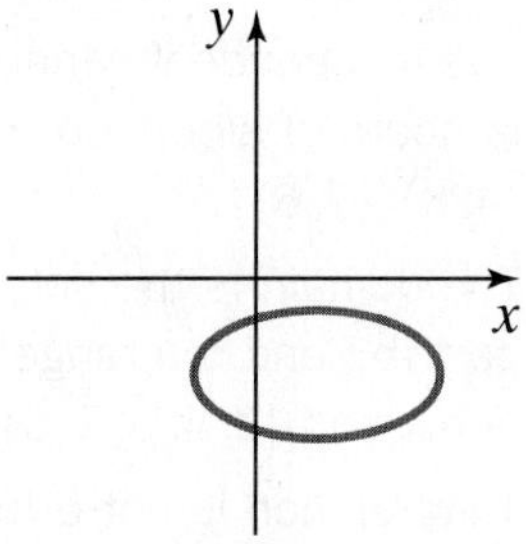

c.

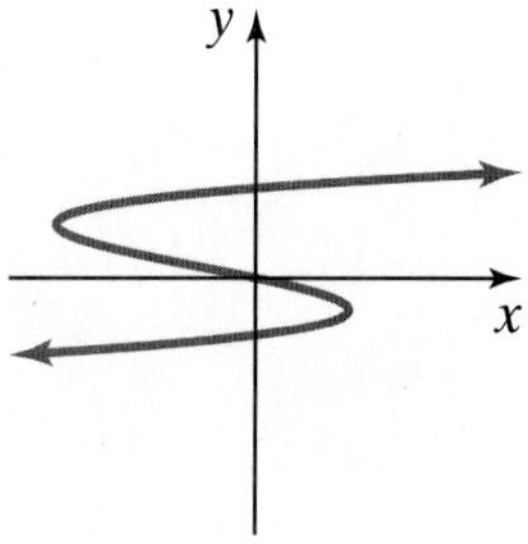

d.

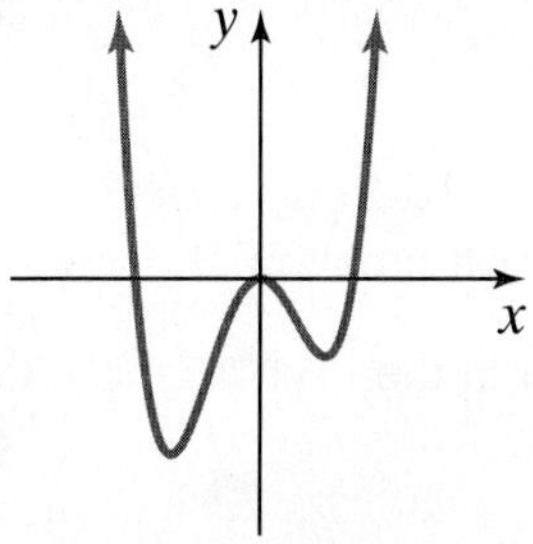

e.

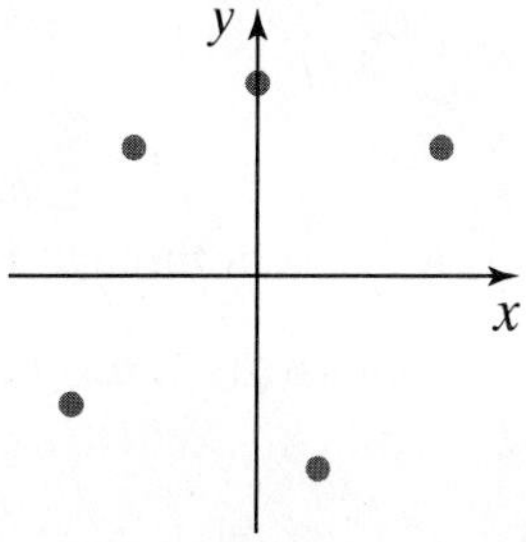

f.

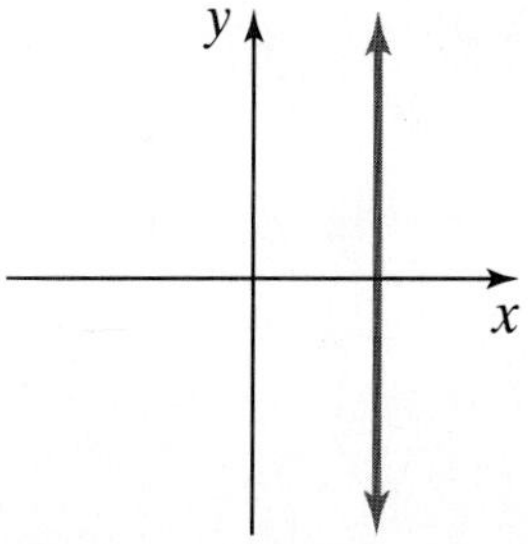

(*eText Screens 17.1-1–17.1-22*)

Solutions Apply the vertical line test to each graph on your own. View the **answers**, or work through this **animation** for complete solutions.

You Try It **Work through this You Try It problem.**

Work Exercises 17–24 in this textbook or in the MyMathLab Study Plan.

How many x- and y-**intercepts** can the graph of a function have? See the **answer**.

OBJECTIVE 5 SOLVE APPLICATION PROBLEMS INVOLVING RELATIONS AND FUNCTIONS

Real-world situations can be described by relations and functions. We use **mathematical models** to describe these situations.

Example 6 Video Entertainment and Sleep

The data in the following table represent the average daily hours of sleep and average daily hours of video entertainment for six students at a local college.

Video Entertainment	Sleep	Video Entertainment	Sleep
8	4	5	7
7	5	4	8
2	9	7	6

a. If a researcher believes the number of hours of video entertainment affects the number of hours of sleep, identify the **independent variable** and the **dependent variable**.

b. What are the ordered pairs for this data?

c. What are the **domain and range**?

d. Is this **relation** a **function**? Explain.

Solutions

a. Since the researcher believes the number of hours of sleep is affected by, or depends on, the number of hours of video entertainment, the independent variable is hours of video entertainment, and the dependent variable is hours of sleep.

b. The independent variable is hours of video entertainment and the dependent variable is hours of sleep, so the corresponding ordered pairs are $\{(8,4),(7,5),(2,9),(5,7),(4,8),(7,6)\}$.

c. The domain is the set of first coordinates (x-coordinates) from the ordered pairs in part (b), and the range is the set of second coordinates (y-coordinates). Therefore, the domain is $\{2,4,5,7,8\}$ and the range is $\{4,5,6,7,8,9\}$.

d. This relation is not a function because one value from the domain, 7, corresponds to more than one value from the range, 5 and 6. This is shown in the **ordered pairs** $(7,5)$ and $(7,6)$.

You Try It **Work through this You Try It problem.**

Work Exercise 25 in this textbook or in the MyMathLab Study Plan.

(eText Screens 17.1-1–17.1-22)

Example 7 High-Speed Internet Access

My video summary The percent of households, y, with high-speed Internet access in 2007 can be modeled by the equation $y = 0.70x + 20.03$, where x is the annual household income (in \$1000s). (*Source:* U.S. Department of Commerce)

a. Identify the **independent** and **dependent variables.**

b. Use the **model** equation to estimate the percent of households in 2007 with high-speed Internet access (to the nearest whole percent) if the annual household income was \$50,000. What point would this correspond to on the graph of the equation?

c. Is the **relation** a **function**? Explain.

d. Determine the **feasible domain.**

Solutions

a. Since the equation is solved for y in terms of x, the independent variable is the annual household income (in \$1000s) and the dependent variable is the percent of households in 2007 with high-speed Internet access.

b.–d. Try to answer these questions on your own. View the **answers**, or watch this **video** for complete solutions to all four parts.

You Try It **Work through this You Try It problem.**

Work Exercises 26–28 in this textbook or in the MyMathLab Study Plan.

17.1 Exercises

In Exercises 1–4, identify the independent and dependent variables.

You Try It

1. $y = 3x^2 - 7x + 5$ **2.** $k = 12v^2$ **3.** $4x + 5y = 18$ **4.** $e = \dfrac{\sqrt{a^2 - b^2}}{a}$

In Exercises 5–8, find the domain and range of each relation.

You Try It

5. $\{(4, -2), (1, -1), (0, 0), (1, 1), (4, 2)\}$

6.

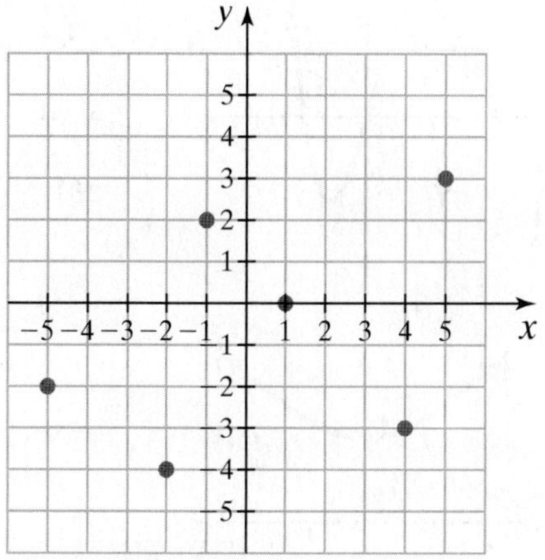

(eText Screens 17.1-1–17.1-22)

You Try It

7.

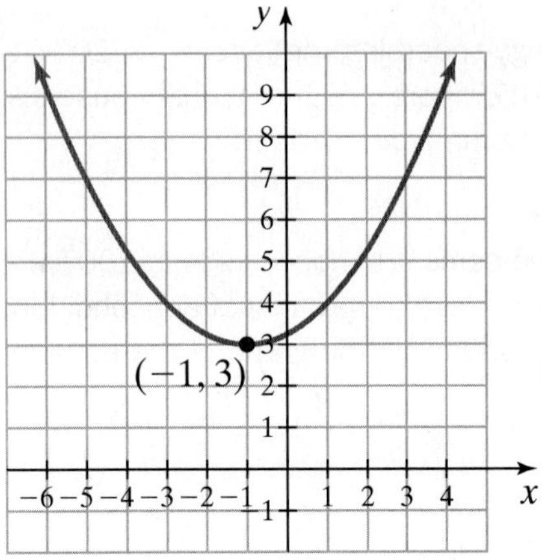

8.

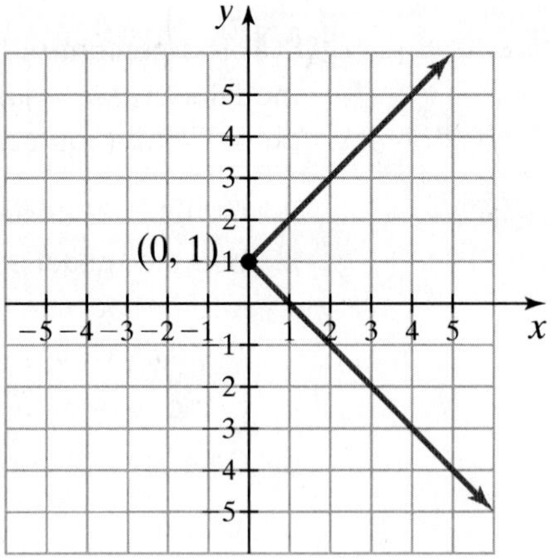

In Exercises 9–16, determine if the relation is a function. Assume x is the independent variable.

You Try It

9. $\{(2, 9), (3, 2), (7, 4), (8, 1), (10, -8)\}$

10. $\{(3, 4), (4, 5), (6, 3), (4, 1), (9, 12)\}$

11. $\{(3, 7), (5, 7), (8, 10), (11, 10), (15, 6)\}$

12. $\{(4, -5), (4, 0), (4, 3), (4, 7), (4, 11)\}$

13. $3x - y = 5$

14. $y = |x + 3| - 1$

15. $|y - 1| + x = 2$

16. $x^2 + y = 1$

In Exercises 17–24, use the vertical line test to determine if each graph is a function.

You Try It

17.

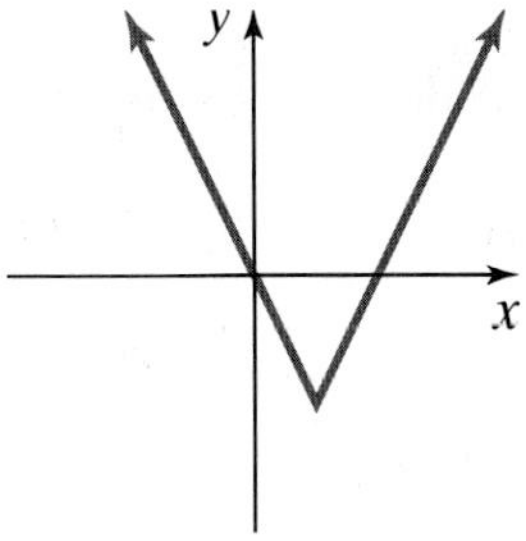

18.

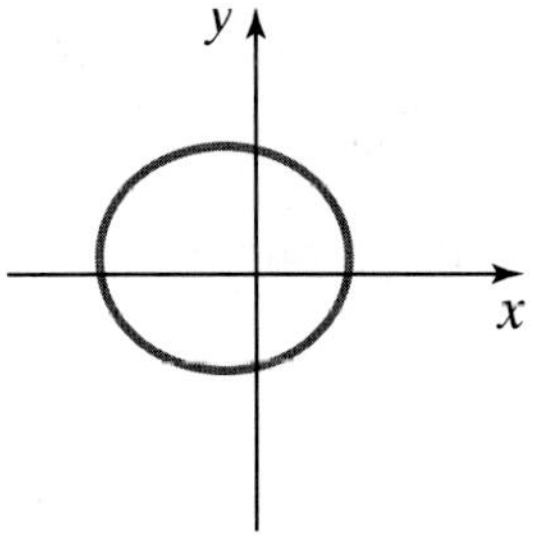

19.

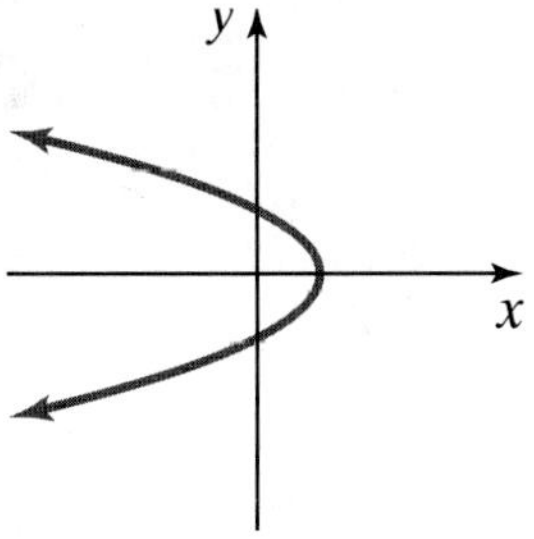

20.

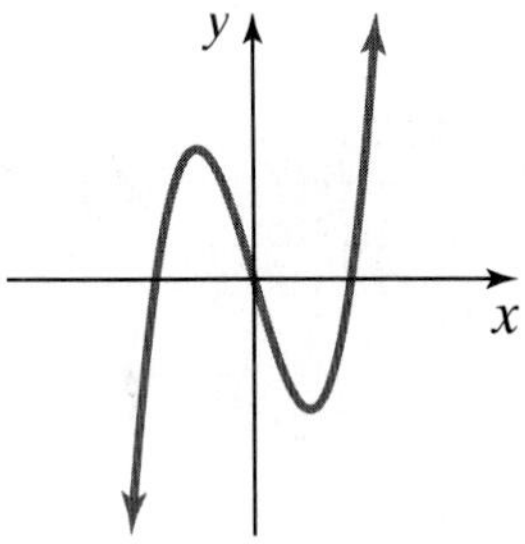

21.

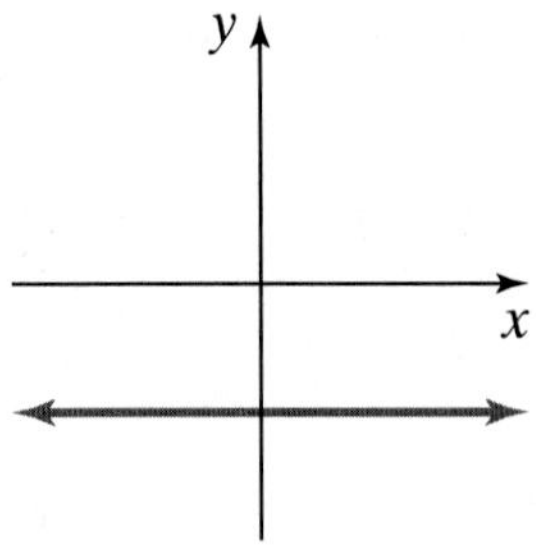

22.

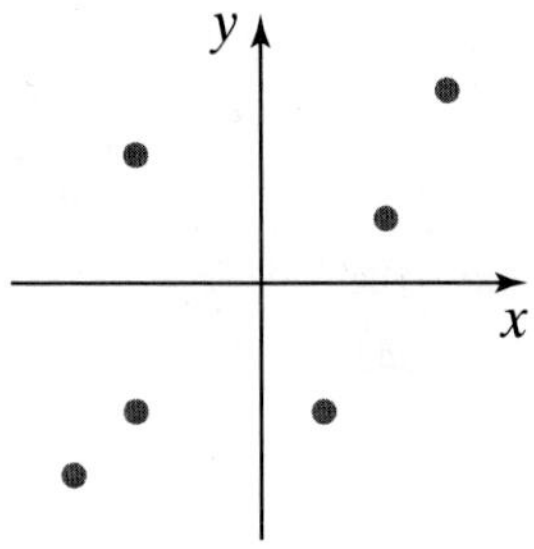

23.

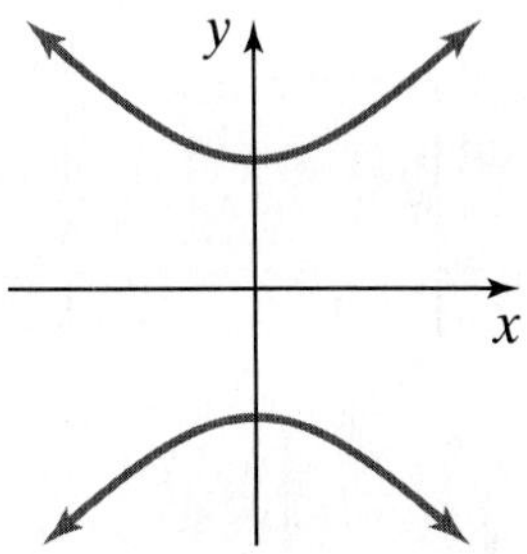

24.

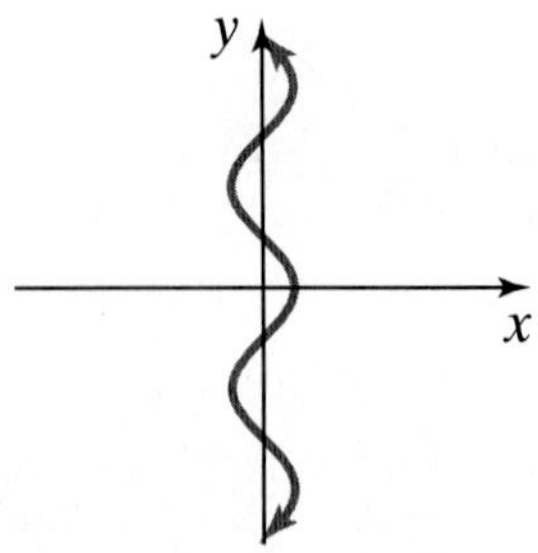

In Exercises 25–28, solve each application problem involving relations and functions.

You Try It

25. Population Data The data in the following table represent the percent of the population with access to clean water and the percent undernourished for several South American countries in 2006.

% population with access to clean water	% undernourished
96	3
91	7
77	15
84	12
92	8

a. If a researcher believes the percent of the population with access to clean water affects the percent undernourished, identify the independent and dependent variables.

b. What are the corresponding ordered pairs for the data?

c. What are the domain and range?

d. Is this relation a function? Explain.

You Try It

26. Medical Expenses The equation $y = 4.09x - 321.96$ relates the amount of out-of-pocket medical expenses (in billions of dollars), x, to the amount paid by insurance (in billions of dollars), y. (*Source:* U.S. Centers for Medicare & Medicaid Services, 2007)

a. Identify the independent and dependent variables.

b. What is the feasible domain?

c. Is the relation a function? Explain.

d. Use the model to estimate the amount paid by insurance if out-of-pocket expenses were $\$270$ billion. What point would this correspond to on the graph of the equation?

e. Solve the equation for x.

f. Use your model to estimate the amount of out-of-pocket expenses if total insurance payments were $\$1$ trillion.

27. Gas Prices The following graph shows the average price per gallon of regular unleaded gasoline in the United States for the years 2002–2007. (*Source:* U.S. Energy Information Administration, 2008)

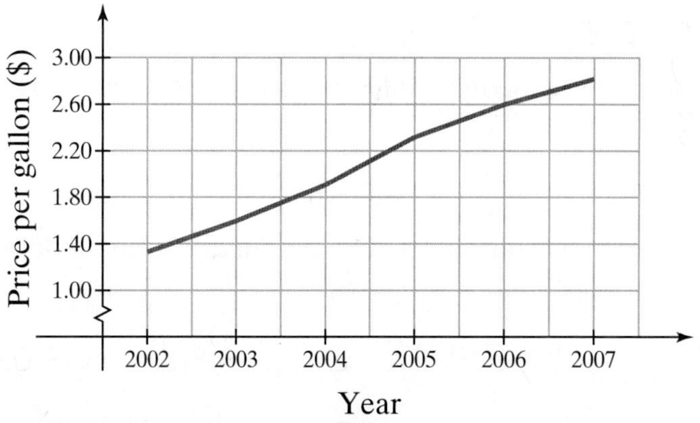

a. Is this graph a function? Explain.

b. Identify the independent and dependent variables.

c. Use the graph to estimate the average price per gallon in 2006.

d. If the data can be modeled by the equation $y = 0.30x + 0.72$, where x is the number of years since 2000 and y is the price per gallon (in dollars), in what year does the model predict that the price per gallon will reach $\$5.00$?

28. College Instructors The percentage of full-time instructors at colleges and universities can be modeled by the equation $y = -0.036x^2 - 0.066x + 61.002$, where x is the number of years since 1990. (*Source:* U.S. National Center for Educational Statistics)

a. Sketch a graph of the equation for the domain $0 \leq x \leq 25$ by plotting points.

b. What is the y-intercept? What does it mean in the context of the problem?

c. Use the model to estimate the percentage of full-time instructors in the year 2030.

d. How would you interpret an x-intercept in this case? Is it reasonable within the context of the problem?

17.2 Function Notation and the Algebra of Functions

THINGS TO KNOW

Before working through this topic, be sure you are familiar with the following concepts:

VIDEO | ANIMATION | INTERACTIVE

You Try It 1. Evaluate Algebraic Expressions (Topic 10.5, **Objective 1**)

You Try It 2. Solve a Formula for a Given Variable (Topic 11.4, **Objective 3**)

You Try It 3. Add Polynomials (Topic 14.3, **Objective 1**)

You Try It 4. Subtract Polynomials (Topic 14.3, **Objective 3**)

You Try It 5. Multiply Two or More Polynomials (Topic 14.4, **Objective 4**)

You Try It 6. Divide Polynomials Using Long Division (Topic 14.7, **Objective 3**)

You Try It 7. Find Restricted Values for Rational Expressions (Topic 16.1, **Objective 2**)

OBJECTIVES

1 Express Equations of Functions Using Function Notation

2 Evaluate Functions

3 Find the Domain of a Polynomial or Rational Function

4 Find the Sum, Difference, Product, and Quotient of Functions

OBJECTIVE 1 EXPRESS EQUATIONS OF FUNCTIONS USING FUNCTION NOTATION

Functions expressed as equations are often named using letters such as f, g, and h. The symbol $f(x)$ is read as "f of x" and is an example of **function notation.** We can use function notation in place of the **dependent variable** in the equation of a function. For example, the function $y = 2x + 3$ may be written as $f(x) = 2x + 3$.

The symbol $f(x)$ represents the value of the dependent variable (output) for a given value of the **independent variable** (input). For $y = f(x)$, we can interpret $f(x)$ as follows: f is the name of the function that relates the independent variable x to the dependent variable y.

Do not confuse function notation with multiplication. $f(x)$ does not mean $f \cdot x$.

When using function notation, any symbol can be used to name the function, and any other symbol can be used to represent the independent variable. For example, consider the function $y = 2x + 3$, which tells us that the value of the dependent variable is obtained by multiplying the value of the independent variable by 2 and then adding 3. We can express this function in many ways:

$f(x) = 2x + 3$	Function name: f	Independent variable: x
$H(t) = 2t + 3$	Function name: H	Independent variable: t
$P(r) = 2r + 3$	Function name: P	Independent variable: r
$\Phi(n) = 2n + 3$	Function name: Φ	Independent variable: n

These four functions are equivalent even though they have different function names and different letters representing the **independent variable**. **Equivalent functions** represent the same set of **ordered pairs**.

When possible, we choose letters for the **variables** to provide meaning. For example, instead of using the variables x and y in our gasoline cost model from **Topic 17.1**, we might use C to represent "cost" and g to represent "gallons of gas." The **function notation** $C(g)$ represents the cost for purchasing g gallons of gas. For this function, g is the independent variable, C is the dependent variable, and $C(g) = 3.24g$ is the **function** that tells us how to find the cost C from the gallons of gas g.

One benefit of using function notation is that it clearly shows the relationship between the independent and dependent variables of an equation. Any equation of a function can be written in function notation using the following procedure:

Expressing Equations of Functions Using Function Notation

Step 1. Choose an appropriate name for the function.

Step 2. Solve the equation for the dependent variable.

Step 3. Replace the dependent variable with equivalent function notation.

When using letters to name functions, the use of lowercase or uppercase matters. For example, f and F are different symbols, so $f(x)$ and $F(x)$ represent different functions. Lowercase and uppercase letters should not be switched within a problem.

Example 1 Expressing Equations of Functions Using Function Notation

Write each function using function notation. Let x be the independent variable and y be the dependent variable.

a. $y = 2x^2 - 4$ **b.** $y - \sqrt{x} = 0$ **c.** $3x + 2y = 6$

Solutions

a. We name the function F. The equation is already solved for y, so we replace y with $F(x)$:

Begin with the original formula: $y = 2x^2 - 4$

Replace y with $F(x)$: $F(x) = 2x^2 - 4$

b. We name the function g. We solve for y and then replace y with $g(x)$:

Begin with the original equation: $y - \sqrt{x} = 0$

Solve for y by adding $\sqrt{x}$ to both sides: $y = \sqrt{x}$

Replace y with $g(x)$: $g(x) = \sqrt{x}$

My video summary

c. Try working through this process yourself. Name the function f. View the **answer**, or watch this **video** for a complete solution to part c.

(eText Screens 17.2-1–17.2-21)

You Try It **Work through this You Try It problem.**

Work Exercises 1–10 in this textbook or in the MyMathLab Study Plan.

OBJECTIVE 2 EVALUATE FUNCTIONS

For $y = f(x)$, the symbol $f(x)$ represents the value of the **dependent variable** y for a given value of the **independent variable** x. For this reason, we call $f(x)$ the **value of the function.** For example, $f(2)$ represents the value of the function f when $x = 2$. When we determine such a function value, we *evaluate the function*.

To **evaluate a function,** we substitute the given value for the independent variable and simplify.

Example 2 Evaluating Functions

My interactive video summary

If $f(x) = 4x - 5$, $g(t) = 3t^2 - 2t + 1$, and $h(r) = \sqrt{r} - 9$, evaluate each of the following.

a. $f(3)$ **b.** $g(-1)$ **c.** $h(16)$ **d.** $f\left(\frac{1}{2}\right)$

Solutions

a. The notation $f(3)$ represents the value of the function f when x is 3. We substitute 3 for x in the function f and simplify:

Substitute 3 for x in the function f: $f(3) = 4(3) - 5$

Simplify: $= 12 - 5$

$= 7$

So, $f(3) = 7$, meaning that the value of f is 7 when x is 3.

b. The notation $g(-1)$ represents the value of the function g when t is -1. We substitute -1 for t in the function g and simplify.

Substitute -1 for t in the function g: $g(-1) = 3(-1)^2 - 2(-1) + 1$

Finish this problem by simplifying the right side. Check your **answer**, or watch this **interactive video** for complete solutions to all four parts.

c. The notation $h(16)$ represents the value of the function h when r is 16. Substitute 16 for r in function h and simplify. Try working through this solution on your own. Check your **answer**, or watch this **interactive video** for complete solutions to all four parts.

d. Try to evaluate $f\left(\frac{1}{2}\right)$ on your own. View the **answer**, or watch this **interactive video** for complete solutions to all four parts.

You Try It **Work through this You Try It problem.**

Work Exercises 11–18 in this textbook or in the MyMathLab Study Plan.

Often **functions** are named by the type of **expression** that defines them. For example, the function $f(x) = 4x - 5$ from **Example 2** is a **linear function** because $4x - 5$ is a **linear expression.** In the same way, $g(t) = 3t^2 - 2t + 1$ is a **quadratic function,** and $h(r) = \sqrt{r} - 9$ is a **square root function.** In Examples 3 and 4, we evaluate a **polynomial function** and a **rational function,** respectively.

(eText Screens 17.2-1–17.2-21)

Example 3 Evaluating a Polynomial Function

My interactive video summary

If $P(x) = 4x^3 - 2x^2 + 8x + 7$, evaluate each of the following.

a. $P(4)$ **b.** $P(-2)$ **c.** $P\left(-\frac{1}{2}\right)$

Solutions

a. Substitute 4 for x in the function P: $P(4) = 4(4)^3 - 2(4)^2 + 8(4) + 7$

Simplify the exponents: $= 4(64) - 2(16) + 8(4) + 7$

Simplify the multiplication: $= 256 - 32 + 32 + 7$

Add and subtract: $= 263$

The value of P is 263 when x is 4.

b. Substitute -2 for x in P: $P(-2) = 4(-2)^3 - 2(-2)^2 + 8(-2) + 7$

Simplify on your own. View the **answer**, or watch this **interactive video** for complete solutions to all three parts.

c. Try to evaluate the function on your own. View the **answer**, or watch this **interactive video** for complete solutions to all three parts.

You Try It **Work through this You Try It problem.**

Work Exercises 19–22 in this textbook or in the MyMathLab Study Plan.

Example 4 Evaluating a Rational Function

My video summary

If $R(x) = \dfrac{5x^2 - 9}{7x + 3}$, evaluate each of the following.

a. $R(1)$ **b.** $R(-3)$

Solutions

a. Substitute 1 for x in the function R: $R(1) = \dfrac{5(1)^2 - 9}{7(1) + 3}$

Simplify the exponent: $= \dfrac{5(1) - 9}{7(1) + 3}$

Simplify the multiplication: $= \dfrac{5 - 9}{7 + 3}$

Add and subtract: $= \dfrac{-4}{10}$

Simplify the fraction: $= -\dfrac{2}{5}$

So, $R(1) = -\frac{2}{5}$. The value of R is $-\frac{2}{5}$ when x is 1.

b. Substitute -3 for x in the function R: $R(-3) = \dfrac{5(-3)^2 - 9}{7(-3) + 3}$

Finish simplifying on your own. See the **answer**, or watch this **video** for complete solutions to both parts.

(eText Screens 17.2-1–17.2-21)

You Try It **Work through this You Try It problem.**

Work Exercises 23 and 24 in this textbook or in the MyMathLab Study Plan.

OBJECTIVE 3 FIND THE DOMAIN OF A POLYNOMIAL OR RATIONAL FUNCTION

In **Topic 17.1**, we defined **domain** as the set of all values for the **independent variable** of the **relation** or **function**. When a function is represented by an equation, the domain is the set of all real numbers for which the function can be evaluated to result in a real number. Consider the **polynomial function** $f(x) = x^3 - 2x + 4$. If we evaluate f for any real number x, the result is a real number, so the domain for f is the set of all real numbers, $\mathbb{R}$ or $(-\infty, \infty)$.

On the other hand consider the **rational function** $g(x) = \frac{x+1}{x-2}$. We cannot evaluate g for $x = 2$ because the result is $g(2) = \frac{2+1}{2-2} = \frac{3}{0}$, which is undefined. So, x cannot be 2, and the domain of g is the set of all real numbers except 2, written in **set-builder notation** as $\{x | x \neq 2\}$. Using **interval notation**, this domain is $(-\infty, 2) \cup (2, \infty)$. Notice that the set-builder notation is less messy, which is typical when stating the domain of rational functions.

We can make the following two generalizations to help with finding the domain of a polynomial or rational function:

- The domain of a polynomial function is the set of all real numbers.
- The domain of a rational function is the set of all real numbers, except those values that cause the denominator to equal zero.

Finding the Domain of a Rational Function

To find the **domain** of a **rational function**, set the **denominator** equal to zero. Then solve the resulting **equation** and exclude the **solutions** from the domain. If the equation has no real solutions, then the domain is the set of all **real numbers**.

Example 5 Finding the Domain of a Function

Find the domain of the given function.

a. $f(x) = 2x^4 - 7x^2 + x - 1$ **b.** $G(x) = \frac{4x+7}{2x-9}$

Solutions

a. f is a **polynomial function**. Evaluating f for any real number x results in a real number. The domain is the set of all real numbers, $\mathbb{R}$ or $(-\infty, \infty)$.

b. G is a rational function. We exclude any values that cause the denominator to be zero.

Set the denominator equal to 0: $2x - 9 = 0$

Add 9 to both sides: $2x = 9$

Divide both sides by 2: $x = \frac{9}{2}$

We must exclude $\frac{9}{2}$, so the domain of G is $\left\{x \middle| x \neq \frac{9}{2}\right\}$.

See **this domain** in interval notation.

You Try It **Work through this You Try It problem.**

Work Exercises 25–28 in this textbook or in the MyMathLab Study Plan.

Example 6 Finding the Domain of a Rational Function

My video summary

Find the domain of $g(x) = \dfrac{x^2 + 2x - 15}{x^2 + 5x - 24}$.

Solution Set the denominator equal to 0: $x^2 + 5x - 24 = 0$

Finish finding the domain by solving this equation and excluding its solutions. View the **answer**, or watch this **video** for a complete solution.

You Try It **Work through this You Try It problem.**

Work Exercises 29–32 in this textbook or in the MyMathLab Study Plan.

Example 7 Finding the Domain of a Rational Function

My video summary

Find the **domain** of each **rational function**.

a. $R(x) = \dfrac{5x - 8}{7}$ **b.** $h(x) = \dfrac{2x - 1}{x^2 + 4}$

Solutions Try to find each domain on your own. View the **answers**, or watch this **video** for detailed solutions.

You Try It **Work through this You Try It problem.**

Work Exercises 33 and 34 in this textbook or in the MyMathLab Study Plan.

OBJECTIVE 4 FIND THE SUM, DIFFERENCE, PRODUCT, AND QUOTIENT OF FUNCTIONS

Two or more **functions** can be combined in the same ways that **algebraic expressions** can be combined, by adding, subtracting, multiplying, and dividing. Consider the functions $f(x) = 2x + 1$ and $g(x) = x^2 - 4$. **Evaluating** these functions when $x = 3$ gives

$$f(3) = 2(3) + 1 = 6 + 1 = 7 \quad \text{and} \quad g(3) = (3)^2 - 4 = 9 - 4 = 5,$$

and the **sum** of these values is

$$f(3) + g(3) = 7 + 5 = 12.$$

If we add the expressions that define $f(x)$ and $g(x)$, we get a third function

$$f(x) + g(x) = (2x + 1) + (x^2 - 4) = x^2 + 2x - 3.$$

This third function is the **sum function** for f and g and is denoted by $(f + g)(x)$. Evaluating this sum function for $x = 3$ gives

$$(f + g)(3) = (3)^2 + 2(3) - 3 = 9 + 6 - 3 = 12.$$

Notice that $f(3) + g(3)$ and $(f + g)(3)$ both resulted in the same value 12, so

$$(f + g)(3) = f(3) + g(3).$$

In fact, $(f + g)(x) = f(x) + g(x)$ for all values of x that are in the **domains** of both f and g. Similarly, $(f - g)(x) = f(x) - g(x)$, $(f \cdot g)(x) = f(x) \cdot g(x)$, and $\left(\frac{f}{g}\right)(x) = \frac{f(x)}{g(x)}$, where $g(x) \neq 0$. Together these four **operations** are known as the **algebra of functions.**

The Algebra of Functions

Let f and g represent two **functions**, then

1. The **sum of f and g** is $(f + g)(x) = f(x) + g(x)$.
2. The **difference of f and g** is $(f - g)(x) = f(x) - g(x)$.
3. The **product of f and g** is $(f \cdot g)(x) = f(x) \cdot g(x)$.
4. The **quotient of f and g** is $\left(\frac{f}{g}\right)(x) = \frac{f(x)}{g(x)}$, provided $g(x) \neq 0$.

Example 8 Finding the Sum and Difference of Functions

My interactive video summary

For $P(x) = x^4 - 9x^2 + 7$ and $Q(x) = 3x^4 - 4x^2 + 2x - 10$, find each of the following.

a. $(P + Q)(x)$ b. $(P - Q)(x)$

Solutions

a.

Define the original problem:	$(P + Q)(x) = P(x) + Q(x)$
Substitute for the function notation:	$= \left(x^4 - 9x^2 + 7\right) + \left(3x^4 - 4x^2 + 2x - 10\right)$
Remove grouping symbols:	$= x^4 - 9x^2 + 7 + 3x^4 - 4x^2 + 2x - 10$
Rearrange terms:	$= x^4 + 3x^4 - 9x^2 - 4x^2 + 2x + 7 - 10$
Combine like terms:	$= 4x^4 - 13x^2 + 2x - 3$

b.

Define the original problem:	$(P - Q)(x) = P(x) - Q(x)$
Substitute for the function notation:	$= \left(x^4 - 9x^2 + 7\right) - \left(3x^4 - 4x^2 + 2x - 10\right)$

Try finishing this problem on your own. See the **answer**, or watch this **interactive video** for complete solutions to both parts.

You Try It **Work through this You Try It problem.**

Work Exercises 35–40 in this textbook or in the MyMathLab Study Plan.

Example 9 Finding the Product of Functions

For $P(x) = 5x - 2$ and $Q(x) = 4x - 9$, find $(P \cdot Q)(x)$.

Solution

Define the original problem:	$(P \cdot Q)(x) = P(x) \cdot Q(x)$
Substitute for the function notation:	$= (5x - 2)(4x - 9)$
Use FOIL to multiply:	$= \overbrace{5x \cdot 4x}^{F} + \overbrace{5x \cdot (-9)}^{O} + \overbrace{(-2) \cdot 4x}^{I} + \overbrace{(-2) \cdot (-9)}^{L}$
Simplify the multiplication:	$= 20x^2 - 45x - 8x + 18$
Combine like terms:	$= 20x^2 - 53x + 18$

(eText Screens 17.2-1–17.2-21)

You Try It **Work through this You Try It problem.**

Work Exercises 41 and 42 in this textbook or in the MyMathLab Study Plan.

Example 10 Finding the Quotient of Functions

My video summary

For $P(x) = 15x^3 + 41x^2 + 4x + 3$ and $Q(x) = 5x + 2$, find $\left(\frac{P}{Q}\right)(x)$. State any values that cannot be included in the **domain** of $\left(\frac{P}{Q}\right)(x)$. (Note that $Q(x)$ cannot be 0.)

Solution

Define the original problem: $\left(\frac{P}{Q}\right)(x) = \frac{P(x)}{Q(x)}$

Substitute for the function notation: $= \frac{15x^3 + 41x^2 + 4x + 3}{5x + 2}$

Complete this problem using **polynomial long division.** To find the values that cannot be included in the domain, set $Q(x)$ equal to 0 and solve. View the **answer**, or watch this **video** for a detailed solution.

You Try It **Work through this You Try It problem.**

Work Exercises 43 and 44 in this textbook or in the MyMathLab Study Plan.

17.2 Exercises

In Exercises 1–10, write each function using function notation. Let x be the independent variable and y be the dependent variable. Use the letter f to name each function.

1. $y = |2x - 5|$

2. $y = 3x^2 + 2x - 5$

3. $y + \sqrt{x} = 1$

4. $2x + y = 3$

You Try It

5. $3x + 4y = 12$

6. $-6x + 18y = 12$

7. $4x^2 - 2y = 10$

8. $3y + 6\sqrt{x - 5} = 0$

9. $\frac{3y - 7}{2} = 3x^2 + 1$

10. $\frac{5y - 8}{3} = \frac{10x^2 + 4}{6}$

In Exercises 11–22, evaluate each function.

You Try It

11. $f(x) = 3x - 5;\ \ f(6)$

12. $h(x) = 2x^2 + 5x - 17;\ \ h(-4)$

13. $F(z) = 2|z - 3| - 5;\ \ F(0)$

14. $T(t) = \frac{5}{6}t + \frac{1}{3};\ \ T(8)$

15. $r(x) = 3 + \sqrt{x - 5};\ \ r(9)$

16. $c(x) = \sqrt{25 - x^2};\ \ c(3)$

17. $\Phi(p) = (p - 1)p^3;\quad \Phi(3)$

18. $R(x) = 8x^2 - 2x + 1;\quad R\left(-\frac{1}{2}\right)$

19. $P(x) = 18x^2 - 6x + 5;\quad P\left(\frac{1}{3}\right)$

20. $Q(t) = t^5 + 3t^4 - 5t^3 - 8t^2 + t - 15;\quad Q(-2)$

You Try It

21. $R(p) = 0.4p^3 - 2.6p^2 - 6.8p + 52;\quad R(10)$

22. $S(x) = 16.4x^4 - 10.6x^2 + 20.1;\quad S(2.5)$

In Exercises 23 and 24, evaluate the given rational function as indicated. If the function value is undefined, state this as your answer.

23. If $f(x) = \dfrac{x^2 - 5x - 14}{5 - x}$, find $f(-2)$, $f(3)$, and $f(5)$.

You Try It

24. If $g(t) = \dfrac{2t^3 - 7}{t^2 + 5}$, find $g(-1)$, $g(0)$, and $g(2)$.

In Exercises 25–34, find the domain of each rational function. Write your answer in set-builder notation.

25. $f(x) = \dfrac{10}{x + 2}$

26. $g(x) = 4x^2 - 3x - 6$

You Try It

27. $f(x) = 5x + 9$

28. $s(t) = \dfrac{3t - 7}{2t}$

29. $h(x) = \dfrac{3x - 7}{5x + 2}$

30. $g(x) = \dfrac{8x}{x^2 + 3x - 88}$

You Try It

31. $R(x) = \dfrac{x + 4}{3x^2 + 10x - 8}$

32. $h(x) = \dfrac{3x - 1}{x^3 - 4x^2 - 12x}$

33. $f(x) = \dfrac{7x - 2}{5}$

34. $R(x) = \dfrac{x + 4}{2x^2 + 1}$

You Try It

In Exercises 35–40, let $P(x) = x^3 - 5x^2 + 2x - 1$, $Q(x) = x^3 - 9x^2 - 12$, and $R(x) = 9x^2 + 4x - 8$. Add or subtract the functions, as indicated.

35. $(P + Q)(x)$

36. $(Q - P)(x)$

37. $(Q + R)(x)$

You Try It

38. $(P - R)(x)$

39. $(Q - R)(x)$

40. $(P + R - Q)(x)$

In Exercises 41 and 42, find $(P \cdot Q)(x)$.

41. $P(x) = 2x - 5;\quad Q(x) = x + 8$

42. $P(x) = x - 4;\quad Q(x) = x^2 + 3x - 5$

You Try It

In Exercises 43 and 44, find $\left(\dfrac{P}{Q}\right)(x)$ and state any values that cannot be included in the domain of $\left(\dfrac{P}{Q}\right)(x)$.

43. $P(x) = 3x^2 - 7x + 5;\quad Q(x) = x - 2$

44. $P(x) = 6x^3 + 11x^2 - 9x - 2;\quad Q(x) = 3x + 1$

You Try It

17.3 Graphs of Functions and Their Applications

THINGS TO KNOW

Before working through this topic, be sure you are familiar with the following concepts:

		VIDEO	ANIMATION	INTERACTIVE
You Try It	1. Read Line Graphs (Topic 12.1, **Objective 1**)	✓		
You Try It	2. Plot Ordered Pairs in the Rectangular Coordinate System (Topic 12.1, **Objective 3**)	✓		
You Try It	3. Graph Linear Equations by Plotting Points (Topic 12.2, **Objective 3**)			✓
You Try It	4. Express Equations of Functions Using Function Notation (Topic 17.2, **Objective 1**)	✓		
You Try It	5. Evaluate Functions (Topic 17.2, **Objective 2**)			✓

OBJECTIVES

1 Graph Simple Functions by Plotting Points

2 Interpret Graphs of Functions

3 Solve Application Problems Involving Functions

OBJECTIVE 1 GRAPH SIMPLE FUNCTIONS BY PLOTTING POINTS

In **Example 2a** of Topic 17.2, we see that $f(3) = 7$. For the **function** f, this means that an input of 3 into the function f results in an output of 7, which corresponds to the **ordered pair** $(3, 7)$. We say that $(3, 7)$ **belongs** to the function f. The **graph of a function** is the graph of all ordered pairs that belong to the function. Since $f(3) = 7$, the point $(3, 7)$ lies on the graph of f. This leads us to the following statement:

Graph of a Function

The **graph of a function** is the graph of all ordered pairs that belong to the function. The point (a, b) lies on the graph of a function f if and only if $f(a) = b$.

In Example 3 of **Topic 12.2**, we graphed the equation $3x - y = 2$ (view the **graph**). Looking at the graph of $3x - y = 2$, we see that the graph is a function because it passes the **vertical line test**. Solving for y, we get $y = 3x - 2$. Replacing y with $f(x)$ allows us to write the equation in **function notation**: $f(x) = 3x - 2$. This is an example of a **linear function**. Because linear functions are also **polynomial functions**, the **domain** of every linear function is the set of all real numbers, $\mathbb{R}$ or $(-\infty, \infty)$ in interval notation.

(eText Screens 17.3-1–17.3-22)

Definition Linear Function

A **linear function** is a function of the form $f(x) = ax + b$, where a and b are real numbers.

To graph linear functions and other simple functions, we use a point-plotting strategy similar to that used to graph linear equations.

Strategy for Graphing Simple Functions by Plotting Points

Step 1. Find several points that **belong** to the function. The exact number of points to find depends on the function being graphed. Try to locate key points such as **intercepts** and **maximum or minimum points.**

Step 2. **Plot** the points found in Step 1.

Step 3. If the points plotted in Step 2 line up, connect them with a straight line; otherwise, connect them with a smooth curve.

Example 1 Graphing Simple Functions by Plotting Points

My animation summary

Graph each function by plotting points.

a. $f(x) = 2x - 1$ **b.** $g(x) = x^2 + 2x - 3$ **c.** $h(x) = 2|x| - 1$

Solutions

a. Step 1. Note that the **domain** of f is the set of all real numbers. We can evaluate $f(x)$ for $x = -2, -1, 0, 1$, and 2. Then we organize the work in Table 1.

x	$y = f(x)$	(x, y)
-2	$f(-2) = 2(-2) - 1 = -5$	$(-2, -5)$
-1	$f(-1) = 2(-1) - 1 = -3$	$(-1, -3)$
0	$f(0) = 2(0) - 1 = -1$	$(0, -1)$
1	$f(1) = 2(1) - 1 = 1$	$(1, 1)$
2	$f(2) = 2(2) - 1 = 3$	$(2, 3)$

Table 1

Step 2. Plot the points found in Table 1.

Step 3. Do the points line up? If so, connect them with a straight line; otherwise, connect with a smooth curve. View the **answer**, or work through this **animation** to see the fully worked solution.

b. Step 1. The **domain** of g is the set of all real numbers. Evaluate $g(x)$ for $x = -4, -3, -2, -1, 0, 1$, and 2 to create a table of ordered pairs that belong to the function. View this **popup** to check your table.

Step 2. Plot the points found in Step 1.

Step 3. Connect them with a line or smooth curve as appropriate. View the **answer**, or work through this **animation** to see the fully worked solution.

c. Create a table of **ordered pairs** that **belong** to the function. Be sure to evaluate the function for negative values of x as well as positive values of x. Plot the ordered pairs and connect them with a line or smooth curve as appropriate. View the **answer**, or work through this **animation** to see the fully worked solution.

(eText Screens 17.3-1–17.3-22)

You Try It **Work through this You Try It problem.**

Work Exercises 1–6 in this textbook or in the MyMathLab Study Plan.

OBJECTIVE 2 INTERPRET GRAPHS OF FUNCTIONS

Graphs of **functions** can be used to show a variety of everyday situations visually. In the next three examples, we use graphs of functions to **model** common situations.

Example 2 Straight-Line Depreciation

The word *depreciation* means "a loss in value." *Straight-line depreciation* is one of several accounting methods allowed by the Internal Revenue Service for deducting losses in the value of equipment as it ages. Claiming a loss in *book value* allows the owner to escape paying taxes on the amount of the loss. The graph in Figure 1 uses straight-line depreciation to depict the book value of a machine as it ages. Use the graph to answer the questions.

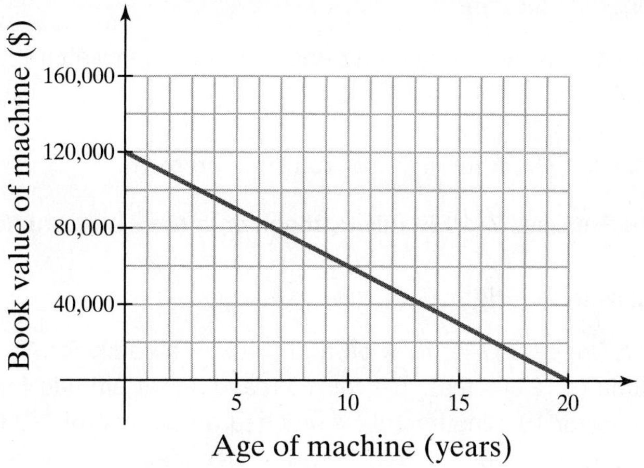

Figure 1
Straight-Line Depreciation

a. What was the **book value** of the machine when it was new?

b. How old will the machine be when it no longer has a book value?

c. What will the book value of the machine be when it is 5 years old?

d. At what age will the machine have lost half of its original book value?

e. By how much does the book value of the machine decrease each year?

Solutions

a. The **graph** contains the point $(0, 120000)$, which represents the book value of the machine when it was new. This book value was $120,000.

b. The **graph** contains the point $(20, 0)$, which shows that the machine will have a book value of $0 (no book value) after 20 years.

c. The **graph** contains the point $(5, 90000)$, which represents the book value of the machine after 5 years. This book value will be $90,000.

d. From part (a), we know the original book value of the machine was $120,000. Half of this is $60,000. The graph contains the point $(10, 60000)$. So, the machine will have lost half of its original book value when it is 10 years old.

e. Since the machine's original value of $120,000 is lost evenly over a 20-year period, the book value decreases by $\dfrac{\$120,000}{20 \text{ years}} = \$6000$ each year.

(eText Screens 17.3-1–17.3-22)

Example 3 Spring Temperatures

My video summary

The graph of the **function** in Figure 2 gives the outside temperatures over one 24-hour period in spring. Use the graph to answer the questions.

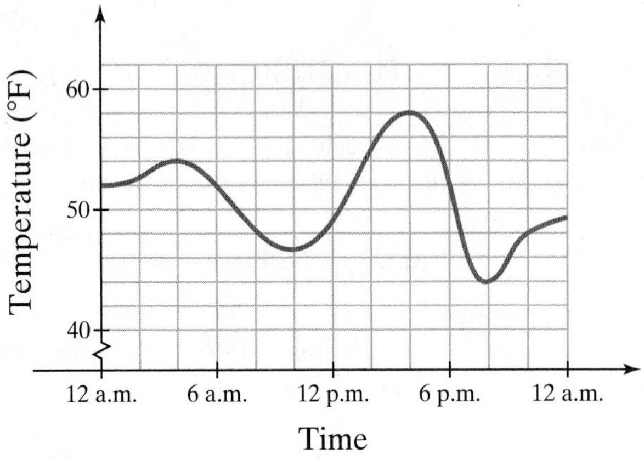

Figure 2
Spring Temperatures

a. Over what time periods was the temperature rising?

b. Over what time periods was the temperature falling?

c. What was the highest temperature for the day? At what time was it reached?

d. What was the lowest temperature for the day shown? At what time was it reached?

e. Over what time period did the temperature decrease most rapidly?

Solution Work through the questions and then watch this **video** to check your answers.

You Try It **Work through this You Try It problem.**

Work Exercises 7–10 in this textbook or in the MyMathLab Study Plan.

Example 4 Flight Altitude

My video summary

A Boeing 757 jet took off and climbed steadily for 20 minutes until it reached an altitude of 18,000 feet. The jet maintained that altitude for 30 minutes. Then it climbed steadily for 10 minutes until it reached an altitude of 26,000 feet. The jet remained at 26,000 feet for 40 minutes. Then it descended steadily for 20 minutes until it reached an altitude of 20,000 feet, where it remained for 30 more minutes. During the final 20 minutes of the flight, the jet descended steadily until it landed at its destination airport. Draw a graph of the 757's altitude as a function of time.

Solution The 757's altitude is a function of time, so the **independent variable**, *time* (in minutes), is represented by the horizontal axis and the **dependent variable**, *altitude*, is represented by the vertical axis. To **draw the graph**, identify key points. Plot these points and connect them. For example, the **ordered pair** $(0, 0)$ represents an altitude of 0 feet (ground) at time 0 minutes (before takeoff). The ordered pair $(20, 18000)$ represents an altitude of 18,000 feet at 20 minutes after takeoff. To show the initial climb of the airplane, we connect the points $(0, 0)$ and $(20, 18000)$ with a straight line. Continue this process. View the **answer** and/or watch this **video** to see the fully worked solution.

You Try It **Work through this You Try It problem.**

Work Exercises 11 and 12 in this textbook or in the MyMathLab Study Plan.

OBJECTIVE 3 SOLVE APPLICATION PROBLEMS INVOLVING FUNCTIONS

Functions are used to **model** a variety of real-world applications in fields of study such as physics, biology, business, and economics. We can develop methods for creating such models ourselves. However, for the next few examples, the models used to solve applications are given.

Example 5 Falling Rock

My interactive video summary

A rock is dropped from the top of a cliff. Its height, h, above the ground, in feet, at t seconds is given by the function, $h(t) = -16t^2 + 900$. Use the model to answer the following questions.

a. Evaluate $h(0)$. What does this value represent?

b. Evaluate $h(2)$. What does this value represent? How far has the rock fallen at this time?

c. Evaluate $h(10)$. Is this possible? Explain.

d. Evaluate $h(7.5)$. Interpret this result.

e. Determine the **feasible domain** and the range that makes sense (or **feasible range**) within the context of the problem.

f. Graph the function.

Solutions

a. $h(0) = -16(0)^2 + 900 = 900$
The height of the rock at $t = 0$ seconds (before the rock is dropped) is 900 feet above the foot of the cliff. Therefore, the height of the cliff is 900 feet.

b. $h(2) = -16(2)^2 + 900 = 836$
Two seconds after being dropped, the rock is 836 feet above the ground. Therefore, the rock has fallen $900 - 836 = 64$ feet.

c. $h(10) = -16(10)^2 + 900 = -700$
A negative height means that the rock has fallen below ground level. Our result shows that the rock would be 700 feet below the ground at the base of the cliff 10 seconds after being dropped. This result is not possible (assuming there is not a 700-foot-deep hole at the base of the cliff). Therefore, $t = 10$ seconds is outside the **feasible domain** of the function.

d. $h(7.5) = -16(7.5)^2 + 900 = 0$
After 7.5 seconds, the rock is 0 feet above the ground, or literally on the ground. Therefore, the rock hits the ground at the foot of the cliff 7.5 seconds after being dropped.

e. The **feasible domain** includes only those domain values that make sense within the context of the application problem. Since t represents time, its values are **non-negative**. From part (d) we know that the rock hits the ground at $t = 7.5$ seconds. This means that function values for times after 7.5 seconds will not make sense (as we saw in part (c)). The feasible domain is $\{t | 0 \le t \le 7.5\}$ or $[0, 7.5]$.

Likewise, the **feasible range** includes only those range values that make sense within the context of the application problem. For example, height is non-negative. From part (a) we know that $h(0) = 900$, which is the largest possible value for h. Therefore, the feasible range is $\{h | 0 \le h \le 900\}$ or $[0, 900]$.

f. **Graph the function** by plotting several points and connecting them with a smooth curve. View the **graph** or watch the **interactive video** for the fully worked solution.

(eText Screens 17.3-1–17.3-22)

Example 6 Rent in Queens, NY

My interactive video summary

The average monthly rent, R, for apartments in Queens, New York, is **modeled** by the **function** $R(a) = 2.2a$, where a is the floor area of the apartment in square feet. Use the model to answer the following questions. (*Source*: Apartments.com, 2010)

a. What is the average monthly rent for apartments in Queens, New York with a floor area of 800 square feet?

b. What is the floor area of an apartment if its rent is \$1430 per month?

c. Determine the **feasible domain** and the **feasible range** of the function.

d. Graph the function.

Solutions

a. $R(800) = 2.2(800) = 1760$
The average monthly rent for an 800-square-foot apartment in Queens is \$1760.

b. In this case, we solve for a when $R(a) = 1430$.

Rewrite the original function:	$R(a) = 2.2a$
Substitute 1430 for $R(a)$:	$1430 = 2.2a$
Divide both sides by 2.2:	$\dfrac{1430}{2.2} = \dfrac{2.2a}{2.2}$
Simplify:	$650 = a$

So, the floor area of an apartment with a monthly rent of \$1430 is 650 square feet.

c. For the application situation to make sense, the floor area of an apartment must not be negative. In a practical sense, it is not likely to find apartments with floor areas of 1 square foot or 100,000 square feet. However, there are no definite limits. Therefore, the **feasible domain** is $\{a|a \geq 0\}$ or $[0, \infty)$.

Also, the problem does not make sense if the monthly rent is negative. As with floor area, you are not likely to find apartments with monthly rents of more than \$1,000,000. However, there is no definite maximum limit. Therefore, the **feasible range** is $\{R|R \geq 0\}$ or $[0, \infty)$.

d. **Graph the function** by plotting points. View the **graph** or watch this **interactive video** for the fully worked solution.

You Try It **Work through this You Try It problem.**

Work Exercises 13–16 in this textbook or in the MyMathLab Study Plan.

17.3 Exercises

In Exercises 1–6, graph each function by plotting points.

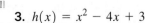

You Try It

1. $f(x) = 2x - 3$

2. $g(x) = -\frac{1}{2}x + 4$

3. $h(x) = x^2 - 4x + 3$

4. $F(x) = |x + 1| - 3$

5. $G(x) = -2|x| + 3$

6. $H(x) = \sqrt{x + 4}$

(eText Screens 17.3-1–17.3-22)

In Exercises 7–10, solve each application problem.

You Try It

7. Text Messaging The graph of the function shown illustrates a text messaging plan offered by AT&T Wireless. Use the graph to answer the following questions.

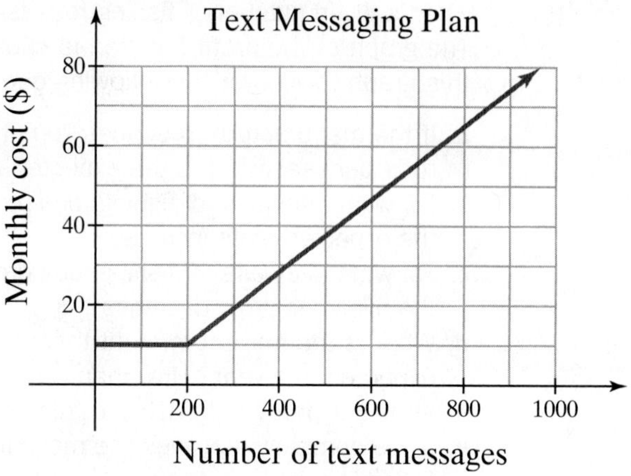

a. Under this plan, what is the monthly cost for the first 200 text messages?
b. After the first 200 text messages, what is the cost per text message under this plan?
c. What is the monthly cost for 600 text messages?
d. What are the feasible domain and the feasible range of this function?
e. A second AT&T Wireless plan offers unlimited texting for \$20 per month. Describe the number of text messages that must be used per month in order to make this second plan a better deal.

8. Straight-Line Depreciation A rancher bought a new truck and will use the truck's depreciation in book value as a tax write-off. The graph of the function shown illustrates the rancher's plan. Use the graph to answer the following questions.

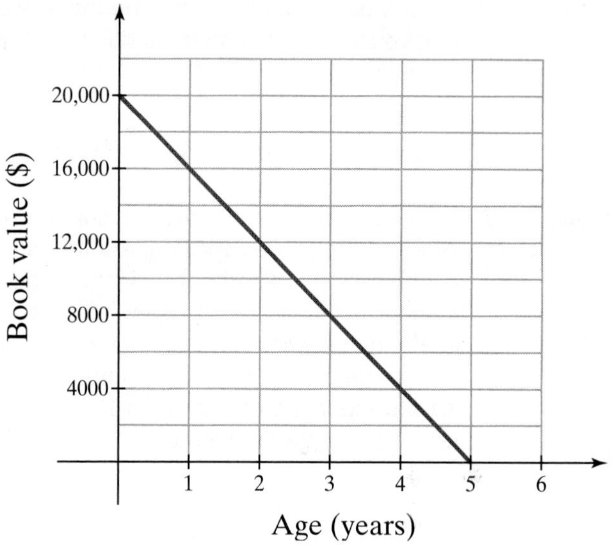

a. What did the rancher pay for the truck?
b. How old will the truck be when its book value is \$0?
c. What are the feasible domain and the feasible range of this function?
d. What will the book value be when the truck is 4 years old?
e. How much money (in book value depreciation) can the rancher write off on taxes each year?

9. Dow Jones Industrial Average The graph of the function shown illustrates the 2010 Dow Jones Industrial Average (DJIA). Each tick mark on the horizontal axis represents the first opening day of the market for that given month. Use the graph to answer the following questions.

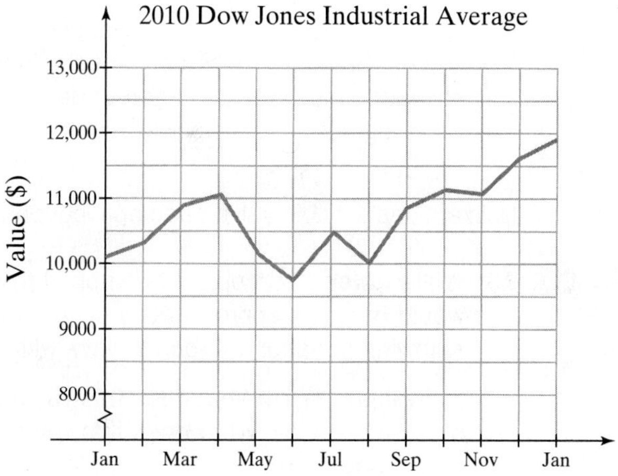

a. Over which month(s) did the DJIA decrease?
b. Over which month(s) did the DJIA increase?
c. Over which month did the DJIA increase most rapidly?
d. Approximate the DJIA's value at the beginning of 2010.
e. Approximate its value at the end of 2010.
f. Approximately how much did the DJIA's value increase over 2010?
g. Approximate the average gain per month.

10. **Fishing Rod Production** A fishing rod manufacturer developed a model for the profit expected when a given number of fishing rods is produced. The graph of the profit function is shown. Use this graph to answer the following questions.

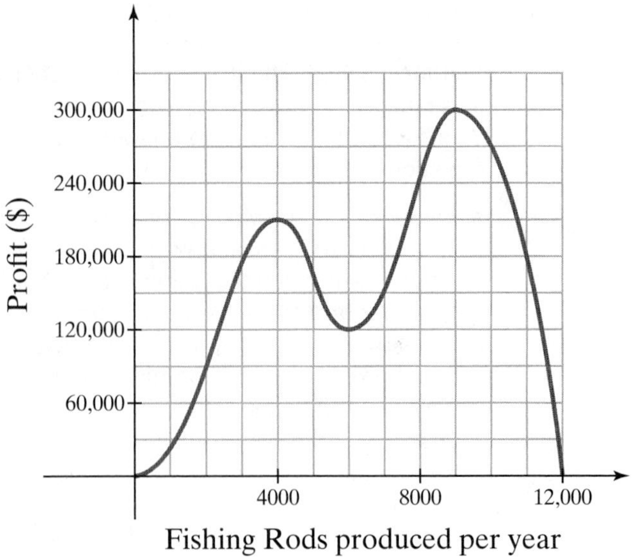

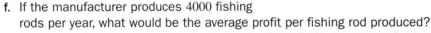

   a. If the manufacturer produces 7000 fishing rods per year, what is the expected profit?
   b. For what interval(s) of fishing rods produced will the expected profit increase?
   c. For what interval(s) of fishing rods produced will the expected profit decrease?
   d. What is the maximum profit that can be expected in a year? How many fishing rods should it produce to achieve this profit?
   e. If the company achieves the maximum profit found in part (d), what would be the average profit per fishing rod produced?
   f. If the manufacturer produces 4000 fishing rods per year, what would be the average profit per fishing rod produced?
   g. If you were the company's chief operating officer (CEO), how many fishing rods would you have produced? Explain why you gave this answer.

In Exercises 11–12, draw a graph of a function that represents each application situation.

You Try It

11. **Exercise** One morning, Karen took a walk. She left her home and walked 4 blocks in 4 minutes at a steady speed. Then she realized that she did not have her cell phone. Being on call for work, she jogged home in 2 minutes to get it. Just as she arrived home, her cell phone rang, so Karen remained at home for 3 minutes during the call. Now pressed for time, she decided to jog instead of walk. She jogged 10 blocks in 5 minutes when her cell phone rang again. She then stopped for 4 minutes to take the call. Now being even more pressed for time, Karen ran 9 blocks back towards her home in 3 minutes. To cool down, she walked the last block home in 1 minute. Draw a graph of Karen's distance from home (in blocks) as a function of time (in minutes).

12. **Stock Price** Bill bought stock in a company for \$10 per share. After his purchase, the price remained constant for two days. At that time, a negative report was released about the company, so the price declined steadily for three days until it had lost a total of \$6 per share. The price remained steady for four days. Then, a positive report was released about the company, so the price rose sharply for two days, gaining \$8 per share. The price continued gaining gradually for four more days, ultimately rising \$1 more per share. At that point, the price began to decline slowly for six days, losing \$2 per share. Then, Bill sold the stock. Draw a graph of the stock's value (in dollars per share) as a function of the time (in days).

In Exercises 13–16, solve each application problem.

13. **Willis Tower** If an object is dropped from the roof of the Willis Tower, its height, h, in meters, at t seconds would be given approximately by the function $h(t) = -4.9t^2 + 442.225$. Use the function to answer the following questions. (*Source*: www.willistower.com).

   a. Evaluate $h(0)$. What does this value represent?
   b. Evaluate $h(5)$. What does this value represent?
   c. Evaluate $h(10)$. Is this possible? Explain.
   d. Evaluate $h(9.5)$. What does this value represent?
   e. Determine the feasible domain and the feasible range of the function in the context of this problem.
   f. Graph the function by plotting points.

You Try It

14. Car Payments Car companies occasionally offer qualified buyers an incentive of 0% financing and no money down. Ethan just purchased a new automobile under such an agreement. The amount of money, A, that he will still owe the company after n monthly payments have been made is given by the function, $A(n) = 34{,}515 - 575.25n$. Use the model to answer the following questions:

a. Evaluate $A(0)$. What does this value represent?
b. Evaluate $A(24)$. What does this value represent?
c. Evaluate $A(72)$. Is this possible? Explain.
d. After 3 years of making payments, how much will Ethan still owe on the car?
e. How long will it take Ethan to pay off the car?
f. Determine the feasible domain and the feasible range of the function in the context of the problem.
g. Graph the function by plotting points.

15. Gas Mileage The amount of gas, G (in gallons), that remains in the tank of a 2011 Toyota Prius after it has been driven m miles, starting with a full tank, is given by the function, $G(m) = \frac{571.2 - m}{48}$. (*Source*: www.hybridcars.com)

a. Evaluate $G(100)$. What does this value represent?
b. Evaluate $G(200)$. What does this value represent?
c. How much gas does a full tank hold?
d. How much gas will remain in the tank after it has been driven 400 miles, starting with a full tank?
e. After filling up, how far can the Prius travel before it runs out of gas?
f. Determine the feasible domain and the feasible range of the function in the context of this problem.
g. Graph the function by plotting points.

16. Area of a Circle The area A of a circle with circumference C is given by the function $A(C) = \frac{C^2}{4\pi}$. Use this function to answer the following questions:

a. Evaluate $A(4\pi)$. What does this value represent?
b. Evaluate $A(12)$. What does this value represent?
c. Determine the area of a circle with a circumference of 16π (in inches).
d. Determine the feasible domain and the feasible range of the function in the context of this problem.
e. Graph the function by plotting points. [Hint: To make graphing easier, substitute multiples of π for C.]

MODULE EIGHTEEN

Radicals and Rational Exponents

MODULE EIGHTEEN CONTENTS

18.1 Radical Expressions

THINGS TO KNOW

Before working through this topic, be sure you are familiar with the following concepts:

		VIDEO	ANIMATION	INTERACTIVE
You Try It	1. FInd the Absolute Value of a Real Number (Topic 10.1, **Objective 4**)	▤		
You Try It	2. Use the Order of Operations to Evaluate Numeric Expressions Exponents and Radicals (Topic 10.4, **Objective 2**)	▤		

OBJECTIVES

1 Find Square Roots of Perfect Squares

2 Approximate Square Roots

3 Simplify Radical Expressions of the Form $\sqrt{a^2}$

4 Find Cube Roots

5 Find and Approximate nth Roots

OBJECTIVE 1 FIND SQUARE ROOTS OF PERFECT SQUARES

We have seen earlier that *squaring* a number means to multiply the number by itself. For example:

The square of 3 is $3^2 = 3 \cdot 3 = 9$

The square of -3 is $(-3)^2 = (-3)(-3) = 9$.

The square of $\frac{1}{5}$ is $\left(\frac{1}{5}\right)^2 = \frac{1}{5} \cdot \frac{1}{5} = \frac{1}{25}$.

The opposite operation of squaring is to find the **square root**. The square root of a **non-negative** real number a is a **real number** b that, when **squared**, results in a. So, b is a square root of a if $b^2 = a$.

Every positive real number has two square roots: one positive and one negative. For example, 3 and -3 are both square roots of 9 because

$$3^2 = 9 \quad \text{and} \quad (-3)^2 = 9$$

3 is the *positive*, or *principal*, *square root* of 9, and -3 is the *negative square root* of 9.

We use the **radical sign** $\sqrt{\ }$ to denote principal square roots. We place a negative sign in front of a radical sign, $-\sqrt{\ }$, to denote negative square roots. For example, $\sqrt{4}$ represents the principal square root of 4, whereas $-\sqrt{4}$ represents the negative square root of 4. So, we write $\sqrt{4} = 2$ and $-\sqrt{4} = -2$.

> **Definition Principal and Negative Square Roots**
>
> A non-negative real number b is the **principal square root** of a non-negative real number a, denoted as $b = \sqrt{a}$, if $b^2 = a$.
>
> A negative real number b is the **negative square root** of a non-negative real number a, denoted as $b = -\sqrt{a}$, if $b^2 = a$.

The **expression** $\sqrt{4}$ is called a *radical expression*. A **radical expression** is an expression that contains a **radical sign**. The expression beneath the radical sign is called the **radicand**.

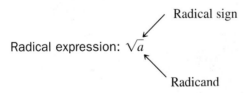

Radical sign

Radical expression: $\sqrt{a}$

Radicand

When the **radicand** is a **perfect square**, the **square root** will simplify to a **rational number**.

Example 1 Finding Square Roots

My video summary

Evaluate.

a. $\sqrt{64}$ b. $-\sqrt{169}$ c. $\sqrt{-100}$

d. $\sqrt{\frac{9}{25}}$ e. $\sqrt{0.81}$ f. $\sqrt{0}$

(eText Screens 18.1-1–18.1-16)

Solutions

a. We need to find the **principal square root** of 64. Since $8^2 = 64$, $\sqrt{64} = 8$.

b. We need to find the **negative square root** of 169. Since $(-13)^2 = 169$, $-\sqrt{169} = -13$.

c. There is no **real number** that can be **squared** to result in -100 because the square of a real number will always be non-negative. So, $\sqrt{-100}$ is not a real number.

d.–f. Try to evaluate these square roots on your own. View the **answers**, or watch this **video** to check your work.

CAUTION The square root of a negative number is not a real number.

You Try It **Work through this You Try It problem.**

Work Exercises 1–4 in this textbook or in the MyMathLab Study Plan.

OBJECTIVE 2 APPROXIMATE SQUARE ROOTS

In **Example 1**, we saw that a **square root** simplifies to a **rational number** if the **radicand** is a **perfect square**. But what happens when the radicand is not a perfect square? In this case, the square root is an **irrational number**.

Consider $\sqrt{12}$. The radicand 12 is not a perfect square, so $\sqrt{12}$ is an irrational number. For such **radical expressions**, we can find decimal approximations using a calculator. Before doing this, consider what might be a **reasonable** result. Notice that 12 is between the perfect squares 9 and 16.

$$9 < 12 < 16$$

So, the **principal square root** of 12 should be between the principal square roots of 9 and 16.

$$\sqrt{9} < \sqrt{12} < \sqrt{16} \quad \text{or} \quad 3 < \sqrt{12} < 4$$

Figure 1 shows the TI-84 Plus calculator display for $\sqrt{12}$. Rounding to three decimal places, we have $\sqrt{12} \approx 3.464$. This approximation is between 3 and 4 as expected.

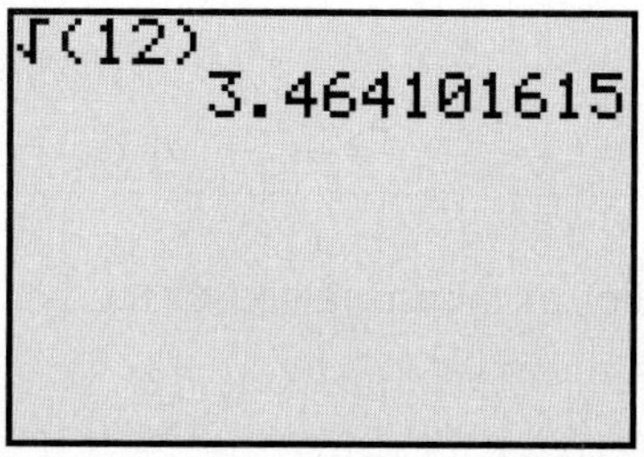

Figure 1 TI-84 Plus Display Approximating $\sqrt{12}$

Example 2 Approximating Square Roots

Use a calculator to approximate each **square root**. Round to three decimal places. Check that the answer is **reasonable**.

a. $\sqrt{5}$ **b.** $\sqrt{45}$ **c.** $\sqrt{103}$

Solutions

a. The **radicand** 5 is between the **perfect squares** $2^2 = 4$ and $3^2 = 9$, so $\sqrt{4} < \sqrt{5} < \sqrt{9}$ or $2 < \sqrt{5} < 3$. The square root is between 2 and 3. From Figure 2, we get $\sqrt{5} \approx 2.236$, which is between 2 and 3.

b. 45 is between $6^2 = 36$ and $7^2 = 49$, so $\sqrt{36} < \sqrt{45} < \sqrt{49}$ or $6 < \sqrt{45} < 7$. Figure 2 shows us $\sqrt{45} \approx 6.708$, which is between 6 and 7.

c. 103 is between $10^2 = 100$ and $11^2 = 121$, so $\sqrt{100} < \sqrt{103} < \sqrt{121}$ or $10 < \sqrt{103} < 11$. In Figure 2, we see $\sqrt{103} \approx 10.149$, which is reasonable.

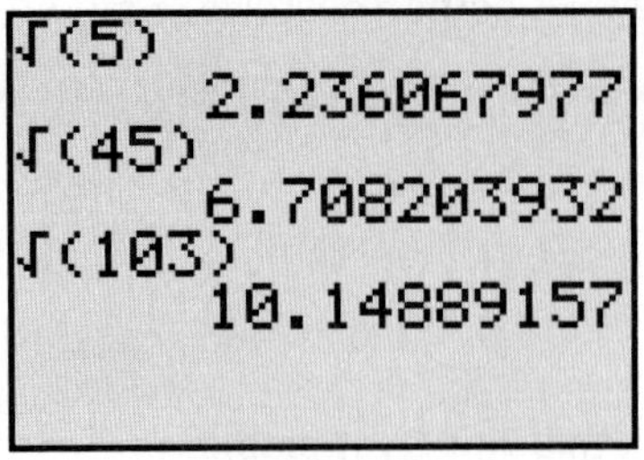

Figure 2 TI-84 Plus Display for Estimating $\sqrt{5}$, $\sqrt{45}$, and $\sqrt{103}$

You Try It **Work through this You Try It problem.**

Work Exercises 5–6 in this textbook or in the MyMathLab Study Plan.

OBJECTIVE 3 SIMPLIFY RADICAL EXPRESSIONS OF THE FORM $\sqrt{a^2}$

Let's now consider **square roots** with **variables** in the **radicand**. A common misconception is to think that $\sqrt{a^2} = a$, but this is not necessarily true. To see why, substitute $a = -5$ in the expression $\sqrt{a^2}$ and simplify:

$$\begin{aligned} &\text{Substitute } -5 \text{ for } a\text{:} && \sqrt{a^2} = \sqrt{(-5)^2} \\ &\text{Simplify } (-5)^2\text{:} && \quad\;\; = \sqrt{25} \\ &\text{Find the principal square root of 25:} && \quad\;\; = 5 \end{aligned}$$

The final result is not -5 but $|-5| = 5$. This illustrates the following square root property for simplifying **radical expressions** of the form $\sqrt{a^2}$.

> **Simplifying Radical Expressions of the Form $\sqrt{a^2}$**
>
> For any **real number** a,
>
> $$\sqrt{a^2} = |a|.$$

When taking the square root of a **base** raised to the second **power**, the result will be the **absolute value** of the base.

Example 3 Simplifying Radical Expressions of the Form $\sqrt{a^2}$

My video summary

Simplify.

a. $\sqrt{(-12)^2}$ **b.** $\sqrt{(2x-5)^2}$ **c.** $\sqrt{100x^2}$

d. $\sqrt{x^2 + 12x + 36}$ **e.** $\sqrt{9x^4}$ **f.** $\sqrt{y^6}$

Solutions

a. $\sqrt{(-12)^2} = |-12| = 12$

b. $\sqrt{(2x-5)^2} = |2x-5|$. Since $2x - 5$ could be negative, the **absolute value** symbol is required to ensure a non-negative result.

c. Since $100x^2 = (10x)^2$, then $\sqrt{100x^2} = \sqrt{(10x)^2} = |10x|$ or $10|x|$.

d.–f. Try to simplify each square root on your own by first writing the **radicand** as a **base** raised to the second **power**. Include absolute value notation if necessary. View the **answers**, or watch this **video** to check your work.

You Try It **Work through this You Try It problem.**

Work Exercises 7–14 in this textbook or in the MyMathLab Study Plan.

OBJECTIVE 4 FIND CUBE ROOTS

We can apply the process of finding **square roots** to other types of roots. For example, the **cube root** of a **real number** a is a real number b that, when **cubed**, results in a. So, b is the cube root of a if $b^3 = a$. For example, 2 is the cube root of 8 because $2^3 = 8$. To write the cube root of a, we use the notation $\sqrt[3]{a}$. The 3 in this **radical expression** indicates a cube root instead of a square root. Using this notation, we write $\sqrt[3]{8} = 2$.

Definition Cube Roots

A real number b is the **cube root** of a real number a, denoted as $b = \sqrt[3]{a}$, if $b^3 = a$.

Recall that the cube of a negative number is a negative number. Unlike square roots, cube roots can have negative numbers in the **radicand**. For example, $\sqrt[3]{-64} = -4$ because $(-4)^3 = -64$. Every real number has one real cube root. If the radicand is positive, the cube root will be positive. If the radicand is negative, the cube root will be negative. Therefore, **absolute value** is not used when simplifying radical expressions of the form $\sqrt[3]{a^3}$.

Simplifying Radical Expressions of the Form $\sqrt[3]{a^3}$

For any real number a,

$$\sqrt[3]{a^3} = a.$$

Example 4 Finding Cube Roots

My video summary

Simplify.

a. $\sqrt[3]{125}$ **b.** $\sqrt[3]{-1000}$ **c.** $\sqrt[3]{x^{15}}$

d. $\sqrt[3]{0.064}$ **e.** $\sqrt[3]{\frac{8}{27}}$ **f.** $\sqrt[3]{-64y^9}$

Solutions

a. $\sqrt[3]{125} = \sqrt[3]{5^3} = 5$

b. $\sqrt[3]{-1000} = \sqrt[3]{(-10)^3} = -10$

c. $\sqrt[3]{x^{15}} = \sqrt[3]{(x^5)^3} = x^5$

d.–f. Try to simplify each **cube root** on your own by first writing each **radicand** as a **base** raised to the third **power**. View this **popup**, or watch this **video** to check your work.

CAUTION: When simplifying expressions of the form $\sqrt[3]{a^3}$, do not use absolute value symbols. Doing so may lead to an incorrect result. For example, $\sqrt[3]{(-5)^3} = -5$, not $|-5| = 5$.

(eText Screens 18.1-1–18.1-16)

You Try It **Work through this You Try It problem.**

Work Exercises 15–20 in this textbook or in the MyMathLab Study Plan.

OBJECTIVE 5 FIND AND APPROXIMATE nth ROOTS

We know that 2 is a **square root** of 4 because $2^2 = 4$ and 2 is a **cube root** of 8 because $2^3 = 8$. In fact, 2 is a **4th root** of 16 because $2^4 = 16$, and 2 is a **5th root** of 32 because $2^5 = 32$. We denote each of these roots using **radical signs** as follows: $\sqrt{4} = 2$, $\sqrt[3]{8} = 2$, $\sqrt[4]{16} = 2$, and $\sqrt[5]{32} = 2$. These are called *nth roots*.

> **Definition Principal nth Roots**
>
> If a and b are **real numbers** and n is an **integer** such that $n \geq 2$, then b is the **principal nth root** of a, denoted as $b = \sqrt[n]{a}$, if $b^n = a$.
>
> **Note:** b must be non-negative when n is even.

In the notation $\sqrt[n]{a}$, n is called the **index** of the **radical expression**, and it indicates the type of root. For example, a cube root has an index of $n = 3$. If no **index** is shown, then it is understood to be $n = 2$ for a square root.

If n is an odd integer, then the **radicand** a can be any real number. However, if n is an even integer, then the radicand a must be **non-negative**. This leads to the following rule for simplifying radical expressions of the form $\sqrt[n]{a^n}$.

> **Simplifying Radical Expressions of the Form $\sqrt[n]{a^n}$**
>
> If n is an integer such that $n \geq 2$ and a is a real number, then
>
> $$\sqrt[n]{a^n} = a \text{ if } n \text{ is odd.}$$
> $$\sqrt[n]{a^n} = |a| \text{ if } n \text{ is even.}$$

Example 5 Finding nth Roots

My video summary

Simplify.

a. $\sqrt[4]{81}$ **b.** $\sqrt[5]{-32}$ **c.** $\sqrt[6]{\dfrac{1}{64}}$

d. $\sqrt[5]{x^{15}}$ **e.** $\sqrt[6]{(x-7)^6}$ **f.** $\sqrt[4]{-1}$

Solutions

a. $\sqrt[4]{81} = \sqrt[4]{3^4} = |3| = 3$

b. $\sqrt[5]{-32} = \sqrt[5]{(-2)^5} = -2$

c. $\sqrt[6]{\dfrac{1}{64}} = \sqrt[6]{\left(\dfrac{1}{2}\right)^6} = \left|\dfrac{1}{2}\right| = \dfrac{1}{2}$

d.–f. Try to simplify each **nth root** on your own by first writing each **radicand** as a **base** raised to the nth **power**. View this **popup**, or watch this **video** to check your work.

CAUTION: When simplifying expressions of the form $\sqrt[n]{a^n}$, use absolute value symbols when n is even but not when n is odd.

You Try It **Work through this You Try It problem.**

Work Exercises 21–26 in this textbook or in the MyMathLab Study Plan.

In **Example 5** we saw that $\sqrt[4]{81} = 3$ because $81 = 3^4$. Since 81 is a **perfect 4th power**, its **4th root** is a **rational number**. But what if the **radicand** of an **nth root** is not a **perfect nth power**? Such expressions represent **irrational numbers**. As with square roots, we can use a calculator to approximate these types of nth roots.

Example 6 Approximating nth Roots

Use a calculator to approximate each root. Round to three decimal places. Check that the answer is **reasonable**.

a. $\sqrt[3]{6}$ **b.** $\sqrt[4]{200}$ **c.** $\sqrt[5]{154}$

Solutions

a. The radicand 6 is between the **perfect cubes** $1^3 = 1$ and $2^3 = 8$, so $\sqrt[3]{1} < \sqrt[3]{6} < \sqrt[3]{8}$ or $1 < \sqrt[3]{6} < 2$. From Figure 3, we get $\sqrt[3]{6} \approx 1.817$, which is between 1 and 2.

b. 200 is between $3^4 = 81$ and $4^4 = 256$, so $3 < \sqrt[4]{200} < 4$. In Figure 3, we see that $\sqrt[4]{200} \approx 3.761$, which is between 3 and 4.

c. 154 is between $2^5 = 32$ and $3^5 = 243$, so $2 < \sqrt[5]{154} < 3$. Figure 3 shows us that $\sqrt[5]{154} \approx 2.738$, which is reasonable.

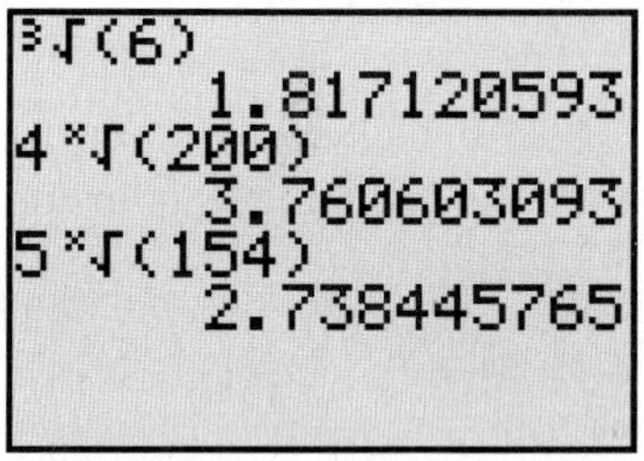

Figure 3 TI-84 Plus Display for Estimating $\sqrt[3]{6}$, $\sqrt[4]{200}$, and $\sqrt[5]{154}$

You Try It **Work through this You Try It problem.**

Work Exercises 27–30 in this textbook or in the MyMathLab Study Plan.

18.1 Exercises

In Exercises 1–4, evaluate each square root. If the answer is not a real number, state so.

You Try It

1. $\sqrt{100}$ **2.** $-\sqrt{49}$ **3.** $\sqrt{\dfrac{9}{121}}$ **4.** $\sqrt{0.09}$

In Exercises 5–6, use a calculator to approximate each square root. Round to three decimal places. Check that the answer is reasonable.

You Try It

5. $\sqrt{55}$ **6.** $\sqrt{637}$

(eText Screens 18.2-1–18.2-14)

You Try It

In Exercises 7–14, simplify. Include the absolute value symbol if necessary.

7. $\sqrt{(-17)^2}$ 8. $\sqrt{(8x)^2}$ 9. $\sqrt{(6x-5)^2}$

10. $\sqrt{16x^2}$ 11. $\sqrt{81x^4}$ 12. $\sqrt{w^{14}}$

13. $-\sqrt{121x^{10}}$ 14. $\sqrt{x^2+16x+64}$

In Exercises 15–26, simplify. Include the absolute value symbol if necessary. If the answer is not a real number, state so.

You Try It

15. $\sqrt[3]{343}$ 16. $\sqrt[3]{-125}$ 17. $\sqrt[3]{-\dfrac{27}{64}}$

18. $\sqrt[3]{0.216}$ 19. $\sqrt[3]{w^{21}}$ 20. $\sqrt[3]{-8x^{15}}$

You Try It

21. $\sqrt[5]{1024}$ 22. $-\sqrt[4]{256}$ 23. $\sqrt[4]{\dfrac{16}{625}}$

24. $\sqrt[5]{-243}$ 25. $\sqrt[4]{x^{20}}$ 26. $\sqrt[6]{(x-4)^6}$

In Exercises 27–30, use a calculator to approximate each root. Round to three decimal places. Check that the answer is reasonable.

27. $\sqrt[3]{42}$ 28. $\sqrt[4]{232}$ 29. $\sqrt[5]{50}$ 30. $\sqrt[6]{348}$

You Try It

18.2 Radical Functions

THINGS TO KNOW

Before working through this topic, be sure you are familiar with the following concepts:

VIDEO ANIMATION INTERACTIVE

You Try It 1. Find the Domain and Range of a Relation (Topic 17.1, **Objective 2**)

You Try It 2. Evaluate Functions (Topic 17.2, **Objective 2**)

You Try It 3. Graph Simple Functions by Plotting Points (Topic 17.3, **Objective 1**)

You Try It 4. Find Square Roots of Perfect Squares (Topic 18.1, **Objective 1**)

You Try It 5. Approximate Square Roots (Topic 18.1, **Objective 2**)

You Try It

6. Simplify Radical Expressions of the Form $\sqrt{a^2}$
(Topic 18.1, **Objective 3**)

You Try It

7. Find Cube Roots
(Topic 18.1, **Objective 4**)

OBJECTIVES

1 Evaluate Radical Functions

2 Find the Domain of a Radical Function

3 Graph Functions That Contain Square Roots or Cube Roots

OBJECTIVE 1 EVALUATE RADICAL FUNCTIONS

A **radical function** is a **function** that contains one or more **radical expressions.** We evaluate radical functions in the same way that we **evaluated functions** in **Topic 17.2.** Substitute the given value for the **variable** and simplify using the **order of operations.**

Example 1 Evaluating Radical Functions

My interactive video summary

For the **radical functions** $f(x) = \sqrt{2x - 5}$, $g(x) = \sqrt[3]{5x + 9}$, and $h(x) = -3\sqrt[4]{x} + 2$, evaluate the following.

a. $f(15)$ **b.** $g(-2)$ **c.** $h(625)$

d. $g\left(-\frac{1}{5}\right)$ **e.** $f(0.5)$ **f.** $h(1)$

Solutions

a. Recall that **radical signs** serve as **grouping symbols** as well as indicating the type of root. So, once we substitute the value for the variable, we simplify the **radicand** before finding the root.

Substitute 15 for x in the function f: $f(15) = \sqrt{2(15) - 5}$

Multiply: $= \sqrt{30 - 5}$

Subtract: $= \sqrt{25}$

Evaluate the square root: $= 5$ because $5^2 = 25$

b. Substitute -2 for x in the function g: $g(-2) = \sqrt[3]{5(-2) + 9}$

Multiply: $= \sqrt[3]{-10 + 9}$

Add: $= \sqrt[3]{-1}$

Evaluate the cube root: $= -1$ because $(-1)^3 = -1$

c. Substitute 625 for x in the function h: $h(625) = -3\sqrt[4]{625} + 2$

Evaluate the 4th root: $= -3(5) + 2$ because $5^4 = 625$

Multiply: $= -15 + 2$

Add: $= -13$

d.–f. Try to evaluate each **radical function** on your own. View the **answers** or watch this **interactive video** to check your work.

(eText Screens 18.2-1–18.2-14)

You Try It **Work through this You Try It problem.**

Work Exercises 1–6 in this textbook or in the MyMathLab Study Plan.

OBJECTIVE 2 FIND THE DOMAIN OF A RADICAL FUNCTION

We know that the **radicand** of a **radical expression** must be **non-negative** if the **index** of the radical expression is even. For example, $\sqrt[4]{-16}$ is not a **real number**. However, if the index of the radical expression is odd, then the radicand can be any real number. We use this information to find the **domain** of **radical functions.**

To find the domain of the function $f(x) = \sqrt{2x - 5}$ from **Example 1**, we recognize that the radicand of a **square root** must be non-negative. Setting the radicand greater than or equal to zero and solving will give the domain of this function:

Set the radicand greater than or equal to 0: $2x - 5 \geq 0$

Add 5: $2x \geq 5$

Divide by 2: $x \geq \dfrac{5}{2}$

The domain of $f(x) = \sqrt{2x - 5}$ is $\left\{x \middle| x \geq \dfrac{5}{2}\right\}$ or, in **interval notation,** $\left[\dfrac{5}{2}, \infty\right)$.

To find the domain of $g(x) = \sqrt[3]{5x + 9}$, we recognize that the radicand of a **cube root** can be any real number. Since there are no restrictions for the radicand, the domain of $g(x) = \sqrt[3]{5x + 9}$ is the set of all real numbers.

Guideline to Finding the Domain of a Radical Function

If the **index** of a **radical function** is even, the **radicand** must be greater than or equal to zero.

If the index of a radical function is odd, the radicand can be any real number.

Example 2 Finding the Domain of a Radical Function

My interactive video summary

Find the **domain** for each radical function.

a. $F(x) = \sqrt[4]{12 - 4x}$ **b.** $h(x) = \sqrt[5]{3x + 5}$ **c.** $G(x) = \sqrt[6]{5x + 7}$

Solutions

a. The **index** is 4, which is even, so the radicand must be non-negative. We set the radicand greater than or equal to zero and solve:

Set the radicand greater than or equal to 0: $12 - 4x \geq 0$

Subtract 12 from both sides of the inequality: $-4x \geq -12$

Divide both sides by -4, reversing the inequality symbol: $x \leq 3$

The domain is $\{x | x \leq 3\}$ or, in **interval notation,** $(-\infty, 3]$.

b.–c. Try to find the domain for each radical function on your own. Check your **answers,** or watch this **interactive video** for complete solutions.

(eText Screens 18.2-1–18.2-14)

You Try It Work through this You Try It problem.

Work Exercises 7–12 in this textbook or in the MyMathLab Study Plan.

OBJECTIVE 3 GRAPH FUNCTIONS THAT CONTAIN SQUARE ROOTS OR CUBE ROOTS

In **Topic 17.3** we graphed simple functions by plotting points. We use the same **strategy** to graph the **square root function** $f(x) = \sqrt{x}$.

First we find several **ordered pairs** that **belong** to the function. The **domain** of the **square root function** is $\{x|x \geq 0\}$, or in interval notation $[0, \infty)$, so we **evaluate** the function for non-negative values (Table 1). Then we plot the points and connect them with a smooth curve (Figure 4).

x	$y = f(x) = \sqrt{x}$	(x, y)
0	$f(0) = \sqrt{0} = 0$	$(0, 0)$
1	$f(1) = \sqrt{1} = 1$	$(1, 1)$
2	$f(2) = \sqrt{2} \approx 1.414$	$(2, \sqrt{2})$
4	$f(4) = \sqrt{4} = 2$	$(4, 2)$
6	$f(6) = \sqrt{6} \approx 2.449$	$(6, \sqrt{6})$
9	$f(9) = \sqrt{9} = 3$	$(9, 3)$

Table 1

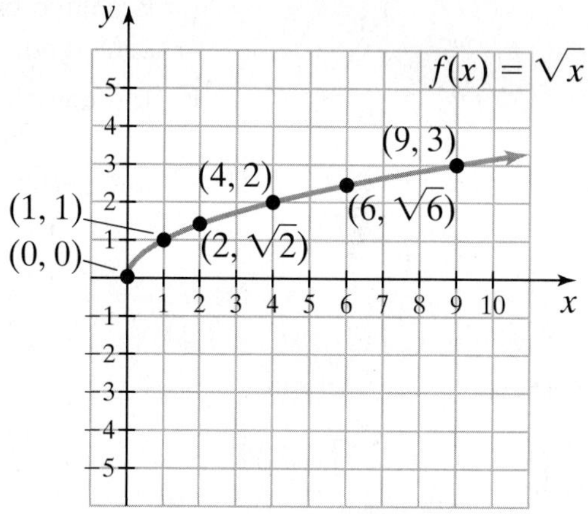

Figure 4

Other functions involving **square roots** will have a graph similar to the square root function, provided that the **radicand** is a **linear expression**.

Example 3 Graphing Functions That Contain Square Roots

My interactive video summary

Graph each function. Compare each graph to that of the **square root function**.

a. $F(x) = \sqrt{x + 1}$ **b.** $g(x) = \sqrt{x} + 1$ **c.** $h(x) = -\sqrt{x}$

Solutions

a. The **domain** of $F(x) = \sqrt{x + 1}$ is $\{x|x \geq -1\}$, or using **interval notation** $[-1, \infty)$. (Do you see **why**?) We **evaluate** the function for values of x greater than or equal to -1.

To make the computations work out nicely, we choose values for x that will make the **radicand** a **perfect square**. Let $x = -1, 0, 3,$ and 8 (**Table 2**). Plotting the points and connecting them with a smooth curve gives the graph shown in **Figure 5**.

x	$y = F(x) = \sqrt{x+1}$	(x, y)
-1	$F(-1) = \sqrt{-1+1} = \sqrt{0} = 0$	$(-1, 0)$
0	$F(0) = \sqrt{0+1} = \sqrt{1} = 1$	$(0, 1)$
3	$F(3) = \sqrt{3+1} = \sqrt{4} = 2$	$(3, 2)$
8	$F(8) = \sqrt{8+1} = \sqrt{9} = 3$	$(8, 3)$

Table 2

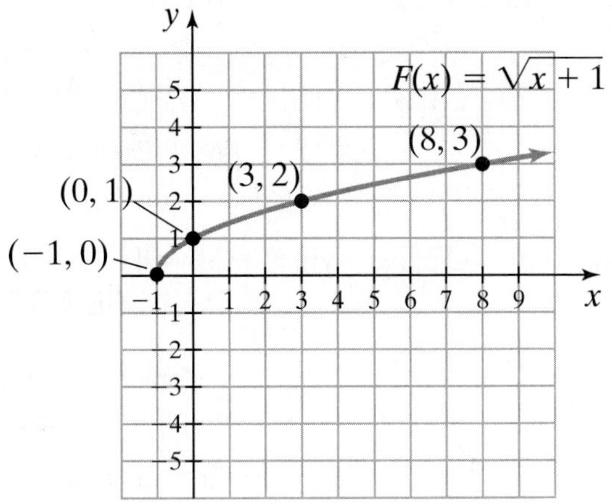

Figure 5

The graph of $F(x) = \sqrt{x+1}$ looks like the graph of the **square root function**, but it is shifted one unit to the left. View this **popup** to see the two graphs together on the same grid.

b. The **domain** of $g(x) = \sqrt{x} + 1$ is $\{x|x \geq 0\}$, or using **interval notation** $[0, \infty)$. We **evaluate** the function for non-negative values of x. For easier computations, we substitute **perfect squares** for x (Table 3).

x	$y = g(x) = \sqrt{x} + 1$	(x, y)
0	$g(0) = \sqrt{0} + 1 = 0 + 1 = 1$	$(0, 1)$
1	$g(1) = \sqrt{1} + 1 = 1 + 1 = 2$	$(1, 2)$
4	$g(4) = \sqrt{4} + 1 = 2 + 1 = 3$	$(4, 3)$
9	$g(9) = \sqrt{9} + 1 = 3 + 1 = 4$	$(9, 4)$

Table 3

Plot the points and connect them with a smooth curve. Next, compare this graph to the graph of the **square root function**. Check your **answer**, or watch this **interactive video** for a complete solution.

c. Try to work this problem on your own. Check your **answer**, or watch this **interactive video** for a complete solution.

You Try It **Work through this You Try It problem.**

Work Exercises 13–15 in this textbook or in the MyMathLab Study Plan.

Now let's graph the **cube root function** $f(x) = \sqrt[3]{x}$. The **domain** of the **cube root function** is the set of all real numbers $\mathbb{R}$, or using **interval notation** $(-\infty, \infty)$. We **evaluate** the function for both positive and negative values. For easier computations, we substitute **perfect cubes** for x (**Table 4**). Then we plot the points and connect them with a smooth curve (**Figure 6**).

x	$y = f(x) = \sqrt[3]{x}$	(x, y)
-8	$f(-8) = \sqrt[3]{-8} = -2$	$(-8, -2)$
-1	$f(-1) = \sqrt[3]{-1} = -1$	$(-1, -1)$
0	$f(0) = \sqrt[3]{0} = 0$	$(0, 0)$
1	$f(1) = \sqrt[3]{1} = 1$	$(1, 1)$
8	$f(8) = \sqrt[3]{8} = 2$	$(8, 2)$

Table 4

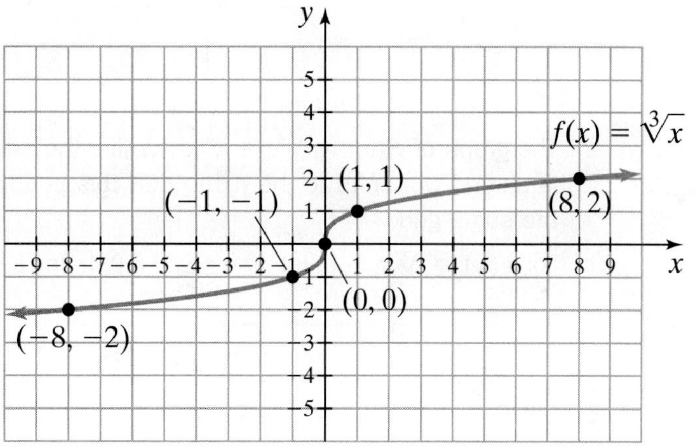

Figure 6

As with functions involving **square roots**, functions involving **cube roots** will have a graph similar to the cube root function if the **radicand** is a **linear expression**.

Example 4 Graphing Functions That Contain Cube Roots

My interactive video summary

Graph each function. Compare each graph to that of the **cube root function**.

a. $F(x) = \sqrt[3]{x - 2}$ **b.** $g(x) = \sqrt[3]{x} - 2$ **c.** $h(x) = -\sqrt[3]{x}$

Solutions The **domain** of each function is the set of all real numbers $\mathbb{R}$ or $(-\infty, \infty)$, so we **evaluate** each function for positive and negative values. For easier computations, we choose values for x that make the **radicand** a **perfect cube**.

a. Let $x = -6, 1, 2, 3$, and 10 (Table 5). Plotting the points and connecting them with a smooth curve gives the graph shown in **Figure 7**.

x	$y = F(x) = \sqrt[3]{x - 2}$	(x, y)
-6	$F(-6) = \sqrt[3]{-6 - 2} = \sqrt[3]{-8} = -2$	$(-6, -2)$
1	$F(1) = \sqrt[3]{1 - 2} = \sqrt[3]{-1} = -1$	$(1, -1)$
2	$F(2) = \sqrt[3]{2 - 2} = \sqrt[3]{0} = 0$	$(2, 0)$
3	$F(3) = \sqrt[3]{3 - 2} = \sqrt[3]{1} = 1$	$(3, 1)$
10	$F(10) = \sqrt[3]{10 - 2} = \sqrt[3]{8} = 2$	$(10, 2)$

Table 5

(eText Screens 18.2-1–18.2-14)

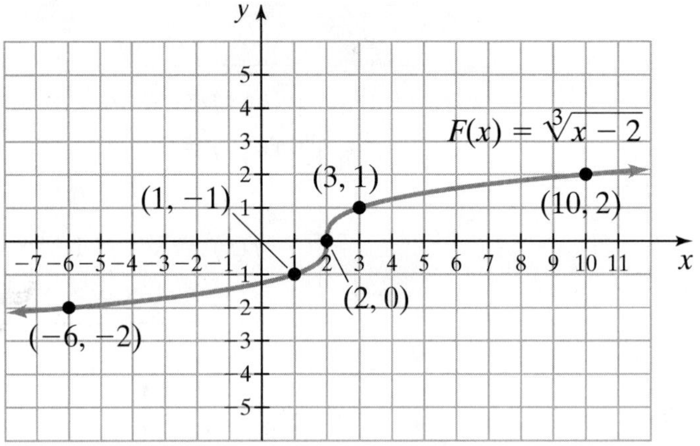

Figure 7

The graph of $F(x) = \sqrt[3]{x-2}$ looks like the graph of the **cube root function**, but it is shifted two units to the right. View this **popup** to see the two graphs together on the same grid.

b. For $g(x) = \sqrt[3]{x} - 2$, let $x = -8, -1, 0, 1,$ and 8 (see Table 6).

x	$y = g(x) = \sqrt[3]{x} - 2$	(x, y)
-8	$F(-8) = \sqrt[3]{-8} - 2 = -2 - 2 = -4$	$(-8, -4)$
-1	$F(-1) = \sqrt[3]{-1} - 2 = -1 - 2 = -3$	$(-1, -3)$
0	$F(0) = \sqrt[3]{0} - 2 = 0 - 2 = -2$	$(0, -2)$
1	$F(1) = \sqrt[3]{1} - 2 = 1 - 2 = -1$	$(1, -1)$
8	$F(8) = \sqrt[3]{8} - 2 = 2 - 2 = 0$	$(8, 0)$

Table 6

Plot the points, connect them with a smooth curve, and compare the graph to the graph of the **cube root function**. Check your **answer**, or watch this **interactive video** for a complete solution.

c. Try to work this problem on your own. Check your **answer**, or watch this **interactive video** for a complete solution.

You Try It **Work through this You Try It problem.**

Work Exercises 16–18 in this textbook or in the MyMathLab Study Plan.

18.2 Exercises

You Try It

In Exercises 1–6, evaluate each radical function. If the function value is not a real number, state so.

1. If $f(x) = \sqrt{x-7}$, find $f(88)$.

2. If $F(x) = \sqrt{9x+10}$, find $F\left(\frac{2}{3}\right)$.

3. If $g(x) = \sqrt[3]{x+5}$, find $g(3)$.

4. If $G(x) = \sqrt[3]{4x-5}$, find $G(-5.5)$.

5. If $h(x) = \sqrt[4]{2x} - 3$, find $h(8)$.

6. If $H(x) = 2\sqrt[4]{x}$, find $H(1)$.

(eText Screens 18.3-1–18.3-32)

You Try It

In Exercises 7–12, find the domain of each radical function.

7. $f(x) = \sqrt{x + 4}$ **8.** $F(x) = \sqrt{6 - 2x}$ **9.** $g(x) = \sqrt[3]{2x + 1}$

10. $G(x) = \sqrt[4]{2x + 5}$ **11.** $h(x) = \sqrt[4]{x} - 3$ **12.** $H(x) = \sqrt[5]{6x + 7}$

You Try It

In Exercises 13–18, graph each radical function. Compare the graph to that of the square root function or cube root function.

13. $f(x) = \sqrt{x - 2}$ **14.** $F(x) = \sqrt{x} - 3$ **15.** $g(x) = 2\sqrt{x}$

You Try It

16. $G(x) = \sqrt[3]{x + 1}$ **17.** $h(x) = \sqrt[3]{x} + 2$ **18.** $H(x) = -2\sqrt[3]{x}$

18.3 Rational Exponents and Simplifying Radical Expressions

THINGS TO KNOW

Before working through this topic, be sure you are familiar with the following concepts:

	VIDEO	ANIMATION	INTERACTIVE
You Try It 1. Simplify Exponential Expressions Using the Product Rule (Topic 14.1, **Objective 1**)	▣		
You Try It 2. Simplify Exponential Expressions Using the Quotient Rule (Topic 14.1, **Objective 2**)	▣		
You Try It 3. Use the Power-to-Power Rule (Topic 14.1, **Objective 4**)	▣		
You Try It 4. Use the Product-to-Power Rule (Topic 14.1, **Objective 5**)	▣		
You Try It 5. Use the Quotient-to-Power Rule (Topic 14.1, **Objective 6**)	▣		
You Try It 6. Simplify Exponential Expressions Using a Combination of Rules (Topic 14.1, **Objective 7**)			▣
You Try It 7. Use the Negative-Power Rule (Topic 14.6, **Objective 1**)	▣		
You Try It 8. Find Square Roots of Perfect Squares (Topic 18.1, **Objective 1**)	▣		
You Try It 9. Simplify Radical Expressions of the Form $\sqrt{a^2}$ (Topic 18.1, **Objective 3**)	▣		
You Try It 10. Find and Approximate nth Roots (Topic 18.1, **Objective 5**)	▣		

(eText Screens 18.3-1–18.3-32)

OBJECTIVES

1 Use the Definition for Rational Exponents of the Form $a^{\frac{1}{n}}$

2 Use the Definition for Rational Exponents of the Form $a^{\frac{m}{n}}$

3 Simplify Exponential Expressions Involving Rational Exponents

4 Use Rational Exponents to Simplify Radical Expressions

5 Simplify Radical Expressions Using the Product Rule

6 Simplify Radical Expressions Using the Quotient Rule

OBJECTIVE 1 USE THE DEFINITION FOR RATIONAL EXPONENTS OF THE FORM $a^{\frac{1}{n}}$

In **Topic 14.1** we defined an **exponential expression** as a **constant** or **algebraic expression** that is raised to a **power**. So far, we have only considered exponential expressions with **integer** powers. But what if the power is a **rational number** such as $\frac{1}{2}$? Suppose a and b are non-negative real numbers such that

$$b = a^{\frac{1}{2}}.$$

Squaring the expressions on both sides of the equal sign results in

$$(b)^2 = \left(a^{\frac{1}{2}}\right)^2.$$

Using the **power-to-power rule** for exponents, we multiply the powers in the expression on the right. This gives

$$b^2 = \left(a^{\frac{1}{2}}\right)^2 = a^{\frac{1}{2}\cdot 2} = a^1 = a.$$

We find that $b = a^{\frac{1}{2}}$ is a non-negative **real number** such that $b^2 = a$. However, this statement is similar to the definition for the **principal square root** of a: $b = \sqrt{a}$ if $b^2 = a$. So, we can conclude that

$$a^{\frac{1}{2}} = \sqrt{a}.$$

This idea can be used to define a **rational exponent of the form** $a^{\frac{1}{n}}$.

> **Definition Rational Exponent of the Form $a^{\frac{1}{n}}$**
>
> If n is an **integer** such that $n \geq 2$ and if $\sqrt[n]{a}$ is a **real number**, then $a^{\frac{1}{n}} = \sqrt[n]{a}$.

The **denominator** n of the **rational exponent** is the **index** of the **root**. The **base** a of the **exponential expression** is the **radicand** of the root. If n is odd, then a can be any real number. If n is even, then a must be **non-negative**.

Example 1 Converting Exponential Expressions to Radical Expressions

Write each exponential expression as a **radical expression**. **Simplify** if possible.

a. $25^{\frac{1}{2}}$ **b.** $(-64x^3)^{\frac{1}{3}}$ **c.** $-100^{\frac{1}{2}}$ **d.** $(-81)^{\frac{1}{4}}$ **e.** $(7x^3y)^{\frac{1}{5}}$

My video summary

Solutions

a. The index of the root is 2, the denominator of the rational exponent. So, the radical expression is a **square root**. The radicand is 25.

$$25^{\frac{1}{2}} = \sqrt{25} = 5$$

b. From the denominator of the rational exponent, the index is 3. So, we have a **cube root**. The radicand is $-64x^3$.

$$(-64x^3)^{\frac{1}{3}} = \sqrt[3]{-64x^3} = \sqrt[3]{(-4x)^3} = -4x$$

c. As in part (a), the **radical expression** is a **square root**. The negative sign is not part of the **radicand** because it is not part of the **base**. We can tell this because there are no grouping symbols around -100.

$$-100^{\frac{1}{2}} = -\sqrt{100} = -10$$

d.–e. Try to convert each expression on your own. View this popup to check your answers, or watch this **video** for the complete solutions.

You Try It **Work through this You Try It problem.**

Work Exercises 1–6 in this textbook or in the MyMathLab Study Plan.

Example 2 Converting Radical Expressions to Exponential Expressions

My video summary

Write each radical expression as an **exponential expression**.

a. $\sqrt{5y}$ b. $\sqrt[3]{7x^2y}$ c. $\sqrt[4]{\dfrac{2m}{3n}}$

Solutions

a. The radical expression is a square root, so the **index** is 2 and the **denominator** of the **rational exponent** is 2. The base of the exponential expression is the radicand $5y$.

$$\sqrt{5y} = (5y)^{\frac{1}{2}}$$

b. The **index** is 3 and the **radicand** is $7x^2y$, so the **denominator** of the **rational exponent** is 3 and the **base** of the **exponential expression** is $7x^2y$. Finish writing the exponential expression. Check your answer, or watch this **video** for the complete solution.

c. Try to convert this expression on your own. Check your **answer**, or watch this **video** for the complete solution.

You Try It **Work through this You Try It problem.**

Work Exercises 13–15 in this textbook or in the MyMathLab Study Plan.

OBJECTIVE 2 USE THE DEFINITION FOR RATIONAL EXPONENTS OF THE FORM $a^{\frac{m}{n}}$

What if the **numerator** of the **rational exponent** is not 1? For example, consider the **exponential expression** $a^{\frac{2}{3}}$. From the **power-to-power rule** for exponents, we write

$$a^{\frac{2}{3}} = a^{\frac{1}{3}\cdot 2} = \left(a^{\frac{1}{3}}\right)^2 = \left(\sqrt[3]{a}\right)^2 \quad \text{or} \quad a^{\frac{2}{3}} = a^{2\cdot\frac{1}{3}} = (a^2)^{\frac{1}{3}} = \sqrt[3]{a^2}.$$

(eText Screens 18.3-1–18.3-32)

This idea can be used to define a rational exponent of the form $a^{\frac{m}{n}}$.

Definition Rational Exponent of the Form $a^{\frac{m}{n}}$

If $\frac{m}{n}$ is a **rational number** in **lowest terms**, m and n are integers such that $n \geq 2$, and $\sqrt[n]{a}$ is a **real number**, then

$$a^{\frac{m}{n}} = \left(\sqrt[n]{a}\right)^m = \sqrt[n]{a^m}.$$

To **simplify** $a^{\frac{m}{n}}$ using the form $\left(\sqrt[n]{a}\right)^m$, we find the **root** first and the **power** second. Using the form $\sqrt[n]{a^m}$, we find the power first and the root second. Both forms may be used, but the form $\left(\sqrt[n]{a}\right)^m$ is usually easier because it involves smaller numbers.

Note: For the rest of this module, we assume that all variable **factors** of the **radicand** of a **radical expression** with an even **index** will be **non-negative** real numbers. Likewise, all variable factors of the **base** of an **exponential expression** containing a rational exponent with an even **denominator** will be non-negative **real numbers**. This assumption allows us to avoid using **absolute value** symbols when simplifying radical expressions.

Example 3 Converting Exponential Expressions to Radical Expressions

My video summary

Write each exponential expression as a radical expression. **Simplify** if possible.

a. $16^{\frac{3}{2}}$ **b.** $\left(\frac{y^3}{1000}\right)^{\frac{2}{3}}$ **c.** $-81^{\frac{3}{4}}$ **d.** $(-36)^{\frac{5}{2}}$ **e.** $(x^2y)^{\frac{2}{5}}$

Solutions

a. We use the form $a^{\frac{m}{n}} = \left(\sqrt[n]{a}\right)^m$. Since the denominator of the **rational exponent** is 2, the radical expression is a **square root**.

Begin with the original exponential expression:	$16^{\frac{3}{2}}$
Use $a^{\frac{m}{n}} = \left(\sqrt[n]{a}\right)^m$ to rewrite as a radical expression:	$= (\sqrt{16})^3$
Simplify $\sqrt{16}$:	$= (4)^3$
Simplify:	$= 64$

b. $\left(\frac{y^3}{1000}\right)^{\frac{2}{3}} = \left(\sqrt[3]{\frac{y^3}{1000}}\right)^2 = \left(\sqrt[3]{\left(\frac{y}{10}\right)^3}\right)^2 = \left(\frac{y}{10}\right)^2 = \frac{y^2}{100}$

c. Note that the negative sign is not part of the **base**, so it goes in front of the **radical expression**.

$$-81^{\frac{3}{4}} = -\left(\sqrt[4]{81}\right)^3 = -(3)^3 = -27$$

d.–e. Try to convert each expression on your own. Check your **answers**, or watch this **video** for the complete solutions.

You Try It Work through this You Try It problem.

Work Exercises 7–12 in this textbook or in the MyMathLab Study Plan.

Example 4 Converting Radical Expressions to Exponential Expressions

My video summary

Write each radical expression as an **exponential expression**.

a. $\sqrt[8]{x^5}$ b. $\left(\sqrt[5]{2ab^2}\right)^3$ c. $\sqrt[4]{(10x)^3}$

Solutions

a. The **index** is 8, so the **denominator** of the **rational exponent** is 8. The **radicand** is a **power** of 5, so the **numerator** is 5.

$$\sqrt[8]{x^5} = x^{\frac{5}{8}}$$

b. The index is 5, so the denominator of the rational exponent is 5. The expression is raised to a power of 3, so the numerator is 3. Finish writing the exponential expression. Check your **answer**, or watch this **video** for the complete solution.

c. Try to convert this expression on your own. Check your **answer**, or watch this **video** for the complete solution.

You Try It **Work through this You Try It problem.**

Work Exercises 16–18 in this textbook or in the MyMathLab Study Plan.

If a **rational exponent** is negative, we can first use the **negative-power rule** from **Topic 14.6** to rewrite the expression with a positive exponent.

Example 5 Using the Negative-Power Rule with Negative Rational Exponents

My video summary

Write each **exponential expression** with positive exponents. **Simplify** if possible.

a. $1000^{-\frac{1}{3}}$ b. $\dfrac{1}{81^{-\frac{1}{4}}}$ c. $125^{-\frac{2}{3}}$ d. $\dfrac{1}{8^{-\frac{4}{3}}}$ e. $(-25)^{-\frac{3}{2}}$

Solutions

a. Begin with the original exponential expression: $1000^{-\frac{1}{3}}$

Use the negative-power rule $a^{-n} = \dfrac{1}{a^n}$: $= \dfrac{1}{1000^{\frac{1}{3}}}$

Rewrite the rational exponent as a radical expression: $= \dfrac{1}{\sqrt[3]{1000}}$

Evaluate the root: $= \dfrac{1}{10}$

b. Begin with the original exponential expression: $\dfrac{1}{81^{-\frac{1}{4}}}$

Use the negative-power rule $\dfrac{1}{a^{-n}} = a^n$: $= 81^{\frac{1}{4}}$

Rewrite the rational exponent as a radical expression: $= \sqrt[4]{81}$

Evaluate the root: $= 3$

c.–e. Try to rewrite and simplify each expression on your own. View this **popup** to check your answers, or watch this **video** for complete solutions.

You Try It **Work through this You Try It problem.**

Work Exercises 19–21 in this textbook or in the MyMathLab Study Plan.

OBJECTIVE 3 SIMPLIFY EXPONENTIAL EXPRESSIONS INVOLVING RATIONAL EXPONENTS

In **Example 5** we used the **negative-power rule** for exponents to rewrite and simplify **exponential expressions** containing negative **rational** exponents. In **Topics 14.1** and **14.6**, we used several other **rules** to **simplify exponential expressions** involving **integer** exponents. We can use these rules to simplify expressions involving rational exponents. For convenience, we repeat them here.

Rules for Exponents

Product Rule:	$a^m \cdot a^n = a^{m+n}$
Quotient Rule:	$\dfrac{a^m}{a^n} = a^{m-n} \quad (a \neq 0)$
Zero-Power Rule:	$a^0 = 1 \quad (a \neq 0)$
Negative-Power Rule:	$a^{-n} = \dfrac{1}{a^n}$ or $\dfrac{1}{a^{-n}} = a^n \quad (a \neq 0)$
Power-to-Power Rule:	$(a^m)^n = a^{m \cdot n}$
Product-to-Power Rule:	$(ab)^n = a^n b^n$
Quotient-to-Power Rule:	$\left(\dfrac{a}{b}\right)^n = \dfrac{a^n}{b^n} \quad (b \neq 0)$

Recall that an exponential expression is **simplified** when

- No parentheses or **grouping symbols** are present.
- No zero or **negative exponents** are present.
- No **powers** are raised to powers.
- Each **base** occurs only once.

Example 6 Simplifying Expressions Involving Rational Exponents

My interactive video summary

Use the **rules for exponents** to **simplify** each expression. Assume all **variables** represent **non-negative** values.

a. $x^{\frac{3}{8}} \cdot x^{\frac{1}{6}}$ **b.** $\dfrac{49^{\frac{7}{10}}}{49^{\frac{1}{5}}}$ **c.** $\left(64^{\frac{4}{9}}\right)^{\frac{3}{2}}$ **d.** $\left(32x^{\frac{5}{6}}y^{\frac{10}{9}}\right)^{\frac{3}{5}}$

e. $\left(\dfrac{125x^{\frac{5}{4}}}{y^{\frac{7}{8}}z^{\frac{9}{4}}}\right)^{\frac{4}{3}}, y \neq 0, z \neq 0$ **f.** $\left(4x^{\frac{1}{6}}y^{\frac{3}{4}}\right)^2\left(3x^{\frac{5}{9}}y^{-\frac{3}{2}}\right), y \neq 0$

(eText Screens 18.3-1–18.3-32)

Solutions

a. Begin with the original expression: $x^{\frac{3}{8}} \cdot x^{\frac{1}{6}}$

Use the **product rule** for exponents: $= x^{\frac{3}{8}+\frac{1}{6}}$

Add: $= x^{\frac{13}{24}}$ ← $\boxed{\frac{3}{8}+\frac{1}{6}=\frac{9}{24}+\frac{4}{24}=\frac{13}{24}}$

b. Begin with the original expression: $\dfrac{49^{\frac{7}{10}}}{49^{\frac{1}{5}}}$

Use the **quotient rule** for exponents: $= 49^{\frac{7}{10}-\frac{1}{5}}$

Subtract: $= 49^{\frac{1}{2}}$ ← $\boxed{\frac{7}{10}-\frac{1}{5}=\frac{7}{10}-\frac{2}{10}=\frac{5}{10}=\frac{1}{2}}$

Use $a^{\frac{1}{n}} = \sqrt[n]{a}$ to rewrite as a radical expression: $= \sqrt{49}$

Simplify: $= 7$

c. Begin with the original expression: $\left(64^{\frac{4}{9}}\right)^{\frac{3}{2}}$

Use the **power-to-power rule** for exponents: $= 64^{\frac{4}{9}\cdot\frac{3}{2}}$

Multiply: $= 64^{\frac{2}{3}}$ ← $\boxed{\frac{4}{9}\cdot\frac{3}{2}=\frac{\overset{2}{\cancel{4}}}{\underset{3}{\cancel{9}}}\cdot\frac{\overset{1}{\cancel{3}}}{\underset{1}{\cancel{2}}}=\frac{2}{3}}$

Use $a^{\frac{m}{n}} = \left(\sqrt[n]{a}\right)^m$ to rewrite as a radical expression: $= \left(\sqrt[3]{64}\right)^2$

Simplify $\sqrt[3]{64}$: $= (4)^2$

Simplify: $= 16$

d.–f. Try to simplify each expression on your own. Check your answers, or watch this **interactive video** for complete solutions to all six parts.

You Try It **Work through this You Try It problem.**

Work Exercises 22–28 in this textbook or in the MyMathLab Study Plan.

OBJECTIVE 4 USE RATIONAL EXPONENTS TO SIMPLIFY RADICAL EXPRESSIONS

Some **radical expressions** can be simplified by first writing them with **rational exponents.** We can use the following process.

> **Using Rational Exponents to Simplify Radical Expressions**
>
> **Step 1.** Convert each radical expression to an **exponential expression** with rational exponents.
>
> **Step 2.** Simplify by writing fractions in **lowest terms** or using the **rules of exponents,** as necessary.
>
> **Step 3.** Convert any remaining rational exponents back to a radical expression.

(eText Screens 18.3-1–18.3-32)

Example 7 Simplifying Radical Expressions

My interactive video summary

Use rational exponents to simplify each radical expression. Assume all **variables** represent **non-negative** values.

a. $\sqrt[6]{y}\cdot\sqrt[3]{y}$ b. $\sqrt{\sqrt[3]{x}}$ c. $\sqrt[6]{x^4}$

d. $\sqrt[8]{25x^2y^6}$ e. $\sqrt[4]{49}$ f. $\dfrac{\sqrt[3]{x}}{\sqrt[4]{x}},\ x \neq 0$

Solutions

a. Rewrite each radical using rational exponents: $\sqrt[6]{y}\cdot\sqrt[3]{y} = y^{\frac{1}{6}}\cdot y^{\frac{1}{3}}$

Use the product rule for exponents: $= y^{\frac{1}{6}+\frac{1}{3}}$

Add: $= y^{\frac{1}{2}}$ ← $\frac{1}{6}+\frac{1}{3}=\frac{1}{6}+\frac{2}{6}=\frac{3}{6}=\frac{1}{2}$

Convert back to a **radical expression**: $= \sqrt{y}$

b. Rewrite the **radicand** $\sqrt[3]{x}$ using a rational exponent: $\sqrt{\sqrt[3]{x}} = \sqrt{x^{\frac{1}{3}}}$

Rewrite the square root using a rational exponent: $= \left(x^{\frac{1}{3}}\right)^{\frac{1}{2}}$

Use the **power-to-power** rule for exponents: $= x^{\frac{1}{3}\cdot\frac{1}{2}}$

Multiply: $= x^{\frac{1}{6}}$

Convert back to a radical expression: $= \sqrt[6]{x}$

c. Convert to a rational exponent: $\sqrt[6]{x^4} = x^{\frac{4}{6}}$

Write the rational exponent in **lowest terms**: $= x^{\frac{2}{3}}$

Convert back to a radical expression: $= \sqrt[3]{x^2}$

d.–f. Try to simplify each **radical expression** on your own. **View this popup to** check your answers, or watch this **interactive video** for complete solutions.

You Try It **Work through this You Try It problem.**

Work Exercises 29–34 in this textbook or in the MyMathLab Study Plan.

OBJECTIVE 5 SIMPLIFY RADICAL EXPRESSIONS USING THE PRODUCT RULE

We can develop a *product rule for radicals* that is similar to the **product-to-power rule for exponents.** Look at the **radical expression** $\sqrt[n]{ab}$.

Rewrite the radical as a rational exponent: $\sqrt[n]{ab} = (ab)^{\frac{1}{n}}$

Apply the product-to-power rule for exponents: $= a^{\frac{1}{n}}b^{\frac{1}{n}}$

Convert back to radical expressions: $= \sqrt[n]{a}\sqrt[n]{b}$

This **product rule for radicals** works in both directions: $\sqrt[n]{ab} = \sqrt[n]{a}\sqrt[n]{b}$ and $\sqrt[n]{a}\sqrt[n]{b} = \sqrt[n]{ab}$
So, we can use the rule to multiply radicals and to simplify radicals.

Product Rule for Radicals

If $\sqrt[n]{a}$ and $\sqrt[n]{b}$ are real numbers, then $\sqrt[n]{a}\sqrt[n]{b} = \sqrt[n]{ab}$.

CAUTION: The **index** on each radical must be the same in order to use the product rule for radicals.

(eText Screens 18.3-1–18.3-32)

Example 8 Using the Product Rule to Multiply Radicals

Multiply. Assume all **variables** represent **non-negative** values.

a. $\sqrt{2}\cdot\sqrt{5}$ **b.** $\sqrt{2}\cdot\sqrt{18}$ **c.** $\sqrt[5]{4x^2}\cdot\sqrt[5]{7y^3}$ **d.** $\sqrt[3]{2}\cdot\sqrt[3]{4}$

Solutions In each case, we use the **product rule for radicals** and then **simplify**.

a. $$\overbrace{\sqrt{2}\cdot\sqrt{5}}^{\text{Original expression}} = \overbrace{\sqrt{2\cdot 5}}^{\text{Product rule for radicals}} = \overbrace{\sqrt{10}}^{\text{Simplify the radicand}}$$

b. $$\overbrace{\sqrt{2}\cdot\sqrt{18}}^{\text{Original expression}} = \overbrace{\sqrt{2\cdot 18}}^{\text{Product rule for radicals}} = \overbrace{\sqrt{36}}^{\text{Simplify the radicand}} = \overbrace{6}^{\text{Evaluate}}$$

c. $$\overbrace{\sqrt[5]{4x^2}\cdot\sqrt[5]{7y^3}}^{\text{Original expression}} = \overbrace{\sqrt[5]{4x^2\cdot 7y^3}}^{\text{Product rule for radicals}} = \overbrace{\sqrt[5]{28x^2y^3}}^{\text{Simplify the radicand}}$$

d. $$\overbrace{\sqrt[3]{2}\cdot\sqrt[3]{4}}^{\text{Original expression}} = \overbrace{\sqrt[3]{2\cdot 4}}^{\text{Product rule for radicals}} = \overbrace{\sqrt[3]{8}}^{\text{Simplify the radicand}} = \overbrace{2}^{\text{Evaluate}}$$

You Try It **Work through this You Try It problem.**

Work Exercises 35–38 in this textbook or in the MyMathLab Study Plan.

In **Topic 16.1** we simplified **radical expressions** of the form $\sqrt[n]{a^n}$. But when the **radicand** was not a **perfect** n**th power**, we approximated the **irrational number**. Now, we can use the **product rule** to simplify such radicals. A radical of the form $\sqrt[n]{a}$ is **simplified** if the radicand a has no **factors** that are perfect nth powers (other than 1 or -1). For example, $\sqrt{12}$ is not simplified because the **perfect square** 4 is a factor of the radicand 12. We simplify $\sqrt{12}$ as follows:

Factor 12 into the product of 4 and 3: $\sqrt{12} = \sqrt{4\cdot 3}$

Use the product rule for radicals: $= \sqrt{4}\cdot\sqrt{3}$

Evaluate $\sqrt{4}$: $= 2\sqrt{3}$

Since the radicand 3 has no perfect-square factors other than 1, this final result is simplified.

We can use the following process to simply **radical expressions** of the form $\sqrt[n]{a}$.

Using the Product Rule to Simplify Radical Expressions of the Form $\sqrt[n]{a}$

Step 1. Write the **radicand** as a **product** of two **factors**, one being the largest possible **perfect** n**th power**.

Step 2. Use the **product rule for radicals** to take the n**th root** of each factor.

Step 3. Simplify the nth root of the perfect nth power.

Example 9 Using the Product Rule to Simplify Radicals

Use the **product rule** to **simplify**. Assume all **variables** represent non-negative values.

a. $\sqrt{700}$ **b.** $\sqrt[3]{40}$ **c.** $\sqrt[4]{x^8y^5}$ **d.** $\sqrt{50x^4y^3}$

(eText Screens 18.3-1–18.3-32)

Solutions We follow the **three-step process.**

a. The largest factor of 700 that is a **perfect square** is 100.

Begin with the original expression: $\sqrt{700}$

Factor 700 into the product $100 \cdot 7$: $= \sqrt{100 \cdot 7}$

Use the product rule for radicals: $= \sqrt{100} \cdot \sqrt{7}$

Evaluate $\sqrt{100}$: $= 10\sqrt{7}$

b. The largest factor of 40 that is a **perfect cube** is 8.

Begin with the original expression: $\sqrt[3]{40}$

Factor 40 into the product $8 \cdot 5$: $= \sqrt[3]{8 \cdot 5}$

Use the product rule for radicals: $= \sqrt[3]{8} \cdot \sqrt[3]{5}$

Evaluate $\sqrt[3]{8}$: $= 2\sqrt[3]{5}$

c. Note that x^8 is a **perfect 4th power** because $x^8 = (x^2)^4$. Also, $y^5 = y^4 \cdot y$. So, the largest factor of x^8y^5 that is a perfect 4th power is x^8y^4.

Begin with the original expression: $\sqrt[4]{x^8y^5}$

Factor x^8y^5 into the product $x^8y^4 \cdot y$: $= \sqrt[4]{x^8y^4 \cdot y}$

Use the product rule for radicals: $= \sqrt[4]{x^8y^4} \cdot \sqrt[4]{y}$

Simplify $\sqrt[4]{x^8y^4}$: $= x^2y\sqrt[4]{y}$

My video summary

d. Try to simplify this **radical expression** on your own. Check your answer, or watch this **video** for a complete solution.

You Try It **Work through this You Try It problem.**

Work Exercises 39–44 in this textbook or in the MyMathLab Study Plan.

Example 10 Using the Product Rule to Multiply and Simplify Radicals

My interactive video summary

Multiply and **simplify**. Assume all **variables** represent **non-negative** values.

a. $3\sqrt{10} \cdot 7\sqrt{2}$ b. $2\sqrt[3]{4} \cdot 5\sqrt[3]{6}$ c. $\sqrt[4]{18x^3} \cdot \sqrt[4]{45x^2}$

Solutions First, use the **product rule** to multiply **radicals**. Then, simplify using the **three-step process.**

a.

Begin with the original expression: $3\sqrt{10} \cdot 7\sqrt{2}$

Rearrange factors to group radicals together: $= 3 \cdot 7 \cdot \sqrt{10} \cdot \sqrt{2}$

Use the product rule for radicals: $= 3 \cdot 7 \cdot \sqrt{10 \cdot 2}$

Multiply $3 \cdot 7$; multiply $10 \cdot 2$: $= 21\sqrt{20}$

4 is the largest factor of 20 that is a perfect square: $= 21\sqrt{4 \cdot 5}$

Use the product rule for radicals: $= 21\sqrt{4} \cdot \sqrt{5}$

Simplify $\sqrt{4}$: $= 21 \cdot 2\sqrt{5}$

Multiply $21 \cdot 2$: $= 42\sqrt{5}$

b.–c. Try to work these problems on your own. View this popup to check your **answers**, or watch this **interactive video** for complete solutions.

(eText Screens 18.3-1–18.3-32)

You Try It **Work through this You Try It problem.**

Work Exercises 45–50 in this textbook or in the MyMathLab Study Plan.

CAUTION Be careful not to confuse an **exponent** with the **index** of a **radical**, or vice versa. For example, consider the product of x^3 and $\sqrt{y}$, written as $x^3\sqrt{y}$. Now, consider the product of x and $\sqrt[3]{y}$, written as $x\sqrt[3]{y}$. Can you see how one expression might be confused with the other? When using such expressions, it is very important to write neatly. Consider using a multiplication symbol to make the expressions as clear as possible: $x^3 \cdot \sqrt{y}$ or $x \cdot \sqrt[3]{y}$.

OBJECTIVE 6 SIMPLIFY RADICAL EXPRESSIONS USING THE QUOTIENT RULE

Next, we develop a *quotient rule for radicals*. Consider the **radical expression** $\sqrt[n]{\dfrac{a}{b}}$.

Rewrite the radical as a rational exponent: $\sqrt[n]{\dfrac{a}{b}} = \left(\dfrac{a}{b}\right)^{\frac{1}{n}}$

Apply the **quotient-to-power rule for exponents:** $= \dfrac{a^{\frac{1}{n}}}{b^{\frac{1}{n}}}$

Convert back to radical expressions: $= \dfrac{\sqrt[n]{a}}{\sqrt[n]{b}}$

We can use the **quotient rule for radicals** to simplify radical expressions involving fractions.

Quotient Rule for Radicals

If $\sqrt[n]{a}$ and $\sqrt[n]{b}$ are **real numbers** and $b \neq 0$, then $\sqrt[n]{\dfrac{a}{b}} = \dfrac{\sqrt[n]{a}}{\sqrt[n]{b}}$.

CAUTION The **index** on each radical must be the same in order to use the quotient rule for radicals.

Like the **product rule**, the **quotient rule** works in both directions:

$$\sqrt[n]{\frac{a}{b}} = \frac{\sqrt[n]{a}}{\sqrt[n]{b}} \quad \text{and} \quad \frac{\sqrt[n]{a}}{\sqrt[n]{b}} = \sqrt[n]{\frac{a}{b}}$$

We now add more requirements for a radical expression to be **simplified**.

Simplified Radical Expression

For a **radical expression** to be **simplified**, it must meet the following three conditions:

Condition 1. The **radicand** has no **factor** that is a **perfect power** of the **index** of the radical.

Condition 2. The radicand contains no fractions or **negative exponents.**

Condition 3. No **denominator** contains a radical.

(eText Screens 18.3-1–18.3-32)

As we saw in **Example 9**, we can use the **product rule for radicals** to help resolve issues with Condition 1. We use the **quotient rule for radicals** to resolve issues with Conditions 2 and 3.

Example 11 Using the Quotient Rule to Simplify Radicals

Use the quotient rule to **simplify**. Assume all **variables** represent **non-negative** values.

a. $\sqrt{\dfrac{25}{64}}$ **b.** $\sqrt[3]{\dfrac{16x^5}{27}}$ **c.** $\sqrt[4]{\dfrac{5x^4}{81}}$ **d.** $\sqrt{\dfrac{5x^7}{45x}}, x \neq 0$

Solutions In each case, we use the **quotient rule for radicals.**

a. Original expression = Quotient rule for radicals = Evaluate each square root

$$\sqrt{\frac{25}{64}} = \frac{\sqrt{25}}{\sqrt{64}} = \frac{5}{8}$$

b. Original expression = Quotient rule for radicals = Write as product using perfect-cube factors = Product rule for radicals = Simplify the cube roots

$$\sqrt[3]{\frac{16x^5}{27}} = \frac{\sqrt[3]{16x^5}}{\sqrt[3]{27}} = \frac{\sqrt[3]{8x^3 \cdot 2x^2}}{\sqrt[3]{27}} = \frac{\sqrt[3]{8x^3} \cdot \sqrt[3]{2x^2}}{\sqrt[3]{27}} = \frac{2x\sqrt[3]{2x^2}}{3}$$

c. Original expression = Quotient rule for radicals = Product rule for radicals = Simplify the 4th roots

$$\sqrt[4]{\frac{5x^4}{81}} = \frac{\sqrt[4]{5x^4}}{\sqrt[4]{81}} = \frac{\sqrt[4]{x^4} \cdot \sqrt[4]{5}}{\sqrt[4]{81}} = \frac{x\sqrt[4]{5}}{3}$$

d. Here we simplify the fraction before using the **quotient rule for radicals.**

Original expression = Simplify the fraction = Quotient rule for radicals = Simplify the square roots

$$\sqrt{\frac{5x^7}{45x}} = \sqrt{\frac{x^6}{9}} = \frac{\sqrt{x^6}}{\sqrt{9}} = \frac{x^3}{3}$$

You Try It **Work through this You Try It problem.**

Work Exercises 51–56 in this textbook or in the MyMathLab Study Plan.

In Example 11, we used the quotient rule for radicals to remove fractions from the radicand. In Example 12, we will use the quotient rule to remove **radicals** from a **denominator.**

Example 12 Using the Quotient Rule to Simplify Radicals

My interactive video summary

Use the **quotient rule** to **simplify**. Assume all **variables** represent positive numbers.

a. $\dfrac{\sqrt{240x^3}}{\sqrt{15x}}$ **b.** $\dfrac{\sqrt[3]{-500z^2}}{\sqrt[3]{4z^{-1}}}$ **c.** $\dfrac{\sqrt{150m^9}}{\sqrt{3m}}$ **d.** $\dfrac{\sqrt{45x^5y^{-3}}}{\sqrt{20xy^{-1}}}$

(eText Screens 18.3-1–18.3-32)

Solutions In each case, we use the **quotient rule for radicals.**

a. Begin with the original expression: $\frac{\sqrt{240x^3}}{\sqrt{15x}}$

Use the quotient rule for radicals: $= \sqrt{\frac{240x^3}{15x}}$

Simplify the radicand: $= \sqrt{16x^2}$

Simplify: $= 4x$

b. Begin with the original expression: $\frac{\sqrt[3]{-500z^2}}{\sqrt[3]{4z^{-1}}}$

Use the quotient rule for radicals: $= \sqrt[3]{\frac{-500z^2}{4z^{-1}}}$

Divide factors and subtract exponents in radicand: $= \sqrt[3]{\frac{-500}{4} \cdot z^{2-(-1)}}$

Simplify radicand: $= \sqrt[3]{-125z^3}$

Simplify the cube root: $= -5z$

c.–d. Try to simplify each radical expression on your own. View this popup to check your **answers**, or watch this **interactive video** for complete solutions.

You Try It **Work through this You Try It problem.**

Work Exercises 57–62 in this textbook or in the MyMathLab Study Plan.

18.3 Exercises

In Exercises 1–12, write each exponential expression as a radical expression. Simplify if possible.

You Try It

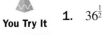

1. $36^{\frac{1}{2}}$ **2.** $(-125)^{\frac{1}{3}}$ **3.** $-81^{\frac{1}{4}}$

4. $(27x^3)^{\frac{1}{3}}$ **5.** $(7xy)^{\frac{1}{5}}$ **6.** $(13xy^2)^{\frac{1}{3}}$

You Try It **7.** $100^{\frac{3}{2}}$ **8.** $-8^{\frac{4}{3}}$ **9.** $(-16)^{\frac{3}{4}}$

10. $\left(\frac{x^3}{64}\right)^{\frac{2}{3}}$ **11.** $\left(\frac{1}{8}\right)^{\frac{5}{3}}$ **12.** $\left(xy^2\right)^{\frac{3}{7}}$

In Exercises 13–18, write each radical expression as an exponential expression.

You Try It

13. $\sqrt{10}$ **14.** $\sqrt[3]{7m}$ **15.** $\sqrt[5]{\frac{2x}{y}}$

You Try It **16.** $\sqrt[7]{x^4}$ **17.** $\left(\sqrt[5]{3xy^3}\right)^4$ **18.** $\sqrt[9]{(2xy)^4}$

In Exercises 19–21, write each exponential expression with positive exponents. Simplify if possible.

You Try It **19.** $25^{-\frac{1}{2}}$ **20.** $\frac{1}{625^{-\frac{3}{4}}}$ **21.** $\left(\frac{27}{8}\right)^{-\frac{2}{3}}$

For Exercises 22–62, assume that all variables represent non-negative values.

In Exercises 22–28, use the rules for exponents to simplify. Write final answers in exponential form when necessary.

You Try It

22. $x^{\frac{1}{3}} \cdot x^{\frac{1}{2}}$

23. $9^{\frac{3}{10}} \cdot 9^{\frac{1}{5}}$

24. $\dfrac{36^{\frac{3}{4}}}{36^{\frac{1}{4}}}$

25. $\left(x^{\frac{5}{3}}\right)^{\frac{9}{10}}$

26. $\left(9m^4n^{-\frac{3}{2}}\right)^{\frac{1}{2}}$

27. $(32x^5y^{-10})^{\frac{1}{5}}\left(xy^{-\frac{1}{2}}\right)$

28. $\left(\dfrac{16x^{\frac{2}{3}}}{81x^{\frac{5}{4}}y^{\frac{2}{3}}}\right)^{\frac{3}{4}}$

You Try It

In Exercises 29–34, use rational exponents to simplify each radical expression. Write final answers in radical form when necessary.

29. $\sqrt[5]{x^2} \cdot \sqrt[4]{x}$

30. $\sqrt[4]{\sqrt[3]{x^2}}$

31. $\sqrt[6]{x^3}$

32. $\dfrac{\sqrt{5}}{\sqrt[6]{5}}$

33. $\sqrt[8]{81}$

34. $\sqrt[10]{49x^6y^8}$

In Exercises 35–38, multiply.

You Try It

35. $\sqrt[4]{5} \cdot \sqrt[4]{7}$

36. $\sqrt{3} \cdot \sqrt{12}$

37. $\sqrt[3]{5x^2} \cdot \sqrt[3]{25x}$

38. $\sqrt[4]{5x} \cdot \sqrt[4]{15y^3}$

You Try It

In Exercises 39–44, use the product rule to simplify.

39. $\sqrt{72}$

40. $\sqrt[3]{56}$

41. $\sqrt{48x^3}$

42. $\sqrt[3]{-125x^6}$

43. $\sqrt{a^{15}}$

44. $\sqrt[4]{x^5y^8z^9}$

You Try It

In Exercises 45–50, multiply and simplify.

45. $\sqrt{6} \cdot \sqrt{30}$

46. $4\sqrt{3x} \cdot 7\sqrt{6x}$

47. $\sqrt{20x^3y} \cdot \sqrt{18xy^4}$

48. $\sqrt[3]{9x^2} \cdot \sqrt[3]{6x^2}$

49. $2x\sqrt[3]{5y^2} \cdot 3x\,\sqrt[3]{25y}$

50. $\sqrt[4]{8x^2} \cdot \sqrt[4]{6x^3}$

In Exercises 51–62, use the quotient rule to simplify.

You Try It

51. $\sqrt{\dfrac{49}{81}}$

52. $\sqrt{\dfrac{12x^3}{25y^2}}$

53. $\sqrt[3]{\dfrac{11x^3}{8}}$

54. $\sqrt[3]{\dfrac{16x^9}{125y^6}}$

55. $\sqrt[4]{\dfrac{21x^5}{16}}$

56. $\sqrt{\dfrac{6x^5}{54x}}$

You Try It

57. $\dfrac{\sqrt{45}}{\sqrt{5}}$

58. $\dfrac{\sqrt{96x^6y^8}}{\sqrt{12x^2y^5}}$

59. $\dfrac{\sqrt{120x^9}}{\sqrt{3x^{-1}}}$

60. $\dfrac{\sqrt[3]{5000}}{\sqrt[3]{5}}$

61. $\dfrac{\sqrt[3]{-72m}}{\sqrt[3]{3m^{-2}}}$

62. $\dfrac{\sqrt{3x^5y^3}}{\sqrt{48xy^5}}$

18.4 Operations with Radicals

THINGS TO KNOW

Before working through this topic, be sure you are familiar with the following concepts:

		VIDEO	ANIMATION	INTERACTIVE
You Try It	1. Find Square Roots of Perfect Squares (Topic 18.1, **Objective 1**)	✓		
You Try It	2. Find Cube Roots (Topic 18.1, **Objective 4**)	✓		
You Try It	3. Find and Approximate nth Roots (Topic 18.1, **Objective 5**)	✓		
You Try It	4. Simplify Radical Expressions Using the Product Rule (Topic 18.3, **Objective 5**)	✓		
You Try It	5. Simplify Radical Expressions Using the Quotient Rule (Topic 18.3, **Objective 6**)			✓

OBJECTIVES

1 Add and Subtract Radical Expressions

2 Multiply Radical Expressions

3 Rationalize Denominators of Radical Expressions

OBJECTIVE 1 ADD AND SUBTRACT RADICAL EXPRESSIONS

As we learned in Topic 14.2, when simplifying **polynomial expressions** we **combine like terms**. Recall that **like terms** have the same **variables** raised to the same corresponding **exponents.** For example, $3x^2y^3$ and $5x^2y^3$ are like terms. We can add or subtract these terms using the reverse of the **distributive property**.

$$3x^2y^3 + 5x^2y^3 = (3 + 5)x^2y^3 = 8x^2y^3$$
$$3x^2y^3 - 5x^2y^3 = (3 - 5)x^2y^3 = -2x^2y^3$$

$4xy^3$ and $2x^2y^2$ are not like terms, so we cannot add or subtract them. They have the same variables, but the variables in each term do not have the same corresponding exponents. This idea is true with **radicals** as well. We can only add and subtract **like radicals.**

Definition Like Radicals

Two radicals are **like radicals** if they have the same **index** and the same **radicand.**

In **Figure 8**, $\sqrt{3}$ and $4\sqrt{3}$ are **like radicals** because they have the same **index** 2 and the same **radicand** 3. The terms $\sqrt[3]{5}$ and $\sqrt[3]{3}$ are **not** like radicals because the radicands are different. Likewise, $\sqrt{5}$ and $\sqrt[3]{5}$ are **not** like radicals because the indices are different.

(*eText Screens 18.4-1–18.4-25*)

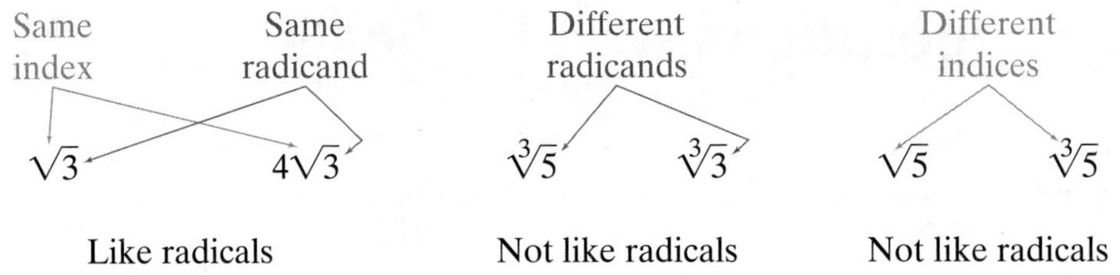

Figure 8

We add and subtract like radicals in the same way that we add and subtract **like terms:** reverse the **distributive property** to factor out the like radical and then simplify.

Example 1 Adding and Subtracting Radical Expressions

Add or subtract.

a. $\sqrt{11} + 6\sqrt{11}$ **b.** $7\sqrt{3} - 5\sqrt[4]{3}$ **c.** $\sqrt[3]{\frac{5}{8}} + 2\sqrt[3]{5}$

Solutions

a. Begin with the original expression: $\sqrt{11} + 6\sqrt{11}$

Identify like radicals: $= \sqrt{11} + 6\sqrt{11}$

Reverse distributive property: $= (1 + 6)\sqrt{11}$

Simplify: $= 7\sqrt{11}$

b. The **indices** are different, so $\sqrt{3}$ and $\sqrt[4]{3}$ are not **like radicals**. We cannot simplify $7\sqrt{3} - 5\sqrt[4]{3}$ further since $\sqrt{3}$ and $\sqrt[4]{3}$ are not like radicals.

c. Begin with the original expression: $\sqrt[3]{\frac{5}{8}} + 2\sqrt[3]{5}$

Quotient rule: $= \dfrac{\sqrt[3]{5}}{\sqrt[3]{8}} + 2\sqrt[3]{5}$

Simplify: $= \dfrac{\sqrt[3]{5}}{2} + 2\sqrt[3]{5}$

Identify like radicals: $= \dfrac{\sqrt[3]{5}}{2} + 2\sqrt[3]{5}$

Reverse distributive property: $= \left(\dfrac{1}{2} + 2\right)\sqrt[3]{5}$

Simplify: $= \dfrac{5}{2}\sqrt[3]{5}$ or $\dfrac{5\sqrt[3]{5}}{2}$

You Try It **Work through this You Try It problem.**

Work Exercises 1–4 in this textbook or in the MyMathLab Study Plan.

Example 2 Adding and Subtracting Radical Expressions

Add or subtract. Assume **variables** represent non-negative values.

a. $15\sqrt[3]{4x^2} - 9\sqrt[3]{4x^2}$ **b.** $2\sqrt{3x} + \sqrt{\frac{x}{4}}$

(eText Screens 18.4-1–18.4-25)

Solutions

a. Begin with the original expression: $15\sqrt[3]{4x^2} - 9\sqrt[3]{4x^2}$

Identify like radicals: $= 15\sqrt[3]{4x^2} - 9\sqrt[3]{4x^2}$

Reverse distributive property: $= (15 - 9)\sqrt[3]{4x^2}$

Simplify: $= 6\sqrt[3]{4x^2}$

b. Begin with the original expression: $2\sqrt{3x} + \sqrt{\dfrac{x}{4}}$

Quotient rule: $= 2\sqrt{3x} + \dfrac{\sqrt{x}}{\sqrt{4}}$

Simplify: $= 2\sqrt{3x} + \dfrac{\sqrt{x}}{2}$

The **radicands** $3x$ and x are different, so $2\sqrt{3x}$ and $\dfrac{\sqrt{x}}{2}$ are not **like radicals**. We cannot simplify further.

You Try It **Work through this You Try It problem.**

Work Exercises 5–8 in this textbook or in the MyMathLab Study Plan.

CAUTION Sometimes it is necessary to **simplify radicals** first before adding or subtracting.

Example 3 Adding and Subtracting Radical Expressions

Add or subtract.

a. $\sqrt{54} + 6\sqrt{72} - 3\sqrt{24}$

b. $\sqrt[3]{24} - \sqrt[3]{192} + 4\sqrt[3]{250}$

Solutions

a. Begin with the original expression: $\sqrt{54} + 6\sqrt{72} - 3\sqrt{24}$

Factor: $= \sqrt{9\cdot 6} + 6\sqrt{36\cdot 2} - 3\sqrt{4\cdot 6}$

Product rule: $= \sqrt{9}\cdot\sqrt{6} + 6\sqrt{36}\cdot\sqrt{2} - 3\sqrt{4}\cdot\sqrt{6}$

Simplify radicals: $= 3\sqrt{6} + 6\cdot 6\sqrt{2} - 3\cdot 2\sqrt{6}$

Multiply factors: $= 3\sqrt{6} + 36\sqrt{2} - 6\sqrt{6}$

Collect like radicals: $= (3 - 6)\sqrt{6} + 36\sqrt{2}$

Simplify: $= -3\sqrt{6} + 36\sqrt{2}$

My video summary

b. Work the problem on your own. Check your **answer**, or watch this **video** for a detailed solution.

You Try It **Work through this You Try It problem.**

Work Exercises 9–12 in this textbook or in the MyMathLab Study Plan.

Example 4 Adding and Subtracting Radical Expressions

Add or subtract. Assume **variables** represent non-negative values.

a. $\sqrt[3]{27m^5n^4} + 2mn\sqrt[3]{m^2n} - m\sqrt[3]{m^2n^4}$

b. $2a\sqrt{16ab^3} + 4\sqrt{9a^2b} - 5\sqrt{4a^3b^3}$

My video summary

Solutions

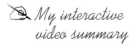

a. Original expression: $\sqrt[3]{27m^5n^4} + 2mn\sqrt[3]{m^2n} - m\sqrt[3]{m^2n^4}$

Factor: $= \sqrt[3]{27m^3n^3 \cdot m^2n} + 2mn\sqrt[3]{m^2n} - m\sqrt[3]{n^3 \cdot m^2n}$

Product rule: $= \sqrt[3]{27m^3n^3} \cdot \sqrt[3]{m^2n} + 2mn\sqrt[3]{m^2n} - m\sqrt[3]{n^3} \cdot \sqrt[3]{m^2n}$

Simplify radicals: $= 3mn\sqrt[3]{m^2n} + 2mn\sqrt[3]{m^2n} - mn\sqrt[3]{m^2n}$

Collect like radicals: $= (3 + 2 - 1)mn\sqrt[3]{m^2n}$

Simplify: $= 4mn\sqrt[3]{m^2n}$

b. Work the problem on your own. Check your **answer**, or watch this **video** for a detailed solution.

You Try It **Work through this You Try It problem.**

Work Exercises 13–16 in this textbook or in the MyMathLab Study Plan.

Example 5 Adding and Subtracting Radical Expressions

My interactive video summary

Add or subtract. Assume **variables** represent non-negative values.

a. $\dfrac{\sqrt{45}}{6x} - \dfrac{4\sqrt{20}}{5x}$ b. $\dfrac{\sqrt[4]{a^5}}{3} + \dfrac{a\sqrt[4]{a}}{12}$

c. $\dfrac{3x^3\sqrt{24x^3y^3}}{2x\sqrt{3x^2y}} - \dfrac{x^2\sqrt{10xy^4}}{\sqrt{5y^2}}$

Solutions Try to work these problems on your own. Remember to find a **common denominator** before adding or subtracting **fractions**. View this **popup** to check your answers, or watch this **interactive video** for detailed solutions.

You Try It **Work through this You Try It problem.**

Work Exercises 17–20 in this textbook or in the MyMathLab Study Plan.

OBJECTIVE 2 MULTIPLY RADICAL EXPRESSIONS

In **Topic 18.3** we saw how to multiply two **radical expressions** when each had only one **term**. Here we extend this idea to radical expressions with more than one term.

To multiply radical expressions, we follow the same approach as when multiplying **polynomial expressions**. We use the **distributive property** to multiply each term in the first expression by each term in the second. Then we simplify the resulting products and combine **like terms** and **like radicals**.

Example 6 Multiplying Radical Expressions

Multiply. Assume **variables** represent non-negative values.

a. $5\sqrt{2x}\left(3\sqrt{2x} - \sqrt{3}\right)$ b. $\sqrt[3]{2n^2}\left(\sqrt[3]{4n} + \sqrt[3]{5n}\right)$

Solutions

a. Begin with the original expression: $5\sqrt{2x}\left(3\sqrt{2x} - \sqrt{3}\right)$

Distributive property: $= 5\sqrt{2x}\cdot 3\sqrt{2x} - 5\sqrt{2x}\cdot\sqrt{3}$

Rearrange factors: $= 5\cdot 3\cdot\sqrt{2x}\cdot\sqrt{2x} - 5\cdot\sqrt{2x}\cdot\sqrt{3}$

Multiply: $= 15\sqrt{4x^2} - 5\sqrt{6x}$

Simplify radical: $= 15\cdot 2x - 5\sqrt{6x}$

Simplify: $= 30x - 5\sqrt{6x}$

My video summary

b. Try this problem on your own. Check your **answer**, or watch this **video** for a detailed solution.

You Try It **Work through this You Try It problem.**

Work Exercises 21–24 in this textbook or in the MyMathLab Study Plan.

Example 7 Multiplying Radical Expressions

Multiply. Assume **variables** represent non-negative values.

a. $\left(7\sqrt{2} - 2\sqrt{3}\right)\left(\sqrt{2} - 5\right)$ b. $\left(\sqrt{m} - 4\right)\left(3\sqrt{m} + 7\right)$

Solutions

a. Use the **FOIL** method to multiply the **radical expressions**. The two first **terms** are $7\sqrt{2}$ and $\sqrt{2}$. The two outside terms are $7\sqrt{2}$ and -5. The two inside terms are $-2\sqrt{3}$ and $\sqrt{2}$. The two last terms are $-2\sqrt{3}$ and -5.

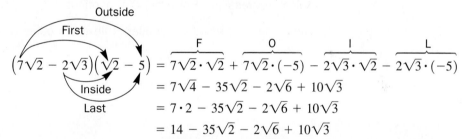

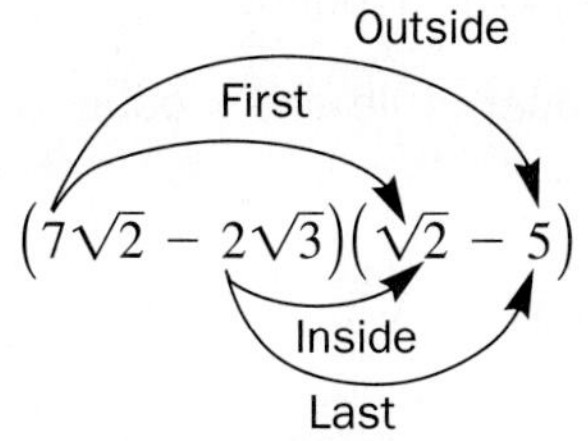

$$\left(7\sqrt{2} - 2\sqrt{3}\right)\left(\sqrt{2} - 5\right) = \overbrace{7\sqrt{2}\cdot\sqrt{2}}^{F} + \overbrace{7\sqrt{2}\cdot(-5)}^{O} - \overbrace{2\sqrt{3}\cdot\sqrt{2}}^{I} - \overbrace{2\sqrt{3}\cdot(-5)}^{L}$$
$$= 7\sqrt{4} - 35\sqrt{2} - 2\sqrt{6} + 10\sqrt{3}$$
$$= 7\cdot 2 - 35\sqrt{2} - 2\sqrt{6} + 10\sqrt{3}$$
$$= 14 - 35\sqrt{2} - 2\sqrt{6} + 10\sqrt{3}$$

My video summary

b. Try this problem on your own. Check your **answer**, or watch this **video** for a detailed solution.

You Try It **Work through this You Try It problem.**

Work Exercises 25–30 in this textbook or in the MyMathLab Study Plan.

Example 8 Using Special Products to Multiply Radical Expressions

Multiply. Assume **variables** represent non-negative values.

a. $\left(\sqrt{y} + 3\right)\left(\sqrt{y} - 3\right)$ b. $\left(3\sqrt{x} - 2\right)^2$

Solutions

a. $(\sqrt{y}+3)(\sqrt{y}-3)$ is a **product of conjugates**. We use the rule for $(A+B)(A-B)$ with $A=\sqrt{y}$ and $B=3$.

Write the product of conjugates rule: $(A+B)(A-B) = A^2 - B^2$

Substitute $\sqrt{y}$ for A and 3 for B: $\left(\sqrt{y}+3\right)\left(\sqrt{y}-3\right) = \left(\sqrt{y}\right)^2 - (3)^2$

Simplify: $= y - 9$

My video summary

b. This **expression** has the form of the **square of a binomial difference** $(A-B)^2$. Use the **special product** rule $(A-B)^2 = A^2 - 2AB + B^2$ with $A = 3\sqrt{x}$ and $B = 2$. Finish working this problem on your own. Check your answer, or watch this **video** for a detailed solution.

You Try It **Work through this You Try It problem.**

Work Exercises 31–36 in this textbook or in the MyMathLab Study Plan.

In Example 8, the product of **conjugates** involving **square roots** resulted in an expression without any **radicals**. This result is true in general and occurs because the product of conjugates equals the **difference of two squares**. We will make use of this result in the next objective.

OBJECTIVE 3 RATIONALIZE DENOMINATORS OF RADICAL EXPRESSIONS

In **Topic 18.3** we saw that a **simplified radical expression** has no **radicals** in the **denominator**. We used the **quotient rule** to eliminate radicals from the denominator of a radical expression. However, the quotient rule is not always enough. For example, simplifying $\sqrt{\frac{3}{32}}$ using the quotient rule and **product rule** yields the following:

$$\sqrt{\frac{3}{32}} = \underbrace{\frac{\sqrt{3}}{\sqrt{32}}}_{\text{Quotient rule}} = \underbrace{\frac{\sqrt{3}}{\sqrt{16}\cdot\sqrt{2}}}_{\text{Product rule}} = \frac{\sqrt{3}}{4\sqrt{2}}$$

We still have a radical in the denominator. Although there is nothing wrong with this, it is common practice to remove radicals from the denominator so that the denominator is a **rational number**. This process is called **rationalizing the denominator**.

Rationalizing a Denominator with One Term

To rationalize a denominator with a single radical of **index** n, multiply the **numerator** and denominator by a radical of index n so that the radicand in the denominator is a **perfect** n**th power**.

Multiplying the numerator and denominator by the same radical is equivalent to multiplying by 1 since $\frac{\sqrt{2}}{\sqrt{2}} = 1$, $\frac{\sqrt{11}}{\sqrt{11}} = 1$, $\frac{\sqrt{x}}{\sqrt{x}} = 1$, and so on. We are simply writing an **equivalent expression** without radicals in the denominator. This idea is like writing **equivalent fractions** with a **least common denominator**.

(eText Screens 18.4-1–18.4-25)

Example 9 Rationalizing Denominators with Square Roots

Rationalize the denominator.

a. $\dfrac{\sqrt{5}}{\sqrt{3}}$ b. $\sqrt{\dfrac{2}{5x}}$

Solutions

a. Since the **denominator** contains a **square root**, we multiply the **numerator** and denominator by a square root so that the **radicand** in the denominator is a **perfect square.** $3 \cdot 3 = 3^2 = 9$ is a perfect square, so we multiply the numerator and denominator by $\sqrt{3}$.

$$\overbrace{\frac{\sqrt{5}}{\sqrt{3}}}^{\text{Original expression}} = \frac{\sqrt{5}}{\sqrt{3}} \cdot \overbrace{\frac{\sqrt{3}}{\sqrt{3}}}^{\text{Multiplying by 1}} = \overbrace{\frac{\sqrt{15}}{\sqrt{9}}}^{\text{Product rule}} = \overbrace{\frac{\sqrt{15}}{3}}^{\text{Simplify}}$$

b. Using the **quotient rule**, we write $\sqrt{\dfrac{2}{5x}} = \dfrac{\sqrt{2}}{\sqrt{5x}}$. Since the denominator contains a square root, we multiply the numerator and denominator by a **square root** so that the **radicand** in the denominator is a **perfect square.** $5x \cdot 5x = (5x)^2 = 25x^2$ is a perfect square, so we multiply the numerator and denominator by $\sqrt{5x}$.

$$\overbrace{\sqrt{\frac{2}{5x}}}^{\text{Original expression}} = \overbrace{\frac{\sqrt{2}}{\sqrt{5x}}}^{\text{Quotient rule}} = \frac{\sqrt{2}}{\sqrt{5x}} \cdot \overbrace{\frac{\sqrt{5x}}{\sqrt{5x}}}^{\text{Multiplying by 1}} = \overbrace{\frac{\sqrt{10x}}{\sqrt{25x^2}}}^{\text{Product rule}} = \overbrace{\frac{\sqrt{10x}}{5x}}^{\text{Simplify}}$$

You Try It **Work through this You Try It problem.**

Work Exercises 37–40 in this textbook or in the MyMathLab Study Plan.

Why would we want to **rationalize a denominator**? View this popup to find out.

Example 10 Rationalizing Denominators with Cube Roots or Fourth Roots

Rationalize the denominator.

a. $\sqrt[3]{\dfrac{11}{25x}}$ b. $\dfrac{\sqrt[4]{7x}}{\sqrt[4]{27y^2}}$

Solutions

a. Using the **quotient rule**, we write $\sqrt[3]{\dfrac{11}{25x}} = \dfrac{\sqrt[3]{11}}{\sqrt[3]{25x}}$. For **roots** greater than 2, it is helpful to write the **radicand** of the denominator in **exponential form.**

$$\sqrt[3]{25x} = \sqrt[3]{5^2x^1}$$

Since the denominator contains a **cube root**, we multiply the numerator and denominator by a cube root so that the radicand in the denominator is a **perfect cube**. To do this, we need the **factors** in the radicand to have

exponents of 3 or **multiples** of 3. We have two 5's and one x, so we need one more 5 and two more x's to get $5^2x^1 \cdot 5^1x^2 = 5^3x^3 = (5x)^3$, which is a perfect cube. Therefore, we multiply the numerator and denominator by $\sqrt[3]{5^1x^2} = \sqrt[3]{5x^2}$.

$$\underbrace{\sqrt[3]{\frac{11}{25x}}}_{\text{Original expression}} = \underbrace{\frac{\sqrt[3]{11}}{\sqrt[3]{25x}}}_{\text{Quotient rule}} = \frac{\sqrt[3]{11}}{\sqrt[3]{25x}} \cdot \underbrace{\frac{\sqrt[3]{5x^2}}{\sqrt[3]{5x^2}}}_{\text{Multiplying by 1}} = \underbrace{\frac{\sqrt[3]{55x^2}}{\sqrt[3]{125x^3}}}_{\text{Product rule}} = \underbrace{\frac{\sqrt[3]{55x^2}}{5x}}_{\text{Simplify}}$$

My video summary

b. Try this problem on your own. Check your **answer**, or watch this **video** for a detailed solution.

You Try It **Work through this You Try It problem.**

Work Exercises 41 and 42 in this textbook or in the MyMathLab Study Plan.

CAUTION

Though not required, it is best to **rationalize the denominator** last when **simplifying radical expressions** since the **radicand** in the denominator will then involve simpler expressions.

Example 11 Rationalizing Denominators

My video summary

Simplify each expression first and then rationalize the denominator.

a. $\sqrt{\frac{3x}{50}}$ **b.** $\frac{\sqrt{18x}}{\sqrt{27xy}}$ **c.** $\sqrt[3]{\frac{-4x^5}{16y^5}}$

Solutions Try these problems on your own. View this **popup** to check your answers, or watch this **video** for detailed solutions.

You Try It **Work through this You Try It problem.**

Work Exercises 43 and 44 in this textbook or in the MyMathLab Study Plan.

The goal of **rationalizing a denominator** is to write an equivalent expression without **radicals** in the denominator. Earlier in this topic, we noted that multiplying **conjugates** involving **square roots** results in an expression without any radicals. So, we can use the **product of conjugates** to rationalize the denominator of a radical expression whose **denominator** contains two **terms** involving one or more square roots.

Would we ever want to rationalize the **numerator**? View this popup to find out.

Rationalizing a Denominator with Two Terms

To rationalize a denominator with two terms involving one or more square roots, multiply the numerator and denominator by the conjugate of the denominator.

Example 12 Rationalizing Denominators with Two Terms

Rationalize the denominator.

a. $\frac{2}{\sqrt{3}+5}$ **b.** $\frac{7}{3\sqrt{x}-4}$ **c.** $\frac{\sqrt{y}-3}{\sqrt{y}+2}$

(eText Screens 18.4-1–18.4-25)

My video summary

Solutions

a. Since the **denominator** has two **terms** and involves a **square root**, we multiply the **numerator** and denominator by the **conjugate** of the denominator $\sqrt{3} - 5$.

Multiply numerator and denominator by $\sqrt{3} - 5$: $\quad \dfrac{2}{\sqrt{3}+5} = \dfrac{2}{\sqrt{3}+5} \cdot \dfrac{\sqrt{3}-5}{\sqrt{3}-5}$

Multiply numerators and multiply denominators: $\quad = \dfrac{2(\sqrt{3}-5)}{(\sqrt{3})^2 - (5)^2}$

Simplify the denominator: $\quad = \dfrac{2(\sqrt{3}-5)}{3-25}$

$= \dfrac{2(\sqrt{3}-5)}{-22}$

Divide out a common factor of 2: $\quad = \dfrac{{}^{1}\cancel{2}(\sqrt{3}-5)}{-\cancel{22}_{11}}$

Simplify: $\quad = -\dfrac{\sqrt{3}-5}{11}$ or $\dfrac{5-\sqrt{3}}{11}$

b. Since the **denominator** has two **terms** and involves a **square root**, we multiply the **numerator** and denominator by the **conjugate** of the denominator, $3\sqrt{x} + 4$.

$$\frac{7}{3\sqrt{x}-4} = \underbrace{\frac{7}{3\sqrt{x}-4} \cdot \frac{3\sqrt{x}+4}{3\sqrt{x}+4}}_{\text{Multiply numerator and denominator by } 3\sqrt{x}+4} = \underbrace{\frac{7(3\sqrt{x}+4)}{(3\sqrt{x})^2 - (4)^2}}_{\text{Multiply numerators and multiply denominators}} = \underbrace{\frac{7(3\sqrt{x}+4)}{9x-16}}_{\text{Simplify}} \text{ or } \underbrace{\frac{21\sqrt{x}+28}{9x-16}}_{\text{Distribute}}$$

We can leave the result in factored form, or we can distribute in the final step. Leaving the numerator factored until the end helps with dividing out **common factors**, if they are present.

c. Work this problem on your own. Check your answer, or watch this **video** for a detailed solution.

You Try It **Work through this You Try It problem.**

Work Exercises 45–50 in this textbook or in the MyMathLab Study Plan.

18.4 Exercises

In Exercises 1–50, assume that all variables represent non-negative values and be sure to simplify your answer.

In Exercises 1–20, add or subtract, if possible.

You Try It

1. $4\sqrt{7} + 9\sqrt{7}$

2. $\sqrt[3]{\dfrac{32}{27}} - \sqrt[3]{108}$

3. $3\sqrt{2} + 5\sqrt{6}$

4. $3\sqrt{7} - 4\sqrt[3]{6} + 2\sqrt{7} - 8\sqrt[3]{6}$

You Try It

5. $\sqrt[3]{6x} - 7\sqrt[3]{6x}$

6. $\dfrac{4\sqrt{3x}}{9} + \dfrac{2\sqrt{3x}}{15} - \dfrac{7\sqrt{3x}}{6}$

7. $10\sqrt{5} - 3\sqrt{x} - 4\sqrt{5} + 9\sqrt{x}$

8. $6\sqrt{a} + 2\sqrt{b} - a\sqrt{3}$

You Try It

9. $\sqrt{45} - \sqrt{80}$

10. $3\sqrt[3]{32} + \sqrt[3]{500}$

11. $5\sqrt{12} - \sqrt{75} + 3\sqrt{20}$

12. $6\sqrt[3]{54} + 2\sqrt{8} - 4\sqrt[3]{16}$

You Try It

13. $2\sqrt[4]{81x^5} - 3\sqrt[4]{x^5}$

14. $6\sqrt{20x^3} + 4x\sqrt{45x}$

15. $3a^2\sqrt{ab^3} + \sqrt{16a^5b^3} - 5b\sqrt{a^5b}$

16. $\sqrt[3]{81m^3n} - 4\sqrt[3]{3m^3n} + 3m\sqrt[3]{162n}$

You Try It

17. $\dfrac{\sqrt{72}}{3y} + \dfrac{5\sqrt{18}}{4y}$

18. $\dfrac{\sqrt[3]{8a^4}}{4a} - \dfrac{\sqrt[3]{27a}}{7}$

19. $\dfrac{\sqrt[4]{81a^7}}{3} - \dfrac{5a\sqrt[4]{16a^3}}{2}$

20. $\dfrac{5x^2\sqrt{128x^5y^3}}{\sqrt{64x^4y}} + \dfrac{2xy\sqrt{18x^3y}}{3\sqrt{y}}$

In Exercises 21–36, multiply.

You Try It

21. $\sqrt{5}(2x + \sqrt{3})$

22. $\sqrt{2x}(\sqrt{6x} - 5)$

23. $2\sqrt{3m}(m - 3\sqrt{5m})$

24. $\sqrt[3]{x}\left(\sqrt[3]{54x^2} - \sqrt[3]{x}\right)$

You Try It

25. $(8\sqrt{3} - 5)(\sqrt{3} - 1)$

26. $(3 + \sqrt{5})(2 + \sqrt{7})$

27. $(\sqrt{x} + 4)(\sqrt{x} + 6)$

28. $(4 - 2\sqrt{3x})(5 + \sqrt{3x})$

You Try It

29. $(3 - \sqrt{x})(4 - \sqrt{y})$

30. $\left(\sqrt[3]{y} + 1\right)\left(\sqrt[3]{y} - 2\right)$

31. $(\sqrt{7z} + 3)(\sqrt{7z} - 3)$

32. $(\sqrt{5} + 3y)^2$

33. $(\sqrt{x} + 3)(3 - \sqrt{x})$

34. $(\sqrt{3a} - \sqrt{2b})(\sqrt{3a} + \sqrt{2b})$

35. $(2\sqrt{x} + 5)^2$

36. $(2\sqrt{m} - 3\sqrt{n})^2$

In Exercises 37–50, rationalize the denominator.

37. $\dfrac{4}{\sqrt{6}}$

38. $\dfrac{\sqrt{3}}{\sqrt{5}}$

You Try It 39. $\sqrt{\dfrac{9}{2x}}$

40. $\dfrac{5}{\sqrt{11x}}$

You Try It 41. $\sqrt[3]{\dfrac{9}{4x^2}}$

42. $\dfrac{\sqrt[4]{3b^5}}{\sqrt[4]{25a}}$

You Try It

43. $\dfrac{8}{\sqrt{12x^3y^4}}$

44. $\dfrac{\sqrt[3]{2x^2}}{\sqrt[3]{36y^5}}$

45. $\dfrac{5}{2 - \sqrt{3}}$

46. $\dfrac{-4x}{5 + \sqrt{6}}$

You Try It

47. $\dfrac{-3}{\sqrt{x} + 1}$

48. $\dfrac{\sqrt{a}}{3\sqrt{a} - \sqrt{b}}$

49. $\dfrac{\sqrt{m} - 4}{\sqrt{m} - 7}$

50. $\dfrac{5\sqrt{3} + 2\sqrt{6}}{4\sqrt{6} - \sqrt{3}}$

(eText Screens 18.5-1–18.5-26)

18.5 Radical Equations and Models

THINGS TO KNOW

Before working through this topic, be sure you are familiar with the following concepts:

		VIDEO	ANIMATION	INTERACTIVE
You Try It	1. Solve Linear Equations Using Both Properties of Equality (Topic 11.1, **Objective 5**)	✓		
You Try It	2. Solve a Formula for a Given Variable (Topic 11.4, **Objective 3**)	✓		
You Try It	3. Use Geometric Formulas to Solve Applications (Topic 11.4, **Objective 4**)	✓		
You Try It	4. Use the Power-to-Power Rule (Topic 14.1, **Objective 4**)	✓		
You Try It	5. Solve Quadratic Equations by Factoring (Topic 15.7, **Objective 1**)			✓
You Try It	6. Multiply Radical Expressions (Topic 18.4, **Objective 2**)	✓		

OBJECTIVES

1 Solve Equations Involving One Radical Expression

2 Solve Equations Involving Two Radical Expressions

3 Use Radical Equations and Models to Solve Application Problems

OBJECTIVE 1 SOLVE EQUATIONS INVOLVING ONE RADICAL EXPRESSION

Recall that when **solving** an **equation in one variable**, we find all values of the **variable** that make the equation true. All of these values together form the **solution set** of the equation. Previously, we have solved **linear equations**, **polynomial equations**, and **rational equations**. Now we learn how to solve *radical equations*.

> **Definition Radical Equation**
>
> A **radical equation** is an equation that contains at least one **radical expression** with a variable in the **radicand**.

Some examples of radical equations are $\sqrt{2x+1}=3$, $\sqrt[3]{5x-4}+7=10$, $\sqrt[4]{19x-2}=2\sqrt[4]{x+1}$, and $\sqrt{x+9}-\sqrt{x-6}=3$.

The first two equations contain one radical expression, whereas the last two equations contain two radical expressions.

(eText Screens 18.5-1–18.5-26)

CAUTION

Not all equations that contain radical expressions are radical equations. If no radical expression contains a variable in the radicand, then the equation is not a radical equation. For example, $2x + 1 = \sqrt{3}$ is not a radical equation.

The key to solving a radical equation is to eliminate all of the radicals from the equation. To do this, we can use the following **theorem**:

> **Theorem The Power Principle of Equality**
>
> If A and B represent **algebraic expressions** and n is a positive **real number**, then any **solution** to the **equation** $A = B$ is also a solution to the equation $A^n = B^n$.

Consider the equation $\sqrt{x} = 5$, which has one **radical expression** $\sqrt{x}$. The radical expression $\sqrt{x}$ is an **isolated radical expression** because it stands alone on one side of the equal sign. An isolated radical can be eliminated from an equation by raising both sides of the equation to the power of the **index**. In this case, we **square** both sides of the equation because the radical expression is a **square root**.

Begin with the original equation: $\sqrt{x} = 5$

Square both sides: $\left(\sqrt{x}\right)^2 = (5)^2$

Simplify: $x = 25$

Checking this answer, we substitute 25 for x in the original equation:

Substitute 25 for x: $\sqrt{25} \stackrel{?}{=} 5$

Simplify: $5 = 5$ True

Since $x = 25$ **satisfies** the original equation, the **solution set** is $\{25\}$.

Sometimes when we raise both sides of an **equation** to an even **power**, the new equation will have **solutions** that are not solutions to the original equation. Such "solutions" are called **extraneous solutions**. They must be identified and excluded from the **solution set**. View this **popup** for a more in-depth explanation of extraneous solutions.

We can use the following steps to **solve** equations involving one **radical expression**.

> **Solving Equations Involving One Radical Expression**
>
> **Step 1.** **Isolate the radical expression.** Use the **properties of equality** to get the radical expression by itself on one side of the equal sign.
>
> **Step 2.** Eliminate the radical. Identify the **index** of the radical expression and raise both sides of the equation to the index power.
>
> **Step 3.** Solve the resulting equation.
>
> **Step 4.** Check each solution from Step 3 in the original equation. Disregard any **extraneous solutions**.

TIP

Extraneous solutions do not result from raising both sides of an equation to an odd power, but it is still good practice to check your answers anyway.

CAUTION

When solving **radical equations**, answers can fail to check for two reasons: An answer may be an extraneous solution, or an error may have been made while solving. When an answer is disregarded as an extraneous solution, it is important to make sure that no errors were made while solving.

Example 1 Solving an Equation Involving One Radical Expression

Solve $\sqrt{3x-2}+6=11$.

Solution We follow the **four-step process.**

Step 1. **Isolate** the **radical expression** by subtracting 6 from both sides.

Begin with the original equation:	$\sqrt{3x-2}+6=11$
Subtract 6 from both sides:	$\sqrt{3x-2}+6-6=11-6$
Simplify:	$\sqrt{3x-2}=5$

Step 2. The radical expression is a **square root**, so we **square** both sides.

Square both sides:	$\left(\sqrt{3x-2}\right)^2=(5)^2$
Simplify:	$3x-2=25$

Step 3. Solve the resulting **linear equation.**

Add 2 to both sides:	$3x=27$
Divide both sides by 3:	$x=9$

Step 4. Check $x=9$ in the original equation.

Substitute 9 for x:	$\sqrt{3(9)-2}+6\stackrel{?}{=}11$
Simplify beneath the radical:	$\sqrt{25}+6\stackrel{?}{=}11$
Evaluate the square root:	$5+6\stackrel{?}{=}11$
Add:	$11=11$ True

The answer checks in the original equation, so the **solution set** is $\{9\}$.

You Try It **Work through this You Try It problem.**

Work Exercises 1–8 in this textbook or in the MyMathLab Study Plan.

CAUTION When solving a **radical equation**, it is important to **isolate the radical** before raising both sides to the **index** power. If this is not done, then the **radical expression** will not be eliminated. View this **popup** to see an example.

Example 2 Solving an Equation Involving One Radical Expression

Solve $\sqrt{2x-1}+9=6$.

Solution

Step 1.	Begin with the original equation:	$\sqrt{2x-1}+9=6$
	Subtract 9 from both sides:	$\sqrt{2x-1}+9-9=6-9$
	Simplify:	$\sqrt{2x-1}=-3$
Step 2.	Square both sides:	$\left(\sqrt{2x-1}\right)^2=(-3)^2$
	Simplify:	$2x-1=9$
Step 3.	Add 1 to both sides:	$2x=10$
	Divide both sides by 2:	$x=5$

Step 4. Check $x = 5$ in the original equation: $\sqrt{2(5) - 1} + 9 \stackrel{?}{=} 6$

Simplify beneath the radical: $\sqrt{9} + 9 \stackrel{?}{=} 6$

Evaluate the square root: $3 + 9 \stackrel{?}{=} 6$

Add: $12 = 6$ False

The answer $x = 5$ does not check in the original equation. It is an **extraneous solution.** Since this is the only possible **solution**, the equation has no real solution. The **solution set** is $\{\}$ or $\varnothing$.

Note: We might have noticed sooner that this equation would have no real solution. In step 1, when we **isolated the radical**, the resulting equation was $\sqrt{2x - 1} = -3$. This equation states that a **principal square root** equals a negative value, but this is impossible since principal square roots must be non-negative. So, this equation has no real solution.

You Try It **Work through this You Try It problem.**

Work Exercises 9 and 10 in this textbook or in the MyMathLab Study Plan.

Example 3 Solving an Equation Involving One Radical Expression

My video summary

Solve $\sqrt{3x + 7} - x = 1$.

Solution We follow the **four-step process.**

Step 1. Isolate the radical expression by adding x to both sides.

Begin with the original equation: $\sqrt{3x + 7} - x = 1$

Add x to both sides: $\sqrt{3x + 7} - x + x = 1 + x$

Simplify each side: $\sqrt{3x + 7} = x + 1$

Step 2. The **radical expression** is a **square root**, so we **square** both sides of the equation to eliminate the radical.

Square both sides: $\left(\sqrt{3x + 7}\right)^2 = (x + 1)^2$

Simplify each side: $3x + 7 = x^2 + 2x + 1$

Try to solve this **quadratic equation** on your own by using the **four-step process** from **Topic 15.7.** Then finish solving the original equation by checking for **extraneous solutions.** Check your answer, or watch this **video** for the complete solution.

You Try It **Work through this You Try It problem.**

Work Exercises 11–16 in this textbook or in the MyMathLab Study Plan.

So far, all of the **radical equations** solved have involved square roots. In Example 4, we look at an equation involving a **cube root.**

Example 4 Solving an Equation Involving One Radical Expression

Solve $\sqrt[3]{3x^2 + 23x + 10} = 12$.

(eText Screens 18.5-1–18.5-26)

My video summary

Solution

Step 1. Begin with the original equation: $\sqrt[3]{3x^2 + 23x} + 10 = 12$

Subtract 10 from both sides: $\sqrt[3]{3x^2 + 23x} + 10 - 10 = 12 - 10$

Simplify: $\sqrt[3]{3x^2 + 23x} = 2$

Step 2. The **radical expression** is a **cube root**, so we **cube** both sides of the equation to eliminate the radical.

Cube both sides: $\left(\sqrt[3]{3x^2 + 23x}\right)^3 = (2)^3$

Simplify: $3x^2 + 23x = 8$

The result is a **quadratic equation**. Try to finish solving this equation on your own. Check your answer, or watch this **video** for the complete solution.

You Try It **Work through this You Try It problem.**

Work Exercises 17–20 in this textbook or in the MyMathLab Study Plan.

Recall from **Topic 18.3** that $a^{\frac{1}{n}} = \sqrt[n]{a}$. Sometimes equations containing one **rational exponent** can be solved using the same process for solving equations with one **radical expression**. We solve such an equation in Example 5.

Example 5 Solving an Equation Involving a Rational Exponent

My video summary

Solve $(x^2 - 9)^{\frac{1}{4}} + 3 = 5$.

Solution This equation is equivalent to $\sqrt[4]{x^2 - 9} + 3 = 5$. Try solving it on your own by using the **four-step process**. Check your answer, or watch this **video** for a complete solution.

You Try It **Work through this You Try It problem.**

Work Exercises 21–24 in this textbook or in the MyMathLab Study Plan.

OBJECTIVE 2 SOLVE EQUATIONS INVOLVING TWO RADICAL EXPRESSIONS

Solving an equation involving two **radical expressions** is similar to solving an equation involving one radical expression. However, it may be necessary to repeat the process for eliminating a radical in order to eliminate both radicals from the equation.

Solving Equations Involving Two Radical Expressions

Step 1. **Isolate** one of the radical expressions. Use the **properties of equality** to get a radical expression by itself on one side of the equal sign.

Step 2. Eliminate the radical from the isolated radical expression. Identify the **index** of the isolated radical expression and raise both sides of the equation to this index **power**.

Step 3. If all the radicals have been eliminated, then solve the resulting equation. Otherwise, repeat Steps 1 and 2.

Step 4. Check each solution from Step 3 in the original equation. Disregard any **extraneous solutions**.

Example 6 Solving an Equation Involving Two Radical Expressions

My video summary

Solve $\sqrt{x+9} - \sqrt{x} = 1$.

Solution We follow the **four-step process.**

Step 1. We choose to isolate $\sqrt{x+9}$.

Begin with the original equation: $\sqrt{x+9} - \sqrt{x} = 1$

Add $\sqrt{x}$ to both sides: $\sqrt{x+9} - \sqrt{x} + \sqrt{x} = 1 + \sqrt{x}$

Simplify: $\sqrt{x+9} = 1 + \sqrt{x}$

Step 2. The isolated **radical expression** is a **square root**, so we **square** both sides.

Square both sides: $\left(\sqrt{x+9}\right)^2 = \left(1 + \sqrt{x}\right)^2$

The right side has the form of the **square of a binomial sum** $(A+B)^2$, so we use the **special product rule** $(A+B)^2 = A^2 + 2AB + B^2$ with $A = 1$ and $B = \sqrt{x}$.

Use $(A+B)^2 = A^2 + 2AB + B^2$ on the right side: $\left(\sqrt{x+9}\right)^2 = (1)^2 + 2(1)\left(\sqrt{x}\right) + \left(\sqrt{x}\right)^2$

Simplify: $x + 9 = 1 + 2\sqrt{x} + x$

Step 3. The equation still contains a radical expression, so we repeat Steps 1 and 2. Try to finish solving this equation on your own. Remember to check for **extraneous solutions.** Check your answer, or watch this **video** for the complete solution.

You Try It **Work through this You Try It problem.**

Work Exercises 25–27 in this textbook or in the MyMathLab Study Plan.

Example 7 Solving an Equation Involving Two Radical Expressions

My video summary

Solve $\sqrt{2x+3} + \sqrt{x-2} = 4$.

Solution Try solving this equation on your own by using the **four-step process.** Check your answer, or watch this **video** for the complete solution.

You Try It **Work through this You Try It problem.**

Work Exercises 28–30 in this textbook or in the MyMathLab Study Plan.

Example 8 Solving an Equation Involving Two Radical Expressions

My video summary

Solve $\sqrt[3]{2x^2 - 9} + \sqrt[3]{3x - 11} = 0$.

Solution We follow the **four-step process.**

Step 1. Begin with the original equation: $\sqrt[3]{2x^2 - 9} + \sqrt[3]{3x - 11} = 0$

Subtract $\sqrt[3]{3x - 11}$ from both sides: $\sqrt[3]{2x^2 - 9} = -\sqrt[3]{3x - 11}$

Step 2. The two **radical expressions** are **cube roots**, so we **cube** both sides.

Cube both sides: $\left(\sqrt[3]{2x^2 - 9}\right)^3 = \left(-\sqrt[3]{3x - 11}\right)^3$

Simplify: $2x^2 - 9 = -(3x - 11)$

Distribute the negative: $2x^2 - 9 = -3x + 11$

Step 3. Both radicals are eliminated. Try to finish solving on your own.

Step 4. Check your **answer**, or watch this **video** for the complete solution.

You Try It **Work through this You Try It problem.**

Work Exercises 31–34 in this textbook or in the MyMathLab Study Plan.

OBJECTIVE 3 USE RADICAL EQUATIONS AND MODELS TO SOLVE APPLICATION PROBLEMS

Often real-world situations are **modeled** by **formulas** that involve **radical expressions.** For example, the formula $t = \frac{\sqrt{d}}{4}$ gives the time t, in seconds, that it takes a **free-falling** object to fall a distance of d feet.

In **Topic 11.4** we learned to solve a formula for a given **variable.** If the given variable is beneath a radical, then we must remove the radical by **isolating the radical** and raising both sides of the equation to the appropriate **power.**

Example 9 Solving a Formula for a Variable Beneath a Radical

Solve each formula for the given variable.

a. Free-falling object: $t = \frac{\sqrt{d}}{4}$ for d.

b. Radius of a **sphere**: $r = \sqrt[3]{\frac{3V}{4\pi}}$ for V.

Solutions

a.

Begin with the original formula: $t = \frac{\sqrt{d}}{4}$

Multiply both sides by 4 to isolate the radical: $4(t) = 4\left(\frac{\sqrt{d}}{4}\right)$

Simplify: $4t = \sqrt{d}$

Square both sides: $(4t)^2 = \left(\sqrt{d}\right)^2$

Simplify: $16t^2 = d$

Solving for d, the formula for a free-falling object is $d = 16t^2$.

My video summary

b. The **variable** V is already contained within the **isolated radical expression.** The radical expression is a **cube root**, so we eliminate the radical by **cubing** both sides. Try to finish solving the formula for V. Check your answer, or watch this **video** for a complete solution.

You Try It **Work through this You Try It problem.**

Work Exercises 35–40 in this textbook or in the MyMathLab Study Plan.

We can use **radical equations** and **models** to solve a variety of application problems in many different disciplines. In Example 10, a radical equation is used to model the readability of written text.

(eText Screens 18.5-1–18.5-26)

Example 10 Assessing the Readability of Written Text

My video summary

A **SMOG** grade for written text is a minimum reading grade level G that a reader must possess in order to fully understand the written text being graded. If w is the number of words that have three or more syllables in a sample of 30 sentences from a given text, then the SMOG grade for that text is given by the **formula** $G = \sqrt{w} + 3$. Use the SMOG grade formula to answer the following questions. (*Source:* readabilityformulas.com).

a. If a sample of 30 sentences contains 18 words with three or more syllables, then what is the SMOG grade for the text? If necessary, round to a **whole number** for the grade level.

b. If a text must have a tenth-grade reading level, then how many words with three or more syllables would be needed in the sample of 30 sentences?

Solutions

a. There are 18 words with three or more syllables, so we substitute 18 for w in the given **formula**.

Begin with the original formula: $G = \sqrt{w} + 3$

Substitute 18 for w: $G = \sqrt{18} + 3$

Approximate the square root: $G \approx 4.24 + 3$

Simplify: $G \approx 7.24$

Rounding to the nearest **whole number**, we get $G = 7$. So, the **SMOG** grade for this text is a seventh-grade reading level.

b. The text must have a tenth-grade reading level, so we substitute 10 for G in the formula and solve for w.

Begin with the original formula: $G = \sqrt{w} + 3$

Substitute 10 for G: $10 = \sqrt{w} + 3$

Try to finish solving this problem on your own. Check your answer, or watch this **video** for a complete solution.

You Try It **Work through this You Try It problem.**

Work Exercises 41–44 in this textbook or in the MyMathLab Study Plan.

Example 11 Punting a Football

An important component of a good punt in football is **hang time**, which is the length of time that the punted ball remains in the air. If wind resistance is ignored, the relationship between the hang time t, in seconds, and the vertical height h, in feet, that the ball reaches can be **modeled** by the **formula** $t = \frac{\sqrt{h}}{2}$.

Use this formula to answer the following questions.

a. If the average hang time for an NFL punt is 4.6 seconds, then what is the vertical height for an average NFL punt? Round to the nearest foot.

b. Cowboys Stadium in Arlington, Texas, has a huge high-definition screen centered over most of the football field. The bottom of the screen is 90 feet above the field. What hang time would result in the ball hitting the screen? Round to the nearest hundredth of a second.

(eText Screens 18.5-1–18.5-26)

Solutions

a. We substitute the hang time 4.6 for t in the formula and solve for h.

Begin with the original formula: $t = \dfrac{\sqrt{h}}{2}$

Substitute 4.6 for t: $4.6 = \dfrac{\sqrt{h}}{2}$

Multiply both sides by 2: $9.2 = \sqrt{h}$

Square both sides: $(9.2)^2 = (\sqrt{h})^2$

Simplify: $84.64 = h$

Rounding, the average NFL punt reaches a vertical height of about 85 feet.

b. We substitute the vertical height, 90 feet, for h and **simplify**.

Begin with the original formula: $t = \dfrac{\sqrt{h}}{2}$

Substitute 90 for h: $t = \dfrac{\sqrt{90}}{2}$

Approximate the square root: $t \approx \dfrac{9.4868}{2}$

Divide: $t \approx 4.7434$

Rounding, a hang time of about 4.74 seconds will result in a punt that hits the screen.

You Try It **Work through this You Try It problem.**

Work Exercises 45 and 46 in this textbook or in the MyMathLab Study Plan.

18.5 Exercises

In Exercises 1–34, solve each radical equation.

You Try It

1. $\sqrt{x + 15} = 3$

2. $\sqrt{2x - 1} = 3$

3. $\sqrt{r} - 3 = -1$

4. $3\sqrt{m} + 5 = 7$

You Try It

5. $\sqrt{z + 5} + 2 = 9$

6. $\sqrt{1 - x} + 1 = 4$

You Try It

7. $\sqrt{4p + 1} - 3 = 2$

8. $5\sqrt{w - 9} - 2 = 8$

9. $\sqrt{q + 10} = -2$

10. $\sqrt{2x - 1} + 6 = 1$

11. $\sqrt{2m + 1} - m = -1$

12. $p - \sqrt{4p + 9} + 3 = 0$

You Try It

13. $\sqrt{32 - 4x} - x = 0$

14. $\sqrt{7z + 2} + 3z = 2$

15. $2\sqrt{4y + 1} + 3y = 4y + 4$

16. $2q + 1 = \sqrt{q^3 + 17}$

You Try It

17. $\sqrt[3]{2x - 1} = -2$

18. $\sqrt[3]{7 - 5w} + 1 = 4$

19. $\sqrt[3]{2n^2 + 15n} + 1 = 4$

20. $\sqrt[4]{x^2 - 6x} - 2 = 0$

21. $(x-5)^{\frac{1}{2}} - 1 = 4$

22. $(2y+3)^{\frac{1}{3}} + 7 = 9$

23. $(2w+5)^{\frac{1}{4}} + 1 = 4$

24. $(p^2 - 19)^{\frac{1}{4}} + 2 = 5$

You Try It

25. $\sqrt{x+15} - \sqrt{x} = 3$

26. $\sqrt{x} - 1 = \sqrt{x-5}$

27. $\sqrt{x-64} + 8 = \sqrt{x}$

28. $\sqrt{6-x} + \sqrt{5x+6} = 6$

You Try It

29. $\sqrt{3x+1} - \sqrt{x+4} = 1$

30. $\sqrt{6+5x} + \sqrt{3x+4} = 2$

31. $\sqrt{7x-4} = \sqrt{4x+11}$

32. $\sqrt[3]{4y-3} = \sqrt[3]{6y+9}$

You Try It

33. $\sqrt[3]{2x^2+5x} = \sqrt[3]{2x+14}$

34. $\sqrt[4]{6w+1} = \sqrt[4]{2w+17}$

In Exercises 35–40, solve each formula for the given variable.

You Try It

35. $T = \dfrac{\pi\sqrt{2L}}{4}$ for L

36. $d = k\sqrt[3]{E}$ for E

37. $r = \sqrt[3]{\dfrac{3V}{\pi h}}$ for h

38. $V = \sqrt{\dfrac{FR}{m}}$ for m

39. $A = P\sqrt{1+r}$ for r

40. $T = \left(\dfrac{LH^2}{25}\right)^{\frac{1}{4}}$ for L

You Try It

In Exercises 41–46, use the given model to solve each application problem.

41. **SMOG Grade** Recall from Example 10 that the formula $G = \sqrt{w} + 3$ gives the minimum reading grade level G needed to fully understand written text containing w words with three or more syllables in a sample of 30 sentences.

a. If a sample of 30 sentences contains 51 words with three or more syllables, then what is the SMOG grade for the text? If necessary, round your answer to a whole number to find the grade level.

b. If a text must have a ninth-grade reading level, then how many words with three or more syllables are needed in a sample of 30 sentences?

42. **Measure of Leanness** The *ponderal index* is a measure of the "leanness" of a person. A person who is h inches tall and weighs w pounds has a ponderal index I given by $I = \dfrac{h}{\sqrt[3]{w}}$.

a. Compute the ponderal index for a person who is 75 inches tall and weighs 190 pounds. Round to the nearest hundredth.

b. What is a man's weight if he is 70 inches tall and has a ponderal index of 12.35? Round to the nearest whole number.

43. **Body Surface Area** The Mosteller formula is used in the medical field to estimate a person's body surface area. The formula is $A = \dfrac{\sqrt{hw}}{60}$, where A is body surface area in square meters, h is height in centimeters, and w is weight in kilograms.

a. Compute the body surface area of a person who is 178 cm tall and weighs 90.8 kg. Round to the nearest hundredth.

b. If a woman is 165 cm tall and has body surface area of 1.72 m^2, how much does she weigh? Round to the nearest tenth.

44. Skid Marks Under certain road conditions, the length of a skid mark S, in feet, is related to the velocity v, in miles per hour, by the formula $v = \sqrt{10S}$. Assuming these same road conditions, answer the following:

a. Compute a car's velocity if it leaves a skid mark of 360 feet.

b. If a car is traveling at 50 miles per hour when it skids, what will be the length of the skid mark?

You Try It

45. Hang Time Recall from Example 11 that hang time t, in seconds, and the vertical height h, in feet, can be modeled by $t = \frac{\sqrt{h}}{2}$. This same formula can be used to model an athlete's hang time when jumping.

a. If LeBron James has a vertical leap of 3.7 feet, what is his hang time? Round to the nearest hundredth.

b. When Mark "Wild Thing" Wilson of the Harlem Globetrotters slam-dunked a regulation basketball on a 12-foot rim in front of an Indianapolis crowd, his hang time for the shot was approximately 1.16 seconds. What was the vertical distance of his jump? Round to the nearest tenth.

46. Distance to the Horizon From a boat, the distance d, in miles, that a person can see to the horizon is modeled by the formula $d = \frac{3\sqrt{h}}{2}$, where h is the height, in feet, of eye level above the sea.

a. From his ship, how far can a sailor see to the horizon if his eye level is 36 feet above the sea?

b. How high is the eye level of a sailor who can see 12 miles to the horizon?

18.6 Complex Numbers

THINGS TO KNOW

Before working through this topic, be sure you are familiar with the following concepts:

		VIDEO	ANIMATION	INTERACTIVE
You Try It	1. Find Square Roots of Perfect Squares (Topic 18.1, **Objective 1**)	▦		
You Try It	2. Simplify Radical Expressions Using the Product Rule (Topic 18.3, **Objective 5**)	▦		▦

OBJECTIVES

1 Simplify Powers of i

2 Add and Subtract Complex Numbers

3 Multiply Complex Numbers

4 Divide Complex Numbers

5 Simplify Radicals with Negative Radicands

OBJECTIVE 1 SIMPLIFY POWERS OF i

So far we have learned about **real number** solutions to **equations**, but not every equation has real number **solutions.** For example, consider the equation $x^2 + 1 = 0$. We can check if a value is a solution by substituting the value for x and seeing if a true statement results. Is $x = -1$ a solution to this equation?

Begin with the original equation: $x^2 + 1 = 0$

Substitute -1 for x: $(-1)^2 + 1 \stackrel{?}{=} 0$

Simplify: $1 + 1 \stackrel{?}{=} 0$

$2 = 0$ False

Because $2 = 0$ is not a true statement, $x = -1$ is not a solution to the equation $x^2 + 1 = 0$. In fact, this equation has no real solution. View this **popup** to see why.

To find the solution to equations such as $x^2 + 1 = 0$, we introduce a new number called the **imaginary unit** i.

> **Definition Imaginary Unit i**
>
> The **imaginary unit** i is defined as
>
> $$i = \sqrt{-1}, \text{ where } i^2 = -1.$$

When working with the **imaginary unit**, we will encounter various powers of i. Let's consider some powers of i and look for patterns that will help us simplify them.

$i^1 = \underbrace{i = \sqrt{-1}}_{\text{Defined}}$

$i^2 = \underbrace{-1}_{\text{Defined}}$

$i^3 = \underbrace{i^2 \cdot i}_{\text{Product rule for exponents}} = \underbrace{(-1)}_{i^2 = -1} \cdot i = -i$

$i^4 = \underbrace{i^2 \cdot i^2}_{\text{Product rule for exponents}} = \underbrace{(-1)}_{i^2 = -1} \cdot \underbrace{(-1)}_{i^2 = -1} = 1$

$i^5 = \underbrace{i^4 \cdot i}_{\text{Product rule for exponents}} = \underbrace{(1)}_{i^4 = 1} \cdot i = i$

$i^6 = \underbrace{i^4 \cdot i^2}_{\text{Product rule for exponents}} = \underbrace{(1)}_{i^4 = 1} \cdot \underbrace{(-1)}_{i^2 = -1} = -1$

$i^7 = \underbrace{i^4 \cdot i^3}_{\text{Product rule for exponents}} = \underbrace{(1)}_{i^4 = 1} \cdot \underbrace{(-i)}_{i^3 = -i} = -i$

$i^8 = \underbrace{i^4 \cdot i^4}_{\text{Product rule for exponents}} = \underbrace{(1)}_{i^4 = 1} \cdot \underbrace{(1)}_{i^4 = 1} = 1$

$\vdots$

Notice that the powers of i follow the pattern $i, -1, -i, 1$. Based on this pattern, what is the value of i^0? View this **popup** to find out. We can use the following procedure to simplify powers of i.

> **Simplifying i^n for $n > 4$**
>
> **Step 1.** Divide n by 4 and find the **remainder** r.
>
> **Step 2.** Replace the **exponent** (power) on i by the remainder, $i^n = i^r$.
>
> **Step 3.** Use the results $i^0 = 1$, $i^1 = i$, $i^2 = -1$, and $i^3 = -i$ to simplify if necessary.

(eText Screens 18.6-1–18.6-24)

Example 1 Simplifying Powers of i

Simplify.

My video summary

a. i^{17} **b.** i^{60} **c.** i^{39} **d.** $-i^{90}$ **e.** $i^{14} + i^{29}$

Solutions

a. **Step 1.** Divide the exponent by 4 and find the remainder:

$$\begin{array}{r} 4 \\ 4\overline{)17} \longleftarrow \text{Exponent, } n \\ \underline{16} \\ 1 \longleftarrow \text{Remainder, } r \end{array}$$

Step 2. Replace n by r: $\overbrace{i^{17}}^{r} = \overbrace{i^{1}}^{n}$

Step 3. Simplify if necessary: $i^{17} = i^{1} = i$

b. Divide the **exponent** 60 by 4 and find the **remainder**.

$$\begin{array}{r} 15 \\ 4\overline{)60} \longleftarrow \text{Exponent, } n \\ \underline{4} \\ 20 \\ \underline{20} \\ 0 \longleftarrow \text{Remainder, } r \end{array}$$

Replace n by r and simplify if necessary.

$$\overbrace{i^{60}}^{n} = \overbrace{i^{0}}^{r} = 1$$

c.–e. Try to simplify these **powers of** i on your own. Check your **answers**, or watch this **video** to see detailed solutions.

You Try It **Work through this You Try It problem.**

Work Exercises 1–5 in this textbook or in the MyMathLab Study Plan.

With the **imaginary unit**, we can now expand our number system from the set of real numbers to the set of **complex numbers.**

> **Complex Numbers**
>
> The set of all numbers of the form
>
> $$a + bi,$$
>
> where a and b are **real numbers** and i is the **imaginary unit**, is called the set of **complex numbers**. The number a is called the **real part**, and the number b is called the **imaginary part**.

If $b = 0$, then the complex number is a purely real number. If $a = 0$, then the complex number is a purely **imaginary number. Figure 9** illustrates the relationships between complex numbers.

(eText Screens 18.6-1–18.6-24)

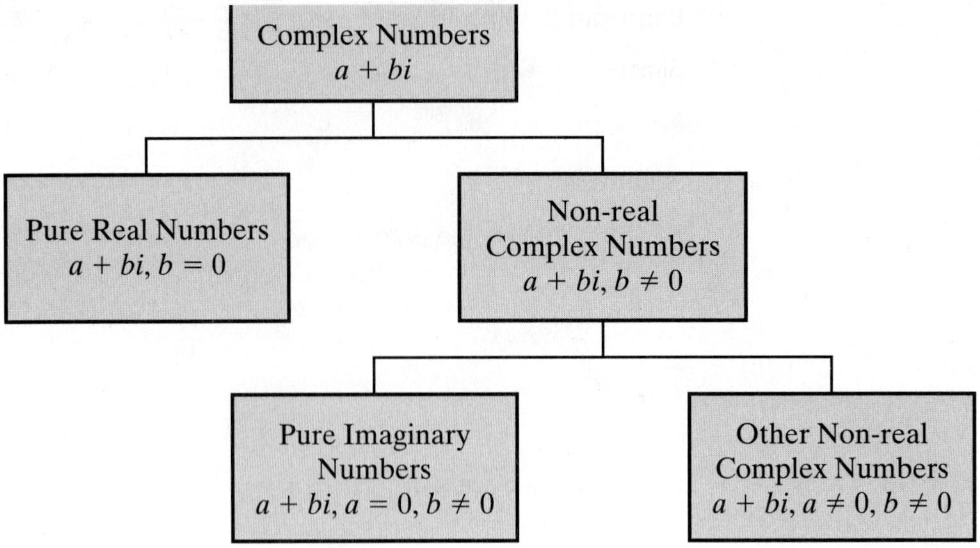

Figure 9 The Complex Number System

Figure 9 shows us that all **real numbers** are **complex numbers**, but not all complex numbers are real numbers. This distinction is important, particularly when solving equations. For example, $x^2 + 1 = 0$ has no *real* solutions, but it does have two *complex* solutions.

A complex number of the form $a + bi$ is written in **standard form**, and is the typical way to write complex numbers. Below are examples of complex numbers written in **standard form**.

	Standard form		Real part		Imaginary part	
Real Number:	7	$=$	7 (Real part, $a = 7$)	$+$	0 (Imaginary part, $b = 0$)	i
Imaginary Number:	$3i$	$=$	0 (Real part, $a = 0$)	$+$	3 (Imaginary part, $b = 3$)	i
Non-real Complex Number:	$4 + 9i$	$=$	4 (Real part, $a = 4$)	$+$	9 (Imaginary part, $b = 9$)	i
	$\frac{1}{3} + \frac{2}{3}i$	$=$	$\frac{1}{3}$ (Real part, $a = \frac{1}{3}$)	$+$	$\frac{2}{3}$ (Imaginary part, $b = \frac{2}{3}$)	i

OBJECTIVE 2 ADD AND SUBTRACT COMPLEX NUMBERS

To add or subtract **complex numbers**, we combine the **real parts** and combine the **imaginary parts**.

Adding and Subtracting Complex Numbers

To add complex numbers, add the real parts and add the imaginary parts.

$$(a + bi) + (c + di) = (a + c) + (b + d)i$$

To subtract complex numbers, subtract the real parts and subtract the imaginary parts.

$$(a + bi) - (c + di) = (a - c) + (b - d)i$$

Example 2 Adding and Subtracting Complex Numbers

Perform the indicated operations.

My video summary

a. $(3 + 5i) + (2 - 7i)$ b. $(3 + 5i) - (2 - 7i)$

c. $(-3 - 4i) + (2 - i) - (3 + 7i)$

Solutions

a. Original expression: $(3 + 5i) + (2 - 7i)$

Remove parentheses: $= 3 + 5i + 2 - 7i$

Collect like terms: $= \underbrace{3 + 2}_{\text{Real}} + \underbrace{5i - 7i}_{\text{Imaginary}}$

Combine real parts: $= \overbrace{5}^{3+2} + 5i - 7i$

Combine imaginary parts: $= 5 \overbrace{- 2i}^{5i - 7i}$

b. Original expression: $(3 + 5i) - (2 - 7i)$

Change to add the opposite: $= (3 + 5i) + (-2 + 7i)$

Remove parentheses: $= 3 + 5i - 2 + 7i$

Finish working the problem on your own. Check your answer, or watch this **video** to see the full solution.

c. Try to work the problem on your own. Check your answer, or watch this **video** for a detailed solution.

You Try It **Work through this You Try It problem.**

Work Exercises 6–13 in this textbook or in the MyMathLab Study Plan.

OBJECTIVE 3 MULTIPLY COMPLEX NUMBERS

When multiplying **complex numbers**, we use the **distributive property** and the **FOIL method** just as when multiplying **polynomials**. Remember that $i^2 = -1$ when simplifying.

Example 3 Multiplying Complex Numbers

Multiply.

a. $-4i(3 - 8i)$ b. $(3 \cdot 4i)(6 + 11i)$

Solutions

a. Multiply using the distributive property.

Original expression: $-4i(3 - 8i)$

Distribute: $= -4i \cdot 3 + (-4i)(-8i)$

Multiply: $= -12i + 32i^2$

Replace i^2 with -1: $= -12i + 32\underbrace{(-1)}_{i^2 = -1}$

Simplify: $= -12i - 32$

Write in standard form: $= -32 - 12i$

b. Multiply using the **FOIL method.**

$$(3-4i)(6+11i) = \overset{\text{F}}{\overline{3\cdot 6}} + \overset{\text{O}}{\overline{3\cdot 11i}} - \overset{\text{I}}{\overline{4i\cdot 6}} - \overset{\text{L}}{\overline{4i\cdot 11i}}$$

$$= 18 + \underbrace{33i - 24i}_{\text{Collect like terms}} - 44i^2$$

$$= 18 + 9i - \underbrace{44i^2}_{i^2=-1}$$

$$= 18 + 9i - 44(-1)$$

$$= \underbrace{18 + 9i + 44}_{\text{Collect like terms}}$$

$$= 62 + 9i$$

Example 4 Multiplying Complex Numbers

My video summary

Multiply $(4-3i)(7+5i)$.

Solution Work the problem on your own. Check your answer, or watch this **video** for a detailed solution.

You Try It **Work through this You Try It problem.**

Work Exercises 14–19 in this textbook or in the MyMathLab Study Plan.

Example 5 Special Products Involving Complex Numbers

Multiply.

a. $(4+2i)^2$ **b.** $\left(\sqrt{3}-5i\right)^2$

Solutions

a. We recognize this expression as the square of a **binomial sum**, where $A = 4$ and $B = 2i$. Use the **special product rule** to multiply.

Square of a binomial sum rule: $(A+B)^2 = A^2 + 2AB + B^2$

Substitute 4 for A and $2i$ for B: $(\underset{A}{\underline{4}} + \underset{B}{\underline{2i}})^2 = \underset{A^2}{\underline{(4)^2}} + \underset{2AB}{\underline{2(4)(2i)}} + \underset{B^2}{\underline{(2i)^2}}$

Simplify: $= 16 + 16i + 4i^2$

Replace i^2 with -1: $= 16 + 16i + 4(-1)$

Multiply: $= 16 + 16i - 4$

Simplify: $= 12 + 16i$

We can also find this product using the **FOIL method.**

Begin with the original expression: $(4+2i)^2$

Write the expression as a product: $= (4+2i)(4+2i)$

Multiply using the FOIL method: $= \overset{\text{F}}{\overline{4\cdot 4}} + \overset{\text{O}}{\overline{4\cdot 2i}} + \overset{\text{I}}{\overline{2i\cdot 4}} + \overset{\text{L}}{\overline{2i\cdot 2i}}$

Simplify: $= 16 + 8i + 8i + 4i^2$

Collect like terms: $= 16 + 16i + 4i^2$

At this point we would continue simplifying as before to find $(4+2i)^2 = 12 + 16i$.

My video summary

b. Try to work the problem on your own. Check your answer, or watch this **video** for a detailed solution.

You Try It **Work through this You Try It problem.**

Work Exercises 20 and 21 in this textbook or in the MyMathLab Study Plan.

In **Topic 14.5** we learned that a **binomial sum** and a **binomial difference** made from the same two **terms** are **conjugates** of each other. We also saw that the **product of conjugates** resulted in the **difference of two squares**. These results extend to our discussion of **complex numbers** as follows.

> **Complex Conjugates**
>
> The complex numbers $(a + bi)$ and $(a - bi)$ are called **complex conjugates** of each other. A complex conjugate is obtained by changing the sign of the imaginary part in a complex number. Also, $(a + bi)(a - bi) = a^2 + b^2$.

Notice that the product of complex conjugates is a *sum* of two **squares** rather than a difference and is always a **real number**. View this **popup** to see why.

Example 6 Multiplying Complex Conjugates

Multiply $(-2 + 7i)(-2 - 7i)$.

Solution Since the two **complex numbers** are **conjugates**, we find the product using the result for the **product of complex conjugates**.

Identifying $a = -2$ and $b = 7$ we get

$$\overbrace{(-2 + 7i)(-2 - 7i)}^{(a+bi)(a-bi)} = \overbrace{(-2)^2}^{a^2} + \overbrace{(7)^2}^{b^2} = 4 + 49 = 53.$$
$$(a + bi)(a - bi)$$

We can also find the same result using the **FOIL method**. View this **popup** to see how.

You Try It **Work through this You Try It problem.**

Work Exercises 22–26 in this textbook or in the MyMathLab Study Plan.

OBJECTIVE 4 DIVIDE COMPLEX NUMBERS

When dividing **complex numbers**, the goal is to eliminate the imaginary part from the denominator and to express the **quotient** in **standard form**, $a + bi$. To do this, we multiply the numerator and denominator by the **complex conjugate** of the denominator.

Example 7 Dividing Complex Numbers

Divide. Write the quotient in standard form.

$$\frac{1 - 3i}{5 - 2i}$$

(eText Screens 18.6-1–18.6-24)

My video summary

Solution The **denominator** is $5 - 2i$, so its **complex conjugate** is $5 + 2i$. We multiply both the **numerator** and denominator by the complex conjugate and simplify to **standard form**.

Multiply numerator and denominator by $5 + 2i$: $\frac{1 - 3i}{5 - 2i} = \frac{1 - 3i}{5 - 2i} \cdot \frac{5 + 2i}{5 + 2i}$

Multiply numerators and multiply denominators: (remember that $(a + bi)(a - bi) = a^2 + b^2$) $= \frac{5 + 2i - 15i - 6i^2}{(5)^2 + (2)^2}$

Simplify exponents: (remember that $i^2 = -1$) $= \frac{5 + 2i - 15i - 6(-1)}{25 + 4}$

Simplify the numerator and denominator: $= \frac{5 - 13i + 6}{29}$

$= \frac{11 - 13i}{29}$

Write in standard form, $a + bi$: $= \frac{11}{29} - \frac{13}{29}i$

You Try It **Work through this You Try It problem.**

Work Exercises 27–30 in this textbook or in the MyMathLab Study Plan.

Note: Remember that multiplying the numerator and denominator of an expression by the same quantity is the same as multiplying the expression by 1.

Example 8 Dividing Complex Numbers

My video summary

Divide. Write the quotient in **standard form**.

$$\frac{5 + 7i}{2i}$$

Solution First, we multiply the **numerator** and **denominator** by the **complex conjugate** of the denominator. The denominator is $2i = 0 + 2i$, so its complex conjugate is $0 - 2i = -2i$.

$$\frac{5 + 7i}{2i} = \underbrace{\frac{5 + 7i}{2i} \cdot \frac{-2i}{-2i}}_{\text{Multiply numerator and denominator by } -2i} = \underbrace{\frac{-10i - 14i^2}{-4i^2}}_{\text{Multiply numerators and multiply denominators}}$$

Finish simplifying on your own, and write the answer in standard form. Check your answer, or watch this **video** for the complete solution.

You Try It **Work through this You Try It problem.**

Work Exercise 31 in this textbook or in the MyMathLab Study Plan.

In Example 8, we can get the same result if we multiply the numerator and denominator by $2i$. View this **popup** to find out why.

(eText Screens 18.6-1–18.6-24)

OBJECTIVE 5 SIMPLIFY RADICALS WITH NEGATIVE RADICANDS

In the next module, we will solve equations involving solutions with radicals having a negative **radicand**. Thus, we must first learn how to simplify a **radical** with a negative radicand such as $\sqrt{-49}$. By remembering that $\sqrt{-1} = i$, we can use the following rule to simplify this expression.

> **Square Root of a Negative Number**
>
> For any positive real number a,
>
> $$\sqrt{-a} = \sqrt{-1} \cdot \sqrt{a} = i\sqrt{a}.$$

So, $\sqrt{-49} = \sqrt{-1} \cdot \sqrt{49} = i \cdot 7 = 7i$.

At this point, you might want to review how to simplify radicals using the **product rule** in **Topic 18.3**.

Example 9 Simplifying a Square Root with a Negative Radicand

My video summary

Simplify.

a. $\sqrt{-81}$ b. $\sqrt{-48}$ c. $\sqrt{-108}$

Solutions

a. $\sqrt{-81} = \underbrace{\sqrt{-1}}_{i} \cdot \sqrt{81} = i \cdot 9 = 9i$

b. $\sqrt{-48} = \underbrace{\sqrt{-1}}_{i} \cdot \sqrt{48} = i \cdot \sqrt{16} \cdot \sqrt{3}$

$= i \cdot 4\sqrt{3} = 4i\sqrt{3}$

c. Try this problem on your own. Check your answer, or watch this **video** for a detailed solution.

You Try It **Work through this You Try It problem.**

Work Exercise 32 in this textbook or in the MyMathLab Study Plan.

CAUTION

When simplifying or performing operations involving **radicals** with a negative **radicand** and an even **index**, it is important to first write the numbers in terms of the **imaginary unit** i if possible.

The property $\sqrt{a} \cdot \sqrt{b} = \sqrt{ab}$ is only true when $a \geq 0$ and $b \geq 0$ so that $\sqrt{a}$ and $\sqrt{b}$ are **real numbers**. This property does not apply to non-real numbers. To find the correct answer if a or b are negative, we must first write each number in terms of the imaginary unit i.

$$\sqrt{-3} \cdot \sqrt{-12} = \sqrt{(-3)(-12)} = \sqrt{36} = 6$$

$$\sqrt{-3} \cdot \sqrt{-12} = \underbrace{\sqrt{-1}}_{i} \cdot \sqrt{3} \cdot \underbrace{\sqrt{-1}}_{i} \cdot \sqrt{12} = i\sqrt{3} \cdot i\sqrt{12} = \underbrace{i^2}_{i^2 = -1}\sqrt{36} = -6$$

We can use a graphing calculator to check the result. See **Figure 10**.

(eText Screens 18.6-1–18.6-24)

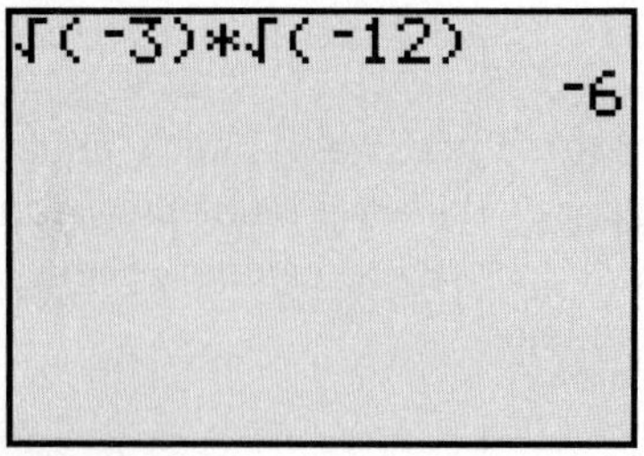

Figure 10

Notice that in order to get the correct answer, we had to first write each number in terms of the **imaginary unit** i.

Example 10 Simplifying Expressions with Negative Radicands

My video summary

Simplify.

a. $\sqrt{-8} + \sqrt{-18}$

b. $\sqrt{-8} \cdot \sqrt{-18}$

c. $\dfrac{6 + \sqrt{(6)^2 - 4(2)(5)}}{2}$

d. $\dfrac{4 - \sqrt{-12}}{4}$

Solutions

a. $\sqrt{-8} + \sqrt{-18} = \underbrace{\sqrt{-1}}_{i} \cdot \sqrt{8} + \underbrace{\sqrt{-1}}_{i} \cdot \sqrt{18}$

$= i \cdot 2\sqrt{2} + i \cdot 3\sqrt{2}$

$= 2i\sqrt{2} + 3i\sqrt{2}$

$= 5i\sqrt{2}$

b. $\sqrt{-8} \cdot \sqrt{-18} = \left(\underbrace{\sqrt{-1}}_{i} \cdot \sqrt{8}\right) \cdot \left(\underbrace{\sqrt{-1}}_{i} \cdot \sqrt{18}\right)$

$= (i \cdot 2\sqrt{2}) \cdot (i \cdot 3\sqrt{2})$

$= \underbrace{i^2}_{i^2 = -1} \cdot 6\sqrt{4}$

$= -1 \cdot 6 \cdot 2$

$= -12$

c.–d. Try these problems on your own. Check your answers, or watch this **video** for detailed solutions.

You Try It **Work through this You Try It problem.**

Work Exercises 33–40 in this textbook or in the MyMathLab Study Plan.

18.6 Exercises

You Try It

In Exercises 1–5, write each power of i as i, -1, $-i$, or 1.

1. i^{41} **2.** i^{28} **3.** $-i^{19}$ **4.** $(-i)^7$ **5.** $i^{22} + i^{13}$

(eText Screens 18.6-1–18.6-24)

You Try It

In Exercises 6–13, find the sum or difference. Write each answer in standard form, $a + bi$.

6. $(3 - 2i) + (-7 + 9i)$

7. $(3 - 2i) - (-7 + 9i)$

8. $i - (1 + i)$

9. $5 + (2 - 3i)$

10. $(2 + 5i) - (2 - 5i)$

11. $(2 + 5i) + (2 - 5i)$

12. $[(-1 + 8i) - (3 - 4i)] + (9 - 4i)$

13. $(6 + 3i) - [(2 + 4i) + (5 - 2i)]$

In Exercises 14–21, perform the indicated operations. Write each answer in standard form.

You Try It

14. $3i(7i)$

15. $2i(4 - 3i)$

16. $-i(1 - i)$

17. $(3 - 2i)(6 + i)$

18. $(-2 - i)(3 - 4i)$

19. $(5 + i)(2 + 3i)$

You Try It

20. $(2 + 7i)^2$

21. $(6 - 2i)^2$

In Exercises 22–26, find the product of the complex number and its conjugate.

You Try It

22. $5 - 2i$

23. $1 - i$

24. $\frac{1}{2} - 3i$

25. $\sqrt{5} + i$

26. $4i$

You Try It

In Exercises 27–31, write each quotient in standard form.

27. $\frac{2 - i}{3 + 4i}$

28. $\frac{1}{2 - i}$

29. $\frac{3i}{2 + 2i}$

You Try It

30. $\frac{5 + i}{5 - i}$

31. $\frac{2 - 3i}{5i}$

You Try It

In Exercises 32–40, write each expression in standard form.

32. $\sqrt{-320}$

33. $\sqrt{-36} - \sqrt{49}$

34. $\sqrt{-1} + 3 - \sqrt{-64}$

You Try It

35. $\sqrt{-2} \cdot \sqrt{-18}$

36. $\left(\sqrt{-8}\right)^2$

37. $\left(i\sqrt{-4}\right)^2$

38. $\frac{-4 - \sqrt{-20}}{2}$

39. $\frac{-3 - \sqrt{-81}}{6}$

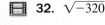

40. $\frac{4 + \sqrt{-8}}{4}$

MODULE NINETEEN

Quadratic Equations and Functions; Circles

MODULE NINETEEN CONTENTS

19.1 Solving Quadratic Equations

THINGS TO KNOW

Before working through this topic, be sure you are familiar with the following concepts:	VIDEO	ANIMATION	INTERACTIVE
You Try It — **1.** Solve Quadratic Equations by Factoring (Topic 15.7, **Objective 1**)	■		
You Try It — **2.** Find Square Roots of Perfect Squares (Topic 18.1, **Objective 1**)	■		
You Try It — **3.** Simplify Square Roots of the Form $\sqrt{a^2}$ (Topic 18.1, **Objective 3**)	■		
You Try It — **4.** Simplify Square Roots Using the Product Rule (Topic 18.3, **Objective 5**)	■		■
You Try It — **5.** Simplify Square Roots Using the Quotient Rule (Topic 18.3, **Objective 6**)			■
You Try It — **6.** Simplify Radicals with Negative Radicands (Topic 18.6, **Objective 5**)	■		

OBJECTIVES

1 Solve Quadratic Equations Using the Square Root Property

2 Solve Quadratic Equations by Completing the Square

(eText Screens 19.1-1–19.1-28)

3 Solve Quadratic Equations Using the Quadratic Formula

4 Use the Discriminant to Determine the Number and Type of Solutions to a Quadratic Equation

5 Solve Equations That Are Quadratic in Form

OBJECTIVE 1 SOLVE QUADRATIC EQUATIONS USING THE SQUARE ROOT PROPERTY

Recall that a **quadratic equation** is an equation that can be written in the standard form $ax^2 + bx + c = 0$ where a, b, and c are **real numbers** and $a \neq 0$. In **Topic 15.7**, we solved quadratic equations by **factoring** and then using the **zero product property**. For example, we can solve the equation $x^2 = 100$ as follows:

Write the original equation:	$x^2 = 100$
Subtract 100 from both sides:	$x^2 - 100 = 0$
Factor the **difference of squares**:	$(x - 10)(x + 10) = 0$
Set each **factor** equal to 0:	$x - 10 = 0$ or $x + 10 = 0$
Solve each resulting equation:	$x = 10$ or $x = -10$

Both values **check**, so the **solution set** is $\{-10, 10\}$.

Some quadratic equations are not easy to solve by factoring, and other quadratic equations cannot be solved by factoring at all. Therefore, in this module, we discuss other methods for solving quadratic equations, beginning with the *square root property*.

To understand this property, look again at the equation $x^2 = 100$. Notice that the two solutions, -10 and 10, are the negative and positive **square roots** of the number 100. Looking at a similar equation, $x^2 = 36$, do you see that the solutions will be the negative and positive square roots of 36? Therefore, $x = -\sqrt{36} = -6$ or $x = \sqrt{36} = 6$. This illustrates the **square root property**.

My video summary

Square Root Property

If u is an **algebraic expression** and k is a **real number**, then $u^2 = k$ is equivalent to $u = -\sqrt{k}$ or $u = \sqrt{k}$. Equivalently, if $u^2 = k$, then $u = \pm\sqrt{k}$.

Watch this **video** for further explanation of the square root property.

Note: If $k \geq 0$, then the solutions to the equation $u^2 = k$ will be **real numbers**. However, if $k < 0$, then the solutions to $u^2 = k$ will be non-real **complex numbers**.

Example 1 Using the Square Root Property

My video summary

Use the square root property to solve each **quadratic equation**. Write each answer in **simplest form**.

a. $x^2 = 144$ **b.** $x^2 = 48$

Solutions For each equation, we apply the square root property and simplify. Work through each of the following, or watch this **video** for fully worked solutions.

a.

Write the original equation:	$x^2 = 144$
Apply the square root property:	$x = \pm\sqrt{144}$
Evaluate:	$x = \pm 12$

These results **check**, so the **solution set** is $\{-12, 12\}$.

b. Write the original equation: $x^2 = 48$

Apply the **square root property**: $x = \pm\sqrt{48}$

Factor the **radicand**: $x = \pm\sqrt{16 \cdot 3}$

Simplify: $x = \pm 4\sqrt{3}$

View the **check**. The solution set is $\{-4\sqrt{3}, 4\sqrt{3}\}$.

You Try It **Work through this You Try It problem.**

Work Exercises 1–4 in this textbook or in the MyMathLab Study Plan.

CAUTION

When solving equations of the form $u^2 = k$, we often simply say that we are taking the square root of both sides. However, because the **square root** of a number yields only one value, the **principal root**, many students forget to include the $\pm$. Remember that applying the square root property for $k \neq 0$ will result in *two* values: the positive and negative square roots of k.

To solve quadratic equations using the square root property, we use the following guidelines.

Solving Quadratic Equations Using the Square Root Property

Step 1. Write the equation in the form $u^2 = k$ to **isolate** the quantity being squared.

Step 2. Apply the **square root property**.

Step 3. Solve the resulting equations.

Step 4. Check the **solutions** in the original equation.

Example 2 Solving a Quadratic Equation Using the Square Root Property

Solve.

a. $x^2 - 16 = 0$ **b.** $2x^2 + 72 = 0$ **c.** $(x - 1)^2 = 9$ **d.** $2(x + 1)^2 - 17 = 23$

Solutions

a. **Isolate** x^2, then apply the square root property.

Original equation: $x^2 - 16 = 0$

Add 16 to both sides: $x^2 = 16 \leftarrow$ Square is isolated

Apply the square root property: $x = \pm\sqrt{16}$

Simplify the radical: $x = \pm 4$

The **solution set** is $\{-4, 4\}$. The check is left to you.

b. **Isolate** x^2, then apply the **square root property**.

Original equation: $2x^2 + 72 = 0$

Subtract 72 from both sides: $2x^2 = -72$

Divide both sides by 2: $x^2 = -36 \leftarrow$ Square is isolated

Apply the square root property: $x = \pm\sqrt{-36}$

Simplify the **radical** with a negative **radicand**: $x = \pm\sqrt{-1} \cdot \sqrt{36}$

$x = \pm 6i$

The solution set is $\{-6i, 6i\}$. The check is left to you.

My video summary

c. In this case, we are squaring an **algebraic expression**. However, this will not change our process. We still isolate the square and apply the square root property.

Original equation: $(x - 1)^2 = 9 \leftarrow$ Square is isolated

Apply the square root property: $x - 1 = \pm\sqrt{9}$

Try to finish solving this equation on your own. View the **answer**, or watch this **video** for the complete solution.

My video summary

d. Try to work this problem on your own. Remember to **isolate** the square first, then apply the **square root property**. Simplify **radicals** if possible. View the **answer**, or watch this **video** for the complete solution.

You Try It **Work through this You Try It problem.**

Work Exercises 5–10 in this textbook or in the MyMathLab Study Plan.

In Example 2c, the **quadratic equation** $(x - 1)^2 = 9$ can be solved using the square root property because the left side of the equation is a **perfect square** and the right side is a **constant**. But what about quadratic equations such as $x^2 - 5x + 3 = 0$? Even if we get the square term by itself

$$x^2 = 5x - 3$$

we cannot apply the square root property because the right-hand side is not a constant. However, every quadratic equation can be written in the form $(x - h)^2 = k$ (as in Example 2c) by using a method known as **completing the square**.

OBJECTIVE 2 SOLVE QUADRATIC EQUATIONS BY COMPLETING THE SQUARE

Consider the following **perfect square trinomials**:

$$x^2 + 2x + 1 = (x + 1)^2 \qquad x^2 - 6x + 9 = (x - 3)^2 \qquad x^2 - 7x + \frac{49}{4} = \left(x - \frac{7}{2}\right)^2$$

$$\left(\frac{1}{2}\cdot 2\right)^2 = 1 \qquad \left(\frac{1}{2}\cdot(-6)\right)^2 = 9 \qquad \left(\frac{1}{2}\cdot(-7)\right)^2 = \frac{49}{4}$$

In each case, notice the relationship between the **coefficient of the linear term** (x-term) and the **constant** term. The constant term of a perfect square trinomial is equal to the square of $\frac{1}{2}$ the **linear coefficient**.

To *complete the square* means to add an appropriate constant so that a binomial of the form $x^2 + bx$ becomes a perfect square trinomial. The appropriate constant is the square of half the linear coefficient, $\left(\frac{1}{2}\cdot b\right)^2$. For example, to complete the square given $x^2 + 10x$, we add $\left(\frac{1}{2}\cdot 10\right)^2 = 5^2 = 25$ so we can write

$$x^2 + 10x + 25 = (x + 5)^2.$$

Example 3 Completing the Square

What number must be added to make the **binomial** a **perfect square trinomial**?

a. $x^2 - 12x$ **b.** $x^2 + 5x$ **c.** $x^2 - \frac{3}{2}x$

Solutions

a. The **linear coefficient** is -12, so we must add $\left(\frac{1}{2}(-12)\right)^2 = (-6)^2 = 36$ to **complete the square.** Thus, the expression $x^2 - 12x + 36$ is a perfect square trinomial and $x^2 - 12x + 36 = (x - 6)^2$.

b. The linear coefficient is 5, so we must add $\left(\frac{1}{2}\cdot 5\right)^2 = \left(\frac{5}{2}\right)^2 = \frac{25}{4}$ to complete the square: $x^2 + 5x + \frac{25}{4} = \left(x + \frac{5}{2}\right)^2$.

My video summary

c. Try to work this problem on your own. View the **answer,** or watch this **video** for the complete solution.

You Try It **Work through this You Try It problem.**

Work Exercises 11–14 in this textbook or in the MyMathLab Study Plan.

When writing **the perfect square trinomial** as a **binomial** squared, note that the first term of the binomial is x (the variable) and the second term is $\frac{1}{2}$ the **linear coefficient** from the trinomial. Consider the perfect square trinomials we saw earlier:

$$x^2 + 2x + 1 = (x + 1)^2 \qquad \left(\frac{1}{2}\cdot 2\right) = 1$$

$$x^2 - 6x + 9 = (x - 3)^2 \qquad \left(\frac{1}{2}\cdot(-6)\right) = -3$$

$$x^2 - 7x + \frac{49}{4} = \left(x - \frac{7}{2}\right)^2 \qquad \left(\frac{1}{2}\cdot(-7)\right) = -\frac{7}{2}$$

To solve a **quadratic equation** of the form $ax^2 + bx + c = 0$ by **completing the square,** where a, b, and c are real numbers, and $a \neq 0$, use the following guidelines:

Solving $ax^2 + bx + c = 0, a \neq 0$, by Completing the Square

Step 1. If $a \neq 1$, divide both sides of the equation by a.

Step 2. Move all constants to the right-hand side.

Step 3. Find $\frac{1}{2}$ times the coefficient of the x-term, **square** it, and add the result to both sides of the equation.

Step 4. The left-hand side is now a **perfect square.** Rewrite it as a binomial squared.

Step 5. Use the **square root property** and solve for x.

Example 4 Solving a Quadratic Equation by Completing the Square

Solve $x^2 - 8x + 2 = 0$ by **completing the square.**

Solution

Step 1. The **leading coefficient** is 1, so we proceed to Step 2.

Step 2. Move all constants to the right-hand side.

$$x^2 - 8x = -2$$

Step 3. Multiply $\frac{1}{2}$ times the coefficient of the x-term, square the result, and add this to both sides of the equation.

$$\left(\frac{1}{2}\cdot(-8)\right)^2 = (-4)^2 = 16 \quad \rightarrow \quad x^2 - 8x + 16 = -2 + 16$$

$$x^2 - 8x + 16 = 14$$

Step 4. Rewrite the left-hand side as a perfect square: $(x - 4)^2 = 14$

Step 5. Use the **square root property** and solve for x.

Use the square root property: $x - 4 = \pm\sqrt{14}$

Add 4 to both sides: $x = 4 \pm \sqrt{14}$

The solution set is $\{4 - \sqrt{14}, 4 + \sqrt{14}\}$.

Example 5 Solving a Quadratic Equation by Completing the Square

Solve $2x^2 - 10x - 6 = 0$ by **completing the square.**

Solution

Step 1. Original equation: $2x^2 - 10x - 6 = 0$

Divide both sides by 2: $\frac{2x^2}{2} - \frac{10x}{2} - \frac{6}{2} = \frac{0}{2}$

Simplify: $x^2 - 5x - 3 = 0$

Step 2. Add 3 to both sides: $x^2 - 5x = 3$

Step 3. $\left(\frac{1}{2}\cdot(-5)\right)^2 = \left(-\frac{5}{2}\right)^2 = \frac{25}{4} \quad \rightarrow \quad x^2 - 5x + \frac{25}{4} = 3 + \frac{25}{4}$

$$x^2 - 5x + \frac{25}{4} = \frac{12}{4} + \frac{25}{4}$$

$$x^2 - 5x + \frac{25}{4} = \frac{37}{4}$$

Step 4. Write the left-hand side as a **binomial** squared: $\left(x - \frac{5}{2}\right)^2 = \frac{37}{4}$

Step 5. Use the **square root property:** $x - \frac{5}{2} = \pm\sqrt{\frac{37}{4}}$

Apply $\sqrt{\frac{a}{b}} = \frac{\sqrt{a}}{\sqrt{b}}$: $x - \frac{5}{2} = \pm\frac{\sqrt{37}}{\sqrt{4}}$

Simplify: $x - \frac{5}{2} = \pm\frac{\sqrt{37}}{2}$

Add $\frac{5}{2}$ to both sides and simplify: $x = \frac{5}{2} \pm \frac{\sqrt{37}}{2} = \frac{5 \pm \sqrt{37}}{2}$

The solution set is $\left\{\frac{5 - \sqrt{37}}{2}, \frac{5 + \sqrt{37}}{2}\right\}$.

(eText Screens 19.1-1–19.1-28)

Example 6 Solving a Quadratic Equation by Completing the Square

My video summary

Solve $3x^2 - 18x + 19 = 0$ by **completing the square.**

Solution Try solving this equation on your own. View the **answer**, or watch this **video** for the complete solution.

You Try It **Work through this You Try It problem.**

Work Exercises 15–20 in this textbook or in the MyMathLab Study Plan.

OBJECTIVE 3 SOLVE QUADRATIC EQUATIONS USING THE QUADRATIC FORMULA

My animation summary

We can solve any **quadratic equation** by **completing the square**. However, this process can be very time consuming. If we solve the general quadratic equation $ax^2 + bx + c = 0$ by completing the square, where a, b, and c are real numbers, and $a \neq 0$, we obtain a useful result known as the **quadratic formula.** The quadratic formula can be used to solve any quadratic equation and is often less time-consuming than completing the square. Work through this **animation** to see how to derive the quadratic formula by using completing the square to solve the general quadratic equation.

Quadratic Formula

The **solutions** to the quadratic equation $ax^2 + bx + c = 0$, $a \neq 0$, are given by the following formula:

$$x = \frac{-b \pm \sqrt{b^2 - 4ac}}{2a}$$

CAUTION

Remember to write the quadratic equation in standard form before identifying the coefficients a, b, and c.

Example 7 Solving a Quadratic Equation Using the Quadratic Formula

Solve $2x^2 - 3x = 2$ using the **quadratic formula.**

Solution First, we write the equation in **standard form.**

$$2x^2 - 3x - 2 = 0$$

Next, we identify the **coefficients.**

$$2x^2 - 3x - 2 = 0$$

$$a = 2 \quad b = -3 \quad c = -2$$

Substitute 2 for a, -3 for b, and -2 for c in the **quadratic formula.**

Quadratic formula: $x = \dfrac{-b \pm \sqrt{b^2 - 4ac}}{2a}$

Substitute values for $a, b,$ and c: $= \dfrac{-(-3) \pm \sqrt{(-3)^2 - 4(2)(-2)}}{2(2)}$

Simplify: $= \dfrac{3 \pm \sqrt{9 + 16}}{4}$

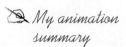

$= \dfrac{3 \pm \sqrt{25}}{4} = \dfrac{3 \pm 5}{4}$

(*eText Screens 19.1-1–19.1-28*)

There are two **solutions** to the equation, one from the + sign and one from the − sign.

$$x = \frac{3+5}{4} = \frac{8}{4} = 2 \quad \text{or} \quad x = \frac{3-5}{4} = \frac{-2}{4} = -\frac{1}{2}$$

The **solution set** is $\left\{-\frac{1}{2}, 2\right\}$.

Example 8 Solving a Quadratic Equation Using the Quadratic Formula

My video summary

Solve $3x^2 + 2x - 2 = 0$ using the **quadratic formula.**

Solution The equation is in **standard form**, so we begin by identifying the **coefficients.**

$$3x^2 + 2x - 2 = 0$$

$$a = 3 \quad b = 2 \quad c = -2$$

Substitute 3 for a, 2 for b, and −2 for c in the quadratic formula.

Quadratic formula: $x = \dfrac{-b \pm \sqrt{b^2 - 4ac}}{2a}$

Substitute values for a, b, and c: $= \dfrac{-(2) \pm \sqrt{(2)^2 - 4(3)(-2)}}{2(3)}$

Simplify: $= \dfrac{-2 \pm \sqrt{4 + 24}}{6}$

Finish simplifying to find the two **solutions.** Remember to **simplify the radical** first and then divide out any **common factors.** View the **answer**, or watch this **video** for the complete solution.

Example 9 Solving a Quadratic Equation Using the Quadratic Formula

Solve $4x^2 = x - 6$ using the **quadratic formula.**

Solution First, we write the equation in **standard form.**

$$4x^2 - x + 6 = 0$$

Next we identify the **coefficients.**

$$4x^2 - 1x + 6 = 0$$

$$a = 4 \quad b = -1 \quad c = 6$$

Substitute 4 for a, −1 for b, and 6 for c in the quadratic formula.

$$x = \frac{-b \pm \sqrt{b^2 - 4ac}}{2a} = \frac{-(-1) \pm \sqrt{(-1)^2 - 4(4)(6)}}{2(4)} = \frac{1 \pm \sqrt{1 - 96}}{8} = \frac{1 \pm \sqrt{-95}}{8}$$

The **radicand** is negative, so the solutions will be non-real **complex numbers.**

$$x = \frac{1 \pm \sqrt{-1} \cdot \sqrt{95}}{8} = \frac{1 \pm i\sqrt{95}}{8}$$

The solution set, in **standard form** $a + bi$, is $\left\{\frac{1}{8} - \frac{\sqrt{95}}{8}i, \frac{1}{8} + \frac{\sqrt{95}}{8}i\right\}$.

(eText Screens 19.1-1–19.1-28)

Example 10 Solving a Quadratic Equation Using the Quadratic Formula

My video summary

Solve $14x^2 - 5x = 5x^2 + 7x - 4$ using the **quadratic formula**.

Solution Try to solve this equation on your own. Remember to write the equation in **standard form** first. View the answer, or watch this **video** for the complete solution.

You Try It **Work through this You Try It problem.**

Work Exercises 21–26 in this textbook or in the MyMathLab Study Plan.

OBJECTIVE 4 USE THE DISCRIMINANT TO DETERMINE THE NUMBER AND TYPE OF SOLUTIONS TO A QUADRATIC EQUATION

In Example 9, the **quadratic equation** $4x^2 = x - 6$ had two non-real solutions. The solutions were non-real because the expression $b^2 - 4ac$ under the **radical** was a negative number. Given a quadratic equation of the form $ax^2 + bx + c = 0$, the expression $b^2 - 4ac$ is called the **discriminant.** Knowing the value of the discriminant can help us determine the number and type of **solutions** to a quadratic equation.

Discriminant

Given a quadratic equation $ax^2 + bx + c = 0$, $a \neq 0$, the expression $D = b^2 - 4ac$ is called the **discriminant.**

If $D > 0$, then the quadratic equation has two real solutions.
If $D < 0$, then the quadratic equation has two non-real solutions.
If $D = 0$, then the quadratic equation has exactly one real solution.

Example 11 Using the Discriminant

My video summary

Use the **discriminant** to determine the number and type of solutions to each of the following **quadratic equations**.

a. $3x^2 + 2x + 2 = 0$ **b.** $4x^2 + 1 = 4x$

Solutions

a. The equation $3x^2 + 2x + 2 = 0$ is in **standard form** with $a = 3$, $b = 2$, and $c = 2$.

$$D = (2)^2 - 4(3)(2) = 4 - 24 = -20 < 0$$

Because the discriminant is less than zero, there are two non-real solutions.

b. The equation $4x^2 + 1 = 4x$ is not written in **standard form**. To write the equation in standard form, subtract $4x$ from both sides:

$$4x^2 - 4x + 1 = 0$$

Now the equation is in standard form with $a = 4$, $b = -4$, and $c = 1$.

$$D = (-4)^2 - 4(4)(1) = 16 - 16 = 0$$

Because the discriminant is zero, there is exactly one real solution.

You Try It **Work through this You Try It problem.**

Work Exercises 27–30 in this textbook or in the MyMathLab Study Plan.

OBJECTIVE 5 SOLVE EQUATIONS THAT ARE QUADRATIC IN FORM

We have now seen several different techniques for solving **quadratic equations.** View this **popup box** for a summary along with some advantages and disadvantages for each approach.

Sometimes equations that are not quadratic can be changed into a **quadratic equation** by using **substitution**. Equations of this type are said to be **quadratic in form** because they have the form $au^2 + bu + c = 0,\ a \neq 0$, after appropriate substitutions are made. Table 1 shows how to write such equations in quadratic form.

Original Equation		Make an Appropriate Substitution		New Equation Is a Quadratic
$2x^4 - 11x^2 + 12 = 0$	Determine the proper substitution $\longrightarrow$	Let $u = x^2$, then $u^2 = x^4$.	New equation is quadratic $\longrightarrow$	$2u^2 - 11u + 12 = 0$
$\left(\frac{1}{x-2}\right)^2 + \frac{2}{x-2} - 15 = 0$	Determine the proper substitution $\longrightarrow$	Let $u = \frac{1}{x-2}$, then $u^2 = \left(\frac{1}{x-2}\right)^2$.	New equation is quadratic $\longrightarrow$	$u^2 + 2u - 15 = 0$
$x^{2/3} - 9x^{1/3} + 8 = 0$	Determine the proper substitution $\longrightarrow$	Let $u = x^{1/3}$, then $u^2 = \left(x^{1/3}\right)^2 = x^{2/3}$.	New equation is quadratic $\longrightarrow$	$u^2 - 9u + 8 = 0$
$3x^{-2} - 5x^{-1} - 2 = 0$	Determine the proper substitution $\longrightarrow$	Let $u = x^{-1} = \frac{1}{x}$, then $u^2 = \left(x^{-1}\right)^2 = x^{-2}$.	New equation is quadratic $\longrightarrow$	$3u^2 - 5u - 2 = 0$

Table 1

In Example 12, we solve each of the equations listed in Table 1.

Example 12 Solving Equations That Are Quadratic in Form

My interactive video summary

Solve each equation.

a. $2x^4 - 11x^2 + 12 = 0$

b. $\left(\frac{1}{x-2}\right)^2 + \frac{2}{x-2} - 15 = 0$

c. $x^{\frac{2}{3}} - 9x^{\frac{1}{3}} + 8 = 0$

d. $3x^{-2} - 5x^{-1} - 2 = 0$

Solutions

a. To solve the equation $2x^4 - 11x^2 + 12 = 0$, we look for an appropriate **substitution.**

The middle term contains x^2, and we can rewrite the first term as $2\left(x^2\right)^2$ using **rules for exponents.** Therefore, an appropriate substitution would be $u = x^2$. Letting $u = x^2$, we get $u^2 = \left(x^2\right)^2 = x^4$. Substituting u for x^2 and u^2 for x^4, our equation becomes

$$2u^2 - 11u + 12 = 0,$$

which is a **quadratic equation,** in terms of u. We can solve this quadratic equation for u by using any of the approaches previously discussed. For this example, we will solve the quadratic equation by factoring.

$$\text{Quadratic equation:}\quad 2u^2 - 11u + 12 = 0$$
$$\text{Factor:}\quad (2u - 3)(u - 4) = 0$$

We now apply the **zero product property** and solve the resulting equations for u.

$$2u - 3 = 0 \quad \text{or} \quad u - 4 = 0$$
$$2u = 3 \qquad\qquad u = 4$$
$$u = \frac{3}{2}$$

(eText Screens 19.1-1–19.1-28)

Here we may be tempted to form a **solution set**. However, we have only solved for u. Since the original equation involved the variable x, we still need to solve for x. Because we said that $u = x^2$, we get

$$x^2 = \frac{3}{2} \quad \text{or} \quad x^2 = 4.$$

We solve these two equations for x by applying the **square root property**.

$$x^2 = \frac{3}{2} \qquad\qquad \text{or} \quad x^2 = 4$$

$$x = \underbrace{\pm\sqrt{\frac{3}{2}}}_{\text{Rationalize the denominator}} = \pm\frac{\sqrt{6}}{2} \qquad\qquad x = \pm\sqrt{4} = \pm 2$$

The solution set is $\left\{\pm 2, \pm\frac{\sqrt{6}}{2}\right\}$.

b.–d. Try to solve these equations on your own. As a hint, look back at the suggested **substitutions** given earlier in **Table 1**. View the **answers**, or watch this **interactive video** to see complete solutions.

You Try It **Work through this You Try It problem.**

Work Exercises 31–37 in this textbook or in the MyMathLab Study Plan.

19.1 Exercises

In Exercises 1–10, solve each equation using the square root property.

You Try It

1. $x^2 = 256$ **2.** $y^2 = 80$ **3.** $m^2 = \frac{25}{64}$ **4.** $p^2 = \frac{20}{49}$

5. $x^2 - 64 = 0$ **6.** $x^2 + 64 = 0$ **7.** $3x^2 = 72$ **8.** $(x + 2)^2 - 9 = 0$

You Try It

9. $(2x + 1)^2 + 4 = 0$ **10.** $3(x - 4)^2 + 2 = 8$

In Exercises 11–14, decide what number must be added to each binomial to make a perfect square trinomial.

You Try It **11.** $x^2 - 8x$ **12.** $x^2 + 10x$ **13.** $x^2 - 7x$ **14.** $x^2 + \frac{5}{3}x$

In Exercises 15–20, solve each quadratic equation by completing the square.

You Try It **15.** $x^2 - 8x - 2 = 0$ **16.** $x^2 + 7x + 14 = 0$ **17.** $2x^2 + 8 = -6x$

18. $4z^2 + 10 = 16z$ **19.** $3x^2 = 7 - 24x$ **20.** $3x^2 + 5x + 12 = 0$

(eText Screens 19.1-1–19.1-28)

In Exercises 21–26, solve each quadratic equation using the quadratic formula.

21. $3x^2 + 8x - 3 = 0$ **22.** $x^2 - 8x - 2 = 0$ **23.** $4x^2 - x + 8 = 0$

You Try It

24. $3x^2 = 1 + 4x$ **25.** $9x^2 - 6x = -1$ **26.** $5x^2 + 3x + 1 = 0$

In Exercises 27–30, use the discriminant to determine the number and type of the solutions to each quadratic equation. Do not solve the equations.

You Try It

27. $x^2 + 3x + 1 = 0$ **28.** $4x^2 + 4x + 1 = 0$

29. $2x^2 + x = 5$ **30.** $3x^2 + \sqrt{12}x + 4 = 0$

In Exercises 31–37, solve the equation after making an appropriate substitution.

You Try It

31. $x^4 - 6x^2 + 8 = 0$ **32.** $(13x - 1)^2 - 2(13x - 1) - 3 = 0$

33. $2x^{\frac{2}{3}} - 5x^{\frac{1}{3}} + 2 = 0$ **34.** $x^6 - x^3 = 6$

35. $\sqrt{x} - 3\sqrt[4]{x} - 4 = 0$ **36.** $3\left(\dfrac{1}{x-1}\right)^2 - \dfrac{5}{x-1} - 2 = 0$

37. $2x^{-2} - 3x^{-1} - 2 = 0$

19.2 Quadratic Functions and Their Graphs

THINGS TO KNOW

Before working through this topic, be sure you are familiar with the following concepts:

	Concept	VIDEO	ANIMATION	INTERACTIVE
You Try It	1. Find x- and y-Intercepts (Topic 12.2, **Objective 4**)	▣		
You Try It	2. Solve Quadratic Equations by Factoring (Topic 15.7, **Objective 1**)	▣		
You Try It	3. Find the Domain and Range of a Relation (Topic 17.1, **Objective 2**)			▣
You Try It	4. Solve Quadratic Equations by Completing the Square (Topic 19.1, **Objective 2**)	▣		
You Try It	5. Solve Quadratic Equations Using the Quadratic Formula (Topic 19.1, **Objective 3**)	▣		

OBJECTIVES

1 Identify the Characteristics of a Quadratic Function from Its Graph

2 Graph Quadratic Functions by Using Translations

3 Graph Quadratic Functions of the Form $f(x) = a(x - h)^2 + k$

4 Find the Vertex of a Quadratic Function by Completing the Square

5 Graph Quadratic Functions of the Form $f(x) = ax^2 + bx + c$ by Completing the Square

6 Find the Vertex of a Quadratic Function by Using the Vertex Formula

7 Graph Quadratic Functions of the Form $f(x) = ax^2 + bx + c$ by Using the Vertex Formula

OBJECTIVE 1 IDENTIFY THE CHARACTERISTICS OF A QUADRATIC FUNCTION FROM ITS GRAPH

In the previous topic, we learned how to solve **quadratic equations.** In this topic, we learn about *graphing quadratic functions*.

Definition Quadratic Function

A **quadratic function** is a second-degree **polynomial function** of the form $f(x) = ax^2 + bx + c$, where a, b, and c are real numbers and $a \neq 0$. Every quadratic function has a "u-shaped" graph called a **parabola.**

The function $f(x) = x^2$ is a quadratic function with $a = 1$, $b = 0$, and $c = 0$. Its graph is shown in Figure 1a. The function $g(x) = -x^2$ is a quadratic function with $a = -1$, $b = 0$, and $c = 0$. Its graph is shown in Figure 1b. Notice that both graphs are **parabolas** and have the characteristic "u-shape."

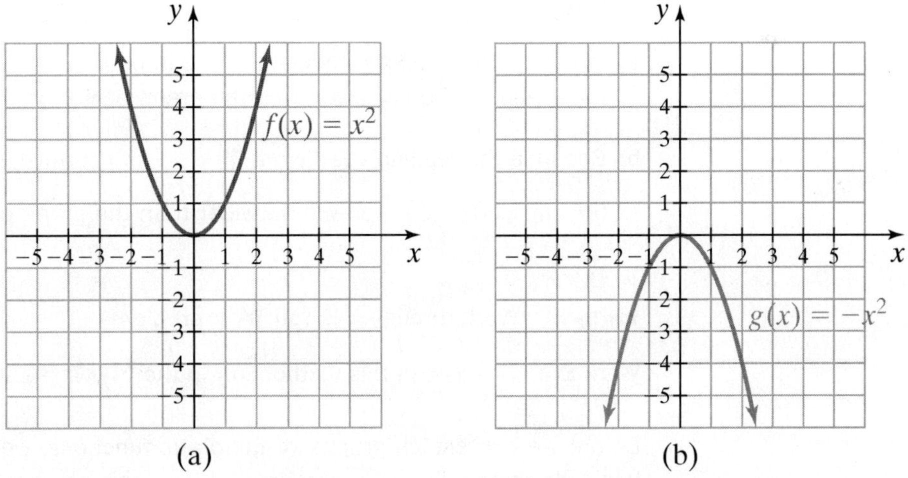

Figure 1
Graphs of $f(x) = x^2$ and $g(x) = -x^2$

A **parabola** either opens up (called **concave up**) or opens down (called **concave down**) depending on the **leading coefficient,** a. If $a > 0$, as in Figure 1a, the parabola will "open up." If $a < 0$, as in Figure 1b, the parabola will "open down."

The **leading coefficient** also affects the shape of the **parabola**. The function $f(x) = 2x^2$ is a **quadratic function** with $a = 2$, $b = 0$, and $c = 0$. Its graph is shown in Figure 2a. The function $g(x) = \frac{1}{2}x^2$ is a quadratic function with $a = \frac{1}{2}$, $b = 0$, and $c = 0$. Its graph is shown in Figure 2b. The graph of $y = x^2$ is shown in gray.

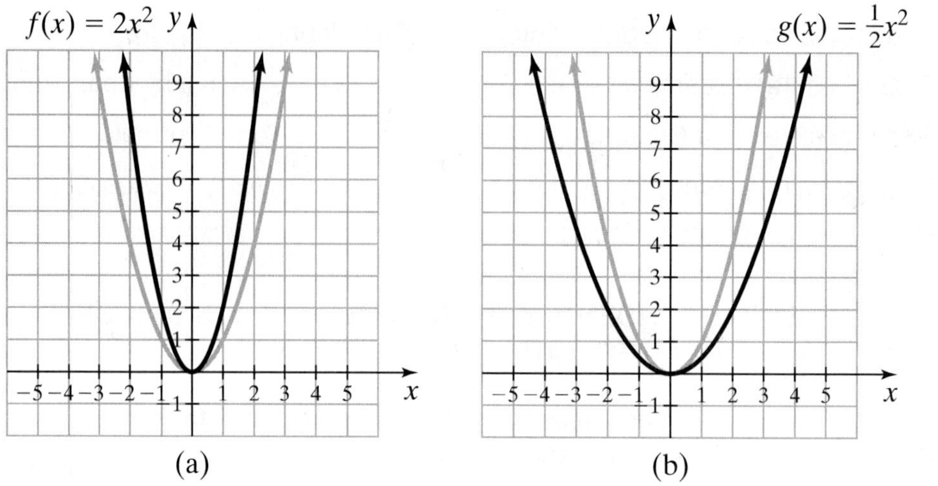

Figure 2

If $|a| > 1$, as in Figure 2a, the graph will be narrower than the graph of $y = x^2$. If $0 < |a| < 1$, as in Figure 2b, the graph will be wider than the graph of $y = x^2$.

Example 1 Determining the Shape of the Graph of a Quadratic Function

Without graphing, determine if the graph of each **quadratic function** opens up or down. Also determine if the graph will be wider or narrower than the graph of $y = x^2$.

a. $f(x) = 2x^2 - 3x + 5$ **b.** $g(x) = -\frac{2}{3}x^2 + 5x - 7$

Solutions

a. Because the **leading coefficient** $a = 2$ is positive, the graph will open up. Since $|a| > 1$, the graph will be narrower than the graph of $y = x^2$.

b. Because the **leading coefficient** $a = -\frac{2}{3}$ is negative, the graph will open down. Since $0 < |a| < 1$, the graph will be wider than the graph of $y = x^2$.

You Try It **Work through this You Try It problem.**

Work Exercises 1–4 in this textbook or in the MyMathLab Study Plan.

Before we can sketch graphs of **quadratic functions**, we must be able to identify the five basic characteristics of a **parabola**: *vertex, axis of symmetry, y-intercept, x-intercepts,* and *domain and range.*

Every parabola has a **vertex.** If the parabola "opens up," the vertex (h, k) is the lowest point on the graph and the function will have a **minimum value.** The minimum value is the smallest possible value for the function and is given by the y-coordinate of the vertex, k. If the parabola "opens down," the vertex (h, k) is the highest point on the graph and the function will have a **maximum value.** The maximum value is the largest possible value for the function and is given by the y-coordinate of the vertex, k. In either case, the y-coordinate of the vertex indicates the maximum or minimum value and the x-coordinate indicates *where* this occurs.

The **axis of symmetry** is an imaginary **vertical line** that passes through the vertex and divides the graph into two mirror images. Points on the graph that are the same horizontal distance from the axis of symmetry will have the same y-coordinate.

The **domain** of every quadratic function is all real numbers $(-\infty, \infty)$. The **range** is determined by k, the y-coordinate of the vertex, and whether the graph opens up or down. If the graph opens up, the range is all real numbers greater than or equal to k, or $[k, \infty)$. If the graph opens down, the range is all real numbers less than or equal to k, or $(-\infty, k]$.

The graph of every quadratic function $f(x)$ crosses the y-axis, so every parabola has a y-**intercept**. However, it may or may not have any x-**intercepts**. We find these intercepts in the usual way.

We summarize the five basic characteristics with the following:

My animation summary

1. **Vertex**
2. **Axis of symmetry**
3. y**-intercept**
4. x**-intercept(s) or real zeros**
5. **Domain and range**

Work through this animation to get a detailed description of each characteristic.

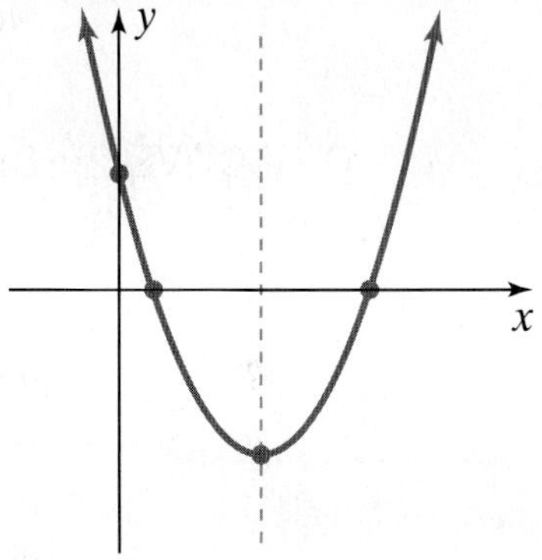

Example 2 Finding the Characteristics of a Quadratic Function from Its Graph

Use the given graph of a **quadratic function** to find the following:

a. **Vertex** **b.** **Axis of symmetry** **c.** y**-intercept**
d. x**-intercept(s)** **e.** **Domain and range**

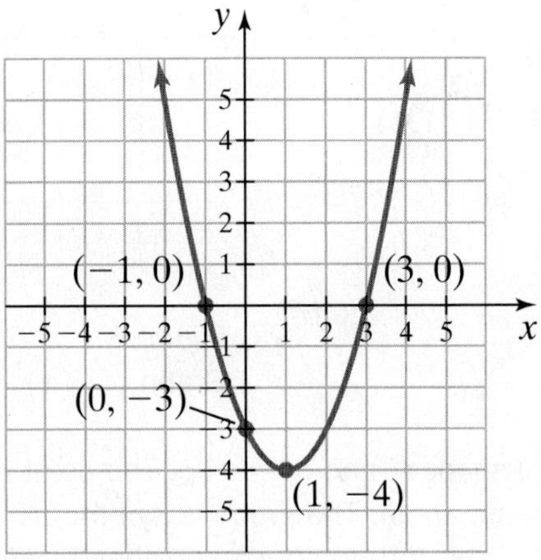

Solutions

a. Because the graph "opens up," the vertex is the lowest point on the graph: $(h, k) = (1, -4)$.

b. The **axis of symmetry** is the vertical line that passes through the **vertex**. Since the x-coordinate of the vertex is $h = 1$, the equation of the axis of symmetry is $x = 1$.

c. The **y-intercept** is the y-coordinate of the point where the graph crosses the y-axis. The graph crosses the y-axis at the point $(0, -3)$, so the y-intercept is -3.

d. The **x-intercepts** are the x-coordinates of points where the graph crosses or touches the x-axis. The graph crosses the x-axis at the points $(-1, 0)$ and $(3, 0)$, so the x-intercepts are -1 and 3.

e. The **domain** of every quadratic function is all real numbers, or $(-\infty, \infty)$. To determine the **range**, first note that the graph has a minimum value because it opens up. The minimum value is the y-coordinate of the vertex, $k = -4$. So, the range is all real numbers greater than or equal to -4, or $[-4, \infty)$ in interval notation.

You Try It **Work through this You Try It problem.**

Work Exercises 5 and 6 in this textbook or in the MyMathLab Study Plan.

OBJECTIVE 2 GRAPH QUADRATIC FUNCTIONS BY USING TRANSLATIONS

In mathematics, a **translation** is when every point on a graph is shifted the same distance in the same direction. We now examine translations of parabolas involving vertical or horizontal shifts.

In **Topic 17.3**, we **graphed** simple functions by plotting points. We can use the same technique to graph the basic **quadratic function** $f(x) = x^2$. Table 2 gives some **ordered pairs**, and Figure 3 shows the resulting graph.

x	$y = x^2$	(x, y)
-3	$(-3)^2 = 9$	$(-3, 9)$
-2	$(-2)^2 = 4$	$(-2, 4)$
-1	$(-1)^2 = 1$	$(-1, 1)$
0	$(0)^2 = 0$	$(0, 0)$
1	$(1)^2 = 1$	$(1, 1)$
2	$(2)^2 = 4$	$(2, 4)$
3	$(3)^2 = 9$	$(3, 9)$

Table 2

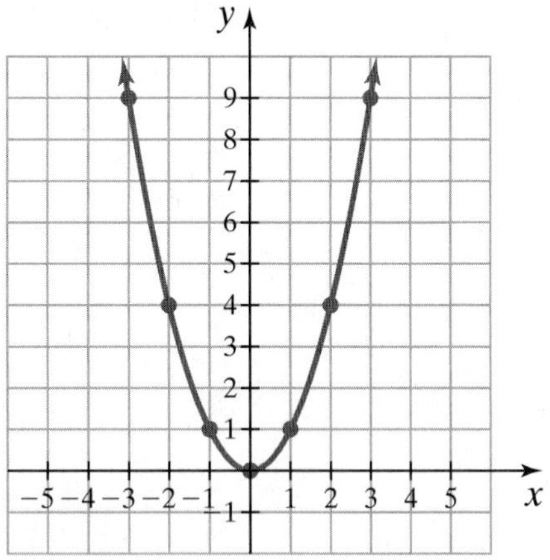

Figure 3
Graph of $f(x) = x^2$

Notice that the graph is "u-shaped" and opens up. The point $(0, 0)$, the lowest point on the graph, is the **vertex** of this **parabola**. The graph is **symmetric** about the vertical line $x = 0$. The **domain** is $(-\infty, \infty)$. From the graph we see that y can be any real number greater than or equal to 0, so the **range** is $[0, \infty)$. Similarly, we can use point plotting to graph other **quadratic functions**.

Example 3 Graphing a Quadratic Function with a Vertical Shift

Sketch the graph of $g(x) = x^2 + 2$. Compare the graph to the graph of $f(x) = x^2$.

Solution Table 3 shows that for every value of x, the y-coordinate of the function g is always 2 greater than the y-coordinate for f. The two functions are graphed in Figure 4. The graph of $g(x) = x^2 + 2$ is exactly the same as the graph of $f(x) = x^2$, except the graph of g is shifted *up* two units.

The point $(0, 2)$, now the lowest point on the graph, is the vertex of this parabola. The graph of g is symmetric about the vertical line $x = 0$. The domain is $(-\infty, \infty)$. From the graph, we see that y can be any real number greater than or equal to 2, so the **range** is $[2, \infty)$. Notice how the vertical shift affected the y-coordinate of the **vertex** and the range, but did not affect the x-coordinate of the vertex.

x	$f(x) = x^2$	$g(x) = x^2 + 2$
-2	$(-2)^2 = 4$	$(-2)^2 + 2 = 4 + 2 = 6$
-1	$(-1)^2 = 1$	$(-1)^2 + 2 = 1 + 2 = 3$
0	$(0)^2 = 0$	$(0)^2 + 2 = 0 + 2 = 2$
1	$(1)^2 = 1$	$(1)^2 + 2 = 1 + 2 = 3$
2	$(2)^2 = 4$	$(2)^2 + 2 = 4 + 2 = 6$

Table 3

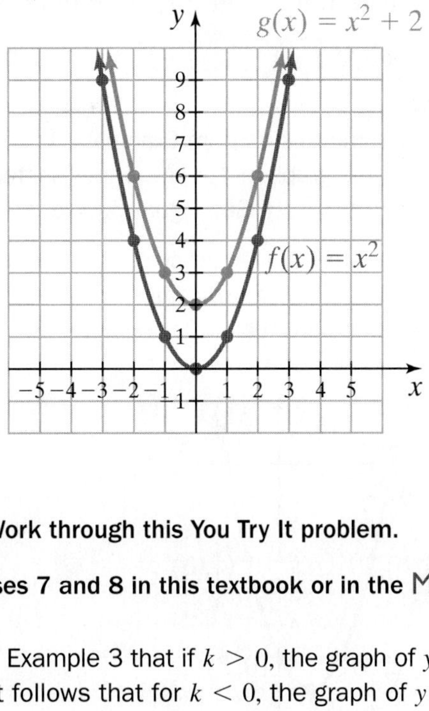

Figure 4
Graph of $g(x) = x^2 + 2$

You Try It **Work through this You Try It problem.**

Work Exercises 7 and 8 in this textbook or in the MyMathLab Study Plan.

We see from Example 3 that if $k > 0$, the graph of $y = x^2 + k$ is the graph of $y = x^2$ shifted *up* k units. It follows that for $k < 0$, the graph of $y = x^2 + k$ is the graph of $y = x^2$ shifted *down* k units.

> **Vertical Shifts of Quadratic Functions**
>
> The graph of $y = x^2 + k$ is a **parabola** with the same shape as the graph of $y = x^2$, but it is shifted vertically $|k|$ units. The graph is shifted up if $k > 0$ and down if $k < 0$. The **vertex** of the parabola is $(0, k)$.

(eText Screens 19.2-1–19.2-36)

Example 4 Graphing a Quadratic Function with a Horizontal Shift

My video summary

Sketch the graph of $g(x) = (x - 2)^2$. Compare this graph to the graph of $f(x) = x^2$.

Solution Construct a table of values to find points on the graph. Use the points to sketch the graph. See this popup box for a sample table of values and to check your graph. Watch this **video** for a detailed solution. Confirm that the graph of $g(x) = (x - 2)^2$ is the same as the graph of $f(x) = x^2$ but shifted two units to the right.

You Try It **Work through this You Try It problem.**

Work Exercises 9 and 10 in this textbook or in the MyMathLab Study Plan.

We see from Example 4 that if $h > 0$, the graph of $y = (x - h)^2$ is the graph of $y = x^2$ shifted *right* h units. It follows that for $h < 0$, the graph of $y = (x - h)^2$ is the graph of $y = x^2$ shifted *left* h units.

Horizontal Shifts of Quadratic Functions

The graph of $y = (x - h)^2$ is a **parabola** with the same shape as the graph of $y = x^2$, but it is shifted horizontally $|h|$ units. The graph is shifted right if $h > 0$ and left if $h < 0$. The **vertex** of the parabola is $(h, 0)$.

Example 5 Graphing a Quadratic Function with a Horizontal and Vertical Shift

Sketch the graph of $g(x) = (x + 1)^2 + 2$. Compare this graph to the graph of $f(x) = x^2$.

Solution Writing $g(x) = (x + 1)^2 + 2 = (x - (-1))^2 + 2$ we see that $h = -1$ and $k = 2$. Based on the results from Examples 3 and 4, the graph of $g(x) = (x + 1)^2 + 2$ will be the same as the graph of $f(x) = x^2$ but shifted left one unit (because $h = -1$) and up two units (because $k = 2$). See Figure 5. See this popup box for a sample table of values.

The point $(-1, 2)$, now the lowest point on the graph, is the **vertex** of this **parabola**. The graph of g is **symmetric** about the vertical line $x = -1$. The **domain** is $(-\infty, \infty)$. From the graph, we see that y can be any real number greater than or equal to 2, so the **range** is $[2, \infty)$.

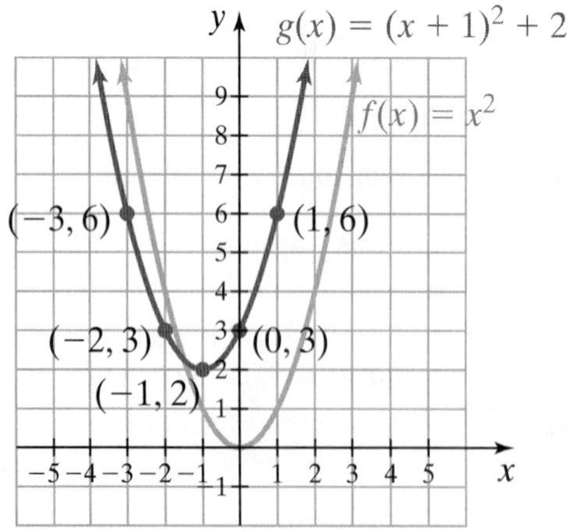

Figure 5
Graph of $g(x) = (x + 1)^2 + 2$

You Try It **Work through this You Try It problem.**

Work Exercises 11 and 12 in this textbook or in the MyMathLab Study Plan.

OBJECTIVE 3 GRAPH QUADRATIC FUNCTIONS OF THE FORM $f(x) = a(x - h)^2 + k$

The **quadratic function** $f(x) = \frac{1}{2}(x - 3)^2 - 2$ has **vertex** $(3, -2)$ and is written in *standard form*.

> **Standard Form of a Quadratic Function**
>
> A quadratic function is in **standard form** if it is written as $f(x) = a(x - h)^2 + k$. The graph is a **parabola** with vertex (h, k). The parabola "opens up" if $a > 0$ or "opens down" if $a < 0$.

Since we can easily determine the coordinates of the vertex (h, k) when written this way, standard form is sometimes called **vertex form.**

Example 6 Graphing a Quadratic Function in the Form $f(x) = a(x - h)^2 + k$

Given the **quadratic function** $f(x) = \frac{1}{2}(x + 3)^2 - 2$, answer the following:

a. What are the coordinates of the **vertex**?

b. Does the graph "open up" or "open down"?

c. What is the equation of the **axis of symmetry**?

d. Find any **x-intercepts.**

e. Find the **y-intercept.**

f. Sketch the graph.

g. State the **domain** and **range** in **interval notation.**

Solutions

a. First we change $x + 3$ to $x - (-3)$ to write the function in **standard form.** Then we find the **vertex** by determining the values for h and k.

$$f(x) = \frac{1}{2}(x + 3)^2 - 2 \quad \rightarrow \quad f(x) = \frac{1}{2}(\overbrace{x - (-3)}^{x + 3})^2 + (-2)$$

$$a = \frac{1}{2} \qquad h = -3 \qquad k = -2$$

$$f(x) = a\,(x - h)^2 + k$$

Because $h = -3$ and $k = -2$, the vertex for this **parabola** is $(-3, -2)$.

b. Since the **leading coefficient** is $a = \frac{1}{2} > 0$, the parabola opens up.

c. The x-coordinate of the vertex is -3, so the equation of the **axis of symmetry** is $x = -3$.

d. We determine the x-**intercepts** (if any) by finding the real solutions to the equation $f(x) = 0$.

Write the original function:	$f(x) = \frac{1}{2}(x+3)^2 - 2$
Set $f(x)$ equal to 0:	$\frac{1}{2}(x+3)^2 - 2 = 0$
Add 2 to both sides:	$\frac{1}{2}(x+3)^2 = 2$
Multiply both sides by 2:	$(x+3)^2 = 4$
Square root property:	$x + 3 = \pm 2$
Subtract 3 from both sides:	$x = -3 \pm 2$

The x-intercepts are $-3 - 2 = -5$ and $-3 + 2 = -1$.

e. The y-**intercept** is found by evaluating $f(0)$.

$$f(0) = \frac{1}{2}(0+3)^2 - 2$$
$$= \frac{1}{2}(3)^2 - 2 = \frac{9}{2} - 2 = \frac{5}{2}$$

The y-intercept is $\frac{5}{2}$.

f. Before sketching a graph of the function, let's summarize what we know: The **vertex** is $(h, k) = (-3, -2)$; the graph opens up; the equation of the **axis of symmetry** is $x = -3$; there are two x-**intercepts**, -5 and -1; the y-**intercept** is $\frac{5}{2}$. From the vertex and intercepts, we can plot the points $(-3, -2)$, $(-5, 0)$, $(-1, 0)$, and $\left(0, \frac{5}{2}\right)$. The y-intercept lies three units to the right of the axis of symmetry. Because of the symmetry of the graph, we can move three units to left of the axis of symmetry to obtain another point on the graph: $\left(-6, \frac{5}{2}\right)$. Using this information, we can sketch the **parabola**, as shown in Figure 6.

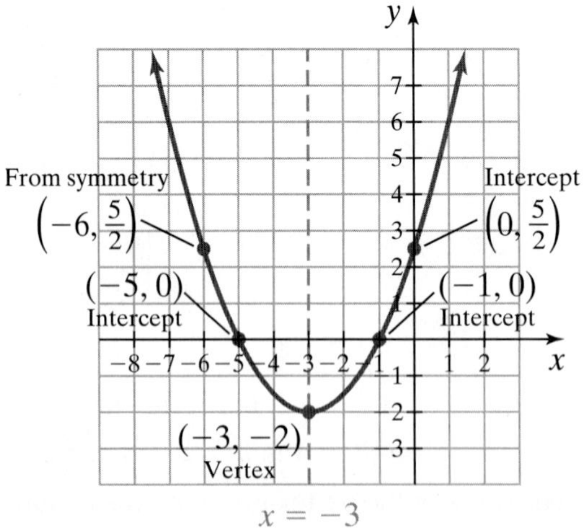

Figure 6

Graph of $f(x) = \frac{1}{2}(x+3)^2 - 2$

g. The **domain** of every **quadratic function** is all real numbers $(-\infty, \infty)$. Because the parabola opens up, the graph has a **minimum** value at the vertex. The minimum value is the y-coordinate of the vertex, $k = -2$. So, the **range** is all real numbers greater than or equal to -2, or $[-2, \infty)$ in **interval notation**.

Example 7 Graphing a Quadratic Function in the Form $f(x) = a(x - h)^2 + k$

My video summary

Given that the **quadratic function** $f(x) = -(x - 2)^2 - 4$ is in **standard form**, answer the following:

a. What are the coordinates of the **vertex**?

b. Does the graph "open up" or "open down"?

c. What is the equation of the **axis of symmetry**?

d. Find any **x-intercepts**.

e. Find the **y-intercept**.

f. Sketch the graph.

g. State the **domain** and **range** in **interval notation**.

Solutions Try to answer the questions on your own. View the answers, or watch this video to see complete solutions.

You Try It **Work through this You Try It problem.**

Work Exercises 13–18 in this textbook or in the MyMathLab Study Plan.

OBJECTIVE 4 FIND THE VERTEX OF A QUADRATIC FUNCTION BY COMPLETING THE SQUARE

In **Topic 17.1** we saw how to use completing the square to **solve a quadratic equation**. The process involved adding the same quantity to both sides of the equation. In this topic, we use **completing the square** to write a quadratic function in **standard form**. Since we are dealing with a **function** instead of an **equation**, we take a slightly different approach. Rather than adding the same quantity to both sides, we will *add and subtract* the same quantity, which is the same as adding 0 to the function.

Writing $f(x) = ax^2 + bx + c$ in Standard Form by Completing the Square

Step 1. Group the variable terms together within parentheses.

Step 2. If $a \neq 1$, factor a out of the **variable terms**.

Step 3. Take half the **coefficient** of the x-term inside the parentheses, **square** it, and *add* it *inside* the parentheses. Multiply this value by a, then *subtract* from c.

Step 4. The expression inside the parentheses is now a **perfect square**. Rewrite it as a **binomial** squared and simplify the **constant** term outside of the parentheses.

Example 8 Writing a Quadratic Function in Standard Form

My video summary

Write the function $f(x) = 2x^2 - 8x + 7$ in **standard form** and find the **vertex**.

Solution Watch this **video**, or use the following steps to **write the function in standard form** by completing the square.

Step 1. $f(x) = (2x^2 - 8x) + 7$

Step 2. $f(x) = 2(x^2 - 4x) + 7$

↑ Factor out a, the coefficient of x^2, from the variable terms.

Step 3. $\left(\frac{1}{2}(-4)\right)^2 = (-2)^2 = 4$

Coefficient of x inside parentheses — Add inside parentheses

We are really adding $2(4) = 8$ to the function. So we must subtract 8 as well.

$$f(x) = 2(x^2 - 4x + 4) + 7 - 8$$

CAUTION: Notice that although we added 4 inside the parentheses, we had to subtract 8 on the outside. Don't forget that a **distributes** to each term inside the parentheses so we must take it into account when determining what to subtract.

Step 4. $f(x) = \underbrace{2(x^2 - 4x + 4)}_{\text{Perfect square trinomial}} + \underbrace{7 - 8}_{\text{Simplify}} \Rightarrow f(x) = 2(x - 2)^2 - 1$

Now we compare the function to the **standard form** $f(x) = a(x - h)^2 + k$, to determine the **vertex**. The vertex is $(h, k) = (2, -1)$.

You Try It **Work through this You Try It problem.**

Work Exercises 19–22 in this textbook or in the MyMathLab Study Plan.

Example 9 Writing a Quadratic Function in Standard Form

My video summary

Write the function $f(x) = -3x^2 - 24x$ in standard form and find the vertex.

Solution The **constant** term is $c = 0$ so we can write $f(x) = -3x^2 - 24x + 0$. Complete the square to **write the function in standard form** and determine the vertex. View the answer, or watch this **video** for the complete solution.

You Try It **Work through this You Try It problem.**

Work Exercise 23 in this textbook or in the MyMathLab Study Plan.

OBJECTIVE 5 GRAPH QUADRATIC FUNCTIONS OF THE FORM $f(x) = ax^2 + bx + c$ BY COMPLETING THE SQUARE

Once we have the function in **standard form**, we can sketch its graph.

Example 10 Graphing a Quadratic Function in the Form $f(x) = ax^2 + bx + c$

My video summary

Rewrite the **quadratic function** $f(x) = 2x^2 - 4x - 3$ in standard form, and then answer the following:

a. What are the coordinates of the **vertex**?
b. Does the graph "open up" or "open down"?
c. What is the equation of the **axis of symmetry**?
d. Find any **x-intercepts**.
e. Find **the y-intercept**.
f. Sketch the graph.
g. State the **domain** and **range** in **interval notation**.

(*eText Screens 19.2-1–19.2-36*)

Solutions Write the function in **standard form** by completing the square then try answering the questions as in **Example 6**. View the answers, or watch this **video** to see each part worked out in detail.

You Try It **Work through this You Try It problem.**

Work Exercises 24–29 in this textbook or in the MyMathLab Study Plan.

OBJECTIVE 6 FIND THE VERTEX OF A QUADRATIC FUNCTION BY USING THE VERTEX FORMULA

My video summary

Just as we saw how the **quadratic formula** comes from solving a **general quadratic equation** by completing the square, we can use completing the square to establish a **formula** for the **vertex**. Work through this **video** to verify that the function $f(x) = ax^2 + bx + c$ is equivalent to $f(x) = a\left(x + \frac{b}{2a}\right)^2 + c - \frac{b^2}{4a}$, then compare this to the standard form $f(x) = a(x - h)^2 + k$.

$$f(x) = a\left(x - \left(-\frac{b}{2a}\right)\right)^2 + c - \frac{b^2}{4a}$$

$$\downarrow \qquad\qquad \downarrow$$

$$f(x) = a(x \quad - \quad h)^2 \quad + \quad k$$

We can see that the **coordinates** of the vertex must be $\left(-\frac{b}{2a}, c - \frac{b^2}{4a}\right)$.

It is not really necessary to memorize the formulas for both coordinates. Since we can find the value of the **y-coordinate** by evaluating the function at the **x-coordinate**, it is common to write the **vertex** as $\left(-\frac{b}{2a}, f\left(-\frac{b}{2a}\right)\right)$.

Formula for the Vertex of a Parabola

Given a **quadratic function** of the form $f(x) = ax^2 + bx + c,\ a \neq 0$, the vertex of the **parabola** is given by

$$(h, k) = \left(-\frac{b}{2a}, f\left(-\frac{b}{2a}\right)\right).$$

The **axis of symmetry** is the vertical line $x = -\frac{b}{2a}$.

Example 11 Using the Vertex Formula

My video summary

Use the **vertex formula** to find the **vertex** for each **quadratic function**.

a. $f(x) = 3x^2 - 12x - 4$ **b.** $f(x) = -\frac{1}{2}x^2 - 10x + 5$

Solutions

a. We start by identifying the **coefficients** of the function: $a = 3$, $b = -12$, and $c = -4$

Using the vertex formula, the x-coordinate is $h = -\frac{b}{2a} = -\frac{(-12)}{2(3)} = \frac{12}{6} = 2$.

We find the y-coordinate by evaluating $k = f(h) = f(2)$.

$$f(2) = 3(2)^2 - 12(2) - 4$$
$$= 12 - 24 - 4 = -16$$

The vertex is $(h, k) = (2, -16)$.

b. Determine the vertex on your own. View the answer, or watch this **video** to see the complete solution.

You Try It **Work through this You Try It problem.**

Work Exercises 30–33 in this textbook or in the MyMathLab Study Plan.

OBJECTIVE 7 GRAPH QUADRATIC FUNCTIONS OF THE FORM $f(x) = ax^2 + bx + c$ BY USING THE VERTEX FORMULA

My video summary

Example 12 Graphing a Quadratic Function Using the Vertex Formula

Given the quadratic function $f(x) = -2x^2 - 4x + 5$, answer the following:

a. What are the coordinates of the **vertex**?

b. Does the graph "open up" or "open down"?

c. What is the equation of the **axis of symmetry**?

d. Find any **x-intercepts**.

e. Find the **y-intercept**.

f. Sketch the graph.

g. State the **domain** and **range** in **interval notation**.

Solutions Find the vertex by using the **vertex formula**. Then try to answer the remaining questions as in **Example 6**. View the answers, or watch this **video** to see complete solutions.

You Try It **Work through this You Try It problem.**

Work Exercises 34–39 in this textbook or in the MyMathLab Study Plan.

19.2 Exercises

In Exercises 1–4, without graphing, determine if the graph of each quadratic function opens up or down. Also determine if the graph will be wider or narrower than the graph of $y = x^2$.

You Try It

1. $f(x) = \frac{1}{4}x^2 - 3$

2. $f(x) = -2x^2 + 4x + 1$

3. $f(x) = -4 + 5x^2 - 2x$

4. $f(x) = -\frac{2}{3}x^2 + x + 7$

In Exercises 5–6, use the given graph of a quadratic function to find the following:

a. Vertex **b.** Axis of symmetry **c.** y-intercept
d. x-intercept(s) **e.** Domain and range

You Try It

5. 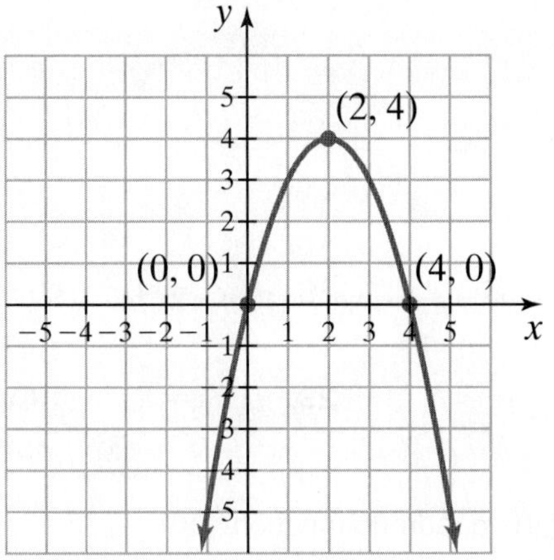

6. 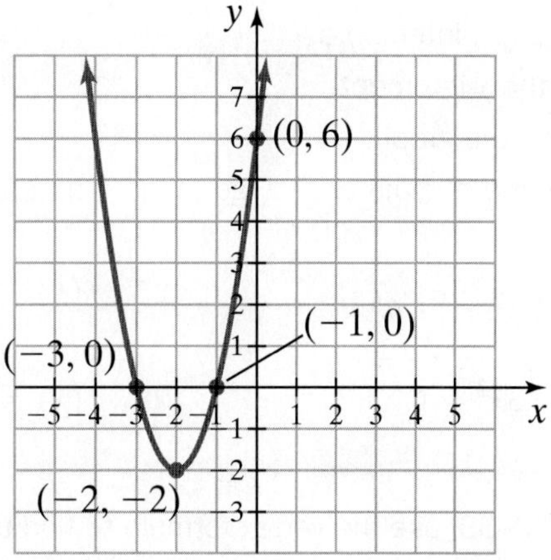

In Exercises 7–12, sketch the graph of each quadratic function by using translations (vertical or horizontal shifts). Compare each graph to the graph of $y = x^2$.

You Try It

7. $f(x) = x^2 - 4$

8. $f(x) = x^2 + \frac{3}{2}$

You Try It

9. $f(x) = (x + 3)^2$

10. $f(x) = (x - 4)^2$

11. $f(x) = (x - 1)^2 - 2$

12. $f(x) = (x + 2)^2 - 3$

You Try It

In Exercises 13–18, use the given quadratic function to answer the following:

a. What are the coordinates of the vertex?
b. Does the graph "open up" or "open down"?
c. What is the equation of the axis of symmetry?
d. Find any x-intercepts.
e. Find the y-intercept.
f. Sketch the graph.
g. State the domain and range in interval notation.

13. $f(x) = (x - 2)^2 - 4$

14. $f(x) = -(x + 1)^2 - 9$

15. $f(x) = -2(x - 3)^2 + 2$

You Try It

16. $f(x) = \frac{1}{2}(x + 2)^2 + 2$

17. $f(x) = -\frac{1}{4}(x - 4)^2 + 2$

18. $f(x) = 3\left(x + \frac{1}{3}\right)^2 - 4$

In Exercises 19–23, write the function in standard form and find the vertex.

19. $f(x) = x^2 + 8x + 9$

20. $f(x) = 3x^2 - 6x + 2$

21. $f(x) = -2x^2 - 12x - 10$

You Try It

22. $f(x) = \frac{1}{2}x^2 + 6x + 1$

23. $f(x) = x^2 + 10x$

You Try It

(eText Screens 19.3-1–19.3-37)

In Exercises 24–29, rewrite the quadratic function in standard form and then answer the following:

a. Determine the vertex by completing the square.
b. Does the graph "open up" or "open down"?
c. What is the equation of the axis of symmetry?
d. Find any x-intercepts.
e. Find the y-intercept.
f. Sketch the graph.
g. State the domain and range in interval notation.

You Try It

24. $f(x) = x^2 + 6x - 7$ **25.** $f(x) = -x^2 - 3x + 4$ **26.** $f(x) = 4x^2 - 7x + 8$

27. $f(x) = -3x^2 - 6x + 1$ **28.** $f(x) = \frac{1}{4}x^2 - 2x + 1$ **29.** $f(x) = 2x^2 - 16x$

In Exercises 30–33, use the vertex formula to find the vertex of the quadratic function.

You Try It

30. $f(x) = x^2 + 5x - 3$ **31.** $f(x) = -x^2 + 2x - 5$

32. $f(x) = -4x^2 + 12x$ **33.** $f(x) = \frac{1}{2}x^2 + 7x + 4$

In Exercises 34–39, use the quadratic function to answer the following:

a. Determine the vertex by using the vertex formula.
b. Does the graph "open up" or "open down"?
c. What is the equation of the axis of symmetry?
d. Find any x-intercepts.
e. Find the y-intercept.
f. Sketch the graph.
g. State the domain and range in interval notation.

You Try It

34. $f(x) = x^2 - 4x - 60$ **35.** $f(x) = 3x^2 + 6x - 4$ **36.** $f(x) = -x^2 + 2x - 6$

37. $f(x) = \frac{1}{2}x^2 + 6x + 1$ **38.** $f(x) = -3x^2 + 7x + 5$ **39.** $f(x) = x^2 - 8x$

19.3 Applications and Modeling of Quadratic Functions

THINGS TO KNOW

Before working through this topic, be sure you are familiar with the following concepts:

VIDEO ANIMATION INTERACTIVE

You Try It

1. Use the Problem-Solving Strategy to Solve Direct Translation Problems (Topic 11.3, **Objective 2**)

You Try It

2. Use Linear Equations to Solve Applications (Topic 12.4, **Objective 8**)

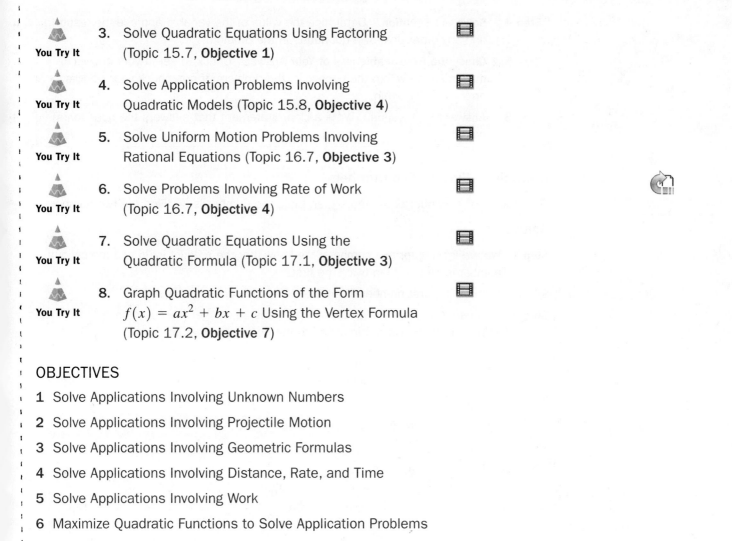

You Try It 3. Solve Quadratic Equations Using Factoring (Topic 15.7, **Objective 1**)

You Try It 4. Solve Application Problems Involving Quadratic Models (Topic 15.8, **Objective 4**)

You Try It 5. Solve Uniform Motion Problems Involving Rational Equations (Topic 16.7, **Objective 3**)

You Try It 6. Solve Problems Involving Rate of Work (Topic 16.7, **Objective 4**)

You Try It 7. Solve Quadratic Equations Using the Quadratic Formula (Topic 17.1, **Objective 3**)

You Try It 8. Graph Quadratic Functions of the Form $f(x) = ax^2 + bx + c$ Using the Vertex Formula (Topic 17.2, **Objective 7**)

OBJECTIVES

1 Solve Applications Involving Unknown Numbers

2 Solve Applications Involving Projectile Motion

3 Solve Applications Involving Geometric Formulas

4 Solve Applications Involving Distance, Rate, and Time

5 Solve Applications Involving Work

6 Maximize Quadratic Functions to Solve Application Problems

7 Minimize Quadratic Functions to Solve Application Problems

OBJECTIVE 1 SOLVE APPLICATIONS INVOLVING UNKNOWN NUMBERS

In **Topic 11.3**, we learned how to solve application problems involving **linear equations**. In this topic, we follow the same six-step process to solve application problems involving quadratic equations.

Problem-Solving Strategy for Applications

Step 1. **Define the Problem.** Read the problem carefully, or multiple times if necessary. Identify what you are trying to find and determine what information is available to help you find it.

Step 2. **Assign Variables.** Choose a variable to assign to an unknown quantity in the problem. If other unknown quantities exist, express them in terms of the selected variable.

Step 3. **Translate into an Equation.** Use the relationships among the known and unknown quantities to form an equation.

continued

Step 4. **Solve the Equation.** Determine the value of the variable and use the result to find any other unknown quantities in the problem.

Step 5. **Check the Reasonableness of Your Answer.** Check to see if your answer makes sense within the context of the problem. If not, check your work for errors and try again.

Step 6. **Answer the Question.** Write a clear statement that answers the question(s) posed.

Example 1 Finding Two Numbers

The **product** of a number and 1 more than twice the number is 36. Find the two numbers.

Solution

Step 1. We are looking for two numbers. We know that the product is 36 and the second number is 1 more than twice the first.

Step 2. Let x = the first number. Then $2x + 1$ = the second number.

Step 3. Because the product of the two numbers is 36, we get the equation $x(2x + 1) = 36$, which simplifies to the **quadratic equation** $2x^2 + x - 36 = 0$.

Step 4. Solve:

Original equation: $2x^2 + x - 36 = 0$

Factor: $(2x + 9)(x - 4) = 0$

Apply the **zero product property** and solve for x:

$$2x + 9 = 0 \quad \text{or} \quad x - 4 = 0$$
$$2x = -9 \qquad\qquad x = 4$$
$$x = -\frac{9}{2}$$

Step 5. If $x = -\frac{9}{2}$, then the other number is $2\left(-\frac{9}{2}\right) + 1 = -8$. Since $\left(-\frac{9}{2}\right)(-8) = 36$, this result is reasonable.

If $x = 4$, then the other number is $2(4) + 1 = 9$. Since $(4)(9) = 36$, this result is also reasonable.

Step 6. The two numbers are $-\frac{9}{2}$ and -8, or 4 and 9.

You Try It **Work through this You Try It problem.**

Work Exercises 1 and 2 in this textbook or in the MyMathLab Study Plan.

Example 2 Finding Consecutive Even Integers

Three consecutive *positive* even integers are such that the square of the third is 20 less than the sum of the squares of the first two. Find the positive integers.

Solution

Step 1. We are looking for three consecutive positive even integers. We know that the square of the third is 20 less than the sum of the squares of the first two.

Step 2. Let x = the first positive even integer. Then $x + 2$ = the second, and $x + 4$ = the third.

Step 3. Because the square of the third is 20 less than the sum of the **squares** of the first two, we get the equation

$$\underbrace{(x+4)^2}_{\text{Square of the third}} = \underbrace{\underbrace{x^2}_{\text{Square of the first}} + \underbrace{(x+2)^2}_{\text{Square of the second}}}_{\text{Sum}} \underbrace{-\ 20}_{\text{20 less than the sum}}.$$

Step 4. Solve:

Original equation:	$(x+4)^2 = x^2 + (x+2)^2 - 20$
Expand:	$x^2 + 8x + 16 = x^2 + x^2 + 4x + 4 - 20$
Combine like terms:	$x^2 + 8x + 16 = 2x^2 + 4x - 16$
Write in standard form:	$0 = x^2 - 4x - 32$
Factor:	$0 = (x-8)(x+4)$

Apply the **zero product property** and solve for x:

$$x - 8 = 0 \quad \text{or} \quad x + 4 = 0$$
$$x = 8 \qquad\qquad x = -4$$

Step 5. Since the consecutive even integers must be positive, we discard the negative solution. The only feasible solution remaining is $x = 8$ as the first positive even integer which would give 10 and 12 as the next two consecutive even integers. Since $8^2 + 10^2 - 20 = 12^2$, this result is reasonable.

Step 6. The three consecutive positive even integers are 8, 10, and 12.

You Try It **Work through this You Try It problem.**

Work Exercises 3 and 4 in this textbook or in the MyMathLab Study Plan.

OBJECTIVE 2 SOLVE APPLICATIONS INVOLVING PROJECTILE MOTION

An object launched, thrown, or shot vertically into the air with an initial velocity of v_0 meters per second (m/s) from an initial height of h_0 meters above the ground can be **modeled** by the equation $h = -4.9t^2 + v_0t + h_0$. The variable h is the height above the ground (in meters) of the object (also known as a *projectile*) t seconds after its departure.

Example 3 Launch a Toy Rocket

A toy rocket is launched at an initial velocity of 14.7 m/s from a platform that sits 49 meters above the ground. The height h of the rocket above the ground at any time t seconds after launch is given by the equation $h = -4.9t^2 + 14.7t + 49$. When will the rocket hit the ground?

Solution

Step 1. We want to find the time when $h = 0$ (the rocket hits the ground) using the equation $h = -4.9t^2 + 14.7t + 49$.

Step 2. We know that h = height in meters and t = time in seconds.

Step 3. Using the equation $h = -4.9t^2 + 14.7t + 49$, set $h = 0$ and solve for t.

Step 4. Solve:

Original equation:	$0 = -4.9t^2 + 14.7t + 49$
Divide both sides by -4.9:	$0 = t^2 - 3t - 10$
Factor:	$0 = (t-5)(t+2)$

Apply the **zero product property** and solve for t:

$$t - 5 = 0 \quad \text{or} \quad t + 2 = 0$$
$$t = 5 \qquad\qquad t = -2$$

Step 5. Because t represents the time (in seconds) after launch, its value cannot be negative. So, $t = -2$ seconds does not make sense. The only reasonable solution is $t = 5$ seconds.

Step 6. The rocket will hit the ground 5 seconds after launch.

You Try It **Work through this You Try It problem.**

Work Exercises 5 and 6 in this textbook or in the MyMathLab Study Plan.

OBJECTIVE 3 SOLVE APPLICATIONS INVOLVING GEOMETRIC FORMULAS

Now let's explore application problems that involve geometric formulas and **quadratic equations.**

Example 4 Dimensions of a Rectangle

My interactive video summary

The length of a rectangle is 6 inches less than four times the width. Find the dimensions of the rectangle if the **area** of the rectangle is 54 square inches.

Solution Work through this **interactive video** to check that the dimensions of the rectangle are 12 inches by 4.5 inches.

You Try It **Work through this You Try It problem.**

Work Exercises 7 and 8 in this textbook or in the MyMathLab Study Plan.

Example 5 Width of a High-Definition Television

Shayna bought a new 40-inch high-definition television. If the length of Shayna's television is 8 inches longer than the width, find the width of the television.

Solution

Step 1. We want to determine the width of the television. We know that the length is 8 inches more than the width, and we assume that the television is rectangular. The size of a television is the length of its ***diagonal***, which in this case is 40 inches.

Step 2. Let w = width of the television. Then $w + 8$ = length of the television.

Step 3. We can create a **quadratic equation** using the **Pythagorean theorem**, $a^2 + b^2 = c^2$.

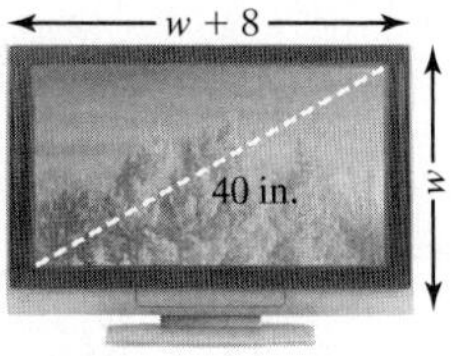

$$w^2 + (w + 8)^2 = 40^2$$

Step 4. Solve:

Original equation:	$w^2 + (w + 8)^2 = 40^2$
Square the binomial:	$w^2 + w^2 + 16w + 64 = 1600$
Combine like terms:	$2w^2 + 16w - 1536 = 0$
Divide both sides by 2:	$w^2 + 8w - 768 = 0$
Factor:	$(w - 24)(w + 32) = 0$

Apply the **zero product property** and solve for w:

$$w - 24 = 0 \quad \text{or} \quad w + 32 = 0$$
$$w = 24 \qquad\qquad w = -32$$

Step 5. Because w represents the width of the television, its value cannot be negative. Thus, $w = -32$ does not make sense. The only reasonable solution is $w = 24$ inches.

Step 6. The width of the television is 24 inches.

You Try It Work through this You Try It problem.

Work Exercises 9 and 10 in this textbook or in the MyMathLab Study Plan.

OBJECTIVE 4 SOLVE APPLICATIONS INVOLVING DISTANCE, RATE, AND TIME

In the next example, we use quadratic equations to solve an application problem involving distance, rate, and time.

Example 6 Speed of an Airplane

My video summary

Kevin flew his new Cessna O-2A airplane from Jonesburg to Mountainview, a distance of 2560 miles. The average speed for the return trip was 64 mph faster than the average outbound speed. If the total flying time for the round trip was 18 hours, what was the plane's average speed on the outbound trip from Jonesburg to Mountainview?

Solution

Step 1. We are asked to find the average outbound speed of the plane from Jonesburg to Mountainview. We know that the distance traveled in each direction is 2560 miles and the total time of the trip is 18 hours. We also know that the speed on the return trip is 64 mph more than on the outbound trip. Because the problem involves distance, rate, and time, we will need the **distance formula** $d = r \cdot t$.

Step 2. Let r = speed of plane on the outbound trip. Then $r + 64$ = speed of plane on the return trip from Mountainview to Jonesburg.

Step 3. Since the total time of the trip was 18 hours, we write

$$\text{time}_{\text{outbound}} + \text{time}_{\text{return}} = \text{time}_{\text{total}}$$
$$\text{time}_{\text{outbound}} + \text{time}_{\text{return}} = 18$$

Because $\text{distance} = \text{rate} \cdot \text{time}$, we also know that $\text{time} = \dfrac{\text{distance}}{\text{rate}}$.

Using the given distance and rate expressions, we get the following equation:

$$\frac{2560}{r} + \frac{2560}{r + 64} = 18$$

Solve this equation on your own. Be sure to check the reasonableness of your answer. View the answer, or watch this **video** for the complete solution.

You Try It Work through this You Try It problem.

Work Exercises 11 and 12 in this textbook or in the MyMathLab Study Plan.

(eText Screens 19.3-1–19.3-37)

OBJECTIVE 5 SOLVE APPLICATIONS INVOLVING WORK

We can also use quadratic equations to solve application problems involving work.

Example 7 Monthly Sales Reports

My video summary

Dawn can finish the monthly sales reports in 2 hours less time than it takes Adam. Working together, they were able to finish the sales reports in 8 hours. How long does it take each person to finish the monthly sales reports alone? (Round to the nearest minute.)

Solution

Step 1. We must find the time it takes for each person to finish the sales reports alone. We know that it takes Dawn 2 hours less than Adam to do the job alone, and it takes them 8 hours to do the job together. Because this problem involves combining **rates of work**, we use the equation

$$\frac{1}{t_1} + \frac{1}{t_2} = \frac{1}{t}.$$

Step 2. Let t_1 = the time it takes Dawn to complete the reports alone (in hours). Then $t_2 = t_1 + 2$ = the time it takes Adam to complete the reports alone. We know that it takes $t = 8$ hours to complete the job when they work together.

Step 3. Substituting the expressions for the individual times and 8 for the total time, we get the following equation:

$$\frac{1}{t_1} + \frac{1}{t_1 + 2} = \frac{1}{8}$$

Solve this equation for t_1. Confirm that it will take Dawn approximately 15 hours and 4 minutes to complete the reports on her own and it will take Adam 17 hours and 4 minutes to complete the same job alone. Be sure to check that the sum of their rates is equal to $\frac{1}{8}$ (within rounding error). To see this problem worked out in detail, watch this **video**.

You Try It **Work through this You Try It problem.**

Work Exercises 13 and 14 in this textbook or in the MyMathLab Study Plan.

OBJECTIVE 6 MAXIMIZE QUADRATIC FUNCTIONS TO SOLVE APPLICATION PROBLEMS

Sometimes with application problems involving **functions**, we often need to find the *maximum* or *minimum* value of the function. For example, a builder with a fixed amount of fencing may wish to maximize the area enclosed, or an economist may want to minimize a **cost** function or maximize a **profit** function. **Quadratic functions** are relatively easy to maximize or minimize because we know a formula for finding the **coordinates** of the **vertex**. Recall that if $f(x) = ax^2 + bx + c$, $a \neq 0$, we know that the coordinates of the vertex are $\left(-\frac{b}{2a}, f\left(-\frac{b}{2a}\right)\right)$. In addition, we know that if $a > 0$, the **parabola** opens *up* and the function has a **minimum** value at the vertex. If $a < 0$, the parabola opens *down* and the function has a **maximum** value at the vertex. See **Figure 7**.

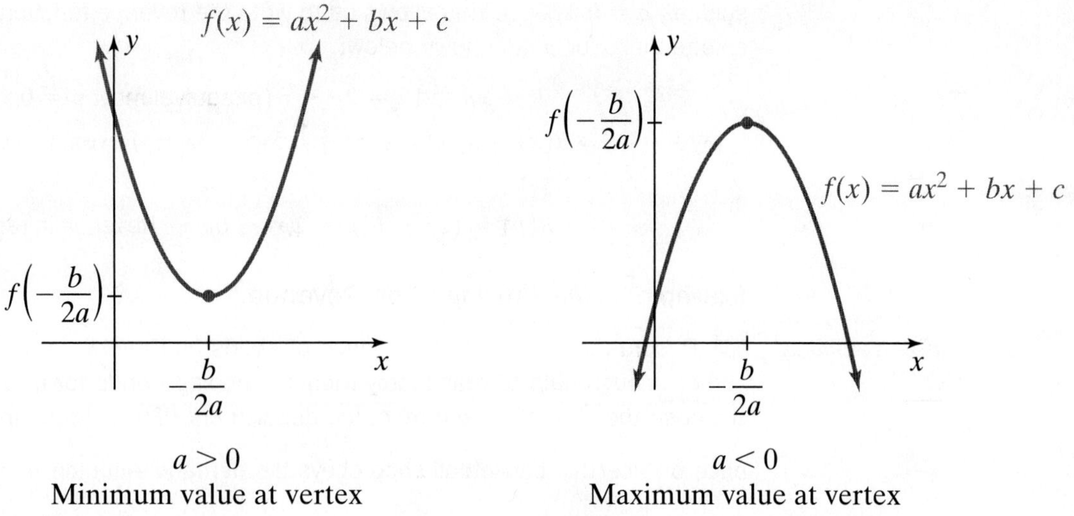

Figure 7

Projectile Motion

Example 8 Launching a Toy Rocket

A toy rocket is launched with an initial **velocity** of 44.1 meters per second from a platform located 1 meter above the ground. The height h of the object above the ground at any time t seconds after launch is given by the function $h(t) = -4.9t^2 + 44.1t + 1$. How long after launch did it take the rocket to reach its maximum height? What is the rocket's **maximum** height?

Solution Because the function is **quadratic** with $a = -4.9 < 0$, we know that the graph is a **parabola** that opens down. So, the function has a maximum value at the **vertex**. The t-coordinate of the vertex is $t = -\frac{b}{2a} = -\frac{44.1}{2(-4.9)} = \frac{-44.1}{-9.8} = 4.5$ seconds. Therefore, the rocket reaches its maximum height at 4.5 seconds after launch. The rocket's maximum height is

$$h(4.5) = -4.9(4.5)^2 + 44.1(4.5) + 1 = 100.225 \text{ meters.}$$

You Try It **Work through this You Try It problem.**

Work Exercises 15–17 in this textbook or in the MyMathLab Study Plan.

Maximize Revenue and Profit

Revenue is the dollar amount received from selling x items at a price of p dollars per item. **Total revenue** is found by multiplying the number of units sold by the price per unit, $R = xp$. For example, if a child sells 50 cups of lemonade at a price of \$0.25 per cup, then the revenue generated is

$$R = \underbrace{(50)}_{x}\underbrace{(0.25)}_{p} = \$12.50.$$

A **demand equation** relates the quantity sold to the price per item, and we can write this equation as quantity in terms of price, such as $x = 2p - 6$, or as price in terms of quantity, such as $p = 0.5x + 3$. This allows us to write the revenue function in terms of a single variable, either x or p, as shown below:

$$R = xp \text{ and } x = 2p - 6 \text{ (or equivalently, } p = 0.5x + 3)$$

$$R(x) = x(0.5x + 3) = 0.5x^2 + 3x \leftarrow \text{Revenue in terms of } x$$

or

$$R(p) = (2p - 6)p = 2p^2 - 6p \leftarrow \text{Revenue in terms of } p$$

Example 9 Maximizing Shoe Revenue

My video summary

Records can be kept on the price of shoes and the number of pairs sold in order to gather enough data to reasonably **model** shopping trends for a particular type of shoe. Suppose the marketing and research department of a shoe company determined the price of a certain basketball shoe obeys the **demand equation** $p = -\frac{1}{50}x + 110$.

a. According to the demand equation, how much should the shoes sell for if 500 pairs of shoes are sold? 1200 pairs of shoes?

b. What is the **revenue** if 500 pairs of shoes are sold? 1200 pairs of shoes?

c. How many pairs of shoes should be sold in order to maximize revenue? What is the **maximum** revenue?

d. What price should be charged in order to maximize revenue?

Solutions

a. If 500 pairs of shoes are sold, the price should be

$$p(500) = -\frac{1}{50}(500) + 110 = -10 + 110 = \$100.$$

If 1200 pairs of shoes are sold, the price should be

$$p(1200) = -\frac{1}{50}(1200) + 110 = -24 + 110 = \$86.$$

b. Because $R = xp$, we can substitute the **demand equation,** $p = -\frac{1}{50}x + 110$, for p to get R as a function of x.

$$R(x) = x\underbrace{\left(-\frac{1}{50}x + 110\right)}_{p} \quad \text{or} \quad R(x) = -\frac{1}{50}x^2 + 110x$$

The **revenue** generated by selling 500 pairs of shoes is

$$\begin{aligned} R(500) &= 500\left(-\frac{1}{50}(500) + 110\right) \\ &= 500(-10 + 110) \\ &= 500(100) \\ &= \$50{,}000. \end{aligned}$$

The revenue from selling 1200 pairs of shoes is

$$\begin{aligned} R(1200) &= 1200\left(-\frac{1}{50}(1200) + 110\right) \\ &= 1200(-24 + 110) \\ &= 1200(86) \\ &= \$103{,}200. \end{aligned}$$

c. $R(x) = -\frac{1}{50}x^2 + 110x$ is a **quadratic function** with $a = -\frac{1}{50} < 0$ and $b = 110.$

Because $a < 0$, the function has a **maximum** value at the **vertex**. Therefore, the value of x that produces the maximum **revenue** is

$$x = -\frac{b}{2a} = -\frac{110}{2\left(-\frac{1}{50}\right)} = \underbrace{\frac{-110}{-\frac{1}{25}} = \frac{-110}{1}\cdot\left(-\frac{25}{1}\right)}_{\text{Change to multiplication by the reciprocal}} = 2750 \text{ pairs of shoes.}$$

The maximum revenue is

$$\begin{aligned} R(2750) &= 2750\left(-\frac{1}{50}(2750) + 110\right) \\ &= 2750(55) \\ &= \$151{,}250. \end{aligned}$$

d. Using the **demand equation**, the price that should be charged to maximize revenue when selling 2750 pairs of shoes is $p(2750) = -\frac{1}{50}(2750) + 110 = \$55.$

You Try It **Work through this You Try It problem.**

Work Exercises 18 and 19 in this textbook or in the MyMathLab Study Plan.

Example 10 Maximizing Profit

My video summary

To sell x waterproof CD alarm clocks, WaterTime, LLC, has determined that the price in dollars must be $p = 250 - 2x$, which is the **demand equation**. Each clock costs \$2 to produce, with fixed costs of \$4000 per month, producing the **cost** function $C(x) = 2x + 4000$.

a. Express the **revenue** R as a function of x.

b. Express the **profit** P as a function of x.

c. Find the value of x that maximizes profit. What is the **maximum** profit?

d. What is the price of the alarm clock that will maximize profit?

Solutions

a. The equation for revenue is $R = xp$. So, we substitute the demand equation $p = 250 - 2x$ for p to obtain the function:

$$R(x) = x\underbrace{(250 - 2x)}_{p} \quad \text{or} \quad R(x) = -2x^2 + 250x$$

(eText Screens 19.3-1–19.3-37)

b. **Profit** is equal to **revenue** minus **cost**.

$$\begin{aligned} P(x) &= R(x) - C(x) \\ &= -2x^2 + 250x - (2x + 4000) \\ &= -2x^2 + 250x - 2x - 4000 \\ &= -2x^2 + 248x - 4000 \end{aligned}$$

c.–d. Try to finish the remaining parts on your own. Find the **vertex** to answer the questions in part (c), and use the **demand equation** to answer part (d). View the **answers**, or watch this **video** for complete solutions.

You Try It **Work through this You Try It problem.**

Work Exercises 20–23 in this textbook or in the MyMathLab Study Plan.

Maximize Area

Example 11 Maximizing Area

My video summary

Suppose you are asked to build a fence around a rectangular field that borders a river. You have 3000 feet of fencing available, but no fencing is required along the river. Find the dimensions of the field that maximizes the enclosed **area**. What is the **maximum** area?

Figure 8

Solution From Figure 8, let x = width of the field and y = length of the field. Since the field is rectangular, we find the area by multiplying length by width:

$$A = xy$$

Ideally, we want the area to depend on only one variable. So, we need to write x in terms of y, or write y in terms of x. The 3000 feet of fencing will be used for two widths and one length, giving us the equation

$$2x + y = 3000.$$

Solving this equation for y gives $y = -2x + 3000$. We can substitute $-2x + 3000$ for y to get the area as a **function** of x alone.

$$A(x) = x\underbrace{(-2x + 3000)}_{y} \quad \text{or} \quad A(x) = -2x^2 + 3000x$$

(eText Screens 10.3-1–10.3-37)

Our area function is **quadratic** with $a = -2 < 0$. Therefore, the function has a **maximum** value at its **vertex**. Try to finish working this problem on your own. View the **answers**, or watch this **video** for a complete solution.

You Try It **Work through this You Try It problem.**

Work Exercises 24 and 25 in this textbook or in the MyMathLab Study Plan.

OBJECTIVE 7 MINIMIZE QUADRATIC FUNCTIONS TO SOLVE APPLICATION PROBLEMS

Let's now consider some situations in which we **minimize** a **function**.

Example 12 Minimizing Costs

A fabric manufacturer has daily production costs of $C = 7000 - 100x + 0.5x^2$, where C is the total **cost** (in dollars) and x is the number of units produced. How many units should be produced each day in order to minimize costs? What is the minimum daily cost?

Solution Writing the cost function in standard form, we have

$$C(x) = 0.5x^2 - 100x + 7000.$$

This is a **quadratic function** with $a = 0.5 > 0$, $b = -100$, and $c = 7000$. Because $a > 0$, the graph of the function is a **parabola** that *opens up*, and the function has a minimum value at the **vertex**. The x-coordinate of the **vertex** tells us the number of units to produce in order to minimize costs, and the y-coordinate tells us the minimum cost. To find these coordinates, we use the **vertex formula**

$$\left(-\frac{b}{2a}, f\left(-\frac{b}{2a}\right)\right).$$

$$x = -\frac{(-100)}{2(0.5)} = \frac{100}{1} = 100 \quad \text{and} \quad \begin{aligned} C(100) &= 0.5(100)^2 - 100(100) + 7000 \\ &= 0.5(10{,}000) - 100(100) + 7000 \\ &= 5000 - 10{,}000 + 7000 \\ &= 2000 \end{aligned}$$

The manufacturer should produce 100 units each day to minimize costs. The minimum daily cost will be \$2000.

You Try It **Work through this You Try It problem.**

Work Exercises 26 and 27 in this textbook or in the MyMathLab Study Plan.

Example 13 Exchange Rates

Between January 1, 2007, and February 1, 2009, the exchange rate between Canadian dollars and U.S. dollars can be approximated by $D(x) = 0.0017x^2 - 0.0397x + 1.2194$, where D is the number of Canadian dollars for one U.S. dollar x months after January 1, 2007. For the given time period, determine the month in which the exchange rate was a **minimum**. What was the minimum exchange rate? Round your answer to three decimal places. (*Source*: U.S. Federal Reserve, St. Louis, MO)

(eText Screens 19.3-1–19.3-37)

Solution $D(x) = 0.0017x^2 - 0.0397x + 1.2194$ is a **quadratic function** with $a = 0.0017 > 0$, $b = -0.0397$, and $c = 1.2194$. Because $a > 0$, the graph of the function is a **parabola** that *opens up*, and the function has a **minimum** value at the **vertex**. So, the value of x that produces the minimum exchange rate is

$$x = -\frac{b}{2a} = -\frac{(-0.0397)}{2(0.0017)} \approx 11.676.$$

How do we interpret this value for x? Since x represents the number of months after January 1, 2007, $x = 1$ indicates February 1, 2007, $x = 2$ means March 1, 2007, and so on. So, $x = 11$ represents December 1, 2007, and $x = 12$ indicates January 1, 2008. Since $11 < 11.676 < 12$, the minimum exchange rate occurred in December 2007 and was approximately

$$D(11.676) = 0.0017(11.676)^2 - 0.0397(11.676) + 1.2194$$
$$\approx \$0.988 \text{ Canadian dollars for one U.S. dollar.}$$

You Try It **Work through this You Try It problem.**

Work Exercises 28 and 29 in this textbook or in the MyMathLab Study Plan.

In economics, **long run average cost (LRAC)**, is the total long run cost divided by the number of units produced. Companies sometimes use the *LRAC* to determine if they should merge resources such as sales regions or franchises. If the *LRAC* is decreasing, then it would be best for the company to merge resources, provided that the merger does not cause the *LRAC* to increase. If the *LRAC* is increasing, a merger may not be beneficial since it would lead to higher costs. This idea is illustrated in the next example.

Example 14 Merging Resources

My video summary

An account rep in one territory oversees $N = 20$ accounts and a second account rep in a nearby territory manages $N = 8$ accounts. The **long run average cost** function for their industry is $C = N^2 - 70N + 1400$.

a. Determine the long run average cost for $N = 20$ accounts and $N = 8$ accounts.

b. What number of accounts minimizes the long run average cost? What is the **minimum** long run average cost?

c. Should the two territories be merged into a single territory?

Solutions Try to work this problem on your own. View the **answers**, or watch this **video** for detailed solutions.

You Try It **Work through this You Try It problem.**

Work Exercises 30 and 31 in this textbook or in the MyMathLab Study Plan.

19.3 Exercises

You Try It

1. **Finding a Number** The product of some negative number and 5 less than three times that number is 12. Find the number.

2. **Finding a Number** The square of a number plus the number is 132. What is the number?

You Try It

3. **Finding Integers** Three consecutive odd integers are such that the square of the third integer is 15 more than the sum of the squares of the first two. Find the integers.

4. **Finding a Number** The sum of the square of a number and the square of 7 more than the number is 169. What is the number?

You Try It

5. **Rocket Launch** A toy rocket is launched from a platform 2.8 meters above the ground in such a way that its height, h (in meters), after t seconds is given by the equation $h = -4.9t^2 + 18.9t + 2.8$. How long will it take for the rocket to hit the ground?

6. **Projectile Motion** Shawn threw a rock straight up from a cliff that was 24 feet above the water. If the height of the rock h, in feet, after t seconds is given by the equation $h = -16t^2 + 20t + 24$, how long will it take for the rock to hit the water?

You Try It

7. **Dimensions of a Rectangle** The length of a rectangle is 1 cm less than three times the width. If the area of the rectangle is $30\,\text{cm}^2$, find the dimensions of the rectangle.

8. **Dimensions of a Rectangle** The length of a rectangle is 1 inch less than twice the width. If the diagonal is 2 inches more than the length, find the dimensions of the rectangle.

You Try It

9. **Loading Ramp** A loading ramp in a steel yard has a horizontal run that is 26 feet longer than its vertical rise. If the ramp is 30 feet long, what is the vertical rise, rounded to the nearest hundredth of a foot?

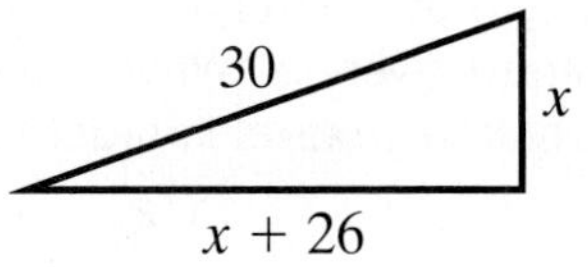

10. **Building a Walkway** A 35- by 20-foot rectangular swimming pool is surrounded by a walkway of uniform width. If the total area of the walkway is $434\,\text{ft}^2$, how wide is the walkway?

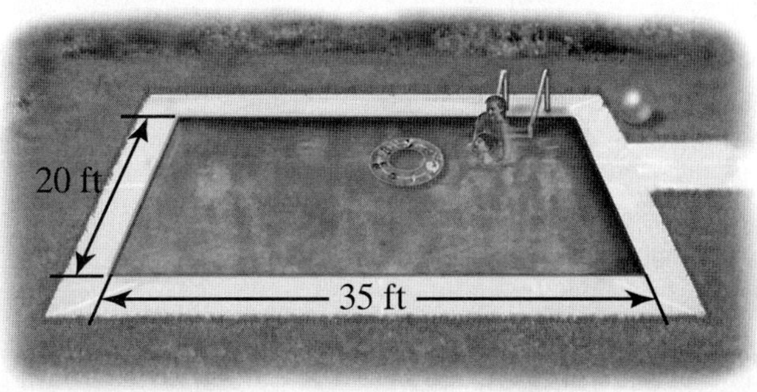

You Try It

11. **Speed of a Boat** Logan rowed her boat upstream a distance of 9 miles and then rowed back to the starting point. The total time of the trip was 10 hours. If the rate of the current was 4 mph, find the average speed of the boat in still water.

12. **Speed of a Car** Imogene's car traveled 280 miles averaging a certain speed. If the car had gone 5 mph faster, the trip would have taken 1 hour less. Find the average speed.

You Try It

13. **Mowing Lawns** Twin brothers, Billy and Bobby, can mow their grandparent's lawn together in 56 minutes. Billy could mow the lawn by himself in 15 minutes less time than it would take Bobby. How long would it take Bobby to mow the lawn by himself?

14. **Working Together** Jeff and Kirk can build a 75-foot retaining wall together in 12 hours. Because Jeff has more experience, he could build the wall himself 4 hours quicker than Kirk. How long would it take Kirk (to the nearest minute) to build the wall by himself?

(eText Screens 19.3-1–19.3-37)

You Try It

15. Maximum Height A baseball player swings and hits a pop fly straight up in the air to the catcher. The height of the baseball in meters t seconds after it is hit is given by the quadratic function $h(t) = -4.9t^2 + 34.3t + 1$. How long does it take for the baseball to reach its maximum height? What is the baseball's maximum height?

16. Maximum Height An object is launched vertically in the air at 36.75 meters per second from a platform 10 meters above the ground. The height of the object above the ground (in meters) t seconds after launch is given by $h(t) = -4.9t^2 + 36.75t + 10$. How long does it take the object to reach its maximum height? What is the object's maximum height?

17. Maximum Height of a Toy Rocket A toy rocket is shot vertically into the air from a launching pad 5 feet above the ground with an initial velocity of 112 feet per second. The height h, in feet, of the rocket above the ground at t seconds after launch is given by the function $h(t) = -16t^2 + 112t + 5$. How long will it take the rocket to reach its maximum height? What is the maximum height?

You Try It

18. Maximum Revenue The price p and the quantity x sold of a small flatscreen TV obeys the demand equation $p = -0.15x + 300$.

a. How much should be charged for the TV if there are 50 TVs in stock?
b. What quantity x will maximize revenue? What is the maximum revenue?
c. What price should be charged per TV in order to maximize revenue?

19. Maximum Revenue The dollar price for a barrel of oil sold at a certain oil refinery tends to obey the demand equation $p = -\frac{1}{10}x + 72$, where x is the number of barrels of oil on hand (in millions).

a. How much should be charged per barrel of oil if there are 4 million barrels on hand?
b. What quantity x will maximize revenue? What is the maximum revenue?
c. What price should be charged per barrel in order to maximize revenue?

In Exercises 20–23, use the fact that profit is defined as revenue minus cost, or $P(x) = R(x) - C(x)$.

You Try It

20. Maximum Profit Rite-Cut riding lawnmowers obey the demand equation $p = -\frac{1}{20}x + 1000$. The cost of producing x lawnmowers is given by the function $C(x) = 100x + 5000$.

a. Express the revenue R as a function of x.
b. Express the profit P as a function of x.
c. Find the value of x that maximizes profit. What is the maximum profit?
d. What price should be charged per lawnmower to maximize profit?

21. Maximum Profit The CarryItAll minivan, a popular vehicle among soccer moms, obeys the demand equation $p = -\frac{1}{40}x + 8000$. The cost of producing x vans is given by the function $C(x) = 4000x + 20{,}000$.

a. Express the revenue R as a function of x.
b. Express the profit P as a function of x.
c. Find the value of x that maximizes profit. What is the maximum profit?
d. What price should be charged to maximize profit?

22. Maximum Profit Silver Scooter, Inc. finds that it costs \$200 to produce each motorized scooter and that the fixed costs are \$1500 per day yielding the cost function $C(x) = 200x + 1500$. The price is given by $p = 600 - 5x$, where p is the price in dollars at which exactly x scooters will be sold. Find the quantity of scooters that Silver Scooter, Inc. should produce and the price it should charge to maximize profit. Find the maximum profit.

23. Maximum Profit Amy, the owner of Amy's Pottery, can produce china pitchers at a cost of \$5 each. She estimates her price function to be $p = 17 - 5x$, where p is the price at which exactly x pitchers will be sold per week. Find the number of pitchers that she should produce and the price that she should charge in order to maximize profit. Also find the maximum profit.

You Try It

24. Maximum Area A farmer has 1800 feet of fencing available to enclose a rectangular area bordering a river. If no fencing is required along the river, find the dimensions of the fence that will maximize the area. What is the maximum area?

25. Maximum Area Jim wants to build a rectangular parking lot along a busy street but has only 2500 feet of fencing available. If no fencing is required along the street, find the maximum area of the parking lot.

You Try It

26. Minimizing Cost A manufacturer of denim jeans has daily production costs of $C = 0.2x^2 - 90x + 10{,}700$, where C is the total cost (in dollars) and x is the number of jeans produced. How many jeans should be produced each day in order to minimize costs? What is the minimum daily cost?

27. Minimizing Loss Function Quality control analysts often use a *quadratic loss function* to determine costs associated with failing to meet product specifications. For a certain machine part, the loss function relating cost to diameter is $L = 4x^2 - 4x + 2.5$, where L is the total loss (in dollars) per part and x is the diameter of the machined part. Find the diameter that minimizes total loss per part. What is the minimum total loss?

You Try It

28. Minimum Exchange Rate Between January 1, 1999, and January 1, 2009, the exchange rate between U.S. dollars and euros can be approximated by $D(x) = 0.003x^2 - 0.029x + 0.932$, where D is the number of U.S. dollars for one euro x years after January 1, 1999. For the given time period, find the year in which the exchange rate was at a minimum. What was the minimum exchange rate? Round your answer to three decimal places. (*Source:* U.S. Federal Reserve, St. Louis, MO)

29. Minimum Bankruptcies Based on data from the Federal Judiciary, the number of business bankruptcy filings per year can be approximated by the model $B(x) = 3.9x^2 - 50.8x + 192.9$, where x is the number of years since 2000 and B is the number of business bankruptcy filings (in 1000s). During what year was the number of filings at a minimum? What was the approximate minimum number of filings to the nearest thousand?

You Try It

30. Merging Account Territories An account manager in one territory oversees $N = 14$ accounts, and a second account manager in a nearby territory manages $N = 18$ accounts. The long run average cost function for their industry is $C = 484 - 44N + N^2$.

a. Find the long run average cost for $N = 14$ accounts and for $N = 18$ accounts.

b. How many accounts will minimize the long run average cost? What is the minimum long run average cost?

c. Should the two territories be merged into a single territory?

31. Bank Merger One bank services $N = 150$ customers, and a second bank services $N = 80$ customers. The long run average cost function for their industry is $C = N^2 - 700N + 110{,}000$.

a. Find the long run average cost for $N = 150$ customers and for $N = 80$ customers.

b. What number of customers will minimize the long run average cost? What is the minimum long run average cost?

c. Should the two banks merge?

19.4 Circles

THINGS TO KNOW

Before working through this topic, be sure you are familiar with the following concepts:

	VIDEO	ANIMATION	INTERACTIVE
You Try It 1. Plot Ordered Pairs in the Rectangular Coordinate System (Topic 12.1, **Objective 3**)	✓		
You Try It 2. Simplify Radical Expressions Using the Product Rule (Topic 18.3, **Objective 5**)	✓		✓

3. Solve Quadratic Equations Using the Square Root Property (Topic 19.1, **Objective 1**)

4. Solve Quadratic Equations by Completing the Square (Topic 19.1, **Objective 2**)

OBJECTIVES

1 Find the Distance Between Two Points

2 Find the Midpoint of a Line Segment

3 Write the Standard Form of an Equation of a Circle

4 Sketch the Graph of a Circle Given in Standard Form

5 Write the General Form of a Circle in Standard Form and Sketch Its Graph

OBJECTIVE 1 FIND THE DISTANCE BETWEEN TWO POINTS

Recall that the **Pythagorean theorem** states that the sum of the **squares** of the lengths of the two **legs** in a **right triangle** is equal to the square of the length of the **hypotenuse**, or $a^2 + b^2 = c^2$. See Figure 9.

My video summary

We can use the Pythagorean theorem to develop a **formula** for finding the distance between two points in a **plane**. See the steps, or watch this **video** for an explanation.

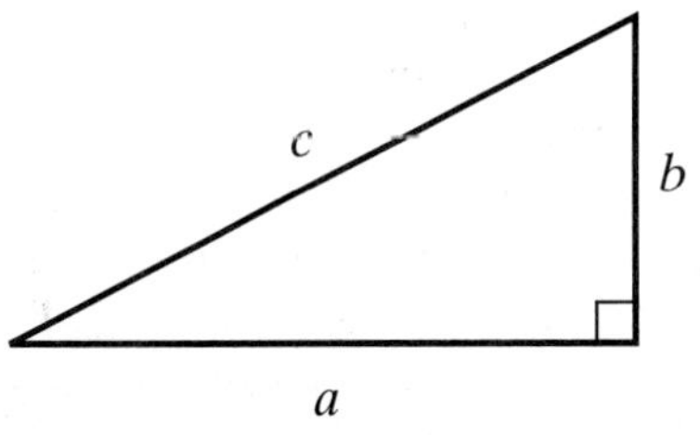

Figure 9
Pythagorean theorem $a^2 + b^2 = c^2$

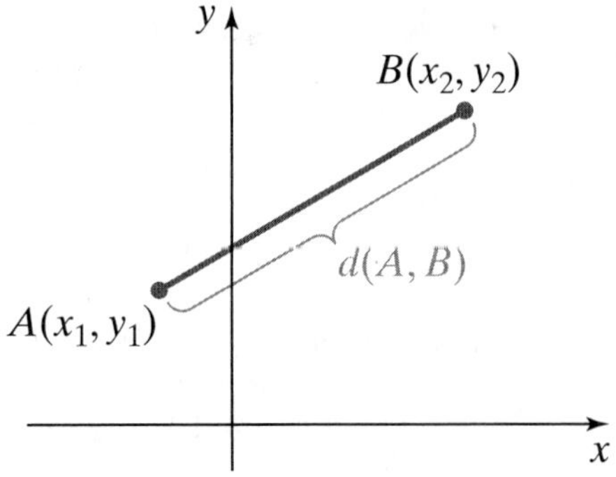

Figure 10

To find the distance $d(A, B)$ between the **points** $A(x_1, y_1)$ and $B(x_2, y_2)$ (see **Figure 10**), we use the **distance formula.**

> **Distance Formula**
>
> The distance $d(A, B)$ between two points (x_1, y_1) and (x_2, y_2) is given by
>
> $$d(A, B) = \sqrt{(x_2 - x_1)^2 + (y_2 - y_1)^2}.$$

CAUTION When using the distance formula, it does not matter which point is (x_1, y_1) and which is (x_2, y_2).

(eText Screens 19.4-1–19.4-29)

Example 1 Using the Distance Formula

Find the distance $d(A, B)$ between points $(-1, 5)$ and $(4, -5)$.

Solution Let $(x_1, y_1) = (-1, 5)$ and $(x_2, y_2) = (4, -5)$ and use the **distance formula.**

Distance formula:	$d(A, B) = \sqrt{(x_2 - x_1)^2 + (y_2 - y_1)^2}$
Substitute the coordinates:	$= \sqrt{(4 - (-1))^2 + (-5 - 5)^2}$
Change to addition:	$= \sqrt{(4 + 1)^2 + (-5 - 5)^2}$
Combine like terms:	$= \sqrt{(5)^2 + (-10)^2}$
Square:	$= \sqrt{25 + 100}$
Add:	$= \sqrt{125}$
Simplify the radical:	$= 5\sqrt{5}$

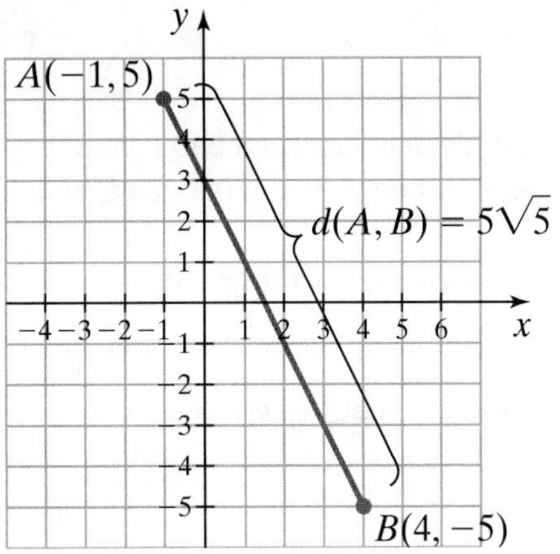

You Try It **Work through this You Try It problem.**

Work Exercises 1–8 in this textbook or in the MyMathLab Study Plan.

OBJECTIVE 2 FIND THE MIDPOINT OF A LINE SEGMENT

The **midpoint** of a line segment is the point that lies exactly halfway between the endpoints of the line segment. The **distance formula** can be used to develop a formula for finding the midpoint. The **coordinates** of the midpoint are found by **averaging** the x**-coordinates** of the endpoints and averaging the y**-coordinates** of the endpoints.

Midpoint of a Line Segment

The **midpoint** of the line segment from $A(x_1, y_1)$ to $B(x_2, y_2)$ is the point with coordinates

$$\left(\frac{x_1 + x_2}{2}, \frac{y_1 + y_2}{2}\right).$$

CAUTION The distance formula involves finding the **difference** between coordinates, but the midpoint formula involves finding the **sum** of the coordinates.

(eText Screens 19.4-1–19.4-29)

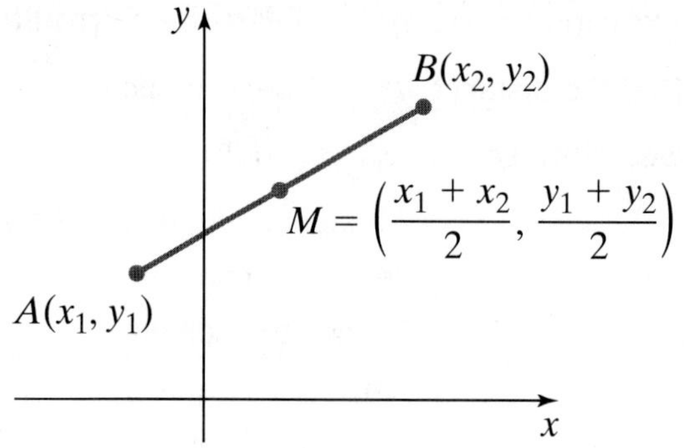

Figure 11

Because the **midpoint** lies halfway between the endpoints of a line segment (see Figure 11), the distance from the midpoint to either endpoint is half the distance between the endpoints.

$$d(A, M) = d(M, B) = \frac{d(A, B)}{2}$$

CAUTION The **distance between two** points is a number, but the midpoint of a line segment is a point. When writing a midpoint, be sure to use an **ordered pair.**

Example 2 Finding the Midpoint of a Line Segment

Find the **midpoint** of the line segment with endpoints $(-3, 2)$ and $(4, 6)$.

Solution Let $(x_1, y_1) = (-3, 2)$ and $(x_2, y_2) = (4, 6)$ and use the midpoint formula.

Midpoint formula: $\left(\frac{x_1 + x_2}{2}, \frac{y_1 + y_2}{2}\right)$

Substitute the coordinates: $= \left(\frac{-3 + 4}{2}, \frac{2 + 6}{2}\right)$

Simplify: $= \left(\frac{1}{2}, \frac{8}{2}\right)$

$= \left(\frac{1}{2}, 4\right)$

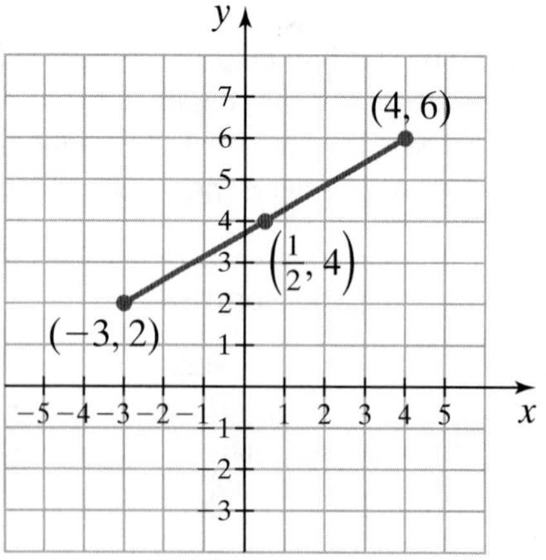

Figure 12

You Try It **Work through this You Try It problem.**

Work Exercises 9–16 in this textbook or in the MyMathLab Study Plan.

OBJECTIVE 3 WRITE THE STANDARD FORM OF AN EQUATION OF A CIRCLE

A **circle** is the set of all points (x, y) in the **coordinate plane** that are a fixed distance r from a fixed point (h, k). The fixed distance r is called the **radius** of the circle, and the fixed point (h, k) is called the **center** of the circle. See **Figure 13**.

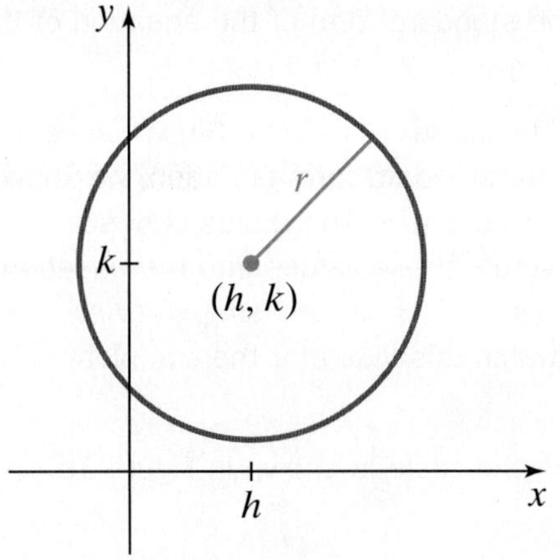

Figure 13
Circle with center (h, k) and radius, r

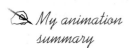

My animation summary

The **distance formula** can be used to develop the standard form of the equation of a circle. Watch this **animation** to see how this is done.

> **Standard Form of the Equation of a Circle**
>
> The **standard form of the equation of a circle** with center (h, k) and radius r is
>
> $$(x - h)^2 + (y - k)^2 = r^2.$$

Example 3 Writing the Standard Form of the Equation of a Circle

Write the standard form of the equation of a **circle** with center $(0, 0)$ and radius 3.

Solution The **center** is $(0, 0)$. Since the center is represented by (h, k) in our standard form equation, we know that $h = 0$ and $k = 0$. The **radius** is 3, so we have $r = 3$. Substitute these values into the equation.

$$\left(x - (0)\right)^2 + \left(y - (0)\right)^2 = 3^2$$
$$x^2 + y^2 = 9$$

The graph of the circle is shown in Figure 14.

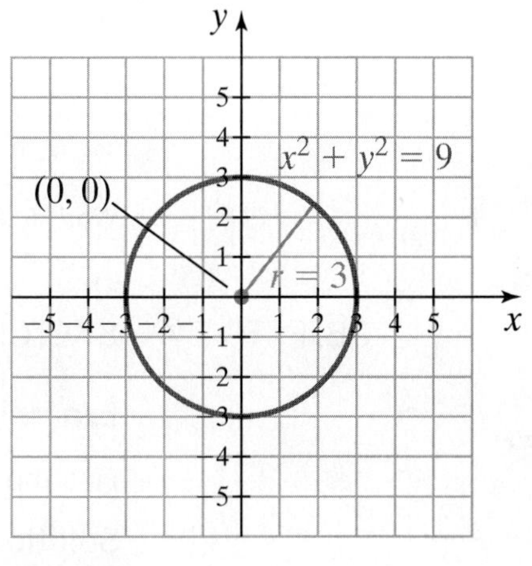

Figure 14

You Try It **Work through this You Try It problem.**

Work Exercises 17 and 18 in this textbook or in the MyMathLab Study Plan.

From Example 3, we see that the equation of a **circle** centered at the **origin** takes on a special form. The **standard form** of the equation of a circle centered at the origin with **radius** r is

$$x^2 + y^2 = r^2.$$

Example 4 Writing the Standard Form of the Equation of a Circle

My video summary

Write the standard form of the equation of the circle with center $(-2, 3)$ and radius 6.

Solution The **center** is $(-2, 3)$. Since the center is (h, k) in our standard form equation, we know that $h = -2$ and $k = 3$. The radius is 6, so $r = 6$. Substitute these values into the equation $(x - h)^2 + (y - k)^2 = r^2$ and simplify. View the answer, or watch this **video** for the complete solution.

The graph of the circle is shown in Figure 15.

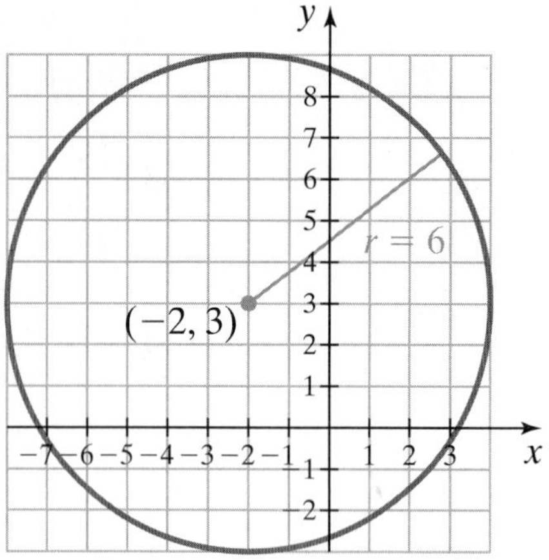

Figure 15

You Try It **Work through this You Try It problem.**

Work Exercises 19–22 in this textbook or in the MyMathLab Study Plan.

Example 5 Writing the Standard Form of the Equation of a Circle

My video summary

Write the standard form of the equation of the **circle** with **center** $(0, -4)$ and **radius** $r = \sqrt{5}$.

Solution Determine h, k, and r. Then substitute these values into the **standard form** of the equation of a circle and simplify. View the answer, or watch this **video** for the complete solution.

You Try It **Work through this You Try It problem.**

Work Exercises 23 and 24 in this textbook or in the MyMathLab Study Plan.

OBJECTIVE 4 SKETCH THE GRAPH OF A CIRCLE GIVEN IN STANDARD FORM

Example 6 Sketching the Graph of a Circle

Find the **center** and **radius** of the **circle** $x^2 + y^2 = 4$ and sketch its graph.

Solution We start by finding the center (h, k) and radius, r. The equation $x^2 + y^2 = 4$ has the form $(x - 0)^2 + (y - 0)^2 = 2^2$. Therefore, the circle is centered at the **origin**, $(h, k) = (0, 0)$, and has radius $r = 2$.

To sketch the graph of this circle, locate the center and then plot a few **points** on the circle. The easiest points to plot are located two units left and right from the center and two units up and down from the center.

$$\overbrace{(0 - \underbrace{2}_{r}, 0)}^{\text{2 units left}} \rightarrow (-2, 0) \qquad \overbrace{(0 + \underbrace{2}_{r}, 0)}^{\text{2 units right}} \rightarrow (2, 0)$$

$$\overbrace{(0, 0 + \underbrace{2}_{r})}^{\text{2 units up}} \rightarrow (0, 2) \qquad \overbrace{(0, 0 - \underbrace{2}_{r})}^{\text{2 units down}} \rightarrow (0, -2)$$

Complete the graph by drawing the **circle** through these points, as shown in Figure 16.

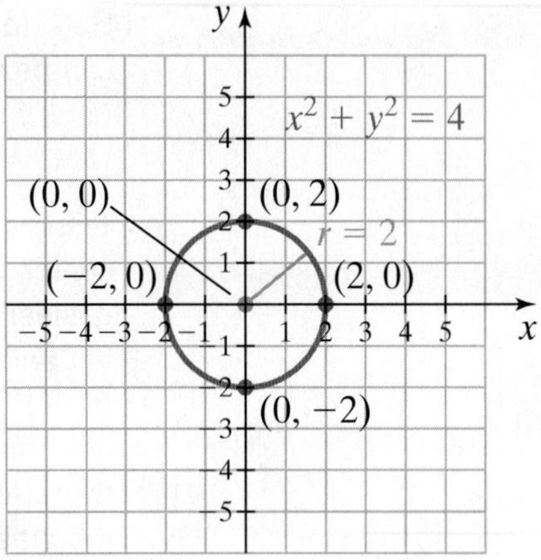

Figure 16
Graph of $x^2 + y^2 = 4$

You Try It **Work through this You Try It problem.**

Work Exercises 25 and 26 in this textbook or in the MyMathLab Study Plan.

Example 7 Sketching the Graph of a Circle

Find the **center** and **radius** of the **circle** $(x + 1)^2 + (y - 4)^2 = 25$ and sketch its graph.

Solution The equation $(x + 1)^2 + (y - 4)^2 = 25$ can be written as $(x - (-1))^2 + (y - 4)^2 = 5^2$. Comparing this equation to the **standard form** of the equation of a circle, we find that the circle has center $(h, k) = (-1, 4)$ and radius $r = 5$.

Locate the center, then plot a few points on the circle. Start with the points that are five units left and right of the center and five units up and down from the center.

$$\overbrace{(-1 - \underbrace{5}_{r}, 4)}^{\text{5 units left}} \rightarrow (-6, 4) \qquad \overbrace{(-1 + \underbrace{5}_{r}, 4)}^{\text{5 units right}} \rightarrow (4, 4)$$

$$\overbrace{(-1, 4 + \underbrace{5}_{r})}^{\text{5 units up}} \rightarrow (-1, 9) \qquad \overbrace{(-1, 4 - \underbrace{5}_{r})}^{\text{5 units down}} \rightarrow (-1, -1)$$

Use these points to sketch the graph of the circle. View the answer.

You Try It **Work through this You Try It problem.**

Work Exercises 27–30 in this textbook or in the MyMathLab Study Plan.

Recall that each point on the graph of an equation is an **ordered pair solution** to the equation. We can improve our sketch by finding additional points. For example, we may want to find the **intercepts** (if any) and then plot those corresponding points.

Example 8 Sketching the Graph of a Circle

Find the **center** and the **radius**, and sketch the graph of the **circle** $(x - 1)^2 + (y + 2)^2 = 9$. Also find any intercepts.

(eText Screens 19.4-1–19.4-29)

My video summary

Solution The equation $(x - 1)^2 + (y + 2)^2 = 9$ can be written in **standard form** as $(x - 1)^2 + (y - (-2))^2 = 3^2$ with center $(h, k) = (1, -2)$ and radius $r = 3$.

To sketch the graph of this circle, locate the center, then plot a few points on the circle. The easiest points to plot are located three units left and right from the center, $(-2, -2)$ and $(4, -2)$ respectively, and three units up and down from the center, $(1, 1)$ and $(1, -5)$ respectively. After plotting these four points, complete the graph by drawing the circle through the points as shown in Figure 17.

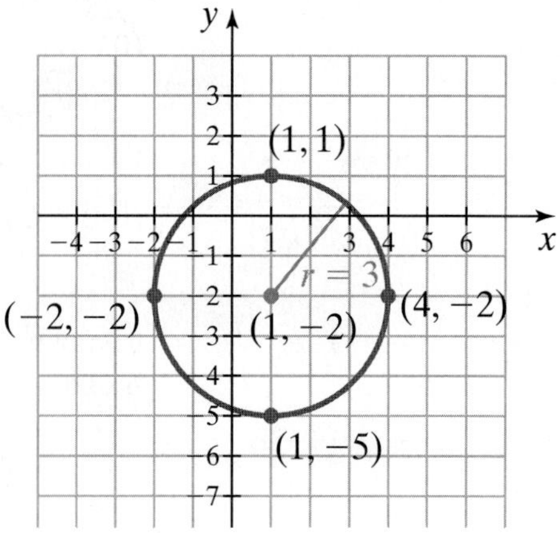

Figure 17
Graph of $(x - 1)^2 + (y + 2)^2 = 9$

To find the x-**intercepts**, set $y = 0$ and solve the resulting **quadratic equation** for x. To find the y-**intercepts**, set $x = 0$ and solve the resulting quadratic equation for y. View the answer, or watch this **video** for the complete solution.

You Try It **Work through this You Try It problem.**

Work Exercises 31 and 32 in this textbook or in the MyMathLab Study Plan.

OBJECTIVE 5 WRITE THE GENERAL FORM OF A CIRCLE IN STANDARD FORM AND SKETCH ITS GRAPH

Since a **circle** is completely defined by its **center** and **radius**, we only need these two pieces of information to write the equation of a circle or to sketch its **graph**. We saw in Examples 3–5 that given the center and radius, we can write the **standard form** of an equation of a circle. Then we used this standard form in Examples 6–8 to find the center and radius and then to sketch the graph of a circle.

Not all equations of circles are given in standard form. Consider the circle $(x + 3)^2 + (y - 1)^2 = 49$. This equation is in standard form, so we can quickly see that the center is $(h, k) = (-3, 1)$ and the radius is $r = \sqrt{49} = 7$. However, let's see what happens if we expand the equation.

Original equation:	$(x + 3)^2 + (y - 1)^2 = 49$
Expand to remove parentheses:	$x^2 + 6x + 9 + y^2 - 2y + 1 = 49$
Collect like terms:	$x^2 + 6x + y^2 - 2y + 10 = 49$
Subtract 49 from both sides:	$x^2 + 6x + y^2 - 2y - 39 = 0$
Rearrange terms:	$x^2 + y^2 + 6x - 2y - 39 = 0$

The equation $x^2 + y^2 + 6x - 2y - 39 = 0$ is written in **general form**. In this form, the **center** and **radius** are less obvious.

General Form of the Equation of a Circle

The **general form of the equation of a circle** is

$$Ax^2 + By^2 + Cx + Dy + E = 0,$$

where A, B, C, D, and E are **real numbers** and $A = B \neq 0$.

(eText Screens 19.4-1–19.4-29)

To find the center and radius of a circle from its equation in general form, we need to **complete the square** in both x and y to write the equation in **standard form.** Before working through Example 9, you might want to review completing the square from **Topic 17.1.**

Example 9 Writing the General Form of a Circle in Standard Form and Sketching Its Graph

Write the equation $x^2 + y^2 + 10x + 9 = 0$ in **standard form.** Find the **center, radius,** and **intercepts,** then sketch the graph.

Solution The equation contains a **linear term** in x, but no linear term in y. So, we only need to **complete the square** in x. Start by rearranging terms to group the x terms together and move any constants to the right-hand side of the equation.

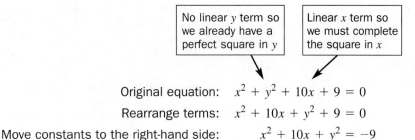

No linear y term so we already have a perfect square in y

Linear x term so we must complete the square in x

Original equation: $x^2 + y^2 + 10x + 9 = 0$

Rearrange terms: $x^2 + 10x + y^2 + 9 = 0$

Move constants to the right-hand side: $x^2 + 10x + y^2 = -9$

To complete the square in x, we divide the **coefficient** of the linear term by 2, **square** the result, and add this to both sides of the equation. Since $\left(\frac{1}{2}\cdot 10\right)^2 = 5^2 = 25$, we need to add 25 to both sides, and then factor the left-hand side.

Complete the square in x: $x^2 + 10x + 25 + y^2 = -9 + 25$

$\left(\frac{1}{2}\cdot 10\right)^2 = 25$

Factor the left-hand side: $(x + 5)^2 + y^2 = 16$

Now the equation is written in **standard form.** Comparing to $(x - h)^2 + (y - k)^2 = r^2$, the **center** is $(h, k) = (-5, 0)$, and the **radius** is $r = \sqrt{16} = 4$.

To find the x-**intercepts,** we let $y = 0$ and solve for x.

Standard form: $(x + 5)^2 + y^2 = 16$

Substitute 0 for y: $(x + 5)^2 + (0)^2 = 16$

Simplify: $(x + 5)^2 = 16$

Square root property: $x + 5 = \pm\sqrt{16}$

Simplify: $x + 5 = \pm 4$

Subtract 5 from both sides: $x = -5 \pm 4$

The x-intercepts are $-5 - 4 = -9$ and $-5 + 4 = -1$. So, the points $(-9, 0)$ and $(-1, 0)$ are on the graph.

To find the y-**intercepts,** we let $x = 0$ and solve for y.

Original equation: $x^2 + y^2 + 10x + 9 = 0$

Substitute 0 for x: $(0)^2 + y^2 + 10(0) + 9 = 0$

Simplify: $y^2 + 9 = 0$

Subtract 9 from both sides: $y^2 = -9$

There is no **real number** whose **square** is -9, so the graph has no y-intercepts. The graph of the circle is shown in Figure 18.

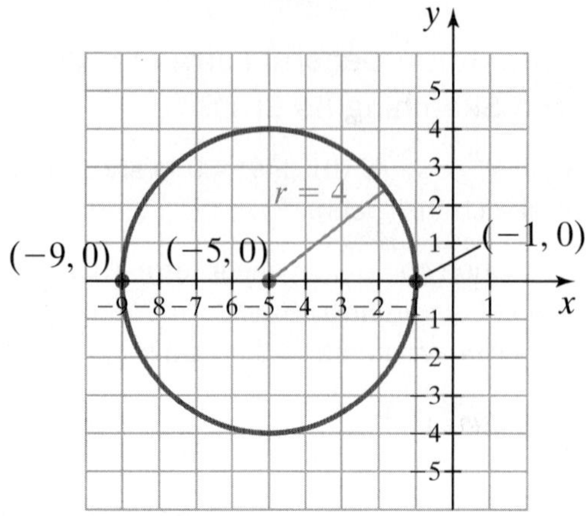

Figure 18
Graph of $x^2 + y^2 + 10x + 9 = 0$

You Try It **Work through this You Try It problem.**

Work Exercises 33 and 34 in this textbook or in the MyMathLab Study Plan.

Example 10 Writing the General Form of a Circle in Standard Form and Sketching Its Graph

My video summary

Write the equation $x^2 + y^2 - 8x + 6y + 16 = 0$ in **standard form**; find the **center**, **radius**, and **intercepts**, and sketch the graph.

Solution The equation contains **linear terms** in both x and y, so we will need to **complete the square** in both x and y. We rearrange the terms, complete the square, and move any **constants** to the right-hand side of the equation.

Rearrange the terms: $x^2 - 8x \quad + y^2 + 6y \quad = -16$

Complete the square in x and y. Remember to add 16 and 9 to both sides: $x^2 - 8x + 16 + y^2 + 6y + 9 = -16 + 16 + 9$

$\left(\frac{1}{2}\cdot(-8)\right)^2 = 16 \qquad \left(\frac{1}{2}\cdot 6\right)^2 = 9$

Factor the left: $(x - 4)^2 + (y + 3)^2 = 9$

Now we have converted the **general form** of the circle into **standard form** with **center** $(h, k) = (4, -3)$ and **radius** $r = 3$. Watch this **video** to verify that there are no y-**intercepts**. The only x-**intercept** is $x = 4$. The graph of the circle is shown in Figure 19.

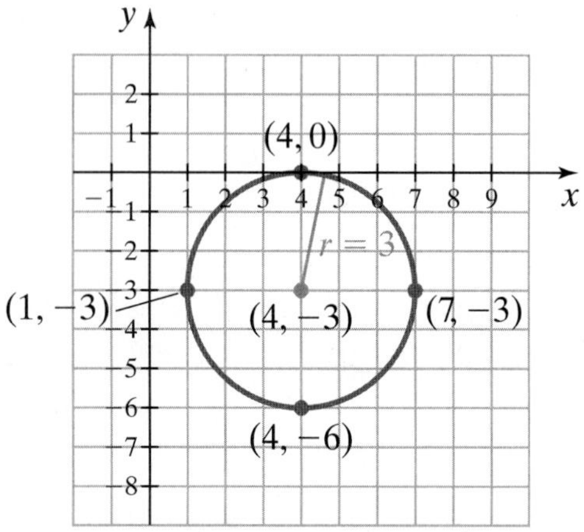

Figure 19
Graph of $x^2 + y^2 - 8x + 6y + 16 = 0$

(eText Screens 19.4-1–19.4-29)

You Try It **Work through this You Try It problem.**

Work Exercises 35–40 in this textbook or in the MyMathLab Study Plan.

Example 11 Writing the General Form of a Circle in Standard Form, Where $A \neq 1$ and $B \neq 1$

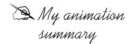

My animation summary

Write the equation $4x^2 + 4y^2 + 4x - 8y + 1 = 0$ in **standard form.** Find the **center, radius,** and **intercepts.** Then sketch the graph.

Solution Work through this **animation** to verify that this equation is equivalent to the following equation in standard form:

$$\left(x + \frac{1}{2}\right)^2 + (y - 1)^2 = 1; \text{center } (h, k) = \left(-\frac{1}{2}, 1\right) \text{ and } r = \sqrt{1} = 1$$

Now work through the **animation** again to verify that the one x-**intercept** is found at $\left(-\frac{1}{2}, 0\right)$, and the y-**intercepts** are found at $\left(0, 1 - \frac{\sqrt{3}}{2}\right)$ and $\left(0, 1 + \frac{\sqrt{3}}{2}\right)$. The graph of the circle is shown in **Figure 20**.

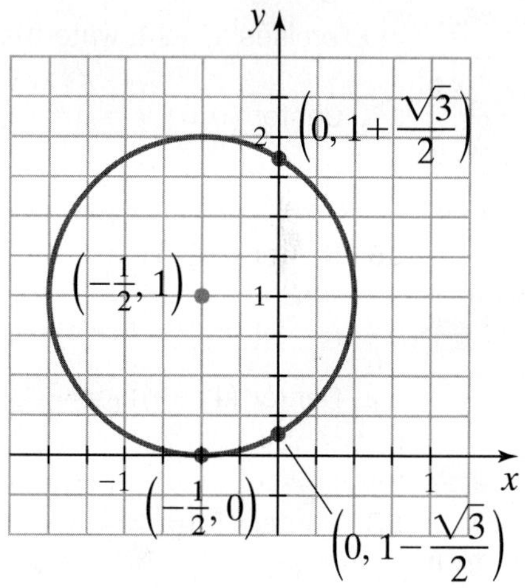

Figure 20
Graph of $4x^2 + 4y^2 + 4x - 8y + 1 = 0$

You Try It **Work through this You Try It problem.**

Work Exercises 41–44 in this textbook or in the MyMathLab Study Plan.

CAUTION

Every equation of a **circle** can be written in **general form.** However, not every equation of the form $Ax^2 + By^2 + Cx + Dy + E = 0$ is the equation of a circle. For example, the equation $2x^2 + 2y^2 + 1 = 0$ has no graph, and the graph of $x^2 + y^2 + 2x - 4y + 5 = 0$ is a single point.

19.4 Exercises

In Exercises 1–8, find the distance $d(A, B)$ between points A and B.

1. $A(2, 7); B(5, 11)$ **2.** $A(1, 5); B(-2, 1)$

You Try It **3.** $A(3, 5); B(-2, -2)$ **4.** $A(2, -4); B(-3, 6)$

5. $A\left(\frac{2}{3}, 5\right), B\left(-1, \frac{1}{2}\right)$

6. $A(0, -\sqrt{2}); B(\sqrt{3}, 0)$

7. $A(2, -3); B(2, 5)$

8. $A(-1, 4); B(5, 4)$

In Exercises 9–16, find the midpoint of the line segment with endpoints A and B.

9. $A(3, 7); B(5, 9)$

10. $A(2, -5); B(4, 1)$

You Try It **11.** $A(-1, 4); B(-2, -2)$

12. $A(-3, 0); B(0, 7)$

13. $A(0, 1); B\left(-3, \frac{1}{2}\right)$

14. $A(\sqrt{27}, 4); B(\sqrt{3}, 2)$

15. $A(1, 3); B(1, -5)$

16. $A(4, -2); B(-3, -2)$

In Exercises 17–24, write the standard form of the equation of each circle described.

You Try It **17.** Center $(0, 0)$, $r = 1$

18. Center $(0, 0)$, $r = \sqrt{3}$

You Try It **19.** Center $(-2, 3)$, $r = 4$

20. Center $(0, 4)$, $r = 8$

21. Center $(1, -4)$, $r = \frac{3}{4}$

22. Center $\left(\frac{1}{5}, 3\right)$, $r = 2$

You Try It **23.** Center $(3, 0)$, $r = \sqrt{2}$

24. Center $\left(-\frac{1}{4}, -\frac{1}{3}\right)$, $r = \frac{3}{4}$

In Exercises 25–30, find the center and radius of each circle and sketch its graph.

You Try It **25.** $x^2 + y^2 = 9$

26. $(x + 3)^2 + y^2 = 4$

You Try It **27.** $(x - 1)^2 + (y + 5)^2 = 16$

28. $(x + 2)^2 + (y + 4)^2 = 36$

29. $\left(x - \frac{1}{4}\right)^2 + \left(y + \frac{1}{2}\right)^2 = 4$

30. $(x + 4)^2 + (y - 5)^2 = 10$

In Exercises 31–32, find the center, radius, and intercepts of each circle and then sketch its graph.

You Try It **31.** $(x - 4)^2 + (y + 7)^2 = 12$

32. $(x + 1)^2 + (y - 3)^2 = 20$

In Exercises 33–44, write the equation of each circle in standard form. Find the center, radius, and intercepts. Then sketch the graph.

You Try It **33.** $x^2 + y^2 + 6x + 5 = 0$

34. $x^2 + y^2 + 2y - 8 = 0$

You Try It **35.** $x^2 + y^2 + 2x - 4y + 1 = 0$

36. $x^2 + y^2 - 10x + 6y + 18 = 0$

(eText Screens 19.5-1–19.5-21)

37. $x^2 + y^2 - 4x - 8y + 19 = 0$

38. $x^2 + y^2 - 6x - 12y - 5 = 0$

You Try It

39. $x^2 + y^2 - 3x - y - \frac{1}{2} = 0$

40. $x^2 + y^2 + \frac{2}{3}x - \frac{1}{2}y - \frac{7}{18} = 0$

41. $2x^2 + 2y^2 - 4x + 8y + 2 = 0$

42. $16x^2 + 16y^2 - 16x - 8y - 11 = 0$

43. $144x^2 + 144y^2 - 72x - 96y - 551 = 0$

44. $36x^2 + 36y^2 + 12x + 72y - 35 = 0$

19.5 Polynomial and Rational Inequalities

THINGS TO KNOW

Before working through this topic, be sure you are familiar with the following concepts:

		VIDEO	ANIMATION	INTERACTIVE
You Try It	1. Factor by Grouping (Topic 15.1, **Objective 4**)			✓
You Try It	2. Factor Trinomials of the Form $x^2 + bx + c$ (Topic 15.2, **Objective 1**)	✓	✓	✓
You Try It	3. Factor Trinomials of the Form $ax^2 + bx + c$ Using Trial and Error (Topic 15.3, **Objective 1**)	✓	✓	✓
You Try It	4. Factor Trinomials of the Form $ax^2 + bx + c$ Using the ac Method (Topic 15.4, **Objective 1**)	✓	✓	✓
You Try It	5. Solve Polynomial Equations by Factoring (Topic 15.7, **Objective 2**)	✓		

OBJECTIVES

1 Solve Polynomial Inequalities

2 Solve Rational Inequalities

OBJECTIVE 1 SOLVE POLYNOMIAL INEQUALITIES

In Topic 11.7, we learned how to solve linear inequalities. In this objective, we learn how to solve **polynomial inequalities.**

Definition Polynomial Inequality

A **polynomial inequality** is an inequality that can be written as

$$P(x) < 0, P(x) > 0, P(x) \le 0, \text{or } P(x) \ge 0,$$

where $P(x)$ is a **polynomial function.**

(eText Screens 19.5-1–19.5-21)

When solving polynomial inequalities, **x-intercepts** will play an important part. To understand why, consider the polynomial functions graphed in Figure 21.

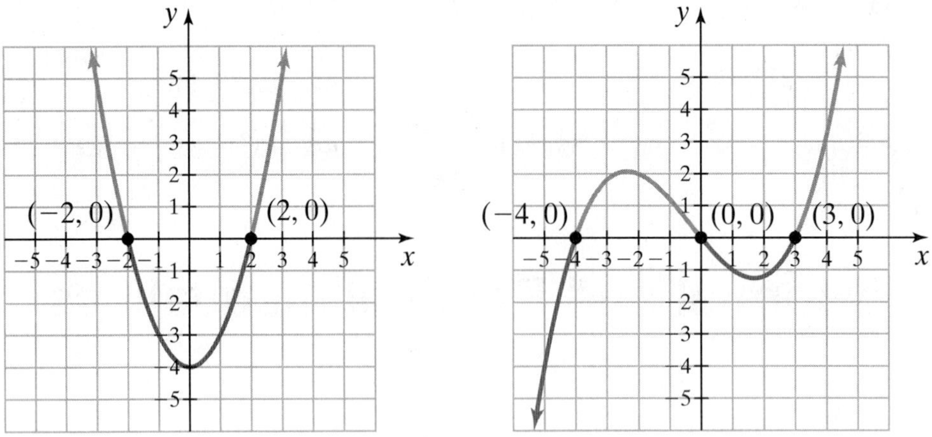

(a) $f(x) = (x + 2)(x - 2)$ (b) $g(x) = 0.1x(x + 4)(x - 3)$ **Figure 21**

From Figure 21, we see that for some x values, the value of the polynomial function will be positive; for others, it will be negative, and for still other x values, the resulting value of the polynomial function will be zero. The x values where the function value is zero are the **x-intercepts**, which are also **boundary points** that divide the **real number line** into **intervals.** Notice that in each interval, the function is always positive or always negative. This is true for all **polynomials** and is central to our approach to solving **polynomial inequalities**.

Example 1 Solving a Polynomial Inequality

My video summary

Solve $x^3 - 3x^2 + 2x \geq 0$.

Solution Work through the following solution or watch this **video**.

First, **factor** the left-hand side to get $x(x - 1)(x - 2) \geq 0$. Second, find all real values of x that make the left-hand side *equal to* zero. These values are the **boundary points.** To find these boundary points, set the **factored polynomial** on the left equal to zero and solve for x.

Set the factored polynomial equal to 0: $x(x - 1)(x - 2) = 0$

Use the **zero product property**: $x = 0$ or $x - 1 = 0$ or $x - 2 = 0$

The boundary points are $x = 0$, $x = 1$, and $x = 2$.

Next, plot each boundary point on a **number line.** Because the expression $x(x - 1)(x - 2)$ is equal to zero at our three boundary points, we use a solid circle ● at each boundary point to indicate that the inequality $x(x - 1)(x - 2) \geq 0$ is satisfied at these points. (**Note**: If the inequality had been a **strict inequality,** such as $>$ or $<$, we would have used an open circle ○ to represent that the boundary points were *not* part of the solution. We will look at this in Example 2.) Notice that in Figure 22, we have naturally divided the number line into four **intervals.**

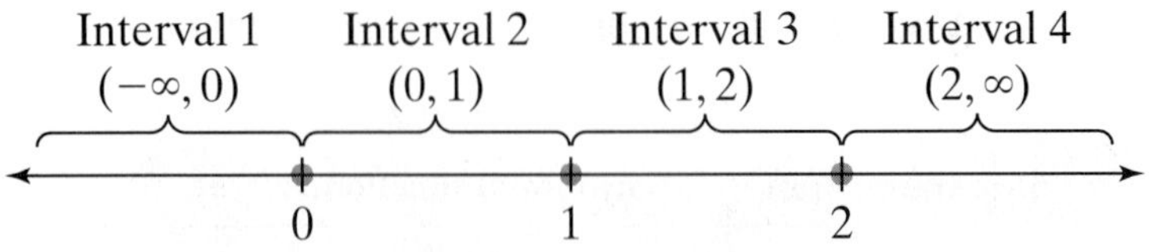

Figure 22

(eText Screens 19.5-1–19.5-21)

The expression $x(x-1)(x-2)$ is equal to zero *only* at the three **boundary points:** $0, 1,$ and 2. So, in any of the four intervals shown in **Figure 22**, the expression $x(x-1)(x-2)$ must be either always *positive* or always *negative* throughout the entire interval. To check whether this expression is positive or negative on each interval, pick a number called a **test value** from each interval. The test value can be any point in the interval but not a boundary point. Possible test values are plotted in Figure 23.

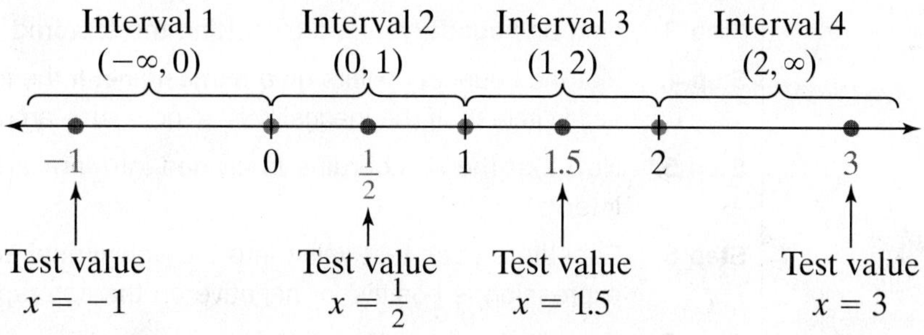

Figure 23

Substitute the **test value** into the expression $x(x-1)(x-2)$ and check to see if it yields a positive or negative value.

Interval	Test Value	Substitute Test Value into $x(x-1)(x-2)$	Comment
1. $(-\infty, 0)$	$x=-1$	$(-1)(-1-1)(-1-2) \Rightarrow (-)(-)(-) = -$	Expression is negative on $(-\infty, 0)$
2. $(0,1)$	$x=\frac{1}{2}$	$\left(\frac{1}{2}\right)\left(\frac{1}{2}-1\right)\left(\frac{1}{2}-2\right) \Rightarrow (+)(-)(-) = +$	Expression is positive on $(0, 1)$
3. $(1,2)$	$x=1.5$	$(1.5)(1.5-1)(1.5-2) \Rightarrow (+)(+)(-) = -$	Expression is negative on $(1, 2)$
4. $(2,\infty)$	$x=3$	$(3)(3-1)(3-2) \Rightarrow (+)(+)(+) = +$	Expression is positive on $(2, \infty)$

If the expression $x(x-1)(x-2)$ is positive on an **interval**, we place a "+" above the interval on the **number line**. If the expression $x(x-1)(x-2)$ is negative, we place a "−" above the interval. See Figure 24.

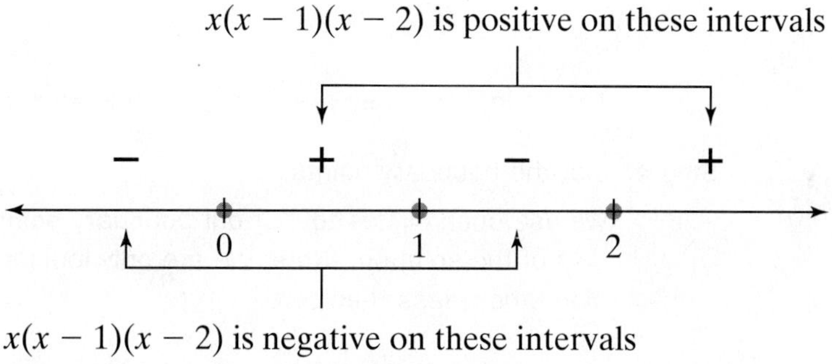

Figure 24

In Figure 24, we see that $x(x-1)(x-2)$ is greater than zero on the intervals $(0,1)$ and $(2,\infty)$ and is equal to zero at the **boundary points** $x=0$, $x=1$, and $x=2$. So, the **solution set** for the inequality $x(x-1)(x-2) \geq 0$ is the interval $[0,1] \cup [2,\infty)$.

You Try It **Work through this You Try It problem.**

Work Exercises 1–6 in this textbook or in the MyMathLab Study Plan.

Example 1 illustrates the following approach for solving a factorable **polynomial inequality**.

Solving Polynomial Inequalities

Step 1. Move all **terms** to one side of the inequality, leaving zero on the other side.

Step 2. **Factor** the nonzero side of the inequality.

Step 3. Find all **boundary points** by setting the **factored polynomial** equal to zero.

Step 4. Plot the boundary points on a **number line.** If the inequality is $\leq$ or $\geq$, use a solid circle ●. If the inequality is $<$ or $>$, use an open circle ○.

Step 5. Now that the number line is divided into **intervals**, pick a **test value** from each interval.

Step 6. Substitute each test value into the **polynomial** and determine whether the expression is positive or negative on the corresponding interval.

Step 7. Find the intervals that **satisfy** the inequality.

Example 2 Solving a Polynomial Inequality

Solve $x^2 + 5x < 3 - x^2$.

Solution We use the steps for **solving polynomial inequalities.**

Step 1. Move all terms to one side of the inequality.

$$2x^2 + 5x - 3 < 0$$

Step 2. Factor.

$$(2x - 1)(x + 3) < 0$$

Step 3. Find **boundary points** by setting the factored polynomial equal to zero.

$$(2x - 1)(x + 3) = 0$$

Zero product property: $2x - 1 = 0$ or $x + 3 = 0$

$$2x = 1 \qquad x = -3$$

$$x = \frac{1}{2}$$

The boundary points are $x = -3$ and $x = \frac{1}{2}$.

Step 4. Plot the boundary points.

We use open circles to plot our **boundary points** because these points are *not* part of the solution. (**Note**: We are only looking for values that make the expression strictly less than zero.)

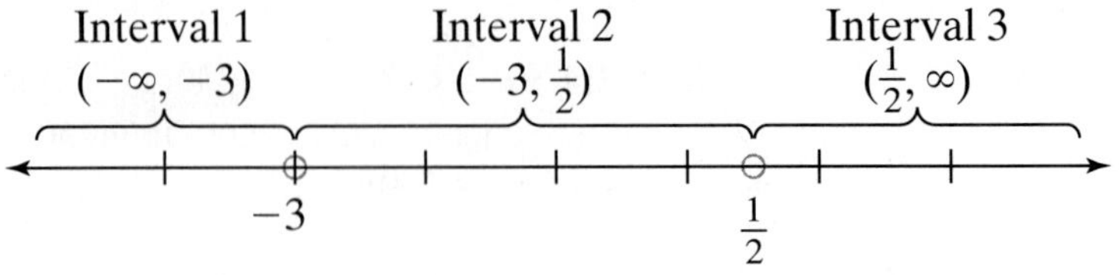

(eText Screens 19.5-1–19.5-21)

Step 5. We have three **intervals**. Pick a **test value** from each interval.

Interval 1: Test value $x = -4$

Interval 2: Test value $x = 0$

Interval 3: Test value $x = 1$

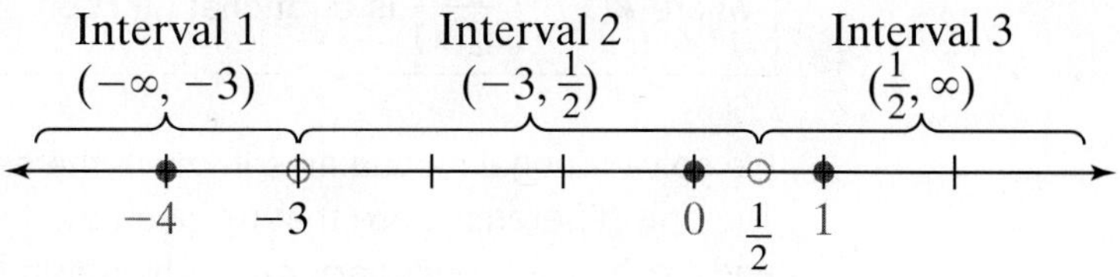

Step 6. Substitute each test value into the polynomial.

$x = -4$: $(2(-4) - 1)((-4) + 3) \Rightarrow (-)(-) = +$,
expression is positive on $(-\infty, -3)$.

$x = 0$: $(2(0) - 1)((0) + 3) \Rightarrow (-)(+) = -$,
expression is negative on $\left(-3, \frac{1}{2}\right)$.

$x = 1$: $(2(1) - 1)((1) + 3) \Rightarrow (+)(+) = +$,
expression is positive on $\left(\frac{1}{2}, \infty\right)$.

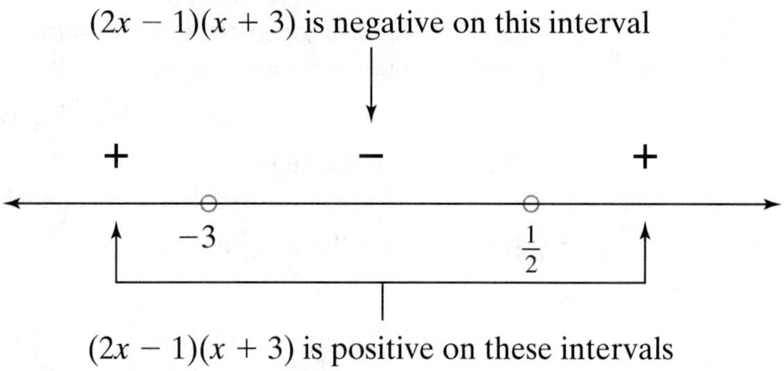

Step 7. Find the **intervals** that **satisfy** the inequality.

Because we want x-values that result in a polynomial value less than zero (negative values), the **solution set** is the **interval** $\left(-3, \frac{1}{2}\right)$.

You Try It **Work through this You Try It problem.**

Work Exercises 7–10 in this textbook or in the MyMathLab Study Plan.

OBJECTIVE 2 SOLVE RATIONAL INEQUALITIES

We now extend our technique for solving **polynomial inequalities** to solve **rational inequalities.**

(eText Screens 19.5-1–19.5-21)

Definition Rational Inequality

A **rational inequality** is an inequality that can be written as

$$R(x) < 0, R(x) > 0, R(x) \le 0, \text{or } R(x) \ge 0,$$

where $R(x) = \dfrac{P(x)}{Q(x)}$ is a **rational function** and $P(x)$ and $Q(x)$ are **polynomial functions.**

We solve rational inequalities in much the same way that we solve polynomial inequalities. The only differences are that we need the left-hand side to be a single **rational expression** and we find the **boundary points** by setting both the **numerator** and **denominator** equal to zero.

Solving Rational Inequalities

Step 1. Move all **terms** to one side of the inequality, leaving zero on the other side.

Step 2. Combine the terms into a single **rational expression** and factor the nonzero side of the inequality (both the **numerator** and **denominator**).

Step 3. Find all **boundary points** by setting the **factored polynomials** in the numerator and denominator equal to zero.

Step 4. Plot the boundary points on a **number line.** If the inequality is $\le$ or $\ge$, use a solid circle ● for the boundary points from the numerator. If the inequality is $<$ or $>$, use an open circle ○. Boundary points from the denominator always have an open circle. If the same boundary point is obtained from both the numerator and denominator, then it is plotted with an open circle.

Step 5. Pick a **test value** from each interval.

Step 6. Substitute each test value into the rational expression and determine whether the expression is positive or negative on the corresponding interval.

Step 7. Find the intervals that **satisfy** the inequality.

Example 3 Solving a Rational Inequality

My video summary

Solve $\dfrac{x-4}{x+1} \ge 0.$

Solution Using the steps for **solving a rational inequality**, work through the following solution, or watch this **video**.

Because the left-hand side of the inequality is already a single **rational expression** in completely factored form, we can skip Steps 1 and 2 and start with Step 3.

Step 3. Find the **boundary points** by setting the **factored polynomials** in the **numerator** and the **denominator** equal to zero.

Numerator: $x - 4 = 0$, so $x = 4$ is a boundary point.

Denominator: $x + 1 = 0$, so $x = -1$ is a boundary point.

Step 4. Plot the boundary points.

Since the inequality is a *greater than or equal to* inequality, we use a closed circle ● to represent the boundary point $x = 4$. Because we cannot divide by zero, we use an open circle ○ to represent the boundary point from the denominator, $x = -1$.

(eText Screens 19.5-1–19.5-21)

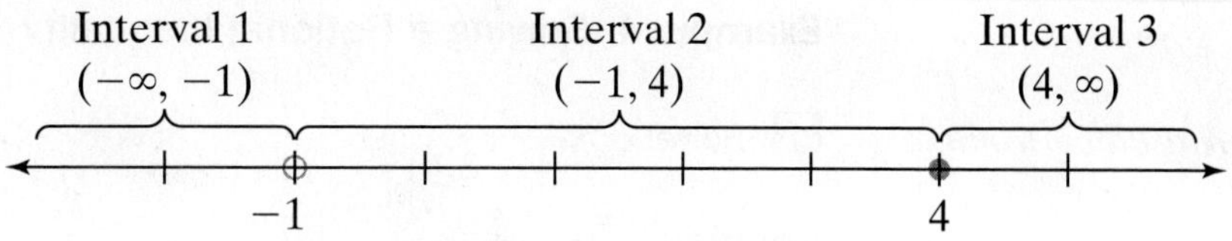

Step 5. Pick a **test value** from each interval.

Interval 1: Test value $x = -2$

Interval 2: Test value $x = 0$

Interval 3: Test value $x = 5$

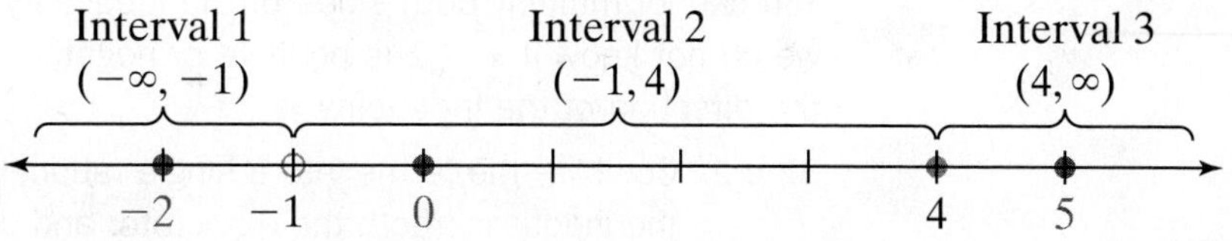

Step 6. Substitute each **test value** into the rational expression, and determine whether the expression is positive or negative on the corresponding interval.

$$x = -2: \frac{(-2 - 4)}{(-2 + 1))} \Rightarrow \frac{(-)}{(-)} = +,$$

expression is positive on the interval $(-\infty, -1)$.

$$x = 0: \frac{(-0 - 4)}{(0 + 1)} \Rightarrow \frac{(-)}{(+)} = -,$$

expression is negative on the interval $(-1, 4)$.

$$x = 5: \frac{(5 - 4)}{(5 + 1)} \Rightarrow \frac{(+)}{(+)} = +,$$

expression is positive on the interval $(4, \infty)$.

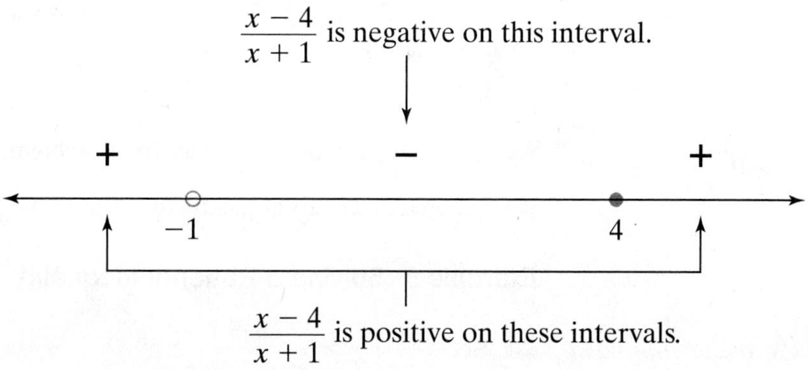

Step 7. Find the intervals that **satisfy** the inequality.

We look for x values that make the **rational expression** greater than or equal to zero. From the last step, we see that the **solution set** to the inequality is $(-\infty, -1) \cup [4, \infty)$.

Note that 4 is included as a solution but not -1 because $x = -1$ makes the **denominator** equal to zero.

You Try It **Work through this You Try It problem.**

Work Exercises 11–14 in this textbook or in the MyMathLab Study Plan.

(eText Screens 19.5-1–19.5-21)

Example 4 Solving a Rational Inequality

My video summary

Solve $x > \frac{3}{x-2}$.

Solution

Step 1. Subtract $\frac{3}{x-2}$ from both sides of the inequality.

$$x - \frac{3}{x-2} > 0$$

CAUTION

You cannot multiply both sides of the inequality by $x - 2$ to **clear the fraction** because we do not know if $x - 2$ is positive or negative. So, we are not sure if we need to reverse the direction of the inequality.

Step 2. Combine the terms into a single rational expression and factor the nonzero side of the inequality (both the numerator and denominator).

We combine the terms by finding a **common denominator**.

Rewrite the inequality: $x - \frac{3}{x-2} > 0$

Rewrite using a common denominator of $x - 2$: $\frac{x(x-2)}{x-2} - \frac{3}{x-2} > 0$

Combine terms: $\frac{x(x-2) - 3}{x-2} > 0$

Distribute: $\frac{x^2 - 2x - 3}{x-2} > 0$

Factor: $\frac{(x-3)(x+1)}{x-2} > 0$

Finish Steps 3–7 on your own. View the **answer**, or watch this **video** to see the entire solution.

You Try It **Work through this You Try It problem.**

Work Exercises 15–20 in this textbook or in the MyMathLab Study Plan.

Example 5 Solving a Rational Inequality

My video summary

Solve $\frac{x+1}{x-2} > \frac{7x+1}{x^2+x-6}$.

Solution Complete the **seven-step process** to solve the rational inequality. View the **answer**, or watch this **video** for the complete solution.

You Try It **Work through this You Try It problem.**

Work Exercises 21 and 22 in this textbook or in the MyMathLab Study Plan.

(eText Screens 19.5-1–19.5-21)

19.5 Exercises

In Exercises 1–10, solve each polynomial inequality. Express each solution using interval notation.

You Try It

1. $(x - 1)(x + 3) \geq 0$

2. $x^2 + 4x - 21 < 0$

3. $x(3x + 2) \leq 0$

4. $2x^2 - 4x > 0$

5. $x^2 \leq 1$

6. $(x - 1)(x + 4)(x - 3) \geq 0$

You Try It

7. $3x^2 + x < 3x + 1$

8. $2x^3 > 24x - 2x^2$

9. $x^3 + x^2 - x \leq 1$

10. $x^3 \geq -2x^2 - x$

In Exercises 11–22, solve each rational inequality. Express each solution using interval notation.

You Try It

11. $\dfrac{x + 3}{x - 1} \leq 0$

12. $\dfrac{2 - x}{3x + 9} \geq 0$

13. $\dfrac{x}{x - 1} > 0$

14. $\dfrac{x^2 - 9}{x + 2} \leq 0$

You Try It

15. $\dfrac{4}{x + 1} \geq 2$

16. $\dfrac{x + 5}{2x - 3} > 1$

17. $\dfrac{x^2 - 8}{x + 4} \geq x$

18. $\dfrac{x}{2 - x} \leq \dfrac{1}{x}$

19. $\dfrac{x - 1}{x - 2} \geq \dfrac{x + 2}{x + 3}$

20. $\dfrac{x - 1}{x + 1} + \dfrac{x + 1}{x - 1} \leq \dfrac{x + 5}{x^2 - 1}$

You Try It

21. $\dfrac{x + 1}{x + 4} < \dfrac{-11x - 17}{x^2 - x - 20}$

22. $\dfrac{x - 8}{x^2 + 5x + 4} < \dfrac{-21x}{(x + 1)(x - 3)(x + 4)}$

MODULE TWENTY

Exponential and Logarithmic Functions and Equations

MODULE TWENTY CONTENTS

20.1 Transformations of Functions

THINGS TO KNOW

Before working through this topic, be sure you are familiar with the following concepts:

VIDEO ANIMATION INTERACTIVE

You Try It 1. Graph Linear Equations by Plotting Points (Topic 12.2, **Objective 3**)

You Try It 2. Find the Domain and Range of a Relation (Topic 17.1, **Objective 2**)

You Try It 3. Graph Simple Functions by Plotting Points (Topic 17.3, **Objective 1**)

You Try It 4. Graph Functions That Contain Square Roots or Cube Roots (Topic 18.2, **Objective 3**)

You Try It 5. Graph Quadratic Functions by Using Translations (Topic 19.2, **Objective 2**)

(eText Screens 20.1-1–20.1-39)

OBJECTIVES

1 Use Vertical Shifts to Graph Functions

2 Use Horizontal Shifts to Graph Functions

3 Use Reflections to Graph Functions

4 Use Vertical Stretches and Compressions to Graph Functions

5 Use Horizontal Stretches and Compressions to Graph Functions

6 Use Combinations of Transformations to Graph Functions

In this topic, we learn how to sketch the graphs of new functions using the graphs of known functions. Starting with the graph of a known function, we "transform" it into a new function by applying various transformations. Before we begin our discussion about transformations, it is critical that you know the graphs of some basic functions that have been explored earlier in this text. Take a moment to review these summaries of some basic functions. Select a function name to review its properties.

REVIEW OF THE BASIC FUNCTIONS

Select a function to review its graph.

The **identity function** $f(x) = x$

The **absolute value function** $f(x) = |x|$

The **square function** $f(x) = x^2$

The **cube function** $f(x) = x^3$

The **square root function** $f(x) = \sqrt{x}$

The **cube root function** $f(x) = \sqrt[3]{x}$

OBJECTIVE 1 USE VERTICAL SHIFTS TO GRAPH FUNCTIONS

Example 1 Vertically Shifting a Function

My video summary

Sketch the graphs of $f(x) = |x|$ and $g(x) = |x| + 2$.

Solution Table 1 shows that for every value of x, the y-coordinate of the function g is always two greater than the y-coordinate of the function f. The two functions are sketched in **Figure 1**. The graph of $g(x) = |x| + 2$ is exactly the same as the graph of $f(x) = |x|$, except the graph of g is shifted *up* two units.

x	$f(x) = \|x\|$	$g(x) = \|x\| + 2$
−3	3	5
−2	2	4
−1	1	3
0	0	2
1	1	3
2	2	4
3	3	5

Table 1

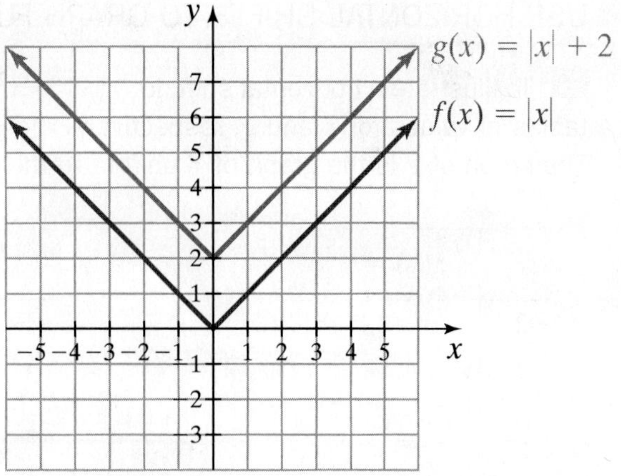

Figure 1

We see from Example 1 that if c is a positive number, then $y = f(x) + c$ is the graph of f shifted *up* c units. It follows that for $c > 0$, the graph of $y = f(x) - c$ is the graph of f shifted *down* c units.

Vertical Shifts of Functions

If c is a positive real number,

The graph of $y = f(x) + c$ is obtained by shifting the graph of $y = f(x)$ vertically upward c units.

The graph of $y = f(x) - c$ is obtained by shifting the graph of $y = f(x)$ vertically downward c units.

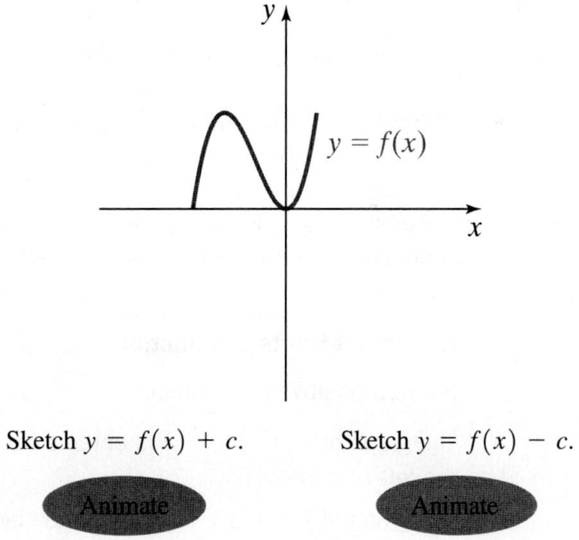

You Try It **Work through this You Try It problem.**

Work Exercises 1–7 in this textbook or in the MyMathLab Study Plan.

OBJECTIVE 2 USE HORIZONTAL SHIFTS TO GRAPH FUNCTIONS

My video summary

To illustrate a horizontal shift, let $f(x) = x^2$ and $g(x) = (x + 2)^2$. Tables 2 and 3 show tables of values for f and g, respectively. The graphs of f and g are sketched in Figure 2. The graph of g is the graph of f shifted to the *left* two units.

x	$f(x) = x^2$
−2	4
−1	1
0	0
1	1
2	4

Table 2

x	$g(x) = (x + 2)^2$
−4	4
−3	1
−2	0
−1	1
0	4

Table 3

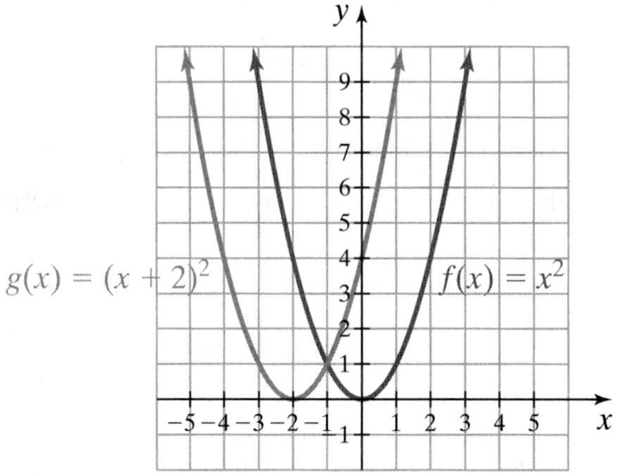

Figure 2

It follows that if c is a positive number, then $y = f(x + c)$ is the graph of f shifted to the *left* c units. For $c > 0$, the graph of $y = f(x - c)$ is the graph of f shifted to the *right* c units. At first glance, it appears that the rule for horizontal shifts is the opposite of what seems natural. Substituting $x + c$ for x causes the graph of $y = f(x)$ to be shifted to the left, whereas substituting $x - c$ for x causes the graph to shift to the right c units.

Horizontal Shifts of Functions

If c is a positive real number,

The graph of $y = f(x + c)$ is obtained by shifting the graph of $y = f(x)$ horizontally to the left c units.

The graph of $y = f(x - c)$ is obtained by shifting the graph of $y = f(x)$ horizontally to the right c units.

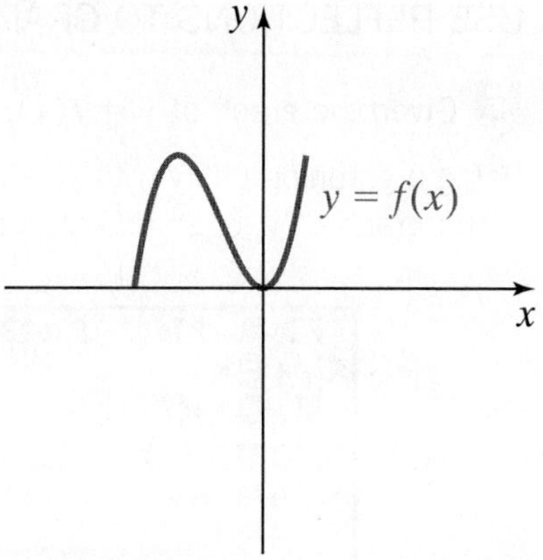

Sketch $y = f(x + c)$. Sketch $y = f(x - c)$.

Animate Animate

You Try It **Work through this You Try It problem.**

Work Exercises 8–14 in this textbook or in the MyMathLab Study Plan.

Example 2 Combining Horizontal and Vertical Shifts

My animation summary

Use the graph of $y = x^3$ to sketch the graph of $g(x) = (x - 1)^3 + 2$.

Solution The graph of g is obtained by shifting the graph of the basic function $y = x^3$ first horizontally to the right one unit and then vertically upward two units. When doing a problem with multiple transformations, it is good practice to always perform the vertical transformation last. Select the animate button to see how to sketch the graph of $g(x) = (x - 1)^3 + 2$.

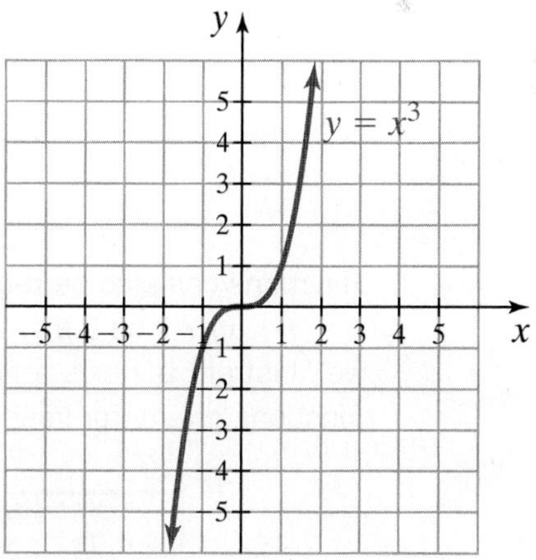

Animate

You Try It **Work through this You Try It problem.**

Work Exercises 15–21 in this textbook or in the MyMathLab Study Plan.

(eText Screens 20.1-1–20.1-39)

OBJECTIVE 3 USE REFLECTIONS TO GRAPH FUNCTIONS

My video summary

Given the graph of $y = f(x)$, what does the graph of $y = -f(x)$ look like?

Using a graphing utility with $y_1 = x^2$ and $y_2 = -x^2$, we can see that the graph of $y_2 = -x^2$ is the graph of $y_1 = x^2$ reflected about the ***x*-axis.**

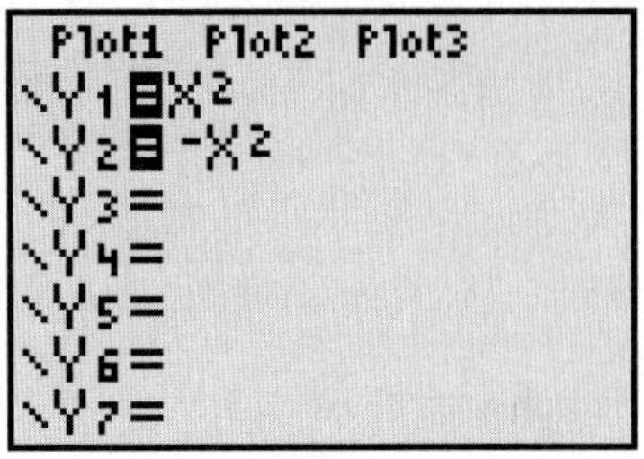

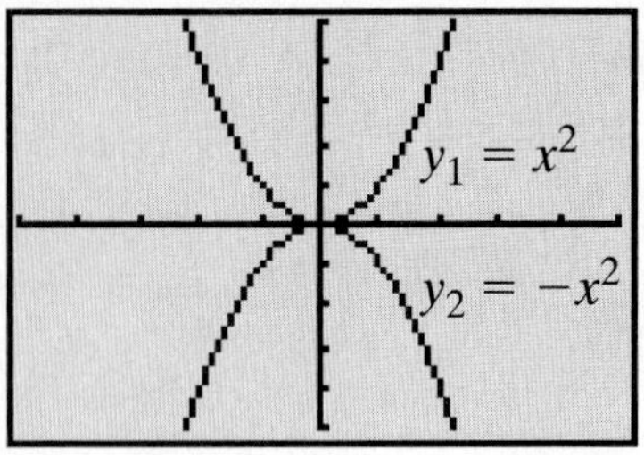

Reflections of Functions about the x-Axis

The graph of $y = -f(x)$ is obtained by reflecting the graph of $y = f(x)$ about the x-axis.

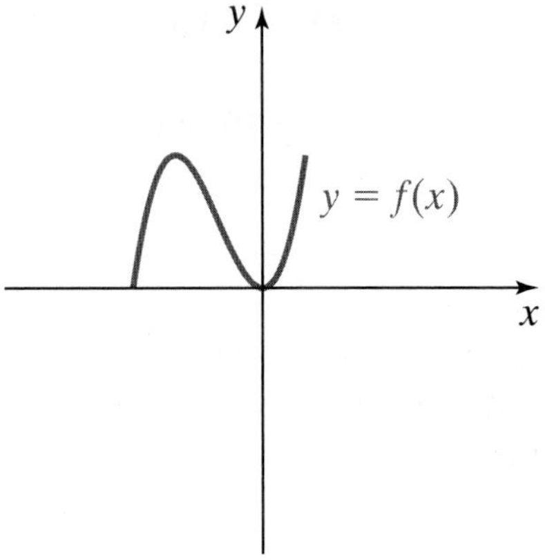

Sketch $y = -f(x)$.

Animate

Functions can also be reflected about the y-axis. Given the graph of $y = f(x)$, the graph of $y = f(-x)$ will be the graph of $y = f(x)$ reflected about the y-axis. Using a graphing utility, we illustrate a y-axis reflection by letting $y_1 = \sqrt{x}$ and $y_2 = \sqrt{-x}$. You can see that the functions are mirror images of each other about the y-axis.

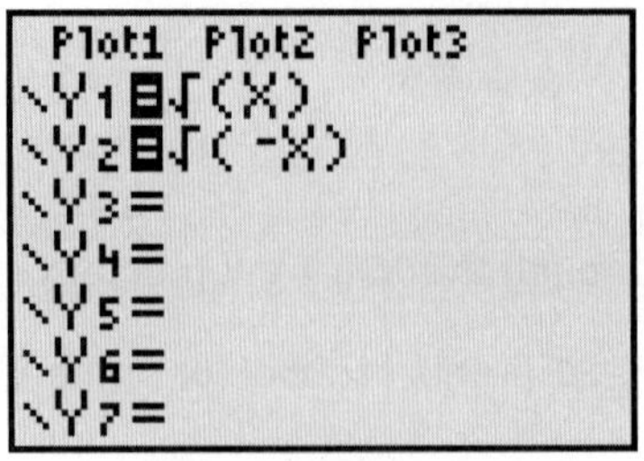

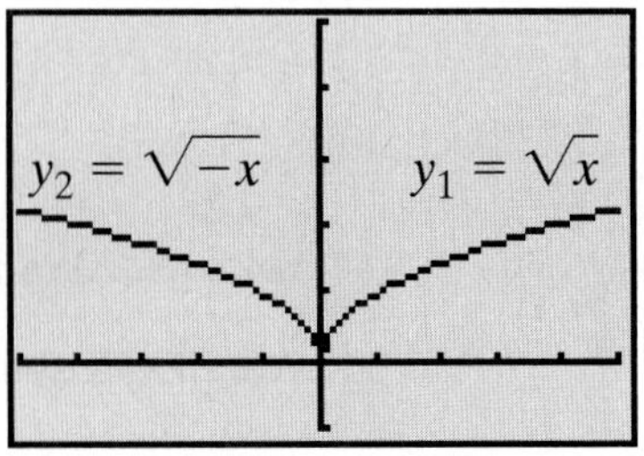

Reflections of Functions about the y-Axis

The graph of $y = f(-x)$ is obtained by reflecting the graph of $y = f(x)$ about the y-axis.

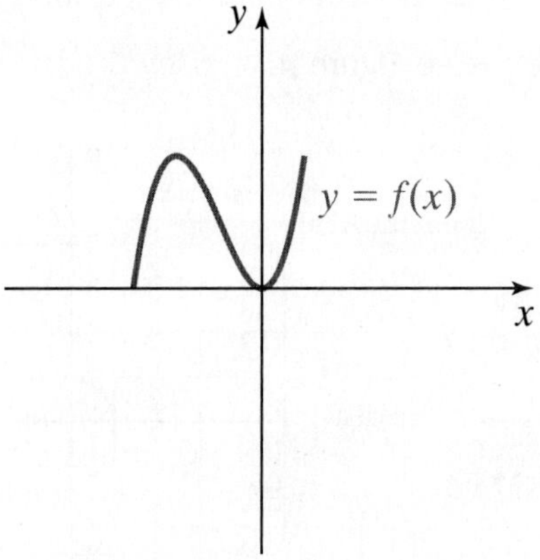

Sketch $y = f(-x)$.

Animate

Example 3 Sketching Functions Using Reflections and Shifts

Use the graph of the basic function $y = \sqrt[3]{x}$ to sketch each graph.

a. $g(x) = -\sqrt[3]{x} - 2$ b. $h(x) = \sqrt[3]{1 - x}$.

Solutions

a. Starting with the graph of $y = \sqrt[3]{x}$, we can obtain the graph of $g(x) = -\sqrt[3]{x} - 2$ by performing two transformations:

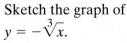

1. Reflect about the x-axis.
2. Vertically shift down two units.

See Figure 3.

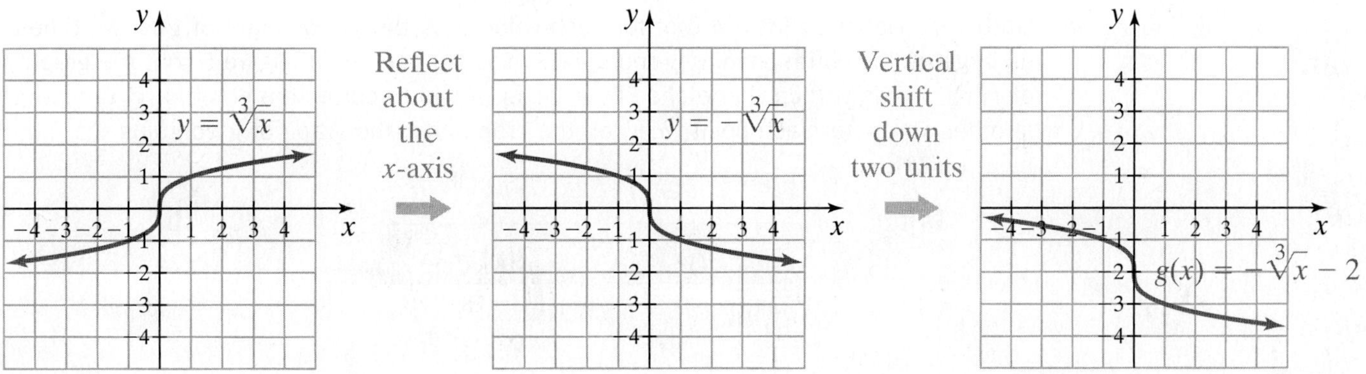

Start with the graph of the basic function $y = \sqrt[3]{x}$.

Sketch the graph of $y = -\sqrt[3]{x}$.

Sketch the graph of $g(x) = -\sqrt[3]{x} - 2$.

Figure 3

b. Starting with the graph of $y = \sqrt[3]{x}$, we can obtain the graph of $h(x) = \sqrt[3]{1 - x}$ by performing two transformations:

1. Horizontally shift left one unit.
2. Reflect about the y-axis.

See Figure 4.

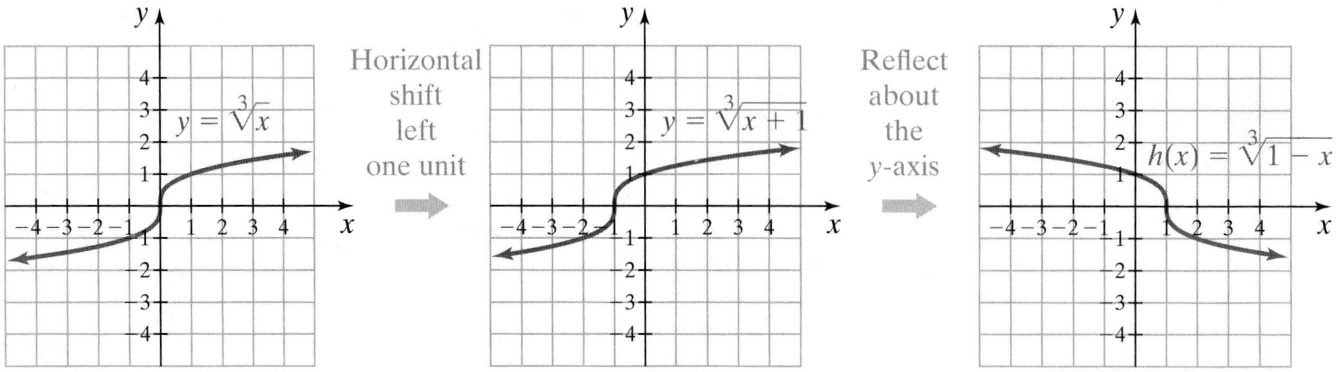

Start with the graph of the basic function $y = \sqrt[3]{x}$.

Sketch the graph of $y = \sqrt[3]{x + 1}$.

Replace x with $-x$, and sketch the graph of $h(x) = \sqrt[3]{-x + 1} = \sqrt[3]{1 - x}$.

Figure 4

You Try It **Work through this You Try It problem.**

Work Exercises 22–33 in this textbook or in the MyMathLab Study Plan.

OBJECTIVE 4 USE VERTICAL STRETCHES AND COMPRESSIONS TO GRAPH FUNCTIONS

Example 4 Vertically Stretching and Compressing

My video summary Use the graph of $f(x) = x^2$ to sketch the graph of $g(x) = 2x^2$.

Solution Notice in Table 4 that for each value of x, the y-coordinate of g is two times as large as the corresponding y-coordinate of f. We can see in Figure 5 that the graph of $f(x) = x^2$ is vertically stretched by a factor of two to obtain the graph of $g(x) = 2x^2$. In other words, for each point (a, b) on the graph of f, the graph of g contains the point $(a, 2b)$.

x	$f(x) = x^2$	$g(x) = 2x^2$
−2	4	8
−1	1	2
0	0	0
1	1	2
2	4	8

Table 4

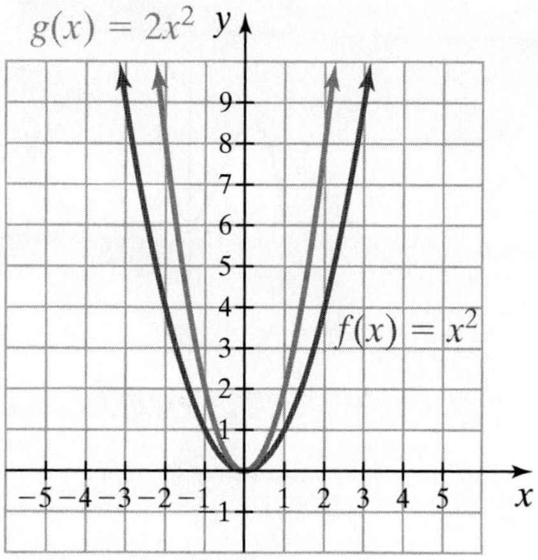

Figure 5

It follows from Example 4 that if $a > 1$, the graph of $y = af(x)$ is a **vertical stretch** of the graph of $y = f(x)$ and is obtained by multiplying each y-coordinate on the graph of f by a factor of a. If $0 < a < 1$, then the graph of $y = af(x)$ is a **vertical compression** of the graph of $y = f(x)$. Table 5 and Figure 6 show the relationship between the graphs of the functions $f(x) = x^2$ and $h(x) = \frac{1}{2}x^2$.

x	$f(x) = x^2$	$h(x) = \frac{1}{2}x^2$
−2	4	2
−1	1	$\frac{1}{2}$
0	0	0
1	1	$\frac{1}{2}$
2	4	2

Table 5

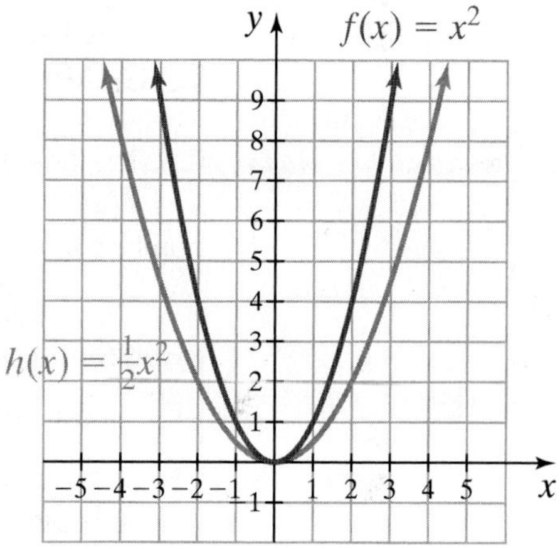

Figure 6

Vertical Stretches and Compressions of Functions

Suppose a is a positive real number:

The graph of $y = af(x)$ is obtained by multiplying each y-coordinate of $y = f(x)$ by a. If $a > 1$, the graph of $y = af(x)$ is a vertical stretch of the graph of $y = f(x)$. If $0 < a < 1$, the graph of $y = af(x)$ is a vertical compression of the graph of $y = f(x)$.

(cText Screens 20.1-1–20.1-39)

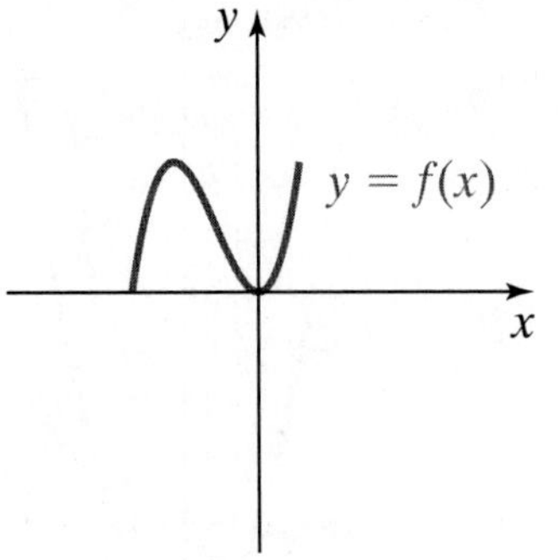

Sketch $y = af(x)$.
$(a > 1)$

Animate

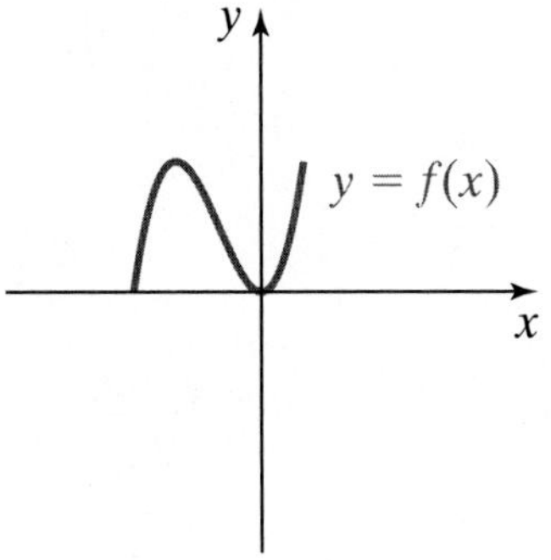

Sketch $y = af(x)$.
$(0 < a < 1)$

Animate

You Try It **Work through this You Try It problem.**

Work Exercises 34–40 in this textbook or in the MyMathLab Study Plan.

OBJECTIVE 5 USE HORIZONTAL STRETCHES AND COMPRESSIONS TO GRAPH FUNCTIONS

My video summary

The final transformation to discuss is a horizontal stretch or compression. A function, $y = f(x)$, will be horizontally stretched or compressed when x is multiplied by a positive number, $a \neq 1$, to obtain the new function, $y = f(ax)$.

> **Horizontal Stretches and Compressions of Functions**
>
> If a is a positive real number,
>
> For $a > 1$, the graph of $y = f(ax)$ is obtained by dividing each x-coordinate of $y = f(x)$ by a. The resultant graph is a horizontal compression.
>
> For $0 < a < 1$, the graph of $y = f(ax)$ is obtained by dividing each x-coordinate of $y = f(x)$ by a. The resultant graph is a horizontal stretch.

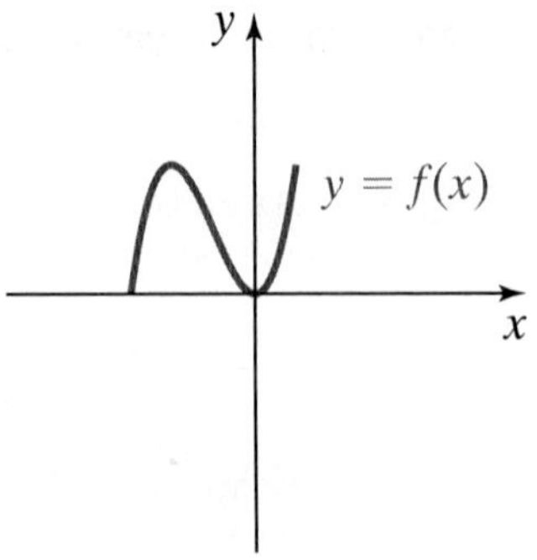

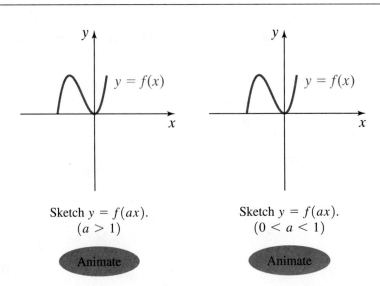

Sketch $y = f(ax)$.
$(a > 1)$

Animate

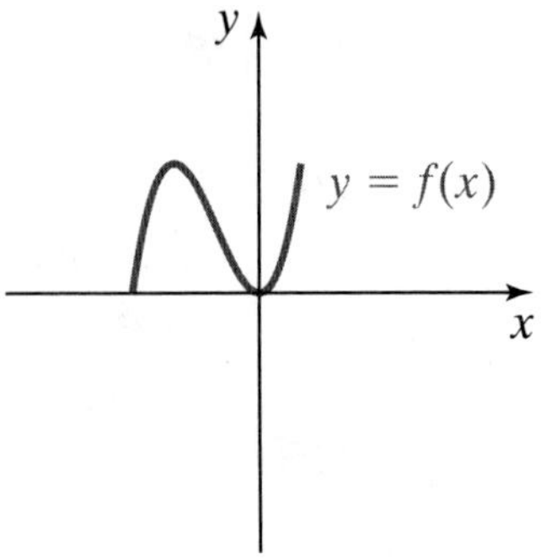

Sketch $y = f(ax)$.
$(0 < a < 1)$

Animate

Example 5 Horizontally Stretching and Compressing

Use the graph of $f(x) = \sqrt{x}$ to sketch the graphs of $g(x) = \sqrt{4x}$ and $h(x) = \sqrt{\frac{1}{4}x}$.

Solution The graph of $f(x) = \sqrt{x}$ contains the ordered pairs $(0, 0), (1, 1), (4, 2)$. To sketch the graph of $g(x) = \sqrt{4x}$, we must divide each previous x-coordinate by 4. Therefore, the ordered pairs $(0, 0), \left(\frac{1}{4}, 1\right), (1, 2)$ must lie on the graph of g. You can see that the graph of g is a horizontal compression of the graph of f. See Figure 7.

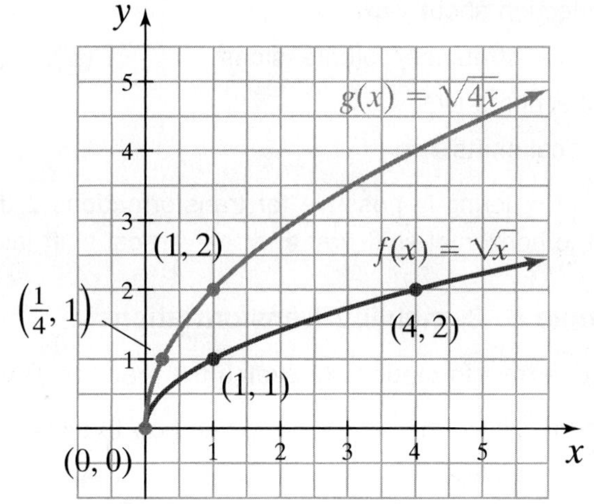

Figure 7
Graphs of $f(x) = \sqrt{x}$ and $g(x) = \sqrt{4x}$

Similarly, to sketch the graph of $h(x) = \sqrt{\frac{1}{4}x}$, we divide the x-coordinates of the ordered pairs of f by $\frac{1}{4}$ to get the ordered pairs $(0, 0), (4, 1), (16, 2)$. You can see that the graph of h is a horizontal stretch of the graph of f. See Figure 8.

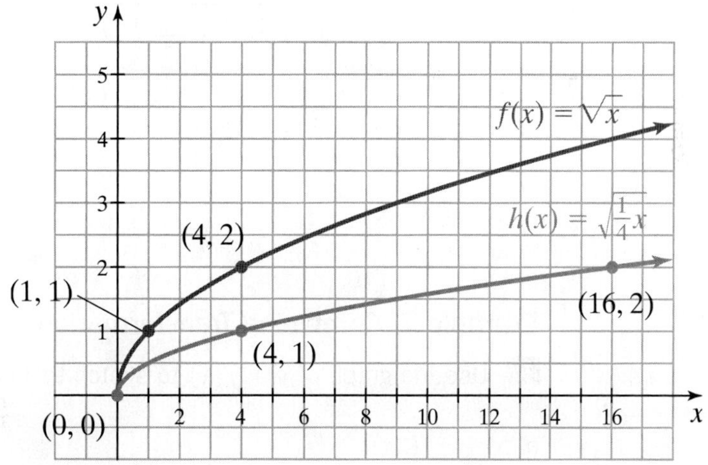

Figure 8
Graphs of $f(x) = \sqrt{x}$ and $g(x) = \sqrt{\frac{1}{4}x}$

You Try It **Work through this You Try It problem.**

Work Exercises 41–45 in this textbook or in the MyMathLab Study Plan.

OBJECTIVE 6 USE COMBINATIONS OF TRANSFORMATIONS TO GRAPH FUNCTIONS

You may encounter functions that combine many (if not all) of the transformations discussed in this topic. When sketching a function that involves multiple transformations, it is important to follow a certain "order of operations." Following is the order in which each transformation is performed in this text:

1. Horizontal shifts
2. Horizontal stretches/compressions
3. Reflection about y-axis
4. Vertical stretches/compressions
5. Reflection about x-axis
6. Vertical shifts

Different ordering is possible for transformations 2 through 5, but you should always perform the horizontal shift first and the vertical shift last.

Example 6 Combining Transformations

My animation summary

Use transformations to sketch the graph of $f(x) = -2(x + 3)^2 - 1$.

Solution Watch the **animation** to see how to sketch the function $f(x) = -2(x + 3)^2 - 1$ as seen in Figure 9.

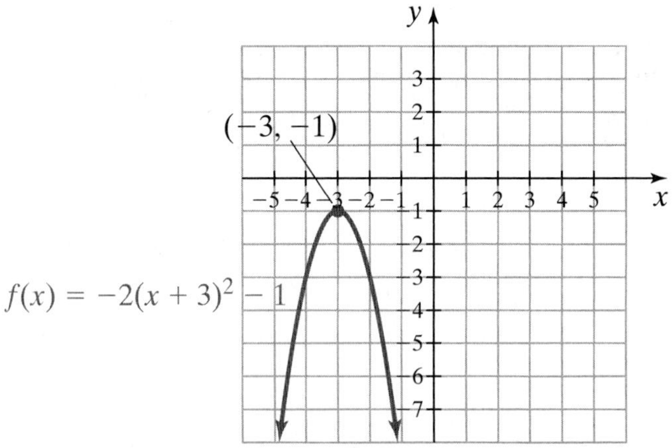

Figure 9
Graph of $f(x) = -2(x + 3)^2 - 1$

Example 7 Combining Transformations

My interactive video summary

Use the graph of $y = f(x)$ to sketch each of the following functions.

a. $y = -f(2x)$

b. $y = 2f(x - 3) - 1$

c. $y = -\frac{1}{2}f(2 - x) + 3$

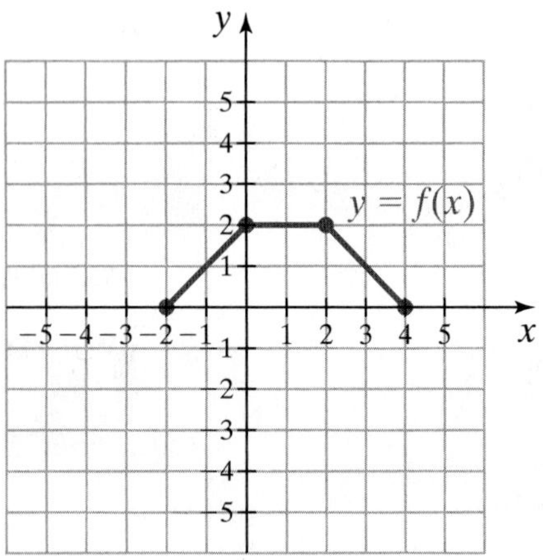

(*eText Screens 20.1-1–20.1-39*)

Solutions Watch the **interactive video** to see any one of the solutions worked out in detail.

a. The graph of $y = -f(2x)$ can be obtained from the graph of $y = f(x)$ using two transformations: (1) a horizontal compression and (2) a reflection about the x-axis. The resultant graph is shown in Figure 10.

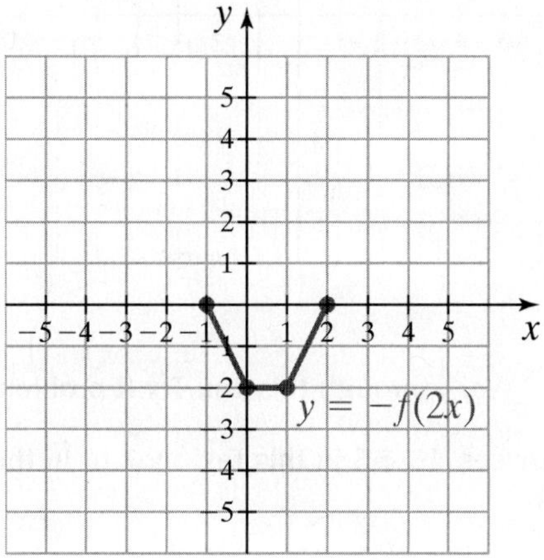

Figure 10
Graph of $y = -f(2x)$

b. The graph of $y = 2f(x - 3) - 1$ can be obtained from the graph of $y = f(x)$ using three transformations: (1) a horizontal shift to the right three units, (2) a vertical stretch by a factor of 2, and (3) a vertical shift down one unit. The resultant graph is shown in Figure 11.

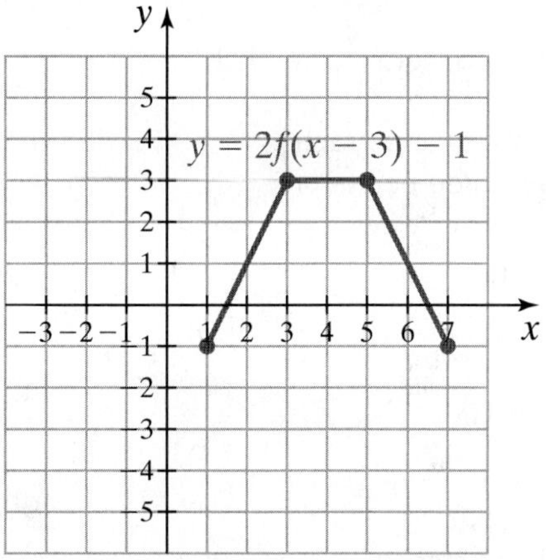

Figure 11
Graph of $y = 2f(x - 3) - 1$

c. The graph of $y = -\frac{1}{2}f(2 - x) + 3$ can be obtained from the graph of $y = f(x)$ using five transformations: (1) a horizontal shift to the left two units, (2) a reflection about the y-axis, (3) a vertical compression by a factor of $\frac{1}{2}$, (4) a reflection about the x-axis, and (5) a vertical shift up three units. The resultant graph is shown in Figure 12.

(eText Screens 20.1-1–20.1-39)

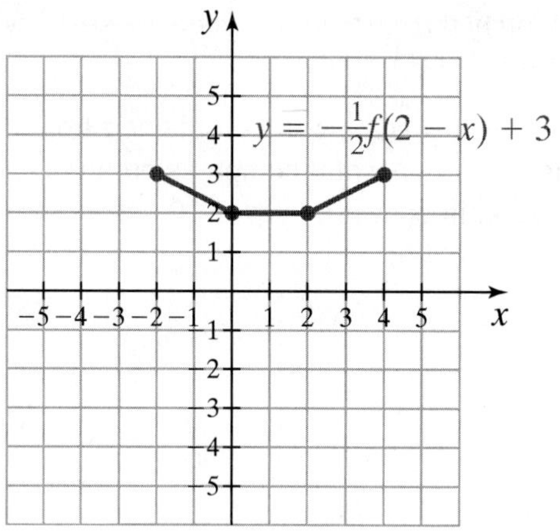

Figure 12

Graph of $y = -\frac{1}{2}f(2 - x) + 3$

You Try It **Work through this You Try It problem.**

Work Exercises 46–55 in this textbook or in the MyMathLabStudy Plan

20.1 Exercises

In Exercises 1–5, use the graph of a known basic function and vertical shifting to sketch each function.

You Try It

1. $f(x) = x^2 - 1$ **2.** $y = \sqrt{x} + 2$ **3.** $h(x) = \sqrt[3]{x} - 2$

4. $y = |x| - 3$ **5.** $g(x) = x^3 + 1$

6. Use the graph of $y = f(x)$ to sketch the graph of $y = f(x) - 1$. Label at least three points on the new graph.

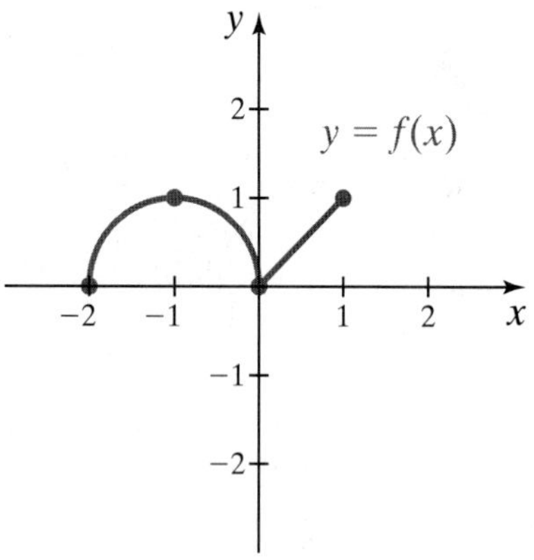

7. Use the graph of $y = f(x)$ to sketch the graph of $y = f(x) + 2$. Label at least three points on the new graph.

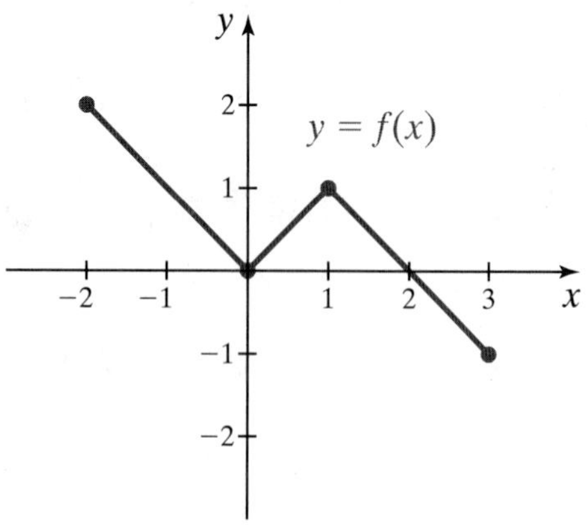

In Exercises 8–12, use the graph of a known basic function and horizontal shifts to sketch each function.

You Try It

8. $f(x) = \sqrt[3]{x - 2}$ **9.** $y = \sqrt{x - 4}$ **10.** $h(x) = (x + 1)^3$

11. $k(x) = |x - 1|$ **12.** $y = (x - 3)^2$

13. Use the graph of $y = f(x)$ to sketch the graph of $y = f(x - 2)$. Label at least three points on the new graph.

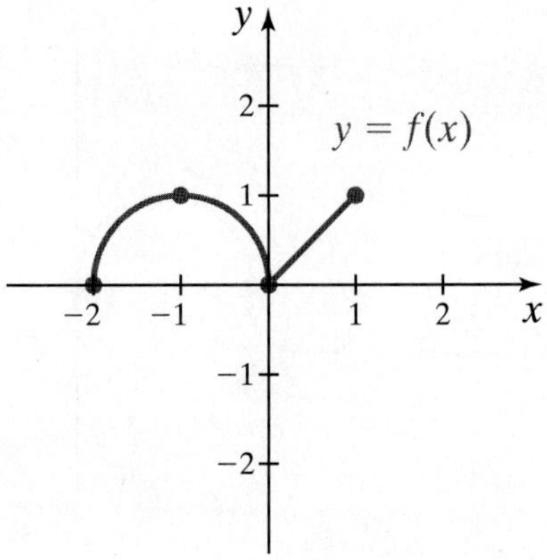

14. Use the graph of $y = f(x)$ to sketch the graph of $y = f(x + 2)$. Label at least three points on the new graph.

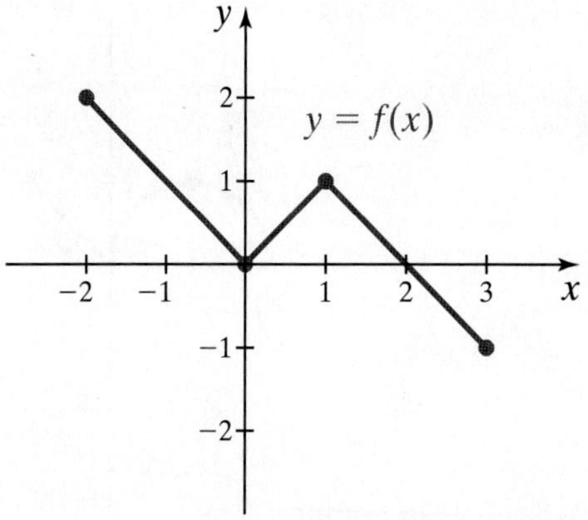

In Exercises 15–19, use the graph of a known basic function and a combination of horizontal and vertical shifts to sketch each function.

You Try It

15. $y = (x + 1)^2 - 2$

16. $f(x) = (x - 3)^2 + 1$

17. $y = \sqrt{x + 3} + 2$

18. $f(x) = |x + 2| + 2$

19. $y = \sqrt[3]{x + 1} - 1$

20. Use the graph of $y = f(x)$ to sketch the graph of $y = f(x - 2) - 1$. Label at least three points on the new graph.

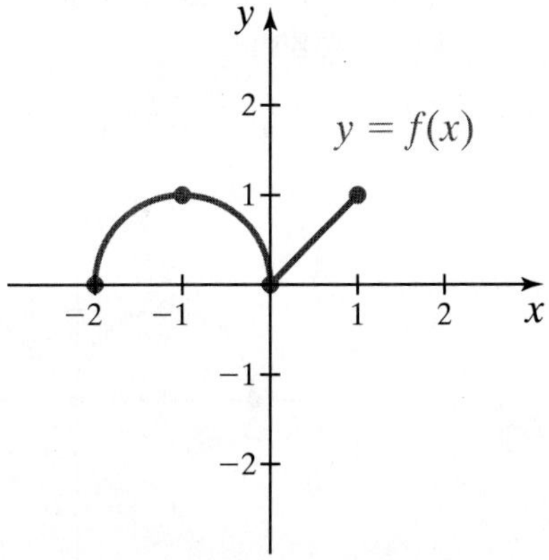

21. Use the graph of $y = f(x)$ to sketch the graph of $y = f(x + 1) + 2$. Label at least three points on the new graph.

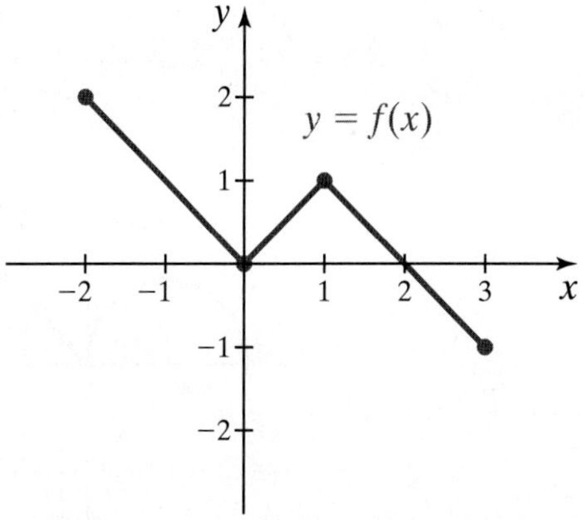

In Exercises 22–31, use the graph of a known basic function and a combination of horizontal shifts, reflections, and vertical shifts to sketch each function.

You Try It

22. $g(x) = -x^2 - 2$

23. $h(x) = \sqrt{2 - x}$

24. $f(x) = |-1 - x|$

25. $h(x) = \sqrt[3]{-x - 2}$

26. $g(x) = -x^3 + 1$

27. $h(x) = -\sqrt[3]{x} + 2$

28. $g(x) = (3 - x)^2$

29. $h(x) = -\sqrt{x} - 1$

30. $f(x) = -|x| + 1$

31. $g(x) = (1 - x)^3$

32. Use the graph of $y = f(x)$ to sketch the graph of $y = -f(x)$. Label at least three points on the new graph.

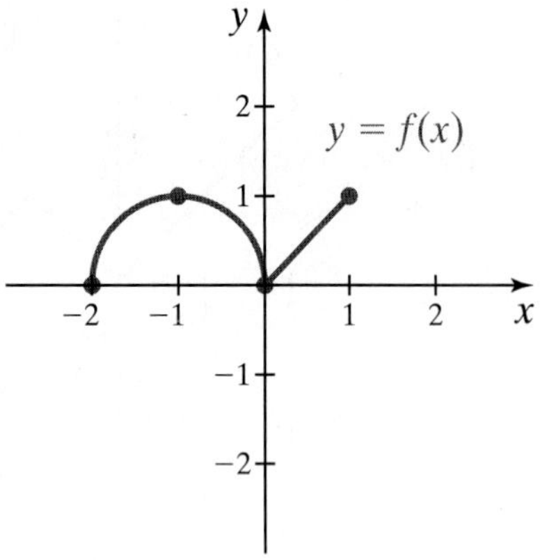

33. Use the graph of $y = f(x)$ to sketch the graph of $y = f(-x)$. Label at least three points on the new graph.

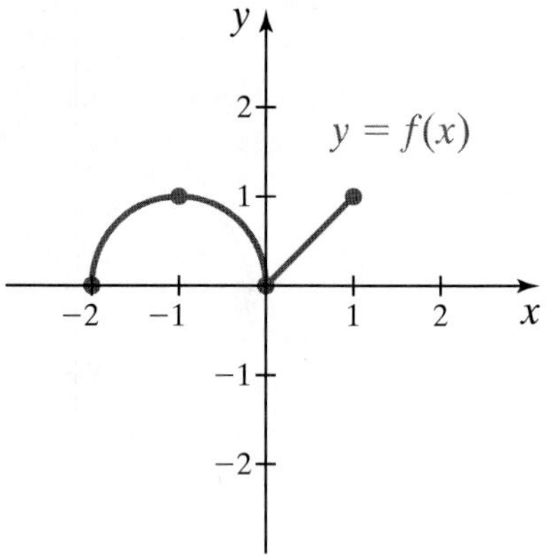

In Exercises 34–38, use the graph of a known basic function and a vertical stretch or vertical compression to sketch each function.

You Try It

34. $f(x) = \frac{1}{4}|x|$

35. $g(x) = 2\sqrt{x}$

36. $f(x) = \frac{1}{3}x^3$

37. $f(x) = \frac{1}{2}\sqrt[3]{x}$

38. $g(x) = 3x^2$

39. Use the graph of $y = f(x)$ to sketch the graph of $y = 3f(x)$. Label at least three points on the new graph.

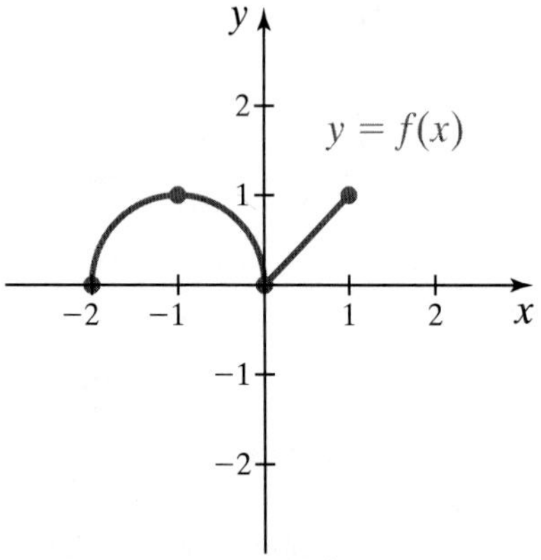

40. Use the graph of $y = f(x)$ to sketch the graph of $y = \frac{1}{2}f(x)$. Label at least three points on the new graph.

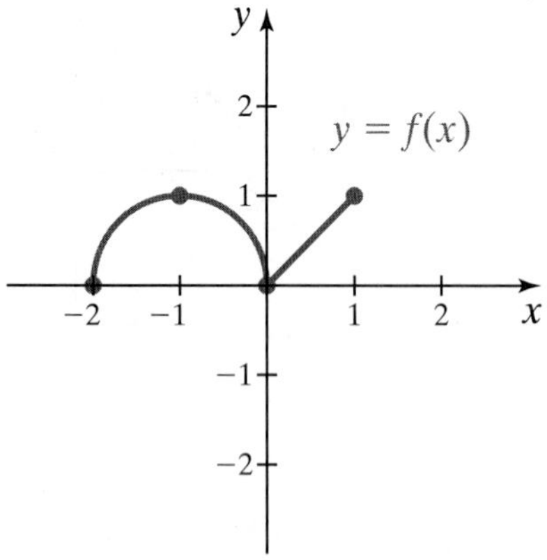

In Exercises 41–43, use the graph of a known basic function and a horizontal stretch or horizontal compression to sketch each function.

You Try It

41. $y = \left|\frac{1}{4}x\right|$

42. $f(x) = \sqrt{2x}$

43. $g(x) = \sqrt[3]{3x}$

44. Use the graph of $y = f(x)$ to sketch the graph of $y = f(2x)$. Label at least three points on the new graph.

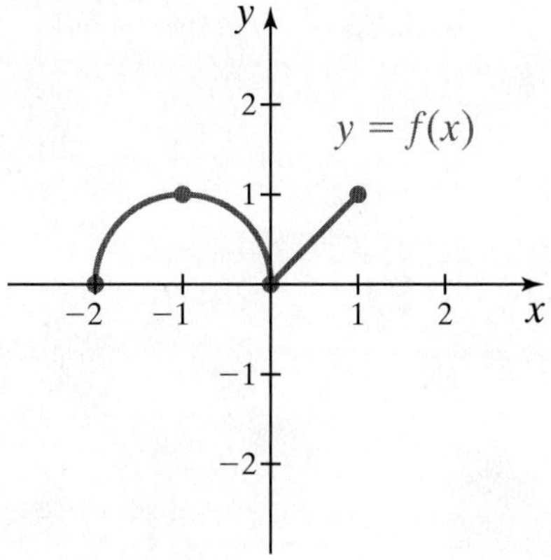

45. Use the graph of $y = f(x)$ to sketch the graph of $y = f(\frac{1}{2}x)$. Label at least three points on the new graph.

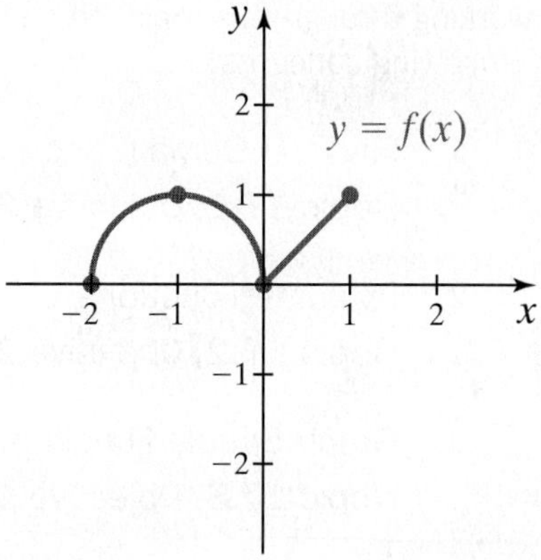

In Exercises 46–50, use the graph of a known basic function and a combination of transformations to sketch each function.

You Try It

46. $f(x) = -(x - 2)^2 + 3$

47. $g(x) = \frac{1}{2}|x + 1| - 1$

48. $f(x) = 2\sqrt[3]{x} - 1$

49. $g(x) = -\frac{1}{2}(2 - x)^3 + 1$

50. $h(x) = 2\sqrt{4 - x} + 5$

In Exercises 51–55, use the graph of $y = f(x)$ to sketch each function. Label at least three points on each graph.

51. $y = -f(-x) - 1$

52. $y = \frac{1}{2}f(2 - x)$

53. $y = -2f(x + 1) + 2$

54. $y = 3 - 3f(x + 3)$

55. $y = -f(1 - x) - 2$

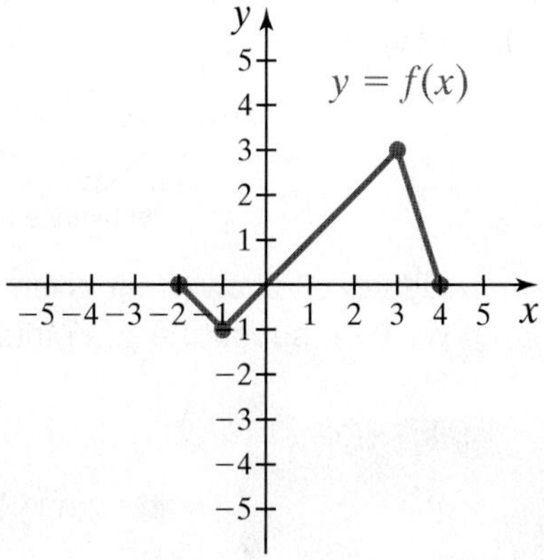

20.2 Composite and Inverse Functions

THINGS TO KNOW

Before working through this topic, be sure you are familiar with the following concepts:	VIDEO	ANIMATION	INTERACTIVE
You Try It — 1. Find the Domain and Range of a Relation (Topic 17.1, **Objective 2**)			✓
You Try It — 2. Evaluate Functions (Topic 17.2, **Objective 2**)	✓		✓
You Try It — 3. Graph Simple Functions by Plotting Points (Topic 17.3, **Objective 1**)		✓	

OBJECTIVES

1 Form and Evaluate Composite Functions

2 Determine the Domain of Composite Functions

3 Determine If a Function Is One-to-One Using the Horizontal Line Test

4 Verify Inverse Functions

5 Sketch the Graphs of Inverse Functions

6 Find the Inverse of a One-to-One Function

OBJECTIVE 1 FORM AND EVALUATE COMPOSITE FUNCTIONS

My video summary

Consider the functions $f(x) = x^2$ and $g(x) = 2x + 1$. How could we find $f(g(x))$? To find $f(g(x))$, we substitute $g(x)$ for x in the function f to get

$$f(g(x)) = f(2x + 1) = (2x + 1)^2.$$

Substitute $g(x)$ into f

The diagram in Figure 13 shows that given a number x, we first apply it to the function g to obtain $g(x)$. We then substitute $g(x)$ into f to get the result.

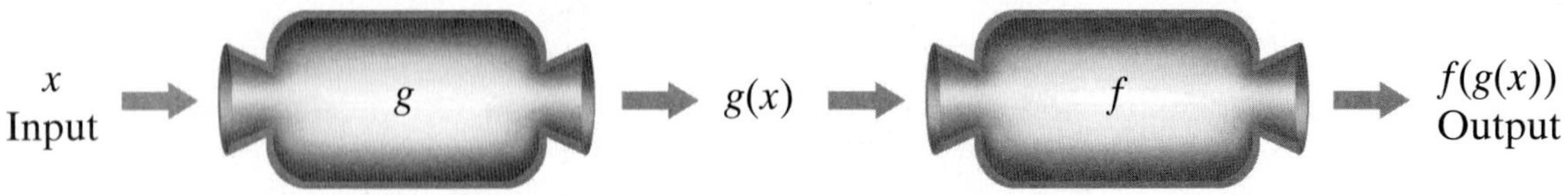

Figure 13 Composition of f and g

The function $f(g(x))$ is called a *composite function* because one function is "composed" of another function.

Definition Composite Function

Given functions f and g, the **composite function,** $f \circ g$, (also called the **composition of f and g**) is defined by

$$(f \circ g)(x) = f(g(x)),$$

provided $g(x)$ is in the domain of f.

CAUTION

The composition of f and g does not equal the product of f and g:

$$(f \circ g)(x) \neq f(x)g(x).$$

Also, the composition of f and g does not necessarily equal the composition of g and f, although this equality does exist for certain pairs of functions.

Example 1 Forming and Evaluating Composite Functions

My interactive video summary

Let $f(x) = 4x + 1, g(x) = \dfrac{x}{x-2}$ and $h(x) = \sqrt{x+3}$.

a. Find the function $f \circ g$.

b. Find the function $g \circ h$.

c. Find the function $h \circ f \circ g$.

d. Evaluate $(f \circ g)(4)$, or state that it is undefined.

e. Evaluate $(g \circ h)(1)$, or state that it is undefined.

f. Evaluate $(h \circ f \circ g)(6)$, or state that it is undefined.

Solutions Work through the **interactive video** to verify the following:

a. $(f \circ g)(x) = 4\left(\dfrac{x}{x-2}\right) + 1 = \dfrac{5x-2}{x-2}$

b. $(g \circ h)(x) = \dfrac{\sqrt{x+3}}{\sqrt{x+3}-2}$

c. $(h \circ f \circ g)(x) = \sqrt{\dfrac{5x-2}{x-2} + 3} = \sqrt{\dfrac{8x-8}{x-2}}$

d. $(f \circ g)(4) = 9$

e. $(g \circ h)(1)$ is undefined

f. $(h \circ f \circ g)(6) = \sqrt{10}$

You Try It **Work through this You Try It problem.**

Work Exercises 1–16 in this textbook or in the MyMathLab Study Plan.

(eText Screens 20.2-1–20.2-34)

Example 2 Evaluating Composite Functions Using a Graph

My interactive video summary

Use the graph to evaluate each expression:

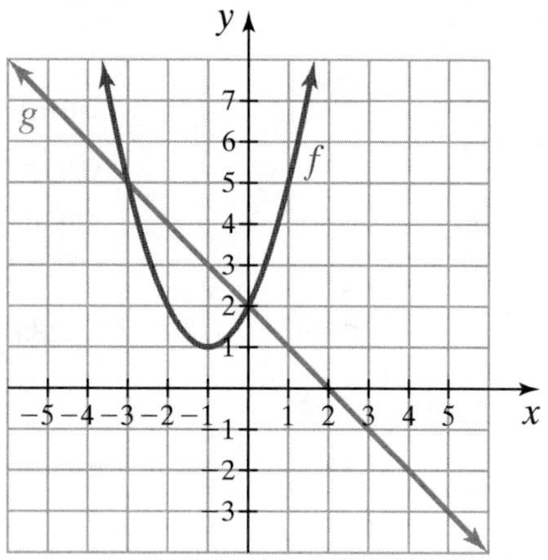

a. $(f \circ g)(4)$

b. $(g \circ f)(-3)$

c. $(f \circ f)(-1)$

d. $(g \circ g)(4)$

e. $(f \circ g \circ f)(1)$

Solutions Work through the **interactive video** to verify the following:

a. $(f \circ g)(4) = 2$ b. $(g \circ f)(-3) = -3$ c. $(f \circ f)(-1) = 5$

d. $(g \circ g)(4) = 4$ e. $(f \circ g \circ f)(1) = 5$

You Try It **Work through this You Try It problem.**

Work Exercises 17 and 18 in this textbook or in the MyMathLab Study Plan.

OBJECTIVE 2 DETERMINE THE DOMAIN OF COMPOSITE FUNCTIONS

My video summary

Suppose f and g are functions. For a number x to be in the domain of $f \circ g$, x must be in the domain of g *and* $g(x)$ must be in domain of f. Follow these two steps to find the domain of $f \circ g$:

1. Find the domain of g.
2. Exclude from the domain of g all values of x for which $g(x)$ is not in the domain of f.

Example 3 Finding the Domain of a Composite Function

My interactive video summary

Let $f(x) = \dfrac{-10}{x-4}$ and $g(x) = \sqrt{5-x}$.

a. Find the domain of $f \circ g$.

b. Find the domain of $g \circ f$.

Solutions

a. First, form the composite function $(f \circ g)(x) = \dfrac{-10}{\sqrt{5-x}-4}$.

To find the domain of $f \circ g$, we follow these two steps:

Step 1. Find the domain of g.

The domain of $g(x) = \sqrt{5-x}$ is $(-\infty, 5]$. The domain of $f \circ g$ cannot contain any values of x that are not in this interval. In other words, the domain of $f \circ g$ is a **subset** of $(-\infty, 5]$.

Step 2. Exclude from the domain of g all values of x for which $g(x)$ is not in the domain of f.

All real numbers except 4 are in the domain of f. This implies that $g(x)$ cannot equal 4 because $g(x)$ equal to 4 would make the denominator of f, $x - 4$, equal to zero. Thus, we must exclude all values of x such that $g(x) = 4$.

Substitute $\sqrt{5-x}$ for $g(x)$: $\quad g(x) = 4$

Square both sides: $\quad \sqrt{5-x} = 4$

$$5 - x = 16$$

Solve for x: $\quad x = -11$

We must *exclude* $x = -11$ from the domain of $f \circ g$. Therefore, the domain of $f \circ g$ is all values of x less than 5 such that $x \neq -11$, or the interval $(-\infty, -11) \cup (-11, 5]$. You may want to view the **interactive video** for a more detailed solution.

b. You should carefully work through the **interactive video** to verify that the domain of $(g \circ f)(x) = \sqrt{5 + \dfrac{10}{x-4}} = \sqrt{\dfrac{5x-10}{x-4}}$ is the interval $(-\infty, 2] \cup (4, \infty)$.

You Try It **Work through this You Try It problem.**

Work Exercises 19–24 in this textbook or in the MyMathLab Study Plan.

OBJECTIVE 3 DETERMINE IF A FUNCTION IS ONE-TO-ONE USING THE HORIZONTAL LINE TEST

Later in this topic, we examine a process for finding the inverse of a function, but keep the following in mind: We can find the inverse of many functions using this process, but that inverse will not always be a function. In this text, we are only interested in inverses that are functions, so we first develop a test to determine whether a function has an inverse *function*. When the word *inverse* is used throughout the remainder of this topic, we assume that we are referring to the inverse that is a function.

My video summary

First, we must define the concept of **one-to-one** functions.

Definition One-to-One Function

A function f is **one-to-one** if for any values $a \neq b$ in the domain of f, $f(a) \neq f(b)$.

This definition suggests that a function is one-to-one if for any two *different* input values (domain values), the corresponding output values (range values) must be different. An alternate definition says that if two range values are the same, $f(u) = f(v)$, then the domain values must be the same; that is, $u = v$.

Alternate Definition of a One-to-One Function

A function f is **one-to-one** if for any two range values $f(u)$ and $f(v)$, $f(u) = f(v)$ implies that $u = v$.

(eText Screens 20.2-1–20.2-34)

The function sketched in Figure 14a is one-to-one because for any two distinct x-values in the domain $(a \neq b)$, the function values or range values are not equal $(f(a) \neq f(b))$. In Figure 14b, we see that the function $y = g(x)$ is *not* one-to-one because we can easily find two different domain values that correspond to the same range value.

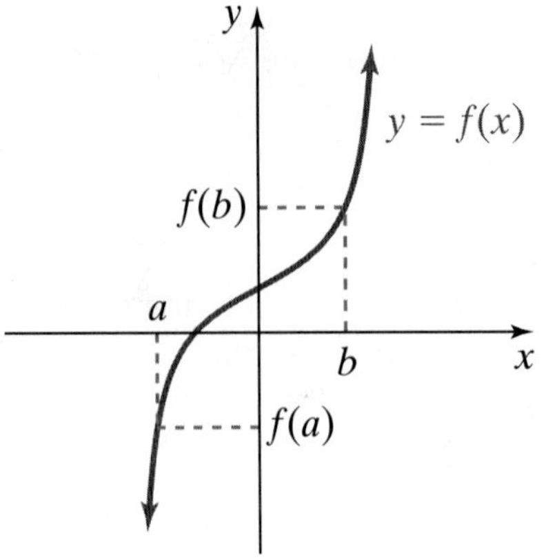

Figure 14a
An example of a one-to-one function

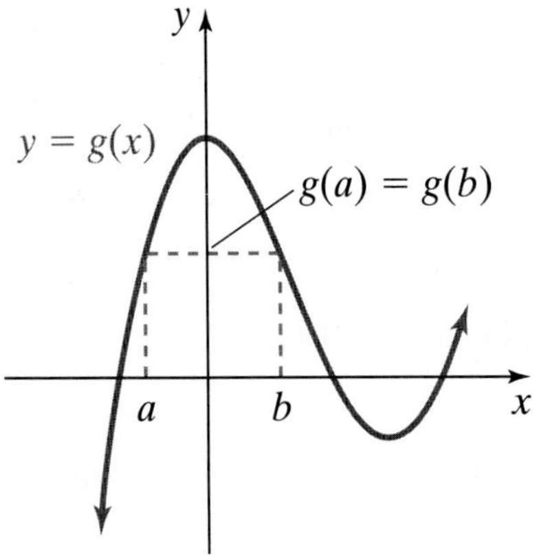

Figure 14b
An example of a function that is not one-to-one

My video summary

Notice in **Figure 15** that the horizontal lines intersect the graph of $y = f(x)$ in at most one place, whereas we can find many horizontal lines that intersect the graph of $y = g(x)$ more than once. This gives us a visual example of how we can use horizontal lines to help us determine from the graph whether a function is one-to-one. Using horizontal lines to determine whether the graph of a function is one-to-one is known as the *horizontal line test.*

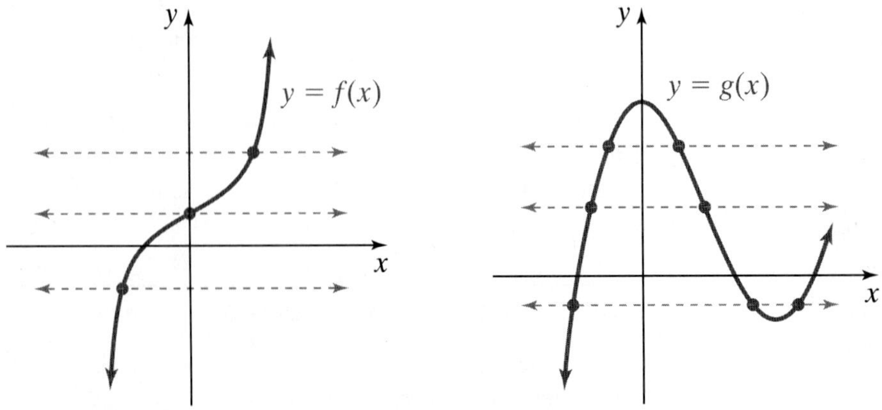

Figure 15 Drawing horizontal lines can help determine whether a graph represents a one-to-one function.

(eText Screens 20.2-1–20.2-34)

> **Horizontal Line Test**
>
> If every horizontal line intersects the graph of a function f at most once, then f is one-to-one.

Example 4 Determining Whether a Function Is One-to-One

Determine whether each function is one-to-one.

a.

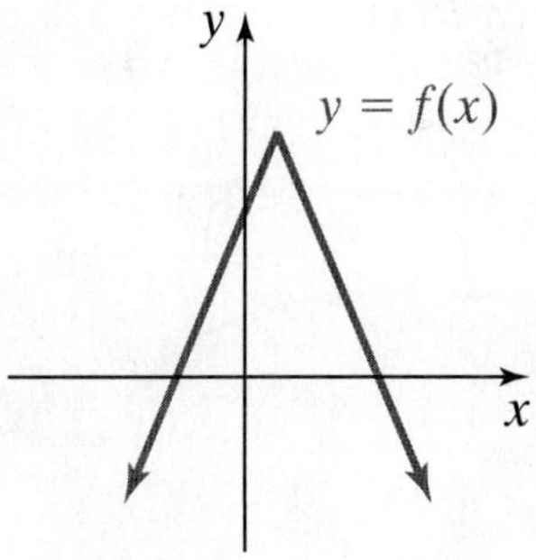

b.

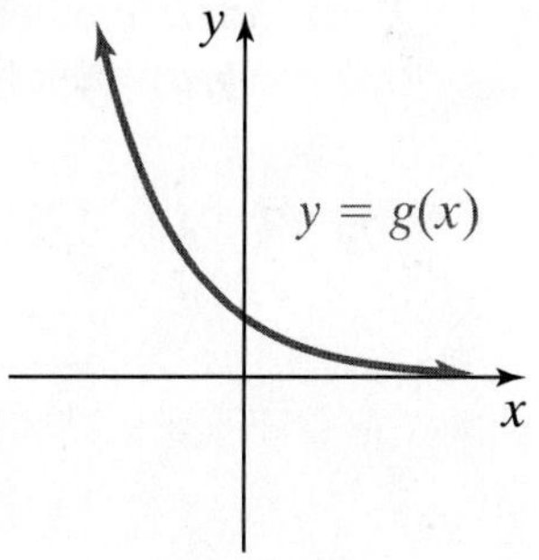

c. $f(x) = x^2 + 1, x \le 0$

d. $f(x) = \begin{cases} 2x + 4 \text{ for } x \le -1 \\ 2x - 6 \text{ for } x \ge 4 \end{cases}$

Solutions The functions in parts b and c are one-to-one, whereas the functions in parts a and d are not one-to-one. Watch the **animation** to verify.

You Try It **Work through this You Try It problem.**

Work Exercises 25–39 in this textbook or in the MyMathLab Study Plan.

OBJECTIVE 4 VERIFY INVERSE FUNCTIONS

My video summary

We are now ready to ask ourselves the question, "Why should we be concerned with one-to-one functions?"

Answer: Every one-to-one function has an inverse function!

To illustrate an inverse function, let's consider the function $F(C) = \frac{9}{5}C + 32$. This function is used to convert a given temperature in degrees Celsius into its equivalent temperature in degrees Fahrenheit. For example,

$$F(25) = \frac{9}{5}(25) + 32 = 45 + 32 = 77.$$

Thus, a temperature of 25°C corresponds to a temperature of 77°F. To convert a temperature of 77°F back into 25°C, we need a different function. The function that is used to accomplish this task is $C(F) = \frac{5}{9}(F - 32)$.

$$\begin{aligned} C(77) &= \frac{5}{9}(77 - 32) \\ &= \frac{5}{9}(45) \\ &= 25 \end{aligned}$$

The function C is the *inverse* of the function F. In essence, these functions perform opposite actions (they "undo" each other). The first function converted 25°C into 77°F, whereas the second function converted 77°F back into 25°C.

(eText Screens 20.2-1–20.2-34)

Definition Inverse Function

Let f be a one-to-one function with domain A and range B. Then f^{-1} is the **inverse function of f** with domain B and range A. Furthermore, if $f(a) = b$, then $f^{-1}(b) = a$.

According to the definition of an inverse function, the domain of f is exactly the same as the range of f^{-1}, and the range of f is the same as the domain of f^{-1}. Figure 16 illustrates that if the function f assigns a number a to b, then the inverse function will assign b back to a. In other words, if the point (a, b) is an ordered pair on the graph of f, then the point (b, a) must be on the graph of f^{-1}.

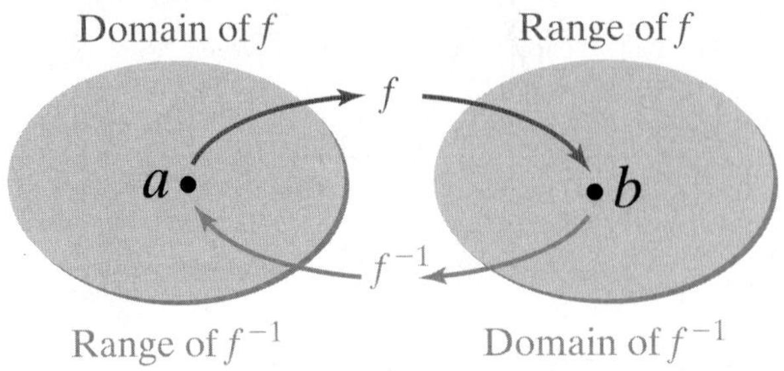

Figure 16

CAUTION Do not confuse f^{-1} with $\frac{1}{f(x)}$. The negative 1 in f^{-1} is *not* an exponent!

My video summary

As with our opening example using an inverse function to convert a Fahrenheit temperature back into a Celsius temperature, inverse functions "undo" each other. For example, it can be shown that if $f(x) = x^3$, then the inverse of f is $f^{-1}(x) = \sqrt[3]{x}$. Note that

$$f(2) = (2)^3 = 8 \quad \text{and} \quad f^{-1}(8) = \sqrt[3]{8} = 2.$$

The function f takes the number 2 to 8, whereas f^{-1} takes 8 back to 2. Observe what happens if we look at the composition of f and f^{-1} and the composition of f^{-1} and f at specified values:

$$\underset{\uparrow\ x\text{-value}}{(f \circ f^{-1})(8)} = f(f^{-1}(8)) = f(2) = \underset{\uparrow\ \text{Same as } x\text{-value}}{8}$$

$$\underset{\uparrow\ x\text{-value}}{(f^{-1} \circ f)(2)} = f^{-1}(f(2)) = f^{-1}(8) = \underset{\uparrow\ \text{Same as } x\text{-value}}{2}$$

Because of the "undoing" nature of inverse functions, we get the following **composition cancellation equations:**

Composition Cancellation Equations

$f(f^{-1}(x)) = x$ for all x in the domain of f^{-1}

and $f^{-1}(f(x)) = x$ for all x in the domain of f

(eText Screens 20.2-1–20.2-34)

These cancellation equations can be used to show whether two functions are inverses of each other. We can see from our example that if $f(x) = x^3$ and $f^{-1}(x) = \sqrt[3]{x}$, then

$$f(f^{-1}(x)) = f(\sqrt[3]{x}) = (\sqrt[3]{x})^3 = x \text{ and } f^{-1}(f(x)) = f^{-1}(x^3) = \sqrt[3]{x^3} = x.$$

Example 5 Verify Inverse Functions

My interactive video summary

Show that $f(x) = \dfrac{x}{2x+3}$ and $g(x) = \dfrac{3x}{1-2x}$ are inverse functions using the composition cancellation equations.

Solution To show that f and g are inverses of each other, we must show that $(f \circ g)(x) = x$ and $(g \circ f)(x) = x$. Work through the **interactive video** to verify that both composition cancellation equations are satisfied.

You Try It **Work through this You Try It problem.**

Work Exercises 40–44 in this textbook or in the MyMathLab Study Plan.

OBJECTIVE 5 SKETCH THE GRAPHS OF INVERSE FUNCTIONS

If f is a one-to-one function, then we know that it must have an inverse function, f^{-1}. Given the graph of a one-to-one function f, we can obtain the graph of f^{-1} by simply interchanging the coordinates of each ordered pair that lies on the graph of f. In other words, for any point (a, b) on the graph of f, the point (b, a) must lie on the graph of f^{-1}. Notice in Figure 17 that the points (a, b) and (b, a) are symmetric about the line $y = x$. Therefore, the graph of f^{-1} is a reflection of the graph of f about the line $y = x$. Figure 18 shows the graph of $f(x) = x^3$ and $f^{-1}(x) = \sqrt[3]{x}$. You can see that if the functions have any points in common, they must lie along the line $y = x$.

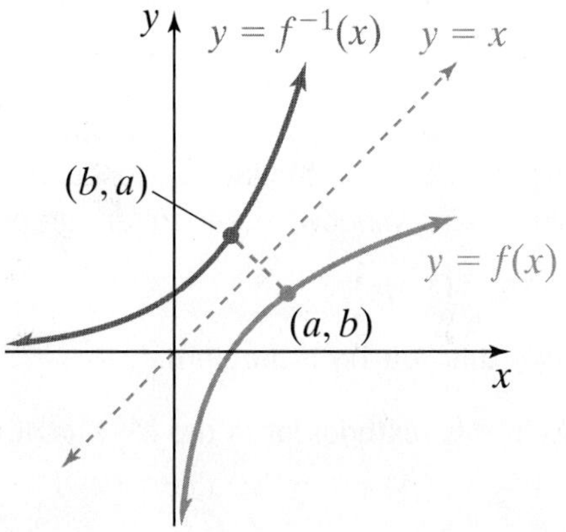

Figure 17 Graph of a one-to-one function and its inverse

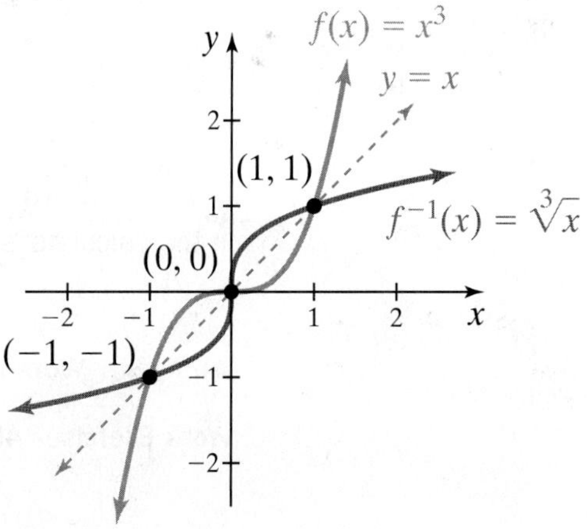

Figure 18 Graph of $f(x) = x^3$ and $f^{-1}(x) = \sqrt[3]{x}$

Example 6 Sketch the Graph of a One-to-One Function and Its Inverse

Sketch the graph of $f(x) = x^2 + 1, x \le 0$, and its inverse. Also state the domain and range of f and f^{-1}.

My animation summary

Solution The graphs of f and f^{-1} are sketched in Figure 19. Notice how the graph of f^{-1} is a reflection of the graph of f about the line $y = x$. Also notice that the domain of f is the same as the range of f^{-1}. Likewise, the domain of f^{-1} is equivalent to the range of f. View the **animation** to see exactly how to sketch f and f^{-1}.

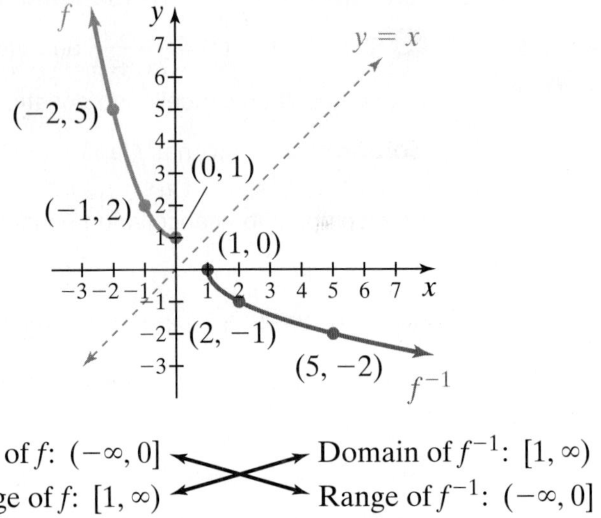

Figure 19 Graph of $f(x) = x^2 + 1, x \le 0$, and its inverse

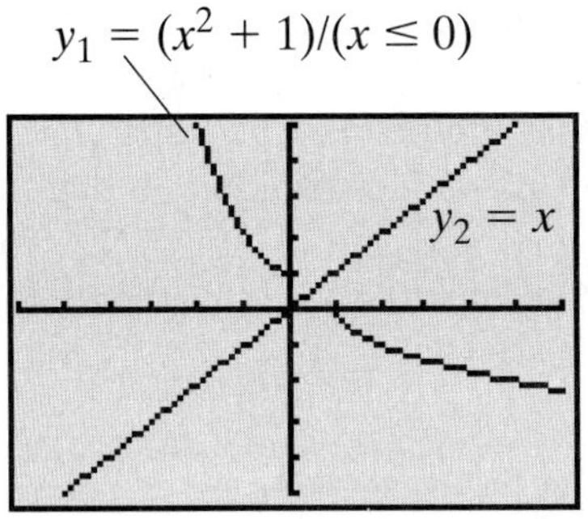

Using a TI-84Plus, we can sketch the functions from Example 6 by letting $y_1 = (x^2 + 1)/(x \le 0)$. We can draw the inverse function by typing the command **DrawInv Y_1** in the calculator's main viewing window. The resultant graphs are shown in the figure.

You Try It **Work through this You Try It problem.**

Work Exercises 45–50 in this textbook or in the MyMathLab Study Plan.

OBJECTIVE 6 FIND THE INVERSE OF A ONE-TO-ONE FUNCTION

My video summary

We are now ready to find the inverse of a one-to-one function algebraically. We know that if a point (x, y) is on the graph of a one-to-one function, then the point (y, x) is on the graph of its inverse function. We can use this information to develop a process for finding the inverse of a function algebraically simply by switching x and y in the original function to produce its inverse function.

(*eText Screens 20.2-1–20.2-34*)

We use as a motivating example the function $f(x) = x^2 + 1, x \le 0$, discussed in Example 6. To find the inverse of a one-to-one function, we follow the four-step process outlined next.

Step 1. Change $f(x)$ to y: $y = x^2 + 1$

Step 2. Interchange x and y: $x = y^2 + 1$

Step 3. Solve for y: $x - 1 = y^2$

$$\pm\sqrt{x - 1} = y$$

(Because the domain of f is $(-\infty, 0]$, the range of f^{-1} must be $(-\infty, 0]$. Therefore, we must use the negative square root or $y = -\sqrt{x - 1}$.)

Step 4. Change y to $f^{-1}(x)$: $f^{-1}(x) = -\sqrt{x - 1}$

Thus, the inverse of $f(x) = x^2 + 1, x \le 0$, is $f^{-1}(x) = -\sqrt{x - 1}$.

Example 7 Find the Inverse of a Function

My animation summary

Find the inverse of the function $f(x) = \dfrac{2x}{1 - 5x}$, and state the domain and range of f and f^{-1}.

Solution Work through the **animation**, and follow the four-step process to verify that $f^{-1}(x) = \dfrac{x}{5x + 2}$. The domain of f is $\left(-\infty, \frac{1}{5}\right) \cup \left(\frac{1}{5}, \infty\right)$, whereas the domain of f^{-1} is $\left(-\infty, -\frac{2}{5}\right) \cup \left(-\frac{2}{5}, \infty\right)$. Because the range of f must be the domain of f^{-1} and the range of f^{-1} must be the domain of f, we get the following result:

Domain of f: $\left(-\infty, \dfrac{1}{5}\right) \cup \left(\dfrac{1}{5}, \infty\right)$ ↔ Range of f^{-1}: $\left(-\infty, \dfrac{1}{5}\right) \cup \left(\dfrac{1}{5}, \infty\right)$

Range of f: $\left(-\infty, -\dfrac{2}{5}\right) \cup \left(-\dfrac{2}{5}, \infty\right)$ ↔ Domain of f^{-1}: $\left(-\infty, -\dfrac{2}{5}\right) \cup \left(-\dfrac{2}{5}, \infty\right)$

You Try It **Work through this You Try It problem.**

Work Exercises 51–59 in this textbook or in the MyMathLab Study Plan.

Inverse Function Summary

1. f^{-1} exists if and only if the function f is one-to-one.
2. The domain of f is the same as the range of f^{-1}, and the range of f is the same as the domain of f^{-1}.
3. To verify that two one-to-one functions, f and g, are inverses of each other, we must use the composition cancellation equations to show that $f(g(x)) = g(f(x)) = x$.
4. The graph of f^{-1} is a reflection of the graph of f about the line $y = x$. That is, for any point (a, b) that lies on the graph of f, the point (b, a) must lie on the graph of f^{-1}.
5. To find the inverse of a one-to-one function, replace $f(x)$ with y, interchange the variables x and y, and solve for y. This is the function $f^{-1}(x)$.

20.2 Exercises

In Exercises 1–8, let $f(x) = 3x + 1$, $g(x) = \frac{2}{x+1}$, and $h(x) = \sqrt{x+3}$.

You Try It

1. Find the function $f \circ g$.

2. Find the function $g \circ f$.

3. Find the function $f \circ h$.

4. Find the function $g \circ h$.

5. Find the function $h \circ f$.

6. Find the function $h \circ g$.

7. Find the function $f \circ f$.

8. Find the function $g \circ g$.

In Exercises 9–16, evaluate the following composite functions given that $f(x) = 3x + 1$, $g(x) = \frac{2}{x+1}$, and $h(x) = \sqrt{x+3}$.

9. $(f \circ g)(0)$ **10.** $(f \circ h)(6)$ **11.** $(g \circ f)(1)$ **12.** $(g \circ h)(-2)$

13. $(h \circ f)(0)$ **14.** $(h \circ g)(3)$ **15.** $(f \circ f)(-1)$ **16.** $(g \circ g)(4)$

In Exercises 17 and 18, use the graph to evaluate each expression.

You Try It

17. **a.** $(f \circ g)(1)$ **b.** $(g \circ f)(-1)$ **c.** $(g \circ g)(0)$ **d.** $(f \circ f)(1)$

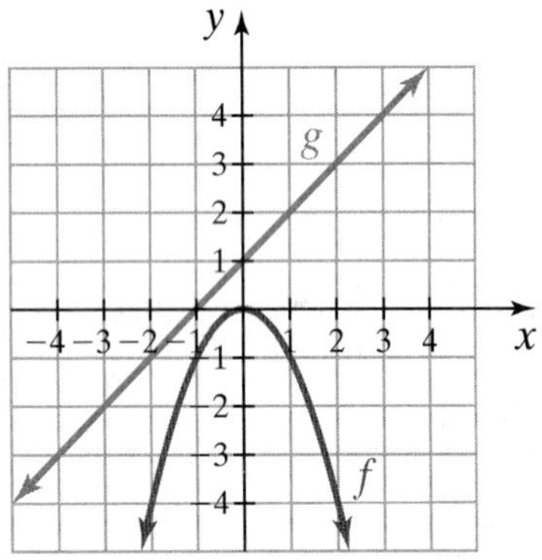

18. **a.** $(f \circ g)(1)$ **b.** $(g \circ f)(-1)$ **c.** $(g \circ g)(0)$ **d.** $(f \circ f)(1)$

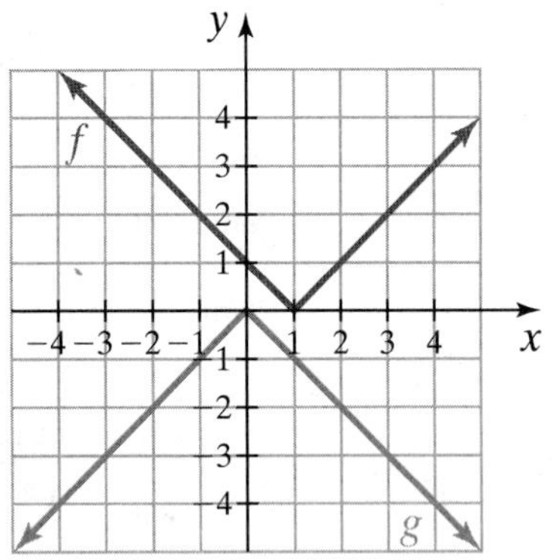

In Exercises 19–24, find the domain of $(f \circ g)(x)$ and $(g \circ f)(x)$.

You Try It

19. $f(x) = x^2, g(x) = 2x - 1$

20. $f(x) = 3x - 5, g(x) = 2x^2 + 1$

21. $f(x) = x^2, g(x) = \sqrt{x}$

22. $f(x) = \frac{1}{x}, g(x) = x^2 - 4$

23. $f(x) = \frac{3}{x+1}, g(x) = \frac{x}{x-2}$

24. $f(x) = \frac{2x}{x-3}, g(x) = \frac{x+1}{x-1}$

In Exercises 25–39, determine whether each function is one-to-one.

You Try It

25. $f(x) = 3x - 1$

26. $f(x) = 2x^2$

27. $f(x) = (x-1)^2, x \geq 1$

28. $f(x) = (x-1)^2, x \geq -1$

29. $f(x) = \frac{1}{x} - 2$

30. $f(x) = 4\sqrt{x}$

31. $f(x) = -2|x|$ **32.** $f(x) = 2$ **33.** $f(x) = (x + 1)^3 - 2$

34.

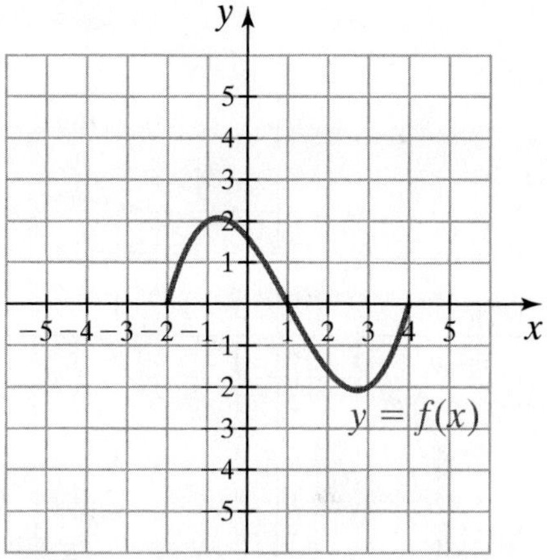

35.

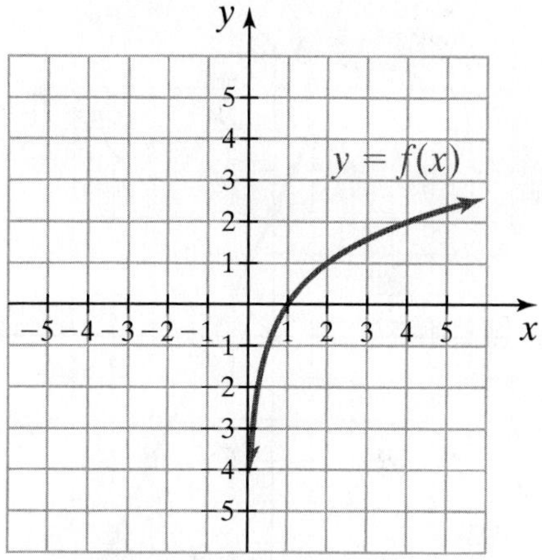

36.

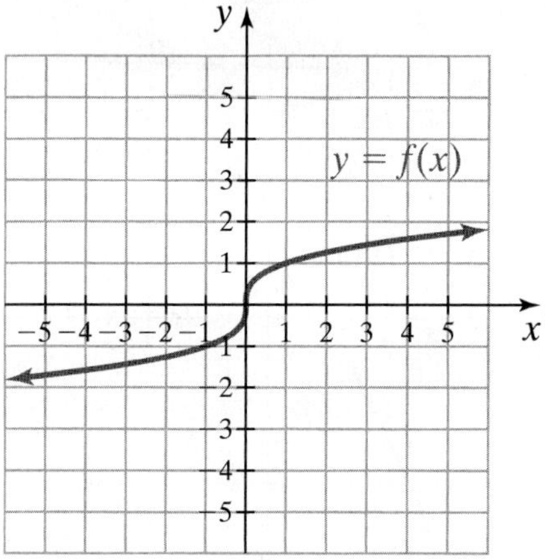

37.

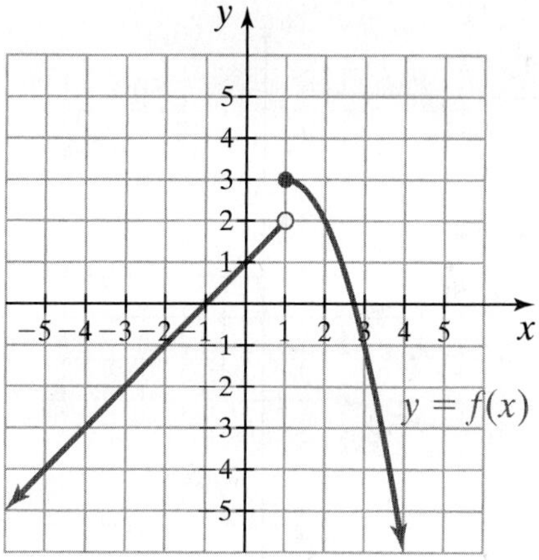

38.

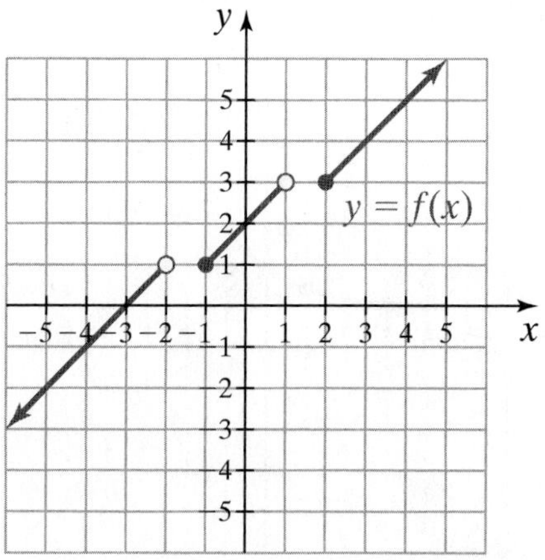

39.

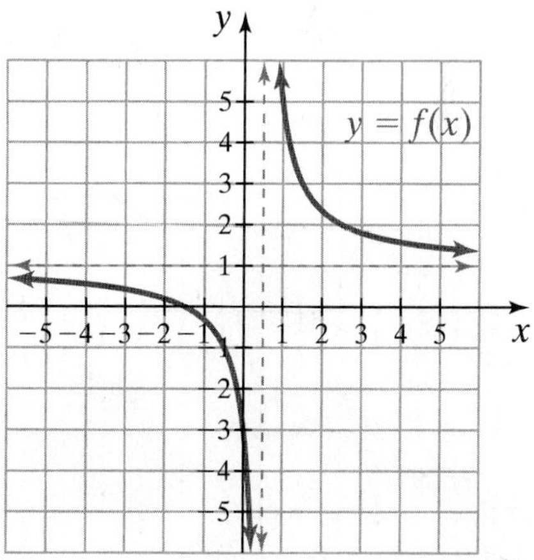

In Exercises 40–44, use the composition cancellation equations to verify that f and g are inverse functions.

You Try It

40. $f(x) = \frac{3}{2}x - 4$ and $g(x) = \frac{2x + 8}{3}$

41. $f(x) = (x - 1)^2, x \geq 1$, and $g(x) = \sqrt{x} + 1$

42. $f(x) = \frac{7}{x + 1}$ and $g(x) = \frac{7 - x}{x}$

43. $f(x) = \frac{x}{5 + 3x}$ and $g(x) = \frac{5x}{1 - 3x}$

44. $f(x) = 2\sqrt[3]{x - 1} + 3$ and $g(x) = \frac{(x - 3)^3}{8} + 1$

In Exercises 45–50, use the graph of f to sketch the graph of f^{-1}. Use the graphs to determine the domain and range of each function.

You Try It

45.

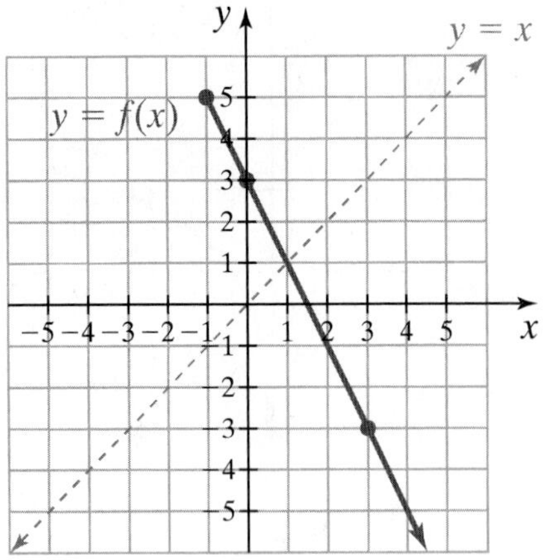

46.

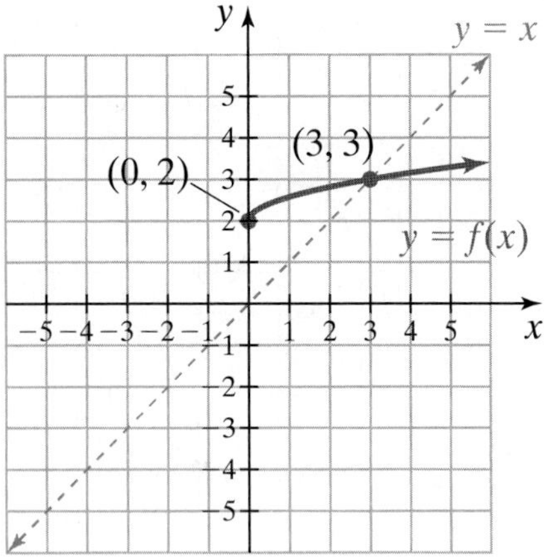

47.

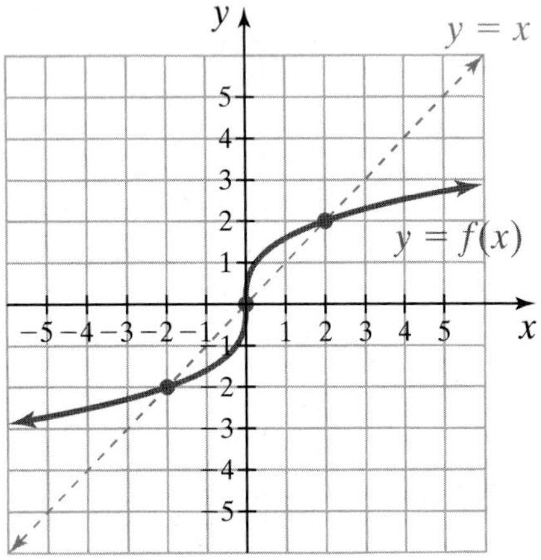

48.

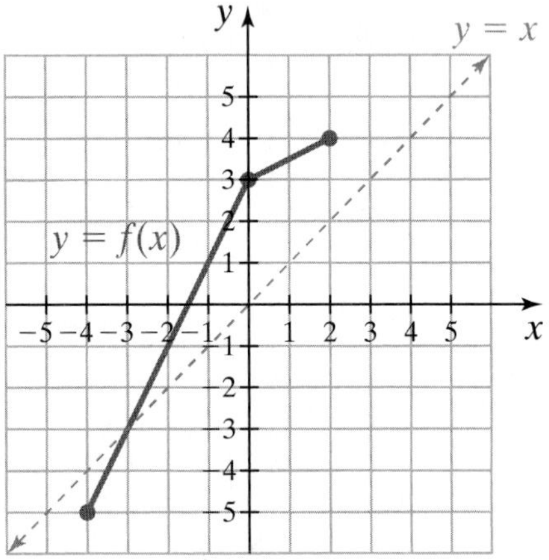

49.

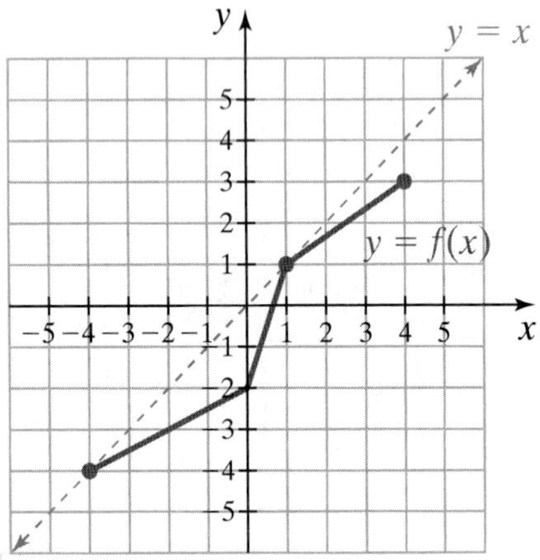

50.

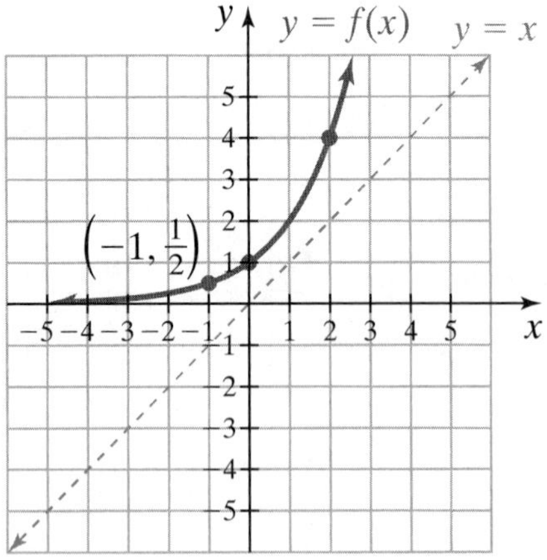

In Exercises 51–59, write an equation for the inverse function, and then state the domain and range of f and f^{-1}.

You Try It

51. $f(x) = \frac{1}{3}x - 5$

52. $f(x) = \frac{3x + 9}{7}$

53. $f(x) = \sqrt[3]{2x - 3}$

54. $f(x) = 1 - \sqrt[5]{x + 4}$

55. $f(x) = -x^2 - 2, x \geq 0$

56. $f(x) = (x + 3)^2 - 5, x \leq -3$

57. $f(x) = \frac{3}{x}$

58. $f(x) = \frac{1 - x}{2x}$

59. $f(x) = \frac{8x - 1}{7 - 5x}$

(eText Screens 20.3-1–20.3-32)

20.3 Exponential Functions

THINGS TO KNOW

Before working through this topic, be sure that you are familiar with the following concepts:

VIDEO ANIMATION INTERACTIVE

You Try It 1. Use Combinations of Transformations to Graph Functions (Topic 20.1, **Objective 6**)

You Try It 2. Determine If a Function Is One-to-One Using the Horizontal Line Test (Topic 20.2, **Objective 3**)

OBJECTIVES

1 Use the Characteristics of Exponential Functions

2 Sketch the Graphs of Exponential Functions Using Transformations

3 Solve Exponential Equations by Relating the Bases

4 Solve Applications of Exponential Functions

OBJECTIVE 1 USE THE CHARACTERISTICS OF EXPONENTIAL FUNCTIONS

Many natural phenomena and real-life applications can be modeled using exponential functions. Before we define the exponential function, it is important to remember how to manipulate exponential expressions because this skill is necessary when solving certain equations involving exponents. In **Topic 19.3**, expressions of the form b^r were evaluated for **rational numbers** r. For example,

$$3^2 = 9, \quad 4^{-2} = \frac{1}{4^2} = \frac{1}{16}, \quad \text{and} \quad 27^{-2/3} = \frac{1}{27^{2/3}} = \frac{1}{(\sqrt[3]{27})^2} = \frac{1}{(3)^2} = \frac{1}{9}.$$

In this topic, we extend the meaning of b^r to include all **real** values of r by defining the exponential function $f(x) = b^x$.

Definition Exponential Function

An **exponential function** is a function of the form $f(x) = b^x$, where x is any real number and $b > 0$ such that $b \neq 1$.

The constant, b, is called the base of the exponential function.

Notice in the definition that the base, b, must be positive and must not equal 1. If $b = 1$, then the function $f(x) = 1^x$ is equal to 1 for all x and is hence equivalent to the constant function $f(x) = 1$. If b were negative, then $f(x) = b^x$ would not be defined for all real values of x. For example, if $b = -4$, then $f\left(\frac{1}{2}\right) = (-4)^{1/2} = \sqrt{-4} = 2i$, which is *not* a positive real number.

(eText Screens 20.3-1–20.3-32)

My video summary

Before we generalize the graph of $f(x) = b^x$, we create a table of values and sketch the graph of $y = b^x$ for $b = 2, 3, \frac{1}{2}$, and $\frac{1}{3}$. See Table 6.

Table 6

x	$y = 2^x$	$y = 3^x$	$y = \left(\frac{1}{2}\right)^x$	$y = \left(\frac{1}{3}\right)^x$
-2	$2^{-2} = \frac{1}{2^2} = \frac{1}{4}$	$3^{-2} = \frac{1}{3^2} = \frac{1}{9}$	$\left(\frac{1}{2}\right)^{-2} = 2^2 = 4$	$\left(\frac{1}{3}\right)^{-2} = 3^2 = 9$
-1	$2^{-1} = \frac{1}{2^1} = \frac{1}{2}$	$3^{-1} = \frac{1}{3^1} = \frac{1}{3}$	$\left(\frac{1}{2}\right)^{-1} = 2^1 = 2$	$\left(\frac{1}{3}\right)^{-1} = 3^1 = 3$
0	$2^0 = 1$	$3^0 = 1$	$\left(\frac{1}{2}\right)^0 = 1$	$\left(\frac{1}{3}\right)^0 = 1$
1	$2^1 = 2$	$3^1 = 3$	$\left(\frac{1}{2}\right)^1 = \frac{1}{2}$	$\left(\frac{1}{3}\right)^1 = \frac{1}{3}$
2	$2^2 = 4$	$3^2 = 9$	$\left(\frac{1}{2}\right)^2 = \frac{1}{4}$	$\left(\frac{1}{3}\right)^2 = \frac{1}{9}$

You can see from the graphs sketched in Figure 20 that all four functions intersect the y-axis at the point $(0, 1)$. This is true because $b^0 = 1$ for all nonzero values of b. For values of $b > 1$, the graph of $y = b^x$ increases rapidly as the values of x approach positive infinity ($b^x \to \infty$ as $x \to \infty$). In fact, the larger the base, the faster the graph will grow. Also, for $b > 1$, the graph of $y = b^x$ decreases quickly, approaching 0 as the values of x approach negative infinity ($b^x \to 0$ as $x \to -\infty$). Thus, the x-axis (the line $y = 0$) is a **horizontal asymptote**.

However, for $0 < b < 1$, the graph decreases quickly, approaching the horizontal asymptote $y = 0$ as the values of x approach positive infinity ($b^x \to 0$ as $x \to \infty$), whereas the graph increases rapidly as the values of x approach negative infinity ($b^x \to \infty$ as $x \to -\infty$). The preceding statements, along with some other characteristics of the graphs of exponential functions, are outlined on the following page.

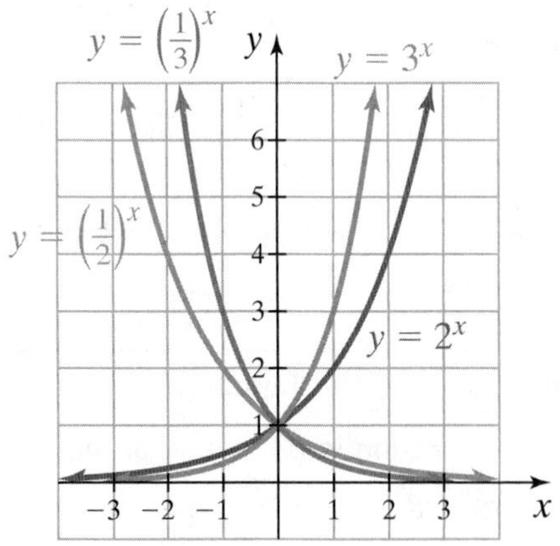

Figure 20
Graphs of $y = 2^x$, $y = 3^x$, $y = \left(\frac{1}{2}\right)^x$, and $y = \left(\frac{1}{3}\right)^x$

(eText Screens 20.3-1–20.3-32)

Characteristics of Exponential Functions

For $b > 0, b \neq 1$, the exponential function with base b is defined by $f(x) = b^x$.

The domain of $f(x) = b^x$ is $(-\infty, \infty)$, and the range is $(0, \infty)$. The graph of $f(x) = b^x$ has one of the following two shapes:

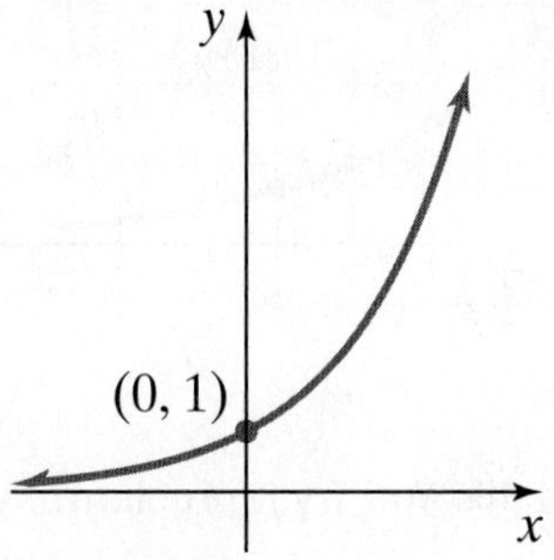

$f(x) = b^x, b > 1$

The graph intersects the y-axis at $(0, 1)$.

$b^x \to \infty$ as $x \to \infty$

$b^x \to 0$ as $x \to -\infty$

The line $y = 0$ is a **horizontal asymptote**.

The function is **one-to-one**.

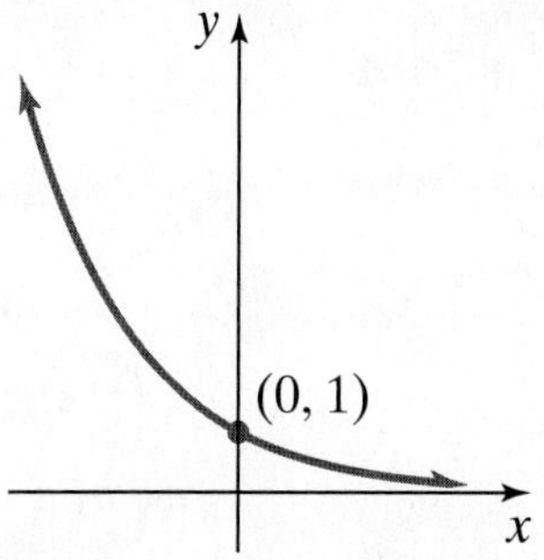

$f(x) = b^x, 0 < b < 1$

The graph intersects the y-axis at $(0, 1)$.

$b^x \to 0$ as $x \to \infty$

$b^x \to \infty$ as $x \to -\infty$

The line $y = 0$ is a **horizontal asymptote**.

The function is **one-to-one**.

Example 1 Sketching the Graph of an Exponential Function

My video summary

Sketch the graph of $f(x) = \left(\frac{2}{3}\right)^x$.

Solution Because the base of the exponential function is $\frac{2}{3}$, which is between 0 and 1, the graph must approach the x-axis as the value of x approaches positive infinity. The graph intersects the y-axis at $(0, 1)$. We can find a few more points by choosing some negative and positive values of x:

$$f(-2) = \left(\frac{2}{3}\right)^{-2} = \left(\frac{3}{2}\right)^2 = \frac{9}{4}$$

$$f(-1) = \left(\frac{2}{3}\right)^{-1} = \left(\frac{3}{2}\right)^1 = \frac{3}{2}$$

$$f(1) = \left(\frac{2}{3}\right)^1 = \frac{2}{3}$$

$$f(2) = \left(\frac{2}{3}\right)^2 = \frac{4}{9}$$

We can complete the graph by connecting the points with a smooth curve. See **Figure 21**.

(eText Screens 20.3-1–20.3-32)

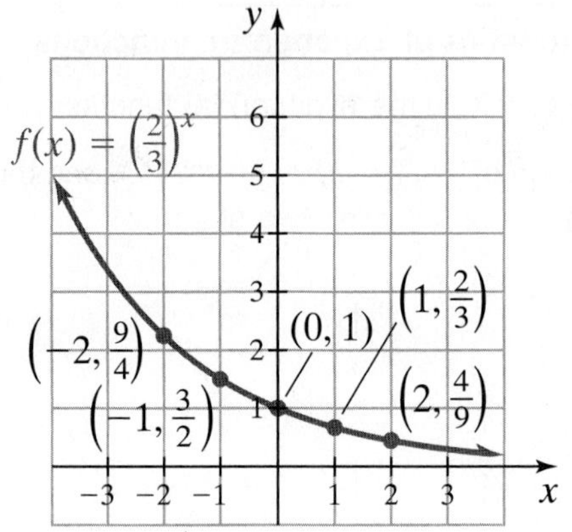

Figure 21
Graph of $f(x) = \left(\frac{2}{3}\right)^x$

You Try It **Work through this You Try It problem.**

Work Exercises 1–5 in this textbook or in the MyMathLab Study Plan.

Example 2 Determining an Exponential Function Given the Graph

Find the exponential function $f(x) = b^x$ whose graph is given as follows.

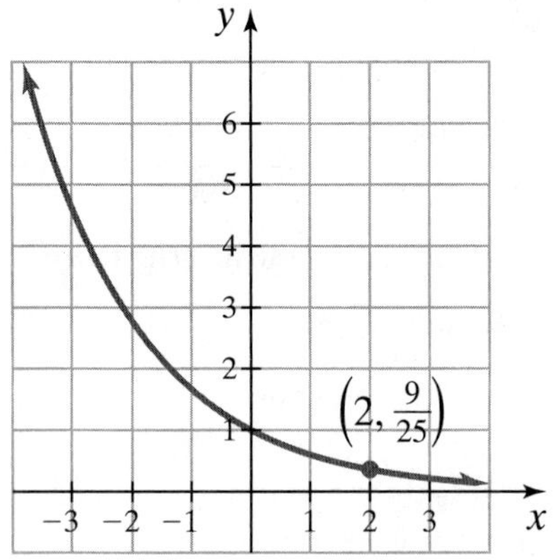

Solution From the point $\left(2, \frac{9}{25}\right)$, we see that $f(2) = \frac{9}{25}$. Thus,

Write the exponential function $f(x) = b^x$:	$f(x) = b^x$
Evaluate $f(2)$:	$f(2) = b^2$
The graph contains the point $\left(2, \dfrac{9}{25}\right)$:	$f(2) = \dfrac{9}{25}$
Equate the two expressions for $f(2)$:	$b^2 = \dfrac{9}{25}$

Therefore, we are looking for a constant b such that $b^2 = \frac{9}{25}$. Using the **square root property**, we get

$$\sqrt{b^2} = \pm\sqrt{\frac{9}{25}}$$

$$b = \pm\frac{3}{5}.$$

By definition of an exponential function, $b > 0$; thus, $b = \frac{3}{5}$. Therefore, this is the graph of $f(x) = \left(\frac{3}{5}\right)^x$.

(eText Screens 20.3-1–20.3-32)

You Try It **Work through this You Try It problem.**

Work Exercises 6–12 in this textbook or in the MyMathLab Study Plan.

OBJECTIVE 2 SKETCH THE GRAPHS OF EXPONENTIAL FUNCTIONS USING TRANSFORMATIONS

Often we can use the **transformation techniques** that are discussed in Topic 20.1 to sketch the graph of exponential functions. For example, the graph of $f(x) = 3^x - 1$ can be obtained by vertically shifting the graph of $y = 3^x$ down one unit. You can see in Figure 22 that the y-intercept of $f(x) = 3^x - 1$ is $(0, 0)$ and the **horizontal asymptote** is the line $y = -1$.

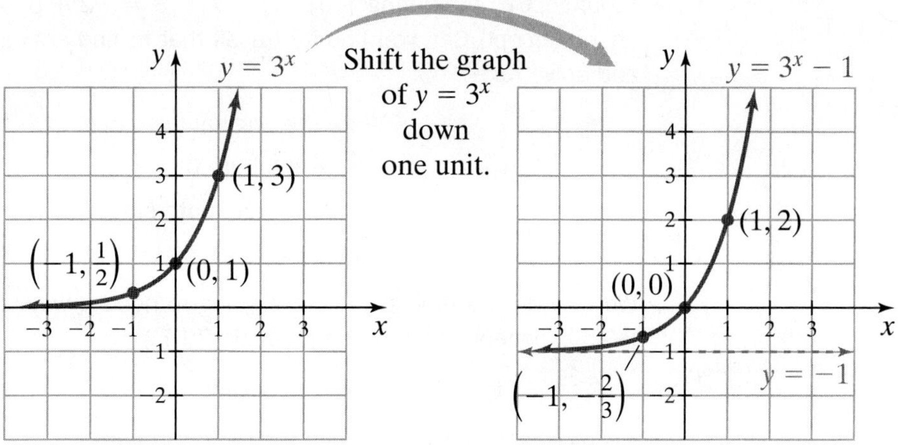

Figure 22
The graph of $f(x) = 3^x - 1$ can be obtained by vertically shifting the graph of $y = 3^x$ down one unit.

Example 3 Using Transformations to Sketch an Exponential Function

My video summary

Use transformations to sketch the graph of $f(x) = -2^{x+1} + 3$.

Solution Starting with the graph of $y = 2^x$, we can obtain the graph of $f(x) = -2^{x+1} + 3$ through a series of three transformations:

1. Horizontally shift the graph of $y = 2^x$ to the left one unit, producing the graph of $y = 2^{x+1}$.
2. Reflect the graph of $y = 2^{x+1}$ about the x-axis, producing the graph of $y = -2^{x+1}$.
3. Vertically shift the graph of $y = -2^{x+1}$ up three units, producing the graph of $f(x) = -2^{x+1} + 3$.

The graph of $f(x) = -2^{x+1} + 3$ is shown in **Figure 23**. Watch the **video** to see every step.

(eText Screens 20.3-1–20.3-32)

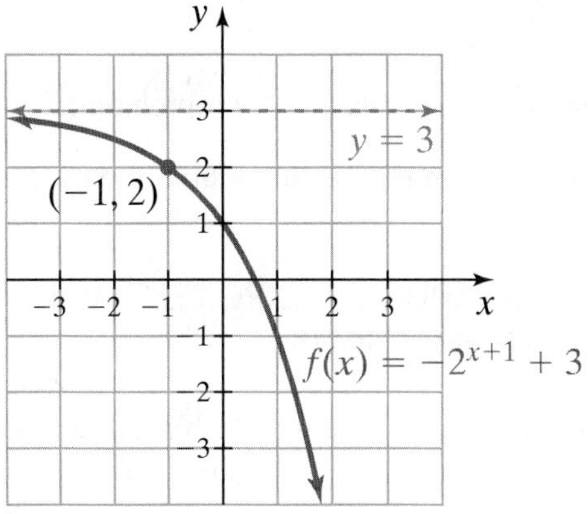

Figure 23
Graph of $f(x) = -2^{x+1} + 3$

Notice that the graph of $f(x) = -2^{x+1} + 3$ in Figure 23 has a y-intercept. We can find the y-intercept by evaluating $f(0) = -2^{0+1} + 3 = -2 + 3 = 1$. Also notice that the graph has an x-intercept. Can you find it? Recall that to find an x-intercept, we need to set $f(x) = 0$ and solve for x.

Write the original function:	$f(x) = -2^{x+1} + 3$
Substitute 0 for $f(x)$:	$0 = -2^{x+1} + 3$
Subtract 3 from both sides:	$-3 = -2^{x+1}$
Multiply both sides by -1:	$3 = 2^{x+1}$

We now have a problem! We are going to need a way to solve for a variable that appears in an exponent. (Stay tuned for Topic 20.7.)

You Try It **Work through this You Try It problem.**

Work Exercises 13–20 in this textbook or in the MyMathLab Study Plan.

OBJECTIVE 3 SOLVE EXPONENTIAL EQUATIONS BY RELATING THE BASES

One important property of all exponential functions is that they are **one-to-one functions.** You may want to review **Topic 20.2**, which discusses one-to-one functions in detail. The function $f(x) = b^x$ is one-to-one because the graph of f passes the **horizontal line test.** In Topic 20.2, the alternate definition of one-to-one is stated as follows:

A function f is one-to-one if for any two range values $f(u)$ and $f(v)$, $f(u) = f(v)$ implies that $u = v$.

Using this definition and letting $f(x) = b^x$, we can say that if $b^u = b^v$, then $u = v$. In other words, if the bases of an exponential equation of the form $b^u = b^v$ are the same, then the exponents must be the same. Solving exponential equations with this property is known as the **method of relating the bases** for solving exponential equations.

Method of Relating the Bases

If an exponential equation can be written in the form $b^u = b^v$, then $u = v$.

(eText Screens 20.3-1–20.3-32)

Example 4 Using the Method of Relating the Bases to Solve Exponential Equations

My animation summary

Solve the following equations.

a. $8 = \frac{1}{16^x}$ b. $\frac{1}{27^x} = \left(\sqrt[4]{3}\right)^{x-2}$

Solutions

a. Work through the **animation** to see how to obtain a solution of $x = -\frac{3}{4}$.

b. Work through the **animation** to see how to obtain a solution of $x = \frac{2}{13}$.

You Try It **Work through this You Try It problem.**

Work Exercises 21–30 in this textbook or in the MyMathLab Study Plan.

OBJECTIVE 4 SOLVE APPLICATIONS OF EXPONENTIAL FUNCTIONS

Exponential functions are used to describe many real-life situations and natural phenomena. We now look at some examples.

Example 5 Learn to Hit a 3-Wood on a Golf Driving Range

Most golfers find that their golf skills improve dramatically at first and then level off rather quickly. For example, suppose that the distance (in yards) that a typical beginning golfer can hit a 3-wood after t weeks of practice on the driving range is given by the exponential function $d(t) = 225 - 100(2.7)^{-0.7t}$. This function has been developed after many years of gathering data on beginning golfers.

How far can a typical beginning golfer initially hit a 3-wood? How far can a typical beginning golfer hit a 3-wood after 1 week of practice on the driving range? After 5 weeks? After 9 weeks? Round to the nearest hundredth yard.

Solution Initially, when $t = 0$, $d(0) = 225 - 100(2.7)^0 = 225 - 100 = 125$ yards. Therefore, a typical beginning golfer can hit a 3-wood 125 yards.

After 1 week of practice on the driving range, a typical beginning golfer can hit a 3-wood $d(1) = 225 - 100(2.7)^{-0.7(1)} \approx 175.11$ yards. After 5 weeks of practice, $d(5) = 225 - 100(2.7)^{-0.7(5)} \approx 221.91$ yards. After 9 weeks of practice, $d(9) = 225 - 100(2.7)^{-0.7(9)} \approx 224.81$ yards.

Using a graphing utility, we can sketch the graph of $d(t) = 225 - 100(2.7)^{-0.7t}$. You can see from the graph in **Figure 24** that the distance increases rather quickly and then tapers off toward a **horizontal asymptote** of 225 yards.

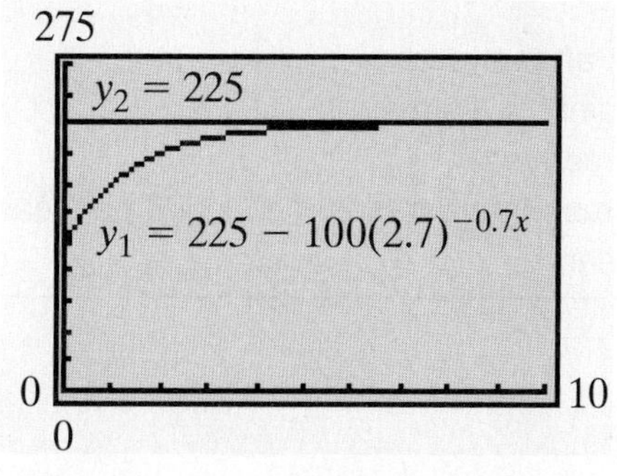

Figure 24
A TI-84 Plus was used to sketch the function $d(t) = 225 - 100(2.7)^{-0.7t}$ and the **horizontal asymptote** $y = 225$.

(eText Screens 20.3-1–20.3-32)

You Try It **Work through this You Try It problem.**

Work Exercises 31–34 in this textbook or in the MyMathLab Study Plan.

Compound Interest

A real-life application of exponential functions is the concept of **compound interest**, or interest that is paid on *both principal and interest*. First, we take a look at how **simple interest** is accrued. If an investment of P dollars is invested at r percent annually (written as a decimal) using simple interest, then the interest earned after 1 year is Pr dollars. Adding this interest to the original investment yields a total amount, A, of

$$A = \underbrace{P}_{\substack{\text{Original}\\\text{investment}}} + \underbrace{Pr}_{\substack{\text{Interest}\\\text{earned}}} = P(1 + r).$$

If this amount is reinvested at the same interest rate, then the total amount after 2 years becomes

$$A = \underbrace{P(1 + r)}_{\substack{\text{Total investment}\\\text{after 1 year}}} + \underbrace{P(1 + r)r}_{\substack{\text{Interest}\\\text{earned}}} = P(1 + r)(1 + r) = P(1 + r)^2.$$

Reinvesting this amount for a third year gives an amount of $P(1 + r)^3$. Continuing this process for k years, we can see that the amount becomes $A = P(1 + r)^k$. This is an exponential function with base $1 + r$.

We can now modify this formula to obtain another formula that will model interest that is compounded periodically throughout the year(s). When interest is compounded periodically, then k no longer represents the number of years but rather the number of pay periods. If interest is paid n times per year for t years, then $k = nt$ pay periods. Thus, in the formula $A = P(1 + r)^k$, we substitute nt for k and get $A = P(1 + r)^{nt}$.

In the earlier simple interest model, the variable r was used to represent annual interest. In the periodically compounded interest model being developed here with n pay periods per year, the interest rate per pay period is no longer r but rather $\frac{r}{n}$. Thus, in the formula $A = P(1 + r)^{nt}$, we replace r with $\frac{r}{n}$ and get the periodic compound interest formula $A = P\left(1 + \frac{r}{n}\right)^{nt}$.

Periodic Compound Interest Formula

Periodic compound interest can be calculated using the formula

$$A = P\left(1 + \frac{r}{n}\right)^{nt},$$

where

A = Total amount after t years
P = Principal (original investment)
r = Interest rate per year
n = Number of times interest is compounded per year
t = Number of years

(eText Screens 20.3-1–20.3-32)

Example 6 Calculating Compound Interest

Which investment will yield the most money after 25 years?

Investment A: $\$12{,}000$ invested at 3% compounded monthly
Investment B: $\$10{,}000$ invested at 3.9% compounded quarterly

Solution Investment A: $P = 12{,}000, r = 0.03, n = 12, t = 25$:

$$A = 12{,}000\left(1 + \frac{0.03}{12}\right)^{12(25)} \approx \$25{,}380.23$$

Investment B: $P = 10{,}000, r = 0.039, n = 4, t = 25$:

$$A = 10{,}000\left(1 + \frac{0.039}{4}\right)^{4(25)} \approx \$26{,}386.77$$

Investment B will yield the most money after 25 years.

You Try It **Work through this You Try It problem.**

Work Exercises 35–37 in this textbook or in the MyMathLab Study Plan.

Present Value

Sometimes investors want to know how much money to invest now to reach a certain investment goal in the future. This amount of money, P, is known as the **present value** of A dollars. To find a formula for present value, start with the formula for periodic compound interest and solve the formula for P:

Use the periodic compound interest formula: $A = P\left(1 + \frac{r}{n}\right)^{nt}$

Divide both sides by $\left(1 + \frac{r}{n}\right)^{nt}$: $\frac{A}{\left(1 + \frac{r}{n}\right)^{nt}} = \frac{P\cancel{\left(1 + \frac{r}{n}\right)^{nt}}}{\cancel{\left(1 + \frac{r}{n}\right)^{nt}}}$

Rewrite $\frac{1}{\left(1 + \frac{r}{n}\right)^{nt}}$ as $\left(1 + \frac{r}{n}\right)^{-nt}$: $A\left(1 + \frac{r}{n}\right)^{-nt} = P$

The formula $P = A\left(1 + \frac{r}{n}\right)^{-nt}$ is known as the **present value formula.**

Present Value Formula

Present value can be calculated using the formula

$$P = A\left(1 + \frac{r}{n}\right)^{-nt},$$

where

$P =$ Principal (original investment)
$A =$ Total amount after t years
$r =$ Interest rate per year
$n =$ Number of times interest is compounded per year
$t =$ Number of years

(eText Screens 20.3-1–20.3-32)

Example 7 Determining Present Value

Find the present value of $8000 if interest is paid at a rate of 5.6% compounded quarterly for 7 years. Round to the nearest cent.

Solution Using the present value formula $P = A\left(1 + \frac{r}{n}\right)^{-nt}$ with $A = \$8000$, $r = 0.056$, $n = 4$, and $t = 7$, we get

$$P = A\left(1 + \frac{r}{n}\right)^{-nt}$$

$$P = 8000\left(1 + \frac{0.056}{4}\right)^{-(4)(7)} \approx 5420.35.$$

Therefore, the present value of $8000 in 7 years at 5.6% compounded quarterly is $5420.35.

You Try It **Work through this You Try It problem.**

Work Exercises 38–40 in this textbook or in the MyMathLab Study Plan.

20.3 Exercises

In Exercises 1–5, sketch the graph of each exponential function. Label the y-intercept and at least two other points on the graph using both positive and negative values of x.

You Try It

1. $f(x) = 4^x$ **2.** $f(x) = \left(\frac{1}{4}\right)^x$ **3.** $f(x) = \left(\frac{3}{2}\right)^x$

4. $f(x) = (.4)^x$ **5.** $f(x) = (2.7)^x$

In Exercises 6–12, determine the correct exponential function of the form $f(x) = b^x$ whose graph is given.

You Try It

6.

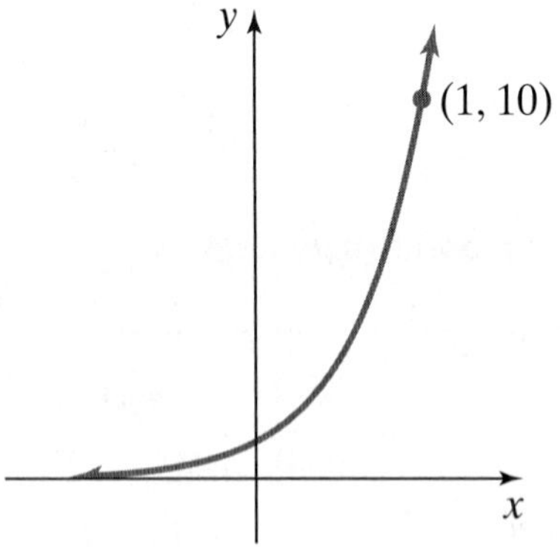

7.

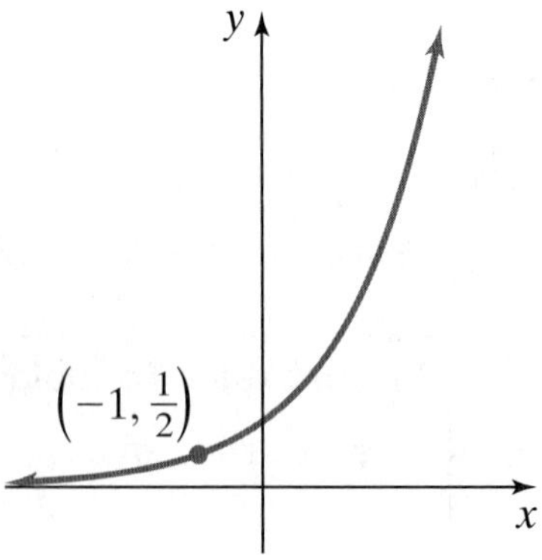

(eText Screens 20.3-1–20.3-32)

8.

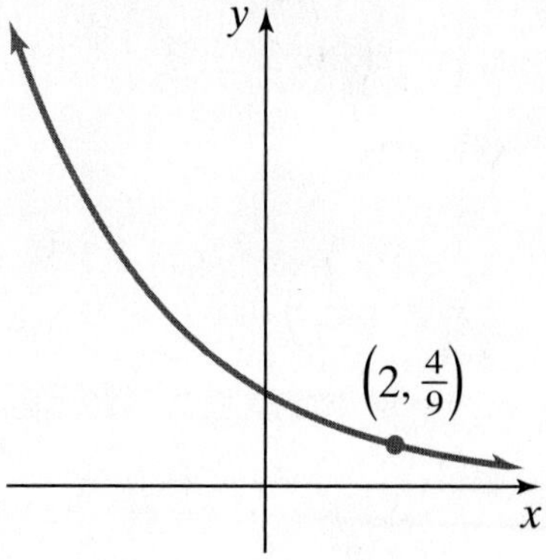

9.

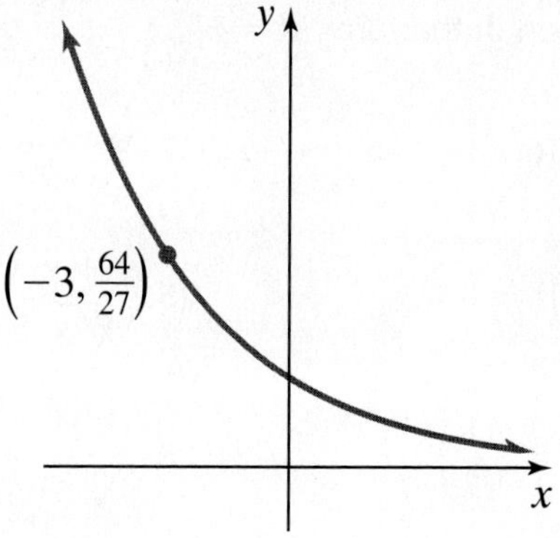

10.

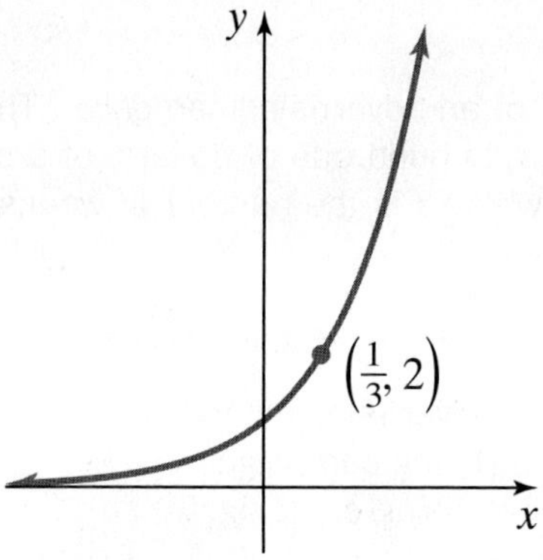

11.

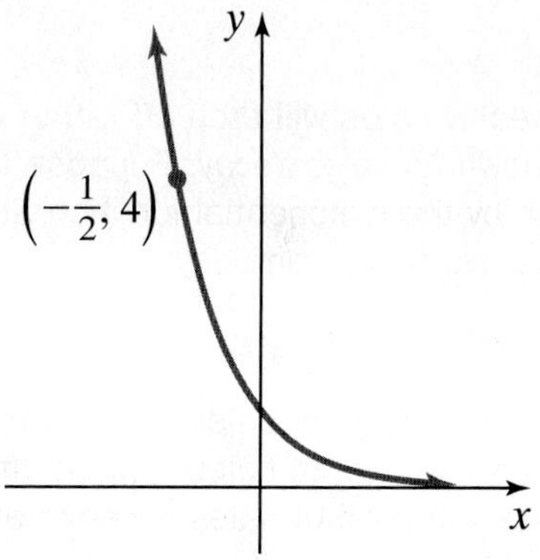

12.

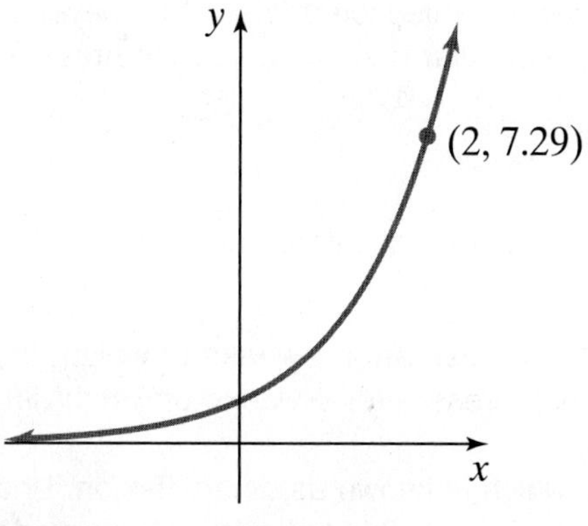

In Exercises 13–20, use the graph of $y = 2^x$ or $y = 3^x$ and transformations to sketch each exponential function. Determine the domain and range. Also, determine the y-intercept, and find the equation of the horizontal asymptote.

13. $f(x) = 2^{x-1}$ **14.** $f(x) = 3^x - 1$ **15.** $f(x) = -3^{x+2}$

You Try It

16. $f(x) = -2^{x+1} - 1$ **17.** $f(x) = \left(\frac{1}{2}\right)^{x+1}$ **18.** $f(x) = \left(\frac{1}{3}\right)^x - 3$

19. $f(x) = 2^{-x} + 1$ **20.** $f(x) = 3^{1-x} - 2$

(eText Screens 20.3-1–20.3-32)

In Exercises 21–30, solve each exponential equation using the method of "relating the bases" by first rewriting the equation in the form $b^u = b^v$.

You Try It

21. $2^x = 16$ **22.** $3^{x-1} = \frac{1}{9}$ **23.** $\sqrt{5} = 25^x$

24. $\left(\sqrt[3]{3}\right)^x = 9$ **25.** $\frac{1}{\sqrt[5]{8}} = 2^x$ **26.** $\frac{9}{\sqrt[4]{3}} = \left(\frac{1}{27}\right)^x$

27. $(49)^x = \left(\frac{1}{7}\right)^{x-1}$ **28.** $\frac{125}{\sqrt[3]{5^x}} = \left(\frac{1}{25^x}\right)$

29. $\frac{3^{x^2}}{9^x} = 27$ **30.** $2^{x^3} = \frac{4^x}{2^{-x^2}}$

You Try It

31. Typically, weekly sales will drop off rather quickly after the end of an advertising campaign. This drop in sales is known as *sales decay*. Suppose that the gross sales S, in hundreds of dollars, of a certain product is given by the exponential function $S(t) = 3000(1.5^{-0.3t})$, where t is the number of weeks after the end of the advertising campaign.

Answer the following questions, rounding each answer to the nearest whole number:

a. What was the level of sales immediately after the end of the advertising campaign when $t = 0$?
b. What was the level of sales 1 week after the end of the advertising campaign?
c. What was the level of sales 5 weeks after the end of the advertising campaign?

32. Most people who start a serious weight-lifting regimen initially notice a rapid increase in the maximum amount of weight that they can bench press. After a few weeks, this increase starts to level off. The following function models the maximum weight, w, that a particular person can bench press in pounds at the end of t weeks of working out.

$$w(t) = 250 - 120(2.7)^{-0.3t}$$

Answer the following questions, rounding each answer to the nearest whole number:

a. What is the maximum weight that this person can bench press initially?
b. What is the maximum weight that this person can bench press after 3 weeks of weight lifting?
c. What is the maximum weight that this person can bench press after 7 weeks of weight lifting?

33. *Escherichia coli* bacteria reproduce by simple cell division, which is known as binary fission. Under ideal conditions, a population of *E. coli* bacteria can double every 20 minutes. This behavior can be modeled by the exponential function $N(t) = N_0(2^{0.05t})$, where t is in minutes and N_0 is the initial number of *E. coli* bacteria.

Answer the following questions, rounding each answer to the nearest bacteria:

a. If the initial number of *E. coli* bacteria is five, how many bacteria will be present in 3 hours?
b. If the initial number of *E. coli* bacteria is eight, how many bacteria will be present in 3 hours?
c. If the initial number of *E. coli* bacteria is eight, how many bacteria will be present in 10 hours?

34. A wildlife-management research team noticed that a certain forest had no rabbits, so they decided to introduce a rabbit population into the forest for the first time. The rabbit population will be controlled by wolves and other predators. This rabbit population can be modeled by the function $R(t) = \frac{960}{0.6 + 23.4(2.7)^{-0.045t}}$, where t is the number of weeks after the research team first introduced the rabbits into the forest.

(eText Screens 20.4-1–20.4-20)

Answer the following questions, rounding each answer to the nearest whole number:

a. How many rabbits did the wildlife-management research team bring into the forest?
b. How many rabbits can be expected after 10 weeks?
c. How many rabbits can be expected after the first year?
d. What is the expected rabbit population after 4 years? 5 years? What can the expected rabbit population approach as time goes on?

Use the **periodic compound interest formula** to solve Exercises 35–37.

You Try It

35. Suppose that $9000 is invested at 3.5% compounded quarterly. Find the total amount of this investment after 10 years. Round to the nearest cent.

36. Suppose that you have $5000 to invest. Which investment yields the greater return over a 10-year period: 7.35% compounded daily or 7.4% compounded quarterly?

37. Which investment yields the greatest return?
Investment A: $4000 invested for 5 years compounded semiannually (twice per year) at 8%
Investment B: $5000 invested for 4 years compounded quarterly at 4.5%

Use the **present value formula** to solve Exercises 38–40.

You Try It

38. Find the present value of $10,000 if interest is paid at a rate of 4.5% compounded semiannually for 12 years. Round to the nearest cent.

39. Find the present value of $1,000,000 if interest is paid at a rate of 9.5% compounded monthly for 8 years. Round to the nearest cent.

40. How much money would you have to invest at 10% compounded semiannually so that the total investment had a value of $2205 after 1 year? Round to the nearest cent.

20.4 The Natural Exponential Function

THINGS TO KNOW

Before working through this topic, be sure you are familiar with the following concepts:

	VIDEO	ANIMATION	INTERACTIVE
You Try It **1.** Sketch the Graphs of Exponential Functions Using Transformations (Topic 20.3, **Objective 2**)	✓		
You Try It **2.** Solve Exponential Equations by Relating the Bases (Topic 20.3, **Objective 3**)		✓	
You Try It **3.** Solve Applications of Exponential Functions (Topic 20.3, **Objective 4**)			

(eText Screens 20.4-1–20.4-20)

OBJECTIVES

1 Use the Characteristics of the Natural Exponential Function

2 Sketch the Graphs of Natural Exponential Functions Using Transformations

3 Solve Natural Exponential Equations by Relating the Bases

4 Solve Applications of the Natural Exponential Function

OBJECTIVE 1 USE THE CHARACTERISTICS OF THE NATURAL EXPONENTIAL FUNCTION

My video summary

We learned in the previous topic that any positive number b, where $b \neq 1$, can be used as the base of an exponential function. However, there is one number that appears as the base in exponential applications more than any other number. This number is called the **natural base** and is symbolized using the letter e. The number e is an irrational number that is defined as the value of the expression $\left(1 + \frac{1}{n}\right)^n$ as n approaches infinity. Table 7 shows the values of the expression $\left(1 + \frac{1}{n}\right)^n$ for increasingly large values of n.

n	$\left(1 + \frac{1}{n}\right)^n$
1	2
2	2.25
10	2.5937424601
100	2.7048138294
1,000	2.7169239322
10,000	2.7181459268
100,000	2.7182682372
1,000,000	2.7182804693
10,000,000	2.7182816925
100,000,000	2.7182818149

Table 7

You can see from Table 7 that as the values of n get large, the value e (rounded to six decimal places) is 2.718282. The function $f(x) = e^x$ is called the **natural exponential function**. Because $2 < e < 3$, it follows that the graph of $f(x) = e^x$ lies between the graph of $y = 2^x$ and $y = 3^x$, as seen in Figure 25.

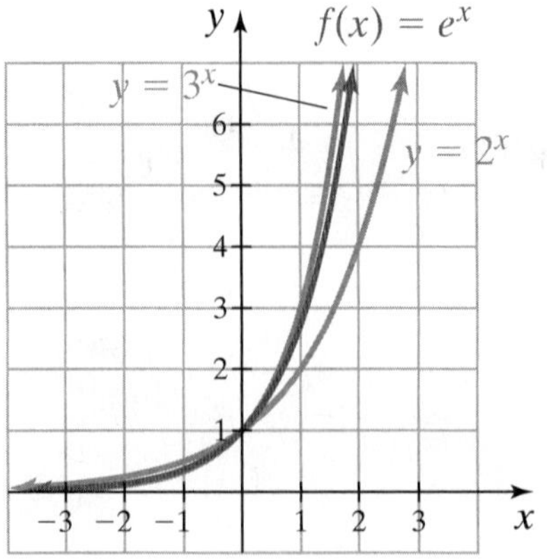

Figure 25
Graph of the natural exponential function $f(x) = e^x$

(eText Screens 20.4-1–20.4-20)

Characteristics of the Natural Exponential Function

The natural exponential function is the exponential function with base e and is defined as $f(x) = e^x$.

The domain of $f(x) = e^x$ is $(-\infty, \infty)$, and the range is $(0, \infty)$. The graph of $f(x) = e^x$ and some of its characteristics are stated next.

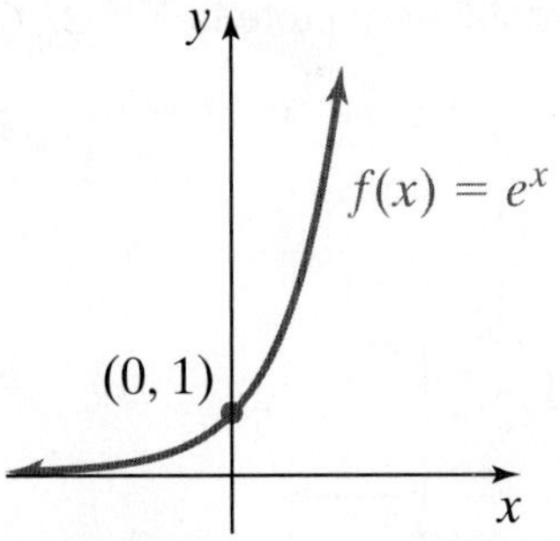

- The graph of $f(x) = e^x$ intersects the y-axis at $(0, 1)$.
- $e^x \to \infty$ as $x \to \infty$
- $e^x \to 0$ as $x \to -\infty$
- The line $y = 0$ is a **horizontal asymptote.**
- The function $f(x) = e^x$ is **one-to-one.**

It is important that you are able to use your calculator to evaluate various powers of e. Most calculators have an $\boxed{e^x}$ key. Find this special key on your calculator and evaluate the expressions in the following example.

Example 1 Evaluating the Natural Exponential Function

Evaluate each expression correctly to six decimal places.

a. e^2 **b.** $e^{-0.534}$ **c.** $1000e^{0.013}$

Solutions Using the $\boxed{e^x}$ key on a calculator, we get

a. $e^2 \approx 7.389056$

b. $e^{-0.534} \approx 0.586255$

c. $1000e^{0.013} \approx 1013.084867$

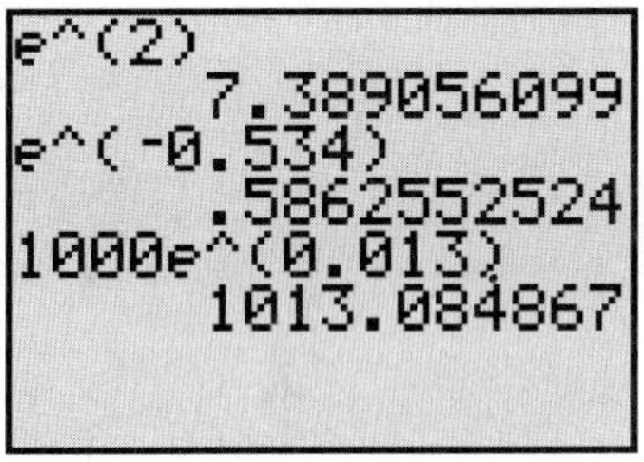

Screen Shot from TI-84 Plus

You Try It **Work through this You Try It problem.**

Work Exercises 1–5 in this textbook or in the MyMathLab Study Plan.

OBJECTIVE 2 SKETCH THE GRAPHS OF NATURAL EXPONENTIAL FUNCTIONS USING TRANSFORMATIONS

Again, we can use the **transformation techniques** that are discussed in **Topic 20.1** to sketch variations of the natural exponential function.

Example 2 Using Transformations to Sketch Natural Exponential Functions

My video summary

Use transformations to sketch the graph of $f(x) = -e^x + 2$. Determine the domain, range, and y-intercept, and find the equation of any asymptotes.

Solution We can sketch the graph of $f(x) = -e^x + 2$ through a series of the following two transformations.

(eText Screens 20.4-1–20.4-20)

Start with the graph of $y = e^x$.

1. Reflect the graph of $y = e^x$ about the x-axis, producing the graph of $y = -e^x$.
2. Vertically shift the graph of $y = -e^x$ up two units, producing the graph of $f(x) = -e^x + 2$.

The graph of $f(x) = -e^x + 2$ is shown in Figure 26. Watch the **video** to see each step. The domain of f is the interval $(-\infty, \infty)$. The range of f is the interval $(-\infty, 2)$. The y-intercept is 1, and the equation of the horizontal asymptote is $y = 2$.

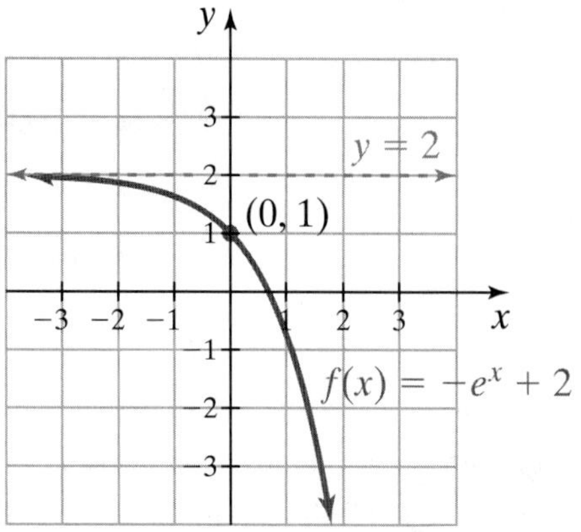

Figure 26
Graph of $f(x) = -e^x + 2$

You Try It **Work through this You Try It problem.**

Work Exercises 6–10 in this textbook or in the MyMathLab Study Plan.

OBJECTIVE 3 SOLVE NATURAL EXPONENTIAL EQUATIONS BY RELATING THE BASES

Recall the **method of relating the bases** for solving exponential equations from **Topic 20.3.** If we can write an exponential equation in the form of $b^u = b^v$, then $u = v$. This method for solving exponential equations certainly holds true for the natural base as illustrated in the following example.

Example 3 Using the Method of Relating the Bases to Solve Natural Exponential Equations

My video summary

Use the method of relating the bases to solve each exponential equation:

a. $e^{3x-1} = \dfrac{1}{\sqrt{e}}$

b. $\dfrac{e^{x^2}}{e^{10}} = (e^x)^3$

Solutions Work through the **interactive video** to verify that the solutions are as follows:

a. $x = \dfrac{1}{6}$

b. $x = -2$ or $x = 5$

You Try It **Work through this You Try It problem.**

Work Exercises 11–18 in this textbook or in the MyMathLab Study Plan.

OBJECTIVE 4 SOLVE APPLICATIONS OF THE NATURAL EXPONENTIAL FUNCTION

Continuous Compound Interest

My animation summary

Recall the **periodic compound interest formula** that is introduced in **Topic 20.3.** Some banks use **continuous compounding**; that is, they compound the interest every fraction of a second every day! If we start with the formula for periodic compound interest, $A = P\left(1 + \frac{r}{n}\right)^{nt}$, and let n (the number of times the interest is compounded each year) approach infinity, we can derive the formula $A = Pe^{rt}$, which is the formula for continuous compound interest. Work through this **animation** to see exactly how this formula is derived.

> **Continuous Compound Interest Formula**
>
> Continuous compound interest can be calculated using the formula
>
> $$A = Pe^{rt},$$
>
> where
>
> A = Total amount after t years
> P = Principal
> r = Interest rate per year
> t = Number of years

Example 4 Calculating Continuous Compound Interest

How much money would be in an account after 5 years if an original investment of $6000 was compounded continuously at 4.5%? Compare this amount to the same investment that was compounded daily. Round to the nearest cent.

Solution First, the amount after 5 years compounded continuously is $A = Pe^{rt} = 6000e^{0.045(5)} \approx \$7513.94$. The same investment compounded daily yields an amount of $A = P\left(1 + \frac{r}{n}\right)^{nt} = 6000\left(1 + \frac{0.045}{365}\right)^{365(5)} \approx \$7513.83$. Continuous compound interest yields only $0.11 more interest after 5 years!

You Try It **Work through this You Try It problem.**

Work Exercises 19–21 in this textbook or in the MyMathLab Study Plan.

Present Value

Recall that the present value P is the amount of money to be invested now to obtain A dollars in the future. To find a formula for present value on money that is compounded continuously, we start with the formula for continuous compound interest and solve for P.

Write the continuous compound interest formula: $A = Pe^{rt}$

Divide both sides by e^{rt}: $\frac{A}{e^{rt}} = \frac{P\cancel{e^{rt}}}{\cancel{e^{rt}}}$

Rewrite $\frac{1}{e^{rt}}$ as e^{-rt}: $Ae^{-rt} = P$

(eText Screens 20.4-1–20.4-20)

Present Value Formula

The present value of A dollars after t years of continuous compound interest, with interest rate r, is given by the formula

$$P = Ae^{-rt}.$$

Example 5 Calculating Present Value

Find the present value of \$18,000 if interest is paid at a rate of 8% compounded continuously for 20 years. Round to the nearest cent.

Solution Using the present value formula $P = Ae^{-rt}$ with $A = \$18{,}000$, $r = 0.08$, and $t = 20$, we get

$$P = Ae^{-rt}$$
$$P = (18{,}000)e^{-(0.08)(20)} \approx \$3634.14.$$

You Try It **Work through this You Try It problem.**

Work Exercises 22 and 23 in this textbook or in the MyMathLab Study Plan.

Exponential Growth Model

You have probably heard that some populations grow exponentially. Most populations grow at a rate proportional to the size of the population. In other words, the larger the population, the faster the population grows. With this in mind, it can be shown in a more advanced math course that the mathematical model that can describe population growth is given by the function $P(t) = P_0e^{kt}$.

Exponential Growth

A model that describes the population, P, after a certain time, t, is

$$P(t) = P_0e^{kt},$$

where $P_0 = P(0)$ is the initial population and $k > 0$ is a constant called the **relative growth rate.** (Note: k may be given as a percent.)

The graph of the exponential growth model is shown in Figure 27. Notice that the graph has a y-intercept of P_0.

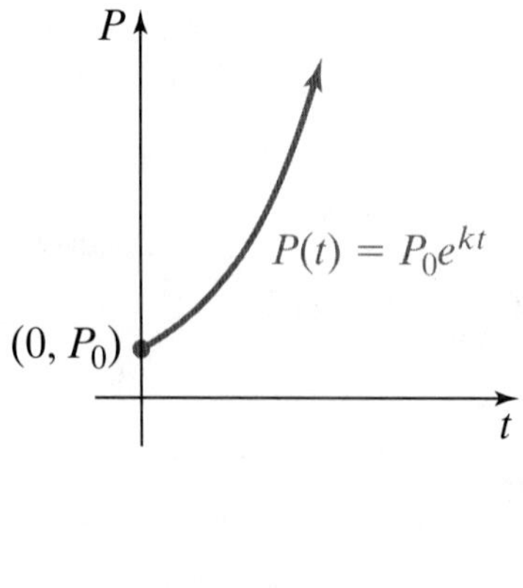

Figure 27
Graph of the exponential growth model $P(t) = P_0e^{kt}$

Example 6 Population Growth

My video summary

The population of a small town follows the exponential growth model $P(t) = 900e^{0.015t}$, where t is the number of years after 1900.

Answer the following questions, rounding each answer to the nearest whole number:

a. What was the population of this town in 1900?
b. What was the population of this town in 1950?
c. Use this model to predict the population of this town in 2012.

Solutions

a. The initial population was $P(0) = 900e^{0.015(0)} = 900.$

b. Because 1950 is 50 years after 1900, we must evaluate $P(50)$. $P(50) = 900e^{0.015(50)} \approx 1905.$

c. In the year 2012, we can predict that the population will be $P(112) = 900e^{0.015(112)} \approx 4829.$

Example 7 Determining the Initial Population

My video summary

Twenty years ago, the State of Idaho Fish and Game Department introduced a new breed of wolf into a certain Idaho forest. The current wolf population in this forest is now estimated at 825, with a relative growth rate of 12%.

Answer the following questions, rounding each answer to the nearest whole number:

a. How many wolves did the Idaho Fish and Game Department initially introduce into this forest?
b. How many wolves can be expected after another 20 years?

Solutions

a. The relative growth rate is 0.12, so we use the exponential growth model $P(t) = P_0e^{0.12t}$. Because $P(20) = 825$, we get

Substitute 20 for t: $P(20) = P_0e^{0.12(20)}$

Substitute 825 for $P(20)$: $825 = P_0e^{0.12(20)}$

Solve for P_0: $P_0 = \dfrac{825}{e^{0.12(20)}} \approx 75.$

Therefore, the Idaho Fish and Game Department initially introduced 75 wolves into the forest.

b. Because $P_0 = 75$, we can use the exponential growth model $P(t) = 75e^{0.12t}$. In another 20 years, the value of t will be 40. Thus, we must evaluate $P(40)$.

$$P(40) = 75e^{0.12(40)} \approx 9113$$

Therefore, we can expect the wolf population to be approximately 9113 in another 20 years.

You Try It **Work through this You Try It problem.**

Work Exercises 24–27 in this textbook or in the MyMathLab Study Plan.

(eText Screens 20.4-1–20.4-20)

20.4 Exercises

In Exercises 1–5, use a calculator to approximate each exponential expression to six decimal places.

You Try It

1. e^3 2. $e^{-0.2}$ 3. $e^{1/3}$ 4. $100e^{-.123}$ 5. $\sqrt{2}e^{\pi}$

In Exercises 6–10, use transformations to sketch each exponential function. Determine the domain and range. Also, label the y-intercept, and find the equation of the horizontal asymptote.

You Try It

6. $f(x) = e^{x-1}$ 7. $f(x) = e^x - 1$ 8. $f(x) = -e^{x+2}$ 9. $f(x) = -e^{x+1} - 1$ 10. $f(x) = e^{-x} - 2$

In Exercises 11–18, solve each exponential equation using the method of relating the bases by first rewriting the equation in the form $e^u = e^v$.

You Try It

11. $e^x = \dfrac{1}{e^2}$ 12. $e^{5x+2} = \sqrt[3]{e}$ 13. $\dfrac{1}{e^x} = \dfrac{\sqrt{e}}{e^{1-x}}$ 14. $(e^{x^2})^2 = e^8$

15. $e^{x^2} = (e^x) \cdot e^{12}$ 16. $e^{x^2} = \dfrac{e^3}{(e^x)^5}$ 17. $\dfrac{e^{x^3}}{e^x} = \dfrac{e^{2x^2}}{e^2}$ 18. $e^{x^3} = \dfrac{(e^{2x^2})^2 \cdot e^x}{e^4}$

You Try It

19. An original investment of \$6000 earns 6.25% interest compounded continuously. What will the investment be worth in 2 years? in 20 years? Round to the nearest cent.

20. How much more will an investment of \$10,000 earning 5.5% compounded continuously for 9 years earn compared to the same investment at 5.5% compounded quarterly for 9 years? Round to the nearest cent.

21. Suppose your great-great grandfather invested \$500 earning 6.5% interest compounded continuously 100 years ago. How much would his investment be worth today? Round to the nearest cent.

You Try It

22. Find the present value of \$16,000 if interest is paid at a rate of 4.5% compounded continuously for 10 years. Round to the nearest cent.

23. Which has the lower present value: (a) \$20,000 if interest is paid at a rate of 5.18% compounded continuously for 2 years or (b) \$25,000 if interest is paid at a rate of 3.8% compounded continuously for 30 months?

You Try It

24. The population of a rural city follows the exponential growth model $P(t) = 2000e^{0.035t}$, where t is the number of years after 1995.
    a. What was the population of this city in 1995?
    b. What is the relative growth rate as a percent?
    c. Use this model to approximate the population in 2030, rounding to the nearest whole number.

25. The relative growth rate of a certain bacteria colony is 25%. Suppose there are 10 bacteria initially.
    a. Find a function that describes the population of the bacteria after t hours.
    b. How many bacteria should we expect after 1 day? Round to the nearest whole number.

26. In 2006, the population of a certain American city was 18,221. If the relative growth rate has been 6% since 1986, what was the population of this city in 1986? Round to the nearest whole number.

27. In 1970, a wildlife resource management team introduced a certain rabbit species into a forest for the first time. In 2004, the rabbit population had grown to 7183. The relative growth rate for this rabbit species is 20%.

    Answer the following questions, rounding each answer to the nearest whole number:
    a. How many rabbits did the wildlife resource management team introduce into the forest in 1970?
    b. How many rabbits can be expected in the year 2025?

20.5 Logarithmic Functions

THINGS TO KNOW

Before working through this topic, be sure you are familiar with the following concepts:

VIDEO | ANIMATION | INTERACTIVE

You Try It 1. Solve Polynomial Inequalities (Topic 19.5, **Objective 1**)

You Try It 2. Solve Rational Inequalities (Topic 19.5, **Objective 2**)

You Try It 3. Determine If a Function Is One-to-One Using the Horizontal Line Test (Topic 20.2, **Objective 3**)

You Try It 4. Verify Inverse Functions (Topic 20.2, **Objective 4**)

You Try It 5. Sketch the Graphs of Inverse Functions (Topic 20.2, **Objective 5**)

You Try It 6. Find the Inverse of a One-to-One Function (Topic 20.2, **Objective 6**)

You Try It 7. Use the Characteristics of Exponential Functions (Topic 20.3, **Objective 1**)

You Try It 8. Sketch the Graphs of Exponential Functions Using Transformations (Topic 20.3, **Objective 2**)

You Try It 9. Solve Exponential Equations by Relating the Bases (Topic 20.3, **Objective 3**)

OBJECTIVES

1 Use the Definition of a Logarithmic Function

2 Evaluate Logarithmic Expressions

3 Use the Properties of Logarithms

4 Use the Common and Natural Logarithms

5 Use the Characteristics of Logarithmic Functions

6 Sketch the Graphs of Logarithmic Functions Using Transformations

7 Find the Domain of Logarithmic Functions

(eText Screens 20.5-1–20.5-28)

OBJECTIVE 1 USE THE DEFINITION OF A LOGARITHMIC FUNCTION

My video summary

Every exponential function of the form $f(x) = b^x$, where $b > 0$ and $b \neq 1$, is **one-to-one** and thus has an **inverse function**. (You may want to refer to **Topic 20.2** to review one-to-one functions and inverse functions.) Remember, given the graph of a one-to-one function f, the graph of the inverse function is a reflection about the line $y = x$. That is, for any point (a, b) that lies on the graph of f, the point (b, a) must lie on the graph of f^{-1}. In other words, the graph of f^{-1} can be obtained by simply switching the x and y coordinates of the ordered pairs of $f(x) = b^x$. Watch this **video** to see how to sketch the graph of $f(x) = b^x$ and its inverse.

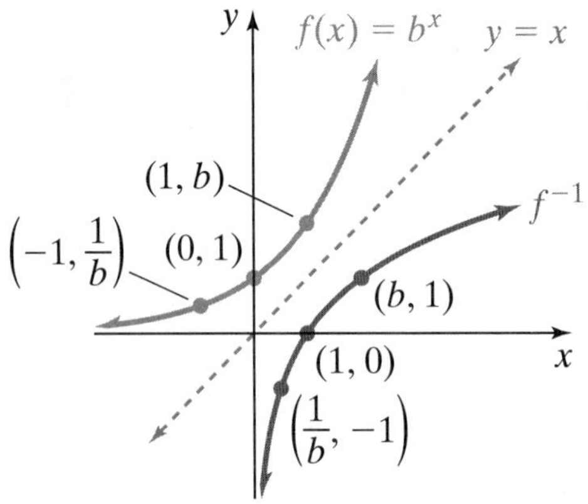

Figure 28
Graph of $f(x) = b^x, b > 1$, and its inverse function

The graphs of f and f^{-1} are sketched in Figure 28, but what is the equation of the inverse of $f(x) = b^x$? To find the equation of f^{-1}, we follow the **four-step process** for finding inverse functions that is discussed in Topic 20.2.

Step 1. Change $f(x)$ to y: $y = b^x$

Step 2. Interchange x and y: $x = b^y$

Step 3. Solve for y: ??

Before we can solve for y, we must introduce the following definition:

Definition Logarithmic Function

For $x > 0, b > 0$, and $b \neq 1$, the **logarithmic function** with base b is defined by

$$y = \log_b x \quad \text{if and only if} \quad x = b^y.$$

The equation $y = \log_b x$ is said to be in **logarithmic form,** whereas the equation $x = b^y$ is in **exponential form**. We can now continue to find the inverse of $f(x) = b^x$ by completing Steps 3 and 4.

Step 3. Solve for y: $x = b^y$ can be written as $y = \log_b x$

Step 4. Change y to $f^{-1}(x)$: $f^{-1}(x) = \log_b x$

In general, if $f(x) = b^x$ for $b > 0$ and $b \neq 1$, then the inverse function is $f^{-1}(x) = \log_b x$. For example, the inverse of $f(x) = 2^x$ is $f^{-1}(x) = \log_2 x$, which is read as "the log base 2 of x." We revisit the graphs of logarithmic functions later on in this topic, but first it is very important to understand the definition of the logarithmic function and practice how to go back and forth writing exponential equations as logarithmic equations and vice versa.

(eText Screens 20.5-1–20.5-28)

Example 1 Changing from Exponential Form to Logarithmic Form

My video summary

Write each exponential equation as an equation involving a logarithm.

a. $2^3 = 8$ b. $5^{-2} = \frac{1}{25}$ c. $1.1^M = z$

Solutions We use the fact that the equation $x = b^y$ is equivalent to the equation $y = \log_b x$.

a. $2^3 = 8$ is equivalent to $\log_2 8 = 3$.

b. $5^{-2} = \frac{1}{25}$ is equivalent to $\log_5 \frac{1}{25} = -2$.

c. $1.1^M = z$ is equivalent to $\log_{1.1} z = M$.

Note that the exponent of the original (exponential) equation ends up by itself on the right side of the second (logarithmic) equation. Therefore, a logarithmic expression can be thought of as describing the exponent of a certain exponential equation.

Watch the **video** to see this example worked out in more detail.

You Try It **Work through this You Try It problem.**

Work Exercises 1–6 in this textbook or in the MyMathLab Study Plan.

Example 2 Changing from Logarithmic Form to Exponential Form

My video summary

Write each logarithmic equation as an equation involving an exponent.

a. $\log_3 81 = 4$ b. $\log_4 16 = y$ c. $\log_{3/5} x = 2$

Solutions We use the fact that the equation $y = \log_b x$ is equivalent to the equation $x = b^y$.

a. $\log_3 81 = 4$ is equivalent to $3^4 = 81$.

b. $\log_4 16 = y$ is equivalent to $4^y = 16$.

c. $\log_{3/5} x = 2$ is equivalent to $\left(\frac{3}{5}\right)^2 = x$.

Watch the **video** to see this example worked out in more detail.

You Try It **Work through this You Try It problem.**

Work Exercises 7–11 in this textbook or in the MyMathLab study plan.

OBJECTIVE 2 EVALUATE LOGARITHMIC EXPRESSIONS

Because a logarithmic expression represents the exponent of an exponential equation, it is possible to evaluate many logarithms by inspection or by creating the corresponding exponential equation. Remember that the expression $\log_b x$ is the exponent to which b must be raised in order get x. For example, suppose we are to evaluate the expression $\log_4 64$. To evaluate this expression, we must ask ourselves, "4 raised to what power is 64?" Because $4^3 = 64$, we conclude that $\log_4 64 = 3$. For some logarithmic expressions, it is often convenient to create an exponential equation and use the **method of relating the bases** for solving exponential equations. For more complex logarithmic expressions, additional techniques are required. These techniques are discussed in Topics 20.6 and 20.7.

(eText Screens 20.5-1–20.5-28)

Example 3 Evaluating Logarithmic Expressions

My interactive video summary

Evaluate each logarithm:

a. $\log_5 25$ **b.** $\log_3 \frac{1}{27}$ **c.** $\log_{\sqrt{2}} \frac{1}{4}$

Solutions

a. To evaluate $\log_5 25$, we must ask, "5 raised to what exponent is 25?" Because $5^2 = 25, \log_5 25 = 2$.

b. The expression $\log_3 \frac{1}{27}$ requires more analysis. In this case, we ask, "3 raised to what exponent is $\frac{1}{27}$?" Suppose we let y equal this exponent. Then $3^y = \frac{1}{27}$. To solve for y, we can use the **method of relating the bases** for solving exponential equations.

Write the exponential equation: $3^y = \dfrac{1}{27}$

Rewrite 27 as 3^3: $3^y = \dfrac{1}{3^3}$

Use $\dfrac{1}{b^n} = b^{-n}$: $3^y = 3^{-3}$

Use the method of relating the bases: $y = -3$
(If $b^u = b^v$, then $u = v$.)

Thus, $\log_3 \frac{1}{27} = -3$.

c. Watch the **interactive video** to verify that $\log_{\sqrt{2}} \dfrac{1}{4} = -4$ and to see each solution worked out in detail.

You Try It **Work through this You Try It problem.**

Work Exercises 12–18 in this textbook or in the MyMathLab Study Plan.

OBJECTIVE 3 USE THE PROPERTIES OF LOGARITHMS

Because $b^1 = b$ for any real number b, we can use the definition of the logarithmic function ($y = \log_b x$ if and only if $x = b^y$) to rewrite this expression as $\log_b b = 1$. Similarly, because $b^0 = 1$ for any real number b, we can rewrite this expression as $\log_b 1 = 0$. These two general properties are summarized as follows.

General Properties of Logarithms

For $b > 0$ and $b \neq 1$,

1. $\log_b b = 1$ and

2. $\log_b 1 = 0$.

In **Topic 20.2**, we saw that a function f and its inverse function f^{-1} satisfy the following two composition cancellation equations:

$$f(f^{-1}(x)) = x \text{ for all } x \text{ in the domain of } f^{-1} \text{ and}$$
$$f^{-1}(f(x)) = x \text{ for all } x \text{ in the domain of } f.$$

(eText Screens 20.5-1–20.5-28)

If $f(x) = b^x$, then $f^{-1}(x) = \log_b x$. Applying the two composition cancellation equations we get

$$f(f^{-1}(x)) = b^{\log_b x} = x \text{ and}$$
$$f^{-1}(f(x)) = \log_b b^x = x.$$

Cancellation Properties of Exponentials and Logarithms

For $b > 0$ and $b \neq 1$,

1. $b^{\log_b x} = x$ and
2. $\log_b b^x = x$.

Example 4 Using the Properties of Logarithms

Use the properties of logarithms to evaluate each expression.

a. $\log_3 3^4$ **b.** $\log_{12} 12$ **c.** $7^{\log_7 13}$ **d.** $\log_8 1$

Solutions

a. By the second cancellation property, $\log_3 3^4 = 4$.

b. Because $\log_b b = 1$ for all $b > 0$ and $b \neq 1$, $\log_{12} 12 = 1$.

c. By the first cancellation property, $7^{\log_7 13} = 13$.

d. Because $\log_b 1 = 0$ for all $b > 0$ and $b \neq 1$, $\log_8 1 = 0$.

You Try It **Work through this You Try It problem.**

Work Exercises 19–26 in this textbook or in the MyMathLab Study Plan.

OBJECTIVE 4 USE THE COMMON AND NATURAL LOGARITHMS

There are two bases that are used more frequently than any other base. They are base 10 and base e. (Refer to **Topic 20.4** to review the natural base e.) Because our counting system is based on the number 10, the base 10 logarithm is known as the **common logarithm.** Instead of using the notation $\log_{10} x$ to denote the common logarithm, it is usually abbreviated without the subscript 10 as simply $\log x$. The base e logarithm is called the **natural logarithm** and is abbreviated as $\ln x$ instead of $\log_e x$. Most scientific calculators are equipped with a [log] key and a [ln] key. We can apply the definition of the logarithmic function for the base 10 and for the base e logarithm as follows.

Definition Common Logarithmic Function

For $x > 0$, the **common logarithmic function** is defined by

$$y = \log x \quad \text{if and only if} \quad x = 10^y.$$

Definition Natural Logarithmic Function

For $x > 0$, the **natural logarithmic function** is defined by

$$y = \ln x \quad \text{if and only if} \quad x = e^y.$$

(eText Screens 20.5-1–20.5-28)

Example 5 Changing from Exponential Form to Logarithmic Form

My video summary

Write each exponential equation as an equation involving a common logarithm or natural logarithm.

a. $e^0 = 1$ b. $10^{-2} = \frac{1}{100}$ c. $e^K = w$

Solutions

a. $e^0 = 1$ is equivalent to $\ln 1 = 0$.

b. $10^{-2} = \frac{1}{100}$ is equivalent to $\log\left(\frac{1}{100}\right) = -2$.

c. $e^K = w$ is equivalent to $\ln w = K$.

Watch the **video** to see this example worked out in more detail.

You Try It **Work through this You Try It problem.**

Work Exercises 27–31 in this textbook or in the MyMathLab Study Plan.

Example 6 Changing from Logarithmic Form to Exponential Form

My video summary

Write each logarithmic equation as an equation involving an exponent.

a. $\log 10 = 1$ b. $\ln 20 = Z$ c. $\log(x - 1) = T$

Solutions

a. $\log 10 = 1$ is equivalent to $10^1 = 10$.

b. $\ln 20 = Z$ is equivalent to $e^Z = 20$.

c. $\log(x - 1) = T$ is equivalent to $10^T = x - 1$.

Watch the **video** to see this example worked out in more detail.

You Try It **Work through this You Try It problem.**

Work Exercises 32–35 in this textbook or in the MyMathLab Study Plan.

Example 7 Evaluating Common and Natural Logarithmic Expressions

My video summary

Evaluate each expression without the use of a calculator.

a. $\log 100$ b. $\ln\sqrt{e}$ c. $e^{\ln 51}$ d. $\log 1$

Solutions

a. $\log 100 = 2$ because $10^2 = 100$ by the **definition of the logarithmic function** or $\log 100 = \log 10^2 = 2$ by **cancellation property** (2).

b. $\ln\sqrt{e} = \ln e^{1/2} = \frac{1}{2}$ by **cancellation property** (2).

c. $e^{\ln 51} = 51$ by **cancellation property** (1).

d. $\log 1 = 0$ by **general property** (1).

Watch the **video** to see this example worked out in more detail.

You Try It **Work through this You Try It problem.**

Work Exercises 36–43 in this textbook or in the MyMathLab Study Plan.

(eText Screens 20.5-1–20.5-28)

OBJECTIVE 5 USE THE CHARACTERISTICS OF LOGARITHMIC FUNCTIONS

To sketch the graph of a logarithmic function of the form $f(x) = \log_b x$, where $b > 0$ and $b \neq 1$, follow these three steps:

Step 1. Start with the graph of the exponential function $y = b^x$, labeling several ordered pairs.

Step 2. Because $f(x) = \log_b x$ is the inverse of $y = b^x$, we can find several points on the graph of $f(x) = \log_b x$ by reversing the coordinates of the ordered pairs of $y = b^x$.

Step 3. Plot the ordered pairs from Step 2, and complete the graph of $f(x) = \log_b x$ by connecting the ordered pairs with a smooth curve. The graph of $f(x) = \log_b x$ is a reflection of the graph of $y = b^x$ about the line $y = x$.

Example 8 Sketching the Graph of a Logarithmic Function

My video summary

Sketch the graph of $f(x) = \log_3 x$.

Solution

Step 1. The graph of $y = 3^x$ passes through the points $\left(-1, \frac{1}{3}\right)$, $(0, 1)$, and $(1, 3)$. See Figure 29.

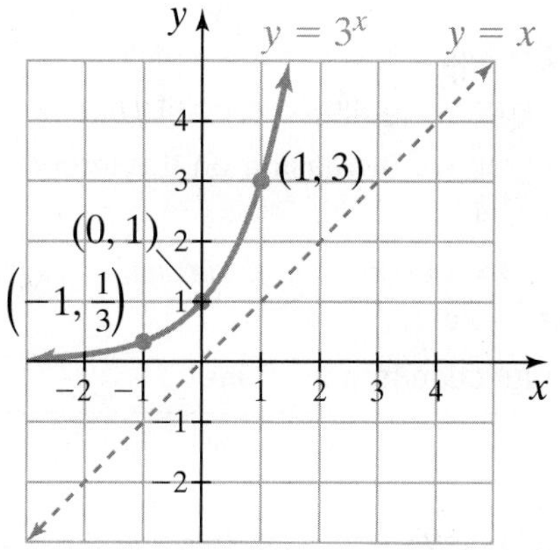

Figure 29
Graph of $y = 3^x$

Step 2. We reverse the three ordered pairs from Step 1 to obtain the following three points: $\left(\frac{1}{3}, -1\right)$, $(1, 0)$, and $(3, 1)$.

Step 3. Plot the points $\left(\frac{1}{3}, -1\right)$, $(1, 0)$, and $(3, 1)$, and connect them with a smooth curve to obtain the graph of $f(x) = \log_3 x$. See Figure 30.

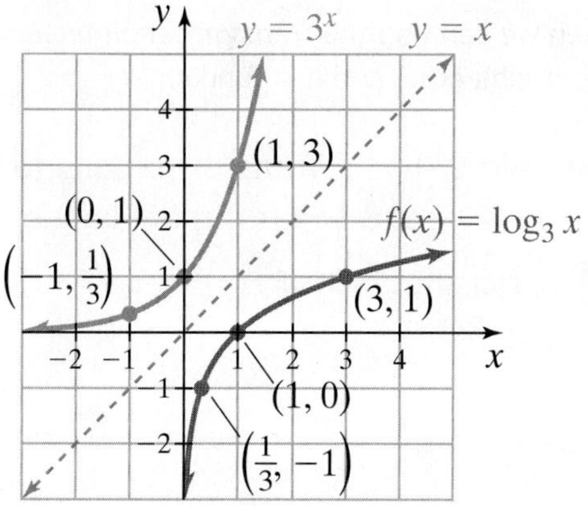

Figure 30
Graphs of $y = 3^x$ and $f(x) = \log_3 x$

(eText Screens 20.5-1–20.5-28)

Notice in Figure 30 that the y-axis is a **vertical asymptote** of the graph of $f(x) = \log_3 x$. Every logarithmic function of the form $y = \log_b x$, where $b > 0$ and $b \neq 1$ has a vertical asymptote at the y-axis. The graphs and the characteristics of logarithmic functions are outlined as follows.

Characteristics of Logarithmic Functions

For $b > 0, b \neq 1$, the logarithmic function with base b is defined by $y = \log_b x$. The domain of $f(x) = \log_b x$ is $(0, \infty)$, and the range is $(-\infty, \infty)$. The graph of $f(x) = \log_b x$ has one of the following two shapes.

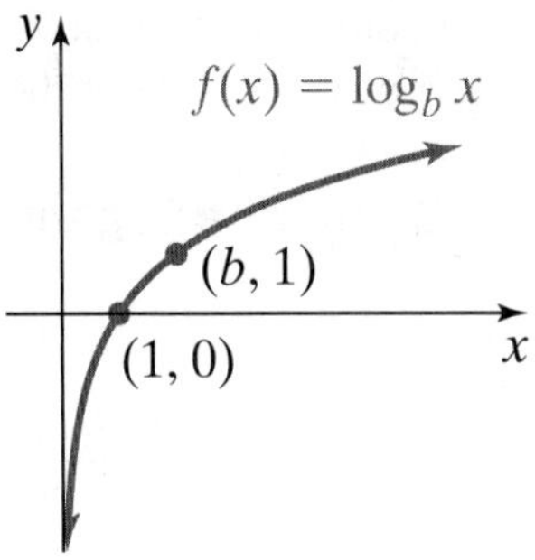

$f(x) = \log_b x, b > 1$

The graph intersects the x-axis at $(1, 0)$.

The graph contains the point $(b, 1)$.

The graph is **increasing** on the interval $(0, \infty)$.

The y-axis $(x = 0)$ is a **vertical asymptote.**

The function is **one-to-one.**

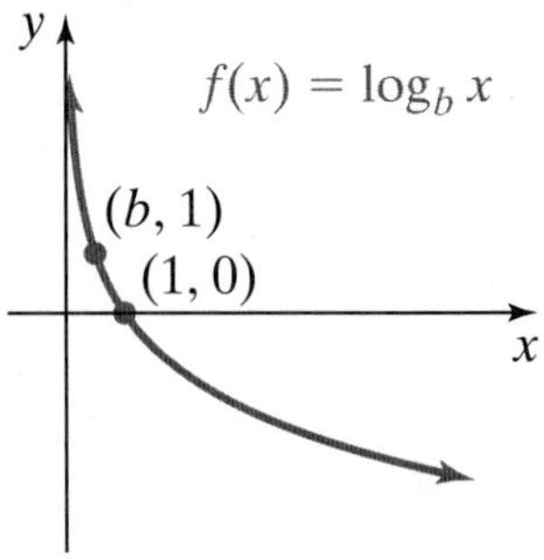

$f(x) = \log_b x, 0 < b < 1$

The graph intersects the x-axis at $(1, 0)$.

The graph contains the point $(b, 1)$.

The graph is **decreasing** on the interval $(0, \infty)$.

The y-axis $(x = 0)$ is a **vertical asymptote.**

The function is **one-to-one.**

You Try It **Work through this You Try It problem.**

Work Exercises 44 and 45 in this textbook or in the MyMathLab Study Plan.

OBJECTIVE 6 SKETCH THE GRAPHS OF LOGARITHMIC FUNCTIONS USING TRANSFORMATIONS

Often we can use the **transformation techniques** that are discussed in **Topic 20.1** to sketch the graphs of logarithmic functions.

Example 9 Using Transformations to Sketch the Graph of a Logarithmic Function

Sketch the graph of $f(x) = -\ln(x + 2) - 1$.

My video summary

Solution Recall that the function $y = \ln x$ has a base of e, where $2 < e < 3$. This means that the graph of $y = \ln x$ is increasing on the interval $(0, \infty)$ and contains the points $(1, 0)$ and $(e, 1)$. Starting with the graph of $y = \ln x$, we can obtain the graph of $f(x) = -\ln(x + 2) - 1$ through the following series of transformations:

1. Shift the graph of $y = \ln x$ horizontally to the left two units to obtain the graph of $y = \ln(x + 2)$.
2. Reflect the graph of $y = \ln(x + 2)$ about the x-axis to obtain the graph of $y = -\ln(x + 2)$.
3. Shift the graph of $y = -\ln(x + 2)$ vertically down one unit to obtain the final graph of $f(x) = -\ln(x + 2) - 1$.

   The graph of $f(x) = -\ln(x + 2) - 1$ is sketched in Figure 31. You can see from the graph that the domain of $f(x) = -\ln(x + 2) - 1$ is $(-2, \infty)$. The vertical asymptote is $x = -2$, and the x-intercept is $\left(\frac{1}{e} - 2, 0\right)$. Watch the **video** to see each step worked out in detail.

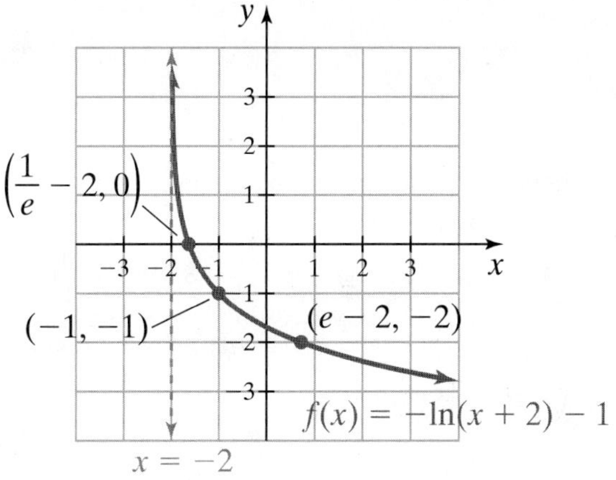

Figure 31
Graph of $f(x) = -\ln(x + 2) - 1$

You Try It **Work through this You Try It problem.**

Work Exercises 46–51 in this textbook or in the MyMathLab Study Plan.

OBJECTIVE 7 FIND THE DOMAIN OF LOGARITHMIC FUNCTIONS

In Example 9, we sketched the function $f(x) = -\ln(x + 2) - 1$ and observed that the domain was $(-2, \infty)$. We do not have to sketch the graph of a logarithmic function to determine the domain. The domain of a logarithmic function consists of all values of x for which the **argument** of the logarithm is greater than zero. In other words, if $f(x) = \log_b[g(x)]$, then the domain of f can be found by solving the inequality $g(x) > 0$. For example, given the function $f(x) = -\ln(x + 2) - 1$ from Example 9, we can determine the domain by solving the linear inequality $x + 2 > 0$. Solving this inequality for x, we obtain $x > -2$. Thus, the domain of $f(x) = -\ln(x + 2) - 1$ is $(-2, \infty)$.

Example 10 is a bit more challenging because the argument of the logarithm is a **rational expression.**

Example 10 Finding the Domain of a Logarithmic Function with a Rational Argument

My interactive video summary

Find the domain of $f(x) = \log_5\left(\frac{2x-1}{x+3}\right)$.

Solution To find the domain of f, we must find all values of x for which the argument $\frac{2x-1}{x+3}$ is greater than zero. That is, you must solve the rational inequality $\left(\frac{2x-1}{x+3}\right) > 0$. See **Topic 19.5** if you need help remembering how to solve this inequality. By the techniques discussed in Topic 19.5, we find that the solution to $\left(\frac{2x-1}{x+3}\right) > 0$ is $x < -3$ or $x > \frac{1}{2}$. Therefore, the domain of $f(x) = \log_5\left(\frac{2x-1}{x+3}\right)$ in set notation is $\{x | x < -3 \text{ or } x > \frac{1}{2}\}$. In interval notation, the domain is $(-\infty, -3) \cup \left(\frac{1}{2}, \infty\right)$. Watch the **interactive video** to see this problem worked out in detail.

You Try It **Work through this You Try It problem.**

Work Exercises 52–58 in this textbook or in the MyMathLab Study Plan.

20.5 Exercises

In Exercises 1–6, write each exponential equation as an equation involving a logarithm.

You Try It

1. $3^2 = 9$ **2.** $16^{1/2} = 4$ **3.** $2^{-3} = \frac{1}{8}$

4. $\sqrt{2}^{\pi} = W$ **5.** $\left(\frac{1}{3}\right)^t = 27$ **6.** $7^{5k} = L$

In Exercises 7–11, write each logarithmic equation as an exponential equation.

You Try It

7. $\log_5 1 = 0$ **8.** $\log_7 343 = 3$ **9.** $\log_{\sqrt{2}} 8 = 6$

10. $\log_4 K = L$ **11.** $\log_a (x - 1) = 3$

In Exercises 12–18, evaluate each logarithm without the use of a calculator.

You Try It

12. $\log_2 8$ **13.** $\log_6 \sqrt{6}$ **14.** $\log_3 \frac{1}{9}$

15. $\log_{\sqrt{5}} 25$ **16.** $\log_4\left(\frac{1}{\sqrt[5]{64}}\right)$ **17.** $\log_{1/7}\sqrt[3]{7}$

18. $\log_{0.1} 100$

In Exercises 19–26, use the properties of logarithms to evaluate each expression without the use of a calculator.

You Try It

19. $2^{\log_2 11}$ **20.** $\log_4 4$ **21.** $\log_9 1$

22. $\log_7 7^{-3}$ **23.** $\log_a a, a > 1$ **24.** $5^{\log_5 M}, M > 0$

25. $\log_y 1, y > 0$ **26.** $\log_x x^{20}, x > 1$

In Exercises 27–31, write each exponential equation as an equation involving a common logarithm or a natural logarithm.

You Try It

27. $10^3 = 1000$
28. $e^{-1} = \frac{1}{e}$
29. $e^k = 2$
30. $10^e = M$
31. $e^{10} = Z$

In Exercises 32–35, write each logarithmic equation as an exponential equation.

You Try It

32. $\ln 1 = 0$
33. $\log(1{,}000{,}000) = 6$
34. $\log K = L$
35. $\ln Z = 4$

In Exercises 36–43, evaluate each expression without the use of a calculator, and then verify your answer using a calculator.

You Try It

36. $\log 10{,}000$
37. $\log\left(\frac{1}{1000}\right)$
38. $\ln 1$
39. $\ln\sqrt[3]{e^2}$
40. $10^{\log e}$
41. $e^{\ln 49}$
42. $\log 10^6$
43. $\ln e + \ln e^3$

In Exercises 44–51, sketch each logarithmic function. Label at least two points on the graph, and determine the domain and the equation of any vertical asymptotes.

You Try It

44. $h(x) = \log_4 x$
45. $g(x) = \log_{\frac{1}{3}} x$

You Try It

46. $f(x) = \log_2(x) - 1$
47. $f(x) = \log_5(x - 1)$
48. $f(x) = -\ln(x)$
49. $y = \log_{1/2}(x + 1) + 2$
50. $y = \log_3(1 - x)$
51. $h(x) = -\frac{1}{2}\log_3(x + 3) + 1$

In Exercises 52–58, find the domain of each logarithmic function.

You Try It

52. $f(x) = \log(-x)$
53. $f(x) = \log_{1/4}(2x + 6)$
54. $f(x) = \ln(1 - 3x)$
55. $f(x) = \log_2(x^2 - 9)$
56. $f(x) = \log_7(x^2 - x - 20)$
57. $f(x) = \ln\left(\frac{x + 5}{x - 8}\right)$
58. $f(x) = \log\left(\frac{x^2 - x - 6}{x + 10}\right)$

(eText Screens 20.6-1–20.6-20)

20.6 Properties of Logarithms

THINGS TO KNOW

Before working through this topic, be sure you are familiar with the following concepts:

		VIDEO	ANIMATION	INTERACTIVE
You Try It	1. Solve Exponential Equations by Relating the Bases (Topic 20.3, **Objective 3**)		✓	
You Try It	2. Change from Exponential Form to Logarithmic Form (Topic 20.5, **Objective 1**)	✓		
You Try It	3. Change from Logarithmic Form to Exponential Form (Topic 20.5, **Objective 1**)	✓		
You Try It	4. Evaluate Logarithmic Expressions (Topic 20.5, **Objective 2**)			✓
You Try It	5. Use the Common and Natural Logarithms (Topic 20.5, **Objective 4**)	✓		
You Try It	6. Find the Domain of Logarithmic Functions (Topic 20.5, **Objective 7**)			✓

OBJECTIVES

1 Use the Product Rule, Quotient Rule, and Power Rule for Logarithms

2 Expand and Condense Logarithmic Expressions

3 Solve Logarithmic Equations Using the Logarithm Property of Equality

4 Use the Change of Base Formula

OBJECTIVE 1 USE THE PRODUCT RULE, QUOTIENT RULE, AND POWER RULE FOR LOGARITHMS

In this topic, we learn how to manipulate logarithmic expressions using properties of logarithms. Understanding how to use these properties will help us solve exponential and logarithmic equations that are encountered in the next topic. Recall from **Topic 20.5** the **general properties** and **cancellation properties** of logarithms. We now look at three additional properties of logarithms.

My video summary

Properties of Logarithms

If $b > 0, b \neq 1$, u and v represent positive numbers and r is any real number, then

$\log_b uv = \log_b u + \log_b v$ Product rule for logarithms

$\log_b \frac{u}{v} = \log_b u - \log_b v$ Quotient rule for logarithms

$\log_b u^r = r \log_b u$ Power rule for logarithms

(eText Screens 20.6-1–20.6-20)

To prove the product rule and quotient rule for logarithms, we use properties of exponents and the **method of relating the bases** to solve exponential equations. The power rule for logarithms is a direct result of the product rule. Watch the videos linked above to see proofs of these properties.

Example 1 Using the Product Rule

My video summary

Use the product rule for logarithms to expand each expression. Assume $x > 0$.

a. $\ln(5x)$ **b.** $\log_2(8x)$

Solutions

a. Use the product rule for logarithms: $\ln(5x) = \ln 5 + \ln x$

b. Use the product rule for logarithms: $\log_2(8x) = \log_2 8 + \log_2 x$

Use the **definition of the logarithmic function** to rewrite $\log_2 8$ as 3 because $2^3 = 8$: $\log_2(8x) = 3 + \log_2 x$

CAUTION $\log_b(u + v)$ is *not* equivalent to $\log_b u + \log_b v$.

You Try It **Work through this You Try It problem.**

Example 2 Using the Quotient Rule

My video summary

Use the quotient rule for logarithms to expand each expression. Assume $x > 0$.

a. $\log_5\left(\frac{12}{x}\right)$ **b.** $\ln\left(\frac{x}{e^5}\right)$

Solutions

a. Use the quotient rule for logarithms: $\log_5\left(\frac{12}{x}\right) = \log_5 12 - \log_5 x$

b. Use the quotient rule for logarithms: $\ln\left(\frac{x}{e^5}\right) = \ln x - \ln e^5$

Use **cancellation property (2)** to rewrite $\ln e^5$ as 5: $= \ln x - 5$

CAUTION $\log_b(u - v)$ is *not* equivalent to $\log_b u - \log_b v$, and $\frac{\log_b u}{\log_b v}$ is *not* equivalent to $\log_b u - \log_b v$.

You Try It **Work through this You Try It problem.**

Example 3 Using the Power Rule

Use the power rule for logarithms to rewrite each expression. Assume $x > 0$.

a. $\log 6^3$ **b.** $\log_{1/2}\sqrt[4]{x}$

(eText Screens 20.6-1–20.6-20)

My video summary

Solutions

a. Use the power rule for logarithms: $\log 6^3 = 3\log 6$

b. Rewrite the fourth root of x using a rational exponent: $\log_{1/2}\sqrt[4]{x} = \log_{1/2} x^{1/4}$

Use the power rule for logarithms: $= \frac{1}{4}\log_{1/2} x$

The process of using the power rule to simplify a logarithmic expression is often casually referred to as "bringing down the exponent."

CAUTION $(\log_b u)^r$ is *not* equivalent to $r \log_b u$.

You Try It **Work through this You Try It problem.**

Work Exercises 1–10 in this textbook or in the MyMathLab Study Plan.

OBJECTIVE 2 EXPAND AND CONDENSE LOGARITHMIC EXPRESSIONS

Sometimes it is necessary to combine several properties of logarithms to expand a logarithmic expression into the sum and/or difference of logarithms or to condense several logarithms into a single logarithm.

Example 4 Expanding a Logarithmic Expression

My interactive video summary

Use properties of logarithms to expand each logarithmic expression as much as possible.

a. $\log_7\left(49x^3\sqrt[5]{y^2}\right)$ b. $\ln\left(\frac{(x^2-4)}{9e^{x^3}}\right)$

Solutions

a. Write the original expression: $\log_7\left(49x^3\sqrt[5]{y^2}\right)$

Use the product rule: $= \log_7 49 + \log_7 x^3\sqrt[5]{y^2}$

Use the product rule again: $= \log_7 49 + \log_7 x^3 + \log_7\sqrt[5]{y^2}$

Rewrite $\sqrt[5]{y^2}$ using a rational exponent: $= \log_7 49 + \log_7 x^3 + \log_7 y^{2/5}$

Rewrite $\log_7 49$ as 2 and use the power rule: $= 2 + 3\log_7 x + \frac{2}{5}\log_7 y$

b. Factor the expression in the numerator: $\ln\left(\frac{(x^2-4)}{9e^{x^3}}\right) = \ln\left(\frac{(x-2)(x+2)}{9e^{x^3}}\right)$

Use the quotient rule: $= \ln(x-2)(x+2) - \ln 9e^{x^3}$

Use the product rule twice: $= \ln(x-2) + \ln(x+2) - \left[\ln 9 + \ln e^{x^3}\right]$

Use cancellation property (2) to rewrite $\ln e^{x^3}$ as x^3: $= \ln(x-2) + \ln(x+2) - \left[\ln 9 + x^3\right]$

Simplify: $= \ln(x-2) + \ln(x+2) - \ln 9 - x^3$

Watch the **interactive video** to see this example worked out in detail.

(eText Screens 20.6-1–20.6-20)

You Try It **Work through this You Try It problem.**

Work Exercises 11–20 in this textbook or in the MyMathLab Study Plan.

Example 5 Condensing a Logarithmic Expression

My interactive video summary

Use properties of logarithms to rewrite each expression as a single logarithm.

a. $\frac{1}{2}\log(x-1) - 3\log z + \log 5$ **b.** $\frac{1}{3}(\log_3 x - 2\log_3 y) + \log_3 10$

Solutions

a. Write the original expression: $\frac{1}{2}\log(x-1) - 3\log z + \log 5$

Use the power rule twice: $= \log(x-1)^{1/2} - \log z^3 + \log 5$

Use the quotient rule: $= \log\frac{(x-1)^{1/2}}{z^3} + \log 5$

Use the product rule: $= \log\frac{5(x-1)^{1/2}}{z^3}$ or $\log\frac{5\sqrt{x-1}}{z^3}$

b. Write the original expression: $\frac{1}{3}(\log_3 x - 2\log_3 y) + \log_3 10$

Use the power rule: $= \frac{1}{3}(\log_3 x - \log_3 y^2) + \log_3 10$

Use the quotient rule: $= \frac{1}{3}\log_3\frac{x}{y^2} + \log_3 10$

Use the power rule: $= \log_3\left(\frac{x}{y^2}\right)^{1/3} + \log_3 10$

Use the product rule: $= \log_3\left[10\left(\frac{x}{y^2}\right)^{1/3}\right]$ or $\log_3\left[10\sqrt[3]{\frac{x}{y^2}}\right]$

Watch the **interactive video** to see this example worked out in detail.

You Try It **Work through this You Try It problem.**

Work Exercises 21–30 in this textbook or in the MyMathLab Study Plan.

OBJECTIVE 3 SOLVE LOGARITHMIC EQUATIONS USING THE LOGARITHM PROPERTY OF EQUALITY

Remember that all logarithmic functions of the form $f(x) = \log_b x$ for $b > 0$ and $b \neq 1$ are **one-to-one.** In **Topic 20.2**, the alternate definition of **one-to-one** stated that

> A function f is one-to-one if for any two range values $f(u)$ and $f(v)$, $f(u) = f(v)$ implies that $u = v$.

Using this definition and letting $f(x) = \log_b x$, we can say that if $\log_b u = \log_b v$, then $u = v$. In other words, if the bases of a logarithmic equation of the form $\log_b u = \log_b v$ are equal, then the **arguments** must be equal. This is known as the **logarithm property of equality.**

(eText Screens 20.6-1–20.6-20)

> **Logarithm Property of Equality**
>
> If a logarithmic equation can be written in the form $\log_b u = \log_b v$, then $u = v$. Furthermore, if $u = v$, then $\log_b u = \log_b v$.

The second statement of the logarithm property of equality says that if we start with the equation $u = v$, then we can rewrite the equation as $\log_b u = \log_b v$. This process is often casually referred to as "taking the log of both sides."

Example 6 Using the Logarithm Property of Equality to Solve Logarithmic Equations

My interactive video summary

Solve the following equations.

a. $\log_7 (x - 1) = \log_7 12$ **b.** $2 \ln x = \ln 16$

Solutions

a. Because the base of each logarithm is 7, we can use the logarithm property of equality to eliminate the logarithms.

Write the original equation: $\log_7 (x - 1) = \log_7 12$

If $\log_b u = \log_b v$, then $u = v$: $(x - 1) = 12$

Solve for x: $x = 13$

b. Write the original expression: $2 \ln x = \ln 16$

Use the power rule: $\ln x^2 = \ln 16$

If $\log_b u = \log_b v$, then $u = v$: $x^2 = 16$

Use the square root property: $x = \pm 4$

The domain of $\ln x$ is $x > 0$; this implies that $x = -4$ is an **extraneous solution**, and hence, we must discard it. Therefore, this equation has only one solution, $x = 4$.

You Try It **Work through this You Try It problem.**

Work Exercises 31–36 in this textbook or in the MyMathLab Study Plan.

CAUTION: Remember to check potential solutions in the *original* equation before forming the solution set.

OBJECTIVE 4 USE THE CHANGE OF BASE FORMULA

Most scientific calculators are equipped with a log key and a ln key to evaluate common logarithms and natural logarithms. But how do we use a calculator to evaluate logarithmic expressions having a base other than 10 or e? The answer is to use the following **change of base formula**.

> **Change of Base Formula**
>
> For any positive base $b \neq 1$ and for any positive real number u, then
>
> $$\log_b u = \frac{\log_a u}{\log_a b},$$
>
> where a is any positive number such that $a \neq 1$.

(eText Screens 20.6-1–20.6-20)

My video summary

Watch this **video** for a proof of the change of base formula.

The change of base formula allows us to change the base of a logarithmic expression into a ratio of two logarithms using any base we choose. For example, suppose we are given the logarithmic expression $\log_3 10$. We can use the change of base formula to write this logarithm as a logarithm involving any base we choose:

$$\log_3 10 = \frac{\log_7 10}{\log_7 3} \quad \text{or} \quad \log_3 10 = \frac{\log_2 10}{\log_2 3} \quad \text{or} \quad \log_3 10 = \frac{\log 10}{\log 3} \quad \text{or} \quad \log_3 10 = \frac{\ln 10}{\ln 3}$$

In each of the previous four cases, we introduced a new base (7, 2, 10, and e, respectively). However, if we want to use a calculator to get a numerical approximation of $\log_3 10$, then it really only makes sense to change $\log_3 10$ into an expression involving base 10 or base e because these are the only two bases most calculators can handle.

Note: $\log_3 10 = \frac{\log 10}{\log 3} \approx 2.0959$ or $\log_3 10 = \frac{\ln 10}{\ln 3} \approx 2.0959$

Example 7 Using the Change of Base Formula

Approximate the following expressions. Round each to four decimal places.

a. $\log_9 200$ **b.** $\log_{\sqrt{3}} \pi$

Solutions

a. $\log_9 200 = \frac{\log 200}{\log 9} \approx 2.4114$ or $\log_9 200 = \frac{\ln 200}{\ln 9} \approx 2.4114$

b. $\log_{\sqrt{3}} \pi = \frac{\log \pi}{\log \sqrt{3}} \approx 2.0840$ or $\log_{\sqrt{3}} \pi = \frac{\ln \pi}{\ln \sqrt{3}} \approx 2.0840$

You Try It **Work through this You Try It problem.**

Work Exercises 37–40 in this textbook or in the MyMathLab Study Plan.

Example 8 Using the Change of Base Formula and Properties of Logarithms

My video summary

Use the change of base formula and the properties of logarithms to rewrite as a single logarithm involving base 2.

$$\log_4 x + 3\log_2 y$$

Solution To use properties of logarithms, the base of each logarithmic expression must be the same. We use the change of base formula to rewrite $\log_4 x$ as a logarithmic expression involving base 2:

Use the change of base formula $\log_4 x = \frac{\log_2 x}{\log_2 4}$: $\log_4 x + 3\log_2 y = \frac{\log_2 x}{\log_2 4} + 3\log_2 y$

Rewrite $\log_2 4$ as 2 because $2^2 = 4$: $= \frac{\log_2 x}{2} + 3\log_2 y$

Rewrite $\frac{\log_2 x}{2}$ as $\frac{1}{2}\log_2 x$: $= \frac{1}{2}\log_2 x + 3\log_2 y$

Use the power rule: $= \log_2 x^{1/2} + \log_2 y^3$

Use the product rule: $= \log_2 x^{1/2} y^3$ or $\log_2 \sqrt{x}\, y^3$

(eText Screens 20.6-1–20.6-20)

Therefore, the expression $\log_4 x + 3\log_2 y$ is equivalent to $\log_2 \sqrt{x}y^3$. Note that we could have chosen to rewrite the original expression as a single logarithm involving base 4. Watch the **video** to see that the expression $\log_4 x + 3\log_2 y$ is also equivalent to $\log_4 xy^6$.

You Try It **Work through this You Try It problem.**

Work Exercises 41–44 in this textbook or in the MyMathLab Study Plan.

Example 9 Using the Change of Base Formula to Solve Logarithmic Equations

My video summary

Use the change of base formula and the properties of logarithms to solve the equation

$$2\log_3 x = \log_9 16.$$

Solution Watch the **video** to see how the change of base formula and the power rule for logarithms can be used to solve this equation.

You Try It **Work through this You Try It problem.**

Work Exercises 45–48 in this textbook or in the MyMathLab Study Plan.

20.6 Exercises

In Exercises 1–10, use the product rule, quotient rule, or power rule to expand each logarithmic expression. Wherever possible, evaluate logarithmic expressions.

You Try It

1. $\log_4(xy)$ **2.** $\log\left(\frac{9}{t}\right)$ **3.** $\log_5 y^3$ **4.** $\log_3(27w)$ **5.** $\ln 5e^2$

6. $\log_9 \sqrt[4]{k}$ **7.** $\log 100P$ **8.** $\ln\left(\frac{e^5}{r}\right)$ **9.** $\log_{\sqrt{2}} 8x$ **10.** $\log_2\left(\frac{M}{32}\right)$

In Exercises 11–20, use the properties of logarithms to expand each logarithmic expression. Wherever possible, evaluate logarithmic expressions.

You Try It

11. $\log_7 x^2y^3$ **12.** $\ln \frac{a^2b^3}{c^4}$ **13.** $\log \frac{\sqrt{x}}{10y^3}$ **14.** $\log_3 9(x^2 - 25)$

15. $\log_2 \sqrt{4xy}$ **16.** $\log_5 \frac{\sqrt{5x^5}}{\sqrt[3]{25y^4}}$ **17.** $\ln \frac{\sqrt[5]{ez}}{\sqrt{x-1}}$ **18.** $\log_3 \sqrt[4]{\frac{x^3y^5}{9}}$

19. $\ln\left[\frac{x+1}{(x^2-1)^3}\right]^{2/3}$ **20.** $\log \frac{(10x)^3\sqrt{x-4}}{(x^2-16)^5}$

In Exercises 21–30, use properties of logarithms to rewrite each expression as a single logarithm. Wherever possible, evaluate logarithmic expressions.

You Try It

21. $\log_b A + \log_b C$ **22.** $\log_4 M - \log_4 N$

23. $2\log_8 x + \frac{1}{3}\log_8 y$ **24.** $\log 20 + \log 5$

25. $\log_2 80 - \log_2 5$

26. $\ln\sqrt{x} - \frac{1}{3}\ln x + \ln\sqrt[4]{x}$

27. $\log_5(x-2) + \log_5(x+2)$

28. $\log_9(x^2 - 5x + 6) - \log_9(x^2 - 4) + \log_9(x + 2)$

29. $\log(x-3) + 2\log(x+3) - \log(x^3 + x^2 - 9x - 9)$

30. $\frac{1}{2}[\ln(x-1)^2 - \ln(2x^2 - x - 1)^4] + 2\ln(2x+1)$

In Exercises 31–36, use the properties of logarithms and the logarithm property of equality to solve each logarithmic equation.

You Try It

31. $\log_3(2x+1) = \log_3 11$

32. $\log_{11}\sqrt{x} = \log_{11} 6$

33. $2\log(x+5) = \log 12 + \log 3$

34. $\ln 5 + \ln x = \ln 7 + \ln(3x - 2)$

35. $\log_7(x+6) - \log_7(x+2) = \log_7 x$

36. $\log_2(x+3) + \log_2(x-4) = \log_2(x+12)$

In Exercises 37–40, use the change of base formula and a calculator to approximate each logarithmic expression. Round your answers to four decimal places.

You Try It

37. $\log_4 51$

38. $\log_7 0.8$

39. $\log_{1/5} 72$

40. $\log_{\sqrt{7}} 100$

In Exercises 41–44, use the change of base formula and the properties of logarithms to rewrite each expression as a single logarithm in the indicated base.

You Try It

41. $\log_3 x + 4\log_9 w$, base 3

42. $\log_5 x + \log_{1/5} x^3$, base 5

43. $\log_{16} x^4 + \log_8 y^3 + \log_4 w^2$, base 2

44. $\log_{e^2} x^5 + \log_{e^3} x^6 + \log_{e^4} x^{12}$, base e

In Exercises 45–48, solve each logarithmic equation.

You Try It

45. $\log_2 x = \log_4 25$

46. $\log_{1/3} x = \log_3 20$

47. $\log_5 x = \log_{\sqrt{5}} 6$

48. $2\ln x = \log_{e^3} 125$

20.7 Exponential and Logarithmic Equations

THINGS TO KNOW

Before working through this topic, be sure you are familiar with the following concepts: VIDEO ANIMATION INTERACTIVE

You Try It — 1. Solve Exponential Equations by Relating the Bases (Topic 20.3, **Objective 3**)

You Try It — 2. Change from Exponential to Logarithmic Form (Topic 20.5, **Objective 1**)

You Try It — 3. Change from Logarithmic to Exponential Form (Topic 20.5, **Objective 1**)

(eText Screens 20.7-1–20.7-20)

You Try It 4. Use the Properties of Logarithms (Topic 20.5, **Objective 3**)

You Try It 5. Expand and Condense Logarithmic Expressions (Topic 20.6, **Objective 2**)

You Try It 6. Solve Logarithmic Equations Using the Logarithm Property of Equality (Topic 20.6, **Objective 3**)

OBJECTIVES

1 Solve Exponential Equations

2 Solve Logarithmic Equations

In this topic, we learn how to solve exponential and logarithmic equations. The techniques and strategies learned in this topic help us solve applied problems that are discussed in **Topic 20.8**. We start by developing a strategy to solve exponential equations.

OBJECTIVE 1 SOLVE EXPONENTIAL EQUATIONS

We have already solved exponential equations using the **method of relating the bases.** For example, we can solve the equation $4^{x+3} = \frac{1}{2}$ by converting the base on both sides of the equation to base 2. View this **popup box** to see how to solve this equation.

But suppose we are given an exponential equation in which the bases cannot be related, such as $2^{x+1} = 3$. Remember in **Topic 20.3, Example 3**, we wanted to find the one x-intercept of the graph of $f(x) = -2^{x+1} + 3$ (Figure 32).

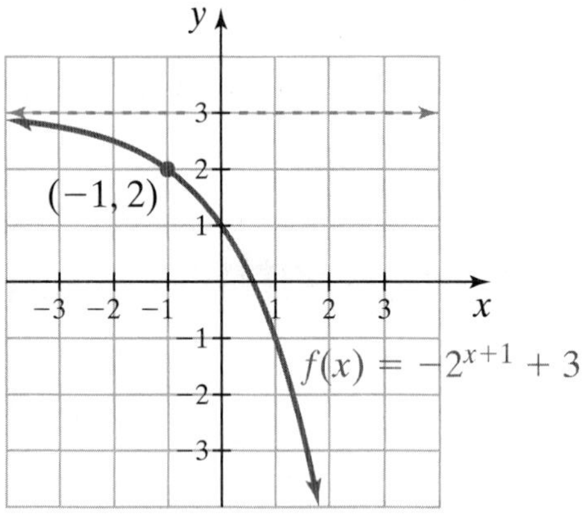

Figure 32
Graph of $f(x) = -2^{x+1} + 3$

To find the x-intercept of $f(x) = -2^{x+1} + 3$, we need to set $f(x) = 0$ and solve for x.

$$f(x) = -2^{x+1} + 3$$
$$0 = -2^{x+1} + 3$$
$$2^{x+1} = 3$$

In **Topic 20.3**, we could not solve this equation for x because we had not yet defined the logarithm. We can now use some properties of logarithms to solve this equation. Recall the following logarithmic properties.

If $u = v$, then $\log_b u = \log_b v$.	Logarithm property of equality
$\log_b u^r = r \log_b u$	Power rule for logarithms

We can use these two properties to solve the equation $2^{x+1} = 3$ and thus determine the x-intercept of $f(x) = -2^{x+1} + 3$. We solve the equation $2^{x+1} = 3$ in Example 1.

Example 1 Solving an Exponential Equation

My video summary

Solve $2^{x+1} = 3$.

Solution

Write the original equation: $2^{x+1} = 3$

Use the **logarithm property of equality:** $\ln 2^{x+1} = \ln 3$

Use the power rule for logarithms: $(x + 1)\ln 2 = \ln 3$

Use the distributive property: $x \ln 2 + \ln 2 = \ln 3$

Subtract $\ln 2$ from both sides: $x \ln 2 = \ln 3 - \ln 2$

Divide both sides by $\ln 2$: $x = \dfrac{\ln 3 - \ln 2}{\ln 2}$

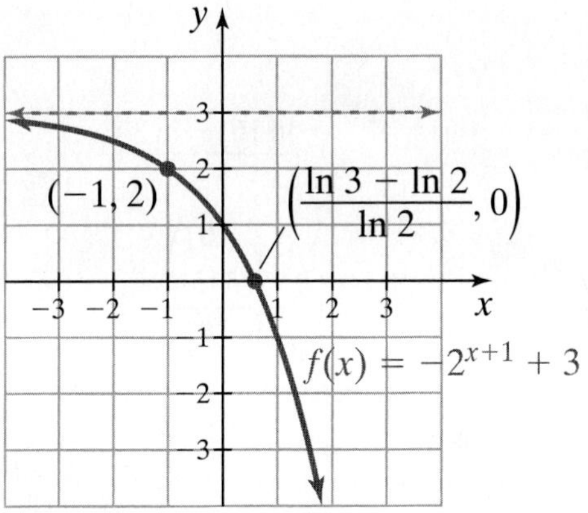

Figure 33 Graph of $f(x) = -2^{x+1} + 3$

The solution to Example 1 verifies that the x-intercept of $f(x) = -2^{x+1} + 3$ is $x = \dfrac{\ln 3 - \ln 2}{\ln 2} \approx 0.5850$. See Figure 33.

When we cannot easily relate the bases of an exponential equation, as in Example 1, we use logarithms and their properties to solve them. The methods used to solve exponential equations are outlined as follows.

Solving Exponential Equations

- If the equation can be written in the form $b^u = b^v$, then solve the equation $u = v$.
- If the equation cannot easily be written in the form $b^u = b^v$,
  1. Use the logarithm property of equality to "take the log of both sides" (typically using base 10 or base e).
  2. Use the product rule of logarithms to "bring down" any exponents.
  3. Solve for the given variable.

Example 2 Solving Exponential Equations

My interactive video summary

Solve each equation. For **part b**, round to four decimal places.

a. $3^{x-1} = \left(\dfrac{1}{27}\right)^{2x+1}$ **b.** $7^{x+3} = 4^{2-x}$

Solutions

a. Watch the **interactive video**, or view this **popup box** to see that the solution is $x = -\dfrac{2}{7}$.

b. We cannot easily use the method of relating the bases because we cannot easily write both 7 and 4 using a common base. Therefore, we use logarithms to solve.

Write the original equation: $7^{x+3} = 4^{2+x}$

Use the **logarithm property of equality**: $\ln 7^{x+3} = \ln 4^{2-x}$

Use the power rule for logarithms: $(x+3)\ln 7 = (2-x)\ln 4$

Use the distributive property: $x\ln 7 + 3\ln 7 = 2\ln 4 - x\ln 4$

Add $x\ln 4$ to both sides, and subtract $3\ln 7$ from both sides: $x\ln 7 + x\ln 4 = 2\ln 4 - 3\ln 7$

Factor out an x from the left-hand side: $x(\ln 7 + \ln 4) = 2\ln 4 - 3\ln 7$

Divide both sides by $\ln 7 + \ln 4$: $x = \dfrac{2\ln 4 - 3\ln 7}{\ln 7 + \ln 4}$

Use the power rule for logarithms in the numerator, and use the product rule for logarithms in the denominator: $= \dfrac{\ln 16 - \ln 343}{\ln 28}$

Use the quotient rule for logarithms to rewrite $\ln 16 - \ln 343$ as $\ln\left(\dfrac{16}{343}\right)$: $= \dfrac{\ln\left(\dfrac{16}{343}\right)}{\ln 28}$

Use a calculator to round to four decimal places: ≈ -0.9199

You Try It **Work through this You Try It problem.**

Work Exercises 1–10 in this textbook or in the MyMathLab Study Plan.

Example 3 Solving Exponential Equations Involving the Natural Exponential Function

My interactive video summary

Solve each equation. Round to four decimal places.

a. $25e^{x-5} = 17$ **b.** $e^{2x-1} \cdot e^{x+4} = 11$

Solutions

a. Isolate the exponential term on the left by dividing both sides of the equation by 25.

Write the original equation: $25e^{x-5} = 17$

Divide both sides by 25: $e^{x-5} = \dfrac{17}{25}$

Use the **logarithm property of equality**: $\ln e^{x-5} = \ln\dfrac{17}{25}$

Use **cancellation property (2)** to rewrite $\ln e^{x-5}$ as $x - 5$: $x - 5 = \ln\dfrac{17}{25}$

Add 5 to both sides: $x = \ln\dfrac{17}{25} + 5$

Use a calculator to round to four decimal places: ≈ 4.6143

b. Write the original equation: $e^{2x-1} \cdot e^{x+4} = 11$

Use $b^m \cdot b^n = b^{m+n}$: $e^{(2x-1)+(x+4)} = 11$

Combine like terms in the exponent: $e^{3x+3} = 11$

Use the **logarithm property of equality**: $\ln e^{3x+3} = \ln 11$

Use **cancellation property (2)** to rewrite $\ln e^{3x+3}$ as $3x + 3$: $3x + 3 = \ln 11$

Subtract 3 from both sides: $3x = \ln 11 - 3$

Divide both sides by 3: $x = \dfrac{\ln 11 - 3}{3}$

Use a calculator to round to four decimal places: ≈ -0.2007

Watch the **interactive video** to see the solutions to this example worked out in detail.

You Try It **Work through this You Try It problem.**

Work Exercises 11–15 in this textbook or in the MyMathLab Study Plan.

OBJECTIVE 2 SOLVE LOGARITHMIC EQUATIONS

We now turn our attention to solving logarithmic equations. In **Topic 20.6**, we learned how to solve certain logarithmic equations by using the **logarithm property of equality**. That is, if we can write a logarithmic equation in the form $\log_b u = \log_b v$, then $u = v$. Before we look at an example, let's review three of the properties of logarithms.

Properties of Logarithms

If $b > 0, b \neq 1, u$ and v represent positive numbers and r is any real number, then

$\log_b uv = \log_b u + \log_b v$	Product rule for logarithms
$\log_b \dfrac{u}{v} = \log_b u - \log_b v$	Quotient rule for logarithms
$\log_b u^r = r \log_b u$	Power rule for logarithms

Example 4 Solving a Logarithmic Equation Using the Logarithm Property of Equality

My video summary

Solve $2 \log_5 (x - 1) = \log_5 64$.

Solution We can use the power rule for logarithms and the logarithmic property of equality to solve.

Write the original equation: $2 \log_5 (x - 1) = \log_5 64$

Use the power rule: $\log_5 (x - 1)^2 = \log_5 64$

Use the logarithm property of equality: $(x - 1)^2 = 64$

Use the square root property: $x - 1 = \pm 8$

Solve for x: $x = 1 \pm 8$

Simplify: $x = 9$ or $x = -7$

(eText Screens 20.7-1–20.7-20)

Recall that the domain of a logarithmic function must contain only positive numbers; thus, $x - 1$ must be positive. Therefore, the solution of $x = -7$ must be discarded. The only solution is $x = 9$. You may want to review how to determine the domain of a logarithmic function, which was discussed in **Topic 20.5**.

You Try It **Work through this You Try It problem.**

Work Exercises 16–19 in this textbook or in the MyMathLab Study Plan.

CAUTION

When solving logarithmic equations, it is important to always verify the solutions. Logarithmic equations often lead to **extraneous solutions**, as in Example 4.

When a logarithmic equation cannot be written in the form $\log_b u = \log_b v$, we adhere to the steps outlined as follows:

Solving Logarithmic Equations

1. Determine the domain of the variable.
2. Use properties of logarithms to combine all logarithms, and write as a single logarithm, if needed.
3. Eliminate the logarithm by rewriting the equation in exponential form. View this **popup box** to review how to change from logarithmic form to exponential form, which was discussed in **Topic 20.5**.
4. Solve for the given variable.
5. Check for any extraneous solutions. Verify that each solution is in the domain of the variable.

Example 5 Solving a Logarithmic Equation

My video summary

Solve $\log_4(2x - 1) = 2$.

Solution The domain of the variable in this equation is the solution to the inequality $2x - 1 > 0$ or $x > \frac{1}{2}$. Thus, our solution must be greater than $\frac{1}{2}$. Because the equation involves a single logarithm, we can proceed to the third step.

Write the original equation:	$\log_4(2x - 1) = 2$
Rewrite in exponential form:	$4^2 = 2x - 1$
Simplify:	$16 = 2x - 1$
Add 1 to both sides:	$17 = 2x$
Divide by 2:	$x = \frac{17}{2}$

Because the solution satisfies the inequality, there are no extraneous solutions. We can verify the solution by substituting $x = \frac{17}{2}$ into the original equation.

Check:	Write the original equation:	$\log_4(2x - 1) = 2$
	Substitute $x = \frac{17}{2}$:	$\log_4\left(2\left(\frac{17}{2}\right) - 1\right) \stackrel{?}{=} 2$
	Simplify:	$\log_4(17 - 1) \stackrel{?}{=} 2$
	This is a true statement because $4^2 = 16$:	$\log_4(16) = 2$

(eText Screens 20.7-1–20.7-20)

You Try It **Work through this You Try It problem.**

Work Exercises 20–24 in this textbook or in the MyMathLab Study Plan.

Example 6 Solving a Logarithmic Equation

My interactive video summary

Solve $\log_2(x + 10) + \log_2(x + 6) = 5$.

Solution The domain of the variable in this equation is the solution to the compound inequality $x + 10 > 0$ and $x + 6 > 0$. The solution to this compound inequality is $x > -6$. (You may want to review compound inequalities from **Topic 2.8**.)

Write the original equation:	$\log_2(x + 10) + \log_2(x + 6) = 5$
Use the product rule:	$\log_2(x + 10)(x + 6) = 5$
Rewrite in exponential form:	$(x + 10)(x + 6) = 2^5$
Simplify:	$x^2 + 16x + 60 = 32$
Subtract 32 from both sides:	$x^2 + 16x + 28 = 0$
Factor:	$(x + 14)(x + 2) = 0$
Use the **zero-product property** to solve:	$x = -14$ or $x = -2$

Because the domain of the variable is $x > -6$, we must *exclude* the solution $x = -14$. Therefore, the only solution to this logarithmic equation is $x = -2$. Work through the **interactive video** to see this solution worked out in detail.

You Try It **Work through this You Try It problem.**

Work Exercises 25–30 in this textbook or in the MyMathLab Study Plan.

Example 7 Solving a Logarithmic Equation

Solve $\ln(x - 4) - \ln(x - 5) = 2$. Round to four decimal places.

Solution The domain of the variable is the solution to the compound inequality $x - 4 > 0$ and $x - 5 > 0$. The solution to this compound inequality is $x > 5$. (You may want to review compound inequalities from **Topic 2.8**.)

Write the original equation:	$\ln(x - 4) - \ln(x - 5) = 2$
Use the quotient rule:	$\ln\left(\frac{x - 4}{x - 5}\right) = 2$
Rewrite in exponential form:	$e^2 = \frac{x - 4}{x - 5}$
Multiply both sides by $x - 5$:	$e^2(x - 5) = x - 4$
Use the distributive property:	$e^2x - 5e^2 = x - 4$
Add $5e^2$ to both sides and subract x from both sides:	$e^2x - x = 5e^2 - 4$
Factor out an x from the left-hand side:	$x(e^2 - 1) = 5e^2 - 4$
Solve for x. Use a calculator to round to four decimal places:	$x = \frac{5e^2 - 4}{e^2 - 1} \approx 5.1565$

(eText Screens 20.7-1–20.7-20)

We approximate the exact answer $x = \dfrac{5e^2 - 4}{e^2 - 1}$ in order to verify that the solution is in the domain of the variable. In this example, we see that 5.1565 is clearly greater than 5. In some cases, we may need to use the exact answer to verify a solution to a logarithmic equation. See this **verification** for Example 7.

You Try It **Work through this You Try It problem.**

Work Exercises 31 and 32 in this textbook or in the MyMathLab Study Plan.

20.7 Exercises

In Exercises 1–15, solve each exponential equation. For **irrational solutions**, round to four decimal places.

You Try It

1. $3^x = 5$ **2.** $2^{x/3} = 19$ **3.** $4^{x^2-2x} = 64$ **4.** $3^{x+7} = -20$

5. $(1.52)^{-3x/7} = 11$ **6.** $\left(\dfrac{1}{5}\right)^{x-1} = 25^x$ **7.** $8^{4x-7} = 11^{5+x}$ **8.** $3(9)^{x-1} = (81)^{2x+1}$

9. $(3.14)^x = \pi^{1-2x}$ **10.** $7(2 - 10^{4x-2}) = 8$ **11.** $e^x = 2$ **12.** $150e^{x-4} = 5$

You Try It

13. $e^{x-3} \cdot e^{3x+7} = 24$ **14.** $2(e^{x-1})^2 \cdot e^{3-x} = 80$ **15.** $8e^{-x/3} \cdot e^x = 1$

In Exercises 16–30, solve each logarithmic equation.

You Try It

16. $\log_4(x + 1) = \log_4(6x - 5)$ **17.** $\log_3(x^2 - 21) = \log_3 4x$

18. $2\log_5(3 - x) - \log_5 2 = \log_5 18$ **19.** $2\ln x - \ln(2x - 3) = \ln 2x - \ln(x - 1)$

You Try It

20. $\log_2(4x - 7) = 3$ **21.** $\log(1 - 5x) = 2$

22. $\log_3(2x - 5) = -2$ **23.** $\log_x 3 = -1$ **24.** $\log_{x/2} 16 = 2$

You Try It

25. $\log_2(x - 2) + \log_2(x + 2) = 5$ **26.** $\log_7(x + 9) + \log_7(x + 15) = 1$

27. $\log_2(3x + 1) - \log_2(x - 2) = 2$ **28.** $\log_4(x - 7) + \log_4 x = \dfrac{3}{2}$

29. $\ln 3 + \ln\left(x^2 + \dfrac{2x}{3}\right) = 0$ **30.** $\log_2(x - 4) + \log_2(x + 6) = 2 + \log_2 x$

In Exercises 31 and 32, solve each logarithmic equation. Round to four decimal places.

You Try It

31. $\ln x - \ln(x + 6) = 1$ **32.** $\ln(x + 3) - \ln(x - 2) = 4$

(eText Screens 20.8-1–20.8-29)

20.8 Applications of Exponential and Logarithmic Functions

THINGS TO KNOW

Before working through this topic, be sure you are familiar with the following concepts:

		VIDEO	ANIMATION	INTERACTIVE
You Try It	1. Solve Applications of Exponential Functions (Topic 20.3, **Objective 4**)			
You Try It	2. Solve Applications of the Natural Exponential Function (Topic 20.4, **Objective 4**)	▣		
You Try It	3. Use the Properties of Logarithms (Topic 20.5, **Objective 3**)			

OBJECTIVES

1 Solve Compound Interest Applications

2 Solve Exponential Growth and Decay Applications

3 Solve Logistic Growth Applications

4 Use Newton's Law of Cooling

We have seen that exponential functions appear in a wide variety of settings, including biology, chemistry, physics, and business. In this topic, we revisit some applications that are discussed previously in this module and then introduce several new applications. The difference between the applications presented earlier and the applications presented in this topic is that we are now equipped with the tools necessary to solve for variables that appear as exponents. We start with applications involving **compound interest**. You may want to review periodic compound interest from **Topic 20.3** and continuous compound interest from **Topic 20.4** before proceeding.

OBJECTIVE 1 SOLVE COMPOUND INTEREST APPLICATIONS

In **Topics 20.3 and 20.4**, the formula for compound interest and continuous compound interest were defined as follows.

Compound Interest Formulas

Periodic Compound Interest Formula

$$A = P\left(1 + \frac{r}{n}\right)^{nt}$$

Continuous Compound Interest Formula

$$A = Pe^{rt},$$

continued

where

A = Total amount after t years

P = Principal (original investment)

r = Interest rate per year

n = Number of times interest is compounded per year

t = Number of years

Example 1 Finding the Doubling Time

My video summary

How long will it take (in years and months) for an investment to double if it earns 7.5% compounded monthly?

Solution We use the periodic compound interest formula $A = P\left(1 + \frac{r}{n}\right)^{nt}$ with $r = 0.075$ and $n = 12$ and solve for t. Notice that the principal is not given. As it turns out, any value of P will suffice. If the principal is P, then the amount needed to double the investment is We now have all of the information necessary to solve for t:

Use the periodic compound interest formula: $A = P\left(1 + \dfrac{r}{n}\right)^{nt}$

Substitute the appropriate values: $2P = P\left(1 + \dfrac{0.075}{12}\right)^{12t}$

Divide both sides by P: $2 = \left(1 + \dfrac{0.075}{12}\right)^{12t}$

Simplify within the parentheses: $2 = (1.00625)^{12t}$

Use the **logarithm property of equality**: $\ln 2 = \ln (1.00625)^{12t}$

Use the power rule: $\log_b u^r = r \log_b u$: $\ln 2 = 12t \ln (1.00625)$

Divide both sides by $12 \ln (1.00625)$: $t = \dfrac{\ln 2}{12 \ln (1.00625)}$

Round to two decimal places: $t \approx 9.27$ years

Note that $0.27 \text{ years} = 0.27 \text{ years} \times \frac{12 \text{ months}}{1 \text{ year}} = 3.24 \text{ months}$. Because the interest is compounded at the end of each month, the investment will not double until 9 years and 4 months.

Example 2 Continuous Compound Interest

My video summary

Suppose an investment of \$5000 compounded continuously grew to an amount of \$5130.50 in 6 months. Find the interest rate, and then determine how long it will take for the investment to grow to \$6000. Round the interest rate to the nearest hundredth of a percent and the time to the nearest hundredth of a year.

Solution Because the investment is compounded continuously, we use the formula $A = Pe^{rt}$.

We are given that $P = 5000$, so $A = 5000e^{rt}$. In 6 months, or when $t = 0.5$ years, we know that $A = 5130.50$. Substituting these values into the compound interest formula will enable us to solve for r:

Substitute the appropriate values: $5130.50 = 5000e^{r(0.5)}$

Divide by 5000: $\dfrac{5130.50}{5000} = e^{0.5r}$

(eText Screens 20.8-1–20.8-29)

Use the **logarithm property of equality:** $\ln\left(\dfrac{5130.50}{5000}\right) = \ln e^{0.5r}$

Use **cancellation property (2)** to rewrite $\ln e^{0.5r}$ as $0.5r$: $\ln\left(\dfrac{5130.50}{5000}\right) = 0.5r$

Divide by 0.5: $r = \dfrac{\ln\left(\dfrac{5130.50}{5000}\right)}{0.5} \approx 0.05153$

Therefore, the interest rate is 5.15%. To find the time that it takes for the investment to grow to \$6000, we use the formula $A = Pe^{rt}$, with $A = 6000$, $P = 5000$, and $r = 0.0515$, and solve for t. Watch the **video** to verify that it will take approximately 3.54 years.

You Try It **Work through this You Try It problem.**

Work Exercises 1–5 in this textbook or in the MyMathLab Study Plan.

OBJECTIVE 2 SOLVE EXPONENTIAL GROWTH AND DECAY APPLICATIONS

In **Topic 20.4**, the exponential growth model was introduced. This model was used when a population grows at a rate proportional to the size of its current population. This model is often called the uninhibited growth model. We review this exponential growth model and sketch the graph in Figure 34.

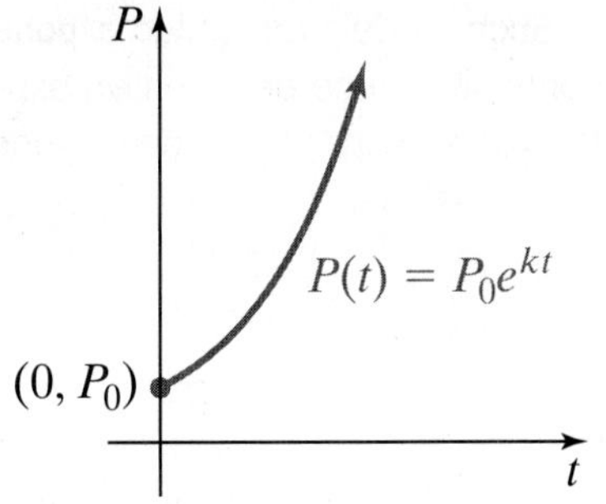

Figure 34
Graph of $P(t) = P_0e^{kt}$ for $k > 0$

Exponential Growth

A model that describes the exponential uninhibited growth of a population, P, after a certain time, t, is

$$P(t) = P_0e^{kt},$$

where $P_0 = P(0)$ is the initial population and $k > 0$ is a constant called the **relative growth rate.** (*Note:* k is sometimes given as a percent.)

Example 3 Population Growth

The population of a small town grows at a rate proportional to its current size. In 1900, the population was 900. In 1920, the population had grown to 1600. What was the population of this town in 1950? Round to the nearest whole number.

My video summary

Solution Using the model $P(t) = P_0 e^{kt}$, we must first determine the constants P_0 and k. The initial population was 900 in 1900 so $P_0 = 900$. Therefore, $P(t) = 900e^{kt}$. To find k, we use the fact that in 1920, or when $t = 20$, the population was 1600; thus,

Substitute $P(20) = 1600$: $P(20) = 900e^{k(20)} = 1600$

$$900e^{20k} = 1600$$

Divide by 900 and simplify: $e^{20k} = \frac{16}{9}$

Use the **logarithm property of equality**: $\ln e^{20k} = \ln \frac{16}{9}$

Use **cancellation property (2)** to rewrite $\ln e^{20k}$ as $20k$: $20k = \ln \frac{16}{9}$

Divide by 20: $k = \frac{\ln\left(\frac{16}{9}\right)}{20}$

The function that models the population of this town at any time t is given by $P(t) = 900e^{\frac{\ln(16/9)}{20}t}$. To determine the population in 1950, or when $t = 50$, we evaluate $P(50)$:

$$P(50) = 900e^{\frac{\ln(16/9)}{20}(50)} \approx 3793$$

Exponential Decay

Some populations exhibit *negative exponential growth*. In other words, the population, quantity, or amount *decreases* over time. Such models are called **exponential decay** models. The only difference between an exponential growth model and an exponential decay model is that the constant, k, is less than zero. See Figure 35.

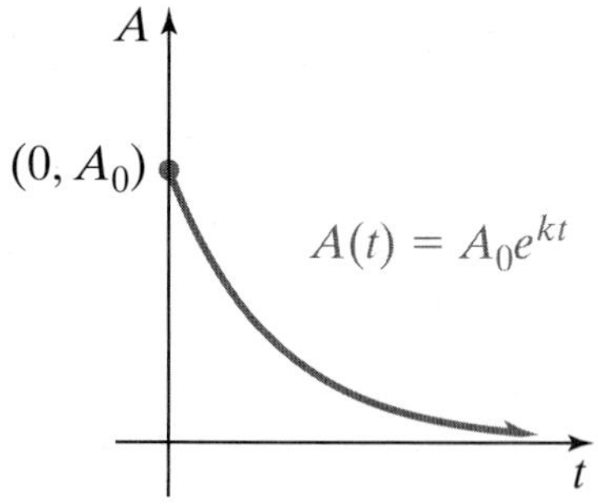

Figure 35
Graph of $A(t) = A_0 e^{kt}$ for $k < 0$

> **Exponential Decay**
>
> A model that describes the exponential decay of a population, quantity, or amount A, after a certain time, t, is
>
> $$A(t) = A_0 e^{kt},$$
>
> where $A_0 = A(0)$ is the initial quantity and $k < 0$ is a constant called the **relative decay constant**. (*Note:* k is sometimes given as a percent.)

Half-Life

My animation summary

Every radioactive element has a half-life, which is the required time for a given quantity of that element to decay to half of its original mass. For example, the half-life of Cesium-137 is 30 years. Thus, it takes 30 years for any amount of Cesium-137 to decay to $\frac{1}{2}$ of its original mass. It takes an additional 30 years to decay to $\frac{1}{4}$ of its original mass and so on. See Figure 36, and view the **animation** that illustrates the half-life of Cesium-137.

(eText Screens 20.8-1–20.8-29)

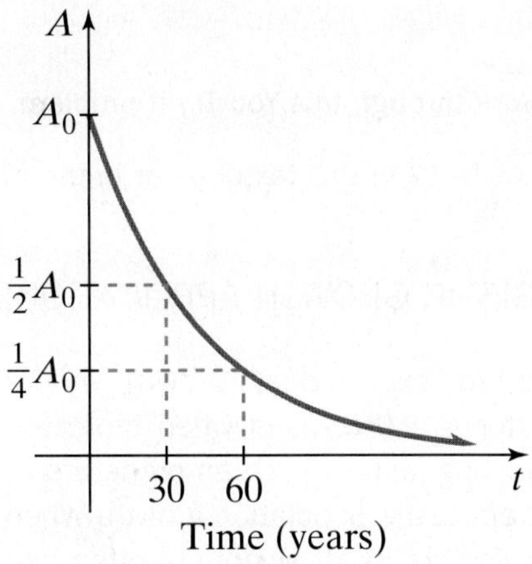

Figure 36
Half-life of Cesium-137

Example 4 Radioactive Decay

My video summary

Suppose that a meteorite is found containing 4% of its original Krypton-99. If the half-life of Krypton-99 is years, how old is the meteorite? Round to the nearest year.

Solution We use the formula $A(t) = A_0e^{kt}$, where A_0 is the original amount of Krypton-99. We first must find the constant k. To find k, we use the fact that the half-life of Krypton-99 is 80 years. Therefore, $A(80) = \frac{1}{2}A_0$. Because $A(80) = A_0e^{k(80)}$, we can set $\frac{1}{2}A_0 = A_0e^{k(80)}$ and solve for k.

$$\frac{1}{2}A_0 = A_0e^{k(80)}$$

Divide both sides by A_0: $$\frac{1}{2} = e^{80k}$$

Use the **logarithm property of equality**: $$\ln\frac{1}{2} = \ln e^{80k}$$

Use **cancellation property (2)** to rewrite $\ln e^{80k}$ as $80k$: $$\ln\frac{1}{2} = 80k$$

Divide both sides by 80: $$\frac{\ln\frac{1}{2}}{80} = k$$

$\ln\frac{1}{2} = \ln 1 - \ln 2 = 0 - \ln 2 = -\ln 2$: $$\frac{-\ln 2}{80} = k$$

Now that we know $k = \frac{-\ln 2}{80}$, our function becomes $A(t) = A_0e^{\frac{-\ln 2}{80}t}$. To find out the age of the meteorite, we set $A(t) = 0.04A_0$ because the meteorite now contains 4% of the original amount of Krypton-99.

Substitute $0.04A_0$ for $A(t)$: $$0.04A_0 = A_0e^{\frac{-\ln 2}{80}t}$$

Divide both sides by A_0: $$0.04 = e^{\frac{-\ln 2}{80}t}$$

Use the **logarithm property of equality**: $$\ln 0.04 = \ln e^{\frac{-\ln 2}{80}t}$$

Use **cancellation property (2)** to rewrite $\ln e^{\frac{-\ln 2}{80}t}$ as $\frac{-\ln 2}{80}t$: $$\ln 0.04 = \frac{-\ln 2}{80}t$$

Divide both sides by $\frac{-\ln 2}{80}$: $$\frac{\ln 0.04}{\frac{-\ln 2}{80}} = t \approx 372 \text{ years}$$

The meteorite is about 372 years old.

You Try It **Work through this You Try It problem.**

Work Exercises 6–12 in this textbook or in the MyMathLab Study Plan.

OBJECTIVE 3 SOLVE LOGISTIC GROWTH APPLICATIONS

The uninhibited exponential growth model $P(t) = P_0e^{kt}$ for $k > 0$ is used when there are no outside limiting factors such as predators or disease that affect the population growth. When such outside factors exist, scientists often use a **logistic model** to describe the population growth. One such logistic model is described and sketched in Figure 37.

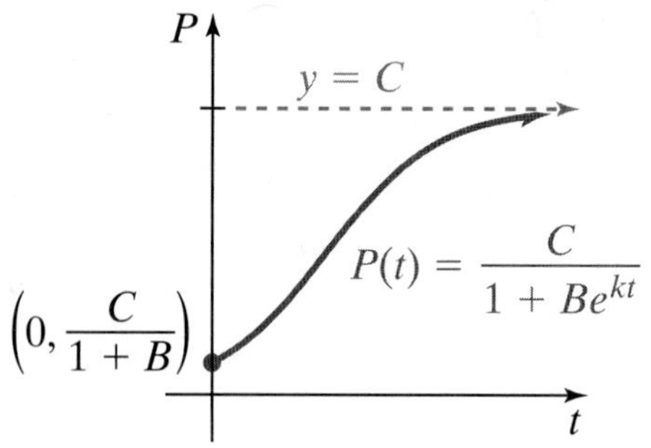

Figure 37 Graph of $P(t) = \dfrac{C}{1 + Be^{kt}}$

> **Logistic Growth**
>
> A model that describes the logistic growth of a population P at any time t is given by the function
>
> $$P(t) = \frac{C}{1 + Be^{kt}},$$
>
> where B, C, and k are constants with $C > 0$ and $k < 0$.

The number C is called the **carrying capacity**. In the logistic model, the population will approach the value of the carrying capacity over time but never exceed it. You can see in the graph sketched in Figure 37 that the graph of the logistic growth model approaches the **horizontal asymptote** $y = C$.

Example 5 Logistic Growth

My interactive video summary

Ten goldfish were introduced into a small pond. Because of limited food, space, and oxygen, the carrying capacity of the pond is 400 goldfish. The goldfish population at any time t, in days, is modeled by the logistic growth function $F(t) = \frac{C}{1 + Be^{kt}}$. If 30 goldfish are in the pond after 20 days,

a. Find B.

b. Find k.

c. When will the pond contain 250 goldfish? Round to the nearest whole number.

(eText Screens 20.8-1–20.8-29)

Solutions

a. The carrying capacity is 400; thus, $C = 400$. Also, initially (at $t = 0$) there were 10 goldfish, so $F(0) = 10$. Therefore,

Substitute $C = 400$ and $F(0) = 10$: $$10 = \frac{400}{1 + Be^{k(0)}}$$

Evaluate $e^0 = 1$: $$10 = \frac{400}{1 + B}$$

Multiply both sides by $1 + B$: $$10 + 10B = 400$$

Solve for B: $$B = 39$$

b. Use the function $F(t) = \dfrac{400}{1 + 39ekt}$ and the fact that $F(20) = 30$ (there are 30 goldfish after 20 days) to solve for k.

Substitute $F(20) = 30$: $$30 = \frac{400}{1 + 39e^{k(20)}}$$

Multiply both sides by $1 + 39e^{20k}$: $$30(1 + 39e^{20k}) = 400$$

Use the distributive property: $$30 + 1170e^{20k} = 400$$

Subtract 30 from both sides: $$1170e^{20k} = 370$$

Divide both sides by 1170: $$e^{20k} = \frac{370}{1170}$$

Simplify: $$e^{20k} = \frac{37}{117}$$

Use the **logarithm property of equality**: $$\ln e^{20k} = \ln \frac{37}{117}$$

Use **cancellation property (2)** to rewrite $\ln e^{20k}$ as $20k$: $$20k = \ln \frac{37}{117}$$

Divide both sides by 20: $$k = \frac{\ln \frac{37}{117}}{20}$$

c. Use the function $F(t) = \dfrac{400}{1 + 39e^{\frac{\ln(37/117)}{20}t}}$, and then find t when $F(t) = 250$.

By repeating the exact same process as in **part (b)**, we find that it will take approximately 73 days until there are 250 goldfish in the pond. Watch the **interactive video** to verify the solution.

You Try It **Work through this You Try It problem.**

Work Exercises 13–15 in this textbook or in the MyMathLab Study Plan.

OBJECTIVE 4 USE NEWTON'S LAW OF COOLING

Newton's law of cooling states that the temperature of an object changes at a rate proportional to the difference between its temperature and that of its surroundings. It can be shown in a more advanced course that the function describing Newton's law of cooling is given by the following.

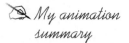

My animation summary

Newton's Law of Cooling

The temperature T of an object at any time t is given by

$$T(t) = S + (T_0 - S)e^{kt},$$

where T_0 is the original temperature of the object, S is the constant temperature of the surroundings, and k is the cooling constant.

View the **animation** to see how this function behaves.

Example 6 Newton's Law of Cooling

Suppose that the temperature of a cup of hot tea obeys Newton's law of cooling. If the tea has a temperature of 200°F when it is initially poured and 1 minute later has cooled to 189°F in a room that maintains a constant temperature of 69°F, determine when the tea reaches a temperature of 146°F. Round to the nearest minute.

My video summary

Solution We start using the formula for Newton's law of cooling with $T_0 = 200$ and $S = 69$.

Use Newton's law of cooling formula: $T(t) = S + (T_0 - S)e^{kt}$

Substitute $T_0 = 200$ and $S = 69$: $T(t) = 69 + (200 - 69)e^{kt}$

Simplify: $T(t) = 69 + 131e^{kt}$

We now proceed to find k. The object cools to 189°F in 1 minute; thus, $T(1) = 189$. Therefore,

Substitute $T(1) = 189$: $189 = 69 + 131e^{k(1)}$

Subtract 69 from both sides: $120 = 131e^{k}$

Divide both sides by 131: $\dfrac{120}{131} = e^{k}$

Use the **logarithm property of equality**: $\ln\dfrac{120}{131} = \ln e^{k}$

Use **cancellation property (2)** to rewrite $\ln e^{k}$ as k: $\ln\dfrac{120}{131} = k$

Now that we know the cooling constant $k = \ln\frac{120}{131}$, we can use the function $T(t) = 69 + 131e^{\left(\ln\frac{120}{131}\right)t}$ and determine the value of t when $T(t) = 146$.

Set $T(t) = 146$: $146 = 69 + 131e^{\left(\ln\frac{120}{131}\right)t}$

Subtract 69 from both sides: $77 = 131e^{\left(\ln\frac{120}{131}\right)t}$

Divide both sides by 131: $\dfrac{77}{131} = e^{\left(\ln\frac{120}{131}\right)t}$

Use the **logarithm property of equality**: $\ln\dfrac{77}{131} = \ln e^{\left(\ln\frac{120}{131}\right)t}$

Use **cancellation property (2)** to rewrite $\ln e^{\left(\ln\frac{120}{131}\right)t}$ as $\left(\ln\dfrac{120}{131}\right)t$: $\ln\dfrac{77}{131} = \left(\ln\dfrac{120}{131}\right)t$

Divide both sides by $\ln\dfrac{120}{131}$: $\dfrac{\ln\frac{77}{131}}{\ln\frac{120}{131}} = t$

Use a calculator to approximate the time rounded to the nearest minute: $t \approx 6$ minutes

So, it takes approximately 6 minutes for the tea to cool to 146°F.

(eText Screens 20.8-1–20.8-29)

The graph of $T(t) = 69 + 131e^{\left(\ln \frac{120}{131}\right)t}$, which describes the temperature of the tea t minutes after being poured, was created using a graphing utility. Note that the line $y = 69$, which represents the temperature of the surroundings, is a **horizontal asymptote**.

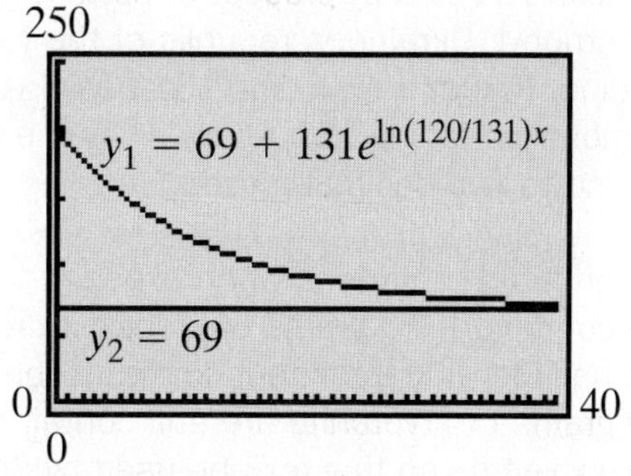

You Try It **Work through this You Try It problem.**

Work Exercises 16–18 in this textbook or in the MyMathLab Study Plan.

20.8 Exercises

You Try It

1. Jimmy invests $15,000 in an account that pays 6.25% compounded quarterly. How long (in years and months) will it take for his investment to reach $20,000?

2. How long (in years and months) will it take for an investment to double at 9% compounded monthly?

3. How long will it take for an investment to triple if it is compounded continuously at 8%? Round to two decimal places.

4. What is the interest rate necessary for an investment to quadruple after 8 years of continuous compound interest? (Round to the nearest hundredth of a percent.)

5. Marsha and Jan both invested money on March 1, 2005. Marsha invested $5000 at Bank A, where the interest was compounded quarterly. Jan invested $3000 at Bank B, where the interest was compounded continuously. On March 1, 2007, Marsha had a balance of $5468.12, whereas Jan had a balance of $3289.09. What was the interest rate at each bank? (Round to the nearest tenth of a percent.)

You Try It

6. The population of Adamsville grew from 9000 to 15,000 in 6 years. Assuming uninhibited exponential growth, what is the expected population in an additional 4 years? Round to the nearest whole number.

7. During a research experiment, it was found that the number of bacteria in a culture grew at a rate proportional to its size. At 8:00 AM, there were 2000 bacteria present in the culture. At noon, the number of bacteria grew to 2400. How many bacteria will there be at midnight? Round to the nearest whole number.

8. A skull cleaning factory cleans animal skulls such as deer, buffalo, and other types of animal skulls using flesh-eating beetles to clean the skulls. The factory owner started with only 10 adult beetles. After 40 days, the beetle population grew to 30 adult beetles. Assuming uninhibited exponential growth, how long did it take before the beetle population reached 10,000 beetles? Round to the nearest whole number.

9. The population of a Midwest industrial town decreased from 210,000 to 205,000 in just 3 years. Assuming negative exponential growth and that this trend continues, what will the population be after an additional 3 years? Round to the nearest whole number.

10. A certain radioactive isotope is leaked into a small stream. Three hundred days after the leak, 2% of the original amount of the substance remained. Determine the half-life of this radioactive isotope. Round to the nearest whole number.

11. Radioactive Iodine-131 is a by-product of certain nuclear reactors. On April 26, 1986, one of the nuclear reactors in Chernobyl, Ukraine, a republic of the former Soviet Union, experienced a massive release of radioactive iodine. Fortunately, Iodine-131 has a very short half-life of 8 days. Estimate the percentage of the original amount of Iodine-131 released by the Chernobyl explosion on May 1, 1986, 5 days after the explosion. Round to two decimal places.

12. Superman is rendered powerless when exposed to 50 or more grams of kryptonite. A 500-year-old rock that originally contained 300 grams of kryptonite was recently stolen from a rock museum by Superman's enemies. The half-life of kryptonite is known to be 200 years.
   a. How many grams of kryptonite are still contained in the stolen rock? Round to two decimal places.
   b. For how many years can this rock be used by Superman's enemies to render him powerless? Round to the nearest whole number.

You Try It

13. The logistic growth model $H(t) = \frac{6000}{1 + 2e^{-0.65t}}$ represents the number of families that own a home in a certain small (but growing) Idaho city t years after 1980.
   a. What is the maximum number of families that will own a home in this city?
   b. How many families owned a home in 1980?
   c. In what year did 5920 families own a home?

14. The number of students that hear a rumor on a small college campus t days after the rumor starts is modeled by the logistic function $R(t) = \frac{3000}{1 + Be^{kt}}$. Determine the following if 8 students initially heard the rumor and 100 students heard the rumor after 1 day.
   a. What is the carrying capacity for the number of students who will hear the rumor?
   b. Find B.
   c. Find k.
   d. How long will it take for 2900 students to hear the rumor?

15. In 1999, 1500 runners entered the inaugural Run-for-Your-Life marathon in Joppetown, USA. In 2005, 21,500 runners entered the race. Because of the limited number of hotels, restaurants, and portable toilets in the area, the carrying capacity for the number of racers is 61,500. The number of racers at any time, t, in years, can be modeled by the logistic function $P(t) = \frac{C}{1 + Be^{kt}}$.
   a. What is the value of C?
   b. Find B.
   c. Find k.
   d. In what year should at least 49,500 runners be expected to run in the race? Round to the nearest year.

You Try It

16. Estabon poured himself a hot beverage that had a temperature of 198°F and then set it on the kitchen table to cool. The temperature of the kitchen was a constant 75°F. If the drink cooled to 180°F in 2 minutes, how long will it take for the drink to cool to 100°F?

17. Police arrive at a murder scene at 1:00 AM and immediately record the body's temperature, which was 92°F. At 2:30 AM, after thoroughly inspecting and fingerprinting the area, they again took the temperature of the body, which had dropped to 85°F. The temperature of the crime scene has remained at a constant 60°F. Determine when the person was murdered. (Assume that the victim was healthy at the time of death. That is, assume that the temperature of the body at the time of death was 98.6°F.)

18. Jodi poured herself a cold soda that had an initial temperature of 40°F and immediately went outside to sunbathe where the temperature was a steady 99°F. After 5 minutes, the temperature of the soda was 47°F. Jodi had to run back into the house to answer the phone. What is the expected temperature of the soda after an additional 10 minutes?

MODULE TWENTY-ONE

Conic Sections

MODULE TWENTY-ONE CONTENTS

Introduction to Conic Sections

In this module, we focus on the geometric study of conic sections. Conic sections (or conics) are formed when a plane intersects a pair of **right circular cones**. The surface of the cones comprises the set of all line segments that intersect the outer edges of the circular bases of the cones and pass through a fixed point. The fixed point is called the **vertex** of the cone, and the line segments are called the **elements**. See Figure 1.

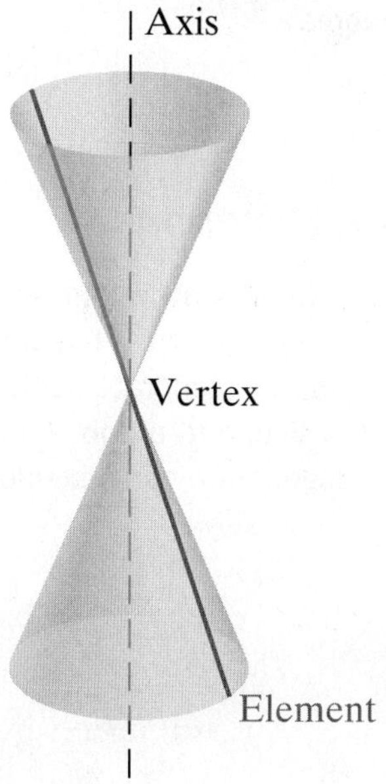

Figure 1
Pair of right circular cones

(eText Screens 21.0-1–21.0-4)

When a plane intersects a right circular cone, a conic section is formed. The four **conic sections** that can be formed are **circles, ellipses, parabolas,** and **hyperbolas**. Watch the following four animations to see how each conic section is formed.

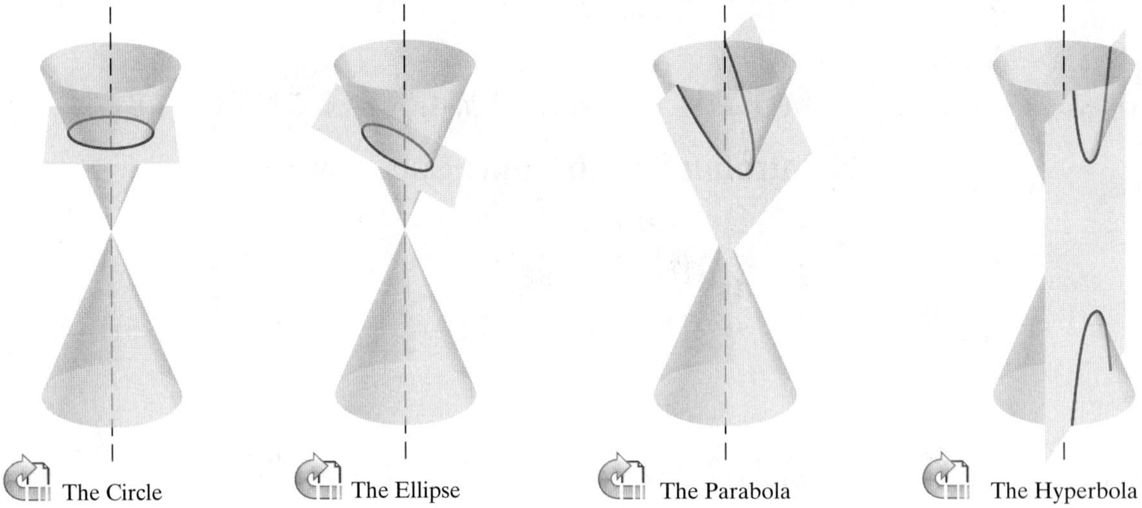

My video summary

Because we studied circles in **Topic 19.4**, they are not covered again in this module. It may, however, be useful to review circles before going on. Watch this **video** to review how to write the equation of a circle in standard form by completing the square. We also need to be able to complete the square to write each of the other conic sections in standard form. These conic sections are introduced in the following order:

Topic 21.1 The Parabola

Topic 21.2 The Ellipse

Topic 21.3 The Hyperbola

DEGENERATE CONIC SECTIONS

It is worth noting that when the circle, ellipse, parabola, or hyperbola is formed, the intersecting plane does not pass through the vertex of the cones. When a plane does intersect the vertex of the cones, a degenerate conic section is formed. The three **degenerate conic sections** are a **point**, a **line**, and a **pair of intersecting lines**. See **Figure 2**. We do not concern ourselves with degenerate conic sections in this module.

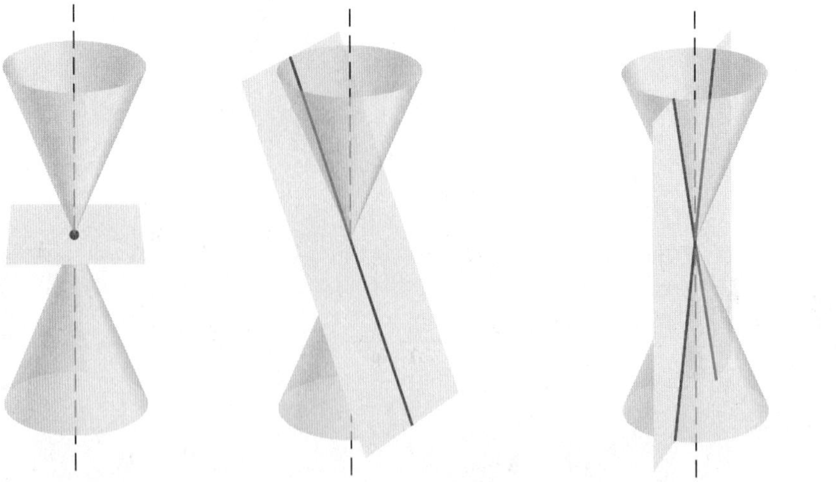

Figure 2 Degenerate conic sections

(eText Screens 21.1-1–21.1-22)

21.1 The Parabola

THINGS TO KNOW

Before working through this topic, be sure you are familiar with the following concepts:

VIDEO ANIMATION INTERACTIVE

You Try It **1.** Write the Equation of a Line Given Its Slope and y-intercept (Topic 12.4, **Objective 3**)

You Try It **2.** Write the Equation of a Line Given Its Slope and a Point on the Line (Topic 12.4, **Objective 4**)

You Try It **3.** Write the Equation of a Line Given Two Points (Topic 12.4, **Objective 5**)

You Try It **4.** Find the Distance Between Two Points (Topic 19.4, **Objective 1**)

You Try It **5.** Write the General Form of a Circle in Standard Form and Sketch Its Graph (Topic 19.4, **Objective 5**)

OBJECTIVES

1 Work with the Equation of a Parabola with a Vertical Axis of Symmetry

2 Work with the Equation of a Parabola with a Horizontal Axis of Symmetry

3 Find the Equation of a Parabola Given Information about the Graph

4 Complete the Square to Find the Equation of a Parabola in Standard Form

5 Solve Applications Involving Parabolas

OBJECTIVE 1 WORK WITH THE EQUATION OF A PARABOLA WITH A VERTICAL AXIS OF SYMMETRY

In Topic 19.2, we studied quadratic functions of the form $f(x) = ax^2 + bx + c, a \neq 0$. We learned that every quadratic function has a U-shaped graph called a *parabola*. You may want to review the different characteristics of a parabola. Work through the following animation, and view each characteristic listed below for a detailed description.

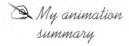

My animation summary

Characteristics of a Parabola

1. **Vertex**

2. **Axis of symmetry**

3. **y-Intercept**

4. **x-Intercept(s) or real zeros**

5. **Domain and range**

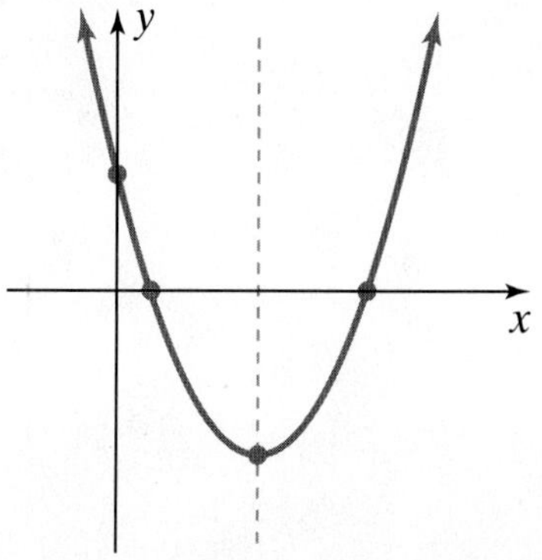

My animation summary

In **Module 19.2**, we studied quadratic functions and parabolas from an algebraic point of view. We now look at parabolas from a geometric perspective. We saw in the introduction to this module that when a plane is parallel to an **element** of the cone, the plane will intersect the cone in a parabola. (Watch the animation.)

The set of points that define the parabola formed by the intersection described previously is stated in the following geometric definition of the parabola.

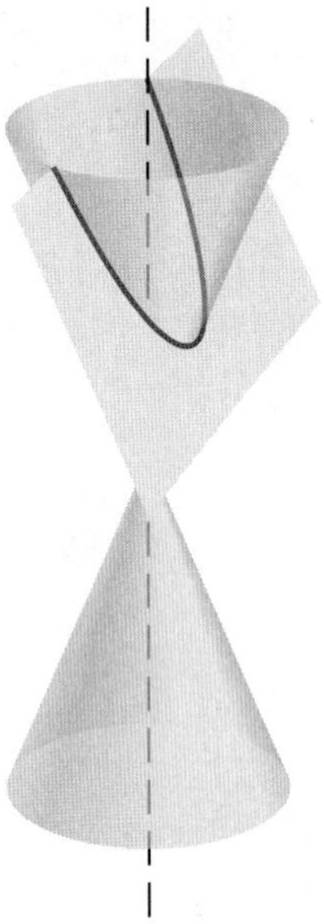

The Parabola

Geometric Definition of the Parabola

A **parabola** is the set of all points in a plane equidistant from a fixed point F and a fixed line D. The fixed point is called the **focus**, and the fixed line is called the **directrix**.

My video summary

Watch the **video** to see how to sketch the parabola seen in Figure 3 using the geometric definition of the parabola.

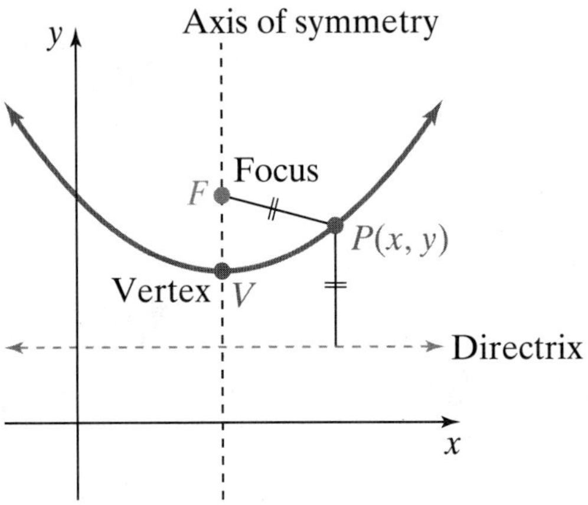

Figure 3
The distance from any point P to the focus is the same as the distance from point P to the directrix.

In Figure 3, we can see that for any point $P(x, y)$ that lies on the graph of the parabola, the distance from point P to the focus is exactly the same as the distance from point P to the directrix. Similarly, because the vertex, V, lies on the graph of the parabola, the distance from V to the focus must also be the same as the distance from V to the directrix. Therefore, if the distance from V to F is p units, then the distance from V to the directrix is also p units. If the coordinates of the vertex in Figure 3 are (h, k), then the coordinates of the focus must be $(h, k + p)$ and the equation of the directrix is $y = k - p$. We can use this information and the fact that the distance from $P(x, y)$ to the focus is equal to the distance from $P(x, y)$ to the directrix to derive the equation of a parabola.

Equation of a Parabola in Standard Form with a Vertical Axis of Symmetry

The equation of a parabola with a vertical axis of symmetry is $(x - h)^2 = 4p(y - k)$,

where

the vertex is $V(h, k)$, $|p|$ = distance from the vertex to the focus = distance from the vertex to the directrix, the focus is $F(h, k + p)$, and the equation of the directrix is $y = k - p$.

The parabola opens *upward* if $p > 0$ or *downward* if $p < 0$.

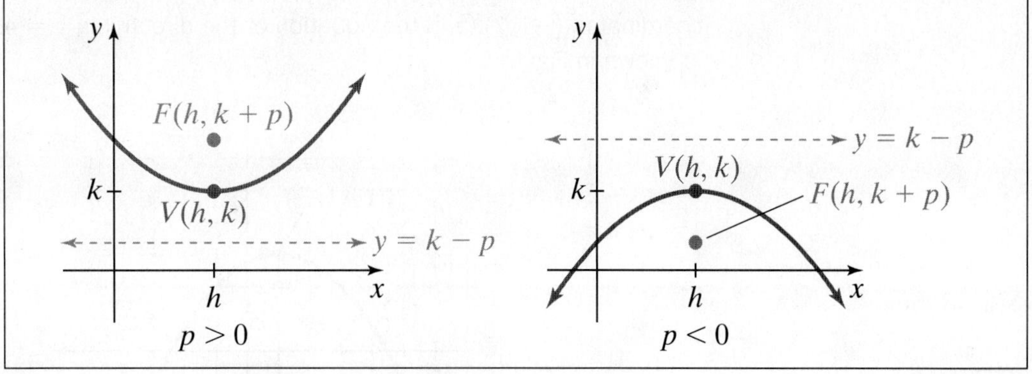

Example 1 Finding the Vertex, Focus, and Directrix of a Parabola and Sketching Its Graph

Find the vertex, focus, and directrix of the parabola $x^2 = 8y$ and sketch its graph.

Solution Notice that we can rewrite the equation $x^2 = 8y$ as $(x - 0)^2 = 8(y - 0)$. We can now compare the equation $(x - 0)^2 = 8(y - 0)$ to the standard form equation $(x - h)^2 = 4p(y - k)$ to see that $h = 0$ and $k = 0$; hence, the vertex is at the origin $(0, 0)$. To find the focus and directrix, we need to find p.

$$4p = 8$$

Divide both sides by 4: $\quad p = 2$

Because the value of p is positive, the parabola opens upward and the focus is located two units vertically *above* the vertex, whereas the directrix is the horizontal line located two units vertically *below* the vertex. The focus has coordinates $(0, 2)$, and the equation of the directrix is $y = -2$. The graph is shown in Figure 4.

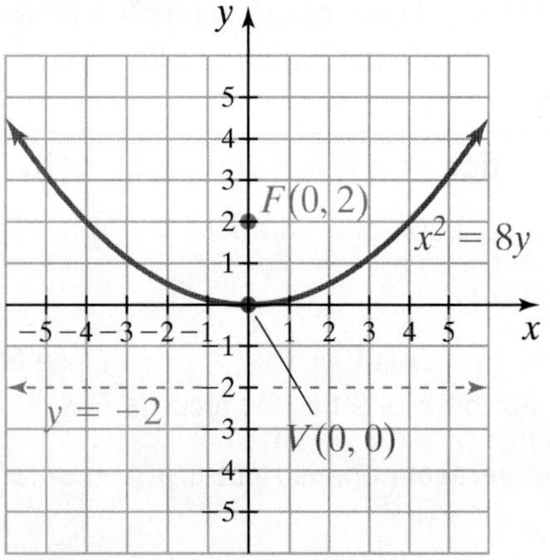

Figure 4

Example 2 Finding the Vertex, Focus, and Directrix of a Parabola and Sketching Its Graph

My video summary

Find the vertex, focus, and directrix of the parabola $-(x + 1)^2 = 4(y - 3)$ and sketch its graph.

(eText Screens 21.1-1–21.1-22)

Solution Watch the **video** to verify that the vertex has coordinates $(-1, 3)$, the focus has coordinates $(-1, 2)$, and the equation of the directrix is $y = 4$. The graph of this parabola is shown in Figure 5.

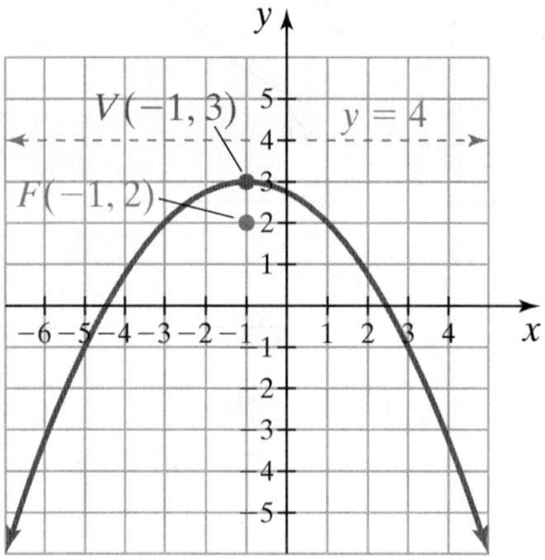

$-(x + 1)^2 = 4(y - 3)$

Figure 5

You Try It **Work through this You Try It problem.**

Work Exercises 1–5 in this textbook or in the MyMathLab Study Plan.

OBJECTIVE 2 WORK WITH THE EQUATION OF A PARABOLA WITH A HORIZONTAL AXIS OF SYMMETRY

In Examples 1 and 2, the graphs of both parabolas had vertical axes of symmetry. The graph of a parabola could also have a horizontal axis of symmetry and open "sideways." We derive the standard form of the parabola with a horizontal axis of symmetry in much the same way as we did with the parabola with a vertical axis of symmetry.

Equation of a Parabola in Standard Form with a Horizontal Axis of Symmetry

The equation of a parabola with a horizontal axis of symmetry is $(y - k)^2 = 4p(x - h)$, where

the vertex is $V(h, k)$, $|p|$ = distance from the vertex to the focus = distance from the vertex to the directrix, the focus is $F(h + p, k)$, and the equation of the directrix is $x = h - p$.

The **parabola opens *right*** if $p > 0$ **or *left* if** $p < 0$.

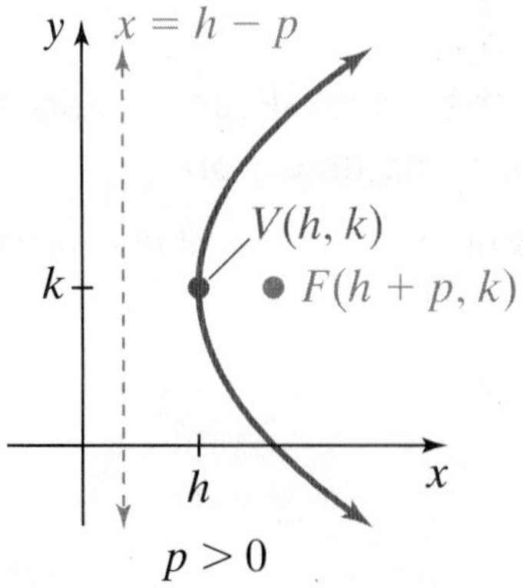

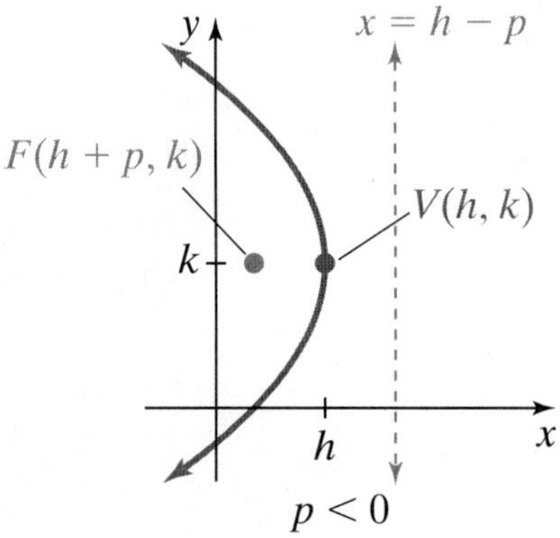

Example 3 Finding the Vertex, Focus, and Directrix of a Parabola and Sketching Its Graph

My video summary

Find the vertex, focus, and directrix of the parabola $(y - 3)^2 = 8(x + 2)$ and sketch its graph.

Solution Watch the video to verify that the vertex has coordinates $(-2, 3)$, the focus has coordinates $(0, 3)$, and the equation of the directrix is $x = -4$. The graph of this parabola is shown in Figure 6.

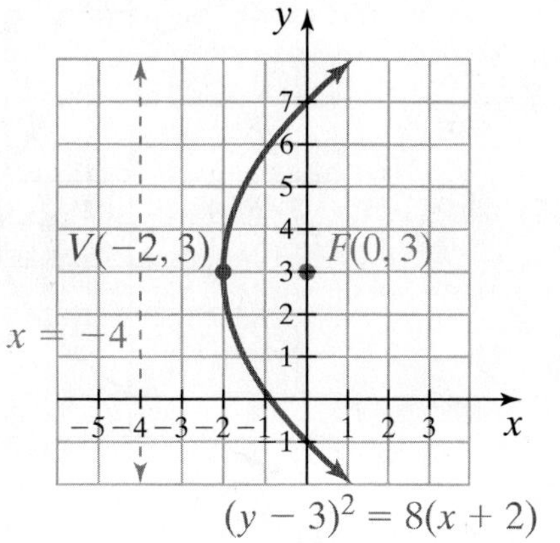

Figure 6

You Try It **Work through this You Try It problem.**

Work Exercises 6–10 in this textbook or in the MyMathLab Study Plan.

OBJECTIVE 3 FIND THE EQUATION OF A PARABOLA GIVEN INFORMATION ABOUT THE GRAPH

It is often necessary to determine the equation of a parabola given certain information. It is always useful to first determine whether the parabola has a vertical axis of symmetry or a horizontal axis of symmetry. Try to work through Examples 4 and 5. Then watch the solutions in the corresponding videos to determine whether you are correct.

Example 4 Finding the Equation of a Parabola

My video summary

Find the standard form of the equation of the parabola with focus $\left(-3, \frac{5}{2}\right)$ and directrix $y = \frac{11}{2}$.

Solution Watch the video to see that the equation of this parabola is $(x + 3)^2 = -6(y - 4)$. The graph is shown in Figure 7.

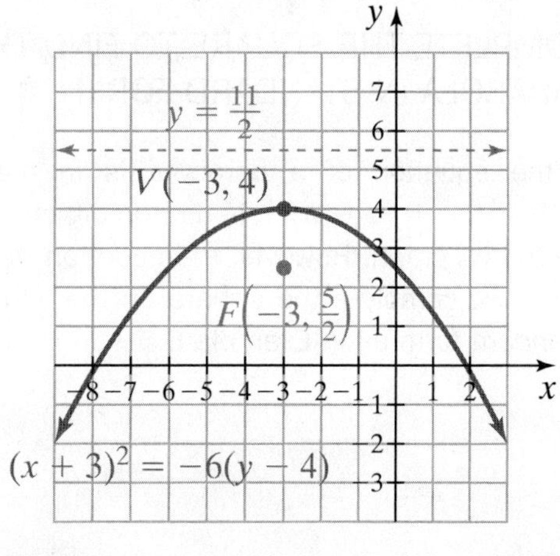

Figure 7

(eText Screens 21.1-1–21.1-22)

Example 5 Finding the Equation of a Parabola

My video summary

Find the standard form of the equation of the parabola with focus $(4, -2)$ and vertex $\left(\frac{13}{2}, -2\right)$.

Solution Watch the **video** to see that the equation of this parabola is $(y + 2)^2 = -10\left(x - \frac{13}{2}\right)$. The graph is shown in Figure 8.

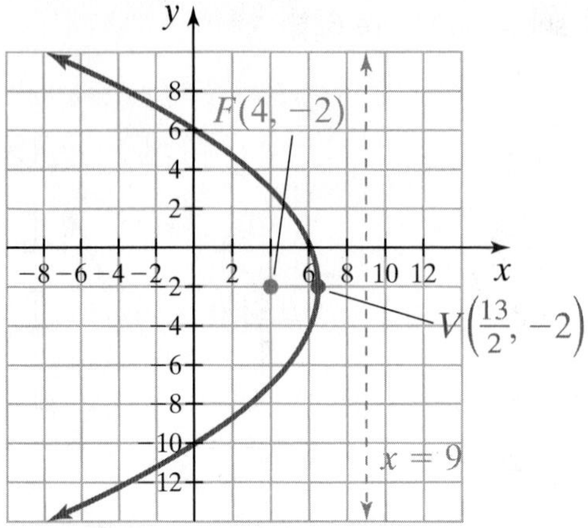

$(y + 2)^2 = -10\left(x - \frac{13}{2}\right)$

Figure 8

We can use a graphing utility to graph the parabola from Example 5 by solving the equation for y.

$$(y + 2)^2 = -10\left(x - \frac{13}{2}\right)$$
$$y + 2 = \pm\sqrt{-10\left(x - \frac{13}{2}\right)}$$
$$y = -2 \pm \sqrt{-10\left(x - \frac{13}{2}\right)}$$

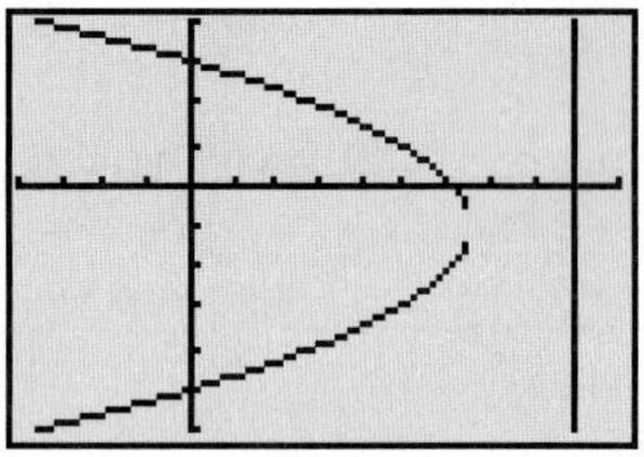

Figure 9

Using $y_1 = -2 + \sqrt{-10\left(x - \frac{13}{2}\right)}$, and $y_2 = -2 - \sqrt{-10\left(x - \frac{13}{2}\right)}$, we obtain the graph seen in Figure 9. *Note:* The directrix was created using the calculator's DRAW feature.

You Try It **Work through this You Try It problem.**

Work Exercises 11–19 in this textbook or in the MyMathLab Study Plan.

OBJECTIVE 4 COMPLETE THE SQUARE TO FIND THE EQUATION OF A PARABOLA IN STANDARD FORM

If the equation of a parabola is in the standard form of $(x - h)^2 = 4p(y - k)$ or $(y - k)^2 = 4p(x - h)$, it is not too difficult to determine the vertex, focus, and directrix and sketch its graph. However, the equation might not be given in standard form. If this is the case, we complete the square on the variable that is squared to rewrite the equation in standard form as in Example 6.

(eText Screens 21.1-1–21.1-22)

Example 6 Writing the Equation of a Parabola in Standard Form by Completing the Square

My video summary

Find the vertex, focus, and directrix and sketch the graph of the parabola $x^2 - 8x + 12y = -52$.

Solution Because x is squared, we will complete the square on the variable x.

$$x^2 - 8x + 12y = -52$$
$$x^2 - 8x = -12y - 52$$
$$x^2 - 8x + 16 = -12y - 52 + 16$$
$$(x - 4)^2 = -12y - 36$$
$$(x - 4)^2 = -12(y + 3)$$

The equation is now in standard form with vertex $(4, -3)$ and $4p = -12$ so $p = -3$. The parabola must open down because the variable x is squared and $p < 0$. The focus is located three units below the vertex, whereas the directrix is three units above the vertex. Thus, the focus has coordinates $(4, -6)$, and the equation of the directrix is $y = 0$ or the x-axis. You can watch the **video** to see this solution worked out in detail. The graph is shown in Figure 10.

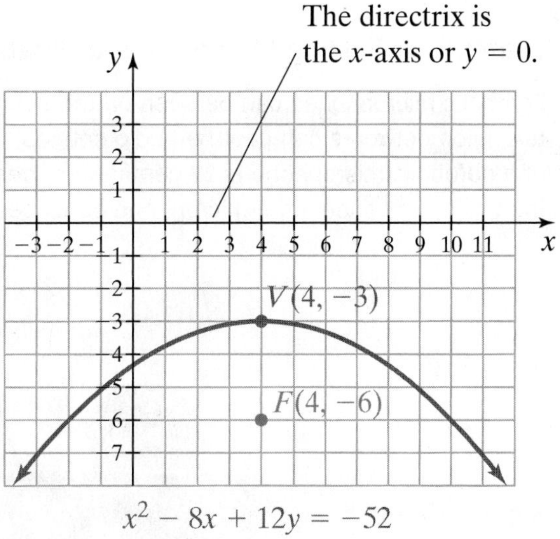

Figure 10

You Try It Work through this You Try It problem.

Work Exercises 20–23 in this textbook or in the MyMathLab Study Plan.

OBJECTIVE 5 SOLVE APPLICATIONS INVOLVING PARABOLAS

The Romans were one of the first civilizations to use the engineering properties of parabolic structures in their creation of **arch bridges**. The cables of many suspension bridges, such as the **Golden Gate Bridge** in San Francisco, span from tower to tower in the shape of a parabola.

Parabolic surfaces are used in the manufacture of many satellite dishes, search lights, car headlights, telescopes, lamps, heaters, and other objects. This is because parabolic surfaces have the property that incoming rays of light or radio waves traveling parallel to the axis of symmetry of a parabolic reflector or receiver will reflect off the parabolic surface and travel directly toward the antenna that is placed at the focus. See Figure 11. When a light source such as the headlight of a car is placed at the focus of a parabolic reflector, the light reflects off the surface outward, producing a narrow beam of light and thus maximizing the output of illumination.

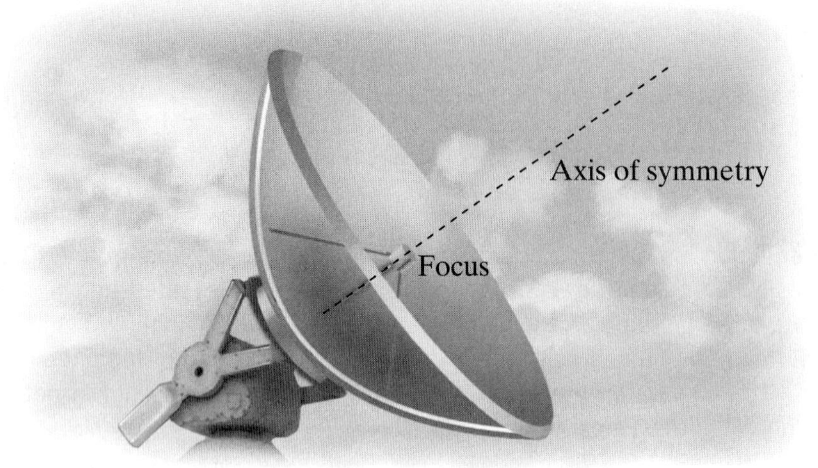

Figure 11
Incoming rays, parallel to the axis of symmetry, reflect off the parabolic surface toward the antenna placed at the focus.

Example 7 Finding the Focus of a Parabolic Microphone

Parabolic microphones can be seen on the sidelines of professional sporting events so that television networks can capture audio sounds from the players on the field. If the surface of a parabolic microphone is 27 centimeters deep and has a diameter of 72 centimeters at the top, where should the microphone be placed relative to the vertex of the parabola?

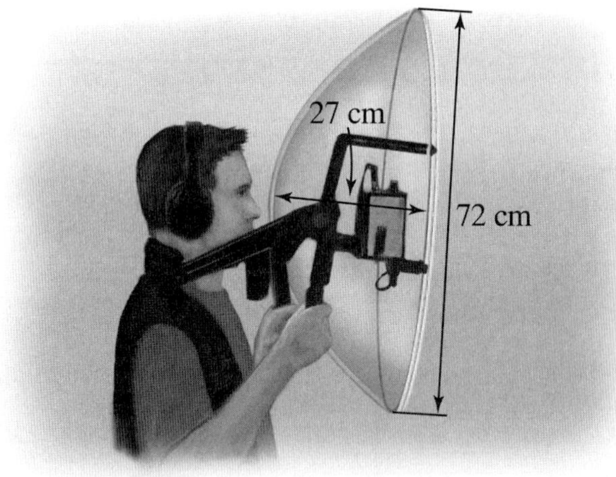

Solution We can draw a parabola with the vertex at the origin representing the center cross section of the parabolic microphone. The equation of this parabola in standard form is $x^2 = 4py$. Substitute the point $(36, 27)$ into the equation to get

$$x^2 = 4py$$
$$(36)^2 = 4p(27)$$
$$1296 = 108p$$
$$p = 12.$$

The microphone must be placed 12 centimeters from the vertex.

(eText Screens 21.1-1–21.1-22)

You Try It Work through this You Try It problem.

Work Exercises 24–28 in this textbook or in the MyMathLab Study Plan.

21.1 Exercises

In Exercises 1–10, determine the vertex, focus, and directrix of the parabola and sketch its graph.

You Try It

1. $x^2 = 16y$

2. $x^2 = -8y$

3. $(x - 1)^2 = -12(y - 4)$

4. $(x + 3)^2 = 6(y - 1)$

5. $(x + 2)^2 = 5(y + 6)$

6. $y^2 = 4x$

You Try It

7. $y^2 = -8x$

8. $(y - 5)^2 = -4(x - 2)$

9. $(y + 3)^2 = 20(x - 4)$

10. $(y + 4)^2 = 9(x + 3)$

In Exercises 11–19, find the equation in standard form of the parabola described.

You Try It

11. The focus has coordinates $(2, 0)$, and the equation of the directrix is $x = -2$.

12. The focus has coordinates $\left(0, -\frac{1}{2}\right)$, and the equation of the directrix is $y = \frac{1}{2}$.

13. The focus has coordinates $(3, -5)$, and the equation of the directrix is $y = -1$.

14. The focus has coordinates $(2, 4)$, and the equation of the directrix is $x = -4$.

15. The vertex has coordinates $\left(-\frac{11}{4}, -2\right)$, and the focus has coordinates $(-3, -2)$.

16. The vertex has coordinates $\left(4, -\frac{1}{4}\right)$, and the focus has coordinates $\left(4, \frac{1}{4}\right)$.

17. The vertex has coordinates $(-3, 4)$, and the equation of the directrix is $x = -7$.

18. Find the equations of the two parabolas in standard form that have a horizontal axis of symmetry and a focus at the point $(0, 4)$, and that pass through the origin.

19. Find the equations of the two parabolas in standard form that have a vertical axis of symmetry and a focus at the point $\left(2, \frac{3}{2}\right)$, and that pass through the origin.

In Exercises 20–23, determine the vertex, focus, and directrix of the parabola and sketch its graph.

20. $x^2 + 10x = 5y - 10$

21. $y^2 - 12y + 6x + 30 = 0$

You Try It

22. $x^2 - 2x = -y + 1$

23. $y^2 + x + 6y = -10$

You Try It

24. A parabolic eavesdropping device is used by a CIA agent to record terrorist conversations. The parabolic surface measures 120 centimeters in diameter at the top and is 90 centimeters deep at its center. How far from the vertex should the microphone be located?

(eText Screens 21.2-1–21.2-26)

25. A parabolic space heater is 18 inches in diameter and 8 inches deep. How far from the vertex should the heat source be located to maximize the heating output?

26. A large NASA parabolic satellite dish is 22 feet across and has a receiver located at the focus 4 feet from its base. The satellite dish should be how deep?

27. A parabolic arch bridge spans 160 feet at the base and is 40 feet above the water at the center. Find the equation of the parabola if the vertex is placed at the point $(0, 40)$. Can a sailboat that is 35 feet tall fit under the bridge 30 feet from the center?

28. The cable between two 40-meter towers of a suspension bridge is in the shape of a parabola that just touches the bridge halfway between the towers. The two towers are 100 meters apart. Vertical cables are spaced every 10 meters along the bridge. What are the lengths of the vertical cables located 30 meters from the center of the bridge?

21.2 The Ellipse

THINGS TO KNOW

Before working through this topic, be sure you are familiar with the following concepts:

		VIDEO	ANIMATION	INTERACTIVE
You Try It	1. Find the Distance Between Two Points (Topic 17.4, **Objective 1**)			
You Try It	2. Complete the Square to Find the Equation of a Parabola in Standard Form (Topic 21.1, **Objective 4**)	▤		

(eText Screens 21.2-1–21.2-26)

OBJECTIVES

1 Sketch the Graph of an Ellipse

2 Find the Equation of an Ellipse Given Information about the Graph

3 Complete the Square to Find the Equation of an Ellipse in Standard Form

4 Solve Applications Involving Ellipses

My animation summary

When a plane intersects a **right circular cone** at an angle between 0 and 90 degrees to the axis of the cone, the conic section formed is an ellipse. (Watch this **animation**.)

The set of points in the plane that defines an ellipse formed by the intersection of a plane and a cone described previously is stated in the following geometric definition.

Geometric Definition of the Ellipse

An **ellipse** is the set of all points in a plane, the sum of whose distances from two fixed points is a positive constant. The two fixed points, F_1 and F_2, are called the *foci*.

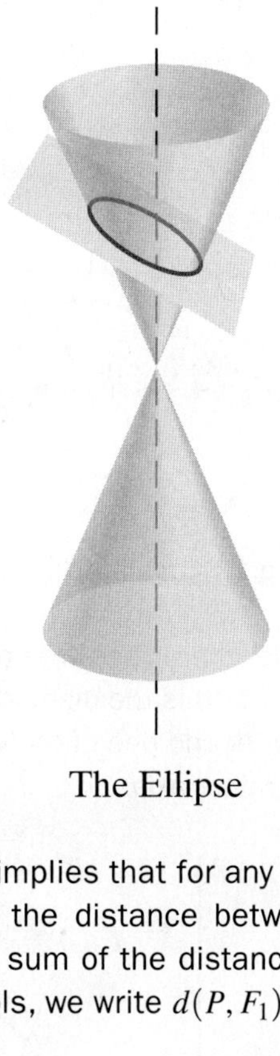

The Ellipse

The previous geometric definition implies that for any two points P and Q that lie on the graph of the ellipse, the sum of the distance between P and F_1 plus the distance between P and F_2 is equal to the sum of the distance between Q and F_1 plus the distance between Q and F_2. In symbols, we write $d(P, F_1) + d(P, F_2) = d(Q, F_1) + d(Q, F_2)$. See **Figure 12**.

(eText Screens 21.2-1–21.2-26)

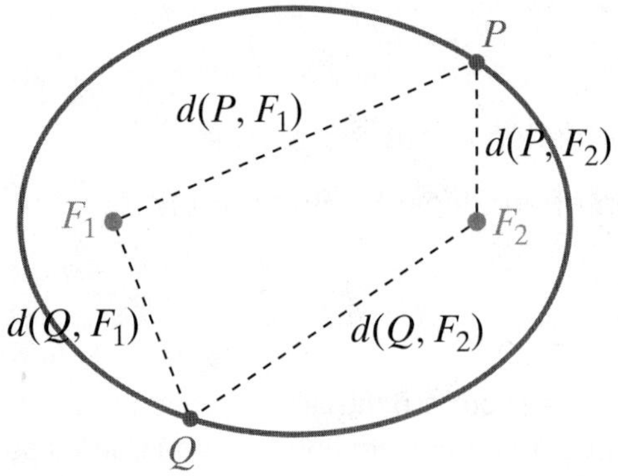

Figure 12
For any points P and Q on an ellipse,
$d(P, F_1) + d(P, F_2) = d(Q, F_1) + d(Q, F_2)$.

OBJECTIVE 1 SKETCH THE GRAPH OF AN ELLIPSE

My video summary

An ellipse has two **axes of symmetry**. The longer axis, which is the line segment that connects the two vertices, is called the **major axis**. The foci are always located along the major axis. The shorter axis is called the **minor axis**. It is the line segment perpendicular to the major axis that passes through the center having endpoints that lie on the ellipse. Watch the **video** to see how to sketch the two ellipses seen in **Figure 13**. In **Figure 13a**, the ellipse has a **horizontal major axis**. The ellipse in **Figure 13b** has a **vertical major axis**.

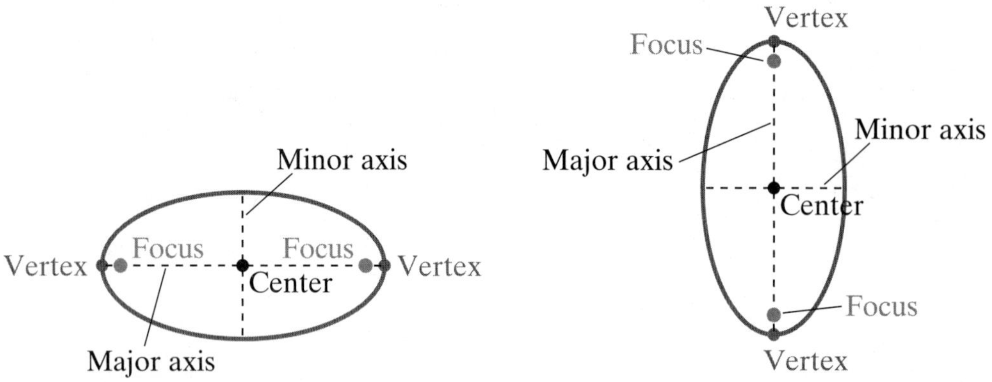

Figure 13 (a) Ellipse with horizontal major axis (b) Ellipse with vertical major axis

Consider the ellipse with a **horizontal major axis** centered at (h, k), as shown in Figure 14, where $c > 0$ is the distance between the center and a focus, $a > 0$ is the distance between the center and one of the vertices, and $b > 0$ is the distance from the center to an endpoint of the **minor axis**.

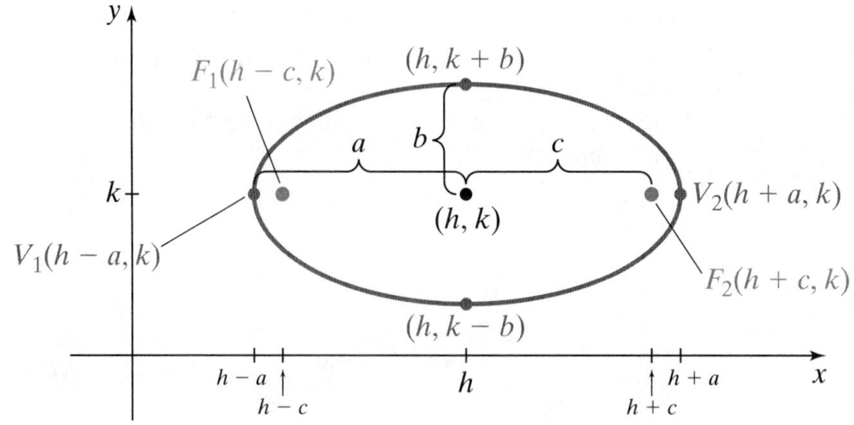

Figure 14

(eText Screens 21.2-1–21.2-26)

Before we can derive the equation of this ellipse, we must establish the following two facts:

Fact 1: The sum of the distances from any point on the ellipse to the two foci is $2a$.

Fact 2: $b^2 = a^2 - c^2$, or equivalently, $c^2 = a^2 - b^2$.

Once we have established these two facts, we can derive the equation of the ellipse. Click **here** to see this derivation. We now state the two standard equations of an ellipse.

Equation of an Ellipse in Standard Form with Center (h, k)

HORIZONTAL MAJOR AXIS

$$\frac{(x-h)^2}{a^2} + \frac{(y-k)^2}{b^2} = 1$$

- $a > b > 0$
- Foci: $F_1(h - c, k)$ and $F_2(h + c, k)$
- Vertices: $V_1(h - a, k)$ and $V_2(h + a, k)$
- Endpoints of minor axis: $(h, k - b)$ and $(h, k + b)$
- $c^2 = a^2 - b^2$

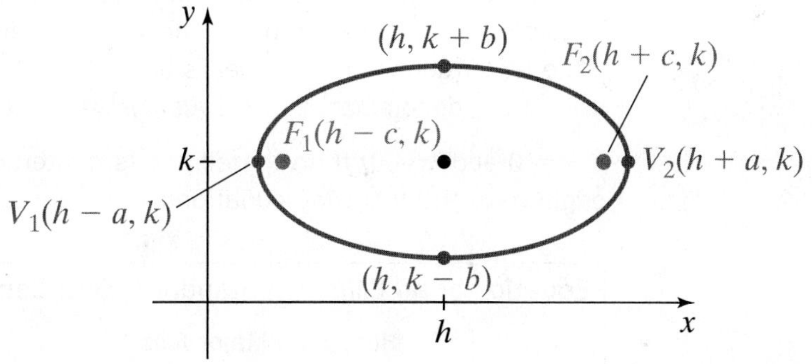

Equation of an Ellipse in Standard Form with Center (h, k)

VERTICAL MAJOR AXIS

$$\frac{(x-h)^2}{b^2} + \frac{(y-k)^2}{a^2} = 1$$

- $a > b > 0$
- Foci: $F_1(h, k - c)$ and $F_2(h, k + c)$
- Vertices: $V_1(h, k - a)$ and $V_2(h, k + a)$
- Endpoints of minor axis: $(h - b, k)$ and $(h + b, k)$
- $c^2 = a^2 - b^2$

continued

(eText Screens 21.2-1–21.2-26)

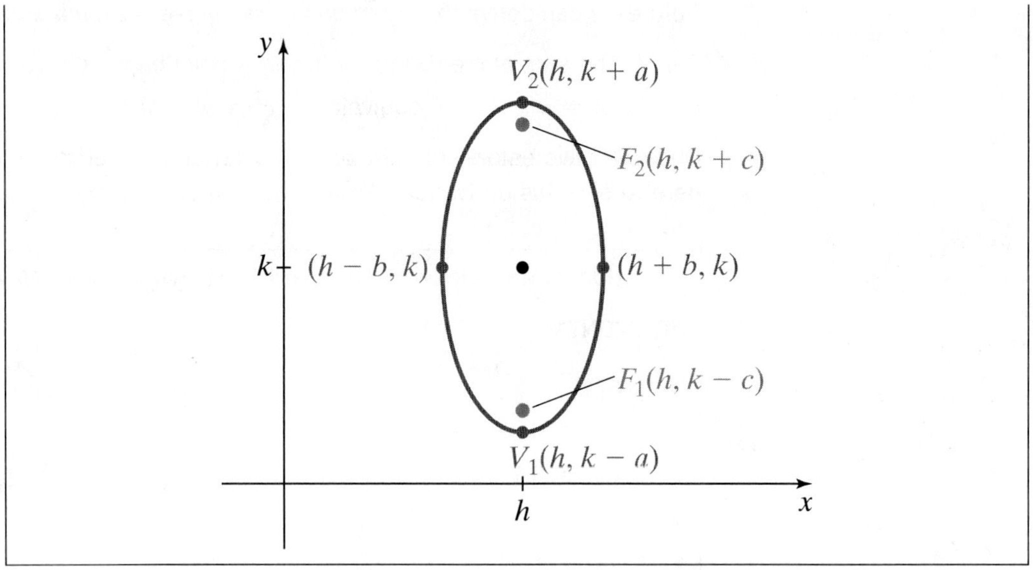

Notice that the two equations are identical except for the placement of a^2. If the equation of an ellipse is in standard form, we can quickly determine whether the ellipse has a horizontal major axis or a vertical major axis by looking at the denominator. If the larger denominator, a^2, appears under the x-term, then the ellipse has a horizontal major axis. If the larger denominator appears under the y-term, then the ellipse has a vertical major axis. If the denominators are equal $(a^2 = b^2)$, then the ellipse is a circle!

If $h = 0$ and $k = 0$, then the ellipse is centered at the **origin**. Ellipses centered at the origin have the following equations.

Equation of an Ellipse in Standard Form Centered at the Origin

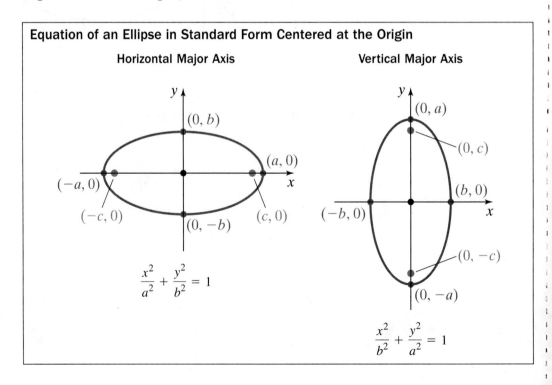

Horizontal Major Axis

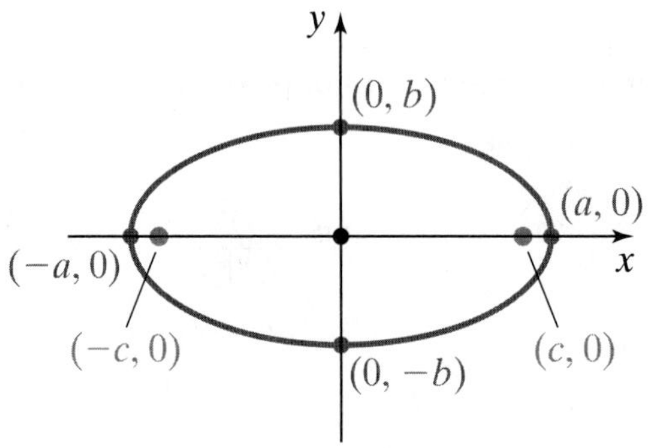

$$\frac{x^2}{a^2} + \frac{y^2}{b^2} = 1$$

Vertical Major Axis

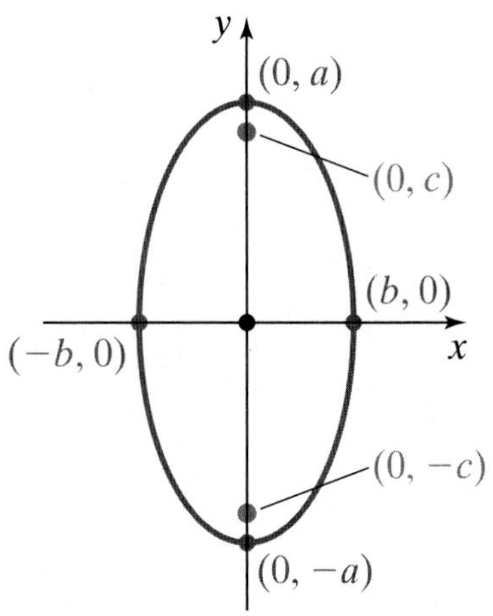

$$\frac{x^2}{b^2} + \frac{y^2}{a^2} = 1$$

(eText Screens 21.2-1–21.2-26)

Example 1 Sketching the Graph of an Ellipse Centered at the Origin

My video summary

Sketch the graph of the ellipse $\frac{x^2}{25} + \frac{y^2}{4} = 1$, and label the center, foci, and vertices.

Solution Watch the **video** to see how to sketch the ellipse shown in Figure 15.

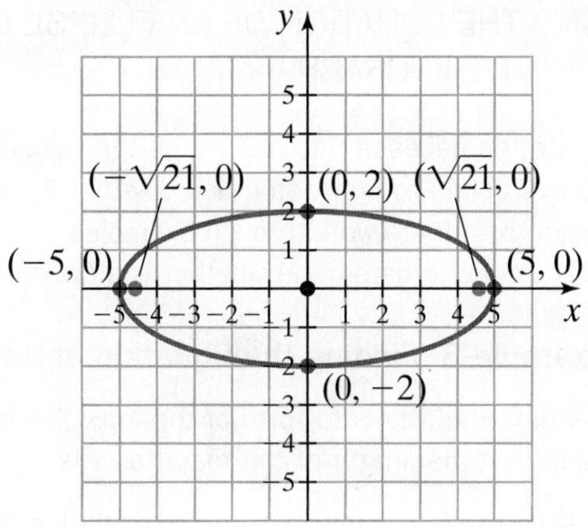

Figure 15

You Try It Work through this You Try It problem.

Work Exercises 1–4 in this textbook or in the MyMathLab Study Plan.

Example 2 Sketching the Graph of an Ellipse

My video summary

Sketch the graph of the ellipse $\frac{(x+2)^2}{20} + \frac{(y-3)^2}{36} = 1$, and label the center, foci, and vertices.

Solution Note that the larger denominator appears under the y-term. This indicates that the ellipse has a vertical major axis. Watch the **video** to verify that the center of the ellipse is $(-2, 3)$. The foci have coordinates $(-2, -1)$ and $(-2, 7)$, the vertices have coordinates $(-2, -3)$ and $(-2, 9)$, and the coordinates of the minor axis are $(-2 - 2\sqrt{5}, 3)$ and $(-2 + 2\sqrt{5}, 3)$. The graph is shown in Figure 16.

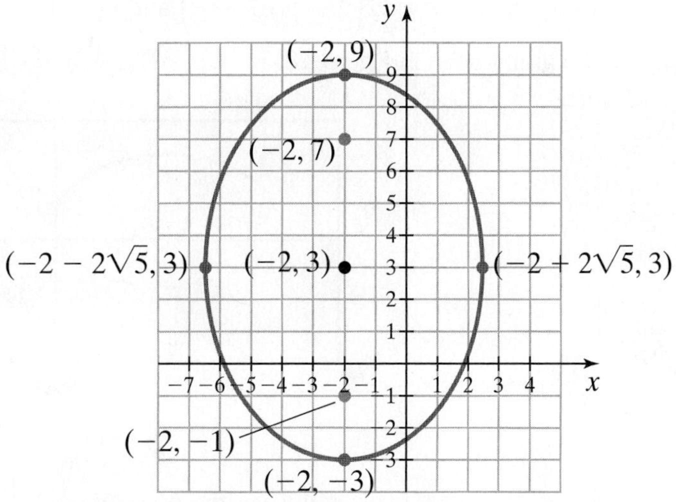

Figure 16

(eText Screens 21.2-1–21.2-26)

You Try It Work through this You Try It problem.

Work Exercises 5–8 in this textbook or in the MyMathLab Study Plan.

OBJECTIVE 2 FIND THE EQUATION OF AN ELLIPSE GIVEN INFORMATION ABOUT THE GRAPH

It is often necessary to determine the equation of an ellipse given certain information. It is always useful to first determine whether the ellipse has a horizontal major axis or a vertical major axis. Try to work through Examples 3 and 4. Then watch the solutions to the corresponding videos to determine whether you are correct.

Example 3 Finding the Equation of an Ellipse

My video summary

Find the standard form of the equation of the ellipse with foci at $(-6, 1)$ and $(-2, 1)$ such that the length of the major axis is eight units.

Solution Watch the **video** to verify that this is an ellipse centered at $(-4, 1)$ with a horizontal major axis such that $a = 4$ and $b = \sqrt{12}$. The equation in standard form is $\frac{(x+4)^2}{4^2} + \frac{(y-1)^2}{(\sqrt{12})^2} = 1$ or $\frac{(x+4)^2}{16} + \frac{(y-1)^2}{12} = 1$. The graph of this ellipse is shown in **Figure 17**.

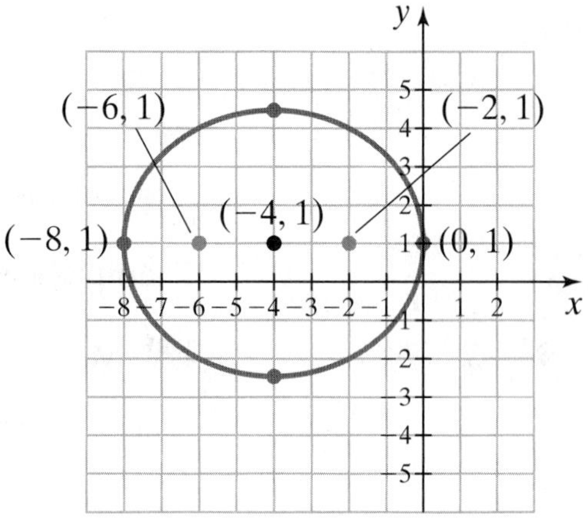

Figure 17 $\frac{(x+4)^2}{16} + \frac{(y-1)^2}{12} = 1$

We can use a graphing utility to graph the ellipse from Example 3 by solving the equation for y:

$$y = 1 \pm \sqrt{12\left(1 - \frac{(x+4)^2}{16}\right)}$$

(View this **popup** box to see how to solve for y.)

Using $y_1 = 1 + \sqrt{12\left(1 - \frac{(x+4)^2}{16}\right)}$ and $y_2 = 1 - \sqrt{12\left(1 - \frac{(x+4)^2}{16}\right)}$, we obtain the graph seen in Figure 18.

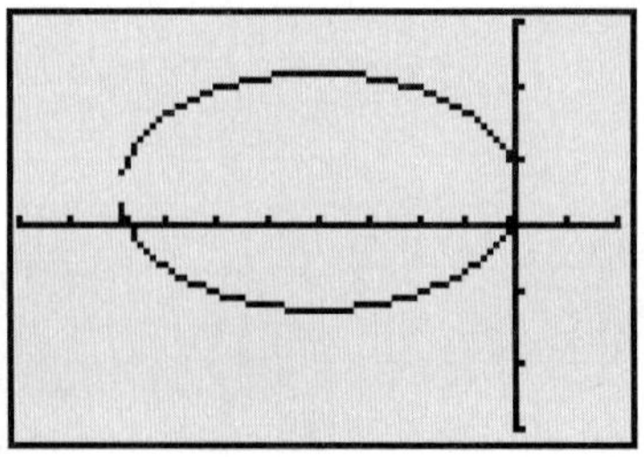

Figure 18

You Try It Work through this You Try It problem.

Work Exercises 9–18 in this textbook or in the MyMathLab Study Plan.

(eText Screens 21.2-1–21.2-26)

Example 4 Finding the Equation of an Ellipse

My video summary

Determine the equation of the ellipse with foci located at $(0, 6)$ and $(0, -6)$ that passes through the point $(-5, 6)$.

Solution To find the equation, we can use **Fact 1** and **Fact 2** of ellipses. Watch the **video** to see how we can use these two facts to determine that the equation of this ellipse is $\frac{x^2}{45} + \frac{y^2}{81} = 1$.

You Try It **Work through this You Try It problem.**

Work Exercises 19 and 20 in this textbook or in the MyMathLab Study Plan.

OBJECTIVE 3 COMPLETE THE SQUARE TO FIND THE EQUATION OF AN ELLIPSE IN STANDARD FORM

If the equation of an ellipse is in the standard form of $\frac{(x-h)^2}{a^2} + \frac{(y-k)^2}{b^2} = 1$ or $\frac{(x-h)^2}{b^2} + \frac{(y-k)^2}{a^2} = 1$, it is not too difficult to determine the center and foci and sketch its graph. However, the equation might not be given in standard form. If this is the case, we complete the square on both variables to rewrite the equation in standard form as in Example 5.

Example 5 Writing the Equation of an Ellipse in Standard Form by Completing the Square

My video summary

Find the center and foci and sketch the ellipse $36x^2 + 20y^2 + 144x - 120y - 396 = 0$.

Solution Rearrange the terms, leaving some room to complete the square and move any constants to the right-hand side:

Rearrange the terms: $36x^2 + 144x \quad + 20y^2 - 120y \qquad = 396$

Then factor and complete the square.

Factor out 36 and 20: $36(x^2 + 4x \quad) + 20(y^2 - 6y \quad) = 396$

Complete the square on x and y. Remember to add $36 \cdot 4 = 144$ and $20 \cdot 9 = 180$ to the right side: $36(x^2 + 4x + 4) + 20(y^2 - 6y + 9) = 396 + 144 + 180$

Factor the left side, and simplify the right side: $36(x+2)^2 + 20(y-3)^2 = 720$

Divide both sides by 720: $\frac{36(x+2)^2}{720} + \frac{20(y-3)^2}{720} = \frac{720}{720}$

Simplify: $\frac{(x+2)^2}{20} + \frac{(y-3)^2}{36} = 1$

(eText Screens 21.2-1–21.2-26)

The equation is now in standard form. Watch the **video** to see the solution to this example worked out in detail. Notice that this is the exact same ellipse that we sketched in Example 2. To see how to sketch this ellipse, refer to **Example 2** or watch the last part of this **video**.

You Try It **Work through this You Try It problem.**

Work Exercises 21–24 in this textbook or in the MyMathLab Study Plan.

OBJECTIVE 4 SOLVE APPLICATIONS INVOLVING ELLIPSES

Ellipses have many applications. The planets of our solar system travel around the Sun in an elliptical orbit with the Sun at one focus. Some comets, such as Halley's comet, also travel in elliptical orbits. See **Figure 19**. (Some comets are seen in our solar system only once because they travel in hyperbolic orbits.) We saw in **Topic 21.1** that the parabola had a special reflecting property where incoming rays of light or radio waves traveling parallel to the axis of symmetry of a parabolic reflector or receiver will reflect off the parabolic surface and travel directly toward the antenna that is placed at the focus.

Figure 19 Planets travel around the Sun in an elliptical orbit.

Ellipses have a similar reflecting property. When light or sound waves originate from one focus of an ellipse, the waves will reflect off the surface of the ellipse and travel directly toward the other focus. See **Figure 20**. This reflecting property is used in a medical procedure to break up kidney stones called *sound wave lithotripsy* in which the patient is placed in an elliptical tank with the kidney stone placed at one focus. An ultrasound wave emitter is positioned at the other focus. The sound waves reflect off the walls of the tank directly to the kidney stone, thus obliterating the stone into fragments that are easily passed naturally through the patient's body. See Example 6.

(eText Screens 21.2-1–21.2-26)

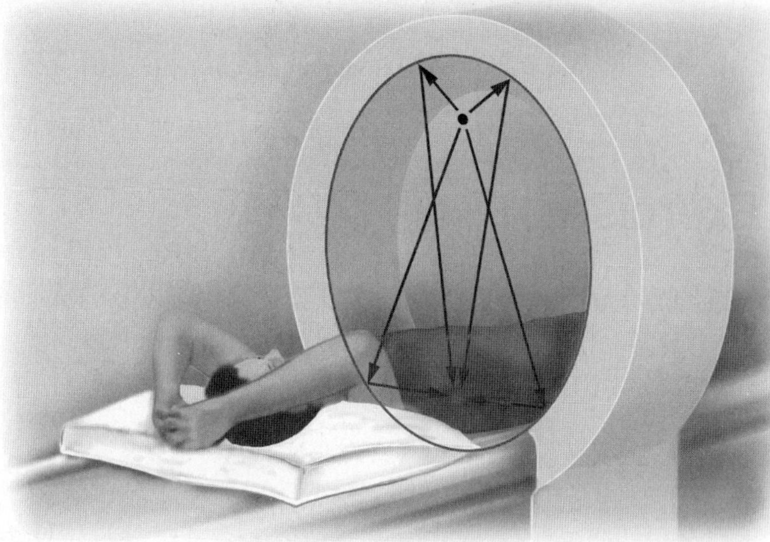

Figure 20 Sound waves or light rays emitted from one focus of an ellipse reflect off the surface directly to the other focus.

Example 6 Position a Patient During Kidney Stone Treatment

A patient is placed in an elliptical tank that is 280 centimeters long and 250 centimeters wide to undergo sound wave lithotripsy treatment for kidney stones. Determine where the sound emitter and the stone should be positioned relative to the center of the ellipse.

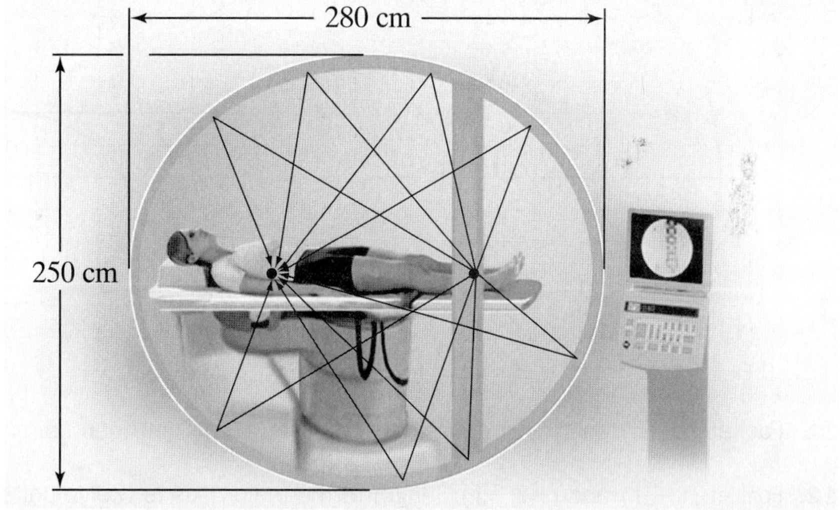

Solution The kidney stone and the sound emitter must be placed at the foci of the ellipse. We know that the major axis is 280 centimeters. Therefore, the vertices must be 140 centimeters from the center. Similarly, the endpoints of the minor axis are located 125 centimeters from the center.

To find c, we use the fact that $c^2 = a^2 - b^2$.

$$c^2 = 140^2 - 125^2$$
$$c^2 = 3975$$
$$c = \sqrt{3975} \approx 63.05$$

The stone and the sound emitter should be positioned approximately 63.05 centimeters from the center of the tank on the major axis.

(eText Screens 21.2-1–21.2-26)

You Try It Work through this You Try It problem.

Work Exercises 25–31 in this textbook or in the MyMathLab Study Plan.

21.2 Exercises

In Exercises 1–8, determine the center, foci, and vertices of the ellipse and sketch its graph.

You Try It **1.** $\frac{x^2}{16} + \frac{y^2}{4} = 1$

2. $\frac{x^2}{12} + \frac{y^2}{25} = 1$

3. $4x^2 + 9y^2 = 36$

4. $20x^2 + 5y^2 = 100$

You Try It **5.** $\frac{(x-2)^2}{36} + \frac{(y-4)^2}{25} = 1$

6. $\frac{(x-1)^2}{9} + \frac{(y+4)^2}{49} = 1$

7. $9(x-5)^2 + 25(y+1)^2 = 225$

8. $(x+7)^2 + 9(y-2)^2 = 9$

In Exercises 9–18, determine the standard equation of each ellipse using the given graph or the stated information.

You Try It **9.**

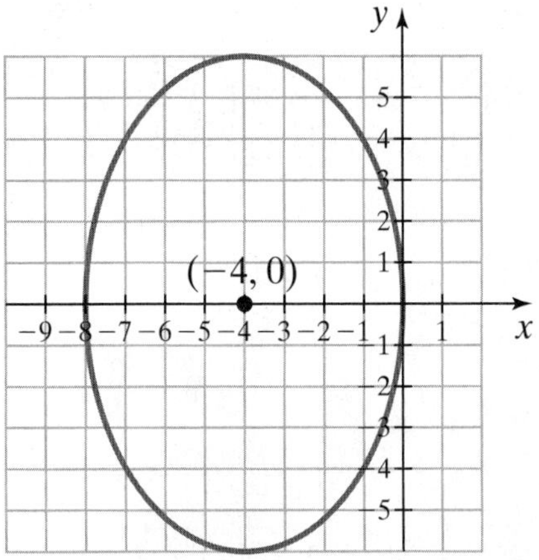

10.

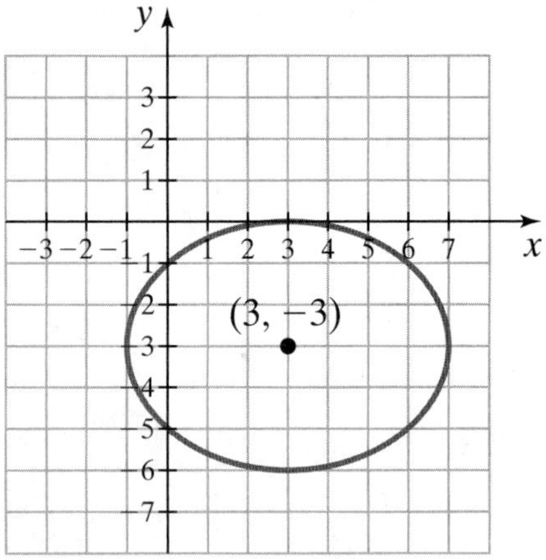

11. Foci at $(-5, 0)$ and $(5, 0)$; length of the major axis is fourteen units.

12. Foci at $(8, -1)$ and $(-2, -1)$; length of the major axis is twelve units.

13. Vertices at $(-6, 10)$ and $(-6, 0)$; length of the minor axis is eight units.

14. Center at $(4, 5)$; vertical minor axis with length sixteen units; $c = 6$.

15. Center at $(-1, 4)$; vertex at $(-1, 8)$; focus at $(-1, 7)$.

16. Vertices at $(2, -7)$ and $(2, 5)$; focus at $(2, 3)$.

17. Center at $(4, 1)$; focus at $(4, 8)$; ellipse passes through the point $(6, 1)$.

18. Center at $(1, 3)$; focus at $(1, 6)$; ellipse passes through the point $(2, 3)$.

In Exercises 19 and 20, determine the standard equation of each ellipse by first using **Fact 1** to find a and then by using **Fact 2** to find b.

You Try It **19.** The foci have coordinates $(-2, 0)$ and $(2, 0)$. The ellipse contains the point $(2, 3)$.

20. The foci have coordinates $(0, -8)$ and $(0, 8)$. The ellipse contains the point $(-2, 6)$.

In Exercises 21–24, complete the square to write each equation in the form $\frac{(x-h)^2}{a^2} + \frac{(y-k)^2}{b^2} = 1$ or $\frac{(x-h)^2}{b^2} + \frac{(y-k)^2}{a^2} = 1$. Determine the center, foci, and vertices of the ellipse and sketch its graph.

You Try It **21.** $16x^2 + 20y^2 + 64x - 40y - 236 = 0$

22. $9y^2 + 16x^2 + 224x + 54y + 721 = 0$

23. $x^2 + 4x + 16y^2 - 32y + 4 = 0$

24. $50x^2 + y^2 + 20y = 0$

You Try It **25.** An elliptical arch railroad tunnel 16 feet high at the center and 30 feet wide is cut through the side of a mountain. Find the equation of the ellipse if a vertex is represented by the point $(0, 16)$.

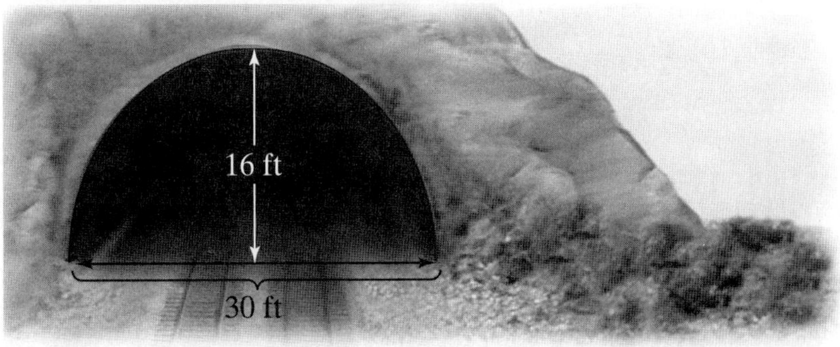

26. A rectangular playing field lies in the interior of an elliptical track that is 50 yards wide and 140 yards long. What is the width of the rectangular playing field if the width is located 10 yards from either vertex?

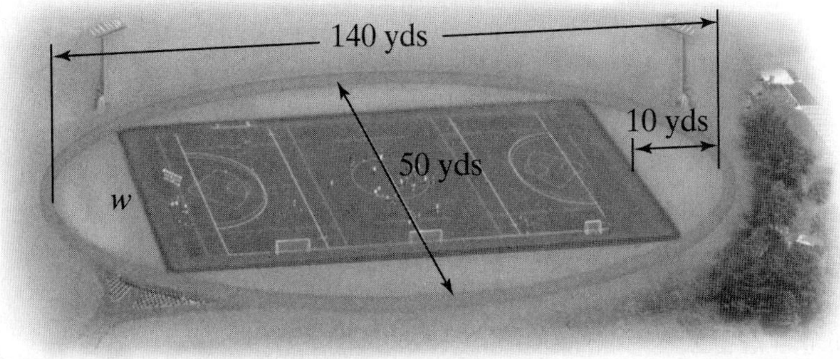

27. A 78-inch by 36-inch door contains a decorative glass elliptical pattern. One vertex of the ellipse is located 9 inches from the top of the door. The other vertex is located 27 inches from the bottom of the door. The endpoints of the minor axis are located 9 inches from each side. Find the equation of the elliptical pattern. (Assume that the center of the elliptical pattern is at the origin.)

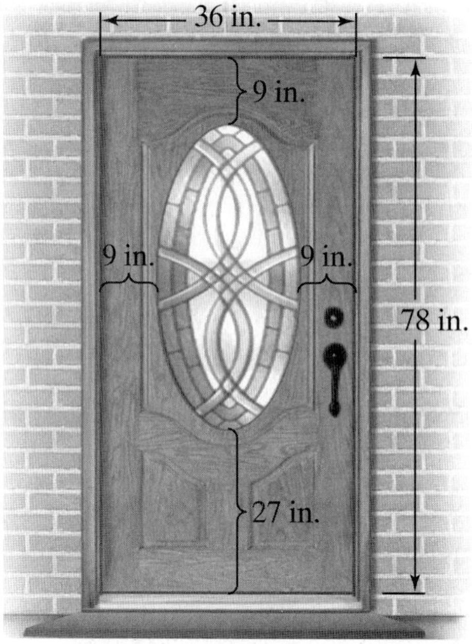

28. A patient is placed in an elliptical tank that is 180 centimeters long and 170 centimeters wide to undergo sound wave lithotripsy treatment for kidney stones. Determine where the sound emitter and the stone should be positioned relative to the center of the tank. (Round to the nearest hundredth of a centimeter.)

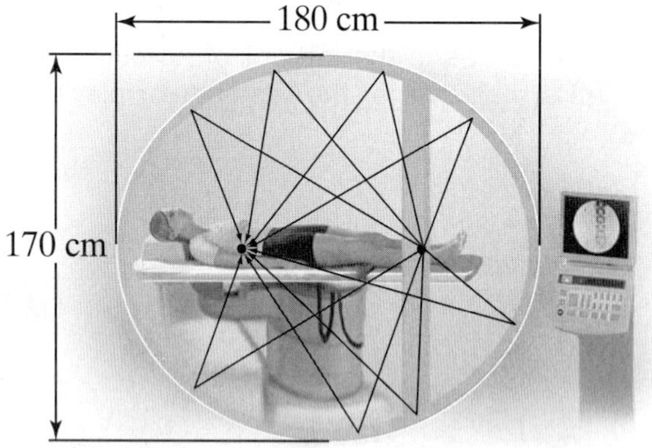

29. A window is constructed with the top half of an ellipse on top of a square. The square portion of the window has a 36-inch base. If the window is 50 inches tall at its highest point, find the height, h, of the window 14 inches from the center of the base. (Round the height to the nearest hundredth of an inch.)

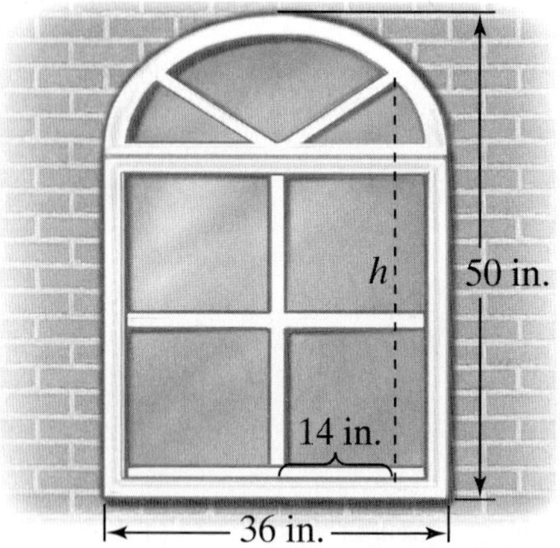

30. A government spy satellite is in an elliptical orbit around the Earth with the center of the Earth at a focus. If the satellite is 150 miles from the surface of the Earth at one vertex of the orbit and 500 miles from the surface of the Earth at the other vertex, find the equation of the elliptical orbit. (Assume that the Earth is a sphere with a diameter of 8000 miles.)

31. An elliptical arch bridge spans 160 feet. The elliptical arch has a maximum height of 40 feet. What is the height of the arch at a distance of 10 feet from the center?

21.3 The Hyperbola

THINGS TO KNOW

Before working through this topic, be sure you are familiar with the following concepts:

		VIDEO	ANIMATION	INTERACTIVE
You Try It	1. Write the Equation of a Line Given Its Slope and y-intercept (Topic 12.4, **Objective 3**)			
You Try It	2. Write the Equation of a Line Given Its Slope and a Point on the Line (Topic 12.4, **Objective 4**)			✓
You Try It	3. Write the Equation of a Line Given Two Points (Topic 12.4, **Objective 5**)	✓		
You Try It	4. Find the Distance Between Two Points (Topic 19.4, **Objective 1**)			
You Try It	5. Complete the Square to Find the Equation of an Ellipse in Standard Form (Topic 21.2, **Objective 3**)	✓		

OBJECTIVES

1 Sketch the Graph of a Hyperbola

2 Find the Equation of a Hyperbola in Standard Form

3 Complete the Square to Find the Equation of a Hyperbola in Standard Form

4 Solve Applications Involving Hyperbolas

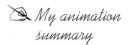

My animation summary

When a plane intersects two **right circular cones** at the same time, the conic section formed is a hyperbola. (Watch this animation.)

The set of points in the plane that defines a hyperbola formed by the intersection of a plane and the cones described previously is stated in the following geometric definition.

Geometric Definition of the Hyperbola

A **hyperbola** is the set of all points in a plane, the difference of whose distances from two fixed points is a positive constant. The two fixed points, F_1 and F_2, are called the foci.

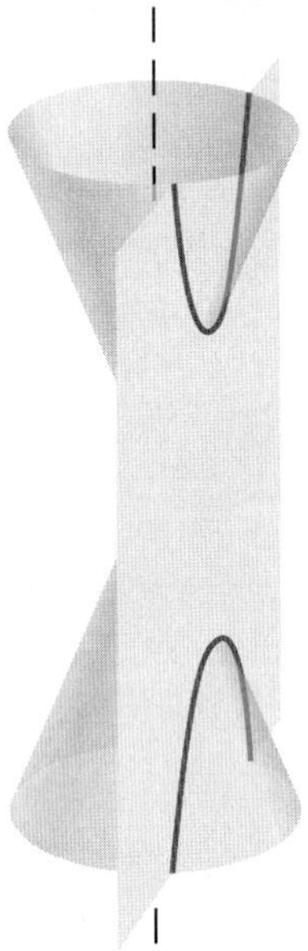

The Hyperbola

Notice that the previous geometric definition is very similar to the **geometric definition of the ellipse**. Recall that for a point to lie on the graph of an ellipse, the **sum** of the distances from the point to the two foci is constant. For a point to lie on the graph of a hyperbola, the **difference** between the distances from the point to the two foci is constant. Because subtraction is not **commutative**, we consider the absolute value of the difference in the distances between a point on the hyperbola and the foci to ensure that the constant is positive. Thus, for any two points P and Q that lie on the graph of a hyperbola, $|d(P, F_1) - d(P, F_2)| = |d(Q, F_1) - d(Q, F_2)|$. See Figure 21.

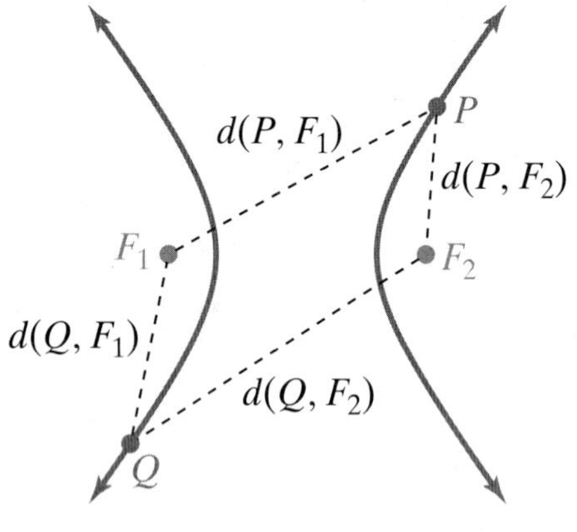

Figure 21
For any points P and Q that lie on the graph of a hyperbola, $|d(P, F_1) - d(P, F_2)| = |d(Q, F_1) - d(Q, F_2)|$.

OBJECTIVE 1 SKETCH THE GRAPH OF A HYPERBOLA

My video summary

The graph of a hyperbola has two branches. These branches look somewhat like parabolas, but they are certainly not because the branches do not satisfy the **geometric definition of the parabola**. Every hyperbola has a center, two vertices, and two foci. The vertices are located at the endpoints of an invisible line segment called the **transverse axis.** The transverse axis is either parallel to the x-axis (horizontal transverse axis) or parallel to the y-axis (vertical transverse axis). The center of a hyperbola is located midway between the two vertices (or two foci). The hyperbola has another invisible line segment called the **conjugate axis** that passes through the center and lies perpendicular to the transverse axis. Each branch of the hyperbola approaches (but never intersects) a pair of lines called *asymptotes*.

(eText Screens 21.3-1–21.3-29)

A **reference rectangle** is typically used as a guide to help sketch the asymptotes. The reference rectangle is a rectangle whose midpoints of each side are the vertices of the hyperbola or the endpoints of the conjugate axis. The asymptotes pass diagonally through opposite corners of the reference rectangle. Watch the **video** to see how to sketch the graphs of the two hyperbolas shown in Figure 22.

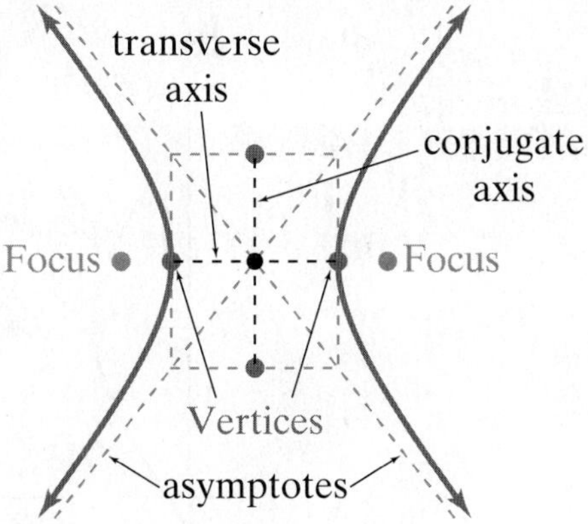

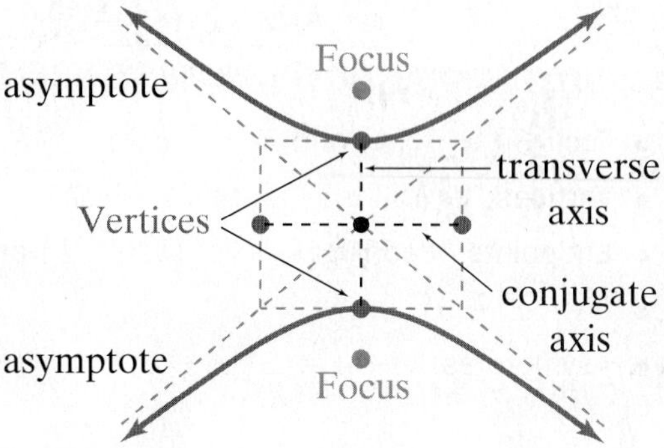

Figure 22 (a) Hyperbola with a horizontal transverse axis

(b) Hyperbola with a vertical transverse axis

To derive the equation of a hyperbola, consider a hyperbola with a horizontal transverse axis centered at (h, k). If the distance between the center and either vertex is $a > 0$, then the coordinates of the vertices are $V_1(h - a, k)$ and $V_2(h + a, k)$, and the length of the transverse axis is equal to $2a$. If $c > 0$ is the distance between the center and either foci, then the foci have coordinates $F_1(h - c, k)$ and $F_2(h + c, k)$. See Figure 23.

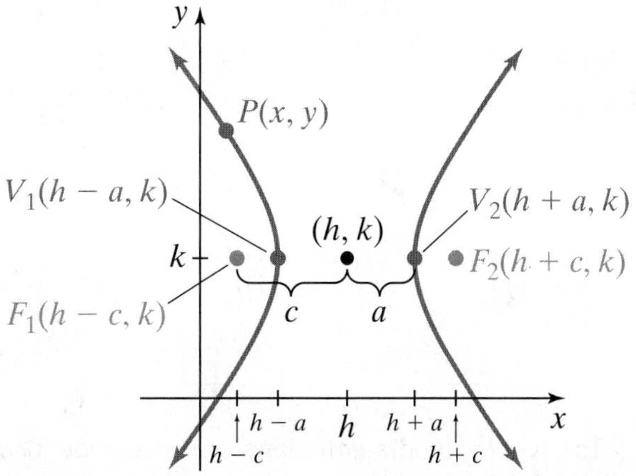

Figure 23

By the geometric definition of a hyperbola, we know that for any point $P(x, y)$ that lies on the hyperbola, the difference of the distances from P to the two foci is a constant. It can be shown that this constant is equal to $2a$, the length of the transverse axis. We now state this fact and denote it as Fact 1 for Hyperbolas.

> **Fact 1 for Hyperbolas**
>
> Given the foci of a hyperbola F_1 and F_2 and any point P that lies on the graph of the hyperbola, the difference of the distance between P and the foci is equal to $2a$. In other words, $|d(P, F_1) - d(P, F_2)| = 2a$. The constant $2a$ represents the length of the transverse axis.

(eText Screens 21.3-1–21.3-29)

Once we know that the constant stated in the geometric definition is equal to $2a$, we can derive the equation of a hyperbola. Click **here** to see this derivation. We now state the two standard equations of hyperbolas.

Equation of a Hyperbola in Standard Form with Center (h, k)

Horizontal Transverse Axis

$$\frac{(x-h)^2}{a^2} - \frac{(y-k)^2}{b^2} = 1$$

- Foci: $F_1(h-c, k)$ and $F_2(h+c, k)$
- Vertices: $V_1(h-a, k)$ and $V_2(h+a, k)$
- Endpoints of conjugate axis: $(h, k-b)$ and $(h, k+b)$
- $b^2 = c^2 - a^2$
- Asymptotes: $y - k = \pm\frac{b}{a}(x-h)$

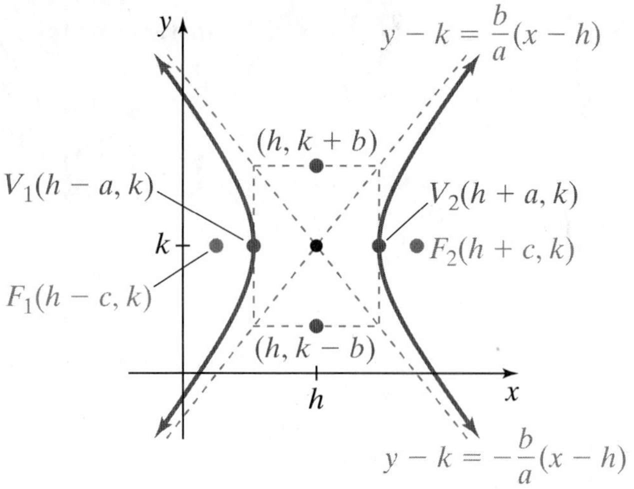

Equation of a Hyperbola in Standard Form with Center (h, k)

Vertical Transverse Axis

$$\frac{(y-k)^2}{a^2} - \frac{(x-h)^2}{b^2} = 1$$

- Foci: $F_1(h, k-c)$ and $F_2(h, k+c)$
- Vertices: $V_1(h, k-a)$ and $V_2(h, k+a)$
- Endpoints of conjugate axis: $(h-b, k)$ and $(h+b, k)$
- $b^2 = c^2 - a^2$
- Asymptotes: $y - k = \pm\frac{a}{b}(x-h)$

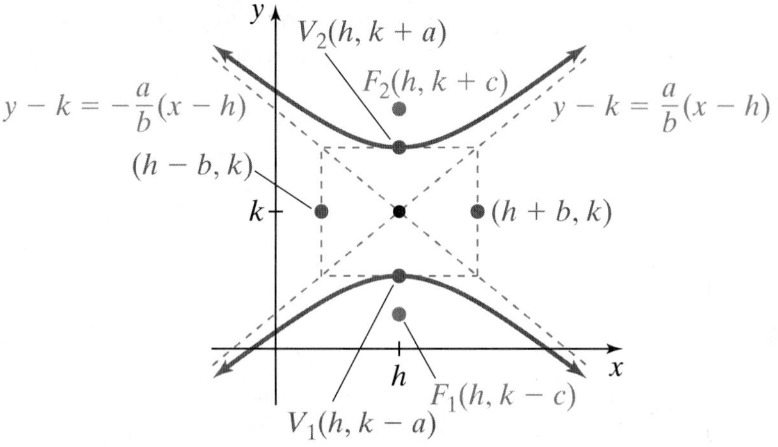

The two hyperbola equations are nearly identical except for two differences. The x-terms are positive and the y-terms are negative in the equation with a horizontal transverse axis. The signs are reversed in the equation with a vertical transverse axis. Also note that a^2 appears in the denominator of the positive squared term in each equation. We can derive the equations of the asymptotes by using the **point-slope form of the equation of a line.**

Note: If $h = 0$ and $k = 0$, then the hyperbola is centered at the **origin.** Hyperbolas centered at the origin have the following equations.

Standard Equations of a Hyperbola with the Center at the Origin

$$\frac{x^2}{a^2} - \frac{y^2}{b^2} = 1 \qquad \frac{y^2}{a^2} - \frac{x^2}{b^2} = 1$$

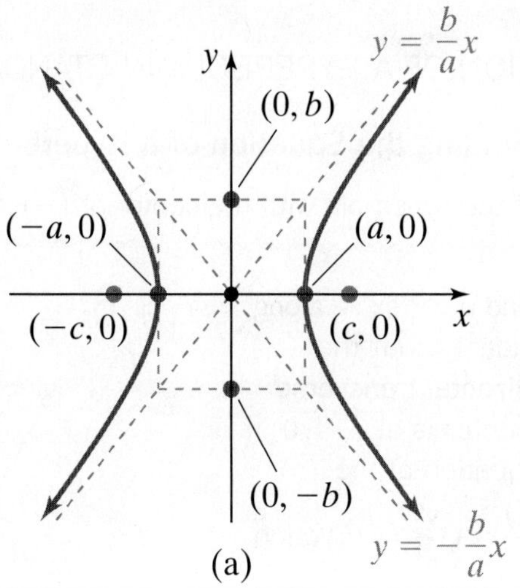

(a)

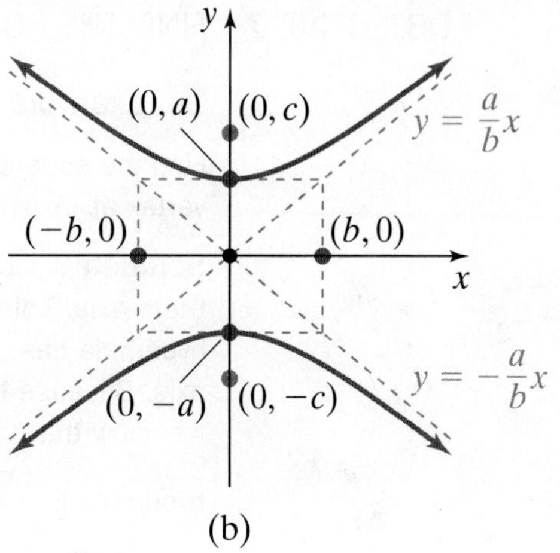

(b)

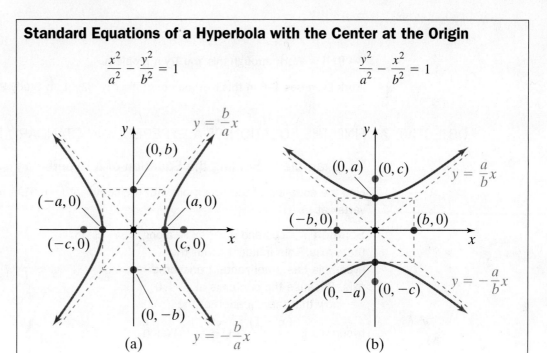

Example 1 Sketching the Graph of a Hyperbola in Standard Form

My interactive video summary

Sketch the following hyperbolas. Determine the center, transverse axis, vertices, and foci, and find the equations of the asymptotes.

a. $\frac{(y-4)^2}{36} - \frac{(x+5)^2}{9} = 1$ **b.** $25x^2 - 16y^2 = 400$

Solution Watch the **interactive video** to see how to sketch each hyperbola shown in Figure 24(a) and 24(b).

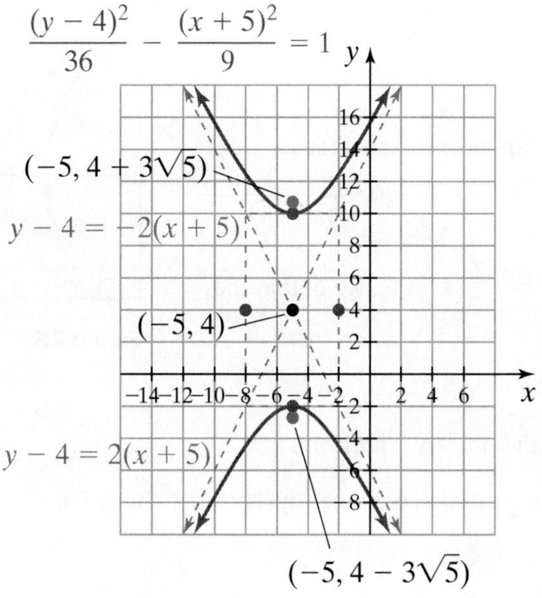

Figure 24(a)

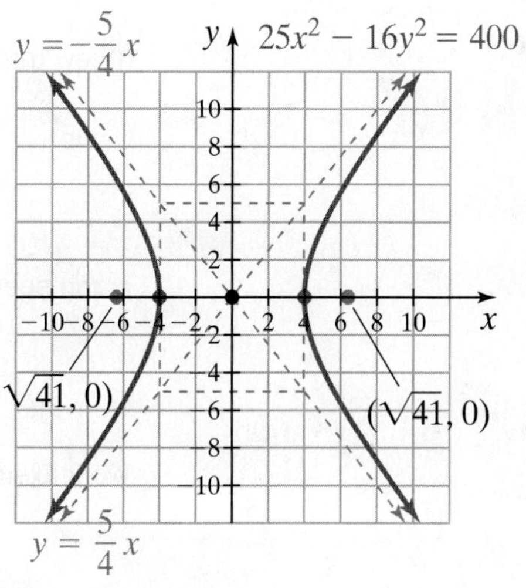

Figure 24(b)

(eText Screens 21.3-1–21.3-29)

You Try It Work through this You Try It problem.

Work Exercises 1–6 in this textbook or in the MyMathLab Study Plan.

OBJECTIVE 2 FIND THE EQUATION OF A HYPERBOLA IN STANDARD FORM

My video summary

Example 2 Finding the Equation of a Hyperbola

Find the equation of the hyperbola with the center at $(-1, 0)$, a focus at $(-11, 0)$, and a vertex at $(5, 0)$.

Solution A focus and a vertex lie along the x-axis. This indicates that the hyperbola has a horizontal transverse axis. Because the center is at $(-1, 0)$, we know that the equation of the hyperbola is $\frac{(x+1)^2}{a^2} - \frac{y^2}{b^2} = 1$. Watch the **video** to verify that the equation in standard form is $\frac{(x+1)^2}{36} - \frac{y^2}{64} = 1$. The equations of the asymptotes are $y = -\frac{4}{3}(x+1)$ and $y = \frac{4}{3}(x+1)$. The graph of this hyperbola is shown in Figure 25.

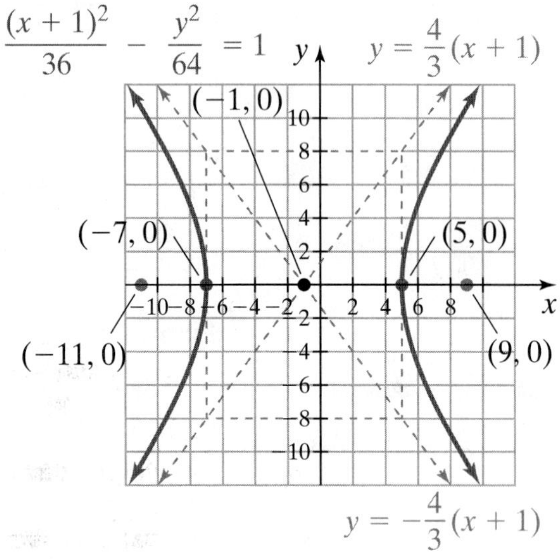

Figure 25

We can use a graphing utility to graph the hyperbola from Example 2 by solving the equation for y:

$$y = \pm\frac{4}{3}\sqrt{(x+1)^2 - 36}$$

((View this **popup** box to see how to solve for y.)

Using $y_1 = \frac{4}{3}\sqrt{(x+1)^2 - 36}$, $y_2 = -\frac{4}{3}\sqrt{(x+1)^2 - 36}$, $y_3 = -\frac{4}{3}(x+1)$, and $y_4 = -\frac{4}{3}(x+1)$, we obtain the graph seen in Figure 26.

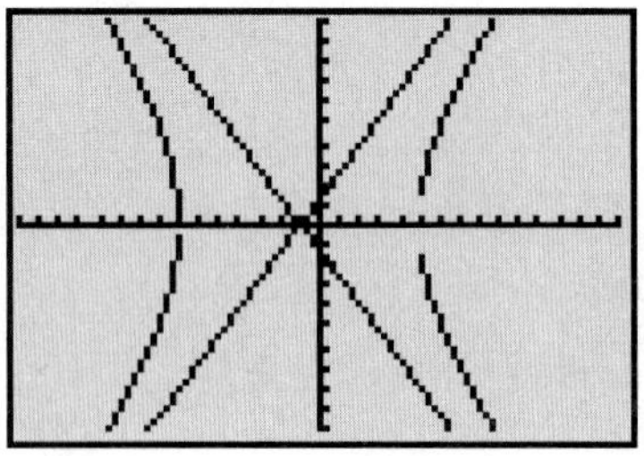

Figure 26

You Try It Work through this You Try It problem.

Work Exercises 7–14 in this textbook or in the MyMathLab Study Plan.

(eText Screens 21.3-1–21.3-29)

OBJECTIVE 3 COMPLETE THE SQUARE TO FIND THE EQUATION OF A HYPERBOLA IN STANDARD FORM

If the equation of a hyperbola is in the standard form of $\dfrac{(x-h)^2}{a^2} - \dfrac{(y-k)^2}{b^2} = 1$ or $\dfrac{(y-k)^2}{a^2} - \dfrac{(x-h)^2}{b^2} = 1$, then it is not too difficult to determine the center, vertices, foci, and asymptotes and sketch its graph. However, the equation might not be given in standard form. If this is the case, we may need to complete the square on both variables as in Example 3.

Example 3 Writing the Equation of a Hyperbola in Standard Form by Completing the Square

My video summary

Find the center, vertices, foci, and equations of asymptotes and sketch the hyperbola $12x^2 - 4y^2 - 72x - 16y + 140 = 0$.

Solution Rearrange the terms leaving some room to complete the square, and move any constants to the right-hand side:

Rearrange the terms: $12x^2 - 72x \qquad - 4y^2 - 16y \qquad = -140$

Then factor and complete the square.

Factor out 12 and -4: $12(x^2 - 6x \quad) - 4(y^2 + 4y \quad) = -140$

Complete the square on x and y. Remember to add $12 \cdot 9 = 108$ and $-4 \cdot 4 = -16$ to the right side: $12(x^2 - 6x + 9) - 4(y^2 + 4y + 4) = -140 + 108 - 16$

Factor the left side, and simplify the right side: $12(x-3)^2 - 4(y+2)^2 = -48$

Divide both sides by -48: $\dfrac{12(x-3)^2}{-48} - \dfrac{4(y+2)^2}{-48} = \dfrac{-48}{-48}$

Simplify: $-\dfrac{(x-3)^2}{4} + \dfrac{(y+2)^2}{12} = 1$

Rewrite the equation: $\dfrac{(y+2)^2}{12} - \dfrac{(x-3)^2}{4} = 1$

The equation is now in standard form. Watch the **video** to see this example worked out in detail. You should verify that this is the equation of a hyperbola with a vertical transverse axis with center $(3, -2)$. The vertices have coordinates $(3, -2 - 2\sqrt{3})$ and $(3, -2 + 2\sqrt{3})$. The foci have coordinates $(3, -6)$ and $(3, 2)$. The equations of the asymptotes are $y + 2 = -\sqrt{3}(x - 3)$ and $y + 2 = \sqrt{3}(x - 3)$.

You Try It **Work through this You Try It problem.**

Work Exercises 15–20 in this textbook or in the MyMathLab Study Plan.

OBJECTIVE 4 SOLVE APPLICATIONS INVOLVING HYPERBOLAS

Hyperbolas have many applications. We saw in **Topic 21.2** that the planets in our solar system and some comets, such as Halley's comet, travel through our solar system in elliptical orbits. However, some comets are seen only once in our solar system because they travel through the solar system on the **path of a hyperbola** with the Sun at a focus. On October 14, 1947, Chuck Yeager became the first person to break the sound barrier. As an airplane moves faster than the speed of sound, a **cone-shaped shock wave** is produced. The cone intersects the ground in the shape of a hyperbola. When two rocks are simultaneously tossed into a calm pool of water, ripples move outward in the form of **concentric circles**. These circles intersect in points that form a hyperbola. Hyperbolas can be used to locate ships by sending radio signals simultaneously from radio transmitters placed at some fixed distance apart. A device measures the difference in the time it takes the radio signals to reach the ship. The equation of a hyperbola can then be determined to describe the current path of the ship. If three transmitters are used, two hyperbolic equations can be determined. The precise location of the ship can be determined by finding the intersection of the two hyperbolas. This system of locating ships is known as long-range navigation or LORAN. See **Example 4**.

Example 4 Use a Hyperbola to Locate a Ship

One transmitting station is located 100 miles due east from another transmitting station. Each station simultaneously sends out a radio signal. The signal from the west tower is received by a ship $\frac{1600}{3}$ microseconds after the signal from the east tower. If the radio signal travels at 0.18 miles per microsecond, find the equation of the hyperbola on which the ship is presently located.

Solution Start by plotting the two foci of the hyperbola at points $F_1(-50, 0)$ and $F_2(50, 0)$. These two points represent the position of the two transmitting towers. Note that $c = 50$. Because the hyperbola is centered at the origin with a horizontal transverse axis, the equation must be of the form $\frac{x^2}{a^2} - \frac{y^2}{b^2} = 1$.

The difference in the distances from the two transmitters to the ship is $\left(\frac{1600}{3}\text{ microseconds}\right) \cdot \left(0.18\,\frac{\text{miles}}{\text{microsecond}}\right) = 96$ miles. This distance represents the constant stated in the **Fact 1 for Hyperbolas**.

Therefore, $2a = 96$, so $a = 48$ or $a^2 = 2304$. To find b^2, we use the fact that $b^2 = c^2 - a^2$.

$$b^2 = c^2 - a^2$$
$$b^2 = 50^2 - 48^2$$
$$b^2 = 196$$

We now substitute the values of $a^2 = 2304$ and $b^2 = 196$ into the previous equation to obtain the equation $\frac{x^2}{2304} - \frac{y^2}{196} = 1$.

You Try It **Work through this You Try It problem.**

Work Exercises 21–24 in this textbook or in the MyMathLab Study Plan.

21.3 Exercises

In Exercises 1–6, determine the center, transverse axis, vertices, foci, and the equations of the asymptotes and sketch the hyperbola.

You Try It

1. $\frac{x^2}{16} - \frac{y^2}{9} = 1$

2. $\frac{y^2}{9} - \frac{x^2}{16} = 1$

3. $\frac{(y-4)^2}{25} - \frac{(x-2)^2}{36} = 1$

4. $\frac{(x+1)^2}{9} - \frac{(y+3)^2}{49} = 1$

5. $20x^2 - 5y^2 = 100$

6. $20(x-1)^2 - 16(y-3)^2 = -320$

In Exercises 7–12, determine the standard equation of the hyperbola with the given characteristics and sketch the graph.

You Try It

7. The center is at $(0, 0)$, a focus is at $(5, 0)$, and a vertex is at $(3, 0)$.

8. The center is at $(0, 0)$, a focus is at $(0, 10)$, and a vertex is at $(0, -6)$.

9. The center is at $(4, -4)$, a focus is at $(6, -4)$, and a vertex is at $(5, -4)$.

10. The center is at $(-6, -1)$, a focus is at $(-6, -9)$, and a vertex is at $(-6, -5)$.

11. The foci are at $(9, 3)$ and $(9, 9)$; the vertex is at $(9, 8)$.

12. The vertices are at $(-2, -3)$ and $(10, -3)$; an asymptote has equation $y + 3 = \frac{7}{6}(x - 4)$.

In Exercises 13 and 14, determine the equation of the hyperbola with the given characteristics and sketch the graph. *Hint:* $|d(P, F_1) - d(P, F_2)| = 2a$.

13. The hyperbola has foci with coordinates $F_1(0, -6)$ and $F_2(0, 6)$ and passes through the point $P(8, 10)$.

14. The hyperbola has foci with coordinates $F_1(-6, 1)$ and $F_2(10, 1)$ and passes through the point $P(10, 13)$.

In Exercises 15–20, complete the square to write each equation in the form $\frac{(x-h)^2}{a^2} - \frac{(y-k)^2}{b^2} = 1$ or $\frac{(y-k)^2}{a^2} - \frac{(x-h)^2}{b^2} = 1$. Determine the center, vertices, foci, endpoints of the conjugate axis, and the equations of the asymptotes of the hyperbola, and sketch its graph.

You Try It

15. $x^2 - y^2 - 4x + 2y - 1 = 0$

16. $x^2 - y^2 + 8x - 6y + 9 = 0$

17. $y^2 - 9x^2 - 12y - 36x - 9 = 0$

18. $x^2 - 16y^2 + 10x + 64y - 55 = 0$

19. $25y^2 - 144x^2 + 1728x + 400y + 16 = 0$

20. $49y^2 - 576x^2 - 98y - 1152x - 28{,}751 = 0$

(eText Screens 21.3-1–21.3-29)

You Try It

21. A light on a wall produces a shadow in the shape of a hyperbola. If the distance between the two vertices of the hyperbola is 14 inches and the distance between the two foci is 16 inches, find the equation of the hyperbola.

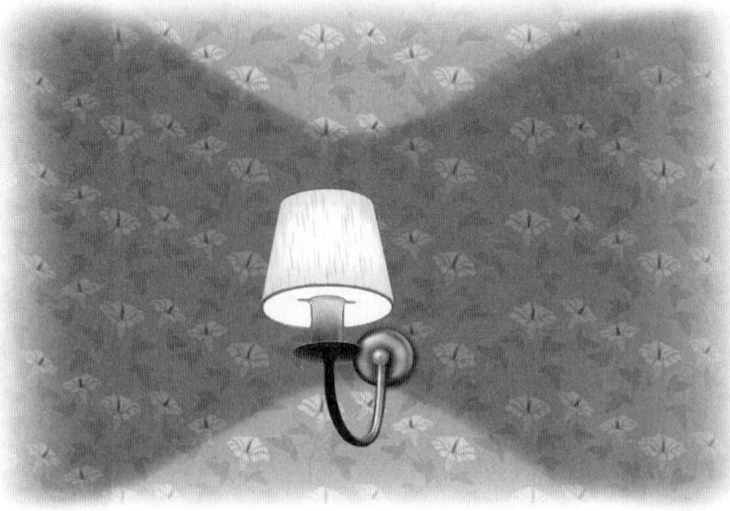

22. This figure shows the hyperbolic orbit of a comet with the center of the Sun positioned at a focus, 100 million miles from the origin. (The units are in millions of miles.) The comet will be 50 million miles from the center of the Sun at its nearest point during the orbit. Find the equation of the hyperbola describing the comet's orbit.

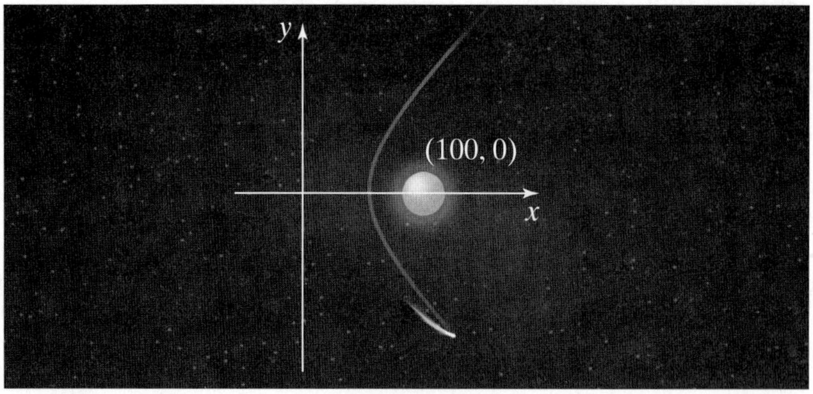

23. A nuclear power plant has a large cooling tower with sides curved in the shape of a hyperbola. The radius of the base of the tower is 60 meters. The radius of the top of the tower is 50 meters. The sides of the tower are 60 meters apart at the closest point located 90 meters above the ground.

a. Find the equation of the hyperbola that describes the sides of the cooling tower. (Assume that the center is at the origin.)

b. Determine the height of the tower. (Round your answer to the nearest meter.)

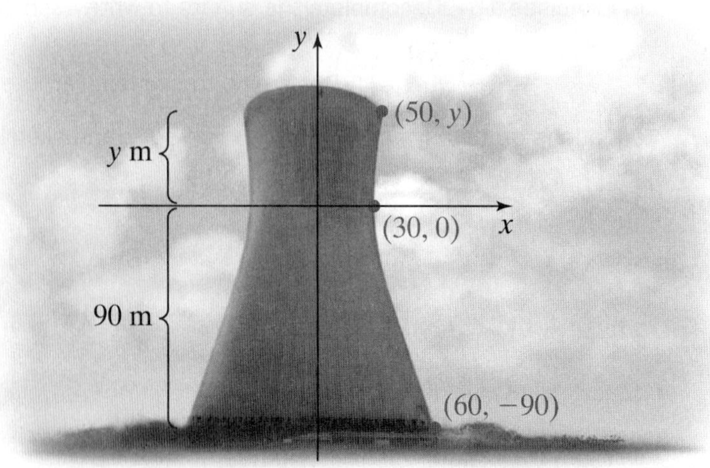

24. One transmitting station is located 80 miles north of another transmitting station. Each station simultaneously sends out a radio signal. The signal from the north station is received 200 microseconds after the signal from the south station. If the radio signal travels at 0.18 miles per microsecond, find the equation of the hyperbola on which the ship is presently located.

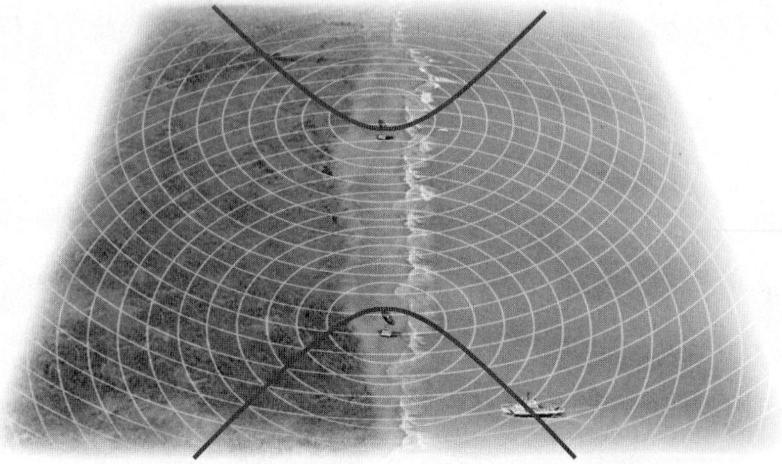

MODULE TWENTY-TWO

Sequences and Series

MODULE TWENTY-TWO CONTENTS

22.1 Introduction to Sequences and Series

THINGS TO KNOW

Before working through this topic, be sure you are familiar with the following concept:

		VIDEO	ANIMATION	INTERACTIVE
You Try It	1. Determine If Graphs Are Functions (Topic 17.1, **Objective 4**)			

OBJECTIVES

1 Write the Terms of a Sequence

2 Write the Terms of a Recursive Sequence

3 Write the General Term for a Given Sequence

4 Compute Partial Sums of a Series

5 Determine the Sum of a Finite Series Written in Summation Notation

6 Write a Series Using Summation Notation

OBJECTIVE 1 WRITE THE TERMS OF A SEQUENCE

Consider the function $f(n) = 2n - 1$, where n is a **natural number**. The graph of this function consists of infinitely many ordered pairs of the form $(n, 2n - 1)$, where $n \geq 1$. Therefore, the ordered pairs that lie on the graph of this function are $(1, 1), (2, 3), (3, 5), (4, 7)$, and so on. A portion of the graph of this function can be seen in Figure 1.

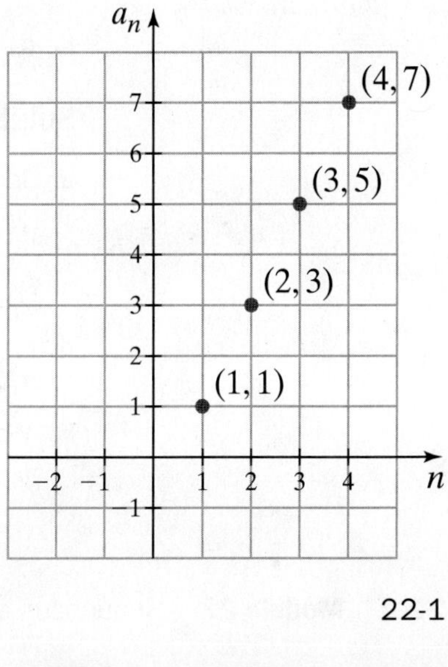

Figure 1
A portion of the graph of the function $f(n) = 2n - 1$, where n is a natural number.

Clearly, the graph seen in **Figure 1** is a function because the graph passes the **vertical line test**. Any function whose domain is the set of natural numbers is called an **infinite sequence**. Instead of using the conventional function notation $f(n)$ to name a sequence, we will use a subscript notation, such as a_n (read as "a sub n"), to name our sequences. We now formally define a sequence.

Definition Sequence

A **finite sequence** is a function whose domain is the finite set $\{1, 2, 3, \ldots, n\}$, where n is a **natural number**.

An **infinite sequence** is a function whose domain is the set of all natural numbers.

The range values of a sequence are called the **terms** of the sequence.

We now rename the sequence $f(n) = 2n - 1$ as $a_n = 2n - 1$. The first four terms of this sequence are $a_1 = 1, a_2 = 3, a_3 = 5$, and $a_4 = 7$.

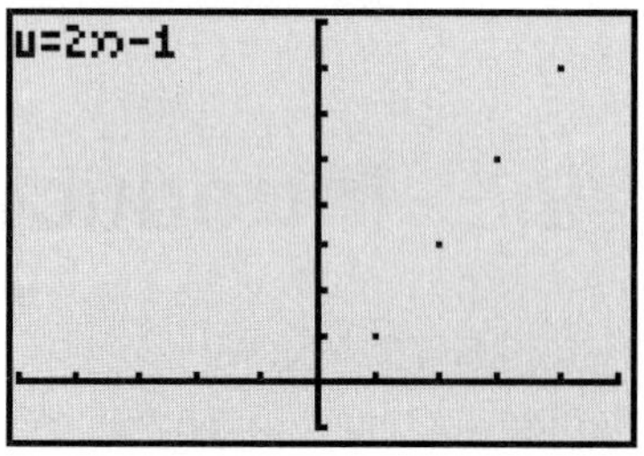

A graphing utility set to *sequence mode* can be used to sketch the graph of a sequence. The figure on the left shows a portion of the graph of the sequence $a_n = 2n - 1$.

Before we start to find the terms of a sequence it is important to introduce factorial notation.

Definition The Factorial of a Non-Negative Integer

The **factorial** of a non-negative integer n, denoted as $n!$, is the product of all positive integers less than or equal to n. Thus, $n! = n(n-1) \cdot \cdots \cdot 3 \cdot 2 \cdot 1$.

Note: By definition, we say that zero factorial is equal to 1 or $0! = 1$.

Some examples of factorial notation are $5! = 5 \cdot 4 \cdot 3 \cdot 2 \cdot 1 = 120$ and $8! = 8 \cdot 7 \cdot 6 \cdot 5 \cdot 4 \cdot 3 \cdot 2 \cdot 1 = 40{,}320$. As you can see, the value of $n!$ gets large rather quickly. In fact, the value of $13!$ is over 6 billion, which is a rough estimate of the population of Earth! Example 1c illustrates how factorial notation can be used to define a sequence.

Example 1 Writing the Terms of a Sequence

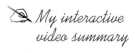

My interactive video summary

Write the first four terms of each sequence whose nth term is given.

a. $a_n = 2n - 1$ **b.** $b_n = n^2 - 1$ **c.** $c_n = \dfrac{3^n}{(n-1)!}$ **d.** $d_n = (-1)^n 2^{n-1}$

Solutions

a. To find the first four terms of the sequence, we evaluate $a_n = 2n - 1$ when n is $1, 2, 3,$ and 4.

$$a_1 = 2(1) - 1 = 2 - 1 = 1, a_2 = 2(2) - 1 = 4 - 1 = 3,$$
$$a_3 = 2(3) - 1 = 6 - 1 = 5, \text{ and } a_4 = 2(4) - 1 = 8 - 1 = 7$$

Therefore, the first four terms of the sequence $a_n = 2n - 1$ are $1, 3, 5,$ and 7.

b. Work through the **interactive video** to verify that the first four terms of the sequence $b_n = n^2 - 1$ are $0, 3, 8,$ and 15.

c. To find the first four terms of the sequence $c_n = \dfrac{3^n}{(n-1)!}$, we evaluate c_n when n is $1, 2, 3$, and 4.

$$c_1 = \frac{3^1}{(1-1)!} = \frac{3}{0!} = \frac{3}{1} = 3, \qquad c_2 = \frac{3^2}{(2-1)!} = \frac{9}{1!} = \frac{9}{1} = 9,$$

$$c_3 = \frac{3^3}{(3-1)!} = \frac{27}{2!} = \frac{27}{2}, \quad \text{and} \quad c_4 = \frac{3^4}{(4-1)!} = \frac{81}{3!} = \frac{81}{6} = \frac{27}{2}$$

Therefore, the first four terms of the sequence $c_n = \dfrac{3^n}{(n-1)!}$ are

$$3, 9, \tfrac{27}{2}, \text{and } \tfrac{27}{2}.$$

d. Work through the **interactive video** to verify that the first four terms of the sequence $d_n = (-1)^n 2^{n-1}$ are $-1, 2, -4$, and 8.

Note: The sequence $d_n = (-1)^n 2^{n-1}$ is an example of an **alternating sequence** because the successive terms alternate in sign.

You Try It **Work through this You Try It problem.**

Work Exercises 1–10 in this textbook or in the MyMathLab Study Plan.

OBJECTIVE 2 WRITE THE TERMS OF A RECURSIVE SEQUENCE

Some sequences are defined recursively. A **recursive sequence** is a sequence in which each term is defined using one or more of its previous terms. Typically, the first term of a recursive sequence is given, followed by the formula for the nth term of the sequence. The following example illustrates two recursive sequences.

My interactive video summary

Example 2 Writing the Terms of a Recursive Sequence

Write the first four terms of each of the following recursive sequences.

a. $a_1 = -3, a_n = 5a_{n-1} - 1$ for $n \geq 2$

b. $b_1 = 2, b_n = \dfrac{(-1)^{n-1} n}{b_{n-1}}$ for $n \geq 2$

Solutions

a. The first four terms of this recursive sequence are $-3, -16, -81$, and -406. Work through this **interactive video** to verify.

b. The first four terms of this recursive sequence are $2, -1, -3$, and $\frac{4}{3}$. Work through this **interactive video** to verify.

Arguably the most famous recursively defined sequence is the **Fibonacci sequence** named after the 13th-century Italian mathematician **Leonardo of Pisa**, also known as Fibonacci. The Fibonacci sequence is defined in Example 3.

Example 3 Writing the Terms of the Fibonacci Sequence

The Fibonacci sequence is defined recursively by $a_n = a_{n-1} + a_{n-2}$, where $a_1 = 1$ and $a_2 = 1$. Write the first eight terms of the Fibonacci sequence.

Solution We are given that $a_1 = 1$ and $a_2 = 1$. We use the recursive formula $a_n = a_{n-1} + a_{n-2}$ to find the next six terms starting with $n = 3$.

$$a_3 = a_2 + a_1 = 1 + 1 = 2$$
$$a_4 = a_3 + a_2 = 2 + 1 = 3$$
$$a_5 = a_4 + a_3 = 3 + 2 = 5$$
$$a_6 = a_5 + a_4 = 5 + 3 = 8$$
$$a_7 = a_6 + a_5 = 8 + 5 = 13$$
$$a_8 = a_7 + a_6 = 13 + 8 = 21$$

You can see in Example 3 that given the first two terms, each term of the Fibonacci sequence is the sum of the preceding two terms. We now write the first 12 terms of the Fibonacci sequence.

$$1, 1, 2, 3, 5, 8, 13, 21, 34, 55, 89, 144, \ldots$$

The numbers of this sequence are known as **Fibonacci numbers**. The Fibonacci sequence and Fibonacci numbers occur in many natural phenomena such as the spiral formation of seeds of various plants, the number of petals of a flower, and the formation of the branches of a tree. See **Exercise 44**.

You Try It **Work through this You Try It problem.**

Work Exercises 11–14 in this textbook or in the MyMathLab Study Plan.

OBJECTIVE 3 WRITE THE GENERAL TERM FOR A GIVEN SEQUENCE

Sometimes the first several terms of a sequence are given without listing the nth term. When this occurs, we must try to determine a pattern and use deductive reasoning to establish a rule that describes the general term, or nth term, of the sequence. Example 4 illustrates two such sequences.

Example 4 Finding the General Term of a Sequence

My video summary

Write a formula for the nth term of each infinite sequence, then use this formula to find the eighth term of the sequence.

a. $\frac{1}{1}, \frac{1}{2}, \frac{1}{3}, \frac{1}{4}, \frac{1}{5}, \ldots$ **b.** $-\frac{2}{1}, \frac{4}{2}, -\frac{8}{6}, \frac{16}{24}, -\frac{32}{120}, \ldots$

Solutions

a. The nth term of the sequence is $a_n = \frac{1}{n}$. Thus, the eighth term of this sequence is $a_8 = \frac{1}{8}$.

b. For this sequence, notice that the first term is negative and that terms alternate in sign. We can therefore represent the sign of each term as $(-1)^n$. Also, notice that the numerators are successive powers of 2. We now have the following pattern:

$$\begin{array}{ccccc} a_1 & a_2 & a_3 & a_4 & a_5 \\ \downarrow & \downarrow & \downarrow & \downarrow & \downarrow \\ -\frac{2}{1}, & \frac{4}{2}, & -\frac{8}{6}, & \frac{16}{24}, & -\frac{32}{120}, \ldots \\ \downarrow & \downarrow & \downarrow & \downarrow & \downarrow \\ \frac{(-1)^1 2^1}{1}, & \frac{(-1)^2 2^2}{2}, & \frac{(-1)^3 2^3}{6}, & \frac{(-1)^4 2^4}{24}, & \frac{(-1)^5 2^5}{120}, \ldots \end{array}$$

Finally, if we factor each successive denominator we get $1 = 1, 2 = 2 \cdot 1, 6 = 3 \cdot 2 \cdot 1, 24 = 4 \cdot 3 \cdot 2 \cdot 1$, and $120 = 5 \cdot 4 \cdot 3 \cdot 2 \cdot 1$. This suggests that the denominator of the nth term can be represented by $n!$

Therefore, the nth term of the sequence is $a_n = \dfrac{(-1)^n 2^n}{n!}$. The eighth term of this sequence is

$$a_8 = \frac{(-1)^8 2^8}{8!} = \frac{256}{40{,}320} = \frac{2}{315}.$$

If you would like to see this solution worked out in detail, watch this **video**.

You Try It **Work through this You Try It problem.**

Work Exercises 15–22 in this textbook or in the MyMathLab Study Plan.

OBJECTIVE 4 COMPUTE PARTIAL SUMS OF A SERIES

Suppose that we wanted to find the sum of the first four terms of the sequence $a_n = 2n - 1$. From Example 1 we saw that the first four terms of this sequence were $a_1 = 1, a_2 = 3, a_3 = 5$, and $a_4 = 7$. Therefore, the sum of the first four terms is $a_1 + a_2 + a_3 + a_4 = 1 + 3 + 5 + 7 = 16$. The expression $1 + 3 + 5 + 7$ is called a **series**.

> **Definition Series**
>
> Let $a_1, a_2, a_3, \ldots$ be a sequence. The expression of the form $a_1 + a_2 + a_3 + \cdots + a_n$ is called a **finite series.**
>
> The expression of the form $a_1 + a_2 + a_3 + \cdots + a_n + a_{n+1} + \cdots$ is called an **infinite series.**
>
> The sum of the first n terms of a series is called the nth **partial sum** of the series and is denoted as S_n.

For the series $1 + 3 + 5 + 7 + 9 + \cdots + 2n - 1$, the first five partial sums are as follows:

$$S_1 = 1$$
$$S_2 = 1 + 3 = 4$$
$$S_3 = 1 + 3 + 5 = 9$$
$$S_4 = 1 + 3 + 5 + 7 = 16$$
$$S_5 = 1 + 3 + 5 + 7 + 9 = 25$$

It appears that $S_n = n^2$. In fact, it can be shown that for any positive integer n, the sum of the series $1 + 3 + 5 + 7 + 9 + \cdots + 2n - 1$ is equal to n^2.

Example 5 Computing Partial Sums of a Series

Given the general term of each sequence, find the indicated partial sum.

a. $a_n = \dfrac{1}{n}$, find S_3. **b.** $b_n = (-1)^n 2^{n-1}$, find S_5.

Solutions

a. The first three terms are $a_1 = 1, a_2 = \frac{1}{2}$, and $a_3 = \frac{1}{3}$. Therefore, the partial sum, S_3, is $S_3 = 1 + \frac{1}{2} + \frac{1}{3} = \frac{11}{6}$.

b. The first five terms are $b_1 = -1, b_2 = 2, b_3 = -4, b_4 = 8$, and $b_5 = -16$. Therefore, the partial sum, S_5, is $S_5 = -1 + 2 + (-4) + 8 + (-16) = -11$.

You Try It **Work through this You Try It problem.**

Work Exercises 23–28 in this textbook or in the MyMathLab Study Plan.

OBJECTIVE 5 DETERMINE THE SUM OF A FINITE SERIES WRITTEN IN SUMMATION NOTATION

My video summary

Writing out an entire finite series of the form $a_1 + a_2 + a_3 + \cdots + a_n$ can be quite tedious, especially if n is fairly large. Fortunately, there is a convenient way to express a finite series using a short-hand notation called *summation notation* (also called *sigma notation*). This notation involves the use of the uppercase Greek letter sigma, which is written as Σ.

Definition Summation Notation

If $a_1, a_2, a_3, \ldots$ is a sequence, then the finite series $a_1 + a_2 + a_3 + \cdots + a_n$ can be written in **summation notation** as $\sum_{i=1}^{n} a_i$. The infinite series $a_{n+1} + \cdots$ can be written as $\sum_{i=1}^{\infty} a_i$.

The variable i is called the **index of summation.** The number 1 is the **lower limit of summation** and n is the **upper limit of summation.**

The lower limit of summation, $i = 1$, below the sigma tells us which term to start with. The upper limit of summation, n, that appears above the sigma tells us which term of the sequence will be the last term to add. There is nothing special about the letter i to represent the index of summation. We will often use different letters such as j or k. Also, it is not necessary for the lower limit of summation to start at 1. In Examples 6b and 6c, the lower limits of summation are 2 and 0, respectively.

Example 6 Determining the Sum of a Series Written in Summation Notation

My interactive video summary

Find the sum of each finite series.

a. $\sum_{i=1}^{5} i^2$ b. $\sum_{j=2}^{5} \frac{j-1}{j+1}$ c. $\sum_{k=0}^{6} \frac{1}{k!}$

(Round the sum to three decimal places.)

Solutions

a. $\sum_{i=1}^{5} i^2 = 1^2 + 2^2 + 3^2 + 4^2 + 5^2 = 1 + 4 + 9 + 16 + 25 = 55$

b. $$\sum_{j=2}^{5} \frac{j-1}{j+1} = \frac{2-1}{2+1} + \frac{3-1}{3+1} + \frac{4-1}{4+1} + \frac{5-1}{5+1}$$
$$= \frac{1}{3} + \frac{2}{4} + \frac{3}{5} + \frac{4}{6} = \frac{21}{10}$$

c. $\displaystyle\sum_{k=0}^{6}\frac{1}{k!} = \frac{1}{0!} + \frac{1}{1!} + \frac{1}{2!} + \frac{1}{3!} + \frac{1}{4!} + \frac{1}{5!} + \frac{1}{6!}$

$$= 1 + 1 + \frac{1}{2} + \frac{1}{6} + \frac{1}{24} + \frac{1}{120} + \frac{1}{720} = \frac{1957}{720} \approx 2.718$$

Work through this **interactive video** to see these solutions worked out in detail.

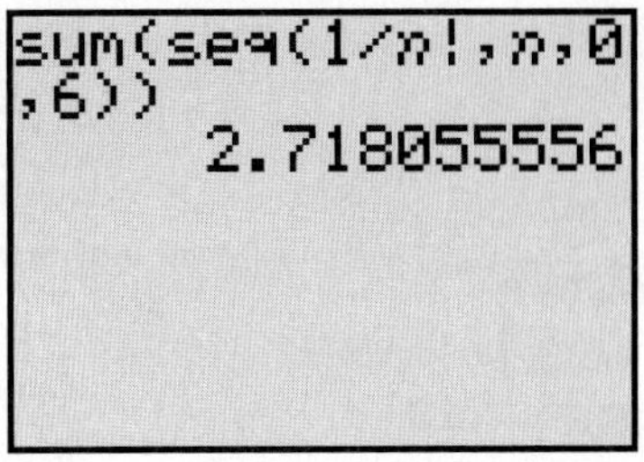

Using a TI-84 Plus, we can calculate the sum obtained in Example 6c. Notice that this number is a good approximation of the **number** e. In fact, it can be shown that the exact value of e is $e = \displaystyle\sum_{n=0}^{\infty}\frac{1}{n!}$.

You Try It **Work through this You Try It problem.**

Work Exercises 29–35 in this textbook or in the MyMathLab Study Plan.

OBJECTIVE 6 WRITE A SERIES USING SUMMATION NOTATION

Given the first several terms of a series, it is important to be able to rewrite the series using summation notation as in Example 7.

Example 7 Writing a Series Using Summation Notation

Rewrite each series using summation notation. Use 1 as the lower limit of summation.

a. $2 + 4 + 6 + 8 + 10 + 12$ **b.** $1 + 2 + 6 + 24 + 120 + 720 + \cdots + 3{,}628{,}800$

Solutions

a. This series is the sum of six terms. Therefore, the lower limit of summation is 1 and the upper limit of summation is 6. Each term is a successive multiple of 2. So, one possible series is $\displaystyle\sum_{i=1}^{6} 2i$.

b. Notice that $1 = 1!, 2 = 2!, 6 = 3!, 24 = 4!, 120 = 5!, 720 = 6!$, and $3{,}628{,}800 = 10!$ Thus, a possible series is $\displaystyle\sum_{n=1}^{10} n!$

You Try It **Work through this You Try It problem.**

Work Exercises 36–44 in this textbook or in the MyMathLab Study Plan.

22.1 Exercises

In Exercises 1–10, write the first four terms of each sequence.

You Try It

1. $a_n = 3n + 1$ **2.** $a_n = 4^n$ **3.** $a_n = \dfrac{4n}{n + 3}$ **4.** $a_n = (-4)^n$

5. $a_n = 5(n + 2)!$ **6.** $a_n = \dfrac{n^3}{(n + 1)!}$ **7.** $a_n = (-1)^n(5n)$

8. $a_n = \dfrac{3^n}{(-1)^{n+1} + 5}$ **9.** $a_n = \dfrac{(-1)^n}{(n + 5)(n + 6)}$ **10.** $a_n = \dfrac{(-1)^n(3)^{2n+1}}{(2n + 1)!}$

In Exercises 11–14, write the first four terms of each recursive sequence.

You Try It

11. $a_1 = 7, a_n = 3 + a_{n-1}$ **for** $n \geq 2$ **12.** $a_1 = -1, a_n = n - a_{n-1}$ for $n \geq 2$

13. $a_1 = 6, a_n = \dfrac{a_{n-1}}{n^2}$ **for** $n \geq 2$ **14.** $a_1 = -4, a_n = 1 - \dfrac{1}{a_{n-1}}$ for $n \geq 2$

In Exercises 15–22, write a formula for the general term, or nth term, for the given sequence. Then find the indicated term.

You Try It

15. $-1, 1, 3, 5, 7, \ldots; a_{11}.$ **16.** $\dfrac{1}{5}, \dfrac{2}{6}, \dfrac{3}{7}, \dfrac{4}{8}, \dfrac{5}{9}, \ldots; a_8.$

17. $1 \cdot 6, 2 \cdot 7, 3 \cdot 8, 4 \cdot 9, \ldots; a_7.$ **18.** $-2, 4, -8, 16, \ldots; a_7.$

19. $\dfrac{2}{5}, \dfrac{2}{25}, \dfrac{2}{125}, \dfrac{2}{625}, \ldots; a_6.$ **20.** $-6, 12, -24, 48, -96, \ldots; a_9.$

21. $-6, 24, -120, 720, \ldots; a_5.$ **22.** $\dfrac{3}{2}, \dfrac{9}{6}, \dfrac{27}{24}, \dfrac{81}{120}, \ldots; a_5.$

In Exercises 23–25, the first several terms of a sequence are given. Find the indicated partial sum.

You Try It

23. $2, 4, 6, 8, 10, \ldots; S_4$ **24.** $3, -6, 9, -12, 15, -18, \ldots; S_9$

25. $\dfrac{1}{2}, -\dfrac{1}{4}, \dfrac{1}{8}, -\dfrac{1}{16}, \ldots; S_5$

In Exercises 26–28, the general term of a sequence is given. Find the indicated partial sum.

26. $a_n = 3n + 8; S_6$ **27.** $a_n = (-1)^n \cdot (4n); S_6$

28. $a_1 = 4, a_n = a_{n-1} - 8$ **for** $n \geq 2; S_8$

In Exercises 29–35, find the sum of each series.

You Try It

29. $\sum_{i=1}^{9} i$ **30.** $\sum_{i=1}^{6} (4i + 3)$ **31.** $\sum_{i=1}^{7} i(i + 2)$ **32.** $\sum_{i=1}^{21} 7$

33. $\sum_{j=0}^{5} (j + 4)^2$ **34.** $\sum_{k=2}^{7} \dfrac{k!}{(k - 2)!}$ **35.** $\sum_{j=0}^{5} (j + 4)^2$

In Exercises 36–43, rewrite each series using summation notation. Use 1 as the lower limit of summation.

You Try It

36. $1 + 2 + 3 + \cdots + 29$

37. $5 + 10 + 15 + \cdots + 50$

38. $1^2 + 2^2 + 3^2 + \cdots + 11^2$

39. $\frac{4}{5} + \frac{5}{6} + \frac{6}{7} + \cdots + \frac{12}{13}$

40. $2 + (-4) + 8 + (-16) + \cdots + (-256)$

41. $-\frac{1}{9} + \frac{1}{18} - \frac{1}{27} + \cdots + \frac{1}{54}$

42. $5 + \frac{5^2}{2} + \frac{5^3}{3} + \cdots + \frac{5^n}{n}$

43. $1 + 7 + \frac{7^2}{2!} + \frac{7^3}{3!} + \frac{7^4}{4!} + \cdots + \frac{7^n}{n!}$

44. The figure below shows the progression of the branching of a tree during each stage of development. Notice that the number of branches formed during a given stage is a Fibonacci number. Assuming that this branching pattern continues, how many branches will form during the 10th stage of development?

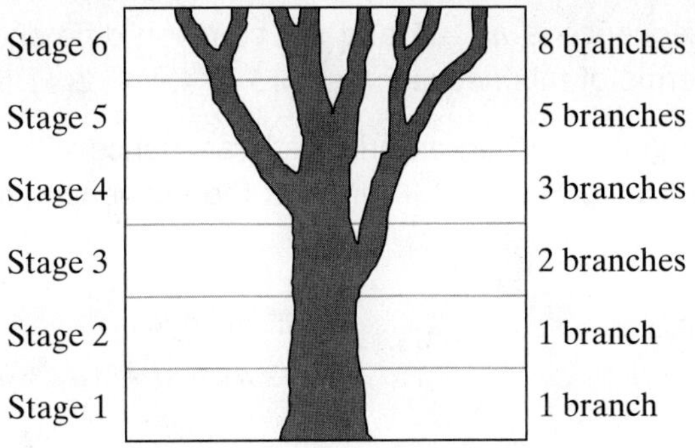

22.2 Arithmetic Sequences and Series

THINGS TO KNOW

Before working through this topic, be sure you are familiar with the following concepts:

VIDEO ANIMATION INTERACTIVE

You Try It

1. Solve Systems of Linear Equations by Substitution (Topic 13.2, **Objective 1**)

You Try It

2. Solve Systems of Linear Equations by Elimination (Topic 13.3, **Objective 1**)

You Try It

3. Determine the Sum of a Finite Series Written in Summation Notation (Topic 22.1, **Objective 5**)

OBJECTIVES

1. Determine If a Sequence Is Arithmetic
2. Find the General Term or a Specific Term of an Arithmetic Sequence
3. Compute the nth Partial Sum of an Arithmetic Series
4. Solve Applications of Arithmetic Sequences and Series

OBJECTIVE 1 DETERMINE IF A SEQUENCE IS ARITHMETIC

In this topic, we will work exclusively with a specific type of sequence known as an **arithmetic sequence**. A sequence is arithmetic if the difference in any two successive terms is constant. For example, the sequence

$$5, 9, 13, 17, \ldots$$

is arithmetic because the difference of any two successive terms is 4. The first term of this sequence is $a_1 = 5$ and the common difference is $d = 4$. Notice that we can rewrite the terms of this sequence as $5, 5 + 4, 5 + 2(4), 5 + 3(4), \ldots$

In general, given an arithmetic sequence with a first term of a_1 and a common difference of d, then the first n terms of the sequence are as follows:

$$a_1$$

$$a_2 = a_1 + d$$

$$a_3 = a_2 + d = \underbrace{(a_1 + d)}_{a_2} + d = a_1 + 2d$$

$$a_4 = a_3 + d = \underbrace{(a_1 + 2d)}_{a_3} + d = a_1 + 3d$$

$$\vdots$$

$$a_n = a_1 + (n - 1)d$$

> **Definition Arithmetic Sequence**
>
> An **arithmetic sequence** is a sequence of the form $a_1, a_1 + d, a_1 + 2d, a_1 + 3d, a_1 + 4d, \ldots,$ where a_1 is the first term of the sequence and d is the common difference. The general term, or nth term, of an arithmetic sequence has the form $a_n = a_1 + (n - 1)d$.

Example 1 Determining If a Sequence Is Arithmetic

My interactive video summary

For each of the following sequences, determine if it is arithmetic. If the sequence is arithmetic, find the common difference.

a. $1, 4, 7, 10, 13, \ldots$

b. $b_n = n^2 - n$

c. $a_n = -2n + 7$

d. $a_1 = 14, a_n = 3 + a_{n-1}$

Solutions Watch this **interactive video** to verify that the sequences in parts (a), (c), and (d) are arithmetic. The sequence in part (b) is not arithmetic. Notice that the arithmetic sequence in part (d) is a **recursive sequence**.

It is worth noting that every arithmetic sequence is a linear function whose domain is the natural numbers. A portion of the graphs of the arithmetic sequences from Example 1a and Example 1c are seen in **Figure 2**. Notice that the ordered pairs of each sequence are **collinear**.

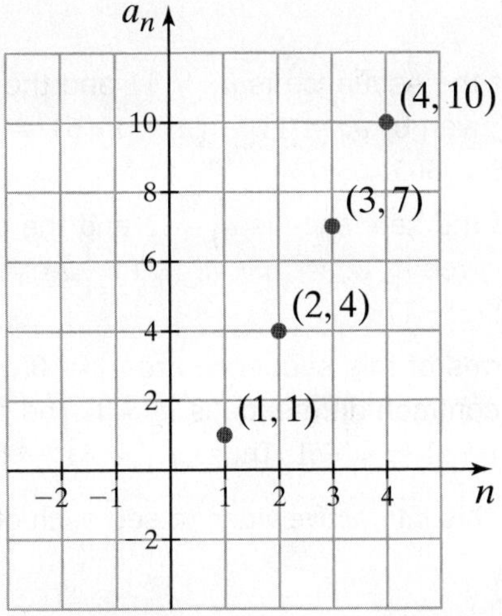

(a) A portion of the graph of the sequence $1, 4, 7, 10, 13, \ldots$

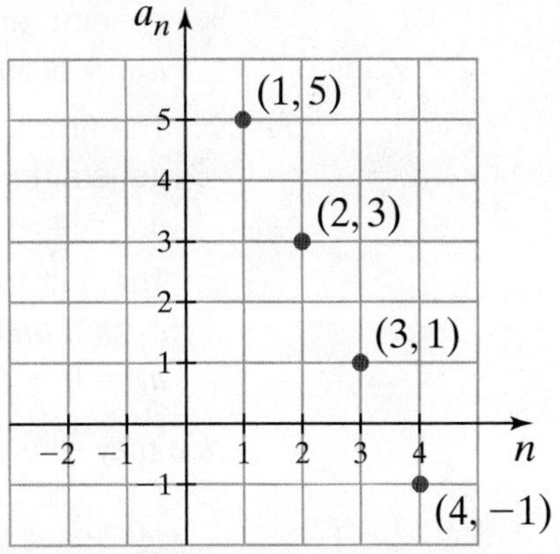

(b) A portion of the graph of the sequence $a_n = -2n + 7$

Figure 2 The graph of every arithmetic sequence is represented by a set of ordered pairs that lies on a straight line.

Note: When the common difference of an arithmetic sequence is positive, the terms of the sequence *increase* and the graph is represented by a set of ordered pairs that lie along a line with positive slope. When the common difference of an arithmetic sequence is negative, the terms of the sequence *decrease* and the graph is represented by a set of ordered pairs that lie along a line with negative slope.

You Try It **Work through this You Try It problem.**

Work Exercises 1–6 in this textbook or in the MyMathLab Study Plan.

OBJECTIVE 2 FIND THE GENERAL TERM OR A SPECIFIC TERM OF AN ARITHMETIC SEQUENCE

By the definition of an arithmetic sequence, the general term of an arithmetic sequence has the form $a_n = a_1 + (n - 1)d$. We can use this formula to find any term of an arithmetic sequence.

Example 2 Finding the General Term of an Arithmetic Sequence

My interactive video summary

Find the general term of each arithmetic sequence, then find the indicated term of the sequence. (In part (c), only a portion of the graph is given. Assume that the domain of this sequence is all natural numbers.)

a. $11, 17, 23, 29, 35, \ldots; a_{50}$

b. $2, 0, -2, -4, -6, \ldots; a_{90}$

c. Find a_{31}.

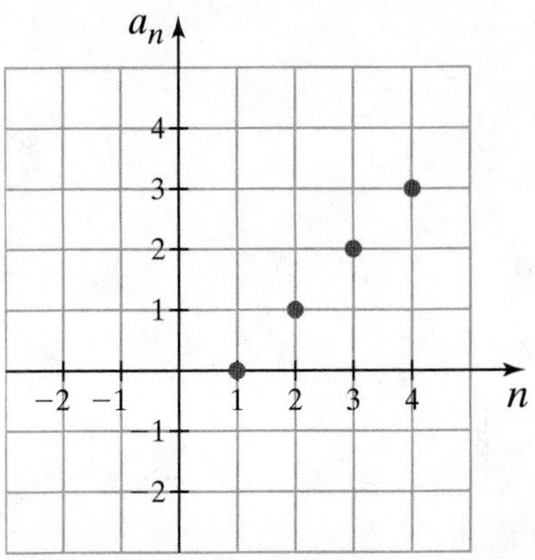

(eText Screens 22.2-1–22.2-22)

Solutions

a. The first term of the sequence is $a_1 = 11$ and the common difference is $d = 6$. The general term is given by $a_n = 11 + (n - 1)(6) = 11 + 6n - 6 = 6n + 5$. Therefore, $a_{50} = 6(50) + 5 = 305$.

b. The first term of the sequence is $a_1 = 2$ and the common difference is $d = -2$. The general term is given by $a_n = 2 + (n - 1)(-2) = 2 - 2n + 2 = 4 - 2n$. Therefore, $a_{90} = 4 - 2(90) = -176$.

c. The first four terms of this sequence are $a_1 = 0, a_2 = 1, a_3 = 2$, and $a_4 = 3$. Thus, $a_1 = 0$ and the common difference is $d = 1$. The general term is given by $a_n = 0 + (n - 1)(1) = n - 1$. Thus, $a_{31} = 31 - 1 = 30$.

You may also watch this **interactive video** to see each of these solutions worked out in detail.

You Try It **Work through this You Try It problem.**

Work Exercises 7–12 in this textbook or in the MyMathLab Study Plan.

Example 3 Finding a Specific Term of an Arithmetic Sequence

My interactive video summary

a. Given an arithmetic sequence with $d = -4$ and $a_3 = 14$, find a_{50}.

b. Given an arithmetic sequence with $a_4 = 12$ and $a_{15} = -10$, find a_{41}.

Solutions

a. We are given that $d = -4$ and $a_3 = 14$. We can use this information to solve for a_1.

Use the formula for the general term of an arithmetic sequence:	$a_n = a_1 + (n - 1)d$
Substitute $n = 3$ and $d = -4$:	$a_3 = a_1 + (3 - 1)(-4)$

Simplifying, we get $a_3 = a_1 - 8$. We can now substitute $a_3 = 14$ to solve for a_1.

Start with the formula for a_3:	$a_3 = a_1 - 8$
Substitute $a_3 = 14$:	$14 = a_1 - 8$
Add 8 to both sides:	$22 = a_1$

Using the formula $a_n = a_1 + (n - 1)d$ with $a_1 = 22$ and $d = -4$, we can simplify to get $a_n = 26 - 4n$. Therefore, $a_{50} = 26 - 4(50) = -174$. Watch this **interactive video** to see every step of this solution.

b. Using the fact that $a_n = a_1 + (n - 1)d$, we get $a_4 = a_1 + (4 - 1)d = 12$ and $a_{15} = a_1 + (15 - 1)d = -10$. This gives us the following system of linear equations:

$$\begin{cases} a_1 + 3d = 12 \\ a_1 + 14d = -10 \end{cases}$$

Using the substitution method or elimination method to solve this system, we get $a_1 = 18$ and $d = -2$. (Watch this **interactive video** to see how to solve this system using either method.) Using the formula $a_n = a_1 + (n - 1)d$ with $a_1 = 18$ and $d = -2$, we can find the general term.

Use the formula for the general term of an arithmetic sequence:	$a_n = a_1 + (n - 1)d$
Substitute $a_1 = 18$ and $d = -2$:	$= 18 + (n - 1)(-2)$
Use the **distributive property**:	$= 18 - 2n + 2$
Simplify:	$= 20 - 2n$

The general term is $a_n = 20 - 2n$. Therefore, $a_{41} = 20 - 2(41) = -62$. Watch this **interactive video** to see this entire solution worked out in detail.

You Try It **Work through this You Try It problem.**

Work Exercises 13–18 in this textbook or in the MyMathLab Study Plan.

OBJECTIVE 3 COMPUTE THE nth PARTIAL SUM OF AN ARITHMETIC SERIES

If $a_1, a_2, a_3, \ldots$ is an arithmetic sequence, then the expression $a_1 + a_2 + a_3 + \cdots + a_n + a_{n+1} + \cdots$ is called an **infinite arithmetic series** and can be written using summation notation as $\sum_{i=1}^{\infty} a_i$. Recall that the sum of the first n terms of a series is called the **nth partial sum** of the series and is given by $S_n = a_1 + a_2 + a_3 + \cdots + a_n$. We can also represent the nth partial sum using summation notation as $S_n = \sum_{i=1}^{n} a_i$. As you can see, the nth partial sum is simply the sum of a finite arithmetic series. Fortunately, there is a convenient formula for computing the nth partial sum of an arithmetic series.

My video summary

Formula for the nth Partial Sum of an Arithmetic Series

The sum of the first n terms of an arithmetic series is called the **nth partial sum** of the series and is given by $S_n = \sum_{i=1}^{n} a_i = a_1 + a_2 + a_3 + \cdots + a_n$. This sum can be computed using the formula $S_n = \dfrac{n(a_1 + a_n)}{2}$.

Watch this **video** to see the derivation of this formula.

Note: The nth partial sum of an arithmetic series is simply the sum of a finite arithmetic series. An arithmetic series *must* be **finite** in order to compute the sum. This is not true for some other types of series. You will see how to find the sum of a special type of infinite series in **Topic 22.3**.

Example 4 Finding the Sum of an Arithmetic Series

My interactive video summary

Find the sum of each arithmetic series.

a. $\sum_{i=1}^{20} (2i - 11)$ **b.** $-5 + (-1) + 3 + 7 + \cdots + 39$

Solutions

a. We can use the formula $S_{20} = \dfrac{20(a_1 + a_{20})}{2}$ to compute the sum of the first 20 terms of this series.

Substitute $i = 1$ in the formula $2i - 11$ to find a_1: $a_1 = 2(1) - 11 = -9$

Substitute $i = 20$ in the formula $2i - 11$ to find a_{20}: $a_{20} = 2(20) - 11 = 29$

We now substitute $a_1 = -9$ and $a_{20} = 29$ into the formula $S_{20} = \frac{20(a_1 + a_{20})}{2}$.

Use the formula for 20th partial sum of an arithmetic series: $S_{20} = \frac{20(a_1 + a_{20})}{2}$

Substitute $a_1 = -9$ and $a_{20} = 29$: $= \frac{20(-9 + 29)}{2}$

Simplify: $= 200$

Therefore, $\sum_{i=1}^{20}(2i - 11) = 200$. You may also watch this **interactive video** to see this solution worked out in detail.

b. Work through the **interactive video** to verify that the sum of this arithmetic series is 204.

You Try It **Work through this You Try It problem.**

Work Exercises 19–27 in this textbook or in the MyMathLab Study Plan.

OBJECTIVE 4 SOLVE APPLICATIONS OF ARITHMETIC SEQUENCES AND SERIES

Example 5 Selling Newspaper Subscriptions

A local newspaper has hired teenagers to go door-to-door to try to solicit new subscribers. The teenagers receive \$2 for selling the first subscription. For each additional subscription sold, the newspaper will pay the teenagers 10 cents more than what was paid for the previous subscription. How much will the teenagers get paid for selling the 100th subscription? How much money will the teenagers earn by selling 100 subscriptions?

Solution The amount of money earned by selling one newspaper subscription can be represented by $a_1 = 2$. The money earned by selling the second subscription is $a_2 = 2.10$. The money earned by selling the third subscription is $a_3 = 2.20$. We see that the amount of money earned by selling n newspaper subscriptions is an arithmetic sequence with $a_1 = 2$ and $d = 0.10$. This sequence is defined by

$$a_n = 2 + (n - 1)(0.10) = 2 + (0.10)n - 0.10 = 0.10n + 1.90.$$

The cash earned by selling the 100th subscription is the 100th term of this sequence, or $a_{100} = 0.10(100) + 1.90 = 11.90$. Therefore, the teenagers are paid \$11.90 for selling the 100th subscription.

To find the total amount earned by selling 100 subscriptions, we must find the sum of the series $\sum_{i=1}^{100}[(0.10)i + 1.90]$.

Using the formula $S_n = \frac{n(a_1 + a_n)}{2}$ with $n = 100$, $a_1 = 2$, and $a_{100} = 11.90$, we get $S_{100} = \frac{100(a_1 + a_{100})}{2} = \frac{100(2 + 11.90)}{2} = 50(13.90) = 695$. Thus, the teenagers will be paid \$695 if they sell 100 subscriptions.

Example 6 Seats in a Theater

My video summary

A large multiplex movie house has many theaters. The smallest theater has only 12 rows. There are six seats in the first row. Each row has two seats more than the previous row. How many total seats are there in this theater?

Solution Try solving this word problem on your own. When you are done, watch this video to see if you are correct, then work through the following "You Try It" problem.

You Try It **Work through this You Try It problem.**

Work Exercises 28–35 in this textbook or in the MyMathLab Study Plan.

22.2 Exercises

In Exercises 1–6, determine if the sequence is arithmetic. If the sequence is arithmetic, find the common difference.

You Try It

1. $8, 14, 20, 26, 32, \ldots$

2. $8, 11, 13, 16, 18, \ldots$

3. $a_n = \dfrac{3n + 1}{2}$

4. $a_n = n(n + 1)$

5. $a_1 = 8, a_n = 2 + a_{n-1}$

6. $a_1 = 5, a_n = 3a_{n-1} + 1$

In Exercises 7–12, find the general term of each arithmetic sequence and then find the indicated term of the sequence. If the sequence is represented by a graph, assume that the domain of the sequence is all natural numbers.

You Try It

7. $2, 7, 12, 17, \ldots; a_{10}$

8. $5, 1, -3, -7, \ldots; a_{31}$

9. $\frac{3}{2}, 3, \frac{9}{2}, 6, \frac{15}{2}, \ldots; a_{50}$

10. $5.0, 3.8, 2.6, 1.4, \ldots; a_{29}$

11. Find a_{17}.

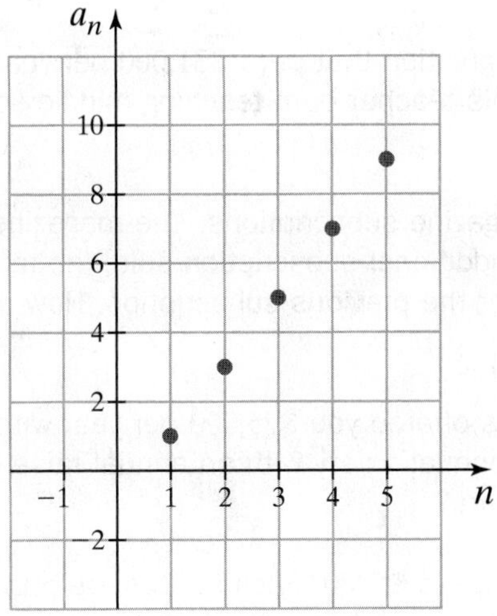

12. Find a_{11}.

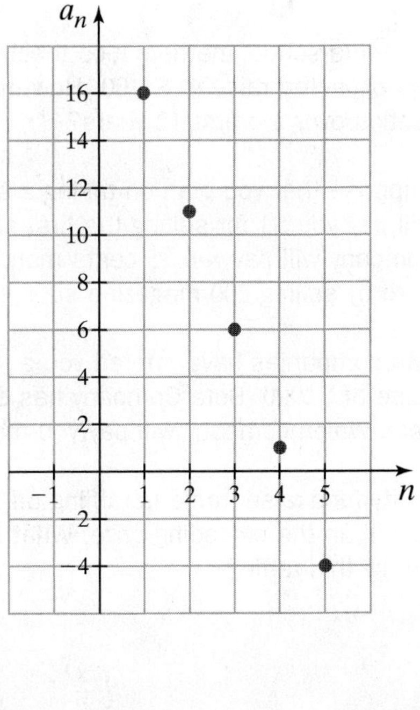

You Try It

13. Given an arithmetic sequence with $d = 3$ and $a_8 = 5$, find a_{30}.

14. Given an arithmetic sequence with $d = -5$ and $a_7 = 11$, find a_{22}.

15. Given an arithmetic sequence with $a_5 = 4$ and $a_{22} = 55$, find a_{36}.

16. Given an arithmetic sequence with $a_6 = 4$ and $a_{20} = -52$, find a_{33}.

17. Given an arithmetic sequence with $a_{16} = 30$ and $a_{30} = 65$, find a_9.

18. Given an arithmetic sequence with $a_8 = -6$ and $a_{19} = -\frac{45}{2}$, find a_{34}.

In Exercises 19–27, find the indicated sum.

You Try It

19. $\sum_{i=1}^{80} i$

20. $\sum_{j=1}^{10} (3j + 7)$

21. $7 + 10 + 13 + 16 + \cdots + 118$

22. $-14 + (-9) + (-4) + 1 + \cdots + 101$

23. $\sum_{i=3}^{14} (-7 - 9i)$

24. $1 + 11 + 21 + 31 + \cdots + a_{102}$

25. $6 + 14 + 22 + 30 + \cdots + (8n - 2)$

26. Find the sum of the first 100 odd integers.

27. Find the sum of the first 100 even positive integers.

You Try It

28. A large multiplex movie house has many theaters. The largest theater has 40 rows. There are 12 seats in the first row. Each row has four seats more than the previous row. How many total seats are there in this theater?

29. A stack of logs has 47 logs on the bottom layer. Each subsequent layer has nine fewer logs than the previous layer. If the top layer has two logs, how many total logs are there in the pile?

30. A middle school mathematics teacher accepts a teaching position that pays $31,000 per year. Each year, the expected raise is $1100. How much total money will this teacher earn teaching middle school mathematics over the first 12 years?

31. Suppose that you plan on taking a summer job selling magazine subscriptions. The magazine company will pay you $1 for selling the first subscription. For each additional subscription sold, the magazine company will pay you 15 cents more than what was paid for the previous subscription. How much will you earn by selling 200 magazine subscriptions?

32. Two companies have offered you a job. Alpha Company has offered you $35,000 per year with an annual raise of $2000. Beta Company has offered you a $46,000 annual salary with an annual raise of $800 per year. Which company will pay you more over the first 10 years?

33. A city fund-raiser raffle is raffling off 25 cash prizes. First prize is $5000. Each successive prize is $200 less than the preceding prize. What is the value of the 25th prize? What is the total amount of cash given out by this raffle?

34. Larry's Luxury Rental Car Company rents luxury cars for up to 18 days. The price is \$300 for the first day, with the rental fee decreasing \$7 for each additional day. How much will it cost to rent a luxury car for 18 days?

35. A ball thrown straight up in the air travels 48 inches in the first tenth of a second. In the next tenth of a second, the ball travels 44 inches. After each additional tenth of a second, the ball travels 4 inches less than it did during the preceding tenth of a second. How long will it take before the ball starts coming back down? What is the total distance that the ball has traveled when it has reached its maximum height?

22.3 Geometric Sequences and Series

THINGS TO KNOW

Before working through this topic, be sure you are familiar with the following concepts:

	VIDEO	ANIMATION	INTERACTIVE
You Try It **1.** Solve Applications of Exponential Functions (Topic 20.3, **Objective 4**)			
You Try It **2.** Find the General Term or a Specific Term of an Arithmetic Sequence (Topic 22.2, **Objective 2**)			▶
You Try It **3.** Compute the nth Partial Sum of an Arithmetic Series (Topic 22.2, **Objective 3**)			▶

OBJECTIVES

1 Write the Terms of a Geometric Sequence

2 Determine If a Sequence Is Geometric

3 Find the General Term or a Specific Term of a Geometric Sequence

4 Compute the nth Partial Sum of a Geometric Series

5 Determine If an Infinite Geometric Series Converges or Diverges

6 Solve Applications of Geometric Sequences and Series

(eText Screens 22.3-1–22.3-34)

OBJECTIVE 1 WRITE THE TERMS OF A GEOMETRIC SEQUENCE

Suppose that you have agreed to work for Donald Trump on a particular job for 21 days. Mr. Trump gives you two choices of payment. You can be paid \$100 for the first day and an additional \$50 per day for each subsequent day. Or, you can choose to be paid 1 penny for the first day with your pay doubling each subsequent day. Which method of payment would you choose? (We will revisit this question later in this topic. See **Example 7**.) Notice that each payment method can be represented by a sequence:

Payment Method 1: $100, 150, 200, 250, 300, \ldots$

Payment Method 2: $0.01, 0.02, 0.04, 0.08, 0.16, \ldots$

The first method of payment is an **arithmetic sequence** with $a_1 = 100$ and $d = 50$. The second method of payment is an example of a **geometric sequence**. Each term of this geometric sequence can be obtained by multiplying the previous term by 2. The number 2 in this case is called the **common ratio**. We can obtain this common ratio by dividing any term of the sequence (except the first term) by the previous term. That is, $r = \frac{a_2}{a_1} = \frac{a_3}{a_2} = \cdots = \frac{a_{n+1}}{a_n}$. The first term of the payment method 2 sequence is $a_1 = 0.01$ and the common ratio is $r = 2$.

Notice that we can rewrite the terms of this sequence as $0.01, (0.01)(2), (0.01)(2^2), (0.01)(2^3), \ldots$. In general, given a geometric sequence with a first term of a_1 and a common ratio of r, then the first n terms of the sequence are

$$\begin{aligned}
&a_1\\
&a_2 = a_1 r\\
&a_3 = a_2 r = \underbrace{(a_1 r)}_{a^2} r = a_1 r^2\\
&a_4 = a_3 r = \underbrace{(a_1 r^2)}_{a^3} r = a_1 r^3\\
&\vdots\\
&a_n = a_1 r^{n-1}.
\end{aligned}$$

> **Definition Geometric Sequence**
>
> A **geometric sequence** is a sequence of the form $a_1, a_1 r, a_1 r^2, a_1 r^3, a_1 r^4, \ldots$, where a_1 is the first term of the sequence and r is the common ratio such that $r = \frac{a_2}{a_1} = \frac{a_3}{a_2} = \cdots = \frac{a_{n+1}}{a_n}$ for all $n \geq 1$. The general term, or nth term, of a geometric sequence has the form $a_n = a_1 r^{n-1}$.

A portion of the two sequences representing the two payment methods are shown in **Figure 3**. The sequence representing payment method 1 (Figure 3a) is arithmetic. The ordered pairs of this sequence are **collinear**. The sequence representing payment method 2 (Figure 3b) is geometric. Notice that the ordered pairs of this sequence do not lie along a common line but, rather, lie on an **exponential curve**.

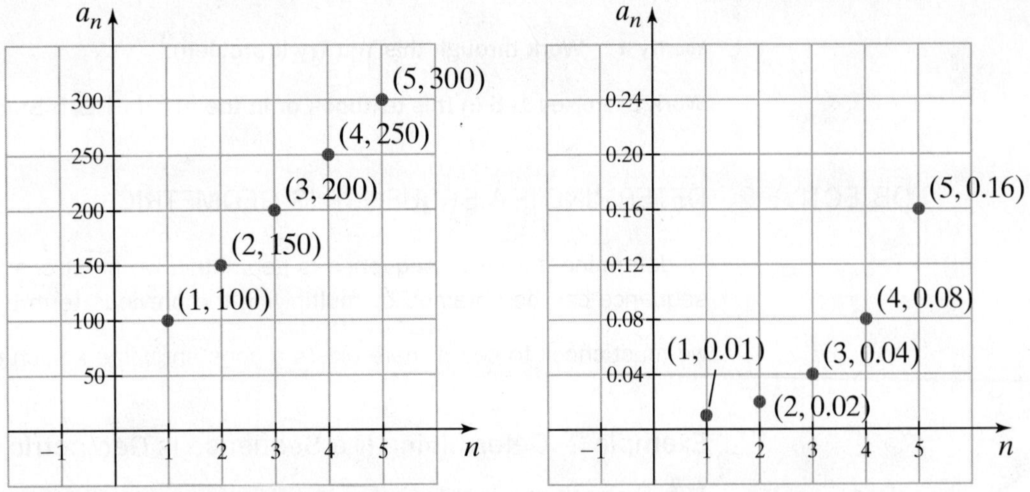

(a)
A portion of the graph of the sequence
100, 150, 200, 250, 300, . . .

(b)
A portion of the graph of the sequence
0.01, 0.02, 0.04, 0.08, 0.16, . . .

Figure 3 The graph of every arithmetic sequence is represented by a set of ordered pairs that lies on a straight line. The graph of a geometric sequence with $r > 0$ is represented by a set of ordered pairs that lies on an exponential curve.

If the first term and the common ratio of a geometric sequence are known, then we can determine the second term by multiplying the first term by the common ratio. The third term can be found by multiplying the second term by the common ratio. We continue this process to find the subsequent terms of the sequence.

Example 1 Writing a Geometric Sequence

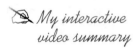

My interactive video summary

a. Write the first five terms of the geometric sequence having a first term of 2 and a common ratio of 3.

b. Write the first five terms of the geometric sequence such that $a_1 = -4$ and $a_n = -5a_{n-1}$ for $n \geq 2$.

Solutions

a. We are given that the first term is $a_1 = 2$ and the common ratio is $r = 3$. The second term is $a_2 = 2 \cdot 3 = 6$. The third term is $a_3 = 6 \cdot 3 = 18$. The fourth term is $a_4 = 18 \cdot 3 = 54$. The fifth term is $a_5 = 54 \cdot 3 = 162$. Therefore, the first five terms of this sequence are 2, 6, 18, 54, and 162.

b. This sequence is defined **recursively**. The first term is $a_1 = -4$. To find a_2, substitute the value of 2 for n in the formula $a_n = -5a_{n-1}$ and simplify:

Given recursive formula:	$a_n = -5a_{n-1}$
Substitute:	$a_2 = -5a_{2-1}$
Simplify:	$= -5a_1$
Substitute $a_1 = -4$:	$= -5(-4)$
Multiply:	$= 20$

Therefore, $a_2 = 20$. We can follow this same procedure to find $a_3, a_4,$ and a_5. You should verify that $a_3 = -100, a_4 = 500,$ and $a_5 = -2500$ on your own, then watch this **interactive video** to see if you are correct.

(eText Screens 22.3-1–22.3-34)

You Try It **Work through this You Try It problem.**

Work Exercises 1–5 in this textbook or in the MyMathLab Study Plan.

OBJECTIVE 2 DETERMINE IF A SEQUENCE IS GEOMETRIC

To determine if a given sequence is geometric, we must check to see if each term of the sequence can be obtained by multiplying the previous term by a common ratio r. That is, we must check to see if there exists a constant value r such that $r = \frac{a_{n+1}}{a_n}$ for any $n \geq 1$.

Example 2 Determining If a Sequence Is Geometric

My interactive video summary

For each of the following sequences, determine if it is geometric. If the sequence is geometric, find the common ratio.

a. $2, 4, 6, 8, 10, \ldots$

b. $\frac{2}{3}, \frac{4}{9}, \frac{8}{27}, \frac{16}{81}, \frac{32}{243}, \ldots$

c. $12, -6, 3, -\frac{3}{2}, \frac{3}{4}, \ldots$

Solutions

a. For this sequence, $\frac{a_2}{a_1} = \frac{4}{2} = 2$ and $\frac{a_3}{a_2} = \frac{6}{4} = \frac{3}{2}$. Since $\frac{a_2}{a_1} \neq \frac{a_3}{a_2}$, there does not exist a common ratio. Hence, this sequence is not geometric. (Note that this sequence is an **arithmetic sequence**.)

b. For this sequence, $\frac{a_2}{a_1} = \frac{\frac{4}{9}}{\frac{2}{3}} = \frac{4}{9} \cdot \frac{3}{2} = \frac{2}{3}$. Note that each term of this sequence (other than the first term) can be obtained by multiplying the previous term by $\frac{2}{3}$.

Therefore, this sequence is geometric with a common ratio of $\frac{2}{3}$.

c. Try to determine if this sequence is geometric. Work through the **interactive video** to see if you are correct.

You Try It **Work through this You Try It problem.**

Work Exercises 6–10 in this textbook or in the MyMathLab Study Plan.

OBJECTIVE 3 FIND THE GENERAL TERM OR A SPECIFIC TERM OF A GEOMETRIC SEQUENCE

By the definition of a geometric sequence, the general term, or nth term, of a geometric sequence has the form $a_n = a_1 r^{n-1}$. We can use this formula to find any term of a given geometric sequence.

Example 3 Finding the General Term of a Geometric Sequence

Find the general term of each geometric sequence.

a. $12, -6, 3, -\frac{3}{2}, \frac{3}{4}, \ldots$

b. $\frac{2}{3}, \frac{2}{9}, \frac{2}{27}, \frac{2}{81}, \frac{2}{243}, \ldots$

Solutions

a. The first term of the sequence is $a_1 = 12$ and the common ratio is $r = \frac{a_2}{a_1} = \frac{-6}{12} = -\frac{1}{2}$. Therefore, $a_n = 12\left(-\frac{1}{2}\right)^{n-1}$.

b. The first term of the sequence is $a_1 = \frac{2}{3}$ and the common ratio is $r = \frac{a_2}{a_1} = \frac{\frac{2}{9}}{\frac{2}{3}} = \frac{1}{3}$.

Therefore, $a_n = \left(\frac{2}{3}\right)\left(\frac{1}{3}\right)^{n-1}$.

Example 4 Finding a Specific Term of a Geometric Sequence

My interactive video summary

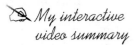

a. Find the seventh term of the geometric sequence whose first term is 2 and whose common ratio is -3.

b. Given a geometric sequence such that $a_6 = 16$ and $a_9 = 2$, find a_{13}.

Solutions

a. We can use the formula $a_n = a_1 r^{n-1}$ with $a_1 = 2$ and $r = -3$ to find the general term. The general term is $a_n = 2(-3)^{n-1}$. Therefore, $a_7 = 2(-3)^{7-1} = 2(-3)^6 = 2(729) = 1458$.

b. Since $a_6 = 16$, we can substitute $n = 6$ into the formula $a_n = a_1 r^{n-1}$ to get $a_6 = a_1 r^5 = 16$. Similarly, we can substitute $n = 9$ into the formula $a_n = a_1 r^{n-1}$ to get $a_9 = a_1 r^8 = 2$. This gives the following two equations.

$$(1)\quad a_1 r^5 = 16$$
$$(2)\quad a_1 r^8 = 2$$

Divide both sides of equation (1) by r^5 to get $a_1 = \frac{16}{r^5}$. Now, substitute $a_1 = \frac{16}{r^5}$ into equation (2) and solve for a_1.

Start with equation (2): $a_1 r^8 = 2$

Substitute $a_1 = \frac{16}{r^5}$: $\left(\frac{16}{r^5}\right) r^8 = 2$

$\frac{r^8}{r^5} = r^3$: $16r^3 = 2$

Divide both sides by 16: $r^3 = \frac{1}{8}$

Take the cube root of both sides: $r = \frac{1}{2}$

Now substitute $r = \frac{1}{2}$ into equation (1) to solve for a_1.

Start with equation (1):	$a_1 r^5 = 16$
Substitute $r = \frac{1}{2}$:	$a_1\left(\frac{1}{2}\right)^5 = 16$
$\left(\frac{1}{2}\right)^5 = \frac{1}{32}$:	$a_1\left(\frac{1}{32}\right) = 16$
Multiply both sides by 32:	$a_1 = 512$

We can now use the formula $a_n = a_1 r^{n-1}$ with $a_1 = 512$ and $r = \frac{1}{2}$ to find the general term. The general term is $a_n = 512\left(\frac{1}{2}\right)^{n-1}$. Therefore,

$$a_{13} = 512\left(\frac{1}{2}\right)^{13-1} = 512\left(\frac{1}{2}\right)^{12} = \frac{512}{2^{12}} = \frac{512}{4096} = \frac{1}{8}.$$

You may wish to work through this **interactive video** to see these solutions worked out in detail.

You Try It **Work through this You Try It problem.**

Work Exercises 11–18 in this textbook or in the MyMathLab Study Plan.

OBJECTIVE 4 COMPUTE THE nth PARTIAL SUM OF A GEOMETRIC SERIES

My video summary

If $a_1, a_1r, a_1r^2, a_1r^3, \ldots$ is a geometric sequence, then the expression $a_1 + a_1r + a_1r^2 + a_1r^3 + \cdots + a_1r^{n-1} + \cdots$ is called an **infinite geometric series** and can be written in summation notation as $\sum_{i=1}^{\infty} a_1 r^{i-1}$. Recall that the sum of the first n terms of a series is called the **nth partial sum** of the series and is given by $S_n = a_1 + a_1r + a_1r^2 + a_1r^3 + \cdots + a_1r^{n-1}$. We can also represent the nth partial sum using summation notation as $S_n = \sum_{i=1}^{n} a_1 r^{i-1}$. As you can see, this nth partial sum is simply the sum of a finite geometric series. Fortunately, there is a convenient formula for computing the nth partial sum of a geometric series.

Formula for the nth Partial Sum of a Geometric Series

The sum of the first n terms of a geometric series is called the **nth partial sum** of the series and is given by

$$S_n = \sum_{i=1}^{n} a_1 r^{i-1} = a_1 + a_1r + a_1r^2 + a_1r^3 + \cdots + a_1r^{n-1}.$$

This sum can be computed using the formula $S_n = \dfrac{a_1(1 - r^n)}{1 - r}$ for $r \neq 1$.

Watch this video to see the derivation of this formula.

Example 5 Computing the *n*th Partial Sum of a Geometric Series

My interactive video summary

a. Find the sum of the series $\sum_{i=1}^{15} 5(-2)^{i-1}$.

b. Find the seventh partial sum of the geometric series $8 + 6 + \frac{9}{2} + \frac{27}{8} + \cdots$.

Solutions

a. Using the formula $S_n = \frac{a_1(1 - r^n)}{1 - r}$ with $n = 15, a_1 = 5$, and $r = -2$, we get

$S_{15} = \frac{5(1 - (-2)^{15})}{1 - (-2)} = \frac{5(1 - (-32{,}768))}{3} = \frac{5(32{,}769)}{3} = \frac{163{,}845}{3} = 54{,}615$. You may also watch this **interactive video** to see this solution worked out in detail.

b. Work through the **interactive video** to verify that the seventh partial sum of this series is $S_7 = \frac{14{,}197}{512}$.

You Try It **Work through this You Try It problem.**

Work Exercises 19–23 in this textbook or in the MyMathLab Study Plan.

OBJECTIVE 5 DETERMINE IF AN INFINITE GEOMETRIC SERIES CONVERGES OR DIVERGES

My video summary

Consider the infinite geometric series $\sum_{n=1}^{\infty} a_1 r^{n-1} = a_1 + a_1 r + a_1 r^2 + \cdots + a_1 r^{n-1} + \cdots$. Is it possible for a series of this form to have a finite sum? Is it possible to add infinitely many terms and get a finite sum? The answer is **yes**, it is possible, but it depends on the value of r. Before we determine the value(s) of r for which an infinite geometric series has a finite sum, let's look at an example. Consider the following geometric series

$$\frac{1}{2} + \frac{1}{3} + \frac{2}{9} + \frac{4}{27} + \frac{8}{81} + \cdots$$

Note that $a_1 = \frac{1}{2}$ and $r = \frac{2}{3}$. We can use the formula $S_n = \frac{a_1(1 - r^n)}{1 - r}$ to find the nth partial sum for any value of n of our choosing. The nth partial sums for $n = 5, 10, 20$, and 40 are given in **Table 1** as well as the value of r^n.

n	$S_n = \frac{a_1(1 - r^n)}{1 - r} = \frac{\left(\frac{1}{2}\right)\left(1 - \left(\frac{2}{3}\right)^n\right)}{1 - \frac{2}{3}}$	$r^n = \left(\frac{2}{3}\right)^n$
5	1.3024691	0.1316872
10	1.4739877	0.0173415
20	1.4995489	0.0003007
40	1.4999999	0.0000001

Table 1

Looking at Table 1, it appears that as n increases, the value of S_n is getting closer to $1.5 = \frac{3}{2}$. Also notice that as n increases, the value of $r^n = \left(\frac{2}{3}\right)^n$ is getting closer to zero. In fact, for any value of r between -1 and 1, the value of r^n will always approach 0 as n approaches infinity. We say, "For values of r between -1 and 1, r^n approaches zero as n approaches infinity" and write: For $|r| < 1, r^n \to 0$ as $n \to \infty$. Thus, if $|r| < 1$, $S_n = \frac{a_1(1 - r^n)}{1 - r} \approx \frac{a_1(1 - 0)}{1 - r} = \frac{a_1}{1 - r}$ for large values of n. Therefore, given an infinite geometric series with $|r| < 1$, the sum of the series is given by $S = \frac{a_1}{1 - r}$.

Note: A formal proof of this formula requires calculus.

Formula for the Sum of an Infinite Geometric Series

Let $\sum_{n=1}^{\infty} a_1 r^{n-1} = a_1 + a_1 r + a_1 r^2 + a_1 r^3 + \cdots + a_1 r^{n-1} + \cdots$ be an infinite geometric series. If $|r| < 1$, then the sum of the series is given by $S = \frac{a_1}{1 - r}$.

Note that if $|r| < 1$, then the infinite geometric series has a finite sum and is said to **converge**. If $|r| \geq 1$, then the infinite geometric series does not have a finite sum and the series is said to **diverge**.

Example 6 Determining If an Infinite Geometric Series Converges or Diverges

My interactive video summary

Determine whether each of the following series converges or diverges. If the series converges, find the sum.

a. $\sum_{n=1}^{\infty} \frac{1}{2}\left(\frac{2}{3}\right)^{n-1}$

b. $3 - \frac{6}{5} + \frac{12}{25} - \frac{24}{125} + \cdots$

c. $12 + 18 + 27 + \frac{81}{2} + \frac{243}{4} + \cdots$

Solutions

a. This is an infinite geometric series with $|r| = \left|\frac{2}{3}\right| < 1$. Since $|r| < 1$, the infinite series must converge and, thus, must have a finite sum. The sum of the series is

$$S = \frac{a_1}{1 - r} = \frac{\frac{1}{2}}{1 - \frac{2}{3}} = \frac{\frac{1}{2}}{\frac{1}{3}} = \frac{1}{2} \cdot \frac{3}{1} = \frac{3}{2}.$$

b. For this infinite geometric series, the common ratio is

$r = \frac{-\frac{6}{5}}{3} = -\frac{6}{5} \cdot \frac{1}{3} = -\frac{2}{5}$. Since $|r| = \left|-\frac{2}{5}\right| = \frac{2}{5} < 1$, the infinite series converges.

The sum is $S = \frac{a_1}{1 - r} = \frac{3}{1 - \left(-\frac{2}{5}\right)} = \frac{3}{1 + \frac{2}{5}} = \frac{3}{\frac{7}{5}} = 3 \cdot \frac{5}{7} = \frac{15}{7}$.

c. This infinite series diverges. Do you know why? Work through the **interactive video** to see why this series diverges.

(eText Screens 22.3-1–22.3-34)

You Try It **Work through this You Try It problem.**

Work Exercises 24–28 in this textbook or in the MyMathLab Study Plan.

OBJECTIVE 6 SOLVE APPLICATIONS OF GEOMETRIC SEQUENCES AND SERIES

In Example 7, we revisit the question that was presented at the beginning of this topic.

Example 7 Choosing a Payment Method

Suppose that you have agreed to work for Donald Trump on a particular job for 21 days. Mr. Trump gives you two choices of payment. You can be paid $\$100$ for the first day and an additional $\$50$ per day for each subsequent day. Or, you can choose to be paid 1 penny for the first day with your pay doubling each subsequent day. Which method of payment yields the most income?

Solution Each payment method can be represented by a sequence.

Payment method 1: $100, 150, 200, 250, \ldots$

Payment method 2: $0.01, 0.02, 0.04, 0.08, \ldots$

To find out which method of payment will yield the greatest income, we must find the sum of the first 21 terms of each sequence.

Payment method 1 is an **arithmetic sequence** with $a_1 = 100$ and $d = 50$. Using the **formula for the general term of an arithmetic sequence** we get $a_n = 100 + (n - 1)50 = 100 + 50n - 50 = 50n + 50$. Note that $a_{21} = 50(21) + 50 = 1100$. Using the **formula for the nth partial sum of an arithmetic series** with $n = 21$, we get

$$S_{21} = \frac{n(a_1 + a_{21})}{2} = \frac{21(100 + 1100)}{2} = \$12{,}600.$$

Payment method 2 is a geometric sequence with $a_1 = 0.01$ and $r = 2$. Using the **formula for the nth partial sum of a geometric series** with $n = 21$, we get $S_{21} = \dfrac{a_1\left(1 - r^{21}\right)}{1 - r} = \dfrac{(0.01)\left(1 - 2^{21}\right)}{1 - 2} = \$20{,}971.51$. Clearly, payment method 2 is the better choice.

Example 8 Total Amount Given to a Local Charity

A local charity received $\$8500$ in charitable contributions during the month of January. Because of a struggling economy, it is projected that contributions will decline each month to 95% of the previous month's contributions. What are the expected contributions for the month of October? What is the total expected contributions that this charity can expect at the end of the year?

Solution The monthly contributions can be represented by a geometric sequence with $a_1 = 8500$ and $r = 0.95$. Thus, the contributions for the nth month is given by $a_n = 8500(0.95)^{n-1}$. The expected contributions for October, or when $n = 10$, are $a_{10} = 8500(0.95)^{10-1} \approx 5357.12$. Thus, the contributions for the month of October are expected to be about $\$5357.12$.

(eText Screens 22.3-1–22.3-34)

The total contributions for the year can be written as the following finite geometric series:

$$8500 + (8500)(0.95) + (8500)(0.95)^2 + \cdots + (8500)(0.95)^{11} = \sum_{i=1}^{12} 8500(0.95)^{i-1}$$

Using the **formula for the nth partial sum of a geometric series** with $n = 12$, we get $S_{12} = \dfrac{a_1(1 - r^{12})}{1 - r} = \dfrac{8500(1 - 0.95^{12})}{1 - 0.95} \approx 78{,}138.79$. Therefore, the charity can expect about $\$78{,}138.79$ in donations for the year.

You Try It **Work through this You Try It problem.**

Work Exercises 29–33 in this textbook or in the MyMathLab Study Plan.

Example 9 Expressing a Repeating Decimal as a Ratio of Two Integers

My interactive video summary

Every repeating decimal number is a **rational number** and can therefore be represented by the quotient of two integers. Write each of the following repeating decimal numbers as the quotient of two integers.

a. $0.\overline{4}$ **b.** $0.2\overline{13}$

Solutions

a. We can rewrite $0.\overline{4}$ as $0.44444\ldots = \dfrac{4}{10} + \dfrac{4}{100} + \dfrac{4}{1{,}000} + \dfrac{4}{10{,}000} + \dfrac{4}{100{,}000} + \cdots.$

This is an infinite geometric series with $a_1 = \dfrac{4}{10}$ and $r = \dfrac{1}{10}$. Because $|r| = \left|\dfrac{1}{10}\right| < 1$, we know that the series converges. Using the formula $S = \dfrac{a_1}{1 - r}$, we see that

$$0.\overline{4} = \frac{\frac{4}{10}}{1 - \frac{1}{10}} = \frac{\frac{4}{10}}{\frac{9}{10}} = \frac{4}{9}.$$

b. Carefully work through the **interactive video** to see that $0.2\overline{13} = \dfrac{211}{990}$.

You Try It **Work through this You Try It problem.**

Work Exercises 34 and 35 in this textbook or in the MyMathLab Study Plan.

Annuities

You Try It

In **Topic 20.3** we derived a formula for **periodic compound interest**. This formula is used to determine the future value of a *one-time* investment. (Use the You Try It icon to see a periodic compound interest practice exercise.) Suppose that instead of investing one lump sum, you wish to invest equal amounts of money at steady intervals. An investment of equal amounts deposited at equal time intervals is called an **annuity**. If these equal deposits are made at the end of a compound period, the annuity is called an **ordinary annuity**.

For example, suppose that you want to invest $\$P$ at the end of each payment period at an annual rate r, in decimal form. Then the interest rate per payment period is $i = \dfrac{r}{\text{number of payment periods per year}}$. We now summarize the total amount of the ordinary annuity after the first k payment periods.

End of 1st payment period: $\underbrace{P}_{\text{1st payment}}$

End of 2nd payment period: $\underbrace{P}_{\text{1st payment}} + \underbrace{Pi}_{\substack{\text{Interest}\\\text{earned on}\\\text{1st payment}}} + \underbrace{P}_{\text{2nd payment}} = \underbrace{P(1+i)}_{\substack{\text{Total amount}\\\text{of 1st payment}}} + \underbrace{P}_{\text{2nd payment}}$

End of 3rd payment period:

$$\underbrace{P(1+i)}_{\substack{\text{Amount of}\\\text{1st payment}}} + \underbrace{Pi(1+i)}_{\substack{\text{Interest earned}\\\text{on amount of}\\\text{1st payment}}} + \underbrace{P}_{\text{2nd payment}} + \underbrace{Pi}_{\substack{\text{Interest}\\\text{earned on}\\\text{2nd payment}}} + \underbrace{P}_{\text{3rd payment}} = \underbrace{P(1+i)^2}_{\substack{\text{Total amount}\\\text{of 1st payment}}} + \underbrace{P(1+i)}_{\substack{\text{Total amount}\\\text{of 2nd payment}}} + \underbrace{P}_{\text{3rd payment}}$$

End of kth payment period: $\underbrace{P(1+i)^{k-1}}_{\substack{\text{Total amount}\\\text{of 1st payment}}} + \underbrace{P(1+i)^{k-2}}_{\substack{\text{Total amount}\\\text{of 2nd payment}}} + \cdots + \underbrace{P(1+i)}_{\substack{\text{Total amount}\\\text{of } (k-1)\text{st}\\\text{payment}}} + \underbrace{P}_{k\text{th payment}}$

The total amount of the ordinary annuity after k payment periods is

$$A = P + P(1 + i) + \cdots + P(1 + i)^{k-2} + P(1 + i)^{k-1}.$$

This is a finite geometric series with $a_1 = P$ and a common ratio of $(1 + i)$. Thus, the amount of the annuity after the kth payment is

$$A = \frac{P\left(1 - (1 + i)^k\right)}{1 - (1 + i)} = \frac{P\left(1 - (1 + i)^k\right)}{-i} = \frac{P\left((1 + i)^k - 1\right)}{i}.$$

Amount of an Ordinary Annuity after the kth Payment

The total amount of an ordinary annuity after the kth payment is given by the formula

$$A = \frac{P\left((1 + i)^k - 1\right)}{i},$$

where

A = Total amount of annuity after k payments

P = Deposit amount at the end of each payment period

i = Interest rate per payment period

Example 10 Finding the Amount of an Ordinary Annuity

My video summary

Chie and Ben decided to save for their newborn son Jack's college education. They decided to invest \$200 every 3 months in an investment earning 8% interest compounded quarterly. How much is this investment worth after 18 years?

Solution This is an ordinary annuity with $P = \$200$ and $i = \frac{0.08}{4} = 0.02$. What is k? See if you can determine k and use the formula $A = \frac{P((1 + i)^k - 1)}{i}$ to determine the total amount of this annuity. When you are done, watch this **video** to see if you are correct.

You Try It **Work through this You Try It problem.**

Work Exercises 36–38 in this textbook or in the MyMathLab Study Plan.

22.3 Exercises

In Exercises 1–5, write the first five terms of the geometric sequence with the given information.

You Try It

1. The first term is 8 and the common ratio is 2.
2. The first term is 162 and the common ratio is $\frac{1}{3}$.
3. The first term is 25 and the common ratio is $-\frac{1}{5}$.
4. $a_n = 7a_{n-1}; a_1 = 3$
5. $a_n = -2a_{n-1}; a_1 = -4$

In Exercises 6–10, determine if the sequence is geometric. If the sequence is geometric, find the common ratio.

You Try It

6. $4, 24, 144, 864, \ldots$
7. $-2, 2, -2, 2, \ldots$
8. $-3, 1, -1, -3, \ldots$
9. $2, -\dfrac{10}{3}, \dfrac{50}{9}, -\dfrac{250}{27}, \ldots$
10. $7.236, -5.7888, 4.63104, -3.704832, \ldots$

You Try It

11. Determine the general term of the sequence $3, 6, 12, 24, \ldots.$
12. Determine the general term of the sequence $\frac{1}{2}, \frac{1}{8}, \frac{1}{32}, \frac{1}{128}, \ldots.$
13. Determine the general term of the sequence $\frac{1}{5}, -\frac{2}{15}, \frac{4}{45}, -\frac{8}{135}, \ldots.$

You Try It

14. Find the seventh term of the geometric sequence whose first term is 5 and whose common ratio is 4.
15. Find the sixth term of the geometric sequence whose first term is 3804 and whose common ratio is $-\frac{1}{4}$.
16. Find the 11th term of the geometric sequence \$5000, \$5050, \$5100.50, ...
17. Given a geometric sequence such that $a_4 = 108$ and $a_7 = 2916$, find a_{10}.
18. Given a geometric sequence such that $a_3 = 16$ and $a_8 = -\frac{1}{2}$, find a_{11}.

In Exercises 19–23, find the sum of each geometric series.

You Try It

19. $\sum_{i=1}^{8} 7(-4)^{i-1}$
20. $\sum_{i=1}^{13} 2\left(\dfrac{3}{5}\right)^{i-1}$
21. $\sum_{i=1}^{10} 4(1.05)^{i-1}$
22. $2 + \dfrac{2}{3} + \dfrac{2}{9} + \dfrac{2}{27} + \cdots + \dfrac{2}{729}$
23. $1 - \dfrac{1}{2} + \dfrac{1}{4} - \dfrac{1}{8} + \cdots - \dfrac{1}{128}$

In Exercises 24–28, determine if each infinite geometric series converges or diverges. If the series converges, find the sum.

You Try It

24. $-1 + \frac{1}{10} - \frac{1}{100} + \frac{1}{1000} - \cdots$

25. $343 + 49 + 7 + 1 + \cdots$

26. $\sum_{i=1}^{\infty} 7\left(\frac{1}{4}\right)^{i-1}$

27. $\sum_{i=1}^{\infty} \frac{1}{5}(3)^{i-1}$

28. $0.5 - 0.05 + 0.005 - 0.0005 + 0.00005 - \cdots$

29. Warren wanted to save money to purchase a new car. He started by saving $1 on the first of January. On the first of February, he saved $3. On the first of March he saved $9. So, on the first day of each month, he wanted to save three times as much as he did on the first day of the previous month. If Warren continues his savings pattern, how much will he need to save on the first day of September?

30. Suppose that you have accepted a job for 2 weeks that will pay $0.07 for the first day, $0.14 for the second day, $0.28 for the third day, and so on. What will your total earnings be after 2 weeks?

31. Mary has accepted a teaching job that pays $25,000 for the first year. According to the Teacher's Union, Mary will get guaranteed salary increases of 4 percent per year. If Mary plans to teach for 30 years, what will be her total salary earnings?

32. A child is given an initial push on a rope swing. On the first swing, the rope swings through an arc of 12 feet. On each successive swing, the length of the arc is 80% of the previous length. After 10 swings, what is the total length the rope will have swung? When the child stops swinging, what is the total length the rope will have swung?

33. Randy dropped a rubber ball from his apartment window from a height of 50 feet. The ball always bounces $\frac{3}{5}$ of the distance fallen. How far has the ball traveled once it is done bouncing?

You Try It

34. Rewrite the number $0.\overline{7}$ as the quotient of two integers.

35. Rewrite the number $0.\overline{325}$ as the quotient of two integers.

You Try It

36. Kip contributes $200 every month to his $401(k)$. What will the value of Kip's $401(k)$ be in 10 years if the yearly rate of return is assumed to be 12% compounded monthly?

37. Mark and Lisa decide to invest $500 every 3 months in an education IRA to save for their son Beau's college education. What will the value of the IRA be after 10 years if the yearly assumed rate of return is 8% compounded quarterly?

38. Marv and Cindy decide to build a new home in 10 years. They will need $80,000 to purchase the lot of their dreams. How much should they save each month in an account that has an assumed yearly rate of return of 7% compounded monthly?

22.4 The Binomial Theorem

THINGS TO KNOW

Before working through this topic, be sure you are familiar with the following concepts:

		VIDEO	ANIMATION	INTERACTIVE
You Try It	1. Multiply Two Binomials (Topic 14.4, Objective 3)			
You Try It	2. Multiply Two or More Polynomials (Topic 14.4, Objective 4)			

OBJECTIVES

1 Expand Binomials Raised to a Power Using Pascal's Triangle

2 Evaluate Binomial Coefficients

3 Expand Binomials Raised to a Power Using the Binomial Theorem

4 Find a Particular Term or a Particular Coefficient of a Binomial Expansion

OBJECTIVE 1 EXPAND BINOMIALS RAISED TO A POWER USING PASCAL'S TRIANGLE

My video summary

In this topic, we will focus on expanding algebraic expressions of the form $(a + b)^n$, where n is an integer greater than or equal to zero. Because $(a + b)$ is a binomial, we call the expansion of $(a + b)^n$ a *binomial expansion*. Consider the expansion of $(a + b)^4$.

$$\begin{aligned}(a+b)^4 &= \underbrace{(a+b)(a+b)}\cdot\underbrace{(a+b)(a+b)} \\ &= (a^2+2ab+b^2)(a^2+2ab+b^2) \\ &= a^4+2a^3b+a^2b^2+2a^3b+4a^2b^2+2ab^3+a^2b^2+2ab^3+b^4 \\ &= a^4+4a^3b+6a^2b^2+4ab^3+b^4\end{aligned}$$

Although the expansion of $(a + b)^4$ using the method above is not too complicated, it would not be desirable to use this method to expand $(a + b)^n$ for large values of n.

The goal in this topic is to try to develop a method for expanding expressions of the form $(a + b)^n$ without actually performing all of the multiplication. We start by studying the expanded forms of $(a + b)^n$ for $n = 0, 1, 2, 3, 4,$ and 5.

$$\begin{aligned}
n = 0&: (a+b)^0 = 1 \\
n = 1&: (a+b)^1 = 1a + 1b \\
n = 2&: (a+b)^2 = 1a^2 + 2ab + 1b^2 \\
n = 3&: (a+b)^3 = 1a^3 + 3a^2b + 3ab^2 + 1b^3 \\
n = 4&: (a+b)^4 = 1a^4 + 4a^3b + 6a^2b^2 + 4ab^3 + 1b^4 \\
n = 5&: (a+b)^5 = 1a^5 + 5a^4b + 10a^3b^2 + 10a^2b^3 + 5ab^4 + 1b^5
\end{aligned}$$

The coefficients of each expansion are highlighted in red. These coefficients are known as **binomial coefficients**. Before we determine a pattern for these coefficients, let's first observe the exponent pattern. Notice in each expansion of $(a + b)^n$, there are always $n + 1$ terms. The sum of the exponents of each term is always equal to n. Also note that the first term is always a^n $\left(\text{or } a^nb^0\right)$ and the last term is always b^n $\left(\text{or } a^0b^n\right)$. As we look at the terms of each expansion from left to right, the exponent of the first variable decreases by 1 and the exponent of the second variable increases by 1. Thus, the exponent pattern of the variables of each expansion is $a^n b^0, a^{n-1}b^1, a^{n-2}b^2, a^{n-3}b^3, \ldots, a^1 b^{n-1}, a^0 b^n$. The pattern for the binomial coefficients is less obvious. To see the pattern for the coefficients, we start by rewriting the six expansions of $(a + b)^n$ again, but this time we only write the coefficients. See Figure 4.

$n = 0$: 1
$n = 1$: 1 1
$n = 2$: 1 2 1
$n = 3$: 1 3 3 1
$n = 4$: 1 4 6 4 1
$n = 5$: 1 5 10 10 5 1

Figure 4
The coefficients of the expansions of $(a + b)^n$, also called Pascal's triangle

Notice that the coefficients in Figure 4 form a "triangle." This triangle is known as **Pascal's triangle**, named after the French mathematician, **Blaise Pascal**. The first and last number of each row of Pascal's triangle is 1. Every other number is equal to the sum of the two numbers directly above it. We can now write the next row of Pascal's triangle, which is the row corresponding to $n = 6$.

$n = 5$: 1 5 10 10 5 1
$n = 6$: 1 6 15 20 15 6 1

This new row of Pascal's triangle represents the coefficients of the expansion of $(a + b)^6$. Therefore,

$$\begin{aligned}(a + b)^6 &= a^6b^0 + 6a^5b^1 + 15a^4b^2 + 20a^3b^3 + 15a^2b^4 + 6a^1b^5 + a^0b^6 \\ &= a^6 + 6a^5b + 15a^4b^2 + 20a^3b^3 + 15a^2b^4 + 6ab^5 + b^6.\end{aligned}$$

Notice the pattern of the exponents of each variable. The exponent of variable a starts with 6 and decreases by 1 for each successive term until it equals 0. The exponent of variable b starts with 0 and increases by 1 each term until it equals 6.

See if you can create Pascal's triangle for values of n up to 10. View this **popup box** to see if you are correct.

Example 1 Using Pascal's Triangle to Expand a Binomial Raised to a Power

My interactive video summary

Use Pascal's triangle to expand each binomial.

a. $(x + 2)^4$ **b.** $(x - 3)^5$ **c.** $(2x - 3y)^3$

Solutions

a. We start by looking at the row of Pascal's triangle corresponding with $n = 4$. We see that this row is 1 4 6 4 1. Using these coefficients and the exponent pattern we get

$$\begin{aligned}(x + 2)^4 &= 1\left(x^4 \cdot 2^0\right) + 4\left(x^3 \cdot 2^1\right) + 6\left(x^2 \cdot 2^2\right) + 4\left(x^1 \cdot 2^3\right) + 1\left(x^0 \cdot 2^4\right) \\ &= x^4 + 8x^3 + 24x^2 + 32x + 16.\end{aligned}$$

b. The row of Pascal's triangle corresponding with $n = 5$ is 1 5 10 10 5 1. Using these coefficients and the exponent pattern we get

$$\begin{aligned}(x - 3)^5 &= 1\left(x^5 \cdot (-3)^0\right) + 5\left(x^1 \cdot (-3)^1\right) + 10\left(x^3 \cdot (-3)^2\right) \\ &\quad + 10\left(x^2 \cdot (-3)^3\right) + 5\left(x \cdot (-3)^4\right) + 1\left(x^0 \cdot (-3)^5\right) \\ &= x^5 - 15x^4 + 90x^3 - 270x^2 + 405x - 243.\end{aligned}$$

(eText Screens 22.4-1–22.4-18)

c. See if you can use Pascal's triangle to show that $(2x - 3y)^3 = 8x^3 - 36x^2y + 54xy^2 - 27y^3$. Work through the **interactive video** to see the solution.

Note: The terms of the expansion of the form $(a - b)^n$ will *always* alternate in sign with the sign of the first term being positive.

You Try It **Work through this You Try It problem.**

Work Exercises 1–6 in this textbook or in the MyMathLab Study Plan.

OBJECTIVE 2 EVALUATE BINOMIAL COEFFICIENTS

Although Pascal's triangle is useful for determining the binomial coefficients of $(a + b)^n$ for fairly small values of n, it is not that useful for large values of n. For example, to find the binomial coefficients of the expansion of $(a + b)^{50}$ using Pascal's triangle, we would need to write the first 51 rows of the triangle to determine the coefficients. Fortunately, there is a convenient formula for the binomial coefficients. This formula requires the use of **factorials**. Recall, $n! = n \cdot (n - 1) \cdot (n - 2) \cdot \cdots \cdot 3 \cdot 2 \cdot 1$ and $0! = 1$. To establish a formula for the binomial coefficients, let's take another look at the expansion for $(a + b)^6$.

$$(a + b)^6 = 1a^6b^0 + 6a^5b^1 + 15a^4b^2 + 20a^3b^3 + 15a^2b^4 + 6a^1b^5 + 1a^0b^6$$

Table 2 shows the relationship between the variable parts of the expansion of the form $a^{n-r}\, b^r$ and the corresponding coefficients. Select any of the coefficients in Table 2 to verify that the formula using factorial notation is true.

Variables	Coefficient	Variables	Coefficient
a^6b^0	$1 = \dfrac{6!}{0! \cdot 6!}$	a^2b^4	$15 = \dfrac{6!}{4! \cdot 2!}$
a^5b^1	$6 = \dfrac{6!}{1! \cdot 5!}$	a^1b^5	$6 = \dfrac{6!}{5! \cdot 1!}$
a^4b^2	$15 = \dfrac{6!}{2! \cdot 4!}$	a^0b^6	$1 = \dfrac{6!}{6! \cdot 0!}$
a^3b^3	$20 = \dfrac{6!}{3! \cdot 3!}$		

Table 2

You can see in Table 2 that for each pair of variables of the form $a^{n-r}b^r$, the corresponding binomial coefficient is of the form $\dfrac{n!}{r! \cdot (n - r)!}$. We will use the shorthand notation $\dbinom{n}{r}$, read as "n choose r," to denote a binomial coefficient.

Formula for a Binomial Coefficient

For non-negative integers n and r with $n \geq r$, the coefficient of the expansion of $(a + b)^n$ whose variable part is $a^{n-r}b^r$ is given by

$$\binom{n}{r} = \frac{n!}{r! \cdot (n - r)!}.$$

(eText Screens 22.4-1–22.4-18)

Example 2 Evaluating Binomial Coefficients

Evaluate each of the following binomial coefficients.

a. $\binom{5}{3}$ b. $\binom{4}{1}$ c. $\binom{12}{8}$

Solutions

a. $\binom{5}{3} = \dfrac{5!}{3!(5-3)!} = \dfrac{5!}{3!\cdot 2!} = \dfrac{5\cdot 4\cdot \cancel{3!}}{\cancel{3!}\cdot 2\cdot 1} = \dfrac{20}{2} = 10$

b. $\binom{4}{1} = \dfrac{4!}{1!(4-1)!} = \dfrac{4!}{1!\cdot 3!} = \dfrac{4\cdot \cancel{3!}}{1\cdot \cancel{3!}} = \dfrac{4}{1} = 4$

c. $\binom{12}{8} = \dfrac{12!}{8!(12-8)!} = \dfrac{12!}{8!\cdot 4!} = \dfrac{12\cdot 11\cdot 10\cdot 9\cdot \cancel{8!}}{\cancel{8!}\cdot 4\cdot 3\cdot 2\cdot 1} = \dfrac{11{,}880}{24} = 495$

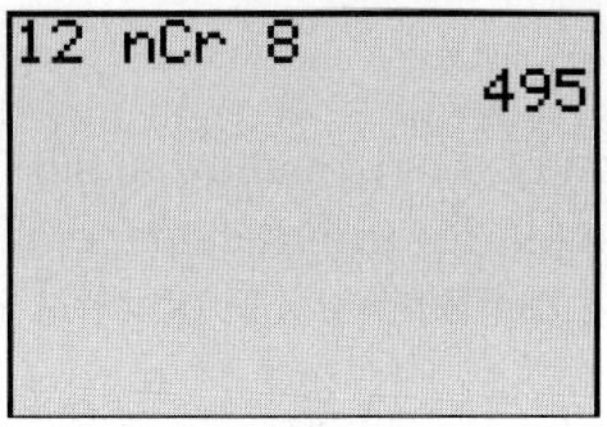

A calculator can be used to compute binomial coefficients. Typically, the key $\boxed{nCr}$ is used. The figure on the left shows the computation of $\binom{12}{8}$ using a graphing utility.

You Try It **Work through this You Try It problem.**

Work Exercises 7–10 in this textbook or in the MyMathLab Study Plan.

OBJECTIVE 3 EXPAND BINOMIALS RAISED TO A POWER USING THE BINOMIAL THEOREM

Now that we know how to compute a binomial coefficient, we can state the Binomial Theorem.

> **Binomial Theorem**
>
> If n is a positive integer then,
>
> $$(a+b)^n = \binom{n}{0}a^n + \binom{n}{1}a^{n-1}b + \binom{n}{2}a^{n-2}b^2 + \cdots + \binom{n}{n}b^n$$
> $$= \sum_{i=0}^{n}\binom{n}{i}a^{n-i}b^i.$$

Example 3 Using the Binomial Theorem to Expand a Binomial Raised to a Power

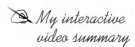

My interactive video summary

Use the Binomial Theorem to expand each binomial.

a. $(x-1)^8$ b. $(\sqrt{x}+y^2)^5$

Solutions

a. Work through the **interactive video** to verify that $(x-1)^8 = x^8 - 8x^7 + 28x^6 - 56x^5 + 70x^4 - 56x^3 + 28x^2 - 8x + 1$ using the Binomial Theorem.

b. The expansion of $(\sqrt{x} + y^2)^5$ is as follows.

$$\begin{aligned}(\sqrt{x} + y^2)^5 &= \binom{5}{0}(\sqrt{x})^5 + \binom{5}{1}(\sqrt{x})^4(y^2) + \binom{5}{2}(\sqrt{x})^3(y^2)^2 + \binom{5}{3}(\sqrt{x})^2(y^2)^3 + \binom{5}{4}(\sqrt{x})(y^2)^4 + \binom{5}{5}(y^2)^5 \\ &= 1\cdot(\sqrt{x})^5 + 5\cdot(\sqrt{x})^4(y^2) + 10\cdot(\sqrt{x})^3(y^2)^2 + 10\cdot(\sqrt{x})^2(y^2)^3 + 5\cdot(\sqrt{x})(y^2)^4 + 1\cdot(y^2)^5 \\ &= x^2\sqrt{x} + 5x^2y^2 + 10x\sqrt{x}y^4 + 10xy^6 + 5\sqrt{x}y^8 + y^{10}\end{aligned}$$

You Try It **Work through this You Try It problem.**

Work Exercises 11–17 in this textbook or in the MyMathLab Study Plan.

OBJECTIVE 4 FIND A PARTICULAR TERM OR A PARTICULAR COEFFICIENT OF A BINOMIAL EXPANSION

We may want to find a particular term of a binomial expansion. Fortunately, we can use the Binomial Theorem to develop a formula for a particular term. We start by writing out the first several terms of $(a + b)^n$ using the Binomial Theorem.

$$(a + b)^n = \binom{n}{0}a^n + \binom{n}{1}a^{n-1}b + \binom{n}{2}a^{n-2}b^2 + \binom{n}{3}a^{n-3}b^3 + \cdots + \binom{n}{n}b^n$$

The first term is $\binom{n}{0}a^n$. The second term is $\binom{n}{1}a^{n-1}b$. The third term is $\binom{n}{2}a^{n-2}b^2$. Following this pattern, we can see that the formula for the $(r + 1)$st term (for $r \geq 0$) is given by $\binom{n}{r}a^{n-r}b^r$.

Formula for the $(r + 1)$st Term of a Binomial Expansion

If n is a positive integer and if $r \geq 0$, then the $(r + 1)$st term of the expansion of $(a + b)^n$ is given by

$$\binom{n}{r}a^{n-r}b^r = \frac{n!}{r!\cdot(n-r)!}a^{n-r}b^r.$$

Example 4 Finding a Particular Term of a Binomial Expansion

My video summary

Find the third term of the expansion of $(2x - 3)^{10}$.

Solution Since we want to find the third term of this expansion, we will use the formula for the $(r + 1)$st term, which is equal to $\binom{n}{r}a^{n-r}b^r$ for $r = 2, n = 10, a = 2x$, and $b = -3$. Therefore, the third term is

$$\binom{10}{2}(2x)^{10-2}(-3)^2 = 45(256x^8)(9) = 103{,}680x^8.$$

Watch this **video** to see every step of this solution.

(eText Screens 22.4-1–22.4-18)

Example 5 Finding a Particular Coefficient of a Binomial Expansion

My video summary

Find the coefficient of x^7 in the expansion of $(x + 4)^{11}$.

Solution The formula for the $(r + 1)$st term of the expansion of $(x + 4)^{11}$ is given by the formula $\binom{11}{r}x^{11-r}4^r$. The term containing x^7 occurs when $11 - r = 7$. Solving this equation for r we get $r = 4$. Therefore, the term involving x^7 is $\binom{11}{4}x^7 4^4$. Simplifying this expression we get $84{,}480x^7$. Thus, the coefficient of x^7 is 84,480. Watch this **video** to see this solution worked out in detail.

You Try It **Work through this You Try It problem.**

Work Exercises 18–25 in this textbook or in the MyMathLab Study Plan.

22.4 Exercises

In Exercises 1–6, use Pascal's triangle to expand each binomial.

You Try It

1. $(m + n)^6$ **2.** $(x + 5)^4$ **3.** $(x - y)^7$

4. $(x - 3)^5$ **5.** $(2x + 3y)^6$ **6.** $(3x^2 - 4y^3)^4$

In Exercises 7–10, evaluate each binomial coefficient.

You Try It

7. $\binom{7}{1}$ **8.** $\binom{10}{4}$ **9.** $\binom{7}{7}$ **10.** $\binom{23}{3}$

In Exercises 11–17, use the Binomial Theorem to expand each binomial.

You Try It

11. $(x + 2)^7$ **12.** $(x - 3)^6$ **13.** $(4x + 1)^5$

14. $(x + 3y)^4$ **15.** $(5x - 3y)^5$ **16.** $(x^4 + y^5)^6$

17. $(\sqrt{x} - \sqrt{2})^4$

You Try It

18. Find the sixth term of the expansion of $(x + 4)^9$.

19. Find the fifth term of the expansion of $(a - b)^8$.

20. Find the third term of the expansion of $(3c - d)^7$.

21. Find the seventh term of the expansion of $(3x + 2)^{10}$.

You Try It

22. Find the coefficient of x^5 in the expansion of $(x - 3)^{11}$.

23. Find the coefficient of x^4 in the expansion of $(4x + 1)^{12}$.

24. Find the coefficient of x^0 in the expansion of $\left(x^2 + \frac{1}{x}\right)^{12}$.

25. Find the coefficient of x^{10} in the expansion of $\left(x - \frac{3}{\sqrt{x}}\right)^{19}$.

MODULE TWENTY-THREE

Additional Topics

MODULE TWENTY-THREE CONTENTS

23.1 Synthetic Division

THINGS TO KNOW

Before working through this topic, be sure you are familiar with the following concepts:

		VIDEO	ANIMATION	INTERACTIVE
You Try It	1. Divide a Polynomial by a Monomial (Topic 14.7, **Objective 2**)	▤		
You Try It	2. Divide Polynomials Using Long Division (Topic 14.7, **Objective 3**)	▤		

OBJECTIVE

1 Divide a Polynomial by a Binomial Using Synthetic Division

OBJECTIVE 1 DIVIDE A POLYNOMIAL BY A BINOMIAL USING SYNTHETIC DIVISION

In **Topic 14.7**, we divided **polynomials** using **long division**. If we divide a polynomial by a **binomial** of the form $x - c$, then a shortcut method called **synthetic division** can be used instead. Synthetic division removes the repetition involved with long division. If we use zeros as placeholders for missing **powers** and line up **like terms**, we only need to work with the **coefficients** because the position of the coefficient will determine the power of the term.

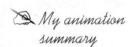

My animation summary

Watch this **animation** to see how synthetic division works as compared to polynomial long division.

We illustrate the **synthetic division** process in the following example.

Example 1 Dividing Polynomials Using Synthetic Division

Divide $2x^4 + 9x^3 - 12x + 1$ by $x + 5$ using synthetic division.

Solution

Step 1. Write both polynomials in **standard form**. Both polynomials are already in standard form, so we move on to Step 2.

Step 2. Rewrite the divisor as a binomial of the form $x - c$. Since $x + 5 = x - (-5)$, $c = -5$.

Step 3. Write down c and the coefficients of the dividend. If there are powers "missing," insert a 0 coefficient as a placeholder for the missing term(s).

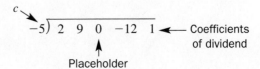

$$\overset{c}{-5)}\ 2 \quad 9 \quad \underset{\uparrow \text{ Placeholder}}{0} \quad -12 \quad 1 \leftarrow \text{Coefficients of dividend}$$

Step 4. Leave some space under the coefficients, draw a horizontal line, and drop down the **leading coefficient** under the line.

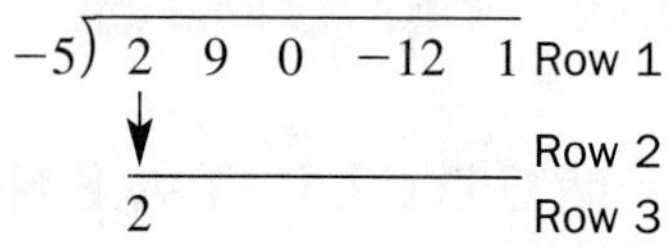

Step 5. Multiply c by the entry in Row 3 and put the result in the next position to the right in Row 2. Add the values from Rows 1 and 2 to get the next entry in Row 3.

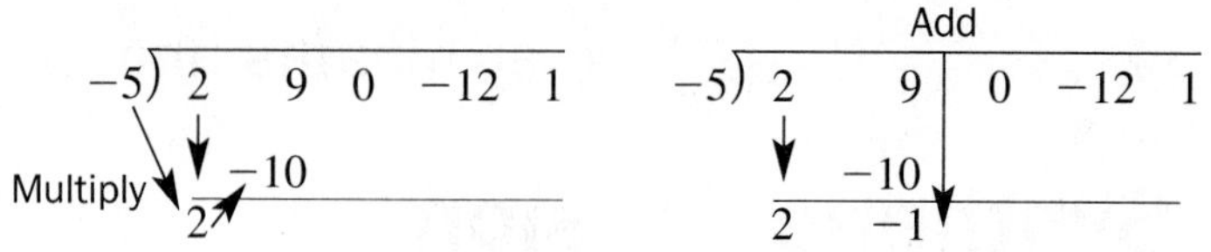

Step 6. Continue with Step 5 until Row 3 is complete.

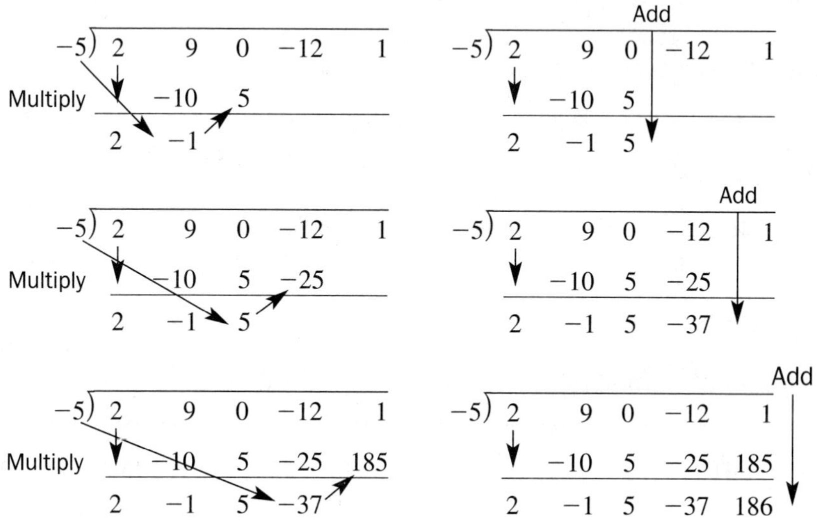

The **quotient** and **remainder** are determined by looking at the completed Row 3 of the **synthetic division**. Because the **divisor** is a **polynomial** of **degree** 1, the quotient will have a degree that is one less than the **dividend**. Since the dividend has a degree of 4, the quotient will have a degree of 3. The last entry in the last row is the remainder. The remaining entries in that row are the coefficients of the quotient when written in **standard form**.

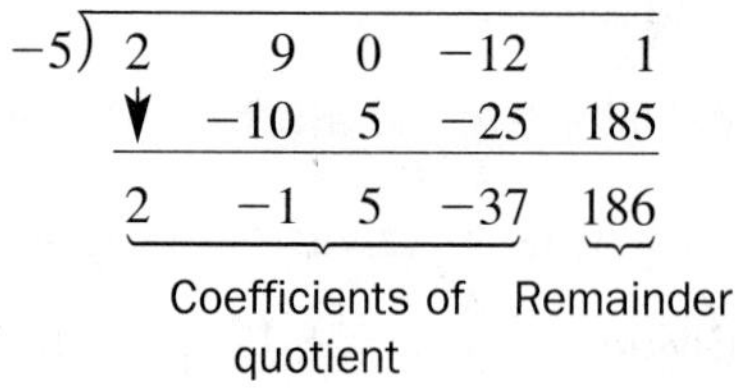

Here, the quotient is $2x^3 - 1x^2 + 5x - 37$, and the remainder is 186. So,

$$\frac{2x^4 + 9x^3 - 12x + 1}{x + 5} = 2x^3 - x^2 + 5x - 37 + \frac{186}{x + 5}.$$

View this **popup box** to see how to solve this same problem using **long division**.

Example 2 Dividing Polynomials Using Synthetic Division

My video summary

Divide $2x^4 - 3x^2 + 5x - 30$ by $x - 2$ using synthetic division.

Solution The **divisor** is a **binomial** of the form $x - c$ where $c = 2$. We start the **synthetic division** process as follows:

$$\begin{array}{r|rrrrrl} 2 & 2 & 0 & -3 & 5 & -30 & \\ & \downarrow & & & & & \leftarrow \text{Products} \\ \hline & 2 & & & & & \leftarrow \text{Sums} \end{array}$$

Complete this process on your own. View the **answer**, or watch this **video** for a detailed solution.

You Try It **Work through this You Try It problem.**

Work Exercises 1–8 in this textbook or in the MyMathLab Study Plan.

Example 3 Dividing Polynomials Using Synthetic Division

My video summary

Divide $4x^3 - 8x^2 + 7x - 4$ by $x - \frac{1}{2}$ using synthetic division.

Solution The **divisor** is a **binomial** of the form $x - c$ with $c = \frac{1}{2}$. Complete the **synthetic division** on your own. View the **answer**, or watch this **video** for the full solution.

You Try It **Work through this You Try It problem.**

Work Exercises 9–10 in this textbook or in the MyMathLab Study Plan.

CAUTION

Synthetic division can only be used when dividing a **polynomial** by a binomial of the form $x - c$.

23.1 Exercises

In Exercises 1–10, divide using synthetic division.

You Try It

1. $x^2 - x - 17$ divided by $x - 5$

2. $x^2 + \frac{16}{3}x - 4$ divided by $x + 6$

3. $(4x^3 + 5x - 7) \div (x + 2)$

4. $(3x^3 + 11x^2 + 2x + 31) \div (x + 4)$

5. $\dfrac{x^3 - 12x + 9}{x - 3}$

6. $\dfrac{7x^4 - 12x^3 - 8x^2 + 7x + 2}{x - 2}$

7. $\dfrac{x^4 + 5x^2 - 41}{x - 2}$

8. $\dfrac{x^4 + 15x^2 - 16}{x + 1}$

You Try It

9. $\dfrac{15x^3 + 14x^2 - 43x + 14}{x - \frac{2}{5}}$

10. $\dfrac{3x^3 - 20x^2 - x - 7}{x + \frac{1}{3}}$

23.2 Solving Systems of Linear Equations Using Matrices

THINGS TO KNOW

Before working through this topic, be sure you are familiar with the following concepts:

VIDEO ANIMATION INTERACTIVE

You Try It 1. Solve Systems of Linear Equations by Substitution (Topic 13.2, **Objective 1**)

You Try It 2. Solve Special Systems by Substitution (Topic 13.2, **Objective 2**)

You Try It 3. Solve Systems of Linear Equations by Elimination (Topic 13.3, **Objective 1**)

You Try It 4. Solve Special Systems by Elimination (Topic 13.3, **Objective 2**)

You Try It 5. Solve Systems of Linear Equations in Three Variables (Topic 13.6, **Objective 2**)

OBJECTIVES

1 Write an Augmented Matrix

2 Solve Systems of Two Equations Using Matrices

3 Solve Systems of Three Equations Using Matrices

OBJECTIVE 1 WRITE AN AUGMENTED MATRIX

In this topic we look at how to solve **systems of linear equations** by using a **matrix**. A matrix is a rectangular array of numbers arranged in rows and columns. The following are examples of **matrices** (the plural of *matrix*).

$$\begin{bmatrix} 3 & 7 \\ -2 & 5 \end{bmatrix} \quad \begin{bmatrix} 0 & 3 & 1 \\ 2 & -4 & 1 \end{bmatrix} \quad \begin{bmatrix} 1 & 0 & 0 \\ 0 & 1 & 0 \\ 0 & 0 & 1 \end{bmatrix} \quad \begin{bmatrix} 1 & 0 & -1 & 5 \\ 0 & 1 & 4 & -3 \\ 0 & 0 & 1 & \frac{1}{2} \end{bmatrix}$$

The size of a matrix is determined by the number of rows and columns. The following matrix is a 2×3 (read *two by three*) matrix because it has 2 rows and 3 columns.

$$\begin{matrix} \text{row 1} \rightarrow \\ \text{row 2} \rightarrow \end{matrix} \begin{bmatrix} 4 & 9 & -6 \\ 3 & -1 & 2 \end{bmatrix}$$

column 1 ↑ column 3

column 2

(eText Screens 23.2-1–23.2-26)

When solving systems of equations using **elimination**, we wrote the equations in the system in **standard form** lining up the **variables**. Doing so gave a similar rectangular structure to our system. We then used the coefficients of the equations to eliminate variables because the solution to a system depends on the **coefficients** of the equations, not on the variables.

Once we have a system written in **standard form**, we can simplify the **elimination** process by first using the coefficients of the equations to form an **augmented matrix** that can be used to solve the system. Consider the following illustration.

Coefficients of x — Coefficients of y — Equal signs — Constants

$$\begin{array}{l} \text{Equation 1} \\ \text{Equation 2} \end{array} \begin{cases} 3x + 2y = 11 \\ 8x - 9y = 4 \end{cases} \longrightarrow \begin{array}{l} \text{Row 1} \\ \text{Row 2} \end{array} \left[\begin{array}{cc|c} 3 & 2 & 11 \\ 8 & -9 & 4 \end{array}\right]$$

System in standard form — Augmented matrix

An augmented matrix for a system of equations in standard form is a **matrix** consisting of the **coefficients** of the variables and the constants from the right-hand side of the equations, separated by a vertical line. There is a column of coefficients for each variable and an additional column for the constants. Each row in the matrix corresponds to an equation in the system. Therefore, systems of two equations in two variables will produce a 2×3 augmented matrix, whereas systems of three equations in three variables will produce a 3×4 augmented matrix.

Example 1 Writing an Augmented Matrix

Write the corresponding **augmented matrix** for each system of equations.

a. $\begin{cases} -2x + 7y = 9 \\ 8x + 3y = 0 \end{cases}$ **b.** $\begin{cases} 6x - 3y = 11 \\ y = 2x - 5 \end{cases}$ **c.** $\begin{cases} 5x - 2y + z = 18 \\ x + 3z = -5 \\ 3y = 9 + z \end{cases}$

Solutions

a. The equations are both in **standard form**, so we can immediately write the augmented matrix.

$$\begin{cases} -2x + 7y = 9 \\ 8x + 3y = 0 \end{cases} \rightarrow \left[\begin{array}{cc|c} -2 & 7 & 9 \\ 8 & 3 & 0 \end{array}\right]$$

b. The first equation is in standard form, but the second is not. We can put it in standard form by subtracting $2x$ from both sides.

$$\underset{\text{Original system}}{\begin{cases} 6x - 3y = 11 \\ y = 2x - 5 \end{cases}} \rightarrow \underset{\text{Standard form}}{\begin{cases} 6x - 3y = 11 \\ -2x + 1y = -5 \end{cases}} \rightarrow \underset{\text{Augmented matrix}}{\left[\begin{array}{cc|c} 6 & -3 & 11 \\ -2 & 1 & -5 \end{array}\right]}$$

My video summary

c. Start by writing the equations in **standard form**. If a variable is missing from an equation, it may be helpful to include the variable with a 0 coefficient.

$$\underset{\text{Original system}}{\begin{cases} 5x - 2y + z = 18 \\ x + 3z = -5 \\ 3y = 9 + z \end{cases}} \rightarrow \underset{\text{Standard form}}{\begin{cases} 5x - 2y + 1z = 18 \\ 1x + 0y + 3z = -5 \\ 0x + 3y - 1z = 9 \end{cases}} \quad \begin{array}{l} \leftarrow +z = +1z \\ \leftarrow x = 1x \\ \leftarrow -z = -1z \end{array}$$

Now write the **augmented matrix**. View the **answer**, or watch this **video** for a complete solution.

(eText Screens 23.2-1–23.2-26)

Notice that in the standard form we wrote $+z$ as $+1z$, x as $1x$, and $-z$ as $-1z$. This was to help make sure we entered the correct coefficient in the augmented matrix.

You Try It **Work through this You Try It problem.**

Work Exercises 1–6 in this textbook or in the MyMathLab Study Plan.

OBJECTIVE 2 SOLVE SYSTEMS OF TWO EQUATIONS USING MATRICES

To solve a **system of equations** using a **matrix** means to find an equivalent matrix from which the solution can be easily obtained. Two matrices are **equivalent matrices** if they represent systems that have the same solution set. We obtain equivalent matrices by applying **row operations**.

> **Row Operations**
>
> Applying any of the following operations to a matrix will result in an equivalent matrix.
>
> **1.** Interchange any two rows.
>
> **2.** Multiply the elements of a row by the same non-zero constant.
>
> **3.** Add a multiple of one row to another row.

Note that these operations are the same as those applied to equations of a system when solving the system by the **elimination** method. We will use the following notation to describe the row operations.

Notation	**Meaning**
$R_i \Leftrightarrow R_j$	Interchange Rows i and j
$kR_i:\ \text{New } R_i$	k times Row i becomes the new Row i
$kR_i + R_j:\ \text{New } R_j$	k times Row i plus Row j becomes the new Row j

View this **popup** to see an example of each row operation.

For a **system of two equations**, we apply **row operations** until we obtain a **matrix** of the form

$$\left[\begin{array}{cc|c} 1 & a & b \\ 0 & 1 & c \end{array}\right].$$

This is called the **row-echelon form** of the matrix. In this form, there are 1's along the diagonal from the upper left to lower right, and 0's below the diagonal. From there we obtain a system of equations that can be solved by **back-substitution**. Consider the following.

Coefficients of y — Equal signs
Coefficients of x — Constants

$$\left[\begin{array}{cc|c} 1 & -3 & 7 \\ 0 & 1 & -2 \end{array}\right]$$

Augmented matrix in row-echelon form

$$\begin{cases} 1x - 3y = 7 \\ 0x + 1y = -2 \end{cases} \longrightarrow \begin{cases} x - 3y = 7 \\ y = -2 \end{cases}$$

Corresponding system

The second equation in the system tells us that $y = -2$. To find the value for x, we can substitute -2 for y in the first equation and solve for x.

$$\begin{aligned} \text{Original equation:} \quad & x - 3y = 7 \\ \text{Substitute } -2 \text{ for } y\text{:} \quad & x - 3(-2) = 7 \\ \text{Multiply:} \quad & x + 6 = 7 \\ \text{Subtract 6 from both sides:} \quad & x = 1 \end{aligned}$$

So the **solution** to the system is the ordered pair $(1, -2)$.

Example 2 Solving a System of Two Equations Using Matrices

Use **matrices** to solve each of the following systems.

a. $\begin{cases} 2x - 5y = 26 \\ 3x + 2y = 1 \end{cases}$ **b.** $\begin{cases} y = 3x + 2 \\ 6x + \dfrac{1}{2}y = 6 \end{cases}$

Solutions

a. Both equations are in **standard form**, so we can write the **augmented matrix**.

$$\begin{cases} 2x - 5y = 26 \\ 3x + 2y = 1 \end{cases} \rightarrow \left[\begin{array}{cc|c} 2 & -5 & 26 \\ 3 & 2 & 1 \end{array}\right]$$

We now apply **row operations** to put the matrix in **row-echelon form**. Our first step is to obtain a 1 in row 1, column 1. We do this by multiplying each entry in the first row by $\frac{1}{2}$.

$$\left[\begin{array}{cc|c} 2 & -5 & 26 \\ 3 & 2 & 1 \end{array}\right] \xrightarrow{\frac{1}{2}R_1} \left[\begin{array}{cc|c} \frac{1}{2}(2) & \frac{1}{2}(-5) & \frac{1}{2}(26) \\ 3 & 2 & 1 \end{array}\right] = \left[\begin{array}{cc|c} 1 & -\frac{5}{2} & 13 \\ 3 & 2 & 1 \end{array}\right]$$

Next we use row 1 to get a 0 in row 2, column 1. We do this by multiplying the entries in row 1 by -3 and adding the results to the entries in row 2.

$$\left[\begin{array}{cc|c} 1 & -\frac{5}{2} & 13 \\ 3 & 2 & 1 \end{array}\right] \xrightarrow{-3R_1 + R_2} \left[\begin{array}{cc|c} 1 & -\frac{5}{2} & 13 \\ -3(1) + 3 & -3\left(-\frac{5}{2}\right) + 2 & -3(13) + 1 \end{array}\right] = \left[\begin{array}{cc|c} 1 & -\frac{5}{2} & 13 \\ 0 & \frac{19}{2} & -38 \end{array}\right]$$

Lastly, we get a 1 in row 2, column 2. We do this by multiplying the entries in row 2 by $\frac{2}{19}$.

$$\left[\begin{array}{cc|c} 1 & -\frac{5}{2} & 13 \\ 0 & \frac{19}{2} & -38 \end{array}\right] \xrightarrow{\frac{2}{19}R_2} \left[\begin{array}{cc|c} 1 & -\frac{5}{2} & 13 \\ \frac{2}{19}(0) & \frac{2}{19}\left(\frac{19}{2}\right) & \frac{2}{19}(-38) \end{array}\right] = \left[\begin{array}{cc|c} 1 & -\frac{5}{2} & 13 \\ 0 & 1 & -4 \end{array}\right]$$

The corresponding **equivalent system** is

$$\begin{cases} x - \dfrac{5}{2}y = 13 \\ y = -4 \end{cases}$$

The second equation tells us that $y = -4$. To find the value for x, we can **substitute** -4 for y in the first equation and solve for x.

$$\begin{aligned} \text{First equation:} \quad & x - \frac{5}{2}y = 13 \\ \text{Substitute } -4 \text{ for } y\text{:} \quad & x - \frac{5}{2}(-4) = 13 \\ \text{Multiply:} \quad & x + 10 = 13 \\ \text{Subtract 10 from both sides:} \quad & x = 3 \end{aligned}$$

The **solution** to the system is the ordered pair $(3, -4)$.

(eText Screens 23.2-1–23.2-26)

My video summary

b. Start by writing the equations in the system in **standard form.**

$$\begin{cases} y = 3x + 2 \\ 6x + \frac{1}{2}y = 6 \end{cases} \rightarrow \begin{cases} -3x + y = 2 \\ 6x + \frac{1}{2}y = 6 \end{cases}$$

Now write the corresponding **augmented matrix.**

$$\begin{cases} -3x + 1y = 2 \\ 6x + \frac{1}{2}y = 6 \end{cases} \rightarrow \left[\begin{array}{cc|c} -3 & 1 & 2 \\ 6 & \frac{1}{2} & 6 \end{array}\right]$$

Use **row operations** to solve the system on your own. View the **answer**, or watch this **video** for a complete solution.

CAUTION

Recall that a system of equations may have one solution, no solution, or an infinite number of solutions.

You Try It **Work through this You Try It problem.**

Work Exercises 7–11 in this textbook or in the MyMathLab Study Plan.

Example 3 Solving a System of Two Equations Using Matrices

My video summary

Use **matrices** to solve the following system.

$$\begin{cases} 3x + \frac{1}{2}y = 4 \\ 2y = -12x + 16 \end{cases}$$

Solution Work through the following, or watch this **video** for the complete solution. Start by writing the equations in the system in **standard form.**

$$\begin{cases} 3x + \frac{1}{2}y = 4 \\ 2y = -12x + 16 \end{cases} \rightarrow \begin{cases} 3x + \frac{1}{2}y = 4 \\ 12x + 2y = 16 \end{cases}$$

Now write the corresponding **augmented matrix.**

$$\begin{cases} 3x + \frac{1}{2}y = 4 \\ 12x + 2y = 16 \end{cases} \rightarrow \left[\begin{array}{cc|c} 3 & \frac{1}{2} & 4 \\ 12 & 2 & 16 \end{array}\right]$$

Apply **row operations** to put the matrix in **row-echelon form.**

$$\left[\begin{array}{cc|c} 3 & \frac{1}{2} & 4 \\ 12 & 2 & 16 \end{array}\right] \xrightarrow{\frac{1}{3}R_1} \left[\begin{array}{cc|c} 1 & \frac{1}{6} & \frac{4}{3} \\ 12 & 2 & 16 \end{array}\right] \xrightarrow{-12R_1 + R_2} \left[\begin{array}{cc|c} 1 & \frac{1}{6} & \frac{4}{3} \\ 0 & 0 & 0 \end{array}\right]$$

Notice that the last row is all zeros. This corresponds to the equation $0 = 0$, which is an **identity.**

The second equation in the system is just a multiple of the first equation, so it provides us with no new information. Therefore, this system is **dependent** and **consistent** and has an infinite number of solutions. Both equations in the system describe the same line so we can use either to obtain the **slope-intercept form** of the line.

Second equation: $2y = -12x + 16$

Divide both sides by 2: $\dfrac{2y}{2} = \dfrac{-12x + 16}{2}$

Simplify: $y = -6x + 8$

The solution set is $\{(x, y) | y = -6x + 8\}$

When solving a **system of two equations** using **matrices**, if we obtain a row of all zeros (as in this example), the system will have an infinite number of solutions. However, if we obtain a row that is all zeros to the left of the vertical bar, but nonzero to the right, then the system has no solution. Consider the following.

$$\left[\begin{array}{cc|c} 1 & 3 & 8 \\ 0 & 0 & 2 \end{array}\right]$$

The second row corresponds to the equation $0x + 0y = 2$, or $0 = 2$. Since this statement is false, the system has no solution.

You Try It **Work through this You Try It problem.**

Work Exercises 12–18 in this textbook or in the MyMathLab Study Plan.

OBJECTIVE 3 SOLVE SYSTEMS OF THREE EQUATIONS USING MATRICES

For a **system of three equations**, we again apply **row operations** until we obtain a **matrix** in the following **row-echelon form**.

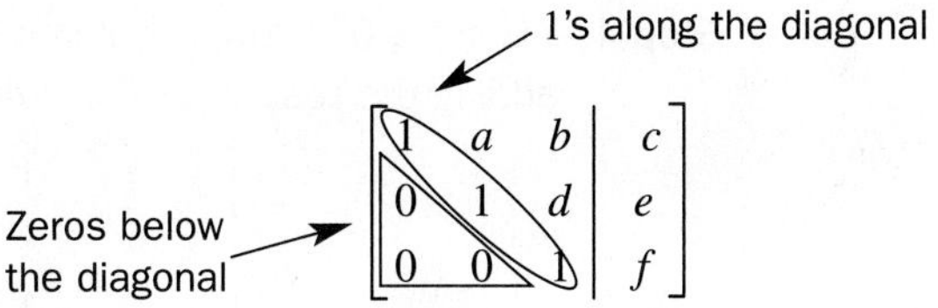

As with two equations, this form has 1's along the diagonal from the upper left to lower right, and 0's below the diagonal. In general, we can use the following steps to write the row-echelon form.

Writing Row-Echelon Form (Three Rows)

Step 1. Get a 1 in the row 1, column 1 entry.

Step 2. Use row 1 to obtain 0's in column 1 below the diagonal.

Step 3. Get a 1 in the row 2, column 2 entry.

Step 4. Use row 2 to obtain 0's in column 2 below the diagonal.

Step 5. Get a 1 in the row 3, column 3 entry.

Example 4 Solving a System of Three Equations Using Matrices

Use **matrices** to solve each of the following systems.

a. $\begin{cases} 2x + y - 3z = -3 \\ x + 2y + z = 4 \\ -3x + y - 4z = 1 \end{cases}$

b. $\begin{cases} 2x + 2y + z = 1 \\ 5x + 2y = 30 + 3z \\ 3x + 4z = -11 \end{cases}$

(eText Screens 23.2-1–23.2-26)

Solutions

a. All three equations are in **standard form**, so we can write the **augmented matrix**.

$$\begin{cases} 2x + 1y - 3z = -3 \\ 1x + 2y + 1z = 4 \\ -3x + 1y - 4z = 1 \end{cases} : \left[\begin{array}{ccc|c} 2 & 1 & -3 & -3 \\ 1 & 2 & 1 & 4 \\ -3 & 1 & -4 & 1 \end{array}\right]$$

We now apply **row operations** to put the matrix in **row-echelon form**.

Step 1. We obtain a 1 in row 1, column 1 by interchanging row 1 and row 2.

$$\left[\begin{array}{ccc|c} 2 & 1 & -3 & -3 \\ 1 & 2 & 1 & 4 \\ -3 & 1 & -4 & 1 \end{array}\right] \xrightarrow{R_1 \Leftrightarrow R_2} \left[\begin{array}{ccc|c} 1 & 2 & 1 & 4 \\ 2 & 1 & -3 & -3 \\ -3 & 1 & -4 & 1 \end{array}\right]$$

Step 2. We get a 0 in row 2, column 1 by multiplying the entries in row 1 by -2 and adding the results to the entries in row 2. We get a 0 in row 3, column 1 by multiplying the entries in row 1 by 3 and adding the results to the entries in row 3.

$$\left[\begin{array}{ccc|c} 1 & 2 & 1 & 4 \\ 2 & 1 & -3 & -3 \\ -3 & 1 & -4 & 1 \end{array}\right] \begin{array}{c} \xrightarrow{-2R_1+R_2} \\ \xrightarrow{3R_1+R_3} \end{array} \left[\begin{array}{ccc|c} 1 & 2 & 1 & 4 \\ 0 & -3 & -5 & -11 \\ 0 & 7 & -1 & 13 \end{array}\right]$$

Step 3. We get a 1 in row 2, column 2 by multiplying the entries in row 2 by $-\frac{1}{3}$.

$$\left[\begin{array}{ccc|c} 1 & 2 & 1 & 4 \\ 0 & -3 & -5 & -11 \\ 0 & 7 & -1 & 13 \end{array}\right] \xrightarrow{-\frac{1}{3}R_2} \left[\begin{array}{ccc|c} 1 & 2 & 1 & 4 \\ 0 & 1 & \frac{5}{3} & \frac{11}{3} \\ 0 & 7 & -1 & 13 \end{array}\right]$$

Step 4. We get a 0 in row 3, column 2 by multiplying the entries in row 2 by -7 and adding the results to the entries in row 3.

$$\left[\begin{array}{ccc|c} 1 & 2 & 1 & 4 \\ 0 & 1 & \frac{5}{3} & \frac{11}{3} \\ 0 & 7 & -1 & 13 \end{array}\right] \xrightarrow{-7R_2+R_3} \left[\begin{array}{ccc|c} 1 & 2 & 1 & 4 \\ 0 & 1 & \frac{5}{3} & \frac{11}{3} \\ 0 & 0 & -\frac{38}{3} & -\frac{38}{3} \end{array}\right]$$

Step 5. We get a 1 in row 3, column 3 by multiplying the entries in row 3 by $-\frac{3}{38}$.

$$\left[\begin{array}{ccc|c} 1 & 2 & 1 & 4 \\ 0 & 1 & \frac{5}{3} & \frac{11}{3} \\ 0 & 0 & -\frac{38}{3} & -\frac{38}{3} \end{array}\right] \xrightarrow{-\frac{3}{38}R_3} \left[\begin{array}{ccc|c} 1 & 2 & 1 & 4 \\ 0 & 1 & \frac{5}{3} & \frac{11}{3} \\ 0 & 0 & 1 & 1 \end{array}\right]$$

The corresponding **equivalent system** is

$$\begin{cases} x + 2y + z = 4 \\ y + \frac{5}{3}z = \frac{11}{3} \\ z = 1 \end{cases}$$

The third equation tells us that $z = 1$. To find the value for y, we can **substitute** 1 for z in the second equation and solve for y.

Second equation: $y + \frac{5}{3}z = \frac{11}{3}$

Substitute 1 for z: $y + \frac{5}{3}(1) = \frac{11}{3}$

Multiply: $y + \frac{5}{3} = \frac{11}{3}$

Subtract $\frac{5}{3}$ from both sides: $y = \frac{6}{3} = 2$

We now know that $y = 2$ and $z = 1$. We can find the value of x by **substituting** these values into the first equation and solving for x.

First equation: $x + 2y + z = 4$

Substitute 2 for y and 1 for z: $x + 2(2) + (1) = 4$

Multiply: $x + 4 + 1 = 4$

Combine like terms: $x + 5 = 4$

Subtract 5 from both sides: $x = -1$

The **solution** to the system is the ordered triple $(-1, 2, 1)$.

My video summary

b. Start by writing the equations in the system in **standard form.**

$$\begin{cases} 2x + 2y + z = 1 \\ 5x + 2y = 30 + 3z \\ 3x + 4z = -11 \end{cases} \rightarrow \begin{cases} 2x + 2y + z = 1 \\ 5x + 2y - 3z = 30 \\ 3x + 0y + 4z = -11 \end{cases}$$

Now write the corresponding **augmented matrix.**

$$\begin{cases} 2x + 2y + 1z = 1 \\ 5x + 2y - 3z = 30 \\ 3x + 0y + 4z = -11 \end{cases} \rightarrow \left[\begin{array}{ccc|c} 2 & 2 & 1 & 1 \\ 5 & 2 & -3 & 30 \\ 3 & 0 & 4 & -11 \end{array}\right]$$

Use **row operations** to solve the system on your own. View the **answer**, or watch this **video** for a complete solution.

You Try It **Work through this You Try It problem.**

Work Exercises 19–21 in this textbook or in the MyMathLab Study Plan.

Example 5 Solving a Speical System of Three Equations Using Matrices

My video summary

Use **matrices** to solve the following system.

$$\begin{cases} x + 2y - 3z = 5 \\ -2x - 3y + 4z = -6 \\ 2x + 4y - 6z = 10 \end{cases}$$

Solution Work through the following, or watch this **video** for the complete solution.

All three equations are in **standard form,** so we can write the **augmented matrix.**

$$\begin{cases} 1x + 2y - 3z = 5 \\ -2x - 3y + 4z = -6 \\ 2x + 4y - 6z = 10 \end{cases} \rightarrow \left[\begin{array}{ccc|c} 1 & 2 & -3 & 5 \\ -2 & -3 & 4 & -6 \\ 2 & 4 & -6 & 10 \end{array}\right]$$

We now apply **row operations** to put the matrix in **row-echelon form.**

Step 1. There is already a 1 in row 1, column 1.

Step 2. We get a 0 in row 2, column 1 by multiplying the entries in row 1 by 2 and adding the results to the entries in row 2. We get a 0 in row 3, column 1 by multiplying the entries in row 1 by -2 and adding the results to the entries in row 3.

$$\left[\begin{array}{rrr|r} 1 & 2 & -3 & 5 \\ -2 & -3 & 4 & -6 \\ 2 & 4 & -6 & 10 \end{array}\right] \xrightarrow[-2R_1+R_3]{2R_1+R_2} \left[\begin{array}{rrr|r} 1 & 2 & -3 & 5 \\ 0 & 1 & -2 & 4 \\ 0 & 0 & 0 & 0 \end{array}\right]$$

Notice that the last row is all zeros. This corresponds to the **identity** $0 = 0$ and indicates the system may have an infinite number of solutions.

The corresponding **equivalent system** is

$$\begin{cases} x + 2y - 3z = 5 \\ \quad\;\; y - 2z = 4. \\ \qquad\quad 0 = 0 \end{cases}$$

The third equation is an identity so provides no information about the solution. The second equation can be solved for y in terms of z.

Second equation: $y - 2z = 4$

Add $2z$ to both sides: $y = 2z + 4$

We now know that $y = 2z + 4$. We can **substitute** $2z + 4$ for y in the first equation, then solve the equation for x in terms of z.

First equation: $x + 2y - 3z = 5$

Substitute $2z + 4$ for y: $x + 2(2z + 4) - 3z = 5$

Distribute: $x + 4z + 8 - 3z = 5$

Combine like terms: $x + z + 8 = 5$

Subtract 8 from both sides: $x + z = -3$

Subtract z from both sides: $x = -z - 3$

Having expressed both x and y in terms of the third variable z, the system has an infinite number of solutions. We can express the **solutions** to the system as the **ordered triple** $(-z - 3, 2z + 4, z)$, where z is any real number.

You Try It **Work through this You Try It problem.**

Work Exercises 22–30 in this textbook or in the MyMathLab Study Plan.

CAUTION

Unlike the two equation case, if we obtain a row of all zeros when solving a system of three equations, there is not necessarily an infinite number of solutions. We must make sure that no other row yields a **contradiction**. Consider the following.

$$\left[\begin{array}{ccc|c} 1 & 3 & 0 & 2 \\ 0 & 0 & 0 & 3 \\ 0 & 0 & 0 & 0 \end{array}\right]$$

The third row corresponds to the equation $0 = 0$, which is an **identity**. However, the second row corresponds to the equation $0 = 3$, which is a contradiction. If any row yields a contradiction, the system has no solution. Therefore, the system represented by this **matrix** has no solution.

(eText Screens 23.2-1–23.2-26)

23.2 Exercises

In Exercises 1–6, write the corresponding augmented matrix for each system.

You Try It

1. $\begin{cases} 2x + 3y = 5 \\ 5x - 2y = 1 \end{cases}$

2. $\begin{cases} 5x - y = 3 \\ 2y = x + 1 \end{cases}$

3. $\begin{cases} \frac{1}{2}x - \frac{1}{3}y = 0 \\ \frac{2}{7}x + \frac{5}{9}y = 2 \end{cases}$

4. $\begin{cases} 2x + 3y - 2z = 7 \\ 5x - 2y + z = 4 \\ 3x + 2y - 4z = 1 \end{cases}$

5. $\begin{cases} 7x - 2y + 5z = 3 \\ 5y + 2z = -2 \\ 3x - 2z = 1 \end{cases}$

6. $\begin{cases} 2x = 1 - 3y \\ 5y = 2 + 2z \\ 3z = 2x - 5 \end{cases}$

In Exercises 7–30, solve each system of linear equations using matrices.

You Try It

7. $\begin{cases} x + 3y = -6 \\ 3x + 2y = -4 \end{cases}$

8. $\begin{cases} x - 5y = 6 \\ 5x - y = 6 \end{cases}$

9. $\begin{cases} 3x + 2y = 4 \\ 6x + 5y = 7 \end{cases}$

10. $\begin{cases} \frac{1}{2}x + 2y = 8 \\ 3x - \frac{1}{3}y = 11 \end{cases}$

11. $\begin{cases} 2x - \frac{1}{2}y = 11 \\ 4x + 3y = 14 \end{cases}$

12. $\begin{cases} 3x - 2y = -1 \\ -6x + 4y = 2 \end{cases}$

You Try It

13. $\begin{cases} 2x + 3y = 2 \\ 4x - 6y = 0 \end{cases}$

14. $\begin{cases} 3x - 2y = 3 \\ 6x + 6y = 1 \end{cases}$

15. $\begin{cases} 2x + \frac{1}{2}y = 3 \\ y = 7 - 4x \end{cases}$

16. $\begin{cases} 3y = 13 + 2x \\ 6x + 2y = -6 \end{cases}$

17. $\begin{cases} \frac{1}{2}x + \frac{1}{3}y = 3 \\ x + y = 7 \end{cases}$

18. $\begin{cases} 3x + 6y = 3 \\ 2y = 1 - x \end{cases}$

You Try It

19. $\begin{cases} 2x - 3y + 4z = 4 \\ 6x + 4y - 2z = 0 \\ 4x - 2y - 3z = 11 \end{cases}$

20. $\begin{cases} x + 2y + 3z = -13 \\ 2x + y - 2z = 5 \\ 3x - 4y + z = 27 \end{cases}$

21. $\begin{cases} x + 2y + z = 9 \\ 2x - y + 3z = 5 \\ 3x + 2y - z = 7 \end{cases}$

You Try It

22. $\begin{cases} x + 2y + 2z = 3 \\ 2x + y - 2z = -3 \\ 7x + 8y + 2z = 3 \end{cases}$

23. $\begin{cases} \frac{1}{2}x + y - 2z = -3 \\ 3x - \frac{1}{3}y + z = 15 \\ 2x - 5y + \frac{1}{4}z = -6 \end{cases}$

24. $\begin{cases} x + y + z = 0 \\ x + 2y + 2z = -1 \\ 2x + 2y - z = -3 \end{cases}$

25. $\begin{cases} x + 2y + z = 4 \\ 2x + y + 2z = 2 \\ x + 7y + 5z = 14 \end{cases}$

26. $\begin{cases} x + y + 2z = 2 \\ 3x - y - z = 1 \\ 5x + y + 3z = 3 \end{cases}$

27. $\begin{cases} x - z = 1 \\ -x + 2y - 3z = 12 \\ 2x - 4y = -6 \end{cases}$

28. $\begin{cases} 3x + 2y - 2z = -1 \\ x - 2y + 2z = 2 \\ x - y + 2z = 0 \end{cases}$

29. $\begin{cases} -2x - y + z = 7 \\ 4x - 6y + 8z = -4 \\ x - 2y + 3z = 0 \end{cases}$

30. $\begin{cases} 4x + 2y + 3z = 9 \\ 3x - y + z = 8 \\ x + y + z = 2 \end{cases}$

23.3 Determinants and Cramer's Rule

THINGS TO KNOW

Before working through this topic, be sure you are familiar with the following concepts:

		VIDEO	ANIMATION	INTERACTIVE
You Try It	1. Solve Systems of Linear Equations by Elimination (Topic 13.3, **Objective 1**)	▤		
You Try It	2. Solve Systems of Linear Equations in Three Variables (Topic 13.6, **Objective 2**)	▤		
You Try It	3. Write an Augmented Matrix (Topic 23.2, **Objective 1**)	▤		
You Try It	4. Solve Systems of Two Equations Using Matrices (Topic 23.2, **Objective 2**)	▤		
You Try It	5. Solve Systems of Three Equations Using Matrices (Topic 23.2, **Objective 3**)	▤		

OBJECTIVES

1 Evaluate a 2×2 Determinant

2 Use Cramer's Rule to Solve a System of Linear Equations in Two Variables

3 Evaluate a 3×3 Determinant

4 Use Cramer's Rule to Solve a System of Linear Equations in Three Variables

OBJECTIVE 1 EVALUATE A 2×2 DETERMINANT

A **square matrix** has an equal number of rows and columns. For example, matrices A and B shown below are square matrices. A is a 2×2 matrix, while B is a 3×3 matrix.

$$A = \begin{bmatrix} 2 & -3 \\ 1 & 4 \end{bmatrix} \qquad B = \begin{bmatrix} 2 & 1 & -1 \\ 3 & 6 & -2 \\ 5 & 4 & 1 \end{bmatrix}$$

Every square matrix has a **real number** associated with it called its *determinant*. We will first define the determinant of a 2×2 matrix.

Definition Determinant of a 2×2 Matrix

The **determinant** of a 2×2 matrix $\begin{bmatrix} a & b \\ c & d \end{bmatrix}$ is denoted by $\begin{vmatrix} a & b \\ c & d \end{vmatrix}$ and is defined by

$$\begin{vmatrix} a & b \\ c & d \end{vmatrix} = ad - cb.$$

(eText Screens 23.3-1–23.3-21)

To evaluate a 2×2 determinant, we find the **difference** between the **products** of the two diagonals. This is illustrated in the following diagram:

$$\begin{vmatrix} a & b \\ c & d \end{vmatrix} = ad - cb$$

CAUTION Pay close attention to the notation for matrices and **determinants.** $\begin{bmatrix} a & b \\ c & d \end{bmatrix}$ is a matrix, while $\begin{vmatrix} a & b \\ c & d \end{vmatrix}$ is the determinant of $\begin{bmatrix} a & b \\ c & d \end{bmatrix}$.

Example 1 Evaluating 2×2 Determinants

Evaluate each determinant.

a. $\begin{vmatrix} 5 & 2 \\ 8 & 7 \end{vmatrix}$ **b.** $\begin{vmatrix} -9 & 3 \\ -4 & 2 \end{vmatrix}$

Solutions

a. $\begin{vmatrix} 5 & 2 \\ 8 & 7 \end{vmatrix} = 5 \cdot 7 - 8 \cdot 2 = 35 - 16 = 19$

b. $\begin{vmatrix} -9 & 3 \\ -4 & 2 \end{vmatrix} = (-9)(2) - (-4)(3) = -18 + 12 = -6$

You Try It **Work through this You Try It problem.**

Work Exercises 1–4 in this textbook or in the MyMathLab Study Plan.

OBJECTIVE 2 USE CRAMER'S RULE TO SOLVE A SYSTEM OF LINEAR EQUATIONS IN TWO VARIABLES

Consider the following **system of linear equations in two variables:**

$$\begin{cases} ax + by = m \\ cx + dy = n \end{cases},$$

where a, b, c, and d are numerical **coefficients,** and m and n are **constants.**

Cramer's Rule for solving a system of linear equations in two variables uses the following three **determinants** that are defined by the coefficients and constants from the equations.

x-coefficients

y-coefficients

$$D = \begin{vmatrix} a & b \\ c & d \end{vmatrix}, D_x = \begin{vmatrix} m & b \\ n & d \end{vmatrix}, \text{ and } D_y = \begin{vmatrix} a & m \\ c & n \end{vmatrix}$$

Constants

Cramer's Rule for Solving Systems of Linear Equations in Two Variables

For the system of linear equations $\begin{cases} ax + by = m \\ cx + dy = n \end{cases}$, the solution is given by

$$x = \frac{D_x}{D} = \frac{\begin{vmatrix} m & b \\ n & d \end{vmatrix}}{\begin{vmatrix} a & b \\ c & d \end{vmatrix}} \quad \text{and} \quad y = \frac{D_y}{D} = \frac{\begin{vmatrix} a & m \\ c & n \end{vmatrix}}{\begin{vmatrix} a & b \\ c & d \end{vmatrix}} \quad \text{provided} \quad D = \begin{vmatrix} a & b \\ c & d \end{vmatrix} \neq 0.$$

My video summary

Watch this **video** to see Cramer's Rule derived.

Example 2 Using Cramer's Rule to Solve a Linear System in Two Variables

Use Cramer's Rule to solve each **system**.

a. $\begin{cases} 3x + 2y = 12 \\ 4x - 5y = -7 \end{cases}$ b. $\begin{cases} 6x + y = -2 \\ 9x - 2y = 11 \end{cases}$

Solutions

a. Begin by finding D, D_x, and D_y.

$$D = \begin{vmatrix} 3 & 2 \\ 4 & -5 \end{vmatrix} = 3(-5) - 4(2) = -15 - 8 = -23$$

$$D_x = \begin{vmatrix} 12 & 2 \\ -7 & -5 \end{vmatrix} = 12(-5) - (-7)(2) = -60 + 14 = -46$$

$$D_y = \begin{vmatrix} 3 & 12 \\ 4 & -7 \end{vmatrix} = 3(-7) - 4(12) = -21 - 48 = -69$$

Then $x = \frac{D_x}{D} = \frac{-46}{-23} = 2$ and $y = \frac{D_y}{D} = \frac{-69}{-23} = 3$. So, the **solution to the system** is the ordered pair $(2, 3)$. View the **check.**

My video summary

b. Try solving this system on your own. View the **answer,** or watch this **video** for a complete solution.

If $D = 0$, then $x = \frac{D_x}{D}$ and $y = \frac{D_y}{D}$ are not **real numbers,** so **Cramer's Rule** does not apply. However, $D = 0$ tells us that the **system** is either **inconsistent** or **dependent.**

Cramer's Rule with Inconsistent and Dependent Systems

For the system of linear equations $\begin{cases} ax + by = m \\ cx + dy = n \end{cases}$, if $D = \begin{vmatrix} a & b \\ c & d \end{vmatrix} = 0$, then the system is either inconsistent or dependent.

1. If $D = \begin{vmatrix} a & b \\ c & d \end{vmatrix} = 0$ and at least one of the determinants $D_x = \begin{vmatrix} m & b \\ n & d \end{vmatrix}$ or $D_y = \begin{vmatrix} a & m \\ c & n \end{vmatrix}$ is not 0, then the system is inconsistent and the solution set is $\varnothing$.
2. If $D = \begin{vmatrix} a & b \\ c & d \end{vmatrix} = 0$ and the determinants $D_x = \begin{vmatrix} m & b \\ n & d \end{vmatrix}$ and $D_y = \begin{vmatrix} a & m \\ c & n \end{vmatrix}$ are both 0, then the system is dependent and has infinitely many solutions.

Example 3 Using Cramer's Rule to Solve a Linear System in Two Variables

Use **Cramer's Rule** to solve each **system**.

a. $\begin{cases} 2x - 5y = 7 \\ -4x + 10y = 11 \end{cases}$

b. $\begin{cases} 6x + 10y = -50 \\ -9x - 15y = 75 \end{cases}$

Solutions

a. Because $D = \begin{vmatrix} 2 & -5 \\ -4 & 10 \end{vmatrix} = 2(10) - (-4)(-5) = 20 - 20 = 0$, the system is either **inconsistent** or **dependent**. Because $D_x = \begin{vmatrix} 7 & -5 \\ 11 & 10 \end{vmatrix} = 7(10) - 11(-5) = 70 + 55 = 125 \neq 0$, the system is inconsistent. The solution set is $\varnothing$.

My video summary

b. Try solving this system on your own. View the **answer**, or watch this **video** for a complete solution.

You Try It **Work through this You Try It problem.**

Work Exercises 5–12 in this textbook or in the MyMathLab Study Plan.

OBJECTIVE 3 EVALUATE A 3×3 DETERMINANT

For every entry of a **3×3 determinant**, there is a 2×2 **determinant** associated with it called its **minor**. The minor is the 2×2 determinant obtained by deleting the row and the column in which the entry appears. For example, consider the following 3×3 determinant.

$$\begin{vmatrix} a_1 & b_1 & c_1 \\ a_2 & b_2 & c_2 \\ a_3 & b_3 & c_3 \end{vmatrix}$$

Entry a_1 is located in the first row and first column of the determinant. To find the minor associated with a_1, delete the first row and the first column to obtain a 2×2 determinant.

$$\begin{vmatrix} \not{a_1} & \not{b_1} & \not{c_1} \\ \not{a_2} & b_2 & c_2 \\ \not{a_3} & b_3 & c_3 \end{vmatrix}$$

The minor associated with a_1 is $\begin{vmatrix} b_2 & c_2 \\ b_3 & c_3 \end{vmatrix}$.

Similarly, the entry b_1 is located in the first row and second column. To find the minor associated with b_1, delete the first row and second column.

$$\begin{vmatrix} \not{a_1} & \not{b_1} & \not{c_1} \\ a_2 & \not{b_2} & c_2 \\ a_3 & \not{b_3} & c_3 \end{vmatrix}$$

The minor associated with b_1 is $\begin{vmatrix} a_2 & c_2 \\ a_3 & c_3 \end{vmatrix}$.

To evaluate a 3×3 determinant, we use a technique called **expansion by minors**. In the definition below, we expand across the first row.

Definition Determinant of a 3×3 Matrix

The **determinant** of a 3×3 matrix $\begin{bmatrix} a_1 & b_1 & c_1 \\ a_2 & b_2 & c_2 \\ a_3 & b_3 & c_3 \end{bmatrix}$ is denoted by $\begin{vmatrix} a_1 & b_1 & c_1 \\ a_2 & b_2 & c_2 \\ a_3 & b_3 & c_3 \end{vmatrix}$ and is defined by

Subtract Add

$$\begin{vmatrix} a_1 & b_1 & c_1 \\ a_2 & b_2 & c_2 \\ a_3 & b_3 & c_3 \end{vmatrix} = a_1\begin{vmatrix} b_2 & c_2 \\ b_3 & c_3 \end{vmatrix} - b_1\begin{vmatrix} a_2 & c_2 \\ a_3 & c_3 \end{vmatrix} + c_1\begin{vmatrix} a_2 & b_2 \\ a_3 & b_3 \end{vmatrix}.$$

Example 4 Evaluating 3×3 Determinants

Evaluate the determinant $\begin{vmatrix} 3 & 1 & -2 \\ -1 & 2 & 4 \\ 0 & 5 & 1 \end{vmatrix}$.

Solution To expand across the first row, we begin by finding the minor associated with each entry in the first row, which are $3, 1$, and -2.

$\begin{vmatrix} 3 & 1 & -2 \\ -1 & 2 & 4 \\ 0 & 5 & 1 \end{vmatrix}$ The minor associated with 3 is $\begin{vmatrix} 2 & 4 \\ 5 & 1 \end{vmatrix}$.

$\begin{vmatrix} 3 & 1 & -2 \\ -1 & 2 & 4 \\ 0 & 5 & 1 \end{vmatrix}$ The minor associated with 1 is $\begin{vmatrix} -1 & 4 \\ 0 & 1 \end{vmatrix}$.

$\begin{vmatrix} 3 & 1 & -2 \\ -1 & 2 & 4 \\ 0 & 5 & 1 \end{vmatrix}$ The minor associated with -2 is $\begin{vmatrix} -1 & 2 \\ 0 & 5 \end{vmatrix}$.

So,
$$\begin{aligned}
\begin{vmatrix} 3 & 1 & -2 \\ -1 & 2 & 4 \\ 0 & 5 & 1 \end{vmatrix} &= 3\begin{vmatrix} 2 & 4 \\ 5 & 1 \end{vmatrix} - 1\begin{vmatrix} -1 & 4 \\ 0 & 1 \end{vmatrix} + (-2)\begin{vmatrix} -1 & 2 \\ 0 & 5 \end{vmatrix} \\
&= 3(2\cdot 1 - 5\cdot 4) - 1(-1\cdot 1 - 0\cdot 4) + (-2)(-1\cdot 5 - 0\cdot 2) \\
&= 3(2 - 20) - 1(-1 - 0) + (-2)(-5 - 0) \\
&= 3(-18) - 1(-1) + (-2)(-5) \\
&= -54 + 1 + 10 \\
&= -43
\end{aligned}$$

You Try It **Work through this You Try It problem.**

Work Exercise 13 in this textbook or in the MyMathLab Study Plan.

In Example 4, we evaluated the **determinant** by expanding the minors across the first row. Actually, we can find the value of a determinant by expanding the minors across any row or down any column. However, we must be careful to properly add or subtract the product of each row or column entry and its minor.

If the **sum** of the row number and column number of an entry is **even**, then the **product** of the entry and its minor is "added." If the sum of the row number and column number of an entry is **odd**, then the product is "subtracted." These additions and subtractions are summarized below for 3×3 determinants:

$$\begin{vmatrix} + & - & + \\ - & + & - \\ + & - & + \end{vmatrix}.$$

When calculating determinants, it is typically easier to expand by minors for rows or columns that contain zeros. In Example 5, we evaluate a determinant by expanding the minors for different rows and columns.

Example 5 Evaluating 3×3 Determinants

My interactive video summary

Evaluate the **determinant** by expanding the minors for the given row or column.

$$\begin{vmatrix} 2 & 4 & -3 \\ 3 & 1 & 0 \\ -1 & -2 & 5 \end{vmatrix}$$

a. First column **b.** Second row **c.** Third column

Solutions

a. The entries down the first column are 2, 3, and -1. The corresponding minors are:

$\begin{vmatrix} 2 & 4 & -3 \\ 3 & 1 & 0 \\ -1 & -2 & 5 \end{vmatrix}$ The minor associated with 2 is $\begin{vmatrix} 1 & 0 \\ -2 & 5 \end{vmatrix}$.

$\begin{vmatrix} 2 & 4 & -3 \\ 3 & 1 & 0 \\ -1 & -2 & 5 \end{vmatrix}$ The minor associated with 3 is $\begin{vmatrix} 4 & -3 \\ -2 & 5 \end{vmatrix}$.

$\begin{vmatrix} 2 & 4 & -3 \\ 3 & 1 & 0 \\ -1 & -2 & 5 \end{vmatrix}$ The minor associated with -1 is $\begin{vmatrix} 4 & -3 \\ 1 & 0 \end{vmatrix}$.

The additions and subtractions that correspond to the first-column entries are $+, -$, and $+$.

$$\text{So, } \begin{vmatrix} 2 & 4 & -3 \\ 3 & 1 & 0 \\ -1 & -2 & 5 \end{vmatrix} = 2\begin{vmatrix} 1 & 0 \\ -2 & 5 \end{vmatrix} - 3\begin{vmatrix} 4 & -3 \\ -2 & 5 \end{vmatrix} + (-1)\begin{vmatrix} 4 & -3 \\ 1 & 0 \end{vmatrix}$$
$$= 2(5 - 0) - 3(20 - 6) + (-1)(0 + 3)$$
$$= 10 - 42 - 3$$
$$= -35$$

b. The entries in the second row are $3, 1$, and 0, and the corresponding minors are $\begin{vmatrix} 4 & -3 \\ -2 & 5 \end{vmatrix}$, $\begin{vmatrix} 2 & -3 \\ -1 & 5 \end{vmatrix}$, and $\begin{vmatrix} 2 & 4 \\ -1 & -2 \end{vmatrix}$. The additions and subtractions that correspond to the second-row entries are $-, +$, and $-$.

$$\text{So, } \begin{vmatrix} 2 & 4 & -3 \\ 3 & 1 & 0 \\ -1 & -2 & 5 \end{vmatrix} = -3\begin{vmatrix} 4 & -3 \\ -2 & 5 \end{vmatrix} + 1\begin{vmatrix} 2 & -3 \\ -1 & 5 \end{vmatrix} - 0\begin{vmatrix} 2 & 4 \\ -1 & -2 \end{vmatrix}.$$

Finish simplifying on your own to obtain the value -35, or watch this **interactive video** for complete solutions to all three parts.

c. Try working this expansion on your own to obtain the value -35, or watch this **interactive video** for complete solutions to all three parts.

You Try It **Work through this You Try It problem.**

Work Exercises 14–16 in this textbook or in the MyMathLab Study Plan.

OBJECTIVE 4 USE CRAMER'S RULE TO SOLVE A SYSTEM OF LINEAR EQUATIONS IN THREE VARIABLES

We can use **Cramer's Rule** to solve a **system of linear equation in three variables** using 3×3 **determinants** defined by the **coefficients** and **constants** from the equations.

Cramer's Rule for Solving Systems of Linear Equations in Three Variables

For the system of linear equations $\begin{cases} a_1x + b_1y + c_1z = n_1 \\ a_2x + b_2y + c_2z = n_2 \\ a_3x + b_3y + c_3z = n_3 \end{cases}$, the solution is given by

$$x = \frac{D_x}{D}, \quad y = \frac{D_y}{D} \quad \text{and} \quad z = \frac{D_z}{D}$$

where

$$D = \begin{vmatrix} a_1 & b_1 & c_1 \\ a_2 & b_2 & c_2 \\ a_3 & b_3 & c_3 \end{vmatrix}, \qquad D_x = \begin{vmatrix} n_1 & b_1 & c_1 \\ n_2 & b_2 & c_2 \\ n_3 & b_3 & c_3 \end{vmatrix},$$

$$D_y = \begin{vmatrix} a_1 & n_1 & c_1 \\ a_2 & n_2 & c_2 \\ a_3 & n_3 & c_3 \end{vmatrix}, \quad \text{and} \quad D_z = \begin{vmatrix} a_1 & b_1 & n_1 \\ a_2 & b_2 & n_2 \\ a_3 & b_3 & n_3 \end{vmatrix},$$

provided $D \neq 0$.

Example 6 Using Cramer's Rule to Solve a Linear System in Three Variables

Use **Cramer's Rule** to solve each **system**.

a. $\begin{cases} x - y + 4z = -5 \\ 6x - 2y + 5z = 1 \\ 3x + 2y + 2z = 10 \end{cases}$ **b.** $\begin{cases} x + y + z = 3 \\ 2x - 3y = 14 \\ 4y + 5z = -3 \end{cases}$

Solutions

a. Begin by finding $D, D_x, D_y,$ and D_z.

$$D = \begin{vmatrix} 1 & -1 & 4 \\ 6 & -2 & 5 \\ 3 & 2 & 2 \end{vmatrix} = 1\begin{vmatrix} -2 & 5 \\ 2 & 2 \end{vmatrix} - (-1)\begin{vmatrix} 6 & 5 \\ 3 & 2 \end{vmatrix} + 4\begin{vmatrix} 6 & -2 \\ 3 & 2 \end{vmatrix}$$

$$= 1(-4 - 10) - (-1)(12 - 15) + 4(12 - (-6))$$

$$= -14 - 3 + 72$$

$$= 55$$

$$D_x = \begin{vmatrix} -5 & -1 & 4 \\ 1 & -2 & 5 \\ 10 & 2 & 2 \end{vmatrix} = -5\begin{vmatrix} -2 & 5 \\ 2 & 2 \end{vmatrix} - (-1)\begin{vmatrix} 1 & 5 \\ 10 & 2 \end{vmatrix} + 4\begin{vmatrix} 1 & -2 \\ 10 & 2 \end{vmatrix}$$

$$= -5(-4 - 10) - (-1)(2 - 50) + 4(2 - (-20))$$

$$= 70 - 48 + 88$$

$$= 110$$

$$D_y = \begin{vmatrix} 1 & -5 & 4 \\ 6 & 1 & 5 \\ 3 & 10 & 2 \end{vmatrix} = 1\begin{vmatrix} 1 & 5 \\ 10 & 2 \end{vmatrix} - (-5)\begin{vmatrix} 6 & 5 \\ 3 & 2 \end{vmatrix} + 4\begin{vmatrix} 6 & 1 \\ 3 & 10 \end{vmatrix}$$

$$= 1(2 - 50) - (-5)(12 - 15) + 4(60 - 3)$$

$$= -48 - 15 + 228$$

$$= 165$$

$$D_z = \begin{vmatrix} 1 & -1 & -5 \\ 6 & -2 & 1 \\ 3 & 2 & 10 \end{vmatrix} = 1\begin{vmatrix} -2 & 1 \\ 2 & 10 \end{vmatrix} - (-1)\begin{vmatrix} 6 & 1 \\ 3 & 10 \end{vmatrix} + (-5)\begin{vmatrix} 6 & -2 \\ 3 & 2 \end{vmatrix}$$

$$= 1(-20 - 2) - (-1)(60 - 3) + (-5)(12 - (-6))$$

$$= -22 + 57 - 90$$

$$= -55$$

Then $x = \frac{D_x}{D} = \frac{110}{55} = 2$, $y = \frac{D_y}{D} = \frac{165}{55} = 3$, and $z = \frac{D_z}{D} = \frac{-55}{55} = -1$. So, the **ordered-triple solution** is $(2, 3, -1)$. View the **check**.

My video summary

b. Try solving this system on your own. View the **answer**, or watch this **video** for a complete solution.

Like with the two-variable case, if $D = 0$, then $x = \frac{D_x}{D}$, $y = \frac{D_y}{D}$, and $z = \frac{D_z}{D}$ are not **real numbers**, and **Cramer's Rule** does not apply. But, $D = 0$ tells us that the **system** is either **inconsistent** or **dependent.**

Cramer's Rule with Inconsistent and Dependent Systems in Three Variables

If $D = 0$, then the system is either inconsistent or dependent.

1. If $D = 0$ and at least one of the determinants D_x, D_y, or D_z is not 0, then the system is inconsistent and the solution set is $\varnothing$.
2. If $D = 0$ and the determinants D_x, D_y, and D_z are all 0, then the system is dependent and has infinitely many solutions. The solutions to these systems can be found using the methods from **Topic 13.6** or **Topic 23.2**.

You Try It **Work through this You Try It problem.**

Work Exercises 17–22 in this textbook or in the MyMathLab Study Plan.

23.3 Exercises

In Exercises 1–4, evaluate each 2×2 determinant.

You Try It

1. $\begin{vmatrix} 7 & 3 \\ 8 & 6 \end{vmatrix}$ **2.** $\begin{vmatrix} 9 & -7 \\ 4 & -2 \end{vmatrix}$ **3.** $\begin{vmatrix} 3 & -6 \\ -2 & 4 \end{vmatrix}$ **4.** $\begin{vmatrix} 10 & 1 \\ 6 & -5 \end{vmatrix}$

In Exercises 5–12 use Cramer's Rule to solve each system of linear equations in two variables.

You Try It

5. $\begin{cases} x + y = 3 \\ -2x + y = 0 \end{cases}$ **6.** $\begin{cases} 2x - y = 1 \\ -3x + 3y = 0 \end{cases}$ **7.** $\begin{cases} 4x - y = -4 \\ -3x + 3y = -6 \end{cases}$

8. $\begin{cases} 10x - 15y = 12 \\ -4x + 6y = 9 \end{cases}$ **9.** $\begin{cases} 2x - 3y = 0 \\ 4x + 9y = 5 \end{cases}$ **10.** $\begin{cases} 3x + 4y = -2 \\ 6x - 3y = 7 \end{cases}$

11. $\begin{cases} -9x + 3y = -15 \\ 6x - 2y = 10 \end{cases}$ **12.** $\begin{cases} 4x + 5y = 51 \\ 3x - 2y = -2 \end{cases}$

In Exercises 13–16, evaluate each 3×3 determinant.

You Try It

13. $\begin{vmatrix} 7 & -3 & 8 \\ 1 & -5 & -2 \\ 3 & 7 & 4 \end{vmatrix}$ **14.** $\begin{vmatrix} 2 & 5 & 1 \\ 4 & 0 & 3 \\ -6 & 1 & 0 \end{vmatrix}$

You Try It

15. $\begin{vmatrix} 4 & 0 & 7 \\ -1 & 3 & -1 \\ 5 & 2 & 6 \end{vmatrix}$ **16.** $\begin{vmatrix} -5 & -1 & -2 \\ 8 & 10 & 0 \\ -7 & 5 & 0 \end{vmatrix}$

In Exercises 17–22 use Cramer's Rule to solve each system of linear equations in three variables.

You Try It

17. $\begin{cases} x + y - z = 4 \\ 5x - y + 3z = 6 \\ x + y - 5z = 8 \end{cases}$ **18.** $\begin{cases} 3x + 3y + 4z = 6 \\ 4x - y + z = 8 \\ -x + y - 3z = -2 \end{cases}$

19. $\begin{cases} -2x + 5y + 5z = -4 \\ 5x + 6y + 5z = -23 \\ -3x + 4y + 5z = -3 \end{cases}$ **20.** $\begin{cases} x - z = -1 \\ 2y - z = 8 \\ 2x + 2y = 0 \end{cases}$

21. $\begin{cases} -x + 2y + z = 2 \\ 3x + y - z = -4 \\ x + 5y + z = 5 \end{cases}$ **22.** $\begin{cases} 4x - 3y + z = 4 \\ 10x - 9y = 8 \\ 6y + 2z = 0 \end{cases}$

Appendix A.4 Squares and Square Roots

n	n^2	$\sqrt{n}$	n	n^2	$\sqrt{n}$	n	n^2	$\sqrt{n}$	n	n^2	$\sqrt{n}$
1	1	1.000	26	676	5.099	51	2601	7.141	76	5776	8.718
2	4	1.414	27	729	5.196	52	2704	7.211	77	5929	8.775
3	9	1.732	28	784	5.292	53	2809	7.280	78	6084	8.832
4	16	2.000	29	841	5.385	54	2916	7.348	79	6241	8.888
5	25	2.236	30	900	5.477	55	3025	7.416	80	6400	8.944
6	36	2.449	31	961	5.568	56	3136	7.483	81	6561	9.000
7	49	2.646	32	1024	5.657	57	3249	7.550	82	6724	9.055
8	64	2.828	33	1089	5.745	58	3364	7.616	83	6889	9.110
9	81	3.000	34	1156	5.831	59	3481	7.681	84	7056	9.165
10	100	3.162	35	1225	5.916	60	3600	7.746	85	7225	9.220
11	121	3.317	36	1296	6.000	61	3721	7.810	86	7396	9.274
12	144	3.464	37	1369	6.083	62	3844	7.874	87	7569	9.327
13	169	3.606	38	1444	6.164	63	3969	7.937	88	7744	9.381
14	196	3.742	39	1521	6.245	64	4096	8.000	89	7921	9.434
15	225	3.873	40	1600	6.325	65	4225	8.062	90	8100	9.487
16	256	4.000	41	1681	6.403	66	4356	8.124	91	8281	9.539
17	289	4.123	42	1764	6.481	67	4489	8.185	92	8464	9.592
18	324	4.243	43	1849	6.557	68	4624	8.246	93	8649	9.644
19	361	4.359	44	1936	6.633	69	4761	8.307	94	8836	9.695
20	400	4.472	45	2025	6.708	70	4900	8.367	95	9025	9.747
21	441	4.583	46	2116	6.782	71	5041	8.426	96	9216	9.798
22	484	4.690	47	2209	6.856	72	5184	8.485	97	9409	9.849
23	529	4.796	48	2304	6.928	73	5329	8.544	98	9604	9.899
24	576	4.899	49	2401	7.000	74	5476	8.602	99	9801	9.950
25	625	5.000	50	2500	7.071	75	5625	8.660	100	10,000	10.000

Appendix A.5 Measurement Conversions and Temperature Formulas

U.S. CUSTOMARY (AMERICAN) UNITS

Length

12 inches (in.) = 1 foot (ft)

3 feet = 1 yard (yd)

5280 feet = 1 mile (mi)

Capacity

8 fluid ounces (fl oz) = 1 cup (c)

2 cups = 1 pint (pt)

2 pints = 1 quart (qt)

4 quarts = 1 gallon (gal)

Weight

16 ounces (oz) = 1 pound (lb)

2000 pounds = 1 ton

Time

60 seconds (s or sec) = 1 minute (min)

60 minutes = 1 hour (hr)

24 hours = 1 day (d)

7 days = 1 week (wk)

METRIC UNITS

Length

Meter (m) is the base unit.

1 kilometer (km) = 1000 meters (m)	1 m = $\frac{1}{1000}$ km (or 0.001 km)
1 hectometer (hm) = 100 m	1 m = $\frac{1}{100}$ hm (or 0.01 hm)
1 dekameter (dam) = 10 m	1 m = $\frac{1}{10}$ dam (or 0.1 dam)
1 decimeter (dm) = $\frac{1}{10}$ m (or 0.1 m)	1 m = 10 dm
1 centimeter (cm) = $\frac{1}{100}$ m (or 0.01 m)	1 m = 100 cm
1 millimeter (mm) = $\frac{1}{1000}$ m or (0.001 m)	1 m = 1000 mm

Mass

Gram (g) is the base unit.

1 kilogram (kg) = 1000 grams (g)	1 g = $\frac{1}{1000}$ kg (or 0.001 kg)
1 hectogram (hg) = 100 g	1 g = $\frac{1}{100}$ hg (or 0.01 hg)
1 dekagram (dag) = 10 g	1 g = $\frac{1}{10}$ dag (or 0.1 dag)
1 decigram (dg) = $\frac{1}{10}$ g (or 0.1 g)	1 g = 10 dg

1 centigram (cg) $= \frac{1}{100}$ g (or 0.01 g)	1 g = 100 cg
1 milligram (mg) $= \frac{1}{1000}$ g or (0.001 g)	1 g = 1000 mg

Capacity

Liter (L) is the base unit.

1 kiloliter (kL) = 1000 liters (L)	1 L $= \frac{1}{1000}$ kL (or 0.001 kL)
1 hectoliter (hL) = 100 L	1 L $= \frac{1}{100}$ hL (or 0.01 hL)
1 dekaliter (daL) = 10 L	1 L $= \frac{1}{10}$ daL (or 0.1 daL)
1 deciliter (dL) $= \frac{1}{10}$ L (or 0.1 L)	1 L = 10 dL
1 centiliter (cL) $= \frac{1}{100}$ L (or 0.01 L)	1 L = 100 cL
1 milliliter (mL) $= \frac{1}{1000}$ L or (0.001 L)	1 L = 1000 mL

AMERICAN AND METRIC CONVERSION FACTS

Length

1 in. = 2.54 cm	
1 ft ≈ 0.305 m	1 m ≈ 3.281 ft
1 yd ≈ 0.914 m	1 m ≈ 1.094 yd
1 mi ≈ 1.609 km	1 km ≈ 0.621 mi

Weight and Mass

1 kg ≈ 2.20 lb	1 lb ≈ 0.45 kg
1 g ≈ 0.035 oz	1 oz ≈ 28.35 g

Capacity

1 fl oz ≈ 29.57 mL	1 L ≈ 33.81 fl oz
1 c ≈ 237 mL	1 L ≈ 4.23 c
1 pt ≈ 0.47 L	1 L ≈ 2.11 pt
1 qt ≈ 0.95 L	1 L ≈ 1.06 qt
1 gal ≈ 3.79 L	1 L ≈ 0.26 gal

TEMPERATURE CONVERSION FORMULAS

Celsius to Fahrenheit: $F = \frac{9}{5}C + 32$

Fahrenheit to Celsius: $C = \frac{5}{9}(F - 32)$

APPENDIX B

Geometric Formulas

<table>
<tr><th>Figure</th><th>Formulas</th><th>Figure</th><th>Formulas</th></tr>
<tr><td>Square
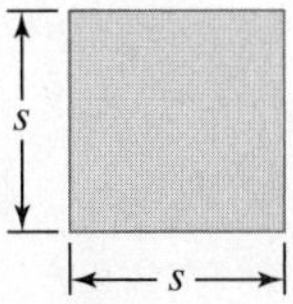</td><td>Perimeter: $P = 4s$
Area: $A = s^2$</td><td>Triangle
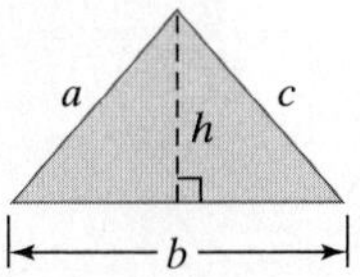</td><td>Perimeter: $P = a + b + c$
Area: $A = \frac{1}{2}bh$</td></tr>
<tr><td>Rectangle
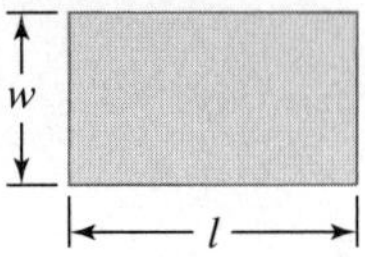</td><td>Perimeter: $P = 2l + 2w$
Area: $A = lw$</td><td>Trapezoid
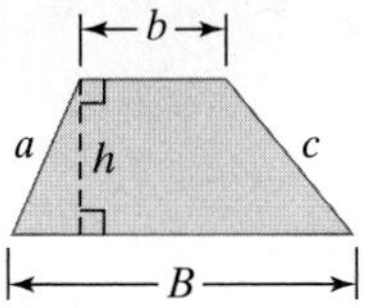</td><td>Perimeter: $P = a + b + c + B$
Area: $A = \frac{1}{2}h(b + B)$</td></tr>
<tr><td>Parallelogram
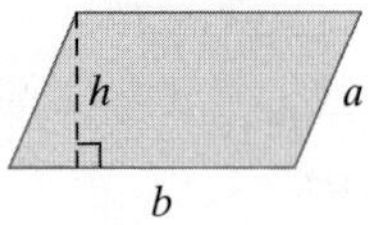</td><td>Perimeter: $P = 2a + 2b$
Area: $A = bh$</td><td>Circle
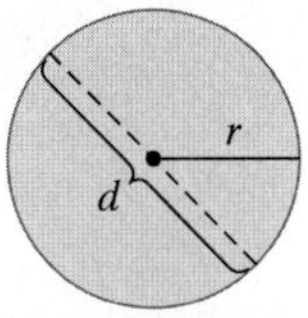</td><td>Circumference: $C = 2\pi r = \pi d$
Area: $A = \pi r^2$</td></tr>
<tr><td>Cube
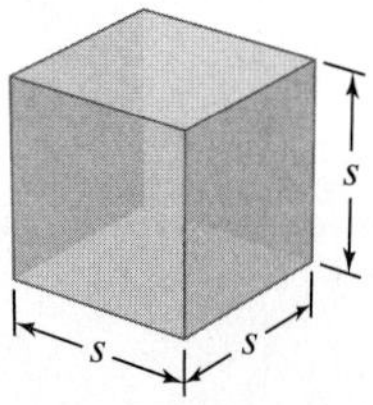</td><td>Volume: $V = s^3$
Surface Area: $SA = 6s^2$</td><td>Sphere
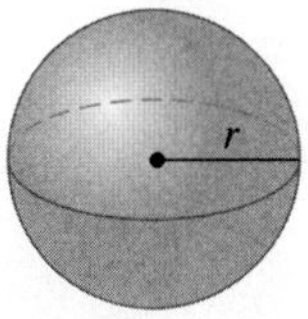</td><td>Volume: $V = \frac{4}{3}\pi r^3$
Surface Area: $SA = 4\pi r^2$</td></tr>
<tr><td>Rectangular Solid
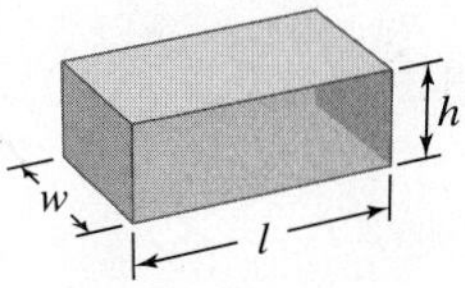</td><td>Volume: $V = lwh$
Surface Area:
$SA = 2lw + 2lh + 2wh$</td><td>Right Cone
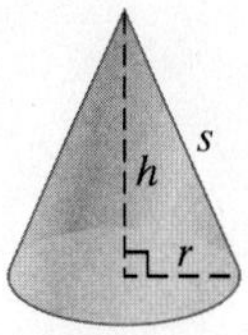</td><td>Volume: $V = \frac{1}{3}\pi r^2 h$
Surface Area:
$SA = \pi rs + \pi r^2 = \pi r\sqrt{r^2 + h^2} + \pi r^2$</td></tr>
<tr><td>Right Circular Cylinder
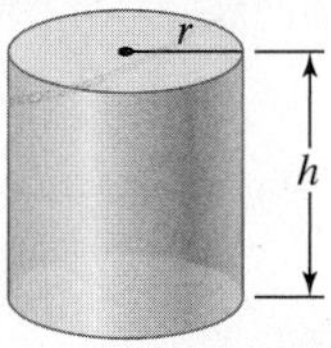</td><td>Volume: $V = \pi r^2 h$
Surface Area:
$SA = 2\pi r^2 + 2\pi rh$</td><td>Pyramid
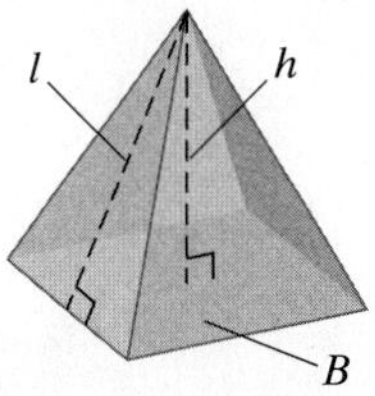</td><td>Volume:
$V = \frac{1}{3}Bh$, where B is the area of the base.
Surface Area (of a Regular Pyramid):
$SA = B + \frac{1}{2}Pl$, where P is the perimeter of the base, and l is the slant height.</td></tr>
</table>

Note: The constant π is often approximated by 3.14 or $\frac{22}{7}$.

Geometric Formulas

Glossary

4th root A real number b is the 4th root of a real number a provided that $b^4 = a$. Every positive real number a has two real 4th roots, one positive, $\sqrt[4]{a}$ (the principal 4th root), and one negative, $-\sqrt[4]{a}$.

5th root A real number b is the 5th root of a real number a provided that $b^5 = a$.

Absolute value The absolute value of a number a, written as $|a|$, is the distance from 0 to a on the number line.

Acute angle An angle that measures between 0° and 90°.

Addends Quantities being added in an addition problem.

Addition The process of combining two (or more) numbers into a single number called the *sum.* The numbers being added are called *terms,* or *addends.* A $+$ is used to indicate addition.

Addition method See *Elimination Method*.

Addition property of equality Let a, b, and c be numbers or algebraic expressions. Then $a = b$, and $a + c = b + c$ are equivalent equations. Also, because subtraction is defined in terms of addition, $a = b$ and $a - c = b - c$ are equivalent equations.

Addition property of inequality Let a, b, and c be real numbers. If $a < b$, then $a + c < b + c$ and $a - c < b - c$. Adding or subtracting the same quantity to/from both sides of an inequality results in an equivalent inequality. The inequality symbol $<$ can be replaced with $>, \leq, \geq$, or $\neq$.

Additive identity The number 0 is called the *additive identity* because if 0 is added to any number, the *sum* is the same identical number.

Additive inverse The opposite of a real number is also called the *additive inverse* of the number because the sum of a number and its opposite is zero.

Adjacent angles Two angles that share a common vertex and a common side are adjacent angles provided that they do not overlap.

Algebraic equation A statement that two quantities are equal where one or both quantities contain at least one variable.

Algebraic expression A variable or combination of variables, constants, operations, and grouping symbols.

Alternate exterior angles A pair of angles created by two parallel lines and a transversal that are located on the outside of each parallel line and on opposite sides of the transversal.

Alternate interior angles A pair of angles created by two parallel lines and a transversal that are located on the inside of each parallel line and on opposite sides of the transversal.

Alternating sequence A sequence in which the terms alternate in sign.

Amount of decrease The amount of decrease can be found by subtracting the new value from the original value. That is, Amount of decrease $=$ Original value $-$ New value.

Amount of increase The amount of increase can be found by subtracting the original value from the new value. That is, Amount of increase $=$ New value $-$ Original value.

Angle A figure formed by two rays that share a common endpoint.

Angle-Side-Angle (ASA) property of congruence If two angles of one triangle have the same measures as the corresponding angles of another triangle and if the corresponding sides between each pair of angles have equal lengths, then the two triangles are congruent.

Annually Happening once per year. In terms of compound interest, compounding *annually* means to compound interest once per year. Interest can also be compounded over other time intervals, such as *semiannually* (twice per year), *quarterly* (4 times per year), *monthly* (12 times per year), or *daily* (365 times per year).

Annuity An investment of equal amounts deposited at equal time intervals is called an *annuity*.

Apex See *Right cone*.

Area The amount of surface contained within the sides of a closed two-dimensional figure. Area is measured in square units such as square inches or square centimeters.

Area of a circle If a circle has radius r, then its area A is given by the formula $A = \pi r^2$. The constant π is often approximated by the decimal 3.14 (or in some cases, the fraction $\frac{22}{7}$).

Area of a rectangle If a rectangle has length l and width w, then its area A is given by the formula $A = l \cdot w$.

Area of a square If a square has sides of length s, then its area A is given by the formula $A = s \cdot s = s^2$.

Area of a triangle If a triangle has base b and height h, then its area A is given by the formula $A = \frac{1}{2}bh$.

Arithmetic sequence A sequence is arithmetic if the difference in any two successive terms is constant. The nth term of the sequence is given by $a_n = a_1 + (n - 1)d$, where a_1 is the first term and d is the constant difference.

Associative property of addition Changing the grouping of terms does not affect the sum. That is, for any numbers, a, b, and c, $(a + b) + c = a + (b + c)$.

Associative property of multiplication Changing the grouping of factors does not affect the product. That is, for any numbers a, b, and c, $(a \cdot b) \cdot c = a \cdot (b \cdot c)$.

Augmented matrix An augmented matrix for a system of equations in standard form is a matrix consisting of the coefficients of the variables and the constants from the right-hand side of the equations, separated by a vertical line.

Average A measure of central tendency found by adding the data values and then dividing by the number of values. Also called the *mean.*

Average rate of change The change in the value of a dependent variable divided by the corresponding change in the independent variable.

Axis of symmetry In the graph of a quadratic equation, the axis of symmetry is an imaginary vertical line that passes through the vertex of the graph and divides the graph into two mirror images.

Back substitution The process of substituting known values for one or more variables back into an original equation to find values of remaining variables.

Bar graph Uses vertical or horizontal bars to display information visually. Each bar represents a different category of data. Longer bars indicate larger values and shorter bars indicate smaller values, so bar graphs are useful for making comparisons among the categories.

Bar notation for repeating decimals Used to show the pattern of digits that repeat within a repeating decimal. A bar (horizontal line) is placed above the digit or pattern of digits that repeat.

Base of a rectangular solid (box) The area of the bottom rectangle.

Base of a right circular cylinder The area of the bottom circle.

Base of a triangle The length of the side perpendicular to the height of the triangle.

Base of an exponential expression The number, variable, or expression that is raised to a power. For example, 3 is the base in 3^5.

Base unit A unit of measurement (e.g., meter) that has other units named in terms of it (e.g., millimeter).

Binomial A simplified polynomial with two terms.

Binomial coefficients The coefficients in the expansion of $(a + b)^n$. The coefficient of the term with variable part $a^{n-r}b^r$ is given by $\binom{n}{r} = \frac{n!}{r!(n-r)!}$.

Binomial difference A binomial in which one term is subtracted from the other such as $A - B$.

Binomial sum A binomial in which the two terms are added such as $A + B$.

Borrow To take a unit from the next largest place value in the minuend in order to make a number larger than what is being subtracted.

Boundary line When graphing a linear inequality, the *boundary line* is the graph of the corresponding linear equation. The boundary line divides the coordinate plane into two half-planes, only one of which contains solutions to the inequality. The line is dashed if the inequality is strict and points on the line are not part of the solution set. The line is solid if the inequality is non-strict and points on the line are included in the solution set.

Boundary point (polynomial inequality) A boundary point for a polynomial inequality $P(x) < 0$ is a value for x such that $P(x) = 0$. These are the x-intercepts for the graph of the polynomial. *Note:* The inequality symbol $<$ can be replaced by $>, \leq,$ or $\geq$.

Boundary point (rational inequality) A boundary point for a rational inequality $R(x) < 0$ is a value for x such that $R(x) = 0$ or $R(x)$ is undefined. These are values that make either the numerator or denominator equal zero. *Note:* The inequality symbol $<$ can be replaced by $>, \leq,$ or $\geq$.

Branch See *Tree diagram*.

Capacity The maximum amount of a substance that can be put in a container.

Carry To move a digit from one column of numbers to a column with a higher place value.

Cartesian coordinate system See *Rectangular coordinate system*

Cartesian plane The plane represented by a rectangular coordinate system, also known as the *xy-plane*.

Celsius scale A scale of temperature where 0°C is the freezing point of water (read "zero degrees Celsius") and 100°C is the boiling point of water. The raised circle indicates "degrees" and the "*C*" indicates we are using the *Celsius scale.*

Center of a circle The point within the circle, (h, k), that is the same distance from all points that lie on the circle.

Centimeter (cm) A measure of length in the metric system equal to one hundredth of a meter.

Change of base formula For any positive base $b \neq 1$ and for any positive real number u, $\log_b u = \frac{\log_a u}{\log_a b}$, where a is any positive real number such that $a \neq 1$.

Circle The set of all points (x, y) in the coordinate plane that are a fixed distance, r, from a fixed point (h, k). The fixed distance, r, is called the radius of the circle, and the fixed point (h, k) is called the center of the circle.

Circle graph A pie-shaped graph in which each category is represented by a sector, or slice, of the circle. The size of the sector depends on how much of the total is made up by the category. Also called a *pie graph* or *pie chart,* a circle graph is typically used to show how categories compare to a total.

Circumference The distance around the circle. Its value is found by using either $C = 2\pi r$ or $C = \pi d$, where r is the radius of the circle and d is the diameter. The constant π is often approximated by the decimal 3.14 (or in some cases, the fraction $\frac{22}{7}$).

Class A category, denoted by a range of values, used to organize and summarize data values into a table or histogram.

Class frequency The number, or count, of data values that fall within a given class interval.

Class interval See *Histogram.*

Clear the decimals To *clear the decimals* from an equation, multiply both sides of the equation by an appropriate power of 10, such as 10, 100, or 1000. Determine which power of 10 to use by finding the largest number of decimal places in any of the constants.

Clear the fractions To *clear the fractions* from an equation, multiply both sides of the equation by the LCD of all the fractions in the equation.

Coefficient (or numerical coefficient) The numeric factor of a term. For example, the coefficient in $4x^2$ is 4, the coefficient in $-3y$ is -3, and the coefficient in 13 is 13 (constant coefficient).

Coefficient of a monomial The coefficient of a monomial is the constant factor.

Coinciding Lines Lines that have the same slope and the same y-intercept.

Collect like terms To rearrange terms in an algebraic expression so that like terms are grouped together.

Combine like terms To add or subtract the like terms. To do this, reverse the distributive property to add or subtract the coefficients of the terms. The variable part remains unchanged.

Commission A percent of the sales that is paid to a salesperson as a wage.

Commission formula A percent equation for computing a sales commission, where the commission is the amount, the commission rate is the percent, and the sales is the base. That is, Commission = Commission rate · Sales.

Commission rate Commission is earned or paid as a percent of the sales. This percent is called the *commission rate*.

Common factor A factor common to two or more numbers. For example, 6 is a factor of both 30 and 42 because $30 = 6 \cdot 5$ and $42 = 6 \cdot 7$.

Common logarithm The base 10 logarithm, $\log_{10}x$, usually denoted as $\log x$. If no base is shown for a logarithm, it is understood to be the common logarithm.

Common logarithmic function For $x > 0$, the common logarithmic function is defined by: $y = \log x$ if and only if $x = 10^y$.

Common multiple A common multiple of two real numbers is a quantity that is evenly divisible by both numbers. For example, 24 is a common multiple of 6 and 12 because it is evenly divisible by both 6 and 12.

Common prime factor A prime number that is a factor of each integer in a group of integers is called a common prime factor of the group of integers. For example, $6 = 2 \cdot 3$, $9 = 3 \cdot 3$, and $15 = 3 \cdot 5$, so 3 is a common prime factor of 6, 9, and 15.

Common ratio The ratio of successive terms in a geometric sequence.

Common variable factors for a GCF If a variable is a common factor of a group of monomials, then the lowest power on that variable in the group will be the power on the variable in the GCF.

Commutative property of addition The order of the terms in addition does not affect the sum. That is, for any numbers a and b, $a + b = b + a$.

Commutative property of multiplication The order of the factors in multiplication does not affect the product. That is, for any numbers a and b, $a \cdot b = b \cdot a$.

Compare real numbers To compare real numbers means to determine which number is larger than the other, or if the two numbers are equal.

Comparison line graph Two or more line graphs plotted on the same set of axes. Each line graph represents a different quantity that varies over time.

Complementary angles Two angles are complementary angles, or *complements,* if the sum of their measures is 90°.

Complements See *Complementary angles.*

Complete the square To complete the square means to add an appropriate constant so that a binomial becomes a perfect square trinomial. The appropriate constant is the square of half the linear coefficient.

Complex conjugates The complex numbers $(a + bi)$ and $(a - bi)$ are called *complex conjugates* of each other. A complex conjugate is obtained by changing the sign of the imaginary part in a complex number.

Complex fraction A fraction in which the numerator and/or denominator contain(s) fractions.

Complex numbers The set of all numbers of the form $a + bi$, where a and b are real numbers and i is the imaginary unit, is called the set of complex numbers. The number a is called the *real part*, and the number b is called the *imaginary part.*

Complex rational expression A rational expression in which the numerator and/or denominator contain(s) rational expressions. Also called a *complex fraction.*

Composite function Given functions f and g, the composite function, $f \circ g$ (also called the *composition of* f and g) is defined by $(f \circ g)(x) = f(g(x))$, provided $g(x)$ is in the domain of f.

Composite number A whole number greater than 1 that is not prime.

Compound inequality Two inequalities that are joined together using the words *and* or *or*.

Compound interest Interest that is paid or earned both on the original principal and on the interest that has been paid previously.

Compound interest formula The overall amount A in an account is given by the formula $A = P\left(1 + \frac{r}{n}\right)^{n \cdot t}$ where P is the principal, r is the annual interest rate written in decimal form, t is the length of time in years, and n is the number of times compounded per year.

Concave down A graph, or part of a graph, that curves down.

Concave up A graph, or part of a graph, that curves up.

Concentration The concentration of a solution refers to the strength of the solution. The higher the concentration, the stronger the solution. Concentration of a component in a solution is determined by the formula:

$$\text{Concentration} = \frac{\text{Amount of component}}{\text{Amount of solution}}.$$

Congruence symbol (≅) Used to state that two triangles are congruent.

Congruent Two geometric figures are congruent if they have the exact same shape and size.

Conic section A geometric figure formed when a plane intersects a right circular cone.

Conjugate axis An invisible line segment that passes through the center of a hyperbola and lies perpendicular to the transverse axis.

Conjugates Two expressions of the form $u + v$ and $u - v$ are conjugates of each other. The conjugate of a binomial is obtained by changing the sign of the second term in the binomial.

Consecutive even integers Even integers that appear next to each other in an ordered list of all even integers. For example, 6, 8, and 10 are consecutive even integers.

Consecutive integers Integers that appear next to each other in an ordered list of all integers. For example, 3, 4, and 5 are consecutive integers.

Consecutive odd integers Odd integers that appear next to each other in an ordered list of all odd integers. For example, 5, 7, and 9 are consecutive odd integers.

Consistent system A system of equations or inequalities that has at least one solution.

Constant A number. The value of a constant never changes. For example, the numbers 10, −6, and 0 are constants.

Constant of variation A fixed non-zero number, k, that relates two quantities that vary directly or indirectly. It is the ratio of the two variables in direct variation $k = \frac{y}{x}$, and the product of the two variables in indirect variation $(k = y \cdot x)$. Also called the *proportionality constant.*

Constant term A term without any variables.

Continuous compounding Compounding interest at every instant, or every fraction of a second every day.

Continuous compound interest formula Continuous compound interest can be calculated using the formula $A = Pe^{rt}$, where A = total amount after t years; P = principal (original investment); r = interest rate per year, and t = number of years.

Contradiction An equation for which no value of the variable can make the equation true. It has no solution, and its solution set is the empty { } or null set ∅.

Converge To approach a fixed value. For example, an infinite geometric series with $|r| < 1$ has a finite sum so is said to *converge*.

Coordinate axes Two perpendicular real number lines that divide the coordinate plane into four quadrants. The horizontal coordinate axis is called the *x-axis* and the vertical coordinate axis is called the *y-axis*.

Coordinate plane The plane represented by a rectangular coordinate system, also known as the *Cartesian plane* or *xy-plane*.

Coordinates The x- and y-coordinates of a point or ordered pair. The coordinates provide the position of the point in the coordinate plane.

Corresponding angles Pairs of angles created by two parallel lines and a transversal that are located in similar positions along each of the parallel lines.

Corresponding angles in congruent triangles Angles that match up in two or more congruent triangles. That is, corresponding angles are located in the same positions relative to the shape of the triangles, and have the same angle measures.

Corresponding angles in similar triangles Angles that match up in two or more similar triangles. That is, corresponding angles are located in the same positions relative to the shape of the triangles, and have the same angle measures.

Corresponding sides in congruent triangles Sides that match up in two or more congruent triangles. That is, corresponding sides are located in the same positions relative to the shape of the triangles, and have the same lengths.

Corresponding sides in similar triangles Sides that are located opposite two corresponding angles.

Counting number A counting number, or *natural number*, is an element of the set $N = \{1, 2, 3, 4, 5, \ldots\}$.

Cramer's Rule A method for solving systems of linear equations by using determinants.

Cross multiplying See *Cross product*.

Cross product For two fractions that are equivalent in a proportion, if the denominator of one fraction is multiplied by the numerator of the other, the result is a cross product.

Cube A rectangular solid in which all six faces are squares of equal size.

Cube root function $f(x) = \sqrt[3]{x}$. The domain of the cube root function is the set of all real numbers $\mathbb{R}$, or using interval notation, $(-\infty, \infty)$.

Cube roots The cube root of a real number a is a real number b that, when cubed, results in a. That is, $b = \sqrt[3]{a}$, if $b^3 = a$.

Cubic centimeter (cm^3) A cube with 1-centimeter edges.

Cubic inch (in^3) A cube with 1-inch edges.

Cubic units Units used to measure volume. See *Volume*.

Cup (c) A unit of measure for volume equal to 8 fluid ounces.

Daily Happening once each day (365 times each year), as with compound interest. See *Annually*.

Day (d) A unit of measure for time equal to 24 hours.

Decimal approximation A rounded decimal used to estimate an exact value.

Decimal as percent To write a decimal as a percent, multiply the decimal by 100%. Multiplying by 100% has two results: (1) the decimal point is moved two places to the right; and (2) the percent symbol (%) is attached.

Decimal number (or Decimal) A number that contains a decimal point. The position of each digit in a decimal number determines its place value. Digits appearing in front of the decimal point form the whole-number part of the number, and digits appearing behind the decimal point form the decimal part (or fractional part) of the number.

Decimal part (or fractional part) Digits appearing behind the decimal point form the *decimal part* (or *fractional part*) of the number.

Decimal place A position of a digit to the right of the decimal point.

Decimal point Separates the whole-number part and fractional part of a decimal number.

Degenerate conic section A conic section formed when a plane intersects the vertex of the cones.

Degree of a monomial The degree of a monomial is the sum of the exponents on the variables.

Degree of a polynomial The degree of a polynomial is the largest degree of its terms. For example, the degree of $10x^5 - 3x^4 + x^3 - 2x^2 + 8x + 9$, is 5, the largest degree of any term in the polynomial.

Degrees Angles are measured in *degrees*. The symbol used to represent degrees is a small raised circle, °. A measure of 360 degrees, or 360°, is equal to one full revolution.

Demand equation A demand equation is an equation that relates quantity sold (demanded), x. to price, p. When solved for x it indicates the quantity that will be demanded for a given price. When solved for p it indicates the price that will be charged for a given demand.

Denominator The expression below the fraction bar in a fraction or quotient.

Dependent system A system of equations where one equation is a multiple of another, yielding an infinite number of solutions. In two variable systems, the graph of a dependent system is coinciding lines.

Dependent variable When solving for a variable, that variable is called the *dependent variable* because its value depends on the value(s) of the remaining variable(s). Any remaining variables are called *independent variables* because we are free to select their values.

Descending order Terms are listed so that the first term has the largest degree and the exponents on the variable decrease from left to right.

Determinant A value computed for a square matrix that can be used to solve systems of linear equations.

Devine formula A popular formula in the healthcare profession that uses a person's height to compute his or her ideal body weight. The formula is $w = 110 + 5.06(h - 60)$ for men and $w = 100.1 + 5.06(h - 60)$ for women, where w is ideal body weight, in pounds, and h is height, in inches.

Diagonal A diagonal of a rectangle is a line that extends between two opposite corners of the rectangle.

Diameter The diameter of a circle is the distance across the circle and through its center. The length of the diameter is two times the length of the radius.

Difference The result of subtracting two numbers. The number being subtracted is called the *subtrahend*, and the number being subtracted from is called the *minuend*.

Digits The symbols 0, 1, 2, 3, 4, 5, 6, 7, 8, and 9.

Directly proportional A relationship between two variables where one variable is a constant multiple of the other. The constant value is called the *constant of variation*.

Directrix A parabola is the set of all points in a plane equidistant from a fixed point F and a fixed line D. The fixed point is called the *focus*, and the fixed line is called the *directrix*.

Direct variation For an equation of the form $y = kx$, we say that y varies directly with x, or y is directly proportional to x. The nonzero constant k is called the *constant of variation* or the *proportionality constant*.

Discount The amount by which the original price is decreased when merchandise is placed on sale.

Discount formula A percent equation for computing discounts, where the discount is the amount, the discount rate is the percent, and original price is the base. That is, $\text{Discount} = \text{Discount rate} \cdot \text{Original price}$.

Discount rate The percent by which the original price is decreased when merchandise is placed on sale.

Discriminant Given a quadratic equation $ax^2 + bx + c = 0, a \neq 0$, the expression $D = b^2 - 4ac$ is called the *discriminant*. If $D > 0$, then the quadratic equation has two real solutions. If $D < 0$, then the quadratic equation has two non-real solutions. If $D = 0$, then the quadratic equation has exactly one real solution.

Distance formula The distance $d(A, B)$ between two points (x_1, y_1) and (x_2, y_2) is given by $d(A, B) = \sqrt{(x_2 - x_1)^2 + (y_2 - y_1)^2}$.

Distributive property Multiplication distributes over addition (or subtraction). That is, for any numbers a, b, and c, $a \cdot (b + c) = a \cdot b + a \cdot c$ or $a \cdot (b - c) = a \cdot b - a \cdot c$. Similarly, $(b + c)a = ba + ca$ and $(b - c)a = ba - ca$.

Diverge To have no limit. For example, an infinite geometric series with $|r| \geq 1$ does not approach a finite sum so is said to *diverge*.

Dividend The quantity that you divide into in a division problem. This is also the numerator of a fraction.

Divide out To simplify a fraction, we can divide out common factors by dividing both the numerator and denominator by the common factors.

Divides evenly If no remainder results when dividing a dividend by a divisor, then the divisor *divides evenly* into the dividend.

Divides exactly If no remainder results when dividing a dividend by a divisor, then the divisor *divides exactly* into the dividend.

Divisibility tests Special rules that can be used to determine if certain numbers are factors of a given number.

Divisible If no remainder results when dividing a dividend by a divisor, the dividend is *divisible* by the divisor and the divisor is a *factor* of the dividend.

Division The process of separating a quantity into groups of equal size. The number being divided is called the dividend, whereas the number that divides is the divisor. The result of division is the quotient. If a and b represent two numbers and $b \neq 0$, then $a \div b = \frac{a}{b} = a \cdot \frac{1}{b}$.

Division properties of 0 For any number a, except 0, $0 \div a = 0$. Division by 0 is not possible. That is, $a \div 0$ is undefined.

Division properties of 1 Any number, except 0, divided by itself is 1. That is, for any number a, except 0, $a \div a = 1$. Any number divided by 1 is itself. That is, for any number a, $a \div 1 = a$.

Division property of equality See *Multiplication property of equality*.

Division shortcut In a division problem, an equal number of trailing zeros can be removed from the dividend and divisor without changing the quotient.

Divisor The quantity "divided by" in a division problem. This is also the denominator of a fraction.

Domain The domain of a relation is the set of all possible values for the independent variable. These are the first coordinates in a set of ordered pairs and are also known as *input values*.

Double bar graph A bar graph that contains two bars for each category.

Double-negative rule If a is a real number, then $-(-a) = a$.

Edge of a solid A line segment where two faces touch (or intersect).

Elements of a cone Line segments that intersect the outer edges of the circular bases of a pair of right circular cones and pass through a fixed point. The fixed point is called the *vertex* of the cone.

Elements of a set The objects in a set. Also called the *members of a set*.

Elimination method A method for solving a system of linear equations in two variables that involves adding two equations together in a way that will eliminate one of the variables.

Ellipse The set of all points in a plane, the sum of whose distances from two fixed points is a positive constant. The two fixed points, F_1 and F_2, are called the foci.

Ellipsis notation for repeating decimals Uses three dots (...) behind a digit that repeats or a pattern of repeating digits to indicate that the pattern continues forever.

Empirical probability An estimated probability of an event computed by dividing the number of times the event occurred (or is observed) by the number of trials. That is,

$$\text{Estimated probability of an event} = \frac{\text{Number of times the event is observed}}{\text{Number of trials of the experiment}}.$$

Endpoint A point where a ray or line segment ends.

Equally likely outcomes Outcomes that have the same chance, or probability, of occurring.

Equal sign An equal sign, =, is used in equations to state that two quantities are equal.

Equation A statement that two quantities are equal. An equation is a *numeric equation* if it contains only numeric expressions and no variables. An equation is an *algebraic equation* if it contains one or more variables.

Equation in one variable An equation that contains a single variable.

Equilateral triangle A triangle with three equal sides and three equal angles.

Equivalent Two quantities that are the same, such as when a fraction and a decimal represent the same value. For example $\frac{1}{2}$ and 0.5 are equivalent.

Equivalent equations Two or more equations with the same solution set.

Equivalent fractions Two or more fractions that represent the same value.

Equivalent functions Two functions that represent the same set of ordered pairs.

Equivalent inequalities Two or more inequalities that have the same solution set.

Equivalent matrices Two or more matrices that represent systems that have the same solution set.

Evaluate a formula To find the value of the isolated variable by substituting values for all other variables in the formula.

Evaluate a function To determine the value of a function for a given value of the independent variable by substituting the value for the independent variable into the function definition and simplifying the resulting numeric expression using the order of operations.

Evaluate algebraic expressions To substitute given values for the variables and simplify the resulting numeric expression using the order of operations.

Evaluate an exponential expression To multiply out the expression to a single numeric value.

Event A combination of possible outcomes for a probability experiment.

Expanded form of a number Shows the number as the addition of the value of each digit in the number. For example, $127 = 100 + 20 + 7$.

Expansion by minor A technique used to evaluate determinants of 3×3 matrices, and larger square matrices.

Experiment Any task or process in which the result is not known in advance. Some examples are flipping a coin, drawing a name from a hat, or rolling a pair of dice. Also called a *probability experiment*.

Exponent A superscripted number that tells how many times a base expression is multiplied by itself. Also called *power*.

Exponential decay A model that describes the exponential decay (decrease) of a population, quantity, or amount A, after a certain time, t, is $A(t) = A_0 e^{kt}$, where $A_0 = A(0)$ is the initial quantity and $k < 0$ is a constant called the *relative decay rate*.

Exponential expression A constant or algebraic expression that is raised to a power. The constant or algebraic expression makes up the base, and the power is the exponent on the base.

Exponential form Involving the use of exponents. For example, the expression x^3 is in exponential form (as opposed to $x \cdot x \cdot x$) and the equation $4^x = 50$ is in exponential form (as opposed to $x = \log_4 50$).

Exponential function A function of the form $f(x) = b^x$ where x is any real number and $b > 0$ such that $b \neq 1$. The constant, b, is called the base of the exponential function.

Exponential growth A model that describes the exponential increase of a population, quantity, or amount A, after a certain time, t, is $A(t) = A_0e^{kt}$, where $A_0 = A(0)$ is the initial quantity and $k > 0$ is a constant called the *relative growth rate*.

Exponent notation a^n represents the product of n factors of the number a. a is the *base*, and n is the *exponent*, or *power*.

Exponent of 1 An exponent of 1 is usually not written because a base number raised to a power of 1 is itself. If a variable has no exponent, then the exponent is understood to be 1 (e.g., $x^1 = x$).

Extraneous solutions A solution to a new equation that occurs during the solving process but is not a solution to the original equation.

Extreme value A value that is much larger or much smaller than the other data values in a list.

Faces The sides of a three dimensional object such as a rectangular solid.

Factor In multiplication, each number or expression being multiplied. For example, in the expression $2 \cdot 3$, the factors are 2 and 3. In the expression $x(y + 1)$, the factors are x and $y + 1$.

Factor a polynomial To write a polynomial as an equivalent expression that is the product of polynomials.

Factor a whole number To write the nonzero whole number as the product of two or more whole numbers.

Factored completely A whole number is *factored completely* if it is written as the product of all prime numbers. A polynomial is *factored completely* if it is written as the product of all prime polynomials.

Factored form When a polynomial is written as an equivalent expression that is the product of polynomials.

Factorial The *factorial* of a non-negative integer n, denoted as $n!$, is the product of all positive integers less than or equal to n. Thus, $n! = n(n - 1) \ldots \ldots 3 \cdot 2 \cdot 1$.

Factoring Writing a polynomial as the product of two or more different polynomials of lesser degree.

Factoring by grouping A method of factoring where terms of the original polynomial are grouped together to look for common factors.

Factor over the integers To write a polynomial as a product of polynomial factors where all coefficients in both the original polynomial and its factors are integers.

Factor tree A branching diagram used to show the prime factors of a number.

Fahrenheit scale A scale of temperature where 32°F is the freezing point of water (read "thirty-two degrees Fahrenheit") and 212°F is the boiling point of water. (read "two hundred twelve degrees Fahrenheit"). The raised circle indicates "degrees" and the "F" indicates we are using the *Fahrenheit scale*.

False equation An equation that is never true.

Feasible domain The set of values for the independent variable that make sense, or are feasible, in the context of the application.

Feasible solution A solution that satisfies the conditions of an application problem and makes sense within the context of the problem.

Fibonacci numbers The numbers that appear in the Fibonacci sequence: $1, 1, 2, 3, 5, 8, \ldots$

Fibonacci sequence A famous sequence named after the 13th-century Italian mathematician Leonardo of Pisa, also known as Fibonacci. The sequence is defined recursively as $a_n = a_{n-1} + a_{n-2}$, where $a_1 = 1$ and $a_2 = 1$.

Finite sequence A function whose domain is the finite set $\{1, 2, 3, \ldots, n\}$ where is n a natural number.

Finite series The sum of a fixed, or finite, number of terms in a sequence.

Finite set A set with a fixed, or finite, number of elements.

First-degree equation A polynomial equation in which the largest exponent on a variable is 1.

Fluid ounce (fl oz) A commonly used American unit of capacity. 8 fluid ounces is equivalent to 1 cup.

Focus of a parabola A *parabola* is the set of all points in a plane equidistant from a fixed point F and a fixed line D. The fixed point is called the *focus*, and the fixed line is called the *directrix*.

FOIL method FOIL is the acronym for a method of multiplying binomials; that is, to multiply the two First terms, the two Outside terms, the two Inside terms, and the two Last terms, then add the resulting products together.

Foot (ft) A common American unit of length equal to 12 inches. About the length from the elbow to wrist in an adult arm.

Formula An equation that describes the relationship between two or more variables.

Fraction A number of the form $\frac{a}{b}$, where a and b are numbers and $b \neq 0$. a is called the *numerator*, and b is called the *denominator*.

Fraction bar A horizontal line used to separate the numerator (or top expression) and the denominator (or bottom expression) in a fraction.

Fraction part of a mixed number For a mixed number having the form $a\frac{b}{c}$, the proper fraction $\frac{b}{c}$ is called its fraction part.

Frequency See *Class frequency*.

Frequency table A table showing the number, or count, of data values that fall within each class interval of a frequency distribution.

Function A relation in which each value in the domain corresponds to exactly one value in the range.

Function notation Notation used to name a function and identify the independent variable. An example of function notation is $f(x)$, read "f of x." For $y = f(x)$, f is the name of the function that relates the independent variable x to the dependent variable y.

Fundamental counting principle The number of outcomes for a series of tasks can be found by multiplying the number of ways each task can be completed. For example, if there are n_1 ways to do task 1 and n_2 ways to do task 2, then there are $n_1 \times n_2$ ways to complete the two tasks. This result extends to any number of tasks.

Gallon (gal) A common American unit of volume equal to four quarts.

General equation for percent An equation used to solve percent problems that relates an amount to a base, usually expressed as $\text{Percent} \cdot \text{Base} = \text{Amount}$.

General form of the equation of a circle $Ax^2 + By^2 + Cx + Dy + E = 0$, where A, B, C, D, and E are real numbers and $A = B \neq 0$.

Geometric sequence A sequence is geometric if the ratio of any two successive terms is constant. The nth term of the sequence is given by $a_n = a_1 \cdot r^{n-1}$, where a_1 is the first term and r is the constant ratio.

Golden ratio The ratio that, according to ancient Greeks, is most pleasing to the eye. It can be roughly approximated by the fraction $\frac{81}{50}$ and appears frequently in nature, art, and architecture.

Gram (g) The base unit of mass in the metric system, about the weight of a paperclip.

Graph an ordered pair To place a point at the ordered pair's location on the coordinate plane.

Graph a number To place a point, or solid circle, (●) at the number's location on the number line.

Graph of a function The graph of all ordered pairs that belong to the function. The point (a, b) lies on the graph of a function f if and only if $f(a) = b$.

Graph of an equation in two variables A visual display of all ordered pair solutions to the equation.

Graph of a system of linear inequalities in two variables The intersection of the graphs of each inequality in the system.

Greatest common factor (GCF) The largest factor common to all terms in a list. The GCF can include both numeric factors, such as 2, and variable factors, such as x.

Greatest common factor (GCF) of a group of integers The largest integer that divides evenly into each integer in the group.

Greatest common factor (GCF) of a group of monomials The monomial with the largest coefficient and highest degree that divides each monomial evenly.

Greatest common factor (GCF) of a polynomial The monomial with the largest coefficient and highest degree that divides each term of the polynomial evenly.

Grouping symbols Mathematical symbols used to group operations so they are treated as a single quantity. When simplifying, all operations within grouping symbols must be done first. Some examples of grouping symbols are parentheses (), brackets [ ], and braces { }. Operators such as absolute value | |; fraction bars, –; and radicals, $\sqrt{\ }$ are also treated as grouping symbols.

Half-plane One half of the coordinate plane, separated from the other half by a boundary line.

Hang time Length of time that an object, such as a punted football or jumping athlete, remains in the air.

Height of a triangle The perpendicular distance from a vertex of the triangle to the side opposite the vertex.

Histogram A special type of bar graph in which the bars touch, and the width of the bars has meaning. Each bar in a histogram represents a class, or category, that is a range of values called a *class interval*. The height of each bar is determined by the number of data values that fall within the corresponding class interval. There are no gaps between the bars of a histogram, unless a class has a frequency of 0, and there is no overlap in the class intervals.

Horizontal line The graph of a linear equation in two variables whose points all have the same y-coordinate, b, is a horizontal line with the equation $y = b$.

Horizontal line test If every horizontal line intersects the graph of a function f at most once, then f is one-to-one.

Hour (h) A common unit of measure for time equal to 60 minutes.

Hyperbola The set of all points in a plane, the difference of whose distances from two fixed points is a positive constant. The two fixed points, F_1 and F_2 are called the foci.

Hypotenuse The longest side of a right triangle. It is the side opposite the right angle.

Identity An equation that is always true for all defined values of its variable.

Identity property of addition If 0 is added to any given number, then the sum is the given number. That is, for any number $a, a + 0 = a$ and $0 + a = a$.

Identity property of multiplication If 1 is multiplied by any number, then the product is the original number. That is, for any number $a, a \cdot 1 = a$ and $1 \cdot a = a$.

Imaginary part of a complex number For a complex number in the standard form $a + bi$, the number b is called the *imaginary part of the complex number.*

Imaginary unit i $i = \sqrt{-1}$ where $i^2 = -1$.

Improper fraction A fraction in which the numerator and denominator are both whole numbers and the numerator is greater than or equal to the denominator. An improper fraction has a value greater than or equal to 1.

Inch (in.) An American unit of length equal to one twelfth of a foot, about the diameter of a U.S. quarter dollar coin.

Inclusive Means to include the extremes as well as everything between. For example, the set of numbers between 4 and 7, *inclusive,* includes 4 and 7 along with all numbers in between them.

Inconsistent system A system of equations or inequalities that has no solution. When solving an inconsistent system, a contradiction (false statement) will result.

Independent system of two equations in two variables A system of two equations in two variables that has either exactly one solution or no solution. The graphs of two lines in the system either intersect or are parallel.

Independent variable A variable for which we can select values. These are also called *input variables* because they represent the set of values that we can *input* into a relation.

Indeterminate A quantity whose value cannot be determined. For example, if a zero dividend is divided by a divisor of 0, the result is indeterminate. That is, $0 \div 0$ is indeterminate.

Index of a radical expression In the radical expression $\sqrt[n]{a}$, n is called the index, and indicates the type of root.

Index of summation A letter used to indicate the terms of a sum.

Inequality A statement that two quantities are not strictly equal to each other.

Inequality symbols Symbols used to indicate two quantities are not equal. For example, the symbol $>$ means greater than; the symbol $<$ means less than.

Infinite arithmetic series The sum of the terms of an infinite arithmetic sequence.

Infinite geometric series The sum of the terms of an infinite geometric sequence.

Infinite series The sum of the terms of an infinite sequence.

Infinite sequence A function whose domain is the set of all natural numbers.

Infinite set A set with an unlimited number of elements.

Integers If the positive whole numbers, their related negative numbers, and zero are grouped together, they form the *integers*. That is, $\ldots, -5, -4, -3, -2, -1, 0, 1, 2, 3, 4, 5, \ldots$.

Intercept A point where a graph crosses or touches a coordinate axis.

Interest When we borrow money, we pay an extra amount of money called *interest* for this privilege. Similarly, if we lend money to someone else, the payment we receive for this generosity is also called *interest*.

Interest rate The amount of interest paid or earned is a percent of the principal over a given time interval. The percent is called the *interest rate*.

Intersecting lines Two lines that cross or touch at a single point on the same plane.

Intersection For any two sets A and B, the intersection of A and B is given by $A \cap B$ and represents the elements that are common to set A and set B.

Interval notation An interval is a set of numbers that lie between a given lower and upper bound. The bounds indicate the end points of the interval. For example, the interval $[3, 9]$ is the set of all values between 3 and 9, including 3 and 9. The interval $(4, 11)$ is the set of all values between 4 and 11, but not including 4 or 11. The interval $(4, \infty)$ is the set of all values greater than 4.

Inverse function Let f be a one-to-one function with domain A and range B. Then f^{-1} is the inverse function of f with domain B and range A. Furthermore, if $f(a) = b$, then $f^{-1}(b) = a$.

Inversely proportional (inverse variation) For equations of the form $y = \frac{k}{x}$ or $y = k\frac{1}{x}$, we say that y varies inversely with x, or y is inversely proportional to x. The constant k is called the constant of variation or the proportionality constant.

Irrational number An irrational number is a real number that is not a rational number, so it cannot be written as the quotient $\frac{p}{q}$ of two integers p and q with $q \neq 0$. The decimal form of an an irrational number will not terminate and will not repeat.

Isolate a quantity A quantity in an equation is isolated if it is by itself on one side of the equation and is the only occurrence of its type in the equation.

Isolate a radical expression To get the radical expression by itself on one side of the equation using the properties of equality.

Isolate a variable To use properties of equality and simplification to get the variable by itself on one side of the equation and not appear on the other side.

Isosceles triangle A triangle that has two sides with equal lengths and two angles with equal measures.

Leading coefficient of a polynomial in one variable For a polynomial in one variable written in standard form, the coefficient of the first term (the term with the highest degree).

Least common denominator (LCD) The smallest number or expression that is divisible by all the denominators in a group of fractions.

Least common multiple The smallest multiple that is evenly divisible by all the numbers or expressions in a list of numbers or expressions.

Legs The two sides of a right triangle that form the right angle.

Length The distance from one end of an object to another along its largest dimension.

Like fractions Two or more fractions that have the same denominator. For example, the fractions $\frac{2}{15}$ and $\frac{8}{15}$ are like fractions because they have a common denominator of 15.

Like radicals Two or more radical expressions that have both the same index and the same radicand.

Like terms Two or more terms that contain the exact same variable factors. Their variables must match, and any exponents on the variables must match. The coefficients may differ.

Line A straight arrangement of points that extend forever in two directions.

Linear coefficient The coefficient of the linear term in a simplified polynomial.

Linear equation in one variable An equation that can be written in the form $ax + b = c$, where x is a variable and a, b, and c are any numbers such that $a \neq 0$. A linear equation is also called *a first-degree equation* because the exponent on the variable is understood to be 1.

Linear equation in three variables An equation that can be written in the standard form $Ax + By + Cz = D$ where A, B, C, and D are real numbers, and A, B, and C are not all equal to 0.

Linear equation in two variables An equation that can be written in the standard form $Ax + By = C$, where A, B, and C are real numbers, and A and B are not both equal to 0.

Linear function A function of the form $f(x) = ax + b$ where a and b are real numbers and $a \neq 0$.

Linear inequality in one variable An inequality that can be written in the form $ax + b < c$, where a, b and c are real numbers and $a \neq 0$. The inequality symbol $<$ can be replaced with $>, \leq, \leq$, or $\neq$.

Linear inequality in two variables An inequality that can be written in the standard form $Ax + By < C$ where A, B, and C are real numbers, and A and B are not both equal to zero. The inequality symbol $<$ can be replaced with $>, \leq, \leq$, or $\neq$.

Line graph A graph consisting of a series of points that are connected by line segments, often used to show how the value of a variable changes over time.

Line segment (segment) A section of a line that has two endpoints.

Liter (L) The base unit of capacity, or volume, in the metric system. A cube that measures 10 centimeters on each side has the capacity of 1 liter.

Logarithmic form The equation $y = \log_b x$ is in *logarithmic form.*

Logarithm property of equality If a logarithmic equation can be written in the form $\log_b u = \log_b v$, then $u = v$. Furthermore, if $u = v$, then $\log_b u = \log_b v$.

Logistic model A model that describes the logistic growth of a population P at any time t is given by the function $P(t) = \frac{C}{1 + Be^{kt}}$ where B, C, and k are constants with $C > 0$ and $k > 0$.

Long division A process for dividing two numbers that uses a sequence of steps involving division, multiplication, subtraction, and dropping down the next digit. The steps are repeated until the remainder can no longer be divided.

Long division symbol The symbol $\overline{)\ \ }$ used to separate the dividend from the divisor in a long division problem.

Long run average cost (LRAC) The total long run cost divided by the number of units produced.

Lower bound The largest value that is less than or equal to all values in an interval. No values in the interval can be less than the lower bound. For example, the interval $[6, 9]$ has a lower bound of 6. No value in the interval is less than 6.

Lower limit of summation The value that indicates the number of the first term of a sequence to be included in a series.

Lowest terms (or Simplest form) A fraction is written in *lowest terms*, or *simplest form*, if the numerator and denominator have no common factors other than 1.

Main fraction bar The fraction bar that separates the numerator and denominator of a complex fraction.

Major axis In an ellipse, the longer axis which is the line segment that connects the two vertices.

Markup An amount added by a merchant to the cost of an item to obtain the selling price of the item.

Mass A measure related to the amount of matter within an object. The term mass is often used interchangeably with *weight* (which is fine in most practical situations). However, an object's weight is affected by gravitational forces, whereas an object's mass is not affected by gravity.

Mathematical expression A statement containing symbols (including numbers) and/or mathematical operations (such as addition or subtraction).

Mathematical model Uses the language of mathematics to describe a problem. Typically the model is an equation that describes a relationship within an application.

Mathematical symbol A character that represents a mathematical relation or operation.

Matrices The plural of *matrix*.

Matrix A rectangular array of numbers arranged in rows and columns.

Maximum The largest value in the range of the function. It occurs at the highest point on the graph of the function.

Mean A measure of central tendency found by adding a list of data values and then dividing by the number of values. Also called the *average*.

Measure of central tendency A measure that indicates the "middle" of a list of data.

Median The middle value in a list of ordered data. If there is an even number of data values, the median is the average of the two middle values. The median is a measure of central tendency that is not greatly affected by a few extreme data values.

Meter (m) The base unit of length in the metric system. About the height of a door knob.

Method of relating the bases If an exponential equation can be written in the form $b^u = b^v$, then $u = v$.

Metric system (International system of units) An international system of measurement (e.g., meters, kilograms, liters) used in science, medicine, and most manufacturing settings.

Midpoint of a line segment The point that lies exactly halfway between the endpoints of a line segment. The midpoint of the line segment from $A(x_1, y_1)$ to $B(x_2, y_2)$ is the point with coordinates $\left(\frac{x_1 + x_2}{2}, \frac{y_1 + y_2}{2}\right)$.

Mile (mi) An American unit of length equal to 5280 feet.

Minimum The smallest value in the range of the function. It occurs at the lowest point on the graph of the function.

Minor rational expressions The rational expressions within the numerator and denominator of a complex rational expression. Also called *minor fractions*.

Minuend The quantity being subtracted *from* in a subtraction (difference) problem.

Minute (min) A common unit for the measure of time equal to 60 seconds.

Mixed number A number of the form $a\frac{b}{c}$, where a is a nonzero whole number and $\frac{b}{c}$ is a proper fraction. $a\frac{b}{c}$ means $a + \frac{b}{c}$, but the + symbol is not written. a is the *whole-number part* and $\frac{b}{c}$ is the *fraction part* of the mixed number.

Mode A measure of central tendency, the mode of a list of data is the data value that appears most frequently. A list of data may have no mode, one mode, or more than one mode.

Model A representation used to illustrate something that cannot be seen easily (for example, a DNA model). See also *Mathematical model*.

Monomial A simplified term in which all variables are raised to non-negative integer powers and no variables appear in any denominator.

Monthly Happening once each month (12 times each year), as with compound interest. See *Annually*.

Multiple The product of two whole number factors. A number that can be divided by a second number without a remainder is said to be a *multiple* of the second number.

Multiplication The simplified process of writing repeated addition. Multiplication symbols include a multiplication sign ($\times$), an asterisk, a multiplication dot ($\cdot$), or parentheses.

Multiplication property of equality Let a, b, and c be numbers or algebraic expressions with $c \neq 0$. Then, $a = b$ and $a \cdot c = b \cdot c$ are equivalent equations. Also, because division is defined in terms of multiplication, $a = b$ and $\frac{a}{c} = \frac{b}{c}$ are equivalent equations.

Multiplication property of inequality Let a, b, and c real numbers. If $a < b$ and $c > 0$, then $ac < bc$ and $\frac{a}{c} < \frac{b}{c}$. Multiplying or dividing both sides of an inequality by a positive number c results in an equivalent inequality. If $a < b$ and $c < 0$, then $ac > bc$ and $\frac{a}{c} > \frac{b}{c}$. Multiplying or dividing both sides of an inequality by a negative number c, and switching the direction of the inequality, results in an equivalent inequality.

Multiplication property of zero For any number a, $0 \cdot a = 0$ and $a \cdot 0 = 0$.

Multiplicative identity The number 1 is called the *multiplicative identity* because if 1 is multiplied by any number, the *product* is the same identical number.

Multiplicative identity property If 1 is multiplied by any number, then the product is the original number. That is, for any number a, $a \cdot 1 = a$ and $1 \cdot a = a$.

Multiplicative inverse The reciprocal of a real number is called the *multiplicative inverse* of the number because the product of a number and its reciprocal is 1.

Natural base There is one number that appears as the base in exponential applications more than any other number. This number is called the *natural base* and is symbolized using the letter e. ($e = 2.7182818284\ldots$)

Natural exponential function The function $f(x) = e^x$.

Natural logarithm The base e logarithm, typically denoted by ln (rather than $\log_e$).

Natural logarithmic function For $x > 0$, the natural logarithmic function is defined by: $y = \ln x$ if and only if $x = e^y$.

Natural numbers Also called *counting numbers*, the elements of the set $\mathbb{N} = \{1, 2, 3, 4, 5, \ldots\}$.

Negative exponent An exponent, or power, whose value is less than zero.

Negative number A number less than zero.

Negative-power rule To remove a negative exponent, switch the location of the base (numerator or denominator) and change the exponent to be positive. That is, $a^{-n} = \frac{1}{a^n}$ and $\frac{1}{a^{-n}} = a^n (a \neq 0)$.

Negative sign (−) The symbol used to indicate that a number is negative, or less than zero. Also used to indicate the opposite of a number.

Negative slope The slope of a line that slants downward, or falls, from left to right.

Negative square root A negative real number is the *negative square root* of a non-negative real number a, denoted as $b = -\sqrt{a}$, if $b^2 = a$.

Nested grouping symbols Grouping symbols within grouping symbols.

Newton's law of cooling The temperature T of an object at any time t is given by $T(t) = S + (T_0 - S)e^{kt}$, where T_0 is the original temperature of the object, S is the constant temperature of the surroundings, and k is the cooling constant.

Nonlinear equation An equation that is not a linear equation.

Non-negative A real number, x, is non-negative if it is 0 or larger. That is, $x \geq 0$.

Non-simplified expression An expression that can be written in a simpler form by combining like terms, removing grouping symbols, and so forth.

Non-strict inequality An inequality that contains one or both of the following inequality symbols: $\leq, \geq$. The possibility of equality is included in a non-strict inequality.

Nonzero A real number is nonzero if it is not equal to 0. All positive and negative numbers are nonzero.

***n*th partial sum** The sum of the first n terms of a series, denoted by S_n.

Null set A set that contains no elements. Also called the *empty set*. Denoted by $\varnothing$ or $\{\ \}$.

Number line A visual representation of all positive and negative numbers and zero.

Numerator The expression above the fraction bar in a fraction or quotient.

Numeric equation A statement of equality between two numeric expressions. A numeric equation does not contain variables.

Numeric expression A combination of numbers and arithmetic operations.

Obtuse angle An angle that measures between 90° and 180°, not inclusive.

One-to-one function A function f is one-to-one if for any values $a \neq b$ in the domain of f, $f(a) \neq f(b)$.

One-to-one function (Alternate definition) A function f is one-to-one if for any two range values $f(u)$ and $f(v), f(u) = f(v)$ implies that $u = v$.

Operation A mathematical action that is performed by following specific rules. Addition, subtraction, multiplication, and division are all operations.

Opposite of an expression If a negative sign appears in front of an expression within parentheses, removing the parentheses and changing the sign of each term in the expression results in the opposite of the expression.

Opposite of a polynomial Is formed by changing the signs on all terms of the original polynomial.

Opposite reciprocals Two nonzero numbers whose product equals -1. This occurs when the two numbers have opposite signs and their absolute values are reciprocals.

Opposites Two numbers whose graphs are located the same distance from a on the number line but lie on opposite sides of 0. The sum of opposites is 0. Also called *additive inverses*.

Ordered A list of data values is *ordered* if it is written in increasing value from smallest to largest.

Ordered pair Each position, or point, on the coordinate plane can be identified using an *ordered pair* of numbers in the form (x, y). The first number x is called the *x-coordinate* or *abscissa* and indicates the point's horizontal position. The second number y is called the *y-coordinate* or *ordinate* and indicates the point's vertical position.

Ordered pair solution An *ordered pair solution* to an equation in two variables, written as (x, y) gives values for the variables that, when substituted in the equation, result in a true statement. For example, (2, 5) is an ordered pair solution to the equation $2x + y = 9$ because $2(2)+(5) = 4 + 5 = 9$.

Ordered triple Each position, or point, in space (three dimensions) can be identified using an ordered triple of numbers in the form (x, y, z).

Ordered-triple solution An *ordered-triple solution* to an equation in three variables, written as (x, y, z) gives values for the variables that, when substituted in the equation, result in a true statement.

Order of operations (1) Evaluate operations within parentheses (or other grouping symbols) first, starting with the innermost set and working out; (2) work from left to right and evaluate any exponential expressions as they occur; (3) work from left to right and perform any multiplication or division operations as they occur; and (4) work from left to right and perform any addition or subtraction operations as they occur.

Ordinary annuity An annuity in which equal deposits are made at the end of a compound period.

Origin (Real number line) The point 0 is called the origin of a real number line. Numbers located to the left of the origin are negative numbers, and numbers located to the right of the origin are positive numbers.

Origin (Rectangular coordinate system) The point at which the x- and y-axes intersect in the rectangular coordinate system. The origin is represented by the ordered pair (0,0).

Original price The price of merchandise before any discount is subtracted or any tax is added. In applications of the percent equation involving percent increase or percent decrease, the *original price* is the base.

Ounce (oz) An American unit of weight equal to one sixteenth of a pound.

Outcome (of an experiment) Any possible result of a probability experiment.

Overall amount for loan or investment Whether borrowing or investing, the overall amount of money involved results from adding the principal and the interest. That is, Overall amount = Principal + Interest, or $A = P + I$.

Paired data Data represented as ordered pairs.

Parabola The set of all points in a plane equidistant from a fixed point F and a fixed line D. The fixed point is called the *focus*, and the fixed line is called the *directrix*.

Parallel lines Two lines that lie on the same plane and never cross (or touch).

Parallelogram A four-sided polygon with two pairs of parallel sides. Its opposite sides are equal in length and its opposite angles are equal in measure.

Pentagon A five-sided polygon.

Percent "Per hundred" or "out of 100." A percent is denoted by the percent symbol, %.

Percent decrease The amount of decrease expressed as a percentage of the original value. That is,

$$\text{Percent decrease} = \frac{\text{Amount of decrease}}{\text{Original value}}.$$

Percent increase The amount of increase expressed as a percentage of the original value. That is,

$$\text{Percent decrease} = \frac{\text{Amount of increase}}{\text{Original value}}.$$

Percent of change The amount of change expressed as a percent of the original amount.

Percent proportion $\frac{\text{Amount}}{\text{Base}} = \frac{\text{Percent Number}}{100}$. The amount is the part compared to the whole, the base is the number after *of*, and the percent number is the number before the percent symbol, %.

Perfect 4th power A quantity that can be written as the 4th power of another quantity of the same type.

Perfect cube A quantity that can be written as the cube of another quantity of the same type.

Perfect square A number whose square root is a whole number or fraction (including decimals that can be written in fraction form).

Perfect square trinomial A trinomial that can be written as the square of a binomial sum, $(A + B)^2$, or the square of a binomial difference, $(A - B)^2$.

Perimeter The distance around a figure. For a polygon, the perimeter is the sum of the lengths of the sides of the polygon.

Perimeter of a parallelogram If a parallelogram has parallel sides of length a, and parallel base sides of length b, then its perimeter P is given by the formula $P = 2a + 2b$.

Perimeter of a rectangle If a rectangle has length l and width w, then its perimeter P is given by the formula $P = 2l + 2w$.

Perimeter of a square If a square has sides of length s, then its perimeter P is given by the formula $P = 4s$.

Perimeter of a trapezoid If a trapezoid has sides of length a and c, and bases b (small base) and B (large base), then its perimeter P is given by the formula $P = a + b + c + B$.

Perimeter of a triangle If a triangle has sides of length a, b, and c, then its perimeter P is given by the formula $P = a + b + c$.

Periodic compound interest formula $A = P\left(1 + \frac{r}{n}\right)^{nt}$ where $A =$ total amount after t years; $P =$ principal (original investment); $r =$ interest rate per year (as a decimal), and $n =$ number of times interest is compounded per year.

Periods Groups of three digits separated by commas that are used to express whole numbers with many digits. The first five periods from right to left are *ones*, *thousands*, *millions*, *billions*, and *trillions*.

Perpendicular Two lines or flat surfaces are *perpendicular* if they meet (or intersect) at a right angle (90°).

Pint (pt) An American unit of volume equal to two cups.

Place value The value of the position where a digit lies in a number, such as ones, tens, hundreds, and tenths, hundredths, thousandths, and so on.

Plane figure A two-dimensional (flat) figure that has length and width but zero thickness (e.g., rectangles, triangles, and circles).

Plot See *Graph a number* and *Graph an ordered pair*.

Plus or minus symbol The symbol $\pm$ is read as "plus or minus" and can be used to indicate two values, one positive and one negative. For example, $x = \pm 6$ is read as "x equals plus or minus six" and means $x = 6$ or $x = -6$. The symbol can also be used to indicate addition and subtraction. For example $x = 5 \pm 3$ is read "x equals five plus or minus three." It means $x = 5 + 3 = 8$ or $x = 5 - 3 = 2$.

Point A location on a number line, on a plane, or in space. Usually represented by a dot.

Point-slope form Given the slope m of a line and a point (x_1, y_1) on the line, the *point-slope form* of the equation of the line is given by $y - y_1 = m(x - x_1)$.

Polygon A closed plane figure made up of connected line segments. Each line segment is called a *side* of the polygon.

Polynomial A polynomial in x is a monomial or a finite sum of monomials.

Polynomial equation A statement that two polynomials are equal to each other.

Polynomial function A function of the form $P(x) = a_n x^n + a_{n-1}x^{n-1} + \cdots + a_1 x + a_0$ where each coefficient, a_i, is a real number and n is a whole number.

Polynomial inequality An inequality that can be written as $P(x) < 0, P(x) > 0, P(x) \leq 0$, or $P(x) \geq 0$, where $P(x)$ is a polynomial function.

Polynomial in one variable A polynomial containing one variable.

Polynomial in several variables A polynomial containing two or more variables.

Polynomial long division A process for dividing two polynomials that uses a sequence of steps involving division, multiplication, subtraction, and dropping down the next term. The steps are repeated until the remainder can no longer be divided.

Positive number A number that is greater than zero.

Positive sign $(+)$ The symbol used to indicate that a number is positive, or greater than zero.

Positive slope The slope of a line that slants upward, or rises, from left to right.

Positive square root A non-negative real number b is the *positive square root* of a non-negative real number a, denoted as $b = \sqrt{a}$, if $b^2 = a$. Also called *principal square root*.

Pound (lb) An American unit of weight equal to 16 ounces.

Power A superscripted number that tells how many times a base number is multiplied by itself. Also called *exponent*.

Power principle of equality If A and B represent algebraic expressions and n is a positive real number, then any solution to the equation $A = B$ is also a solution to the equation $A^n = B^n$.

Power-to-power rule When an exponential expression is raised to a power, multiply the exponents, that is, $(a^m)^n = a^{m \cdot n}$.

Present value formula Present value can be calculated using the formula $P = A\left(1 + \frac{r}{n}\right)^{-nt}$, where $P =$ principal (original investment); $A =$ total amount after t years; $r =$ interest rate per year (as a decimal), $n =$ number of times interest is compounded per year.

Prime factorization The expression of a composite number as the product of its prime factors. Every composite number has a unique prime factorization.

Prime number A whole number greater than 1 whose only whole number factors are 1 and itself.

Prime polynomial A polynomial whose only polynomial factors with integer coefficients are 1 and itself.

Principal The amount of money that is borrowed or invested.

Principal nth root If a and b are real numbers and n is an integer such that $n \geq 2$, then b is the principal nth root of a, denoted as $b = \sqrt[n]{a}$, if $b^n = a$. Note that b must be non-negative when n is even.

Principal square root A non-negative real number b is the principal square root of a non-negative real number a, denoted as $b = \sqrt{a}$, if $b^2 = a$. Also called *positive square root*.

Probability The chance or likelihood that an event will occur. Probability can be expressed as a fraction, decimal, or percent.

Probability experiment (or Experiment) Any task or process in which the result is not known in advance. Some examples are flipping a coin, drawing a name from a hat, or rolling a pair of dice.

Product The result of multiplication.

Product of complex conjugates The product of complex conjugates is equal to the sum of the squares of the real and imaginary parts. That is, $(a + bi)(a - bi) = a^2 + b^2$.

Product rule for exponents When multiplying exponential expressions with the same base, add the exponents and keep the common base. That is, $a^m \cdot a^n = a^{m+n}$.

Product rule for radicals If $\sqrt[n]{a}$ and $\sqrt[n]{b}$ are real numbers, then $\sqrt[n]{a}\,\sqrt[n]{b} = \sqrt[n]{ab}$.

Product-to-power rule When raising a product to a power, raise each factor of the base to the common exponent. That is, $(ab)^n = a^n b^n$.

Proper fraction A fraction in which the numerator and denominator are both whole numbers and the numerator is smaller than the denominator. A proper fraction has a value less than 1.

Property of equivalent fractions If a, b, and c are numbers, then $\frac{a}{b} = \frac{a \cdot c}{b \cdot c}$ and $\frac{a}{b} = \frac{a \div c}{b \div c}$ as long as b and c are not equal to 0.

Proportion An equation stating that two ratios (or rates) are equal, such as $\frac{a}{b} = \frac{c}{d}$, where $b \neq 0$ and $d \neq 0$.

Proportionality constant See *Constant of variation*.

Purchase price The price of merchandise before the sales tax is added.

Pyramid A solid containing a base that is a polygon and sides that are triangles sharing a common vertex.

Pythagorean Theorem For right triangles, the sum of the squares of the lengths of the legs of the triangle equals the square of the length of the hypotenuse, that is, $a^2 + b^2 = c^2$.

Quadrants The four regions into which the x-axis and the y-axis divide the rectangular coordinate system.

Quadratic equation in one variable An equation that can be written in the standard form $ax^2 + bx + c = 0$, where a, b, and c are real numbers and $a \neq 0$.

Quadratic formula The solutions to the quadratic equation $ax^2 + bx + c = 0, a \neq 0$, are given by the formula

$$x = \frac{-b \pm \sqrt{b^2 - 4ac}}{2a}.$$

Quadratic function A second-degree polynomial function of the form $f(x) = ax^2 + bx + c$, where a, b, and c are real numbers and $a \neq 0$. Every quadratic function has a "u-shaped" graph called a *parabola*.

Quadrilateral A four-sided polygon with the measures of the angles adding to 360°.

Quart (qt) An American unit of volume equal to two pints.

Quarterly Happening once each quarter (four times each year). See *Annually*.

Quotient The result of division.

Quotient rule for exponents When dividing exponential expressions with the same nonzero base, subtract the denominator exponent from the numerator exponent and keep the common base. That is, $\frac{a^m}{a^n} = a^{m-n}\,(a \neq 0)$.

Quotient rule for radicals If $\sqrt[n]{a}$ and $\sqrt[n]{b}$ are real numbers and $b \neq 0$, then $\sqrt[n]{\frac{a}{b}} = \frac{\sqrt[n]{a}}{\sqrt[n]{b}}$

Quotient-to-power rule When raising a quotient to a power, raise both the numerator and denominator to the common exponent. That is, $\left(\frac{a}{b}\right)^n = \frac{a^n}{b^n}(b \neq 0)$.

Radical equation An equation that contains at least one radical expression with a variable in the radicand.

Radical expression An expression that contains a radical sign.

Radical function A function that contains one or more radical expressions.

Radical sign ($\sqrt{\ }$) The symbol used to indicate roots of real numbers.

Radicand The expression beneath the radical sign.

Radius The fixed distance, r, from the center of the circle, (h, k), to any point that lies on the circle. The length of the radius is half the length of the diameter.

Range The set of all values for the dependent variable of a relation. These are the second coordinates in a set of ordered pairs and are also known as *output values*.

Rate A special type of ratio in which quantities of different types are compared.

Rate of change The ratio of the vertical change to the horizontal change when moving from one point on the graph to another. The slope of the line connecting the two points is equal to the rate of change.

Rate of work The number of jobs that can be completed in a given unit of time. If one job can be completed in t units of time, then the rate of work is given by $\frac{1}{t}$.

Ratio A comparison of two quantities, usually in the form of a quotient. There are three common notations used to express ratios: colon (e.g., 7:12); *to* (e.g., 7 to 12); and a fraction (quotient) (e.g., $\frac{7}{12}$).

Rational equation A statement in which two rational expressions are set equal to each other.

Rational exponent of the form $a^{m/n}$ If m/n is a rational number in lowest terms, m and n are integers such that $n \geq 2$, and $\sqrt[n]{a}$ is a real number, then $a^{m/n} = (\sqrt[n]{a})^m = \sqrt[n]{a^m}$.

Rational exponent of the form $a^{1/n}$ If n is an integer such that $n \geq 2$, and if $\sqrt[n]{a}$ is a real number, then $a^{1/n} = \sqrt[n]{a}$.

Rational expression An expression that can be written as the quotient $\frac{P}{Q}$ of two polynomials P and Q as long as $Q \neq 0$.

Rational function A function of the form $R(x) = P(x)/Q(x)$ where $P(x)$ and $Q(x)$ are polynomial functions with $Q(x) \neq 0$.

Rational inequality An inequality that can be written as $R(x) < 0$, $R(x) > 0$, $R(x) \leq 0$, or $R(x) \geq 0$, where $R(x) = \frac{P(x)}{Q(x)}$ is a rational function.

Rationalizing a denominator The process of removing radicals so that the denominator contains only a rational number.

Rational number A number that can be written as a fraction $\frac{p}{q}$, where p and q are integers and $q \neq 0$.

Ray A section of a line that has one endpoint and continues without end in one direction.

Real number A real number is any number that is either rational or irrational. Combining the set of rational numbers with the set of irrational numbers forms the set of real numbers, represented by $\mathbb{R}$.

Real number line A graph that represents the set of all real numbers, also known simply as the *number line*.

Real part of a complex number For a complex number in the standard form $a + bi$, the number a is called the *real part of the complex number*.

Reciprocals Two nonzero numbers whose product is 1. Also called *multiplicative inverses*.

Rectangle A four-sided polygon in which the sides meet at right angles. Its opposite sides are equal in length.

Rectangular coordinate system A system that consists of two number lines, one horizontal and one vertical. The horizontal number line is the x-axis and the vertical number line is the y-axis. The two axes intersect at a point called the *origin*. The plane represented by this system is called the *coordinate plane*. The axes divide the plane into four regions called *quadrants*. Also called the *Cartesian coordinate system*.

Rectangular solid A solid that consists of six rectangular surfaces called its *faces*. A line segment where two faces touch (or intersect) is an *edge* of the solid. A corner point where three faces touch (or intersect) is a *vertex*. Opposite faces have the exact same shape and size.

Recursive sequence A sequence in which each term is defined using one or more of its previous terms. Typically, the first term of a recursive sequence is given, followed by the formula for the nth term of the sequence.

Reference rectangle A rectangle used as a guide when sketching the asymptotes of the graph of a hyperbola. The reference rectangle is a rectangle whose midpoints of each side are the vertices of the hyperbola or the endpoints of the conjugate axis. The asymptotes pass diagonally through opposite corners of the reference rectangle.

Regular hexagon A regular six-sided polygon.

Regular polygon A polygon whose sides are all of equal length and whose angles are all of equal measure.

Regular pyramid A pyramid that has a base that is a regular polygon and is perpendicular to the height. Its triangle sides are congruent.

Relation The correspondence between two sets of numbers that can be represented by a set of ordered pairs.

Relative decay rate See *Exponential decay*.

Relative frequency The percent of data values that fall within a given class interval.

Relative frequency table A table showing the percent of data values that fall within each class interval.

Relative growth rate See *Exponential growth*.

Remainder The amount left over after division when there are no more digits to drop down and the divisor cannot divide into the dividend any further.

Repeating decimal A decimal number in which a fixed number of digits form a pattern that repeats without end. A repeating decimal is indicated by placing a bar over the pattern of digits that repeat or by writing three dots (...), called an *ellipsis*, behind the established pattern. A repeating decimal can always be written as a fraction.

Restricted value For an algebraic expression, any value that makes the expression undefined.

Restricted value for rational expressions Any value that causes the denominator of the rational expression to equal zero.

Rhombus A parallelogram with four equal sides.

Right angle An angle with a measure of 90 degrees.

Right circular cylinder A solid that consists of two circles of equal size and a curved rectangular side. When positioned upright, the circles align one directly above the other so that each is perpendicular to the height.

Right circular cone A solid that contains a circular base and a curved surface with one vertex called the *apex*. If positioned upward, the *apex* (or tip) aligns directly above the center of the circle so that the circle is perpendicular to the height.

Right triangle A triangle with a $90°$ angle, or right angle. The *hypotenuse* of a right triangle is the side opposite the right angle and is the longest of the three sides. The other two sides are called the *legs* of the triangle.

Rise See *Slope*.

Round To approximate a number to a desired place value by replacing the number with an estimate that has fewer nonzero digits.

Rounding down Keeping the digit in the desired place value the same and placing zeros in all remaining place values to the right.

Rounding up Increasing the digit in the desired place value by 1 and placing zeros in all remaining place values to the right.

Row-echelon form A matrix form in which there are 1's along the diagonal from the upper left to lower right, and 0's below the diagonal.

Row operation An operation on a row of a matrix that results in an equivalent matrix. Applying any of the following operations to a matrix will result in an equivalent matrix. 1. Interchange any two rows. 2. Multiply the elements of a row by the same nonzero constant. 3. Add a multiple of one row to another row.

Run See *Slope*.

Sales tax An amount charged by a state or local government when a purchase is made. Typically, the sales tax is a percent of the purchase price.

Sales tax rate Sales tax is charged as a percent of the price of the merchandise purchased. This percent is the *tax rate*.

Satisfy A value satisfies an equation or inequality if, when substituted for the variable, it results in a true statement.

Scalene triangle A triangle with three different side lengths and three different angle measures.

Scatter plot Paired data graphed as points on a coordinate plane.

Scientific notation A number is written in *scientific notation* if it has the form $a \times 10^n$, where a is a real number, such that $1 \le a < 10$, and n is an integer.

Second (s or sec) A common unit for the measure of time. 60 seconds is equivalent to 1 minute.

Semiannually Happening twice each year. See *Annually*.

Series The sum of the terms of a sequence.

Set-builder notation A way of expressing a set of values for a variable that gives the name of the variable and the condition(s) that must be met, separated by a vertical line.

Side Each line segment of a polygon.

Side of an angle Each of the two rays that form an angle.

Side-Angle-Side (SAS) property of congruence If the lengths of two sides of one triangle equal the lengths of two sides of another triangle and if the corresponding angles between each pair of sides have the same measure, then the two triangles are congruent.

Side-Side-Side (SSS) property of congruence If the lengths of the three sides of one triangle equal the lengths of the sides of another triangle, then the two triangles are congruent.

Similarity symbol $(\sim)$ A symbol used to state that two triangles are similar.

Similar triangles Two or more triangles that have the same shape but not necessarily the same size.

Simple interest Interest that is paid or earned on only the original principal. It is computed using the formula $\text{Simple interest} = \text{Principal} \cdot \text{Rate} \cdot \text{Time}$, or $I = P \cdot r \cdot t$.

Simplified In the simplest form possible.

Simplified algebraic expression An algebraic expression is simplified if grouping symbols are removed and like terms have been combined.

Simplified exponential expressions An exponential expression is simplified if (1) there are no parentheses or other grouping symbols, (2) there are no zero or negative exponents, (3) there are no powers raised to powers, and (4) each base occurs only once.

Simplified polynomial A polynomial is a simplified polynomial if all of its terms are simplified and none of its terms are like terms.

Simplified term A term containing a single numeric factor and none of the variable factors can be combined using the rules for exponents.

Simplify a complex fraction Write an equivalent fraction in which the numerator and denominator contain no minor fractions and have no common factors other than 1.

Simplify a complex rational expression Write an equivalent rational expression of the form $\frac{P}{Q}$ such that P and Q are polynomials with no common factors other than 1.

Simplify a ratio Write an equivalent ratio by dividing out all common factors and common units.

Slant height of a regular pyramid The length of a line from the apex of the pyramid drawn perpendicular to an edge of the base of the pyramid.

Slant height of a right cone The length of a line drawn from the apex of the cone to any point on the edge of its base.

Slope The ratio of the vertical change in y, or *rise*, to the horizontal change in x, or *run*, when moving from one point to another point along a straight line.

Slope formula Given two points, (x_1, y_1) and (x_2, y_2), on the graph of a line, the slope m of the line containing the two points is given by the formula

$$m = \frac{\text{Change in } y}{\text{Change in } x} = \frac{\text{Rise}}{\text{Run}} = \frac{y_2 - y_1}{x_2 - x_1}, \text{ where } x_1 \neq x_2.$$

Slope-intercept form A linear equation in two variables of the form $y = mx + b$ is written in slope-intercept form, where m is the slope of the line and b is the y-intercept.

Slope of a horizontal line All horizontal lines (which have equations of the form $y = b$) have slope 0.

Slope of a vertical line All vertical lines (which have equations of the form $x = a$) have undefined slope.

Solid A three-dimensional figure that is bounded by surfaces. Solids have length, width, and thickness (or height).

Solution A value that, when substituted for a variable, makes the equation or inequality true.

Solution region of a system of linear inequalities The common region that all solution sets of the inequalities in a system of linear inequalities share.

Solution set The collection of *all* solutions to an equation or inequality.

Solution to a linear inequality in two variables An ordered pair that, when substituted for the variables, makes the inequality true.

Solution to an equation in one variable A value that, when substituted for the variable, makes the equation true.

Solution to an equation in two variables An ordered pair of values that, when substituted for the variables, makes the equation true.

Solution to a system of linear equations (three variables) An ordered triple that, when substituted for the variables, makes all equations in the system true.

Solution to a system of linear equations (two variables) An ordered pair that, when substituted for the variables, makes all equations in the system true.

Solution to a system of linear inequalities (two variables) An ordered pair that, when substituted for the variables, satisfies all inequalities in the system.

Solve To solve an equation or inequality means to find its solution set, or the set of all values that make the equation or inequality true.

Solve a proportion To find an unknown quantity that makes a proportion true, usually by first cross-multiplying.

Solve for a variable An equation is solved for a variable if the variable is by itself on one side of the equation and is the only time that the variable occurs in the equation.

Sphere A solid that consists of all points in three dimensions that are located at an equal distance from a fixed point called the *center*. A *radius* extends from the center to the surface of a sphere. A *diameter* extends from one side of the sphere to the other, through the center.

Square A rectangle with four equal sides.

Square centimeter A square with 1-centimeter sides.

Square inch A square with 1-inch sides.

Square matrix A matrix with an equal number of rows and columns.

Square pyramid A regular pyramid with a square base.

Square root For $a > 0$, $\sqrt{a}$ is the positive or principal square root of a. That is, $\sqrt{a} = b$ only if $b^2 = a$ and $b > 0$. Also, $\sqrt{0} = 0$.

Square root function A function involving a square root. The basic square root function is $f(x) = \sqrt{x}$.

Square root of a negative number For any positive real number a, $\sqrt{-a} = \sqrt{-1} \cdot \sqrt{a} = i\sqrt{a}$. The square root of a negative number is not a real number.

Square root property If u is an algebraic expression and k is a real number, then $u^2 = k$ is equivalent to the compound equality $u = -\sqrt{k}$ or $u = \sqrt{k}$. Equivalently, if $u^2 = k$ then $u = \pm\sqrt{k}$.

Square units Units used to measure area. See *Area* or *Surface area*.

Squaring property of equality If A and B represent algebraic expressions, then any solution to the equation $A = B$ is also a solution to the equation $A^2 = B^2$.

Standard form of a complex number A complex number in standard form is written as $a + bi$ where a and b are real numbers and i is the imaginary unit.

Standard form of a number When a number is written using digits.

Standard form of a quadratic function A quadratic function is in standard form if it is written as $f(x) = a(x - h)^2 + k$. The graph is a parabola with vertex (h, k). The parabola "opens up" if $a > 0$ or "opens down" if $a < 0$. This form is also called the *vertex form*.

Standard form of the equation of a circle The standard form of the equation of a circle with center (h, k) and radius r is $(x - h)^2 + (y - k)^2 = r^2$.

Straight angle An angle that measures 180°.

Strict inequality An inequality containing one or more of the following inequality symbols: $<, >, \neq$. There is no possibility of equality in a strict inequality.

Subscript notation Use of subscripts to distinguish between different terms or quantities.

Substitution Substitution is a process in which a value or algebraic expression is substituted for a variable, or vice versa.

Substitution method A method for solving a system of linear equations in two variables that involves solving one of two equations for one variable, substituting the resulting expression into the other equation, and then solving for the remaining variable.

Subtraction Taking one number away from another to find the difference of the two numbers. The *subtrahend* is the number being subtracted and the *minuend* is the number being subtracted from.

Subtraction property of equality See *Addition property of equality*.

Subtrahend The number being subtracted in a subtraction (difference) problem.

Sum The result of adding two real numbers. The numbers being added are called *terms*, or *addends*.

Sum and difference of two terms rule (Product of conjugates rule) The product of the sum and difference of two terms (conjugates) is equal to the difference of the squares of the two terms. That is, $(A + B)(A - B) = A^2 - B^2$.

Summation notation A short-hand notation that uses the Greek letter sigma, Σ, to express a series. For example, the finite series $a_1 + a_2 + a_3 + \ldots + a_n$ can be written as $\sum_{i=1}^{n} a_i$. The infinite series $a_1 + a_2 + \ldots + a_n + a_{n+1} + \ldots$ can be written as $\sum_{i=1}^{\infty} a_i$.

Sum of the angles of a triangle The measures of the three angles in a triangle add to 180 degrees.

Supplementary angles Two angles are supplementary angles, or *supplements*, if the sum of their measures is 180°.

Supplements See *Supplementary angles*.

Surface area The total area of the surface on all sides of a solid. Surface area is measured in *square units*.

Surface area of a cone The surface area *(SA)* of a cone is given by the formula $SA = \pi rs + \pi r^2$, where r is the radius, and $s = \sqrt{r^2 + h^2}$ is the slant height.

Surface area of a cube If a cube has sides of length s, then its surface area *(SA)* is given by the formula $SA = 6s^2$.

Surface area of a rectangular solid If a rectangular solid has length l, width w, and height h, then its surface area *(SA)* is given by the formula $SA = 2lw + 2lh + 2wh$.

Surface area of a regular pyramid The surface area *(SA)* of a regular pyramid is found by the formula $SA = B + \frac{1}{2}Pl$, where B is the area of the base, P is the perimeter of the base, and l is the slant height.

Surface area of a right circular cylinder If a right circular cylinder has radius r and height h, then its surface area *(SA)* is given by the formula $SA = 2\pi r^2 + 2\pi rh$.

Surface area of a sphere The surface area *(SA)* of a sphere is found by the formula $SA = 4\pi r^2$, where r is the radius.

Symmetric A two-dimensional graph is symmetric if it has mirror images (reflections) on opposite sides of a dividing line.

Synthetic division A shortcut method for dividing a polynomial by a binomial of the form $x - c$. Synthetic division removes the repetition involved with long division by considering only the coefficients involved.

System of linear equations in three variables A collection of linear equations in three variables considered together. A solution to the system is an ordered triple that, when substituted for the variables, makes all equations in the system true.

System of linear equations in two variables A collection of linear equations in two variables considered together. A solution to the system is an ordered pair that, when substituted for the variables, makes all equations in the system true.

System of linear inequalities in two variables A collection of linear inequalities in two variables considered together. A solution to the system is an ordered pair that, when substituted for the variables, makes all inequalities in the system true.

Table A series of rows and columns used to organize and display information.

Term Each quantity being added in a numeric or algebraic expression. Because subtraction is defined by adding the opposite, a subtracted quantity is a "negative" term. A term that contains one or more variables is a variable term. A term without any variables is a constant term.

Terminating decimal A decimal number that ends or has a fixed number of digits behind the decimal point.

Terms of the polynomial The monomials that make up the polynomial.

Test point A point used to find the half-plane that contains the ordered pair solutions to a linear inequality in two variables.

Test value After using boundary points to divide the number line into intervals, a test value is a number chosen from an interval to test whether an expression is positive or negative on the interval. *Note:* A test value cannot be a boundary point.

Theoretical probability Computation of the likelihood of an event based on the laws of probability. Theoretical probability takes *all* outcomes into account when finding the probability of an event.

Ton An American unit of weight equal to 2000 pounds.

Total revenue Revenue is the dollar amount received from selling x items at a price of p dollars per item. *Total revenue* is found by multiplying the number of units sold by the price per unit, $R = xp$.

Trailing zeros The zeros that lie to the right of the last nonzero digit in a number.

Transversal line A line that intersects two or more lines at different points.

Transverse axis An invisible line segment passing through the center of a hyperbola that connects the vertices of the hyperbola.

Trapezoid A four-sided polygon with one pair of parallel sides called *bases.*

Tree diagram A visual display used to list the possible outcomes of a series of tasks. Beginning with the first task, a branch is created for each possible choice. Choices for subsequent tasks are branched out from each previous choice.

Trial One instance of a probability experiment.

Triangle A three-sided polygon. The measures of the angles add to 180°.

Trinomial A simplified polynomial with three terms.

True equation An equation that is always true. Also called an *identity*.

U.S. Customary Units system (American Units) A system of measurements (e.g., inches, ounces, gallons) used in the United States.

Undefined quotient If a nonzero dividend is divided by a divisor of 0, the result is an undefined quotient. That is, $a \div 0$ is undefined for $a \neq 0$.

Undefined slope Vertical lines have *undefined slope* because the slope has a denominator of 0 and division by 0 is undefined.

Uniform motion Movement at a constant speed.

Union For any two sets A and B, the *union* of A and B is given by $A \cup B$ and represents all the elements that are either in set A or in set B.

Unit distance The distance from 0 to 1 (the length of one unit) on the number line.

Unit fraction A fraction that is equivalent to 1.

Unit price The cost for one unit of a product. Found by dividing the total price by the number of units. That is, $\text{Unit price} = \dfrac{\text{Total cost}}{\text{Number of units}}$.

Unit rate A rate that has a denominator of 1 unit.

Unlike fractions Fractions with different denominators.

Upper bound The smallest value that is greater than all the values in the interval. No values in the interval can be more than the upper bound. For example, the interval $[6, 9]$ has an upper bound of 9. No value in the interval is more than 9.

Upper limit of summation The value that indicates the subscript of the last term of a sequence to be included in a series.

Value of the function For $y = f(x)$, the symbol $f(x)$ represents the value of the dependent variable y for a given value of the independent variable x. For this reason, we call $f(x)$ the *value of the function*.

Variable A symbol (usually a letter) that is used in place of a numeric value that can change, or vary, depending on the situation.

Variable term A term that contains a variable.

Variation equation An equation used to describe how one quantity changes with respect to one or more additional quantities.

Varies directly See *Direct variation*.

Varies inversely See *Inverse variation*.

Velocity The rate of change of distance with respect to time. The absolute value of velocity is speed.

Vertex form See *Standard form of a quadratic function*.

Vertex formula For a quadratic function $f(x) = ax^2 + bx + c$, the x-coordinate of the vertex is given by $x = -\dfrac{b}{2a}$. The y-coordinate is found by evaluating the function using the x-coordinate. Thus, the coordinates of the vertex are $\left(-\dfrac{b}{2a}, f\left(-\dfrac{b}{2a}\right)\right)$. Alternatively, we can use $f\left(-\dfrac{b}{2a}\right) = c - \dfrac{b^2}{4a}$ to find the y-coordinate.

Vertex of an angle The point where the two sides of an angle meet.

Vertex of a parabola The lowest point (if the parabola opens up) or the highest point (if the parabola opens down) of the graph of the parabola.

Vertex of a solid A corner point where three or more faces touch (or intersect). The tip of a cone is also called a *vertex.*

Vertex of a triangle A corner point of a triangle. A triangle has three vertices.

Vertical angles Two nonadjacent angles formed by two intersecting lines. Vertical angles have equal measures.

Vertical line test If a vertical line intersects (crosses or touches) the graph of a relation at more than one point, then the relation is not a function. If every vertical line intersects the graph of a relation at no more than one point, then the relation is a function.

Vertices The plural of *vertex*.

Volume The amount of the space within a solid. Space is three dimensional, so volume is measured in units that are also three dimensional, *cubic units*.

Volume of a cube The volume (V) of a cube is found by cubing its side length (s). That is, $V = s^3$.

Volume of a rectangular solid The volume (V) of a rectangular solid is found by multiplying the length, l, times the width, w, times the height, h. That is, $V = lwh$.

Volume of a regular pyramid The volume (V) of a regular pyramid is found by the formula $V = \dfrac{1}{3}Bh$, where B is the area of the base, and h is the height.

Volume of a right circular cylinder The volume (V) of a right circular cylinder is found by multiplying the area of the circular base, πr^2, by the height, h. That is, $V = \pi r^2 h$, where r is the radius, and h is the height.

Volume of a right cone The volume (V) of a right cone is found by the formula $V = \dfrac{1}{3}\pi r^2 h$, where r is the radius, and h is the height.

Volume of a sphere The volume (V) of a sphere is found by the formula $V = \frac{4}{3}\pi r^3$, where r is the radius.

Week (wk) A unit of time equal to 7 days.

Weight A measure related to the pull of Earth's gravity on an object.

Whole-number part of a decimal Digits appearing in front of the *decimal point* form the *whole-number part* of the decimal number.

Whole-number part of a mixed number For a mixed number having the form $a\frac{b}{c}$, the whole number a is called its *whole-number part*.

Whole numbers The numbers 0, 1, 2, 3, 4, 5, 6, 7, 8, 9, 10, 11, and so on, with zero (0) being the smallest whole number. There is no largest whole number.

Word form of a number When a number is written using words (as opposed to using digits).

x-axis The horizontal axis in the rectangular coordinate system.

x-coordinate The first number in an ordered pair. Also called the *abscissa*.

x-intercept The x-coordinate of a point where a graph crosses or touches the x-axis.

xy-plane The plane represented by a rectangular coordinate system. Also known as the *Cartesian plane*.

y-axis The vertical axis in the rectangular coordinate system.

y-coordinate The second number in an ordered pair. Also called the *ordinate*.

y-intercept The y-coordinate of a point where a graph crosses or touches the y-axis.

Yard (yd) An American unit of length equal to 3 feet or 36 inches.

Zero product property If A and B are real numbers or algebraic expressions and $A \cdot B = 0$, then $A = 0$ or $B = 0$.

Zero-power rule A nonzero base raised to the 0 power equals 1. That is, $a^0 = 1$ $(a \neq 0)$.

MODULE 1

1.2 Exercises

1. tens **3.** hundred thousands **5.** nine hundred three **7.** twelve thousand, nine hundred thirty-four **9.** seven million, three hundred twenty thousand, six hundred ninety-five **11.** 463 **13.** 11,655,712,931 **15.** 928,350,000 **17.** $700 + 40 + 8$ **19.** $50{,}000 + 2000 + 800 + 6$ **21.** $2{,}000{,}000 + 700{,}000 + 6000 + 500 + 40$ **23.** $>$ **25.** $<$ **27.** 560 **29.** 7000 **31.** 20,000 **33.** 56,000,000 **35. a.** 5 **b.** 6 **c.** Austria, South Korea **d.** Germany, Canada **37. a.** 95 cm **b.** 128 cm **c.** 5 years old **d.** 14 years old **39. a.** 300,382 **b.** 1,360,301 **c.** 1970 **d.** 1980

1.3 Exercises

1. 12 **3.** 59 **5.** 767 **7.** 6938 **9.** 12,309 **11.** 57,598 **13.** 31 **15.** 652 **17.** 1129 **19.** 7525 **21.** 99,525 **23.** $10 + 14$ **25.** $8 + (2 + 16)$ **27.** 22 **29.** 32 **31.** 112 **33.** 718 **35.** 13,056 **37.** 3 **39.** 19 **41.** 33 **43.** 221 **45.** 361 **47.** 8 **49.** 325 **51.** 98 **53.** 457 **55.** 704 **57.** 5952 **59.** 18,316 **61.** 160 **63.** 6600 **65.** 389,000 **67.** 22 m **69.** 56 in. **71.** 34 in. **73.** 203 **75.** 191 **77. a.** \$2900 **b.** \$2840 **79. a.** 2000 stores **b.** 1809 stores **81. a.** 90 ft **b.** 82 ft

1.4 Exercises

1. a. $5 \cdot 9$ **b.** $(5 \cdot 6) \cdot 3$ **3. a.** 19 **b.** 0 **5.** $7 \cdot 8 + 7 \cdot 1$ **7.** $7 \cdot 20 - 7 \cdot 9$ **9.** 150 **11.** 25,515 **13.** 1798 **15.** 9144 **17.** 74,530 **19.** 72,372 **21.** 752,678 **23.** 840,630 **25.** 283,200 **27.** 44,660,000 **29.** 13,889,000,000 **31.** 16 sq ft **33.** 16,740 calories **35.** 675 GB **37.** \$672 **39. a.** \$17,700 **b.** \$18,720 **41.** \$564

1.5 Exercises

1. 6 **3.** 7 **5.** 1 **7.** 0 **9.** 24 **11.** 247 **13.** 478 R 4 **15.** 74 **17.** 43 R 39 **19.** 240 R 15 **21.** 243 R 74 **23.** 706 R 8 **25.** 37 **27.** 90 **29.** 100 **31.** 4000 **33.** \$65,000 **35.** \$367 **37.** 145 pizzas **39.** 5 treats; 5 treats **41.** \$300,000

1.6 Exercises

1. 8^5 **3.** $7^2 \cdot 9^4$ **5.** 16 **7.** 1 **9.** 0 **11.** 72 **13.** 80,000 **15.** 8 **17.** 44 **19.** 17 **21.** 0 **23.** 64 **25.** 22 **27.** 20 **29.** 10 **31.** 31 **33.** 35 **35.** 81 **37.** 5 **39.** 10 **41.** 4 **43.** 2 **45.** 15 **47.** 602 **49.** 27 missions

1.7 Exercises

1. 34 **3.** 52 **5.** 57 **7.** 48 **9.** 8 **11.** 47 **13.** 73 **15.** 1 **17.** 2 **19.** 0 **21.** expression **23.** equation **25.** equation **27.** not a solution **29.** not a solution **31.** solution **33.** not a solution **35.** solution **37.** not a solution **39.** $x - 12$ **41.** $80 - x$ **43.** $\frac{x}{32}$ **45.** $x - 25$ **47.** $4x + 10$ **49.** $x - 12 = 65$ **51.** $\frac{x}{4} = 25$ **53.** $6x + 2 = 50$ **55.** $63 - x = 41$ **57.** $2(x + 1) = x + 8$

MODULE 2

2.1 Exercises

1. 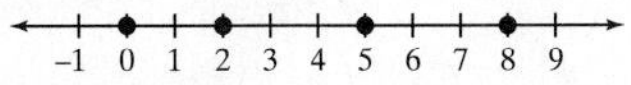

3. 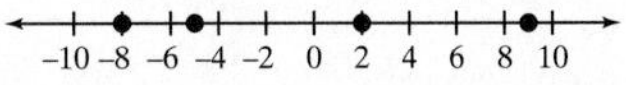

5. $<$ **7.** $>$ **9.** $<$ **11.** 6 **13.** 0 **15.** 78 **17.** 4 **19.** −85 **21.** 13 **23.** −10 **25.** 12 **27.** 20,320 **29.** 2806 **31.** −5314

33.

Day	Water Level (feet)
Sunday	−2
Monday	−1
Tuesday	3
Wednesday	5
Thursday	4
Friday	2
Saturday	−1

a. Tuesday, Wednesday, Thursday, Friday **b.** Wednesday; 5 feet **c.** Sunday; −2 feet **d.** Sunday and Friday; −2 and 2

2.2 Exercises

1. 11 **3.** −4 **5.** −10 **7.** 22 **9.** 18 **11.** 138 **13.** 675 **15.** −2 **17.** 3 **19.** −23 **21.** −8 **23.** 0 **25.** 1 **27.** 0 **29.** 117 **31.** −3 **33.** −59 **35.** 0 **37.** −420 **39.** 0 **41.** −22 **43.** −10 **45.** 6 **47.** $-53 + (-26) = -79$ **49.** $-31 + 19 = -12$ **51.** 18°F **53.** \$639 **55.** 66 points

2.3 Exercises

1. 5 **3.** 11 **5.** −30 **7.** 10 **9.** 9 **11.** 50 **13.** 173 **15.** 0 **17.** 11 **19.** 0 **21.** 27 **23.** 36 **25.** 61 **27.** −36 **29.** $-38 - 12 = -50$ **31.** $-16 - (-37) = 21$ **33.** 71°F **35.** 8 strokes **37.** \$353 **39.** 59 min

2.4 Exercises

1. −28 **3.** 45 **5.** −40 **7.** 72 **9.** 210 **11.** −60 **13.** −168 **15.** 60 **17.** −54 **19.** 36 **21.** −81 **23.** −1000 **25.** 1 **27.** −9 **29.** −3 **31.** 11 **33.** undefined **35.** −7 **37.** $2(-16) = -32$ **39.** $-234 \div 9 = -26$ **41.** $32(-6) = -192$ **43.** $(-25)(-6) = 150$ **45.** −36 degrees **47.** 8 min

2.5 Exercises

1. 45 **3.** 13 **5.** 15 **7.** 8 **9.** −12 **11.** 10 **13.** −18 **15.** 135 **17.** 13 **19.** 180 **21.** −51 **23.** −42 **25.** 25 **27.** 100 **29.** −30 **31.** 1 **33.** 134 **35.** −5 **37.** −103 **39.** −7 **41.** −9 **43.** −13 **45.** −12 **47.** **a.** −41 **b.** 7 **c.** −101 **49.** **a.** −47 **b.** 7 **c.** 257 **51.** 18 **53.** 124 **55.** 85 **57.** −4 **59.** $-9(-4 - x)$ **61.** $-15 - 20x$ **63.** $(x + 7) \div 3$ **65.** $x(9 + x)$ **67.** $2x - x \div (-5)$

2.6 Exercises

1. solution **3.** not a solution **5.** solution **7.** 7 **9.** −6 **11.** −8 **13.** −4 **15.** 5 **17.** 9 **19.** −48 **21.** −55 **23.** 9 **25.** −18 **27.** −3 **29.** 0 **31.** −15 **33.** 0 **35.** 14

MODULE 3

3.1 Exercises

1. $4x, 7y; 4, 7$ **3.** $9z^2, -8z, 6; 9, -8, 6$ **5.** $w^3, -4w^2, -6w, 5; 1, -4, -6, 5$ **7.** $3x$ and $-x$; $9y$ and $8y$ **9.** $7m$ and $-m$; −12 and 8 **11.** $23z$ **13.** $9ab$ **15.** $7p^2 - 11p + 20$ **17.** $15m + 3n$ **19.** $-7t^2 + 20t$ **21.** $7w^2 + 4w$ **23.** $3k^2 - 13k - 12$ **25.** $7ab + 17$ **27.** $56x$ **29.** $-60n$ **31.** $91z^2$ **33.** $-112ab$ **35.** $7x + 21$ **37.** $80w + 8$ **39.** $-12x - 42$ **41.** $-20y + 40$ **43.** $4x + 23$ **45.** $15y - 6$ **47.** $5w + 13$ **49.** $7x - 3n + 6$

3.2 Exercises

1. −5 **3.** −6 **5.** 3 **7.** 18 **9.** 4 **11.** 5 **13.** 80 **15.** −18 **17.** 4 **19.** −9 **21.** 7 **23.** −8 **25.** 16 **27.** −15 **29.** −6 **31.** 6 **33.** 8 **35.** −6 **37.** 8 **39.** 13 **41.** −17 **43.** 4 **45.** −2 **47.** −1 **49.** 5

3.3 Exercises

1. not a linear equation in one variable; not an equation **3.** linear equation in one variable **5.** not a linear equation in one variable; nonlinear **7.** not a linear equation in one variable; two variables **9.** 2 **11.** 1 **13.** −6 **15.** −1 **17.** 9 **19.** 3 **21.** −5 **23.** 3 **25.** 18 **27.** 7 **29.** 0 **31.** −9 **33.** 2 **35.** −13 **37.** −6 **39.** 4 **41.** 15

3.4 Exercises

1. $2x - 12 = x$ **3.** $3(x + 5) = 4(x \div 7)$ **5.** $7 + (8 - x) = 15$ **7.** $(x + 3)(x + 7) = 17$ **9.** $\frac{1}{2}(x + 7) = 2x - 4$ **11.** $17(x - 4) = 3x + 7$ **13.** 3 **15.** −5 **17.** −8 **19.** −1 **21.** 7 **23.** 6 **25.** 4 in. by 9 in. **27.** 6 ft by 14 ft **29.** 26 ft by 34 ft **31.** rectangle: 19 by 12; triangle: 17, 21, 24 **33.** 2 **35.** 3 **37.** 66 cm, 29 cm **39.** deep space rocket: 70 tons; space shuttle: 25 tons **41.** Keegan: 3; Brennan: −2 **43.** small: 110 cal; shot: 27 cal **45.** 17 ft, 19 ft, 29 ft

MODULE 4

4.1 Exercises

1. numerator: 7; denominator: 12 **3.** numerator: −19; denominator: 32 **5.** $\frac{4}{6}$ **7.** $\frac{5}{9}$ **9.** $\frac{3}{5}$ **11.** $\frac{11}{16}$ **13.** $\frac{11}{12}$ **15.** $\frac{25}{89}$ **17.** $\frac{3}{12}; \frac{9}{12}$ **19.** $\frac{395}{443}; \frac{48}{443}$ **21.** proper **23.** improper **25.** $\frac{14}{9}; 1\frac{5}{9}$ **27.** $\frac{16}{3}; 5\frac{1}{3}$ **29.** $1\frac{9}{16}$ **31.**

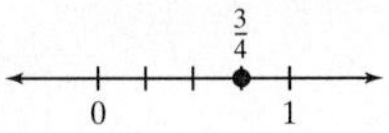

33.

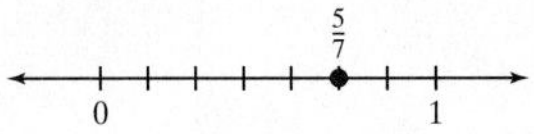

35.

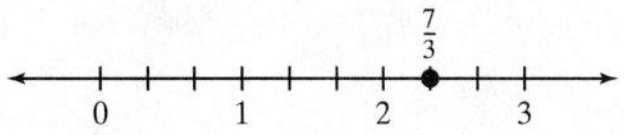

37.

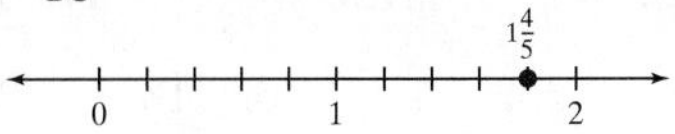

39. $2\frac{2}{3}$ **41.** 13 **43.** 29 **45.** $4\frac{21}{32}$ **47.** $\frac{21}{4}$ **49.** $\frac{97}{10}$ **51.** $\frac{371}{25}$

4.2 Exercises

1. 1, 2, 3, 4, 6, 12 **3.** 1, 19 **5.** 1, 2, 3, 4, 6, 8, 9, 12, 18, 24, 36, 72 **7.** 1, 3, 5, 11, 15, 33, 55, 165 **9.** 1, 2, 3, 4, 6, 8, 12, 24 **11.** $2 \cdot 3$ **13.** prime **15.** $2 \cdot 2 \cdot 2 \cdot 13$ **17.** $3 \cdot 5 \cdot 5$ **19.** $2 \cdot 2 \cdot 2 \cdot 2 \cdot 3 \cdot 3$ **21.** $3 \cdot 3 \cdot 5 \cdot 7$ **23.** $2 \cdot 2 \cdot 3 \cdot 3 \cdot 11$ **25.** 3 **27.** 15 **29.** 140 **31.** $3a^2$ **33.** 21 **35.** m^2n **37.** $6xz$ **39.** equivalent **41.** equivalent **43.** not equivalent **45.** equivalent **47.** $\frac{40}{56}$ **49.** $\frac{1}{3}$ **51.** $\frac{8}{12}$ **53.** $\frac{4}{28}$ **55.** $\frac{6}{2}$ **57.** $\frac{12x}{40}$ **59.** $\frac{1}{2}$ **61.** $\frac{7}{6}$ **63.** $\frac{38}{81}$ **65.** $\frac{7}{16}$ **67.** $-\frac{1}{3}$ **69.** 3 **71.** $\frac{7}{27}$ **73.** $-\frac{2}{3}$ **75.** $\frac{12}{17}$ **77.** $-\frac{15x}{8}$ **79.** $\frac{5x}{11y}$ **81.** $\frac{3x}{2z}$ **83.** $\frac{11xy}{6z}$ **85.** $-\frac{1}{10x^2}$

4.3 Exercises

1. $\frac{20}{63}$ **3.** $\frac{1}{52}$ **5.** $\frac{3}{4}$ **7.** 2 **9.** 14 **11.** $-\frac{5}{12}$ **13.** $\frac{3}{4}$ **15.** 15 **17.** $\frac{y}{x^2}$ **19.** $-\frac{4q}{9}$ **21.** $\frac{16}{49}$ **23.** $\frac{64}{81}$ **25.** $\frac{25}{9}$ **27.** $\frac{4}{7}$ **29.** $\frac{1}{10}$ **31.** $\frac{5}{16}$ **33.** $\frac{2}{7}$ **35.** −12 **37.** $-\frac{2}{3}$ **39.** $\frac{15}{7x}$ **41.** $\frac{w^2x}{y}$ **43.** $-\frac{4}{75q^2}$ **45.** $\frac{4}{9}$ **47.** $\frac{3}{2}$ **49.** $\frac{3p}{4q}$ **51.** 180 acres **53.** 32 servings **55.** 20 tsp **57.** $\frac{15}{4}$ in.^{2} **59.** 27 ft^2 **61.** $\frac{5}{4}$ in.^{2}

4.4 Exercises

1. $\frac{7}{9}$ **3.** $\frac{16}{15}$ **5.** $\frac{3}{11}$ **7.** $\frac{1}{7}$ **9.** 1 **11.** −2 **13.** $\frac{5}{7}$ **15.** $-\frac{2}{5}$ **17.** 1 **19.** $\frac{1}{5}$ **21.** 42 **23.** 40 **25.** 17 **27.** 36 **29.** 270 **31.** 60 **33.** $\frac{29}{24}$ **35.** $-\frac{3}{8}$ **37.** $\frac{31}{28}$ **39.** $\frac{7}{3}$ **41.** $-\frac{1}{10}$ **43.** $\frac{31}{30}$ **45.** $\frac{8x + 3}{25}$ **47.** $\frac{67z}{90}$ **49.** $\frac{23}{14a}$ **51.** $\frac{2}{z}$ **53.** $\frac{36 - 5x}{12x}$ **55.** $\frac{14x^2 + 25}{105x}$ **57.** $\frac{2a + 4}{3}$ **59.** $\frac{5x + 4}{6}$ **61.** $\frac{7}{12}$ **63.** $\frac{3}{10}$ **65.** $1\frac{11}{12}$ c **67.** $\frac{39}{50}$ **69.** $\frac{1}{25}$

4.5 Exercises

1. $\frac{8}{9}$ **3.** $\frac{1}{3}$ **5.** $\frac{5}{9}$ **7.** $\frac{4}{3}$ **9.** $-\frac{6}{25}$ **11.** $\frac{6}{37}$ **13.** $\frac{6}{xy}$ **15.** $\frac{3}{5x+1}$ **17.** $-\frac{5}{9}$ **19.** $\frac{1}{3}$ **21.** $\frac{9}{64}$
23. $-\frac{3}{2}$ **25.** $\frac{1}{2}$ **27.** $-\frac{5}{54}$ **29.** $\frac{11}{12}$ **31.** $-\frac{55}{64}$ **33.** $\frac{5}{19}$ **35.** $\frac{6}{625}$

4.6 Exercises

1. $3\frac{13}{20}$ **3.** $2\frac{14}{15}$ **5.** $8\frac{1}{6}$ **7.** $6\frac{1}{42}$ **9.** $21\frac{5}{7}$ **11.** $\frac{32}{81}$ **13.** $\frac{5}{9}$ **15.** $1\frac{47}{48}$ **17.** $\frac{26}{67}$ **19.** $\frac{81}{100}$
21. $3\frac{4}{7}$ **23.** $7\frac{1}{3}$ **25.** $9\frac{7}{10}$ **27.** $6\frac{1}{4}$ **29.** $6\frac{1}{3}$ **31.** $7\frac{19}{40}$ **33.** $13\frac{7}{10}$ **35.** $5\frac{2}{9}$ **37.** $5\frac{1}{4}$ **39.** $6\frac{17}{24}$
41. $4\frac{17}{80}$ **43.** $4\frac{5}{9}$ **45.** $3\frac{4}{5}$ **47.** $\frac{7}{18}$ **49.** $2\frac{7}{15}$ **51.** $1\frac{11}{20}$ **53.** $17\frac{1}{2}$ in. **55.** $67\frac{17}{24}$ ft^2 **57.** $4\frac{3}{8}$ ft
59. $53\frac{71}{72}$ yd^2

4.7 Exercises

1. solution **3.** solution **5.** not a solution **7.** solution **9.** $\frac{6}{17}$ **11.** $\frac{4}{9}$ **13.** $-\frac{2}{15}$ **15.** $-\frac{11}{18}$ **17.** 18
19. $\frac{3}{4}$ **21.** $-\frac{2}{7}$ **23.** $\frac{5}{84}$ **25.** 2 **27.** $\frac{3}{5}$ **29.** $-\frac{7}{4}$ **31.** $-\frac{3}{16}$ **33.** $-\frac{3}{4}$ **35.** −45 **37.** 18 **39.** 5

MODULE 5

5.1 Exercises

1. tenths **3.** tens **5.** thousands **7.** ones **9.** sixty-three hundredths **11.** forty-three and fifty-nine hundredths
13. seven thousand five hundred sixteen and five hundred thirty-four thousandths **15.** five hundred thirteen ten-thousandths
17. 12.8 **19.** 315.013 **21.** 21.05 **23.** 800.075 **25.** $\frac{57}{100}$ **27.** $19\frac{7}{10}$ **29.** $\frac{5}{8}$ **31.** $2\frac{1}{250}$

33. 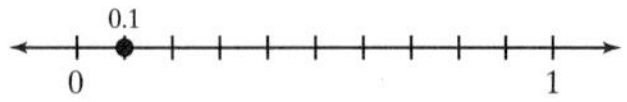

35. 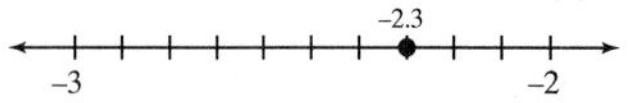

37. 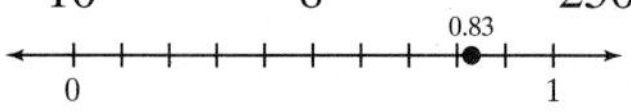

39. 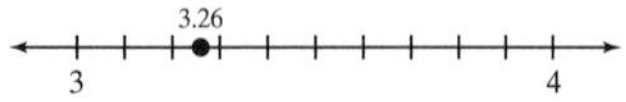

41. < **43.** > **45.** < **47.** > **49.** = **51.** 0.7 **53.** 24.906
55. −0.096 **57.** 132.60 **59.** −240

5.2 Exercises

1. 7.97 **3.** 139.68 **5.** 463.985 **7.** 17.82 **9.** 339.8627 **11.** 539.9627 **13.** 57.748 **15.** 62.0093
17. 53.8291 **19.** 49.0019 **21.** 5.65 **23.** 122.221 **25.** 123.0223 **27.** 7.95 **29.** 5.13 **31.** 44.28
33. 300 **35.** 200 **37.** 170 **39.** 24.82 **41.** −112.63 **43.** 582.942 **45.** −33.22 **47.** 42.1102
49. −13.72 **51.** $-16.531x + 1.32$ **53.** $5.499y - 2.098$ **55.** $44.021b + 24.074$ **57.** 4.64 Mbps **59.** \$84.13
61. \$123.21

5.3 Exercises

1. 34.68 **3.** 106.157 **5.** 306.18 **7.** 31.11244 **9.** 3618.1566 **11.** 56.052012 **13.** −107.415
15. −135.84 **17.** 18.5871 **19.** 0.1431 **21.** 511.7 **23.** 27,284 **25.** 1520 **27.** 0.15472 **29.** −350
31. 2400 **33.** 3000 **35.** 30,000 **37.** 12.0409 **39.** 103.823 **41.** 104.8576 **43.** 26π yd; 81.64 yd
45. 15.2π ft; 47.728 ft **47.** 23,300 gal **49.** 34.8945 USD **51.** \$22.87 **53.** 30.48 cm

5.4 Exercises

1. 0.53 **3.** −4.7 **5.** 1.375 **7.** −7.94 **9.** $0.7\overline{1}$ **11.** $-0.1\overline{06}$ **13.** $2.\overline{7}$ **15.** −7.75 **17.** $1.\overline{8}$
19. 1.85 **21.** $-2.2\overline{3}$ **23.** $-2.541\overline{6}$ **25.** 1.71 **27.** 130 **29.** −0.076 **31.** 17.2 **33.** 0.78913
35. −94.1 **37.** −8640 **39.** 0.003194 **41.** 200 **43.** 11 **45.** 2.74 oz **47.** 75 kg **49.** 28.4 mi/gal
51. 3200 pictures

5.5 Exercises

1. 0.8 **3.** −0.875 **5.** $0.\overline{5}$ **7.** $0.\overline{72}$ **9.** 0.5625 **11.** −3.75 **13.** $-1.3\overline{8}$ **15.** $0.6\overline{8}$ **17.** 6.25
19. −4.78 **21.** 0.7 **23.** −0.471 **25.** 4.13 **27.** 4.667 **29.** < **31.** = **33.** < **35.** <

37. $\frac{5}{16}$, 0.325, $\frac{1}{3}$ **39.** $\frac{17}{2}$, $8.\overline{5}$, 8.56 **41.** 0.486 **43.** 0.25 **45.** 12.695 **47.** 13.865 **49.** 55.89 **51.** 0.76 **53.** 7.425 **55.** 1.8625 **57.** 36π ft^2; 113.04 ft^2 **59.** 81π yd^2; 254.34 yd^2

5.6 Exercises

1. solution **3.** not a solution **5.** solution **7.** 8.8 **9.** 9 **11.** 0.4 **13.** 13 **15.** 5 **17.** 1.4 **19.** 1.5 **21.** −2.6 **23.** −5.5 **25.** 2.5 **27.** −2.1 **29.** 3.1 **31.** 5.1 **33.** 4 **35.** 1.5 **37.** 5.5 **39.** 2.5 **41.** −7.5 **43.** 1147 min **45.** 1.5 h swimming, 4 h jogging **47.** 65 mi

MODULE 6

6.1 Exercises

1. $\frac{6}{17}$ **3.** $\frac{15}{11}$ **5.** $\frac{3}{10}$ **7.** $\frac{11}{6}$ **9.** $\frac{2}{1}$ **11.** $\frac{5}{8}$ **13.** $\frac{1}{5}$ **15.** $\frac{13}{7}$ **17.** $\frac{13}{7}$ **19.** $\frac{18}{11}$ **21.** $\frac{117}{160}$ **23.** $\frac{2}{3}$ **25.** $\frac{63}{40}$; yes **27.** $\frac{\$102}{5\text{ h}}$ **29.** $\frac{257\text{ lb}}{4\text{ ft}^2}$ **31.** $\frac{75\text{ cal}}{14\text{ crackers}}$ **33.** $\frac{9\text{ points}}{5\text{ min}}$ **35.** 40 mi/gal **37.** 45 words/min **39.** 24.3 MB/sec **41.** 463.9 yd/game **43.** \$0.59/cucumber **45.** \$0.17/envelope **47.** pack of 16: \$0.62/battery; pack of 20: \$0.62/battery; pack of 20 is the better deal. **49.** 5-oz can: \$0.30/oz; 20-oz 4-pack: \$0.38/oz; 5-oz can is the better deal.

6.2 Exercises

1. $\frac{10}{15} = \frac{30}{45}$ **3.** $\frac{480\text{ mi}}{15\text{ gal}} = \frac{256\text{ mi}}{8\text{ gal}}$ **5.** $\frac{60}{75} = \frac{4}{5}$ **7.** $\frac{5}{8} = \frac{5}{8}, \frac{65}{104} = \frac{5}{8}$; true **9.** $\frac{12}{20} = \frac{3}{5}, \frac{180}{300} = \frac{3}{5}$; true **11.** false **13.** true **15.** false **17.** false **19.** $\frac{7}{4}$ **21.** 36 **23.** 3 **25.** 2 **27.** 4 **29.** 7.2 **31.** 0.75 **33.** $\frac{3}{8}$ **35.** $8\frac{1}{5}$ **37.** 20.7 **39.** 4.467

6.3 Exercises

1. 15 min **3.** 3125 fish **5.** 600 cal **7.** 8 gal **9.** 17.0 cm **11.** 985.5 h **13.** 8.3 gal **15.** \$106.08 **17.** 73.3 lb **19.** 61 s

6.4 Exercises

1. a. $\angle A$ corresponds to $\angle D$, $\angle B$ corresponds to $\angle E$, $\angle C$ corresponds to $\angle F$, $\overline{AB}$ corresponds to $\overline{DE}$, $\overline{BC}$ corresponds to $\overline{EF}$, $\overline{AC}$ corresponds to $\overline{DF}$ **b.** *DEF* **c.** 50° **d.** 73 mm **3. a.** $\angle X$ corresponds to $\angle O$, $\angle Y$ corresponds to $\angle N$, $\angle Z$ corresponds to $\angle M$, $\overline{XY}$ corresponds to $\overline{ON}$, $\overline{YZ}$ corresponds to $\overline{NM}$, $\overline{XZ}$ corresponds to $\overline{OM}$ **b.** *ONM* **c.** $\angle O$ **d.** $\overline{XZ}$ **5.** SSS **7.** SAS **9.** congruent; ASA **11.** not congruent **13.** congruent; ASA **15. a.** $\angle F$ corresponds to $\angle U$, $\angle G$ corresponds to $\angle T$, $\angle H$ corresponds to $\angle V$, $\overline{FG}$ corresponds to $\overline{UT}$, $\overline{GH}$ corresponds to $\overline{TV}$, $\overline{FH}$ corresponds to $\overline{UV}$ **b.** *UTV* **17. a.** $\angle X$ corresponds to $\angle K$, $\angle Y$ corresponds to $\angle L$, $\angle Z$ corresponds to $\angle J$, $\overline{XY}$ corresponds to $\overline{KL}$, $\overline{YZ}$ corresponds to $\overline{LJ}$, $\overline{XZ}$ corresponds to $\overline{KJ}$ **b.** *KLJ* **19.** 6 ft **21.** 11.25 m **23.** 1.19 km **25.** 80 ft **27.** 115.2 yd **29.** 8.6 ft

6.5 Exercises

1. 7 **3.** 0.5 **5.** $\frac{3}{5}$ **7.** 8.832 **9.** 3.062 **11.** 4.868 **13.** 16 **15.** 41 **17.** $\sqrt{115}$; 10.72 **19.** 46.8 in. **21.** $10\sqrt{3034}$ ft; 551 ft

MODULE 7

7.1 Exercises

1. 0.942 **3.** 0.02 **5.** 2.35 **7.** 0.004 **9.** 73.5% **11.** 6.4% **13.** 91% **15.** 124% **17.** $\frac{13}{100}$ **19.** $\frac{13}{20}$ **21.** $\frac{9}{125}$ **23.** $2\frac{2}{5}$ **25.** $\frac{3}{400}$ **27.** $\frac{47}{900}$ **29.** $\frac{4}{75}$ **31.** 60% **33.** $6\frac{2}{3}$% or $6.\overline{6}$% **35.** $132\frac{1}{2}$% or 132.5% **37.** $781\frac{1}{4}$% or 781.25% **39.** 81.82% **41.** 271.43% **43.** 0.025; $\frac{1}{40}$ **45.** 48%; $\frac{12}{25}$ **47.** $146\frac{2}{3}$% or $146.\overline{6}$%; $1.4\overline{6}$

7.2 Exercises

1. $40\% \cdot x = 25$ **3.** $22.5\% \cdot 160 = x$ **5.** $12\% \cdot x = 17$ **7.** $x = 15\% \cdot 36$ **9.** $22 = 130\% \cdot x$ **11.** $11\% \cdot 94 = x$ **13.** 21 **15.** 13.8 **17.** 150% **19.** 20% **21.** 55 **23.** 30

7.3 Exercises

1. $\frac{30}{b} = \frac{22}{100}$ **3.** $\frac{a}{130} = \frac{19}{100}$ **5.** $\frac{17.5}{80} = \frac{p}{100}$ **7.** $\frac{a}{110} = \frac{16.5}{100}$ **9.** $\frac{18}{b} = \frac{42}{100}$ **11.** $\frac{18.2}{72.6} = \frac{p}{100}$ **13.** 10.5 **15.** 18 **17.** 60% **19.** 160% **21.** 80 **23.** 40

7.4 Exercises

1. 45 million adults **3.** \$6.40 **5.** 895 video games **7.** 36 lb **9.** \$3708 million **11.** 43% **13.** 123% **15.** 35% **17.** 15% **19.** 26% **21.** 13.5% **23.** 25% **25.** 21% **27.** 23.4%

7.5 Exercises

1. a. \$43.92 **b.** \$592.92 **3.** \$4.12; \$54.07 **5.** 7.35% **7.** 6.75% **9.** \$9000 **11.** \$4481.40 **13.** 0.5% **15.** \$4000 **17. a.** \$21.50 **b.** \$64.50 **19.** \$26.99; \$9.00 **21.** 32.5% **23.** 16%

7.6 Exercises

1. \$600 **3.** \$47.50 **5.** \$3587.50 **7.** \$814 **9. a.** \$10,719.14 **b.** \$1719.14 **11. a.** \$6317.39 **b.** \$1492.39 **13.** \$21,733.99

MODULE 8

8.1 Exercises

1. line; $\overleftrightarrow{AB}$ **3.** segment; $\overline{FG}$ **5.** ray; $\overrightarrow{DC}$ **7.** $\angle LON, \angle NOL$ **9.** $\angle DGE, \angle EGD$ **11.** acute **13.** obtuse **15.** right **17.** acute **19.** complementary **21.** neither **23.** complementary **25.** 67° **27.** 139° **29.** 69° **31.** 22° **33.** $m\angle b = 137°, m\angle c = 43°, m\angle d = 137°$ **35. a.** 57° **b.** 54° **37.** $m\angle 1 = 126°, m\angle 2 = 54°, m\angle 3 = 126°, m\angle 4 = 54°, m\angle 5 = 126°, m\angle 7 = 126°, m\angle 8 = 54°$ **39.** 53°

8.2 Exercises

1. 20 ft **3.** 14 yd **5.** 58 in. **7.** 42 cm **9.** 25 km **11.** 21 in. **13.** 11.2 m **15.** 38.8 cm **17.** 13 ft^2 **19.** 27.5 m^2 **21.** 9 $in.^2$ **23.** 74.2 km^2 **25.** 90 $in.^2$ **27.** 30 cm^2 **29.** $125\frac{1}{3}$ yd^2 **31.** 124 $in.^2$ **33.** $C = 10\pi$ in. ≈ 31.4 in.; $A = 25\pi$ $in.^2 \approx 78.5$ $in.^2$ **35.** $C = 12.3\pi$ m ≈ 38.62 m; $A = 37.8225\pi$ $m^2 \approx 118.76$ m^2 **37.** $P = 28$ m; $A = 48$ m^2 **39.** $P = 38$ cm; $A = 74$ cm^2 **41.** $P = 54$ ft; $A = 72$ ft^2 **43.** $P = 44$ m; $A = 84$ m^2 **45.** 15.23 in.; 13.08 $in.^2$ **47.** 26 yd; 84 yd^2 **49.** 4657 cm^2; 248 cm **51.** 17.03 ft^2 **53.** 792 yd^2 **55.** 129,076 ft^2

8.3 Exercises

1. 120 $in.^3$ **3.** 10 ft^3 **5.** 1728 ft^3 **7.** $14{,}000\pi$ mm^3; 43,960 mm^3 **9.** 20π m^3; 62.8 m^3 **11.** 4500π mm^3; 14,130 mm^3 **13.** 96π $in.^3$; 301.44 $in.^3$ **15.** 75 ft^3 **17.** 166 yd^2 **19.** 512 $in.^2$ **21.** 9 $in.^2$ **23.** 20π $in.^2$; 62.8 $in.^2$ **25.** 27π m^2; 84.78 m^2 **27.** 196π mm^2; 615.44 mm^2 **29.** 24π yd^2; 75.36 yd^2 **31.** 175 ft^2 **33.** 3680 $in.^3$ **35.** 128 ft^3 **37.** 7776 cm^2 **39.** 9420 m^3 **41.** 434.67 $in.^3$; 277.45 $in.^2$ **43.** 1808.64 ft^3 **45.** 718.01 $in.^3$ **47.** 42.39 ft^3 **49.** 30,181.68 ft^3

8.4 Exercises

1. 15 ft **3.** 5 ft **5.** 3 mi **7.** 25.5 ft **9.** $9\frac{1}{3}$ yd **11.** 18,480 mi **13.** 330 ft **15.** $2\frac{37}{44}$ mi **17.** 14,520 yd **19.** 12 yd 2 ft **21.** 41 in. **23.** 248 in. **25.** 61.5 m **27.** 1200 cm **29.** 0.654 m **31.** 8.42 m **33.** 41 cm **35.** 5.4 km **37.** 3.025 m **39.** 140 hm **41.** 1.39 dam **43.** 8.3 km **45.** 12,000 m **47.** 139.7 cm **49.** 4.4 m **51.** 14.6 m **53.** 42.2 km **55.** 243.84 cm **57.** 1.5 yd **59.** 15.0 in. **61.** 16.5 cm by 3.8 cm

8.5 Exercises

1. 112 oz **3.** 14 lb **5.** 84 oz **7.** 1.8 tons **9.** 150.4 oz **11.** 17,200 lb **13.** 8 lb **15.** 195 oz **17.** 14 lb 12 oz **19.** 214 oz **21.** 16 g **23.** 12 kg **25.** 6.52 g **27.** 0.084 kg **29.** 0.25 g **31.** 490 g

33. 5 mg **35.** 0.5 g **37.** 3.7 kg **39.** 275 lb **41.** 84.2 kg **43.** 70.9 g **45.** 3600 g **47.** 138.9 g **49.** 17.85 oz

8.6 Exercises

1. 4 c **3.** 100 qt **5.** 48 fl oz **7.** 31 gal **9.** 6 fl oz **11.** 1.5 qt **13.** 4.75 gal **15.** 120 pt **17.** 136 c **19.** 9 qt **21.** 4 pt **23.** 2 qt **25.** 50 qt **27.** 21 gal 2 qt **29.** 19 qt **31.** 21,000 mL **33.** 0.052 L **35.** 0.674 L **37.** 8.75 dL **39.** 91 L **41.** 0.75 L **43.** 1.3 L **45.** 1540.5 mL **47.** 11.8 L **49.** 14.3 L **51.** 61.6 L **53.** 7.4 L **55.** 2.1 qt **57.** 49,400 gal

8.7 Exercises

1. 42 d **3.** 288 hr **5.** 1680 sec **7.** 1800 min **9.** 87.5 d **11.** 270 min **13.** 176 hr **15.** $5\frac{1}{3}$ hr **17.** 11,520 sec **19.** $3\frac{17}{36}$ d **21.** 588 min **23.** 348.0 hr **25.** 113°F **27.** −13°F **29.** 36.7°C **31.** 12.5°C **33.** 23°F **35.** 38.9°C

MODULE 9

9.1 Exercises

1. 24 **3.** 368 **5.** 40 **7.** 5.6 **9.** 214 mi **11.** 4 **13.** 2.3 **15.** no mode **17. a.** mean: 17.81 in.; median: 17.5 in.; mode: 14.9 in. **b.** mean: 20.51 in.; median: 19 in.; mode: 14.9 in. **19. a.** mean: 88.9; median: 90; mode: 90 **b.** mean: 89.4; median: 90; mode: 90

9.2 Exercises

1. a. 175–199 **b.** 150–174 **c.** 18 **d.** 34 **e.** 31 **3. a.** 28% **b.** 13–18 **c.** 27%

5.

Weeks unemployed	Tally	Frequency
Less than 5	||||	4
5–14	~~||||~~	5
15–26	|||	3
27 or more	~~||||~~ |||	8

7.

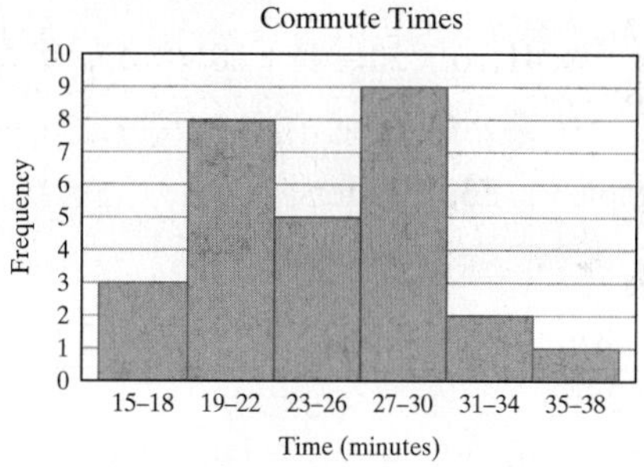

9.

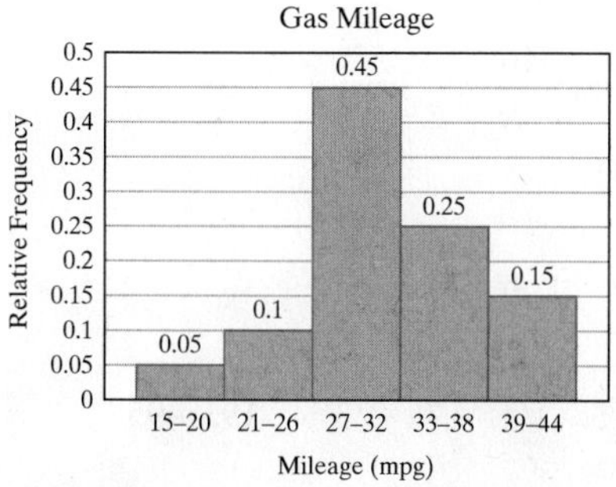

9.3 Exercises

1. 12 **3.** 12 **5.** H, S, FF; H, S, O; H, W, FF; H, W, O; H, T, FF; H, T, O; F, S, FF; F, S, O; F, W, FF; F, W, O; F, T, FF; F, T, O; C, S, FF; C, S, O; C, W, FF; C, W, O; C, T, FF; C, T, O **7.** 272 **9.** 168 **11.** 756 **13.** 4096 **15.** 985,527 **17.** 244,904

9.4 Exercises

1. 0.286 **3.** 0.429 **5.** 0.696 **7.** 0.39 **9.** 0.164 **11.** 0.25 **13.** 0.5 **15.** 0.419

MODULE 10

10.1 Exercises

1. integer and rational number **3.** irrational number **5.** integer and rational number **7.** rational number

9. 

11. 17 **13.** −12.45 **15.** $-4.\overline{27}$ **17.** 12.5 **19.** $\frac{15}{28}$ **21.** >

23. = **25.** = **27.** > **29.** false **31.** true **33.** false **35.** $45 \leq 50$ **37.** $4 \neq 5$
39. ≠ **41.** ≠

10.2 Exercises

1. 15 **3.** −71 **5.** −4.03 **7.** $-\frac{11}{3}$ **9.** $-\frac{46}{15}$ **11.** 4 **13.** −2 **15.** 0 **17.** 0 **19.** 0

21. 6 **23.** $\frac{40}{21}$ **25.** −4 **27.** $-\frac{1}{2}$ **29.** $\frac{11}{8}$ **31.** $-7+5$ **33.** $(3.7+2.1)+8.1$ **35.** $-15+12$

37. $12-(-13)$ **39.** $(2-7)+5$ **41.** $\frac{5}{9}-\frac{2}{3}$ **43.** a loss of 181.46 points, or −181.46 points

45. a difference of 2319 m

10.3 Exercises

1. −6 **3.** 6 **5.** −70 **7.** 936 **9.** $\frac{4}{3}$ **11.** 0 **13.** $\frac{32}{3}$ **15.** −13 **17.** 0 **19.** 18 **21.** −7

23. 16 **25.** $\frac{5}{12}$ **27.** $\frac{4}{3}$ **29.** $\frac{25}{18}$ **31.** 0.15 **33.** −3 **35.** 0.22(18) **37.** $\frac{2}{3}(-20)$

39. $(19-12)\frac{7}{8}$ **41.** $0.7(20-4)+18$ **43.** $\frac{5 \text{ chaperones}}{60 \text{ students}}$ **45.** $\frac{3.76}{5}$ **47.** $\frac{(30+12)}{\frac{7}{8}}$ **49.** \$2.52

51. 39,610 people **53.** 12 liters

10.4 Exercises

1. 256 **3.** $\frac{81}{256}$ **5.** 1 **7.** −81 **9.** $\frac{27}{125}$ **11.** −0.0001 **13.** 10 **15.** −1 **17.** $\frac{7}{4}$ **19.** 1

21. −4 **23.** 11 **25.** 54 **27.** 23 **29.** 4 **31.** $\frac{17}{13}$ **33.** $\frac{3}{2}$ **35.** $\frac{40}{7}$ **37.** 32 **39.** 88

41. 1 **43.** 9 **45.** $\frac{31}{5}$ **47.** 5

10.5 Exercises

1. 20 **3.** 144 **5.** $-\frac{41}{4}$ **7.** 38 **9.** −1 **11.** −4 **13.** $\frac{26}{5}$ **15.** 0 **17.** −23 **19.** −95

21. 29 **23.** 33 **25.** $9+5$ **27.** $x+11$ **29.** $-3(8.3)$ **31.** $3+(6+15)$ **33.** $7(y \cdot 4)$
35. $1.45+(a+b)$ **37.** $(-6+6)+c$ **39.** $5+x$ **41.** $24z$ **43.** $3a+3b$ **45.** $8y-20$ **47.** $32x+12$

49. $-12x-8y$ **51.** $-7+n$ **53.** $\frac{3x}{4}-4$ **55.** $0.3+0.68x$ **57.** $12x+48-60y$ **59.** $15x+3y-21$

61. $4(x+1)$ **63.** $\frac{1}{2}(y+6)$ **65.** Identity Property of Multiplication

67. Inverse Property of Multiplication

10.6 Exercises

1. 2 terms, coefficients: 3 and −5 **3.** 2 terms, coefficients: $-\frac{3}{4}$ and −9 **5.** 3 terms, coefficients: 1, $-\frac{1}{2}$, and 8
7. $2a$ and $-4a$, $-b$ and $2b$ **9.** $6.3m$ and $-8m$, $-4.5n^2$ and $2.2n^2$ **11.** $7z$ **13.** $5x^2 + 4x - 8$
15. $7x + 10x^2$ **17.** $22a - 2b$ **19.** 12.3 **21.** $-1.35c^2$ **23.** $-4x^2 - x - 2$ **25.** $-4x + 15$ **27.** $19x - 8$
29. $7w - 17$ **31.** $22x^2 + 18x + 8$ **33.** $x + 8$ **35.** $\frac{10}{x}+2$ **37.** $\frac{3}{5}x+1$ **39.** $3s + 18$ cm
41. $3n - 1$ satellites **43. a.** $12x + 4$ ft **b.** 64 ft **45. a.** $0.49x + 175$ gallons **b.** 180.88 gallons

MODULE 11

11.1 Exercises

1. No, because the algebraic expression is not equated to a second expression.
3. No, because the algebraic equation is nonlinear (the variable is raised to an exponent other than 1).
5. No, because there are two different variables in the equation. **7.** $x = 3$ is a solution **9.** $z = -4$ is a solution
11. $x = \frac{9}{4}$ is a solution **13.** $a = 5$ **15.** $y = -3$ **17.** $x = 7$ **19.** $x = -\frac{3}{4}$ **21.** $x = 3$ **23.** $m = 5$
25. $w = -1$ **27.** $x = 27$ **29.** $m = 32$ **31.** $z = 14$ **33.** $x = \frac{5}{6}$ **35.** $w = \frac{17}{5}$ **37.** $x = 5$ **39.** $x = 5$
41. $z = 2$ **43.** $x = 8.2$ **45.** $y = 5$ **47.** $b = -2$ **49.** $z = 0$ **51.** $y = -\frac{18}{7}$

11.2 Exercises

1. {−1} **3.** {−3} **5.** {2} **7.** {−2} **9.** {0} **11.** {1} **13.** {−5} **15.** {−6} **17.** {−2}
19. $\left\{\frac{17}{30}\right\}$ **21.** {1} **23.** $\left\{\frac{3}{2}\right\}$ **25.** $\left\{-\frac{11}{5}\right\}$ **27.** $\left\{-\frac{8}{15}\right\}$ **29.** {4} **31.** {7} **33.** {2}
35. {−3} **37.** $\left\{\frac{19}{20}\right\}$ or {0.95} **39.** $\left\{\frac{1073}{230}\right\}$ **41.** { } or ∅ **43.** $\mathbb{R}$ **45.** { } or ∅ **47.** $N = 22$ sliders **49.** $h \approx 57.71$ inches

11.3 Exercises

1. $x + 13 = 38$ **3.** $\frac{x}{4} - 5 = 31$ **5.** $\frac{x+2.7}{6} = \frac{x}{9}$ **7.** $\frac{x}{2} = x - 19$ **9.** 114 **11.** 3 **13.** $-\frac{11}{4}$
15. 11 **17.** 14.4 inches, 57.6 inches **19.** Taipei 101: 508 meters, Burj Khalifa: 828 meters
21. Kareem Abdul-Jabbar: 6 times, Michael Jordan: 5 times, Bill Russell: 5 times **23.** 345, 346 **25.** 57, 59
27. 158, 159 **29.** 235 pounds **31.** security guard: 25 hours, landscaper: 15 hours

11.4 Exercises

1. 350 miles **3.** $w = 110 + 5.06(70 - 60) = 160.6$ pounds **5.** $15 **7.** 19.0% **9.** 12 feet
11. $P = 381$ cm, $A = 7543.8$ cm^2 **13.** $3675 **15.** $24.50 **17.** $3750 **19.** 3.5 hours **21.** 15 feet
23. $r = \frac{C}{2\pi}$ **25.** $y = \frac{C - Ax}{B}$ **27.** $R = \frac{E}{I} - r$ **29.** $C = \frac{5}{9}(F - 32)$ **31.** 54.1 gallons **33.** $940.50
35. 42.6 cubic yards

11.5 Exercises

1. $w = 4$ feet, $l = 22$ feet **3.** 2100 square yards **5.** 15 feet, 20 feet, 25 feet **7.** 5760 cubic inches
9. 20°, 70° **11.** 30° **13.** 34°, 73°, 73° **15.** 50° **17.** 0.5 hour **19.** 148.1 miles

11.6 Exercises

1. 28 **3.** 85% **5.** 3640 **7.** 216 pounds **9.** 21.9% **11.** \$349.93 **13.** 45% **15.** 150% **17.** 8% **19.** \$10,196 **21.** 6.2% **23.** 27.7% **25.** 21.6% **27.** 3.6 liters **29.** 21 cups

11.7 Exercises

1. $\{x \mid x < -3\}$ **3.** $\{z \mid z \geq 0\}$ **5.** 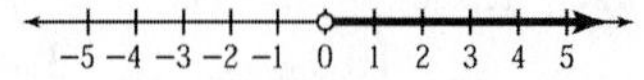

7. 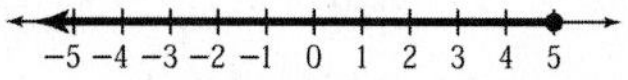

9. 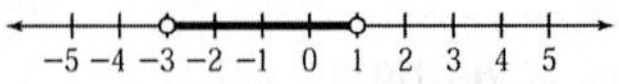

11. $(-5, -2)$ **13.** $(3, 10]$ **15.** 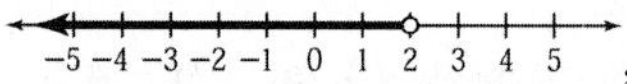

, $(-\infty, 2)$

17. 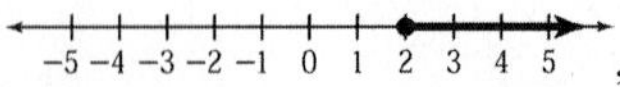

, $[2, \infty)$ **19.** 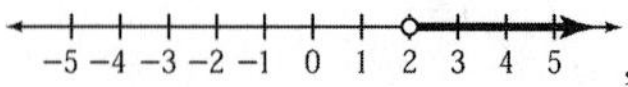

, $(2, \infty)$

21. 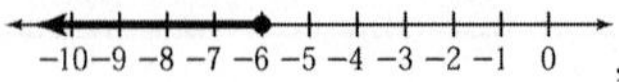

, $(-\infty, -6]$ **23.** 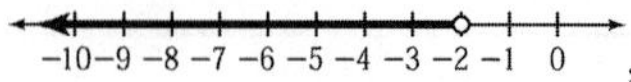

, $(-\infty, -2)$

25. 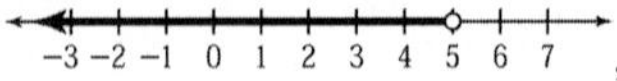

, $(-\infty, 5)$ **27.** 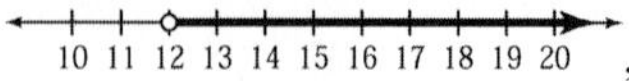

, $(12, \infty)$

29. 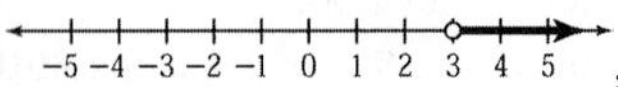

, $(3, \infty)$ **31.** $(2, \infty)$ **33.** $\left(-\frac{9}{8}, \infty\right)$ **35.** $\varnothing$

37. 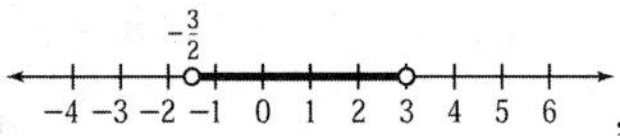

, $\left(-\frac{3}{2}, 3\right)$ **39.** 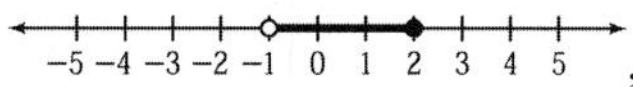

, $(-1, 2]$

41. 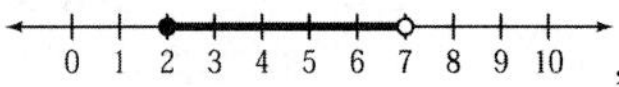

, $[2, 7)$ **43.** 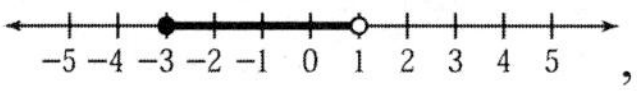

, $[-3, 1)$ **45.** 159 bracelets

47. 105 guests

11.8 Exercises

1. $\{-2, 0, 6\}$ **3.** $\{-6, -4, -2, 0, 1, 2, 4, 5, 6, 9, 15\}$ **5.** $\{x \mid 2 \leq x < 9\}$ **7.** $(-8, 3)$ **9.** $(1, 4]$

11. $(-\infty, -3] \cup (4, \infty)$ **13.** $(-\infty, \infty)$ **15.** $(-\infty, -6)$ **17.** $\{-3, 3\}$ **19.** $\left\{-\frac{5}{3}, -\frac{2}{3}\right\}$ **21.** $\left\{-\frac{3}{7}\right\}$

23. $\left\{0, \frac{3}{2}\right\}$ **25.** $\left[-\frac{1}{3}, \frac{1}{3}\right]$ **27.** $\left(-2, \frac{7}{2}\right)$ **29.** $\left\{\frac{7}{3}\right\}$ **31.** $(-\infty, -3) \cup (3, \infty)$ **33.** $\left(-\infty, \frac{9}{4}\right] \cup \left[\frac{15}{4}, \infty\right)$

35. $\left(-\infty, \frac{4}{9}\right) \cup \left(\frac{4}{9}, \infty\right)$ **37.** $\left[-1, \frac{5}{2}\right]$ **39.** $(-\infty, -8] \cup [-2, \infty)$ **41.** $(-\infty, \infty)$

MODULE 12

12.1 Exercises

1. a. 4 **b.** June **c.** May, 20 **d.** January, February, December **3. a.** *A*: (3, 1); Quadrant I **b.** *B*: (5, −2); Quadrant IV **c.** *C*: (−2, 0); *x*-axis **d.** *D*: (−2, −4); Quadrant III

5-9. 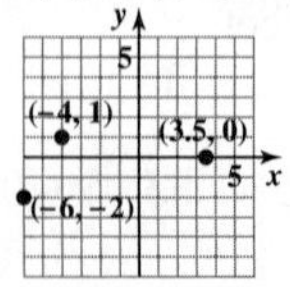

(−4, 1): Quadrant II; (−6, −2): Quadrant III; (3.5, 0): *x*-axis

11. 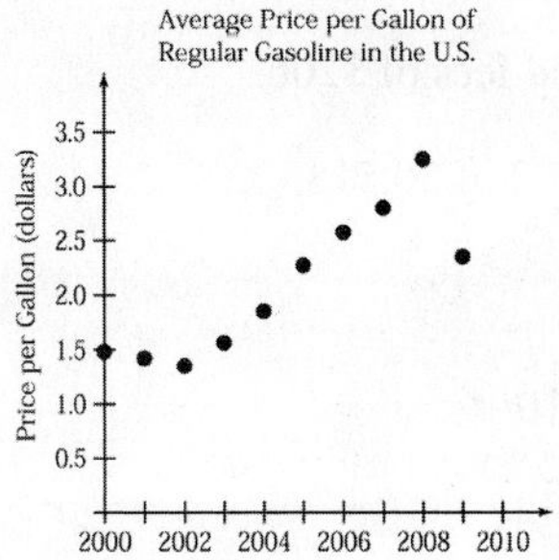

13. 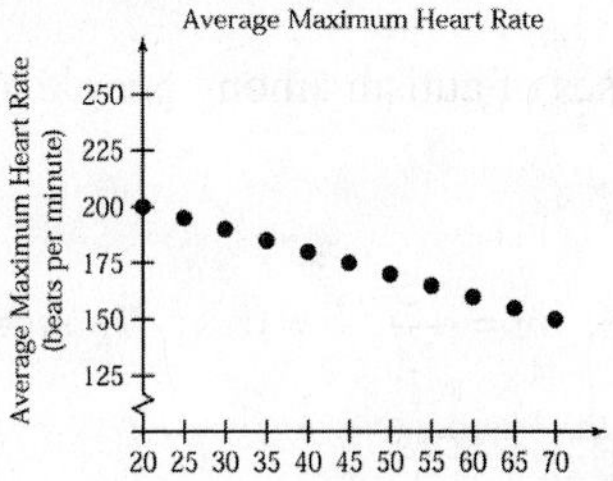

The scatter plot shows that average maximum heart rate steadily decreases as age increases.

12.2 Exercises

1. a. no **b.** yes **c.** yes **3. a.** no **b.** yes **c.** no **5. a.** $x = 0$ **b.** $y = \frac{3}{4}$

7. 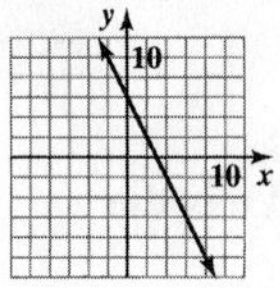
9. 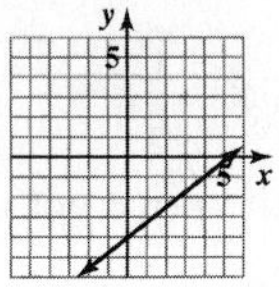
11. 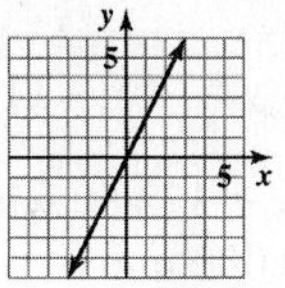
13. 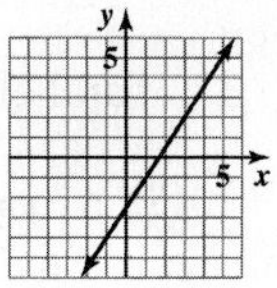
15. 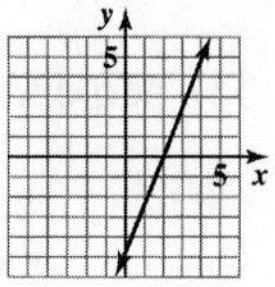

17. the x-intercept is 3, the y-intercept is 2 **19.** the x-intercepts are -2, 0, and 1, the y-intercept is 0

21. the x-intercept is $\frac{1}{2}$, the y-intercept is -2 **23.** 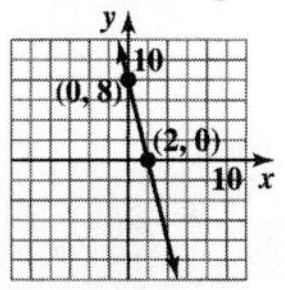

25. 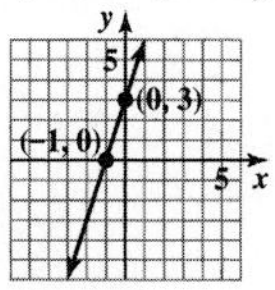

27. 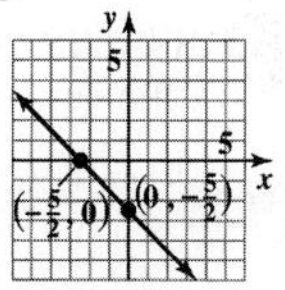

29. 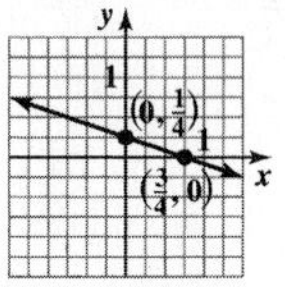
31. 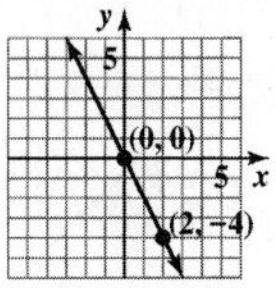

33. 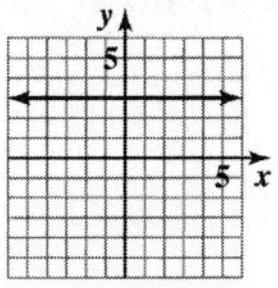
35. 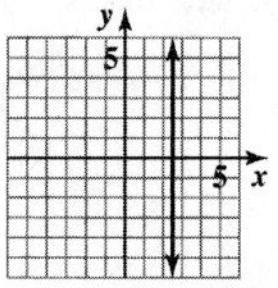

37. a. 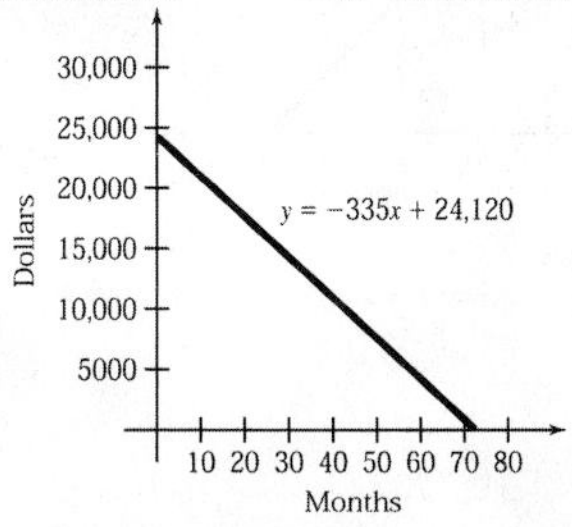

b. 15,745 **c.** Natalie still owes $15,745 toward the loan after 25 months.
d. The y-intercept is $24,120, which is the total cost of the automobile at the time of purchase.
e. The x-intercept represents the number of months that are needed for the entire payment to be made.

12.3 Exercises

1. $m = \frac{3}{4}$ **3.** $m = -\frac{6}{5}$ **5.** $m = \frac{1}{2}$ **7.** $m = \frac{4}{3}$ **9.** $m = 0$ **11.** 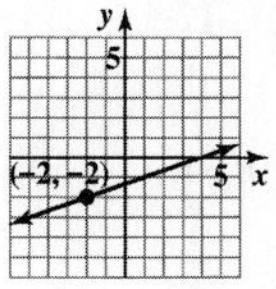

13. 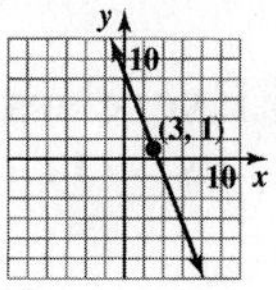

15. 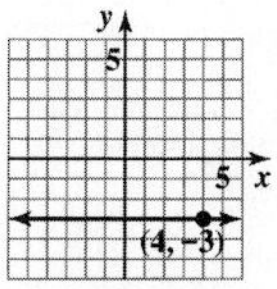

17. $m_2 = \frac{2}{3}$, $m_3 = -\frac{3}{2}$ **19.** $m_2 = 6$, $m_3 = -\frac{1}{6}$ **21.** 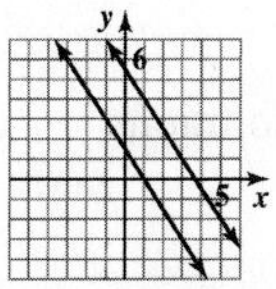
23. 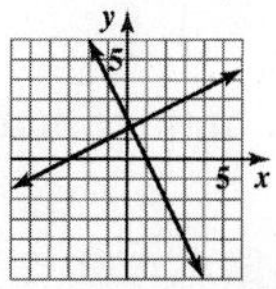

25. grade $= \frac{1}{8} = 0.125$, or 12.5% **27.** $m = 206$, or an annual increase in tuition and fees of \$206.

29. $m \approx 30{,}518$, or an annual increase in the cases of autism among people ages 3 to 22 of 30,518.

12.4 Exercises

1. $m = -3,\ b = 9$ **3.** $m = -\frac{6}{5},\ b = -\frac{4}{5}$ **5.** $m = -\frac{5}{4},\ b = 0$ **7.** $m = 0,\ b = 10$

9. 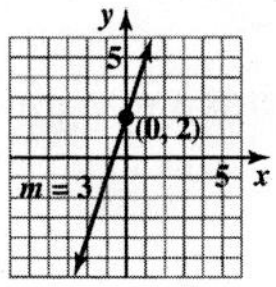

11. 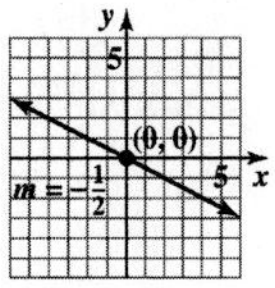

13. 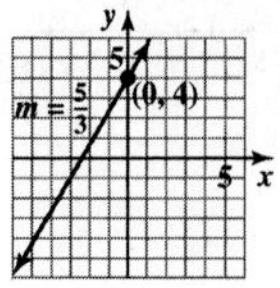

15. $y = 4x + 8$ **17.** $y = \frac{1}{3}x + 4$ **19.** $y = \frac{5}{4}$

21. $y = -3x + 11$ **23.** $y = -\frac{2}{3}x + 9$ **25.** $y = \frac{2}{3}x$ **27.** $y = 1$ **29.** $y = \frac{3}{2}x + 3$ **31.** $y = -\frac{2}{9}x$

33. parallel **35.** perpendicular **37.** parallel **39.** $y = -6x - 11$ **41.** $y = \frac{1}{6}x + \frac{7}{3}$

43. a. $y = 1.3x - 0.65$ **b.** 5.59% **45. a.** $y \approx -1.29x + 372.50$ where x is the price per unit and y is the number of units sold (in 1000s). **b.** approximately 179,000 units

12.5 Exercises

1. a. no **b.** yes **c.** no **3. a.** no **b.** yes **c.** no **5.** 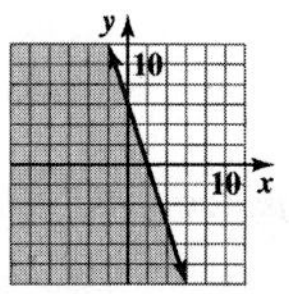

7. 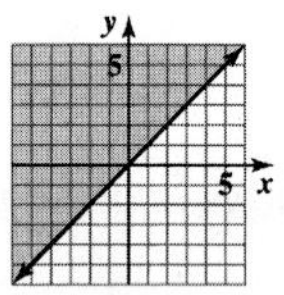

9. 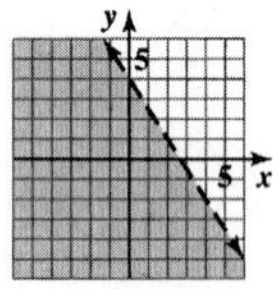

11. 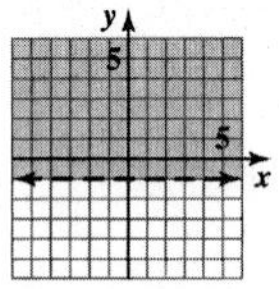

13. a. $450c + 300s \le 90{,}000$ **b.** 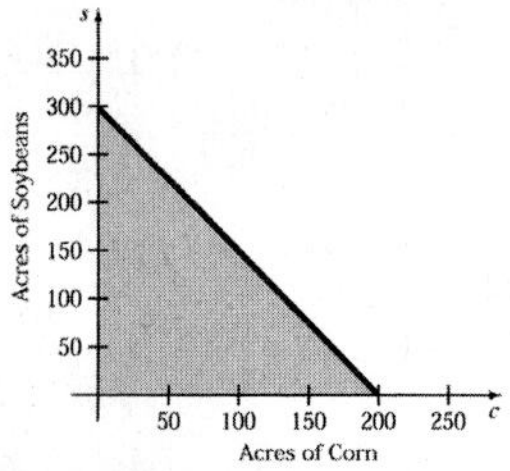

c. no (93,750 > 90,000) **15. a.** $0.10d + 0.25q > 12.00$ **b.** 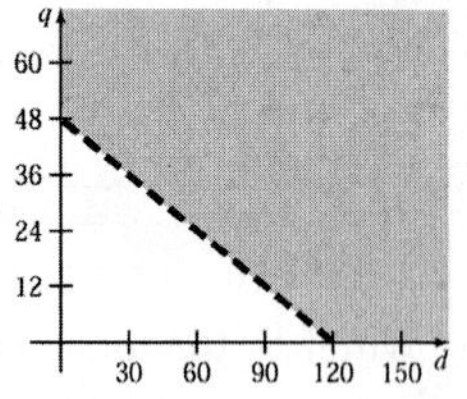

c. yes (14.75 > 12.00)

MODULE 13

13.1 Exercises

1. a. yes **b.** no **3. a.** no **b.** yes **5. a.** yes **b.** no **7.** no solution **9.** no solution

11. infinite number of solutions **13.** 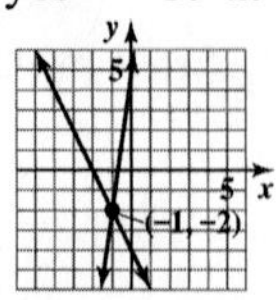

15. 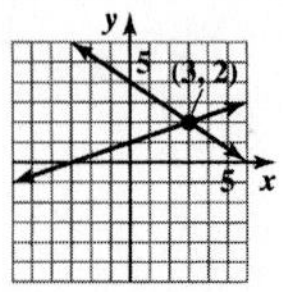

17. 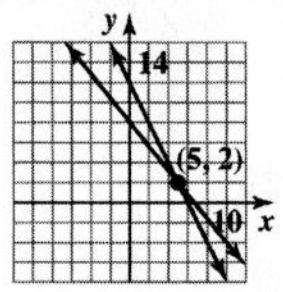

19. 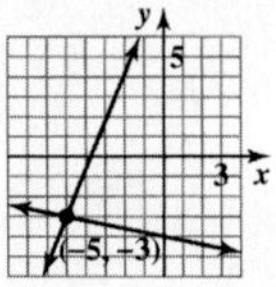

21. 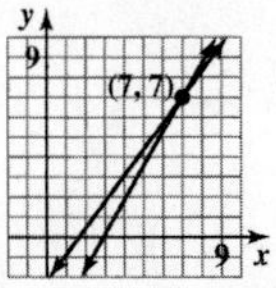

13.2 Exercises

1. (3, 7) **3.** (1, 1) **5.** (2, −3) **7.** (3, −2) **9.** (3, 1) **11.** (−2, −2) **13.** (0, 0) **15.** $\left(-\frac{5}{3}, \frac{26}{9}\right)$

17. (−1, 2) **19.** (2.2, −1.1) **21.** no solution **23.** $\left\{(x, y) \mid y = \frac{1}{3}x + \frac{2}{3}\right\}$ **25.** $\left\{(x, y) \mid y = \frac{1}{3}x + \frac{1}{3}\right\}$

27. $\left\{(x, y) \mid y = \frac{2}{3}x + \frac{1}{3}\right\}$

13.3 Exercises

1. (6, 2) **3.** (7, 2) **5.** (5, 0) **7.** (−2, 2) **9.** (−1, 3) **11.** (6, 5) **13.** (−6, 3) **15.** (−4, 7)

17. (0, 0) **19.** $\left(\frac{1}{2}, -1\right)$ **21.** (6, −4) **23.** $\left\{(x, y) \mid y = \frac{1}{3}x - 2\right\}$ **25.** { } or ∅

13.4 Exercises

1. Clayton: 12 years, Josh: 14 years **3.** Isaiah: 180 pounds, Geoff: 160 pounds
5. Statue of Liberty: 305.5 feet, Gateway Arch: 630 feet **7.** width: 75 cm, length: 124 cm
9. equal sides: 30 cm long, short side: 20 cm long **11.** angle 1: 25°, angle 2: 65°
13. swimming time: 0.5 hr, running time: 1.0 hr **15.** plane speed: 210 miles/hr, wind speed: 14 miles/hr
17. Coke: 161 cans, Dr. Pepper: 110 cans **19.** mango: 1.5 pounds, kiwi: 2.5 pounds
21. 18-carat gold: 140 ounces, 12-carat gold: 210 ounces **23.** 1 glass of milk: 11 carbs, 1 snack bar: 13 carbs
25. 1 cheeseburger with onion: 480 calories, 1 order of french fries: 400 calories

13.5 Exercises

1. a. yes **b.** yes **c.** no **3. a.** yes **b.** no **c.** yes **5.** 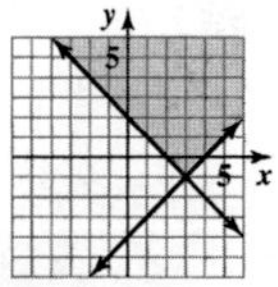

7. 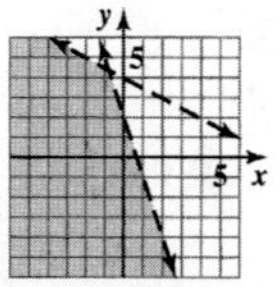

9. 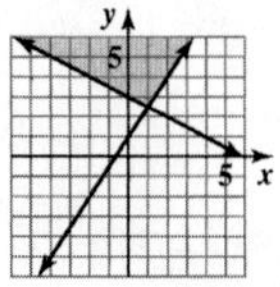

11. 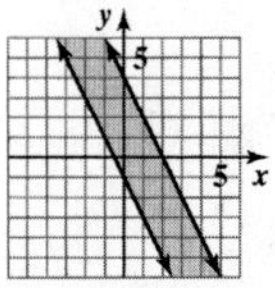

13. 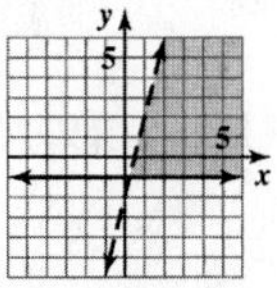

15. 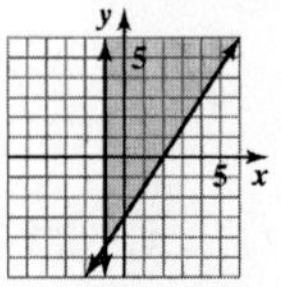

17. 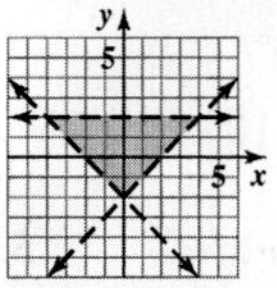

19. a. 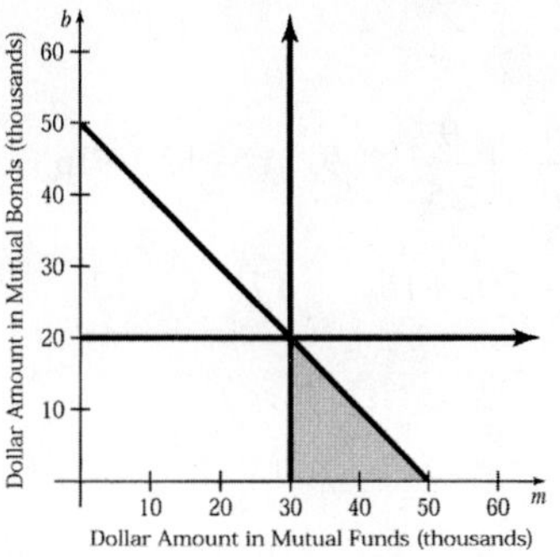

b. no **c.** yes **21. a.** 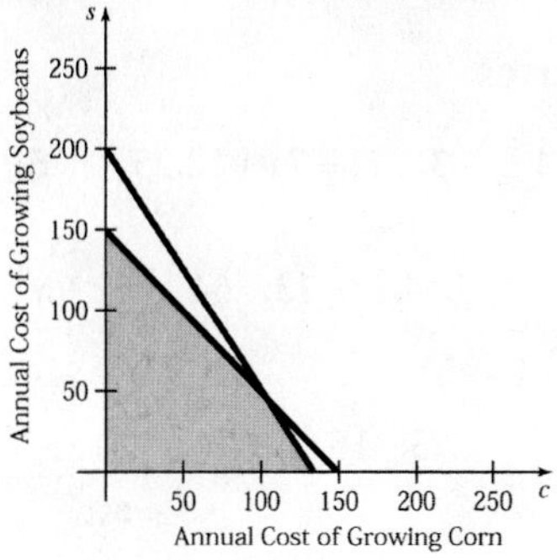

b. yes
c. no

13.6 Exercises

1. yes: $(-1,1,-2)$, no: $(1,-1,2)$ **3.** $(-1,2,3)$ **5.** $\left(\frac{1}{4},-\frac{1}{2},\frac{3}{8}\right)$ **7.** $(-3,0,1)$ **9.** $(29,16,3)$

11. $(5,-2,-4)$ **13.** $\left(\frac{1}{2}y-\frac{1}{2}z-3,y,z\right)$ **15.** no solution **17.** $(2,z-3,z)$

19. Hot dog: \$1.50, Hamburger: \$2.50, Chicken sandwich: \$3.75

21. Adult: 520, Children: 210, Senior citizen: 280 **23.** $y=-\frac{1}{80}x^2+\frac{37}{40}x-\frac{161}{16}$; 5.25 million

MODULE 14

14.1 Exercises

1. 7^6 **3.** w^6 **5.** $30a^{10}$ **7.** $18m^{13}n^6$ **9.** b^4 **11.** $(-10)^3=-1000$ **13.** $4q^{12}$ **15.** x^2y^6z

17. 1 **19.** –1 **21.** 1 **23.** x^{24} **25.** $(-2)^{12}$ **27.** p^6q^6 **29.** $a^{21}b^7$ **31.** $49r^6s^{18}$ **33.** $\frac{a^8}{b^8}$

35. $\frac{t^3}{4^3}=\frac{t^3}{64}$ **37.** $\frac{u^{35}}{2^7}=\frac{u^{35}}{128}$ **39.** x^{26} **41.** $4p^{12}$ **43.** $\frac{x^{12}}{y^7}$

14.2 Exercises

1. trinomial **3.** none of these **5.** none of these **7.** coefficient: 4; degree: 4 **9.** coefficient: 1; degree: 7 **11.** coefficient: –9; degree: 1 **13.** $5x+4$; degree: 1; leading coefficient: 5 **15.** $-3x+9$; degree: 1; leading coefficient: –3 **17.** $4.1x^3+x^2-6.7x+3.8$; degree: 3; leading coefficient: 4.1 **19.** 7 **21.** –7 **23.** 7.44 **25.** 39 **27.** $-5x+5$ **29.** $16x+1$ **31.** $1.3x^2+3.2x-9.1$ **33.** $7x^3-8$

14.3 Exercises

1. $6x+14$ **3.** $3m^2+10$ **5.** $8p^3-2p^2-5p-11$ **7.** $1.7w^2-3.5w-0.9$ **9.** $-7x^2+24$

11. $8z^3-3z^2-4z+13$ **13.** $3y+22$ **15.** $5t^2+2t-4$ **17.** $-5x^3+12x^2-2x$ **19.** $2.5n^2-3.6n+5.9$

14.4 Exercises

1. $6x^8$ **3.** $6y^{14}$ **5.** $\frac{4}{5}x^5$ **7.** $15x^2+10x$ **9.** $3x^3-3x^2+3x$ **11.** $-3x^6-6x^4+21x^3$

13. $x^2+7x+10$ **15.** $y^2-\frac{11}{15}y-\frac{4}{5}$ **17.** $x^2+7x-18$ **19.** $56z^2+27z-18$ **21.** x^3-27

23. $2x^3-x^2+x+1$ **25.** $2x^5+7x^3-3x^2+3x-9$ **27.** $x^5+x^4-2x^3+2x-1$ **29.** $30x^3+25x^2-30x$

14.5 Exercises

1. y^2+4y+4 **3.** $x^2+7x+12.25$ **5.** $\frac{9}{16}z^2+\frac{3}{2}z+1$ **7.** $36x^2+\frac{36}{5}x+\frac{9}{25}$ **9.** $x^2-12x+36$

11. $z^4-8.4z^2+17.64$ **13.** $64-16x+x^2$ or $x^2-16x+64$ **15.** $\frac{25}{4}z^2-20z+16$ **17.** x^2-81 **19.**

$9z^2-16$

21. $4.41-9y^2$ **23.** $16x^4-49$

14.6 Exercises

1. $\frac{1}{y^3}$ **3.** $\frac{6}{w^5}$ **5.** 64 **7.** $\frac{1}{2^3}+\frac{1}{4}=\frac{3}{8}$ **9.** z^5 **11.** $(-3)^4=81$ **13.** $-12t^3$ **15.** $\frac{2^5}{9^2}=\frac{32}{81}$

17. $\frac{32}{x^4}$ **19.** $5t^6$ **21.** $\frac{x^6}{216y^{15}}$ **23.** $\frac{9b^2}{625a^4c^2}$ **25.** $\frac{q^{56}}{p^{16}}$ **27.** $\frac{9a^{27}}{b^{23}}$ **29.** 8.9×10^{10}

31. 4.06×10^{-7} **33.** 2.54×10^4 **35.** 5×10^{-9} **37.** 0.00000297 **39.** 0.0000000008

41. 5,878,000,000,000 **43.** 0.000053 **45.** 6.9×10^{15} **47.** 3×10^5

14.7 Exercises

1. $6x^8$ **3.** $\frac{1}{6}$ **5.** $2x-5$ **7.** $8y^5-4y^3+1$ **9.** $2a-7-\frac{3}{a}$ **11.** $x+12$ **13.** x^2-5x+7

15. $x^2-3x+6-\frac{34}{x+5}$ **17.** $3x+5+\frac{5}{5-2x}$ **19.** $4m^2-8m+21-\frac{49}{m+2}$

14.8 Exercises

1. 3, 7; –4, 4; 12, 2; –5, 0; 7 **3.** 6, 5; –2.5, 5; 7, 8; –4.3, 0; 8 **5.** –7 **7.** $-18a+18b$

9. $-2m^2+16mn+6n$ **11.** $-3x-2xy+2y$ **13.** $3x^3-3xy-4$ **15.** $-6mn^2+m-2n$ **17.** $3x^3y-9x^2y$

19. $3x^3y+3x^2y^2$ **21.** $15x^2+16xy-15y^2$ **23.** $0.6a^3-0.08a^2b^2+9ab-1.2b^3$ **25.** $a^2+6ab+9b^2$

27. $16m^2-8mn+n^2$ **29.** $25x^2-20xy+4y^2$ **31.** $2x^3+5x^2y-5xy^2+y^3$ **33.** $-6x^4+6x^3y-2x^2y+2xy^2$

MODULE 15

15.1 Exercises

1. 7 **3.** 9 **5.** x^2 **7.** a^3b **9.** $6x^5$ **11.** 5 **13.** $18mn^4$ **15.** $6(5x+2)$ **17.** $4w^3(5w-7)$

19. $4y^3(y^2-4y+7)$ **21.** $5ab^3(5b^2-2ab+a^2)$ **23.** $-8x(x-3)$ **25.** $-3(3x^2+9x-4)$

27. $(y-4)(5x+8)$ **29.** $(m-2n)(m-3n)$ **31.** $(y+3)(5x+6)$ **33.** $(3b-4)(3a-2)$

35. $(6x-11y)(x-1)$ **37.** $(x+6)(x^2+7)$ **39.** $(x-5)(x^2+1)$

15.2 Exercises

1. $(x+7)(x+3)$ **3.** $(w+9)(w+8)$ **5.** $(n-7)(n-2)$ **7.** $(p+9)(p-6)$ **9.** prime

11. $(x-18)(x-6)$ **13.** $(x-33)(x+2)$ **15.** $(w+12)(w-8)$ **17.** $(x+2y)(x+4y)$

19. $(a-3b)(a-8b)$ **21.** $5(x+2)(x+8)$ **23.** $2x(x-15)(x+9)$ **25.** $20(m-25)(m-1)$

27. $-(x-3)(x+1)$ **29.** $-(x-5y)(x-2y)$

15.3 Exercises

1. $(3x+1)(x+5)$ **3.** $(5x+2)(x+4)$ **5.** $(3n-7)(2n-1)$ **7.** $(12t-7)(t+1)$ **9.** prime

11. $(10p+1)(2p-1)$ **13.** $(2w-3)(w+13)$ **15.** $(5t-3)(3t-8)$ **17.** $(z-6)(4z+3)$

19. $(2x+3)(6x-1)$ **21.** $(7y-8)(y+1)$ **23.** $(2w-9)(w-3)$ **25.** $(2m-3)(12m+5)$

27. $(3x+y)(3x-8y)$ **29.** $(2a-7b)(a-10b)$

15.4 Exercises

1. $(5x+2)(x+3)$ **3.** $(4t-5)(t+3)$ **5.** $(3p-2)(p-8)$ **7.** $(2w-11)(w-3)$ **9.** $(10y-3)(y+5)$

11. $(2q-1)(9q-2)$ **13.** $(5x+1)(4x+5)$ **15.** $(2t-1)(4t+5)$ **17.** $(z-4)(6z+5)$ **19.** prime

21. $(x+2y)(4x+3y)$ **23.** $(2a+3b)(7a-3b)$ **25.** $2x^2(x+5)(5x-3)$ **27.** $-(m+4)(3m-2)$
29. $-3ab(12a-49+4b)$

15.5 Exercises

1. $(y-4)(y+4)$ **3.** $(x-y)(x+y)$ **5.** $(x^5-2)(x^5+2)$ **7.** $(3y-1)(3y+1)$ **9.** $(4-5m)(4+5m)$
11. $3(2x-3)(2x+3)$ **13.** $5(4x^2+9)$ **15.** $3y(x-2)(x+2)$ **17.** $(x^2+1)(x+1)(x-1)$
19. $(x^2+y^2)(x+y)(x-y)$ **21.** $5(x^4+1)(x^2+1)(x+1)(x-1)$ **23.** $(x-1)^2$ **25.** $(2x+7)^2$
27. prime **29.** $(m-7n)^2$ **31.** $x(5y+7)^2$ **33.** $(z^3+5)^2$ **35.** $(y-5)(y^2+5y+25)$
37. $(2x-1)(4x^2+2x+1)$ **39.** prime **41.** $(xy-3)(x^2y^2+3xy+9)$ **43.** $(x-1)(x^2+x+1)(x^6+x^3+1)$

15.6 Exercises

1. $3(x-2)(x+2)$ **3.** $(a+2b)(a^2-2ab+4b^2)$ **5.** $(x+3)(x-2)$ **7.** $5(z-3)^2$ **9.** $(2x-5)^2$
11. $x(5x-1)(x+3)$ **13.** prime **15.** $(3x+2)^2$ **17.** $-3(p-2q)(2p+q)$ **19.** $9y(1+2y)$
21. $(5m+4n)^2$ **23.** $(b+2a)(b-a)(b+a)$ **25.** $(x-5)(x+4)$ **27.** $3y(2y+1)(8y-3)$
29. $(p-1)(p-9)$

15.7 Exercises

1. $\{-2,1\}$ **3.** $\left\{-\frac{1}{7},\frac{1}{3}\right\}$ **5.** $\{2,3\}$ **7.** $\{-3,-2\}$ **9.** $\{-7,-2\}$ **11.** $\left\{-2,\frac{1}{3}\right\}$ **13.** $\{-5,5\}$
15. $\left\{-\frac{4}{3}\right\}$ **17.** $\left\{-3,\frac{1}{2}\right\}$ **19.** $\{-3,5\}$ **21.** $\{-3,8\}$ **23.** $\left\{\frac{5}{4}\right\}$ **25.** $\{-12,1,3\}$ **27.** $\{-2,7,8\}$
29. $\{-2,0,1\}$ **31.** $\left\{-2,-\frac{3}{7},2\right\}$ **33.** $\left\{-\frac{5}{2},-\frac{1}{3},\frac{5}{2}\right\}$ **35.** $\{1,2,3\}$ **37.** $\left\{-2,-\frac{5}{3},5\right\}$ **39.** $\{-1,1,2\}$

15.8 Exercises

1. The room numbers are 80 and 82. **3.** The numbers are 7, 8, and 9. **5.** 20% of community colleges provide access to grades through mobile devices, and 19% provide access to transcripts. **7.** The width of the frame is 4 inches. **9.** The sail has a base of 8 yards and a height of 12 yards. **11.** The corner post is 7.5 feet tall. **13.** The hypotenuse is 20. **15.** It will take the object 3 seconds. **17.** The ball will hit the ground in 3 seconds. **19.** The city had a population of 7000 in 2006. **21.** There are 42 occupied units.

MODULE 16

16.1 Exercises

1. 4 **3.** $\frac{1}{2}$ **5.** -10 **7.** 1 **9.** $\frac{2}{19}$ **11.** $x \neq 8$ **13.** $t \neq \frac{4}{7}$ **15.** $p \neq -2, 3$ **17.** $y \neq -\frac{5}{3}, 6$
19. No restricted values **21.** $\frac{3}{5}$ **23.** $\frac{6}{y+6}$ **25.** $\frac{z-5}{z+7}$ **27.** $\frac{t+8}{2t-5}$ **29.** $\frac{1}{x-4}$ **31.** $\frac{x-8y}{x-4y}$
33. $\frac{1}{x+2}$ **35.** -1 **37.** $-\frac{w+2}{w+8}$ **39.** $\frac{n^2+7}{n-5}$

16.2 Exercises

1. $\frac{5b}{a}$ **3.** $\frac{z}{y^2}$ **5.** $\frac{5x}{21}$ **7.** $\frac{3}{5}$ **9.** $\frac{1}{(x+2)(x-1)}$ **11.** $\frac{3r(r+4)(r-4)}{r+2}$ **13.** $\frac{x+1}{x-1}$

15. $\frac{(x+1)(3x-2)}{(3x+2)(x-1)}$ **17.** 1 **19.** $\frac{x-2}{x+1}$ **21.** $\frac{x-2}{2(x+2)(x+1)}$ **23.** $-\frac{5x+1}{2x+1}$ **25.** 1 **27.** $\frac{x+3y}{3x-y}$

29. $\frac{y}{x^2+xy+y^2}$ **31.** $\frac{5xy}{2}$ **33.** $\frac{(3x+2)(x+3)}{(x+1)(2x+3)}$ **35.** $\frac{x^2}{21(x-1)}$ **37.** $\frac{1}{3x}$

39. $\frac{(x+2)(x+3)}{(x^2+3x+9)(x^2+2x+4)}$ **41.** $(x+3)(x-1)$ **43.** $\frac{(2x+3)(3x+2)}{(x+2)(x+5)}$ **45.** $\frac{3(h+4)(h-2)}{(7h-2)(6h+1)}$

47. $\frac{(x-y)}{2}$ **49.** $\frac{15x^3y}{16}$ **51.** $\frac{x+4}{x-2}$ **53.** $\frac{x+4}{x-2}$

16.3 Exercises

1. $18x^5$ **3.** $14a^3b^5c^3$ **5.** $12x(2x-1)(3x+1)$ **7.** $-(x-3)(x+2)$ **9.** $-(x-3)(x+1)(x+3)$

11. $(y-3)(y+3)^2$ **13.** $(2x+1)(x-3)(3x-2)$ **15.** $6x^2(3x-2)(x+5)(2x+1)$

17. $(2x+3)(x-1)(3x-2)(x+5)$ **19.** $\frac{9x^2y^2}{15x^4y^5}$ **21.** $\frac{(x+1)(x-1)}{2(x-1)(x+3)}$ **23.** $\frac{(3x-5)(x+2)}{(2x+3)(x+2)}$

25. $\frac{-(x+1)(x-1)(x-2)}{3x(x-1)(x-2)(x-3)}$ **27.** $\frac{5(3x-5)(x-2)}{10(x-3)(x-2)(x+1)}$

16.4 Exercises

1. $\frac{6x}{7}$ **3.** $\frac{a^2+b^2+3}{5c^2}$ **5.** $\frac{x}{3a^3}$ **7.** $\frac{3x+2}{2x+3}$ **9.** $-\frac{4}{x}$ **11.** $\frac{2}{x+3}$ **13.** $\frac{x+3}{x-3}$ **15.** $\frac{x+1}{x+2}$

17. $\frac{13}{6x}$ **19.** $\frac{8x}{(x-2)(x+2)}$ **21.** $\frac{11x^2-5x-10}{10x(x-2)}$ **23.** $\frac{x^2+4x+10}{2x(x+5)}$ **25.** $\frac{1}{x(x-1)(x+1)}$

27. $\frac{2x+1}{2x-5}$ **29.** $\frac{6}{x-3}$ **31.** $\frac{3(2x+3)}{(x-2)(x+2)}$ **33.** $\frac{2x-1}{2x+1}$ **35.** $\frac{x^2}{(x-2)(x-1)(x+1)}$

37. $\frac{5}{x+2}$ **39.** $\frac{x-3}{x-1}$ **41.** $\frac{x^2-2x-1}{x(x+1)(x-1)}$ **43.** $\frac{(2x+3)(x-15)}{(x+6)^2(2x-9)}$ **45.** $\frac{2}{x+2}$ **47.** $\frac{4}{x}$

16.5 Exercises

1. 6 **3.** $\frac{5}{(x-3)(x-1)}$ **5.** $-\frac{x}{2x-3}$ **7.** $\frac{2(x+2)}{3(2x-1)}$ **9.** $\frac{3}{(x+1)(x+4)}$ **11.** $-\frac{3x-13}{5x-18}$ **13.** $\frac{7x+1}{7x-1}$

15. $\frac{1}{(x-6)(x+5)}$ **17.** $\frac{x-4}{4x}$ **19.** $\frac{3}{x-5}$ **21.** $\frac{x-1}{x-7}$ **23.** $\frac{6x}{x^2+9}$ **25.** $\frac{6}{x-y}$

16.6 Exercises

1. No, because $\frac{\sqrt{x^2-4x+1}}{x-5}$ is not a rational expression. **3.** Yes **5.** $\left\{\frac{15}{2}\right\}$ **7.** $\{12\}$ **9.** $\left\{\frac{5}{21}\right\}$

11. $\{-15,-4\}$ **13.** $\{-12\}$ **15.** { } or ∅ **17.** $\{12\}$ **19.** $\{2\}$ **21.** $C=\frac{ET}{R}$ **23.** $s=\frac{x-m}{z}$

25. $x_2=\frac{y_2-y_1+mx_1}{m}$ **27.** $B=\frac{E+D}{R+1}$

16.7 Exercises

1. The inspector can expect to find 240 defective units. **3.** There are 120 grams of fat in a quart of the ice cream. **5.** He will need 2020 more pounds of corn seed. **7.** 15 cm **9.** The lighthouse is 63 meters tall.
11. One circuit has a resistance of 12 ohms and the other circuit has a resistance of 36 ohms.
13. The current's speed was 3 mph. **15.** Mae's walking speed is 1.5 mph.
17. It will take 18 minutes for both copy machines to copy the final exams.
19. It would take Shawn 6 hours to stain the deck by himself. **21.** It will take 450 minutes to fill the tub.

16.8 Exercises

1. a. $y = 3.5x$ **b.** $y = 24.5$ **3. a.** $y = 24x^2$ **b.** $y = 600$ **5. a.** $p = \frac{5}{8}n^3$ **b.** $p = 625$

7. She will be 40 feet below the surface. **9.** 900 gallons could be pumped through the pipe. **11. a.** $y = \frac{600}{x}$

b. $y = 20$ **13. a.** $n = \frac{518.4}{m^2}$ **b.** $p = 32.4$ **15.** The car will be 9 years old.
17. The satellite will weigh approximately 79.01 kg.

MODULE 17

17.1 Exercises

1. independent: x, dependent: y **3.** Answers may vary. independent: x, dependent: y **5.** domain: $\{0,1,4\}$, range: $\{-2,-1,0,1,2\}$ **7.** domain: $(-\infty,\infty)$, range: $[3,\infty)$ **9.** yes **11.** yes **13.** yes **15.** no
17. yes **19.** no **21.** yes **23.** no **25. a.** independent: percent with access to clean water, dependent: percent undernourished **b.** $\{(96,3),(91,7),(77,15),(84,12),(92,8)\}$ **c.** domain: $\{96,91,77,84,92\}$, range: $\{3,7,15,12,8\}$ **d.** yes, it passes the vertical line test **27. a.** yes, it passes the vertical line test **b.** independent: Year, dependent: Price per gallon **c.** \$2.60 **d.** 2015

17.2 Exercises

1. $f(x) = |2x-5|$ **3.** $f(x) = -\sqrt{x}+1$ **5.** $f(x) = -\frac{3}{4}x+3$ **7.** $f(x) = 2x^2-5$ **9.** $f(x) = 2x^2+3$

11. 13 **13.** 1 **15.** 5 **17.** 54 **19.** 5 **21.** 124 **23.** $f(-2) = 0$; $f(3) = -10$; $f(5)$ is undefined

25. $\{x \mid x \neq -2\}$ **27.** $\{x \mid x \neq -2, 5\}$ **29.** $\left\{x \mid x \neq -\frac{2}{5}\right\}$ **31.** $\left\{x \mid x \neq -4, \frac{2}{3}\right\}$ **33.** $\mathbb{R}$

35. $2x^3-14x^2+2x-13$ **37.** $x^3+4x-20$ **39.** x^3-18x^2-4x-4 **41.** $2x^2+11x-40$

43. $3x-1+\frac{3}{x-2}$, $x \neq 2$

17.3 Exercises

1. 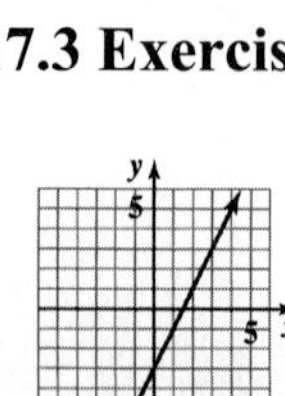

3. 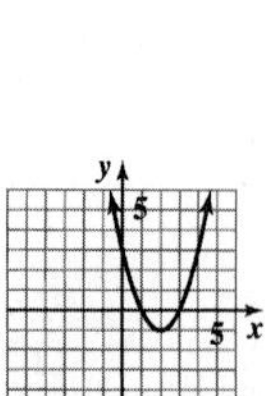

5. 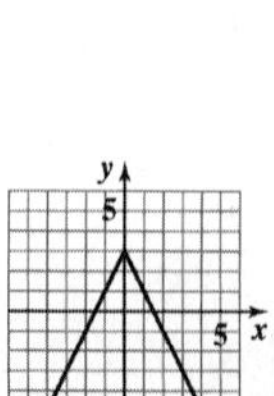

7. a. \$10 **b.** about \$0.09 **c.** about \$46 **d.** domain: $[0,\infty)$, range: $[10,\infty)$ **e.** you must use over 311 text messages (approximately) to make it a better deal **9. a.** March, June, and July
b. February and December **c.** January, April, May, August, September, October, and November
d. September **e.** \$12,600 **f.** \$8600 **g.** \$4000 **h.** about \$333.33

11. 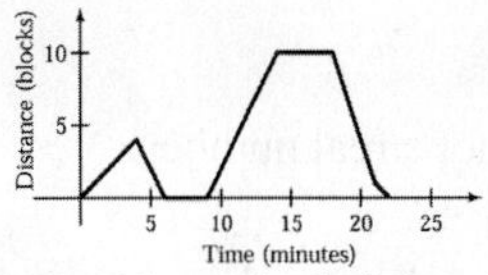

13. a. 442.225 m, the initial height of the object or the height of the building **b.** 319.725 m, the height of the object after 5 seconds **c.** –47.775 m, not possible, the object can't have a negative height **d.** 0 m, when the object hits the ground **e.** domain: $[0, 9.5]$, range: $[0, 442.225]$

f. 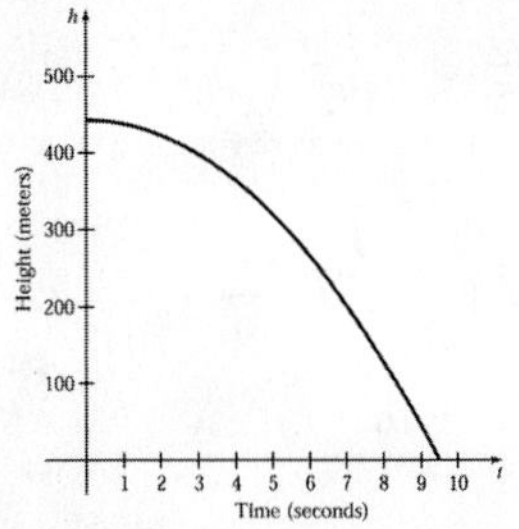

15. a. 9.817; the number of gallons remaining in the tank after driving 100 miles. **b.** 7.733; the number of gallons remaining in the tank after driving 200 miles. **c.** 11.9 gallons **d.** 3.567 gallons **e.** 571.2 miles **f.** domain: $[0, 571.2]$, range: $[0, 11.9]$

g. 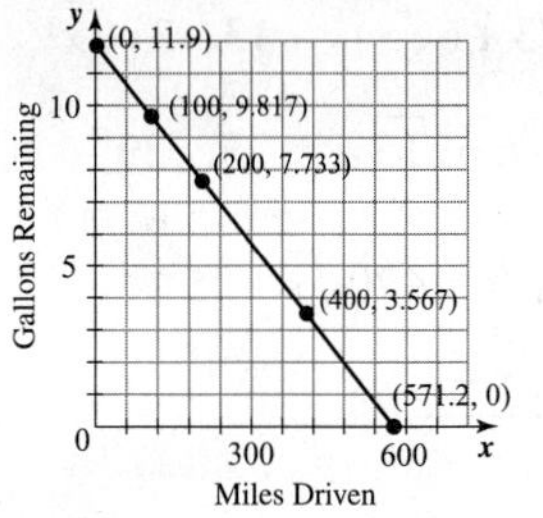

MODULE 18

18.1 Exercises

1. 10 **3.** $\frac{3}{11}$ **5.** 7.416 **7.** 17 **9.** $|6x-5|$ **11.** $9x^2$ **13.** $-11|x^5|$ **15.** 7 **17.** $-\frac{3}{4}$

19. w^7 **21.** 4 **23.** $\frac{2}{5}$ **25.** $|x^5|$ **27.** 3.476 **29.** 2.187

18.2 Exercises

1. 9 **3.** 2 **5.** -1 **7.** $\{x \mid x \geq -4\}$ or $[-4, \infty)$ **9.** $\mathbb{R}$ or $(-\infty, \infty)$ **11.** $\{x \mid x \geq 0\}$ or $[0, \infty)$

13. 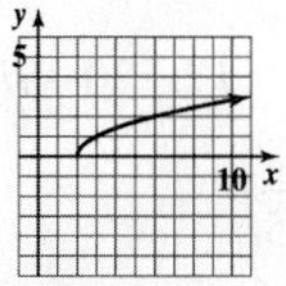 It looks like the graph of the square root function, but it is shifted two units to the right.

15. 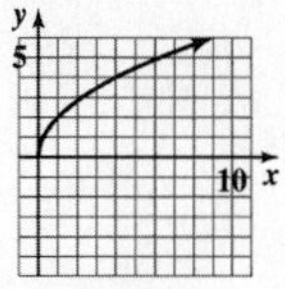 It looks like the graph of the square root function, but it is stretched by a factor of two.

17. 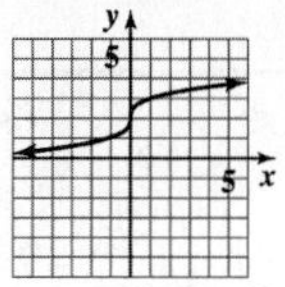 It looks like the graph of the cube root function, but it is shifted two units up.

18.3 Exercises

1. $\sqrt{36}=6$ **3.** $-\sqrt[4]{81}=-3$ **5.** $\sqrt[5]{7xy}$ **7.** $\left(\sqrt{100}\right)^3=1000$ **9.** $\left(\sqrt[4]{-16}\right)^3$ is not a real number.

11. $\left(\sqrt[3]{\frac{1}{8}}\right)^5=\frac{1}{32}$ **13.** $10^{\frac{1}{2}}$ **15.** $\left(\frac{2x}{y}\right)^{\frac{1}{5}}$ **17.** $\left(3xy^3\right)^{\frac{4}{5}}$ **19.** $\frac{1}{25^{\frac{1}{2}}}=\frac{1}{5}$ **21.** $\left(\frac{8}{27}\right)^{\frac{2}{3}}=\frac{4}{9}$

23. $9^{\frac{1}{2}}=3$ **25.** $x^{\frac{3}{2}}$ **27.** $\frac{2x^2}{y^{\frac{5}{2}}}$ **29.** $\sqrt[20]{x^{13}}$ **31.** $\sqrt{x}$ **33.** $\sqrt{3}$ **35.** $\sqrt[4]{35}$ **37.** $5x$

39. $6\sqrt{2}$ **41.** $4x\sqrt{3x}$ **43.** $a^7\sqrt{a}$ **45.** $6\sqrt{5}$ **47.** $6x^2y^2\sqrt{10y}$ **49.** $30x^2y$ **51.** $\frac{7}{9}$

53. $\frac{x\sqrt[3]{11}}{2}$ **55.** $\frac{x\sqrt[4]{21x}}{2}$ **57.** 3 **59.** $2x^5\sqrt{10}$ **61.** $-2m\sqrt[3]{3}$

18.4 Exercises

1. $13\sqrt{7}$ **3.** $3\sqrt{2}+5\sqrt{6}$ **5.** $-6\sqrt[3]{6x}$ **7.** $6\sqrt{5}+6\sqrt{x}$ **9.** $-\sqrt{5}$ **11.** $5\sqrt{3}+6\sqrt{5}$ **13.** $3x\sqrt[4]{x}$

15. $2a^2b\sqrt{ab}$ **17.** $\frac{23\sqrt{2}}{4y}$ **19.** $-4a\sqrt[4]{a^3}$ **21.** $2x\sqrt{5}+\sqrt{15}$ **23.** $2m\sqrt{3m}-6m\sqrt{15}$ **25.** $29-13\sqrt{3}$

27. $x+10\sqrt{x}+24$ **29.** $12-3\sqrt{y}-4\sqrt{x}+\sqrt{xy}$ **31.** $7z-9$ **33.** $9-x$ **35.** $4x+20\sqrt{x}+25$

37. $\frac{2\sqrt{6}}{3}$ **39.** $\frac{3\sqrt{2x}}{2x}$ **41.** $\frac{\sqrt[3]{18x}}{2x}$ **43.** $\frac{4\sqrt{3x}}{3x^2y^2}$ **45.** $10+5\sqrt{3}$ **47.** $\frac{-3\sqrt{x}+3}{x-1}$

49. $\frac{m+3\sqrt{m}-28}{m-49}$

18.5 Exercises

1. $\{-6\}$ **3.** $\{4\}$ **5.** $\{44\}$ **7.** $\{6\}$ **9.** $\{\ \}$ or $\varnothing$ **11.** $\{4\}$ **13.** $\{4\}$ **15.** $\{2, 6\}$

17. $\left\{-\frac{7}{2}\right\}$ **19.** $\left\{-9, \frac{3}{2}\right\}$ **21.** $\{30\}$ **23.** $\{38\}$ **25.** $\{1\}$ **27.** $\{64\}$ **29.** $\{5\}$ **31.** $\{5\}$

33. $\left\{-\frac{7}{2}, 2\right\}$ **35.** $L=\frac{8T^2}{\pi^2}$ **37.** $h=\frac{3V}{\pi r^3}$ **39.** $r=\frac{A^2}{P^2}-1$ or $r=\frac{A^2-P^2}{P^2}$ **41. a.** 10th grade **b.** 36 words **43. a.** 2.12 m^2 **b.** 64.5 kg **45. a.** 0.96 sec **b.** 5.4 ft

18.6 Exercises

1. i **3.** i **5.** $-1+i$ **7.** $10-11i$ **9.** $7-3i$ **11.** 4 **13.** $-1+i$ **15.** $6+8i$ **17.** $20-9i$

19. $7+17i$ **21.** $32-24i$ **23.** 2 **25.** 6 **27.** $\frac{2}{25}-\frac{11}{25}i$ **29.** $\frac{3}{4}+\frac{3}{4}i$ **31.** $-\frac{3}{5}-\frac{2}{5}i$

33. $-7+6i$ **35.** -6 **37.** 4 **39.** $-\frac{1}{2}-\frac{3}{2}i$

MODULE 19

19.1 Exercises

1. $\{-16, 16\}$ **3.** $\left\{-\frac{5}{8}, \frac{5}{8}\right\}$ **5.** $\{-8, 8\}$ **7.** $\left\{-2\sqrt{6}, 2\sqrt{6}\right\}$ **9.** $\{\ \}$ or $\varnothing$ **11.** 16 **13.** $\frac{49}{4}$

15. $\{4-3\sqrt{2},\ 4+3\sqrt{2}\}$ 17. $\left\{\frac{3}{2}-\frac{\sqrt{7}}{2}i,\ \frac{3}{2}+\frac{\sqrt{7}}{2}i\right\}$ 19. $\left\{\frac{-12-\sqrt{165}}{3},\ \frac{-12+\sqrt{165}}{3}\right\}$

21. $\left\{-3,\ \frac{1}{3}\right\}$ 23. $\left\{\frac{1}{8}-\frac{\sqrt{127}}{8}i,\ \frac{1}{8}+\frac{\sqrt{127}}{8}i\right\}$ 25. $\left\{\frac{1}{3}\right\}$ 27. $D=5>0$, so there are two real solutions

29. $D=4>0$, so there are two real solutions 31. $\{-2,\ -\sqrt{2},\ \sqrt{2},\ 2\}$ 33. $\left\{\frac{1}{8},\ 8\right\}$

35. $\{256\}$ 37. $\left\{-2,\ \frac{1}{2}\right\}$

19.2 Exercises

1. up, wider 3. up, narrower 5. a. $(2,4)$ b. $x=2$ c. 0 d. 0 and 4 e. domain: $(-\infty,\infty)$ and range: $(-\infty,4]$ 7. 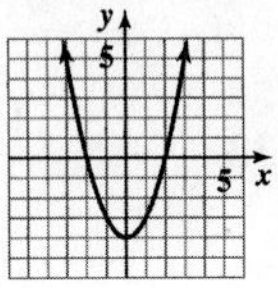 It looks like the graph of $y=x^2$, but shifted down four units.

9. 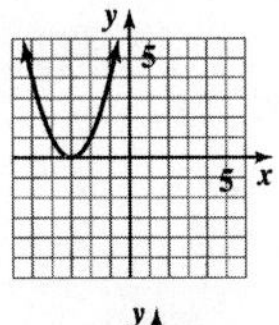 It looks like the graph of $y=x^2$, but shifted to the left three units.

11. 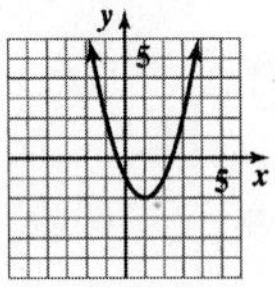 It looks like the graph of $y=x^2$, but shifted to the right one unit and down two units.

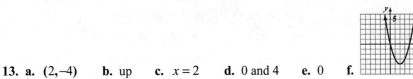

13. a. $(2,-4)$ b. up c. $x=2$ d. 0 and 4 e. 0 f. 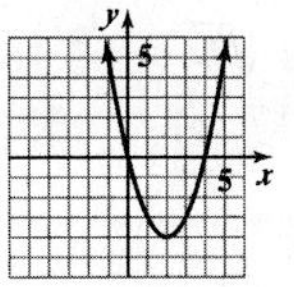 g. domain: $(-\infty,\infty)$ and range: $[-4,\infty)$ 15. a. $(3,2)$ b. down c. $x=3$ d. 2 and 4 e. -16 f. 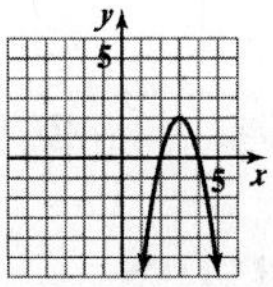
g. domain: $(-\infty,\infty)$ and range: $(-\infty,2]$

17. a. $(4,2)$ b. down c. $x=4$ d. $4-2\sqrt{2}\approx 1.1716$ and $4+2\sqrt{2}\approx 6.8284$ e. -2 f. 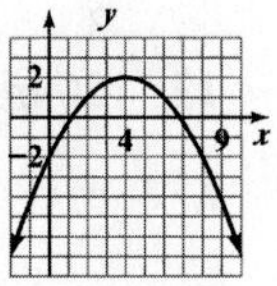
g. domain: $(-\infty,\infty)$ and range: $(-\infty,2]$ 19. $f(x)=(x+4)^2-7$, $(-4,-7)$ 21. $f(x)=-2(x+3)^2+8$, $(-3,8)$

23. $f(x)=(x+5)^2-25$, $(-5,-25)$ 25. $f(x)=-\left(x+\frac{3}{2}\right)^2+\frac{25}{4}$ a. $\left(-\frac{3}{2},\frac{25}{4}\right)$ b. down c. $x=-\frac{3}{2}$

d. −4 and 1 **e.** 4 **f.**

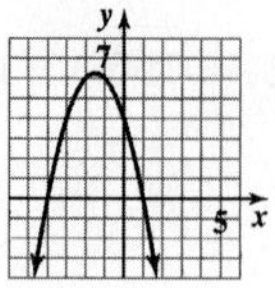

g. domain: $(-\infty,\infty)$ and range: $\left(-\infty,\frac{25}{4}\right]$

27. $f(x)=-3(x+1)^2+4$ **a.** $(-1,4)$ **b.** down **c.** $x=-1$

d. $-1-\frac{2\sqrt{3}}{3}\approx-2.1547$ and $-1+\frac{2\sqrt{3}}{3}\approx0.1547$ **e.** 1 **f.**

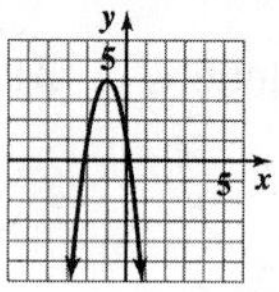

g. domain: $(-\infty,\infty)$ and range: $(-\infty,4$

29. $f(x)=2(x-4)^2-32$ **a.** $(4,-32)$ **b.** up **c.** $x=4$ **d.** 0 and 8 **e.** 0

f.

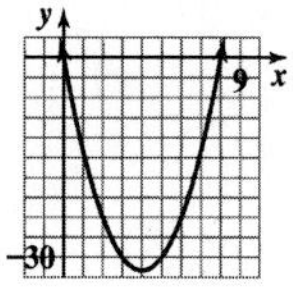

g. domain: $(-\infty,\infty)$ and range: $[-32,\infty)$ **31.** $(1,-4)$ **33.** $\left(-7,-\frac{41}{2}\right)$

35. a. $(-1,-7)$ **b.** up **c.** $x=-1$ **d.** $-1-\frac{\sqrt{21}}{3}\approx-2.5275$ and $-1+\frac{\sqrt{21}}{3}\approx0.5275$ **e.** −4

f.

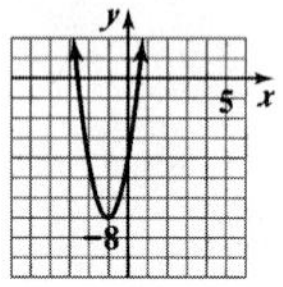

g. domain: $(-\infty,\infty)$ and range: $[-7,\infty)$ **37. a.** $(-6,-17)$ **b.** up **c.** $x=-6$

d. $-6-\sqrt{34}\approx-11.8310$ and $-6+\sqrt{34}\approx-0.1690$ **e.** 1 **f.**

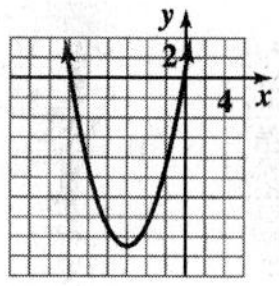

g. domain: $(-\infty,\infty)$ and range: $[-17,\infty)$ **39. a.** $(4,-16)$ **b.** up **c.** $x=4$ **d.** 0 and 8 **e.** 0

f.

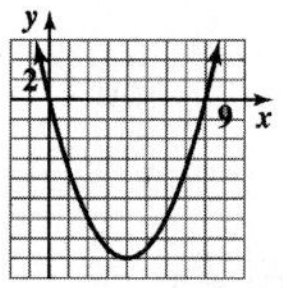

g. domain: $(-\infty,\infty)$ and range: $[-16,\infty)$

19.3 Exercises

1. $-\frac{4}{3}$ **3.** 1, 3, and 5 or 3, 5, and 7 **5.** 4 sec **7.** $\frac{10}{3}$ cm by 9 cm **9.** 3.76 ft **11.** 5 mph **13.** 120 min **15.** 3.5 sec, 61.025 m **17.** 3.5 sec, 201 ft **19. a.** \$71.60 **b.** 360, \$12,960 million **c.** \$36 **21. a.** $R(x)=-\frac{1}{40}x^2+8000x$ **b.** $P(x)=-\frac{1}{40}x^2+4000x-20{,}000$ **c.** 80,000; \$159,980,000 **d.** \$6000 **23.** 1.2; \$11; \$7.20 **25.** 625 ft by 1250 ft; 781,250 ft^2 **27.** $\frac{1}{2}$; \$1.50 **29.** 2006; 27,000 **31. a.** \$27,500; \$60,400 **b.** 350; −\$12,500 **c.** yes

19.4 Exercises

1. 5 **3.** $\sqrt{74}$ **5.** $\frac{\sqrt{829}}{6}$ **7.** 8 **9.** $(4,8)$ **11.** $\left(-\frac{3}{2},1\right)$ **13.** $\left(-\frac{3}{2},\frac{3}{4}\right)$ **15.** $(1,-1)$ **17.** $x^2+y^2=1$ **19.** $(x+2)^2+(y-3)^2=16$ **21.** $(x-1)^2+(y+4)^2=\frac{9}{16}$ **23.** $(x-3)^2+y^2=2$

25. $(h,k)=(0,0)$, $r=3$, 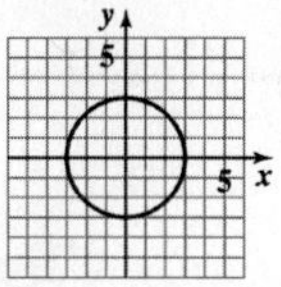 **27.** $(h,k)=(1,-5)$, $r=4$, 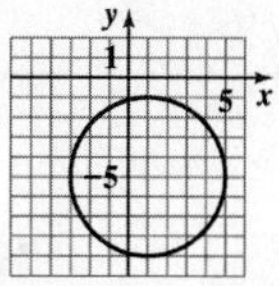 **29.** $(h,k)=\left(\frac{1}{4},-\frac{1}{2}\right)$, $r=2$, 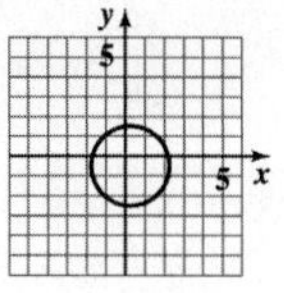 **31.** $(h,k)=(4,-7)$, $r=2\sqrt{3}$, x-int: none, y-int: none, 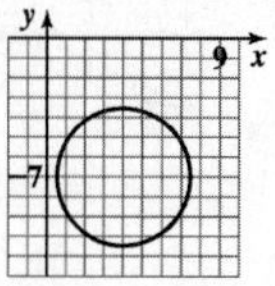

33. $(x+3)^2+y^2=4$, $(h,k)=(-3,0)$, $r=2$, x-int: $x=-5,1$, y-int: none, 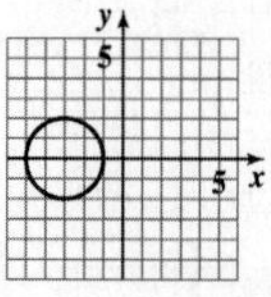

35. $(x+1)^2+(y-2)^2=4$, $(h,k)=(-1,2)$, $r=2$, x-int: $x=-1$, y-int: $y=2\pm\sqrt{3}$, 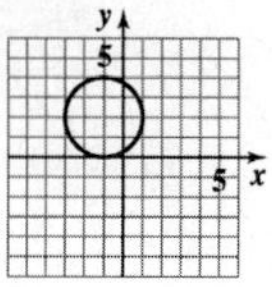

37. $(x-2)^2+(y-4)^2=1$, $(h,k)=(2,4)$, $r=1$, x-int: none, y-int: none, 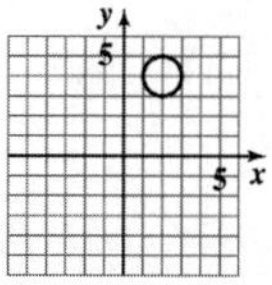

39. $\left(x-\frac{3}{2}\right)^2+\left(y-\frac{1}{2}\right)^2=3$, $(h,k)=\left(\frac{3}{2},\frac{1}{2}\right)$, $r=\sqrt{3}$, x-int: $x=\frac{3\pm\sqrt{11}}{2}$, y-int: $y=\frac{1\pm\sqrt{3}}{2}$, 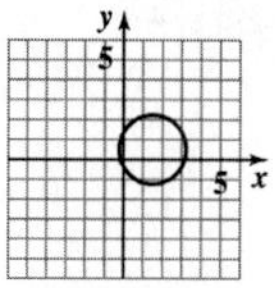

41. $(x-1)^2+(y+2)^2=4$, $(h,k)=(1,-2)$, $r=2$, x-int: $x=1$, y-int: $y=-2\pm\sqrt{3}$, 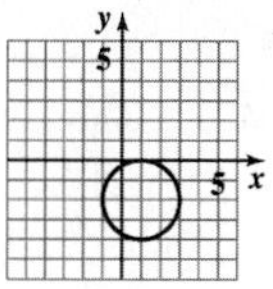

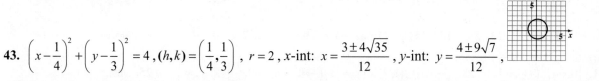

43. $\left(x-\frac{1}{4}\right)^2+\left(y-\frac{1}{3}\right)^2=4$, $(h,k)=\left(\frac{1}{4},\frac{1}{3}\right)$, $r=2$, x-int: $x=\frac{3\pm4\sqrt{35}}{12}$, y-int: $y=\frac{4\pm9\sqrt{7}}{12}$, 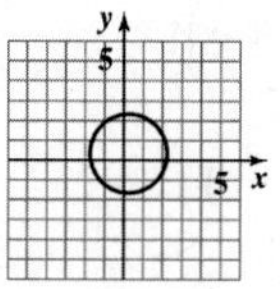

19.5 Exercises

1. $(-\infty,-3]\cup[1,\infty)$ **3.** $\left[-\frac{2}{3},0\right]$ **5.** $[-1,1]$ **7.** $\left(-\frac{1}{3},1\right)$ **9.** $(-\infty,1]$ **11.** $[-3,1)$

13. $(-\infty,0)\cup(1,\infty)$ **15.** $(-1,1]$ **17.** $(-4,-2]$ **19.** $\left(-3,-\frac{1}{2}\right]\cup(2,\infty)$ **21.** $(-3,5)$

MODULE 20

20.1 Exercises

1. 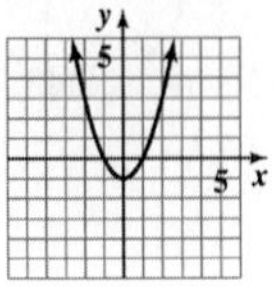
3. 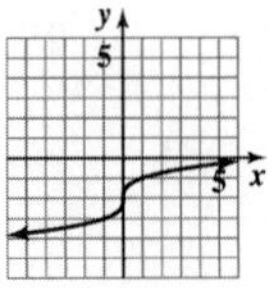
5. 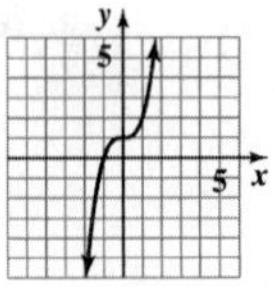
7. 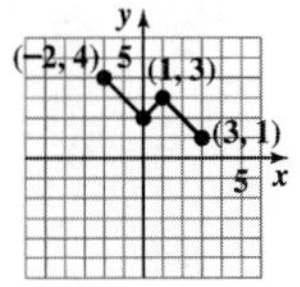

9. 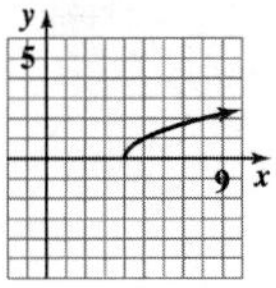
11. 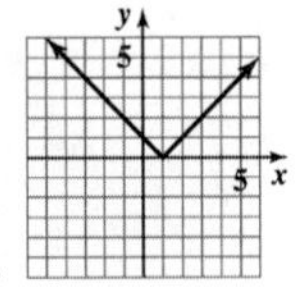
13. 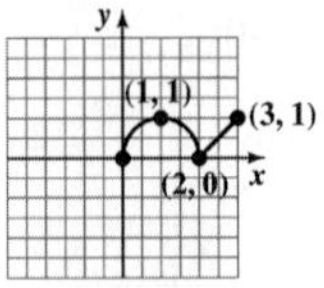

15. 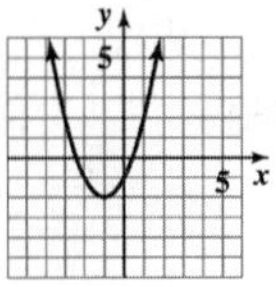
17. 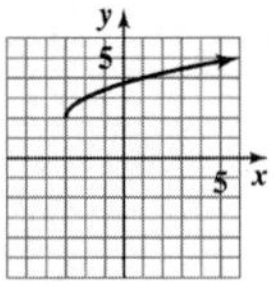
19. 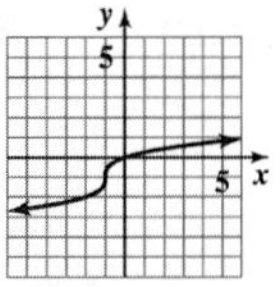
21. 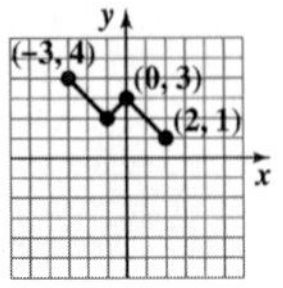

23. 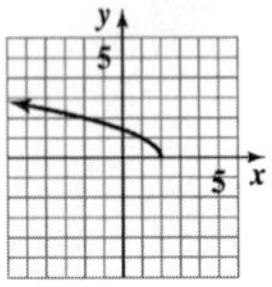
25. 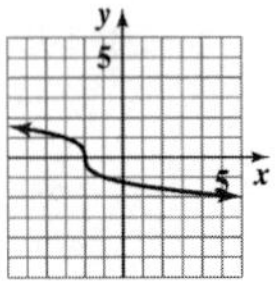
27. 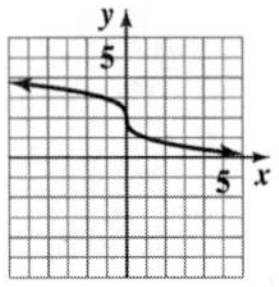
29. 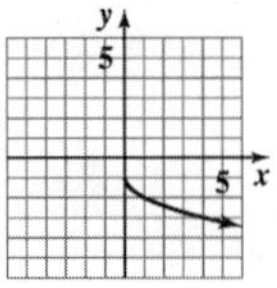
31. 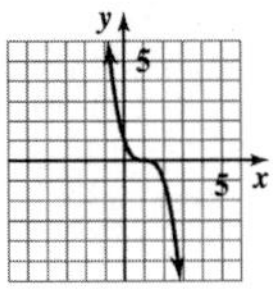
33. 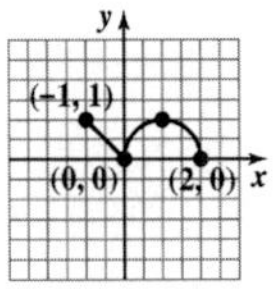

35. 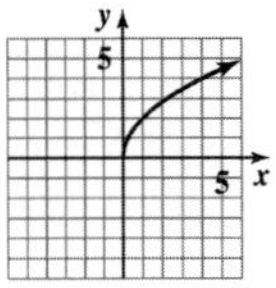
37. 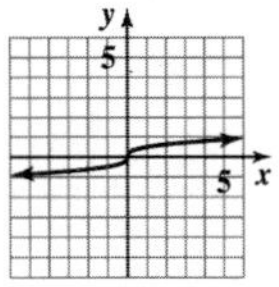
39. 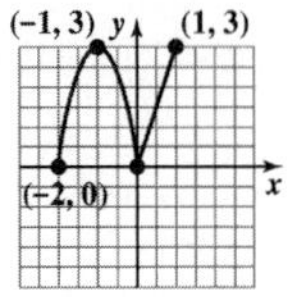

41. 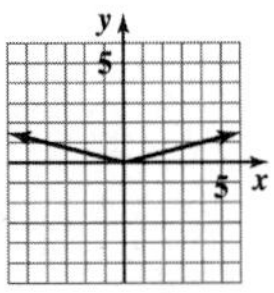
43. 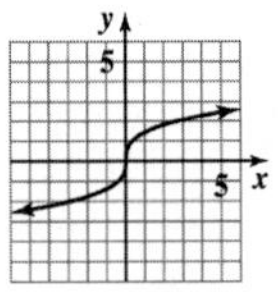
45. 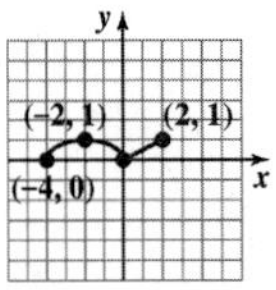

47. 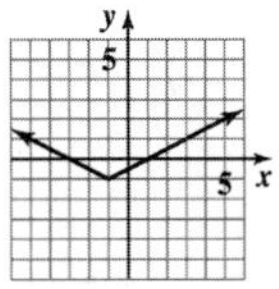
49. 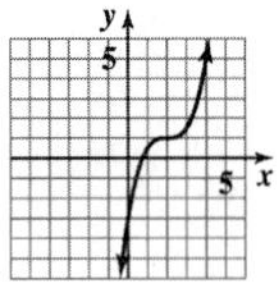
51. 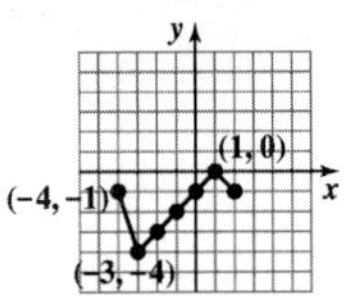

53. 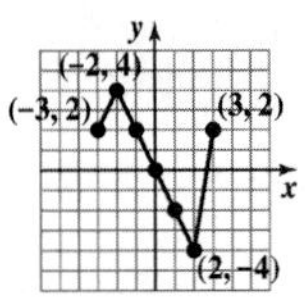

55. 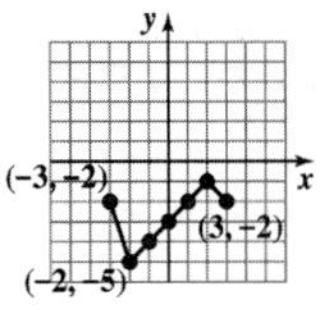

20.2 Exercises

1. $(f \circ g)(x) = \dfrac{x+7}{x+1}$ **3.** $(f \circ h)(x) = 3\sqrt{x+3}+1$ **5.** $(h \circ f)(x) = \sqrt{3x+4}$ **7.** $(f \circ f)(x) = 9x+4$

9. 7 **11.** $\dfrac{2}{5}$ **13.** 2 **15.** −5 **17. a.** −4 **b.** 0 **c.** 2 **d.** −1 **19.** $(-\infty,\infty)$; $(-\infty,\infty)$

21. $[0,\infty)$; $(-\infty,\infty)$ **23.** $(-\infty,1)\cup(1,2)\cup(2,\infty)$; $(-\infty,-1)\cup\left(-1,\frac{1}{2}\right)\cup\left(\frac{1}{2},\infty\right)$ **25.** yes **27.** yes

29. yes **31.** no **33.** yes **35.** yes **37.** no **39.** yes **41.–43.** $(f \circ g)(x) = (g \circ f)(x) = x$

45. 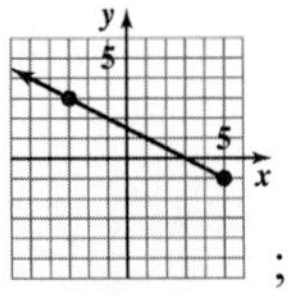 ; domain: $(-\infty,5]$; range: $[-1,\infty)$ **47.** 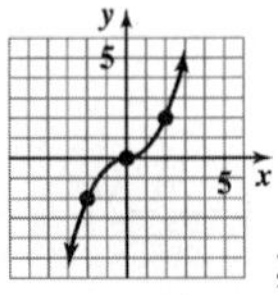 ; domain: $(-\infty,\infty)$; range: $(-\infty,\infty)$

49. 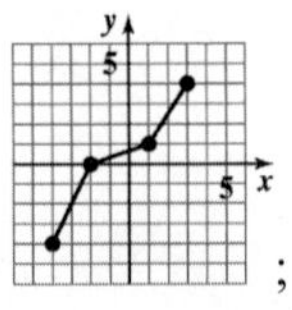 ; domain: $[-4,3]$; range: $[-4,4]$ **51.** $f^{-1}(x) = 3x+15$; domain f = range $f^{-1} = (-\infty,\infty)$; range f = domain $f^{-1} = (-\infty,\infty)$ **53.** $f^{-1}(x) = \dfrac{x^3+3}{2}$; domain f = range $f^{-1} = (-\infty,\infty)$; range f = domain

$f^{-1} = (-\infty,\infty)$ **55.** $f^{-1}(x)=\sqrt{-x-2}$; domain f = range $f^{-1}=[0,\infty)$; range f = domain $f^{-1}=(-\infty,-2]$

57. $f^{-1}(x)=\dfrac{3}{x}$; domain f = range f^{-1} = range f = domain $f^{-1}=(-\infty,0)\cup(0,\infty)$ **59.** $f^{-1}(x)=\dfrac{7x+1}{5x+8}$; domain f = range $f^{-1}=\left(-\infty,\dfrac{7}{5}\right)\cup\left(\dfrac{7}{5},\infty\right)$; range f = domain $f^{-1}=\left(-\infty,-\dfrac{8}{5}\right)\cup\left(-\dfrac{8}{5},\infty\right)$

20.3 Exercises

1. 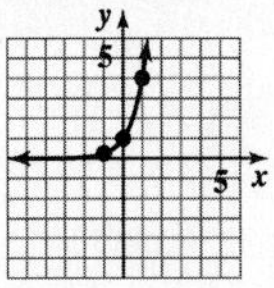 **3.** 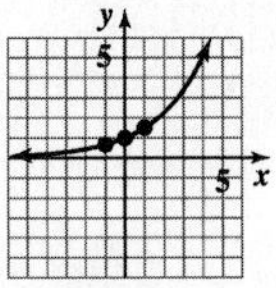 **5.** 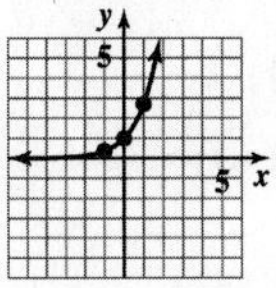 **7.** $f(x)=2^x$ **9.** $f(x)=\left(\dfrac{3}{4}\right)^x$

11. $f(x)=\left(\dfrac{1}{16}\right)^x$ **13.** 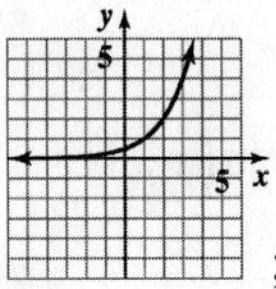 ; domain: $(-\infty,\infty)$; range: $(0,\infty)$; y-intercept: $\dfrac{1}{2}$; $y=0$

15. 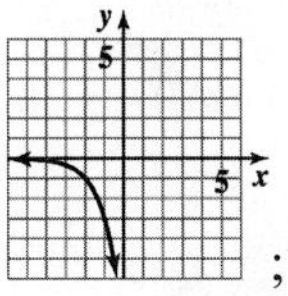 ; domain: $(-\infty,\infty)$; range: $(-\infty,0)$; y-intercept: -9; $y=0$ **17.** 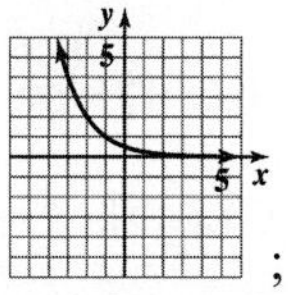 ; domain: $(-\infty,\infty)$, range: $(0,\infty)$; y-intercept: $\dfrac{1}{2}$; $y=0$ **19.** 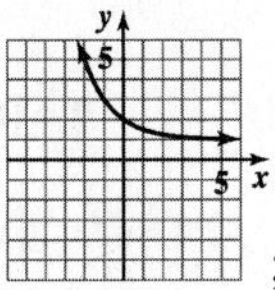 ; domain: $(-\infty,\infty)$; range: $(1,\infty)$; y-intercept: 2; $y=1$

21. $\{4\}$ **23.** $\left\{\dfrac{1}{4}\right\}$ **25.** $\left\{-\dfrac{3}{5}\right\}$ **27.** $\left\{\dfrac{1}{3}\right\}$ **29.** $\{-1, 3\}$ **31. a.** \$300,000 **b.** \$265,640 **c.** \$163,299 **33. a.** 2560 bacteria **b.** 4096 bacteria **c.** 8,589,934,592 bacteria **35.** \$12,752.18 **37.** Investment B **39.** \$469,068.04

20.4 Exercises

1. 20.085537 **3.** 1.395612 **5.** 32.725881 **7.** 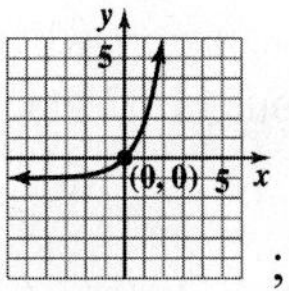

; domain: $(-\infty,\infty)$; range: $(-1,\infty)$; $y=-1$

9. 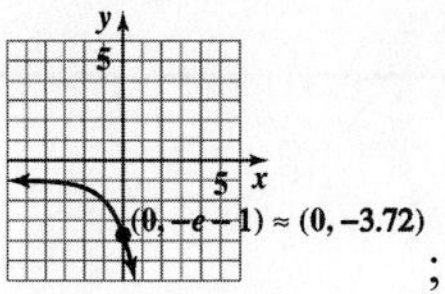

; domain: $(-\infty,\infty)$; range: $(-\infty,-1)$; $y=-1$ **11.** $\{-2\}$ **13.** $\left\{\dfrac{1}{4}\right\}$ **15.** $\{-3,4\}$

17. $\{-1,1,2\}$ **19.** \$6798.89; \$20,942.06 **21.** \$332,570.82 **23.** (a) has the lower present value. **25. a.** $P(t)=10e^{0.25t}$ **b.** 4034 bacteria **27. a.** 8 rabbits **b.** 478,993 rabbits

20.5 Exercises

1. $\log_3 9=2$ **3.** $\log_2 \dfrac{1}{8}=-3$ **5.** $\log_{1/3} 27=t$ **7.** $5^0=1$ **9.** $\sqrt{2}^6=8$ **11.** $a^3=(x-1)$ **13.** $\dfrac{1}{2}$

15. 4 **17.** $-\dfrac{1}{3}$ **19.** 11 **21.** 0 **23.** 1 **25.** 0 **27.** $\log 1000=3$ **29.** $\ln 2=k$ **31.** $\ln Z=10$

33. $10^6 = 1{,}000{,}000$ **35.** $e^4 = Z$ **37.** -3 **39.** $\frac{2}{3}$ **41.** 49 **43.** 4 **45.** 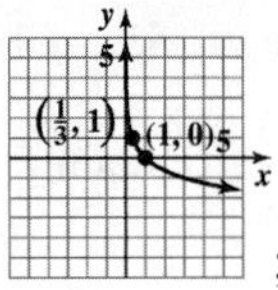

; $(0, \infty)$; $x = 0$

47. 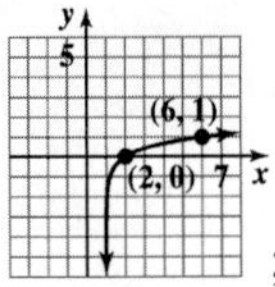

; $(1, \infty)$; $x = 1$ **49.** 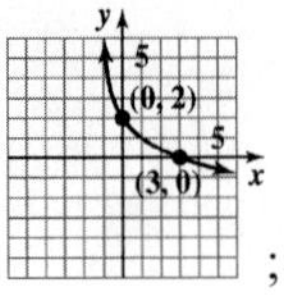

; $(-1, \infty)$; $x = -1$ **51.** 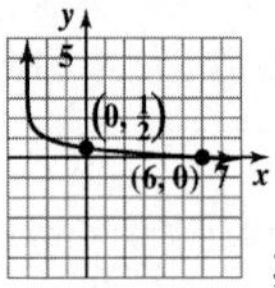

; $(-3, \infty)$; $x = -3$

53. $(-3, \infty)$ **55.** $(-\infty, -3) \cup (3, \infty)$ **57.** $(-\infty, -5) \cup (8, \infty)$

20.6 Exercises

1. $\log_4 x + \log_4 y$ **3.** $3\log_5 y$ **5.** $\ln 5 + 2$ **7.** $2 + \log P$ **9.** $6 + \log_{\sqrt{2}} x$ **11.** $2\log_7 x + 3\log_7 y$

13. $\frac{1}{2}\log x - 1 - 3\log y$ **15.** $1 + \frac{1}{2}\log_2 x + \frac{1}{2}\log_2 y$ **17.** $\frac{1}{5} + \frac{1}{5}\ln z - \frac{1}{2}\ln(x-1)$

19. $-\frac{4}{3}\ln(x+1) - 2\ln(x-1)$ **21.** $\log_b(AC)$ **23.** $\log_8\left(x^2\sqrt[3]{y}\right)$ **25.** 4 **27.** $\log_5\left(x^2 - 4\right)$

29. $\log\left(\frac{x+3}{x+1}\right)$ **31.** $\{5\}$ **33.** $\{1\}$ **35.** $\{2\}$ **37.** 2.8362 **39.** –2.6572 **41.** $\log_3\left(xw^2\right)$

43. $\log_2(xyw)$ **45.** $\{5\}$ **47.** 36

20.7 Exercises

1. {1.4650} **3.** {–1, 3} **5.** {–13.3627} **7.** {4.4841} **9.** {0.3334} **11.** {0.6931}

13. {–0.2055} **15.** {–3.1192} **17.** {7} **19.** {2, 3} **21.** $\left\{-\frac{99}{5}\right\}$ **23.** $\left\{\frac{1}{3}\right\}$ **25.** {6} **27.** {9}

29. $\left\{-1, \frac{1}{3}\right\}$ **31.** { } or ∅

20.8 Exercises

1. about 4 years 8 months **3.** 13.73 years **5.** Bank A: 4.5%, Bank B: 4.6% **7.** 4147 bacteria
9. 200,119 people **11.** 64.84% **13. a.** 6000 families **b.** 2000 families **c.** 1988 **15. a.** 61,500
b. 40 **c.** $-\dfrac{\ln\left(\frac{43}{2}\right)}{6}$ **d.** 2009 **17.** about 11:52 PM

MODULE 21

21.1 Exercises

1. vertex: (0, 0), focus: (0, 4), directrix: $y = -4$, 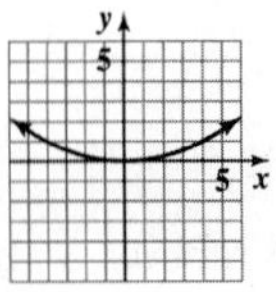

3. vertex: (1, 4), focus: (1, 1),

directrix: $y = 7$, 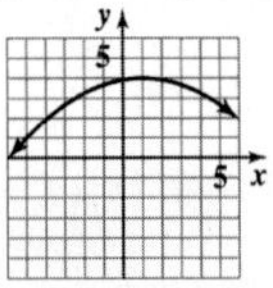

5. vertex: (–2, –6), focus: $\left(-2, -\frac{19}{4}\right)$, directrix: $y = -\frac{29}{4}$, 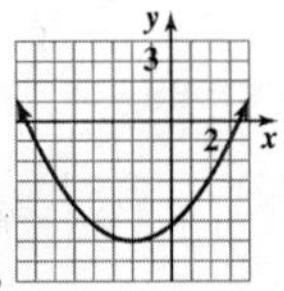

7. vertex: (0, 0), focus: (–2, 0), directrix: $x = 2$, 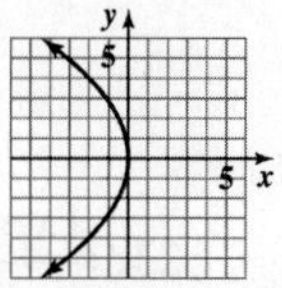 **9.** vertex: (4, –3), focus: (9, –3),

directrix: $x = -1$, 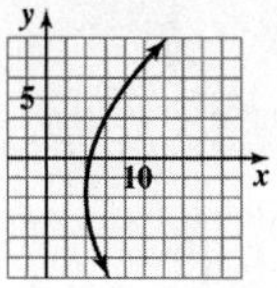 **11.** $y^2 = 8x$ **13.** $(x-3)^2 = -8(y+3)$ **15.** $(y+2)^2 = -\left(x+\frac{11}{4}\right)$

17. $(y-4)^2 = 16(x+3)$ **19.** $(x-2)^2 = -2(y-2)$ and $(x-2)^2 = 8\left(y+\frac{1}{2}\right)$

21. vertex: (1, 6), focus: $\left(-\frac{1}{2}, 6\right)$, directrix: $x = \frac{5}{2}$, 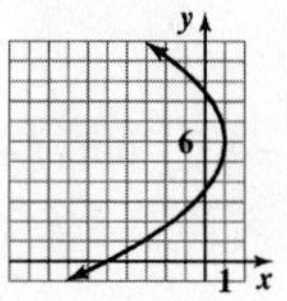 **23.** vertex: (–1, –3), focus: $\left(-\frac{5}{4}, -3\right)$,

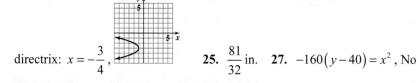

directrix: $x = -\frac{3}{4}$, 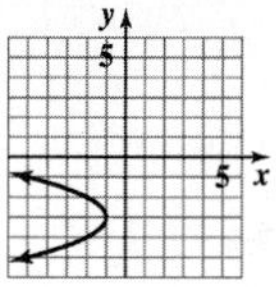 **25.** $\frac{81}{32}$ in. **27.** $-160(y-40) = x^2$, No

21.2 Exercises

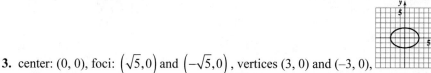

1. center: (0, 0), foci: $\left(\sqrt{12}, 0\right)$ and $\left(-\sqrt{12}, 0\right)$, vertices (4, 0) and (–4, 0), 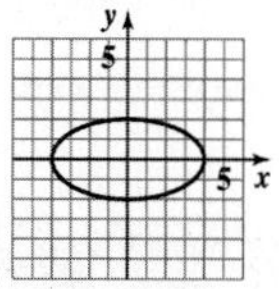

3. center: (0, 0), foci: $\left(\sqrt{5}, 0\right)$ and $\left(-\sqrt{5}, 0\right)$, vertices (3, 0) and (–3, 0), 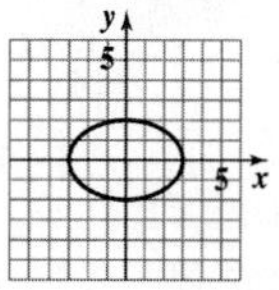

5. center: (2, 4), foci: $\left(2+\sqrt{11}, 4\right)$ and $\left(2-\sqrt{11}, 4\right)$, vertices $(8, 4)$ and $(-4, 4)$, 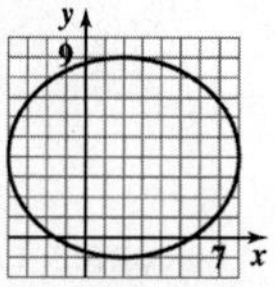

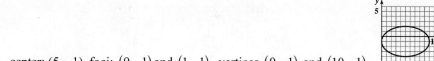

7. center: (5, –1), foci: $(9, -1)$ and $(1, -1)$, vertices $(0, -1)$ and $(10, -1)$, 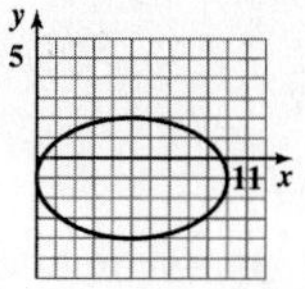

9. $\frac{(x+4)^2}{16} + \frac{y^2}{36} = 1$ **11.** $\frac{x^2}{49} + \frac{y^2}{24} = 1$ **13.** $\frac{(x+6)^2}{16} + \frac{(y-5)^2}{25} = 1$ **15.** $\frac{(x+1)^2}{7} + \frac{(y-4)^2}{16} = 1$

17. $\frac{(x-4)^2}{4} + \frac{(y-1)^2}{53} = 1$ **19.** $\frac{x^2}{16} + \frac{y^2}{12} = 1$ **21.** $\frac{(x+2)^2}{20} + \frac{(y-1)^2}{16} = 1$, center: (–2, 1), foci: (–4, 1) and (0, 1), vertices: $\left(-2-2\sqrt{5},1\right)$ and $\left(-2+2\sqrt{5},1\right)$, 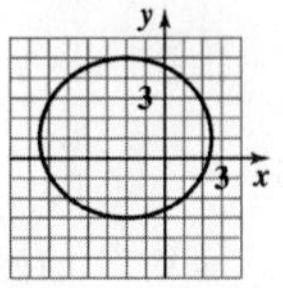 **23.** $\dfrac{\left(x+2\right)^2}{16}+\dfrac{\left(y-1\right)^2}{1}=1$, center: (–2, 1),

foci: $\left(-2-\sqrt{15},1\right)$ and $\left(-2+\sqrt{15},1\right)$, vertices: $\left(-6,1\right)$ and $\left(2,1\right)$, 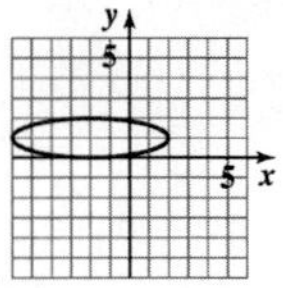 **25.** $\dfrac{x^2}{225}+\dfrac{y^2}{256}=1$

27. about 25.8 yd **29.** 44.80 in. **31.** about 39.69 ft

21.3 Exercises

1. center: (0, 0), transverse axis: $y=0$, vertices: $\left(-4,0\right)$ and $\left(4,0\right)$, foci: $\left(-5,0\right)$ and $\left(5,0\right)$, asymptotes: $y=\pm\dfrac{3}{4}x$,

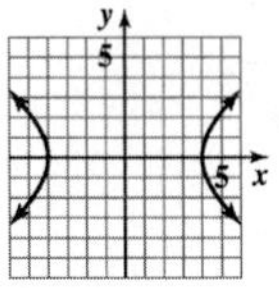

3. center: (2, 4), transverse axis: $x=2$, vertices: $\left(2,-1\right)$ and $\left(2,9\right)$,

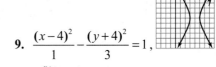

foci: $\left(2,4-\sqrt{61}\right)$ and $\left(2,4+\sqrt{61}\right)$, asymptotes: $y-4=\pm\dfrac{5}{6}(x-2)$, 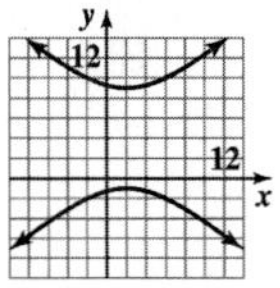

5. center: (0, 0), transverse axis: $y=0$, vertices: $\left(-\sqrt{5},0\right)$ and $\left(\sqrt{5},0\right)$, foci: $\left(-5,0\right)$ and $\left(5,0\right)$, asymptotes:

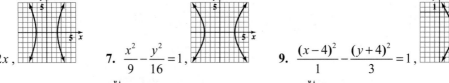

$y=\pm 2x$, 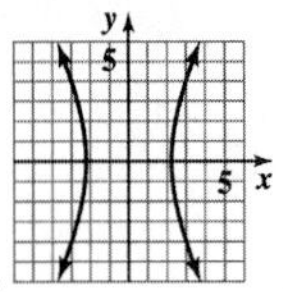 **7.** $\dfrac{x^2}{9}-\dfrac{y^2}{16}=1$, 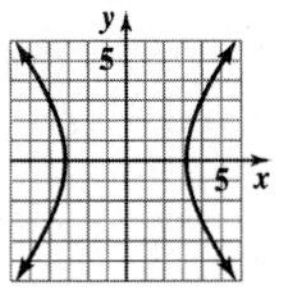 **9.** $\dfrac{(x-4)^2}{1}-\dfrac{(y+4)^2}{3}=1$, 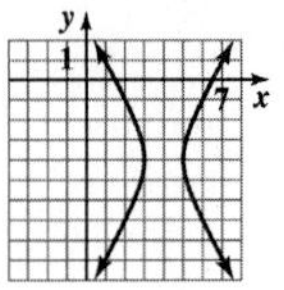

11. $\dfrac{(y-6)^2}{4}-\dfrac{(x-9)^2}{5}=1$, 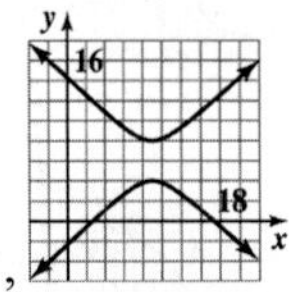 **13.** $\dfrac{y^2}{20}-\dfrac{x^2}{16}=1$, 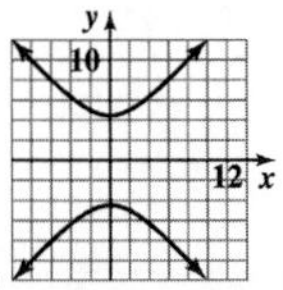

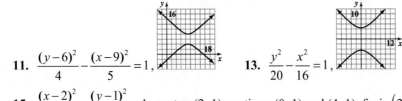

15. $\dfrac{(x-2)^2}{4}-\dfrac{(y-1)^2}{4}=1$, center: (2, 1), vertices: (0, 1) and (4, 1), foci: $\left(2-2\sqrt{2},1\right)$ and $\left(2+2\sqrt{2},1\right)$, endpoints of

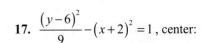

conjugate axis: (2, –1) and (2, 3), asymptotes: $y-1=\pm\left(x-2\right)$, 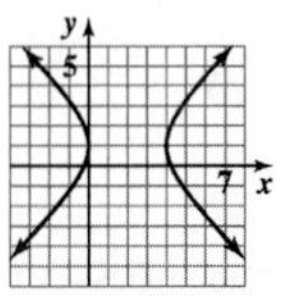 **17.** $\dfrac{\left(y-6\right)^2}{9}-\left(x+2\right)^2=1$, center:

(–2, 6), vertices: $\left(-2,3\right)$ and $\left(-2,9\right)$, foci: $\left(-2,6-\sqrt{10}\right)$ and $\left(-2,6+\sqrt{10}\right)$, endpoints of conjugate axis:

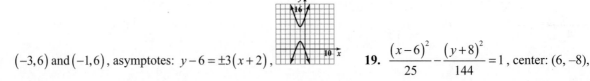

$\left(-3,6\right)$ and $\left(-1,6\right)$, asymptotes: $y-6=\pm 3\left(x+2\right)$, 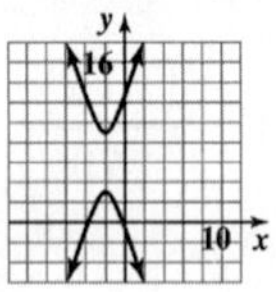 **19.** $\dfrac{\left(x-6\right)^2}{25}-\dfrac{\left(y+8\right)^2}{144}=1$, center: (6, –8), vertices: (1, –8) and (11, –8), foci: $(-7,-8)$ and $(19,-8)$, endpoints of conjugate axis: $(6,-20)$ and $(6,4)$,

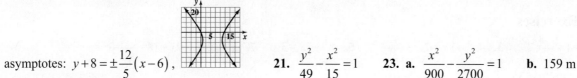

asymptotes: $y+8=\pm\frac{12}{5}(x-6)$,

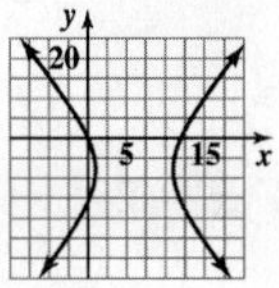

21. $\frac{y^2}{49}-\frac{x^2}{15}=1$ **23. a.** $\frac{x^2}{900}-\frac{y^2}{2700}=1$ **b.** 159 m

MODULE 22

22.1 Exercises

1. 4, 7, 10, 13 **3.** 1, $\frac{8}{5}$, 2, $\frac{16}{7}$ **5.** 30, 120, 600, 3600 **7.** –5, 10, –15, 20 **9.** $-\frac{1}{42}, \frac{1}{56}, -\frac{1}{72}, \frac{1}{90}$

11. 7, 10, 13, 16 **13.** 6, $\frac{3}{2}, \frac{1}{6}, \frac{1}{96}$ **15.** $a_n=2n-3$, $a_{11}=19$ **17.** $a_n=n(n+5)$, $a_7=84$

19. $a_n=\frac{2}{5^n}$, $a_6=\frac{2}{15{,}625}$ **21.** $a_1=-6$, $a_n=(-1)^n(n+2)!$ for $n\geq 2$, $a_5=-5040$ **23.** 20 **25.** $\frac{11}{32}$

27. 12 **29.** 45 **31.** 196 **33.** 271 **35.** 271 **37.** $\sum_{i=1}^{10}5i$ **39.** $\sum_{i=1}^{9}\frac{i+3}{i+4}$ **41.** $\sum_{i=1}^{6}\frac{(-1)^i}{9i}$

43. $\sum_{i=1}^{n+1}\frac{7^{i-1}}{(i-1)!}$

22.2 Exercises

1. yes, 6 **3.** yes, $\frac{3}{2}$ **5.** yes, 2 **7.** $a_n=5n-3$, $a_{10}=47$ **9.** $a_n=\frac{3}{2}n$, $a_{50}=75$

11. $a_n=2n-1$, $a_{17}=33$ **13.** 71 **15.** 97 **17.** $\frac{25}{2}$ **19.** 3240 **21.** 2375 **23.** –1002

25. no solution **27.** 10,100 **29.** 147 **31.** \$3185 **33.** \$200; \$65,000 **35.** 1.3 sec; 312 in.

22.3 Exercises

1. 8, 16, 32, 64, 128 **3.** 25, –5, 1, $-\frac{1}{5}, \frac{1}{25}$ **5.** –4, 8, –16, 32, –64 **7.** yes, –1 **9.** yes, $-\frac{5}{3}$

11. $a_n=3\cdot 2^{n-1}$ **13.** $a_n=\frac{1}{5}\left(-\frac{2}{3}\right)^{n-1}$ **15.** $-\frac{951}{256}$ **17.** 78,732 **19.** –91,749 **21.** ≈ 50.31

23. $\frac{85}{128}$ **25.** converges, $\frac{2401}{6}$ **27.** diverges **29.** \$6561 **31.** $\approx$ \$1,402,123.44 **33.** 200 ft

35. $\frac{322}{990}$ **37.** $\approx$ \$30,200.99

22.4 Exercises

1. $m^6+6m^5n+15m^4n^2+20m^3n^3+15m^2n^4+6mn^5+n^6$

3. $x^7-7x^6y+21x^5y^2-35x^4y^3+35x^3y^4-21x^2y^5+7xy^6-y^7$

5. $64x^6+576x^5y+2160x^4y^2+4320x^3y^3+4860x^2y^4+2916xy^5+729y^6$ **7.** 7 **9.** 1

11. $x^7+14x^6+84x^5+280x^4+560x^3+672x^2+448x+128$

13. $1024x^5+1280x^4+640x^3+160x^2+20x+1$

15. $3125x^5-9375x^4y+11{,}250x^3y^2-6750x^2y^3+2025xy^4-243y^5$ **17.** $x^2-4\sqrt{2}x^{\frac{3}{2}}+12x-8\sqrt{2}x^{\frac{1}{2}}+4$

19. $70a^4b^4$ **21.** $1{,}088{,}640x^4$ **23.** 126,720 **25.** 19,779,228

MODULE 23 Additional Topics

23.1 Exercises

1. $x+4+\frac{3}{x-5}$ **3.** $4x^2-8x+21-\frac{49}{x+2}$ **5.** x^2+3x-3 **7.** $x^3+2x^2+9x+18-\frac{5}{x-2}$

9. $15x^2+20x-35$

23.2 Exercises

1. $\left[\begin{array}{cc|c} 2 & 3 & 5 \\ 5 & -2 & 1 \end{array}\right]$ **3.** $\left[\begin{array}{cc|c} \frac{1}{2} & -\frac{1}{3} & 0 \\ \frac{2}{7} & \frac{5}{9} & 2 \end{array}\right]$ **5.** $\left[\begin{array}{ccc|c} 7 & -2 & 5 & 3 \\ 0 & 5 & 2 & -2 \\ 3 & 0 & -2 & 1 \end{array}\right]$ **7.** $(0, -2)$ **9.** $(2, -1)$

11. $(5, -2)$ **13.** $\left(\frac{1}{2}, \frac{1}{3}\right)$ **15.** $\{\}$ or $\varnothing$ **17.** $(4, 3)$ **19.** $(1, -2, -1)$ **21.** $(1, 3, 2)$

23. $(4, 3, 4)$ **25.** $(0, 2, 0)$ **27.** $\left(-2, \frac{1}{2}, -3\right)$ **29.** $(-3, 0, 1)$

23.3 Exercises

1. 18 **3.** 0 **5.** $(1, 2)$ **7.** $(-2, -4)$ **9.** $\left(\frac{1}{2}, \frac{1}{3}\right)$ **11.** $\{(x, y) \mid y = 3x - 5\}$ **13.** 164 **15.** -39

17. $(2, 1, -1)$ **19.** $(-3, 2, -4)$ **21.** $\varnothing$

Index

B

E

F

G

M

Q

R

T

U

V

W

X

Y

Z